WATER-ROCK INTERACTION

PROCEEDINGS OF THE 9TH INTERNATIONAL SYMPOSIUM ON WATER-ROCK
INTERACTION – WRI-9/TAUPO/NEW ZEALAND/30 MARCH - 3 APRIL 1998

Water-Rock Interaction

Edited by
GREG B. AREHART
Wairakei Research Centre, Institute of Geological and Nuclear Sciences, Taupo, New Zealand

JOHN R. HULSTON
Nuclear Sciences, Institute of Geological and Nuclear Sciences, Lower Hutt, New Zealand

A.A.BALKEMA/ROTTERDAM/BROOKFIELD/1998

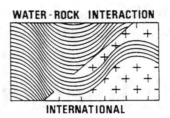

WRI-9

The texts of the various papers in this volume were set individually by typists under the supervision of each of the authors concerned.

Authorization to photocopy items for internal or personal use, or the internal or personal use of specific clients, is granted by A.A.Balkema, Rotterdam, provided that the base fee of US$ 1.50 per copy, plus US$ 0.10 per page is paid directly to Copyright Clearance Center, 222 Rosewood Drive, Danvers, MA 01923, USA. For those organizations that have been granted a photocopy license by CCC, a separate system of payment has been arranged. The fee code for users of the Transactional Reporting Service is: 90 5410 942 4/98 US$ 1.50 + US$ 0.10.

Published by
A.A.Balkema, P.O.Box 1675, 3000 BR Rotterdam, Netherlands (Fax: +31.10.413.5947)
A.A.Balkema Publishers, Old Post Road, Brookfield, VT 05036-9704, USA (Fax: 802.276.3837)

ISBN 90 5410 942 4
© 1998 A.A.Balkema, Rotterdam
Printed in the Netherlands

Table of contents

Preface	XXIII
Organisation	XXV

New Zealand highlight speakers

From basins to mountains and back again: N.Z. basin evolution since 10 Ma *R.G.Allis, R.Funnell & X.Zhan*	3
Hydrothermal alteration in New Zealand geothermal systems *P.R.L.Browne*	11
Chemistry of 3.2 Ga seafloor hydrothermal vent fluids *C.E.J.de Ronde*	19
Fire and water: Physical roles of water in large eruptions at Taupo and Okataina calderas *B.F.Houghton & C.J.N.Wilson*	25
Metal reactions at the water-soil interface *R.G.McLaren*	31
Conditions for rapid large-volume flow *R.H.Sibson*	35
Clean, green and steaming: Environmental geochemistry in New Zealand *J.Webster*	39

1 Surficial systems

The Arno River catchment basin, Tuscany, Italy: Chemical and isotopic composition of water *A.Adorni-Braccesi, L.Bellucci, C.Panichi, G.La Ruffa, F.Podda, G.Cortecci, E.Dinelli, A.Bencini & E.Gimenez Forcada*	47
Trace metals dissolved in the rainwaters on northern Sardinia (Mediterranean Sea) *R.Caboi, C.Ardau, L.Rundeddu & F.Frau*	51
Geochemistry of the Arno River, Italy: Natural and anthropogenic contributions *G.Cortecci, E.Dinelli, F.Lucchini, L.Fanfani, G.La Ruffa, F.Podda, A.Bencini, E.Gimenez Forcada & A.Adorni Braccesi*	55

Geochemistry of riverine particulate and dissolved loads, Darling River Basin, Australia 59
C.E.Martin & M.T.McCulloch

Sea water sulfate addition to a forested catchment: Results after five years of experimental treatment 63
C.-M.Mörth & P.Torssander

$\delta^{34}S$ dynamics in the system bedrock - soil - runoff - atmosphere: Results from the GEOMON network of small catchments, Czech Republic 67
M.Novák

Geological controls on drainage water compositions across a granite-related zoned mineral field, Zeehan, Western Tasmania 71
T.E.Parr & D.R.Cooke

The hydrogeochemistry of thallium in natural waters 75
P.Shand, W.M.Edmunds & J.Ellis

Decalcification and acidification of coastal dune sands in the Netherlands 79
P.J.Stuyfzand

Factor analysis of stream water chemistry following storms in Eastern Pennsylvania and New Jersey, USA 83
B.J.Woodward V, B.Bowen & D.E.Grandstaff

2 Processes involving organic matter

Analysis of environmentally significant organic and inorganic metal species by coupled IC-ICP-MS, GC-ICP-MS and LC-ICP-MS 89
J.R.Brydie, A.P.Gize, P.R.Lythgoe, D.A.Polya, G.Kilpatrick, K.Hall & K.Sajan

Peat-water interactions: South Taupo Wetland, New Zealand 93
C.Chagué-Goff

Groundwater chemistry and water-rock interactions at hydrocarbon storage cavern sites in Korea 97
H.T.Chon, J.U.Lee, S.Y.Oh & H.D.Park

Organic matter maturation as an indicator of hydrothermal processes in sedimentary basins 101
M.Glikson & S.D.Golding

Adsorption of L-alanine monomer, dimer, trimer, tetramer and pentamer by some allophanes 105
H.Hashizume & B.K.G.Theng

Sorption and fractionation of natural organic matter on kaolinite and goethite 109
P.Maurice, K.Namjesnik-Dejanovic, S.Lower, M.Pullin, Y.-P.Chin & G.R.Aiken

Solid phase partitioning of uranium and copper in the presence of HFO and bacteria 115
L.A.Warren & F.G.Ferris

Trace metal/microbial interactions in an Antarctic freshwater system 119
K.S.Webster, J.G.Webster & P.E.Nelson

Influence of autochthonous microorganisms on the migration of redox-sensitive radionuclides 123
A.Winkler, T.Taute, A.Pekdeger, I.Stroetmann & G.Maue

Phosphorus in soils and ground waters of the Indian Ocean atoll islands 127
P.V.Yelpatyevsky & T.N.Lutsenko

Carboxylates in fluid-inclusions in minerals 131
Yishan Zeng & Jiaqi Liu

3 Groundwater quality

Halogen geochemistry of a Middle Jurassic calcareous aquifer in northern France 137
F.Barbecot, C.Marlin, E.Gibert & L.Dever

Modelling of redox conditions and control of trace elements in clayey groundwaters 141
C.Beaucaire, H.Pitsch & C.Boursat

Hydrochemistry in an indurated argillaceous formation (Tournemire tunnel site, France) 145
L.De Windt, J.Cabrera & J.-Y.Boisson

Different approaches to estimate trace element concentrations in groundwaters 149
L.Duro & J.Bruno

An integrated groundwater quality model based on hydrochemical environments 153
J.Griffioen, A.L.Lourens, C.B.M.te Stroet, B.Minnema, M.P.Laeven, P.J.Stuyfzand, C.G.E.M.van Beek & W.Beekman

Linkage between hydrochemistry and geological cover of groundwaters in a Triassic sandstone aquifer (Buntsandstein), SW Germany 157
T.G.Kretzschmar

Groundwater quality variations in the Eocene Bagshot Formation, UK 161
J.M.Macmillan & J.D.Mather

Nitrate loading of shallow ground water, prairie vs cultivated land, northeastern Kansas, USA 165
G.L.Macpherson

Allochthonous ions dissolved in recent and fossil groundwaters: Identification and origins 169
E.Mazor

High-rate denitrification from several electron donors in a schist aquifer 173
H.Pauwels, O.Legendre & J.-C.Foucher

Sorption of fluorescent tracers in a physically and chemically heterogeneous aquifer material 177
T.Ptak & H.Strobel

Influence of eruptive volcanic lithologies on surface and ground water chemical compositions, Lake Taupo, New Zealand 181
M.R.Rosen & L.Coshell

Hydrogeochemical investigations on arsenic contamination of a shallow aquifer 185
Ch.Sommer-von Jarmersted, U.Maiwald, A.Pekdeger & M.Th.Schafmeister

Remediation of high fluoride groundwaters from arid regions using heat-treated soils: A column experiment study in Xinzhou, China 189
Yanxin Wang, Yuan Ximing, Guo Huaming, Wang Hong & Wang Yangen

4 Groundwater general

Trace element hydrogeochemistry of Mt. Etna, Sicily: Insight on water-rock interaction 195
A. Aiuppa, P. Allard, W. D'Alessandro, A. Michel, F. Parello & M. Treuil

Boron isotope geochemistry as a tracer for the evolution of natural aquatic systems 199
S. R. Barth

A small-scale dispersion experiment in a heterogeneous sandy aquifer, Botany Sands aquifer, Sydney, Australia: Tracer movement and interaction with geological material 203
P. Beck & J. Jankowski

Controls on sulfate reduction in a dual porosity aquifer 207
S. H. Bottrell, S. J. Moncaster, J. H. Tellam & J. W. Lloyd

$^{87}Sr/^{86}Sr$ in groundwater as indicators of carbonate dissolution 211
S. S. Dogramaci, A. L. Herczeg & Y. Bone

Trace elements as residence time indicators in groundwaters: The East Midlands Triassic sandstone aquifer, England 215
W. M. Edmunds & P. L. Smedley

Retarded intraparticle diffusion in heterogeneous aquifer material 219
M. Finkel & R. Liedl

The source of stable chlorine isotopic signatures in groundwaters from crystalline shield rocks 223
S. K. Frape, G. Bryant, P. Durance, J. C. Ropchan, J. Doupe, R. Blomqvist, P. Nissinen & J. Kaija

High permeabilities of Quaternary granites in Japan and its implications for mass and heat transfer in a magmatic-hydrothermal system 227
K. Fujimoto, M. Takahashi, N. Doi & O. Kato

A geochemical model for groundwaters of the arid Ti-Tree Basin, Central Australia 231
G. A. Harrington & A. L. Herczeg

Tidal influences on metal concentrations in groundwater, Geelong, Australia 235
S. Horner & T. R. Weaver

Hydrogeochemical processes in a fractured rock aquifer of the Lachlan Fold Belt: Yass, New South Wales, Australia 239
J. Jankowski, R. I. Acworth & S. Shekarforoush

Reverse ion-exchange in a deeply weathered porphyritic dacite fractured aquifer system, Yass, New South Wales, Australia 243
J. Jankowski, R. I. Acworth & S. Shekarforoush

Comparison of oxygen and hydrogen isotopes from two perennial karst springs, Indiana, USA 247
N. C. Krothe

Saline intrusion into an urban sandstone aquifer 251
R. J. Newton, A. P. Barker, S. H. Bottrell & J. H. Tellam

Geochemical processes in two carbonate-free aquifer systems of North Cameroon 255
R. Njitchoua, L. Dever & B. Ngounou-Ngatcha

Geochemical and other porosity types in clay-rich rocks *F.J. Pearson*	259
Adsorption of herbicides by aquifer sediments *J.E. Rae, A. Parker & A.J. Peters*	263
Biogeochemical reactions induced by artificial recharge to carbonate aquifers *K.J. Rattray, A.L. Herczeg & P.J. Dillon*	267
The origin of sodium-bicarbonate groundwaters in a fractured aquifer experiencing magmatic carbon dioxide degassing, the Ballimore region, central New South Wales, Australia *S. Schofield & J. Jankowski*	271
Origin and mobility of arsenic in groundwater from the Pampean Plain, Argentina *P.L. Smedley, H.B. Nicolli, A.J. Barros & J.O. Tullio*	275

5 Sedimentary basins

Formation waters and diagenetic modifications: General trends exhibited by oil fields from the Norwegian shelf – A model for formation waters in oil prone subsiding basins *P. Aagaard & P.K. Egeberg*	281
REE distribution in fine-grained sediments from the Portuguese Atlantic shelf *M.F. Araújo & M.A. Gouveia*	285
Water-rock reactions in evaporite basins: Their role in the formation of potash deposits *C. Ayora, D.I. Cendón, C. Taberner, I. Fanlo, J. García-Veigas & J.J. Pueyo*	289
The origin of the Canadian Shield brines: Freezing or evaporation of seawater? *D.J. Bottomley, A. Katz, A. Starinsky, L.H. Chan, M. Douglas, I.D. Clark, K.G. Raven & D.C. Gregoire*	293
Minor and trace element chemistry and provenance in Alpine glacial meltwaters *G.H. Brown & R. Fuge*	297
Diagenesis of nonmarine sediments in an evolving tectonically-induced rain shadow *C.P. Chamberlain, D. Craw & M. Poage*	301
Neogenesis during thermal stimulation of bitumen, Alberta, Canada *J.S. Dudley & C.H. Moore*	305
Heterogeneity of formation waters within and between oil fields by halogen isotopes *H.G.M. Eggenkamp & M.L. Coleman*	309
Fluids, migration systems and diapirism, East Coast North Island, New Zealand *B.D. Field, R. Funnell, G. Lyon & C.I. Uruski*	313
Petroleum systems of the East Coast region, New Zealand *B.D. Field, R. Funnell, S. Killops, K. Rogers & C.I. Uruski*	317
Chlorite coatings in deeply buried sandstones – Examples from the Norwegian continental shelf *J. Jahren, E. Olsen & K. Bjørlykke*	321

Mechanisms of vertical variations of $\delta^{13}C(CH_4)$ value in sediments M.O.Jędrysek	325
Reservoir heterogeneity due to fault related, rock-water interaction M.Lee	329
Transformation of diatomite into porcelanite and opaline chert under the influence of an andesite intrusion in the Miocene Iwaya Formation, Japan E.Nakata, M.Chigira & M.Watanabe	333
Salt springs and structural setting of the Marchean Adriatic foredeep, Central Italy T.Nanni & P.Vivalda	337
Deuterium content and salinity of brines, Filitelnic gas-field, Transylvanian Basin, Romania D.C.Papp	341
Dolomitization of Ekofisk Oil Field reservoir chalk by injected seawater R.Petrovich & A.-A.Hamouda	345
Temporal fluctuations of syntectonic fluids in the Cascadia accretionary wedge J.C.Sample, C.D.Coathe & K.D.McKeegan	349
Surface characterization of biotite reacted with acid solution H.Seyama, A.Tanaka, J.Sato, M.Tsurumi & M.Soma	353
Brines in Siberian Platform: Geochemical and isotopic evidence for water-rock interaction S.L.Shvartsev	357
Reservoir connectivity determined from produced water chemistry, Standard Draw-Echo Springs gas field, Wyoming, USA L.K.Smith & R.C.Surdam	361
Geochemistry of waters from two adjoining basins in Hungary I.Varsányi, J.M.Matray & L.Ó.Kovács	365
Do stable isotopes and fluid inclusions allow to constrain the origin and timing of dolomitization in deeply buried carbonate reservoirs? Example of the Pinda Formation, Angola F.R.Walgenwitz, H.Eichenseer & P.Biondi	369
Evidence of Proterozoic primary $CaCO_3$ precipitation from the McArthur Group of northern Australia P.R.Winefield & P.McGoldrick	373

6 Weathering

The use of strontium isotopes in weathering studies D.C.Bain	379
Granitoid weathering in the laboratory: Chemical and Sr isotope perspectives on mineral dissolution rates T.D.Bullen, A.F.White, D.V.Vivit & M.S.Schulz	383

The field dissolution rate of feldspar in a Pennsylvania (USA) spodsol as measured by atomic force microscopy   387
M.A.Nugent, P.Maurice & S.L.Brantley

Degradation processes of trachytes in monument façades, Azores, Portugal   391
M.I.Prudêncio, J.C.Waerenborgh, M.A.Gouveia, M.J.Trindade, E.Alves, M.A.Sequeira Braga, C.A.Alves, M.O.Figueiredo & T.Silva

Laboratory studies of the chemical weathering of rock from the English Lake District   395
R.Stidson, J.Hamilton-Taylor & E.Tipping

Comparisons of short-term and long-term chemical weathering rates in granitoid regoliths   399
A.F.White & D.A.Stonestrom

7 Metamorphism

Devolatilization in a siliceous dolomite, petrologic and stable isotope systematics   405
R.Abart

Metamorphic fluid flow at marble-schist boundaries, Corsica, France   409
I.S.Buick & I.Cartwright

Shear zone-related hydrothermal alteration in Proterozoic rocks in Finland   413
A.Lindberg & M.Siitari-Kauppi

Hydrocarbon gases and fluid evolution in very low-grade metamorphic terranes: A case study from the Central Swiss Alps   417
M.Mazurek, H.N.Waber & A.Gautschi

Chemical zonation of contact metamorphic garnet: A record of fluid-rock interaction, Juneau gold belt, SE Alaska   421
H.H.Stowell & T.Menard

Low-grade oceanic metamorphism and tectonic thickening of the oceanic crust from the Eltanin Fracture Zone (Pacific ocean)   425
I.A.Tararin

Stable isotope studies of calcite from very-low grade metamorphic greywacke terranes of the North Island, New Zealand   429
S.Woldemichael

8 Magma – Water interaction

Shallow magmatic degassing: Processes and PTX constraints for paleo-fluids associated with the Ngatamariki diorite intrusion, New Zealand   435
B.W.Christenson, C.P.Wood & G.B.Arehart

The Gorely Volcano Crater Lake: New data on structure and water chemistry   439
Yu.O.Egorov, G.M.Gavrilenko, A.B.Osipenko & L.G.Osipenko

Gas-water interaction at Mammoth Mountain volcano, California, USA   443
W.C.Evans, M.L.Sorey, R.L.Michel, B.M.Kennedy & L.J.Hainsworth

Budget and sources of volatiles discharging at Kudryavy Volcano, Kurile Islands, Russia — 447
T.P.Fischer, S.N.Williams, Y.Sano & M.A.Korzhinski

Sulfur isotopes in rocks from the Katla Volcanic Centre – With implications for Iceland mantle heterogeneities? — 451
L.W.Hildebrand & P.Torssander

Changes in Cl concentrations and isotope values of hot spring waters at Kuju volcano, Japan, prior to the 1995 eruptive activity — 455
R.Itoi, T.Kai, M.Fukuda & I.Kita

Fumarole gas geochemistry in estimating subsurface temperatures at Hengill in Southwestern Iceland — 459
G.Ívarsson

Sulfur and oxygen isotopic variations of dissolved sulfate in Crater Lake, Mt. Ruapehu, New Zealand — 463
M.Kusakabe & B.Takano

Kinetics of postmagmatic clay mineral crystallization in lava flows — 467
A.Mas, P.Dudoignon, D.Proust & F.Schenato

Hydrothermal system evolution induced by magma degassing: The case of Vulcano — 471
P.M.Nuccio, A.Paonita & F.Sortino

Magma degassing and geochemical detection of its ascent — 475
P.M.Nuccio & M.Valenza

Variability of volcanic gases by trace element-determination in volcanic sulphur — 479
H.Puchelt, U.Kramar, B.Spettel & H.H.Schock

Characterization of a magmatic/meteoric transition zone at the Kakkonda geothermal system, northeast Japan — 483
M.Sasaki, K.Fujimoto, T.Sawaki, H.Tsukamoto, H.Muraoka, M.Sasada, T.Ohtani, M.Yagi, M.Kurosawa, N.Doi, O.Kato, K.Kasai, R.Komatsu & Y.Muramatsu

The Joule-Thomson expansion of CO_2 and H_2O in geothermal and volcanic processes — 487
D.M.Sirkis, G.C.Ulmer, D.E.Grandstaff & N.P.Flynn

Carbon dioxide and helium emissions from a reservoir of magmatic gas beneath Mammoth Mountain, California, USA — 491
M.L.Sorey, W.C.Evans, C.D.Farrar & B.M.Kennedy

Modeling the interaction of magmatic gases with water at active volcanoes — 495
R.B.Symonds & T.M.Gerlach

Magmatic sulfur content of the 1995-1996 Ruapehu eruptions, New Zealand — 499
T.Thordarson, C.P.Wood & B.F.Houghton

D/H composition of water from Neogene magmatites in the East Carpathians, Romania — 503
I.Ureche, D.C.Papp & V.Feurdean

Fluid-magmatic differentiation of the low-water granitic melts as a possible result of cavitation — 507
G.A.Valuy

9 Ore deposits

Effects of fluid flow and temperature variations on lead mineralization in the Southeast Missouri ore district — 513
M.S.Appold & G.Garven

Isotopic signature of hydrothermal sulfates from Carlin-type ore deposits — 517
G.B.Arehart

Zircon-fluid interaction in the Bayan Obo REE-Nb-Fe ore deposit, Inner Mongolia, China — 521
L.S.Campbell

Thermal and geochemical evolution of La Guitarra epithermal deposit, Temascaltepec, Mexico — 525
A.Camprubí, À.Canals, E.Cardellach, Z.D.Sharp & R.M.Prol-Ledesma

Regional-scale fluid flow and origins of Pb-Zn-Ag mineralisation at Broken Hill, Australia: Constraints from oxygen isotope geochemistry — 529
I.Cartwright

Fluidization, metallogenic mechanism and type of the Bankuan gold deposit, China — 533
Y.J.Chen, H.Y.Chen, H.H.Wang, X.Li, S.X.Hu, S.G.Fu & C.Y.Jin

Water-rich quartz and adularia veins of the Hishikari epithermal Au-Ag deposit, southern Kyushu, Japan — 537
K.Faure, Y.Matsuhisa, H.Metsugi & C.Mizota

Behaviour of Re-Os, Sm-Nd, and U-Pb systematics in hydrothermal ores — 541
R.Frei, Th.F.Nägler, R.Schönberg & J.D.Kramers

Chemistry of hydrothermal zircon: Investigating timing and nature of water-rock interaction — 545
P.W.O.Hoskin, P.D.Kinny & D.Wyborn

Fluid migration-reaction model of Zijinshan epithermal deposit as traced by variation of oxygen isotope compositions of altered wall rocks — 549
R.Hua & J.Hu

Mineralogical, sulfur isotope and fluid inclusion studies of gold mineralization, Bendigo, Victoria, Australia — 553
X.Li, P.Jackson, P.A.Kitto & Y.Jia

Morphology of pyrite and marcasite at the Golden Cross mine, New Zealand — 557
J.L.Mauk, P.W.O.Hoskin & R.R.Seal, II

Variation of carbon and oxygen isotopes in the alteration halo to the Lady Loretta deposit: Implications for exploration and ore genesis — 561
P.McGoldrick, P.Kitto & R.Large

Approaching equilibrium from the hot and cold sides in the FeS_2-FeS-Fe_3O_4-H_2S-CO_2-CH_4 system in light of fluid inclusion gas analysis — 565
D.I.Norman, B.A.Chomiak & J.N.Moore

A hybrid origin for porphyritic magmas sourcing mineralising fluids — 569
M.G.Rowland & J.J.Wilkinson

Alkaline leaching of uranium ore from the North Bohemian Cretaceous, Czech Republic 575
P. Štrof, P. Ira, J. Emmer, J. Novák, L. Gomboš & T. Pačes

Geochemical studies of the Kujieertai uranium deposit in Yili Basin, northwest China 579
Z. Sun, W. Shi, X. Li & J. Liu

Application of isotope studies of Australian groundwaters to mineral exploration: 583
The Abra Prospect, Western Australia
D. J. Whitford, A. S. Andrew, G. R. Carr & A. M. Giblin

Spectral characterisation of the hydrothermal alteration at Hishikari, Japan 587
K. Yang, J. F. Huntington & K. M. Scott

Sulfur-isotope geochemistry of Chinkuashih copper-gold deposits, Taiwan: Preliminary 591
results
H. W. Yeh, L. P. Tan & M. Kusakabe

Hydrogen and oxygen isotopes of water-rock interaction in Dalongshan uranium deposit, 595
Anhui province, China
J. P. Zhai, H. F. Ling & K. Hu

10 Geothermal fluids and gases

Boron isotopes in geothermal and ground waters in New Zealand 601
J. K. Aggarwal

Geochemistry of natural waters in Skagafjördur, N-Iceland: I. Chemistry 605
A. Andrésdóttir, S. Arnórsson & Á. E. Sveinbjörnsdóttir

Organic gas in Öxarfjördur, NE Iceland 609
H. Ármannsson, M. Ólafsson, G. Ó. Fridleifsson, W. G. Darling & T. Laier

Gas chemistry of the Krafla Geothermal Field, Iceland 613
S. Arnórsson, Th. Fridriksson & I. Gunnarsson

Precious metals in deep geothermal fluids at the Ohaaki geothermal field 617
K. L. Brown & J. G. Webster

New data on the chemical composition of waters in the Paratunka hydrothermal system, 621
Kamchatka
O. V. Chudaev, V. A. Chudaeva, P. Shand & W. M. Edmunds

Thermal fluids and scalings in the geothermal power plant of Kizildere, Turkey 625
L. B. Giese, A. Pekdeger & E. Dahms

Correlations between B/Cl ratios and other chemical and isotopic components of Taupo 629
Volcanic Zone, NZ geothermal fluids – Evidence for water-rock interaction as the major
source of boron and gas
J. R. Hulston

Chemical and isotopic features of gas manifestations at Phlegrean Fields and Ischia island, 633
Italy
S. Inguaggiato & G. Pecoraino

Fluid chemistry and water-rock interaction in a CO_2-rich geothermal area, Northern Portugal 637
J.M.Marques, L.Aires-Barros, R.C.Graça, M.J.Matias & M.J.Basto

Sulfur redox chemistry and the origin of thiosulfate in hydrothermal waters of Yellowstone National Park 641
D.K.Nordstrom, Y.Xu, M.A.A.Schoonen, K.M.Cunningham & J.W.Ball

Hydrogeochemical and isotope geochemical features of the thermal waters of Kızıldere, Salavatlı, and Germencik in the rift zone of the Büyük Menderes, western Anatolia, Turkey: Preliminary studies 645
N.Özgür, A.Pekdeger, M.Wolf, W.Stichler, K.P.Seiler & M.Satir

Precious and base metal deposition in an active hydrothermal system, La Primavera, Mexico 649
R.M.Prol-Ledesma, R.Lozano-Sta.Cruz, E.Alcalá-Montiel, V.A.Cruz-Casas, S.Hernández-Lombardini, F.Juárez-Sánchez, A.Canals & E.Cardellach

Geochemistry of natural waters in Skagafjördur, N-Iceland: II. Isotopes 653
Á.E.Sveinbjörnsdóttir, S.Arnórsson, J.Heinemeier & E.Boaretto

Gas geochemistry in the Yangbajing geothermal field, Tibet 657
Zhao Ping, Jin Jian, Zhang Haizheng, Duo Ji & Liang Tingli

11 Geothermal general

Pliocene to present-day water-rock interaction processes at 3.5 km depth within a 3.8 Ma old Larderello monzogranite 663
G.Cavarretta & M.Puxeddu

Geothermal system in Tapi rift basin, Northern Deccan Province, India 667
D.Chandrasekharam & S.R.Prasad

Thermal and chemical evolution of the Tiwi Geothermal System, Philippines 671
J.N.Moore, T.S.Powell, C.J.Bruton, D.I.Norman & M.T.Heizler

Low-temperature alteration of basalts from the Tangihua Complex, New Zealand 675
K.N.Nicholson & P.M.Black

A new type of hydrothermal alteration at the Kizildere geothermal field in the rift zone of the Büyük Menderes, western Anatolia, Turkey 679
N.Özgür, M.Vogel & A.Pekdeger

I-S series in geothermal fields: Comparison with diagenetic I-S series 683
P.Patrier, H.Traineau, P.Papanagiotou, E.Turgné & D.Beaufort

Geothermal resource development at Tattapani in Madhaya Pradesh, India 687
S.K.Sharma & J.Tikku

Illite, illite-smectite and smectite occurrences in the Broadlands-Ohaaki geothermal system and their implications for clay mineral geothermometry 691
S.F.Simmons & P.R.L.Browne

Gas behavior at some geothermal fields in Japan, revealed by Laser Raman Microprobe analysis of fluid inclusions 695
S.Taguchi, H.Takagi, H.Maeda, K.Sanada, M.Hayashi, M.Sasada, T.Sawaki, T.Uchida & T.Fujino

Chemical stability of the hydrothermal silicates at the Los Azufres geothermal field, Mexico 697
I.S.Torres-Alvarado

Evaluation of geothermal activity using thermally stimulated and radiation storage processes of quartz 701
N.Tsuchiya, T.Suzuki & K.Nakatsuka

Water-rock interaction at the boundary of Wairakei geothermal field 705
C.P.Wood

12 Oceanic

Alteration of basalts from the Ninetyeast Ridge, Indian Ocean (ODP data) 711
A.V.Artamonov, V.B.Kurnosov & B.P.Zolotarev

Modelling the halmyrolytic formation of palygorskite from serpentinite 715
C.M.Destrigneville, A.M.Karpoff & D.Charpentier

The underwater eruption in the Academia Nauk caldera (Kamchatka) and its consequences 719
S.M.Fazlullin, S.V.Ushakov, R.A.Shuvalov, A.G.Nikolaeva, E.G.Lupikina & M.Aoki

Halide systematics in sedimentary hydrothermal systems, Escanaba Trough – ODP Leg 169 723
J.M.Gieskes, C.Mahn, R.James & J.Ishibashi

Helium and carbon isotopes in submarine gases from the Aeolian arc, Southern Italy 727
S.Inguaggiato & F.Italiano

Fluid chemistry of sediment-rich hydrothermal systems on the continental margin and the mid-oceanic ridge 731
J.Ishibashi, U.Tsunogai, T.Gamo & H.Chiba

Fluid chemistry of seafloor magmatic hydrothermal system in the Manus Basin, PNG 735
J.Ishibashi, H.Takahashi, T.Gamo, K.Okamura, T.Yamanaka, H.Chiba, J.-L.Charlou & K.Shitashima

Alkali element and B geochemistry of sedimented hydrothermal systems 739
R.H.James & M.R.Palmer

Formation of clay minerals in the sedimentary sequence of middle valley, Juan de Fuca Ridge – ODP Leg 169 743
K.S.Lackschewitz, R.Botz, D.Garbe-Schönberg, P.Stoffers, K.Horz, A.Singer & D.Ackermand

Hydrothermal basalt alteration at the surface of the TAG active mound, MAR26°N 747
H.Masuda, M.Nakamura, K.Tanaka, H.Chiba, T.Gamo & K.Fujioka

Feeder zones of massive sulfide deposits: Constraints from Bent Hill, Juan de Fuca Ridge – ODP Leg 169 751
P.Nehlig & L.Marquez

Trace elements in hydrothermal fluids at the Manus Basin, Papua New Guinea *K.Shitashima, T.Gamo, K.Okamura & J.Ishibashi*	755
Mineralogy and chemical composition of clay minerals, TAG hydrothermal mound *A.A.Sturz, M.J.T.Itoh & S.E.Smith*	759

13 *Fluids and tectonics*

Rock-exchanged fluid oxygen isotope ratios in active collisional mountain belts, Pakistan and New Zealand *D.Craw, P.O.Koons, C.P.Chamberlain & M.Poage*	765
Underpressured paleofluids and future fluid flow in the host rocks of a planned radioactive waste repository *L.W.Diamond*	769
Soil gas emissions and tectonics in volcanic areas of Italy and Hawaii *S.Gurrieri, S.De Gregorio, I.S.Diliberto, S.Giammanco & M.Valenza*	773
Mineral-water interactions and stress: Pressure solution of halite aggregates *R.Hellmann, J.P.Gratier & T.Chen*	777
Fluids and faults: The chemistry, origin and interactions of fluids associated with the San Andreas fault system, California, USA *Y.K.Kharaka, J.J.Thordsen, W.C.Evans & B.M.Kennedy*	781
Fluid flow during folding and thrusting in carbonates: 2-D patterns of Sr and O isotope alteration *A.M.McCaig & J.G.Kirby*	785
The formation of albite veins in high-pressure terrains: Examples from Corsica and Zermatt-Saas, Switzerland *J.A.Miller, I.Cartwright & A.C.Barnicoat*	789
Lateral variations in mylonite thickness as influenced by fluid/rock interactions in a shear zone in Africa *U.Ring*	793
Subsurface horst features beneath the geothermal reservoirs in Taupo Volcanic Zone (TVZ) *S.Tamanyu*	797
Ar/Ar dating and uplift rate of hydrothermal minerals in the Southern Alps, New Zealand *D.A.H.Teagle, C.M.Hall, S.C.Cox & D.Craw*	801
Trap integrity and fluid migration: Coupled mechanical/fluid flow models *P.Upton, K.Baxter & G.W.O'Brien*	805
Monitoring of thermal and mineral waters in the frame of READINESS *H.Woith, C.Milkereit, J.Zschau, U.Maiwald & A.Pekdeger*	809
Large temperature fluctuations recorded in veins in the Victory gold deposit, Western Australia: A consequence of episodic fluid influx during progressive deformation? *Y.Xu & J.M.Palin*	813

14 Experimental

Pitzer specific ion interaction parameters for Ag-Cl from solubility measurements in the system AgCl-HCl-H_2O to 275°C 819
J.J.Bao & D.A.Polya

Dissolution of sanidine up to 300°C near equilibrium at approximately neutral pH 823
G.Berger, D.Beaufort & J.-C.Lacharpagne

Stable isotope exchange equilibria and kinetics in mineral-fluid systems 827
D.R.Cole, L.R.Riciputi, J.Horita & T.Chacko

Solubility and potentiometric studies of REE complexation with simple carboxylate (acetate, oxalate) ligands from 25° to 80°C 831
R.Ding, C.H.Gammons & S.A.Wood

Two-dimensional measurement of natural radioactivity of rocks by photostimulated luminescence 835
M.Hareyama, N.Tsuchiya & M.Takebe

Leaching experiments with acid cation-exchange resin as a new tool to estimate element availabilities in geological samples 839
W.Irber, P.Möller & W.Bach

Quantitative analysis of high density fluid inclusions 843
P.Knoll, M.Pressl, R.Abart & R.A.Kaindl

Modified set-up for column experiments to improve the comparability of water-rock interaction data: Column cap and hydraulic control system 847
D.Lazik

Solubility of Platinum in aqueous fluids buffered by manganese oxides 851
G.G.Likhoidov, L.P.Plyusnina & J.A.Scheka

Semiquantitative measurements of CO_2 gas in liquid-rich inclusions by laser Raman microspectroscopy 855
S.Maeda, S.Taguchi, H.Takagi, K.Sanada, M.Hayashi, M.Sasada, T.Sawaki, T.Fujino & T.Uchida

An autoradiographic method for studying irradiation-induced luminescence in feldspars 859
M.Siitari-Kauppi, S.Pinnioja & A.Lindberg

A Raman spectroscopic study of thio-arsenite and arsenite species in low-temperature aqueous solutions 863
S.A.Wood, C.D.Tait & D.R.Janecky

Study of electrical conductivity of H_2O at 0.21-4.18 GPa and 20-350°C 867
H.Zheng, H.Xie, Y.Xu, M.Song, J.Guo & Y.Zhang

15 Modelling

Enhancements to the geochemical model PHREEQC – ID transport and reaction kinetics 873
C.A.J.Appelo & D.L.Parkhurst

Physicochemical model of water-atmosphere-coal system 877
O.V.Avchenko

Modeling the metasomatism of marbles 881
V.N.Balashov & B.W.D.Yardley

Forward modelling of complex water evolution – Soda waters in Northland, New Zealand 885
F.May

Chemical and isotopic features and flow path modelling of thermal fluids of the Abano system, Italy 889
C.Panichi, L.Bellucci, S.Caliro, F.Gherardi, G.Volpi, G.Magro & M.Pennisi

Trace element speciation in hydrotherms due to the influence of CO_2 on genetic 'silicate rock-thermal fluid' processes 893
E.N.Pentcheva, L.Van't Dack & R.Gijbels

PHOX: Automated calculation of mineral stability and aqueous species predominance fields in Eh (or log (fO_2) or pε)-pH space 897
D.A.Polya

The reaction between ferrous iron and Mn-oxides in a transport system: Column experiment and solute transport modeling 901
D.Postma & C.A.J.Appelo

Hydrogeochemical processes at the fracture/matrix boundary of fractured sandstones 905
M.Sauter & R.Liedl

Calculations of fluid-ternary solid solution equilibria: An application of the Wilson equation to fluid-$(Fe,Mn,Mg)TiO_3$ equilibria 909
Y.Shibue

Competitive pool growth model and numerical simulation for morphological diversity of hot-spring mineral deposits 913
H.Shigeno

16 Mineral surfaces

Colloidal interactions of precipitated Zn carbonates with clay minerals 919
H.B.Bradl

'Natural' schwertmannite formed in a lake from waters draining pyritic deposits 923
C.W.Childs, K.Inoue, C.Mizota, M.Soma & B.K.G.Theng

Attachment features between an aerobic *Pseudomonas sp.* bacteria and hematite observed with atomic-force microscopy 927
J.Forsythe, P.Maurice & L.Hersman

Influence of temperature on the sorption isotherm of potassium on a montmorillonite 931
E.C.Gaucher, L.Claude, H.Pitsch & J.Ly

Mineral surfaces and the sorption of bacteria in groundwater 935
J.S.Herman, A.L.Mills & E.P.Knapp

Lead adsorption onto aquifer gravel using batch experiments and XPS 939
C.Hinton & M.E.Close

Water-rock interaction and sorption of redox-sensitive elements: Experiments on olivine 943
and uranium
J.Suksi, M.Upero, A.Adriaens & K.-H.Hellmuth

Arsenic removal from geothermal bore waters: The effect of mono-silicic acid 947
P.J.Swedlund & J.G.Webster

Trace metal adsorption onto acid mine drainage iron oxide 951
J.G.Webster, P.J.Swedlund & K.S.Webster

Kinetics of calcite precipitation: Molar measurements and molecular descriptions 955
P.Zuddas, G.De Giudici & A.Mucci

17 Waste storage and disposal

Geochemical modelling of groundwater/bentonite interaction for waste disposal systems 961
D.Arcos, J.Bruno & L.Duro

Uranium series isotopic data of fracture infill materials from the potential underground 965
laboratory site in the Vienne granitoids, France
J.Casanova & J.-F.Aranyossy

Influence of mine watering on groundwater quality at Monteponi, Sardinia, Italy 969
R.Cidu & L.Fanfani

Ferricrete provides record of natural acid drainage, New World District, Montana 973
G.Furniss & N.W.Hinman

Landfill leachate – Chalk rock interactions: The fate of nitrogen and sulphur species 977
N.C.Ingrey & J.D.Mather

Hydrogeochemical characteristics of deep groundwater in Korea for geological disposal 981
of radioactive waste
J.U.Lee, H.T.Chon & Y.W.John

Alteration of cold crucible melter Ti/Zr-based ceramics 985
G.Leturcq, G.Berger, T.Advocat & A.Bonnetier

Remediation of a sandstone aquifer following chemical mining of uranium in the Stráž 989
deposit, Czech Republic
J.Novák, R.Smetana & J.Šlosar

Molecular characterization of manganese oxides and trace metals in stream sediments 993
from a mining-contaminated site
P.A.O'Day, K.E.Geiger & C.C.Fuller

Determination of background chemistry of water at mining and milling sites, Salt Lake 997
Valley, Utah, USA
D.D.Runnells, D.P.Dupon, R.L.Jones & D.J.Cline

Attenuation of leachate contaminants in an engineered wetland 1001
M.Sartaj & L.Fernandes

Capillary barriers for the surface sealings of landfills 1005
N. von der Hude & F. Huppert

Flocculation of metal-rich colloids in a stream affected by mine drainage 1009
P. Zuddas, F. Podda & A. Lay

Uranium mobility in surface waters draining mineralized areas in the western US 1013
R. B. Wanty, W. R. Miller, R. A. Zielinski, G. S. Plumlee, D. J. Bove, F. E. Lichte, A. L. Meier & K. S. Smith

Author index 1017

Preface

The 9th International Symposium on Water-Rock Interaction (WRI-9) sponsored by the International Association of Geochemistry and Cosmochemistry (IAGC) and the New Zealand Institute of Geological and Nuclear Sciences was held in Taupo, New Zealand, March 30-April 3, 1998. Nearly 300 manuscripts were submitted by scientists from 27 countries for presentation in both oral and poster symposia. Following technical review and revision, 241 of these papers were accepted and are included in this volume.

The range of topics falling under the umbrella of Water-Rock Interaction continues to broaden, as indicated by the session topics:
A. Experimental and modelling;
B. Groundwater;
E. Geothermal;
F. Sedimentary basins;
G. Oceanic drilling programme/Groundwater;
H. Waste storage and disposal/Processes involving organic matter;
M. Surficial systems/Mineral surfaces/Weathering;
N. Ore deposits;
O. Magma-water interaction;
P. Fluids and tectonics/Metamorphism.

A special emphasis on poster presentations was made at the symposium and six poster sessions: C, D, I, J, K and L were arranged to cover all the above topics. Traditional geochemical areas are well-represented and range from low- and ambient-temperature environments through groundwater and hydrothermal waters to the interaction of waters with magmas at very high temperatures. These areas were explored in sessions on Groundwater, Geothermal, Ore deposits, a special session on the Ocean drilling programme, and Magma-water interaction. To mark the founding work in New Zealand on water-rock experiments at high temperature by Ellis and Mahon the conference started with a session on Experimental and modelling. The Groundwater session was the largest at the meeting and the Sedimentary basins session was substantial with new data on basinal brines, diagenesis, fluid evolution and hydrocarbon formation and migration. Fluids and tectonics was a new session and together with metamorphism covered both chemical and physical processes. Combined sessions on Waste storage and disposal/Processes involving organic material and Surficial systems/Mineral surfaces/Weathering completed the symposium.

There is expanding interest in integrated studies of water-rock systems, and several sessions were devoted to exploring some of these possibilities. New Zealand is one of the best places in the world to study water-rock interaction. To provide an insight into New Zealand work we scheduled New Zealand highlight speakers to commence many of the sessions. They spanned a continuum of water-rock interaction topics and stretched the traditional boundaries of this field.

All of the submitted manuscripts were reviewed by at least two reviewers and returned to the authors with suggested modifications. In a few cases, manuscripts were modified slightly by us as editors to meet formatting requirements; hopefully this process did not change the import of the manuscripts. We are greatly indebted to our numerous colleagues who provided, often at very short notice, careful and extensive reviews of the manuscripts. Many of these colleagues reviewed a significant number of manuscripts. It would not have been possible to do a thorough job without them. In addition to the editors, reviewers of manuscripts were:

J. Aggarwal,	J. Gamble,	J. Palin,	S. Tamanyu,
M. Brandiss,	D. Hayba,	M. Rattenbury,	R. Wanty,
K. Brown,	A. Herczeg,	M. Rosen,	J. Webster,
P. R. L. Browne,	C. Kendall,	J. Scott,	C. P. Wood,
C. Chagué-Goff,	Y. Kharaka,	R. Seal,	S. Wood,
B. Christenson,	R. Kettler,	D. Sheppard,	A. White,
S. Dogramacı,	J. Laranthal,	S. Simmons,	P. White,
C. de Ronde,	C. Lind,	M. Sorey,	H. Yeh,
M. Edmunds,	G. Lyon,	M. Stewart,	R. Zierenberg,
W. C. Evans,	E. Mazor,	N. Sturchio,	R. Zielinski.
B. Field,	E. Mroczek,	B. Symonds,	

Initially, we had hoped to minimise the paperwork and postage and save both time and cost by accepting electronic submissions. While this worked well for a very few papers, the majority arrived in somewhat less than optimal condition. In the end, it was necessary to request paper copies for the majority of the papers submitted. After this educational experience with electronic manuscripts we have learnt a great deal about how to put together such a publication. We are greatly indebted to three people without whom this volume would never have been completed. Christine Toms, Diane Tilyard, and Jackie Duncan went far beyond the normal call of duty in assembling, coordinating, formatting, and organising not only the papers in this volume, but also the myriad details of registration and other logistics. They all heaved a sigh of relief when the final package went off to the printer!

In the final analysis, we hope that this volume represents the latest word in research into the everbroadening field of Water-Rock Interaction. The papers herein should be of interest to a wide range of researchers and applied scientists. We hope that in reading through these proceedings the reader will find both the most recent applications to their own fields, and also some new and different ideas and approaches from other disciplines that will stimulate their thinking in a new and different way.

Greg B. Arehart
John R. Hulston
Editors, WRI-9 Proceedings

Brian W. Robinson
Secretary-General, WRI-9

Organisation

INTERNATIONAL ASSOCIATION OF GEOCHEMISTRY AND COSMOCHEMISTRY (IAGC)

Executive Committee (IAGC)
President: Gunter Faure, USA
Vice-President: Eric Galimov, Russia
Secretary: Mel Gascoyne, Canada
Treasurer: David Long, USA
Past-President: Hitoshi Sakai, Japan

Council Members:
John Gurney, South Africa
Russell Harmon, USA
Jochen Hoefs, Germany
Petr Jakes, Czechoslovakia
Gero Kurat, Austria
Marc Javoy, France
Malcolm McCulloch, Australia
N. Sobolev, Russia
K. Subbarao, India
Yishan Zeng, China
Ron Fuge (Ex-Officio), UK

WORKING GROUP ON WATER-ROCK INTERACTION

Executive Committee
Chairman: W.M.(Mike) Edmunds, UK

Members:
Tomas Paces, Czechoslovakia
Yves Tardy, France
Brian Hitchon, Canada
Hitoshi Sakai, Japan
Halldor Armannsson, Iceland
Yousif Kharaka, USA
Oleg Chudaev, Russia
Brian Robinson, New Zealand

WRI-9 ORGANISING COMMITTEE

Honorary Chairmen
Dr A.J.Ellis
W.A.J.Mahon, DesignPower NZ Limited

Secretary General
Dr B.W.Robinson, IGNS, Wairakei Research Centre

Scientific programme
Prof. P.M.Black, University of Auckland
Ass. Prof. P.R.L.Browne, University of Auckland
Dr S.F.Simmons, University of Auckland
Dr B.W.Christenson, IGNS, Wairakei Research Centre
Dr J.Webster, ESR, Auckland
Dr M.R.Rosen, IGNS, Wairakei Research Centre

Field trips
Dr S.F.Simmons, University of Auckland
Dr J.Webster, ESR, Auckland
Dr M.R.Rosen, IGNS, Wairakei Research Centre
Dr D.Craw, University of Otago
Dr B.F.Houghton, IGNS, Wairakei Research Centre
B.J.Scott, IGNS, Wairakei Research Centre

Sponsorship
Dr D.S.Sheppard, IGNS, Nuclear Sciences

Treasurer
Jackie Duncan, IGNS, Wairakei Research Centre

Conference administration
The Conference Company, Auckland

The Organising Committee are grateful to the following major sponsors who are supporting the WRI-9 Symposium:

Institute of Geological and Nuclear Sciences
Air New Zealand

New Zealand highlight speakers

From basins to mountains and back again: N.Z. basin evolution since 10 Ma

R.G. Allis, R. Funnell & X. Zhan
Institute of Geological and Nuclear Sciences, Lower Hutt, New Zealand

ABSTRACT: Increasing convergence on the New Zealand plate boundary since 10 Ma has caused rapid evolution of its basins. Exhumation of the landward portions of basins is providing sediments to prograding shelves and causing accelerated burial in basins offshore. The deformation is perturbing both thermal and pore pressure regimes, with implications for petroleum generation, migration and entrapment. Away from the active plate boundary zone, the heat flow is between 50 - 60 mW/m^2, and pore pressure is close to hydrostatic to ca. 4 km depth. Within the plate boundary zone, heat flows range from 40 to more than 100 mW/m^2, and pore pressure gradients range between hydrostatic and lithostatic to depths as shallow as 1 km.

1. INTRODUCTION

Most of New Zealand's prospective petroleum-bearing basins began to form during Late Cretaceous rifting and breakup of the Gondwana continent. By ca. 40 Ma, most of the NZ continent had subsided below sea level. However, during the Late Neogene the region was affected by initially transform motion on the Australian-Pacific plate boundary, and during the last 10 Ma by an increasing component of compression (King and Thrasher, 1996; Sutherland, 1995). This has resulted in rapid uplift of the Southern Alps adjacent to the Alpine Fault in the South Island, southward prolongation of the Hikurangi subduction zone beneath the North Island, and rifting in the Taupo Volcanic Zone (TVZ; Fig. 1). Active transpressional tectonics now dominate the frontal arc basins of the East Coast and offshore Fiordland, the Taranaki Basin has been affected by its foreland setting, and the Great South and Northland Basins remain largely on passive margins.

The purpose of this paper is to highlight the rapidity of basin evolution in N.Z. because of its location on an active plate boundary. Basins are being destroyed as they are uplifted and eroded, whereas new basins are forming and deepening due to sediments accumulating on prograding continental shelves. This paper reviews some of the thermal and fluid flow signatures of this deformation. The thermal regime has important implications for the

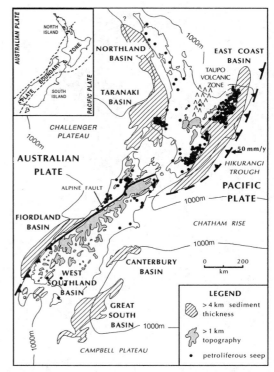

Fig. 1. Relationship between major sedimentary basins (>4 km depth) and the present plate boundary zone. Dots are petroliferous seeps and stains.

depth and timing of oil generation from potential source rocks, whereas the fluid flow and pressure regime influences oil migration and its entrapment in reservoirs.

2. THERMAL REGIME

The present day heat flow of N.Z. is shown in Fig.2 (based partly on unpublished data from GNS). Most of the thermal data are from completion reports of petroleum exploration wells and are from depths of 1 to 5 km. Conductive heat flows have been derived using the methods of Funnell et al., (1996). In a few wells at relatively shallow depth, the data have been corrected for the thermal effects of surrounding topography. Apart from the TVZ, hot springs make little contribution to the total heat flow. The location of boiling springs is also shown on Fig. 2. In the TVZ, the heat flow is predominantly convective in the form of plumes of hot water and steam in geothermal fields.

The total heat output here is around 4000 MW, or equivalent to an average heat flow of around 700 mW/m^2 (Bibby et al., 1995).

The average heat flow furthest from the active plate boundary zone is between 50 and 60 mW/m^2, being best defined in the western Taranaki Basin, and in the southern Great South Basin. This value is consistent with mature continental crust which has cooled since the last thermal disturbance, a rifting event during the Late Cretaceous. Heat flow systematically outside the 50 - 60 mW/m^2 range has probably been affected by Late Neogene tectonism. High heat flow (>80 mW/m^2) extends northwards from the TVZ into the western Lau-Havre Trough, and also through the Coromandel Peninsula into eastern Northland. There is a close correlation between high heat flow and the occurrence of Pliocene-Recent rhyolitic volcanism. In the southeast South Island, high heat flow coincides with the Late Miocene Dunedin basaltic volcanic complex (Funnell and Allis, 1996). Elevated heat flow also occurs in part of the northwest South Island, along the Southern Alps, and in central Fiordland (Shi et al., 1996; Townend, in press). These anomalies are due to the effects of uplift and erosion during the last 10 Ma causing elevation of the isotherms.

Heat flow is depressed beneath the southern North Island, apparently because of the cooling effects of the underlying subduction zone and associated crustal accretion. The effects of the cooling can be detected as far west as the southern Taranaki basin, where the oil window has been depressed by about 1 km compared to northern Taranaki (Armstrong et al., 1996).

2.1 Southern Alps

The most extreme example of the thermal effects of rapid exhumation occurs in the central Southern Alps (Fig. 3). The cross-section shows the thermal regime modelled by Shi et al., (1996) based on the known uplift and erosion rates across the Southern Alps, and the convergence history inferred from plate tectonic considerations. Here, Pacific Plate continental crust is deflected upwards against Australian plate continental crust over a distance of around 50 km. Prior to convergence, the crust was probably covered by several km of sediments, as is presently the case in northern Canterbury and the Chatham Rise. Mid-lower crustal material is now being continually exhumed in the zone of maximum uplift, where uplift and erosion rates probably

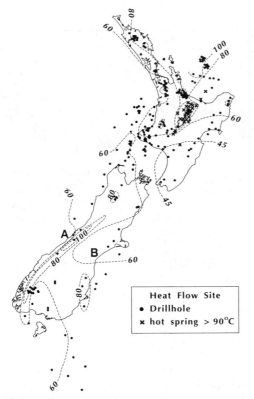

Fig. 2. Heat flow of New Zealand. Dots are data points; squares are geothermal fields. A-B is line of cross section in Fig. 3. Contours in mW/m^2.

exceed 10 km/Ma. Temperatures of at least 200°C are present at less than 5 km depth. Allis and Shi (1995) have pointed out that there may be a discrepancy between thermal regime inferred from young zircon fission track ages (ca 250°C at 10 km depth, and 1 Ma ages; Kamp and Tippett, 1993), and the evidence of fluid inclusions suggesting temperatures of 250°C at around 4 km depth (Craw, 1997).

The seismicity beneath the Southern Alps suggests that Pacific plate upper mantle is descending beneath the Southern Alps as an incipient subduction zone. Although elevated temperatures occur in the upper crust, temperatures are likely to be depressed in the root zone which is building up beneath the Southern Alps (Fig.3).

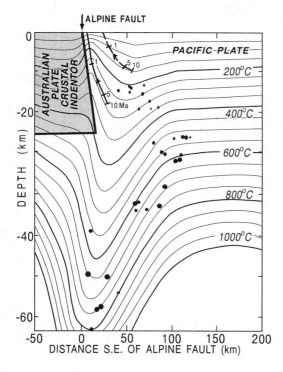

Fig. 3. Inferred present-day isotherms in the crust and upper mantle beneath the central Southern Alps assuming the Australian plate acts as rigid indentor (Allis and Shi, 1995). Two uplift paths for rock presently exposed at the surface are shown as curving lines with arrows; ticks mark the exhumation history in Ma. The solid uplift path coincides with the zone of maximum uplift rate (10 mm/y). Dots are earthquake focii, with size proportional to magnitude (Reyners, 1987).

2.2 North Island

The amount of shortening in the North Island landmass has been much less than in the Southern Alps (ca. 60 km of crustal shortening during the last 10 Ma), with the resulting basin inversion and erosion less than 3 km (Fig. 4). The map of cumulative erosion in North Island basins was derived from a study of uplifted mudstone compaction trends, and uplifted coal rank trends (Armstrong et al., in prep.).

There are at least three distinct episodes of erosion affecting the contours in Fig. 4. In the north (Waikato Basin), a N-S axis of uplift has resulted in up to 2 km of erosion parallel to the coastline, along the axis of a former depocentre indicated by coal ranks of Eocene coal measures. The age of this uplift is uncertain but post-dates 10 Ma. Some of the exhumation may have been synchronous with Pliocene to Recent rifting both to the west in the offshore northern Taranaki basin, and to the east in the TVZ.

In the onshore Taranaki and Wanganui Basins, maximum basin inversion coincides with high topography adjacent to the southern end of the TVZ. The exhumation is also uplifting Pleistocene marine sediments around the southern Taranaki-Wanganui coastline, indicating that it is a recent phenomena. The semi-circular shape of the erosion contours appears to be consistent with thermally induced thinning lithosphere adjacent to the southern TVZ.

The southern Taranaki Basin near the northern end of the South Island has had its own distinctive exhumation history. Between 10 and 5 Ma, the area was uplifted and eroded, probably as the northernmost expression of increasing compression on the Alpine Fault, and accelerated exhumation of the Southern Alps. In the last 5 Ma, the area has subsided below sea level, presumably because compression has eased here with the southward prolongation of the Hikurangi subduction zone and the associated development of the Marlborough transform fault system in the northeast South Island.

Fig.5 shows four examples of the wide variation in basin development across the North Island since 10 Ma, and the effects on the thermal regime. In the offshore northern Taranaki Basin (Ariki-1), high sedimentation rates in the Miocene coincided with crustal extension associated with volcanism. Subsidence continued through the Quaternary as a result of progradation of the continental shelf and associated crustal loading. In the onshore northern Taranaki Basin (Kaimiro-1) sediments infilled a

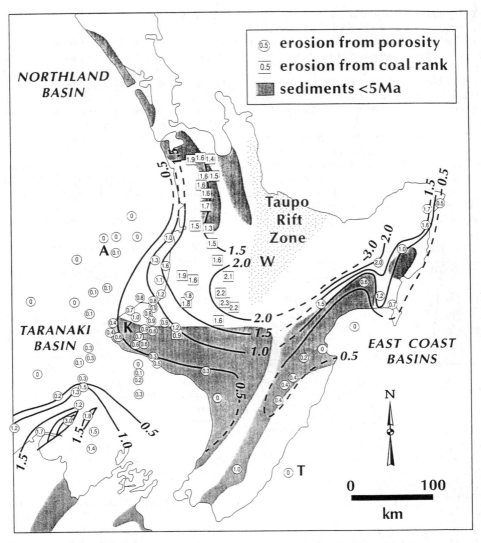

Fig. 4. Late Neogene exhumation (km) of North Island basins based on uplifted porosity and coal rank trends. There are insufficient data to identify faulted offsets in the contours of the East Coast. A, K, W, T, are geohistory sites modelled in Fig. 5.

prominent Miocene foredeep, modified by late phase extension. Basement subsidence and sedimentation continued until Quaternary uplift resulted in erosion over most of the Taranaki Peninsula. Recent volcanism (Mt Taranaki) has added to the topography, caused a 20°C increase in sediment temperature, and initiated oil generation in Eocene source rocks (Allis et al., 1996).

In the TVZ, Early Quaternary andesites lie directly on Mesozoic basement, indicating that the region was previously a topographic high. The region was probably part of the axial ranges behind Miocene forearc basins to the northeast. Rifting and foundering of the region has culminated in extensive rhyolite volcanism, especially since 0.3 Ma, indicative of partial melting of the lower crust. Widespread lakebeds in the last 0.1 Ma (Huka Fm.) could indicate that the region is now at its lowest elevation since 10 Ma.

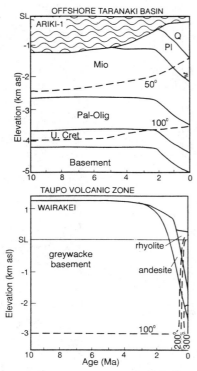
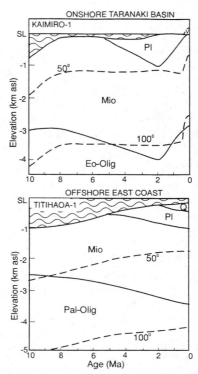

Fig. 5. Examples of four different geohistories within 300 km of each other in the North Island (Fig. 4 for locations).

The offshore East Coast basins are characterised by progressive thickening through both increased sedimentation with time and overthrusting (Fig.5). This is due to the increasing exhumation of the NZ landmass since 10 Ma, and eastern basins moving landward on a growing accretionary wedge.

3. PRESSURE REGIME

The pressure and fluid flow regimes in NZ's basins are strongly dependent on their recent tectonic history (Allis et al., 1997; Fig. 6). Most wells in the Great South and Canterbury Basins show hydrostatic pressure profiles to between 4 and 5 km depth. This is due to their passive margin setting since ca. 50 Ma. In the Taranaki Basin, most wells exhibit overpressuring by ca. 3 km depth, and in East Coast Basins, over-pressuring may begin at less than 1 km depth. The deepest wells in the TVZ extend to less than 3 km depth, so the deep pressure regime is unknown. Wells in geothermal fields typically show pressures between hot and cold hydrostatic, due to their location in plumes of hot water rising from groundwater circulating through hot rock at depth.

Allis et al. (1997) investigated the causes of over-pressuring by modelling the geohistories using a 1-D finite element code developed by Person et al. (1993). Some of the results from this modelling are shown in Fig. 7. Important assumptions are the relationships between porosity and depth, and permeability and porosity, for the characteristic lithologies. With normal compaction, porosity decreases exponentially from $50 \pm 5\%$ at the surface to less than 10% at 5 km depth; permeability decreases according to a power relationship by at least 6 orders of magnitude from its near-surface value. The vertical pressure regime is particularly sensitive to the mudstone permeability relationship, because it is the lowest permeability layers which most impede the compaction-driven outflow. This can be seen from the two models for Ariki-1 (Fig. 7). Mudstone permeabilities are about 10% less than that for muddy siltstone for any given porosity. The predicted present-day pressure profile at 4 - 5 km depth in Ariki-1 is 25 MPa (250 bar) lower with the muddy siltstone permeability relationship. In this well, and much of the northern offshore Taranaki Basin, the development of overpressures below 3 km depth is mostly a consequence of rapid sedimentation since 2 Ma.

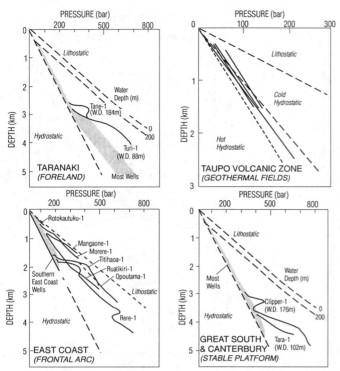

Fig. 6. Pore pressure trends in four areas with distinctive recent tectonic histories.

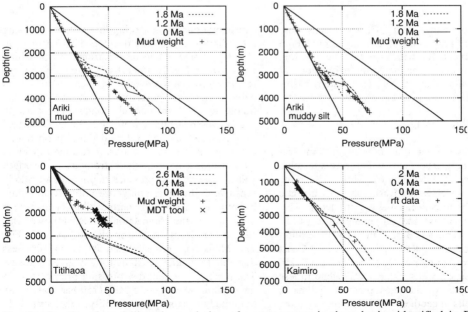

Fig. 7. Results of modelling the evolution of pore pressure in three basins identified in Fig. 5. Overpressuring is sensitive to rapid sedimentation and the permeability of mudstones (Ariki-1), the timing of sedimentation ceasing and exhumation beginning (Kaimiro-1), and excess fluid upflow from depth in an accretionary prism (Titihaoa-1); rft is repeat formation test pressure data.

In the onshore Taranaki basin, Quaternary exhumation has allowed partial recovery of the earlier compaction-driven overpressures (Fig. 7, Kaimiro-1). The modelling shows that substantial pressure recovery occurs on a time-scale of 1 m.y., and therefore fossil overpressuring caused by rapid uplift is unlikely to survive on a geological time-scale. This model highlights the importance of having accurate geohistories, especially over the last few million years, for quantitative interpretation of the pressure regime. Webster and Adams (1996) have suggested that overpressures in the onshore Taranaki basin may also be caused by gas generation at depth.

The third geohistory modelled in Fig. 7 is from the offshore East Coast Basin. Here, despite using a relatively low permeability relationship for mudstone, it is not possible to match the observed overpressures at 2 km depth. The simplest explanation is that high fluid pressures at shallow depth are due to excessive fluid upflow from greater depth. This would be consistent with the basin's setting on an active accretionary prism. The East Coast is notable for the number of petroliferous, saline, and gas seeps (Fig. 1), some of which are attributed to fluid upflow from the underlying subduction zone (Giggenbach et al., 1993). Extensive gas hydrates offshore from the East Coast (Townend, 1997) may be the submarine equivalent of the gas seepages onshore.

REFERENCES

Allis, R.G., P.A.Armstrong & R.H.Funnell 1996. Implications of a high heat flow anomaly around New Plymouth, North Island, New Zealand. *NZJGG, 38,* 121-130.

Allis, R.G., R.H.Funnell & X.Zhan 1997. Fluid pressure trends in basins astride New Zealand's plate boundary zone. In Hendry et al., *Geofluids II,* Extended Abstracts, 214-217, Queens Univ. Belfast.

Allis, R.G. & Y.Shi, 1995. New insights to temperature and pressure beneath the central Southern Alps, New Zealand. *NZJGG, 38,* 585-592.

Armstrong, P.A., D.S.Chapman, R.H.Funnell, R.G.Allis & P.J.J.Kamp 1996. Thermal modelling and hydrocarbon generation in an active-margin basin: the Taranaki Basin, New Zealand. *AAPG Bull., 80,* 1216-1241.

Bibby, H.M., G.Caldwell, F.J.Davey & T.H.Webb 1995. Geophysical evidence of the structure of the Taupo Volcanic Zone and its hydrothermal circulation. *JVGR, 68,* 29-58.

Craw, D. 1997. Fluid inclusion evidence for geothermal structure beneath the Southern Alps, New Zealand. *NZJGG, 40,* 43-52.

Funnell, R.H. & R.G.Allis 1996. Hydrocarbon maturation potential of offshore Canterbury and Great South Basins. *Proc. N.Z. Petroleum Conference, 1,* 22-31, Ministry of Commerce, N.Z.

Funnell, R.H., D.S.Chapman, R.G. Allis & P.A. Armstrong 1996. Thermal state of the Taranaki Basin, New Zealand. Journal of Geophysical Research, 101, 25197-25515.

Giggenbach, W.F., Y.Sano & H.Wakita, 1993. Isotopic composition of heium, CO_2, and CH_4 contents in gases produced along the New Zealand part of a convergent plate boundary. *G.C. Acta, 57,* 3427-3455.

Kamp, P.J.J. & J.M.Tippett, 1993. Dynamics of Pacific plate crust in the South Island (N.Z.) zone of oblique continent-continent convergence. *J.G.R., 98,* 16105-16118.

King, P. & G.Thrasher 1996. Cretaceous-Cenozoic geology and petroleum systems of the Taranaki Basin, New Zealand. *IGNS Mono. 13,* pp. 244, Lower Hutt, N.Z.

Person, M., D.Tonpin, J.Weick, & P.Eadington, 1993. Paleo-2D: a finite element programme for simulating two-dimensional groundwater flow, heat transfer, and petroleum generation within compacting sedimentary basins. Draft, pers. comm. 8/24/93.

Reyners, M. 1987. Subcrustal earthquakes in the central South Island, New Zealand, and the root of the Southern Alps. *Geology, 15,* 1168-1171.

Shi, Y., R.G.Allis, & F.Davey 1996. Thermal modelling of the Southern Alps, New Zealand. *PAGEOPH, 146,* 469-501.

Sutherland, R. 1995. The Australian-Pacific boundary and Cenozoic plate motions in the SW Pacific: some constraints from Geosat data. *Tectonics, 14,* 819-831.

Townend, J. 1997. Estimates of conductive heat flow through bottom-simulating reflectors on the Hikurangi and southwest Fiordland continental margins. *Marine Geology, 141,* 209-220.

Townend, J. *in press.* Heat flow through the West Coast, New Zealand, NZJGG.

Webster M. & S.Adams, Geopressures, hydrocarbon generation and migration, onshore Taranaki. *Petroleum Exploration News in New Zealand, 47,* 13-22.

Hydrothermal alteration in New Zealand geothermal systems

P.R.L. Browne
The Geothermal Institute and Geology Department, The University of Auckland, New Zealand

ABSTRACT: Since 1950 petrographic studies of the hydrothermal alteration of cores and cuttings recovered from about 200 exploration wells drilled into 9 active geothermal fields in the Taupo Volcanic Zone and at the Ngawha field in Northland have been used to assess the hydrologies of their reservoirs. Concurrently with providing engineers with this practical information, geologists and geochemists have also treated active geothermal fields as natural, open laboratories where fluid-rock interactions occur. This has allowed us to recognise some of the parameters that control these interactions. For example, thermally sensitive minerals include zeolites, other calc-silicates and some clays, whereas the identities of the replacement feldspars reflects the amount of fluids that have moved through the rocks (a relationship first recognised by Alfred Steiner of the NZ Geological Survey). However, the greatest advances have been made by combining the results of petrographic observations, fluid chemistry and thermodynamic considerations.

1 INTRODUCTION

The potential for using geothermal energy to generate power in the Taupo Volcanic Zone (TVZ) was recognised long before the Second World War (e.g. Grange, 1937; Bruce and Shorland, 1933). However, it was only after the war, when there were severe power shortages, that serious exploration began. With much foresight the New Zealand wartime Government had instructed an engineer, Lt. Francis Tuck, attached to the New Zealand Division in Italy to report on the use of geothermal power at Larderello in Tuscany. The power station there had been operating long before the start of World War II but had been destroyed by the retreating German army in 1944. Tuck's report and the boldness of some politicians of the day meant that drilling at the Wairakei field began with the famous D line in 1950 (Grindley, 1965). No one knew what to expect and the predicted cross-sections of that date depicted magma at shallow depths and only dry steam discharging from the wells, as occurs at Larderello. In the event, the drillholes discharged a mixture of steam and hot water, derived from a single liquid-phase, so that Wairakei became the first liquid-dominated system to be prospected by drilling. In November 1958 it began to generate power (Grindley, 1965).

The novelty of the system meant that much new technology, including methods of making measurements, had to be developed; this was largely done by the pioneer scientists and engineers of the then Department of Scientific and Industrial Research (DSIR) and the Public Works Department. It was quickly realised that application of the methods used in oil field exploration just did not work in hot geothermal fields hosted by volcanic rocks. As a result, cores were needed, and these had to be examined petrographically to learn about the nature of the Wairakei reservoir. These cores were used to determine not only the lithology, stratigraphy and hence structure of the field (Grindley, 1965; Healy, 1956), but also to identify the solid products of fluid/rock interactions expressed by their hydrothermal alteration.

2 DEVELOPMENT OF GEOTHERMAL PETROLOGY

During the 1950's and early 1960's geothermal petrology developed very quickly. There were a number of reasons for this and they include some that were administrative, but others that resulted from the talents of the scientists and engineers that became involved and, thirdly, because of the nature of the reservoir rocks and the thermal fluids themselves. These included:

1. Geothermal science was almost the exclusive preserve of the DSIR which had a small administrative staff at its Head Office that co-ordinated the geothermal work of scientists in its constituent divisions, notably the New Zealand Geological Survey, Chemistry, Applied Maths, Physics and Engineering

Divisions and the Institute of Nuclear Sciences. There were regular co-ordination meetings that fostered the free flow of data and ideas between all the DSIR scientists themselves and engineers with the Ministry of Works (MOW, the successor of the Public Works Department), and the Department of Electricity.

2. The study of geothermal petrology has greatly benefited from the involvement of a distinguished group of fluid geochemists who were all well aware of the importance of the rocks through which the thermal fluids they studied passed and the mineralogical significance of fluid/rock interactions. Notable among the pioneers were Stuart Wilson, Tony Mahon, Dick Glover and Jim Ellis; the frequent worldwide citations of their work today testify to its quality and importance. These scientists clearly relished their pioneering roles and recognised it even at the time they did it. From the mid 1970's the geothermal geochemistry mantle also covered younger scientists who made important studies of the stable isotopic composition of hydrothermal minerals (e.g. Blattner, 1975; Robinson, 1978). In this period Werner Giggenbach began to make his outstanding contributions to our collective understanding of the processes of fluid/rock interaction (e.g. Giggenbach, 1980, 1984) by placing them on a more secure footing thermodynamically. Jeff Hedenquist and Dick Henley also made major impacts on the geothermal scene, among their other contributions, by pointing out the close similarities that alteration and mineralisation in active geothermal fields have with many epithermal ore deposits. Hedenquist and Stewart (1985) first drew the alteration petrologists' attention to the presence of subsurface CO_2-rich and bicarbonate waters in some geothermal reservoirs.

3. The reservoir rocks themselves were ideal to study petrographically and the MOW was generous in the number of cores that it recovered. The rocks that were encountered were all young volcanics or else volcanically derived. Silicic pyroclastic rocks predominated but there were also minor amounts of andesite at Wairakei and most of that field is covered by fine-grained lacustrine sediments locally up to 300 m thick. However, the primary constituents of the volcanic rocks are fairly simple, typically including devitrified or fresh glass, quartz, andesine ± biotite ± hypersthene ± augite ± titanomagnetite ± ilmenite with trace amounts of apatite and zircon. All of these, except quartz, apatite and zircon, are unstable in the geothermal environment so they are readily pseudomorphed by hydrothermal minerals. These changes are easily seen petrographically so that there is seldom any difficulty for a petrographer in deciding whether a mineral is primary or secondary; this is not the case incidentally (e.g. at the Ngawha field in Northland) where the reservoir rocks are the Mesozoic and Permian greywackes and argillites that form the basement of the North Island. These rocks have been subjected to low-grade metamorphism which produced an assemblage of secondary minerals that are mostly also stable in a geothermal environment so that distinguishing between those that are geothermal and those that are metamorphic (or detrital) is not usually an easy task.

The main rationale for taking cores was that they are samples of the geothermal reservoir itself so that a knowledge of their lithology can help reveal the reservoir stratigraphy, structure and hydrology. Another purpose was to determine the hydrothermal alteration mineralogy and to interpret it in terms of reservoir conditions. It is this second aspect which concerns us here. Taxpayer money was not spent taking cores for purely scientific purposes; there had to be results that were helpful to the reservoir engineers. This was indeed the case, but much of the scientific side of fluid-rock interaction study was piggy-backed onto the requirements and expectations of MOW engineers. It is worth considering briefly here the practical uses of hydrothermal alteration petrology as applied to assessing geothermal reservoirs.

3 PRACTICAL APPLICATIONS OF HYDROTHERMAL ALTERATION PETROGRAPHY IN GEOTHERMAL INVESTIGATIONS.

Alfred Steiner, a refugee from Czechoslovakia, worked as the geothermal petrologist at the New Zealand Geological Survey from 1947 until his retirement in 1966. He was a very able man with a microscope and also learnt X-Ray Diffractometry when this technique became available at the Survey in the mid 1960's. Steiner laid the basis for geothermal petrology in this country. He published a series of papers on subsurface hydrothermal alteration at Wairakei (Steiner, 1953; 1955; 1968; 1970), culminating in his Bulletin (published in 1977) and also on the Waiotapu field where 7 wells were drilled in 1956 (Steiner, 1963). However, Steiner also wrote very many reports on individual wells for MOW and his fellow scientists. In these it is clear that he recognised a relationship between alteration mineralogy and permeable "fault fissures" and also between temperature, expressed in terms of rank, and mineralogy. This originally empirical relationship was followed and extended by Browne (1970) for the Broadlands-Ohaaki field and in the Philippines by Reyes (e.g. 1990). Elders (e.g. 1977) and others at UC Riverside described similar relationships for the geothermal systems of the Imperial Valley, California and for Cerro Prieto in Baja California, Mexico where the reservoir rocks are mostly fluviatile sediments.

For the New Zealand systems the following conclusions could be drawn from the alteration mineralogy to add to a picture of reservoir conditions.

(a) Downwell temperatures can be predicted. For example, mordenite is a low temperature phase (stable below 110°C) but wairakite usually indicates temperatures above 210°C; epidote is also a good temperature indicator, as indeed are most calc-silicates. For an unknown reason the abundance and crystal sizes of epidote seems to increase with reservoir temperature although this relationship has not been quantified. Clay minerals were also used as temperature indicators. Because wells are drilled using cold water and mud as bit lubricants they do not achieve thermal stability until some months after being drilled. The mineralogical interpretations of subsurface temperatures, however, can be used immediately to plan future drilling. Since 1975, the results of fluid inclusion geothermometry have been compared with the measured bore tempeatures and those deduced from the alteration mineralogy. This has allowed us to recognise places in a geothermal reservoir that have heated, cooled or been thermally stable since the inclusions formed (Browne et al, 1976).

(b) Locating adequate permeability is always a problem in geothermal exploration and this is usually more difficult to do than to find hot rock. The hydrothermal mineralogy of cores and cuttings was used both in New Zealand and the Philippines (e.g. Browne, 1970; Reyes 1990) to:

(i) predict well output qualitatively;
(ii) determine the hydrological function of a particular lithological unit, e.g. whether it functions as an aquifer or an aquitard;
(iii) locate permeable zones;
(iv) determine the optimum depth to set the production casing and to decide whether or not to deepen a particular well;
(v) help recognise where "well damage" may have occurred through cracks becoming sealed with drilling mud.

The most permeability-sensitive minerals are the feldspars. With increasing permeability, andesine is replaced by albite, then albite plus adularia then adularia alone. There is little primary K-feldspar present in the volcanic rocks of the TVZ, so virtually all K-feldspar seen is secondary. Incidentally, alkaline volcanic rocks in the Olkaria geothermal field in Kenya contain primary sanidine and orthoclase phenocrysts that are also replaced by adularia in permeable zones (Browne,1984). This merely involves a slight change in the structure of the feldspar, not in its composition; a small but perhaps interesting point.

(c) The location of some boiling zones (past or present) can be predicted from the occurrence of vein adularia, quartz and bladed calcite (Browne, 1978; Simmons and Christenson, 1994).

(d) The sites of corrosive acid fluids can be determined from the presence of alunite, pyrophyllite, diaspore and perhaps dickite in the cores and cuttings recovered from a well. These are, fortunately, rare in the New Zealand fields but Reyes (1990) has shown how this relationship is used in the Philippines. With hindsight it should have also been possible for NZ petrologists to identify depths where casing corrosive CO_2-rich and bicarbonate waters occur from the presence of smectite plus siderite or calcite in the cores and cuttings. However, this relationship was not recognised until such fluids had first caused their damage (Hedenquist,1990; Hedenquist and Stewart, 1985).

4 GEOTHERMAL SYSTEMS AS NATURAL LABORATORIES

Most people at this meeting will be more interested in the hydrothermal alteration of geothermal rocks for the insights they provide about fluid/rock interactions than in the practical applications of geothermal petrology described above.

The latter has merely been a consequence of the former and results from our viewing active geothermal systems as natural large scale laboratories. It is true we cannot control the physical parameters prevailing in a reservoir, and such systems are open, but we can, at least, measure some of them. Temperature and pressure measurements are made regularly down wells, even if some of the results are not always easy to interpret, and the discharged fluids are routinely analysed and the compositions of the deep fluids calculated (e.g. Ellis and Mahon, 1977). Subsurface permeabilities and well outputs can be measured or estimated and present fluid flow paths established to some extent. This information can be applied to understanding the principles of fluid/rock interactions and the significance of a wide range of minerals that occur in active geothermal fields world-wide. A few general points emerge from the studies.

1. The geothermal minerals are not an exotic assemblage. Most occur also in low-grade metamorphic rocks and/or epithermal ore deposits. There are very few exceptions to

this: e.g. wairakite, first described by Steiner (1955) and Coombs (1955), is a common mineral in geothermal fields world-wide, but is rare elsewhere.

2. Equilibrium is not always or everywhere achieved between fluids and rocks and between minerals. This is not a particularly upsetting situation as what we commonly see in thin sections are reactions frozen in progress. It is usually just as (or more) important to recognise the <u>direction</u> of a reaction and this can be usually determined by looking at textural relations in thin sections.

3. The dissolved gases are very important in controlling the mineralogy that develops at a particular geothermal field. Where the amounts of dissolved gases are low, then the alteration is dominated by calc-silicates (epidote, wairakite, prehnite) but in gassy systems calcite is the main calcium-bearing phase (Browne and Ellis, 1970). This is an important practical point because much attention needs to be paid, when designing a power station, to the amount of non-condensible gases that separate together with the steam from the thermal fluid.

4. Some progress has been made in understanding the processes of subsurface boiling and its mineralogical signatures. Ascending liquid with high concentrations of dissolved CO_2 gas (I know this phrase sounds absurd but that's how it's expressed) first boils deeper than do less gassy fluids. Boiling is a self-limiting process and ceases locally as gases are lost from a liquid and cooling also occurs. Instead, in many geothermal reservoirs the locations of boiling constantly move in all four dimensions, depending also upon the local "plumbing" and how this changes through mineral deposition. Boiling can cause the deposition of vein adularia (because boiling results in an increase in the pH of the residual liquid), quartz (through cooling) and bladed calcite (from loss of CO_2 with the separated vapour). For an as yet unknown reason (but see Tulloch,1982) the calcite is bladed (called "angel wing" in epithermal deposits of the Coromandel District by the old time miners); when cooling occurs this mineral is replaced by quartz but the bladed morphology is commonly preserved.

5. Ore grade concentration of gold and silver have deposited in several geothermal fields and in some pipework through which thermal fluids move (Weissberg, 1969); Weissberg et al, 1979; Brown, 1986). In addition, minor concentrations of base metal sulfides occur in cores and cuttings from the deeper parts of some reservoirs. Much has been learnt about the processes of metal transport and deposition from these occurrences (Henley, Hedenquist and Roberts, 1986).

6. The lithology of the host rocks themselves has very little effect on the identities of the hydrothermal minerals that form (Browne, 1989), a rather surprising but welcome conclusion perhaps, but one allowing ready transfer from New Zealand to elsewhere of experiences gained in interpreting alteration mineralogy.

7. Transfer of constituents between solid and liquid phases is a very real process which can be quantified (Bogie and Browne, 1979; Appleyard et al, submitted). One surprising result, however, is that the much greater abundance of quartz in many altered samples is deceptive in that rock analysis and application of the Gresens equation shows that the silica needed to produce this added quartz derives very locally, from the primary minerals present in the host rock. By contrast, potassium and sodium are highly mobile and both can be either added or depleted depending upon local permeability, the volume of fluid that has moved through the rocks and some other factors.

8. The hydrothermal minerals that occur in veins reflect the nature of the fluids circulating through their host rocks since these minerals deposit directly from such fluids. As expected, quartz, adularia, calcite, epidote and wairakite are common vein minerals but iron-rich chlorite is also. This is perhaps surprising given that the iron contents of the typical neutral pH thermal fluids are less than 0.3 mg/kg (Browne and Ellis, 1970) meaning that chlorite saturation is achieved even at low Fe concentrations.

9. Hydrothermal alteration that occurs by the replacement of the primary phases has received less study than the veins but is volumetrically very much more important. Some true replacement reactions occur with concurrent dissolution of the primary phases and deposition of the secondary minerals; this can be recognised where the original cleavages, zoning and even twinning characteristics survive within the new host. In some places, however, it is more likely that the primary phase is first dissolved entirely, or partly so, before a hydrothermal mineral deposits within the space created (i.e. this is not replacement in the literal sense). Some of the replacement reactions are quite complex and are not easily represented by simple equations; for example, the rather common one whereby andesine is partly replaced by chlorite. Volume preserving equations are also more realistic representations

of replacement reactions than are conventional stoichiometric ones.

10. The deposition of hydrothermal minerals can occur very quickly. This is especially obvious for carbonates where pipes can become blocked with calcite within only a few weeks; however, even calc-silicates can deposit at a measurable pace. For example, Browne et al (1989) showed that in one of the Ngatamariki wells individual wairakite crystals are growing at average rates of 190 μm^3 per day, prehnite from 0.2 to 0.5 μm^3 per day, and epidote at 0.2 μm^3 per day.

5 POSSIBLE FUTURE DIRECTIONS

The fast pace of geothermal drilling in New Zealand from 1950 to about 1980 meant that an enormous number of core and cutting samples from 9 geothermal fields were recovered. Studies of their alteration petrology was largely made on a reconnaissance level. It is only in the past 10 years or so that more detailed mineralogical studies have been made using these samples; fortunately such studies could be done because the samples of cores and cuttings were kept, thanks initially to MWD. Much more work still remains to be done using them and it seems likely to me that detailed mineralogical studies will continue on them, using old samples but with modern techniques and applying new ideas. For example, the initial petrological studies made on the Waiotapu cores (Steiner, 1963) were followed in 1983 by Hedenquist (Hedenquist and Browne, 1989) using new methods and then by Simmons in 1989. Similarly, Browne and Ellis's (1970) work on fluids and cores from Broadlands - Ohaaki field was followed by Lonker et. al. (1990), Hedenquist (1990) and Simmons and Browne (1990). More studies will undoubtedly follow as long as the cores and cuttings are not discarded as a cost saving measure.

Much more remains to be done, even without cores, by looking at surface alteration (there is no complete map showing the distribution of surface hydrothermal alteration of the Wairakei field, for example).

One other fruitful direction of study is likely to be the duration of geothermal systems and how these change during their lifetimes. Isotope studies and simple old-fashioned petrography will reveal some answers provided, of course, that the right questions are asked. The geothermal systems in the Taupo Volcanic Zone have a wide range of ages: the Waimangu field was born only late last century but thermal activity at Ohakuri has been dead for over 40,000 years. However, the Te Kopia field has undergone a turbulent history for the past 120,000 years or so, probably due to its location astride the Paeroa Fault. Surficial rocks on its upthrown side that are now being altered by steam, first reacted with neutral pH waters at least 200 m below ground surface (Bignall and Browne, 1994). More recently, Arehart et al (1997) were able to date near fresh and altered igneous rocks from the Ngatamariki and Rotokawa fields. A dioritic pluton at the former has a 800 ka age. This pluton generated a geothermal system (or systems) that has been active for several thousand years but more remains to be learnt about the evolution of thermal activity at both these fields.

The time is now ripe to make studies of the compositions of individual minerals, first field-wide and then on a regional basis. This has already begun, for example, by looking at the distribution and compositions of clay minerals (Harvey and Browne, 1991; Chi Ma, et al, 1992; Simmons and Browne, 1990) but we know little about how and, more importantly, why, secondary plagioclases vary in composition, for example. The same remark also applies to other minerals showing solid solution characteristics, such as those of the epidote group and the chlorites. Detailed studies made using thin section petrography, the electron microprobe and the scanning electron microprobe will, I expect, identify many minerals missed during the initial petrographic studies.

The manner in which replacement reactions actually occur is not really understood and offers fruitful opportunities for careful study and, perhaps simple controlled laboratory experiments. The geothermal systems themselves, however, will remain outstanding natural laboratories to check mineralogical geothermometers, to learn about the stabilities of minerals in a natural environment and to recognise the processes that control their genesis.

6 REFERENCES

AREHART, G.B., WOOD, C.P., CHRISTENSON, B.W., BROWNE, P.R.L. and FOLAND, K.A. (1997). Timing of volcanism and geothermal activity at Ngatamariki and Rotokawa, New Zealand. Proceedings of the 19th New Zealand Geothermal Workshop, 117-122.

APPLEYARD, E.P., BROWNE, P.R.L. and A.G. REYES (submitted) Composition, volume, density and mass balance relationships of hydrothermally altered rocks from geothermal settings. Applied Geochemistry.

BIGNALL, G. and BROWNE, P.R.L. Surface hydrothermal alteration and evolution of the Te Kopia thermal area, Orakei-korako-Te Kopia geothermal systems, Taupo Volcanic Zone, New Zealand. Geothermics, 645-658, 1994.

BLATTNER, P. (1975). Oxygen isotopic composition of fissure-grown quartz, adularia, and calcite from Broadland geothermal field, New Zealand: with an appendix on quartz-K-feldspar-calcite-muscovite oxygen isotope geothermometers. American Journal of Science 275(7), 785-800.

BOGIE, I. and P.R.L. BROWNE (1979). Geochemistry of hydrothermal alteration of the Ohaaki Rhyolite, Broadlands geothermal field. p. 326-330. In: Proceedings of the New Zealand Geothermal Workshop 1979. Geothermal Institute, University of Auckland, Auckland.

BROWNE, P.R.L. (1970). Hydrothermal alteration as an aid in investigating geothermal fields. Geothermics, 2, 564-570.

BROWNE, P.R.L. (1971). Mineralisation in the Broadlands geothermal field, Taupo Volcanic Zone, New Zealand. Society of Mining Geology of Japan, Special Issue 2, 64-75.

BROWNE, P.R.L. (1978). Hydrothermal alteration in active geothermal fields. Annual review of earth and planetary sciences, 6, 229-250.

BROWNE, P.R.L. (1984). Subsurface stratigraphy and hydrothermal alteration of the eastern section of the Olkaria geothermal field, Kenya. Proc. 6th NZ Geothermal Workshop, 33-41.

BROWNE, P.R.L. (1989). Contrasting alteration styles of andesitic and rhyolitic rocks in geothermal fields of the Taupo Volcanic Zone. Proceeding of the 11th NZ Geothermal Workshop, 111-116.

BROWNE, P.R.L., COURTNEY, S.F. and C.P. WOOD (1989). Rates of formation of calc-silicate minerals in drillhole casing, Ngatamariki Geothermal Field, New Zealand. American Mineralogist, 74, 759-763.

BROWNE, P.R.L. and A.J. ELLIS (1970). The Ohaaki-Broadlands hydrothermal area, New Zealand: mineralogy and related geochemistry. American Journal of Science, 269, 97-131.

BROWNE, P.R.L., ROEDDER, E. and A. WODZICKI (1976). A comparison of past and present geothermal waters, from a study of fluid inclusions, Broadlands geothermal field, New Zealand. p. 140-149. In: Cadek, J., Paces, T. (eds). Proceedings: International Symposium on Water-Rock Interaction, Czechoslovakia 1974. 463 p.

BRUCE, J.A. and F.B. SHORLAND (1933). Utilisation of natural heat resources in thermal regions. NZ Journal of Agriculture, 47, 29-32.

CHI MA, BROWNE, P.R.L. and C.C. HARVEY (1991). Crystallinity of subsurface clay minerals in the Te Mihi sector of the Wairakei Geothermal System. Proc. 14th NZ Geothermal Workshop, 345-350.

COOMBS, D.S. (1955). X-ray observations on wairakite and non-cubic analcime. Mineral Mag. 30(230), 699-708.

ELDERS, W.A. (1977). Petrology and a practical tool in geothermal studies. Trans. Geothermal Resources Council, 1, 85-87.

ELLIS, A.J. and W.A.J. MAHON (1977). Chemistry and geothermal systems. Academic Press, New York, p. 393.

GIGGENBACH, W.F. (1980). Geothermal gas equilibria. Geochimica et cosmochimica acta 44(12), 2021-2032.

GIGGENBACH, W.F. (1984). Mass transfer in hydrothermal alteration systems - a conceptual approach. Geochimica et Cosmochimica acta 48, 2693-2711.

GRANGE, L.I. (1937). The geology of the Rotorua-Taupo subdivision. New Zealand Geological Survey Bulletin, 37, pp. 138.

GRINDLEY, G.W. (1960). Sheet 8, Taupo. Geological Map of New Zealand, 1:250,000. DSIR.

GRINDLEY, G.W. (1963). Geology and Structure of Waiotapu geothermal field. In: Waiotapu Geothermal Field. DSIR Bulletin, 155, 10-25.

GRINDLEY, G.W. (1965). The geology, structure and exploitation of the Wairakei geothermal field, Taupo, New Zealand. NZ Geol. Survey Bulletin, 75, pp. 135.

HARVEY, C.C. and BROWNE, P.R.L. (1991). Mixed layer clay geothermometry in the Wairakei geothermal field, New Zealand. Clays and Clay Minerals, 39, 614-621.

HEALY, J. (1956). Preliminary account of hydrothermal conditions at Wairakei, New Zealand. Pac.Sci.Congr., 8th Manila, Proc, 2, 214-227.

HEDENQUIST, J.W. (1986). Geothermal systems in the Taupo Volcanic Zone: their characteristics and relation to volcanism and mineralisation. p. 134-168. In: Smith, I.E.M. (ed.). Late Cenozoic volcanism in New Zealand. 371 p. Bulletin / Royal Society of New Zealand 23.

HEDENQUIST, J.W. (1990). The thermal and geochemical structure of the Broadlands-Ohaaki geothermal system, New Zealand. Geothermics, 19, 151-185.

HEDENQUIST, J.W. and P.R.L. BROWNE (1989). The evolution of the Waiotapu geothermal system, New Zealand, based on the chemical and isotopic composition of its fluids, minerals and rocks. Geochimica et cosmochimica acta 53, 2235-2257.

HEDENQUIST, J.W. and M.K. STEWART (1985). Natural CO_2-rich steam-heated waters at Broadlands, New Zealand: their chemistry, distribution and corrosive nature. Trans. Geothermal Resources Council, 9, 245-250.

HENLEY, R.W., HEDENQUIST, J.W. and P.J. ROBERTS (1986). Guide to the active epithermal (geothermal) systems and precious metal deposits of New Zealand. Monograph Series of Mineral Deposits, Gebruder Borntraeger, pp. 211.

LONKER, S.W., FITZGERALD, J.D., HEDENQUIST, J.W. and J.L. WALSHE (1990). Mineral-fluid interactions in the Broadlands-Ohaaki geothermal system, New Zealand. Am.J.Sci., 290, 995-1068.

REYES, A.G. (1990). Petrology of Philippine geothermal systems and the application of alteration mineralogy to their assessment. J.Volcan. and Geothermal Research, 43, 279-309.

ROBINSON, B.W. (1978) (Ed.). Stable isotopes in the earth sciences. DSIR Bulletin 220, pp. 229.

SIMMONS, S.F. and B.W. CHRISTENSON (1994). Origins of calcite in a boiling geothermal system. Am.J.Sci., 294, 361-400.

SIMMONS, S.F. and P.R.L. BROWNE (1990). Shallow and marginal alteration at Ohaaki-Broadlands geothermal system. Proc. 12th NZ Geothermal Workshop, 25-30.

STEINER, A. (1953). Hydrothermal rock alteration at Wairakei, New Zealand. Econ. Geol. 48(1), 1-13.

STEINER, A. (1955). Wairakite, the calcium analogue of analcime, a new zeolite mineral. Mineral Mag. 30(230), 691-698.

STEINER, A. (1963). The rocks penetrated by drillholes in the Waiotapu thermal area, and their hydrothermal alteration in Waiotapy geothermal field. DSIR Bulletin, 155, 26-34.

STEINER, A. (1968). Clay minerals in hydrothermally altered rocks at Wairakei, New Zealand. Clays and Clay Minerals 16(3), 193-213.

STEINER, A. (1970). Genesis of hydrothermal K-feldspar (adularia) in an active geothermal environment at Wairakei, New Zealand. Mineralogical Magazine, 37, 916-922.

STEINER, A. (1977). The Wairakei geothermal area, North Island, New Zealand: its subsurface geology and hydrothermal rock alteration. NZ Geological Survey Bulletin 90, pp. 136.

TULLOCH, A.J. (1982). Mineralogical observations on carbonate scaling in geothermal wells at Kawerau and Broadlands. Proc. Pacific Geothermal Conf. 1982, 131-134. Incorporating the 4th NZ Geothermal Workshop, part 1, 283 p.

WEISSBERG, B.G. (1969). Gold-silver ore-grade precipitates from New Zealand thermal waters. Econ. Geol. 64(1), 95-108.

WEISSBERG, B.G., SEWARD, T.M. and P.R.L. BROWNE (1979). Ore metals in active geothermal systems. p. 738-780. In: Barnes, H.L. (ed.) Geochemistry of Hydrothermal Ore Deposits, 798 p. Wiley: New York.

Chemistry of 3.2 Ga seafloor hydrothermal vent fluids

Cornel E.J.de Ronde
Institute of Geological and Nuclear Sciences, Lower Hutt, New Zealand

ABSTRACT: Ancient (ca. 3.2 Ga) seafloor hydrothermal vents, known as the Ironstone Pods, have been identified in the Barberton greenstone belt of South Africa. Ironstone Pod hydrothermal fluid end-member concentrations (Mg = 0) of various dissolved components derived from bulk fluid inclusion crush-leach experiments, include; Cl (730 mmol/l), Br (2.59), I (0.058), Na (822), NH_4 (11.4), K (21.5), Ca (42.6), and Sr (0.15). This hydrothermal fluid also contains up to 1.07 mol.% CO_2, 0.03 mol.% N_2, 0.02 mol.% CH_4, 262 ppm COS, and minor amounts of C_2-C_4 hydrocarbons. Hydrothermal end-member Ca, Sr and NH_4, in particular, and to a lesser degree K, I and CO_2, commonly plot on, or very close to, modern vent fluid trends. By contrast, end-member Na and Br concentrations are distinct (higher) from modern vent fluids. High I and NH_4 concentrations are consistent with contributions from sediments and/or organic matter.

1 INTRODUCTION

In the 20 years since the discovery of active seafloor venting at the Galapagos Spreading Centre (Corliss et al. 1979), numerous hydrothermal fluid discharges have been analysed from vent sites spanning a variety of tectonic environments, including mid-ocean ridges (MOR), backarc basins and island arcs (e.g., Ishibashi and Urabe 1995). These studies have shown how vent fluids at different sites can have variable compositions. For example, chloride concentrations can range between 230% and 6% of seawater values; gases (e.g., CO_2 and H_2S) and metals (e.g., Fe, Mn, Cu) can also have a wide range in concentration between sites (Von Damm 1990).

Archean greenstone belts commonly show extensive alteration of the supracrustal rocks and in particular for the mafic volcanics. A number of workers have suggested that this pervasive alteration can be attributed to seafloor metasomatism (e.g., de Wit et al. 1987; de Ronde et al. 1994). In the Barberton greenstone belt of South Africa, evidence for related seafloor hydrothermal activity has been found in the form of Cu-Zn-Ba ± Pb ± Au volcanogenic massive sulphides and massive Fe-rich pods, known as the Ironstone Pods (de Wit et al. 1982; de Ronde et al. 1994). The pods are considered to have formed directly on the Barberton seafloor as constructional hydrothermal mounds (de Ronde et al. 1994). Application of fluid inclusion microthermometric and chromatographic techniques has enabled the fluid composition of the pod vent fluids to be determined, and also provides insight into the composition of mid-Archean age seawater (de Ronde et al. 1997).

2 FLUID INCLUSIONS

2.1 *Microthermometry*

Fluid inclusion microscopy studies were performed on quartz considered to be paragenetically related to the dominant hematite and goethite of the Ironstone Pods. Samples included microveinlets and vugs contained within the pods, and stockwork veins located immediately below the pods. Four types of inclusions were noted; (1) dominant type I inclusions are 2-phase aqueous inclusions with salinities in the range 0.7-15.8 wt.% NaCl equiv. and homogenisation temperatures (T_h) of 82-219 °C; type Ib inclusions contain CO_2 (based on the observation of CO_2 clathrates) with salinities <3.5 wt.% NaCl and T_h between 110 and 188 °C, (2) rare type II and IIa inclusions with much higher salinities (~24-30 wt.% $CaCl_2$ equiv.) and generally lower T_h (33-109 °C), (3) 3-phase liquid CO_2-bearing type III inclusions (salinities <1 wt.% NaCl and T_h 120-200 °C), and (4) late, 2-phase aqueous inclusions (not measured). Inclusion types I-III are considered primary or pseudosecondary in origin and type IV secondary.

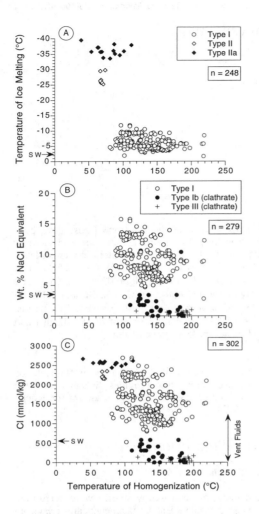

Figure 1. Fluid inclusion microthermometric data.

For additional details on Ironstone Pod fluid inclusion microthermometry, including textural observations, see de Ronde et al. (1994).

Plotting ice melting temperature (T_{mice}) against T_h differentiates type I inclusions from types II and IIa (Fig. 1A). The apparent salinities of types Ib and III inclusions are determined via clathrate melting experiments, and thus cannot be plotted on Figure 1A, but rather are compared with the data for type I inclusions by plotting wt.% NaCl equivalent against T_h (Fig. 1B); the typically lower salinities of types Ib and III inclusions can clearly be seen. In order to compare all the different fluid inclusions on one diagram, the salinities can be converted to mmol/kg Cl and again plotted against T_h (Fig. 1C). Here, the distinctly higher chloride concentrations and lower T_h values of types II and IIa inclusions are still distinguished from the more intermediate chloride concentrations and moderate-to-high T_h values of type I inclusions. The latter can in turn be easily distinguished from the low-chloride inclusions of types Ib and III. A significant proportion of the type I and all of the types Ib and III salinity data in Figure 1C overlap with the highest chlorinities recorded for hydrothermal fluids of modern black smokers (i.e., up to 230% SW values; Von Damm, 1990).

2.2 Fluid inclusion gases

Ironstone Pod fluid inclusion volatiles are dominated by H_2O (98.85-99.90 mol.%), with CO_2 (<0.01 to 1.07%) being the next most abundant species (Table 1). This is consistent with microthermometric results which show that the inclusions are dominantly aqueous, with minor CO_2 locally (e.g., inclusion types Ib and III). Nitrogen (0.01-0.10%) and CH_4 (0.01-0.03%) are also present in the inclusions. Noticeable concentrations of COS occur in many samples (19-262 ppm); detectable hydrocarbons are usually on the order of a few ppm (Table 1).

No evidence is seen for boiling to have occurred at the site of pod formation. However, the low salinity, gas-rich type Ib inclusions and 3-phase liquid CO_2-bearing type III inclusions occurring with variable (but higher) salinity, relatively gas-poor type I inclusions, could be interpreted as evidence for sub-seafloor phase separation. That is, types Ib and III inclusions may represent condensed vapor from a boiled hydrothermal fluid, represented by dominant type I inclusions. However, the very high salinity of fluid inclusion types II and IIa combined with the fact that they do not contain any salts as daughter phases (e.g., NaCl), suggests that the precursor to the hydrothermal fluid was evaporative in origin (de Ronde et al. 1994). Thus the salinity of type I inclusions may have at least two sources. The combination of the highest measured type I T_h value (219 °C) and CO_2 concentration (1.07 mol.%) enables a *minimum* estimate of water depth above the Ironstone Pods to be calculated of 982 m (de Ronde et al. 1997).

2.3 Fluid inclusion anions/cations

Anion/cation data for Ironstone Pod fluid inclusions are given in Table 2. Anions in the inclusions are dominated by Cl, with concentrations of up to 183% that of present seawater (SW), and range between 690 and 999 mmol/l (cf. modern SW at 556 mmol/l). Bromide concentrations are consistent between samples, and are approximately three times higher

Table 1. Gas chromatographic analyses of Ironstone Pod fluid inclusions in quartz (de Ronde et al. 1997).

Mole % (unless otherwise stated)

Sample	N_2^+	CH_4	CO_2	H_2O	COS (ppm)*	C_2H_4 (ppm)	$C_2H_6^+$ (ppm)	C_3H_6 (ppb)	C_3H_8 (ppm)	C_4H_8 (ppb)	C_4H_{10} (ppm)
Ironstone pod quartz											
B-85-3†	0.020	0.024	1.074	98.853	262	1	9	-	9	-	5
B-85-4	0.095	0.031	-	99.868	-	1	20	-	15	-	24
ISP-86-6	0.010	0.006	0.341	99.638	34	<1	2	-	1	-	3
ISP-86-22	0.080	0.023	-	99.877	179	1	12	-	7	-	10
ISP-90-1†	0.051	0.015	0.012	99.903	155	1	15	10	8	20	15
ISP-90-2†	0.051	0.025	0.003	99.902	127	2	24	-	22	-	5
ISP-90-3	0.015	0.021	0.409	99.547	78	1	5	-	-	-	-
ISP-90-4†	0.011	0.021	0.972	98.991	55	<1	3	-	2	-	1
ISP-90-5	0.013	0.012	0.624	99.341	101	<1	4	-	-	-	5
ISP-90-6	0.011	0.021	0.928	99.035	19	1	11	-	8	-	6
ISP-90-7	0.026	0.009	1.033	98.916	153	1	8	-	3	-	3

Notes: Analyses run under isothermal (80 °C) conditions; the crusher is pumped up to 7000 p.s.i. and released, resulting in the appropriate crushing pressure, then released quickly to zero pressure; N_2^+ values are a combination of N_2-CO-Ar-O_2 where N_2 is the dominant gas; similarly, the $C_2H_6^+$ value is a combined C_2H_2/C_2H_6 peak where C_2H_6 is the dominant gas
*, ppm and ppb are in molar units (for additional information on this technique, see Bray et al. 1991)
†, average from two separate analyses; -, not detected; fluid inclusion microthermometry has been performed on several quartz grains from each of samples B-85-3 & 4 (de Ronde et al. 1994), and ISP-90-1, 3, 4 & 5 (de Ronde et al. 1997)

than seawater for a given Cl content. Similarly, SO_4 and I concentrations in the fluid inclusions are fairly consistent between samples (when detected); SO_4 values are about an order-of-magnitude lower, and I two orders-of-magnitude higher than modern seawater, respectively (Table 2).

Cation concentrations in the pod fluid inclusions are dominated by Na, ranging from 557 to 918 mmol/l, or 120-200% seawater values. Ammonium concentrations are two-to-three orders-of-magnitude greater in the inclusions relative to modern seawater. Potassium ranges from almost seawater values to approximately four times greater in the inclusions. Conversely, Mg ranges from seawater concentrations down to values 27% lower. Both Ca and Sr concentrations are more than an order-of-magnitude greater in the inclusions relative to seawater.

The data given in Table 2 show that several of the cation and anion concentrations within the pod fluid inclusions are within an order-of-magnitude of modern seawater values. The exceptions are higher I, NH_4 and Sr values. Both CO_2 and CH_4 in the fluid inclusions have concentrations two-orders-of magnitude greater than seawater (Von Damm 1990). The charge balances for the various fluid inclusion analyses range between 1.1 and 1.6 which are typical for this kind of analytical work on relatively dilute fluids involving large dilution factors in the leaching process (e.g., Channer and Spooner 1993). In addition, there are negative ions, such as bicarbonate, which will be present in the inclusions, but which have not been determined.

3 HYDROTHERMAL END-MEMBER FLUID

In modern seafloor systems, the hydrothermal end-member component is calculated by fitting a least squares line to the data, and extrapolating back to zero magnesium. For example, the pod fluid inclusion chloride data, extrapolated to Mg = 0, gives a Cl concentration of 730 mmol/l, or a concentration 130% higher than modern seawater, but still well within the range recorded for modern vent fluids (see Fig. 1C). Hydrothermal vent fluids are also considered to contain zero sulfate; extrapolation of the pod data to SO_4 = 0 (not shown) gives a similar Cl concentration of 715 mmol/l. The fit of least squares lines to the fluid inclusion data range from quite poor for NH_4 and K, and to a lesser degree I and Na, to quite reasonable for Cl, Br, Ca, and Sr (de Ronde et al. 1997). The quality of fit can to a large degree be attributed to the errors involved in determining the concentrations of the various components, and to natural variability (see Channer and Spooner 1993).

Ironstone Pod fluid inclusion hydrothermal end-member concentrations for the various dissolved components are shown in Figure 2, and are listed in Table 3. This calculated end-member generally plots on, or close to, the trends defined by modern

Table 2. Anion/cation data for Ironstone Pod fluid inclusions in quartz (from de Ronde et al. 1997).

Sample	Anions (mmol/l)				Cations (mmol/l)						Q^+/Q^-
	Cl^-	Br^-	SO_4^{2-}	I^-	Na^+	NH_4^+	K^+	Mg^{2+}	Ca^{2+}	Sr^{2+}	
B-85-3	771	2.81	-	0.06	828	-	13.1	-	-	-	1.1
ISP-86-6	690	1.86	-	0.02	557	11.4	11.2	31.8	130	2.2	1.3
ISP-90-3	856	2.75	2.3	0.07	996	25.4	38.5	14.5	128	-	1.6
ISP-90-4	711	2.06	0.7	0.04	734	10.3	18.3	17.5	130	2.1	1.5
ISP-90-5	857	2.53	2.1	0.05	828	4.6	21.2	25.2	172	2.6	1.5
ISP-90-7	999	2.58	2.0	0.05	918	-	19.5	50.9	219	4.7	1.5
Modern Seawater†	556	0.86	28.7	0.0005	477	<0.01	10.1	54.2	10.5	0.09	1.0

Notes: The data listed represent average results for up to 7 analyses of each sample; the species that were analysed (by ion chromatography) for, but not detected or were blank dominated include; anions F^-, CH_3COO^-, $HCOO^-$, NO_3^- and HPO_4^{2-}, and cations Li^+, Rb^+ and Cs^+; undetermined anions (e.g., HCO_3^-, CO_3^{2-}) considered main reason for charge imbalance (for additional information on this technique, see Channer and Spooner 1993) -, not detected or blank dominated (i.e., the blank peak area was 25% or more of the sample peak area); 2+ cations not detected in B-85-3 due to low yield
†, Anion/cation data for modern seawater from Von Damm (1990) and Holland (1978)

hydrothermal end-member vent fluids, with the exception of Na, Br, and possibly CO_2 (Figs. 2A, E and G). For example, end-member Ca and Sr concentrations are similar between Ironstone Pod and MOR vent fluids, an interesting result considering that the initial seawater has very divergent Ca and Sr concentrations (Figs. 2B and D, and Table 3). Potassium concentrations typically reflect seawater/rock interaction (e.g., Von Damm, 1990) and are again similar between pod and modern vent hydrothermal end-member compositions (Fig. 2F); the slightly lower pod value possibly reflects a different, lower K-bearing (komatiitic) underlying rock sequence. Ironstone Pod hydrothermal end-member NH_4 and I (not shown) concentrations are similar to those of back-arc basin vent sites, although are significantly higher than those from MOR and seamount-hosted vent fields (Fig. 2F). These high concentrations are interpreted to be the result of sediment and/or organic matter contributions to the hydrothermal fluid (cf. You et al. 1991).

The distinctly higher Na concentration for the pod hydrothermal end-member is consistent with a high Cl concentration for these fluids (Fig. 2A) and indicates NaCl is the dominant salt in the hydrothermal fluid.

Carbon dioxide concentrations in pod fluid inclusions range from the highest values for modern vents (e.g., those of Axial, Kasuga, and Loihi seamounts, and the Okinawa Trough vent fields), to concentrations 2-3 times higher. This may be, in part, due to the inferred (back-arc) tectonic setting of the pods (de Ronde and de Wit 1994). That is, the abundance of CO_2 in the inclusions might be attributed to the volatile-rich nature of arc magmas. Modern seafloor hydrothermal systems located in back-arc environments tend to have much higher gas (dominantly CO_2) concentrations dissolved in the hydrothermal fluid than do systems associated with MOR's (e.g., Ishibashi and Urabe 1995).

Of the pod fluid inclusion samples analysed for combined cations and anions, B-85-3 is most like the calculated hydrothermal end-member fluid, as Mg was not detected in the analysis of this sample (Table 2). Sample B-85-3 also contains the highest concentration of CO_2 of all the samples analysed (Table 1) and marks one end of a range of samples containing relatively high CO_2 concentrations. Thus samples containing relatively high concentrations of CO_2 in the fluid inclusions most likely indicate a high proportion of hydrothermal end-member, having mixed to varying degrees with a non-hydrothermal fluid (seawater). A calculated Barberton seawater end-member composition for the inclusions is also given in Table 3.

SUMMARY

Ironstone Pod hydrothermal fluid end-member concentrations for dissolved components such as Ca, Sr and NH_4, in particular, and to a lesser degree K, I and CO_2, commonly plot right on, or very close to, modern vent fluid trends. By contrast, hydrothermal end-member Na and Br concentrations are distinct (higher) from modern vent fluids. High I and NH_4 concentrations in the hydrothermal fluid are consistent with contributions from sediments and/or

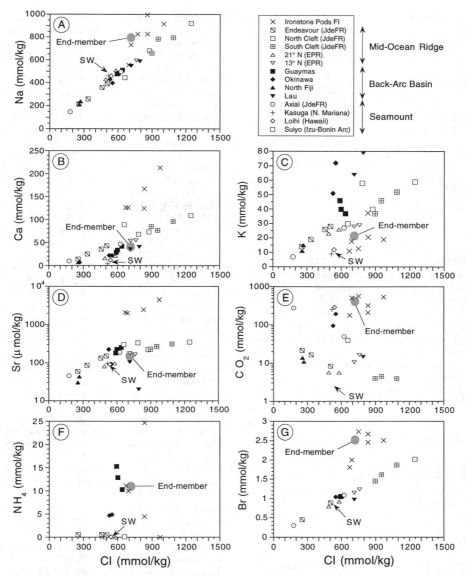

Figure 2. Concentrations of selected major dissolved components versus Cl concentration for the Ironstone Pod fluid inclusions (crosses; Table 2). Also shown are selected end-member hydrothermal vent fluids from modern vent sites to highlight the variable chemistry within these seafloor hydrothermal systems. The grey dot labelled 'End-member' relates to the concentration of that component in the Ironstone Pod hydrothermal fluid, once corrected back to zero Mg (Table 3). The concentration of any one component for modern seawater is given by the black arrow labelled 'SW', where the tip of the arrow points exactly at the appropriate value. Data for most of the vent fields is compiled in de Ronde (1995), with additional data from Von Damm (1990) and You et al. (1994). FI, fluid inclusions; JdeFR, Juan de Fuca Ridge; EPR, East Pacific Rise.

organic matter.

An estimate of the Barberton seawater end-member (Table 3) shows that several components are commonly within an-order-of magnitude of modern seawater values, with the exception of significantly higher I, NH_4, Ca and Sr. Sulfate concentrations are minimum estimates for Barberton seawater, although are consistent with the lower SO_4

Table 3. Ironstone Pod end-member fluids.

Anions (mmol/l)				Cations (mmol/l)							Q^+/Q^-
Cl^-	Br^-	SO_4^{2-}	I^-	Na^+	NH_4^+	K^+	Mg^{2+}	Ca^{2+}	Sr^{2+}		
Hydrothermal end-member											
730	2.59	0*	0.058	822	11.4	21.5	0	42.6	0.15		1.3
Seawater end-member											
920	2.25	2.3†	0.037	789	5.1	18.9	50.9†	232	4.52		1.5

Notes: Hydrothermal end-member calculated by regressing data through Mg = 0; seawater end-member calculated by extrapolating from Mg = 50.9 mmol/l
*, Given a value of 0 although regression to Mg = 0 gives SO_4 = 0.64 mmol/l
†, Highest value measured

concentrations postulated for the Archean oceans (e.g., Grotzinger and Kasting 1993). Fluid inclusion samples containing the greatest seawater component have higher N_2 (up to 0.1 mol.%) and low CO_2, when compared to the hydrothermal end-member fluid. Barberton ambient seawater is considered to have been an evaporitic brine of $NaCl-CaCl_2$ composition during the time of pod deposition. Both the pod hydrothermal and seawater end-members have Br concentrations noticeably higher than those for modern seawater or vent fluids, suggesting that ~3.2 Ga Barberton seawater had a true Br/Cl value higher than today's seawater (Channer et al. 1997).

ACKNOWLEDGMENTS

This paper is a review of recent research done by the author in collaboration with D.M. deR. Channer, K. Faure, C.J. Bray, and E.T.C. Spooner whose input is gratefully acknowledged.

REFERENCES

Bray, C.J., Spooner, E.T.C. & A.V. Thomas 1991. Fluid inclusion volatile analysis by heated crushing, on-line gas chromatography; applications to Archean fluids. J. Geochem. Explor. 42: 167-193.

Channer, D.M. DeR. & E.T.C. Spooner 1993. Combined gas and ion chromatographic analysis of fluid inclusions: applications to Archean granitic pegmatite and Au-quartz vein fluids. Geochim. Cosmochim. Acta 58: 1101-1118.

Channer, D.M. DeR. de Ronde, C.E.J. & E.T.C. Spooner 1997. The Cl--Br--I- composition of ~3.23 Ga modified seawater: implications for the geological evolution of ocean halide chemistry. E.P.S.L. 150: 325-335.

Corliss, J.B., Dymond, J., Gordon, L.I., Edmond, J.M., von Herzen, R.P., Ballard, R.D., Green, K., Williams, D., Bainbridge, A., Crane, K. & T.J.H. van Andel 1979. Submarine thermal springs on the Galapagos Rift. Science 203: 1073-1083.

de Ronde, C.E.J. 1995. Fluid chemistry and isotopic characteristics of seafloor hydrothermal systems and associated VMS deposits: potential for magmatic contributions. In, *Magmas, Fluids, and Ore Deposits* (J.F.H. Thompson, ed.). Mineral. Assoc. Canada Short Course 23: 479-509.

de Ronde, C.E.J. & M.J. de Wit 1994. Tectonic history of the Barberton greenstone belt, South Africa: 490 million years of Archean crustal evolution. Tectonics 13: 983-1005.

de Ronde, C.E.J., de Wit, M.J. & E.T.C. Spooner 1994. Early Archean (>3.2 Ga) Fe-oxide-rich, hydrothermal discharge vents in the Barberton greenstone belt, South Africa. Geol. Soc. Am. Bull. 106: 86-104.

de Ronde, C.E.J., Channer, D.M. deR., Faure, K., Bray, C.J. & E.T.C. Spooner 1997. Fluid chemistry of Archean seafloor hydrothermal vents; implications for the composition of circa 3.2 Ga seawater. Geochem. Cosmochim. Acta 61: 4025-4042.

de Wit, M.J., Hart, R., Martin, A. & P. Abbott, 1982. Archean abiogenic and probable biogenic structures associated with mineralised vent systems and regional metasomatism, with implications for greenstone belt studies. Econ. Geol. 77: 1783-1802.

de Wit, M.J., Hart, R.A. & R.J. Hart 1987. The Jamestown ophiolite complex, Barberton mountain belt: a section through 3.5 Ga oceanic crust. J. Afr. Earth Sci. 5: 681-730.

Grotzinger, J.P. & J.F. Kasting 1993. New constraints on Precambrian ocean composition. J. Geol. 101: 235-243.

Holland, H.D. 1978. *The Chemistry of the Atmosphere and Oceans*. Wiley-Interscience, New York, 351 p.

Ishibashi, J. & T. Urabe 1995. Hydrothermal activity related to arc-backarc magmatism in the Western Pacific. In, *Back-arc Basins: Tectonics and Magmatism* (B. Taylor, ed.), Plenum Pub. Co., New York.

Von Damm, K.L. 1990. Seafloor hydrothermal activity: black smoker chemistry and chimneys. Ann. Reviews Earth Planet. Sci. 18: 173-204.

You, C.-F., Butterfield, D.A., Spivack, A.J., Gieskes, J.M., Gamo, T. & A.J. Campbell 1994. Boron and halide systematics in submarine hydrothermal systems: effects of phase separation and sedimentary contributions. E.P.S.L. 123: 227-238.

Fire and water: Physical roles of water in large eruptions at Taupo and Okataina calderas

B.F. Houghton & C.J.N. Wilson
Institute of Geological and Nuclear Sciences, Wairakei, New Zealand

ABSTRACT: Interaction between magma and water in explosive eruptions occurs to variable extents in many eruptions, illustrated here by the 181AD Taupo and 1886AD Tarawera eruptions. Waters involved can be surface, groundwater, or preheated geothermal waters. Limited water involvement boosts eruptive power but has only subtle effects on deposits (mostly in juvenile clast densities), while large-scale water involvement maintains or increases eruptive power, and has marked effects on deposits, such as increased clast densities, poor sorting, and evidence for flushing of fine ash and contemporaneous water erosion.

1 INTRODUCTION

In contrast to most presentations at this conference, this paper deals only with physical aspects of water:rock interaction, i.e. the involvement of water in large-scale pyroclastic volcanism. Water plays two vital roles in explosive eruptions. First, as the dominant volatile species dissolved in magmas (1-6 wt %), water is exsolved and rapidly decompressed to constitute the driving force for "dry" explosive eruptions. Second, contact between magma and external (sea-, lake-, ground-) water may itself be explosive, with flashing of the water into steam which powers "wet" explosions.

Although magma:water interaction is common at some stage in eruptions at many volcanoes, the most voluminous and violent occurrences are at calderas due to their low topographic relief, frequently large large magma volumes and the presence of large reservoirs of external water in the form of caldera lakes. Here we contrast the influence of external water during the most recent eruptions from two calderas in central Taupo Volcanic Zone (TVZ).

Central TVZ (Fig. 1) is a Quaternary arc undergoing rapid extension (7-18 mm a^{-1}) which accompanies voluminous silicic magmatism and the development of a nested cluster of caldera centres since c. 1.6 Ma (Houghton et al. 1995; Wilson et al. 1995). Since 65 ka silicic volcanism has focused at Taupo and Okataina (Fig. 1a-c). At Taupo, two "wet" caldera-forming events at 1.8 and 26.5 ka are separated by 26 smaller lava-producing and "dry" and "wet" explosive episodes. At Okataina two large dome complexes have formed over this time period in a series of mostly "dry" explosive and extrusive eruptions.

1.1 *Case studies*

The c. 35 km^3 (magma) Taupo 181AD eruption is among the most powerful explosive eruptions in the world during the past 5 ka (Wilson & Walker, 1985). Its products serve as holotypes for the most widely dispersed "dry" (ultraplinian: Walker, 1980) and "wet" fall deposits (phreatoplinian: Self & Sparks, 1978), and low-aspect-ratio (= violently emplaced) ignimbrites (Walker at al., 1980). The eruption consisted of 6 phases (Fig. 1d), characterised by widely varying magma discharge rates and degrees of magma: water interaction, from a series of NE-SW aligned vents close to the eastern margin of Lake Taupo (Smith & Houghton 1995).

The 1886AD eruption of Tarawera, one of the two dome complexes within Okataina caldera, was New Zealand's largest historical eruption. About 0.8 km^3 of basaltic magma was erupted in 5 hours from a 17 km long fissure extending SW across and beyond the dome complex into a low-lying region hosting a large intense geothermal system which was violently unroofed by the eruption. All phases of the eruption were "wet" in the sense of involving external water, but the eruption products vary greatly reflecting variety in both the abundance and the nature (i.e. cold ground water versus superheated geothermal fluid) of the water.

1.2 *Taupo 181AD*

The 181 AD eruption began with relatively weak "wet" explosive activity that produced a locally dispersed, fine grained pumiceous ash fall deposit

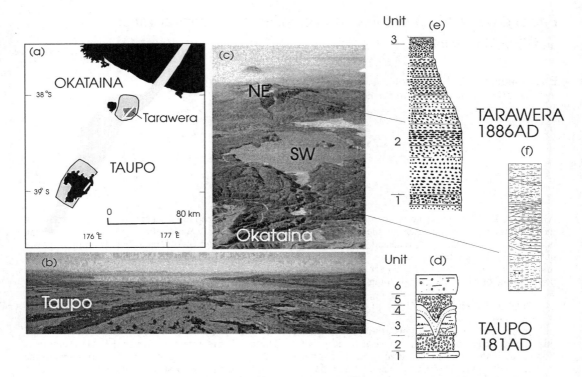

Fig. 1. (a) Taupo and Okataina caldera volcanoes and young TVZ (lightly shaded area) (after Wilson et al., 1995). (b) View of Taupo caldera from the NE; note the caldera lake. (c) View from the SW along the line of the 1886AD eruptive fissure. Its SW part is now largely filled by a lake, while the NE part bisects the Tarawera dome complex. (d) Stratigraphy of the proximal 181AD Taupo products, typically 10-20 m thick. (e) Stratigraphy of NE Tarawera 1886AD products; typical thicknesses are 30-60 m. (f) Stratigraphy of SW Tarawera products; aggregate thicknesses are 5 to 30 m.

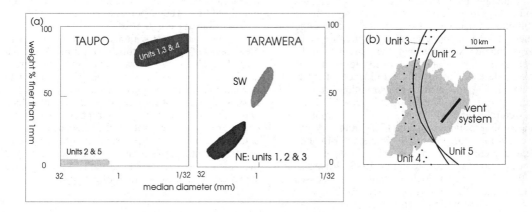

Fig. 2. (a) Comparative data on median diameters and contents of fine ash between "dry" and "wet" units of the Taupo (left) and Tarawera (right) eruptions. See text for discussion. (b) Map of the 3 cm isopachs for Units 2-5 from the Taupo 181AD eruption. Note how the "wet" deposits (Units 3 and 4; dotted lines) were dispersed further upwind (west) than their associated "dry" counterparts (Units 2 and 5).

(Unit 1). This is overlain by a uniform, non-graded, well sorted "dry" widespread pumice fall deposit (Unit 2). Unit 3 is a fine grained pumiceous ash deposit containing some layers rich in aggregates of ash particles (accretionary lapilli) and was erupted from the SW end of the vent system. Unit 3 was then extensively eroded and gullied and this erosion is interpreted as due to eruption of an assemblage rich in liquid water, causing erosion rather than deposition of ash (Walker, 1981). Unit 4 is an exceptionally fine-grained and finely laminated ash with numerous intraformational erosion gullies and an abundance of ash aggregates and vesicular matrix textures. Unit 5 represents an abrupt return to "dry" eruptive conditions producing a uniform coarse-grained pumice fall deposit with subordinate volumes of interbedded coeval ignimbrite close to source. Unit 5 is overlain and truncated by a thin widespread non-welded ignimbrite (Unit 6) which is not discussed further here.

1.3 Tarawera 1886AD

The 1886 AD eruption produced 3 geographically distinct products:
(i) a coarse-grained scoria fall deposit confined to within 400 m of the NE half of the eruptive fissure (Fig. 1e),
(ii) a widespread scoria lapilli bed covering at least 10 000 km^2, and
(iii) a fine-ash rich deposit erupted from the SW half of the eruptive fissure (Fig. 1f).

The proximal scoria fall deposit contains 3 units (Fig. 1). Unit 1 is rich in rhyolitic wall rock fragments and moderate-density juvenile basalt clasts with quenched outer rims. Unit 2 contains red and black, ragged, often fluidal, scoria clasts and is often welded, the welded subunits defining a series of lenses along the fissure. Unit 3 consists of wall rock blocks to ash and subordinate dense juvenile basalt clasts.

1.4 Contrasts between the eruptions

These two eruptions show contrasts in the nature of the magma, the nature of the external water phase and the architecture and hydrology of the vent region. The Taupo eruption involved rhyolite and the 1886 AD Tarawera eruption high-alumina basalt. The water at Taupo was cold lake water contained within a proto-Lake Taupo similar to the modern lake. On the NW half of the Tarawera fissure, magma encountered limited quantities of cold groundwater in fractures in pre-existing rhyolite lava and in pore space in pumiceous rhyolite pyroclastics. To the SW the basalt encountered the Rotomahana geothermal system containing fluid at a reservoir temperature of 230-240 °C.

2 MAGMA/WATER INTERACTION PRIOR TO FRAGMENTATION

Juvenile clasts from the Tarawera eruption contain a striking abundance of cm to sub-mm sized wall rock xenoliths. These xenoliths must have been incorporated prior to the final magma/water interaction leading to explosive fragmentation, and represent a earlier pre-fragmentation stage of disruptive contact with groundwater. Incorporation of the inclusions has important consequences for the rheology of the magma - the decrease in temperature and the increased abundance of solid particles both raising the viscosity of the basalt. We suggest this increase in viscosity contributed to the widespread dispersal of the eruption products by permitting the 1886 basalt to behave in a more viscous fashion more typical of higher-SiO$_2$ magmas. In sharp contrast, rhyolite pumices from the Taupo eruption (and other rhyolitic "wet" deposits) virtually never contain wall-rock lithics; the inherently high viscosity of rhyolite magma seems to preclude bulk interaction with wall rock and groundwater on the scale of single clasts.

3 FRAGMENTATION STYLES AND ERUPTIVE MECHANISMS

The most striking effects of magma/water interaction are seen in the morphology and vesicularity of the resulting pyroclasts and in the grain size characteristics of the deposits (Houghton & Wilson, 1989). Rising magma can encounter external water over a range of depths and in a variety of states, from gas-rich but non-vesiculated, to a vesicle-charged foam, to degassed. Contact with external water quenches the exterior of the resulting clasts and inhibits further vesiculation.

At Taupo there is a clear distinction between the "dry" activity and that involving significant amounts of external water based on grain size alone of the fall products. Poor sorting and an abundance of fine ash distinguishes Units 1, 3 and 4 (where there is independent evidence for the presence of liquid water) from Units 2 and 5. For example, the former typically contain >50 wt % <1 mm ash, while the latter have < 5 wt % (Fig. 2a). This reflects a major contrast in the water:magma ratio, with phases 1, 3 and 4 characterized by abundant liquid water in the eruption causing premature flushing and deposition of the fine ash component.

At Taupo we also recognise three different types of assemblage of juvenile pyroclasts (Fig. 3; Houghton & Wilson, 1989). The "dry" units (2 and

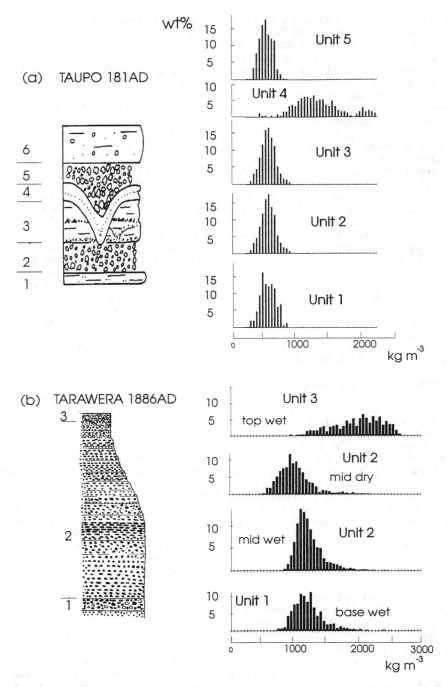

Fig. 3. Representative density data from magmatic clasts for (a) Taupo 181AD and (b) Tarawera 1886AD eruptions. The clast densities were determined on 16-32 mm diameter lapilli by water immersion (Houghton & Wilson, 1989). Equivalent vesicularities for the rhyolite in (a) are 75% (density 600 kg m^{-3}, 50% (density 1200 kg m^{-3}) and 25% (density 1800 kg m^{-3}), and for the basalt in (b) 75% (density 700 kg m^{-3}, 50% (density 1400 kg m^{-3}) and 25% (density 2100 kg m^{-3}).

5) contain strikingly unimodal assemblages of highly vesicular pumice commensurate with disruption of a magmatic foam by bubble growth alone. Deposits of "wet" phases 1 and 3 are also dominated by highly vesicular pumice but contain a "tail" of denser clasts quenched immediately prior to the peak of vesiculation. The external water came into contact with a highly vesicular foam, which was possibly already starting to fragment due to decompression of magmatic volatiles. Phase 4 contains an assemblage of pyroclasts characterised by a broad range of densities without any single sharply defined vesicularity peak, and in large part involved disruption of previously degassed, weakly to non-vesicular magma.

We infer that external water played a role in all phases of the 1886 eruption at Tarawera but the signature of the interaction is subtle, often not reflected in the grain size of the deposits. Proximal fall deposits are generally coarse and fines-poorer (like the "dry" units at Taupo), while parts of Unit 3 and the deposits from the SW end of the fissure are finer grained and richer in fine ash due to incorporation of high contents of pre-existing ash-grade lithic material (Fig. 2a). However, observations of the vesicularity of the basaltic clasts (Fig. 3b) and the abundance of rhyolitic wall-rock lithics are more sensitive measures of eruptive processes. Units 1 and 3 both contain abundant lithics and quenched basaltic bombs which we link to moderate interaction with ground water at times when the magma ascent rate was relatively low. However the distribution of vesicularity within their populations of basaltic clasts is quite different reflecting that the Phase 1 magma was newly arriving at the surface and probably actively vesiculating when it encountered ground water whereas partially degassed, and possibly vent-recycled, magma was erupted during the waning Phase 3.

The Tarawera Unit 2 deposits are more complicated and show clear evidence (e.g. the presence of multiple welded lenses) of being the products of mixing of clasts from numerous point sources along the eruptive fissure. Eruptive styles clearly varied with time and spatially along the fissure with two end-member styles, with identical grain size characteristics but different vesicularity signatures. One extreme is a relatively uniform population of lower-density scoria (mid-dry, Fig. 3b); the other has an equally well defined density peak but at significantly higher density (mid-wet, Fig. 3b).

The contrast between involvement of groundwater and surface water is demonstrated in the differences in abundance of wall rock fragments (lithics) in deposits at Taupo and Tarawera. In explosions involving ground water, disruption and incorporation of host-rock material of the aquifer is inevitable. Thus all the 1886 deposits at Tarawera contain a significant abundance of wall rock inclusions (up to 50 wt % or more of the deposits). In contrast, surface water may flood into a vent system without any country rock being incorporated. Thus in the Taupo "wet" deposits the abundance of wall rock fragments is low (less than 5 wt %) and no greater than in the associated "dry" units.

4 ROLE OF WATER IN PLUME TRANSPORT

The role of water in plume transport is still one of the most poorly explained aspects of wet eruptions, in part because of a lack of historical observed analogues, and in part because of computational complexities. In existing models plume heights depend on the thermal (sensible heat) flux and the proportion of the volatile phase (e.g. Sparks, 1986). Addition of water should in principle reduce the plume height through a reduction in the sensible heat flux and hence cause "wet" deposits to be less-widely dispersed, and may make the plume more unstable and prone to gravitational collapse, generating some kind of pyroclastic density current (surge or flow). However, deposits at Taupo and Tarawera show other factors are involved.

At Taupo, the "wet" deposits of phases 3 and 4 are still dispersed over substantial areas, comparable to those of similar-sized dry eruptions (Self & Sparks, 1978). The only difference is that the wet deposits tend to be more symmetrically dispersed about the vent (e.g. they are the only fall materials from this eruption to occur west of Lake Taupo; Fig. 2b). It is not clear in these cases whether the presence of water causes the plume to expand more rapidly, or whether the plumes were lower in height and thus less-affected by high winds aloft. Data from larger examples from the 26.5 ka Taupo eruption suggest the former is the case (Wilson, 1994). Some proximal plume instability is indicated by local near-source density current deposits in Phase 3, but these have not been seen in other wet Taupo units; in addition, note that both "dry" fall units (phases 2 and 5) were also associated with some degree of plume instability.

At Tarawera, additional fragmentation caused by magma:water interaction, coupled with the additional steam generated, caused the eruption plume from the NW part of the fissure to disperse pyroclasts much more widely than usual for "dry" basaltic events. At the SW end of the fissure, a high plume was generated, but much of the material was erupted with a considerable lateral component due to expansion of the superheated geothermal water and was emplaced as powerful density currents which swept out to >6 km from source. The cold groundwater at the NW end of the fissure served to enhance the power of the eruptive plume while clearly not enough water was incorporated to significantly cool the column (cf. presence of welded deposits). In contrast the superheated geothermal water caused intense fragmentation of the country rock and caused much of the erupted material to be laid down as density current deposits.

5 DEPOSITIONAL PROCESSES DURING "WET" ERUPTIONS

Evidence from Tarawera suggests that limited involvement of external water can leave scant evidence in the deposits, except for the changes in clast vesicularity and abundnaces of country-rock lithics. With greater involvement of water, "wet" deposits begin to show the effects of premature flushing of fine ash due to the presence of liquid water in the plume, resulting in a fining of the mean grain size and poorer sorting (Fig. 2a). Initially, water causes some accretion of fine ash as isolated mud pellets, such as those seen in the otherwise dry, well sorted lapilli fall deposits from the NW end of the Tarawera fissure. With increasing amounts of water, more and more fine ash is flushed from the plume as a range of forms, from loose aggregates, through accretionary lapilli and saturated ash to mud rain (Walker 1981). At its most extreme, so much water was landing during parts of phases 3 and 4 at Taupo that the fall deposition was creating contemporaneous sheet wash, resulting in net erosion rather than deposition.

In deposits emplaced by density currents, the critical factor is whether the water is present as steam or in liquid form, as this controls the cohesion of the accreting deposits. Both dry, superheated low-angle megaripples and steeper, wet, cohesive bedforms occur, the former particularly at SW Tarawera, the latter locally in Unit 3 of the Taupo eruption.

The poorer sorting shown by "wet" deposits discussed here is also shared by otherwise "dry" deposits where localised rain-flushing has occurred due to chance meteorological conditions (Walker, 1981). There is some controversy as to the extent to which the poorer sorting and finer grain size of typical "wet" deposits reflects a genuine reduction in grain size, or are merely an artefact of the premature flushing of fine ash that in "dry" events would be transported further from vent.

6 CONCLUSIONS

Magma:water interaction is probably more widespread than implied in the volcanological literature. Limited interaction, such as at NW Tarawera, leaves only subtle evidence in the form of decreased juvenile clast densities and higher abundances of included lithic fragments. "Wet" deposits are often less well documented, both in New Zealand and elsewhere, for two reasons. First, "wet" deposits are often fine-grained and hence susceptible to rapid erosion, such as the extensive gullying which affected the SW Tarawera deposits soon after 1886. Second, "wet" deposits are non-coherent but retain moisture and thus can become covered by vegetation very rapidly. At many volcanoes the extent and magnitude of magma:water interaction is thus under-estimated.

REFERENCES

Houghton, B.F. & Wilson, C.J.N. 1989. A vesicularity index for pyroclastic deposits. *Bull. Volcanol.* 51: 451-462.

Houghton, B.F., Wilson, C.J.N., McWilliams, M.O., Lanphere, M.A., Weaver, S.D., Briggs R.M. & Pringle, M.S. 1995. Chronology and dynamics of a large silicic magmatic system: central Taupo Volcanic Zone, New Zealand. *Geology* 23: 13-16.

Self, S. & Sparks, R.S.J. 1978. Characteristics of widespread pyroclastic deposits formed by the interaction of silicic magma and water. *Bull. Volcanol.* 41: 196-212.

Smith, R.T. & Houghton, B.F. 1995. Vent migration and changing eruptive style during the 1800a Taupo eruption: new evidence from the Hatepe and Rotongaio phreatoplinian ashes. *Bull. Volcanol.* 57: 432-439.

Sparks, R.S.J. 1986. The dimensions and dynamics of volcanic eruption columns. *Bull. Volcanol.* 46: 3-15.

Walker, G.P.L. 1980. Taupo pumice: product of the most powerful known (ultraplinian) eruption? *J. Volcanol. Geotherm. Res.* 8: 69-94.

Walker, G.P.L. 1981. Characteristics of two phreatoplinian ashes and their water-flushed origins. *J. Volcanol. Geotherm. Res.* 9: 395-407.

Walker, G.P.L., Heming, R.F. & Wilson, C.J.N. 1980. Low-aspect ratio ignimbrites. *Nature* 283: 286-287.

Wilson, C.J.N. 1994. Ash-fall deposits from large-scale phreatomagmatic volcanism: limitations of available eruption-column models. *U.S. Geol. Surv. Bull.* 2047: 93-99.

Wilson, C.J.N. & Walker, G.P.L. 1985. The Taupo eruption, New Zealand. I. General aspects. *Phil. Trans. R. Soc. Lond.* A314: 199-228.

Wilson, C.J.N., Houghton, B.F., McWilliams, M.O., Lanphere, M.A., Weaver, S.D. & Briggs, R.M. 1995. Volcanic and structural evolution of Taupo Volcanic Zone, New Zealand: a review. *J. Volcanol. Geotherm. Res.* 68: 1-28.

Metal reactions at the water-soil interface

R.G. McLaren
Department of Soil Science, Lincoln University, Canterbury, New Zealand

ABSTRACT: Soil contamination by heavy metals can result in subsequent contamination of water bodies, adversely affect soil biota and plant growth and introduce metals into the food chain. The long-term bioavailability and mobility of both native and contaminant metals in the soil depends on their ability to desorb from the soil solid phase into the soil solution. Methods for studying metal desorption from soils are reviewed and recent studies of metal desorption are described. The concentrations of both native and added metals (Cd and Zn) desorbed from soils are shown to be substantially reduced as soil pH increased. Desorbed metal concentrations also decrease with an increase in contact period between contaminant metal and the soil (Cd and Co). Studies of metal desorption kinetics show that metal desorption rates are also decreased by increasing soil pH or contact period.

1 INTRODUCTION

In many countries, including New Zealand, there is increasing interest in the land disposal of wastes as a means of recycling plant nutrients and organic matter, however, the presence of various contaminants in many wastes is seen as a cause for concern. Soil contamination resulting from waste disposal or some other activities can lead to subsequent contamination of surface and groundwater with contaminants such as nutrients, metals and a wide range of organic chemicals. In addition, high contaminant concentrations in the soil, particularly in the case of metals, may have adverse effects on soil biota and plant growth (McLaren & Smith, 1996).

The bioavailability and potential mobility of both nutrient and contaminant ions in the soil depends on their concentration and speciation in the soil solution, and on the soil's ability to release ions from the solid phase to replenish those removed by plant uptake or leaching. In the case of metal ions, it is now generally accepted that solution metal concentrations are most likely to be controlled by sorption-desorption reactions at the surface of both inorganic and organic soil colloidal materials (Swift & McLaren, 1991). This paper reviews some recent research relating to metal reactions at the water-soil interface, in particular concentrating on some of the factors that affect release of sorbed metals back into the soil solution.

2 METAL DESORPTION

2.1 *Methodology*

In comparison with the numerous studies of metal sorption by soils and soil components, there have been relatively few concerned with desorption. In some ways this is a paradox because it is the desorption process that controls the amount and rate of release of metals into the soil solution, and hence their potential bioavailability and mobility. Swift & McLaren (1991) have argued that the term desorption has been misused to describe studies in which sorbed metals have been 'extracted' (rather than desorbed) back into solution using reagents such as acids or complexing agents (e.g. EDTA or DTPA). Displacement of sorbed metals using weak electrolyte solutions would seem much more likely to simulate metal desorption into natural soil solution than the use of acids or synthetic complexing agents. A batch equilibration method to determine desorption of both native and recently sorbed Cu has been described by Hogg et al. (1993). This method involves repeated equilibration of soil samples in metal-free weak electrolyte (0.01M $Ca(NO_3)_2$. Following each equilibration, samples are centrifuged and the desorbed metal is determined in the supernatant equilibrium solution. The soil residue is then resuspended in fresh metal-free electrolyte solution and the process repeated. The resulting desorption data can be expressed as cumulative desorption *versus* number of equilibrations (Figure 1) or in the form of a desorption isotherm. This method has been used successfully to examine desorption of native and applied Zn (Singh et al., 1998) and Cd (Gray et al., 1998) from a range of New Zealand soils.

method has been used to examine the kinetics of Cd and Co desorption from iron and manganese oxides (Backes et al., 1995) and from soil clay fractions (McLaren et al., 1998). The second technique involves equilibration of soil/water suspensions with strips of ion-exchange resin membrane. The resin membrane has a high affinity for metals, and as metal ions are desorbed from the solid phase they are quickly removed from solution by the resin membrane. Thus metal concentrations in the solution remain low, stimulating further desorption. Periodic removal of the resin membrane strips from the soil suspensions and determination of the retained metals enables desorption rates to be determined. This system has been used to examine some of the factors affecting the kinetics of Zn desorption from soils (Singh et al., 1996).

2.2 Effect of pH

The effect of soil pH on the desorption of native and added Cd and Zn has been examined using the repeated batch desorption method (McLaren et al., 1997; Gray et al., 1998). In the case of Cd, soils at different pH values were sampled from a field liming experiment and for zinc, soil pH was adjusted in the laboratory by the addition of HCl or $Ca(OH)_2$. For both metals, there was a dramatic reduction in the amounts of native metal desorbed from soils as pH increased. An example for Cd is shown in Figure 2.

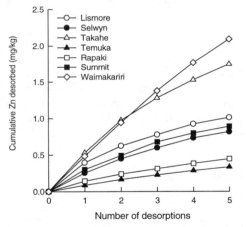

Figure 1. Cumulative desorption of native Zn from some New Zealand soils (Singh et al., 1997)

One of the disadvantages of the repeated equilibration batch desorption method is that it can not be used to study desorption kinetics. For this purpose two other techniques to examine desorption have been developed. The first technique involves continuous pumping of weak electrolyte solution through a cell in which the sample undergoing desorption is retained on a micropore filter. Monitoring of the solution leaving the cell enables the rate of metal desorption to be determined. This

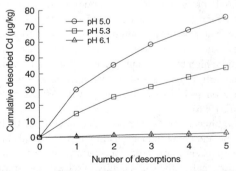

Figure 2. Effect of pH on desorption of native Cd from Te Kowhai silt loam (Gray et al., 1998)

For Cd at low pHs, up to 22% of total soil Cd could be desorbed from some soils but at pHs approaching

6.0, the proportions of total Cd desorbed decreased to much lower values (0-7%). In the case of Zn, the cumulative amounts of native Zn that could be desorbed by 10 equilibrations, even at low pHs, represented very small proportions only (3-5%) of total soil Zn concentrations. Desorption of added Cd and Zn from soils was also decreased substantially by increasing soil pH.

2.3 Effect of contact period

A number of studies have demonstrated that increasing the contact time between soil and sorbed metal decreases the metal's subsequent ability to desorb from the soil. McLaren et al. (1998) showed that for both Cd and Co sorbed by soil clay fractions, increases in sorption period were accompanied by substantial decreases in the proportions of sorbed metal desorbed. Similarly, Gray et al. (1998) have demonstrated a reduction in added Cd desorption from a range of New Zealand soils by increasing the contact period before desorption (Table 1). Such effects have implications for the long-term mobility of metals in soils. For example, in soils contaminated with Cu, Cr and As (in timber preservative), it has been shown that increased contact period between the soil and contaminant metals can reduce leaching of these metals from the soil (McLaren et al., 1994).

Table 1. Effect of contact period on cumulative Cd desorbed from soil (mg Cd/kg soil)

Soil	Contact period (days)	
	5	30
Te Kuiti	2.33	1.87
Waiareka	0.48	0.23
Takahe	2.09	1.77
Temuka	1.50	1.13

2.4 Desorption kinetics

It is considered by some authors (Sparks, 1985) that chemical equilibrium is seldom attained in systems such as soils, and thus the usefulness of equilibrium models such as sorption/desorption isotherms for predictive purposes may be limited unless models also incorporate relevant kinetic data. However, to date, there have been very few studies that have examined soil metal desorption kinetics. The only published studies (Kuo & Mikelsen, 1980 (Zn); Lehman & Harter, 1984 (Cu); Dang et al., 1994 (Zn)) have all used complexing agents (EDTA, DTPA or sodium citrate) to 'desorb' metals from the soil. Although it can be argued that natural complexing agents play an important role in solubilising metals in soil, the use of relatively high concentrations of synthetic complexing agents to stimulate desorption is likely to produce substantially higher rates of desorption than would be likely under normal soil conditions.

Using the continuous desorption technique described above, McLaren et al. (1998) found that desorption kinetics for Cd and Co sorbed onto soil clay fractions were described well by both a two-site first-order model (Figure 3) and a model involving a log-normal distribution of first-order rate constants. Reaction half-lives derived from the models suggested a movement of Cd and Co to slower desorption reactions, with increasing sorption period.

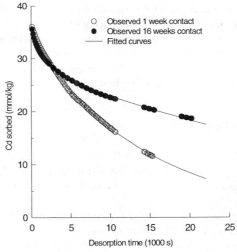

Figure 3. Effect of initial sorption period on desorption of Cd from Wakanui clay (McLaren et al., 1998)

The kinetics of Zn desorption from soil, studied using the ion-exchange resin membrane technique, have also been show to fit a two-site first-order rate model. In this study (McLaren et al., 1997),

Table 2. Fitted desorption reaction half-lives for Zn desorption by ion-exchange resin membrane from Selwyn soil (2-site first-order rate model).

	$t_{0.5}(k_1)$ h	$t_{0.5}(k_2)$ h
pH 4.4	0.03	0.44
pH 4.9	0.07	17.5

desorption reaction half-lives ($t_{0.5}k$) increased substantially with an increase in soil pH (Table 2).

3 CONCLUSIONS

Results such as those described above give clear indications of some of the factors likely to effect the continuing bioavailability and mobility of contaminant metals in soils. However, the research has consisted mainly of laboratory studies involving synthetic or extracted soil constituents, or homogenised soils, maintained at wide solution to soil ratios, quite atypical of natural conditions. The challenge is to show that the observations made in the laboratory do have direct relevance to the field situation. The successful modeling of the long-term fate of metals in the environment certainly requires a sound understanding of the nature and magnitude of any processes leading to the attenuation of metal bioavailability and mobility in soils.

4 REFERENCES

Backes, C.A., McLaren, R.G., Rate, A.W. & Swift, R.S. 1995. Kinetics of cadmium and cobalt desorption from iron and manganese oxides. *Soil Sci. Soc. Am. J.* 59:778-785.

Dang, Y.P., Dalal, R.C., Edwards, D.G. & Tiller, K.G. 1994. Kinetics of zinc desorption from vertisols. *Soil Sci. Soc. Am. J.* 58:1392-1399.

Gray, C.W., McLaren, R.G., Roberts, A.H.C. and Condron, L.M. 1998. Sorption and desorption of cadmium from some New Zealand soil. *Aust. J. Soil Res.* In press.

Hogg, D.S., McLaren, R.G. & Swift, R.S. 1993. Desorption of copper from some New Zealand soils. *Soil Sci. Soc. Am. J.* 57:361-366.

Kuo, S. & Mikkelsen, D.S. 1980. Kinetics of zinc desorption from soils. *Plant Soil* 56:355-364.

Lehman, R.G. & Harter, R.D. 1984. Assessment of copper-soil bond strength by desorption kinetics. *Soil Sci. Soc. Am. J.* 48:769-772.

McLaren, R.G., Backes, C.A., Rate, A.W. & Swift, R.S. 1998. Effect of sorption period on the kinetics of cadmium and cobalt desorption from soil clay fractions. *Soil Sci Soc. Am. J.* In press.

McLaren, R.G., Carey, P.L., Cameron, K.C., Adams, J.A. & Sedcole, J.R. 1994. Effect of soil properties and contact period on the leaching of copper, chromium and arsenic through undisturbed soils. *Trans 15th World Congress of Soil Science, Vol 3a*. 156-169.

McLaren, R.G., Singh, D. & Cameron, K.C. 1997. Influence of pH on the desorption of native and applied zinc from soils. *Extended Abstracts 4th International Conference on the Biogeochemistry of Trace Elements*, 503-504.

McLaren, R.G. & Smith, C.J. 1996. Issues in the disposal of industrial and urban wastes. In R. Naidu, R.S. Kookana, D.P. Oliver, S. Rogers & M. McLaughlin (eds), *Contaminants and the soil environment in the Australasia-Pacific region:*183-212. Dordrecht: Kluwer Academic Publishers.

Singh, D., McLaren, R.G. & Cameron, K.C. 1996. Kinetics of zinc desorption from soils using ion-exchange resin membranes. *Aust. & N.Z. Nat. Soils. Conf.* 2:247-248.

Singh, D., McLaren, R.G. & Cameron, K.C. 1998. Desorption of native and added zinc from a range of New Zealand soils in relation to soil properties. *Aust. J. Soil. Res.* In press.

Sparks, D.L. 1986. Kinetics of ionic reactions in clay minerals and soils. *Adv. Agron.* 38:231-266.

Swift, RS. & McLaren, R.G. 1990. Micronutrient adsorption by soils and soil colloids. In G.H. Bolt, M.F. De Boodt, M.B. Hayes & M.B. McBride (eds), *Interactions at the soil colloid-soil solution interface:*257-292. Dordrecht: Kluwer Academic Publishers.

Conditions for rapid large-volume flow

R.H.Sibson
Department of Geology, University of Otago, Dunedin, New Zealand

ABSTRACT: Because all brittle failure modes are fluid-pressure dependent, flow through stressed crust may be enhanced by a variety of structural conduits 'self-generated' by migrating fluids. Rapid localised flow requires gaping extension fractures often interlinked by shear and extensional-shear fractures within a mesh structure. Activation of such structures as fluid conduits requires the *tensile overpressure* condition ($P_f > \sigma_3$) to be met. At depth this can be achieved only in the absence of throughgoing cohesionless faults that are favourably oriented for reactivation within the prevailing stress field. Failure mode plots demonstrate that in extensional/transtensional regimes, large-volume flow through fault-fracture meshes is possible under hydrostatic fluid pressures in the near-surface to depths dependent on the tensile strength of the more competent material within the rock-mass. In contrast, large volume flow within compressional/transpressional regimes requires fluids overpressured to near-lithostatic values at all depths.

1 INTRODUCTION

Much hydrothermal mineralisation in the form of veining and associated zones of hydrothermal alteration is hosted within brittle fault-fracture meshes comprising varying proportions of pure extension and extensional-shear fractures interlinked by low-displacement shears (Sibson, 1996). These mesh structures may or may not be associated with major throughgoing faults but clearly form important localised conduits for flow of hydrothermal fluids. Three factors, solubility criteria (Henley, 1985), vein textures indicating open-space filling or repeated crack-seal episodes on extension fractures (Fournier, 1985; Robert et al., 1995), and the need for rapid transport between different P-T-X environments to promote instability and localised precipitation, suggest that many hydrothermal deposits are the product of multiple episodes of rapid, large-volume flow through locally gaping fault-fracture meshes. This paper investigates the mechanical circumstances under which such flow systems may develop.

2 STRESS/FLUID-PRESSURE CONTROLS ON BRITTLE FAILURE

Stress conditions for brittle failure in fluid-saturated intact rock under triaxial stress (principal compressive stresses, $\sigma_1 > \sigma_2 > \sigma_3$) with internal fluid pressure, P_f, are represented in Fig. 1 by a composite failure envelope for intact, isotropic rock normalised to tensile strength, T (Secor, 1965). The standard criteria for the three modes of macroscopic failure can all be expressed as functions of fluid pressure in relation to the state of differential stress and the material properties within the rock-mass, principally the tensile strength, T, a measure of rock 'competence' (Sibson, 1996).

When $(\sigma_1 - \sigma_3) > 5.66T$, shear fractures (faults) form in planes containing σ_2 and lying at angles $\theta_c = 0.5 \tan^{-1}(1/\mu_i)$ to the σ_1 direction in accordance with the Coulomb criterion:

$$P_f = \sigma_3 + [4T\sqrt{K} - (\sigma_1 - \sigma_3)]/(K - 1) \qquad (1)$$

where $K = [\sqrt{(1 + \mu_i^2)} + \mu_i]^2$, μ_i being the coefficient of internal friction.

When $5.66T > (\sigma_1 - \sigma_3) > 4T$, extensional-shear fractures may form at $0 < \theta < \theta_c$ to σ_1 according to the Griffith criterion:

$$P_f = \sigma_3 + [8T(\sigma_1 - \sigma_3) - (\sigma_1 - \sigma_3)^2]/16T \qquad (2)$$

When $(\sigma_1 - \sigma_3) < 4T$, extension fractures form in planes perpendicular to σ_3 in accordance with the hydraulic fracture criterion:

$$P_f = \sigma_3 + T \qquad (3)$$

Injection of fluid into an intact rock mass under differential stress ($\sigma_1 - \sigma_3$) can thus induce different modes of brittle failure depending on the ratio of differential stress to tensile strength.

3 FORMATION AND ACTIVATION OF FAULT-FRACTURE MESHES

Fluid infiltration into a stressed intact rock-mass may therefore 'self-generate' a variety of brittle structures which contribute to rock-mass permeability, especially in material of low intrinsic permeability. Provided ($\sigma_1 - \sigma_3$) < 5.66T_{max}, the higher tensile strength material will tend to fail by extensional or extensional-shear fracturing while lower strength material will fail in shear, leading to the formation of mixed-mode fault-fracture meshes. The degree of regularity within a developing fault-fracture mesh reflects the heterogeneity of the rock-mass which, in

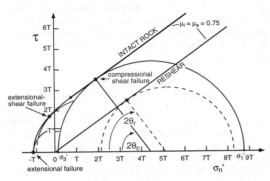

Fig. 1 - Generic Mohr diagram showing a composite Griffith-Coulomb failure envelope for intact rock normalised to tensile strength, T (after Secor, 1965) plus the reshear condition for a cohesionless fault, both with slopes $\mu_i = \mu_s = 0.75$ in the compressional field. Critical stress circles are shown for the three macroscopic modes of brittle failure in intact rock, plus a stress state for reshear of an optimally oriented cohesionless fault.

turn, affects the heterogeneity of the local stress field (Fig. 2). More regular meshes are likely to develop where the stress field is symmetric with respect to layering.

Note, however, that mesh generation or activation will be inhibited by the presence, within the infiltrated rock-mass, of cohesionless faults that are favourably oriented for frictional reactivation within the prevailing stress field. From consideration of the reshear condition for a cohesionless fault (Fig.1):

$$\tau = \mu_s(\sigma_n - P_f) \qquad (4)$$

(where τ and σ_n are respectively the components of shear and normal stress on the fault, and μ_s is the static coefficient of rock friction) it is apparent that the $P_f > \sigma_3$ fluid-pressure condition necessary for extensional or extensional-shear failure cannot be achieved in the presence of throughgoing faults that are favourably oriented for reactivation.

3.1 Brittle failure mode plots

The conditions for different modes of brittle failure represented by equations 1-4 can be transposed to plots of differential stress ($\sigma_1 - \sigma_3$) versus effective vertical stress ($\sigma_v' = (\sigma_v - P_f)$) for compressional ($\sigma_v = \sigma_3$) and extensional ($\sigma_v = \sigma_1$) tectonic regimes (Fig. 3). Failure modes in strike-slip regimes ($\sigma_v = \sigma_2$) may lie anywhere between these end-member situations. Uniform 'Andersonian' stress fields are assumed, without any stress heterogeneities arising from fault stepovers (e.g. Segall & Pollard, 1980) or other forms of irregularity. Transitions between different failure modes can then be equated to depth in the central tabulation for different values of the pore-fluid factor, $\lambda_v = P_f/\sigma_v$. Note how with increasing tensile strength, the failure condition moves further from the line representing the reshear of an optimally oriented, cohesionless fault. For each value of tensile strength, there is a transition with increasing σ_v' from pure extension fracturing through extensional-shear to compressional shear failure in both plots. For most sedimentary rocks long-term tensile strengths lie in the 1-10 MPa range, but crystalline rocks may have strengths of 20 MPa or more (Lockner, 1995).

The plots emphasize the marked contrast between the stress and fluid-pressure levels need to induce brittle failure in compressional and extensional tectonic regimes. At the same σ_v' value, far higher stress levels are needed to induce failure in compressional regimes where only compressional shear failure can occur unless $\sigma_v' < 0$, requiring $P_f > \sigma_v$ and $\lambda_v > 1.0$. However, in extensional regimes both extensional and extensional-shear fracturing may occur at shallow depths under hydrostatic fluid pressure levels ($\lambda_v \sim 0.4$). The plots also define the conditions where, for a specific range of tensile strengths, mixed-mode brittle failure leading to mesh development is likely to occur.

4 CONDITIONS FOR LARGE-VOLUME FLOW

Initial formation or reactivation of a fault-fracture mesh involve the opening or re-opening of hydraulic extension and extensional-shear fractures. This requires fluid pressure, at least locally, to exceed

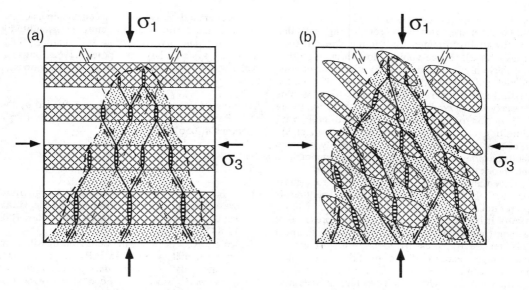

Fig. 2 - Development of fault-fracture meshes by the advance of fluid-pressure fronts (lightly stippled) through a stressed rock-mass (cross-hatched areas represent high-tensile strength competent units): (a) far-field stresses symmetric with respect to layering; (b) irregular competence heterogeneity. Light dashed lines represent orientations of favourably oriented faults which, if throughgoing, would inhibit mesh development.

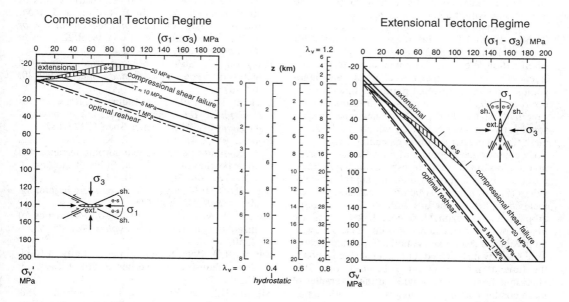

Fig. 3 - Brittle failure mode plots for compressional and extensional tectonic regimes, plotting differential stress at failure $(\sigma_1 - \sigma_3)$ against effective vertical stress σ_v' for various values of tensile strength, T. Expected orientations of extension fractures (ext.), Coulomb shear fractures or faults (sh.), and extensional shear fractures (e-s) are shown in the insets. Fields occupied by the different failure modes are labelled on the plots with the field of extensional-shear failure cross-hatched. Values for σ_v' can be equated to specific depth/fluid-pressure combinations in the central tabulation.

the least compressive stress (i.e. $P_f > \sigma_3$) which can only occur in the absence of throughgoing faults that are favourably oriented for frictional reactivation (Figs. 1 & 2). Large-volume flow through meshes can thus only be maintained: (i) within intact crust devoid of favourably oriented, cohesionless faults; (ii) in regions where existing faults have become *severely misoriented* for frictional reactivation (lying at $\theta > c.55°$ to the σ_1 direction) so that the $P_f > \sigma_3$ condition can still be maintained (Sibson et al., 1988); or (iii) where existing, favourably oriented faults have regained cohesive strength comparable to intact rock through hydrothermal cementation.

Given that bulk permeability of a rock-mass pervaded by a set of parallel planar fractures is proportional to the cube of fracture aperture (Snow, 1968), it is easy to appreciate how the presence of gaping fractures within a fault-fracture mesh under this 'tensile overpressure' condition will greatly enhance permeability. The brittle failure mode plots in Fig. 3 can thus be used to define the stress/fluid-pressure settings where fault-fracture meshes may develop and large-volume flow may occur. In extensional (and transtensional) regimes, flow through fault-fracture meshes may occur under hydrostatic fluid pressures to depths dictated by the maximum tensile strength of upper crustal material. For $T = 5$ MPa, this corresponds to depths of ~1.3 km. The coincidence of this shallow, large-volume flow regime with the boiling horizons of ascending, near-hydrostatically pressured geothermal plumes (Henley, 1985) defines the epizonal environment for mineralisation. Increased tensile strength through silicification, etc., has the potential to deepen the extent of mixed-mode mesh development, and large-volume flow in extensional regimes. Interestingly, Fournier (1991) has suggested a transition to suprahydrostatic fluid pressures near the base of some active geothermal fields which would allow mesh flow to greater depths in extending crust.

By contrast, development of fault-fracture meshes in compressional (and transpressional) regimes requires fluids overpressured to supralithostatic values ($P_f > \sigma_v$), unless structural irregularities create local stress heterogeneities with $\sigma_v < \sigma_3$. A large proportion of mesozonal lode gold deposits comprise meshes of flat-lying extension veins associated with fault-veins on steep reverse faults that demonstrably were severely misoriented for frictional reactivation at the time of mineralisation (Sibson et al., 1988; Hodgson, 1993). They provide further examples of mesh conduits acting as channels for episodic large-volume flow through fault-valve action with the meshes activated by near-lithostatic fluid over-pressures (Cox, 1995; Robert et al., 1995).

In general, flow through fault-fracture meshes is unlikely to be steady-state given the dynamic local interactions between stress and fluid-pressure cycling, strength variations from cementation, and the creation and destruction of fracture permeability by fracturing and hydrothermal self-sealing. Such complex systems have the potential for 'burping' valve activity over a range of different time periods.

Acknowledgements: This work was funded by the NZ Public Good Science Fund through FRST Contract #C05611.

5 REFERENCES

Cox, S.F. 1995. Faulting processes at high fluid pressures: an example from the Wattle Gully Fault, Victoria, Australia. *J. Geophys. Res.* 100: 12,841-12,859.

Fournier, R.O. 1985. The behavior of silica in hydrothermal systems. In B.R. Berger & P.M. Bethke (eds), *Geology and geochemistry of epithermal systems*, Soc. Econ. Geol. Reviews in Economic Geology 2: 45-61.

Fournier, R.O. 1991. The transition from hydrostatic to greater than hydrostatic fluid pressures in presently active hydrothermal systems in crystalline rock. *Geophys. Res. Lett.* 18: 955-958.

Henley, R. W. 1985. The geothermal framework of epithermal deposits. In B.R. Berger & P.M. Bethke (eds), *Geology and geochemistry of epithermal systems*, Soc. Econ. Geol. Reviews in Economic Geology 2: 1-24.

Hodgson, C.I. 1993. Mesothermal lode gold deposits. *Geol. Assoc. Canada Spec. Paper* 40: 635-678.

Lockner, D.A. 1995. Rock failure. In, *Rock physics and phase relations: a handbook of physical constants*, AGU Reference Shelf 3: 127-147.

Robert, F., A-M. Boullier & K. Firdaous 1995. Gold-quartz veins in metamorphic terranes and their bearing on the role of fluids in faulting. *J. Geophys. Res.* 100: 12,861-12,879.

Secor, D. T. 1965. Role of fluid pressure in jointing. *Am. J. Sci.* 263: 633-646.

Segall, P. & D.D. Pollard 1980. Mechanics of discontinuous faults. *J. Geophys. Res.* 85: 4337-4350.

Sibson, R.H. 1996. Structural permeability of fluid-driven fault-fracture meshes. *J. Struct. Geol.* 18: 1031-1042.

Sibson, R.H., F. Robert. & K.H. Poulsen, 1988. High-angle reverse faults, fluid-pressure cycling, and mesothermal gold-quartz deposits. *Geology* 16: 551-555.

Snow, D.T. 1968. Rock fracture spacings, openings, and porosities. *J. Soil. Mech. Found. Div., Proc. Am. Soc. Civil Engrs.* 94: 73-91.

Clean, green and steaming: Environmental geochemistry in New Zealand

Jenny Webster
Institute of Environmental Science and Research (ESR), Auckland, New Zealand

ABSTRACT: The evolution of environmental geochemistry in New Zealand has been guided by the initiative of individual geochemists, the development and refinement of trace metal analytical techniques and changes made to government science funding and environmental legislation. Landmark research and development in geothermal and Antarctic environmental geochemistry between 1960 and 1990 is reviewed. A trend towards interdisciplinary research and problem prevention rather than cure, has been identified for environmental sciences in general. This is represented in environmental geochemistry by recognition of the role of bacteria in many geochemical processes, and by development of predictive geochemical models.

1 INTRODUCTION

Conservation of the natural environment is an issue which has had a consistently high public and political profile in New Zealand (NZ). Although we have had very few large scale industrial environmental disasters, the general quality of surface water and groundwater is beginning to show the effects of land clearance and intensive agriculture. Power generation has also affected river flows and quality. Rivers have been dammed for hydroelectricitic power, and used for cooling water or to receive fluid wastes from coal, gas and geothermal power stations.

I believe the particularly rapid growth of the environmental geochemistry over the last 20 years, has been influenced by three factors:

- The initiative of individual geochemists
- The development of analytical techniques for low level trace metals
- The restructuring of government science funding and environmental legislation.

As it would be somewhat over-ambitious to attempt to provide a complete review of NZ environmental geochemistry here, I have focused on two areas in which NZ geochemists were able to make very significant contributions to the international literature in the 1960s: geothermal geochemistry and Antarctic geochemistry.

Wairakei Geothermal Power Station was commissioned in 1958, ushering in a period of intense geochemical study of geothermal systems and their reservoirs. Working in the first country to develop liquid-dominated geothermal fields, NZ scientists were also the first to report the environmental consequences of discharging spent bore fluid into a natural river system.

In Antarctica, geochemists from Victoria, Massey and Waikato Universities were the first to describe and determine the unusual chemistry of the inland, ice-covered, stratified lakes of the McMurdo Dry Valleys (eg. Wilson & Wellman, 1962). In the course of these investigations lake algae were also "accidentally collected on a nylon line", heralding the first of many interdisciplinary biogeochemical studies on these lakes.

2 EVOLUTION

Environmental geochemistry was pioneered by geochemists involved in research with an entirely different purpose. The geochemical methodology used to study ore-forming processes and to improve geochemical exploration techniques, for example, has proved to be particularly well suited to studying the impacts of mining activity on the environment.

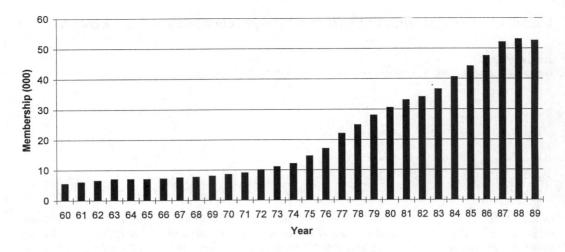

Figure 1. Annual membership of the NZ Royal Forest & Bird Society (courtesy NZRF&B Soc.)

To chose a local example; Weissberg & Wodzicki (1968) determined trace metal concentrations in soils and river sediments near the Tui mine in the Coromandel region, and studied metal dispersion mechanisms. The study was undertaken in the hope that these results would *"...facilitate the interpretation of geochemical prospecting data collected in other, similar areas"*, and was followed by many other stream sediment surveys in the Coromandel region (eg., as recently compiled by Christie & Mitchell, 1992). Twenty years later this same data is being used in surveys of the impact of mineralisation and mining on the Coromandel catchments (eg., Livingston, 1987).

2.1 *Beginnings*

There was a sudden and very significant increase in public environmental awareness in the 1970s. This is reflected in the membership of various environmental NGOs such the NZ Royal Forest & Bird Society (Figure 1), and coincides with the first appearance of what we would now term "environmental geochemistry" publications by NZ scientists in local and international journals.

High levels of mercury were reported for the first time in sediments and trout of the Waikato river (Weissberg and Zobell, 1973; Brook et al., 1976). The Waikato receives all cooling water, condensate and spent bore effluents discharged from the Wairakei Power Station and field, as well as natural discharge from a many other geothermal fields.

Attention was then paid to Hg in other geothermal catchments and those draining Hg mineralisation such as the Puhipuhi cinnabar deposit (Hoggins & Brooks, 1973). Indeed, after the Hg poisoning incident at Minimata Bay in Japan, the levels of Hg in fish or shell fish in general became a subject of some considerable interest, and numerous publications.

Axtmann (1975) conducted the first environmental audit of Wairakei Power Station, and Ellis and Mahon's (1977) landmark text on "Chemistry and Geothermal Systems" recognised some of the water quality issues associated with the disposal of spent bore fluids.

On the Antarctic continent, attention was still focussed on the Dry Valley saline lakes. Research undertaken with the NZARP (NZ Antarctic Research Programme) at this time included chemical, bathymetric and isotopic characterisation of specific lakes (eg. Hendy et al 1977). Climatic histories were established on the basis of historical and recent lake level fluctuations and salinity gradients.

The need to establish base line data for contaminant levels in these inland drainage systems was noted. However, in light of more recent evidence, it would appear that early attempts to measure trace metal concentrations and profiles were unsuccessful. Analytical techniques were not quite up to the task of either the very low level analysis required, or analysis in highly saline fluids, highlighting a severe limitation of environmental geochemistry at this time. Trace metal concentrations could be reliably determined in sediments, or in the tissues of plants and animals, but not in waters unless they had been severely contaminated.

2.2 *Diversification*

The 1980's were a period of significant diversification, largely due to improved analytical technology. With the development and refinement of methods for low level analysis of trace elements in natural waters, more detailed geochemical surveys of groundwater, effluent and surface water quality could be made (eg. Dickson & Hunter, 1981; Williamson, 1986).

The effects of geothermal flows on lake and catchment water quality in the Central North Island region were reported for the first time (eg. Timperley & Vigor-Brown, 1986); a subject which continues to be of interest and debate (eg. Timperley & Huser, 1996; Deely & Sheppard 1996). The development of analytical methods for the low level analysis of As in water, and for the separate determination of As^{III}, As^V and organic As (Aggett & Aspell, 1976), initiated research on the fate and transport of As in rivers receiving As-rich geothermal fluids. Quantitative reduction of As^V to the more toxic As^{III} in the sediments and base waters of some of the hydroelectricity lakes on the Waikato River was reported (Aggett & O'Brien, 1985). During seasonal turn-over higher As concentrations are released into the main body of the river (McLaren & Kim, 1995).

Improved analytical and "clean" sampling capabilities were also put to the test in Antarctica, where NZARP now had a considerably expanded geochemical programme. The development of a photoaccoustic field method for low level Hg determination (Patterson, 1984), for example, facilitated measurement of Hg in air and snow without the attendant problems of sample contamination during transport and analysis in NZ (eg., Dick et al., 1990). Mercury levels were up to an order of magnitude lower than had been previously measured in Antarctica.

Trace metal concentrations could now be reliably determined in lake waters and saline brines (eg. Figure 2). Metal concentration profiles were shown to be unique for each metal, governed by individual metal desorption and sulphide precipitation characteristics (Webster, 1994).

The spectre of acid mine drainage also raised its head for the first time in NZ at this time. Seepage from the base of an abandoned tailings dam at the Tui mine (Cu-Pb-Zn) at Te Aroha had been identified as the cause of high metal levels and acidity in the receiving stream waters (Hendy, 1981). Concerns were raised for the health of aquatic life in the catchment and commercial fisheries off shore, as well as for the grass on the local golf course which used the stream water for irrigation!

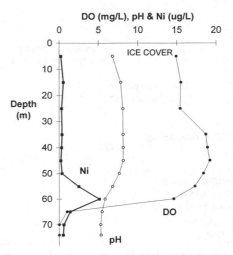

Figure 2. Dissolved Ni profile in Lake Vanda, as determined in 1987 (Webster, 1994), relative to depth profiles of pH and DO.

3 EXTERNAL PRESSURES

With the restructuring of government science and science funding in NZ in the 1990s, and the introduction of a contestable funding pool, funding criteria became more aligned to political and public requirements. Appropriately termed the "Public Good Science Fund", allocations from this fund were preferentially made to research programmes demonstrating applicability and benefit to the "public of NZ". A review of the allocation of PGSF funding between 1991/92 (when the fund was set up) and 1997/98, reveals that, although total funding has not significantly increased over this period, the proportion allocated to environmental research has approximately doubled.

In 1992, the various sections of environmental legislation were combined into the Resource Management Act (RMA). The new Act invested a great deal of responsibility in regional government authorities for the protection of air, water and land environments under their jurisdiction. Consequently the Act increased the stringency of the discharge consent application and award process.

The requirements of the RMA have highlighted some very important gaps in the current knowledge of how contaminants are released, transported and immobilised in natural environments: gaps which can be, and are being, addressed through geochemical research.

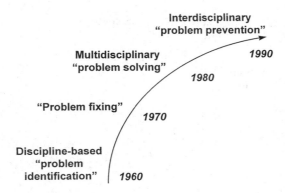

Figure 3. Trends in the field of environmental science (courtesy Prof. John Hay, SEMS).

3.1 *A merging of disciplines*

In a recent address, Prof. John Hay of the School of Environmental and Marine Sciences at the University of Auckland proposed that there were two principal trends in the nature of environmental science: a trend towards interdisciplinary studies, and a movement from problem identification and "fixing" to prevention (Figure 3). Aspects of this trend are evident in environmental geochemistry as we plot the field's evolution into the 1990s.

The need for multidisciplinary study is particularly evident. There has been, for example, increasing attention paid to the role of bacteria in what had previously been considered to be indisputably inorganic chemical processes. Thermophilic bacteria have been implicated in the deposition of silica and calcite sinters in NZ geothermal systems (eg., Jones et al., 1996). Likewise the oxidation of As^{III} to As^{V} in natural waters is catalysed by bacteria, as is methylation of both As and Hg (which forms the toxic bioaccumulative methylmercury species). Geothermal systems have proven to be an ideal environment in which to study the abilities and tolerances of "extremophilic" bacteria.

At the other end of the temperature scale, "extremophilic" bacteria on the Antarctic continent have developed tolerance to extremes of cold, salinity and desiccation (Hawes et al, 1992). Recent studies of these bacteria, including the prolific cyanobacteria, have highlighted their ability to precipitate salts such as barite from undersaturated Antarctic meltwaters (Tazaki et al.,1996). Combined biogeochemical studies of the interactions of these bacteria with natural and contaminant trace metals, in the tradition of the early Antarctic scientists at Lake Vanda, are ongoing.

A survey of the papers to be presented at this conference suggests that this renewed interest in biogeochemistry is by no means restricted to the geothermal and Antarctic bacteria.

3.2 *Prevention not cure...*

The second trend, that of problem prevention rather than cure, is reflected in the PGSF preference for process-oriented, rather than descriptive or survey-oriented research. The need to understand rather than measure is emphasised, so that results can also be used in other environments.

More specifically, there has been increased attention paid to the ability of geochemical computer models to accurately predict the speciation of contaminants. Speciation influences contaminant toxicity, dispersion and therefore degree of environmental impact. Although specifically developed for this purpose, it is widely acknowledged that careful application is required for natural systems. The models generally assume thermodynamic equilibrium, well-characterised solution and surface chemistry, and full representation of all possible reactions in the model database. Clearly there are few natural environments in which all of these assumptions will be justified. However, such models can be successfully used to model individual reactions and processes within a system.

MINTEQA2 (Allison et al., 1991), for example, has been used to model the solubility of metal sulphides in pore waters of contaminated intertidal muds of the Manukau Harbour in Auckland (Williamson et al.,1994), and trace metal adsorption in the acidic seepage from the Tui tailings dam (Webster, 1994). Even when metal behaviour can not be accurately modelled, the discrepancy itself can provide information on the reaction mechanisms. Inaccuracies in modelled As adsorption from Si-rich geothermal fluids, for example, has highlighted the ability of Si to both adsorb to and polymerise on oxide surfaces. In so doing, Si inhibits the adsorption of As, and possibly other contaminants (Swedlund et al., 1996).

4 THE FUTURE

Where to from here? The need for more accurate geochemical models is generally acknowledged to be important. To achieve this, research must be directed towards testing their use, updating and correcting the databases and improving their ability to model some of the more complex (including uniquely NZ) systems. Incorporating the influence of bacteria into these models will certainly be a challenge.

Like other areas of environmental science, environmental geochemistry will continue to be influenced by emerging issues in the biological sciences. For the last ten years, for example, there has been a concerted international effort to develop water quality criteria to protect aquatic life. There is a growing realisation that solution and surface complexation of trace metals decreases their toxicity. Attempts are being made to correct criteria for this, but further data is required (particularly on the toxicity of solution complexes). Sediment toxicity, and the inability of estuaries to simply flush contaminants away, are also emergent issues. Assessing contaminant impacts in terms of load or flux, rather than concentration, may be the next challenge for environmental managers and scientists alike.

In closing I would like note that it has been suggested that geochemistry holds the solution to many global environmental problems, from mineral fertilisers to new energy sources and the use of CO_2-utilizing bacteria to clean fossil fuel emissions (Fyfe, 1996). Something to think about…

Acknowledgements and apologies

Committing a review such as this to paper is inherently fraught with danger. I am bound to have omitted the work of many very capable New Zealand geochemists, and for this I apologise. I would like to acknowledge the influence of: Dr Alan Mann, Prof. Terry Seward, Dr Kirk Nordstrom, Dr Kevin Brown and Prof. Bill Fyfe.

REFERENCES

Aggett J. and Aspell A. (1976) The determination of As(III) and total As by atomic absorption spectroscopy. *Analyst* 101, 341-347.

Aggett J. and O'Brien G.A. (1985) Detailed model for the mobility of arsenic in lacustrine sediments based on measurements in lake Ohakuri. *Environ. Sci. Technology* 19, 231-238.

Allison J.D., Brown D.S., Novo-Gradac K.J. (1991) MINTEQA2/PRODEFA2, a geochemical computer assessment model for environmental systems. *EPA/600/3-91/021* USEPA Publication, Athens, Georgia.

Axtmann R.C. (1975) Environmental impact of a geothermal power plant *Science* 187, 795.

Brooks R.R., Lewis J.R. and Reeves R.D. (1976) Mercury and other heavy metals in trout of central North Island, New Zealand. *NZ Jour. Marine & Freshwater Research* 10, 233-244.

Christie A.B. and Mitchell A.G. (1992) A geochemistry database of the Coromandel Penninsula. *Proc. NZ Aus. Inst. Mining & Metallurgy 26th Annual Conference.*

Deely J.M. and Sheppard D.S. (1996) Whangaehu River, New Zealand: geochemistry of a river discharging from an active crater lake. *Applied Geochemistry* 11, 447 - 460.

Dickson R.J. and Hunter K.A. (1981) Cu and Ni in surface waters of Otago Harbour. *NZ Jour. Marine and Freshwater Research* 15, 475-480

Ellis A.J. and Mahon W.A.J. (1977) *Chemistry and Geothermal Systems* Academic Press, New York.

Fyfe W.S. (1996) Conservation of the Earth's surface: the ultimate biosphere resource. *Plenary presentation at 4th Int. Symp. on the Geochemistry of the Earth's Surface.* Yorkshire, UK.

Hendy C.H., Wilson A.T., Popplewell K.B. and House D.A. (1977) Dating of geochemical events in Lake Bonney, Antarctica and their relation to glacial and climate changes. *NZ Jour. Geology and Geophysics* 20, 1103 - 1122.

Hendy C.A. (1981) The Tui mine - after the miners have left. *NZ Environment* 29, 17-19.

Hoggins F.E. & Brooks R.R. (1973) Natural dispersion of mercury from Puhipuhi, Northland NZ. *NZ Jour. Marine & Freshwater Research* 7, 125-132.

Jones B., Renault R.W. and Rosen M.R. (1996) High temperature (>90°C) calcite precipitation at Waikite Hot Springs, north island, New Zealand. *Jour. Geol. Soc. London* 153, 481-496.

Livingston M.E. (ed) (1987) Preliminary studies of the effects of past mining on the aquatic environment, Coromandel Peninsula. *Water & Soil Misc. Publication No.* 104.

McLaren S.J. and Kim N.D. (1995) Evidence for a seasonal fluctuation of arsenic in NZ longest river and the effect of treatment on concentrations in drinking water. *Jour. Environ. Pollution* 90, 67-73.

Patterson J.E. (1984) A differential photoacoustic mercury detector. *Analytic. Chim. Acta.* 174, 459-468.

Dick, A.L. Sheppard D.S. and Patterson J.E. (1990) Mercury content of Antarctic ice and snow: Initial results. *Atmospheric Environment* 24, 973-978.

Swedlund P.J., Webster J.G. and Miskelly G.M. (1996) Arsenic removal from geothermal bore water: the effect of dissolved silica. *Proc. 18th NZ Geothermal Workshop*, 89-93.

Tazaki K., Webster J., and Fyfe W.S. (1996) Transformational processes of microbial barite to sediments in Antarctica. *Japanese Mineralogical Journal* 12, 63-68.

Timperley M.H. and Huser B.A. (1996) Inflows of geothermal fluid chemicals to the Waikato River catchment, New Zealand. *NZ Jour. Marine and Freshwater Research* 30, 525-535.

Timperley M.H. and Vigor-Brown P.J. (1986) Water chemistry of lakes in the Taupo Volcanic zone. *NZ Jour. Marine and Freshwater Research* 20, 173-184.

Webster J.G. (1994) Trace metal behaviour in oxic and anoxic Ca-Cl brines of the Wright Valley drainage, Antarctica. *Chemical Geology* 112, 255-274.

Webster J.G. (1995) Chemical processes affecting trace metal transport in the Waihou River and estuary, New Zealand. *NZ J Marine and Freshwater Research* 29(4), 539-553.

Weissberg B.G. and Wodzicki A. (1968) Geochemistry of some soils and stream sediments in the vicinity of the Tui mine, Te Aroha. *D.S.I.R. Chemistry Report No.* CD 2104.

Weissberg & Zobell (1973) Geothermal mercury pollution in New Zealand. *Bulletin Environmental Contamination and Toxicology* 9, 148-155.

Williamson R.B. (1985 & 1986) Urban stormwater quality I & II. *NZ Jour. Marine and Freshwater Research* 19, 413-427 & 20, 315-328.

Williamson R.B., Hume T.M. and Mol-Krijen J. (1994) A comparison of the early diagenetic environment in intertidal sands and muds of the Manukau Harbour, New Zealand. *Env. Geology* 24, 254-266.

Wilson A.T. & Wellman H.W. (1962) Lake Vanda: an Antarctic lake. *Nature* 196, 1171-1173.

1 Surficial systems

The Arno River catchment basin, Tuscany, Italy: Chemical and isotopic composition of water

A. Adorni-Braccesi, L. Bellucci & C. Panichi – *Istituto Int. per le Ricerche Geotermiche, CNR, Pisa, Italy*
G. La Ruffa & F. Podda – *Dipartimento di Scienze della Terra, University of Cagliari, Italy*
G. Cortecci & E. Dinelli – *Dip. di Scienze della Terra e Geologico-Ambientali, University of Bologna, Italy*
A. Bencini & E. Gimenez Forcada – *Dipartimento di Scienze della Terra, University of Firenze, Italy*

ABSTRACT: D, ^{18}O and ^{34}S contents, together with chemical concentrations of major components, were measured in 23 sample sites in the Arno river and in 11 main tributaries. Isotopic and chemical variations in the river are very similar, showing a progressive increase in heavy isotopes and solutes downstream of the source to the mouth of the river. The lateral inputs from tributaries appear to be important and three main sub-basins can be identified by simple mass-balance calculations. Seawater intrusion in the river and its relationship with confined and unconfined aquifers are discussed.

1 INTRODUCTION

In recent years, the decline in water quality of many Italian rivers as a result of human activity has been a subject of much concern. Overflows and floods have also occurred frequently, at times with catastrophic consequences, as in 1966 when Florence, Pisa and a large part of Tuscany were flooded. The National Research Council (CNR) of Italy has launched several projects of this type, including an inventory of the areas affected by landslides and floods. The objectives of this project are (1) to clarify the hydrodynamic relationship between the river water, main tributaries and nearby aquifers; (2) to delimit the extent of the seawater intrusion in the river with respect to the coastal aquifers of the basin; and (3) to identify any sources of natural and human pollution in order to assess the environmental risks.

Twelve months into this study of the Arno river basin, the hydrochemical features of the system have already been better defined; the results have also led to an extensive chemical survey of trace elements in the water and bed sediments in order to map anomalous situations that may be related to pollution phenomena (Cortecci et al., this volume).

2 DESCRIPTION OF THE STUDY AREA

The Arno catchment basin drains a total area of 8228 km² with a maximum elevation of 1500 m a.s.l.

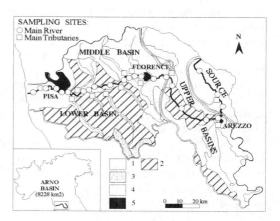

Figure 1. Geological sketch map of the Arno river basin and location of sampling sites. (1 alluvial recent sediments; 2 alluvial consolidated sediments; 3 evaporites; 4 sandstones; 5 quartzites.).

(Fig. 1). The Arno is 242 km long and its flowrate is highest in January (3.4 m³/s 10 km from the source and 195 m³/s near the mouth) and lowest in August (0.3 m³/s 10 km from the source and 10 m³/s near the mouth).

Extreme water flows were recorded during the November 1966 flood, with 312 m³/s 10 km from the source, 4100 m³/s at Florence and 2290 m³/s near The geology is dominated by sedimentary lithologies. Turbiditic sandstone and marls outcrop in the upper course, while the middle part is characterised by marine and lacustrine sediments, shales, marly-calcareous, arenaceous and chaotic clays with scattered sandstone.

The right-hand tributaries (Bisenzio, Ombrone and Usciana) drain turbiditic sandstone and minor limestone, whereas marine and lacustrine clayey and sandy deposits and minor Messinian evaporite outcrops are drained by the left-hand tributaries (Greve, Elsa, Pesa, Egola and Era). The coastal plain consists of alluvia, with gravel and a silty sandy matrix.

The area is heavily industrialised and urbanised, Arezzo, Florence and Pisa being the main cities.

3 SAMPLING METHODS AND ANALYTICAL TECHNIQUES

In November '96 and March '97, 23 surface water samples were collected from source to mouth of the Arno river for isotopic and chemical analyses.

The 11 main tributaries were sampled just before the confluence.

Electrical conductivity, pH and Eh were measured in situ. For isotopic analysis dissolved sulfate was separated by means of an ion-exchange resin, then eluted and precipitated as $BaSO_4$.

Routine procedures were used to measure the isotopic compositions and the values are reported in δ (‰) notation. Analytical precision was ±0.1‰, ±1‰, and ±0.3‰ for $\delta^{18}O$, δD and $\delta^{34}S$, respectively.

4 CHEMICAL AND ISOTOPIC COMPOSITION OF WATER

Very similar isotopic variations can be observed in water samples collected between the source area and the coastal plain in November '96 and March '97, under similar flow rate conditions (Fig. 2). A progressive increase in δ values from -50.1‰ to -35.9‰ for hydrogen and from -7.8‰ to -5.6‰ for oxygen are observed in November and from -50.9‰ to -36.7‰ and from -7.9‰ to -5.9‰ in March. The increase in δD and $\delta^{18}O$ 156 km from the source is probably due to an incomplete mixing at the sampling point between the Arno and its tributary, the latter being enriched in heavy isotopes. The relative enrichments in D and ^{18}O about 25 km upstream of the mouth can be ascribed to seawater contribution.

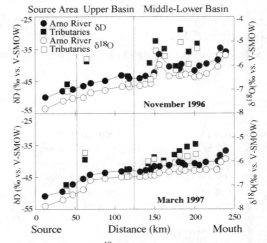

Figure 2. δD and $\delta^{18}O$ variations of the Arno river and main tributaries with respect to the distance from the river source to the sea. (Dashed area indicates seawater intrusion).

Chloride and sulfate concentrations in the Arno river increase from source to mouth following the same trend as the hydrogen and oxygen isotopic compositions (Fig. 3).

Based on the evolutionary trends of the isotopic and chemical data, the Arno catchment basin can be subdivided into three parts: Source Area, Upper Basin and Middle-Lower Basin.

An overall range from -11.0‰ to +16.3‰ can be observed for ^{34}S of dissolved sulfate in the Arno along its entire course (Fig. 3). Most of the depleted values refer to the Source Zone (-11.0‰ to -2.0‰); they correspond to the lowest contents of aqueous sulfate, thus suggesting a supergenic origin of sulfate from oxidation of pyrite in country rocks. Comparable $\delta^{34}S$ values from -7.5‰ to +1.6‰ were observed by Schwarcz and Cortecci (1974) for sulfate in spring waters emerging from sandstone in the uppermost part of the nearby Serchio river Basin. The supergenic source for sulfate ions possibly prevails downstream of Florence, keeping the $\delta^{34}S$ values around -2.5‰.

The increase of $\delta^{34}S$ in the Middle-Lower Basin

probably reflects anthropogenic and natural contributions, both sources providing sulfate enriched in ^{34}S. The $\delta^{34}S$ values as high as +13‰ in the Elsa tributary are consistent with the presence of evaporites in the upper reaches of the watershed (Fig.1).

The $\delta^{34}S$ signature of these evaporites is in the range +17 to + 23‰ (Testa & al., 1996).

The enrichment of 12‰ observed in ^{34}S for samples collected in the vicinity of the mouth clearly indicates mixing with sulfate of seawater.

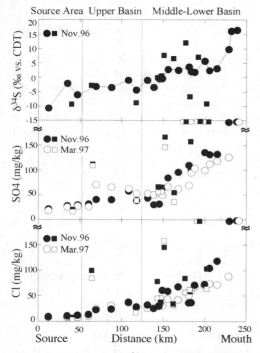

Figure 3. Variations of $\delta^{34}S$ values and chloride and sulfate concentrations in the Arno river and tributaries from the source to the sea. Dashed area indicates seawater intrusion. (In the Middle-Lower Basin some ionic concentrations are out of scale).

5 MASS BALANCES BY MEANS OF CHEMICAL AND ISOTOPIC DATA

Using the $\delta^{18}O$ data, chloride and sulfate contents and the flow rate at the beginning (δ_{in}, C_{in}, Q_{in}) and end (δ_{out}, C_{out}, Q_{out}) of each sub-basin, the mass balance was calculated for the river (Table 1).

Comparison between calculated values ($\delta^{18}O_{calc}$, Cl^-_{calc} and $SO_4^{2-}_{calc}$) for the Arno river and the average contents of the tributaries in each of the three sub-basins indicates that: i) the isotopic variations observed in the river essentially reflect the $\delta^{18}O$ composition of the tributaries, which varies as a function of the differences in the mean elevations of the three basins; ii) the measured and calculated chloride and sulfate concentrations show significant discrepancies.

In the Source Area, the apparent excess of solutes may simply be due to the contribution of a number of small streams having the same isotopic composition, but slightly different salt contents. In the Upper Basin, the inputs from the tributaries cannot explain the observed chemical deficiency of about 70%, and additional contributions of diluted water from the close network of rills must be invoked. Finally, in the Middle-Lower Basin the roughly 35% excess of chloride and sulfate ions in the Arno river probably comes from anthropogenic sources, as indicated by the sulfur isotopic data.

Table 1. Mass balance of the relative contribution of different tributaries to the Arno river, using $\delta^{18}O$, Cl^- and SO_4^{2-} data.

	$\delta^{18}O_{in}$	$\delta^{18}O_{out}$	$\delta^{18}O_{cal}$	$\delta^{18}O_{meanTrib}$	Δ %
SA	-7.80	-7.38	-7.34	-7.10	3.3
UB	-7.38	-6.90	-6.63	-6.30	5.0
M-LB	-6.90	-6.46	-6.52	-5.70	1.4
	Cl_{in}	Cl_{out}	Cl_{calc}	$Cl_{meanTrib}$	Δ %
SA	8.16	11.35	11.68	7.80	33.2
UB	11.35	28.01	37.30	63.83	-71.1
M-LB	28.01	103.90	248.09	168.7	32.0
	SO_{4in}	SO_{4out}	SO_{4calc}	$SO_{4meanTrib}$	Δ %
SA	21.61	29.30	30.10	22.09	26.6
UB	29.30	38.42	43.50	75.65	-73.9
M-LB	38.42	131.12	307.25	187.44	39.0

S.A.=Source Basin (Q_{in}= 3.2 m^3/s ; Q_{out}= 34 m^3/s);
U.B.=Upper Basin (Q_{in}= 34 m^3/s; Q_{out}= 95 m^3/s);
M.-L.B.= Middle-Lower Basin (Q_{in}= 95 m^3/s; Q_{out}= 145 m^3/s).

6 SALT INTRUSION AND RIVER / GROUND-WATER INTERACTION

In the coastal plain near Pisa, seawater intrusion is

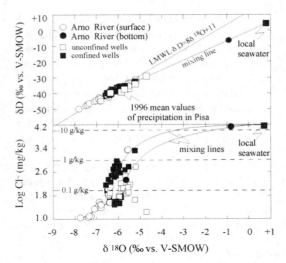

Figure 4. δD, δ^{18}O and chloride content in the Arno river and nearby groundwaters

morphologically constrained by a rise in terrain about 25 km from the mouth of the Arno.

Isotopic and chemical data clearly indicate a mixing between seawater and river (Fig. 4).

Ten kilometres from the sea, the seawater component at shallow levels is as high as 15%, and 80% at deeper levels (Grassi & Rossi, 1997). As regards any interactions between the Arno river and nearby groundwaters, the phreatic and confined aquifers appear to be affected to different degrees by the river (Fig. 4). Mixing processes should be responsible for the deviation from the local meteoric water line (LMWL) of the studied well waters, which show ^{18}O and D enrichments as high as 1.5‰ and 10‰, respectively.

The relationships between Cl and δ^{18}O are in line with the inferred river/groundwater interactions, but they also reveal that isotopically different feeding waters are involved (Fig. 4). The isotopic values of the phreatic aquifer may be related to summer-autumn meteoric inputs and/or to partial evaporation in the unsaturated zone. On the other hand, the confined groundwaters appear to be recharged by waters with an isotopic composition that is very similar to the local precipitation in Pisa in 1996 (-6.5‰ and -41.9‰ for δ^{18}O and δD, respectively), and to the Arno in Pisa in both November '96 and March '97. There are no data available so far to distinguish between these two different sources.

7 CONCLUSIONS

A study of the isotopic and chemical compositions of the waters and solutes associated with the Arno river basin indicates that the watershed can be divided into three parts: the Source Area, with a total surface area of 883 km^2, a mean altitude of 900 m a.s.l., and precipitous morphology; the Upper Basin, covering 3195 km^2 and with a mean altitude of 700 m a.s.l.; and the Middle-Lower Basin, surface area of 4150 km^2, a mean altitude of 200 m a.s.l., with a hilly morphology and two large plains around Florence and Pisa. In each of these areas, the Arno is an admixture of the different water types present in its catchment basin, but mainly rainfall and the main tributaries.

The chemistry of the Arno waters appears to be determined by geochemical and biological processes within the catchment basin (e.g. weathering of minerals, ion exchange and plant uptake). Along the Middle-Lower Basin, chemical data suggest that anthropogenic sources may have a significant influence on the concentrations of some major ions (e.g. Cl and SO$_4$) and of some heavy metals such as Pb, Cr, Zn (see Cortecci & al., this volume).

REFERENCE

Cortecci, G., Dinelli, E., Lucchini, F., Fanfani, L., La Ruffa, G., Podda, Bencini, A., Gimenez Forcada E. & A. Adorni-Braccesi 1998. Geochemistry of the Arno River, Italy: natural and anthropogenic contributions, this volume.

Grassi, S. & S. Rossi 1997. Il cloro delle acque sotterranee della pianura di Pisa. *Atti Soc. Tosc. Sci. Nat., Mem., Serie A.* 103: 1 - 8.

Schwarcz, H.P. & G. Cortecci 1974. Isotopic analysis of spring and stream water sulfate from the Italian Alps and Apennines. *Chem. Geol..* 13: 285 - 294.

Testa, G., Barbieri, M. & G. Cortecci 1996. The origin of Messinian evaporites from Tuscany (Italy) based on new stratigraphic and geochemical data. In *Proc-17th Regional African European Meeting of Sedimentology.* Sfax, Tunisia. March 26-28, 1996: 264 - 265.

Trace metals dissolved in the rainwaters on northern Sardinia (Mediterranean Sea)

R.Caboi, C.Ardau, L.Rundeddu & F.Frau
Dipartimento di Scienze della Terra, Università di Cagliari, Italy

ABSTRACT: Rainwater chemistry in northern Sardinia is little affected by crustal aerosols because of the low influence of sporadic pulses of desert dusts in this area. Most rain events exhibit pH < 5 and the total concentration of most trace metals is almost completely in dissolved form. This can be important in relation to the effect of heavy metals on the marine ecosystem as their solubilities are scarcely reduced by pH-dependent adsorption and precipitation processes in the presence of crustal aerosols.

1 INTRODUCTION

Concentrations of trace elements in rainwater derive from the type of aerosol that has been scavenged from atmosphere. In the Mediterranean area, atmospheric aerosols are supplied mainly by fluxes of marine, anthropogenic and crustal origin. The anthropogenic "background" of European origin is supplied to the atmosphere continuously, while the crustal flux of African origin (Saharan dust) has an intermittent nature, and is related to seasonal climatic changes that control the dispersion of the material introduced in the atmosphere. Meteoric water scavenges a largely variable mix of particulate supplied by the three sources.

The solid state speciation of trace metals in aerosols affects their dissolution in rainwater; heavy metals from anthropogenic-dominated aerosols are more soluble than those from crustal-dominated ones. In the Mediterranean atmosphere, when European anthropogenic aerosol is perturbed by mixing with desert dusts, trace metal chemistry in rainwaters (that scavenged the "mixed" aerosol) is affected by the total (i.e. dissolved + particulate) trace metal composition and by the pH increase. The latter controls trace metal solubility by means of pH-dependent adsorption and precipitation processes (Chester et al. 1996).

The influence of desert dusts is particularly important in southern Sardinia because of its proximity to the African shore. Recent studies (Caboi et al. 1995; Frau et al. 1996) have shown (i) that the dissolved contents of some crustal trace metals in rainwater (Sr and Mn) increase as a consequence of mineral dust dissolution and their solubilities are little affected by the pH increase, (ii) that other crustal trace metals, such as Al and Fe, exhibit an "inert" character, whereas (iii) the anthropogenic trace metal concentrations (Pb, Cu, Zn and Cd) in rainwater solution decrease owing to their negligible release from crustal particulate and to their decreased solubility as the pH increases. This paper aims at studying the behaviour of a set of trace metals (Al, Mn, Sr, Ba, Rb, Co, Ni, V, Mo, Cr, Zn, Cd, Cu and Pb) in northern Sardinia, in an area

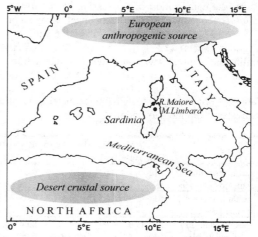

Fig. 1 Location of sampling stations in the Mediterranean Sea; main anthropogenic and crustal sources of atmospheric aerosols.

characterized by a smaller influence of desert particulate on rainwater, it being farther from its source, and by a greater importance of anthropogenic aerosol of European origin (Fig. 1).

2 METHODS

In the period between October 1994 and December 1996, 149 rain events were sampled, and collected daily with a *wet&dry* apparatus in two stations in northern Sardinia, one maritime (RM, Rena Maiore, 65 events, total rainfall: 1347 mm), the other mountain – forestal (ML, M.te Limbara, 84 events, total rainfall: 2319 mm).

Metals were analysed on filtered (0.4 µm, Nuclepore) rainwater samples. After filtration, the samples were acidified at 1% with an HNO_3 Carlo Erba RS Superpure. ICP-MS Perkin-Elmer "ELAN 5000" was used for the following metals analyses: Al, Mn, Mo, V, Cd, Cu, Pb, Ni, Co, Ba, Sr, Rb, Zn and Cr. Major chemical components were analysed by ion-chromatography and AAS.

Summary statistics, cluster and factor analyses were performed by a UNISTAT package on logarithmic and normalized values of the chemical data.

3 RAINWATERS IN THE TWO STATIONS

3.1 *Marine and continental contribution to the salinity of rainwaters*

Assuming Na and Cl in rainwater entirely of marine origin, the contribution of the marine source to the dissolved content of a major chemical component "X" in a rainwater sample "i" is given by: $sX_{i(rain)} = (X/Na)_{sea} * Na_{i(rain)}$. The contribution of the continental source was calculated by subtraction:

Table 1 - Marine (sSal), anthropogenic (aSal) and mineralogic (mSal) contributions to the salinity, and mean pH values in the rainwaters of northern (ML and RM) and southern Sardinia (CC).

	ML 84 events mm 2319 forestal	RM 65 events mm 1347 maritime	CC* 16 events mm 184 maritime
Sal (µeq l^{-1})	697	700	2363
sSal/Sal %	58	69	84
aSal/cSal %	72	77	29
mSal/cSal %	28	23	71
pH	5.26	5.03	5.58

*After Caboi et al. (1995), samples collected in southern Sardinia

$cX_i = tX_i - sX_i$, where tX is the analytical concentration. In Table 1 the mean total salinities (Sal) and the percentages of marine (sSal = ΣsX), anthropogenic (aSal = $H^+ + NH_4^+ + NO_3^- + cSO_4^{2-}$) and mineralogic (mSal = $cCa^{2+} + cMg^{2+} + cK^+ + cHCO_3^-$) contributions to the continental salinity (cSal = aSal+mSal) at the two stations are compared with those of rainwater in the station of Capo Carbonara in southern Sardinia (October 1993 - May 1994). Marine contribution in rainwater is prevalent also inland of the mountain - forestal area of M.te Limbara. The effect of the dissolution of minerals contained in crustal aerosols turns out to be significantly lower in rainwater in northern Sardinia, where the pH is lower and the contribution of anthropogenic origin to the non-marine salinity higher.

3.2 *Trace metals*

Mean compositions of trace elements, standard deviations and detection limits (D.L. = arithm. mean + 3*σ of blank signal) are shown in Table 2.

Table 2 - Summary statistics of trace metal concentrations (µg l^{-1}) and detection limits (D.L.) in the rainwaters from the two stations

		Al	Co	Ba	cRb	sRb	Mn	cSr	sSr	V	Mo	Cr	Ni	Zn	Cu	Cd	Pb
	D.L.	0.1	0.01	0.03	0.01		0.01	0.03		0.01	0.01	0.4	0.03	2.8	0.04	0.02	0.1
arithmetic mean	ML	7	0.04	1.7	0.20	0.05	4.4	2.0	3.1	0.75	0.03	0.5	0.4	39	1.3	0.23	1.1
standard deviation		9	0.03	4.2	0.71	0.08	6.6	3.8	4.9	0.66	0.03	0.6	0.3	63	2.2	0.92	1.1
arithmetic mean	RM	8	0.05	1.5	0.07	0.05	2.1	2.1	3.2	0.68	0.03	0.5	0.6	34	1.1	0.18	1.5
standard deviation		10	0.08	1.5	0.07	0.05	2.7	4.1	3.5	0.65	0.02	0.7	0.5	39	2.3	0.41	1.4
weighted mean	ML	6	0.03	1.4	0.15	0.04	4.4	1.7	2.6	0.70	0.03	0.5	0.4	23	0.9	0.12	1.1
for mm of rain	RM	5	0.03	1.1	0.07	0.05	1.5	2.0	3.1	0.61	0.03	0.5	0.5	24	1.2	0.10	1.4
geometric mean	ML	4	0.03	1.0	0.06	0.02	2.3	0.3	1.2	0.57	0.02	0.4	0.3	16	0.8	0.06	0.7
	RM	5	0.03	1.0	0.05	0.03	1.3	0.5	1.9	0.52	0.02	0.3	0.5	24	0.6	0.07	0.9

Considering the large standard deviations, there are non-significant differences in the mean values of trace elements in the two stations. For Rb and Sr, whose concentrations are conservative in sea water, the marine and continental contributions were calculated as for major elements. A half detection limit value was assumed for concentrations lower than D.L. in calculating the average values; this assumption only affects chromium values, where 65 % were lower than D.L..

4 CLUSTER AND FACTOR ANALYSES

A "cluster analysis" was performed on the complete set of data (using log-values of main continental chemical components as normalized variables) to distinguish homogeneous groups of rainwater. Data elaboration led to distinguish three groups of rain events with different chemical characters (Table 3).

Group 1 is made up of 23 events and presents a salinity prevalently of marine and anthropogenic origin (the latter being 90% of non-marine salinity), and a mean pH value of 4.3. These events can be defined as "anthropogenic rains".

Group 3 comprises 8 events and shows an important crustal contribution to salinity (~ 80% of non-marine salinity) which increases the pH to 6.9. These events are indicated as "crustal rains".

Group 2, that includes most of the sampled events (~80%) and shows chemical characters (pH ~ 5.1, mSal = 24% of non-marine salinity) indicating a low crustal contribution, is between these two "end-members" groups of rains. These events can be defined as "mixed rains".

Cluster analysis has allowed us to distinguish the different types of rain events better than with the distinction in CPR (Crust-Poor Rain) and CRR (Crust-Rich Rain) reported in Frau et al. 1996. Eight events have not been grouped by cluster analysis and can be statistically considered "outliers".

4.1 Anthropogenic and crustal trace metals

Different chemical characters of the three types of events affect trace metal concentrations and the value of enrichment factors (EF_{crust}= $(Me/Al)_{rain}/(Me/Al)_{crust}$ where $(Me/Al)_{rain}$ are the ratios relating to each trace metal in the rainwater, and $(Me/Al)_{crust}$ are their ratios in average crustal material) used to estimate the non-crustal contribution to the dissolved components in the rains, assuming Al concentration entirely from crustal source (Table 4).

The concentration of aluminium is actually higher in "anthropogenic rains" despite its crustal origin because its solubility is reduced by the higher pH in "crustal rains"; thus the EF_{crust} values have poor efficiency in discriminating the origin of trace metals soluble in rainwater. This is particularly clear for Sr in Group 3 with EF_{crust} = 883.

Solid state speciation in aerosols and rainwater pH greatly influence the solubility of anthropogenic trace metals such as Cu, Zn, Pb, and Cd (see e.g., Chester et al. 1996, Caboi et al. 1995). Dissolved concentrations of most crustal trace metals (Rb, Sr, Ba, Mn, Cr) are higher in Group 3 than in Group 1, while most anthropogenic trace metals (Cu, Pb, Zn, Cd, V) show an opposite behaviour. This agrees with the above-mentioned processes that affect trace metal solubility in rainwater.

To better investigate the behaviour of trace metals, principal component factor analysis (R-mode) was performed, considering standardized variables relating to the non-marine concentrations of the major chemical ions in the rainwaters and all trace

Table 3 - Mean pH values, anthropogenic (aSal) and mineralogic (mSal) contributions to total salinity (Sal) in the three groups of rains by cluster analysis.

	Group 1 23 events rain: 409 mm	Group 2 110 events rain: 2888 mm	Group 3 8 events rain: 243mm
pH	4.32	5.12	6.90
Sal (µeq l^{-1})	1095	515	736
cSal (µeq l^{-1})	248	96	438
aSal/cSal %	90	76	23
mSal/cSal %	10	24	77

Table 4 - Geometric means of trace metal concentrations (µg l^{-1}) and EF_{crust} values in Groups 1, 2 and 3

	Al	cRb	cSr	Ba	Mn	Co	V	Cr	Ni	Mo	Cu	Pb	Zn	Cd
Group 1	9.55	0.09	0.21	1.02	1.83	0.04	1.10	0.25	0.65	0.03	1.00	2.69	29.23	0.12
Group 2	3.30	0.04	0.28	0.83	1.53	0.03	0.45	0.34	0.34	0.01	0.64	0.69	17.72	0.06
Group 3	3.55	0.18	12.57	3.20	3.65	0.02	0.61	0.45	0.21	0.03	0.26	0.11	7.21	0.02
Group 1-EF_{crust}	1	6	6	17	18	22	82	25	97	136	227	1218	1670	4216
Group 2-EF_{crust}	1	8	21	39	45	45	98	99	147	150	420	903	2863	6306
Group 3-EF_{crust}	1	31	883	140	99	36	122	124	85	385	156	131	1108	1519

Table 5 - Factor loading matrix (Varimax rot.) of the main components (bold italic font for significant values)

	Factor 1	Factor 2	Factor 3	Factor 4
eigenvalues	6.45	4.35	1.59	1.39
% of var.	30.7	20.7	7.5	6.6
cumul. %	30.7	51.4	58.9	65.5
Al	*0.53*	0.16	*0.65*	-0.04
Ba	0.06	*0.75*	0.21	0.12
cCa	0.06	*0.92*	-0.04	0.21
Cd	-0.09	-0.07	*0.34*	0.21
cHCO_3	-0.14	*0.81*	-0.22	0.23
cK	0.06	0.18	0.12	*0.87*
cMg	*0.40*	*0.60*	-0.13	0.30
Co	0.09	0.27	*0.78*	-0.06
cRb	0.19	*0.32*	0.08	*0.82*
cSO_4	*0.71*	*0.40*	*0.38*	-0.02
cSr	0.02	*0.92*	-0.14	0.13
Cu	0.10	-0.13	0.25	0.01
H^+	*0.80*	-0.25	0.14	-0.24
Mn	0.09	*0.44*	0.12	*0.62*
Mo	*0.75*	0.15	-0.16	0.19
NH_4	*0.59*	0.27	*0.35*	0.26
Ni	*0.42*	-0.07	*0.48*	0.03
NO_3	*0.77*	0.14	*0.32*	0.22
Pb	*0.81*	-0.20	*0.34*	-0.12
V	*0.86*	0.02	0.01	0.23
Zn	0.12	-0.17	*0.59*	0.17

metals (with exception of Cr) of 141 events (excluding outliers). Table 5 presents the loading factor matrix for the four factors which account for most of the variability.

Factor 1, which explains most of the variance, and factor 3 are characterized by higher loading factors of variables linked to the anthropogenic components in rainwaters (H^+, NO_3, cSO_4, NH_4); trace metals linked to factor 1 are mainly of anthropogenic origin and their solubility is clearly pH dependent; V and Mo present a significant linear correlation in mixed types of events ($r^2 = 0.61$) and may both derive from polluting industrial power plants (Abbaticchio et al. 1989). Zn, Cd, Ni and Co, which show higher loads in factor 3, are less affected by H^+ activity. Though Al is entirely of crustal origin, it is linked to these factors only by its higher solubility at a low pH.

Factor 2 explains the solubilisation of the mineral carbonatic fraction of crustal aerosols, while factor 4 is linked to the mineral silicatic fraction and probably to Fe-Mn oxides and hydroxides.

5 CONCLUSIONS

The behaviour of trace metals in rainwaters compared in three different types of events, statistically distinguished by cluster analysis as "anthropogenic rains" (mean pH 4.3) "mixed rains" (mean pH 5.1) and "crustal rains" (mean pH 6.9), can be summarized as follows: i) the low pH in Group 1 promotes the solubilisation of both anthropogenic and crustal trace metals, but the dissolved contents of the latter remain low because the crustal component in the scavenged aerosol is negligible; ii) the high pH in Group 3 reduces the solubilisation of anthropogenic metals, while crustal metals are seemingly much more soluble, but this depends on their abundance in the scavenged aerosol; iii) Al solubility is strongly decreased when the pH increases and its dissolved contents in Group 2 and 3 are clearly lower than in Group 1 although the amount of Al in a crustal aerosol is generally one order of magnitude greater than that in an anthropogenic aerosol; iv) factor analysis points out that Al, Pb, V and Mo are particularly linked to H^+ activity, while solubility of Zn, Cd, Ni and Co seems less affected by rain acidity; v) Ba, Sr, Mn and Rb are clearly linked to major metals of crustal origin such as Ca, Mg and K.

Acknowledgements: This research was supported by the financial contribution of the European Community (PIC-INTERREG 1)

6 REFERENCES

Abbaticchio, P.E. & V. Amicarelli, 1989. Problemi connessi alla corretta valutazione dell'emissione atmosferica di metalli pesanti in Europa. *Inquinamento*, 3, 118-121.

Caboi, R., F. Frau and A. Cristini, 1995. Metal concentrations in rainwater (Sardinia, Western Mediterranean Sea). In "Water-Rock Interaction", *Proceed. of the 8th Internat. Symp. on Water-Rock Interaction - WRI-8, Vladivostok, Russia 15-19 august 1995*, Y.K. Kharaka & O.V. Chudaev (eds.), A.A.Balkema, Rotterdam, 337-340.

Chester, R., M. Nimmo and S. Keyse, 1996. The influence of Saharan and Middle Eastern desert-derived dust on the trace metal composition of Mediterranean aerosols and rainwaters: An overview. In *the Impact of Desert Dust Across the Mediterranean*, S. Guerzoni and R. Chester (eds.), 1996, Kluwer Academic Publishers, Netherlands, 253-273.

Frau, F., R. Caboi and A. Cristini, 1996. The impact of Saharan dust on trace metal solubility in rainwater in Sardinia, Italy. In *The Impact of Desert Dust Across the Mediterranean*, S. Guerzoni and R. Chester (eds.), 1996, Kluwer Academic Publishers, Netherlands, 285-290.

Geochemistry of the Arno River, Italy: Natural and anthropogenic contributions

G.Cortecci, E.Dinelli & F.Lucchini – *Dipartimento di Scienze della Terra e Geologico-Ambientali, University of Bologna, Italy*

L.Fanfani, G.La Ruffa & F.Podda – *Dipartimento di Scienze della Terra, University of Cagliari, Italy*

A.Bencini & E.Gimenez Forcada – *Dipartimento di Scienze della Terra, University of Florence, Italy*

A.Adorni Braccesi – *Istituto Internazionale per le Ricerche Geotermiche, CNR, Pisa, Italy*

ABSTRACT: Waters and bed sediments from the Arno River and its main tributaries were analysed for major and trace elements, basically with the purpose of evaluating the anthropogenic effects on the quality of the water-sediment system. Areal distributions of selected elements are given and anomalous situations clearly related to pollution are described.

1 INTRODUCTION

Within a research project of environmental geochemistry aimed at monitoring the waters and bed sediments of the Arno basin (northern Tuscany) by elemental and isotopic analyses with repeated sampling in different seasons for a term of three years, we report here the first results on chemistry of waters and stream sediments. The aims of the project are to define the hydrological behaviour of the studied rivers and their metal mobilization, in order to prevent and reduce environmental risks. Isotopic data are discussed by Adorni Braccesi et al. (1998) elsewhere in this volume.

A previous study on the Arno River by Bencini & Malesani (1993) revealed close connections between geochemical and hydrologic parameters, systematic changes in the chemical composition from river source to river mouth, and anomalous inputs of some chemical species near urban areas.

The Arno River flows into the Tyrrhenian Sea, after 242 km across heavily industrialized land (textiles, paper-mills, tanneries, mechanical plants) and densely urbanized territory (the main towns are Arezzo, Florence and Pisa), with highly specialized and widespread agricultural activities.

The lithology of the basin (8228 km^2) is dominated by sedimentary rocks (Fig. 1). Sandstones and marls outcrop in the upper part of the basin. In the middle part, near Florence, the rocks are represented by sandstones, shales and calcareous turbidites interbedded in chaotic clays with scattered ophiolitic blocks. Downstream from Florence, the right-hand tributaries drain sandstones and minor cherty limestones, while the left-hand tributaries drain essentially fine-grained marine and lacustrine clayey and sandy deposits and, in the upper sections of the sub-basins, clastics, limestones and minor evaporitic outcrops. The coastal plain corresponds to a graben filled by marine and continental deposits.

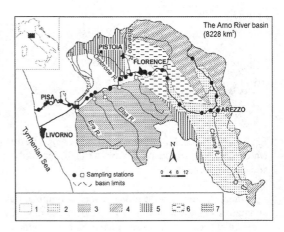

Fig. 1 - Sampling stations and schematic geology of the Arno river basin. 1 - Alluvial deposits; 2 - Pleistocene continental and lacustrine deposits; 3 - Marine Plio-Quaternary deposits; 4 - Cervarola turbiditic sandstones; 5 - Macigno Fm. turbiditic sandstones; 6 - Chaotic shaly rocks with calcareous and ophiolitic olistoliths; 7 - Metamorphic rocks of the Tuscan sequence.

2 SAMPLING AND ANALYSES

The Arno River and eleven tributaries were sampled for water and sediment in November 1996 for a total of twenty-three stations along the main river from source to mouth and one station for each tributary just before the confluence (Fig. 1).

Major cations were analysed by AAS, alkalinity by titration, chloride by argentometry and sulfate by turbidimetry. Trace element were measured by ICP-MS on filtered samples (0.45μm).

The bed sediments were sampled by a small dredge mostly from the centre of the rivers. Major and trace elements were determined in the <80 mesh fraction by X-ray fluorescence spectrometry on pressed powder pellets.

3 RESULTS AND DISCUSSION

3.1 *Chemical composition of waters*

The samples from the Arno exhibit a downstream increase in dissolved contents with a nearly regular change in composition (Fig. 2). The samples from some tributaries deviate significantly from this evolutionary trend due to the different lithologies of the sub-basins or anthropogenic inputs.

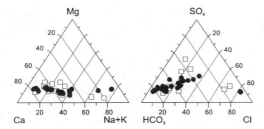

Fig. 2 - Triangular diagram for the Arno basin water samples. Filled circles = Arno River; open squares = tributaries.

Trace element distributions reveal a number of anomalies as shown in Fig. 3 for selected elements. Most of the twenty analysed elements are enriched in the right-side tributaries downstream from Florence, and the main river, that preserves a memory of the anthropic input, partially acts as a diluting container for some tens of kilometres. Interestingly, the Li/Cl ratio may be used as a contamination marker. As shown in Fig. 4, it decreases discontinuously from source to mouth in the Arno River, describing a stepwise path. The

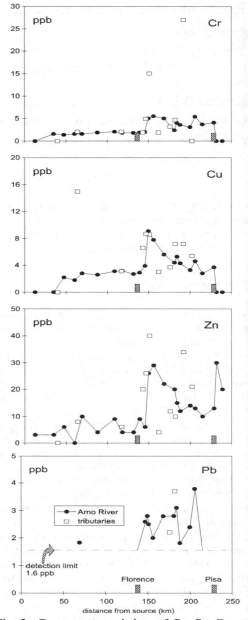

Fig. 3 - Downstream variations of Cr, Cu, Zn and Pb for the Arno River and its tributaries.

steps identify basin areas with increasing anthropic burden, except for the last step that approaches the seawater value. In the tributaries, the Li/Cl ratio depends on the proportions of natural and anthropogenic contributions, the latter being enriched in chloride. It should be pointed out that, except for As and Zn, the contents of metals dissolved in the Arno downstream from Florence and in most tributaries are higher than those observed in the Po River and its most polluted tributary the Lambro river (Pettine et al., 1996).

Strength Elements) and REE are enriched. Elsewhere these elements as well as LILE (Large Ion Lithofile Elements) do not appear appreciably affected by pollutants.

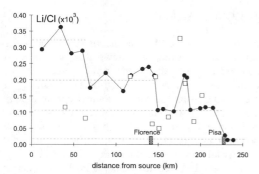

Fig. 4 - Li/Cl downstream variations for the Arno and its tributaries (symbols as in Fig. 2).

3.2 Chemical composition of bedload sediments

The main mineralogy of sediments from the Arno River does not change significantly from source to mouth and consists mainly of plagioclase, clays and quartz (±carbonate minerals). The sediments from the left-hand tributaries show a slightly higher carbonate content.

Accordingly, the major element composition of the sediments from the Arno and its tributaries is fairly homogeneous, and is mostly controlled by the lithologies crossed by the rivers. Exceptions to the observed chemical homogeneity are represented by high P and Mn contents in the sediments from some tributaries and the Arno River in the vicinity of Florence.

In the upper part of the basin the contents of trace elements largely involved in human activities (Cr, Cu, Zn and Pb) match the values observed in the country rocks (van de Kamp & Leake, 1995; Andreozzi et al., 1997). Cu, Zn and Pb increase sharply above background values at, or just before, Florence, and their anomalies can be traced downstream for about 30 km (Fig. 5). Along this tract the river receives highly polluted tributaries (Ombrone and Bisenzio rivers). In the following 60 km, sediment quality improves and no other critical zones are identified. Cr behaves differently and displays only one clear-cut anomaly in the lower part of the river, where also HFSE (High Field

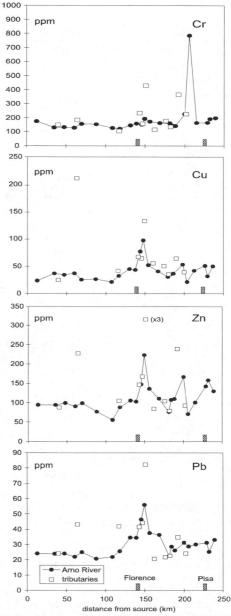

Fig. 5 - Metal content in the bed sediments of the Arno River and its tributaries.

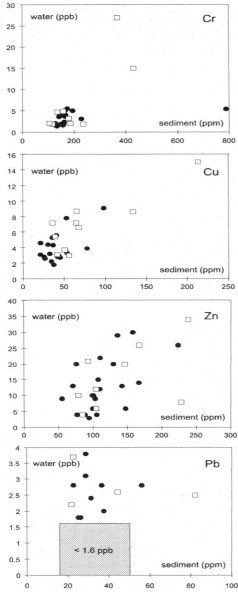

Fig. 6 - Relations between metal concentrations in waters and sediments (symbols as in Fig. 2).

3.3 *Chemical relationships between waters and sediments*

The relations between water and sediment concentrations for Cr, Cu, Zn and Pb (Fig.6) reflect the geochemical behaviour of these elements in surface environments (Adriano, 1986). The low dissolved contents and the lack of correlation for Cr and Pb match the their low aqueous mobility. The satisfactory correlations observed for Cu and Zn support the existence of adsorption and desorption processes between sediment and water.

4 CONCLUSIONS

A first survey of the geochemistry of the Arno River and its tributaries shows that the chemical composition of dissolved solids and bedload sediments is generally affected by the lithology of the sub-basins and locally by anthropogenic inputs. The latter are largely predominant in the middle and lower part of the basin, around and downstream from Florence, where industrial activity and urban settlements are widespread.

The supply of contaminants from the tributaries causes an increase in "anthropogenic" metals in the sediments of the main river Arno, where they are trapped not far from the confluence. This conclusion is supported by the nearly parallel distributions of metals in the sediments and waters along the Arno River.

REFERENCES

Adorni Braccesi, A., Bellucci, L., Panichi, C., La Ruffa, G., Podda, F., Cortecci, G., Dinelli, E., Bencini, A., & E. Gimenez Forcada 1998. The Arno River catchment basin, Italy: chemical and isotopic compositions of waters (this volume).

Adriano, D.C. 1986. *Trace elements in the terrestrial environment*. Springer-Verlag, New York.

Andreozzi, M., Dinelli, E. & F. Tateo 1997. Geochemical and mineralogical criteria for the identification of ash layers in the stratigraphic framework of a foredeep; the Early Miocene Mt. Cervarola sandstones, northern Italy. *Chem. Geol.* 137: 23-39.

Bencini, A. & P. Malesani 1993. *Fiume Arno: acque, sedimenti e biosfera*. Accademia "La Colombaia". Firenze: Leo S. Olschki Editore.

Pettine, M, Bianchi, M., Martinotti, W., Muntau, H., Renoldi, M. & G. Tartari 1996. Contribution of the Lambro River to the total pollutant transport in the Po watershed (Italy). *Sci. Total Environ.* 192: 275-297.

van de Kamp, P.C. & B.E. Leake 1995. Petrology and geochemistry of siliciclastic rocks of mixed feldspathic and ophiolitic provenance in the Northern Apennines, Italy. *Chem. Geol.* 122: 1-20.

Geochemistry of riverine particulate and dissolved loads, Darling River Basin, Australia

C.E. Martin
Research School of Earth Sciences and Department of Geography, The Australian National University, Canberra, A.C.T., Australia

M.T. McCulloch
Research School of Earth Sciences, The Australian National University, Canberra, A.C.T., Australia

ABSTRACT: This paper discusses the geochemistry of riverine particulate and dissolved loads of rivers in the Darling River basin of Australia, with the goals of identifying their sources and assessing the possible impacts phosphate fertilizer usage in the region has had on river chemistry.

1 INTRODUCTION

The geochemistry of river waters serves to monitor both natural rock weathering processes that take place over long time spans, and the comparatively short-lived but often intense perturbations to Earth's surface caused by human activity. In order to understand the global cycles of many trace elements, particularly elements that have low natural abundances in surface waters and are of common use in society, it is important to investigate the extent to which these abundances have been altered from their original state.

The Darling River basin of eastern Australia is a good area to study both natural and human impacted riverine processes. The largest inland river in Australia, the Darling provides water for the agricultural industry in an area where water supply by rains tends to vary dramatically from year to year, as droughts are followed by periodic floods that may fall in restricted regions of the catchment and then slowly transported downsteam. Phosphate fertilizers are used in parts of the drainage basin, and have been suggested as the cause of nuisance algal blooms which have been known to occur in the rivers with increasing frequency. Land usage practices may have been contributing to increasing erosion and thereby suspended sediment transport of P as well. The effects of drought and flooding as well as of human activity on the geochemistry of the river system are the topic of this investigation.

2 SAMPLES AND METHODS

Sampling locations are shown in Figure 1. Water samples were collected from several locations within the Namoi catchment in April 1995 while under extreme drought conditions and when toxic blue-green algal (cyanobacteria) blooms were prevalent. Samples from the Darling river were taken at a single location in the town of Bourke over the period February - August 1996 as a series of floods occurred, notably a 50-year event that peaked in late February. Samples of rocks, soils and reservoir sediment from the Chaffey Reservoir subcatchment of the Namoi, suspended sediment from the Namoi near its confluence with the Darling, and phosphate fertilizers that are sold in the

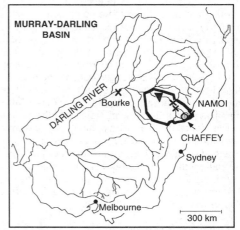

Fig. 1 Map of the Murray-Darling basin (after Douglas et al., 1995), showing geographic reference points and the location of sampling sites (X marks water sample locations, inverted triangle is for location of Namoi suspended sediment, other samples taken in Chaffey Reservoir catchment).

Namoi catchment were also analyzed.

Water samples were filtered to <0.45 μm and acidified after being returned from the field. Sample splits were analyzed for Sr and Nd isotopic compositions by thermal ionization mass spectrometry and trace element concentrations by inductively-coupled plasma mass spectrometry, using methods described elsewhere (Martin & McCulloch, 1997). Bulk solids were totally dissolved or leached in dilute HCl ± H_2O_2 for analysis of Sr and Nd isotopes and trace elements.

3 RESULTS

3.1 *Namoi catchment*

Figure 2 shows the results of the isotopic analysis of rocks, soils, sediment from Chaffey Reservoir, phosphate fertilizers, leachate solutions of some of the bulk samples, and water from the Peel River just upstream of where it enters Chaffey Reservoir.

The two major rock types in Chaffey Reservoir catchment are basalt and metagraywacke, and soils in the catchment are classified as either "basaltic" or "sedimentary" depending on their color and proximity to outcrop (Caitcheon et al., 1995). The rocks have the highest Nd and lowest Sr isotopic composition of the materials in the catchment that were analyzed. The soils have lower Nd and higher Sr isotopic compositions than the rocks with which they are associated in the field. The fertilizers have the lowest Nd and highest Sr isotopic compositions. The Chaffey Reservoir sediment is different from the soils, having a Nd isotopic composition that is similar to the metagraywacke, while its Sr isotopic composition is intermediate between the two rock types. The Peel River water has a slightly higher Sr isotopic composition than the reservoir sediment, but still lies between the values found in the rocks.

The leachates of the rocks and soils should be related to the more leachable or water-soluble component present in these materials. In the case of Nd the leachate solutions have an isotopic composition that is not distinguishable from the bulk samples. For Sr, the leachates of the rocks are both more radiogenic than the bulk sample, whereas for the fertilizer it is identical to the bulk and for the two sedimentary rocks the leachate could be either higher or lower than the bulk sample.

3.2 *Darling River at Bourke*

The floods that took place on the Darling River at Bourke in 1996 were caused by intense rains in different upstream locations following a long period of drought. The largest flood, which took place in late February and peaked at 95 GL/day, had its

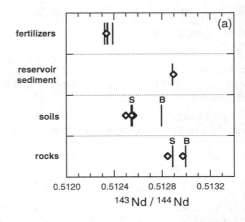

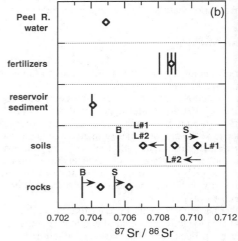

Figure 2. Diagrams showing the (a) Nd and (b) Sr isotopic compositions of different materials in Chaffey Reservoir catchment (Martin & McCulloch, 1997). Bulk samples are shown as bars (label B = basaltic, S = metasedimentary), leachates (L) are shown as open diamonds, and the changes from bulk rock through sequential leachate solution compositions are indicated by arrows.

source (~80% of the water) in the northeastern portion of the Murray-Darling basin (southeast Queensland). The much smaller floods of May (peak flow of 33.3 GL/day) and June (peak flow of 36.3 GL/day) were derived from different sources, the May flood waters coming in substantial proportion from rivers to the south (~65%) within about 150 km of the sampling location, whereas the June flood waters had no component from the south but came from the north (45%) and northeast (55%)

(A. Amos, NSW Department of Land and Water Conservation, written communication, 1996).

Figure 3 shows the variation in Sr isotopic composition and Sr/Ca ratio in the river waters from these different floods. The Sr isotopic composition of river waters depends predominantly on the age and rock types present in the source region, although it will also reflect the relative weathering rates of the constituent minerals. In addition, the Sr/Ca ratios of the waters might be expected to vary depending on the degree of saturation with respect to Ca salts and the amount of cation exchange which had taken place. The May flood, which came from southerly sources, had a $^{87}Sr/^{86}Sr$ that was significantly higher than any of the other Darling river waters, and as the northerly derived waters of the June flood arrived at Bourke, they provided the lowest $^{87}Sr/^{86}Sr$ and highest Sr/Ca ratio waters measured. Relative to the smaller events, the February flood had a relatively constant $^{87}Sr/^{86}Sr$ of about 0.707 but had a large range in Sr/Ca ratio.

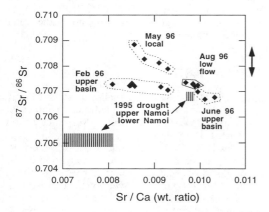

Figure 3. Variation diagram of Sr isotopic composition vs. Sr/Ca ratio of the Darling river sampled at Bourke, NSW, over a period of 7 months in 1996, with floods and their sources indicated. Range of $^{87}Sr/^{86}Sr$ in waters sampled in 1989 (Douglas et al., 1995) (arrow), and fields for results from the Namoi catchment (Martin & McCulloch, 1997) are plotted for comparison.

4 DISCUSSION

The isotopic and trace element compositions of the waters and solids allow us to calculate the mixtures of components that are presently being delivered in solid and dissolved form to rivers from the catchment and to determine how much of it may be from fertilizer application.

4.1 *Chaffey catchment: fertilizer or natural sources?*

The environmental problem of recurrent blue-green algal blooms in the Namoi river, exemplified by Chaffey Reservoir, is generally agreed to be caused by high phosphorus concentrations in the waters that are sustained by high particulate concentrations, but there is debate as to whether these are natural or due to phosphate (± sulfate) fertilizer application. The trace element and isotopic compositions of the natural source materials in the catchment and fertilizers applied there can be compared with that of the reservoir sediment to place limits on the contribution of each potential source.

On the basis of Sr and Nd isotopes, the basaltic rocks must be a significant contributor to the sediment in Chaffey Reservoir, in agreement with previous results based on major element compositions and mineral magnetic data (Caitcheon et al., 1995). All the other naturally occurring materials in the catchment as well as the fertilizers, have Sr and Nd isotopic compositions that could be added to the basalt to provide the bulk composition of the sediment. A good fit for a three-component model, involving the sedimentary soil plus fertilizer as additional components to the basalt, is obtained for most of the elements analyzed in the reservoir sediment, as is shown in Figure 4. It is clear that

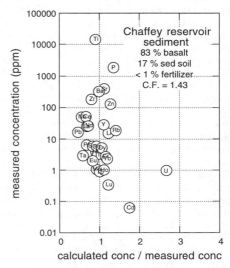

Figure 4. Plot of measured concentration of elements in Chaffey reservoir sediment, plotted versus the ratio of the concentration calculated relative to that measured, for a mixture of approximately 83% basalt with 17% sedimentary soil and < 1% fertilizer and a concentration factor of about 43% (after Martin & McCulloch, 1997).

there has been removal of some of the more mobile chemical constituents during the formation of the reservoir sediment, and this model is consistent with a concentration by about 40% of the relatively immobile constituents in the sediment, such as Ti and the rare earth elements. By this analysis, the high concentration of P in the reservoir sediment is interpreted to be due to natural processes operating in the catchment and not to the introduction of additional P from phosphate fertilizer applications.

4.2 *Darling river: regional rainfall effects?*

Sr isotopic variations on the order of ±0.001 in $^{87}Sr/^{86}Sr$ have been reported for a number of large rivers (e.g. Palmer and Edmond, 1992), similar to those observed in the Darling River, but most researchers have not considered in detail the reasons behind this variability. Usually the Sr system is considered to consist of a low concentration, radiogenic Sr endmember derived from old silicate rocks mixed with a high concentration, relatively nonradiogenic endmember derived from carbonate rocks.

In the case of the Darling river there is an intermediate Sr isotopic composition and concentration component present, that is probably typical of the northeastern drainage basin and contributes some water under all flow conditions. This component is also similar to the Namoi river sampled under drought conditions. The May 1996 flood waters come from a more radiogenic terrain but have a similar Sr/Ca ratio to the February 1996 flood. It is likely that the variations in Sr/Ca ratios observed are due to redissolution by flood waters of Ca-rich salts that had accumulated during the previous drought interval. The observation that the June 1996 flood waters have the highest Sr/Ca ratios suggests that soluble salts ceased to be available for dissolution. On the basis of Sr isotopes, the source of Sr in the February and June floods was the same, consistent with the hydrologic analysis of water sources.

In order to determine the most typical Sr isotopic composition being supplied to this river and by implication other rivers with variable $^{87}Sr/^{86}Sr$, it is important to have an understanding of the relationship between water flux and $^{87}Sr/^{86}Sr$, including the climatology and meteorology of the region. In the case of the Darling River at Bourke, the greatest flux of water in the catchment appears to have an $^{87}Sr/^{86}Sr$ of 0.7072-0.7074.

5 CONCLUSIONS

Variations in the Nd and Sr isotopic compositions and elemental concentrations of riverine particulate and dissolved loads can be used to determine the extent of human influence and the atmospheric/hydrologic influences on transport of elements. In the case of Chaffey Reservoir catchment, it was possible to determine that the dominant source of reservoir sediment was from the rocks in the catchment and that fertilizer was a very minor contributor of most elements. Although fertilizer was shown to be minor in this case, in other locations it may be more important, and should be readily distinguishable with isotopic and elemental tracing. In the case of the Darling River, it was possible to investigate the influence of rainfall location on river chemistry. A global implication of these results is that, if rainfall distribution patterns in a catchment change, the Sr isotopic composition of dissolved Sr from the rivers draining that catchment can be significantly affected.

6 ACKNOWLEDGEMENTS

We thank J. Olley and G. Caitcheon (CSIRO Land and Water) for providing the Namoi catchment solid samples and A. Amos (NSW DLWC, Bourke) for invaluable logistical support in obtaining the water samples from the Bourke site.

REFERENCES

Caitcheon G., Donnelly T., Wallbrink P., & Murray A. (1995) Nutrient and sediment sources in Chaffey Reservoir Catchment. Australian J. Soil Water Cons. 8, 41-49.

Douglas G.B., Gray C.M., Hart B.T., & Beckett R. 1995. A strontium isotopic investigation of the origin of suspended particulate matter (SPM) in the Murray-Darling River system, Australia. *Geochim. Cosmochim. Acta* 59: 3799-3815.

Martin C.E. & McCulloch M.T. 1997. Nd-Sr Isotopic and Trace Element Geochemistry of Dissolved and Particulate Stream Loads and Soils in a Fertilized Catchment. *Geochim. Cosmochim. Acta*, in review.

Palmer M. R. & Edmond J. M. (1992) Controls over the strontium isotope composition of river water. Geochim. Cosmochim. Acta 56, 2099-2111.

Sea water sulfate addition to a forested catchment: Results after five years of experimental treatment

C.-M. Mörth & P.Torssander
Department of Geology and Geochemistry, Stockholm University, Sweden

ABSTRACT: By building a roof over a forested catchment it has been possible control the deposition and to introduce a sulfur tracer to study soil processes. The catchment is located in a polluted area at Lake Gårdsjön in W Sweden. Soil sulfur dynamics and acidification reversal processes have been studied by reducing the sulfate deposition to pre-industrial levels and changing the sulfur isotope composition of deposited sulfate to resemble seawater sulfate. The results show that sulfate fluxes in runoff are controlled by organic sulfur cycling in the O-horizon and sorption processes in the B-horizon. After six years there is evidence for net sulfur mineralization from the upper soil (the O-horizon). This is supported by mass balance calculations where the difference between input and output has been compared to the change in soil sulfate. Acidification reversal therefore seems to be prolonged by desorption of sulfate and by net mineralization of organic sulfur.

1. INTRODUCTION

Acidification has been one of the major issues during the last decades, causing severe damages on ecosystems where the soil buffering capacity is low (Andersson and Olsson, 1985). In view of sulfur emission reductions (Hedin et al., 1992) and the lower load of sulfur deposited, acidification reversibility has been investigated in several catchments (i.e. Wright et al., 1988; Hultberg & Skeffington, in press). The movement of sulfate to and from acidified catchments has been of great interest due to the theory of sulfate as a mobile anion that increases the flux of base cations (Reuss and Johnson, 1986). The currently presented research was aimed to study the processes that regulate the sulfate mass flux in runoff and the sources and sinks of sulfur in catchment.

2. SITE DESCRIPTION

The presented experiment was performed at Lake Gårdsjön in a full-scale catchment manipulation. The catchment is situated on the Swedish west coast (Fig. 1) about 20 km from the sea. Average soil depth is about 40 cm. Norway spruce trees covers the whole catchment. To enable control over the deposition a roof has been built over a whole catchment. The roof covers 6300 m². The natural deposition under the roof is controlled by a sprinkler system that can simulate rain and rain storm events. The chemistry of the sprinkler water is different from natural throughfall and bulk deposition. It has been shown that the throughfall sulfate mass flux balances the runoff sulfate mass flux in the Lake Gårdsjön area (Hultberg and Grennfelt, 1992). The sulfate concentration in the sprinkler water is reduced by about 80% (from about 120 μM to 16 μM). The sprinkler water added is spiked with small amounts of seawater sulfate, which has a sulfur $\delta^{34}S_{SO4}$ value of +19.5‰ and $\delta^{18}O_{SO4}$ of +9.5‰. This enables tracing of soil sulfur processes if the soil sulfur has different isotope compositions (see results).

3. RESULTS

The sulfur concentration and $\delta^{34}S$ values were determined in deposition, runoff, groundwater and soils during one year prior to the start of the experiment and five years of experimental control. Soil samples have been taken both before and during the experiment.

The $\delta^{34}S_{SO4}$ values of the deposition before the roof experiment started were almost identical for bulk deposition (mean value +7.4‰) and throughfall (mean value +7.5‰). When plotting the Cl/SO$_4$ ratio vs the $\delta^{34}S_{SO4}$ value, two sources of sulfur in the

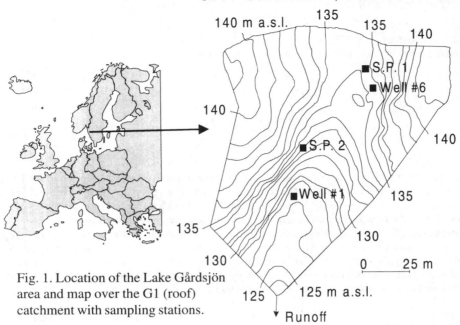

Fig. 1. Location of the Lake Gårdsjön area and map over the G1 (roof) catchment with sampling stations.

atmospheric deposition are identified: antrophogenic sulfur ($\delta^{34}S$ value of about +4.0‰) and seawater spray ($\delta^{34}S$ value of about +20.7‰, Mörth and Torssander, 1992).

The runoff sulfate concentration before the experiment started was stable at around 200 mM (Fig. 2). The $\delta^{34}S_{SO4}$ values showed only small variation. The mean value for that period was +5.45‰ ±0.16 (95% confidence interval). After the reduction of the sulfate input the runoff sulfate concentration has decreased and has been reduced by about 45%. After five years we have not observed any larger differences in $\delta^{34}S_{SO4}$ values at all, Fig. 2 (about +0.5‰ could be identified after two years).

Similarly, the groundwater wells show no response to the change in sulfur source (i.e. change of the isotope composition in the incoming sulfate) and there is a similar decrease of the sulfate concentration as in runoff. Because of the similarities between runoff and groundwater well #1 sampling of the well has stopped.

Two soil profiles (Fig. 1; SP1 and SP2) were analyzed before the experiment started and the upper one (SP1) re-sampled after almost two years of treatment. The B-horizon $\delta^{34}S_{SO4}$ values of adsorbed sulfate were similar between the soil profiles and through time, mean $\delta^{34}S_{SO4}$ values for SP2 and SP1 were +5.35‰ and +5.99‰ respectively, with small spatial variation (Torssander and Mörth, in press). Total sulfur $\delta^{34}S_{\Sigma S}$ values however show large variations with depth. In the upper soil (O-horizon) the mean value for the two soil profiles was +5.80‰. In the B-horizon the mean value was +12.83‰.

4. DISCUSSION

At low temperatures there is only one process that can fractionate the sulfur isotope composition significantly, except for change of source and that process is bacterial sulfate reduction (Krouse and Grinenko, 1991; Van Stempvoort et al., 1992). Hence, fractionation must be considered if appropriate environments are present in forested catchments where this can occur. It has been proven that peat or boggy areas are likely locations for anaerobic reduction (Gorham et al., 1984). The roof catchment is rather unique for boreal forests of the northern-hemisphere because boggy areas are lacking, except for few square meters at the outlet. The average soil depth of about 40 cm and the very shallow groundwaters are also reasons why there is no evidence for bacterial sulfate reduction (Torssander & Mörth, in press).

The sulfur isotope compositions of sulfate in runoff, groundwater and soil extractions (between +5.5‰ and +6.0‰) collected before the experiment started were similar, in contrast to the large variations in throughfall sulfate. This suggests that the inorganic adsorbed sulfate was the source to runoff and that the inorganic sulfate pool was homogeneous in its isotope composition. This implies that sulfate in solution exchanges with adsorbed sulfate and that sulfate adsorption is reversible (Mörth and Torssander, 1995).

The initial lack of response in runoff $\delta^{34}S_{SO4}$ values after changing the $\delta^{34}S_{SO4}$ value in input water is not surprising when considering the large pool of sulfate in the soil (the pool is about 10 to 15 larger than the yearly deposition). This suggests mixing of the sprinkler water sulfur with a large inorganic sulfur reservoir in the catchment. The sulfate pool is sufficiently large to control the runoff sulfur isotope composition and the runoff sulfate mass flux for a long period of time. This will increase the recovery time of the soil to pre acidification mass fluxes of sulfate and will also have a large impact on the mass flux of base cations (Reuss and Johnson, 1986).

The high total sulfur $\delta^{34}S_{\Sigma S}$ values and the large total sulfur concentration in the B-horizon compared to the inorganic sulfate pool in the B-horizon suggest that the organic sulfur pool in the B-horizon at Lake Gårdsjön is large, because inorganic sulfates or sulfides are lacking in the soil. This implies that organic sulfur is accumulating in the B-horizon. The high $\delta^{34}S_{\Sigma S}$ values of around +13‰ implies that this accumulation occurred a long time ago since no heavy sulfur source is present in the area apart from sea spray. The organic sulfur pool in the O-horizon has a $\delta^{34}S_{orgS}$ value (about +6‰) identical to the inorganic sulfate, which reflects atmospheric deposition during later part of the 20th century. This implies that most of the organic sulfur in the B-horizon has an origin from sea spray and accumulated earlier, before the $\delta^{34}S$ values of deposition and O-horizon organic sulfur changed.

Oxygen isotopes in sulfate can be used to determine the fraction of sulfur transformed within a catchment (Krouse and Grinenko, 1991). It is suggested that isotope shifts of the oxygen isotope composition between throughfall and runoff is caused by organic sulfur cycling in the upper soil, the O-horizon (Gélineau et al., 1988; Van Stempvoort et al., 1992). Oxygen isotope measurements in the G1 catchment suggest that around one third of sulfate is transferred through the O-horizon organic sulfur. The $\delta^{18}O_{SO4}$ shift between input water (throughfall and sprinkler water) and runoff is about +3.1‰ (from +9.5‰ to +5.4‰).

However, modeling the runoff $\delta^{34}S_{SO4}$ values,

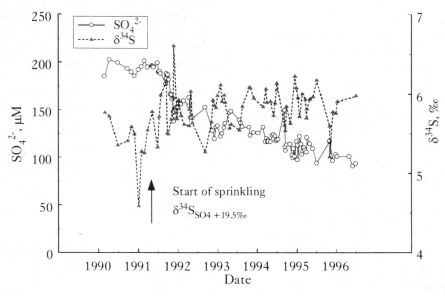

Fig. 2. The runoff SO_4^{2-} concentration and $\delta^{34}S_{SO4}$ values.

using a model that mixes sprinkler sulfate with soil sulfate (Mörth, 1995), suggests that (after six years of treatment) the runoff $\delta^{34}S_{SO_4}$ value should be around +7‰. This in combination with the mass balance results where more sulfate has been transported out than can be accounted for in the soil, indicates net mineralization. Since the total soil sulfur $\delta^{34}S_{\Sigma S}$ values are different between the upper and lower soil the source for this mineralization must be the upper soil. Net mineralization further increases the time needed for an acidification reversal.

5. CONCLUSIONS

Organic sulfur cycling in the O-horizon and sulfate desorption in the B-horizon are the main sulfur flux mechanisms in the catchment. Net sulfur mineralization from the upper soil must occur to explain the difference between the observed soil sulfate pool changes and the runoff sulfate mass flux. This observation is supported by sulfur isotope measurements and modeling.

6. ACKNOWLEDGEMENTS

Klara Hajnal is thanked for preparation of isotope analysis. This project was funded by NFR, The Swedish Natural Science Research Council.

7. REFERENCES

Andersson, F. & B.Olsson, 1985. Lake Gårdsjön; An acid forest lake and its catchment. *Ecological Bulletin 37, Publishing House of the Swedish Research Councils.* Stockholm.

Gélineau, M., R.Carignan. and A.Tessier, 1989. Study of the transit of sulfate in a Canadian shield lake watershed with stable oxygen isotope ratios. *Appl. Geochem.* 4: 195-201.

Gorham, E., S.E.Bayley and D.W.Schindler, 1984. Effects of acid deposition upon peatlands. *Can. J. Fish. Aquat. Sci.* 41: 1256-1268.

Hedin, L.O., L.Granat, G.E.Likens, T.A.Buishand, J.N.Galloway, T.J.Butler and H.Rodhe, 1994. Steep declines in atmospheric base cations of Europe and North America. *Nature* 367: 351-354.

Hultberg, H. & P.Grennfelt, 1992. Sulphur and seasalt deposition as reflected by throughfall and runoff chemistry in forested catchments. *Environ. Pollut.* 75: 215-222.

Hultberg, H. & R.Skeffington, in press. *Experimental reversal of acid rain effects: The Gårdsjön roof project.* John Wiley & Sons. New York.

Krouse, H.R. & V.A.Grinenko, 1991. *Stable isotopes. Natural and anthropogenic sulphur in the environment.* SCOPE 43. John Wiley & Sons. New York.

Mörth, C-M. & P.Torssander, 1995. Sulfur and oxygen isotope ratios in sulfate during an acidification reversal study at Lake Gårdsjön, Western Sweden. *Water Air Soil Pollut.* 79: 261-278.

Mörth, C-M., 1995. *Sulfur isotopes used as a tracer of acidifcation reversal, dispersion of aciud mine drainage and sulfur dynamics in small and large catchments.* Meddelanden från Stockholms Universitets institution för Geologi och Geokemi No 290. Stockholm University. Dept. of Geology and Geochemistry. Stockholm.

Reuss, J.O & D.W.Johnson, 1986. *Acid deposition and the acidification of soils and waters.* Springer-Verlag. New York.

Torssander, P. & C-M.Mörth, in press. Sulfur dynamics in the roof experiment at Lake Gårdsjön deduced from sulfur and oxygen isotope ratios in sulfate. In: *Experimental reversal of acid rain effects: The Gårdsjön project.* Hultberg & Skeffington (Eds). John Wiley & Sons. New York.

Van Stempvoort, D.R., E.J.Reardon and P.Fritz, 1990. Fractionation of sulfur and oxygen isotopes in sulfate by soil sorption. *Geochim. Cosmochim. Acta* 54: 2817-2826.

Van Stempvoort, D.R., P.Fritz and E.J.Reardon, 1992. Sulfate dynamics in upland forest soils, central and southern Ontario, Canada: stable isotope evidence. *Appl. Geochem.* 7: 159-175.

$\delta^{34}S$ dynamics in the system bedrock - soil - runoff - atmosphere: Results from the GEOMON network of small catchments, Czech Republic

M. Novák
Czech Geological Survey, Prague, Czech Republic

ABSTRACT: Sulfur isotope ratios were determined in bedrock, vertical soil profiles, forest floor moss, spruce throughfall, bulk atmospheric deposition in an open area and surface runoff in forested catchments of the Czech national monitoring network GEOMON. Depth-dependent shifts in $\delta^{34}S$ ratios in the soil reflect both mixing of geologic and atmospheric S sources, and S isotope fractionation *in situ*.

1 INTRODUCTION

It is generally accepted that most of the sulfur cycled between soil and forest biomass is derived from the atmosphere rather than from bedrock (Likens et al., 1977). Sulfur enters forested catchments in the form of wet deposition (SO_4^{2-}) and dry deposition (mostly SO_2). It has been shown that the turnover time of S in the catchment is on the order of decades (Mayer et al., 1995). The amount of S made available to plants by weathering of accessory sulfides at their rooting depth is site-specific. It depends on the stage of pedogenesis and S content in the fresh rock.

Under favourable conditions sulfur isotope ratios can be used to identify the depth within the soil profile at which bedrock-derived S becomes the predominant S pool. The main condition is that the isotopic separation between the potential S sources, the atmosphere and the bedrock, is significant.

Here I present a set of 156 determinations of $\delta^{34}S$ ratios in vertical soil profiles. The samples were taken in spruce forest ecosystems of the Czech Republic, representing an area of 60 000 km². The data are discussed in light of isotope signatures of potential S sources and possible mechanisms of redistribution of S isotopes within the soil.

2 METHODOLOGY

2.1 *Sampling*

Vertical soil profiles were sampled in 18 forested catchments belonging to the Czech national monitoring network GEOMON (Fig. 1; Novák et al., 1996). Input-output budgets for major and trace elements in these catchments have been measured for 5 to 15 years. The bedrock was represented by granite, rhyolite and gneiss of a similar composition (12 sites), non-carbonate sedimentary rocks (7 sites) and serpentinite (1 site). The soils were Dystric and Eutric Cambisols (15 sites), Dystric Planosol and Ferro-humic Podzol (1 site each). The average thickness of the soil horizons sampled was 11, 15 and 30 cm for A, B and C, respectively. Some 200 g of soil from each depth interval was dried, sieved (< 2 mm) and homogenized.

Forty grab samples (10 kg in all) of orthogneiss were taken at two sites in the Northern Czech Republic (JEZ, Fig. 1, and Přísečnice, 30 km SW of JEZ), and analyzed for whole-rock $\delta^{34}S$.

Stream baseflow was sampled at 6 sites (LIZ, ALB, HAR, SAL, HAL, SOL; Fig. 1) in mid-September 1996 and analyzed for $\delta^{34}S$.

Spruce throughfall, bulk atmospheric deposition in open area and surface runoff were sampled at 4 sites (LYS, SAL, CER, LES; Fig. 1) in October 1996. Throughfall and bulk deposition were sampled cumulatively while the runoff sample was taken at the end of the 30-day observation period.

2.2 *Analytical techniques*

Total S was extracted from soil samples as $BaSO_4$ using the Eschka's procedure (Novák et al. 1996). S concentration was determined gravimetrically (>0.03 wt. %) and on a LECO SC-132 Sulfur Analyzer (<0.03 wt. %). $BaSO_4$-S was converted to SO_2 at 980°C. Sulfur was extracted from whole-rock samples as ZnS using Johnson-Nishita distillation, converted to Ag_2S and thermally decomposed to SO_2 at 800 °C. Sulfate-S was precipitated from water samples as $BaSO_4$ and converted to SO_2. $\delta^{34}S$ ratios were measured on a Finnigan MAT 251 mass spectrometer.

Fig. 1 Study sites. 1-5 denote areas where bedrock $\delta^{34}S$ ratios were previously studied.

3. RESULTS AND DISCUSSION

3.1 *Isotope signatures of bedrock sulfur*

Both orthogneisses belonging to the same Krušné Mts. crystalline unit contained < 0.005 wt. % S. Johnson-Nishita distillation performed on 100 g of heavy mineral concentrate did not yield enough S for isotope determination. When 200 g of ground whole-rock sample were subjected to the same procedure, small-volume isotope analysis yielded $\delta^{34}S$ ratios of +16.0 and +5.8 ‰ for Přísečnice (SW of JEZ) and JEZ, respectively. These values are high compared to older data by Šmejkal (unpubl.) ranging between -1 to +4 ‰.

Table 1 summarizes older data on $\delta^{34}S$ ratios of bedrock in several areas of the Czech Republic. Geologically, area no. 1 comprises catchments SAL, HAR, ALB and LIZ ($\delta^{34}S$ of -12 to +7, median +4 ‰), and area no. 3 (-1 to +4 ‰) is close to JEZ. The data were obtained mostly during past ore exploration projects. The scatter in $\delta^{34}S$ is considerable even within a single geological unit (e.g, -33 to +38 ‰ in area no. 4 ; Table 1). Isotope analysis of bedrock sulfur in the remaining catchments is currently in progress.

Table 1. $\delta^{34}S$ (‰) of bedrock sulphides in selected areas of the Czech Republic.

Area*	Material analyzed	$\delta^{34}S$ (‰) Median	Range	N**
1	sulfides in granites	0	-5 to +2	15
2	sulfides in paragneiss	+4	-12 to +7	25
3	sulfides in orthogneiss	0	-1 to +4	5
4	sulfides in shales	+8	-33 to +38	35
5	sphalerite, galena	-5	-15 to +3	100

*see Fig. 1; **number of analyses. Analysed by: K. Žák, V. Šmejkal and J. Hladíková, Stable Isotope Laboratory of the Czech Geological Survey, Prague.

3.2 *Use of stream baseflow $\delta^{34}S$ to fingerprint bedrock sulfur*

In 1995 we attempted to estimate average $\delta^{34}S$ of catchment bedrock by analysing stream baseflow in all 18 GEOMON catchments (Novák et al., 1996). In 1996 six sites were re-visited and $\delta^{34}S$ of baseflow compared to the previous year (Table 2). For LIZ, ALB, HAR, SAL and HAL stream baseflow S was always isotopically heavier (higher $\delta^{34}S$) than soil S sampled in triplicate from the C horizon in the lower segment of each catchment. At these sites, $\delta^{34}S$ of baseflow was lower in the second year compared to the first year by 3.1 to 7.7 ‰. The year-to-year variability suggests that baseflow S was not a clear indication of average bedrock $\delta^{34}S$. At SOL $\delta^{34}S$ of baseflow was extremely negative (-13.3 ‰). Still, $\delta^{34}S$ of the soil horizon C remained positive (on average +3 ‰), similar to local atmospheric input (+2 to +5 ‰). Consequently, at this site it was unlikely that the C horizon (depth of 22-55 cm) contained bedrock S.

Table 2. Comparison between $\delta^{34}S$ (‰) of stream baseflow and the C horizon of spruce forest soil.

Site	Stream baseflow		Soil C horizon		
	1994*	1995*	Pit I	Pit II	Pit III
LIZ	12.0	5.2	3.5	1.8	4.7
ALB	7.8	4.7	1.4	4.1	0.9
HAR	12.0	5.0	2.3	3.6	4.7
SAL	14.0	6.3	2.8	5.0	3.6
HAL	14.0	10.3	9.0	2.1	4.8
SOL	-13.3	n.d.	2.2	3.2	3.6

*mid-September

3.3 *Mixing of atmospheric and geologic S in runoff*

During seasons of higher catchment runoff the admixture of atmospheric S in stream water apparently also increases. At LYS, SAL, CER and LES $\delta^{34}S$ of spruce throughfall is lower than $\delta^{34}S$ of bulk atmospheric deposition measured in an open area (Table 3; cf. also Novák et al., 1995). At the catchment level, the average atmospheric $\delta^{34}S$ must be weighted by the forested-to-unforested land ratio. If $\delta^{34}S$ of bedrock falls between those of throughfall and open area deposition (e.g., between +3.5 and +5.1 ‰ for SAL, see chapter 3.1), or, alternatively, if the admixture of bedrock-derived S in runoff is negligible, $\delta^{34}S$ of runoff will also fall between those of throughfall and open area deposition (+4.6 ‰ for SAL, Table 3). This appears to be the case also for LYS. At CER and LES $\delta^{34}S$ of runoff was outside the deposition range indicating that the bedrock component in runoff may be significant. Compared to atmospheric deposition the surface runoff S was isotopically lighter at CER and isotopically heavier at LES. Thus average bedrock S could be <+2.6 ‰ for CER and >+6.4 ‰ for LES (Table 3). The assumptions of such simple mixing are: (1) negligible seasonal variability in $\delta^{34}S$ of input (a lag between input and output might account for different $\delta^{34}S$ measured in deposition and runoff on the same sampling date), and (2) unchanged S isotope signatures of input with increasing/decreasing industrial pollution on the scale of decades.

Table 3. Comparison between $\delta^{34}S$ (‰) of spruce throughfall, atmospheric deposition in open area and surface runoff (October 1996).

Site	Spruce throughfall	Open area deposition	Surface runoff	% of forest*
LYS	3.9	6.3	5.4	70
SAL	3.5	5.1	4.6	100
CER	3.0	3.8	2.6	100
LES	3.2	3.5	6.4	95

*within the catchment

3.4 Vertical $\delta^{34}S$ profiles in soils

Explaining $\delta^{34}S$ ratios in soils and catchment runoff as a simple mixture of atmo- and geologic S may become further complicated if some of the processes occuring within the ecosystem fractionate S isotopes. Only if all S pools resulting from such processes remain *in situ* is the mixing approach still applicable. Table 4 gives S concentrations and $\delta^{34}S$ ratios for three depth levels in the soils of the GEOMON network. All depth levels were randomly sampled in triplicate except for RAZ (Fig. 1) where data from two soil pits are available ([S_{tot}] of 0.14 and 0.10, 0.05 and 0.06, 0.04 and 0.04 wt. % for A, Bv and C, respectively; $\delta^{34}S$ of 2.6 and 3.9, 5.1 and 5.0, 2.5 and 5.9 ‰ for A, Bv and C, respectively). Vertical signals based on average $\delta^{34}S$ at individual sites were discussed by Novák et al. (1996) with respect to local pollution levels. Here a synthesis of the concentration and isotope trends is presented. Histograms in Fig. 2 depict vertical changes in soil S concentration. With an increasing soil depth S concentrations decrease and the range of values narrows. While at JEZ and MOL, sites situated in one of the most polluted region of the world (atmospheric deposition >80 kg S ha^{-1} yr^{-1}), [S_{tot}] in the A horizon is among the highest, [S_{tot}] in B and C is among the lowest (cf. Tab. 4 and Fig. 2). Unlike in the topsoil, in deeper soil large S inputs do not necessarily result in higher S concentration. However, due to increasing soil density the amount of S stored in B and C may be higher than in the topsoil (horizon A). Histograms

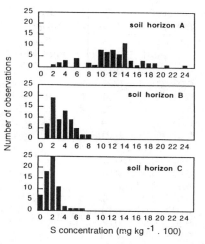

Fig. 2 Histograms for S concentration in soils.

for the $\delta^{34}S$ ratios are given in Fig. 3. From the soil horizon A to B and C the mean $\delta^{34}S$ ratios increased along with the scatter in measured values. There was a statistically significant negative S isotope shift from forest floor moss (data by Buzek, unpubl.) to the A horizon. This shift may be the result of an initial isotope fractionation in the ecosystem: either $\delta^{34}S$ of the topsoil containing decaying spruce litter mirrors average atmospheric input and moss redistributes S isotopes relative to rainfall, or, moss reflects unchanged $\delta^{34}S$ of the rainfall and the topsoil redistributes S isotopes by means of microbial activity or higher-plant assimilation. A later S isotope fractionation occurs along the transport pathway from the horizon A to C. Since no isotope effect is associated with inorganic sulfate adsorption, we have previously suggested that this statistically significant positive isotope shift may be a result of preferential release (revolatilization?) of the lighter isotope ^{32}S during decomposition of organic matter (Novák et al., 1996). Since $\delta^{34}S$ of stream baseflow is on average even higher than that of the C horizon, the scenario of increasing admixture of bedrock S in the upper 50 cm of the soil instead of the second fractionating process remains open. The arguments opposing such scenario can be summarized as follows: (1) baseflow, whose $\delta^{34}S$ was found to vary between years, may reflect a high proportion of S originating in deeper soil horizons themselves characterized by relatively high $\delta^{34}S$, not necessarily high bedrock $\delta^{34}S$; (2) although the territory under study was unevenly covered by isotope analyses of bedrock, the median $\delta^{34}S$ in the existing data from large geological units in the Czech Republic (cf. Table 1, total N of analyses >300) is close to 0 ‰, lower than baseflow values in Fig. 3; and (3) a positive $\delta^{34}S$ shift from the A to C horizon can be present despite negative $\delta^{34}S$ of the bedrock (SOL).

Table 4. Sulphur concentration ([S_{tot}]) and stable isotope composition ($\delta^{34}S$) in three soil horizons.

Site	Bedrock*	Soil	[S_{tot}] (wt %) Pit I	Pit II	Pit III	$\delta^{34}S$ (‰) Pit I	Pit II	Pit III
JEZ	og	A	0.14	0.18	0.15	1.8	1.7	0.6
		Bv	0.03	0.01	0.01	3.0	2.7	0.4
		C	0.01	0.01	0.01	6.5	3.3	1.1
MOL	rh	A	0.24	0.13	0.21	4.4	2.9	2.9
		A/B	0.03	0.03	0.04	n.d.	7.0	3.3
		Bs	0.01	0.01	0.01	3.3	5.1	n.d.
LYS	gr	A	0.16	0.11	0.12	4.1	3.2	2.0
		Bv	0.02	0.02	0.02	0.5	6.0	3.3
		C	0.01	0.02	0.01	1.1	4.9	4.6
PLB	sr	A	0.12	0.14	0.11	3.3	6.0	2.8
		Bv	0.02	0.02	0.02	5.7	3.5	6.2
		C	0.01	0.03	0.01	7.3	4.7	4.6
MLL	rh	A	0.02	0.13	0.06	1.4	1.9	2.5
		Bv	0.01	0.04	0.03	2.1	4.9	5.6
		B/C	0.02	0.02	0.02	0.4	5.0	4.7
HAL	gw	A	0.04	0.10	0.11	2.9	2.4	2.7
		g	0.01	0.05	0.02	5.0	2.9	4.6
		gm	0.02	0.03	0.01	9.0	2.1	4.8
JZD	gr	A	0.15	0.13	0.15	5.0	2.0	-2.0
		A/B	0.06	0.01	0.08	7.2	0.8	6.0
		Bs	0.07	0.01	0.02	4.2	0.7	11.6
ALB	pg	A	0.14	0.14	0.10	0.3	2.6	n.d.
		Bv	0.03	0.04	0.06	-3.0	4.2	3.4
		C	0.02	0.03	0.03	1.4	4.1	0.9
LIZ	pg	A	0.11	0.13	0.12	3.2	1.5	2.8
		Bv	0.05	0.02	0.03	3.5	1.3	3.0
		C	0.03	0.06	0.02	3.5	1.8	4.7
PLP	gr	A	0.08	0.14	0.11	2.7	3.1	2.9
		Bv	0.03	0.04	0.04	5.5	3.7	2.7
		C	0.02	0.02	0.02	5.2	6.2	4.5
HAR	pg	A	0.12	0.14	0.03	2.7	3.7	2.9
		A/B	0.07	0.05	0.02	3.6	2.8	3.8
		Bm	0.01	0.01	0.01	2.3	3.6	4.7
SAL	pg	A	0.10	0.10	0.10	3.0	3.2	3.1
		Bv	0.04	0.04	0.04	2.3	5.0	3.6
		C	0.02	0.02	0.03	2.8	5.0	3.6
SEP	ms	A	0.12	0.06	0.18	3.5	2.4	1.8
		Bs	0.05	0.06	0.07	3.8	2.7	2.1
		B/C	0.03	0.03	0.02	7.1	3.0	1.8
TEP	og	A	0.12	0.14	0.14	2.5	1.5	2.9
		A/B	0.03	0.04	0.03	3.4	3.0	2.9
		B/C	0.02	0.03	0.02	5.0	2.9	10.7
VOD	og	A	0.12	0.19	0.12	2.5	1.4	3.4
		E	0.04	0.05	0.04	3.2	0.8	4.9
		Bs	0.03	0.04	0.03	3.5	1.1	5.1
SOL	sh	A	0.03	0.04	0.04	1.4	2.1	2.0
		A/B	0.02	0.03	0.03	2.5	1.5	2.6
		Bv	0.01	0.02	0.01	2.2	3.2	3.6
CER	sa	A	0.08	0.11	0.09	0.0	2.7	1.9
		Bv	0.08	0.05	0.02	0.8	5.0	3.2
		C	0.03	0.06	0.03	2.9	3.3	9.1

* og - orthogneiss, rh - rhyolite, gr - granite, sr - serpentinite, gw - greywacke, pg - paragneiss, ms - mica shist, sh - shale, sa - sandstone.

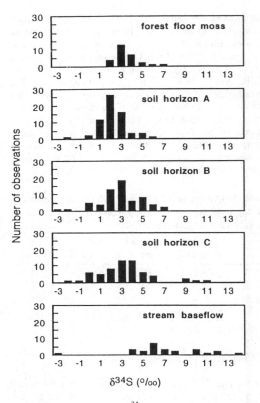

Fig. 3 Histograms for $\delta^{34}S$ ratios in the catchments

ACKNOWLEDGEMENTS

This work was supported by the Czech Granting Agency (205/93/2404) and the European Union (CANIF, ERB IC20 CT960024).

REFERENCES

Likens, G.E., Bormann, F.H., Pierce, R.S., Eaton, J.S. & N.M. Johnson 1977. *Biogeochemistry of a forested ecosystem.* New York: Springer.

Mayer, B., Fritz, P., Prietzel J. & H.R. Krouse 1995. The use of stable S and O isotope ratios for interpreting the mobility of sulfate in aerobic forest soils. *Appl. Geochem.* 10: 161-173.

Novák, M., Bottrell, S.H., Fottová, D., Buzek, F., Groscheová, H. & K. Žák 1996. S isotope signals in forest soils of Central Europe along an air pollution gradient. *Environ. Sci. Technol.* 30: 3473-3476.

Novák, M., Bottrell, S.H., Groscheová, H., Buzek F., & J. Černý 1995. S isotope characteristics of two North Bohemian forest catchments. *Water Air Soil Pollut.* 85: 1641-1646.

Geological controls on drainage water compositions across a granite-related zoned mineral field, Zeehan, Western Tasmania

T.E. Parr & D.R. Cooke
Centre for Ore Deposit Research, Geology Department, University of Tasmania, Hobart, Tas., Australia

ABSTRACT: Mineralogical zonation around the Heemskirk granite, Western Tasmania, is controlling the composition of drainage waters seeping from abandoned mine sites in the Zeehan mineral field. The most contaminated, acid waters (pH <3; Σ trace metals > 1ppm) are discharging from adits in the pyrite zone (proximal to granite). Less contaminated waters are sourced from the intermediate sidero-pyrite zone, and the least polluted waters (pH >5.5; Σ trace metals <1ppm) are derived from the distal siderite zone.

INTRODUCTION

The chemical composition of creeks affected by acid mine drainage (AMD), will be controlled by the local geology, anthropogenic factors and local biological activity. Understanding the geology of an area is therefore essential to understanding the nature and variability of AMD. In this study, we have investigated drainage water compositions from a zoned mineral field in western Tasmania, to identify possible mineralogical and host lithological controls on drainage water composition.

MINING HISTORY

Over 50 mineral occurrences in the Zeehan mineral field, Western Tasmania, were mined for Ag and Pb from 1887 to 1913. The last significant producer in the district was the Oceana Mine, which closed in 1960. The mineral field produced 0.2 Mt of Pb, 0.76 t Ag, 2,754 t Zn and lesser amounts of copper, tin and cadmium (Both & Williams, 1968a). The mines were worked for galena, sphalerite, chalcopyrite and stannite (Both & Williams, 1968b).

ZONED MINERAL FIELD

The Zeehan mineral field is a classic example of a granite-related zoned mineral field. Both and Williams (1968b) defined three mineralogical zones in the Zeehan field. From west to east, they are the Pyrite Zone, Intermediate Zone (Pyrite + Siderite gangue) and Siderite Zone (Figure 1). A fourth zone of cassiterite occurs within the granite (Figure 1).

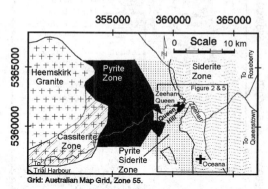

Figure 1 Mineralogical zonation in the Zeehan mineral field (modified from Taylor and Mathison, 1990 after Both & Williams 1968b).

Pyrite is most common in the western part of the field (proximal to granite), and around Queen Hill where geophysical modeling has shown a granite cupola lies close to the surface (Leaman, 1990). Cassiterite is restricted to the Queen Hill area (Figure 1), and within the granite. Sphalerite is found across the Zeehan field, but is more abundant in the western (pyrite) zone. Galena also occurs across the Zeehan field, but the abundance of galena increases from west to east into the Siderite Zone. The Ag content of galena is low in the pyrite zone, increases to a maximum in the intermediate zone and decreases in the siderite zone (Both & Williams, 1968b).

Lithologies

The Heemskirk granite crops out 6 km west of the township of Zeehan (Figure 1) and was emplaced during the late Devonian. Magmatic-hydrothermal fluids released from the granite migrated out along faults and precipitated vein mineralisation across the Zeehan field (Figure 1, Both & Williams, 1968a). Gravity surveys by Leaman (1990) show that the granite dips gently eastward under the Zeehan mineral field, with two local granite highs occurring under the mineral field. One of the granite highs occurs in the study area, under Queen Hill (Figure 2). This granite high has complicated the simple east-west mineral zonation in the Queen Hill area, where a small pyrite zone occurs above the granite high (Figure 1, Both & Williams, 1968b).

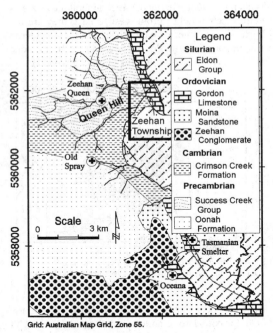

Figure 2 Simplified geological map of part of the Zeehan mineral field showing the distribution of major lithologies.

Detailed descriptions of the geology of the Zeehan district are provided in Everard et al (1992) and Seymour and Calver (1995). Mineralization is hosted by the Precambrian Oonah Formation (interbedded mudstones and siltstones); siliceous mudstones and sandstones of the Success Creek Group, the Crimson Creek Formation (interbedded volcaniclastic turbiditic greywackes and siltstone-mudstones) Zeehan Conglomerate, Moina Sandstone, Gordon Limestone and the Eldon Group (siliceous sandstones, siltstones and mudstones; Figure 2, Everard et al, 1992, Seymour and Calver, 1995).

WATER CHEMISTRY

Drainage waters from twenty sites around the mineral field were sampled four times at three weekly intervals. A total of 19 elements were measured by ICP-EOS (Ag, Al, As, Ca, Cd, Co, Cu, Fe, K, Mg, Mn, Mo, Na, Ni, Pb, Si, Sn, V and Zn), together with field measurements of pH, Eh, conductivity and temperature.

Maximum and minimum concentrations measured for each element are listed in Table 1. Some of the water samples exceeded all limits set by the ANZECC (1992) guidelines, showing that some drainage waters in the Zeehan field are contaminated beyond the recommended guidelines (Table 1).

The major contaminants of AMD in the Zeehan field are Fe, Zn, Sn, Cu and Pb, which is consistent with the ore and gangue mineralogy of the mineral deposits. Not surprisingly, the highest Fe concentrations occur in drainage waters from the pyrite zone, and Sn is elevated proximal to the subsurface granite high at Queen Hill (Figure 2). The lowest pH waters ($\approx$ 2-3) are draining from the Zeehan Queen mine site in the pyrite zone (Figure 1). The highest pH values (up to 7.60) were recorded from below the Oceana mine in the siderite zone (Figure 1). This mine is hosted by the Gordon Limestone, which is buffering drainage waters to alkaline conditions.

DRAINAGE WATER CLASSIFICATION

An AMD classification system has been adopted based on Ficklin et al. (1992) and hereafter referred to as a "Ficklin Plot" (Figure 3 & 4). The classification scheme plots pH against the log of the sum of As, Co, Cu, Cd, Mo, Ni, Pb, Sn, V and Zn in parts per million (Figure 3 & 4).

Table 1: Maximum & minimum values analysed drainage waters in the Zeehan mineral field. ANZECC water quality guidelines are listed for comparative purposes (ANZECC, 1992) BD = below detection limits (generally <0.05 ppm on the ICP-EOS) Ag was below detection in all samples.

Attribute	Maximum	Minimum	Mean	Recommended max. for aquatic ecosystems
	(ppm unless otherwise stated)			
pH	7.6	2.6	4.9	6.5 - 9.0
T (°C)	23.6	6.7	10.5	-
Cond. (µS)	3800	50	692	-
Alkalinity mg CaCO$_3$/l	154.5	BD	11.5	-
SO$_4^{-2}$ (mg/l)	2420.0	1.8	400.7	-
Cl (mg/l)	48.8	10.1	18.9	-
Al	17.7	BD	2.1	0.01
As	4.64	BD	0.2	0.05
Co	0.09	BD	0.0	-
Cu	7.61	BD	0.3	0.005
K	14.8	BD	0.8	-
Fe	198.6	BD	21.7	1
Mg	27.8	0.9	6.8	-
Mn	54.7	BD	3.9	-
Mo	0.12	BD	0.0	-
Na	11.0	4.6	7.7	-
Ni	0.17	BD	0.0	0.02
Pb	3.55	BD	0.4	0.005
Si	10.4	0.4	3.5	-
Sn	10.05	BD	0.6	0.000008
V	0.05	BD	0.0	-
Zn	50.06	BD	6.4	0.01
Ca	213.7	BD	10.9	-
Cd	0.50	BD	0.0	0.002

Mineralogy

From Figure 3, it is obvious that mineralogy is a major control on the composition of the drainage waters in the Zeehan field. Mineral deposits in the pyrite zone produce high acid, high metal to acid, high metal mine waters (Figure 3). Consequently, this mineralogical zone is the source of the most contaminated drainage waters in the Zeehan mineral field. Mine sites in the Siderite zone predominantly produce AMD compositions that are clustered in the near-neutral low metal to acid, low metal zones, and are generally the least contaminated waters (Figure 3). High metal, acid and high metal, near-neutral waters are produced from the intermediate mineral zone (Figure 3), probably due to the varying abundance of pyrite and siderite in this area. Pyrite abundance is most likely the fundamental control on acid generation in the Zeehan mineral field.

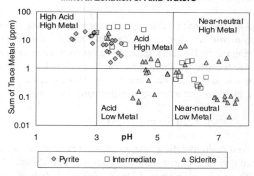

Figure 3 Ficklin plot showing mineralogical controls on AMD in the Zeehan field.

Host lithologies

Individual data points on a Ficklin plot can also be discriminated in terms of lithologies (Figure 4). Each lithology from Zeehan forms a close cluster, ranging from high metal, high acid in the siliceous Success Creek Group down to near-neutral, low metal waters associated with the Gordon Limestone (Figure 4). This distribution relates to the acid-buffering capacity of each lithology. The siliceous units have little to no acid-buffering capacity, whereas the limestone and (to a lesser extent) the volcanic units have a high buffering capacity.

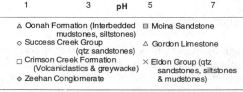

Figure 4 Ficklin plot of data points zoned by lithology.

CONTAMINATION MAP

An acid mine drainage hazard map has been constructed for the Zeehan water catchment (Figure 5). Because drainage water from each sample site has been classified on the basis of pH and trace metal contaminants, these two factors have been combined into a "contamination number" (cn):

$$cn = \Sigma \text{ trace metals}/pH$$

Four contamination categories were defined based on contamination numbers from <5 to >150 (Figure 3). The pyrite zone around Queen Hill is highlighted as a high contamination zone, as are point sources from the Old Spray Mine and the Tasmanian Smelter (Figure 5).

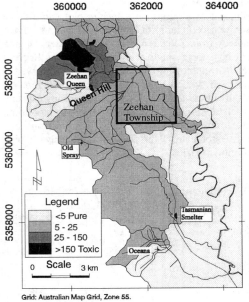

Figure 5 AMD contamination map of the Zeehan Mineral Field.

CONCLUSIONS

High pyrite content is the major factor in generating polluted drainage waters from abandoned mine sites around Zeehan. Siderite does not appear to contribute significantly to acid production. The major contaminants (Fe, Zn, Pb, Sn & Cu) are consistent with the known mineralogies of the mineral field. Some lithologies (eg Gordon Limestone, mafic units in the Crimson Creek Formation) are helping to buffer acidity and reduce contamination in the drainage waters. Sites of highest priority for remediation have been identified based on their contamination numbers (Figure 5).

ACKNOWLEDGMENTS

This study was part of an Honours study by the senior author at the University of Tasmania. The authors gratefully acknowledge Mineral Resources Tasmania and especially Mr Wojtek Grun for their support and funding for this project.

REFERENCES

Australian & New Zealand Environment & Conservation Council, 1992 Australian Water Quality Guidelines for Fresh and Marine Waters, National Water Quality Management Strategy. ANZECC; Australia: p. 2-1 to 2-42.

Both, R.A., & Williams, K.L. 1968a Mineralogical Zoning in the Lead-Zinc ores of the Zeehan Field, Tasmania. J. Geol. Soc. Aust. 15: 121-137.

Both, R.A., & Williams, K.L. 1968b Mineralogical Zoning in the Lead-Zinc ores of the Zeehan Field, Tasmania. J. Geol. Soc. Aust. 15: 217-243.

Everard, J.L., Findlay, R.H., McClenaghan, M.P., Seymour, D. B. & Brown, A.V. 1992 Current systematic regional mapping - new potentials for mineral discovery. Bull. Geo. Surv. Tas. 70:70-87.

Ficklin, W.H., Plumlee, G.S., Smith, K.S. & McHugh, J.B. 1992 Geochemical classification of mine drainages and natural drainages in mineralized areas. *In* Y.K. Kharaka & A. S. Maest (eds) Water-Rock Interaction: 1: 381 - 384. Brookfield: A. A. Balkerra.

Leaman, D.E. 1990 An interpretation form of the Heemskirk Granite. Zeehan. Open File RGC Exploration P/L, EL 42/87 Annual Report for the Period Oct. 1990 – Sept. 1991, App. 6 (Unpub.)

Seymour, D.B. & Calver, C.R. 1995 Explanatory notes for the Time Space Diagram & Stratotectonic elements map of Tasmania. Tas. Geo. Surv. Record 1995/01: 13-22.

Taylor, S. & Mathison, I.J. 1990 Oceana Lead-Zinc-Silver Deposit. *In* F.E. Hughes (eds), Geology and Mineral Deposits of Australia and Papua New Guinea 2 AUSIMM: Parkville: 1253-1256..

The hydrogeochemistry of thallium in natural waters

Paul Shand, W. Mike Edmunds & Jill Ellis
British Geological Survey, Maclean Building, Wallingford, UK

ABSTRACT: Thallium is a highly toxic element, but its occurrence in the hydrological system is poorly documented. Waters from a range of geological terranes representing a variety of water types (groundwaters, thermal waters, surface waters and mine waters) were analysed for thallium. Typical concentrations in surface and ground waters are of the order of $10^{-1} - 10^1$ ng l^{-1} with higher concentrations (a few µg l^{-1}) in waters associated with the oxidation of sulphide minerals. The highest concentrations (up to 99 µg l^{-1}) were found in geothermal waters. High concentrations of thallium, however, are not ubiquitous in thermal waters but strongly dependant on host lithology. Thallium is fractionated from K and Rb during weathering and hydrothermal alteration.

1. INTRODUCTION

Thallium (from the Greek *thallos*: green twig) is one of the most highly toxic elements. Although its chemistry and toxicology are well documented, there are limited data on its occurrence in natural waters.

Thallium is still used in some rodenticides and has a wide range of industrial uses. Large amounts are also released in emission sources from power plants, smelters, cement plants and in runoff from mines.

The water soluble salts of thallium (sulfate, carbonate, acetate) are the most toxic and deaths occur with ingestion of only trace amounts. Thallium substitutes for potassium in many biological processes and is released slowly from the body.

2. GEOCHEMISTRY OF THALLIUM

Thallium exists in two oxidation states, Tl^+ and Tl^{3+}, but Tl^+ generally forms the most stable compounds. It is found as a trace constituent in metal ores, coal, the potassium salts sylvite and carnallite and potassium silicate minerals, showing both chalcophile and lithophile behaviour. The similar behaviour of thallium to potassium and rubidium, due to similar ionic radii, is reflected in similar enrichments in acid igneous rocks, pegmatites and sheet silicates. Marine manganese nodules also have high concentrations of thallium as do rocks with free carbon. Several soluble thallous salts exist, but thallium minerals are extremely rare in the natural environment due to their low temperature stabilities (Elhaddad & Moh, 1993).

Thallium shows little tendency to form inorganic or organic complexes, but biomass uptake may be important and it is a useful biogeochemical pathfinder element in ore prospecting (Warren & Horsky, 1986).

Vink (1993) has shown that the Tl^+ ion is the dominant species in most natural environments in terms of pH and Eh (Figure 1) indicating that thallium

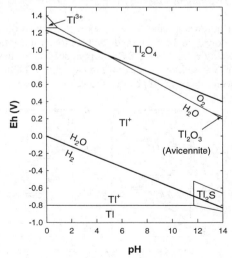

Figure 1. Eh-pH diagram for the system Tl-S-O-H. Activities for dissolved Tl and S species are 10^{-8} m and 10^{-3} m respectively. Modified from Vink (1993).

is theoretically very mobile, supporting the significant dispersion noted around oxidising sulphide bodies (Vink, 1993).

In seawater the concentration of thallium is very low (12 - 16 ng kg^{-1}; Flegal & Patterson). Typical concentrations in fresh waters are also of the order of ng l^{-1} but oilfield waters (Korkisch & Steffan, 1979) may be higher, up to 660 µg l^{-1}. The mobility of thallium was shown to be enhanced in geothermal systems by Dall'aglio et al. (1994) who reported concentrations in the range of µg l^{-1}.

3. SAMPLING AND ANALYTICAL TECHNIQUES

Thallium may be analysed routinely by ICP MS with detection limits of the order of 0.1 - 0.5 µg l^{-1}. The vast majority of natural waters analysed contain thallium at levels lower than this detection limit. In order to determine baseline concentrations, thallium was extracted from 250 ml samples by solvent (4-methylpentan-2-one) extraction similar to the method used by Cidu et al. (1994) for gold. The 3-sigma detection limits achieved ranged from 0.3 to 2.6 ng l^{-1}.

This paper reviews selected data compiled during routine analysis (section 4) and presents data on baseline concentrations (section 5) in waters using the extraction procedure discussed above.

4. ANALYTICAL DATA

4.1 Thallium in soils

Soil porewaters and aqueous extracts from profiles developed on Tertiary sands and Silurian mudstones (Gooddy et al., 1995; BGS unpublished data) have been analysed for a wide range of elements. Thallium concentrations in all profiles showed a decrease from the upper organic horizons downwards with porewater concentrations up to 400 ng l^{-1} in the O/A horizons.

The partition coefficients were high for all horizons (log K_d from 1.4-2.8) for both weakly and strongly bound thallium. The K_d's were generally higher in the organic horizons than in the deeper mineral horizons.

4.2 Thallium in the unsaturated zone

Unsaturated zone profiles have been collected from arid and temperate zones and the porewaters extracted by centrifugation. Thallium concentrations are often above detection limit with concentrations as high as several hundred ng l^{-1}. An example of a profile collected from Tertiary glauconitic sands in England is shown in Figure 2 where it can be seen that thallium is mobilised at pH 4-5 possibly in association with K-rich horizons.

4.3 Thallium in surface waters

Thallium has been studied in surface waters collected from low order upland streams from a wide range of sedimentary, igneous and metamorphic rock types. Most surface waters were below detection limit but high concentrations (up to 490 ng l^{-1}) were found in waters close to auriferous ore bodies.

4.4 Thallium in groundwaters

The majority of groundwaters analysed were also below the detection limits of ICP MS. Maximum concentrations in UK aquifers were found in acid groundwaters (Tl 350 ng l^{-1} pH 4.2,) where sulphide oxidation was occurring. High concentrations (up to 7730 ng l^{-1}) were found in ferruginous springs in northern Sardinia. The springs discharge through altered Tertiary volcanic rocks with the highest

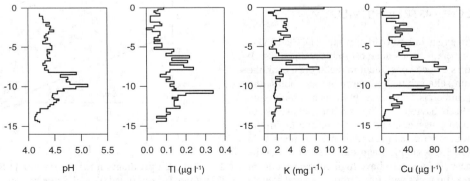

Figure 2. Chemistry of porewaters from an unsaturated zone profile in glauconitic sands.

concentrations in springs close to auriferous veins.

4.5 Thallium in thermal waters

The thermal (T=45°C) mineral waters of Bath in England contained 93 ng l^{-1}, on the threshold of detection. Mine waters from the South Crofty and Wheal Jane tin mines of Cornwall were significantly higher with 19,000 and 3,600 ng l^{-1} respectively. The highest concentrations were found in thermal waters (T=76°C) from Southampton in arkosic sandstone of Permo-Triassic age which contained 99,000 ng l^{-1}.

A large number of thermal waters (T up to 89 °C) from Kamchatka have been analysed. Bedrock geology comprises Neogene-Quaternary calc-alkaline basalts and andesites. Thallium concentrations were relatively low compared with other thermal waters with a maximum concentration of only 500 ng l^{-1}.

5. BASELINE CONCENTRATIONS

More than eighty water samples were collected for the determination of thallium following solvent extraction. These included groundwaters from a range of UK aquifers including the Chalk and sandstones of Tertiary, Permo-Triassic and Cretaceous age. In addition samples were analysed from saline mineral waters in Wales, stream waters from the Devonian Cheviot granite and andesites of England and groundwaters from a high grade metamorphic terrain in Sri Lanka.

Thallium concentrations as high as 34 ng l^{-1} were present in UK sandstone aquifer groundwaters. However, some samples were below detection limit even with preconcentration. The data are shown plotted against pH in Figure 3.

Although there is little correlation overall between thallium and pH, individual rock units show a tendency to higher concentrations with lower pH. These more acid waters are typically found where sulphide oxidation is known to occur. There is little correlation of thallium with Eh and although correlations exist between thallium and K or Rb in individual aquifers, the correlation may be negative as well as positive.

The saline waters of Wales, considered to be 5-10 kyr old contained very little thallium (0.9 - 1.1 ng l^{-1}) and the Cheviot and Sri Lanka samples from hard rock areas were only slightly higher (0.4 - 4 ng l^{-1}).

6. DISCUSSION

Although a high mobility of thallium has been implied by the stability of Tl^+ in most of the Eh-pH space occupied by natural waters, several geochemical processes are likely to limit its mobility. The most important of these is the incorporation of thallium into clays and sheet silicates. Adsorption is also an important processes and likely to be an important mechanism in the fractionation of thallium from K and Rb. The low concentrations typical of most surface and ground waters provide evidence for limited mobility of thallium. Thallium is plotted against T (°C) and K concentration in Figure 4.

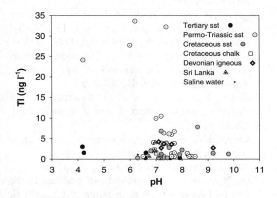

Figure 3. Plot of pH vs. Tl concentration in preconcentrated waters samples.

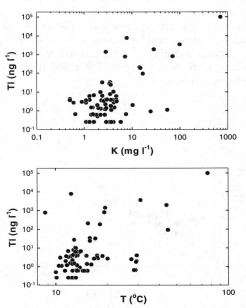

Figure 4. T (°C) and K concentration vs. thallium for complete data set.

There is an overall increase in thallium with T and K concentration but considerable scatter reflects the range of waters analysed. Concentrations of thallium in waters relatively close to ambient T are typically of the order of ng l^{-1}. The thallium is likely to be derived from a variety of sources including biotite, potassium feldspar, sulphide minerals, clays and carbonaceous material. Locally, particularly in the vicinity of oxidising ore bodies, thallium is easily dispersed and may reach concentrations of several µg l^{-1}.

High concentrations of thallium are generally associated with thermal waters. This association is not ubiquitous, however, and appears to be controlled by the rock type and regional tectonic province. Hydrothermal regions where the bedrock is dominated by high-K or acid igneous or sedimentary rocks are likely to have waters with high thallium (e.g. Dall'aglio et al., 1994). Calc-alkaline terrains where bedrock lithology is dominated by basaltic to andesitic volcanism on the other hand lack a primary source of thallium as evidenced by the low concentrations of thallium in the thermal waters of Kamchatka.

There is a relatively good correlation of thallium with rubidium (r=0.6) but limited data are as yet available. Tl/K and Tl/Rb ratios are much lower than in terrestrial rocks probably as a consequence of a lower mobility of Tl or the confinement of Tl to the least favoured minerals for weathering e.g. K-feldspar. The relative mobility of thallium to potassium increases in a consistent manner as indicated by the good correlation between thallium and Tl/K ratio (Figure 5). The stability of the major silicate minerals such as K-feldspar and biotite is strongly controlled by temperature and pH and it appears that the relative mobility of thallium is in some way related to the type and kinetics of alteration.

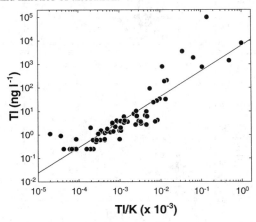

Figure 5. Plot of thallium concentration vs. Tl/K ratio in studied waters.

7. CONCLUSIONS

Thallium concentrations measured in natural waters range from <0.5 to 99,000 ng l^{-1}. Background concentrations in most waters close to ambient temperature are typically in the range $10^{-1} - 10^{1}$ ng l^{-1}, showing that thallium is not a particularly mobile element. Thallium incorporated in sulphide minerals is dispersed relatively easily during oxidation reactions and concentrations may reach several µg l^{-1}. Thermal waters contained the highest concentrations of thallium up to 99 µg l^{-1}. The primary control on the concentrations of thallium is the host rock lithology and mineralogy. Important secondary controls such as adsorption and incorporation into clay minerals strongly limit thallium mobility in most environments. Both solubility and dispersion are enhanced at high temperatures in response to hydrothermal alteration. Tl/K and Tl/Rb ratios are very variable and low in comparison with typical rocks. Thallium, thus appears to be strongly fractionated by weathering and alteration processes.

8. REFERENCES

Cidu, R., Fanfani, L., Shand, P., Edmunds, W.M., Van't dack, L. & Gibjels, R. 1994 The determination of gold at the ultratrace level in natural waters. *Analytica Chimica Acta*, **296**, 295-304.

Dall'aglio, M., Fornaseri, M. & Brondi, M. 1994. New data on thallium in rocks and natural waters from central and southern Italy. Insights into applications. *Miner. Petrogr. Acta*, **38**, 103-112.

Elhaddad, M.A. & Moh, H. 1993. The low temperature synthesis of thallium minerals in aqueous environment in relation to natural occurrences. *N. Jb. Miner. Abh.*, **166**, 9-14.

Flegal, A.R. & Patterson, C.C. 1985. Thallium concentrations is seawater. *Mar. Chem.*, **15**, 327-331.

Gooddy, D.C., Shand, P., Kinniburgh, D.G., & Van Riemsdijk, W.H. 1995 Field-based partition coefficients for trace elements in soil solutions. *Journal of Soil Science*, **46**, 265-285.

Korkisch, J & Steffan, I. 1979. Determination of thallium in natural waters. *Intern. J. Environ. Anal. Chem.*, **6**, 111-118.

Warren, H.V. & Horsky, S.J. 1986. Thallium, a biogeochemical prospecting tool for gold. *J. Geochem. Explor.*, **26**, 215-221.

Vink, B.W. 1993. The behaviour of thallium in the (sub)surface environment in terms of Eh and pH. *Chem. Geol.*, **109**, 119-123.

Decalcification and acidification of coastal dune sands in the Netherlands

P.J. Stuyfzand
Kiwa N.V., Research and Consultancy, Nieuwegein, Netherlands

ABSTRACT: The leaching of calcium carbonate constitutes the main weathering reaction of coastal dune sands, prior to acidification. This leaching was studied on 28 plots along the North Sea coast, by monitoring for 3 years the fluxes and chemistry of bulk precipitation, the upper layer of calcareous groundwater and the drainage water from lysimeters. Dune shrub gave the highest and pine the lowest decalcification rate (250 versus 150 mg $CaCO_3$ m^{-2} d^{-1}), from 7 vegetation types. Dry, atmospheric deposition of acidifying substances was highest for pines, but was counteracted by a less favourable water balance and more accumulation of organic matter. On 6 plots the leaching was favoured by oxidation of organic matter and pyrite in dune sand. The decalcification rate can be calculated with a linear equation, if the decalcification front resides below the root zone. An accelerated $CaCO_3$ leaching is postulated during the initial period when calcium carbonate is still present in the root zone.

1 INTRODUCTION

The coastal dunes of the Netherlands constitute an important catchment area for drinking water supply, and contain important nature reserves. The most prominent weathering reaction in calcareous dunes with a precipitation excess, is the leaching of calcium carbonate. This is followed by acidification, which is of great concern for drinking water supply and nature conservation, for well known reasons.

The decalcification process was studied or noticed in many dune areas in the world mainly through geochemical analysis, amongst others by Salisbury (1925), Ball & Williams (1974), Burges & Drover (1953), Olson (1958), Van Breemen & Protz (1988) and Rozema et al. (1985). In this contribution the fluxes and chemistry of bulk precipitation and shallow groundwater are used to quantify the leaching rate under 7 vegetation covers.

2 SAMPLING

The upper 0.3-3 metres of dune groundwater were sampled for chemical analysis, on the 28 plots shown in Fig.1. Characteristics for each plot are given in Table 1. Every plot was situated within a flat dune terrain with homogeneous vegetation cover, where the upper groundwater derived from rainfall which passed that vegetation. On each plot, on average, about 36 samples were taken in 1980-1983, from 4 sampling wells with short screens (0.01-0.5 m long). The 4 plots near Castricum are composed of huge, 58 years old lysimeters (25 by 25 m, 2.5 m deep), each with a different, mature vegetation cover. Here the drainage water was sampled.

Bulk precipitation was monitored with 8 collectors close to the plots (Fig.1), on a biweekly basis. All samples were filtrated in the field through 0.45 μm filter paper. Water fluxes were measured using standard rain gauges and both small size (1 m^2) and huge (625 m^2) lysimeters.

3 GEOLOGY AND GEOCHEMISTRY

The older dune sands more inland (Fig.1), formed in the period 200-3800 BC, the youger dunes in the period 900-1850 AD. The age of the dune landscape thus increases from the beach inland. In between Egmond and Bergen a transition zone exists, from primarily calcareous younger dunes (>2% $CaCO_3$) in the south to dunes primarily poor in lime (<0.4% $CaCO_3$) in the north. The calcium carbonate is largely present as shell debris, composed of aragonite ($Sr_{0.002}Na_{0.024}Mg_{0.002}Ca(CO_3)_{1.016}$).

4 BULK PRECIPITATION

Mean annual precipitation was 0.8 m. The weighted mean composition of bulk precipitation is given in Table 1. There are only minor differences between the 3 dune areas. The chemical composition is explained by contributions from: sea spray (>97% for Na^+, Cl^- and Mg^{2+}), continental mineral aerosols (>95% for Al, SiO_2), biogenic inputs (DOC, PO_4^{3-}), and anthropogenic pollution (>80% for NO_3^-/NH_4^+, >50% for SO_4^{2-}).

5 COASTAL DUNE GROUNDWATER

The mean composition of shallow groundwater on the 28 plots is presented in Table 1. Their chemical composition is governed mainly by the distance to the coast, vegetation cover, depth of decalcification of dune sand, and thickness of the unsaturated zone (Stuyfzand, 1993). Coastal dune groundwater is characterized by: a high dry deposition of sea spray (depending on vegetation cover and distance to the beach); large-scale N_2-fixation, especially by sea buckthorn (*Hippophaë rhamnoides*); and strong annual and seasonal fluctuations in its chemistry. Natural groundwater recharge varied from 0.62 m/y for bare dunes to 0.14 m/y for dunes covered with pines (Table 1). In addition to the 4 huge lysimeters data from small (1 m²) lysimeters were used to estimate the recharge for plots without lysimeters.

6 DECALCIFICATION AND GROUNDWATER

Most Ca^{2+} in the calcareous groundwaters, derives from $CaCO_3$ dissolution. Other Ca^{2+} sources are sea spray, the weathering of silicate minerals and cation-exchange. The sea spray contribution (mg/l basis) is equal to 0.021·Cl^- in groundwater (0.021 = Ca^{2+}/Cl^- in sea water). This is generally < 5%. The dissolution of Ca-silicates, may contribute ≤2 mg Ca^{2+}/l for each dissolved 6 mg SiO_2/l, in case of congruent anorthite dissolution. We take half this amount as most probable, which is also <5%. Desorbed Ca^{2+} principally derives from earlier carbonate dissolution, and is treated here as such. The mean decalcification flux (Q_{CaCO_3}), in mg $CaCO_3$ m⁻² d⁻¹, then becomes:

$$Q_{CaCO_3} = 2.5 \cdot Ca^{\#} \cdot N \quad (1)$$

with: N = mean groundwater recharge [mm/d]; $Ca^{\#}$ = Ca^{2+}-0.021 Cl^--SiO_2/6 = mean Ca^{2+} concentration (mg/l) of upper groundwater, stemming from $CaCO_3$ dissolution; 2.5 = conversion factor Ca^{2+} to $CaCO_3$.

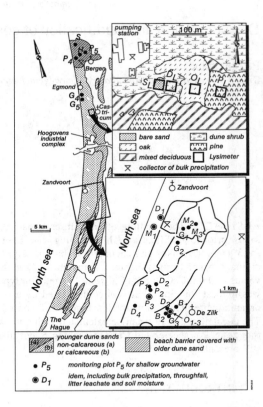

Fig. 1. Location of 28 monitoring plots for the upper dune groundwater and the most important collectors for bulk precipitation. B = bracken (*Pteridium aquilinum*); D = dune shrub (mainly *Hippophaë rhamnoides*); G = (dune) grasses; H = heather (*Calluna vulgaris*); M = mosses; O = oaks (*Quercus robur*); P = pines (*Pinus nigra ssp nigra*); S = bare or scanty vegetation cover.

The main leaching zone is considered here to reside above the water table, which is consistent with geochemical observations in the calcareous dunes. Results of calculation using Eq.1, are presented as bar diagrams for the 24 calcareous plots in Fig.2. It can be concluded that $CaCO_3$ leaching is favoured by (1) passage through dune peat above or close to the water table (B_2, O_1), due to CO_2 and H_2SO_4 release upon oxic decomposition and pyrite oxidation, and (2) sea buckthorn (all D-plots).

Below *Hippophaë rhamnoides* the intercepted atmospheric acids and acid precursor NH_4^+, and biogenic acids (CO_2 and organic acids) seem to be most efficiently consumed by dissolving $CaCO_3$ in the root zone. The strong N_2-fixation in its root nodules is a H^+-indifferent process, but prevents the

Table 1. Mean composition of bulk precipitation and the upper 0.3-3 metres of dune groundwater under various vegetation covers, in the coastal dune area of the Western Netherlands in the period 1980-1983.

monitoring plot	dune peat †	decalc. depth m-surf	MGL§ m-surf	N ‡ m/y	pH	Cl⁻	SO₄²⁻	HCO₃⁻	NO₃⁻	Al³⁺	Na⁺	K⁺	Ca²⁺	Mg²⁺	NH₄⁺	Fe	Mn	SiO₂	DOC
											mg/L								
BULK PRECIPITATION (WEIGHTED MEAN)																			
Plots I	-	-	-	0.80	4.5	16	7	0	3.4	0.10	9	0.4	1	1.1	1.29	0.10	0.02	0.2	1.5
Plots II+III	-	-	-	0.83	4.66	12	7	<1	3.6	-	6.7	0.4	1.3	0.9	1.31	0.04	0.03	-	1.7
Plots IV	-	-	-	0.82	4.4	15	8	0	3.5	0.07	8	0.6	2	1.0	1.21	0.05	0.03	<0.5	1.5
I: MINISCREENS, DUNES POOR IN CALCITE, NORTH OF BERGEN																			
S = scanty	-	13.0	3.1	0.50	4.40	29.0	17.0	<1	10.5	1.92	17.8	1.6	2.0	2.5	0.03	0.24	0.03	10.0	3.4
H = heather	h	20.0	2.1	0.37	4.79	33.4	20.4	5	2.7	0.80	20.1	2.1	6.3	3.0	0.09	2.05	0.04	13.1	9.8
P₄ = pines-4	-	20.0	2.9	0.14	4.80	128	46	5	0.7	0.90	74.8	2.2	9.5	10.8	0.04	0.42	0.09	16.2	10.4
P₅ = pines-5	-	18.0	3.7	0.20	5.31	63.4	36.3	7	1.0	0.60	38.3	2.8	6.8	7.9	0.04	0.87	0.05	23.1	8.3
II: LYSIMETERS, CALCAREOUS DUNES, WEST OF CASTRICUM																			
S_L = bare	-	<0.1	2.2	0.62	7.96	16.1	16.5	116	16.7	<0.01	9.2	2.33	47.4	2.4	0.03	0.03	<0.01	4.4	3.1
D_L = duneshrub	-	<0.1	2.3	0.34	7.42	63.3	50.8	270	35.3	<0.01	33.4	3.28	119	6.4	0.04	0.04	0.03	8.7	7.8
O_L = oaks	-	<0.1	2.3	0.30	7.38	79.1	47.4	276	0.5	<0.01	42.1	4.86	110	5.7	0.03	0.08	0.07	10.8	10.4
P_L = pines	-	<0.1	2.4	0.14	7.41	435	304	388	0.6	<0.01	242	15.4	252	24.7	0.04	0.04	<0.01	13.0	17.2
III: PIEZOMETERS, CALCAREOUS DUNES SOUTH OF EGMOND AAN ZEE																			
G₄ = grasses-4	-	<0.1	0.4	0.25	7.15	115	24.1	330	<0.1	<0.02	67.5	3.56	103	13.8	0.10	2.23	0.16	6.5	20.6
G₅ = grasses-5	-	<0.1	1.1	0.33	7.58	76.5	32.2	202	9.4	<0.02	48.3	2.25	73.3	7.4	0.02	0.05	0.01	4.6	-
IV: MINISCREENS, CALCAREOUS DUNES, SOUTH OF ZANDVOORT AAN ZEE																			
M₁ = mosses-1	-	0.05	1.5	0.52	7.77	58.1	34.3	152	36.1	<0.02	33.9	1.75	66.7	6.3	<0.1	<0.2	<0.08	5.2	2.7
M₂ = mosses-2	-	0.20	1.7	0.49	7.86	22.9	22.9	141	25.6	<0.02	15.7	1.19	56.2	4.1	0.16	0.11	<0.05	4.2	3.3
M₃ = mosses-3	-	0.3	2.5	0.43	7.74	39.5	37.1	176	34.2	<0.02	28.0	1.16	74.0	5.1	<0.10	0.28	<0.05	5.1	4.1
G₁ = grasses-1	-	<0.1	0.7	0.38	7.36	34.9	28.7	315	8.5	<0.02	19.1	0.98	109	7.4	0.45	2.44	0.26	4.8	6.3
G₂ = grasses-2	-	0.3	0.3	0.33	7.31	34.4	20.8	321	<0.2	<0.02	21.2	1.40	104	6.1	0.59	14.21	1.78	9.6	20.8
G₃ = grasses-3	+	1.3	3.1	0.33	6.66	42.0	66.4	287	4.9	<0.02	21.5	2.22	112	6.4	0.37	1.52	0.13	10.1	33.5
B₁ = bracken-1	-	2.0	4.1	0.41	7.76	21.8	30.5	158	14.7	<0.02	12.5	2.06	61	5.2	0.07	<0.2	<0.10	7.3	1.8
B₂ = bracken-2	+	1.3	2.1	0.41	7.20	34.6	128	282	381	<0.02	19.8	0.41	266	3.6	0.05	<0.2	<0.10	5.4	27.9
D₁ = duneshrub	-	<0.1	2.7	0.38	7.77	121	54.6	162	52.4	<0.02	68.1	1.54	89	6.3	0.09	<0.2	<0.08	7.6	1.6
D₂ = duneshrub	-	0.3	2.4	0.38	7.34	41.7	40.3	275	21.9	<0.02	24.6	0.67	104	7.2	0.06	0.02	0.09	5.4	5.5
D₃ = duneshrub	S	1.3	3.7	0.38	7.12	41.2	73.6	268	85.4	<0.02	23.5	0.85	136	5.9	<0.05	<0.3	<0.08	3.6	13.6
D₄ = duneshrub	+	0.3	2.1	0.41	7.15	94.2	77.6	344	0.6	<0.02	53.9	1.64	135	7.6	0.46	3.61	0.15	7.3	8.0
O₁ = oaks-1	S	1.4	3.5	0.31	7.12	86.7	167	298	18.3	<0.02	48.4	2.28	162	8.0	0.22	3.33	0.10	10.4	32.7
O₂ = oaks-2	-	1.5	3.0	0.31	7.58	61.2	67.6	202	7.1	<0.02	36.3	1.69	80.7	10.3	0.07	0.35	<0.09	6.8	5.0
O₃ = oaks-3	-	1.5	3.6	0.31	7.46	83.7	93.1	212	45.6	<0.02	45.6	5.64	109	11.5	<0.05	0.45	<0.09	7.3	18.6
P₁ = pines-1	+	0.4	2.2	0.14	6.90	146	78.1	638	1.4	<0.02	91.0	1.17	220	13.5	0.29	4.04	0.33	6.9	37.6
P₂ = pines-2	-	0.4	1.6	0.14	7.47	107	86	352	<0.2	<0.02	57.6	0.75	130	13.4	<0.1	0.24	0.37	8.5	-
P₃ = pines-3	S	0.1	2.9	0.20	7.26	73.8	76.0	336	4.7	<0.02	42.8	1.45	131	7.4	0.12	0.71	0.23	7.6	15.9

†: interaction above or at the groundwater table: +, S, - = peat clearly present, in traces and not present respectively; h = 0.1 m thick, humic top layer; § = Mean Groundwater Level in metres below land surface; ‡ = if bulk deposition 'gross precipitation', else 'natural groundwater recharge'.

uptake of $H^+ + NO_3^-$, and thereby contributes to the high leaching rate.

Although pines intercept higher amounts of atmospheric acids and acid precursors than all other vegetation types, and probably produce more CO_2 and biogenic acids than sea buckthorn, they rank among the vegetation types with the lowest decalcification rate (Fig.2)! This is explained by: (a) CO_2 losses to the atmosphere through diffusion, due to shortage of water to take up the CO_2 produced and due to precipitation of calcite upon drying of the intensively rooted soil horizons; (b) denitrification yielding bicarbonate, which buffers H^+ by formation of volatile H_2CO_3; and (c) acid buffering by the tree canopy, NO_3^- uptake and neoformed soil organic matter.

7 CALCULATED DECALCIFICATION RATES

It is recognized that $CaCO_3$ dissolution is a very fast process, requiring a few days to reach equilibrium. The width of the transition zone between completely decalcified and practically unleached dune sand generally ranges from 0.05-0.5 m, as observed elsewhere.

The decalcification rate (v_{CaCO3}), i.e. the migration velocity of the schematized sharp lime boundary downwards, then simply becomes, in mm/d:

$$v_{CaCO_3} = \frac{Q_{CaCO_3}}{(1-\varepsilon) \cdot \rho_s \cdot (CaCO_3)_{solid}} \quad (2)$$

with: ε = porosity [-]; ρ_s = density of solids [kg/dm³]; $(CaCO_3)_{solid}$ = content [mg/kg dry weight]; Q_{CaCO3} = mean $CaCO_3$ flux, calculated with Eq.1 [mg $CaCO_3$ m⁻² d⁻¹].

Application of Eq.2 to the calcareous dunes, which contain on average 5 % $CaCO_3$ by dry weight, with ε = 0.4, ρ_s = 2.65 kg/l and the median value for Q_{CaCO3}= 200 mg $CaCO_3$ m⁻² d⁻¹ (Fig.2), yields a decalcification rate of 2 µm/d, or 9 cm/century. This rate compares well with geochemical data from the dunes near Zandvoort (Fig.3).

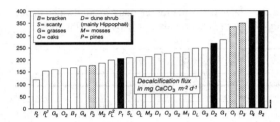

FIG. 2 The daily CaCO$_3$ flux from the decalcification zone in primarily calcareous dunes, arranged in increasing order. Interaction with intercalated dune peat is indicated by black bars (strong) or dots (weak, peat in traces).

Calculations for prolonged periods are only approximate and heavily rely on various assumptions (Stuyfzand, 1993). Nevertheless a linear depletion of the CaCO$_3$ stock is evidenced by the data in Fig.3, and confirmed by data in Fig.4. The data from Schiermonnikoog and Lake Michigan reveal, however, a higher decalcification rate during the initial 20 and 50-100 years, respectively. This appears to be the general picture, because : (1) also the curve for Southport (UK; Salisbury, 1925) can be divided into a high rate during the initial 100 y and a slow constant rate afterwards; and (2) there is other evidence from Burges and Drover (1953) who noticed that 50% was leached in a coastal dune system of New South Wales (Australia) in 50 y and the remaining part required an additional 150 y.

The duration of this initial phase of accelerated decalcification depends mainly on the original lime content (Fig.4), when gross precipitation does not deviate much : 100 y for Southport (6.3% CaCO$_3$), 50-100 y for Lake Michigan (2.6%), 20 y for Schiermonnikoog (1.2%) and probably 5 y for Blakeney Point (0.6%?).

The duration of accelerated leaching thereby becomes about 20 y for each 1.2 % CaCO$_3$ originally present. The rate of CaCO$_3$ loss from the soil is not proportional to the carbonate left at any time, as stated by Olson (1958).

The initial, high leaching rate is explained by: (1) specific characteristics of pioneer vegetations, with their N$_2$-fixation (reducing uptake of atmospheric HNO$_3$) and low

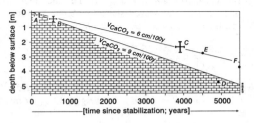

FIG. 3 Measured decalcification depth of dune sand in relation to the time elapsed since stabilization (younger dunes) or dune formation (older dunes), for the primarily calcareous dunes south of Zandvoort.

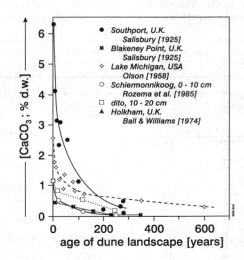

FIG. 4 Progress in decalcification of the upper 10-20 cm mineral soil, for calcareous sand dunes in selected humid areas of the world (rainfall = 700-900 mm/y).

evapotranspiration losses (reducing CO$_2$ losses to the atmosphere); (2) an initial optimal contact of CO$_2$ and organic acids with CaCO$_3$ in a calcareous root zone; and (3) a high uptake and redistribution of Ca^{2+} by calciphilous vegetation.

REFERENCES

Ball, D.F. & W.M. Williams 1974. Soil development on coastal dunes at Holkham, Norfolk, England. *Proc. 10th Intern. Congress of Soil Science (Moscow)* VI: 380-386.

Burges, A. & D.P. Drover 1953. The rate of podzol development in sands of the Woy Woy district, N.S.W. *Australian J. Botany* 1: 83-94.

Olson, J.S. 1958. Rates of succession and soil changes on Southern Lake Michigan sand dunes. *Bot. Gazette* 119: 125-170.

Rozema, J., P. Laan, R. Broekman, W.H.O. Ernst & C.A.J. Appelo 1985. On the lime transition and decalcification in the coastal dunes of the province of north Holland and the island of Schiermonnikoog. *Acta Bot. Neerl.* 34: 393-411.

Salisbury, E.J. 1925. Note on the edaphic succession in some dune soils with special reference to the time factor. *J. Ecol.* 13: 322-328.

Stuyfzand, P.J. 1993. *Hydrochemistry and hydrology of the coastal dune area of the Western Netherlands*. Ph.D Thesis Vrije Univ. Amsterdam, published by Kiwa, ISBN 90-74741-01-0, 366 p.

Van Breemen, N. & R. Protz 1988. Rates of calcium carbonate removal from soils. *Can. J. Soil Sci.* 68: 449-454.

Factor analysis of stream water chemistry following storms in Eastern Pennsylvania and New Jersey, USA

B.J.Woodward, V
I.C.F. Kaiser Engineers and Constructors, Inc., Fairfax, Va., USA

Brad Bowen
P.R.C. Environmental Management, Pa., USA

D.E.Grandstaff
Department of Geology, Temple University, Pa., USA

Abstract: Variations in water chemistry in four small, first-order streams during and following storms were analyzed using R-mode Factor Analyses coupled with hysteresis plots. These analyses indicate that there are three statistically significant factors controlling stream water composition during rainstorms. These factors were subsequently interpreted to represent three physical sources/processes: inputs of rain, rainwater solutes, and throughfall; dissolution of soluble soil salts; and leaching of organic matter. Differences in bedrock type did not greatly affect the number and nature of factors and processes.

1 INTRODUCTION

Before finally reaching streams, rainwater may react with foliage or plant surfaces (throughfall); leaf litter, humus, and land surfaces (overland flow); and/or deeper soil or rock layers (interflow/baseflow). Investigation of variations in stream water chemistry following storms can reveal important information about biogeochemical processes and water:rock interactions. To determine the nature of processes acting during storms and whether variations in bedrock composition and mineralogy would affect these processes, we have investigated changes in stream water chemistry during and after storms in four small (≤ 1 km^2), lithologically distinct drainage basins. Results of this study may be compared with a previous study by Miller and Drever (1977).

2 LOCALITIES AND ANALYSIS

Samples were collected prior to, during, and following storms in small, first-order streams at four different locations in eastern Pennsylvania and New Jersey. The drainage basins were chosen to be individually lithologically homogeneous but so that each had a different bedrock lithology. Geochemical data were collected from drainage basins located in Phoenixville (granodiorite), Uhlerstown (diabase), and Lumberville (shale), in eastern Pennsylvania and Wharton State Forest, New Jersey (unlithified quartz sand).

Water samples were collected in acid-washed polyethylene bottles. The pH, Eh, temperature, alkalinity, TDS, temperature, and discharge were measured in the field. Discharge was measured using a Gurley pygmy flow meter at stream locations having measured cross-sections. In the laboratory, water samples were filtered through a 0.45 µm membrane filter. Ca, Mg, Na, K, Fe, SiO_2, Cl, PO_4, NO_3, SO_4 and trace elements were measured using atomic absorption and ion chromatography. Data were analyzed using R-mode factor analysis (Davis, 1986) using NCSS (Hintze, 1996) and Statistica (Statsoft, 1991). Factor analysis was done using Varimax rotation and robust covariance estimation. Interpretation of factors was aided by examination of hysteresis (concentration *vs.* discharge) diagrams for individual species.

3 CHEMICAL VARIATIONS

Chemical variations were measured for five different storms. Variations were similar at all sites, thus only a single storm, at an unnamed tributary to Pickering Creek, Phoenixville, PA on October 29, 1993, is presented in detail (Fig. 1).

Discharge and species concentrations in Pickering Creek were measured starting about 12 hours before the beginning of the storm (0 hours) and continued for about 91 hours (until 79 hours after the storm began).

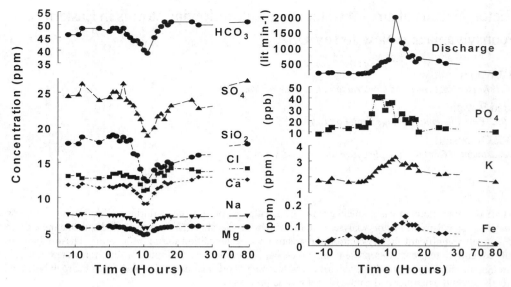

Figure 1. Variations in stream water chemistry and discharge variations during a storm at Phoenixville, PA, October 29, 1993. Note breaks in X axes between 30 and 70 hours and in Y axes.

The storm had a total precipitation amount of 4.3 cm. No further precipitation occurred during the study period. During and following the storm, the discharge increased by a factor of 10, reaching a maximum after 11 hours. Concentrations of Ca, Mg, Na, SiO_2, Cl, and sulfate decreased as discharge increased, whereas K, NO_3, phosphate, and Fe increased. Although most minimum and maximum species concentrations coincided with the maximum discharge at 11 hours, the maximum concentrations for PO_4 and NO_3 occurred at about 8 hours (Figure 1).

The relationship between discharge and TDS is shown in Figure 2. Prior to the storm, the water chemistry was dominated by baseflow and had relatively constant composition. During the first 11 hours after storm initiation, the discharge increased and the TDS concentration was relatively low. After 11 hours, the discharge decreased but TDS remained higher than in the initial period. By approximately 79 hours the TDS and discharge approached the pre-storm values. During the first period (0 to 11 hours), discharge was dominated by overland flow as well as direct precipitation. This short, 11-hour timeframe allowed minimal opportunity for leaching and consequently, the stream had low TDS. During the second period (11 to 79 hours), discharge became increasingly dominated by interflow and baseflow contributions. Within this longer, 68 hour period there was more opportunity for interactions between the infiltrating precipitation and soil particles, contributing to a higher TDS concentration in the stream water and a different chemical composition. Although the TDS decreased, the data points do not lie along or near the theoretical dilution line, even during the initial period. This confirms that rainwater contributions from storm events do not purely dilute baseflow (Miller and

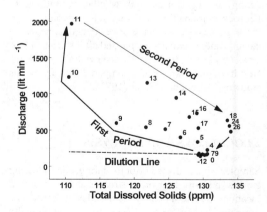

Figure 2. Hysteresis plot of TDS *vs.* Discharge. Numbers near circles indicate sampling times in hours before/after beginning of storm. For discussion of first and second period see text.

Drever, 1977), but that species are added to the water by reactions, even at very short time periods.

4 FACTOR ANALYSIS

Factor analysis is a statistical procedure used to reveal an underlying structure or pattern in a set of multivariate observations (Davis, 1986). In this study, R-mode factor analysis was applied to determine interrelations between the variables of interest (species concentrations, pH, and stream discharge). Using R-mode factor analysis, these variables are reduced or converted into new variables or factors, based on their diversification throughout the storm event. These factors may then be interpreted in terms of physical sources or processes. One of the problems in Factor Analysis is determination of the proper number of factors to be retained. Two criteria are often used, the Kaiser Criterion and the Scree Plot (Hintze, 1996). The Kaiser criterion states that only factors with eigenvalues greater than about 1.0 should be retained. Using the scree plot of eigenvalues, only those factors having high eigenvalues, forming the "cliff" should be retained. Both the scree plot (Figure 3) and Kaiser criterion suggest that three factors should be retained. The three factors, significantly correlated species, and eigenvalues are given in Table 1.

TABLE 1. Significant Factor Loadings after Varimax Rotation.

	Factor I	Factor II	Factor III
Discharge	0.86	0.41	
pH		-0.75	
Na	-0.64	-0.67	
Ca		-0.87	
Mg		-0.78	
K	0.77		0.41
Fe	0.83		
Cl		-0.89	
HCO$_3$		-0.95	
SO$_4$	-0.83		
NO$_3$	-0.78		
PO$_4$			0.73
SiO$_2$	-0.93		
Percent Variation (Cumulative)	47	88	98
Eigenvalue	5.89	5.49	1.27

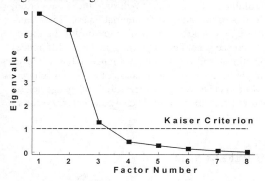

Figure 3. Scree plot of R-mode factor eigenvalues after varimax rotation. The scree plot and Kaiser criterion both indicate three significant factors.

Plots of concentration *vs.* discharge (hysteresis plots) for individual species also aid factor interpretation. For example, at a given discharge, the highest concentrations of Ca (Factor II) occur during the second period, which is dominated by baseflow/interflow (Figure 4). This indicates that Ca (as well as other chemical species in Factor II) is primarily derived from reactions deeper within the soil. In contrast, at given discharge, the highest

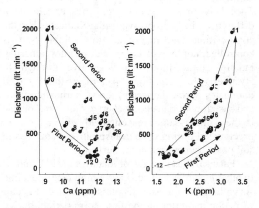

Figure 4. Hysteresis plots for Ca and K *vs.* Discharge. Note the loops are oriented in different directions, i.e. Ca has higher concentrations at given discharge during the second period, whereas K has higher concentrations during the first period. This indicates that the factor containing Ca (Factor II) is dominant, producing the highest concentrations, during the second period, whereas, the factors containing K (Factors I and III), have their greatest effects during the first period.

concentrations of K (Factors I and III) occur during the first period, indicating that it is primarily derived from reactions with throughfall and surface flow.

Factor I has high factor loadings for SO_4, NO_3, K, Fe, and discharge (Table 1). SO_4 and NO_3 may be derived from acid rain or dry deposition on leaf surfaces. K is often the highest concentration cation in throughfall. Fe may be present as colloidal sized (< 0.45 μm) particles entrained by the increased discharge. Thus, Factor I may be interpreted as primarily representing acid rain, dry deposition, and throughfall contributions. Hysteresis plots indicate that Factor I (and III) dominate during the first period.

Significant species correlated with Factor II are: Ca, Mg, Na, Cl, HCO_3, pH, and discharge (Table 1). The difference in sign between discharge and the chemical factors indicates that they are negatively correlated, *i.e.* concentrations decrease as discharge increases. Factor II is relatively constant at the four localities. At Uhlerstown, K also has a high factor loading in Factor II. Factor II is similar to Factor 1 of Miller and Drever (1977), although they did not analyze for anions other than HCO_3. Miller and Drever (1977) interpreted this factor as representing inputs from soluble soil salts. The composition of the proposed soluble soil salt may be determined from plots of Factor I species fluxes *vs.* discharge. The composition of the salt (normalized to Mg = 1) is approximately $Ca_{1.1}Mg_1Na_{1.1}Cl_{1.7}(HCO_3)_{3.6}$. This was confirmed by leaching dry soil with distilled water. The relative uniformity of Factor II, despite substantial differences in bedrock type, may indicate that the composition and nature of the soluble salt is controlled, not by rock composition and weathering, but by solute inputs in vegetative throughfall. Hysteresis plots (*e.g.* Figure 4) indicate that Factor II dominates during the second period, during interflow- and baseflow-dominated discharge. This is reasonable if the soluble salts are contained within the soil, precipitated during evapotranspirative water loss in soil pores, rather than at the surface.

Factor III has high factor loadings for PO_4 and K. In other localities NO_3 also has a high factor loading. As with Factor I, hysteresis plots indicate that Factor III is dominant during the first storm period, during rising discharge, when discharge is dominated by overland flow. Thus, Factor III may represent leaching of nutrients from leaf litter and soil humus by overland flow.

5 CONCLUSIONS

Chemical variations in stream water following storms in eastern Pennsylvania and New Jersey appear to be controlled by three underlying factors. During initial periods, storm waters are influenced by acid rain (SO_4 and NO_3) and throughfall (K) [Factor I], then leaching of nutrients (K, PO_4, NO_3) from humic materials during overland flow [Factor III]. At longer times the composition of discharge is primarily influenced by dissolution of soluble salts in the soil (Ca, Mg, Na, Cl, HCO_3) by interflow and baseflow (Factor II). The nature of these factors appears fairly uniform over four lithologically different sites investigated. The relatively constant nature of Factor II (soluble soil salts), despite large differences in lithology and mineralogy, suggests that the composition of the salts may be largely controlled by biological activity, *e.g.* solute inputs in throughfall, rather than rock composition and weathering.

In a previous study of storm water chemistry, Miller and Drever (1977) also extracted three factors, although two of them had eigenvalues less than 1.0. They interpreted one of the factors, which had high loadings for Mg, SiO_2 and HCO_3, as indicating weathering of ferromagnesian minerals. None of the factors isolated in this study correspond with this ferromagnesian weathering factor, and there is no strong evidence in this study for the effects of primary mineral weathering at any of the four localities. Miller and Drever (1977) also concluded that Factor II (soluble salts) dominated during the first period; results of this study indicate that Factor II is greater during the second period. The reason for the timing difference is not known; it may be due to differences in climate and vegetation, difference in basin size, or might be due to different measurements of discharge. Miller and Drever (1977) measured only gauge height, an indirect measure of discharge which may not be adequate.

6 REFERENCES

Davis, J. C., 1986. *Statistics and Data Analysis in Geology*. Wiley, New York, 646 p.

Hintze, J. L., 1996. *NCSS 6.0*. Kaysville, Utah.

Miller W. R. and Drever, J. I, 1977. Water chemistry of a stream following a storm, Absaroka Mountains, Wyoming. *Geol. Soc. Amer. Bull.*, 88, 286-290.

Statsoft, 1991. *Statistica*. Statsoft Corporation, Tulsa, Oklahoma.

2 Processes involving organic matter

Analysis of environmentally significant organic and inorganic metal species by coupled IC-ICP-MS, GC-ICP-MS and LC-ICP-MS

James R. Brydie, Andrew P. Gize, Paul R. Lythgoe & David A. Polya
Department of Earth Sciences, University of Manchester, UK

Graham Kilpatrick & Keith Hall
LC2 Ltd, Analytical Laboratories, Manchester, UK

Kurijan Sajan
School of Marine Sciences, Cochin University of Science and Technology, India

ABSTRACT: Coupled ion chromatography-ICP-MS, gas chromatography-ICP-MS and liquid chromatography-ICP-MS analytical techniques have been developed and used to analyse trace inorganic and organic metal species from modern and ancient estuarine/marine sediments. Identification and quantitation of arsenate, arsenite, selenate, and selenite in water at the ng/g level has been achieved by IC-ICP-MS. Volatile and semi-volatile organometals, notably alkyl lead and alkyl tin species, have been separated and identified by GC-ICP-MS with sub-picogram detection limits. Organically-bound Ni, V, Fe and Cu have been detected in the bulk metalloporphyrin fraction of a Kupferschiefer extract using LC-ICP-MS. The development of these coupled analytical techniques offers hitherto unobtainable identification, quantitation of organic and inorganic metal species in marine and estuarine environments.

1 INTRODUCTION

The speciation of metals in estuarine and marine sediments and waters determines to a large extent their chemical and biochemical availability and hence their toxicological effects. Identifying and quantifying these inorganic, organic or organometallic species is therefore critically important to understanding these environments, both modern and ancient (particularly ore forming environments such as the Kupferschiefer). Until recently, identification and quantitation of individual metal species (as opposed to total dissolved metal) in complex natural waters at sub-ppm and sub-ppb level has been fraught with difficulties. Whilst ion, gas and liquid chromatographic techniques (IC, GC, LC) are particularly useful for separating ionic, volatile and non-volatile species respectively, the analytical detection limits associated with these methods are higher than can be achieved with inductively coupled plasma mass spectrometry (ICP-MS). We report recent developments in coupling ion, gas and liquid chromatographic separations with ICP-MS to achieve both effective separation of metal species and low analytical detection limits.

2 ANALYSIS OF IONIC SPECIES BY COUPLED IC-ICP-MS

A Dionex QIC ion chromatograph (IC) was interfaced with a PQ/2 ICP-MS, enabling the passing of the aqueous eluant (20 mM NaOH) from the IC into the nebuliser of the ICP-MS. An anion self-regenerating suppresser was used to minimise sodium deposition on the cones of the ICP-MS (Lythgoe et al., 1997). The utility of the technique was demonstrated by the analysis of a mixture of low and high oxidation state As and Se species dissolved in water at concentrations of 10 ng/g. Poor IC separation of arsenite from selenite and selenate species was resolved by IC-ICP-MS analysis scanning at m/z = 75 for As and 82 for Se. (Fig. 1).

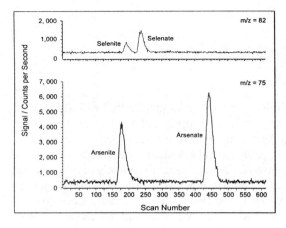

Fig. 1. IC-ICP-MS scans of a synthetic solution containing 10 ng/g each of arsenite, arsenate, selenite and selenate.

IC-ICP-MS analysis of natural waters has proved more problematical, particularly for relatively acidic waters in which metal precipitation may occur during elution with NaOH. The use of a Dionex Onguard H-cartridge to remove Fe, however, has enabled the qualitative analysis of arsenic in acid waters draining pyritic coal spoil.

3 ANALYSIS OF VOLATILE ORGANO-METALS BY COUPLED GC-ICP-MS

Several economically and environmentally significant metals (e.g., Pb, Sn, Hg, Au, Pt, and U) can form volatile and semi-volatile organic species and complexes, such complex formation being particularly pronounced for "soft" metals (Ahrland, 1973). The quantitative roles of these complexes in aqueous solutions is largely unknown. Current research has focussed on one type of organometallic compound, where s-bonds are formed between carbon and the less electropositive, "heavy" metals. These include the environmentally and toxicologically significant alkyl derivatives of lead, tin, mercury, arsenic, and selenium.

Separation of organo-lead compounds was achieved by gas chromatography (GC), with the heated GC capillary column inserted close to the ICP-MS. The advantage of the technique is that neither prior separation nor derivatisation of the organometals is required, thus reducing sample loss by adsorption. Although the organo-leads are eluted in a complex mixture of aliphatic and aromatic compounds, making identification and quantitation difficult by conventional GC and GC-MS techniques, these are dissociated in the ICP. Consequently, detection is by monitoring of the four Pb isotopes alone, which currently has resulted in sub-nanogram level detection limits for tetraethyl lead in petrol (Fig. 2).

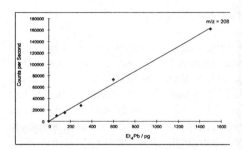

Fig. 2 Calibration curve for GC-ICP-MS analysis for Et_4Pb in petrol dissolved in dichloromethane.

Injected sample volume was 0.1 µl.

We have also used GC-ICP-MS to investigate the effects of natural microbial methylation on distribution and speciation of alkyl lead in estuarine sediments. Jarvie et al. (1975) showed evidence for anaerobic microbial methylation of lead by the conversion of trimethyl lead acetate to tetramethyl lead. The environmental significance is that although the toxicity of inorganic bioavailable lead is relatively low compared to other heavy metals, alkyllead compounds are highly toxic (Fergusson, 1990). The environmental pathways of alkylleads have been investigated. Harrison et al. (1985) showed atmospheric breakdown of tetraalkyllead to tri-and di-alkyllead compounds. In aqueous systems, Harrison et al. (1986) showed that tetraakyllead compounds (R_4Pb) decomposed to the ionised species (R_3Pb+) which were suggested to have an appreciable environmental lifetime (ca. 1-2 yr in Mersey, U.K. estuarine sediments). Although a microbial pathway for metal alkylation has held in the literature (Hg, As, and Se), the evidence for microbial lead alkylation has been challenged. Reisinger et al. (1982) used isotopically labelled methyl donors, and reported the formation of tetraalkylleads only as the result of inorganic reactions, similar to those implicated by Jarvie et al. (1975). In the latter two studies, both organically- and inorganically-bound sulphur was invoked as a reaction intermediary. Lead sediment concentrations in the Mersey increase upstream (ca. 100 µg/g at the mouth, 200-250 µg/g in the upper estuary: Langston, 1986). In intertidal Mersey sediment cores, a colouration difference is evident. The upper sections are light brown, "oxidised" and at about 15 cm depth, become black, "anoxic". Pyrolysis-gas chromatography-mass spectrometry of the humic acids isolated from such cores indicate a humic acid organo-sulphur maximum at the boundary between "oxidised" and "anoxic" sediments (Alsop and Gize, 1995). Similar organo-sulphur maxima have been recorded in British Columbia fjords (Francois, 1987). Preliminary data from shallow oxidised sediments in the Mersey estuary has identified alkylleads. Whilst tetraethyllead was found to be the dominant alkyllead in both these sediments (Fig. 3) and unused fuel (e.g Lythgoe et al., 1997), substantially higher ratios of tetramethyllead to tetraethyllead were found in the sediments It is currently postulated that mixed ligands (e.g. Et_3PbHS) are responsible for environmental persistence of organoleads.

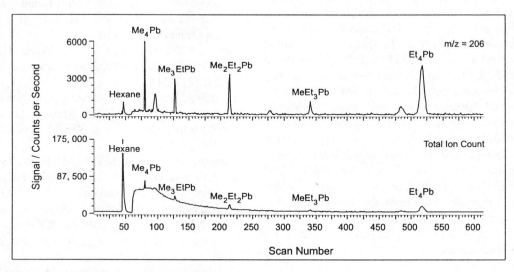

Fig. 3 GC-ICP-MS scan of EDTA/hexane extracted oxidised sediments from the Mersey Estuary. Me = methyl; Et = ethyl;.

4 ANALYSIS OF NON-VOLATILE ORGANOMETALS BY LC-ICP-MS

A GC interface requires sufficient heat to volatilise a compound before transfer to an ICP-MS. Some important organo-metals, especially the metalloporphyrins, are insufficiently volatile to be analysed by gas chromatography. High performance (pressure) liquid chromatography has been the traditional separation technique. The difficulty encountered in coupled LC-ICP-MS was extensive carbon coating of the Ni cones by the solvent (methanol). The method used to solve this problem was based on solvent stripping by solvent diffusion across a permeable membrane, using a pressure gradient induced by an argon counterflow. The bulk metalloporphyrin fraction of a Kupferschiefer extract analysed in this fashion was shown to contain organically-bound Ni, V, Fe and Cu.

CONCLUSIONS

Coupled IC-ICP-MS has been developed and demonstrated to be capable of separating and quantifying redox sensitive species of arsenic and selenium in synthetic and natural waters at the ng/g level. Coupled GC-ICP-MS and LC-ICP-MS provide the means not only to separate volatile and non-volatile organic metal species extracted from sediments, aqueous or non-aqueous fluids but also to quantify the concentrations of these species at sub-µg/g and even sub-ng/g levels. Using a combination of these chromatographic techniques, it is possible to separate and quantify the compounds about which currently there exists the least knowledge of their roles in ore genesis and the environment - the organometals.

ACKNOWLEDGEMENTS

JRB acknowledges the receipt of an NERC PhD studentship. KS acknowledges receipt of a Commonwealth Fellowship.

REFERENCES

Ahrland, S. 1973. Thermodynamics of stepwise formation of metal-ion complexes in aqueous solution. *Struct. Bonding*, 15, 167-188.

Alsop, A.L., and Gize, A.P., 1995, Potential Pt-Au-humic organic-sulphur species interactions: in, Proceedings Third Biennial SGA Meeting, Prague, Aug. 28-31, 1995, 719-721

Fergusson, J.E., 1990, The Heavy Metals: Chemistry, Environmental Impact, and Health Effects, Pergamon Press, 429-457

Francois, R., 1987, A study of sulphur enrichment in the humic fraction of marine sediments during early diagenesis: Geochim. Cosmochim. Acta, 51, 17-27

Harrison, R.M., Hewitt, C.N., and Radojevic, M., 1985, Environmental pathways of alkyllead compounds: 5th International Conference Heavy Metals in the Environment, 1, 82-84

Harrison, R.M., Hewitt, C.N., and Radojevic, M., 1986, Environmental pathways of alkyllead compounds: 6th International Conference Heavy Metals in the Environment, 110-116

Jarvie, A.W., Markall, R.N., and Potter, H.R., 1975, Chemical alkylation of lead: Nature, 255, 217-218

Langston, W.J., 1986, Metals in sediments and benthic organisms in the Mersey Estuary: Estuarine, Coastal, and Shelf Science, 23, 239-261.

Lythgoe, P.R., Kilpatrick, G., Hall, K., Gize, A.P. and Polya, D.A., 1997, Analysis of trace organo-metallics and metallic components by coupled chromatography ICP-MS and laser ablation-ICP-MS. 4th Biennial SGA meeting, Turku, Finland, August 11-13, 1997.

Reisinger, K., Stoeppler, M., and Nürnberg, H.W., 1982, Evidence for the absence of biological methylation of lead in the environment: Nature, 291:228-230

Peat-water interactions: South Taupo Wetland, New Zealand

C. Chagué-Goff
Institute of Geological and Nuclear Sciences Ltd, Lower Hutt, New Zealand

ABSTRACT: Comparison is made between the chemical composition of peat and sediment in two wetlands of the southern shore of Lake Taupo, New Zealand. The elemental composition in the relatively undisturbed Stump Bay Wetland is related to the nature and amount of inorganic input into the wetland, and the Cu and Zn concentrations do not exceed 60 ppm. Trace element distribution in Turangi Wetland reflects the influence of recent anthropogenic activity. Surface Cu and Zn concentrations of up to 110 ppm and 440 ppm, respectively, are attributed to sewage effluent discharge in the last 30 years.

1 INTRODUCTION

South Taupo Wetland lies at the southern end of Lake Taupo, in the central North Island, New Zealand (175°50' E, 38°58' S) (Fig. 1). Much of the area, which includes the Stump Bay Wetland and the Turangi Wetland, is underlain by pumice and greywacke alluvium. Formation of beach ridges has resulted in water ponding and establishment of wetland vegetation in Stump Bay Wetland (Cromarty & Scott, 1996), which is a natural wetland relatively undisturbed by human activity. Turangi Wetland on the other hand has developed in low-lying surfaces of the Tongariro River delta (Warner, 1985) and is now part of a land treatment facility for sewage effluent.

This paper discusses the influence of natural and anthropogenic factors on wetland development and peat chemistry.

2 METHODS

A total of 14 cores were collected in both sites using a stainless-steel McAulay type auger. They were packed in plastic wrap and kept chilled at 4°C before analysis. In the laboratory, cores were sliced into 5 cm intervals. Samples were dried at 80°C until constant weight and ash yield was determined by ashing oven-dried samples at 550°C for 16 hours.

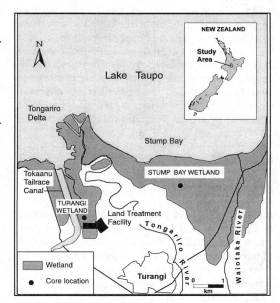

Fig. 1 Location map showing Turangi Wetland and Stump Bay Wetland, which are part of the South Taupo Wetland. The location of the representative core in each site is also shown.

Major and trace elements were determined by X-ray fluorescence.

For the purpose of this paper, the stratigraphy, ash yield and elemental composition (Si, Al, K, Cu and Zn) of one representative core in each site is reported.

3 WETLAND STRATIGRAPHY

3.1 Stump Bay Wetland

Core stratigraphy consists of silty or sandy sediment overlain by a thick peat layer (~ 1.0-1.1 m) with intercalated thin silty layers, suggesting that peat formation has been interrupted by episodic flooding events in the past. The peat layer thins towards the edge of the lake, down to a few cm.

3.2 Turangi Wetland

On the west side of the Tokaanu Tailrace Canal, the muddy mineral substrate is overlain by peat, whereas on the east side of the canal, the inorganic-rich peaty layer is itself overlain by a mud layer. Historical and ^{137}Cs data suggest that flooding of the Tongariro River, possibly enhanced by lake level changes in the past 50 years, has led to an increase in inorganic input in the wetland (Chagué-Goff et al., in press). Organic accumulation has ceased and has been surpassed by detrital inorganic sedimentation.

4 RESULTS

4.1 Stump Bay Wetland

In the representative core, ash yield ranges from 21 to 32% in the peat layers, except in the basal peat, where it reaches 51% (Fig. 2), reflecting the uptake of mineral matter and mixing from root action in the first stages of wetland formation. The inorganic content of the silt-rich layers is around 70%.

Si, K, Al, Cu and Zn concentrations closely follow ash distribution, being low in the peat layers and high in the silty layers (Fig. 2). The Cu and Zn content does not exceed 60 ppm.

4.2 Turangi Wetland

Ash yield is about 70% in the upper muddy layer, decreases to 62% in the inorganic-rich peaty layer and is above 80% in most of the underlying muddy substrate (Fig. 3).

Si, K and Al distribution is fairly similar to the ash distribution. The Cu content is about twice as high in the top 30 cm than in the remainder of the core, reaching up to 110 ppm. Zn concentrations are highest in the top 18 cm of the core, reaching up to 440 ppm.

5 DISCUSSION

The relatively high inorganic content in the peat of Stump Bay Wetland (>20%) is an indicator of the influence of mineral- and nutrient-rich surface and groundwater on the wetland. However, favourable conditions have been maintained for a long enough period to allow peat formation and accumulation. Only temporary episodic flooding has resulted in the deposition of thin silty layers within the peat sequence.

The high ash yield (> 60%) in the sediments of Turangi Wetland on the other hand is indicative of the depositional environment in the Tongariro Delta, which is prone to frequent flooding by the river.

Si, Al and K distributions in both wetland cores suggest that these elements are mainly associated with the inorganic fraction. Their concentration in the peat and sediment is dependent upon the mineral matter abundance and variety, which is mainly related to mineral input by surface water at the time of deposition (Chagué-Goff et al., 1996).

Cu and Zn concentrations in Stump Bay Wetland also reflect their association with the inorganic matter of peat and sediment, in an environment relatively undisturbed by human activity. Similar concentrations and distributions are observed in other cores of the wetland (Chagué-Goff, unpublished data) and thus are most likely to represent natural background levels.

In Turangi Wetland, Cu and Zn distribution can not be solely explained in terms of mineral matter abundance. Cu and Zn contents are noticeably higher at the top of the core than in the remainder of the core and in the core taken in Stump Bay Wetland, suggesting an additional source for these elements. Their increased concentrations have been attributed to sewage effluent discharge, which also has an effect on the vegetation and water quality down gradient of the land treatment facility (Rosen & Chagué-Goff, 1997; Chagué-Goff et al., in press). Further evidence for the anthropogenic input is the spatial distribution of these elements in the wetland. Cu and Zn concentrations in the top sediment layers

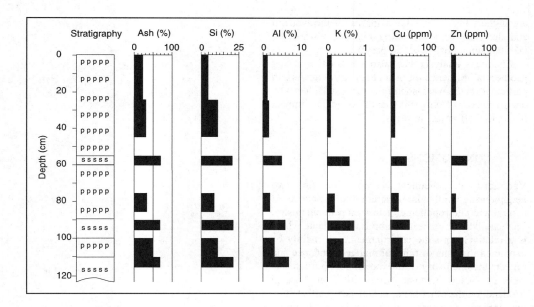

Fig. 2: Stratigraphy, ash yield, Si, Al, K, Cu and Zn concentrations in Stump Bay Wetland (p = peat, s = sediment).

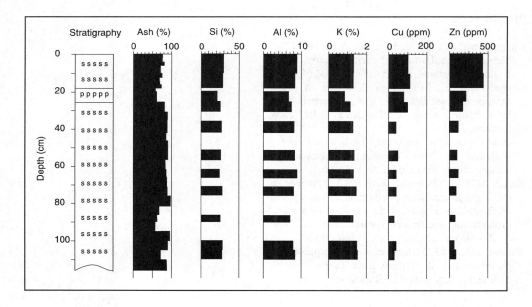

Fig. 3: Stratigraphy, ash yield, Si, Al, K, Cu and Zn concentrations in Turangi Wetland. This core was taken near the point of effluent inflow in the wetland (p = peaty, s = sediment).

are highest near the effluent inflow in the wetland and decrease with increasing distance toward the lake (Chagué-Goff et al., in press).

Chemical analysis has also revealed a strong geothermal influence on peat chemistry (e.g. As), in particular on the west side of the canal, which can be distinguished from the anthropogenic impact (Chagué-Goff et al., in press).

6 CONCLUSIONS

Variations in chemical composition in two neighbouring wetlands illustrate the influence of natural and anthropogenic factors on peat chemistry. Elemental distribution in Stump Bay Wetland, which is a relatively pristine environment, can mainly be explained in terms of mineral matter abundance and reflects the chemistry of the inorganic input into the wetland. Trace element composition in Turangi Wetland, on the other hand, is noticeably affected by the discharge of sewage effluent in the last 30 years, which results in an increase of Zn and Cu in the top layers. However, the effluent does not appear to contribute any Si, K and Al to the wetland.

REFERENCES

Chagué-Goff, C., F.Goodarzi & W.S.Fyfe 1996. Elemental distribution and pyrite occurrence in a freshwater peatland, Alberta. *J. Geol.* 104:649-663.

Chagué-Goff, C., M.R.Rosen & P.Eser in press. Sewage effluent discharge and geothermal input in a natural wetland, Tongariro Delta, New Zealand. *Ecol. Eng.*

Cromarty, P. & D.A.Scott (eds.) 1996. *A directory of wetlands in New Zealand*. Department of Conservation, Wellington, New Zealand.

Rosen, M.R. & C.Chagué-Goff 1997. Influence of the Turangi oxidation ponds land treatment area on the quality of water in the surrounding shallow groundwater aquifer and natural wetlands. *Proc. N. Z. Water & Wastes Assoc. Conf.*, 90-94.

Warner, J.G. 1985. Sedimentary and morphological characteristics of southern Lake Taupo. Unpubl. MA thesis, University of Auckland.

Groundwater chemistry and water-rock interactions at hydrocarbon storage cavern sites in Korea

H.T.Chon, J.U.Lee, S.Y.Oh & H.D.Park
Department of Mineral and Petroleum Engineering, College of Engineering, Seoul National University, Korea

ABSTRACT: Rocks and groundwater in the vicinity of oil and LPG storage cavern sites were sampled and analyzed to investigate the process of water-rock interaction in the water curtain system. The change of chemical compositions against Cl indicates that water-rock interactions occurred from the flow of groundwater at the LPG storage cavern site. The hydrolysis and weathering of aluminosilicate minerals are attributed to water-rock interaction at the LPG storage cavern site by the interpretation of WATEQ4F simulation. Those at the oil storage cavern site were also interpreted by the dissolution of concrete and precipitation of carbonates. From the results of factor analysis, it can be concluded that the main controlling the chemical composition of groundwater are chemical compositions of injection water, water-rock interaction, microbiological effect, dissolution of concrete and contamination from the outside.

1. INTRODUCTION

Several hydrocarbon storage caverns on a large scale have been under construction and operated in Korea. For this study, two sites were selected from those for the storage of oil and LPG. Such hydrocarbon storage caverns have a water curtain system to maintain the internal pressure to prevent them from leaking hydrocarbons. Thus, the chemical composition of groundwater in the water curtain system plays an important role in underground storage of hydrocarbon from the viewpoint of geochemical and microbiological clogging. Because of this importance, they have been monitored periodically. The purpose of this research is to investigate the process of water-rock interaction in the water curtain system using monitored chemical data.

2. GEOLOGY AND GROUNDWATER IN THE WATER CURTAIN SYSTEM

The LPG storage cavern site is located near the western coast of the Korean peninsula. Precambrian metamorphic rocks in the Kyeonggi massif facilitating cavern has been subjected to several periods of tectonic activity. The metamorphic rocks consist of augen gneiss, calcareous-interbedded chlorite banded gneiss, biotite banded gneiss, biotite schist and leucocratic gneiss. The primary minerals are quartz, K-feldspar, chlorite, biotite, and plagioclase, calcite and sericite occurred as accessory minerals. The water curtain system was set up at 90m below depth from ground surface (0m sea level) and the top of the cavern is at 25m below depth from the water curtain system. The injection water is pumped at some distance and supplied to the water curtain system through the pipe line. The flow path of groundwater can be summarized as follows ; injection water → horizontal well and piezometric well in the water curtain system → seepage point at cavern.

The oil storage cavern site is located near the southern coast of the Korean peninsula. The geology around the storage cavern site consists of cretaceous volcanic rocks and micrographic granite. Cretaceous volcanic rocks are composed of andesite tuff, aphanitic andesite and porphyritic andesite. The primary minerals of micrographic granite are orthoclase, quartz and plagioclase, and secondary minerals are hornblende, perthite, microcline, biotite, garnet and opaque mineral. The cavern which also has water curtain system at 0m sea level, is located at -30~-60m sea level. Groundwater and surface water near the cavern site are mixed, filtered by active carbon, and treated by NaOCl for sterilization of bacteria, and then, injected into the water curtain system. Therefore, the flow path of groundwater can be summarized as follows ; treated water → horizontal well in the water curtain system → seepage point at cavern.

3. SAMPLING AND ANALYSIS

Groundwater was sampled at some points along flow path and locally recharged shallow groundwater was also sampled at the upper part of the water curtain system. Sampling was undertaken at the same location both in rainy and dry seasons. Eh, pH, electric conductivity (EC), and temperature were measured in-situ. Cations were analyzed by Inductively Coupled Plasma Atomic Emission Spectrophotometry (ICP-AES) and Atomic Absorption Spectrometer (AAS), alkalinity by titration, and anions by Ion Chromatography (IC) in the laboratory, respectively. Electroneutrality was also calculated to verify the analytical deviation which has been discussed by several scientists (Greenberg et al., 1992, Jackson, 1993, Minear and Keith, 1982).

4. RESULT AND DISCUSSION

4.1 *LPG storage cavern site*

Injection water pumped at 110m depth is Na-Ca-Cl type saline water with a maximum total dissolved solids (TDS) content of 1,204 mg/l. The high salinity can be attributed to the inflow of sea water from the fact that cavern site is located near coastal area. But the molar ratio of Ca/Mg is much greater than 1/2, and the concentration of Si is higher, which indicates that injection water undergoes geochemical evolution locally as well as being mixed with sea water and fresh water recharged.

The chemical composition of groundwater in the horizontal well of water curtain system indicates that initial injection saline water is weakly diluted by typical recharge water (Ca-HCO$_3$ type). As groundwater moves from horizontal well in the water curtain system to cavern, the concentrations of Ca, K, Al, Sr increase, but Si, Mg decrease. When some cations against Cl concentration were plotted on the X-Y diagram, groundwater samples in cavern were appeared on the upper part from the line of simple mixing (Figure 1). This indicates that water-rock interactions occur in cavern.

Hydrogeochemical modeling by WATEQ4F (Ball and Nordstrom, 1991) was performed to obtain the saturation indices of specified minerals which were related to water-rock interaction in cavern. As a result of water-rock interaction, the concentration of Ca increases and this leads to oversaturation and precipitation of calcite. This indicates that not only the dissolution of calcite but also the hydrolysis and weathering of Ca-plagioclase and Ca-Mg minerals occur. But, saturation index of Ca-plagioclase shows the oversaturation state of all groundwater samples in cavern, Ca-Mg minerals such as diopside may be the sources of Ca increase in groundwater. High concentration of K in groundwater will be explained by the dissolution of K-feldspar which is second most abundant mineral. Kaolinite is stable as a weathered mineral of K-feldspar, but thermodynamic equilibrium is reached between groundwater and K-feldspar as groundwater approaches to cavern (Figure 2). Therefore, K concentration in groundwater can be also attributed to the hydrolysis and weathering of biotite, which and oversaturation state of chlorite will explain the increase of K concentration and the decrease of Mg and Si concentration in groundwater.

The results from the factor analysis, which is a popular technique (Drever, 1988), show that the factors controlling the chemical compositions of groundwater are (i) Na-Cl-Mg-Br-Ca-Sr, (ii) pH-alkalinity-Si-K and (iii) NO$_2$-NO$_3$. The first controlling factor explains the salinity of initial injection water. The second controlling factor explains the pH increase controlled by the dissolution of K-feldspar and calcite. The two simple reactions can explain this controlling factor as follows;

$2KAlSi_3O_8 + 9H_2O + 2H^+$
$\rightarrow Al_2Si_2O_5(OH)_4 + 2K^+ + 4H_4SiO_4$ (4.1)

$CaCO_3 + H^+ \rightarrow Ca^{2+} + HCO_3^-$ (4.2)

The third controlling factor explains the consumption of NO$_2$ and NO$_3$ by bacteria in anaerobic environment.

4.2 *Oil storage cavern site*

Treated injection water with a TDS content of 50-221 mg/l is classified as fresh water. The ion concentrations of the horizontal well in the water curtain system are controlled by the mixing ratio of surface water and ground water in injection water rather than by water-rock interaction. For the case of seepage water having strong alkalinity (9.3~12.5 pH) and maximum TDS content of 964 mg/l, it is due to the dissolution of concrete of the cavern. The concentrations of K, Ca, Al, SO_4 increase but the concentrations of Mg, NO_3 decrease as groundwater moves to seepage point.

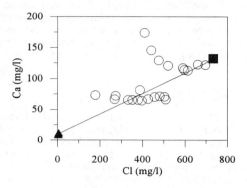

Figure 1. Ca against Cl concentration of groundwaters at LPG storage cavern site. Straight line denotes the simple mixing line. Solid rectangle, solid triangle, and open circle represent injection water, local recharged fresh water and evolved groundwater, respectively.

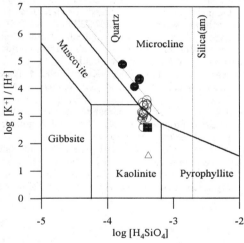

Figure 2. Stability of microcline, muscovite, kaolinite, gibbsite, and amorphous silica in equilibrium with K, H, and H_4SiO_4 in aqueous solution at 25℃, 1atm. The stability field is based on the thermodynamic data from Helgeson (1969) and Nesbitt (1977). Solid rectangle, open circle, solid circle, and triangle represent injection water, horizontal well waters, more evolved waters and local recharged fresh waters, respectively. The dotted line parallel to muscovite and microcline boundary can be produced if thermodynamic properties of muscovite follow other sources.

Table 1. The result of factor analysis of groundwater samples at hydrocarbon storage cavern sites.

(a) LPG storage cavern site

	Factor 1	Factor 2	Factor 3	Percentage of variance (%) [Cum. Perc.]
Na	0.932			
Cl	0.930			
SO_4	0.855			
Mg	0.841			
Br	0.837			37.6
Al	0.788			[37.6]
Ca	0.774	0.424		
Ba	0.766			
Sr	0.679			
Fe	0.674			
pH		0.963		
Alkalinity		0.962		24.6
Si		-0.758		[62.3]
K		0.711		
NO_3			0.955	10.7
NO_2			0.953	[73.0]

(b) oil storage cavern site

	Factor 1	Factor 2	Factor 3	Percentage of variance (%) [Cum. Perc.]
EC	0.987			
Alkalinity	0.981			
Ca	0.973			
K	0.963			51.3
Al	0.955			[51.3]
Sr	0.909			
pH	0.621			
Cl		0.952		
SO_4		0.799		28.4
NO_3		0.793		[79.7]
Na		0.741		
Si			0.862	8.5
Mg			0.716	[88.3]

According to hydrogeochemical modeling by WATEQ4F, Portlandite ($Ca(OH)_2$), formed from the mixing of cement and water, is dissolved continuously, which makes seepage water have a strong alkaline pH value and undersaturated state in the whole pathline (Figure 3). But Mg is precipitated as carbonates and NO_3 is used by bacteria in the anaerobic seepage interval.

The results from the factor analysis show that the factors controlling the chemical composition of groundwater are (i) Conductivity-alkalinity-Ca-K-Al-Sr-pH, (ii) Cl-SO$_4$-NO$_3$-Na and (iii) Si-Mg. The first controlling factor explains the dissolution of concrete composed of CaO, Al$_2$O$_3$, CaSO$_4$, Na$_2$O, K$_2$O and so on. The second controlling factor explains the inflow of pollutants from the outside. Because the oil storage cavern site is located at industrial area, surface water used as injection water is contaminated with NO$_3$, Na, Cl and SO$_4$ by human activity. The third controlling factor explains the local water-rock interaction which is weaker than the dissolution of concrete.

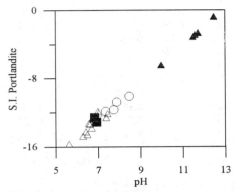

Figure 3. Saturation index (S.I.) of portlandite against pH at the oil storage cavern site. Solid rectangle, solid triangle, open circle, and open triangle represent treated injection water, seepage water, evolved groundwaters in cavern and local recharged groundwater, respectively.

5. CONCLUSION

The characteristics of chemical compositions and the result of hydrogeochemical modeling can be summarized as follows.
 1. In the case of LPG storage cavern site, fresh shallow groundwater mix with saline injection water and the concentrations of metal ions are changed as it moves to the cavern, which means that water-rock interaction has happened. In the case of oil storage cavern site, however, water-rock interaction is weak but interaction between water and concrete in the cavern makes the seepage water strong alkaline.
 2. Using the results of WATEQ4F simulation, it is concluded that the concentrations of dissolved ions are controlled by the hydrolysis and weathering of aluminosilicate minerals at LPG storage cavern site. At oil storage cavern site, the concentrations of dissolved ions are controlled by the dissolution of concrete and the precipitation of the carbonates.

 3. Factor analysis shows that the initial injection water with high salinity, the change of pH by the dissolution of K-feldspar, and the microbiological effect are the factors controlling the chemical characteristics of groundwater at LPG storage cavern site. While, the dissolution of concrete, the intrusion of external polluted water, and the local water-rock interaction are the controlling factors at oil storage cavern site.

REFERENCES

Ball, J.W. and D.K. Nordstrom 1991. User's manual for WATEQ4F, with revised thermodynamic data base and test cases for calculating speciation of minor, trace and redox elements in natural waters. *U.S. Geol. Surv. Open File Rep. 91-183.*
Drever, J.I., 1988, *The geochemistry of natural waters* (2nd ed), New Jersey: Prentice Hall.
Greenberg, A.E., L.S. Clesceri and A.D. Eaton (Eds) 1992. Standard methods for the examination of water and wastewater(18th Eds). Washington DC.: American Public Health Association.
Helgeson, H.C., 1969, Thermodynamics of hydrothermal systems at elevated temperatures and pressures, *Amer. J. Sci.* 267:729-804.
Jackson, G.B. 1993. Applied water and spentwater chemistry - A laboratory manual. N.Y.: Van Nostrand Reinhold.
Minear, R.A. and L.H. Keith 1982. Water analysis vol. 1, 2 and 3. Academic Press Inc.
Nesbitt, H.W., 1977, Estimation of the thermodynamic properties of Na-, Ca- and Mg-beidellites, *Canadian Mineralogist* 15:22-30.
Nordstrom, D.K., J.W. Ball, R.J. Donahoe and D. Whittemore 1989. Groundwater chemistry and water-rock interactions at Stripa. *Geochim. Cosmochim. Acta.* 53. : 1727-1740.

Organic matter maturation as an indicator of hydrothermal processes in sedimentary basins

M. Glikson & S. D. Golding
Department of Earth Sciences, The University of Queensland, Qld, Australia

ABSTRACT: The involvement of hydrothermal processes in ore grade mineralisation render organic maturation studies of primary importance in reconstructing fluid flow pathways (from source rock to reservoir), and evaluation of hydrothermal fluxes in a sedimentary basin. The nature of organic matter, as well as maturation profiles, display significantly different characteristics in sediments subjected to convective/advective heat transfer than those affected solely by conduction as a result of gradual increase in temperature during burial over extended geological time. Typically erratic and irregular maturation profiles as expressed by reflectance of the prevailing organic matter are a distinct feature in regions subjected to rapid high temperature effect. These basins are frequently host to ore deposits associated with migrated hydrocarbons. The latter in the form of pyrobitumen are always of relatively high reflectance (>1.8%) suggesting temperatures of fluids exceeding 150°C. Surges or fluxes of hydrothermal fluids can be detected from multiple generation bitumen populations. Therefore, not only exploration for hydrocarbons, but for ore deposits as well may benefit from organic maturation studies, utilising the irreversible thermal imprint on organics.

1. INTRODUCTION

The role of organic matter in mineralisation has been the subject of numerous studies emphasising the key role of hydrocarbons in ore formation (Ripley et al., 1990; Mauk & Hishima, 1992; and others). In these cases an active role in transport, solubility and sulfidation has been highlighted. The passive role of organic matter in adsorbing and trapping metals has likewise been documented in numerous studies (Brumsack, 1980; Beveridge et al., 1983; Glikson et al., 1985; Ripley et al., 1990). Clearly a distinction has to be made between source rock or primary organic matter and hydrocarbons. The former may act as a sink for metals, but does not play an active role in solubility and transportation. Hydrocarbons on the other hand have been transported with metal-laden fluids and precipitated together, thus sharing a common pathway, and thermal history. The latter may be expressed in the reflectance of the organic matter.

Methods of study: Organic petrology was carried out using coal petrological techniques, electron microscopy and microanalytical techniques as outlined in Glikson & Taylor (1986), isotope geochemistry as in Golding et al. (1996).

2. PROTEROZOIC MOUNT ISA BASIN, LAWN HILL FORMATION, RIVERSLEIGH FOLD ZONE

The organic-rich Riversleigh shales are host to the Ag-Pb-Zn Century deposit. The organic matter can be classified into two main types: (1) *insitu* alginite (source rock), and its residue after oil generation; and (2) bitumens/pyrobitumens. The latter are polymerised heavy oils, the abundance of which is known to characterise active hydrothermal systems (Altebaumer et al., 1981; Simoneit et al., 1992). Immature alginite when observed in TEM is not in itself associated with minerals (Fig. 1) due to lack of affinity between lipidic materials and metals (Glikson & Taylor, 1986). However, alginite residue after oil generation (spent alginite) acts as a trap for physically adsorbed metals (Fig. 2), although no ore grade mineralisation is seen to be associated with spent alginite. On the other hand in ore deposits such as Century, pyrobitumen is the organic component associated with the metals. Clearly the hydrocarbons associated with the ore deposit were transported with the hot Ag-Pb-Zn rich fluids. Reflectance values of the pyrobitumen in the Century orebody proper are 2.0-2.2%, implying fluid temperatures of around 200-250°C (Barker & Pawlewicz, 1986).

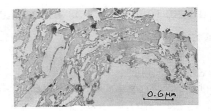

Figure 1: TEM of immature alginite

Altogether 3 main types of bitumen may be distinguished within the Mount Isa Basin sediments: Bitumen referred to as type B in an earlier study (Glikson et al., 1992), typically has lower reflectance than any other organic matter present in the same sequence. Its appearance is as discrete bodies (Fig. 3a) within the source rock and/or overlying sediments, supporting migrated oil as source. As bitumen B is of the lowest maturation in these sediments, it has to represent the latest thermal episode. A second population of bitumens referred to as type C (Glikson et al., 1992) forms globules ranging in size from several to a few hundred micrometers (Fig. 3b&c). Type C consists of an inner core of same or similar reflectance to that of the alginite of the sequence if present, surrounded by an outer shell of significantly higher reflectance pyrobitumen. Whereas the core may represent an early stage of organic maturation and oil generation, the outer shell represents a rapid high temperature thermal episode, probably the main thermal event that affected the basin, and preceded the episode represented by the generation of type B. A third type D, as in Glikson et al. (1992), is a vein bitumen or infill of cavities, a residue from oil-fluid migration (e.g. Century deposit).

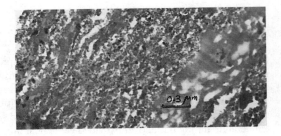

Figure 2: Mature alginite, after oil generation, electron-dense particles = minerals.

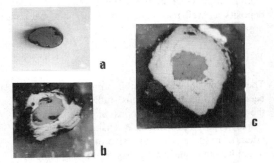

Figure 3: a) Bitumen (Type B), reflected light; b) and c) Bitumen/pyrobitumen (Type C).

Erratic maturation profiles are obtained from plotting the reflectance of the outer rims of the bitumens, whereas when plotting reflectance of cores alone 'normal' maturation profiles such as are common in steadily subsiding basins are obtained (Fig. 4). Erratic maturation profiles have been reported from geothermal systems (Barker, 1991), and active hydrothermal systems such as the Red Sea (Robert, 1988).

3. PERMO-TRIASSIC BOWEN BASIN

The Bowen Basin coals are strongly affected by magmatism and volcanism accompanying the break-up of the supercontinent of Gondwana and irregular maturation profiles within a single seam are not uncommon. As a result, isoreflectance contours show limited lateral continuity due to localised thermal aureoles in proximity to intrusions. The diameter of thermal aureoles may extend over several kms. when associated with

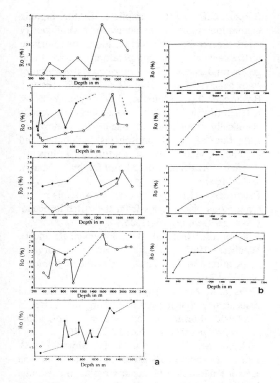

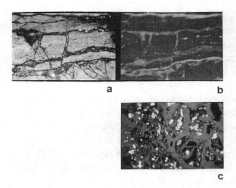

Figure 5. Calcite and bitumen cleat infill in reflected white light (a) and fluorescence model; (b) Pyrite and bitumen in reflected light (c).

Figure 4: a) - reflectance profiles of organic matter in Mount Isa; b) - reflectance of cores in bitumen C.

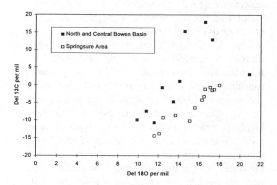

Figure 6: C-O isotope systematics of carbonate veins in Bowen Basin coals.

some of the larger sills. Widespread coking and excessive bitumen confirm rapid heating of the coals to temperatures locally approaching 300°C. Calcite and often pyrite mineralisation is intimately associated with bitumen (Fig. 5).

Carbonate $\delta^{13}C$ and $\delta^{18}O$ values exhibit a wide range of values and are broadly positively correlated (Fig. 6). Such covariation of C and O is to be expected in hydrothermal carbonates deposited over a range of temperatures; however, the slope of the broad correlation trend is too steep to be explained by temperature variation alone. Carbonate deposition from a mixed CO_2- and CH_4- bearing fluid of relatively constant C and O isotope composition best explains the steep correlation trend and extreme range of $\delta^{13}C$ values for the north and central Bowen Basin samples (cf. Golding et al., 1996).

These pervasive carbonate-rich veins in the coals are the product of magmatism-related hydrothermal activity. This interpretation is supported by the unradiogenic $^{87}Sr/^{86}Sr$ isotopic compositions (0.70352 to 0.70506) and C and O isotopic ratios of carbonates that indicate mixing between magmatic and meteoric fluids (Golding et al., 1996).

Carbonates from coals in the Springsure area, south-west Bowen Basin exhibit a more restricted range of $\delta^{13}C$ values and fall on a separate correlation trend (Fig. 6). Calculated fluid compositions at 150 ± 50°C for the Springsure end member carbonate C and O isotope compositions indicate mixing between magmatic and meteoric fluids with a significant input of organic carbon.

The steep slope of both correlation trends suggests that the carbonates were deposited at relatively high temperatures where mineral-fluid oxygen isotope fractionations are smaller. The common observation of coking structures adjacent to carbonate mineralisation supports the high temperature hypothesis.

4. SUMMARY AND CONCLUSIONS

Erratic maturation profiles, as expressed by reflectance of oil generating organic matter are a direct response to uneven and irregular convective or advective heat transfer by short-time span pulses, as distinct from more consistent heat transfer by conduction over long geological time. In contrast, maturation profiles in gradually subsiding basins over long geological time display a consistent gradual increase in reflectance with depth. Convective/advective heat transfer is a highly complex process due to its dependence on a large number of factors, any one of which (e.g. density/ composition of fluid, density of rock, pressure, temperature, driving mechanism etc.) may be subject to variations within the basin, within the rock formation, and within the coalseam or carbonaceous material itself, and thus influence local, as well as regional maturation profiles. Hydrothermal effects on Bowen Basin coals are supported by the isotope geochemistry of carbonate mineralisation.

5. REFERENCES

Altebaumer, F.J., Leithauser, D. & R.G. Schaefer, 1981. Effect of geologically rapid heating on maturation and hydrocarbon generation in Lower Jurassic shales from NW-Germany: *Advances in Organic Geochemistry*: pp 80-86.

Barker, C.E. & M.J. Pawlewicz, 1986. The correlation of vitrinite reflectance with maximum temperature in humic organic matter. In: Buntenbath & Stegema (ed.) *Palaeogeothermics*. Springer Verlag: pp 79-93.

Barker, C.E. 1991. Implications for organic maturation studies of evidence for a geologically rapid increase and stabilisation of vitrinite reflectance at peak temperature: Cerro Pieto geothermal system. *The American Association of Petroleum Geologists Bull.* 75(12): 1852-1863.

Beveridge, T.J., Meloche, J.D., Fyfe, W.S. & R.G.E. Murray, 1983. Diagenesis of metals chemically complexed to bacteria: Laboratory formation of metal phosphates, sulphides and organic condensates in artificial sediments. *Applied Environmental Microbiology*. March: 1094-1108.

Brumsack, H.J. 1980. Geochemistry of Cretaceous black shales from the Atlantic ocean. *Chemical Geology*. 31:1-25.

Glikson, M., Chappell, B.A., Freeman, R. and E. Webber, 1985: Trace element - organic associations in oil shales. *Chemical Geology*. 53:155-174.

Glikson, M. & G.H. Taylor, 1986. Cyanobacterial mats; major contributors to the organic matter in Toolebuc Formation oil shales. *Geol. Soc. Australia Special Publ.* No. 12:276-286.

Glikson, M., Taylor, D. & D. Morris, 1992. Petroleum source rock studies in Proterozoic and Lower Palaeozoic sedimentary basins in Australia, and the maturation path of alginite. *Organic Geochemistry*. 18(6):881-897.

Golding, S.D., Glikson, M., Collerson, K.D., Zhao, J-X, Baublys, K. & J.M. Crossley, 1996. Nature and source of carbonate mineralisation in Bowen Basin coals: Implications for the origin of coalseam gases. In: *Mesozoic of the Eastern Australian Plate Conference,* Beeston (ed.) Geol. Soc. of Australia. No. 46:205-212.

Mauk, J.L. & G.B. Hieshima, 1992. Organic matter and copper mineralisation at White Pine, Michigan, U.S.A. *Chemical Geology* 99:189-211.

Ripley, E.M., Shaffer, N.R. & M.S. Gilstrap, 1990. Distribution and geochemical characteristics of metal enrichment in the New Albany shale (Devonian-Mississippian), Indiana. *Economic Geology*. 85: 1790-1807.

Robert, P. 1988. *Organic Metamorphism and Geothermal History*. Elf-Aquitaine and D. Reidel Publ. 311 pp.

Simonet, B.R.T., Kawka, O.E. & G.M. Wang, 1992. Biological maturation in contemporary hydrothermal systems, alteration of immature organic matter in zero geological time. In: *Biological Markers in Sediments and Petroleum.*

Adsorption of L-alanine monomer, dimer, trimer, tetramer and pentamer by some allophanes

H. Hashizume
National Institute for Research in Inorganic Materials, Tsukuba, Ibaraki, Japan

B. K. G. Theng
Manaaki Whenua-Landcare Research, Palmerston North, New Zealand

ABSTRACT: We have determined the adsorption of mono-, di-, tri-, tetra- and penta-L-alanine by three allophanes with different Al/Si ratios: Kanuma (Al/Si=1.4), Kitakami (Al/Si=1.5), and Te Kuiti (Al/Si=1.6). The extent of adsorption was little influenced by the molecular weight of the amino acids. However, for a given alanine species adsorption tended to increase in the order Kanuma<Kitakami<Te Kuiti, that is, as the Al/Si ratio of the allophane increased. The same trend was previously observed for the adsorption of phosphate by allophanes.

1 INTRODUCTION

Amino acids and peptides are an integral part of the organic matter in soils and sediments where they can survive for thousands of years (Stevenson, 1982; Bada, 1991). Because their long-term persistence is apparently due to close association with clay and mineral surfaces (Curry et al., 1994), the adsorption of amino acids and peptides by crystalline layer silicates (e.g., kaolinite and montmorillonite) has been extensively investigated (Theng, 1974, 1979; Dashman and Stotzky, 1985). In the case of glycine and its oligomers, Greenland et al. (1962) found adsorption to increase with molecular weight.

Relatively little is known, however, about the interactions of amino acids, peptides, and proteins with poorly crystalline (short-range order) minerals like allophane. Milestone (1971) reported that the adsorption of bovine serum albumin by allophane was greater at pH 6 than at pH 7. Similarly, Hashizume and Theng (1996) found adsorption of DL-alanine by allophane to be highest at pH 6. This observation was explained in terms of the charge characteristics of allophane, involving structural (OH)Al(H$_2$O) groups.

Here we investigate the adsorption at pH 6 of L-alanine and its oligomers by three allophanes with different Al/Si ratios in order to clarify the effect of molecular weight on, and the role of surface Al in, the interaction process.

2 EXPERIMENTAL

The allophanes were obtained from pumice beds near Kanuma (Tochigi, Japan), Kitakami (Iwate, Japan), and Te Kuiti (New Zealand). The clay (<2µm e.s.d.), fraction was separated by dispersing the bulk material at pH 3.5 with an ultrasonic probe, sedimenting under gravity, coagulating with 1M NaCl at pH 6, and removing excess electrolyte by dialysis against deionized water. The dialysed samples were dried at 50 °C before use. Reagent grade L-alanine and its oligomers were obtained from Sigma, and used as received.

Adsorption isotherms were determined by equilibrating 200 mg allophane with 15ml solution of the amino acid at 20±1°C. The solutions were made up in 0.004M NaCl containing 0.0001M NaN$_3$ to inhibit microbial growth, and adjusted to pH 6 by addition of 0.1 M HCl or NaOH. After centrifugation, the amino acid concentration in the supernatant solutions was determined using a Shimadzu TOC-5000 analyzer. The amount adsorbed was estimated by difference between the amount initially added and that at equilibrium with the allophane. All measurements were replicated at least twice, blank runs were carried out, and the concentration of the stock amino acid solutions was checked at frequent intervals.

3 RESULTS AND DISCUSSION

The isotherms for the adsorption of L-alanine and its oligomers by Kanuma, Kitakami, and Te Kuiti allophanes are shown in Figs. 1, 2, and 3, respectively. For any one allophane the extent of adsorption over the range of equilibrium concentrations used (<0.003M) does not vary greatly with the molecular weight of the alanine species. This contrasts with the behaviour of glycine and its oligomers (Greenland et al., 1962), and of alkylammonium cations (Theng, 1974) in montmorillonite where adsorption increases in a

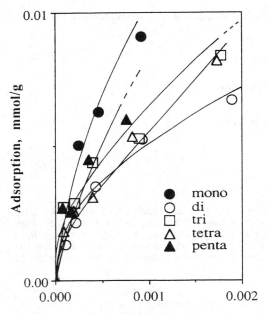

Figure 1. Isotherms for the adsorption of L-alanine, monomer, dimer, trimer, tetramer and pentamer by Kanuma allophane at pH 6.

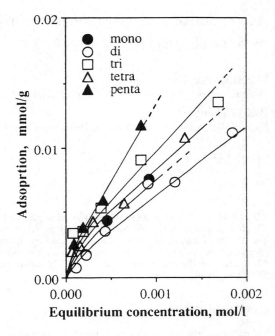

Figure 2. Isotherms for the adsorption of L-alanine, monomer, dimer, trimer, tetramer and pentamer by Kitakami allophane at pH 6.

regular manner with chain length. As the adsorbate molecules in montmorillonite tend to lie flat on the clay surface, the increase in adsorption and surface affinity with molecular size has commonly been ascribed to the incremental contribution of van der Waals interactions.

However, unlike montmorillonite which has a layer structure, the primary (unit) particle of allophane is a hollow spherule with an outer diameter of 3.5 - 5.5 nm. The spherule wall of about 0.7 nm thickness consists of an outer gibbsitic sheet to which $(O_3)Si(OH)$ groups are attached on the inside. Structural defects within the wall give rise to perforations of about 0.3nm in diameter where $(OH)Al(H_2O)$ groups are exposed. These groups become protonated at pH < 6, and lose protons at pH > 6. That is, at pH 6 the net surface charge of allophane is close to zero (Parfitt, 1990) while the alanine species would exist in the zwitterionic form.

Adsorption at pH 6 would therefore be primarily controlled by electrostatic interactions involving the COO^- and NH_3^+ groups of the amino acid and $(OH_2)^+Al(H_2O)$ and $(OH)Al(OH)^-$ groups of allophane, respectively. Since the hydrophobic alkyl groups of alanine and its oligomers would essentially be repelled from the hydrophilic surface (wall) of allophane particles, the alkyl side chains are likely to extend away from the wall. In this conformation van der Waals interactions between amino acid and allophane surface are not important,

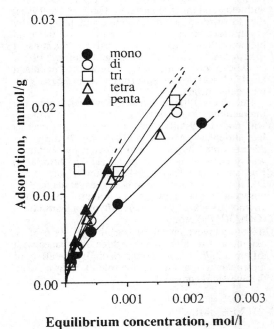

Figure 3. Isotherms for the adsorption of L-alanine, monomer, dimer, trimer, tetramer and pentamer by Kanuma allophane at pH 6.

and the level of adsorption would vary little with adsorbate molecular weight as the isotherms indicate.

The observed order of adsorption, with Kanuma allophane showing the lowest and the Te Kuiti sample the highest uptake, is also consistent with the proposed mechanism since the lower the Al/Si ratio of the allophane the smaller would be the concentration of reactive Al groups. That such groups are involved in the process is supported by phosphate adsorption studies (Theng, et al., 1982; Parfitt, 1990). Phosphate ions are known to interact with $(OH)Al(OH_2)$ groups, and like that of alanine species the adsorption of phosphate by allophanes tends to increase as the Al/Si ratio of the samples increases.

4 ACKNOWLEDGMENTS

We thank Mr. C. Feltham for assistance with using the TOC-5000 analyzer.

5 REFERENCES

Bada, J. L. 1990. Amino acid cosmogeochemistry. *Phil. Trans. R. Soc. Lond.* B333: 349-358.

Curry, G.B., B.K.G. Theng & H. Zheng 1994. Amino acid distribution in a loess-palaeosol sequence near Luochuan, Loess Plateau, China. *Org. Geochem.* 22: 287-298.

Dashman, T. & G. Stotzky 1985. Physical properties of homoionic montmorillonite and kaolinite complexed with amino acids and peptides. *Soil Biol. Biochem..* 17: 189-195.

Greenland, D.J., R.H. Laby & J.P. Quirk 1962. Adsorption of glycine and its di-, tri-, and tetra-peptides by montmorillonite. *Trans. Faraday Soc.* 58: 829-841.

Hashizume H. & B.K.G. Theng 1996. Adsorption of alanine by allophane. *Abstracts, Australian and New Zealand National Soils Conference 1996* : Volume 2, Oral Papers, 119-120.

Milestone, N.B. 1971. Allophane, its structure and possible uses. *New Zealand Inst. Chem.* 35:191-197.

Parfitt, R.L. 1990. Allophane in New Zealand - A review. *Aust. J. Soil Res.* 28:343-360.

Stevenson, F.L. 1982. *Humus Chemistry.* Wiley-Interscience: New York.

Theng, B.K.G. 1974. *The Chemistry of Clay-Organic Reactions.* Adam Hilger: London.

Theng, B.K.G. 1979. *Formation and Properties of Clay-Polymer Complexes.* Elsevier: Amsterdam.

Theng, B.K.G., M. Russell, G.J. Churchman & R.L. Parfitt 1982. Surface properties of allophane, halloysite, and imogolite. *Clays Clay Miner.* 30:143-149.

Sorption and fractionation of natural organic matter on kaolinite and goethite

P. Maurice, K. Namjesnik-Dejanovic, S. Lower & M. Pullin
Water Resources Research Institute, Kent State University, USA

Y.-P. Chin
Department of Geological Sciences, Ohio State University, Ohio, USA

G. R. Aiken
US Geological Survey, WRD, Boulder, Colo., USA

ABSTRACT: High pressure size exclusion chromatography (HPSEC) was used to determine directly the molecular-weight fractionation that occurred upon sorption of a muck fulvic acid (FA) to kaolinite and goethite, at pH 3.7, 22°C, 24hrs. equilibration. FA sorption affinity was found to be much greater for goethite than for kaolinite, with sorption maxima, A_{max} = 0.22 and 0.05 mgC m^{-2}, respectively. Fractionation measurements agreed well with theoretical predictions. At concentrations < A_{max}, for both sorbents, the average molecular weight (M_w) of FA remaining in solution decreased by as much as several hundred daltons, indicating preferential sorption of the higher molecular weight components. At concentrations > A_{max}, solution M_w did not change, most probably because preferential sorption was overwhelmed by excess FA. Polydispersity decreased slightly upon sorption to kaolinite, but there was no significant change for goethite. Our results demonstrate quantitatively the preferential sorption of higher molecular weight components, and show that effects on solution M_w vary as a function of [FA].

1 INTRODUCTION

Natural organic matter (NOM) is an important component of surface- ground- and soil waters, and it plays a key role in mineral growth and dissolution, trace-metal cycling, and the global C budget. Additionally, NOM plays a crucial, though complex, role in the transport of organic and inorganic pollutants through porous media. On the one hand, NOM that is bound to mineral surfaces may remove trace metals, nonpolar organic compounds (NOCs) and other pollutants from the water column. On the other hand, NOM that remains free or in colloidal form within the water column may increase pollutant mobility. Hence, understanding and quantifying the bulk partitioning of NOM between solid and dissolved phases is fundamental to a wide range of pollutant transport phenomena. Moreover, because NOM consists of a variety of molecules with variable structure, hydrophobicity, functionality, and reactivity, we need to understand how NOM fractionates upon sorption to mineral surfaces. For example, preferential sorption of more hydrophobic components may increase NOC retardation through porous media.

A variety of field and laboratory investigations (e.g., Davis and Gloor, 1981; Tipping, 1981; Gu et al., 1995) have suggested that higher molecular weight NOM components sorb preferentially to mineral surfaces, resulting in fractionation of the NOM pool. In general, these studies have relied on indirect or qualitative methods of molecular-weight analysis. In this study, we use high pressure size exclusion chromatography (HPSEC) to measure directly the molecular weight change upon sorption of a muck soil fulvic acid (FA) to kaolinite and goethite. Combination of HPSEC with isotherm analysis, absorbance measurements and dissolved Fe and Al determinations provides an integrated understanding of the sorption and fractionation processes.

2 MATERIALS AND METHODS

2.1. Mineralogic samples

Goethite was synthesized according to the method described by Gu et al. (1994). Following preparation, the sample was freeze-dried. X-ray diffraction (XRD) analysis performed on a Philips XRD 300 Generator using Cu Kα radiation showed that the majority of the material was goethite (γ-FeOOH) with a trace amount of hematite (α-Fe$_2$O$_3$). BET surface area =58.0 m^2/g.

The kaolinite sample used in these experiments was well crystallized standard Kga-1b from the Clay Minerals Society of America. The sample was cleaned by rinsing 100 g of kaolinite approximately 4 times in 1 liter of 1 M NaCl at pH 3, after which, the sample was washed repeatedly in Milli-Q UV water. The kaolinite was then freeze-dried. BET surface area = 12.64 m^2/g.

2.2. NOM sample

The FA sample used in these experiments was

isolated from a muck soil underlying the streambed of McDonalds Branch, a small freshwater wetland in the Pine Barrens region of the Atlantic coastal plain of New Jersey (USA). Surface waters overlying this muck unit typically have pH = 3.6 to 4.2 and dissolved organic carbon (DOC) concentrations = 18 to 30 mgC L^{-1}, but occasionally as high as 100 mgC L^{-1}. The muck sample was NaOH extracted under a N$_2$ atmosphere in the USGS laboratory of George Aiken. Following extraction, the sample was fractionated using XAD-8 resin following the method of Aiken et al. (1992), and freeze-dried. Results of elemental analyses are shown in Table 1. This FA has a high O/C ratio (0.69) relative to IHSS reference Suwannee River FA (O/C = 0.58), which indicates that it is highly polar.

Table 1. Weight % of constituents in the FA sample

C	H	N	S	P	O	Ash
47.6	4.1	0.9	0.4	0.2	44.0	2.4

2.3. Sorption experiments

A stock solution of 45 mgC L^{-1} FA was prepared from freeze-dried isolate and MilliQ-UV water, and allowed to equilibrate in a refrigerator, overnight. Various dilutions were prepared gravimetrically by dilution with 0.01 N NaCl electrolyte solution. These diluted samples were split by placing 36 mL into each of 4 40 mL polypropylene centrifuge tubes, 1 for a control sample (no mineralogic particulates), 1 for goethite-reacted (designated goethite on graphs), and 2 for kaolinite-reacted (kaolinite1 and kaolinite2). 3 mL of kaolinite or goethite suspension were then added to each tube; 3 mL MilliQ water was added to each control. In order to account for differences in specific surface areas, different concentrations of stock suspensions were used: 9.21 g L^{-1} goethite versus 42.26 g L^{-1} kaolinite. Sample pH was adjusted to 3.7 using 0.01 N NaOH/HCl. This pH was chosen as the Muck pore-water pH at the time of sampling. The DOCs of the control samples ranged from 0 to 41 mgC L^{-1}. The centrifuge tubes were capped tightly, and placed in a horizontal position in a 22 ± .2 °C water bath shaker table covered to keep out light. Samples were allowed to equilibrate for 24 hours, after which they were centrifuged under refrigeration at 12,000 rpm for 20 min. Samples for metals analysis were acidified with ultrapure HCl and stored in polyethylene bottles. Samples for DOC, HPSEC and UV-Vis analysis were placed in glass bottles and analyzed within 48 hrs of collection. Samples were refrigerated prior to analysis.

2.4. HPSEC

HPSEC measurements were performed in the method of Chin et al. (1994). Briefly, instrumentation consists of a column which uses a modified silica stationary phase (Waters Protein Pack 125) with pores that exclude colloids of a certain size while allowing small analytes to diffuse in. The solvent delivery system is a Waters 510 pump operated isocratically, and the entire system is connected to a Waters 486 UV/Vis detector. Analytes were monitored at λ =230 nm; higher wavelengths are less sensitive to smaller molecular weight moeities (Chin, et al., 1994). Mobile phases consisted of aqueous solutions buffered to pH = 6.8. Salts were added to both the mobile phase and sample matrix (equivalent of 0.1M NaCl) to ensure that ion exclusion effects were suppressed in the column. The SEC system was calibrated using random coil sodium polystyrenesulfonates (18K, 8K, 5.4K, 1.8K) and acetone. Weight-average molecular weights (M$_w$) were determined using the equations presented in Chin et al. (1994). Aqueous samples were introduced using a 20μL rotary injector (Rheodyne) and did not require any pretreatment.

2.5. Aluminum and Fe concentrations

Al was measured using a flow-injection analysis (FIA) system as described by Sutheimer and Cabaniss (1995). The FIA system works by detecting a fluorescent complex between Al and lumogallion [3-(2,4-dihydroxyphenylazo)-2-hydroxy-5-chloro-benzenesuphonic acid]. The method has no interference from organic complexing agents or from other metals. However, the method does not detect polymeric Al species. Thus, the Al measured during our experiments is strictly the dissolved organic and inorganic monomeric Al. The system was calibrated using a standard curve made by diluting a 1000 ppm AA/ICP calibration standard (Aldrich Chemical Company) in high purity pH 2.0 HCl (Milli-Q water and 99.999% HCl - Aldrich). All standards were made fresh daily using polymeric labware. A 10 μl sample loop was used. The FIA system gave a linear response between 0.05 and 1.0 ppm Al with a precision of 1% at 0.1 ppm. Fe was measured using a Perkin Elmer graphite-furnace AA with an automatic sampler.

2.6. UV-Vis

Absorbance spectra were collected between 200 and 900 nm using a double beam Hitachi U2000 spectrometer. The spectra were collected at 200 nm/min using 1 cm quartz cells with Milli-Q water as the reference. Although synchronous spectra were collected, we report extinction coefficients (ε), which were calculated based only on data at λ = 280 nm.

2.7. Dissolved Organic Carbon

DOC concentrations were measured on a Shimadzu TOC5000 carbon analyzer. Prior to analysis, samples were acidified and purged for 2 min. to remove inorganic C.

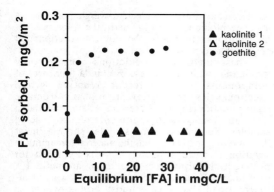

Figure 1. Sorption of FA on kaolinite and goethite

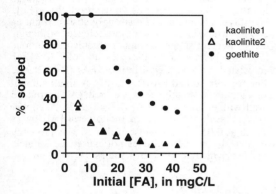

Figure 2. Percent sorption of FA on kaolinite and goethite

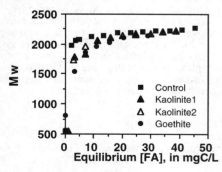

Figure 3. Comparison of weight average molecular weights (M_w) of control and reacted samples

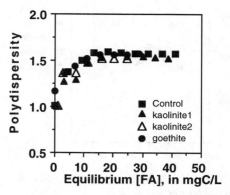

Figure 4. Comparison of polydispersity values of control and reacted samples

3 RESULTS AND DISCUSSION

Figure 1 shows FA sorption isotherms for kaolinite and goethite at pH 3.7. FA has a much higher affinity for goethite than to kaolinite, with sorption maxima, A_{max} = 0.22 and 0.05 mgC m^{-2}, respectively. As shown in Figure 2, this greater sorption density for goethite is reflected in greater % sorption, over the range of initial [FA]. At initial [FA] $\leq$ 10 mgC L^{-1}, there was essentially complete sorption onto goethite.

M_ws of control and reacted samples are shown in Figure 3. The control data show a decrease in M_w with concentration. This trend is probably, for the most part, an artifact of the HPSEC detection system, which is sensitive to concentration. Nevertheless, comparison of control versus reacted sample values shows that at equilibrium [FA] < 15 mgC L^{-1}, solution M_w decreases by as much as several hundred daltons. Note that for sorption onto goethite, at the lowest concentrations, sorption was so complete that the M_w in the equilibrium solution was identical to the MilliQ water blank. The decrease in solution M_w indicates that the higher molecular weight FA components sorb preferentially to both kaolinite and goethite. Despite the different sorption affinities for kaolinite versus goethite, the M_w values for any given equilibrium [FA] tend to be only slightly lower for goethite- than for kaolinite- reacted samples. However, the change in M_w from initial to equilibrium conditions was greater for goethite than for kaolinite because the stronger sorption affinity resulted in greater differences between initial and equilibrium [FA]. Figure 4 shows that the polydispersity values (weight average molecular weight divided by number average molecular weight of goethite-reacted samples were similar to the control samples at any given equilibrium [FA]; polydispersity values for kaolinite-reacted samples were slightly lower.

M_w and polydispersity do not change at higher equilibrium [FA]. Even if some preferential sorption occurs, its effects on solution characteristics may be overwhelmed by excess FA concentrations as A_{max} is exceeded.

The above observations can be compared with predictions based on polydisperse polymer sorption

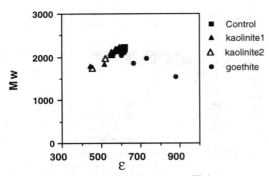

Figure 5. Changes in extinction coefficient, ε upon sorption to kaolinite and goethite

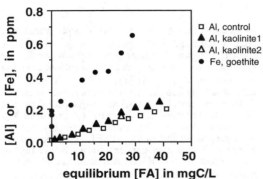

Figure 6. Changes in [Fe] and monomeric [Al] upon sorption to goethite and kaolinite, respectively. Control values for goethite-reacted samples are not plotted; [Fe] in an ~45 mg L^{-1} control = 0.014 ppm.

theory (Cohen-Stuart et al., 1980). In the case of a binary mixture, the following scenario may be expected. At very low concentrations, in the rising portion of the sorption isotherm, both components can be easily accomodated on the surface, so that fractionation will not occur. At somewhat higher concentrations, but < A_{max}, there is not enough available surface to accommodate all components, and the higher molecular-weight component will begin to displace the lower molecular-weight component. As concentration increases to A_{max}, all of the lower molecular-weight molecules are displaced from the surface, the surface is covered by larger molecules, and the solution again contains both components. For complex mixtures, the scenario is similar, but any component with a given molecular weight will be displaced by a component with only slightly higher molecular weight.

Our results agree with these theoretical predictions. Although limitations of the HPSEC technique do not permit us to observe effects at very low concentrations, we do observe the results of displacement of lower-molecular-weight components at concentrations < A_{max} and the presence of distributions similar to the control solutions, at concentrations > A_{max}.

These theoretical predictions consider only molecular weight effects. FA is a mixture of components with different reactive group functionalities and different aliphatic/aromatic characteristics. Figure 5 presents ε data, calculated from absorbance values, for control and reacted samples. Kaolinite-reacted samples show a linear decrease in ε, indicating that the material remaining in solution is more aliphatic. This is evidence for preferential sorption of the more aromatic FA constituents. Goethite-reacted samples show a mirror-image trend, but this is most probably due to interference from Fe in solution, brought about by goethite dissolution.

As shown in Figure 6, there was substantial goethite dissolution, as measured by [Fe]. Hydrolyzed (inorganic) Fe strongly absorbs light at 280nm, which is the cause of interference. Although the y-intercept suggests some inorganic dissolution, [Fe] was enhanced by the presence of FA. [Fe] correlated well with equilibrium [FA] ($R^2 = 0.95$), but not with sorbed FA ($R^2 = 0.40$). A small amount of kaolinite dissolution also occurred; [Al] correlated well with equilibrium [FA] ($R^2 = 0.993$). At higher [FA], dissolution increased monomeric [Al] by as much as 0.05 mg L^{-1}. Most of this Al was probably organically bound. Al does not cause interference, and hence did not effect the ε values.

4 CONCLUSIONS

Results of our experiments demonstrate preferential sorption of higher molecular-weight FA components to kaolinite and goethite. At relatively low [FA], this effect can lead to changes in solution M_w on the order of several hundred daltons. As A_{max} is exceeded, solution M_w does not change, most probably because excess FA overwhelms sorption effects. ε values indicate sorption of more aromatic constituents to kaolinite; ε data for goethite are ambiguous because of Fe-FA complexation. Finally, our results show substantial goethite dissolution and moderate kaolinite dissolution in the presence of muck soil FA.

5 ACKNOWLEDGMENTS

Funding for this research was provided by National Science Foundation grant # EAR-9628461.

6 REFERENCES

Aiken G.R., D.M. McKnight, K.A. Thorn, and M. Thurman, 1992. Isolation of hydrophilic organic acids from water using nonionic macroporous resins. *Org. Geochem.*, 18: 567.
Chin Y.P., G.R. Aiken, E. O'Loughlin, 1994.

Molecular weight, polydispersity, and spectroscopic properties of humic substances, *ES&T*, 28: 1853.

Cohen-Stuart, M.A., J.M.H.M. Scheutjens, and G.J. Fleer, 1980. Polydispersity effects and the interpretation of polymer adsorption isotherms. *J. Polymer Sci.*, 18: 559-573.

Davis, J.A., and R. Gloor, 1981. Adsorption of dissolved organics in lake water by aluminum oxide. Effect of molecular weight. *ES&T*, 15: 1223-1229.

Gu, B., J. Schmitt, Z. Chen, L. Liang, and J.F. McCarthy, 1994. Adsorption and desorption of NOM on iron oxide: Mechanisms and models. *ES&T* 28: 38-46

Gu, B., J. Schmitt, Z. Chen, L. Liang, and J. F. McCarthy, 1995. Adsorption and desorption of different organic matter fractions on iron oxide. *GCA*, 59: 219-229

Sutheimer, S.H., S.E. Cabaniss, 1995. Determination of trace aluminum in natural waters by flow-injection analysis with fluoresence detection of the lumogallion complex. *Analyt. Chim. Acta*, 303: 211- 221.

Tipping, E. 1981. The adsorption of aquatic humic substances by iron oxides. *Geochim. Cosmochim. Acta*, 45: 191-199.

Solid phase partitioning of uranium and copper in the presence of HFO and bacteria

Lesley A. Warren & F. Grant Ferris
Department of Geology, University of Toronto, Ont., Canada

ABSTRACT: The affinity of hydrous ferric oxide (HFO) and HFO formed in association with *B. subtilis*, for Cu and U were determined in the laboratory. HFOs were precipitated alone, or as surficial coatings on *B. subtilis*, and then exposed to either Cu ($10^{-4.5}$M) or U (10^{-4}M) at a pH of 6 and sampled over a 24 hour period for both dissolved and total metal concentrations. Results indicated that solid phase partitioning of both U and Cu occurred in both treatments; although the bacteria - HFO treatment appeared to enhance solid phase partitioning of both metals over that seen in the HFO alone treatment. Solid phase partitioning of Cu showed a steady increase over the time course of the experiment in both treatments; while solid phase U partitioning was lower than that seen for Cu, and remained static in the bacteria - HFO treatment and decreased in the HFO alone treatment.

1 INTRODUCTION

It is widely accepted that mineral surfaces, especially those of HFO play a significant, if not dominant role in regulating the behavior of metals in natural aquatic systems (Benjamin 1983). Such recognition has lead to the development of surface complexation (SCM) and surface precipitation (SPM) models (Dzombak and Morel 1990; Farley et al. 1985) which are increasingly used in modeling reactive transport of contaminants, with some success. At the same time, bacteria are receiving more recognition in the developing conceptual picture of metal cycling as they are known to be equally widespread, abundant and potentially reactive surfaces. Bacteria are responsible for processes that concentrate, disperse, alter or fractionate metals and thus are likely important players in determining metal behavior. While a host of qualitative information has been collected on the ability of bacteria to sorb metals and act as nucleation sites for mineral precipitation (Thompson and Beveridge, 1993) and specifically for Fe oxide formation (Ferris et al. 1989, 1987); very little quantitative information on microbial surface reactions has, as yet been documented. This information would aid greatly in the development of any model including bacterial surfaces as reactive geochemical substrates.

In particular, metal and radionuclide contamination of groundwater systems requires the development of effective remediation strategies. Bioremediation techniques involving bacteria appear very promising as they provide a potentially low cost, effective treatment for removal of contaminants through solid (biological) phase capture and subsequent removal of solution contaminants. We have evaluated the ability of *Bacillus subtilis* (Gram positive bacterium) coated with HFO to sorb uranium and copper from solution and compared these results to those for HFO alone.

2 EXPERIMENTAL METHODS

All experiments were run in the laboratory using a *B. subtilis* cell density equivalent to an optical density of 0.2 (at 600 nm) under non growth conditions. HFO and *B. subtilis* coated with HFO were prepared by titrating a 0.1 mmolar Fe solution ($FeNO_3 \cdot 6H_2O$) alone (HFO treatment) or with bacterial cells in solution (B. subtilis - HFO treatment) to pH 7 with KOH, and gently stirring this for 24 hours. Analysis of dissolved iron levels (ferrozine spectrophotometric assay HACH DR/2000) at the end of this initial HFO formation stage, indicated that most Fe had precipitated as HFO as solution Fe concentrations were below detection. Further, STEM analyses also visually confirmed HFO precipitation had occurred directly on the cell surfaces.

Bacterial cells and the HFO precipitates were then spun down and gently washed with 10 mls of ultrapure H_2O and then exposed to either a 10^{-4}M solution of U (U_T) or a $10^{-4.5}$M solution of Cu (Cu_T). The pH of both solutions for each metal was adjusted to a pH of 6. Experiments were run as a time series for 24 hours over which time the experimental containers were gently shaken to keep cells and HFO particles in solution. Samples were taken for total and dissolved (0.2 μm filtered) analyses (U by NAA SLOWPOKE reactor; and Cu by flame atomic absorption) at time 0, 1, 2, 4, and 24 hours.

3 RESULTS AND DISCUSSION

Results show an enhanced solid phase partitioning of both U and Cu with B. *subtilis* - HFO over HFO alone (Figs. 1,2). However the results for the two elements show different patterns. Solid phase U (U_S) concentrations remained stable over the time course in the B. *subtilis* - HFO treatment; and showed a decreasing trend before returning to the initial value observed in the HFO alone treatment (Fig.1; Table 1).

In contrast, solid phase Cu (Cu_S) increased over the course of the experiment in both treatments; although to higher levels in the B. *subtilis* - HFO treatment (Fig. 2; Table. 1). The higher Cu_S values observed in the bacteria - HFO treatment compared to the HFO treatment, probably reflects sorption of Cu by bacterial surface sites over and above that by the HFO sites and/or stabilization of the oxide sites by the underlying bacterial surface as has been noted by other authors for organic matter (Laxen 1985; Buffle et al, 1984). Copper is well known for its strong affinity for organic matter.

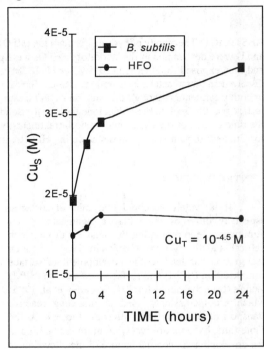

Fig. 2. Solid phase concentrations of Cu (Cu_S; M) over a 24 hour period for B. *subtilis* coated with HFO (B. *subtilis*) and for HFO alone (HFO).

Table 1: The proportion of solid phase copper Cu_S or uranium U_S over 24 hours for the B. *subtilis* - HFO treatment and the HFO alone treatment.

TIME (hour)	B. *subtilis*-HFO %Cu_S	%U_S	HFO %Cu_S	%U_S
0	37	21	47	8
1	51	24	51	5
2	56	23	55	3
4	69	23	54	2
24	76	25	57	11

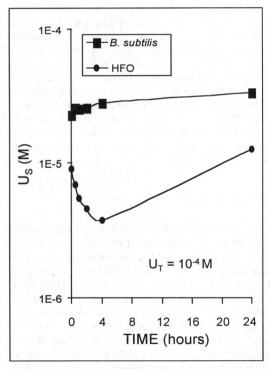

Fig. 1. Solid phase concentrations of U (U_S; M) over a 24 hour period for B. *subtilis* coated with HFO (B. *subtilis*) and for HFO alone (HFO).

Results indicated a higher solid phase partitioning of Cu in the presence of both *B. subtilis* coated with HFO and the HFOs alone than that observed for U (Table 1). One explanation for this result may be that only high energy binding sites on the HFO are capable of binding U, and under our experimental conditions we may have saturated these sites. This explanation is consistent with the observed trend of a static response in U_S concentrations for the *B. subtilis* - HFO over the 24 hours (somewhere around 23-25% of the total uranium in solution). In contrast to the results for the *B. subtilis* coated with HFO, the results for HFO alone indicate an initial decline in U_S from 8% to 3% in the first 4 hours, before returning to 11% after 24 hours. This result, again, highlights the apparent greater reactivity of the bacterially associated HFOs in comparison to those formed abiotically.

4 CONCLUSIONS

Bacterially associated HFOs were demonstrated to enhance solid phase partitioning of both copper and uranium over HFO alone, in laboratory batch experiments. Cu and U solid phase partitioning in the presence of *B. subtilis* coated with HFO showed differing results. Solid phase Cu showed a steady increase over the 24 hour period up to 76% of the total Cu in solution; while U sorption appeared to remain static at the initial solid phase U partitioning of 25%. The results for HFO also showed an increase in solid phase Cu partitioning over the course of the experiment but at a lower proportion from 47 to 54% of the total Cu in solution, than that seen with the *B. subtilis* associated HFO. Uranium sorption to the HFOs was weak, and decreased initially to 3% from 8% observed at the start of the experiment, before returning to 11% after 24 hours. We hypothesize that uranium is bound by higher energy sites on the HFOs and these may have been saturated under the conditions of the experiment.

5 ACKNOWLEDGMENTS

We thank Jennifer Cox, Tracy Fleury and Andrew Wolf for help in the collection and analyses of the data. This research was sponsored by the Environmental Management Science Program of the Office of Environmental Management and Office of Energy Research at the United States Department of Energy under Award # DE-FG07-96ER62317.

6 REFERENCES

Benjamin, M.M. 1983. Adsorption and surface precipitation of metals on amorphous iron oxide. *Environ. Sci. Technol.* 17:686-692.

Buffle, J., A. Tessier and W. Haerdi. 1984. Interpretation of trace metal complexation by aquatic organic matter. In. Kramer, C.J.M. and J.C. Duinker (Eds.) *Complexation of Trace Metals in Natural Waters* The Hague: Junk Publishers pp. 301-316.

Dzombak D.A. and F.M.M. Morel. 1990. *Surface Complexation Modeling: Hydrous Ferric Oxide* New York: Wiley - Interscience 331 pp.

Farley, K.J., D.A. Dzombak and F.M.M. Morel. 1985. A surface precipitation model for the sorption of cations on metal oxides. *J. Colloid Interface Sci.* 106:226-242.

Ferris, F.G., S. Schultze, T.C. Witten, W.S. Fyfe and T.J. Beveridge. 1989. Metal interactions with microbial biofilms in acidic and neutral pH environments. *Appl. Environ. Microbiol.* 55:1249-1257.

Ferris, F.G., W.S. Fyfe and T.J. Beveridge. 1987. Bacteria as nucleation sites for authigenic minerals in a metal-contaminated lake sediment. *Chem. Geol.* 63:225-232.

Laxen, D.P.H. 1985. Trace metal adsorption/coprecipitation on hydrous ferric oxide under realistic conditions. *Water Res.* 19:1229-1236.

Thompson, J.B. and T.J. Beveridge. 1993. Interactions of metal ions with bacterial surfaces and the ensuing development of minerals. In: Rao. S. (Ed.) *Particulate Matter and Aquatic Contaminants*. Boca Raton: Lewis pp.65-104

Trace metal/microbial interactions in an Antarctic freshwater system

K.S. Webster, J.G. Webster & P.E. Nelson
Institute of Environmental Science and Research, Auckland, New Zealand

ABSTRACT: The interaction of cyanobacteria with natural and anthropogenic trace metals in and around Lake Vanda has been examined. Lake Vanda is perennially ice-covered and stratified with respect to temperature, salinity, pH and redox potential. Trace metals in the lake can potentially interact with phytoplankton, which exist at discrete horizons controlled by temperature and light requirements. As yet, there is no evidence of trace metal uptake or precipitation in the most prolific cyanobacteria community just above the anoxic zone, where concentrations of dissolved metals are also highest. Analysis of suspended sediment from the anoxic zone however, suggests that nitrate- or sulfate-reducing bacteria may be involved in trace metal precipitation. At the former site of Vanda Station, there are small areas of trace metal-contaminated soils. In microbial mats taken from these and surrounding areas, Cu, Ni and Zn enrichment shows a strong correlation with Fe, suggesting that these metals are bound to Fe-oxide particulate phases in the mat matrix. Pb, Ag and Cd are also enriched in the mat, but show a poor correlation with Fe.

1 INTRODUCTION

This research was undertaken to increase our understanding of how indigenous Antarctic microbial communities (planktonic and mat) interact with natural and contaminant trace metals in meltwaters and soils. Bacterial communities can interact with heavy metals in a number of ways. In some cases, bacteria may show a toxic response to heavy metals in the environment, others can actively release trace metals bound to sediments or adsorb metals onto cell walls. Biosorption of metals can leach and precipitate metals from solution under chemical conditions which may otherwise not favour metal precipitation (Gadd 1992). Red and green ice and snow algae of the Arctic have demonstrated the ability to concentrate elements such as phosphorus, sulfur, silicon and aluminium from wind-blown aerosols as a nutrient source to enable their survival in adverse conditions (Tazaki et al., 1994). However, despite microbial mats being the dominant life form of continental Antarctica (they occur wherever sufficient moisture occurs on ice-free land), the interaction of Antarctic bacteria with trace metals has not previously been investigated.

1.1 *Lake Vanda*

Lake Vanda is a perennially ice-covered inland lake in the Wright Valley of Victoria Land, Antarctica (77°S). The lake is approximately 8km long, 2 km wide and was 74m deep at its deepest point when measured in mid-summer 1995 and in early 1997. It has no outlet, and lake level has risen steadily over the last 10 years, reflecting an increase in seasonal meltwater inflow from the Onyx River. The rising lake level was a factor in the decision to remove the NZ base, Vanda Station, from the lakeshore in 1994/95. The lake is meromictic, comprising a homogeneous layer of low salinity above 50 m depth, and deeper layer where salinity increases with depth to form a Ca-Cl brine. In 1995, the lake was anoxic below 65m; below this, concentrations of up to 34 mg/kg H_2S and temperatures of up to 17°C were measured (Figure 1). There has been little change since the major ion, thermal and pH profiles were first measured in the 1960's (eg., Angino & Armitage, 1963) as the density stratification prevents the seasonal overturn often observed in freshwater lakes.

1.2 *Trace metals*

Significant interaction between trace metals and microbial life in Lake Vanda could take place both in the main body of the lake, and at the lake margin where lake waters are beginning to flood a small gully contaminated by greywater (washing water) waste residues from the former base. Although extensive remediation of the base site has been undertaken, some areas of soil show elevated trace metal concentrations; up to 23 mg/kg Pb, 64 mg/kg Zn, 40 mg/kg Cu as well as significant levels of other trace metals.

Previous studies have established distribution profiles for selected trace metals in Lake Vanda (Green et al., 1993; Webster, 1994). Trace metals enter the lake in water from the Onyx River, predominately bound to suspended material (inorganic and organic) and are distributed across the lake in the very dilute water layer just below the ice cover. As this particulate matter settles slowly through the oxic layers of the lake, encountering changes in pH and major ion chemistry, trace metals adsorbed onto the surface of oxides are released. Oxides eventually dissolve completely at, or immediately below, the redox boundary (i.e. oxic/anoxic transition), releasing bound trace metals. In the anoxic zone, thiophilic metals react with H_2S to form metal-sulphides. For this study, our sampling has focussed on the region near and just below the redox boundary, where trace metals are predominantly dissolved and bacteria are most active.

1.3 Microbial communities

In Lake Vanda, an algal biomass maximum occurs just above the redox boundary. The position of this horizon corresponds to the level at which autotrophs receive sufficient light for growth and can utilise nutrients available from the anoxic water below. The position of this horizon is usually identified using chlorophyll-*a* concentrations, which showed a peak at 62m depth when measured in January 1997. The phytoplankton in both the upper and lower oxic layers are predominantly cyanobacteria (*Phormidium spp.*) (Vincent & Vincent, 1982). Benthic microbial mats grow on sediments at the base and margins of the lake and are dominated by oscillatorian cyanobacteria which form a mucilaginous matrix 0.5-5mm thick. This matrix acts as a physical barrier to the exchange of solute components, including biologically active gases and nutrients (Hawes et al., 1993).

2 SAMPLING & ANALYTICAL METHODS

Water, sediment and microbial samples were collected from Lake Vanda, the Onyx River and flooded soils of the surrounding area in November/December 1995, and January 1997. Standard trace metal sampling techniques were used in the collection of all samples.

Water samples for chlorophyll-*a*, TOC and VOC were collected and preserved for later analysis. Conductivity, temperature, pH and dissolved oxygen

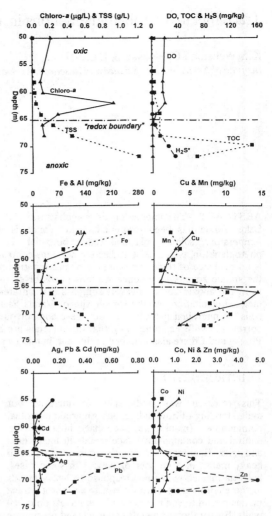

Figure 1 Chlorophyll-*a*, dissolved oxygen (DO), total suspended sediment (TSS) and total organic carbon (TOC) profiles through the deepest part of Lake Vanda, together with trace metal concentrations in suspended sediments.

(DO) were measured immediately on site using standard portable meters, while samples for major and trace element analysis were preserved for analysis by HPIC and ICP-MS or graphite furnace AAS (see Webster et al 1997 for details of sampling and analytical techniques). For the profile through the main body of Lake Vanda, an Aquapac *in situ* Fluorometer was used to locate the phytoplankton-rich horizon just above the redox boundary (at ca. 2m depth). From 56m to the base of the lake at 72m, water samples were collected at 2m intervals.

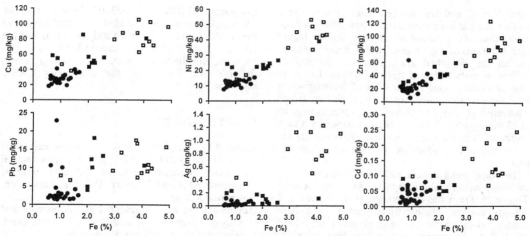

Figure 2 Trace metal concentrations in cyanobacterial mats (☐=1995; ■=1997) and underlying sediments/soils (●=1997) relative to Fe concentration.

Suspended sediments in Lake Vanda were collected onto preweighed 0.22 µm filter membranes by vacuum filtration. After drying (60°C) and reweighing to determine total suspended sediment/phytoplankton concentrations (TSS), these were digested in hot concentrated HNO_3 and analysed for trace metals by ICP-MS.

Samples of cyanobacterial mat and associated soils or sediments, were collected and sealed in airtight HDPE containers for return to New Zealand. Mats were washed and placed in an ultrasonic bath to remove loosely adhering sediment, then dried at 60°C. The sediments and soils were dried, sieved (500µm mesh), and digested in hot concentrated HNO_3 prior to analysis for Fe, Al, Mn, Ni, Cu, Pb & Zn by AAS and for Ag and Cd by ICP-MS.

3 RESULTS & DISCUSSION

3.1 Phytoplankton and trace metals in Lake Vanda

The results from the December 1995 sampling programme indicated that the phytoplankton-rich layer just above the redox boundary in Lake Vanda appeared to be enriched in selected trace metals and oxide phases (Webster et al., 1996). However, when re-sampled on a finer scale (2m intervals) in January 1997, it became evident that trace metal enrichment in the suspended sediment and phytoplankton actually occurred somewhat below the zone of maximum phytoplankton activity at 62m depth (see Figure 1). The Co, Cu, Pb, Ag and Cd content of the suspended sediment for example, increased at the redox boundary, peaked at 66m, then decreased in the anoxic zone. Mn and Ni increased at the redox boundary and remained high in the anoxic zone, while Zn showed peaks at 66m as well as 70m. Enrichment in the anoxic zone, and the absence of a corresponding Fe peak, preclude an oxide association for the metals. There is, however, a dramatic increase in the total concentration of suspended sediments below 66m depth, indicating precipitation or flocculation of solids at this depth in the water column. H_2S was also detectable for the first time at 66m depth. This suggests that the enrichment noted for Cu, Pb, Ag and Cd may be due to formation of particulate sulfide phases at this depth; a reaction likely to be at least facilitated, if not induced, by sulfate-reducing bacteria active in the anoxic zone. The selective precipitation of the sulfides of these metals (compared to those of Fe & Zn for example) is consistent with MINTEQA2 (Allison et al, 1991) modelling of relative sulfide saturation indices. However, sulfide precipitation may also be a detoxification mechanism and hence reflect the relative toxicity of the metals to the bacteria present. At greater depths, the dilution effect of increasing concentrations of organic carbon, detrital silicate minerals and salts in the suspended sediment causes a decrease in suspended sediment trace metal concentrations.

3.2 Trace metal/mat interactions at sediment surfaces

Thick microbial mats, dominated by cyanobacteria, grow in the shallow waters at the margins of the Lake Vanda and the Onyx River. Dried mats also occur in the subaerial sediments of relict lake shorelines. A survey of 24 mats from the Onyx River,

Lake Vanda, and the flooded soils of the Vanda Station site has been completed. Mats collected from the station site appeared to be have slightly higher concentrations of Ag, Pb and Cd than those collected outside the disturbed area. This was particularly evident in mats collected from Greywater Gully, an area historically used for disposal of washing water from the base and where surface soils also contained relatively high concentrations of these metals (e.g. up to 23 mg/kg Pb). However, there is generally a high degree of variability in the trace metal concentration of the mats, both within and outside the station site area.

The trace metal concentrations of the mats were uniformly higher than those of the sediments beneath (Figure 2). From this observation alone, however, it remains unclear whether trace metals have been directly taken up by the phytoplankton or remain attached to inorganic oxide-rich sediments which have been incorporated in the mucilaginous matrix of the mat. Sediment typically comprises between 20 and 90% of the mat samples by weight, and can not be physically or chemically separated from the organic matter prior to digestion. However, the digestion method used dissolves only oxide and carbonate phases. It was noted that the Fe:Al ratio was similar in all the sediments and mats digested, which suggests that the mats do not selectively take up either Fe or Al. Hence the Fe concentration has been taken as indicative of the amount of metal-binding sedimentary oxide present in the mat samples, and mat metal concentrations are plotted versus Fe in Figure 2. A stronger digestion (HNO_3 + HCl) was used on the 1995 mat samples, than in 1997 (HNO_3 only). Generally the mats subjected to the stronger digest showed slightly higher Fe concentrations, suggesting that a portion of the crystalline as well as amorphous oxide phases had been dissolved.

There was a linear correlation between the proportion of Fe dissolved during the digest and Ni, Zn, Cu (and possibly Cd) concentrations in the mat, which suggests that these metals are bound to sediment Fe-oxide in the mat matrix. Other metals, such as Pb and Ag do not show a linear relationship with Fe, and further work is required to determine the way in which these metals are taken up and bound by cyanobacterial mats. Work by Tazaki et al (1997), on a mat collected from the Onyx River as part of this study, has identified precipitation as a sulphate salt (barite: $BaSO_4$) within the bacterial cells as an uptake/binding mechanism for barium. We hope to establish the viability of this mechanism for other metals in the near future.

4. ACKNOWLEDGEMENTS

This study was funded by FRST PGSF Research Contracts: CO 3507 and CO3106 in New Zealand. We would like to acknowledge the collaboration of Profs. Bill Fyfe and Kazue Tazaki, as well as Drs Ian Hawes, Julie Hall, Clive Howard-Williams in this research, and the logistic support of Antarctica NZ, the NZ Air Force and the US Navy VXE-6 squadron.

5. REFERENCES

Allison J.D., D.S. Brown & K.J. Novo-Gradac 1991. MINTEQA2/PRODEFA2, a geochemical computer assessment model for environmental systems. *EPA/600/3-91/021* USEPA Publication, Athens, Georgia.

Angino E.E. & K.B. Armitage 1963. A geochemical study of Lakes Bonney and Vanda, Victoria Land, Antarctica. *J. of Geology.* 71: 9-95.

Gadd G.M. 1992. Microbial Control of Heavy Metal Pollution. In Fry et al (eds), *Microbial Control of Pollution:* 89-112. Cambridge University Press, UK.

Green W.J., D.E. Canfield, Y. Shengsong, K.E. Chave, T.G. Ferdelman & G. Delanois 1993. Metal Transport and Release Processes in Lake Vanda: the Role of Oxide Phases. *Physical and Biogeochemical Processes in Antarctic Lakes*, Antarctic Research Series 59: 145-163.

Hawes I., C. Howard-Williams & R. Pridmore 1993. Environmental Control of Microbial Biomass in the onds of the McMurdo Ice Shelf, Antarctica. *Archiv fur Hydrobiologie* 3: 271-287.

Tazaki K., W.S. Fyfe, S. Iizumi, Y. Sampei, H. Watanabe, M. Goto, Y. Miyake & S. Noda 1994. Aerosol Nutrients for Arctic Ice Algae. *Nat. Geographic Res. & Explor.* 10: 116-131.

Tazaki K., J. Webster & W.S. Fyfe 1997. Transformation Processes of Microbial Barite to Sediments in Antarctica. *Japanese Mineralogical Journal* 12: 63-68.

Vincent W.F. & C.L. Vincent 1982. Factors Controlling Phytoplankton Production in Lake Vanda (77°S). *Can. J. Fisheries & Aquatic Sci.* 39: 1602-1609.

Webster J.G. 1994. Trace Metal Behaviour in Oxic and Anoxic Ca-Cl Brines of the Wright Valley Drainage, Antarctica. *Chemical Geology* 112: 255-274.

Webster J.G., W.S. Fyfe, K.S. Webster, I. Hawes 1996. Trace metal interaction with Antarctic microbial communities. *Proc. 4th Int. Symposium on the Earth's Surface*, Yorkshire, England.

Webster J.G., K.S. Webster, I. Hawes 1997 in press. Trace metal transport in Lake Wilson: a comparison with Lake Vanda. *Proc. Int. Workshop, Polar Desert Ecosystems*, Christchurch, NZ.

Influence of autochthonous microorganisms on the migration of redox-sensitive radionuclides

A. Winkler, T. Taute & A. Pekdeger
Institute for Environmental Geology, Free University of Berlin, Germany

I. Stroetmann & G. Maue
Department of Hygiene, Technical University of Berlin, Germany

ABSTRACT: The mobility of redox-sensitive radionuclides in soils and sediments is affected by biotic parameters. Microorganisms interact with the redox-sensitive elements Technetium (Tc) and Selenium (Se) by adsorption, reduction/oxidation, incorporation and complexation processes. To estimate the influence of autochthonous microorganisms, special procedures for drilling, sampling and carrying out the experiments were applied to avoid microbial contamination.

Sorption experiments with nonsterile surface material and allochthonous microorganisms showed a high immobilization of Tc and Se. This immobilization was not predictable by thermodynamic calculations with computer codes. Former investigations with high fixation of Tc and Se have been done under normal laboratory conditions. These experiments were carried out with groundwater avoiding any microbial contamination. It could be shown that the microbial population of the sediments was rather low and that microbial activity was not high enough for a significant fixation of Tc (colony forming units CFU = $< 10^6$).

1 INTRODUCTION

Technetium (atomic number 43, Tc-99 half-life is 2.13 10^5 years) is the lightest radioelement in the periodic system. Primordial Tc has decayed, but today Tc is produced in nuclear power plants (6% of the fission products). Tc occurs mainly in two oxidation states. Tc(VII) as the pertechnetate anion TcO_4^- is highly mobile and has been used as a tracer in hydrogeologic studies (Moser 1979). Under reducing conditions, Tc(IV) is chemically stable like $TcO(OH)_2$. Tc(IV) compounds have a very low solubility, they are more or less immobile in aqueous systems. Similar properties are known for Se which is mobile as selenate(VI) and selenite (IV) and insoluble as Se(0). The high mobility in oxidizing environments and a long half-life makes these two radionuclides objects of special concern in the term of high active waste (HAW) disposal.

The transport behavior of redox-sensitive elements such as Tc and Se is not only determined by the geo-chemical parameters which can be measured within the macro environment, but to a large extent by micro environments (local redox minimum zones) and microbial processes (Pignolet et al. 1986; Schulte & Scoppa 1987; Henrot 1989; Stroetmann, 1995). Reproducible investigations on the migration behaviour of Tc and Se showed a strong fixation in aerobic environments, which cannot be explained by taking thermodynamical and geochemical aspects into consideration (Winkler et al. 1995).

Microbial activity was detected at depth to 1200 m in crystalline rock with total counts of 10^5-10^7 bacteria/cm^2 (Pedersen 1993). Generally, migration experiments are carried out without considering the influence of microbiological activity. The purpose of these studies was to investigate whether the original indigenous microbial population is able to alter the migration behavior of redox-sensitive elements (like Tc and Se) in a similar way as the allochthonous.

2 METHODS

2.1 Sediments with autochthonous microorganisms

A sterile sampling technique was used to gain Tertiary and Quaternary sediments of various types for experiments with an unaltered autochthonous population. Dry drilling was used to exclude contamination by the drilling fluid. The drilling tools were treated with an open flame and the sampling PE tube was sterilized with formaldehyde gas. The first centimeters of the sediment at the ends of the 1 m long tubes were sampled with sterilized tools. The sediment free parts of the core were filled with sterile filtered inert gas (Ar) and sealed immediately. The tubes were stored in Ar filled PVC casings and sealed gas-tight for transport, storage (12°C) and preparation. The climatic conditions during sampling (5-8°C) made additional cooling at the sampling site unnecessary.

The bore hole was developed as a well and pumped out for some hours until ready for sampling. To keep the chemical alteration of the water at a minimum, an autoclaved stainless steel container and sterilized viton tubing was used. The containers were filled with an inert gas. To hinder the aquifer's gases from degassing, an overpressure equal to the hydrostatic pressure of the formation (2000 hPa) was applied on the containers.

Table 1. Chemical composition of the water used for the column experiments. All values in mg/l except for Cu, Cd, As and Hg (µg/l).

DOC	SO_4^{2-}	Cl^-	HS	NO_2	Na^+	K^+	Ca^{2+}
4,82	102	15	<0,1	<0,1	5,8	1,6	62

Mg^{2+}	Fe^{2+}	Mn^{2+}	Zn^{2+}	Cu^{2+}	Cd^{2+}	As^{2+}	Hg^{2+}
4,2	4,8	0,1	0,08	<0,1	<0,1	<0,1	<0,1

Apart from other sediments, a sand with a relatively high amount of organic carbon was chosen in order to have a sediment sample with a good nutrient supply for microbial life.

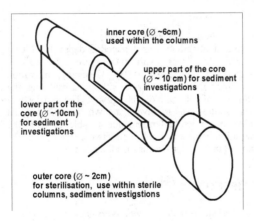

Fig. 1. Sample core splitting for the different purposes of investigation.

2.2 Sterilization and column preparation

The sediments used in the sterile columns (abiotic controls) were sterilized by γ-irradiation (Co-60, 35kGy). This proved to be the best method for sediments (regarding the effectiveness as well as the saving of the physical and chemical properties of the organic and inorganic components (Stroetmann et al. 1993). Other materials and apparatus were autoclaved or treated with formaldehyde gas. Because some samples are from anaerobic aquifers, the columns could not be filled in the sterile bench. The preparation of the columns occurred in an anaerobic glove box, which was sterilized by UV light for some days. Only the inner part of the sediment cores was used for the investigation of the microbial influence on the Tc and Se migration behavior (Fig. 1). The investigations were carried out at groundwater temperature of 9±2°C (Fig. 2) in order to maintain the same microbial population.

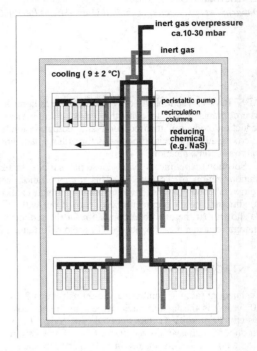

Fig. 2. Refrigerator device for the flow-through-column experiments with a double-inert gas supply system. The columns received an inert gas overpressure of max. 30 mbar (hPc). The boxes containing suction pumps had a constant gas flow.

When samples were taken for the measurements of the sterile columns, sterility was controlled. It could be shown, that the mobile microorganisms had little or no influence on the migration behavior. Therefore, it was necessary to examine the immobile microorganisms on the sediments. It is impossible to take soil or sediment samples from the recirculation columns without the risk of microbial contamination. Therefore, several small columns were built parallel with the same sediment to investigate the development of the immobile microbial population on the sediments within various time steps (Fig. 3).

2.3 Microbial investigations

A number of various microbial investigation techniques were applied: e.g., the counting method of bacteria with epifluorescence microscope using acridinorange (AO; 3,6 tetramethyldiaminoacridine; Fry, 1988). This method is principally suitable for the sediment investigations. Nowadays it is used for quantifying all cells, including non colony forming ones, because living and dead cells become the same color. The counting of the colony-forming units (CFU) is a way to determine the number of the reproducible microorganisms. Six different agars were applied for cultivation of aerobe microorganisms with different nutrient supply, two agars for facultative and obligate anaerobe microorganisms Stroetmann (1995).

experimental setup resulted in a good acceptance by the microorganisms. Compared with an investigation of the microbial community in the beginning, the original species composition remained identical for several (6-7) weeks. It is possible to maintain the physical and chemical parameters and to carry out sorption experiments in order to learn about the influence of the autochthonous microbial flora on the transport behavior of redox-sensitive elements. Sampling, transport and experiments could be done under the aspect of the feasibility of preserving in-situ conditions of an anaerobe aquifer for laboratory experiments with autochthonous microorganisms without a measurable contamination with allochthonous organisms.

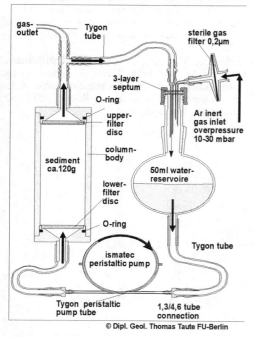

Fig. 3. Scheme of the flow through column devices

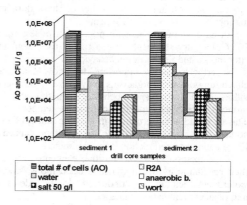

Fig. 4. Comparison of the results of the two cell counting methods AO and CFU per g (ds).

Calculations with the geochemical equilibration code PHREEQE using the CHEMVAL database show that 99% of the Tc is in the heptavalent oxidation state as TcO_4^-.

One problem of time-dependent investigations with parallel columns is a broader variation of the sorption values. A comparison of sterile and nonsterile experiments showed, that there was no significant Tc and Se fixation due to the activity of the autochthonous microorganisms. This is also true for some parameter variations with higher mineralized water (NaCl 50 g/l) and water with pH values of about 8.

The number of CFU in the nonsterile columns was always less than 10^6 CFU/g dry sediment.

3 RESULTS

Before the columns were filled, microbial investigations showed 10^7 cells/g of dry sediment (ds) with acridineorange counting (AO). The colony-forming units (CFU) showed up to some 10^5 cells/g dry sediment (Fig. 4). The difference of about two orders of magnitude between AO and CFU remaind over the whole investigation period.

It could be shown that the parameters of the

4 CONCLUSIONS AND DISCUSSION

The composition of the microbial flora in former investigations with allochthonous bacteria differed in each column. The new experiments showed a constant microbial community over a period of several weeks. There was no fixation of the redox-sensitive elements due to the activity of autochthonous bacteria.

The comparison of sterile and nonsterile recirculation column experiments showed no significant influence of autochthonous microbial activity on the sorption of redox-sensitive radionuclides.

It is not yet possible to determine whether autochthonous microorganisms are able to create reducing micro environments on particle surfaces. All the experiments with allochthonous bacteria have been done under laboratory conditions, e.g., 20°C. These conditions supported a higher metabolism and led to much higher CFU compared with groundwater conditions at about 10°C.

First experiments with a higher CFU of the autochthonous bacteria indicate the ability of these flora to immobilize redox-sensitive elements.

An autochthonous bacteria population of less than 10^6 CFU/g dry sediment proved to be insufficient to alter the Tc and Se migration behavior.

Nevertheless, it is important to know that the allochthonous microorganisms may have a stronger influence on the transport behavior than the autochthonous microorganisms at groundwater temperature. This may lead to uncertainties in the assessment of the contaminant migration because the aquifers' capacity for retention and retardation is overestimated.

If the fixation of redox-sensitive elements exceeds their prediction by thermodynamic calculations, sterile experiments should be conducted, to exclude any microbial contamination influence.

ACKNOWLEDGMENTS

Excellent technical support was given by J. Arnold, M. Jürgens, D. Majerczyk, and D. Lange. This study was financially supported by the German Ministry of Education and Research (BMBF), project numbers 02 E 8050 0, 02 E 8060 0 and 02 E 8060 A8.

REFERENCES

CHEMVAL (1994) CHEMVAL database, available via NEA (Nuclear Energy Agency, Paris).
Fry, C. J. 1988. Determination of Biomass . in B. Austin (ed), *Methods in Aquatic Bacteriology*, John Wiley & Sons Ltd.
Henrot, J. 1989. Bioaccumulation and chemical modification of Tc by soil bacteria. *Health Physics*, 57, 239.
Moser, H. 1979. Isotopenhydrologische Methoden zur Bestimmung der Durchlässigkeit des Grundwasserleiters. *Mitt.Ing.u.Hydrogeologie*, 8: 79-103, Aachen.
Pedersen, K. 1993. The deep subterranean biosphere. *Earth Science Reviews*, 34 S. 243- 260
Pignolet, L., K. Fonsny, F. Auvray & M. Cogneau 1986. Microbial action on technetium fixation on marine sediment. in T.H. Sibley & C. Myttenaere (eds), *Application of distribution coefficients to Radiological Assesment Models*: 361-370, London, New York: Elsevier Applied Sci.
Schulte, E.H. & P. Scoppa 1987. Sources and behaviour of technetium in the environment. *The Science of the total Environment*, 64: 163-179.
Stroetmann, I. 1995. Einfluß mikrobieller Aktivität auf das Migrationsverhalten redoxsensitiver Radionuklide (Technetium und Selen) in Lockergesteinen. Hygiene Berlin, Band 21; W. Dott & H. Rüden (eds).
Stroetmann, I., P.Kämpfer, & W.Dott, 1993. Untersuchungen zur Effizienz von Sterilisationsverfahren an unterschiedlichen Böden. *Zentralblatt für Hygiene und Umweltmedizin*. Vol.195; S.111-120.
Winkler, A., T. Taute, A. Pekdeger, I.Stroetmann, G.Maue, W. Dott 1995. Microbial alteration of the transport behaviour of redox sensitive radionuclides in sediments. *Proceedings WRI-8*: 277-280. Rotterdam: Balkema.

Phosphorus in soils and ground waters of the Indian Ocean atoll islands

P.V. Yelpatyevsky & T.N. Lutsenko
Pacific Institute of Geography, Vladivostok, Russia

ABSTRACT: The distribution of phosphorus has been studied in ground waters on the islands and adjacent shoals on the different-age atoll of the western part of the Indian Ocean. Waters on the islands contain from 10 to 100 times phosphorus than ocean waters. Water dissolved organic matter plays a big role in the mobilization of phosphorus in carbonate-calcium conditions. A discharge of the phosphorus-containing waters into adjacent shoals may increase the productivity of the atoll ecosystem.

1 INTRODUCTION

The atolls are a specific objects with respect to geochemistry because of the bedrock of the island ecosystems are not the usual substrates such as quartz, primary silicates, and clay minerals. These islands are formed by calcium carbonates from the remainders of limestone sea organism skeletons.

Small islands are always the place of nest colonies of marine fish-feeding birds. As a result of their vital activity, the transference of sea biomass from the adjacent sea on the islands occurs, and the mineral components are accumulated. In a wet environment, the most soluble products of vital activity of the colonies are washed out by infiltrate waters but calcium-carbonate substrate fixes phosphorus in the slightly soluble compounds. As a result, the soils and rocks, enriched with calcium phosphates, have been formed. They often use as a natural phosphate fertilizer (island guano) but the extend of deposits in general does not reach commercial value. Fundamental review of G.E. Hutchinson (1950) shows the wide spread of the phosphotization process within the coral islands. Phosphate soils and rocks has been established for the most Micronesian Islands by F.R.Fosberg (1957), and one should agree with the K.A.Rodgers (1994) opinion that phosphates are typical for the low islands of tropical ocean. From our point of view, these formations are at least of Holocene age, and should be considered as soils because they are a result of action of both: biotic (vegetation, ornithofauna) and climatic factors on the mineral substrate (carbonate rocks). At first sight the fact of the phosphorus accumulation in the soils and rocks does not presume its input into surrounding waters. Atoll ecosystems are highly productive, and the land pool of phosphorus can promote that. The phosphorus contribution into the productivity of the atoll ecosystem is possible if soluble phosphorus compounds are able to penetrate through the carbonate rocks up to the surface of the ground waters, and to migrate into surrounding waters. The goal of this study is to estimate phosphorus concentrations in the ground waters of the atoll islands, and to investigate the opportunities of their output beyond the island.

2 OBJECTS AND METHODS

The works were conducted on the islands of the western part of the Indian Ocean. The following low islands were examined : the northern island of the Poivre Atoll, the northern island of the Farghuar Atoll, some of the islands of the Cargados Carajos Atoll, and rising and litified the Assumption Atoll.

On the low islands, water samples were collected from freshly dug soil profile pits. On the Poivre and Farquhar from 6 to 8 soil profiles were located on transects which crossed the islands and reached the beach. Here during a low tide period the samples were also collected from holes up to 2 m in depth. The samples supposedly characterize a leakage of island fresh waters during the low tide. In addition, sea water samples were collected on the reef flat near the edge, and outside the atoll. Each studied small sandkay of the Cargados Carajos atoll was characterized by the only one profile and by one water sample. On the rising and cemented Assumption atoll, the water from karst wells at a depth of 5-6 m was sampled.

At the same day, as collected the water was filtered through 0.23 µm filters, and the phosphorus content was analyzed by colorimetry of blue form of phosphorus-molibdenum complex, with ascorbic acid as a reductant. Total soluble phosphorus was determined after boiling of the samples with potassium persulfate and sulfuric acid. To study the possible relationship of phosphorus with dissolved organic matter (DOM), we used gel-chromatography of one sample on sephadex G-15, and phosphorus was analyzed in separate eluate fractions.

3 RESULTS AND DISCUSSION

The more ancient part of the northern island of the Poivre atoll is formed by partly or completely cemented limestones. Phosphate-carbonate illuvial humus soil with a thickness of humus horizon up to 20 cm locates on the limestone surface. Dark-brown and coffee-brown limestones reach the level of the ground waters. C.G.Piggott (1968) describes these soils as "Jemo Series" after as R.F.Fosberg (1954) marked them. Phosphatizied soils with the horizon of raw acid humus on the hardpan surface have not observed on the Seychelle Islands. It is not clear whether the soils existed or not before the islands have became the coco-plantations. The phosphorus content reaches 12% in these soils, sometimes the high levels of phosphorus in phosphate-carbonate material are traced from the depth of 1.4 -2 m up to ground water level.

The humus-carbonate soils are the younger formations, and they consist of a carbonate sand with low humus content in the upper part. The phosphorus levels in the material are within 0.63 - 0.03 %. The water analysis shows that they are fresh. Chloride contents do not reach 100 mg/l even at a distance of 100 m from the beach. The concentration of phosphorus in soils is different. Under phosphate-carbonate soils, it is 2-3 times higher than those under non-phosphate soils (Fig. 1, Table 1). The variations for soluble phosphorus are less contrast than for the soils. The phosphorus washed out from phosphate-carbonate soils, usually diffuses in ground waters in much larger area than the soil area. Thus, the high level of available nutrient elements can be supported in a large part of the plantation. In the beach hole during the low tide period, the phosphorus concentrations are reduced in high brackish waters with chloride level 6.4 g/l, but they are essentially higher than those for sea water (Table 1).

Table 1. Concentrations of phosphorus (µg/l) and chloride (g/l) in island and sea waters.

No samp.	Place of sampling	P	Cl⁻
	The Poivre atoll		
1	Humus-carbonate soil	570	0,013
2	Phosphate-carbonate soil	1380	0,033
3	The same	1510	0,10
4	Humus-carbonate soil	590	0,087
5	Ground waters on a beach during the low tide	41	6,4
6	Coastal waters of reef flat during the low tide	28	n.an.*
7	The same in bushes of *Talassia*	37	-'-
8	Sea water at the reef edge	14	-"-
9	Sea water mile away from shore	9	-"-
	The Farghuar atoll		
10	Humus-carbonate soil	1350	7,2
11	Phosphate-carbonate soil	1680	2,8
12	Humus-carbonate soil	2020	7,7
13	Phosphate-carbonate soil	1290	3,0
14	Humus-carbonate soil	48	0,62
15	Ground waters on a beach	76	n. an.
16	Lagoon waters near the shore during the low tide	35	-"-
17	Lagoon waters 150 m from the shore	15	-"-

* - not analyzed

The decline of phosphorus concentrations goes non-proportionally to dilution by sea water but to a larger extent. The phosphorus is partly removing from water migration on the way from the sources of the input up to discharge into coastal waters.

During the low tide, the coastal sea water keeps a relatively high phosphorus content and in thickets of marine grass *Thallassia* which is greatly developed on the reef flat near the shore. The phosphorus content here rises possibly due to metabolizm secretions, and the decrease of water exchange speed. At the reef edge, the phosphorus concentrations are still higher than those in the open sea.

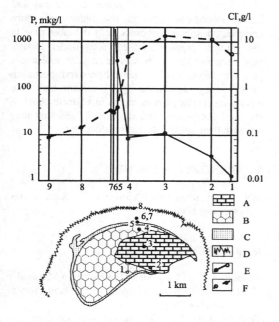

Fig.1. Content of phosphorus and chloride in soil and coastal sea waters in profile across the northern island of Poivre atoll (soil contours in accordance with Piggott, 1968, with author additions). A - phosphate-carbonate soils; B - humus-carbonate soils; C - beach sands; D - coral reef; E - plot of phosphorus content; F - plot of chloride content. Numbers of samples are the same as in Table 1.

Similar studies were conducted on the northern island of the Farquhar atoll which is formed by narrow (0.3 - 0.6 km) accumulations of carbonate material. A small width of the island is an obstacle to the accumulation of fresh water, and as a result the ground waters are brackish (Table 1). The position of the atoll in the Indian Ocean arid zone where the annual sum of atmospheric precipitation is about 1000 mm (Stoddart & Walsh, 1979), also promotes that. The ornithogenous phosphate-carbonate soils locate in the central part of the northern island and fixe the oldest section of the island. Inside a contour of this soils, carbonate material is cemented in a different extent. This shows the various ages of the accumulation. The phosphorus content in the soils is about 10-12 %. As in the previous case, the highest levels of phosphorus were determined under ornithogenous soils. However, the areas with waters enriched with phosphorus exceeds the area of these soils. Waters sampled at the beach of the atoll lagoon, had the oceanic salinity but the phosphorus content remained rather high (Table 1). The phosphorus level in the lagoon water near the shore was as twice lower. In the atoll lagoon, due to the lower speed of the water exchange than in open sea, the phosphorus content is reduced but do not reach the levels for open oligotrophic waters of the tropical ocean.

The waters under the phosphate-carbonate soils and sometimes under the humus-carbonate (non-phosphotized) soils are yellow on both of the islands. This shows the mobility of soil humus, and allows to suggest the phosphorus migration in the composition of organo-mineral complexes. We examined this assumption. One sample was concentrated by several times using the freezing and separated by gel-chromatography on sephadex G-15. In eluate fractions, an absorbance at 420 nm as an index of organic matter content, electroconductivity as an index of electrolyte concentration, and the phosphorus content after decomposition of organic matter by potassium persulphate were determined (Fig. 2).

Dissolved organic matter was sub-divided into three fractions. These are: 1) high-molecular compounds, 2) mid-moleculur compounds , 3) low-molecular compounds going next to electrolyte peak. Phosphorus is washed out only in three successive fractions of mid-molecular DOM. Phosphorus recovery was 86.5% and 58 % of it was removed in the only eluate fraction. There was no phosphorus in fractions containing the main amount of salts. The distribution received may be considered as a witness of the phosphorus relationship with DOM with a molecular weight of several thousands.

There are many karst wells in the completely lithified limestones of the Assumption Island. The island ground waters are characterized by a high chlorinity, which does not depend on the distance from the shore. The average chloride content is 14.6 ± 0.9 g/l (7 samples). The phosphorus content is rather high but monotonous and constitute 145 ± 29 µg/l. The similarity of hydrochemical data is the result of the karst development, and as consequence, an easy water exchange takes place. In spite of complete lithofication of the island and transformation of phosphorus-content components into less soluble compounds, the phosphorus inputs to waters in essential amounts but they are of one order of magnitude lower than those under the younger carbonate-phosphate soils.

On the young islands, the process of phosphate accumulation has only begun. Such islands are not numerous, and some of the islands of the Cargados-Carajos Atoll are an example. The Coco Island is a

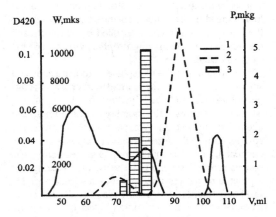

Fig. 2. Distribution of phosphorus and chloride in water under the phosphate-carbonate soil in gel-filtration on sephadex G-15 (the Island Farquar, sample number 11). 1 - absorbance curve - index of organic substances; 2 - conductivity curve - index of electrolytes content; 3 - phosphorus content in separate eluate fractions.

sandkey with the elevation above tidal level about 1.5-2 m. The is island occupied by a nest colony of *Sterna fuscata* with density nest about 1 on 1-2 m^2, and there are *Gigis alba* nest on rare bushes of *Turnefortia argenta*. Humus-carbonate soils on this island are not formed yet, and the indications of soil-formation are only observed up to depth of 30 sm. At a depth of 1.2 m, ground waters have a salinity about normal oceanic but phosphorus contents reach the level of 110 µg/l.

The present density of the colony is much less on other island (in the Cocos Island). Nevertheless, the highly fresh waters with a chloride level of 1.5 g/l contain 235 µg/l of phosphorus. Perhaps it is due to a higher density of nest colonies in the past.

4 CONCLUSION

The phosphorus, accumulates on the islands from the mineral part of the sea biomass on a small area, is included into insoil migration in spite of the richness of the substrate by calcium. In the absence of a bird population, the power of vertical and lateral migrations flow in the older and more complex phosphate-carbonate soils is higher than in the younger humus-carbonate soils under today's nest colonies. The water-soluble organic substances play a mobilizing role in phosphorus transportation under the conditions of calcium-carbonate environment. Phosphorus inputs in adjacent shoal of the atoll ecosystem from the island, and that may increase their eutrophity and productivity.

5 ACKNOWLEDGEMENTS

We would thank Dr. Yaroslav V. Kuzmin for polishing our English.

6 REFERENCES

Fosberg, R.F. 1954. Soils of the Northern Marshall Atolls, with special reference to the Jamo Series. *Soil Science*.78:99-107.
Fosberg, R.F. 1957. Description and occurrence of atoll phosphate rock in Micronesia. *Amer. J. Sci.* 255:584-592.
Hutchinson, G.E. 1950. Survey of existing knowledge of biochemistry: 3. The biochemistry of vertebrate excretion. *Bull. Amer. Mus. Nat. Hist.* 96:1-554.
Piggott C.J. 1968. A Soil Survey of Seychelles. *Technical Bull. Ministry of Overseas Develop.* England. 2:1-87.
Rodgers R.A. 1994. The geochemical role of atoll phosphates on Vaitupu, Nuculaelae, Funafuti and neighboring low islands of Tuvalu, central Pacific. *Mineral. Deposita.* 29:60-68.
Stoddart D.R. & R.P.D. Walsh. 1979. Long-term climatic change in the Western Indian Ocean. *Phil. Trans. R. Soc. London.* 286B:11-23.

Carboxylates in fluid-inclusions in minerals

Yishan Zeng
Department of Geology, Peking University, Beijing, People's Republic of China

Jiaqi Liu
Yichang Institute of Geology and Mineral Resource, CAGS, Hubei, People's Republic of China

ABSTRACT: The carboxylate (formate, acetate, propionate and oxalate) compositions for fluid inclusion leachates from minerals collected from various hydrothermal mineral deposits were determined using ion chromatograph. The minerals studied include quartz, fluorite, barite and a few ore minerals occurred in tungsten, gold and some non-metal mineral deposits. The analysis results showed that short-chain carboxylic acids or carboxylates are common components in hydrothermal ore-forming fluids.

1 INTRODUCTION

It is well known that aqueous carboxylic acids, carbo-xylate ions and metal-carboxylate complexes commonly participate in geochemical processes in sedimentary basin, hydrothermal system and other geological environments. Recently, geochemists has paid attention to the role of carboxylic acids in hydrothermal ore-forming processes. Giordano (1994) reviewed the advances in this field.

Analysis for the aqueous organic compositions of inclusions in minerals will give us some direct evidences on the presence of carboxylates in ore-forming fluids. However, up to now, only a little study result about this aspect is reported in literature. For example, Germanov and Mel'kanovitskaya analyzed leachates of ore-rocks samples (e.g., sample A consists of quartz, sphalerite, galena and other minerals) from polymetallic deposits at the North Caucaso, Pre-USSR by gas chromatography, and detected monofunctional carboxylates ($C_1 - C_6$). They thought that these carboxylates were residue of hydrothermal solution trapped fluid inclusions in minerals. R. McLimans (1977) reported the detection of aqueous organic compounds with molecular weights of 15-80 in fluid inclusions from the Wisconsin Pb-Zn district, USA (see Giordano, 1994).

The present study is aimed at examination of the carboxylic acid or carboxylate compositions of fluid inclusion in minerals by analyzing leachates from mineral-samples from various hydrothermal ore deposits.

2 GEOLOGICAL DESCRIPTION OF SAMPLES

We collected samples from some hydrothermal tungsten, gold and other mineral deposits in China. The majority of samples was gangue minerals, and some associated ore minerals also were studied. The general geological characteristics of these samples are briefly described as follows.

2.1 Samples from tungsten deposits

The Xihuashan tungsten-mineralized field locates in the south part of Jiangxi Province, China. We studied quartz and wolframite samples of different mineralization stage in the Xihuashan-Dangping beryl-wolframite deposits. These quartz-type deposits occur in a biotite-granite intrusive body. The samples from this deposit are:

Sample D-30 and D-30A (quartz) was selected from an early-stage pegmatoid vein. The two sampling points are less than 50 m apart. Fluid inclusions in these minerals are mostly 3-5 μm in diameter and the vapor phase occupies of 10-15 vol.% of the total volume of inclusion at room temperature (V/V+L=10-15%). Sample D-7 (quartz) was from a beryl-wolframite-molebdenite quartz vein. The fluid inclusions have a size of 10-15 μm and V/V+L of 10-15%. Secondary inclusions (T_h=100-150 °C) in a considerable quantity are pre-sent in this sample. D-27 (quartz) was collected from a late-stage barren quartz vein. In this mineral the diameters of most inclusions are 2-5 μm and V/V+L are lower than 5%.

We specially studied samples from single crystal of quartz in druse cavity occurring within mica-rich greisen nearby wolframite-bearing quartz vein. Sample D-6q was from a piece of a quartz crystal. Two-phase fluid inclusions in it are relatively large (mostly 10-20 μm) and commonly have V/V+L of 10-40%. The tabular wolframite crystals intergrowing with the quartz also were studied (sample D-6w). Sample D-33-3 and D-33-6 were from different parts of a large rock crystal (30 cm in length). A few two-phase fluid inclusions (V/V+L=10-30%) of tens micrometers in diameter were observed in the crystal.

2.2 Quartz samples from gold deposits

We studied mineral samples from three hydrothermal quartz vein-type gold deposits in the northwest part of Hebei Province, China.

The Dongping sulfide-free type gold deposits occur within a syenite complex Sample 15 is a piece of milk-white massive quartz from an auriferous quartz vein (the main lode). V/V+L of most of inclusions (5-10 μm in diameter) are 15-20%, and some multiphase (aqueous phase-liquid CO_2-vapor) fluid inclusions were observed in the mineral. Sample Z-1 (quartz) was from a late-stage milk-white barren quartz vein containing barite-chalcedony nodule in the Xiaoyingpan gold mine. Most of inclusions (3-8 μm) in the quartz are liquid ones and some inclusions have V/V+L of 5-10%. We selected two samples from an auriferous quartz vein containing sulfides (pyrite, galena, chalcopyrite, sphalerite and so on) at the Zhang-quanzhuang gold mine, associated quartz (ZQZ-q) and pyrite (ZQZ-p). More than 90% inclusions (3-5 μm in diameter) in quartz are liquid ones (V/V+L <5%). Both the Xiaoyingpan and Zhangquanzhuang gold deposits occur in Archeozoic metamorphic rock series.

2.3 Samples from other mineral deposits

The Lanying epithermal fluorite deposit located near Beijing. Fluorite veins in a considerable amount occur in Jurassic volcanic rocks and quartz-syenite intrusive body. Some galena and quartz are associated with fluorite in these veins. Sample 13 was a pure greenish fluorite crystal. Inclusions (3-15 μm) in the mineral have V/V+L of less than 5-10%.

The Shangpingshan vein-type barite deposits locate in the northeast part of Hebei Province and occur in Cambrian limestone strata. There are some galena, fluorite, sphalerite and quartz in barite veins. Sample S-5 is an early-stage barite. Liquid, vapor-liquid (V/V+L=0-80%) and multiphase inclusions containing daughter mineral (NaCl) were observed in this sample. Sample S-7 is a relatively late-stage barite, and inclusions (mostly 5-25 μm) in the mineral have V/V+L of 0-90%. The transparent cubic crystals of fluorite (S-6) usually occur within druse cavity in barite veins and they are the last minerals in the deposit. Inclusions in the fluorite are relatively small (5-10 μm), their V/V+L are less than 10%. Sample S-8 is galena crystal associated with barite (S-5). All these samples were collected from the same mining area of this deposit.

3 LEACHING EXPERIMENTS

Detail petrographic and microthermometric studies of mineral samples were performed before leaching experiments.

First, the mineral samples were crushed and sieved. Mineral grains (0.5-1.2 mm) were purified by hand-picking under a binocular microscope (mineral purity is higher 99%). In this study we used the method roughly similar to that described by Channer and Spooner (1992). The standard procedure is: 5 g mineral sample was put in 10 ml high pure H_2O_2 and heated at 75 °C for 12 hours. Ore minerals such as galena, pyrite and wolframite were not treated by hot H_2O_2 to avoid possible oxidation instead of soaking these minerals in deionized water at ambient conditions for 3 days. Then samples were rinsed with deionized water until the conductivity of rinsing solution was less than 1 μs/cm and dried at 50-60 °C. The primary aim of cleaning is the removal of any organic contaminants of mineral samples. Hand crushing cleaned mineral grains in an agate mortar by agate pestle and transferring the crushed mineral powder into a glass bottle containing 10 ml H_2O (conductivity ranged from 0.6 to 0.8 μs/cm), then the contents in the bottle were vibrated in a ultrasonic wave cleaning cell for about 10 minutes. The resulting leachates were filtered by 0.2 μm filter, and then analyzed using DX-2000i/sp ion chromatograph linked to a GC-6 integrator as soon as possible (IonPac AS11 anion exchange column, NaOH eluant-solution, H_2SO_4 regenereant-solu-

Table 1 Carboxylates in fluid inclusion leachates from minerals (μg/ml)

Sammple No.	Mineral	T_h(℃)	Salinity	Formate	Acetate	Propionate	Oxalate
D-30	Quartz	360-300	10	1.18	0.20	-	0.14
D-30A	Quartz			0.76	0.81	-	0.13
D-7	Quartz	340-240	7-9	0.29	0.15	-	0.02
	Quartz*			0.05	0.15	-	0.05
D-27	Quartz	200-150	3-5	2.03	0.65	0.08	0.13
D-6q	Quartz	310-150	3-8	1.25	0.45	-	0.04
	Quartz*			0.07	0.47	0.23	0.09
D-6w	Wolframite			0.67	0.38	-	0.17
	Wolframite*			0.24	0.30	-	0.26
D-33-3	Quartz	305-170	2.5-8	1.02	0.47	-	0.03
	Quartz*			0.05	0.50	-	0.10
D-33-6	Quartz	310-170	3-9	1.29	1.01	-	0.04
	Quartz*			0.42	0.51	-	-
15	Quartz	360-285	6-7	0.27	-	-	0.04
	Quartz*			0.17	1.73	0.76	0.19
Z-1	Quartz	220-180	10-11	0.65	0.28	-	0.02
	Quartz*			-	0.46	-	0.08
ZQZ-q	Quartz	230-195	8-10	0.86	-	-	0.01
ZQZ-p	Pyrite			0.46	-	-	0.06
13	Fluorite	150-130	7	1.28	-	0.09	0.01
	Fluorite*			-	0.34	0.31	-
	Fluorite*			0.66	-	0.23	0.20
S-5	Barite	320-90	12-13	0.59	0.17	-	0.03
	Barite*			0.12	-	0.15	0.22
S-7	Barite	270-200	9-13	0.04	-	-	0.08
	Barite*			-	0.16	-	-
S-6	Fluorite	120-90	6-8	1.54	-	0.21	0.01
S-8	Galena			0.75	0.08	-	0.17
	Galena*			0.35	0.24	-	0.25

Notes: Leachates — 5 g mineral + 10 ml water;
  "*"— the sample was analyzed by exclusion ion chromatography (DX-100 ion chromatograph and AS4A-SC anion exchange column);
  "-"— lower than the detection limit;
  T_h — homogenization temperature (℃);
  Salinity — weight equivalent percent of NaCl.

tion). In these experiments all containers contacting with sample and leachates were also treated by H_2O_2 before use.

To check the possible contamination during leaching runs, blank test were carried out. We also performed a few parallel leaching runs for some mineral samples. The determination results for carboxylate compositions in leachates and homogenization temperature of primary inclusions in mineral are listed in Table 1.

4 GEOLOGICAL IMPLICATIONS

With a few exceptions, little secondary inclusions were observed in samples in the present study. Roedder (1979) indicated that primary aqueous inclusions in gangue minerals are generally thought to represent actual preserved samples of ore-forming solutions.

Our mineral samples were collected from mineral deposits occurred in a variety of geological environments and some of them were from different mineralized stages in a deposit. From the semiquantitative data for the carboxylate composition presented in Table 1 it can be concluded that short-chain carboxylic acids or carboxylates are common components in ore-forming fluids. It is worth mentioning that the determined homogenization temperatures of primary inclusions in minerals fall into a wide temperature range (see Table 1). These data indicated that the carboxylic acids in hydrothermal fluids are able to exist even at temperatures higher than 300 °C. The theoretical calculation (see Shock, 1995) and laboratory experiments (e.g., Kawamura et al 1986) has indicated that low molecular weight organic acids can persist at elevated pressures and temperatures. It seems to be necessary that to re-evaluate and fur-ther study the role of carboxylic acids in hydrothermal ore formation.

This project was supported by the National Natural Science Foundation of China (No. 49673193).

REFERENCES

Channer, D.M.& E.T.C. Spooner 1992. Analysis of fluid inclusion leachates from quartz by ion chromatography. *Geochim. Cosmochim. Acta.* 56: 249-259.

Germanov, A.I. & S.G. Mel'kanovitskaya 1975. Organic acids in hydrothermal ores of polymetallic deposits and in groundwater of sedimentary cover. *Dokl. Akad. Nauk, SSSR.* 225: 192-195 (in Russian).

Giordano, T.H. 1994. Metal transport in ore fluids by organic ligand complexation. In E.D. Pittman & M D. Lewan (eds.), *Organic acids in geological processes.* Verlag Berlin Heidelberg.

Kawamura, K., Tannenbaum, E., Huizinga, B.J. & Kaplan, I.R. 1986 Volatile organic acids generated from kerogen during laboratory heating. *Geochem. J.* 20: 51-59.

Roedder, E. 1979. Fluid inclusions as samples of ore fluids. In H.L. Barnes (ed.) *Geochemistry of hydrothermal deposits.* Wiley, New York.

Shock, E.L. 1995. Organic acids in hydrothermal solutions: standard molal thermodynamic properties of caroxylic acids and estimates of dissociat ion constants at high temperature and pressures. *Amer. Sci. J.* 295: 496-580.

3 Groundwater quality

Halogen geochemistry of a Middle Jurassic calcareous aquifer in northern France

F. Barbecot, C. Marlin & L. Dever
Laboratoire d'Hydrologie et de Géochimie Isotopique, Université Paris-Sud, France

E. Gibert
Laboratoire d'Hydrologie et de Géochimie Isotopique, Université Paris-Sud, France (Presently: IAEA, Isotope Hydrology Section, Vienna, Austria)

ABSTRACT: The primary source of F and Br in groundwater from the Middle Jurassic aquifer in Caen area (northern France) is identified to come from water-rock interaction. The increasing content of F along the flow path is explained by cation exchange with clay minerals which induces a loss of Ca. Concurrently with the water-carbonate interaction indicated by the increase of the ^{13}C content of TDIC, the Br/Cl weight ratio of most groundwaters also increases downflow involving a Br source within the water-bearing rock. However, the high Cl content and Br/Cl weight ratio locally observed for some groundwaters at the confined-unconfined limit of the aquifer imply a direct recharge by marine water through Flandrian organic-rich layers, during a high sea level period during the Late Quaternary.

1 INTRODUCTION

The aquifer of the Middle Jurassic constitutes the main aquifer for water supply in the Caen area and is located along the french Channel coast. Downgradient, this aquifer, once confined, is characterized by an increase in the groundwater salinity, which limits the pumping to the unconfined or slightly confined water-bearing formation.

Composed of Middle Jurassic carbonates and marls belonging to the western edge of the Paris Basin, the sedimentary layers thicken and dip down around 1° towards the North-East (Rioult *et al*, 1986). The water flow follows the decreasing gradient which is in agreement with the slope of the sedimentary series (Pascaud *et al*, 1991).

The main aquifer in the study area consists of Middle Jurassic marine deposits. This carbonate multi-layer system is composed of the Bajocian formations (calcareous and oolitic formations of 15 to 20 m thick) and by the Bathonian formations (marly and calcareous layers reaching 100 m in thickness). Eastward, the Dogger aquifer becomes confined under the Callovian marls, with a thickness up to more than 110 m, and under the Flandrian deposits, which is an alternation of peats and marine clays. Locally, the Flandrian deposits directly overlay the Bathonian formations because of the erosion of the Callovian marls. Therefore, this feature may have allowed a direct recharge of the Bathonian aquifer by marine water during high sea level periods and during deposition of the Flandrian Formation. Today, in this region, the annual precipitation averages 710±10 mm falling year round (Data from Meteo-France).

In order to discuss the origin of Cl, Br, F in groundwater, eighteen samples have been sampled along a flow line and analyzed for their chemical compositions (Br, F and major elements) and isotopic content ($\delta^{18}O$, δ^2H, $\delta^{13}C$, $A^{14}C$).

2 DISTRIBUTION AND ORIGIN OF HALIDES

The Middle Jurassic aquifer contains freshwater in the recharge area (TDS less than 750 mg.L^{-1}). The groundwater becomes brackish along major flow paths (TDS up to 1960 mg.L^{-1}). The geochemical facies of groundwater evolves from a Ca-HCO$_3$ type to a Na-HCO$_3$ and Na-Cl/HCO$_3$ type. Three main processes have been proposed to explain such an evolution : (1) a mixing with sea water (up to 4%) (2) precipitation/dissolution reactions with the matrix of the aquifer and (3) cation exchange with clay minerals from the aquifer and the upper formation (Barbecot *et al.*, 1997).

2.1 Fluoride

As soon as the aquifer becomes confined, the fluoride content of groundwater exceeds 2 mg.L^{-1} and reaches 9 mg.L^{-1} in deeper parts of the aquifer. The high F concentration is associated with groundwater characterized by a Na-HCO$_3$ to Na-Cl/HCO$_3$ geochemical facies. As the content of F gradually rises with increasing distance from the outcrop, two main primary sources of F can be invoked (Travi, 1988) : (1) incongruent dissolution of fluorapatite and (2) leaching of marine clay. As the dissolved F content is inversely correlated with the Ca content, the F concentration appears to be controlled by minerals which contain Ca such as fluorite (Fig. 1). Moreover, most groundwaters (F content $\geq$ 0.5 mg.L^{-1}) reach saturation with respect to fluorite. (Plummer et al, 1976).

The groundwaters coming from the unconfined and from the slightly confined part of the aquifer do not seem to be disturbed by any cation exchange. This implies a Ca-Na steady state between water and clay minerals due to a high renewal rate for these groundwaters (Barbecot et al., 1997).

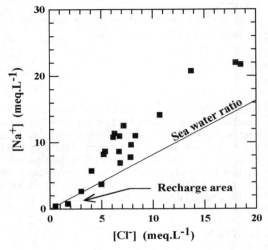

Fig. 2 : Na versus Cl concentrations.

2.2 ^{13}C content of TDIC

Along the main flow path, the groundwater is continuously at saturation with respect to calcite. Concurrently with the loss of Ca due to cation exchange, the pH (from 6.8 to 9.1) and total dissolved inorganic carbon contents (from 320 to 560 mg.L^{-1}) increase due to carbonate mineral dissolution to maintain the geochemical equilibrium with respect to calcite. The enrichment in ^{13}C of total dissolved inorganic carbon (TDIC) is explained both by dissolution of carbonates from the aquifer and carbon exchange between the groundwater and the water-bearing rock through time. Then, the Na/(Ca+Mg) ratio increases with an enrichment in ^{13}C, as shown in figure 3. This reflects the two types of water-rock interactions dominating the geochemical evolution of the groundwater. However, the evolution of the Na/(Ca+Mg) ratio can not be due to a mixing with sea water as the groundwater with the higher Cl contents does not have the highest Na/(Ca+Mg) ratio.

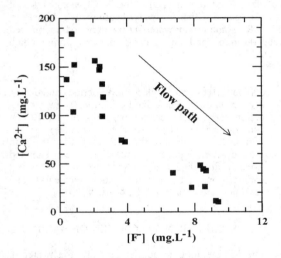

Fig. 1 : Calcium versus fluoride contents.

As the F and Ca contents are controlled by the saturation with respect to flurorite and calcite respectively, the concentrations of Ca, are likely to decrease downgradient by cation exchange on phyllosilicates. Indeed, the aquifer is made up of an intercalation of pure Ca-carbonate layers and calcaro-marly layers. This is in agreement with an excess of Na with respect to a sea-water dilution line (Fig. 2) and with the relative mobility between Na and Ca, Na being known to be more mobile than Ca under such a system (Petruzzelli & Lopez, 1993; Walraevens & Cardenal, 1994).

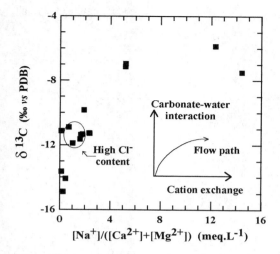
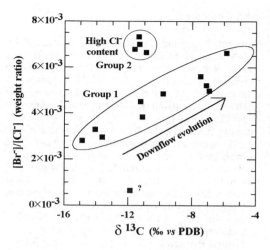

Fig. 3 : δ^{13}C of TDIC *versus* [Na$^+$]/([Ca^{2+}] + [Mg^{2+}]).

Fig. 4 : Br/Cl weight ratio *versus* δ^{13}C of TDIC.

2.3 Bromide

Unlike the F, as the Br content is not controlled by a equilibrium with a mineral phase, it can be used as a sharp tracer of water sources (Edmunds, 1996).

The Br/Cl weight ratio of groundwater varies between 2.8 x 10^{-3} to 7.6 x 10^{-3} from the recharge area to the deep part of the aquifer. However, this variation does not follow the increasing distance from the outcrop. It reaches a maximum when the aquifer is slightly confined under the discontinuous Callovian marls and the Flandrian organic-rich deposits. The evolution of Br/Cl as a function of δ^{13}C characterizes two groups of groundwater (Fig. 4).

The group 1 indicates a good relationship between δ^{13}C and Br/Cl. The bromide enrichment can be ascribed to the interaction with the aquifer along the flow path (1) by a mixing or leaching of a Br-rich water trapped in the aquifer or (2) by dissolved organic matter decomposition (Murphy & Davis, 1989).

Located in the aquifer part slightly confined under the Flandrian peat, the second group corresponds to the highest Br/Cl weight ratio (up to 7.6 x 10^{-3}) and chloride content (up to 660 mg.L^{-1}). The chemical evolution of those waters follows a mixing with a marine water enriched in Br. This local signature in bromide and chloride is induced by a direct recharge to the aquifer, during high sea level, through Flandrian peats. Indeed, the actual outcrop altitude is too high to be overlapped by the last sea transgression.

Some Flandrian peat leaching experiments have confirmed the possible supply of enriched Br/Cl water by such a process leading to a Br/Cl ratio at least ten times higher than the marine one (see also Worden, 1996).

3 CONCLUSION

A comparison of the halogens distribution and ratios in the Bathonian-Bajocian aquifer indicates the main factors controlling the geochemical evolution. The fluoride increase along the flow path is allowed by the loss of calcium occurring by 2Na/Ca exchange with the matrix in a system controlled by saturation with respect to fluorite and calcite. Whereas, the bromide content increases continuously in the groundwater by water-rock interaction. However, locally, the high Br/Cl ratio and high Cl concentrations of some groundwater, resulting from a mixing with sea water, imply an infiltration of marine water through Flandrian Br-rich peats during the last transgression.

ACKNOWLEDGEMENTS

This work is supported the EC project, PALAEAUX (program No ENV4-CT95-0156). The *Laboratoire d'Hydrogéologie* from the University of Avignon is gratefully ackownledged for performing chemical analyses.

REFERENCES

Barbecot, F; Marlin, C., Gibert, E., Dever, L. 1997. Geochemical evolution of a coastal aquifer to a Holocene sea-water intrusion (Dogger aquifer in northern France). *IAEA-SM-349/15*. In press.

Edmunds, W. M. 1996. Bromine geochemistry of British groundwaters. *Mineralogical Magazine* N° 60: 275-284.

Murphy, E. M., Davis, S.N. 1989. Characterization and Isotopic composition of Organic and Inorganic Carbon in the Milk River Aquifer. *Water resources research* 25 (8): 1893-1905.

Pascaud, P., Mauger, D. 1991. Carte hydrogéologique du département du Calvados. Caen, (1/100 00). Orléans : BRGM (eds).

Petruzzelli, D., Lopez, A. 1993. Solid-Phase Characteristics and Ion Exchange Phenomena in Natural Permeable Media. *Migration and Fate of Pollitanys in Soils and Subsoils*. D. a. H. Petruzzelli (eds), F. G. NATO ASI Series, Vol. G 32 : 75-91.

Plummer, L. N., Jones, D. F., Truesdell, A.H., 1976, WateqF, a Fortran IV version of WateqQ, a computer program for calculating equilibrium of natural waters, *U.S. Geochimical Survey Water Res. Invest.* 76(13) : pp 61.

Rioult, M., Coutard, J.P., Helluin, M., Pellerin, J.,Quinejure-Helluin, E., Larsonneur, C., Alain, Y. (1986). *Carte géol. de France*, (1/50 000), feuille de Caen et notice. Orléans : BRGM (eds) : pp 104.

Travi, Y. (1988). Hydrochimie et hydrologie isotopique des aquifères fluorurés du bassin du Sénégal. Origine et conditions de transport du fluor dans les eaux souterraines., *Thesis*, Paris-Sud University : pp 197.

Walraevens, K., Cardenal, J. 1994. Aquifer recharge and exchangeable cations in a Tertiary clay layer (Bartonian clay, Flanders-Belgium). *VM Goldschmidt Conference report*, Edinburgh, Mineralogical Society (London,eds) : 955-957.

Worden, R. H. 1996. Controls on halogen concentrations in sedimentary formation waters. *Mineralogical Magazine* 60: 259-274.

Modelling of redox conditions and control of trace elements in clayey groundwaters

C. Beaucaire, H. Pitsch & C. Boursat
Commissariat à l'Energie Atomique, DCC/DESD/SESD/LIRE, Saclay, Gif sur Yvette, France

ABSTRACT: The study of the belgian Boom clay and its surrounding aquifers shows that the fluid phase speciation, focused on the electroactive constituents Fe, Mn, S, and the accurate knowledge of the mineralogy of the host-rock may assess the main processes of redox control in natural clayey groundwaters. Speciation of other redox sensitive elements (U, Co, Zn) allows identification of mineralogical regulation by oxides or sulfides.

1 INTRODUCTION

The redox conditions of groundwaters are important for the safety assessment of nuclear waste repositories. Speciation, solubility and therefore mobility of several radionuclides depend directly on redox conditions.

The definition of the redox conditions in a natural environment depends on the reactions that may take place between selected host-rock constituents (pyrite, oxide, organic matter...) and the solution. These different components of the rock constitute what Scoot and Morgan (1990) named its oxidative and reductive capacities. The direct measurement of the redox potential is difficult to perform and both the experimental realisation and the signification in terms of equilibria are often subject to discussion. An alternative is to evaluate the redox potential from the existing electro-active species pairs in the water phase or from heterogeneous equilibria between the minerals and the solution.

This work is based on experimental data concerning the Boom Clay formation in north-eastern Belgium. Three kinds of groundwaters have been studied : clay porewater collected directly from piezometers drilled into the host-rock from the underground laboratory located at Mol, and fluids sampled from drill holes in the Neogene and Rupelian aquifers, which over- and underly the Boom clay. The chemical evolution of these fluids has been particularly followed in the framework of the ARCHIMEDE project : a model has been proposed for the regulation of the major components and some trace elements (Beaucaire et al., 1995).

This study is focused on redox dependent solid-solution equilibria. The specific issues discussed are : (1) redox potential measurements, their limit and interpretation; (2) concentration and speciation of electro-active elements (Fe, Mn, S); (3) evaluation of minerals as redox buffers of the groundwaters and (4) modelling and implication on the control of trace elements.

2 REDOX POTENTIAL MEASUREMENT

Several campaigns of E_h measurement have been performed on water samples from the aquifers in deep drill holes whose depths range from -50 to -500 m. We generally observe a drop of the signal even after several sampling cycles without reaching a steady state. This can be corelated to the very low concentration level of electroactive species in the fluid, and measurements are also likely to be perturbed by intrusion of oxygen in the water column and/or through the tubing of the sampling device (Grenthe et al., 1992; Pitsch et al., 1995). Nevertheless, successful E_h measurements were carried out with a stainless steel cell on interstitial water perturbed by heating and γ irradiation, showing that the medium remains quite reducing (Beaufays et al., 1994).

3 GLOBAL CHEMICAL COMPOSITION AND SPECIATION

Interstitial waters are of the sodium-bicarbonate type and have an alkaline pH (8.2). In the Rupelian aquifer, the groundwaters evolve from a Na-HCO$_3$ end-member (here represented by Herentals drill hole) to a Na-Cl one with lower pH (Meer). The major chemical characteristics are summarized in table 1.

In the absence of E_h data, different electro-active constituents have been followed. Among these, S is the most concentrated electroactive one (10^{-5} M) followed by Fe and Mn (10^{-6} M). Uranium (10^{-9} M) is also considered as a good indicator of redox conditions.

Titration of S species has been performed directly in the field by potentiometric measurement with $HgCl_2$ (Boulègue et al., 1978). No traces of sulfide or thiosulfate have been detected in solution. Additional analysis in the laboratory, by ion chromatography and capillary electrophoresis showed that in interstitial waters and aquifer groundwaters, sulfur is essentially under sulfate form with concentrations ranging from 0.01 mM in the interstitial water to 4.7 mM in the more concentrated waters in the aquifers.

Table 1 : Chemical composition of interstitial water and Rupelian aquifer fluids. The concentrations are expressed in mM except for Fe, Mn and U in µM.

	Herentals	Meer	Interstitial water
T (°C)	16.6	21.3	17
pH	8.25	7.74	8.2
E_h (mV)	60	150	-310
HCO_3	10.8	13.2	12.5
SO_4	0.1	4.7	0.05
Cl	5.4	93.7	0.5
H_4SiO_4	0.15	0.21	0.14
Na	15.7	115	12.7
K	0.33	0.96	0.25
Ca	0.15	0.82	0.06
Mg	0.16	1.7	0.12
Fe	0.95	0.4	2.7
Mn	1.6	0.6	1.8
U	0.0017	0.00045	0.0002

We used a computer program (modified version of PHREEQE : Parkhurst et al. 1980) with the MINTEQ data bank for the speciation of the waters and calculation of saturation indexes. Carbonate appears as the main complexing agent in these groundwaters. If we assume that Fe and Mn are essentially under the valence II, $FeCO_3$ and $FeHCO_3^+$ cannot be neglected and represent up to 75% of the total Fe in interstitial fluid. Conversely, Mn is essentially present as Mn^{2+}, only 15% of the total metal is constituted by $MnHCO_3^+$. Interstitial waters are close to saturation with siderite ($FeCO_3$) and rhodocrosite ($MnCO_3$). On the contrary, the fluids of the aquifer became more and more undersaturated with respect to siderite as chloride concentrations increases.

4 HETEROGENEOUS EQUILIBRIA

4.1 Mineralogy of the Boom Clay

The clayey phase represents 60% of the bulk rock. Non clayey-minerals are quartz (20 to 25 %), pyrite, carbonates and some felspars, all present in a few percent.

Pyrite appears as aggregates of euheral crystals and framboids. In oxidised samples, pyrite is partly replaced by anhydrite and ferric sulfate as jarosite if intense oxidation is maintained. Correlatively, this implies a great increase of sulfate concentration in the interstitial solution for the more oxidized samples.

Carbonates are disseminated in the clay and are mainly present as calcite and dolomite. Besides these phases, crystals of siderite and rhodocrosite have been observed. This is in agreement with the modelling of interstitial waters.

According to their low sulfate concentrations, the interstitial fluids can be expected to be saturated with respect to pyrite.

4.2 Modelling

4.2.1 Interstitial fluids

Assuming that the interstitial water is saturated with respect to pyrite, we can calculate the redox potential from the following equilibrium :

$$Fe^{++} + 2SO_4^{--} + 16H^+ + 14e = FeS_2 \downarrow + 8H_2O \quad (1)$$

with $\log K^0_s = -16.17$

Considering that Fe is principally complexed by carbonate, we can calculate the concentration of free metal in solution :

$$Fe^{++} + HCO_3^- = FeCO_3 + H^+ \quad \log K^0 = -5.60 \quad (2)$$

$$Fe^{++} + HCO_3^- = FeHCO_3^+ \quad \log K^0 = 7.83 \quad (3)$$

with $\quad [Fe]_T = [Fe^{++}] + [FeCO_3] + [FeHCO_3^+]$

and $\quad [S]_T = [SO_4^{--}]$

$[S]_T$ and $[Fe]_T$ are the measured concentrations of the total element in solution.

Finally, the redox potential, in mV, is given by:

$$E_h = E_0 + 4.21[\log(Fe^{++}) + 2\log(SO_4^{--}) - 16pH]$$

In the case of the interstitial water, the redox potential is estimated to be -250 mV. For this potential, the solution is saturated with respect to the following minerals (log of indexes of saturation are indicated) :

siderite :	-0.14
rhodocrosite :	0.17
goethite :	-0.06
uraninite :	-0.28

In conclusion, the mineral assemblage formed by goethite-pyrite-siderite can be considered in this kind of system as a redox buffer that fixes respectively the E_h, S and Fe concentrations in solution for a given pCO_2.

4.2.2 Aquifer

In the aquifers, the lithostratigraphy consists of alternative fine clayey sands and silts. Glauconite is also present up to 90% in the Berchem sands (Neogene). Conversely to the Boom Clay, pyrite is rare and has only been observed as an amorphous phase in the Rupelian (Steurbaut, 1992; Beaufays et al., 1992). The presence of oxide and oxihydroxide is generally common in this kind of formation and suggests the possibility of a redox control of the water by these minerals.

Assuming that Fe is essentially present as $FeCO_3$ and $FeHCO_3^+$, we can estimate the concentrations of the free metal in the solution and the redox potential is calculated from the following equilibrium:

$$Fe^{++} + H_2O = FeOOH\downarrow + 2H^+ + e^-$$

with $Log\ K^0_s = 12.53$

According to this equilibrium, the potential becomes more oxidant when the pH decreases. This is the case for the more saline groundwaters (Meer drill-hole), where redox potential reaches -135 mV.

In the Neogene aquifer, where the bicarbonate concentrations are lower and pH more acidic, the potentials are also more oxidant. Correlatively, when the pH decreases, the solutions become more undersaturated with respect to siderite and rhodocrosite.

Modelling of these different fluids keeping saturation with goethite shows that equilibrium with uraninite is achieved in both aquifers (Table 2).

Fluid speciation indicates that U is essentially in the form of the $U(OH)_4$ complex, except in the Meer sample, where $UO_2(CO_3)_3^{2-}$ represents 50% of total U (Table 3). Given the relatively low kinetics of U reduction, this implies that the residence time of groundwaters even in the Neogene aquifer are such that the equilibrium with reduced species of U is achieved.

Co, Ni and Zn are also very sensitive to redox condition because they form very insoluble precipitates with sulfide. Table 2 shows that the interstitial waters are oversaturated with respect to ZnS and CoS. The presence of a high content of organic matter in the interstitial fluid, with complexing capacity towards Co and Zn, can increase the solubility of these metals leading to the apparent oversaturation. In fact, the presence of mixed (Co-Ni-Fe-Cu) sulfur phases has been observed in the Boom clay (Jedwab and Nicaise, 1989).

Table 2 : Indexes of saturation for the main minerals in interstitial waters and in the aquifers.

	Interst. water	Herentals	Meer	Neogene
E_h (mV) calculated	-250	-226	-135	-170
siderite	-0.14	-0.61	-1.32	-1.10
rhodocrosite	0.18	0.10	-1.04	-0.1
pyrite	0	-5.8	-18.1	-6.63
CoS	1.44			
NiS	-2.05			
ZnS	1.75			
uraninite	-0.27	0.69	-0.59	-0.08
$USiO_4$	0.76	1.71	1.19	0.77

Table 3 : Repartition of U species in the aquifers and in the interstitial waters. Concentrations are expressed in nM.

	Interstitial water	Herentals	Meer	Neogene aquifer
$U(OH)_4$	0.2	1.69	0.24	0.17
$UO_2(CO_3)_3^{4-}$	0	0	0.20	0
% U(IV)	100	100	53.7	100

5 CONCLUSION

In the interstitial waters, we have identified a mineral assemblage (goethite-siderite-pyrite) that can be considered as a redox buffer. In the aquifers, on the basis of Fe(II)-FeOOH equilibrium, the evolution of calculated redox potentials appears also consistent with the behavior of the other electroactive species such as U.

From this study we can conclude that :

(1) when E_h data are lacking, speciation of the fluid phase (specially for electro-active species) and

mineralogy of the host rock can provide data for the calculation of the redox potential;

(2) further speciation of trace elements allows identification of mineralogical controls of redox-sensitive constituents (U, Co, Zn).

The next step will be the validation of the computed potential values by E_h measurements in deep drill holes by means of a long distance optode.

6 ACKNOWLEDGMENTS

Authors wish to thank SCK-Mol, Waste & Disposal division, for technical support and fruitful discussion, and EDF, Direction des études et recherches, for financial support.

REFERENCES

Beaucaire C., Toulhoat P. and Pitsch H., 1995. Chemical characterisation and modelling of the interstitial fluid in the Boom clay formation. *Proceedings of the 8th international symposium on water-rock interaction* : 779-782.

Beaufays R., Blommaert W., Bronders P., De Cannière P., Del Marmol P., Henrion P., Monsecour M., Patyn J. and Put M., 1992. Characterisation of the Boom clay and its multilayered hydrogeological environment. *Final report EUR R-2944*.

Beaufays R., De Cannière P., Fonteyne A., Labat S., Meynendonckx L., Noynaert L., Volckaert G., Bruggemen A., Lambrechts M., Vandervoort F., 1994. CERBERUS : a demonstration test to study the near field effects of an HLW canister in an argillaceous formation, *Report EUR 15718 EN*.

Boulègue J., Ciabrini J.P., Fouillac C., Michard G. and Ouzounian G., 1978. Field titrations of dissolved sulfur species in anoxic environments - Geochemistry of Puzzichello waters. *Chemical Geology* : 25, 19-29.

Grenthe I., Stumm W., Laaksuharju M., Nilsson A.C. and Wikberg P., 1992. Redox potentials and redox reactions in deep groundwater systems. *Chemical Geology* : 98, N°1/2; 131-150.

Jedwab J. and Nicaise D. 1989. Etude des constituants mineurs de l'argile de Boom. *Rapport inédit ONDRAF*.

Parkhurst D.L., Thorstenson D.C., Plummer L.N., 1980. PHREEQE : a computer program for geochemical calculations, *USGS Water resources investigations* : 80-96, 210 pp.

Pitsch H., L'Hénoret P., Boursat C., De Cannière P. and Fonteyne A., 1995. Characterization of deep underground fluids-Part II : Redox potential measurements. *Proceedings of the 8th international symposium on water-rock interaction* : 471-475.

Scott M.J. and Morgan J.J., 1990. Energetics and conservative properties of redox systems. In : D.C. Melchior and R.L. Basset (Editors), Chemical modeling of aqueous systems II. *Am. Chem. Soc. Symp. Ser.*: 419, 368-378.

Steurbaut E., 1992. Integrated stratigraphic analysis of lower rupelian deposits (oligocene) in the belgian basin. *Annales de la Société Géologique de Belgique* : T.115 (fascicule 1), 287-306.

Hydrochemistry in an indurated argillaceous formation (Tournemire tunnel site, France)

L. De Windt, J. Cabrera & J.-Y. Boisson
Institut de Protection et de Sûreté Nucléaire, DPRE/SERGD, Fontenay-aux-Roses, France

ABSTRACT: The Tournemire site (a thick claystone layer located between two limestone aquifers) is selected to develop research on the containment properties of indurated argillaceous formations that are considered for high level radioactive waste disposal. Waters sampled in fractures (induced by tectonics) have a claystone signature and provide *in situ* hydrochemical data for this very low water content formation. Porewater and fractures chemical data are modelled with geochemical codes. To characterise the porewater chemistry of Tournemire argilites, leaching/cation exchange method seems to reach quantitative results whereas squeezing on pre-hydrated samples gives qualitative results.

1 INTRODUCTION

Indurated argillaceous formations are considered as potential host media for high level radioactive waste disposal. Their characteristics differ, in many aspects, from plastic clays or crystalline rocks. The French Institute for Protection and Nuclear Safety (IPSN) has selected the Tournemire site (southern France) to develop *in situ* research on the containment properties of such formations (Barbreau & Boisson 1993, De Windt et al 1997). A railway tunnel gives access to a sub-horizontal argilite layer (250 m thick, Toarcian and Domerian formations) located between two karstic aquifers.

Water circulations seem to be of very minor importance in the clayey layer. Some water inflows were noticed in fractures induced by tectonics. A key issue in the site characterisation is thus to adequately characterise the groundwater chemistry and to identify the buffer mechanisms in the matrix and in the fracturing network. In return, interpretation of the hydrochemical data help to better understand water circulations in this low permeability formation. Technically, the characterisation of porewater chemistry is made difficult by the very low water contents of argilites, and different extraction techniques have to be assessed. It is the purpose of this paper to present the main results on the hydrochemistry of the Tournemire argillaceous formation.

2 POREWATER CHEMISTRY

The values of hydraulic conductivity and effective diffusion coefficient are homogeneous along the argilite lithology, with mean values of 10^{-13} m/s and 10^{-12} m^2/s respectively. The total porosity is 10 % with no macro-porosity. Specific surfaces are about 20 m^2/g. With such hydrogeological and porosity conditions, solute transport is extremely slow, even for conservative species, with a substantial diffusive component. However, care has to be taken in such low permeability regimes for which hydraulic tests are at their limits of applicability and/or representativeness. Hydrochemical parameters, Cl$^-$ content for instance, may help to corroborate such hypothesis of very low porewater movements.

A typical mineralogy of the Tournemire argilites is given in Table 1. Since no Cl-bearing mineral phase has been detected, Cl is assumed to occur exclusively in the liquid phase and its concentration can be derived from the analysis of aqueous extracts of the rocks and the measurement of the rock porosity. Porewater Cl$^-$ concentrations of 8 to 11 mmol/l were obtained (free water content) for samples from the centre of the argilite formation. Moderate Cl$^-$ contents in porewater seem to demonstrate very slow but effective solute transfers along the geological ages (dozens of millions years): dilute Ca-HCO$_3^-$ water of the surrounding aquifers having very slowly mixed with the Na-Cl connate porewater. The very low water transfer regime of the Tournemire formation seems thus to be enforced by such data. A detailed profile of the whole argilite layer is now in progress in parallel with further studies of the site paleohydrogeology.

The cation exchange properties of the Tournemire argilites were determined by the Nickel ethylenediamine method (Bayens & Bradbury, 1994). A typical set of cation exchange data is: CEC of 80 meq/kg, Na^+ 18 meq/kg, K^+ 10 meq/kg, Ca^{2+} 33 meq/kg ang Mg^{2+} 14 meq/kg. Characteristics for the exchange population of the Tournemire argilite are the low ratios of the exchangeable Na/K and Na/Ca compared with other marine sedimentary clayey rocks (Pearson & Scholtis, 1995).

Table 1. Mineralogy of the Tournemire argilites.

Mineral phase	Proportion (%)
Illite	15-25
Interstr. illite/smectite	15-20
Kaolinite chlorite	15-25
Quartz	15-20
Calcite	10-15
Dolomite/ankerite	0-2
Pyrite	0-3
Organic carbon	0-1
Others (K-feldspar, albite)	0-5

Due to low water contents, acquisition of the porewater hydrochemistry is difficult and cannot be carried out *in situ*. Two alternatives for porewater analysis, leaching/cation exchange (Bayens & Bradbury, 1994) and leaching/squeezing (Entwisle & Reeder, 1993) methods, are tested. Both techniques were performed in laboratory under anoxic conditions to avoid pyrite oxidation, which is known to deeply disturb the porewater chemical composition.

For very accurate results, the selectivity coefficients should be specifically calculated for the Tournemire argilites from the *in situ* exchanged ion population and the ion concentrations in aqueous extracts. Such data are not yet available for the Tournemire argilites and analogous data of a shale with similar clay mineral composition have been used (Bayens & Bradbury, 1994). In the preliminary calculations performed, the activities of Ca^{2+}, F^- and $SiO_2(aq)$ in the starting solution were fixed by mineral solubility control of calcite, fluorite and chalcedony at a fixed p_{CO_2} of 10^{-2} bar. In all these calculations, electroneutrality was balanced by Cl^-. The calculated porewater is of $Na-Cl-(HCO_3^-)$ water type with a moderate total dissolved solid content (TDS) of around 1300 mg/l (see Table 2). The pH of the porewater is about 7.4. The calculated Cl^- concentration of 9 mmol/l compare fairly well with the experimental values. No porosity consideration was taken into account by the modelling. It is worth noting the consistency between the calculated Cl^- content and the Cl^- content obtained from porosity and leachable experimental data. Due to the presence of organic matter, Fe(II)-carbonates and pyrite, reducing conditions probably prevail but were not considered in this preliminary modelling.

The moisture content of the Tournemire argilites (3.5 %) is too low for extracting porewater by mechanical squeezing. Sufficient deionised water were thus added to disaggregated rocks in order to reach moisture contents in the range 20-30%. Chemical compositions fluids (after 7 days of equilibration) were then obtained by mechanical squeezing. The hydrochemical data are reported in Table 2. The water is of $Na-HCO_3-Cl$ type and rather dilute. The low content in Cl^- might be related to the fact that up to 1000 % moisture content is necessary to extract all the porewater Cl^- by leaching. The rather low content in sulphates might indicate that pyrite oxidation occurred but was minimised. With a relatively simple experimental approach, qualitative results can be reached by squeezing on pre-hydrated Tournemire argilites. Work is in progress to improve the methodology. Nevertheless, in addition to the known problem of ion fractionation during squeezing [Entwisle & Reeder, 1993], dissolution of carbonates and modification of the exchangeable cations probably take place when moisture contents are moved from 3.5 to 30 %. The leaching/cation exchange method seems to be more able to reach accurate results for argillaceous rocks with very low water content (at least for waters with moderate TDS).

3 HYDROCHEMISTRY IN FRACTURES

The fracturing network within the argillaceous formation was essentially generated by the North-South Pyrenean compressional tectonics. The fracturing is mainly sub-vertical (thus potentially favourable to any interconnections with the surrounding aquifers) and shows microfissures, fractures of centimetre scale and faults (fault breccia) of meter scale. Some of the subvertical fracture planes are not fully sealed and show millimetre apertures, whereas others present decimetre scale dilation zones where calcite geodes take places. The paragenesis of the filling minerals is relatively simple: fissures and fractures are filled by calcite with occurrence of cubic pyrite and trapped matrix pieces in minute amounts. Permanent water flows (3 ml/h) and neoformation of sulphate minerals are presently observed in the subvertical planes with calcite geodes. The neoformation of Na-Ca-sulphates is probably related to fluid circulations too, since Na^+ is the main cation found in the waters sampled in fractures. Permanent water flows (10 ml/h) are also noticed in brecciated and/or fractured

Table 2. Hydrochemical data set.

	Modelled matrix porewater (see text)	Squeezed matrix porewater (see text)	Groundwater from fracture (see text)	Surrounding water (Cernon fault)
T (°C)	20	20	~ 13	~ 13
pH	7.4	8.2	8.0 - 8.2	7.4
Eh (mV)	-	-	< - 250	oxidising
p_{CO2} (bar)	~ 10^{-2}	-	~ 10^{-3}	-
TDS (mgr/l)	~ 1300	~ 500	~ 900	~ 500
$[Na]_T$ (mmol/l)	14.1	5.4	12.5	0.11
$[K]_T$ (mmol/l)	1.3	0.09	0.14	0.01
$[Ca]_T$ (mmol/l)	0.87	0.22	0.37	2.3
$[Mg]_T$ (mmol/l)	0.49	0.18	0.29	0.82
$[Sr]_T$ (µmol/l)	-	-	5.7	-
$[Fe]_T$ (µmol/l)	-	-	2.7	< 0.0
$[Si]_T$ (mmol/l)	0.20	0.17	0.11	-
$[Cl^-]_T$ (mmol/l)	9.3	1.2	9.3	0.12
$[F^-]_T$ (mmol/l)	0.22	-	0.23	-
$[HCO_3^-]_T$ (mmol/l)	9.4	4.2	4.4	4.9
$[SO_4^{2-}]_T$ (mmol/l)	-	0.40	0.21	0.13

claystones which are close to (~ 8 m) the interface between claystones and limestones. A mechanical packer covered with *Teflon* and flushed with Ar was set up in a dry borehole. The pH in the packer chamber was 8.2, but 8.0 if directly measured at the water outlets. A minimal value for *in situ* p_{CO2} of $10^{-2.6}$ bar may be estimated from the above two pH data and carbonates aqueous concentrations. This p_{CO2} value has to be considered only as indicative. Work is in progress to better constrain the carbonate system in fractures. The hydrochemical data for the fractures are given in Table 2. The water presents an argilite signature: Na-Cl water type, reducing and slightly alkaline.

The karstic aquifers surrounding the argilite formation yield the same chemistry: oxidising conditions, dilute and Ca-HCO_3^- water. The hydrochemistry of the Cernon fault (see Figure 1) is reported in Table 2 for comparison with the fracture data. The water sampled in fractures clearly differs from limestone water type and is rather close to the argilite porewater type. The Cl^- content for instance is much higher in fracture waters than in carbonated aquifers. On the other hand, weak but significant ^{14}C activities were measured in the fracture waters. These preliminary results suggest slowly inflowing waters (probably from the upper aquifer) whose chemistry was partly buffered by argilite minerals through water-rock interactions and/or diffusive porewater transfers.

Although the origin of the water from the fractured argilites is still not fully understood, it offers the chance to provide *in situ* hydrochemical data about a very low water content argillaceous formation, and consequently to be useful to validate the relatively perturbative laboratory methods for porewater extraction. The mineral saturation indices computed with MINTEQ (Allison *et al*, 1991) are reported in Table 3. The MINTEQ database is more particularly suitable to deal with low temperature mineral equilibriums (Engi 1992, Griffault *et al* 1996). The assumptions made *a priori* for the porewater modelling of the previous sub-section seems valuable: calcite, dolomite, chalcedony and even fluorite are in equilibrium with the water elements. Fluorite was not identified by mineralogical analysis, but it is frequently quoted as the mineral controlling the fluorine content in other marls or shales formations (Bayens & Bradbury 1994, Pearson & Scholtis 1995). Because rather high calcite contents in Mg and Sr are found in argilites and in fracture fillings, magnesite and strontianite saturation indices should be better interpreted in terms of solid solutions with calcite. It is worth mentioning the rather large variations in thermodynamic equilibrium constants for dolomite, magnesite and strontianite with regard to the database (Engi, 1992). Natural water are usually saturated with chalcedony which results from low temperature degradations of alumino-silicates. As discussed in the previous sub-section, the existence of strongly reducing conditions within the argilites is not surprising, but it is here clearly demonstrated by rather high Fe contents and a good agreement between siderite and goethite saturations for an Eh value of -240 mV. This value has to be considered as a maximum, since lower reducing conditions may exist due to the presence of pyrite (Griffault et al, 1996).

Table 3.Fracture hydrochemistry: saturation indices.

Calcite	0.0	Quartz	0.2
Dolomite	- 0.2	Chalcedony	- 0.2
Magnesite	- 0.7	Goethite	0.0
Strontianite	- 1.0	Fluorite	0.0
Siderite	- 0.1	Gypsum	- 2.7

Eh = - 240 mV, pH = 8.1.

4 CONCLUSIONS

Interpretation of the hydrochemical data leads to the hypothesis of very slow water transfers both in the matrix and fractures environments. This assumption is consistent with very weak measured hydraulic conductivity and diffusion coefficient for argilites and very small flow rates in fractures.

Waters sampled in fractures with an argilite signature offer the chance to provide *in situ* hydrochemical data about a very low water content argillaceous formation, and consequently to be useful to validate porewater extraction methods. Qualitative results can be obtained by squeezing on pre-hydrated Tournemire argilite samples. But the leaching/cation exchange method seems to be more able to reach accurate results for such argillaceous rocks with very low water content (at least for waters with moderate TDS).

5 ACKNOWLEDGEMENTS

The preliminary porewater analysis and modelling were performed in collaboration with Drs M. Mazurek & H. N. Waber (Rock-Water Interaction Group, Univ. of Berne, CH) and Drs. M. Cave, S. Reeder & D.C. Entwisle (BGS, Nottingham, UK).

6 REFERENCES

Allison, J. D., Brown, D. S. & Novo-Gradac, K. J. 1991. MINTEQA2/PRODEF2, A Geochemical Assessment Model for Environmental Systems: Version 3.0 User's Manual. EPA/600/3-91/021, U.S. EPA, Athens, GA, 30605.

Barbreau, A. & Boisson, J.-Y. 1993. Caractérisation d'une formation argileuse / synthèse des principaux résultats obtenus à partir du tunnel de Tournemire de janvier 1992 à juin 1993, *Rapport d'avancement n° 1 du contrat CCE-CEA n° FI 2W CT91-0115*: EUR 15736FR.

Bayens, B. & Bradbury, M. H. 1994. Physico-chemical characterisation and calculated in situ porewater chemistries for a low permeability Palfris marls sample from Wellenberg. Wettingen, Switzerland: *NAGRA report*, NTB 94-22.

Engi, M. 1992. Thermodynamic data for minerals: a critical assessment, *The Stability of Clay Minerals (Price G. D. & Ross N. L., eds), The Mineralogical Society Series* **3**: 267-328. London: Chapman & Hall.

De Windt, L., Cabrera, J. & Boisson, J.-Y. 1997. Radioactive waste containment in indurated claystones: comparison between the chemical containment properties of matrix and fractures », Geological Society of London (Special Publication on Chemical Containment of Waste in the Geosphere), submitted for publication.

Entwisle D.C. & Reeder, S. 1993. Extraction of pore fluids from mudrocks for geochemical analysis, *Geoch. of Clay-Pore Fluids Interactions (Manning D. A. C., Hall P. L., Hughes P. R., eds), The Mineralogical Society Series* 4: 365-388. London: Chapman & Hall.

Griffault L., Merceron T., Mossman J.R., Neerdael B., De Canniere P., Beaucaire C., Daumas S., Bianchi A. and Christen R. (1996) Aquisition et régulation de la chimie des eaux en milieu argileux: projet « Archimede-Argiles »: EUR 17454FR.

Pearson Jr, F. J. & Scholtis, A. 1995. Controls on the chemistry of porewater in a marl of very low permeability, *Water-Rock Interaction (Kharaka & Chudaev, eds)*: 35-38. Rotterdam: Balkema.

Different approaches to estimate trace element concentrations in groundwaters

L. Duro & J. Bruno
QuantiSci, CENT Parc Technològic del Vallès, Cerdanyola del Vallès, Spain

ABSTRACT. Evidence from Natural System Studies indicates that trace elements, and consequently also radionuclides, are associated to the dynamic cycling of major geochemical components. In this work, this association is considered and modeled by using two different approaches: (1) solid solution models and (2) simple coprecipitation and codissolution models. The result of the application of these two approaches to data collected in El Berrocal (Spain) is shown and discussed. The trace metals taken into consideration are: Ba, Sr, Mn and U. The results obtained indicate that the behavior of Sr, Mn and Ba is closely related to that of calcium and that U is mainly associated to the cycle of iron.

1. INTRODUCTION

Unless an accidental large spill of contaminants is introduced in the geosphere, the concentration of heavy metals in groundwater is normally below 10^{-6} mole·dm^{-3}. For this reason, heavy metals are normally considered as trace components of "unaltered" geochemical systems. Because of this small content, it is very rare that the concentration of these components is governed by the precipitation of pure solid phases of the metals in question and, for this reason, the association between minor and major components of geochemical systems is so important.

There are two main different approaches to account for the association of trace and major components of groundwater.

a) Sorption approaches to describe surface interactions. The most recent approaches to deal with sorption are Surface Complexation Models (SCM), which treat the mineral surfaces as electron-donor ligands able to coordinate the trace metal. The SCM theory has been extensively developed in the last decade (see for example Dzombak and Morel, 1990). However, due to the large amount of parameters needed for the application of SCM, they are not normally applied to systems presenting a large degree of heterogeneity, such as natural environments.

b) Solid solution approaches. These models describe lattice interactions. The methods existing for calculating the solubility of mixed solids are based on the calculation of the activity of each one of the pure end-members in the bulk of the solid phase. This requires a detailed knowledge of the type of mixed solid and of the mixing capacities of both end-members of the solid solution. The level of knowledge on a natural system is usually not as extensive as to allow the application of surface complexation, neither solid solution, approaches.

We have used two alternative models to describe the interaction between minor and major geochemical components of natural waters: coprecipitation and codissolution approaches. We assume that no analyses of the aqueous concentration of these metals are available and, therefore, we will base our modeling on the knowledge of the major aqueous chemistry and the major and trace chemistry of the minerals.

2. COPRECIPITATION MODEL

Let us assume the coprecipitation of a trace metal, T, with a major metal, M, forming a mixed solid phase, for example $T_xM_{(1-x)}CO_3(s)$. (For the sake of simplicity we assume that both cations, T and M, are divalent). If the trace metal T were forming a pure carbonate phase, $TCO_3(s)$, its solubility equilibrium would be given by:

$$TCO_3(s) \Leftrightarrow T^{2+} + CO_3^{2-} \qquad (eq.1)$$

with a solubility constant defined by:

$$K_{s0} = \frac{[T^{2+}]\gamma_T \cdot [CO_3^{2-}] \cdot \gamma_{CO_3}}{a_{TCO_3(s)}} \quad (eq.2)$$

where γ_T and γ_{CO_3} stand for the activity coefficients for species T^{2+} and CO_3^{2-} respectively.

If the dissolution of a minor amount of $TCO_3(s)$ in the host carbonate phase, $MCO_3(s)$, can be considered ideal, its thermodynamic activity, $a_{TCO_3(s)}$, may be equated to its molar fraction, $\chi_{TCO_3(s)}$, for percentages of $TCO_3(s)$ less than 5%. Therefore, we can define a conditional solubility product, $K_{so}*$, as:

$$K_{so} \cdot a_{TCO_3(s)} \approx K_{so} \cdot \chi_{TCO_3(s)} = K_{so}^* \quad (eq.3)$$

In this approach we are assuming that the activity coefficient of the trace component of the solid solution is 1 and, strictly speaking, it should be calculated for each composition of a solid solution by using the excess-free-energy of mixing (see Glynn et al., 1990).

This is a very simple model that can be used to calculate the aqueous concentration of a trace metal, T, which can be solubility controlled by a mixed $T_xM_{(1-x)}CO_3(s)$ phase. Furthermore, it is simple to integrate this approach in the traditional codes used to perform geochemical calculations. It just requires the multiplication of the solubility constant of the trace solid phase, $TCO_3(s)$, by its molar fraction in the bulk of the solid.

3. CODISSOLUTION MODEL

This model is based on the stoichiometric saturation concept, which was formally defined by Thorstenson and Plummer (1977). According to these authors, the composition of a solid solution may remain invariant during the dissolution of the solid phase due to kinetic restrictions, i.e. the dissolution of a solid phase $T_xM_{(1-x)}CO_3(s)$, will be given by the following equation:

$$T_xM_{(1-x)}CO_3(s) \Rightarrow x\,T^{2+} + (1-x)\,M^{2+} + CO_3^{2-} \quad (eq.4)$$

The solution must be under-saturated with regard to this solid in order to apply this model. In this way, the aqueous concentration of T can be calculated once the aqueous concentration of M is known:

$$[T] = \frac{x}{1-x}[M] \quad (eq.5)$$

4. APPLICATION TO EL BERROCAL

The mineralogical description of the site allows assessment of the degree of association of trace metals with major minerals (Pérez del Villar et al., 1996). Nine water samples (from #0 to #8) were selected from different locations in order to perform the analyses.

The main trace element-major element associations documented in El Berrocal (Pérez del Villar et al., 1996) are: barium and manganese present in calcite at 0.005 and 0.04 mole fractions respectively; uranium associated with iron(III)-oxyhydroxides in a 0.5% molar.

In El Berrocal, it is not possible to reproduce the concentrations of the trace metals considered by assuming equilibration of the groundwater with any of the pure solid phases contained in the geochemical databases. These waters are under-saturated with respect to calcite and at equilibrium or slightly over-saturated with respect to iron(III)-oxyhydroxides. Calcite is the main source of calcium in the groundwaters, and we can assume that the main sources of manganese and barium are the mixed Mn-Ba-Ca carbonates. If we take the saturation index of the major solid as an indication of the approach to use (equilibrium approach, i.e. coprecipitation, or codissolution approach) it seems clear that a codissolution approach will be more appropriate to apply to barium and manganese and that a coprecipitation approach will be more adequate in the case of uranium/iron associations. The results obtained are shown in figures 1 to 4.

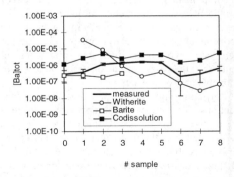

Figure 1. Comparison between calculated and measured barium concentrations (in mole·dm^{-3}).

We can see that equilibrium with barite reproduces very well the measured aqueous barium concentration in sample 0, in the vicinity of a barite dike. The assumption of Ba-CaCO$_3$ congruent

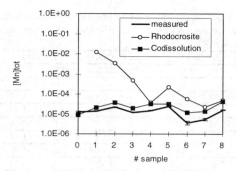

Figure 2. Comparison between calculated and measured manganese concentrations (in mole·dm^{-3}).

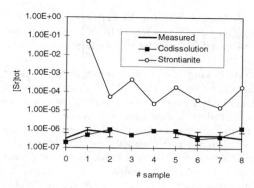

Figure 4. Comparison between calculated and measured strontium concentrations (in mole·dm^{-3}).

codissolution predicts fairly well the observed trend for samples 3 to 8.

Equilibrium with rhodocrosite does not produce satisfactory results, whereas the assumption of Mn-bearing calcite codissolution reproduces the trend and the measured concentrations within a range of ± 0.3 logarithmic units.

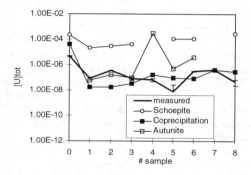

Figure 3. Comparison between calculated and measured uranium concentrations (in mole·dm^{-3}).

Apart from samples 1 and 2 and 3, where autunite has been identified, the rest of the measurements are better reproduced by assuming coprecipitation of uranium with iron(III)-oxyhydroxides.

In spite of the fact that no associations with major geochemical components were reported for strontium in El Berrocal, we applied a codissolution approach by assuming a Sr molar fraction of 10^{-3} in calcite (see figure 4). This figure has been taken from the absence of Sr in the analyses of calcite from el Berrocal and to the detection limit of Sr in EDS technique.

5. DISCUSSION AND CONCLUSIONS

In the models presented, we have based the application of a codissolution model on an undersaturation of the system with respect to the major component of the mixed solid. Nevertheless, this fact is not indicative of an undersaturation with respect to the mixed solid itself. In order to test whether our assumption is valid in the systems studied, we must calculate the state of saturation of the system with regard to the mixed solid phase considered in each case. First of all, we can calculate the state of saturation of the system with respect to the minor component of the solid solution and finally we will show the state of saturation of the system with respect to the solid solution considered.

According to Guggenheim (1937), the activity coefficient of the minor component in a solid solution $T_\chi M_{(1-\chi)}CO_3$ (γ_T) can be estimated from:

$\ln \gamma_T = \chi^2_{MCO3(s)} [a_0 - a_1(3\chi_{TCO3(s)} - \chi_{MCO3(s)})...]$ (eq.6)

where a_0 and a_1 are coefficients to calculate the excess free energy of mixing. Only a_0 is needed if the solution is considered regular (Urusov, 1974).

The conditional solubility constant of each one of the end-members of the solid solution can be calculated by using equation 3, where the activity, a_{TCO3}, is the molar fraction of the solid component in the solid solution times its activity coefficient (calculated as in equation 6).

The following cases were investigated:

a) Sr-bearing calcite in El Berrocal. $\chi_{Sr} = 0.001$; $a_{SrCO3(s)} = 0.001 \cdot \gamma_{Sr}$; $K_{so} = 10^{-9.25}$. Tesoriero and Pankow (1996) determined $a_0 = 5.7$ for Sr-CaCO$_3$ solid solution.

b) Ba-bearing calcite in El Berrocal. $\chi_{Ba} = 0.005$; $a_{BaCO3(s)} = 0.005 \cdot \gamma_{Ba}$; $K_{so} = 10^{-13.35}$. Tesoriero and Pankow. (1996) determined a value of $a_0 = 4.6$ for a Ba-CaCO$_3$ solid solution.

c) Mn-bearing calcite in El Berrocal. $\chi_{Mn} = 0.04$; $a_{MnCO3(s)} = 0.04 \cdot \gamma_{Mn}$; $K_{so} = 10^{-10.41}$. From Pingitore et al. (1988) it is possible to estimate a γ between 5 and 64, depending on the coprecipitation rate.

From the previous analysis, it is possible to say that in all the cases studied here, the system was under-saturated with regard to the minor component of the solid solution (see table 1). The activity of the major component, CaCO$_3$(s), is not greatly affected by forming a solid solution and, therefore, the saturation index of the system with respect to this end-member will not be quantitatively affected.

In order to estimate the saturation state of the aqueous system with respect to the solid solution considered, we can calculate the solubility constant for the mixed solid. According to Thorstenton and Plummer (1977), the solubility constant of a solid solution, $T_\chi M_{(1-\chi)}CO_3$, is given by the following expression:

$$K_{ss} = K_{MCO3(s)}^{(1-\chi)} \cdot K_{TCO3}^{\chi} \cdot (1-\chi)^{(1-\chi)} \cdot \chi^{\chi} \cdot \exp[a_0 \cdot \chi \cdot (1-\chi)]$$

where $K_{MCO3(s)}$ and $K_{TCO3(s)}$ are the solubility constants of the pure end-members of the solid solution. The saturation index of the El Berrocal groundwaters with respect to the mixed solids considered in this analysis are shown in Table 1.

Although a negative saturation index of the major component of a solid solution does not indicate the undersaturation with respect of the solid solution itself, we can see that in the case we are analyzing, this parameter can be very useful, because also the system is under-saturated with respect to the solid solutions considered. This is an important help to estimate the trace metals concentrations in systems where no analytical data is available.

The main conclusion that can be extracted from this work is the importance of considering the association of major and trace elemental cycles when developing an understanding of trace element behavior in geochemical systems.

Although the saturation index of the major component is not sufficient to validate the coprecipitation and codissolution models, it can give clues for the type of approach to use.

The role of other processes, such as sorption and limitation at the source, must also be considered when interpreting the behavior of trace element in natural systems.

Table 1. Saturation index of the groundwaters with regard to the solid solution (SS) and the minor end-member (P)

#	Sr-CaCO$_3$		Mn-CaCO$_3$		Ba-CaCO$_3$	
	SI(P)	SI(SS)	SI(P)	SI(SS)	SI(P)	SI(SS)
1	-4.43	-2.95	-3.02	-2.94	-5.73	-2.94
2	-4.05	-2.20	-2.33	-2.19	-4.72	-2.19
3	-3.66	-1.72	-1.82	-1.71	-3.90	-1.70
4	-2.52	-0.37	-0.60	-0.37	-2.70	-0.36
5	-2.94	-1.15	-1.19	-1.14	-3.53	-1.14
6	-2.60	-1.07	-1.48	-1.07	-3.82	-1.06
7	-2.16	-0.55	-0.85	-0.55	-3.26	-0.54
8	-2.60	-0.41	-0.71	-0.41	-3.22	-0.40

6. REFERENCES

D.A. Dzombak and F.M.M. Morel. 1990. Surface complexation modeling. Hydrous ferric oxide. Wiley. NY. 393 p.

P.D. Glynn, E.J. Reardon, L.N. Plummer and E-Busenberg. 1990. Reaction paths and equilibrium end-points in solid-solution aqueous-solution systems Geochim. et Cosmochim. Acta, 54: 267-282.

E.A. Guggenheim. 1937. Theoretical basis of Raoult's law. Trans. Faraday Soc., 33, 151-159.

L. Pérez del Villar, M. Pelayo, J.S. Cózar, B. De la Cruz, J. Pardillo, P. Rivas, E. Reyes, A. Delgado, R. Núñez, M.T. Crespo, A. Jiménez, A. Quejido and M. Sánchez. 1996. T.R. 3 in El Berrocal Project. Characterization and validation of natural radionuclide migration processes under real conditions on the fissured granitic environment in El Berrocal Project. ECC n1 F12W/CT91/0080. Topical reports Vol. I: Geological studies

N.E. Pingitore, M. Eastman, M. Sandidge, K. Oden, and B. Freiha. 1988. The coprecipitation of manganese(II) with calcite: an experimental study Mar. Chem., 25. 107-120.

A.J. Tesoriero and J.F. Pankow. 1996. Solid solution partitioning of Sr^{2+}, Ba^{2+}, and Ca^{2+} to calcite. Geochim. et Cosmochim. Acta. 60(6): 1053-1064.

D.C. Thorstenson and L.N. Plummer. 1977. Equilibria criteria for two-component solids reacting with fixd composition in an aqueous phase. Example: the magnesian calcites Am. J. Sci., 277, 1203-1223.

V.S. Urusov. 1974. Energetic criteria for solid solution miscibility gap calculations Bull. Mineral., 97, 217-222.

An integrated groundwater quality model based on hydrogeochemical environments

Jasper Griffioen, Aris L. Lourens, Chris B. M. te Stroet & Bennie Minnema
Netherlands Institute of Applied Geoscience TNO, Delft, Netherlands

Marc P. Laeven, Pieter J. Stuyfzand, Cees G. E. M. van Beek & Wouter Beekman
Kiwa N.V. Research and Consultancy, Nieuwegein, Netherlands

ABSTRACT: Models of reactive groundwater transport are needed for prediction of groundwater quality. We present an integrated, hierarchical approach with which both the major groundwater composition and individual trace compounds can be modelled. The approach aims at minimizing the computation time and incorporating mechanistically oriented model parameters as much as possible. The major groundwater chemistry can be modelled with a multiple solute model, when one is solely interested in degradation of organic matter in relation to denitrification, or with a multicomponent geochemical transport model, when the entire major composition has to be modelled. The sorption parameters for trace compounds (either microorganics or trace metal(loid)s) are calculated from mathematical functions that combine literature values for the process parameters (intrinsic binding constants, partition coefficients) with system parameters such as pH, Ca-concentration, organic carbon content. Degradation constants for microorganics are obtained from literature for each specific redox environment.

1 INTRODUCTION

There is a growing need for prediction of groundwater quality. Numerical reactive transport models can be of use for this. The aim of this study is to develop a flexible modelling approach that can be used for prediction of groundwater quality at individual drinking water abstraction locations. The model is currently being developed and this paper presents the general set-up and some aspects of the individual submodels.

2 MODELLING DEMANDS

The quality at drinking water abstraction locations can be threatened by various pollutants. Both the major groundwater quality may deteriorate and individual trace compounds may contaminate groundwater. A simple modelling approach that relies on mechanistic modelling concepts is most wanted, since computation time required is long for numerical transport models and geochemical data of the subsurface are in general scarce.

We have distinguished four types of modelling scenarios (Figure 1). First, denitrification is the relevant issue. Here, one is solely interested in the reactants of the act of denitrification. A multiple solute transport model will be sufficient. Second, the major groundwater quality is the relevant issue, for example the hardness, pH, etc. Both reactants and products need to be considered. A multicomponent, geochemical transport model is required. Third, the fate of individual trace metal(loids) is of

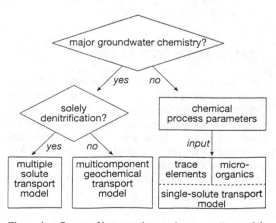

Figure 1: Set-up of integrated groundwater quality model.

interest. A single-solute transport model may tackle this problem when sorption processes are controlling the mobility. One should assure here that other, classical reactions are irrelevant and that the amount of sorption of the trace metal at issue is negligible compared to that of the major ions. If not, one should use the multicomponent geochemical transport model. Last, the fate of microorganics such as pesticides, is of interest. A single-solute model that considers sorption and degradation often suits here.

3 HIERARCHICAL STRUCTURE OF THE INTEGRATED MODEL

Chemical model parameters should preferably be based on mechanistic concepts for the geochemical processes operative instead of empirical formulas. The concept of hydrogeochemical environments is used for this purpose. A hydrogeochemical environment is defined as an underground spatial entity with a unique pore water composition and a unique assemblage of reactive components as the porous matrix. The distinct characteristics refer to both the intensity and capacity factors for geochemical processes during groundwater flow. The model parameters for a trace compound vary for a given study area with the hydrogeochemical environments occurring (Figure 2). The sorption parameters are estimated by mathematical functions that combine capacity parameters obtained from analyses (concentration, contents), with literature values for the intensity parameters (thermodynamic constants, typical sorption capacities per unit content). Also the degradation constants for microorganics strongly depend on the hydrogeochemical environment.

4 SET-UP OF INDIVIDUAL MODELS

The model set-up is based on the finite difference models MODFLOW (McDonald & Harbaugh, 1988) and MT3D96 (Zheng, 1996). The set of individual models are embedded in a Graphical User Interface, which takes care of pre- and postprocessing and invokes the models. The integrity of the whole system and its (intermediate) results is controlled and visualised in a decision support fashion.

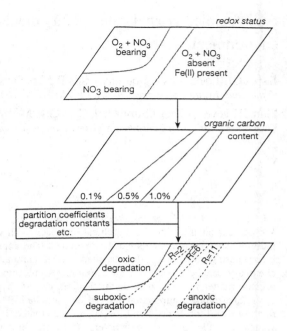

Figure 2: Process parameters for fate of microorganics as determined by hydrogeochemical environments occurring.

4.1 *Major chemistry*

A multicomponent geochemical transport model will be established that is based on PHAST, a coupling between PHREEQC (Parkhurst, 1995) and HST3D (Kipp, 1986).

Degradation of organic matter in relation to denitrification is calculated with a multiple solute version of MT3D96, whereby the implicit coupling between transport and chemistry is set aside. Advective/dispersive transport is calculated with the Eulerian/Lagrangian approach as is done within MT3D96. Degradation is calculated as a redox process among the species involved according to reaction stochiometry. Transport and chemistry are explicitly coupled by the operator split method. One can define an unlimited amount of oxidants and reductants.

The amount of degradation per time step per grid cell, is calculated by explicit analytical solutions of kinetic differential equations and superposition of oxidants and reductants (either dissolved or solid) concentration fields. One can choose either first-order degradation with respect to oxidants (NO_3-limited) or zero-order degradation with respect to organic matter (carbon-limited), for the relevant oxidant/reductant couples. The oxidants O_2, NO_3 and SO_4 are consumed in order of decreasing energy yield. The reactivity of solid organic carbon may decrease with decreasing remaining fraction of organic matter (cf. Middelburg, 1989):

$$\log \mu_t^0 = \log \mu_0^0 - a(1-\alpha)$$

where μ^0 is the zero-order degradation constant, subscript t and 0 refer to time t and time zero, respectively, a is an empirical constant, α is the fraction of organic matter remaining.

4.2 Sorption of trace elements

An input module is constructed that calculates a single-solute isotherm per hydrogeochemical environment for each trace element. Five inorganic trace elements are considered, so far: Zn, Cd, Ni, Cu and As. The amount of sorption of these compounds can be estimated in two ways. First, empirical linear sorption isotherms from literature are used, which are based on concentration units (e.g. $\log K_d = a\, pH + b \log OC + c \log CEC$, where K_d is the distribution coefficient, OC is the organic carbon content and CEC is the cation-exchange capacity.

Second, theoretical (non-linear) isotherms can be constructed from general isotherm type curves (Figure 3). The general isotherm type curves are established for cation-exchange on clay minerals and surface complexation on oxides and organic matter, assuming the sorbents to be additive.

The general isotherm type curves are obtained from PHREEQC calculations for a range of fresh water compositions: 1. dilute precipitation to agriculturally influenced NO_3-rich groundwater, 2. pH from 4 to 7 and 3. one tenth of the maximum tolerable concentration for drinking water for the trace element of interest to ten times that value. A total of 63 compositions is obtained in this way. The range of compositions obtained represents the compositions that will be of practical interest for drinking water supply and for which it can be stated that the trace compounds are still trace metals and not major compounds.

The activity of the free trace metal is calculated in three non-iterative steps by formulas that take into account the major inorganic aqueous complexes and complexation with dissolved organic acids, including competition of Ca, Mg, Fe, Al and other trace metals. Here, we assume that the organic acids can be represented by malonic acid using the method of Morel & Hering (1993). The calculated activity of the free ion is subsequently used to calculate the related sorbed amount for each sorbent present from the general isotherm type curves and the amount of sorbent present. The results are used to calculate a single-solute isotherm between the total sorbed amount and the total concentration of the trace element at issue. This isotherm is used by the single-solute transport model MT3D96 for transport calculations. An isotherm is calculated for each hydrogeo-chemical environment.

4.3 Fate of microorganics

The fate of microorganics, especially pesticides, is assumed to be determined by (reversible) sorption and degradation. A module is created in which the chemical process parameters for fate of microorganics are calculated from spatial data about the redox status and sorption capacity on the one hand, and literature data on organic carbon/water partition coefficients, degradation constants, etc., on the other hand.

Degradation constants are literature values, whereby the values are classified into different redox groups ranging from oxic to methanogenic. Sorption of hydrophobic microorganics is calculated from the organic carbon/water partition coefficient (or octanol/water partition coefficient), DOC concentration and organic carbon content of the porous medium. For hydrophylic microorganics that have acid/base properties, the major sorption process is estimated from sediment analyses, e.g. anion sorption to oxides or partitioning to organic

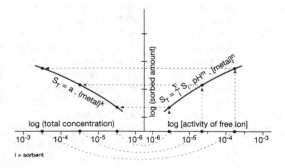

Figure 3: Calculation of single-solute isotherms from general isotherm type curves.

matter, and literature data on sorption are used. The pH-dependant dissociation is here taken into account.

5 DISCUSSION

Many assumptions become included in this integrated model approach, especially for estimating the amount of sorption of the five trace elements considered. The major uncertainty is probably the estimate of the sorption capacity of each sorbent considered: Tessier et al. (1996) and Fuller et al. (1996) concluded that laboratory-derived adsorption data sets may be useful for predicting metal adsorption in the field, whereas Coston et al. (1995) concluded that equating the amount of extracted Fe with ferrihydrite may not account for the observed adsorption.

The alternative is the use of empirical formulas as provided by e.g. Christensen et al. (1996). The empirical formulas are usually obtained by fitting of log-transformed data of batch sorption experiments. A major disadvantage of that approach is that sorption is assumed to be linear and no interference happens among the various dissolved compounds present.

We observed during establishing the general isotherm type curves, for example, that the absence or presence of Cu strongly influences the amount of sorption of Zn and Cd. This effect has not been reported in empirical studies, by statistical data analysis. The major primary variable is in most cases pH. The addition of other primary variables results in little improvement of the regression coefficient, and thus explanation of the variation of the dependent secondary variable by the variations of the independent primary variables.

Our theoretical model approach, however, has made clear that sorption of trace metals can seldom be modelled as a single-solute process, even when one takes into account the specific circumstances of a hydrogeochemical environment. The use of a multi-component geochemical transport model will then be necessary, which implies long computation times. The degree of accuracy is presumably near-identical for that approach since the major uncertainty lies in the availibility of field data.

6 ACKNOWLEDGMENTS

David Parkhurst from the U.S.G.S. is gratefully thanked for making available a test version of PHAST, a coupling of PHREEQC and HST3D.

7 REFERENCES

Christensen, T.H., Lehmann, N., Jackson, T. & Holm, P.E. (1996). Cadmium and nickel distribution coefficients for sandy aquifer material. J. Cont. Hydrol (24), 75-84.

Coston, J.A., Fuller, C.C. & Davis, J.A. (1995). Pb^{2+} and Zn^{2+} adsorption by a natural aluminum- and iron-bearing surface coating on a aquifer sand. Geochim. Cosmochim. Acta (59), 3535-3547.

Fuller, C.C., Davis, J.A., Coston, J.A. & Dixon, E. (1996). Characterization of metal adsorption variability in a sand and gravel aquifer, Cape Cod, Massachusetts, U.S.A. J. Cont. Hydrol. (22), 165-187.

Kipp, K.L. (1986). HST3D: A computer code for simulation of heat and solute transport in three-dimensional ground-water flow systems. U.S.G.S. Water Resources Investigations Report, 86-4095, 517 pp.

McDonald, M.G. & Harbaugh, A.W. (1988). A modular three-dimensional finite-difference ground-water flow model. U.S.G.S. Tech. Water-Res. Inv., Book 6, Chapter A1.

Middelburg, J.J. (1989). A simple rate model for organic matter decomposition in marine sediments. Geochim. Cosmochim. Acta (53), 1577-1581.

Morel, F.M. & Hering, J.G. (1993). Principles and applications of aquatic chemistry. John Wiley & Sons, New York, 588 pp.

Parkhurst, D.L. (1995). User's guide to PHREEQC - a computer program for speciation, reaction-path, advective-transport, and inverse geochemical calculations. U.S. Geol. Surv. Water Res. Inv. 95-4227, 143 pp.

Tessier, A., Fortin, D., Belzile, N., DeVitre, R.R. & Leppard, G.G. (1996). Metal sorption to diagenetic iron and manganese oxyhydroxides and associated organic matter: Narrowing the gap between field and laboratory measurements. Geochim. Cosmochim. Acta (60), 387-404.

Zheng, C. (1996). MT3D96. A modular three-dimensional transport model for simulation of advection, dispersion and chemical reactions of contaminants in ground-water systems. S.S. Papadopulos & Ass., Inc., Bethesda, Maryland, U.S.A.

Linkage between hydrochemistry and geological cover of groundwaters in a Triassic sandstone aquifer (Buntsandstein), SW Germany

T.G. Kretzschmar
JMAS, Junta Municipal de Agua y Saneamiento, Ciudad Juarez, Chihuahua, Mexico

ABSTRACT: The widely extensive Bunter sandstone (Lower Triassic) in SW Germany represents an important aquifer system which can be subdivided into an unconfined zone (Black Forest) and a confined zone covered by carbonates with evaporites and continental beds. As a result of erosion, valley incision and significant long-term groundwater circulation in the unconfined zone, most of the carbonate cement and high TDS formation water of the Bunter has been leached. Thus, present-day groundwater is very low to low-mineralized, contains little hydrogen carbonate, and is devoid of geogenic sulfate and chloride. By contrast, slowly circulating groundwater from the confined zone is highly mineralized. The hydrochemical data indicate a direct correlation between the mineralization and both the age of the waters as well as their distance from the outcropping sandstone and points to a connected flow system supported by the hydraulic data.

1 INTRODUCTION

1.1 Geological setting

The Bunter sandstone (*Buntsandstein*) formation of the Lower Triassic is an important aquifer in Germany. In the South it rests on crystalline rocks or continental red beds of the Lower Permian and crops out in the Black Forest and other areas. Due to the dip of the strata, the sandstone is overlain towards the East by clayey redbeds (uppermost Bunter Fm.) and the carbonate-evaporite sequence of the Middle Triassic (Muschelkalk), as well as younger sedimentary rocks (Fig. 1). As a result of uplift and denudation since the middle Tertiary, the sandstone thickness in Black Forest is more or less reduced by erosion; under younger sediment cover it ranges mostly among 300 and 400 m (Geyer & Gwinner, 1986). At its top the sandstone is sealed by red silty clays and calcareous or dolomitic marls of the Lower Muschelkalk. The latter are overlain by dolomites, sulfate rock, and halite, which is partially removed by subsolution.

1.2 Aquifer Charateristics

The original state of the sandstone after deposition under a semi-arid climate is poorly known. It is assumed that the reddish coatings of the quartz grains

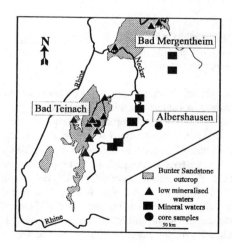

Fig 1: Outcrops of the Bunter Sandstone in SW-Germany as well as simplified locations of the examined springs and wells.

and the presence of carbonate cement in some parts of the sequence mainly result from diagenesis (e.g., Valeton, 1953). Most of the sandstones are cemented by silica and therefore have a very low matrix permeability, although their porosity is still high (10 to 20 %). Thus, the hydraulic permeability of the rock mass ($\sim 10^{-6}$ m/s) is mainly achieved through fractures. In the Black Forest, the sandstone aquifer (unconfined) has been exposed to weathering and groundwater circulation for a long time (1 Ma). Where the exposed sandstone is cut by valleys, groundwater circulation down to the base of the aquifer has leached most of the original carbonate (in the upper part of the sandstone sequence) and diagenetic carbonate cement, particularly so in the middle part of the Bunter sandstone. In this geomorphic setting mainly low TDS waters are present.

By contrast, deep wells both in the Black Forest and in areas where the aquifer is confined have yielded high TDS waters. It is generally assumed that water circulation in this zone is negligible. Only a minor proportion of the high groundwater mineralization can originate from water-sandstone interaction. The source of the high concentrations of chloride and sulfate is not clear but may be (1) the dissolution of syndepositional playa salts or (2) downward directed diffusive flux of salts from the Middle Muschelkalk to the Bunter sandstone.

In contrast to Kretzschmar, 1994 and 1995, where petrographic and thermodynamic results were discussed, the following paper focuses mainly on the hydrochemical composition of the waters.

2 HYDROCHEMISTRY

2.1 Springs (very low TDS)

The results of different spring water analyses are summarized in Table 1 and Fig. 2. They show a great variation in its chemical composition with the tendency to higher concentrations where deeper circulation takes place. The springs represent mainly Ca-Mg-HCO_3-type waters. In the case of anthropogenic influence, high concentrations of NO_3 or Cl (BR1, BR2 and WE1) lead to Na-Cl-type waters. Unaffected springs can show a certain variation in the chemical composition due to the unit of the Bunter they are fed.

Indicators for the penetration depth and residence time of shallow water in the Bunter formation are SiO_2, Li and As (see also Hem, 1985). Whereas As and Li range around the detection limit and can only be detected accurately in the deepest springs (see Fig. 2), the easily detectable silica content closely correlates with the penetration depth.

Table 1: Chemical composition of different spring waters of the lower Bunter (*su*; Kretzschmar, 1995), middle Bunter (*smc1* and *smb*; Hauser, 1994, Kretzschmar, 1995), and the upper Bunter (*so*; Hauser, 1994, Kretzschmar, 1995).

sample	pH	el. cond. µs/cm	Na	Mg	K	Ca	Cl	SO_4	NO_3	HCO_3	Li	As
			mg/l								µg/l	
Mo1	6,93	132	5,4	4,4	1,5	12,7	10,6	4,8	3,0	50,0	n.d.	n.d.
BR1	7,39	243	8	1,5	1,8	6,2	15,5	8,4	5,4	6,1	n.d.	1,5
BR2	4,99	97	12,9	1,2	1,7	7,8	28,6	7,5	2,3	6,1	n.d.	n.d.
B2	7	141	4	2,9	2,6	14,7	7,9	6,6	17,7	34,2	n.d.	n.d.
Gl2	7,09	372	3,9	20,1	1,7	45,8	8,4	18,0	14,8	183	n.d.	1,5
Stollen	6,76	217	4,3	8,2	2,2	21	9,8	4,1	10,4	76,3	0,9	1,0
Kurg.	7	355	4,9	15,1	2,7	39,4	10,0	9,3	15,1	180,0	4,8	3,3

n.d. = not detectable, around the detection limit

2.2 Low TDS wells

The low TDS wells located around the rim of the outcropping Bunter sandstone have a borehole depth of between 15 and 120 m. The chemical composition (Table 2) does not show a great chemical variation. All analyzed samples are of Ca-HCO_3-type and also the variation of each well during a year is quite low, so that the influence of very shallow water can be assumed to be minor. Isotope studies (^{3}H) showed that even the samples with a high nitrate content have residence times of over 20 years (Kretzschmar, 1995).

Table 2: Chemical composition of different low TDS well waters of the Bunter (Kretzschmar, 1995).

sample	pH	el. cond. µs/cm	Na	Mg	K	Ca	Cl	SO_4	NO_3	HCO_3	Li	As
			mg/l								µg/l	
CW1	7,54	411	16,6	18,5	2,8	47,6	24,5	24,3	19,0	186,1	23,7	9,1
NG1	7,1	648	10,9	27,0	2,4	86,8	22,4	53,2	23,3	300,0	16,0	2,2
NG2	7,32	590	15,4	19,8	1,6	81,0	33,7	36,4	19,3	260,0	12,4	2,8
pf6	7,22	676	17,6	30,4	2,8	88,6	51,5	54,5	26,8	292,9	9,7	1,9
HB1	7,25	714	7,0	48,5	5,4	93,2	2,9	156,8	0,8	363,1	45,1	3,5
SU1	7,2	550	5,6	20,5	1,8	83,0	6,8	59,5	11,4	282,0	10,0	2,7
WT7	7,11	861	23,5	28,8	2,9	125,3	68,1	57,3	40,4	356,9	16,6	3,5
WT8	7,11	822	14,8	36,0	2,5	122,8	46,9	73,1	29,6	375,3	16,2	3,0
KH1	7,01	501	5,5	19,0	1,2	70,5	28,4	13,2	30,2	295,9	9,5	3,0

2.3 High TDS wells

The chemical composition of some of the investigated wells bearing high TDS water is shown in Table 3. The mineralization ranges from 5000 to 70000 mg/l TDS (e.g., Carlé, 1976; Frantz, 1990). These waters

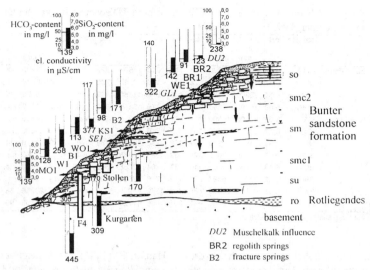

Fig. 2: Simplified cross section of different types of springs in the Bunter in the area of Bad Teinach (after Irouschek, 1990).

contain high concentrations of Na, Cl, and SO_4 (Na-Cl- and to a lesser extent Na-(Ca)-Cl-(SO_4)-type waters). ^{14}C dating indicates that the oldest of these waters are at least 40000 years old. The Lithium content correlates well with the mineralization. However, as the northern high TDS wells of the study area show only small amounts of Arsenic, a correlation between As and mineralization can only be established for the Black Forest. The reason for the differences in the Arsenic concentration has so far not been resolved and needs further investigation. The high TDS wells only present in the covered Bunter (Fig. 1) show a correlation between the mineralization and the distance to the outcropping Bunter, with wells furthest away showing the highest mineralization (Fig. 3).

3 DISCUSSION AND CONCLUSIONS

The chemical long-term evolution of the groundwaters is difficult to assess because of the unknown earlier chemical composition of the formation water in the Bunter sandstone. Processes like the dissolution of syndepositional playa salts and/or diffusive salt transport from overlying evaporites may have also played a role in the system. The absence or presence of geogenic sulfate and chloride can be best explained by a leaching model, assuming that, at some earlier time, chloride and sulfate concentrations in the pore waters were high in addition to relatively large amounts of carbonate cement.

Lithium and Arsenic concentrations increase with increasing distance to the outcropping Bunter. They can be therefore used as a tracer for the distance of the waters to its recharge area. Even for the other earth-alkali and hydrogencarbonate ions a good correlation is given (Fig. 3). The oxidation state shows a reverse correlation with values up to 700 mV in the very low TDS springs and values just above reduction in the high TDS wells. With distance form the outcropping Bunter the Li/Mg-ratios correlate well with the degree of mineralization (Fig. 4). It can be explained as a product of Magnesium/Lithium exchange in minerals (mainly in the carbonate cement), due to direct water-rock interaction.

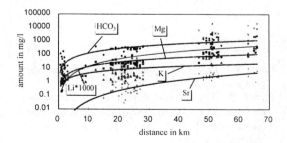

Fig. 3. Relevant ions vs. the distance to the outcropping Bunter.

Table 3: Chemical composition of different high TDS waters of the Bunter (Kretzschmar, 1995).

	sample	pH	el. cond. µs/cm	Na mg/l	Mg	K	Ca	Cl	SO₄	NO₃	HCO₃	Li µg/l	As
Black Forest	CN 1	6,58	30400	8000	118	144	648	14623	3622	n.d.	729	14900	685
	BB1	6,67	7540	2096	28	74	178	2942	706	n.d.	714	6630	n.a.
	LB1	6,61	28200	8130	83	105	676	11080	3469	n.d.	1172	3350	590
	SAWÖ	5,99	10290	5396	376	119	776	2486	2000	n.d.	1687	1510	200
Main/ Tauber	IF2	6,03	55600	19360	375	489	1280	27385	6848	n.d.	2590	20500	n.a.
	IF1	6,12	27200	9860	161	317	616	10369	5222	n.d.	2285	8600	n.a.
	NH1	5,92	46500	12385	227	358	786	16310	4736	n.d.	2270	9470	14
	BÖ1	5,97	10990	2648	129	62	588	3368	2112	n.d.	1522	1880	n.a.
	BÖ2	6,14	13630	2967	110	81	704	3988	1600	n.d.	2072	3720	n.a.
	Bm1	6,2	23000	5396	376	119	776	7090	3648	n.d.	1339	3990	1
	Bm2	6,15	67500	23990	680	389	1272	31905	6285	n.d.	3505	20210	38
	Bm3	5,85	49300	13690	682	287	806	17729	7296	n.d.	3109	4100	99

n.a.= not analyzed, n.d.= not detectable

It can be therefore concluded that the hydrochemical data of the different Bunter waters indicate, at least in the investigated area, a direct correlation between the mineralization and both the age of the waters as well as their distance from the outcropping sandstone.

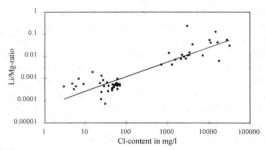

Fig. 4. Li/Mg-ratios (after Schmassmann et al., 1992) against the Cl-content as another conservative element and a representative tracer for the distance from the outcropping Bunter sandstone.

4 REFERENCES

Carlé, W. (1976). Die Mineralwässer von Bad Mergentheim. *Heilbad und Kurort*, 1: 3-23.

Frantz, S. (1990) *Hydrochemische Untersuchungen an Formationswässern des Eyacher Kohlensäuregebietes*. unpubl. dipl. thesis, pp. 105, Geol.-Pal. Institut, Univ. Tübingen.

Geyer, O.F. and M.P.Gwinner (1986). *Die Geologie von Baden-Württemberg* (3. Auflage). E. Schweizerbart'sche Verlagsbuchhandlung: Stuttgart, pp. 472.

Hauser, R. (1994). *Geologie, Deckschichten, Quellen und Bohrkernpetrographie im Mittleren Buntsandstein (Bad Teinach, Nordschwarzwald)*. unpubl. dipl. thesis, pp. 132, Geol.-Pal. Institut Univ. Tübingen.

Hem, J.D. (1985). Study and Interpretation of the Chemical Characteristics of Natural Water. *U.S. Geological Survey Water-Supply Paper*, 2254: 1-263.

Irouschek, T. (1990). *Hydrogeologie und Stoffumsatz im Buntsandstein des Nordschwarzwaldes*. Ph.D. thesis, pp. 139, Geol.-Pal. Institut Univ. Tübingen.

Kharaka, Y.K., W.D.Gunter, P.K.Aggarwal, E.H. Perkins, and J.D.DeBraal, (1988). SOLMINEQ.88, A Computer Program for Geochemical Modeling of Water-Rock Interaction. *U.S. Geological Survey Water-Resources Investigations Report*, 88-4227: 1-207.

Kretzschmar, T. & G.Einsele (1994). Chemical evolution of shallow and deep groundwaters in a Triassic sandstone aquifer (*Buntsandstein*), SW-Germany. In Y.K.Kharaka & O.V.Chudaev (eds), *Water-Rock-Interaction* 8: 453-457. Rotterdam: Balkema.

Kretzschmar, T. (1995). *Hydrochemische, petrographische und thermodynamische Untersuchungen zur Genese tiefer Buntsandsteinwässer in Baden-Württemberg*. Ph.D. thesis, pp.142 Geol.-Pal. Institut Univ. Tübingen.

Schmassmann, H., M.Kullin, and K.Schneemann, (1992). Hydrochemische Synthese Nordschweiz: Buntsandstein-, Perm- und Kristallin-Aquifere. *Nagra Technische Berichte*, 20: 1-493.

Valetom, I. (1953). Petrographie des süddeutschen Hauptbuntsandsteins. *Heidelb. Beitr. Mineral. Petrogr.* 3: 335-379.

Groundwater quality variations in the Eocene Bagshot Formation, UK

J. M. Macmillan & J. D. Mather
Royal Holloway, University of London, Egham, UK

ABSTRACT: Major and trace constituents have been analysed in groundwaters in the Eocene Bagshot Formation using ICP-AES, ion chromatography and colorimetric methods. Samples were obtained from fourteen sites comprising springs, shallow wells and boreholes. Typically pH increases from 4.0 to 6.5 with depth and the composition changes from Ca-Na-SO_4-Cl to Fe-Ca-Na-SO_4-HCO_3-Cl. Groundwater composition is controlled by mineral dissolution and redox reactions modified at surface by atmospheric inputs and at depth by ion exchange processes. Anthropogenic practices such as fertiliser usage and waste disposal are interpreted to have affected many samples.

1. INTRODUCTION

This paper presents a preliminary interpretation of variations in groundwater quality within the Tertiary Bagshot Formation. The study area (Figure 1) is centred around the Bagshot outcrop southwest of London. The natural vegetation consists of heathland and mixed deciduous and coniferous woodland. Although the aquifer is not suitable for potable supply, significant quantities are abstracted for agricultural, horticultural and amenity purposes.

2. GEOLOGY

The Bagshot Formation (Early to Middle Eocene) consists of an unconsolidated to poorly consolidated sandy sequence up to 115 m thick, deposited in estuarine to shallow marine conditions. The lowest unit comprises very fine to fine sands with interbedded muds and mud-clast conglomerates (Goldring et al., 1978). It is succeeded by a middle unit of glauconitic sandy clays and an upper unit of fine sands with some thin clay bands (Curry, 1992). At outcrop, the sands appear yellowish orange to pale grey and, in the subsurface, dark greenish grey to olive grey.

Analysis of cored Lower Bagshot material shows that the average grain size is very fine to fine sand (75 - 212 μm) with typically 20% "fines" (< 75 μm) of which only 1% is clay (< 2 μm). The sand and

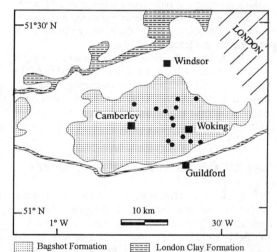

Figure 1   Simplified geological map of the study area. Sample locations are indicated by •

coarse silt fraction is predominantly quartz with subordinate feldspar, glauconite, muscovite, flint fragments, heavy minerals (including pyrite), and lignite. Qualitative XRD analysis of the < 75 μm fraction shows this to include chlorite, kaolinite, illite-mica and some smectite. Apart from iron-speciation, there appears to be no crystalline mineralogical difference between the yellow oxidised sediments and the dark green reduced sediments. The

Table 1 Borehole details and chemical analyses for Bagshot Formation sample sites

Well code	Location (number of analyses)	Sample Source	Land Use	Screen Depths m bgl	Water Depth m bgl	pH	EC μS/cm	Fe_{ToT} mg/l	Al^{3+} mg/l	NO_3^- mg/l	Si mg/l
1	Chertsey (12)	spring	woodland	0	-	4.1	75	<1	6.6	6	8
2	Bracknell (5)	spring	forest	0	-	4.5	80	3	0.4	5	5
3	Chobham (1)	borehole	residential	6 – 14	2	5.0	351	11	0.0	0	14
4	Chobham (3)	borehole	golf course	12 – 42	3	6.7	309	3	0.0	0	8
5	Wentworth (1)	borehole	residential[b]	14 – 26	16	4.9	431	0	0.0	38	8
6	Woking (4)	borehole	horticulture	4 – 20	2	5.5	280	11	0.0	0	12
7	Woking (4)	borehole	golf course	34 – 40	24	6.0	302	22	0.0	0	13
8	Windlesham (4)	well[a]	horticulture	2 – 4	3	4.6	310	10	0.2	0	5
9	Windlesham (1)	borehole	parkland	19 – 29	10	6.2	285	19	0.1	0	17
A	West Hill (1)	borehole	golf course	36 – 56	9	6.2	175	18	0.0	0	10
B	Worplesdon (1)	borehole	golf course	10 – 22	9	5.7	407	15	0.0	0	19
C	Ripley (3)	borehole	residential[c]	7 – 8	3	5.4	557	0	0.1	46	5
D	Ripley (3)	borehole	landfill	11 – 12	1	6.7	910	6	0.1	0	3
E	Chertsey (1)	borehole	test track	<20	13	4.6	250	7	0.2	0	15

m bgl metres below ground level, [a]shallow brick-lined well, [b] mixed residential/golf course, [c] mixed residential/agriculture upgradient from site C

Bagshot Formation is underlain by the London Clay, a thick sequence of blue-grey marine clays. Towards the axis of the London Basin syncline, and locally elsewhere, it is covered by Quaternary fluvial sands and gravels and Recent alluvium.

Spring lines are developed along the basal outcrop with the London Clay and again at the junction of the Upper and Middle units. The overlying superficial deposits are generally in hydraulic continuity with the Bagshot Formation, however, it may be confined locally where the cover is more clay-rich.

3. FIELD AND LABORATORY PROCEDURES

Water samples were obtained from 14 localities (Table 1) in the Bagshot Formation. Electrical conductivity, pH and alkalinity were determined in the field. Unfiltered samples for nitrite and ferrous iron were collected in pyrex bottles and analysed within 3 hours using colorimetric methods. All samples for cation and anion analysis were filtered (0.45mm) and collected in HDPE bottles. The cation samples were acidified with concentrated nitric acid to 1%.

Major and trace cations were determined by ICP-AES whilst major anions were determined by ion chromatography. The Piper diagram (Figure 2) illustrates the relationship between the main anions and cations. Other species not included in figure 2 are listed in table 1.

Percussion drilled sediment cores were obtained from two groundwater monitoring boreholes at a restored landfill site at Ripley. These cores were taken in the Lower Bagshot Beds and include both yellow oxidised material and dark greenish grey reduced sediments. Particle size and XRD analyses were carried out on selected core samples.

4. RESULTS AND DISCUSSION

At the sampled sites pH (Figure 3) increases from around 4.0 – 4.5 at surface to over 6.0 at around 50 m below ground level. Aluminium values are inversely proportional to pH, decreasing from 6.6 to 0.0 mg/l at depth. Silica concentrations, which range from 5 to 19 mg/l, do not show any clear relationship with either pH or depth. The low surface pH arises from the absence of readily soluble minerals to neutralise the impact of acidic rainfall, the decomposition of organic matter and the oxidation of iron compounds particularly pyrite.

The rise in pH with depth indicates that buffering is taking place. In the near surface this is probably controlled by the dissolution of gibbsite, $Al(OH)_3$ and aluminium solubility. The aluminium brought in to solution displaces base cations from ion exchange sites thereby promoting further gibbsite breakdown. Dissolution continues until equilibrium is reached between the groundwater, gibbsite and the exchange complexes (Appelo & Postma, 1994). The relatively high silica concentrations suggest that feldspar and clay mineral dissolution is also occurring. Silicate dissolution buffers porewaters by removing H^+ and releasing base cations and silica to groundwater.

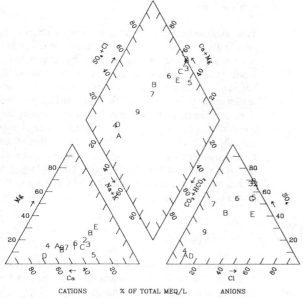

Figure 2 Piper diagram of Bagshot groundwater composition

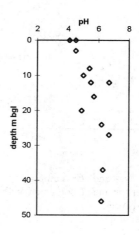

Figure 3 pH depth profile for Bagshot groundwater samples

Iron concentrations range from 0 to > 20 mg/l, the majority is present as the reduced Fe^{2+} form. The Fe^{2+} ion will be oxidised to Fe^{3+} on contact with oxygenated groundwater or air and precipitated as iron oxyhydroxides, hence the yellowish-orange colour of the Bagshot Beds at surface and the dark greenish grey colour in the subsurface. The dissolution of iron (II) silicates, such as chlorite, glauconite and smectite, in general proceeds much faster under anoxic conditions compared with oxic conditions (Appelo & Postma, 1994) and may be the source of the high iron concentrations in many Bagshot groundwaters since these minerals were identified by XRD in core samples.

It is anticipated from theoretical redox considerations that nitrate will be unstable in the presence of high dissolved iron. The data in table 1 indicate that nitrate was detected at only 4 sites where iron concentrations are low or absent. Nitrate, derived from rainfall, dry deposition and nitrogen fixation, is around 5 mg/l at the two springs. Low concentrations of nitrite at site 1 suggest that denitrification is occurring. Exceptionally high nitrate concentrations are present at sites 5 and C (38 and 46 mg/l respectively). Cores taken in the Bagshot section at site C comprise yellowish-orange sands implying oxygenated groundwaters, and hence nitrate is preserved. Oxygenated groundwaters can also be inferred from the borehole log at site 5. The land use at these sites (table 1) could involve significant fertiliser use. The potassium content, a common fertiliser constituent, is also high in these groundwaters which suggests that the nitrate could be derived from fertilisers.

Assuming that chloride behaves conservatively and that the only source term is rainfall (3 mg/l mean concentration, Barrett et al., 1987) implies an evapotranspiration concentration factor of around 5 for the deepest wells and 17 – 25 for the two springs. The latter values are higher than expected. However, if dry deposition and the effects of throughfall (concentration factor of 2-3 for forest trees, Kinniburgh & Edmunds, 1984) were taken into account, the evapotranspiration factor for the wooded spring sites could decrease significantly.

Presuming that rainfall concentrations correspond with average seawater, sodium and chloride concentrations should plot on or close to the seawater dilution line on figure 4. Although a good correlation is present, there are exceptions.

Anomalous groundwater composition, where sodium chloride composition cannot be related to rainfall input, is evident at site D down gradient of a landfill. Here it is likely that landfill leachate modifies

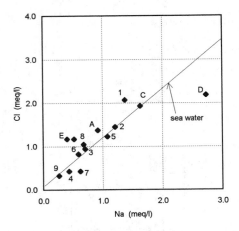

Figure 4 Na and Cl composition of Bagshot Groundwaters in meq/l compared with average sea water composition

groundwater composition since the up gradient well, (site C) does not appear to be affected.

Typically cation abundance in Bagshot groundwaters has the form Ca > Na > Mg > K, however, at site E it is Mg > K > Na > Ca. In view of the land use (vehicle test track within heathland) de-icing chemicals such as rock salt (impure halite with potassium / magnesium salts) could be responsible though this is not confirmed by the low total dissolved solids implied by the EC of 250 µS/cm.

As anticipated, alkalinity concentrations are very low to absent in the springs and near-surface wells with their low pH. The Piper diagram shows a progressive enrichment in bicarbonate in the deeper samples. The assumption that alkalinity can be equated with bicarbonate, however, is probably incorrect since the anoxic decomposition of organic matter produces both bicarbonate and organic acid anions which contribute to alkalinity (Baedecker & Cozzarelli, 1992). From the pH and Piper plot position, the landfill leachate contaminated sample (site D) is superficially comparable with two "deep" uncontaminated groundwaters, sites 4 and A. However, the concentrations of the base cations and main anions are approximately three times higher in D and the organic alkalinity component is unlikely to be the same in the two settings.

5. SUMMARY

The composition of Bagshot groundwaters is principally controlled by dissolution and redox reactions modified at the surface by atmospheric inputs and at depth by ion exchange processes. Nitrate pollution from fertilisers, although widespread, is restricted to the near surface aerobic zone and has no negative impact on water quality for current abstraction purposes. Road salting and landfill leachate, however, have the potential to contaminate the aquifer to a greater extent and may impair water suitability for irrigation. Bagshot groundwaters contain high dissolved iron and are not suitable for potable supply.

ACKNOWLEDGEMENTS

This research is funded by EPSRC and RMC (UK) Ltd. The authors wish to thank the site owners for allowing access to their boreholes.

REFERENCES

Appelo, C.A.J. & D. Postma 1994. *Geochemistry, groundwater and pollution*. Rotterdam: A.A. Balkema.

Barratt et al. 1987. *Acid deposition in the United Kingdom, 1981-1985*. Stevenage: Warren Spring Laboratory

Baedecker, M.J. & I.M. Cozzarelli 1992. The determination and fate of unstable constituents of contaminated groundwater. In S. Lesage & R.E. Jackson (eds), *Groundwater Contamination and Analysis at Hazardous Waste Sites*. New York: Marcel Dekker.

Curry, D. 1992. Tertiary. In P.McL.D. Duff & A.J. Smith (eds), *Geology of England and Wales*: 389 - 411. London: The Geological Society.

Goldring, R., D.W.J. Bosence, & T. Blake 1975. Estuarine sedimentation in the Eocene of southern England. *Sedimentology* 25: 861-876.

Kinniburgh, D.G. & W.M. Edmunds 1984 The susceptibility of UK groundwaters to acid deposition. Report to UK Department of Environment.

Nitrate loading of shallow ground water, prairie vs cultivated land, northeastern Kansas, USA

G.L. Macpherson
University of Kansas, Lawrence, Kans., USA

ABSTRACT: The Konza Prairie NSF Long-Term Ecological Research (LTER) Site is a dissected tallgrass-prairie upland in northeastern Kansas. The study area on Konza described here is located on loess-like alluvium. Two hydrogeologically similar sites were used to compare nitrate loading to ground water for a period of 2.5 years. One site was cultivated (cropped and fertilized) and one was untreated tallgrass prairie. Results show that the unconfined aquifer at both sites is strongly stratified in nitrate content when water levels are high because macropore flow adds higher-nitrate recharge water to the upper part of the aquifer. This recharge water is not enriched in other components that might indicate evapoconcentration in the soil zone. During times of falling water levels, nitrate content increases at both sites, due to mixing of the previously recharged high-nitrate water with the resident ground water.

1 INTRODUCTION

The supply of nitrate to ground water depends upon nitrogen loading rates, assimilation by biota, and oxidation-reduction reactions between nitrate and other forms of nitrogen. Dissolved nitrate in high concentrations can be toxic to human infants and to livestock, and thus is a public health concern. High nitrate concentrations are found mostly in agricultural regions where feedlots or fertilizer application provide point and non-point sources of nitrate (Spalding and Exner, 1993). Other common sources of nitrate include meteoric precipitation and degradation of soil organic matter.

This study examines a shallow, fine-grained unconfined alluvial aquifer within the 34 km^2 Konza Prairie. This site is here called the EPSCoR study area of the Konza LTER. The purpose of the study was to document temporal changes in nitrate concentrations at two geologically and hydrologically similar sites, one under cultivation and one an unplowed prairie. Nitrogen flux to the ground water was hoped to be the only significant variable during the 2.5-year study, and the temporal data reported here are used to examine the processes that result in vertical stratification of nitrate in aquifers.

The reduction of nitrate to nitrogen gases or to ammonium has been used as the explanation for spatial variability of nitrate concentrations in aquifers. Vertical stratification is prominent, and it has long been known that high nitrate concentrations are more likely to be found near the water table rather than deeper in an aquifer. Researchers have used stratification of nitrate concentrations in the saturated zone of an aquifer coupled with coincident stratification of dissolved oxygen (DO) and antithetic concentrations of bicarbonate alkalinity as evidence that denitrification occurs in the upper part of an aquifer (see Geyer et al., 1992, for examples). These observations, along with measurement of denitrification potential and bacterial population studies, all provide permissive evidence that nitrate enters an aquifer through the unsaturated zone and is denitrified with depth as fluid moves vertically through the aquifer.

2 SETTING

Konza is a tallgrass prairie with wooded riparian zones. This dissected limestone upland is underlain by flat-lying, thin (1-2 m) Permian limestones and cherty limestones that alternate with more easily weathered shales (2-3 m). The limestones underlie the uplands and form benches along the relatively steep slopes. Riparian zones along higher order streams are filled with fine-grained material that is presumably reworked loess. This valley fill often contains a chert-gravel lag at the base and is terraced (Smith, 1991). The cultivated site for this study is a terrace that has been under continuous cultivation since 1939 (Dodds *et al.*, 1996). The prairie site, never cultivated, is on the same age terrace.

Two locations were selected for well nests. One site is located within a field under cultivation for various crops (soybeans, wheat). The area around and between the wells was manually seeded with the crop being grown, harvested when the field was harvested, and any chemicals applied to the field were also applied to it. The second site is about 1 km upstream from the cultivated site and is located within a prairie. The prairie site was never irrigated, fertilized, or mowed during the study period. Both sites are approximately 100 m from the major stream draining the Konza.

3   METHODS

Six wells were installed at each site. Each was completed with PVC pipe and a single, PVC screen and each bottom cap was slotted so that the total length open to the aquifer is 30 cm. Deep wells are 10-cm diameter and shallow and intermediate-depth wells are 5-cm diameter. Gravel pack was placed around the screen, and alternating layers of bentonite and cuttings seal the rest of the annular space. Steel well protectors were installed and cemented into place.

Wells were sampled every two weeks beginning August, 1993, except during winter months and during periods of heavy rainfall. Low-yield wells were bailed with a Teflon® bailer and higher-yield wells were pumped with a bladder pump. Water samples were gravity filtered through 0.45 µ Millipore® membrane filters and stored in acid-washed, 250 mL, low-density polyethylene (LDPE) bottles. Anions and cations were determined by ion chromatography with a Dionex 4000i ion chromatograph. Alkalinity was titrated using 0.02 N H_2SO_4. A Hydrolab® (downhole probe) was used on all wells with sufficient water depth to determine *in situ* pH, dissolved oxygen, and redox state.

Annual precipitation was about 148% of mean average precipitation in 1993, 78% of average in 1994, and 119% of average in 1995.

4   TEMPORAL TRENDS

Figure 1 shows the trends in nitrate concentrations through time at the cultivated site. Shown on the plot is a deep (A-2, 10 m), intermediate (A-3, 8 m), and

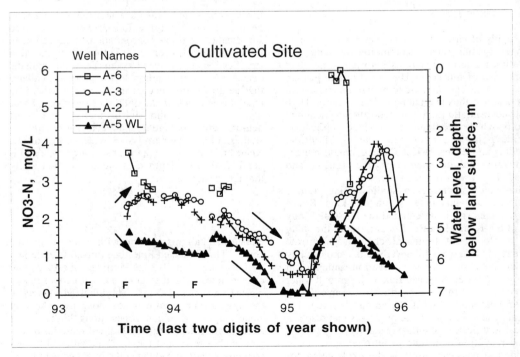

Figure 1: Nitrate in ground water beneath the cultivated EPSCoR site at Konza is stratified when water levels are rising and ground water is better mixed at other times. Depth to water (WL) is shown by the triangles. "F's" show dates of fertilizer application. Representative wells shown. Arrows highlight trends discussed in the text.

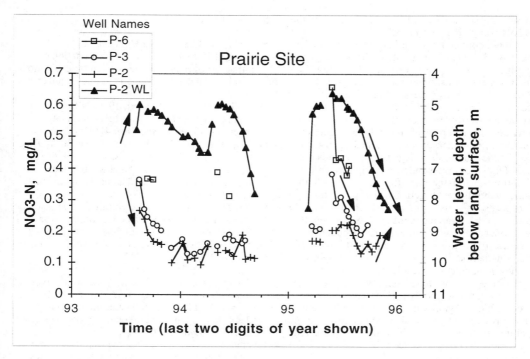

Figure 2: Ground water beneath the prairie EPSCoR site at Konza is stratified in nitrate content when water levels are rising and better mixed at other times. Water level changes are more dramatic at this site than at the cultivated site, and nitrate levels are 10 times lower. All wells were dry during the winter of 1994-1995.

shallow (A-6, 5.5 m) well. Also shown is the depth to water in one of the deep wells at that site. The shallow A-6 well was dry except during periods of the highest water levels. This well typically had higher nitrate than the deeper wells at any sampling time. The intermediate A-3 and deep A-2 wells typically produced water with less nitrate than A-6, with A-3 having higher nitrate sooner than A-2. The highest nitrate concentrations in the shallow A-6 well correspond to the highest water levels at the cultivated site, in the second quarter of 1995. Fertilizers were applied in March and August of 1993 and March of 1994. No fertilizers were applied in 1995.

Figure 2 shows the trends in nitrate concentrations through time at the prairie site. Shown on the plot are the trends for a deep (P-2, 7.8 m), intermediate (P-3, 6.4 m) and shallow (P-6, 4.8 m) well. Although no fertilizer was added to this site, the high nitrate in the shallowest well occurs during times of highest water levels, just as at the cultivated site. At this prairie site, nitrate was almost always higher in the intermediate P-3 well than in the deeper P-2 well.

During most periods, water level directional trends at both sites coincided with nitrate concentration trends. Exceptions are at the very beginning and very end of the monitoring period, when trends were opposite. The opposing trend is much more lengthy at the cultivated site, but the pattern of stratification during high water levels and more homogeneous water chemistry during lower water levels is the same at both sites.

5 DISCUSSION

Vertical stratification of nitrate in aquifers is well documented (e.g., Delwiche and Steyn, 1970; Böttcher *et al.*, 1990; Postma *et al.*, 1991; Smith *et al.*, 1991; Simpkins and Parkin, 1993; Hendry *et al.*, 1983; Geyer *et al.*, 1992). Temporal variations in ground water nitrate are less frequently reported. Schnabel *et al.* (1993) attributed stream water variability in nitrate to variable nitrate contents of a layered aquifer system, as have others doing catchment studies (e.g., Trudgill, 1995). Lowrance (1992) collected temporal data, using means of data to show riparian zone uptake of nitrate lowering nitrate content of shallow ground water. The data reported here reveal information about the process of nitrate loading of a shallow alluvial aquifer.

When the shallowest wells at both sites of the EPSCoR study area first produce water after a wet period, both contain nitrate at levels than are 75% to several 100% higher than nitrate in the deeper wells. The time to recover to a very nearly unstratified ground-water chemistry after such a wet period is on the order of 3 months or less at both sites. After the water level high, again at both the prairie and cultivated sites, there is a short (prairie site) to long (cultivated site) period during which water level directional trends and chemical trends oppose each other. These relationships suggest that two processes occur. First, the correspondence of high water levels and nitrate-chemistry stratification suggests that ground-water is recharged fairly rapidly through macropores at this site. This is not simple entrainment of evapoconcentrated soil water that might occur as water levels rise, because Cl concentrations are nearly constant over the monitoring period. Second, the antithetic response of water level and nitrate concentration suggests that the higher-nitrate shallow water mixes with and elevates the nitrate concentration of the deeper water in a lagged fashion. Again, this phenomenon happens at both cultivated and prairie sites, although it is more apparent (because of the 10 times higher nitrate concentrations) at the cultivated site.

6 SUMMARY AND CONCLUSIONS

Cultivated and prairie sites at the EPSCoR study area of the Konza LTER show similar processes of nitrate loading of the underlying unconfined alluvial aquifer. Although nitrate concentrations are approximately 10 times higher in the cultivated-site ground water than in the prairie-site ground water, both sites show distinct nitrate-concentration stratification when water levels are high. Also at both sites, ground water "recovers" to more homogeneous nitrate content in the aquifer within a period of about 3 months. Rising nitrate concentrations corresponding to falling water levels in the aquifer do not result from evapotranspiration. Instead, mixing of the shallow (high nitrate) and deeper (lower nitrate) ground water results in a net increase in nitrate in the deeper water when recharge volume is large.

7 ACKNOWLEDGMENTS

Research support came from the NSF Experimental Program to Stimulate Competitive Research (EPSCoR) and LTER Program at the Konza Prairie Research Natural Area, Kansas State University, Manhattan, KS 66506. Thanks to W. Simpopo and Marcia Schulmeister for help in sample processing.

8 REFERENCES

Böettcher, J., O.Strebel, S.Vokerlius & H.L. Schmidt 1990. Using isotope fractionation of nitrate-nitrogen and nitrate-oxygen for evaluation of microbial denitrification in a sandy aquifer. *J. Hydrol.* 114:413-424.

Delwiche, C.C. & P.L.Steyn 1970. Nitrogen isotope fractionation in soils and microbial reactions. *Environ. Sci. Tech.* 4(11):929-935.

Dodds, W.K., M.K.Banks, C.S.Clenan, C.W.Rice, D.Sotomayor, E.A.Strauss & W.Yu 1996. Biological properties of soil and subsurface sediments under grassland and cultivation. *Soil Biol. Biochem.* 28(7):837-846.

Geyer, D.J., C.K.Keller, J.L.Smith & D.L. Johnstone 1992. Subsurface fate of nitrate as a function of depth and landscape position in Missouri Flat Creek watershed, USA. *J. Contam. Hydrol.* 11(1992): 127-147.

Hendry, M.J., R.W.Gillham & J.A.Cherry 1983. An integrated approach to hydrogeologic investigations, a case history. *J. Hydrol.* 63: 211-233.

Lowrance, R. 1992. Groundwater nitrate and denitrification in a coastal plain riparian forest: *J. Environ. Qual.* 21: 401-405.

Postma, D., C.Boesen, H.Kristiansen & F.Larsen 1991. Nitrate reduction in an unconfined sandy aquifer-- water chemistry, reduction processes, and geochemical modeling. *Wat. Resour. Res.* 27(8): 2027-2045.

Schnabel, R.R., J.B.Urban & W.J.Gburek 1993. Hydrologic controls in nitrate sulfate, and chloride concentrations. *J. Environ. Qual.* 22(1993): 589-596.

Simpkins, W.W. & T.B.Parkins 1993. Hydrogeology and redox geochemistry of CH4 in a Late Wisconsan till and loess sequence in central Iowa. *Wat. Resour. Res.* 29(11): 3643-3657.

Smith, G. N., 1991, Geomorphology and geomorphic history of Konza Prairie Research Natural Area, Riley and Geary Counties, Kansas: unpublished M.S. thesis, Kansas State University, Manhattan, Kansas.

Smith, R.L., B.L.Howes & J.H.Duff 1991. Denitrification in nitrate-contaminated ground water-- occurrence in steep vertical geochemical gradients. *Geochim. Cosmochim. Acta* 55:1815-1825.

Spalding, R.F. & M.E.Exner 1993. Occurrence of nitrate in groundwater, a review. *J. Environ. Qual.* 22:392-402.

Trudgill, S.T., ed. 1995. Solute modeling in catchment systems. New York, John Wiley & Sons, 473 pp.

Allochthonous ions dissolved in recent and fossil groundwaters: Identification and origins

Emanuel Mazor
Department of Environmental Sciences and Energy Research, Weizmann Institute of Science, Rehovot, Israel

ABSTRACT: Positive linear correlations observed in studied groundwater systems between the concentrations of Cl, Br and other ions are interpreted as indicating an allochtonous origin of these ions. Based on low Cl/Br ratios and abundance of a Ca-Cl component, as well as an isotopic composition of meteoric water, the allochtonous source is identified in certain fossil waters as brine-spray and evaporitic dust. Ions that do not correlate with the Cl concentration seem to reflect water-rock interactions.

1. INTRODUCTION

Ions dissolved in recent and fossil groundwaters can be autochtonous, i.e. originate from water-rock interactions, such as dissolution of halite and gypsum, CO_2-containing water reacting with rock minerals, dolomitization and dedolomitization, ion exchange or membrane filtration. On the other hand, dissolved ions can be allochtonous, i.e. brought in with the recharge water, e.g. air-borne sea-spray and air-borne brine-spray, washed into the groundwater system by the meteoric recharge water. The concentration of the allochtonous ions can vary significantly according to the evapotranspiration intensity prior to deep infiltration of the recharged water (Mazor and George, 1992; Mazor, 1997). The initial concentration of part of the allochtonous ions can be preserved in the water, whereas the abundance of another part is modified by water-rock interactions.

The present communication deals with the identification of the allochtonous ions in recent and fossil groundwaters, in light of a list of case studies. Identification of the allochtonous ions provides information on the environmental conditions that prevailed at the time a studied groundwater was recharged, they serve as markers that help to identify the water-rock interactions that determined the final composition of studied groundwaters.

2. COMPOSITIONAL INDICATORS OF AN ALLOCHTONOUS ORIGIN OF DISSOLVED IONS

2.1 Identification of fossil groundwaters.

Groundwater that has a salinity and composition that are significantly different from the local recent groundwater, has been formed under different environmental conditions and hence is fossil. Such groundwaters are always confined and occur at depths that are greater than the depth of the local base-flow, e.g. formation waters associated with petroleum compounds, encountered in subsidence basins.

2.2 Undersaturation in respect to halite indicates an allochtonous origin of Cl.

Halite is the only potential source of Cl that may be dissolved in water as a result of interaction with common rocks. Groundwater that comes in contact with halite reaches a Cl saturation concentration of ~180 g/l. Chloride is conservative and, hence, the concentration of Cl in groundwater provides a criterion to identify autochtonous Cl, in the cases of saturation concentrations with regards to halite. In all cases of undersaturation with respect to halite, halite is absent from the relevant rock assemblage, and hence, the Cl is external, i.e. washed down by the infiltrating recharge water. Examples are most of the currently formed groundwaters that have a wide range of Cl concentration, but receive the bulk of it from airborne sea-spray.

The kind of cation that balances the Cl provides a further clue to the origin of the Cl. Examples: in the case of halite dissolution the balancing ion is Na, in the case of seawater the other cations are present in known ratios, and in the case of evaporitic brines often a Ca-Cl component is present besides the Na-Cl component.

2.3 Information obtained from the Cl/Br ratio.

The Cl/Br weight ratio in seawater is ~290, in halite it is >3000, and in residual evaporitic brines that precipitated halite, the Br concentration is elevated and the Cl/Br ratios are often in the range of 20 to 200. These distinct categories of the Cl/Br ratio serve as useful indicators of the source of these ions in groundwater, as (a) both ions are conservative, and (b) the ratio is maintained in the case of dilution of a saline water by intermixing with fresh water, often happening in boreholes and pumped wells. Linear correlations often observed between the concentration of Cl and Br in groundwaters collected in a study area provide further evidence that these ions have a common allochtonous source, ruling out in such cases Br contribution from organic matter or Cl contribution from halite.

Chloride and bromide are allochtonous in many fossil groundwaters as is borne out by combinations of the following observations: (a) presence of a Ca-Cl component, (b) Cl/Br < 200, and (c) lack of correlation between type of host rocks and the Cl/Br ratio or Cl concentration of the encountered water.

2.4 Sea-spray identified as the major source of Cl and Br in recent groundwaters.

The most likely external source of Cl is airborne sea-spray, as is confirmed by the Cl/Br ratio that is ~290 in the bulk of recent groundwaters. Further more, in many recent groundwaters other ions preserve a marine abundance ratio (Mazor and George, 1992, Mazor, 1997), as is demonstrated by the example of unconfined groundwaters of the Campaspe Basin, Australia (Fig 1a).

2.5 Brine-spray identified as the source of Cl and Br in many fossil groundwaters.

This is evidenced by the Cl/Br ratios that are in many cases distinctly lower than the marine value, and by the presence of a Ca-Cl component. In most fossil groundwaters the isotopic composition of the water resembles non-evaporated meteoric water, thus supporting the brine-spray model and in many cases ruling out direct entrapment of seawater or brine.

Positive linear correlations to Cl and Br indicate other ions are allochtonous as well. Water samples collected at adjacent wells, or from different depths in the same well, often differ in their salinity, but a positive linear correlation is observed between the concentration of Cl, Br and other ions. This indicates that these ions were brought into the system together with the Cl and Br, and in these cases they did not originate from water-rock interactions. Examples are given in Figs. 1b to 1e from sedimentary basins, and in Fig. 1f for a crystalline terrain.

3. CASE STUDIES

3.1 Campaspe, Australia - unconfined recent groundwaters, sea-spray tagged.

Fig. 1a depicts composition diagrams of unconfined and semi-confined groundwaters, a few hundreds kilometers inland, Campaspe River region, Victoria, Australia (data from Arad and Evans, 1987). Remarkable linear positive correlations are observed between the concentrations of Cl, Br, Na and Mg, in seawater-like relative abundances, indicating these groundwaters are tagged by sea-spray. In addition, a light isotopic composition, plotting on the global meteoric water line, reveals the non-evaporated nature of these groundwaters. The wide range of salinity is the result of different degrees of evapotranspiration prior to recharge during dry periods. Salts accumulating on the surface are washed down by subsequent strong rain events. The preservation of the linear correlations indicates that the involvement in water-rock interaction was negligible, in contrast to ions that do not correlate with Cl and reflect water-rock interactions.

3.2 Mersey Basin, northwest England, Permo-Triassic sandstones

Tellam (1995) conducted a detailed hydrochemical study of 70 to 370 m deep layers of Permo-Triassic sandstones in the Mersey Basin, northwest England. Fresh groundwater is encountered at a shallow depth and below occur confined saline groundwaters (3000 to 192000 mg/l Cl) that were recharged under environmental conditions that were distinctly different from the present ones, indicating the saline waters are fossil. A remarkable linear correlation is seen in Fig. 1b between the concentrations of Cl, Na, Ca, Mg and SO_4, indicating that these ions are allochtonous, i.e. they originated from outside the rock system and were brought in by the recharged water. The saline groundwaters are Na-Cl dominated, the equivalent ratio of Cl/Na is close to 1, and the Cl/Br ratios are > 1500. These observations are convincing indications that the allochtonous source of the dissolved salts was mainly composed of halite dust, or airborne spray of an evaporitic brine that redissolved halite.

3.3 Ontario, Canada, and Michigan, U.S.A., Ordovician carbonates, shales and sandstones.

Fig. 1c depicts the composition of formation waters taped in Ordovician host rocks at Ontario and Michigan (data from Dollar et al., 1991). The composition is entirely different from the local recent fresh groundwater, indicating these are fossil waters. The concentrations of Na, Ca, Mg, and Br reveal a

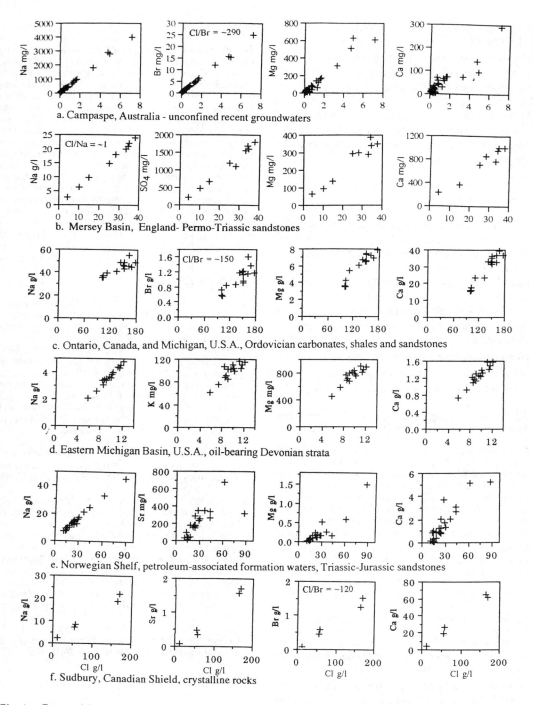

Fig. 1: Composition as a function of Cl concentration. Correlations indicate common allochtonous sources.

linear positive correlation with the concentration of Cl, indicating these ions are of the same allochthonous source of evaporitic brine. The δD values are around -30‰, indicating an origin from a non-evaporated meteoric water. The Cl/Br ratios are in the range of 100 to 187, i.e. reflecting an evaporitic brine affinity, as does the conspicuous Ca-Cl component. Thus an allochtonous source from brine-spray washed down by meteoric water in a dry climate is concluded.

3.4 Eastern Michigan Basin, U.S.A., oil-bearing Devonian strata

Fig. 1d depicts composition diagrams of formation waters from the eastern shallow part of the Michigan Basin (data from Weaver et al., 1995). The linear positive correlations indicate that Cl,, Na, Ca, Mg and K (as well as Sr and Br) had a common allochtonous source, suggested by a Cl/Br ratio of ~130 to have been brine-spray. The linear correlation further indicates that these ions were little involved in secondary water-rock interactions. The isotopic composition of the waters is distinctly light: δD = -83 to -36 ‰ and $\delta^{18}O$ = -9.5 to -5.9 ‰ , revealing an affinity to non-evaporated meteoric water.

3.5 Norwegian Shelf, petroleum-associated formation waters, Triassic-Jurassic sandstones

Composition of formation waters from the Norwegian Shelf, hosted in Upper Triassic - Jurassic sandstones interbedded with shales and mudstone, are presented in Fig. 1e (data from Egeberg and Aagaard, 1989). The Cl/Br ratios are in the range of 140 to 240, well in the limits of evaporitic brines. The concentration of Na, Ca, Mg, and Sr are positively correlated with the Cl concentrations, indicating that these cations are of the same external source of evaporitic origin. The δD values, -46 to -23 ‰, indicate an origin from meteoric water.

3.6 Sudbury, Canadian Shield, crystalline rocks

Saline groundwaters have been encountered in fracture-compartments within many crystalline shields, and Fig. 1f depicts the composition of such waters from the Sudbury mining region, Ontario (data from Frape et al., 1984). The waters resemble those observed in the sedimentary basins - Cl is linearly correlated to Na, Sr, Br and Ca, indicating a common allochtonous origin, and the Cl/Br ratio is ~120.

CONCLUSIONS

The hydrochemical history of groundwaters can be understood in terms of allochtonous and autochtonous sources of the dissolved ions, provided detailed chemical data are available. Positive linear correlations of the concentrations of Cl, Br and other accompanying ions in water samples from different locations within a study area indicate an allochtonous input. In recent groundwaters the latter is mainly sea-spray, whereas in fossil waters it was often brine-spray and evaporitic dust that were washed down by meteoric waters, and direct entrapment of residual brines or seawater are other possibilities. Lack of linear correlations between the concentrations of various ions indicates their origins were decoupled. In such cases two possibilities warrant consideration - supply from more than one kind of allochtonous source, e.g. spray from different types of brines, or contribution from water-rock interactions.

REFERENCES

Arad A. and Evans R. 1987. The hydrology, hydrochemistry and environmental isotopes of the Campaspe River aquifer system, north-central Victoria, Australia. *J. of Hydrology.* 95: 63-86.

Dollar P. S., Frape S. K. and McNutt R. H. 1991. Geochemistry of formation waters, southwestern Ontario, Canada and southern Michigan, U.S.A.: Implications for origin and evolution. *Ontario Geological Survey Open File Report* 5743, 72 p.

Egeberg P. K. and Aagaard P. 1989. Origin and evolution of formation waters from oil fields on the Norwegian shelf. *Applied Geochemistry.* 4: 131-142.

Frape S. K. and Fritz P. 1982. The chemistry and isotopic composition of saline groundwaters from the Sudbury Basin, Ontario. *Canadian J. of Earth Sci.* 19: 645-661.

Mazor E. 1997. Chemical and Isotopic Groundwater Hydrology - the Applied Aspect. Marcel Dekker Inc., New York, pp. 413.

Mazor E. and George R. 1992. Marine airborne salts applied to trace evapotranspiration, local recharge and lateral groundwater flow in Western Australia. *J. of Hydrology.* 139: 63-77.

Tellam J. H. 1995. Hydrochemistry of the saline groundwaters of the lower Mersey Basin Permo-Triassic sandstone aquifer, UK. *J. of Hydrology.* 165: 45-84.

Weaver T. R., Frape S. K. and Cherry J. A. 1995. Recent cross-formational fluid flow and mixing in the shallow Michigan Basin. *GAS Bull.* 107: 697-707.

High-rate denitrification from several electron donors in a schist aquifer

H. Pauwels
BRGM, Research Division, Orléans, France

O. Legendre & J.-C. Foucher
BRGM, National Mining Division, Orléans, France

ABSTRACT: The decrease in nitrate concentration in a schist aquifer in Brittany is partly interpreted as the result of denitrifying processes involving several electron donors - organic matter, pyrite, ferrous iron. At least the two first processes are enhanced by the presence of bacteria. A small-scale tracer test has demonstrated a high rate of autotrophic denitrification, explaining the lack of nitrate in some parts of the aquifer. Despite pyrite oxidation, however, the sulphate concentration in the water remains low due to the precipitation of secondary mineral phases such as jarosite and natroalunite; this is presumed to occur mainly in the network of small fissures, where denitrification is more effective, rather than in the large fissures or fractures.

1 INTRODUCTION

Nitrate contamination of both surface water and groundwater due to intensive agricultural activity is a crucial problem in Brittany (western France). In the Coët Dan catchment which is an area of particularly intensive farming, mean annual nitrate concentrations in the river at the basin outlet have frequently exceeded 60 mg.l^{-1} since 1989 (Cann, 1996), and groundwater contamination with values as high as 200 mg.l^{-1} are observed in the shallow zones (Pauwels, 1996). One process of nitrate removal from groundwater that has been observed in other contexts is natural denitrification (Hiscock et al., 1991). This process has been evidenced in the Coët Dan groundwater (Pauwels, 1994), where it is directed by (a) the mineralization of organic matter, and (b) the oxidation of pyrite (detected in rock samples collected during drilling).

The purpose of the present paper is to discuss the effectiveness of these two types of natural denitrification as determined from groundwater sampling and a tracer test.

2 GEOLOGICAL AND HYDROGEOLOGICAL SETTING

The Coët Dan basin is a small catchment (12 km²) located approximately 70 km southwest of Rennes. It is underlain by Proterozoic (530 Ma) schist, including sandstone, siltstone and claystone, with the top few metres comprising recent alluvium, weathered schist and sandstone from which pyrite has been dissolved.

3 HYDROCHEMICAL BASELINE CONDITIONS

High anthropogenic nitrate (as much as 200 mg.l^{-1}) and chloride concentrations have been measured in the shallow part of the Coët Dan aquifer. In addition, hydrogeological investigations peformed on sampling wells (Martelat et al., 1996) indicate that in places shallow groundwater flux penetrates into the deeper parts of the aquifer, potentially contaminating the underlying groundwater in nitrate and chloride. However, whereas SO_4 and CO_2 concentrations are observed to increase with depth, the nitrate concentrations decrease sharply (Fig. 1).

The decrease in NO_3 and increase in SO_4 and CO_2 concentrations with depth have been partly explained by two denitrifying processes (Pauwels, 1994):

1. Denitrification by the oxidation of pyrite:

$$5\ FeS_2 + 14\ NO_3^- + 4H^+ \rightarrow$$
$$7N_2 + 10\ SO_4^{2-} + 5\ Fe^{2+} + 2\ H_2O \qquad (1)$$

2. Denitrification by the oxidation of organic matter:

$$CH_2O + 4/5\ NO_3^- + 4/5\ H^+ \rightarrow$$
$$2/5\ N_2 + CO_2 + 7/5\ H_2O \qquad (2)$$

Nevertheless, the SO_4 and CO_2 concentrations, both products of the denitrifying processes, remain relatively low (Fig. 1), which could suggest that the two processes are limited.

Samples collected at various periods indicate that denitrifying bacteria are always present in the groundwater, even where nitrates are absent. They include *Thiobacillus denitrificans*, an autotrophic bacterium that uses sulphides, and also heterotrophic bacteria able to reduce nitrate to nitrite and gaseous nitrogen.

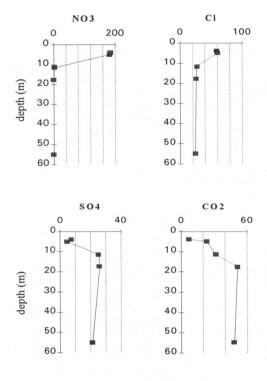

Fig. 1. Variation of nitrate, chloride, sulphate and carbon dioxide (mg.l^{-1}) as a function of depth.

4 SMALL-SCALE TRACER TEST

A small-scale tracer test with the injection of nitrate, accompanied with bromide as a conservative component was carried out between an injection well and a production well lying 15.2 m apart. The injection well was screened between 32 m and 82 m depth so that tracer transfer occurred only in the lower part of the aquifer where natural nitrate concentrations are below the detection limit.

The first indications of both tracers arrived virtually simultaneously only five and half hours after injection, with 73% of the bromide as against 47% of the nitrate being recovered after 10 days. According to the model of the observed breakthrough curves, tracer transfer takes place in a dual-porosity medium. Part of the tracer is transported rapidly (less than 40 hours) along large fissures or fractures (of the low-porosity / high-permeability medium). The other part arrives later, after migrating through the small fissures (high-porosity / low-permeability medium) (Pauwels et al., 1997). Global tracer recovery at 73% for bromides and 47% for nitrates indicates a rapid denitrification process.

5 RESULTS AND DISCUSSION

5.1 *Heterotrophic denitrification*

The total dissolved CO_2 increases with depth partly as the result of organic matter oxidation through nitrate reduction, but concentrations remain low and no carbonate mineral phase which could carry the CO_2 has been detected. Moreover, the DOC content in the surface waters is fairly low, indicating that heterotrophic processes of denitrification must be rather limited.

During the tracer test, the DOC in the injection well was too low to explain the nitrate decline by heterotrophic denitrification.

5.2 *Denitrification with oxidation of mineral compounds*

According to equation (1), denitrification through pyrite oxidation must involve an increase of sulphate concentrations. Actual concentrations in the groundwater vary from 5 mg.l^{-1} in the shallow part of the aquifer, where nitrate concentrations are high, to 30-35 mg.l^{-1} in the underlying part. Investigations by X-ray diffraction revealed the presence of two secondary sulphate minerals, jarosite ($KFe_3(SO_4)_2(OH)_6$) and natroalunite ($NaAl_3(SO_4)_2(OH)_6$), which could explain the limited SO_4 concentrations. This indicates that denitrification through pyrite oxidation is more important than would be expected from the SO_4 concentration.

Because no increase in sulphate concentration as the result of nitrate reduction was observed during the tracer test, precipitation of SO_4 in secondary minerals is also suspected

The precipitation of jarosite implies oxidation of Fe^{2+} to Fe^{3+}. In this case, NO_3^- is the only available electron donor and hence ferrous iron participates in the denitrification:

$$NO_3^- + 5\ Fe^{2+} + 6H^+ \rightarrow 1/2\ N_2 + 5\ Fe^{3+} + 3H_2O \quad (3)$$

Gallionella ferruginea, a bacterium capable of mediating such a reaction, was not detected in the water samples.

Ferric ions are also assumed to be partly removed from the water by precipitation of iron hydroxyde as they are observed in the cutting samples.

5.3 Denitrification kinetics

The rate of denitrification was also determined through the artificial tracer test. Since bromide is considered to be a conservative tracer, it is possible, from the Br concentration of each water sample collected at the wellhead of the production well, to calculate the concentration of nitrate in the produced water if denitrification had not occurred (i.e. $[NO_3]_0$). The nitrate concentrations were therefore plotted as $\ln([NO_3]/[NO_3]_0)$ vs. time from the tracer-injection pulse; in such diagrams, a first-order model is represented by a straight line. In the present case, two fairly linear segments are observed (Fig. 2a,b).

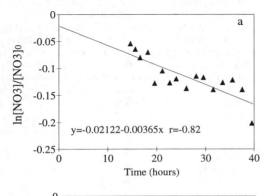

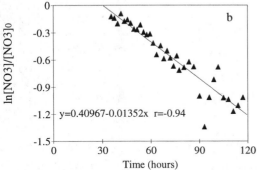

Fig. 2. Plot of the logarithm of nitrate concentration over expected nitrate concentration assuming no denitrification versus time. The solid lines mark the regressions: (a) tracer directly transferred through large fissures, (b) tracer having partly migrated through small fissures

The tagged water arriving in the production well during the first 40 hours, circulated mainly through the high-permeability medium (larger fissures); the denitrification rate determined from the slope gave a half life ($t_{1/2}$) of 7.9 days (Fig. 2a). The tracer arriving at the production well after 40 hours, mainly migrated through the low-permeability medium (smaller fissures); the slope of Figure 2b indicates a higher denitrification rate with a half life ($t_{1/2}$) of only 2.1 days.

A high autotrophic denitrification rate was concluded from the tracer test. By way of comparison, the half life for denitrification with pyrite in the Fuhrberg field, near Hannover, was found to be in the range of 1 to 2.3 years (Frind et al., 1990). During natural circulation in the Coët Dan aquifer, a similar rate to that of the tracer test can be expected. As heterotrophic denitrification is deduced from the chemical composition of the waters, it must occur simultaneously with the autotrophic denitrification, and thus at a high rate.

5.4 Nitrate transfer and denitrification

Nitrate concentrations in the deeper part of the aquifer are variable with time (Fig. 3). At some periods and in some places, concentrations below the limit of detection occur from a depth of 7 m. At other periods, nitrate is detected up to 100 m depth: rapid circulation and a lower denitrification rate in the larger fissures and fractures enable nitrate to reach such depths when the water flux is directed downwards. Nitrate contamination of the aquifer must thus be mainly through water circulation in the larger fissures, since waters are rapidly denitrified in the smaller fissures.

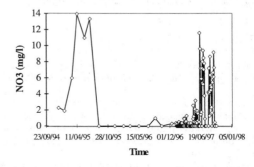

Fig. 3. Nitrate concentration (mg.l^{-1}) vs. time in a well screened over 7 to 16 m depth.

Jarosite and natroalunite have been observed in cuttings and their presence could explain the limited sulphate concentrations in the waters. However, thermodynamic calculations, using EQ3 software (Wolery et al., 1990), based on the chemical composition of the waters indicate that these are undersaturated with respect to both minerals, although saturation of some samples with respect to jarosite has been evidenced by using thermodynamic data of Baron and Palmer (1996). Nevertheless, the denitrification study from the tracer test demonstrated that the kinetics, at least of water-rock interactions, vary according to the size of the fissures. These variations must therefore involve differences in the chemical composition of the waters. Secondary mineral precipitation is probably not uniform in the aquifer and may occur preferentially where SO_4 production is the highest, i.e. in the smaller fissures.

6 CONCLUSIONS:

Three principal conclusions may be drawn from the chemical composition of groundwater and tracer behaviour:
1. Nitrate removal from the aquifer of the Coët Dan basin by denitrification involving three electron donors (organic matter, Fe(II) and pyrite) is rapid.
2. The rate of denitrification through pyrite oxidation is lower in open fractures and large fissures, which allow nitrate to contaminate the aquifer at depth.
3. Denitrification involving pyrite oxidation is more effective than denitrification involving organic matter oxidation: SO_4 and Fe concentrations remain low due to jarosite, natroalunite and iron hydroxide precipitation.

Acknowledgements: This is BRGM contribution N[o]. 97021. The technical equipment used on site was funded by CST-BVRE. The BRGM Translation Department is thanked for editing the English.

7. REFERENCES:

Baron D. and Palmer C.D. 1996 Solubility of jarosite at 4-35°C. *Geochim. Cosmochim. Acta* 60, 185-195.

Cann, Ch. 1996. Variations des teneurs en azote dans quelques cours d'eau bretons. In P. Mérot & A. Jigorel (eds), Hydrologie dans les pays celtiques. *Les Colloques* **79**, 193-201, INRA, Paris 1996.

Frind, E.O., Duynisveld, W.H.M., Strebel, O. & Boettcher, J. 1990. Modeling of multicomonent transport with microbial transformation in groundwater: the Fuhrberg case. *Water Resour. Res.* 26, 1707-1719.

Hiscock, K.M., Lloyd, J.W. & Lerner, D.N. 1991. Review of natural and artificial denitrification of groundwater. *Wat. Res.*, 25 (9), 1099-1111.

Martelat, A., Lachassagne, P. & Pauwels, H. 1996. Dénitrification naturelle dans un aquifère pyriteux du Morbihan (France), caractéristiques hydrogéologiques du système. Actes du 1er Colloque Interceltique d'Hydrologie et de Gestion des Eaux - Bretagne 96 - Rennes 8-11 July, Edition INSA, 23-24.

Pauwels, H. 1994. Natural denitrification in groundwater in the presence of pyrite: preliminary results obtained at Naizin (Brittany, France). *Min. Mag.*, 58A, 696-697.

Pauwels, H. 1996. Preliminary results on chemical variations in a schist aquifer: implications for nitrate transport and denitrification. In P. Mérot &. A. Jigorel (eds.), Hydrologie and les pays celtiques. *Les Colloques* **79**, 111-117, INRA, Paris 1996.

Pauwels, H., Kloppmann, W., Foucher, J-C, Lachassagne, P. & Martelat, A. 1997. Tracer tests applied to nitrate transfer and denitrification studies in a shaly aquifer (Coët-dan basin, Brittany, France). 7th Proceedings on *Water Tracing*, Slovenia, May 1997, In A. Kranjc (ed.) *Tracer Hydrology*, 327-330, Balkema, Rotterdam.

Wolery, T.J., Jackson, K.J., Bourcier, W.L., Bruton, C.J., Viani, B.E., Knauss, K.G. & Delany, J.M. 1990. Current status of the EQ3/6 software package for geochemical modeling. In D.C. Melchior and R.L. Basset (eds), Chemical Modeling of Aqueous Systems II. *American Society Symposium Series, 416*, 104-116, American Chemical Society.

Sorption of fluorescent tracers in a physically and chemically heterogeneous aquifer material

T. Ptak & H. Strobel
Applied Geology, University of Tübingen, Germany

ABSTRACT: Sorption properties of two fluorescent groundwater tracers, Fluoresceine and Rhodamine WT, were investigated for different lithocomponents and grain size fractions of aquifer material from the 'Horkheimer Insel' test site in Germany. While Fluoresceine proved to be practically non-sorbing, Rhodamine WT showed a sorptive behaviour, where the sorption capacity can be described as a function of the grain size. The results can be used for the interpretation of field scale tracer tests and for the description of aquifer properties controlling the transport of reactive solutes.

1 INTRODUCTION

If aquifer transport parameters are needed, e.g. for modelling of flow and transport in groundwater or for the planning of aquifer remediation activities, transport experiments at field scale can be performed. Since in general it is not possible to inject contaminants into groundwater due to law regulations, tracers are used as a substitute. Very often fluorescent tracer dyes are preferred, since they can be detected very easily and at very low concentrations, using a fluorimeter. Most tracer testing was concerned with the application of non-reactive tracers. However, if reactive transport has to be investigated, and if contaminants cannot be used for the transport experiments, reactive (e.g. sorbing) tracers, with a reactive behaviour according to the contaminant investigated, can be used instead.

If two tracers, one non-reactive and one reactive, are injected simultaneously into an aquifer within a tracer experiment, the chemical aquifer properties and the effects of chemical aquifer heterogeneities can be investigated directly. In this differential approach, both tracers experience the same hydraulic conductivity field, i.e. the same physical heterogeneity. Therefore, the difference in transport behaviour of the two tracers (described e.g. by effective retardation factors, second order moments of tracer breakthrough curves etc.) can be attributed to the chemical aquifer properties only.

This differential tracing approach was applied at field scale at the 'Horkheimer Insel' test site in South Germany, using Fluoresceine and Rhodamine WT tracers (Ptak & Schmid, 1996; Kleiner, 1997). For the interpretation of the field tracer experiments, it was necessary to obtain an insight into the sorption properties of the two tracers used.

Fluoresceine ($C_{20}H_{10}Na_2O_5$) and Rhodamine WT ($C_{29}H_{29}N_2O_5Na_2Cl$) are green and orange fluorescent tracer dyes respectively. Both are polar and ionizable. Therefore, empirical equations relating the distribution coefficient (K_d) to the octanol/water partition coefficient (K_{ow}) and to the organic carbon content of the aquifer material (f_{oc}), developed for non-polar chemicals, are not necessarily valid. Adsorption on mineral surfaces (clay, carbonate, quartz) may dominate the sorption of ionic compounds if the organic carbon content is low. Other factors affecting the sorption are pH, the valence and concentration of background ions, the concentration of the dyes itself, temperature, residence time and also the structural characteristics of the dye molecules. The sorption properties of these tracers in natural porous media were investigated e.g. by Sabatini & Austin (1991), Shiau et al. (1993) and Soerens & Sabatini (1994). In the experiments performed so far, bulk samples of aquifer material were used as sorbents. However, aquifer sediments are in general a heterogeneous mixture of different lithocomponents with different grain size fractions (e.g. Grathwohl & Kleineidam, 1995). In order to investigate if and how the sorption of the tracers depends on the lithological and grain size composition of the aquifer material, laboratory batch experiments were performed, described in the following.

2 EXPERIMENTAL APPROACH

Aquifer material samples of about 10 cm thickness from 100 mm diameter drill cores, obtained during construction of monitoring wells, were separated into grain size fractions and, in a petrographical analysis, into lithocomponents.

Then batch sorption experiments were performed for the different lithocomponents as well as for the individual grain size fractions. Sorption isotherms were measured and Freundlich isotherms were fitted (Strobel, 1996).

Figure 1 shows the lithological composition for the different grain size fractions and for the bulk aquifer material.

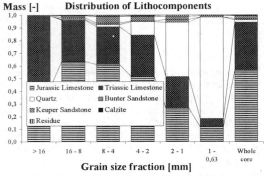

Figure 1. Lithological composition of the 'Horkheimer Insel' aquifer material for different grain size fractions and bulk material.

The Freundlich type partitioning is described by:

$$c_s = K_{Fr} \cdot c_{eq}^{\frac{1}{n}} \quad (1)$$

where c_s [μg l^{-1}] and c_{eq} [μg l^{-1}] are the equilibrium tracer concentrations in the solid and in the aqueous phases respectively. K_{Fr} [l kg^{-1}] and n^{-1} [-] are the Freundlich coefficient and the Freundlich exponent.

3 RESULTS FOR FLUORESCEINE

For the Fluoresceine concentration values at field scale, encountered during the field experiments, no significant sorption of Fluoresceine could be observed. As an example, Figure 2 shows the Fluoresceine concentration values measured in one of the batch experiments as a function of time.

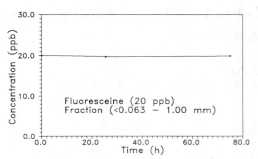

Figure 2. Fluoresceine concentrations as a function of time in one of the batch experiments.

From a practical point of view, it is in most cases admissible to accept Fluoresceine as a quasi ideal tracer (Käss, 1992). Fluoresceine is generally reported to be the least sorptive of all tracer dyes, especially when the organic carbon content of the aquifer material is low. The organic carbon content of the 'Horkheimer Insel' aquifer material can be as low as 0.2 mg g^{-1} (Strobel, 1996).

Following this experience and the experimental results, Fluoresceine can be assumed practically non-sorbing for the 'Horkheimer Insel' aquifer material.

4 RESULTS FOR RHODAMINE WT

Rhodamine WT is known to be significantly more sorptive (e.g. Sabatini & Austin, 1991). Figure 3 shows an example of Rhodamine WT concentrations in one of the batch experiments.

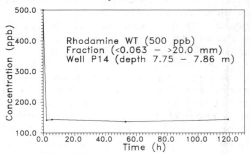

Figure 3. Rhodamine WT concentrations as a function of time in one of the batch experiments.

Within the batch experiments, sorption equilibrium c_{eq} was achieved almost immediately, which is typical for surface sorption, and no kinetic behaviour was observed at the time scale of the field tracer experiments (7 - 14 days).

Figure 4 shows the sorption isotherms of Rhodamine WT measured for the different lithocomponents and the fitted Freundlich sorption isotherms, together with the isotherm parameters.

The petrographical analysis (Figure 1) revealed, that within the samples consisting of Triassic limestone, Jurassic limestone, quartz, Keuper sandstone, Bunter sandstone and calcite the three components quartz, Jurassic and Triassic limestone dominated with about 95 % of the sample mass. Since the sorption properties (Freundlich coefficient K_{Fr}) were practically almost identical for these three components and similar for the remaining 5 % (Keuper sandstone, Bunter sandstone and calcite), it seems not necessary to regard the lithological composition of the aquifer material as a property controlling the sorption process in case of an aquifer material such as at the 'Horkheimer Insel' test site.

On the other hand, in batch experiments performed using the separated grain size fractions of the aquifer material, a relationship could be established between sorption capacity and grain size. As expected, the smaller diameter grain size fractions had a higher sorption capacity. This is typical for a compound such as Rhodamine WT, being sorbed to the surface of the grains. Figure 5 shows the sorption isotherms for Rhodamine WT for the grain size fractions investigated.

From the batch experiments, the following relationship could be obtained:

$$K_{Fri} = 0.132 \cdot d_i^{-1.0} \qquad (2)$$

where K_{Fri} is the Freundlich coefficient [l kg^{-1}] and d_i is the representative diameter [mm] of a grain size fraction i, and:

$$K_{Di} = \frac{1}{n_i} \cdot K_{Fri} \cdot c_{eq}^{\frac{1}{n_i}-1}$$
$$= 0.85 \cdot 0.132 \cdot d_i^{-1.0} \cdot c_{eq}^{0.85-1} \qquad (3)$$

where K_{Di} is the distribution coefficient [l kg^{-1}] at sorption equilibrium, c_{eq} [µg l^{-1}] is the equilibrium concentration and n_i^{-1} [-] is the Freundlich exponent of a grain size fraction i. Then, if the grain size distribution of an aquifer material sample is known, the effective K_D value of the whole sample is:

$$K_D = \sum_{i=1}^{m} x_i \cdot K_{Di} \qquad (4)$$

where m is the number of grain size fractions, x_i [-] is the mass fractional contribution of each grain size fraction to the total sample mass and K_{Di} is the distribution coefficient following equation (3).

5 CONCLUSIONS

Laboratory batch experiments performed for different lithocomponents and grain size fractions of aquifer material samples from the 'Horkheimer Insel' test site showed that there is practically no sorption of Fluoresceine. Rhodamine WT is significantly sorptive, with equilibrium concentrations achieved almost immediately. In case of the 'Horkheimer Insel' aquifer material, sorption capacity of Rhodamine WT can be described as a function of the grain size distribution only, without considering the lithological composition.

The tracers used, Fluoresceine and Rhodamine WT, may be viewed as substitutes to simulate the lower (no sorption) and upper (sorption) limits

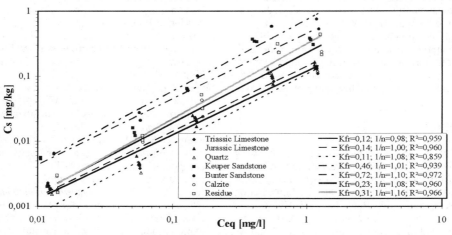

Figure 4. Sorption isotherms of Rhodamine WT for lithocomponents (obtained from the 2 - 4 mm grain size fraction).

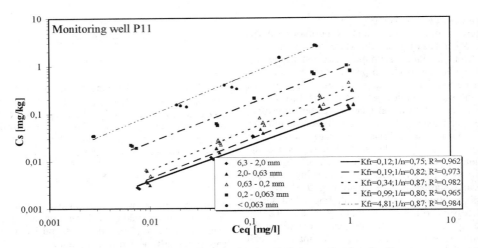

Figure 5. Sorption isotherms of Rhodamine WT for grain size fractions.

describing the transport of reactive compounds, e.g. some pesticides, adsorbing to mineral surfaces.

If the approach of characterizing the sorption properties of some tracers and/or contaminants for the individual aquifer material lithocomponents and grain size fractions is applied, costly laboratory experiments have to be performed only once. Then, if a new site has to be characterized with respect to reactive transport properties, it is only necessary to investigate the lithological and grain size composition, and the effective parameters, controlling reactive transport, can be obtained from the combination of the individual contributions of the lithocomponents or grain size fractions, following e.g. equation (4).

This approach was introduced into a numerical stochastic modelling framework to describe the transport of Rhodamine WT in a physically and chemically heterogeneous aquifer in Ptak (1997).

Future applications are in preparation, especially for the transport of such contaminants in groundwater, which require also the consideration of the lithological composition of the aquifer material.

REFERENCES

Grathwohl, P. & Kleineidam, S. (1995) *Impact of heterogeneous aquifer materials on sorption capacities and sorption dynamics of organic contaminants.* Groundwater Quality: Remediation and Protection, IAHS Publ. No. 225, 79-86.

Käss, W. (1992) *Geohydrologische Markierungstechnik.* Lehrbuch der Hydrogeologie, Band 9, Verlag Gebrüder Borntraeger, Stuttgart, F.R.G.

Kleiner, K. (1997) *Erprobung von Multitracer-Analyseverfahren und Einsatz ausgewählter Tracer im Feldversuch.* Diplomarbeit, Geologisches Institut, Universität Tübingen, F.R.G.

Ptak, T. & Schmid, G. (1996) Dual-tracer transport experiments in a physically and chemically heterogeneous porous aquifer: Effective transport parameters and spatial variability. *J. Hydrol.,* 183(1-2), 117-138.

Ptak, T. (1997) *Evaluation of reactive transport processes in a heterogeneous porous aquifer within a non-parametric numerical stochastic transport modelling framework based on sequential indicator simulation of categorical variables.* Proc. First Europ. Conf. on Geostatistics for Environmental Applications, geoENV 96, 20.-22. November 1996, Lisbon, Portugal, Kluwer Academic Publishers.

Sabatini, D.A. & Austin, T.A. (1991) Characteristics of Rhodamine WT and Fluorescein as adsorbing ground-water tracers. *Ground Water,* 29(3), 341-349.

Shiau, B.J., Sabatini, D.A. & Harwell, J.H. (1993) Influence of Rhodamine WT properties on sorption and transport in subsurface media. *Ground Water,* 31(6), 913-920.

Soerens, T.S. & Sabatini, D.A. (1994) Cosolvency effects on sorption of a semipolar, ionogenic compound (Rhodamine WT) with subsurface materials. *Environ. Sci. Technol.,* 28, 1010-1014.

Strobel, H. (1996) *Sorption eines reaktiven Tracers (Rhodamin WT) in heterogenem Aquifermaterial vom Testfeld "Horkheimer Neckarinsel.* Diplomarbeit, Geologisches Institut, Universität Tübingen, F.R.G.

Influence of eruptive volcanic lithologies on surface and ground water chemical compositions, Lake Taupo, New Zealand

M.R.Rosen
Institute of Geological and Nuclear Sciences, Taupo, New Zealand

L.Coshell
Coshell and Associates, Perth, W.A., Australia

ABSTRACT: Assignment of water samples collected from rivers around Lake Taupo to the lithology through which the water passes, shows clustering into distinct groups related to volcanic lithologies present in the lake basin. The two main volcanic lithologies present in the Taupo Basin are rhyolite and andesite, but water collected from geothermal areas also forms a distinctive grouping. Surface water samples collected from greywackes plot in an intermediate position between rhyolite and andesite. Ground water samples taken from the rhyolite correlate with the surface water samples, particularly for cations, but a higher proportion of HCO_3 in the ground water samples compared with river waters suggests that HCO_3 is added to the ground water through degradation of organic matter in the soil zone. Lake Taupo water plots within the rhyolite field indicating that interaction with rhyolitic rocks controls the chemical composition of Lake Taupo.

1 INTRODUCTION

The major source of ions, particularly sodium (Na) and chloride (Cl), that contribute to the chemical composition of Lake Taupo, a 616 km^2 caldera lake located in the center of the North Island, New Zealand, has been debated for a number of years. This is because the concentrations of Na and Cl in the lake water are 2.4 times higher than their concentrations in the water that flows into the lake (Timperley and Vigor-Brown, 1986). Two hypotheses have been proposed to account for the excess salt in the lake water, 1) contributions from geothermal seeps under the lake and around the margin of the lake (Schouten, 1980), and 2) dissolution of pumice on the bed of Lake Taupo (Timperley and Vigor-Brown, 1985). The first hypothesis can account for excess Na and Cl (both present in geothermal water), but the second hypothesis cannot account for excess Cl (Cl is not present in appreciable amounts in pumice).

Mass balance calculations made by Timperley and Vigor-Brown (1986) demonstrated that the chemical compositions of most lakes in the Taupo Volcanic zone can be characterised by contributions from geothermal steam/water, precipitation, and water-rock interaction in cold spring inflow. According to their calculations, the chemical composition of Lake Taupo is similar to the composition of cold spring water, in which the weathering of pumice by carbonic acid is the major source of dissolved salts.

This paper defines the role of weathering of each major volcanic rock type in the Lake Taupo catchment, and shows that the water chemistry of individual subcatchments around the basin can be characterised by the degree of interaction with the type of volcanic rocks over which the water flows. The data can then be extended to the subsurface to determine the controls on the ionic composition of ground water. These data can be used to differentiate between the two theories for the composition of Lake Taupo water.

2 GEOLOGIC SETTING

Lake Taupo is the largest lake in New Zealand, and is a caldera formed from multiple eruptions of the Taupo volcano over the past 300,000 years. The size of the Lake Taupo Catchment is 2673 km^3 excluding the lake (Schouten et. al., 1981). Most of the rocks in the Taupo Catchment are volcanic in origin; with rhyolites most common to the north and east of the lake and andesites most common to the west and southwest (Figure 1). Southeast of the lake is a large area of greywackes located in the Kaimanawa Mountains. These mountains are the source of many of the streams entering Lake Taupo.

Schouten (1980) showed from river data that where the subsoil cover is derived from rhyolite or greywacke, Na forms the most important cation and the monovalent cations are in excess of the divalent cations. Where the subsoil cover is derived from andesite, Ca is the most important cation and divalent cations are in excess of monovalent cations.

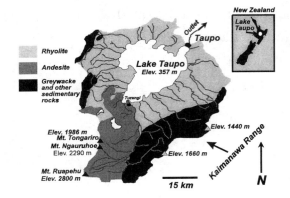

Figure 1. Map of the Taupo Catchment showing major rivers and rock types.

This paper extends these observations to ground water analyses, and geothermally influenced ground water.

3 METHODS

Chemical analyses of rivers and geothermal streams measured by Schouten et al. (1981) were characterised by the lithology over which the water flowed and given different symbols on a piper diagram (Figure 2). Additional analyses of geothermally influenced ground water were taken from Gibbs (1979), cold ground water springs from Timperley (1983) and Rosen, (unpublished data), and average lake water from White et. al. (1983).

Geological and petrophysical examination of rhyolitic rocks in the Waimarino River catchment (10 km northeast of Turangi) were undertaken to determine different types of ground water aquifers, and to quantify the flow characteristics of the deep ignimbrite. Pump-tests were conducted on a ground water monitoring well that penetrated the ignimbrite, and permeability tests were conducted on core material from the ignimbrite.

4 TYPES OF GROUND WATER AQUIFERS

Although ground water accounts for only 5% ± 3% of discharge into Lake Taupo (Schouten, 1980), ground water can be locally significant to the overall discharge pattern to the lake because of the highly permeable nature of the rhyolitic pumices to the northeast of the lake. In addition, deeper ground water discharges to the lake are unquantified, and it is unknown if groundwater leaves Lake Taupo.

Three types of aquifers are present in the Taupo catchment (Rosen et. al., 1995): 1) shallow, high transmissivity aquifers (<10 m) composed of rhyolitic and andesite pumice, 2) shallow to deep (>30 m) river derived sand and gravel aquifers, many containing greywacke boulders and cobbles, with moderate transmissivities that are confined to river valley systems, and 3) deep (>40 m) fractured, welded ignimbrite aquifers with low transmissivities. Minipermeameter and core plug measurements have shown that the matrix permeability is low (Figure 3), but is of the same order of magnitude as calculated from pump-test transmissivities ($T=1.2 \times 10^{-4}$ m^3/min), which include fracture permeability. This indicates that fractures may not be the controlling factor in tranmissivity measurements.

Many aquifers in the southern part of the Taupo basin (both deep and shallow) are low in dissolved oxygen and contain moderately high concentrations of dissolved iron and manganese (>10 mg/L Fe, >1 mg/L Mn), and ammonium (>2 mg/L).

5 DISCUSSION

The piper diagram shown in Figure 2 illustrates the importance of water-rock interaction in determining the ionic composition of both surface and ground waters entering Lake Taupo. Rivers to the south and west of the Taupo catchment that flow through andesitic rocks plot in an area of the diagram where divalent cations are more abundant, and where sulfate is the most common anion. Conversely, rivers that flow through rhyolitic rocks (both ignimbrites and rhyolitic pumice) plot in areas where the monovalent cations are dominant, and bicarbonate is the most abundant anion. Rivers that flow over greywackes plot in a position intermediate to the andesite and rhyolite fields. Geothermal water has a very different composition to the water derived from water-rock interaction. Geothermal water is dominated by Na and Cl, and plots in a similar position to seawater. Cold ground water that is mixed with geothermal fluids, plots in an intermediate position between the geothermal and andesitic/rhyolitic fields. Cold ground water flowing through rhyolitic pumices plot within the same field of relative cation abundances as the surface water samples, but the anions show a higher abundance of bicarbonate. This is because recharge water must pass through the soil zone before it can reach the ground water table. Interaction with plant material and decay of organic matter increases the bicarbonate in solution (Stumm, 1994), and this extra bicarbonate is recharged to the aquifer. In addition, dissolution of pumice also contributes bicarbonate to solution (Timperley and Vigor-Brown, 1985).

Average analyses of Lake Taupo water plot within the rhyolite field, indicating that weathering of rhyolitic rocks is the most important factor

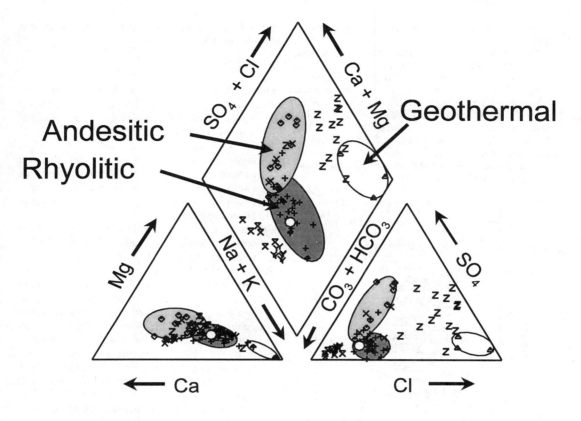

Figure 2. Piper diagram showing major ions in solution of rivers and streams, ground water, lake water, and geothermal water. Symbols are: surface water flowing over rhyolites (+), surface water flowing over andesite (◊), surface water flowing over greywacke (x), streams in geothermal areas (Δ), average Lake Taupo water (**O**), geothermally influenced ground water (z), cold ground water flowing through rhyolite (X,Y). Notice that rivers and streams flowing over rhyolite and andesite fall into different fields, and Lake Taupo water falls generally in the rhyolite field.

controlling the composition of Lake Taupo water.

Mass balance calculations have been carried out by many authors to determine the origin of major ions in Lake Taupo water (Schouten, 1980; Timperley and Vigor-Brown, 1985; 1986). However, these calculations were based on a limited knowledge of the ground water aquifers. The mass of dissolved ions in ground water was determined by subtracting all known surface water masses, assuming there was no loss of ground water from the basin. A total flow of 8 m^3/s was calculated by Timperley and Vigor-Brown (1985), but if the ground water flow from the shallow pumice aquifer near Taupo township is calculated, then this area alone, which captures only 1% of the Taupo basin, can account for 25% of this flow. Although Schouten (1980) stated that "there is no reason to believe that any significant subsurface outflow occurs from Lake Taupo", there have been no studies completed that deomonstrate this. Clearly, more work needs to be done to quantify the ground water input and outflows (if any) to the basin before further mass balance calculations can be attempted.

6 CONCLUSIONS

Water samples collected from rivers around Lake Taupo correlate with the lithology through which the water passes, show clustering into distinct groups related to rhyolite and andesite volcanic lithologies present in the lake basin. Water collected from geothermal areas also form a distinctive grouping. Ground water samples taken from the rhyolite correlate with the surface water samples, particularly for cations. Lake Taupo water plots within the rhyolite field indicating that interaction with rhyolitic rocks controls the chemical composition of Lake Taupo.

Three types of ground water aquifers have been recognised in the Taupo basin; 1) shallow (<10 m), permeable rhyolitic pumice aquifers, 2) shallow to deep (0 - 30 m) alluvial aquifers with intermediate permeabilities associated with river systems, and 3) deep (> 40 m) ignimbrite aquifers containing both fracture and matrix permeability of low transmissivity. The contribution of ground water to the chemical composition of Lake Taupo is uncertain due to the limited number of quantitative measurements of ground water flow.

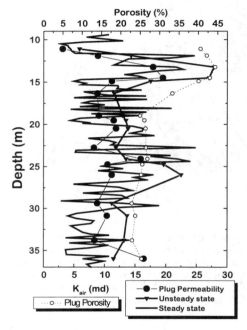

Figure 3. Comparison of permeability measurements from a 30 meter long ignimbrite core from the southeast portion of the rhyolite area. Note that the permeability is greater near the top of the core.

7 ACKNOWLEDGMENTS

We would like to thank James Brown from Core Lab Services for the use of the unsteady state permeability measurements and the core plug measurements. Arnon Arad provided stimulating discussion during his sabbatical leave to GNS in 1994 and helped us to formulate some of the ideas in this manuscript. Funding for this project was provided by the Taupo District Council, Environment Waikato, LIRO, and NSOF. Reviews by W.C. Evans and J.R. Hulston greatly improved the manuscript.

REFERENCES

Gibbs, M.M. 1979. Ground water input to Lake Taupo, New Zealand: Nitrogen and phosphorus inputs from Taupo township. *New Zealand Journal of Science*, 22, 235-243.

Rosen, M.R., A. Arad & L. Coshell. 1995. The importance of groundwater to the chemical composition of Lake Taupo water: A reassessment. The first International Limnogeological Congress, [abstract], Copenhagen, Denmark, p. 115.

Schouten, C.J. 1980. Budget of water and its constituents for Lake Taupo, Hamilton Science Centre Internal Report No 80/53, 41 p.

Schouten, C.J., W. Terzaghi & Y. Gordon 1981. Summaries of water quality and mass transport data for the Lake Taupo catchment, New Zealand. Water & Soil Miscellaneous publications No. 24, National Water and Soil Conservation Organisation, Wellington, 167 p.

Stumm, W. 1994. Aquisition of solutes and regulation of the composition of natural waters. In G. Bidoglio & W Stumm (eds) *Chemistry of Aquatic Systems: Local and Global Perspectives:*1-31. Dordrecht: Kluwer Academic Pubishing.

Timperley, M.H. 1983. Phosphorus in spring waters of the Taupo volcanic zone, North Island, New Zealand. *Chemical Geology*, 38, 287-306.

Timperley, M.H. & R.J. Vigor-Brown 1985. Weathering of pumice in the sediments as a possible source of major ions for the waters of Lake Taupo, New Zealand, *Chemical Geology*, 49, 43-52.

Timperley, M.H. & R.J. Vigor-Brown 1986. Water chemistry of lakes in the Taupo, Volcanic Zone, Zealand. *New Zealand Journal of Marine and Freshwater Research*, 20, 173-183.

White, E. M. Downes, M. Gibbs, L. Kemp, L. Mackenzie & G. Payne 1980. Aspects of the physics, chemistry, and phytoplankton biology of Lake Taupo. *New Zealand Journal of Marine and Freshwater Research*, 14, 139-148.

Hydrogeochemical investigations on arsenic contamination of a shallow aquifer

Ch. Sommer-von Jarmersted, U. Maiwald, A. Pekdeger & M.Th. Schafmeister
FU Berlin, FR Rohstoff- und Umweltgeologie, Germany

ABSTRACT: In a former industrial area a high arsenic contamination with concentrations up to 80,000 mg/kg in soils and 100,000 µg/l in groundwater was detected. In this area sulfuric acid was produced until World War I. Arsenic and several heavy metals are the residues of the roasting process of sulfide ores. Wells for drinking water production downstream of the contaminated site already reveal significant arsenic concentrations in groundwater. A future use of the site as residential area is planned. Therefore a remediation plan has to be designed.

1 INTRODUCTION

Within an area of 500 by 150 m, which was formerly used as production site for sulfuric acid, high contaminations with arsenic and several heavy metals were detected. The sediments in the investigated site consist of sands and gravel. Maximum values of 80,000 mg/kg arsenic were found in sediment samples. Locally, in the center of the pollution the contamination was still detected at a depth of 22 m. Ground water analyses show arsenic concentrations up to 100,000 µg/l in the central part of the area. The maximum geogene background concentration of arsenic in sedimentary aquifers is about 1.5 µg/l (Schleyer & Kerndorff, 1992). Also in groundwater production wells several hundred meters away, significantly increased concentrations were detected during the last two decades.

Besides groundwater flow which leads to the lateral spreading of the contamination another mechanism must have influenced arsenic transport. Since the ground water table lies between 1.7 m and 3 m below the surface a vertical transport to a depth of 22 m is quite astonishing.

In order to design a reliable remediation plan hydrogeological and geochemical investigations were performed which mainly aimed at the following questions:

- Which arsenic bearing mineral phases occur ?
- In which order of magnitude can the arsenic be mobilized?
- Which are the main transport processes to depth?

2 INVESTIGATION METHODS

In order to determine the distribution of arsenic bearing minerals through depth, sediment samples of the different layers were taken from a 15 m deep bore hole in the center of the polluted area. Additionally 38 characteristic samples from drillings distributed over the whole contamination area were selected to characterize the contamination in its lateral extension and homogeneity. Twenty-four samples were chosen from 10 drillings up to 2 m depth, 14 samples were selected out of 40 bore holes up to depth of 6m. A detailed description of the geological setting was possible by these drillings.

Extensive laboratory investigations, including microscopy, REM, EDX, aqua regia digestion and elution analyses, gave information about the various phases of arsenic.

The mobility of arsenic within the area was investigated by a longtime pumping test. The results of these investigations provided the fundamentals for a transport model which is used as a base for the remediation plan.

3 RESULTS

The uppermost 2 m of the contaminated area consist of a slag-bearing debris layer. In the center of pollution, besides the slag, a coloured powdered matrial is observed. These anthropogenic sediments are underlain by fluviatile sands and gravels which can easily be differentiated by their characteristic colours. Six different lithotypes were recognized in the center of pollution. Maximum concentrations up to 80,000 mg/kg are detected within the debris layers. Significant concentrations varying between 2 mg/kg and 60,000 mg/kg are also found in the sand layers. Figure 1 illustrates the spectra of concentrations within the different lithotypes and the eluted amount of arsenic.

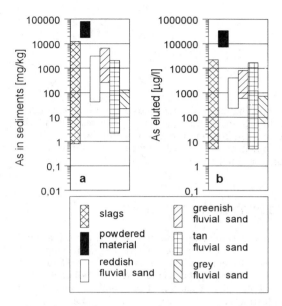

Fig.1 a) Arsenic concentration ranges in different lithotypes, determined by aqua regia digestion. b) water soluble arsenic in different lithotypes, determined by elution.

Drill-cores from the central part of the site usually show high arsenic values still at great depths. Figure 2 shows an example of the vertical arsenic and $CaCO_3$ concentrations in a 15 m deep bore hole. The arsenic contamination in the subsurface of the investigated area shows a very characteristic shape: at the top we find contaminations distributed over the entire site. Downwards the contamination reveals locally cone shaped structures up to several meters below the groundwater table (Fig. 3).

Different arsenic bearing minerals were identified by microscopic, REM and EDX analyses which are typical for different lithotypes. The slags and the powdered material within the debris layers are definitely the suppliers of arsenic. The occurrence of As-S, Fe-S and Fe-Si-Al in different quantities proves the existence of metal sulfides and/or -sulfates and of metal-bearing slags in the debris layers.

The reddish and greenish colors of the fluvial sands are caused by thin coatings on quartz particles.

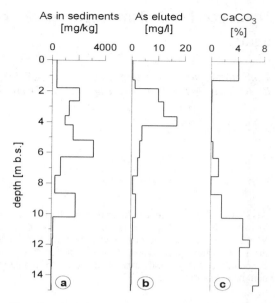

Fig.2 Vertical distribution in a borehole in the center of contamination: a) arsenic in sediments, determined by aqua regia dilution, b) water soluble arsenic, c) $CaCO_3$ content of sediments.

By EDX-analyses they could clearly be identified as secondary mineral phases, like Ferro-oxides and hydroxides with high arsenic contents. Additionally Fe- and Ca-rich arsenates (scorodite type) were found. They are typical weathering products of As- and FeAs-sulfides (Bowell, 1991). Carbonates like calcite, ankerite and siderite were found in the sands as well. At shallow depths (< 6 m) below the debris layers the sediments are strikingly free of carbonate. Below 6 m the content of the Calcite and the other carbonates increases (Fig.2).

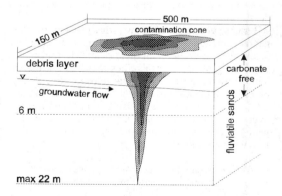

Fig.3 Schematic illustration of the main contamination cone.

The analyses of the sediment elutes reveal again the highest metal concentrations in the debris layers on top. Figure 4 shows the correlation between the

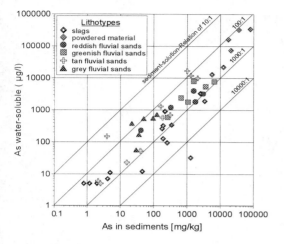

Fig.4 Relationship between arsenic in sediments and water-soluble content in different lithotypes.

solid metal concentration and the eluted arsenic. A broad spectrum of arsenic mobilization can clearly be seen which shows no relationship to different lithotypes.

4 CONCLUSIONS

During decades of sulfuric acid production the residues were widely deposited over the former production site and are responsible for the shallow contamination of the whole area. The deep contamination cone can be explained by heavy liquids, which carried the metals perpendicular to the lateral groundwater movement. Such a liquid might have been e.g. highly concentrated sulfuric acid ('oleum', density 1,84 g/cm^3) which has been oozed into the subsurface by accidental leakage. During the leakage of this liquid the arsenic bearing minerals (mainly oxides and sulfides) were dissolved in the top layers and carried downwards. On its way to depth the acid was increasingly neutralized by carbonates. Arsenic and several heavy metals were precipitated and co-precipitated as oxides, hydroxides, carbonates and sulfates. This processes can be characterized as slowly downward moving reaction fronts. The missing carbonate content in the upper sediments and the analyzed mineralisation in deeper sediments is in good accordance with this thesis. The ferrihydrite in the sediments is of primary interest for the sorption and coprecipitation of arsenates (Davis et al., 1989; Bowell, 1994). EDX-analyses confirm this processes.

The linear relationship between acid soluble and water soluble arsenic concentrations, no matter which arsenic mineral or lithotypes occur, raises the following consequences for a possible remediation scheme for this large area. To prevent further dissolution and vertical transportation of contaminants by precipitation and seepage the shallow contamination in the debris layer has to be covered with impermeable materials.

The main contamination at greater depth which is located several meters within the saturated zone is characterized by high arsenic concentrations of the groundwater. Therefore this zone must be surrounded by a scaling wall to prevent the horizontal transportation by ground water flow.

The results of the mobilization test and the transport modeling must show if a hydraulic clean up of the contamination is possible.

5 REFERENCES

Bowell, R.I. 1991. The mobility of gold in tropical rain forest soils.- PhD thesis, 438 p., University of Southampton.

Bowell, R.I. 1994. Sorption of arsenic by iron oxides and oxyhydroxides in soils. *Appl. Geochem.*, 9, pp. 279-286.

Davis, I.A., C.C.Fuller, B.A. Rea & R.G. Claypool-Frey 1989. Sorption and coprecipitation of arsenate by ferrihydrite.- In: Miles (ed.). Water-Rock-Interaction, pp. 187-189.

Schleyer, R & H. Kerndorff 1992. Die Grundwasserqualität westdeutscher Trinkwasserresourcen, 249 p., VCH Weinheim.

Remediation of high fluoride groundwaters from arid regions using heat-treated soils: A column experiment study in Xinzhou, China

Yanxin Wang, Yuan Ximing & Guo Huaming
China University of Geosciences, Wuhan, People's Republic of China

Wang Hong & Wang Yangen
Water Conservancy Bureau of Shanxi Province, Taiyuan, People's Republic of China

ABSTRACT : Cost-effective remediation of high fluoride groundwaters is urgently needed to eliminate fluorosis which is the most widespread endemic disease in China. Three types of Cenozoic soils have been collected *in situ*, heat-treated and then used as fluoride-removing adsorbents in column experiments on water samples from Xinzhou. Hydraulic conductivity (K) and working exchange capacity (WEC) measurements and fluoride removal experiments under static and dynamic conditions show that one of these soil materials, the Cenozoic red soil after 400 ℃ heat-treatment is a permeable and powerful specific adsorbent for fluoride with K values between 0.05 and 0.11cm/s and WEC between 89.10 and 147.61mg/kg.

1 INTRODUCTION

Fluorosis is the most widespread endemic disease in China. It was roughly estimated that more than 40 million Chinese people are victims of endemic dental fluorosis (Zhen,1992), which is largely related to long-term intake of high fluoride waters. Most of these waters occur in the semi-arid and arid regions of northern China(Shen et al. 1990; Ren et al. 1996). Cost-effective remediation of high fluoride groundwaters is urgently needed for the sustainable social and economic development of these regions. Naturally occurring minerals capable of removing fluoride should be the first choice for remediating this problem from an economic point of view. It is well known that zeolite has long been employed to remove fluoride from industrial waste waters (Maier,1953) and from high fluoride groundwaters (Chai et al. 1992). In the present study, three types of Cenozoic soils were collected from the study area (Xinzhou Basin, Shanxi Province, northwestern China), reworked under different temperatures, and then used to treat high fluoride groundwaters from the study area.

2 SAMPLES AND METHODS

Three types of Cenozoic soils were used as fluoride-removing materials in this study: one is red soil and the other two are loess. Denoted respectively as S_1, S_2 and S_3, these soils are mainly composed of clay minerals and quartz, with up to 6.3% Fe_2O_3 and up to 530 ppm F.

The groundwaters used in our column experiments include three drinking waters and one thermal water, the hydrochemistry of which is summarized in Table 1. These waters are representative of the high fluoride groundwaters widely developed in Xinzhou, a rift basin of the Fengwei Graben (Wang et al. 1993). It can be seen that all the waters are enriched in Sr, SiO_2 and F. Our monitoring data show that their fluoride contents vary with time. For instance, the fluoride concentrations of sample 1 and 3 are respectively 4.0 and 4.6mg/l in November, 1994. In order to increase the permeability and exchange capacity of the soil materials, after being soaked in distilled water for 48h, they were separately put into a Muffle furnace with temperatures between 100 and 1000 ℃, at an interval of 100 ℃. After 2h, they were taken out, ground and sieved. Then each 10g material with grain size 0.25-0.5mm was put into a beaker containing 100ml water sample 2, and the amount of removed fluoride was measured after 48h. This experiment is called " static ".

At the same time, the heat-treated 100g soil materials of different sizes (0.10-0.25mm, 0.25-0.50mm and 0.5-1.0mm) were filled either in glass tubes with a diameter of 2.1cm or in small PVC columns with a diameter of 4.25cm. And the following dynamic experiments were made.

1. Hydraulic conductivity measurement: Using distilled water, the hydraulic conductivities of the soil materials were determined in the PVC column by keeping a constant water head difference over the length of the filled materials and by measuring the discharge at the outlet of the column. The hydraulic conductivity can then be calculated according to Darcy's law.

2. Working exchange capacity measurement: Fixed beds of the soil materials in the PVC column were eluted by standard solutions. Data regarding such elution runs at constant flow rate are collected in breakthrough curves in which the outlet concentration c of fluoride is reported as a function of the effluent volume V. In practice, the area delimited by the two coordinate axes, the value of inlet concentration c_0 and the value of breakthrough volume V_b (the volume of effluent at which outlet fluoride concentration approaches c_0 and then does not change sensibly), represents the mass of the fluoride adsorbed in the soil materials at the breakthrough point. From this, the value Working Exchange Capacity (WEC) for the soil material can be calculated by the formula:

$$WEC = \frac{\int_0^{V_b}(c-c_0)dV}{m_e} \qquad (1)$$

where m_e is the mass of the soil material.

3. Fluoride removal experiment: Fixed beds of the soil materials in the glass or PVC columns were eluted by the high fluoride groundwaters sampled from Xinzhou. The flow rate of the inlet waters was strictly controlled. After each 50ml volume of effluent, the fluoride content and pH values of the effluent were measured. As soon as the concentration of fluoride reached 1.0mg/l (The Chinese national standard value for drinking water), the experiment was stopped and the quantity of removed fluoride calculated.

A larger PVC column, 60cm in length and 19.5cm in diameter, was also designed. 4149g of heat-treated soil S_1 with grain size 0.25-0.50mm and 8402g with grain size 0.5-1.0mm were filled in this column and eluted by the high fluoride groundwaters from Xinzhou. The flow rate of the inlet waters was about $0.01 m^3/s$. Before the effluent's fluoride concentration reached 1.0mg/l, 55L of water sample 2, 145L of water sample 3 and 86L of water sample 4 had been treated.

3 RESULTS AND DISCUSSION

3.1 Hydraulic conductivity

Before heat-treatment, the soil materials are practically impermeable. The soil sample S_1 with grain size greater than 1.0mm has a permeability of only 0.003 cm/s. Heat-treatment effectively

Table 1. Hydrochemistry of high fluoride groundwaters sampled in October, 1995 from Xinzhou (in mg/l except pH and t)*

Sample No.**	1	2	3	4
pH	6.0	6.5	6.5	7.0
t(°C)	12.0	43.0	11.0	12.0
TDS	421	1468	445	198
Na^+	40.2	252.0	34.6	58.0
K^+	2.55	12.6	2.89	1.62
Ca^{2+}	76.5	241.0	87.8	68.9
Mg^{2+}	14.73	3.47	18.60	14.84
Al	<0.005	0.075	<0.005	<0.005
Li	0.004	0.096	0.011	0.004
Sr	0.51	0.91	0.47	0.37
Cl^-	20.9	174.0	34.0	25.8
SO_4^{2-}	22.4	710.0	43.3	35.8
HCO_3^-	250.0	65.8	226.0	247.0
SiO_2	15.3	34.9	15.7	15.9
F^-	1.50	5.00	1.25	1.30

* The Fe and Zn contents are below detection level (0.004mg/l); ** 1-Water supply well of the Beiguoxia Village; 2-Thermal water well of the Duncun Village; 3-Water supply well of the Houboming Village; 4-Tap water of the Sanitation and Antiepidemic Station of the Xinzhou Prefecture.

enhances their hydraulic conductivities (Table 2). Clearly, the K value of S_1 increases with heating temperature.

3.2 Working Exchange Capacity (WEC)

The WEC value of the soil sample S_1, which shows the greatest capacity for removing fluoride, has been measured (Table 3). It can be seen that the smaller the specific surface area, the greater the WEC value. Taking account of hydraulic conductivity, our best choice should be the heat-treated soil material S_1 with grain size 0.25-0.50mm.

Table 2. Hydraulic conductivities of the soil materials

Soil Sample	Heating Temp.(°C)	Grain Size (mm)	K(cm/s)
S_1	400	0.25-0.5	0.05
S_1	400	0.5-1.0	0.11
S_1	500	0.5-1.0	0.12
S_1	600	0.5-1.0	0.17
S_1	700	0.5-1.0	0.20
S_2	400	0.5-1.0	0.28
S_2	500	0.25-0.5	0.09
S_3	400	0.25-0.5	0.04

Table 3. WEC values of S_1

Grain Size(mm)	c_0(mg/l)	V_b(ml)	WEC (mg/Kg)
0.1-0.25	10.5	4000	147.61
0.25-0.50	11.0	3200	113.70
0.5-1.0	10.2	3200	89.10

3.3 Fluoride removal under static and dynamic conditions

Static experiments show that there is a best heating temperature between 400 and 500 °C for the soil material S_1 (Fig.1). Dynamic experiments with S_1 in glass and PVC small columns indicate that there is still a best heating temperature at 400 °C(Fig.1). Therefore, heat-treatment at 400 °C has an optimal

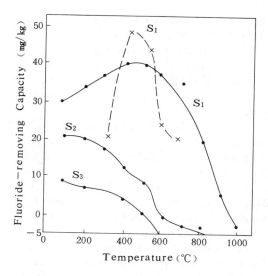

Figure 1. The change of fluoride-removing capacity of soil materials with heating temperature
●-Result of static experiments; ×-Results of dynamic experiments

Table 4 The chemistry of effluent samples of fluoride removal column experiments(in mg/l, the figure in brackets is the concentration before experiment as given in Table 1).

Sample No.	2	4
K^+	58.5(12.6)	10.9(1.62)
Na^+	252.5(252.0)	40.2(58.0)
Ca^{2+}	113.3(241.0)	53.5(68.9)
Mg^{2+}	22.5(3.47)	7.7(14.84)
Al	1.04(0.075)	0.35(<0.005)
Li	0.26(0.096)	0.18(0.004)
Fe^{2+}	0.83(<0.004)	0.14(<0.004)
Sr	0.70(0.91)	0.30(0.37)
Zn	0.02(<0.004)	0.02(<0.004)
Cl^-	180.2(174.0)	27.9(25.8)
SO_4^{2-}	700.0(710.0)	33.3(35.8)
HCO_3^-	25.1(65.8)	220.4(247.0)
H_2SiO_3	59.33(45.4)	49.65(20.7)
F^-	0.05(5.00)	0.28(1.30)

effect on the enhancement of both the permeability and the WEC of the soil material S_1.

The powerful fluoride-removing capacity of S_1 has been proved by the larger column experiment. Two effluent samples were collected during the experiment. The chemistry of the samples is given in Table 4.

It is clear from Table 4 that the fluoride concentration has been lowered below 1.0mg/l. A comparison of hydrochemistry before and after the experiment shows that the concentrations of K^+, Al, Li, Fe^{2+}, Zn and Si are increased, whereas those of Ca^{2+}, Sr and HCO_3^- decreased. It is doubtless that all the increased components originate from the water-soil interactions in the columns.

Therefore heat-treated Cenozoic soil S_1 has been found to be a new type of cost-effective material for remediation of high-fluoride groundwaters from the Xinzhou Basin, where more than 300,000 people are victims of endemic fluorosis.

To reveal the mechanism of the soil material S_1's fluoride-removing capacity, infrared spectrometry and X-ray powder diffraction analysis were made which indicate that more than 60% of it is composed of illite, and the rest of quartz, kaolinite and feldspar. The major difference in mineral composition of the soil before and after heat-treatment is the partial loss of water and transformation of Fe^{2+} into Fe^{3+}. It is believed that most of organic substances in the soil has also been lost due to heat-treatment.

It is well known that for many clay minerals, there are both negative double layers on the crystal surface planes and positive double layers on the marginal surfaces of the clay slices. Heat-treatment at about 400 °C should have caused reasonable breakage of the margins to create more positive double layers. The role of Fe^{3+} is similar to that of Al^{3+} in creating more positive double layers. Among the anions commonly found in groundwaters, except OH^-, F^- has the strongest tendency of being adsorbed by some clay minerals. Therefore, the substantial increase of positive double layers due to heat-treatment should be responsible for the soil S_1's enhancement of fluoride-removing capacity.

Although some preliminary research on the activation and regeneration procedures using acids, alkali, Fe and Al salts has been made (Wang et al. 1997), more experimental work is still needed to understand the selective nature and the activation and regeneration properties of the soil materials used in this study.

ACKNOWLEDGEMENTS

This work was financially supported by the National Natural Science Foundation of China(NSFC No.49772158). Drs John R. Hulston and Michael Rosen are very much appreciated for reviewing the manuscript.

REFERENCES

Chai, H. et al. 1992. *Use of zeolite from Jinyun, China.* Beijing: Geology Press. (in Chinese)

Maier, F.J., 1953. Defluoridation of municipal water supplies. *J. Amer. Water Works Assc.* 45: 879-888.

Ren, F. et al. 1996. Geochemical environment of high fluoride groundwaters in northern China and the relationship between fluorine speciation and endemic fluorosis. *Acta Geoscientia Sinaca.* 17: 85-97 (in Chinese with English abstract)

Shen, Zh. et al. 1990. The characteristics of fluorine in groundwaters of north China and the significance of fluorite-water interaction to fluorine transport. *J. China Univ. Geosci.* 1: 93-97

Wang, Yanxin et al. 1993. A comparative hydrochemical study on mineral waters from the Xinzhou Basin and the Baikal Rift Zone. *Earth Science-J. China Univ. Geosci.* 18: 661-670. (in Chinese with English abstract)

Wang, Yanxin et al. 1997. *Remediation of high fluoride groundwaters using some natural soil-rock materials.* Wuhan: China University of Geosciences. (in Chinese.)

Zhen, B. 1992. *Endemic fluorosis and industrial fluoride pollution.* Beijing: China Environmental Science Press. (in Chinese)

4 Groundwater general

Trace element hydrogeochemistry of Mt. Etna, Sicily: Insight on water-rock interaction

A. Aiuppa & F. Parello – *Istituto di Mineralogia, Petrografia e Geochimica, Palermo, Italy*

P. Allard – *CNRS-CEA, Centre des Faibles Radioactivités, Orme de Merisiers, Gif sur Yvette, France*

W. D'Alessandro – *CNR, Istituto Geochimica dei Fluidi, Palermo, Italy*

A. Michel & M. Treuil – *Groupe des Sciences de la Terre, Lab. Pierre Sue, CEN-CEA Saclay, Gif sur Yvette, France*

ABSTRACT: The processes controlling trace element content in etnean groundwaters have been investigated. The results show that, according to previous studies of major elements and dissolved gases (CO_2 and He), the distribution of trace elements is controlled by water-rock interaction processes that act at low temperature, and in an acid environment favoured by rising magmatic CO_2-rich gases. Only for some samples and for the more volatile elements (As and Sb) can a different process be hypothesized (transport in the rising magmatic gas phase).

1 INTRODUCTION

Mt. Etna is an alkaline volcano with persistent activity. Its edifice grew during the last 500.000 years, inside an area of structural weakness located between the appennino-magrebide chain at N and the iblean foreland at S. Its activity is mainly effusive, along with an intense and continuous degassing from summit craters (35000 t/d CO_2 - Allard et al., 1991). Etnean aquifers are confined in highly permeable fractured lavas, and are limited at depth by impermeable sedimentary formations. ∂D and $\partial^{18}O$ data confirm the meteoric origin of waters, where concentrations and relative ratios of ionic species in solution are controlled by the interaction of a CO_2-rich magmatic gas phase with etnean basalt (Anzà et al., 1989, Allard et al. 1997). The interaction in the system water-gas-basalt acts at low temperatures (<50°C). The only evidences of high temperature processes have been recorded in the area of Paternò at south of the volcano, where a thermal aquifer (T≈150°C - Chiodini et al. 1996) lies inside the sedimentary basement under the etnean volcanites. In this work preliminary results on trace element distribution in etnean groundwaters are presented. Samples have been collected in two peripheral areas of the volcanic edifice (Zafferana in the E sector and Paternò in the SW sector) previously recognized as anomalously degassing.

2 MAJOR ELEMENTS

Hydrogeochemistry of etnean groundwaters has been studied by many authors (Anzà et al. 1989, Allard et al. 1997, D'Alessandro et al. 1997). Despite their purely meteoric origin, many of the analysed waters are highly enriched in bicarbonate and total dissolved carbon. This high content is linked to the gradual conversion of a magmatic CO_2-rich gas phase ($\partial^{13}C(CO_2) \approx -4‰$) during a low temperature interaction with the host basalts. Such a process is confirmed by the high positive correlation of HCO_3 (and also of TDC, pCO_2 and $\partial^{13}C$) with cations (generally Mg≥Na»Ca>K), all increasing in concentration in samples originating from a higher elevation on the volcano and/or having a longer pathflow. According to this fact, the interaction is likely to occur in the more central (elevated) parts of the volcano, rather than at the site of water discharge. The magmatic origin of the interacting gas phase is also confirmed by helium isotope data (6.9±0.2 Ra). A different chemical composition is observed for two samples (VS and ST) in the area of Paternò, which are likely to have interacted at higher temperature (T≈150°C) in the sedimentary basement. They are very saline brines with essentially a Na-Cl composition.

3 TRACE ELEMENTS

Sixteen water samples have been collected and stored in pre-washed nalgene bottles after filtration with 0.45 µm Millipore MF filters and acidification to pH ≈ 2 with ultrapure nitric acid. Samples have been analyzed for 36 trace elements (Li, Al, V, Cr, Mn, Fe, Co, Ni, Cu, Zn, As, Rb, Sr, Mo, Cd, Sb, Cs, Ba, W, Pb, Th, U, REE) by a Fison ICP-MS Plasmaquad PQ2+ at Pierre Sue Laboratory in Saclay, using internal standard method. No preconcentration procedure has been applied to the samples.

3.1 Alkalis and alkaline earths

Following Stumm and Morgan (1981), alkalis and alkaline earths are A-type metal cations, having an inert gas electronic configuration and forming stable complexes with hard anions (e.g. OH-). As a rule, they are present in solution as free ions, so having a high mobility in the aquatic system.

The observed order of abundance in etnean groundwaters is Sr»Rb>Li>Ba»Cs, with average concentrations (calculated excluding more concentrated VS and ST samples) being 434, 36, 25, 13 and 0.26 ppb respectively. This order is due to relative abundance in etnean lavas (Sr>Ba>Rb=Li>Cs) and to the different mobility of the elements during water-rock interaction. To better evaluate the last point, enrichment factors (EF) have been calculated following the equation: $EF = (C_i/C_{ref})_w / (C_i/C_{ref})_r$; where C_i and C_{ref} are the concentrations of the element i and of the reference element, w and r refer respectively to water and rocks. An EF value of 1 indicates that the element i has the same mobility during water rock-interaction as the reference element, here chosen as K. This is the case of Rb, which has the highest average EF (0.93), while Sr, Li and Cs have lower EF values (0.47, 0.43 and 0.37 respectively) and are thus less mobile than K. Secondary minerals responsible for their preferential uptake relative to K could be calcite and zeolites for Sr, magnesian minerals for Li and amorphous silica for Cs. Ba has the lowest EF value (0.02), and this could be explained by his high ionic radius and his tendence (like Cs) to be concentrated in clay minerals. All these minerals have been found in the low temperature alteration assemblages of basaltic rocks in an experimental study by Gislason and Eugster (1987). A different behaviour is observed for VS and ST, where EF values higher than 10 have been found for Li, Cs and Ba. It can be argued that these elements are enriched relative to K by the interaction with the sedimentary substrate, mainly constituted of clay formations.

Fig. 1 and 2 clearly show the high Li/K and Na/K ratios for VS and ST as compared to the same ratios in the basalt (straight line). Water samples show a linear dependence, whose slope is different from that of the rock line (EF<1 for Li and EF>1 for Na). Clearly in both graphs some samples from Paternò sector depart from water-sample line, toward the composition of VS and ST samples. Similar conclusions could be reached by Ba-K and Cs-K graphs. It is thus possible to conclude that the two studied areas (Zafferana and Paternò) show a different composition for alkalis and alkaline earths, as a results of the interaction with the impermeable basement in the Paternò area. This could be due to the different morphology of the substrate itself or to an higher thermal flux allowing the rise of thermal waters having interacted at depth with the clays.

3.2 Transition metals

Transition metal content in etnean groundwater is expected to be controlled by ferromagnesian mineral dissolution. Following Gislason and Eugster (1987), Ti-Fe oxides are stable during basalt-water interaction, being normally oversaturated in the solution (with oversaturation growing with decreasing temperature). Olivine and pyroxene, on the contrary, are generally undersaturated and release Mg, Fe, Mn, Cr, V, Co and Ni to the solution. Fe and Mn solubility in aqueous systems is generally limited by the formation of secondary minerals, and is governed by the pH and the Eh of the solution. Given their low pH, etnean groundwaters generally fall in the stability field of Kaolinite when plotting activitiy diagrams in the system $CaO-MgO-Na_2O-K_2O-SiO_2-Al_2O_3-H_2O$ at 25°C. No Fe and Mn are then lost by Mg-Fe-Mn smectites, and their removal is probably due to the formation of insoluble

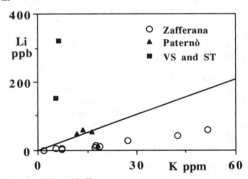

Fig. 1 - Li vs. K diagram.

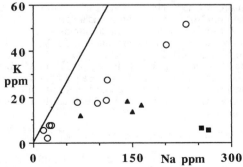

Fig. 2 - K vs. Na diagram.

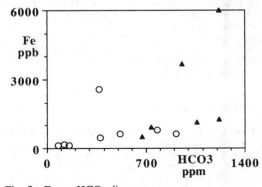

Fig. 3 - Fe vs. HCO_3 diagram.

amorphous Fe-Mn hydroxide. As previously pointed out by Giammanco et al. (1996), Fe/Mn ratios in etnean aquifers are generally higher than the same ratios in volcanic rocks. So the observed order of mobility (Mg»Fe>Mn), given the higher tendency of Fe with respect to Mn to precipitate as hydroxide in oxidizing environment, suggests a water-rock interaction process acting in a reducing environment. These reducing conditions are probably caused by the rising magmatic gases interacting with the aquifers.

As shown in Fig. 3, Fe concentrations correlate well with HCO_3 concentrations, but for three samples an anomalous Fe enrichment is observed, resulting from still more reducing conditions (Fig. 4). In the area of Paternò, where Fe concentrations are the highest (up to 6 ppm), the presence of a CH_4-rich gas phase has been reported (Chiodini et al. 1996). The discharge of these Fe-rich springs is followed by the precipitation of a brown colored Fe-Mn hydroxide. Where these extremely reducing conditions prevail, vanadium mobility is highly limited and its concentrations, generally very high (up to 150 ppb), are at minimum values (Fig. 5-6). Furthermore, this is probably also the result of V uptake in Fe-hydroxide. In more oxidizing environment, V forms more soluble complexes (vanadates) that have the same mobility as chrome (chromates) (Fig. 6).

3.3 Volatile trace metals

In the last years, increasing attention has been paid to the study of the role of high temperature volcanic gases in trasporting and releasing of heavy metals to the atmosphere. Previous studies on Mt. Etna (Buat-Menard and Arnold, (1978); Quisefitz et al. (1982)) revealed that a number of volatile elements (As, Sb, Cu, Zn, Pb, Se, Cd, Hg) are highly enriched in volcanic aerosols emitted by the summit craters relative to bulk magma composition. This high volatility is explained by their tendency to form strong complexes with halogens and sulfur compounds. As, Sb, Cu, Zn, Pb and Cd

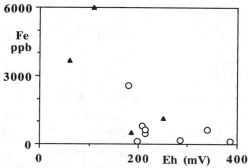

Fig. 4 - Fe vs. Eh diagram.

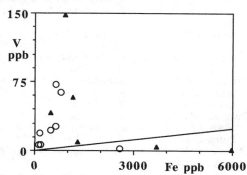

Fig. 5 - V vs. Fe diagram.

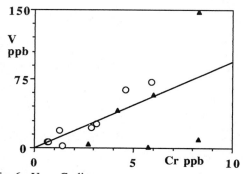

Fig. 6 - V vs. Cr diagram.

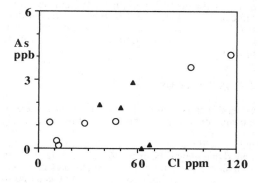

Fig. 7 - As vs. Cl diagram.

concentrations have been measured in etnean aquifers to evaluate their possible interaction with high temperature Cl-rich volcanic gases. With exception of Cd and Pb, whose concentrations (from 20 to 50 ppt) are close to the detection limits, all other elements are well intercorrelated, suggesting a common origin. Moreover, they correlate well with Cl concentration (Fig. 7) and to measured R/R_a ratios (Allard et al., 1997), as should be expected if these metals are brought to the aquifers by the dissolution of acid gases. This hypotesis, however, even if it can't be ruled out, seems to be unlikely, because good correlation are also recorded with HCO_3 (Fig. 8), Mg and transition metals, as it is to be expected if rock dissolution is responsable for As and Sb content in solution. Following this interpretation, the role of magmatic gases is not direct (hot Cl-rich magmatic

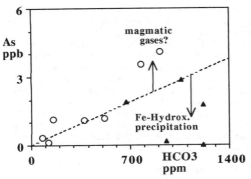

Fig. 8 - As vs. HCO$_3$ diagram.

gases transporting trace metals), but indirect ("cold") CO$_2$-rich magmatic gases leeding to intense rock dissolution). During basalt weathering, ground mass, where incompatible elements such as Cl, As and Sb are preferentially partitioned, is the most instable phase (relative to olivine, pyroxene and plagioclase; Gislason and Eugster, 1987). This could explain the enrichment of volatile trace elements relative to transition metals observed in etnean waters. Only for two more enriched samples in the Zafferana area a different, magmatic (?) origin for As and Sb seems more probable. Finally, it has to be pointed out that, like V, As solubility seems to be limited where extreme reducing conditions prevail (Fe-hydroxide uptake).

4 CONCLUSIONS

Several factors control the trace element content in etnean ground waters. The first limiting factor being rock composition, releasing metals to the solution during acid-attack favoured by CO$_2$ injection by the underlying magma chamber. As an example, the high vanadium concentrations in etnean basalt (up to 300 ppm) explain the high concentrations measured in solution. Basalt dissolution, however, is incongruent, so that the selective dissolution of different phases (groundmass > olivine > pyroxene > plagioclase > Fe-Ti oxide) and the formation of secondary minerals limit the solubility of trace elements. An order of mobility has been deduced calculating enrichment factors (EF) for alkalis, earth alkaline and transition metals. A petrochemical and petrographic study of altered etnean rocks is in progress to test these suggestions from water chemistry. For some elements (Fe, Mn and V) an important control seems to be also the Eh of the solution, defining their oxidation states. For others, like As and Sb, characterized by a high volatility, their direct derivation from the gas phase, even if improbable, can't be ruled out. If this is true, the samples having the highest As/Mg and As/HCO$_3$ (Petrulli and Ponteferro near Zafferana, Acqua Difesa near Paternò) are confirmed as the most affected by deep recharge and also the most important from the point of view of volcanic activity monitoring.

5 REFERENCES

Allard, P., J.Carbonelle, D.Dajlevic, J.Le Bronec, P.Morel, M.C.Robe, J.M.Maurenas, R.Faivre-Pierret, D.Martin, J.C.Sabroux & P.Zettwoog 1991. Eruptive and diffuse emissions of CO$_2$ from Mount Etna. *Nature* 351: 387-391.

Allard, P., P.Jean-Baptiste, W.D'Alessandro, F.Parello, B.Parisi & C.Flehoc 1997. Mantle-derived helium and carbon in groundwaters and gases of Mount Etna, Italy. *Earth Planet. Sci. Lett.*, 148: 501-516.

Anzà, S., G.Dongarrà, S.Giammanco, V.Gottini, S.Hauser & M.Valenza 1989. Geochimica dei fluidi dell'Etna. Le acque sotterranee. *Miner. Petrogr. Acta* 23: 231-251.

Buat-Ménard, P. & M.Arnold 1978. The heavy metals chemistry of atmospheric particulate matter emitted by Mt. Etna volcano. *Geophys. Res. Lett.* 5: 245-248.

Chiodini, G., W.D'Alessandro & F.Parello 1996. Geochemistry of the gases and of the waters discharged by the mud volcanoes of Paternò, Mt.Etna (Italy). *Bull. Volcanol.* 58: 51-58.

D'Alessandro, W., S. De Gregorio, G.Dongarrà, S.Gurrieri, F.Parello & B.Parisi 1997. Chemical and isotopic characterization of the gases of Mount Etna (Italy). *J. Volcanol. Geotherm. Res*, 78: 65-76.

Giammanco, S., M.Valenza, S.Pignato & G.Giammanco 1996. Mg, Mn, Fe and V concentrations in the ground waters of Mount Etna (Sicily), *Water Res.* 30/2: 378-386.

Gislason, S.R. & H.P.Eugster 1987. Meteoric water-basalt interaction: I. A laboratory study. *Geochim. Cosmochim. acta* 51: 2822-2840.

Quisefit, J.P., G.Bergametti & R.Vie Le Sage 1982. Nouvelle évaluation des fluxes particulaires de l'Etna-1980. *C.R. Acad. Sci.* 295 II: 943-945.

Stumm, W. & J.J.Morgan 1981. *Aquatic chemistry*. Wiley-Interscience.

Boron isotope geochemistry as a tracer for the evolution of natural aquatic systems

S.R. Barth
Department of Earth and Space Sciences, State University of New York at Stony Brook, N.Y., USA (Presently: Institut für Geologie und Paläontologie, Universität Tübingen, Germany)

ABSTRACT: Boron isotopic compositions ($\delta^{11}B$) and B concentrations in variably mineralized groundwaters from crystalline basement reservoirs in SW Germany and N Switzerland have been examined by negative thermal ionization mass spectrometry (NTIMS). Basement-hosted groundwaters from the study area are characterized by large variations in $\delta^{11}B$ from -3.5 to +17.6 ‰ (0.2-13.2 mg B/L). Boron isotope systematics, in combination with chemical parameters, are used to distinguish three distinct groundwater types and to constrain their evolution via water-rock interaction and/or admixture of external (sediment-derived) components. Boron isotopes are useful tracers for discerning distinct sources of dissolved salts in basement-hosted groundwaters and, hence, may help to trace the pathway of groundwater migration in the subsurface.

1 INTRODUCTION TO BORON ISOTOPE GEOCHEMISTRY

The potential of boron isotope geochemistry for tracing geochemical processes in a variety of geological environments has been demonstrated by a number of studies which considerably increased over the past ten years (see reviews by Bassett, 1990; Barth, 1993; Palmer & Swihart, 1996; Leeman & Sisson, 1996). This is mainly a consequence of the development and further improvement of the analytical techniques (reviews by Aggarwal & Palmer, 1995; Swihart, 1996).

The relatively large mass difference between the two stable isotopes of boron, ^{10}B and ^{11}B, leads to a wide range of boron isotope variations in nature ($\delta^{11}B \approx 90$ ‰ (Barth, 1993); $\delta^{11}B = \{[(^{11}B/^{10}B)_{Sample} / (^{11}B/^{10}B)_{NIST\ SRM-951\ Standard}] - 1\} \times 10^3$).
Natural waters encompass a range in $\delta^{11}B$ of 76 ‰, i.e. from -16 ‰ to +60 ‰. This is mainly the result of the large variability among diverse natural boron sources and of partitioning reactions during the adsorption of boron accompanied by isotopic fractionation. Such water-rock reactions account for instance for the enrichment in ^{11}B of seawater ($\delta^{11}B$ of +39.5 ‰; Spivack & Edmond, 1987) by ≈ 40-50 ‰ relative to average continental crust. The high $^{11}B/^{10}B$ ratio of seawater is attributed to isotopic fractionation related to differential uptake of the two dominant boron species, $B(OH)_3^0$ and $B(OH)_4^-$, during the adsorption of boron onto active surfaces in the marine environment.

2 BORON ISOTOPE VARIATIONS IN CONTINENTAL AQUATIC SYSTEMS

Boron is an ubiquitous minor or trace constituent in natural aquatic systems, derived from the interaction of fluids with the crust and/or mixing processes among fluids from different reservoirs. Several studies have investigated the boron isotope systematics of natural waters from various sediment-hosted continental settings (e.g. Palmer, 1991; Vengosh et al., 1991a, b, 1994, 1995; Land & Macpherson, 1992; Xiao et al., 1992; Stueber et al., 1993; Moldovanyi et al., 1993).

The boron isotopic composition of sediment-hosted fluids may be utilized as a sensitive tracer for discerning a marine *versus* non-marine origin of the boron source. This is evidenced by isotopically heavy boron isotope values of brines from Australian salt lakes and the Dead Sea ($\delta^{11}B$ of +26 to +60 ‰; Vengosh et al., 1991a, b) relative to distinctly lighter values of brines from salt lakes in Tibet (+1 to +15 ‰; Xiao et al., 1992; Vengosh et al., 1995) and geothermal fluids from the Salton Sea, USA (-3 to -1 ‰; Palmer, 1991). Saline formation waters from several sedimentary basins in the U.S. display a wide range in $\delta^{11}B$ from +12 to +50 ‰, i.e. +18 to +39 ‰ in the Illinois Basin (Stueber et al., 1993), +27 to +50 ‰ in the Gulf Coast Basin (Moldovanyi et al., 1993), and +12 to +40 ‰ in the Gulf of Mexico Basin (Land & Macpherson, 1992). A large part of these variations may result from interactions of the formation waters with sedimentary reservoir rocks.

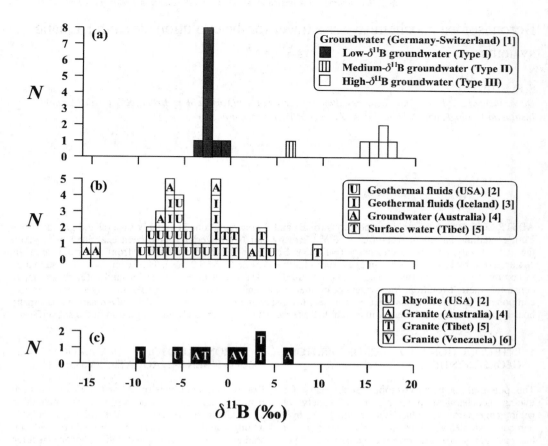

Figure 1. Histograms showing a comparison of $\delta^{11}B$ values determined for (a) groundwaters from crystalline basement reservoirs in SW Germany and N Switzerland [1] with available data reported for (b) natural waters hosted by diverse types of magmatic rocks [2-5], and (c) acidic volcanic and plutonic rocks [2, 4-6]. References: [1] Barth (this study), [2] Palmer & Sturchio (1990), [3] Aggarwal et al. (1992; meteoric water dominated fluids), [4] Vengosh et al. (1991a), [5] Vengosh et al. (1995), [6] Spivack et al. (1987).

Few studies have investigated the boron isotope systematics of natural waters hosted by magmatic reservoir rocks (e.g. Palmer & Sturchio, 1990; Vengosh et al., 1991a, 1995; Aggarwal et al., 1992) (see Fig. 1). Among these, geothermal fluids from the Yellowstone National Park, USA ($\delta^{11}B$ of -9 to +4 ‰) appear to reflect the boron isotopic compositions of related volcanic rocks, i.e. rhyolites with $\delta^{11}B$ values of -10 and -5 ‰ (Palmer & Sturchio, 1990). Similarly, groundwaters from the Great Artesian Basin, Australia ($\delta^{11}B$ of -16 to +2 ‰) and input waters to the Da Qaidam Basin, Tibet (-1 to +10 ‰) show ranges in $\delta^{11}B$ which largely overlap with those of associated plutonic rocks, i.e. granites with $\delta^{11}B$ values of -3 to +7 ‰ and -2 to +4 ‰, respectively (Vengosh et al., 1991a, 1995).

At variance, more variable $\delta^{11}B$ values (-2 to +36 ‰; not shown in Fig. 1) have been reported for brines hosted by granitic and metamorphic basement rocks of the Canadian Precambrian Shield (Bottomley et al., 1994). These variations have been interpreted in terms of infiltration of evaporitic brines representing remnants of marine incursions.

In order to constrain the potential of boron isotopes for tracing the evolution of groundwater via water-rock interaction and/or mixing processes with external fluids, variably mineralized groundwaters from crystalline basement reservoirs in SW Germany and N Switzerland have been investigated. It will be evaluated whether boron isotopes are useful tracers for discerning distinct sources of dissolved salts in basement-hosted groundwaters.

3 BORON ISOTOPE SYSTEMATICS OF GROUNDWATER FROM CRYSTALLINE BASEMENT RESERVOIRS

The results of boron isotope analyses, performed on a series of basement-hosted groundwaters (number of samples $N = 17$), are shown in Figs. 1 and 2, and are reported in Table 1 along with some relevant chemical data. Three dinstinct groundwater types may be distinguished: (1) low-δ^{11}B groundwater (Type I), (2) medium-δ^{11}B groundwater (Type II), and (3) high-δ^{11}B groundwater (Type III).

Low-δ^{11}B groundwaters furnish evidence for a systematic relationship of increasing salinity with slightly decreasing δ^{11}B values. This may reflect the evolution of Type I groundwaters by leaching of crystalline host rocks along the pathway of groundwater flow.

Water-rock interaction may have played an important role in the evolution of the medium-δ^{11}B groundwater (Type II) derived from a low-permeability zone within the crystalline basement. A pronounced degree of mineralization and a Br/Cl (molar) ratio of 0.0045, which is within the range of those determined for Type I groundwaters (0.0021-0.0071), is in agreement with this interpretation.

Admixture of external components by infiltration of sediment-hosted groundwaters, possibly derived from stratigraphically overlying Permian and Lower Triassic sedimentary strata, is considered as a major process in the evolution of high-δ^{11}B groundwaters. Type III groundwaters are characterized by high

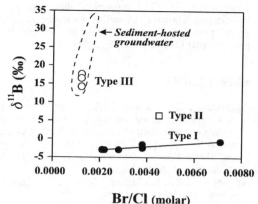

Figure 2. δ^{11}B versus Br/Cl (molar) ratios for groundwaters from crystalline basement reservoirs in SW Germany and N Switzerland.

δ^{11}B values and uniform low Br/Cl (molar) ratios of 0.0012 which are within the range of those determined for saline sediment-hosted groundwaters from overlying Permian and Lower Triassic strata (δ^{11}B = +6.4 to +33.0 ‰, Br/Cl = 0.0010-0.0017) as shown in Fig. 2.

Table 1. Boron isotopic and chemical compositions of basement-hosted groundwaters from SW Germany and N Switzerland.

Water Type	δ^{11}B [1] (‰)	B [1] (mg/L)	Cl [2] (g/L)	TDS [2] (g/L)
I	-3.5 to -0.6	0.2-1.3	0.03-0.2	0.5-1.1
II	+6.4	9.9	3.9	7.6
III	+14.2 to +17.6	0.8-13.2	0.4-3.6	3.3-7.9

[1] Boron isotopic compositions and B concentrations have been measured by negative thermal ionization mass spectrometry (NTIMS) with analytical uncertainties of ± 0.5 ‰ ($2\sigma_{mean}$) and ± 0.4 % (2σ), respectively (see Barth, 1997a, b). Boron isotopic compositions are reported as δ^{11}B values relative to a mean ^{11}B/^{10}B ratio of 4.00125 measured for the NIST SRM-951 standard (Barth, 1997c). Data are from Barth (this study). [2] Data are from Barth (this study) and Pearson et al. (1991).

4 CONCLUSIONS

The boron isotopic signatures of basement-hosted groundwaters which evolved without significant contributions from external sources (such as the Type I and Type II groundwaters) are within the range of δ^{11}B values reported for granitic rocks from a variety of locations world-wide (see Fig. 1). At variance, basement-hosted groundwaters which have evolved by admixture of sediment-derived components (such as the Type III groundwaters) are clearly set apart from the known range of granitic rocks by distinctively higher δ^{11}B values. Boron isotopes, in combination with other chemical parameters, may thus be utilized as sensitive tracers for discerning distinct (internal *versus* external) sources of dissolved salts in basement-hosted groundwaters. Application of boron isotope geochemistry may provide new insights into geochemical processes operating in the Earth's crust and hydrosphere and may help to trace the pathway of groundwater migration in the subsurface.

5 ACKNOWLEDGEMENTS

My thanks are due to MK Stewart and JR Hulston for reviews of the paper, and to GB Arehart for editorial handling. The support of this work by a Habilitation fellowship as well as by a related travel grant provided by the Deutsche Forschungsgemeinschaft DFG is gratefully acknowledged.

6 REFERENCES

Aggarwal, J.K., Palmer, M.R. & Ragnarsdottir, K.V. 1992. Boron isotopic composition of Icelandic hydrothermal systems. In Y.K. Kharaka and A.S. Maest (eds), *Proceed. 7th Intern. Symp. on Water-Rock Interaction WRI-7*, 893-895. Rotterdam: A.A. Balkema.

Aggarwal, J.K. & Palmer, M.R. 1995. Boron isotope analysis: A review. *Analyst* 120: 1301-1307.

Barth, S. 1993. Boron isotope variations in nature: A synthesis. *Geol. Rundsch.* 82: 640-651.

Barth, S. 1997a. Comparison of NTIMS and ICP-OES methods for the determination of boron concentrations in natural fresh and saline waters. *Fresenius J. Anal. Chem.* 358: 854-855.

Barth, S. 1997b. Boron isotopic analysis of natural fresh and saline waters by negative thermal ionization mass spectrometry. *Chem. Geol. (Isot. Geosci. Sect.)* [in press].

Barth, S. 1997c. Application of boron isotopes for tracing sources of anthropogenic contamination in groundwater. *Wat. Res.* [in press].

Bassett, R.L. 1990. A critical evaluation of the available measurements for the stable isotopes of boron. *Appl. Geochem.* 5: 541-554.

Bottomley, D.J., Gregoire, D.C. & Raven, K.G. 1994. Saline groundwaters and brines in the Canadian Shield: Geochemical and isotopic evidence for a residual evaporite brine component. *Geochim. Cosmochim. Acta* 58: 1483-1498.

Land, L.S. & Macpherson, G.L. 1992. Origin of saline formation waters, Cenozoic Section, Gulf of Mexico Sedimentary Basin. *Amer. Assoc. of Petrol. Geol. Bull.* 76: 1344-1362.

Leeman, W.P. & Sisson, V.B. 1996. Geochemistry of boron and its implications for crustal and mantle processes. In E.S. Grew and L.M. Anovitz (eds), Boron: Mineralogy, Petrology and Geochemistry, *Reviews of Mineralogy*, Vol. 33: 645-707. Washington: Mineral. Soc. of America.

Moldovanyi, E.P., Walter, L.M. & Land, L.S. 1993. Strontium, boron, oxygen, and hydrogen isotope geochemistry of brines from basal strata of the Gulf Coast sedimentary basin, USA. *Geochim. Cosmochim. Acta* 57: 2083-2099.

Palmer, M.R. 1991. Boron isotope systematics of hydrothermal fluids and tourmalines: A synthesis. *Chem. Geol. (Isot. Geosci. Sect.)* 94: 111-121.

Palmer, M.R. & Sturchio, N.C. 1990. The boron isotope systematics of the Yellowstone National Park (Wyoming) hydrothermal system: A reconnaissance. *Geochim. Cosmochim. Acta* 54: 2811-2815.

Palmer, M.R. & Swihart, G.H. 1996. Boron isotope geochemistry: An overview. In E.S. Grew and L.M. Anovitz (eds), Boron: Mineralogy, Petrology and Geochemistry, *Reviews of Mineralogy*, Vol. 33: 709-744. Washington: Mineral. Soc. of America.

Pearson, F.J. et al. 1991. Applied Isotope Hydrogeology: A Case Study in Northern Switzerland. 439 pp. Amsterdam: Elsevier B.V.

Spivack, A.J. & Edmond, J.M. 1987. Boron isotope exchange between seawater and the oceanic crust. *Geochim. Cosmochim. Acta* 51: 1033-1043.

Spivack, A.J., Palmer, M.R. & Edmond, J.M. 1987. The sedimentary cycle of the boron isotopes. *Geochim. Cosmochim. Acta* 51: 1939-1949.

Stueber, A.M., Walter, L.M., Huston, T.J. & Pushkar, P. 1993. Formation waters from Mississippian-Pennsylvanian reservoirs, Illinois basin, USA: Chemical and isotopic constraints on evolution and migration. *Geochim. Cosmochim. Acta* 57: 763-784.

Swihart, G.H. 1996. Instrumental techniques for boron isotope analysis. In E.S. Grew and L.M. Anovitz (eds), Boron: Mineralogy, Petrology and Geochemistry, *Reviews of Mineralogy*, Vol. 33: 845-862. Washington: Mineral. Soc. of America.

Vengosh, A., Chivas, A.R., McCulloch, M.T., Starinsky, A. & Kolodny, Y. 1991a. Boron isotope geochemistry of Australian salt lakes. *Geochim. Cosmochim. Acta* 55: 2591-2606.

Vengosh, A., Starinsky, A., Kolodny, Y. & Chivas, A.R. 1991b. Boron isotope geochemistry as a tracer for the evolution of brines and associated hot springs from the Dead Sea, Israel. *Geochim. Cosmochim. Acta* 55: 1689-1695.

Vengosh, A., Starinsky, A., Kolodny, Y. & Chivas, A.R. 1994. Boron isotope geochemistry of thermal spring from the northern Rift Valley, Israel. *J. Hydrol.* 162: 155-169.

Vengosh, A., Chivas, A.R., Starinsky, A., Kolodny, Y., Baozhen, Z. & Pengxi, Z. 1995. Chemical and boron isotope compositions of non-marine brines from the Qaidam Basin, Qinghai, China. *Chem. Geol.* 120: 135-154.

Xiao, Y.-K., Sun, D., Wang, Y., Qi, H. & Jin, L. 1992. Boron isotopic compositions of brine, sediments, and source water in Da Qaidam Lake, Qinghai, China. *Geochim. Cosmochim. Acta* 56: 1561-1568.

A small-scale dispersion experiment in a heterogeneous sandy aquifer, Botany Sands aquifer, Sydney, Australia: Tracer movement and interaction with geological material

Peter Beck & Jerzy Jankowski
Groundwater Centre, Department of Applied Geology, University of New South Wales, Sydney, N.S.W., Australia

ABSTRACT: A small-scale natural gradient tracer experiment was carried out at the East Lakes Experimental Site (ELE) in the Botany Sands aquifer. The experiment was conducted to examine the transport, dispersivity and interaction of aquifer material with several conservative and non-conservative tracers. The behaviour of the non-reactive tracer boron and two reactive tracers lithium and potassium are the focus of this paper. Ten sampling sessions, over a period of 32 days showed that boron migrated at a rate of 0.34 m/day, travelling a total of 10.8 m. At the same time lithium and potassium migrated at average rates of 0.12 m/day and 0.34m/day travelling 3.9 m and 10.8 m respectively. Mean longitudinal dispersivity values of 0.0145, 0.0148 and 0.0243 were calculated for B, Li^+ and K^+. The shape of the plume recorded by these three tracers suggests that aquifer material and hydraulic conductivity strongly affected the transport behaviour of the plume.

1 INTRODUCTION

Recent scientific publications have discussed five comprehensive and unique field test experiments, under natural gradient conditions (Mackay et al. 1986; Moltyaner & Killey 1988; Boggs et al. 1992; Garabedian et al. 1991; LeBlanc et al 1991; Jensen et al. 1993). All of these tests provided detailed information on hydrogeological conditions affecting transport of solutes, gave new data for the understanding of dispersion processes and helped to validate theoretical equations. The conservative tracers used in these studies were either Br^- or Cl^-. In prior investigations the reactive inorganic tracer used was Li^+ (Garabedian et al. 1991; LeBlanc et al 1991). Organic tracers were used at the Borden (Mackay et al. 1986) and Columbus (Boggs et al. 1992) sites. The field experiments discussed were conducted on large distances, ranging from 20 to 282 m (Killey & Moltyaner 1988; LeBlanc et al 1991). Results showed that the transport of tracers was affected by variations in dispersivity and aquifer heterogeneity, as these play an important role in dispersive mixing of tracers, over long distances. The small-scale natural gradient tracer test, discussed here, was undertaken in a heterogeneous sandy aquifer to obtain detailed information on the movement and behaviour of several tracers, in a small portion of the aquifer.

2 SITE DESCRIPTION

The East Lakes Experimental (ELE) site is located in the central part of the Botany Sands aquifer, in the Botany Basin, Sydney. The site was established in 1992 for the detailed study of sorption, dispersion and advection of non-reactive and reactive tracers. A dense three dimensional network of 49 multilevel piezometers has been installed within the upper shallow part of the aquifer on a grid of 7 m by 11 m (Figure 1). A total of 815 sampling points are available, horizontally these points are set out on a one metre grid. In the vertical sense the sampling points are between 150 and 200 mm apart (Figure 2). The close spacing of the sampling points has helped to resolve the sites geology and the vertical resolution available provides an insight into the hydrogeochemical heterogeneity of the aquifer (Evans, 1993; Jankowski et al. 1994, 1997).

The dominant lithology across the site is quartz sand, with sieve analyses indicate these sands are 77 - 95 % of medium grained, 3 % - 10 % fine grained with 2 % - 14 % silty-clay material. Grain size analyses show that the highest clay content exists between 7.5 m and 8.5 m a.s.l. The zone has a higher fine sand content (5 % to 8 %) than the aquifer above and below this layer (3 % to 4 %) (Figure 2). The presence of the fine sand and the silt/clay contribute to a hydraulic conductivity value below

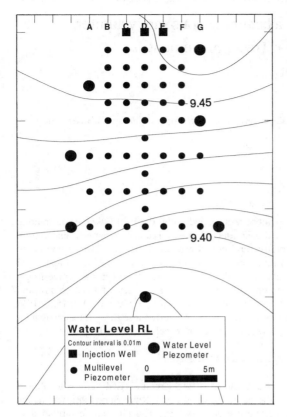

Figure 1. Location of injection wells, multilevel piezometers and average water table position.

6×10^{-5} m/s. This section represents approximately 4 m of the saturated part of the aquifer, with the water levels at the site varying between 9.7 m and 9.3 m a.s.l. (between 1.70 and 1.87 m below ground surface). The hydraulic gradient varies between 0.003 to 0.009 with an average value of 0.007.

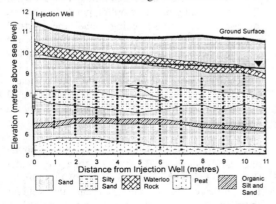

Figure 2. Geological cross-section through the site.

Hydraulic conductivity of the ELE site was estimated from 516 falling head measurements. The mean hydraulic conductivity of all calculated values for the entire site is 1.69×10^{-4} m/s (14.6 m/day), with a maximum of 5.8×10^{-4} m/s (50 m/day) and a minimum of 2.1×10^{-5} m/s (1.8 m/day). The presence of clay in the central part of the aquifer section results in a hydraulic conductivity that is an order of magnitude lower than the surrounding zones. This layer has a profound effect upon solute movement in the aquifer (Evans 1993). Several low hydraulic conductivity lenses exist at the site. These low conductivity zones can be classified as mini aquitards within the aquifer causing the lower part of the section to be semi-confined to confined.

3 BACKGROUND CHEMISTRY

Prior to injection of the tracer 88 groundwater samples, from 5 vertical sections, were collected and analysed to establish the background hydrochemistry of the site. Chemical composition is discussed in detail by Jankowski et al. (1997). The background concentrations of boron lithium and potassium were: boron - max. 0.15 mg/l; av. 0.07 mg/l; min 0.00 mg/l, lithium was not detected (> 0.01 mg/l), potassium - max. 6.04 mg/l; av. 3.74 mg/l; min. 0.10 mg/l. The concentrations taken for the contouring the tracer movement were higher than the maximum background data for all tracers.

4 TRACER APPLICATION AND SAMPLING

A total of 300 litres of solution was injected into three wells (100 l per well) (Figure 1) over a period of 1 hour. The injection wells were constructed of 51 mm ID PVC tubes with a slotted interval of 0.5 m at the depth of 8.1 to 8.6 m below ground surface. Chemical composition of the injected solution was specially selected to meet several objectives. These were to obtain results of longitudinal, horizontal-traverse and vertical-traverse dispersivity values from non-reactive tracers, obtain information about sorption of reactive cations and to obtain data for the behaviour of heavy metals. The solution contained a "cocktail" of ten elements. Three non-reactive B, Br$^-$ and Cl$^-$ (110.07 mg/l, 186.08 mg/l, 741.89 mg/l), two reactive cations, K$^+$ and Li$^+$ (106.93 mg/l, 123.72 mg/l) and five heavy metals Cd^{2+}, Cu^{2+}, Ni^{2+}, Pb^{2+}, and Zn^{2+} (56.19 mg/l, 56.43 mg/l, 54.83 mg/l, 51.27 mg/l 51.40 mg/l). The EC of this solution was 2877 µS/cm and it had a pH of 5.8.

Following injection, ten sampling sessions over a 32 day period were carried out, collecting a total of 2,500 samples. Water samples were withdrawn from PVC mini-piezometers (ID - 5 mm) using peristaltic pumps and 60 ml syringes.

5 RESULTS

The vertical distribution of boron, potassium and lithium plumes (along line E) are shown in Figures 3, 4 and 5 respectively. Boron was expected to be a non-reactive tracer and its behaviour is similar to chloride, which has been used as a tracer in a previous tracer test (Evans 1993). The mean solute velocity in the chloride tracer test was 0.35 m/day. The average velocity for boron tracer on this line is 0.34 m/day. This indicates that the boron tracer behaviour can be considered non-reactive. At day 6 boron had travelled up to 3 m, from the injection well with the maximum concentration (80 mg/l) located between 1 and 2 meters. At day 16 the front of the tracer was at 6 m and had a maximum concentration of 41.1 mg/l (37% of initial concentration). Between day 16 and 24 the tracer plume split in two parts. At day 32 boron showed a very narrow plume between 7 and 11 m and a trailing plume at up to 4 m from injection well (Figure 3). The maximum boron concentration at day 32 was 8 mg/l (7% of initial concentration).

Potassium was expected to be a reactive tracer with

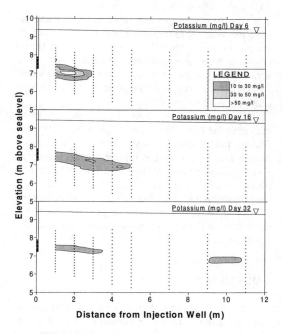

Figure 4. Vertical distribution of potassium tracer.

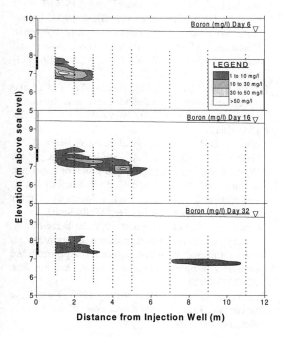

Figure 3. Vertical distribution of boron tracer.

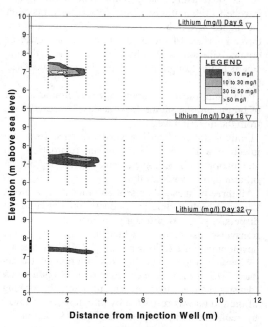

Figure 5. Vertical distribution of lithium tracer.

some sorption and/or ion exchange with silty clay or silty sand. At day 6 K^+ was detected 3 meters from the injection well, with a maximum concentration of 70.8 mg/l (66% of initial concentration).

By day 16 potassium had travelled 5 m from the injection wells (Figure 4). Between day 16 and 20 the potassium plume split in two, and at day 32 an advanced front was sampled between 9 and 10 m with the tail portion only 3.5 m from injection wells. The concentration in the advanced front was 23 mg/l (22% of initial concentration).

Lithium was also expected to be a reactive tracer, matching the behaviour of potassium. However, the lithium plume could not be defined at a distance greater than 4 m from the injection well (Figure 5). Over the 32 days and 10 sampling rounds the lithium plume could only be delineated between the injection well and 4 m. The maximum concentration of Li^+ at day 32 was 3.5 mg/l (3% of initial concentration), located between 2 to 3 m from the injection well.

6 DISCUSSION

The boron tracer travelled at a rate of 0.34 m/day. This movement rate is very similar to that calculated from hydrogeological parameters (0.32 m/day) for quartz sand using Darcy's law. The boron plume travelled preferentially along geological units with the highest hydraulic conductivity. Movement of the lower interface of the plume is on the top edge of organic silt and sand, which has a lower hydraulic conductivity. Longitudinal dispersion was calculated to be 0.0145.

Potassium tracer movement rate was similar to that of boron and the plume front at day 32 was 10.8 m from the injection wells.

Lithium travelled only a small distance from the injection wells showing significant retardation and attenuation with geological material. At day 32 the lithium plume was very thin.

All three plumes are present and mostly flow along zones of high hydraulic conductivity. Longitudinal dispersivity is significantly higher than vertical transverse dispersivity, which is reflected in the development of thin long plumes for all three tracers.

Although initially the injected solution of tracers had 7.75 times higher Electrical Conductivity (density 6 times) than the groundwater, the sinking of the plume was restricted by the presence of lower hydraulic conductivity units. The variation in hydraulic conductivity therefore, restricted the effects of density contrast.

REFERENCES

Boggs, J.M., S.C. Young, L.M. Beard, L.W. Gelhar, K.R. Rehfeldt & E.E. Adams 1992. Field study of dispersion in a heterogenous aquifer. 1. Overview and site description. *Water Resour. Res.* 28: 3281-3291.

Evans, D.J. 1993. A physical and hydrochemical characterisation of a sand aquifer in Sydney, Australia. *MAppSc Project Report*, UNSW.

Garabedian, S.P., D.R. LeBlanc, L.W. Gelhar & M.A. Celia 1991. Large-scale natural gradient tracer test in sand and gravel, Cape Cod, Massachusetts. 2. Analysis of spatial moments for a non-reactive tracer. *Water Resour. Res.* 27: 911-924.

Jankowski, J., R.I. Acworth & D.J. Evans 1994. Detailed hydrogeochemical sampling of an unconsolidated sand aquifer in the Botany Basin, Sydney, Australia: II. Hydrogeology and groundwater chemistry. *Proc. 3rd Int. Symp. Environ. Geochem., Cracow, 12-15 September 1994*: 170-172.

Jankowski, J., R.I. Acworth & P. Beck 1997. Vertical heterogeneity in the Botany Sands aquifer, Sydney, Australia: Implications for chemical variations and contaminant plume delineation. *Proc. 27th Cong. Int. Assoc. Hydrogeol., Nottingham, 21-27 September 1994*: 445-450.

Jensen, K.H., K. Bitsch & P.J. Bjerg 1903. Large-scale dispersion experiments in a sandy aquifer in Denmark: Observed tracer movements and numerical analyses. *Water Resour. Res.* 29: 673-696.

Killey, R.W.D. & G.L. Moltyaner 1988. Twin Lake tracer test: Setting, methodology, and hydraulic conductivity distribution. *Water Resour. Res.* 24: 1585-1612.

LeBlanc, D.R., S.P. Garabedian, K.M. Hess, L.W. Gelhar, R.D. Quadri, K.G. Stollenwerk & W.W. Wood 1991. Large-scale natural gradient tracer test in sand and gravel, Cape Cod, Massachusetts. 1. Experimental design and observed tracer movement. *Water Resour. Res.* 27: 895-910.

Mackay, D.M., D.L. Freyberg, P.V. Roberts & J.A. Cherry 1986. A natural gradient experiment on solute transport in a sand aquifer. 1. Approach and overview of plume movement. *Water Resour. Res.* 22: 2017-2029.

Moltyaner, G.L. & R.W.D. Killey 1988. Twin Lake tracer test. Longitudinal dispersion. *Water Resour. Res.* 24: 1613-1627.

Controls on sulfate reduction in a dual porosity aquifer

S. H. Bottrell
Department of Earth Sciences, University of Leeds, UK

S. J. Moncaster
Department of Earth Sciences, University of Leeds & School of Earth Sciences, University of Birmingham, UK

J. H. Tellam & J. W. Lloyd
School of Earth Sciences, University of Birmingham, UK

ABSTRACT: The factors controlling bacterial sulfate reduction in groundwaters have been investigated by comparing the chemistry and sulfate-sulfur isotopic composition of fissure waters and matrix pore waters from the sulfate reduction zone. Fissure waters have low concentrations (<0.1mM) of sulfate which is strongly enriched in ^{34}S and sulfide (concentrations up to 0.03mM) which is strongly ^{34}S depleted. Pore-waters in the aquifer matrix have markedly different sulfur chemistry, crucially: (1) sulfate concentrations are far higher than in the fissure waters, and (2) there is no sulfur isotopic evidence for the porewaters having been significantly affected by sulfate reduction. The fissure and pore waters have evolved independently, lack of sulfate reduction in the pore waters results from the exclusion of sulfate reducing bacteria from the aquifer matrix blocks because the pore interconnections are too small for them to pass. Following drilling, there was a significant stimulation of sulfate reduction and ground kerogen was released, indicating that lack of organic matter limits sulfate reduction in the fissure waters.

1 INTRODUCTION

Bacterial sulfate reduction can be an important process controlling the chemistry of deep groundwaters. The sulfide produced is toxic, rendering groundwater unsuitable for drinking, and corrosive, limiting potential industrial usage. Dual-porosity aquifers contain groundwater both in open fissures and in pore-spaces in the aquifer matrix in the blocks between fissures. The degree of chemical and physical interaction between these different groundwaters is controlled largely by physical properties of the aquifer, such as the porosity and permeability of the aquifer matrix and the spacing of the fissures defining the matrix blocks. Thus conclusions relating to bacterial sulfate reduction in such a system will also apply to other bacterial processes, such as nitrate reduction which may be important in the attenuation of nitrate pollution in groundwaters.

2 STUDY SITE AND SAMPLING DETAILS

We have studied groundwaters affected by sulfate reduction in the Lincolnshire Limestone aquifer of E. England. This is a classic dual porosity aquifer with most groundwater flow via solutionally-enlarged fissures on joints and bedding planes but high porosity (typically ~25%) in the limestone blocks between. In the outcrop zone pyrite oxidation consumes dissolved oxygen and sulfate concentrations increase. Downdip the limestone is confined between clays and further pyrite oxidation rapidly removes remaining dissolved oxygen (Moncaster et al. 1992). Further downdip in the confined aquifer, zones of nitrate reduction and subsequently sulfate reduction are apparent in the chemistry of the fissure waters sampled from boreholes (Smith-Carrington et al. 1983; Bishop and Lloyd 1990). During this study we sampled fissure waters from boreholes (see Fig. 1) for chemical and sulfur isotopic analysis and also drilled to recover core from the full thickness of the aquifer in the sulfate reducing zone. Drilling was undertaken under the strictest possible conditions, using only local, low-sulfate groundwater stored in nitrogen-purged tanks. Core was frozen in liquid nitrogen within 10 minutes of being brought to surface. Matrix porewaters were extracted by leaching under strictly anoxic conditions and analyzed for chemical and sulfate-sulfur isotopic compositions.

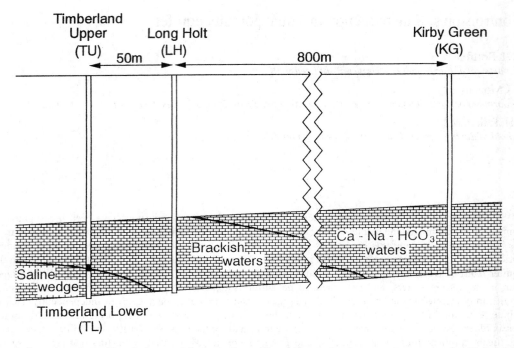

Fig. 1 Location plan of boreholes used for sampling. The sulfate reduction zone is associated with brackish and saline waters where the top surface of the aquifer is at 60m below surface. The Timberland well has a strong salinity gradient around 78m below surface and was packered at this depth to isolate an upper water (with chemistry similar to Long Holt well) and a lower, more saline, water.

3 RESULTS AND DISCUSSION

3.1 EVIDENCE FOR SULFATE REDUCTION IN FISSURE AND PORE WATERS

Up-dip of the sulfate reduction zone sulfate is present at concentrations of ~1mM, but within this zone they fall to <0.1mM. However, sulfide is only present at concentrations <0.03mM. The isotopic effects of sulfate reduction on the dissolved sulfur species are clear, with large enrichments in ^{34}S in sulfate (at +30 to +37$^o/_{oo}$ CDT) while sulfide is strongly depleted in ^{34}S (at -20 to -39$^o/_{oo}$ CDT). Porewaters in the aquifer matrix have markedly different sulfur chemistry. Figure 2 summarizes the porewater data from the Long Holt core. At low Cl concentration (similar to the fissure waters) the porewaters have much higher sulfate than the fissure waters. Other pore waters are richer in Cl and, although SO_4/Cl is lower, sulfate concentrations are still far higher than in the fissure waters. Most of the data lie close to a mixing curve between these two end-member compositions (which will be a hyperbola on this plot since the end-members have different SO_4/Cl ratios) but with some spread toward the fissure-water composition. Pore-water sulfate-sulfur isotopic compositions show a wide range from slightly positive values (+3$^o/_{oo}$ CDT) to quite ^{34}S-depleted values, around -20$^o/_{oo}$ CDT. All of these data are very significantly depleted in ^{34}S compared to the fissure waters.

The porewater data are complicated by the fact that there is chemical and isotopic evidence for porewaters of two different origins, but crucially:

1) porewater sulfate concentrations are far higher than in the fissure waters,

2) there is no sulfur isotopic evidence for the porewaters having been significantly affected by sulfate reduction. The isotopic compositions of sulfate in the porewaters are consistent with the sources inferred from their chemistry i.e. mixtures of

^{34}S-depleted pyrite derived sulfate (high SO$_4$/Cl) and ^{34}S-enriched marine-derived sulfate (low SO$_4$/Cl).

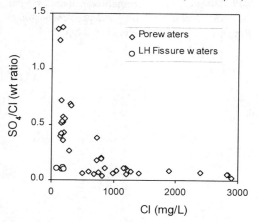

Fig. 2 Recalculated chemical data from porewater extractions of Long Holt core. Data from Long Holt fissure waters are plotted for comparison.

Thus, on the timescales of bacterial sulfate reduction and diffusional exchange at least, the fissure and pore waters have evolved independently. The fissure waters have been profoundly affected by sulfate reducers while the porewaters are unaffected. This is probably due to the exclusion of sulfate reducing bacteria from the aquifer matrix blocks because the pore interconnections are too small for them to pass. This is supported by mercury porosimeter measurements which show critical porethroat size to be <0.5µm even in the limestones with highest porosity and permeability. The bacteria are therefore restricted to the immediate vicinity of the fissures and their walls. This finding is in agreement with similar inferences made from bacterial activity measurements on core material (Fredrickson et al. 1997). Other than historical data relating to groundwater exploitation (sulfide was present ~100 years ago) we cannot know when sulfate reduction took place in the fissure waters, but there has been negligible subsequent modification of porewater compositions by diffusional exchange with the fissure waters.

3.2 STIMULATION OF SULFATE REDUCTION AFTER DRILLING

Following drilling to recover the core, the new well and pre-existing wells in the vicinity were monitored and showed a significant change in the sulfur chemistry of the fissure waters with sulfate being converted to sulfide (see Fig. 3). This affected all three waters sampled in the vicinity of the drilling site.). Initial rates of sulfate reduction associated with this phase were of the order of 0.6µmolL^{-1} hr^{-1}, amongst the highest measured or inferred for sulfate reduction in groundwater (see

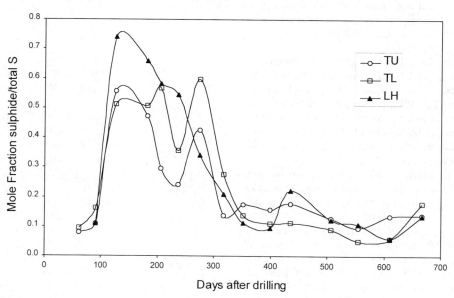

Fig. 3 Mole fraction of total dissolved S present as sulfide during the period after drilling.

Jakobsen and Postma, 1994 and refs therein. This strongly suggests that, even at sulfate concentrations as low as 0.1mM, sulfate availability is not limiting on sulfate reduction in groundwaters. Rather, the limiting factor appears to be the presence of an electron donor which would have been supplied in abundance in the form of labile organic matter during drilling since the kerf on the coring bit will have ground up substantial amounts of the limestone (which contains 0.5 to 3 wt% organic carbon in the form of "kerogen").

4 CONCLUSION

The primary limitation on bacterial sulfate reduction in the fissure water is the availability of organic matter (or other electron donors). Reduction is limited to the fissure waters because bacteria are too large to pass pore throats into the aquifer matrix blocks and utilize the organic matter and sulfate therein. By analogy, the same constraints will apply to nitrate reducing bacteria in this aquifer and we would predict that nitrate attenuation potential will be severely limited by this constraint.

5 ACKNOWLEDGEMENTS

This work was supported by NERC via research grant GR3/7839 and we are endebted to Rob Newton for assistance with analyses and data processing.

6 REFERENCES

Bishop, P.K. and J.W. Lloyd, 1990. Chemical and isotopic evidence for hydrogeochemical processes occurring in the Lincolnshire Limestone. *J. Hydrol.* 121: 293-320.

Frederickson, J.K., J.P. McKinley, B.N. Bjornstad, P.E. Long, D.B. Ringelberg, D.C. White, L.R. Krumholz, J.M. Suflita, F.S. Colwell, R.M. Lehman, T.J. Phelps and T.C. Onstott, 1997. Pore-size constraints on the activity and survival of subsurface bacteria in a Late Cretaceous shale-sandstone sequence, northwestern New Mexico. *Geomicrobiol. J.* 14: 183-202.

Jakobsen, R. and D. Postma, 1994. In situ rates of sulfate reduction in an aquifer (Romo, Denmark) and implications for the reactivity of organic matter. *Geology*, 22: 1103-1106.

Moncaster, S.J., S.H. Bottrell, J.H. Tellam and J.W. Lloyd, 1992. Sulfur isotope ratios as tracers of natural and anthropogenic sulfur in the Lincolnshire Limestone aquifer, eastern England. In: *Water-rock Interaction*, Y.K. Kharaka and A.S. Maest (eds), Balkema, Rotterdam: 813-816.

Smith-Carrington, A.K., L.R. Bridge, A.S. Robertson and S.S.D. Foster, 1983. *The nitrate pollution problem in groundwater supplies from Jurassic limestones in central Lincolnshire.* Report Inst. Geol. Sci., No. 83/3, 22pp.

$^{87}Sr/^{86}Sr$ in groundwater as indicators of carbonate dissolution

S.S. Dogramaci
Department of Geology and Geophysics, University of Adelaide, S.A. & Centre for Groundwater Studies, Australia

A.L. Herczeg
CSIRO, Land and Water & Centre for Groundwater Studies, Australia

Y. Bone
Department of Geology and Geophysics, University of Adelaide, S.A., Australia

ABSTRACT: The $^{87}Sr/^{86}Sr$ ratio of dissolved Sr in groundwater was analysed in order to investigate the effect of carbonate-dissolution on the chemical composition of the groundwater in the regional Murray Group Aquifer in the southwestern Murray Basin, Australia. Groundwater from the south and central part of the Murray Group Aquifer shows a $^{87}Sr/^{86}Sr$ ratio evolution from 0.7097 at the basin margin to a less radiogenic value of 0.7085 further downgradient, which is similar to the calcite $^{87}Sr/^{86}Sr$ ratio in the aquifer matrix. Results from a model for the incongruent dissolution of carbonate minerals in the aquifer matrix closely match the observed $^{87}Sr/^{86}Sr$ and Sr/Ca ratios. The progressive decrease in $^{87}Sr/^{86}Sr$ with a concomitant increase in the Mg/Ca ratio, and the relationship between $^{87}Sr/^{86}Sr$ ratio and $\delta^{13}C_{TDIC}$ composition in the groundwater further supports the occurrence of incongruent dissolution of carbonates in the Murray Group Aquifer.

1 INTRODUCTION

The interaction between carbonate minerals and groundwater in calcareous aquifers may play an important role in the evolution and composition of groundwater chemistry. Because strontium ions can replace calcium ions in carbonate minerals (Faure, 1986), strontium concentrations and their isotopic signatures can be used to identify the source of Sr in the groundwater and the extent of carbonate mineral-groundwater reactions, (Stueber et al., 1984; McNutt, 1987; Collerson et al., 1988; McNutt et al., 1990).

Groundwater in calcareous aquifers, particularly in large basins in arid and semi-arid regions, often have a long residence time. Equilibrium, therefore, may be established between the carbonate minerals and the groundwater. Reactions involving these minerals may occur by the mechanism known as incongruent dissolution, i.e. neomorphism of high Mg-calcite to low Mg-calcite.

Furthermore, carbonate mineral reactions with groundwater can reduce the ^{14}C concentrations independently of radiometric decay, resulting in a dilution of the ^{14}C content of groundwater. To correct groundwater ^{14}C composition for this dilution effect, the precise nature of these reactions must be known.

Our study area extends over 60,000 km^2 of the southwest Murray Basin, and consists of two major aquifers; the Murray Group Aquifer which overlies the Renmark Group Aquifer. These are separated by

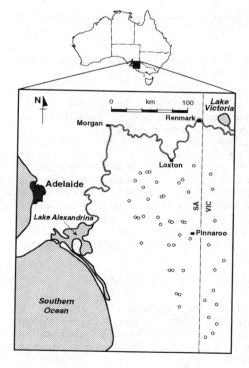

Fig. 1. Location of the study area, southwest of the Murray Basin, and groundwater sampling locations.

a 20-30 m thick aquitard (Fig. 1). Groundwater in the unconfined carbonate Murray Group Aquifer is an important source for domestic, irrigation and stock use in the study area.

2   RESULTS AND DISCUSSION

The $^{87}Sr/^{86}Sr$ ratio of dissolved Sr in the Murray Group Aquifer from the south and central part of the study area decreases with increasing Sr concentration (Fig. 2). It approaches a value of 0.7085 which is equivalent to the average Tertiary limestone $^{87}Sr/^{86}Sr$ ratio.

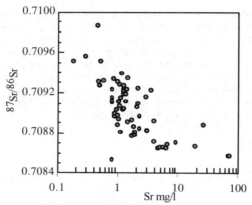

Fig. 2. $^{87}Sr/^{86}Sr$ ratios vs. Sr concentration in the Murray Group Aquifer.

This is due to the mixing of the two end members: (1) water that derives Sr from calcareous minerals in the Tertiary limestone ($^{87}Sr/^{86}Sr$ ratio of ~ 0.7085) and (2) water that derives Sr from trace amounts of silicates ($^{87}Sr/^{86}Sr$ ratio of ~ 0.7110). The observed $^{87}Sr/^{86}Sr$ ratio in the groundwater represents a mixture of these two end members depending on their relative proportions of Sr.

Dissolution of the calcitic minerals in the Murray Group Aquifer can initially occur as a result of reaction with soil water charged with high pCO_2 (~$10^{-1.5}$ to $10^{-2.5}$ atm., Dighton and Allison, 1985) as follows:

$$CO_2 + H_2O + CaCO_3 \rightarrow Ca^{2+} + 2HCO_3^- \quad (1)$$

This results in increasing Ca, Mg and Sr concentrations relative to that of the infiltrating rainwater and a change in the Sr/Ca ratio depending on the Sr/Ca ratios of the calcitic minerals and initial infiltrating rainwater. The Sr/Ca ratio progressively increases from 0.004 for the fresh groundwater, to 0.08 for the more saline groundwater in the Murray Group Aquifer, which is an order of magnitude higher than that in the calcitic minerals (Fig. 3). It appears therefore that dissolution of the calcitic minerals alone cannot explain the high Sr/Ca ratio in these saline groundwaters. We suggest the most probable mechanism that results in the current $^{87}Sr/^{86}Sr$ and Sr/Ca ratio relationship is incongruent dissolution of the calcitic minerals. The generalised form of incongruent dissolution is shown below:

$$Ca^{2+} + (Ca_{1-X}Mg_X)CO_2 \rightarrow CaCO_3 + Mg^{2+} \quad (2)$$

The calcitic minerals of the aquifers reflect all the components present in sea water from which these minerals precipitated (Tucker & Wright 1990). Mg and to a lesser extent, Sr, can replace Ca in calcitic minerals. The concentration of Mg commonly ranges from 1 to 16% (rarely, much higher) and Sr concentration is <1% in calcitic minerals. When the concentration of Mg in calcitic minerals exceed 1%, it can affect its solubility behavior. Generally high Mg-calcite is characterised by higher solubility relative to low Mg-calcite (Plummer & Mackenzie, 1974; Tucker & Wright, 1990). High Mg-calcite dissolves when it comes into contact with Mg-poor groundwater, releasing Mg and Sr into the groundwater, thereby causing an increase in the groundwater Mg/Ca ratio.

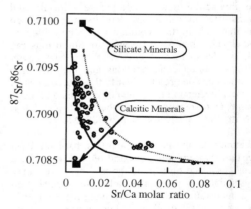

Fig. 3. The $^{87}Sr/^{86}Sr$ and Sr/Ca ratios of the carbonate and silicate mineral of the aquifer matrix along with the computed and observed variation of Sr/Ca and $^{87}Sr/^{86}Sr$ ratio in the groundwater of the Murray Group Aquifer. Note that the modeled solid and dashed lines represent two hypothetical initial Sr/Ca ratios of 0.001 and 0.01 in groundwater.

The increase in Sr concentration in groundwater during incongruent dissolution, is due to its low partition coefficient in calcite (0.057) (Katz et al., 1972). Because the change in Ca concentration during incongruent dissolution of carbonates is small (Wigley et al., 1978), the addition of Sr to groundwater results in an increase in the Sr/Ca ratio.

Incongruent dissolution of carbonates progressively modifies the $^{87}Sr/^{86}Sr$ ratio in groundwater, and eventually approaches that of the calcitic minerals. This occurs in two steps: (1) dissolution of calcitic minerals, causing the addition of Sr to groundwater and modification of the $^{87}Sr/^{86}Sr$ ratio, and (2) reprecipitation of calcitic minerals, thereby removing some of the Sr from the groundwater without changing the $^{87}Sr/^{86}Sr$ ratio. If this process continues, the $^{87}Sr/^{86}Sr$ ratio in the groundwater will ultimately be similar to that of the calcitic minerals of the aquifer matrix.

Starinsky et al. (1983) presented a model for incongruent dissolution of Mg-calcite to explain the Sr/Ca and $^{87}Sr/^{86}Sr$ ratios in oil field brines, and suggested that the final $^{87}Sr/^{86}Sr$ ratio of the brine depends exclusively on the $^{87}Sr/^{86}Sr$ of reacting minerals. The assumptions for this analytical model are:

(1) incongruent dissolution will not affect the concentration of Ca in groundwater, assuming the amount of Ca added by Mg-calcite dissolution is equal to the amount of Ca removed by calcite precipitation.
(2) incongruent dissolution occurs in a closed system, with much more Ca in the aquifer rock than in the groundwater.

From these two assumptions the change in Sr/Ca and $^{87}Sr/^{86}Sr$ ratios is mathematically treated by mass conservation of Ca in the system. The final equation used to determine the $^{87}Sr/^{86}Sr$ ratio of the final groundwater during the dissolution reprecipitation reaction according to Starinsky et al. (1983) is:

$$(\frac{^{87}Sr}{^{86}Sr})_L = \frac{(\frac{Sr}{Ca})_{Li}(\frac{^{87}Sr}{^{86}Sr})_{Li} \exp(-K_{Sr}^C \cdot (\frac{Ca_S}{Ca_L})) + \frac{1}{K_{Sr}^C}(\frac{Sr}{Ca})_R(\frac{^{87}Sr}{^{86}Sr})_R (1-\exp(-K_{Sr}^C(\frac{Ca_S}{Ca_L})))}{(\frac{Sr}{Ca})_{Li} \exp(-K_{Sr}^C \cdot \frac{Ca_S}{Ca_L}) + \frac{1}{K_{Sr}^C}(\frac{Sr}{Ca})_R (1-\exp(-K_{Sr}^C \frac{Ca_S}{Ca_L}))} \quad (3)$$

where $(^{87}Sr/^{86}Sr)_L$ and $(^{87}Sr/^{86}Sr)_R$ are the strontium isotopic ratios of groundwater and carbonate rocks respectively, the $(Sr/Ca)_{Li}$ and $(Sr/Ca)_R$ are initial ratios in groundwater and carbonate rocks before incongruent dissolution, the (Ca_S/Ca_L) is the ratio of the amount of Ca dissolved from the carbonate rocks to its amount in the initial solution, and K_{Sr}^C is the partition coefficient between calcite and water.

The Sr/Ca ratio used in our application of the model were measured values in the calcitic minerals of the Murray Group Aquifer matrix (0.005), and in the groundwater. The calcite $^{87}Sr/^{86}Sr$ ratio is the average of 10 measurements.

The modeled curves correspond to the observed data and suggests that incongruent dissolution of calcitic minerals can adequately predict the current Sr/Ca and $^{87}Sr/^{86}Sr$ ratio evolution of the groundwater in the Murray Group Aquifer. The occurrence of incongruent dissolution is further supported by the increase in Mg/Ca ratio and the concomitant decrease in the $^{87}Sr/^{86}Sr$ ratio (Fig. 4). The relationship of the Mg/Ca and $^{87}Sr/^{86}Sr$ ratios suggests that dissolution of calcitic minerals results in the addition of Mg relative to Ca, and relatively low $^{87}Sr/^{86}Sr$ ratios in the groundwater, which is in accord with the above model.

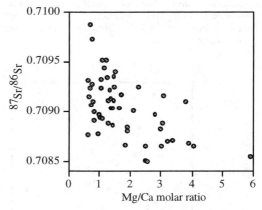

Fig. 4. The observed relationship between $^{87}Sr/^{86}Sr$ and Mg/Ca ratios.

Incongruent dissolution will also change the $d^{13}C$ composition of the total dissolved inorganic carbon (TDIC) of groundwater by the addition of enriched $d^{13}C$ from these Tertiary carbonates. Smith et al. (1975) showed that continuous dissolution and reprecipitation of calcitic minerals within the London Basin Aquifer is responsible for the enrichment of $\delta^{13}C_{TDIC}$ from -13‰ to -1‰ along a flow path of 8km. Wigley (1975) and Wigley et al. (1978, 1979) presented a more quantitative discussion on the evolution of $\delta^{13}C_{TDIC}$ in groundwater with special emphasis on incongruent dissolution and dedolomatisation. During incongruent dissolution of Mg-calcite, the $\delta^{13}C_{TDIC}$ will not only reflect the Mg-calcite source, but also the isotopic fractionation during the low Mg-calcite precipitation.

The $\delta^{13}C_{TDIC}$ and the $^{87}Sr/^{86}Sr$ ratio relationship (Fig. 5) shows that there is a general trend of enrichment in $\delta^{13}C_{TDIC}$ with a decrease in the $^{87}Sr/^{86}Sr$ ratio. The more enriched $\delta^{13}C_{TDIC}$ groundwaters have low $^{87}Sr/^{86}Sr$ ratios and are similar to those in the aquifer matrix. The progressive enrichment of $\delta^{13}C_{TDIC}$, combined with a decrease of the $^{87}Sr/^{86}Sr$ ratio only occurs as a result of the incongruent dissolution of the calcitic minerals.

As a result of the above mentioned reaction, the ^{14}C activity of groundwater decreases due to progressive addition of carbonate from the Tertiary aquifer matrix, which is devoid of ^{14}C. Accordingly, this should be taken into account when correcting the measured ^{14}C of the groundwater for age determination. If only the dissolution process is considered when estimating the initial ^{14}C of groundwater (according to equation 1), the result would be an over-estimation of the groundwater age. The most appropriate ^{14}C correction model is that of Wigley et al. (1978), which is based on the carbonate mass transfer during incongruent dissolution, taking into account the ^{13}C isotope balance between the carbonate rocks and groundwater.

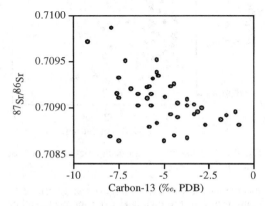

Fig. 5. $^{87}Sr/^{86}Sr$ ratio vs. $\delta^{13}C_{TDIC}$ in the groundwater from the Murray Group Aquifer.

3 CONCLUSIONS

The $^{87}Sr/^{86}Sr$ ratio, in combination with Sr, Ca, Mg concentrations and $\delta^{13}C_{TDIC}$ composition, provide powerful constraints on the nature of calcitic mineral-groundwater reactions in the limestones of the Murray Group Aquifer, southwestern Murray Basin.

Incongruent dissolution of Mg-calcite in the aquifer is shown to occur by a model of $^{87}Sr/^{86}Sr$, Sr/Ca and Mg/Ca ratio variations. This is supported by the $\delta^{13}C_{TDIC}$ composition of groundwater, which becomes increasingly more enriched in ^{13}C, as the $^{87}Sr/^{86}Sr$ ratio progressively decreases to that of the calictic minerals. The effect of incongruent dissolution therefore, must be considered when determining ^{14}C ages in regional calcareous aquifers.

4 REFERENCES

Dighton, J.C., & G.B. Allison. 1985. Carbon dioxide profiles of nine sites in the Murray Basin.*Technical Memorandum, Division of Soils CSIRO Australia*, No. 34.

Faure, G. 1986. Principles of Isotope Geology. *John Wiley & Sons*, Inc. 589 pp.

Katz, A., E. Sass, & A. Starinsky. 1972. Strontium behavior in the aragonite-calcite transformation: an experimental study at 40-98 ∞C. *Geochim. Cosmochim. Acta.*, 36, 481-496.

Plummer, L.N., H.L. Vacher, F.T. Mackenzie, O.P. Bricker, & L.S. Land.1976. Hydrogeochemistry of Bermuda: A case history of groundwater-water diagenesis of biocalcarnites. *Geol. Soc. Amer. Bull.* 87, 1301-1316.

Plummer, L.N., & F.T. Mackenzie. 1974. Predicting mineral solubility from rate data: application to the dissolution of magnesian calcite. *Am. J. Sci.* 274, 61-83.

Smith, D.B., R.A. Downing, R.A. Monkhouse, & F.J. Pearson. 1975. Stable carbon and oxygen isotope ratios of groundwater from the Chalk and Lincolnshire Limestone. *Nature*, 257, 783-784.

Starinsky, A., M. Bielski, B. Lazar, G. Steinitz, & M. Raab.1983. Strontium isotope evidence on the history of oilfield brines, Mediterranean Coastal Plain, Israel. *Geochim Cosmochim Acta*, 47, 687-695.

Stueber, M. A., P. Pushker, & A.E. Hetherington. 1984. A Strontium isotopic study of Smackover brines and associated solids, southern Arkansas. *Geochim. Cosmochim. Acta*, 48, 1637-1649.

Tucker, M.E., & V.P. Wright. 1990. Carbonate Sedimentology. *Backwell Scientific Publications.* 482 pp.

Wigley, T. M. L. (1975) Carbon 14 dating of Groundwater From Closed and Open Systems. *Water Resour. Res.* 11, 324-328.

Wigley, T. M. L., L.N. Plummer, & F.J. Pearson, Jr. 1978. Mass transfer and Carbon Isotope Evolution in Natural Water Systems. *Geochim. Cosmochim. Acta* . 42, 1117-1139.

Wigley, T. M. L., L.N. Plummer, & F.J. Pearson, Jr. 1979. Errata: *Geochim. Cosmochim. Acta*. 43, 1393.

Trace elements as residence time indicators in groundwaters: The East Midlands Triassic sandstone aquifer, England

W. Mike Edmunds & Pauline L. Smedley
British Geological Survey, Crowmarsh Gifford, Wallingford, UK

ABSTRACT: The increase of several trace elements along flow lines in the red-bed Triassic Sandstone aquifer of the East Midlands, England is described. Using well-calibrated radiocarbon analyses, the age dependence of this increase is examined and six parameters - Li, Rb, Sr, Mn, Mo and $\delta^{13}C$ are used to derive a "chemical" timescale with which to determine age relationships in groundwaters for which no radiocarbon analyses are available. This approach provides a means of investigating the residence time distribution of groundwater on a wider range of groundwaters than normal and offers one way of extending the radiocarbon age scale.

1 INTRODUCTION

Hydrogeochemical information for natural groundwaters can be considered essentially in two categories 1) those parameters which are inert and therefore serve as indicators or tracers of input conditions and 2) those which are reactive and which record the processes taking place as a result of water-rock interaction. Some reactive tracers may be attenuated during flow from the input source, whilst others increase in concentration with time, unless or until their concentration is limited by mineral saturation or other geochemical controls.

The objective of the present paper is to investigate the possibility that time-dependent water-rock reactions may give rise to "linear" build up of reaction products (especially trace elements) that can be used to estimate groundwater residence times up to and beyond the range of radiocarbon. This hypothesis is examined using data obtained from the East Midlands Sherwood Sandstone aquifer which has been the site of international investigations of the use of groundwater dating techniques (Andrews et al. 1984). Radiocarbon ages, which have been verified in previous studies are used here to explore the time-dependent increase in concentration of various trace elements as water moves down gradient.

2 HYDROGEOLOGY

The Sherwood Sandstone of the East Midlands (120 to 300m thick) is a red-bed sandstone thought to be mainly of fluviatile origin. The lower half of the aquifer consists of red or pink uncemented sandstone with minor argillaceous partings. The percentage quartz/feldspar/lithic clasts/clay (mainly muscovite, illite, biotite, chlorite) is typically 71/12/14/2 %. The upper half of the sandstone consists of cross-bedded and in parts pebbly red or pink sandstone. Detrital dolomite and a non-marine calcite cement (around 1-2%) also occur, the small amounts of carbonate tending to dominate the hydrogeochemistry. The sandstone may be considered as a single hydraulic unit although the local differences in stratigraphy and facies may result in minor differences in vertical and horizontal permeability as well as transmissivity. Various studies have been carried out on the hydrogeochemistry, including isotope geochemistry (Bath et al. 1979; Edmunds et al. 1982; Edmunds et al. 1994; Andrews et al. 1994).

3 METHODS AND ANALYSIS

The East Midlands aquifer was first sampled in 1975 (Edmunds et al. 1982) when 31 samples were collected from 26 different sites. During the present investigation a total of 45 samples were collected from 35 different boreholes (Fig.1); some sites sampled in 1975 had been abandoned whilst several new sources were available. Well-head analysis at each site included temperature, specific electrical conductance (SEC at 25°C), alkalinity (as HCO_3), pH, Eh and dissolved oxygen (DO).

Filtered (0.45 µm) samples were collected for Cl. (and other anion analysis) by colorimetry. Filtered and acidified (1% v/v HNO_3) samples were collected

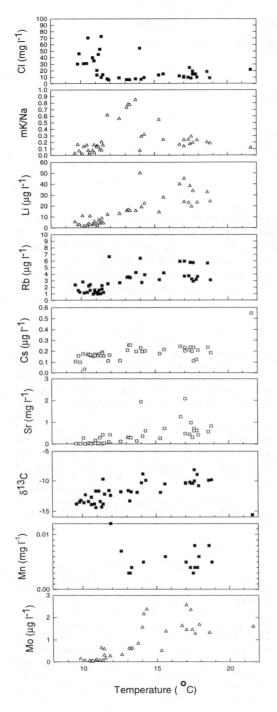

Fig. 1. Down-gradient profiles of selected parameters in the East Midlands aquifer; groundwater temperature is used as a proxy for distance.

in acid-washed polyethylene bottles for major cation, SO_4 and trace element analysis by ICP-OES and ICP-MS. Carbon-13 was measured by mass spectrometry.

4 RESULTS

Full analyses of the 1992 samples are presented elsewhere (Edmunds and Smedley in prep), along with results for the non-reactive tracers, but they confirm the earlier geochemical model for the aquifer (Bath et al. 1979; Edmunds et al. 1982). The only detectable changes in the chemistry of the waters since 1975 are found in the unconfined aquifer, where the concentrations of some indicators of pollution (eg DOC, NO_3) have increased. There is evidence that palaeowaters are found beneath modern polluted waters in the unconfined aquifer. A well-defined redox boundary occurs in the aquifer some 4km beneath confining beds yet oxygen has persisted under natural flow conditions for several thousands of years and the aquifer has a low capacity for *in situ* reduction of contaminants. Chloride (Fig. 1) and sulphate at outcrop have relatively high concentrations due to pollution but at depth, in waters of pre-industrial origin, concentrations of chloride (as well as sulphate and the other halogens) are low and represent modern or palaeo-atmospheric inputs or glacial melt water (Edmunds et al. 1982; Andrews et al. 1994). Chloride concentrations remain below 20 mg l^{-1} over most of the confined aquifer.

Over 30 trace elements have been measured in the 1992 campaign and those considered as the basis for the present paper and which illustrate the Down gradient evolution are shown in Fig.1. In these plots the concentrations are plotted against temperature as an analogue of distance and/or depth.

5 DISCUSSION

5.1 *The age of groundwater: radiocarbon evidence*

The radiocarbon results corrected using the WATEQ-ISOTOP program (Bath et al. 1979) and supported by an international intercomparison study (Andrews et al. 1983) show that the groundwater increases in age along the west to east flow direction with maximum ages in excess of 35 k yr BP. During the present study groundwater from one site (Grove 2) was reanalysed for carbon-14 and the result (20.8 pmc) compares well with that obtained (18.0 pmc) in the 1975 study. On the basis of this and the other hydrogeochemical evidence it is clear that there have been no major changes in the aquifer in the 15-year period and the original carbon-14 corrected ages are used as the basis for the present interpretation. The radiocarbon results suggest a mean groundwater (particle) flow velocity of 0.7 m yr^{-1} under the natural (pre-development) gradient. Supporting evidence for the groundwater age comes also from the oxygen and hydrogen stable isotope ratios as well as from the ratios of the noble gases (Bath et al. 1979).

5.2 Reactive tracers of residence time

The reactions of minor carbonate minerals may be followed using pH, HCO$_3$, Ca and Mg as well as some trace elements, especially Sr and the stable isotope ratio, δ^{13}C. It has been shown (Edmunds et al. 1982) that the ratio of Ca:Mg:HCO$_3$ varies significantly across the aquifer and that the proportions of each reflect the reaction of different amounts of dolomite or calcite that contribute to the composition of the groundwaters. Since the groundwaters are saturated with respect to calcite and/or dolomite increases in major ions do not show promise as indicators of residence time. However the progressive down-gradient enrichment in ^{13}C as well as Sr may be used as a residence time indicators. Barium cannot be used since barite saturation is rapidly achieved. Strontium is not solubility limited and the down-gradient increase is due initially to the incongruent reactions of carbonate minerals then at depth to the dissolution of gypsum. The δ^{13}C becomes more positive (from -14 to around -9 ‰) along the flow direction. This is because the calcite, which has a δ^{13}C of around -7 ‰ is more soluble than the dolomite, but the dolomite has a δ^{13}C near to zero (marine origin) and dominates the reactions over the long-term.

Silicon remains near-constant at around 3.5 mg l^{-1} Si across the aquifer which indicates the rapid dissolution of silicate minerals or the mobilisation of silica from exchange complexes, buffered at low temperatures probably by chalcedony. Other changes in the hydrogeochemistry however, imply that silicate minerals continue to react during down-gradient flow. The molar Na/Cl ratio increases across the aquifer and is a good indicator of reaction in the silicate system either resulting from mineral dissolution or as a result of cation exchange. However this change is not linear and is masked by pollutant sources near outcrop. Potassium which is low at outcrop increases significantly across the aquifer, and is buffered at concentrations of K around 7 mg/l (mK/Na ratios approaching 1). These high values are related to reaction of K-feldspar in the arkosic sandstone.

The trace alkali metals lithium, rubidium and caesium all show progressive increases in concentration down-gradient. The initial concentrations are very low and unlike major ions are not influenced by the pollution (Fig. 1). Lithium increases regularly from around 1 µg l^{-1} at outcrop to around 40 µg l^{-1} in the fresh waters at depth. Similar increases can be seen in rubidium (from 1 to around 6

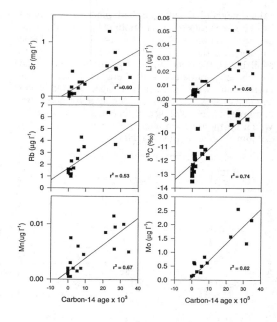

Fig. 2. Radiocarbon ages plotted against concentrations of Sr, Li, Rb, Mn, Mo and δ^{13}C.

µg l^{-1}) and to a lesser extent in caesium (from 0.1 to around 0.25 µg l^{-1}). These increases are consistent with the relative exchange capacities of the alkali metals (Cs>Rb>Li). The source of these elements are likely to be clays associated with the red-bed formations, although the K-feldspar dissolution may also may be a source of Rb. In high-temperature waters, lithium and Li/Na increase in a linear manner with increasing temperature (Ellis and Wilson 1960; Fouillac and Michard 1981) suggesting that Li is released as a lattice impurity. At low temperatures the same process is taking place at a much slower rate (Edmunds et al. 1986) and there is a strong dependence on the lithological Li/Na ratio. Lithium therefore is an ideal low-temperature residence time indicator.

Other trace element increases in this aquifer have also been considered. Some elements such as As, U and Sb show a time-dependent increase in the aerobic section of the aquifer (Smedley and Edmunds in prep.) but are influenced by redox reactions at depth. Two metals (Mn and Mo) are not affected by the redox reactions over the pH-Eh range in this aquifer and increase smoothly, most probably by desorption from the oxide surfaces.

5.3 The 'chemical' ages of the groundwater

Six candidates are selected as residence time indicators - Li, Rb, Sr, Mn, Mo and δ^{13}C. These are unaffected by solubility controls to limit upper concentrations or values, neither have they reached plateaux controlled by redox reactions or by partitioning between the solid and aqueous phases. These are plotted in Fig. 2 against ^{14}C ages for those sites where both sets of data are available. Little or no impact from pollution is found for these parameters and there is a reasonable correlation with radiocarbon age. Groundwater 'chemical ages' were then calculated for the more comprehensive 1992 set of chemical analyses for which no radiocarbon data were available using the combined linear regression data.

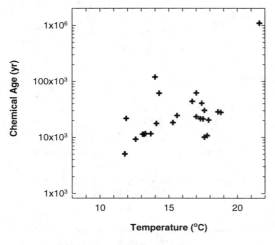

Fig. 3. Chemical ages (see text) plotted against groundwater temperature (as a proxy for distance)

The "chemical ages" for the larger set of data have been plotted in Fig. 3 against temperature as a measure of distance and evolution as in Fig. 1. Results shown are only for the pre-industrial (unpolluted) waters. The majority of waters increase linearly in age with distance and an age range up to 160 k yr for low salinity (Cl) waters is now indicated. Outliers younger or older than the main linear trend are considered to relate to areas of the aquifer with higher or lower transmissivities The saline waters in the deepest part of the aquifer give an age in excess of 10^6 yr. The extrapolated age of the fresh waters may indicate the timespan for enhanced groundwater circulation as a result of glaciation and/or lowered sea levels during the late Pleistocene.

Similar chemical ageing trends are found in other aquifers and although the same indicators used in this study are likely to be applicable, other chemical or isotopic indicators may also apply, depending on lithology and mineralogy. In each case the relationship between concentration and age using radiocarbon (or other age indicator) needs to be established. In most groundwater investigations it is probable that on grounds of cost only a few radiocarbon analyses will be possible. Therefore the present approach offers a means of examining the residence time distribution more closely using chemical data which can be determined on all sites in the aquifer.

6 REFERENCES

Andrews, J.N., Balderer, W., Bath, A.H., Clausen, H.B., Evans, G.V., Florkowski, T., Goldbrunner, J.E. Ivanovich, M., Loosli, H., Zojer, H. (1984) Environmental isotope studies on two aquifer systems: A comparison of groundwater dating methods. *Isotope Hydrology, 1983*. IAEA Vienna.

Andrews, J.N., Edmunds, W.M., Smedley,P.L,Fontes, J-Ch., Fifield, L.K. & Allan, G.L. (1994). Chlorine-36 as a palaeoclimatic indicator: the east Midlands Triassic aquifer (UK). *Earth and Planet. Sci. Lett.* 122: 159-171.

Bath, A.H., Edmunds, W.M. and Andrews, J.N.(1979) Palaeoclimatic trends deduced from the hydrochemistry of a Triassic sandstone aquifer, United Kingdom. *Isotope Hydrology 1978, Vol.II.* 545-568. IAEA. Vienna.

Downing, R.A., Edmunds, W.M. and Gale, I.N. (1987) Regional groundwater flow in sedimentary basins in the UK. *Geol. Soc. Special Publ. No.34*, 105-125. (Geological Society, London.)

Edmunds, W.M., Bath, A.H. & Miles. D.L. (1982) Hydrochemical evolution of the East Midlands Triassic sandstone aquifer, England. *Geochim. Cosmochim. Acta*, 46: 2069-2081.

Edmunds, W M, Cook, J M & Miles, D L. (1986). Lithium mobility and cycling in dilute continental waters. *Proc. 5th Int. Symp. Water-Rock Interaction*. Reykjavik, Iceland. 183-187.

Edmunds, W.M. & Smedley, P.L. (in prep) . Residence time indicators in groundwater: the East Midlands Triassic aquifer.

Ellis, A.J & Wilson, S.H. (1960). The geochemistry of alkali metal ions in the Wairakei hydrothermal system. *New Zealand J.Sci.*, 3: 593-617.

Fouillac, C. & Michard, G. (1981). Sodium/Lithium ratio in water applied to geothermometry of geothermal reservoirs. *Geothermics*, 10: 55-70.

Smedley, P.L & Edmunds, W.M. (in prep). Redox processes and trace element behaviour in the East Midlands Triassic Sandstone aquifer.

Retarded intraparticle diffusion in heterogeneous aquifer material

M. Finkel & R. Liedl
Applied Geology, University of Tübingen, Germany

ABSTRACT: An analytical model of intraparticle diffusion in heterogeneous aquifer material is presented which allows for transient boundary conditions. This model has been developed in order to predict the mass transfer of PAH between the mobile phase and the intraparticle pores for conditions which are similar to field-scale scenarios. The modelling approach is designed to simulate small-scale diffusion processes within a more comprehensive computer code for modelling reactive transport of organics in groundwater. The diffusion model has been verified against another analytical approach and a well-known numerical method. It also could be validated successfully with respect to data from laboratory experiments.

1 INTRODUCTION

Transport of dissolved polycyclic hydrocarbons (PAH) in non-consolidated aquifers is highly affected by reaction kinetics, i. e. slow sorption processes that lead to a considerably retarded movement of PAH plumes and to a long persistence in the subsurface (Grathwohl, 1992).

The different mechanisms contributing to slow sorption have been summarized by Pignatello & Xing (1996) who point out that diffusive processes are responsible for the kinetic sorption behaviour of organic compounds (Fig. 1). This statement is supported by the results of various lab and field experiments (Miller & Pedit, 1992, Grathwohl et al., 1994, Harmon & Roberts, 1994, Pedit & Miller, 1994, Grathwohl & Kleineidam, 1995). The experiments also show that mass transfer due to slow sorption is highly dependent upon the physico-chemical heterogeneity of the aquifer material which is determined by the lithological composition and the grain size distribution.

It is obvious that computer models for PAH transport in aquifers have to take into account these experimental findings. In particular, a module has to be provided simulating intraparticle diffusion for field-scale conditions. This cannot be achieved by simply using analytical approaches (Carslaw & Jaeger, 1959, Crank, 1975) which are restricted to a rather limited range of application (homogeneous aquifers, steady-state boundary conditions).

In order to overcome these restrictions several numerical models of intraparticle diffusion have been developed (Wu & Gschwend, 1988, Pedit & Miller, 1995). However, if these models are intended to be employed at the field scale enormous demands of computer resources have to be expected due to the high number of grid cells which are necessary in order to resolve the diffusive processes at the grain scale.

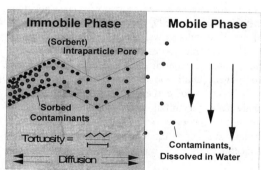

Fig.1: Schematic view of the retarded intraparticle diffusion process (from Grathwohl, 1994)

This paper presents an analytical model of intraparticle diffusion in heterogeneous aquifer material allowing for transient boundary conditions. This model has been developed in order to predict the mass transfer of PAH between the mobile phase and the intraparticle pores for conditions which are similar to field-scale scenarios. The modelling approach is designed to serve as part of a more complex computer code for simulating reactive transport of organics in groundwater.

2 MODELLING APPROACH

The intraparticle diffusion model presented here is based upon the following assumptions (Fig. 1):
- The aquifer material consists of a finite number of different lithological components and grain size classes.
- Grains are approximated as porous spheres. The intraparticle pores are filled with immobile water.
- PAH transport within the grains can be described by radial diffusion.
- PAH transport along intraparticle pores is retarded due to equilibrium sorption onto the solid phase. This process is modelled by a linear sorption isotherm.
- Sorptive sites are evenly spread within the grains.

Retarded intraparticle diffusion within spherical grains is modelled by Fick's second law

$$\frac{\partial c_{jk}}{\partial t} = \frac{D_{app,j}}{r^2} \frac{\partial}{\partial r}\left(r^2 \frac{\partial c_{jk}}{\partial r}\right) \quad (1)$$

where t is time, r is the radial coordinate, c_{jk} is the concentration of dissolved PAH and the indices j, k represent the different lithological components and grain size classes, respectively.. The apparent diffusion coefficient $D_{app,j}$ accounts for the tortuosity of the flow paths as well as for linear retardation of PAH in the micropores (Fig. 1). It can be obtained from batch experiments for each lithological component (Grathwohl & Kleineidam, 1995).

In order to solve the diffusion equation (Eq. 1) analytically the initial concentration is assumed to be uniform, i. e. $c_{jk}(r,0) = c^0$, and the boundary values are approximated by a step function of concentrations

$$c_{jk}(R_k, t) = c^l \quad \text{if} \quad t_{l-1} < t < t_l \quad (2)$$

where R_k denotes the grain size radius for class k and the changes in concentration are numbered by the superscript l. This type of boundary (Fig. 2) allows for simulating transient processes so that the modelling approach presented here may be applied to time-variant reactive transport problems.

The diffusion model formulated above can be solved analytically for the concentrations $c_{jk}(r,t)$.

Integrating with respect to r yields the mass of the contaminant in a (j,k)-grain as a function of time

$$M_{jk}(t) = \frac{8}{\pi} R_k^3 \left(\varepsilon_j + (1-\varepsilon_j)\rho_j K_{d,j}\right) \sum_{n=1}^{\infty} \frac{1}{n^2} [c^l - \sum_{m=1}^{l} (c^m - c^{m-1}) e^{-D_{app,j} n^2 \pi^2 (t - t_{m-1})/R_k^2}] \quad (3)$$

for $t_{l-1} < t \leq t_l$ with ρ_j dry solid density, ε_j intraparticle porosity and $K_{d,j}$ partition coefficient. Using this result the total contaminant mass in the grains at time t is obtained as

$$M(t) = \sum_{j,k} \frac{3 f_{jk} \rho_b V}{4\pi R_k^3 (1-\varepsilon_j)\rho_j} M_{jk}(t) \quad (4)$$

with f_{jk} mass fraction of (j,k)-grains, ρ_b bulk density, V volume of grid cell (or batch reactor).

Similarly, the total mass flow rate through the surfaces of the grains is given by

$$F(t) = \sum_{j,k} \frac{3 f_{jk} \rho_b V}{4\pi R_k^3 (1-\varepsilon_j)\rho_j} F_{jk}(t) \quad (5)$$

where

$$F_{jk}(t) = 4\pi R_k^2 \left(\varepsilon_j + (1-\varepsilon_j)\rho_j K_{d,j}\right) D_{app,j} \frac{\partial c_{jk}}{\partial r}(R_k, t)$$

$$= 8\pi R_k \left(\varepsilon_j + (1-\varepsilon_j)\rho_j K_{d,j}\right) \cdot \quad (6)$$

$$\cdot \sum_{n=1}^{\infty} \sum_{m=1}^{l} (c^m - c^{m-1}) e^{-D_{app,j} n^2 \pi^2 (t - t_{m-1})/R_k^2}$$

represents the mass flow rates through the surfaces of the (j,k)-grains according to Fick's first law. It should be mentioned that negative values of F_{jk} in Eq. (5) indicate that solute is taken up by the grains.

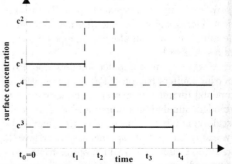

Fig. 2: Schematic example of time-dependent boundary conditions

3 MODEL VERIFICATION

The model has been verified for a closed batch system, where an analytical solution is available for homogeneous aquifer material (Crank, 1975). The decrease in concentration of mobile PAH in the batch reactor is simulated by a step function according to Eq. (2). The concentrations c^l have been determined from a mass balance for the entire batch system accounting for the decreasing PAH mass outside the grains. Both solutions are nearly identical for solid-to-solution ratios varying by a factor of 50 (Fig. 3).

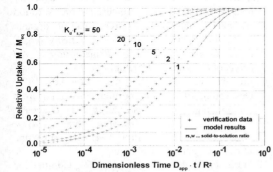

Fig. 3: Model verification against analytical data

Further evidence of the correctness of the model is given by comparisons with results of numerical simulations of Wu & Gschwend (1988). Fig. 4 illustrates the impact of particle size distribution on the relative uptake of PAH. The sample consists of four different grain size classes with equal mass fractions (25%). It can be seen that it is not possible to represent the PAH uptake of a heterogeneous sample by employing an averaged radius, e. g. the geometric mean, unless the range of the grain size distribution is small.

Fig. 5 shows the uptake characteristics for the constituents of size distribution no. 2. As could be expected, all relative uptakes finally approach a value of 0.25 according to the uniform mass fraction distribution of the sample (Fig. 4). It has to be noted, however, that the relative uptakes of the classes 1-3 temporarily increase beyond 0.25. This is due to the fact that the uptake of class 4 (large grain size) is comparatively slow because of the small surface-to-volume ratio of large grains. As a consequence, class 4 initially is not "seen" by the solute which preferentially diffuses into the smaller grains (large surface-to-volume ratio). Later on, slow sorptive uptake by the large grains continues so that solute is partly released from the classes 1-3 and still sorbed by class 4 until equilibrium is obtained.

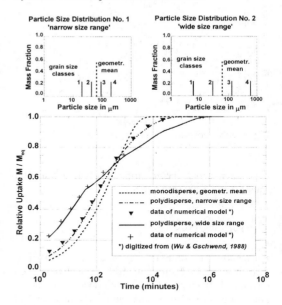

Fig. 4: Model verification against data from numerical simulations

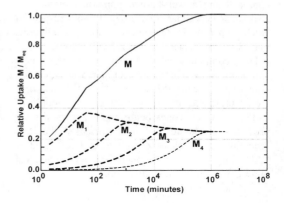

Fig. 5: Diffusive uptake for each grain size class

4 MODEL VALIDATION

Model validation has been performed with respect to long-term batch experiments (Rügner et al., 1997) investigating the diffusive uptake of phenanthrene by quartz and Keuper sandstone ("Stubensandstein"). Fig. 6 indicates a good agreement of lab data and model results until the

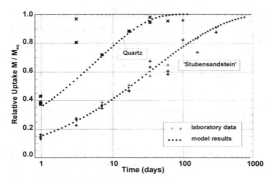

Fig. 6: Slow sorption of phenanthrene

experiments had been stopped after about 60d (quartz) and 300d (sandstone).

5 CONCLUSIONS

An analytical solution for retarded intraparticle diffusion of PAH in spherical grains has been developed. It accounts for non-uniform lithological and grain size distributions and uses time-variant surface concentrations as boundary conditions. The model has been verified against another analytical approach and a well-known numerical method. It also could be validated successfully with respect to batch experiments. Therefore, model predictions may also be expected to yield reliable results.

The computational efficiency of this new approach will have to be compared to common numerical schemes. It is expected that the analytical model presented above may serve as a module of general transport codes being applied to field-scale problems. It should also be emphasised that this modelling approach is, of course, not restricted to PAH but may be applied to any chemical undergoing intraparticle diffusion.

6 ACKNOWLEDGEMENTS

The research project was funded by the State of Baden-Württemberg (PWAB research programme), Germany.

7 REFERENCES

Carslaw, H. S. & J. C. Jaeger 1959. *Conduction of heat in solids*, 2nd ed. Oxford University Press.

Crank, J. 1975. *The mathematics of diffusion.* Oxford: Clarendon.

Grathwohl, P. 1992. Persistence of contaminants in soils and sediments due to diffusion-controlled de-sorption. *Int. Symp. on Environmental Contamination in Central and Eastern Europe*: 604-606. Budapest.

Grathwohl, P. 1994. Persistenz organischer Schadstoffe in Boden und Grundwasser - können einmal entstandene Untergrundverunreinigungen wieder beseitigt werden? In J. Matschullat & G. Müller (eds.) *Geowissenschaften und Umwelt*: 263-273. Berlin: Springer.

Grathwohl, P. & S. Kleineidam 1995. Impact of heterogeneous aquifer materials on sorption capaci-ties and sorption dynamics of organic contaminants. In K. Kovar & J. Krasny (eds.), *Groundwater quality: Remediation and protection*: 79-86. IAHS Publ. no. 225.

Grathwohl, P., W. Pyka & P. Merkel 1994. Desorption of organic pollutants (PAHs) from contaminated aquifer material. In T. H. Dracos & F. Stauffer (eds.), *Transport and reactive processes in aquifers,* (Proc. IAHR/AIRH Symp., Zürich). Rotterdam: Balkema.

Harmon, T. C. & P. V. Roberts 1994. Comparison of intraparticle sorption and desorption rates for a halogenated alkene in a sandy aquifer material. *Env. Sci. Technol.* 28(9): 1650-1660.

Miller, C. T. & J. A. Pedit 1992. Use of a reactive surface-diffusion model to describe apparent sorption-desorption hysteresis and abiotoc degradation of lindane in a subsurface model. *Env. Sci. Technol.* 26(7): 1417-1427.

Pedit, J. A., & C. T. Miller 1994. Heterogeneous sorption processes in subsurface systems, 1, Model formulations and application. *Env. Sci. Technol.* 28(12): 2094-2104.

Pedit, J. A., & C. T. Miller 1995. Heterogeneous sorption processes in subsurface systems, 2, Diffusion modeling approaches. *Environ. Sci. Technol.* 29: 1766-1772.

Pignatello, J. J., & B. Xing 1996. Mechanisms of slow sorption of organic chemicals to natural particles. *Environ. Sci. Technol.* 30(1): 1-11.

Rügner H., S. Kleineidam & P. Grathwohl 1997. Sorptionsverhalten organischer Schadstoffe in heterogenem Aquifermaterial am Beispiel des Phenanthrens. *Grundwasser*, 2(3),133-138.

Wu, S. C., & P. M. Gschwend 1988. Numerical modeling of sorption kinetics of organic compounds to soil and sediment particles. *Water Resour. Res.* 24 (8): 1373-1383.

The source of stable chlorine isotopic signatures in groundwaters from crystalline shield rocks

S.K.Frape, G.Bryant, P.Durance, J.C.Ropchan & J.Doupe
Department of Earth Sciences, University of Waterloo, Ont., Canada

R.Blomqvist, P.Nissinen & J.Kaija
Geological Survey of Finland, Espoo, Finland

ABSTRACT: Detailed analyses of stable chlorine isotopic signatures of saline waters and brines at Canadian and Fennoscandian research sites shows a wide range of isotopic signatures. Limited rock and mineral isotopic data shows that the leaching of chlorine bearing minerals could influence the $\delta^{37}Cl$ groundwater signature at many sites. Mixed with the rock isotopic signature at some sites in Finland is the overprinting of glacial and paleo Baltic seawaters. Therefore multiple isotopic signatures and sources of chloride in Shield environments is normal, rather than systems with a single chlorine source.

1 INTRODUCTION

The origin of high salinity fluids in crystalline shield environments is still very much in debate. Surprisingly researchers studying these highly saline fluids often choose to support one source or one process for the origin of the salinity. The end members most often supported are (i) marine or marine derived salinity (Bottomley et al., 1994); (ii) dissolution and leaching reactions involving water and the host rock (Frape and Fritz, 1987; McNutt et al., 1990). A variety of isotopic ratios have been used to assess and support the different origins. However many of the isotopic ratios used are cationic species which have been shown to be easily modified by host rock over-printing. More recently attempts have been made to assess the evolution and possible origin of the major component of salinity, chloride, using the stable chlorine isotopic signatures of shield brines (Kaufmann, et al., 1987; Frape et al., 1996).

The possibility of multiple end member signatures and sources of salinity at sites in crystalline shield rocks is difficult to assess, unless differences in chloride signatures can be shown to exist. The research described here started with reconnaissance scale studies of groundwaters across the Canadian and Fennoscandian Shields. At some sites in Finland and at the Stripa site, in Sweden more complete studies were possible because of previous detailed sampling. As well, rock and mineral phases are being analyzed as a comparison to the fluids in order to assess processes like mixing and diffusion versus dissolution of host rock components.

2 SAMPLING AND METHODS

Groundwater samples were obtained from exploration boreholes and open flowing fractures in mines as well as surface exploration boreholes. In more detailed studies hydrogeological instrumentation allowed us to obtain samples from discrete isolated intervals in many boreholes. The samples are filtered, but for Cl isotopes no further field treatment is carried out.

Analytical methodology used by the University of Waterloo for stable chlorine isotopes parallels the methods outlined by Long et al. (1993). Whole rock chlorine isotopic analyses involves dissolution of the rock to free the chloride. The free Cl is then treated in the same manner as an aqueous sample. Precision of the methodology is better than $\pm\ 0.15$‰.

3 RESULT AND DISCUSSION

Shield brines can have total dissolved solid contents of over 300 gl^{-1} (Frape and Fritz, 1987). Canadian Shield brines are dominantly Ca-Na-Cl in composition. Fennoscandian Shield brines have a much more variable composition with both Ca-Na-Cl and Na-Ca-Cl fluids often found within the same site (Nurmi et al., 1988).

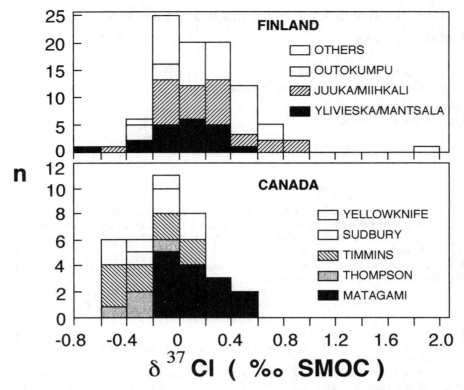

Figure 1. Distribution of stable chlorine isotopic signatures of saline groundwaters from the Canadian and Fennoscandian Shields.

3.1 $\delta^{37}Cl$ Signatures for Shield Brines

Figure 1 shows the isotopic distributions of stable chlorine isotopes at numerous sites across the two major areas of study. The two Shields have similar isotopic distributions although the Canadian Shield in general has a bias toward depleted values and the Finnish samples a bias toward enriched values.

Several other observations can be made from the data set. Although the mean values for data sets presented are close to SMOC (0 ‰), the data sets show a wide divergence away from present day seawater.

Groundwater values from the Stripa site, Sweden are not included on Figure 1. They have a very wide range of values compared to saline groundwaters found at any other Shield site. The most interesting aspect of the sample distribution is the relationship of the isotopic signature with borehole depth and borehole orientation (Figure 2). Water samples from the vertical and deep V_1 and V_2 boreholes have a trend toward more positive values. Water samples from shallow wells, the shallow M3 borehole and the sub-horizontally drilled N1 and E1 boreholes have more depleted isotopic values. The simplest explanation for the isotopic distributions at the site are a deep positive end member ($\sim +1.2$ ‰) mixing

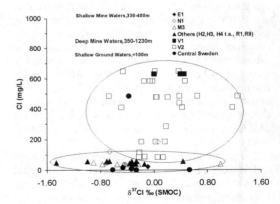

Figure 2. Stable chlorine isotopic signatures versus chloride concentration for groundwaters from the crystalline bedrock of the Stripa area, Sweden.

in proportions with a surficial or shallow end member with a negative isotopic signature (~ - 0.8 ‰). The orientation of fractures and boreholes would tend to favour the explanation, but the source of the chloride in both end members is still an unknown.

3.2 Rock leaching and whole rock chlorine isotopic signatures

Two studies are ongoing at this time, but provide initial insight into one of the possible end member signatures. At the Finnish site designated Juuka/Miihkali the groundwaters have $\delta^{37}Cl$ signatures ranging up to + 1.1 ‰ in the deep, Na-Ca-Cl. Leaching experiments and whole rock dissolution samples gave $\delta^{37}Cl$ values from + 1.3 to 1.5 ‰. The similar nature of the rock values and the most concentrated brines would support a strong rock control of the brine values. In the case of the Juuka site the serpentine mineralization contain considerable chlorine in the structure of secondary Mg, Fe hydroxy-chlorine minerals. These secondary minerals are believed to be the major source of the rock bound chlorine at this site.

A second study, comparing the chloride bearing mineral sodalite, from three different host rocks and locations is summarized in Table 1.

Table 1. $\delta^{37}Cl$ values for selected sodalite samples ($\delta^{37}Cl$ precision ± 0.12 ‰)

Location	Host Rocks	Age (Ma)	$\delta^{37}Cl$ (SMOC)
1) Bancroft (Canada)	pegmatites	1160	+0.02 n=20
2) Mt-Ste-Hillaire (Canada)	syenite diorite	122	-0.06‰ n=17
3) Namibia (Africa)	carbonatite dykes	749	-0.11‰ n=5

Although the minerals come from different sites, of different ages and in one case a different continent; the $\delta^{37}Cl$ values are virtually identical and similar to the world wide sea water signature. Analyses on biotites and other Cl bearing minerals at several of our sites show variable ranges of positive and negative isotopic values. The initial results from mineral analyses would suggest that long term water rock dissolution reactions have the potential to overprint distinctive rock isotopic signatures on the waters.

3.3 Other suspected sources of chlorine in the Fennoscandian Shield

Nurmi et al. (1988) presented strong evidence, based on major ion analyses, for the intrusion of proto Baltic seawater into the shallow portions of some sites in Finland. The Baltic Sea is low salinity compared to ocean water. The $\delta^{37}Cl$ signature of present day Baltic Sea water is consistently -0.2 ‰. The signature is very similar to salt samples analyzed from Poland and northern Germany and values reported by Eggenkamp et al. (1995) for Zechstein evaporites (range -0.58 to 0.00‰). It is safe to assume that throughout time the Baltic has received a considerable portion of its chloride load from this evaporitic source.

Figure 3 provides some insight into additional chloride sources in Finnish groundwaters. First if we refer to the JuMi (Juuka) site. As discussed previously, the rock/mineral $\delta^{37}Cl$ signature at this site ranges from +1.3 to +1.5 ‰. Many of the fluids found at the site are quite dilute (< 1000 mg l^{-1} Cl); but have enriched $\delta^{37}Cl$ signatures. The authors believe that fresh glacial melt water, lacking in significant chloride penetrated many of the Finnish sites during post glacial times. The mixing trend would follow a line approximated by the 3a line in Figure 3. The intrusion of fresh $\delta^{18}O$ depleted water would result in a lowering of Cl concentration and little change in $\delta^{37}Cl$ isotopic signature. Note that the JuMi point shown as the concentrated end member on the figure represents four samples at depths below 700m at the site. The diagram is also coded for the $\delta^{18}O$ signature of the water component of each sample. The least concentrated samples, which are also usually the shallower samples at each site tend to be the most isotopically depleted in oxygen-18. Glacial melt waters and probably paleo-Baltic waters of 10-20000 years ago in this region would have had a much more negative isotopic signature than present day Baltic waters ($\delta^{18}O$ < -20 ‰ versus present of -8 ‰). Several additional trends of the diagram (1 and 3b) appear to suggest a colder climate dilute glacial or Baltic water infiltrated many of the shallow portions of our Finnish sites and mixed with the original rock $\delta^{37}Cl$ signatures.

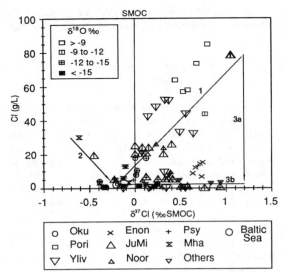

Figure 3. The $\delta^{37}Cl$ signature versus Cl⁻ concentration coded for $\delta^{18}O$ signature for groundwaters from a number of research sites on the Fennoscandian Shield. (Depth <300m) Numbers and lines represent possible mixing scenarios (modified after Frape et al., 1996).

4 SUMMARY

Stable chlorine isotope analyses of Shield groundwaters provides an additional means for in assessing the origin and evolution of the salinity of these fluids. Initial isotopic results for chlorine bound in mineral phases at one site suggests that the associated brines derived a substantial portion of their $\delta^{37}Cl$ signature from the rocks. Ongoing chlorine isotopic studies similar to those reported for strontium isotopes (McNutt et al., 1990) will most likely confirm the relationship of halides sourced in rock and mineral phases as reported by Kamineni (1987) and Kamineni et al. (1992).

Additionally a dilute but extensive component of Baltic seawater with a distinct $\delta^{37}Cl$ signature can be found at many Finnish sites and possibly occurs in the shallow waters and vertical fracture systems in Stripa, Sweden. This finding would confirm that the rock mass can be open to penetration and rapid mixing of allocthonous surficial waters during some events such as deglaciation. Further study and use of stable chlorine isotopes will reveal the possible rates at which the rock mass can reassert its isotopic signature over externally sources salinity.

ACKNOWLEDGEMENTS

The National Science and Engineering Research Council of Canada, Geological Survey of Finland, Canadian and Finnish mining companies, Heide Flatt and Bob Drimmie Waterloo.

REFERENCES

Bottomley, D.J., D.C. Gregoire & K.G. Raven 1994. Saline groundwaters and brines in the Canadian Shield: Geochemical and isotopic evidence for a residual evaporite brine component. *Geochim. Cosmochim. Acta* 58:1483-1498.

Eggenkamp, H.G.M., R. Krevlen & A.F. Koster van Groos 1995. Chlorine stable isotope fractionation in evaporites. *Geochim. Cosmochim Acta* 59:5169-5175.

Frape, S.K. & P. Fritz 1987. Geochemical trends for groundwaters from the Canadian Shield. In P. Fritz & S.K. Frape (eds) *Geol. Assoc. Can. Spec. Paper* 33:211-223.

Frape, S.K., G. Bryant, R. Blomqvist & T. Ruskeeniemi 1996. Evidence from stable chlorine isotopes for multiple sources of chloride in groundwaters from crystalline shield environments. In *IAEA Proceedings* SM-336:19-30.

Kamineni, D.C. 1987. Holgen-bearing minerals in plutonic rocks: a possible source of chlorine in saline groundwater in the Canadian Shield. In P. Fritz & S.K. Frape (eds) *Geol. Assoc. Canada. Spec. Paper* 33:69-79.

Kamineni, D.C., M. Gascoyne, T.W. Melnyk, S.K. Frape & R. Blomqvist 1992. Cl and Br in mafic and ultra mafic rocks: significance for the origin of salinity in groundwater. In Y. Kharaka & A. Maest (eds.) *WRI-7*:801-804.

Kaufmann, R.S., S.K. Frape, P. Fritz & H. Bentley 1987. Chlorine stable isotope composition of Canadian Shield brines. In P. Fritz & S.K. Frape (eds.) *Geol. Assoc. Canada. Spec. Paper* 33:89-94.

McNutt, R.H., S.K. Frape, P. Fritz, M.G. Jones & I.M. MacDonald 1990. The $^{87}Sr/^{86}Sr$ values of Canadian Shield brines and fracture minerals with applications to groundwater mixing, fracture history and geochronology. *Geochim. Cosmochim. Acta* 54:205-215.

Nurmi, P.A., I.T. Kukkonen & P.W. Lahermo 1988. Geochemistry and origin of saline groundwaters in the Fennoscandian Shield. *Appl. Geochem* 3:185-203.

High permeabilities of Quaternary granites in Japan and its implications for mass and heat transfer in a magmatic-hydrothermal system

K. Fujimoto & M. Takahashi
Geological Survey of Japan, Tsukuba, Japan

N. Doi & O. Kato
JMC Geothermal Engineering Co. Ltd, Takizawa, Japan

ABSTRACT: The permeability of Quaternary unaltered granite from deep geothermal wells in Japan is higher than 100 μd, which is one order of magnitude higher than the reported laboratory permeability of various granitic rocks. Mean crack-width, which ranges from 0.12 to 1.30 μm, has a positive correlation with the measured permeability. The granite is considered to be permeable just after the solidification and subsequent alteration, and/or compaction reduces the permeability and crack-width. Grain boundaries may be important as a migration path for degassing and circulation of hydrothermal fluid in the magmatic-hydrothermal stage.

1 INTRODUCTION

Transport of water in a granite is one of the essential processes in heat and mass transfer in the magmatic-hydrothermal stage. Thus, permeability of the granite in this stage is a fundamental parameter, but no data has been obtained yet due to lack of appropriate samples. Recent studies revealed that young granitic intrusives underlie some geothermal fields such as Kakkonda in Japan, the Geysers in USA, and Tongonan in Philippine, and that they are a heat source (Yasukawa et al., 1994).

In Kakkonda geothermal area, Japan, some deep wells penetrated Quaternary granites which have temperatures greater than 350°C and are accompanied by contact metamorphism (e.g., Kato and Doi, 1993; Uchida et al., 1996). This geothermal system is an active magmatic-hydrothermal system, and therefore, very important, because studies have been on fossil systems such as porphyry copper deposits.

We have analyzed the permeability, porosity, and pore size of this hot granite to elucidate the migration path of fluids, and the implications for heat and mass transfer in the magmatic-hydrothermal stage.

2 EXPERIMENTAL AND RESULTS

2.1 Samples

The six samples used in this study were from young granitic rocks in Japan, which are listed in Table 1. All of them are medium-grained granodiorite or tonalite, which exhibit shallow intrusive features. Five are core samples (>1500 m depth) from geothermal areas.

Four of the samples (1, 2, 3, and 4) are core samples from the hot deep zone (>320 °C, 1740 to 2764 m depth) in the Kakkonda geothermal area, where a geothermal power plant (80 MW) is operating. The Quaternary unaltered granodiorite (samples 1 and 2) is probably the youngest intrusive in the area. Samples 3 and 4 are also from Quaternary intrusive rocks, however, they are probably older than the granodiorite at Kakkonda, as they have been weakly altered and metamorphosed as indicated by biotitization of hornblende and recrystallized biotite.

Table 1. Brief description of the samples. Age data are from Kanisawa et al. (1994) for samples 1 and 2, NEDO (1992) for sample 5 and Harayama (1992) for sample 6.

No.	Locality/Well-depth	Rock facies	Alteration/Metamorphism	Age(Ma)	Temperature
1	Kakkonda/well 21-2568.5m	granodiorite	no alteration	0.34-0.07	>400°C
2	Kakkonda/well 13-2346.4m	granodiorite	no alteration	0.34-0.07	>350°C
3	Kakkonda/well 5-1740.3m	tonalite	weak alteration/metamorphism	Quaternary	>320°C
4	Kakkonda/well 20-2764.7m	tonalite	weak alteration/metamorphism	Quaternary	>350°C
5	Nyuto/TZ7-1493.5m	granodiorite	alteration	~1.7	~140°C
6	Takidani / surface	granodiorite	weak alteration	~1	

Table 2. Pressure conditions and permeability of core samples. Pc is the confining pressure, Pp is the pore pressure, and Pe is the effective confining pressure (Pe = Pc - Pp). 2a and 2b are different specimens.

No.	Pc (MPa)	Pp (MPa)	Pe (MPa)	Permeability (µd)
1	5	3	2	370
2a	5	3	2	116
2b	5	3	2	73.7
	5	3	2	83.7
	40	10	30	10.5
	40	20	20	14.4
	40	25	15	20.1
	40	25	15	21.5
3	5	3	2	2.5
4	5	3	2	11.7
	40	38	2	19.5
	40	30	10	5.7
	40	20	20	2.27
	40	10	30	0.73
5	5	3	2	0.86
	40	10	30	0.10
	40	20	20	0.19
	40	30	10	0.37
	40	35	5	0.69
6	5	3	2	7.4

Table 3. Permeability (Pe = 2 MPa), porosity, and mean crack-width.

No.	Permeability (µd)	Porosity (%)	Mean crack-width (µm)
1	370	3.27	1.3
2	97	2.17	0.92
3	2.5	2.32	0.28
4	11.7	1.31	0.28
5	0.86	1.64	0.12
6	7.4	1.25	0.26

The subsurface geology and alteration of the Kakkonda geothermal area were described by Kato and Doi (1993), and petrology and geochemistry of the Quaternary granite were presented by Kanisawa et al. (1994).

The granodiorite sample 5 also comes from a geothermal well at 1500 m depth and temperature of 140°C in the Nyuto hot spring area. It is moderately altered and mafic minerals are nearly completely changed to secondary minerals such as chlorite, epidote, calcite and quartz. The age is reported to be about 1.7 Ma (NEDO, 1992). The granodiorite sample 6, the only surface sample, is from the Takidani granodiorite which is the youngest exposed granite in the world, with the age of about 1.0 Ma (Harayama, 1994). It appears unaltered in hand-specimen and in thin section.

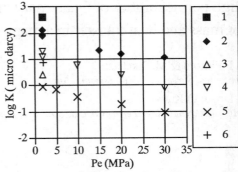

Figure 1. Permeability (K) versus effective confining pressure (Pe). Pe = Pc - Pp. The legend number is the sample number listed in Table 1.

2.2 Permeability

Permeability of core samples was measured by the Transition Pulse method (Brace et al., 1968), following the method described by Takahashi et al. (1990). After the pore pressure equilibrated within the specimen, the fluid pressure was raised instantaneously by a few bars at one end of the specimen. The pore pressure then decreased exponentially with time as water flowed through the specimen. The permeability can be calculated from the pore pressure decay rate. This method has superior precision and independent control of pore pressure (Pp) and confining pressure (Pc), compared to other methods. The latter is important, as the pressure release effect can be minimized for core samples.

The specimens were cylindrical in shape, with a 30 mm radius and 30 mm length, without any apparent cracks. Length of the specimens was parallel to the original core, for samples 1 to 5.

The temperature was kept at 22 °C, within 1 °C fluctuation. First, the Pc is set at 5 MPa and Pp at 3 MPa for the measurement at the effective confining pressure (Pe = Pc - Pp) at 2 MPa. After that, the Pc is increased up to 40 MPa slowly, and then the Pp is raised slowly to measure permeability at several Pc values. The pressure conditions and measured permeability are listed in Table 2. Forty MPa, nearly the maximum Pc of our apparatus, corresponds to lithostatic pressure of about 1.6 km depth, if the rock density is 2.5 g/cc.

The measured permeability, when the Pe is at 2 MPa, ranges from 0.86 to 370 µd (Table 3). Quaternary unaltered Kakkonda granodiorite (samples 1 and 2) have permeabilities higher than 100 µd, which is more than one order of magnitude higher than the permeabilities of the other samples.

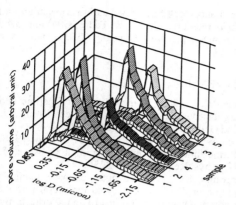

Figure 2. Histograms of crack-width (D). Samples are arranged in the order of permeability.

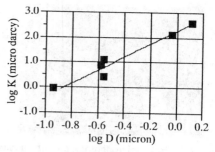

Figure 3. Permeability (K) when the Pe is at 2 MPa versus mean crack-width (D). The slope of the fitted line is 2.63.

The permeability decreases about one order of magnitude as the Pe is increased up to 30 MPa (Fig. 1 and Table 2).

2.3 Porosity and crack-width

Porosity and crack-width distribution were measured by a mercury intrusion porosimeter (Carlo Erba Porosimeter 2000) following the method of Lin et al (1995). We measured the volume of the intruded mercury with a capacitance system as the intrusion pressure was increased up to 200 MPa. The porosity was obtained from the cumulative volume of the intruded mercury. Pore-size estimates are based on the relation between the pore-volume versus intrusion pressure, as the mercury can be intruded into narrower pore spaces at higher pressures. Assuming the pore spaces to be platy cracks, the crack-width can easily be calculated from the intrusion pressure, the contact angle and the surface tension, to get a histogram of the crack-width (Table 3 and Fig. 2).

Porosity ranges from 1.3 to 3.3 %. Quaternary unaltered granodiorite samples, which are permeable as mentioned before, show relatively large porosities of 2 %, whereas, the other samples have porosities lower than 2%, except the tonalite sample 3. Mean crack-width, which ranges from 0.12 to 1.30 μm, has a positive correlation with the measured permeability (Fig. 3).

3. DISCUSSIONS AND CONCLUSIONS

The permeability values obtained in this study are relatively high compared to other well known granites, such as Westerly granite or Barre granite, which range from 0.001 to a few tens micro darcies (Brace, 1980). In particular, the permeability of the Quaternary unaltered granodiorite from Kakkonda geothermal area (samples 1 and 2), exceeds 100 μd, which is comparable to relatively impermeable sandstone.

Recently, Morrow and Lockner (1997) pointed out the difficulty in obtaining meaningful permeability result after core retrieval. It is possible that the values for permeability have some uncertainty, however, in this case, the Quaternary unaltered granodiorite should be more permeable than the altered and older tonalite. Both should have nearly the same high permeability if core retrieval seriously affects samples, because they were collected from nearly the same temperature and pressure conditions. In fact, the granodiorite is an order of magnitude more permeable than the tonalite.

Which penetration path is then responsible for the high permeability? Scanning electron microscope (SEM) observations of the fractured surface of the unaltered granodiorite from Kakkonda area show that the grain boundaries are open and contain only a little precipitate (Figs. 4 and 5). The opening widths are at micron scales and concordant with the measured mean crack-width. The mineral surfaces show euhedral growth and some dissolution etch pits exist, which indicates that the grain boundaries were filled with hydrothermal fluid. Blue dyed epoxy impregnated thin sections of the specimens also demonstrate that the grain boundaries are open and interconnected (Takahashi et al., 1992). These results suggest that the grain boundaries are an effective migration path for fluids.

However, the grain boundaries of samples 3 to 6 are not as open as in samples 1 and 2. The alteration minerals (sericite and epidote) were precipitated in the grain boundaries of sample 5, the granodiorite from Nyuto hot spring area (Fig. 6).

Our results indicate that grain boundaries can play a major role in the degassing and circulation of hydrothermal fluid during the magmatic-hydrothermal stage. The grain boundaries, however, will be narrowed due to the subsequent precipitation of secondary minerals, recrystallization and compaction which reduce permeability and pore size.

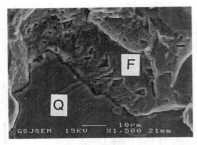

Figure 4. SEM microphotograph of unaltered Quaternary granite of Kakkonda geothermal area, Japan (sample 1). The lower part is a fractured surface of quartz (Q) and the upper part is a feldspar grain surface with dissolution etch pits (F). The grain boundary is open and no mineral is precipitated.

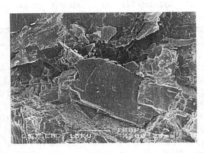

Figure 5. SEM microphotograph of the same sample as Fig. 4 (sample 1). The euhedral shape and clear surface indicates the existence of free space.

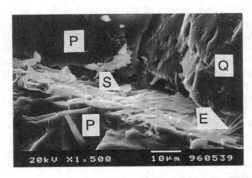

Figure 6. SEM microphotograph of a granodiorite sample from Nyuto hot spring area, Japan (sample 5). The grain boundary between quartz(Q) and plagioclase(P) is open, but epidote(E) and sericite (S) are precipitated.

ACKNOWLEDGMENTS

Japan Metal & Chemicals Co. Ltd., Tohoku Geothermal Energy Co. Ltd. and Dr. S. Harayama (GSJ) provided samples. Dr. W. Lin (Dia Consultants Co. Ltd.) helped us in permeability and porosity measurement. Dr. K. Faure (GSJ) improved the manuscript. Mr. Oowada (GSJ) made thin sections.

REFERENCES

Brace, W.F. 1980. Permeability of crystalline and argillaceous rocks. Int. J. Rock Mech. Min. Sci. and Geomech. Abstr., 17: 241-251.

Brace, W.F., J.B.Walsh, and W.T. Frangos 1968. Permeability of granite under high pressure. J. Geophys. Res., 73: 2225-2236.

Harayama, S 1992. Youngest exposed granitoid pluton on Earth; Cooling and rapid uplift of the Pliocene-Quaternary Takidani Granodiorite in the Japan Alps, Central Japan. Geology, 20: 657-660.

Kanisawa,S., N.Doi, O.Kato, and K.Ishikawa 1994. Quaternary Kakkonda granite underlying the Kakkonda geothermal field, northeast Japan. J. Petrol. Mineral. Econ. Geol., 89: 390-407. JE

Kato, O. and N.Doi 1993. Neo-granitic pluton and later hydrothermal alteration at the Kakkonda geothermal field, Japan. *Proc. 15th NZ Geothermal Workshop*, 155-161.

Lin, W., M.Takahashi, and N.Sugita 1995. Change of microcrack width induced by temperature increase in Inada granite. J. Jpn. Soc. Engineer. Geol. 36: 300-304. JE

Morrow, CA and D.A. Lockner 1997. Permeability and porosity of the Illinois UPH 3 drillhole granite and a comparison with other deep drillhole rocks. J.G.R. 102: 3067-3075.

NEDO 1992 *Report of geothermal development survey project, No. 27 East Tazawa-ko lake area*, 1011p. New Energy Development Organization, Japan. JE

Takahashi,M., A.Hirata, and H.Koide 1990 Effect of confining pressure and pore pressure on permeability of Inada granite. J. Jpn. Soc. Engineer. Geol. 31: 105-114. JE

Takahashi, M., Z.Xue, A.Oowada and Y.Ishijima 1992. On the study of interconnected pore network using blue dyed epoxy impregnated thin section. J. Jpn. Soc. Engineer. Geol. 33: 294-306. JE

Uchida, T, M.Yagi, M.Sasaki, M.Kamenosono, N.Doi, and S.Miyazaki 1996. Investigation of deep-seated geothermal reservoir by NEDO's 4,000 m well in Kakkonda, Japan. *Abst. 8th Intl. Symp. Observation of the continental crust through drilling.* 55-59.

Yasukawa, K, M.Yagi, H.Muraoka and K.Hisatani 1994. Current status of exploration and exploitation of deep-seated geothermal resources. *Energy* 225: 24-36. JE

(JE: in Japanese with English abstract)

A geochemical model for groundwaters of the arid Ti-Tree Basin, Central Australia

G.A. Harrington & A.L. Herczeg
Centre for Groundwater Studies and CSIRO Land and Water, Glen Osmond, Australia

ABSTRACT: A geochemical model has been developed to elucidate the mechanisms responsible for the spatial variation of major ion concentrations in the groundwaters of the Ti-Tree Basin, Northern Territory, central Australia. Groundwaters in this arid basin are relatively fresh (TDS range from 500-1500 mg/l), and are of Na-HCO_3 type. Stable isotopes of water indicate that the groundwaters are derived from rainfall, with a bias towards large, monsoonal events. The geochemical model suggests that the groundwaters initially acquire solutes by dissolution of gypsum and carbonates, as well as weathering of silicates. Subsequently, concentration of the recharge water by evapotranspiration increases TDS concentration with concomitant increases in Na and Cl relative to other ions due to cation exchange, carbonate and silica precipitation and reverse-weathering to form cation-rich clay minerals.

1 INTRODUCTION

The Ti-Tree Basin is located 200 km north of Alice Springs in the Northern Territory of central Australia. It occupies about 5000 km² of flat sandy plain. The only topographical features are the uplifted basin margins and several ephemeral rivers. The Basin contains a thick (up to 300 m) sequence of Cainozoic sediments which yield high quality groundwater (TDS values often < 1000 mg/l). The region is arid, and potential evapotranspiration exceeds average rainfall by almost 3000 mm per annum. Water from underground aquifers is relied upon extensively for domestic and agricultural purposes, as well as the irrigation of several large horticultural developments that were established after a groundwater resource investigation in the mid-1980's (McDonald, 1988).

There are two main issues in the Ti-Tree Basin that need to be addressed, namely groundwater sustainability and groundwater quality. A thorough understanding of both of these is vital to ensure long-term settlement and horticultural activity in the Basin. This paper focuses on the water quality aspect by interpreting preliminary hydrochemical and stable isotope data to elucidate the mechanisms responsible for spatial variations in major ion concentrations across the Ti-Tree Basin.

The authors wish to acknowledge the collaboration and technical support provided by the Water Resources Branch of the Northern Territory Department of Lands, Planning and Environment, Alice Springs.

2 RESULTS AND DISCUSSION

2.1 General Description

A total of 32 boreholes were sampled during August-November 1996 and analysed for major ion chemistry and stable isotopes (^{2}H, ^{18}O and ^{13}C). Chemical analyses of several typical groundwater samples from the Ti-Tree Basin are presented in Table 1. Salinity generally ranges from 500 to 1500 mg/l, with a few samples up to 3000 mg/l. The more dilute waters are Na-HCO_3 type. As the waters become more saline there is a change towards a Na-Cl type.

2.2 Major Ion Trends

For the relatively fresh (TDS < 1000 mg/l) groundwaters, sulphate and bicarbonate ions dominate over chloride (Fig. 1). This indicates that their chemistry is derived primarily by water-rock interactions, because sulphate and bicarbonate concentrations in rainfall of the arid interior are negligible compared to chloride (Keywood, 1995). Conversely, the more saline waters (TDS > 1000 mg/l) are dominated by chloride, thus suggesting a

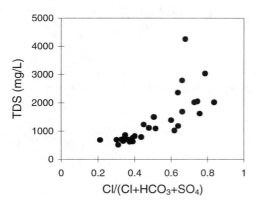

Fig. 1. TDS concentration versus Chloride relative to total anion concentration, Ti-Tree Basin.

more significant component of their chemistry is controlled by evapotranspiration. By plotting major ion/chloride ratios versus chloride, it is possible to identify whether changes in groundwater chemistry are a result of mixing between two or more waters with different major ion/chloride ratios, water-rock interactions or evapotranspiration. It is assumed that chloride is derived only from rainfall, and that there is no dissolution or precipitation of chloride-bearing minerals in the Ti-Tree Basin. If evapotranspiration was the only mechanism by which groundwater chemistry changes, then the major ion/chloride versus chloride plots would be horizontal, as all ion concentrations would increase by the same amount. Therefore the hydrochemical trends observed in the Ti-Tree Basin (Fig. 2(a-f)) must be derived by mixing and/or water-rock interactions, as well as evapotranspiration. As chloride concentration increases, there is an appreciable decrease in sodium (Fig. 2(a)), bicarbonate (Fig. 2(b)), and dissolved silica (Fig. 2(c)) concentrations relative to chloride in the groundwater. It is also evident that both sodium (Fig. 2(d)) and calcium (Fig. 2(e)) increase relative to bicarbonate as chloride increases. Calcium and sulphate concentrations increase in equimolar quantities (Fig. 2(f)) suggesting that dissolution of gypsum may be an important process during evolution of the groundwater chemistry.

2.3 Stable Isotopes, δ^2H and $\delta^{18}O$

The stable isotopic composition of Ti-Tree Basin groundwater samples, expressed as delta values relative to Standard Mean Ocean Water (IAEA, Vienna), range from -63‰ to -50‰ and -9‰ to -5‰ for deuterium and oxygen-18 respectively.

Monthly rainfall samples collected at Alice Springs (200 kilometres south of the study area) over a 20 year period (IAEA, 1969-1990) plot on a Local Meteoric Water Line (LMWL):

$$\delta^2H = 6.86\delta^{18}O + 4.48 \quad (1)$$

as shown in Figure 3. The δ^2H-$\delta^{18}O$ relationship is similar to that obtained by Calf et al. (1991), both of which exhibit a slope less than the value of 8 for the world Meteoric Water Line (MWL) (Craig, 1961). The long-term amount-weighted mean isotopic composition of Alice Springs rainfall is $\delta^2H=-23.5‰$ and $\delta^{18}O=-4.7‰$ (Fig. 3).

Groundwater samples from the Ti-Tree Basin plot below the Local Meteoric Water Line (Fig. 3), indicating that recharge waters must have been evaporated prior to entering the saturated zone. The stable isotopic composition of all groundwaters can be approximated by the following relationship:

$$\delta^2H = 3.55\delta^{18}O - 31.09 \quad (2)$$

The slope of equation (2) suggests that evaporation of recharge waters probably took place in the soil or vadose zone rather than from a free-surface (Allison, 1982). This would imply that groundwater recharge is predominantly diffuse in the areas where these samples have been obtained.

2.4 Carbon-13

There are several processes within the vadose and saturated zones of the Ti-Tree Basin groundwater system that may influence the carbon-13 composition of the dissolved inorganic carbon (DIC). These include: dissolution or precipitation of carbonate minerals, respiration from plant roots, and oxidation of organic matter.

Table 1. Chemical analyses of groundwater from selected boreholes, Ti-Tree Basin

Bore RN	TDS (mg/l)	pH	Na^+ (mg/l)	K^+ (mg/l)	Ca^{2+} (mg/l)	Mg^{2+} (mg/l)	Cl^- (mg/l)	HCO_3^- (mg/l)	SO_4^{2-} (mg/l)	NO_3^- (mg/l)	log P_{CO2}
10676	687	6.8	143	21	35	24	71	405	79	34	-1.18
11767	859	6.8	171	25	46	36	149	408	110	38	-1.19
5532	1110	7.1	245	38	52	44	265	389	170	77	-1.52
14249	762	7.4	125	29	49	30	132	306	79	82	-1.89
13542	1500	7.2	243	43	88	82	363	472	220	110	-1.55
14691	647	7.7	102	23	39	34	98	293	57	100	-2.19

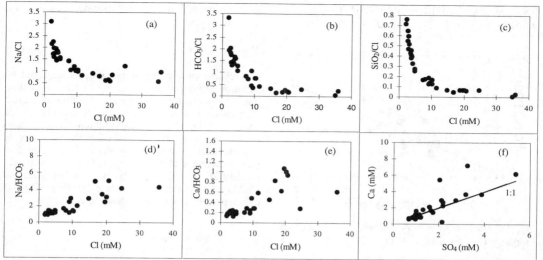

Fig. 2. (a)-(f) Hydrochemical plots of Na/Cl, HCO$_3$/Cl, SiO$_2$/Cl, Na/HCO$_3$ and Ca/HCO$_3$ mole ratios versus Cl; and Ca versus SO$_4$, Ti-Tree Basin groundwater samples.

A plot of $\delta^{13}C$ versus the reciprocal of DIC enables the relative importance of each process to be investigated (Fig. 4). This plot indicates that the groundwater $\delta^{13}C$ composition is a result of different degrees of mixing between two end-members: one with a relatively enriched $\delta^{13}C$ composition of ~-5‰, and the other with a more depleted $\delta^{13}C$ composition of ~-14‰. As DIC concentration increases, there is an addition of the more-depleted ^{13}C end-member.

2.5 Chemical Model

A simple model has been proposed to explain the origin of the major ionic species and hydrochemical trends (Fig. 2) observed in the Ti-Tree Basin. Initially, all waters were normalised up to a chloride concentration of 10mM. For example, suppose a particular water sample had a chloride concentration of 5mM, then all ionic concentrations of that sample would be multiplied by factor of two. Doing this enabled the relative comparison and interpretation of different water analyses without the effects of evapotranspiration.

Sulphate ions are accounted for by assuming they originated by dissolution of gypsum (Fig. 2f):

$$CaSO_4 \cdot 2H_2O \leftrightarrow Ca^{2+} + SO_4^{2-} + 2H_2O \quad (3)$$

and a corresponding number of moles of calcium were subtracted from the normalised analyses. The excess (or deficit) of calcium ions (after accounting for dissolution of gypsum) and magnesium ions are assumed to have originated by the dissolution (or precipitation) of carbonate minerals:

$$Ca_xMg_{(1-x)}CO_3 + CO_2 + H_2O \leftrightarrow XCa^{2+} + (1-X)Mg^{2+} + 2HCO_3^- \quad (4)$$

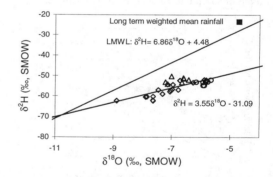

Fig. 3. Stable isotope plot of Ti-Tree Basin groundwaters. LWML represents the Local Meteoric Water Line.

Groundwater samples plotted on a stability field diagram with respect to various silicate minerals indicate that the waters are in equilibrium with Na-smectite. It is therefore proposed that all bicarbonate ions not accounted for by dissolution of calcite, and all sodium ions (except those derived from rainfall) are derived by the action of CO$_2$ weathering secondary silicate minerals such as Na-smectite:

$$3Na_{0.33}Al_{2.33}Si_{3.67}O_{10}(OH)_2 + CO_2 + 4.5H_2O \leftrightarrow$$
$$3.5Al_2Si_{2.4}O_{5.8}(OH)_4 + Na^+ + HCO_3^- + 2.6SiO_2 \quad (5)$$

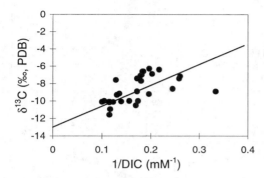

Fig. 4. Carbon-13 versus reciprocal of Dissolved Inorganic Carbon (DIC = HCO_3 + CO_2).

Figure 5 shows the results of the model described above, by plotting the residual normalised sodium concentration (after accounting for silicate weathering) versus residual normalised bicarbonate concentration (after accounting for calcite dissolution/precipitation and silicate weathering). The "excess" of modelled Na concentrations above the 1:1 line is assumed to be the fraction of meteoric input of sodium to each groundwater sample.

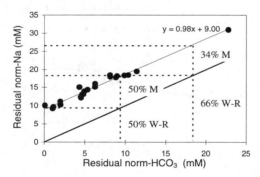

Figure 5. Model concentrations showing fraction of meteoric (M) and water-rock (W-R) derived sodium.

Although the model described above only accounts for the evolution of relatively fresh groundwaters (Cl <10mM), several mechanisms can be suggested for the origin of the more saline waters. It is proposed that the relatively dilute waters become concentrated by evapotranspiration (ET) in the unsaturated zone, or are mixed with waters in which chloride is the more dominant anion (Fig. 2(a-c)). Concentration by ET in the unsaturated zone results in both exchange of sodium ions for calcium ions on clay-mineral surfaces:

$$2Na\text{-}X + Ca^{2+} \leftrightarrow Ca\text{-}X + 2Na^+ \quad (6)$$

and reverse-weathering of silicates, as indicated by equation 5 proceeding from right to left. Providing the reverse-weathering process is more dominant than cation exchange, there will be an addition of sodium (Fig. 2(d)) and calcium (Fig. 2(e)) concentration relative to bicarbonate as chloride concentration increases. It is worthwhile noting that cation exchange alone will not affect SiO_2 or HCO_3^- concentrations (cf. Fig. 2(b-c)). Furthermore, the Ca:SO_4 ratio would be less than 1:1 (Fig. (2(f)) if cation exchange was a dominant process.

3 CONCLUSIONS

A geochemical model describing the evolution of groundwater chemistry in the Ti-Tree Basin of central Australia suggests that (1) more dilute waters (Cl < 10mM) are primarily derived by dissolution of gypsum, dissolution or precipitation of carbonate minerals, and weathering of silicates; and (2) the more saline waters (TDS > 1000 mg/l) have a chemical composition that reflects significant evapotranspiration during recharge, with removal of cations and bicarbonate ions relative to chloride, resulting in a Na-Cl type water.

4 REFERENCES

Allison, G.B. 1982. The relationship between ^{18}O and deuterium in water in sand columns undergoing evaporation. *J. Hydrol.* 55:163-169.

Calf, G.E., P.S. McDonald & G. Jacobson 1991. Recharge mechanism and groundwater age in the Ti-Tree Basin, Northern Territory. *Aust. J. Earth Sc.* 38:299-306.

Craig, H. 1961. Isotopic variations in meteoric waters. *Science* 133:1702-1703.

International Atomic Energy Agency. 1969-1990. World Survey of Isotope Concentration in Precipitation. *Environ. Isotope Data* No.1-10.

Keywood, M. 1995. Origins and sources of atmospheric precipitation from Australia: chlorine-36 and major-element chemistry. *PhD thesis, Australian National University.* 207pp.

McDonald, P.S. 1988. Groundwater Studies, Ti-Tree Basin, 1984-1988. *Northern Territory Power and Water Authority Report* No. 1/90.

Tidal influences on metal concentrations in groundwater, Geelong, Australia

S. Horner & T. R. Weaver
VIEPS, School of Earth Sciences, The University of Melbourne, Parkville, Vic., Australia

ABSTRACT: Fluctuations in groundwater elevations and water chemistry with tidal cycle were measured in a tidally-influenced aquifer between a landfill and a coastal wetlands at Limelighters Lagoon, Victoria. These fluctuations were related to tidal stage, distance inland, and aquifer heterogeneity. Over a single tidal cycle, aqueous concentrations of copper, lead and zinc varied significantly with time up to 70 m inland (eg. 0.029-0.12 mg Cu/L), indicating that metals were being transferred to and from aqueous solution as the tide progressed. Major ion chemistry varied mainly with distance from shore rather than tidal stage indicating that the changing metal concentrations were related to short-term variations in local geochemical conditions rather than the input of more saline or fresher water as the tidal cycle progressed. Variations in metal concentrations were sufficient that individual metals met, or exceeded, water quality guidelines depending on the stage of the tidal-cycle stage at which groundwater samples were collected.

1 INTRODUCTION

Much research has been performed on major-ion and heavy-metal behaviour in aqueous systems (eg. Hahne & Kroontje, 1973; Forstner et al., 1989) and on the physical response of aquifer systems to tidal fluctuations (eg. Lanyon et al., 1982; Nielsen, 1990; Erskine, 1992). Roxburgh (1985) presents one of the few studies that combines these approaches, describing how major-ion concentrations vary in response to tidal fluctuations in a tidally-influenced aquifer. Here, we address the potential effects of tidally-driven fluctuations in groundwater elevations and chemistry on the transfer of trace metals between solid and aqueous phases in a tidally-influenced aquifer.

Fluctuations in water levels and chemistry with tidal stage were monitored in a heterogeneous unconsolidated coastal aquifer between a landfill and an internationally-significant coastal wetland, Limeburners Lagoon, near Geelong, Victoria. Groundwater elevations, water quality parameters, trace metals, and major ions were measured at 21 piezometers in 8 nests at 2-3 hourly intervals over 12-hour tidal cycles.

1.1 Geology and hydrogeology

The study area is located approximately 5 km northeast of Geelong, Victoria, at Limeburners Lagoon, a shallow tidal arm of Corio Bay (Figure 1). Approximately 150 m west of Limeburners Lagoon lies the Bell Road Landfill which receives a mixture of domestic and industrial waste.

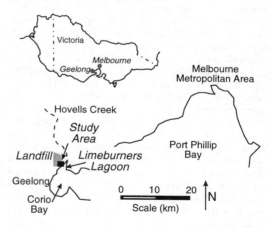

Figure 1: Location of the study area and Limeburners Lagoon, Geelong (after Horner, 1997)

The aquifer between Limeburners Lagoon and the Bell Road Landfill is the Tertiary Moorabool Viaduct Formation, which consists of fine grained sands with minor zones of gravel, clay and limestone (Bowler, 1963). This formation is overlain by basalts of the Late Pliocene Newer Volcanics which range in thickness from about 0.5 m near the shore to over 7 m at the landfill site. Throughout the region, the Moorabool Viaduct Formation is underlain by the Tertiary Batesford-Fyansford Formations (Bowler, 1963) of which the Fyansford Formation is considered to act as a basal aquitard.

In the study area, the water table occurs in the Moorabool Viaduct Formation, the regional groundwater flow direction is to the southeast, toward the shores of Limeburners Lagoon, and the hydraulic conductivity of the aquifer is estimated to range from 2×10^{-3} to 3.6 m/d (2.9×10^{-8} to 4.2×10^{-5} m/s) (Ellis, 1992). At the Bell Road Landfill site total dissolved solids (TDS) contents in groundwater in the Moorabool Viaduct Formation range from 2800 to 12000 mg/L (Ellis, 1992). The TDS of brackish surface water from Limeburners Lagoon is approximately 12000 mg/L.

1.2 Instrumentation and groundwater sampling

Twenty-one bores were installed in 8 piezometer nests over a distance of approximately 130 m from the shore of the lagoon towards the southeastern boundary of the landfill (Figure 2). The 5 piezometers nests closest to the shore (nests 4-8) were spaced at 10 m intervals up to a distance of 40 m from the shore. Nests 1-3 were separated by greater distances. Piezometers in nests 1-3 were installed using a hollow-stem auger and were screened over the bottom 1 m. Piezometers in nests 4-8 were installed by hand auger and were screened over the bottom 30 cm. Shallow piezometers (A wells) were screened near the water table; deeper piezometers in the nests were screened 1-2 m lower (C wells). B wells were completed at intermediate depths. All shallow piezometers were screened in the Moorabool Viaduct Formation. Piezometers 1-3C were screened in the Fyansford Formation.

Three groundwater monitoring and sampling rounds were conducted. Sampling run I involved collection of water-quality parameters at 2-hourly intervals over an entire tidal cycle. Rising-head tests were conducted in each piezometer to measure hydraulic conductivity (Table 1). Sampling run II involved collection of samples for copper, lead, zinc, major ions, nitrate and ammonia analyses, and water-quality parameters at low tide, mid-flood tide, high tide, mid-ebb tide and low tide (3-hourly intervals). Electrical conductivity (EC), pH, Eh, dissolved oxygen (DO), and temperature were measured in the field. All samples for cation analysis were filtered through 0.45 µm cellulose acetate filters and acidified with nitric acid in the field. Heavy metals and other cations were analysed by ICPMS and anions were analysed by ion chromatography at AMDEL Laboratories, Sydney.

Eh, pH, DO and EC meters and probes were calibrated and standardised in the field. Variations in groundwater elevations with tidal stage were measured during sampling run III.

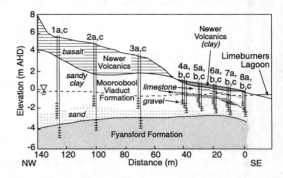

Figure 2: Cross-section showing locations of piezometers (after Horner, 1997).

2 TIDAL INFLUENCES ON GROUND-WATER QUALITY

Groundwater monitoring indicated that, although changes in groundwater elevations and major-ion concentrations were relatively minor, significant variations in water-quality parameters and heavy-metal concentrations occurred with tidal stage.

2.1 The saline-fresh water interface

The average tidal range in Limeburners Lagoon is approximately 0.80 m, and the maximum change in groundwater elevation associated with tidal stage measured during sampling run III was 12.5 cm (Well 7C, Table 1). The largest changes in water levels occurred in the deeper, C level wells in the Moorabool Viaduct Formation. Changes of <5 cm were measured inland at wells 1C and 2C in the Fyansford Formation. Water level fluctuations in piezometers closest to the shore (8A and 8C) were lower than in more distal piezometers (Table 1) and were out of phase with tidal stage, indicating the heterogeneous nature of the aquifer.

Water-quality and major-ion analyses indicate the extent to which saline (brackish) water from the lagoon penetrated the Moorabool Viaduct Formation aquifer during a tidal cycle. Chloride concentrations decreased landward in both the shallow (A) and deeper (C) piezometers from about 15000 mg/L at the shore to a background value of about 3000 mg/L landward of nest 5, indicating that brackish water from Limeburners Lagoon reaches up to 30 m inland in this aquifer. The effect of heterogeneity on salt water intrusion in this aquifer is clear. At nest 6,

Table 1: Measured time lags (high tide vs maximum change in groundwater elevation), maximum change in groundwater elevation over entire tidal cycle, and hydraulic conductivity values measured at piezometers.

	8A	7A	6A	5A	4A	3A	2A	1A
Measured time lag (hr)	4	nm	0	2	2	nm	nm	2
Max. groundwater elevation change (cm)	2.0	1.0	7.0	2.0	9.5	1.5	1.0	3.0
Hydraulic conductivity (m/s)	9.86e-8	1.32e-7	5.72e-8	1.21e-7	4.63e-6	1.15e-5	1.09e-5	4.83e-6
Distance inland (m)	0	10	20	30	40	70	100	130
	8C	7C	6C	5C	4C	3C	2C	1C
Measured time lag (hr)	4	0	2	2	2	0	4	2
Max. groundwater elevation change (cm)	1.7	12.5	8.0	4.0	3.0	1.0	3.0	2.0
Hydraulic conductivity (m/s)	9.86e-9	3.61e-8	5.06e-7	3.55e-5	1.67e-5	5.37e-7	6.79e-7	5.81e-7
Distance inland (m)	0	10	20	30	40	70	100	130

nm = time lag for maximum change in groundwater elevation to occur not measured

groundwater from the shallow well (6A) is more saline (7300 mg Cl⁻/L) than groundwater from the deeper well (6C, 3600 mg Cl⁻/L).

2.2 Water-quality parameters and major ions

Eh and pH varied with both distance inland and tidal-cycle stage. Eh values were highest (170-250 mV) in groundwater from piezometer nests 1-3, 70 to 130 m away from the shore, and lowest (50-150 mV) in groundwater from nests closer to shore (0-40 m). The maximum variation in Eh with tidal stage occurred in piezometers closest to shore (eg. 8C, Eh=80 mV, mid-flood tide; Eh=170 mV, high tide). Over an entire tidal cycle, pH values declined by up to 2 pH units near shore (eg. 7.0 to 4.6, 6A); further inland, pH values fluctuated by 1 pH unit over the same tidal cycle (eg. between 5.4 and 6.3, 1A).

Major cation concentrations in groundwater from A and C level wells declined inland from the saline water interface, reaching background values (eg. Ca~95 mg/L) at about 30 m inland (Figure 3). Nitrate and bicarbonate concentrations displayed different trends from major cation and chloride concentrations. Bicarbonate concentrations were highest (550-650 mg/L) in groundwater from wells 4-6A, and decreased in groundwater from wells 1-4A and 8A (~300 mg/L). Bicarbonate concentrations in C level wells decreased from 344 mg/L in well 7C to between 240 and 300 mg/L in groundwater from all other C wells (Figure 3). Nitrate concentrations were highest (1.0-1.5 mg as N/L) in groundwater from wells 1A to 3A, furthest from shore and closest to the landfill, and were <0.1 mg/L in shallow groundwater shoreward of well 5A. Nitrate concentrations varied significantly with distance in deeper wells, from <0.1 mg/L in groundwater from wells 1C, 7C, and 8C, to between 1 and 2 mg/L in groundwater from wells 4C, 5C, and 6C (Figure 3). None of the major-ion and nitrate concentrations varied significantly with tidal stage.

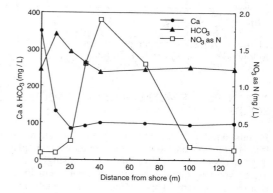

Figure 3: Concentrations (mg/L) of Ca^{2+}, HCO_3^-, NO_3^- (as N) in groundwater with distance inland (C level wells, high tide).

2.3 Copper, lead and zinc concentrations

Unlike the changes identified in major-ion chemistry with distance, concentrations of copper, lead and zinc varied significantly with tidal stage. In general, lead, copper and zinc concentrations were low at times other than high tide. Metal concentrations increased at high tide, with maximum increases in concentration occurring in the deeper wells. The instances of groundwater containing dissolved copper at concentrations greater than 40 µg/L (the Dutch B guideline concentration is 50 µg/L) increased at high tide (Figures 4a and 4b).

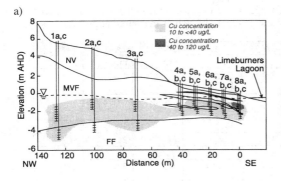

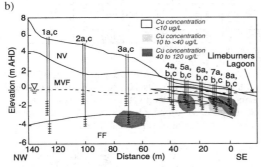

Figure 4: Aqueous copper concentrations (µg/L) at a) low tide (0 hours) and b) high tide (+6 hours)

3 TRANSFER OF METALS BETWEEN AQUIFER SOLIDS AND GROUNDWATER

In this aquifer, large changes were identified in trace-metal concentrations, redox potential and pH in groundwater up to 70 m inland over a single tidal cycle whereas major-ion concentrations remained relatively constant. This indicates that metals are most probably transferred into and out of solution as a result of the changing Eh-pH environment. The tidal cycle is the dominant mechanism leading to the large variations in geochemical conditions with time at the near- to mid-shore piezometers as groundwater with different geochemical characteristics is driven progressively into different portions of the aquifer. For example, as high tide is reached, pockets of more reducing, more acidic water move inland from the near-shore environment, and more saline, oxidised lagoon water replaces the near-shore water. Metals are then transferred to, or from, the solid phase as the surrounding aquifer matrix and the recently-placed groundwater attempt to approach equilibrium (eg. Figures 4a and b).

In tidally-influenced aquifers, the constantly changing hydraulic conditions mean that groundwater-flow paths are always changing. In a geologically and geochemically heterogeneous aquifer such as the Moorabool Viaduct Formation, groundwater is, therefore, continually being exposed to different sediment types and matrix chemistry. Under these conditions, trace metals will be transferred to or from solution in an attempt to reach a new equilibrium between the aquifer matrix and groundwater as the groundwater enters a geochemically different portion of the aquifer.

4 REFERENCES

Bowler, J.M., 1963. Tertiary stratigraphy and sedimentation in the Geelong-Maude area. *Proc. Royal Soc. Vic.*, No.76, Pt.1-2, pp.69-136.

Ellis, D., 1992. *A hydrogeological assessment of the Bell Road Landfill, Corio, Victoria*. Unpublished M.App.Sc., Univ. New South Wales: Australia.

Erskine, A.D., 1991. The effect of tidal fluctuations on a coastal aquifer in the UK. *Ground Water*, v.29, no.4, pp.556-562.

Forstner, U., Ahlf, W. and Calmano, W., 1989. Studies on the transfer of heavy metals between sedimentary phases with a multi chamber device: combined effects of salinity and redox potential. *Marine Chem.*, v.28, pp.145-158.

Hahne, H.C.H. and W. Kroontje, 1973. Significance of pH and chloride concentration on behaviour of heavy metal pollutants: mercury (II), cadmium(II), zinc(II) and lead(II). *J. Env. Qual.*, v.2, no.4, pp.440-450.

Horner, S., 1997. *Physical and chemical influences of tidal activity on groundwater in a coastal aquifer: implications for major ion and trace metal behaviour*. Unpublished MSc Thesis, University of Melbourne: Australia.

Lanyon, J.A, I.I. Eliot and D.J. Clarke, 1982. Groundwater level variation during semidiurnal spring tidal cycles on a sandy beach. *Aust. J. Mar. Freshwater Res.*, v.33, pp.377-400.

Nielsen, P., 1990. Tidal dynamics of the water table in beaches. *Water Resources Res.*, v.26, no.9, pp.2127-2134.

Roxburgh, I.S., 1985. Groundwater chemistry change due to tidal fluctuations in the Stonehouse Brewery well, Plymouth, England. *Water Resources Research*, v.4, no.6, pp.1235-1247.

5 ACKNOWLEDGMENTS

ARC Small Grant No. S04947383 to TRW funded this research. The City of Greater Geelong provided access to the Bell Road Landfill and reports.

Hydrogeochemical processes in a fractured rock aquifer of the Lachlan Fold Belt: Yass, New South Wales, Australia

Jerzy Jankowski & Seyed Shekarforoush
Groundwater Centre, Department of Applied Geology, University of New South Wales, Sydney, N.S.W., Australia

R. Ian Acworth
Groundwater Centre, Water Research Laboratory, University of New South Wales, Manly Vale, N.S.W., Australia

ABSTRACT: The computer program NETPATH was used to model chemical and stable isotopic data from groundwaters sampled in the Dicks and Williams Creek catchments, which are located near Yass, New South Wales. The purpose was to investigate the significance of several hydrogeochemical processes occurring in this fractured rock aquifer. Water-rock interaction with fracture-filling minerals is shown to produce chemically distinct groundwater types that reflect the mineralogy of the aquifer. The modelling of three endpoint groundwaters supports the authors' hydrogeochemical interpretation that the dominant processes in this system are oxidation-reduction, weathering and dissolution of the aquifer matrix. In the recharge zone, the oxidation of pyrite, the dissolution of chlorite and calcite, and the precipitation of goethite are the dominant processes. The chemistry of discharging waters is controlled by the dissolution of goethite in the presence of reduced sulphur producing FeS.

1 INTRODUCTION

The hydrogeochemistry of an early Palaeozoic fractured aquifer system in the Southern Tablelands of New South Wales was investigated to delineate the pervasive geochemical processes that dominate water evolution in this greenschist grade, regionally metamorphosed, aquifer system. The study area encompasses the Dicks and Williams Creek catchments and is located some 30 km southeast of Yass, within the Lachlan Fold Belt of southeastern Australia. Several workers (Jankowski & Acworth 1993; Acworth et al. 1997; Lawson 1989) have studied dryland salinisation processes in the Dicks and Williams Creek catchments. Hydrogeological studies of these catchments have shown that the chemical composition of groundwaters is intimately related to aquifer mineralogy (Jankowski & Acworth 1993; Jankowski et al. 1994).

2 GEOLOGY

The Dicks and Williams Creek catchments consist of fractured, quartz-rich Ordovician meta-sediments overlain in the valleys and localised depressions by unconsolidated clay-rich Quaternary sediments. The Ordovician sediments are of deep marine origin, and have been associated with deposition within the Monaro Basin. Lithologically, these sediments consist of interbedded shale, siltstone and sandstone. X-ray diffraction and thin section analyses indicate that the strata consist of quartz grains in a matrix of chlorite, biotite, sericite, and goethite replacing pyrite and calcareous material. Deep weathering in the Tertiary has resulted in the accumulation of clay minerals within fracture zones (Lawson 1989). The mineralogical content of the clay fraction is formed from 50% chlorite and kaolinite, 30% illite and 20% expandable lattice clay minerals. Silica remobilisation has produced northerly trending quartz dykes in arenaceous units throughout the area. Three major north-south striking faults run through the study area, with the Sullivan Line Fault (Abell 1991) lying directly between the Dicks and Williams Creek catchments. The two catchments are geologically distinct. The Dicks Creek catchment is part of the Pittman formation and consists of quartz-rich sandstone, siltstone and shale with minor black shale and calcareous sandstone (Lawson 1989). In the Williams Creek area, pyrite-bearing shale is the dominant lithology, this strata is referred to as the Acton Shale. This lithological and mineralogical variation is reflected in the chemistry of the groundwaters. A total of 95 samples from 25 bores (with total depths ranging from 27 to 102 m), and 11 samples from shallow and surface waters were analysed. The fractured aquifer is semi-confined on

Table 1. Average chemical data from Dicks Creek and Williams Creek catchments.

Zone	n	EC µS/cm	TDS mg/l	pH	CO_2 mg/l	O_2 mg/l	Na^+ mg/l	K^+ mg/l	Ca^{2+} mg/l	Mg^{2+} mg/l	Fe^T mg/l	HCO_3^- mg/l	SO_4^{2-} mg/l	Cl^- mg/l	S^{2-} mg/l	SiO_2 mg/l
RZ_w	11	55	45	6.05	8.3	3.6	8.4	1.1	0.3	1.6	0.4	11.2	3.5	10.8	ND	7.8
RZ_{gw}	12	584	361	5.76	106.4	2.3	39.4	1.0	8.1	41.2	4.0	60.8	144.2	48.3	0.19	12.6
DCI_{gw}	50	1346	975	6.46	133.3	0.2	104.6	2.0	54.8	84.1	6.9	294.9	247.0	153.8	1.17	27.1
$DCII_{gw}$	15	679	568	6.73	64.1	0.2	79.3	1.1	15.5	40.1	9.6	215.9	140.1	36.3	0.56	32.6
WC_{gw}	18	2370	1647	5.42	216.8	1.1	138.2	3.1	31.8	234.2	7.3	116.0	887.4	205.9	0.25	21.7

RZ_w - Recharge waters; RZ_{gw} - Recharge groundwaters; DCI_{gw} - Dicks Creek 1; $DCII_{gw}$ - Dicks Creek 2; WC_{gw} - Williams Creek

the middle to lower slopes, and on the floors of valleys, by debris-flow deposits (Acworth et al. 1997).

3 GROUNDWATER CHEMISTRY

Oxygen-18 and deuterium isotopic data indicate that groundwaters in the aquifer system are of meteoric origin, and that these waters are transmitted rapidly through the fracture system. Different hydrogeochemical evolution pathways are observed in this aquifer system. These pathways depend mainly upon geology and mineralogy and residence time, and are irrespective of the elevation of the bores or the depth of the aquifer interval.

Five chemical types of groundwaters were selected for modelling (RZ_w, RZ_{gw}, DCI_{gw}, $DCII_{gw}$ & WC_{gw}). Selection was based upon the location of the borehole along flow path (recharge-discharge zones) and the sampled groundwaters chemical composition (Table 1). Fresh meteoric Recharge Waters (RZ_w) rapidly evolve into the initial water used in all models. Recharge Zone groundwaters – (RZ_{gw}) occur in bores adjacent to the exposed bedrock with aquifer intervals between 80-100 m. They are fresh Mg-Na-SO_4-Cl waters, slightly acidic, oxidised and of low salinity. The chemical composition of the RZ_{gw} results from the rapid alteration of Recharge Waters (RZ_w) through water-interaction within fractures in both fresh and weathered rock.

The chemical composition of Dicks Creek I (DCI_{gw}) and Dicks Creek II ($DCII_{gw}$) groundwaters result from a series of complex water-rock interaction reactions in the presence of organic material. These groundwaters are typical of discharge waters and often flowing under artesian pressures from a depth of up to 77 m. They are more saline, are depleted in dissolved oxygen, and show pH buffering between the H_2CO_3/HCO_3 couple.

Chemical classification of these waters characterises DCI_{gw} and $DCII_{gw}$ as Mg-Na-SO_4-HCO_3-Cl and Na-Mg-HCO_3-SO_4 respectively. The Williams Creek groundwaters (WC_{gw}) has the highest salinity, are acidic, slightly oxic waters giving them the appearance of being the most juvenile waters. Magnesium and sulphate comprise up to 70% of the total ions in WC_{gw} samples.

Chemical evolution of these waters commences in the recharge zone where H^+ is generated by the oxidation of pyrite and used for dissolution and weathering reactions. Rainfall charged with soil zone CO_2 attack soluble salts, carbonates and aluminosilicates. Both groundwater systems are closed to CO_{2gas} mass transfer from the surface, but are open to CO_{2gas} production or consumption within the aquifer through the oxidation of organic matter and the associated SO_4 reduction, $CaCO_3$ dissolution, and the weathering of aluminosilicates.

Figures 1a and 1b show that there is an excess of Na over Cl in $DCII_{gw}$ (Na/Cl ratio 3.4) this is due to ion exchange on clays lining fractures (Cerling et al. 1989). The flux of chloride from rainwater is insignificant (around 1 mg/l Cl), and positive heads negate the overlying colluvium as a Cl source. All of the chloride in the groundwater can be accounted for by water-rock interaction with the rocks matrix (where Cl is stored in pore spaces as soluble salts in pores or as Cl bearing minerals on grain boundaries) as these sediments are of marine origin. Longer residency time and increasing distance along flow path elevate Cl concentrations (DCI_{gw} and WC_{gw}). Sodium is delivered from the dissolution of soluble salts and is released by hydrolyses of albite. The latter, is probably of marginal significance, as feldspars constitute a small portion of the whole rock mineralogy.

Calcium originates from the dissolution of calcite in the rocks, particularly calcareous sandstone. All groundwaters are undersaturated with respect to

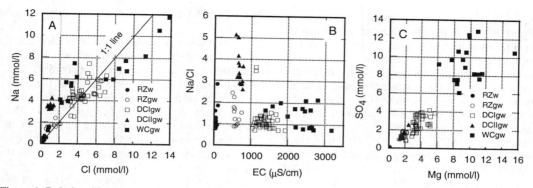

Figure 1. Relationships between Na versus Cl (A), Na/Cl versus EC (B) and SO$_4$ versus Mg (C).

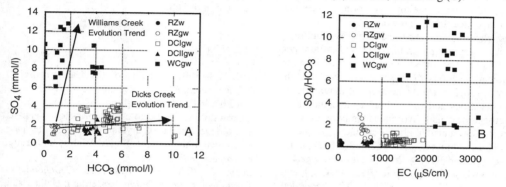

Figure 2. Relationships between SO$_4$ versus HCO$_3$ (A) and SO$_4$/HCO$_3$ versus EC (B).

calcite (log IAP/K$_{cal}$ = 0.00 to -5.56). Some Ca can be released by the alteration of anorthite, but the most likely source is ion exchange. Magnesium is delivered from the dissolution of chlorite, which pseudomorphs biotite and is abundant in these greenschist grade metamorphic rocks. Dolomite is not present in rock samples and was not found in thin section or XRD analyses. Potassium is delivered from two sources, the alteration of biotite and the weathering of K-feldspar. Little potassium is present in the waters as the illitic clays, that line fractures, retard it.

Sulphate is the major anion in the groundwater system. Gypsum can be discounted as a source as it was not identified through petrographic or XRD analyses. Sulphate originates from the oxidation of pyrite in the oxic waters of the recharge zone, producing large amounts of H$^+$ and Fe^{2+}. The later is oxidised to Fe^{3+} and removed from solution. The Fe^{3+} precipitates either as orange lepidocrocite or as the amorphous iron oxide ferrihydrite, and is later transformed to haematite or goethite. Both minerals are supersaturated in the oxic Williams Creek groundwaters. The H$^+$ produced from this reaction is consumed by the dissolution of carbonates and aluminosilicates. In the Dicks Creek catchment SO$_4$ is reduced to S^{2-} through the oxidation of organic matter (Table 1) and precipitates as FeS rather than FeS$_2$ (both supersaturated), removing any remnant Fe^{2+}. The FeS is less stable, but due to reaction kinetics precipitates more rapid than pyrite (Appelo & Postma 1993). Black FeS was found in all of the bores where reducing anoxic waters flowed under artesian pressure.

4 HYDROGEOCHEMICAL PROCESSES

The sequence of hydrogeochemical reactions from recharge to discharge zones in Williams and Dicks Creek catchments are best described as two separate hydrochemical systems. Flowing groundwaters from Williams Creek have evolved in a fracture system containing traces of organic matter and abundant disseminated pyrite and chlorite. Oxidation of pyrite produces SO$_4$ and generates H$^+$ which attacks chlorite producing Mg-SO$_4$ rich waters (Figure 1c). Artesian waters of Dicks Creek are rich in HCO$_3$

(Figure 2a, b) with lower Mg and SO_4 concentrations (Figure 1c). Two geological variants could explain the difference in water types. Either there is less pyrite in the Dicks Creek catchment, or the same amount of pyrite is oxidised in both catchments, producing a high SO_4 concentration, and sufficient organic matter is available in the Dicks Creek strata to mediate SO_4 to S^{2-}. The H^+ produced from the oxidation of FeS_2 reacts with CO_2 released from the reduction of SO_4 producing H_2CO_3, which dissociates to HCO_3, keeping carbonate equilibrium with pH being buffered between 6.4 and 6.7. As the reaction progresses the consumption of H^+ reduces the groundwaters acidity, slowing the dissolution of silicate minerals, mainly chlorite, keeping magnesium concentrations lower.

To test the hydrogeochemical processes outlined above, the computer program NETPATH (Plummer et al. 1991) has been used to calculate mass balance of chemical reactions. The selection of phases is based on geological and mineralogical assumptions, field observations and the authors understanding of the hydrogeology and hydrogeochemistry of each catchment. Ten plausible phases are $CO_{2(gas)}$, calcite, NaCl, chlorite, biotite, goethite, FeS_2, FeS, albite (RZ_{gw} -DCI_{gw} and RZ_{gw} - WC_{gw}), kaolinite (RZ_{gw} - $DCII_{gw}$ and RZ_{gw} - WC_{gw}), illite (RZ_{gw} - DCI_{gw}), and Ca/Na-exchange (RZ_{gw} - $DCII_{gw}$). The modelled mass transfer is supported by ^{13}C and $^{34}S_{SO4}$ isotopic data and is as follow:

From RZ_{gw} to DCI_{gw}:
RZ_{gw}+6.37$CO_{2(gas)}$+1.36$CaCO_3$+4.04NaCl+0.53 Chlorite+0.66Biotite+4.32Goethite+7.77FeS_2+0.15 Albite=>1.06Illite+12.98FeS+DCI_{gw}.

The isotopic composition calculated by NETPATH of the final water is -17.5‰ for $\delta^{13}C$ and 8.8‰ for $\delta^{34}S_{SO4}$, which is consistent with the observed data of -17.4‰ and 7.2‰ respectively.

From RZ_{gw} to $DCII_{gw}$:
RZ_{gw}+2.8$CO_{2(gas)}$+1.56$CaCO_3$+0.72NaCl+0.32 Chlorite+0.00Biotite+2.80Goethite+4.08FeS_2+1.18 Ca/Na-exchange=>0.27Kaolinite+6.72FeS+$DCII_{gw}$.

The isotopic composition calculated by NETPATH of the final water has a $\delta^{13}C$ of -14.3‰ and a $\delta^{34}S_{SO4}$ of 8.4‰, which is consistent with observed values of -16.0‰ and 5.4‰ respectively.

From RZ_{gw} to WC_{gw}:
RZ_{gw}+14.94$CO_{2(gas)}$+0.79$CaCO_3$+5.51NaCl+1.90 Chlorite+0.05Biotite+18.36Goethite+27.53FeS_2+0.14Albite=>3.03Kaolinite+45.84FeS+WC_{gw}.

The isotopic composition calculated by NETPATH of the final water gives a $\delta^{13}C$ of -19.9‰ and a $\delta^{34}S_{SO4}$ of 9.3‰, which is consistent with the observed data: -18.6‰ and 7.4‰ respectively.

5 CONCLUSIONS

The mass balance modelling of the three chemically different discharge groundwaters shows that the most important reactions, which satisfy $\delta^{13}C$ and $\delta^{34}S_{SO4}$ isotopic data, are the oxidation of FeS_2, the formation of goethite and the dissolution of chlorite by H^+ which results from FeS_2 oxidation. The concentration of carbon species in the system is primarily a function of the $CaCO_3$ dissolution, the weathering of aluminosilicates, and the outgassing of CO_2 in the presence of H^+ derived from the pyrite oxidation. In the discharge zones the reduction of goethite in the presence of S^{2-} encourages the precipitation of FeS.

REFERENCES

Abell, R.S. 1991. Geology of the Canberra 1:100 000 sheet area. *BMR Geol. Geophys. Bull.* 233.

Acworth, R.I., A. Broughton, C. Nicoll & J. Jankowski 1997. The role of debris-flow deposits in the development of dryland salinity in the Yass River catchment, New South Wales, Australia. *Hydrogeol. J.* 5: 22-36.

Appelo, C.A.J. & D. Postma 1993. *Geochemistry, groundwater and pollution*. Rotterdam: Balkema.

Cerling, T.E., B.L. Pederson & K.L. Von Damm 1989. Sodium-calcium ion exchange in the weathering of shales: Implications for global weathering budgets. *Geology* 17: 552-554.

Jankowski, J., & R.I. Acworth 1993. The hydrogeochemistry of groundwater in fractured bedrock aquifers beneath dryland salinity occurrences at Yass, NSW. *AGSO J. Aust. Geol. Geophys.* 14: 279-285.

Jankowski, J., R.I. Acworth, R.P. Schneider & B.R. Chadwick 1994. Hydrogeochemical and microbiological processes in fractured bedrock aquifers developed in the Palaeozoic metasediments of south-east Australia. *Proc. 3rd Int. Symp. Environ. Geochem., Cracow, 12-15 September 1994*: 176-178.

Lawson, S.J. 1989. Dryland salinity project: Williams Creek drilling completion report. *Tech. Rep. Dep. Water Resour.* TS 88.017.

Plummer, L.N., E.C. Prestemon & D.L. Parkhurst 1991. *An interactive code (NETPATH) for modelling NET geochemical reactions along a flow PATH*. USGS Water Res. Invest. Rep. 91-4078.

Reverse ion-exchange in a deeply weathered porphyritic dacite fractured aquifer system, Yass, New South Wales, Australia

Jerzy Jankowski & Seyed Shekarforoush
Groundwater Centre, Department of Applied Geology, University of New South Wales, Sydney, N.S.W., Australia

R. Ian Acworth
Groundwater Centre, Water Research Laboratory, University of New South Wales, Manly Vale, N.S.W., Australia

ABSTRACT: The hydrogeochemistry and the evolution of groundwater flowing through a fractured aquifer within the Spring Creek catchment near Yass, NSW has been developed from newly available data. The main lithology in the area is a porphyritic dacite, which has been deeply weathered into montmorillonitic clays. The porphyritic dacite erupted in a submarine setting, accordingly all of the Cl in these waters is attributed to the dissolution of halite. Chemical classification of samples from the Spring Creek catchment show that fresh meteoric waters are Na-Cl type, these evolve into Mg-Na-HCO_3 type recharge groundwaters, and then into Mg-Ca-Cl-(HCO_3) type discharge waters. Discharge groundwaters are characterised by a Na:Cl ratio of 0.24. This deficiency is attributed to reverse ion exchange reactions of Na for Mg and Ca on exchange sites of high cation exchange capacity montmorillonitic clays. Mass balance modelling indicates that the dissolution of NaCl supplies Cl to aquatic system and reverse ion exchange removes Na and delivers both Mg and Ca into these groundwaters.

1 INTRODUCTION

Sixteen deep bores ranging from 11 to 102 m have been intermittently sampled over the last few years from a porphyritic dacite fractured aquifer system, 20-30 km north of Yass, New South Wales (Lawson 1989: Lertsirivorakul 1990; Jankowski & Acworth 1993).

All of the rocks in the catchment are highly fractured, with macro and micro-fractures and quartz veins providing conduits for groundwater flow (White 1988; Lawson 1989). These fractures follow the north-south regional structural trend. Groundwater recharge occurs in elevated areas where bedrock crops out, and flows through fractures into valleys where colluvial cover provides an aquifer seal allowing these waters to discharge under artesian pressure. This catchment is severely affected by dryland salinity where groundwater from the fractured aquifer discharges to the surface.

This paper describes the hydrogeochemical evolution of groundwaters in the Spring Creek catchment through mass balance modelling. The computer program NETPATH (Plummer et al. 1991) was used as a modelling tool and shows how water-rock interaction in fractures and veins controls groundwater chemistry.

2 GEOLOGY

The study area is located in the southern part of the Lachlan Fold Belt of southeastern Australia, within the Cowra-Yass Synclinorial Zone (Cas 1983). Regional east-west compression has given the rocks of the Lachlan Fold Belt a northerly trend. In the study area, the Early Silurian Mundoonen Sandstone is unconformably overlain by Middle Silurian porphyritic dacite of the Douro Volcanic Series. Submarine conditions prevailed during the eruption of the Douro Volcanic Series. This sequence of interbedded porphyritic dacite and rhyodacite has a total thickness of about 2300m. Intense weathering during Tertiary times has produced a deep weathering profile.

The mineralogical composition of the porphyritic dacite is similar to andesite. Petrographic and XRF analyses indicate that the phenocrysts and groundmass in samples from the Douro Volcanic Series consist of quartz, feldspars and biotite, with disseminated pyrite present in fractures and quartz veins. Minor calcite, dolomite, haematite and goethite, and traces of gypsum and halite are found in fresh and weathered rocks. Weathering has altered biotite to chlorite, and feldspar to sericite (White 1988). Clay comprises up to 30% of fresh rock

Table 1. Average chemical composition of recharge waters (R_w), recharge groundwaters (R_{gw}) and discharge groundwaters (D_{gw}).

Zone	n	EC µS/cm	TDS mg/l	pH	Na mmol/l	K mmol/l	Ca mmol/l	Mg mmol/l	HCO$_3$ mmol/l	Cl mmol/l	SO$_4$ mmol/l	SiO$_2$ mmol/l	Na/Cl
R_w	9	132	100	5.94	0.62	0.032	0.04	0.21	0.53	0.62	0.03	0.33	1.00
R_{gw}	2	707	573	6.68	3.18	0.025	0.59	1.69	5.43	1.66	0.12	0.42	1.92
D_{gw}	9	2383	1522	6.57	3.85	0.075	4.45	5.19	7.84	16.07	0.40	0.51	0.24

(Lertsirivorakul 1990), and XRD clay fraction analyses indicate that the dacitic rocks are largely altered to montmorillonite, illite, chlorite and kaolinite.

3 GROUNDWATER CHEMISTRY

Since these rocks were deposited under marine conditions it is reasonable to attribute all of the Cl in the groundwaters to halite dissolution. All groundwaters have low SO$_4$ concentrations, showing that neither gypsum nor pyrite influence groundwater chemistry.

Fresh, oxidising, recharge waters (R_w) were sampled from shallow piezometers and creeks, these are Na-Cl type with appreciable Mg and HCO$_3$ concentrations. These waters are fresh (130 µS/cm EC; TDS - 100 mg/l) and have a pH of around 5.9 (Table 1). As the recharge waters sink through the deep weathering profile into the fracture system they evolve into recharge groundwaters (R_{gw}) which are more alkaline (pH of 6.7), Mg-Na-HCO$_3$ type waters (EC of 700 µS/cm; TDS - 570 mg/l). Recharge groundwaters were sampled from deep bores in elevated areas, these waters evolve rapidly down gradient into Mg-Ca-Cl-(HCO$_3$) discharge groundwaters (D_{gw}). Discharge groundwaters flow under artesian pressure in boreholes on the lower slopes and in valley floors. They have an average EC of 2000 - 2600 µS/cm (TDS - 1400 to 1700 mg/l) and pH values of 6.0 - 6.7 (Table 1).

4 HYDROGEOCHEMICAL PROCESSES

Infiltrating groundwaters are rich in CO$_2$, which dissociates into H$_2$CO$_3$ and HCO$_3$, generating hydrogen ions, which react with the weathered rock. The dissolution of calcite, chlorite, albite and halite in the weathered rock of the unsaturated zone provide a source for the major ions in the $R_{(gw)}$. This weathering produces kaolinite, which is subsequently altered to montmorillonite, which is the dominant fracture filling clay mineral. Recharge groundwaters have Na:Cl ratios greater than unity (Figure 1a, b), with $R_{(gw)}$ waters having an average Na:Cl ratio of 1.9.

Two chemical reactions are interpreted to contribute for the Na excess over Cl to aquatic systems; (1) the weathering of Na-aluminosilicates (Meybeck 1987) and, (2) ion exchange reaction (Cerling et al. 1989). The rate of groundwater flow through the fracture network in the Spring Creek catchment reduces the potential impact of albite dissolution, as albite weathers too slowly to contribute significant quantities of Na. This reaction tends to produce waters with pH values that are neutral or alkaline, which may explain the slight increase in pH from 5.9 in $R_{(w)}$ to 6.7 in $R_{(gw)}$. The most feasible reaction that explains the Na excess in $R_{(gw)}$ is ion exchange on clay derived from the weathering of dacite, as follows:

$$Ca^{2+} (Mg^{2+}) + Na_2\text{-clay} \Rightarrow 2Na^+ + Ca (Mg)\text{-clay}$$

From the recharge zone to the discharge zone the groundwaters interact with the aquifer matrix. The Cl concentrations increase 10-fold with a very small increase of Na, the loss on average of 14.4 mmol/l of Na can only be attributed to clay mediated reverse ion exchange (Table 1). The reverse ion exchange reaction requires the presence of clay and associated exchangeable Ca or Mg, and the flowing water to have a higher concentration of Na than the clay to drive a reaction as the system attempts to regain equilibrium. This process is well documented and has been identified as a dominant process where seawater intrudes coastal aquifers containing Ca-rich clays (Howard & Lloyd 1984). In the Spring Creek catchment Na-Cl rich waters are transmitted through an aquifer where Ca and Mg are the dominant ions on the exchange sites of clays. As the water attempts to reach equilibrium Na is lost and Ca and Mg

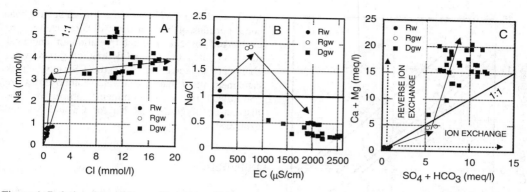

Figure 1. Relationships between Na versus Cl (A), Na/Cl versus EC (B) and Ca+Mg versus SO_4+HCO_3 (C).

are incorporated into the groundwater, producing Ca-(Mg)-Cl type waters.

The ratio of Ca + Mg versus $SO_4 + HCO_3$ (Figure 1c) will be close to unity if the dissolution of calcite, dolomite and gypsum are the dominant reactions in a system. Ion exchange will tend to shift the trend lines to the right due to an excess of $SO_4 + HCO_3$ (Cerling et al. 1989; Fisher & Mullican 1997). This process is clearly seen in the recharge groundwaters (R_{gw}). The hydrochemical trend on Figure 1c for Spring Creek groundwater is shifted to the left, showing a large excess of Ca + Mg over $SO_4 + HCO_3$, which can be only balanced by reverse ion exchange.

$2Na^+ + Ca\ (Mg)\text{-clay} \Rightarrow Na_2\text{-clay} + Ca^{2+}\ (Mg^{2+})$

When the water and exchanging clay minerals reach equilibrium with respect to sodium, calcium and magnesium ion exchange processes cease. This system is not in a state of equilibrium, as there clear evidence of ongoing ion exchange. This disequilibrium suggests that production of kaolinite, its transformation into Ca (Mg)-montmorillonite, and the dissolution of NaCl are occurring simultaneously.

The chemical data were plotted on aluminosilicate stability diagrams to identify the equilibrium conditions that prevail in the system. These diagrams show that all groundwaters are in equilibrium with kaolinite on the Na stability diagram. On Ca and Mg stability diagrams, all discharge groundwaters are in equilibrium with Ca and Mg-montmorillonite. Therefore, groundwater chemical compositions are controlled by the equilibrium between cation concentration in the groundwater and Na, Ca and Mg-montmorillonite.

5 MASS BALANCE MODELLING

The mass balance between recharge and discharge groundwaters was solved using the computer program NETPATH (Plummer et al. 1991). Two models were found using the following 7 phases: $CaCO_3$, CO_{2gas}, NaCl, K-feldspar, Na/Ca and Na/Mg-exchange and kaolinite. In each exchange reaction, the species listed first exchanges to the clay and the second species is released into solution. The mass transfer is as follows:

MODEL 1:
R_{gw} + 0.48 $CaCO_3$ + 3.67 CO_{2gas} + 14.44 NaCl + 0.05 K-feldspar $\Rightarrow$ 3.38 Na/Ca-exchange + 3.50 Na/Mg-exchange + 0.03 Kaolinite + D_{gw}

MODEL 2:
R_{gw} + 0.10 Chlorite + 4.15 CO_{2gas} + 14.44 NaCl + 0.05 K-feldspar $\Rightarrow$ 3.86 Na/Ca-exchange + 3.02 Na/Mg-exchange + 0.18 Kaolinite + D_{gw}

6 DISCUSSION

A plot of Na – Cl vs Ca + Mg - SO_4 - HCO_3 (Figure 2) allows us to test the model that reverse ion exchange is the dominant reaction in the aquifer system. By subtracting chloride from sodium (remembering that Cl is a conservative parameter, and assuming all Cl comes from halite dissolution), waters that are not influenced by ion exchange will plot close to zero on this axis. By adding calcium and magnesium and subtracting sulphate and bicarbonate, the dissolution of calcite, dolomite and gypsum will also return values of zero if dissolution is congruent and no ion exchange occurs. If ion exchange is the dominant process in the system the waters should form a line with slope of -1.

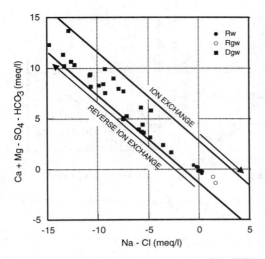

Figure 2. Relationship between Ca+Mg-SO$_4$-HCO$_3$ versus Na-Cl in Spring Creek groundwaters.

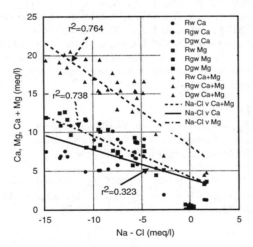

Figure 3. Relationships between Ca, Mg and Ca+Mg versus Na-Cl in Spring Creek groundwaters.

Linear regression on the data in Figure 2 indicates that the data fit onto a line that has a gradient that is very close to unity (-1.04), with a goodness of fit (r^2 = 0.91). This provides supporting evidence that Ca, Mg and Na concentrations are interrelated through reverse ion exchange. However, it is still unclear which ion (Ca or Mg) controls this hydrochemical reaction. To discriminate between these sources plots of Ca versus Na - Cl, Mg versus Na - Cl and Ca + Mg versus Na - Cl were developed (Figure 3). The correlation coefficients between the X and Y-axes show that the strongest correlation (r^2 - 0.76) is for the exchange of Na for Ca + Mg, indicating that equal portions of both of these divalent cations are exchanged for Na. The interpretation of chemical data and mass balance modelling of groundwater chemical compositions in the Spring Creek catchment show that groundwater compositions are controlled by two dominant processes: the dissolution of NaCl, and Na for Ca and Mg ion exchange.

REFERENCES

Cas, R.A.F. 1983. A review of the palaeogeographic and tectonic development of the Palaeozoic Lachlan Fold Belt of southeastern Australia. *J. Geol. Soc. Aust. Spec. Pub.* 10, 104 pp.

Cerling, T.E., B.L. Pederson & K.L. Von Damm 1989. Sodium-calcium ion exchange in the weathering of shales: Implications for global weathering budgets. *Geology* 17: 552-554.

Fisher, R.S. & W.F. Mullican II 1997. Hydrochemical evolution of sodium-sulfate and sodium-chloride groundwater beneath the northern Chihuahuan Desert, Trans-Pecos, Texas, USA. *Hydrogeol. J.* 5: 4-16

Howard, K.W.F. & J.W. Lloyd 1984. Major ion characterisation of coastal saline ground waters. *Ground Water* 21:429-437.

Jankowski, J. & R.I. Acworth 1993. The hydrogeochemistry of groundwater in fractured bedrock aquifers beneath dryland salinity occurrences at Yass, NSW: *AGSO J. Aust. Geol. Geophys.* 14: 279-285.

Lawson, S. 1989. Dryland salinity project: Begalia drilling completion report. *Tech. Rep. Dep. Water Resour.* TS 88.018.

Lertsirivorakul, R. 1990. Hydrogeology and dryland salinity of Begalia catchment, Yass, New South Wales. *MAppSc Project Report*, UNSW.

Meybeck, M. 1987. Global chemical weathering of surficial rocks estimated from river dissolved loads. *Am. J. Sci.* 287:401-428.

Plummer, L.N., E.C. Prestemon & D.L. Parkhurst 1991. *An interactive code (NETPATH) for modelling NET geochemical reactions along a flow PATH*. USGS Water Res. Invest. Rep. 91-4078.

White, R. 1988. The origin of dryland salinity at the Begalia catchment, Yass. *BSc (Hons.)* UNSW.

Comparison of oxygen and hydrogen isotopes from two perennial karst springs, Indiana, USA

Noel C. Krothe
Department of Geological Sciences, Indiana University, Bloomington, Ind., USA

ABSTRACT: Oxygen and deuterium isotopes were compared to decipher aquifer recharge, flow and storage characteristics of two large karst aquifer systems. The distance between the springs is only one mile (1.6km) but each represents flow from a different aquifer. Orangeville Rise receives recharge from diffuse sources on a large sinkhole plain with no prominent perennial sinking streams. Lost River Rise is recharged from the sinkhole plain but also from a large perennial sinking stream the Lost River. Although the recharge to the aquifer is different the amount of prestorm versus storm water is similar across a storm event. The storm peak of the hydrograph at Lost River Rise has a larger rain water component than Orangeville Rise. Pre-storm water contributes 75 to 80% of the total discharge over the monitoring period. However the distribution of the storm water across the hydrograph varies significantly with Orangeville Rise and Lost River Rise having 12 to 15% and up to 40% rainwater respectively at peak discharge.

1. Introduction

Carbonate aquifers in karst terranes are highly complex due to heterogeneities within the aquifer system. Fundamental properties of water recharge, groundwater flow and subsurface water storage can be extremely variable, even within a single carbonate aquifer. Recharge can range along a continuum from concentrated to dispersed, with sinking streams representing the most concentrated form of recharge and intergranular infiltration representing the most dispersed form. Water transmission within the subsurface can occur rapidly by means of conduit flow, or by diffuse flow, where water seeps slowly along small joints and fractures, or between individual grains of aquifer material. The volume of water stored within the vadose (unsaturated) and phreatic (saturated) zones is a function of both the aquifer's primary and its secondary porosity.

This study compares the short-term storm response of oxygen and deuterium isotopes at two major karst springs approximately 1 mile (1.6km) apart but receiving discharge from two distinct aquifers (Figure 1). The Orangeville Rise recharge is diffuse in nature receiving precipitation through a myriad of sinkholes and other solutional features on

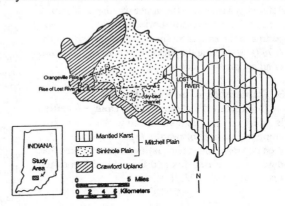

Figure 1. Study area and physiographic provinces within the upper Lost River Basin. Dashed line shows result of tracer tests.

the Mitchell Plain but no perennial streams. The Lost River Rise receives recharge from the sinkhole plain but also from a large perennial sinking stream the Lost River (Figure 1).

Orangeville Rise and Rise of Lost River are perennial springs located in south central Indiana (Figure 1) within two physiographic provinces, the Mitchell Plain and the Crawford Upland.

2. Conceptual Model of Groundwater Storage

A model illustrating water storage compartments in the vicinity of Orangeville Rise is shown in Figure 2 *(Lakey & Krothe, 1996)*.

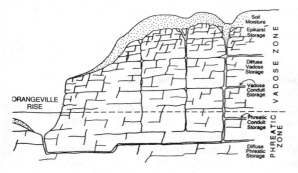

Figure 2. Model for water storage compartments of the vadose and phreatic zones in the vicinity of Orangeville Rise and Lost River Rise.

Water which passes through the soil zone may reside for some time in epikarst storage. Epikarst is defined as a region in the upper weathered layers of rock at the base of soil horizons, lying above the permanently saturated (phreatic) zone. *Williams (1983)* was among the first to describe the formation of an epikarst-surface.

3. Stable Isotopes in Water Studies

If it is assumed that discharge at Orangeville Rise is derived by simple mixing between two components: prestorm water Q_{ps} and rainwater Q_R, then two mass balance equations can be written that describe the water flux and the isotope flux as the rise *(Fritz et al., 1976)*:

$$Q_M = Q_r + Q_{ps} \quad (1)$$

$$Q_m \delta_M = Q_R \delta_R + Q_{ps} \delta_{ps} \quad (2)$$

where Q_M is the measured discharge at any instant in time and is defined by the sum of the two components. Delta notation represents the ^{18}O or deuterium (D) composition of the instantaneously measured discharge (δ_M), the rainwater (δ_R), or prestorm water (δ_{ps}). By combining equations (1) and (2) a third equation can be written which identifies the rainwater contribution to discharge at any instant in time, in terms of the isotopic composition of the measured discharge, prestorm water, and rainwater.

$$Q_R = Q_M \frac{(\delta_M - \delta_{ps})}{(\delta_R - \delta_{ps})} \quad (3)$$

To use (3) as a means of separating discharge into rainwater and prestorm water components, the isotopic composition of rainwater must be significantly different than that of prestorm water, and the prestorm water should have an identifiable and uniform isotopic composition throughout the basin (Table 1). Ideally, isotopic composition of rainfall should be constant through both time and areal distribution.

Table 1. Selected Lost River Chemical Data from Domestic Wells Located in the Lost River Basin (*n*=26)

Parameter	Range	Average
Specific Conductance, µS	408-1985	668
Calcium, mg L^{-1}	70-388	113
Sulfate, mg L^{-1}	13.9-976	125
δD, *‰	-43 to -46	-44
δ^{18}O, *‰	-6.4 to -6.8	-6.6

*SMOW = δD and δ^{18}O standard

4. Field Investigation

The field investigation was conducted in two stages. During the first stage, water was sampled at Orangeville Rise and Lost River Rise on a biweekly schedule to establish base flow chemical characteristics. It was critical during this stage to determine the base flow or prestorm isotopic composition to be used in the mass balance equations. Water from domestic wells was also sampled during this stage of the investigation to establish isotopic composition of groundwater within the basin (Table 1). A rating curve was also defined for Orangeville Rise.

The second stage of sampling identified short-term variability of water chemistry associated with changed in discharge brought on a storm in the Lost River basin: in October, corresponding to the driest time of the year. Sampling frequency during the second stage of monitoring was keyed to the rate of change in discharge; every 2 to 4 hours during the time of rapid discharge increase, peak flow and initial discharge recession, and less frequently during later stages when discharge decreased less rapidly.

5. Results

5.1 Biweekly Sampling

Results from the biweekly sampling showed that base

flow discharge at Orangeville Rise, in the 3 months prior to October storm monitoring, had an isotopic composition of -40‰ and -6.2‰ for δD and δ¹⁸O, respectively. The base flow discharge at the rise of Lost River had a nearly identical isotopic composition of -40‰ δD and -6.4 δD respectively. These values were used as the isotopic signature of prestorm water in hydrograph separations of spring discharge for the October storm.

Isotopic composition of groundwater was found to be quite uniform even though major-ion chemistry in the basin was variable due to the presence of gypsum (Table 1). The average δD and δ¹⁸O of groundwater was found to be -44‰ and -6.6‰, respectively, very similiar to the base flow of the spings.

6. Hydrograph Separation

As a means of investigating components of discharge, storm hydrographs were separated into pre-storm and rainwater components. Separations were performed using both deuterium and oxygen isotopic data as a means of checking for consistency in the data.

Separations of the October storm having a δD of -56‰ and a δ¹⁸O of -8.2‰ for the Orangeville Rise are shown in Figure 3 along with the equation used to solve for instantaneous rainwater discharge.

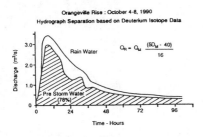

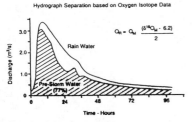

Figure 3. Hydrograph separation of discharge from Orangeville Rise into prestorm and rainwater components.

Separations based on the two sets of isotopic data identify slightly different amounts of prestorm and rainwater at any instant on the hydrograph; however, they reveal very similar trends in the contribution of prestorm and rainwater to total instantaneous discharge and identify similar quantities of prestorm water over the 4 days of monitoring. At peak flow, rainwater accounted for approximately 15% of the instantaneous discharge. The greatest percentage of rainwater at any time on the hydrograph occurred 18 hours after peak discharge, where it made up 40 to 45% of the total. Separation revealed that a secondary pulse of prestorm water arrived just after the time of maximum rainwater contribution. For the entire 4 days of monitoring, rainwater made up 20 to 25% of the total discharge at Orangeville Rise.

The October storm hydrograph separation for Rise of Lost River is shown in Figure 4.

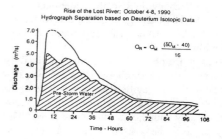

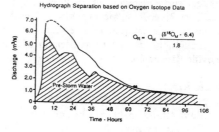

Figure 4. Hydrograph separation of discharge from Lost River Rise into prestorm and rainwater components.

Results based on oxygen and deuterium isotopic data reveal that 75% of discharge over the five days of monitoring was derived from pre-storm water. While the separations based on the two data sets differ during rising discharge, the plots are quite similar throughout the remainder of the hydrograph, with increases and decreases in the rainwater component mirroring one another. Rainwater comprised approximately 30% of the instantaneous discharge at peak flow, while four hours into discharge recession, rainwater made up 38% of the instantaneous discharge, the greatest proportion of rainwater discharge at any point on the hydrograph. Separations reveal pulses of pre-storm water during the first 36 hour of discharge recession, and a small proportion of rainwater contributed to discharge throughout recession.

7. Conclusions for the Upper Lost River Basin

At both Rises, pre-storm water contributed 75 to 80% of the total discharge over the monitoring periods, and separations of storm hydrographs in each case showed pulses of pre-storm water during discharge recession.

Discharge at both Rises increased very rapidly after the October storm, and reached peak discharge at approximately 11 hours after the storm. At Orangeville Rise, rainwater made up 12 to 15% of total discharge at peak flow, whereas at Rise of the Lost River, peak flow contained up to 40% rainwater.

Based on isotopic data, rainwater provided the largest proportion to discharge 18 to 24 hours after peak flow at Orangeville Rise, but at Rise of the Lost River, maximum rainwater contribution occurred only four hours after peak flow. This outcome may be partially derived from the intermittent surface stream which flows directly into the Lost River Rise, but may also reflect the longer period of time required for new rainwater to travel through the carbonate aquifers(s) of the Orangeville Rise basin.

8. References

Fritz, P., J.A. Cherry, K.U. Weyer, and M. Sklash, Storm run-off analysis using environmental isotopes and major ions, *Interpretation of Environmental Isotope and Hydrochemical Data in Groundwater Hydrology,* IAEA, Vienna, 111-130, 1976.

Lakey, B.L. and N.C. Krothe, Stable isotopic variation of storm discharge from a perennial spring, Indiana, *Water Resource Research*, 32, 721-731, 1996.

Williams, P.W., The role of the subcutaneous zone in Karst Hydrology, *J. Hydrol.*, 61, 45-67, 1983.

Saline intrusion into an urban sandstone aquifer

Robert J. Newton, Andrew P. Barker & Simon H. Bottrell
School of Earth Sciences, University of Leeds, UK

John H. Tellam
School of Earth Sciences, University of Birmingham, UK

ABSTRACT: A saline-freshwater mixing zone has been sampled by taking advantage of rail tunnels beneath the Mersey Estuary and Liverpool city centre. SO_4 in the saline tunnel waters below the estuary is shown to be enriched in ^{18}O by 1.5‰ but without any significant change in $\delta^{34}S$, despite showing a decrease in SO_4 concentration and a decline in SO_4^{2-}/Cl^-. This is the result of a tidally-driven estuarine cycle of bacterial SO_4 reduction at high tide, followed by draining of some pore water with ^{34}S enriched SO_4 and reoxidation of part of the newly formed monosulphides at low tide. A major low permeability fault separates these Cl rich waters from freshwaters on the landward side of the fault. The saline waters on the estuary side of the fault are affected by mixing with meteoric waters and calcite dissolution. On the landward side of the fault the freshwaters have been modified by ion exchange induced by flushing of an earlier saline intrusion.

1 INTRODUCTION

Liverpool has been an industrial centre since the 1840's and abstraction of groundwater from the Permo-Triassic sandstone aquifer was the principal source of water for most of its industries. Abstraction peaked in the 1970's when 34% of the recharge to the aquifer was from intrusion of saline water from the Mersey Estuary and the Manchester Ship Canal (Howard, 1987). The rail tunnels beneath the city centre (the Loop Line) were constructed at this time and the subsequent rise in groundwater levels as abstraction declined, has caused a major problem with water ingress. The rail tunnel beneath the estuary itself connecting Liverpool and Birkenhead (the Mersey Tunnel), was constructed in 1886 and incorporates a second unlined tunnel beneath the main tunnel leading to pumped sumps at either end, keeping the main tunnel largely dry.

The tunnels are cut in the Permo-Triassic Sherwood Sandstone Group at the northern edge of the Cheshire Basin and rest unconformably on the Westphalian Coal Measures of the Upper Carboniferous. The Loop Line tunnels intersect two faults; the Castle Street Fault (CSF) and the Kingsway Fault, both of which run N-S parallel to the estuary.

Samples were collected from the estuary (×12, covering the tidal range), from the Irish Sea (×1), from the Mersey Tunnel (×3), from the Loop Line (×26) and from the pumping station at Birkenhead (×1). These were analysed for major cations and anions, $\delta^{34}S$ and $\delta^{18}O$ in SO_4, $\delta^{13}C$ in dissolved inorganic carbon and $^{87}Sr/^{86}Sr$. In addition, four dissaggregated samples of the Wilmslow Sandstone were leached for 24 hours in a 2% acetic acid solution to dissolve the calcite cement and extract other forms of readily soluble strontium. $\delta^{34}S$ is expressed relative to the CDT standard, $\delta^{18}O$ relative to the SMOW standard and $\delta^{13}C$ relative to the PDB standard.

2 RESULTS

The main chemical features of the groundwaters and estuary waters are summarised in Fig. 1 (Cl^- and NO_3^-) and Fig. 2 (Na^+, K^+, Ca^{2+} and SO_4^{2-}).

All groundwaters on the estuary side of the CSF are close to calcite saturation. Three samples were collected from beneath the estuary itself. The sample marked A on Fig. 2 was collected from the centre of the Mersey Tunnel and is the most saline having $[Cl^-]$ = 15380 mg/L, close to the mean $[Cl^-]$ of the estuary (15250 mg/L). The two other samples collected beneath the estuary (group B on Fig. 2) have lower [Cl⁻] but higher [Ca] than those measured in the sample from the centre of the tunnel, but have a similar depletion in SO_4.

The groundwaters to the landward side of the CSF can be divided into two groups. The five samples nearest the fault have [Cl⁻] in the range 245-1070 mg/L. Samples further inland from the fault have [Cl⁻] in the range 44 to 207 mg/L with a mean of 106 mg/L (n = 18; σ_{n-1} = 43). All waters to the east of the fault are characterised by the presence of NO_3 (Fig. 1) and are undersaturated with respect to calcite (saturation index for calcite in the range -0.6 to -3.7).

Two estuary samples gave a mean $^{87}Sr/^{86}Sr$ ratio of 0.70924, which was identical to the sample from the centre of the tunnel. The $^{87}Sr/^{86}Sr$ ratio of groundwaters became progressively more radiogenic with distance from the estuary. The highest ratio, 0.71147, was measured in the sample furthest inland. The acetic acid leachates of the Wilmslow Sandstone give values for the abundance of readily soluble Sr in the range 0.07 to 2.34 mg/kg. The $^{87}Sr/^{86}Sr$ of the leachates varied between 0.71049 and 0.71120.

The ^{34}S and ^{18}O in SO_4 data for the estuary and groundwaters are summarised in Fig. 3. The seawater sample from the Irish Sea gave $\delta^{34}S$ = +21.0‰ and $\delta^{18}O$ = +8.8‰.

The estuary waters had $\delta^{13}C$ in the range -5.6‰ to -5.2‰ with a mean of -5.4‰ (n = 3, σ_{n-1} = 0.21). Within the tunnels the samples had values between -24.1‰ and -16.5‰, the $\delta^{13}C$ becoming more negative as [Cl⁻] decreased.

3 DISCUSSION

The contrasting chemistries on either side of the CSF demonstrates that the fault acts as a barrier to groundwater flow (see Fig. 1 for e.g.). This division provides a basis for discussing the results.

3.1 WATERS ON THE ESTUARY SIDE OF THE CSF

Much of the chemistry of the groundwaters can be explained by simple mixing of infiltrating estuary water with fresh groundwater (Fig. 2a and 2b). However all groundwaters on the estuary side of the CSF have lower [SO_4] than would be expected from mixing (Fig. 2d) and all the samples with the exception of the one taken from the centre of the tunnel are enriched in Ca relative to calculated mixing concentrations (Fig. 2c).

The sample collected from the centre of the

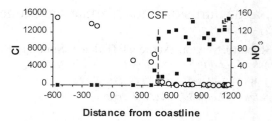

Fig. 1. Variations in Cl (open circles) and NO3 (filled squares) with distance. All concentrations in mg/L.

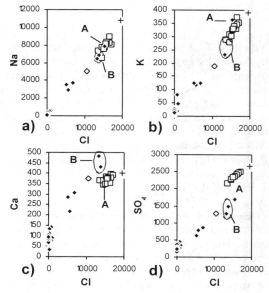

Fig. 2. Plots of Na, K, Ca, and SO_4 vs. Cl in groundwaters (filled diamonds) and estuary waters (open squares). All values are in mg/L. The open diamond represents the pump station and the cross represents seawater. See text for explanation of A and B.

tunnel has similar [Ca] to the estuary (A on Fig. 2), but is depleted in SO_4. This section of the tunnel is roofed only by Quaternary sediments, so this water has not been in contact with the sandstone aquifer. Therefore the Ca enrichment occurs within the aquifer, whereas the SO_4 depletion occurs before the water reaches the sandstone.

The two most likely causes of Ca enrichment are mineral dissolution or ion exchange. The sandstone is known to contain 0-20% calcite cement. The estuary floor is a zone of intense biological activity and waters passing through these sediments will become considerably enriched in biogenic CO_2. So despite the saturation of the estuary with respect to calcite, calcite dissolution in the sandstone is possible. Evidence for the biological addition of CO_2 ($\delta^{13}C \sim -25‰$) is available in the isotopic composition of HCO_3^- dissolved in the groundwaters infiltrating into the tunnels ($\delta^{13}C = -16.5‰$), which is enriched in ^{12}C compared with bicarbonate in the estuary ($\delta^{13}C = -5.4‰$). Rough calculations suggest that if the aquifer contained as little as 0.5% calcite then the dissolution of this could account for the observed increase.

Using pumping rates given by Rushton *et al* (1988) we can calculate the approximate volume of water that has passed through the aquifer in the region of the tunnels. We have calculated this to be 1.6×10^{12} L over the lifetime of the tunnels. Combining this figure with ion exchange capacity measurements from Carlyle (1991) shows that more than enough water (by three orders of magnitude) has passed through the sandstones to saturate the exchange sites with Na. Therefore calcite dissolution rather than ion exchange is controlling the cation chemistry of these waters. This is supported by a slight trend to increased $^{87}Sr/^{86}Sr$ away from the tunnel centre. As the sandstone leaching experiments show, this would be expected from calcite dissolution.

The concentration of SO_4 drops by about 25% between the estuary and the tunnels (Fig. 2d). The environment of the estuary floor sediments is ideal for bacterial SO_4 reduction to occur and this would seem the most likely explanation for the observed drop in concentration. However, this process produces residual SO_4 which is heavier (enriched in both ^{18}O and ^{34}S) than the source, as light sulphide is fixed in the sediment. By contrast, the $\delta^{34}S$ of the SO_4 in waters from the estuary and from the tunnels immediately below are practically identical (means +21.0‰ c.f. +21.2‰) (Fig. 3b), but interestingly the ^{18}O isotopic composition of the groundwater SO_4 shows a mean positive shift of 1.5‰ (Fig. 3a). The scenario that we envisage to produce these observed values begins with SO_4 reduction at high tide producing residual heavy SO_4 and light sulphide (probably monosulphide). As the tide drops, large areas of mud will be exposed to the atmosphere and the newly formed monosulphide will begin to oxidise back to SO_4 by a combination of spontaneous chemical and bacterially mediated reaction pathways. These reactions will be buffered at approximately pH 8 by the saline water and will therefore not involve Fe^{3+}. This limits possible oxidants to water and atmospheric oxygen. Atmospheric oxygen has a $\delta^{18}O$ of +23.8 and during incorporation into SO_4 undergoes a fractionation of -8.7‰. If most of the oxygen incorporated into the SO_4 came from this source then this would produce SO_4 with a $\delta^{18}O$ of $\sim +15‰$. The $\delta^{34}S$ of this SO_4 would be very light compared to the estuary water. The $\delta^{18}O$ of this SO_4 would also be made heavier by partial bacterially mediated equilibration between water oxygen and SO_4 oxygen. During low tide pore water will drain back to the estuary. This will contain mainly residual SO_4 and a small amount of SO_4 from the reoxidation of the sulphide. The SO_4-S left in the porewater as the tide begins to rise will be a mixture of light S from sulphide oxidation plus a reduced amount of heavy residual S from the initial SO_4 reduction, resulting in no net change in $\delta^{34}S$. The two pools of SO_4-O will both be ^{18}O enriched giving the observed positive shift in $\delta^{18}O$.

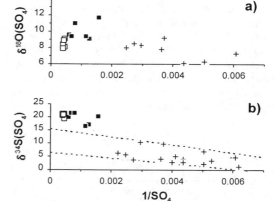

Fig. 3. $\delta^{34}S$ and $\delta^{18}O$ of SO_4 plotted against reciprocal SO_4 concentration (mg/L). Crosses represent freshwaters on the landward side of the CSF, filled squares represent groundwaters on the estuary side of the CSF and open squares represent estuary waters.

3.2 GROUNDWATER ON THE LANDWARD SIDE OF THE CSF.

These groundwaters are all undersaturated with respect to calcite and contain high [NO_3^-], the result of anthropogenic pollution and characteristic of recent recharge.

Samples collected from near the fault have somewhat higher [Cl$^-$] compared with those further away. These waters have a major cation ratio ([Na]+[K])/([Ca]+[Mg]) similar to the much more saline groundwaters on the estuary side of the fault and their $^{87}Sr/^{86}Sr$ ratio is also similar. The SO_4 $\delta^{34}S$ of these waters (Fig. 3b) suggests that these features do not result from present day landward flow of saline water across the fault. Earlier saline intrusion possibly caused by high industrial abstraction rates, would have resulted in equilibration of the saline water with the ion exchange sites within the sandstone. This saline water was flushed by fresh groundwaters and 'reverse' ion exchange is now occurring. Evidence for this argument comes from the Sr isotope ratio of groundwaters on this side of the fault. Comparing the Sr isotope ratio with excess Ca and Sr in the groundwater (Fig. 4) shows two distinct trends. Fresh waters away from the fault have Sr which is isotopically identical to that in the aquifer carbonate. The waters nearest the fault on the landward side, however, have far less radiogenic strontium, similar to waters on the estuary side. This group of waters defines a mixture between estuary water strontium and an end-member associated with excess Ca and Sr with $^{87}Sr/^{86}Sr$ of 0.7098. This is far less radiogenic than the calcites in the Wilmslow sandstone ($^{87}Sr/^{86}Sr$ = 0.71049 to 0.71120) and may be related to calcites in the Chester Pebble Beds which are far less clay-rich and would therefore have produced less radiogenic Sr during diagenesis.

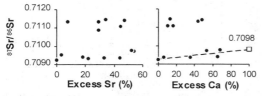

Fig. 4. $^{87}Sr/^{86}Sr$ variations with excess Ca and Sr. Excess is defined as the Ca or Sr present above that predicted by simple mixing.

The S isotope data from these waters (Fig. 3b) show a mixing line between a SO_4 source with $\delta^{34}S$ of +7 to +16‰ and one with a $\delta^{34}S$ of ~+2‰. The first range is consistent with values from gypsum or gypsum derived building materials. The second value could be from a number of sources such as sewage effluent (0 to +4‰, van Dover et al, 1992) or SO_4 leached from bricks and concrete (+0.4 to +2.2‰, Tellam et al, 1994).

4 CONCLUSIONS

Waters ingressing to the Liverpool railway tunnels are characterised by two very different evolutionary pathways resulting in groundwaters with very different chemistries. These waters are separated by a fault with a low hydraulic conductivity. Groundwaters on the estuary side of the fault have a chemistry that is predominantly controlled by processes that occur during marine intrusion, notably SO_4 reduction and calcite dissolution and further modified by mixing with fresh groundwater. On the landward side of the fault the chemistry results from recharge in an urban environment.

The relatively low SO_4 concentration in groundwaters beneath the estuary is due to bacterial SO_4 reduction in the sediments at the base of the estuary. The $\delta^{34}S$ of the SO_4 shows no net change due to the loss of heavy residual SO_4 and the addition of light SO_4 from sulphide oxidation. This process also leads to enrichment in ^{18}O relative to SO_4 in the estuary via the addition of heavy O from the atmosphere and bacterially mediated isotopic exchange with water in a cycle of repeated SO_4 reduction and sulphide oxidation. A particularly important conclusion of this study is that sulphur isotope studies in isolation cannot always be used to identify SO_4 reduction in the natural environment.

The sulphur isotopic composition of dissolved SO_4 in the fresh groundwaters appears to be dominated by two sources. One source is gypsum, either from the glacial till or leached from gypsum-derived building materials. The other source is more difficult to identify and may be from other building materials, leaks from sewers, or pyrite oxidation in the tills.

REFERENCES

Carlyle, H.F. 1991. The hydrochemical recognition of ion exchange during seawater intrusion at Widness, Merseyside, UK. Unpub. Ph.D. thesis, Univ. of Birmingham

Howard, K.W.F. 1987. Beneficial aspects of sea-water intrusion. *Groundwater* **25**, 398-406

Rushton, K.R., Kawecki, M.W. and Brassington, F.C. 1988. Groundwater model of conditions in Liverpool sandstone aquifer. *J. Inst. Wat. Env. Mang.* **2**, 67-84

Tellam, J.H. 1994. The groundwater chemistry of the Lower Mersey Basin Permo-Triassic sandstone aquifer system, UK: 1980 and pre-industrialisation-urbanisation. *J. Hydrol.* **161**, 287-325

van Dover, C., Grassie, J.F., Fry, B., Garritt, R.H. and Starczak, V.R. 1992. Stable isotope evidence for entry of sewage-derived organic material into a deep-sea food web. *Nature.* **360**, 153-156

Geochemical processes in two carbonate-free aquifer systems of North Cameroon

R. Njitchoua & L. Dever
Laboratoire d'Hydrologie et de Géochimie Isotopique, Université de Paris-Sud, France

B. Ngounou-Ngatcha
Centre de Recherches Hydrologiques, Yaoundé, Cameroon

ABSTRACT: The present paper examines the geochemical processes controlling the groundwater chemistry in the Upper Cretaceous sandstone aquifer, Benue Basin, and the North-Diamaré plain alluvial aquifer, both in northern Cameroon. Groundwaters from both systems have evolved under open system conditions with respect to biogenic CO_2. As inferred from thermodynamic data and mineral stability diagrams, all groundwaters are saturated to supersaturated with respect to quartz, chalcedony and amorphous silica, and undersaturated with respect to calcite. Groundwater has acquired its major-ion chemistry through equilibrium reactions with silicates. The main chemical reactions involved in these equilibrium processes are (1) carbonic weathering of silicate minerals, (2) cation exchange reactions between water and montmorillonites, and (3) to a lesser extent, transformation of Na-montmorillonite into kaolinite.

1 INTRODUCTION

This paper discusses the major processes controlling the chemical composition of groundwater from two carbonate-free aquifer systems in northern Cameroon, namely, the Upper Cretaceous sandstone aquifer of the Benue Basin, and the alluvial aquifer of North-Diamaré piedmont plain, at the southwestern border of the Great Chad Basin (Figure 1). Groundwater in these areas is the only available water supply.

The Upper Cretaceous sandstones series, also known as the Garoua sandstone aquifer (GSA), consist of feldspath sandstones interbedded with clayey levels (Tillement, 1972; Maurin & Guiraud, 1990). The alluvial deposits in the North-Diamaré plain are of Quaternary age, and made up of sandy clays, clayey sands, coarse sands and gravels (Tillement, 1972). In both aquifers, groundwater generally occurs under unconfined conditions. However, in the Garoua sandstone aquifer, the occurence of discontinuous clayey layers within the bulk of sandstones can locally generate a confinement.

As recently reported by Njitchoua et al. (1997) and Njitchoua & Ngounou-Ngatcha (1997), the $\delta^{18}O$ values of groundwater from both aquifers vary within -0.27 and -4.86 ‰ relative to V-SMOW, and are consistent with a recharge mechanism through direct infiltration of meteoric water and/or lateral inflow of river water.

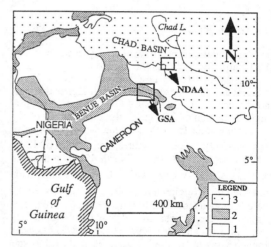

Figure 1. Geographical locations of the areas under study. GSB and NDAA refer to Garoua Sandstone aquifer and to North-Diamaré alluvial aquifer, respectively. 1 = Crystalline rocks (Precambrian); 2 = Sandstones (Cretaceous); 3 = alluvial deposits (Quaternary).

Chemically, groundwater from the GSA is acidic (5.0≤ pH≤6.5) and less mineralized (50-300 mg/L TDS), whereas groundwater from the NDAA shows neutral values of pH (6.6-7.4) and TDS (300-500 mg/L). The dominant ionic species in groundwater from both aquifers are Ca^{2+}, Na^+ and HCO_3^-. The

Na^+/Ca^{2+} (in meq/L) ratios vary widely, being on average 0.74 for GSA groundwater, and 1.78 for NDAA groundwater. The concentrations of aqueous silica vary within 10-100 mg/L for the GSA water, and 84-120 mg/L for the NDAA waters.

The main objective of this paper will be to determine the mineral assemblage as well as the major chemical reactions that might have controlled the composition of groundwater from the GSA and the NDAA. The methods used are based on the interpretation of (1) thermodynamic data such as the partial pressure of CO_2 gas, and the saturation index of groundwater with respect to mineral phases; (2) partial equilibrium diagrams. Moreover, the $\delta^{13}C$ measurements will be also used to determine the origin of CO_2 gas, which is the major factor that governs the water-rock interaction processes.

2. RESULTS AND DISCUSSION

The values of saturation index, partial pressure of CO_2 (pCO_2) and activity of dominant aqueous species have been computed using the WATEQF program (Plummer et al., 1976).

2.1. Origin of CO_2 gas

The calculated pCO_2 values vary from $10^{-2.5}$ to $10^{-1.5}$ atm for the Garoua sandstone aquifer, and from $10^{-2.0}$ to $10^{-1.5}$ atm for the North-Diamaré alluvial aquifer. These values, much higher than the atmospheric partial pressure ($10^{-3.5}$ atm), suggest a high production of CO_2 gas in the soil zone.

Groundwater from the Garoua sandstone aquifer has relatively low $\delta^{13}C$ values, varying from -12 to -20 ‰ versus PDB. In the North-Diamaré alluvial aquifer, similarly negative $\delta^{13}C$ values have been reported by Ketchemen (1992). Such low $\delta^{13}C$ values suggest that, although the study areas are adjacent to the "Cameroon Volcanic Line" where CO_2 gas with ^{13}C contents within -0.7 and -8 ‰ has been found (Kling et al., 1989; Kusakabe et al., 1989), the CO_2 in the GSA and NDAA systems is of biogenic origin. This finding is supported by Figure 2 showing the ^{13}C-pH. relationship, which indicates that DIC in groundwaters results from an equilibrium with a CO_2 gas with ^{13}C contents between -16 and -20 ‰, consistent with a vegetal cover dominated by C4 and CAM plants.

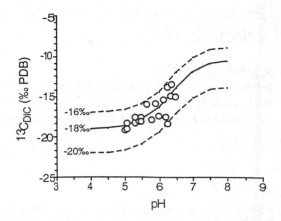

Figure 2. $\delta^{13}C$-pH relationship for groundwater from the Garoua sandstone aquifer. The curves represents the isotopic maturation line for different values of $\delta^{13}C$ of CO_2.

2.2. Mineralization processes

As can be seen in Figure 3, groundwater from the GSA and NDAA are saturated to supersaturated with respect to quartz, chalcedony and amophous silica (Figure 3A), and undersaturated with respect to calcite (Figure 3B). On the other hand, in Fig. 4 showing the mineral stabilities in terms of major-cation/proton activities versus dissolved silica activities (in log scale), groundwater from the GSA plots in the stability field of kaolinite (Figs. 4A-B), whereas groundwater from NDAA plots on the equilibrium straight line between kaolinite and Na-montmorillonite (Figure 4A) and in the field of stability of Ca-montmorillonite (Figure 4B).

As inferred from the above observations, the composition of groundwater from the GSA and NDAA systems appears to result from the interaction between CO_2-rich water and silicate minerals.

For the GSA, the mineral assemblage that may have equilibrated with groundwater includes, at least: quartz, chalcedony, amorphous silica, plagioclases, K-feldspar and kaolinite. The observed major-ion species may be here determined by the carbonic acid weathering of silicates, as indicated by the following incongruent dissolution of albite for example

$$2NaAlSi_3O_8 + 11H_2O + 2CO_2 = 2Na^+ + 2HCO_3^-$$
albite
$$+ 4H_4SiO_4 + Al_2Si_2O_5(OH)_4 \quad (1)$$
kaolinite

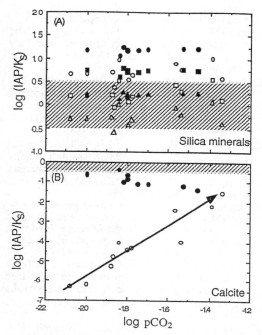

Figure 3. Relationships between saturation index (Log [IAP/K_s]) and pCO_2 for groundwater from the Garoua sandstone aquifer (open symbols) and the North-Diamaré alluvial aquifer (full symbols). The dashed zones represent the state of saturation, owing to the analytical errors. A - Silica minerals: quartz = circles, chalcedony = squares, amorphous silica = triangles. B - Calcite

This equation shows that the increase in soil CO_2 gas may lead to increased concentrations of Na^+, HCO_3^- and dissolved silica, simultaneously with the formation of kaolinite, as shown by the trend in Figure 3B indicating an increase in Ca^{2+} and HCO_3^- which are the main species controlling the saturation index of calcite.

For these groundwaters, the relatively low values of rNa^+/rCa^{2+} may be attributed to the high alterability of Ca-feldspars relative to Na-bearing minerals. The concentrations of dissolved silica are controlled by the solubilities of silica minerals, such as quartz, chalcedony, and mostly amorphous silica. Moreover, the weak TDS contents observed for groundwater in this system are to be attributed: (1) either to a relatively short time of contact between water and the mineral assemblage; almost all the measured radiocarbon activities for these waters vary between 115 and 90 percent of modern carbon, corresponding to apparent ages between modern and less than 1 ka B.P.; or (2) to the lack of sufficient soluble cation bearing minerals within the host rock which might thus be essentially composed of silica minerals. The latter assumption is in good accord with relatively low pH values measured for groundwater from this system.

In the NDAA, on the contrary, processes appear to be more complex. The mechanisms that can explain the observed major-ion composition of groundwater may likely include the acid hydrolysis of calcium and sodium bearing minerals and cation exchange between water and clay minerals. The fact that groundwater from this system is plotted either on the kaolinite/sodium-montmorillonite line, or in the field of stability of calcium-montmorillonites, suggests that sodium- and calcium-montmorillonites are the major weathering products that may have equilibrated with solution. The chemical reactions are expected to have occured in the following order:

(1) chemical weathering of sodium and calcium feldspars leading to the supply of Ca^{2+}, Na^+, HCO_3^- and dissolved silica in solution, and the formation of Na-montmorillonites and kaolinite to a lesser extent.

(2) cation exchange reactions between sodium of montmorillonites and calcium of water may follow, leading to the occurence of Ca-montmorillonites and Na-rich water. This hypothesis is consistent with the plot of data points in the stability field of Ca-montmorillonites (Figure 4B), and the lack of relationship between pCO_2 and saturation index with respect to calcite (3B).

(3) equilibrium reactions between kaolinite and Na-montmorillonites, as inferred from the plot of groundwater representing points on the kaolinite/Na-montmorillonite equilibrium line (Figure 4A).

Mechanisms (2) and (3) may explain the excess of Na^+ prior to Ca^{2+} in solution. The occurence of Ca-montmorillonite within the mineral assemblage may also explain the low concentrations of Ca^{2+} in water, and therefore the undersaturation state of water with respect to calcite. Since the saturation with respect to calcite is partly related to Ca^{2+}, the small amount of Ca^{2+} in solution due to cation exchange reactions may explain the decrease in saturation index values with increasing pCO_2 as seen in Figure 3B.

3. CONCLUDING SUMMARY

The chemical composition of groundwater from the Garoua sandstone aquifer and the North-Diamaré plain groundwater system generally results from the interaction between CO_2-rich recharging water with silicate minerals forming the host rock.

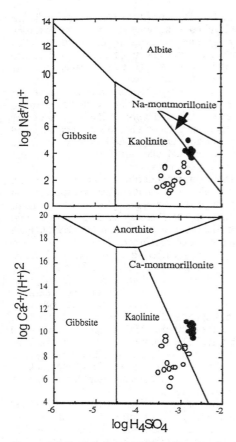

Figure 4. Mineral phase stability diagrams in terms of major-cation/proton activities versus dissolved silica activities (in log scale) for groundwater from the Garoua sandstone aquifer (open circles) and the North-Diamaré plain alluvial aquifer (full circles).

In the Garoua sandstone aquifer, the reaction involved is simply the incongruent dissolution of primary silicates to form kaolinite which is expected to be the sole weathering product. In the North-Diamaré plain aquifer, on the contrary, the mineralization processes appear to be more complex, involving acid dissolution of silicates, cation exchange reactions, and to a lesser extent, equilibrium reactions among clay minerals.

REFERENCES

Ketchemen, B. 1992. Hydrogéologie du Grand Yaéré, Extrême Nord du Cameroun. Synthèse hydrogéologique et étude de la recharge par les isotopes de l'environnement. *Thèse de 3ème cycle, Université de Dakar, Sénégal*: 172 p.

Kling, G.W., M.L. Tuttle & W.C. Evans 1989. The evolution of thermal structure and water chemistry in Lake Nyos. *J. Volcano. Geotherm. Res.* 39: 151-165.

Kusakabe, M., T. Ohsumi & S. Aramaki 1989. The Lake Nyos gas disaster: chemical and isotopic evidence in waters and dissolved gases from three Cameroonian lakes, Nyos, Monoun and Wum. *J. Volcano. Geotherm. Res.* 39: 167-185.

Mook, W.G. 1980. Carbon-14 in hydrogeological studies. *In Handbook of environmental isotope geochemistry*, vol. 1, the terrestrial environment, Fritz, A. & J.-Ch. Fontes (Editors), Amsterdam, the Netherlands, Elsevier: 49-74.

Njitchoua, R. & B. Ngounou-Ngatcha 1997. Hydrogeochemistry and environmental isotopic investigations of the North Diamaré Piedmont Plain, Extreme-North of Cameroon. *J. Af. Earth Sci.* (under press)

Njitchoua, R., L. Dever, J.-Ch. Fontes & E. Naah 1997. Geochemistry, origin and recharge mechanisms of groundwaters from the Garoua Sandstone Aquifer, Northern Cameroon. *J. hydrol.* 190, 1-2: 123-140.

Plummer, L.N., B.F. Jones & A.H. Truesdell 1976. WATEQF: a Fortran IV version of WATEQ, a computer program for calculating chemical equilibrium of natural waters. *U.S. Geol. Surv. Water Resources. Invest. Rep.* 76-13.

Tillement, B. 1972. Hydrogéologie du Nord-Cameroun. Rapport 6: 294 p. *Direction des Mines et de la Géologie.* Yaoundé: Cameroon.

Geochemical and other porosity types in clay-rich rocks

F.J. Pearson
Paul Scherrer Institut, Villigen PSI, Switzerland

ABSTRACT: The concept of geochemical porosity is introduced. It is the proportion of the bulk volume of a saturated material containing water available for water-rock reactions and solute transport. It must be known for geochemical and transport modelling, and when determining pore water chemistry in rock too impermeable for conventional water sampling. It is to be distinguished conceptually from total and water-content porosity, and from advective and diffusive transport porosity. In coarse-grained rock, all types of porosity may have nearly the same value, so it is not necessary to distinguish among them. In five clay-rich materials where transport is by diffusion, geochemical and diffusion porosities for water molecules are the same as the water-content porosities. For solutes, which do not have access to interlayer water of clay minerals or surface-sorbed water, these porosities are only one-third to one-half the water-content porosities.

1 INTRODUCTION

Porosity is a rock property as important to geochemists as to hydrologists and reservoir engineers. Total porosity is the proportion of the bulk volume of a material occupied by interstices. Geochemical porosity is the proportion containing the fluid in which water-rock reactions and solute transport occur. Other types include isolated vs. connected, and fracture vs. matrix porosity, and effective porosity for advective and diffusive transport. If more than one fluid phase is present, the proportions of the interstitial space occupied by each must also be known. Only water-saturated material will be considered here.

In material such as well-sorted sands, numerical values of the various types of porosity are virtually the same, so there is little reason to distinguish among them. However, in other materials different types of porosities may not have the same values.

A fractured crystalline rock may contain fluid in isolated inclusions, in connected interstices among mineral grains, and in fractures. The total porosity is that of all three environments, while the effective porosity for fluid flow and advective solute transport is that of the fractures. The interstitial matrix porosity is available for diffusive but not advective solute transport.

Clay-rich materials contain several types of water, each with different effects on aqueous geochemistry and solute transport properties (Pearson *in press*). To understand pore-water chemistry it is necessary to distinguish between the bound water associated with mineral surfaces - clay-mineral interlayer water and Donnan layer water - and the free water that is relevant to water-rock reactions and solute transport. Together, these volumes comprise the connected porosity of the material.

The free water, or geochemical porosity for a substance, is the focus of this paper. It must be known for calculations of water-rock reactions and reactive transport, and to translate the results of leaching studies to *in situ* pore water compositions.

2 TYPES OF POROSITY

Geochemical porosity is best described within the context of other porosity types.

2.1 *Physical Porosity*

Total porosity, n_{tot} the ratio of void volume to total volume, is found from sample bulk and grain densities. Values for connected physical porosity are often based on sample impregnation techniques such as mercury injection, or on measurements of water content. Because water content porosities, n_{wc}, describe connected rather than total porosity:

$$n_{wc} \leq n_{tot}.$$

Water content values of clay-rich rocks depend on the drying conditions. It is not clear whether water in clay interlayer positions, usually part of measured water-losses, should be included in the total porosity. As shown below, such water is accessible to diffusing water molecules, but not to solutes.

2.2 Transport Porosity

Transport porosities relate velocities of solutes in a fluid to its flux and, with other parameters, the rates of diffusion of substances in a porous medium to their rates of diffusion in free water.

Fluid flux - volume per time and cross-sectional area - has units of velocity and is referred to as the Darcy velocity. Its ratio to the average linear (tracer) velocity is the advective transport porosity, n_{adv}^i, sometimes known as the effective porosity. Isolated pores and pores open only at one end to the connected pore system (dead-end pores) are not part of the advective transport porosity. Thus, generally:

$$n_{adv}^i \leq n_{wc} \leq n_{tot}.$$

Diffusive transport through porous media is described by equations analogous to Fick's laws of free-water diffusion. Fick's second law becomes:

$$D_{eff}^i \frac{\partial^2 C^i}{\partial x^2} = \alpha^i \frac{\partial C^i}{\partial t}. \quad (1)$$

D_{eff}^i, the effective diffusion coefficient for i:

$$D_{eff}^i = n_{diff-area}^i \cdot G \cdot D_0^i. \quad (2)$$

D_0^i is the free water diffusion coefficient (Li and Gregory 1974), and G the pore geometry factor. $n_{diff-area}^i$ is the area across which diffusion occurs in a porous medium and is analogous to n_{adv}^i. G and $n_{diff-area}^i$ are determined by the properties of the medium and the diffusing substance i.

α^i in (1) is the capacity of the medium for i:

$$\alpha^i = n_{diff-vol}^i \cdot R^i. \quad (3)$$

$n_{diff-vol}^i$, the diffusion-accessible porosity, is the volume of fluid into which i can diffuse, and R^i is its retardation factor, representing solute-medium interactions of all types.

$n_{diff-area}^i$ and $n_{diff-vol}^i$ may differ in media such as fractured rock in which the porosity comprises both a through-transport porosity and a significant volume of dead-end pores (Bradbury and Green 1985). In spite of this, one commonly refers only to a single diffusion porosity, n_{diff}^i (e.g. Bourke and others 1993; De Cannière and others 1996).

(1) is often written in terms of the apparent diffusion coefficient:

$$D_{app}^i = \frac{D_{eff}^i}{\alpha^i}. \quad (4)$$

Given a single diffusion porosity, the apparent diffusion coefficient of a non-reactive substance ($R^i = 1$) is independent of porosity so that G, the pore geometry factor, can be found experimentally.

Through-diffusion experiments carried out in cells providing boundary conditions that permit closed form solutions to (1) lead to both D_{eff}^i and α^i, from which $n_{diff-area}^i$ and $n_{diff-vol}^i$ can be found (Bourke and others 1993; Bradbury and Green 1985). Such experiments show that for the diffusion of water molecules themselves in clay-rich material:

$$n_{adv}^{H_2O} < n_{diff}^{H_2O} \approx n_{wc} \leq n_{tot},$$

while for the diffusion of solutes:

$$n_{adv}^i < n_{diff}^i < n_{wc} \leq n_{tot}.$$

n_{diff}^i for non-reactive solutes can also be measured with the out-diffusion or leaching method used for geochemical porosity and described below.

2.3 Geochemical Porosity

Geochemical porosity, $n_{geochem}$, is used in geochemical and solute transport modelling. Diffusive and advective transport move mass, so geochemical porosity will be similar to advective porosity in systems in which transport is principally by advection, and similar to diffusion porosity in systems in which transport is principally by diffusion. Clay-rich rocks are more likely to be of the latter type than of the former, so for them:

$$n_{geochem} \approx n_{Diff} \leq n_{wc} < n_{tot}.$$

Geochemical porosity is also needed to characterise pore water chemistry in material like clay-rich rock that too little water to be sampled conventionally. This chemistry can be found by modelling equilibrium between water and reactive formation minerals. Such calculations require knowledge of the reactive minerals present, and of specific formation properties such as cation-exchange (Appelo and Postma 1993; Baeyens and Bradbury 1994; Pearson and Scholtis 1995). The concentrations of solutes that are not controlled by mineral-water reactions must also be known for such calculations.

Non-reactive solute concentrations in pore water,

C^i_{fluid}, can be found from the density and geochemical porosity, and the solute concentrations in the bulk material, C^i_{bulk}, which can be determined by aqueous leaching of a disaggregated sample:

$$C^i_{fluid} = C^i_{bulk} \cdot \rho_{bulk} \cdot \frac{1}{n^i_{geochem}} \quad (5)$$

This is the inverse of the out-diffusion method used to determine the diffusion porosity n^i_{diff} when C^i_{fluid} is known (Bourke and others 1993).

3. RELATIVE VALUES OF GEOCHEMICAL AND OTHER TYPES OF POROSITY

Data on the London clay, a clay-rich Canadian glacial till, the Palfris Marl in central Switzerland, the Opalinus clay in north-western Switzerland, and the Boom Clay in Belgium illustrate values of geochemical and other types of porosity in clay-rich rocks (Table 1). The last three units are being studied as potential host rocks for radioactive waste repositories. The Boom Clay results also show the influence of pore-water salinity on diffusion and geochemical porosity.

3.1 London Clay

Bourke and others (1993) describe through-diffusion and out-diffusion experiments with HDO, HTO and I⁻. Diffusion porosities for both HDO and HTO are about 0.6, the same as the water-content porosity. Thus, diffusing HDO and HTO have access to all the water that can be extracted by drying, including not only free water, but water bound in sorbed surface layers and clay interlayers.

Through-diffusion and out-diffusion porosities for I⁻ are 0.2 and 0.3, vs. 0.6 for the HDO, HTO, and water content porosities. Thus, I⁻ has access only to one-third to one-half of the water content, presumably because the size and negative charge of the ion exclude it from very small pores and from the vicinity of negatively charged mineral surfaces.

3.2 Clay-Rich Glacial Till

Van der Kamp and others (1996) used an incremental out-diffusion (radial diffusion) method to determine geochemical porosity. For HDO, geochemical and water content porosities are the same, while for Cl⁻, the former is only about 0.5 that of the latter. Again, diffusing water molecules have access to the entire water content of the material, while Cl⁻ has access to only half.

3.3 Boom Clay

De Cannière and others (1996) summarise the transport properties of the Boom Clay including through-diffusion porosities for HTO, Br⁻, and I⁻. The diffusing substances were measured as tracers in NaCl solutions of ionic strengths of 0.02 M, about that of natural pore water, and 1 M.

The diffusion porosity for HTO in the 0.02 M solution can be taken as the water content porosity. The relative porosities of Br⁻ and I⁻ in 0.02 M solution are virtually the same: 0.40 and 0.43 respectively. In the 1 M solution, the HTO porosity is greater than in the 0.02 M solution by a factor of 1.26. The relative I⁻ porosity at 1 M is 0.70.

The higher I⁻ porosity in the 1 M solution is consistent with a decrease in Donnan layer thickness with increasing salinity, allowing a greater volume of free water able to contain reactive solutes.

3.4 Opalinus Clay

Data on this unit are from as yet unpublished reports of the Mt. Terri Project (Gautschi and others 1993).

Geochemical porosities range from 0.09 to 0.11 and were calculated using (5) with C^i_{fluid} values from water squeezed from core at high pressures (Entwisle and Reeder 1993), and C^i_{bulk} from aqueous leaching. Water content porosities range from 0.15 to 0.20, and vary sympathetically with the geochemical porosities leading to smaller variability in the porosity ratios than in the porosities themselves.

3.5 Palfris Marl

Mercury-injection porosities of undeformed clay-rich marl samples were 0.029, while water-content porosities of similar samples were 0.032 (Nagra 1997). The Cl⁻ content of the bulk rock, from aqueous leaching, was 0.70 mmol/kg. One contaminated ground-water sample could be corrected indicating a pore water Cl⁻ content of 0.2 M. Using (5), these give a geochemical porosity of 0.009, leading to a porosity ratio of 0.3±0.1.

4 SUMMARY AND CONCLUSIONS

Geochemical porosity is used in modelling water-rock reactions and reactive solute transport. In five clay-rich rocks, where transport is by diffusion, geochemical and diffusion porosities for water molecules are the same and equal the water-content porosities.

Table 1: Comparison of geochemical to other porosity values for several clay-rich materials.

Material and Source of Data	Geochemical Porosity and Measurement technique	Ratio of Geochemical to Other Porosity
London Clay	Through-Diffusion	
	HDO 0.61 ± 0.05	1.02
	HTO 0.59 ± 0.03	0.98
Bourke and others (1993)	I⁻ 0.21 ± 0.06	0.35
	Out-Diffusion	
	HDO 0.59 ± 0.05	0.98
	HTO 0.62 ± 0.06	1.03
	I⁻ 0.32 ± 0.06	0.53
Glacial Till van der Kamp and others (1996)	Incremental Out-Diffusion (Radial Diffusion)	
	HDO 0.33	1.03 ± 0.04
	Cl⁻ 0.17	0.53 ± 0.08
Boom Clay	Through Diffusion; Various Cl⁻ Solns	Based on HTO at 0.02 M Cl⁻
De Cannière and others (1996)	HTO: 0.02 M Cl⁻ 0.46 ± 0.02	
	1.0 M Cl⁻ 0.59 ± 0.07	1.
	Br⁻ : 0.02 M Cl⁻ 0.18 ± 0.03	1.26
	I⁻ : 0.02 M Cl⁻ 0.19 ± 0.01	0.40
	1.0 M Cl⁻ 0.30 ± 0.09	0.43
		0.70
Opalinus Clay (See text)	Cl⁻ Leaching vs. Squeezing 0.09 to 0.11	0.54 ± 0.04
Palfris marl (See Text)	Cl⁻ Leaching vs. Borehole Sample 0.009	0.30 ± 0.1

The geochemical and diffusion porosities for solutes, which do not have access to clay interlayer or surface-sorbed water, are only one-third to one-half the water-content porosities, and vary with the salinity of the pore water.

5   REFERENCES

Appelo, C. A. J. and Postma, D. 1993. *Geochemistry, Groundwater and Pollution*. Rotterdam: Balkema.

Baeyens, B. and Bradbury, M. H. 1994. *Physico-chemical characterisation and calculated in situ porewater chemistries for a low permeability Palfris marl sample from Wellenberg*. Wettingen, Switzerland: Nagra Technical Report **94-22**.

Bourke, P. J., Jefferies, N. L., Lever, D. A. and Lineham, T. R. 1993. Mass transfer mechanisms in compacted clays. *In*: Manning and others 1993. 331-350.

Bradbury, M. H. and Green, A. 1985. Measurement of important parameters determining aqueous phase diffusion rates through crystalline rock matrices. *Journal of Hydrology*, **82**, 39-55.

De Cannière, P., Moors, H., Lolievier, P., De Preter, P. and Put, M. 1996. *Laboratory and in situ Migration Experiments in the Boom Clay*. Luxembourg, European Commission Report **EUR 16927**.

Entwisle, D. C. and Reeder, S. 1993. New apparatus for pore fluid extraction from mudrocks for geochemical analysis. *In*: Manning and others 1993. 365-388.

Gautschi, A., Ross, C. and Scholtis, A. 1993. Pore water-groundwater relationships in Jurassic shales and limestones of northern Switzerland. *In*: Manning and others 1993. 412-422.

Li, Yuan-Hui and Gregory, Sandra 1974. Diffusion of ions in sea water and in deep-sea sediments. *Geochimica et Cosmochimica Acta* **38** 701-714.

Manning, D. A. C., Hall, P. L. and Hughes, C. R., (eds.) 1993. *Geochemistry of Clay-Pore Fluid Interactions*, London: Chapman and Hall.

Nagra 1997. *Schlussbericht zu den geologischen Oberflächenuntersuchungen am Wellenberg*. Wettingen, Switzerland: Nagra Technischer Bericht **96-01**.

Pearson, F. J. in press. What is the porosity of a mudrock? *In*: Aplin, A. C., Fleet, A. and Macquaker, J. (eds.) *Proceedings of a Meeting on Mudrocks at the Basin Scale: Properties, Controls and Behaviour*. London: Geological Society Special Publication.

Pearson, F. J. and Scholtis, A. 1995. Controls on the chemistry of pore water in a marl of very low permeability. *In:* Kharaka, Y. K. and Chudaev, O. V. (eds.) *Proceedings of the 8th international symposium on water-rock, Water-Rock Interaction-WRI-8, Vladivostok, Russia*, Rotterdam: Balkema, 35-38.

Van der Kamp, G., Van Stempvoort, D. R., Wassenaar, L. I. 1996. The radial diffusion method. *Water Resources Research* **32** 1815-1822.

Adsorption of herbicides by aquifer sediments

J.E. Rae, A. Parker & A.J. Peters
Postgraduate Research Institute for Sedimentology, University of Reading, UK

ABSTRACT: This study aims to determine the extent of adsorption of common herbicides onto aquifer sands, and make a preliminary assessment of the key controls on this adsorption. A range of mineralogical, chemical and physical properties have been determined on depth intervals of one core of Tertiary sands from the London Basin aquifer. Batch experiments have been carried out on samples from the same depth intervals in order to assess the adsorption. Triazine herbicides exhibited an unexpected degree of adsorption which may be related to protonation of the compounds in the low-pH porewaters. The phenylureas exhibited adsorption characteristics which were correlated with surface area, ion-exchange capacity, silt content and carbon content of the sediment.

1 INTRODUCTION

The transport and ultimate fate of herbicides in the environment is controlled to a large extent by the process of sorption, as this determines their mobility in the system and also their availability for chemical and biodegradation (Karickhoff, 1984, Weber *et al.*, 1993; Murphy & Zachara, 1995). Although adsorption of herbicides onto soils has been extensively studied, the extent of adsorption in sedimentary environments and the processes operating are not well known.

2 MATERIALS AND METHODS

A sediment core was obtained from the Tertiary Thanet Sands which form part of the major London Basin aquifer system in the region of London, UK. Dry percussion drilling at the Shortlands site (SHL) yielded sections of approximately 1 m length. The core sections were halved longitudinally and the centre channel of one half of each section was sampled. The sediment was then placed in pore-fluid extraction cups and centrifuged at 6000 rpm for 30 minutes in a refrigerated centrifuge at 10° C. The pore-fluid was collected for major-ion, pH and conductivity determinations (see below). The sediment was then transferred to a drying bowl and dried in air at 40° C for 24 hours. It was disaggregated by gentle use of a mortar and pestle, sieved to <2 mm, and any aggregations that remained visible to the naked eye were further gently disaggregated. Sub-samples of the original material were taken to determine physical and chemical characteristics of the sediment, as described below.

Herbicide standards were made by dissolving 10 mg each of atrazine, isoproturon, linuron and diuron in 25 ml High Performance Liquid Chromatography (HPLC) grade methanol to give seven stock standards of 400 µg ml^{-1}. Ten millimolar CaCl$_2$ (aq) was prepared by diluting 20 ml of 1 M CaCl$_2$ (Analar grade) in 2 litres of ultra high-quality (UHQ) water. The herbicide solutions were diluted in 10 mM CaCl$_2$ (aq) to make mixed solutions of 600, 480, 360, 240, 120 and 60 ng ml^{-1} for use in batch equilibration experiments.

Batch equilibration was achieved by mixing 15 g of dried sediment with herbicide solution at the six concentration levels in 50 ml PTFE centrifuge tubes. The equilibration was performed in triplicate. Headspace was eliminated from the tubes to minimise loss due to herbicide volatilisation and/or oxidation. Control tubes were prepared at each concentration level with no sediment present to monitor loss of herbicide. Two millilitres of each solution were stored in HPLC vials to use as HPLC calibration standards. The filled centrifuge tubes were placed in a temperature-controlled rotary tumbler at 8 (±1)°C. The tubes were then rotated end-over-end, 4 times per minute, for a 24-hour period.

Following equilibration, the tubes were removed from the tumbler and centrifuged at 3000 rpm for 15 mins. Three millilitres of supernatant were removed from each tube with a gas-tight syringe and filtered through 0.2 µm PTFE membrane syringe filter directly into a vial ready for analysis by HPLC. The pH of the remaining solution at equilibrium (pH$_{eqm}$) was determined using an electrode measurement.

Herbicide concentrations in batch solutions were determined by HPLC using a Dionex liquid chromatograph system and autosampler. The analytical column was a 250 mm x 4.6 mm, 5 μm "Apex 1" reverse-phase cyanopropyl column (Jones Chromatography), used in conjunction with a 20 mm pre-column of the same packing material. Isocratic elution using acetonitrile: methanol:water (1:7:12) at a flow rate of 1 ml min^{-1} was used. Detection was achieved with UV absorbance at wavelength 211 nm. This combination was found to separate all seven compounds in approximately 12 minutes.

Pore-fluids were analysed for the following major ions using a combination of inductively-coupled plasma atomic-emission spectrometry (ICP-AES), (ARL 35000) for Mg^{2+}, Ca^{2+} and total Fe, atomic absorption (AAS) (Perkin Elmer 3030) for Na^+ and K^+, and ion chromatography (IC) (Dionex 2000 i) for Cl^-, SO_4^{2-} and NO_3^{2-}. The specific conductivity and pH of the pore-fluid were measured using appropriate electrodes. Sub-samples of the core material were used to determine: water content by drying; particle size analysis by laser granulometry (Coulter LS 130); whole rock and clay mineralogy by x-ray diffraction (XRD) (Sietronics SIE 112 microprocessor-controlled Philips PW 1050 system); specific surface area by N_2-BET adsorption (Coulter SA 1300); free iron oxide/hydroxide by citrate-dithionite digestion; cation exchange capacity using $CaCl_2$/$MgSO_4$ and triethanalamine and organic and total carbon by CHN analyser (Perkin Elmer 2240B).

3 RESULTS AND DISCUSSION

The variation in sediment properties with depth in the core is illustrated in Figure 1, and Figure 2 shows the relationships between sediment properties and adsorption coefficient (K_d) for sediment and porewater in the case of isoproturon. The other phenylureas exhibited similar trends.

The batch adsorption experiments yielded data which were used to construct linear isotherms from which the adsorption coefficients K_d were calculated. No data are available for atrazine below 11.2 m since adsorption here was very high, resulting in solution concentrations below detection. In general, the extent of adsorption of the herbicides by the overlying fluvial sediment (< 11.2 m depth) was relatively low, whereas adsorption by the Thanet Sands aquifer sediments was comparatively high.

A number of aquifer properties are potentially important in controlling herbicide sorption. These include the carbon content, particularly organic carbon (OC), pore-fluid pH, surface area (SA) and cation exchange capacity (CEC) (Mustafa & Gamar,

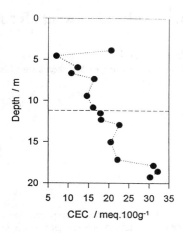

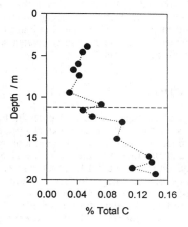

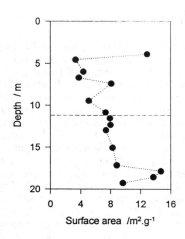

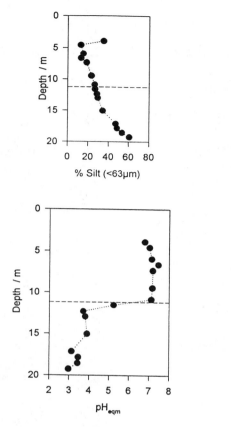

Figure 1. Variation in sediment properties with depth.

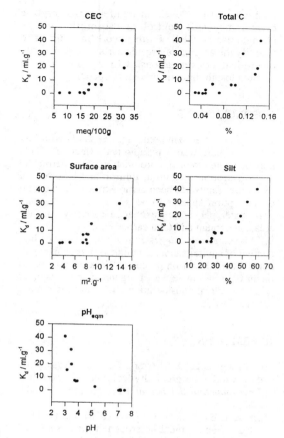

Figure 2. Variation in adsorption coefficient (K_d) for isoproturon with sediment properties: cation exchange capacity (CEC); total carbon; surface area, silt content and pore-water pH.

1972; Schwarzenbach & Westall, 1981; Karickhoff, 1984, Barber et al., 1992).

The key control on the high level of atrazine adsorption below 11.2 m at this site may well be pore-water pH. At these depths, pH values of around 3.5 could lead to protonation of atrazine, thus enhancing its adsorption characteristics (Armstrong & Chesters, 1968).

Linear correlation analysis for the three phenylurea herbicides suggests that the extent of adsorption is related to five aquifer properties (Table 1).

Figure 1 illustrates the correlation of K_d values for isoproturon with the five sediment properties. (Graphs for linuron and diuron are very similar in appearance.) The remaining sediment and pore-water properties measured appear to exert no significant influence on adsorption of these compounds. It is tentatively suggested that the correlations given may be an artefact of one key controlling factor; that is the organic carbon content.

Table 1. Linear correlation coefficient for three phenylurea herbicides

Herbicide	pH	Cation exchange capacity	Surface area	% Silt	Total carbon
Diuron	-0.80	-0.84	0.55	0.92	0.91
Linuron	-0.81	0.64	0.30	0.80	0.88
Isoproturon	-0.75	0.86	0.70	0.94	0.83

Although at low levels in aquifer sands, organic carbon may still control the adsorption of the phenylureas studied. Further work in progress to quantify the amount of organic carbon, together with theoretical considerations, will provide a more detailed insight into the processes operating.

4 CONCLUSIONS

1. Uptake of atrazine onto aquifer sands with low pH (~3.5) pore-waters is rapid and effective. This may be related to *in situ* protonation of the herbicide.
2. Adsorption of diuron, linuron and isoproturon by aquifer sands has been demonstrated. The extent of adsorption appears to be related to five factors: pore-water pH, cation exchange capacity, surface area, silt content and total carbon content.
3. It is tentatively suggested that the organic carbon content of aquifer sands, although very low, may still control the adsorption process. Work in progress to quantify these low levels of organic carbon, together with theoretical considerations, will provide a more detailed insight.

REFERENCES

Armstrong, D.E. & Chesters, G. 1968. Adsorption catalyzed chemical hydrolysis of atrazine. *Environmental Science and Technology*. 2: 683-687.

Barber, L.B., Thurmar, E.M. & Runnells, D.D. 1992. Geochemical heterogeneity in a sand and gravel aquifer: effect of sediment mineralogy and particle size on the sorption of chorobenzines. *Journal of Contaminant Hydrology*. 9: 35-54.

Karickhoff, S. W. 1984. Organic pollutant sorption in aquifer systems. *Journal of Hydraulic Engineering*. 110: 707-735.

Murphy, E.M. & Zachara, J.M. 1995. The role of sorbed humic substances on the distribution of organic and inorganic contaminants in groundwater. *Geoderma*. 67: 103-124.

Mustafa, M.A. & Gamer, Y. 1972. Adsorption and desorption of diuron as a function of soil properties. *Soil Science Society of America*. 36: 561-565

Schwarzenbach, R.P. & Westall, J. 1981. Transport of non-polar organic compounds from surface water to groundwater. Laboratory Sorption Studies. *Environmental Science and Technology*. 15: 1360-1367.

Weber, J.B., Best, J.A. & Gonese, J.U. 1993. Bioavailability and bioactivity of sorbed organic chemicals in sorption and degradation of pesticides and organic chemicals in soil. *SSSA Special Publication No. 32*. SSSA: Madison, WI, USA.

Biogeochemical reactions induced by artificial recharge to carbonate aquifers

K.J. Rattray
Centre for Groundwater Studies & Flinders University of South Australia, S.A., Australia

A.L. Herczeg & P.J. Dillon
Centre for Groundwater Studies & CSIRO Land and Water, Glen Osmond, S.A., Australia

ABSTRACT: Injection of surface water into ground water for subsequent reuse is becoming a very attractive option for regions such as South Australia that have limited surface resources and have dry summers. The interaction between two dissimilar water types needs to be better understood because of potential reactions such as Fe or Mn hydroxide dissolution/precipitation, adsorption/desorption of trace metals, dissolution or precipitation of carbonate matrix, and changes in water quality due to cation exchange and redox reactions. This study examines mass transfer reactions during the process of artificial recharge to carbonate aquifers in South Australia, where the receiving water quality varies from place to place. Proton generation from organic matter oxidation and sulphide mineral oxidation, catalysed by bacterial processes during injection of surface waters, initiates carbonate dissolution and cation exchange. Some months after injection ceases, sulphate reduction, cation exchange and carbonate precipitation dominate. Where there are large salinity differences between injected and receiving waters, cation exchange and carbonate precipitation are important.

1 INTRODUCTION

Aquifer storage and recovery (ASR) is a process that involves artificial recharge of surface runoff or storm water into an aquifer, for subsequent reuse by extraction in drier months. It is becoming an increasingly attractive option in places where surface storages are not viable or are expensive, and in areas with dry summers and winter dominated rainfall.

One of the primary issues facing the viability of such schemes involves uncertainties when oxygenated surface waters, that may have substantial amounts of dissolved or particulate organic matter (i.e., high BOD) are injected into suboxic or anaerobic groundwaters that usually have higher salinity. This paper describes some results from three ASR experimental sites near metropolitan Adelaide, South Australia.

There are a number of practical implications to this work. For example, there is the potential for the transmissivity of the aquifer to be adversely affected through the precipitation of inorganic minerals such as Fe-oxyhydroxides or carbonate minerals or through the swelling and dispersion of clays as a result of cation exchange reactions. Biological growths, stimulated by the addition of organic matter to the system, may also result in significant clogging around the injection well and contribute to inorganic mineral precipitation. In addition the water quality of the recovered water may be adversely affected by reactions such as the reductive dissolution of Fe and Mn oxides, H_2S production and methanogenisis, that may result from the anoxification of the groundwaters.

A number of biogeochemical reactions may be induced by addition of water containing oxygen and labile organic matter (dissolved and particulate) into anoxic or suboxic carbonate aquifers (Table 1). Acidity generated from CO_2 production and/or sulphide mineral oxidation (SMO) can induce a number of geochemical reactions including carbonate dissolution and silicate hydrolysis. Generation of reducing conditions through high organic matter loadings promote Fe, Mn and SO_4 reduction, and even methane fermentation. Other reactions not directly related to the microbiologically mediated sequence, may be caused by mixing of two dissimilar water types through the perturbation of ion activities that may result in cation exchange and mineral dissolution or precipitation.

A substantial body of work exists on microbiological controls on biogeochemical processes (Chapelle and McMahon, 1991) as well as studies on pollution from organic plumes and landfills (eg., Back & Baedecker, 1989; Bennett et al., 1993; Herczeg et al., 1993). Specific studies on geochemical processes during artificial recharge ASR are relatively few (eg. Wood & Bassett, 1975).

Table 1: List of some reactions that may occur during ASR: Organic matter oxidation via (1) oxygen, (2) Fe(III) and (3) SO$_4$ reduction. Pyrite oxidation via oxygen (4a) and Fe(III) reduction (4b). Carbonate dissolution (5) and cation exchange (6).

$$O_2 + CH_2O = CO_2 + H_2O \quad (1)$$
$$FeOOH + CH_2O = Fe^{2+} + HCO_3 \quad (2)$$
$$SO_4^{2-} + 2CH_2O = H_2S + 2HCO_3^- \quad (3)$$
$$15/4 O_2 + FeS_2 = 2SO_4 + Fe(OH)_3 + 4H^+ \quad (4a)$$
$$14Fe^{3+} + FeS_2 + 8H_2O = $$
$$\quad = 2SO_4^{2-} + 16H^+ + 15Fe^{2+} \quad (4b)$$
$$CO_2 + H_2O + CaCO_3 = 2HCO_3 + Ca^{2+} \quad (5)$$
$$X\text{-}Ca^{2+} + 2Na^+ = X\text{-}Na_2 + Ca^{2+} \quad (6)$$

2 SITE DESCRIPTION AND METHODOLOGY

Three experimental sites were investigated as part of this study (Andrews Farm (AF), Clayton (C) and Strathalbyn (S)), all located within approximately 150 km of the Adelaide metropolitan area. The receiving aquifers are all Tertiary carbonate sand aquifers. AF is confined, S is semi-confined and C is unconfined and karstic. At AF the ambient groundwater is a Na–Cl–HCO$_3^-$ type and at S and C the ambient groundwaters are Na-Cl types. The salinity of these waters varies from "fresh" at AF (TDS ~2,150 mg/L), through brackish at S (TDS ~6,300 mg/L) to quite saline at C (TDS ~30,000 mg/L). Injected water at AF is storm water, ponded prior to injection in an adjacent wetland, with a low TDS. At S and C the injected water is from Lake Alexandrina (a large freshwater lake fed by the River Murray). Dissolved oxygen (DO) concentrations are zero, or very low (S) in all receiving groundwaters, whereas the injected waters tend to be high in DO.

Groundwaters from each of the injection wells and observation boreholes were sampled during and after injection of surface waters into these systems. The sampling interval varied from weekly to over two months, with the sampling interval generally shorter during and immediately following injection. At the time of writing, no significant groundwater pumping (ie., the recovery phase of ASR) had taken place, however, long term experimental pumping will take place at AF and C in the near future. This study therefore focuses on biogeochemical effects induced by the injection phase.

Field measurements of dissolved oxygen, pH, Eh, EC and temperature were made and laboratory analyses conducted for the major ions as well as Fe, Mn, DOC and TOC. In some cases stable isotope measurements were also made for the total inorganic carbon and sulphate in the system. For each set of analyses ion concentrations were calculated for a hypothetically conservatively mixed solution using mass balance calculations. Chloride was assumed to behave conservatively in these systems, therefore the measured end-member Cl$^-$ concentrations, were used to determine the mixing ratio between the injected water and the ambient groundwater.

3 RESULTS AND DISCUSSION

3.1 Mass balance calculations

Concentrations of major and minor ions from injection wells and observation holes show large variations from site to site, as well as variations between various observation holes at each site and over the weeks to months after injection. Rather than display the raw measured data, we have normalised the data to account for conservative mixing (as given by end member Cl concentrations) and plotted mass transfer of bicarbonate, Na+K, Ca+Mg and SO$_4$ at one observation borehole at each of the surface water injection sites as a function of time (Figures 1-2 a&b). For example, a HCO$_3$ 'excess' corresponds to a measured HCO$_3$ concentration in the ground water that is greater than that calculated from conservative mixing. A negative number corresponds to removal of HCO$_3$ compared with conservative mixing.

Calculated excess concentrations for Andrews Farm all indicate net transfer of ions into the solution as a result of mixing of storm water and mains water with the groundwater (Fig.1). Injection occurred from late June to mid Aug 1995 which resulted in substantial addition of HCO$_3$ and Ca+Mg into solution, and to a lesser extent Na and SO$_4$. The increase in HCO$_3$ and Ca indicates calcite dissolution, caused by generation of CO$_2$ by oxidation of organic material plus a small amount of pyrite oxidation (sect. 3.2). Net transfer of Ca and HCO$_3$ occurred over the following 7 months. After about mid Feb 1996, the excess concentrations decreased and net fluxes approached zero by 26 July 1996 when further significant injection of surface water took place.

Major element fluxes for Strathalbyn (Fig. 2a) and Clayton (Fig. 2b) are different from those at Andrews Farm. There is a large amount of Na and HCO$_3$ apparently removed from the waters following injection at Strathalbyn. This is believed to be due to cation exchange reactions between the injected surface waters and the ambient groundwaters at the freshwater-saltwater interface, resulting in the removal of Na and the addition of Ca (Sayles and Mangelsdorf, 1977). The addition of large amounts of Ca results in subsequent calcite precipitation and hence the removal of HCO$_3$. Similar reactions have been observed for calcite dissolution where the removal of Ca through cation exchange has resulted in the dissolution of calcite in the aquifer (eg. Baier and Wesner, 1971; Chapelle, 1983). Quantitative calculations for these processes, that also take sulphate reduction into consideration, give a very good match between calculated excess HCO$_3$ and measured excess HCO$_3$ (Rattray et al., 1996). There is some removal of SO$_4$ at Strathalbyn which is indicative of sulphate reduction.

Groundwater compositions at Clayton (Fig. 2b)

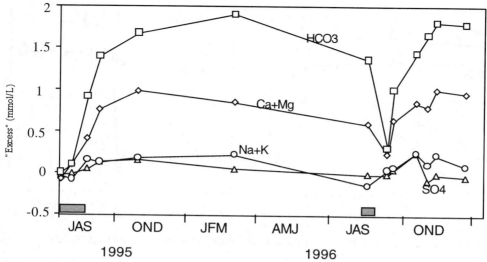

Fig. 1 "Excess" concentrations of HCO3, Ca+Mg, Na and SO4 over an 18 month period at Andrews Farm from mid 1995 to the end of 1996. The stippled rectangles represent injection event and their approximate magnitude.

appear to be dominated by release of Ca + Mg and removal of Na from the solution. Mass transfer of bicarbonate and sulphate is much lower than the cations, suggesting the release of Ca occurs via cation exchange, with Na replacing Ca at exchange sites in the system. Similar observations have been made in seawater intrusion studies where the high ionic strength of the intruding seawater and the predominance of Na in the saline solution drive cation exchange in the opposite direction to that usually observed in dilute solutions, where exchangers generally have a higher selectivity for Ca (Stuyfzand, 1985 cited in Appelo & Postma, 1994). Negative values for SO4 again indicate sulphate reduction takes place.

3.2 Sulphur isotope data

Negative SO4 fluxes at Andrews Farm indicate sulphate reduction, and this is confirmed by sulphur isotope data which show variations of >16‰ over a one year cycle. Native groundwaters have $\delta^{34}S$ of ~+10‰ prior to injection, $\delta^{34}S$ values rapidly decrease to –6‰ after injection, then increase over the next 300 days back towards towards +10‰. This relationship is inversely correlated with calculated SO4 excess concentrations, in that decreasing $\delta^{34}S$ corresponds to increased SO4 excess concentrations, whereas the increasing trend of $\delta^{34}S$ concentration corresponds to sulphate reduction (i.e., negative "excess" values).

3.3 Comparison of the three sites

The three ASR experimental sites all had about the same concentrations of dissolved oxygen and dissolved and particulate organic matter, but vastly different amounts of water injected. That is, the stress on Andrews Farm (total of 240 ML injected) was much greater than the other two sites. The other differences are that the aquifer matrix at AF had much higher reduced sulphur mineral content than the other two sites.

We focussed here on the geochemical effects induced by the biogeochemical reactions (i,.e., the end-products) rather than defining the nature of the microbiologically catalysed reactions themselves. Essentially the reaction sequence can be divided into two distinct phases; (1) during or immediately following injection there is an acid producing phase, exemplified by CO2 production, sulphide oxidation and carbonate dissolution. (2) further anaerobic organic matter oxidation accompanied by mobilisation of Fe, sulphate reduction and carbonate precipitation.

Although there is apparent dissolution of carbonate matrix at Andrews farm inferred from the mass balance calculations, the amount of calcite dissolved due to that process alone is <0.001% of the total rock matrix due to the dominance of rock mass over that of solutes dissolved in water. However, an increase in transmissivity may occur especially if the calcite acts as cement to bind minerals such as quartz, which may be mobilised especially during pumping. Changes in aquifer transmissivity could also be affected by the amount of particulate matter injected, and reactions involving Fe/Mn oxyhydroxides.

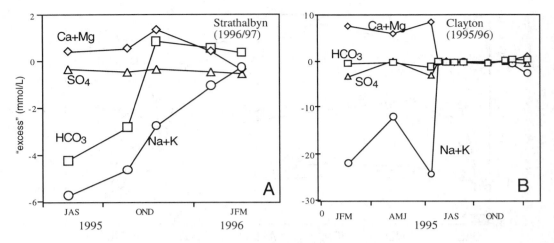

Figure 2a and 2b. Excess concentrations of major ions for Strathalbyn (A) and Clayton (B) ASR sites.

4 CONCLUSIONS

This study shows that a number of biogeochemical reactions are induced in carbonate aquifers upon the injection of oxygenated, organically loaded surface waters and that these reactions are interrelated. At each of the three field sites, the amount of oxidisable organic matter must have exceeded the amount of oxygen injected because there was apparent anoxification of the ground water. That is, net sulphate reduction when integrated over the time scale of recharge and recovery. The oxidation of organic matter and sulphide minerals resulted in dissolution of calcite at Andrews Farm with subsequent carbonate precipitation and cation exchange. Net addition of Ca and Mg to groundwaters may be caused by cation exchange where fresh water is injected into saline or brackish water.

5 ACKNOWLEDGEMENTS

We are grateful to the Mines and Energy Dept SA for logistical and field assistance, and Paul Pavelic at CSIRO for general collaboration on the ASR project.

6 REFERENCES

Appelo, C.A.J. & D.Postma 1994. *Geochemistry, Groundwater and Pollution.* A.A. Balkema, Rotterdam. 536 pp.

Back, W. & M.J.Baedecker 1989. Chemical hydrology in natural and contaminated environments. *J. Hydrol.* 106:1-28.

Bair, D.C. & G.M.Wesner 1971. Reclaimed waste water for groundwater recharge. *Water Resour. Bull.* 7(5): 991-1001.

Bennett, P.C., D.E.Siegel, M.J.Baedecker, & M.F. Hult 1993. Crude oil in a shallow sand and gravel aquifer 1. Hydrogeology and inorganic geochemistry. *Appl. Geochem.* 8: 529-549.

Chapelle, F.H. & P.B.McMahon 1991. Geochemistry of dissolved inorganic carbon in a Coastal Plain Aquifer: Sulphate from confining beds as an oxidant in microbial CO_2 production. *J. Hydrol.* 127: 85-108.

Herczeg, A.L., S.R.Richardson, & P.J.Dillon 1993. Importance of methanogenesis for organic carbon mineralisation in groundwater contaminated by liquid effluent. *Appl. Geochem.* 6: 533-542.

Rattray, K.J., P.J.Dillon, A.L.Herczeg & P. Pavelic, 1996. Geochemical processes in aquifers receiving injected surface water. Centre for Groundwater Studies Report No. 65 ISBN No. 1 875753 15X.

Sayles, F.L. & P.C.Mangelsdorf 1977. The equilibration of clay minerals with seawater: exchange reactions. *Geochim. Cosmochim. Acta* 41: 951-960.

Wood, W.W. & R.L.Bassett 1975. Water quality changes related to the development of anaerobic conditions during artificial recharge. *Water Resour. Res.* 11: 553-558.

The origin of sodium-bicarbonate groundwaters in a fractured aquifer experiencing magmatic carbon dioxide degassing, the Ballimore region, central New South Wales, Australia

Shane Schofield & Jerzy Jankowski
Groundwater Centre, Department of Applied Geology, University of New South Wales, Sydney, N.S.W., Australia

ABSTRACT: Ninety-five percent pure Na-HCO$_3$ groundwaters occur in a confined, fractured, Permian to Jurassic aquifer in the Ballimore region, central New South Wales. These waters range in salinity from 3,000 to 8,800 mg/l, have CO$_{2(aq)}$ concentrations above 1 g/l, log P$_{CO2}$ 0.22 (P$_{CO2}$ = 1.7 atm), and exhibit strong H$_2$CO$_3$/HCO$_3$ couple pH buffering. Geological interpretation and geochemical modelling suggests that Late Miocene magmatic activity (ca. 12 Ma) is intimately related to the genesis of these unique waters. Chemical and isotopic data indicate that these groundwaters are of meteoric origin and that an external source of CO$_2$ is present.

1 INTRODUCTION

Several combinations of hydrogeochemical processes and water-rock interaction reactions have been used to explain the evolution of Na-HCO$_3$ groundwaters. Some Na-HCO$_3$ groundwaters originate from the dissolution of CaCO$_3$ in the presence of biogenic CO$_2$, accompanied by Ca for Na ion-exchange (Foster 1950; Chapelle & Knobel 1985). Blake (1989) and Herczeg et al. (1991) explain the presence of Na-HCO$_3$ rich groundwaters through three different reactions: the reaction of Na with kaolinite to form Na-beidellite and H$^+$, the dissolution of CaMg(CO$_3$)$_2$ by H$^+$ which release Ca, Mg and HCO$_3$ ions to solution, then the exchange of Ca and Mg for Na. Similar processes are thought to be responsible for Na-HCO$_3$ groundwaters of the Great Artesian Basin of Australia (Habermehl 1983). Balfe & Carmichael (1980) and Baker et al. (1995) have identified a relationship between Na-HCO$_3$ rich groundwaters in the Bowen-Gunnedah-Sydney basin system and the occurrence of dawsonite (NaAlCO$_3$(OH)$_2$)). All of these Na-HCO$_3$ groundwaters have pH values above 7, often above 8, and contain little CO$_2$. Several Na-HCO$_3$ groundwaters with high CO$_2$ concentrations and pH values of less than 7 are discussed in the literature, but all these occur in tectonically active areas (Irvin & Barnes 1975, 1980; Barnes 1985; Evans et al. 1986).

This paper will discuss the origin of effervescent, slightly acidic (pH 6.1-6.9), Na-HCO$_3$ groundwater in a tectonically stable area.

2 GEOLOGY AND HYDROGEOLOGY

In the Ballimore region basement rocks consist of dominantly marine, Ordovician to Carboniferous igneous and metasedimentary strata of the Lachlan Fold Belt, which are metamorphosed to greenschist grade and have a well developed north-south foliation. These rocks are unconformably overlain by terrestrial, Permian to Jurassic sediments of the Gunnedah and Surat Basins.

Outcrop relationships and geophysical data suggest that a series of northwest trending extensional horsts and grabens span this region. Regional cross sections tied by borehole control show thickening of the sedimentary units across faults, indicating that extensional tectonics since the Early Permian have repeatedly initiated motion on the faults that bound these horsts and grabens. The sedimentary sequence consists of Early Permian polymictic conglomerate, siltstone and mudstone, Late Permian quartzose sandstone interbedded with coal and shale, Triassic quartz-lithic sandstone overlain by a lacustrine shale with lithic sandstone interbeds, Middle Jurassic interbedded lithic sandstone, shale and coal and Late Jurassic quartzose sandstone. The Late Jurassic quartzose sandstone has high porosity and permeability.

Regional cross sections indicate that this unit thickens appreciably into localised depocentres where Late Jurassic syndepositional faulting has accommodated the accumulation of up to 200 m (drilled thickness) of sediment.

Dating of Tertiary igneous and volcanic rocks along the east Australian plate margin shows that volcanic rocks become younger (53 to 0 Ma) from far north Queensland to southwestern Victoria (Sutherland 1981). The northerly movement of the Australian plate over one or several stationary hotspot(s) is thought to be responsible for this magmatism (Wellman & McDougall 1974; Sutherland 1981). Wellman & McDougall (1974) dated Miocene volcanic rocks from the Ballimore region, using K-Ar methods, at 14-17 Ma to the north, 14 Ma to the east and 12 Ma to the west.

X-ray fluorescence analyses of core from the sedimentary sequence have failed to detect any sodium rich units. However, both basement rocks and Miocene intrusive and volcanic samples contain significant quantities of sodium. X-Ray Fluorescence analysis of probable Late Miocene samples from the Ballimore region confirm the presence of compositionally intermediate, sodium rich (5.3 to 7.3 wt% Na_2O) plutonic and extrusive rocks. Geochemically, these rocks are classified as being syenite (intrusive) or trachyte (extrusive).

For over a century, $Na-HCO_3$ type waters have flowed from boreholes in the Ballimore region. These waters are highly effervescent, with CO_2 concentrations in excess of 1 g/l and CO_2 partial pressures that are one thousand times higher than atmospheric levels.

Groundwater flow in the Ballimore region is controlled by deep-seated faults and fractures. Borehole lithological correlations show there is no link between the age of the geological unit within which the aquifer is intersected, and the occurrence of $Na-HCO_3$ type groundwater. Groundwater sampled from adjacent boreholes, where the aquifer is intersected in geological units of different ages, have similar isotopic compositions, chemistry and water levels. Detailed observation of visual porosity in thin-sections indicate that there is a rapid loss in primary porosity with depth. The loss in porosity is due to the precipitation of quartz overgrowths, authigenic kaolinite, siderite and calcite in the matrix of the sandy units.

3 HYDROGEOCHEMISTRY

Eight hydrogeochemical groups are recognised and average analyses are shown in Table 1. The fresh groundwaters (groups 1 and 2) are typical recharge waters sampled from Late Jurassic quartzose sandstone. Saline groundwaters (group 3) are found within thin Quaternary alluvial deposits and at the upper interface of the fractured aquifer system. Figures 1a and 1b illustrate the effects of simple mixing between Na-Cl and $Na-HCO_3$ water types.

The dominant ionic species in the soda water groups are Na and HCO_3, with these two ions comprising 80-95% of the ions in solution (Figure 1c). The concentration of other major ions is low, with Ca and Mg concentrations less than 3 mmol/l.

Strongly negative redox conditions and low sulphate concentrations indicate that oxidation of organic matter has removed sulphate from solution, producing 1-2 mg/l of HS^-. The Cl concentration in the soda waters is low, with a plot of Na versus Cl and Cl versus HCO_3 showing distinctive saline and soda trends (Figures 1a and 2a). The ratio of alkalinity to DIC is determined by the processes that contribute CO_2 to the system. In Figure 2b the DIC of each soda water group varies independently of alkalinity, indicating that the aquifer system is open with respect to CO_2 gas. The soda waters do not lie along the trends attributed to the dissolution of calcite, SO_4 reduction, or methanogenesis, showing that CO_2 comes from an alternate source.

In the simplest sense $\delta^{13}C$ data indicates whether the carbon in the groundwater system is derived from a

Table 1. Average chemical compositions of different groundwaters from the Ballimore region.

Type	N°	TDS	CO_2	Na	K	Ca	Mg	HCO_3	SO_4	Cl	DIC	P_{CO2}	$\delta^{13}C$
		mg/l	mg/l					mmol/l				atm	‰ PDB
Fresh 1	3	442	89	1.9	0.1	0.7	1.1	4.5	0.1	1.1	15.2	-0.54	-14.10
Fresh 2	3	339	111	3.6	0.4	0.1	0.3	1.3	0.1	3.3	4.4	-1.08	-17.18
Saline	3	7914	151	88.8	0.2	4.8	16.9	10.1	7.7	109.1	16.6	-0.77	-16.27
Soda 1	5	8360	1058	90.2	1.2	1.2	2.0	95.6	0.0	8.4	117.4	-0.26	-0.66
Soda 2	15	5208	953	51.3	2.4	1.8	1.8	61.6	0.0	1.5	107.6	0.10	-1.71
Soda 3	8	4881	1059	45.9	2.0	2.6	2.3	57.7	0.0	1.6	109.4	0.14	-0.40
Soda 4	3	4091	939	40.6	2.2	1.8	1.9	47.1	0.2	1.4	50.0	-0.66	-0.21
Soda 5	9	3418	984	28.9	2.2	3.0	2.7	37.9	0.0	4.4	100.3	0.22	-2.07

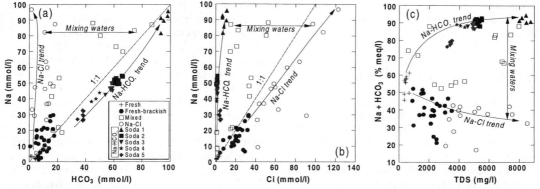

Figure 1. (a) Na vs. HCO_3; (b) Cl vs Na, (a) & (b) show mixing and evolutionary trends of Na-Cl and Na-HCO_3 waters; (c) TDS vs. Na+HCO_3, showing the dominance of Na and HCO_3 ions in the soda waters.

heavy inorganic source (>–10‰), or a light biogenic source (<–10‰). The measured isotopic ratios quoted in Table 1 emanate from intermixing of several carbon sources: soil zone CO_2 (biogenic), calcite dissolution (inorganic) and CO_2 resulting from the oxidation of organic matter (biogenic).

The geochemical processes controlling carbon species in this aquifer system are shown by the overlapping regions on the $\delta^{13}C$ versus DIC^{-1} plot (Figure 2c). The groundwaters in Region I are fresh to saline, have low to moderate concentrations of HCO_3 derived from a mixed biogenic and inorganic carbon source ($\delta^{13}C_{DIC}$ values of -22 to -5‰). Region II comprises waters that stem from the mixing of the three end-members fresh, saline and soda. These waters have $\delta^{13}C_{DIC}$ ratios that range from 0 to -15‰ with the heavier carbons reflecting a higher proportion of soda water in the mix. Region III contains the soda waters. These HCO_3 rich waters have a $\delta^{13}C_{DIC}$ range of -5 to +2.5‰, indicating an solely inorganic carbon source. There is no overlap between regions I and III. This shows that the origin of carbon in the soda waters is independent of the carbon in the shallow groundwater system.

Since the system is open to the generation/addition of CO_2, and $\delta^{13}C$ data identify this carbon as being of inorganic origin, the CO_2 must come from either carbonate dissolution, silicate weathering or a mantle derived CO_2 source. Carbonate dissolution can be eliminated, as there are no Ca-CO_3 type waters in the area. Silicate dissolution would be relatively slow and would produce mixed inorganic-biogenic $\delta^{13}C$ values as light soil zone CO_2 would be needed to drive the silicate weathering reaction. Both of the above reactions would elevate the pH of the groundwater through the consumption of H_2CO_3. This leaves a magmatic/mantle derived CO_2 source. Such a source would have the appropriate carbon signature (Evans et al. 1986), would be capable of continually providing CO_2 and would maintain pH equilibrium in the water at the H_2CO_3/HCO_3 buffering position.

Carbon-14 analyses show that Na-HCO_3 waters contain less than 2% of modern carbon. This

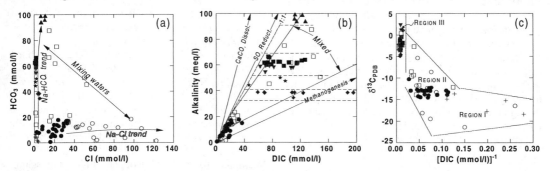

Figure 2. (a) Cl vs. HCO_3: shows low Cl concentration in the soda waters; (b) DIC vs. Alkalinity: shows CO_2 in the soda waters form independently of conventional groundwater CO_2 sources; (c) DIC vs. $\delta^{13}C_{PDB}$: shows that heavier $\delta^{13}C$‰ values are related to higher DIC concentrations.

indicates either that the water is very old, or that the ^{14}C in recharging meteoric waters is swamped by a "dead" carbon source.

4 HYDROGEOCHEMICAL MODELLING

The hydrogeochemical modelling code NETPATH (Plummer et al. 1991) has been used to calculate a mass-balance model which describes this groundwater system. Average analyses of the fresh groundwaters sampled from the Late Jurassic quartzose sandstone and the most concentrated Na-HCO_3 type water, Fresh 1 and Soda 1 (Table 1) were used as end member waters. The following mass-balance model, which is supported by $\delta^{13}C$ data, describes the chemical evolution of the Na-HCO_3 type waters:

Fresh (1)+101.5$CO_{2(gas)}$+81.6Albite+7.4NaCl+1.1K-mica+0.5$CaCO_3$+0.2Chlorite+0.1$FeCO_3$+0.1CH_2O
$\Rightarrow$ Soda (1)+124.6Kaolinite+0.03 FeS_2

Modelling suggests that the flux of 101.5 mmol of CO_2 gas and the dissolution of 81.6 mmol of albite are the two most important reactions in the aquifer system.

5 ORIGIN OF Na-HCO_3 GROUNDWATERS

The origin of Na-HCO_3 groundwaters can be best explained by Na-feldspar weathering reactions driven by CO_2 derived from an inorganic CO_2 source. Sodium rich intrusive rocks provide both a CO_2 and a Na source. Meteoric water pooling in the quartz sandstone filled Late Jurassic depocentres provides a fresh water reservoir, and a pressure drive to the system. This fresh water migrates along faults and fractures in the sedimentary sequence, until it reaches an area where CO_2 is stored. This reservoir could reside in a structural or stratigraphic trap, either in fractures or in zones of preserved porosity. The water would readily mix with the CO_2 producing aggressive H_2CO_3 rich fluids. The aggressive nature of the fluids would corrode Na-feldspar present in one of the sodic units. The resulting Na-HCO_3 rich water could then by pushed through the fracture network by CO_2 gas expansion, enter the bore and discharge at the bore head as effervescent Na-HCO_3 type water.

REFERENCES

Baker, J.C., G.P. Bai, P.J. Hamilton, S.Z. Golding & J.B. Keene 1995. Continental-scale magmatic carbon dioxide seepage recorded by dawsonite in the Bowen-Gunnedah-Sydney Basin system, Eastern Australia. *J. Sed. Res.* A65(3): 522-530.

Balfe, P.E. & D.C. Carmichael 1980. An occurrence of dawsonite in the Bowen Basin. *Qld. Gov. Mining J.* 81: 519-522.

Barnes, I. 1985. Mineral-water reactions in metamorphism and volcanism. *Chem. Geol.* 49: 21-29.

Blake, R. 1989. The origin of high sodium bicarbonate waters in the Otway Basin, Victoria, Australia. *Proc. 6 Int. Symp. Water-Rock Int., Malvern, 3-8 August 1989*: 83-85. Rotterdam: Balkema.

Chapelle, F.H. & L.L. Knobel 1985. Stable carbon isotopes of HCO_3 in the Aquia Aquifer, Maryland: Evidence for an isotopically heavy source of CO_2. *Ground Water* 23: 592-599.

Evans, W.C., T.S. Presser & I. Barnes 1986. Selected soda springs of Colorado and their origins. *USGS Water-Supply Paper 2310*: 45-52.

Foster, M.D. 1950. The origin of high sodium bicarbonate waters in the Atlantic and Gulf Coastal Plains. *Geochim. Cosmochim. Acta* 1: 33-48.

Habermehl, M.A. 1983. Hydrogeology and hydrochemistry of the Great Artesian Basin, Australia. *Proc. Int. Conf. Groundwater & Man, Sydney, 5-9 December 1983, Aust. Water. Resour. Counc. Conf. Ser.* 8(3): 83-98.

Herczeg, A.L., T. Torgersen, A.R. Chivas & M.A. Habermehl 1991. Geochemistry of groundwaters from the Great Artesian Basin, Australia. *J. Hydrol.* 126: 225-245.

Irvin, W.P. & I. Barnes 1975. Effect of geologic structure and metamorphic fluids on seismic behaviour of the San Andreas fault system in central and northern California. *Geology* 3: 713-716.

Irvin, W.P. & I. Barnes 1980. Tectonic relations of carbon dioxide discharges and earthquakes. *J. Geophys. Res.* 85(B6): 3115-3121.

Plummer, L.N., E.C. Prestemon & D.L. Parkhurst 1991. *An interactive code (NETPATH) for modelling NET geochemical reactions along a flow PATH.* USGS Water Res. Invest. Rep. 91-4078.

Sutherland, F.L. 1981. Migration in relation to possible tectonic and regional controls in Eastern Australian Volcanism. *J. Volc. & Geophys. Res.* 9: 181-213.

Wellman, P. & I. McDougall 1974. Potassium-Argon on the Cainozoic volcanic rocks of New South Wales. *J. Geol. Soc. Aust.* 21: 247-272.

Origin and mobility of arsenic in groundwater from the Pampean Plain, Argentina

P.L. Smedley
British Geological Survey, Wallingford, UK

H.B. Nicolli & A.J. Barros
Instituto de Geoquímica, San Miguel, Argentine

J.O. Tullio
Dirección de Aguas, Olascoaga, Santa Rosa, Argentine

ABSTRACT: Groundwaters from Tertiary and Quaternary loess deposits of northern La Pampa Province, Argentina have high concentrations of arsenic (range <0.004 mg l^{-1} to 5.3 mg l^{-1}) as well as other trace elements including F, V, U, B and Mo. Salinities are highly variable (SEC range 0.77–17.5 mS cm^{-1}) and are mainly generated by high degrees of evaporation. The groundwaters are uniformly oxidising and the arsenic is dominated by arsenate. Highest concentrations are associated with high pH and HCO$_3$ and As also closely correlates positively with V and F. There is a weaker positive correlation with B and Be. The arsenic occurrence in solution is not related to oxidation of sulphide minerals but is more likely to be derived by dissolution from pyroclastic debris in the loess deposits, glass shards being a possible source. Mobility of arsenate in the oxidising conditions is enhanced where high pH and HCO$_3$ prevent sorption onto surfaces of Fe and Al oxide in the aquifer sediments.

1. INTRODUCTION

Groundwaters from the Chaco-Pampean Plain of Argentina have long been known to contain high concentrations of As as well as other potentially toxic trace elements. Nicolli et al. (1989) reported As concentrations between 19 µg l^{-1} and 3.8 mg l^{-1} in groundwaters from the Province of Córdoba and concentrations up to 720 µg l^{-1} were noted in the Pampean Carcarañá river basin (Nicolli and Merino, 1997). High As concentrations are also problematic in groundwaters from neighbouring Chile (Cáceres et al., 1992). The health effects of arsenic in the drinking water in Argentina are not well-established but in parts of Córdoba Province, chronic arsenic disease is known as 'Bell Ville disease' after the town where prevalence is highest. Typical symptoms include skin-pigmentation disorders, hyperkeratosis and skin cancer. Statistics for the prevalence of skin and bladder cancer in are reported to be higher in parts of Argentina than in control areas (Astolfi et al., 1981; Hopenhayn-Rich et al., 1996).

In this investigation, 109 groundwater samples have been collected from the Eduardo Castex area of La Pampa Province in central Argentina (Figure 1) in order to determine the distribution of arsenic and related trace elements in the groundwaters and to establish the processes involved in mobilisation. Results and some preliminary interpretations are presented in this account.

2. REGIONAL SETTING

La Pampa Province in central Argentina experiences a semi-arid temperate climate with an average annual rainfall of 780 mm and a mean annual temperature of 16°C. The area investigated lies in the northern part of the Province, between latitudes 35°20' S and 36° S and longitudes 64° W and 65°10' W and covers an area of 110 km x 70 km (Figure 1).

The region comprises a largely flat-lying plain although the regional slope is from west (elevation 310 m) to east (145 m). There is little surface drainage in the region but ephemeral ponds exist in localised

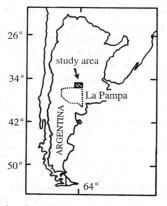

Figure 1. Location map showing study area and outline of La Pampa Province.

depressions at times of highest rainfall and experience high rates of evaporation and consequent deposition of salt during the dry season.

Land-use is predominantly agricultural, with cattle-rearing being dominant and grain crops also of regional importance.

3. GEOLOGY AND HYDROGEOLOGY

The Pampean region comprises a thick sequence (generally around 200–300 m) of Tertiary and Quaternary loess sediments of mainly silt or fine-sand grade. These are wind-blown and reworked sediments with a high proportion of fine volcanic detritus and are of rhyolitic composition. Sediments from the region have yet to be investigated in detail but Pampean loess typically comprises mainly sodic plagioclase (20–60%), quartz (20–30%) and glass shards (15–50%), with minor amounts of muscovite, calcite and biotite and clay minerals dominated by illite (Nicolli et al., 1989; Sayago, 1995). Compositions are relatively uniform across the region (Sayago, 1995). Discrete thin (5–15 cm thick) layers of dacitic/rhyolitic volcanic ash are visible in road cuttings at shallow depths within the Quaternary deposits.

Subsurface layers of calcrete, often laterally extensive, are also abundant at the Tertiary/Quaternary interface.

Groundwater use is mainly for agriculture (cattle, irrigation) and for potable supply. In the major towns, groundwater is treated by reverse osmosis to remove dissolved salts and As before use for potable water. In the rural areas, this is not possible and water is typically used without pre-treatment.

Regional groundwater flow is easterly following the topographic gradient but smaller-scale flow patterns are controlled by localised topography. Depressions form discharge areas which fill during rainfall periods (mainly summer rains) and evaporate during dry periods. Salt encrustations form around the periphery of such depressions following prolonged evaporation. Water levels vary according to topographic variation, being as deep as 120 m below ground level in the west and as shallow as 3–5 m below ground level in the eastern part of the study area. Pumping depresses the water level locally.

Borehole depth is variable depending on the regional topography. Boreholes in the west (Ing. Foster area) reach up to 140 m below ground level. Those in the east may be as shallow as 10 m or less.

4. REGIONAL HYDROGEOCHEMISTRY

Groundwaters from La Pampa have highly variable salinity, SEC values ranging from 0.77–17.5 mS cm^{-1}. The salinity is generated by evaporation and the most saline groundwaters are commonly restricted to

Table 1. Statistical summary of chemical data for groundwaters from La Pampa.

	Units	Min	Max	Median	n
Well depth	m	5.4	140	29.1	103
Gw level	m	2.1	129	13.1	74
Temp'ture	°C	16.1	29.1	19.8	106
pH		6.99	8.66	7.85	107
Eh	mV	131	492	325	102
DO	mg l^{-1}	0.8	9.9	6.1	105
SEC	µS cm^{-1}	773	17520	2610	107
Ca	mg l^{-1}	1.55	599	21.0	107
Mg	mg l^{-1}	2.08	521	21.8	107
Na	mg l^{-1}	122	3101	541	107
K	mg l^{-1}	3.15	70.6	12.0	106
Cl	mg l^{-1}	8.5	4580	205	107
SO$_4$	mg l^{-1}	6.78	3171	285	107
HCO$_3$	mg l^{-1}	195	1440	650	107
NO$_3$-N	mg l^{-1}	<0.23	144	9.01	107
NH$_4$-N	mg l^{-1}	<0.02	0.14	<0.02	82
Si	mg l^{-1}	22.1	39.2	29.6	107
As(III)	mg l^{-1}	<0.004	0.215	<0.004	107
As$_{total}$	mg l^{-1}	<0.004	5.28	0.14	107
Ba	µg l^{-1}	5	259	36	107
Sr	mg l^{-1}	0.07	13.3	0.60	107
V	mg l^{-1}	0.02	5.43	0.54	107
B	mg l^{-1}	0.46	13.8	2.96	107
Be	µg l^{-1}	<0.02	0.16	0.034	78
P$_{total}$	mg l^{-1}	<0.2	0.67	<0.2	107
F	mg l^{-1}	0.034	29.2	3.78	107
Br	mg l^{-1}	0.055	11.6	0.65	42
I	mg l^{-1}	0.02	0.73	0.12	107
Li	µg l^{-1}	5.7	117	24.6	78
Al	µg l^{-1}	2.9	991	17.8	78
Cr	µg l^{-1}	0.43	10.4	2.17	78
Fe	mg l^{-1}	<0.006	1.16	0.06	107
Mn	µg l^{-1}	0.05	78	2.01	78
Co	µg l^{-1}	<0.090	1.38	0.13	78
Ni	µg l^{-1}	<0.020	14.3	0.75	78
Cu	µg l^{-1}	0.4	89	7.7	78
Zn	µg l^{-1}	2.5	1438	60	78
Rb	µg l^{-1}	1.5	29	4.8	78
Y	µg l^{-1}	<0.007	0.36	0.043	78
Mo	µg l^{-1}	2.7	991	61	78
Cd	µg l^{-1}	<0.02	2.7	0.13	78
Sb	µg l^{-1}	<0.10	0.92	0.10	78
Cs	µg l^{-1}	<0.006	0.29	0.034	78
La	µg l^{-1}	<0.024	15.4	0.024	78
Tl	µg l^{-1}	<0.007	0.14	0.012	78
Pb	µg l^{-1}	<0.060	13.8	0.50	78
Bi	µg l^{-1}	<0.024	0.74	<0.024	78
U	µg l^{-1}	6.2	248	31	78

shallow parts of the aquifer.

The groundwaters are universally oxidising with high dissolved-oxygen concentrations (0.8–9.9 mg l^{-1}) and redox potentials (Eh measured on-site 131–492 mV; Table 1). Nitrate-N concentrations are also notably high in the groundwaters with a maximum value observed at 144 mg l^{-1}. The nitrate most likely derives from agricultural pollutants and concentrations are highest where water levels are shallow. Pollution may have made a further contribution of Cl and SO$_4$ to the groundwater.

Major-element compositions are controlled by reaction of silicate minerals and by carbonate equilibria. Sodium is the dominant cation in the Pampean groundwaters and HCO$_3$ is usually the dominant anion, though Cl and SO$_4$ are also important in the more saline samples. Sodium derives mainly by dissolution of sodic plagioclase in the sediments, but is further concentrated by evaporative losses. Bicarbonate concentrations and pH reflect soil-zone and carbonate reactions.

Concentrations of As reach up to 5.3 mg l^{-1} in the groundwaters, more than 500 times the recently revised WHO recommended limit for As in drinking water. Measured As(III) concentrations are a small proportion of the total As present and since organic As forms such as MMAA and DMAA are usually negligible in groundwaters (e.g. Chen et al., 1994), As(V), arsenate, is taken to be the dominant species.

The regional distribution of As in the groundwaters is shown in Figure 2. There is no apparent regional trend in As concentrations and concentrations are highly variable on a local scale, reflecting depth variations and major-element chemical composition.

Dissolved As concentrations correlate closely with HCO$_3$ concentrations and pH (Figure 3). Enhanced mobility of arsenate at high pH has often been noted in natural waters, the relationship arising from the degree of sorption of the species onto Fe oxide (e.g. goethite, ferrihydrite). Relationships between dissolved As concentration and HCO$_3$ have also been reported (e.g. Smedley et al., 1996).

Arsenic in water is commonly taken to be the result of oxidation of sulphide minerals, notably pyrite, in aquifer sediments. If this were the major source of the As, a correlation might be expected between the two solutes. However, Figure 3 shows that no correlation exists between As and SO$_4$ in the Pampean groundwaters. Indeed, the SO$_4$ is thought to be dominated by evaporative concentration of recharge water rather than by significant oxidation of sulphide minerals. Stable-sulphur-isotopic compositions ($\delta^{34}S_{SO4}$) of selected samples are also rather enriched (close to +8 ‰) to have resulted from direct oxidation of pyrite in the aquifer.

Significant correlations exist between As and both F and V in the groundwaters, which themselves reach very high concentrations (F range 0.03–29 mg l^{-1}; V range 0.02–5.4 mg l^{-1} Figure 3). Weaker positive correlations also occur with B and Be (0.5–14 mg l^{-1} and <0.02–0.16 µg l^{-1} respectively). The groundwaters also have notably high concentrations of U and Mo (Table 1). All these trace elements are known to be enriched in volcanic materials and to be mobile under oxidising conditions (e.g. vanadate, molybdate, borate). Many form anionic complexes in alkaline solution (F commonly forming complexes with B, Be, Fe(III) and Al). The trace elements are likely to be preferentially mobilised under conditions where pH (and HCO$_3$) are highest, these being controlled by carbonate reactions. The source of these trace elements in the Pampean loess sediments has not yet been established but volcanic ash and/or glass shards are a likely potential source.

Sorption of As and other anionic species onto Fe oxide and Al oxide is also likely to be at a minimum under conditions of high pH and high HCO$_3$, perhaps as a result of HCO$_3^-$ – HAsO$_4^{2-}$ competition.

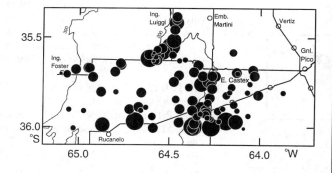 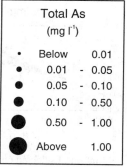

Figure 2. Regional distribution of total As in the groundwaters of La Pampa.

5. SUMMARY

Oxidising groundwaters from northern La Pampa in Argentina have high concentrations of As which relate closely to high pH and HCO_3 concentrations. The major-element chemistry of the groundwaters is controlled by carbonate reactions and evaporation. Arsenic concentrations correlate closely with F and V and to a lesser extent with B and Be. Concentrations of Mo and U are also high in the groundwaters. These elements are thought to have derived from volcanic glass in the loess and are at their most mobile under alkaline and oxidising conditions.

Although the occurrence of As in groundwater has commonly been associated with reducing groundwaters (e.g. Chen et al., 1994), this investigation shows that high concentrations can be maintained in solution under highly oxidising conditions. Although Fe- and Al- oxide minerals are present in the loess sediments, sorption is likely to be restricted in the groundwaters with highest As concentrations under the prevailing conditions of high pH and HCO_3.

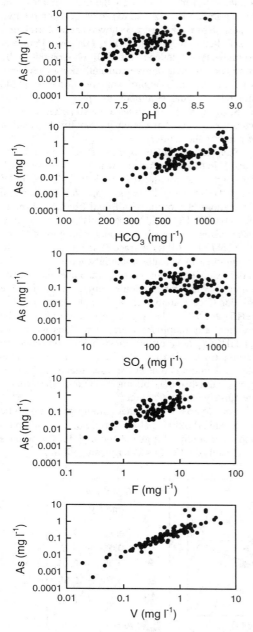

Figure 3. Variation of selected chemical parameters with As concentration in the Pampean groundwaters.

6. REFERENCES

Astolfi, E A N, Maccagno, A, García Fernández, J C, Vaccaro, R and Stimola, R. 1981. Relation between arsenic in drinking water and skin cancer. *Biol. Trace Elem. Res.*, 3, 133-143.

Cáceres, L, Gruttner, V E and Contreras, N. 1992. Water cycling in arid regions: Chilean case. *Ambio*, 21, 138-144.

Chen, S-L, Dzeng, S R, Yang, M_H, Chiu, K-H, Shieh, G-M and Wai, C M. 1996. Arsenic species in groundwaters of the Blackfoot disease area, Taiwan. *Environ. Sci. Technol.*, 28, 877-881.

Hopenhayn-Rich C, Biggs, M L, Fuchs, A, Bergoglio, R E, Nicolli, H B and Smith, A H. 1996. Bladder cancer mortality associated with arsenic in drinking water in Argentina. *Epidem.* 7, 117-124.

Nicolli, H B and Merino, M H. 1997. High contents of F, As and Se in groundwater of the Carcarañá river basin, Argentine Pampean Plain, *Environ. Geol.*, In press.

Nicolli, H B, Suriano, J M, Gómez Peral, M A, Ferpozzi, L H and Baleani, O A. 1989. Groundwater contamination with arsenic and other trace elements in an area of the Pampa, Province of Córdoba, Argentina. *Environ. Geol. Water Sci.*, 14, 3-16.

Sayago, J M. 1995. The Argentine neotropical loess: a review. *Q. Sci. Rev.*, 14, 755-766.

Smedley, P L, Edmunds, W M and Pelig-Ba, K B. 1996. Mobility of arsenic in groundwater in the Obuasi area of Ghana. *In*: Environ. Geochem. Health, *eds*: Appleton, JD, Fuge, R & McCall, G J H., Geol. Soc. Special Publ. No 113, 163-181.

5 Sedimentary basins

Formation waters and diagenetic modifications: General trends exhibited by oil fields from the Norwegian shelf – A model for formation waters in oil prone subsiding basins

P. Aagaard
Department of Geology, University of Oslo, Blindern, Norway

P. K. Egeberg
Department of Chemistry, Agder College, Kristiansand, Norway

ABSTRACT: The main characteristics of oil field formation waters from the Norwegian shelf have been reviewed and discussed. The original pore waters were derived by mixing of two end members: a residual brine and meteoric water. Two main processes, apart from the mixing, have changed the chemical composition: 1) early brine modification by dolomitization of calcite and precipitation of K-feldspar, and 2) burial diagenesis with recrystallization of calcite, albitization, chlorite precipitation and at greater depth illitization and recrystallization of quartz. The burial diagenesis is also reflected in changes in $\delta^{18}O$ towards depth, leading to more positive values. The basins offshore Norway exhibit a P_{CO2} which is 2-3 orders of magnitude higher than those of hydrothermal systems in Iceland and New Zealand for the same temperature.

1 INTRODUCTION

The origin and evolution of formation waters from oil fields on the Norwegian shelf, were first compiled and discussed by Egeberg and Aagaard (1989). Later, North Sea formation waters from the UK sector have been presented in compilations of Warren and Smalley (1994) and Bjørlykke et al. (1995). The sedimentary basins offshore Norway and UK thus constitute large natural laboratories, where considerable data on formation waters and mineral reactions have been recorded. The purpose of the present communication, however, is not to review all those data, but to summarize some main characteristics of the formation water systems on the Norwegian shelf, which we believe are general features of subsiding oil prone sedimentary basins.

2 BRINE COMPONENTS

Worden (1996) recently reviewed the halogen content of reported sedimentary formation waters, and concluded that all chloride and bromide data reflected a marine influence and, in Mesozoic and Paleozoic reservoirs, specially the presence of evaporates. This agrees well with the model of Egeberg and Aagaard (1989) for formation waters on the Norwegian shelf, as a mixing of a brine end member and a meteoric end member. The chloride/bromide data corresponded to an endmember of an evaporation residue with 6 m chloride concentration. The corresponding theoretical Na/Cl ratio of 0.755 (Carpenter, 1978) was identical to the slope of the linear Na vs Cl curve. Isotopic composition (δ^2H, $\delta^{18}O$) of such an evaporation residue (Pierre et al., 1984) were also in

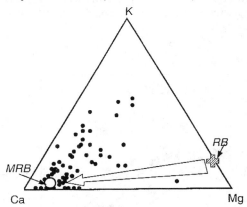

Figure 1. Relative proportions of Mg, Ca and K in oil field formation waters on the Norwegian shelf. RB and MRB denote residual and modified residual brine respectively. Arrow indicate changes from RB to MRB accompanying early dolomitization and K-feldspar precipitation. The deviation from the MRB composition is caused by later diagenetic reactions during burial.

composition is caused by later diagenetic reactions during burial.
accord with formation water data. The Mg, Ca and K content of the brine end member were not the same as the residual brine, but had been modified by early/contemporaneous dolomitization and K-feldspar formation, leading to the end member Ca brine with negligible content of K (Figure 1). The dilute end member for formation waters on the Norwegian shelf, was identified as meteoric water.

Figure 1 demonstrates considerable change in the relative proportions of Mg, Ca and K components during mixing with dilute end member and during later diagenetic reactions.

3 DIAGENESIS

The Jurassic clastic reservoirs offshore Norway are dominated by arkoses and subarkoses with minor occurrences of sublitharenites/litharenites (Bjørlykke et al., 1992), e.g. the Brent Group sandstones normally have a quartz content between 40-90%, up to 15-20% kaolinite, a variable albite content, and a K-feldspar content decreasing from about 10% at shallow burial to below detection limit at 3.5-4 km depth. Calcite is present as cement in most sandstones, siderite is a frequent early cement, while the late carbonate cements are ankerites and ferroan calcite. The other late diagenetic minerals are quartz, albite, illite and minor chlorite, reflecting the original

development towards heavier oxygen composition with burial.

clastic mineral assemblage. Major diagenetic reactions on a mass basis include recrystallization of carbonates, albitization of K-feldspar (more or less contamporaneous with illitization in mudstones), later illitization of kaolinites and quartz precipitation (with silica sourced from dissolving quartz and other mineral reactions).

The oxygen isotopic composition of the formation waters with burial (Figure 2) also reflects those diagenetic changes. There is a definite trend of formation water oxygen to become heavier with depth. Egeberg and Aaagard (1989) attributed this shift mainly to recrystallization of early formed calcite and to illitization and chloritization of kaolinite and smectites. Other data from the North Sea given by Warren and Smalley (1993) also exhibit the same trend. This trend is in general accord with previous closed system model calculations by Dickinson (1986), done to simulate recrystallization of calcite and quartz and authigenic illite formation during burial diagenesis.

The large variations in K concentrations depicted in Figure 1 may contradict a control of illite precipitation. However, previous equilibrium calculations describing phase relations in North Sea Jurassic sandstones and comparisons with formation water chemistry (Aagaard et al., 1992, Bjørlykke et al., 1995), have shown illite to be stable. A noteworthy feature of subsiding basins such as the North Sea, is the ability to extract K for illite growth, while uplifted basins often contain excess of K (Bjørlykke et al., 1995).

4 CARBON DIOXIDE TREND

Smith and Ehrenberg (1989) presented data on partial pressure of carbon dioxide from hydrocarbon reservoirs both on the Norwegian shelf and the Gulf Coast. These basins exhibited similar trends with exponential increase towards depth. Smith and Ehrenberg argued that the sources of CO_2 were inorganic reactions between feldspars, clay minerals and carbonates. An inorganic origin was also advocated by Hutcheon and Abercrombie (1989) to explain high CO_2 in the Ventura field. Carothers and Kharaka (1980) on the other hand, interpreted carbon isotopic composition of bicarbonate in the

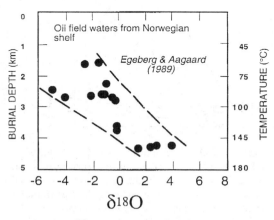

Figure 2. $\delta^{18}O$ composition of formation waters from oil fields on the Norwegian shelf, depicting a

inorganic sources. Whatever the sources, the transport mechanism of CO_2 in the basins is also critical. In this regard, the high solubility in oil may thus be an important factor.

It can also be concluded that the high concentrations of CO_2 in sedimentary basins constrain to a large extent diagenetic reactions during burial. A comparison with P_{CO2} from hydrothermal systems on Iceland and New Zealand (Figure 3) reveal that P_{CO2} in oil prone sedimentary basins are 2-3 orders of magnitude higher for the same temperature.

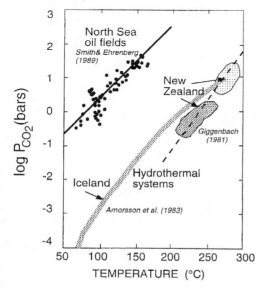

Figure 3. Trend of partial pressure of carbon dioxide found in hydrocarbon reservoirs on the Norwegian shelf (Smith and Ehrenberg, 1989) compared with data from hydrothermal systems (Giggenbach, 1981; Arnorsson et al., 1983).

5 CONCLUDING REMARKS

Hydrocarbon reservoir waters from the Norwegian shelf are charaterized by:
- Having a brine component of a modified evaporation residue and a dilute endmember (which is meteoric in oil fields offshore Norway)
- Oxygen isotopic trend towards depth consistent with the diagenetic reactions recorded in the sediments
- High P_{CO2} versus temperature trend far exceeding those of hydrothermal systems in Iceland and New Zealand.

These features are also common to other sedimentary basins, and set formation waters in subsiding, oil prone basins apart from most other water rocks reaction environments.

6 REFERENCES

Aagaard, P., Egeberg, P.K. and Jahren, J.S., 1992, North Sea clastic diagenesis and formation water constraints: In Kharaka & Maest, Water Rock Interaction WRI-7, Park City, p. 1147-1152.

Arnorsson, S., Gunnlaugsson, E., and Svavarsson, H., 1983, The chemistry of geothermal waters in Iceland. II Mineral equilibria and independent varables controlling water compositions: Geochim. Cosmochim. Acta, v. 47, p. 547-566.

Bjørlykke, K., Nedkvitne, T., Ramm, M., and Saigal, G.C., 1992, Diagenetic processes in the Brent Group (Middle Jurassic) reservoirs of the North Sea - An overview: in Morton, A.C., Hazeldine, R.S., Giles, M.R. and Brown, S. (eds), Geology of the Brent Group, Geol. Soc. Spec. Publ. No 61, p. 263-287.

Bjørlykke, K., Aagaard, P., Egeberg, P.K., and Simmons, S.P., 1995, Geochemical constraints from formation water analyses from the North Sea and the Gulf Coast Basins on quartz, feldspar and illite precipitation in reservoir rocks: in Cubitt, J.M. and England, W.A (eds), The Geochemistry of Reservoirs, Geol Soc. Spec.Publ. No 86, p. 33-50.

Carothers, W.W. and Kharaka, Y.K., 1980, Stable carbon isotopes of HCO3- in oil-field waters - implications for the origin of CO_2: Geochim. Cosmochim. Acta, v. 44, p. 323-332.

Carpenter, A.B.1978, Origin and chemical evolution of brines in sedimentary basins: Oklahoma Geological Survey Circular 79, p. 60-77.

Dickinson, W.W., 1986, Modeling of Oxygen Isotopes in Formation Waters - Review of the Principles with Examples: in Hitchon, B., Bachu, S., and Sauveplane, C.M., 1986, Hydrogology of Sedimentary Basins: Application to Exploration and Exploitation, Proc. 3rd Can./Amer. Conf. on Hydrogeology, p. 221-229.

Egeberg, P.K. and Aagaard, P., 1989, Origin and evolution of formation waters from oil fields on the Norwegian Shelf: Applied Geochemistry, v. 4, p. 131-142.

Giggenbach, W.F., 1981, Geothermal mineral equilibria: Geochim. Cosmochim. Acta, v. 45, p. 393-410.

Hutcheon, I. and Abercrombie, H.J., 1989, The role of silicate hydrolysis in the origin of CO_2 in sedimentary basins: in Miles, D.L. (ed) Water Rock Interaction WRI-6, Malvern, p. 321-324.

James, A.T., 1990, Correlation of Reservoired Gases Using the Carbon Isotopic Composition of Wet Gas Components: AAPG Bull., v. 74, p. 1441-1458.

Land, L.S. and Macpherson, G.L., 1992, Origin of Saline Formation Waters, Cenozoic Section, Gulf of Mexico Sedimentary Basin: AAPG Bull., v. 76, p. 1344-1362.

Pierre, C., Ortlieb, L., and Person, A., 1984, Supratidal evaporite dolomite at Ojo de Liebre Lagoon: mineralogical and isotopic arguments for primary crystallization: J. Sed. Petr., v.54, p. 1049-1061.

Smith, J.T, and Ehrenberg, S.N., 1989, Correlation of carbon dioxide abundance with temperature in clastic hydrocarbon reservoirs: relationship to inorganic chemical equilibrium: Marine and Petroleum Geology, v. 6, p. 129-135.

Warren, E.A. and Smalley, P.C., 1993, The chemical composition of North Sea formation waters: a review of their heterogeneity and potential applications: in Parker, J.R.(ed) Petroleum Geology of Northwest Europe, Proc. 4th Conf., p. 1347-1352.

Warren, E.A. and Smalley, P.C., 1994, North Sea Formation Waters Atlas; Geol. Soc. Memoir No 15.

Worden, R.H., 1996, Controls on halogen concentrations in sedimentary formation waters: Mineralogical Magazine, v. 60, p. 259-274.

REE distribution in fine-grained sediments from the Portuguese Atlantic shelf

M.F.Araújo & M.A.Gouveia
Department of Chemistry, ITN, Sacavém, Portugal

ABSTRACT: The importance of the trace element geochemistry, particularly the REE distribution has been recognized to be an important tool to characterize the origin of sediments and the corresponding sedimentation processes. REE have been determined in sediments collected at two fine silt-clayed deposits located at the Northwestern Portuguese shelf and at the main estuaries that drain the adjacent coastal area. We are discussing the shelf sediments to assess terrigeneous input pattern of REE. The shale-normalized patterns seem to indicate distinct pathways of the different river borne materials when related with the fine sediments deposited at the shelf. Therefore, while for the system Douro estuary/silt-clayed deposit, these patterns are identical, in the case of the system Minho estuary/silt-clayed deposit, a relation between the river borne materials and the deposited sediments can not be established.

INTRODUCTION

The Northwestern Portuguese margin is characterized by a strong fluvial input, since five major rivers drain to this region. As a consequence of the high energy environment, the cover of the shelf is essentially constituted by sands. However, at the mid shelf, at about a depth of 100 m, two important fine grained deposits lay off the main estuaries on the region: Minho and Douro (Fig. 1). The "Douro" deposit is limited by Cretaceous and Paleocene outcrops on the west (Musellec, 1974), while the "Minho" deposit lays into an "open" sea without relieves.

The general circulation over the West Iberian continental margin is dominated by a poleward flow all over the year. In summer, this flow can be attenuated by a southward current in the upper 50-100m (Haynes & Barton, 1990).

Previously published studies have been pointing out a probable terrigeneous origin of these fine grained sedimentary deposits (Araújo et al., 1997; Drago et al., 1994). Radiocarbon dating and accumulation rates, determined respectively by the ^{14}C and ^{210}Pb techniques of sediment cores collected at the Douro deposit, show that this deposit is a recent and still active formation (Carvalho & Ramos,

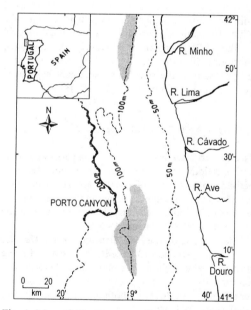

Fig. 1. Map of Northwestern Portuguese shelf with the two fine grained sedimentary deposits.

1989; Drago et al, 1994). The "Minho" formation is an older deposit and lower accumulation rates have been measured.

In this paper we make an attempt to infer the provenance of the fine grained sedimentary deposits by comparing the REE shale-normalized patterns of marine surficial sediments collected in both deposits and at the main estuaries that drain the adjacent coastal zone.

SAMPLING AND METHODS

Twelve surficial sediments were collected at the two fine grained sedimentary deposits (nine samples at the "Douro" deposit and three at the "Minho") with a "Smith-McIntyre" grab during the scientific cruise GEOMAR92 (November 1992) organized by the Portuguese Hydrographic Institute. Sediments were also collected with a grab, at the estuaries during the scientific campaign SEDIMINHO I/93 (February 1993). Subsamples were taken from the central part of the grab to avoid contaminations, stored in plastic boxes and deep frozen prior to being dried. About 50 g of this material was ground and passed through a plastic sieve to obtain a grain size <64 μm. Fractions between 0.2 and 0.3 g of the dried homogenized sediment were weighed in polyethylene vials for INAA. Irradiations were carried out in the core grid of the RPI reactor (Sacavém). A complete description of our procedure, as well as the accuracy and precision of the overall measurement have been given previously (Gouveia et al., 1992).

RESULTS AND DISCUSSION

Usually to facilitate data comparison, REE concentrations are represented using normalization methods to a chosen reference material. In the case of sediments, it has been a common practice, to use a set of REE concentrations in an average shale composition (P. Henderson, 1984).

Thus, results on the REE distribution (La, Ce, Nd, Sm, Eu, Tb, Yb and Lu) have been normalized using the set of values of the shale abundance (Haskins and Haskins, 1966), used in many studies of the marine geochemistry of REE in both sediment and water (Sholkovitz, 1990).

The shale normalized REE distribution of sediments from the "Minho" deposit is represented in Fig. 2. The range of variation of the shale normalized patterns of the REE distribution for nine sediment samples collected from the "Douro" deposit is shown in Fig. 3.

It is noteworthy the uniformity of the patterns, since no significant variabilities of the REE distributions could be detected for the two sedimentary environments. Our results are also highly comparable to values previously published by Sholkovitz (1990) for shelf and slope sediments with variable origins, indicating a constant REE distribution in sediments, as suggested by Taylor & McLennan, 1985.

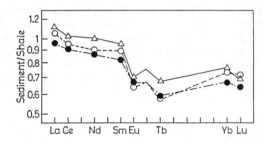

Fig. 2. Shale normalized REE distribution in surficial sediments from the "Minho" deposit.

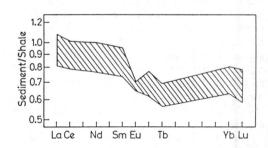

Fig. 3. Shale normalized REE distribution in surficial sediments from the "Douro" deposit.

In general, REE patterns are uniformly flat and similar, and show a negative Eu anomaly throughout each sedimentary deposit. According to Sholkovitz (1990), marine sediments are similar to shales, in spite of a preferential and more rapid removal of LREE from the oceans. This can be explained by the increasing stability of the solvated heavier HREE, forming stronger complexes with carbonate ions, and consequently being retained in solution.

Although the REE contents in both deposits can be highly comparable to shales, fine sediments deposited at the northern formation are apparently slightly enriched in LREE.

If we assume that these deposits have their origin in the river borne material, it should be expect that the geochemical patterns of the fluvial sediments could be recognized in the sedimentary deposits.

Since the REE abundances are highly grain size dependent (Taylor & McLennan, 1985) we have tried to compare the shelf sediments with the finer grain sized estuarine sediments. In Fig. 4 and Fig.5 are represented the shale normalized patterns from sediments from Minho and Douro estuaries.

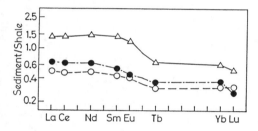

Fig. 4. Shale normalized REE distribution in surficial sediments from Minho estuary.

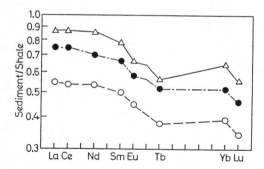

Fig. 5. Shale normalized REE distribution in surficial sediments from Douro estuary.

The total REE abundances for the Douro sediments are lower (Fig. 5), although they present a pattern similar to the shelf sediments (Fig.3). It should be noticed that these are fine surficial sediments, and a more direct relation would be expected with the particulated suspended matter. Shale-normalized patterns of the sediments from Minho estuary exhibit large variations on the total REE and a clear enrichment on the LREE. However, Minho estuarine sediments are in general coarser than all the other sediments.

In order to understand the REE fractionation, we have calculated the $(La/Yb)_n$ ratios for the fine sediments collected at sedimentary deposits and for sediments collected in the estuaries. The four fine grained sediment samples collected at Douro estuary present similar distribution in LREE and HREE, with $(La/Yb)_n$ ranging from 1.36 up to 1.43. They also present striking similarities with the nine samples collected all over the "Douro" sedimentary deposit, with $(La/Yb)_n$ ranging from 1.23 up to 1.32. The same comparison made for sediments collected at the "Minho" deposit and at the Minho estuary showed that although all over the deposit the ratio $(La/Yb)_n$ is quite constant, varying between 1.43 and 1.47, at the Minho estuary is highly variable.

Generally, all studied sediments present an enrichment in LREE when compared to shales. However, previous work (Elderfield et al.,1990) have shown that nearshore sediments and fluvial suspended matter, often exhibit REE patterns that are also LREE enriched.

In summary, the shale normalized patterns of the sedimentary formations are remarkable similar exhibiting REE patterns comparable to typical shales, although LREE enriched. This enrichment has been found in sediments with rather different origins and it has been attributed to the finer grain size of the sediments. There is a close relation between the Douro estuary/deposit system, although we did not found a relative enrichment in LREE in the marine environment. However estuarine sediments have been collected in a mixing area, with high salinity values. We could not find a direct relation between sediments from the Minho deposit/estuary, due to the large variabilities for the shale normalized patterns (and grain size) of the estuarine sediments.

ACKNOWLEDGMENTS

The authors thank the Captain and crew of the *R.V. Almeida Carvalho* for their assistance during the scientific cruise GEOMAR92, organized by the Portuguese Hydrographic Institute, during which sediments were collected. We are indebted to our colleagues from DISEPLA group that have participated in the expeditions GEOMAR92 and

SEDIMINHO I/93 for all the cooperation. Thanks are also due to the staff of the Reactor Department for their assistance with the neutron irradiations. Part of this work was funded by a JNICT research contract.

REFERENCES

Araújo M.F., J.-M. Jouanneau P.Valério, P.Paiva, A.Gouveia, O.Weber & J.M.A. Dias 1997. Geochemical signatures of the main West Atlantique European estuaries in the mud fields of the shelf. Exemples of Tagus, Douro, Minho and Gironde. Submitted to *Oceanologica Acta*.

Carvalho, F.P. & L. A. Ramos 1989. Lead 210 chronology in marine sediments from the Northern continental margin of Portugal. *Actas do II Congresso sobre a Qualidade do Ambiente*, Lisboa, A143-A151.

Drago, T., J.-M. Jouanneau, J.M.A. Dias, R. Prud'Homme, S.A. Kuehl & A.M.M. Soares 1994. La vasière Ouest-Douro et le piegeage des sédiments estuariens récents, *GAIA*, 9, 53-58.

Elderfield, H., R. Upstill-Goddard & E.R. Sholkovitz 1990. The rare earth elements in rivers, estuaries and coastal seas and their significance to the composition of ocean water. *Geochim. Cosmochim. Acta,* 54, 971-991.

Gouveia, M.A., M.I. Prudêncio, I. Morgado & J.M.P. Cabral 1992. New data on the GSJ reference rocks JB-1a and JG-1a by instrumental neutron activation analysis. *J.Radioanal. Chem.*, 158 (1), 115-120.

Haskin M.A. & L.A. Haskin 1966. Rare earths in European shales: a redetermination. *Science*, 154:507-509.

Haynes, R. & E.D. Barton 1990. A poleward flow along the Atlantic coast of the Iberian Peninsula. *J. Geophys. Res.,* 95:679-691.

Henderson, P. 1984. *Rare Earth Element Geochemistry*. P. Henderson (ed.) ch.1, Amsterdam (The Netherlands): Elsevier.

Musellec, P. 1974. Géologie du plateau continental portugais au Nord du Cap Carvoeiro. Thèse 3ème Cycle, Univ. Rennes, 74p.

Sholkovitz, E.R. 1990. Rare-earth elements in marine sediments and geochemical standards. *Chem. Geology*, 88:333-347.

Taylor, S.R. & S.M. McLennan 1985. The Continental Crust: its Composition and Evolution. Blackwell Scientific Publications.

Water-rock reactions in evaporite basins: Their role in the formation of potash deposits

C.Ayora, D.I.Cendón & C.Taberner – *Institut de Ciències de la Terra, CSIC, Barcelona, Spain*
I.Fanlo – *Cristalografía y Mineralogía, Departamento Ciències de la Tierra, Universidad de Zaragoza, Spain*
J.García-Veigas – *Serveis Científico-Tècnics, Universidad de Barcelona, Spain*
J.J.Pueyo – *Departamento de Geoquímica, Universidad de Barcelona, Spain*

ABSTRACT: The chemical evolution of several potash-forming and potash-free evaporite basins has been reconstructed using mineral associations, solute content analysed in fluid inclusions and numerical simulation of evaporation scenarios. The sulphate depletion of brine is responsible for the type of potash deposit, K-sulphates or sylvite. Both, dolomitization and the addition of a Ca-rich solution into the basin account for the ratios between the major solutes analysed. Early diagenetic replacement of anhydrite by polyhalite could explain the variable and depleted amount of potassium found in fluid inclusions. These processes took place to different extents in the basins of the same age, and within the same basin, and therefore, cannot be attributed to global changes in the oceans composition.

1 INTRODUCTION

Traditionally the origin of potash deposits (chlorides and sulphates) has been controversial. The last stages of evaporation of present day marine water lead to precipitation of magnesium and potassium sulphates and carnallite, whereas sylvite never precipitates. However, most of the potash deposits in the Phanerozoic sedimentary record consist of sylvite and carnallite. Two main hypotheses were proposed to explain this contradiction: 1) the sylvite is not formed from seawater, and 2) the ocean composition was different to that of the present day. A diagenetic origin for sylvite has also been proposed (Braitsch, 1971). The presence of sedimentary textures has unequivocally established the syn-depositional origin for the sylvite (Lowenstein and Spencer, 1990). A sulphate depletion in the evaporitic brine is needed to explain the formation of sedimentary sylvite. The scarcity of sulphate causes the lack of precipitation of minerals, such as polyhalite, which consume potassium, and prevent the brine reaching sylvite saturation. Recently, Kovalevich (1991) and Hardie (1996) proposed that the concentration of major solutes in the ocean has experienced secular variation during the last 600 My due to fluctuations in the hydrothermal convective flux of water in the ocean ridges. The interaction of water with basic igneous rocks would consume Mg and Na, and supply Ca and K to the ocean. The increase of Ca would account for the precipitation of gypsum/anhydrite and the depletion of oceanic sulphate.

This hypothesis assumes that the solute concentration is not significantly modified by water-rock interactions taking place in the basin during evaporation. In order to investigate the existence of such processes we have reconstructed the chemical evolution of several evaporite basins, based on: 1) mineralogy and petrography of the evaporite sequences, and 2) solute content of brines trapped in halite fluid inclusions.

2 METHODOLOGY

Five marine evaporite sequences have been studied: The Lorraine basin (France), forms part of the larger German Basin that formed during the Upper Triassic. The evaporitic sequence studied is about 75 m thick, and consists of halite, sulphates and carbonates, potash salts are not present. The Lorca basin (Spain), was a marginal basin of the western Mediterranean during the Messinian. The lower part of this sequence is formed by halite with minor sulphate. No potash minerals are present. The Central Sicily basin (Italy), was part of the central Mediterranean Messinian basins. The marine sequence has a salt member with kainite beds in the middle of it. The South Pyrenean foreland basin (Spain), was an elongated east-west trough. Two main depocenters developed during the Upper Eocene: the Catalonian depocenter (Suria), and the Navarra depocenter (Subiza). The marine sequence contains sulphates, halite and potash facies (sylvite and carnallite).

To ensure the primary origin of the brine, several primary fluid inclusions, were selected for microanalysis. The electrolyte composition of the fluid inclusions was determined by direct X-ray microanalysis of frozen fluid inclusions (Cryo-SEM-EDS), according to the methodology described and improved on in Ayora et al. (1994a). This method allows the quantitative analyses of a representative group of inclusions in the same crystal with sizes larger than 15 μm. In each fluid inclusion the Na, Mg, K, Cl and S (expressed as SO_4) contents were measured. Though values of precision error varied for each element and analytical process, they were always lower than 10%. The consistency of the results was checked against the charge balance and the NaCl saturation index.

The brines evolution and the amount of mineral formed in the five basins were calculated using the conceptual model of a hydrologically open basin. The main principle supporting the calculations is the conservation of the volume of the brine and the conservation of the mass of each solute. The basin is assumed to have evolved with a constant volume of water. Thus, the evaporated volume of water was replaced by an equal volume of inflow water, and complete mixing was assumed to have occurred. The inflow water may be either seawater, continental water or mixtures of both depending on the geometry of the basin. See Ayora et al. (1994 a,b) for a detailed explanation. Mineral precipitation was controlled by thermodynamic equilibrium between the minerals and the brine which has a constant temperature of 25°C. The thermodynamic calculations are based on the Pitzer ion-ion interaction model and use the parameters given by Harvie et al. (1984).

3 RESULTS AND DISCUSSION

3.1 Sulphate depletion

In our first estimation a brine was simulated to evaporate with present day seawater as the only recharge into the system. From the lack of Mg-minerals it follows that the amount of Mg can be considered quite constant and used as a reasonable indicator of the degree of evaporation. The prediction did not match the analysed solute content.

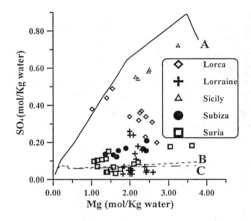

Figure 1.- Calculated evolution of SO_4 content in respect to Mg (mol/Kg water), for A : evaporation of present day seawater, B : dolomitization modelled as substitution of 0.018 mol/Kg water Mg by Ca, C: Additional inflow of 1mol/Kg of $CaCl_2$ solution. The mean SO_4 content in fluid inclusions is also represented for each basin (see legend).

The most relevant feature is that the SO_4 content recorded in the fluid inclusions is variable and much lower than predicted (Figure 1). The predicted values for Cl were lower than observed, whereas those of Na were higher (Figure 2). Two main hypotheses have been suggested to explain the depleted values of SO_4 in the analyses: 1) bacterial sulphate reduction, 2) the addition of Ca into the basin and removal of SO_4 by precipitating Ca-sulphates. Simulations of bacterial reduction of SO_4 to H_2S result in a different solute distribution to those observed in the fluid inclusions (Ayora et al., 1994b). Moreover, the $\delta^{18}O$ and $\delta^{34}S$ of the sulphates interbedded in the sequences are fairly constant, and don't match the ratio of isotopic enrichment expected from sulphate reduction.

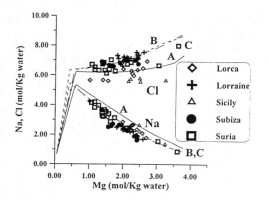

Figure 2.- Calculated evolution of Na and Cl content, with respect to Mg (mol/Kg water) for A, B and C (see Figure 1).

Alternatively, two processes leading to the addition of Ca from a source external to the basin were considered: 1) dolomitization, modelled by means of the substitution of Mg by Ca in the recharge; and 2) the additional inflow of a hypothetical solution 1 mol/kg of $CaCl_2$, which represents a diagenetic or hydrothermal source. With respect to pure seawater evaporation, the excess of Ca causes the precipitation of Ca-sulphates and a depletion in the amount of dissolved sulphate. In the case of dolomitization, the consumption of Mg makes the plot of the rest of the solutes shift to the left in Fig . 2. In the case of the addition of $CaCl_2$, the increase in Cl in the basin enhances the formation of halite and a corresponding decrease in Na occurs. From the chemical point of view, both processes lead to an identical result: the perfect match between the predicted and analysed proportions of solutes (Figure 2).

To match the solute content recorded in fluid inclusions, a different amount of dolomitization or $CaCl_2$-brine inflow is needed for each studied basin. Taking into account the amount of evaporite minerals accumulated in each sequence, a number of 1 m deep basins are needed for evaporation, and a total amount of Ca-excess is broadly calculated (Table 1). The mass of dolomite involved should be taken as a base for a further, more detailed evaluation in the field.

Table 1. Characteristics of the studied basins relate to the amount of sulphate depletion.

	thick. sequen. studied (m)	$QCaCl_2$/ Qinflow	Ca excess (x10^4 molCa /m^2)	dolomite required (m^3/m^2 basin)
Lorca	95	0.016-14	10	6.43
Lorrain	75	0.019-12	11.4	7.33
Sicily	160	0.005	6.4	4.12
Subiza	20	0.015	1.65	1.06
Suria	90	0.018-15	12.96	8.34

The solute content recorded in fluid inclusions is consistent with the mineral associations observed. In all cases the mineral sequence consists of halite and minor anhydrite. The depletion of sulphate explains the undersaturation of brine in glauberite and polyhalite, and the lack of these minerals as primary phases. The solute proportion is closer to seawater evaporation in the case of Sicily, where Mg-K-sulphates consistently form most of the potash deposits. The degree of sulphate depletion is significant in the southpyrenean basin where the mined seams consist of sylvite.

3.2 Variation in potassium

The K content analysed in fluid inclusions is variable within the same basin, and is either higher or lower than predicted by the evaporation of present day seawater (Figure 3). Evaporation with an additional $CaCl_2$-rich inflow not only delays the precipitation of sylvite, but follows the same rate as the seawater plot (Figure 3). The calculations assuming dolomitization, however, involve a decrease in Mg, and a relative increase in the rate of K accumulation (Figure 3).

As far as we know no realistic process leading to a K increase in the brine, with respect to that predicted from evaporation, can be suggested. Two processes, however, may be suggested as responsible for the K decrease in the brine: 1) the K adsorption by clay minerals, and 2) the reaction of the brine with previously formed Ca-sulphate to form polyhalite. All the calculations gave K concentrations sorbed in clays at least one order of magnitude lower than the variations recorded in fluid inclusions.

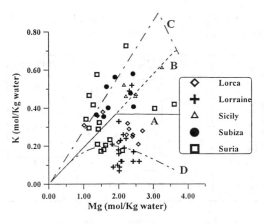

Figure 3.- Calculated evolution of K content with respect to Mg (mol/Kg water) for A,B and C (see Figure 1), D: Replacement of 5% of Ca-sulphates to form polyhalite.

In all the basins studied polyhalite was observed to grow, replacing previous anhydrite aggregates along the boundaries of halite grains. This process takes place according to the reaction:
$4CaSO_4 + 2K^+ + Mg^{2+} + 2H_2O = K_2MgCa_2(SO_4)_4 \cdot 2H_2O + 2Ca^{2+}$

The replacement of just 5% of the mass of anhydrite formed within the halite is required to predict the lower K concentrations analysed (Figure 3). Although the model calculates that replacement is continuous, the variable K content indicates that this is not so.

The basins with lower and more variable K contents, such as Lorraine, and Lorca, are those where textures show a shallow brine depth. It is expected that the smaller the brine's volume the stronger the modifications due to water-rock reactions will be. Following the calculations, the rest of the basins could reach potash mineral saturation if the basin's restriction causes the brine to reach evaporation stages that are advanced enough.

4 CONCLUSIONS

The depletion in sulphate analysed in fluid inclusions is consistent with the mineralogy of the potash deposits included in the sequence. The depletion in sulphate, and the relative proportion of the major solutes, can be explained either by dolomitization or by the addition of a Ca-rich brine into the basin. Dolomitization predicts K values close to the higher values analysed. The K depletion may be explained using early diagenetic reactions of the brine with previously formed anhydrite giving polyhalite. As no process leading to an increase in the K content of the brine is suggested, dolomitization appears to better explain the K contents of fluid inclusions.

Instead of secular variations in the ocean's composition, sulphate depletion is attributed to processes of water-rock interaction which took place within the evaporite basin. This is supported by these observations: 1) two basins of the same age display very different amounts of sulphate depletion: Messinian of Sicily and Lorca; 2) two sub-basins from the same basin show different amounts of sulphate depletion: Suria and Subiza; and 3) the same basin can experience different degrees of sulphate depletion during its evolution, as shown by the initial and advanced stages of Lorca basin. The variable K content, even for samples closely located, confirms the importance of water-rock interactions within the basin.

ACKNOWLEDGEMENTS

This paper has been supported by the Spanish Government DGCYT-PB93-0165

REFERENCES

Ayora C., García-Veigas J., and Pueyo J.J. (1994a) *Geochim. Cosmochim. Acta*, 58: 43-55.
Ayora C., García-Veigas J., and Pueyo J.J. (1994b) *Geochim. Cosmochim. Acta*, 58: 3379-3394.
Braitsch O. (1971) *Salt deposits, their origin and composition*, Springer-Verlag, 232 pp.
Hardie L.A. (1996) *Geology,* 24: 279-283.
Kovalevich V.M. (1990) *Geochemistry International* 26: 20-27
Lowenstein T.K. and Spencer R. (1990) *Amer. J. Sci.* 290: 1-42

The origin of the Canadian Shield brines: Freezing or evaporation of seawater?

D.J. Bottomley – *Atomic Energy Control Board, Ottawa, Ont., Canada*
A. Katz & A. Starinsky – *Institute of Earth Sciences, The Hebrew University of Jerusalem, Israel*
L.H. Chan – *Department of Geology and Geophysics, Louisiana State University, Baton Rouge, La., USA*
M. Douglas & I.D. Clark – *Ottawa-Carleton Geoscience Center, University of Ottawa, Ont., Canada*
K.G. Raven – *Intera Consultants Ltd, Ottawa, Ont., Canada*
D.C. Gregoire – *Geological Survey of Canada, Ottawa, Ont., Canada*

ABSTRACT: New chemical and isotopic (δ^6Li and ^{87}Sr/^{86}Sr) data of deep seated calcium chloride brines from Yellowknife, N.W.T., is reported and discussed.

The δ^6Li and Li/Br ratios constrain the origin of the brines to seawater. The composition and salinity of the brines can be explained by freezing or evaporation of seawater, followed by brine-mineral interaction. In both cases, the Ca-chloridic nature was dictated mainly by dolomitization of marine sediments. In the evaporitic scenario, SO$_4$ was removed as gypsum, and the low Na/Cl ratio (0.28) was attained by halite precipitation followed by feldspar albitization. A cryogenic mineral precipitation sequence (mirabilite→hydrohalite), and subsequent sericitization, would lead to identical results. While direct geological evidence to resolve between the models is still missing, cryogenesis is favored by the Na-Cl-Br relationships.

The elevated ^{87}Sr/^{86}Sr ratios in the brines (up to 0.7147) are explained by albitization or sericitization in the evaporitic and cryogenic models, respectively.

1 INTRODUCTION

Hypersaline brines are common in deep mines across the Canadian Precambrian Shield, but their origin has remained enigmatic. Their salinities may reach 300g/L and they are calcium-chloridic (Ca>(SO$_4$+ HCO$_3$), in equivalents). Some of the most concentrated brines occur in the Miramar Con gold mine, Yellowknife volcanic belt, in the southwestern part of the Archean Slave structural province, N.W.T., Canada. The mine is situated in shear zones in 2.6-2.7 Ga old metabasalts. Certain chemical and isotopic characteristics of the Yellowknife brines (e.g. their non-marine radiogenic ^{87}Sr/^{86}Sr ratio) led earlier investigators to propose that most, if not all the solutes in these liquids were inherited from the crystalline basement minerals via brine-rock interaction (e.g. Frape et al., 1984, Frape and Fritz, 1987, Kamineni et al., 1992).

Recently, both Herut et al. (1990) and Bottomley et al. (1994) proposed seawater as a possible source for the major brine solutes, but differed about the concentration mechanism which gave birth to the original brines. Whereas Herut and co-workers (1990) demonstrated that freezing of seawater plus some modification of the brine by water-rock interaction may explain both the salinity and chemical composition of the brines including the radiogenic Sr, Bottomley and his group advocated an evaporitic concentration process, rather than seawater freezing, to achieve the same.

Lately, we had the opportunity to re-sample and analyze the Miramar Con mine at depths up to 1600m and to gather new chemical and isotopic information relevant to the origin of the apparently most saline "end-member" of these liquids studied sofar. Moreover, the development of precise Li-isotope measurement techniques (Chan, 1987) is providing now a new tool to constrain the source of the solutes in the Canadian fluids.

2 RESULTS AND DISCUSSION

Chemical analyses of the nine most concentrated Yellowknife brines (>100g Cl/L) are given in Table 1 along with Sr and Li isotopic ratios. The solute/chloride concentration ratios display the following enrichment factors with respect to seawater: Mg (0.023), K (0.076), SO$_4$ (0.08), Na (0.32), Li (2.56), Br (2.65), Sr (18.3) and Ca (19.8).

The two ion pairs, Sr-Ca and Li-Br, show almost identical enrichments, and their ratios are marine. SO_4 and Mg are almost complete depleted, and Na is significantly so, whereas Ca is extremely enriched.

The $^{87}Sr/^{86}Sr$ ratios achieve a maximum of 0.71475. The δ^6Li values are very similar to that of modern seawater (-32.3‰) or slightly heavier (up to about -40‰). The $\delta^{34}S$ values (not shown in the Table) are similar to that of modern seawater or slightly heavier.

3 THE MARINE ORIGIN OF THE BRINES

The new δ^6Li data provide, for the first time, direct evidence that the Canadian Shield brines are marine in origin. This is not only indicated by the striking similarity between the respective δ^6Li values, but also by the absence of crustal rocks, let alone the local metabasalts, which could have donated the appropriate Li to the brines. The Li/Br ratios in the brines strongly support the same conclusion.

4 SEAWATER TO BRINE TRANSFORMATION

The seawater→brine transformation could take place either by evaporation or freezing. The resolution between the two options is hampered by the lack of direct geological evidence. Stratigraphic considerations allow to extend the now-eroded evaporitic Devonian record several hundred kilometers east of the present day erosion contact (Qing, 1991). Large-scale freezing of seawater is common in subarctic regions off the Canadian coast and was the more so during the ice ages. Its potential for brine production has been discussed and shown by Bein and Arad (1992). Evaporation of seawater at Yellowknife during the Devonian would simplify the problem of long distance brine migration involved in a cryogenic origin, but raises the question how did such concentrated brines survive, undiluted, for hundreds of millions years within intensely fractured zones. Also, whereas the Yellowknife brines are chemically and isotopically similar to other deep-seated fluids found across the crystalline Precambrian shield (Herut et al., 1990), they are clearly distinct from ancient evaporitic fluids and formation waters in the Alberta and Michigan basins (e.g. Wilson and Long, 1993).

5 BRINE EVOLUTION AND INTERACTION

Mass balance calculations coupled with marine-evaporitic mineral sequence information (Raab, 1996) and with cryogenic mineral paragenesis (Nelson and Thompson, 1954; Herut et al., 1990) show both processes to be feasible. In the cryogenic basin, SO_4 is precipitated along with an equivalent mass of Na as mirabilite ($Na_2SO_4 \cdot 10H_2O$), followed by hydrohalite ($NaCl \cdot 2H_2O$) until the characteristic low Na/Cl and Br/Cl ratios of these brines (≈ 0.28 and 4.25×10^{-3}, respectively) are attained. Subsequently, the dense brine infiltrates through the basinal sediments and dolomitizes $CaCO_3$ minerals (aragonite and calcite) therein. This step is the main Mg↔Ca exchange reaction which turns the brine into a typical calcium-chloridic solution. Finally, during migration through the silicate basement additional Mg↔Ca, as well as K↔Ca exchange take place between the brine and host rocks, in the form of chloritization and sericitization. Thereby, the original marine $^{87}Sr/^{86}Sr$ ratios are elevated to their present values.

Table 1. Chemical composition and isotope data for hypersaline (>100 g Cl/L) Yellowknife brines. Ion concentrations are in meq/kg solution unless otherwise specified.

Sample	2931	4080	3553	3425	4010	4012	3424	3554	4011
Mg	15.7	18.8	15.5	15.8	15.2	13.6	19.8	21.8	18.8
Sr	16.3	16.8	22.7	22.5	23.7	23.8	24.3	25.6	26.6
Ca	2108	2144	2940	3032	3038	3069	3087	3155	3266
Na	848	839	1103	1097	1108	1112	1136	1173	1175
Li (μeq/kg)	360	351	530	506	460	477	437	476	441
K	4.0	3.9	5.6	5.5	6.1	5.9	5.6	6.2	6.4
SO4	43.4	25.5	0.2	53.2	0.0	0.0	53.0	0.2	0.0
Cl	2989	3043	3817	4069	4095	4104	4170	4173	4372
Br	11.8	12.1	16.3	16.9	16.1	16.3	16.8	17.5	.317
δ^6Li	-36.3		-35.4	-32.3			-34.7	-32.1	
$^{87}Sr/^{86}Sr$	0.714383		0.714745	0.714740			0.714305	0.714233	

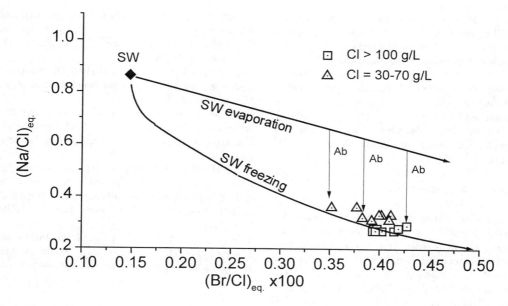

Fig.1 The position of the Yellowknife brines on a seawater Na/Cl-Br/Cl evolution diagram. "Ab" arrows denote albitization of plagioclase.

The major stages in the marine evaporitic evolution model include concentration of the brine beyond halite saturation until the present Br/Cl ratio of $\approx 4.25 \times 10^{-3}$ is obtained. Dolomitization of $CaCO_3$, accompanied by the removal of SO_4 as gypsum, enriches the brine in Ca, and depletes it in Mg. The lowering of the Na/Cl ratio from that achieved during evaporation (≈ 0.58) to that measured in the brine (≈ 0.28) is brought about by plagioclase albitization, which complements also the Ca balance therein. Here again, the elevated $^{87}Sr/^{86}Sr$ ratios are provided by brine-silicate interaction.
In the two models, dolomitization of marine $CaCO_3$ is a major source of Ca and sink of Mg. Hence both are compatible with the Sr/Ca ratio of the Yellowknife brines (which happens to be similar to that of modern seawater) as well as with their Li/Br ratio and δ^6Li values. Neither Li nor Br are involved in this reaction to any significant extent.

6 THE Na-Cl-Br RATIOS AND CRYOGENESIS

Herut et al. (1990) have demonstrated that the evaporation and freezing paths of seawater can be resolved on a Na/Cl-Br/Cl diagram. Figure 1 represents such a plot for the data in Table 1. All the nine brines (dot-centered rectangles) plot almost exactly along the freezing path. Although albitization could have produced similar results, the fact that, within experimental error, all of these brines reached the freezing line and did not cross it is striking. The same Figure contains eight additional samples (dot centered triangles) which have been diluted at some later stage after infiltration. Even these data points are located close to the freezing line and indicate, that no NaCl mineral was in contact with the brines during their dilution.

7 CONCLUSIONS

1. δ^6Li, Li/Br ratios and mass balance calculations demonstrate, that the Canadian Precambrian shield brines occurring at Yellowknife, N.W.T., are marine in origin.
2. Seawater freezing and seawater evaporation could equally well account for the concentration of the brines.
3. In both models, the main Ca↔Mg exchange reaction is dolomitization of marine $CaCO_3$. The main differences between the evaporitic and cryogenic evolution paths are related to the depletion

of Na and SO$_4$. In cryogenesis, this is achieved by removal of mirabilite and hydrohalite, as contrasted to the removal of halite and gypsum, followed by subsurface albitization, in the evaporitic path.

4. Although not ruling-out a marine evaporitic evolution, the position of the brines on a Na/Cl-Br/Cl diagram lend strong support in favor of the cryogenic model.

5. It is likely that comparable brines ubiquitous in the crystalline shield of Canada share a similar origin and evolution.

8 ACKNOWLEDGEMENTS

The Miramar Mining Corporation is thanked for allowing this study at the Con Mine and to their engineering and geology staff for assistance with underground access to sampling sites. We also wish to thank A. Agranat, E. Kasher, S. Meiri (geochemical laboratory, Hebrew University), J. Loop (University of Ottawa geochemical laboratory), T. Blanchard (geochemical laboratory, LSU) for their invaluable technical assistance, and to A. Sweezey and R. Timlin for assistance with the mine water sampling. This study was funded in part by the Atomic Energy Control Board, Canada. Li and Sr isotope work was supported by NSF Grant EAR 9506390 to L. H. C.

9 REFERENCES

Bottomley, D., D.C. Gregoire and K.G. Raven 1994. Saline groundwaters and brines in the Canadian Shield: Geochemical and isotopic evidence for a residual evaporite brine component. *Geochim. et Cosmochim. Acta* 58: 1483-1498.

Chan, L.H., 1987. Lithium isotope analysis by thermal ionization mass spectrometry of lithium tetraborate. *Analytical Chemistry* 59: 2662-2665.

Frape, S.K. & P. Fritz. 1987. Geochemical trends for groundwaters from the Canadian Shield. *Geological Association of Canada special Paper* 33: 211-223.

Frape, S.K., P. Fritz and H. McNutt 1984. The role of water-rock interaction in the chemical evolution of groundwaters from the Canadian Shield. *Geochim. et Cosmochim. Acta* 48: 1617-1627.

Herut, B., A. Starinsky, A. Katz and A.Bein 1990. The role of seawater freezing in the formation of subsurface brines. *Geochim. et Cosmochim. Acta* 54: 13-21.

Kamineni, D.C., M. Gascoyne, T.W. Melnyk, S.K. Frape and R. Blomqvist 1992. Cl and Br in mafic and ultramafic rocks: Significance for the origin of salinity in groundwater. *Proc. 7th Intl Symp. on Water-Rock interaction, Park City, Utah, July 13-18, 1992.* vol.1: 801-804.

Nelson, K.H. & T.G. Thompson 1954. Deposition of salts from sea water by frigid concentration. *Jour. Marine Res.* 13(2): 166-182.

Raab, M. 1996. The evolution of evaporites and brines along the Dead Sea Rift. *Ph.D. dissertation.* The Hebrew University of Jerusalem (in Hebrew, English abstract).

Qing, H. 1991. Petrography and geochemistry of Middle Devonian Presqu'ile dolomite, Pine Point N.W.T. and adjacent subsurface, and northeastern British Columbia. *Ph.D. dissertation*, McGill University.

Wilson, T.P. & D.T. Long 1993. Geochemistry and isotope chemistry of Michigan Basin brines: Devonian formations. *Applied Geochemistry* 8: 81-100.

Minor and trace element chemistry and provenance in Alpine glacial meltwaters

G.H.Brown & R.Fuge
Centre for Glaciology and Institute of Geography and Earth Sciences, University of Wales, Aberystwyth, UK

ABSTRACT: The use of major ion chemistry of bulk melt waters to define solute provenance and routing has been found to be somewhat equivocal, since dissolved species are frequently derived from more than one source. It is possible that trace and minor element data may allow more precise identification of solute provenance, though few data exist on trace and minor metals in glacial meltwaters. Meltwaters draining Haut Glacier d'Arolla, Valais, Switzerland, are dominated by the Ca^{2+} cation with HCO_3^- and SO_4^{2-} being the major anions. Data for trace metals determined by ICP-MS reveal strong correlations for Ca and Sr and Mg and Sr and it is likely that all 3 elements derive mainly from carbonates. Iron, Ni, and Co could be derived from pyrite or ferromagnesian minerals while metals such as Al and Ti are thought to be released from alumino-silicates. Uranium, most likely transported as a carbonate complex, probably derives from accessory minerals such as allanite and apatite.

1 INTRODUCTION

Recent studies of variations in major dissolved ions transported by glacial meltwaters have added significantly to our understanding of chemical weathering rates and processes in glacial environments (*e.g.* Tranter *et al.*, 1993; Brown *et al.*, 1994a,b; 1996; Sharp, 1991; Sharp *et al.*, 1995). While these studies are invaluable to both hydrological and geochemical interpretations in glaciated regions, solute provenance and routing is often equivocal, since dissolved species are frequently derived from more than one source. For example, SO_4^{2-} may be derived from both atmospheric (*e.g.* acid sulphate aerosol) and lithogenic (*e.g.* pyrite, gypsum) sources.

Variations in the major ion chemistry of glacial runoff have been exploited to separate the solute load into atmospheric, snowpack and crustal sources (*e.g.* Sharp *et al.*, 1995; Brown *et al.*, 1996a). However, to date little attempt has been made to utilise variations in trace and minor elements in these waters to deconvolute the solute load. It is possible that concentration data and inter-element ratios of some of the minor and trace metals may allow more precise identification of solute provenance.

In the present study concentration data for the minor and trace metals Al, Ba, Co, Cu, Fe, Mn, Ni, Pb, Rb, Ti, U and Zn has been obtained for bulk meltwaters draining Haut Glacier d'Arolla, Valais, Switzerland and an attempt made to link their chemistry with solute sources.

2 THE STUDY

Haut Glacier d'Arolla is situated at the head of the Val d'Hérens, Valais, Switzerland. The altitude of the glacier ranges from ~2560-3500 m and covers ~6.3 km^2 of a ~12 km^2 catchment. The maximum length of the glacier is ~4.2 km. Between 2 and 5 portals contribute meltwater to the bulk runoff (Sharp *et al.*, 1993).

2.1 *Geology of the study area*

The study area is underlain by the Arolla Series of the Dent Blanche nappe (Mazurek, 1986), the predominant lithology being the Arolla gneisses which are collectively a variable series of metagranites. A part of the glaciated area is underlain by gabbro and amphibolite (Brown, 1991).

Mineralogy of the bedrock is extremely varied, consisting of quartz, feldspar, olivine, pyroxene, mica, amphibole, cordierite, hematite, magnetite and talc (Tranter *et al.*, 1997). Mazurek (1986) lists allanite, apatite, sphene and zircon as accessory minerals in some of the bedrocks. In addition, carbonates (up to 0.58%) and sulphides (up to

0.71%) occur in trace quantities (Tranter et al., 1997). However, Brown et al. (1996a) record concentrations of carbonate up to 12% and pyrite up to 5% in fine (< 5 mm) morainic material.

2.2 Methodology

Bulk meltwater samples were collected hourly/bi-hourly ~100 m from the glacier snout across the 24 hour period 17-18 August 1996. The waters were vacuum filtered through 0.45µm cellulose nitrate membranes, transferred to pre-contaminated plastic bottles and acidified with HNO_3.

Analyses were performed by ICP-MS using Rh as an internal standard and a single external standard containing 100 µg l^{-1} of each element being determined.

3 RESULTS AND DISCUSSION

Table 1 gives the concentration range of some major ions in the bulk melt water.

Table 1. Concentration of major ions in bulk meltwater draining Haut Glacier d'Arolla during the 1989-90 meltseasons (concentrations in µeq l^{-1}).

Ion	Minimum	Maximum
Ca^{2+}	140	470
Mg^{2+}	15	52
Na^+	5.2	36
K^+	3.3	19
HCO_3^-	180	460
SO_4^{2-}	20	340
NO_3^-	0.00	36
Cl^-	0.85	6.7

Calcium is the dominant cation in these meltwaters, reflecting the fact that carbonate dissolution is a major source of solute in the subglacial hydrological system (Tranter et al., 1993; Brown et al., 1996a). It has been demonstrated that carbonate-derived Ca^{2+} is the dominant cation in other catchments which, in common with that in the study area, are dominated by igneous and metamorphic bedrock (Raiswell, 1984; Drever and Hurcomb, 1986). Much of the Mg^{2+} in meltwaters is also thought to derive from carbonate weathering (Brown et al., 1996a), with some possibly derived from ferromagnesian minerals. The alkali metals possibly derive from feldspars with some Na being of marine origin (Sharp et al., 1995; Brown et al., 1996a).

While carbonate breakdown appears to be the major subglacial weathering process, the oxidation of pyrite is also of major significance, as evidenced by the relatively high concentration of SO_4^{2-} in the meltwaters.

3.1 Minor and trace metals

The concentration ranges of minor and trace metals in the bulk meltwaters during 17-18 August 1996 at Haut Glacier d'Arolla are given in Table 2. From these data, it is apparent that concentrations of many of the trace metals determined in the meltwaters are generally in the range found for average river water (Brown and Fuge, submitted). Aluminium and Fe are somewhat more concentrated than average surface water (Langmuir, 1997).

Table 2. Trace and minor metals in bulk meltwaters draining Haut Glacier d'Arolla, 17-18 August 1996 (in µg l^{-1}).

Metal	Minimum	Maximum
Al	93	175
Ti	3.1	9.0
Fe	334	650
Mn	3.7	7.0
Sr	8.9	21
Ba	0.2	2.3
Rb	0.1	0.9
Co	0.1	0.4
Cu	0.2	2.3
Ni	0.6	1.6
Pb	0.0	0.9
Zn	0.2	3.5
U	0.3	1.1

It has been suggested that coupled carbonate dissolution and pyrite oxidation is the dominant chemical weathering process in the distributed facet of the subglacial hydrological system (Tranter et al., 1993). It is of interest to determine which of the trace metals also derive from these minerals. In an attempt to provide this information correlation coefficients have been computed and these are given in Table 3. It should be noted that as the samples were collected throughout a 24 hour discharge cycle some of the co-variance will be due to the inverse and inter-dependent relationship between bulk discharge and dissolved metal concentrations. In addition, the relatively small number of samples (n = 20) may also impose limitations on interpretations of solute provenance. However, within these limitations some conclusions can be drawn.

Strontium along with Ca and Mg is an alkaline earth element and commonly replaces Ca in crystal lattices. In the meltwaters these 3 elements show very strong inter-correlation and it is suggested that Sr, along with Ca and Mg, derives mainly from the breakdown of carbonate minerals (Goldschmidt, 1954; Holland, 1978). The remaining alkaline earth element Ba, however, shows no significant correlation with Ca or Mg and possibly derives from a silicate mineral such as K-feldspar (Puchelt, 1972).

Table 3. Correlation coefficients for Ca, Mg, and Fe with the other metals and SO_4. n = 20.

	Ca	Mg	Fe
Mg	0.99	-	-
Fe	0.71	0.71	-
Al	0.31	0.33	0.76
Ti	0.60	0.63	0.61
Mn	0.33	0.35	0.76
Sr	0.96	0.97	0.62
Ba	0.44	0.43	0.05
Rb	0.54	0.54	0.44
Co	0.57	0.60	0.46
Cu	0.16	0.21	0.05
Ni	0.74	0.73	0.49
Pb	0.03	0.03	0.30
Zn	0.27	0.32	0.38
U	0.87	0.89	0.38
SO_4^{2-}	0.73	0.75	0.17

The fairly clear correlation of U with Ca and Mg is probably not directly related to its source. It is more likely to reflect the transport of U as a carbonate complex, the carbonate being released during the weathering of carbonates (Levinson, 1980; Krauskopf and Bird, 1996). Uranium is possibly derived from minerals such as allanite and apatite which have all been listed as accessory minerals in the bedrocks (Mazurek, 1986).

The oxidation of pyrite is an important subglacial weathering process in the distributed system (Tranter et al., 1993), and is reflected in the high SO_4 content of the meltwaters. It is then of interest that no significant correlation of Fe and SO_4 is recorded in the present study (see below). There is, however, a weak correlation of Fe with Ca and Mg.

Metals such as Ni and Co are frequently found to be relatively concentrated in pyrite (Levinson, 1980). It is possible then that these metals could derive in part from the weathering of pyrite. However, neither of these metals correlates significantly with Fe in the meltwater. It is likely that as the meltwaters are alkaline, pH ranges between 7.7 and 9.1, that Fe in solution will be relatively immobile as the pH of hydrolysis of Fe^{3+} is 2.3 and that of Fe^{2+} is 5.5. The pH of hydrolysis for Ni^{2+} and Co^{2+}, however, is 6.7 and 6.8 respectively (Levinson, 1980), which could mean that they remain in solution longer than Fe. In addition, Ni and Co may be released from ferromagnesian minerals.

Lead, Zn and Cu are chalcophile elements and as such could also be derived during sulphide oxidation. It is equally likely that they derive from silicate minerals, Pb possibly from feldspar and Zn and Cu from ferromagnesian minerals.

Manganese shows a weak correlation with Fe and this might imply a common source. However, Mn and Fe show similar behaviour in the weathering environment, being strongly influenced by Eh. In addition, under alkaline conditions both would be expected to be insoluble.

Of the trace and minor elements considered in the present study, Al and Ti are geochemically strongly associated with silicate minerals and it would seem likely that that they are derived from the breakdown of such silicates. Rubidium also probably derives from silicates, where it is found replacing K.

4 CONCLUSIONS

From this limited study it is apparent that minor and trace elements in glacial meltwaters are at or near the concentrations found in average river waters.

It seems likely that minor and trace metal analyses could in conjunction with data for major ions give valuable additional information on solute provenance for glacial meltwaters.

5 REFERENCES

Brown, G.H. 1991. Solute provenance and transport pathways in alpine glacial environments. *Unpublished Ph.D. Thesis, University of Southampton.*

Brown, G.H., M.J. Sharp, M. Tranter, A.M. Gurnell, and P.W. Nienow. 1994a. The impact of post-mixing chemical reactions on the major ion chemistry of bulk meltwaters draining the Haut Glacier d'Arolla, Valais, Switzerland, *Hydrological Processes.* 8: 465-480.

Brown, G.H., M. Tranter, M.J. Sharp, T.D. Davies and S. Tsiouris. 1994b. Dissolved Oxygen Variations in Alpine glacial meltwaters, *Earth Surface Processes and Landforms.* 19: 247-253.

Brown, G.H., M. Tranter and M. Sharp. 1996a. Subglacial chemical erosion - seasonal variations in solute provenance, Haut Glacier d'Arolla, Switzerland, *Annals of Glaciology.* 22: 25-31.

Brown, G.H., M. Tranter and M.J. Sharp. 1996b. Experimental investigations of the weathering of suspended sediment by Alpine glacial meltwaters, *Hydrological Processes.* 10: 579-598.

Brown, G.H. and R. Fuge. Submitted. Trace element chemistry of glacial meltwaters in an Alpine headwater catchment, *Headwater Control IV: Hydrology, Water Resources and Ecology in*

Headwaters, Meran/Merano, Italy, 20-23 April 1998, IAHS Publication.

Drever, J.I. and D.R. Hurcomb. 1986. Neutralisation of atmospheric acidity by chemical weathering in an alpine drainage basin in the North Cascade Mountains. *Geology.* 14: 221-224.

Goldschmidt, V.M. 1954. *Geochemistry.* Oxford University Press. London.

Holland, H.D. 1978. *The Chemistry of the atmosphere and oceans*, John Wiley and Sons, New York.

Krauskopf, K.B. and D.K. Bird. 1996. *Introduction to geochemistry (2nd edition).* McGraw-Hill, London.

Langmuir, D. 1997. *Aqueous Environmental Geochemistry.* Prentice Hall, New Jersey.

Levinson, A.A. 1980. *Introduction to Exploration Geochemistry* 2nd edition. Applied Publishers, Illinois.

Mazurek, M. 1986. Structural evolution and metamorphism of the Dent Blanche nappe and the Combin zone west of Zermatt (Switzerland), *Eclogae Geologicae Helvetiae.* 79: 41-56.

Puchelt, H. 1972. *Handbook of Geochemistry vol. II.* (K.H. Wedepohl - ed.). Springer. Berlin.

Raiswell, R. 1984. Chemical models of solute acquisition in glacial meltwaters. Journal of Glaciology. 30: 49-57.

Sharp, M.J. 1991. Hydrological inferences from meltwater quality data - the unfulfilled potential. *Proceedings of the British Hydrological Society National Symposium, Southampton, 16-18 September 1991*: 5.1-5.8.

Sharp, M., K.S. Richards, N. Arnold, I.C. Willis, P. Nienow, and J-L Tison. 1993. Geometry, bed topography and drainage system structure of the Haut Glacier d'Arolla, Switzerland. Earth Surface Processes and Landforms. 18: 557-571.

Sharp, M., M. Tranter, G.H. Brown and M. Skidmore. 1995. Rates of chemical denudation and CO_2 drawdown in a glacier-covered alpine catchment, *Geology.* 23: 61-64.

Tranter, M., G.H. Brown, R. Raiswell, M.J. Sharp, and A.M. Gurnell. 1993. A conceptual model of solute acquisition by Alpine glacial meltwaters. *Journal of Glaciology.* 39: 573-581.

Tranter, M., M. Sharp, G.H. Brown, I.C. Willis, B.P. Hubbard, M.K. Nielsen, C.C. Smart, S. Gordon, M. Tulley, and H.H. Lamb. 1997. Variability in the chemical composition of in-situ subglacial meltwaters, *Hydrological Processes.* 11: 59-77.

Diagenesis of nonmarine sediments in an evolving tectonically-induced rain shadow

C.P.Chamberlain & M.Poage
Department of Earth Sciences, Dartmouth College, Hanover, N.H., USA

D.Craw
Geology Department, University of Otago, Dunedin, New Zealand

ABSTRACT: Immature nonmarine sediments formed progressively during rise of the Southern Alps, New Zealand, at the Pacific-Australian plate boundary. Sedimentary sequences have been variably water-saturated and reduced and oxidized by groundwater during uplift. Groundwater percolation has resulted in diagenesis, including authigenic clay formation in a complex set of mineralogical transformations. Authigenic kaolinite records the oxygen isotope signature of the groundwater, and hence the rain water, at time of alteration. Kaolinites have become isotopically lighter by at least 5 per mil between Miocene and present, reflecting the observed development of a rain shadow to the east of the Southern Alps.

1 INTRODUCTION

Diagenesis in nonmarine sedimentary sequences results from passage of groundwater through the sediments. The mineralogical transformations which occur depend mainly on the chemistry of the water and how this water interacts with the primary material. Chemically mature sediments show little evidence of diagenesis, but immature sediments are distinctly more reactive. Immature nonmarine sediments typically form in tectonically active environments where rapid physical erosion and deposition prevent significant chemical weathering. Rise of mountains in tectonically active areas can have a dramatic effect on a region's climate by interrupting airstreams. Changes in climate can then affect the groundwater which causes diagenesis. There are therefore, complex inter-relationships between tectonic processes, immature nonmarine sedimentation, diagenesis, and climate change.

1.1 Nonmarine sediments, southern New Zealand

This study describes diagenetic mineralogical transformations in late Tertiary to Recent sediments in southern New Zealand. The sediments have formed at various stages during development of the current Pacific-Australian plate boundary. Deposition was sufficiently removed from the main uplift zone that the sediments have been preserved, although most have been deformed by subsequent tectonism. All the sedimentary sequences were deposited to the east of the main range, the Southern Alps (Fig. 1).

The Southern Alps interrupt a strong moist westerly airstream, causing high rainfalls on the western slopes and a pronounced rain shadow to the east of the mountains (Fig. 1).

Miocene sediments contain some immature clasts of underlying Mesozoic metasedimentary basement. These latter clasts, and the upper 5-20 m of basement unconformably beneath the sediments, have been diagenetically altered to clay minerals.

Similar alteration has occurred in Pliocene-Recent immature molasse deposits and alluvial fans derived from rising mountain ranges to the east of the Southern Alps (Fig. 1). Alteration continued in these sediments after they were deformed and uplifted, until they were so cemented and uplifted that groundwater no longer flowed through them.

2 AUTHIGENIC CLAY MINERALS

Authigenic clay minerals have formed by direct replacement of primary minerals, and in pore spaces after local elemental mobility during degradation of primary minerals in adjacent clasts. Two distinct clay mineral transformation paths have been identified, according to the oxidation state of the geochemical system (Craw, 1994). The samples for this study were obtained in the general vicinity of Naseby (Fig. 1).

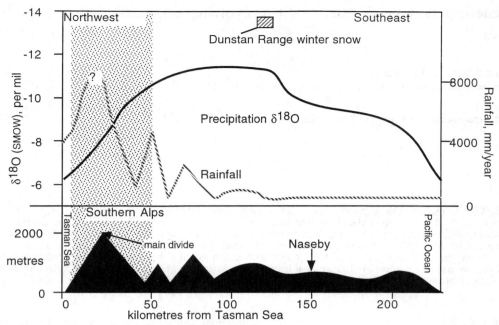

Fig. 1: Sketch topographic profile (black) across southern New Zealand. The main Southern Alps are indicated with stipple. Profiles for rainfall (striped line, right-hand ordinate) and rain oxygen isotopic ratio (heavy black line; calculated from Stewart et al., 1983) across the mountains are shown also.

2.1 Non-oxidized alteration.

Alteration is pronounced where groundwater interacted with primary schist or greywacke minerals, in clasts or underlying basement. The alteration caused breakdown of muscovite and chlorite to smectites and kaolinite (Fig. 2). Chlorite broke down to green ferrous/ferric iron smectite which locally also developed kaolinite interlayers (Craw, 1984). Muscovite developed minor illite interlayering on the submicron scale during incipient alteration, resulting in broadening of the 10Å X-ray diffraction peak. Illite then broke down to white or pale brown smectite. The clay mineral suite is dominated by smectites, and kaolinite was formed only after extreme degradation of the rock, when albite also altered to kaolinite (Craw, 1984; 1994).

A wide range of smectite mineral compositions has formed in these rocks (Craw et al., 1995; Fig. 3), with different compositions of smectite interleaved on the micron scale or smaller. These smectite minerals are initially interleaved with primary phyllosilicate minerals on the micron scale, and have similar compositions to their host mineral. Green smectites from chlorite are low in silica and potassium, and high in iron, whereas white smectites from muscovite are high in silica, aluminium, and potassium (Fig. 3 A, B). With increasing alteration these two end-member smectite compositions evolve towards an intermediate compositional range with pale greenish-brown colour (Fig. 3A, B). Authigenic pyrite occurs intergrown with smectites and kaolinite in both weakly and strongly altered rocks.

2.2 Oxidized alteration.

This type of alteration occurred where oxygenated groundwater flowed through the rock and resulted in overall oxidation of the original rock. The effect is noticeable in the quartz-dominated coal measures only where permeable quartz gravels have been silicified, with minor included goethite. Where oxidized water has interacted with primary rock, the main alteration reactions involved breakdown of albite to kaolinite, and some muscovite to illite and then to kaolinite (Fig. 2). Chlorite altered initially to iron smectite, but this was transformed to kaolinite and goethite after only mild alteration (Fig. 2). Iron and manganese-bearing carbonate accompanies kaolinite and goethite in highly altered rocks.

Oxidation of rocks initially altered along the non-oxidized transformation path (Fig. 2) has occurred during uplift and erosion of some altered rocks. This oxidation process caused transformation of smectites to kaolinite and goethite, oxidation of pyrite to goethite, and kaolinitization of remaining albite. Further degradation of muscovite to illite and then kaolinite also occurred.

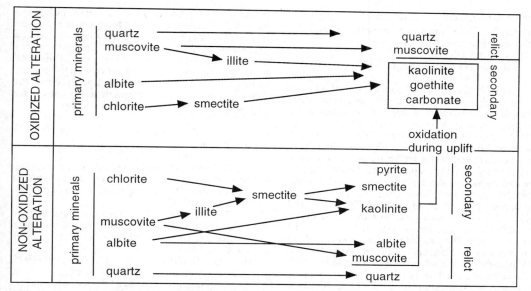

Fig. 2: Summary of mineralogical transformations which occurred during nonmarine diagenesis of immature clasts and basement. Non-oxidized alteration has different products from oxidized alteration, but the two processes converge with oxidation during uplift.

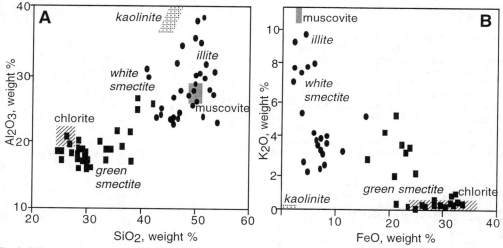

Fig. 3: Microprobe analyses of authigenic clay minerals formed during diagenesis of immature clasts and basement. White smectites (dots) and kaolinite are derived from muscovite, and green smectites (squares) and kaolinite are derived from chlorite or biotite.

3 OXYGEN ISOTOPIC ANALYSES

3.1 *Isotopic rain shadow*

The isotopic composition of groundwater reflects that of the meteoric water from which it is derived. This signature is in turn passed on to clay minerals which are formed authigenically during nonmarine diagenesis. If the isotopic composition of meteoric water changes with time, so too will the isotopic composition of the clay minerals. The dramatic rain shadow which has developed to the east of the Southern Alps (above; Fig. 1) is reflected in the oxygen isotopic composition of the rain water (Fig. 1). This is because the very high rainfalls on the western side of the mountains extracts heavy isotopes preferentially, and atmospheric water which reaches the east side is depleted in these isotopes. Meteoric water in the Southern Alps rain shadow is 3-5 per mil depleted in ^{18}O compared to expected values for equivalent elevations (Fig. 1).

3.2 *Clay mineral isotopic ratios*

We have obtained a preliminary set of oxygen isotopic analyses for authigenic kaolinites from three geological situations: altered metasedimentary basement beneath Miocene sediments; unoxidized altered water-saturated Pliocene molasse sediments; and altered molasse and fan sediments which have been uplifted and subjected to oxidized alteration. Ranges of data for each situation are narrow (Fig. 4). The ages of alteration are poorly known, but realistic estimates have been made (Fig. 4). The data are compared (Fig. 4) to present precipitation, and projected Miocene precipitation based on deduced Miocene climate and the isotopic ratios for similar climate in New Zealand today (Stewart et al., 1983). The resultant diagram (Fig. 4) depicts the development of the Southern Alps rain shadow which must have occurred in the late Pliocene or early Quaternary. This is presumably the timing of the rise of the Southern Alps as a major mountain chain. More accurate dating is required to better define the timing of this uplift.

4 CONCLUSIONS

Chemically immature components of Cenozoic terrestrial sediments, and their underlying basement, have undergone a complex sequence of clay mineral formation and recrystallization in the South Island of New Zealand during burial and uplift induced by the rise of the Southern Alps. Under non-oxidizing conditions, the clay fraction is dominated by smectites which ultimately transform to kaolinite. Under oxidizing conditions, the clay fraction is dominated by kaolinite and goethite. Kaolinite and goethite have also formed from the non-oxidized assemblage when the diagenetically altered sediments were uplifted and oxidized. Diagenetic clays took on the oxygen isotopic signature of the altering meteoric-sourced groundwater, and retained this signature to the present unless overprinting clay mineral transformatons imposed a new groundwater signature during uplift. Clay minerals thus record the development of isotopically light meteoric water in the rain shadow to the east of the Southern Alps.

REFERENCES

Craw, D. 1984. Ferrous-iron-bearing vermiculite-smectite series formed during alteration of chlorite to kaolinite, Otago Schist, New Zealand. *Clay Mins 19*: 509-520.

Craw, D. 1994. Contrasting alteration mineralogy at an unconformity beneath auriferous terrestrial sediments, central Otago, New Zealand. *Sed. Geol.* 92: 17-30.

Craw , D., D.W.Smith, & J.H.Youngson 1995. Formation of authigenic Fe2+-bearing smectite-vermiculite during terrestrial diagenesis, southern New Zealand. *N.Z.J.Geol.Geophys.38*: 151-158.

Stewart, M.K., M.A.Cox, M.R.James & G.L. Lyon, 1983. *Deuterium in New Zealand rivers and streams.* Institute of Nuclear Sciences Report INS-R-320, Wellington, New Zealand, 32 pp.

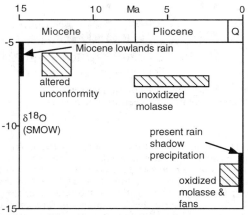

Fig. 4: Variation of authigenic clay mineral oxygen isotope ratio with estimated time of alteration.

Neogenesis during thermal stimulation of bitumen, Alberta, Canada

J.S. Dudley
Imperial Oil Resources Limited, Calgary, Alb., Canada

C.H. Moore
C.H. Moore Research and Applications, Bloomington, Ind., USA

ABSTRACT: Large reserves of bitumen occur at Cold Lake in the Lower Cretaceous Clearwater Formation. The mineralogical immaturity of this volcanic litharenite makes it prone to "operation-induced" clastic diagenesis (neogenesis) from the injection of 300°C steam as part of the cyclic steam stimulation (CSS) recovery process. Neogenesis has long been recognized as occurring during CSS and proposed to damage the reservoir through neogenetic clay growth. Its actual impact on production and performance of CSS has been less certain. Integration of diverse methodologies and data, including post-steam core analysis, coupled reaction-fluid flow simulation, pertinent field data, and the regional stratigraphic/sedimentologic framework, has shown that a more complex multi-mechanistic process involving mineral dissolution, fines migration, in addition to neogenetic growth can impact performance by impeding fluid flow.

1 INTRODUCTION

The Clearwater Formation at Cold Lake, Alberta contains one of the major bitumen resources in the world. The highly viscous nature of the bitumen in this subsurface reservoir necessitates thermal recovery processes such as cyclic steam stimulation (CSS). CSS involves both injection of 300°C steam and production in the same wellbore. The mineralogical immaturity of the Clearwater volcanic litharenite makes it prone to "operation-induced" clastic diagenesis (neogenesis) as a result of steam (hot water) - reservoir interaction. The term "neogenesis" refers to any of a number of steam-induced changes in a reservoir due to steam stimulation. These changes can include consolidation /cementation, mineral growth and dissolution, changes in permeability, gas evolution, and evolution of water chemistry from that injected to that produced. Such changes can potentially affect performance in a variety of ways including reduced fluid flow, scaling of wellbore and near wellbore, casing failures, gas drive from inorganic sources, reduced compaction drive and variable sand production. This study focuses on reduced fluid flow, and tests whether steam-induced growth of permeability-impairing minerals causes decreased fluid flow to the wellbore.

The role of neogenesis in causing reduced fluid flow is a priority due to past unexplained CSS performance with low oil and total fluid to steam ratios. This performance had not been predicted and there was concern that similar poor performance might be encountered in future development areas.

2 PREVIOUS WORK

The potential for neogenetic changes to impact steam stimulation performance by impairing permeability was noted some time ago (e.g., Sedimentology Research Group, 1981), but the actual impact on performance had yet to be demonstrated.

Past studies of post-steam core document the growth of neogenetic clay and propose this as a possible reservoir damage mechanism (Sedimentology Research Group, 1981; Fialka et al., 1993).

Core-flood experiments have shown that Clearwater mineral-water reactions include volcanic rock fragment and carbonate dissolution resulting in both production of fines and growth of smectites and carbonates (Huang & Longo, 1994; Mok et al., 1995). Measured resultant changes in permeability have been variable, including some increases, as well as drastic drops of up to 98% of the initial value (e.g., Kirk et al., 1987).

3 DATA

This study utilizes newly described post-steam core, numerical simulation of mineral reactions, and regional mineral distribution related to stratigraphy to understand how cyclic steam stimulation-reservoir interaction might impede CSS well performance.

3.1 Post-steam core

Initial, pre-steam permeabilities of the predominantly unconsolidated reservoir sands are on the order of 1-3 Darcies. Post-steam core is commonly consolidated with permeabilities having decreased by only a few tenths of a Darcy. However, the most drastic drop in permeability observed in core taken after thermal stimulation has been an order of magnitude to as low as .3 Darcies from a pre-steam value of 2-3 Darcies. This damage is caused by fines migration due to a disruption of early diagenetic rims of berthierine on the sand grains. Petrography and scanning electron microscopy reveal berthierine fragments lodged in pore throats against intact berthierine rims. Disruption of the berthierine rims appears to be due to a combination of volcanic rock fragment dissolution and possible differential thermal expansion of the sand grain and the clay (Lin, 1985). Though post-steam core available from the area experiencing impaired fluid flow does not demonstrate as severe a permeability decrease as that described above, similar disruption of berthierine is observed.

3.2 Simulation

Calibration of the REACTRAN computer programme (Dudley & Moore, 1992; Mok et al., 1995) has lead to the ability to simulate field scale coupled reaction-fluid flow. Such simulations predict the amount and
distribution of mineral growth and dissolution associated with steaming and the potential impact upon permeability. Field scale simulations were conducted on the lithologies of the area in which poorer fluid flow had been experienced. These simulations predict that volcanic rock fragment and carbonate dissolution leads to growth of clays distributed close to the wellbore and the horizon of steam injection. The permeability drop predicted is not overly severe (a 1 Darcy decrease from an initial value of 3 Darcies), but may well be a minimum due to simplifying assumptions such as spherical grains and no migration of the neogenetic clay which may, in fact, be particularly prominent near the wellbore.

3.3 Siderite and scale

Post-steam core, experiments, and simulations show that siderite commonly undergoes dissolution upon steaming and is, therefore, the most likely cause of calcite scale precipitation in the Cold Lake thermal operation. This is further supported by field data showing a correlation between bottom hole pump failures due to scale, and the occurrence of siderite-replaced clay clasts. These clasts are formed by the incision of a paleovalley into sideritized muds and may, therefore, be predictable from the regional stratigraphic model. The area of problematic CSS fluid flow is located within the trend of calcite scale-induced pump failures and associated siderite clasts.

4 DISCUSSION

4.1 Combined mechanisms

The area that suffered from poor fluid flow is located where a number of the above damaging phenomena occur together. Its mineralogy is conducive to the growth of new clays near the wellbore where their impact on fluid flow would be greatest. Fines migration is likely to be extreme due to a combination of sources. The neogenetic clays themselves would likely migrate toward the nearby wellbore. Berthierine occurs near the top of the perforation zones and appears to have been disrupted in nearby post-steam core. In addition, the area is on trend with the occurrence of siderite-replaced clasts and bottom hole pump failures. Siderite-replaced clasts are concentrated in lower portion of the perforation zone. The expected dissolution of siderite clasts would release other fines such as quartz and clay, as well as leading to calcite scale precipitation near the wellbore. Some of the problem wells showed fines in the produced emulsions and calcite scale on a pulled liner suggesting a relationship to their poorer performance.

Post-steam core from the problem area is currently being studied to further test this hypothesis, but may not be a completely valid test, in that it was obtained 10 metres away from the producing wellbore and may, therefore, not be representative of the most significant damage location.

4.2 A new model

Rather than a single mechanism, such as neogenetic clay growth, a multi-mechanistic process involving mineral dissolution, fines migration, and neogenetic mineral growth, may be required for neogenesis to impede fluid flow during cyclic steam stimulation. It is the combination of otherwise non-damaging mechanisms that is significant and not the individual mechanisms. The multi-mechanistic nature of the neogenetic process presents the opportunity to prevent fluid flow problems by avoiding some, but not necessarily all, of the contributing mechanisms by either selective drilling locations and/or completion intervals.

5 CONCLUSIONS

1) Neogenetic clay growth alone may not significantly impact CSS fluid flow in mineralogical suites currently being exploited.

2) A combination of mechanisms including mineral dissolution, physical mineral disruption, fines migration, scale precipitation and neogenetic clay growth is likely required to impede fluid flow of CSS.

3) Formation damage and poor CSS performance may be prevented by avoiding one or more of the damaging mechanisms through prudent placement of wells and/or completion intervals.

6 ACKNOWLEDGEMENTS

Thanks to Imperial Oil Resources Limited for permission to publish these results. We wish to acknowledge the helpful support of F.J. Longstaffe and U. Mok (University of Western Ontario) and W.D. Gunter (Alberta Research Council).

7 REFERENCES

Dudley, J.S. and Moore, C.H. 1992. Computer modelling of steam flood experiments. *Proceedings of the 7th International Water-Rock Interaction Conference, Park City* : 223-1226.

Fialka, B.N., McClanahan, R.K., Robb, G.A. and Longstaffe, F.J. 1993. The evaluation of cyclic steam stimulation of an oil sand reservoir using post-steam core analysis. *Journal of Canadian Petroleum Geology*, v. 32, no. 2: 56-62.

Huang, W.L. & Longo, J.M. 1994. Experimental studies of silicate-carbonate reactions - II Applications to steam flooding of oil sands. *Applied Geochemistry*, 9, no 5: 523-532.

Kirk, J.S., Bird, G.W., & Longstaffe, F.J. 1987. Laboratory study of the effects of steam condensate flooding in the Clearwater Formation: high temperature flow experiments. *Bulletin of Canadian Petroleum Geology*, 35, no. 1: 34-47.

Lin, F.-C. 1985. Clay-coating reduction of permeability during oil sand testing. *Clays and Clay Minerals*, 33, no. 1: 76-78.

Mok, U., Wiwchar, B., Gunter, W.D., Moore, C.H., Dudley, J.S., and Longstaffe, F.J. 1995. The spatial mineral distribution in a flow system: experiment and computer simulations. *Proceedings of the 8th Int'l. Water-Rock Interaction Conference, Vladivostok, Russia, August* : 41-743.

Sedimentology Research Group 1981. The effects of in situ steam injection on Cold Lake oil sands. *Bulletin of Canadian Petroleum Geology*, 29: 447-478.

Heterogeneity of formation waters within and between oil fields by halogen isotopes

H.G.M. Eggenkamp & M.L. Coleman
Postgraduate Research Institute for Sedimentology, The University of Reading, Whiteknights, UK

ABSTRACT: Chlorine isotope data have been measured on produced waters, oil-zone waters and mudstones from North Sea area oil-fields. Large chlorine isotope variations exist between fields, within fields, and between different water types in fields. Produced waters are mixtures between water that has dissolved evaporite and water affected by water-rock interaction or membrane filtration. Oil-zone waters range from rock-interacted types that may have meteoric or marine origins. Mudstones are complex: in the upper 2Km $\delta^{37}Cl$ decreases with depth. In deeper samples no systematic variation is found, and within one field a range of over 2‰ occurs at one depth. Bromine isotopes have been measured on one set of samples. Cl and Br isotope compositions correlate negatively, indicating a different fractionation mechanism. Br isotopes are a potentially powerful tool in assessing formation waters and oil-reservoirs.

1 INTRODUCTION

Our knowledge about chemical heterogeneity in formation waters related to North Sea oil occurrences has improved considerably in the past decade. It is known by now that very large variations in chemical composition are found not only between fields, but also within fields (Coleman 1992, Warren & Smalley 1995, Thurlow & Coleman 1997). Variations were explained by different histories of the various waters found within and between fields (Thurlow & Coleman 1997).

Chlorine is generally the most common anion in formation waters. The element is a very conservative one which indicates that it can be used as a tracer for physical processes. Cl has two stable isotopes (^{35}Cl and ^{37}Cl), and the ratio between them can be used to refine conclusions based purely on Cl concentrations.

We have measured Cl isotope compositions in a large variety of formation waters relative to oil in and around the North Sea. It was found that extremely large variations in Cl isotope compositions were found. While normally values are within 2‰ of the standard (ocean water chloride), Cl isotope compositions in formation waters can be as negative as -5‰. We will try to assess these variations in relationship to potential processes and experiments are being done in our laboratory to find indications that could explain the found variations.

Cl isotopes are measured in a) aquifer formation waters, b) water extracted from dry oil (oil zone water), and c) salt residue extracted from mudstones.

Bromine is also a common element in formation waters. The bromide concentration, and the Cl/Br ratio can give very useful information about sources of the sample (e.g. Matray et al. 1989). Like Cl, Br also has two stable isotopes (^{79}Br and ^{81}Br). We expected that the Br isotope variation could also be useful in determining sources and processes that took place in formation waters. For this reason we recently developed a method which made it possible to measure Br isotope compositions in formation waters (Eggenkamp & Coleman 1997a).

2 METHODS

Cl isotope variations have been measured using the method described by Eggenkamp (1994). An amount of liquid sample, containing between 20 and 100 μL Cl⁻ is mixed with KNO_3 to get a high ionic strength, and a buffer ($Na_2HPO_4 \cdot 2H_2O$/Citric acid), to reach a constant low pH. To this a $AgNO_3$ solution is added to precipitate AgCl. This is filtered over a glass fibre filter, and reacted with CH_3I in a sealed evacuated borosilicate glass capsule. After 2 days at 80 °C the CH_3I has reacted with AgCl to form AgI and CH_3Cl. The CH_3Cl and CH_3I are then separated by gas chromatography, and the pure CH_3Cl is measured on a VG SIRA 12 dual inlet mass spectrometer.

Cl isotopes of oil and mudstones cannot be determined directly. The Cl is extracted from the oil using a method recently developed in our laboratory (Thurlow & Coleman 1997). The Cl in the mudrocks is extracted by mixing rock chips with water and shaking the mixture for 24 hours. The rock disaggregates and Cl is dissolved in the water. In both cases the Cl isotope composition can be measured as described above.

The method for measuring bromine isotope compositions in formation waters (Eggenkamp &

Coleman 1997a) is a combination of classical methods (Friedheim & Mayer 1892; Willey & Taylor 1978) is based on the difference in redox behaviour of Cl and Br, which makes it possible to separate the two elements very efficiently in a two step process.

3 RESULTS AND DISCUSSION

3.1 *Produced water*

Figure 1 shows Cl isotope data of the produced water samples from three different oil fields in (or near) the North Sea.

Samples from Forties (Central North Sea) and the Westland (Southern, Netherlands) show a very clear positive correlation between Cl concentration and $\delta^{37}Cl$. This indicates probably a two end member mixing between a high concentration end member with a $\delta^{37}Cl$ close to 0‰. The high Cl is probably due to dissolving Permian Salts which are present under the southern and central North Sea. The $\delta^{37}Cl$ also points into this direction as Eggenkamp et al. (1995) showed that $\delta^{37}Cl$ values in the evaporites are close to 0‰ or slightly negative. The Oseberg samples (Northern North Sea, Ziegler & Coleman 1998) have no high concentration end member because no evaporite deposits are present (or at least shown) under the northern North Sea. Here the end member with the high $\delta^{37}Cl$ seems to be caused by contamination with sea water (added in plot for comparison).

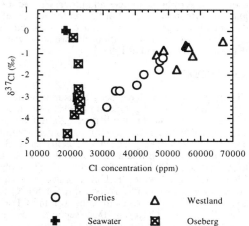

Figure 1. Cl concentrations and $\delta^{37}Cl$ of aquifer formation waters in various North Sea oil fields.

The other end member, which appears to be common to all fields, has a low Cl concentration (appr. 20000 ppm) combined with an extremely low $\delta^{37}Cl$ of approximately -5‰. These are among the most negative $\delta^{37}Cl$ values ever found in natural samples. Only in water samples, sampled below convergent margins (Ransom et al. 1995) were even more negative values found.

It is not yet known what process produces these extremely low $\delta^{37}Cl$ values. Proposed mechanisms are diffusion (Desaulniers et al. 1986, Eggenkamp et al. 1994), membrane filtration (Campbell 1985, Phillips & Bentley 1987) and mineral reactions (Ransom et al. 1995). Diffusion cannot be the cause of these low values, as the Cl concentrations are too high to cause these low $\delta^{37}Cl$ values (see e.g. Eggenkamp & Coleman 1997b). Both other processes are not yet very well understood. Membrane filtration seems to produce conflicts between theory (Phillips & Bentley 1987) and practice (Campbell 1985), while mineral reactions seem to need illogically large fractionations (Ransom et al. 1995). Much research seems to be necessary to solve these problems, but it seems reasonable to expect that the values are caused by reactions between Cl in solution and surrounding rocks. It seems likely that in some cases this negative $\delta^{37}Cl$ end member composition represents oil-zone water coproduced with the more saline aquifer.

3.2 *Oil zone water*

It was expected that water extracted from (dry) oil (oil zone water, Figure 2) had $\delta^{37}Cl$ values that would be comparable to those of the aquifer waters. It seems this is not always the case. It was however noticed before that clear variations between aquifer and oil zone waters from individual fields exist (Thurlow & Coleman 1997). In the case of Forties Field the oil zone water affects produced water composition but it is not clear if the same process resulted in the other low $\delta^{37}Cl$ compositions.

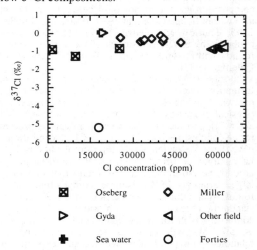

Figure 2. Cl concentrations and $\delta^{37}Cl$ of oil zone waters in various North Sea oil fields.

Thurlow & Coleman (1997) concluded that oil-zone water is fossilised formation water, trapped by the oil when this came into place, and as such reflecting water of low salinity from before salinisation by evaporites. In general these samples seem to reflect mixtures between a slightly (but significantly) negative end member and sea or meteoric water. Because oil-zone water is extracted from oil using distilled de-ionised water, it is impossible to determine oxygen and hydrogen isotopes on oil zone waters, which would have made it easier to determine origin of those waters. However, our recent data make it unlikely that oil-zone waters are fossil formation water.

3.3 Cl extracted from mudstones

It could be reasonable to assume that (at least part of) the water present in oil fields comes from the same rocks as the oil. For this reason we have taken a series of mudstones, leached the salt content from them, and measured the Cl isotope composition of this. This Cl represents the residual Cl, left after expulsion of the rest of the Cl when the muds were compacted. The water would then be transported into the reservoir, but the mechanism is not yet clear. In Figure 3 we show the $\delta^{37}Cl$ relative to the burial depth of the mudstone samples. Samples are divided into two groups, samples from the UK mainland, and from the UK continental shelf. The latter samples are generally more deeply buried. The shallowest samples show positive $\delta^{37}Cl$ values, which is probably due to interactions between Cl and the clay. The heavier isotope seems to be bound better to the clay minerals at shallower burials. Cl isotope fractionation may be dependent on the pressure at which it appears, and the heavier isotope seems to be bound more strongly to the clay.

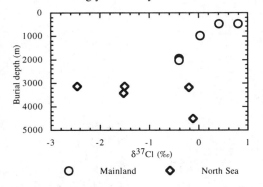

Figure 3. Relationship between burial depth and $\delta^{37}Cl$ for mudstones from the UK mainland and the UK continental shelf.

At greater depths lower $\delta^{37}Cl$ are found. These results agree with experiments by J.L. Loomis (unpublished data) who found that fluids expelled from clay-water mixtures show lower $\delta^{37}Cl$ at lower pressures, indicating that residual $\delta^{37}Cl$ must be positive, while at higher pressures the $\delta^{37}Cl$ start to rise again, which should lead to lower residual $\delta^{37}Cl$ in the clay. Samples that have seen much deeper burial show a much more complex behaviour, as is shown in four samples from the Miller oil-field, which were sampled at nearly the same depth (~3200 metres), and show a 2.5‰ variation. It is not clear what could be the reason for this behaviour.

3.4 Bromine isotopes

Bromine is also an important element in formation water, so we developed a method to be able to measure Br isotopes in formation waters. Our first set of samples was that from the Oseberg field, as this shows a very large range of $\delta^{37}Cl$ values. It was expected that Br isotope values would show the same sign of fractionation as the Cl isotopes, but as can be seen in figure 4, the $\delta^{81}Br$ values are positive while $\delta^{37}Cl$ values are (very) negative.

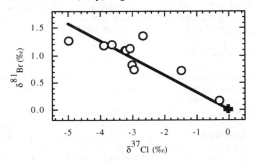

Figure 4. Relationship between Cl and Br isotope compositions in aquifer formation waters from the Oseberg field (Norwegian CS)

This indicates that the process for fractionating Br isotopes must be a different one from the one fractionating Cl isotopes, which might be because of the different redox behaviour of Br (much easier to oxidise), different behaviour in biological processes (easier to incorporate in organic compounds by organisms), and different behaviour when precipitating salt from a brine (Eggenkamp 1995; Eggenkamp & Coleman 1997a).

Although it is not yet known what caused the positive $\delta^{81}Br$, values the combination of Cl and Br isotopes indicate that the main process in the Oseberg Field is a two end member mixing between sea water and aquifer water with a $\delta^{37}Cl$ of approximately -5‰, and a $\delta^{81}Br$ of approximately +1.5‰. As the correlation is not perfect (r^2 = -0.75) it is also likely that other processes are taking place.

4 CONCLUSIONS

Results show that large variations in both halogen concentrations, and halogen isotope compositions exist between, and also within North Sea oil fields. Halogen isotopes add important information to our knowledge of these samples as they add an extra tool to distinguish between sources or end-members from data collections. Also, especially Cl, which is a very conservative element is able to indicate physical processes which otherwise, with other isotopes or plain chemistry could not be recognised. It is also likely that Cl isotope compositions are influenced by reactions between solutions and rocks, especially clay- mineral-rich mudstones. This is still very much under investigation and we expect to produce more conclusive results in the near future.

ACKNOWLEDGEMENTS

Part of the research presented in this paper was sponsored by BP, which also made available formation water and oil samples. Norsk Hydro provided oil and water samples under a CEC funded project (JOU2-0182). The current work is supported by a grant from the UK Natural Environment Research council to MLC (GR3/10360). Dr. J.H.S. MacQuaker (Manchester) provided the mudstones used in this study. We would like to thank Dr. Y.K. Kharaka for reviewing this manuscript.

REFERENCES

Campbell, D.J. 1985. *Fractionation of stable chlorine isotopes during transport through semipermeable membranes.* 103 pp. Tucson: M.Sc. Thesis.

Coleman, M.L. 1992. Water composition variation within one formation. In Y.K. Kharaka and A.S. Maest (eds), *Proceedings of the 7th International Symposium on Water-Rock Interaction:* 1109-1112. Rotterdam: Balkema.

Desaulniers, D.E., R.S. Kaufmann, J.A. Cherry & H.W. Bentley 1986. ^{35}Cl-^{37}Cl variations in a diffusion-controlled groundwater system. *Geochim. Cosmochim. Acta* 50:1757-1764.

Eggenkamp, H.G.M. 1994. $\delta^{37}Cl$: The geochemistry of chlorine isotopes. *Geol. Ultrai.* 116, 150 pp. Utrecht: Ph.D. Thesis.

Eggenkamp, H.G.M. 1995. Bromine stable isotope fractionation: Experimental determination on evaporites. *Terra Nova 7 Abstr. Suppl.* 1:331.

Eggenkamp, H.G.M. & M.L. Coleman 1997a. Bromine isotope compositions in natural samples: A novel application of a classical method. *Terra Nova 9 Abstr. Suppl.* 1:443.

Eggenkamp H.G.M. & M.L. Coleman 1997b. Comparison of chlorine and bromine isotope fractionation in diffusion: Geochemical consequences. In *Seventh Annual V.M. Goldschmidt Conference* p. 64. LPI Contribution No. 921. Houston: Lunar and Planetary Institute.

Eggenkamp, H.G.M., J.J. Middelburg & R. Kreulen 1994. Preferential diffusion of ^{35}Cl relative to ^{37}Cl in sediments of Kau Bay, Halmahera, Indonesia. *Chem. Geol (Isot. Geosc. Section)* 116:317-325.

Eggenkamp, H.G.M., R. Kreulen & A.F. Koster van Groos 1995. Chlorine stable isotope fractionation in evaporites. *Geochim. Cosmochim. Acta* 59:5169-5175.

Friedheim, C. & R.J. Mayer 1892. Über die quantitative trennung und Bestimmung von Chlor, Brom und Jod. *Z. anor. Chem.* 1:407-422.

Matray, J.M., A. Meunier, M. Thomas & J.C. Fontes 1989. Les eaux de formation du trias et du Dogger du Bassin Parisien: Histoire et effets diagénétiques sur les réservoirs. *Bull. Centres Rech. Explor.-Prod. Elf-Aquitaine* 13:483-504.

Phillips, F.M. & H.W. Bentley 1987. Isotopic fractionation during ion filtration. I: Theory. *Geochim. Cosmochim. Acta* 51:683-695.

Ransom, B., A.J. Spivack & M. Kastner 1995. Stable Cl isotopes in subduction-zone pore waters: Implications for fluid-rock reactions and the cycling of chlorine. *Geology* 23:715-718.

Thurlow, J.E. & M.L. Coleman 1997. Heterogeneity and evolution of formation waters in the oilfields of the Central North Sea. In J.P. Hendry et al. (eds), *Geofluids II:* 307-310. Belfast: The Queen's University of Belfast.

Warren, E.A. & P.C. Smalley 1994. (eds) *North Sea Formation Waters Atlas.* Geological Society of London Memoir No. 15. 208 pp.

Willey, J.F. & J.W. Taylor 1978. Capacitive integration to produce high precision isotope ratio measurements on methyl chloride and bromide. *Anal. Chem.* 50:1930-1933.

Ziegler, K & M.L. Coleman 1998. Chemical and isotopic variations of formation waters from the Oseberg field (Brent group, Norwegian North Sea): implications for palaeohydrodynamics. *Appl. Geoch.* 13:in press.

Fluids, migration systems and diapirism, East Coast North Island, New Zealand

B.D.Field, R.Funnell, G.Lyon & C.I.Uruski
Institute of Geological and Nuclear Sciences, Lower Hutt, New Zealand

ABSTRACT: Understanding fluid migration systems is especially important for companies exploring for hydrocarbons in the East Coast region because the best potential hydrocarbon source rocks (Late Cretaceous-Paleocene) are separated from the best reservoir rocks (Neogene) by a thick section of mudstones with good fluid seal potential. The Cretaceous-Cenozoic rocks the East Coast region of New Zealand comprise three main packets: a Cretaceous sandstone-mudstone succession, a Late Cretaceous-Paleogene mudstone-dominated unit, and a Neogene sandstone-mudstone-limestone succession deposited on the Hikurangi subduction margin. High sedimentation rates in the Neogene and vertical fluid migration from subducted sediments have contributed to over-pressured formation fluids, causing mud volcanoes, mud diapirs and problems for drillers. Avenues for fluid movement include faults, joints, lateral connectivity of adjacent clastic facies (e.g., slope channel conduits) and across unconformities at which low-permeability units have been removed by erosion, and bouyancy-created conduits (e.g., diapirism).

1 GEOLOGICAL SETTING

The East Coast region lies on the Neogene transpressive Hikurangi subduction margin. The region consists of three main sedimentary packets: a mainly sandstone and mudstone Cretaceous packet; a mudstone-dominated middle packet that contains latest Cretaceous to Paleocene hydrocarbon source rocks and Eocene smectitic mudstones and Oligocene calcareous mudstones; and a Neogene packet of sandstone, mudstone and bioclastic limestone with good reservoir potential. Total sediment thicknesses range to over 6 km and sedimentation rates were commonly very high. Over-pressuring of formation fluids is common: mud volcanoes are widespread and high fluid pressures are encountered in exploration wells. The region has New Zealand's highest concentration of hydrocarbon seeps (McLernon, 1972, Francis, 1995), proving generation, expulsion, and migration from Cretaceous-Paleocene source rocks through the Paleogene mudstone packet have occurred (Figure 1).

Petroleum exploration plays fall into two categories: sub-Eocene, and Neogene. Neogene plays require migration pathways through the Paleogene mud packet.

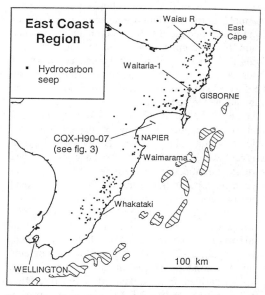

Figure 1: Locality map of the East Coast region, North Island, showing hydrocarbon seeps (dots) and areas of bottom-simulating seismic reflectors (inferred to be gas hydrate accumulations, hatch pattern; after Field & Uruski et al., in press).

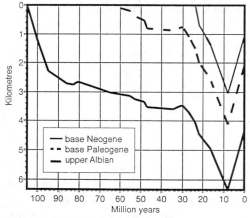

Figure 2: Subsidence plot, Rere-1.

2 FLUIDS

Fluid volumes in the region, which covers an area of about 70,000 km² are difficult to estimate for any particular time. The Cretaceous and Neogene packets are generally more than 2000 m thick; the "middle" unit, of Late Cretaceous to Oligocene age is commonly only a few hundred metres thick. Loading and porosity reduction due to burial were probably "normal" until late Oligocene to Neogene times (Figure 2), when rapid sedimentation and subsidence would have trapped large volumes of formation fluids and contributed to accelerated expulsion of fluids. Allis et al. (1997) inferred overpressuring on the East Coast was due largely to fluid migration from subducted sediments.

Some of the fluids discharged from springs and mud volcanoes permit study of the evolution of fluids within an active accretionary prism (Giggenbach et al., 1995). Waters have intermediate to high salinities (up to 26,000 mg Cl/kg) and oxygen isotopic ratios enriched relative to local groundwaters. The chemistry indicates close attainment of a water-rock equilibrium at 85±25 °C. The gas isotopic compositions indicate a range of fluid origins from microbial to thermally mature (Lyon et al., 1992). CH_4, N_2, Ar & He contents indicate a mixture of crustal and meteoric gases possibly derived from subducted sediments. High $^3He/^4He$ values (up to 3.3 R_a) around warm springs, indicate the presence of mantle helium in central East Coast, presumably moving up along the subducting plate boundary. Lewis & Marshall (1996) found depleted ^{13}C concentrations in cements associated with submarine fluid vents and inferred

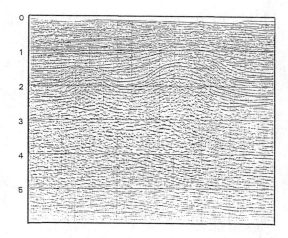

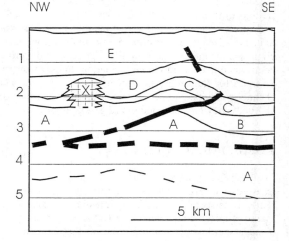

Figure 3: Seismic line CQX-H90-07, showing diapiric structure (X). Depths are in seconds TWT. A=?Cretaceous; B=Paleogene; C=Middle Miocene; D=Late Miocene; E=post-Miocene.

the exhalative hydrocarbons had a thermogenic origin.

3 DIAPIRISM

Katz & Wood (1980) reported shale diapirs from seismic reflection lines from the region. Figure 3 shows a diapiric structure in Hawkes Bay that intruded Late Miocene sediments but not the Pliocene succession. Diapirism might have commenced even earlier in some places, depending on local rates of Miocene sedimentary loading.

4 MUD VOLCANOES

Mud volcanoes are common on the East Coast (Ridd, 1970). Some have erupted violently, throwing mud and rocks 90 m into the air. Pettinga (1996) recorded a mud volcano (near gas seeps near Waimarama) that discharged about 10,000 m^3 of mud within hours of its birth in 1994. Nelson & Healy (1984) recorded pockmark-like depressions on the seabed offshore from Gisborne that they thought might be evidence for submarine mud volcanism or fluid vents.

5 MIGRATION PATHWAYS

5.1 Faults

The transpressive plate boundary setting of the region has led to widespread faulting during most of Neogene time and continuing today. Many hydrocarbon seeps occur near faults, suggesting faults facilitate migration of Cretaceous-Paleocene fluids through to Neogene reservoir rocks and the surface. Dean et al. (1976) referred to "diapiric walls" of mobile mudstone rising up fault zones. Kvenvolden & Pettinga (1989) described two gas seeps in southern Hawke's Bay that they inferred were sourced through migration up faults and melange zones.

5.2 Fractures

Fluid migration through fractures is likely in the Late Cretaceous-Paleocene Whangai Formation, a widespread siliceous shale at the base of the middle packet. Some Miocene limestones also have well developed, closely-spaced fractures and appear to have good permeability.

5.3 Porosity & permeability

Porosities and permeabilities are higher for Neogene units (Field, 1995; Field & Uruski et al., in press). The Plio-Pleistocene limestones of the East Coast (Beu, 1995) have high porosities, and permeabilities of over 10,000 mD (though most are less than 100mD).

5.4 Facies pathways

Sedimentary facies may provide good conduits for fluids. There are significant unconformities, many of regional extent, overlain by paralic sands. These units, which cut across older deposits, allow migration between different stratigraphic levels in the region; some hydrocarbon seeps occur at unconformities. Also, Neogene turbidite units locally have well preserved channel systems (e.g., at Whakataki coast & Waiau River). Such channels could link sand-rich basin floor fan successions with upper bathyal to shelf sediments, both across the SW-NE structural grain of the region and parallel to it (due to axial feeding of slope basins).

6 HAZARDS & PROBLEMS

Sudden gas release could pose a significant hazard to shipping through decreasing water density. Two "disturbances of the sea" off East Cape in 1877 were inferred by Gregg (1958) to have been due to submarine igneous activity, though Ridd (1970) thought submarine mud volcanoes were a more likely cause. The lack of mention in historic records of discolouration of the water by mud suggests the discharge was mainly gas - possibly gas hydrate phase change triggered by the earthquake recorded at the time. Gas hydrates appear to be common in the East Coast region (see modelling by Townend in press). Tsunamis generated from rapid submarine expulsion from diapirs have been reported near Gisborne by de Lange (1997).

Less catastrophic problems include mud pollution associated with fluid expulsion (both onland and offshore), the high erodability of weakly compacted Neogene mudstones, and difficulties experienced during drilling. The Waitaria-1 petroleum exploration well was abandoned largely because of over-pressured Miocene mudstones.

7 CONCLUSIONS

The "middle packet", mainly of Eocene mudstones, is a critical factor in determining migration pathways for crustal fluids and Cretaceous-Paleocene-sourced hydrocarbons moving into porous Neogene formations. Migration to the surface has occurred over most of (at least) the onshore part of the region and pathways include faults, permeable formations, facies-controlled conduits and unconformities. Overpressured and smectitic mudstones have caused problems in some exploration wells. Further studies and modelling of these mudstones will provide important insights into the fluid dynamics of the plate margin and its hydrocarbon resources.

REFERENCES

Allis, R., R.Funnell & X.Zhan 1997. Fluid pressure trends in basins astride New Zealand's plate boundary zone. In J.P.Hendry, P.F.Parnell, A.H.Ruffell & R.H.Worden (eds), *Contributions to the 2nd Int. Conf. on Fluid Evolution, Migration and Interaction in Sedimentary Basins and Orogenic Belts, Ireland, March 10-14, 1997*:214-217.

Beu, A.G. 1995. Pliocene limestones and their scallops. Lithostratigraphy, Pectinid biostratigraphy and paleogeography of eastern North Island late Neogene limestone. Inst.Geol.Nucl.Sci. monograph 10†.

Campbell, S. & T.R.Healy 1984. Pockmark-like structures on the Poverty Bay sea bed - possible evidence for submarine mud volcanism. *N.Z.J.Geol.Geophys.*27: 225-230.

Dean,P., P.J.Hill & A.D.Milne 1976. A review of the hydrocarbon potential of the PPL 38002 area, East Coast, North Island. N.Z. openfile petrol. rept 670‡.

de Lange, W.P. 1997. Tsunami hazard associated with marl diapirism off Poverty Bay, New Zealand. Abstract. *Geol.Soc.N.Z.misc.publ.*95A:49.

Field, B.D. 1995. Reservoir potential of the East Coast oil and gas province. *Petrol.Expl.N.Z. News*45:11-19.

Field, B.D. 1997. East Coast study brings new insights. *Petrol.Expl.N.Z.News*51:8-21.

Field, B.D., C.I. Uruski and 11 others in press. Cretaceous-Cenozoic geology and petroleum systems of the East Coast region, New Zealand. Inst. Geol.Nucl.Sci.monograph 19†.

Francis, D.A. 1995. Oil and gas seeps of northern and central East Coast basin. *Petrol.Expl. N.Z.News*44:21-27.

Gregg, D.R. 1958. Reports of a submarine eruption off New Zealand in 1877. *N.Z.J.Geol.Geophys.*1:459-460.

Giggenbach, W.F., M.K.Stewart, Y.Sano, R.L.Goguel & G.L.Lyon 1995. Isotopic and chemical composition of waters and gases from the East Coast accretionary prism, New Zealand. In: "Isotope and geochemical techniques applied to geothermal investigations". IAEA- TECDOC-788œ:209-231.

Katz, H.R. & R.A.Wood 1980. Submerged margin east of the North Island, New Zealand and its petroleum potential. *UN ESCAP, CCOP/SOPAC Tech Bull.*3:221-235.

Lewis, K. & B.A.Marshall 1996. Seep faunas and other indicators of methane-rich dewatering on New Zealand convergent margins. *N.Z.J.Geol.Geophys.*39:181-200.

Lyon, G.L., D.Francis & W.F. Giggenbach 1992. The stable isotope composition of some East Coast natural gases. Proc.1991 NZ Oil Expln Conf.2:310-319. Ministry of Commerce.

McLernon, C.R. 1972. Indications of petroleum in New Zealand. N.Z. openfile petrol. rept 839‡.

Pettinga, J.R. 1996. Mud volcano eruption, emergent accretionary margin, southern Hawke's Bay. Abstract. *Geol.Soc.N.Z.misc.publ.*91A:142.

Pettinga, J.R. & K.A. Kvenvolden 1989. Hydrocarbon gas seeps of the convergent margin, North Island, New Zealand. *Mar.Petrol.Geol.*6:2-8.

Ridd, M.F. 1970. Mud volcanoes in New Zealand. *Am.Assoc. Petrol.Geol. Bull.*54:601-616.

Townend, J. in press. Estimates of conductive heat flow through bottom-simulating reflectors on the Hikurangi and southwest Fiordland continental slopes. *Mar.Geol.*

† Available from Institute of Geological & Nuclear Sciences, Box 30368, Lower Hutt, N.Z.

‡ Available from Energy and Resources Division, Ministry of Commerce, Box 1473, Wellington.

Petroleum systems of the East Coast region, New Zealand

B.D.Field, R.Funnell, S.Killops, K.Rogers & C.I.Uruski
Institute of Geological and Nuclear Sciences, Lower Hutt, New Zealand

ABSTRACT: Comprehensive analysis of the Cretaceous-Cenozoic development of the East Coast region of New Zealand indicates significant oil and gas generation occurred from Cretaceous times to the present (Field & Uruski et al., in press). Very large structural closures exist and there are thick (in places stacked) potential reservoir units. The region is highly prospective, though still classed as "frontier".

1 GEOLOGICAL SETTING

The East Coast region lies on the transpressive Hikurangi subduction margin (Figure 1), where subduction began in the Early Miocene. Parts of the margin are well exposed onshore. The pre-Miocene history of the region comprises a mid Cretaceous subduction system, a Gondwana-breakup rifting phase, a comparatively quiescent "drift" phase and a late Paleogene subduction zone to the northwest of Northland and East Cape. The lithological succession is dominated by sandstone and mudstone, though packets of conglomerate, breccia, and pelagic and coarse bioclastic limestones also occur. The region has had a long and varied history of tectonism, and extensional, compressional and wrench structures are all present.

2 SOURCE ROCKS

The most likely source rock in the deeper parts of the region is the Paleocene marine Waipawa Formation "Black Shale", a brownish black mudstone with TOC values of around 2-6%, a high sulfur content and hydrogen indices ranging from about 50 to 500 kilograms of hydrocarbons per tonne of organic carbon. It has both oil and gas potential. The Waipawa Formation is widespread in outcrop and probably extends offshore - if so, it is estimated to have a volume of over 500km^3. The Cretaceous-Paleocene Whangai Formation, a widespread, marine siliceous mudstone underlying the Waipawa Formation, also has generative potential, as do some

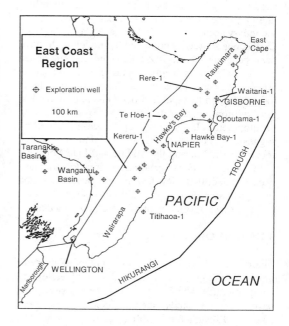

Figure 1: Locality map of the East Coast region. Only two wells have been drilled offshore (both encountered significant gas) yet the offshore part of the region has many large structural closures.

earlier Cretaceous rocks, particularly those with significant detrital terrestrial organic matter (e.g., Glenburn Formation).

3 MATURATION & EXPULSION

The East Coast region has New Zealand's highest concentration of seeps, proving both hydrocarbon generation and expulsion have occurred (McLernon, 1972, Francis, 1995). Significant gas was found in Hawke Bay-1 (Heffer et al., 1976), Te Hoe-1 (Dobbie & Carter, 1990) and Titihaoa-1 (Biros et al., 1995; Uruski, 1997).

Extensive source rock data are now available (Rogers, 1995; Elgar, 1997; Field & Uruski et al., in press). Maturation modelling using BASSIM (basin simulation software developed at IGNS and University of Utah), with new kinetics data, indicates considerable variability in the timings of generation and expulsion. In some places, any Cretaceous marine source rocks would probably have expelled most of their oil before many of the Neogene potential reservoir sandstones were deposited and before Neogene trap formation. Any oils from Cretaceous source rocks in these areas are therefore likely to be found in Cretaceous plays or as secondary accumulations in Neogene plays. Gas from Cretaceous sources is predicted by modelling to have been generated during the Paleogene in some areas, but most was generated during the Neogene. Paleocene source rocks are predicted in most places to have started to expel hydrocarbons in only the last few million years, after main reservoir and trap development. In some deeper areas modelling suggests expulsion of oil from Paleocene source rocks occurred during the Middle and Late Miocene.

In general, most generation appears to have occurred in only the last few million years, after (and in some cases during) deposition of reservoir units and the main development of structural closures.

4 RESERVOIR UNITS

There is a wide variety of types of potential reservoir units, including alluvial fan and incised valley fill conglomerates, shelf and paralic sands, slope and basin floor fan sands, shelfal bioclastic limestones (Beu, 1995) and units with fracture porosity. The reservoir characteristics of these are generally not well known, though reconnaissance work has provided some porosity-permeability values (Field, 1995). The best potential reservoir units are Neogene, with permeabilities of some paralic sands ranging above several hundred millidarcies and of flysch sands around 15-25 mD (Field & Uruski et al. in press). Pliocene bioclastic limestones have porosities of generally 10-40% and permeabilities ranging up to over 10,000 millidarcies (Beu, 1995).

Stratigraphic units with reservoir potential occur at several levels in most parts of the region, so any one exploration well is likely to penetrate stacked reservoirs and multiple plays.

5 TRAPS

Tectonism in Cretaceous and Neogene times has produced a range of both ages and types of structural closures. Plate motions and the partitioning of associated deformation have produced trap structures derived from normal, reverse, thrust and wrench fault systems, sometimes contemporaneously in different parts of the region, and the reactivation of structures is common. The main periods of structuring were mid Cretaceous and Neogene. Younger plays are more likely to be preserved, as the region is still tectonically active.

Many structural closures are of similar size to those of the giant Maui Gas Field of the Taranaki Basin (King & Thrasher 1996; Field & Uruski et al. in press) and the prospective parts of the East Coast region are extensive.

The interplay of tectonism and eustasy has enhanced the potential for stratigraphic traps, with paralic and neritic sandstones and limestones occurring between units of bathyal mudstone.

The Eocene to Oligocene succession is dominated by mudstones and could form good seals for pre-Eocene plays. Neogene plays are probably sealed by Miocene bathyal mudstones or mudstone units that occur between the porous Pliocene limestones.

6 PETROLEUM SYSTEMS

Improved knowledge of the source rocks, burial history, thermal regimes, reservoir units and structures in the region allows the refinement of petroleum systems models (after Magoon & Dow, 1994) for parts of the region or specific plays. Since the source rocks lie, in most places, below a thick section of mud-rich Eocene-Oligocene formations, plays will fall into two categories: sub-Eocene and Neogene. Neogene plays will require migration pathways through the Eocene mud units. The modelling of fluid flow up fault zones may therefore be critical to understanding migration pathways and their tectonic control, and hence petroleum systems, in the East Coast region.

REFERENCES

Beu, A.G. 1995. Pliocene limestones and their scallops. Lithostratigraphy, Pectinid biostratigraphy and paleogeography of eastern North Island late Neogene limestone. Inst.Geol.Nucl.Sci. monograph 10†.

Biros, D., R. Cuevas & W. Moehl 1995. PPL38318 Offshore East Coast basin, North Island, New Zealand. Titihaoa-1 geologic operations summary. N.Z.petrol.rept 2081‡.

Dobbie, W. & M.J. Carter 1990. Te Hoe-1 well completion report. N.Z.petrol.rept 1835‡.

Elgar, N. 1997. Petroleum geology and geochemistry of oils and possible source rocks of the southern East Coast Basin, New Zealand. Unpubl. PhD, Victoria Univ. of Wellington.

Field, B.D. 1995. Reservoir potential of the East Coast oil and gas province. *Petrol. Expl.N.Z.News*45:11-19.

Field, B.D. 1997. East Coast study brings new insights. *Petrol.Expl.N.Z.News*51:18-21.

Field, B.D., C.I. Uruski. and 11 others in press. Cretaceous-Cenozoic geology and petroleum systems of the East Coast region, New Zealand. Inst.Geol.Nucl.Sci. monograph 19†.

Francis, D.A. 1995. Oil and gas seeps of northern and central East Coast basin. *Petrol. Expl.N.Z.News*44:21-27.

Heffer, K., A. Milne & C. Simpson 1976. Well completion report - Hawkes Bay-1. N.Z.petrol.rept 667‡.

King, P.R. & G.P. Thrasher 1996. Cretaceous-Cenozoic geology and petroleum systems of the Taranaki Basin, New Zealand. Inst.Geol.Nucl.Sci. monograph 13†.

Magoon, L.B. & W.G. Dow 1994. The petroleum system. In L.B. Magoon & W.G. Dow (eds), *The petroleum system - from source to trap:* 3-23. Tulsa: AAPG.

McLernon, C.R. 1972. Indications of petroleum in New Zealand. N.Z. openfile petrol. rept 839‡.

Rogers K.M. 1995. Oil seeps and potential source rocks of the northern East Coast Basin, New Zealand: a geochemical study: Unpubl. PhD, Victoria Univ. of Wellington.

Uruski, C.I. 1997. Titihaoa-1. *Petrol.Expl.N.Z. News*52 (December 1997).

† Available from the Institute of Geological and Nuclear Sciences Ltd, Box 30368, Lower Hutt.

‡ Available from Energy & Resources Division, Ministry of Commerce, Box 1473, Wellington.

Chlorite coatings in deeply buried sandstones – Examples from the Norwegian continental shelf

J.Jahren, E.Olsen & K.Bjørlykke
Department of Geology, University of Oslo, Blindern, Norway

ABSTRACT: Subarkosic sandstones from two wells from the Smorbukk Field in the Haltenbanken area offshore mid Norway, have been studied. Cores range from 3148 to 4884 m depth and cover sections from early Jurassic to late Cretaceous. Special emphasis was put on the relationship between quartz cementation and chlorite coating but other diagenetic processes are also described. Shallow marine sandstone intervals in the lower Jurassic Tilje Formation (well 1. 4640-4847 m), and middle Jurassic Garn Formation (well 1. 4415-4432 m, well 2. 4172-4224 m) contained pervasive chlorite coatings. Chlorite coatings occur in almost all Garn samples from well 2, less than half the Tilje samples from well 1 and not at all in the Garn Formation from well 1. These iron-rich chlorite coatings cover detrital quartz grains and retard both the quartz cementation and the porosity reduction expected in these sandstones. The sandstones studied have therefore higher than normal porosities.

1 INTRODUCTION

The porosity preserving effects of authigenic chlorite coatings on detrital quartz grains have been recognised by many authors (Heald and Anderegg 1960, Pittman and Lumsten 1968, Tillman and Almond 1979, Houseknecht and Hathon 1987, Pittman et al. 1992, Ehrenberg 1993, etc). Pervasive chlorite coatings limits the area available for quartz cementation (Heald 1955, Pittman et al.1992), and since diffusion of silica is slow, the chemical compaction (dissolution of quartz at grain to grain contacts and quartz cementation) also slows down. Quartz cementation has traditionally been viewed as a stress driven process (Sorby 1863, Rutter 1983) enhanced by the presence of clay lamina (Heald 1955, Weyl 1959, and others) forming stylolites but the actual process of quartz cementation is mainly a function of temperature (Bjørkum 1996). This paper presents new observations of the relationship between quartz cementation, porosity preservation and chlorite coating in deeply buried subarkosic sandstones.

2 METHODS AND MATERIALS

Core descriptions of a total of 425 m of sandstone cores taken from two Haltenbanken wells located in the Smørbukk Field (Ehrenberg et al. 1992) were conducted at GECO in Stavanger. Sands from the Jurassic-Cretaceous section comprising the Tilje Formation, Tofte Formation, Ror Formation, Ile Formation, Not Formation, Garn Formation, Lange Formation and the Lysing Formation were described. Optical microscopy, X-ray Diffraction (XRD), X-ray Fluorescence (XRF) and Scanning Electron Microscopy (SEM) were conducted at the Department of Geology, University of Oslo, and Transmission Electron Microscopy (TEM) were performed at the Department of Physics, University of Oslo. 129 XRD samples were analysed using a Philips PW 170 Diffractometer, XRF analyses were conducted on 31 main element and 20 trace element samples using a Philips PW 2400 Spectrometer, 92 thin sections were studied optically and SEM of 24 thin sections and 48 stubs were performed in a JEOL 840. TEM analyses were done in a JEOL 2000 Fx.

3 RESULTS AND DISCUSSION

Stratigraphy, sedimentary facies and diagenetic aspects in sandstones from the Haltenbanken area has been described by Bjørlykke et al. 1986, Heum et al. 1986, Ehrenberg et al. 1992, among others. The

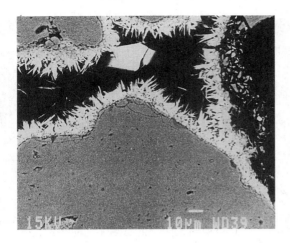

Figure 1. SEM images of grain coating chlorites, a) backscatter photomicrograph of a polished thin section showing the radial oriented nature of chlorite coatings. The sample is from the Tilje Formation (well 1, 4673 m burial depth), and b) photomicrograph of prismatic quartz crystals growing through the chlorite coating (Tilje Formation, well 1, 4844 m burial depth).

sandstones studied comprise subarkoses and quartz arenites with variable grain sizes and sorting. The clastic quartz content is 60% (point counting results) on average in the studied sandstones. The chlorite content in both the Tilje Formation and the Garn Formation is on average 4% (optical thin section point count results indicate a grain coating chlorite content between 0.3 and 7% for the samples studied). Chlorite is always found in the coarse-grained sandy units of the shallow marine Tilje Formation but some medium to fine-grained sands contain only minor amounts of chlorite. The shallow marine Garn Formation shows chlorite coating in one of the wells only (well 2). The chlorite is found both as ooids and late diagenetic coatings (figures 1a and 1b). The amount of ooids is always higher in the Tilje Formation than in the Garn Formation. The presence of ooids indicate a shallow marine environment with appreciable amounts of iron supply indicating proximity to an estuary environment.

The grain coating chlorites show little compositional variation and belong to the iron-rich chlorite variety of chamosite). The average composition of the Tilje Formation authigenic chlorite was $Mg_{0.75}Fe_{3.31}Al_{1.70}(Al_{1.21}Si_{2.79})O_{10}(OH)_8$ while the average composition of the Garn Formation chlorite was $Mg_{0.88}Fe_{2.89}Al_{1.88}(Al_{1.17}Si_{2.83})O_{10}(OH)_8$. The average porosity in the Garn and the Tilje formations is about 10% (3-25%). Chlorite coated intervals are 3-5 times more porous than non coated intervals (figure 2). Quartz cement is unevenly distributed and range from 0 to about 25% of the total rock. Sandstones without chlorite coating show the highest content of quartz cement. High mica content and the formation of stylolites is always associated with high quartz cement values. Textural evidence indicate that quartz cementation is a late diagenetic process post-dating the formation of kaolinite, siderite, calcite cement and chlorite coatings also found as authigenic phases in the sandstones. Chlorite coatings post-date the other pre-quartz cement diagenetic minerals. The formation of quartz cement, illite, dolomite and albite are contemporaneous late diagenetic processes.

The role played by coatings in porosity preservation is clearly documented in figure 1b, quartz cement morphology is very different from non chlorite coated surfaces. Authigenic quartz growth on non-coated surfaces are continuous and covers the detrital quartz grain entirely. Authigenic quartz on coated surfaces are typically formed as prismatic crystals (figure 1b) growing in optical continuity with the detrital grain beneath but unable to grow laterally across the chlorite coating. These observations are the key to understanding the porosity preserving nature (figure 2) of any coating, limited availability of nucleation surface and difficulty for the cementing material to grow laterally across coated surfaces (Jahren 1993). Since quartz diffusion out of the system (10-100 m) is slow the system will be nearly isochemical (Bjørlykke and Egeberg, 1993). When the area available for nucleation and growth of quartz cement is very small the pore water will be supersaturated with respect to quartz.

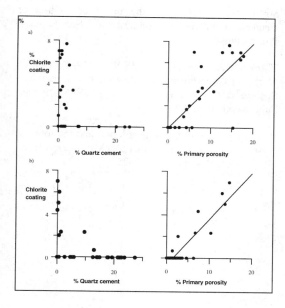

Figure 2. Grain coating chlorite content vs. quartz cement and primary porosity, a) Well 2 b) Well 1. Quartz cement content is highest in non-coated samples. The preserved primary porosity values show a positive correlation with increasing chlorite coating content.

The quartz supersaturation will approach the supersaturation created at the grain to grain contact by the differential stress-driven pressure solution process. Further quartz dissolution at grain to grain contacts will be effectively balanced if overgrowing the coating require a higher supersaturation than provided by the pressure-solution process. Effective redistribution of silica will therefore temporarily be stopped or slowed down in coated systems compared to non-coated systems.

Chlorite coatings on quartz grains probably depends on the presence of a precursor mineral which dissolves and provides the necessary ions for chlorite formation. This may be linked to a particular verdine facies where iron-rich minerals precipitate at the ocean floor (Ehrenberg 1993). The fact that the deltaic Brent Group of the North Sea does normally not show any development of chlorite coatings suggest that this verdine facies does not develop in deltaic environments but in more estuarine-tidal settings.

4 CONCLUDING REMARKS

Two wells from the Haltenbanken area situated offshore Norway have been studied. Sandy sections from Jurassic reservoir units illustrate the important part chlorite coatings play in porosity preservation in deeply buried sandstones. Porosity is commonly rapidly lost below 4000 m burial in regions with geothermal gradients > 3°C/100 m like the Haltenbanken area. In this study, porosity values higher than 25% and very little quartz cement has been found in samples buried to more than 4800 m depth. The reason for this is pervasive chlorite coatings limiting the available space for accommodation of quartz cement derived from differential pressure driven dissolution of detrital quartz grains at stylolites. The dissolution process will be slow due to this lack of accommodation space and primary porosity will be preserved.

This study shows that a link between the occurrence of chlorite coatings, precursor minerals and depositional environment can be developed. Predictive assessments of the distribution of chlorite coatings in the subsurface must be based on facies and provenance.

ACKNOWLEDGMENT

This research was supported by Saga Petroleum. The samples studied was provided by Den Norske Stats Oljeselskap (STATOIL). The manuscript benefitted from constructive reviews by R. Wanty and J. R. Hulstone.

REFERENCES

Bjørkum, P.A. (1996) How important is pressure in causing dissolution of quartz in sandstones?, Jour. Sed. Res., 66, 147-154.

Bjørlykke, K., Aagaard, P., Dypvik, H., Hastings, D.S., and Harper, A.S. (1986) Diagenesis and reservoir properties of Jurassic sandstones from the Halten-banken area, Offshore mid Norway, In: Habitat of hydrocarbons on the Norwegian continental shelf (Eds. A.M. Spencer et al.) Graham and Trotman, 275-286.

Bjørlykke, K., and Egeberg, P.K. (1993) Quartz cementation in sedimentary basins. AAPG Bull. 77, 1538-1548.

Ehrenberg, S.N. (1990) Relationship between diagenesis and reservoir quality in sandstones of the Garn Formation, Haltenbanken, Mid-Norwegian continental shelf. AAPG Bull. 74, 1538-1558.

Ehrenberg, S.N. (1993) Preservation of anomalously high porosity in deeply buried sandstones by grain-coating chlorite. Examples from the Norwegian continental shelf. AAPG Bull. 77, 1260-1286.

Ehrenberg, S.N., Gjerstad, H.M., and Hadler-Jacobsen, F. (1992) A gas condensate fault trap in the Haltenbanken Province, offshore Mid Norway. AAPG Memoir, 54, 323-348

Heald, M.T. (1955) Stylolites in sandstones. Jour. Geol., 63, 101-114.

Heald, M.T., and Anderegg, R.C., (1960) Differential cementation in the Tuscarora sandstone. Jour. Sed. Petrol., 30, 568-577.

Heum, K., Dalland, AQ., and Meisingset, K.K., (1986) Habitat of hydrocarbons at Haltenbanken (PVT-modelling as a predictive tool in hydrocarbon exploration). In: Spencer, A.M. (Ed.). Habitat of Hydrocarbons on the Norwegian Continental Shelf. Nor. Pet. Soc. Graham and Trotman, 259-274.

Housknecht, D.W., and Hathon, L.A. (1987) Relationships among thermal maturity, sandstone diagenesis and reservoir quality in Pennsylvanian strata of the Arkoma basin. AAPG Bull. 71, 568-569.

Jahren, J.S. (1993) Microcrystalline quartz coatings in sandstones: A scanning electron microscopy study. In: Karlson, G. (Ed), Extended abstracts from the 45 annual meeting of the Scandinavian Society for Electron Microscopy. 111-112.

Pittman, E.D., and Lumsden, D.N. (1968) Relationships between chlorite coatings on quartz grains and porosity, Spiro sand, Oklahoma. Jour. Sed. Pet., 38, 668-670.

Pittman, E.D., Larese, R.E., and Heald, M.T. (1992) Clay coats: occurrence and relevance to preservation of porosity in sandstones. SEPM Special Publication. 47, 241-255.

Rutter, E.H. (1983) Pressure solution in nature, theory and experiment. Geol. Soc. London Jour. 140, 725-740.

Sorby, H.C., (1863) On the direct correlation of mechanical and chemical forces. Royal Soc. London, Proceedings, 12, 538-550.

Tillman, R.W., and Almond, W.R. (1979) Diagenesis of the Frontier Formation offshore bar sandstones. Sparehead Ranch Field. Wyoming. SEPM Special Publication 26, 337-378.

Weyl, P.K. (1959) Pressure solution and the force of crystallisation - A phenomenological theory. Jour. Geophysical Res., 64, 2001-2025.

Mechanisms of vertical variations of $\delta^{13}C(CH_4)$ value in sediments

Mariusz O. Jędrysek
Laboratory of Isotope Geology and Biogeochemistry, University of Wrocław, Poland

ABSTRACT: These observations concern bubble methane from recent freshwater sediments in Lake Moszne (E Poland). The $\delta^{13}C(CH_4)$ value varied widely downward in vertical profile of about 3 meters, from about -4.5‰/m (late summer 1993) to about +2.5‰/m (late winter). Likewise, the production rate gradually decreased downwards ceasing at depth of about 1-3 meters. These vertical variations apparently are due not to temperature or oxidation effect, but rather to higher gradient of downward decrease of production rate *via* acetic acid fermentation than that *via* the CO_2-H_2 pathway. The downward increase of $\delta^{13}C(CH_4)$ value in winter, late autumn and, at greater depths, in late summer may be a consequence of ^{13}C isotope enrichment of the residual pool of precursors of methane (predominantly CO_2).

1 INTRODUCTION

Considerable attention has been directed towards quantifying tropospheric methane sources and their isotopic composition, and preparing mass fluxes and isotopic budget. On the other hand, as conditions of aquatic environment are recorded during early diagenesis, understanding of methanogenesis is crucial for understanding of $\delta^{13}C$ and $\delta^{18}O$ isotope profiles (Jędrysek 1997). It is well known that mechanisms of biogenic methane production (dominantly CO_2 reduction or acetic acid fermentation) can result in a wide range of $\delta^{13}C(CH_4)$ value and the high values corresponded to higher contribution of the acetate process (e.g. Sugimoto and Wada 1993, Jędrysek 1995, 1997). The factor limiting acetate dissimilation is the availability of acetate. When acetate production rate is very high but other microorganisms outcomplete the methanogens, little of it is converted to methane. If hydrogen is produced in substantial amounts by bacteria and dissociation of water, and likewise the CO_2 concentration in water increases, it may enhance methanogenesis *via* the CO_2/H_2 reduction pathway in the uppermost horizons of sediments.

The objective of this several year study was to obtain new data on CH_4 generated from lake sediments with respect to spatial and temporal factors to enable better understanding of the mechanisms of methanogenesis and supposed oxidation of methane, and to provide new data for isotope mass balances of atmospheric greenhouse gases. Methane samples reported in this work represents natural freshwater conditions of Lake Moszne (E Poland).

2 RESULTS and DISCUSSION

Sampling technique (agitation of sediments) and analytical procedure (methane oxidation and MS analysis) have been described earlier (Jędrysek 1995, 1997).

Both gross acetate production rate and acetate concentration in sediments are highest in the surficial zone of 0-2 cm and decrease several times downcore at 8-10 cm, although, the variation of acetate oxidation rate constants with depth has yet to be better clarified (Michelson *et al.* 1989). Acetate turnover is higher near the surface, but it is unsure if acetate dissimilation is significantly more important for methanogenesis than CO_2/H_2 reduction pathway, because H_2 production is highest near the surface as well. The input of fresh organic matter into sediments, at the end of summer, could enhance the production of acetic acid and provide higher $\delta^{13}C(CH_4)$ values. Consequently, the late summer/early autumn methane could show a more positive $\delta^{13}C$ value than in the overall annual cycle. In fact, highest temperatures of sediments (Fig. 1),

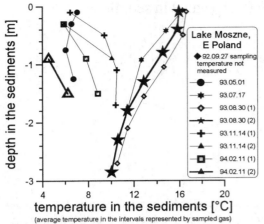

Fig. 1. Temperature variations in vertical profiles in the Lake Moszne sediments (E Poland) - each point corresponds to the gas sampling interval.

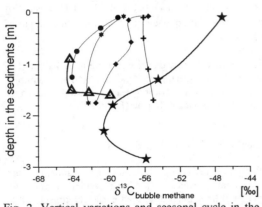

Fig. 2. Vertical variations and seasonal cycle in the vertical variations in $\delta^{13}C(CH_4)$ value in Lake Moszne sediments.

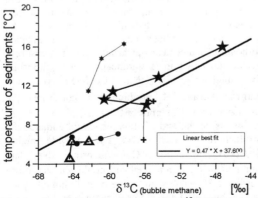

Fig. 3. Correlation in the system: $\delta^{13}C(CH_4)$ and temperature of sediments in Lake Moszne sediments.

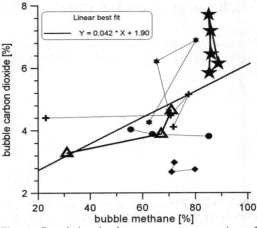

Fig. 4. Correlation in the system: concentration of bubble CH_4 - CO_2 in Lake Moszne sediments.

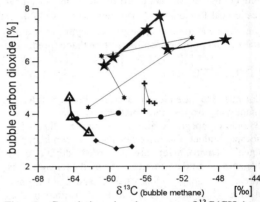

Fig. 5. Correlation in the system: $\delta^{13}C(CH_4)$ - concentration of bubble CO_2 in Lake Moszne sediments.

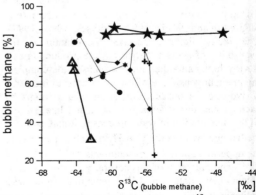

Fig. 6. Correlation in the system: $\delta^{13}C(CH_4)$ - concentration of bubble CH_4 in Lake Moszne sediments.

$\delta^{13}C(CH_4)$ values (Fig. 2), CO_2 and CH_4 concentrations (Fig. 4, 5, 6) and also the highest methane production were observed (also ebullitive flux measurements) during late summer and early autumn (Jędrysek 1997). After the acetate released in the shallower sediments was exhausted through methane production, methanogenesis could still proceed in the deeper layer, using CO_2 derived from organic and inorganic sources. In general, a slight positive correlation ($R^2=0.35$) in the temperature of sediments-$\delta^{13}C(CH_4)$ system (Fig. 3) may be suggestive for direct role of temperature in the process of methanogenesis. However, this is not always very true. For example, in tropical conditions, where vertical and annual variations in temperature of sediments is vanishingly low the $\delta^{13}C(CH_4)$ value is very low and strongly decreases with increasing depth (Jędrysek 1997).

Likewise, the lowest $\delta^{13}C(CH_4)$ value but negative $\delta^{13}C(CH_4)$ - depth correlation have been found for temperate climate conditions, especially in colder seasons or deeper parts of sediments. In Lake Moszne, some profiles e.g. 94.02.11 and deeper parts of 93.11.14 and 93.08.30 profiles, show in general negative $\delta^{13}C(CH_4)$-depth correlations (Fig. 2). The seasonal variation in vertical profiles of the $\delta^{13}C(CH_4)$ value in the Lake Moszne (Fig. 2) not always correlated well with the corresponding seasonal variation of temperature in sediments (Figs 1, 2, 3). The downward increase in temperature corresponds to downward decrease in $\delta^{13}C(CH_4)$ value in the uppermost part of the 93.11.14 $\delta^{13}C(CH_4)$ profile. In the deeper part of the same profile the downward decrease in temperature corresponds to downward increase in $\delta^{13}C(CH_4)$ value - the temperature and $\delta^{13}C(CH_4)$ value varies to the opposite directions. On the other hand, the downward increase in the temperature of sediments in 94.02.11 corresponds to downward increase in $\delta^{13}C(CH_4)$ value - the temperature and $\delta^{13}C(CH_4)$ value varies in the same direction. The 93.05.01 temperature profile show, in the interval from 0 to about 0.8 m, a downward decrease and below ca 0.8 m a downward increase of temperature, but the $\delta^{13}C(CH_4)$ value in this profile show only downward decrease, both above and below 0.8m - temperature and $\delta^{13}C(CH_4)$ value varies independently. These observations may suggest that temperature seems to be not the factor directly responsible for isotope signature of methane, consequently, other mechanism than temperature variation, probably govern the isotope pattern observed.

At higher temperature of summer the rate of decomposition of organic matter is apparently higher than during colder seasons. The Lake Moszne sediments are composed mostly of organic detritus. The temperature of the Lake Moszne sediments varied from 4.52°C during winter to 16.31°C during summer. Thus, in the surficial regions CO_2 and acetic acid, and in the deeper regions mostly CO_2, are apparently produced most efficiently at the end of summer when in the whole profile temperature is the highest. Therefore, the porewater during this season is saturated with respect to methane precursors (Figs. 4, 5, 6). It was observed by Chanton et al. (1989) that due to temperature-controlled solubility changes and temperature-dependent diffusion transport of methane, inventories of sedimentary gas bubbles was several times higher during summer than in winter. Thus, the observed more intensive bubbles production, higher $\delta^{13}C(CH_4)$ value and also higher CO_2 and CH_4 concentration in summer than in winter, resulted not only from more intensive bacterial activity at higher temperature but also from limited diffusion at higher temperature.

On the other hand, when temperature decreases, the concentration of methane precursors decreases being accompanied by significant isotope effect, i.e. the residual CO_2-acetic acid pool becomes gradually enriched in heavy isotopes - closed system effect. This pool is especially limited in the deepest part of sediments, hence the highest ^{13}C-enrichment was observed in the deepest parts of the winter and late autumn profiles. This model may also explain the lower gradients of the $\delta^{13}C(CH_4)$ value decrease at greater depths (93.05.01, 93.07.17 and 93.08.30). Likewise, the surprising increase of the $\delta^{13}C(CH_4)$ value in the deepest part (below 3 m) of the 93.08.30 profile may be explained by limited production of CO_2 at this depth, by active methanogenesis and by lower diffusion. Consequently, the deeper stratum CO_2 gets isotopically heavy as more CH_4 produced, so CH_4 gets heavy too - closed system isotope effect. Moreover, some extremely enriched CO_2 may diffuse upwards causing ^{13}C-enrichment of the carbon pool of methane precursors in overlying levels.

In anoxic marine sediments, methane production from bicarbonate was found to occur below the uppermost 2 cm (sampled interval 0-28 cm), whereas significant methane production from acetate only occurred at depths below 10 cm where sulfate was exhausted (Crill and Martens 1986). An implicit assumption could be made that acetate diffusion within natural environment was negligible, so that all the acetate produced within a given stratum was immediately consumed by reaction such as CH_4 fermentation, sulfate reduction or sorption (Michelson *et al.* 1989). Partial consumption of methane by oxidation can significantly shift hydrogen and carbon isotopic ratios positively in the residual methane (e.g. Barker and Fritz 1981, Coleman *et al.* 1981). Therefore, $\delta^{13}C(CH_4)$ value should be also higher when CO_2 concentration increase - this is the case (Fig. 5). Thus, one may conclude that the phenomenon of general decrease of $\delta^{13}C(CH_4)$ with increasing depth in sediments (Fig. 2) was possibly caused either by active methane consumption close to the surface region and/or by higher contribution of acetic acid fermentation in the surface parts and a relatively greater contribution by the CO_2/H_2 pathway in deeper parts of sediments. The latter process might be emphasized here, since the positive depth-$\delta^{13}C$ correlation in sediments possibly could not be a result of varying degrees of bacterial oxidation of trapped methane, because anoxic conditions were observed just several centimeters below the water-sediment interface, and the sulfate concentration in pore waters was close to zero. Moreover, in general higher concentration of CH_4 corresponds to higher concentration of CO_2 (Fig. 4). An opposite trend is expected when active oxidation produce CO_2. Likewise, methane concentration do not correlate to $\delta^{13}C(CH_4)$ value (Fig. 6), which prove that oxidation rate of methane is negligible as compared to the production rate.

In later stage of diagenesis, acetate and CO_2 may originate from different compounds showing different isotope ratios - supposedly enriched in heavy carbon isotopes. Thus, the vertical variation in $\delta^{13}C(CH_4)$ can be regarded as reflecting not only acetate/carbon dioxide pathways, oxidation and kinetic enrichment in ^{13}C of residual precursors of methane, but also the isotope characteristics of precursors of acetate and carbon dioxide. Further studies will be needed on this point. It could be an important basis to develop new tools for paleoenvironmental reconstruction based on vertical variations of carbon isotope composition of carbon-bearing compounds of sediments (organic matter, carbonates).

3 ACKNOWLEDGMENTS

This study was supported by the International Atomic Energy Agency (grant no. POL9400) and University of Wrocław grant no. 2022/W/ING/96-20.

4 REFERENCES

BARKER J.F. and FRITZ P., 1981, Carbon isotope fractionation during microbial methane oxidation., *Nature*, **293**, 289-291.

CHANTON J.P., MARTENS C.S., and KELLY C.A., 1989, Gas transport from methane-saturated, tidal freshwater and wetland sediments., *Limnol.Oceanogr.*, **34**, 807-819.

COLEMAN D.D., RISATTI J.B. and SCHOEL M., 1981, Fractionation of carbon and hydrogen isotopes by methane-oxidizing bacteria., *Geochim. Cosmochim. Acta*, **45**, 1033-1037.

CRILL P.M. and MARTENS C.S., 1986, Methane production from bicarbonate and acetate in an anoxic marine sediment., *Geochim.Cosmochim.Acta*, **50**, 2089-2097.

JĘDRYSEK M.O., 1995, Carbon isotope evidence for diurnal variations in methanogenesis in freshwater lake sediments., *Geochim. Cosmochim. Acta*, **59**, 557-561.

JĘDRYSEK M.O., 1997, Spatial and temporal variations in carbon isotope ratio of early-diagenetic methane from freshwater sediments: methanogenic pathways., *Acta Universitatis Vratislaviensis - Prace Geol.-Miner.*, (vol. **LXIII**, No 1997, pp. 1-110).

MICHELSON A.R., JACOBSON M.E., SCRANTON and MACKIN J.E, 1989, Modeling of distribution of acetate in anoxic estuarine sediments., *Limnol.Oceanogr.*, **34**, 747-757.

SUGIMOTO A. and WADA E., 1993, Carbon isotopic composition of bacterial methane in a soil incubation experiment: Contributions of acetate and CO_2/H_2., *Geochim. Cosmochim. Acta*, **57**, 4015-4027.

Reservoir heterogeneity due to fault related, rock-water interaction

Mingchou Lee
Mobil Exploration and Producing Technical Center, Dallas, Tex., USA

ABSTRACT: There are three main producing sandstones of the study field within the Upper Rotliegend (Lower Permian) in northwest Germany. The Havel is mostly located in the water zone, but it has the best reservoir quality. The upper two units, the Wustrow and Dethlingen, are heterogeneous, independent of facies. Detrital feldspar of the Wustrow and Dethlingen were more altered in samples near the fault-transfer zone than in samples near the graben center. Four alteration zones where cement compositions change gradually are recognized. Hot $CaCl_2$-rich saline fluids were present during cementation. During two episodes, peaking at around 205 and 180 MA, diagenetic fluids entered the reservoir through fault-transfer zones, and altered nearby sandstones. Away from the entry zones, sandstones in the paleo-HC zones were protected from further alteration. The Havel suffered only limited alteration because the sands were sealed from lateral fluid movement by conglomerates near the fluid entry zone.

1 INTRODUCTION

A variety of factors including sedimentary facies, sorting, grain size, compaction, pressure dissolution, cementation, structural/tectonic influences can control heterogeneity of reservoir quality. To better understand the control of diagenetic processes, quantitative information related to the timing and conditions of cementation has been collected using fluid inclusions data, stable isotope geochemistry and K/Ar geochronology of clays. Sibson et al., (1975), Sibson (1990, 1996) addressed fluid flow along faults using active fault fluid-flow models. Knipe (1993) discussed the possible influence of fault zone processes and diagenesis on fluid flow. Gaupp et al. (1993) proposed a fault-diagenesis model in which the influx of acidic fluids from the adjacent Carboniferous Coal Measures into Permian Rotliegend sandstones caused intensive feldspar and carbonate dissolution and major dickite/kaolinite cementation a few hundred meters from the graben boundary fault. Extensive illite cementation occurred in an outer zone away from boundary faults. Because few quantitative analyses were available, evaluation of the various fault related diagenesis models has been difficult.

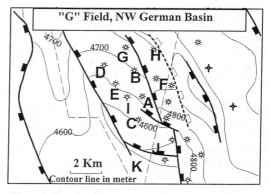

Figure 1. Location map, showing the top of the Wustrow.

2 PROBLEM STATEMENT

The Permian Rotliegend sandstone is the main reservoir unit of many hydrocarbon fields in the Southern North Sea, Netherlands, and northwest Germany. In the studied field, the G field (Figure

1), the main reservoir units, the Wustrow, Dethlingen, and Havel, consist largely of aeolian dune and associated facies. Reservoir heterogeneity exists at all scales, from the field scale to the single well scale. In general, the Dethlingen sandstones show inferior reservoir quality to the Wustrow except in the uppermost Dethlingen section of the A well. Some parts of the Wustrow sandstones in this field have poor reservoir quality, whereas other parts have excellent quality. The Havel dune facies has the best reservoir quality, with some thin beds showing extensive alterations. The multiple trends of the porosity and permeability (ϕ and k) correlation suggest complicated geological controls.

The questions that this study addresses are: "What geological processes, other than facies distribution, controlled the observed reservoir heterogeneity, and where are the reservoirs with optimal quality located?"

3 METHODOLOGY

This study uses an iterative methodology. First, we used geological data including logs, facies, petrography, and lab core porosity/permeability to construct a preliminary geological model. Based on this model, we selected a set of core samples to test such model using advanced analyses (i.e. stable isotope, age dating, fluid inclusion, etc) for the timing, the conditions, and spatial distribution of diagenetic processes. Such quantitative data allow us to critically evaluate the validity of the model, and to modify the model when deviation from the predicted outcome is found. The preliminary model we built was similar to Gaupp's model but with an added effect of early hydrocarbon emplacement on protecting reservoir quality. A set of samples was selected from 9 wells for testing the validity of the model.

4 DATA AND SAMPLE ANALYSIS

The degree of detrital feldspar (both K-feldspar and plagioclase) and volcanic rock-fragments alteration varies systematically. The alteration is more intensive in the C, D, E, and I wells where detrital feldspar and lithic grains were extensively altered, and diagenetic illite and pore-filling quartz are

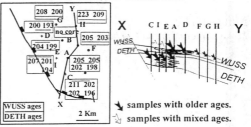

Figure 2. Schematic cross section, showing the location of the occurrence of diagenetic illites formed during the first episode.

abundant. In the F, G and H wells, the dominant clay in the Wustrow sandstones is early grain-coating Mg-Fe chlorite formed after compaction. Minor amounts of illitic clays grew after the chlorite grain coats. In the Dethlingen sandstones, the early chlorite was replaced by a moderate amount of illite. In the field, two kinds of illite polytype, 1M(cis) and 1M(trans), are found in the samples. In general, samples from both the Wustrow and the Dethlingen in the C, D, E, and I wells have larger amounts of 1M(cis) illite, whereas 1M(trans) is dominant in samples from the F, G, and H wells. A spatial zonation of the distribution of diagenetic minerals exists. The sequence of mineral zones, moving away from the boundary fault, are: platy grain-coating, 1M(cis)-rich illite + pore-filling quartz + calcite (moderate)---> platy/fibrous illite + pore-filling quartz + calcite (abundant) ---> chlorite grain-coating + fibrous, mostly 1M(trans) illite (minor) + calcite (minor).

The homogenization temperatures of the primary aqueous inclusions in the quartz and calcite cements range from 105 °C to 157 °C. The eutectic temperatures for the inclusions range from -53°C to -64°C, and the majority of the melting temperatures are approximately -30°C to -21 °C. The data suggest that the fluids trapped are very saline, $CaCl_2$-rich fluids.

For the Wustrow and the Dethlingen samples, a distinct bimodal distribution of K/Ar illite ages exists, i.e., the 225 to 190 Ma and the 185 to 160 Ma groups (the uncertainty in age data is typically 5 m.y.). The "older" age group peaks at

205-200 Ma, and the "younger" age group peaks at 175-170 Ma. The Havel ages fall in 210 to 180 Ma. These kinds of age ranges are similar to what are observed in the other Rotliegend fields (Lee, unpublished data). Stable oxygen isotopic ratios of chlorite-free clay fractions are analyzed. The δO^{18} values of diagenetic clays in this field fall into a narrow range of 18.2 to 15.8 permil. The range of δO^{18} values of the clays with older K/Ar ages is 18.2 to 17.6 permil, and the range of δO^{18} values of the younger clays is 17.8 to 16.7 permil. The average δO^{18} value of the older clays is about 0.5 permil higher than the younger clays. The burial depth of the reservoirs during the two episodes was between 2.5 to 3 km.

5 RESULTS AND DISCUSSION

Integration of facies, lab ϕ and k data, petrography, geochemistry, burial history, and structural setting shows that intensive rock/water interactions occurred during two major episodes of fault-related diagenesis, peaking at around 205 and 180 Ma. In Figure 2, the large arrows pointing the C, I, E, and A wells represent the extensive diagenetic alteration that occurred in these sites during the first (older) episode. The evidence for the first event, that impacted the Wustrow in the F, G and H wells, is given by the minor amount of hairy illite that grew after the grain-coating clays, shown by the small arrows. It appears that the diagenetic front penetrated the Wustrow in these wells. Based on these old ages, we conclude that no additional growth of illite had occurred in these rocks since the first episode. The data also show that the duration of this episode of diagenetic alteration is relatively short and reservoirs throughout the graben were influenced almost at the same time. The alteration that occurred during the second event is most extensive in the Dethlingen of the D well and the Wustrow of the C and I wells, where detrital feldspar were completely or nearly completely removed, and thick grain coating 1M(cis) illite and pore-filling quartz are common. The majority of illite diagenesis in the Dethlingen in the B, F, G, and H wells, including the formation of hairy 1M(trans) illite, occurred during the last stage of this episode.

The evidences presented above indicate strong structural and facies influences on water/rock interactions (Figure 3). The active diagenetic fluids entered the Rotliegend reservoirs of the "G" Field via the EW-trending fault transfer zone where two main NNW trending boundary normal faults are connected. Reservoirs in both the hanging wall and the footwall (Zwingmann, 1995) were affected.

Sedimentological evidence indicates that during the time of Havel deposition, the transfer zone was a relay-ramp and the sediments were transported into the graben over the ramp northward. Alluvial fan sediments were very well developed near the ramp, whereas the dune facies was deposited in the interior of the graben. During the Dethlingen and Wustrow time, the topographic relief was minimal, and the dune facies was deposited across the graben faults. Reactivation of the graben faults had dissected the Rotliegend. The Havel suffered only limited alteration because the sands were sealed from lateral fluid movement by conglomerates near the fluid entry zone.

6 SOURCE OF FLUIDS

Fluid inclusions data obtained in this study show that high salinity, $CaCl_2$-rich brines existed during quartz and carbonate cementation. After applying a pressure correction, the trapping temperatures of inclusion formation during the first diagenesis episode are between 140° to 170°C, much higher than what would be expected assuming a

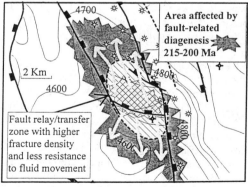

Figure 3, Revised 2D diagenetic model, showing strong structural influence on water/rock interaction.

geothermal gradient of 30° to 35°C/km. During Early Jurassic time, active rifting occurred on the both sides of the Atlantic Ocean. Similar high temperature and Late Triassic to Early Jurassic diagenetic events are found in the Upper Triassic sands in the Paris Basin (Clauer et al., 1995), in Triassic sandstones in the Irish Sea (Bushell, 1986), in Triassic shales of the Moroccan Atlantic Margin (Huon et al., 1993), and in Upper Triassic clastics in the Newark Basin (Steckler et al., 1993). Hardie (1990) suggested that modern rift and strike-slip systems are zones of upwelling hydrothermal $CaCl_2$ brines rich in Na, K, and Ca, but low in Mg, and with high concentrations of Fe, Mn, Pb, Zn Cu, Ba and other trace elements. The fluids are driven upward by density instabilities of thermal origin. $CaCl_2$-rich hydrothermal fluids might have existed in this tectonically active region and promoted major diagenetic alteration in various clastic sediments. The elevated temperatures, the large volume of fluid involved (to remove and precipitate minerals), and highly evolved nature of formation fluids ($\delta O^{18} \sim 10$ permil) suggested that convective fluid circulating systems might have existed during the time of the main diagenetic episodes.

7 ACKNOWLEDGMENTS

I want to acknowledge with gratitude the support of Mobil E&P Technical Center and Mobil Erdgas-Erdoel Gmbh. I appreciate greatly the analytical support by J. Lewandowski, K. Kronmueller, J.T. Edwards, J. Higgins, F. Roof, N. Houghton, J. Aronson, C. Elliott, L. Abel, T. Tsui, D. Ulmer-Scholle, and R. Reynolds.

8 REFERENCES

Bushell, T.P. (1986) Reservoir geology of the Morecambe Field. *In* J. Brooks, J. Gofs, and B. van Hoorne (Eds.). Habitat of Paleozoic Gas in North West Europe Geol. Soc. Special Publication, 23.

Clauer, N., J.R. O'Neil, and S. Furlan (1995) Clay minerals as records of temperature conditions and duration of thermal anomalies in the Paris Basin, France. Clay Minerals, 30, 1-13.

Gaupp, R., Matter, A., Platt, J., Ramseyer, K., and Walzebuck, J. (1993) Diagenesis and fluid evolution of deeply buried Permian (Rotliegend) reservoirs, Northwest Germany. AAPG 77, 1111-1128

Hardie, L.A. (1990) The roles of rifting and hydrothermal CaCl2 brines in the origin of potash evaporites: An hypothesis. Am. Jour. Sci. 290, 43-106.

Huon, S., J. Cornee, A. Pique, N. Rais, N. Clauer, N. Liewig, and R. Zayane (1993) Mise en evidence au Maroc d'evenements thermiques d'age triasico-liasique lies a l'ouverture de l'Atlantique. Bull. Soc. geol. France, 2, 165-176.

Knipe, R.J. (1993) The influence of fault zone processes and diagenesis on fluid flow. *In* Horbury and Robinson (eds.) "Diagenesis and basin development". AAPG Studies, 36, 135-151.

Sibson, R.H. (1990) Conditions for fault-valve behaviour. *In* R.J. Knipe and E.H. Rutter (eds.) "Deformation mechanisms, rheology and tectonics", Geol. Soc. Special Publication, 54, 15-28.

Sibson, R.H., J.M. Moore, and A.H. Rankin (1975) Seismic pumping - A hydrothermal fluid transport mechanism. J. Geol. Soc. London, 131, 653-859.

Sibson, R.H. (1996) Structural permeability of fluid-driven fault-fracture meshes. J. Struc. Geol. 18, 1031-1042.

Steckler, M.S., G.D. Karner, G.I. Omar, and B.P. Kohn (1993) Pattern of hydrothermal circulation within the Newark Basin from Fission-Track Analysis. Geology, 21, 735-738.

Zwingmann, H. (1995) Study of the conditions of gas emplacement in sandstone reservoirs (Rotliegende of Germany). mineralogical, morphological, geochemical and isotopical aspects. Ph.D. thesis, U. of Strasbourg, France, 197p..

Transformation of diatomite into porcelanite and opaline chert under the influence of an andesite intrusion in the Miocene Iwaya Formation, Japan

E. Nakata
Central Research Institute of Electric Power Industry, Chiba, Japan

M. Chigira
Disaster Prevention Research Institute, Kyoto University, Japan

M. Watanabe
Central Research Service Corporation, Chiba, Japan

ABSTRACT: We found two alteration types of diatomite intruded by andesite dike. Type I: A porcelainte layer occurs along andesite-siliceous rock contact with a width of about 20cm–20m, and a chert layer consisting of amorphous silica (opal-A_G) and tridymite-rich opal-CT occurs between the porcelanite and diatomite with a width of about 5cm–20m. Chert layer is made by the compaction of smectite-rich diatomite and precipitation of opal-A_G and tridymite-rich opal-CT from silica-supersaturated hydrothermal solution came from andesite. Type II: Diatomite is only slightly compacted near andesite dike almost without mineral alteration, probably because little hydrothermal solution infiltrated into diatomite.

1 INTRODUCTION

Opal-A, which is major mineral of diatomite, is known to transform into quartz through a metastable phase of opal-CT at progressive diagenesis (Murata and Nakata, 1974). Kastner and Siever (1993), also reported the formation of opal-CT, quartz, and smectite in diatomite intruded by basalt. Although mineralogical changes of siliceous rocks have been studied, change of physical and hydraulic properties with the mineralogical changes has scarcely been studied.

We investigated the alteration mechanism of diatomite of the Miocene Iwaya Formation intruded by andesite in Akita Prefecture, northeast Japan. In this paper we discuss the formative mechanism of opaline chert and change of physical properties, such as porosity, permeability, and pore-size distribution, with mineralogical change of silica minerals around andesite dike. Some results have already been reported (Chigira, et al., 1995).

2 GEOLOGICAL SETTING

The Iwaya Formation consists mostly of a diatomite with intercalations of thin acid or basic tuff beds and a 4 to 4.8 m-thick pumice tuff bed. The age of the Iwaya Formation is estimated to be from 15 or 12 Ma to 10.6 or 8.9 Ma from fossils (Yamamoto et al.,1994).

Andesite dikes intruded into diatomite with widths from 4 m to 400 m. The K-Ar age of one andesite dike was 9.5 ± 0.5 Ma, suggesting that the andesite dikes intruded into diatomite nearly contemporaneous with the latter's deposition.

3 OCCURRENCE OF SILICEOUS ROCKS NEAR ANDESITE DIKE

We found nine contacts of siliceous rock and andesite dikes, and one contact of siliceous rock and pumice tuff in the study area. The siliceous rocks near the contacts with andesite dikes were classified into two types.

Type I: A brownish gray or dark gray porcelanite layer occurs along the contact with andesite. Thin layers of black soft siliceous rock is often intercalated between the porcelanite layer and andesite. A very hard blackish vitreous chert layer occurs along the periphery of the porcelanite layer. The porcelanite and chert are alteration products of diatomite. Smectite occurs in and around the chert layers. Thickness of porcelanite and chert layers range from a few centimeters to about 20 m for each.

Type II: Almost diatomite. Diatomite color, however, change into blackish gray within several centimeters from andesite. Chert layers do not occur along around the andesite. Silica mineral alteration is hardly, with one exception where porcelanite layer with only a few centimeters occurs along the contact with andesite.

We drilled two holes, A and B. Hole A penetrated type I contact between siliceous rocks and an andesite dike with a thickness of 400 m. Hole B penetrated type II contact between siliceous rock and an andesite dike with a thickness of 30 m (or 200 m).

Drill hole A, also penetrated alteration zones on and beneath a gently-dipping pumice tuff with 4.8 m thickness that was intruded by the andesite dike too. The alteration type of the siliceous rocks in contact with the pumice tuff was of type I, suggesting that the porcelanite and chert layers were made by the interaction of diatomite and hydrothermal solution infiltrated into diatomite from the pumice tuff.

Drill hole B, on the other hand, only a thin porcelanite layer occurred along the contact with andesite.

4 ALTERATION NEAR HEAT SOURCE

4.1 Mineralogy

Silica minerals identified by XRD, SEM and FE-SEM were opal-A, opal-A_G, opal-CT, and quartz. Alteration zones of the contacts of type I are divided into three zones of silica minerals (Figure 1).

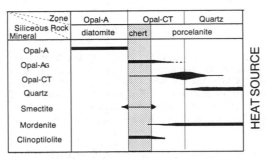

Figure 1. Alteration zones and mineral assemblages observed in the diatomite intruded by andesite.

The opal-A zone consists of diatomite, which is composed largely of diatom tests. The opal-CT zone consists of chert composed of opal-A_G and opal-CT, in addition to porcelanite composed of opal-CT, except for the presence of authigenic quartz near the andesite. The quartz zone consists of porcelanite composed of authigenic quartz and cristobalite-rich opal-CT. As the relative amount of authigenic quartz with internal standard $BaCO_3$ increases, opal-CT decreases toward the andesite with a decrease in distance from the chert. Smectite is present in chert and in siliceous rocks near the chert in hole A, and in diatomite that occurs about 5m from the intrusive contact in hole B.

Opal-CT is usually characterized by d(4.1 Å) spacings. They changed as follows in the drilled core A (Figure 2). The d(4.1Å) decreased from 4.123Å of chert to 4.050Å of porcelanite near andesite, also decreased from 4.106Å of chert to 4.056Å of porcelanite near the pumice tuff. Similar d(4.1Å) change was observed for the samples obtained from other outcrops: the change occurred within an interval of several centimeters to maximum 40 m in this study area. Once the d(4.1Å) became around 4.06Å, relative amount of quartz increased sharply and the d(4.1Å) was constant at about 4.06Å in the quartz zone. The d(4.1Å) decrease is due to the increase of low-cristobalite (101) and decrease in low-tridymite (002) in opal-CT.

Mizutani (1977) showed that the rate of d(4.1Å) decrease is definitely dependent on the reaction temperature, being faster at higher temperatures. We infer that large thermal gradient occurred around andesite and the pumice tuff, and the opal-A_G and tridymite-rich opal-CT occurred at lower temperatures than the temperatures of the place where cristobalite rich opal-CT and quartz occurred. No authigenic quartz were identified around the pumice tuff, probably because the temperature there was lower than near andesite.

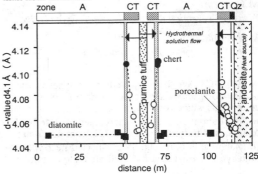

Figure 2. The distribution of d(4.1Å) spacing on the hole A. ●; chert, ○; porcelanite, ■; diatomite

4.2 Temperature inferred from zeolite minerals

Authigenic zeolite minerals, mordenite, clinoptilolite, and phillipsite, were identified with silica minerals. From these minerals, we can roughly estimate the alteration temperature. Mordenite was found to coexist with opal-CT and/or quartz in the chert and porcelanite. Clinoptilolite was found with opal-A_G and tridymite-rich opal-CT in the chert and the porcelanite near the chert layer at the hole A. Phillipsite occurred in an intercalated thin acid tuff in the opal-CT zone. These zeolites are known to be made by low-temperature hydrothermal alteration (less than 150 ℃) in field (Hawkins, 1978). Mordenite has been synthesized at the most high-temperature (380℃) in these zeolites (Ames and Sand, 1958). Clinoptilolite is known to coexists with opal-CT and clay at DSDP porcelanite samples (Thein and von Rad, 1987)

From the facts described above, we infer that the alteration zones were made under relatively low temperatures in comparison with andesitic magma temperatures.

4.3 Morphology of silica minerals

SEM analysis clarified that chert consisted of spherical particles of amorphous silica with a diameter of $0.5\,\mu m$ (opal-A_G; Florke et al., 1991) and opal-CT lepispheres (Wise and Kelts, 1972) with a diameter of $0.5{\sim}2\,\mu m$ which are aggregated small hexagonal blade-shaped crystals. Chert has a large

Figure 3. SEM photograph showing amorphous silica (opal-A_G) in the chert. Original pore space had been filled up by the precipitated opal-A_G.

Figure 4. SEM photograph showing smooth surface opal-CT lepisheres d(4.1Å) = 4.052Å. The diameter of opal-CT decrease with progressive alteration.

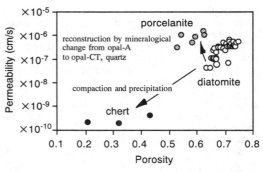

Figure 5. Porosity and permeability of Siliceous rocks.

amount of opal-A_G (Figure 3) and opal-CT lepispheres.

Porcelanite of the opal-CT zone consisted of opal-CT lepispheres. The diameter of the lepishperes decreased from about 1~2 μm near the chert to about 0.5 μm when approaching to the pumice tuff or andesite. With decreasing the diameter, the opal-CT lepisphere changed their surface structure from so-called tridyimite-type twining (Florke, et al., 1976) to smooth cristobalite rich opal-CT lepisphere (Figure 4). Authigenic quartz grains were euhedral.

The amorphous silica-like material and smooth surface lepispheres discovered in our study were similar to those noticed by earlier researches (Pollard and Weaver, 1973 : Oehler, 1975).

5 CHANGE OF PHYSICAL PROPERTIES

With the mineralogical changes stated above, porosity, permeability, and pore-size distribution also changed. Porosity decreased from diatomite to chert, and permeability also was the smallest for the chert (10^{-10} cm/s order). In contrast to the porosity difference between porcelanite and diatomite, however, porcelanite was more permeable than the diatomite (Figure 5). Pore-size distribution of diatomite and porcelanite except for the porcelanite near the chert was unimodal, with a peak between 0.4 μm and 0.8 μm. Pore-size distribution of the chert also was unimodal with a peak between 0.01 μm and 0.04 μm. The porcelanite near the chert had bimodal pore-size distribution, each peak being between 0.01–0.04 and 0.4–0.8 μm.

The porosities, permeabilities, and the pore-size distributions of diatomite and porcelanite, indicate that pores connected to each other increased by the alteration from diatomite to porcelanite.

6 THE FORMATION OF IMPERMEABLE CHERT NEAR HEAT SOURCE

The formation of a chert layer is closely related to the mechanism of the hydrothermal alteration of diatomite by magma intrusion at the study area. The occurrence of chert layers in the study area indicates that they were made along their precursors, compacted zones of smectite-bearing diatomite; smectite formed by the heat probably softened the diatomite which in turn was compacted to make the preferred orientation of smectite and diatom frustules (Figure 6).

The compaction and resultant preferred orientation of diatom frustules and smectite decreases the

Figure 6. SEM photograph showing compacted diatom tests in diatomite near the chert.

permeability perpendicular to the compacted zone of diatomite and causes a gradient of pore pressure, effective stress, and temperature across it.

Temperature, effective stress, and fluid pressure near andesite are shown schematically in Figure 7 which was drawn by the consideration on heat transfer by advection or conduction and on fluid flow. Hydrothermal solution flowing away from andesite decreases its temperature to the most when it crosses the precursor above stated and also increases the mean pore velocity to the most extent there, probably being highly supersaturated with amorphous silica. Amorphous silica, thus precipitates along the precursor from the hydrothermal solution (Chigira and Watanabe, 1994).

An impermeable chert layer make a semiclosed system between the layer and andesite. Therefore, temperature between the layer and andesite would be kept higher than outer zone, which is preferable for the transformation of opal-A to opal-CT or quartz.

When hydrothermal solution does not infiltrate into diatomite in large amount, temperature, effective stress, and pore pressure do not have large gradients even though smectite occurred in diatomite, so a chert layer does not form. However, diatomite is heated by conduction and it is compacted as far as tens of meters away from the contact with andesite, so permeability decrease with the compaction.

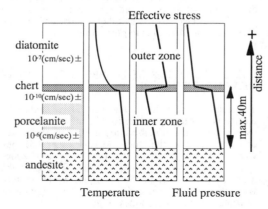

Figure 7. Conception of formation of type I around heat source. Smectite rich parts were compacted with increase in the effective stress. The solution supersaturated into amorphous silica with increase in the mean pore velocity at the compacted zone.

REFERENCES

Ames, L. L. and Sand, L. B. 1958. Hydrothermal Synthesis of Wairakite and Calcium-Mordenite. *Amer. Mine.*, 43, 476-480.

Chigira, M. and Watanabe, M. 1994. Silica Precipitation Behavior in a Flow Field with negative temperature gradients. *Jour. Geophys. Res.*, 99, 15539-15548.

Chigira, M., Nakata, E. and Watanabe, M. 1995. Self-sealing of Rock-Water systems by Silica precipitation. in Proceedings of the 8th International Symposium on Water-Rock Interaction, edited by Kharaka, Y. and Chudaev, O., 73-77.

Florke, O. W., Hollmann, R., von Rad, U., and Rosch, H. 1976. Intergrowth and Twinning in Opal-CT Lepispheres. *Contrib. Mineral. Petrol.*, 58, 235-242.

Florke, O. W., Martin, G.B., Bochum, R., and Wirth, R. 1991. Nomenclature of Micro- and Non-crystalline Silica Minerals, based on Structure and Microstructure. *Neues Jahrbuch Miner. Abh.*, 163, 19-42.

Jones, J. B. and Segnit, E. R. 1971. The Nature of Opal I.Nomenclature and Constituent Phase. *Jour. Society Australia*, 18, 57-68.

Hawkins, D. B. 1978. Hydrothermal Synthesis of Clinoptilolite and Comments on The Assemblage Phillipsite-Clinoptilolite-Mordenite. *Natural Zeolites.*, 337-343.

Kastner, M. and Siever, R. 1993. Siliceous Sediments of the Guaymas Basin: The Effect of High Thermal Gradients on Diagenesis. *Jour. Geology*, 91, 629-941.

Mizutani, S. 1977. Progressive Ordering of Cristobalite Silica in the Early Stage of Diagenesis. *Contrib. Mineral. Petrol.*, 61, 129-140.

Murata, K. J. and Nakata, J. K. 1974. Cristobalitic stage in the diagenesis of diatomaceous shale. *Science*, 184, 567-568.

Oehler, J. H. 1975. Origin and Distribution of Silica Lepispheres in Porcelanite from the Monterey Formation of California. *Jour. Sedimentary Petrology*, 45, 252-257.

Pollard, C. O. and Weaver, C. E. 1973. Opaline Spheres: Loosely-Packed Aggregates from Silica Nodule in Diatomaceous Miocene Fuller's Earth. *Jour. Sedimentary Petrology*, 43, 1072-1076.

Thein, J. and von Rad, U. 1987. Silica Diagensis in Continental Rise and Slope Sediments of Eastern North America (Site 603 and 605, Leg93: Sites 612 and 613, Leg95). *Initial. Reports, DSDP*, 501-525.

Wise, S. W. and Kelts, K. B. 1972. Inferred Diagenetic History of a Weakly Silicified Deep Sea Chalk. *Trans. Gulf Coast Assoc. Geol. Societies*, 22, 177-203.

Yamamoto, S., Irino, T., Tada, R., and Iijima, A. 1994. Reconstruction of a Miocene volcanic seamount and the sedimentation depth for siliceous shale of the Onnagawa Formation in the Fujisato area, northern Akita, *Jour. Geol. Soc. Japan*, 100, 557-573.

Salt springs and structural setting of the Marchean Adriatic foredeep, Central Italy

T. Nanni & P. Vivalda
Dipartimento di Scienze dei Materiali e della Terra, Università degli Studi di Ancona, Italy

ABSTRACT: In the present work the relationship between salt springs emerging from adriatic foredeep deposits and structural setting is examined. The genesis of the salt waters is discussed and it is demonstrated that the presence of salt waters is an indicative phenomenon of an active tectonic regime.

1 INTRODUCTION

Many localities of the Marche region have springs with mineralised waters. In this work the salt waters of marchean foredeep are examined. The salt springs emerge from plio-pleistocenic and messinian deposits. Many springs are also located in the continental deposits of alluvial plains (Fig. 1). On the surface, the salt waters emerging from plio-pleistocenic deposits generate large morphological forms very similar to mud vulcanoes.

2 STRUCTURAL SETTING OF THE CENTRAL ADRIATIC FOREDEEP

The strucural setting of the central adriatic foredeep in Central Italy is represented by folds bordered by thrusts connected with Apennines tectogenesis (Fig. 1). The messinian and plio-pleistocenic pre-orogenic deposits, which outcrop in proximity to the umbro-marchean ridge and the adriatic coast, are strongly tectonised by thrusts and faults with apennine and antiapennine direction.

The structural setting of plio-pleistocenic post-orogenic sequence is characterised by a monocline style toward the Adriatic sea and interrupted by apennine and transversal faults. Seismic profiles (Calamita, 1990; Ori et al., 1991) prove the presence of thrusts buried by the plio-pleistocenic sequence in which, generally, the faults are connected with the buried thrusts.

The messinian deposits of the marchean sequence are formed by arenaceous bodies, marly clays, tripoli, gypsum-arenites, bituminous clays and gypsum of the "Gessoso-solfifera" Formation. The plio-pleistocenic sequence is formed by marly clays with alternance of sandstone bodies (Fig. 1).

3 SALT WATERS

The analyses of physical and chemical parameters permit a subdivision of the salt waters in to three groups with different salinity. Every group includes waters emerging either from the plio-pleistocenic or from the messinian deposits (Tab. 1, Fig. 2).

Variable dilution by vadose waters causes the different salinity of the salt waters. The vadose waters may flow to considerable depth by circulating through aranaceous bodies in the plio-pleistocenic clays and through the fractures.

The waters with higher salinity (A and B groups) emerge in the main thrust fronts and are associated with the faults. This situation has been observed both in the inner part of Marche foredeep, and in the vicinity of adriatic coast (Fig. 1, 3). The notable values of ammonium ion and the modest contents of NO_2 suggest a groundwater circulation that prevents mixing with oxygenated surface waters. Isotopic data of saline waters prove an age older than 40.000 years and an uniform distribution in ^{18}O and ^{2}H (Chiarle et al., 1992).

The springs with lower salinity (Group C) are present principally in the centre-oriental foredeep. Many springs emerge from continental deposits and the salt waters are mixed with groundwaters of the alluvial aquifers. Furthermore, in these areas the plio-pleistocenic sequence has a monocline style

Fig.1. Geolithological scheme of the marchean adriatic foredeep. 1. Alluvial terraced deposits (Pleistocene-Holocene); 2. Clays and marly clays (post-orogenic Plio-pleistocene); 3-4. Marly clays (pre-orogenic lower and middle p.p. Pliocene; 5-6-7. Marly clays (5), bituminous clays and gypsum of messinian evaporites (6) and marly clays with arenaceous bodies (7); 8. Marly limestones, marls and clayey marls (Serravallian-Tortonian); 9. "Marnoso-arenacea" Formation; 10. Meso-cenozoic limestone ridge; 11.Salt springs; 12. Sulphureous springs; 13. Salt waters present in the alluvial aquifers; 14. Faults.

toward the Adriatic sea and covers the pre-orogenic deposits characterised by buried thrusts (Fig. 4).

Table 1. Summary of chemical composition in the salt waters of marchean foredeep (Electrical conductivity in mS, Element concentration in mg/l)

N°	Cond.	Na	K	Mg	Sr	Cl	SO4	Br
A								
110	\	22950.00	254.00	1788.00	\	45285.00	10.10	\
25	237.0	38746.00	180.00	3126.00	165.30	90454.00	640.60	\
40B	76.0	18500.00	114.00	701.00	26.44	29244.00	0.01	\
69B	50.0	30680.00	265.00	1384.00	178.60	58412.00	6.25	\
94A	208.0	54270.00	317.00	2671.00	325.00	106080.00	32.00	1970.00
94C	226.0	56570.00	387.00	3173.00	374.80	119090.00	140.00	2430.00
B								
40A	40.0	9420.00	76.80	443.60	17.96	14587.00	0.01	\
45	24.3	9424.00	202.80	540.00	10.68	16763.00	0.89	89.32
51	9.2	3263.00	43.20	140.10	7.27	5399.00	\	25.80
75	33.8	9602.00	63.20	542.10	34.00	17872.00	3100.00	\
90A	21.2	5820.00	142.00	321.70	16.17	9877.00	0.01	\
90B	14.8	4030.00	102.00	184.60	8.95	6347.00	0.01	\
C								
7	1.32	264.80	19.10	83.30	2.97	354.50	358.00	1.67
109	3.9	1289.00	37.20	63.90	1.89	1231.00	0.01	\
20C	6.8	2874.00	49.60	152.40	2.08	699.00	5074.00	\
22	3.1	631.20	1.76	53.50	2.23	716.70	430.90	2.72
27	0.5	29.30	6.69	38.30	0.75	30.03	16.04	0.10
50D	4.0	961.90	7.99	4.55	0.12	334.40	2.00	\

Here the salt waters are strongly diluted by vadose waters circulating in the arenaceous units, in contact with salt pliocenic waters in the faults areas.

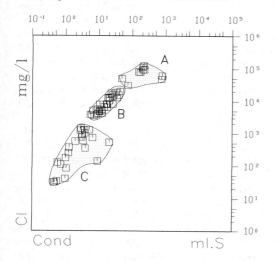

Fig. 2. Electrical conductivity-Cl relationships in the salt waters of the marchean foredeep. The three groups of waters with different salinity are indicated.

The waters emerging from messinian deposits originated from marine waters subjected to evaporation up to a saline concentration, similar to seawater when gypsum begins to precipitate. The brines remain trapped in the clayey deposits and isolated from gypseous bodies of messinian sequence. These events recurred, as demonstrated by the presence of salt waters in different levels of the sequence.

The origin of the plio-pleistocenic salt waters is not associated with the evaporation of marine waters. Plio-pleistocenic sediments are typical of a deep basin, and waters originated from marine waters trapped in the sediments and subjected to ultrafiltration processes through the clayey membrane, with the formation of the brines. Some enrichment of salinity may derive from mixing with the underlying messinian waters, lifting along fractures. The presence of free gas (CH_4, H_2S, CO_2 etc.), always associated with salt waters, permits the lifting of the waters (Chiarle et al., 1992).

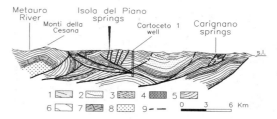

Fig. 3. Geological cross-section through marchean foredeep in northern area. 1. Plio-pleistocenic post-orogenic deposits; 2-3. Lower and middle p.p. Pliocene pre-orogenic deposits; 4. Messinian deposits; 5-6-7. Umbro-marchean mesocenozoic sequence; 8. "Anidriti di Burano" Formation; 9. Thrusts and faults.

The common origin from a basal brine of all the salt waters is confirmed by analysing Cl, Na and Br ions (Fig. 5 and Tab. 1).

The chemical data generally indicate a significant enrichment in Sr and Li. Certainly the enrichment derives from reactions with clay minerals. In the waters characterised by lower salinity there are enrichments in K and in Mg, while in the waters with higher salinity there is a depletion. The various percentages of K are produced by interactions with clayey sediments crossed during the lifting. In the springs in proximity to the limestone ridge, the enrichment in Mg may derive from waters circulating in the ridge and diffused through faults and joints. The enrichment of salt waters in ^{18}O and ^{13}C indicates a contribution of groundwaters from limestone aquifers (Chiarle et al., 1992).

Sometimes the marchean salt waters are characterised by the presence of sulphates.

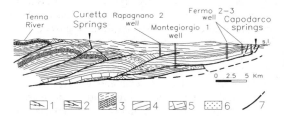

Fig. 4. Geological cross-section through marchean foredeep in the southern area. 1-2. Plio-pleistocenic post-orogenic deposits; 3. Lower and middle Pliocene and messinian pre-orogenic deposits; 4-5-6. Umbro-marchean mesocenozoic sequence; 7. Thrusts and faults.

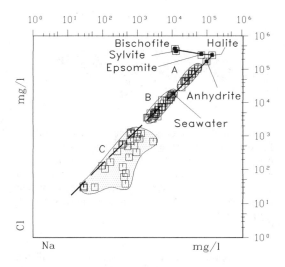

Fig. 5. Cl-Na relationships in the salt waters of marchean foredeep. The plots relatively fall close to the seawater evaporation-dilution trajectory considered as the dilution line of basal brine.

They are typical of some springs in the northern part of the region where messinian deposits, with the greatest thickness, outcrop or are at shallow depth. The considerable presence of sulphates is connected with the lifting of the messinian waters mixed with the pliocenic ones.

In conclusion, the correlation between salt springs and tectonic characters of the discharge areas demonstrates that the springs with higher salinity emerge in the main thrusts involving post-orogenic plio-pleistocenic sequence. A similar compressive tectonic regime, without the presence of arenaceous units, reduces groundwater circulation and prevents mixing with vadose waters, and supports the squeezing out of the salt waters. The deep movement connected with tectonic stress, prevents the contact with messinian deposits.

In the areas with compressive structures buried by plio-pleistocenic sediments and characterised by a monocline style with many faults, the waters with lower salinity are present. In these situations the salt waters are diluted by vadose waters circulating in the arenaceous bodies. Most of the studied springs are located in areas with more or less evident tectonic compression. Similar processes were described by Unrh et al., (1992) in western California, where perennial saline springs typically emerge along west-vergent thrust faults.

In the present study a strong relationship between salt waters and tectonic stress is demonstrated. This may suggest that the thrusts, especially in the coastal area, have recently been or are still active.

(Work funded by GNDCI of CNR; pubbl. n. 1657)

4 REFERENCES

Calamita F. 1990. Thrusts and fold-related structures in the Umbria-Marche Apennines (Central Italy). *Annales Tectonicae*. 4, 83-177.

Chiarle M., Nanni T., Sacchi E. & Zuppi G.M. 1992. Saline formation waters in Pliocene clays of Po valley, Italy: tectonic significance. *Proceedings of the international Symposium on Water-Rock Interaction WRI-7*. Park City (USA), 13-18 July 1992.

Ori G.G., Serafini G., Visentini G., Ricci Lucchi F., Casnedi R., Colalongo M.L. & Mosna S. 1991. The Plio-pleistocene adriatic foredeep (Marche and Abruzzo, Italy): an integrated approach to surface and subsurface geology. *3 nd. E.A.P.G. Conf. Adriatic foredeep field trip guidebook*. Firenze, AGIP spa, p.p.85.

Unruh J.R., Davisson M.L., Criss R.E. & Moores E.M. 1992. Implications of perennial saline springs for abnormally high fluid pressures and active thrusting in western California. *Geology*, v.20, 431-434.

Deuterium content and salinity of brines, Filitelnic gas-field, Transylvanian Basin, Romania

D.C. Papp
Geological Institute of Romania, Cluj-Napoca Branch, Romania

ABSTRACT: Based on 19 wellhead water samples collected from the Filitelnic gas-field, Transylvanian Basin (Romania), our study suggests that the correlation of isotopic and salinity data of brines uplifted by producing wells, represents an indicator of the mixing phenomena which appear in the methane-deposits water system. The analytical data of selected samples show the deuterium and NaCl content of water. Five situations have been delimited.

1 INTRODUCTION

During extractive process, the methane-deposits water system contains a mixture of diverse water types: (1) "connate" formation water; (2) meteoric ground water; and (3) condensation water. These mix phenomena are very evident by the correlation of isotopic and salinity data. The importance of this kind of study consists in that they offer a possibility to follow the dynamic regime of the uplifted waters in every well, representing an indicator of rational exploitation of a gas-field.

Our study, based on 19 wellhead water samples collected from Filitelnic gas-field represents only a model, due of course, to the fact that the exploitation parameters change practically continuously. Analysis of hydrogen isotopic composition of collected waters were performed at the Institute of Isotopic and Molecular Technology, Cluj-Napoca.

2 REGIONAL AND GEOLOGICAL SETTING

Representing one of the most important gas-field of the Transylvanian Basin, Filitelnic structure lies at about 28km SSE from Tg. Mures town.

In this area, covering sedimentary rocks of the Transylvanian Basin are Senonian, Helvetian, Badenian, Sarmatian, and Pliocene in age, having a maximum thickness of 4533 meters. Only Badenian, Buglovian and Sarmatian deposits contain hydrocarbon gases.

Fig. 1 Location sketch map of the Filitelnic gas-field

The structural arrangement of the stratigraphic levels is different, forming two main domes, one in the western part of the structure, the other in the southern one.

The collector strata consist of sands, sandy marls, and sandstones, separated by impermeable intercalation of marls and gritty marls, forming 14 producing complexes. They are saturated with free gases - CH_4 (99.5%), N_2, C_2H_6, C_3H_4, and "connate" formation water (22.23 - 130.62 g/l NaCl). These values have been obtained at the early in the production history of the gas-field.

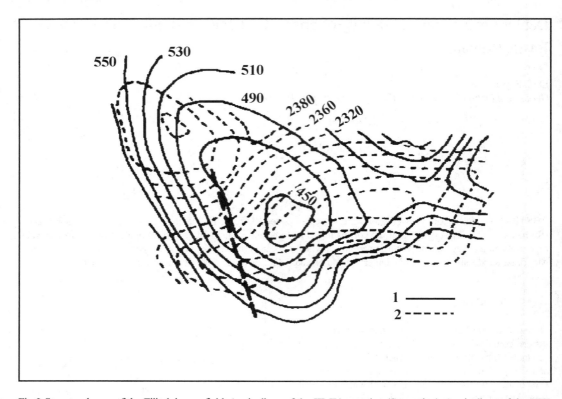

Fig.2 Structural map of the Filitelnic gas-field. 1 = isolines of the III-IV complex (Sarmatian); 2 = isolines of the XIII complex (Badenian).

3 SAMPLE COLLECTION AND ANALYTICAL DATA

In our study, the samples come from the wells situated especially to the margins of the structures, near the water/gas limits, which uplift salted waters (brine). It is important to note that none of the water sample represent true "connate" formation water because during the extractive process, the condensation water (produced by the condensation of the water vapours that saturated the gas in deposit conditions) appears anyhow in each of the wells.

The analytical data of selected samples are given in Table I, listed in order of decreasing depth. The deuterium content is relative to SMOW standard, and the salinity is expressed in NaCl content of water (g/l).

4 RESULTS AND CONCLUSIONS

To interpret the results, the following situations can be delimited (Fig.3):

1# high δD and high salinity = formation water - well with inundation tendency;

2# high δD and lower salinity = mixture of formation water and condensation water - well with cone-up tendency;

3# high δD and low salinity = condensation water - well with high productivity;

4# low δD and low salinity = mixture of condensation water and meteoric ground water - well with possible technical defects;

5# low δD and high salinity = mixture of formation water and meteoric ground water - well with possible technical defects;

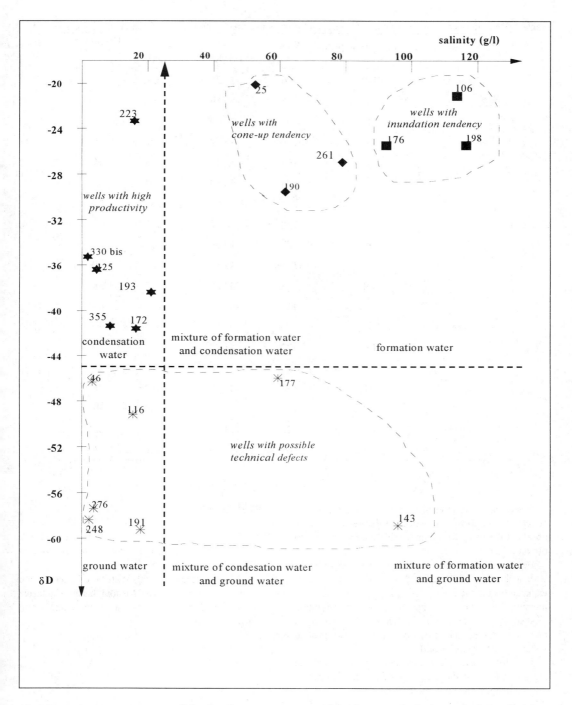

Fig.3 Deuterium content versus salinity for diverse water types uplifted by some producing wells from Filitelnic gas-field. Points are identified by well name and type.

Table 1. Deuterium content and salinity of brines uplifted by some producing wells from Filitelnic gas-field.

Age	Productive complex	Well	δD (‰ SMOW)	Salinity (g/l)	Diagnosis
Sarmatian	III-IV	106	-21.25	114.80	1#
	IV	125	-36.66	4.62	3#
		190	-29.60	61.79	2#
		198	-25.42	116.67	1#
	V	191	-59.13	17.18	4#
		193	-38.59	21.94	3#
		261	-27.03	79.88	2#
		248	-58.17	2.35	4#
	VI	25	-19.96	52.70	2#
		330bis	-35.69	1.60	3#
	VII	143	-58.80	86.23	5#
		116	-49.11	15.22	4#
Buglovian	VIII	276	-57.84	4.23	4#
	IX	172	-41.79	16.40	3#
	X	46	-46.16	2.62	4#
	XI	177	-45.97	59.52	4#
	XII	176	-25.42	92.04	1#
Badenian	XIII	223	-23.50	16.76	3#
	XIV	355	-41.47	7.66	3#

It is of particular interest to see to which degree our conclusions confirm in practice. In this respect it may note that the wells with high productivity according to our study are so indeed.

Because we obtained a high percentage of wells with technical defects (36.8 %), as a supplementary explanation for the lower dD and lower salinity, shale membrane filtration may be taken into consideration (Fisher, 1990). Nonetheless, in two wells in which we have determined the presence of meteoric ground water, breaks were detected.

We also consider that our model study offers an important tool for the diagnosis of the phenomena of well inundation and cone-up tendency.

5 REFERENCES

Blaga, L. 1968. Isotopic exchange of hydrogen between fluids of the hydrocarbon fields. *Isotopenpraxis*. 4: 178-183.

Fisher, J.B., Boles, J.R. (1990) Water rock interaction in Tertiary sandstones, San Joaquin basin, California, U.S.A.: Diagenetic controls on water composition. *Chemical Geol.* 82: 83-101.

… Water-Rock Interaction, Arehart & Hulston (eds) © 1998 Balkema, Rotterdam, ISBN 90 5410 942 4

Dolomitization of Ekofisk Oil Field reservoir chalk by injected seawater

R. Petrovich
Geoscience Branch, Research and Services Division, Phillips Petroleum Company, Bartlesville, Okla., USA

A.-A. Hamouda
Reservoir Management Branch, Exploration and Production, Phillips Petroleum Company Norway, Stavanger Office, Tananger, Norway

ABSTRACT: Seawater that is being injected into the fractured chalk reservoir of the Ekofisk Oil Field is reacting with it at 131°C and 28-34 MPa and being produced again, after some months, from a few wells. Changes in its composition, modeled with SOLMINEQ.88, show that very slow dolomitization is occurring in the reservoir and the precipitated dolomite is very disordered. Also, brine chemistry shows that the very sparse indigenous dolomite in the Ekofisk reservoir chalk is partially ordered. Our observations provide direct evidence of dolomitization of $CaCO_3$ rocks by seawater, such as is inferred to occur within the endo-upwellings of atolls.

1 DOLOMITIZATION OF $CaCO_3$ ROCKS BY SEA WATER AND THE EKOFISK WATERFLOOD

The question whether seawater can cause dolomitization of $CaCO_3$ sediments and rocks and if so, under what conditions has been debated for a long time (*e.g.*, Machel and Mountjoy, 1986). The reason for doubt is that although dolomite, and not calcite, is stable in contact with seawater even at sea-bottom temperatures, dolomitization of calcite by unadulterated seawater has not been achieved in the laboratory, apparently because of the inhibiting effect of sulfate ions (Baker and Kastner, 1981; Morrow and Ricketts, 1988; Brady *et al.*, 1996). Injection of seawater into the fractured chalk reservoir of the Ekofisk Oil Field (Brewster *et al.*, 1986; Sulak *et al.*, 1990) has provided an opportunity to find out whether the chemistry of seawater that has been flowing through the chalk for some months at 131°C and at pressures in the 28-34 MPa range has been changed in a way that indicates dolomitization.

2 METHOD OF STUDY

Three wells that are producing or have produced water of an unmistakable seawater chemistry have been identified. The first of these wells, 2/4 B-12, has been producing seawater since the Fall of 1994, the second, 2/4 C-6A, had produced seawater from February 1990 until the Fall of 1994, and the third, 2/4 C-23, has produced seawater intermittently since its completion. In all three cases, seawater is contaminated by minor amounts of formation brine that is being mixed with it within the formation and/or the well. We obtained the nominal initial (*i.e.*, pre-reaction) compositions of the seawater-like waters by adding to seawater enough locally produced brine to match the observed [Cl⁻]; no other conservative dissolved species could be used because the waters had been analyzed for engineering purposes. The changes in concentrations of dissolved ionic species were obtained as the differences between the actual concentrations in the mean produced waters and nominal initial concentrations in the same waters.

3 THE OBSERVED PATTERN OF CHANGES IN WATER CHEMISTRY

The changes in the molalities of Ca^{2+}, Mg^{2+}, K^+, SO_4^{2-}, and HCO_3^- ions that were undergone by the mean slightly contaminated seawaters produced from Wells 2/4-B12, 2/4-C6A and 2/4-C23 with their passage through the Ekofisk chalk reservoir are summarized in Figure 1. In all three cases the molalities of Ca^{2+} and HCO_3^- have increased, and the molalities of Mg^{2+}, K^+ and SO_4^{2-} have decreased by significant amounts. Whatever its cause (see below), the decline in SO_4^{2-} molality is a convenient progress variable for the evolution of injected seawater: the changes in Ca^{2+}, Mg^{2+}, and K^+ molalities are strongly correlated with it. The very significant increase in HCO_3^- molality is

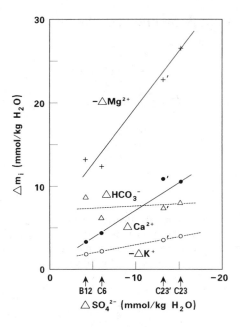

Figure 1. Changes in molalities of Mg^{2+}, Ca^{2+}, K^+ and HCO_3^- ions as functions of the change in SO_4^{2-} molality in slightly contaminated seawaters produced from wells 2/4 B-12, 2/4 C-6A, and 2/4 C-23 (C23 assumes contamination with the highest-salinity brine, C23' contamination with a repeatedly produced medium-salinity brine).

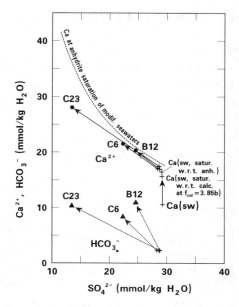

Figure 2. Evolution of the chemical compositions of the slightly modified seawaters produced from wells 2/4 B-12, 2/4 C-6, and 2/4 C-23, treated as if initially mixed with their contaminating brines. Superposed projections on the ($[SO_4^{2-}]$, $[Ca^{2+}]$) plane and the ($[SO_4^{2-}]$, $[HCO_3^-]$) plane. Crosses denote initial or intermediate positions of the representative point of the water, with *sw* denoting seawater, *satur. w. r. t.* denoting "saturated with respect to"; full circles and triangles denote the final positions of the representative point. Anhydrite saturation curve calculated using SOLMINEQ.88 with extended Debye-Hückel activity coefficients.

relatively uniform and to the first approximation is not correlated with the others.

4 DOLOMITIZATION AND OTHER REACTIONS IN THE CHALK RESERVOIR

Dissolution of calcite to achieve metastable equilibrium between seawater, calcite, and CO_2 gas is clearly buffered not only by the overwhelming excess of calcite, but also by the CO_2 from the hydrocarbon phases, which explains the high and relatively uniform HCO_3^- molalities of the produced seawaters. The net increases in Ca^{2+} molality are within a factor of two of the amounts predicted by the water-rock interaction software package SOLMINEQ.88 of Kharaka and co-workers (1988) for the present reservoir conditions. As seen in Figure 2, it appears that dissolution of calcite is also being buffered by precipitation of anhydrite, but the evidence is not clear-cut (sulfur isotopic analyses, now in progress, will show whether sulfate is being removed mainly by anhydrite precipitation or by bacterial reduction of sulfate).

Although some loss of Mg^{2+} to illitization and chloritization of kaolinite and illite-smectite is likely (the chalk contains about 1 percent of clay minerals and K^+ is clearly being lost by illitization), so much Mg^{2+} is being lost that the conclusion that calcite dissolution is being accompanied by precipitation of a dolomite or a magnesian calcite seems to be unavoidable. If the considered seawaters are in equilibrium with both calcite and a dolomite of a uniform state of order, the familiar relation

$$a_{Ca^{2+}}/a_{Mg^{2+}} = K_{calcite}^2/K_{dolomite} \tag{1}$$

must be satisfied, where $a_{Ca^{2+}}$ is the activity of Ca^{2+}, $a_{Mg^{2+}}$ is the activity of Mg^{2+}, $K_{calcite}$ is the solubility product of calcite, and $K_{dolomite}$ is the solubility product of dolomite of that particular state of order. In Figure 3, the representative points of the considered seawaters were plotted on a $\log(a_{Ca2+}/a_{Mg2+})$,T-plot of Morrow and coworkers (1994), which includes these authors' and Baker and Kastner's (1981) experimental points and several calcite-dolomite stability-field boundaries, with calcite stable to the right and dolomite to the left. In viewing this plot, one should be aware that while the calculated stability field boundaries differ greatly even for a specified state of dolomite order, both boundary 1 for completely disordered dolomite and boundary 5 for completely ordered dolomite are based on the highly accurate enthalpy measurements of Navrotsky and Capobianco (1987). The representative points of Ekofisk produced seawaters fall on the boundary between the stability fields of calcite and a very disordered dolomite, comparable to the dolomite that was obtained by Baker and Kastner (1981) by reaction between calcite and sulfate-free chloride solutions at 200°C. In view of the fact that all three seawaters were slightly contaminated by brines somewhere along their paths, it is significant that the representative points of seawaters from wells 2/4-B12 and 2/4-C6A, whose chemistry remained quite steady during many months of sampling and whose flow into the said wells is well understood, coincide and lie right on the most reliable boundary between the stability fields of calcite and completely disordered dolomite. Seawater produced from well 2/4-C23 gives a somewhat higher $\log(a_{Ca2+}/a_{Mg2+})$ ratio, but the implications of this difference are not clear because the chemistry of water produced from that well fluctuates from the slightly contaminated seawater to a medium-salinity brine and the local flow field is poorly understood, so that mixing in the well could be affecting the results. However, the general implications of the observed pattern are clear: in the parts of the Ekofisk chalk reservoir that are being flooded by seawater, calcite is dissolving and a very disordered dolomite is precipitating. Fortunately, the process is very slow because of the relatively low seawater/rock ratios.

5 SOME GEOLOGICAL IMPLICATIONS

The above approach can also be used to test the potential of SOLMINEQ.88 with the current solubility constants to calculate saturation of formation waters with respect to dolomite. Very sparse dolomite rhombohedra, tens of micrometers across, occur here and there in different Ekofisk chalk samples from cores obtained before the waterflood (a pattern of oc-

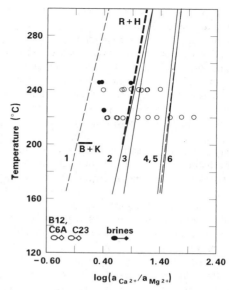

Figure 3. Representative points of Ekofisk reservoir brines and of seawaters produced from wells 2/4-B12, 2/4-C6A, and 2/4-C23, with activities calculated with SOLMINEQ.88 using extended Debye-Hückel activity coefficients (circles) and Pitzer ion-interaction activity coefficients (diamonds), superposed on Morrow and coworkers' (1994) $\log(a_{Ca2+}/a_{Mg2+})$,T-plot (redrafted) of (a) the results of their own and Baker and Kastner's (1981) calcite dolomitization experiments and (b) different extrapolated and calculated boundaries between the calcite and dolomite stability fields, all at the pressure of saturated water vapor. Morrow and coworkers' calcite converted to dolomite is indicated by full circles, their calcite not converted to dolomite by empty circles, Baker and Kastner's (1981) calcite converted to dolomite by the B+K heavy line. The heavy dashed line marked R+H is the calcite-dolomite stability field boundary extrapolated by Morrow and coworkers from Rosenberg and Holland's (1964) high-temperature data and the thin lines are calcite-dolomite stability field boundaries calculated by different authors (referenced by Morrow and coworkers) for completely disordered dolomite (1, 2), partially ordered dolomite (3), and completely ordered dolomite (4, 5, 6). Ekofisk data points are for 27.6 MPa, but if their a_{Ca2+}/a_{Mg2+} values were corrected for the effect of the pressure difference, they would be displaced to the left by only 0.02 log units.

currence similar to those in the neighboring Eldfisk and Tor reservoir chalks, documented by Maliva and Dickson (1994)); thus formation brines in at least some parts of the Ekofisk reservoir are saturated with respect to dolomite. Moreover, a_{Ca2+}/a_{Mg2+} ratios of uncontaminated, Ba^{2+}-rich formation brines produced from six Ekofisk wells are highly consistent: values calculated using extended Debye-Hückel activity coefficients fall within the range 4.0 ± 0.5, values calculated using Pitzer activity coefficients within the range 6.5 ± 0.5. When the representative points of these brines are plotted on Morrow and coworkers' $\log(a_{Ca2+}/a_{Mg2+})$,T- plot (Fig. 3), they fall on the boundary between the stability fields of calcite and a partially disordered dolomite, not of calcite and fully ordered dolomite. We do not know how typical this is of formation brines in deep sedimentary basins, but it is clear that in calculating equilibria between formation brines and reservoir rocks, the state of order of dolomites needs to be considered very carefully.

There may not be many settings in which seawater naturally invades carbonate rocks at temperatures above 100°C, but one comes readily to mind: that of endo-upwellings, the convective rising of seawater through the carbonate edifices of atolls, driven by the residual heat of the underlying volcanoes (Rougerie and Wauthy, 1986). Our observations confirm that at temperatures attainable in such edifices, dolomitization of limestones by convecting seawater (Aharon et al., 1987) does not require a drastic reduction of SO_4^{2-} molality.

ACKNOWLEDGEMENTS: The authors thank the managements of Phillips Petroleum Company and its Ekofisk partners, namely, Fina Exploration Norway SCA, Norsk Agip A/S, Elf Petroleum Norge A/S, Norsk Hydro a.s., Total Norge A.S., Den Norske Stats Oljeselskap, and Saga Petroleum ASA, for their permission to publish this paper.

REFERENCES

Aharon, P., R.A.Socki & L.Chan 1987. Dolomitization of atolls by sea water convection flow: test of a hypothesis at Niue, South Pacific. *J.Geol.*95: 187-203.
Baker, P.A. & M.Kastner 1981. Constraints on the formation of sedimentary dolomite. *Science*213: 214-216.
Brady, P.V., J.L.Krumhansl & H.W.Papenguth 1996. Surface complexation clues to dolomite growth. *Geochim.Cosmochim.Acta*60:727-731.
Brewster, J., J.Dangerfield & H.Farrell 1986. The geology and geophysics of the Ekofisk Field waterflood. *Mar.Petrol.Geol.*3:139-169.
Kharaka, Y.K., W.D.Gunter, P.K.Aggarwal, E.H. Perkins & J.D.DeBraal 1988. SOLMINEQ.88: a computer program for geochemical modeling of water-rock interactions. *U.S. Geol. Survey Water-Resources Investig. Rept.* 88-4227, 420 pp.
Machel, H.G. & E.W.Mountjoy 1986. Chemistry and environments of dolomitization - a reappraisal. *Earth-Sci.Rev.*23:175-222.
Maliva, R.G. & J.A.D.Dickson 1994. Origin and environment of formation of late diagenetic dolomite in Cretaceous/Tertiary chalk, North Sea Central Graben. *Geol.Mag.*131:609-617.
Morrow, D.W., B.L.Gorham & J.N.Y.Wong 1994. Dolomite-calcite equilibrium at 220 to 240°C at saturation vapor pressure: experimental data. *Geochim.Cosmochim.Acta*58:169-177.
Morrow, D.W. & B.D.Ricketts 1988. Experimental investigation of sulfate inhibition of dolomite and its mineral analogues. In V.Shukla & P.A. Baker (eds.) *Sedimentology and Geochemistry of Dolostones*:25-38. Tulsa, S.E.P.M.
Navrotsky, A. & C.Capobianco 1987. Enthalpies of formation of dolomite and of magnesian calcites. *Amer.Mineral.*72:782-787.
Rosenberg, P.E. & H.D.Holland 1964. Calcite-dolomite-magnesite stability relations in solutions at elevated temperatures. *Science*145:700-701.
Rougerie, F. & B.Wauthy 1986. Le concept d'endo-upwelling dans le fonctionnement des atolls-oasis. *Oceanol.Acta*9:133-148.
Sulak, R.M., G.R.Nossa & D.A.Thompson 1990. Ekofisk Field enhanced recovery. In A.T.Buller et al. (eds.) *North Sea Oil and Gas Reserves - II* : 281-295. London: Graham & Trotman.

Note added in October 1997: Sulfate in seawater-like waters (*i.e.*, waters with [Cl⁻] ranging from 587 to 638 mmol/L) that were produced from five Ekofisk Field wells and contained 24.9, 18.7, 14.6, 4.6, and 1.3 mmol/L of sulfate gave $\delta^{34}S$ values (relative to Cañon Diablo troilite) of +18.1, +20.6, +20.7, +19.0, and +19.3 permil, respectively. The lack of significant deviation of these $\delta^{34}S$ values from the seawater sulfate value of +20.0 permil shows that the loss of sulfate from the injected seawater is due not to the bacterial reduction of sulfate, but to the precipitation of anhydrite and minor barite.

Temporal fluctuations of syntectonic fluids in the Cascadia accretionary wedge

J.C. Sample
Department of Geological Sciences, California State University, Long Beach, Calif., USA

C. D. Coathe & K. D. McKeegan
Department of Earth and Space Sciences, University of California, Los Angeles, Calif., USA

ABSTRACT: Isotopic values of syntectonic, carbonate cements and veins from the Cascadia accretionary wedge record the history of fluid flow during recent subduction. For many of the carbonates, low $\delta^{18}O_{PDB}$ values and high $^{87}Sr/^{86}Sr$ ratios suggest a deep, warm source for fluids. Variations of $\delta^{18}O_{PDB}$ within individual veins indicate cyclical pulsing of warm fluids from depth up fluid conduits. The pulsing may be related to seismically induced fluid flow.

1 SIGNIFICANCE OF FLUID EXPULSION IN ACCRETIONARY WEDGES

Subduction and accretion of sediment at convergent margins causes fluid expulsion that influences a variety of organic and inorganic processes. Fluids transfer heat and mass, and they cause precipitation or dissolution of mineral phases, thereby influencing the physical properties of the sediments during deformation (COSOD II, 1987; Kastner et al., 1991). Cements and veins precipitated during tectonically induced fluid expulsion provide a record of evolving fluid characteristics and sources during subduction. This paper presents results from isotopic studies of carbonate cements and veins formed during Quaternary time in the Cascadia accretionary wedge.

The Cascadia accretionary wedge includes accreted clastic sediments largely derived from the Columbia River and the Oregon coastal mountains (Kulm and Scheidegger, 1979), which mantle the incoming plate to thicknesses of 4 km (MacKay et al., 1992). The age of oceanic basement varies between 7-9 Ma in the study area. The isotopic data presented here are derived from samples collected in a northern area of strike-slip faulting and a southern area dominated by landward-verging thrust faults and associated folds (Fig. 1). Active faults occur in both areas, and strike-slip faults in the northern area appear to link with the decollement at about 3.5 km depth (Goldfinger et al., 1992; Sample et al., 1993; Sample and Kopf, 1995). Fluid seepage is indicated at the faults by the presence of benthic, chemosynthetic communities and buildups of carbonate deposits at the seafloor (Moore et al., 1991; Sample et al., 1993; Carson et al., 1994).

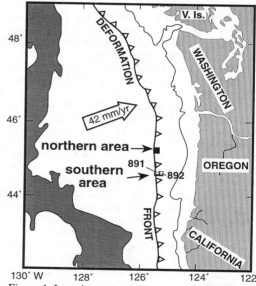

Figure 1. Location map of study area. ODP Sites 891 and 892 shown in southern area. Subduction rate from DeMets et al. (1994).

2 SAMPLE DESCRIPTIONS AND ANALYTICAL TECHNIQUES

Analyzed samples were collected with the submersible *Alvin*, and by drilling at Ocean Drilling Program (ODP) Sites 891 and 892. *Alvin* samples vary from

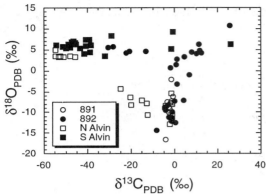

Figure 2. Oxygen and carbon isotopes of carbonate cements and veins from the Cascadia margin.

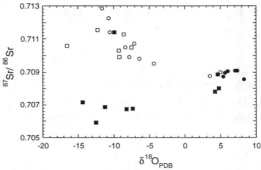

Figure 3. Oxygen and strontium isotopes of carbonate veins and cements from the Cascadia margin.

weakly to strongly lithified sandstone and mudstone, which contain tectonic structures such as veins, tension fractures, brecciation, shear fractures with slickensides, grain-scale cataclasis and stress solution. Carbonate cements and veins typically compose from 10% to 80% of the sample and have dominant compositions of calcite, Mg-calcite, Ca-dolomite, and Mn-calcite. Dolomite and aragonite are rare. ODP samples are mainly poorly lithified sands, silts, and muds. Deformation structures are carbonate-filled veins, and deformation bands in fine-grained material. Carbonate contents range from 1 to 15 % in non-concretionary samples, and cements comprise calcite and Mg-calcite.

Carbonate compositions and contents were determined by standard X-ray diffraction and electron microprobe techniques summarized elsewhere (Sample et al., 1993; Sample and Kopf, 1995; Sample and Reid, in press). Carbonates were analyzed for O, C, Sr, and Nd isotopic characteristics, but only O and Sr data are discussed here. Isotopic techniques for analysis of bulk samples are summarized in Sample and Kopf (1995) and Sample and Reid (in press). In situ oxygen isotopic analysis was accomplished with a Cameca ims 1270 secondary ionization mass spectrometer located in the UCLA Keck Laboratory for Isotope Geochemistry, employing a Cs^+ primary ion beam source, and a mass-resolving power of 3500. Instrument mass fractionation was corrected by repeat analysis of a calcite standard mounted on the same thick section as each calcite vein sample. Typical precision is 1 ‰ (1σ). Transects were completed across individual veins, using a spot size of 30 μm.

3 ISOTOPIC RESULTS FROM BULK ANALYSES

Oxygen isotopic values measured in Cascadia bulk samples span a range much greater than observed at other accretionary margins. $\delta^{18}O_{PDB}$ values range from a low of -16.5‰ at ODP Site 891 to a high of +10.7 at Site 892 (Fig. 2). Low values are associated with strike-slip faults in the northern area, and with thrust faults and a zone of broken formation in Site 892 (Sample et al., 1993; Sample and Kopf, 1995). Site 891, which penetrates the frontal thrust of the accretionary wedge, has low $\delta^{18}O_{PDB}$ values in carbonates that are distributed widely throughout the section. The low oxygen isotopic values are not in oxygen-isotope equilibrium with current pore waters, whose $\delta^{18}O_{SMOW}$ values range from -1.3‰ to +0.3‰ (Kastner et al., 1995), and whose estimated current downhole temperatures reach of maximum of 25°C (Westbrook et al., 1994).

$^{87}Sr/^{86}Sr$ ratios along the margin also span a very large range (Fig. 3). In the northern area, at Site 891, and at Site 892, the values range from 0.7088 to 0.7128, 0.7103 to 0.7113, and 0.7058 to 0.7089. respectively. The predominance of low ratios at Site 892 suggest derivation of Sr from water-rock interactions within the accretionary wedge. Common, high ratios near the frontal thrust regions in the northern (*Alvin*) and southern (Site 891) areas suggest the presence of radiogenic fluids derived from the decollement. A deep, decollement source for the radiogenic fluids is further supported by the strong correlation between low $\delta^{18}O$ values and high $^{87}Sr/^{86}Sr$ ratios at the northern, strike-slip zone (Fig. 4).

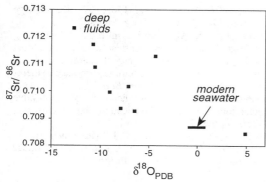

Figure 4. Correlation between strontium and oxygen isotopes for samples from the northern area.

4 ISOTOPIC RESULTS FROM MICRODRILLING AND ION MICROPROBE ANALYSES OF VEINS

More detailed subsampling of matrix and veins by microdrilling has revealed large differences in oxygen isotopic values. Carbonate veins vary by up to 10‰ relative to the carbonate cement of their sedimentary host (Sample and Reid, in press). The nature and details of some of these variations have been investigated with ion microprobe measurement of veins. The samples chosen for analysis are from an active fault found 106 metres below seafloor at Site 892. The fault is imaged on a seismic reflection profile and links with a fault scarp at the seafloor. A reversal of seismic-reflection wave polarity in the fault zone suggests that it is an active conduit for fluid flow (Sample and Kopf, 1995). $\delta^{18}O_{PDB}$ values for two large veins and a number of small veins from one sample (146 892D 10X 5, 98-100 cm) range from -10 ‰ to -2 ‰, with a mean value of -8 ‰ for 30 analyses. The mean of the ion probe data is identical (within error) to the mean of four analyses obtained from microdrilled material and processed by traditional phosphoric acid dissolution techniques. A representative transect for one of the veins is shown in Figure 5. $\delta^{18}O_{PDB}$ values fluctuate by 6 ‰ over a distance of about 1.5 mm. This indicates significant temporal variations in either the temperature or isotopic composition of the fluids from which the calcite precipitated during the growth of the vein. The position of the greatest variation in isotopic values indicates that the largest fluctuations in fluid characteristics occurred during growth of the center of the vein.

5 CONNECTION BETWEEN ISOTOPIC RESULTS AND CASCADIA TECTONICS

The isotopic results show that many Cascadia carbonates have very low $\delta^{18}O_{PDB}$ values and very high $^{87}Sr/^{86}Sr$ ratios relative to modern seawater. The isotopic values of oxygen require some combination of higher paleotemperatures during precipitation and strong ^{18}O depletions of the pore fluid in the past, resulting from open-system behavior. It is unlikely that changing the paleofluid composition by water

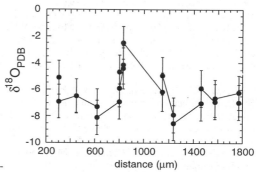

Figure 5. Results of ion microprobe traverse across crack-seal vein formed along active thrust at Site 892.

rock interaction alone can account for much of the observed depletion in ^{18}O. Rather the isotopic data can be interpreted as evidence for deep fluid sources within the accretionary wedge or from the decollement, coupled with precipitation of the veins at high temperatures. Paleotemperatures calculated from the oxygen isotopes of samples from Site 892 range up to 90°C, or 80°C higher than measured downhole temperatures. The magnitude of the temperature anomaly requires rapid upward migration of fluids (rates of 10^6 to 10^7 mm/a; Sample, 1996) in order to sustain the temperature for precipitation of veins with low $\delta^{18}O_{PDB}$ values. Rapid migration suggests a mechanism of seismically induced fluid migration (e.g., Sibson, 1992). The fluctuations of isotopic values within individual veins suggests several cycles of fluid pulses are recorded during the growth of the vein.

6 ACKNOWLEDGMENTS

Research of JCS supported by NSF grants OCE-9020919 and OCE-9633330. The production of this publication was partially funded by California State University, Long Beach.

7 REFERENCES

Carson, B., E. Seke, V. Paskevich, & M. L. Holmes 1994. Fluid expulsion sites on the Cascadia accretionary prism: Mapping diagenetic deposits with processed GLORIA imagery. *J. Geophys. Res.* 99: 11,959-11,970.

COSOD II, 1987. Report of the Second Conference on Scientific Ocean Drilling, COSOD II: Strasbourg, France, European Science Foundation, p. 142.

DeMets, C., R.G. Gordon, D.F. Argus, & S. Stein 1994. Effects of recent revisions to the geomagnetic reversal time scale on estimates of current plate motions. *Geophys. Res. Lett..* 21: 2191-2194.

Goldfinger, C., L.D. Kulm, R.S. Yeats, B. Appelgate, M.E. MacKay, & G.F. Moore, 1992. Transverse structural trends along the Oregon convergent margin: Implications for Cascadia earthquake potential and crustal rotations. *Geology.* 20:141-144.

Kastner, M., H. Elderfield, & J. B. Martin, 1991. Fluids in convergent margins: What do we know about their composition, origin, role in diagenesis and importance for oceanic chemical fluxes. *Phil. Trans. Roy. Soc. Lond., Ser. A.* 335:275-288.

Kastner, M., J. Sample, M. Whiticar, & M. Hovland, 1995. Geochemical evidence for fluid flow and diagenesis at the Cascadia convergent margin, In G. Westbrook, B. Carson, & R. J. Musgrave, (eds), *Proc. ODP, Sci. Results.* 146: 375-384. College Station, TX: Ocean Drilling Program.

Kulm, L. D., & K. F. Scheidegger, 1979. Quaternary sedimentation on the tectonically active Oregon continental slope, In L. J. Doyle, & O. H. Pilkey, (eds), *Geology of Continental Slope:*:247-263.

MacKay, M.E., G.F. Moore, G.R. Cochrane, J.C. Moore, & L.D. Kulm, 1992. Landward vergence and oblique structural trends in the Oregon margin accretionary prism: Implications and effect on fluid flow. *Earth Planet. Sci. Lett.* 109: 477-491.

Moore, J. C., K. M. Brown, F. Horath, G. Cochrane, M. Mackay, & G. Moore, 1991. Plumbing accretionary prisms: Effects of permeability variations. *Phil. Trans. Roy. Soc. Lond., Ser. A.* 335:275-288.

Sample, J. C., 1996. Isotopic evidence from authigenic carbonates for rapid upward flow in accretionary wedges. *Geology.* 24:897-900.

Sample, J. C., M. R. Reid, H. J. Tobin, & J. C. Moore, 1993. Carbonate cements indicate channeled fluid flow along a zone of vertical faults at the deformation front of the Cascadia accretionary wedge. *Geology.* 21:507-510.

Sample, J. C., & A. Kopf, 1995. Geochemistry of syntectonic carbonate cements and veins from the Oregon margin (ODP Leg 146): Implications for the hydrogeologic evolution of the accretionary wedge, In B. Carson, G. Westbrook, R. Musgrave, (eds), *Proc. ODP, Sci. Results,* 146: 137-148. College Station, TX, (Ocean Drilling Program).

Sample, J.C. & M.R. Reid, in press. Contrasting hydrogeologic regimes along strike-slip and thrust faults in the Oregon convergent margin: evidence from the chemistry ff syntectonic carbonate cements and veins. *Geol. Soc. Amer. Bull.*

Sibson, R. H., 1992. Implications of fault-valve behaviour for rupture nucleation and recurrence. *Tectonophys.* 211:283-293.

Westbrook, G., B. Carson, R.J. Musgrave, et al., 1994. *Proc. ODP, Init. Repts., 146 (Pt. 1)*: College Station, TX (Ocean Drilling Program).

Surface characterization of biotite reacted with acid solution

H. Seyama & A. Tanaka
National Institute for Environmental Studies, Tsukuba, Ibaraki, Japan

J. Sato & M. Tsurumi
Hirosaki University, Aomori, Japan

M. Soma
University of Shizuoka, Japan

ABSTRACT: As a model of the chemical weathering of silicate minerals by acidic solution, surface alteration of acid-leached mica (biotite) was examined by secondary ion mass spectrometry and X-ray photoelectron spectroscopy. Iron, magnesium, potassium and aluminum were selectively leached during acid dissolution (0.05 mol l^{-1} H$_2$SO$_4$), resulting in the formation of an altered layer rich in Si (SiO$_2 \cdot n$H$_2$O) on biotite surface. After acid dissolution for one week, the thickness of altered surface layer was estimated to be about 100 nm.

1 INTRODUCTION

The natural process of rock and mineral weathering is a major research topic in earth and environmental sciences. Surface analytical techniques such as secondary ion mass spectrometry (SIMS) and X-ray photoelectron spectroscopy (XPS) are useful tools to get clues on the mechanism of chemical weathering (Mogk, 1990; Shotyk & Metson, 1994) which is essentially a surface process. In spite of the previous studies on dissolution of silicate minerals (Drever, 1985; Colman & Dethier, 1986; Mogk, 1990; Shotyk & Metson, 1994), there is still a lack of knowledge concerning the growth of surface altered layer on silicate minerals during chemical weathering. As a part of systematic studies for elucidating the mechanisms of silicate mineral weathering, we report a study on surface alteration of acid-leached mica examined by SIMS and XPS.

2 EXPERIMENTAL

2.1 Sample

Biotite (Fukushima, Japan) was used as a mica sample in this study. The composition of biotite determined by chemical analysis was (K$_{0.91}$Na$_{0.01}$)(Fe$_{2.61}$Mg$_{0.03}$Mn$_{0.09}$Ti$_{0.03}$Al$_{0.24}$)Al$_{1.22}$Si$_{2.78}$O$_{10}$(OH,F)$_2$. A thin section of biotite was cleaved (about 100 mg) and stirred in 0.05 mol l^{-1} H$_2$SO$_4$ (50 ml) for periods of between 3 h and 10 days. After acid dissolution, the biotite sample was washed with deionized water and dried in a desiccator.

2.2 Secondary ion mass spectrometry (SIMS)

Secondary ion mass spectrometry is the mass spectrometry of atomic species which are emitted when a solid surface is bombarded by an energetic primary ion beam. Most of the particles sputtered away by primary ion bombardment are uncharged, but a small fraction consists of atomic or molecular ions (secondary ions). Secondary ion intensities recorded as a function of time provide a depth profile of the elements in the sample. Positive secondary ion depth profiling was performed using a CAMECA IMS4f instrument. The biotite samples were coated with a layer of gold to reduce sample charging. Depth profiles were obtained using a 17 keV O$^-$ primary ion beam with a typical current of 100 nA. The O$^-$ beam was rastered over an area of 250 × 250 μm^2. Positive secondary ions were collected from a circular area of 60 μm diameter with mass resolution of about 250. In order to compensate the negative sample charging due to O$^-$ bombardment, the secondary accelerating voltage was automatically adjusted by monitoring ^{28}Si$^+$ intensity during depth profiling.

2.3 X-Ray photoelectron Spectroscopy (XPS)

X-Ray photoelectron spectroscopy measures kinetic energy distribution of electrons (photoelectrons) emitted by X-ray irradiation from the core levels of elements constituting a solid sample. The determination of core electron binding energy, defined as the difference between the energy of primary photon and the kinetic energy of photoelectron, allows in principle the identification of elements. Because the inelastic mean free path of photoelectron in solids is short, typically in the order of nm, XPS is surface sensitive. X-Ray photoelectron spectra were recorded on a Vacuum Generators ESCALAB 5 instrument using Al K_α and Mg K_α X-ray sources (12 kV and 10 mA) with an analyzer pass energy of 50 eV. Electron energies were determined relative to the Au $4f_{7/2}$ binding energy (84.0 eV) of a vacuum-evaporated gold film on the sample.

3 RESULTS AND DISCUSSION

Mica has a layered aluminosilicate structure comprising SiO_4^{4-} tetrahedra linked to form a flat sheet. Divalent cations such as Fe^{2+} and Mg^{2+} occupy the octahedral site in the aluminosilicate layer. The major divalent cation in biotite is Fe^{2+}. Some of the Si^{4+} ions in the tetrahedral site are replaced with Al^{3+} ions. As indicated in the chemical composition, a minor portion of Al^{3+} ion also occupies the octahedral site in this biotite sample.

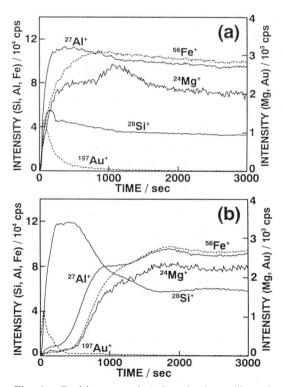

Fig. 1. Positive secondary ion depth profiles of biotite (a) before and (b) after acid dissolution for one week.

The positive secondary ion depth profiles of biotite samples before and after acid dissolution are shown in Fig. 1. The depth profile of acid-leached biotite provides evidence for the formation of an altered surface layer. The secondary ion intensities of constituent elements of the unleached biotite increased rapidly with decreasing $^{197}Au^+$ intensity during the first few hundred seconds of depth profiling. On the other hand, a lag in the rise of $^{56}Fe^+$ and $^{24}Mg^+$ intensities, i.e., an altered surface layer depleted of Fe and Mg, was observed in the depth profile of acid-leached biotite. The $^{56}Fe^+$ and $^{24}Mg^+$ profiles of the reacted biotite were similar to each other. Aluminum was also removed from the mineral surface, although the depth of Al depletion is less than that of the divalent cations, as shown in Fig. 1b, indicating that Fe^{2+} and Mg^{2+} are more susceptible to leaching by acid solution than Al^{3+}. Most (84%) of the Al^{3+} ions in the biotite substitute for Si^{4+} ions, and hence the change of $^{27}Al^+$ intensity with measurement time in Fig. 1 mainly shows the profile of Al^{3+} ions in the tetrahedral site. In addition, Fig. 1b shows that the $^{28}Si^+$ intensity rose sharply at the beginning of SIMS analysis and then decreased to a value representative of the bulk. These results of depth profiling indicate the selective leaching of Fe^{2+}, Mg^{2+} and Al^{3+} ions in the aluminosilicate layer of mica, i.e., the formation of an altered surface layer residually enriched in Si during acid dissolution.

The K^+ ion is held between the aluminosilicate layers and would also be expected to be leached by acid treatment. However the depth profile of $^{41}K^+$ secondary ion could not provide clear evidence for the surface depletion of K in the biotite after acid dissolution. The $^{41}K^+$ intensity of the surface was often higher than that of the bulk for both unleached and acid-leached samples. Alkali metal is known to become mobile in the sample during the SIMS analysis by sample charging (Benninghoven et al., 1987). Thus the surface enrichment of K may be due to charge induced migration of K^+ under O^- bombardment.

After acid dissolution for one week, the thickness of the altered surface layer was estimated based on the sputter rate. The experimentally determined sputter rate of O^- primary ion beam was 0.07 nm sec^{-1} 100nA^{-1} for a silicate glass. Assuming the same sputter rate for both altered surface layer and bulk phase in the biotite, the thickness of altered surface layer was calculated to be about 100 nm. It is noteworthy that the total dissolved aluminosilicate layer, deduced from the concentration of elements in the acid solution after the treatment, was thicker than the altered layer. Similar thick altered layers as deep as several tens of nm or more have been observed in previous studies on the chemical weathering of labradorite feldspar (Muir et al., 1989; Muir & Nesbitt, 1991; 1992).

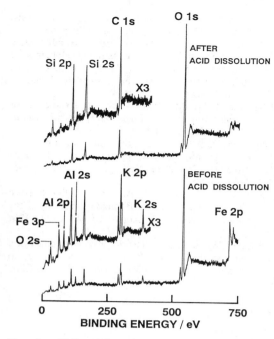

Fig. 2. X-Ray photoelectron spectra of biotite before and after acid dissolution for one week.

The formation of an altered surface layer was confirmed by the X-ray photoelectron spectra of biotite samples before and after acid dissolution (Fig. 2). Table 1 shows the surface composition of biotite samples determined by XPS. The atomic composition relative to Si was calculated from the area intensities of the photoelectron spectra on the basis of the relative sensitivity of each line. The relative sensitivities were experimentally determined from the relative spectral intensities of compounds of known chemical composition such as metal halides and zeolites (Seyama & Soma, 1988). The relative atomic abundances of major cations (Si^{4+}, Al^{3+}, Fe^{2+} and K^+) in the untreated biotite (reaction time = 0) were consistent with the elemental (bulk) composition determined by chemical analysis of the original sample. Magnesium concentration was too low to detect by XPS. On the other hand, the relative abundance of oxygen (O/Si = 5.47) in the untreated biotite was 27% larger than the bulk value (O/Si = 4.32). The excess of surface oxygen atoms was presumably attributable to surface OH^- ion and/or adsorbed H_2O. We have already reported the consistency of the relative abundances of cations and excess oxygen in the surface layer of freshly powdered silicate minerals (Seyama & Soma, 1985; 1988). There are two kinds of Al^{3+} ion in the biotite, *i.e.*, tetrahedrally (major) and octahedrally (minor) coordinated Al^{3+}. However the Al 2p line of the untreated biotite showed apparently a single peak without a spectral structure indicating presence of the two Al species.

Table 1. Surface compositions of biotite samples determined by XPS.

Element (line)	Reaction time (h)				
	0	3	24	168	240
Si 2s	1	1	1	1	1
Al 2p	0.58	0.32	0.19	0.12	0.06
Fe 2p	0.96	0.51	0.27	0.19	0.09
K 2p	0.32	0.14	0.10	0.05	0.01
O 1s	5.47	4.53	4.01	3.78	3.60

Values are atomic concentrations relative to Si (atomic ratio).

As shown in Table 1, the relative surface abundances of Al, Fe and K in the acid-leached biotite decreased with increasing reaction time, indicating the selective leaching of Al, Fe and K during acid dissolution. The relative abundance of oxygen in biotite also decreased after the treatment, suggesting the formation of a leached layer rich in Si at the mineral surface during acid dissolution which is consistent with the result of SIMS depth profiling.

The chemical shifts of photoelectron binding energies of the biotite samples are listed in Table 2. The binding energies of Al 2p and K $2p_{3/2}$ remained unchanged by acid dissolution within the limits of experimental error. The observed Fe $2p_{3/2}$ binding energies of acid-treated biotite samples were higher than that of the untreated biotite (freshly cleaved sample) which corresponds to Fe^{2+} ion in silicate minerals (Seyama & Soma, 1987). It is likely that Fe^{2+} ions in the surface layer of the acid-treated samples were partially oxidized to Fe^{3+} during dissolution reaction and/or drying in air after acid dissolution.

Table 2. Photoelectron binding energies of biotite samples.

Element (line)	Reaction time (h)				
	0	3	24	168	240
Si 2s	153.3	153.5	154.0	154.1	154.6
Al 2p	74.1	74.1	74.2	74.2	74.0
Fe $2p_{3/2}$	710.8	711.9	711.5	711.3	711.2
K $2p_{3/2}$	293.3	293.1	293.5	293.4	293.3
O 1s	531.3	531.6	532.4	532.5	533.1

The Si 2s and O 1s photoelectron binding energies of unleached biotite were 153.3 and 531.3 eV, respectively. These photoelectron spectra were broadened and shifted to higher binding energies during acid dissolution. Both Si 2s and O 1s binding energies of the biotite samples acid-treated for more than one week (Si 2s = 154.1-154.6 eV and O 1s = 532.5-533.1 eV) were comparable to those of silicon dioxides, *i.e.*, quartz (Si 2s = 154.4 eV and O 1s = 532.6 eV) and silica gel (Si 2s = 154.6 eV and O 1s = 533.1 eV), suggesting the formation of $SiO_2 \cdot nH_2O$ layer on the mineral surface. A similar hydrated silicon dioxide layer has also been observed on acid-leached olivine (Seyama et al., 1996).

4. CONCLUSION

Although the dissolution conditions used in this study were drastic compared with those during natural weathering, the experimental results reflect the possible changes occurring in the surface chemical composition of biotite during chemical weathering. This work supports the following dissolution process for trioctahedral mica in acidic solution. At the initial stage, divalent cations (Fe^{2+} and Mg^{2+}) in the octahedral site and K^+ ion in the interlayer cation site are selectively removed. This is followed by leaching of Al^{3+} ion in the tetrahedral site, resulting in breakdown of the aluminosilicate layer of mica and the formation of an altered layer rich in Si ($SiO_2 \cdot nH_2O$) at the mica surface during acid dissolution. The release of Si from the top surface of the altered layer into solution is the slowest reaction in the dissolution process. The overall dissolution of mica proceeds gradually through these steps, the thickness of the altered layer depending on the rate of Si dissolution.

REFERENCES

Benninghoven A., F. G. Rüdenauer & H. W. Werner 1987. *Secondary ion mass spectrometry, basic concepts, instrumental aspects, applications and trends*: 895-899. New York: Wiley.

Colman, S. M. & D. P. Dethier, (ed.) 1986. *Rates of chemical weathering of rocks and minerals*. Orland: Academic Press.

Drever, J. I. (ed.) 1985. *The Chemistry of Weathering*. Dordrecht: Reidel,

Mogk, D. W. 1990. Application of Auger electron spectroscopy to studies of chemical weathering. *Rev. Geophys.* **28**: 337-356.

Muir, I. J., G. M. Bancroft & H. W. Nesbitt 1989. Characteristics of altered labradorite surfaces by SIMS and XPS. *Geochim. Cosmochim. Acta* **53**: 1235-1241.

Muir, I. J. & H. W. Nesbitt 1991. Effects of aqueous cations on the dissolution of labradorite feldspar. *Geochim. Cosmochim. Acta* **55**: 3181-3189.

Muir, I. J. & H. W. Nesbitt 1992. Controls on differential leaching of calcium and aluminum from labradorite in dilute electrolyte solutions. *Geochim. Cosmochim. Acta* **56**: 3979-3985.

Seyama, H. & M. Soma 1985. Bonding-state characterization of the constituent elements of silicate minerals by X-ray photoelectron spectroscopy. *J. Chem. Soc., Faraday Trans. 1* **81**: 485-495.

Seyama, H. & M. Soma 1987. Fe 2p spectra of silicate minerals. *J. Electron Spectrosc. Relat. Phenom.* **42**: 97-101.

Seyama, H. & M. Soma 1988. Application of X-ray photoelectron spectroscopy to the study of silicate minerals. *Res. Rep. Natl. Inst. Environ. Stud., Japan*, No. 111.

Seyama, H., M. Soma & A. Tanaka 1996. Surface characterization of acid-leached olivines by X-ray photoelectron spectroscopy. *Chem. Geol.* **129**: 209-216.

Shotyk, W. & J. B. Metson 1994. Secondary ion mass spectrometry (SIMS) and its application to chemical weathering. *Rev. Geophys.* **32**: 197-220.

Brines in Siberian Platform: Geochemical and isotopic evidence for water-rock interaction

S.L.Shvartsev
Tomsk Department of the United Institute of Geology, Geophysics and Mineralogy SB RAS, Russia

ABSTRACT: Analyses and the calculation of the saturation states of hypersaline brines of Ca-Na-Cl and Na-Ca-Cl types from 14 localities on the Siberian Platform show, that the water-rock system is in an equilibrium-nonequilibrium state and demonstrate the utility of the $^{87}Sr/^{86}Sr$ ratio in the study of brines system. The isotopic ratios range from 0,70943±13 to 0,72010±11 and have obtained their singnatures by water-rock interactions, primary involving the feldspars.

1 INTRODUCTION

Strong calcium-chloride brines occur widely in sedimentary basins containing halite associations, but their origin remains acutely debatable, though with a tendency for the views to come together. Most researchers agree that the brines are preserved sea water from various stages of evaporative concentration substantially altered by interaction with the rocks during diagenesis and catagenesis (Land, 1995, Shvartsev, Bukaty, 1996, Shvartsev, 1995). But the detailed mechanisms, scales, and nature of these interactions remain debatable.

We have studied the geochemistry of hypersaline brines of the Siberian Platform more than twenty five years. That allowed us to form the new conception of its forming and genesis. According to our idea, under conditions in the Earth's crust, the water-rock system is everywhere equilibrium-nonequilibrium and hence, is capable of continuous interaction and development during a long geological period. Each stage of interaction corresponds to a characteristic assemblage of secondary minerals and type of water geochemistry. Water not only gives rise to secondary minerals, but simultaneously concentrates mobile elements in the solution (Shvartsev, 1991). To explain this statement in a more detailed way, we focus on the example of brines of the west part of the Siberian Platform (Fig.1).

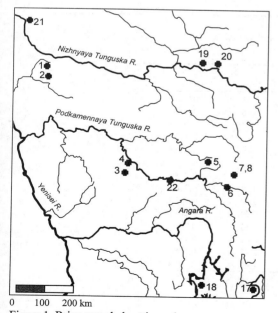
Figure 1. Brine sample location scheme

2 BRINE GEOCHEMISTRY

The studied brines of the Siberian Platform are in terrigenous, salt-bearing carbonate deposits of the middle-late Precambrian and lower-middle Palaezoic including traps and intrusions, there has been universal development of strong brines. These brines are grouped according to salinity into low-concentrated, concentrated, and superconcentrated ones; compositionally they are divided into Cl-Na, Na-Ca, and Ca types. Concentrated brines contain very high values of Ca, Mg, Na, Sr, Br, and other elements (Table 1).

Table 1. Chemical composition of the studied brines, g/l

Sample number	Locality	Age of sediments	K^+	Na^+	Ca^{2+}	Mg^{2+}	Sr^{2+}	Cl^-	Br^-	SO_4^{2-}	Total salinity
1	Tanachiskaja well, 7	$Є_1$ us	18,0	64,3	50,7	7,41	1,90	225	3,08	0,30	372
2	Tanachiskaja well, 2	$Є_{1-2}$ kst	14,9	53,9	63,3	6,23	2,08	234	-	0,29	386
3	Jurubchenskaja well, 69	Rf	-	48,3	17,6	7,01	0,43	152	-	0,71	226
4	Kujumbinskaja well, 4	$Є_1$ us	12,0	35,7	74,7	10,5	-	238	4,83	0,18	375
5	Chambinskaja well, 114	V_{vn}	-	48,1	49,2	11,9	-	194	4,15	0,10	308
6	Elohtinskaja well, 2	V_{vn}	1,96	44,5	24,0	10,8	0,50	138	2,63	2,65	224
7	Teterinskaja well, 104	$Є_1$ us	0,23	19,7	95,2	12,2	-	255	-	0,13	407
8	Teterinskaja well, 104	$VЄ_{tt}$	23,6	18,5	100	15,4	3,53	268	6,87	0,13	425
17	Omoloi well, 13i	$Є_1$ an	6,26	1,64	138	42,2	12,3	40,6	8,24	0,01	626
18	Bratsk well, 8	$Є_1$ mt	12,2	11,3	55,7	14,2	5,27	222	4,91	0,13	357
19	Spring, the Tutonchana River	P_1	0,35	20,7	12,1	0,01	0,30	53,4	0,28	0,59	87,5
20	Nizhnyaja Tunguska well, 44	P_2	0,56	86,4	31,7	2,23	1,64	195	1,43	0,31	318
21	Bol'shoi Porog well, 3	P_1	1,17	57,3	15,4	0,01	0,34	116	0,31	0,67	191
22	Spring below Oskoba settlement	$Є_1$	0,17	31,7	0,16	0,47	0,05	50,4	0,04	4,01	87,1
Ratio of elements concentration in brines relative to initial sea water			2,75	0,39	130	0,66	-	1,04	5,55	0,04	0,95

Table 2 Genetic ratios and isotopic composition of brines and sea water of the corresponding geological epoch

Sample number	Cl/Br	Ca/Cl	rNa/rCl	$^{87}Sr/^{86}Sr$	1/Sr, g/l	$^{87}Sr/^{86}Sr$ of the sea water
1	73,0	0,22	0,44	0,71031	0,53	0,7080
2	-	0,27	0,36	0,71150	0,48	0,7088
3	-	0,12	0,49	0,70943	2,33	0,7060
4	49,2	0,31	0,23	0,72010	-	0,7080
5	46,7	0,25	0,38	0,71546	-	0,7068
6	52,5	0,17	0,50	0,71809	2,00	0,7068
7	-	0,37	0,12	0,70969	-	0,7080
8	39,0	0,37	0,11	0,70943	0,28	0,7065
17	49,2	0,39	0,06	0,7187	0,08	0,7085
18	45,3	0,25	0,08	0,7160	0,19	0,7075
19	193,4	0,23	0,60	0,7139	3,33	0,7080
20	136,7	0,16	0,68	0,7132	0,61	0,7070
21	374,3	0,13	0,76	0,7131	2,93	0,7080
22	1145	0,003	0,97	0,7110	19,2	0,7088

Studied waters strongly differ in genesis and degree of metamorphization. For example, based on values of genetic ratios (Table 2), brines of sample 22 unambiguosly correspond to the infiltration type; i.e., it originated from meteoric water. Samples 20, 21 and 3 are also indicative of infiltration waters, which, however, were partially mixed with ancient sedimentation waters. The rest of samples of sedimentation waters have different metamorphization degrees.

First of all, we revise the degree of discrepancy between the examined brines and sea water at the corresponding stage of evaporative concentration (Table 1). As the resulting data show, in these case, as in other (Shvartsev, Bukaty, 1995), the natural brines are > 100 times enriched with calcium, 5,5 times enriched with bromide, and almost about 3 times enriched with potassium. At the same time, they are extremly depleted in sulphate and in a lesser degree in sodium and magnesium. Such profound changes in the composition from the initial sedimentary waters, testify to the great role of host rocks in the formation of the modern brines.

In order to reveal mechanisms of elemental concentration in underground waters, we revise the equilibrium of brines under study with calcite, gypsum, dolomite, anorthite - the leading calcium minerals in the rocks (Shvartsev, Bukaty, 1996), and kaolinite, montmorillonite, illite etc. (Shvartsev 1991, 1995). As it was shown, in natural conditions the contradictory equilibrium-nonequilibrium system is formed, in which one part of minerals (anorthite, gypsum, calcite) is dissolved; the others (dolomite, kaolinite, montmorillonite, illite) are formed. It is impossible for the solution to equilibrate with anorthite, which therefore dissolves always at rate and serves as a source of calcium. The latter may pass completely into dolomite, and then there will be no obvious changes in the solution composition, or else it may partially accumulate in solution. This depends on the geochemical medium.

3 THE $^{87}Sr/^{86}Sr$ RATIOS OF THE STRONG BRINES

The use of Sr isotopes in brine studies has been discussed by McNutt et. al. (1990), Fritz et. al. (1987) and many others. They have shown, that the isotopic ratio of saline waters are sensitive tracers for identifying different water masses and genesis.

Strontium isotopes in the brines have been studied of the Siberian Platform by Pinneker and Shvartsev (1996). Additional data obtained recently are used in this work. The isotopic strontium composition ofstudied brines is shown in Table 2. From the table we notice that values of $^{87}Sr/^{86}Sr$ increase from infiltration to sedimentation waters and with an

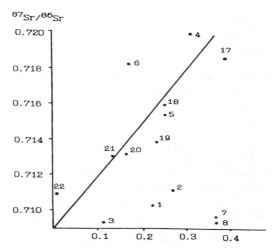

Figure 2. Sr isotopic composition versus Ca/Cl ratio of brines.

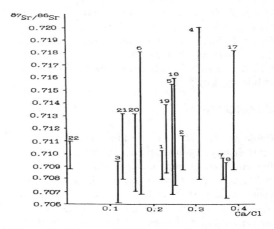

Figure 3. Changes of Sr isotopes in the brines (high value) relatively sea water corresponding geological epoch (low value)

increasing its metamorphization degree. This is clearly shown in Fig.2, where the metamorphization degree of the waters is expressed in terms of Ca/Cl ratios.

Comparative isotope analysis shows that studied brines have a considerably higher radiogenic strontium than sea water (Koepnick et.al., 1985, Gavshin et. al., 1994) of the corresponding geological epoch. This is true for infiltration and sedimentation brines, but the mentioned difference not always increases with the degree of brine metamorphization.

Sample 1 for example represent the fossil water, which is strong metamorphised (Table 2), but the isotopic value is not high. The same situation is with the samples 7 and 8. Therefore we confirm the conclusion of McNutt et. al. (1990), that sometimes even an "old" connate waters, with high TDS may have a low $^{87}Sr/^{86}Sr$ value.

A strong influence of rock on the strontium isotopic composition of brines is observed in sample 4, 6, 17, 18 and 15 because they are connate waters that have interacted with terrigenous and volcanogenic rocks for a long time. The enrichment of sedimentation waters by radiogenic strontium isotope is correlated with their metamorphization degree (Figs. 2 and 3). This conclusion is very important because it is indicative of the continuous interaction of buried brines with water-enclosing rocks and the slow accumulation of radiogenic strontium by dissolution of primarily feldspars. However, the absence of any positive correlation between $^{87}Sr/^{86}Sr$ and 1/Sr (Table 2) suggest the different degree of brine-rock interaction. Such interpretation of obtained data is in good agreement with the notion of equilibrium-nonequilibrium state of brine-rock system.

4. CONCLUSION

Consequently in the sedimentary basins a contradictory equilibrium-nonequlibrium system is formed in which some minerals (anorthite, gypsum and others) are dissolved; the others (dolomite, kaolinite etc.) are formed. A part of calcium during this interaction is continuously concentrated in aqueous solution, formed a calcium type of brine. This takes place over geologic time and results in the enrichment of many elements in the aqueous solution. It is confirmed by Sr isotopes that values are considerably higher than sea water corresponding geological epoch. The $^{87}Sr/^{86}Sr$ isotope ratio in the brines of the Siberian Platform is a sensitive indication of their interaction with rock and degree of dilution of sedimentation brines by infiltration waters. The revealed relationship between the concentration of radiogenic strontium in brines and the degree of their metamophization serve as a firm basis for the development of mechanisms for the geological evolution of connate brines.

REFERENCES

Fritz P. and Frape S.K. (Eds), 1987 Saline waters and gases in crystalline rocks. Geolog. Assoc. Canada Spec. Pape, 33

Gavshin V.M., Ponomarchuk B.A. Nikitin I.A., and Razvorotneva L.I. 1994. Unusual correlation the strontium isotopes in the Riphean carbonate rocks.Doklady Akademii Nauk: 335, N 1: 77-80.(in Russian)

Koepnick R.B., Burke W.N., Denison R.E. et al. 1985. Construction of the sea water $^{86}Sr/^{87}Sr$ curve for the Cenozoic and Cretaceous: Supportin data. Chem Geol., 58,N 1/2: 55-80

Land L.S. 1995. Na-Ca-Cl saline formation waters, frio formation (oligocine), south Texas, USA: Products of diagenesis. Geochim.and Cosmoch. Acta, 59, N 11: 2163-2174

McNutt R.H., Frape S.K., Fritz P. et al., 1990. $^{87}Sr/^{86}Sr$ values of Canadian Shield brines and fracture minerals with applications to ground water mixing, fracture history, and geochronology. Geoch. et Cosmoch. Acta, 54, N 1: 205-215

Pinneker E.V., Shvartsev S.L. 1996, Strontium isotopy in the brines of the Siberian Platform. Transact. of the Russ. Acad. of Sci. Earth Sci.Sect., 351, N8: 1310-1312

Shvartsev S.L. 1991. Interaction of water with aluminosilicate rocks. Review. Sov.Geol. and Geophys. N12:13-37.

Shvartsev S.L.1995. Equilibrium-nonequilibrium state of the water-rock system. Water-rock interact., Kharaka and Maest (Eds.) Rotterdam:Balkema: 751-754.

Shvartsev S.L., Bukaty M.B. 1996. Evolution of $CaCl_2$ brine of Tungussky Basin (Siberian Platform): The role of water-rock interaction. Transact of the Russ. Acad of Sci. Earth Sci. Sect., 344, N7: 246-250.

Reservoir connectivity determined from produced water chemistry, Standard Draw-Echo Springs gas field, Wyoming, USA

L. K. Smith & R. C. Surdam
Institute for Energy Research, University of Wyoming, Laramie, Wyo., USA

ABSTRACT: Composition of produced water indicates connectivity between the upper and lower Almond Formation at Standard Draw-Echo Springs field. Wells in the sweetspot (highest gas production) are completed in the upper Almond, an offshore marine bar. However, these wells produce relatively fresh, bicarbonate dominated water typical of the lower Almond. Geochemical modeling indicates that fractures, which are present throughout the field, are most likely to be open in the sweetspot. The modeling shows that calcite, a common fracture-filling cement, is unstable only in that portion of the field. Another fracture network connecting the Almond with deeper zones is also indicated by the produced water. Throughout the basin, Almond water with anomalously high total dissolved solids produces from wells adjacent to major surface-mapped lineaments. Thus produced water chemistry in the Almond Formation can be used to assess reservoir and fluid connectivity at a range of scales.

1 INTRODUCTION

At Standard Draw-Echo Springs field in Wyoming, numerous wells produce more gas than has been estimated from reserve calculations for the upper Almond Formation. Iverson and Surdam (1995) hypothesized that the additional gas was coming from the lower Almond, producing through upper Almond perforations via fractures connecting the two zones. Both open and cemented fractures are present in core from the field. Also, major surface lineaments cross the field.

This study offers support for local and regional fracture connectivity using the chemistry of the produced water.

2 ZONAL DIFFERENCES IN FORMATION WATER CHEMISTRY

Basinwide, the Almond Formation produced water chemistry exhibits differences between the upper and lower zones. For example, the lower Almond is generally less saline. This is to be expected because the upper Almond is of marine origin, whereas the lower Almond was deposited in a coastal plain and estuary with numerous thin coal beds. There is greater scatter in the upper Almond salinity values which we interpret as resulting from mixing with lower Almond water. Figure 1 shows that there are also compositional differences. The upper Almond water is chloride-dominated whereas the lower Almond water tends to be more bicarbonate-rich.

3 CONNECTIVITY BETWEEN UPPER AND LOWER ALMOND

Although trends in Almond Formation produced water chemistry tend to follow original depositional environment, the composition of water producing from any given well at Standard Draw-Echo Springs field may bear little relation to the zone in which the well was completed. For ex-

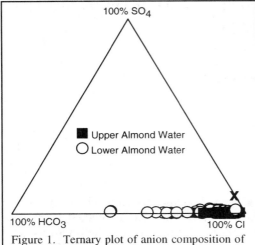

Figure 1. Ternary plot of anion composition of formation water in the Almond Formation, Green River Basin. X = seawater.

ample, the 226F well in the southern sweet spot (Figure 2) is completed only in the upper Almond. However, water producing from that well and other wells in the area is relatively fresh (Figure 2). These low total dissolved solids (TDS) are not due to condensation of water vapor producing with the gas stream. A correction has been applied to each sample to account for that (McKetta and Wehe, 1957). Also, there is no relation between chloride content and the oxygen isotopes of the samples (Smith and Surdam, 1997) which is what would be expected if the fresher waters were due to water vapor condensation (Poulson et al., 1996). The gas:water production ratio (GWR) also indicates that the fresher water is the result of lower Almond water producing from upper Almond perforations. If fresh water was due to condensation of water vapor, then wells with high GWRs (low water production) would have low TDS (all the water from vapor) and wells with low GWRs (high water production) would have high TDS (all the water from more saline formation water). However, the opposite trend exists at Standard Draw-Echo Springs field. Wells with low GWR produce water with the lowest TDS (Figure 3). The bicarbonate domi-

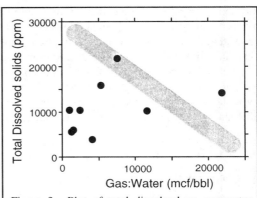

Figure 3. Plot of total dissolved vs. gas:water ratio (TDS vs. GWR). Grey line represents the trend that would expected if the fresher waters were the result of dilution by dewpoint water.

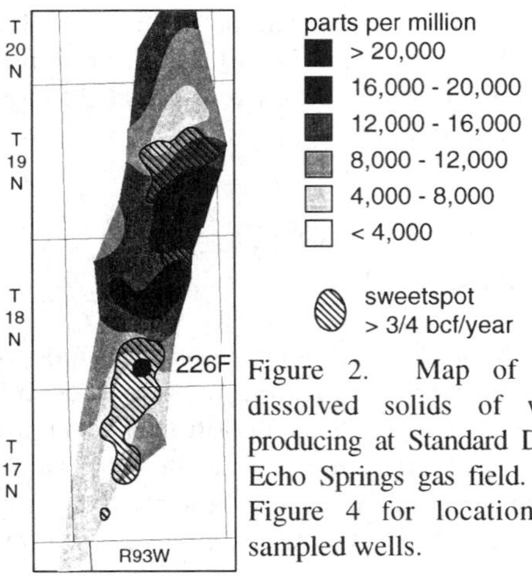

nance of the fresher water is another indication that it originates in the lower Almond (Figure 4).

A set of nine water analyses taken over a two year period for a single well provide additional evidence. This well was completed in the lower Almond. Over the two-year period, the produced water freshened through time, indicative of lower Almond coal de-watering caused by production and pressure depletion. This would explain fresh lower Almond water producing from the upper Almond in other wells after some pressure depletion in the upper Almond had occurred.

Water chemistry at Standard Draw-Echo Springs field also indicates which areas of the field have better fracture connectivity due to lack of fracture cement. Calcite is the most common fracture-fill-

Figure 2. Map of total dissolved solids of water producing at Standard Draw-Echo Springs gas field. See Figure 4 for locations of sampled wells.

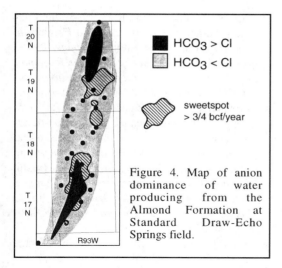

Figure 4. Map of anion dominance of water producing from the Almond Formation at Standard Draw-Echo Springs field.

ing cement (Dunn et al., 1995; Dunn and Humphreys, 1996). The calcite saturation index was calculated for each water sample then mapped to determine the relationship to sweet spots. An area where calcite is unstable almost exactly corresponds with the southern sweet spot, whereas calcite is stable over the rest of the field (Figure 5). This is not due to higher CO_2 content of produced gas in that part of the field because it is relatively constant throughout the field. Undersaturation with respect to calcite means that calcite as a matrix cement is probably less abundant there than in the rest of the field and that fractures are more likely to be open. Similar patterns were observed for kaolinite and illite (Smith and Surdam, 1997). This helps explain the fact that TDS and bicarbonate (Figures 2 and 4) do not correspond with the northern sweet spot as well as they do with the southern sweet spot. Lower Almond water cannot produce through upper Almond perforations in the northern area because fractures there are more likely to be occluded by cements.

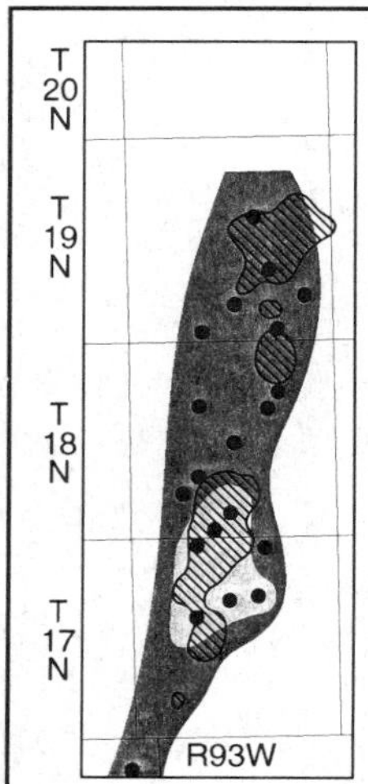

4 CONNECTIVITY WITH DEEPER ZONES

Almond Formation produced water chemistry also indicates possible connection with deeper zones in one portion of the field as well as elsewhere in the basin. One portion of the field has water with higher total dissolved solids (>20,000ppm, Figure 2). A major linear feature has been mapped from surface expression in that part of the field (Jaworowski et al., 1995). Also, those higher TDS waters have Na:K ratios that are more similar to Niobrara Formation water than to Almond Formation water elsewhere in the field. The Niobrara is approximately 1200 m below the Almond.

The water chemistry also suggests that other major lineaments in the Washakie Basin are, in places, open and connecting the Almond with deeper formations. Figure 6 shows a map of TDS overlain with major lineaments. The TDS anomalies (highlighted in dark grey) all correspond to major lineaments.

Figure 5. Map showing areas of Standard Draw-Echo Springs field where the produced waters are either oversaturated or undersaturated with respect to calcite.

5 CONCLUSIONS

Connectivity between the upper and lower Almond is indicated by water with a lower Almond signature (fresh, bicarbonate-dominated) producing from wells completed in the upper Almond. Connectivity is further indicated by chemical modeling of the waters which indicates a paucity of fracture cements in the gas production sweetspot. Connectivity between the Almond Formation and deeper zones is indicated by the association of anomalously high TDS waters producing only from those wells on or adjacent to major lineaments in the basin. Almond Formation produced water chemistry can be used to evaluate reservoir connectivity at a range of scales.

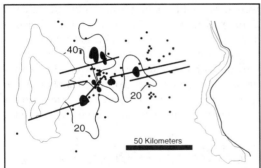

Figure 6. Map of the Green River Basin showing correspondence of anomalous TDS in produced water (highlighted in dark gray) with major lineaments in the basin (from Jaworowski et al., 1995). Light grey areas are Almond Fm outcrops. Contour interval = 20,000 ppm.

ACKNOWLEDGMENTS

The funding for this research was provided by the Gas Research Institute. Amoco Corp. provided regional water data for the study. R. Billingsley and L. Evans, both from Amoco, provided helpful discussion, and P. Lee, also from Amoco, helped in the collection of new water samples. Reviews by P. White and J.R. Hulston improved the final version of this manuscript.

REFERENCES

Dunn, T.L., Aguado, B., Humphreys, J., and Surdam, R.C., 1995, Cements and in-situ widths of natural fractures, Almond Formation, Green River Basin, Wyoming *in* Jones, R. (ed) Resources of Southwestern Wyoming: Wyoming Geological Association 1995 Field Conference Guidebook, p. 255-270.

Dunn, T.L., and Humphreys, J., 1996, Natural fracture permeability and cement distributions in anomalously pressured gas reservoirs: Proceedings of Reservoir Characterization Symposium sponsored by Rocky Mountain Association of Geologists and Denver Geophysical Society, Denver, Colorado, September 13, 1996.

Iverson, W.P., and Surdam, R.C., 1995, Tight gas sand production from the Almond Formation, Washakie Basin, Wyoming: *SPE paper 29559 presented at the SPE Rocky Mountain Regional/Low-Permeability Reservoirs Symposium, Denver, CO, March 20-22, 1995*, p. 163-176.

Jaworowski, C., Simon, R., and Surdam, R., 1995, Chapter 3 - The significance of northeast-trending lineaments, regional fractures, and digital geologic data to oil and gas production within the east-central Greater Green River Basin, Wyoming In *Natural Fractures and Lineaments of the East-Central Greater Green River Basin: Gas Research Institute Topical Report No. GRI-95/0306*, p. 5-22.

McKetta, J.J., and Wehe, A.H., 1958, Chart for the water content of natural gases: *Petroleum Refiner*, v. 37, no. 8, p. 153-154.

Poulson, S.R., Homoto, H., and Ross, T.P., 1996, Stable isotope geochemistry of waters and gases (CO_2, CH_4) from the overpressured Morganza and Moore-Sams fields, Louisiana Gulf Coast: *Applied Geochemistry*, v. 10, p. 407-417.

Smith, L.K., and Surdam, L.K., 1997, Local and regional Almond Formation reservoir connectivity determined from produced water chemistry, Washakie Basin, Wyoming: (abs) AAPG 1997 Annual Convention, Dallas, TX, Official Program, p.A109.

Geochemistry of waters from two adjoining basins in Hungary

I. Varsányi
Department of Mineralogy, Geochemistry and Petrology, University of Szeged, Hungary

J. M. Matray
Département Stockage, ANTEA, Groupe BRGM, Orléans, France

L. Ó. Kovács
Hungarian Geological Survey, Budapest, Hungary

ABSTRACT: Waters from the Quaternary, Pliocene and Pontian layers in two adjacent areas in the Pannonian Basin (southeast Hungary) were studied. Water stable isotope data suggest that the period of infiltration within the whole studied depth interval (25-2500 m) was between 70,000 and 10,000 years BP in the Szeged area, and it was older than 70,000 years in the Körös area. Water flow in the Körös area was therefore much more restricted than in the Szeged area. The general evolution of chemical and isotopic characteristics with depth is essentially explained by transformation of organic matter.

1 INTRODUCTION

The Pannonian Basin is a large (100,000 km^2), non-uniform, multilayer flow system formed during the Neogene period. It is generally suggested that two flow systems exist: the Quaternary layers, and deeper horizons down to about 2500 m. The bottom of the deeper flow system is the boundary between the Pontian and Pannonian stages. Under this boundary occur practically stagnant NaCl-NaHCO$_3$ type waters. The Pliocene and Pontian layers are characterised by a NaHCO$_3$ water type (Erdélyi, 1979). In the Quaternary, both Ca-Mg(HCO$_3$)$_2$ and NaHCO$_3$ type waters exist. The study area includes two depressions separated by a high (Fig.1, Fig.2). The depression located westward of the high is named Szeged area, and the eastern one is named Körös area. The goal of this study is to draw conclusions on the hydrology of these adjacent depressions by using hydrochemistry and isotope geochemistry.

2 GEOLOGICAL BACKGROUND

The Paleozoic-Mesozoic basement is overlain by hundreds to thousands of metres of Miocene, Pliocene and Quaternary sediments. During the whole Neogene period, subsidence occurred at different rates and in different places of the basin, thus explaining the formation of various subbasins. The prePannonian (lower to middle Miocene) sediments formed in a marine to brackish environment. The Pannonian (M$_3$Pa) in the late Miocene was characterised by lake deposits followed by lacustrine and deltaic sedimentations in the Pontian (M$_3$Po). During the Pliocene (Pl), sedimentation became more and more fluvial whereas the Quaternary (Q) was purely fluvial. The mineralogical composition of the sediment from the upper Miocene to the Pleistocene is characterised by quartz, feldspar, micas, calcite, dolomite, illite/smectite mixed layers, chlorite and kaolinite.

3 METHODS

Fifty samples were collected at the well head of producing drinking water and geothermal wells with perforation depths ranging from 26 to 2473 m. In situ measurements, laboratory major and trace element analyses and several isotopic techniques were used to get insight into the geochemistry of formation waters. Alkalinity, chemical oxygen demand (COD) and Cl$^-$ were determined by titration, and Br$^-$, F$^-$, NH$_4^+$, NO$_2^-$, NO$_3^-$, SO$_4^{2-}$, SiO$_2$, and HBO$_2$ by spectrophotometry in the unfiltered samples. Samples reserved for determining calcium, magnesium, sodium, potassium, iron, and manganese were measured by AAS techniques with a Perkin-Elmer Zeeman 5000 equipment. Total dissolved organic carbon was analysed with a Beckman analyser. These analyses were made at the Public Health Institute, Budapest, Hungary. Acetate was determined with HPLC technique at the Institute of Biology, Szeged, Hungary. All isotopic analyses were performed by mass spectrometry in the BRGM laboratories in Orléans, France. Water stable isotopes

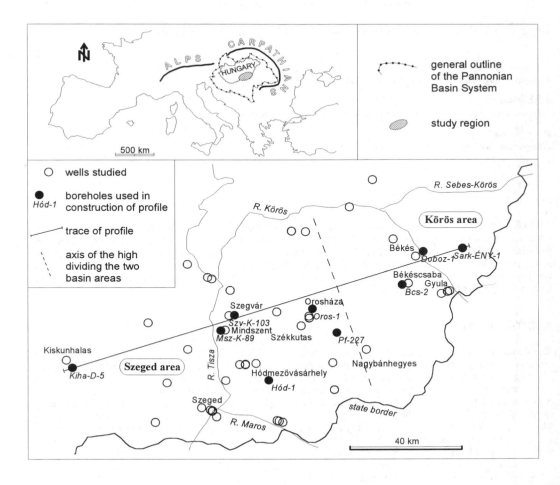

Figure 1. Location of the studied wells

were measured by equilibration techniques on undistilled samples with an accuracy of 0.8‰ for deuterium and 0.1‰ for oxygen-18. Stable carbon isotope ratios from total dissolved inorganic carbon were measured on CO_2 gas extracted with H_3PO_4. The δ values are reported relative to PDB standard with a precision of ±0.1‰.

4 DISCUSSION

Most of the waters are $NaHCO_3$ type. However, some samples from the shallowest layers display an intermediate type between $NaHCO_3$ and $Ca-Mg(HCO_3)_2$. Acetate was found in 11 waters with concentrations as high as 850 mg/l, i.e. in waters where acetate contributes up to 1/3 of the total alkalinity. Fig.2 shows that at equal depths, waters from the Körös area have sodium concentrations and alkalinities at least 10 mmol/l greater than in the Szeged area. This difference can reach 30 mmol/l in the deepest samples. The waters in the Körös area contain dissolved organic matter in much higher concentrations than in the Szeged area (Fig.2).

The isotopic composition of waters is given in Fig.3. The present-day local meteoric water line (LMWL) was determined by Deák et al. (1987) in a village located 100 km north of the studied area.

In the Szeged area, two groups can be distinguished. Waters from the Quaternary, Pliocene and Pontian layers plot exactly on this line thus indicating their meteoric origin.

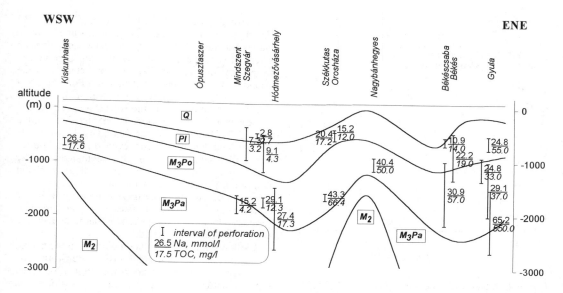

Figure 2. Geological profile across the study region with values of sodium and TOC

Waters produced from the boundary between the Pannonian and Pontian show combined deuterium and oxygen-18 'shifts'. The line through these five samples intersects the LMWL at values of −90±5‰ in δ^2H and −12±1‰ in δ^{18}O, indicating the connection of these five samples with paleometeoric waters. The isotope divergencies shown by these five samples suggest a mixing between this paleometeoric end-member and an oilfield water collected in the same area (Mátyás, 1996).

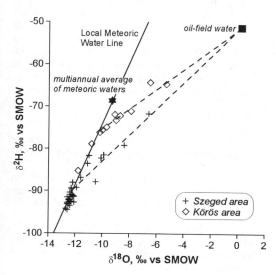

Figure 3. Trends in waters' stable isotope values

The suggested paleometeoric waters are depleted in heavy isotopes compared to modern meteoric waters thus implying recharge under lower temperatures (Dansgaard, 1964). It is assumed that the drop in temperature reflects a colder climate during water infiltration. The fact that δ^2H and δ^{18}O values of samples from the Szeged area are very homogeneous despite differences in depth (550 to 2500 m) strongly suggests that infiltration occurred during the same and probably the last cold period. The position of several samples on the LMWL can be explained by a mixture between the paleometeoric source and the present day meteoric water (Varsányi et al., 1997).

In the Körös area, samples produced from the Pliocene layers plot on the LMWL. Waters from the deeper (Pontian) reservoirs show isotope deviations from the LMWL. The line through these latter intersects the LMWL at about −76‰ in δ^2H and −10‰ in δ^{18}O. These values suggest the infiltration of paleometeoric water at a warmer period than that evidenced in the Szeged area. One sample from the Körös area plots between the two paleometeoric waters. This position suggests infiltration of water between the colder and warmer paleorecharge periods.

In Fig.4, the NH_4^+ contents are compared to the δ^{13}C values. The ammonium concentration which is generally linked to organics correlates well with the δ^{13}C values up to values of 10 and 20 mg/l for the samples from the Szeged and Körös areas, respectively. In both areas, the δ^{13}C values determined on the total dissolved inorganic carbon increase with increasing NH_4^+ concentrations.

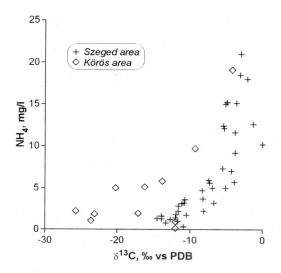

Figure 4. Relationships between NH_4 and $\delta^{13}C$ values

Fig.4 displays two parallel trends, each representing a different area. For the Szeged area, the samples with more than 10 mg/l of ammonium have the highest $\delta^{13}C$ values independently of their NH_4^+ concentrations. These waters are the deepest and warmest and have the highest concentrations in dissolved organic material. The correlation between the NH_4^+ contents and the $\delta^{13}C$ values suggests that ammonium and CO_2 were produced by transformation of organic matter. During the transformation of organic matter, the $\delta^{13}C$ of the dissolved CO_2 increases up to a constant value. The lighter isotopes originate from weaker and more easily broken bonds. This explains that the CO_2 released from sediment organic matter (SOM) becomes enriched in carbon-13 as the decay of SOM proceeds. The separation of samples into two groups in Fig.4 may be explained either by the difference in the type of organic matter or in the reaction path in the two adjacent areas.

5 CONCLUSIONS

Hydrological conclusions can be drawn from chemical and isotopic data determined in formation waters from the Quaternary, Pliocene and Pontian layers. Water stable isotopes indicate different infiltration periods in the two areas. In the Szeged area, the small range of δ^2H and $\delta^{18}O$ values in waters from different depths (25-2500 m) suggests that infiltration within the whole depth interval occurred in the same cold period. According to pollen records in Europe, the last cold period took place roughly between 70,000 and 12,000 years BP (Guiot et al., 1989) and this is the assumed age of waters in the Quaternary, Pliocene and Pontian layers. It is supposed that several samples produced from the deepest studied layers, close to the boundary between the Pannonian and Pontian, are a mixture of paleometeoric water with older oilfield waters. The infiltration in the Körös area occurred in a warmer and earlier period, i.e. earlier than 70,000 years BP. Waters from the boundary between the Pontian and Pannonian and even from the Pontian layers represent a mixture of paleometeoric water with these old oilfield waters. These different infiltration periods suggest that the sediments in the Szeged area are more permeable then those in the Körös area. The different ages of waters are also supported by differences in concentration of chemical components in the two areas.

Carbon stable isotope data indicate that the source of bicarbonate in these waters is the transformation of organic matter. The type of the original sediment organic matter or the pathway of the transformation must have been different in the two areas.

ACKNOWLEDGEMENTS : This work is the result of a collaborative project between Hungary and France. It was financed by the Hungarian Research Fund (OTKA project N° T 017208) and BRGM (Research Division: project N° P40).

REFERENCES

Dansgaard, W., 1964. Stable isotopes in precipitation. *Tellus*, v.16: 436-468.

Deák, J., Stute, M., Rudolph, J. and Sonntag, C., 1987. Determination of the flow regime of quaternary and Pliocene layers in the Great Hungarian Plain (Hungary) by D, ^{18}O, ^{14}C, and noble gas measurements. International symposium on the use of isotopes techniques in water resources development, IAEA, Vienna, Austria.

Erdélyi, M., 1979. Hydrodynamics of the Hungarian Basin Proc. No.18. Budapest VITUKI.

Guiot, J., Pons, A., De Beaulieu, J. L. and Reille, M., 1989. A 140,000-year continental climate reconstruction from two European pollen records. *Nature* 338: 309-313.

Mátyás, J., 1996. Stable isotopic mass balance in sandstone-shale couplets: An example from the Neogene Pannonian basin. *Földtani Közlöny*, v. 126, 1: 77-88.

Varsányi, I., Matray, J.M. and Ó.Kovács, L., 1997. Geochemistry of formation waters in the Pannonian Basin (Southeast Hungary). *Chemical Geology*, v.140: 89-106.

Do stable isotopes and fluid inclusions allow to constrain the origin and timing of dolomitization in deeply buried carbonate reservoirs? Example of the Pinda Formation, Angola

F.R. Walgenwitz, H. Eichenseer & P. Biondi
Elf Exploration Production, CSTJF, L1/37, Pau, France

ABSTRACT: Oxygen isotope compositions and homogenization temperatures of fluid inclusions measured in massively dolomitized Albian carbonate reservoirs of the Angola margin, vary according to the present-day burial depths and reservoir temperatures. It is argued that fluid inclusions and stable isotopes are reset during Tertiary burial, and reflect the conditions of recrystallization of pre-existing dolostones, early diagenetic in origin.

1 INTRODUCTION

Current thinking is that burial dolomitization is extensive, although considerable uncertainties remain concerning the origin of parental fluids. Mattes and Mountjoy (1980), Zenger (1983), Qing and Montjoy (1992), Mountjoy and Amthor (1994) have suggested massive replacement of limestones at burial depths of 1km or more, which poses obvious problems concerning the source of magnesium and its transport over distances sometimes of tens or hundreds of kilometers (Amthor et al., 1993).

In many case studies, stable isotopes and fluid inclusions are the main geochemical tools used to support the conclusion that massive replacement dolomitization may occur in deep subsurface environments. However, dolomite can be metastable and potentially prone to dissolution and recrystallization at various stages of the burial history (Mazzulo, 1992). Thus, oxygen isotopes and fluid inclusions can be reset and are not suitable to evidence the conditions of dolomitization (Kupecz and Land, 1994). Such processes have clearly been documented in Paleozoic dolostones. Massively dolomitized Albian mixed ramp reservoirs, offsfore Angola, currently buried at depth ranging from 2000m to 3500m during the Tertiary, provide the opportunity to further support this conclusion.

2 GEOLOGICAL CONTEXT

The oil-prone mixed ramp series of the Albian Pinda Group, offshore Angola, reach a thickness of more than 600m and developed during the earliest spreading stages of the South Atlantic. The carbonates deposits are homogeneously dolomitized over several hundreds of meters. Rapid burial occurred during the Tertiary. However, due to a difference in the rate of subsidence since the Late Miocene, reservoir depth and temperature increase from the East (2000m, 120°C) to the West (3500m, 160°C).

3 PETROGRAPHY

Dolostones contain both microrhombs ($\approx$50mm) substituting grains (ooids and oncoids), and coarse crystalline dolospar mosaics (200-300mm ; Fig.1). Both microrhombs and mosaics are replacement dolomites. Mosaics typically show cloudy cores and clear rims. Microprobe analyses show that all types of dolomite are almost stoichiometric. Cloudy cores are extremely rich in aqueous and oil fluid inclusions (Fig.2), and show a blotchy red cathodoluminescence. SEM observations in back-scattered electron mode, also show tiny solid inclusions of calcite. Clear rims do not contain fluid inclusions and show several growth zones in cathodoluminescence.

These rims are developed when intercrystalline or moldic porosity is present, and thus considered as cements. They occur in the oil and water legs of reservoirs.

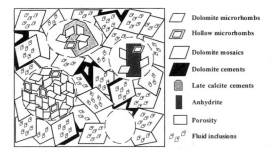

Fig.1: Main features of dolomite petrography

The siliciclastic components include detrital quartz, K-Feldspars and plagioclases. Both the detrital plagioclases and K-Feldspars were more or less intensely recrystallized during burial, as shown by the numerous petroleum and aqueous fluid inclusions (Fig.2). K-Feldspars are overgrown by thick limpid rims (50 to 100 mm) of authigenic adularia, whose K-Ar ages of 98±16Ma (Girard, 1988) are close to the Albian time of deposition. Similarly with the limpid dolomite cements, the adularia overgrowths do not contain fluid inclusions.

Anhydrite is frequent, either as nodules (of depositional origin), or as cements in fractures and moldic porosity. The main types of porosity are moldic porosity (by dissolution of the non-dolomitized fraction of the ooid and oncoid grains), and intercrystalline porosity.

4 STABLE ISOTOPE COMPOSITION

In every particular reservoir, the $\delta^{18}O$ values are grouped in a rather narrow interval ($\Delta\delta^{18}O=3$ to 4‰, Fig.3), with an ^{18}O-increase in the depositional facies rich in anhydrite.

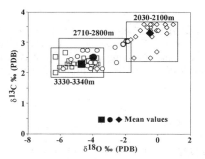

Fig.3: Plot of $\delta^{18}O$ against $\delta^{13}C$ of some selected reservoirs at increasing depth.

The overall $\delta^{18}O$ variation in all wells ranges from +1‰ to -6‰PDB, but the mean $\delta^{18}O$ values vary from -0.3‰ to -4.7‰ according to the depth of reservoirs (Fig.3 and 4). The $\delta^{13}C$ values are less variable (+2 to +3.5‰PDB), and statistically decrease with the present-day reservoir depth as well (Fig.3).

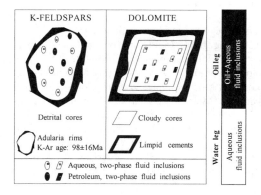

Fig.2: Fluid inclusion distribution in dolomite and K-Feldspars, and in reservoirs according to the present-day fluids.

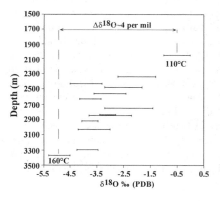

Fig.4 : Plot of mean $\delta^{18}O$ values against depth.

5 FLUID INCLUSION DATA

Two-phase aqueous inclusions (liquid+gas) are observed in the recrystallized feldspars, cloudy cores of dolomite and anhydrite cements, both in the oil and water legs of reservoirs (Fig.2). Petroleum inclusions are restricted to the oil legs of reservoirs.

Mean homogenization temperatures of aqueous inclusions are similar in all types of host minerals, and range from 110°C to 150°C according to the reservoir depth (Fig.5). They are only 10 to 20°C lower than the present-day reservoir temperatures. Salinities estimated from ice-melting temperatures range from 8 to 21 weight per cent equivalent NaCl.

Due to different PVT and compressibility characteristics, the cogenetic petroleum inclusions homogenize at temperatures about 20 to 30°C lower than the associated aqueous inclusions (Fig.5).

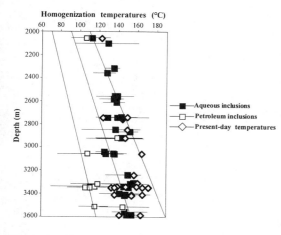

Fig.5: Plot of fluid inclusion homogenization temperatures and reservoir temperatures versus depth.

6 DISCUSSION

The ≈4‰ difference in average oxygen compositions between the shallowest and the deepest reservoirs (Fig.4), is consistent with the value expected from the isotopic fractionation of dolomite in the temperature range indicated by homogenization temperatures of aqueous fluid inclusions (i.e. 110 to 150°C). According to basin modeling studies, such homogenization temperatures would indicate that fluid inclusions (both oil and aqueous) were entrapped during the Late Miocene. Can we thus conclude that replacement dolomitization occurred at depths close to the present-day reservoir depth ?. There are at least two observations inconsistent with this interpretation:

- oil fluid inclusions in the cloudy cores of dolomites and the recrystallized feldspars are restricted to the oil legs of reservoirs (Fig.2). This clearly demonstrates they were entrapped during hydrocarbon migration since the Late Miocene. Massive dolomitization coeval with hydrocarbon filling seems quite unlikely.
- the limpid dolomite cements postdate the cloudy dolomite cores. If fluid inclusions in the cloudy cores indicate the timing and conditions of replacement dolomitization, it results that dolomite cements would postdate hydrocarbon migration and would have grown in oil-saturated reservoirs. This conclusion is unrealistic if we take into consideration the low residual water saturation and cement volume.

A key observation is actually given by the K-feldspars. As previously mentioned, they are composed of detrital cores, recrystallized at great depth during the Late Miocene. However, their authigenic adularia rims have formed during the Albian, soon after deposition. This observation clearly indicate that detrital cores have been selectively recrystallized, probably due to a difference in chemical composition between the detrital and authigenic components. Authigenic adularia, formed in the miscibility gap of feldspars at low temperatures, is a pure potassic end-member, more stable in the diagenesis realm than detrital feldspars derived from high grade metamorphic basements of Pan-African age (≈500Ma).

It is proposed that chemical zoning in dolomite crystals may also account for the selective recrystallization of cloudy cores, as in K-feldspars. As already stated by Land et al. (1975) and Sibley (1982), chemical differences exist between cloudy centers and clear rims which could indicate that clear rims precipitated from more dilute solutions. Therefore, it could be inferred that cloudy cores in the Pinda dolomites selectively recrystallized, either according to a greater excess in Ca content or a lower degree of crystal order, or both. The tiny inclusions of calcite observed in cloudy cores, could be interpreted as evidences of former Ca-excess. Our observations also suggest that centers of rhombs, as well as detrital feldspar cores, are still accessible to pore fluids even after the formation of the clear epitaxic rims and cements. In this assumption, both fluid inclusions and stable isotopes are reset and relevant to the re-equilibration of pre-existing dolostones, formed at shallower depth and probably during an early, sub-depositional diagenetic stage. Dolomite fabrics observed in the Pinda dolomites are common to many other dolomite reservoirs all over the world. The original information given by similar diagenetic features in feldspars and dolomite crystals of mixed carbonate-siliclastic deposits, allows to directly evidence the process of selective recrystallization. It demonstrates that oxygen isotopes and fluid inclusions may lead to erroneous interpretation of the timing and origin of massive dolomitization in deeply buried carbonate deposits.

REFERENCES

Amthor, J., Mountjoy, E. & Machel, H. 1993. Subsurface dolomites in Upper Devonian Leduc Formation buildups, central part of Rimbey-Meadowbrook reef trends, Alberta, *Canada Bull. Canad. Petrol. Geol.* 41, 2 : 164-185.

Girard, J.P., Aronson, J. & Savin S. 1988. Separation, K/Ar dating and 18O/16O ratio measurements of diagenetic K-Feldspar overgrowths : An example from the Lower Cretaceous arkoses of the Angola margin. *Geochim. Cosmochim. Acta* 52 : 2207-2214.

Kupecz, J. & Land, L. 1994. Progressive recrystallization and stabilization of early-stage dolomite : Lower Ordovocian Ellenburger Group, west Texas. In *Dolomites, Purser, Zenger & Tucker Eds* : 255-282.

Land, L., Salem, M. & Morrow, D. 1975. Paleohydrology of ancient dolomites : geochemical evidence. *Am. Assoc. Petroleum Geologists Bull.* 59 : 1602-1625.

Mattes, B. & Montjoye, E. 1980. Burial dolomitization of the Upper Devonian Miette Buildup, Jasper National park. In *Concepts and Models of Dolomitization*, Spec. Publ. Soc. Econ. Paleont. Mineral. 28 : 259-299.

Mountjoy, E. & Amthor, J. 1994. Has burial dolomitization come of age ? Some answers from the Western Canada Sedimentary Basin. In *Dolomites, Purser, Zenger &Tucker Eds* : 203-230.

Qing, H. & Mountjoy, E. 1992. Large-scale fluid flow in the Middle Devonian Presqu'ile barrier, Western Canada Sedimentary Basin. *Geology* 20 : 903-906.

Mazzullo, S. 1992. Geochemical and neomorphic alteration of dolomites, a review. In *Carbonates and Evaporites* 7 : 21-37.

Sibley, D. 1982. The origin of common dolomite fabrics : clues from the Pliocene. *Jour. Sed. Petrology* 52, 4 : 1087-1100.

Zenger, D. 1983. Burial dolomitization in the Lost Burro Formation (Devonian), east-central California, and the significance of late diagenetic dolomitization. *Geology* 11 : 519-522.

Evidence of Proterozoic primary CaCO$_3$ precipitation from the McArthur Group of northern Australia

Peter R. Winefield & Peter McGoldrick
CODES, University of Tasmania, Hobart, Tas., Australia

ABSTRACT: Radiating fans of acicular 'Coxco needles' are the characteristic feature of the Coxco Dolomite Member (≈1640Ma), a subunit of the Teena Dolomite within the Palaeoproterozoic McArthur Group. In the past, 'Coxco needles' have been variably interpreted as dolomitic pseudomorphs after aragonite, gypsum or trona. More recently, gypsum and an emergent brine pool depositional setting has been the favoured interpretation. New work reported here has found that crystal morphology, geochemistry, petrographical and sedimentological relationships are more consistent with a subaqueously deposited aragonitic precursor. The widespread occurrence of 'Coxco needles' at a confined stratigraphic interval is thought to be a function of a subtle change in the HCO$_3^-$ concentration within the water body during deposition of the Teena Dolomite. Although elevated atmospheric CO$_2$ during the Proterozoic supports increased carbonate precipitation (including aragonitic fans), it can not satisfactorily explain the apparently synchronous precipitation of Coxco fans. Therefore, changes in the bathymetry of the McArthur Basin coincident with deposition of a broadly transgressive sequence is inferred to have triggered the widespread chronostratigraphic precipitation of carbonate (i.e aragonite fans) within a variety of lithofacies.

1 INTRODUCTION

The Palaeoproterozoic McArthur Basin is located along the southern and western margins of the Gulf of Carpenteria in northern Australia. The McArthur Group is best exposed in the southern McArthur Basin and is divided into the lower Umbolooga and the upper Batten Subgroups (Pietsch et al., 1991). The Teena Dolomite is part of the Umbolooga Subgroup and directly underlies the Barney Creek Formation, host to the large Pb-Zn-Ag deposit at McArthur River. The upper unit of the Teena Dolomite is defined as the Coxco Dolomite Member of which a characteristic feature is the presence of colloquially named 'Coxco needles', which are radiating clusters of acicular crystal casts (<10cm). Similar radiating fans have also been recognised in stratigraphically equivalent sequences in the Victoria Basin on the western edge of the Northern Territory and in the Tennant Creek area approximately 500km southwest of the McArthur River Mine.

Several precursor minerals have been suggested for these needles. Brown et al. (1969) interpreted the shape and habit as being indicative of aragonite pseudomorphs, while Walker et al. (1977) argued they were gypsum pseudomorphs, based largely on a study of interfacial angles. A third alternative interpretation, proposed by Jackson et al. (1987), was that at least some of the crystal casts were morphologically similar to trona described from the Eocene Green River Formation (USA).

2 METHODOLOGY

As part of the measurement of regional sections through the Teena Dolomite and Barney Creek Formation, a large number of samples containing Coxco needles were collected and field relationships noted. Drillcore from the McArthur River area was also utilised as a source of additional, unweathered sample material. Crystal morphology was examined in handspecimen, while several polished thin-sections were prepared for petrographic analysis. Electron microprobe and stable isotope analysis was also conducted on a number of Coxco needle samples and initial results are compared here with analyses of fibrous marine cement, micritic sediment and late dolospar sampled from the Teena Dolomite.

3 COXCO NEEDLE MORPHOLOGY

Coxco needles are acicular pseudomorphs, now dolomite, that form distinctive bottom-nucleated rosettes or fans, either in isolation or as intergrown bundles forming distinct layers in massive, crystalline dolostone and planar laminated dololutite (Fig. 1).

Figure 1 Radiating Coxco needle fans forming distinct layers in laminated dolomitic siltstone.

They are also observed in direct association with cyclic, 'fibrous' microbialites and within massive dolostone interbedded with carbonaceous shale and dolomitic siltstone. Coxco needle fans grow exclusively upward and are commonly draped by finely laminated dolomicrite with some laminae pinching out against individual fans.

Individual 'needles' average 5cm in length with a height to width ratio of approximately 10:1. Needle terminations are commonly blocky and feathery, although pointed examples are also observed. In cross-section, Coxco needles appear pseudo-hexagonal with crystal casts generally having irregular six-sided forms. Under plain light, needles are made up of a mosaic of generally irregular, equant dolospar cement. Crystal boundaries commonly cross-cut the relict Coxco needle structures, although there are a number of examples where there is a sharp boundary between the irregular, equant dolospar cement, which defines the needle shape, and the finer crystalline dolomicritic matrix.

4 GEOCHEMISTRY

Comparison of Fe and Mn compositions of late coarser dolospar (av. Fe 12100ppm; av. Mn 4080ppm) and fibrous marine cements (av. Fe 7000ppm; av. Mn 4080ppm) with the results for Coxco needle dolospar (av. Fe 3620ppm; av. Mn 675ppm) shows that the Fe and Mn contents of Coxco needles are more comparable to fibrous marine cements than the late dolospar. Sr values were generally below the detection limits of the microprobe. Figure 2 illustrates that the $\delta^{13}C$ and $\delta^{18}O$ values of Coxco needles are slightly heavier than the late dolospar field but plot within the field for fibrous marine cements and micritic sediment.

5 DISCUSSION

The sequence through the middle McArthur Group, which includes the Teena Dolomite, is broadly transgressive and the distinct absence of desiccation features (e.g. mud cracks, peritidal or lacustrine tepee structures), evaporite pseudomorphs (after anhydrite, gypsum, or halite) and supra-tidal lithofacies (e.g. microbially laminated dolostone with evidence of emergence and coarser rippled layers; brecciation associated with post-depositional leaching of evaporites) supports an entirely subaqueous depositional environment for the Coxco Dolomite Member.

The fan-like geometry, pseudo-hexagonal cross-sections, acicular nature of individual Coxco needles and feathery to blocky crystal terminations are all consistent with primary aragonite precipitation (Loucks and Folk, 1976; Mazzullo, 1980). In contrast, subaqueous bottom-nucleated gypsum commonly has 'swallow-tail' or 'spear-like' terminations (Warren, 1996) dissimilar to morphologies exhibited by Coxco needles. The irregular coarse equant mosaic of pseudomorphing dolospar is also consistent with several criteria for identifying primary aragonite cements in ancient carbonate sequences (Sandberg, 1985).

Excellent fabric preservation of Coxco needles implies very early dolomitisation of the original mineral. Dolomitisation of gypsum would require a solid volume loss or the movement of large quantities of pore-water through the rock (Pierre and Rouchy, 1988). The absence of brecciation or solution-collapse phenomena, or evidence of significant pore-water movement (e.g. vughs etc.) would appear to discount the latter process in this case. The dolomitisation of aragonite would in contrast involve relatively little volume loss and most likely would not result in sediment disruption or brecciation.

Exclusively upright growth, sediment onlapping and draping relationships and the presence of Coxco needles in beds generally lacking in detrital carbonate all indicate that they grew either as positive relief features just below the sediment-water interface or nucleated directly at the sediment surface in a subaqueous environment. The lack of any sedimentary inclusions preserved within the pseudomorphing dolomite and the available geochemical evidence would also discount a wholly subsurface or diagenetic origin for the Coxco needles.

6 SIGNIFICANCE

The presence of Coxco needles throughout a variety of chronostratigraphically equivalent lithofacies over a very large area (>100,000km^2) indicates that their precipitation was a function of a subtle but widespread change in the chemistry of the water body. If the precursor mineral to the Coxco needles was aragonite, then this change in water chemistry could be interpreted as an increase in the ratio of HCO_3^- to Ca^{2+} so that $CaCO_3$ precipitation was favoured.

A number of mechanisms could have affected the concentration of HCO_3^- during deposition of the Teena Dolomite in the southern McArthur Basin. Grotzinger (1989) invoked atmospheric conditions to explain the large volume of carbonate precipitates (including aragonite fans) in late Archaean/early Proterozoic carbonate sequences. This is supported by the likelihood that the Archaean/early Proterozoic atmosphere contained more CO_2 than in the Phanerozoic. This would have increased the HCO_3^- to Ca^{2+} ratio thereby favouring precipitation of carbonate. The increased rate of carbonate production is inferred by Grotzinger (1989) to have exhausted the concentration of Ca^{2+} and therefore only very limited calcium sulphate would have been precipitated. The change to conditions more akin to the Phanerozoic is thought to be recorded by the first appearance of bedded or massive gypsum in the McArthur Basin at stratigraphic levels well below the Teena Dolomite. The increased uptake of Ca^{2+} by $CaCO_3$ precipitation would have increased the ratio of Mg/Ca and provides a possible mechanism for the early dolomitisation of carbonate precipitates. The identification of aragonite fans in the Teena Dolomite would appear to complicate this model. A return to atmospheric conditions similar to that invoked in the late Archaean/Early Proterozoic is supported by the recognition of chronostratigraphically equivalent Coxco lithofacies in the Victoria Basin and in the Tennant Creek area. However, there is no evidence of aragonite pseudomorphs or increased carbonate precipitation from stratigraphically equivalent carbonate sequences from the Mt Isa Basin in western Queensland, Australia. A higher degree of fabric destruction related to silicification and surficial weathering in addition to relatively poor exposure might account for the absence of any evidence pointing to increased precipitation of carbonate.

A second mechanism that could have affected carbonate precipitation is tectonically induced changes in sea-level. Kennedy (1996) suggested that transgression and flooding (related to deglaciation) of carbonate shelves by bicarbonate-charged deep water caused the subsequent rapid precipitation of $CaCO_3$ recorded by the deposition of Neoproterozoic cap dolostones in Australia. This demonstrates that changes in the bathymetry of a water body can influence that precipitation of carbonate cements through the upwelling of HCO_3^- charged fluids onto carbonate platforms. Loutit et al. (1994) inferred from Apparent Polar Wander Path (APWP) data that a major tectonic event was coincident with the deposition of the Teena to Reward Dolomite stratigraphic interval in the McArthur Basin. A related change in the bathymetry of the southern McArthur Basin and the upwelling of HCO_3^- charged basin waters during the deposition of the Teena Dolomite could, therefore, possibly account for the precipitation of aragonitic Coxco needles over a large area.

7 CONCLUSIONS

The fan-like geometry, acicular nature of individual crystals, pseudo-hexagonal cross-sections, blocky or feathery terminations, excellent fabric preservation and irregular coarse mosaic of pseudomorphing dolospar cement are all supportive of an aragonitic precursor for the Coxco needles. Sedimentological

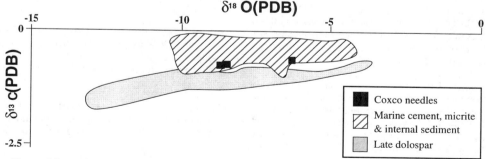

Figure 2 Isotopic compositions of Coxco needles compared to fields for fibrous cement, micrite, internal sediment and late dolospar from the Teena Dolomite.

relationships and their exclusively upright growth suggests that the Coxco fans grew as either positive relief structures immediately below the sediment-water interface or nucleated directly on it. This is supported by their occurrence in beds lacking detrital carbonate and by the similarities in the isotopic and elemental composition of Coxco needle dolospar, fibrous marine cements and micritic sediments sampled from the Teena Dolomite.

The occurrence of Coxco needles in a number of stratigraphically equivalent lithofacies over a large lateral area is thought to be a function of a subtle change in the HCO_3^- concentration of basin waters. It is likely that although bedded or massive gypsum is recognised within lower McArthur Basin stratigraphy, the CO_2 content of the atmosphere was still elevated during the Palaeoproterozoic compared to the Phanerozoic. This would support the development of Coxco needle lithofacies over a large area (>100,000km^2) but it can not sufficiently explain their synchronous precipitation. Therefore, changes in the bathymetry of the McArthur Basin recorded by a broadly transgressive sequence is favoured as the trigger mechanism for widespread chronostratigraphic carbonate (i.e. aragonite fan) precipitation within a variety of Coxco Dolomite Member lithofacies.

8 ACKNOWLEDGMENTS

This work forms part of a PhD study on the Proterozoic McArthur Group carbonates of northern Australia. Funding comes from an Australian Commonwealth Postgraduate Research Award and the AMIRA P384A Proterozoic Sediment-Hosted Base Metal Deposits research project. The authors are indebted to Steve Abbott and Peter Beier (NTGS) for information regarding Coxco needle lithofacies in the Victoria Basin and Tennant Creek area respectively. Comments by Ralph Bottrill, Stuart Bull, David Rawlings, and John Warren are gratefully acknowledged.

9 REFERENCES

Brown, M.C., Claxton, C.W., and Plumb, K.A., 1969. The Proterozoic Barney Creek Formation and some associated units of the McArthur Group, Northern Territory. Bureau of Mineral Resources, Australia, *BMR Record 1969/145*

Grotzinger, J.P., 1989. Facies and evolution of Precambrian carbonate depositional systems: emergence of the modern platform archetype. In Crevello, P.D., Wilson, J.L, Sarg, J.F. and Read, J.R., eds, *Controls on Carbonate Platform and Basin Development*, SEPM Special Publication No. 44, p79-106.

Jackson, M.J., Muir, M.D, and Plumb, K.A., 1987. Geology of the southern McArthur Basin, Northern Territory. Bureau of Mineral Resources, Australia, *BMR Bulletin 220*.

Kennedy, M.J., 1996. Stratigraphy, sedimentology, and isotopic geochemistry of Australian Neoproterozoic Post-glacial Cap Dolostones: deglaciation, $\delta^{13}C$ excursions, and carbonate precipitation. *Journal of Sedimentary Research*, v66, no.6, p1050-1064

Loucks, R.G. & Folk, R.L., 1976. Fanlike rays of former aragonite in Permian Capitean Reef pisolite. *Journal of Sedimentary Petrology*, v46, p483-485

Loutit, T.S., Wyborn, L.A.I, Hinman, M.C., and Idnurm, M., 1994. Palaeomagnetic, tectonic, magmatic and mineralisation events in the Proterozoic of northern Australia. *AusIMM Annual Conference Proceedings*, Darwin, p123-156

Mazzullo, S. J., 1980. Calcite pseudospar replacive of marine acicular aragonite, and implications for aragonite cement diagenesis. *Journal of Sedimentary Petrology*, v50, p409-422

Pietsch, B.A., Rawlings, D.J., Creaser, P.M., Kruse, P.D., Ahmad, P., Fererczi, P.A. and Findhammer, T.L.R., 1991. *Bauhinia Downs 1:250000 Geological Map Series*. NTGS Explanatory Notes SE53-3

Pierre, C., & Rouchy, J.M., 1988. The carbonate replacements after sulphate evaporites in the middle Miocene of Egypt. *Journal of Sedimentary Petrology*, v58, p446-456.

Sandberg, P., 1985. Aragonite cements and their occurrence in ancient limestones. In Schneidermann, N and Harris, P.M., eds, *Carbonate Cements*, SEPM Special Publication No. 36, p33-57

Walker, R.N., Muir, M.D., Diver, W.L., Williams, N., and Wilkins, N., 1977. Evidence of major sulphate evaporite deposits in the Proterozoic McArthur Group, Northern Territory, Australia. *Nature*, v265, p526-529

Warren, J.K., 1996. Evaporites, brines and basemetals: what is an evaporite? Defining the rock matrix. *Australian Journal of Earth Sciences*, v43, p115-132

6 Weathering

The use of strontium isotopes in weathering studies

D.C. Bain
The Macaulay Land Use Research Institute, Craigiebuckler, Aberdeen, UK

ABSTRACT: The uses of $^{87}Sr/^{86}Sr$ ratios in weathering studies have been summarized, indicating their strengths and weaknesses. Catchment weathering rates for calcium can be calculated provided the strontium isotopic ratio in the input and output are sufficiently different but the accuracy is severely limited by problems in obtaining reliable ratios for the weathering-derived strontium. The sources of strontium can be determined for different components of ecosystems from their isotopic ratios. In small catchments, the streamwater strontium isotopic composition can (1) provide information on relative mineral weathering rates; (2) indicate which minerals are weathering; (3) be used to monitor changes in weathering rates and processes due to environmental change as the $^{87}Sr/^{86}Sr$ ratio is very constant regardless of flow rate and can be measured with very high precision. Strontium isotope ratios have also been used to detect increases in acidification over time from determinations in tree rings and mussel shells.

1 INTRODUCTION

The use of strontium isotope ratios for estimations of weathering and atmospheric inputs to ecosystems was first reported by Gosz et al., 1983 and Graustein & Armstrong, 1983. These authors exploited the fact that measurable variations in $^{87}Sr/^{86}Sr$ ratios in different components such as rainwater, bedrock, soil, vegetation and streamwater are due entirely to mixing of strontium derived from these components. As strontium and calcium behave similarly geochemically, the weathering rate for calcium in a catchment has been calculated from calcium input-output chemistry using $^{87}Sr/^{86}Sr$ ratios when these ratios in rainwater and streamwater were sufficiently different (Jacks et al., 1989).

Problems of acidification of soils and waters due to anthropogenic deposition have been studied in great detail in the past decade and there has been increased use of strontium isotope methods to address these problems, particularly in relation to mineral weathering and weathering rates. This paper attempts to summarise the strengths and weaknesses of the application of strontium isotopes in such studies, quoting selected papers for illustration.

2 CALCIUM WEATHERING BUDGETS

The method for calculating calcium weathering rates on a catchment scale requires not only the input and output values for calcium but also reliable figures for the $^{87}Sr/^{86}Sr$ ratio in the input and output plus the $^{87}Sr/^{86}Sr$ ratio of the strontium released by weathering. The method can only be used if the ratios in the input and output are sufficiently different, and if reliable ratios can be obtained for the weathering-derived strontium. It has been applied by a number of workers (e.g. Jacks et al., 1989; Miller et al., 1993) and is very dependent on assessing the contributions from weathering in soils. The soil component has been assessed in some examples from the strontium isotope ratio in the solution obtained by dissolving the soil in HF, but it has been shown that differential weatherability of the soil minerals results in soil digestion being unreliable for estimating the strontium weathering component (Bain &

Bacon, 1994; Bailey et al., 1996). One attempted solution to this problem was to measure the strontium isotope ratio in solutions derived from extracting the mineral soil with organic acids such as citric and salicylic acids, the rationale being that only the more easily weathered minerals release strontium (Wickman & Jacks, 1992). This approach is an improvement on total soil digestion, particularly if organic acid extraction is preceded by an extraction with NH_4Cl to remove exchangeable strontium so that only the strontium weathering component is measured.

The precision of measuring $^{87}Sr/^{86}Sr$ is very high and the differences in $^{87}Sr/^{86}Sr$ ratios measured for catchment studies can be very small. It is important, therefore, to constrain all the factors involved as much as possible in order to measure the weathering rate as accurately as possible. As pulses of acidification occur during rain events which cause a sharp increase in stream flow rate, Bain et al. (1997) investigated the effect of flow rate on the $^{87}Sr/^{86}Sr$ ratio of the streamwater in a small catchment located in andesite bedrock and found that the ratio did not vary significantly, even in samples taken at four-hourly intervals.

3 ECOLOGICAL STUDIES

The variations in the $^{87}Sr/^{86}Sr$ ratios in the various components of ecosystems are not due to fractionation by biological processes but entirely to mixing of strontium derived from the different components with different isotopic composition. Graustein & Armstrong (1983) first used $^{87}Sr/^{86}Sr$ ratios in this way to conclude that 75% of the strontium in the vegetation in New Mexico catchments was derived from atmospheric transport and 25% from weathering of the underlying bedrock. They pointed out that the precision of these estimates was limited by uncertainties about the isotopic composition of strontium released by weathering.

In a later, much more detailed study of a forest ecosystem, Miller et al. (1993) used $^{87}Sr/^{86}Sr$ ratios as a tracer of cation sources in streamwater and were able to deduce that mineral weathering reactions contribute about 70% and soil cation-exchange reactions about 30% of annual strontium exports. They were also able to show that 50-60% of the strontium in the organic-soil-horizon and vegetation pools has an atmospheric origin so that any reduction in atmospheric cation inputs combined with strong acid anion inputs may result in significant depletion of this cation reservoir.

The importance of atmospheric deposition as a source of calcium was also found by Wickman & Jacks (1993) using $^{87}Sr/^{86}Sr$ ratios in atmospheric deposition, pine twigs and salicylic acid leachates of rocks for four pine stands growing on areas of exposed bedrock where the exchangeable pool of base cations was very small. At four sites, only 30% of the strontium (and so calcium) in wood biomass was received from weathering.

4 FINGERPRINTING APPLICATIONS

Strontium dissolved in streamwater is derived from the dissolution of minerals in the catchment and from atmospheric deposition. If the minerals in the rock and soil have different $^{87}Sr/^{86}Sr$ ratios, the ratio in the streamwater will differ from that in the bulk rock or soil because of differences in the rates at which minerals release strontium. The $^{87}Sr/^{86}Sr$ ratio of the strontium released to solution may be higher (e.g. Goldich & Gast, 1966) or lower (e.g. Brass, 1975) than that of the bulk rock, but the streamwater ratio is usually similar to the isotopic composition of the catchment bedrock (Jones & Faure, 1978). Reported values of $^{87}Sr/^{86}Sr$ for streamwater range from 0.7036 for water draining young volcanic rocks to 0.7384 for streams draining old (>1000 my) rocks, and the average $^{87}Sr/^{86}Sr$ ratio for global runoff from major river systems is 0.7119 (Palmer & Edmond, 1989).

Comparison of strontium isotope ratios in rainwater and streamwater in catchments can sometimes yield surprising, and revealing, results. For example, in a small catchment in andesitic lavas of Devonian age, Bain & Bacon (1994) found the $^{87}Sr/^{86}Sr$ ratio in streamwater to be 0.7080, lower than that in the rainwater which was 0.7098, even though the $^{87}Sr/^{86}Sr$ ratios in the HF-digested whole soils were 0.7197 or higher. The explanation for this apparent anomaly was that the strontium in the streamwater was derived from a source containing non-radiogenic strontium, most probably plagioclase feldspar. Thus the strontium isotope ratio appeared to "fingerprint" the dissolved strontium in the stream and act as a tracer for the origin of the solutes in the streamwater.

Strontium isotope ratios were also used by Douglas et al. (1995), not in streamwater but in suspended particulate matter in the Murray-Darling River system, to indicate that weathering of plagioclase produces an unradiogenic strontium pool which dominates the colloidal and dissolved fractions.

Strontium isotopes were used in a novel way by Wallander et al. (1997) and Wickman & Wallander (1997) in experiments to establish the importance of several ectomycorrhizal fungi for the uptake of phosphorus from apatite and potassium from biotite or microcline. Needles from seedlings from different treatments were analysed for their strontium isotope composition and, knowing the strontium ratios in the mineral components and the amounts of these components and their strontium contents, it was possible to calculate the fraction of strontium derived from weathering of the particular minerals. This is a useful application of the use of strontium isotopes as tracers for mineral weathering.

5 TEMPORAL CHANGES

Because of possible increases in weathering rates in recent years due to acidification, attempts have been made to study the variation in $^{87}Sr/^{86}Sr$ ratios with time.

One approach was to measure the ratios in tree rings and this showed a decreasing isotope ratio over 50 years and was attributed to a deterioration in the soil due to biomass consumption and acid deposition although changes in the strontium isotope ratio in the deposition and in the strontium released by weathering could not be ruled out (Wickman, 1992).

An analogous approach involved measuring the strontium isotope ratio in the shell of a mussel which lives in freshwater streams and records changes in chemical composition of the water in the growth segments of its aragonite shell. The results showed a lowering of $^{87}Sr/^{86}Sr$ ratios over the period 1901 to 1980 suggesting ongoing acidification in the catchments containing the streams with the mussels (Åberg et al., 1995). These authors suggested the possibility of using museum collections for the study of environmental changes in historical times.

It has been suggested that the measurement of the strontium isotopic composition of streamwater in small catchments not only provides a useful method for evaluating relative mineral weathering rates without the complications of secondary mineral formation and biological uptake of cations inherent in major element studies, but is also a useful method for determining how these rates change with time (Blum et al., 1994). The $^{87}Sr/^{86}Sr$ ratio can be measured with very high precision and small catchments where the streamwater ratio is very consistent regardless of flow rate could be used for long-term monitoring of the ratio to detect any change in weathering processes due to climatic change (Bain et al., 1997).

6 CONCLUSIONS

The use of $^{87}Sr/^{86}Sr$ ratios can provide important information in weathering studies, some of it unobtainable by other methods. If the ratios in the input and output are sufficiently different, weathering rates for calcium on a catchment scale can be calculated but the major limitation on the accuracy of this method is in obtaining reliable ratios for the weathering-derived strontium. Sources of strontium in different components of ecosystems can be established and quantified from their isotopic ratios. In favourable circumstances, the strontium isotope ratio in streamwaters can indicate which minerals in a catchment are weathering and determining the composition of the streamwater. The strontium isotopic composition of streamwater can also provide information on relative mineral weathering rates, and because of the high precision of the isotope measurement and the constancy of the ratio regardless of flow rate, the $^{87}Sr/^{86}Sr$ ratio in streamwater in small catchments could be used to monitor changes in weathering rates and processes due to environmental change.

7 ACKNOWLEDGEMENT

This work was funded by the Scottish Office Agriculture, Environment and Fisheries Department.

REFERENCES

Åberg, G., T. Wickman & H. Mutvei 1995. Strontium isotope ratios in mussel shells as indicators of acidification. *Ambio* 24:265-268.

Bailey, S.W., J.W. Hambreck, C.T. Driscoll & H.E. Gaudette 1996. Calcium inputs and transport in a base-poor forest ecosystem as interpreted by Sr isotopes. *Water Resources Res.* 32:707-719.

Bain, D.C. & J.R. Bacon 1994. Strontium isotopes as indicators of mineral weathering in catchments. *Catena* 22:201-214.

Bain, D.C., A.J. Midwood & J.D. Miller 1997. Strontium isotope ratios in streams and the effect of flow rate in relation to weathering in catchments. *Catena* (submitted).

Blum, J.D., Y. Erel & K. Brown 1994. $^{87}Sr/^{86}Sr$ ratios of Sierra Nevada stream waters: Implications for relative mineral weathering rates. *Geochim. Cosmochim. Acta* 58:5019-5025.

Brass, G.W. 1975. The effect of weathering on the distribution of strontium isotopes in weathering profiles. *Geochim. Cosmochim. Acta* 39:1647-1653.

Douglas, G.B., C.M. Gray, B.T. Hart & R. Beckett 1995. A strontium isotopic investigation of the origin of suspended particulate matter (SPM) in the Murray-Darling River system, Australia. *Geochim. Cosmochim. Acta* 59:3799-3815.

Goldich, S.S. & P.W. Gast 1966. Effects of weathering on the Rb-Sr and K-Ar ages of biotite from the Morton Gneiss, Minnesota. *Earth Plant. Sci. Lett.* 1:372-375.

Gosz, J.R., D.G. Brookins & D.I. Moore 1983. Using strontium isotope ratios to estimate inputs to ecosystems. *Bioscience* 33:23-30.

Graustein, W.C. & R.L. Armstrong 1983. The use of strontium-87/strontium-86 ratios to measure atmospheric transport into forested catchments. *Science* 219:289-292.

Jacks, G., G. Åberg & P.J. Hamilton 1989. Calcium budgets for catchments as interpreted by strontium isotopes. *Nordic Hydrol.* 20:85-96.

Jones, L.M. & G. Faure 1978. A study of the strontium isotopes in lakes and surficial deposits of the ice-free valleys, southern Victoria Land, Antarctica. *Chem. Geol.* 22:107-120.

Miller, E.K., J.D. Blum & A.J. Friedland 1993. Determination of soil exchangeable-cation loss and weathering rates using Sr isotopes. *Nature* 362:438-441.

Palmer, M.R. & J.M. Edmond 1989. The strontium budget of the modern ocean. *Earth Planet. Sci. Lett.* 92:11-26.

Wallander, H., T. Wickman & G. Jacks 1997. Apatite as a P source in mycorrhizal and non-mycorrhizal *Pinus sylvestris* seedlings. *Plant & Soil* (submitted).

Wickman, T. 1992. Sr isotopes and element concentrations of tree rings - a mirror of acidification? *Lundqua Report* 34:334-337.

Wickman, T. & G. Jacks 1992. Strontium isotopes in weathering budgeting. Pp. 611-614 in: *water-Rock Interaction* vol. 1 (Y.K. Kharaka & A.S. Maest, editor). Balkema, Rotterdam.

Wickman, T. & G. Jacks 1993. Base cation nutrition for pine stands on lithic soils near Stockholm, Sweden, *Appl. Geochem. Suppl. Issue* No. 2:199-202.

Wickman, T. & H. Wallander 1997. Biotite or microcline as a potassium source in ectomycorrhizal and non-mycorrhizal *Pinus sylvestris* seedlings. *Plant & Soil* (submitted).

Granitoid weathering in the laboratory: Chemical and Sr isotope perspectives on mineral dissolution rates

T.D. Bullen, A.F. White, D.V. Vivit & M.S. Schulz
US Geological Survey, Menlo Park, Calif., USA

ABSTRACT: Long-term flow-through column experiments using four fresh granitoids and their weathered equivalents are being conducted to determine dissolution rates of granitoid minerals and their influence on solute chemistry and $^{87}Sr/^{86}Sr$. In general, column effluents attained steady-state compositions after 1 to 12 months of reaction, and have low Ca/Si and high $^{87}Sr/^{86}Sr$ and Rb/K relative to initial column effluents. The chemical data suggest that the steady-state compositions are controlled largely by plagioclase and biotite dissolution, whereas initial compositions were strongly influenced by calcite and K-feldspar dissolution. Relative biotite:plagioclase mass weathering rates, calculated for the fresh granitoids using the steady-state $^{87}Sr/^{86}Sr$ values, vary from 4 to 66 and are 1.4-4.5 times less than those calculated using K/Na ratios. Relative rates calculated for the weathered granitoids show similar variability, but are generally less due to the diminished reactivity of their altered biotites. These results demonstrate that silicate mineral weathering rates calculated using $^{87}Sr/^{86}Sr$ differ from those using chemical fluxes, and must be viewed with caution. Moreover, the role of calcite must be considered when using Sr isotopes to determine silicate mineral weathering rates in freshly exposed granitoid terrains.

1 INTRODUCTION

Granitoids encompass a relatively simple multi-mineralic system, consisting mainly of feldspars, amphiboles, micas and quartz. Determination of mineral reaction rates during granitoid dissolution should involve a straightforward deconvolution of chemical fluxes using mass balance relations (e.g., Garrels and Mackenzie, 1967). However, assumptions must often be made about the relative reactivity of the silicate minerals (e.g. high reactivity of biotite, low reactivity of K-feldspar relative to plagioclase) (cf. Blum et al., 1993). Moreover, reactive trace phases such as calcite are often ignored in the mass balance calculations, but clearly can have a profound influence on granitoid dissolution chemistry (Clow et al, 1997).

Sr isotopes are often used to further constrain the calculation of granitoid mineral weathering rates. The power of Sr isotope methods results from the fact that each reactive mineral is likely to have a distinct $^{87}Sr/^{86}Sr$ value. Sr in the solute load resulting from the dissolution of two or more minerals will have an intermediate $^{87}Sr/^{86}Sr$ value reflecting the proportional contribution of each mineral. The fundamental assumption of this approach is that Sr is released congruently from each mineral (i.e., at the same rate as stoichiometric constituents). However, evidence that Sr may be preferentially released from minerals such as biotite and K-feldspar suggests that the use of Sr isotopes for assessment of mineral weathering rates may be problematic (Bullen et al., 1997).

In order to rigorously quantify both mineral weathering rates and the relative mass transport of Sr during granitoid weathering, we are carrying out long-term dissolution experiments using four fresh granitoids and their weathered equivalents. In this paper we use the chemistry and $^{87}Sr/^{86}Sr$ of reaction column effluents to assess relative mineral weathering rates after one year of reaction time.

2 TECHNIQUES

Fresh and weathered granitoid samples were collected from four well-characterized research watersheds (Loch Vale, CO, Yosemite, CA, Panola Mt., GA, and Luquillo, Puerto Rico). Samples were crushed and wet sieved to produce approximately 750 g of the 0.25-0.85 mm size fraction. Each sample was placed in a quartz glass column through which distilled, deionized water, saturated at 5% atm. CO_2 and atm. O_2 was passed at an average flow rate of 10 ml/hr.

Temperature was maintained at approximately 20°C in a controlled laboratory. Aggregate effluent samples were taken weekly during the first half year and bi-weekly thereafter for analysis of chemistry and $^{87}Sr/^{86}Sr$. Pure mineral separates were obtained from each granitoid using heavy liquid, dye-staining and hand-picking methods. Minerals were dissolved in sealed Teflon beakers using Teflon-distilled HF, HCl and HNO_3. Major- and trace-cation contents of effluents and mineral digests were analyzed using a Perkin-Elmer/Sciex Elan 6000 inductively-coupled plasma mass spectrometer. The precision of the chemical analyses is approximately 3% or better. Sr was separated from the effluents and mineral digests using standard cation exchange techniques. $^{87}Sr/^{86}Sr$ was analyzed using a Finnigan MAT 261 mass spectrometer in two-collector dynamic mode. $^{87}Sr/^{86}Sr$ values are precise at the 95% confidence level to 0.00004 or better for the column effluents, and 0.00002 or better for the mineral digests. Details of these procedures are given in Bullen et al. (1997).

3 RESULTS

The relative weight proportions of silicate minerals, as well as Sr concentrations and $^{87}Sr/^{86}Sr$ values obtained for the mineral separates from the fresh granitoids are given in Table 1. In general, the relative proportions of minerals are highly variable between sites. On the other hand, the relative proportion of plagioclase to biotite is fairly constant, differing by less than a factor of 1.6. This should allow meaningful comparisons of relative plagioclase and biotite weathering rates to be made among the sites.

$^{87}Sr/^{86}Sr$ and Ca/Si in effluents from the fresh granitoids, and $^{87}Sr/^{86}Sr$ in effluents from the weathered granitoids are plotted in Figure 1 as a function of reaction time. In general, $^{87}Sr/^{86}Sr$ in the effluents increases from initial to apparent steady state (i.e. relatively constant) values over times ranging from 1000 to 6000 hours. The exception is $^{87}Sr/^{86}Sr$ in effluent from the Yosemite fresh granite, which has progressively increased but has not reached a steady-state value after nearly 9000 hours of reaction time. Another persistent feature of the data is that at steady state, $^{87}Sr/^{86}Sr$ in effluent from each fresh granitoid is greater than that in effluent from its weathered equivalent. In detail, variations of $^{87}Sr/^{86}Sr$ in the effluents differ among the granitoids. For example, $^{87}Sr/^{86}Sr$ in effluent from the Loch Vale fresh granite initially increases, then levels for nearly 2000 hours prior to increasing to the steady-state value. In contrast, $^{87}Sr/^{86}Sr$ in effluent from the Loch Vale weathered granite initially remains constant, then increases to the steady-state value.

Ca/Si in effluents from the fresh granitoids initially increases over a short time period to maximum values, then falls persistently at various rates. In each case, the low initial Ca/Si ratios are due to exceptionally high Si concentrations, most likely the result of leaching of Si from the surfaces of the freshly ground silicate minerals. In detail, there is an apparent inverse correlation between Ca/Si and $^{87}Sr/^{86}Sr$ of the effluents, the pattern being most clear for the Panola granite. In this case, $^{87}Sr/^{86}Sr$ remains low while Ca/Si is high, then increases abruptly when Ca/Si begins to decrease. This pattern essentially holds for effluents from the other fresh and weathered granitoids.

With increasing reaction time, Rb/K in effluents from the fresh and weathered granitoids generally increases from values similar to that in K-feldspar to values similar to that in biotite, as shown in Figure 2 for the Yosemite fresh granite. Because K-feldspar and biotite are the only granitoid minerals (other than muscovite) that contain significant Rb and K, and because K is a major stoichiometric constituent of each mineral, the variations of the Rb/K ratio probably reflect changes in the reaction rates of each mineral.

Table 1. Chemical and Sr isotope data for silicate mineral separates from fresh granitoids

	PLAGIOCLASE	K-FELDSPAR	BIOTITE	HORNBLENDE	MUSCOVITE	QUARTZ
YOSEMITE	.36 / 1100 / 0.7065	.15 / 750 / 0.7075	.08 / 9 / 0.8053	.05 / 75 / 0.7069	-	.35
LUQUILLO	.55 / 560 / 0.7041	(not separated)	.10 / 17 / 0.7827	.14 / 80 / 0.7042	-	.20
LOCH VALE	.20 / 175 / 0.7422	.45 / 190 / 0.8662	.04 / 8 / 4.8339	-	-	.30
PANOLA	.35 / 750 / 0.7081	.20 / 600 / 0.7160	.10 / 11 / 1.2730	.02 / 55 / 0.7098	.05 / 21 / 0.7548	.27

Numbers listed in the order: relative weight proportions / Sr concentrations (in ppm) / $^{87}Sr/^{86}Sr$ values. Numbers for quartz are relative weight proportions. "-" means not present.

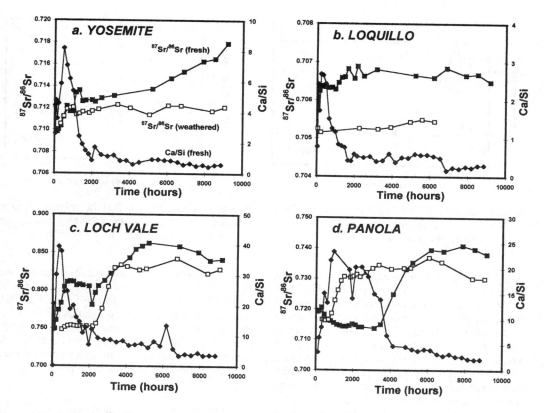

Figure 1. Variations of $^{87}Sr/^{86}Sr$ and Ca/Si in granitoid column effluents. Closed squares are $^{87}Sr/^{86}Sr$ of effluents from fresh granitoid, open squares are $^{87}Sr/^{86}Sr$ of effluents from weathered granitoids, closed diamonds are Ca/Si in effluents from fresh granitoids.

4 DISCUSSION

The persistent high Ca/Si and low Rb/K of the early effluents relative to the steady-state values is an intriguing result of these column experiments. First, we attribute the high Ca/Si to preferential calcite dissolution that overwhelms Ca contributions from other minerals such as plagioclase and hornblende. Cathode luminescence miscroscopy has revealed both inter- and intra-granular calcite in each of the granotoid samples. Moreover, the total amount of Ca leached from each granite during the period when Ca/Si is high is equal to or less than the amount of Ca determined to be in calcite based on coulometric titration of dissolved inorganic carbon from each granitoid. Second, we attribute the low Rb/K to preferential K-feldspar dissolution that overwhelms K and Rb contributions from the micas. Together the Sr-rich calcite and K-feldspar impart a relatively low $^{87}Sr/^{86}Sr$ to the initial column effluents.

As the mass flux from calcite and K-feldspar decreases, $^{87}Sr/^{86}Sr$ increases toward the steady-state value that should reflect a greater relative role of plagioclase and biotite dissolution. The similarity of Rb/K in later steady-state effluents to that in biotite (Fig. 2) suggests that most of the K and Rb at that point are derived from biotite. On the other hand, Ca/(Ca+Na) in later steady-state effluents is generally greater than that in plagioclase, as shown in Figure 3. This implies that calcite and/or some other Ca-rich mineral such as fluorite is dissolving from the granitoids after Sr isotope steady-state is reached, although calcite appears to have been totally leached from the Luquillo and Loch Vale weathered samples.

Regardless, if we assume that Sr at steady-state is derived from plagioclase and biotite dissolution only, we can determine the relative proportion of Sr derived from each mineral and thus calculate their relative reaction rates (cf. Blum et al., 1993). Similarly, if we assume that at steady state all Na is derived from

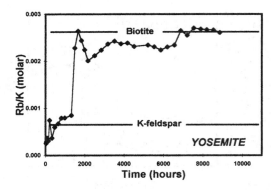

Figure 2. Variation of Rb/K in effluent from the fresh Yosemite granite, compared to values for K-feldspar and biotite separates

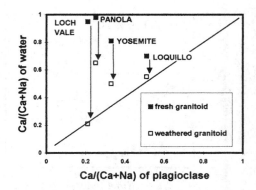

Figure 3. Molar Ca/(Ca+Na) in column effluent waters compared to values for plagioclase; line across diagram represents 1:1 correlation

plagioclase and all K from biotite, we can likewise determine the relative reaction rates of each mineral.

Results for the steady-state column effluents are shown in Table 2. Calculated rates on a mass basis for the fresh granitoids using Sr isotopes are lower than those calculated using K/Na, by factors of 1.4 to 4.5. Assuming that $^{87}Sr/^{86}Sr$ of calcite is less than that in the steady-state effluents, consideration of minor residual calcite in the Sr isotope determination would increase the calculated relative rate, thereby reducing the discrepancy somewhat. Calculated rates for the weathered granitoids are generally less than those for the fresh granitoids, indicating that biotite becomes less reactive as weathering proceeds. Regardless, the variability of ratios among the granitoids is surprising considering their relatively constant plagioclase:biotite abundance ratios, and probably reflects differences in reactive mineral surface areas.

Table 2. Biotite:plagioclase relative weathering rate

	K/Na	Sr ISOTOPES
YOSEMITE	90.4 (52.9)	66.0 (21.7)
LUQUILLO	29.1 (27.3)	6.5 (3.6)
LOCH VALE	15.3 (15.3)	4.2 (2.5)
PANOLA	44.1 (28.4)	14.6 (12.4)

(Values for weathered samples in parentheses)

5 IMPLICATIONS

The results of these experiments clearly demonstrate that estimates of granitoid mineral weathering rates using Sr isotopes can be substantially different from those based on chemical fluxes, particularly for fresh rocks. Furthermore, dissolution of K-feldspar and trace amounts of calcite may significantly influence the Ca, K and Sr fluxes during weathering of freshly exposed granitoid terrains. We maintain that Sr isotopes can provide useful information on weathering rates, but should be applied with caution.

REFERENCES

Blum, J.D., Erel, Y. and Brown, E., 1993, $^{87}Sr/^{86}Sr$ ratios of Sierra Nevada stream waters: implications for relative mineral weathering rates. *Geochim. Cosmochim. Acta* 57: 5019-5025.

Bullen, T.D., White, A.F., Blum, A.E., Harden, J. and Schulz, M.S., 1997, Chemical weathering of a soil chronosequence on granitoid alluvium: II. Mineralogic and isotopic constraints on the behavior of strontium. *Geochim. Cosmochim. Acta* 61, 2: 291-306.

Clow, D.W., Mast, M.A., Bullen, T.D. and Turk, J.T., 1997, $^{87}Sr/^{86}Sr$ as a tracer of mineral weathering reactions and calcium sources in an alpine/subalpine watershed, Loch Vale, Colorado. *Water Resources Res.* 33, 6: 1335-1351.

Garrels, R.M. and Mackenzie, F.T., 1967, Origin of the chemical compositions of some springs and lakes. In *Equilibrium Concepts in Natural Water Systems* (ed. R.F. Gould): 222-242. ACS.

The field dissolution rate of feldspar in a Pennsylvania (USA) spodsol as measured by atomic force microscopy

M. A. Nugent
Department of Geosciences, Penn State University, University Park, Pa., USA (Presently: Department of Geosciences, SUNY at Stony Brook, N.Y., USA)

P. Maurice
Department of Geology, Kent State University, Ohio, USA

S. L. Brantley
Department of Geosciences, Penn State University, University Park, Pa., USA

ABSTRACT: A dissolution rate is calculated for a field-weathered peristeritic feldspar based on AFM images before and after weathering. The rate based on the development of microtexture (10^{-18} mol$_{feldspar}$/cm^2/sec) is ~2 orders of magnitude slower than a rate calculated from a minimum overall surface retreat (10^{-16} mol$_{feldspar}$/cm^2/sec). However, because AFM tip-sample interactions may underestimate the depth and width of weathering features, the rate based on peristerite microtopography may underestimate the actual rate. The calculated rates are comparable to other field rates.

1. INTRODUCTION

Weathering rates of feldspar measured in the laboratory are up to 3 orders of magnitude faster than field rates (White, 1995). Such a large discrepancy suggests that either differences in approaches taken in rate calculations are significant, or that field conditions, and variability in field conditions, are discrepant from those used in laboratory experiments.

In this study, atomic force microscopy (AFM) is used to investigate the microtopography of peristerite feldspar (albite and oligoclase exsolution lamellae) weathered in a soil for 3 years, and to calculate a dissolution rate. AFM provides nanometer-scale imaging and does not require sample coating procedures. However, AFM imaging presents a variety of challenges, and hence is best used in combination with other microscopic techniques.

2. METHODS
2.1 *Sample preparation and site characterization*

Tablets of peristerite from Quebec ($Ab_{10}An_{90}$), determined by microprobe analysis) were prepared by cutting approximately along (001) cleavage planes, polishing to 0.25 µm, and ultra-sonicating in acetone between polishings. Three tablets were mixed with soil from a site in State College, PA. This mixture was placed in a plastic bottle riddled with holes, buried at the site, and removed after 3 yrs. Samples were ultra-sonicated ~20 min. in acetone to remove the reddish coloring associated with burial. Polished, unweathered samples (blanks) were also prepared.

Samples were buried in the B horizon of a pine-tree wooded spodsol, at a site removed from anthropogenic input (with the exception of acid rain). State College receives an annual average precipitation of 101 cm (pH, 4.3). The B horizon consists predominantly of residual quartz, with minor kaolinite and iron oxides, and is overlain by an organic-rich O horizon. Porewater pH varied from 4.9 to 5.6, and averaged 5.1 in summer and 5.6 in winter over summer 1996, and winter, 1997. Soil porewater saturation states were calculated using SOLMINEQ (Kharaka et al., 1988). Porewaters were under- and super-saturated with respect to all feldspars and kaolinite, respectively. These calculations were made without consideration for organically-bound Al.

2.2 *Sample Analysis*

Samples were imaged using a Digital Instruments Nanoscope III AFM with a binocular microscope attachment (see Maurice, 1996). Weathered samples were not imaged before burial; instead, separate unweathered samples were used. Images were collected in air using TMAFM. Etched Si tips 125 µm long used during imaging have a nominal tip radius of curvature of 5-10 nm, although the radius of curvature may increase during imaging. Data were collected in >20 areas on weathered and unweathered samples, at scan speeds <1 Hz, at a wide range of scan angles, and with frequent tip changes. Amplitude and height images were collected simultaneously. 'Height' images allow for measurement of relief, while amplitude images, which provide essentially the first derivative of height data, highlight changes in relief and are easier to interpret for rough surfaces.

3. RESULTS AND DISCUSSION
3.1 *Microtopography*
Polished, unweathered blanks
Ubiquitous, cross-cutting grooves of random orientation, observed under AFM, are interpreted as polish-

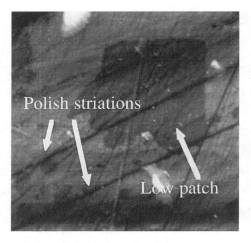

Figure 1. TMAFM height image of polished, unweathered blank showing a low patch, polish striations, and some surface litter. Image is 4x4 μm.

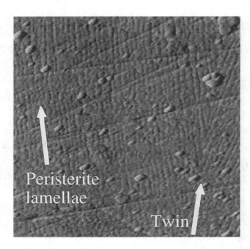

Figure 3. TMAFM 1x1 μm amplitude image of weathered peristerite. Albite (010) twin is 14° to peristerite lamellae. Particles may be tip artifacts.

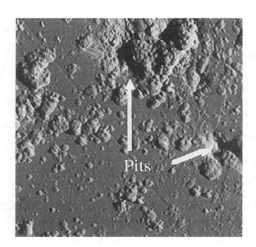

Figure 2. TMAFM amplitude image of weathered, pitted persiterite, partially covered with a natural coating. Image is 10x10 μm.

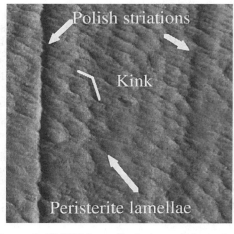

Figure 4. TMAFM amplitude image of weathered peristerite, showing features developed on lamellae. Image is 400 nm on a side. Lamellae run NW-SE.

ing striations (Figure 1). The apparent depth of polishing striations is typically ~3-4 nm. Observed low patches (Figure 1) are interpreted as polishing rip-up of peristerite. Relief between the surface and low patches is ~5nm (maximum, 11 nm). Pits ranging in size from nms to μms were also observed. There was no evidence of twinning or exsolution lamellae on the blanks.

Polished, weathered samples
After 3 years of weathering, the surface is partially covered (>35% coverage, based on several 10x10 μm areas) by a patchy coating (Figure 2). This coating is made of particles ranging in size from 10's to 1000's of nm. Pits imaged on the weathered surface have the size and shape of pits imaged on unreacted samples, suggesting pre-existing pits are not necessarily the locus of weathering. After 3 years of weathering, polish lines, some of which have broadened and shallowed, are still visible under AFM. Low patches imaged on the unweathered sample were not imaged on the weathered sample.

After 3 years of weathering, microtextures directly related to bulk peristerite microfeatures (albite (010) twins oriented 14° to peristerite lamellae) are resolvable under AFM (Figure 3). This microtexture was not observed on unweathered samples. Lamellae grooves are from 0.7 to 1.5 nm deep, and are the deepest linear feature developed due to weathering (Figures 3, 4). Higher resolution imaging reveals that these grooves are curved, instead of straight (Figure 4). Linear features on lamellae ridges (Figure 4) are more shallow than lamellae grooves.

3.2 *Interpretation of the lamellae grooves: AFM tip-sample interaction*

There are two possible explanations for the grooves in Figure 3. First, the grooves may be phase boundaries, implying dissolution along phase boundaries is faster than on either phase. Second, the grooves may be due to preferential dissolution of either albite or oligoclase lamellae. Laboratory dissolution rates predict oligoclase dissolves faster than albite (Blum and Stillings, 1995). If the grooves represent a phase and not a phase boundary, this suggests the grooves are oligoclase. We note that albite and oligoclase lamellae of most An_{10} peristerites are of similar thickness (Korekawa et al., 1970), suggesting the same is true for the An_{10} peristerite used in this study.

Based on cross section analyses (Figure 5) performed by the AFM software across the lamellae, 'high' ridges appear thicker than 'low' grooves. We might then assume the grooves are phase boundaries; however, this interpretation is complicated by the manner in which the AFM tip and sample interact during imaging of features on the order of 10's of nm. Blum and Eberl (1992) report that, for vertical features < 30 nm high, resolution is limited by the tip radius of curvature. Hence, low-relief features may appear broadened by as much as 25% when imaging in contact mode (Wilson et al., 1996), suggesting the ridges associated with peristerite lamellae may appear wider, and the grooves, thinner, than they actually are. In addition, the depth of the grooves may be underestimated due to the finite width of the tip (Figure 5). Interpretation of AFM microtopography on the order of nms is best done in conjunction with other techniques, such as TEM, which can help to definitively resolve AFM tip-sample convolutions.

3.3 *Dissolution rate calculation*

Two dissolution rates can be estimated. Because low patches were absent on weathered samples, we assume the feldspar ledges around them dissolved during weathering. Therefore surface retreat of 11 nm during weathering corresponds to a dissolution rate of 10^{-16} $mol_{feldspar}/cm^2/sec$.

A second dissolution rate is calculated by attributing the surface area increase upon weathering to the development of lamellae grooves. The surface area on a polished, unweathered peristerite, measured by the AFM software, is 1.007 $\mu m^2/\mu m^2$ (average of 7 measurements, normalized to planar surface area). Surface area measurements, made on the 3 year sample in 14 areas with no pits and no coating, average 1.016 $\mu m^2/\mu m^2$. Therefore, the surface area increase upon weathering is 0.009 $\mu m^2/\mu m^2$. This surface area increase is only reflective of areas without significant microtopography, such as pits and microcracks.

First, we confirm that the surface area increase is described by the development of peristerite lamellae grooves. Assuming 5 nm wide, 1 nm deep, straight, V-shaped lamellae grooves (based on AFM observations), the change in surface area (0.009 $\mu m^2/\mu m^2$) yields a groove frequency of 23 μm^{-1}. This underpredicts the observed frequency of ~32 μm^{-1}, suggesting grooves may vary in frequency, width, or depth. Alternatively, the discrepancy between calculated and observed frequency may be related to smoothing of polishing features upon weathering. Nevertheless, this calculation suggests the increase in surface area is largely attributable to lamellae grooves. Therefore, assuming lamellae grooves are 5 nm wide, 1 nm deep, straight, V-shaped, and separated by the observed 26 nm intervals (for a total repeat distance of 31 nm), we calculate a normalized surface area of 1.013 $\mu m^2/\mu m^2$, and a dissolution rate of 10^{-18} $mol_{feldspar}/cm^2/sec$.

3.4 *Implications for weathering*

Presumably, peristerite dissolution proceeds simultaneously at many surface sites on the polished samples: phase boundaries, ledges, albite and oligoclase lamellae, pre-existing and newly formed pits, crystallographic sites, defects, and microcracks outcropping on the surface. The overall dissolution rate is the

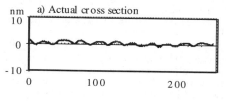

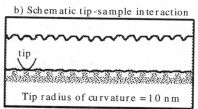

Figure 5. a) Cross section analysis of peristerite microtexture, as measured by AFM. b) Schematic representation of AFM-tip sample interaction (same horizontal and vertical scale as cross section).

Table 1. Dissolution rates for albite, oligoclase, and peristerite

	Field Rate (Geo. SA)	Lab Rate (BET S.A.)
	$mol_{feldspar}/cm^2/sec$	
Albite	none available	8×10^{-17} [1]
Oligoclase	8×10^{-17} [2] - 3×10^{-19} [3]	1×10^{-16} [4]
This study	1×10^{-16} - 9×10^{-19} (pH 5-5.6)	none available

[1] pH 5, Knauss and Wolery, 1986, [2] pH 5, Paces, 1983
[3] pH 5.8, Schnoor, 1990, [4] pH 5, Oxburgh et al., 1994

sum of the rates of these processes. By identifying two major changes due to weathering (low patches, lamellae grooves), we identify two contributors to dissolution.

The 'low patch' dissolution rate (10^{-16} $mol_{feldspar}/cm^2/sec$) is ~2 orders of magnitude faster than the 'groove' rate (10^{-18} $mol_{feldspar}/cm^2/sec$). The low patch rate, representative of ledge retreat, may not represent a long-term dissolution mechanism for the polished surface.

The "groove" rate only includes the volume lost due to the development of the grooves. If the surface defining the ridges was also retreating due to dissolution, then the overall rate would be larger than the groove rate. In addition, because of AFM tip-sample interactions, the true depth and width of grooves may be greater than that imaged. Therefore, the actual groove rate may be faster than the calculated rate. For example, if albite and oligoclase lamellae are both 15.5 nm wide (based on the observed repeat distance, 31 nm), box-shaped corrugations 1 and 3 nm deep correspond to dissolution rates of $10^{-17.3}$ and $10^{-16.7}$ $mol_{feldspar}/cm^2/sec$, respectively. To more accurately quantify these rates, TEM observations of lamellae widths and a better measure of lamellae depth are required.

If the grooves represent phase boundaries, our estimated rate (10^{-18} $mol_{feldspar}/cm^2/sec$) is the rate of peristerite dissolution as driven by the strain along phase boundaries. However, if the grooves are oligoclase lamellae, then using the albite laboratory rate (Table 1), our estimates reveal the ratio of oligoclase to albite dissolution rates is 1.01.

We note that our rates are comparable to field rates calculated using geometric surface area (Table 1). The groove rate is ~2 orders of magnitude slower than lab rates. As would be expected, the low patch rate, based on ledge migration, is significantly faster than the groove rate. While confirmation of lamellae depth is required to further constrain the groove rate, we note that AFM measurements provide us with a novel way of calculating feldspar dissolution rates which are comparable to rates calculated by other methods.

4. REFERENCES

Blum, A.E. and D.D. Eberl 1992 Determination of clay particle thickness and morphology using scanning force microscopy. In: *Proc. of the 7th Int. Symp. on Water Rock Interaction.* Y.K. Kharaka and A.E. Maest (eds), Vol. 1, 133-136.

Blum, A.E. and L.L. Stillings 1995 Feldspar dissolution kinetics. In: *Chemical Weathering Rates of Silicate Minerals,* A.F. White and S.L. Brantley (eds), 291-351.

Kharaka, Y.K., W.D. Gunter, P.K. Aggarwal, E. Perkins, and J.D. Debraal 1988 SOLMINEQ 88: A computer program for geochemical modeling of water-rock interactions. U.S.G.S. Water Resources Invest. Report 88-4427.

Knauss, K., and T. Wolery 1986 Dependence of albite dissolution kinetics on pH and time at 25 and 70°C. *Geochimica et Cosmochimica Acta* 50, 2481-2497.

Korekawa, M., H.-U. Nissen, and D. Philipp 1970 X-ray and electron microscope studies of a sodium rich low plagioclase. *Zeitschrift fur Kristallographie* 131, 418-436.

Maurice, P. 1996 Applications of atomic force microscopy in environmental colloid and surface chemistry. *Colloids and Surfaces A* 107, 57-75.

Oxburgh, R., T. Drever, and Y. Sun 1994 Mechanism of plagioclase dissolution in acid solution at 25°C. *Geochimica et Cosmochimica Acta* 58, 661-669.

Paces, T. 1983 Rate constants of dissolution derived from the measurement of mass balance in hydrologic catchments. *Geochimica et Cosmochimica Acta* 47, 1855-1863.

Schnoor, J. 1990 Kinetics of chemical weathering: A comparison of laboratory and field rates. In: *Aquatic Chemical Kinetics,* W. Stumm (ed), 475-504.

White, A.F. 1995 Chemical weathering rates of silicate minerals in soils. In: *Chemical Weathering Rates of Silicate Minerals*, A.F. White and S.L. Brantley (eds), 407-461.

Wilson, D.L., K. Kump, S. Eppell, and R. Marchant 1996 Morphological restoration of AFM images. *Langmuir* 11, 265-272.

Degradation processes of trachytes in monument façades, Azores, Portugal

M.I.Prudêncio, J.C.Waerenborgh, M.A.Gouveia, M.J.Trindade & E.Alves
Instituto Tecnológico e Nuclear, Sacavém, Portugal

M.A.Sequeira Braga & C.A.Alves
Universidade do Minho, Braga, Portugal

M.O.Figueiredo & T.Silva
Instituto de Investigação Científica Tropical, Lisboa, Portugal

ABSTRACT: Petrographic, mineralogical and chemical analyses confirmed the provenance of the stones used as building materials of the main façade of the Ribeira Grande church (Azores, Portugal). In spite of the serious state of decay of the monument stones most of the chemical differences found seem to be partially inherited from the quarry. However a significant increase in the CaO content in the monument stones is observed certainly due to the use of mortars, cements and lime as building or restoration materials. A depletion of Eu is also found in those samples probably due to preferential dissolution of plagioclase. A major difference between the quarry and monument environments is deduced from the study of the Fe forms. A much higher oxidation degree is observed in the samples from the church stones (where Fe is mainly present as αFe_2O_3) as compared to any of the rock samples including the most weathered ones.

1 INTRODUCTION

Scientific research oriented to the conservation of monuments must pass by a first step of understanding and quantification of the causes and mechanisms responsible for the degradation of stone. In order to avoid the indiscriminate use of standard-treatments in rocks of different nature and exposed to different environments, it is not acceptable to suggest any treatment if the causes of deterioration are not well known.

The stones used as building materials are subjected to various drastic changes, including their extraction of the quarry and placing in the building (internal or external exposures). Also to be taken into account are the climatic and microclimatic conditions, as well as the interaction between the stone and other building and restoration materials (mortars and cements) used since the construction of the monument. Furthermore the monument façades are frequently coated with lime.

Volcanic rocks are the typical building materials used in Azores islands and serious cases of stone degradation are found in churches built of trachytes and basalts. These monuments are under an intense maritime influence.

The volcanic rocks studied belong to the Fogo volcanic complex, which is located in the central part of the S. Miguel Island (Fig. 1). The area under study correspond to two quarries located in the north slope of the volcanic complex, where thick trachytic lava flows occur subjacent to pyroclastic material of recent eruptions (França, 1993).

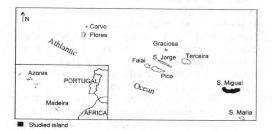

Fig. 1. Location map of the studied area.

In this work results are also presented for the main façade (built of trachyte) of the Misericórdia / Espírito Santo Church, Ribeira Grande, S. Miguel Island. Petrographic, mineralogical and chemical analyses of stone monuments and quarries were made in order to confirm the origin of the stones used in the construction of the church and to compare the degradation processes of the stone in natural outcrops and in the monument.

2 EXPERIMENTAL

Eight stones from the façade of the church (RG-1→8) and seven rock samples from two quarries (RS from Ribeira do Salto and PF-1→6 from Porto Formoso) used for the construction of the church according to historical documents were studied. PF-1 and PF-4 are the most weathered PF quarry samples and PF-6 corresponds to the least weathered according to a macroscopic observation.

Chemical and mineralogical analyses were made by using the X-ray fluorescence spectrometry, instrumental neutron activation analysis and X-ray diffraction (XRD). Mössbauer spectroscopy was used for the identification of Fe oxides and the chemical characterization of Fe forms. Mössbauer spectra were analyzed using the non-linear least-squares method of Stone (1967).

3 RESULTS AND DISCUSSION

The outcrops of trachytes in the Porto Formoso (PF) quarry are light gray, fine to medium-grained textures with phenocrysts of feldspars. These phenocrysts are partially brownie.

The trachytes of Ribeira do Salto (RS) are dark gray, fine-grained with feldspars phenocrysts and with vacuoles.

Petrographic analyses of PF quarry samples revealed a microlitic to trachytic texture. The phenocrysts are mainly of potassium feldspar, plagioclase and microphenocrysts of amphibole (hornblende) and opaque minerals. Alkaline feldspars, amphibole, pyroxene, biotite, opaque minerals, apatite and zircon compose the matrix. The phenocrysts fractures and the matrix become brownie in the early stages of weathering.

The RS quarry sample has an intergranular texture. The phenocrysts are plagioclases, alkaline feldspars, olivine and pyroxene. Olivine and opaque minerals also occur as microphenocrysts. The matrix is composed of feldspars, olivine, opaque minerals and apatite.

The macroscopic and microscopic observations of PF quarry and the monument stones of the main façade showed similar characteristics.

The mean values of the major elements contents in the monument stones and the two quarries are given in Table 1.

The chemical results plotted on the TAS diagram (Middlemost, 1994) indicate a trachytic composition for the PF quarry and the monument stones. RS quarry sample is a trachyandesite (Fig. 2). It should be noted that this classification is approximated since some of the samples are more or less weathered.

Trace elements contents confirm the similarity of the church stones to the PF quarry samples. RS sample (latite) has higher contents of Fe, Sc, Cr, Co, and a positive Eu anomaly (Saunders 1984). The REE patterns of the monument stones and the PF samples (trachytes) show a negative Eu anomaly (Fig. 3 and 4).

Table 1. Mean values of the major elements contents (%) in the monument (RG) and quarries (PF and RS) samples.

	RG	PF	RS
SiO_2	64.9	66.5	56.7
Al_2O_3	16.5	16.2	16.3
Fe_2O_3	4.13	3.49	7.27
MgO	0.39	0.32	2.62
CaO	1.33	0.49	5.31
Na_2O	5.12	5.20	4.21
K_2O	6.19	6.12	4.96
TiO_2	0.69	0.79	2.08
P_2O_5	0.14	0.086	0.41
MnO	0.16	0.18	0.13
L.O.I.	0.38	0.25	0.03

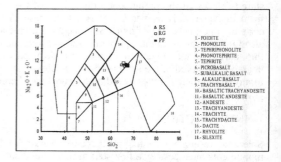

Fig. 2. Chemical results plotted in the TAS diagram (Middlemost, 1994).

The CaO content in the monument stones is higher than in the PF quarry (Table 1), which can be explained by the interaction of the stones with the building materials.

A significant positive Ce anomaly is observed in the more weathered PF quarry samples analyzed, which suggests oxidation and preferential retention of this element.

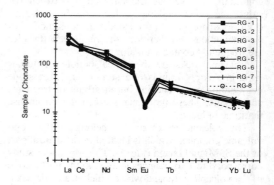

Fig. 3. REE patterns of the monument stone samples.

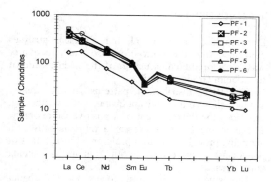

Fig. 4. REE patterns of PF quarry samples.

The monument stones show a larger negative Eu anomaly (Eu/Eu*=0.18-0.24) than the PF quarry samples (Eu/Eu*=0.46-0.76).

Mössbauer spectra were taken of Porto Formoso samples (rock which, as seen above is most probably the one used as building material for the church façade) and of two church stone samples. The hyperfine parameters estimated from the analysis of the spectra taken at room temperature and at 80 K are in Table 2. Typical spectra are shown in Fig. 5. Assignment of Fe phases to the different contributions to the spectra was done comparing these parameters with published data (e.g. Mitra 1992 and in the case of oxides and hydroxides Bowen *et al* 1993 and references therein).

Mössbauer spectra of PF-6 and PF-2 samples are very similar confirming the chemical and mineralogical analyses. Both spectra obtained with the sample at 293 K were fitted with 3 sextets and 4 quadrupole doublets. Two sextets are assigned to magnetite, Fe_3O_4, and the third one with larger hyperfine field corresponds to hematite, αFe_2O_3; as far as the quadrupole doublets are concerned 1 is due to Fe^{3+} in the silicate structure and/or in superparamagnetic oxides; 2 are attributed to Fe^{2+} in biotites and/or amphiboles; the resolution of these spectra does not allow a clear separation of the quadrupole doublets assigned to these silicates certainly due to the presence of several Fe containing phases with partially overlapping contributions to the spectra. The isomer shift of the 4th Fe^{2+} doublet is significantly lower than those found for Fe^{2+} in the silicates; it is typical of Fe^{2+} in oxide spinels, in this case probably ulvospinel since this is an accessory mineral which may be found in trachytes forming solid solution with magnetite (Haggerty 1976).

PF-1 is quite different from the previous samples: no hematite is detected, magnetite seems to be strongly oxidized to maghemite γFe_2O_3. The Fe^{2+} doublet assigned to ulvospinel in the PF-2 and PF-6

Table 2. Hyperfine parameters of Fe estimated from the Mössbauer spectra.

	δ (mm/s)	ε (mm/s) or Δ (mm/s)	B_{hf} (T)	I (%)
PF-6	**293 K**			
Fe_3O_4	0.27	-0.06	46.8	3
	0.67	-0.012	43.8	6
αFe_2O_3	0.36	-0.22	50.4	3
Fe^{3+}	0.43	0.50	-	28
Fe^{2+} silicate	1.17	2.76	-	14
	1.12	2.26	-	32
Fe^{2+} spinel	0.92	1.02	-	14
PF-1	**293 K**			
γFe_2O_3	0.29	0.04	48.0	29
Fe^{3+}	0.39	0.62	-	38
Fe^{2+} silicate	1.12	2.65	-	22
Fe^{2+} spinel	0.93	1.03	-	11
PF-1	**80 K**			
γFe_2O_3	0.38	-0.04	50.0	28
$\alpha FeOOH$	0.47	-0.25	36.8	9
Fe^{3+}	0.50	0.57	-	32
Fe^{2+} silicate	1.27	2.94	-	20
Fe^{2+} spinel	1.03	1.90	-	10
RG-8	**293 K**			
αFe_2O_3	0.37	-0.22	50.9	66
Fe^{3+}	0.39	0.56	-	23
Fe^{2+} silicate	1.11	2.72	-	7
	1.10	1.98	-	4
RG-2	**293 K**			
αFe_2O_3	0.375	-0.23	50.4	57
Fe^{3+}	0.37	0.60	-	43

δ - isomer shift relative to metallic α-Fe at 293 K; Δ - quadrupole splitting measured in the paramagnetic state; $\varepsilon = (e^2 V_{zz} Q / 4)(3\cos^2\theta - 1)$ quadrupole shift calculated from $(\phi_1 + \phi_6 - \phi_2 - \phi_5)/2$ where ϕ_n is the shift of the nth line of the sextet due to quadrupole coupling. B_{hf} - magnetic hyperfine field. I - relative areas. Estimated errors are $\pm$ 0.5 T for B_{hf}, $\pm$ 1% for I, $\pm$ 0.03 mm/s for δ, Δ and ε.

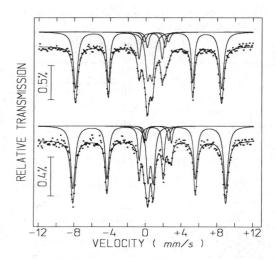

Fig. 5. Mössbauer spectra of RG-8 sample taken at 293 K (top) and 80 K (bottom). Calculated function is plotted on the experimental points. The fitted Fe^{3+} and Fe^{2+} doublets and the α-Fe_2O_3 sextet are also plotted, slightly shifted.

samples is also observed. In the 80 K spectrum besides the broad sextet assigned to γFe_2O_3 the spectrum could only be properly fitted if a second sextet with parameters typical of goethite, $\alpha FeOOH$, was considered (Table 2). In environmental samples isomorphous substitutions of Fe and/or defects and decrease in particle size are commonly observed and may explain why this hydroxide in PF-1 is not magnetically ordered at room temperature; therefore its contribution to the 293 K spectrum overlaps the Fe^{3+} quadrupole doublet and accordingly the area of this doublet is within experimental error approximately the sum of the goethite and Fe^{3+} doublet contributions in the 80 K spectrum. $\alpha FeOOH$ and γFe_2O_3 are not clearly observed in the powder X-ray diffractograms. Mössbauer spectroscopy of the PF-1 sample at 4.2 K with the application of an external field is now in progress hoping to unambiguously identify these oxides.

The spectra of the church stone samples which were already analyzed show much higher oxidation degrees than those observed for the weathered rock samples of the PF quarry. In RG-2 all the Fe present is in the Fe^{3+} form, 57 % of which as αFe_2O_3; in the RG-8 spectra (Fig. 5) the relative area of the hematite contribution is 66 % but 11 % of the total Fe is still found in the form of Fe^{2+} in the silicates structures (biotite and/or amphibole; there is no evidence of a doublet attributable to ulvospinel).

4 CONCLUSIONS

Petrographic, mineralogical and chemical analyses confirmed that the Porto Formoso trachyte quarry was used for the construction of the main façade of the Ribeira Grande Church.

The REE distribution patterns in the quarry samples showed a significant change with weathering, especially the development of a positive Ce anomaly. The monument stones present a more negative Eu anomaly, probably due to a preferential dissolution of plagioclase associated with intense leaching conditions.

The study of the iron forms by Mössbauer spectroscopy in both quarry and monument samples showed a much higher oxidation degree or even a total oxidation of this element in the building stones. In the weathered samples of the quarry maghemite (oxidation of primary Fe_3O_4) and goethite occur. In the church samples analyzed most of the Fe is present as αFe_2O_3. This suggests the leaching of intermediate alteration products and a more advanced alteration degree of the monument samples as compared to the more weathered quarry samples.

The use of trachytic rocks in buildings appears to give rise to serious problems as far as their maintenance is concerned. The results obtained so far appear to indicate significant differences in the chemical content of some elements and in the mineralogical composition between the monument and the quarry samples. Particularly differences in the oxidation-reduction conditions need to be further studied.

5 REFERENCES

Bowen L. H., De Grave E. & Vandenberghe R. E. 1993. in *Mössbauer Spectroscopy Applied to Magnetism and Materials Science*. G. J. Long & F. Grandjean (eds.) ch.4, N. York: Plenum Press.

França, Z. 1993. Contribuição para o estudo dos xenólitos sieníticos do arquipélago dos Açores. Universidade dos Açores.

Haggerty, S. E. 1976. In *Reviews in Mineralogy: Oxide Minerals*. D. Rumble III (ed.) ch.4, Chelsea (Michigan, USA): BookCrafters Inc.

Middlemost, E.A.K. 1994. Naming materials in the magma / igneous rock system. *Earth-Science Reviews* 37: 215-224.

Mitra, S. 1992. *Applied Mössbauer Spectroscopy*. Oxford: Pergamon Press.

Saunders, A. D. 1984. *Rare Earth Element Geochemistry*. P. Hendersson (ed.) ch.6, Amsterdam (The Netherlands): Elsevier.

Stone, A. J. 1967. Least squares fitting of Mossbauer spectra appendix to Bancroft G. M., Maddock A. G., Ong W. K., Prince R. H., Stone A. J. *J. Chem. Soc. (a)*, 1966.

Laboratory studies of the chemical weathering of rock from the English Lake District

R. Stidson & E. Tipping
Institute of Freshwater Ecology, Windermere Laboratory, Ambleside, UK

J. Hamilton-Taylor
Institute of Environmental and Biological Sciences, Lancaster University, UK

ABSTRACT: This study examines the weathering of whole, mixed mineral rocks in the laboratory. Dissolution experiments were carried out in batch type reactors at 10°C in the pH range 2.5 to 6.5. The solutions were analysed for pH, silicon, aluminium, sodium, potassium, magnesium and calcium. The behaviour of these rocks is quantitatively similar to results obtained with pure minerals. The dissolution rate decreases in a linear fashion within the pH range studied. There is an initial fast release of base cations, and with time dissolution begins to reach steady state. Dissolution is not stoichiometric. Fulvic acid was found significantly to increase the rate of dissolution of aluminium at pH 5.5 and 6.5.

1 INTRODUCTION

In the catchment of the River Duddon in the English Lake District, streamwater compositions vary substantially (Table 1). It is presumed that this is due to different weathering rates from the underlying rocks of the Borrowdale Volcanic Series, an Ordovician continental-arc volcanic field comprising predominantly basaltic andesite to andesite lava flows. In the weathering reactions, hydrogen ions are removed from solution in exchange for base cations (calcium, magnesium, potassium and sodium) and aluminium.

In this work, a laboratory weathering study was carried out on rock from the area, to attempt to understand these differences. Although much work has been undertaken on the study of the dissolution of single minerals (e.g. Chou & Wollast, 1984), and this information has been used in weathering models (e.g. Sverdrup, 1990), little has been done on studying the weathering of whole multi-mineral rocks in the laboratory, and then applying these results to the field.

This paper reports the results of work from the first rock sample, and it is planned to extend this work to other rock types in the area to build up a fuller picture.

Table 1. Analysis of two Duddon streamwaters, Gaitscale Gill (GG) and Hardknott Gill (HG), at base flow conditions. Summer 1997. Concentrations in µM.

Stream	pH	Na	K	Ca	Mg	Al	Si
GG	5.0	157	5	22	18	4.5	30
HG	7.4	194	10	130	55	0.3	40

2 METHODS

2.1 Rock analysis

The samples used in this experiment were taken from a single large rock collected from the Hardknott Gill catchment. Analysis by optical microscopy, XRF and XRD determined that it was a basaltic rock that had been metamorphosed under low-grade hydrous conditions to a feldspar - chlorite - quartz assemblage. Table 2 shows the whole rock chemical composition and the average composition of rock containing a similar quantity of SiO_2 from the area (Beddoe-Stephens et al., 1995). Comparatively, rock HK1 has a low CaO composition and slightly low K_2O value which are probably a result of hydrothermal leaching. The Fe_2O_3 composition is slightly higher than average.

Table 2. Whole rock chemical composition in wt % of rock used in this work (HK1) and average composition for a rock containing a similar quantity of SiO_2 from the area (BS).

	HK1	BS
SiO_2	49.3	52.0
Al_2O_3	16.1	15.6
Fe_2O_3	13.9	9.53
CaO	1.55	8.40
MgO	10.0	10.0
MnO	0.49	0.28

	HK1	BS
Na_2O	2.48	2.00
K_2O	0.46	1.60
TiO_2	0.97	1.00
P_2O_5	0.20	0.22
Bal	4.48	-

2.2 Preparation of rock samples

The rock was first broken into pieces of ca. 10cm diameter with a sledgehammer. These pieces were then fragmented, using a jaw cutter, to mimic natural fracturing. These smaller cube-shaped pieces were sieved to select pieces in the 8 - 16 mm size range. Only pieces where all the surfaces consisted of fresh rock were used. These rocks were then submerged in distilled water and subjected to ultrasound for 15 minutes. This process was repeated up to eight times using fresh distilled water and is estimated to remove over 90% of fine surface particles created by the cutting process, based on light scattering experiments.

Each of the rocks was of a cubic shape. Taking the average side length, the geometric surface area can be estimated to be 2.3×10^{-4} m^2g^{-1}. Due to larger pieces of rock being used, gas adsorption methods used by other researchers to measure surface area are inapplicable here, however, this can be taken into account as far as possible by using previously published surface roughness factors.

2.3 Dissolution Experiments

Each experiment was conducted on 10 pieces of rock, with a combined mass of 38 (±0.1) g. The rocks were submerged in $120cm^3$ of solution in $250cm^3$ conical flasks, which were covered in clear plastic film. For experiments in the pH range 2.5 to 4, the solutions were replaced daily for a total of 13 days. At pH 5.5 and 6.5 the solutions were replaced weekly for a total of 10 weeks. The replaced solutions were then taken for analysis. This frequent replacing of solutions was intended to minimise any over saturation and precipitation of secondary phases. Experiments were conducted in the absence of light in an incubator at 10°C, the flasks being gently agitated.

Experiments were carried out using unbuffered solutions (made with nitric acid) and the buffer 2-[N-Morpholino]ethanesulfonic acid (MES), which has been shown to have negligible or low metal binding (Good et al., 1966). Repetition of some experiments, using different buffer concentrations, was carried out to check for buffer-metal binding. All experimental solutions contained 0.01M lithium nitrate to maintain constant ionic strength. Experiments were also done to study the effect of fulvic acid. A summary is shown in Table 3.

Table 3. Compositions of solutions for dissolution experiments. FA represents Fulvic Acid.

pH	solution	conc mM	FA mg l^{-1}
2.5	HNO_3	3.3	none
3.0	HNO_3	1	none
3.5	HNO_3	0.33	none
4.0	HNO_3	0.1	none
5.5	MES	1, 10	none
6.5	MES	1, 10	none
3.0	HNO_3	1	10
4.0	HNO_3	0.1	10
5.5	MES	1	10
5.5	MES	1	30

2.4 Analysis

The solution pH was measured using a glass electrode, dissolved aluminium colorimetrically using pyrocatechol violet, and soluble reactive silicon by reaction with molybdate and then reduction to silicomolybdenum blue. Base cations were measured using atomic absorption spectroscopy.

3 RESULTS

Fig. 1 shows the results for aluminium, silicon, sodium and magnesium at pH 2.5. The rate of dissolution decreases with time. For sodium and magnesium the initial decrease is very rapid. For potassium and calcium the results were similar in both trend and magnitude to those of sodium and magnesium respectively. It is interesting to note that magnesium and calcium dissolved in similar amounts, although magnesium is present in much larger quantities in the original rock. The results at higher pH values were similar but the concentrations fell progressively with increase in pH. It is not certain if steady state was fully reached.

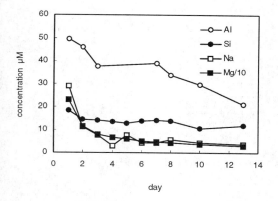

Fig. 1. Concentrations in solution of aluminium, silicon, sodium and magnesium at pH 2.5.

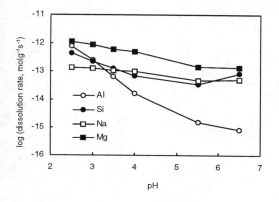

Fig. 2. 'Steady-state' dissolution rate of aluminium, silicon, sodium and magnesium.

Fig. 2 shows the results for the unbuffered solutions and MES (at the lower concentration) at the end of the experiment, where dissolution had started to reach steady state.

For the base cations and aluminium, increasing the concentration of MES buffer from 10^{-3}M to 10^{-2}M has an almost negligible effect at pH 5.5 and a small increase in dissolution rate at pH 6.5. A greater effect was shown with silicon. Its concentration in solution was increased by a factor of 2.8 at pH 5.5 and 4.4 at pH 6.5.

For the base cations the overall effect on dissolution by adding fulvic acid to the solution can be considered negligible. For silicon there is a slight increase, but the increase in dissolved aluminium at higher pH is substantial. At pH 5.5, 10mgl^{-1} of fulvic acid increases the dissolved concentration by a factor of 3.9 and 30mgl^{-1} by a factor of 11.3.

4 DISCUSSION

Fig. 1 shows the initial high dissolution rates and then the decrease to values approaching steady state. Although it is possible that some of the initial rapid dissolution is due to small fine surface particles created by the cutting process and not removed by the cleaning, it is likely to be predominantly due to the presence of a thin surface layer with a non-stoichiometric composition (Chou & Wollast, 1984).

The results in Fig. 2 show that the behaviour of whole rocks is qualitatively similar to that of pure minerals. Weathering rates calculated per m^2 of the rock are of similar magnitude generally to published studies for pure minerals. Assuming 10 as a value of surface roughness (based on a typical surface roughness for feldspars of 9 (Blum, 1994), quartz of 1.6 (Nickel, 1973) and chlorite of 64 (Sverdrup, 1990)), rates for this rock fall in the range of $10^{-12.4}$ to $10^{-8.3}$ mol m^{-2} s^{-1}. A review article (Blum & Stillings, 1995) gives a summary of dissolution rates of plagioclase feldspars which at pH 3 range from $10^{-11.5}$ to 10^{-6} mol m^{-2} s^{-1}. Sverdrup's (1990) review gives rates of between 10^{-13} and $10^{-11.5}$ for quartz dissolution and of 10^{-10} for chlorite dissolution at pH 3.

Dissolution is non-stoichiometric with respect to rock composition. Previous studies on single pure minerals have reported steady state stoichiometric dissolution (eg Stillings & Brantley, 1995). However, this rock contains a mixture of phases, so although individual phases may be dissolving stoichiometrically, this would not give rise to overall stoichiometric dissolution for the whole rock. There are also likely to be small impurity phases present, which will further affect the dissolution stoichiometry.

This work on a mixed-mineral rock complements pure phase work but is more directly relevant to the field situation. Currently, more weathering experiments are being set up on rocks from the other sub-catchment being studied, in which the weathering rate appears lower.

The intended extension of this work is to incorporate the rates of rock dissolution into a computer model (Tipping, 1996) which simulates streamwater chemistry on the basis of soil-water chemical interactions and water flows in upland catchments with acid soils.

5 REFERENCES

Beddoe-Stephens, B., M.G. Petterson, D.Millward, & G.F. Marriner, 1995. Geochemical variation and magmatic cyclicity within an Ordovician continental-arc volcanic field: the lower Borrrowdale Volcanic Group, English Lake District. *J. Volc. Geoth. Res.* 65:81-110.

Blum, A.E. 1994. Feldspars in weathering. In: I. Parsons (ed), *Feldspars and Their Reactions*:595-629. Netherlands: Kluwer Academic Pub.

Blum, A.E. & L.L. Stillings, 1995. Feldspar dissolution kinetics. In: A.F. White & S.L. Brantley (eds), *Chemical Weathering Rates of Silicate Minerals*. Mineral. Soc. Am., Rev. Mineral. 31:291-351.

Chou, L & R. Wollast, 1984. Study of the weathering of albite at room temperature and pressure with a fluidized bed reactor. *Geochim. Cosmochim. Acta.* 48:2205-2217.

Good, N.E., G.D. Winget, W. Winter, T.N. Connolly, S. Izawa & R.M.M. Singh, 1966. Hydrogen ion buffers for biological research. *Biochem.* 5:467-477.

Nickel, E. 1973. Experimental dissolution of light and heavy minerals in comparison with weathering and intrastitial solution. *Contr. Sediment.* 1:1-68.

Stillings, L.L. & S.L. Brantley, 1995. Feldspar dissolution at 25°C and pH 3: Reaction stoichiometry and the effect of cations. *Geochim. Cosmochim. Acta.* 59:1483-1496.

Sverdrup, H.U. 1990. *The Kinetics of Base Cation Release due to Chemical Weathering*. Sweden: Lund University Press.

Tipping, E. 1996. CHUM: a hydrochemical model for upland catchments. *J. Hydrol.* 174:305-330.

Comparisons of short-term and long-term chemical weathering rates in granitoid regoliths

A.F.White & D.A.Stonestrom
US Geological Survey, Menlo Park, Calif., USA

ABSTRACT: Chemical weathering rates in three deeply weathered granitoid profiles are evaluated based on SiO_2 fluxes. Aqueous fluxes below the zone of seasonal variations are estimated from mean-annual pore-water SiO_2 concentrations of 0.1 to 1.5 mM and average hydraulic-flux densities of 6×10^{-9} to 14×10^{-9} m s^{-1} for quasi-steady state portions of the profiles. Corresponding fluid-residence times in the profiles ranged from 10 to 44 years. Total solid state SiO_2 losses, based on chemical and volumetric changes, ranged from 19 to 110 kmoles SiO_2 m^{-2} as a result of 200 to 375 ka of chemical weathering. Extrapolation of contemporary solute fluxes to comparable time periods reproduced SiO_2 losses to about an order of magnitude. A strong positive correlation exists between moisture content and rates of chemical weathering.

INTRODUCTION

This study compares weathering rates based on contemporary solute fluxes to weathering rates based on cumulative solid-state chemical changes for three well-characterized regoliths. Contemporary solute fluxes are based on measured hydraulic conductivities, pressure heads, and pore-water chemistries. Cumulative chemical changes are calculated by mass-balance using measured volume fractions of inert elements. The extent of agreement produced by these two approaches is a gauge of our ability to characterize chemical weathering rates in soils and regoliths (White, 1995).

2 METHODS

The change in mass $^*\Delta M_j$ of component j in a volume of regolith caused by chemical weathering over the fluid residence time $^*t$ (s) can be calculated by integration of the product of c_j, the increase in the component's concentration in pore water over that in precipitation (moles m^{-3}), and q_h, the hydraulic flux density (m s^{-1})

$$^*\Delta M_j = -\int_0^{^*t} c_j q_h \, dt \qquad (1.1)$$

For conditions common to most regoliths, the hydraulic flux density is given by

$$q_h = -K_m \nabla H \qquad (1.2)$$

where K_m is the hydraulic conductivity (m s^{-1}) and ∇H is the hydraulic gradient (m m^{-1}). A special case of Darcy's law concerns quasi-steady, unsaturated zone flow under conditions of either constant pressure head or spatial inflection of pressure head within a vertical section of a regolith. In such a case $dH/dz = 1$ (eq. 1.2) and the hydraulic-flux density becomes $q_h = -K_m$.

Movement of pore-water solutes as described by Eqns. 1.1 and 1.2 produces extremely small changes in regolith chemistry over hydrologic residence times. On the time scale of regolith development, however, solid-state mass losses become significant and can be calculated as

$$\Delta M_j = \left(\rho_s \frac{C_{j,p}}{m_j} 10^6 \right) \int_{z=0}^{z=d} \tau_j \, dz \qquad (1.3)$$

where d is the total profile depth (m), ρ_s is the solid-phase density (g m^{-3}) and m_j is the atomic weight of component j (g mole^{-1}). The mass transfer coefficient τ_j is defined as (White, 1995)

$$\tau_j = \frac{C_{j,w}}{C_{j,p}} \frac{C_{i,p}}{C_{i,w}} \quad (1.4)$$

where $C_{i,p}$ and $C_{i,w}$ are the protolith and regolith compositions of a inert component i such as Ti, Zr and Nb or quartz. When $\tau_j = -1$, component j is completely lost during weathering. If $\tau_j = 0$, component j is immobile and affected only by internal, closed-system chemical processes.

The corresponding weathering rates per unit of surface of regolith (moles m^{-2} s^{-1}) based on eqns 1.1 and 1.3 are

$$*R = *\Delta M_j / *t \quad \text{and} \quad R = \Delta M_j / t \quad (1.5)$$

where t is the age of the regolith since the inception of weathering.

3. REGOLITH CHARACTERIZATION

The above methods were applied to regolith profiles in the Rio-Icacos watershed of Puerto Rico (White, 1995), the Panola Mountain watershed of Georgia (Hooper et al., 1990), and the Merced-River watershed of California (White et al., 1995). The deeply weathered profiles developed from similar grantoid protoliths with similar physical properties, mineralogies, and chemistries. The regoliths are relatively old (200-375 ka), allowing sufficient time for strong weathering signatures to develop. The regoliths are thick (4-10 m) which minimizes evapotranspiration, bioturbation and the extraction of solutes by plant roots. The Rio Icacos site is tropical (22 °C, 4200 mm precipitation yr^{-1}). The Panola site is subtropical (16°C, 1240 mm precipitation yr^{-1}). The Riverbank site is Mediterranean (6°C, 300 mm precipitation yr^{-1}).

Ceramic-cup suction samplers collected pore waters at each sites periodically over 1 to 2 years. Average pore SiO$_2$ concentrations increase in the order Rio Icacos < Panola < Riverbank (Fig. 1) which corresponds to decreasing precipitation. As expected for weathering-dominated regimes, SiO$_2$ increases with depth at each site. SiO$_2$ is not a significant component of precipitation, is minimally impacted by nutrient cycling in vegetation, and is not strongly affected by ion-exchange processes.

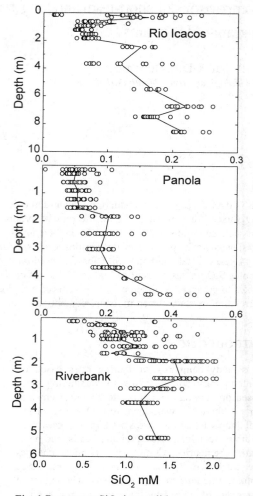

Fig. 1 Pore water SiO$_2$ in regoliths.

Pressure heads were determined using porous-cup tensiometers equipped with analog vacuum gauges. Gravimetric water contents were converted to soil-moisture saturations using measured bulk densities and an assumed particle density (2.65 g cm^{-3}). Saturated hydraulic conductivities K_{sat} were measured using a falling-head permeameter. Unsaturated hydraulic conductivities K_{unsat} were measured using the steady-state centrifuge method (Nimmo et al., 1992).

As shown for the Riverbank regolith at 2.1 m, conductivities decreased from 3.5 10^{-6} m s^{-1} at 100% saturation to 8.9 10^{-10} m s^{-1} at 63% (Fig. 2). Conductivities in the Rio Icacos saprolite at 0.7 m depth decreased from 5.5 10^{-6} m s^{-1} at 100% moisture saturation to 5.0 10^{-9} m s^{-1} at 72% saturation.

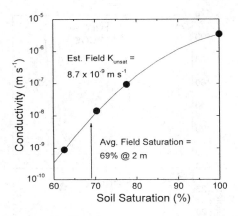
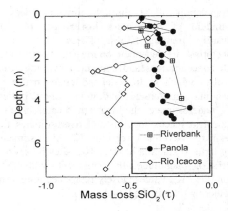

Fig. 2 Experimental hydraulic conductivities for Riverbank regolith. Arrow is extrapolated field value.

Fig. 3 Mass losses of SiO2 from regoliths as calculated from eqn. 1.4.

Saturated hydraulic conductivities of Panola regolith were measured on cores taken from between 0.78 and 2.35 m. Conductivities increased with depth from 1.3×10^{-9} m s^{-1} in the upper soil to 5.0×10^{-6} m s^{-1} in the deeper saprolite. The large range of experimental conductivities means that SiO$_2$ fluxes are strongly dependent on moisture content. Conductivities under actual field conditions were estimated by extrapolation of approximately steady state moisture conditions onto the experimental conductivity curves (Fig. 2).

Protolith mineralogies of all three sites correspond to granitoid compositions. The principal mineralogical change during weathering is a depletion of plagioclase and K-feldspar which manifests in a corresponding loss of SiO$_2$ The mass-transfer coefficients, expressed as τ_{SiO2}, are less than zero for all three regoliths indicating a net loss of SiO$_2$ from the profiles (Fig. 3). SiO$_2$ losses are greatest in the Rio Icacos regolith, exceeding 50% in basal portions of the saprolite. This is indicative of more aggressive weathering of a tropical environment.

4. DISCUSSION

Short term weathering SiO$_2$ losses are compared to long-term losses from the three regoliths in Table 1. As expected, $*\Delta M_{SiO2} \ll \Delta M_{SiO2}$. This is because fluid residence times are much shorter than regolith ages, that is, $*t \ll t$. The respective short-term and long-term weathering rates, $*R$ and R, are calculated by dividing $*\Delta M_{SiO2}$ and ΔM_{SiO2} by $*t$ and t, respectively. Long term rates increase with increasing precipitation and temperature at the three sites. Agreement between $*R$ and R for the Panola regolith are close (0.17 moles m^2 y^{-1}). $*R$ for the Riverbank regolith (0.4 moles m^2 y^{-1}) is about five times the long term rate R (0.08 moles m^2 y^{-1}). $*R$ for the Rio Icacos regolith (0.04 moles m^2 y^{-1}) is approximately an order of magnitude less than R (0.56 moles m^2 y^{-1}).

The sensitivity of estimated weathering rates to different hydraulic-flux densities is shown graphically by plotting ΔM_{SiO2} versus the duration of weathering t (fig. 4). Solid points are equivalent to the long-term weathering rate R for each profile.

Table 1 Hydraulic-flux densities, SiO$_2$ mass losses, and weathering rates.

	Moisture Saturation (%)	Hydraulic Gradient ∇H (m m^{-1})	Hydraulic Flux Density q_h (m s^{-1})	Fluid Residence *t (yr)	Mass Loss $*\Delta M_{SiO2}$ (moles m^{-2})	Regolith Age ^t (yr)	Mass Loss ΔM_{SiO2} (moles m^{-2})	Rate $*R_{SiO2}$ moles m^{-2} yr^{-1}	Rate R_{SiO2} moles m^{-2} yr^{-1}
Rio Icacos	73	1.1	6.1E-09	44	1.8	2.0E+05	1.1E+05	0.04	0.56
Panola	100	1.0	1.4E-08	10	1.7	3.5E+05	2.6E+04	0.17	0.17
Riverbank	69	1.0	8.6E-09	19	7.0	2.5E+05	1.9E+04	0.37	0.08

Extrapolation of *t through time produces a continuum of SiO_2 mass losses that plot as upwardly trending lines. These "weathering trajectories" are dependent on measured pore-water SiO_2 concentrations and computed hydraulic-flux densities which in turn depend on estimates of prevailing unsaturated hydraulic conductivities. K_{unsat} underestimates the long term weathering rate in the Rio Icacos regolith and overestimates the rate in the Riverbank regolith. K_{sat} overestimates long term weathering at these two sites by several orders of magnitude. In agreement, with field saturation, K_{sat} values closely reproduce long term weathering rates in the Panola regolith.

5. CONCLUSION

This study presents quantitative methods for evaluating short term and long term weathering rates based on current solute fluxes and long term changes in regolith chemistries. Results indicate that comparison of rates is most sensitive to estimates of hydraulic-flux densities which can vary by several orders of magnitude depending on the degree of regolith moisture saturation. The dependence of long term weathering on annual precipitation at the three sites indicates such fluxes strongly impact weathering reactions. Possible past changes in climate may have affected precipitation and the degree of saturation and hence the apparent weathering rates.

REFERENCES

Hooper, R.P., Christophersen, N. & Peters, N.E. 1990. Modeling stream water chemistry as a mixture of soilwater end members-Application to Panola Mountain. J. Hydrol. 116: 321-343.

Nimmo, J. R., Askin, K. C., and Mello, K. A., 1992. Improved apparatus for measuring conductivity at low water content. Soil Science Amer. J., 56: 1788-1761.

White, A. F., 1995. Chemical weathering rates of silicate minerals in soils. Reviews Mineral. 31: 405-461.

White, A. F. , Blum, A. E., Schulz, M. S. , Bullen, T. D., Harden, J. W. and Peterson, M. L. 1995. Chemical weathering rates in a soil chronosequence on granitic alluvium. Geochim. Cosmochim. Acta 60: 2533-2550.

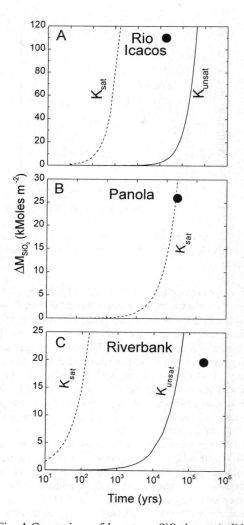

Fig. 4 Comparison of long-term SiO_2 losses (solid points) with projected SiO2 losses based on current pore water compositions and hydraulic flux densities (lines)

7 Metamorphism

Devolatilization in a siliceous dolomite, petrologic and stable isotope systematics

R. Abart
Institute of Mineralogy-Crystallography and Petrology, Karl-Franzens-University, Graz, Austria

ABSTRACT: The evolution of the pore fluid of siliceous carbonates during progressive metamorphism may shed light on the hydraulic rock properties at elevated pressures and temperatures. In this case study petrographic and stable isotope data are employed to reconstruct the history of the pore fluid of a two to six meters thick dolomite oolite layer of the uppermost Werfen formation in the Monzoni contact aureole (South Tyrol, N-Italy). Comparison of the observations and model predictions indicates that the dolomite oolite acted as an aquitard. Less than about 16.5 moles of H_2O per 1000 cm^3 of the original dolomite and essentially no carbon-bearing species infiltrated into the dolomite oolite during contact metamorphism.

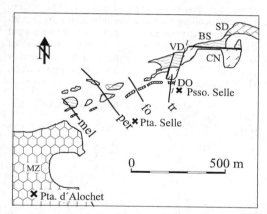

Fig. 1. Geological sketch map of the eastern Monzoni intrusive complex. Only rocks of Mid Triassic age are shown: Monzoni intrusives - MZ, dolomite oolite of the uppermost Werfen formation - DO, Contrin limestone - CN, Buchenstein formation - BS, Schlern dolomite - SD, vulcanic dikes - VD. Solid lines indicate isograds taken from Inderst (1987): First occurence of tremolite - tr, forsterite - fo, periclase - per, melilite - mel.

1 GEOLOGIC SETTING

The Monzoni intrusive complex is a small (approximately 5 km^2), Mid Triassic (217-230 Ma) batholith in the western central Dolomites (South Tyrol, N Italy). The intrusives are dominated by pyroxenite, gabbro, monzogabbro, diorite, and monzodiorite. The country rocks comprise a stratigraphic sequence of Permo-Triassic rocks. The lithologic units strike at a high angle to the intrusive contact and can be followed all across the contact aureole (see Fig. 1).

2 CONTACT METAMORPHISM

The Monzoni complex is a shallow intrusion with an intrusion depth of less than 3 km and solidus temperatures of 1170 to 1195°C (Masch and Huckenholz, 1993). Temperatures were less than 200°C in the Permo-Triassic country rocks prior to intrusion. The intrusion produced an approximately 1000 meter wide contact aureole which is well documented by newly formed mineral parageneses in the Permo-Triassic country rocks (Inderst, 1987; Masch & Heuss-Assbichler, 1991). Maximum temperatures were about 700°C at the intrusive contact as is manifest from the formation of periclase + calcite from dolomite at $X_{CO2} = 0.4$ (Inderst, 1987).

2.1 *The dolomite oolite of the Werfen formation*

A two to six meter thick massive dolomite oolite forms a prominent layer within the uppermost Werfen formation. It can be followed from outside the contact aureole up to the intrusive contact. The dolomite is quite pure. Its composition may be described in the CaO - MgO - SiO_2 - (H_2O - CO_2) model system with the molar proportions being 48:45:3, respectively. The unmetamorphosed dolomite oolite is comprised of dolomite (93 vol. %), calcite (4 vol. %), and quartz (3 vol. %). The following reactions, which produced talc, tremolite, forsterite, and periclase, have been identified with increasing metamorphic grade::

(1) 3 dolomite + 4 quartz + 1 H_2O = 1 talc + 3 calcite + 3 CO_2
(2) 2 talc + 3 calcite = 1 dolomite + 1 tremolite + 1 H_2O + 1 CO_2
(3) 1 tremolite + 11 dolomite = 8 forsterite + 13 calcite + 1 H_2O + 9 CO_2
(4) 1 dolomite = 1 calcite + 1 periclase + 1 CO_2

A T-X_{CO2} diagram for the system CaO - MgO - SiO_2 - H_2O - CO_2 is shown in Fig. 2.

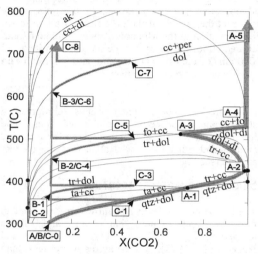

Fig. 2. T-X_{CO2} diagram for the system CaO - MgO - SiO_2 - H_2O - CO_2 at P = 500 bars calculated with data from Holland and Powell (1989): Quartz (qtz), talc (ta), tremolite (tr), forsterite (fo), calcite (cc), dolomite (do), periclase (per), and diopside (dio). Only the equilibria relevant for the composition of the dolomite oolite are labelled. Heavy gray line: fluid evolution path for perfect internal buffering, thin gray line: perfect external buffering, intermediate thickness line: hybrid scenario.

During contact metamorphism, mineral reactions may have controlled the composition of the pore fluid i.e. "internal buffering" (Greenwood, 1975). Alternatively the fluid composition may have been controlled by the introduction of volatile species from an external source, e.g. from the crystallizing pluton - "external buffering". Three possible fluid evolution paths are illustrated in Fig. 2. The initial composition of the pore fluid is assumed to have been X_{CO2} = 0.1, and the porosity is assumed to have been constant at 1 %. The heavy gray line indicates the fluid evolution for the scenario of perfect internal buffering. In this case the paragenetic and fluid evolution is determined by the initial compositions and the proportions of the rock and the pore fluid. A second scenario, perfect external buffering, is indicated by the thin gray line. This scenario implies infinitely rapid transport of an aqueous fluid from an external source into the reacting rock. In this scenario the composition of the pore fluid is kept constant at X_{CO2} = 0.1 by successive infiltration of water regardless of the composition and amount of the internally produced fluid. A third scenario, illustrated by the intermediate thickness gray line, allows for internal buffering of the initially water-rich fluid up to X_{CO2} = 0.5. As soon as this composition is reached infiltration of water drives the buffering reaction to completion and resets the fluid composition to X_{CO2} = 0.1. When the equilibrium temperature of the next devolatilization reaction is reached, internal buffering starts again. This path is referred to as the "hybrid scenario".

Tab. 1. Mineral modes in cm^3 per 1000 cm^3 of the initial dolomite ooltie for perfect internal buffering. Column headings refer to the labels in Fig. 2, path A. Total solid volume (V_S), moles of fluid released during reaction (F exp.), atomic fractions of oxygen and carbon that remain in the rock after reaction (F_O, F_C), oxygen (SMOW) and carbon (PDB) isotope compositions.

	A-0/1	A-1	A-1/2	A-2	A-2/3	A-3	A-3/4	A-5
T°(C)	384	385	426	426	512	512	529	730
X_{CO2}	0,72	0,72	0,99	0,99	0,69	0,69	0,98	1,00
qtz	13	11	7	0	0	0	0	0
ta	18	0	0	0	0	0	0	0
tr	0	20	27	26	9	0	0	0
fo	0	0	0	0	0	2	47	47
cc	52	46	49	48	42	41	118	569
do	903	903	896	887	890	889	788	0
per	0	0	0	0	0	0	0	138
di	0	0	0	11	27	34	0	0
V_S	985	981	978	972	968	966	953	754
F exp.	0,49	0,29	0,25	0,38	0,20	0,19	1,12	12,30
F_O	,993	,996	,997	,993	,998	,998	,977	,720
F_C	,987	,995	,995	,990	,998	,997	,964	,558
$\delta^{18}O$	27,0	27,0	26,9	26,9	26,9	26,9	26,7	24,8
$\delta^{13}C$	0,5	0,5	0,4	0,4	0,4	0,4	0,3	-1,0

The changes in mineral modes, the amount of fluid produced by the reactions, and the amount of water involved in external buffering are listed for the three different scenarios in Tab. 1-3. The pore-fluid evolution provides insight into the hydraulic behaviour of the dolomite oolite during contact metamorphism. Internal buffering implies that the dolomite oolite was completely impermeable and did not interact with external fluid at all. In contrast, the external buffering scenario implies high permeability and introduction of 140 moles of H_2O per 1000 cm^3 of the original dolomite. The hybrid scenario requires introduction of 16.5 moles of H_2O into 1000 cm^3 of the original rock.

Perfect internal buffering would have produced diopside. This mineral does not occur in the dolomite oolite, and perfect internal buffering can be excluded. The mineral successions predicted for the external buffering- and the hybrid scenarios are both compatible with the petrographic observations. Independent information from the stable isotope systematics is used to further constrain the evolution of the dolomite-oolite pore-fluid.

Tab. 2. Mineral modes for perfect external buffering - fluid evolution path B in Fig. 2. Abbreviations as in Tab. 1. Number of moles of H_2O that equilibrated with the rock (H_2O eq), atom equivalent instantaneous fluid/rock ratio for oxygen (F/R_O).

	B-0	B-1	B-2	B-3
T°(C)	300	367	452	600
X_{CO2}	0,1	0,1	0,1	0,1
qtz	0	0	0	0
ta	37	0	0	0
tr	0	37	0	0
fo	0	0	47	47
cc	68	53	118	569
do	875	884	788	0
per	0	0	0	138
V_S	980	974	953	616
H_2O eq	7,56	1,08	10,76	110,11
F_C	0,973	0,995	0,957	0,558
F/R_O	0,084	0,012	0,123	2,177
$\delta^{18}O$	25,1	24,8	22,4	9,6
$\delta^{13}C$	0,4	0,4	0,3	-1,1

Tab. 3. Mineral modes for the hybrid scenario - fluid evolution path C in Fig. 2. Abbreviations as in previous tables.

	C-0/1	C1/2	C-2/3	C-3/4	C-4/5	C-5/6	C-6/7	C-7/8
T°(C)	357	386	386	386	502	610	685	685
X_{CO2}	0,50	0,10	0,30	0,10	0,50	0,10	0,50	0,10
qtz	19	0	0	0	0	0	0	0
ta	9	37	0	0	0	0	0	0
tr	0	0	37	37	28	0	0	0
fo	0	0	0	0	5	24	24	24
cc	45	68	53	53	60	85	88	564
do	915	875	884	884	873	836	832	0
per	0	0	0	0	0	0	1	146
V_S	988	980	974	974	967	945	944	734
H_2Oeq	0,00	0,00	1,92	0,00	0,48	0,00	0,86	13,13
F_C	0,993	0,993	0,995	0,995	0,995	0,983	0,998	0,542
F/R_O	0,000	0,012	0,000	0,018	0,000	0,023	0,000	0,233
$\delta^{18}O$	27,0	26,6	26,8	26,6	26,6	26,3	26,2	21,7
$\delta^{13}C$	0,5	0,4	0,4	0,4	0,4	0,4	0,4	-1,0

3 STABLE ISOTOPE SYSTEMATICS

The bulk carbonate carbon and oxygen isotope compositions of the dolomite oolite at different distances from the intrusive contact and model predictions for the three scenarios are shown in Fig. 3. The compositions in the unmetamorphosed dolomite oolite are $\delta^{13}C$(PDB) = 0 to 0.5 per mil and $\delta^{18}O$(SMOW) = 26.5 to 27.5 per mil. Within the contact aureole $\delta^{18}O$ is shifted from the original values to about 25 per mil at a distance of 500 to 600 meters from the intrusive contact. $\delta^{13}C$ is shifted down to $\delta^{13}C$ = -1.25 to -0.75 per mil at 350 to 400 meters from the intrusive contact.

3.1 Isotopic effects of devolatilization in open and closed systems

At the temperatures of contact metamorphism the CO_2 produced during decarbonation is enriched in the heavy isotopes ^{13}C and ^{18}O with respect to the residual rocks. If the isotopically heavy CO_2 is lost from the system, this causes the residual rock to become successively depleted in the heavy isotope. If the fluid is lost continuously as it is produced, this process is referred to as "Rayleigh distillation". The extent of the isotopic shift produced during devolatilization depends on the fluid-rock equilibrium fractionation factor and on the fraction of the isotopic element under consideration removed from the system. The isotopic shifts resulting from Rayleigh distillation were calculated using the expression given by Valley (1986). The fluid-rock equilibrium fractionation factors were calculated from the fractionation factors given by O´Neil et al. (1969) and by Bottinga (1968). The fraction of the element that remains in the rock after devolatilization was derived from reaction stoichiometries, mass balance constraints, and the constant 1 % porosity assumption (see Tab. 1-3). For the externally derived fluid an oxygen isotope composition of $\delta^{18}O$ = 10 per mil in equilibrium with the Monzoni intrusves was assumed. The isotopic effect of infiltration was modelled according to Taylor (1974).

Model calculations for the external buffering scenario predict much larger $\delta^{18}O$ shifts than observed and perfect external buffering can be safely excluded. The observed oxygen isotope values are somewhat lower than those predicted for the internal buffering scenario and somewhat higher than expected for the hybrid scenario. The largest shift in $\delta^{13}C$ is associated with periclase formation (see Fig. 3). A 50 % loss of the total carbon is associated with this reaction and the observed shift is exactly what is expected for Rayleigh distillation. Calcite-dolomite thermometry (Inderst, 1987) indicates that periclase formation occurred at 700°C and X_{CO2} = 0.4. This suggests limited introduction of a water-rich fluid and supports the hybrid scenario. Model predictions for this scenario agree well with the observed carbon isotope shifts, but they overestimate the oxygen isotope shifts associated with periclase formation (see Fig.3). The corresponding fluid to rock ratio of 16.5 moles H_2O per 1000 cm^3 of the original dolomite oolite is hence regarded as a maximum estimate for the amount of external fluid infiltrated

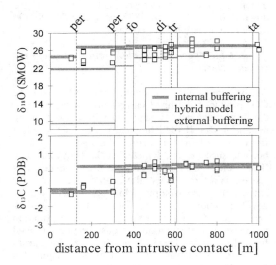

Fig. 3. Carbon and oxygen-isotope systematics of the dolomite oolite (bulk carbonates); first occurrences of index minerals for external buffering and hybrid scenario - solid vertical lines, internal buffering - dashed vertical lines; isotopic trends predicted for the three fluid evolution paths indicated by gray bars.

4 CONCLUSIONS

The succession of mineral zones and the carbon- and oxygen isotope systematics of the two to six meter thick dolomite oolite from the uppermost Werfen formation in the eastern Monzoni contact aureole indicate limited infiltration of external fluid during contact metamorphism. Despite its low thickness the dolomite oolite layer appears to have been rather impermeable. The paragenetic and oxygen isotope evolution of the dolomite indicate infiltration of less than 16.5 moles of a water-rich fluid per 1000 cm^3 of the original rock. The carbon isotope systematics can be readily explained by Rayleigh distillation during devolatilization reactions.

5 AKNOWLEDGEMENTS

This study was funded by the Austrian National Bank - Jubiläumsfonds, grant Nr. 5428. L. Masch, S. Heuss-Assbichler and P. Brack are thanked for their contributions to this work.

6 REFERENCES

Bottinga, Y 1968. Calculation of fractionation factors for carbon and oxygen exchange in the system calcite - carbon dioxide - water. J. Phys. Chem., 72, 800-808.

Greenwood, H.J. 1975. Buffering of pore fluids by mineral reactions. *Am. Journ. Sci.*, 275, 573-593.

Holland T.J.R. & Powell, R. (1989) An enlarged and updated internally consistent thermodynamic dataset with uncertainties and correlations: The system K_2O - Na_2O - CaO - MgO - MnO - FeO - Fe_2O_3 - Al_2O_3 - TiO_2 - SiO_2 - C - H_2 - H_2O. Journ. met. Geol., 8, 89-124.

Inderst, B. 1987. Mischkristalle und fluide Phase bei Hochtemperatur Dekarbonatisierungsreaktionen in zwei Thermoaureolen. unpublished PhD thesis, University of Munich, 215 p..

Masch, L. & Heuss-Assbichler, S. 1991. Decarbonation reactions in siliceous dolomites and impure limestones. In G. Voll, J. Töpel, D.R.M. Pattison, F. Seifert (eds), *Equilibrium and kinetics in contact metamorphism*, 211-229.

Masch, L. & Huckenholz, H.G. 1993. Der Intrusivkörper von Monzoni und seine thermometamorphe Aureole. Beihefte zu European Journal of Mineralogy, 5/2, 81-135.

O´Neil, J.R., Clayton, R.N., and T.K.Mayeda, 1969. Oxygen isotope fractionations in divalent metal carbonates. J. Chem. Phys., 51, 5547-5558.

Taylor, H.P. Jr. 1974. The application of oxygen and hydrogen isotope studies to problems of hydrothermal alteration and ore deposition. *Econ. Geology*, 69, 843-883.

Valley, J.W. 1986. Stable isotope geochemistry of metamorphic rocks. In Valley, J.M., Taylor, H.P. Jr., O´Neil, J.R. (eds), *Stable isotopes in high-temperature geological processes*. Rev. Mineral., 16, 441-490.

Metamorphic fluid flow at marble-schist boundaries, Corsica, France

I.S. Buick
School of Earth Sciences, La Trobe University, Bundoora, Vic., Australia

I. Cartwright
Department of Earth Sciences, Monash University, Clayton, Vic., Australia

ABSTRACT: Interlayered marbles, metabasalts and micaceous schists from the Schistes Lustrés Nappe, Corsica (France) experienced high-pressure/low-temperature metamorphism and subsequent retrogression at greenschist facies during the Alpine orogeny. Oxygen isotope profiles across metre to tens of centimetre-wide marble bands are characterised by high $\delta^{18}O(Cc)$ values in marble cores, consistent with little alteration from Mesozoic marine limestone precursors, and steep gradients to lower values at contacts with surrounding rock types (metabasalts and/or micaschists). Carbon isotope values are little reset in these boundary layers. The oxygen isotope profiles are dominated by diffusion and could have developed over timescales of 10^2-10^6 years. Some profiles may also involve a small component of across strike fluid flow with time-integrated fluid fluxes of less than 0.5 m^3/m^2.

1 INTRODUCTION

In collisional orogens fluids may be evolved at various times during subduction and subsequent exhumation. Some fluid escapes from the subducting slab at relatively high crustal levels and interacts with the rocks in the accretionary wedge. However, much of the fluid is probably liberated at depth during high-pressure/low-temperature metamorphism (HP/LT) under blueschist- or eclogite-facies conditions (Peacock, 1993), or during subsequent exhumation under greenschist-facies conditions. In this paper we constrain syn-metamorphic fluid flow in the Schistes Lustrés Nappe, a structurally coherent HP/LT metamorphic terrain in Corsica (France) using stable isotope profiles across marble-schist and marble-metabasalt contacts.

2 GENERAL GEOLOGY

Alpine Corsica (Fig. 1) contains a series of nappes that were emplaced during the collision of the European continental plate with a Tethyan subduction complex during the late Cretaceous and early Eocene (Gibbons et al., 1986). The structurally lowest (Schistes Lustrés) nappe, which itself can be further divided into several thrust-bound units, was emplaced onto Hercynian European continental basement. It contains structurally-intercalated metasediments (schists, marbles, quartzites; the Schistes Lustrés Unit), the dismembered relicts of a Tethyan ophiolite (the Upper and Lower Ophiolite Units), and thrust slices of obducted European continental basement.

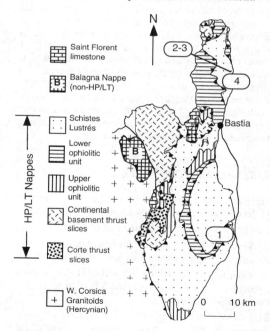

Figure 1: Geological map of Alpine Corsica, France. Traverses (in ovals) are discussed in the text.

The Schistes Lustrés Nappe was metamorphosed at high pressures, most typically in the blueschist facies (P ~ 8 kbar, T ~ 400°C), although relict eclogite-facies assemblages (P ~ 12 kbar, T ~ 530 °C) are also locally preserved (Caron et al., 1981; Gibbons et al., 1986). Major compressional structures were reactivated as extensional shear zones and faults during the late Oligo-Miocene (Jolivet et al., 1991). Extension was associated with a non-pervasive greenschist-facies event (P ~ 5 kbar, T ~ 450°C), and locally extensive veining (Miller et al., 1998; this volume).

3 SAMPLE LOCALITIES

Stable isotope profiles were measured across interlayered calcite-rich marble and metabasalts and/or calcite-bearing micaceous schists (schistes lustrés *sensu stricto*) at the following localities (Fig. 1): a) Sant'Andrea di Cotone quarry (traverse 1); b) Santa Catalina quarry, Cap Corse (traverses 2 and 3); and c) at a road cut several kilometres to the south of Santa Catalina quarry (traverse 4). Marble (calcite + quartz + phengite + lawsonite + glaucophane) from traverse 1 is interlayered with metabasites (glaucophane + lawsonite + garnet + phengite ± relict jadeitic pyroxene), and rare jadeite-bearing acid gneisses that record blueschist-facies conditions (Caron et al., 1981). Marbles from traverses 2 to 4 (calcite + quartz + phengite + chlorite ± graphite ± lawsonite ± glaucophane) are interlayered with metabasites (quartz + epidote + chlorite + actinolite + albite ± glaucophane ± calcite ± lawsonite) and micaschists (quartz + phengite + calcite + albite ± carpholite) that have been extensively, but not completely retrogressed under greenschist-facies conditions. Marble at traverse 1 dips steeply, whereas at traverses 2 to 4 marble layers are sub-horizontal due to folding associated with the greenschist-facies overprint.

4 STABLE ISOTOPE PROFILES

The isotope data show the following general features:
1) the cores of marble bands greater in width than ~1-2 m generally contain calcite with $\delta^{18}O(Cc)$ = 23 to 26 ‰ and $\delta^{13}C(Cc)$ = 0 to 2 ‰ (e.g Fig. 2a.);
2) calcite from metabasites and micaschists has considerably lower $\delta^{18}O(Cc)$ and $\delta^{13}C(Cc)$ values (typically 12 to 17 ‰ and 0 to -3 ‰, respectively) than in adjacent marbles;
3) $\delta^{18}O(Cc)$ values in the marbles decrease from unreset values in cores to values close to those of the adjacent metabasites and micaceous schists at marble contacts (Figs 2a and 3), whereas carbon isotopes show smaller or no appreciable shifts over the same distance (Fig. 2b). The $\delta^{18}O$ values of the metabasites/schists locally increase towards the marble contact, resulting in sigmoidal boundary layers that are commonly tens of centimetes to, at most, 2-3 metres wide (Figs 2a and 3);
4) for an individual marble band, the mid-points of the isotopic boundary layers are generally located at the contact between marble and silicate rock, or on the marble (inboard) side of these contacts. Moreover, the widths of the boundary layers for oxygen isotope values commonly differ at either contact with silicate rock, resulting in asymmetric profiles. This is especially true for the thinner (< 0.5-1.0 metre wide) marble bands. These marble bands, which do not contain inner plateau regions of high, constant $\delta^{18}O(Cc)$, instead show asymmetrically peaked oxygen isotope profiles (Figs 4 and 5); and
5) individual marble bands show near-linear trends on a $\delta^{18}O(Cc)$ v. $\delta^{13}C(Cc)$ diagram; if data from adjacent schists or metabasalts are included these trends are "L"-shaped.

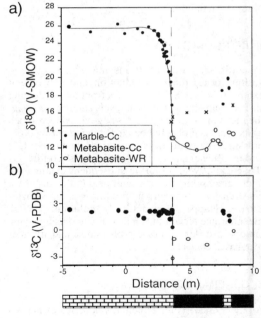

Figure 2. Traverse across marble and blueschist-facies metabasite, Sant'Andrea di Cotone (Fig. 1).

The calcite from the marbles cores has an isotopic composition consistent with unaltered Mesozoic marine carbonate elsewhere in the Alpine chain (Bickle & Baker, 1990), and can not have been significantly affected by metamorphic fluid/rock interaction. The $\delta^{18}O(WR)$ values of metabasalts (typically ~11-14 ‰; e.g. Fig. 2) are too high to represent unaltered igneous values (~5-6‰). Metabasalts away from high-^{18}O marble layers on Corsica commonly show such high values, which we interpret as reflecting low-temperature sea-floor hydrothermal alteration that has been preserved through Alpine metamorphism (cf. Matthews & Schliestedt, 1984). The stable isotopic composition of calcite from the micaschists and metabasites is also consistent with that observed in other HP/LT terrains (Ganor et al., 1989) and is generally, but not always, in equilibrium with the $\delta^{18}O$(WR) values.

range of porosities between 10^{-3} and 10^{-6}, the boundary layers at traverses 1 and 2 could have been formed within periods of 5×10^2 to 10^6 years, which is similar to that estimated in studies of diffusion-dominated boundary layers elsewhere (Ganor et al., 1989; Cartwright & Valley, 1991) and is reasonable given the timescale of metamorphic events (~10^6-10^7 years) during the Alpine orogeny.

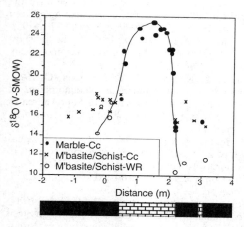

Figure 3. Traverse 2 across marble and retrogressed metabasite and micaschist, Santa Catalina quarry (Fig. 1).

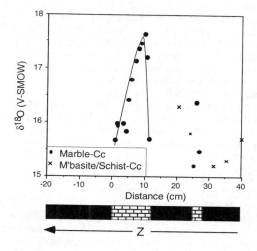

Figure 4. Traverse 3 across marble and partially retrogressed metabasite and micaschist, Cap Corse (Fig. 1).

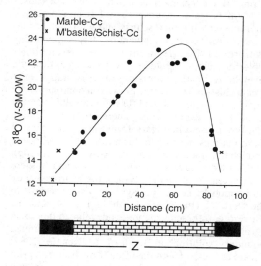

Figure 5. Traverse 4 across marble and partially retrogressed metabasite and micaschist, Cap Corse (Fig. 1).

5 THE GEOMETRY OF FLUID FLOW

The contact between marbles and micaschists or metabasalts represents a sharp initial discontinuity in $\delta^{18}O$ across which subsequent isotopic exchange has taken place, as shown by the boundary layer oxygen isotope gradients. There is little evidence of new mineral assemblages developed in the marbles across these boundary layers. The shape of the isotopic profiles is qualitatively similar to that which results from combined advection-dispersion between two semi-infinite bodies in a system in which local isotopic equilibrium has been maintained (e.g. Bickle & McKenzie, 1987; Bickle & Baker, 1990; Ganor et al., 1989; Cartwright & Valley, 1991).

The large shifts in oxygen isotope values and very small shifts in carbon isotope values across the boundary layers is consistent with isotopic exchange occurring in the presence of a water-rich fluid (Bickle & Baker, 1990). Assuming a diffusivity for oxygen in water of ~10^{-8} m^{2}s^{-1} (Bickle & Baker, 1990) and a

If diffusion was accompanied by a component of fluid flow across strike, the advective distances, given by the position of the mid-point of the profiles in the boundary layers, varies from several centimetres (Fig. 4) to several tens of centimetres (Fig. 2a). For the displacement of oxygen isotope discontinuities across strike, the time-integrated fluid flux into the marble layers calculated from equations in Bickle & McKenzie (1987) and Bickle & Baker (1990) is very small (< 0.1-0.5 m^3/m^2) and is comparable to those calculated from other outcrop-scale studies (e.g. Bickle & Baker, 1990; Ganor et al., 1989). Alternatively, the mid points of the boundary layers could mark the edges of channels oriented along the marble-schist/metabasalt contacts, along which water-rich fluid flowed. The isotopic composition of this fluid could have been either a) in isotopic equilibrium with the silicate rocks, or b) intermediate between that of the silicate rocks and marble, and the initially sharp channel boundaries would have subsequently been quickly smoothed out by diffusion.

6 THE TIMING OF FLUID FLOW

Metabasites, marbles and acid gneisses at Sant-Andrea di Cotone (traverse 1) contain high-pressure mineralogies with only minimal evidence for subsequent retrogression (Caron et al., 1981). Therefore, we suggest that the boundary layer at the marble/metabasite contact at this locality developed during HP/LT metamorphism in the late Cretaceous.

Marbles from traverses 2 to 4 (Cap Corse) are interlayered with mica schists and with metabasites whose blueschist-facies mineralogy has been variably overprinted by greenschist-facies assemblages (chlorite + albite + epidote). At these localities, and elsewhere in the Schistes Lustrés nappe, the greenschist-facies overprint was accompanied by significant channelled fluid flow to form albite (Miller et al., 1998) vein systems. Therefore, it is possible that the isotopic boundary layers at the margins of these marbles developed during the Oligo-Miocene greenschist-facies overprint. However, although profiles from traverses 3 and 4 (Figs 4 and 5) were both taken from shallowly-dipping marbles, they show opposite senses of asymmetry with respect to the vertical (the Z direction in Figs 4 and 5). Therefore, if these asymmetries developed due to an across strike component of advection (the pinned boundary solution of Bickle & McKenzie, 1987; Bickle & Baker, 1990) then fluid flow occurred prior to the main syn-greenschist-facies folding event that resulted in the current sub-horizontal orientation of these marbles.

7 ACKNOWLEDGEMENTS

ISB and IC gratefully acknowledge an ARC Australian Research Fellowship and QE II Fellowship, respectively. This research was funded by ARC Large Grant No. A39531124 to IC and ISB. Jodie Millier is thanked for help in the field and Mary Jane for help with stable isotope analyses. Mark Brandriss provided a thorough and helpful review.

8 REFERENCES

Bickle, M.J. & Baker, J. 1990. Advective-diffusive transport of isotopic fronts: an example from Naxos, Greece. *Earth Planet Sci. Lett.* 97, 78-93.

Bickle, M.J. & McKenzie, D.M. 1987. The transport of heat and matter by fluids during metamorphism. *Contrib. Mineral. Petrol.* 95, 384-392.

Caron, J.M., Kienast, J.R. & Triboulet, C. 1981. high-pressure--low-temperature metamorphism and polyphase alpine deformation at Sant'Andrea di Cotone (Eastern Corsica, France). *Tectonophys.* 78, 419-451.

Cartwright, I. & Valley, J.W. 1991. Steep oxygen-isotope gradients at marble-metagranite contacts in the northwest Adirondack Mountains, New York, USA: products of fluid-hosted diffusion. *Earth Planet Sci. Lett.* 107, 148-163.

Ganor, J., Matthews, A. & Paldor, N. 1989. Constraints on effective diffusivity during oxygen isotope exchange at a marble-schist contact, Sifnos (Cyclades), Greece. *Earth Planet Sci. Lett.* 94, 208-216.

Gibbons, W., Waters, C. & Warburton, J. 1986. The blueschist facies *schistes lustrés* of Alpine Corsica: A review. *Geol. Soc. Am. Mem.* 164, 301-311.

Jolivet, L., Daniel, J.-M. & Fournier, M. 1991. Geometry and kinematics of extension in Alpine Corsica. *Earth Planet Sci. Let t.* 104, 278-291.

Matthews, A. & Schliestedt, M. 1984. Evolution of the blueschist and greenschist facies rocks of Sifnos, Cyclades, Greece. *Contrib. Mineral. Petrol.* 95, 384-392.

Miller, J.A., Cartwright, I. & Barnicoat, A. 1998. The formation of albite veins in high pressure terrains (this volume).

Peacock, S.M. 1993.The importance of blueschist → eclogite dehydration reactions in subducting oceanic crust. *Geol. Soc. Amer. Bull.* 105, 684-694.

Shear zone-related hydrothermal alteration in Proterozoic rocks in Finland

A. Lindberg
Geological Survey of Finland, Espoo, Finland

M. Siitari-Kauppi
Department of Radiochemistry, University of Helsinki, Finland

ABSTRACT: The bedrock of Southern Finland consists mainly of Proterozoic, highly metamorphosed gneisses, both sedimentary and volcanic in origin. Synorogenic migmatite granites are common, as are late-orogenic granitoids varying in composition from granites to tonalites in what is known as the granitoid complex of central Finland (1900 - 1800 Ma). In the Syyry area, Sievi, the potential of the bedrock as a repository for high level nuclear waste was studied with the aid of several deep (300 - 1000 m) drill holes, one of which intersected a thick (60 m) brittle-ductile shear zone. Mylonitization was followed by recrystallization, after which the rock underwent jointing in different directions. The result of the hydrothermal alteration was that narrow (10 - 50 mm), sharp alteration zones have formed around fissures. Porosity is high in the alteration zone being 2 - 5 times that of the unaffected rock matrix. Several porosity profiles were measured and visualized with the ^{14}C-PMMA method.

1 GENERAL GEOLOGY

The main rock types of the Syyry area are acidic plutonic rocks, tonalite, granodiorite and quartz diorite, but in the vicinity there are also volcanic and sedimentary schists and gneisses. Southwest of Syyry there is a large formation of volcanic rocks, mainly andesitic and dacitic porphyry, agglomerate and amphibolite, and to the southeast a zone of mica schist. The basic plutonic rocks (peridotite, gabbro and diorite) grade into quartz diorite, tonalite and granodiorite without sharp contacts. Granite occurs only in small amounts (Anttila et al. 1993, Salli 1967).

The stratigraphy of the area (Salli 1967) from oldest to youngest (supracrustal versus plutonic) is as follows: arkose schists with some volcanic material - ultramafic plutonic rocks and gabbro-diorite; greywacke schists - quartz diorite, granodiorite; porphyroblastic mica schists - granites; volcanic and subvolcanic material - various dyke rocks.

The dominant rock type in outcrops as well as in drill cores from the Syyry site, is tonalite, which contains abundant inclusions (xenoliths) of various schist and plutonic rock types. Aplite, quartz, epidote and pegmatite occur as intersecting dykes and veins. Mica gneiss occurs in the outcrops of the study area as inclusions only, but immediately to the south there is a fairly large greywacke-like mica gneiss formation which, according to magnetic measurements, continues into the study area (Anttila et al. 1993).

The main folding of the area was isoclinal and the dips of schistosity are therefore mostly steep or vertical. The orientation of the fold axis undulates gently from northwest to southeast. The most intensive fracturing is restricted to the regional fracture zones surrounding a block covering 12 x 25 km in the Syyry study area (Anttila et al. 1993). The rocks in the area underwent three significant plastic deformation phases followed by three mainly brittle deformation phases (Kärki 1991).

This study provides information about the last deformation phases and adjacent weak metamorphic processes - the zeolite facies.

2 SAMPLE DESCRIPTION

The most intensively fractured zone in drill core SY7 covers the core length 182 - 207 m, where the rock is mostly crushed and the less fractured parts are macroscopically porous. The pore diameter is 2

- 3 mm and total porosity around 10%. Two core samples, 161 m and 242 m, were studied in detail. Their dimensions are: length around 20 cm and diameter 56 mm. The cores were sawed into two halves to get an even surface (Figs 1 and 2); both have very distinct fissures with alteration zones and an unaltered, "fresh" rock matrix, which enables hydrothermal alteration to be compared with fairly intact rock.

The first sample, 161 m, comes from a contact of tonalite and greywacke - mica gneiss (Fig. 1). Thin (0.5 - 1.0 mm) fissures cut both rock types and are visible due to the presence of a reddish alteration zone (5 - 15 mm wide). These fissures are filled with laumontite, but XRD analyses showed some leonhardite, too. The main minerals of mica gneiss are plagioclase, microcline, quartz and biotite, but there are also some zonal plagioclase phenocrysts. Tonalite contains plagioclase, quartz and biotite, and variable amounts of hornblende. Grain size of the tonalite is approximately 3 - 4 mm and that of mica gneiss 0.1 mm. The minerals of mica gneiss are altered only in the vicinity of fissures, but, in tonalite, plagioclase is slightly sericitized and hornblende is partly altered into biotite and biotite into chlorite.

The other sample, 242 m (Fig. 2), is tonalite mineralogically similar to sample 161 m described above. It contains two different kinds of fracture: at the end of the sample (the left end in Fig. 2) there is a permeable open fissure surface with calcite and analcime layers; polyphase jointing is nicely visible as the parallel occurence of calcite-analcime fillings and the visible alteration zone is approximately 20 mm wide on both sides of the open fissure. There is another fissure (similar to that in sample 161 m) 11 cm apart from the permeable on; it is cemented by laumontite and surrounded by a 10-mm-wide alteration zone.

3 CHEMICAL ANALYSIS

Several plagioclase grains were selected for EDS analyses, because microscopic examination showed that plagioclase was the most systematically altered mineral around fissures and we assumed that some series of alteration intensity would also be found. From the mica gneiss section of sample 161 m we selected a set of three plagioclase grains along a 10-mm line leading straight out from the fissure filled with a 1-mm-thick laumontite layer. This fissure is closed (unpermeable) in terms of conventional determination. The results are listed in Table 1, point 2 being nearest the fissure and the point 4 outermost. Point 1 is a fresh plagioclase grain from the mica gneiss matrix. All points analysed are plotted on Fig. 1.

Similar analyses were made on tonalite, points 5 - 8. Plagioclase grains 5 and 6 represent the near-fissure effect, as grains 7 and 8 are beyond the influence of the laumontite fissures. Here the CaO/Na_2O ratio of altered grains is at the same as in the mica gneiss. The relatively fresh plagioclase grains in both samples have the same composition.

Plagioclase analyses were also made on two polished thin sections of tonalite sample 242 m. Points 9 and 10 are near (10 - 20 mm) the strongly altered end of the sample and are microscopically thoroughly sericitized. Points 11 and 12 are from the optically unaltered middle part of the sample (50 - 60 mm from the altered end) and sericitization of these plagioclase grains does not differ from that of the normal tonalite matrix (Fig. 2). However, plagioclase 11 does not seem chemically belong to either group, and is situated between them.

The results show clearly that the optically slightly altered (normal) plagioclase grains (1, 7, 8, (11), 12 in Table 1) also have different SiO_2/Al_2O_3 ratios from those of the altered grains, which lack aluminium.

Table 1. Compositions of plagioclase grains measured by microprobe (EDS), the means of 2 - 3 measurements (results in weight %).

No	SiO_2	Al_2O_3	CaO	Na_2O	CaO/Na_2O
1	57.7	26.3	8.68	6.31	1.38
2	69.1	19.2	0.30	11.16	0.03
3	63.5	23.0	4.52	8.67	0.52
4	62.8	22.9	4.78	8.51	0.56
5	68.5	19.4	5.75	10.88	0.53
6	68.0	20.0	0.83	10.66	0.08
7	58.1	26.3	8.90	6.31	1.41
8	58.1	26.2	8.49	6.44	1.32
9	68.7	18.9	0.09	11.28	0.01
10	60.1	24.5	6.48	7.50	0.86
11	59.6	25.3	7.57	7.02	1.08
12	57.8	26.2	8.62	6.35	1.36

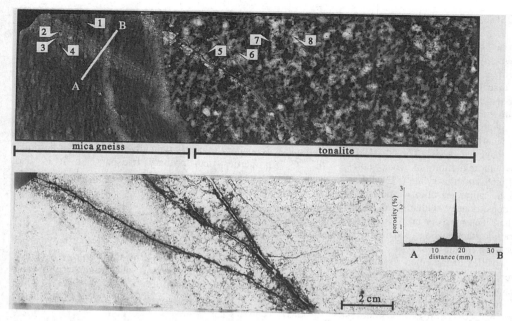

Figure 1. Photograph (top) and autoradiogram (bottom) of sample 161 m. EDS analyses are from points 1 - 8 (Table 1) and porosity profile from line A - B.

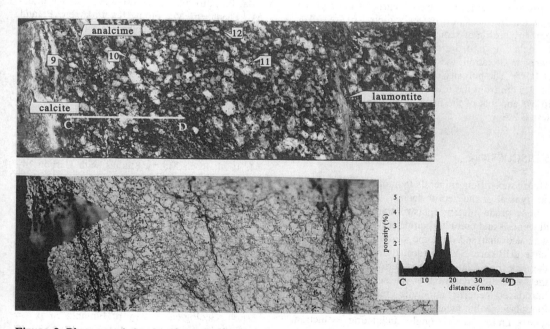

Figure 2. Photograph (top) and autoradiogram (bottom) of sample 242 m. EDS analyses are from points 9 - 12 (Table 1) and porosity profile from line C - D.

4 POROSITY MEASUREMENTS

The porosity of the samples was determined with the ^{14}C-polymethylmethacrylate (^{14}C-PMMA) method (Hellmuth et al. 1993, 1994), which enables the spatial distribution of the pore space and the heterogeneities of porosity to be studied on submillimetric to centimetric scales. Features limited in size by the range of ^{14}C beta radiation can be measured afterwards with autoradiography and digital image analysis. The ^{14}C-PMMA method involves impregnation of rock cores with ^{14}C labelled methylmethacrylate (^{14}C-MMA) in vacuum, irradiation polymerization, autoradiography and optical densitometry with digital image processing techniques. The porosity profiles can be scanned on the autoradiographs using rectangular frames across water-bearing fissures.

The porosities of the filled fissures in sample 161 m ranged from 2% to 5%. The increased porosity was observed to distances of 3 to 7 mm from the fissure, being 0.4% in mica gneiss. A zone of uniform increased porosity was not seen in the tonalite part of sample 161 m, but autoradiographs revealed a few areas at distances of 3 to 5 mm from the fissure in which porosities were 0.5%. The porosities of unaltered, fresh matrix of mica gneiss and tonalite were 0.1 % in sample 161 m. The porosity profile of sample 242 m shows that porosity is higher next to the conducting fracture owing to feldspar alteration, the range there being from 0.4% to 0.9%. The porosity of the filled fissure at a depth of 1.5 cm from the fracture surface ranged from 2% to 6% and that of the fresh tonalite in sample 242 m was 0.2%.

5 DISCUSSION

The fissure-filling minerals laumontite and analcime are typical of the zeolite facies, which presents the lowest grade metamorphism, transitional between diagenesis and prehnite-pumpelleyite facies (prehnite was also analysed on some samples taken from the same drill core, SY7). Warm fluids clearly affected the rock for a relatively short period during and after the last brittle deformation phase. Winkler (1967) introduced the term laumontite-prehnite-quartz facies to replace zeolite facies. In the Syyry samples all the quartz grains have strongly undulating extinction, showing recrystallization at high pressure. Quartz is thus older than laumontite.

Zeolitization in the Syyry bedrock is most likely a replacement reaction, in which the needed Ca derives from the plagioclase surrounding fissures. The Na needed for analcime may originate from the small mafic intrusions in the vicinity.

This kind of alteration is of great, and for the present partly unknown significance to the disposal of radioactive waste. While an increase in porosity means new paths for the diffusion of radionuclides, on the other hand the greatly increased internal surface area significantly enhances sorption possibilities in comparison to intact rock matrix. Secondary minerals typically possess better sorption properties than primary feldspars, so it is also necessary to study the alteration of bedrock outside the permeable fractures.

6 ACKNOWLEDGEMENTS

This work was supported by the Ministry of Trade and Industry of Finland and Radioactivity and Nuclear Safety Authority.

7 REFERENCES

Anttila, P., Kuivamäki, A., Lindberg, A., Kurimo, M., Paananen, M., Front, K., Pitkänen, P. and Kärki A. (1993). The Geology of the Syyry Area, Summary Report. Nuclear Waste Commission of Finnish Power Companies, Report YJT-93-19, 40 p.

Hellmuth, K.H., Siitari-Kauppi, M. and Lindberg, A. 1993. Study of Porosity and Migration Pathways in Crystalline Rock by Impregnation with ^{14}C-polymethylmethacrylate. Journal of Contaminant Hydrology, 13: 403-418

Hellmuth, K.H., Lukkarinen, S. and Siitari-Kauppi, M. 1994. Rock Matrix Studies with Carbon-14-Polymethylmethacrylate (PMMA); Method Development and Applications. Isotopenpraxis Environ. Health Stud., 30: 47-60

Kärki, A. (1991). Structural interpretation of Syyry in Sievi. Kivitieto Oy. TVO/Site investigations, Work report 91-06, 53 p. (in Finnish).

Salli, I. (1967). Pre-quaternary rocks of Lestijärvi and Reisjärvi map sheet areas. Explanation to the maps of pre-quaternary rocks. Map sheets 2341 and 2343. Geological map of Finland 1:100000. Espoo, Geological Survey of Finland, 37 p. (in Finnish, English summary).

Winkler, H.G.F. (1967). Petrogenesis of metamorphic rocks. Springer-Verlag, New York, 237 p.

Hydrocarbon gases and fluid evolution in very low-grade metamorphic terranes: A case study from the Central Swiss Alps

M. Mazurek & H. N. Waber
GGWW, Institutes of Geology and of Mineralogy and Petrology, University of Bern, Switzerland

A. Gautschi
Nagra, Wettingen, Switzerland

ABSTRACT: In a marl formation that was subjected to low-temperature metamorphism, free gases and groundwaters were sampled and compared with the bulk and isotopic compositions of fluid inclusions. Results indicate different sources for gas (derived thermally from kerogen) and water (which represents a mixture of connate and metamorphic water). No indications exist for later interactions of the rocks with meteoric waters.

1 INTRODUCTION

Most investigations of hydrocarbon reservoirs in geologic formations relate to environments with high fracture or matrix porosities and permeabilities. Only in recent years has more attention been paid to the study of fluids in shales and other rock types so far regarded as seals. This paper investigates the evolution of hydrocarbon gases and aqueous fluids in a very low-permeability marl formation, which usually would be considered as a gas seal. The opportunity to conduct such a study is related to a project focused at the site investigation for a repository for low- and intermediate level radioactive waste conducted by Nagra (National Cooperative for the Disposal of Radioactive Waste) at Wellenberg, Central Swiss Alps. A remarkable dataset from field observations, borehole tests and laboratory studies has been collected to characterize rocks and fluids, their past evolution and their present interactions at Wellenberg (Nagra 1997). A total of 7000 m of pre-Quaternary rocks has been cored in 7 boreholes to date, and a large number of groundwater and gas samples have been collected. Extensive hydraulic testing was performed in all boreholes, comprising packer tests, fluid logging and long-term monitoring.

2 GEOLOGIC FRAMEWORK

The Palfris formation (Early Cretaceous) is a marl that is being studied as a potential rock unit to host and isolate waste mainly because of its very low permeability. It is one of the major décollement horizons of the Helvetic Alps (which consist of a layered sequence of clays/marls and limestones). While the sedimentary thickness is only ca. 200 m, the Palfris marls at Wellenberg have been accumulated by tectonic processes (folding, imbrication) to a large mass exceeding 1000 m in thickness. A number of Alpine vein mineralizations occur, and a relative time sequence can be established on the basis of cross-cutting relationships.

During the late Alpine orogenic movements that culminated in the emplacement of the Helvetic nappe pile some 20 Ma b.p., all formations were subjected to a very low-temperature metamorphism (Wellenberg is located close to the Alpine kaolinite → pyrophyllite isograd, cf. Frey 1987). Fluid inclusions indicate temperatures of about 200-250 °C and the existence of a two-phase fluid regime consisting of a methane-rich gas and an aqueous phase (Diamond 1998). The results of vitrinite reflectance analyses show that all the samples underwent an advanced stage of catagenesis with R_0 values between 1.5-2%, i.e. above the oil generation window. Visual kerogen analysis shows complete absence of fluorescence (and therefore absence of aromatic hydrocarbons). The original kerogen type is II or II-III.

Present-day groundwaters and gases reside mainly in the microporous rock matrix (porosity 1-3 vol%) and, quantitatively less important, in Alpine fluid inclusions and in brittle faults that evolved during post-Alpine uplift. Water/rock interactions during and after brittle deformation occurred but resulted in only very limited mineralization of fracture walls.

3 FLUID REGIME AND GAS SAMPLING

Different types of groundwaters are observed in the

marl. The central part of the marl body contains a Na-Cl water (salinity 0.2 molal) rich in dissolved methane and is interpreted as connate seawater diluted and modified by dehydration reactions during metamorphism (Pearson & Scholtis 1995, Diamond 1998). Druses occur in the veins, and these are filled with a methane-rich gas. Water flow, if any at all, occurs in fractures but not in the highly impermeable rock matrix.

A study by Gautschi et al. (1990) dealing with samples from the Palfris formation indicated that gas-rich fluid inclusions are very abundant in calcite veins, and the isotopic signatures of these hydrocarbon gases are consistent with a thermal, i.e. low-grade metamorphic, origin by cracking of kerogen. For the present study, gases were sampled in the following settings:

1. Free gas sampled at the well-head and down-hole
2. Gas dissolved in groundwater (water sampled at formation pressure and degassed in the laboratory)
3. Gas trapped in fluid inclusions, mainly in vein-filling calcite (liberated by cold crushing or thermal decrepitation of rock samples).

Whether the free gases represent gas dissolved in the groundwater under formation pressures and temperatures or an *in-situ* free gas phase is not clear.

4 RESULTS

4.1 Rock matrix and veins (fluid inclusions)

Gas content: The gas contents per unit rock volume vary over 3 orders of magnitude. Most of the gas volume trapped in the rock originates from fluid inclusions in calcite veins (mean: 1270 cm^3 [STP] per dm^3 of rock), while the rock matrix (mean: 290) contains substantially less gas.

Bulk gas composition: The gas phase in fluid inclusions is dominated by methane with CO_2 and higher hydrocarbons as minor constituents. Other gases, such as N_2 and H_2, have been analyzed but are all below detection. No differences between vein and whole-rock samples can be identified.

Isotopic gas composition: As shown in Table 1, fluid inclusions have average methane compositions of $\delta^{13}C = -34.4$ ‰ and $\delta^2H = -107$ ‰. No systematic discrimination is observed between veins and whole-rock materials or between different vein generations. Ethane and propane have slightly heavier compositions of C.

4.2 Gas compositions in groundwaters and in the free gases

Groundwater samples taken down-hole and degassed

Table 1. Average bulk and isotopic compositions of different types of gas. Full ranges of bulk compositions are given in brackets. Isotopic compositions of C and O are expressed in ‰ relative to PDB, those of H relative to SMOW.

	rock matrix and veins (fluid inclusions)	free well-head gas	total gas dissolved in groundwater	free gas (down-hole)
meth. / (eth.+ prop.)	189 (4-545)	2547 (82-11309)	1585 (873-1937)	420 (249-474)
CO_2, vol% @ STP	1.47 (0-3.77)	0.24 (0-0.86)	0.72 (0-1.81)	0.05 (0-0.10)
methane, vol% @ STP	95.5 (76.1-99.8)	99.6 (98.7-99.96)	99.2 (98.1-100)	99.7 (99.6-99.8)
ethane, vol% @ STP	2.0 (0.2-16.1)	0.16 (0-1.21)	0.05 (0-0.11)	0.24 (0.20-0.36)
propane, vol% @ STP	0.54 (0-3.07)	0.005 (0-0.019)	0.001 (0-0.003)	0.007 (0-0.037)
butane, vol% @ STP	0.27 (0-1.57)	0.001 (0-0.004)	b.d.	0.003 (0-0.02)
$\delta^{13}C$ (meth.)	-34.4	-37.7	-33.6	-31.7
δ^2H (meth.)	-107	-148	-137	-137
$\delta^{13}C$ (eth.)	-21.9	-24.0	-25.5	-24.0
$\delta^{13}C$ (prop.)	-22.6	-25.5	-26.4	-24.6

in the laboratory indicate that methane is close to saturation. The gas composition in groundwaters is dominated by methane and compositionally similar to the gas phase present in fluid inclusions (Table 1). All isotopic gas compositions listed in Table 1 are within the ranges defined by the fluid inclusions, suggesting a common genetic history.

4.3 Isotopic composition of groundwaters and of aqueous fluid inclusions

Only very small amounts of the Na-Cl groundwater could be sampled in the boreholes, and the degree of contamination by drilling fluids is high. The stable isotopic composition of the Na-Cl groundwater plots on a line with a slope of 1.8 to the right of the global meteoric water line. The correction of the measured isotopic compositions to *in-situ*, contamination-free conditions yields values of $\delta^2H = -40$ to -10 ‰$_{SMOW}$, which also agrees with data obtained by vacuum distillation of porewater from rock matrix samples.

The isotopic composition of H from aqueous fluid inclusions varies within a wide range between -45 and +3 ‰, with an average at -19 ‰ (Table 2).

4.4 Bulk composition of fluid inclusions

The ratio of water and methane in fluid inclusions was measured by thermal decrepitation of vein quartz and calcite, and results are given in Table 2. The relative proportion of H_2O in the total fluid varies widely between 53 and 94 mol%, and the proportion

Table 2. Bulk composition of fluids derived from thermal decrepitation of vein calcite and quartz, together with the isotopic composition of H in water from aqueous inclusions.

host mineral	methane, mol%	CO_2, mol%	water, mol%	δ^2H of water, ‰ SMOW
vein calcite	4.9	1.5	93.5	-23
vein calcite	4.5	1.5	94.0	-27
vein calcite	22.1	1.0	76.9	-23
vein calcite	12.0	3.6	84.3	1
vein calcite	13.7	2.7	83.6	3
vein calcite	18.9	0.9	80.2	2
vein calcite	11.2	1.1	87.6	-27
vein calcite	14.1	1.4	84.5	-31
vein calcite	29.6	1.8	68.6	-29
vein calcite	43.9	3.5	52.6	-20
vein calcite	44.0	1.5	54.5	-20
vein calcite	10.8	1.0	88.2	-9
vein calcite	6.3	0.8	92.9	-5
vein calcite	21.3	1.6	77.1	-15
vein quartz	12.6	2.4	85.0	-45
vein quartz	18.9	1.3	79.8	-39
vein quartz	12.6	0.9	86.5	-17

of methane is 5 - 44 mol%. The inhomogeneous distribution of gas-rich and aqueous inclusions on a scale of centimeters is consistent with microthermometric observations that show small-scale heterogeneity of the ratio of gas-rich and aqueous fluid inclusions (Diamond 1998).

5 DISCUSSION

5.1 Gas provenance

Table 1 shows that the compositions of all types of gas overlap widely. The similarity of the bulk and isotopic compositions of gases in fluid inclusions and the free formation gas suggests a common origin from the Alpine metamorphic event (see also Gautschi et al. 1990). Ballentine et al. (1994) make the same conclusion on the basis of the identity of the methane / $^{40}Ar_{rad}$ ratios in all types of gas.

A plot discriminating different processes of hydrocarbon gas generation (Whiticar et al. 1986) is shown in Figure 1a. The combination of high $\delta^{13}C$ and high δ^2H of methane is indicative of a thermal origin for all gases during Alpine metamorphism, while bacterial effects can be excluded.

The compositions of fluid inclusions show a wide range of variability with respect to relative proportions of gas species as well as to the gas/water ratios. Such variability is best explained in terms of local production and buffering, while external sources flushing the formation would be expected to produce a more uniform pattern of compositions.

There is a trend for very argillaceous rock matrix samples to have the relatively highest $\delta^{13}C$ values of methane in fluid inclusions. In these samples, clastic (i.e. terrestrial) organic source material dominates over the marine component, and this material is known to have higher δ values, which is consistent with the isotopic data in fluid inclusions. This is another argument for local production and buffering of hydrocarbon gases.

From worldwide observations in several gas fields, Faber (1987) developed an empirical relation between the isotopic composition of C in hydrocarbons and vitrinite reflectance of the source rocks, and so $\delta^{13}C$ of methane, ethane and propane can be used as indirect monitors of thermal maturity. Figure 1b shows that the Wellenberg data can be related to marine kerogen type II/III source materials (with possible admixtures of terrestrial type III materials, i.e. continental input), and most analyses are related to values of vitrinite reflectance above

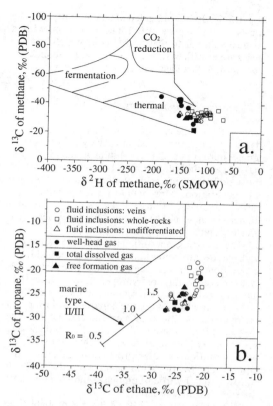

Figure 1. Diagnostic plots of hydrocarbon gas provenance and conditions of generation (references see text).

$R_0=1.5$ %. It is concluded that the source of the gases is an overmature rock formation with a maturity similar to that of the rocks which host the gases today ($R_0 = 1.5 - 2$ %), and so these data are consistent with a local source.

5.2 Sources of Alpine fluids

Isotope thermometers can be applied on the available dataset. C in calcite - methane pairs yields consistent temperatures of ca. 200°C (with measured $\delta^{13}C_{calcite} = 0\text{-}1‰$ and the calibration of Bottinga 1969), which most probably represents the conditions of Alpine gas generation. Moreover, this temperature is a further argument for in-situ gas generation because higher values would be expected for gas generated at greater depth.

The isotopic compositions of C and H in methane and calcite are remarkably constant and indicate generation from one single source. In contrast, the isotopic composition of hydrogen in water from fluid inclusions shows a wide scatter (Table 2), and there is no correlation with the smaller variations in isotopic compositions of methane and calcite. The measured fractionations of ^2H between methane and water are >74 ‰ in all cases but two (Tables 1, 2), which indicates disequilibrium because the equilibrium fractionation never exceeds 74 ‰. It follows that hydrocarbon gases and water have different in-situ sources.

The wide scatter of δ^2H of water is probably due to the mixing of two end-member fluids. Assuming that processes predating Alpine metamorphism had no significant effects on porewater compositions (geological evidence suggesting such processes is lacking), these end-members could be the connate porewater (higher δ^2H) and the Alpine metamorphic water (lower δ^2H). Dilution of connate water by dehydration reactions during metamorphism is consistent with the salinities (about half that of seawater) of both water sampled in fluid inclusions and groundwater samples taken down-hole.

6 CONCLUSIONS

Three sources of fluids were identified in the Palfris formation:
1. Connate water, i.e. Cretaceous seawater
2. Water from metamorphic dehydration of clay minerals
3. Methane-rich hydrocarbon gas derived from thermal decomposition of kerogen.

No indications exist for interactions of the rocks with meteoric water, neither from isotopic nor from mineralogic evidence. The similarity of the chemical and isotopic characteristics of Alpine fluids trapped in fluid inclusions compared to groundwater and free or dissolved gas are indicative of a common genetic history of these fluids. The conclusion is that the central, very low-permeability part of the Palfris formation contains exclusively Alpine fluids.

7 REFERENCES

Ballentine, C.J., Mazurek, M. & Gautschi, A. 1994. Thermal constraints on crustal rare gas release and migration: Evidence from Alpine fluid inclusions. *Geochim. Cosmochim. Acta* 58:4333-4348.

Bottinga, Y. 1969. Calculated fractionation factors for carbon and hydrogen isotope exchange in the system calcite-carbon dioxide-graphite-methane-hydrogen-water vapor. *Geochim. Cosmochim. Acta* 33:49-64.

Diamond, L.W. 1998. Underpressured paleofluids and future fluid flow in a planned radwaste host rock. *Proc. 9th Int. Symp. Water-Rock Interaction (this volume)*.

Faber, E. 1987. Zur Isotopengeochemie gasförmiger Kohlenwasserstoffe. *Erdöl Erdgas Kohle* 103, 210-218.

Frey, M. 1987. Very low-grade metamorphism of clastic sedimentary rocks. In: Frey, M. (ed.) *Low-grade metamorphism: 9-58*, Glasgow: Blackie.

Gautschi, A., Faber, E., Meyer, J., Mullis, J., Schenker, F. & Ballentine, C. 1990. Hydrocarbon and noble gases in fluid inclusions of Alpine calcite veins – Implications for hydrocarbon exploration. *Bull. Swiss Assoc. of Petroleum Geol. and Eng.* 56:13-36.

Nagra 1997. Schlussbericht zu den geologischen Oberflächenuntersuchungen am Wellenberg. *Nagra Technical Report* NTB 96-01, Wettingen, Switzerland.

Pearson, F.J. & Scholtis, A. 1995. Controls on the chemistry of pore water in a marl of very low permeability. *Proc. 8th Int. Symp. Water-Rock Interaction:*35-38.

Whiticar, M.J., Faber, E. & Schoell, M. 1986. Biogenic methane formation in marine and freshwater environments: CO_2 reduction vs. acetate fermentation - Isotope evidence. *Geochim. Cosmochim. Acta* 50:693-709.

Chemical zonation of contact metamorphic garnet: A record of fluid-rock interaction, Juneau gold belt, SE Alaska

H.H. Stowell
Department of Geology, University of Alabama, Tuscaloosa, Ala., USA

T. Menard
Department of Geology and Geophysics, University of Calgary, Alb., Canada

ABSTRACT: Metamorphic aureoles in the Grand Island diorite complex contain garnet with highly variable major- and trace-element zonation. Garnets in staurolite-bearing pelite have high Grs - low Y cores, low Grs - high Y midsections, and outer sections of increasing Grs and decreasing Y toward the rims. Garnets from clinozoisite-bearing calcic pelite have oscillatory zonation in Grs, Y, and Yb. Grains adjacent to veins have three peaks in Grs; grains further from veins have two. Garnet zoning may be explained by episodic interaction with a Ca-bearing fluid, with or without mobility of other components, producing episodic growth of garnet, plagioclase, and epidote. Subsequent to initial garnet growth, garnet zone and lower grade minerals grew during several fluid-infiltration pulses. Fluid flow pulses or changes in fluid compositions may be related to episodic plutonism in the complex followed by emplacement of later plutons to the east and/or influx of regional metamorphic fluids from below. Late fluid-infiltration may be related to mineralization in the Juneau gold belt.

1 INTRODUCTION

The chemical zonation of garnet from contact metamorphic rocks can provide information about the temporal variability of fluid-rock interaction (*e.g.*, Stowell *et al.* 1996). Oscillatory major- and trace-element zoning in garnet from contact metamorphic aureoles in the Grand Island diorite complex, SE Alaska, suggest several episodes of fluid infiltration and metasomatism during garnet zone metamorphism.

2 GEOLOGIC SETTING

The Grand Island diorite complex south of Juneau, Alaska, consists of diorite and quartz diorite plutons that intruded greenschist facies rocks in the western metamorphic belt of the Coast plutonic-metamorphic complex. The igneous assemblage epidote + garnet + hornblende and thermobarometry of the contact-metamorphic rocks indicate crystallization at >6 kbars (Zen & Hammarstrom, 1984; Inman, 1992, Stowell & Crawford, submitted). Garnet-biotite granodiorite from the Grand Island diorite complex has been dated at 94 Ma (U/Pb zircon; Brew *et al.*, 1992). To the north, along strike from the complex, are the largest mines in the Juneau gold belt, the foremost lode-gold-producing district in Alaska. The age of mineralization is estimated at ~55 Ma, when crustal stresses changed in response to shifting motions of plate in the northeast Pacific (Goldfarb *et al.*, 1991). Mineralization may have resulted from rapid metamorphic dewatering and focusing of fluid flow along faults (Goldfarb *et al.*, 1991). The Treadwell deposit, near Juneau, is hosted by diorite similar in age (91±2 Ma; Gehrels, submitted) and mineralogy to the Grand Island diorite complex; other deposits are in metamorphic rocks.

The country rocks around the Grand Island diorite complex include pelite, calcic pelite, marble, amphibolite, and layers and pods of calc-silicate minerals. These rocks were affected by the following metamorphic events (Brew *et al.*, 1992; Stowell & Crawford, submitted): a low-grade regional event that produced chlorite-zone mineral assemblages, contact metamorphism around the diorite plutons, and finally, low-grade metamorphism associated with crustal thickening or emplacement of plutons to the east.

Contact metamorphism caused by intrusion of plutons in the Grand Island diorite complex produced narrow, 20-200 m aureoles with peak metamorphic grade increasing from chlorite-zone outside the aureoles up to staurolite + kyanite zone adjacent to the plutons (Stowell & Inman, 1991; Inman, 1992).

3 METHODS

Major- and trace-element chemical compositions of garnet and other minerals were determined using wavelength-dispersion spectrometry on a JEOL 8600 electron probe microanalyzer. Operating conditions were 20 kV accelerating potential, 50 nA beam current, 60 s count time, and 2-15-μm beam diameter. Standards and ZAF methods were used to calculate compositions (Doyle & Chambers, 1981).

4 GARNET COMPOSITIONAL ZONING

4.1 Garnet in pelite

Staurolite-zone pelite samples were collected ca. 10 m from diorite plutons on Stephens Passage (e.g., sample 8d). The matrix assemblage is staurolite + garnet + biotite + plagioclase + muscovite + quartz + graphite + retrograde chlorite. The staurolite was partially replaced by biotite and small, euhedral crystals of garnet. Chlorite and muscovite partially replaced and pseudomorphed staurolite, garnet, and biotite, and overgrew tectonite fabrics.

Compositional zoning divides the garnet into three zones. The garnet cores have a high Grs content (X_{Grs}=0.22), middle portions have less (0.06<X_{Grs}<0.11), and outer zones of increasing Grs (to X_{Grs}=0.16) at the rim (Fig. 1). The sharp decrease of Grs from the core to the mid-region could reflect resorption of garnet during production of staurolite or loss of epidote from the assemblage. Garnet resorption, however, is unlikely because the garnet core is euhedral. Although there are too few inclusions to demonstrate a change of assemblage, the later interpretation is preferred. Y and Yb concentrations are very low in the Grs-rich core, increase sharply outside the core, and decrease in the outer zone, near the rim (Figs. 1 & 2).

Figure 2. Y X-ray map of garnet grain in Figure 1. High concentrations are shown in light grays. Grain is 500 microns in diameter.

The major- and trace-element zonation suggests that the garnet core grew from Ca-rich, and Y- and Yb-poor solid phases (such as calcite) or that the core grew in equilibrium with a fluid of this composition. The rim, by contrast, may have grown in an assemblage lacking minerals as rich in calcium. Metasomatism during the growth of garnet cannot be dismissed, but the major- and trace-element zonation are compatible with growth without significant changes in bulk rock composition.

4.2 Garnet in calcic pelite

Samples of calcic pelitic schists, were collected within 25 m of a diorite pluton about 6 km from sample 8d. The peak metamorphic assemblage in these samples was garnet + biotite + plagioclase + epidote + quartz + muscovite + titanite, with minor tourmaline, apatite, and zircon. Grains of garnet in schist are small, 0.2 - 0.5 mm, contain inclusions of quartz and epidote, and some have been partially replaced by chlorite. Allanite-rich epidote inclusions adjacent to euhedral cores of high-Grs garnet are inferred to have grown contemporaneously with cores of allanite found in matrix epidote. Pyrrhotite, chalcopyrite, and sphalerite replaced the rim zones of the peak metamorphic minerals but do not occur as inclusions; thus, the sulfides probably were added to the rock during retrograde alteration. Similar sulfide textures and assemblages also occur locally in the plutons of the Grand Island diorite complex. The retrograde alteration may correlate with more extensive alteration and gold mineralization in the Treadwell deposit, 20 km to the north along strike.

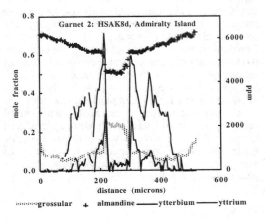

Figure 1. Garnet compositional zoning, pelite (152 analyses).

Garnet has oscillatory zoning in Grs (X_{Grs}=0.10-0.35) and antipathetic zoning in Alm, Y, and Yb (Figs. 3 & 4). Sps decreases and Fe/(Fe+Mg) increases from the core to the rim. Grains of garnet are locally more abundant in selvages adjacent to quartz veins and have three peaks in Grs content, those grains more than 2 m from the veins have only two peaks.

sericite during later retrograde metamorphism (<400°C), indicates later fluid infiltration. Thus, there were four or five fluid infiltration events.

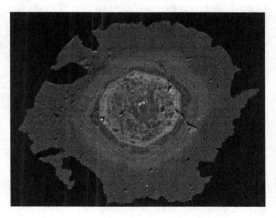

Figure 4. Y X-ray map of garnet grain from Figure 3. High concentrations are shown in light grays. Grain is 500 microns in diameter.

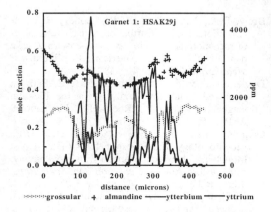

Figure 3. Garnet compositional zoning, calcic pelite (100 analyses).

Plagioclase is zoned from oligoclase to albite, from core to rim. It has equilibrium textures with retrograde minerals; therefore, all the plagioclase currently in the samples likely postdates the formation of garnet.

Na, Y, Yb, and Zr are negatively correlated with Grs and positively correlated with Mn in the garnet. Na in garnet may mirror Na in plagioclase (Hickmott & Spear, 1992). In this case, Grs in garnet and anorthite in plagioclase (subsequently replaced or resorbed) may have increased and decreased together. Consequently, the Y, Yb, and Zr zonation may reflect simultaneous growth of epidote and high Grs garnet alternating with epidote consumption during growth of low Grs garnet. Ti concentrations in the garnet, on the other hand, are positively correlated with Grs, which may reflect consumption of ilmenite or rutile to liberate Ti during growth of high-Grs garnet, or metasomatic addition of Ti.

Matrix epidote contains ca. 10 μm, subhedral cores of zoned, REE-rich epidote (0.2-0.3 mole fraction pistacite) that are generally zoned with increasing La and decreasing Sm and Eu from the center to the rim. A 1 μm wide ring, ca. 3 μm from the center of the grains, has high Ce concentrations that may reflect epidote consumption and concentration of REE at the grain edges. The rims of matrix epidote are unzoned clinozoisite.

Peak metamorphic minerals in the schist were extensively retrograded to clinozoisite + albite + muscovite + sulfide. Local growth of chlorite and

4.3 *Garnet in veins*

Andradite-rich garnet in quartz veins, from location 29 discussed above, is less strongly zoned than any garnet grains from the schist, but also shows numerous oscillations in Grs content (X_{Grs}=0.40-0.58); Adr is antipathetic (X_{Adr}= 0.04-0.17). These garnets contain lower concentrations of Y and Yb (<700 ppm), higher concentrations of Ti, and similar concentrations of Zr than garnet in the schist. The low concentrations of Y and Yb in Adr-rich garnet in veins compared to garnet in schist suggests that: 1) these elements were less abundant in the Ca-rich vein-forming fluid than in the schist, or 2) these elements were strongly partitioned into garnet from the schist.

5 DISCUSSION

Garnet compositional zoning could have been produced by metasomatism, or changes in pressure and temperature. The oscillatory zoning, observed in calcic pelite, would require large changes in temperature (>200°C at constant pressure) or pressure (> 2 kbar). Such large episodic changes of pressure or temperature seem unreasonable for a contact-metamorphic aureole with no other physical evidence for such changes. Large temperature changes would drive a sequence of reactions not observed in the Grand Island diorite complex. Therefore, the patterns of compositional zoning in the garnet are not the result of changes in pressure and temperature. Changes of mineral assemblage can also affect the compositional zoning of minerals, but this is an unlikely explanation for the oscillatory zoning, because it would require the improbable cyclic addition and loss of at least one mineral.

The simplest interpretation is that the sharp increases of Grs and Ti in garnet reflect Ca-metasomatic events that drove metamorphic reactions, and produced garnet (with high Grs content), plagioclase (consuming Na), and epidote (consuming Y and Yb). Garnet growth may have been rapid, as suggested by the inclusion of tiny grains of allanite-rich epidote preferentially in Grs-rich zones. Garnet growth continued with decreasing Grs content after each event. Ca metasomatism is supported by the Adr-rich garnet in veins.

Comparison of garnet from within veins, adjacent to veins, and far from veins provides information about the nature and extent of fluid flux. Adr-rich garnet within the veins is the most Ca-rich of the three and contains the lowest concentrations of Y and Yb. Garnet in vein selvages has three peaks in Grs concentration suggesting that these samples were effected by three phases of Ca-metasomatism during garnet growth. Garnet further from these veins has only two peaks in Grs content suggesting that these samples were affected by fewer phases of Ca-metasomatism during garnet growth.

Fluid flow and metasomatism likely varied temporally in response to successive intrusive events. In addition, variation in fluid flow and metasomatism between sample locations may be related to variation in the number of overlapping aureoles, or may be related to fluid flow through different rock types. Samples from a sequence that contains interbedded calc-silicates, experienced Ca-metasomatism during garnet growth and during retrograde metamorphism; samples from a sequence lacking interbedded calcic rocks did not experience Ca-metasomatism.

6 CONCLUSION

Major, minor, and trace element garnet chemistry from matrix, vein selvage, and vein assemblages is compatible with Ca-metasomatism with little or no trace element mobility during contact metamorphism. The chemical zonation of garnet from these assemblages is inferred to result from infiltration of several pulses of Ca-rich fluid.

7 ACKNOWLEDGMENTS

Partial support for the research was provided by the U.S. Geological Survey and Oak Ridge Associated Universities. M. Bersch provided invaluable help with analyses obtained from the JEOL electron probe microanalyzer at the University of Alabama.

REFERENCES

Brew, D.A., Himmelberg, G.R., Loney, R.A., & A.B. Ford 1992. Distribution and characteristics of metamorphic belts in the south-eastern Alaska part of the North American Cordillera. *J. Metamorph. Geol.* 10: 465-482.

Doyle, J.H., & W.F. Chambers 1981. ZAF80: an improved quantitative analysis for the Flextran language systems. Rockwell International RFP3215.

Gehrels, G.E. submitted. Geology and U/Pb geochronology of the western flank of the Coast Mountains between Juneau and Skagway, southeastern Alaska. *Geol. Soc. Am. Spec. Pap.*

Goldfarb, R.J., Snee, L.W., Miller, L.D., & R.J. Newberry 1991. Rapid dewatering of the crust deduced from ages of mesothermal gold deposits. *Nature* 54: 296-298.

Hickmott, D.D. & F.S. Spear 1992. Major and trace element zoning in metamorphic garnets from western Massachusetts. *J. Petrol.* 33: 965-1005.

Inman, K. 1992. Thermobarometric constraints on the emplacement conditions of the epidote-bearing Grand Island pluton near Juneau, SE Alaska. *M.S. thesis*. Univ. Alabama: 193.

Stowell, H.H., & M.L. Crawford. submitted. Metamorphic history of the western Coast plutonic-metamorphic complex, western British Columbia and southeast Alaska. *Geol. Soc. Am. Spec. Pap.*

Stowell, H.H., & K. Inman 1991. Comparative thermobarometry of contact metamorphic rocks: Grand Island pluton near Juneau, SE AK. *Geol. Soc. Am. Abstr. Prog.* 23: A445.

Stowell, H.H., Menard, T. & C.K. Ridgway 1996. Ca-metasomatism and chemical zonation of garnet in contact metamorphic aureoles, Juneau gold belt, southeastern Alaska. *Can. Min.*, 34, 6: 1195-1209.

Zen, E-A., & J.M. Hammarstrom 1984. Magmatic epidote and its petrologic significance. *Geology* 12: 515-518.

Low-grade oceanic metamorphism and tectonic thickening of the oceanic crust from the Eltanin Fracture Zone (Pacific ocean)

I.A.Tararin
Far East Geological Institute, Vladivostok, Russia

ABSTRACT: Chlorite-epidote-amphibole schist, ferrogabbro-diorite, plagiogranite, and metabasalt of the ophiolite assemblage, dredged from the seamount of the South-Pacific Basin from the north-west propagation of the Eltanin Fracture Zone, underwent metamorphism in two stages. The earliest metamorphism resulted in growth of actinolite, ferro-actinolite, chlorite, albite, and epidote and took place under low greenschist facies conditions (T=340-380°C, P<1 kb as indicated by the amphibole chemistry) corresponding to low-grade sea-floor metamorphism. These low-grade metamorphic assemblages were overprinted by high-pressure metamorphism of lower amphibolite facies (T=520-570°C, P=4-5 kb), presumably resulting from the tectonic thickening of the oceanic crust during its migration north-west from the East Pacific Rise. During this metamorphic episode ferro-tschermakite and ferro-pargasite formed, overprinting the earlier actinolite and ferro-actinolite.

1 INTRODUCTION

According to modern ideas (Miyashiro, 1973; Coleman, 1977; Dobretsov, 1995; Dobretsov et al. 1989, 1994; Humpris and Thompson, 1978; Silant'ev, 1984, 1995; Kawachata et al. 1987; Kurnosov, 1986), sea-floor metamorphism of the oceanic crust shows some specific features which distinguish it from continental crustal metamorphism. The differences are first of all seen in P-T conditions of metamorphism and its geochemical peculiarities. As Miyashiro (1973) emphasized, the presence of calcic plagioclase and amphibole assemblage is an evidence of sea-floor metamorphism testifying low P_S and P_{CO_2}. Even in intensely metamorphosed basalts, igneous textures are still preserved and this, combined with textural and mineralogical evidence, are typical features of sea-floor metamorphism. Oceanic metabasalts actually show a wider range of chemical composition than fresh basalts as a result of metasomatism accompanying hydrothermal metamorphism. Chemical changes seem to indicate that metabasalts of the oceanic crust become progressively Na-enriched and Ca-depleted as it is reacted at progressively higher temperature with the sea water that follows the spilitic trend of metamorphism (Cann, 1969; Vallance, 1974; Silant'ev, 1984). In contrast with the continents, in the oceanic crust the rocks formed under high pressure are rare. They are sampled from subduction zones at convergent plate margins, where metamorphism in the blueschist and eclogite facies (T=250-500°C, P=8-15 kb; Dobretsov, 1995; Korikovskii, 1995) took place in the downgoing oceanic plate. High pressure metamorphism characterizes also the fracture zones, where it is related to the tectonic thickening of the oceanic crust (Pushcharovsky, 1995; Pushcharovsky et al. 1991) or displacements of crustal and mantle blocks on low-angle overthrusts of decollement faults (Karson and Dick, 1984; Silant'ev, 1995).

2 GEOLOGICAL SUMMARY

An example of metamorphic rocks that have undergone low-grade sea-floor metamorphism of greenschist facies followed by high-pressure metamorphism of lower amphibolite facies is the metamorphic sequence of the crustal assemblage dredged from the seamount more than 2500 m high in the South-Pacific Basin on the north-west part of the Eltanin Fracture Zone (Pushchin et al. 1990). Chlorite-epidote-amphibole schist, ferrogabbro-diorite, plagiogranite, metadolerite, and metabasalt comparable with the upper part of an ophiolite section have been recovered on dredges N10-21 (lat 31°03.1'S, long 164°49.5'W, depth 4900-4800 m) and N10-24 (lat 31°04.6'S, long 164°49.8'W, depth 4100-3800 m). Metamorphic rocks have been collected in the form of small (up to 3-4 cm) angular and sub-rounded fragments of massive texture, forming the substrate of ferromanganese nodules and crusts commonly averaging 1-2 cm in thickness.

Chlorite-epidote-amphibole schists consist mainly of acid plagioclase (from An_{4-7} to An_{15-17} and even An_{29-40}) and green amphibole. Chlorite, epidote, and quartz are subordinate. Accessory minerals are magnetite, ilmenite, sphene, pyrite, and apatite. Rarely the relics of brown hornblende are found to coat and replace clinopyroxene, and also occur as independent crystals. Green amphiboles occur as aggregates of randomly-oriented fine-grained crystals, which together with chlorite and ilmenite have replaced igneous pyroxenes.

The amphibole grains are variably zoned from blue green and pale green cores to dark green rims. This zoning pattern and the compositions of successive amphiboles record the evolution of pressure-temperature conditions experienced by the rocks. Electron microprobe (Table 1) and textural data show that the amphibole crystals consist of actinolite or ferro-actinolite cores surrounded by more Al-rich rims (ferro-tschermakite or ferro-pargasite). Change of mineral assemblage observed under the microscope, and wide compositional variations of amphibole from metamorphic rocks, suggest two stages of metamorphism. The earliest metamorphism resulted in growth of albite, chlorite, epidote, and actinolite or ferro-actinolite under lower greenschist facies conditions (T=340-380°C, P<1 kb, using the empirical amphibole geothermobarometer of Mishkin (Mishkin, 1994; Tararin et al. 1995) corresponding to conditions of sea-floor metamorphism. The ferro-tschermakite and ferro-pargasite mantles of amphibole crystals are calculated to have formed at pressures of 3.9-4.8 kb and temperatures of 520-570°C (Table 1), in the second stage of metamorphism under lower amphibolite facies conditions. Experimental amphibole-plagioclase calibrations (Plyusnina, 1982) give much higher pressures of the superimposed metamorphism. So for Sample N10-21-3, where high-Al ferro-tschermakite associated with plagioclase $An_{29.3}$, calculated temperature is 540°C and calculated pressure is 8 kb. Similar P-T parameters (T=510°C, P>8 kb) have

Table 1. Representative microprobe analyses of amphiboles from metamorphic rocks of the NW part of the Fltanin Fracture Zone

	N10-21-6		N10-21-5					N10-21-3		N10-21-1	
	1	2	1	2	3	4	5	1	2	1	2
SiO_2	52.44	52.01	51.54	53.81	52.54	39.92	40.10	48.75	39.67	52.47	51.64
TiO_2	0.19	0.08	0.03	0.15	0.00	0.04	0.10	0.00	0.00	0.00	0.01
Al_2O_3	2.38	2.06	3.17	1.51	1.84	17.24	16.62	4.70	15.40	1.88	2.61
FeO	19.64	21.10	19.88	18.09	22.95	23.39	22.66	27.31	27.37	22.53	19.85
MnO	0.15	0.21	0.09	0.09	0.23	0.05	0.09	0.25	0.30	0.33	0.28
MgO	11.37	10.17	10.99	12.62	9.23	4.70	5.04	6.29	3.34	9.44	10.62
CaO	11.71	11.76	12.17	12.12	11.81	11.15	11.34	11.55	10.67	11.73	12.14
Na_2O	0.27	0.06	0.38	0.04	0.15	1.97	2.01	0.43	1.90	0.07	0.07
K_2O	0.11	0.12	0.08	0.06	0.09	0.49	0.46	0.11	0.28	0.09	0.12
Total	98.26	97.57	98.33	98.49	98.85	98.95	98.42	99.39	98.93	98.54	97.34
Si	7.665	7.732	7.589	7.799	7.773	5.965	6.078	7.318	6.005	7.761	7.668
Ti	0.025	0.007	0.003	0.017	0.000	0.004	0.009	0.000	0.000	0.000	0.001
Al^{IV}	0.335	0.268	0.411	0.201	0.277	2.035	1.922	0.682	1.995	0.239	0.312
Al^{VI}	0.083	0.100	0.132	0.062	0.100	1.007	0.975	0.147	0.751	0.095	0.137
Fe^{3+}	0.279	0.175	0.312	0.119	0.100	0.789	0.679	0.654	1.167	0.169	0.198
Fe^{2+}	2.120	2.441	2.135	2.073	2.734	2.145	2.289	2.763	2.292	2.614	2.274
Mn	0.018	0.027	0.008	0.008	0.029	0.004	0.008	0.032	0.036	0.041	0.034
Mg	2.476	2.251	2.411	2.721	2.036	1.050	1.139	1.404	0.753	2.080	2.355
Ca	1.835	1.872	1.922	1.884	1.868	1.787	1.840	1.856	1.731	1.859	1.939
Na	0.077	0.021	0.103	0.014	0.044	0.569	0.586	0.119	0.564	0.025	0.025
K	0.019	0.021	0.014	0.010	0.016	0.088	0.084	0.020	0.051	0.016	0.021
X_{Mg}	0.539	0.480	0.530	0.568	0.425	0.329	0.332	0.337	0.247	0.443	0.509
T,°C	415	350	375	340	350	550	575	430	520	350	360
P, kbar	<1	<1	<1	<1	<1	4.8	4.7	1.3	3.9	<1	<1
Amphibole	Act	fAct	Act	Act	fActHb	fTsch	fParg	fAct	fTsch	fAct	Act

Note: Sample N10-21-6, 21-5, 21-3 - chlorite-epidote-amphibole schist; N10-21-1 - plagiogranite. Temperature and pressure are determined using amphibole geothermobarometer (Mishkin, 1994; Tararin et al. 1995). Fe^{3+} is calculated using $Fe^{3+}=Al^{IV}-Al^{VI}-2Ti-Na(A)-K(A)+Na(M_4)$. Amphibole after Leake (1978): Act - actinolite, fAct - ferro-actinolite, fActHb - ferro-actinolitic hornblende, fTsch - ferro-tschermakite, fParg - ferro-pargasite. $X_{Mg}=Mg/(Mg+Fe^{2+})$.

been calculated for Sample N10-21-5, in which ferro-tschermakite and ferro-pargasite amphibole are associated with plagioclase $An_{14.2}$. Plagiogranite shows only the first stage of metamorphism, sea-floor metamorphism under the conditions of greenschist facies (T=350-360°C, P<1 kb; Table 1, Sample N10-21-1). This rock consists of plagioclase An_{14-2} (40-50%), quartz (40-50%), K-feldspar, pale green actinolite, chlorite, and epidote. Sphene, ilmenite (MnO=2.5 wt %), and apatite are common accessory minerals. The rocks have low K_2O and FeO, high Na/K ratio, low Ni, Co, Cr contents, and REE patterns (Pushchin et al. 1990) that compare closely with typical oceanic plagiogranite.

3 DISCUSSION AND CONCLUSION

Data obtained from mineral chemistry, and using the empirical amphibole geothermobarometer of Mishkin (Mishkin, 1994; Tararin et al. 1995) and experimental amphibole-plagioclase calibrations of Plyusnina (1982), indicate that metamorphic rocks of the ocean floor from the north-west part of the Eltanin Fracture Zone underwent low-grade sea-floor and superimposed high-pressure metamorphism, testifying to polymetamorphism of the rocks of the oceanic crust. Against the background of retrograde metamorphism that volcanic and plutonic rocks of the oceanic crust usually undergo, and that correspond to the conditions of zeolite, prehnite-pumpellyite, and greenschist facies of low pressure, there is observed the prograde metamorphism which overprinted on earlier low-grade assemblage. In an example from the north-west part of the Eltanin Fracture Zone (Pacific ocean) early conditions calculated as <1 kb, 340-380°C evolved upgrade to 4-5 kb, 520-570°C, implying concurrently rising temperature and pressure that affected the zonation pattern in amphiboles of the metamorphic rocks. During high-pressure metamorphism, high-Al amphibole+acid plagioclase+epidote assemblage was formed, which overprinted the earlier assemblage. Actinolite (or ferro-actinolite) cores of amphibole were jacketed by ferro-tschermakite of ferro-pargasite rims. Such variability of metamorphic conditions presumably result from the tectonic thickening of the oceanic crust, or displacement of different blocks along decollement faults. In the Eltanin Fracture Zone, oceanic crust thickening and high-pressure metamorphism are caused by its movement to the north-west from the East-Pacific Rise. This motion is consistent with the general age progression of volcanic rocks along the Louisville Ridge as required by the hot-spot hypothesis, from the present-day location of Louisville hotspot (50°26'S, 139°09'W, immediately north of the Eltanin Fracture Zone) to Mokuku Guyot (Hawkins et al. 1987; Lonsdale, 1988; Watts et al. 1988; Ballance et al. 1989).

Thus, within the ocean plate can be recognized different geodynamic regimes of metamorphism: low-grade sea-floor metamorphism caused by the rock and sea water interaction, and high-pressure metamorphism resulting from tectonic movements and thickening of oceanic crust along major transform fractures or decollement faults.

4 REFERENCES

Ballance, P.F., J.A.Barron, C.D.Blome, D.Bukry, P.A.Cawood, G.C.H.Chapronierre, R.Frisch, R.H.Herzer, C.S.Nelson, P.Quinterno, H.Ryan, D.W.Scholl, A.J.Stevenson, D.G.Tappin, D.G., and T.L.Vallier, 1989. Late Cretaceous pelagic sediments, volcanic ash and biotas from near Louisville hotspot, Pacific Plate, paleolatitude ~42°S. *Paleogeogr., Paleoclimat., Paleoecol.* 71: 281-299.

Cann, J.R. 1969. Spilites from the Carlsberg Ridge, Indian Ocean. *J. Petrol.* 10: 1-19.

Coleman, R.G. 1977. *Ophiolites*. Berlin - N.Y.: Springer-Verlag. 229.

Dobretsov, N.L. 1995. Tectonics and metamorphism: Problems of relationship. *Petrology* 3: 4-23 (In Russian).

Dobretsov, N.L. (ed), 1985. *Rifean-Paleozoic ophiolites of North Eurasia*. Novosibirsk: Nauka. 200 (In Russian).

Dobretsov, N.L., V.A.Simonov, and V.Yu.Kolobov, 1994. Formation of Oceanic Lithosphere in Mid-Atlantic Slow-Spreading Ridges. *Petrology* 2: 363-379 (In Russian).

Dobretsov, N.L., N.V.Sobolev, V.S.Shatsky, et al., 1989. *Eclogites and glaucophane schists of the folded terranes*. Novosibirsk: Nauka. 234 (In Russian).

Hawkins, J., P.Lonsdale, and R.Batiza, (1987). Petrologic evolution of the Louisville seamount chain. In: B.H.Keating, P.Fruer, R.Batiza, and G.W.Boehlert (eds), *Seamounts, Islands and Atolls* (Geophys. Monogr. Ser., 43). Am. Geophys. Union: 235-254. Wash., D.C.

Humphris, S.E., & G.Thompson 1978. Hydrothermal alteration of oceanic basalts by seawater. *Geochim. Cosmochim. Acta* 42: 107-125.

Karson, J.A. & H.J.B.Dick 1984. Deformed and metamorphosed oceanic crust on the Mid-Atlantic Ridge. *Ofioliti* 9: 279-302.

Kawahata, H., M.Kusakabe, Y.Kikuchi, 1987. Strontium, oxygen, and hydrogen isotope geochemistry of hydrothermally altered and weathered rocks in DSDP Hole 504B, Costa Rica Rift. *Earth Planet. Sci. Lett.* 85: 343-355.

Korikovskii, S.P. 1995. Contrasting Models for Prograde-Retrograde Metamorphic Evolution of Phanerozoic Foldbelts in Collision and Subduction Zones. *Petrology* 3: 45-63 (In Russian).

Kurnosov, V B. 1986. *Hydrothermal alterations of basalts in the Pacific ocean and metal-bearing deposits (using data of deep-sea drilling)*. M.: Nauka. 252 (In Russian).

Leake, B.E. 1978. Nomenclature of amphiboles. *Canad. Mineral.* 16: 501-520.

Lonsdale. P. 1988. Geography and History of the Louisville hotspot chain in the southwest Pacific. *J. Geophys. Res.* 93: 3078-3104.

Mishkin, M.A. 1994. On the origin of the metamorphic rocks of the Bering Sea floor. *Dokl. Akad. Nauk Russia* 338: 641-644 (In Russian).

Miyashiro, A. 1973. *Metamorphism and metamorphic belts*. London: George Allen and Unwin Ltd. 492.

Plyusnina, L.P. 1982. Geothermometry and geobarometry of plagioclase-hornblende bearing assemblages. *Contrib. Mineral. Petrol.* 80: 140-146.

Pushchin, I.K., I.A.Tararin, V.G.Sakhno, R.A.Oktyabrsky, O.V.Chudaev, 1990. Highs of ophiolitic assemblage of the southwestern Pacific to north-east of New Zealand. *Dokl. Akad. Nauk SSSR* 315: 945-949 (In Russian).

Pushcharovsky, Yu.M. 1995. On Three Paradigms in Geology. *Geotectonica* 1: 4-11 (In Russian).

Pushcharovsky, Yu.M., Yu.N.Raznitsin, S.D.Sokolov, 1991. The tectonic crust layering in recent oceans and their paleoanalogus. In: R.G.Garetsky (ed), *Geodynamics and tectonosphere evolution*: 97-112. M.: Nauka (In Russian).

Silant'ev, S.A. 1984. *Metamorphic rocks of the Atlantic ocean floor*. M.: Nauka. 104 (In Russian).

Silant'ev, S.A. 1995. Metamorphism of the Contemporary Oceanic Basins. *Petrology* 3: 24-36 (In Russian).

Tararin, I.A., E.P.Lelikov, M.A.Mishkin, and V.M.Chubarov, (1995). Metamorphic rocks in the Philippine Sea. In: *Geology and Geophysics of the Philippine Sea*, H.Tokuyama et al (eds), Terra Sci.Publlish.Company, Tokyo, 329-356.

Vallance, T.G. 1974. Spilitic degradation of a tholeiitic basalt. *J. Petrol.* 15: 79-96.

Watts, A.B., J.K.Weissel, R.A.Duncan, and R.L.Larson, 1988. The origin of the Louisville Ridge and its relationship to the Eltanin Fracture Zone System. *J. Geophys. Res.* 93: 3051-3077.

Stable isotope studies of calcite from very-low grade metamorphic greywacke terranes of the North Island, New Zealand

Solomon Woldemichael
Geology Department, University of Auckland, New Zealand

ABSTRACT: The stable isotope compositions of 30 calcites from greywacke basement terranes of northern and central North Island were measured. Their $\delta^{18}O$ values (7.9 to 26.6 ‰) suggest they were in equilibrium with ^{18}O-rich metamorphic fluids ($\delta^{18}O$ 0±4 to 18±2 ‰). These results indicate that the pore fluid has evolved through water/rock interactions by oxygen isotope exchange with the host sediments. The $\delta^{13}C$ values of calcite (-24.0 to -2.7 ‰) suggest that the calcite depositing fluid contained carbon derived from a low-^{13}C source, possibly organic material in sediments.

1 INTRODUCTION

As a secondary mineral, calcite may form from a bicarbonate-rich fluid either slowly, as fluids circulate, or by sudden release of carbon dioxide during boiling. The most important factor by which the oxygen isotope composition of the calcite may be controlled is fluid composition. The final calcite-fluid fractionation (α_{Ca-H2O}) depends only upon temperature (O'Neil et al. 1969). The carbon isotope composition of calcite, however, can be expected to reflect the origin of the carbon.

This study is part of my PhD studies on the alteration processes of the North Island greywackes. This paper describes the oxygen and carbon isotopic composition of calcite from these rocks. Among the questions addressed are: what are the origins of calcite in the greywackes and the nature of the fluids that deposited these.

2 GEOLOGICAL BACKGROUND

The greywacke suite of rocks of the northern and central North Island (Figure 1) comprise incomplete remnants of forearc basin (Murihiku terrane), trench slope basin (Manaia Hill terrane) and accretionary complexes (Bay of Islands and Omahuta terranes), developed during the late Permian to mid Cretaceous in a subduction system that extended along the eastern margin of Australian- Antarctic Gondwanaland. Summaries of the nature and

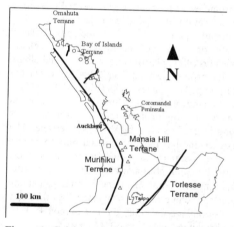

Figure 1: Calcite sample locations. Inferred extent of the major greywacke terranes of the northern and central North Island are shown by bold line. Symbols are: squares: Murihiku terrane; triangles: Manaia Hill terrane; circles: Bay of Islands terrane.

relationship between these terranes have been given in Black (1997).

Lithologically these terranes consist mainly of metasediments ranging in particle size from fine silt to coarse sand. Most show poor sorting, and grains are subrounded to angular in most samples. The detrital mineralogy is relatively simple, consisting mostly of plagioclase, quartz and different types of rock fragments. The relative proportions of these constituents were controlled by the tectonic settings of their sedimentary basins, and differ between terranes. Murihiku terrane sediments are quartz-poor and dominantly of lithic volcanogenic type. The lithic fragments are mainly andesitic. Manaia Hill terrane sediments are also dominantly lithic volcanogenic greywackes, but contain more clasts of sedimentary origin and packets of arkosic and quartz-rich sandstones. The Bay of Islands terrane contains mainly clastic grains of plagioclase and quartz, with subordinate rock fragments. Omahuta terrane sediments are dominantly lithic volcanogenic and have less quartz than the Bay of Island terrane.

3 REGIONAL METAMORPHISM

The Manaia Hill, Bay of Islands and Omahuta terranes underwent an early Cretaceous metamorphism which increased in grade from zeolite facies in the east to prehnite-pumpellyite and then pumpellyite-actinolite facies in the west. The metamorphic event probably was a thermal relaxation after the cessation of subduction (Black et al. 1993). The metamorphic minerals include quartz, laumontite, prehnite, pumpellyite, actinolite, calcite, phyllosilicates (illite, chlorite and smectite) and albite. Peak metamorphism ranged from 200-300° C for the prehnite-pumpellyite facies rocks (Liou et al. 1985).

The metamorphism in Murihiku terrane is in contrast to the regional pattern of east to west increasing in grade shown by the other terrane. The Murihiku terrane has metamorphosed only to zeolite facies, and contains a low grade assemblage dominated by laumontite and clay minerals (chlorite, illite and smectite), but including calcite, heulandite and quartz. Temperatures suggested by the clay mineral assemblage generally did not exceed 150° C, and may have been as low as 60° C.

4 SAMPLES AND ANALYTICAL TECHNIQUE

The oxygen and carbon isotope composition of the calcite are measured on the CO_2 gas extracted using methods similar to that of McCrea (1950).

The oxygen and carbon isotope composition of the calcite are reported in the familiar δ expression as parts per thousand (‰). The standards are (SMOW) oxygen and (PDB) for carbon isotopes respectively. The errors on individual analysis of both oxygen and carbon are not more than 0.1 ‰.

Calcite sample localities are shown in Figure 1. Calcite was petrographically identified in most thin section irrespectively of grain size or metamorphic grade of the rocks. Calcite occurs almost in all cases as vein filling but some times as replacement of clastic grains or ground mass. Calcite is usually intergrowing with other metamorphic minerals. A total of 30 samples from 21 localities representing Murihiku, Manaia Hill and Bay of Islands terranes were selected for this study.

5 RESULTS

The isotopic compositions of the calcites vary widely. $\delta^{18}O$ values range from 7.9 ‰ to 26.6 ‰ and $\delta^{13}C$ values range from -24.0 ‰ to -2.7 ‰ (Figure 2).

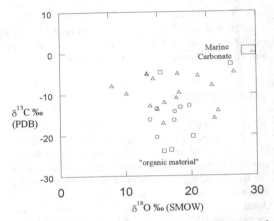

Figure 2. Carbon and oxygen isotope composition of calcites. Symbols as for Figure 1.

None of the calcites have $\delta^{13}C$ values (0±2 ‰) that suggest a direct source of CO_2 from the marine carbonates (Figure 2). Most of the calcites show a tendency towards low $\delta^{13}C$ values. These carbon isotope compositions reflect crystallisation of calcite from formation waters containing carbon derived from low-^{13}C sources, perhaps from CO_2 generated by thermal decarboxylation of organic material. The organic fragments present in the greywackes are the most likely candidate for the source of such carbon.

6 ISOTOPE COMPOSITION OF THE CALCITE-DEPOSITING WATER

The water from which the calcite formed must have been in oxygen isotope equilibrium with the calcite, although the calcite isotope composition might have been altered due to further interaction with fluid. In order to calculate the isotopic composition of the water with which the calcite last equilibrated we need to know the approximate calcite formation temperatures or the last alteration temperature. My petrographic studies show those calcites in the greywackes of the North Island are intergrown with most metamorphic minerals, suggesting that the calcite formation temperature was within the range of the peak metamorphic temperature.

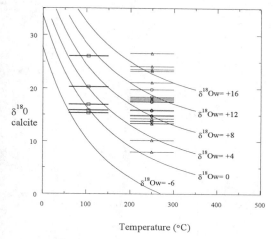

Figure 3. Calcite $\delta^{18}O$ values versus inferred temperature range. Curves represent the composition of calcite in equilibrium with water of constant $\delta^{18}O_w$ values as a function of fractionation temperature (O'Neil et al. 1969). Symbols as for Figure 1 and the lines show the range of the $\delta^{18}O_w$ values accounting for the uncertainties in temperature.

For the purpose of estimating the isotope composition of the calcite forming fluid, the temperature of isotope exchange is assumed to be the same as indicated by the metamorphic mineral assemblages. In Murihiku terrane rocks, the calcite formation temperature probably was very similar to that indicated by the clay mineral assemblage (60-150° C). A reasonable temperature range for Manaia Hill and Bay of Island terrane calcite might range from 200 to 300° C. The fluid/mineral O-isotope exchange reaction was approximated by the system calcite-H₂O and the fractionation factor given by O'Neil et al. (1969) used to calculate the water oxygen isotope composition.

The oxygen isotope composition of the calcite and these temperature range (60 to 150° C for Murihiku, 200 to 300° C for Manaia Hill and Bay of Islands terranes) have been plotted (Figure 3) to illustrate the possible range of $\delta^{18}O_w$ of the calcite forming fluid. In general, results suggest that crystallisation of most calcites at the metamorphic conditions require water which had an oxygen isotope composition different from that of oceanic or meteoric origin. The inferred oxygen isotope composition of metamorphic fluid strongly indicates equilibrium with an ^{18}O-rich metamorphic fluid. The results also suggest the oxygen isotope composition of such metamorphic fluid to be very variable.

Crystallisation of Murihiku terrane calcites at diagenetic temperatures (60-150° C) required a fluid with average $\delta^{18}O_w$ composition ranging between 0±4 and 10±4 ‰.

The fluid involved in the formation of the calcites from the Bay of Islands terrane, is inferred to have relatively narrow $\delta^{18}O_w$ range than the other terranes, ranging from 7±1 to 12±1 ‰. The calcite from the Manaia Hill terrane suggests very variable oxygen isotope composition for their mineralising fluid than Bay of Island terrane. For the Manaia Hill terrane, most of the inferred $\delta^{18}O_w$ values range between 6±2 and 12±2 ‰, but a mineralising fluid with $\delta^{18}O_w$ values of as high as 18±2 ‰ and as low as 1±2 ‰ is also inferred.

7 DISCUSSION AND CONCLUSION

The greywackes were deposited in marine environment. The major source of water in the metamorphic fluids was probably ocean water, originally trapped in the pore space of Murihiku, Manaia Hill and Bay of Islands sediments. As the formation temperature rose due to burial, the pore fluid is likely to have undergone a positive oxygen isotope shift and change in chemical composition due to interaction with the host sediments. The calcites were precipitated from these ^{18}O rich fluids probably after the pore fluid was expelled from the pore space, perhaps during the early Cretaceous as the sedimentary sequence approached its maximum burial depth. The calcites may form either due to a sudden drop in pressure or slowly as the fluid circulated. With the current information it is not possible to distinguish between the two possible mechanisms.

Although the calculated $\delta^{18}O_w$ values has been successful in showing the ^{18}O enrichment trend, it is only approximation solution to the metamorphic condition and some of the scattered values probably reflect the uncertainties associated with the assumptions. Primarily that the calcite crystallisation temperature probably differ significantly from the temperature derived from the metamorphic mineral assemblages and in the worst situation calcite and the metamorphic fluid may not be in equilibrium at any temperature. Secondly, even if the calcites were at isotope equilibrium with the fluid at the peak metamorphic temperature, perhaps they may continue to attain oxygen isotope equilibrium as the temperature fulls until it reached a temperature below which the calcite become isotopicaly isolated and its $\delta^{18}O_w$ values remains unchanged to surface temperature.

8 ACKNOWLEDGMENTS

I thank Professors P.M. Black and P.R.L. Browne for theirs reviews of the manuscript. I also acknowledge M. Brandriss for comments contributing to significant improvements of this paper.

Isotopic analyses were carried out at the Nuclear Sciences, Institute of Geological and Nuclear Sciences. I thank Dr P. Blattner for his generous assistance.

9 REFERENCES

Black, P.M. 1997. Origin and assemblage of terranes in dynamic convergent plate margin environment: Mesozoic basement terranes, New Zealand. Terrane Dynamics 97, Extended Abstract.

Black, P.M., Clark, A.S.B. and Hawke, A.A. 1993. Diagenesis and very low-grade metamorphism of volcanoclastic environments, North Island, New Zealand: Murihiku and Waipapa terranes. Journal of metamorphic geology, 11:429-435.

Liou, J.G., Maruyama, S., and Cho, M. 1985. Phase equilibria and mineral parageneses of metabasites in low-grade metamorphism. Mineralogical magazine 49: 321-333.

McCrea, J.M. 1950. On the isotopic chemistry of carbonates and a paleotemperature scale. The journal of chemical physics, 18:849-857.

O'Neil, J.R., Clayton, R.N. and Mayeda, T.K. 1969. Oxygen isotope fractionation in di-valent metal carbonate. Journal of chemical physics 51: 5547-5558.

8 Magma – Water interaction

Shallow magmatic degassing: Processes and PTX constraints for paleo-fluids associated with the Ngatamariki diorite intrusion, New Zealand

B.W.Christenson, C.P.Wood & G.B.Arehart
Wairakei Research Centre, Institute of Geological and Nuclear Sciences, New Zealand

ABSTRACT: A study of the alteration halo surrounding a diorite intrusion in the Ngatamariki geothermal system is being conducted to ascertain the nature of the reservoir fluids extant at the time of diorite emplacement. Preliminary results show that hot, highly saline (>30 wt % NaCl) fluids were in equilibrium with a low density, volatile-rich phase within, and immediately adjacent to the diorite, and that these fluids interacted with convecting meteoric fluids during formation of the extensive alteration halo. Temperatures in this system range from 250 °C in the alteration halo to > 500 °C in the diorite, and evidence points to the existence of repeated pressure transients across the brittle-plastic transition during subsolidus cooling of the intrusion.

1 INTRODUCTION

A diorite intrusion intersected by geothermal well Ngatamariki NM4 is the only in situ plutonic rock yet encountered in the TVZ. Alteration within and adjacent to the diorite is extensive and distinctive, and suggests that the intrusion served as a convective heat source for the hydrothermal system active at the time. Petrographic, stable isotope and fluid inclusion analyses of alteration associated with this intrusion provide clear insights into both the nature of the paleo-fluids and the reservoir processes adjacent to the pluton, and how shallow, degassing magma bodies evolve through time.

2 GEOLOGY & HYDROTHERMAL PETROLOGY OF NM4

The reservoir stratigraphy of the Ngatamariki field consists of a sequence of interbedded rhyolites, andesites and silicic pyroclastic rocks, and in this respect is similar to the adjacent Rotokawa field (Browne et al., 1992). A feature which sets the Ngatamariki field apart from other explored systems in the TVZ, however, is the presence of a diorite intrusion, intersected between 2640-2750 m in NM4 (Fig. 1). Although the diorite is relatively old (ca. 700 ka, Arehart et al., 1997) and does not serve as a present-day heat-source for the field, there is an

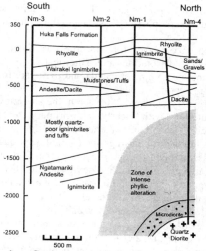

Fig. 1 Geologic cross section through the 4 exploration wells at Ngatamariki.

extensive alteration halo associated with the intrusion, suggesting that it interacted strongly with hydrothermal fluids in the past.

There is an abrupt increase in alteration rank and intensity beneath the Wairakei Ignimbrite (ie. below ca 750 m depth), suggesting that convection driven by this intrusion had probably ceased by the time this unit was deposited (ca. 330 ka bp). Between 750 and 1700 m depth, the alteration is characterised by an

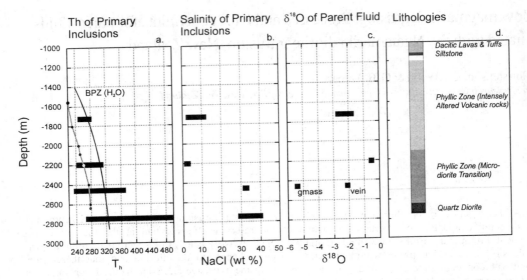

Fig. 2 Phyllic zone fluid inclusion and mineral stable isotope data in NM4. a.) Th ranges compared to present day measured temperatures and boiling point temperatures. b.) Salinities (wt % NaCl). c.) $\delta^{18}O$(SMOW) for waters in equilibrium with analysed mineral phase. d.) Stratigraphic column.

illite-rich, mixed-layer illite-smectite clay, quartz and pyrite, with dissolution and silicification textures predominating. Veins of calcite and pyrite are found at 1000 m depth.

Between 1700 m and 2200 depth, the alteration is very intense, with nearly all primary textures destroyed. Groundmass alteration over this interval consists largely of coarse grained sericite and quartz, with lesser amounts of chlorite and pyrite. Late-stage illite is also present, often replacing sericite. Quartz-pyrite veining, including minor galena and sphalerite, is prevalent over this interval, although sericite-anhydrite veining is observed at 2208 m.

The exact boundary of the pluton is difficult to ascertain owing to the alteration intensity over the transition interval, but a contact between a coarse diorite and what is identified as microdiorite is observable at 2459 m, and the finer-grained variant appears to extend through to 2208 m depth. The coarse diorite at 2749 m consists of up to 60 % plagioclase (oligoclase), 25-30 % amphibole and 5 % magnetite, with interstitial quartz comprising the remaining 5-10 of the rock. A later, green fibrous amphibole grows from the earlier phase, and also occurs in late quartz veins and late interstitial quartz. Oligoclase is variably replaced by albite and epidote, and epidote is also found intergrown with late (ie.

interstitial) quartz. Equilibrium temperatures in excess of 350 °C are indicated for the amphibole-bearing assemblage.

At least two hydrothermal alteration events are represented in the cores. Over much of the phyllic alteration interval, fine grained illite is observed overprinting the earlier, coarser-grained sericite. An early assemblage consisting of quartz + sericite + anhydrite + apatite is locally overprinted by quartz + wairakite + epidote + sphene + chlorite ± calcite. Calcite is observed locally replacing epidote, making this the latest stage alteration phase encountered. The earliest assemblage is similar to that commonly observed in phyllic alteration zones in porphyry copper deposits, whereas the latter is more typical of TVZ-type geothermal systems.

Correlation of the diorite to possibly coeval volcanic rocks higher in the sequence is difficult owing to the altered state of the overlying volcanic units. However, given the age constraints provided by Arehart et al. (1997) for the pluton and the Ngatamariki andesite, and the 330 ka age of the little-altered Wairakei Ignimbrite (Wilson et al., 1995), it would appear that the dacite found between 900-1090 m (Fig.1) is the most likely eruptive coeval to the diorite, which would imply the pluton intruded to within 2 km of the ground surface.

3 FLUID INCLUSION AND MINERAL STABLE ISOTOPE DATA

Preliminary results of fluid inclusion and mineral stable isotope analyses from three core intervals in the phyllic zone and one from the diorite proper are summarised in Fig. 2. Quartz is the most prevalent fluid inclusion host mineral, and occurs in veins, vug fillings and as recrystallised groundmass. Anhydrite from anhydrite/sericite veins in core from 2208 m have also provided rare, but usable inclusions.

There are at least three assemblages of fluid inclusions present in hydrothermal alteration products in the diorite and microdiorite. One assemblage contains high salinity, liquid-dominated inclusions (with halite daughter minerals). These are intimately associated with low salinity (ie. daughter-free) vapour dominated inclusions, and as such, are evidence of fluid immiscibility in the hydrothermal system. Secondary inclusions are relatively common in crystals containing the high salinity fluids, and whilst highly variable vapour-liquid ratios in these inclusions imply entrapment of boiling fluid, the salinities are markedly lower than in the primary inclusions. It is assumed that this assemblage of inclusions represents post-magmatic hydrothermal activity. A third population of inclusions is found in quartz that is demonstrably later than those described above. These inclusions are predominantly liquid-dominated, and are devoid of daughter minerals. This assemblage is presently interpreted to be related to the later hydrothermal over-print evident in the hydrothermal alteration phases described above.

There is a large range of homogenisation temperatures (T_h) recorded in the high salinity inclusions in the diorite and microdiorite (Fig 2a), with the higher end of the range clearly reflecting conditions extant within the cooling pluton. The lower end of the Th range is thought to represent, at least in part, cooling of the heat source, although pressure fluctuations in the hydrothermal environment also contribute to variations in the observed Th values.

Salinities of primary inclusion fluids are shown in Fig. 2b. Inclusion compositions in the diorite (2459 and 2749 m) range from 28 to 41 weight per cent NaCl. Whereas inclusions in vein quartz from 1725 m range from 2 to 12 wt. % NaCl, those in anhydrite from anhydrite-sericite veins from 2208 m have relatively low salinities (<2 wt % NaCl).

$\delta^{18}O$ signatures of the fluid are plotted in Fig. 2c. These values derive from fractionation factors for quartz-water (Clayton et al., 1972) and anhydrite-water (Lloyd, 1968) equilibria, and temperatures are constrained by the relevant T_h values. Calculated compositions for fracture-bound waters range between -0.5 to -2.8 (SMOW). Although these values are 2 to 3 per mil heavier than the present day thermal waters from the adjacent Ohaaki system (Hedenquist, 1990), they are considerably lighter than the +8 to +10 per mil values representative of andesitic water (eg. Giggenbach, 1992). Similarly, δD_{water} from matrix sericite in core from 2208 m suggests a composition, at -38 ± 5 per mil (fractionation factor of Gilg & Sheppard, 1996), only marginally heavier than present-day Ohaaki fluids (-38 to -42; Hedenquist, 1990).

Mosaic quartz from the recrystallised groundmass from the microdiorite at 2459 m yields a derived water $\delta^{18}O$ composition significantly lighter (-5.5) than that for the adjacent vein (Fig. 2c). Given that this value is rather similar to present-day Ohaaki reservoir compositions, we interpret the mosaic quartz to represent recrystallisation from later, meteoric-dominated fluids. Overall, the mineral stable isotope data point to mixing between magmatic fluids and meteoric waters.

4 RESERVOIR PARAMETERS AND PROCESSES IN THE INTRUSIVE ENVIRONMENT

Fluid pressures in the subsolidus Ngatamariki diorite would have been bound by lithostatic loading on the high end, and hydrostatic loading on the low end. If we assumed, as mentioned previously, that the dacite eruptives intersected at 900 m (Fig. 1) were coeval to the diorite, the depth to the intrusion represented at 2749 m would have been ca. 1800 m. For an average rock density of 2.7 g/cm^3, this equates to ca. 480 bar lithostatic pressure. Hydrostatic reservoir pressures at this level would have been approximately 180 bars. However, the actual pressure conditions at any one location in the hydrothermal system in and adjacent to the diorite (ie. lithostatic, hydrostatic, or intermediate value) would have been governed by the rheological properties of the reservoir medium.

The entire observed range of fluid inclusion salinities from 2749 m (Fig. 2) is observable over relatively small distances ($\approx$ few hundred microns) within single quartz crystals. Noting that there is an inverse relationship between Th and salinity amongst these inclusions, it seems that dilution of the brine by meteoric water can be effectively ruled out as a cause. Rather, the compositional variations are better explained by the occurrence of pressure transients in the hydrothermal environment.

It has long been recognised that pressure variations within the magma-ambient environment are a natural consequence of shallow (3-4 km) pluton emplacement, where in-situ crystallization leads to saturation of the magma with respect to H$_2$O and other volatile species, and a net increase in molar volume of the system (eg. Burnham, 1979). At constant volume, this leads to increased fluid pressures, and ultimately hydraulic fracturing of the rocks confining the H_2O-saturated carapace. Evidence of such fracturing at Ngatamariki is found in the large number of quartz veins and segregations throughout the diorite and enclosing microdiorite.

Pressure transients can markedly affect the composition of the magmatic fluid. For example, an aqueous fluid with a bulk salinity of ca. 10 wt % NaCl, at 450 °C, and confining pressure of ca. 330 bars consists of two immiscible phases, including a high density brine (salinity of 28 wt % NaCl) and a low density, volatile-rich vapour (ca. 0.2 wt % NaCl). A pressure drop of just 30 bars can increase the salinity of the high-density brine from 28 to 40 wt % NaCl, whereas the salinity of the vapour decreases to < 0.1 wt % NaCl.

This process can explain not only the observed salinity variations on the sub-millimeter scale in the Ngatamariki diorite, but more importantly, it also represents a mechanism by which heat and mass was transferred across the pluton-reservoir interface. Owing to density and viscosity differences in the two fluid phases, and relative permeability effects under high vapour/liquid ratios, the low density, volatile rich phase is likely to be more mobile during these events and will be compositionally similar to that released from passively degassing volcanoes (eg. Shinohara, 1995). This vapour interacted with convecting, meteoric-dominated fluids and led to the formation of the observed phyllic alteration assemblage via reaction path processes analogous to those discussed by Christenson and Wood (1993).

The intermediate salinities observed in vein quartz from 1725 m are best explained by flushing of the higher salinity brine by convecting, meteoric-dominated hydrothermal fluid from its source in the diorite. Given that permeability within the diorite would have remained low until it had cooled to below ca. 350 °C (eg. Fournier, 1991, Hayba and Ingebritsen, 1997), it is unlikely that meteoric infiltration and dispersal of the brine occurred until late in the sequence of events. This view is supported by the limited distribution of similar high salinity fluids in the granitic heat-source of the present-day Kakkonda system (Kasai et al., 1996).

REFERENCES

Arehart GB, Wood CP, Christenson BW, Browne PRL, Foland, K (1997) Timing of volcanism and geothermal activity at Ngatamariki and Rotokawa, New Zealand. Proc. 19th NZ Geoth. Workshop, p. 117-122.

Browne PRL, Graham IJ, Parker RJ and Wood CP (1992) Subsurface andesite lavas and plutonic rocks in the Rotokawa and Ngatamariki geothermal systems, Taupo Volcanic Zone, New Zealand. Jour Volc Geoth Res 51:199-215.

Burnham CW (1979) Magmas and hydrothermal fluids. In "Geochemistry of Hydrothermal Ore Deposits", HL Barnes ed., J. Wiley & Sons, New York.

Christenson BW and Wood CP (1993) Evolution of a vent-hosted hydrothermal system beneath Ruapehu Crater Lake, New Zealand. Bull Volcanol 55:547-565.

Clayton RN, O'Neil JR and Mayeda TK (1972) Oxygen isotope exchange between quartz and water. Jour Geophys Res 77:3057-3067.

Fournier, RO (1991) The transition from hydrostatic to greater than hydrostatic fluid pressures in presently active continental hydrothermal systems in crystalline rock. Geoph Res Let 18:955-988.

Giggenbach WF (1992) Isotopic shift in waters from geothermal and volcanic systems along convergent plate boundaries and their origin. Earth Planet Sci Lett 113:495-510.

Gilg HA and Sheppard SMF (1996) Hydrogen isotope fractionation between kaolinite and water revisited. Geochim Cosmchim Acta 60:529-533.

Hayba DO and Ingebritsen SE (1997) Multiphase groundwater flow near cooling plutons. Jour Geoph Res 102:12,235-12,252.

Hedenquist JW (1990) The thermal and geochemical structure of the Broadlands-Ohaaki geothermal system, New Zealand. Geothermics 19:151-185.

Kasai K, Sakagawa Y and Miyazaki S (1996) Supersaline brine obtained from quaternary Kakkonda Granite by the NEDO deep geothermal well WD-1A in the Kakkonda Geothermal Field, Japan. GRC Trans. 20:623-629.

Lloyd RM (1968) Oxygen isotope behavior in the sulfate-water system. Jour Geophys Res 73:6099-6110.

Shinohara H (1994) Exsolution of immiscible vapor and liquid phases from a crystallisng silicate melt: Implications for chlorine and metal transport. Geochim Cosmchim Acta 58:5215-5221.

Sourirajan S and Kennedy GC (1962) The system NaCl at elevated temperatures and pressures. Am Jour Sci 260:115-141.

Wilson CJN, Houghton BF, McWilliams MO, Lanphere MA, Weaver SD and Briggs RM (1995) Volcanic and structural evolution of Taupo Volcanic Zone, New Zealand: a review. Jour Volc Geoth Res 68:1-28.

The Gorely Volcano Crater Lake: New data on structure and water chemistry

Yu.O.Egorov
Institute of Volcanic Geology and Geochemistry, Petropavlovsk-Kamchatsky, Russia

G.M.Gavrilenko, A.B.Osipenko & L.G.Osipenko
Institute of Volcanology, Petropavlovsk-Kamchatsky, Russia

ABSTRACT: New data on structure, water chemistry and current state of the acid lake in the active crater of the Gorely Volcano, South Kamchatka are presented. Physical and chemical parameters of studied water are similar to those of other volcanic acid lakes but differ in (1) enhanced cation/anion weight ratio and (2) Fe/Al>1. The reason for these peculiarities is the post-eruptive water-rock interaction combined with magma degassing. It is expected that study of dynamics of thermal/chemical variations in the lake waters will allow us to predict future eruptions of the Gorely Volcano.

1 INTRODUCTION

Crater lakes hosted by active volcanoes are the surface expression of high-level geothermal systems and act as condensers for fluids and gases released by shallow magma bodies. Variations in inputs of magmatic heat and volatiles are generally reflected in the temperature and chemistry of the lake. The purpose of this paper is to present results of the first thermal/chemical water study for the Gorely Volcano acid crater lake (GVCL), South Kamchatka. This study is specially aimed at developing a better conceptual model of the GVCL using new geochemical data.

2 GORELY VOLCANO

Gorely volcano (1829 m) is located 75 km SW from Petropavlovsk - Kamchatsky. Gorely, dissimilar to other Kamchatka volcanoes, looks like a ridge with gentle slopes, formed by superimposition of several stratovolcanic cones and numerous flank cinder cones. Four large craters with small nest-shaped craters located inside them are situated at the volcano summit (Fig.1).

About ten of the Gorely Volcano eruptions are known during the historical time. The last one occurred in mid - 1980s and was characterized by weak phreato-magmatic activity accompanied by volcanic ash and pyroclastic ejections up to 5 km high. Currently, the Gorely is active again. The most intensive fumarolic activity is concentrated in the central crater of the volcano and in a few sites nearby.

The Central crater of the Gorely Volcano has a "telescopic" structure with steep walls up to 100 m high composed of layered pyroclastics and lavas. A well-shaped funnel with a lake on the bottom is located in the central part of the crater. The walls of the funnel are up to 50-60 m high and rockfalls are frequent. Fumarole discharges emanate from vents observed on even grounds at the walls of funnel and near water surface. The crater is filled by fumarolic gases.

3 CRATER LAKE

Sitting atop the Gorely volcanic massif, fresh and acid crater lakes are poorly studied. Originally, the acid crater lake in the Central crater was reported in August, 1960 and thereafter in 1978-1979, 1982, and 1984, (Kirsanov and Ozerov, 1983; Fedotov et al., 1985). The lake disappeared before the 1985 eruption and the well-shaped funnel up to 100 m deep was formed in its place. Bright red fluorescence was observed in July-November, 1986 at the bottom of the funnel. It is suggested that the temperature was about 900°C based on color scale. These phenomena were caused by magma degassing in channel under shallow-deep condition (Ivanov et al., 1988).

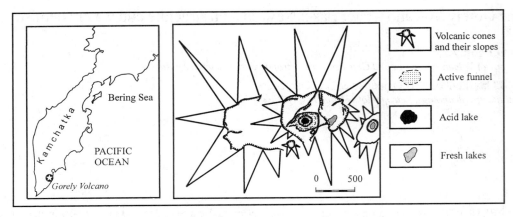

Fig.1 Schematic map of the Gorely Volcano

It is unknown when the lake in the Central crater of the Gorely volcano was reformed after the last eruption in 1986. Our visual observations confirmed that from 1992 until present the bottom of the active crater is filled by the lake (Gavrilenko et al., 1996).

The lake is located on the bottom of the active funnel and has a mean diameter ca.60-70 m. Taking in consideration the steepness of the funnel walls and visual observation of the funnel during the period of lake absence, it can be expected that depth of the GVCL is several tens of meters. The measurements during sampling showed that the temperature of water is $+37^{\circ}C$ and pH=1.13. As the result of chemical processes, lake's water is vigorously convecting and has bright blue color which is typical for acd crater lakes (e.g. Delmelle & Bernard, 1994). It is determined by formation of colloidal sulfur resulted in H_2S oxidation by SO_2 and O_2.

4 WATER CHEMISTRY

Representative chemical analyses of acid and fresh-water lakes of the Gorely Volcano are given in Table 1. The chemical composition of the lake waters is typical of active volcanic crater lakes (e.g. Rowe et al., 1992), being essentially a mixture of sulfuric and hydrochloric acids (Fig.2). As a rule, such lakes are composed of meteoric water accumulated in craters of active volcanoes. Acid volcanic gases dissolve in lake waters and decrease the pH. GVCL waters have high acidities (pH = 1.1-1.2). High concentrations of SO_4, Cl and F reflect injection at the lake bottom of magmatic volatiles (SO_2, H_2S, HCl and HF). Similar anions contents were also measured in the lake waters of Poas (Rowe et al., 1992); Kawah Ijen (Delmelle and Bernard, 1994) and Ruapehu (Christenson, 1994) (Fig.2a). High SO_4/Cl ratio in the GVCL waters may be attributed to elevated volcanic activity (Menyailov, 1975; Takano and Watanuki, 1990).

Synchronously, acidic alteration of wallrock and volcano-clastic debris washed into the lake increased the metal contents of the lake's waters. Major elements, such as Al, Fe, Ca, Mg, Na, and K, are dominated among the cations. These elements differ from typical acid crater lakes in elemental ratios only.

Table 1. Chemical composition of the Gorely Volcano crater lakes water
(Concentrations are in milligrams per liter)

	Na^+	K^+	Ca^{2+}	Mg^{2+}	Al^{3+}	Fe^{3+}	Fe^{2+}	Cl^-	SO_4^{2-}	HSO_4^-	TDS *	pH	T,°C
Acid lake	460	77	561	778	243	240	1335	355	5760	8906	22.7	1.1	37
Fresh lake	1.0	0	6.5	0.5	---	0.0	2.0	1.5	10	---	0.03	7.0	8

Note: * - Total dissolved solid content in g/l

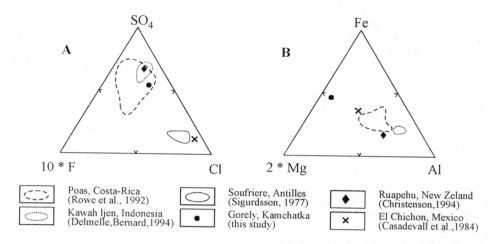

Fig.2 Anion (A) and cation (B) composition of crater lakes water.

5 DISCUSSION

It is anticipated that GVCL waters chemically are similar to those from typical crater lakes during elevated volcanic activity. Basically, this is true for anion composition of the GVCL water (Fig.2a). At the same time, GVCL water is characterized by cation composition (Fig.2b) illustrated the specificity of processes which occurred in hydrothermal system of the Gorely Volcano. Two main peculiarities are as follows: (1) GVCL has enhanced cation/anion weight ratio; (2) Fe/Al ratio > 1. Both peculiarities presumably are inherent in acid crater lakes with cyclic behaviour and correlate with increasing of volcanic activity (Gavrilenko et al., 1996). These values in acid lake during quiescent periods are significantly below (e.g. Delmelle & Bernard, 1994).

The reasons for the peculiarities mentioned above are the duration of water-rock interaction and differences in dynamics of metal weathering as a result of exchange reactions. During the initial stage of acid alteration K, Na, Ca, Mg, and Fe were dissolved. The cation/anion ratio in this stage is relatively high > 0.1 (in the case of GVCL > 0.2). At a later time cation reserve of volcanics is depleted and anion input continues coupled with acid volcanic gases. As a result cation/anion weight ratio decreases to < 0.1.

Fe is the most mobile component during period of elevated magmatic degassing be cause of its high content in high-temperature volcanic gases. As a cosequence Fe to a greater extent than Al came in acid water of lakes on the initial stage of hydrothermal alteration of «fresh» volcanics (Fe/Al > 1). When the active phase in volcano development was terminated and the fluid input decreased, acid alteration became dominant. During this stage Fe and Al completely came in solution with subsequent decreasing of Fe/Al ratio attained <1. This corresponds to that in enclosing volcanics.

The proposed mechanism of chemical alteration explains main peculiarities of GVCL water chemistry. The lake was formed several years after the volcano eruption and, consequently, the walls and the bottom of crater are composed of fresh volcanics. Also, strong fumarolic activity occurs in the lake caused Fe-dominated input. As a result, in the GVCL water cation/anion weight ratio is more than 0.2, and Fe/Al ratio is much more than 1 (Tab.1).

6 CONCLUSIONS

Data on the activity of the Gorely Volcano, South Kamchatka confirm the cyclicity of the GVCL evolution. Cyclic behaviour is reflected in crater lake occurrences during inter-eruptive periods, and disappearances of the lake prior to, or during eruptive activity. It is suggested that phreato-magmatic character of the Gorely Volcano eruptions caused by lake water penetration in magma channel.

Because a lake covers the most active area within the crater, changes in the level and chemistry of lake waters can be expected to reflect variations in the activity of the volcano. It is expected that study of the dynamics of thermal/chemical variations in the GVCL, will allow us to predict the future increasing of the Gorely Volcano activity.

ACKNOWLEDGEMENTS

This study was supported by INTAS Foundation (grant INTAS 94-3129).

REFERENCES

Casadevall, T.J., S.De la Cruz-Reyna, W.I.Rose, S.Bagley, D.L.Finnegan, and W.Zoller, 1984. Crater lake and post-eruption hydrothermal activity, El Chichon Volcano, Mexico. *J.Volc. Geoth. Res.* **23:** 169-191.

Christenson, B. 1994. Convection and stratification in Ruapehu Crater Lake, New Zealand: Implications for Lake Nyos-type gas release eruptions. *Geochem. J.* **28:** 185-197.

Delmelle, P. & A.Bernard 1994. Geochemistry, mineralogy, and chemical modeling of the acid crater lake of Kawah Ijen Volcano, Indonesia. *Geochim. Cosmochim. Acta* **58:** 2445-2460.

Fedotov, S.A., B.V.Ivanov and V.N.Dvigalo, 1985. Activity of volcanoes of the Kamchatka and the Kurile Islands in 1984. *Volcanol. and Seismol.* No.5: 3-23 (in Russian).

Gavrilenko, G.M., A.B.Osipenko, Yu.O.Egorov and M.G.Gavrilenko, 1996. Lake-phantom in the crater of the Gorely Volcano. *IWGCL Newsletter.* **9:** 9-10

Ivanov, B.V., V.A.Droznin and V.A.Vakin, 1988. The 1985 Gorely Volcano Eruption. *Volcanol. and Seismol.* No.4: 93-98 (in Russian).

Kirsanov, I.T. & A.Yu.Ozerov 1983. Products composition and energetic effect of the 1980-1981 Gorely Volcano Eruption. *Volcanol. and Seismol.* No.1: 25-42 (in Russian).

Menyailov, I.A. 1975. Prediction of eruptions using changes in composition of volcanic gases. *Bull. Volcanol.* **39:** 112-125.

Rowe, G.L., S.Ohsawa, B.Takano, S.L.Brantley, J.F.Fernandez and J.Barquero, 1992. Using crater lake chemistry to predict volcanic activity at Poas Volcano, Costa-Rica. *Bull. Volcanol.* **54:** 494-503.

Sigurdsson, H. 1977. Chemistry of the crater lake during the 1971-1972 Soufriere eruption. *J.Volc. Geoth. Res.* **2:** 165-186.

Takano, B. & K.Watanuki 1990. Monitoring of volcanic eruptions at Yugama crater lake by aqueous sulfur oxyanions. *J.Volc. Geoth. Res.* **40:** 71-87.

Gas-water interaction at Mammoth Mountain volcano, California, USA

W.C. Evans, M.L. Sorey & R.L. Michel
US Geological Survey, Menlo Park, Calif., USA

B.M. Kennedy
Lawrence Berkeley National Laboratory, Calif., USA

L.J. Hainsworth
Lawrence Livermore National Laboratory, Calif., USA

ABSTRACT: The large diffuse flux of cold CO_2 issuing from soils on Mammoth Mountain since ~1989 has been estimated to total ≥400 t/d. This study has revealed a flux of ~30-50 t/d of CO_2 dissolved in cold, dilute ground water; virtually every cold spring on the flanks of Mammoth Mountain contains anomalous CO_2, with dissolved inorganic carbon (DIC) concentrations reaching 45 mmol/kg. Gas absorption and transport by cold water is thus an important process here. Numerical modeling of gas absorption indicates that preferential CO_2 loss to solution can explain anomalous compositional variations among soil-gas samples, apart from which the soil gas in all areas of diffuse emissions displays a striking uniformity. Some level of dissolved CO_2 flux predates 1989, indicating that diffuse leakage of gas from depth is a long-term phenomenon at this volcano.

1 INTRODUCTION

Geophysical techniques detected a dike intrusion in 1989 beneath Mammoth Mountain (last active with phreatic eruptions ~700 yBP). The first geochemical response to the dike was an increase in flow and $^3He/^4He$ ratio in fumarole MMF, the only notable steam vent on the mountain (Sorey et al., 1993). Within one year of intrusion, however, a huge flux (≥400 t/d) of cold CO_2 began escaping through soils around the flanks, eventually killing ~60 ha of forest (Farrar et al., 1995; Rahn et al., 1996). Compositional and isotopic similarities between MMF and soil gas in tree-kill areas clearly pointed to a common gas source, although a few soil-gas samples had some anomalous characteristics that required further study. Based on the magnitude and homogeneity of the diffuse flux, Sorey et al. (this volume) propose that the gas is being released from a high-pressure gas reservoir that predates and was spiked, or perhaps breached, by the 1989 dike.

High ratios of $^{40}Ar/^{36}Ar$ (~400), $CO_2/^{36}Ar$ (~10^7), and $N_2/Ar_{(total)}$ (~250) in MMF and in the most CO_2-rich soil-gas samples indicate limited interaction with deeply circulating, thermal or non-thermal, meteoric waters rich in atmospheric ^{36}Ar. However, the fact that the diffuse gas flux occurs at low temperatures and over wide areas of the mountain provides ample opportunity for interaction with shallow groundwater or infiltrating snowmelt. This paper documents the impact of water on these cold gas emissions.

2 THE GASES

Compositional and isotopic data for a few of the 60+ gas samples collected to-date are shown in Table 1. The MMF analysis is representative of eight samples collected from this fumarole during the past four years, v543 and BP-1 are two of the most CO_2-rich soil gases collected thus far, and CH12S is a sample of gas bubbling from one of the cold springs near the northern base of the mountain (map in Sorey et al., this volume). MMSA-1 represents dissolved gas in water from a nearby well, with the values expressed in percent of the total dissolved gas (20.8 mmol/kg).

The similarity between MMF gas and the diffuse emissions is best exemplified by v543, which is nearly identical in CO_2, He, N_2, CH_4, and all isotopes. Both samples have N_2/Ar ratios much greater than air (83.6) due to a dominant component of non-atmospheric N_2. The more reactive species H_2 and H_2S are probably scrubbed from the gas stream by various mechanisms in the shallow, low-temperature environment beneath v543.

Two samples of soil air from forested locations outside the tree-kill areas are also shown. BK-1 is our most CO_2-rich background sample (2.45%), but based on a ^{14}C almost equal to that of atmospheric CO_2, contains only biogenic C from normal soil respiration processes. Most other background sites in healthy forest contain ~0.25-2.0% CO_2; but many, like BK-2, show some component of deep, abiogenic C. Using ^{13}C and ^{14}C, the soil-gas samples at Mammoth Mountain form a mixing line

Table 1. Gas composition in vol-% and isotopic data (delta values in ‰). "na" means not analyzed.

Sample	MMF	v543	BP-1	CH12S	MMSA-1	BK-1	BK-2	BP$_{calc}$
	fumarole	soil gas		spring	well	background soil gas		model
He	0.0015	0.0015	0.0030	0.0029	0.00004	0.0006	0.0005	0.0030
H$_2$	0.0322	<0.0002	<0.0002	<0.0002	<0.00001	<0.0002	0.0004	nm
Ar	0.0046	0.0039	0.0194	0.1515	0.075	0.941	0.931	0.0201
O$_2$	0.0629	0.0518	0.1238	2.890	1.11	18.00	20.49	0.1224
N$_2$	1.155	1.146	2.568	14.33	3.17	78.61	77.89	2.492
CH$_4$	0.0019	0.0017	0.0011	0.0004	<0.0005	0.0009	<0.0002	nm
CO$_2$	98.70	98.80	97.28	82.63	95.64	2.448	0.684	97.36
H$_2$S	0.0461	<0.0005	<0.0005	<0.0005	<0.05	<0.0005	<0.0005	nm
N$_2$/Ar	249	297	133	94.5	42.3	83.6	83.7	124
δ^{13}C-CO$_2$	-4.6	-4.8	-4.5	-4.9	-5.9	-20.0	-14.9	-3.7
^{14}C	na	na	0.33	1.9	4.0	99.2	67.5	0.25
δ^{15}N-N$_2$	1.48	1.44	0.99	na	na	na	na	1.38
^{3}He/^{4}He	5.7	5.5	4.8	na	4.5	na	na	5.7
^{40}Ar/^{36}Ar	405	370	316	na	na	na	na	347

^{14}C is in percent of atmospheric value in year of sample collection. ^{3}He/^{4}He is relative to the ratio in air.

between biogenic CO$_2$ (with ^{14}C = present day atmosphere and δ^{13}C = -20‰) and abiogenic CO$_2$ (with ^{14}C = 0 and δ^{13}C = -4.5‰), as shown in Sorey et al. (this volume).

In fact, the complete chemical and isotopic composition of most soil-gas samples can be characterized as mixtures between v543 and air, after allowing for some diffusional fractionation in δ^{13}C and δ^{15}N due to gas transport through the soil. Samples that deviate markedly from this two-component mixing scheme (like BP-1) show primarily an apparent enrichment in He concentration (Fig. 1). Most such samples were collected in wet soils either near the shore of Horseshoe Lake or underneath heavy snow cover. This suggests that water is preferentially removing the CO$_2$, which is about 200 times more soluble than He at 0°C.

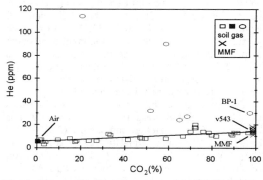

Fig. 1. He *vs.* CO$_2$ in soil gases and in MMF (1994-1997). Filled symbols (n=5) are from healthy forest and plot near air; open symbols are from tree-kill areas. Ellipses are samples collected from wet soils.

3 THE WATERS

Samples were obtained from 10 springs and three wells on the flanks of Mammoth Mountain (Table 2). Except for one slightly warm spring (SLS) at 18°C, water temperatures were ≤10°C showing no anomalous thermal component. Tritium levels were 6-18 TU indicating average flow times of ~1-25 years. Maximum flow paths (summit to discharge point) are ~2-4 km except for one feature (BS) which is 15 km away but is thought to receive some flow from Mammoth Mountain (Sorey et al., in prep.). Spring discharges range from <1 to >10 kg/s, with the combined water flow from some spring groups exceeding 50 kg/s. The total discharge of the sampled features (excluding BS) is estimated at 400 kg/s, and represents nearly half of the precipitation that falls on Mammoth Mountain.

Three cold springs (RMRS, LS, HS) located off Mammoth Mountain were also sampled to constrain the regional background water composition. This comparison is problematic however, because those drainages contain somewhat different soil and vegetation types, as well as outcrops of marine carbonate rocks in the RMRS and LS cases.

Fig. 2 shows the ^{13}C - ^{14}C plot for the DIC in the waters. For comparison, the soil-gas samples in Table 1 are also plotted, and the soil-gas biogenic-abiogenic mixing line is drawn in from Sorey et al. (this volume). The general trend of the DIC in the waters lies close to the mixing line of the CO$_2$ in the soils. Two of the springs (TLS and MMSA-3) on Mammoth Mountain contain mainly biogenic C and thus help define the background. All other waters contain a substantial fraction of abiogenic C. In the case of the regional background sites, RMRS and LS, this could derive from direct fluid interaction with carbonate rocks; but the features on Mammoth

Table 2. Spring (final letter - S) and well samples.

Feature	pH	DIC mmol/kg	^{14}C	pCO$_2$ matm
Regional background sites:				
LS	8.87	0.72	57	0.05
RMRS	7.14	0.75	75	2.08
HS	6.4	0.73	na	6.68
Mammoth Mountain sites:				
TLS	6.68	0.69	91	4.17
MMSA-3S	6.02	2.03	100	24.9
BS	7.23	2.20	37.5	5.65
LB-1	6.97	2.60	26.3	9.88
MINS	7.08	2.90	39.3	9.03
SLS	6.02	9.41	9.8	157
VSS	5.76	15.87	2.8	214
MLS	5.59	16.24	2.0	202
CH15S	5.94	18.02	9.8	231
MMSA-2B	5.79	18.52	4.6	280
MMSA-1	5.43	21.68	3.9	309
RMCS	5.48	26.97	1.2	408
LBCS	5.42	29.44	na	435
CH12S	5.20	45.37	1.9	601

^{14}C = percent atmospheric value in year of sampling.

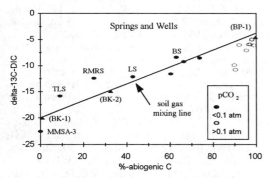

Fig. 2. Waters (ellipses) from Table 2 and soil gases (triangles) from Table 1 (%-abiogenic C = 100 - ^{14}C).

Mountain, which has no carbonate exposures, presumably contain CO$_2$ gas from depth.

Using the ^{14}C and DIC values in Table 2, the concentration of biogenic DIC in the waters can be estimated and ranges from 0.3 mmol/kg in RMCS to 2.0 mmol/kg in MMSA-3, with an average of 0.8 mmol/kg. Cold water (0-5°C) could acquire this much biogenic DIC during recharge through normally respiring soils containing 0.55-4.4% CO$_2$, a range typical of these forest soils. Meanwhile, the three spring groups with the highest DIC also have the highest flow rates and discharge ~25 t/d of abiogenic CO$_2$. The remaining features, along with other possible seepage into streams, probably transport an additional 5-25 t/d of abiogenic CO$_2$.

4 GAS-WATER INTERACTION

For modeling the interaction as the CO$_2$-rich gas from depth encounters cold ground water or infiltrating snowmelt, the deep parent-gas mole fractions (x_i) are computed by correcting MMF (or v543) for air and are: 0.990 CO$_2$, 0.0092 N$_2$, 18 ppm Ar, 15 ppm He, and no O$_2$. The trace reactive species CH$_4$, H$_2$, and H$_2$S are not included in the model because reactions governing their removal in soils are complex. The partial pressure of each species at depth, P_i, is then Px_i where P is total pressure. Infiltrating water is assumed to be saturated with atmospheric gases. The partial pressures p_i (Henry law pressures) of the gases in infiltrating water are then P_{atm} times 0.781 N$_2$, 0.209 O$_2$, 0.00933 Ar, 0.00036 CO$_2$, and 5.2 ppm He, where atmospheric pressure $P_{atm} \cong 0.7$ atm at tree-kill elevation.

The gas-water interaction is modeled using a simple equation of gas exchange across an interface (see e.g., Broecker, 1974):

$$F_i/F_j = S_iD_i(P_i - p_i)/S_jD_j(P_j - p_j) \quad (1)$$

where the relative fluxes F of two species i and j across the interface are given by the ratio of their solubilities S, diffusion coefficients in water D, and pressure difference at the interface. We assume that both recharge and gas-water interaction take place at 0°C (e.g., infiltrating snowmelt), fixing all S_i, D_i, and p_i values, which are virtually independent of P. Calculated F_i/F_j values at any low temperature (<25°C) show that infiltrating water will be a large sink for CO$_2$, a small sink for He, and a source for N$_2$, O$_2$, and Ar. The gas exchange is run as a series of steps, with the gas composition modified after each step. Because the gas/water ratio is unknown, the dissolved gas composition is held constant. Thus P, which sets all P_i, is the only adjustable variable.

The BP-1 sample is used to illustrate results from a typical run. BP$_{calc}$ (Table 1) represents the model results for gas-water interaction at P = 1.6 atm. This composition results from loss of 52% of the CO$_2$ and 3% of the He from the parent gas composition to solution in water, while 24% of the N$_2$, 82% of the Ar, and all of the O$_2$ in BP$_{calc}$ are added to the gas from the water. In this particular run, soil respiration was also simulated by then removing 64% of the O$_2$ and replacing it mole for mole with CO$_2$ containing ^{14}C. Given the uncertainties in the temperature of interaction, that the gas/water ratio is unknown, and that equation 1 does not account for any physical properties of the soil or bedrock medium where the interaction takes place, BP$_{calc}$ is a reasonable match to BP-1. Similar modeling of other "He-enriched" soil-gas samples (Fig. 1) shows that all such compositions can be explained by interaction of the same parent gas with cold water at pressures of 1.2 - 1.6 atm, followed by air dilution.

Gas-water interaction can also be studied qualitatively using results from the water sampling. The CH12S water has absorbed a sufficient amount of the deep gas during underground flow, that at issuance, the pressure drop allows degassing. However, only a fraction of the CO_2 exsolves because of its moderate solubility. In this sense, the CH12S gas bubbles (Table 1) resemble a residual gas like BP-1, He-enriched due to the much lower solubility of He. The well sample MMSA-1 was collected under pressure without degassing. Gas dissolved in this fluid (Table 1) is proportionately low in He. Nevertheless, the fluid is enriched in He by a factor of four relative to atmospheric saturation, and the He isotopes show a definite magmatic signature. The CH12S and MMSA-1 gases represent both products of the gas-water interaction modeled above: a residual gas phase that is He-rich, and a liquid phase that is He-poor, relative to the parent-gas composition.

The water data also support the inference from modeling that gas-water interaction takes place at low pressures, just above ambient. Interaction at high pressures would create fluids with pCO_2 values much higher than those in Table 2. Of course such fluids could then lose CO_2 by silicate hydrolysis or by degassing through the overlying unsaturated zone prior to discharge. This latter possibility is of particular concern because many of the springs are effervescent. However, for the low pH (low HCO_3^-) fluids, the former process is minor, and the fact that such fluids (i.e., CH12S and MMSA-1) still contain the relatively insoluble species from recharge (including O_2) seems to preclude substantial prior degassing. Taken together, the gas and water data suggest that water flows fairly rapidly (years) from recharge points on the mountain to discharge points near the base, encountering the rising gas along shallow flow paths, and dissolving and transporting about 10% of the total CO_2 flux.

5 IMPLICATIONS AND FUTURE WORK

Numerical modeling suggests that intersample variations in tree-kill area soil-gas compositions can generally be attributed to cold-water interaction with preferential removal of CO_2 from a common parent gas source. The large flux of dissolved CO_2 in cold ground waters shows that such a process does in fact occur and on a scale sufficient to influence the soil-gas emissions. It is still unclear whether the dissolved flux is additive to the diffuse soil flux or is simply a part of the diffuse flux that is removed primarily during spring snowmelt when many of the He-enriched samples were collected. In contrast, most soil-gas flux surveying was done in the dry summer when water infiltration is at a minimum.

A more important question concerns the magnitude of the dissolved flux prior to dike intrusion. Pre-1989 information is available for only a few of the ground waters (VSS, MMSA-1 and 2B). These features were CO_2-rich well before 1989, providing additional evidence for long-term gas leakage from a deep reservoir beneath Mammoth Mountain. Information on the other springs will be obtained by the novel technique of checking ^{14}C in the rings of trees that overhang the springs (e.g., Hainsworth et al. 1995) for evidence of pre-1989 effervescence of abiogenic CO_2. If the long-term (pre-1989) CO_2 discharge was as high as the present 30-50 t/d rate, it may be necessary to invoke continuous thermal destruction of carbonate rocks at depth to supply much of the observed CO_2 at this volcano.

REFERENCES

Broecker, W.S. 1974. *Chemical Oceanography*. New York: Harcourt Brace Jovanovich.

Farrar, C.D., M.L. Sorey, W.C. Evans, J.F. Howle, B.D. Kerr, B.M. Kennedy, C.-Y. King & J.R. Southon 1995. Forest-killing diffuse CO_2 emission at Mammoth Mountain as a sign of magmatic unrest. *Nature* 376: 675-678.

Hainsworth, L.J., C.D. Farrar & J.R. Southon 1995. Magmatic CO_2 record in pine needles and growth rings at Mammoth Mountain, California. *Eos Trans. AGU* 76: F688.

Rahn, T.A., J.E. Fessenden & M. Wahlen 1996. Flux chamber measurements of anomalous CO_2 emission from the flanks of Mammoth Mountain, California. *Geophys. Res. Lett.* 23: 1861-1864.

Sorey, M.L., B.M. Kennedy, W.C. Evans, C.D. Farrar & G.A. Suemnicht 1993. Helium isotope and gas discharge variations associated with crustal unrest in Long Valley caldera, California, 1989-1992. *J. Geophys. Res.* 98: 15,871-15,889.

Sorey, M.L., W.C. Evans, C.D. Farrar & B.M. Kennedy 1998. Carbon dioxide and helium emissions from a reservoir of magmatic gas beneath Mammoth Mountain, California, USA. (this volume).

Sorey, M.L., W.C. Evans, B.M. Kennedy, C.D. Farrar, L.J. Hainsworth & B. Hausback 1998. Carbon dioxide and helium emissions from a reservoir of magmatic gas beneath Mammoth Mountain, California, USA. (submitted to *J. Geophys. Res.*).

Budget and sources of volatiles discharging at Kudryavy Volcano, Kurile Islands, Russia

Tobias P. Fischer & Stanley N. Williams
Department of Geology, Arizona State University, Tampa, Ariz., USA

Yuji Sano
Department of Earth and Planetary Sciences, Hiroshima University, Japan

Mikhail A. Korzhinski
Institute of Experimental Mineralogy Chernogolovka, Moscow District, Russia

ABSTRACT: Geochemical studies concerning fluxes of sedimentary material and fluids from the slab into zones of arc magma generation demonstrated the significance of sediment recycling in arc magmas. SO_2 flux and fumarole gas compositions allow derivation of a volatile budget for Kudryavy. SO_2 flux is 73 t/d. $^3He/^4He$ corresponds to 84% mantle origin and a flux of 2200 mol/a mantle He. 1.0 Mt/a of mantle material is required to generate this flux. This mass produces a flux of 0.025 mol/a ^3He and 50 Mmol/a of mantle CO_2. Using C-isotopic data, 12% of CO_2 are derived from mantle, 67% from carbonate in subducted, altered oceanic crust, 21% from subducted sediments. The flux of 280 Mmol/a of carbonate derived CO_2 requires 0.41 Mt/a of oceanic crust, and 0.35 Mt/a of sedimentary material. Mantle-Nitrogen contributes 2% of the total N_2 flux of 5.4 Mmol/a. Assuming N derived from subducted sediments, its concentration there is 460 mg/kg.

1 INTRODUCTION

The contribution of volatiles such as CO_2, SO_2, and H_2O from volcanoes to the atmosphere has received considerable attention due to their potential climatic impact and need for comparison to anthropogenic fluxes. Recycling of volatiles from the atmosphere back to the mantle and volcanoes occurs at subduction zones and zones of arc magma generation. The release of volatiles, mostly water, from subducted sediments is responsible for hydration of overlying peridotitic mantle wedge and generation of arc magmatism. The amount of subducted and released volatiles must be balanced with the amount released from individual volcanoes, at least in the long term. The purpose of this communication is to investigate this hypothesis for one single arc volcano and compare the amount of volatiles potentially released due to subduction and the amount measured at the volcano.

Kudryavy, located on the northeastern shore of Iturup island, is a basaltic andesite cone which last erupted in 1883. On top of the cone are fumarole fields covering an area of 300 x 70 m. Vent temperatures reach a maximum of 920°C (in August 1995). Kudryavy is the site of a unique Re-sulfide mineralization (Korzhinsky et al., 1994), but its gas chemistry is typical for high temperature subduction zone volcanoes (Taran et al., 1995). Kudryavy is part of the Kurile island arc along which formation of accretionary prisms is rare and nearly 100% of sea floor sediments are subducted (von Huene and Scholl, 1991).

2 ISOTOPIC COMPOSITION OF He AND C

Using the chemical composition of gas samples collected in August 1995 and eliminating effects of air contamination, the average composition of the parent gases feeding Kudryavy fumaroles can be derived and is shown in Table 1. The isotopic compositions of C, He, and Ne are shown in Table 2. The focus here is on the composition of non-reactive gases (He, Ar, N_2, CO_2), He and C isotopes, and the flux of volatiles using SO_2 flux data.

Table 1. Composition of parent gas feeding Kudryavy high-temperature fumaroles in mmol/mol. -: not available

	H_2O	CO_2	St	SO_2	H_2S	HCl	HF	He	H_2	Ar	O_2	N_2	N_{2exc}
parent	968	13.8	13.8	11.5	2.3	3.5	-	0.00009	-	0.0005	-	0.19	0.18
flux Mmol/a	29180	416	416	3.47	69	106	-	0.00270	-	0.0150	-	5.7	5.40

Table 2. Isotopic composition of C (‰ PDB), He (RA), and Ne and total amounts of ^{3}He and ^{4}He

T (°C)	δ^{13}C-CO$_2$	^{3}He/^{4}He	^{3}He/^{20}Ne	^{3}He/^{4}He$_{corr}$	CO$_2$/^{3}He	^{3}He pmol/mol	^{4}He mmol/mol
187	-5.2	6.65	127	6.66	9.3x10^9	1.68	0.000179
480	-6.6						
690	-7.6	6.61	14.8	6.73	16.5x10^9	0.75	0.00008
710	-7.3	3.10	0.508	6.63	5.3x10^9	2.13	0.00023
920	-7.3	6.35	4.49	6.76	8.7x10^9	1.32	0.00014
Flux (mol/a)						0.025	2700

The ^{3}He/^{4}He$_{corr}$ have a narrow range of 6.63 - 6.76 R$_A$, independent of temperature. The average ^{3}He/^{4}He ratio of 6.7 R$_A$ is consistent with values for other Kuril-Kamchatka arc volcanoes and implies a mantle He fraction of approximately 84% (Poreda and Craig, 1989).

Absolute concentrations of ^{3}He, in mmol/mol, can be obtained by

$$x_{^3He} = 14 \times 10^{-6} \, x_{He} R_A \quad (1)$$

where x_{He} is the amount of He in the gas sample.

The amount of ^{4}He in mmol/mol can be obtained by
$$x_{^4He} = x_{He} - x_{^3He} \quad (2)$$
Results are shown in Table 2.

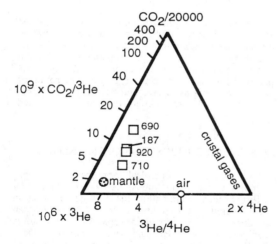

Fig. 1. Relative amounts of CO$_2$, ^{3}He, ^{4}He, and gas temperatures in °C.

The average δ^{13}C value of CO$_2$ in the high temperature gases is 7.2‰ (relative to PDB). The relative amounts of CO$_2$, ^{3}He and ^{4}He may be used to identify the sources of these gases. Based on the composition of gases from MORB glasses, a best ratio of CO$_2$/^{3}He of 2x10^9 is derived for mantle gases (Marty and Jambon, 1987). The CO$_2$/^{4}He ratios for gases derived from crustal sources in New Zealand range from 200 to 50000 (Giggenbach et al., 1993). Figure 1 shows the relative amounts of CO$_2$, ^{3}He and ^{4}He of Kudryavy gases. The CO$_2$/^{3}He ratios range from 5.3-16.5x10^9 (Table 2). The addition of CO$_2$ to the mantle-derived MORB CO$_2$ may be the result of the decomposition of marine sediments on the subducting slab carried down to the zones of arc magma generation as proposed by Marty and Giggenbach (1990), and Varekamp et al. (1992).

Recently Sano and Marty (1995) quantitatively determined the sources of C in fumarole gases from island arc volcanoes using the CO$_2$/^{3}He ratio in combination with the δ^{13}C value of CO$_2$ to distinguish between sedimentary, limestone, and MORB carbon. Results for Kudryavy are shown in Table 3.

The source of non-atmospheric N$_2$ can also be evaluated using the ^{3}He, and ^{4}He contents of the gases (Giggenbach et al., 1993). Figure 2 shows relative abundances of N$_2$(excess) (corrected for air contamination), ^{3}He, and ^{4}He, indicating primarily a sedimentary/metasedimentary source for N$_2$.

Table 3. Flux and average relative CO$_2$ contributions from mantle, limestone and sediments.

	Mantle (%)	Limestone (%)	Sediment (%)
Average (%)	12.5	67	21
Flux (Mmol/a)	50	280	86

3 VOLATILE FLUXES

The SO$_2$ flux was measured by the COSPEC (Stoiber et al., 1983). Total SO$_2$ flux is 73±15 t/d, or 416 Mmol/a. The high temperature, medium temperature and low temperature field contribute 52±9 t/d, 14±4 t/d and 7±4 t/d, respectively.

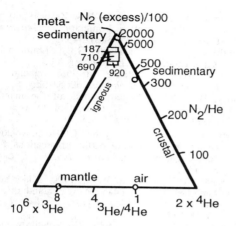

Fig. 2. Relative amounts of N_2, ^3He and ^4He

Using the SO_2 flux and the parent gas feeding Kudryavy fumaroles, the flux of gas species from the volcano can be calculated according to

$$F_i = x_i\, F_{SO_2}/x\, SO_2 = 30.1 x_i \quad (3)$$

where F_i is the flux of species i in Mmol/a and x_i is the concentration of gas species i in mmol/mol. For the flux calculations, the amount of total S (St) is used, not that of SO_2 alone. The fluxes of the gas species are shown in Table 1, 2, and 3. The He flux is 2700 Mmol/a with an average $^3He/^4He$ ratio of 6.9 R_A, and assuming a mantle $^3He/^4He$ ratio of 8 R_A, approximately 84% are of mantle origin and 16% are radiogenic. Therefore, 2200 mol/a of the total He are of mantle origin and 500 mol/a are from other sources. The flux of ^3He from the mantle is

$$F_{^3He} = 8.0\, R_A\, F_{He} = 0.025 \text{ mol/a} \quad (4)$$

This is considerably smaller than the ^3He flux of 0.34 mol/a. measured at White Island with the same methodology (Marty and Giggenbach, 1990). In order to obtain estimates of the amount of mantle material required to produce this He flux, information on the He concentrations of the mantle are required. Sarda and Graham (1990) propose a concentration of 2200 mol/Mt based on analyses of Mid-Atlantic popping rocks. Using this value as being representative of the He concentration in the mantle, 1.0 Mt/a of mantle material is required to produce the observed He flux of 2200 mol/a. At a $^3He/^4He$ ratio of the mantle of 8 $R_A = 11.2 \times 10^{-6}$, the ^3He concentration in the mantle is 0.025 mol/Mt.

The origin of radiogenic ^4He in fumarole gases is more problematic because of the large amounts of crustal rocks and the very efficient extraction mechanism required to produce the radiogenic He flux of 500 mol/a. The most likely explanation for the presence of radiogenic He is the contribution of He from old crustal rocks and that produced from magma aging.

According to Table 1, the flux of CO_2 is 416 Mmol/a. Using C-isotopes of CO_2 in combination with $^3He/^4He$ ratios (Table 2), the relative contributions of MORB CO_2, marine limestone CO_2 and sedimentary CO_2 to the total flux can be calculated. The results are shown in Table 3. Since the flux of CO_2 and the concentration of CO_2 in the mantle were calculated from the accepted value of $CO_2/^3He$ of 2×10^9, the amount of material required to produce the mantle CO_2 flux is the same as that required to produce the mantle derived ^3He and ^4He flux. The CO_2 concentration in the mantle then becomes 50 Mmol/Mt, or 0.22 wt%. This is well below the CO_2 concentration of 1.4 % in the mantle reported by Pineau and Javoy (1983).

The proportion of CO_2 derived from subducted carbonate is 67%, corresponding to a flux of 280 Mmol/a. The carbonate contents of deep sea sediments near the Kurile trench is near zero (Rea et al., 1993), suggesting that subducted sedimentary carbonates contribute only little to the carbonate derived CO_2 flux. Staudigel et al. (1989) estimate the carbonate content in the upper 500m of altered oceanic crust to be 30 000 mg/kg. Using these values and assuming complete devolatilization, 0.41 Mt/a of hydrothermally altered oceanic crust would have to be subducted below Kudryavy to produce the observed carbonate-derived CO_2 flux. In order to estimate the amount of altered crust subducted below Kudryavy, an average spacing of the volcanoes in the Kuriles of 26 km (Voldavetz and Piip, 1959) and complete devolatilization of a 500 m thick carbonate layer is assumed. At a subduction rate of 0.08 m/a (von Huene and Scholl, 1991), the mass of altered oceanic crust subducted is 2.6 Mt/a, 6.3 times the amount required. Even if only a fraction of the CO_2 devolatilizes, the carbonate-derived CO_2 is sufficient to supply the amount measured at the volcano.

The decomposition of organic matter in sedimentary rocks contributes 87 Mmol/a to the measured CO_2 flux. VonHuene and Scholl (1991) classify the Kurile continental margin as a Type 2, indicating that no accretionary prisms are formed and almost 100% of the seafloor sediments get subducted. Rea et al (1993) report a organic carbon content of deep sea sediments of 1000 - 5000 mg/kg. Using the average concentration of 3000 mg/kg, the mass of sediments required to produce the observed CO_2 flux is 0.34 Mt/a. Assuming a solid thickness of the sedimentary pile at the Kurile trench to be

240m (von Huene and Scholl, 1991), the mass of sediments subducted along 26 km of trench is 1.00 Mt/a or almost three times that required to produce the measured CO_2 flux. Therefore, subduction of altered oceanic crust and sediments are well able to account for the excess, non-mantle CO_2 flux of Kudryavy.

The flux of N_2 from Kudryavy is 5.4 Mmol/a (Table 1). Mantle N contents are 2.8 mg/kg, or 0.1 Mmol/Mt (Marty, 1986). The amount of mantle material required to produce the N_2 flux is the same as that required to produce the ^{3}He and mantle CO_2 fluxes, 1.0 Mt/a. Using the mantle N concentration of 0.1 Mmol/Mt, the mantle N_2 flux is 0.1 mol/a, or < 2% of total N_2 discharge. The most likely source for the excess, non-mantle N_2 is subducted sedimentary material. Assuming that the N_2 flux is produced from the 0.34 Mt/a of subducted sediments generating the organic CO_2 flux, the concentration of N_2 in these sediments is required to be 460 mg/kg. The total N content in bulk sediment in the northern Kurile trench was found to be 3000 mg/kg on average (Rea et al., 1993). This suggests that the amount of N_2 supplied from subducted sediments is sufficient to produce measured N_2 flux at Kudryavy.

4 CONCLUSIONS

The model used for comparing the fluxes of material subducted below Kudryavy to those released from the volcano, assumes a steady-state release of volatiles from the zones of arc magma generation to the volcano. Despite its limitations, it allows to establish a budget of inert volatiles released from the volcano and an assessment of their sources. Kudryavy is well suited for application of this model because it has had long-term, high temperature activity without eruptions and is located on a trench without an accretionary prism. Estimations of the budget of volatiles supplied to and released from arc volcanoes has significant implications for evaluating the Earth's volatile budget and element cycling in subduction zones.

ACKNOWLEDGMENTS
This work was supported by NSF International Programs (SNW and TPF) and NASA Earth Systems Science Graduate Fellowship (TPF). We thank our Russian colleagues, especially K. Smulovich, and G. Steinberg for logistical support, G. Lyons, W. Giggenbach, and B. Christensen for generous help with gas analyses at IGNS, New Zealand. This work has benefited from extensive discussions with W. Giggenbach. We thank B. Symonds and an anonymous reviewer for comments on an earlier version of the manuscript.

REFERENCES

Giggenbach, W.F., Sano,Y., & Wakita, H., 1993. Isotopes of H_2, and CO_2 and CH_4 contents in gases produced along the New Zealand part of a convergent plate boundary. *Geochim. Cosmochim. Acta* 35: 3427-3455.

Korzhinsky, M.A., Tkachenko, S.I., Shmulovich, K.I., Taran, Y.A. & Steinberg, G.S. , 1994. Discovery of a pure rhenium mineral at Kudryavy volcano. *Nature* 369: 51-52.

Marty, B. 1995. Nitrogen content of the mantle inferred from N2/Ar correlation in oceanic basalts. *Nature* 377: 326-329.

Marty, B. & Jambon, A., 1987. C/^{3}He in volatile fluxes from the solid earth: Implications for carbon geodynamics. *Earth Planet. Sci. Lett.* 83: 16-26.

Marty, B. & Giggenbach, W.F., 1990. Major and rare gases at White Island, New Zealand: origin and flux of volatiles. *Geophys. Res. Lett.* 17(3): 247-250.

Pineau, F. & Javoy, M., 1983. The volatile record of a 'popping' rock from the Mid-Atlantic Ridge at 15°N: concentrations and isotopic composition (abs.), *Terra Cognita* 6: 191.

Poreda, R. & Craig, H., 1989. Helium isotope ratios in Circum-Pacific volcanic arcs. *Nature* 338: 473-478.

Rea, D.K., Basov, I.A., Janecek, T.R. & Palmer-Julson, A., 1993. Site 881, *Proc. ODP Init. Repts.* 145, pp. 37.

Sano, Y. & Marty, B., 1995. Origin of carbon in fumarolic gas from island arcs. *Chem. Geol.* 119: 265-274.

Sarda, P. & Graham, D., 1990. Mid-ocean ridge popping rocks: implications for degassing at ridge crests. *Earth and Planet Sci. Let.* 97: 268-289.

Staudigel, H., Hart, S.R., Schmincke, H.-U., & Smith, B.M., 1989. Cretaceous ocean crust at DSDP Sites 417 and 418: Carbon uptake from weathering versus loss by magmatic outgassing. *Geochim. Cosmochim. Acta* 53: 3091-3094.

Stoiber, R.E., Malinconico, L.L. & Williams, S.N., 1983. Use of the correlation spectrometer at volcanoes. in: *Forecasting volcanic events*, H. Tazieff and J. Sabroux, eds. 1, pp. 425-444, Elsevier, New York.

Taran, Y.A., Hedenquist, J.W., Korzhinsky, M.A., Tkachenko, S.I. & Shmulovich, K.I., 1995. Geochemistry of magmatic gases from Kudryavy volcano, Iturup, Kurile Islands. *Geochim. Cosmochim. Acta* 59: 1749-1761.

Varekamp, J.C., Kreulen, R., Poorter, R.P.E. & Van Bergen, M.J., 1992. Carbon sources in arc volcanism, with implications for the carbon cycle. *Terra Nova* 4: 363-373.

von Huene, R. & Scholl, D.W., 1991. Observations at convergent margins concerning sediment subduction, subduction erosion, and the growth of continental crust. *Rev. Geophys.* 29: 279-316.

Sulfur isotopes in rocks from the Katla Volcanic Centre – With implications for Iceland mantle heterogeneities?

L.W.Hildebrand & P.Torssander
Department of Geology and Geochemistry, Stockholm University, Sweden

ABSTRACT: S concentrations and isotopic compositions of ΣS have been analysed in a rock suite from the Katla Volcanic Center, SE Iceland. The $\delta^{34}S$ values for the Fe-Ti basalts are between -1.8 and +1.9 ‰, with the lowest $\delta^{34}S$ values in the rocks with the highest S concentration (600-1500ppm). The $\delta^{34}S$ variation is slightly larger than the $\delta^{34}S$ range from -2.0 to +1.0 ‰ determined previously in Icelandic basalts (Torssander, 1989). The Katla rhyolites have more positive $\delta^{34}S$ values, from +0.9 to +2.4 ‰. All the Katla rocks are sulfate deficient implying that the $\delta^{34}S$ distribution cannot be explained by crustal assimilation, fractional crystallisation, partial melting or degassing processes. The basalt S isotope composition may therefore indicate heterogeneities in the source area.

1 INTRODUCTION

Iceland defines a chemical production anomaly which has formed a plateau across the Mid Atlantic Ridge between the Reykjanes Ridge and the Kolbeinsey Ridge. S isotope ratios can be used to determine the mantle complexity beneath Iceland. Previous S-isotope studies in Iceland have covered a larger area with sparse sampling from several volcanic areas. This study focuses on the Katla Volcanic Centre. Using these new results, we hope to gain insight on the observed S isotope variation in Icelandic rocks.

2 BACKGROUND

The active volcanic zones of Iceland (Fig.1) include some 20 recently active volcanic systems. The recent volcanic activity is dominated by the Western and Eastern Rift Zones but is also occurring in off-rift volcanic systems. The Eastern Rift is assumed to be propagating towards the south (Oskarsson et al. 1982). While the rift volcanism is exclusively tholeiitic, the off-rift and propagating rift volcanism is characterised by alkalic or transitional rocks.

Katla volcanic system covers an area of about 700 km² but 2/3 of it is covered by the Myrdalsjökull ice-sheet. The age of the system is estimated at <200 kyrs. The Katla central volcano is characterised by its bimodal volcanism, erupting mostly basalts, and <10% silicic ejecta. A geothermal system is located inside the caldera (~10 km in diameter) and confined to the volcanic centre (Björnsson et al. 1994). The volcanic outcrops consist mostly of hyaloclastites but interglacial compact lava flows cap the hyaloclastites closer to the centre.

The Katla basalts, transitional on an alkali scale, are all fairly evolved (<6.7% MgO), and homogeneous, but a distinction can be made between the most primitive (> 6.0 % MgO, rare), the most frequent group (6.7-4.2% MgO) and the "evolved" FeTi-basalts (4.2-3.4% MgO). High Fe and Ti contents characterise all the basalts and clear Fe and Ti enrichments are observed due to the

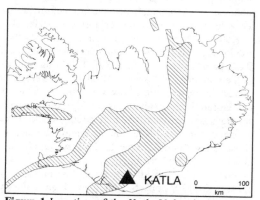

Figure 1 Location of the Katla Volcanic Centre.

extensive olivine fractionation before onset of Fe-Ti oxide fractionation (evolved basalts).

The Katla rhyolites, confined mostly to the volcanic centre, constitute a chemically homogeneous group (67-72% SiO_2) and show tendencies of fractionation (e.g. Zr depletion in the most evolved samples). Trace element studies favour the regional Fe-Ti basalt as the parent for the Katla rhyolites.

Fourteen rock samples from the Katla volcanic system were analysed for S concentration and $\delta^{34}S$. The samples were mostly basalts with some of differentiation. The other samples consist of three rhyolites and one intermediate sample. The S concentration and $\delta^{34}S$ have been determined with the "Kiba" - reaction technique and IRMS (Ueda & Sakai, 1983). All samples were measured for $\delta^{34}S$ in total S since the sulfate content is 5% or less. The accuracy of the $\delta^{34}S$ measurements is better than ±2‰.

3 RESULTS AND DISCUSSION

3.1 S concentration

The S concentration in the Katla basalts is generally 115-885 ppm (average 456 ppm) except for one sample (1500 ppm). This is lower than the estimate for juvenile S concentrations for MORB (800-1000 ppm; Chaussidon et al. 1989). As most of the samples are glassy hyaloclastites, eruption was subglacial or submarine. Moore and Schilling (1973) concluded that S is not lost during degassing below water depths of 200-500 m but explosive basaltic eruptions from depths of 2000 m indicate that degassing can occur at greater depths (Gill et al. 1990). The water depth during Katla eruptions is unknown, but was probably mostly shallow (<1000 m) since decreasing lithostatic pressure during ice unloading causes vigorous crustal movements due to rapid isostatic rebound; hence, the production rate for the volcanoes are much higher during interglacials.

Katla basalts contain on average more S than previous analyses of Icelandic tholeiites and alkali-basalts (Torssander, 1989). This could be due to their higher Fe content (12.6-16.8% FeO) that prevail S saturation. However, the variable S concentration in basalts of relatively homogeneous composition indicates that degassing could have occurred, but to what extent is somewhat unclear. However, we note that the less glassy samples are also the most depleted in S.

The rhyolite obsidian/pitchstone samples only contain 31-172 ppm S, but their high chloride content (1147-1410 ppm) would eventually contradict any vigorous degassing since Cl degasses more efficiently due to its higher diffusion rate (Baker et al. 1996).

3.2 S isotopes

The $\delta^{34}S$ of the basalts is between -1.8 and +1.9‰. This variation is slightly larger than previously found in Icelandic basalts (Torssander, 1989). This is similar to $\delta^{34}S$ values for MORB, which suggest that the depleted mantle vary little in its S isotope composition. The average $\delta^{34}S$ of total S in ocean-floor basalts varies from +0.3 to +0.7‰ (Grinenko et al. 1975; Sakai et al. 1984).

Even though degassing has occurred there is no correlation between S content and $\delta^{34}S$. Rapid degassing of SO_2 from ascending magma results in chemical and isotopic disequilibrium between sulfate and sulfide in the melt and between melt and vapor, with sulfide being more strongly partitioned into the melt. This was first identified in basalts from Kilauea Volcano, Hawaii (Sakai et al. 1982). The curious fact that lighter S would be enriched in the melt during degassing arises from the predominance of sulfide over sulfate in the melt and because SO_2 is enriched in heavier S relative to the melt. However, any correlation between $\delta^{34}S$ and S concentration is not seen for the Katla basalts (fig. 2). The most negative $\delta^{34}S$ value was determined for the basalt having the highest S concentration of 1500 ppm, which is large enough to suggest undegassed. Heavier S was found in the Katla basalts with lower S concentrations, contradictory to what would be expected from effervescence processes. Previous studies in Iceland have suggested degassing to be the main cause for the S isotope distribution in Icelandic rocks (Torssander, 1989), but it is now apparent that the most negative $\delta^{34}S$ values have been found for the rocks with highest S concentration. Thus, degassing cannot be the only cause for the overall large variation in $\delta^{34}S$ of Icelandic rock, at the low sulfate/sulfide ratio encountered.

The three more evolved Fe-Ti basalt samples have a slightly higher average $\delta^{34}S$ value (+0.7 ‰) compared with the other Fe-Ti basalts ($\delta^{34}S$=-0.6 ‰). This observed $\delta^{34}S$ increase parallels an assumed magmatic differentiation, as the Katla suite shows enrichment in Fe and Ti, followed by depletion consistent with Fe-Ti oxide fractionation. Fe-Ti oxides commonly have inclusions of sulfide globules as a result of supersaturation and crystallisation of sulfides. This explanation is not

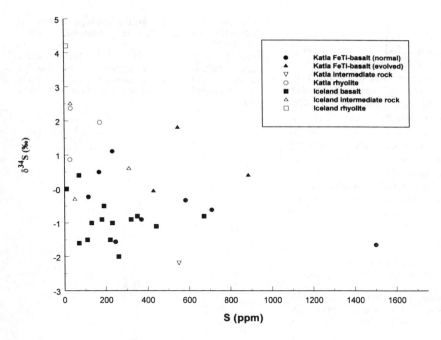

Figure 2 S concentration and isotope composition in Icelandic rocks.

satisfying for the $\delta^{34}S$ variation among the Katla basalts as this process requires a corresponding decrease in Fe and S (as seen for the Galapagos Rift suite, (Davis et al. 1991) and no sulfide globules are observed in any of the samples. Furthermore, the low sulfate/sulfide ratio suggests that this process would not increase the $\delta^{34}S$ value of the basalt to a large extent. Thus, the overall $\delta^{34}S$ variation seen for the Katla and Icelandic basalts/andesites (-2.2 to +1.9 ‰) is inconsistent with simple igneous processes. It suggests that the $\delta^{34}S$ values are related to differences in the S isotope composition in the source region. S isotope variations in the mantle was measured by analysing micro-sulfide inclusions in liquid globules or grains in various minerals or rocks of mantle origin and found to be heterogeneous (Chaussidon et al. 1989).

The $\delta^{34}S$ for the Katla rhyolites range between +0.9 and +2.4‰. Compared to the basalts they are relatively enriched in $\delta^{34}S$, a phenomenon also observed in previous studies (e.g Krafla; Torssander, 1989). This could partly be caused by late stage oxidative degassing that could enrich the melt in $\delta^{34}S$.

Enrichment in $\delta^{34}S$ is known to be achieved by magma-seawater interaction ($\delta^{34}S$ of seawater sulfate is 21‰). However, Icelandic rhyolites have low $\delta^{18}O$ caused by meteoric water-rock interaction ($\delta^{18}O$ in Katla is +3.9 to 6.2‰), indicating that seawater is probably not involved in the hydrothermal systems. Heavy S with another origin than seawater is concurred by the fact that even rhyolites far from the coastline (Krafla) are enriched in $\delta^{34}S$ (Torssander, 1989).

Trace element studies of the rhyolites (e.g. Rb) suggest that there has been no crustal contamination/assimilation.

The proposed kinetic S isotope effects of thermal decomposition of pyrite to cause an enrichment of $\delta^{34}S$ (Kajiwara et al. 1981) is difficult to apply to the Katla rhyolites. High temperature (600°C) experimental studies on pyrite decomposition to pyrrhotite and native S show that the pyrrhotite dominant solid residue is enriched by up to 4.5‰ over the initial material. It is difficult to envision that the solid pyrrhotite residue would be present in the resulting silicic melt without the elemental S. Degassing of H_2S from shallow level magma chambers or late stage oxidative degassing could possibly enrich the rhyolites and intermediate rocks in $\delta^{34}S$.

4 CONCLUSIONS

The Katla Fe-Ti transitional basalts show higher S content and larger range in $\delta^{34}S$ values than previous analyses of Icelandic basalts. SO_2 degassing may occur during eruption of most of these rocks, but as rocks with higher S content have more negative $\delta^{34}S$ values, degassing cannot cause the large variation in isotope composition observed. Furthermore, partial melting, crustal assimilation and fractional crystallisation processes cannot cause the observed large variation either, especially in light of the low sulfate/sulfide ratios. Therefore, the variation in $\delta^{34}S$ may be due to heterogeneities in the source region.

5 REFERENCES

Baker, L. L.& M. J. Rutherford 1996. Sulfur diffusion in rhyolite melts. *Contrib. Mineral. Petrol.* 123:335-344.

Björnsson, H., F. Palsson. M.T. Gudmundsson 1994. The topography of the Katla caldera beneath the ice cap Myrdalsjökull, South-Iceland. *EOS, Trans. AGU* 75:321.

Chaussidon, M. F., F. Albarede and S. M. F. Sheppard 1989. Sulfur isotope variations in the mantle from ion microprobe analyses of micro-sulphide inclusions. *Earth. Planet. Sci. Lett.* 92: 144-156.

Davis, A. S., D. A. Clague, M. J. Schulz. J. R. Hein. 1991. Low sulfur content in submarine lavas: An unreliable indicator of subaerial eruption. *Geology* 19:750-753.

Gill J., Torssander P., Lapierre H., Taylor R., Kaiho K., Koyama M., Kusakabe M, Aitchison J., Cisowski S., Dadey K., Fujioka K., Klaus A., Lovell M., Marsaglia K., Pezard P., Taylor B. and Tazaki K. 1990. Explosive deep water basalt in the Sumisu backarc rift. *Science*, 248, 1214-1217.

Grinenko, V. A., L. V. Dmitrev, A. A. Migdisov, A. Y. Sharaskin 1975. Sulfur contents and isotope composition for igneous and metamorphic rocks from Mid-Ocean Ridges. *Geochem. Int.* 12:132-137.

Moore, J. G. & J.-G. Schilling. 1973 Vesicles, water, and sulfur in Reykjanes Ridge basalts. *Contrib. Mineral. Petrol.* 41:105-118.

Oskarsson, N., G. E. Sigvaldasson, S. Steinthorsson 1982. A dynamic model of rift zone petrogenesis and the regional petrology of Iceland, *J. Petrol.* 23:28-74.

Sakai, H., T. J. Casadevall, J. G. Moore 1982. Chemistry and isotope ratios of sulfur in basalts and volcanic gases at Kilauea volcano, Hawaii. *Geochim. Cosmochim. Acta* 46:729-738

Sakai H, DesMarais DJ, Ueda A and Moore JG 1984. Concentrations and isotope ratios of carbon, nitrogen and sulfur in oceanfloor basalts. *Geochim. Cosmochim. Acta* 48:2433-2441

Torssander, P. 1989. Sulfur isotope ratios of Icelandic rocks. *Contrib. Mineral. Petrol.* 102:18-23

Ueda, A. and Sakai, H. 1983. Simultaneous determination of the concentration and isotope ratios of sulfate- and sulfide-sulfur and carbonate carbon in geological samples. *Geochem. J.* 17:185-196.

Changes in Cl concentrations and isotope values of hot spring waters at Kuju volcano, Japan, prior to the 1995 eruptive activity

R. Itoi, T. Kai & M. Fukuda
Geothermal Research Center, Kyushu University, Kasuga, Japan

I. Kita
Research Institute of Materials and Resources, Mining College, Akita University, Japan

ABSTRACT: Hot spring waters discharging in the fumarolic area at Kuju volcano have been sampled and analyzed for chemical component and stable isotopes regularly from 1992 to 1996. During this period, eruptive activity started on October 11, 1995 and new craters were formed 200m south of the present fumarolic area. Concentrations of Cl and isotope values of hot spring waters show an appreciable increase starting a few months before the beginning of the eruption. These chemical and isotopic data together with discharge behavior of hot springs suggest an increase in volcanic gas outflow from andesitic magma, probably caused by an intrusion of magma at depth.

1 INTRODUCTION

Kuju volcano is located in the central part of Kyushu Island, Japan. Eruptive activity started on October 11, 1995, about 250 years after the last eruption. This activity was accompanied by a small amount of volcanic ash discharge in October and December, 1995, followed by continuous discharging volcanic gases from newly formed craters. The craters, located about 200 m south of the present fumarolic area, are on the northern flank of Mt.Hossho, one of the volcanic cones of Kuju volcano. Most fumaroles exceed 200°C (Ehara, 1992); maximum temperature was 508°C in 1960 (Mizutani et al., 1986). There are several hot springs located within and surrounding the fumarolic area. These hot spring waters are formed by mixing volcanic gases with ground water (Ehara et al., 1981; Mizutani et al., 1986). Therefore, changes in contribution of volcanic gases to hot spring waters might be detected by sequential sampling and analyzing waters. This information may also provide an estimation of activities of degassing from magma and of magma itself at depth.

Increases in Cl concentration and isotope values were found in waters sampled within the fumarolic area before the beginning of eruptive activity. Moreover, one hot spring stopped discharging before the eruption and two other hot springs also gradually decreased their flow rates and ceased discharging after the eruption. These changes in water chemistry and discharging behavior of hot spring may be due to increases in

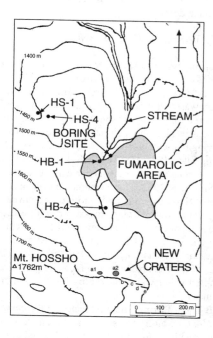

Figure 1 Location map of sampling points (HS-1, HS-4, HB-1, Boring Site, and HB-4), fumarolic area, and newly formed craters during 1995 eruption at Kuju volcano.

supply and temperature of volcanic gases a few months before the beginning of eruptive activity, suggesting

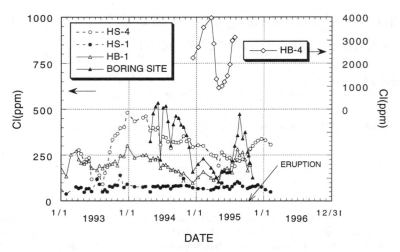

Figure 2 Concentration of Cl with time for hot spring waters at Kuju volcano. Eruptive activity started on October 11, 1995, indicated by arrow.

direct outflow of gases from a possible intrusion of magma at depth.

2 FEATURES OF HOT SPRINGS

Hot spring water flows out from the banks as well as the bottom of the stream running northward in the central part of the fumarolic area. Mixing of low TDS stream water with hot spring water causes significant dilution during rainy season; thus, waters discharging from places high above the stream were sampled. Discharging hot waters have been sampled at sites HB-1, Boring Site, HB-4, HS-1 and HS-4 (Figure 1) with an interval of two weeks except once a month during winter time since 1992. These waters are acid sulfate except for HB-4 which is acid chloride, and have pHs of 1-2. Temperatures were measured with a thermocouple. Continuous measurement of temperature has been conducted at HS-1 from April 1995. An automatic rain gage has been installed 1km north(1283 m a.s.l.) of the fumarolic area to measure precipitation. Sampled water was analyzed by ion cromatography for anions.

HB-1 and Boring Site are located in the central fumarolic area along the stream and their flowrate are rather small(~0.5-1 l/min). HB-4 is located in the upper stream and discharges maximum flowrate of ~3 l/min from fractures possibly connected to the most active and hottest fumaroles(Mizutani et al., 1986). The other two hot springs, HS-1 and HS-4, are located on the peripheral of the fumarolic area and their flowrate is 60-120 l/min. Temperatures are 40-55°C for HS-1 and HS-4, and 80-95°C for HB-1, Boring Site and HB-4. Sulfate concentrations for HS-1 and HS-4 are relatively stable at 820 to 2500 ppm whereas those of HB-1 and Boring Site ranges widely from 2300 to11500 ppm. HB-4 ranges from 2300 to 3400 ppm.

The discharge rate for HB-4 started decreasing in early June 1995, whereas the other hot springs showed increased flowrates, and finally stopped discharging at the end of July 1995. HB-1 and Boring Site also stopped discharging within a month after the eruption. HS-1 and HS-4 have continued discharging.

3 RESULTS AND DISCUSSION

Figure 2 shows Cl concentration changes with time during 1993-1996. All vents but HB-4 have Cl concentration below 500 ppm. HB-4, however, has a maximum Cl concentration of 3970 ppm, which is consistent with those measured for fumarolic condensate of the highest temperature fumaroles(Mizutani et al., 1986).

Values for HB-1, Boring Site, and HS-4 show a continuous decrease during 1994 in spite of fluctuations at the Boring Site. Then, they show a slight increase in early 1995 which is consistent with HB-4. On the other hand, values of HS-4 continue decreasing. The former then increase during May, 1995. These increases suggest that an increase in the outflow or HCl content of the volcanic gases. These decreasing and increasing trends for Cl during 1994-1995 are consistent with estimated discharge rate of HCl gas from the fumarolic area on the base of rain chemistry(Itoi et al., 1997). On the other hand, values of HS-1 remain almost constant and change in Cl concentration of HS-4 seems to

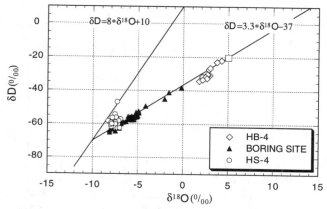

Figure 3 Plot of δD and $\delta^{18}O$ for hot spring waters. Isotope values of possible andesitic magma water($\delta D=-20‰$ and $\delta^{18}O=5‰$) is indicated by open square. A mixing line between meteoric water and magmatic water, and a line for meteoric water are also shown.

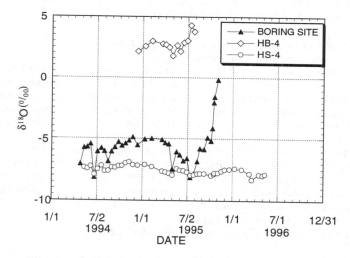

Figure 4 Changes in $\delta^{18}O$ with time for hot spring waters of HB-4, BoringSite, and HS-4.

be delayed compared with those of HB-1, Boring Site and HB-4.

Stable isotope values for samples from HB-4, Boring Site, and HS-4 are plotted in Figure 3 along with the possible composition of magmatic water. Values of HB-4 are plot close to the source of magmatic water, indicating that HB-4 has a large magmatic gas component. Moreover, four water samples collected during June and July plot on the mixing line, very close to the values of magmatic water. This suggests that some waters contain nearly 100% condensed magmatic gas. Most values of Boring Site also plotted on the mixing line and four samples show a trend of increasing in $\delta^{18}O$ from -5 to 0 ‰ during the period of October to November(Figure 4). These four samples were collected after the eruption. On the other hand, values of HS-4 plot within a small range and do not show any significant change.

Figure 4 shows changes in $\delta^{18}O$ with time for HB-4, Boring Site, and HS-4. Increases at HB-4 occur during May to June and during July for Boring Site. These increases in $\delta^{18}O$ suggest that the mixing ratio of volcanic gases rich in steam with high isotopic values developed first at places where most active fumaroles are located, then proceed outwards. This may be due to a rise in pressure at depth resulting in increaseing

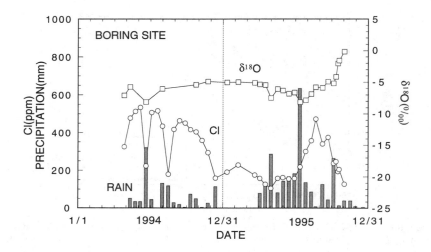

Figure 5 Changes in Cl concentration, $\delta^{18}O$ of waters at Boring Site, and precipitation measured 1km north of the fumarolic area with time.

flow of volcanic gas to the surface. The increase in $\delta^{18}O$ for HB-4 is rather small compared with that of Boring Site. This is because the pre-eruptive $\delta^{18}O$ for HB-4 is already close to magmatic water. No significant change in values of HS-4 was found in spite of apparent change in Cl concentration with time shown in Figure 2. This may be because that resident time after mixing volcanic gas with ground water is rather long compared with those of HB-4 and Boring Site. This allows hot spring water to react and equilibrate with host rocks isotopically. The relatively low temperature of HS-4 also suggests that mixing of ground water may contribute to the low isotope values. On the other hand, residential times for the HB-4 and Boring Site waters are short enough to retain their magmatic signature.

Figure 5 shows changes in Cl concentration and $\delta^{18}O$ with time for the Boring Site together with precipitation data measured 1km north of the fumarolic area. The precipitation data represents accumulated precipitation for the two weeks before the date of data collection. Sharp drops of Cl in June and August, 1994, and for $\delta^{18}O$ in June, 1994, occurred in a synchronous manner. These drops correspond to periods of relatively high precipitation, indicating dilution of hot spring water with infiltrated rain water both in low Cl concentration and isotope values. Neither Cl nor $\delta^{18}O$ show any significant drop during 1995 in spite of relatively frequent and large precipitation compared with those during 1994. No reasonable explanation has been found for the decreasing Cl concentration and increasing $\delta^{18}O$ starting August 1995. Further investigations on this matter are required.

4 CONCLUSIONS

Cl concentrations, and $\delta^{18}O$ values of hot spring waters discharging in the fumarolic area at Kuju volcano show appreciable increases with time starting a few months before the beginning of eruptive activity in 1995. These hot springs are formed by mixing volcanic gases with ground water, and indicate the contribution of volcanic gases to the hot spring waters. Accordingly, increases in $\delta^{18}O$ for HB-4 and the Boring Site suggest increases in the direct supply of magmatic gas associated with the eruptive activity that started on October 11, 1995.

REFERENCES

Ehara, S., Yuhara, K. and Noda, T. 1981. Hydrothermal system and the origin of the volcanic gases of Kuju-Iwoyama volcano, Japan, deduced from heat discharge, water discharge and volcanic gas emission data. *Kazan* 26:35-56(in Japanese)

Ehara, S. 1992. Study on thermal processes beneath Kuju volcano, Japan A step to volcano energy utilization. *GRC Trans.* 16:153-156

Itoi,. R., Kai, T. and Fukuda, F. 1997. Volcanic activity of Kuju volcano and its effects on acidification of rain in the local area. *Proc. of Int. Cong. of Acid Snow and Rain*, 1997

Mizutani, Y., Hayashi, S. and Sugiura, T. 1986. Chemical and isotopic compositions of fumarolic gases from Kuju-Iwoyama, Kyushu, Japan. *Geochemical J.*20:273-285

Fumarole gas geochemistry in estimating subsurface temperatures at Hengill in Southwestern Iceland

Gretar Ívarsson
Reykjavík District Heating, Iceland

Abstract: A regional study on fumarole gas geochemistry (CO_2, H_2S, H_2, N_2, CH_4) has been conducted on three geographically connected geothermal fields in the Hengill area in Southwestern Iceland to determine the subsurface temperatures. The highest temperatures are consistently found to be at the now utilised Nesjavellir field, while progressively lower temperatures are found at the Ölkelduháls field and the Hveragerði field. Gas geochemistry and resistivity measurements suggest that the Nesjavellir field and Ölkelduháls field are not connected at subsurface levels and that temperatures of up to 300°C can be expected at the latter location. While one experimental drillhole at Ölkelduháls has failed to reach predicted temperatures, the field is expected to be a possible future site for both electric production and space heating. Continued research and experimental drilling is anticipated in the next few years.

1 INTRODUCTION

The Nesjavellir power plant, located 30 km east of the capital Reykjavík (Fig. 1), was commissioned in 1990 and originally designed to produce a maximum of 400 MW_t for space heating purposes, with the possibility of parallel electric production. Presently some 150 MW_t have been installed, and about 80 MW_t are produced annually, which amounts to 25% of the energy needed to sustain space heating in the larger Reykjavík area (150,000 people), the difference being produced from 3 low-temperature fields (70-130°C) within and around Reykjavík.

Recent growth of industry in Iceland, employing large quantities of electric energy, has resulted in a shortage of electricity on the national power grid. A redesigned plan for the Nesjavellir plant includes a maximum production of 206 MW_t and 60 MW_e. Two turbines are being installed at Nesjavellir, estimated to commence production in 1998. As the reserve energy will now be used for electric production, research on new unexploited geothermal areas has been accelerated. The Ölkelduháls region, located 6 km southeast of the Nesjavellir field (Fig. 1), is considered promising in that respect and has experienced extensive research since the early 1980's. Research indicates that it is a viable future site for exploitation, that subsurface temperatures may reach 300°C and that it is not connected to the Nesjavellir field, in spite of the fact that surface geothermal manifestations are more or less continuous between the two areas. One 1100 m deep research/production well was drilled in late 1994, but further drilling has been suspended, pending results from land ownership disputes.

In this paper a regional study of fumaroles in the Hengill volcanic field is described. The gas chemistry is used to evaluate subsurface temperatures, and the findings are intergrated with geological and geophysical results.

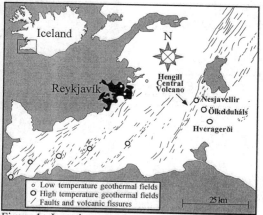

Figure 1: Location map

2 GEOLOGICAL SETTINGS

The field area is dominated by the active Hengill central volcano and it's northeast - southwest trending fis-

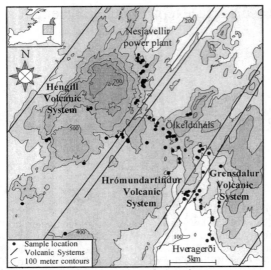

Figure 2: Field area and sample locations

sure swarm (Figs. 1 and 2). The central volcano is irregularly shaped, formed mostly in subglacial eruptions during the last glacial period, but recent lava flows are common both northeast and southwest of Hengill in the adjoining fissure swarm. The Nesjavellir geothermal field and power plant are located on northeastern perimeter of the central volcano, close to the most recent eruption site in that area, which occurred some 2000 years ago. Subparallel to the Hengill volcanic system towards the east are two older volcanic systems (dormant or extinct), which apparently have drifted eastward from the volcanic zone with time. The younger one is identified as the Hrómundartindur system and the older one is called the Grensdalur system. The Ölkelduháls geothermal field belongs to the younger of the two systems, but intense surface geothermal activity characterises the whole area. The most recent eruptive activity in the Ölkelduháls area occurred at the end of the last glacial period. The regional fault trend is parallel to the fissure swarm. (northeast - southwest), but a series of faults trending almost perpendicular to the regional trend (northwest - southeast) are observed to connect the Hengill system to the Grensdalur system through the Ölkelduháls geothermal field. Most of the geothermal activity outside the Hengill system occurs along this series of faults. The whole area is seismically active, but an unusually intense seismic swarm has centered around the Ölkelduháls and Grensdalur area since the summer of 1994. The swarm, which continues to this date, consists mostly of minor quakes, but the larger ones have registered between 4 and 5 on the Richter scale. No systematic change has been observed in the fumarole gas chemistry following this increased seismic activity, but some minor change in surface activity have occurred. Vertical movement of over 2 cm has suggested to some that minor magma injections are occurring underneath the Ölkelduháls area at a depth of 6 km (Sigmundsson et.al 1997).

3 GAS GEOCHEMISTRY

During the last few years, nearly 100 fumalole gas samples have been collected in the field area, including many duplicates to monitor changes due to increased seismic activity. Figure 2 shows a topographic map of the field area with sample locations. Also shown are the three volcanic systems and their associated geothermal fields; the Hengill geothermal field, which includes the Nesjavellir field; the Ölkelduháls geothermal field, which incorporates geothermal manifestations from both the Hrómundartinur volcanic system and partly the Grensdalur volcanic system; and the Hveragerði geothermal field, which is considered to represent horizontal subsurface flowage from the other fields which are topographically higher.

The concentration of various gases in fumarole steam are considered to be in equilibrium with mineral buffers at any given temperature. Five gas geothermometers based on the concentrations and ratios of H_2, H_2S and CO_2 have been developed (Arnórsson and Gunnlaugsson 1985) and are used in this study. A detailed account of sampling methods and chemical analyses has been presented elsewhere (Ívarsson 1996). Experience has demonstrated that their applications has proven to be a reliable method in estimating subsur-

Table 1. The average concentration and calculated temperatures for the selected geothermal fields.

Concentration mmole/kg steam	Nesjavellir 30 samples	Ölkelduháls 31 samples	Hveragerði 19 samples
H_2	50	4	3
N_2	12	4	6
CH_4	1,5	0,1	0,1
O_2+Ar	0,2	0,2	0,7
H_2S	50	9	4
CO_2	475	330	100

Temperature °C	Nesjavellir 30 samples	Ölkelduháls 31 samples	Hveragerði 19 samples
H_2S/H_2	300	287	289
CO_2/H_2	308	283	291
H_2	308	287	282
H_2S	317	286	273
CO_2	305	296	262
Average °C	308	288	279

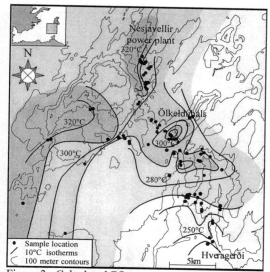

Figure 3: Calculated CO_2 temperatures

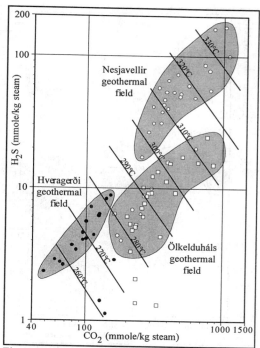

Figure 4: CO_2 versus H_2S

face temperatures in high-temperature geothermal systems in Iceland.

Table 1 summarises the analytical results from the three geothermal areas. As expected the now utilised geothermal field at Nesjavellir, located at the active central volcano of Hengill, provides the highest calculated temperatures, in excess of 300°C. The Hveragerði field, which is used for space heating purposes in the local community, produces the lowest average temperature or about 280°C. Intermediate between the two, both in geographical location and in geological age, is the field at Ölkelduháls, which has a calculated average temperature of 288°C.

Figure 3 shows the calculated CO_2 temperature drawn as isotherms. The field at Ölkelduháls is well confined with calculated CO_2 temperatures above 320°C. Another less well defined high occurs slightly to the southeast, while belonging geologically in part to the Grensdalur volcanic system, it is geographically connected to the Ölkelduháls field, as well as showing geochemical similarities. High CO_2 temperatures occur at the location of the Nesjavellir power plant and the 320°C temperature isotherm is observed to stretch towards the south, where extensive surface geothermal manifestations occur. Southwest of Hengill isolated fumaroles give CO_2 temperatures in excess of 320°C, but the lack of any substantial surface activity precludes any detailed mapping of the subsurface temperatures.

Comparing the relative concentration ratios of CO_2 and H_2S in the steam of the three different geothermal areas (Fig. 4) it is evident that at comparable CO_2 concentrations the Nesjavellir field contains greater quantities of H_2S than the Ölkelduháls field. An independent study on atmospheric gas concentrations of various geothermal fields in Iceland (Ívarsson et.al 1993)

demonstrated that the Nesjavellir field was releasing more H_2S into the atmosphere than any other field in Iceland. This difference in the two fields suggests that the fields are not connected and have different heat sources. Similarly, comparative concentrations of H_2S between the Ölkelduháls and Hveragerði field, at different CO_2 concentrations, suggests that the latter field might be the result of lateral flow from the Ölkelduháls field, accompanied by boiling and degassing.

In 1994 the first and so far only hole was drilled into the Ölkelduháls field. Originally estimated to reach 2000 meters, drilling was suspended at 1100 meters when a promising aquifer was encountered. Sampling and analyses of the drillhole steam later revealed a considerable compositional difference between it and steam produced from surrounding fumaroles. Temperatures in the hole reached only 200°C, far less than expected. The obvious conclusion is that the hole never penetrated the lower parts of the geothermal system. Resistivity measurements had indicated that slightly west of the drillsite was an area where cold groundwater possibly invaded the geothermal system. This may also have contributed to the lack of success.

4 RESISTIVITY AND TEMPERATURE

Resistivity measurements were done at Nesjavellir in 1986 (Árnason et.al. 1987) and in the Ölkelduháls area

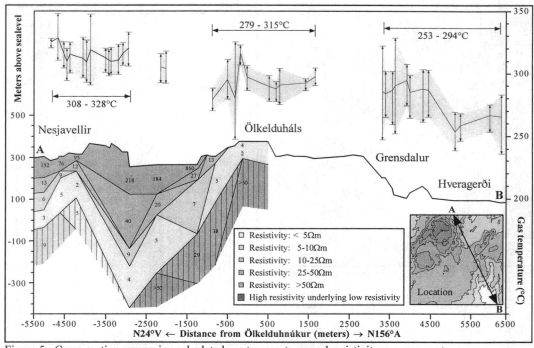

Figure 5: Cross section comparing calculated gas temperatures and resistivity measurements

in 1992 (Árnason 1993). Figure 5 demonstrates a comparison between resistivity measurements and calculated subsurface temperatures based on fumarole gas geochemistry. Generally speaking high surface resistivity is interpreted as relatively fresh surface bedrock, while decreasing resistivity with depth is considered to reflect increased subsurface temperatures. High resistivity underlying a low resistivity cover again suggests that the chlorite-facies has been reached, where boiling can occur in the bedrock and temperatures above 240°C can be expected. A relatively good correlation is observed between the two parameters, where usually the greatest intensity of surface geothermal manifestations and highest calculated fumarole gas temperatures follow locations where both low resistivity layers, as well as where high resistivity layers below lower resistivity, ascend towards the surface. Unfortunately resistivity measurements have not yet been extended to the southeastern part of the field area, so no comparison can be made there. The data so far indicates that the Nesjavellir field and the Ölkelduháls field are not connected at depth, therefore allowing both fields to be independently utilized for power production.

5 REFERENCE

Arnórsson, S. & E. Gunnlaugsson 1985. New gas geothermometers for geothermal exploration - calibration and application. *Geochim. Cosmochim. Acta* 47, 1307-1325.

Árnason, K. 1993. Geothermal activity in the Ölkelduháls area. Resistivity measurements 1991 - 1992. *National Energy Authority*. Report in Icelandic. OS-93037/JHD-10, 82 p.

Árnason, K., G.I. Haraldsson, G.V. Johnsen, G. Þorbergsson, G.P. Hersir, K. Sæmundsson, L.S. Georgsson, S.T. Rögnvaldsson & S.P. Snorrason 1987. Nesjavellir - Ölkelduháls. Surface exploration. *National Energy Authority*. Report in Icelandic. OS-87018/JHD-02, 112 p.

Ívarsson, G. 1996. Geothermal gas in the Hengill region. Collection and analyses. *Reykjavík District Heating Service*. Report in Icelandic. 1996-01, 42 p.

Ívarsson, G., M.Á. Sigurgeirsson, E. Gunnlaugsson, K. H. Sigurðsson & H. Kristmannsdóttir 1993. Geothermal gas in the atmosphere. The concentration of H_2S, SO_2 and Hg from selected geothermal areas. *National Energy Authority*. Report in Icelandic. OS-93074/JHD-16, 69 p.

Sigmundsson, F., P. Einarsson, S.T. Rögnvaldsson, G.R. Fougler, K.M. Hodgkinson & G. Þorbergsson 1997. The 1994-1995 seismicity and deformation at the Hengill triple junction, Iceland: Triggering of earthquakes by minor magma injection in a zone of horizontal shear stress. *J. Geophys. Res*. in press.

Sulfur and oxygen isotopic variations of dissolved sulfate in Crater Lake, Mt. Ruapehu, New Zealand

Minoru Kusakabe
Institute for Study of the Earth's Interior, Okayama University, Misasa, Japan

Bokuichiro Takano
Chemistry Department, Graduate School of Arts and Sciences, University of Tokyo at Komaba, Japan

ABSTRACT: Sulfur and oxygen isotopic compositions of dissolved bisulfate (HSO_4^-) were measured for samples collected from 1982 to 1995 at Crater Lake, Mt. Ruapehu, NZ. The $\delta^{34}S_{HSO_4^-}$ values are fairly constant at around +18‰ throughout the period, and are consistent with experimentally determined sulfur isotope fractionation between HSO_4^- and reduced sulfur ($S°$ and H_2S) during disproportionation reaction of gaseous SO_2 with an assumed $\delta^{34}S$ of 10‰ in a magma-hydrothermal system beneath Crater Lake. The HSO_4^--rich hydrothermal fluid is mixed with downward-percolating lake water to form a lower temperature hot water region at the upper part of the magma-hydrothermal system. The temperature of the hot water system is estimated to be 140±30°C based on ^{18}O fractionation between HSO_4^- and lake water. Oxygen isotopic covariance between HSO_4^- and lake water gives strong evidence that lake water circulates beneath the lake floor. The kinetics of sulfur isotopic equilibration between HSO_4^- and $S°$ suggests that the residence time of HSO_4^- in the hot water system is 1-2 years.

1 DISPROPORTIONATION REACTION OF SO_2

Through the study of active crater lakes, it is possible to understand the dynamics of the magma-hydrothermal systems that lie below these lakes (Special Issue on Geochemistry of Crater Lakes, 1994). Magmatic gases are composed of H_2O, CO_2, SO_2, HCl, H_2S, etc. which condense into the system to make it highly acidic, and thereby promote rock-water interaction within the system. In a magma-hydrothermal system, SO_2 will react with hydrothermal water to form sulfuric acid and elemental sulfur (or H_2S) as given below:

$$3SO_2 + 3H_2O \rightarrow 2HSO_4^- + S° + H_2O + 2H^+ \quad (1)$$

$$4SO_2 + 4H_2O \rightarrow 3HSO_4^- + H_2S + 3H^+ \quad (2)$$

Disproportionation of SO_2 is often invoked to explain the highly acidic nature of hydrothermal water (e.g., Kiyosu and Kurahashi, 1983). Reaction (1) takes place under high total sulfur concentration and plays an important role in active crater lakes, as these lakes often contain high concentration of sulfate and molten sulfur (Christenson and Wood, 1993, Rowe, 1994, Ohsawa et al., 1995).

It is known known that there is a large sulfur isotope fractionation between HSO_4^- and $S°$ through reaction (1) (Robinson, 1973; Kusakabe and Komoda, 1992). Figure 1 shows how fast the sulfur isotope fractionation ($1000\ \ln\alpha_{HSO_4^- - S°}$) approaches equilibrium at different temperatures after the SO_2 disproportionation reaction. It should be noted that a

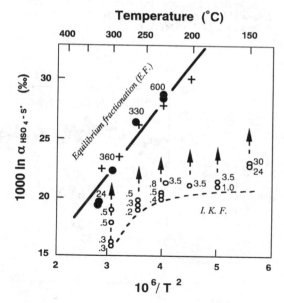

Fig. 1 Sulfur isotope fractionation between HSO_4^- and $S°$ produced by disproportionation reaction of SO_2. I.K.F. stands for the initial kinetic fractionation, which gradually approaches the equilibrium fractionation (E.F.) with time. The numbers by the data points indicate time (hr) after initiation of the reaction.

463

large fractionation (16-20‰) is attained initially (initial kinetic fractionation, IKF), followed by a gradual approach to the equilibrium fractionation (EF) at any given temperature. Under the assumption that the isotopic equilibration obeys the first order reaction, with respect to the isotopic distance between EF and IKF and to the total sulfur concentration in the system, one can calculate the time required for any disequilibrium HSO_4^--$S°$ pair to reach equilibrium under a given temperature and total sulfur concentration. For example, it will take 1-20 days for a HSO_4^--$S°$ pair to reach equilibrium, fractionation of approximately 26‰, after SO_2 is injected into a 250°C hydrothermal liquid with total sulfur concentrations of 0.03-0.3 molal.

2 $\delta^{34}S_{HSO_4^-}$ OF CRATER LAKE, RUAPEHU

The sulfur and oxygen isotopic ratios of dissolved bisulfate (HSO_4^-) have been measured for samples collected from Crater Lake, Mt. Ruapehu between 1982-1995 (Fig. 2). The $\delta^{34}S_{HSO_4^-}$ values were high at about 18‰ and had a tendency to increase only slightly from 1982 to the end of 1994. They varied irregularly during 1984-1989, and started to decrease in 1995 as the volcanic activity increased. The $\delta^{18}O_{HSO_4^-}$ values, on the other hand, showed considerable decrease between 1982 and 1985, followed by a smaller degree of variation with a mean value of ca. 14‰ (SMOW).

There is only one $\delta^{34}S$ value (-6‰) available so far for elemental sulfur and that sample was floating on the lake's surface (collected in Feb. 1991). Assuming that this value represents the $\delta^{34}S$ of elemental sulfur in the lake, and that it does not change with time, the isotopic fractionation between HSO_4^- and $S°$ ($\Delta_{HSO_4^--S°} = \delta^{34}S_{HSO_4} - \delta^{34}S_{S°}$) is 23.5±1.0‰. This is consistent with the view that HSO_4^- and $S°$ were formed by the disproportionation reaction of SO_2 having $\delta^{34}S$ = ca. 10‰ at T<300°C in a magma-hydrothermal system that lies beneath the lake. Direct injection of magmatic SO_2 into the lake water is probably limited because it would produce a HSO_4^--$S°$ pair with a $\Delta_{HSO_4^--S°}$ of 21‰ or so (IKF, Fig. 1) for lake water temperatures of <60°C (Christenson, 1994), and because the $\Delta_{HSO_4^--S°}$ value would be "quenched" considering the slow sulfur isotope exchange at that temperature (Kusakabe and Komoda, 1992). The observed sulfur isotopic fractionation $\Delta_{HSO_4^--S°}$ of 23.5‰ suggests that either the HSO_4^--$S°$ pair was equilibrated at 250-300°C, in the deeper part of the magma-hydrothermal system, or the HSO_4^--$S°$ pair which had an IKF value lower than 20‰ shifted to an equilibrium value at lower temperatures (i.e., leading to higher $\Delta_{HSO_4^--S°}$) in the shallower part of the system. The narrow range of $\delta^{34}S_{HSO_4^-}$ values may indicate that the disproportionation reaction took place in an environment with relatively constant physical conditions, favoring the former option.

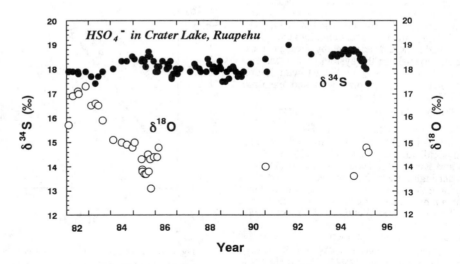

Fig. 2 Variation with time (1982-1995) in $\delta^{34}S$ and $\delta^{18}O$ of HSO_4^- dissolved in Crater Lake, Ruapehu, New Zealand.

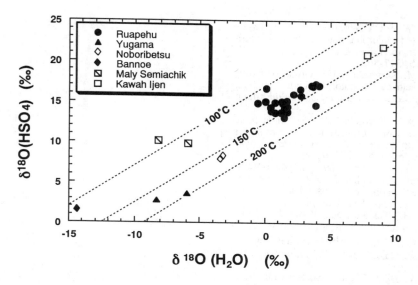

Fig. 3. Relationship between $\delta^{18}O$ values of dissolved sulfate and water from some active crater lakes. Note a positive correlation between them. Dotted lines indicate the equilibrium relationship (Mizutani and Rafter, 1969) at the temperatures indicated.

3 $\delta^{18}O_{HSO_4^-}$ OF CRATER LAKE, RUAPEHU

Once the HSO_4^- ion is formed, it will exchange its oxygen very quickly with the surrounding water under acid hydrothermal conditions, at a rate 2-3 orders of magnitude greater (Chiba and Sakai, 1985) than the sulfur isotope exchange mentioned above. At temperatures higher than 200°C oxygen isotopic equilibration between HSO_4^- and water is almost instantaneous. Thus, the oxygen isotopic ratio of dissolved HSO_4^- is probably controlled by that of the surrounding water. The $\delta^{18}O$ of HSO_4^- and lake water were found to covary with a relatively constant oxygen isotopic fractionation between them (Fig. 3). This suggests that, at Ruapehu, HSO_4^- underwent oxygen isotopic exchange with water at 140±30°C (Mizutani and Rafter, 1969). It is interesting to note that this feature can also be found in other active crater lakes like Kawah Ijen, etc. (Fig. 3), giving strong evidence that lake water circulates through a shallow part of the hydrothermal system below the crater lake floor. At temperatures below 150°C, sulfur isotopic exchange between HSO_4^- and S° becomes slow, and lack of HSO_4^- with high $\delta^{34}S$, (e.g., higher than 23‰) resulting from low-T equilibration, suggests that the residence time of HSO_4^- in the hot water system is less than 1-2 years.

4 $\delta^{34}S$ VERSUS VOLCANIC ACTIVITY

A clear decrease in $\delta^{34}S$ values of dissolved HSO_4^- was observed for samples collected in 1995 as volcanic activity intensified (Fig. 2). The decrease may reflect either the introduction of fresh, magmatic SO_2 with lower $\delta^{34}S$, resulting in the formation of HSO_4^- with low $\delta^{34}S$, or the addition of HSO_4^- produced from oxidation of vent-filling sulfur (remobilization). If the vent-filling sulfur is exposed to water vapor at the time of increased volcanic activity, SO_2 (and H_2S) is produced and admixed with magmatic SO_2. These processes may have contributed to the lowering of $\delta^{34}S$ of HSO_4^- in the lake water, although isotopic fractionation during remobilization of sulfur needs to be carefully examined.

REFERENCES

Chiba, H.& H.Sakai 1985. Oxygen isotope exchange rate between dissolved sulfate and water at hydro-thermal temperatures. *Geochim. Cosmochim. Acta*, 49: 993-1000.

Christenson, B.W. 1994. Convection and stratification in Ruapehu Crater Lake, New Zealand: Implications for Lake Nyos-type gas release eruptions. *Geochem. J.*, 28, 185-197.

Christenson, B.W. & C.P.Wood 1993. Evolution of a vent-hosted hydrothermal system beneath Ruapehu Crater Lake, New Zealand. *Bull. Volcanol.* 55, 547-565.

Kiyosu, Y. & M. Kurahashi 1983. Origin of sulfur species in acid sulfate-chloride thermal waters, northeastern Japan. *Geochim. Cosmochim. Acta,* 47: 1237-1245.

Kusakabe, M. & Y. Komoda 1992. Sulfur isotopic effects in the disproportionation reaction of sulfur dioxidenat hydrothermal temperatures. *Rept. Geol. Surv. Japan,* No. 279: 93-96.

Mizutani, Y & T.A. Rafter 1969. Oxygen isotopic composition of sulphates- Part 3. Oxygen isotopic fractionation in the bisulphate ion-water system. *N. Z. Jour. Sci.* 12:54-59.

Ohsawa, S., B. Takano, M. Kusakabe 1993. Variation in volcanic activity of Kusatsu-Shirane volcano inferrred from $\delta^{34}S$ in sulfate from the Yugama crater lake. *Bull. Volcanol. Soc. Jpn,* 38:95-99.

Robinson, B.W. 1973. Sulphur isotope equilibrium during sulphur hydrolysis at high temperatures. *Earth Planet. Sci. Lett.*, 18, 443-450.

Rowe, G.L. 1994 Oxygen, hydrogen, and sulfur isotope systematics of the crater lake system of Poas Volcano, Costa Rica. *Geochem. J.*, 28, 263-287.

Special Issue on "Geochemistry of Crater Lakes": *Geochem. J.* 28, No. 3, 1994 (Kusakabe, M., ed.)

Kinetics of postmagmatic clay mineral crystallization in lava flows

A. Mas, P. Dudoignon & D. Proust
Ecole Supérieure d'Ingénieurs de Poitiers, Section Génie Civil, UMR 6532, Université de Poitiers, France

F. Schenato
UFRGS-CP Geociencias, Porto Alegre, Brazil

ABSTRACT: The clay minerals crystallized in the mesostasis of basaltic rocks show a vertical zonation of their mineralogy through the thick basaltic units: saponite in outer zones to chlorite/saponite mixed layer in inner part of the flows. The cooling rates, calculated for thick flows (45m to 30m, Parana Basin, Brazil) and thin flows (3.60m to 1.60m, Mururoa Atoll, French Polynesia), permits us to determine the associated clay minerals crystallinity and the crystallization kinetics. Protosaponite is characterized by a very low number of unit cells in coherent scattering array along Z ($1<N<2$ when $v < 10^{-6}$ °C/s). Saponite is characterized by an increasing number of unit cells in coherent scattering array along Z, with a decreasing cooling rate ($N>2$ when $v>10^{-7}$°C/s).

1 INTRODUCTION

The dominant clay minerals crystallized in the basaltic flows are saponite, chlorite/saponite mixed layers and, locally, potassic phases such as celadonite or glauconite. As a function of cooling rates, these clay minerals, which crystallize during the postmagmatic stage, present a vertical zonation of their crystallochemical characteristics through each lava flow (Dudoignon and al., 1988, 1992, 1994). This paper presents the crystallochemical evolution of random chlorite/saponite mixed layer to pure saponite in flows of various thickness (45 m to 1.60 m). The main topic of the work is the correlation of the cooling rates with the chemical compositions and the crystallinity of the newformed clay minerals.

2 METHODS

Two volcanic sites were studied: thick basaltic flows were sampled in the Parana Basin (Brazil) and thin basaltic flows in the Mururoa volcanic pile (French Polynesia). The samples from Mururoa Atoll were taken from three basaltic units of the inclined Fuschia drill hole (Table 1; Bardintzeff et al., 1986). The thick basaltic flows were sampled in the Trizz and Estancia Velha quarries (Table 1; Schenato, 1997).

Table 1: Location and flow type of studied samples: (1) tholeiitic basalt, (2) oceanitic basalt, (3) hawaiite.

flow type	thickness	Geological site
aerial (1)	45 m	Estancia Velha
aerial (1)	30 m	Trizz
subaerial (2)	3.60 m	Mururoa
submarine (2)	1.70 m	Mururoa
dyke (3)	1.60 m	Mururoa

The mineralogical identification includes Microprobe analysis of thin sections, X-ray diffraction on the $< 2\mu m$ fraction (XRD), mathematical XRD decomposition (Decomp software; Lanson and Besson, 1992) and XRD modelisation (Newmod software), thermo-differential analysis TDA and I.R. spectrometry. In order to distinguish the clay minerals or mixed layer assemblages we used the M^+, 4Si, $3R^2$ triangular diagram: $M^+ = 2Ca + Na + K$, $4Si = Si / 4$, $3R^2 = (Fe^{2+} + Fe^{3+} + Mn) / 3$ (Meunier et al., 1991).

The cooling rates of the lava flows have been successively calculated using the Jaeger equations (Carshaw & Jaeger, 1959; Jaeger, 1961, 1968).

3 RESULTS

3.1 *Petrographic observations*

In a general way the lava flows present two distinct parts: (1) the peripheric zones (top and base) characterized by a few centimeters thick quench structure and (2) an inner and holocrystallized part which is the most dominant.

In all these basalts or oceanitic basalts the inner and holocrystallized parts of the flows are composed of altered olivine (2-10%) and pyroxene (10-20%) phenocrysts, plagioclases (An 60-70%), titanomagnetite microcrystals and the micro to cryptocrystalline mesostasis (10-20%). The mesostasis represents the late postmagmatic stage of crystallization with K feldspar, quartz, apatite and green to brown clay matrix. The altered olivines are generally replaced by a green cryptocrystalline clay matrix. Zeolites are locally associated with the clay minerals in vesicles and vein infillings in the subaerial environment.

In the flow periphery, quench zones are 2 to 6 centimeters thick. Microscopic observations of the quench zones reveal an abundant subglassy dark mesostasis (up to 60%), with disseminated phenocrysts of altered olivine, pyroxene and plagioclase. S.E.M. observations reveal that this mesostasis is composed of micro to cryptocrystalline plagioclase, K-feldspar, opaques and dissiminated clays. In submarine lava flows, these outer quench zones are yellow and show a replacement of the glassy mesostasis by saponite along the seawater contact. In these microsites of seawater/rock interaction, the zeolites, with dominant phillipsite, are associated with saponite precipitations. The crystallization temperatures of the saponites sampled in the peripheral zones of thin flows from Mururoa were estimated around 80°C from the $\delta^{18}O$ analyses (Dudoignon et al., 1992).

3.2 *Crystallochemistry of clay minerals*

The microprobe analysis of the clay matrices display three domains in the M^+, 4Si, $3R^2$ triangle: (1) the saponite group, (2) the mixed layer chlorite/saponite group shifted to the chlorite domain and (3) the celadonite group shifted to the M^+ - 4Si line (Fig.1).

In every type of lava flow, the saponites were analyzed in the peripheric zones of the flows (top and bottom). The chemical compositions shift to the chlorite/saponite mixed layer domain, when they have been carried out towards the central part of subaerial flow (Fig.1b). In the inner part of the submarine lava flows the clay minerals are saponite, locally associated with celadonite crystallizations (Fig.1a). In the inner part of the dyke and thick lava flows, the chemical compositions of clay minerals show the two distinct domains of saponite and celadonite.

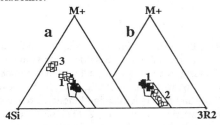

Fig.1: Location of the clay mineral assemblages in the M^+ - 4Si - $3R^2$ triangle; a=1.70m submarine lava flow, b=3.60m subaerial lava flow; 1=saponite, 2=chlorite/saponite mixed layer, 3=celadonite; white squares=inner part and black squares=peripheric zone

The XRD diagrams performed on the thick lava flows have sufficient quality to be decomposed with DECOMP software. In the periphery of these flows only one saponite component is needed to fit correctly the experimental diffractogram. The (001) reflection is located at 17.00 Å after glycol saturation on oriented preparation. In inner parts of the flows, the decomposition of the X-ray diagram needs two components: one dominant pure saponite (001 reflection at 17.20 Å) and one random chlorite/saponite mixed layer (001 reflection at 16.50 Å after glycol treatment).

In the thin lava flows (Mururoa Atoll) only the XRD diagrams performed in the base of subaerial and submarine lava flow have sufficient quality to be decomposed; they are pure saponite (001 reflection at 17.10 Å after glycol treatment).

In the inner parts of these thin flows and in the dyke the very low amplitude of the diagrams may be explained by the very poor crystallinity of the material.

Using NEWMOD software, the superposition between calculated and experimental patterns needs very low number (N) of unit cells in coherent scattering array along Z. This N value varies from the mathematical minimum 1 in the inner part of the thinnest basaltic unit (1.60 m) to 8 in the thickest one (45 m).

4 THE COOLING RATES IN THE 350-40°C TEMPERATURE RANGE

Each basaltic unit shows a decrease of the calculated cooling rates from the top to the bottom which is explained by the heat retention of the country rock (Fig.2). In a general way the scale of cooling rates decreases when the unit thickness increases. The cooling rates, in the center of the subaerial flows (45 m, 30 m, 3.60 m) for the chlorite to chlorite/saponite mixed layer crystallization temperature range (350°C - 200°C), are respectively 1.70 10^{-7} °C/s, 3.70 10^{-7} °C/s, and 2.6 10^{-5}°C/s, and in the saponite crystallization temperature range (110°C-80°C), they are respectively: 3.3 10^{-8} °C/s, 7.3 10^{-8} °C/s, and 6.1 10^{-6} °C/s.

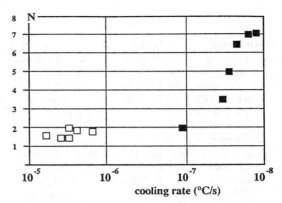

Fig.3: Representation of the N values versus cooling rates. White squares: thin flows (Mururoa), and black squares: thick flows (Estancia Velha and Trizz)

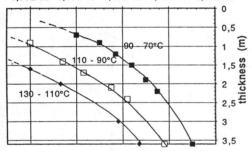

Fig.2: Vertical evolution of the calculated cooling rates (°C/s) from inner part to the base of the thin subaerial flow (3.60m), in the 130-110°C, 110-90°C and 90-70°C temperature ranges.

5 CORRELATION OF COOLING RATES WITH NUMBER OF COHERENT UNIT CELLS (N).

The following correlation is made for the clay minerals formed during the late postmagmatic stage (110°C-80°C) in inner and lower part of each unit (Fig.3). The upper part of the lava flow is not taken into consideration because of a doubt about eventual late alteration (weathering or seawater weathering) during the period preceding the next flow. These saponitic assemblages display two groups; (1) the clay minerals crystallized in the 10^{-7} - 10^{-8} °C/s cooling rate range and (2) clay minerals crystallized in the 10^{-5} - 10^{-7}°C/s cooling rate range. The first shows an increasing N (up to 8 with decreasing cooling rate). The second shows very small N values restrained to the lowest limit (between 1 and 2).

6 DISCUSSION - CONCLUSION

In the thick subaerial lava flows (30, 45m), the crystallized clay minerals in the mesostasis present a vertical evolution of their chemical composition which follows with their XRD crystallographic identification (N up to 8): saponite (base contact) to mixed layer chlorite/saponite (inner part). This evolution is identifiable with the X-ray diffraction, as long as the cooling rates are < 10^{-7}°C/s.

In the thin flows, the cooling rates go up to 10^{-5} °C/s. Under these conditions, the clay mineral cristallinity degrees are very low (1<N<2). Clay identifications necessitate a complementary method of analysis (TDA).

As a result, these minerals can be called protosaponite and proto-chlorite/saponite mixed layer, the same way the protoceladonite was already described in the submarine environment (Mevel, 1980; Dudoignon et al., 1988).

The precipitation of chlorite/saponite mixed layer, prior to the saponite, at higher temperature and thus higher cooling rates could explain why these phases can not be clearly identified using XRD in these thin lava flows.

7 REFERENCES

Bardintzeff, J.M., J.Demange, and A.Gachon, 1986. Petrology of the volcanic bedrock of Mururoa Atoll (Tuamotu archipelago, French Polynesia). *J.Vol. Geoth. Res.* 28: 55-83.

Carshaw, H.S. & J.C.Jaeger, 1959. *Conduction of Heat in solids*, 2nd Ed., Oxford University Press, New York.

Dudoignon, P., A.Meunier, D.Beaufort, A.Gachon, and D.Buigues, 1988. Hydrothermal alteration at Mururoa Atoll. *Chem. Geol.* 76: 385-401.

Dudoignon, P., C.Destrigneville, A.Gachon, D.Buigues, and B.Ledesert, 1992. Mécanisme des altérations hydrothermales associées aux formations volcaniques de l'atoll de Mururoa. *C.R. Acad. Sci. Paris*, t.314, série II: 1043-1049.

Dudoignon, P., G.Porel, and C.Guy, 1994. Pétrographie et quantification des phases secondaires dans les coulées basaltiques sous-marines et aériennes de l'atoll de Mururoa: mesure des surfaces d'échange par traitement d'images. *C. R. Acad. Sci. Paris*, t.319, série II: 775-781.

Jaeger, J.C. 1961. The cooling of Irregulary Shaped Igneous Bodies. *Am. J. Sci.* 259: 721-734.

Jaeger, J.C. 1968. *Cooling and Solidification of Igneous Rocks, Basalts: the Poldervaart Treatise on Rocks of Basaltic Composition, vol.2*, Wiley and sons, 503-536.

Lanson, B. & G.Besson, 1992. Characterization of the end of smectite to illite transformation, decomposition of X-ray patterns. *Clays and Clay Mins.* 40: 40-52.

Meunier, A., A.Inoue, and D.Beaufort, 1991. Chemiographic analysis of trioctahedral smectite to chlorite conversion series from the Okyu Caldera, Japan. *Clays and Clay Mins.* 39: 409-415.

Mevel, C. 1980. Mineralogy and chemistry of secondary phases in low-temperature altered basalts from Deep Sea Drilling Project, legs 51,52 and 53. *Init. Rep. Deep Sea Drill. Proj.* 51-53, part2: 1229-1317.

Schenato, F. 1997. *Alteraçao pos-magmatica de um derrame basaltico da Bacia do Parana (Regiao de Estancia Vehla, RS, Brasil): processos de resfriamento e vesiculaçao*. Thesis. Universidade Federal do Rio Grande do Sul et Université de Poitiers.

Hydrothermal system evolution induced by magma degassing: The case of Vulcano

P.M. Nuccio & A. Paonita
Istituto di Mineralogia, Petrografia e Geochimica, Università di Palermo, Italy

F. Sortino
Istituto di Geochimica dei Fluidi, CNR, Palermo, Italy

ABSTRACT: A model of the evolution of a hydrothermal system in relation to magma degassing is here proposed. We have been able to identify increased vaporization of brine, resulting salt deposition and re-mobilization by vapor phase. The generation and expansion of a central vapor monophase zone clearly has important implications on magma-brine interaction and in the phreato-magmatic explosion hazard assessment.

1 INTRODUCTION

The influence of magma degassing processes on the chemical-physical conditions of hydrothermal systems are of great importance to asses the possibility of water-magma interaction, and consequently the hazard of phreatic explosions.

Here, we propose a model of hydrothermal brine evolution in relation to magma degassing, applying it to the volcanic system of Vulcano, where of fumaroles have shown relevant geocgemical changes, together with important modifications of temperature and mass discharge (Capasso et al. 1997).

The chemical composition of fumarolic gases of Vulcano crater have been interpreted as the result of mixing processes between magmatic fluids and vapors separated from an hydrothermal system (Carapezza et al., 1981 ; Cioni & D'Amore, 1984; Chiodini et al., 1993, 1995; Capasso et al., 1997). According to Carapezza et al. (1981) the evolution of the deep hydrothermal system beneath Vulcano Island is regulated by thermal energy and mass balances. They also predicted that a central vapor monophase zone can be generated after a thermal energy input from the depth. Recently, Nuccio et al. (1997a,b) developed a rigorous geochemical model describing this mixing process. The physical and chemical parameters of the hydrothermal system were evaluated on the basis of the fumarolic gases chemistry and of the equation of state for the H_2O-CO_2-NaCl system given by Bowers and Helgeson (1983). The mixing fraction of magmatic gas and its CO_2 content was computed by a set of mass and energy balance equations (Fig.1). According to these results, CO_2 in magmatic gases displays temporal variations, characterised by concentration peaks.

2 METHODS

Since we are able to calculate the CO_2 concentrations in the two mixing components (Nuccio et al., 1997 a,b), we can evaluate the concentration of other species to assess their origin. We have used binary correlation diagram methods, in which the concentration of a specie in various fumaroles was plotted versus the corresponding CO_2 concentration. If the specie derives from two sources and its behavior is conservative, then the samples will display a linear correlation. By extrapolation of this trend to the CO_2 concentration of the magmatic and hydrothermal endmembers, we can estimate the corresponding concentration of generic species. We extensively applied this method for several minor and trace elements at Vulcano, so to distinguish magmatic and hydrothermal species.

CO_2 correlation diagrams clearly show that He (Fig.2), N_2 and Hg mainly are components of magmatic gases. Further, magmatic genesis of He is already comfirmed on the basis of $^3He/^4He$ isotope ratios (Italiano and Nuccio, 1996). In contrast, there is a general prevalence of halogens (HCl, HF), total sulfur (Stot) and the majority of trace elements (Se, As, Sb, Cu, Au, Pb) in the hydrothermal endmember (Fig. 3, 4). The primary origin of these species can be found in waters feeding the brine (having diverse possible genesis), but also in the igneous rocks where the hydrothermal system is located.

Nevertheless, we can equally classify these species as hydrothermal because their migration happens via the hydrothermal gaseous component.

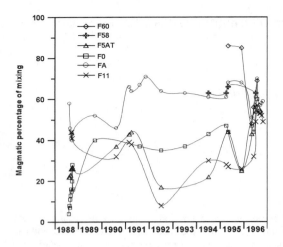
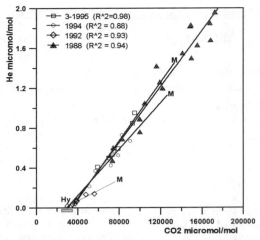

Fig.1: Magmatic percentages of mixing for various fumaroles, according to Nuccio et al. (1997). The comparable mixing fractions, showed during 1988 and 1996 by several fumaroles, correspond to two main volcanic crises.

Fig.2: He vs. CO_2 correlation diagram (Nuccio et al. 1997). M and Hy refer to magmatic and hydrothermal gases respectively. Magmatic concentration of He appears very variable during time, while there is negligible He content in hydrothermal gases.

3 MAGMA DEGASSING AND MODEL OF HYDROTHERMAL SYSTEM EVOLUTION

Magma degassing at depth normally occurs by diffusion, but water oversaturation of magma can substantially results as a consequence of two process: magma depressurization during magma uprising towards the surface, and magma crystallization which increases concentrations of dissolved volatiles in melt. In either case, exsolution process cause vapor separation and, according to the Henry's law, the partition of dissolved volatiles from melt to bubbles (Nuccio & Valenza, 1993). Therefore, effusive bubble degassing takes place and the melt will become progressively more depleted in volatiles that are preferentially partioned into bubbles (i.e., He, CO_2, N_2 etc.).

In response to the diffusive or bubble degassing, the composition of fumarolic gases will show significant variations, governed by the partition process.

Effusive bubble degassing increases fluid and heat fluxes to the surface, greatly affecting the boiling brine (Fig.6a). An input of magmatic heat can completely vaporizes a central portion of brine, enlarging the mixing zone between magmatic and hydrothermal gases (Fig.6b). Then, the boiling brine expands outward from the central monophase zone, depositing the dissolved salts. The hydrothermal vapors separated from the brine, can re-mobilize deposited salt, and excess of HCl, HF, S, etc. can be produced by solid - vapor reactions between hydrothermal gases and rock mineral (Chiodini et al. 1993; Symonds et al., 1993).

If the heat input is sufficient, the migration of the brine boiling interface outward from the central vapor-monophase zone presumably causes almost all fumaroles to be fed by the enlarged, compositionally uniform mixing zone. Consequently, we cannot obtain good fit lines in our correlation diagrams, because it is difficult to estimate the composition of the endmembers by the extrapolation of the fits.

4 THE CASE OF VULCANO

Since 1988, the He and N_2 temporal variation diagrams have displayed concentration peaks which are perfectly correlated to CO_2 peaks (Nuccio et al. 1997a,b). According to their low solubility in magmatic melts (Carroll and Webster, 1994) and to the degassing model of Nuccio and Valenza (1993), these peaks suggest volatile exsolutions are caused by magma ascent.

The concentrations of HCl, HF and sulfur in the fumarolic gases decreased during 1988-96 period (Capasso et al., 1997). The progressive exhausting of this species can be caused by volatilization from solid phases.

HCl correlates negatively with CO_2 during 1992/93 (Fig.3), suggesting the hydrothermal gases mobilized salts deposited during the centrifugal migration of the brine boiling interface after magma energy input (Fig.6b). We think that magmatic HCL

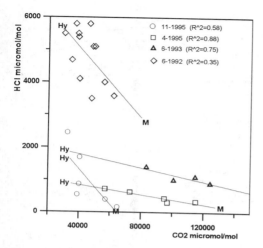
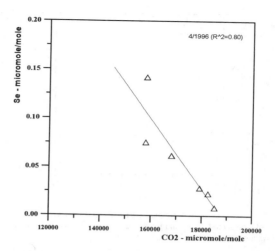

Fig.3: HCl vs. CO$_2$ correlation diagram. M and Hy are as in Fig.2. The 1992 and 1993 fits are caused by the schematic situation described in Fig.6b, while 4/1995 fit is related to situation in Fig.6a.

Fig.4: Se is anticorrelated with CO$_2$, suggesting it mainly comes from the hydrothermal system. M and Hy are as in Fig.2.

would display a positive correlation between HCL vs. CO$_2$.

Salt deposits were completely exhausted in 1994 and during the first half of 1995, according to the low HCl contents of gas discharges (Fig.3). During this period, the true HCl content in hydrothermal gases is probably low (<0.1vol%) and similar to the magmatic one.

At the end of 1995, there are new negative correlations beetween HCl and CO$_2$ (Fig.3), suggesting a new increased vaporization of the brine portion nearest to mixing zone, with a new salt deposition (Fig.6b).

The fit lines of HF and S contents vs. CO$_2$ agree with this interpretation, suggesting a similar origin and behavior of these species. HF concentration in magmatic gas appears negligible, while magmatic gas appears to have an appreciable quantity of sulfur content (0.1-0.2 vol%).

Mobilization during rising of fumarolic gases from mixing zone to surface could cause HCl, HF, Stot, etc. concentration would not be related to CO$_2$. Modest mobilizations of halogens and sulfur along the shallow portion of fumarolic conduits, after the mixing, are probable and they can explain why the correlations are no good in a few cases. Nevertheless, the correlations among some trace elements, sulfur and halogens (Fig.5) are equally good also in this cases. This suggests that trace elements are mainly carried in fumarolic gases as chloride and sulfides.

Generally, we observed a great compositional difference among various fumaroles of Vulcano crater, so that the calculated mixing percentages of magmatic component show a large variability among the various fumaroles sampled (Fig.1). This consents to have good correlations among the different species vs. CO$_2$, and to obtain reliable concentration values for the endmembers.

As we foresee above, during the main volcanic crises (i.e., in 1988 and 1996), mixing percentages of all fumaroles vary towards much more homogeneous values (Fig.1), according to their similar chemical composition. In this case, it becomes very difficult to estimate the composition of the endmembers by the extrapolation of the fits, and to identify the mobilization processes.

5 CONCLUSIONS

Chemical and thermal variations of the fumarolic fluids at Vulcano have been studied on the basis of previous models (Nuccio et al., 1997a,b ; Nuccio and Valenza, 1993) and used to conclude the followings:

1) Main episodes of magma degassing lead to advective transfer of large amount of thermal energy to the above hydrothermal system.

2) Consequently, increased vaporization of the hydrothermal brine can cause a centrifugal migration of the brine boiling interface and an extension of the central vapor mixing zone, with a simultaneous salt deposition.

3) Hydrothermal vapor mobilizes the salt deposits, growing rich in some species (halogens, sulfur, metals), while the mix with the magmatic gas in an enlarged mixing zone allows a thermal and chemical homogenization of the gaseous mixture.

By our model, we should be able to detect the

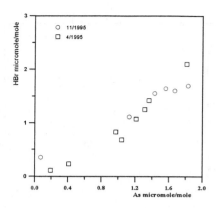

Fig.5: As is directly correlated with HBr and HCl and HF. At Vulcano island, it is mainly released from the hydrothermal system together with Se, Sb, Cu, Au, Pb.

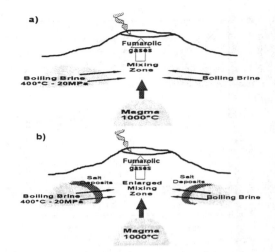

Fig.6: Evolution of hydrothermal brine. It is notable that the enlargement of the mixing zone as a consequence of magmatic input (b) and the progressive working out of salt deposits until the exhausting (a) can be considered as a cyclic condition.

evolution of a hydrothermal system related to magma degassing, as this should appreciably modify the slope of species (halogens, sulfur, metals) vs. CO_2 fit lines, and homogenize fumarolic gases composition, by increasing brine vaporization. This would lead to an enlargement of the central vapor-monophase zone, reducing the probability of magma-brine interaction and the phreato-magmatic explosion hazard.

6 REFERENCES

Bowers T.S., Helgeson H.C., 1983. Calculation of the thermodynamic and geochemical consequences of non ideal mixing in the system H_2O-CO_2-NaCl on the phase relation in geologic systems: Equation of state for H_2O-CO_2-NaCl fluids at high pressures and temperatures. *Geochim. Cosmochim. Acta* 47, 1247-1275.

Capasso G., Inguaggiato S., Nuccio P.M., Pecoraino G., Sortino F. 1997. Change in the composition of La Fossa crater fumaroles. *Acta Volcan.*, in press.

Capasso G., Favara R., Inguaggiato S., 1997. Chemical features and isotopic gaseous manifestation on Vulcano Island (Aeolian Island): an interpretative model of fluid circulation. *Geochim. Cosmochim. Acta*, in press.

Carapezza M., Nuccio P.M., Valenza M., 1981. Genesis and evolution of the fumaroles of Vulcano (Aeolian Island, Italy): a Geochemical Model. *Bull. Volcanol.* 44-3.

Carroll M.R., Webster J.D. 1994. Solubilities of sulfur, noble gases, nitrogen, chlorine, and fluorine in magmas. In: Carroll M.R.& Halloway J.R. (eds.), *Volatiles in magmas* Reviews in Mineralogy, vol. 30. 231-271.

Chiodini G., Cioni R., Marini L., 1993. Reactions governing the chemistry of crater fumaroles from Vulcano Island, Italy, and implications for volcanic surveillance. *Appl. Geochem. 8, 357-371.*

Chiodini G., Cioni R., Marini L., Panichi C., 1995. Origin of fumarolic fluids of Vulcano Island, Italy and implications for volcanic surveillance. *Bull. Volcanol.* 57, 99-110.

Cioni R., D'Amore F., 1984. A genetic model for the fumaroles of Vulcano Island (Sicily, Italy). *Geothermics* 13, 375-384.

Italiano F., Nuccio P.M. 1996. Variazioni del rapporto isotopico dell'elio alle fumarole di Vulcano. *Comunicazione alla Commissione Grandi Rischi*, Roma, 5 Luglio 1996.

Nuccio P.M., Paonita A., Sortino F. 1997,a. Volcanic unrest at Vulcano (Aeolian islands, Italy): Evidence of new degassing magma. IAVCEI97 General Assembly, Puerto Vallarta, Mexico. *IAVCEI abstracts volume*, p. 43.

Nuccio P.M., Paonita A., Sortino F. 1997,b. Processi di miscelazione tra gas magmatici ed idrotermali. In: La Volpe, Nuccio, Pivitera & Sbrana (eds.) *Working Group final relation of G.N.V. Progetto Vulcano* G.N.V., in press.

Nuccio P.M., Valenza M., 1993. Modification of geochemical parameters during magma ascent: the case of Vulcano, Aeolian Island. IGF-CNR pub.

Symonds R.B., Reed M.H. 1993. Calculation of multicomponent chemical equilibria in gas-solid-liquid systems: calculation methods, thermo-chemical data and applications to studies of high temperature volcanic gases with examples from Mount St. Helens. *Am. J. Sci.* 293, 758-864.

Magma degassing and geochemical detection of its ascent

P.M. Nuccio & M. Valenza
Istituto di Mineralogia, Petrografia e Geochimica, Università di Palermo, Italy

ABSTRACT: A geochemical model of magma ascent is proposed to interpret the observable variations of geochemical parameters in monitored volcanoes. The use of helium isotope data, volcanic gas composition and volatile solubility constants in the melt can lead to the detection and computation of magma movements. The geochemical changes are sensitive tracers of magma ascent suitable for forecasting volcanic eruptions.

1 INTRODUCTION

In the last few years, considerable effort has been expended in the identification of eruption precursors and fully empirical approaches have been used (Giggenbach, 1975; Menyailov, 1975; Hirabayashi et al., 1982; Martini et al., 1991). Unfortunately the same change of monitored geochemical parameter may be a precursor at one volcano but not at another (Naughton et al., 1975). More promising results, of general application, can be reached when the observed variations are interpreted in terms of volcanic processes: the aim of this paper is to develop this approach.

The premise of this study is that volcanic eruption is the ultimate expression of magma ascent, which is accompanied by a decrease of pressure As magma rises towards the surface, degassing occurs; it depends on many factors such as melt composition, magma ascent rates and other parameters regulating solubility of different volatiles in the melt. Processes preceding a volcanic eruption develop over the course of months or years and normally lead to the release of magmatic volatiles, which are able to reach the surface some time in advance of the magma itself. Here, we evaluate geochemical processes related to magma ascent, the effects of which can be observed at the surface by regular sampling of the volcanic gases and therefore can be used in detection of magma movements and in forecasting of eruptions.

2 MAGMA VESICULATION

Volatiles are dissolved in magma in different concentrations, related to source characteristics, evolution, contamination and degassing history of magma itself. At static pressure, magma outgassing normally occurs by diffusion. As magma rises through volcanic conduit volatile solubility decreases, until the magma reaches volatile saturation and it begins to exsolve gas into bubbles.

2.1 Bubble growth and ascent

A review of theories regarding bubble growth in magmas is reported in Sparks (1978). The amount of separated vapor depends on degree of melt supersaturation and, in turn, on the depressurization (ascent) of the magma.

The velocity at which the bubbles rise by buoyancy towards the magma top is obtained from Stokes' law. As a consequence of the melt viscosity, the bubbles remain trapped in the magma for a time that is a function of their radius and is inversely proportional to the viscosity itself.

2.2 Exsolution and fractionation of volatiles

The separation of a gas phase and the nucleation of bubbles implies that all the volatiles are distributed between the gas and the melt, irrespective of their degree of saturation. This distribution can be described, for components showing activities in both phases close to the unity, by Henry's law:

$$K_{h,g} = P_g / \chi_g \qquad (1)$$

where P_g is partial pressure of the volatile in the gas phase, χ_g is molar fraction of non-ionized volatile dissolved in the melt and $K_{h,g}$ is Henry's constant for the component g, which is a function of T and melt composition.

The concentration ratio of two volatiles dissolved in the melt is modified in the bubble under closed-system conditions according to:

$$(\chi_1/\chi_2)_b = (\chi_1/\chi_2)_m \, (K_{h,1}/K_{h,2}) \qquad (2)$$

where: $(\chi_1/\chi_2)_b$ is the molar fraction ratio of two volatiles in the bubble, $(\chi_1/\chi_2)_m$ is their initial molar ratio in the magma and $(K_{h,1}/K_{h,2})$ is the ratio of their respective Henry's law constants.

Therefore, a principal feature of magmatic vesiculation is volatile partition and fractionation. Taking into consideration the data available in the literature, the bubbles developed in a water-saturated silicic melt would be enriched in CO_2, He, N_2 and depleted in H_2O with respect to the melt. As a consequence of different gas solubilities, magma degassing leads to enrichment of magma in more soluble species (Jambon et al., 1986; Carroll and Draper, 1994) and a progressive increase of ratios between the more soluble volatile and the less soluble one (i.e., H_2O/CO_2, He/Ar, He/N_2). However, magma could also degas under open-system conditions, developing a progressive fractionation of volatiles, that could be reasonably modeled as a Rayleigh distillation process (Marty, 1995).

2.3 Hydrogen front

Silicic magmas generally attain water saturation at relatively shallow depth (1-2 km), before they erupt. The exsolution of water shifts the equilibria of dissolution reactions, with the consequence that more moles of molecular water could be present in the magma (melt + vapor) and a larger amount of hydrogen would be generated, according to the dissociation reaction of water at high temperature:

$$H_2O = H_2 + {}^1\!/_2 O_2 \qquad (3)$$

H_2 diffuses in silica melt almost a million times faster than O_2 (Carmichael et al., 1974) and therefore, a transient of water exsolution and its subsequent separation in the bubbles would be characterized by generation of a *hydrogen front* moving towards the surface, while the melt would experience a temporary increase in the fO_2. It is worth noting that, in a steady state regime of diffusive degassing, this faster diffusion of H_2 generates an oxygen excess in the magma column that is compensated by an auto-reduction process caused by the H_2 coming from the deeper portions of the magma itself (Sato and Valenza, 1980).

2.4 Kinetic mass fractionation

Isotopic mass fractionation can be prominent when a limited exchange is involved and equilibrium is not reached (Kaneoka, 1980). For instance, when volatiles are exsolved quickly into bubbles they can experience isotope mass fractionation. On the other hand, if the bubbles are not removed from magma, re-equilibration with volatile dissolved in the melt can prevail after time.

In isotope mass fractionation, the lighter isotope is enriched in the gas separated into bubbles and depleted in the residual gas dissolved in the melt.

During a single-step fractionation process the maximum enrichment possible is very close to $(m_2/m_1)^{1/2}$, where m_2 and m_1 are the respective masses of isotopes. Therefore, kinetic fractionation is generally significant only in very light molecules, having a relatively high isotope mass ratio. Helium is probably the best candidate for showing the effects of mass fractionation, also because it is not reactive. The relation linking the residual fraction F of the helium, its isotope ratio (R_i), before the separation of the bubbles, and its final isotope ratio (R_f), can be expressed as follows:

$$F_{He}^{(k-1)} = \frac{R_f}{R_i} \qquad (4)$$

3 RESIDUAL VOLATILES AND ESTIMATION OF MAGMA ASCENT

A volatile saturated magma will separate a gas phase as a consequence of decompression or magma ascent. We mathematically obtained the relation linking the fraction of water (F_{H2O}), remaining in the melt after its exsolution, to the corresponding fraction of helium through the *analogue H_2O-Henry's constant*/He Henry's constant ratio:

$$F_{H2O} = \frac{1}{\left(\dfrac{1}{F_{He}} - 1\right)\dfrac{K_{h,H2O}}{K_{h,He}} + 1} \qquad (5)$$

This relation can be used to evaluate magma ascent, by means of the Henry's constant ratio and the helium residual fraction (F_{He}). Analogous relation can be also applied to CO_2-saturated mafic melts.

Now, we assume that the gas phase is pure water having an ideal behaviour and H_2O solubility in melt can be described by a regular solution model. Then, under isothermal condition the fraction of the water that remains dissolved in the melt is related to the variation of pressure (Mueller and Saxena, 1974):

$$F_{(H2O)} = \frac{a_{H2O}^m}{a_{H2O}^0} \approx \frac{\chi_{H2O}^m}{\chi_{H2O}^0} = \frac{1}{P^0} P \exp\left[\frac{-V_{H2O}^m(P-P^0)}{RT}\right] \quad (6)$$

where V_{H2O}^m and a_{H2O}^m are partial molar volume and activity of H_2O in melt, and superscripts "0" refer to some reference equilibrium condition which serves as lower integration limit.

Moreover, for relatively small ΔP values the exponential term in the (6) can be neglected and the $F_{(H2O)}$ value approximated to the pressure ratio P/P_0, which is also equal to the depth ratio H/H_0. Therefore, the upward movement of magma can be easily inferred if the initial depth is known independently (i.e.- by volcanological or geophysical evaluations) :

$$F_{(H2O)} \approx \frac{P}{P_o} \approx \frac{H}{H_o} \quad (7)$$

4 VARIATIONS OF MASS OUTPUT AND GAS COMPOSITION

The diffusive flux of a volatile species from magma is controlled by the first Fick's law:

$$\Phi_{z \text{ (volatile)}} = -D_z^{melt} \frac{dC_z^{melt}}{dx} \quad (8)$$

where $\Phi_{z(volatile)}$ is the flux of the volatile species, D_z is its diffusion coefficient in the melt and dC_z/dx is its concentration gradient at the top of the magma chamber.

Depressurization of a saturated melt as a consequence of magma ascent causes the exsolution of a vapor phase, the partition of volatiles between melt and bubbles, and decrease of volatile concentration in the melt, producing a corresponding variation in volatile fluxes.

The gas outputs will show marked decreases, related to diffusive magma outgassing and preferential partition of volatiles into bubbles, with consequent modification of gas concentration gradients (Fig.1). In contrast, the output of water would remain almost constant, as its concentration gradient is not appreciably modified for a limited pressure extent, being controlled by its saturation curve.

The advective flux of magma volatiles towards the surface will increases only when bubbles will be able to leave the top of magma column. The following relationship links gas concentrations to volatile flux:

$$C_z = \frac{\Phi_z}{\sum_o^n \Phi_n} \times 100 \quad (9)$$

Consequently, the transient volatile output, caused by bubble separation, will be accompanied by transient compositions of released volatiles according to their partition. This will be characterized by a decrease of gas concentration, directly related to its vapor/melt distribution constant, and a consequent increase of water concentration. The opposite will occur during the subsequent phase of bubble outgassing. Taking into account that silicic magmas display higher viscosity, we expect a larger time interval between the initial phase of transient characterized by decreasing fluxes and gas concentrations, and the period when both gas concentrations and fluxes are maximized.

The bubble discharge could last until the bubbles having smaller diameters can reach the magma top. The excess of discharged gases is related to both the volume and displacement of the magma.

A peculiar case occurs when bubbles coalesce into a viscous foam, after extensive vesiculation of a silicic magma, which prevents the release of the trapped gases. Fumarolic gases will change their chemical compositions according to their partition in a vapor phase, and both steam and gas outputs will decline, but the subsequent phase of output increase of gases could be missed. Physical effects of magma buoyancy can be presumably detected from the surface, because this process may take place only at relatively shallow levels.

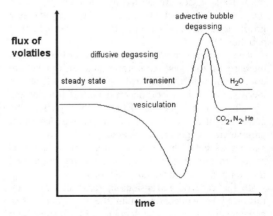

Figure 1. Variation of volatile flux according to magma degassing regime. After magma vesiculation, diffusive outgassing will shows an output decrease of gases preferentially partitioned into bubbles. Subsequent advective bubble outgassing will cause an output increase of all volatiles.

5 INTERPRETATION OF GEOCHEMICAL CHANGES

Each variation in geochemical parameters could be considered as a potential indicator of process occurring beneath the surface in a volcanic system. These can be related to the circulation of the fluids, to the fluid/rock interaction, and so on. Sometimes two or more of these processes could occur together, making it difficult to recognize the one is driving the volcanic evolution. Among these, magma ascent certainly is one of the most important to be detected, especially because of civil defense implications. As magma approaches to the surface normally it supersaturates with volatiles, undergoes separation of a vapor phase and degases.

When we consider magma degassing processes, we must verify the consistency of this hypothesis with the variations observed in the other parameters.

Decreasing trends of some components (e.g. gas/steam ratio, He, etc.) in the fumarolic gases are compatible with progressive depletion of more volatile components during magma degassing. We can reasonably aspect that this process normally develops on time scale, at least, of a year. Higher rates (moth or days) of decreasing gas concentrations are suggestive of other process.

Exsolution processes in the melt, following magma ascent, could be supported by a substantially unchanged steam output and corroborated by a decrease of the $^3He/^4He$ isotope ratio, as helium isotopes can be directly related to a magma source and are very sensitive to the kinetic fractionation. During this phase, the gas concentration changes in the volcanic gases are settled by equation (2), while the variation of helium isotope ratio is inversely proportional to the fraction of the exsolved helium in equation (4).

The above hypothesis may be verified by an increase of the gas/steam ratio and of the helium concentration in the fumarolic gases, as well as the increases of helium isotope ratio and of the outputs of all volatile species. During this phase, the amplitude of the variations of intensive geochemical parameters is generally related to the extent of magma ascent. The excess output of volatiles is related to both the extent of magma migration and the volume of the supersaturated melt.

In the case of a transient increase of gas/steam ratio and He concentration in the fumarolic gases, accompanied by a general decrease of both steam and gas outputs, in absence of other obvious explanations, the possibility of bubbles coalescing into a foam must be considered.

The quality of the final results and their confidence level critically depend on:
1) The accuracy of the physico-chemical constants used for the computations described above.
2) The number of the available geochemical parameters.
3) The frequency of the measurements during the evolutionary event.

6 REFERENCES

Carmichael I.S.E., Turner F.J., Verhoogen J. 1974. Igneous petrology. McGraw-Hill ed., New York.

Carroll M.R., Draper D.S. 1994. Noble gases trace elements in magmatic processess. Chem. Geol. 117, 37-56.

Giggenbach W.F. 1975. Variations in the carbon, sulfur and chlorine contents of volcanic gases discharged from White Island, New Zealand. Bull. Volcanol. 39, 15-17.

Hirabayashi J., Ossaka J. and Ozawa T. 1982. Relationship between volcanic activity and chemical composition of volcanic gases - A case stuty on the Sakurajima Volcano. Geochem. J. 16, 11-21.

Jambon A., Weber H.W. and Braun O. 1986. Solubility of He, Ne, Ar, Kr, and Xe in a basalt malt in the range 1250-1600°C. - Geochemical implications. Geochim Cosmochim. Acta 50, 401-408.

Kaneoka I. 1980. Rare gas isotope mass fractionation: an indicator of gas transport into or from magma. Earth Planet. Sci. Lett. 48, 284-292.

Martini M., Giannini L., Capaccioni B. 1991. Geochemical and seismic precursors of volcanic activity. Acta Vulcanol. 1, 7-11.

Marty B. 1995. Nitrogen content of the mantle inferred from N2-ar correlation in oceanic basalts. Nature 377, 326-329.

Menyailov I. 1975. Prediction of eruptions using changes in composition of volcanic gases. Bull. Volcanol. 39, 112-125.

Mueller R.F., Saxena S.K. 1974. Chemical Petrology. Springer-Verlag. New York.

Naughton J.J., Finlayson J.B. and Lewis W.A. 1975. Some results from recent chemical studies at Kilauea Volcano, Hawaii. Bull. Volcanol. 39, 64-69.

Sato M. and Valenza M. 1980. Oxygen fugacities of the layered series of Skaergaard intrusion, East Greenland. Am. J. Sci. 280-A,134-158.

Sparks R.S.J. 1978. The dynamics of bubble formation and growth in magmas: a review and analysis. J. Volcanol. Geother. Res. 3, 1-37.

Variability of volcanic gases by trace element-determination in volcanic sulphur

H. Puchelt & U. Kramar
Institute for Petrography and Geochemistry, University of Karlsruhe, Germany

B. Spettel
Max Planck Institute for Chemistry, Mainz, Germany

H. H. Schock
Mineralogical Institute, University of Tübingen, Germany

ABSTRACT: Trace elements in volcanic gases have been analysed in their solphatara products i.e. from elemental sulphur by instrumental neutron activation analysis in more than 150 samples from Italian, Greek, North American, Middle or South American and Asian volcanoes. Elements found include arsenic, antimony, selenium, tellurium, mercury, and bromine. The concentrations occasionally went up to over 1000 ppm.

Investigation of volcanic emanations have been analysed since the studies of Allen (1922) and Allen and Zies (1923). Due to difficulties both in sampling and analyses trace components have only occasionally been mentioned. In 1963 chapter K (volcanic emanation) of „Data of geochemistry" was published by White and Waring, giving only one analysis for „yellow native sulphur" containing 76 ppm of selenium by Neumann van Padang (1951). Papike et al (1992) gave his presidential address to the geological society of America mentioning the importance of trace element investigations on volcanic emanations using the unique example of the valley of 10000 smokes in Alaska.

Because analysis of elemental sulphur is difficult the only possibilities are instrumental methods which require neither solution or decomposition of the samples. We used instrumental neutron-activation analysis with irradiation in the research-reactor FR 2 in Karlsuhe nuclear centre at the beginning and the Triga reactor at the cancer research centre at Heidelberg and Mainz recently. In the first case approximately 100 mg of the samples were transferred to quartz ampoules of highest purity, evacuated and sealed in order to avoid losses due to heating during the one day irradiation. Today we place the samples in small PE capsules for the other reactortype is water-cooled. After 3-6 days of cooling of the samples they are measured on intrinsic Ge detector sand an anticompton spectrometer in our institutes for 1000 or 2000 sec.

resp. Most of the samples gave simple spectra with the γ-lines of Se-75, As-76, J-131 from Te-131, Sb-124, Hg 203 and Br-82. For calculation the γ-lines given in table 1 are used. Element concentrations were calculated by

Table 1
Table of usable γ-lines

Element	Irradiation Product	T1/2	Gamma Energy	- Interf.
As	As-76	1,1d	559,1 keV	
			657,04 keV	
Se	Se-75	120,4d	264,65 keV	
			400,64 keV	
Te	J-131	8,05d	364,49 keV	
Sb	Sb-124	60,3d	1691,00 keV	
			602,7 keV	Cs 134*
Hg	Hg-203	46,6d	279,19 keV	Se 75**
Br	Br-82	35,3h	776 keV	
			554 keV	
			619 keV	

*604,7 keV
**276,52 keV

comparison with single-element or fluxmonitor standards. The analyses showed, that the Se concentrations changed even within one volcanic area over a range of several magnitudes i.e. from 0.5 to 1700 ppm. Arsenic changes from 0.05 to 265000 ppm. Tellurium was found between <1 up to 775 ppm. Mercury reached 12 ppm.

In the contrast bacterogenic sulphur has according to literature <10 ppm Se, <1 ppm Te,<0,5 ppm As and < 0,1 ppm Hg.

Table 2
Reactions for formation of volcanic sulphur

$2H_2S + SO_2$	⇔	$3S + 2H_2O$
$H_2S + O$	⇔	$S + H_2O$
SO_2	⇔	$S + O_2$
$3H_2SO_3$	⇔	$2H_2SO_4 + H_2O + S$

Assuming that volcanic sulphur (s. Table 2) can form from stable volatile compounds only the compounds of table 3 have to be considered. If in the centre of the solphatara at high temperature the various compounds have certain concentrations it is logic that during temperature-decrease and possible admixture of atmospheric gases the different components decrease in their concentration in the formed sulphur. Data for one example from Santorini in the Aegean Sea (Nea Kameni Island - Figure 1) are listed in table 4. Other examples (Nisyros Aegean Sea) are discussed in the proposed talk.

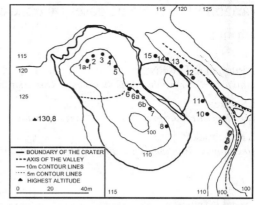

Position of sampled solfataras on Nea Kameni (Santorini)

Fig. 1

A clear relation of trace elements in sulphur and sulphur-isotopes was not found so far. Differences of trace element concentrations in volcanic sulphur are most probably developed during the ascent of the gases to the surface

Table 3
Physical data of volcanic gases.

Element Boilingp.	Hydride Boilingp.	Chloride Boilingp.	Bromide Boilingp.	Oxide Boilingp.
S 444,6°C	H_2S -60,28°C	SCl_2 decomp 59°C S_2Cl_2 138°C SCl_4 decomp.	S_2Br_2 54°C	SO_2 -10°C
As 613°C*	H_3As -55°C decomp 230°C	$AsCl_3$ 130,2°C	$AsBr_3$ 221°C	As_2O_3 193°C As_2O_5 decomp. 315°C
Sb 1750°C*	-17°C	$SbCl_5$ 79°C	$SbBr_3$ 280°C	Sb_2O_3 1550C°
Se 684,7°C	H_2Se -41,5°C	$SeCl_4$ decomp. 288°C	$SeBr_4$ decomp. 75°C Se_2Br_2 decomp. 227°C	SeO_2 340 - 350°C SeO_3 decomp 180°C
Te 1327°C	H_2Te -2°C	$TeCl_2$ 327°C $TeCl_4$ 392°C	$TeBr_2$ 339°C $TeBr_4$ decomp 421°c	TeO_2 1245°C TeO_3 decomp 395°C
Hg 357,3		Hg_2Cl_2 383°C* $HgCl_2$ 276°C	Hg_2Br_2 345°C* $HgBr_2$ 325°C	Hg_2O decomp 100°C HgO decomp 450°C

* sublimation

Table 4
Trace elements in volcanic sulphur [ppm]

Sample	As	Sb	Se	Te	$\delta^{34}S$
Nea Kameni (Greece)					
STN 1a	2,2	1,2	43,6		±0,0‰
STN 1b	0,10	0,27	27,7		+1,4‰
STN 1c			92,4		
STN 1d	1,30	0,83	139,6		±0,0‰
STN 2a	0,29	0,15	23,77		+1,9‰
STN 3a			121,9		+6,5‰
STN 4a	0,55	0,32	301,1	2,90	+3,3‰
STN 5a	0,14	0,12	23,4		+6,8‰
S5	0,54	0,34	141,9		+6,8‰
STS 7a	0,65	0,20	10,04		+9,9‰
STS 8a					+6,3‰
STS 9a	16,9		2,39		+2,1‰
S 9a	0,35	0,15	29,6		+2,1‰
S 10a	0,90	0,078	8,71		+6,0‰
S 11a$_1$		0,30	0,76		+3,2‰
S 12a	0,13	0,34	1,67		+5,5‰
S 13a	52,2	0,17	0,49	1,05	+4,8‰
S 14a	0,28	0,070	1,89		+6,0‰
S 15a	0,79	0,34	6,46		+4,4‰
Hawaii (Big Island) USA					
1	0,22	0,39	0,51		
2		0,43	0,37		
3		0,27	4,97		
Papandajan (Java) Indonesia					
	61,5	2,44	124,8	64,5	
Valley of the 10000 smokes (Alaska) USA					
Fumar. 135	1,65	3,06	5,03		
Long Fissure 135	39,4	6,80	8,30		
Vent 6	81800	0,44	0,46		
Near Baked Mnt. cmp	239,0	0,238	1170	100,9	
32(2) 12.08.19	4,66	0,22	541,5	32,3	
Sapl. Y	243000	0,90	278,8		
NE slope Baked Mnt. 12.08.1919	265000	0,70	183,7		
Volcano (Italy)					
Big Crater	8,6	0,19	38,4	2,8	
Gangue in Main Crater	50,9	0,47	115,8	5,2	
Soltatara (Italy)					
SE corner	<1	2,57	6,13	52,8	
NE end of crater	<1	<0,5	3,1	10	
S corner	0,14	0,52	1,82	4,36	
close to center	12,87	0,48	1,69		
La Corona/Santiaguito (Guatemala)					
about 3" below surface 05.03.1940	2860	0,71	371,6	200,6	
melted S near surf. 40 A5	4570	0,62	517,4	374,0	
about 3" below surface 05.03.1940	5240	0,74	1720	774	

REFERENCES.

Allen, E.T., 1922. Chemical aspects of volcanismwith a collection of the analyses of volcanic gases. *Franklin Institute Jour.*, 193, 29-80.

Allen,E.T. & Zies,E.G. 1923, A chemical study of fumaroles of the Katmai region.National Geographic society, *Contrib. Tech. Papers Katmai Series 2*, 75-155.

Karpov, G.A. & Fazlullin, S. M. 1995. The creation and prolonged existence of the zone of molten native sulfur at the bottom of thermal lake within the volcanic-hydrothermal system (Uzon caldera, Kamchatka), *Water-Rock Interaction, (Kharaka & Chudaev, eds.)* 307-313.

Neumann van Padang,M. 1951, Catalogue of the active volcanoes of the world, including solfatara fields; Pt.1, Indonesia: *Internat. Volcanol. Assoc. Naples*, 271 p.

Papike, J. J.; Keith, T. E. C.; Spilde, M. N.; Galbreath, K. C.; Shearer, C. K. & Laul, J.C. 1991. Geochemistry and mineralogy of fumarolic deposits, Valley of Ten Thousand Smokes,Alaska: Bulk chemical and mineralogical evcolution of dacite-rich protolith. *American Mineralogist 76*, 1662-1673.

Puchelt, H; Hoefs, J, & Nielsen, H. 1971, Sulfur Isotope Investigations of the Aegean Volcanoes, *Acta the International Scientific Congress on the Volcano of the Thera (Greece)*, 303-317

Sturchio,N.C. Williams,S.N. Sano, Y.(1993) The hydrothermal system of Volcan Puracé, Colombia. *Bull.Volcanolog.55*, 289-296

Symonds, R. B.; Rose, W. I.; Gerlach, T. M.; Briggs, P. H. & Russel S. H., 1990, Evaluation of gases, condensates, and SO_2 emissinsfrom Augustine volcano, Alaska: the degassing of a Cl-rich volcanis system., *Bull. Volcanol 52*, 355-374

Toutain, J-P.; Baubron, J-C.; Le Bronec, J.; Alard, P.; Briole, PP.; Marty, B.; Miele, G.; Tedesco, D. & Luongo,G., 1992, Continuos monitoring of distal gas emanations at Vulcano, southern Italy.

White, D. E. & Waring, G. A. (1963) Volcanic Emanations, Data of Geochemistry, sixth edition Chapter K, *Geologoical Suirvey, professional Paper* 440-K, 27 pgs.

Characterization of a magmatic/meteoric transition zone at the Kakkonda geothermal system, northeast Japan

M. Sasaki, K. Fujimoto, T. Sawaki, H. Tsukamoto, H. Muraoka, M. Sasada, T. Ohtani, M. Yagi, M. Kurosawa, N. Doi, O. Kato, K. Kasai, R. Komatsu & Y. Muramatsu
Japan

ABSTRACT: Petrological and geochemical features of a transition zone between magmatic and meteoric environments have been investigated at the Kakkonda geothermal system. Fluid inclusions from hydrothermal veins suggest mixing of saline brine (magmatic) with low salinity fluid (meteoric). The cathodoluminescence study reveals that deformation/fracturing mechanism might be different under low and high temperature conditions. Comparing the Pb-Zn-Cu ratios in the rocks, Pb and Zn are enriched in the shallow reservoir and Cu gradually increases with depth toward the granitic rocks. Inclusion fluids from a quartz vein in a granitic rock contain metal elements (Pb, Zn and Cu) as well as major elements (Na, K, Ca. Fe, Mn and Cl).

1 INTRODUCTION

Deep magmatic and shallow meteoric environments in a magma-related hydrothermal system have contrasting characteristics in fluid (magmatic / meteoric), deformation mode (ductile / brittle), and fluid pressure (lithostatic / hydrostatic). Characterization of the magmatic-meteoric transition is important for exploitation of deep geothermal resources as well as understanding the generation of epithermal-type and porphyry-type ore deposits. In this study, petrological characteristics at a transition zone and polymetallic mineralization features have been investigated at the Kakkonda geothermal system, Japan. Samples used in this study were collected from NEDO's well WD-1 (Uchida et al., 1996) and JMC's wells No. 13, 18, 19 and 20.

2 KAKKONDA GEOTHERMAL SYSTEM

In the Kakkonda geothermal system, some deep wells were drilled through a thermally metamorphic aureole and into a Quaternary granitic pluton which is a heat source of the system (Kato and Doi, 1993; Uchida et al., 1996). Temperature profiles in Figure 1 reveal the existence of a thermal conduction zone below 3,200 m depth, whilst there are two different hydrothermally connected geothermal reservoirs, which are shallow (< 1,500 m) and deep (> 1,500 m) reservoirs (Ikeuchi et al., 1996). Fluids discharged from both reservoirs have low salinities (< 1 wt.%) and are meteoric in origin, whilst Na-K-Ca-Fe-type brine with a high salinity over 19 wt.% is trapped inside the granitic pluton of the thermal conduction zone and is considered to be magmatic in origin (Kasai et al., 1996).

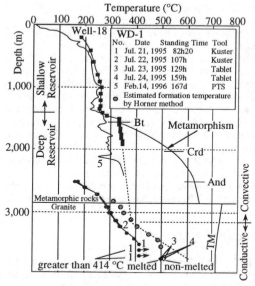

Figure 1. Temperature profiles for the well WD-1 and Well-18 (modified after Uchida et al., 1996). Metamorphism: an estimated metamorphic temperature (after Enami, 1997), Bt: biotite isograd, Crd: cordierite isograd, And: andalusite isograd, TM: ternary minimum of granitic melt.

Table 1. Hydrothermal vein mineral assemblages. Classification of zones with vertical depths are after Kato and Sato (1995).

Zones / Minerals	Shallow Reservoir			Deep Reservoir				In Granite
	Mnt zone	M/S zone	Ser zone	Bt zone	Crd zone	Ath zone	And zone	
Quartz	────────	────────	────────	────────	────────	────────	────────	----
Calcite	────────	────────	────────					
Anhydrite				────────	────────	────────	────────	—
Stillbite	────────							
Laumontite	────────	--						
Wairakite	────────	------						
Prehnite		────────	----					
Epidote			-- ────────					—
K-feldspar			────────	────────				
Albite				────────	────────	────────	────────	
Chlorite			---- ────────	────────	────────	────────		
Sericite		--		--			--	
Actinolite						-----	-- -	--
Biotite							-- -	
Sphalerite			—					
Chalcopyrite			—					
Pyrite	------	------	------	------	------	------	------	------
Specularite				────────	────────	────────	────────	----
Magnetite				--	------	------		
Pyrrhotite					-----	-----		

Abbreviation of zones; Mnt: montmorillonite, M/S: montmorillonite/sericite mixed-layer clay mineral, Ser: sericite, Bt: biotite, Crd: cordierite, Ath: anthophyllite, And: andalusite.

3 HYDROTHERMAL VEIN MINERALS

Based on a petrological study of metamorphic and hydrothermally altered rocks, seven mineral-assemblage zones have been identified by Kato and Sato (1995). Hydrothermal vein-mineral assemblages also change with depth (Table 1). The Ca, K and Na-bearing alumino-silicate minerals occur at different depths; Ca-bearing minerals (zeolite, prehnite and epidote) in the shallow reservoir, a K-bearing mineral (K-feldspar) at around the boundary of two reservoirs, and a Na-bearing mineral (albite) in the deep reservoir. Calcite and anhydrite are common in the shallow and deep reservoirs, respectively, and fill center parts of veins. Sphalerite and chalcopyrite occur predominantly in the sericite zone. Specularite usually occurs in open fractures, whilst magnetite occurs in sealed veins

4 FLUID INCLUSIONS

Fluid inclusions are useful to elucidate the geochemical evolution of hydrothermal fluid. In the Kakkonda geothermal system, liquid-rich and vapor-rich two-phase inclusions and polyphase inclusions are observed. Polyphase inclusions are predominantly observed in the deep reservoir and inside the granitic pluton (Komatsu and Muramatsu, 1994). Daughter minerals identified in polyphase inclusions are halite, sylvite, Fe-bearing chloride and Ca(?)-bearing chloride. Microthermometries of fluid inclusions for vein minerals which were collected from the deep reservoir are shown in Figure 2. The variation in salinity suggests the mixing of low and high salinity fluids, which are considered to be meteoric and magmatic, respectively.

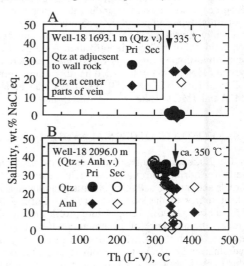

Figure 2. Microthermometry on fluid inclusions from a quartz vein (A) and a quartz-anhydrite vein (B) in the deep reservoir. Arrows indicate present formation temperatures.

Considering the wide variation in salinity for anhydrite (Fig. 2B), precipitation of anhydrite may be an indicator for the mixing front in the Kakkonda geothermal system. Chemical analysis of each fluid inclusion has been conducted by micro LA-ICP-MS. Major components in polyphase inclusions from a quartz vein in a granitic rock are Na, K, Ca, Mn, Fe and Cl with minor amounts of B, Cu, Zn, Pb and Ba, and trace amounts of Li, Mg, Al, Rb and Sr.

5 DEFORMATION MODE OF GRANITE

Mode of deformation must be taken into account as a controlling factor for development of fractures at deep and high-temperature portions of a hydrothermal system. Three spot cores of granitic rocks collected at 2,938, 3,230 and 3,729 m depths of the well WD-1 are investigated, using a cathodoluminescence (CL) luminoscope.

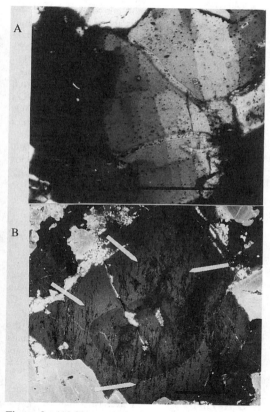

Figure 3. (A) Photograph of quartz grains in granitic rocks from 3,230m depth of WD-1. Fluid inclusion healing planes are coincident with wavy extinction. (B) CL Photograph of quartz grains in granitic rocks, showing existence of many healed fractures parallel to c-axis of host quartz grains. Scale bar: 1 mm.

Some of the quartz grains in the granitic rocks show weak wavy extinction. Fluid inclusions sometimes line along subgrain boundaries in such quartz grains (Fig. 3A), indicating a close relation between formation of fluid inclusions and deformation of quartz. In order to investigate effect of water to deformation of the quartz grains, water contents in quartz grains were analyzed (excluding visible fluid inclusions), using micro FT-IR spectroscopy. The water contents, however, range from 40 to 150 ppm, close to those for typical quartz in granitic rocks.

In the 2,938 m depth sample from the convection zone, fluid inclusion healing planes in quartz grains are sometimes identified, whilst they are rarely identified in the 3,230 and 3,729 m depth samples from the conduction zone. The CL study, however, revealed existence of many healed fractures in the quartz grains from the 3,230 m depth sample. The healed fractures observed by the CL are often parallel to c-axis of quartz grains (Fig. 3B), indicating that the fracturing under high temperatures might be controlled by physical properties of quartz grains such as thermal expansion. In the 3,729 m depth sample, fluid inclusions are relatively abundant and large in size at margin of quartz grains comparing to center of them. Healed fractures are only observed in the center of quartz grains by the CL, whilst no luminescence was observed at the margin. No luminescence at margin of quartz grains may be attributed to recrystalization of quartz crystals through interaction between quartz grains and inclusion fluids.

6 IMPLICATION FOR ORE FORMING PROCESS

All the ore forming processes, e.g. metal acquisition, transportation and deposition, are observed in the Kakkonda geothermal system. The rocks from the shallow reservoir (Fig. 4A and Fig. 4B) are relatively rich in Pb and Zn, whilst those from the deep reservoir and the granitic rocks are relatively rich in Cu (Fig. 4C and 4D).

The metal concentrations in each fluid inclusion, analyzed by micro LA-ICP-MS, are also plotted on Figure 4D. Vapor-rich inclusions are relatively enriched in Cu compared to polyphase inclusions.

The Pb-Zn-Cu ratios of polyphase inclusions are similar to the brine collected from the bottom of the well WD-1, and are slightly enriched in Zn, compared with experimentally determined solubility data reported in Hemley et al. (1992). The difference in metal ratios between vapor-rich and polyphase inclusions may indicate difference in distribution coefficient for each metal element. One of other characteristics revealed by the analysis is an enrichment of B in vapor-rich inclusions rather than polyphase inclusions.

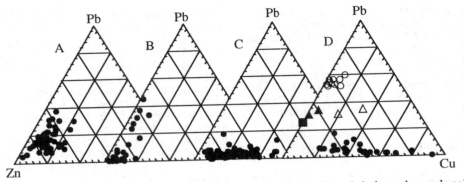

Figure 4. Pb-Zn-Cu ratios (ppm) of rock samples and inclusion fluids. Closed circle: rock sample collected at 20 m depth intervals (Muraoka, 1997) in the well WD-1 (A: 0-1,000 m, B: 1,000-1,500 m, C: 1,500-3,000 m, D: 3,000-3,729 m), closed triangle and open triangle: inclusion fluids in polyphase and vapor-rich inclusions from a quartz vein in a granitic rock, respectively, open circle: experimentally determined solubility data (Hemley et al., 1992), closed square: brine collected at the bottom of the well WD-1 (NEDO, 1996).

7 SUMMARY

In the Kakkonda geothermal system, the deepest depth where meteoric water infiltrates is considered about 3,200 m depth, based on temperature profiles (Ikeuchi et al., 1996; Fig. 1). This study indicates the difference in deformation mode at 2,938 m and below 3,230 m depths. The formation temperature at 3,230 m depth is about 380°C. Homogenization temperatures reported from many epithermal-type deposits seldom exceed a temperature of 400°C. The temperature of about 400 °C may be critical for the change in deformation mode at magma-related hydrothermal system, resulting in mixing between magmatic and meteoric fluids. We hope that further study on the Kakkonda geothermal system will elucidate the change in fluid pressures (lithostatic/hydrostatic) at the transition zone.

ACKNOWLEDGMENTS

New Energy and Industrial Technology Development Organization (NEDO), Japan Metal & Chemicals Co. Ltd. and Tohoku Geothermal Energy Co. Ltd. provided samples.

REFERENCES

Enami,M. 1997. Kakkonda geothermal field and contact metamorphism. *Abst. Japan Earth Planet. Sci. Joint Meeting*: 225. (in Japanese)

Hemley,J.J., G.L.Cygan, J.B.Fein, G.R.Robinson and W.M.D'Angelo 1992. Hydrothermal ore-forming processes in the light of studies in rock-buffered systems: I. iron-copper-zinc-lead sulfide solubility relations. *Econ. Geol.* 87: 1-22.

Ikeuchi,K., R.Komatsu, N.Doi, Y.Sakagawa, M.Sasaki, H.Kamenosono and T.Uchida 1996. Bottom of hydrothermal convection found by temperature measurements above 500°C and fluid inclusion study of WD-1 in the Kakkonda geothermal field, Japan. *Geothermal Resources Council Trans.* 20: 609-616.

Kasai,K., Y.Sakagawa, S.Miyazaki, M.Sasaki, T.Uchida 1996. Supersaline brine obtained from Quaternary Kakkonda granite by the NEDO's deep geothermal well WD-1a in the Kakkonda geothermal field, Japan. *Geothermal Resources Council Trans.* 20: 623-629.

Kato,O. and N.Doi 1993. Neo-granitic pluton and later hydrothermal alteration at the Kakkonda geothermal field, Japan. *Proc. 15th NZ Geothermal Workshop*: 155-161.

Kato.O. and K.Sato 1995. Development of deep-seated geothermal reservoir bringing the Quaternary granite into focus in the Kakkonda geothermal field, northeast Japan. *Resource Geology* 45: 131-144. (in Japanese with English abstract)

Komatsu,R. and Y.Muramatsu 1994. Fluid inclusion study of the deep reservoir at the Kakkonda geothermal field, Japan. *Proc. 16th NZ Geothermal Workshop*: 91-96.

Muraoka, H. 1997. Petrochemical profiling of the aureole of the Kakkonda Granite using cuttings samples along the WD-1a, northeast Japan. *Geothermal Resources Council Trans.* 21: 309-316.

NEDO 1996. Report on Deep-Seated Resources Survey in FY 1995: 271-272. (in Japanese)

Uchida,T., K.Akaku, M.Sasaki, H.Kamenosono, N.Doi, and S.Miyazaki 1996. Recent progress of NEDO's "Deep-Seated Geothermal Resources Survey" project. *Geothermal Resources Council Trans.* 20: 643-648.

… Water-Rock Interaction, Arehart & Hulston (eds) © 1998 Balkema, Rotterdam, ISBN 90 5410 942 4

The Joule-Thomson expansion of CO_2 and H_2O in geothermal and volcanic processes

D. M. Sirkis
US Army Corps of Engineers, Philadelphia, Pa., USA

G. C. Ulmer, D. E. Grandstaff & N. P. Flynn
Department of Geology, Temple University, Philadelphia, Pa., USA

ABSTRACT: In kimberlites and other volcanoes whose gas or fluid phase is CO_2-rich, it has been generally assumed that gas emission during eruption/emplacement would produce a large cooling effect due to gas free-expansion. However, during protracted eruptions, gas emission may be more analogous to an isenthalpic Joule-Thomson expansion. A review of the chemical engineering data and use of various equations of state discloses that CO_2 expansions from the geothermal gradient (or hotter) with P > 600 bars will produce a heating effect. Under similar expansion conditions H_2O will produce a cooling effect. The implications for volcanic- and geothermal environments rich in CO_2 are discussed.

1 THERMODYNAMIC BACKGROUND

During volcanic and hydrothermal (geyser-like) eruptions, gases are emitted from within the earth. Such gas emanations are commonly modeled as expansions into free space (into a vacuum). If the process is adiabatic (dQ = 0), for an ideal gas there is no change in internal energy ($\Delta E = 0$), and the change in gas temperature during expansion is given by

$$T_2/T_1 = (P_2/P_1)^{R/C_p}$$

where T is the temperature (K), P the pressure, C_p the (average) heat capacity at constant pressure, R the gas constant, and 1 and 2 are the initial and final conditions.

Where P_2 is less than P_1, then T_2 will be less than T_1. If P_1 is in the kilobar[1] range and P_2 is essentially 1 bar, then, even if T_1 is at magmatic temperatures, T_2 may be very low. Thus, McGetchin (1968) proposed that during kimberlite eruptions the gases might even become cryogenically cooled (to 254 K).

However, if the eruption of gas is protracted, the model of expansion into free space may not be valid. During protracted eruptions, a steady state system may evolve in which gas at high pressure is emitted through a restrictive orifice or vent system or through a semi-permeable fluidized bed into a lower pressure regime, creating a steady state pressure drop. This case is essentially identical to the classical experiment of Joule and Thomson (Lord Kelvin) in which gases are passed through an orifice, porous plug, or vent, from a reservoir in which the pressure is maintained at pressure (P_1) into a second reservoir having a lower, but constant pressure (P_2). Such processes are essentially isenthalpic ($\Delta H = 0$) and adiabatic. For a real gas, the change in gas temperature with change in pressure under such conditions is governed by the Joule-Thomson (JT) effect and Joule-Thomson coefficient (μ_{JT}). The value of the Joule-Thomson coefficient is given by the equation

$$\mu_{JT} = (dT/dP)_H = 1/C_p (T(dV/dT)_p - V) = V/C_p (\alpha T - 1)$$

in which V is the volume, and α the coefficient of thermal expansion. At low pressures $\alpha T > 1$ and μ_{JT} is positive. Therefore, upon decrease in pressure, the temperature of a real gas also decreases.

The Joule-Thomson effect is well known and extensively used in refrigeration and gas liquification. However, at high pressures, αT may be less than 1 and μ_{JT} negative. Thus a Joule-Thomson expansion (JTE) from high pressure will cause the temperature of such real gases to increase.

The value of μ_{JT} is a function of T and P. The locus of points at which $\mu_{JT} = 0$ is the inversion curve. The curve may be plotted in terms of T and P, or to

---

[1] Most previous literature and data for treatment of critical and reduced pressures and temperatures are in units of bars and degrees K. Therefore, we have retained use of bars (rather than Pascals) for ease of comparison with previous results.

facilitate comparisons between gases according to the principal of corresponding states, in terms of reduced temperature ($T_r = T/T_c$) and pressure ($P_r = P/P_c$), where T_c and P_c are the critical temperature and pressure of the gas (T_c for CO_2 = 304 K and for H_2O = 648 K and P_c for CO_2 = 73 bars and for H_2O = 219.5 bars). By plotting in reduced P and T, one may represent the inversion curve for a given gas by a parabola for each desired equation of state (EOS) that may be of interest. Figure 1 shows for CO_2 the comparison of various EOS calculations. Inside the curve $\mu_{JT} > 0$ and JTE leads to cooling; outside the curve $\mu_{JT} < 0$ and JTE leads to heating.

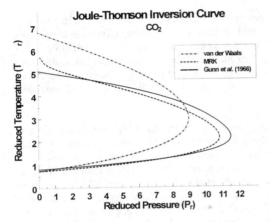

Figure 1. Joule-Thomson Inversion Curves for CO_2 as calculated with vdW and MRK EOS and compared with a regression curve based on a compilation of measured J-T inversion conditions for a variety of gases (Gunn et al., 1966). Although the minimum temperature is often depicted as occurring at zero pressure (as above), the minimum P and T conditions along the inversion curve occur at the intersection of the inversion curve with the vapor-liquid curve.

Predictions from van der Waals (vdW) and Modified Redlich-Kwong (MRK) may be compared with the equation of Gunn et al. (1966), which summarizes, in terms of reduced pressure and temperature, results of 89 inversion measurements. This comparison shows that the MRK EOS is superior to the vdW for prediction of inversion conditions. Juris and Wenzel (1972) compared predictions from several EOS (including MRK, vdW, Dieterici, Virial, Berthelot, Beattie-Bridgeman, Benedict-Webb-Rubin, and Martin-Hou EOS) with actual measured values and with the summary regression equation of Gunn et al. (1966); they concluded that, although not perfect, the MRK EOS provided a better prediction of inversion conditions than did other EOS tested.

2 IMPLICATIONS TO GEOLOGICAL SYSTEMS

Inversion curves predicted by the MRK EOS for CO_2, H_2O, and the mixture (0.5CO_2 + 0.5H_2O) are given in Figure 2, which is plotted in normal P and T. As in Figure 1 the areas inside the inversion curves are the conditions of cooling during JTE and the areas outside the inversion parabolas are the areas of heating during JTE. Also plotted is the continental geothermal gradient (GTG).

2.1 Geothermal/Hydrothemal

Geothermal systems and/or hydrothermal solutions typically originate more shallowly than do volcanic processes. As shown in Figure 2, the JTE of water vapor from the P and T values of the GTG, would produce a heating effect as would the JTE of CO_2 vapor at any T above ~485 K on the GTG. However, if a geothermal field (like Yellowstone) or hydrothermal vent (black smoker) is associated with T values a few hundred degrees above the local GTG, then JTE of H_2O vapor will produce a cooling effect, while the JTE of CO_2 will produce heating.

2.2 Volcanic

Any volcanism is by definition hotter than the local GTG, so that the JTE of volcanic H_2O vapor will be

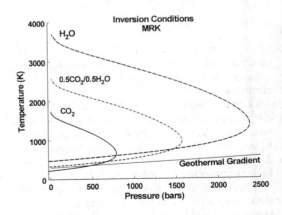

Figure 2. IP curves in normal P and T space for mixture of CO_2 and H_2O obtained from MRK EOS. Recall by comparison to Figure 1 that inside these parabolas, cooling effects would pertain upon expansion and outside these parabolas, heating effects would pertain upon expansion. See text for discussion.

expected to produce cooling, whereas the JTE of volcanic CO_2 will likely be a heating.

2.3 Specific Examples

To return to T_r and P_r allows one, *via* the treatment of corresponding states, to produce a generalized plot for all nearly-non-polar gases (Andrews, 1971). This type of treatment is shown in Figure 3 to which we have added isenthalpes, sub-horizontal lines of equal enthalpy (H1, H2, H3). Adiabatic cooling of a magmatic fluid would follow the general trend of these isenthalps as a trajectory through P_r-T_r space. At a P_r equivalent to 5 km, Figure 3 shows, as bold vertical lines, ranges of T_r from volcanic (1473 K) down to 573 K at 5 km for both CO_2 and H_2O.

Specifically, if one starts at any T_r at 5 km along the CO_2 vertical line in Figure 3 and expands along an isenthalpe to a surface P and T, expansion will initially produce heating (negative slope portion of the

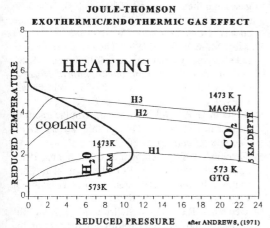

Figure 3. Corresponding states treatment of the JTE expansion effects on CO_2 and H_2O upon expanding from a P_{tot} equivalent to a depth of 5 km. See text for discussion

isenthalpe). By comparison the expansion from this depth from any T_r on the H_2O vertical line along an isenthalpe will produce only cooling (positive slope).

3 MAGNITUDE OF THE JTE EFFECTS

The importance of the JTE effects can best be appreciated from an example: expanding from 1.5 kbar and 1200 K by free expansion with H_2O produces a final T of 229 K and with CO_2 a final T of 345 K. JTE by comparison, under the same conditions, with water comes to 1070 K and CO_2 to 1220 K. It may seem that the JTE CO_2 heating is almost negligible, but the final temperature is 150K greater than that of the JTE of H_2O, and 900 to 1000 K hotter than either CO_2 or H_2O in free expansion.

4 GEOLOGIC APPLICATIONS

4.1 Kimberlites

The last 5 km of kimberlitic emplacement has been the subject of many discussions: *e.g.*, McGethchin and Ullrich (1973) modeled the Cane Valley, Utah, diatreme. They concluded that with a starting magmatic T of 1173 K, the explosive expansion of the CO_2-rich material would have reached an ejection velocity at the surface of MACH 3.8 and a free space expansion T of 254 K, thereby having cooled below ambient surface T by about 40 K. Their model predicted the kimberlite would have already been cooled to 934 K at about 1 km below the surface. According to Wyllie and Haas (1965) the solidus of the CO_2-rich kimberlitic material is 914 K or more. Therefore either the diatreme would suffer a 'thermal death' or the kimberlitic emplacement and advance in the last 1 km would have been only by brecciation and fluidized bed expansion and injection, while the expansion continued to cool the material.

Anyone who has used CO_2 from a laboratory tank will recall that the released gas causes the needle valve to condense moisture and eventually freeze. Note from Figures 2 or 3, that at the valve, the T is ~293 K and the tank pressure above liquid CO_2 is 73 bars so that the conditions are within the CO_2 inversion curve. However, as the P and T of any kimberlite is outside of the cooling region, a JTE treatment at the Cane Valley Diatreme rules out cooling.

To independently test the idea of temperature changes for natural CO_2 expansions, Sirkis *et al.* (1992) examined the magnetic properties of diatremes near Oka, Quebec, and found, using the paleomagnetic conglomerate test, that the xenoliths and autoliths of these diatremes had all exceeded the Curie temperature (>803 K). This indicates that such a cooling had not happened. If such had been the case, it would have quenched at least the autoliths with their original sedimentary magnetization randomly oriented from their 'cold' brecciation- inclusion into the cooling fluidized bed. Thus, no data were found to support any type of expansion-cooling of the solids. Rather, the evidence is that the kimberlitic magma was hot enough to heat the autoliths to above their Curie temperatures.

Further data on the emplacement of kimberlitic pipes has come from the work of Flynn *et al.* (1996)

who examined the wall rock alterations of two kimberlites that cross-cut coal and shale in Pennsylvania, USA, and one that cross-cut bedded salt in New York, USA. Some 12 mineral assemblages could be used for geothermometric estimates and the conclusion was that the emplacement temperature ranged from a minimum ~ 733 K (coking of coal) to less than 1094 K (lack of anhydrous, sylvite-free, halite melting or phlogopite dehydration). These data substantiate that no major expansion- cooling of these CO_2-rich (as high as 14 wt%) kimberlites occurred during emplacement.

4.2 Lake Nyos, Cameroon

The release of large amounts of CO_2 from the volcanic Lake Nyos in 1986 and the subsequent death of more than 1700 people has focused much attention on the exact mechanism responsible for the CO_2 release with theories as to whether this gas was all volcanic or some was biogenic arising from the sediments of the caldera lake. The data indicate that as much as 100 million cubic meters poured out of the lake's surface in just 2 hours. By examination of Figure 2 or 3 it is obvious that the P and T range likely for the gases expanding from the bottom of the lake, even if there were initial bacterial heating or heat of solution, would have resulted in a cooling JTE effect. If the gas was however not 'in residence' in the strata, but rather a volcanic 'leak' from under the lake, it would have produced a JTE heating in addition to the higher temperature from the magma chamber. Evans et al. (1995) rule out the volcanic release idea on the basis that the heat flow into the bottom water is too low to indicate volcanic involvement. However, the possibility that some on-going JTE cooling effect is involved does not seem to have been considered, and may be supported by the idea of Kusabe and Sano (1992), who claim that there is cold, high pressure CO_2 not far below the lake bottom.

4.3 Non-Terrestrial Example

It is important to realize that not all gases will behave as water with a JTE cooling effect. Each gas should be specifically checked in the physical chemistry and chemical engineering data bases.

For example, Soderblom et al. (1990) suggested that solar-heated frozen nitrogen expands and that the expansion was a possible driving force for Triton geysers observed during the Voyager 2 Flyby of Neptune. The authors claimed an isentropic decompressional cooling of 4 K for the nitrogen involved. Such expansion-produced cooling releases enthalpy as an additional heating feed-back loop to further the energetics of their process beyond just the solar-induced thawing-expansion. Even if a JTE occurred, μ_{JT} for nitrogen (Roebuck and Osterburg, 1935) also supports their model, as N_2 has a positive value (cooling effect) at one bar from 0 to 573 K.

REFERENCES

Andrews, F. C. 1971. *Thermodynamics*. Wiley-Interscience. New York.

Evans, W.C., G.W.Kling, & M.L. Tuttle, 1995. Lake Nyos, Cameroon: Examining the 1986 gas release from the system standpoint. WRI-8, 303-306.

Flynn, N.P., G.C. Ulmer & D.P. Gold, 1996. EMP data and the temperature of emplacement of some peridotite dikes. *Geol Soc Amer Abst Vol*, Northeast Section #1340, 55. (Buffalo, NY)

Gunn, R. D., P. L. Chueh, & J. M. Prausnitz, 1966. Inversion temperatures and pressures for cryogenic gases and their mixtures. *Cryogenics* 6, 324-329.

Juris, K. & L.A. Wenzel, 1972. A study of inversion curves, *AIChE Journal* : 184, 684-688.

Kusabe, M. & Y. Sano, 1992. The origin of gases in Lake Nyos, Cameroon. In S.J. Freeth, K.M. Onuoha & C.S. Ofoegbu (eds.) *Natural Hazards in West &Central Africa:* 83-95. Wiesbaden: Vieweg

McGetchin, T.R. 1968 Source and emplacement of kimberlites at Moses Dike, Utah. *Trans. Amer. Geophys. Union (EOS)* 50, 345.

McGetchin,T.R.,Y.S. Nikhanj &A.A. Chodos, 1973. Carbonatite-kimberlite Relations in the Cane Valley Diatreme, San Juan county, Utah. *J. Geophys Res.* 78:1854-1869.

McGetchin, T.R., and G.W. Ullrich, 1973. Xenoliths in Maars and Diatremes with Inferences for the Moon, Mars, and Venus. *J. Geophys Res.* 78:1833-1853.

Roebuck, J.R. & H.Osterburg 1935 The Joule-Thomson Effect in Nitrogen. *Phys. Rev.*: 48, 450-2.

Sirkis, D., G.C. Ulmer, D.E. Grandstaff, J. Castro, & D.P. Gold 1992 Testing a model of diatreme emplacement at Oka, Quebec, using rock magnetism.*Trans. Amer. Geophys. Union (EOS)*: 74, 103.

Soderblom, L.A., S.W.Kieffer, T.L. Becker, R.H. Brown, A.F. Cook, C.J. Hansen, T.V.Jognson, R.L. Kirk & E.M. Shoemaker, 1990. Triton's geyser-like plumes: discovery and basic characterization. *Science* 250:410-415.

Wyllie, P.J. & J.L.Haas, 1965. System $CaO-SiO_2-O_2-H_2O$ at 1 kbar. *Geochim. Cosmochim Acta* 20: 8, 896.

Carbon dioxide and helium emissions from a reservoir of magmatic gas beneath Mammoth Mountain, California, USA

M.L.Sorey & W.C.Evans
United States Geological Survey, Menlo Park, Calif., USA

C.D.Farrar
United States Geological Survey, Carnelian Bay, Calif., USA

B.M.Kennedy
Lawrence Berkeley National Laboratory, Calif., USA

ABSTRACT: Carbon dioxide and helium with isotopic compositions indicative of a magmatic source are discharging at anomalous rates from soils on the flanks of Mammoth Mountain, on the southwestern rim of the Long Valley caldera in eastern California. The gas is released from fumaroles and through more diffuse emissions from normal-temperature soils; it also leaves the mountain dissolved in ground water. Diffuse CO_2 emissions and high CO_2 concentrations in soils have been responsible for tree kills over a total area of ~60 ha. We estimate that the total diffuse gas flux from the surface of the mountain is ~500 tons/day, and that an additional ~40 tons/day of CO_2 are dissolved in ground water that flows away from the mountain. Isotopic and chemical analyses of soil and fumarolic gas demonstrate a remarkable homogeneity in composition, suggesting that the CO_2 and associated helium and excess nitrogen may be derived from a common gas reservoir supplied by magmatic degassing and thermal metamorphism of carbonate rocks.

1 INTRODUCTION

Mammoth Mountain, a dacitic volcano with a history of volcanism over the past 50,000-200,000 years, lies on the southwestern rim of the 760,000 year-old Long Valley caldera (LVC) in eastern California (Fig. 1). Mammoth Mountain also forms the southern end of the Inyo Craters volcanic chain that has produced intermittent rhyolitic and phreatic eruptions over the last 40,000 years, the most recent about 600 years ago (Bailey, 1989; Miller, 1985). Anomalous seismic activity occurred beneath Mammoth Mountain in 1989 with a six-month period of earthquake swarms that included ~four M~3 earthquakes (Hill et al., 1990). In spite of its modest energy release, the 1989 Mammoth Mountain swarm is noteworthy because of its duration and evidence that it was accompanied by a magmatic intrusion in the upper crust beneath the mountain (Hill, 1996; Hill et al., 1990).

The 1989 seismicity and associated deformation at Mammoth Mountain was accompanied and followed by remarkable changes in gas discharge at the surface. These changes include increases in temperature, flow, and $^3He/^4He$ ratios in steam vent MMF (Mammoth Mountain Fumarole) on the north side of the mountain (Sorey et al., 1993); and the development of tree kill areas induced by an enormous, diffuse flux of cold CO_2 out through the soil on the flanks of the mountain (Farrar et al., 1995). Although not recognized and tied to tree

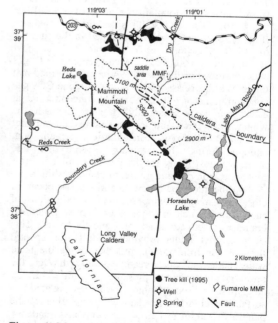

Figure 1. Map of the Mammoth Mountain region on the southwestern edge of Long Valley caldera, showing areas of tree kill caused by high [CO_2], fumarole MMF, and springs and wells containing dissolved carbon that is isotopically similar to that discharging in MMF.

mortality until 1994, the diffuse CO_2 flux has been variously estimated to total 1200 t/d by Farrar et al. (1995) and 400 t/d by Rahn et al. (1996). We report here on the distribution and rate of diffuse gas discharge and on the results of geochemical investigations of fumarolic gas, soil gas, and ground waters on Mammoth Mountain. With these data we present a conceptual model for the gas source(s) beneath this area.

2 VARIATIONS IN $^3He/^4He$ IN FUMAROLE MMF

Periodic determinations of $^3He/^4He$ from MMF fumarole gas provide a time history of variations in the magmatic component of helium discharging from Mammoth Mountain. Helium isotopic ratios have ranged from ~5 to ~7 R_A (times ratio in air) since September 1989, compared with pre-1989 values near 4 R_A (Fig. 2). Corresponding increases in rates of fluid mass and heat from this vent have also been found (Sorey, et al., 1993).

3 AREAS AND RATES OF DIFFUSE CO_2 DISCHARGE

The most obvious signs of diffuse CO_2 discharge from Mammoth Mountain are areas of tree kill on the north, west, and south sides of the mountain (Fig. 1). Such areas were first noted in the summer of 1990; the total area of tree kill is now approximately 60 ha. Measurements of gas concentrations made since 1994 have confirmed that the trees kills are associated with root-zone CO_2 gas concentrations of ~30-90%. CO_2 concentrations at each area vary seasonally (highest beneath the winter snow pack and lowest during late summer and fall). However, concentrations remain high enough during the growing season that tree mortality most likely results from a combination of root zone asphyxia and nutrient uptake inhibition.

For the most part, the tree-kill areas occur in close proximity to mapped faults or extensions of such faults, which in some cases form well-defined boundaries between dead trees and live trees. This suggests that faults provide conduits for upward flow of gas from depth, and that above the local water table the gas spreads laterally through the unsaturated zone by pressure or gravity-driven advection. Topographic variations related to surface-water drainage channels also influence the pattern of lateral gas flow, resulting in a tendency for CO_2 concentrations, and presumably gas discharge rates, to be higher in the dry creek channels than in adjacent regions. In this way, the flow pattern for cold, relatively dense CO_2 resembles that of groundwater flow in shallow

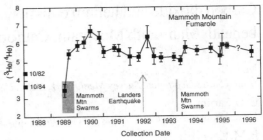

Figure 2. Temporal variations in the air-corrected helium isotopic composition (R/R_A) for fumarole MMF (error bars represent 1σ uncertainties), periods of significant seismic activity beneath Mammoth Mountain (stippled bars), and the timing of the M-7.3 Landers earthquake.

water-table situations.

Anomalous CO_2 concentrations have been detected in only a few areas of limited extent above the tree line on Mammoth Mountain, including (1) the saddle area below the summit (~3,000 m elevation) and under the gondola, (2) the area around fumarole MMF, and (3) parts of the summit ridge. CO_2 concentrations of over 95% have been measured in the fault controlled saddle area in vaults that access water lines for snow-making operations at the Mammoth Mountain Ski Area.

It is difficult to accurately determine the total diffusive gas flux from Mammoth Mountain because of the variability in gas flow rates within each discharge area and the steep, roadless terrain over much of the mountain. The first published estimate of 1,200 t/d (Farrar et al., 1995) was based on extrapolation of a single gas flow rate measurement of 5,800 g m^{-2} d^{-1} over a tree-kill area of 20 ha. Subsequent measurements reported by Rahn et al. (1996) suggested that the average gas flow rate within each tree-kill area was closer to 950 g m^{-2} d^{-1}; this rate was applied to an estimated total tree kill area of 42 ha to obtain a total flux estimate of 400 t/d. Our own set of ~250 measurements made during the summer of 1996 show a range in the average value of gas flow rate at each discharge area of 73 to 2,950 g m^{-2} d^{-1}, yielding an estimated total gas flux of 510 t/d with an uncertainty of ±20%.

4 SOIL-GAS COMPOSITION

Compositions of soil gas in the tree-kill areas, in the saddle area beneath the summit ridge, and in fumarole MMF are remarkably similar, except for the absence of reactive species (e.g. H_2S, CH_4, H_2) in the cold soil gas. None of the gases contains

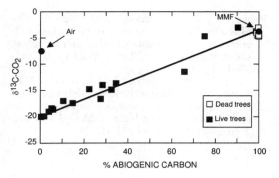

Figure 3. Plot of $\delta^{13}C$ of CO_2 vs. the fraction of CO_2 from abiogenic sources in soil gas and fumarole MMF (determined from $\delta^{14}C$ values).

high-temperature species such as SO_2 or CO. The similarity in composition among the samples from each of these sites is indicated by comparable ratios of He/CO_2 and N_2/Ar and values of $^3He/^4He$ and $\delta^{13}C$-CO_2.

Similarities in isotopic composition are seen in a plot of ^{13}C vs ^{14}C (fig. 3), showing a mixing line between shallow biogenic ($\delta^{13}C = -20‰$) and deep, abiogenic ($\delta^{13}C = -4$ to $-5‰$) end members. Although the heavy end-member $\delta^{13}C$ values are similar to MORB values, they could conceivably also result from carbon sources in the metasedimentary rocks that crop out around the rim of Long Valley caldera. However, we take the helium isotopic values of 4-6 R_A as indications that much of the CO_2 derived from depth is of magmatic origin.

5 CO_2 DISSOLVED IN GROUNDWATER

Water samples collected from cold springs and wells around the flanks of Mammoth Mountain show evidence of widespread discharge of CO_2 with a carbon isotopic composition similar to MMF and tree-kill area soil gas. Dissolved CO_2 with heavy $\delta^{13}C$ and depleted $\delta^{14}C$ values occurs at each of the sites shown in Fig. 1. In addition, helium isotopic ratios of 4.2 and 4.5 R_A have been measured in samples from two wells on the north side of the mountain. In contrast, these groundwaters are relatively dilute and contain tritium contents of 6-18 T.U., indicating that they have not circulated to any significant depth.

The groundwater gas data indicate that the deep-source gas with its magmatic component has been discharging from Mammoth Mountain since before 1989. However, our estimate of the dissolved gas flux (~40 t/d) is much smaller than the present-day diffusive gas flux.

6 CONCEPTUAL MODEL

Key elements of a conceptual model that accounts for the gas discharge at Mammoth Mountain include (1) a reservoir of low-temperature, high-pressure gas, underlain by a zone of high-temperature liquid and bounded by a low-permeability seal, (2) an underlying zone of recent magmatic intrusion, and (3) fault conduits for gas transport through the low-permeability seal and upward to areas of anomalous gas discharge at the land surface. The gas reservoir would contain a mixture of crustal and magmatic gases; the underlying liquid reservoir serves to dissolve high-temperature magmatic gases like SO_2. The edifice outside these reservoirs is characterized by (1) low temperatures, (2) an absence of groundwater except for shallow outflow zones that carry away precipitation and snowmelt, and (3) local areas of hot rock that might serve to boil small amounts of local meteoric water to drive steam vents like MMF.

Chivas et al. (1987) have described an example of a high-pressure gas pocket - a 240 bar, 92°C reservoir drilled into at 2.5 km depth near a Recent maar volcano in Australia. We hypothesize a process for the gradual formation of "cold" gas pockets or caps above hydrothermal fluids that is similar to that described by Giggenbach et al. (1991), in which the gas reservoir is a relict hydrothermal system, supplied with heat and volatiles by episodic dike intrusions below and self-sealed by mineral deposition and acid alteration in a boiling zone that now caps off the gas reservoir.

A reservoir of high-pressure gas, as opposed to degassing of individual intrusions, is favored for Mammoth Mountain because of the homogeneity in composition of the gas discharging over widely spaced areas and the large quantity of gas that apparently has been discharging for at least the past 7 years. We would, however, call upon the 1989 intrusive event and associated seismicity to have increased the rate of gas flow from the deep reservoir by increasing the permeability of the reservoir seal. There is seismic evidence for the presence of a reservoir of compressible gas beneath the mountain from analyses of earthquake data that indicate anomalously low V_p/V_s ratios, similar to the situation at the Geysers geothermal field in northern California (Foulger et al., 1995; Julian et al., 1996). The extent of the seismically anomalous zone beneath Mammoth Mountain is on the order of 20 km³.

The quantities of gas stored in a pressurized (say 5-50 bars) reservoir of this size could easily supply the amount of CO_2 discharged from Mammoth Mountain since 1989, whereas degassing of a single small magmatic intrusion could not. Gas transport from the reservoir to the land surface most likely occurs along fault conduits of limited areal extent. This would be consistent with the relatively rapid transport rates suggested by the timing of the onset of the increased gas discharge in 1989-90 and the need to minimize the opportunity for the gas to encounter and dissolve in deeply circulating groundwater.

7 REFERENCES

Bailey, R.A. 1989. Geologic map of Long Valley caldera, Mono-Inyo Craters volcanic chain, and vicinity, Eastern California. *U.S. Geol. Surv. Map* I-1933, 2 sheets and 11 p.

Chivas, A.R., I. Barnes, W.C. Evans, J.E. Lupton, & J.O. Stone, J.O. 1987. Liquid carbon dioxide of magmatic origin and its role in volcanic eruptions. *Nature* 326 (6113): 587-589.

Farrar, C.D., M.L. Sorey, W.C. Evans, J.F. Howle, B.D. Kerr, B.M. Kennedy, C.-Y. King, & J.R. Southon 1995. Forest-killing diffuse CO_2 emissions at Mammoth Mountain as a sign of magmatic unrest. *Nature* 376 (6542): 675-678.

Foulger, G.R., A.M. Pitt, B.R. Julian, & D.P. Hill 1995. Three-dimensional structure of Mammoth Mountain, Long Valley caldera, from seismic tomography. *EOS Trans. AGU* 76 (46): F351.

Giggenbach, W.F., Y. Sano, & H. U. Schmincke 1991. CO_2-rich gases from Lakes Nyos and Monoun, Cameroon; Laacher See, Germany; Dieng, Indonesia; and Mt. Gambeir, Australia - variations on a common theme. *J. Volc. and Geother. Res.* 45: 311-323.

Hill, D.P. 1996, Earthquakes and carbon dioxide beneath Mammoth Mountain, California. *Seismol. Res. Lett.* 67(1): 8-15.

Hill, D.P., W.I. Ellsworth, M.J.S. Johnston, J.O. Langbein, D.H. Oppenheimer, A.M. Pitt, R.P. Reasonberg, M.L. Sorey, & S.R. McNutt 1990. The 1989 earthquake swarm beneath Mammoth Mountain, California: an initial look at the 4 May through 30 September activity. *Bull. Seismol. Soc. Am.* 80: 325-339.

Julian, B.R., A. Ross, G.R. Foulger, & J.R. Evans 1996. Three-dimensional image of a geothermal reservoir: The Geysers, California. *Geophys. Res. Lett.* 23 (6): 685-688.

Miller, C.D. 1985. Holocene eruptions at the Inyo volcanic chain, California: implications for possible eruptions in Long Valley. *Geology*, 13: 14-17.

Rahn, T.A., J.E. Fessenden, & M. Wahlen 1996. Flux chamber measurements of anomalous CO_2 emission from the flanks of Mammoth Mountain, California. *Geophys. Res. Lett.* 23 (14): 1861-1864.

Sorey, M.L., B.M. Kennedy, W.C. Evans, C.D. Farrar, & G.A. Suemnicht 1993. Helium isotope and gas discharge variations associated with crustal unrest in Long Valley caldera, California. *J. Geophys. Res.* 98: 15,871-15,889.

Modeling the interaction of magmatic gases with water at active volcanoes

Robert B. Symonds & Terrence M. Gerlach
US Geological Survey, Cascades Volcano Observatory, Vancouver, Wash., USA

ABSTRACT: Despite the abundance of $SO_{2(g)}$ in magmatic gases, precursory increases in magmatic $SO_{2(g)}$ are not always observed prior to volcanic eruption. This may occur because many terrestrial volcanoes contain abundant ground or surface water that scrubs magmatic gases until a dry pathway to the surface is established. To better understand the evolution of discharged gas compositions during the dry-out process, we investigated the mixing of a typical high-temperature (915°C) magmatic gas from Merapi volcano, Indonesia with cold (25°C) air-saturated water using thermochemical equilibrium programs. These calculations demonstrate that scrubbing may mask $SO_{2(g)}$ and $HCl_{(g)}$ degassing, although $CO_{2(g)}$, $H_2S_{(g)}$, and $HF_{(g)}$ are affected less by the scrubbing process. Therefore scientists should be wary of low $SO_{2(g)}$ fluxes from restless volcanoes in pre- and early eruption stages and should monitor both scrubbed and non-scrubbed gases to detect magmatic degassing and to monitor the dry-out process.

1 INTRODUCTION

Along with seismic and geodetic monitoring, gas studies play an important role in monitoring potentially restless volcanoes. The main purpose for monitoring gases from a re-awakening volcano is to help detect the intrusion of magma. The dominant magmatic gases are $H_2O_{(g)}$, $CO_{2(g)}$, and $SO_{2(g)}$ followed by lesser amounts of $H_{2(g)}$, $H_2S_{(g)}$, $HCl_{(g)}$, $HF_{(g)}$, $CO_{(g)}$, $S_{2(g)}$, $COS_{(g)}$, and rare gases (Symonds et al., 1994). Hence, the obvious gas precursors of magmatic intrusion are increased fluxes (or concentrations) of $CO_{2(g)}$ and $SO_{2(g)}$, given that $H_2O_{(g)}$ may derive from magmatic or meteoric sources. Of these, the $SO_{2(g)}$ flux is most commonly measured because it can be determined easily with the correlation spectrometer (COSPEC).

Certainly some eruptions are preceded by the expected large increase in $SO_{2(g)}$ emission rates as was the case prior to the 1992 eruptions of Mount Pinatubo (Daag et al., 1996). In contrast, before and after the 1992 eruptions of Mount Spurr, there were generally very low emission rates of $SO_{2(g)}$ despite the high emission rates of $SO_{2(g)}$ during eruptions and persistently elevated seismic activity (Doukas and Gerlach, 1994). Doukas and Gerlach (1994) also observed elevated $CO_{2(g)}$ emissions and significant $H_2S_{(g)}$ odor in the plume during the repose periods between eruptions. They argue that scrubbing of the magmatic gases by cold water within the volcanic edifice reduced the $SO_{2(g)}$ emissions. Apparently, $CO_{2(g)}$ and $H_2S_{(g)}$ are less affected by scrubbing. The purpose of this paper is to investigate scrubbing using a thermochemical modeling approach.

2 NUMERICAL MODELING

We envision scrubbing in active volcanoes to involve the chemical interaction of cold or hydrothermal water with rising magmatic gases. The acidic waters formed by this process may also react with the wall rock, although we consider water-rock reactions to be less significant in the short time-frame of precursory degassing.

To investigate this process, we modeled the thermochemical mixing of a high-temperature magmatic gas with 25°C air-saturated water (ASW). The gas selected for the modeling is a typical high-temperature (915°C) magmatic gas from Merapi Volcano, Indonesia (sample 79-2; Le Guern et al., 1982). This sample contains 88.87% $H_2O_{(g)}$, 7.07% $CO_{2(g)}$, 1.54% $H_{2(g)}$, 1.15% $SO_{2(g)}$, 1.12% $H_2S_{(g)}$, 0.59% $HCl_{(g)}$, 0.16% $CO_{(g)}$, 0.08% $S_{2(g)}$, and 0.04% $HF_{(g)}$ (Le Guern et al., 1982). The modeling was conducted by titrating (numerically) incremental amounts of the 915°C gas into 1 kg of 25°C ASW (with 4.5×10^{-4} mols $N_{2(aq)}$ and 2.4×10^{-4} mols $O_{2(aq)}$) at 0.1 MPa constant pressure. From 0.0002 to 100 kg of gas were added to the ASW. Temperature of the mixture was calculated using an enthalpy balance equation whereby the enthalpy of the mixture equals the enthalpy of the added components. The calculations span from liquid-only to liquid-plus-gas to gas-only over the 25° to 900°C temperature range. To model the liquid and liquid-plus-gas region, we used program CHILLER (Reed, 1982) for modeling aqueous-gas-solid systems constrained mostly by thermochemical data from SUPCRT92 (Johnson et al., 1992). The gas-only region was modeled with program GASWORKS and its GASTHERM thermochemical data for modeling high-temperature gas-solid systems (Symonds and Reed, 1993). For these calculations, we assumed that the gases behave ideally.

Fig. 1 shows the temperature of the mixture over the course of the calculations. The calculations proceed from left to right on the diagram as gas is added to ASW. When <0.008 kg of gas is added to 1 kg of ASW, all the gas dissolves as ions in solution. As the calculations proceed from 0.008 to 1.7 kg of gas added, the mixture consists of both liquid and gas, and the percentage of gas increases from 0.003 to >97% (Fig. 2).

The pH of the aqueous solution decreases from ~3.5 at the start of the calculations (0.0002 kg of gas added) to about -0.5 just before the solution boils completely (1.7 kg of gas added). The dominant aqueous species of C is $H_2CO_{3(aq)}$ due to the low pH of the solution. $Cl^-_{(aq)}$ is the dominant aqueous species of Cl throughout the aqueous and aqueous-plus-gas regions of the calculations, although $HCl_{(aq)}$ becomes a significant minor species just before the solution boils to dryness. The dominant aqueous species of F are $F^-_{(aq)}$ (<0.003 kg of gas added) and $HF_{(aq)}$ (0.003 - 1.7 kg of gas added). The most abundant aqueous species of S are $SO_4^{2-}{}_{(aq)}$ (<0.009 kg of gas added), $H_2S_{(aq)}$ (0.009 - 0.1 kg of gas added), and $HSO_4^-{}_{(aq)}$ (0.2 - 1.7 kg of gas added). Other aqueous species of S, including polythionates, are insignificant. $NH_4^+{}_{(aq)}$ is the main aqueous species of N, which, of course, derives entirely from the ASW.

Solid or liquid native sulfur is predicted to form when between 0.0008 and 1.9 kg of gas are added. When >1.9 kg of gas are added, the mixture exceeds the saturation temperature of sulfur and S only exists in the gaseous state. Production of native sulfur by

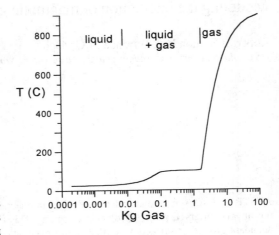

Figure 1. Diagram showing the resulting mixture temperature for reacting (numerically) the 915°C Merapi volcanic gas with 25°C ASW. Horizontal axis shows the kg of gas added to 1 kg of ASW.

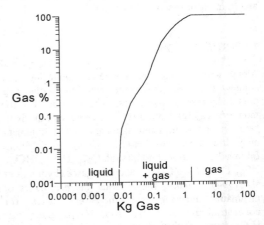

Figure 2. Diagram showing the resulting percentage of gas in the mixture for the Merapi gas-ASW mixing calculations. See caption for Fig. 1.

injection of volcanic gas into water is a common occurrence at many of the world's volcanic crater lakes.

A separate gas phase forms when ≥0.008 kg of gas are added (Fig. 2). At the onset of gas formation at 0.008 kg of gas added, the gas phase is dominated by $CO_{2(g)}$, $H_2O_{(g)}$, $H_2S_{(g)}$, and $N_{2(g)}$ (Fig. 3). As the calculations proceed across the liquid-plus-gas

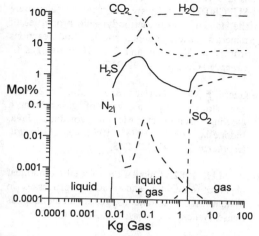

Figure 3. Diagram showing the mol% of the major (>1%) and selected minor (<1%) gas species for the Merapi gas-ASW mixing calculations. See caption for Fig. 1.

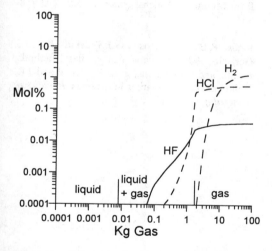

Figure 4. Diagram showing the mol% of minor (<1%) gas species for the Merapi gas-ASW mixing calculations. See caption for Fig. 1.

region, the mol% of $H_2O_{(g)}$ increases, the mol% of $CO_{2(g)}$ decreases, the mol% of $H_2S_{(g)}$ increases and then decreases in concert with $H_2S_{(aq)}$, and the mol% of $N_{2(g)}$ shows a net decrease with $N_{2(aq)}$ (Fig. 3). Minor amounts (>0.0001 mol%) of $HF_{(g)}$ and $HCl_{(g)}$ form when ≥0.06 kg and ≥0.3 kg of gas are added, respectively (Fig. 4). However, magmatic $SO_{2(g)}$ is scrubbed entirely throughout the liquid-plus-gas region (Fig. 3).

When the liquid phase boils completely at 1.8 kg of gas added, the concentrations of the easily scrubbed magmatic species, $SO_{2(g)}$ and $HCl_{(g)}$, increase by several orders of magnitude (Fig. 3-4). $H_{2(g)}$ also becomes a significant minor species when >2 kg of gas are added to the mixture. As still more gas is added, the concentrations (Figs. 3-4) and ratios (Fig. 5) of individual gas species and the temperature of the gas mixture (Fig. 2) approach magmatic values.

As suggested above, the drying out of the system across the liquid-plus-gas region is characterized by the progression from a $CO_{2(g)}$-$H_2O_{(g)}$-$H_2S_{(g)}$ gas to a $H_2O_{(g)}$-$CO_{2(g)}$-$H_2S_{(g)}$ gas. This drying out process is also characterized by dramatic increases in the $H_2O_{(g)}/CO_{2(g)}$ and $HCl_{(g)}/HF_{(g)}$ ratios (Fig. 5).

In summary, these calculations suggest that (1) under wet conditions (water/gas >10^2 at 0.1 MPa), no gas precursors are predicted; (2) under drying conditions (water/gas ~10^2 - ~10^1 at 0.1 MPa) the predicted gas precursors are elevated concentrations and fluxes of $CO_{2(g)}$ and $H_2S_{(g)}$, small concentrations and fluxes of $SO_{2(g)}$, and $HF_{(g)} \gg HCl_{(g)}$; and (3) only nearly dry systems will vent significant $SO_{2(g)}$.

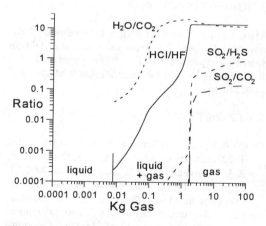

Figure 5. Diagram showing the ratios of selected gas species for the Merapi gas-ASW mixing calculations. See caption for Fig. 1.

3 CONCLUSIONS

This study allows us to reach the following conclusions:

(1) Scrubbing of magmatic gases by water may mask $SO_{2(g)}$ degassing.

(2) Low $SO_{2(g)}$ fluxes from restless volcanoes during the pre- and early eruption stages may indicate scrubbing rather than the absence of magmatic degassing.

(3) We recommend monitoring $CO_{2(g)}$, $SO_{2(g)}$, $H_2S_{(g)}$, $HF_{(g)}$, and $HCl_{(g)}$ in pre- and early eruption stages to help detect both non-scrubbed and scrubbed magmatic gases and to monitor the dry-out process.

(4) Many geothermal systems and some inactive volcanic systems have persistent $H_2S_{(g)}$ odors, probably from the degassing of older magma, that generally do not indicate the threat of an impending magmatic eruption. However, the presence of intense and persistent $H_2S_{(g)}$ odors without formerly significant $H_2S_{(g)}$ degassing may indicate new shallow magma degassing with the possibility of an eruption, especially if there are other precursory phenomenon.

(5) Scrubbing may be widespread as most volcanoes are probably wet.

ACKNOWLEDGMENTS

Mark Reed provided us with a beta version of the new SOLTHERM data base and valuable insight on running program CHILLER. This paper benefited from reviews by Jake Lowenstern, Larry Mastin, and Tom Murray.

REFERENCES

Daag, A.S., B.S. Tubianosa, C.G. Newhall, N.M. Tuñgol, D. Javier, M.T. Dolan, P.J. Delos Reyes, R.A. Arboleda, M.M.L. Martinez, & M.T.M. Regalado 1996. Monitoring sulfur dioxide emissions at Mount Pinatubo, in C.G. Newhall and R.S. Pungongbayan (eds.) *Fire and mud: eruptions and lahars of Mount Pinatubo, Philippines*:409-414. Quezon City:Philippine Institute of Volcanology and Seismology and Seattle:University of Washington Press.

Doukas, M.P. & T.M. Gerlach 1994. Results of volcanic gas monitoring during the 1992 eruption at Mt. Spurr, in T.E.C. Keith (ed), The 1992 eruptions of Crater Peak Vent, Mount Spurr Volcano, Alaska. *U.S. Geol. Surv. Bull.* 2139:47-57.

Johnson, J.W., E.H. Oelkers & H.C. Helgeson 1992. SUPCRT92: a software package for calculating the standard molal thermodynamic properties of minerals, gases, aqueous species, and reactions from 1 to 5000 bar and 0 to 1000 degrees C. *Comp. in Geosci.* 18:899-947.

Le Guern, F., T.M. Gerlach, & A. Nohl 1982. Field gas chromatograph analyses of gases from a glowing dome at Merapi volcano, Java, Indonesia, 1977, 1978, 1979. *J. Volcanol. Geotherm. Res.* 47:223-245.

Reed, M.H. 1982. Calculation of multicomponent chemical equilibria and reaction processes in systems involving minerals, gases and an aqueous phase. *Geochim. Cosmochim. Acta* 46:513-528.

Symonds, R.B. & M.H. Reed 1993. Calculation of multicomponent chemical equilibria in gas-solid-liquid systems: Calculation methods, thermochemical data and applications to studies of high-temperature volcanic gases with examples from Mount St. Helens. *Amer. J. Sci.* 293:758-864.

Symonds, R.B., W.I. Rose, G.J.S. Bluth & T.M. Gerlach, 1994. Volcanic-gas studies: methods, results, and applications, in M.R. Carrol and J.R. Holloway (eds.), Volatiles in magmas. *Reviews in Mineralogy* 30:1-66.

Magmatic sulfur content of the 1995-1996 Ruapehu eruptions, New Zealand

T. Thordarson
Institute of Geological and Nuclear Sciences, Wairakei Research Centre, New Zealand & CSIRO, Magmatic Ore Deposit Group, Division of Exploration and Mining, Australia

C. P. Wood & B. F. Houghton
Institute of Geological and Nuclear Sciences, Wairakei Research Centre, New Zealand

ABSTRACT: Two significant eruption episodes occurred at the Ruapehu volcano in New Zealand between September 1995 and July 1996 during a period characterised by intermittent small explosive eruptions. Juvenile ejecta are porphyritic (25 to 42 vol.%) medium-K calc-alkaline andesite with ~58 wt.% SiO_2. Dacitic glass compositions, represented by microlite-free groundmass phase and inclusions in phenocrysts, indicate a complex melt evolution, where processes of crystal fractionation and magma mixing were important. The degassed groundmass contains ~65 to ~225 ppm S, whereas the S-content of the inclusions ranges from ~250 ppm to ~865 ppm. The total atmospheric SO_2-venting by the activity at Ruapehu is estimated as 266,000 tons.

Table 1. Main events of the 1995-96 Ruapehu eruption episodes, including a sample list and data distribution.

Eruption	Description	Samples	Data distribution[1]
18/9/1995	single explosion and lahar	Ru57	grm=14; incls.=17
23/91995	<1 min eruption with three explosions and 4 lahars	Ru71, Ru74, TRU7	grm=53; incls.=49
11/10/1995	8 hour-long sustained explosive eruption and lahars	Ru73, TRU1, TRU3, TRU4, TRU5	grm=100; incls.=72
14/10/1995	5 hour-long sustained explosive eruption		
17/6/1996	12 hour-long sustained eruption	P53738	grm=19; incls.=18

[1] grm., groundmass glass; incls., glass inclusions

1. INTRODUCTION

Ruapehu volcano, at the southern end of the Taupo Volcanic Zone, is one of the most active andesite cone volcanoes in the world. After 18 years dormancy, activity resumed at Ruapehu in 1995 with sporadic small explosive eruptions through the lake-filled summit crater between 18 September and end of October. A similar eruption episode began ~8 months later on 17 June lasting to late July 1996. The most significant events are listed in Table 1. The total magma volume emitted by these eruptions is about 0.1 km³ (Nairn et al., 1996; Houghton et al., 1996), with discrete events probably not exceeding ~0.02 km³. The largest events were followed by strong atmospheric SO_2 outgassing, exceeding 15,000 tonnes/day in 1995 and 10,000 tonnes/day in 1996.

In this study, we present major element and sulfur analysis of glass inclusions in phenocrysts and of microlite-free groundmass glass (defined as containing <5 vol.% of microlites) from juvenile 1995 and 1996 ejecta. We also discuss the implications of the data for estimating atmospheric venting of SO_2.

2. METHODS

This study includes 11 samples, from eruptions on 18 September, 23 September, and 11 October in 1995 and the eruption on 17 June 1996. These data represent analyses of 156 glass inclusions from phenocryst phases and 186 spot-analysis of microlite-free groundmass glass (Table 1).

Microprobe analyses were obtained at University of Hawaii following the analytical procedures outlined in Thordarson et al. (1996). Sulfur was analysed by using the CSIRO-trace routine of Robinson and Graham (1992) and the estimated detection limit for S is 30 ppm.

3. PETROLOGY

Juvenile 1995-96 scoria and scoria bombs are porphyritic medium-K calc-alkaline andesite (Nairn et al. 1996; Nakagawa et al., 1996). The phenocryst phases are plagioclase (An_{48-86}), clinopyroxene ($En_{40}Fs_{20}Wo_{40}$-$En_{54}Fs_5Wo_{41}$) and orthopyroxene ($En_{62}Fs_{35}Wo_3$-$En_{83}Fs_{14}Wo_3$) with modes ranging from 25 to 42 vol.%. The vesicular (>60 vol.%) groundmass consists of glass and microlites (predominantly plagioclase). Microlite content of the groundmass ranges from <1 vol.% to almost 100 vol.%, with typical values between 35-70 vol.%. Microlite-free glass is occasionally found in quenched outermost selvages of scoria bombs, as groundmass in lapilli-size scoria clasts, and in 0.1-1.0 mm wide cloaks around glomerophyric clusters.

Whole-rock compositions of the 1995-96 Ruapehu eruptives cluster tightly around ~58 wt.% SiO_2, becoming slightly more mafic with time (Fig. 1). Hybrid magma of 60.2-64.9 wt.% SiO_2 was also erupted on 23 September.

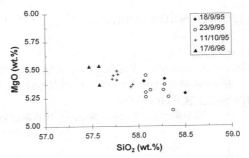

Figure 1. MgO-SiO_2 variation diagram of whole-rock compositions of 1995-96 Ruapehu eruptives.

4. GROUNDMASS GLASS CHEMISTRY

Most of the microlite-free groundmass glass is dacite, with SiO_2 contents from 61.0 to 69.6 wt.% (Fig. 2a). One sample, TRU7 from 23 September, has more evolved groundmass glass compositions (72.7-75.3 wt.% SiO_2). The groundmass glass (i.e. melt) compositions define broad, near-continuous linear trends on variation diagrams (Fig. 2a). Melt compositions become slightly more mafic with time with most evolved melt erupted on 23 September.

Sulfur concentrations in the groundmass glass range from 60 to 225 ppm with an average value of 115±33 ppm (Fig. 2b; Table 2). No correlation was found between eruption date and S-content of the melt, but the least evolved melts have the highest S-values.

5. GLASS INCLUSION CHEMISTRY

The glass inclusions have a similar compositional range and trend as the groundmass glass (i.e. 61.1-71.3 wt.% SiO_2). No correlation is found between phenocryst and inclusion chemistry, but all inclusions trapped in the same phenocryst are

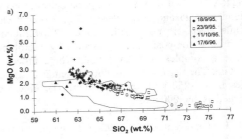

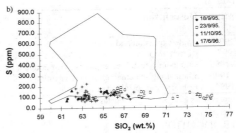

Figure 2. Composition of groundmass glass in 1995-96 Ruapehu products. Solid lines outline the compositional field of glass inclusions Compare with Figure 3.

Table 2. Average sulfur content (ppm) in groundmass glass and glass inclusions from the 1995-96 eruptives

Eruption	S-grm	S-incl.	Max S-incl
18/9/95	126 (26)	229 (59)	332
23/9/95	119 (32)	211 (85)	473
11/10/95	112 (35)	287 (167)	866
17/6/96	106 (39)	274 (152)	492
average of all	115 (33)	256 (133)	

grm, groundmass glass; incl., glass inclusions. Number in brackets are standard deviation. Last column reports maximum measured sulfur content in inclusions from each eruption chemically identical. The glass inclusions generally have low MgO (<2.5 wt.%), and lack the high MgO values found in some of the groundmass glass (Fig. 3a). There is a marked compositional difference between the groundmass glass and glass inclusions from the same events. The compositional range of the glass inclusions on 18 and 23 September is greater than of the groundmass glass, with a shift towards more mafic compositions. The opposite is found in the 11 October 1995 and 17 June 1996 eruptives, where the inclusion data are shifted towards more silicic compositions and also show a spread towards lower MgO values (compare Figs. 2 and 3). It appears that inclusion compositions from the 11 October and 17 June eruptives are shifted towards the melt composition of the magmas erupted on 23 September and 11 October, respectively.

Sulfur concentrations in glass inclusions range from 81 ppm to 866 ppm, with ~84% of the measurements between 100 to 350 ppm (Fig. 3b; Table 2). All four events include inclusions with S-content greater than maximum groundmass values (i.e. 225 ppm S) and 15% of analysed inclusions have high S-content (>350 ppm, Fig 3b). Most inclusions are in phenocrysts with intermediate core compositions (e.g. pyroxene, Mg#=66-79). However, 8 of the sulfur-rich (S>350 ppm) inclusions are hosted by Mg-rich clinopyroxene (Mg# =87-91). Despite a large scatter in the sulfur data (Fig. 3b), it is readily apparent that: a) the baseline S-content of the glass inclusions is ~100 ppm, or similar to the average S-content of the groundmass glass, b) regardless of major element composition, the range in measured S-values in the inclusions is at least 200 ppm, c) the average S-content of inclusions is higher than that of the groundmass glass at equivalent major element compositions, and d) there is no obvious correlation between S-content and major element composition of the glass inclusions.

6. DISCUSSION

The 1995-96 melt compositions (Fig. 3) clearly indicate that fractionation through crystallisation of pyroxene and plagioclase was an important process in their evolution. However, detailed petrologic studies (Nakagawa et al., 1996) suggest a more complex evolution, involving sequential injection and re-mixing of high-temperature (~1100°C) parental magmas with crystal-rich low-temperature (>1000°C) derivatives.

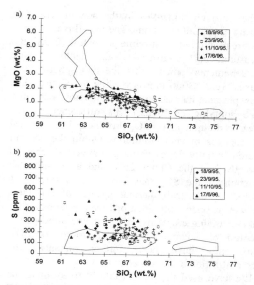

Figure 3. Composition of glass inclusions in 1995-96 Ruapehu products: (a) MgO-SiO$_2$ variation diagram; (b) S-SiO$_2$ variation diagram. Solid lines outline the compositional field of groundmass glass.

The groundmass glass represents degassed melts containing ~100 ppm S. This value is approaching the sulfur capacity of 10-60 ppm S for dacite melts at one atmospheric pressure (Katsura and Nagashima, 1974; Carroll and Webster, 1994). The large range of S-content in the glass inclusions (i.e., 100-870 ppm), suggests that the phenocrysts trapping the inclusions were growing in an actively degassing melt. We propose that the groundmass and the inclusions with low S-content represent partly degassed melts residing in the upper part of the conduit system (<1 km depth; e.g., Stix et al., 1993; Kazahaya et al., 1994). The inclusions with high S-content may represent fresh undegassed high-temperature magma derived from a deeper source (i.e. a crustal magma chamber?). If this is correct, then the deeper level magmas injected into the upper part of the conduit system contain dacite melts spanning a fairly large compositional range (Fig. 3b).

Magma degassing at Ruapehu can be viewed as having occurred in two stages, 1) instantaneous release of volatiles during eruptions which are an unknown quantity because gas-fluxes were not measured from actively erupting ash-rich plumes and 2) as more prolonged degassing of magma residing in the upper part of the conduit system, either through fumaroles in the summit crater or during open vent situations. The SO_2-fluxes of stage 2 were monitored by airborne gas plume (COSPEC) measurements (Nairn et al., 1996). These measurements indicate that the atmospheric SO_2-fluxes were <3000 tons/day while the Crater Lake was present, increasing to >10,000 tons/day after lake-emptying eruptions. This difference in the measured atmospheric SO_2-fluxes can be explained by changes in vent conditions within the summit crater. Initially the escaping gases were discharged directly into the Crater Lake and a considerable fraction of the SO_2 may have been immediately stored in the hydrothermal system of the lake. However, open vent situations were established by lake-emptying eruptions, allowing free atmospheric venting of magmatic SO_2. Furthermore, the SO_2-flux may have been further enhanced by remobilization of the stored SO_2.

We have used the data (Table 2) to estimate the atmospheric venting of SO_2 by the Ruapehu eruptions using the following assumptions: a) the difference between the mean S-content of the inclusions (256 ppm S) and the groundmass (115 ppm S) is taken to represent the amount of SO_2 released at the point of magma fragmentation and eruption, and b) the difference between the maximum S-value of the inclusions (i.e. 866 ppm) and the mean S-content of groundmass is taken as the total sulfur contribution of this magma to the Ruapehu system during the 1995 and 1996 eruption episodes. The former assumption indicates that the melt released ~140 ppm S upon eruption. This is equivalent to 0.7 kg SO_2 per m^3 of melt, assuming melt density of 2500 kg/m^3. After correcting for phenocryst modes, the total volume of the erupted melt is ~0.07 km^3. Thus, the cumulative SO_2 emission by the eruptive events during the 1995 and 1996 episodes is estimated as 49,000 tons. The latter assumption indicates the total contribution of the melt to the gas phase was ~750 ppm S or 3.8 kg SO_2 per m^3 of melt. Using the melt volume given above these values indicate that the overall SO_2 emission from the 1995-96 activity at Ruapehu was at least 266,000 tons. This is a minimum estimate for the total SO_2-flux because it does account for possible contribution from unerupted magma.

ACKNOWLEDGMENTS

Support for this work was provided by the New Zealand FRST Ruapehu contingency contract C05529. We also thank Tari Mattox and an anonymous reviewer for constructive reviews.

REFERENCES

Houghton B. F., Nairn I. A., Scott B. J. Thordarson T., Wilson C. J. N., 1996, Processes and products of the 1995-96 eruptions of Ruapehu, New Zealand. In: Morrisson B., Landis C., and Tulloch A., *Geological Society of New Zealand 1996 Annual Conference programme and abstracts*, p 93.

Katsura T. and Nagashima S., 1974. Solubility of sulfur in some magmas at 1 atmosphere. *Geochim. Cosmochim. Acta*, 38: 517-531.

Kazahaya K. and Shinohara H., 1994. Excessive degassing of Izu-Oshima volcano: magma convection in a conduit. *Bull. Volc.*, 56: 207-216.

Nairn, I. A. and Ruapehu Surveillance Group., 1996, Volcanic eruption at a New Zealand ski resort prompts reevaluation of hazards: *Eos*, 77, 189-191.

Nakagawa M., Wada K., Wood P., Thordarson T., and Gamble J., 1996: Petrology of the juvenile ejecta from the 1995 eruption at Mt. Ruapehu, New Zealand. *Eos*, 77, 2: p 127.

Robinson, B. W. and Graham, J., 1992. Advances in electron microprobe trace-element analysis. *J. Comput.-Assist. Microscopy*, 4: 263-265.

Stix J., Zapata G. J. A., Calvache V. M., Cortes J. G. P., Fischer T. P., Gomez M. D., Narvaez M. L., Ordonez V. M., Ortega E. A. Torres C. R., Williams S. N., 1993. A model of degassing at Galeras Volcano, Columbia, 1988-1993. *Geology*, 21: 963-967.

Thordarson, Th., Self, S., Óskarsson, N. and Hulsebosch, T., 1995. Sulfur, chlorine, and fluorine degassing and atmospheric SO_2-mass-loading by the 1783-84 Laki (Skaftár Fires) eruption in Iceland. *Bull. Volc.*, 58: 205-255.

D/H composition of water from Neogene magmatites in the East Carpathians, Romania

I. Ureche & D.C. Papp
Geological Institute of Romania, Cluj-Napoca Branch, Romania

V. Feurdean
Institute of Isotopic and Molecular Technology, Cluj-Napoca, Romania

ABSTRACT: Based on hydrogen isotope composition of water extracted from some igneous rocks of Neogene age, located in the East Carpathians (Bargau and Gurghio Mountains), our study highlights the interaction between magmatites and water of meteoric or seawater origin. The δD (SMOW) values range from -51.04‰ (magmatic water) to -82.86‰ (meteoric water). The values between these limits are considered to characterise the mixture of different water sources.

1 INTRODUCTION

Numerous studies have documented the interaction between magma or igneous rocks and meteoric ground water (Taylor, 1974; Magaritz and Taylor, 1976; Venneman et al., 1993). Most interactions seem to occur within the upper parts of intrusive complexes and in the surrounding rocks, by means of large scale hydrothermal convective systems, generated by the heat supplied by cooling igneous rocks.

The purpose of this study is to present a case of water/rock interaction, occurring in the magmatites of Neogene age, located in the subvolcanic zone of the East Carpathians - Bargau Mountains. The magmatites consist of intermediate, calc-alkaline rocks with a great variety of mineralogical and petrographical types, from basaltic andesites to rhyolites, and their subvolcanic correspondents.

2 REGIONAL SETTING

In this study we report the hydrogen isotopic data of water contained in rock samples localised in the Magura Cornii and Magura Arsente intrusive structures (north-western part of the Bargau Mountains). Each of these two intrusive structures, represented by laccoliths with sills and dykes, develops over approximately 8km^2, and cross-cut Paleogene sedimentary deposits.

3 PETROLOGY AND GEOCHEMISTRY OF THE IGNEOUS ROCKS

The Magura Cornii intrusive structure consists of

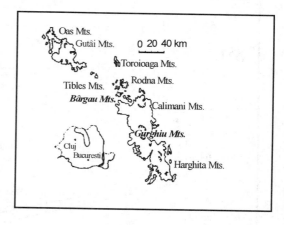

Fig 1. Location map of the Bargau and Gurghiu Mountains

Pannonian (9.7 Ma) hornblende ± pyroxene ± biotite andesites, and microdiorites characterised by the following average contents: SiO_2 - 57.8%, K_2O - 2.4%, Rb/Sr - 0.146, $^{87}Sr/^{86}Sr$ - 0.708.

The magmatites from the Magura Arsente intrusive structure are represented by Pannonian (9.2 Ma) hornblende - pyroxene ± biotite microdiorites. The geochemical features of the studies examples are as follows (average contents):

Table 1. Hydrogen isotopic composition of water from some igneous rocks from Bargau Mountain as compared with Gurghiu Mountains igneous rocks - Ostoros volcanic structure (Papp, 1996)

Sample	Rock type	Location	δD ‰(SMOW)	Presumptive origin of water
1	andesite	Magura Cornii	-53.96	magmatic water
2	microdiorite	Magura Arsente	-56.69	mixture of magmatic and meteoric water
3	contact breccia	Magura Arsente	-67.16	mixture of meteoric and magmatic water
4	igneous xenolith	Magura Arsente	-82.86	meteoric water
5	andesite	Ostoros	-61.38	mixture of magmatic and meteoric water
6	intrusive breccia	Ostoros	-51.04	magmatic water
7	contact breccia	Ostoros	-77.75	meteoric water
8	volcanic tuff	Ostoros	-62.25	mixture of seawater and meteoric water

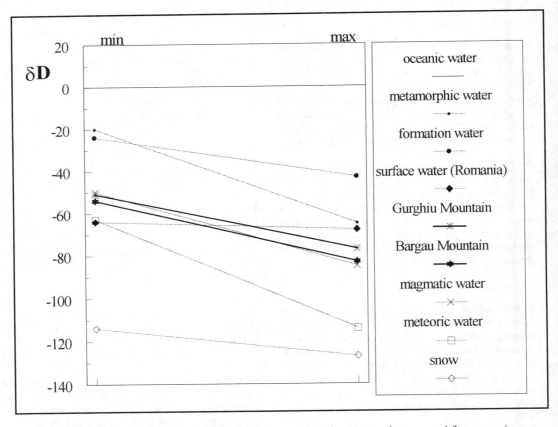

Fig.1 Range of hydrogen isotopic composition - δD (‰ SMOW) - of water samples extracted from some igneous rocks from Bargau and Gurghiu Mountains as compared with diverse water types.

SiO$_2$ - 54.5 %, K$_2$O - 1.87 %, Rb/Sr - 0.176, ^{87}Sr/^{86}Sr - 0.705. Detailed petrological and major and trace elements geochemical studies are reported in Ureche et al., 1995.

4 ANALYTICAL METHODS

Hydrogen analyses were carried out in the Institute of Isotopic and Molecular Technology Cluj - Napoca, with a double - collection Thompson THN 200D mass spectrometer. Analytical results principally refer to fluid inclusion waters, extracted from rock samples by heating at 650°C under vacuum. The hygroscopic water expelled below 110°C was eliminated. All data are expressed in conventional d notation as per - mil deviation of D/H ratios with respect to the SMOW standard. Analyses are given in Table 1.

5 DISCUSSIONS AND CONCLUSIONS

At the present stage of this study, it seems that the intrusive structures under study have been in interaction with meteoric ground waters. It is interesting to compare our case to that of the other segments of the East Carpathian Neogene volcanic chain, in order to determine if the interaction was a local phenomenon, or if it was general for such rocks. The only available data are those from the Ostoros volcanic structure - Gurghiu Mountains (Papp, 1996) (Fig. 1). The hydrogen isotopic results, given in Table 1., suggest that in comparison with the Bargau Mountain intrusive structures, in the Ostoros volcanic structure, the interaction between magmatites and water from meteoric or seawater sources, is more evident. This fact can be explained by the extrusive character of magmatic activity from this area, against the exclusively intrusive one from the Bargau Mountain.

With the presently available isotopic data it is difficult to constrain accurately the conditions of water/rock interaction. A detailed knowledge of the conditions under which the whole process has occurred implies the establishment of an "isotopic map" of all magmatic structures, both for the hydrogen and oxygen isotopes.

REFERENCES

Magaritz, M., Taylor H.P. 1976 Oxygen and Hydrogen isotope studies along 500km traverse across the Coast Range batholith and its country rocks. Central British Columbia. *Can. J. Earth Sci.* 13: 1514-1536.

Papp, D. 1996 Originea fluidelor naturale in lumina studiului izotopilor hidrogenului si oxigenului, referat. *arh. Univ. "Babes-Bolyai"* Cluj-Napoca: 32-59.

Taylor, H.P. Jr. 1973 The application of oxygen and hydrogen isotope studies to problems of hydrothermal and ore deposition. *Econ. Geol.* 69: 843-883.

Vennemann, T.W., Muntean J.L. 1993 Stable isotope evidence for magmatic fluids in the Pueblo Viejo epithermal acid sulphate Au-Ag deposit, Dominican Republic, *Econ. Geol.* 88: 55-74.

Ureche, I., Papp, D., Nitoi, E. 1995 Neogene magmatites from the Rodna and Bargau Mountains (East Carphatians): some petrographical and petrochemical peculiarities, *Rom. J. Mineral.* 77: 48-49.

Fluid-magmatic differentiation of the low-water granitic melts as a possible result of cavitation

G.A.Valuy
Far East Geological Institute, Far Eastern Branch, Russian Academy of Sciences, Vladivostok, Russia

ABSTRACT: The main feature of fine-grained enclaves (autoliths) in granitoids are described (geological position, chemical and mineralogical compositions, etc.) as a result of protracted studies of shallow granite intrusions outcropping on the coast of the Sea of Japan, Russian Far East. A new model of formation of these is proposed. Autoliths appear in granite melt with low water contents and form during the process of degassing as a result of dynamic pressure fall (i.e. cavitation) under the small depth conditions.

1 INTRODUCTION

The problem of formation of fine-grained enclaves in granitoids is the subject of some investigations (Lacroix, 1893; Didier, 1973;). The author encountered this problem while studying shallow granites along the Sea of Japan coast in the Russian Primorye (fig.1). Beautiful coastal outcrops made it possible to obtain abundant geological, mineralogical and petrochemical data on enclaves. The genesis of these enclaves is a great enigma and the purpose of this investigation.

2 CHARACTERISTICS OF THE ENCLAVES

The enclaves are 10-100 cm in diameter and have a specific magmatic texture that has not been observed in any other intrusions, including those of the same composition. This texture is fine to medium grain size and formed by prismatic crystals of plagioclase with xenomorphic quartz, K-feldspar, and biotite between them. In granophyric granites, it is microspherulitic.

Fig.1 Location of studed granite massifs of the Eastern Sichote-Alin volcanic Belt.

Fig. 2. Microgranodiorite enclaves in granite of the Oprichninsky massif.

Fig. 3. Enclaves of fine-crystalline rocks with white rims in granodiorite of the Uspensky massif.

Table 1. Major elements (wt.%) of granites and their autoliths from the Oprichninsky massif.

Sample	SiO_2	TiO_2	Al_2O_3	Fe_2O_3	FeO	MgO	MnO	CaO	Na_2O	K_2O	P_2O_5	LOI
220a*	75.78	0.17	12.38	0.33	1.72	0.38	0.05	0.66	3.97	4.30		0.00
220b$_2$	75.22	0.17	12.12	0.86	1.72	0.47	0.11	0.79	3.87	3.81		0.22
220b$_1$	72.42	0.29	12.76	0.53	3.27	0.57	0.08	0.79	3.97	3.60		0.02
220g*	65.92	0.46	15.06	0.08	5.32	1.70	0.21	2.49	4.80	2.56		0.35
220e	75.32	0.06	11.99	0.04	2.75	0.28	0.03	0.66	3.99	4.10		0.00
220k	72.54	0.23	13.02	0.40	2.62	0.57	0.08	0.66	4.50	3.32		0.00
220p$_3$*	66.96	0.40	15.56	0.51	3.91	0.75	0.18	1.83	4.60	3.05		0.45
220n	74.00	0.17	12.38	0.61	2.62	0.47	0.08	0.79	3.98	3.25		0.11
220o*	74.54	0.21	12.76	0.39	2.24	0.57	0.08	0.92	3.93	3.90		0.05
231p	74.32	0.27	12.72	0.77	2.29	0.52	0.08	0.87	4.16	3.46	trace	0.08
231k*	68.46	0.41	15.11	1.61	2.42	0.73	0.10	2.01	4.90	2.49	0.62	0.028
231m*	61.90	0.88	16.43	3.09	3.79	1.87	0.22	3.47	5.68	2.56		0.23
233*	64.56	0.82	15.90	1.73	3.41	1.64	0.23	2.54	4.97	3.00		0.53
233a	72.44	0.37	12.72	0.96	3.36	0.53	0.07	0.49	4.59	4.20	0.22	0.00
233b*	67.00	0.55	16.43	0.63	3.06	1.07	0.07	2.15	5.20	2.88	0.22	0.24
233c	73.74	0.25	12.06	2.97	1.31	0.31	0.06	0.74	4.36	4.12		0.05
244*	61.42	0.96	17.38	0.10	5.83	2.14	0.20	3.49	5.61	1.67	0.40	0.40
244a	73.26	0.39	11.92	2.24	2.77	0.23	0.07	0.61	3.99	4.12		0.00
264	75.70	0.12	12.12	0.19	1.56	0.21	0.14	0.70	3.90	3.50	0.62	0.42
265	73.26	0.29	12.22	0.76	2.34	0.33	0.11	1.15	4.97	3.37	0.47	0.08
267	74.06	0.28	12.45	3.17	1.07	0.52	0.07	0.73	4.06	3.44	0.11	0.44

*- Enclave, others - host granite. Samples' location (distance from the intrusion contact): 220 - 100 m; 231 - 1200 m; 244 (mid-crystalline granites of aplite and granite textures with enclaves) - 4 km; 264-267 - coarse-crystalline granites lacking any enclaves from the central part of intrusion; blank cell - not determined.

Table 2. Petrophysical characteristics of granitoids and their enclaves

Rock	Density (g/cm3)		Porosity (n, %)		Magnetic susceptibility c 10-6CGS		Remanent magnetization In·10-6 CGS	
	X	N	X	N	X	N	X	N
Oprichninsky massif								
Granite	2.58	30	1.0	8	500	110	200	20
Enclave	2.56	8	2.8	8	1100	138	1100	8
Olginsky and Vladimirovsky massifs								
Granodiorite	2.62	30	0.8	8	750	24	100	15
Enclave	2.58	8	2.9	8	1200	21	1200	8
Valentinovsky massif								
Granite	2.60	15	-	-	380	15	-	-
Enclave	2.58	16	-	-	450	16	-	-
Uspensky and Zapovedny massifs								
Granite	2.60	15	-	-	15	24	1	8
Enclave	2.56	10	-	-	25	12	5	8

X - group mean, N - number of determinations.
Analyst: E.Ya. Dubinchik (VSEGEI, Sankt-Peterburg).

The enclaves are often zoned with relatively coarse cores and fine-crystallized rims resembling quenching zones in dikes. In some enclaves, plagioclase prisms in rims are oriented along the contacts.

The enclaves of different texture are often observed in one outcrop, with finer ones crossing contacts between medium grain size enclaves and the host granites. Composition of the enclaves and host rocks correlates with each other (e.g. enclaves are granodiorite in granites, quartz-diorite in granodiorite). However variation of enclave composition in host rocks of the same composition may be marked even in one outcrop (e.g. varying from granite to granodiorite). Difference between host rock and enclave composition increases from the massif edges to the core and from the shallower to deeper intrusions (Table 1). Enclaves in granites are always larger than those in granodiorites and diorites (Figs. 2 and 3).

Mineralogy of the enclaves is homogenous: 40-70% zoned plagioclase, 10-20% quartz, 1-3% (up to 17% in the microspherulitic varieties) K-feldspar, 4-20% biotite, and 3-10% hornblende. In comparison with the host granites, enclaves are richer in apatite, magnetite, sphene and allanite. The enclaves are more porous and less dense relative to rocks of the same composition, but different in position (Table 2).

The problem of the enclaves' origin is enigmatic. Some researchers believe that they are autoliths, forming at either early or late stages of crystallization, but from the same magma as the host granites. Other researchers consider the enclaves as xenoliths. In our opinion, the enclaves are explained best by the hypothesis of liquid immiscibility resulting from cavities of fluid-poor melt, when it was occupying the magmatic chamber.

3 DISCUSSION

When melts intrude into shallow magma chambers, intensive gas emission occurs as a result of pressure loss. Under these unstable dynamic conditions gas bubbles increase rapidly in size, absorbing volatile magma components, but when they get into regions of high pressure gas bubbles disappear or cave in. In hydrodynamics this process is called "cavitation" (Knapp et al., 1974;). Cavitation is like boiling, but birth and increase in the size of gas bubbles under cavitation it is caused by dynamic pressure loss. When bubbles cave in very high pressure are developed. The pressure rise waves in surrounding liquid may reach several thousand of bars (Knapp et al., 1974). Incidentally when cavitation occurs in liquid with low saturated vapour pressure and low dissolved gas content ("dry" melts) the inside of a bubble can absorb only small part of the full energy and practically all of the energy of the cave-in process is dissipated in the pressurizing enveloping liquid. If the quantity of dissolved gas is large enough, the cavity absorbs a considerable part of the cave-in energy, and the maximum cave-in pressure decreases. Probably because of this autoliths often occur in shallow granites and are absent in deep-seated granites. This pressure is sufficient to cause destruction of some materials, when cavitation occurs near their surface. At a distance from the solid surface the pressure wave inside liquid volume when cave-in occurs causes mass formation of crystallization centres under higher temperature conditions (Flemings, 1977). For granitic melts these will be nuclei of plagioclase and of hornblende. This formation of numerous crystallization nuclei explains the fine-grained structure of enclaves. But the enclaves are not cavitation bubble itself, these later are filled with melt, but a volume of melt, enriched in fluids after cave-in, that was effected by increased cave-in pressure or a shock wave. Although cavitation is a fast process, it entails finite time interval and has several stages called cavern formation. Their growth in size results from the momentary pressure rise and also the diffusion of cations such as Na, Ca, Mg and Fe^{2+}. Gases and fluids after cave-in return to the melt, resulting in volatile enrichment in part of melt. According to some investigators (Knapp et al., 1974;) diffusion velocity under cavitation conditions surpasses usual diffusion velocities in melts by several orders of magnitude.

Water as a prevailing component of granite fluids lower the viscosity and surface tension of the melt some ten times or more (Epelbaum et al., 1973), while enrichment in Ca, Mg, Fe^{2+} heightens those parameters many times, but evidently surface tension rise under ΣRO influence is greater than lowering it under influence of fluids, otherwise this part of the melt would not form separate drop-like enclaves. Besides, in subliquidus melts viscosity is influenced by gas bubbles and by crystalline phases, and viscosity is increased many times (Persikov, 1984). As a result some volume of the melt undergoes viscosity change and thus two liquids with different physical properties are formed - and they must have border betwen them. Surface tension on this border evidently defines plagioclase crystal orientation along contact of drop-like enclave, that is observed in some enclaves of the Oprichninsky massif (Valuy, 1979). Besides, due to curvilinearity of the border additional (Laplacicum) pressure appears $\Delta P = 2\sigma/R$, that also can influence the crystalline phase orientation.

When the drop of the second melt appears in the original magmatic melt, they are in dynamic equilibrium, and drop form is defined by conditions of minimum full surface energy. If surface energy is isotropic, the equilibrium form of the second melt will not be spherical. If surface energy is a function of orientation, the equilibrium form will not be spherical, as is often observed. The bubbles are seldom insulated from each other, and they usually grow and cave in near to each other or to other surfaces. The destruction of gas filled cavitie on cave in may form a cloud of bubbles in the course of next expansion. The larger the bubbles grow, the longer the time interval of the cave in, the more energy is released, and the greater the pressure peak, and all that leads to formation of autoliths of greater dimensions.

4 CONCLUSION

This paper does not provide strict proof of cavitation participation of fine-grained enclaves in granitoids, but I believe that such an explanation is nearer to nature in complex phenomena of crystallization and gas extraction from "dry" magmatic flows at small depths by process, which is observed in liquids, without using any casual factors, such as different magmatic flows mixing together. Incidentally experemented mixing of two liquids shows, that drop-like structure are absent, but that rather long layers and subhorizontal cells are formed (Huppert and Turner, 1984). Most direct proof of cavitation reality is perhaps the higher content of H_2 in fluid phase of autoliths as compared with granites (Valuy and Gantimurova, 1985), because it is known that under cavitations condition some gas ionization occurs, their optical glow (Knapp et al., 1974; and others), and probably water dissociation with some H_2 emission.

I think that cavitation explains such autolith properties, as presence in a single outcrop enclaves of various dimensions and compositions, and change of their dimensions and composition with change of distance from contact and the depth of the massif. The grain size of autolith also depends on composition: medium grains size is larger and more basic, as they were formed in the course of caving-in of bubbles of greater size, that held greater quantity of fluids. The presence of zonal enclaves and "enclaves inside enclaves" (fine-grained enclaves within more midium-grained enclaves) show the multiple stages of bubble cave-in.

5 REFERENCES

Didier J. 1973. Granites and their enclaves. In *Development in Petrology*. Vol.3. Amsterdam, London, New York, 393 p.

Epelbaum M.B., I.V. Balashov and T.P. Salova 1973. Surface tension of acidic magmatic melts at high parameters. *Geochemistry*, 3: 461-464.

Flemings M. 1977. *Solidification processing*. Moscow, Mir, 360 p.

Huppert H.E. & I.S. Turner 1984. Convection with double diffusion. In *Sovremennaya gidrodinamika. Uspehi i problemy*. Moscow, Mir, p. 413-453

Knapp P., J. Daily and F. Hammit 1974. *Cavitation*. Moscow, Mir, 687 p.

Lacroix A. 1893. *Les enclaves des roches volcaniques*. Mascon, 770 p.

Persikov E.S. 1984. *Viscosity of magmatic melts*. Moscow. Nauka, 159 p.

Valuy G. 1979. *Feldspars and crystallisation conditions of granitoids*. Moscow, Nauka, 146 p.

Valuy G.A. & T.P. Gantimurova 1985. On the question of liquid immiscibility in granitic melts. In *Magma and magmatic fluids*. Chernogolovka, p. 28-31.

9 Ore deposits

Effects of fluid flow and temperature variations on lead mineralization in the Southeast Missouri ore district

Martin S. Appold & Grant Garven
Department of Earth and Planetary Sciences, John Hopkins University, Baltimore, Md., USA

ABSTRACT: Numerical modeling was carried out on the physical, thermal, and chemical behavior of a Late Paleozoic topographically-driven groundwater system in the Ozark region of the central USA, and its relationship to Mississippi Valley-type ore formation in the Southeast Missouri Pb-Zn-Cu district was explored. The results indicate that if ore-forming brines had transported both metals and reduced sulfur at concentrations near saturation with respect to major sulfide minerals such as galena, then mineralization would have been concentrated in the Southeast Missouri district due to regional maximizations in temperature change and fluid velocity occurring there.

1 INTRODUCTION

Mississippi Valley-type (MVT) deposits have long been inferred to be precipitates of sedimentary brines. The brines that formed the Southeast Missouri MVT deposits appear to have been part of a giant groundwater system that circulated throughout much of the present-day Ozark region (Fig. 1). Paleomagnetic studies (e.g. Wu and Beales, 1981) date this circulation to be Late Pennsylvanian to Permian, and suggest a link with Alleghanian-age tectonism along the Appalachian-Ouachita tectonic belt of eastern North America. Pore fluids in Ozark-area rocks were probably set in motion by a variety of geologic processes associated with this tectonism. However, the strongest mechanism for fluid flow would have been the topographic gradient created during uplift of the Arkoma foreland basin (Garven et al., 1993). This uplift would have created a strong episode of south-to-north fluid movement, with transport of heat and solute from deeply buried sediments in the south, into the stable mid-continent region of MVT mineralization further north. As fluids moved north into areas of shallower stratigraphic cover, they would have cooled, possibly depositing metal sulfides and other minerals. The present contribution examines temperature variations associated with the Late Paleozoic topographically-driven flow system in the Ozark region, and explores how these variations could have controlled galena mineralization patterns in the Viburnum Trend of the Southeast Missouri ore district.

2 FLUID FLOW AND HEAT TRANSPORT

Groundwater flow and temperature patterns were computed along a 700 km long profile extending

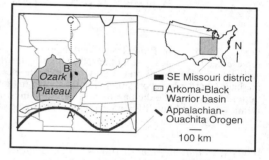

Fig. 1 Map of central USA, showing the location of the Southeast Missouri district (Viburnum Trend and Old Lead Belt) in relation to the Arkoma-Black Warrior foreland basin and the trend of the model profiles, A-B and A-C.

from the Arkoma Basin, through the Viburnum Trend subdistrict of the Southeast Missouri district, and ending in east central Iowa. The profile was constructed from isopach and structure contour data (Imes, 1989; 1990), and from erosional unloading data (Beaumont et al., 1987). Solutions to the equations for variable density flow and heat transport were found using the finite element code, RST2D (Raffensperger and Garven, 1995). Boundary conditions consisted of no flow at the base and right margin of the grid, constant head at the left margin, and water table conditions along the top. A constant heat flux of 60 mWm^{-2} was assigned to the base, and an isotherm of 20° C to the top boundary. Hydraulic conductivities were constrained by data from Imes (1989, 1990). Model values (in m/yr) assigned were

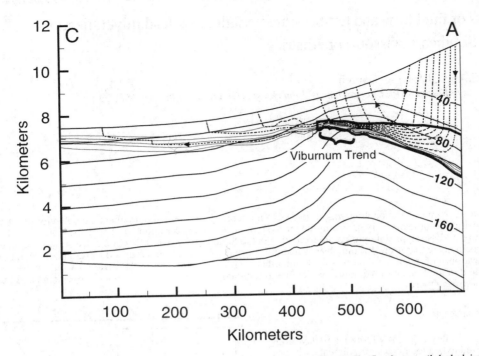

Fig. 2 Steady-state fluid flow and temperature field on the profile A-C. Isotherms (labeled in °C) are represented by the heavy solid lines, flowlines by the dashed pattern, and hydrostratigraphic boundaries by light solid lines. Arrowheads on the flowlines indicate the direction of flow. A restricted portion of the grid used in the later reactive transport simulations is shown in the bold outline. Hydrostratigraphic units represented from bottom to top include the basement, Lamotte Sandstone, Bonneterre Formation, St. Francois Confining Unit, Ozark Aquifer, and undifferentiated Upper Paleozoic sediments.

0.001 for the basement, 100 for the Lamotte Sandstone, 20 for the Bonneterre Formation, 0.004 for the St. Francois Confining Unit, 50 for the Ozark Aquifer, and 2 for the undifferentiated Upper Paleozoic sediments. A solid phase thermal conductivity of 2.0 $Wm^{-1}K^{-1}$ was assigned to the undifferentiated Upper Paleozoic sediments, 3.0 $Wm^{-1}K^{-1}$ to the remaining sedimentary units, and 3.25 $Wm^{-1}K^{-1}$ to the Precambrian basement.

Figure 2 shows the steady-state flowlines and temperature field subject to the conditions above. Recharge occurs over the highest elevations of the profile, while discharge is strongest over the regional crest in the basement (i.e. over the Ozark Dome). Flow is focused mostly through the Ozark Aquifer, due to its combined high permeability and thickness. Lower mass flow rates but comparable average linear velocities exist in the St. Francois Aquifer (Lamotte + Bonneterre). Maximum calculated average linear velocities reach about 7 m/yr in both the Ozark Aquifer and Lamotte Sandstone along the southern flank of the Ozark Dome. Very little flow takes place in the basement or in the deepest sediments of the Arkoma Basin (the southernmost 40-50 km of the profile): average linear velocities in the basement are on the order of only 10^{-5} to 10^{-4} m/yr. Any significant mass transfer in the basement would be expected to have been restricted to fracture zones and the more porous, weathered unconformities. The low average linear velocities of 10^{-3} to 10^{-2} m/yr in the portion of the St. Francois Aquifer in the Arkoma Basin are significant because they indicate that solute is not easily transported out of the deep foreland basin. This has bearing on the geographic location of sources of ore-forming constituents and suggests that the ore constituents were acquired over a longer or different part of the flow path.

The temperature field is strongly affected by fluid flow. In the southernmost section of the profile, where recharge is occurring, isotherms are deflected strongly downward from the initial conductive state. In the discharge region over the Ozark Dome, isotherms are deflected strongly upward. The resultant effect is that the isotherms largely parallel the respective contacts between the basement, Lamotte Ss., Bonneterre Fm., and St. Francois Confining Unit over the southern 150 km of the profile. This means that temperature changes very

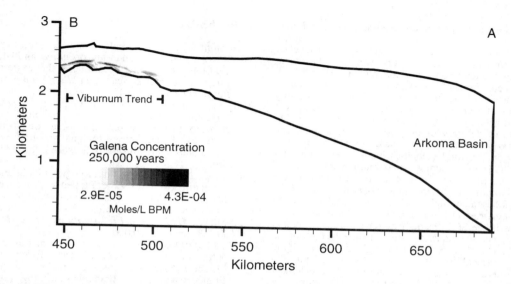

Fig. 3 Galena concentration in moles per liter bulk porous medium after 250,000 years of simulation time under the groundwater advection and thermal patterns summarized in Figure 2 and in the text. The profile, A-B, shown here was extracted from the larger profile, A-C, in Figure 2 and is delineated there in bold outline.

little in the fluid as it traverses the initial portion of the St. Francois Aquifer, the probable conduit for the ore fluids, until the region of the Southeast Missouri district is reached, where a 20-25° C cooling gradient is predicted over the approximate length of the Viburnum Trend. Temperatures predicted here range from about 80° to 105° C, coinciding with the lower to middle range of fluid inclusion homogenization temperatures observed in the subdistrict, and falling within the allowable temperature gradient determined (Rowan and Leach ,1989). The model temperatures shown are most sensitive to three factors: the basal heat flux and the thermal and hydraulic conductivities of the hydrostratigraphic units. However, within the uncertainties of these factors, sensitivity analyses showed the existence of a modest temperature gradient within the Viburnum Trend spanning the lower to middle range of fluid inclusion homogenization temperatures, and the near absence of temperature variations along the flow path south of the trend to be robust results.

3 METAL TRANSPORT AND DEPOSITION

Fluid migrating in the Ozark region during the Late Paleozoic was responsible for transporting not only heat but a diverse suite of chemical constituents. These chemical constituents were, at least in part, the building blocks for the Southeast Missouri ores. Their precipitation from solution could in principle have been caused by a variety of mechanisms. Examined here is the role that regional temperature variations would have had on a fluid initially saturated with respect to the major sulfide ore minerals.

Reactive transport simulations were carried out over a restricted portion of the original grid (cf. Fig. 2) with RST2D, in order to conserve CPU time. Results of the regional-scale simulation were used to define the thermal and hydraulic boundary conditions on the smaller grid. The model maintains chemical equilibrium at each time step, and for initial conditions used the major element concentrations of the "regional brine" of Plumlee et al. (1994). The St. Francois Aquifer south of the Viburnum Trend was assumed to be the source region for the metals and reduced sulfur. The fluid in this part of the aquifer was maintained at equilibrium with respect to galena so that it would behave as a constant source of metals and sulfur. Equimolar amounts of Pb and S were also assumed in order to maximize the concentrations of these elements in the fluid. This resulted in Pb and S concentrations of about 2.0 and 0.3 ppm, respectively (equivalent to about 10^{-5} M). Elsewhere in the profile, the fluid was depleted in metals and sulfur but otherwise retained the same geochemistry. In all, 17 chemical components defining 39 aqueous species and 14 mineral phases were treated in the simulation.

Space limitations restrict us to show only the galena concentration after 250,000 years of reactive flow (Fig. 3), but which is nonetheless representative of the mineralization patterns that develop very early in the simulations. Significant enrichment occurs in the region of the Viburnum Trend. That this should be the case is attributable to the fact that the most cooling occurs here. Isotherms largely parallel the direction of fluid flow south of the trend, providing little opportunity for sulfide minerals to precipitate there.

Within the Viburnum Trend, a second level of enrichment in galena occurs over local basement highs and towards the northern end of the subdistrict. This enrichment has less to do with relative changes in the size of the temperature gradient, which is fairly constant over the trend, than with relative changes in fluid velocity. Higher fluid velocities lead to higher rates of delivery of ore-forming constituents and allow for more rapid accumulation of ore. Regionally, fluid velocities increase in a south-to-north fashion, reaching a maximum over the crest of the Ozark Dome, near the northern end of the Viburnum Trend. Enrichment over the crests of the local basement highs within the trend is also related to increases in fluid velocity. In the model, fluid flow in the St. Francois Aquifer is forced through a narrower interval as it passes between the crest of the impermeable local basement high and the overlying low permeability St. Francois Confining Unit. To conserve mass, the groundwater must be accelerated through this interval, leading to a local maximum in fluid velocity and galena concentration.

4 CONCLUSIONS

The numerical modeling presented here shows that topography generated in the Ozark region during the Late Paleozoic as a consequence of uplift of the Arkoma-Black Warrior foreland basin would have led to vigorous groundwater advection throughout the region. Average linear groundwater velocities could easily have reached magnitudes of meters per year in the Ozark and St. Francois Aquifers, and reach maxima near the crest of the Ozark Dome where the Southeast Missouri deposits are located. This groundwater system was an important vehicle for heat transport in the region. Predicted steady-state temperatures over the Viburnum Trend subdistrict range from about 80-105° C, decreasing towards the north. Temperatures in the St. Francois Aquifer south of the Viburnum Trend remain relatively constant near the maximum of 105° C encountered in the trend, as calculated isotherms largely parallel the fluid flow lines in this region.

These results imply, and the reactive transport simulations confirm, that if brines in the St. Francois Aquifer had been saturated with respect to galena, then galena would have been regionally concentrated in the Southeast Missouri district due to cooling. Secondary levels of enrichment in galena mineralization occur towards the northern end of the Viburnum Trend and over the crests of local basement highs as a result of local fluid velocity increases there. The results indicate the primary role that fluid velocity plays in the localization of ore in a cooling hydrothermal system, and suggests that other geologic features promoting higher fluid velocities, such as breccia zones or other high permeability facies, would also be enriched in mineralization.

5 REFERENCES

Beaumont, C., G.M. Quinlan, and J. Hamilton, 1987. The Alleghanian orogeny and its relationship to the evolution of the eastern interior, North America. *Canadian Society of Petroleum Geologists* 12: 425-455.

Garven, G., S. Ge, M.A. Person, and D.A. Sverjensky, 1993. Genesis of stratabound ore deposits in the midcontinent basins of North America. I. The role of regional groundwater flow. *American Journal of Science* 293: 497-568.

Imes, J.L., 1989. Major geohydrologic units in and adjacent to the Ozark Plateaus Province, Missouri, Arkansas, Kansas, and Oklahoma--basement confining unit. *U.S. Geological Survey Hydrologic Investigations Atlas HA-711-B*: scale 1:750,000.

Imes, J.L., 1990. Major geohydrologic units in and adjacent to the Ozark Plateaus Province, Missouri, Arkansas, Kansas, and Oklahoma--St. Francois aquifer, St. Francois confining unit, Ozark aquifer. *U.S. Geological Survey Hydrologic Investigations Atlas HA-711-C, D, E*: scale 1:750,000.

Ohle, E.L. and P.E. Gerdemann, 1989. Recent exploration history in Southeast Missouri. In: *Mississippi Valley-type Mineralization of the Viburnum Trend*, R.D. Hagni & J.R.M. Coveney (ed), Society of Economic Geologists Guidebook Series 5: 1-11.

Plumlee, G.S., D.L. Leach, A.H. Hofstra, G.P. Landis, E.L. Rowan, and J.G. Viets, 1994. Chemical reaction path modeling of ore deposition in Mississippi Valley-type deposits of the Ozark region, U. S. midcontinent. *Economic Geology* 89: 1361-1383.

Raffensperger, J.P. and G. Garven, 1995. The formation of unconformity-type uranium deposits 2. Coupled hydrochemical modeling: *American Journal of Science* 295: 639-696.

Rowan, E.L. and D.L. Leach, 1989. Constraints from fluid inclusions on sulfide precipitation mechanisms and ore fluid migration in the Viburnum Trend lead district, Missouri. *Economic Geology* 84: 1948-1965.

Wu, Y. and F. Beales, 1981. A reconnaissance study by paleomagnetic methods of the age of mineralization along the Viburnum Trend, Southeast Missouri. *Economic Geology* 76: 1879-1894.

Isotopic signature of hydrothermal sulfates from Carlin-type ore deposits

Greg B. Arehart
Department of Geological Sciences, University of Nevada-Reno, Nev., USA

ABSTRACT

Stable isotopes from sulfates of three origins in Carlin-type deposits (CTDs) from western North America fall into distinct clusters and can be utilized to discriminate between barite of Paleozoic (sedimentary exhalative) origin, barite of hydrothermal origin related to gold mineralization, and supergene alunite and jarosite of weathering origin. Although the latter are obvious from field relations, the distinction between hydrothermal and sedex barite is not always so clear. Sedex sulfate sulfur was probably reduced and mobilized during gold mineralization, equilibrated with hydrothermal fluids, and then precipitated as sulfate during the late stages of mineralization, with a distinctly different sulfur and oxygen isotope signature. Isotopic signatures of sulfates may, therefore, be utilized as a mineral exploration tool for CTDs.

1 INTRODUCTION

Sedimentary rock-hosted disseminated gold deposits, also known as Carlin-type gold deposits (CTDs), are among the most economically important types of ore deposits currently being sought and mined in the US. In excess of 100 million ounces of gold reserves have been identified in northern Nevada, the area hosting the largest proportion of these ore deposits (Teal and Jackson, 1997). Although high-grade portions of these deposits, such as at Mercur (Utah) and Getchell (Nevada), have been mined since the last century (Joralemon, 1951; Jewell and Parry, 1987), they were not recognized as a class until the discovery and development of Carlin as a bulk-tonnage mine between 1962 and 1965 (Hausen and Kerr, 1968).

Figure 1 shows the location of many of the major CTDs in western U.S.A. Although additional deposits which may be Carlin-type or similar deposits are known from locations as diverse as southern China (Ashley et al., 1991); southeast Asia (Garwin et al., 1995); and possibly Peru (Alvarez A. and Noble, 1988), the majority of deposits for which there are more than descriptive data available are located in the Great Basin of western North America.

2 DEPOSIT PARAGENESIS

Carlin-type deposits are commonly hosted in silty carbonates to calcareous siltstones and shales, although a number of other rock types may host ore locally (Arehart, 1996). Gold in Carlin-type deposits is closely associated with hydrothermally-generated arsenian pyrite, pyrite, and locally arsenopyrite (Wells and Mullens, 1973; Arehart et al., 1993a, b); lesser but significant amounts of gold occur in jasperoidal quartz and minor amounts are associated with phyllosilicates and carbonaceous matter (Hausen and Kerr, 1968; Bakken and Einaudi, 1986; Hofstra et al., 1988). Temporally late in the gold stage are arsenic sulfides (realgar and orpiment), that slightly postdate the dominant alteration minerals such as kaolinite, sericite and quartz. Penecontemporaneously or somewhat later in the paragenetic sequence are barite, stibnite, and late calcite, commonly as open fracture fillings in silicified rocks. Late oxide minerals related to weathering, including the sulfates alunite and jarosite, are common above the primary ores (Radtke, 1985; Arehart et al., 1992).

Figure 1. Location map of CTDs in western North America.

3 STABLE ISOTOPE DATA

Sulfur isotope values reported from CTDs span a wide range, from -30 to +20‰ in sulfides and 0 to >35‰ in sulfates (Figure 2). It has long been observed that chemical and optical zoning may exist in many sulfides, and it has been shown recently that many sulfides are isotopically zoned as well (Eldridge et al., 1994). In CTDs, this isotopic zoning correlates with chemical zoning of As and Au in the pyrite (Arehart et al., 1993a, b). Pyrite having the highest $\delta^{34}S$ values also contains the highest concentrations of gold. These zones also contain high concentrations of As (but not all high-As zones contain Au). Calculated and measured $\delta^{34}S$ values of gold-bearing pyrite are near +20‰, whereas all non-ore pyrite is significantly lighter (Arehart et al., 1993b, Hofstra, 1997). Extremely low sulfur isotope values are represented by the latest stage marcasite or pyrite, which have $\delta^{34}S$ values as low as -30‰ (Arehart et al., 1993b).

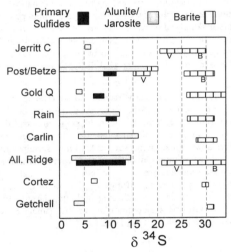

Figure 2. Sulfur isotope values of sulfides and sulfates in CTDs of the Great Basin. V=vein barite, B=sedex barite.

Sulfur and oxygen isotope values of sulfates that are spatially associated with the ore also are highly variable and fall into three distinct groups (Figures 2, 3). Some of these values are clearly related to primary sedimentary sulfate, which is abundant in the Paleozoic section in the Great Basin; the isotopic composition of these sulfates falls directly on the seawater evolution curve (Claypool et al., 1980), as would be expected. Other sulfates are clearly related to postore processes (e.g. alunite, Arehart et al., 1992) and have inherited their sulfur (and consequently their $\delta^{34}S$ values) through the process of near-quantitative oxidation of sulfides in the near-surface environment (Field, 1966; see Figure 2). When these non-hydrothermal sulfates are excluded from the data, the remaining sulfates (barite) have $\delta^{34}S$ values ranging between +15 and +25‰ and $\delta^{18}O$ values near zero (Figure 3).

Although this third group of sulfates are clearly hydrothermal, in most cases barite veins are later than the bulk of gold mineralization. Commonly, barite forms late vein-filling crystals along with quartz, stibnite, and calcite, typically growing in open fractures.

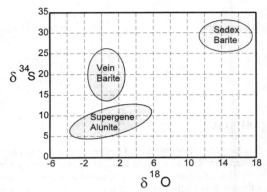

Figure 3. Sulfur and oxygen isotope values for sulfate minerals in CTDs.

4 ISOTOPIC COMPOSITION OF HYDROTHERMAL WATER

Oxygen isotope values for the hydrothermal fluid(s) must be inferred (calculated) from measurements made on ore-related alteration phases. The most common alteration phase on which oxygen measurements have been made is quartz (jasperoid). Jasperoid $\delta^{18}O$ values range widely between deposits and even within deposits (O'Neil and Bailey, 1979; Holland et al., 1988; Hofstra et al., 1988; Arehart, 1996). At Jerritt Canyon, there is a strong correlation between $\delta^{18}O$ of jasperoid and gold grade (Hofstra et al., 1988), and Holland et al. (1988) documented higher $\delta^{18}O$ values of jasperoids from ore deposits vs. non-economic prospects. Most investigators have interpreted the most positive oxygen values as being representative of the ore-bearing fluid (or at least the most exchanged fluid). Based on measured $\delta^{18}O$ values, temperatures between 200 and 250°C, and equilibrium fractionation between quartz and water, the calculated oxygen isotope composition of the fluids responsible for several CTDs ranges from -10 to +10‰.

5 ORIGIN OF HYDROTHERMAL SULFIDES AND SULFATES

There are several potential sources of sulfur for the hydrothermal fluids which created CTDs; these include sedimentary rock-hosted sulfides and/or sulfates; igneous rock-hosted sulfides (no significant sulfates are known from these rocks in Nevada), and sulfur from potential magmatic fluids. Both sedimentary and igneous rock-hosted sulfides have $\delta^{34}S$ values (ca. -5 to +9‰) that are significantly lower than the ore sulfides. No simple mechanism is available that can shift these lower values to the higher ones recorded in ore-stage sulfides. Similarly, magmatic sulfur generally has values of 0±5‰, and is not easily shifted to higher values. The most reasonable source for hydrothermal sulfur in CTDs of the Great Basin is through reduction of sulfate sulfur (Arehart et al., 1993b).

The ore-generating fluid initially had $\delta^{34}S$ values near +20‰ (note that gold deposition occurred very early in the paragenesis) but evolved through time to lower values, as recorded in latest-stage marcasite (-30‰). A most reasonable mechanism that explains this shift is Rayleigh fractionation of the initial fluid during ore depositional processes. A hypothetical model of the evolution of hydrothermal sulfur is shown in Figure 4.

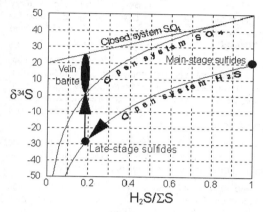

Figure 4. Schematic Rayleigh fractionation of main-stage sulfides to produce late-stage sulfides, followed by oxidation of ore fluids to produce late-stage barite.

If Rayleigh fractionation occurs because of oxidation of H_2S in either open or closed systems, resultant $\delta^{34}S$ values of sulfate in the fluid also will decrease through time. Depending upon the relative effectiveness of equilibrium vs. non-equilibrium processes, the $\delta^{34}S$ value of sulfate in solution will fall between these two extremes. Therefore, any sulfate minerals that are deposited will have $\delta^{34}S$ values between 0 and 25 ‰, if formed from the same or similar solutions which deposited latest-stage sulfide having $\delta^{34}S$ values of -30 ‰ (Figure 4).

In fact, this process is probably what has been recorded in hydrothermal barite from CTDs. Late-stage barite from the Post/Betze orebody was deposited in the same veins and breccias as light sulfides, and is depleted in ^{34}S relative to early non-vein barite (Arehart et al., 1993b). Although this barite could have formed as a result of quantitative oxidation of the original hydrothermal solutions, this alternative is considered unlikely because of the significant difference in their positions in the paragenetic sequence. In addition, the available evidence indicates that barite was not stable at this point in the evolution of the hydrothermal system. Barite deposited early in the paragenetic sequence (essentially remobilized sedex barite) has $\delta^{34}S$ values in excess of 25‰. Similar trends in $\delta^{34}S$ values of barite are observed at Alligator Ridge (Ilchik, 1990), Gold Quarry, and Rain (Howe et al., 1995).

The late hydrothermal barite also has oxygen isotopic compositions consistent with derivation from late hydrothermal fluids, which were probably closer to unexchanged meteoric water in character than were the main-stage fluids. Early gold-rich fluids were more exchanged with the wallrocks through which they traveled; there is a positive correlation between oxygen isotope values and gold grade in jasperoids (Hofstra et al., 1988; Holland et al., 1988). In contrast, water-rock ratios for the latest stage fluids were most likely very high and therefore more meteoric in character. In addition, many models for CTD ore deposition (Hofstra et al., 1988; Kuehn and Rose, 1995; Arehart, 1996) implicate fluid mixing in these deposits; such late fluid mixture(s) also are likely to have a significant non-exchanged component. Presuming that these late fluids are represented by the lower values of fluid compositions as recorded by jasperoidal quartz (-10 to -5‰), then late barite deposited at 225°C should have $\delta^{18}O$ values between -4‰ and +1‰, which is in agreement with measured values (Figure 3).

6 IMPLICATIONS FOR MINERAL EXPLORATION

The presence of two isotopically distinct types of barite could provide important clues to CTD mineralization in the Great Basin. Although in many cases there are textural differences between sedex and hydrothermal barite that allow identification, there are some samples for which the textural criteria may not apply. In these cases, isotopic measurements provide an important discriminator between the two types of barite. Because barite mineralization may extend beyond easily recognizable rock alteration in some CTDs, stable isotopes could be useful in identifying hydrothermal systems that may host gold. Whether this type of relationship holds for CTDs in other parts of the world is as yet undetermined as there are insufficient isotopic data available.

7 SUMMARY

Stable isotopic measurements on sulfates associated with CTDs yield three distinct groups of samples (sedex barite, hydrothermal barite, and supergene alunite and jarosite) that are easily discriminated on the basis of S and O isotopes. Available data are consistent with deposition of sedex barite from Paleozoic ocean waters during primary sedimentation. Reduction of this sedex barite during hydrothermal circulation yielded an ore fluid having $\delta^{34}S$ values of approximately +20‰ from which primary gold-bearing pyrite was deposited. Late sulfides and vein barite were deposited from a hydrothermal fluid that had undergone Rayleigh fractionation (sulfides) and late-stage oxidation (barite). Supergene sulfates (alunite, jarosite) formed as the result of weathering of the primary sulfide ores.

8 REFERENCES

Alvarez-A., A. and Noble, D.C., 1988, Sedimentary rock-hosted disseminated precious metal mineralization at Purisima Concepcion, Yauricocha district, central Peru: *Economic Geology* 83: 1368-1378.

Arehart, G.B., 1996, Origin and characteristics of sediment-hosted gold deposits: A review; *Ore Geology Reviews* 11: 383-403.

Arehart, G.B., Kesler, S.E., O'Neil, J.R., and Foland, K.A., 1992, Evidence for the Supergene Origin of Alunite in Sediment-Hosted Micron Gold Deposits, Nevada; *Economic Geology* 87: 263-270.

Arehart, G.B., Chryssoulis, S.L. and Kesler, S.E., 1993a, Gold and arsenic in iron sulfides from sediment-hosted disseminated gold deposits: Implication for depositional processes; *Economic Geology* 88: 171-185.

Arehart, G.B., Eldridge, C.S., Chryssoulis, S.L., Kesler, S.E. and O'Neil, J.R., 1993b, Ion microprobe determination of gold, arsenic, and sulfur isotopes in sulfides from the Post/Betze sediment-hosted disseminated gold deposit, Nevada; *Geochimica et Cosmochimica Acta* 57: 1505-1520.

Ashley, R.P., Cunningham, C.G., Bostick, N.H., Dean, W.E. and Chou, I.-M., 1991, Geology and geochemistry of three sedimentary-rock-hosted disseminated gold deposits in Guizhou Province, People's Republic of China; *Ore Geology Reviews* 6: 133-151.

Bakken, B.M., and Einaudi, M.T., 1986, Spatial and temporal relation between wall rock alteration and gold mineralization, main pit, Carlin gold mine, Nevada, U.S.A., *in* Macdonald, A.J. (ed.); *Proceedings of Gold '86, an International Symposium on the Geology of Gold:* Toronto, Konsult International, Inc., p. 388-403.

Claypool, G.E., Holser, W.T., Kaplan, I.R., Sakai, H. and Zak, I., 1980, The age curves of sulfur and oxygen isotopes in marine sulfate and their mutual interpretation; *Chemical Geology* 28: 199-260.

Eldridge, C.S., Aoki, M., Arehart, G.B., Danti, K., Hedenquist, J.W. and McKibben, M.A., 1994, Rapid redox shifts in gold-depositing hydrothermal systems: the SHRIMP sulfur isotopic evidence for boiling; *Abstracts of the Eighth International Conference on Geochronology, Cosmochronology, and Isotope Geology*, USGS Circular 1107: 90.

Field, C.W., 1966, Sulfur isotopic method for discriminating between sulfates of hypogene and supergene origin; *Economic Geology* 61: 1428-1435.

Garwin, S.L., Hendri, D. and Lauricella, P.F., 1995, The geology of the Mesel Deposit, North Sulawesi, Indonesia; *Proceedings of the PACRIM '95 Conference*: 221-226.

Hausen, D.F. and Kerr, P.F., 1968, Fine gold occurrence at Carlin, Nevada *in* Ridge, J.D. (eds) *Ore Deposits of the United States, 1933-1967, Volume 1*; New York, AIME: 908-940.

Hofstra, A.H., 1997, Isotopic composition of sulfur in Carlin-type gold deposits: Implications for genetic models; *Society of Economic Geologists Guidebook Series*, vol. 28: 119-129.

Hofstra, A.H., Northrop, H.R., Rye, R.O., Landis, G.P. and Birak, D.J., 1988, Origin of sediment-hosted disseminated gold deposits by fluid mixing - Evidence from jasperoids in the Jerritt Canyon gold district, Nevada, USA; *Proceedings of Bicentennial Gold '88*: 284-289.

Holland, P.T., Beaty, D.W. and Snow, G.G., 1988, Comparative elemental and oxygen isotope geochemistry of jasperoid in the northern Great Basin: Evidence for distinctive fluid evolution in gold-producing hydrothermal systems; *Economic Geology* 83: 1401-1423.

Howe, S.S., Theodore, T.G. and Arehart, G.B., 1995, Sulfur and oxygen isotopic composition of vein barite from the Marigold mine and surrounding area, north-central Nevada: Applications to gold exploration; *Geology and Ore Deposits of the American Cordillera, Program with Abstracts*, no page number.

Ilchik, R.P., 1990, Geology and geochemistry of the Vantage gold deposits, Alligator Ridge-Bald Mountain mining district; Nevada: *Economic Geology* 85: 50-75.

Jewell, P.W. and Parry, W.T., 1987, Geology and hydrothermal alteration of the Mercur gold deposit, Utah; *Economic Geology* 82: 1958-1966.

Joralemon, P., 1951, The occurrence of gold at the Getchell mine, Nevada; *Economic Geology* 46: 267-310.

O'Neil, J.R. and Bailey, G.B., 1979, Stable isotope investigation of gold-bearing jasperoid in the central Drum Mountains, Utah; *Economic Geology* 74: 852-859.

Radtke, A.S., 1985, Geology of the Carlin gold deposit, Nevada; US Geological Survey Professional Paper 1267, 124 pgs.

Wells, J.D., and Mullens, T.E., 1973, Gold-bearing arsenian pyrite determined by microprobe analysis, Cortez and Carlin gold mines, Nevada; *Economic Geology* 68: 187-201.

Teal, L. and Jackson, M., 1997, Geologic overview of the Carlin Trend gold deposits and descriptions of recent deep discoveries; *Society of Economic Geologists Guidebook Series*, vol. 28: 3-37.

Zircon-fluid interaction in the Bayan Obo REE-Nb-Fe ore deposit, Inner Mongolia, China

Linda S. Campbell
School of Geological Sciences, Kingston University, Kingston upon Thames, UK

ABSTRACT: The Bayan Obo REE-Nb-Fe ore deposit in China, of disputed origin, is strongly enriched in the light rare earth elements (LREE). The rare earth phases are dominated by the minerals bastnaesite, huanghoite, monazite and aeschynite. Prominant gangue phases are apatite, magnetite, fluorite, carbonates, aegirine, alkali amphiboles and baryte. Rarely seen, however, are zircon grains. Typically 200-600 microns in length, Bayan Obo zircons display a prominant sieve texture in outer zones and oscillatory growth zonation in the core regions, defined by fluctuations in Hf substitution for Zr (≤ 7 weight % HfO_2). The presence of xenotime veinlets cutting through zircon, together with the coarse, sieve texture of many grains (also containing LREE mineral inclusions), suggests reaction of zircon with the mineralizing fluids. Theoretical modelling of probable reactions indicates an acidic fluid enriched in F. Further evidence for the nature of the mineralizing fluids is provided by abundant carbonate dissolution textures with replacement by fluorite.

1. INTRODUCTION

Zircon is well known to be resistant to decomposition under a wide range of geological conditions. Although evidence for the mobility of Zr and other high field strength elements in crustal fluids is growing, (e.g. Gieré, 1990, Rubin *et al.*, 1989, Rubin *et al.*, 1993), there are virtually no reports of the aqueous *dissolution* of zircon in natural systems. However, Wayne and Sinha (1988) report on leaching of trace elements from mylonitized zircon fragments, and indicate that this was facilitated by metamictization in U and Th rich zones in the original grains. Nevertheless, it is generally thought that Zr mobility is most common in F-rich, alkalic systems (Rubin *et al.*, 1993).

Heavy rare earth elements (HREE) that substitute for Zr have extremely slow diffusion rates through zircon, even at high activation energies (Cherniak *et al.*, 1997). The present study provides evidence for the hydrothermal breakdown of zircon by F-rich mineralizing fluids, releasing HREE for the local precipitation of xenotime in an otherwise strongly LREE-enriched environment.

The origin of the Bayan Obo REE deposit is highly contentious, (Le Bas *et al.*, 1992, Wang *et al.*, 1994, Campbell and Henderson, *in press*), with arguments for an igneous *versus* sedimentary host rock (dolomite), and source of disputed mineralizing fluids, still unresolved. The present study supports a hydrothermal, F-rich fluid model for the mineralization, and adds constraints to the composition of that fluid.

2. CHARACTER AND COMPOSITION OF ZIRCON

The occurrence of zircon in the main mineralized areas at Bayan Obo is extremely rare, and only about 15 grains have been found in polished sections by the present author. Zircon from Bayan Obo has not previously been reported, and the original environment of crystallization remains uncertain. In Figure 1, a typical zircon grain displays an oscillatory-zoned euhedral core enriched in Hf, and a sieve-textured outer region containing inclusions of bastnaesite, baryte (both appearing white) and aegirine (appearing black). The two veinlets, one running through the centre of the grain, are composed of xenotime, $(Y,HREE)PO_4$.

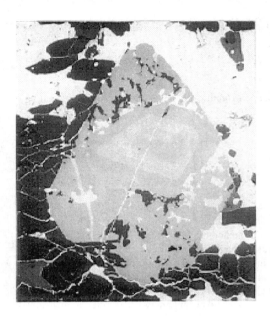

Figure 1. Backscattered electron image of zoned zircon. (Long axis, ~700 microns.)

Partial analyses by electron microprobe (Cameca SX50) are given in Table 1 for three spots.

Table 1. Partial zircon analyses by electron microprobe.

Oxide %	Centre	Inner rim	Sieve textured area
P_2O_5	0.06	0.23	0.07
SiO_2	31.92	31.94	31.49
ZrO_2	62.31	65.35	65.79
HfO_2	7.02	1.77	1.70
Y_2O_3	n.d.	0.37	0.12
FeO	n.d.	n.d.	0.87
Total	101.31	99.67	100.04

Thorium, U and the REE were occasionally detected close to their limits of detection, around 0.05-0.2 oxide weight percent. The concentrations of alpha-particle emitters is therefore probably insufficient for significant metamictization to have occurred which would have contributed to the susceptibility for dissolution, as suggested in Wayne and Sinha, (1988).

3. CARBONATE DISSOLUTION TEXTURES

Further evidence as to the nature of the mineralizing fluids at Bayan Obo can be found in the numerous examples of carbonate dissolution resulting in fluorite precipitation, according to the reaction:

(1) $CaCO_3 + 2HF = CaF_2 + CO_2 + H_2O$

Fluted calcite grain boundaries and preserved "islands" around less soluble magnetite and amphiboles provide the best textural evidence. In addition, some of the resultant CO_2 and water were trapped as 2-phase (L-V) and 3-phase (L-L-V) fluid inclusions in fluorite and apatite.

4. THEORETICAL EXAMINATION OF REACTIONS

Two possible reactions for zircon dissolution have been considered:

(2) $ZrSiO_4 + 8HF(aq) = ZrF_4 + SiF_4(g) + 4H_2O$

(3) $ZrSiO_4 + 4HF(aq) = ZrO_2 + SiF_4(g) + 2H_2O$

In reaction (3), baddeleyite (ZrO_2) is precipitated, but no evidence of this has been found in the Bayan Obo specimens, so reaction (2) is favoured. Calculation of the Gibb's free energy for these reactions indicates that zircon will dissolve readily at ambient to high temperatures. This is represented in Figure 2, where the preference for fluoride over all other halide ligands in these reactions is apparent. The calculations take no account of pressure or reaction kinetics.

5. DISCUSSION

The presence of xenotime, an HREE mineral, in such a strongly LREE-enriched geochemical environment, is indicative of zircon as the source of the HREE. Since diffusion coefficients of HREE in zircon are extremely small (Cherniak et al., 1997), the only way in which the HREE could become available for xenotime precipitation would be by breakdown of zircon. It is proposed that F-rich ore fluid-zircon reaction occurred at conditions unsuitable for the formation of baddeleyite, ZrO_2. Xenotime precipitated along minor fractures through the original zircons, and new, sieve textured zircon re-precipitated as F was removed from solution with the precipitation of LREE fluorocarbonates and fluorite.

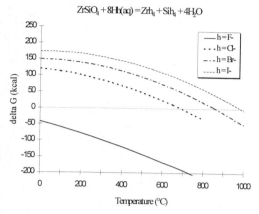

Figure 2. Gibb's free energy of the reaction of zircon with different halide acids. (Reaction (2)).

6. CONCLUSIONS

This study provides a further example of the hydrothermal mobility of Zr, and more significantly, the capability of F-enriched solutions to dissolve zircon, theoretically at ambient to high temperatures. The implications for radiometric dating of zircons are therefore highly significant. The xenotime occurrence and close association of sieve textured zircon with rare earth mineral inclusions at Bayan Obo provides unusual evidence for the nature of the mineralizing fluids.

ACKNOWLEDGEMENTS

Professor Zhang Peishan and colleagues at Academia Sinica, Beijing, are thanked for their helpful discussions and fieldwork arrangements and the co-operation of The Baotou Iron and Steel Company is gratefully acknowledged. The specimens were collected by Paul Henderson in 1990. Terry Williams and John Spratt (The Natural History Museum, London) assisted with the analysis of zircon.

REFERENCES

Campbell, L. S. and Henderson, P. 1997. Apatite paragenesis in the Bayan Obo REE-Nb-Fe ore deposit, Inner Mongolia, China. *Lithos,* (in press).

Cherniak, D. J., Hanchar, J. M. and Watson, E. B. 1997. Rare-earth diffusion in zircon. *Chem. Geol.,* 134: 289-301.

Gieré, R. 1990. Hydrothermal mobility of Ti, Zr and REE: examples from the Bergell and Adamello contact aureoles (Italy). *Terra Nova,* 2: 60-67.

Le Bas, M. J., Keller, J., Tao Kejie, Wall, F., Williams, C. T. and Zhang Peishan. 1992. Carbonatite dykes at Bayan Obo, Inner Mongolia, China. *Mineralogy and Petrology,* 46: 195-228.

Rubin, J. N., Henry, C. D. and Price, J. G. 1989. Hydrothermal zircons and zircon overgrowths, Sierra Blanca Peaks, Texas. *Amer. Mineral.,* 74: 865-869.

Rubin, J. N., Henry, C. D. and Price, J. G. 1993. The mobility of zirconium and other "immobile" elements during hydrothermal alteration. *Chem. Geol.,* 110: 29-47.

Wang, J., Tatsumoto, M., Li, X., Premo., W. R. and Chao, E. C. T. 1994. A precise ^{232}Th-^{208}Pb chronology of fine-grained monazite: Age of the Bayan Obo REE-Fe-Nb ore deposit, China. *Geochimica et Cosmochimica Acta,* 58: 3155-3169.

Wayne, D. M. and Sinha, A. K. 1988. Physical and chemical response of zircons to deformation. *Contrib. Min. Petrol.,* 98: 109-121.

Thermal and geochemical evolution of La Guitarra epithermal deposit, Temascaltepec, Mexico

Antoni Camprubí & Àngels Canals – *Departament de Cristal·lografia, Mineralogia i Dipòsits Minerals, Universitat de Barcelona, Spain*

Esteve Cardellach – *Departament de Geologia, Universitat Autònoma de Barcelona, Spain*

Zachary D. Sharp – *Institut de Minéralogie, Université de Lausanne, Switzerland*

Rosa María Prol-Ledesma – *Instituto de Geología, Universidad Nacional Autónoma de México, Mexico*

ABSTRACT: La Guitarra is a polymetallic and multi-stage, low-sulfidation epithermal vein system. The Ag-Au ore is mainly displayed in silica bands. Stage-I fluids show salinities from 5 to 12 wt. % NaCl eq.; T from 162 to 212°C; $\delta^{18}O_w$ from +1.2 to +5.2‰; X_{FeS} in sphalerite from 0 to 25%. Stage-II (Ag-bearing stage): salinity from 4 to 6 wt. % NaCl eq.; T from 161 to 210°C; $\delta^{18}O_w$ from -2.0 to +6.0‰; X_{FeS} in sphalerite from 0 to 16%. Time-space distributions of the above variables convey to magmatic and meteoric fluid recognition. Coupled with mineralogical evidences of boiling, those characteristics point to a mixing-boiling model for ore deposition.

1 INTRODUCTION

The Temascaltepec mining district is located in the centre of Mexico (Fig. 1), belonging to the Taxco-Guanajuato silver belt. It is made up by a set of epithermal-type veins apparently related to the latest hydrothermal events of the Mexican Ignimbritic Belt (MIB). In this area the MIB is partly overlain by the Trans-Mexican Volcanic Belt (TMVB).

The local geology of the area is *(1)* the Tierra Caliente metamorphic complex of the Nahuatl Terrane, pre-Albian in age; *(2)* a late Laramide granitic stock of calc-alkaline affinity (middle Eocene), which is the host rock of La Guitarra system; *(3)* Oligocene-upper Eocene rhyolites, rhyolitic tuffs and ignimbrites of the MIB; *(4)* TMVB basalts and andesites, from late Miocene to present.

The epithermal deposits of the Temascaltepec district belong to the *adularia-sericite* or *low-sulfidation* type. La Guitarra system contains polymetallic and multi-stage Ag-Au-bearing veins, striking NW-SE, with thicknesses that can reach up to 15 m, thinning with depth down to 5 m. The average grades are 6 ppm for Au and 345 ppm for Ag. Three main mineralisation stages (I, II and III) with characteristic mineralogical and textural features, were defined and described by Camprubí et al. (1997a and b). Stage II is divided into four substages (A, B, C and D). Substage IIB is the main carrier of Ag-Au ore, displayed in silica bands with hydraulic brecciation.

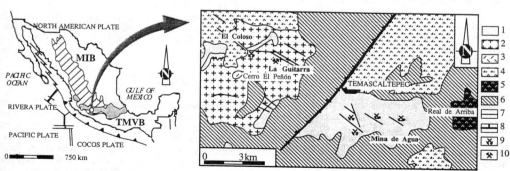

Fig. 1. Geological map of the Temascaltepec district. Key: (1) Tierra Caliente metamorphic complex (pre-Albian), (2) Late-Laramidic granites (Middle Eocene), (3) rhyolitic neck (MIB), (4) rhyolitic tuffs and ignimbrites (MIB), (5) andesites (Miocene) (6) basalts and andesites (Plio-Quaternary, TMVB), (7) epithermal veins, (8) Temascaltepec River normal fault, (9) abandoned mines, (10) active mines. MIB = Mexican Ignimbritic Belt; TMVB = Trans-Mexican Volcanic Belt.

2 ALTERATION PATTERNS

The alteration envelope around La Guitarra system and neighbouring areas is represented by sericitic, argillic and propylitic assemblages. Propylitic alteration is the most external zone and consists of quartz, chlorite, calcite, pyrite and epidote. In the upper part of the system the argillic alteration is the dominant type with kaolinite, montmorillonite and quartz as typomorphs. This alteration becomes argillic in the vicinities of the veins, also associated with narrow silicification bands. Quartz-sericite haloes are developed around the veins but the intensity of the alteration is strong only on vein contact. They consist of quartz, sericite and minor montmorillonite.

3 FLUID INCLUSIONS

Samples for fluid inclusion studies were collected from different mineralisation stages and levels of the mining works in order to determine time and space variations of fluid temperatures and composition. Sampling covers a 400 m vertical and 1200 m horizontal section of the vein. Primary fluid inclusions in quartz (I, IIA and IIB stages), sphalerite and fluorite (both from stage I) found in growth bands or clusters were studied. All inclusions are biphasic at room temperature (liquid+vapour), with vapour to liquid ratios between 0.05 and 0.15 and sizes from 5 to 90 µm. Homogenisation occurred to the liquid phase. Microthermometric and Raman analyses did not show the presence of other gases than water vapour.

Stage I: The widest range of salinities was recorded in this stage: from 5 up to 12 wt.% NaCl eq., for a Th variation of 162 to 212°C. Despite this wide range of salinities, fluid density remains constant throughout this stage, Th variations being compensated with changes in salinities. In stage I a decrease in temperature and salinity was observed from early to late bands, as well as within single zoned fluorite crystals. An upwards Th decrease of 30°C was measured on samples with an elevation difference of 100 m. NaCl is the dominant solute; however hydrohalite melting temperatures in this stage yield lower values than in other stages, corresponding to $NaCl/CaCl_2$ ratios up to 0.9. The highest content in $CaCl_2$ is found in samples with the highest salinity.

Stage IIA: A fairly constant Th was measured throughout this stage, with a variation from 178 to 188°C, except for the samples from the uppermost mining level and the NW part of the system (between 144 and 160°C). These Th ranges do not have the same correlation pattern with salinity observed in stage I. On the contrary, quite different salinities were obtained (8 to 5 wt.% NaCl eq.) for a constant Th value (179°C). On the other hand, we found salinities about 4 wt.% NaCl eq. for Th values between 144 and 188°C.

Stage IIB: An upwards Th decrease trend, as well as a time dependence, was observed. Quartz from early and latest bands collected in the upper levels show the lowest Th range (171 to 161°C). The largest Th variations were recorded in the lower part of the system (210 to 164°C). In contrast with the previous stages, salinity is nearly constant throughout this stage (4 to 6. wt.% NaCl eq.), and fluid inclusions in early bands are slightly more saline than in the latest.

4 MINERAL GEOTHERMOMETRY

The complex mineral association allowed for the use of three independent geothermometers: 1) pyrostilpnite and pyrargyrite equilibrium at 192±5°C (Keighin & Honea 1969); 2) xanthoconite-proustite equilibrium, at 192±10°C (Hall 1966); and 3) stephanite and argentite +pyrargyrite equilibrium, at 197±5°C (Keighin & Honea 1969). The first two pairs were found in stages I and III, and the whole assemblage is widespread in stage IIB. Coupling mineral and fluid inclusion geothermometry (Fig. 2) displays the waning trend of the hydrothermal system.

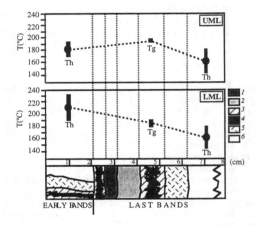

Fig. 2. Temperature variations for stage IIB in the upper mining levels (UML) and in the lower (LML), from fluid inclusions (Th) and mineral geothermometers (Tg). Key: (1) opal, (2) chalcedony, (3) microcrystalline quartz, (4) sulfide-sulfosalt mineralisation, (5) medium-grained quartz, (6) drusic amethyst-quartz.

5 OXYGEN ISOTOPE STUDY

$\delta^{18}O$ values were measured for 34 quartz samples from all stages at different elevations of the system. The $\delta^{18}O$ of water coexisting with quartz was calculated using the equation of Clayton et al. (1972), with the temperatures obtained from homogenisation of fluid inclusions and mineral geothermometry. Oxygen isotopic composition of the fluids range between -2.0 and +6.0‰, pointing respectively to the presence of meteoric and magmatic or evolved meteoric fluids.

The $\delta^{18}O$ values of stage I quartz range from +14.8 to +19.0‰; for a temperature range of 162 to 212°C, the calculated $\delta^{18}O_w$ varies between +1.2 and +5.2‰, the maximum value coinciding with the highest Th.

The $\delta^{18}O_{qz}$ values measured for stage IIA samples is between +15.6 and +19.2‰; at Th between 144 and 188°C, calculated $\delta^{18}O_w$ values range from +0.1 to +5.8‰. Stage IIB quartz has $\delta^{18}O$ values ranging from +10.6 to +18.7‰ and $\delta^{18}O_w$ between -2.0 and +6.0‰ for a Th range of 161-210°C (Fig. 3).

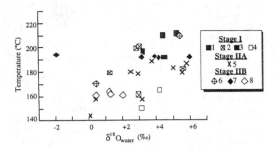

Fig. 3. Temperature-$\delta^{18}O_w$ correlation in stages I, IIA and IIB. Obtained from: (1) sphalerite; (2) fluorite; (3 and 7) mineral geothermometres; (4) quartz; (5 and 8) quartz from last bands; (6) quartz from early bands. Contemporarity is assumed for quartz and the other minerals.

In stage IIB, the lowest $\delta^{18}O_w$ are usually found in the upper mining levels, whose range (-2.0 to +6.0‰) is wider than in the lower levels (between +1.2 and +5.4‰). $\delta^{18}O_w$ is higher in early bands than in the later ones. Stage IID has the highest $\delta^{18}O_{qz}$ values (+23.5‰). Both $\delta^{18}O_{qz}$ and $\delta^{18}O_w$ ranges widen from stage to stage. Systematic isotopic analysis of silica bands in a stage IIB sample (Fig. 4) shows a compositional gap between early and latest bands, coinciding with brecciation: from +18.7 to 16.8‰ and from +17.1 to +15.5‰, respectively. Assuming the thermal evolution illustrated in Fig. 2, $\delta^{18}O_w$ shows a greater change of composition: from +6.3 to +5.2‰ in early bands, and from +3.5 to +0.4‰ in the latest.

BAND	$\delta^{18}O_{qz}$ (‰)	T(°C)	$\delta^{18}O_w$ (‰)
A	+16.8	220	+5.8
B	+16.8	210	+5.2
C	+18.7	187-197	+5.6/+6.3
D	+16.8	187-197	+3.7/+4.4
E	+17.1	180	+3.5
F	+16.3	170	+2.0
G	+15.5	160	+0.4

Fig. 4. $\delta^{18}O_{qz}$ and $\delta^{18}O_w$ in a sample of stage IIB from the upper levels. In italics, temperatures inferred from Fig. 2. Key: (1) host rock, (2) opal, (3) white chalcedony, (4) ore minerals, (5) microcrystalline quartz, (6) chalcedony and grey microcrystalline quartz, and (7) drusic quartz.

6 FLUID EVOLUTION AND PRECIPITATION MECHANISMS

The variation range of $\delta^{18}O_w$ values may be interpreted as the result of mixing of meteoric water with magmatic fluids. The positive correlation of salinity with $\delta^{18}O_w$ also supports this hypothesis. Thus, in the upper levels of the system, dominant light $\delta^{18}O_w$ values indicate a greater influence of meteoric waters. In addition, the progressive decrease in time of fluid salinity (Fig. 5) and FeS content in sphalerite (X_{FeS}), suggests an increasing influence of meteoric water. It is significant that, in stage IIB, $\delta^{18}O_w$ systematically drops from early to latest bands (Figs. 3 to 5). Otherwise, the range of $\delta^{18}O_w$ extreme values widens from the lowest part of the system to the upper, which demonstrates that the meteoric water influence is weaker in the deepest part of the system.

The high X_{FeS} in La Guitarra is common in many other low-sulfidation epithermal deposits: e.g. in Topía (Loucks et al. 1988), Fresnillo (Gemmell et al. 1988) and Creede (Hayba et al. 1985) were found up to 24% molar FeS in sphalerite. Gemmell et al. (1988) sustain that magmatic fluids, more reductive, are responsible for high X_{FeS} in sphalerites, and that its decreasing is due to dilution by meteoric water.

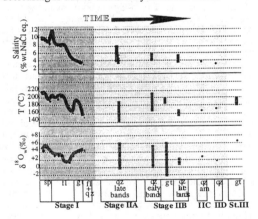

Fig. 5. Time evolution of salinity, temperature and $\delta^{18}O_w$ of mineralising fluids in La Guitarra. Sp=sphalerite, fl=fluorite, gt=mineral geothermometres, qz=quartz, am=amethyst.

Mixing of fluids with contrasting isotope signatures may be inferred for stage I where the highest $\delta^{18}O_w$, the highest Th, and the more saline fluids are found. Mixing is the main precipitation mechanism in stage I but local presence of bladed calcite coupled with silver minerals suggests that restricted episodic boiling would be also present.

During the stage IIB, the lack of significant salinity variations would indicate that it was under the domain of a single fluid. Thus the general Th decreasing (Fig. 2) would be explained by a simple conductive cooling. This effect is considered as determinant in several epithermal deposits (Matsuhisa 1986), but it is not common in active geothermal systems (Truesdell et al. 1977). In La Guitarra it is valid in its lowest part, but the rest of the system can not be explained by means of this effect. The systematic presence of mineral geothermometers in ore bands at any height, indicates temperatures above those found in early and latest bands of the upper part of the system. Moreover, the bands richer in Ag-Au assemblage display the higher $\delta^{18}O_w$. These facts point to specific hydrothermal pulses from which the ore assemblage precipitated, although it is more difficult to detect them in depth.

Boiling is manifested in stages I, IIA and IIB by the presence of bladed calcite and adularia as well as by the characteristic brecciation of stage II. However, boiling alone cannot explain the wide range of $\delta^{18}O_w$ values for stage IIB, because it does not induce significant fractionation in the reported temperature limits (Matsuhisa 1986). Hence, mixing of fluids would justify $\delta^{18}O_w$ variations in this stage.

7 CONCLUSIONS

Temperatures and fluid compositions (145-220°C and 4.0-12% wt. NaCl eq.) in La Guitarra are similar to those reported in many epithermal silver deposits of Mexico, from the Taxco-Guanajuato metallogenetic province (González-Partida 1981, Mango et al. 1991) or in the Fresnillo district (Simmons et al. 1988), where Th range between 140 and 300°C, and salinities up to 15% wt. NaCl eq.

Variations in X_{FeS} (up to 25%), Th, salinities and $\delta^{18}O_w$ (between +1.2 and +5.2‰) demonstrate that during stage I occurred a dilution of early fluids (magmatic) coupled with a temperature decreasing and an increasing variability of $\delta^{18}O_w$, due to meteoric water incursion. Ag minerals appear to be associated with an episode of boiling related to a hydrothermal pulse of magmatic influence, as they coincide with the precipitation of bladed calcite and the maximum X_{FeS} variability. During stage IIB, the relative decreasing of salinity would indicate that the stage I end member fluid of magmatic origin experienced a significative salinity change, but maintaining its isotopic signature. Though the mineralogy of stage IIA also evidences boiling, it does not contain ore mineralisation. Ore precipitation in stage IIB is due to (1) boiling as episodic separation of vapour at different heights, from specific hydrothermal pulses; and (2) mixing of fluids. The overall increasing variability of $\delta^{18}O_w$ in time suggests that the system was progresively being more influenced by meteoric water, particularly in its upper part. Meteoric influx would be favoured by hydraulic brecciation due to boiling in stage IIB.

ACKNOWLEDGEMENTS

Financial support for this study was received from EU through contract CI1*-CT94-0075. Logistic assistance during field work was provided by Luismin S.A. de C.V.

REFERENCES

Camprubí, A., Canals, À., Cardellach, E. & Prol-Ledesma, R.M. 1997a. Low-sulfidation epithermal Ag-Au deposits of La Guitarra (Temascaltepec, Mexico). *Research and explora-tion—where do they meet?* A.A. Balkema Publ. 165-168.

Camprubí, A., Canals, À., Cardellach, E. & Prol-Ledesma, R.M. 1997b. Características del sistema epitermal Ag-Au La Guitarra (Temascaltepec, estado de México). *Mem. XXII Conv. Nal. AIMMGM.* 15 p. (In Spanish)

Clayton, R.N., O'Neal, J.R. & Mayeda, T.K. 1972. Oxygen isotope exchange between quartz and water. *Jour. Geophys. Res.* 77: 3057-3067.

Gemmell, J.B., Simmons, S.F. & Zantop, H. 1988. The Santo Niño silver-lead-zinc vein, Fresnillo district, Zacatecas, Mexico: part I. Structure, vein stratigraphy, and mineralogy. *Econ. Geol.* 83 (8): 1619-1641.

González-Partida, E. 1981. *La province filonienne Au-Ag de Taxco-Guanajuato (Méxique).* PhD Dissertation, Institut National Polytechnique de Lorraine. 234 p. (In French)

Hall, H.T. 1966. *The systems Ag-As-S, Ag-Sb-S, and Ag-Bi-S: Phase relations and mineralogical significance.* PhD Disser-tation, Brown University.

Hayba, D.O., Bethke, P.M., Heald, P. & Foley, N.K. 1985. Geologic, mineralogic, and geochemical characteristics of volcanic-hosted epithermal precious-metal deposits. In: B.R. Berger & P.M. Bethke (eds.), *Geology and geochemistry of epithermal systems*, vol.2. Revs. in Econ. Geol. P. 129-167.

Keighin, C.W. & Honea, R.M. 1969. The system Ag-Sb-S from 600°C to 200°C. *Mineral. Deposita* 4: 153-171.

Loucks, R.R., Lemish, J. & Damon, P.E. 1988. Polymetallic epithermal fissure vein mineralization, Topia, Durango, Mexico: Part I. District geology, geochronology, hydrother-mal alteration, and vein mineralogy. *Econ. Geol.* 83 (8): 1499-1528.

Mango, H., Zantop, H. & Oreskes, N. 1991. A fluid inclusion and isotope study of the Rayas Ag-Au-Cu-Pb-Zn mine, Guanajuato, Mexico. *Econ. Geol.* 86: 1554-1561.

Matsuhisa, Y. 1986. Effect of mixing and boiling of fluids on isotopic compositions of quartz and calcite from epithermal deposits. *Mining Geol.* 36 (6): 487-493.

Simmons, S.F., Gemmell, J.B. & Sawkins, F.J. 1988. The Santo Niño silver-lead-zinc vein, Fresnillo District, Zacatecas, Mexico: Part II. Physical and chemical nature of ore-forming solutions. *Econ. Geol.* 83 (8): 1619-1641.

Truesdell, A.H., Nathenson, M. & Rye, R.O. 1977. The effects of subsurface boiling and dilution on the isotopic composition of Yellowstone thermal waters. *Jour. Geophys. Res.* 82: 3694-3704.

# Regional-scale fluid flow and origins of Pb-Zn-Ag mineralisation at Broken Hill, Australia: Constraints from oxygen isotope geochemistry

I. Cartwright
Department of Earth Sciences, Monash University, Clayton, Vic., Australia

ABSTRACT: Metasedimentary and metavolcanic rocks at Broken Hill, Australia, show regional-scale lowering of $\delta^{18}O$ values from 12-14‰ to 9-10‰ with values as low as 7‰ within a few hundreds of metres of the Pb-Zn-Ag orebodies. The association of low $\delta^{18}O$ values with the orebodies suggests a link with mineralisation. The $\delta^{18}O$ values are similar to those recorded in exhalative deposits where convective circulation of ocean water has occurred. However, the overall scale of resetting (tens of kilometres) is much larger than that around typical exhalative deposits and has dimensions similar to those of large hydrothermal systems. Mineralisation may have occurred at hydrothermal vents at or close to the ocean floor producing the localised low $\delta^{18}O$ values, with later hydrothermal circulation causing larger-scale resetting of oxygen isotopes. Preservation of the $\delta^{18}O$ values through regional metamorphism implies limited pervasive later fluid-rock interaction in this terrain.

1 INTRODUCTION

The Broken Hill orebody consists of Pb-Zn-Ag sulphides that are associated with quartz + gahnite + garnet layers. Although studied for over a century, there is no consensus as to the origins of the orebody, mainly due to the granulite facies metamorphism and deformation that affected the region. Mineralisation models have included: exhalative, syndiagenetic, and metamorphic. Here, new oxygen isotope data are presented that suggests that the orebody is exhalative and was formed as part of an extensive hydrothermal fluid flow system that persisted post its formation.

1.1 Local geology

The Broken Hill inlier comprises deformed and metamorphosed Proterozoic rocks of the Willyama Supergroup that are commonly interpreted as detrital sediments with minor carbonates and interlayered rhyolitic to basic volcanic rocks (eg Willis et al., 1983; Stevens et al., 1988; Page & Laing, 1992), although there are disagreements over the relative amount of volcanic material present (eg Wright et al., 1987). Most Pb-Zn mineralisation occurs within the Broken Hill Group that comprises pelites, calcareous rocks, and rhyodacitic volcanics.

The Broken Hill region underwent metamorphism at ~1.6 Ga (Page & Laing, 1992). Temperatures increased southward, with andalusite + muscovite (~500 °C), sillimanite + muscovite (580-680 °C), sillimanite + K-feldspar (680-760 °C) and two-pyroxene (760-800 °C) metamorphic zones defined (Phillips, 1980: Fig. 1). At higher grades, pelitic and quartzofeldspathic rocks underwent partial melting, with subsequent retrogression fluxed by fluids exsolved from crystallising melts. Later shear zones also caused locally-intense retrogression.

2 BROKEN HILL OREBODY

The main Pb-Zn-Ag orebody occurs within the Hores Gneiss. Smaller Pb-Zn orebodies occur within other units of the Broken Hill Group. The orebody comprises stacked lenses of galena + sphalerite with rhodonite + fluorite + quartz + garnet + calcite. The orebody is associated with quartz + gahnite + garnet rocks that have been interpreted as exhalatives (eg Stevens et al., 1988), skarns (Ehlers et al., 1996), and garnet sandstones (Wright et al., 1981).

2.1 Origins of the Broken Hill orebody

Most workers have considered that the orebody is pre-metamorphic. The association of mineralisation with probable volcanic rocks has been used to suggest a volcanic exhalative origin (eg Stanton, 1976). However, some features of the orebody (the large size, low Cu content, absence of a well-defined footwall feeder) led Phillips *et al.*, (1985) to propose a sediment-hosted exhalative origin. Wright *et al.*, (1987) proposed that the orebody was formed by brines in compacting sediments during diagenesis, while Sawkins (1984) envisaged mineralisation by basinal-scale flow of brines, similar to the model of Rye & Williams (1981) for mineralisation at HYC. In contrast to the pre-metamorphic models, Ehlers *et al.* (1996) proposed a metamorphic origin for the orebody from Pb-Pb model ages. That interpretation was questioned by (Sun *et al.*, 1996) who interpreted the same Pb-Pb data as indicating a 1600 Ma reworking of 1690 Ma ores. It is clear that there is still much uncertainty regarding the genesis of the Broken Hill orebody.

from north to south, and on an outcrop scale $\delta^{18}O$ values generally vary by only 1-2‰. The quartz + gahnite + garnet rocks that are associated with the Pb-Zn orebodies have $\delta^{18}O$ values of 9-10‰. The $\delta^{18}O$ values of pelites and volcanics in the high-grade region (7-10‰) cannot be explained by closed-system resetting, and indicate fluid flow at some time. However, in complex terrains such as Broken Hill, fluid flow may have occurred prior to, during or after regional metamorphism.

3.1 *Post-peak metamorphic fluid flow*

Many Broken Hill rocks underwent minor retrogression during cooling from the peak of regional metamorphism; however, this probably did not cause large-scale $\delta^{18}O$ resetting, because: a) there is no correlation between the degree of retrogression and $\delta^{18}O$ values; and b) minerals in unretrogressed rocks have ^{18}O fractionations close to those expected for the conditions of regional metamorphism (Fig. 2).

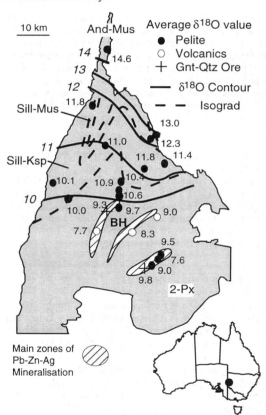

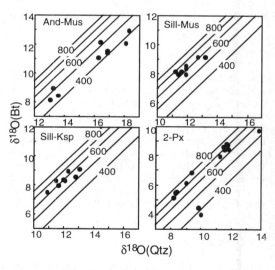

Figure 2: $\delta^{18}O$ values of biotite and quartz from pelites and metavolcanics. Lines show expected fractionations at different temperatures (°C).

Figure 1. Map of Willyama outcrop showing average $\delta^{18}O$ values and the location of the Pb-Zn orebodies.

3.2 *Regional metamorphic fluid flow*

3 OXYGEN ISOTOPE DATA

Figure 1 shows average $\delta^{18}O$ values for individual outcrops (based on 2-15 samples per outcrop). The $\delta^{18}O$ values of metasediments and volcanics decrease

The decrease in $\delta^{18}O$ values corresponds broadly to increasing metamorphic grade (Figs 1,3). Lowering of $\delta^{18}O$ values of metasediments due to up-temperature fluid flow during regional metamorphism has been documented in a few terrains (eg Cartwright *et al.*, 1995). However, such fluid flow systems probably only operate in terrains with relatively simple structures where fluids can be channelled parallel to lithological strike. Broken Hill is structurally complex and metamorphic fluids would have had to flow across lithological strike and across structures, which is unlikely.

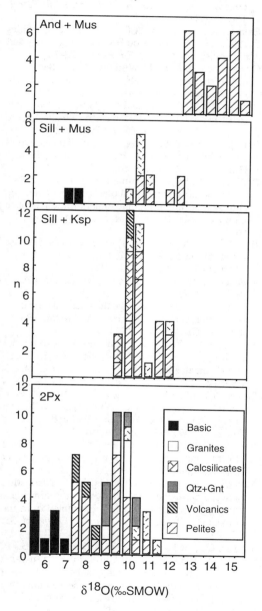

Figure 3. Variation in $\delta^{18}O$ values with metamorphic grade

3.3 *Contact metamorphic fluid flow*

Granitic rocks at Broken Hill that may be intrusive (eg the Alma Gneiss: Vernon & Williams, 1981) have similar $\delta^{18}O$ values (8-10‰) to many of adjacent metasediments and volcanics in the south of the area. However, these rocks account for <20% of the outcrop and are unlikely to have provided sufficient fluid to reset $\delta^{18}O$ values over such a wide area.

4 RELICT HYDROTHERMAL SYSTEM ?

The metasediments and metavolcanics with the lowest $\delta^{18}O$ values (7-8‰) are close to the from the Pb-Zn orebodies (Fig. 1), suggesting some link with mineralisation. Exhalative deposits, such as volcanic hosted massive sulphide (VHMS) deposits at Kuroko show a zonation in oxygen isotope values. Rocks close to the ore body have $\delta^{18}O$ values of 6-8‰ (similar to the lowest $\delta^{18}O$ values at Broken Hill), while rocks away from the orebody have higher $\delta^{18}O$ values. This isotopic zonation reflects the temperature of alteration by infiltrating ocean water around the hydrothermal vent.

4.1 *Application to Broken Hill*

Exhalative origins for Pb-Zn-Ag mineralisation at Broken Hill have previously been proposed (eg Stanton, 1976), and base metal mineralisation in such settings is common. The quartz + gahnite + garnet rocks may also have formed by hydrothermal activity as did siliceous ores at Kuroko (both sets of rocks have $\delta^{18}O$ values of ~10‰). The U-Pb model ages for the Broken Hill orebody (1675 Ma: Sun et al., 1996) may be the age of hydrothermal activity, and intrusive bodies like the Alma Gneiss, may have provided the heat source for hydrothermal activity.

Given the temperature of mineralisation of similar exhalative deposits, the hydrothermal fluids would probably be dominated by ocean water. $\delta^{34}S$ values of Pb-Zn deposits at Broken Hill are -4 to +4‰ (Gustafson & Williams, 1981), similar to other Proterozoic exhalative massive sulphides, but different to the sediment-hosted HYC and Mount Isa deposits which have a broad range of positive $\delta^{34}S$ values (Gustafson & Williams, 1981).

4.2 *Large-Scale Hydrothermal Activity*

The overall scale of alteration at Broken Hill (tens of kilometres; Fig. 1) is much larger than that in exhalative deposits. One way to reconcile this difference is shown in Fig. 4. The orebody was produced at or close to the seafloor near hydrothermal vents. Following ore formation the overlying sediments were deposited. Hydrothermal circulation then continued or resumed, forming a broad area of $\delta^{18}O$ resetting. The age difference between the Broken Hill and Sundown Groups is not large, making it plausible that these two stages are related and were driven by the same heat source.

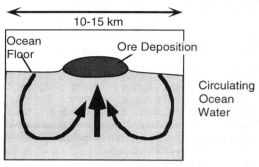

Stage 1: Formation of Exhalitive Ore Deposit

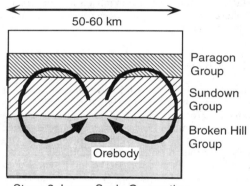

Stage 2: Large-Scale Convection

Figure 4. Conceptual model for mineralisation and later large-scale hydrothermal activity at Broken Hill

4.3 *Comparison with other Australian deposits*

Major base metal deposits formed in eastern Australia during the Proterozoic. U-Pb model ages of HYC and Mount Isa that lie some 1500 km northeast of Broken Hill are 1640 and 1653 Ma, respectively (Sun *et al.*, 1996). HYC has undergone little metamorphism and is almost certainly a sediment-hosted deposit probably formed during diagenesis in a basinal environment with fluids possibly from a nearby exhalative source (Rye & Williams, 1981).

Mineralisation at this time may have involved the formation of both exhalative (Broken Hill) and diagenetic (Mount Isa and HYC) deposits by igneous-hydrothermal processes.

REFERENCES

Cartwright, I., J.K. Vry & M. Sandiford 1995. Changes in stable isotope ratios of metapelites and marbles during regional metamorphism, Mount Lofty Ranges, South Australia: Implications for crustal scale fluid flow *Contrib. Mineral. Petrol.* 120: 292-310.

Ehlers, K., J. Foster, A.P. Nutman & D. Giles 1996. New constraints on Broken Hill geology and mineralisation. *CODES Sp. Pub.* 1: 73-76.

Gustafson, L.B. & N. Williams 1981. Sediment-hosted stratiform deposits of copper, lead, and zinc *Econ. Geol.* 75th anniversary volume: 485-627.

Page, R.W. & W.P. Laing 1992. Felsic metavolcanic rocks related to the Broken Hill Pb-Zn-Ag ore body, Australia. *Econ. Geol.* 87: 2138-2168

Phillips, G.N. 1980. Water activity changes across an amphibolite-granulite facies transition, Broken Hill. *Contrib. Mineral. Petrol.* 75: 377-386

Phillips, G.N., N.L. Archibold & V.J. Wall 1985. Metamorphosed high-Fe tholeiites: their alteration and relationship to sulphide mineralization, Broken Hill, Australia. *Trans. Geol. Soc. S. Africa* 88: 49-59.

Rye, D.M. & N. Williams 1981. Studies of the base metal sulfide deposits at McArthur River, Northern Teeritory, Australia: III. The stable isotope geochemistry of the H.Y.C., Ridge and Cooley deposits. *Econ. Geol.* 76: 1-26

Sawkins, F.J. 1984. Anorogenic felsic magmatism, rift sedimentation and giant Proterozoic Pb-Zn deposits. *Geology* 12: 451-454.

Stanton, R.L. 1976. Petrochemical studies of the ore environment at Broken Hill, New South Wales: environmental synthesis. *Trans. Inst. Min. Metal.* B85: 121-223.

Stevens, B.J.P., R.G. Barnes, E.E. Brown, W.J. Stroud & I.L. Willis 1988. The Willyama Supergroup in the Broken Hill and Euriowe Blocks, NSW. *Precamb. Res.* 40/41: 297-327.

Sun, S., G.R. Carr & R.W. Page 1996. A continued effort to improve lead-isotope model ages. *AGSO Research Newsletter* 24: 19-20.

Vernon, R.H. & P.F. Williams 1981. Distinction between intrusive and extrusive or sedimentary parentage for felsic gneisses. Examples from the Broken Hill Block, NSW. *Austr. J. Earth Sci.* 35: 379-388.

Willis, I.L., R.E. Brown, W.J. Stroud & B.J.P. Stevens 1983. The early Proterozoic Willyama Supergroup: stratigraphic subdivision and interpretation of high- to low-grade metamorphic rocks in the Broken Hill Block, New South Wales. *J. Geol. Soc. Austr.* 30: 195-224.

Wright, J.V. R.C. Haydon & G.W. McConachy 1987. Sedimentary model for the giant Broken Hill Pb-Zn deposit, Australia. *Geology* 15: 598-602.

Fluidization, metallogenic mechanism and type of the Bankuan gold deposit, China

Y.J.Chen, H.Y.Chen, H.H.Wang & X.Li
Department of Geology, Peking University, Beijing, People's Republic of China

S.X.Hu, S.G.Fu & C.Y.Jin
Department of Earth Science, Nanjing University, People's Republic of China

ABSTRACT: Auriferous quartz veins in the Bankuan gold deposit occur in the interlayer broken zone of the basal conglomerate of the Tietonggou Group and at the unconformity between the Tietonggou Group and the crystalline basement. The composition of fluid inclusions indicates that the nature and composition of ore-forming fluids changed drastically soon after they reached the Tietonggou Group from the crystalline basement, resulting in gold precipitation. Therefore the Bankuan deposit is assigned to the conglomerate or unconformity strata-bound type deposit. 137 thermometric data are concentrated in three ranges 400-340°C, 320-220°C and 200-160°C, representing a three-stage metallogenesis.

1 GEOLOGY OF THE BANKUAN GOLD DEPOSIT

The Bankuan gold deposit lies near the Bankuan village, Shanxian County, Henan Province. It has been regarded as a Rand type gold deposit by some geologists (Qin, 1989; Wei & Lv, 1996). The author's comprehensive study does not support this point of view, but suggests it should be a conglomerate or unconformity strata-bound gold deposit (Chen & Fu, 1992).

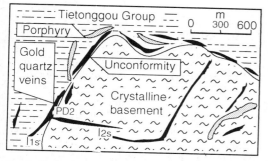

Fig.1 Geological sketch map of the Bankuan gold deposit, Henan Province, China

The Bankuan gold deposit is situated in the contact zone between the Tietonggou Group and the underlying crystalline basement in the Xiaoshan terrane. The crystalline basement, developed before 2200 Ma, was comprised of migmatites developed from greenstone, tonalite-trondhjemite-granodiorite plutons, and greenschist-facies bimodal volcanic rocks (Chen & Zhao, 1997). High angle schistosity and gneissosity were well developed in the crystalline basement. The crystalline basement is unconformably covered by the Tietonggou Group, which was slightly metamorphosed and had been accumulated during the late Paleoproterozoic (2150-1850 Ma) (Chen & Zhao, 1997). With thin interlayers of carbonaceous phyllite, the Tietonggou Group consists mainly of foliated quartzitic conglomerates and quartzose sandstones, showing the features of intramontane molasses. Interbedded sliding faults are well developed along the carbonaceous phyllite in the Tietonggou Group. Auriferous quartz veins are developed best at the unconformity between the Tietonggou Group and the crystalline basement and at the interbedded sliding faults along the carbonaceous phyllite interlayers within the lower portion of the Tietonggou Group (Fig.1).

According to the texture and structure of ores, the characteristics and associations of minerals, wall-rock alterations and other factors, the mineralization may be divided into three stages. These three stages are from early to late characterized respectively by coarse pyrite-bearing quartz veins, disseminated stockworks of poly-metallic sulfides, and thin stockworks of coarse quartz-carbonate occasionally with cubic pyrite.

2 GEOCHEMISTRY OF FLUID INCLUSIONS OF THE BANKUAN DEPOSIT

2.1 Thermometry of inclusions

Fig.2 is the histogram of 137 homogenization temperatures of fluid inclusions. Three peaks appear in the ranges of 400-340°C, 320-220°C and 200-160°C. They correspond to three mineralizing stages. It is also noticed that the early stage mineralization was overprinted by late stage mineralization, as indicated by the positive variation of temperature of inclusions from the core to the margin of a crystal; temperatures falling in different ranges; the shapes of inclusions changing from regular (elliptic, oval) to irregular; their sizes becoming from small to large; the ratios of gas/liquid changing from high to low; the distribution changing from insular to linear.

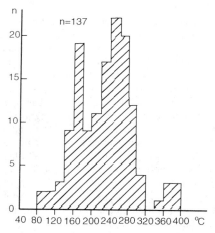

Fig. 2 Histogram of homogeneous temperatures of fluid inclusions in quartz from the Bankuan gold deposit.

2.2 Chemical composition of fluid inclusions

The sample localities are illustrated in Fig.3. As viewed from decrepitation temperatures (300-260°C), inclusion fluids from these samples represent those of the middle mineralizing stage. As shown in Table 1, samples from the crystalline basement (Nos. 6~9) are characterized by higher pH, ρ, Na^+, Ca^{2+}, Mg^{2+}, Cl^-, HCO_3^-, SO_4^{2-}, ΣM^+, ΣM^- and $\Sigma M^\pm$, but lower Eh, F/Cl, CO_2/H_2O and KN/MC than those from the Tietonggou Group (Nos. 1~5). In addition, all the above indices for sample 8810, collected from the unconformity, are just intermediate between the two groups, especially, between samples 8821 and 8823. These phenomena show that from deep to shallow the fluid changed from basic/reducing to acidic/oxidizing, and that the ore-forming materials and ore-forming fluid should be derived largely from the deep and would be partially removed and precipitated during their upward transportation. Due to significant differences in abundance of ions for samples collected from above and below the unconformity, the behavior of the ore-forming fluid should be changed suddenly at the unconformity. It is such a change that created a favorable environment for the precipitation of some ore-forming elements.

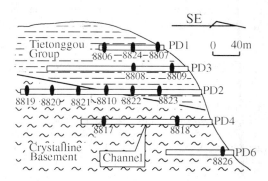

Fig.3 Cross section showing the sample localities of vein I of the Bankuan gold deposit, Henan, China

3 DISCUSSIONS

3.1 Physiochemical variation of ore-forming fluid and its metallogenic significance at unconformity

Current prospecting and mining practice has proved that the best gold mineralization is produced at the unconformity. For instance, along vein I, which penetrated both the Tietonggou Group and the crystalline basement, channel PD2 was excavated at the unconformity, providing the highest grade ores (reaches more than 300 g/t). The physiochemical variations of the fluid at the unconformity and their metallogenetic significance are described in detail as follows:

(1) At the unconformity, the drop of ions in fluid inclusions clearly indicates that a large quantity of

ions had been precipitated, with the involvement of K^+ and Na^+ in K-feldsparization, albitization, sericitization, etc. and of Fe^{2+}, Mg^{2+}, Ca^{2+}, etc. in chloritization, pyritization, carbonatization, etc.. All these alterations are typical indicators of gold mineralization. In fact, gold is a siderophile element and is usually co-precipitated with Fe^{2+}. During fluidization, gold was transported in the form of coordination complexes, with K^+ and Na^+ being the most ideal coordinating cations. When the coordinating cations reacted with the wallrock, the coordination anions would be destroyed, resulting in the removal of the coordination ligands out from the central coordination ion — Au^+ or Au^{3+}, which precipitated as Au^0 simultaneously.

(2) The drastic decrease of Cl^- and HCO_3^- appeared to be the result of the formation of sericite, chlorite, carbonates (e.g. ferric dolomite) and apatite through the reaction of the anions with the wallrocks. In the detection of fluid inclusions, HS^-, S^{2-} and other sulfur species are oxidized into SO_4^{2-}, the calculated concentration of SO_4^{2-} represents the bulk of sulfur species including SO_4^{2-}, HS^-, S^{2-}, etc.. As for reducing solution, the concentration of SO_4^{2-} generally represents that of S^{2-} and/or HS^-, and for oxidizing fluid, it represents the most part of SO_4^{2-}. A sudden decrease in measured SO_4^{2-} from reducing to oxidizing fluids is a clear indication of a decrease in HS^- and/or S^{2-}, implying the precipitation of sulfides in huge amounts. Because gold can combine with S^{2-} and Cl^- to form stable coordination anions and then be transported in the fluid, the precipitation of ligands such as S^{2-} and Cl^- must have led to the separation of gold from the coordination anions, followed by precipitation of gold from the fluid. As a result, gold mineralization would take place.

Table 1 Composition of fluid inclusions in quartz from the Bankuan gold deposit

No.	1	2	3	4	5	6	7	8	9
Sample location	8824 cover	8807 cover	8809 cover	8823 cover	8810 unconformity	8821 basement	8818 basement	8826 basement	8817 basement
K^+	3.32	4.66	4.55	3.73	3.73	5.94	5.58	3.00	4.09
Na^+	4.74	7.04	8.00	7.09	9.13	16.36	5.46	7.79	8.87
Ca^{2+}	2.53	2.04	3.29	3.36	5.30	42.54	48.52	84.32	86.44
Mg^{2+}	0.63	0.57	2.04	0.37	0.79	0.61	1.09	2.50	1.71
HCO_3^-	1.74	2.73	13.33	2.24	5.11	24.00	33.72	68.73	97.53
F^-	0.27	0.30	0.85	0.77	0.23	0.40	0.23	0.38	1.06
Cl^-	6.01	1.05	5.49	14.00	4.71	51.39	30.57	18.88	7.67
SO_4^{2-}	11.38	7.84	15.52	12.32	25.04	11.88	22.56	32.37	17.39
CO	2.96	6.53	4.79	3.28	3.30	4.68	2.51	2.38	1.95
CH_4	0.021	-	-	0.031	-	0.075	-	0.056	-
CO_2	1.17	1.20	1.30	1.25	1.04	1.04	1.29	1.06	1.21
H_2O	9.91	13.90	14.08	12.37	13.21	13.44	12.84	11.38	15.11
ΣM^+	11.22	14.31	17.88	14.55	18.95	65.45	60.65	97.61	101.11
ΣM^-	19.40	11.92	35.19	29.33	35.09	87.67	87.08	120.37	123.65
ΣM^0	14.06	21.63	20.17	16.93	17.55	19.24	16.64	14.88	18.27
$\Sigma M^\pm$	30.62	26.23	53.07	43.88	54.04	153.12	147.74	217.98	224.76
pH	4.70	4.81	5.17	4.66	4.92	6.06	7.49	7.44	7.65
Eh	176.95	170.44	149.14	179.31	163.34	96.49	11.89	14.85	2.43
ρ	24	36	31	32	58	69	63	115	54
F/Cl	0.045	0.286	0.155	0.055	0.049	0.008	0.008	0.020	0.138
CO_2/H_2O	0.12	0.086	0.092	0.10	0.079	0.077	0.10	0.093	0.080
K/Na	0.70	0.66	0.57	0.53	0.41	0.36	1.02	0.39	0.46
KN/MC	2.55	4.48	2.35	2.90	2.11	0.52	0.22	0.12	0.15
T(℃)	270	280	280	265	260	268	300	300	265

Data calculated from detections completed in Department of Earth Sciences, Nanjing University, using gas & liquid chromatography. All the analyses was taken for solutions obtained from mass explosion of quartz. 10^{-6} for ions (including cations and anions) in inclusion; 10^{-4} for gaseous components (molecules) in quartz; mV for Eh; μΩ/cm for conductivity (ρ); ΣM^+, ΣM^-, ΣM^0, and $\Sigma M^\pm$ representing respectively the sums of cations, anions, gaseous components and ions; KN/MC=(K+Na)/(Mg+Ca).

3.2 Mineralization mechanism around unconformity and carbonaceous phyllite beds

The reasons why the unconformity is a factor capable of changing the nature of the fluid are presented as follows:

(1) When the fluid ascended from the crystalline basement to the Tietonggou Group, the original water-rock balance was destroyed because the nature of the wallrocks changed. Subsequently, the behavior of the fluid could vary and some components such as Fe, Mg, Ca, etc. would react with the wallrocks (the Tietonggou Group), leading to their precipitation in the form of pyrite, ferric dolomite, etc..

(2) High angle schistosity and gneissosity of the crystalline basement make the ore-forming fluid move upward easily. At the unconformity, the ascending fluid was hampered by the gently dipping strata of the Tietonggou Group. The accumulation, deceleration and directional change of the moving fluid could lead to precipitation of the components transported in the fluid.

(3) As the unconformity is more sizable than schistosity or gneissosity, the fluid would lay its transported components down when it was moved from schistosity or gneissosity to the unconformity as if a river flew into a sea.

Changes in the nature of fluids would take a period of time (no matter how short it would be), and the movement of the fluids would possess inertia. For these reasons, favorable gold mineralization is highly expected to be produced in the carbonaceous phyllite interbeds within the Tietonggou Group. In the reducing environment along the carbonaceous beds, Au^+ and Au^{3+} could be reduced into Au^0 and precipitated from the fluid. Moreover, strong adsorption of gold on the carbonaceous matter would also promote the reduction and precipitation of gold.

3.3 Genetic type of the Bankuan gold deposit

On the basis of the above data, the Bankuan gold deposit can be easily distinguished from the Witwatersrand gold deposit. It should be assigned to the unconformity or conglomerate strata-bound type instead of the conglomerate type (or Rand type) (Bache, 1987). In the following are presented several lines of evidence for this conclusion:

(1) The ore bodies were emplaced at the unconformity between the conglomerate formation of the Tietonggou Group and its underlying crystalline rocks or at the carbonaceous phyllite interbeds within the conglomerate formation. Their strata-bound features are distinct.

(2) The gold abundance of the Tietonggou Formation was estimated up to 2.1×10^{-9}, which is the highest value in the area, while its underlying crystalline basement and overlying Xiong'er Group only contain as much as 1.0×10^{-9} and 0.6×10^{-9}, respectively(Chen & Fu, 1992).

(3) The sulfur and lead isotope data indicate that at least part of the ore-forming material was derived from the conglomerate formation of the Tietonggou Group (Chen & Fu, 1992).

(4) Drastic change in the behavior of ore-forming fluid should occur at the unconformity between the Tietonggou Formation and the crystalline basement. It is such a drastic change that led to gold deposition at the unconformity or in the carbonaceous phyllite interbeds within the Tietonggou Group. Therefore, the emplacement of the ore-bodies is entirely compatible with the emplacement conditions for strata-bound deposits as proposed by Tu (1989).

REFERENCES

Bache, J.J. 1987. *World Gold Deposits*. London: North Oxford Academic.

Chen, Y.J. & S.G.Fu 1992. *Gold Mineralization in West Henan*. Beijing: China Seismology Press. (in Chinese with English abstract).

Chen, Y.J. & Y.C.Zhao 1997. Geochemical characteristics and evolution of REE in the Early Precambrian sediments: evidence from the southern margin of the North China Craton. *Episodes*, 20:109-116.

Qin, G.Q. 1989. *Gold Deposits in Xiaoshan Terrain*. Zhengzhou: Report of Qinling-Bashan Program (in Chinese, unpublished).

Tu, G.Z. 1989. *Geochemistry of Strata-bound Deposits in China (3)*. Beijing: China Science Press . (in Chinese).

Wei, Y.F. & Y.J.Lv 1996. *Gold Deposits of China*. Beijing: China Geological Press.

Water-rich quartz and adularia veins of the Hishikari epithermal Au-Ag deposit, southern Kyushu, Japan

K. Faure & Y. Matsuhisa
Department of Geochemistry, Geological Survey of Japan, Tsukuba, Japan

H. Metsugi
Metal Mining Agency of Japan, Tokyo, Japan

C. Mizota
Department of Agriculture and Forestry, Iwate University, Morioka, Japan

ABSTRACT: Thermogravimetric, manometric, and Fourier transformation infra-red analyses indicate that quartz and adularia from epithermal veins contain orders of magnitude more water than would be expected in crystalline non-hydrous minerals. The water is contained within structural sites such that it is released only at specific thermal energies, and not merely loosely bound to mineral surfaces or trapped within fluid inclusions. Petrographic studies reveal that the veins contain minor amounts of chalcedonic quartz, and that quartz and adularia contain abundant mineral inclusions, possibly clay minerals. The chalcedonic quartz and mineral inclusions are most likely the source of the water. Results from this study show that chemical and isotopic analyses of water extracted from whole-rocks, ostensibly from fluid inclusions, should be interpreted with caution. Ubiquitous recrystallisation and replacement textures in the Hishikari veins attest to a complex history that may hamper the determination of the original composition of fluids and origin of metals.

1 INTRODUCTION

The Hishikari gold deposits are classified as a low-sulfidation epithermal deposits, based on the scheme proposed by Hedenquist (1986). Ore reserves are estimated to be 1.4 million metric tons with an average gold grade of 70 g/metric ton in the Honko ore zone, and ~ 2 million metric tons at 25 g/metric ton in the Yamada ore zone of the Hishikari deposits (Izawa et al., 1990). Veins cross-cut shales and sandstones of the Shimanto Supergroup, and the Hishikari Lower Andesites which unconformably overly the sediments (Fig. 1). Gold mineralisation in the veins occur predominantly near or below the unconformity, but can also occur above the unconformity. K-Ar ages indicate that mineralisation of the Hishikari deposit began at about 1.25 Ma, lasted ~ 0.6 m.y. and ended by 0.66 Ma (Izawa et al., 1993). Petrogenetic studies, specifically of the gold-bearing stage, have been hampered because the fluid inclusions are too small for microthermometry, the grain-size of the quartz and adularia is too small for adequate mineral separation and the concentration of suitable hydrous minerals is too low for stable isotope analysis (Matsuhisa and Aoki, 1994). In the course of trying to resolve the latter issue, it was discovered that quartz and adularia, the predominant minerals in the Hishikari veins, contained an order of magnitude more water than would be expected for non-hydrous

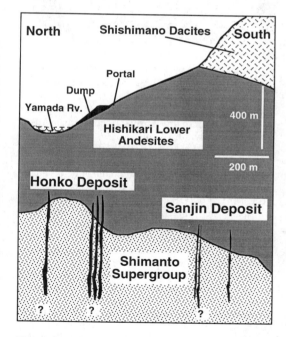

Figure 1. Schematic cross-section of the Hishikari gold deposits.

minerals. Here we report the results of our findings and tentative conclusions.

2. VEIN MORPHOLOGY AND MINERALOGY

In the simplest case, veins (termed mono-axial or simple veins, Fig. 2) consist of several paired symmetrical bands, with the oldest pair on the outside (towards wallrock) and the youngest in the center. However, most veins consist of a complex combination of mono-axial veins, termed multi-axial veins, which can be cross-cutting, divergent or parallel. Multiple thin layers within mono-axial units, in addition to several stages of parallel imbedded mono-axial units, usually make it difficult to unravel the geological relationships within a vein. For this reason, we will only discuss mono-axial vein systems in this study.

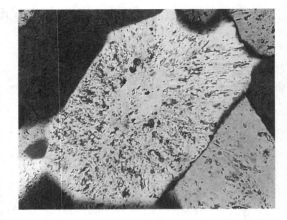

Figure 3. Photograph (plane polarised light) of mineral inclusions (dusty, elongated material) in quartz. Scale is 0.19 mm across photograph.

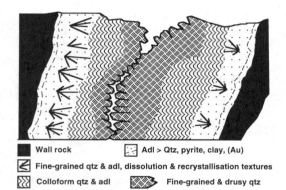

Figure 2. Sketch of a mono-axial vein (Keisen 40 m Level) from the Hishikari epithermal Au-Ag deposits. Vein width 14 cm.

Gangue minerals consist predominantly of quartz and adularia (KAlSi$_3$O$_8$), and rare calcite, truscottite [(Ca, Mn)$_2$Si$_4$O$_9$(OH)$_2$], smectite and kaolinite. The principal ore minerals are pyrite, marcasite, chalcopyrite, argentite, electrum and naumannite. The adularia:quartz ratio usually decreases from the outer portions (100-80%) of a mono-axial vein towards the center which is invariably characterized by drusy quartz (Fig. 2). A columnar adularia band (~5 mm thick), if present, occurs on the outer edges of a mono-axial vein, with the crystal c-axes perpendicular to the wall-rock contact. Adularia and quartz in the middle of mono-axial veins are usually very fine grained (1μm to ~ 200μm), typically with a colloform appearance in hand-specimen, and exhibit a range of re-crystallisation textures (Fig. 2). Thin-section studies reveal that quartz is sometimes chalcedonic, but X-ray diffraction analyses indicate that the quartz has peak positions and relative peak heights indistinguishable from pure and well crystallized quartz. Hot NaOH dissolution of quartz-rich samples show that up to about 15% of the mineral matter can be removed, which indicates that some of the material is not completely crystalline. Recrystallisation and replacement textures, and textures indicative of an amorphous precursor (Saunders, 1990) are ubiquitous in all the Hishikari veins, particularly in the colloform zone (Fig. 2),.

Quartz, more so than adularia, invariably contains numerous mineral inclusions, which range from very small (<10μm) anhedral to large (~0.5 mm) prismatic rod-shaped inclusions (Fig. 3). Positive identification of the mineral inclusions is in progress. In one vein sample, acicular truscottite is clearly replaced by quartz, adularia and calcite. The quartz and adularia that replaced the truscottite, psuedomorph the acicular texture and contain mineral inclusions that appear to be petrographically similar to the mineral inclusions that are common in the quartz and adularia. The mineral inclusions are usually aligned parallel to the acicular textures within the quartz and adularia. It appears, therefore, that the mineral inclusions may represent reaction products of the replacement (and recrystallisation) process that occurred in the Hishikari veins.

Across mono-axial veins, Au-rich portions are usually within the outer, adularia-pyrite-rich bands ("ginguro"). Vertically along the veins, the concentration of Au and Ag is uniform, but at a discrete level within the mine (bonanza zone) the concentration is particularly high. The bonanza zone which has Au and Ag concentrations at least 10 times greater than outside the bonanza zone is often, but not always, associated with the unconformable contact between the sediments of the Shimanto Supergroup and the overlying Lower Hishikari Andesites. Analysis of veins and smectite separated from veins, indicates that Au is associated, but not exclusively, with argentite, electrum, and portions of veins which have an adularia:quartz ratio greater than 0.5. In addition, one "early stage" smectite (<5 μm size fraction) which has been analyzed, contains 2000 ppm Au, whereas "late stage" smectite-kaolinite contains no Au.

3. TGA AND FTIR ANALYSES

Thermogravimetric analysis (TGA) and manometric measurements of the Hishikari quartz-adularia reveal that they contain anomalous quantities of water (Figs. 4 and 5). Hishikari quartz-adularia samples crushed to micron-size powder (hence no remaining fluid inclusions) can contain up to 1 wt.% H_2O (excluding adsorbed water), compared to 0.01 and 0.1 wt.% H_2O for granules of typical quartz or feldspar, depending on the size and number of fluid inclusions (Fig. 4). Numerous TGA data indicate five general stages of volatile release: 20-120°C (adsorbed water); 120 to 300°C; 450 to 750°C; 750 to >1200°C and, at the α/β transition, ~570°C (Fig. 4).

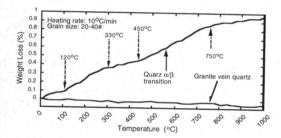

Figure 4. A typical TGA (atmospheric conditions) of Hishikari vein material. Granite quartz is also plotted for comparison. Temperatures plotted on the curve are inflection points observed from numerous TGA data of Hishikari vein samples.

Extraction of water from vein samples, which have been crushed to a fine powder, in a vacuum line supports the findings obtained from TGA (Fig. 5). Extraction of water at different temperatures, for periods much longer than the duration of volatile outgassing, highlights the point that the water is not merely adsorbed onto the mineral surfaces but that a specific thermal input is required to release the water (Fig. 5). Duplication of water yields are accurate to within one decimal place. Cryogenic purification and manometric measurements of volatiles released at

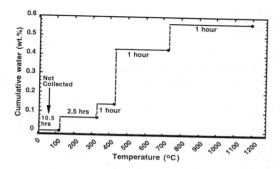

Figure 5. Extraction of water from powdered Hishikari vein sample. Extractions were done in a vacuum line and measured manometrically after extraction time and temperature ranges as indicated in the figure.

different temperatures confirms that the volatiles are predominantly H_2O and that ~35 wt.% of total H_2O is released just before or at the α/β transition, ~35 wt.% after the α/β transition up to 1200°C and ~30 wt.% between 120 and 450°C. A small proportion of hydrogen (<5 wt.% of the total water), not as H_2O, is released at temperatures greater than 600°C. Quartz-rich samples tend to have higher amounts of total water (~ 0.5 to 1 wt.%) and adularia has lesser amounts (~ 0.3 to 0.5 wt.% H_2O; Fig. 6).

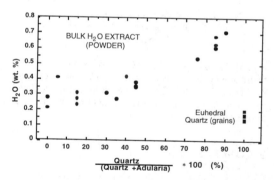

Figure 6. Bulk water extracted after 2 hours of outgassing at 120°C from powdered Hishikari vein samples. Euhedral quartz (Fig. 2) were extracted as grains.

Fourier transformation infra-red (FTIR) analysis support the presence of large amounts of water and that hydrogen is predominantly in the form of "molecular water" (O-H vibrations) rather than as "silanole group water" (Si-O-H and/or Al-O-H vibrations). The ratio of "molecular" to "silanole group water" range from 70 to 100%, with an average of 92%.

4. D/H RATIO OF WATER EXTRACTED FROM QUARTZ AND ADULARIA

D/H ratios and yields of the H_2O released at different temperatures vary substantially. Water extracted from quartz and adularia powders from ambient to 120°C increases in D/H ratio, and is considered to be adsorbed water. At temperatures between 120° (~-80‰) to 570°C (~ -100‰) δD values of water decrease, and then slightly increase again by ~ 5‰ at higher temperatures (Fig. 7). δD values of a small proportion of hydrogen (< 0.01 wt.%) not extracted as H_2O, at temperatures > 600°C, have values lower than -225‰ (possibly hydrogen from "silanole group water" or free hydrogen). Reproducible isotopic values were attained only if the samples were powders and hydrogen, not as H_2O, was excluded from the extraction.

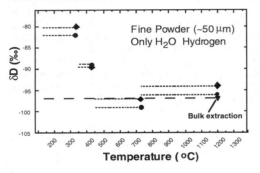

Figure 7. δD values (SMOW) of water extracted from one Hishikari vein sample. δD values of bulk extraction (120 to 1200°C) are similar to calculated weighted averages for the extractions at different temperature intervals.

5. CONCLUSIONS

Quartz and adularia in the Hishikari epithermal veins contain anomalous proportions of water. Excluding adsorbed water and water present in fluid inclusions, the vein quartz and adularia powder can contain up to 1 wt.% water. We have determined that the water extracted from the Hishikari veins is located in specific structural sites that can only be released from the minerals by the input of a particular thermal energy. We propose that the water is present predominantly in chalcedonic quartz as micro "clusters" of H_2O that was trapped during rapid growth of the starting colloidal material (hydrous) mineral. In addition, water is present in mineral inclusions that are ubiquitous in the quartz and adularia of the Hishikari veins. The mineral inclusions are most likely hydrous reaction products of replacement and recrystallisation. Although it is not possible at present to make any meaningful conclusions from the hydrogen isotope compositions of the water, it is however possible to obtain reproducible values. An important finding of this study shows that the results of isotopic or chemical composition analyses of water, ostensibly extracted from fluid inclusions in minerals, should be interpreted with caution, because the water may originate from another source. Ubiquitous recrystallisation and replacement textures suggest that the veins had a complex history, which may obstruct the determination of the original fluid(s) composition and, hence, the ultimate source of the gold.

6. REFERENCES

Hedenquist, J., 1986, Geothermal systems in Taupo Volcanic Zone: their characteristics and relation to volcanism and mineralisation: Bulletin of the Royal Society of New Zealand, v. 23, p. 134-168.

Izawa, E., et al., 1990, The Hishikari deposit; high grade epithermal veins in Quarternary volcanics of southern Kyushu, Japan: Journal of Geochemical Exploration, v. 36, p. 1-56.

Izawa, E., Kurihara, M., and Itaya, T., 1993, K-Ar ages and the initial Ar isotopic ratio of adularia-quartz veins from the Hishikari gold deposit, Japan., in Shikazano, N., Naito, K., and Izawa, E., (eds), High grade epithermal gold mineralization, the Hishikari deposit, Volume No. 14: Resource Geology Special Issue: Tokyo, Komiyama Printing Co., p. 63-69.

Matsuhisa, Y., and Aoki, M., 1994, Temperature and oxygen isotope variations during formation of the Hishikari Epithermal gold-silver veins, Southern Kyushu, Japan: Economic Geology, v. 89, p. 1608-1613.

Saunders, J.A., 1990, Colloidal transport of gold and silica in epithermal precious-metal systems: evidence from the Sleeper deposit, Nevada: Geology, v. 18, p. 757-760.

Behaviour of Re-Os, Sm-Nd, and U-Pb systematics in hydrothermal ores

R. Frei
Gruppe Isotopengeologie, Min.-Pet. Institut, Universität Bern, Switzerland (Presently: Geologisk Institut, Københavns Universitet, Denmark)

Th. F. Nägler, R. Schönberg & J. D. Kramers
Gruppe Isotopengeologie, Min.-Pet. Institut, Universität Bern, Switzerland

ABSTRACT: Isotope analyses of hydrothermal ore minerals from the Harare-Shamva greenstone belt (Zimbabwe) indicate that: The Re-Os systematics in molybdenite and scheelite (RAN deposit) remained almost closed since their deposition 2.6 Ga ago. Sm-Nd data of scheelite and powellitic scheelite (alteration phase) show a similar retentivity. In contrast, U-Pb systematics of molybdenite, wolframite and scheelite have been affected probably by Proterozoic hydrothermal activity and (sub-) recent weathering. Further, Re-Os data from paragenetically younger sulfides, especially pyrrhotite, show strong indications for (sub-) recent Re-gain (Kimberley section). Coupled with the observed Re-loss in powellitic scheelite, a redistribution of Re during oxidising conditions is indicated. Only Fe-poor arsenopyrite remains less affected. Re mobility in sulfides represents a major drawback to the application of Re-Os to crustal whole-rocks, as the budget of the latter is supposed to be dominated by these phases.

1 INTRODUCTION

The stability of the Re-Os system in sulfides and other ore minerals relative to rock-water interaction is an important point of concern, as they are assumed to dominate the Re and Os budgets of the continental crust. Yet, it is not well constrained. For example, McCandless et al. (1993) have shown that molybdenite may be susceptible to Re-loss through supergene alteration processes. Suzuki et al. (1996) suggested that the Re-Os closure temperature for vein molybdenite is roughly 500°C and that the system within this mineral is unlikely to be disturbed by later thermal events. Here we present isotope results of various base metal sulfides and scheelite from a hydrothermal system in the Harare-Shamva greenstone belt (Zimbabwe). The aim was to trace the behaviour of several isotope systems in an ore mineral paragenesis typical of a crustal hydrothermal system.

1.1 Geological setting

The RAN-Kimberley deposits comprise structurally controlled gold mineralization associated with monzonite-granodiorite plutonism in the Harare-Shamva greenstone belt, Zimbabwe. The Harare-Shamva terrain is characterised by a c. 2.7 Ga bimodal volcanic assemblage of the Bulawayan Supergroup unconformably overlain by c. 2.67 Ga clastic Shamvaian Supergroup sedimentary sequence (Jelsma et al, 1996; Vinyu, 1994). This sequence has been intruded by both syn- and post-kinematic granitoids. The Kimberley-RAN deposits are hosted by metagreywacke belonging to the Shamvaian Supergroup and by the 2649 ± 6 Ma old Bindura granodiorite (Jelsma et al., 1996), respectively. In a recent study of the Kimberley deposit, Frei & Pettke (1996) proposed a model in which a partial remobilisation of a scheelite - tourmaline bearing Archean (2.647 ±9 Ma) precursor during Proterozoic hydrothermal activity, led to concomitant gold deposition. Indications from Pb-isotopes of dravite point to a more primitive fluid source for the Archean event, most probably related to the contemporaneous emplacement of post-kinematic granitoids. The observed evolution of the fluid from one with a more primitive Pb isotope signature toward one with a more radiogenic one, along with progressive mineral crystallisation in the Proterozoic mineral succession, has been used to support a multi-stage hydrothermal event (1.95 to 2.02 Ga.). The enrichment in crustal Pb thereby is most likely due to the interaction with the older metasediments of the Shamvaian Supergroup (Frei & Pettke, 1996).

1.2 Samples

RAN: Samples are from molybdenite- and scheelite-bearing quartz veins which transect the Bindura host granodiorite. Molybdenite occurs closely associated with blue-fluorescing scheelite and often is included in the latter. Wolframite is subordinate and is spatially related to the occurrence of molybdenite. Yellow-fluorescing powellite is a patchy alteration phase of molybdenite-bearing scheelite. Dravitic tourmaline occurs as fine needles within quartz.

Kimberley: Sample KIM-1 is taken from a cataclastically deformed part of the shear zone. Two generations of arsenopyrite are present: The texturally older one occurs as clasts within a laminated gold-bearing ore paragenesis consisting of arsenopyrite, chalcopyrite and pyrrhotite. The younger sulfides were deposited as cementations around the older clastic ore. The existence of two generations of arsenopyrite may be correlated with the two different generations of tourmaline in the Kimberley reefs (2647 ± 9 Ma and 2021 ± 17 Ma, Frei & Pettke, 1996). Biotite is massively developed on country rock fragments within the cataclastically reworked shear zone and is present throughout the matrix. Sample KIM-2 is an ore sample taken within a relatively undeformed quartz vein. 'Young' arsenopyrite occurs as laminations in quartz-rich matrix together with minor scheelite and chalcopyrite. Biotite is developed along the contact zone between the vein and the host Shamvaian metagreywacke. KIM-81: Scheelite-bearing quartz vein within an undeformed section of the shear zone. Gold occurs as small blebs within quartz. Arsenopyrite is a minor phase.

2 RESULTS AND DISCUSSION

2.1 U-Pb dating

U-Pb ages of molybdenite and wolframite from the RAN mine, both occurring as inclusions in blue fluorescing scheelite are all significantly in excess of the maximal possible age of 2646 Ma (emplacement of the host granodiorite). $^{207}Pb/^{206}Pb$ ages range between 2.30 - 2.57 Ga and cannot be correlated with the regional geological history. The data indicate a deficit of c. 60% of U for molybdenite and c. 80% U for wolframite. The $^{207}Pb/^{206}Pb$ ages preclude continuous U diffusion from these ore minerals.

Thus we interpret the data to be affected by the well dated early Proterozoic thermal overprint of the Kimberley - RAN area (Frei & Pettke, 1996).

2.2 Pb stepwise leaching (PbSL)

Bulk MoS_2 yielded radiogenic Pb isotopic compositions with $^{206}Pb/^{204}Pb$ of ~68 (RAN-59a) and 171 (RAN-61). A number of leaching experiments aimed at generating Pb isochrons were performed on MoS_2 and powellite separates from the RAN mine. The technique followed Frei & Kamber (1995). Leachates and the residue from sample RAN-61 define a hyperbolic array in the $^{206}Pb/^{204}Pb$ vs. $^{207}Pb/^{204}Pb$ Pb diagram. A disturbed leaching pattern is also exhibited by data from powellite from sample RAN-59a. It is characterised by decreasingly radiogenic Pb from step to step and by an approach of the final two steps towards the 2.65 Ga reference isochron.

We explain these data as open system behaviour of the U-Pb system, in line with the results from U-Pb bulk data of the MoS_2 sample. Further, the residue, assumed to be the most pristine portion of the molybdenite, plots on the 2.65 Ga isochron defined by dravitic tourmaline in veins from the Kimberley mine (Frei & Pettke, 1996). PbSL data of powellite imply that the alteration of originally Archean molybdenite-bearing scheelite most probably took place during the Proterozoic overprint.

2.3 Sm-Nd

The four samples from RAN mine closely scatter around a 2.65 Ga isochron. Positive initial ε_{Nd} (0.8-2.2) are similar to published initial values (0.5-2.8) from the Bindura monzonite (Jelsma et al., 1996), thus they are suggestive (like the Pb isotope data) of a close genetic relationship of the Archean W-Mo mineralisation with the post-tectonic granitoids in this area. The ε_{Nd} value of sample KIM-81 (-1.4) from Kimberley at 2.6 Ga is within the range of the Shamvaian host greywackes at this age (-3.1 to +1.1, Jelsma et al., 1996). At 2.0 Ga the respective value would be unrealistically high, (+12.9).

Thus we conclude that the Sm-Nd systematics of scheelite from RAN and Kimberley mines was only marginally affected by the Proterozoic overprint and concomitant "powellitisation" of pre-existing molybdenite-bearing scheelite.

2.4 Re-Os

The two duplicate analyses of molybdenite from the RAN mine show reproducible Re concentrations around 2.83 ppm, whereas the Os contents in sample RAN-61b (79 ppb) are lower than in sample RAN-61a (86 ppb). Re-Os ages range between 2590 - 2604 Ma. The different Re-Os isotopic signatures of molybdenite aliquots can be explained by variable contamination by minute relicts of scheelite which escaped hand picking. All molybdenite data, including the two data points from scheelite of the same sample, define an isochron (Fig. 1) with an age of 2591 ± 65 Ma (MSWD = 0.14, $^{187}Os/^{188}Os$ initial

~2). The age is in line with the cross-cutting relationship between the RAN-veins and the host granodiorite (2649 ± 6 Ma, Jelsma et al., 1996).

These findings substantiates the conclusions from the other isotope systems, that the W-Mo ore mineral paragenesis has formed during an Archean mineralising event. Results also suggest that the Re-Os isotopic system in molybdenite and scheelite was resistant in these phases towards a later alteration. This is in vast contrast to the U-Pb system in the same sample. The isochron initial supports a crustal origin for Os in the Mo-W ore minerals from the RAN deposit.

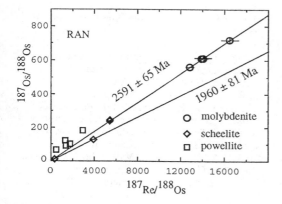

Figure 1. Re-Os isochron diagram from RAN-mine samples. Molybdenite and scheelite from sample RAN-61 define an isochron of 2591 ± 65 Ma with an initial Os isotopic value of +2. Molybdenite-bearing scheelite and scheelite from RAN-60 define a tie line indicating an age of 1.96 Ga, compatible with the age of the Proterozoic overprint. Powellite data lie consistently to left of the Archean isochron, indicating preferential loss of Re during a later stage. Re and Os were analysed following Nägler & Frei (1997) and Frei et al. (sub).

Powellite from the RAN mine is characterised by high Re concentrations and we suggest that it was formed at the expense of molybdenite-bearing scheelite during an alteration event. Data points of powellite (RAN-59a,b) lie consistently to the left of the 2586 Ma molybdenite-scheelite reference isochron (Fig. 1).

This, leading to much too old and geologically meaningless Re-Os ages is interpreted as the result of a major disturbance of the Re-Os system in this mineral, very similar to that shown for the U-Pb system. The disturbance most probably reflects Re-loss.

Scheelite and molybdenite-bearing scheelite from sample RAN-60 do not plot on the RAN-61 scheelite-molybdenite reference isochron either but rather define an own trend line indicating an age of 1960 ± 81 Ma (Fig. 1,2). The initial $^{187}Os/^{188}Os$-intercept is around +2 for this paragenesis, very similar to that defined by the RAN-61 scheelite-molybdenite reference line.

The following interpretations are possible: (i) If scheelite from sample RAN-60 pertained to an initial Archean paragenesis (as indicated by sample RAN-61), then the RAN-60 data would imply open system behaviour with respect to Os and the age defined by the tie line would not have geological relevance. (ii) Molybdenite-bearing scheelite from sample RAN-60 was formed or fully recrystallised during a mineralisation event in the Proterozoic. The 1.96 Ga age defined by the tie line should then be interpreted as a growth age of scheelite. We prefer (ii) as the age defined by the tie line corresponds exactly to the Proterozoic mineralisation age for base metal sulfides in cataclastically reactivated shear zones in the Kimberley mine (Frei & Pettke, 1996). The coincidence in the initial Os isotopic value of both lines indicates a common source of the mineralising fluids. The summary of Re-Os isotope data from the RAN veins support a partial resetting of an Archean Mo-W precursor mineralisation. Re-Os ages of molybdenite and scheelite are still broadly indicative of the Archean mineralisation event following the emplacement of the Bindura granodiorite, and a later Proterozoic one at around 2 Ga.

Re-Os data of sulfides and scheelite from Kimberley are plotted in Fig. 2. The Os isotopic composition of duplicates and triplicates of arsenopyrite, chalcopyrite, pyrrhotite and mixed sulfides (pyrrhotite, arsenopyrite, chalcopyrite) from a cataclastically deformed shear zone (KIM-1), and of scheelite, arsenopyrite and chalcopyrite from an undeformed portion (KIM-2) within a shear zone show $^{187}Os/^{188}Os$ values from +2.05 - +4.74. These values are in the range of those typical of average crustal rocks. Pyrrhotite within the cataclastically reactivated shear zone is characterised by the lowest $^{187}Os/^{188}Os$ (+2.1 to +2.4). Except for pyrrhotite and the mixed sulfide composites, the paragenetically young sulfides from both samples define a quasi linear trend which is slightly off the 1.96 Ga isochron defined by molybdenite-bearing scheelite from the RAN mine. Duplicate analyses of pyrrhotite and mixed sulfides, together with a scheelite sample (KIM-81) from a scheelite-rich quartz vein belonging to the undeformed young ore mineral paragenesis, branch off the trend line towards high Re/Os values. Older clastic arsenopyrite (KIM-1c) is clearly distinguishable from paragenetically younger arsenopyrite from the same sample (Fig. 2), by its lower Re/Os ratio. The reproducible data points from the older generation lie on the 2.6 Ga molybdenite-scheelite isochron from RAN mine (Fig. 2) which therefore supports the presence of an initial Archean mineralisation in the Kimberley shear zones also.

Neither of the two investigated young ore mineral parageneses (KIM-1, 2) have yielded geologically meaningful age constraints but rather indicate a disturbance with respect to Re (data points shifted to the right of the 1.96 reference isochron, Fig.2). This effect is more pronouncedly observed for Fe-rich ore minerals (e.g., pyrrhotite) than for Fe-poor phases. Furthermore, PbSL data of pyrrhotite, together with bulk arsenopyrite and quartz Pb isotope data (Frei & Pettke, 1996) from the same sample (KIM-2), define an isochron of 1936 ± 26 Ma, which has been interpreted as a Proterozoic mineralisation age.

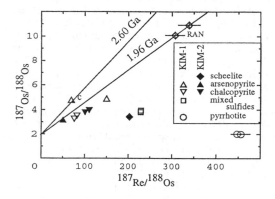

Figure 2. Re-Os isochron diagram from the Kimberley mine samples. Two isochrons from the RAN mine are plotted for comparison. Clastic arsenopyrite (c), is a texturally older generation of arsenopyrite (see text).

The apparent Re-gain of pyrrhotite in particular from the Proterozoic ore mineral paragenesis must have taken place at some time very much later than the deposition of these minerals because the Os isotopic signature of pyrrhotite is still very close to the assumed initial value of +2. As we observe an apparent Re-loss in powellitic scheelite from RAN mine (Fig.1) it is reasonable to assume that Re was redistributed during a (sub-)recent period (weathering environment), in which an alteration fluid percolated through the RAN and Kimberley shear system. If a similar effect is assumed for the U-Pb system in molybdenite, wolframite, scheelite and powellite from RAN mine, then it has to be postulated that only U was noticeably mobile during weathering, whereas Pb was more pronouncedly affected earlier during the Proterozoic overprint. Re-Os isotope data suggest that pyrrhotite was capable of preferentially taking up excess quantities of Re from the alteration fluids.

3 ACKNOWLEDGEMENTS

The authors wish to thank Mr. King, manager of the mines, and his staff, as well as Th. Pettke for collaboration. SNF grant No. 20-40442.94 to JDK, is greatly acknowledged.

4 REFERENCES

Frei, R. & Kamber, B.S. 1995. Single mineral Pb-Pb dating. *Earth Planet. Sci. Lett.* 129: 261-268

Frei, R., & Pettke, T. 1996. Mono-sample Pb-Pb dating of pyrrhotite and tourmaline: Proterozoic vs. Archean intracratonic gold mineralization in Zimbabwe. *Geology*, 24: p: 823-826.

Frei, R., Nägler, Th.F., Schönberg, R. & Kramers, J.D. submitted. Re-Os, Sm-Nd, U-Pb and stepwise Pb leaching isotope systematics in vein-type gold mineralization, genetic tracing and age constraints of crustal hydrothermal acitivity.

Jelsma, H.A., Vinyu, M.L., Valbracht, P.J., Davis, G.R. 1996. Constraints on Archaean crustal evolution of the Zimbabwe craton: a U-Pb zircon, Sm-Nd and Pb-Pb whole rock isotope study. *Contrib. Mineral .Petrol.*, 124: 55-70.

McCandless, T.E., Ruiz, J., & Campbell, A.R. 1993. Rhenium behavior in molybdenite in hypogene and near-surface environments: Implications for Re-Os geochronometry. *Geochim. Cosmochim. Acta*, 57: 889-905.

Nägler, Th.F., & Frei, R. 1997. 'Plug' in' Os distillation. *Schweiz. Min. Petrogr. Mitt*, 77: 123-127.

Suzuki, K., Shimizu, H., & Masuda, A. 1996. Re-Os dating of molybdenites from ore deposits in Japan: Implication for the closure temperature of the Re-Os system for molybdenite and the cooling history of molybdenum ore deposits. *Geochim. Cosmochim. Acta*, 60: 3151-3159.

Vinyu, M.L., Frei, R., & Jelsma, H.L. 1996. Timing between granitoid emplacement and associated gold mieralization; examples from the ca. 2.7 Ga Harare-Shamva greenstone belt, northern Zimbabwe. *Can. J. Earth Sci.* 33: 981-992.

Chemistry of hydrothermal zircon: Investigating timing and nature of water-rock interaction

P.W.O. Hoskin
Research School of Earth Sciences, Australian National University, Canberra, A.C.T., Australia

P.D. Kinny
School of Applied Geology, Curtin University of Technology, Perth, W.A., Australia

D. Wyborn
Department of Geology, Australian National University, Canberra, A.C.T., Australia

ABSTRACT: Hydrothermal and igneous zircon from altered granitoid rocks from the Lachlan Fold Belt, Australia, provide U-Pb isotope data that accurately dates the time of water-rock interaction. Trace element data acquired from the same zircons indirectly provides evidence that the metasomatising aqueous-fluid was enriched in Zr, REE, Th, U, Nb, Ta, Hf, Mo, W and Au. This, and other literature examples demonstrate the usefulness of hydrothermal zircon as a tool for investigating the timing and nature of water-rock interaction as far back as the Archaean.

1 INTRODUCTION

The interaction of aqueous fluids with rocks is an important process in the Earth's crust involving the redistribution of elements and the promotion of breakdown or stabilisation of some minerals. The timing of this fluid-rock interaction is often difficult to determine.

Important constraints on the timing of mineralisation and the refining of proposed models for mineral deposition can be made if the fluid-rock interaction can be directly and reliably dated. Recently, Yeats et al. (1996) proposed a two-stage model for volcanic-hosted VMS and lode gold mineralisation at Mount Gibson in the Yilgarn Craton, WA, based on U-Pb ages from hydrothermal zircon.

In such cases where zircon crystallisation is a product of water-related metasomatism, a direct, accurate, and precise age of the water-rock interaction can be determined. The occurrence of hydrothermal zircon is now being noted more frequently (e.g. Coleman and Erd, 1961; Claoué-Long et al., 1990; Kerrich and King, 1993; Rubin et al., 1993), as is pervasive Zr-metasomatism by aqueous-fluids (e.g. Drummond et al., 1986; Aja et al., 1996; Yeats et al., 1996).

Indirect evidence for the chemical nature of the fluid may be gained by analysis of the fluid-formed zircon. This investigation presents precise U-Pb age determinations and multi-element analyses for zircon formed during water-rock interaction in shallow-level crustal rocks.

2 GEOLOGICAL AND SAMPLE SETTING

The Boggy Plain zoned pluton is a mid-Palaeozoic concentrically zoned, high level, I-type pluton (Wyborn, 1983). The pluton is located in the south-eastern part of the Lachlan Fold Belt, eastern Australia, and forms part of the extensive Boggy Plain Supersuite (BPS) which is recognised as having potential economic significance for Cu, Au, W and Bi (Wyborn et al., 1987).

The Lachlan Fold Belt (LFB) is an elongated belt along the eastern continental margin of Australia, covering an area of ~300,000 km^2, consisting mainly of Ordovician marine sediments intruded by granitoids, and with minor associated volcanics. The LFB has been extensively studied, with particular focus on the origin and nature of the granitoid rocks that comprise a little over 20% of the belt (by area). Although the Palaeozoic tectonic setting of the LFB and the nature of the dominant processes causing compositional variation within most granite suites remains contentious (e.g. Chappell, 1996; 1997; Keay et al., 1997), the genesis of the BPS is comparatively well understood.

Compositional variation in the BPS is ascribed to fractional-crystallisation of originally crystal-poor magmas (Wyborn et al., 1987), rather than fractionation by restite separation of crystal-laden magmas as proposed for other granitoid suites of the LFB (Chappell et al., 1987).

The Boggy Plain zoned pluton is concentrically zoned from marginal diorites, through granodiorite, to a central aplite, and covers a range in SiO_2 from 50–75 wt%.

High temperature (>500 °C) water-rock interaction within the pluton has produced low-grade disseminated W mineralisation in the central aplite and locally within the adjacent adamellite. Although the water-rock interaction is not macroscopically evident, the alteration or replacement of magmatic minerals (apatite, titanite, magnetite) and the deposition of hydrothermal minerals (scheelite, ilmenite, rutile, zircon, yttrobetafite) has occurred pervasively throughout the whole-rock. Metasomatic enrichment is evident in whole-rock analyses of the aplite, where W, Mo, and the REE in particular, are elevated above concentration levels calculated for fractional-crystallisation alone (Wyborn, 1983).

3 METHODS

Zircon from two whole rock samples from the Boggy Plain zoned pluton (BP11, inner adamellite; BP42, central aplite) previously noted to contain texturally and chemically unusual zircon—believed to be of hydrothermal origin (Wyborn, 1983)—were separated by heavy-liquid techniques, mounted in epoxy-resin and sectioned. The zircons were examined by transmitted and reflected light microscopy, and were imaged with a SEM in back-scattered electron and cathodoluminescence modes to identify internal structural and chemical features of individual crystals.

The examinations, above, reveal that many crystals are composites, showing at least two generations of crystal growth. The first generation is characterised by fine oscillatory zonation and is clear in transmitted light. Most zircons from BP11 (inner adamellite) are only of this generation, which is typical of zircon that has crystallised from a silicate melt. A second generation, that generally mantles the first (but not always), reveals no internal structure. This zircon type is usually 'spongy' and brown in transmitted light. Spongy textures have been reported for hydrothermal zircon by other workers (e.g. Wayne and Shina, 1992).

U-Pb isotope analyses on these samples requires an *in situ* technique due to the composite nature of the crystals. *In situ* SIMS analyses by SHRIMP I at the ANU was performed using standard operating procedures.

Minor (Y, Hf) and trace element analyses were performed by Laser Ablation ICPMS (LA-ICPMS) methods using the methodology and instrumental techniques outlined in Hoskin (in press). SIMS determination of Hf was used as an internal standard for LA-ICPMS data reduction; external standardisation was performed against the NIST glass 610. Trace element data was acquired from the identical U-Pb isotope analytical sites.

4 RESULTS

Hydrothermal zircon U-Pb isotope analyses yield concordant apparent ages that overlap the field of concordant data from magmatic zircons from the

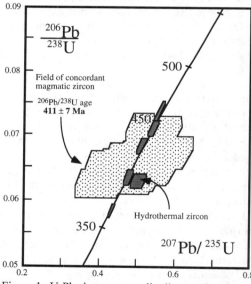

Figure 1. U-Pb zircon concordia diagram for zircons from the Boggy Plain zoned pluton.

whole Boggy Plain zoned pluton (Kinny, unpublished data). Problems with the calibration of Pb/U isotope ratios, caused by the sometimes extremely high U concentrations in the hydrothermal zircon, produce some scattering of apparent ages.

Table 1. Zircon chemistry (ppm) for BP11 (inner adamellite) and BP42 (aplite), Boggy Plain zoned pluton

Elements	P	Sc	Ti	V	Mn	Y	Nb	Mo	Ce	Gd	Yb	Hf (wt%)	Ta	W	Au	Th	U
Igneous																	
BP11 4.1	622	152	33	2	6	1781	4		52.4	31	559	3.1	2.0			290	423
BP11 5.1	892	251	75	2	4	2799	4		54.4	60	846	3.1	2.2			355	470
BP42 1.1	756	187	<73	<6	<10	1669	2		35	35	490	2.8	1.3			164	219
Hydrothermal																	
BP11.1.1	1709	244	69	7	14	2540	5		63.2	55	754	3.2	2.0			290	410
BP11.2.1	3454	467	137	32	<16	6285	25		975	386	1555	3.1	4.4			933	1929
BP11.3.1	3482	141	<56	<5	68	1677	4		172	36	524	3.1	1.7			319	444
BP11.7.1	1676	196	<95	<8	<16	1927	4		62.4	43	574	3.1	2.1			260	360
BP11.A.1	5247	224	213	9	70	2977	9		240	100	805	2.6	2.3			443	864
BP11.B.1	5162	688	<981	<86	237	4843	<48		490	211	1023	3.1	<8			361	681
BP11.C.1	697	217				2285	19		94.3	69	664	3.6	2.7			321	549
BP42.2.2	3415	702	<130	<16	<31	9171	23	<8	123	137	2912	4.0	17	1.3	<2	4384	10125
BP42.3.2	2221	604	115	9	29	8473	62	5	245	183	2480	4.2	19	25	4.9	4758	11411
BP42.4.1	1442	495	205	<24	<41	6818	60	<2	154	151	2063	3.9	2.3	14	<3	1102	1393
BP42.5.1	771	283	77	4	<7	4396	11	5	90	115	1143	3.4	1.9	4.0	1.5	443	539
BP42.6.1	3059	164	<134	<18	<32	16424	191	<8	181	268	4807	4.9	46	7.1	1.9	6054	12999

All data collected by LA-ICPMS analysis except for internal standard element (Hf) by SIMS. A space denotes 'not analysed for'.

Clearly, however, the age of formation of the hydrothermal zircon cannot be resolved from the age of igneous zircon crystallisation at 411 ±7 Ma.

Minor and trace element concentrations for igneous and hydrothermal zircon are given in Table 1. Only three REEs are listed: Ce, Gd and Yb. The chondrite normalised REE pattens (for BP42, central aplite) are presented in Figure 2.

Igneous zircon from both the adamellite and the aplite have strongly HREE-enriched patterns with a prominent Ce anomaly (Ce/Ce* = 32–77). Similarly, hydrothermal zircon is also HREE-enriched, but the overall shape of the pattern is different due to an enrichment in the LREE by about two orders of magnitude compared to the igneous zircon. The Ce anomaly is not as prominent (Ce/Ce* = 0.06–4.2), although the Eu anomaly is larger. Hydrothermal zircon is also more enriched in the HREE (Figure 2) and in other incompatible elements with respect to the igneous zircon from the same sample (Figure 3) including Nb, Ta, Th and U. Gold, Mo and W were detected in some analyses, but these elements are near or at the limit of detection.

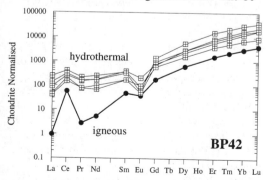

Figure 2. Chondrite normalised REE patterns for hydrothermal and average igneous zircon (BP42).

DISCUSSION

Occurrences of 'spongy' zircon commonly mantling magmatic zircon, and its association with incipient W mineralisation within the Boggy Plain zoned pluton is strong evidence for crystallisation during hydrothermal activity. Despite some scattering of U-Pb age data, the timing of water-rock interaction is unresolved from the time of igneous crystallisation at 411 ±7 Ma. This suggests that the origin of the mineralising fluid may be related to the final stages of magmatic crystallisation. In contrast, hydrothermal zircon from the Yilgarn Craton, WA, records Au mineralisation at 2627 ±13 Ma, 300 Ma after formation of the host rock, from a fluid apparently related to a later magmatic event (Yeats et al., 1996).

The enrichment of the hydrothermal zircon with respect to igneous zircon from the same sample in the REE (including Y, Sc), Nb, Ta, Hf, Th and U is

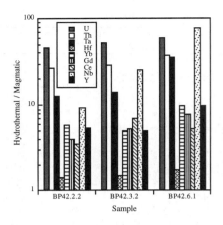

Figure 3. Some hydrothermal zircon trace elements normalised to igneous zircon concentrations from the same sample.

probably indirect evidence that the aqueous-fluid was likewise enriched in these elements. The presence of Mo, W and Au in some analyses must reflect relatively high concentrations of these elements in the fluid. These elements are highly incompatible in the zircon structure (having too large ionic radii for the $^{IV}Si^{4+}$ site) and are probably present in sub-microscopic fluid inclusions.

REFERENCES

Aja, S.U., Wood, S.A. & Williams-Jones, A.E. 1996. The aqueous geochemistry of Zr and the solubility of some Zr-bearing minerals. *Appl. Geochem.*, 10: 603–620.

Chappell, B.W. 1996. Magma mixing and production of compositional variation within granite suites: Evidence from the granites of southeastern Australia. *J. Pet.*, 37: 449–470.

Chappell, B.W. 1997. Compositional variation within granite suites of the Lachlan Fold Belt: its causes and implications for the physical state of magma. *Trans. Roy. Soc. Edin. Earth. Sci.*, 88: 159–170.

Chappell, B.W., White, A.J.R. & Wyborn, D. 1987. The importance of residual source material (restite) in granite petrogenesis. *J. Pet.*, 28: 1111–1138.

Claoué-Long, J.C., King, R.W. & Kerrich, R. 1990. Archaean hydrothermal zircon in the Abitibi greenstone belt: constraints on the timing of gold mineralisation. *Earth Planet. Sci. Let.*, 98: 109–128.

Coleman, R.G. & Erd, R.C. 1961. Hydrozircon from the Wind River Formation, Wyoming. *USGS Sur. Res.*, 1961: C297–300.

Drummond, M.S., Ragland, P.C. & Wesolowski, D. 1986. An example of trondhjemite genesis by means of alkali metasomatism: Rockford Granite, Alabama Appalachians. *Contrib. Min. Pet.*, 93: 98–113.

Hoskin, P.W.O. Minor and trace element analysis of natural $ZrSiO_4$ (zircon) by SIMS and laser ablation ICPMS: A consideration of two broadly competitive techniques. *J. Trace Microprobe Tech.*, in press.

Keay, S., Collins, W.J. & McCulloch, M.T. 1997. A three-component Sr-Nd isotopic mixing model for granitoid genesis, Lachlan fold belt, eastern Australia. *Geology*, 25: 307–310.

Kerrich, R. & King, R. 1993. Hydrothermal zircon and baddeleyite in Val-d'Or Archean mesothermal gold deposits: characteristics, compositions, and fluid-inclusion properties, with implications for timing of primary gold mineralisation. *Can. J. Earth Sci.*, 30: 2334–2351.

Rubin, J.N., Henry, C.D. & Price, J.G. 1993. The mobility of zirconium and other "immobile" elements during hydrothermal alteration. *Chem. Geol.*, 110: 29–47.

Wayne, D.M. & Sinha, A.K. 1992. Stability of zircon U-Pb systematics in a greenschist-grade mylonite: An example from the Rockfish Valley fault zone, central Virginia, USA. *J. Geol.*, 100: 593–603.

Wyborn, D. 1983. *Fractionation processes in the Boggy Plain zoned pluton.* Unpub. Ph.D thesis, Aust. Nat. Uni., 301p.

Wyborn, D., Turner, B.S. & Chappell, B.W. 1987. The Boggy Plain Supersuite: A distinctive belt of I-type igneous rocks of potential economic significance in the Lachlan Fold Belt. *Aust. J. Earth Sci.*, 34: 21–43.

Yeats, C.J., McNaughton, N.J. & Groves, D.I. 1996. SHRIMP U-Pb geochronological constraints on Archean Volcanic-hosted massive sulfide and lode gold mineralisation at Mount Gibson, Yilgarn Craton, Western Australia. *Econ. Geol.*, 91: 1354–1371.

Fluid migration-reaction model of Zijinshan epithermal deposit as traced by variation of oxygen isotope compositions of altered wall rocks

Renmin Hua & Jinhua Hu
State Key Laboratory on Research of Mineral Deposits, Nanjing University, People's Republic of China

ABSTRACT: Based on the variation of oxygen isotope compositions of altered wall rocks from different sites and depths in Zijinshan epithermal Au-Cu deposit, together with data from oxygen isotope composition of ore solution and fluid inclusion studies, a four-stage model of fluid migration-reaction for this deposit is established. An inversion calculation is conducted and is in good accordance with the measured values.

1 INTRODUCTION

Convection or convective circulation of fluid has long been considered as the major ore forming mechanism of a hydrothermal ore deposit (Barnes, 1979). Studies on modern geothermal systems and related epithermal deposits provide strong evidence that meteoric water penetrated and circulated through fractures and rock pores and was heated by high geothermal gradient and/or magmatic heat source, and evolved into an ore-forming solution(Spycher, 1989). In the deeper portion of such geothermal system, however, a magmatic water dominated fluid system might exist, which probably resulted in porphyry copper mineralisation (White et al, 1990). In the pathway of convection, fluid-rock interaction, including oxygen isotope exchange, took place. Therefore, the variation of oxygen isotope compositions of altered wall rocks can be used as a tracer of fluid convection (Taylor, 1974; Hua et al, 1995).

2 GEOLOGY OF ZIJINSHAN DEPOSIT

Zijinshan epithermal Au-Cu deposit of Fujian province is located at the inland zone of coast area in southeast China and tectonically at the transitional zone of Wuyi uplift and Southwest Fujian depression. The northwest trending fault system is the major structural control of the area, through which the Cretaceous subvolcanic dacite porphyry intruded into Jurassic granitic rocks. Crypto-explosive breccia as well as hydrothermal breccia are well developed at the top and peripheral contact of dacite porphyry, to form a northwestward zone of breccia veins, which is spatially closely related to gold mineralisation. The granitic wall rocks were widely altered by sericitization, dickitization, alunitization and silicification. The shallow portion of the deposit is centered by strong silicification alteration and gold ore veins, whereas porphyry copper mineralisation developed in the deep portion.

Zijinshan deposit is considered as the first example of acid-sulfate type epithermal deposit in mainland China (Zhang et al, 1991). Because of its special geological background, unique metallogenic feature, and great economic potential, it attracts not only geological researchers but also many mining companies. Studies on geology, geochemistry and ore genesis have been carried out since the late 1980s (Zhang et al, 1992; Wang et al, 1992; Shi et al, 1993; Zhang et al, 1996), which also include simple discussion of source and characteristics of ore solution. It is generally accepted that the source of ore solution is meteoric water.

Based on these previous studies, the present paper made a further study on the convection (migration)-reaction mechanism of ore solution, which is traced by variation of oxygen isotope compositions of altered rocks.

3 DETERMINATION OF OXYGEN ISOTOPE COMPOSITIONS

37 rock samples from different sites and depths of Zijinshan deposit were selected for determination of oxygen isotope compositions. The analyses were conducted on a MAT 252 gas-ratio mass spectrometer at the State Key Laboratory on Research of Mineral Deposits, Nanjing University.

The rock samples are divided into two groups: the first group of 21 samples are collected from outcrops and shallow cores of drill holes at scattered localities to show the horizontal variation of the oxygen isotope compositions; another group consists of 16 samples of cores of different depths from two drill holes. The results of analyses of the two groups are listed in Tables 1 and 2, respectively.

Table 1. Oxygen isotope compositions of near-surface rocks in Zijinshan deposit

No	sample location	alteration	$\delta^{18}O$(‰)
Z01	line4, tunnel2	qtz+alu	11.85
Z03	line4, tunnel2	s.sil	10.75
Z51	ZK301, 70m	s.sil	11.32
Z52	ZK307, 98m	s.sil	12.68
Z53	ZK001,126m	s.sil	12.06
Z54	ZK006,100m	qtz+alu	12.34
Z55	ZK010,120m	qtz+alu	9.31
Z56	ZK709, 108m	s.sil	12.33
Z70	ZK804,100m	qtz+dic	8.13
Z71	ZK801, 0m	s.sil	10.63
Z72	ZK411, 107m	qtz+alu	9.54
Z75	ZK1906, 100m	qtz+alu	11.12
Z76	ZK1905, 5m	qtz+alu	9.95
Z77	ZK1904,100m	qtz+alu+dic	9.96
Z78	ZK2702, 40m	qtz+alu	6.76
Z79	ZK1201, 101m	s.sil	10.81
Z80	ZK1202, 34m	qtz+alu	7.44
Z81	ZK1105, 20m	qtz+alu	10.36
Z82	ZK1109, 100m	qtz+alu	9.55
Z83	ZK1118, 20m	qtz+alu+dic	9.85
Z87	ZK1207, 22m	qtz+dic	8.69

abbrev: qtz=quartz, alu=alunite, ser=sericite, dic=dickite, s.sil= strong silicification.

Table 2. Oxygen isotope compositions of samples from two drill holes in Zijinshan deposit

No	sample location	alteration	$\delta^{18}O$(‰)
Z57	ZK312, 0m	qtz+alu	9.24
Z59	ZK312, 140m	qtz+alu	11.06
Z61	ZK312, 298m	qtz+alu	9.83
Z63	ZK312, 460m	qtz+alu	9.81
Z65	ZK312, 620m	qtz+alu	10.58
Z67	ZK312, 790m	qtz+alu	10.00
Z69	ZK312, 950m	qtz+alu+dic	10.68
Z88	ZK1117, 0m	s.sil	10.58
Z90	ZK1117, 150m	s.sil	12.94
Z92	ZK1117, 303m	qtz+alu	10.62
Z94	ZK1117, 470m	qtz+alu	9.86
Z96	ZK1117, 640m	qtz+ser	9.75
Z98	ZK1117, 800m	qtz+ser	10.32
Z100	ZK1117, 970m	qtz+ser	9.09
Z102	ZK1117, 1138m	qtz+ser	9.18
Z104	ZK1117, 1268m	qtz+ser	9.90

Besides the above rock samples, four quartz samples were applied for fluid inclusion study on a THMSE 600 heating/cooling stage, as well as the analyses of hydrogen and oxygen isotope compositions of fluid inclusion waters. Table 3 shows the analytical data for these 4 quartz samples (Z11, Z19, Z34 and Z36). Combining with data from other authors (Zhang et al, 1992; Wang et al, 1992), the values of $\delta^{18}O$ for ore solution varies from -1.63‰ to +6.90‰, which can be generally attributed as the results of isotope exchange by interaction between meteoric water ($\delta^{18}O$ =-9.0‰) and granitic wall rocks ($\delta^{18}O$=11.5‰).

Table 3. Hydrogen and oxygen isotope compositions (‰) of ore solution of Zijinshan deposit

No	mineral	$\delta^{18}O_{qua}$	$\delta^{18}O_{H2O}$	δD_{H2O}	T(°C)
Z11	qtz	12.85	6.90	-50.2	350
Z19	qtz	10.16	4.95	-51.2	378
Z34	qtz	9.69	-1.63	-48.9	218
Z36	qtz	10.93	2.20	-66.3	270
Zj385	qtz	10.8	1.16		250
Zj53	qtz			-76	
Zj64	qtz			-60	
Zj296	qtz			-64	
D2-2	qtz	13.99	5.31	-60.9	256
D2-3	qtz	13.02	1.47	-60.0	202
Z-218	qtz	13.94	1.57	-66.1	190

4 VARIATION OF OXYGEN ISOTOPE COMPOSITIONS OF ALTERED ROCKS

It is distinct from Table 1 that rocks with higher $\delta^{18}O$ values are spatially in accordance with strong silicification zone where gold ore bodies are concentrated, and hence this high $\delta^{18}O$ center is in correspondence with gold mineralisation center.

The vertical variation of $\delta^{18}O$ values (Table 2) is more irregular as a whole. Roughly there is a trend of upward increasing of $\delta^{18}O$ values. Higher $\delta^{18}O$ values are generally in shallow portions of the drill holes. The highest values appear at a depth of about 150 m below surface, which are about 2‰ higher than those at surface. It is noticeable that gold ores also occur in this shallow sector (mainly 80 to 200 m below surface) with highest $\delta^{18}O$ values.

The above-mentioned variation of $\delta^{18}O$ values of altered rocks provided following information on the migration-reaction process of fluid.

1. Most samples outside the strong silicification zone (also gold mineralising center) and in the deeper portions have $\delta^{18}O$ values lower than the initial value of granite(11.5 ‰), revealing that they had deposited in equilibrium with meteoric water of low $\delta^{18}O$ value.

2. The higher $\delta^{18}O$ values in strong silicification zone also resulted from water-rock exchange of oxygen isotope. Differing from the former, however, this reaction-exchange took place in discharge stage

of the fluid convection, and the fluid itself was already the product of previous water-rock interaction in recharge stage.

3. The distinct decreasing of $\delta^{18}O$ values at surface relative to those at about 150 m depth can be explained by later isotope exchange between surface meteoric water and altered rocks above underground water table.

Combining $\delta^{18}O$ variation with the geological, geochemical and structural features of the deposit, the present study proposed a metallogenic model in regard to fluid convection for Zijinshan Au-Cu deposit, which mainly includes following four stages. In the first stage, hydrothermal activity related to dacite porphyry resulted in extensive alteration (quartz-sericitization). Porphyry copper mineralisation occurred in deeper part on the southeast of Zijinshan deposit. The second stage started when downgoing meteoric water penetrated along fractures and rock pores in peripheral regions. Water-rock interaction took place (dickitization), through which the fluid could extract gold and copper, and evolved into an ore-bearing solution. A reservoir of such solution might be exist in certain deep aquifer. The reactivation of fault system caused fluid upflowing from reservoir which initiated the third stage of discharge. Different from the second stage, the fluid migrated mainly in a restricted area where fractures and breccia most developed. Water-rock interaction took place together with temperature and pressure decreasing, resulting in copper and gold ore deposition. The last stage was the post-mineralisation interaction between fresh meteoric water and mineralised rocks, causing reduction of $\delta^{18}O$ values of rocks and leaching of gold in surface area, associated with secondary enrichment of gold in a depth from 80 to 200m.

5 INVERSION CALCULATION OF OXYGEN ISOTOPE VARIATION BY WATER-ROCK INTERACTION

The water-rock exchange of oxygen isotope is based on following equation (Taylor, 1974):

$$W\delta^{18}O^i_w + R\delta^{18}O^i_r = W\delta^{18}O^f_w + R\delta^{18}O^f_r \quad (1)$$

and then:

$$W/R = (\delta^{18}O^f_r - \delta^{18}O^i_r)/[\delta^{18}O^i_w - (\delta^{18}O^f_r - \Delta)] \quad (2)$$

In this paper, W/R ratio is shown in weight unit instead of atomic unit. For granitic wall rocks of Zijinshan area, the following equation is deduced:

$$W/R = 0.5(\delta^{18}O^f_r - \delta^{18}O^i_r)/[\delta^{18}O^i_w - (\delta^{18}O^f_r - \Delta)] \quad (3)$$

An inversion calculation is to evaluate $\delta^{18}O^f_r$ value by known W/R ratio, $\delta^{18}O^i_w$, $\delta^{18}O^i_r$ and Δ, that is:

$$\delta^{18}O^f_r = [W/R (\delta^{18}O^i_w + \Delta) + 0.5 \delta^{18}O^i_r]/(0.5 + W/R) \quad (4)$$

The calculation also follows the four stages as described in above text.

In the first stage, $\delta^{18}O^i_w$ = 10.0 ‰ (magmatic water), $\delta^{18}O^i_r$ for granitic wall rock is 11.5‰; W/R ratio is supposed as 0.1 and temperature 350°C. The calculated $\delta^{18}O^f_r$ value is 11.81‰ by using equation 4, slightly higher than the initial value.

The second stage is recharge and downgoing of meteoric water ($\delta^{18}O^i_w$ = -9.0 ‰) reacting with wall rocks ($\delta^{18}O^i_r$ = 11.81 ‰). Temperature increased from about 100°C in surface to higher than 300°C at depths. W/R ratio was generally higher in surface area and lowering downward. Three sections are divided for simple calculation of water-rock interaction in this stage, as shown in Table 4.

Table 4. Variation of $\delta^{18}O$ values (‰) by reaction between downgoing meteoric water and wall rocks

T (°C)	Δ	W/R	$\delta^{18}O^i_w$	$\delta^{18}O^i_r$	$\delta^{18}O^f_w$	$\delta^{18}O^f_r$
100	15.73	1	-9.0	11.81	-7.31	8.42
200	8.45	0.5	-7.31	11.81	-1.98	6.48
300	4.63	0.1	-1.98	11.81	5.65	10.28

It is obvious that the calculated final $\delta^{18}O$ value of water is quite different from its initial value. In temperatures of 200 °C to 300 °C, the calculated $\delta^{18}O^f_w$ is in the range of -1.98‰ to 5.65‰, very similar to measured $\delta^{18}O_{H2O}$ values in Table 3 (-1.63 ‰ to 6.90 ‰).

Fluid-rock interaction of the third stage was under decreasing temperatures from 350 °C to 200 °C, and higher W/R ratio due to the high porosity of pathway. Table 5 shows the result of $\delta^{18}O^f_r$ calculation, which is generally in accordance with measured $\delta^{18}O$ values of samples from two drill holes (Table 2).

Table 5. Variation of $\delta^{18}O$ values (‰) by reaction between upgoing ore solution and wall rocks

T (°C)	Δ	W/R	$\delta^{18}O^i_w$	$\delta^{18}O^i_r$	$\delta^{18}O^f_w$	$\delta^{18}O^f_r$
350	3.37	2	5.65	11.81	6.21	9.58
300	4.63	2	5.65	11.81	5.96	10.59
250	6.26	2	5.65	11.81	5.63	11.89
200	8.45	2	5.65	11.81	5.19	13.64

Water-rock interaction at the last stage resulted in the decreasing of $\delta^{18}O^f_r$ values in near surface section and modified the trend of upward increasing of $\delta^{18}O^f_r$ by previous fluid-rock exchange. When $\delta^{18}O^i_w$ = -9.0 ‰, T= 100 °C, W/R=1, and $\delta^{18}O^i_r$ = 13.64‰ (from Table 5), the calculated $\delta^{18}O^f_r$ is 9.03 ‰, corresponding to the average of measured values.

6 CONCLUSION

A four-staged fluid migration-reaction model for Zijinshan epithermal Au-Cu deposit is established based mainly on the variation of oxygen isotope compositions of altered wall rocks of different spatial distribution. It includes an early alteration by deep-sourced magmatic water, a major ore-forming process of meteoric water convection (recharge and discharge stages), and a late stage meteoric reaction in near surface area. An inversion calculation of simulating the conditions of these four stages is conducted, which is coincident with the proposed model.

7 ACKNOWLEDGEMENTS

This study was supported by Grant 49372108 from China National Natural Science Foundation. The authors thank geologists from Zijinshan Gold Mine and the 4th Geological Team of Fujian for their kind assistance.

8 REFERENCES

Barnes, H.L. (ed) 1979. *Geochemistry of Hydrothermal Ore Deposits,* 2nd edition. New York: Wiley-Intersciences

Hua, R., P. Wu & K. Chen 1995. Two-stage water-rock interaction indicated by variation of oxygen isotope compositions of altered rocks in the Yinshan polymetallic deposit, Jiangxi, China. In Kharaka & Chudaev (eds), *Water-Rock Interaction:* 187 - 190. Rotterdam: Balkema.

Shi, L., W. Peng & R.Huang 1993. Metallogenetic series of subvolcanic porphyry copper deposit of Zijinshan, Fujian. In *Proceedings of 5th Conference on Mineral Deposits*:366 - 369. Beijing: Geol. Pub. House.

Spycher, N.F. & M.Reed 1989. Evolution of a Broadlands-type epithermal ore fluid along alternative P-T paths. *Econ. Geol.* 84:328 - 359.

Taylor, H. P. Jr 1974. The application of oxygen and hydrogen isotope studies to problems of hydrothermal alteration and ore deposition. *Econ. Geol.* 69: 843 - 883.

Wang, W., H. Zhao & J. Chen 1992. Geology and ore genesis of Zijinshan Au-Cu deposit. In Z.Lu & K. Tao (eds), *Monograph of Volcanic Geology and Resources in Coastal Region of Southeast China*: 83 - 91. Beijing: Geol. Pub. House.

White, N.C. & J.W. Hedenquist 1990. Epithermal environmemts and styles of mineralisation. *Jour. Geochem. Explor.*36: 445 - 475.

Zhang, D., D. Li & Y. Zhao 1991. Zijinshan deposit: the first example of Quartz-alunite type epithermal Cu-Au deposit in mainland China. *Geol. Review.* 37(6): 481 - 491.

Zhang, D., D. Li & Y. Zhao 1992. *Alteration and Mineralisation zoning of Zijinshan Cu-Au Deposit.* Beijing: Geol. Pub. House.

Zhang, D., D. Li & Y. Zhao 1996. Wuzhiqilong deposit: a transformed top zone of porphyry copper mineralisation. *Mineral Deposits.* 15 (2): 109 - 122.

Mineralogical, sulfur isotope and fluid inclusion studies of gold mineralization, Bendigo, Victoria, Australia

Xia Li & Peter Jackson
School of Earth Sciences, La Trobe University, Bundoora, Vic., Australia

Paul A. Kitto
CODES, University of Tasmania, Hobart, Tas., Australia

Y. Jia
Geology Department, Changchun University of Earth Sciences, Jilin, People's Republic of China

ABSTRACT: The alteration halo in Bendigo is at least 150m wide around the core of the Nell Gwynne anticline. Gold occurrences correspond to higher concentration of CH_4 and CO_2 during the whole process of the fluid evolution. The homogenization temperatures in the mineralised veins have a range of 220-400°C, whereas those in the barren veins are 145-300°C. Fluid immiscibility, including boiling, is responsible for gold deposition in the CH_4-CO_2-H_2O-$NaCl$ system. CH_4-H_2O fluid inclusions exclusively occur in gold-rich laminated veins, spurs and breccia veins rather than in the barren veins. Salinities range from 1.5% to 13.5 wt% equivalent NaCl. The near-primordial values of the sulfur obtained from the ore veins and the highly altered zone may indicate a magmatic source or derivation from the metamorphic terrain.

1. GEOLOGICAL REVIEW

The Bendigo ore field, Australia's second largest goldfield, is located in central Victoria, 152km NNW of Melbourne. It is hosted by the lower to middle Ordovician turbidite sedimentary rocks in the Bendigo-Ballarat Zone, 5-8km west of the major Whitelaw fault and 10 km NNE of the middle Devonian Harcourt Batholith. It consists of numerous deposits distributed mainly along parallel-trending anticlinal domes separated by 100 to 400 meters. The field has produced approximately 22 million ounces of gold. 17 million ounces of this total came from reef mining, which is approximately 53% of the primary gold produced in Victoria. Mineralisation is mainly restricted to narrow, structurally controlled fault-and fold-related dilatant zones and wider, disseminated and veinlet-hosted mineralisation within a Lower Ordovician turbidite sequence.

2. MINERALOGICAL STUDY

Primary alteration haloes present in metasandstone units in the Nell Gwynne anticlinal dome, contain: hydrothermal phengite (3-10%), chlorite (3-5%), carbonates (2-15%, maximum-50%) and sulfides (1-5%). Alteration mineral haloes extends farther above the mineralised zone in anticlinal areas than above unmineralised synclinal areas. The zone of hydrothermal phengite and pyrite-chalcopyrite (minor)-sphalerite (minor) extends farthest above the mineralised saddle reef positions, more than 150m; chlorite, to 130m; arsenopyrite, to 60m; and carbonates, to 50-80 m (Figure 1).

The turbidite sequence has been metamorphosed to lower greenschist facies resulting in the formation of metamorphic sericite and very minor chlorite. As a result of increased Si^{4+} accompanied by substitution of Fe^{2+} and Mg^{2+} for Al^{3+}, detrital and metamorphic mica tend to be hydrothermal phengite. Hydrothermal chlorite is characterised by enrichment in Al^{3+} and Si^{4+}, with depletion of Fe^{2+} and Mg^{2+} compared with metamorphic chlorite. The source for Fe^{2+}, Mg^{2+}, Ca^{2+} and Al^{3+} is believed to be from both hydrothermal fluids and sediments, as is evident from the occurrence of extensive disseminated siderite, sideroplesite, ankerite, pyrite and arsenopyrite. Ca^{2+} and Fe^{2+} (Mg^{2+}) were produced by decomposition of feldspar and chlorite, respectively. Si^{4+} was probably derived from ore fluids and (or) the dissolution of pre-existing silicates. Based on the studies of bulk-rock compositional variation (Li et al., 1997) and mass-balance calculations (Gao & Kwak, 1997), it is suggested that CO_2, S, Au, As, and K were introduced during the hydrothermal alteration and mineralisation.

3. SULFUR ISOTOPES

Approximately 100 sulfur isotope values for pyrite were obtained using a laser ablation sulfur isotope microprobe in University of Tasmania. The total range of $\delta^{34}S$ values of all sulfide occurrences in the Nell Gwynne and Deborah anticlinal domes is

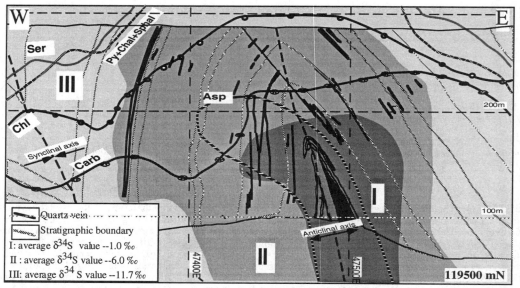

Figure 1. The distribution of hydrothermal minerals and the presentation of sulfur isotope zoning across the Nell Gwynne anticline, Bendigo, Australia.

from -23.5 to +19.3‰. Coarse, disseminated pyrite crystals hosted by sandstone or siltstone have a wide range of $\delta^{34}S$ from -7.1 to +19.3‰. The $\delta^{34}S$ values of framboid-like pyrite aggregate in black shales range from -23.5 to -2.4‰, which reflects typical biogenic origins. Pyrite crystals from laminated quartz veins and a 1 cm wide quartz veinlet have relatively narrow sulfur compositions ranging from -2.1 to +11.1‰.

The sulfur isotope compositional variations define haloes around the saddle reef position in the Nell Gwynne anticlinal dome (Figure 1). The saddle reef is surrounded by an inner zone of low $\delta^{34}S$ values (I), an intermediate zone of medium $\delta^{34}S$ values (II) and an outer background zone (III). In the core of the anticline $\delta^{34}S$ values range from -3.1 to + 5.5‰ with an average of +1.0‰, which is probably representative of main-stage sulfide mineralisation. $\delta^{34}S$ values between +3.3 ‰ and +8.0‰ with an average of + 6.0 ‰ extend from the previous area to a distance of 50 m from the central zone of the anticline (II). The sulfur isotope values represent the mixing of sulfur derived from host metasedimentary rocks and hydrothermal fluid solutions. The presence of the sulfur isotope trend confirms the ore-forming fluid to be from a deep-seated source with the ore fluids having moved upwards in conduits along the axial planes of anticlines rather than being of local origin.

There is an increasing trend of $\delta^{34}S$ values in the range of +7.5 to +18.0‰ with an average of +11.7‰ in sandstones located over 100m away from the mineralised zone (III). The heavy sulfur-rich values in the outer zone are interpreted to indicate background values of diagenetic pyrite as represented by the values found in the accompanying shales.

There is a similarity in sulfur isotope compositions of the strongest alteration zone (I) and sulfides from gold-mineralised veins. The $\delta^{34}S$ values for pyrite in laminated veins in the adjacent Deborah anticlinal dome range from -6.7 to +9.8‰, with an average value of +1.9‰ (n=27), whereas those of later spurs and breccia veins are in the narrow range of -4.7 to +2.7‰ with an average of -0.04 ‰ (n=6) and from -1.4 to +4.8‰ with an average of 3.1‰ (n=5), respectively. The $\delta^{34}S$ values of sulfides from veins and adjacent wall rocks were also reported: +1.4 - +7.5 ‰ (Gao and Kwak, 1995), +0.7 - +5.3 ‰, (Cox et al., 1995) and +1.0 - +3.0 ‰ (Stüwe et al., 1988) from the Wattle Gully mine in the BBZ. The obviously lower $\delta^{34}S$ values for pyrite from the Nell Gwynne and Deborah anticlines than those indicated above in the Wattle Gully mine, result from the abundant biogenic sulfur (in black shales) buffering.

Sphalerite (-7.4‰) and galena (-1.4‰) in sulfide-rich quartz veins yield a formation temperature of 349°C for the vein, based on the experimental data of Ohmoto and Rye (1979), and assuming isotopic

Types	I(V_{CH_4}+L_{H_2O})	II(L_{CO_2}±V_{CO_2}+L_{H_2O})	III (V_{H_2O}+L_{H_2O})			IV(V_{H_2O}+L_{H_2O}+S)
			III$_a$	III$_b$	III$_c$	
Geometry of inclusions		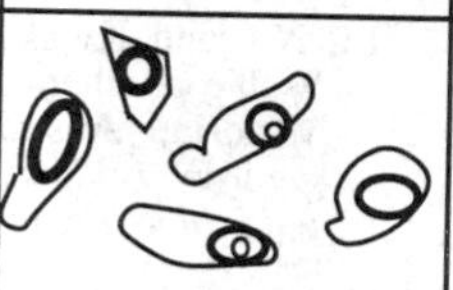		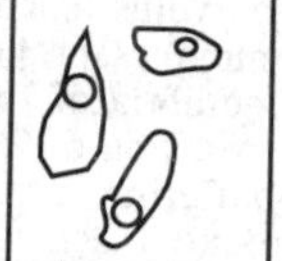	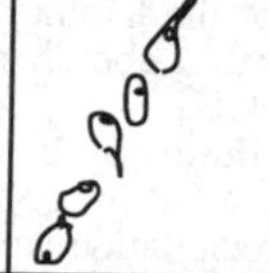	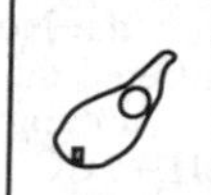
Degree of filling	0.5-0.8	0.4-0.95	0.20-0.6	0.8-0.9	0.9-0.95	0.8-0.9
Size (μm)	5-20	5-20	10-20	5-20	5-15	15
T_{mCO_2} (°C)	—	-56.6 to -60	—	—	—	—
T_{mclath} (°C)	?	4.5 to 11.9	—	—	—	—
T_{mice} (°C)	-1.5 to -8.5	—	-1.2 to -6.9	-3 to -6	-1.2 to 10.49	—
$T_{hCO_2(L)}$ (°C)	—	14.1 to 31.1	—	—	—	—
$T_{h\,total}$ (L/V °C)	317-402 (L)	220-400 (L)	220-381 (V)	205-378 (L)	145-200 (L)	—
Bulk composition	X_{CH_4}=0-6%	X_{CH_4}=0-9% X_{CO_2}=0-38%				

Figure 2. Summary of microthermometric data of fluid inclusions in Central Deborah Mine, Bendigo.

equilibrium. This temperature is consistent with fluid inclusion homogenisation temperatures (Figure 2) obtained from quartz and sphalerite in the laminated, spur and breccia veins.

4. FLUID INCLUSIONS

Three main types of fluid inclusions have been identified at room temperature in this study (Figure 2): type I, two-phase, primary V_{CH_4} + L_{H_2O} fluid inclusion; type II, two / three - phase, primary/pseudosecondary L_{CO_2} ± V_{CO_2} + L_{H_2O} fluid inclusions, and type III, V_{H_2O} + L_{H_2O} inclusions grouping into three clusters (III$_a$: primary / pseudoseconary, vapour-rich, with V/L = 70-90/10-30; III$_b$: primary / pseudosecondary, liquid-rich, with V/L = 15 - 25 / 75 - 85; III$_c$: secondary, liquid - rich, with V/L = 10 - 15 / 85 - 90).

The carbonaceous phase compositions were obtained using microthermometric data (T_{mCO_2} and T_{hCO_2}) by Chaixmeca freezing-heating stage and then confirmed by Laser Raman Spectroscopy. Type I inclusions coexist with primary, type III$_b$ inclusions (V/L=25/75) and have T_h ranging from 317-402°C, whereas type II and type IIIb inclusions with relatively lower V/L ratio occur together with T_h values of 220-400 °C. It is inferred that at least two discrete immiscibility events occurred during the formation of the veins. Boiling (H_2O liquid-vapour immiscibility) was observed in the laminated veins on the basis of type III$_a$ and type III$_b$ inclusions coexisting and homogenising at the same temperatures. Type III$_b$ inclusions in late, barren quartz veinlets have a low T_h range from 200 to 250°C. Type III$_c$ inclusions are usually irregular in shape, and infrequently show necking-down phenomena, and have the lowest T_h range of 145 - 200 °C.

Fluids responsible for gold mineralisation in Au-bearing quartz veins at Bendigo are of the CH_4- CO_2- H_2O - NaCl system, while those of barren quartz (both massive veins, and small veinlets) are characterised by CO_2 - H_2O - NaCl and H_2O - NaCl systems. The carbonaceous phase compositions of type I, II inclusions range from pure CH_4 and mixed CH_4 and CO_2, up to pure CO_2 phases. Minor hydrocarbons, such as C_2H_4 or C_2H_6 occur in the carbonaceous phase of different inclusions, while no H_2S, N_2 nor other gases were detected.

Salinities estimated from clathrate melting temperature for type II inclusions and final melting temperature of ice for type I, III, inclusions are between 1.6 and 13.5wt % equivalent NaCl. The values in the late stage barren veins are slightly higher than those in the early stage mineralised veins. Gold precipitation in quartz veins was related to CH_4 - CO_2 - H_2O - NaCl compositional inclusions having the higher temperatures (T_h= 350 - 400 °C) of the total range (220 to 400°C). The estimated homogenisation temperatures in the barren veins range from 205 to 300 °C. The bulk compositions of ore forming fluids are X_{CH_4} = 0 - 0.11, X_{CO_2} = 0 - 0.38 and X_{H_2O} = 0.6 - 0.97. Higher X_{H_2O} values and a corresponding decrease in X_{CO_2} values (X_{CH_4}: trace) exist in non-ore forming fluids.

5. DISCUSSION AND CONCLUSIONS

Graphite associated with visible gold occurring along the margins of the laminated veins (such as Kingsley reef, the North Deborah mine) is proposed to have buffered the CH_4 - CO_2 equilibria of the early stage ore-forming fluids, by the reaction of $2C + 2 H_2O = CH_4 + CO_2$. Wall-rock buffering also resulted in carbonate precipitation around black slate relicts in quartz veins by the reaction of CO_2 + (Fe^{2+}, Mg^{2+}, Ca^{2+} from wall rock) $\Leftrightarrow$ (Fe, Mg, Ca) CO_3 (siderite, sideroplesite, ankerite, calcite), which removed CO_2 from fluids and increased the amount of CH_4 present. Thus, the CH_4 is considered as a product derived from fluid-wall rock interaction. During the whole process of wall rock and fluid interaction the fluids became more reduced and the wall rock more oxidised.

The occurrence of non-aqueous CH_4 and CO_2 fluid inclusions and intense carbonate alteration indicate that low salinity, reduced, high CH_4 and CO_2 fluids accompanied Bendigo gold mineralisation. Fluid immiscibility, including boiling, of fluids represented by these inclusions having compositions in the CH_4-CO_2 - H_2O - NaCl system occurred and led to gold deposition in mesothermal gold mineralisation. The source of mineralising fluids is interpreted to be derived from dehydration and devolatilization reactions during the prograde metamorphism at depth below the Whitelaw Fault. The near-primordial values of the sulfur obtained from the ore veins and the highly altered zone (I) may indicate a magmatic source or derivation from the metamorphic terrain (such as greenstones), assuming re-equilibration.

ACKNOWLEDGEMENTS

The authors are indebted to Dr. Teunis A. P. Kwak and Dr. Dennis Arne for their helpful advices on some aspects of our manuscript.

REFERENCES

Cox, S.F., Sun, S.S., Etheridge, M.A., Wall, V.J. and Potter, T.F. 1995. Structural and geochemical controls on the development of Turbidite-hosted gold quartz vein deposits, Wattle Gully Mine, Central Victoria, Australia. *Economic Geology*. 90: 1722-1746.

Gao, Z.L. and Kwak, T.A.P., 1995. Turbidite-hosted gold deposits in the Bendigo-Ballarat and Melbourne zones, Australia. I. geology, mineralisation, stable isotopes, and implications for exploration. *International Geology Reviews*. 37: 910-944.

Gao, Z. L. and T. A. P. Kwak, 1997. The geochemistry of wall rock alteration in turbidite-hosted gold vein deposits, Central Victoria, Australia. *Journal of Geochemical Exploration. 1511 (in press)*.

Li, X., and Kwak, T.A.P. and Brown, R., 1997. Wall rock alteration in the Bendigo gold ore field, Victoria, Australia; Use in exploration. *Ore geology reviews (special issue for mesothermal gold deposits, in press)*.

Ohmoto, H. and Rye, R.O., 1979. Isotopes of sulfur and carbon. In: Barnes, H.L. (ed.), *Geochemistry of hydrothermal ore deposits*. Wiley, New York, pp. 509-567.

Stüwe, K., Keays, R.R. and Andrew, A., 1988. Wall rock alteration around gold quartz reefs at the Wattle Gully mine, Ballarat Slate Belt, central Victoria. In: Goode, A.D.T., & Bosma, L.I., (Compilers), *Bicentennial Gold '88. Geol. Soc. Aust. Abstr.* 22: 478-480.

Morphology of pyrite and marcasite at the Golden Cross mine, New Zealand

J.L. Mauk
Geology Department, The University of Auckland, New Zealand

P.W.O. Hoskin
Research School of Earth Sciences, Australian National University, Canberra, A.C.T., Australia

R.R. Seal, II
US Geological Survey, Reston, Va., USA

ABSTRACT: The Golden Cross epithermal gold deposit contains both pyrite and marcasite, that display systematic variations in their morphology, which presumably relate to temperature, water/rock ratios and FeS_2 supersaturation. Pyrite shows a sequence from cubes to octahedra to pyritohedra to dendritic growth, whereas marcasite shows a parallel sequence from tabular to bladed aggregates to euhedral miscellaneous morphologies. Samples with high gold grades contain dominantly pyritohedra, anhedral pyrite and/or euhedral miscellaneous marcasite morphologies. Low grade samples contain mostly pyrite cubes and/or tabular marcasite.

1 INTRODUCTION

Pyrite, the isometric form of FeS_2, is the most common sulfide mineral in the Earth's crust, and commonly occurs in many ore deposits. Marcasite is the orthorhombic dimorph of pyrite, and is also quite common, although not nearly to the extent of pyrite.

Pyrite can occur in a wide variety of habits. Dana (1903) listed eighty-five forms, Sunagawa (1957) reported seventy-five forms from Japan, and Goldschmidt (1920) illustrated over one hundred and eighty forms. However, most pyrite crystals occur as either cubes, octahedra or pyritohedra. The crystal habits of marcasite are less well-documented, but common forms include tabular, prismatic and cockscomb aggregate.

This study was designed to: (1) investigate the changes in the morphology of pyrite and marcasite within a single epithermal system; (2) correlate changes in pyrite morphology with corresponding changes in marcasite morphology; and (3) relate morphologic changes to physical and chemical conditions where pyrite and marcasite were deposited.

2 GEOLOGICAL SETTING

Golden Cross is a low sulfidation epithermal deposit located in the Hauraki Goldfield of the north island of New Zealand (Figure 1; Christie and Brathwaite, 1986). The deposit is hosted by Miocene to Pliocene calc-alkaline volcanic rocks that formed westward of a former west-dipping subduction zone (Skinner, 1986; Brathwaite and Christie, 1996). Regional basement consists of Upper Jurassic greywacke that crops out in the northern portion of the Coromandel Peninsula, but is progressively downfaulted to the south by a series of east-northeast-trending faults, and is not exposed in the Golden Cross region (Brathwaite and Christie, 1996).

Production at the Empire zone of the Golden Cross mine began in January 1992, from both an open pit and underground workings, and will likely terminate in 1997. Total production from the mine will exceed 500,000 ounces of Au.

Hydrothermal mineral zoning in the deposit consists of a central vein-rich core that contains quartz with lesser adularia. Chlorite and illite occur in the deeper (>200 m) portions of the deposit, whereas interstratified illite-smectite occurs in the shallower (<200 m) portions (de Ronde and Blattner, 1988; Simpson et al., 1995). Estimated temperatures of formation of the mineralisation range from 140°C in the shallow portion of the system to 225°C in the deeper portion (de Ronde and Blattner, 1988; Simpson et al., 1995).

3 METHODS

Eleven whole-rock samples were analysed from five main rock types: (1) ore veins; (2) late, barren veins; (3) breccias; (4) country rock; and (5) massive sulfide. Samples were prepared for study by hand crushing with a porcelain mortar and pestle. Samples were sieved, then separated using di-iodomethane. The heavy fraction was hand-picked to remove electrum and composite grains containing silicates, then subjected to HF acid digestion for periods of 3 hours to one day. Subsequent hand-

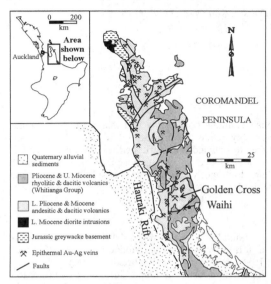

Figure 1. Generalised geologic map of the Coromandel Peninsula, showing the location of the Golden Cross mine (after Skinner, 1986).

picking produced FeS_2 separates that were greater than 95% pure. The first 200 to 300 FeS_2 grains were counted, and their morphologies recorded to provide a statistical basis for comparing changes in FeS_2 morphology among different rock types. Note that the hand crushing increased the abundance of anhedral pyrite and marcasite grains.

FeS_2 separates were analysed for Au by INAA at the SLOWPOKE Reactor Facility at the University of Toronto using methods outlined in Hancock et al. (1991). Whole rock samples were analysed by fire assay for Au at Australian Laboratory Services.

XRD analyses of FeS_2 separates were used to determine the unit cell parameters of pyrite and marcasite, using least squares data refinement.

Pyrite and marcasite separates were prepared, based on morphology, and analysed for their sulfur isotope compositions by conventional techniques using reaction with Cu_2O at 1050°C at the U.S. Geological Survey in Reston, VA.

4 RESULTS

Table 1 shows morphologies for pyrite and marcasite in eleven samples from the Golden Cross deposit, along with the Au values (in ppm) for FeS_2 concentrates and whole rock (WR) samples. Whole rock Au assays correlate roughly with pyrite morphology. Cubic pyrite is most abundant in samples with relatively low Au whole rock values: the late, barren vein, one breccia sample (45820), the country rock sample, and the massive sulfide material. In contrast, samples with relatively high concentrations of pyritohedra and anhedral pyrite generally have whole rock values of greater than 3 ppm Au.

Compared to other marcasite morphologies, the relative abundance of tabular marcasite is highest in three of the low grade samples: the barren vein, the low-grade breccia (45820), and the massive sulfide material. These samples also contain relatively high concentrations of cubic pyrite. Miscellaneous euhedral marcasite morphologies occur in four relatively high grade samples: three ore veins, and one breccia (45816). Two of the ore vein samples contain relatively high concentrations of pyritohedra, whereas the other ore vein sample (45813) and the breccia sample (45816) contain only anhedral pyrite. Taken together, these data indicate that tabular marcasite correlates with cubic pyrite, whereas miscellaneous euhedral marcasite morphologies correlate with pyritohedra and anhedral pyrite.

No systematic variation exists between unit cell parameters of pyrite and marcasite and the Au content of the FeS_2 concentrates.

Nineteen stable isotope measurements revealed a range of $\delta^{34}S$ values from 1.0 to 4.8‰, with an average value of 2.7‰. There is no clear isotopic difference between pyrite and marcasite, or among any of the FeS_2 morphologies.

5 DISCUSSION

Mechanisms of pyrite and marcasite nucleation and growth are temperature dependant and vary with degree of FeS and FeS_2 supersaturation, and with pH (Allen et al., 1912, 1914; Murowchick and Barnes, 1986, 1987; Schoonen and Barnes, 1991a, b, c). Experimental studies of temperature and FeS_2-supersaturation variations as controlling factors upon pyrite habit between 250 and 500°C, found that at a mean temperature of 250°C and low FeS_2-supersaturation pyrite needles were produced (Murowchick and Barnes, 1987). As the temperature and/or degree of FeS_2-supersaturation increased up to 450°C, the sequence cube to octahedron to pyritohedron was produced. Dendritic growth of pyrite occurred at extremely high degrees of FeS_2-supersaturation.

Although maximum experimental temperatures employed by Murowchick and Barnes (1987) were

Table 1. Pyrite and marcasite morphology statistics, in percent, and geochemical data. Rounding causes the apparent totals of the morphology data to range between 99 and 101%.

Sample No.	45812	45815	03008	45813	45814	45816	45820	45811	45810	45818	45819
Rock Type*	o.v.	o.v.	o.v.	o.v.	b.v.	bx	bx	bx	bx	Rx	ms
Cubic py	1	20	2	0	68	0	32	0	10	35	31
Octahedral py	1	1	0	0	1	0	0	0	1	0	2
Pyritohedra	34	36	0	0	8	0	6	0	67	0	0
Anhedral py	40	29	40	56	19	74	57	35	20	65	57
Tabular mc	0	9	2	11	3	13	5	6	2	0	10
Bladed mc**	22	0	57	18	0	0	0	60	0	0	1
Misc mc	3	5	0	15	0	12	0	0	0	0	0
FeS_2 Au (ppm)	790	2200	78	18	n.d.	3.0	41	5.5	120	n.d.	n.d.
WR Au (ppm)	42	59	3.4	9.9	1.7	82	0.3	3.5	29	n.d.	0.01

*o.v. = ore vein; b.v. = late, barren vein; bx = breccias; Rx = country rock; ms = massive sulfide; py = pyrite; mc = marcasite, Misc mc = miscellaneous euhedral marcasite morphologies
** includes cockscomb aggregates
WR = whole rock
n.d. = not detected

well above those of epithermal deposits, the variation in habit with increasing temperature and/or FeS_2-supersaturation occurs in some natural hydrothermal systems. In a study of Japanese hydrothermal vein and replacement deposits, Sunagawa (1957) determined that both pyrite habit and abundance varied from highly altered host rock that had seen an "abundant supply" of the mineralising fluid, to rocks that had not been as intensely altered. A similar correlation between pyrite habit and ore grade was found by Bush et al. (1960) in the Tintic district, Utah, where barren rock contained cubic pyrite and productive rock contained pyritohedra.

FeS_2 habit in the Golden Cross deposit varies from pyrite cubes and tabular marcasite in hydrothermally altered country rock, through to pyritohedra and miscellaneous euhedral marcasite forms in ore veins. Temperatures in the altered country rock during geothermal activity presumably were lower than those in the veins (de Ronde and Blattner, 1988), suggesting that the changes in pyrite habit from cube through to pyritohedra, and the marcasite habit variation from tabular to miscellaneous euhedral forms, correlate with increasing temperature, and perhaps with increased water/rock ratios and FeS_2-supersaturation. The correlation between ore grade and pyrite habit at Golden Cross is similar to that described from the Tintic district (Bush et al., 1960).

The general pyrite sequence of cube to octahedron to pyritohedron to dendritic growth reported from the experimental work by Murowchick and Barnes (1987) occurs at Golden Cross. Based on mutual occurrence, the marcasite sequence that appears to correspond to the pyrite sequence comprises tabular to bladed aggregates to miscellaneous euhedral morphologies, that presumably reflects increasing temperature, fluid flow, and FeS_2-supersaturation.

Experimental studies by Murowchick and Barnes (1986) found that at temperatures up to 240 °C pyrite forms from solutions at pH > 5, and marcasite forms at pH < 5. The alternating and intergrown occurrence of pyrite and marcasite at Golden Cross may therefore represent precipitation from a fluid in which pH fluctuated about 5. However, equilibrium alteration mineral assemblages of calcite, quartz, and adularia, suggest a neutral to alkaline fluid pH of approximately 7.

These conflicting indications highlight important questions. Did marcasite precipitate during relatively brief periods of acidic fluid introduction, events whereby Au solubility in a bisulfide complex is reduced, causing deposition of Au? Can marcasite precipitate from near neutral fluids, in spite of current experimental evidence to the contrary? Further experimental work and studies of natural systems may help to highlight additional factors controlling marcasite and pyrite occurrence and distribution.

The lack of systematic variation between unit cell parameters of pyrite and marcasite and the Au content of the FeS_2 concentrates indicates that Au is not present in solid solution in pyrite or marcasite, but rather occurs as discrete inclusions of electrum.

Sulfur isotope data from this study show a narrow range, despite the wide variation in FeS_2 morphologies that were analysed. These results are consistent with previous studies at Golden Cross (de

Ronde and Blattner, 1988) and the Hauraki Goldfield (Christie and Robinson, 1992), that demonstrated a narrow range of sulfur isotope values (-2.2 to 4.1‰). These data have been interpreted as reflecting a magmatic sulfur source, related to the volcanic rocks of the Coromandel Peninsula (Christie and Robinson, 1992).

6 CONCLUSIONS

Although our sample population is not large, the data reported herein nonetheless show changes in the morphology of pyrite and marcasite in the Golden Cross epithermal system. The inferred sequence for pyrite from cubes to octahedra to pyritohedra to dendritic growth, presumably reflects not only temperature, but water-rock ratios and FeS_2 supersaturation. Marcasite shows a corresponding sequence from tabular to bladed aggregates to euhedral miscellaneous morphologies.

7 ACKNOWLEDGMENTS

We thank Coeur Gold New Zealand, Ltd. for financial support and access to the Golden Cross deposit. We thank Louise Cotterall for drafting the figure, and John Slack and Dan Hayba for helpful reviews.

8 REFERENCES

Allen, E.T., Creenshaw, J.L., & Larsen E.S. 1912. The mineral sulfides of iron. *Am. J. Sci.* 33: 169-236.

Allen, E.T., Creenshaw, J.L., & Merwin, H.E. 1914. Effect of temparature and acidity on the formation of marcasite (FeS_2) and wurtzite (ZnS). *Am. J. Sci.* 38: 393-421.

Appleman, D.E. & Evans, H.T. 1973. Indexing and least-squares refinement of powder diffraction data: U.S. Geol. Surv., Job 9214, Computer Contrib. 20. *U.S. Nat. Tech. Infor. Service*, PB2-16188.

Brathwaite, R.L. & Christie, A.B. 1996. *Geology of the Waihi Area*, Lower Hutt: Institute of Geological and Nuclear Sciences Ltd.

Bush, J.B., Cook, D.R., Lovering, T.S., & Morris, H.T. 1960, The Chief Oxide-Burgin area discoveries, East Tintic district, Utah; A case history - Pt. 1, U.S.G.S. studies, and exploration. *Econ. Geol.* 55: 1116-1147.

Christie, A.B. & Brathwaite, R.L. 1986. Epithermal gold-silver and porphyry copper deposits of the Hauraki Goldfield - A review. In R.W.Henley, J.W.Hedenquist & P.J.Roberts, *Guide to the Active Epithermal (Geothermal) Systems and Precious Metal Deposits of New Zealand: Monograph Series on Mineral Deposits.* 26: 129-145.

Christie, A.B. & Robinson, B.W. 1992. Regional sulfur isotope studies of Au-Ag-Pb-Zn-Cu deposits in the Hauraki Goldfield, south Auckland. *New Zealand J. Geol. Geophys.* 35: 145-150.

Dana, E.S. 1903. *The system of mneralogy of James Dwight Dana.* New York: Wiley.

de Ronde, C.E.J. & Blattner, P. 1988. Hydrothermal alteration, stable isotopes, and fluid inclusions of the Golden Cross epithermal gold-silver deposit, Waihi, New Zealand. *Econ. Geol.* 83: 895-917.

Goldschmidt, V. 1920. *Atlas der Kristallformen, Bd. 6, markasit-pyrit, plates 101-144*: Heidelberg: Carl Winters Universitäbuchhandlung.

Hancock, R.G.V., Pavlish, L.A., Farquhar, R.M., Salloum, R., Fox, W.A. & Wilson, G.C. 1991. Distinguishing European trade copper and northeastern North American native copper. *Archeometry* 33: 69-86.

Murowchick, J.B. & Barnes, H.L. 1986. Marcasite precipitation from hydrothermal solutions. *Geochim. Cosmochim. Acta.* 50: 2615-2629.

Murowchick, J.B. & Barnes, H.L. 1987. Effects of temperature and degree of supersaturation on pyrite morphology. *Amer. Mineral.* 72: 1241-1250.

Schoonen, M.A.A. & Barnes, H.L. 1991a. Reactions forming pyrite and marcasite from solution: I. Nucleation of FeS_2 below 100°C. *Geochim. Cosmochim. Acta.* 55: 1495-1504.

Schoonen, M.A.A. & Barnes, H.L. 1991b. Reactions forming pyrite and marcasite from solution: II. Via FeS precursors below 100°C. *Geochim. Cosmochim. Acta.* 55: 1505-1514.

Schoonen, M.A.A. & Barnes, H.L. 1991c. Reactions forming pyrite and marcasite from solution: III. Hydrothermal processes. *Geochim. Cosmochim. Acta.* 55: 3491-3504.

Simpson, M.P., Simmons, S.F., Mauk, J.L., & McOnie, A. 1995. The three dimensional distribution of hydrothermal minerals at the Golden Cross epithermal Au-Ag deposit, Waihi, New Zealand. *The Australasian Institute of Mining and Metallurgy Publication Series.* 9/95: 551-556

Skinner, D.N.B. 1986. Neogene volcanism in the Hauraki Volcanic Region. *Royal Society of New Zealand Bulletin.* 23: 20-47

Sunagawa, I. 1957. Variation in the crystal habit of pyrite. *Geological Survey of Japan Report.* 175: 1-47.

Variation of carbon and oxygen isotopes in the alteration halo to the Lady Loretta deposit: Implications for exploration and ore genesis

Peter McGoldrick, Paul Kitto & Ross Large
CODES, University of Tasmania, Hobart, Tas., Australia

ABSTRACT: Carbon and oxygen isotope analyses of carbonates were measured on twenty-nine sediment samples within the geochemical halo of the Lady Loretta stratiform sediment-hosted Zn-Pb-Ag deposit. Previous research indicates that the sediments in the inner halo are siderite-bearing and pass outwards to ankerite-bearing and then dolomite-bearing further from the deposit.

This isotope study shows that the inner zone siderites are distinctly heavier in oxygen isotopes ($\delta^{18}O_{sid}$ = 23.2 to 28.3‰) and lighter in carbon ($\delta^{13}C_{sid}$ = −1.9 to −3.2‰) than the outer zone dolomites ($\delta^{18}O_{dol}$ = 16.5 to 18.2‰ and $\delta^{13}C_{dol}$ = −0.9 to −1.5‰). Equilibrium fluid calculations suggest that the siderite halo formed at temperatures near 100°C from a fluid with $\delta^{18}O$ = +6‰ and $\delta^{13}C$ = −6‰, having the characteristics of an evolved basinal brine. In contrast, the outer dolomite sediments formed in equilibrium with a lower temperature (~50°C) fluid, with $\delta^{18}O$ = −11‰, $\delta^{13}C$ = −5‰, suggesting either a meteoric or evaporated seawater origin.

The data also suggest that C/O isotopes on carbonates have the potential to be an exploration indicator of favourable stratigraphy for Zn–Pb–Ag mineralisation. For example, within a given district, carbonate sediments which show the heaviest $\delta^{18}O$ values and lightest $\delta^{13}C$ values may indicate proximity to mineralisation.

1 INTRODUCTION

The Lady Loretta deposit is a small, but high-grade, example of the important stratiform sediment-hosted Zn-Pb-Ag deposits in Palaeoproterozoic sediments in north west Queensland. It has an ore reserve of 8.3 Mt of 18.4% Zn, 8.5% Pb and 125 g/t Ag (Hancock and Purvis, 1990).

Direct evidence from fluid inclusions for ore fluid temperatures is not available in most stratiform sediment-hosted Zn-Pb deposits because of their fine grain size. However, O isotopes in ore-related carbonates may provide indirect evidence of ore fluid temperatures.

Studies by Carr (1984) and Large & McGoldrick (1997) have demonstrated that the alteration halo at the Lady Loretta deposit exhibits a change in the dominant carbonate chemistry from the inner core to the outer extremity, varying from siderite (inner zone) to ankerite to dolomite. The siderite zone forms an envelope which extends about 50 m into the footwall and hangingwall sediments and along strike for at least 500 m. This is surrounded by ankerite-bearing sediments, with minor dolomite patches, which extend approximately a further 100 m across strike beyond the siderite zone. Beyond the ankerite halo, the sediments contain dolomite with low Fe and Mn content.

Isotopic studies were undertaken on the carbonate-bearing sediment surrounding Lady Loretta to determine whether any variations in $\delta^{18}O$ or $\delta^{13}C$ existed which (a) may provide information on the fluids responsible for the formation of the alteration halo, and/or (b) may be useful as vectors to ore.

2 SAMPLING AND ANALYSES

The samples used in this study are from surface and underground drill holes which intersected the host carbonate-bearing sediments over a stratigraphic interval of 300 m enclosing the deposit.

Isotopic analyses were undertaken at the Central Science Laboratory, University of Tasmania, using the technique of McCrea (1950). Carbonate content of samples analysed varied from 10 to 70 wt %. A variety of different acid reaction methods were employed on siderite-bearing sediments, with temperatures varying from 25° to 75°C, and times of 2 to 6 days. Constant results were measured on 15 mg samples reacted in H_3PO_4 for 3 days at 50°C, and for 6 days at 25°C. The isotopic composition of C and O in extracted CO_2 were obtained using a Micromass 602D stable-isotope mass spectrometer. Results are expressed in δ (‰) notation relative to standard mean ocean water (SMOW) for oxygen, and Pee Dee belemnite (PDB) for carbon. The precision of analyses for both C and O is ± 0.1‰. Samples with high sulfide content produced significant H$_2$S contamination, and did not give consistent results. These analyses were rejected. Useable analyses were obtained from twenty-nine samples.

3 RESULTS AND DISCUSSION

The carbon and oxygen isotopic values are plotted on Figure 1 to demonstrate: (i) internal isotopic variations, (ii) comparison with previous data sets from the McArthur basin, and (iii) variations with respect to proximity to ore.

3.1 *Internal Variation*

The C/O covariance diagram for Lady Loretta carbonates, Figure 1a, illustrates a consistent pattern of δ^{13}C and δ^{18}O values dependent on the carbonate type. Dolomites from the background dolomitic sediments outside the alteration halo form a tight grouping with $\delta^{13}C_{dol}$ values varying from –0.9 to –1.5‰ and $\delta^{18}O_{dol}$ values varying from 16.5 to 18.2‰. Ankerites from the outer alteration halo show a similar oxygen isotope range (15.9 to 17.9‰) but are slightly lighter in carbon isotope composition with $\delta^{13}C_{ank}$ values varying from –1.6 to –2.4‰. Siderites from the inner alteration halo surrounding the lead–zinc orebody exhibit much heavier oxygen isotope values ($\delta^{18}O_{sid}$ = 23.2 to 28.3‰), and carbon isotope values that overlap the ankerite field ($\delta^{13}C_{sid}$ = –1.9 to –3.2‰).

3.2 *Comparison with McArthur Basin data*

In Figure 1a, a comparison of the Lady Loretta data is made with data on carbonates from the McArthur Basin (Donnelly & Jackson, 1988; Rye & Williams, 1981; Large & Bull, unpublished data). The Lady Loretta dolomite and ankerite samples have significantly lower δ^{18}O values than sedimentary and hydrothermal carbonates from the McArthur Basin and the giant HYC Zn-Pb-Ag deposit. However, the Lady Loretta siderites overlap with the field of carbonates (ankerite) from the HYC stratiform ore zone; both data sets being heavier in oxygen isotopes and lighter in carbon isotopes than their host sedimentary dolomites. From an exploration view point these two data sets suggest that carbonates with heavy δ^{18}O and light δ^{13}C are more likely to be related to the halo of a stratiform Zn-Pb deposit.

3.3 *Proximity to ore deposit*

Figure 1b plots of variation in δ^{18}O and δ^{13}C in carbonates relative to vertical stratigraphic distance from the Lady Loretta ore horizon. A good linear correlation exists in the ankerite halo between δ^{18}O and distance from ore, showing a correlation coefficient of r^2 = 0.56. No correlation exists between δ^{13}C and distance from ores (r^2 = 0.03).

4 GENETIC CONSIDERATIONS

The marked difference (6–9‰) in δ^{18}O values between the inner halo siderites and background dolomites at Lady Loretta cannot be explained by differences in fractionation factors for dolomite–water and siderite–water. If HCO$_3^-$ is the dominant dissolved C species, then two different fluids must have been involved in carbonate formation. Modelling calculations outlined in the next section, together with geological evidence for an early diagenetic formation of the dolomite, indicate Lady Loretta dolomite formed in equilibrium with low temperature (~ 50°C) meteoric waters (brackish), or extremely evaporated seawaters with $\delta^{18}O_{fluid}$ ~ +11‰, whereas the inner halo siderites (spatially associated with base metal mineralisation) formed in equilibrium with hotter fluids (~100°C) with a distinctly heavy oxygen isotope signature (δ^{18}O ~ +6‰) characteristic of evolved basinal brines.

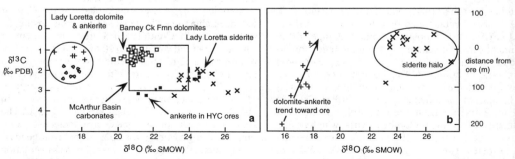

Figure 1 (a) C/O isotope data from Lady Loretta carbonates compared to carbonates from the McArthur Basin. (b) The variation of δ^{18}O with distance from the orebody for sideritic and dolomite/ankerite samples.

4.1 Equilibrium fluid calculations

Theoretical curves for the isotopic composition of dolomites and siderites precipitated from fluids which contain HCO_3^- as the dominant dissolved carbon species are shown in Figure 2. The dominant dissolved carbon bearing species in seawater is thought to be HCO_3^- and it is present in alkaline environments at temperatures less than 150°C (Ohmoto, 1986; Rollinson, 1993). Two sets of curves are shown on the C/O covariance plot for fluids in equilibrium with dolomite-HCO_3^- and siderite-HCO_3^-, respectively. Each pair of theoretical curves in Figure 2 encompass the most likely range in isotopic mineral compositions derived from published experimental data (Deines et al., 1974; Friedman & ONeil, 1977; Ohmoto & Rye, 1979; Land, 1983; Carothers et al., 1988). Based on the Lady Loretta study by Carr (1981) a palaeo-temperature of ~50°C has been estimated for dolomite deposition. From Figure 2 we interpret that the Lady Loretta dolomite must have formed in equilibrium with a fluid having $\delta^{18}O_{fluid}$ and $\delta^{13}C_{fluid}$ values of –11‰ and –5‰ respectively. The low $\delta^{18}O_{fluid}$ values for Lady Loretta dolomite may reflect a brackish to meteoric water influence within a shallow shelf-like environment, where waters are known to be alkaline, and strongly reducing near the sediment/seawater interface (Carr, 1981; Kharaka et al., 1977; Hudson, 1978; Longstaffe, 1987; Gao et al., 1992). Experimental and natural studies of extreme evaporation (e.g. halite saturation) indicate that residual liquids may become depleted in $\delta^{18}O$ (Sofer & Gat, 1975; Holser, 1979; Longstaffe, 1987). This interpretation is consistent with geological evidence for sulfate evaporite horizons in the ore sequence (Dunster, 1997). The $\delta^{13}C_{fluid}$ values of –5‰ are lighter than typical marine carbonates (Longstaffe, 1987) and may reflect the mixing of HCO_3^- produced by microbial degradation of organic matter from sulfate-reducing bacteria and from methogenic bacteria (Carothers & Kharaka, 1980; Bailey et al., 1973; Tissot et al., 1974).

Ankerite deposition from the same fluid, responsible for dolomite deposition, would require lighter $\delta^{13}C_{fluid}$ values close to –6‰ at 50°C (Fig. 2). Furthermore, the C/O covariance plot in Figure 2 indicates that siderite alteration was not related to, or in equilibrium with, dolomite deposition. We suggest that in the siderite envelope that surrounds the Lady Loretta orebody, siderite deposition was associated with an ore fluid having $\delta^{18}O_{fluid}$ and $\delta^{13}C_{fluid}$ values of +6‰ and –6‰, respectively at temperatures near 100°C. Ore fluids with such an isotopic composition are most closely associated with evolved connate fluids, or possibly magmatic hydrothermal brines (Fig. 9, Ohmoto & Rye, 1979).

4.2 Mechanics of halo formation

Although the isotopic data prove that a distinctive heavy $\delta^{18}O$ fluid was involved in the formation of the Lady Loretta halo, there is no evidence from the isotopes to determine whether the halo formed by exhalation and carbonate alteration syn-diagenetically on the basin floor, or due to later replacement of dolomite by siderite deeper in the sediment pile. Both mechanisms are possible based on the C/O isotope variations.

5 CONCLUSIONS

5.1 Genetic implications

The C/O isotope data on the Lady Loretta carbonates indicate that a distinctly heavy $\delta^{18}O$ fluid was involved in the formation of the deposit. The data suggest that the siderite halo to the orebody formed at temperatures near 100°C from a fluid with $\delta^{18}O = +6‰$ and $\delta^{13}C = -6‰$. Evolved, high salinity, basinal brines are the most likely source for the ore fluid based on this isotopic evidence, and other geological considerations.

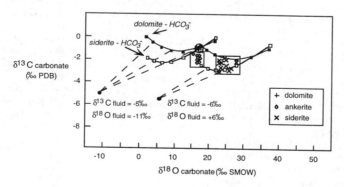

Figure 2 C/O isotope covariance diagram for Lady Loretta carbonates together with fluid equilibrium curves for dolomite-HCO_3^- and siderite-HCO_3^-.

5.2 *Exploration implications*

Two features of the isotope data have important implications for exploration:
1. Within a given district, carbonate sediments which exhibit the heaviest $\delta^{18}O$ values and lightest $\delta^{13}C$ values of the district data set are most likely related to the halo of a lead–zinc deposit.
2. Within the ankerite halo of a deposit, $\delta^{18}O$ can be used as a vector to ore. The $\delta^{18}O$ values increase with proximity to the deposit.

6 ACKNOWLEDGMENTS

This work formed part of ARC/AMIRA project P384 and the sponsor companies are thanked for their support, in particular the former owners of Lady Loretta (Pancontinental Mining).

7 REFERENCES

Bailey, N.J.L., Krouse, H. R., Evans, C.R. & Rogers, M.A., 1973: Alteration of crude oil by waters and bacteria - evidence from geochemical and isotope studies. *American Association of Petroleum Geologists Bulletin* 57: 1276–1290.

Carr, G.R., 1981: The mineralogy, petrology and geochemistry of the zinc–lead–silver ores and host sedimentary rocks at Lady Loretta, northwest Queensland. *Unpubl. PhD thesis, University of Wollongong.* 431 pp.

Carothers, W.W. & Kharaka, Y.K., 1980: Stable carbon isotopes of HCO_3^- in oil-field waters - implications for the origin of CO_2. *Geochem. Cosmochim. Acta* 44: 323–332.

Carothers, W.W., Adami, L.H. & Rosenbauer, R.J., 1988: Experimental oxygen isotope fractionation between siderite-water and phosphoric acid liberated CO_2-siderite. *Geochim. Cosmochim. Acta* 52: 2445–2450.

Deines, P., Langmuir, D. & Harmon, R.S., 1974: Stable carbon isotope ratios and the existence of a gas phase in the evolution of carbonate ground water. *Geochem. Cosmochim. Acta* 52: 727–737.

Donnelly, T.H. & Jackson, M.J., 1988: Sedimentology and geochemistry of a mid-Proterozoic lacustrine unit from Northern Australia. *Sed. Geol.*, 58: 145–169.

Dunster, J.N., 1977: The Lady Loretta Formation: sedimentology and stratiform sediment-hosted base metal mineralisation. *Unpubl. PhD thesis, University of Tasmania.* 293 pp.

Dutton, S.P. & Land, L.S., 1985: Meteoric burial diagenesis of Pennsylvanian arkosic sandstones, southwestern Anadarko Basin, Texas. *American Assoc. Petroleum Geologists Bulletin* 69: 22–38.

Friedman, I. & O'Neil, J.R., 1977: Data of geochemistry: Compilation of stable isotope fractionation factors of geochemical interest. *U.S. Geological Survey Professional Paper* 440-kk: 12.

Gao, G., Land, L. & Folk, R., 1992: Meteoric modification of early dolomite and late dolomitization by basinal fluids, Upper Arbuckle Group, Slick Hills, Southwestern Oklahoma. *American Assoc. Petroleum Geologists Bulletin* 76: 1649–1664.

Holser, W.T., 1979: Trace elements and isotopes in evaporites. *In* R. G. Burns, (ed.): *MARINE MINERALS. Reviews in Mineralogy*: 295–346.

Hudson, J. D., 1978: Concretions, isotopes, and the diagenetic history of the Oxford Clay (Jurassic) of central England. *Sedimentology* 25: 339–370.

Kharaka, Y. K., Lico, M.S., Wright, V.A., & Carothers, W.W., 1977: Geochemistry of formation waters from Pleasant Bayou No. 2 well and adjacent areas in coastal Texas. *Proc. Geopressured-Geothermal Energy Conf., 4th, Austin, Texas*: 168–193.

Land, L.S., 1983: The application of stable isotopes to studies of the origin of dolomite and to problems of diagenesis of clastic sediments. *Soc. Econ. Paleo. Mineral. Short Course* 4.1-4: 22.

Large, R.R. & McGoldrick, P., 1997: Alteration halos associated with stratiform sediment hosted ("Sedex") Pb-Zn-Ag deposits; Development of an alternative exploration approach. *Abstracts of GSA SGEG Meeting, Canberra, Jan 30-31.*

Longstaffe, F. J. 1987: Stable isotope studies of diagenetic processes. *In* T. K. Kyser (ed.): *Stable Isotope Geochemistry of Low Temperature Fluids. Mineralogical Association of Canada* 13: 187-257.

McCrea, L. M., 1950: The isotope chemistry of carbonates and a paleotemperature scale. *Jour. Chemistry Physics* 18: 849-857.

Ohmoto, H. & Rye, R.O., 1979: Isotopes of sulfur and carbon. *In* H. L. Barnes (ed.): Geochemistry of Hydrothermal Ore Deposits. New York, Wiley Intersci.: 509-567.

Rye, D. M. & Williams, N., 1981: Studies of the base metal sulfide deposits at McArthur River, Northern Territory, Australia: III. The stable isotope geochemistry of the H.Y.C., Ridge and Cooley deposits: *Econ. Geol.* 76: 1–26.

Sofer, Z. and Gat, J. R., 1975: The isotopic composition of evaporating brines: effect of the isotope activity ratio in saline solutions. *Earth and Planetary Science Letters* 26: 179–186.

Tissot, B., Durand, B., Espitalie, J. & Combaz, A., 1974: Influence of nature and diagenesis of organic matter in formation of petroleum. *American Assoc. Petroleum Petroleum Geologists Bulletin* 58: 499–506.

Approaching equilibrium from the hot and cold sides in the FeS_2-FeS-Fe_3O_4-H_2S-CO_2-CH_4 system in light of fluid inclusion gas analysis

D.I. Norman & B.A. Chomiak
New Mexico Tech, N.Mex., USA

J.N. Moore
EGI, University of Utah, Utah, USA

ABSTRACT: Water-rock reactions in Broadlands, Tiwi, and The Geysers geothermal systems were studied by fluid inclusion analysis. Inclusions are rich in CH_4 which is attributed to fluid interaction with organic-rich sediments. Broadlands and Tiwi inclusions have less H_2S than is calculated for fluid equilibrium with local pyrite-bearing wall rocks. The Geysers fluid inclusions, and steam, have greater than equilibrium concentrations of H_2S. We attribute the high H_2S to slow kinetics in fluids recharging the systems - cool ground waters (Broadlands and Tiwi) or hot magmatic volatiles (The Geysers). A slope of 0.08 on CO_2/CH_4 vs. H_2S plots suggests that the reaction $2Fe_3O_4 + 12H_2S + CO_2 = 6FeS_2 + 10H_2O + CH_4$ controls sulfidation and pyrite destruction reactions.

1 INTRODUCTION

Analyses of fluid inclusions from active geothermal systems indicate that they contain gaseous species derived from magma, the atmosphere, and the crust (eg. Norman et al. 1996). Meteoric waters and evolved (meteoric) waters enter a convecting geothermal system from cooler regions of the crust; magmatic fluids flux upward starting at near magmatic temperatures. Regardless of source, the gas chemistry of the fluids should evolve in response to changes in temperature and by interaction with wall rocks. Our goal was to determine how and if geothermal fluids from different sources attained equilibrium with wall rocks. Here we report on water-rock reactions in the CH_4-CO_2-H_2S-H_2O-Fe_3O_4-FeS_2-FeS system inferred from analyses of fluid inclusion volatiles. Inclusions studied are from Broadlands, New Zealand; Tiwi, the Philippines; and The Geysers, California. To constrain the interpretations, analyses were performed on splits of samples studied by microthermometry, and inclusion analyses were compared to analyses of present geothemal waters at The Geysers. Fluid source was estimated from N_2-Ar-CH_4 ratios.

2 THE GEOTHERMAL SYSTEMS

The Broadlands, Tiwi, and The Geysers geothermal systems are similar in that basement rocks contain carbonaceous, sulfide-bearing, marine sediments. At Broadlands the marine sediments are just below most of the production wells, which penetrate altered and pyrite-pyrrhotite mineralized volcanics. The Tiwi geothermal production wells are in pyrite-mineralized volcanics. The Geysers wall rocks are pyrite-pyrrhotite bearing marine sediments.

The Tiwi geothermal system is complex. Above 1800 m depth the reservoir is recharged by shallow ground waters; below, geothermal waters are mixtures of sea water and ground waters. Fluid inclusion studies indicate a similar division in the past. Th values are 190° to 330° C (Moore et al. 1997a).

The Geysers geothermal system produces 240° C steam. Fluid inclusion analyses indicate that the geothermal system was water-dominated in the past with paleo temperatures up to 440° C (Moore & Gunderson 1995).

3 METHODS

Volatiles in fluid inclusions are measured by quadrupole mass spectrometry. All analyses are by the crush-fast-scan (CFS) method (details in Norman et al. 1996). The CFS method can measure at maximum the volatiles contained in a 30 micron, liquid-filled inclusion, hence analyses are of single inclusions, or a small number of inclusions.

A quartz sample from Broadlands well BR12 was obtained from Patrick Browne, Geothermal Institute, New Zealand. Fifteen aliquots from one crystal were analyzed yielding 81 analyses. Twenty-eight primary inclusions in the crystal have an average Th = 262.1° C and Tm = -0.3° C.

Splits of The Geysers calcite and quartz samples from drill cores previously studied by fluid inclusion microthermometry (Moore & Gunderson 1995) were analyzed yielding 128 CFS analyses. Tiwi samples are from the Matalibong-25 (Mat25) drill core and were studied petrographically and by fluid inclusion microthermometry (Moore et al. 1997a). Fourteen samples from depths of 1113 to 2428 m were analyzed yielding 94 analyses.

Thermodynamic calculations were done using SUPCRT92. We assumed that fluids were pure water, and that the inclusion analyses represent the amount of gaseous species in the geothermal liquid. Inclusion salinities range up to that of sea water. Calculated amounts of aqueous H_2S in equilibrium with sulfide minerals will overestimate the amount in sea water-like fluids by up to 40%. Analysis of fluid inclusions trapped under boiling conditions generally results in analyses having excess gas species (Norman et al. 1996). The errors in the two assumptions tend to cancel one another out.

4 RESULTS

The fluid inclusion analyses of the three systems plot differently on N_2-Ar-CH_4 diagrams (Fig. 1). Broadlands analyses indicate a mixture of gaseous species derived primarily from magmatic and crustal sources. (We differentiate meteoric fluids that plot near air saturated water (ASW) on a N_2-Ar-CH_4 diagram from fluids that have similar N_2/Ar ratios but plot near the CH_4 corner. The former are called meteoric fluids, the latter evolved fluids.)

Tiwi inclusion volatiles represent magmatic and meteoric fluids that have interacted with crust to varying degrees. N_2/Ar ratios of less than 20 are the result of N_2 loss (as NH_3) during illite alteration. The Geysers fluid inclusion and steam analyses plot as mixtures of meteoric, evolved, and magmatic fluids. The Geysers fluids with N_2/Ar < 20 have additions of radiogenic Ar that lower the ratio.

In Fig. 2 Broadlands analyses plot in the magnetite field because Th values indicate trapping at about 260° C. Broadlands fluids also are more

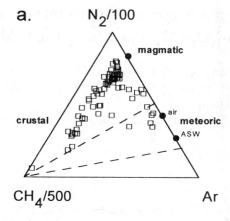

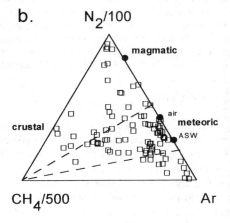

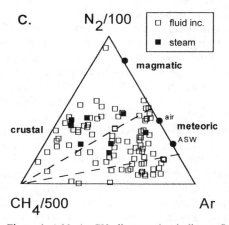

Figure 1. A N_2-Ar-CH_4 diagram that indicates fluid sources. The dashed lines indicate the minimum and maximum N_2/Ar ratios of meteoric waters trapped in fluid inclusions under boiling conditions. The Geysers analyses include steam (Truesdell et al., 1987).
a) Broadlands; b) Tiwi; c) The Geysers.

reduced than the present gas chemistry (Henley & Ellis, 1983), which is far from equilibrium with pyrite and pyrrhotite in the bed rock. Tiwi analyses exhibit higher concentrations of H_2S than Broadlands inclusion fluids, but like Broadlands analyses, they plot more in the magnetite field than expected from Th values. In contrast, The Geysers inclusion fluids and steam plot mostly in the pyrite field given temperatures greater than 240° C.

A few analyses obtained for each geothermal system have $CO_2/CH_4 < 0.1$; some ratios are near 0.01. Calculation shows that boiling cannot produce a fluid with so much more CH_4 than CO_2; neither can necking down of the fluid inclusions. Ratios of CO_2/CH_4 of 0.1 to 0.01 are typical of natural gas, and suggest that methane-rich fluids were formed by pyrolysis of carbonaceous matter.

5 INTERPRETATION

Fluid inclusion analyses indicate that past Broadlands and Tiwi geothermal fluids were mostly undersaturated in H_2S (Fig. 2). The low amounts of fluid inclusion H_2S might be argued to be the result of in situ H_2S oxidation, but the data do not support this contention. The Geysers analyses are similar to present steam concentrations, hence they do not appear to have lost H_2S by *in situ* oxidation. Maximum H_2S-FeS_2 equilibrium temperatures for Broadlands and Tiwi analyses are 250° C and 330° C respectively (Fig. 2), which are about the maximum respective Th values. This cannot be a coincidence.

How H_2S-FeS_2 equilibrium temperatures in Fig. 2a vary from Th is consistent with what we know about the geothermal systems (Elder, 1981). Cool fluids, mainly of meteoric origin, enter the geothermal systems on the margins. Our data indicate that equilibrium was sluggish. The Geysers was a water-dominated system at the time inclusions were trapped. Figure 2c indicates the Geysers fluids were at equilibrium with the wall rock, or possibly like present fluids, oversaturated with H_2S. This is only possible if high temperature, presumably magmatic fluids, entered the reservoir and kinetics prevented the system from attaining equilibrium.

Figure 2 shows two different processes. Broadlands and Tiwi fluids are undersaturated in H_2S.

Fluids should react to replace pyrite with magnetite, thereby increasing H_2S concentrations until equilibrium amounts are reached. Fluid chemistry

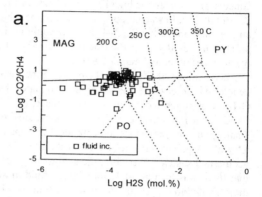

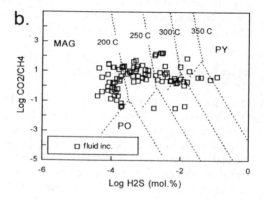

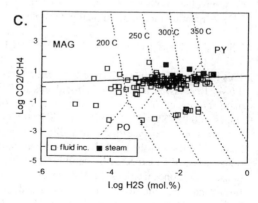

Figure 2. A log CO_2/CH_4 vs. log H_2S diagram. Stability fields for pyrite, pyrrhotite and magnetite are shown for several temperatures. See text for explanation of regression lines. a) Broadlands; b) Tiwi; c) The Geysers.

would move to the right on Fig 2. as this occurs, and approach equilibrium from the low H_2S or cold side. The Geysers fluids should sulfidize wall rock, thereby losing H_2S and approaching equilibrium concentrations from the high H_2S or hot side of Fig. 2. Fluids in all systems appear to be approaching an area just above the Fe_3O_4-FeS_2-FeO equilibrium point. This could well be the true intersection of the pyrite, pyrrhotite, and magnetite fields.

It is unlikely that the gas chemistry of the three geothermal reservoirs was at times dominated by CH_4 as is suggested by a few analyses (Fig. 2). Fluid inclusion analyses of single inclusions or small groups of inclusions indicate local gas chemistry at a small increment of time in the life of the geothermal system. Therefore, the analyses indicate that there was locally a CH_4-dominated fluid in all three geothermal systems. Methane in time would mix with reservoir fluids with the effect of lowering the oxidation state of the solutions. The low oxidation state allows the occurrence of pyrrhotite.

6 DISCUSSION & CONCLUSION

Our data indicate that Broadlands, Tiwi, and The Geysers inclusion fluids should have interacted with iron minerals and pyrite in wall rocks. A detailed stable isotopic study of sulfur in these geothermal systems is needed to resolve this question. However, the reaction between pyrite, magnetite and H_2S is an oxidation-reduction reaction that must involve another reaction pair such as CO_2-CH_4. A reaction thermodynamically possible is:

$$2 Fe_3O_4 + 12H_2S + CO_2 = 6FeS_2 + 10H_2O + CH_4.$$

This reaction plots on Fig 2. with a slope of 1/12 or 0.083. Regression analysis of Broadlands and the main group of The Geysers analyses indicates slopes of 0.08 (SD= 0.13) and 0.083 (SD= 0.038).

Broadlands analyses indicate additions of magmatic volatiles, but the fluids have lower concentrations of H_2S than expected by equilibrium calculations. Low concentrations of H_2S in Broadlands fluids may be due to a magma degassed with respect to H_2S, hence, there is little H_2S. Well BR12 on the margin of the geothermal field may be anomalous.

The Geysers system is steam-dominated because it lacks recharge. However, the gas analyses indicate that the geothermal system was never in equilibrium with wall rocks as one would expect if the geothermal system was sealed at temperatures of 240° to 260° C for an extended period. Fluids with the highest concentrations of H_2S have magmatic N_2/Ar ratios, strongly indicating an input of magmatic volatiles into The Geysers geothermal system. This is consistent with He isotope analyses that show a strong input of mantle fluids into The Geysers geothermal system (Moore et al. 1997b).

REFERENCES

Elder, J.W. 1981. Geothermal systems. Academic Press, London: 508 p.

Henley, R.W. and Ellis, A.J. 1983. Geothermal systems ancient and modern, a geochemical review. Earth Science Reviews 19: 1-50.

Moore, J.N. and Gunderson, R.P. 1995. Fluid inclusions and isotope systematics of an evolving magmatic-hydrothermal system. Geochimica et Cosmochimica Acta 59: 3887-3907.

Moore, J.N., Norman, D.I., Kennedy, B.M. and Adams, M.C. 1997a. Origin and chemical evolution of The Geysers, California, hydrothermal fluids: Implications from fluid inclusion gas compositions. Geothermal Resources Council Transactions 21: 635-641.

Moore, J.N., Powell, T.S., Norman, D.I. and Johnson, G.W. 1997b. Hydrothermal alteration and fluid-inclusions systematics of the reservoir rocks in Matlibong-25, Tiwi, Philippines. Proceedings 22nd Workshop on Geothermal Reservoir Engineering: 447-455.

Norman, D.I., Moore, J.N., Yonaka, B. and Musgrave J.A. 1996. Gaseous species in fluid inclusions: A tracer of fluids and an indicator of fluid processes. Proceedings of the 21st Workshop of Geothermal Reservoir Engineering: 233-240.

Truesdell, A.H., Haizlep, J.R., Box, W.T. Jr. and D'Amore, F. 1987. Field wide chemical and isotopic gradients in steam from The Geysers. Proceedings 11th Workshop on Geothermal Reservoir Engineering: 241-246.

A hybrid origin for porphyritic magmas sourcing mineralising fluids

M.G. Rowland & J.J. Wilkinson
Department of Geology, Imperial College, London, UK

ABSTRACT: This contribution presents results of a study of the hypogene evolution of the porphyry copper deposit of Quebrada Blanca, Chile. The hydrothermal evolution of the deposit and the characteristics of the porphyritic lithology associated with mineralisation are used to show that Quebrada Blanca can be considered representative of Chilean porphyry copper deposits. Evidence is presented suggesting that the porphyritic lithology is the product of interaction between a felsic melt and a volatile-saturated mafic melt. The presence of a source of volatiles and metals other than the felsic magmas traditionally associated with porphyry deposits has important implications for the genetic models for this deposit type.

1. INTRODUCTION

The orthomagmatic model for the formation of porphyry deposits (Burnham, 1979), in which a volatile phase exsolved from a crystallising felsic melt scavenges ore elements, fits well porphyry-associated tin-tungsten and molybdenum deposits. Both are associated with highly differentiated portions of melt (Manning, 1988, Dingwell & Scarfe, 1983) and display unidirectional solidification textures (USTs) consistent with second boiling (Kirkham & Sinclair, 1988). Porphyry copper deposits are associated with less fractionated, more oxidised suites (Blevin & Chappel, 1992). This contribution presents evidence that a mafic magma and its associated volatile phase was a key factor in the development of mineralisation within a porphyry copper system. The importance of this factor (over and above that hypothesized by Sillitoe (1997) and references therein) calls into question the relevance of the orthomagmatic model to the genesis of such deposits.

2. GEOLOGY OF QUEBRADA BLANCA

The porphyry copper deposit of Quebrada Blanca (Hunt et al., 1983) is located in the Chilean altiplano at 21°00S, 68°48W. Along with the nearby deposit of Collahuasi, Quebrada Blanca marks the northern extent of the major deposits of the Late Eocene-Early Oligocene belt (Sillitoe, 1988) that include El Salvador and Chuquicamata.

2.1 Lithologies, petrography and field relations

The deposit is centred on a 2 km by 5 km body of quartz monzonite with a granodiorite core. The quartz monzonite has a seriate texture and consists of euhedral plagioclase and hornblende (now replaced by biotite), subhedral alkali feldspar and anhedral quartz. The granodiorite is medium grained and weakly porphyritic, with phenocrysts of altered plagioclase and corroded biotite.

The intrusion is cut by two groups of dykes. Coarse, porphyritic dykes are located at all levels of exposure, and have a phenocryst assemblage comprising several generations of euhedral to subhedral, growth-zoned plagioclase, euhedral biotite, an amphibole pseudomorphed by biotite, and oxyhornblende exhibiting magnetite exsolution (Figure 1a). Quartz ocelli (Figure 1b), sometimes consisting of microcrystalline quartz aggregates, are present but vary in size and frequency. They contain fluid inclusions. Accessory sphene has been replaced by magnetite. Mafic dykes are seen in deeper exposures below the base of supergene enrichment and have basaltic-andesite to dioritic compositions. Mafic material is also seen as enclaves with and without felsic reaction rims. Finer-grained samples have an intergranular texture, with plagioclase laths enclosing anhedral to subhedral quartz, biotite, and amphibole. In coarser-grained samples, poikilitic quartz partially encloses biotite laths, amphibole and oligoclase. The amphibole has altered partly to

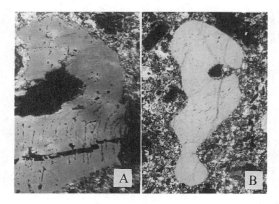

Figure 1. (a) Oxyhornblende exsolving magnetite (b) Quartz ocella. Both approximately 5mm long axis.

Figure 2. Contact between granodiorite and mafic material. Note entrainment of granodiorite phenocrysts.

Figure 3. Backveining of a mafic dyke by granodioritic material. Core 6 cm across in all photographs.

oxyhornblende and displays magnetite exsolution. Magnetite is very abundant within the mafic material and chalcopyrite forms clots.

Dyke contacts, both porphyritic and mafic, tend to be faulted/brecciated in the quartz monzonite but are more fluid within the granodiorite core. This is particularly true of the mafic material, the contacts of which may be convolute to flame-like. Occasionally, phenocrysts derived from the granodiorite may become entrained within the mafic dykes and aligned parallel to flow (Figure 2). Backveining of the dykes by the granodiorite is also observed (Figure 3). Porphyritic dykes occasionally incorporate mafic clasts in addition to fragments of quartz monzonite and granodiorite, but cross-cutting relationships between the dykes are not seen.

2.2 Hydrothermal evolution

Details of the veins at Quebrada Blanca, their assemblages, alteration halos and any other salient points, are summarised in the paragenesis (Table 1).

Table 1. Parageneses of veins at Quebrada Blanca and at El Salvador (Gustafson & Hunt, 1975, Gustafson & Quiroga, 1995) detailing vein characteristics.

STAGE	Vein characteristics: gangue and opaque assemblages, halos.	
	Quebrada Blanca	El Salvador, Chile
EARLY	Biotite + Magnetite ± quartz ± *potassic halo* Only noted at depth. Discontinuous and diffuse	Biotite ± Magnetite ± Quartz ± *albitic halo* Only noted at depth.
		Quartz ± K-feldspar ± Anhydrite ± Sulphides + *potassic halo*
	Quartz + K-feldspar ± Anhydrite ± Magnetite ± Sulphides ± *potassic or sericitic halo*	(Magnetite noted in veins transitional to veins with molybdenite)
	Quartz ± *Potassic/sericitic halo*	Quartz ± Anhydrite ± Molybdenite *Halos generally absent*
	Quartz + Molybdenite + Chalcopyrite± Anhydrite *Halos generally absent*	Sulphide + sericite + Biotite ± Quartz *Potassic or sericitic halos*
LATE	Pyrite > Quartz *Strong sericitic to argillic halo*	Pyrite + Quartz *Very strong sericitic to argillic halos*

3. DISCUSSION

3.1 Can Quebrada Blanca be considered representative of Chilean Porphyry Coppers?

The presence of a porphyry with a phenocryst assemblage comprising plagioclase, biotite and hornblende has been documented for the deposits of El Salvador (Gustafson & Hunt, 1975) and El Teniente (Camus, 1975). An assemblage as above but lacking biotite has also been reported for El Abra (Ambrus, 1977). In each case, quartz ocelli have been noted, as they have at Chuquicamata (Lopez, 1939). Both the porphyry at El Teniente and that at El Abra are clearly related to hypogene mineralisation. Fluid inclusion data from the quartz ocelli within the porphyry at Quebrada Blanca suggest that this lithology was the source of early mineralising fluids: inclusions within the ocelli are very similar to those within early veins with anhydrite and sulfides (Figure 4). Both sets could be interpreted as the separation of a single fluid into

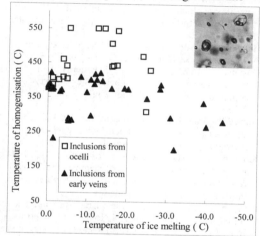

Figure 4. Graph of temperatures of first melting vs. ice melting for inclusions from quartz ocelli and early veins with anhydrite and k-feldspar. Inset: Primary fluid inclusion array indicating phase separation.

a volatile-rich phase and a volatile-poor, saline phase. Petrographic evidence supports this process (Figure 4 inset). A comparison of the vein paragenesis of Quebrada Blanca with that documented at El Salvador (Table 1) suggests that hydrothermal evolution at Quebrada Blanca is also typical of porphyry copper deposits.

3.2 A hybrid origin for the porphyry at Quebrada Blanca

Convolute to flame-like contacts, entrainment and flow alignment of phenocrysts derived from the granodiorite within the mafic dykes and backveining of the dykes by the granodiorite all suggest contemporaneity of the two melts. Furthermore, poikilitic quartz enclosing oligoclase laths and abundant but irregularly/poorly developed biotite laths are textures characteristic of dominantly liquid introduced into a felsic melt (Vernon, 1991).

The contemporaneity of mafic and granodioritic melts creates the possibility of interaction between them and the formation of a hybrid magma. The porphyry may represent such a hybrid since its phenocryst assemblage could have been formed by mixing the early crystallising phases of the quartz monzonite and the mafic material together. Oxyhornblende is seen within the mafic material while an amphibole that is always replaced by biotite is seen within the quartz monzonite. Both the mafic material and the porphyry contain biotite with the same high order birefringence. The plagioclase phenocrysts form three distinct groups, possibly representing a group of phenocrysts derived from each of the parent magmas, mafic and felsic, and one representing primary crystallisation within the hybrid. Complex zoning of the phenocrysts supports the idea of varying melt compositions with time. Quartz is not an early crystallising phase in either of the mixing magmas and it seems unlikely that the ocelli were inherited from either of the parents. The resemblance of the quartz to liquid droplets (Figure 1b) may suggest that the quartz was actually a silica-rich fluid immiscible with the silicate melt. The occurrence of ocelli composed of microcrystals is consistent with recrystallisation from a quenched metastable precursor.

3.3 Volatile saturation of the dioritic melt

Three lines of evidence point to the mafic material being saturated with respect to a volatile phase. First, olivine is not seen within the mafic material although magnetite is. Magnetite replaces olivine as a liquidus phase in low magnesium, high aluminium basalts at low pressures when the melt is saturated with respect to water (Johnson et al., 1994). These basalts are the most common type of basalt in arc settings and could be a parent to the mafic material. Sisson & Lange (1993) estimated that exsolution in most subduction-related basalts begins at depths greater than 6km, so a basalt emplaced at 2-3km, the depth of formation of porphyry deposits, would likely have a coexisting volatile phase.

Figure 5. An early biotite-magnetite vein cutting granodiorite. Note diffuse boundaries.

Second, a generation of veins predates veins associated with hypogene mineralisation and shows a strong spatial relationship with the mafic material (see Table 1). These veins show mineralogical affinities with the mafic material (consisting of biotite and magnetite) and are extremely diffuse within the granodiorite (see Figure 5), pointing to a similar time of formation as the dykes. They could actually represent the mafic-associated volatile phase.

Third, the sphene within the porphyry is partially replaced by magnetite, suggesting that sphene is not stable within the hybrid, possibly as a result of an increase in oxidation state. Calc-alkaline magmas are usually buffered at a constant oxidation state during crystallisation, but may become more oxidised on mixing with a more mafic magma and its coexisting volatile phase (Matthews et al., 1994).

4. CONCLUSIONS

Although the porphyritic lithology associated with mineralisation and the evolution of the hydrothermal system at Quebrada Blanca can be considered representative of Chilean porphyry copper deposits, the magmatic processes responsible for the production of the lithology are distinct from those outlined in the traditionally accepted genetic model. Invasive mafic magmas may supply volatiles and metals to crystallising felsic material in the form of their associated melt volatile phase. Components of the volatile phase may be concentrated in hybrid lithologies which form as a result of the interaction of the two melts. These lithologies probably exsolve their volatile contents as they themselves are intruded at higher levels within the felsic magma chamber. This sequence of events would explain the relatively oxidised state of granites associated with porphyry copper mineralisation and their less differentiated nature: large degrees of fractionation within the felsic melt are not required to produce the concentrations of copper seen as the copper is not derived from the felsic melt.

5. REFERENCES

Ambrus, J. 1977. Geology of the El Abra porphyry copper deposit, Chile. *Econ. Geol.* 72: 1062-1085.

Blevin, P.L. & Chappell, B.W. 1992. The role of magma sources, oxidation states and fractionation in determining the granite metallogeny of eastern Australia. *Trans. Royal Soc. Edin.* 83: 305-316.

Burnham, C.W. 1979. Volatiles in magmas. In H.Barnes, (ed) *Geochemistry of hydrothermal ore deposits*, 2nd Edition, Wiley.

Camus, F. 1975. Geology of the El Teniente orebody with emphasis on wall-rock alteration. *Econ. Geol.* 70: 1341-1372.

Dingwell, D.B., Scarfe, C.M. 1983. Major element partitioning in the system haplogranite-HF-H2O - implications for leucogranites and high silica rhyolites. *EOS* 64: 342.

Gustafson, L.B. & Hunt, J.P. 1975. The porphyry copper deposit at El Salvador, Chile. *Econ. Geol.* 70: 857-912.

Gustafson, L.B. & Quiroga, J. 1995. Patterns of mineralisation and alteration below the porphyry copper orebody at El Salvador, Chile. *Econ. Geol.* 90: 2-16.

Hunt, J.P., Bratt, J.A., & Marquardt, J.C. 1983. Quebrada Blanca, Chile: An enriched porphyry copper deposit. *Mining Engineering* 35: 636-644.

Johnson, M.C., Anderson, A.T. & Rutherford, M.J. 1994. Pre-eruptive volatile contents of magmas. In M. Caroll & J. Holloway, (eds) *Volatiles in magmas*. Reviews in mineralogy 30, Min. Soc. America.

Kirkham & Sinclair, 1988. Comb quartz layers in felsic intrusions and their relationship to porphyry deposits. In R. Taylor & D. Strong, (eds) *Recent advances in the geology of granite related ore deposits*. CIMM Spec. Vol. 39: 50-71.

Lopez, V.M. 1939. The primary mineralisation at Chuquicamata, Chile. *Econ. Geol.* 34: 674-711.

Manning, D.A.C. 1988. Late stage granitic rocks in SW England and SE Asia. In R. Taylor & D. Strong, (eds) *Recent advances in the geology of granite related ore deposits*. CIMM Spec. Vol. 39: 50-71.

Matthews, S., Jones, A., & Beard, A. 1994. Buffering of melt oxygen fugacity by sulphur redox reactions in calc-alkaline magmas. *J. Geol. Soc. London.* 151: 815-823.

Sillitoe, R.H. 1988. Epochs of intrusion-related copper mineralisation in the Andes. *J. of South American Earth Sci.* 1: 89-108.

Sillitoe, R.H. 1997. Characteristics and controls of the largest porphyry copper-gold and epithermal gold deposits in the circum-Pacific region. *Australian J. Earth Sci.* 44: 373-388.

Sisson, T.W. & Lange, G.D. 1993. Water in basalt and basaltic-andesite glass inclusions from subduction related volcanics. *Earth and Planetary Sci. Letts.* 117: 619-635.

Vernon, R.H. 1991. Interpretation of microstructures in microgranitoid enclaves, in Didier & Barbarin (eds) *Enclaves and granite petrology*, Elsevier: 277-291.

Alkaline leaching of uranium ore from the North Bohemian Cretaceous, Czech Republic

P. Štrof, P. Ira & J. Emmer
AMM s.r.o., Liberec, Czech Republic

J. Novák & L. Gomboš
DIAMO s.p., Stráž pod Ralskem, Czech Republic

T. Pačes
Czech Geological Survey, Praha, Czech Republic

ABSTRACT: The thermodynamic model AMMFLOW was designed to simulate carbonate leaching of uranium ore from the North Bohemian Cretaceous sandstones using peroxide as oxidizing agent. Results of modelling are compared with laboratory experiments with disintegrated rock (particle size <1mm) and microstructure grains (10-30mm). Results of modelling represent water-rock interaction in a hypothetical cube of intact rock with 200mm side. Obtained results predict a very poor recovery of uranium using carbonate and peroxide solutions for leaching. This is especially true for rock containing pyrite.

1 INTRODUCTION

A new kinetic flow model has been developed to describe evolution of subsurface dissolutions in the Straz uranium deposit (North Bohemian Cretaceous basin, Czech Republic). The linear-flow-diffusive model AMMFLOW represents water-rock interaction where the solution equilibrium and adsorption equilibrium are maintained, while interaction between solution and solid phases and oxidation-reduction processes are treated as irreversible kinetic processes (Štrof et al. 1996). This chemical code is designed in such a way that it is specific to the conditions of the Straz deposit and performs very fast the kinetic computations so that it can be used as a subprogram in large-scale numerical transport models.

1.1 Partial thermodynamic equilibrium in solution:
A representative set of 21 master chemical components are defined after statistical evaluation of groundwater analyses. The components are: SO_4^{-2}, Al^{+3}, NH_4^+, Fe^{+2}, Fe^{+3}, Ca^{+2}, Mg^{+2}, Na^+, K^+, Mn^{+2}, NO_3^-, SiO_2, CO_3^{-2}, $O_{2(aq)}$, $N_{2(aq)}$, F^-, U^{+4}, UO_2^{+2}, $H_3AsO_3^0$, $H_3AsO_4^0$ and Be^{+2}. The 106 ion species are defined for this set of master components to describe thermodynamic equilibrium in solution over the whole range of pH. Equilibrium is represented by individual mass-action laws

$$K_i = \prod_{j=1}^{N} a_j^{v_j} \quad (1)$$

where K_i = equilibrium constant of i-th chemical reaction (temperature dependant), N = total number of species (106), a_j = activity of j-th species and v_j = stoichiometric reaction coefficient ($v<0$ for initial species, $v>0$ for products and $v=0$ for species, which do not take part in the i-th reaction). Activity coefficients γ_j are calculated using modified Debye-Hückel's law (Parkhurst et al. 1980):

$$\log \gamma_j = \frac{-A Z_j^2 \sqrt{I}}{1 + B r_j \sqrt{I}} + C_j Z_j^2 I \quad (2)$$

where I = ion strength of solution, Z_j = ion charge, m_j = molality of j-th species, A,B = empirical constant, r_j = effective diameter of j-th ion and C_j = individual species empirical constant. The equations of mass-action together with the equations of mass balance and electrical neutrality represent a system of nonlinear equations. It is reduced mathematically to three rational equations and after that solved by the iteration simplex method developed by Nelder and Mead (1964). This makes the computation independent of estimated initial conditions and shortens the iteration loop.

1.2 Interaction kinetics between solution and solid phases:
Major minerals are selected after mineralogical analysis of core specimens from drill holes. The current version of AMMFLOW calculates with 26 solid phases (minerals): kaolinite, gibbsite, hydroalunite, goethite, hydrojarosite, siderite, pyrite, alumina-ammonium-sulfate, gypsum, calcite, magnesite, dolomite, dialogite, quartz, uraninite, ruthefordine and amorphous phases

$Al(OH)_3$, $Fe(OH)_3$, $FeOOH$, $Fe(OH)_2$, $Ca(OH)_2$, $Mg(OH)_2$, $Mn(OH)_2$, SiO_2 and UO_2. Basic kinetic equations for dissolution, nucleation and crystal growth or precipitation of amorphous phases are designed to comply with the transition state theory (Nagy and Lasaga 1993, Steefel and Cappellen 1990, or Oelkers et al. 1994). The rate constants are compiled from literature and they are supplemented with our experimental estimates to reach the explicit kinetic equations. Dissolution or crystal growth or amorphous phase precipitation:

$$\frac{dM}{dt} = -k_{dis,pre} M_S \left(a_{H^+}^{n_H^+} + B a_{OH^-}^{n_{OH}^-} \right) \left[1 - e^{-n_{1,dis,pre} \left| \frac{\Delta G_r}{RT} \right|} \right]^{n_{2,dis,pre}} \quad (3)$$

Crystal nucleation:

$$\frac{dM}{dt} = k_{nuc} \left[1 - e^{-n_{1,nuc} \left| \frac{\Delta G_r}{RT} \right|} \right]^{n_{2,nuc}} \quad (4)$$

where M = "concentration of mineral" (i.e. ratio of mineral mass to volume of reacting water, M_S = effective mineral surface, k_{xx} = rate constant, a_{H^+} = activity of protons, a_{OH^-} = activity of OH^-, n_{H^+}, n_{OH^-}, B, $n_{1,xx}$ and $n_{2,xx}$ = empirical coefficients and t = time. Gibbs free energy of reaction, ΔG_r, is represented by saturation index I_s:

$$\frac{\Delta G}{RT} = I_s = \ln \left(\frac{Q}{K_{eq}} \right) \quad (5)$$

where K_{eq} = temperature dependent equilibrium constant, Q = ion activity product of a mineral dissolution or precipitation in a nonequilibrium state, R = universal gas constant and T = absolute temperature.

1.3 Kinetics of oxidation-reduction processes: Six oxidation-reduction reactions are considered: $O_{2(g)}$-H_2O, Fe^{+3}-Fe^{+2}, NO_3^--$N_{2(g)}$, $N_{2(g)}$-NH_4^+, UO_2^{+2}-U^{+4} and As^{+5}-As^{+3}. Kinetics of oxidation-reduction process is described formally with the same equation as in case of the mineral dissolution:

$$\frac{dOR}{dt} = -k_{s,OR} a_{H^+}^{n_H^+} \left[1 - e^{-n_{1,s} \left| \frac{\Delta G_r}{RT} \right|} \right]^{n_{2,s}} \quad (6)$$

where OR = either the oxidized form or reduced form of a chemical component, $k_{s,OR}$ = rate constant of oxidation (reduction), and $n_{1,s}$, $n_{2,s}$ = empiric coefficient.

1.4 Cation adsorption on solid surface: Adsorption of 12 major cations is modeled. They are: H^+, Al^{+3}, NH_4^+, Fe^{+2}, Fe^{+3}, Ca^{+2}, Mg^{+2}, Na^+, K^+, Mn^{+2}, UO_2^{+2} and Be^{+2}. General sorption mechanism on alumosilicate or hydroxide surface is characterized by equation:

$$Me^{+v} + v S-OH \rightarrow (S-O)_v \equiv Me + v H^+ \quad (7)$$

where Me^{+v} = cation with charge v, S-OH = single binding position on mineral surface and the symbol $\equiv$ represents cation binding. The adsorption equilibrium is represented by simultaneous solution of the following set of equations:

$$\frac{X_i \, a_{H^+}^v}{M_i \, v \, \Delta X^v} = A_i \; ; \quad \sum_i X_i + \Delta X = X \quad (8)$$

where X_i = active position on surface occupied by i-th cation, v = charge of adsorbed cation, M_i = solution concentration of i-th cation, A_i = adsorption affinity of i-th cation, ΔX = unoccupied positions and X = total adsorption capacity [eq/kg of solid].

1.5 Diffusion: A linear flow is described by Darcy's law and proceeds inside a "channel". The channel is placed in a "particle". The "particle" represents porous surroundings through which diffusion proceeds without flow of water. The diffusive transport is described by the second Fick's law:

$$\frac{\partial M(x,t)}{\partial t} = D \frac{\partial^2 M(x,t)}{\partial x^2} \quad (9)$$

where M(x,t) = concentration, t = time a D = diffusion coefficient. The differential equation is solved using finite difference method in which the differential equation is expressed by

$$\Delta M = \frac{(M_{p-1} - M_p) D_k \Delta t}{\Delta x} \quad (10)$$

where M_p = concentration into element p, D_k = diffusive coefficient, Δt = time step a Δx = element length along x coordinate of the diffusive path inside the "particle".

2 EXPERIMENTAL

Laboratory experiments with carbonate leaching of uranium ore were performed on sedimentary rock of the "Hvezdov" layer of the Bohemian Cretaceous basin (Trojáček and Gomboš 1995). Diffusive leaching tests were made with disintegrated rock (particles <1mm) and microstructure intact rock grains (particles 10-30mm). Samples were pretreated by distilled water to remove oxidized layers on uranium ore and pyrite mineralization. The pretreated samples were placed in carbonate solution with concentration 1.5 g/l CO_3^{-2} (volume ratio 1:10) containing peroxide as oxidizing agent (1ml 3% H_2O_2 per 100ml of solution). Leaching solution was periodically withdrawn and replaced by original carbonate and peroxide solution.

3 RESULTS AND CONCLUSIONS

Alkaline leaching of uranium ore dissolves uraninite (UO_2) and released U^{+4} is immediately oxidized to the UO_2^{+2}. Uranium oxidation takes place only if a sufficient amount of oxidizing agent is present. In the experiments, concentration of peroxide was in the range of 5 to 350 mg/l and greatly fluctuated in analysed leachate samples. We modelled the leaching with average concentration of peroxide to be 200 mg/l. This is a sufficient amount for effective oxidation of uranium dioxide. The dissolution of uranium ore is accompanied by pyrite and alumosilicates dissolution. These processes are incorporated also in the computation. The concentrations of major chemical components in the solid phase was 1% by weight of Al in kaolinite and 0.8% by weight of Fe in pyrite. The carbonate ion concentration was maintained at 1.5 g/l.

Modelling shows that the dissolution of uraninite in Straz deposit is probably the result of dissolution of three hypothetical forms of UO_2 with different dissolution rate constants. In the current model only two forms of uraninite are considered. The third form of UO_2 is practically unreactive in the reported time range and into the model was incorporated as equivalent unreactive part of uraninite. The unreactive form of uraninite represents about 20 to 30% of total content of uranium and reduces recovery by carbonate leaching to 70-80%. The pH dependence factor of uraninite dissolution is defined by $\{a_H^{nH} + B \cdot a_{OH}^{nOH}\}$ in equation (3), where B = function (n_{H+}, n_{OH-}, pH_{min}). The empirical coefficients in (3) were adjusted according to our experiments and literature data (Bruno et al. 1991). The empirical coefficients for pH dependence of UO_2 dissolution are listed in Table 1.

Table 1. pH depend. coefficients of UO_2 dissolution.

solid phase	n_{H+}	n_{OH-}	pH_{min}
Uraninite$_{1,2}$	0.6	0.2	6.5

Results of modelling for four disintegrated samples v45873, v45874, v46177 and v46183 with particles <1mm are in Table 2. Laboratory and computation results are compared for sample v45873 in Figure 1. Results of modelling for the microstructure samples with particles 10 to 30 mm are shown in Table 3. The comparison of laboratory and computation results for sample v45873 is in Figure 2.

Table 2. Computed parameters for particles <1mm.

	Input to model				Output values			
Sample	Size [mm]	Weight [g]	Input U [mg]	Input U [weight %]	Uraninite$_1$ [weight %]	Uraninite$_2$ [weight %]	$k_{DISS,1}$ [day^{-1}]	$k_{DISS,2}$ [day^{-1}]
v45873	<1	20	5.96	0.03	0.02	0.01	1500	13
v45874	<1	20	9.8	0.049	0.03	0.019	1400	11
v46177	<1	20	9.05	0.045	0.03	0.015	1400	17
v46183	<1	20	1.02	0.005	0.003	0.002	15000	2

Table 3. Computed parameters for particles 10-30 mm.

	Input to model				Output values			
Sample	Size [mm]	Weight [g]	Input U [mg]	Input U [weight %]	Uraninite$_1$ [weight %]	Uraninite$_2$ [weight %]	$k_{DISS,1}$ [day^{-1}]	$k_{DISS,2}$ [day^{-1}]
v45873-a	10.88	22.23	4.67	0.021	0.013	0.008	1700	19
v45873-b	21.66	24.71	15.26	0.062	0.037	0.025	1700	19
v45874	22.69	28.14	89.72	0.32	0.19	0.13	1400	19
v46177-a	11.12	20.47	5.09	0.025	0.015	0.01	1400	17
v46177-b	19.48	18.37	15.49	0.084	0.055	0.029	1400	17
v46183	16.34	20.78	0.55	0.0026	0.002	0.0006	14000	2

Diffusive coefficients for both uranium forms U^{+4} and UO_2^{+2} were set to 0.1 cm^2/day. Results of AMMFLOW computation for all particle sizes show relatively small variation of the rate constants of uraninite dissolution. The rate constants for the faster dissolving uraninite form are in the range of 1400 to 1700 day^{-1}. The rate constants for the slowly dissolving uraninite are in the range of 11 to 19 day^{-1}. Exception is the sample v46183, which represents concentrated uranium ore with quartz grains about 2cm.

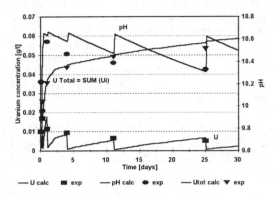

Figure 1. Alkaline leaching of uraninite for sample v45873 with particles <1mm.

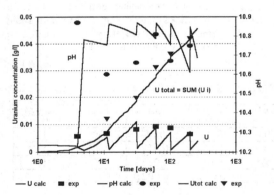

Figure 2. Alkaline leaching of uraninite for sample v45873 with particles 10.9mm.

A model representing dissolution of a hypothetical cube of intact rock with 200mm side was computed. Carbonate solution with 1.5 g/l of CO_3^{-2} and 200mg/l of peroxide as oxidizing agent was used. The third unreactive form of uraninite which comprises 30% of the sample was subtracted from the total concentration of uraninite in rock. About 20% recovery of uranium was reached after 15 year of simulated leaching. The major reason for the low recovery of uranium by alkaline leaching is the presence of pyrite in ore. Pyrite oxidation at the surface of the hypothetical cube consumes oxidizing agent so that the transfer of peroxide to the rock core is inhibited. The peroxide concentration is high in a thin surface layer of sample and decreases rapidly to the center of the cube. Uraninite leaching is therefore inhibited by saturated concentration of U^{+4} in solution. Carbonate leaching of uranium ore in the Straz deposit is therefore very ineffective. Experimental and modelling representation of this process is in progress. Acid leaching is more rapid and the results indicate much higher efficiency in recovering uranium from the rock.

REFERENCES

Bruno J., Casas I. and Puigdomenech I. 1991: The kinetics of dissolution of UO_2 under reducing conditions and the influence of an oxidized surface layer (UO_{2+x}): Application of a continuous flow-through reactor; Geochim. Cosmochim. Acta, 55, 647-658.

Nagy K.L. and Lasaga A.C. 1993: Simultaneous precipitation kinetics of kaolinite and gibbsite at 80°C and pH 3; Geochim. Cosmochim. Acta, 57, 4329-4335.

Nelder J.A. and Mead R. 1964: A simplex method for function minimization; Comput. J., 7, 308-313.

Oelkers E.H., Schott J. and Devidal J-L. 1994: The effect of aluminum, pH, and chemical affinity on the rates of aluminosilicate dissolution reactions; Geochim. Cosmochim. Acta, 58, 2011-2024.

Parkhurst D.K, Thorstenson D.C. and Plummer L.N. 1980: PHREEQE - A computer program for geochemical calculations; U.S. Geol. Survey Water-Resources Investigations Report 80-96.

Štrof P., Ira P., Emmer J. 1996: Development and application of mathematical models for solution of physical-chemical relations in uranium leaching; AMM, Liberec, Czech Republic, in Czech only.

Trojáček J. and Gomboš L. 1995: The verification of possibility of carbonate leaching on to bed Hvezdov; DIAMO, Straz, Czech Republic, in Czech only.

Steefel C.I. and Cappellen P. 1990: A new kinetic approach to modeling water-rock interaction: The role of nucleation, precursors, and Ostwald ripening; Geochim. Cosmochim. Acta, 54, 2657-2677.

Geochemical studies of the Kujieertai uranium deposit in Yili Basin, northwest China

Sun Zhanxue, Shi Weijun, Li Xueli & Liu Jinhui
East China Geological Institute, Linchuan, Jiangxi, People's Republic of China

ABSTRACT: Mineralogy, lithochemistry and hydrochemistry of the inter-layer oxidation zone in the Kujieertai uranium deposit, Xinjiang Province, northwest China, are discussed in this paper. The inter-layer oxidation zone can be divided into four sub-zones — the fully oxidized, the partially oxidized, the uranium ore and the original rock. These sub-zones are characterized by different mineralogical and chemical compositions of rocks, different hydrochemitry of groundwaters, and different colors of rocks as well as various geochemical parameters. Of them, the uranium ore sub-zone is indicated by the highest content of U, Ge, Re, Mo, Se, organic carbon and S^{2-}, the lowest total of iron of rocks, and lowest $\Delta Eh_{W,U}$, highest saturation index and uranium content of groundwaters.

1 INTRODUCTION

For the past six years, the uranium exploration activities in China have been concentrated on in-situ leachable sandstone type uranium deposits. The Kujieertai uranium deposit is one of the biggest such uranium deposits in northwest China (Chen, 1995). Because uranium mineralization in sandstone formations usually occurs in the roll-front of inter-layer oxidation zone (Granger, et al, 1969; Spirakis, 1979; IAEA, 1983; Dahlkamp, 1993), it is necessary to find out the geochemical characteristics of inter-layer oxidation zones for locating uranium ore bodies. The lithochemistry, hydrochemistry and mineralogy of oxidation zones in the Kujieertai uranium deposit, Yili Basin, Xinjiang Province, northwest China, are discussed in the paper.

2 GEOLOGICAL SETTING

The Kujieertai uranium deposit is located in the southern part of the Yili Basin, the large Mesozoic down-warded basin in Xinjiang Province, northwest China. The ore deposit is in sandstone of the middle to upper Jurassic Shuixigou Formation (J_{2-3sh}). Jurassic alluvial and pluvial fans deposited in the basin constitute the major part of the Shuixigou Formation.

2.1 Rocks

The host rock for the uranium ore bodies is mainly lithic sandstone of which lithoclast makes up to 20-40 percent. In addition, sandy conglomerate, siltstone and mudstone can also be seen in the host rocks in the deposit. The host rocks are in EW-trending extension with a low dip of 2 to 18°.

2.2 Uranium mineralization

In the deposit, uranium mineralization is mainly located along the redox front and is completely controlled by inter-layer oxidation zones. Pitchblende is the major uranium mineral, and coffinite and uranium black are secondary. Uranium-bearing ilmenite and land asphalt have also been found in the deposit. In addition, some uranium exists in adsorption and dispersion states. The age of uranium mineralization ranges from 1 Ma to 25 Ma (Chen, 1995; Zhu, 1996).

2.3 Hydrogeology

There are eight aquifers (Aquifer 1 to Aquifer 8) in the sedimentary strata of the basin. Except the top aquifer (Aquifer 1), the rest (Aquifer 2 to Aquifer 8) are confined. Aquifer 5 which consists of fine-coarse grained sandstone with permeability coefficient of 0.4-0.6 m/d is the major host for the uranium ore bodies in the area. The upper and lower confining beds of the uranium-host aquifer are mudstone with the thickness of 8-9 meters.

3 MINERALOGY AND LITHOCHEMISTRY OF INTER-LAYER OXIDATION ZONE

3.1 Mineralogy

Different mineral assemblage and various colors of rocks in the region can be seen due to difference in alteration (oxidation) of host rocks.

The fully oxidized sub-zone is indicated by red hematitization of sandstone with the destruction of organic matter. In the sub-zone, clay minerals such as kaolinite and illite are dyed by limonite and hematite into red brown.

The partially oxidized sub-zone is characterized by brown, gray and white sandstone with rich goethite and powdered limonite.

The uranium ore sub-zone is demonstrated by coexistence of pyrite, limonite, U-bearing bitumen of Judea, pitchblende and uranium black.

The reduced sub-zone (the original rock sub-zone or the unaltered rock sub-zone) is mainly characterized by gray and dark gray sandstone with pyrite, chlorite and carbon debris.

3.2 Lithochemistry

The chemical composition and iron phase analysis of rocks in different zones (from fully oxidized rocks to unaltered rocks) are listed in Table 1 and Table 2. Meanwhile, some environmental parameters such as Δ Eh, organic carbon (Corg) and S^{2-} are shown in Figure 1.

Noted that Δ Eh can be expressed as follows (Shi, 1991):

Table 1. Chemical composition of rocks in the inter-layer oxidation zone of the Kujieertai uranium deposit

Composition	Fully oxidized sub-zone	Partially oxidized sub-zone	Uranium ore sub-zone	Original rock sub-zone
SiO_2(%)	81.73	78.12	88.88	70.46
Al_2O_3(%)	8.66	11.46	4.74	13.93
Fe_2O_3(%)	2.67	1.69	1.56	2.34
CaO(%)	0.17	0.16	0.13	1.82
MgO(%)	0.37	0.28	0.01	0.85
MnO(%)	0.04	0.03	0.01	0.06
TiO_2(%)	0.26	0.51	0.16	0.69
P_2O_5(%)	0.08	0.07	0.04	0.11
K_2O(%)	2.53	2.95	1.73	2.68
Na_2O(%)	0.15	0.15	0.14	0.21
FeO(%)	0.47	1.09	1.50	1.68
U(ppm)	8.65	15.17	3100	11.2
Th(ppm)	6.60	8.62	6.3	11.3
Se(ppm)	0.39	0.18	10.45	0.97
Re(ppm)	0.00	0.20	1.23	0.10
Mo(ppm)	1.75	0.80	4.60	2.07
V(ppm)	33.5	70.5	17.63	89.33
Sc(ppm)	7.4	17.5	2.57	13.17
Ga(ppm)	17.0	22.0	5.27	22.33
In(ppm)	0.10	0.11	<0.1	1.17
Ge(ppm)	0.83	0.43	7.07	0.62

Table 2. The iron phase analysis of the inter-layer oxidation zone in the Kujieertai uranium deposit

Iron phase	Fully oxidized sub-one	Partially oxidized sub-zone	Uranium ore sub-zone	Original rock sub-zone
Magnetic Fe	0.060	0.033	0.097	0.006
Fe in carbonates	0.034	5.7	0.11	0.17
Fe in hematite & limonite	1.6	0.52	0.10	0.46
Fe in silicates	0.59	1.25	0.54	1.47
Fe in pyrrhotite	0.08	0.08	0.52	0.075
Fe in pyrite	0.009	0.25	0.06	0.185
Total of Fe	2.5	7.8	1.45	2.23

$$\Delta \text{ Eh} = \text{Eh}_S - \text{Eh}_O \qquad (1.1)$$

where Δ Eh is a parameter to express the reduction capability of rocks, Eh_O is the redox potential (Eh_O) of alkaline solution of the oxidizer as $KMnO_4$, and Eh_S is the equilibrium Eh value of the solution with the measured rock sample.

As indicated by Table 1, Table 2 and Figure 1, the fully oxidized sub-zone is characterized by the highest content of hematite and limonite, the lowest content of organic carbon, pyrite, and Δ Eh value, and lower concentration of Re, Mo, V, Sc, Ga, In, Ge and other associated elements; the

Table 3. Hydrochemistry of some water samples in the Kujieertai uranium deposit

No.	Zonation	HCO_3^-	Cl^-	SO_4^{2-}	$K^+ + Na^+$	Mg^{2+}	Ca^{2+}	Fe^{2+}	Fe^{3+}	U	pH	Eh	TDS
2402	Oxidized sub-zone	129.36	88.59	209.46	113.18	25.62	69.04	0.10	0.90	3E-4	8.30	60	0.57
3016	U ore sub-zone	152.31	69.77	186.16	84.18	24.18	54.17	0.10	0.30	4E-3	7.45	-79	0.49
2416	Original rock sub-zone	171.47	97.13	201.73	91.31	24.54	75.35	0.10	0.25	2E-3	7.4	-29	0.58

Note: Conc. units for ions are mg/L; Eh — mV; TDS — g/L.

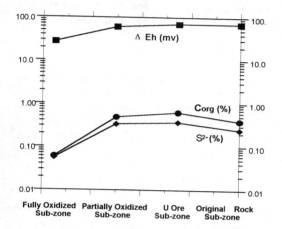

Figure 1. Some geochemical characteristics of the inter-layer oxidation zone in the Kujieertai uranium deposit

partially oxidized sub-zone by lower content of hematite and limonite, intermediate Δ Eh value, higher content of organic carbon, S^{2-}, Sc, Ga and Ge ; and the uranium ore sub-zone by the highest content of U, Ge, Re, Mo, Se, and iron in pyrrhotite, the lowest total of iron, higher Δ Eh value, and the highest content of organic carbon and S^{2-} respectively. The original rock (unaltered rock) sub-zone is similar to the uranium ore sub-zone but has lower content of organic carbon and S^{2-}, and has much lower uranium concentration.

4 HYDROGEOCHEMISTRY OF THE INTER-LAYER OXIDATION ZONE

4.1 Hydrochemistry

The hydrochemical data of some water samples in the Kujieertai uranium deposit are presented in Table 3.

Table 3 shows that the groundwater in the deposit is of fresh $HCO_3 \cdot SO_4 — Ca \cdot Na$ type with weak alkaline pH value.

4.2 Redox condition and hydrogeochemical characteristics

Using geochemical code MINTEQA2 (Jerry, 1991), the saturation index, the critical Eh value ($Eh_{C,U}$) and $\Delta Eh_{W,U}$ have been calculated. Here, $\Delta Eh_{W,U}$ can be expressed as (Shi, 1989; 1991):

$$\Delta Eh_{W,U} = Eh_W - Eh_{C,U} \qquad (4.1)$$

where $\Delta Eh_{W,U}$ is a parameter to express the reduction capability of water for uranium precipitation, Eh_W is the redox potential (Eh_O) of water solution, and $Eh_{C,U}$ is the critical redox potential value of uranium. Uranium can precipitate from the solution if $\Delta Eh_{W,U}$ is less than zero.

The results of the above geochemical modeling are listed in Table 4.

Table 4. Uranium species, saturation index (SI) of pitchblende, $Eh_{C,U}$ and $Eh_{W,U}$ of groundwater in the Kujieertai uranium deposit.

No.	U	Species (%)		SI	Ehw (mV)	$Eh_{C,U}$ (mV)	$\Delta Eh_{W,U}$ (mV)
	UO_2CO_3	$UO_2(CO_3)_2^{2-}$	$UO_2(CO_3)_3^{4-}$				
2402	0.0	29.1	70.9	-5.78	60	2.40	57.6
3016	2.1	80.6	17.2	1.8	-79	59.4	-138.4
2416	2.1	76.4	21.5	0.53	-29	78.4	-107.4

As demonstrated by Table 3 and Table 4, the oxidized sub-zone is characterized by positive $\Delta Eh_{W,U}$, negative saturation index, the lowest uranium content and Fe^{2+}/Fe^{3+} ratio; the uranium ore sub-zone by negative $\Delta Eh_{W,U}$, positive saturation index, the highest uranium content and Fe^{2+}/Fe^{3+} ratio; and the original sub-zone (the reduced sub-zone) by negative $\Delta Eh_{W,U}$, and lower saturation index, lower uranium content and higher Fe^{2+}/Fe^{3+} ratio respectively.

For these different sub-zones, uranium species of groundwater are predominantly $UO_2(CO3)_2^{2-}$ and $UO_2(CO3)_3^{4-}$.

5 CONCLUSIONS

The following conclusions can be reached:

1) The inter-layer oxidation zone of the Kujieertai uranium deposit can be divided into four sub-zones: the fully oxidized, the partially oxidized, the uranium ore and the original rock ones.

2) The fully oxidized sub-zone is indicated by red hematitization of sandstone, and is characterized by the highest content of hematite and limonite, the lowest content of organic carbon, pyrite and Δ Eh value, and lower concentration of Re, Mo, Sc, V, In and other associated elements of the rocks; and positive Δ $Eh_{w,u}$, negative saturation index, the lowest uranium content and Fe^{2+}/Fe^{3+} ratio of the groundwaters.

3) The partially oxidized sub-zone is demonstrated by brown, gray and white sandstones with rich goethite and powdered limonite. Compared with the fully oxidized sub-zone, the sub-zone is characterized by lower redox potential value (Eh_w) and uranium concentration of groundwaters, higher Δ $Eh_{w,u}$, and higher uranium content of rocks as well as higher Fe^{2+}/Fe^{3+} ratio of groundwaters. Furthermore, the sub-zone shows lower content of hematite and limonite, intermediate Δ Eh value, higher content of organic carbon, S^{2-}, V, Ga and other associated elements.

4) The uranium ore sub-zone is characterized by coexistence of pyrite, limonite, U-bearing land asphalt, pitchblende and uranium black; by the highest content of U, Ge, Re, Mo, Se, and iron in pyrrhotite, organic carbon and S^{2-} as well, the lowest total of iron, high Δ Eh value of rocks, and by negative Δ $Eh_{W,U}$, positive saturation index, the highest uranium content and Fe^{2+}/Fe^{3+} ratio of groundwaters.

5) The original rock sub-zone is suggested by gray and dark gray sandstones with pyrite, chlorite and carbon debris. The original rock (unaltered rock) sub-zone is similar to the uranium ore sub-zone but has less content of organic carbon and S^{2-}, and has much less uranium concentration in the rocks. For the groundwater solution in the sub-zone, it often shows negative Δ $Eh_{w,u}$, and lower saturation index and uranium content as well as higher Fe^{2+}/Fe^{3+} ratio.

6) Uranium species in the groundwater solution for the four different sub-zone of the inter-layer oxidation zone in the deposit are predominantly $UO_2(CO3)_2^{2-}$ and $UO_2(CO3)_3^{4-}$.

6 ACKNOWLEDGMENTS

This study was supported by the National Nuclear Science Foundation under the project No. Y7196R1802. Many thanks are due to staff members of Geological Team 216, Northwest China Geological Bureau for their kindest help to our field work.

7 REFERENCES

Chen, D.S. 1995. The mechanism and the model of uranium mineralization of the interlayer oxidation zone in Yili Basin. The Scientific Report of the Beijing Institute of Uranium Geology.

Dahlkamp, F.J. 1993. Uranium ore deposits. Springer-Verlag, Berlin: 250-324.

Granger, H.C. & C.G.Warren. 1969. Unstable sulfur compounds and the origin of roll-type uranium deposit, Econ. Geol. 64: 160-171

Jerry, D. 1991. Manual of MINTEQA2. EPA, USA.

Shi, W.J. 1989. The neutralizing-reduction-minerali- zation of uranium. In D.l. Miles (ed), Water-Rock Interaction: 621-624. Rotterdam: Balkema.

Shi, W.J. 1991. Hydrogeochemistry of uranium. Beijing: China Atomic Energy Press.

Spirakis, C.S. 1979. Interpretation of thermo-luminescenece patterns around a Wyoming roll-type uranium deposit, US Geol. Survey. Open-File Rep. 77-640.

Zhou, J.C. 1996. The inter-layer oxidation zones in the Kujieertai uranium deposit, Xinjiang, NW China. East China Geological Institute: 1-66.

Application of isotope studies of Australian groundwaters to mineral exploration: The Abra Prospect, Western Australia

D.J.Whitford & A.S.Andrew
CSIRO Division of Petroleum Resources, North Ryde, N.S.W., Australia

G.R.Carr & A.M.Giblin
CSIRO Division of Exploration and Mining, North Ryde, N.S.W., Australia

ABSTRACT: The Abra Pb-Ag-Cu-Au Prospect in the Proterozoic Bangemall Basin (WA) was chosen to evaluate the potential application of Pb, S and Sr isotope studies of groundwaters to mineral exploration. Robust techniques have been developed for sampling, and field-based pre-concentration techniques using ion exchange resins in permeable sachets have been demonstrated for Pb, Sr, and in some circumstances, S. Waters sampled from 43 drill holes over an area of about 50 km^2 are broadly similar with respect to major solutes. Pb isotope ratios range from those characteristic of the mineralisation to more radiogenic values more distant from known sulfides. δ^{34}S values less than 14‰ (CDT) probably represent background sulfate from aerosol fallout whereas higher values up to 30‰ may reflect interaction with primary sulfides and sulfates. Measured ^{87}Sr/^{86}Sr ratios vary from 0.725 - 0.814 with most values concentrated between 0.73 - 0.74. Highest values are found in samples closest to mineralisation.

1 INTRODUCTION

Hydrogeochemical methods are becoming an increasingly important tool in Australian mineral exploration, particularly in areas of poor exposure, deep weathering and where transported overburden obscures the underlying geology (Giblin, 1997). As exploration companies target ore bodies with no direct geochemical manifestation at the earth's surface, techniques are required that can be used to assess and rank targets that have been established, mainly through the use of geophysics. Insofar as the chemistry of groundwaters reflect sub-surface geology, hydrogeochemistry is a powerful and cost-effective technique.

The isotopic composition of S, Pb and Sr extracted from groundwaters provides direct information about the source or sources of these elements; S and Pb are direct ore indicators, and Sr yields important information about subsurface lithologies and alteration related to mineralisation. The isotopic compositions in groundwaters are unaffected by mineral precipitation, evaporation and dilution and thus provide information that is complementary to that obtainable from major and trace element abundances.

Research sponsored by the Australian Minerals Industry Research Association (AMIRA) has been directed at:

- defining the conditions under which concealed mineralization can be detected from the isotopic composition of S, Pb and Sr in groundwaters,
- assessing the scale of isotope haloes in a variety of environments and deposit types, and
- optimising the integration of isotopic techniques with conventional hydrogeochemistry.

The Abra Prospect was one of several test sites (Figure 1) chosen to include a variety of deposit types, tectonic settings, climatic and topographic environments, and groundwater chemistry (Andrew et al., in press).

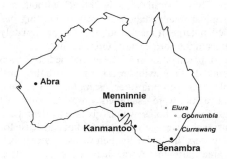

Figure 1 Location of the Abra Prospect together with other test sites.

Mineralisation at Abra comprises a Pb-Ag stringer zone enriched in Cu and Au at its base, overlain by a stratiform Pb-Ag zone. Barite is abundant throughout. The deposit occurs within a sequence of medium to fine grained sedimentary rocks in the Proterozoic Jillawarra mineralised belt of the central Bangemall Basin (Figure 2). Lead model ages suggest a mineralisation age of about 1600 Ma.

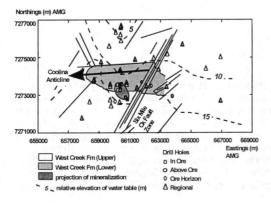

Figure 2 Simplified geological map of the Abra prospect also showing the relative elevation (m) of the water table, the distribution of sampling sites and the surface projection of the mineralised zone.

Abra represents a good case history study for the application of hydrogeochemical exploration methods. Following its discovery in 1981, the prospect and surrounding areas have been explored in some detail. The relatively simple geological setting permits a study focused on the dispersion of ore - related signatures without the additional complication imposed by a complex geological setting. The mineralization is "blind" without any geochemical manifestation at surface. Drilling has revealed a large mineralized body, approximately 200 m from the surface at its closest point.

The mineralization has been defined by deep diamond drilling with most of the holes still suitable for groundwater sampling. Regional groundwaters were sampled from a suite of percussion drill holes which gives excellent two-dimensional coverage of up to 50 km^2, at least for groundwaters immediately below the water table. The area around Abra has undergone deep weathering in places although elsewhere, there is reasonable outcrop, particularly of cherts and silicified siltstones and arenites.

2 BACKGROUND

Hydrology

In the early stages of mineral exploration, detailed groundwater flow patterns and the distribution of aquifers is generally unknown. Groundwater movement is typically in fractured bedrock aquifers. Different aquifers can be recognised from chemical parameters such as Eh and pH measured in the field as well as from laboratory-based measurements of major and trace element abundances. The relative elevation of the water table can be used to make first order inferences about flow directions (Figure 2). Lack of knowledge of local hydrology is not a major hurdle. The occurrence of ore signatures in groundwaters provides genuine incentive to explore further despite uncertainties in the source of that signature.

Target isotope signatures

Lead isotopes have been used widely in mineral exploration in Australia and there is now a substantial metallogenic database which can be used to define target isotope ratios in any geological province (Carr et al., 1995).

Similarly most sulfide mineralization has a characteristic S isotopic signature that is easily distinguishable from "background" S derived from aerosol input. The aerosol input has been shown to vary systematically in isotopic composition across large areas of the continent (Chivas et al., 1991), and local background values can be estimated.

Boundary conditions on background values of $^{87}Sr/^{86}Sr$ can often be established from regional geochronological databases. Strontium isotopic indicators of alteration related to mineralization are defined by relative increases in $^{87}Sr/^{86}Sr$ reflecting potassic alteration, rather than any absolute diagnostic signature.

3 SAMPLING METHODS

Uncased percussion or rotary air blast (RAB) exploration holes provide the best sites for groundwater isotope sampling although diamond drill holes, water bores, surface waters and natural seeps can also be used. Cased holes may be appropriate but there is the possibility of Pb contamination from plasticiser in the polyvinyl chloride (PVC) commonly used for casing. Contamination of diamond drill holes and water bores with anthropogenic Pb is a difficult but not insuperable problem. Bacterial reduction of sulfate either in the drill hole or in the local aquifer can render sulfur isotope data unusable due to the selective and unquantifiable removal of ^{32}S by the bacteria. Sites where sulfate reduction is a problem are readily identifiable from the presence of H_2S.

Samples were collected with a 900mL sampling tube bailer fitted with one-way rubber valves and stainless steel anti-fouling end fittings. Samples were generally taken 5m below the water table. In selected holes, sampling to depths of 300m was achieved using simple field-portable equipment built around a hand-cranked mechanical winder using stainless steel wire over a tripod-mounted pulley.

Sachets fabricated from fine nylon screen and filled with chelex® ion exchange resin were used to collect Pb and Sr from groundwaters to depths of 650m; an anion resin was used for S.

4 RESULTS

The waters can be divided into four groups according to the relationship of the sampled drill hole to mineralisation: within ore, above ore, ore horizon beyond the mineralised zone, and regional samples that might be considered as background.

The major element chemistry of the waters is summarised in Figure 3. All analysed samples are chemically similar and are characterised by low total dissolved salts when compared with many Australian groundwaters. Waters from drill holes that have penetrated ore have the greatest variability. The most striking feature of the trace element abundances in the Abra groundwaters is the enrichment in Ba. Waters from the mineralised zone have Ba concentrations up to 10,200 µg/L. Even background samples, more than 1 km from mineralisation and in the stratigraphically overlying unit have values >100 µg/L. Waters from the mineralised zone are also sporadically enriched in Cs, Mn and the chalcophile elements.

Lead isotope ratios of the groundwaters range from values characteristic of the mineralization to more radiogenic values. Ore values are concentrated near known mineralisation. Contamination with Pb derived from PVC casing is a significant problem.

Sulfur isotope values vary between 11-30‰ (CDT). Values less than 14‰, characteristic of the group most distant from known mineralisation, probably represent background sulfate from aerosol fallout. Higher values may reflect interaction with primary sulfides and sulfates. There is no evidence of bacterial activity in the sampled holes.

Measured $^{87}Sr/^{86}Sr$ ratios vary from 0.725 - 0.814 with most values concentrated between 0.73 - 0.74. Highest values are found in waters from the mineralised zone with other groups indistinguishable. Significant variations were observed in depth profiles from the deeper holes. All measured $^{87}Sr/^{86}Sr$ ratios are higher than those observed in the barite associated with mineralisation.

The spatial variations in the isotopic composition of S, Pb and Sr are shown in Figure 4. Ore-related isotopic signatures of S can be detected up to 6 km from known mineralisation but the lack of an intervening dispersion plume makes the significance ambiguous.

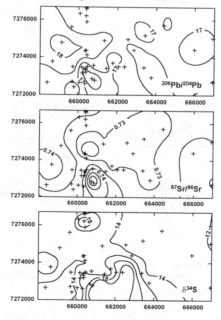

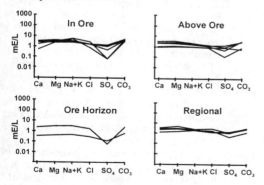

Figure 3 Schoeller plots of major solute concentrations (mE/L) in selected representative groundwaters from Abra.

Figure 4 Contour maps showing the distribution of S, Pb and Sr isotopic compositions in groundwaters (5m below water table) around the Abra prospect.

There are few simple correlations between major and trace element abundances on the one hand and isotopic composition on the other. Barium, which appears to be related to mineralisation shows a negative correlation with $^{206}Pb/^{204}Pb$ and weak positive correlations with $\delta^{34}S$ and $^{87}Sr/^{86}Sr$ (Figure 5). There are no correlations between isotopic compositions and abundance levels of Pb, S and Sr.

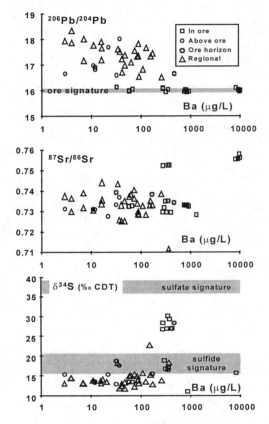

Figure 5 Variation of Pb, S and Sr isotopic composition with Ba contents in Abra groundwaters (5m below water table).

5 DISCUSSION

The isotopic composition of S and Pb in groundwaters indicate the presence of mineralisation at Abra. They are direct and easily interpretable indicators in contrast to elements such as Ba and Mn, which are often characteristic of mineralised environments, but are nevertheless indirect indicators. Lead isotopes provide the best discriminant; "ore" signatures were detected in all holes that penetrated sulfides despite subsequent blockage of holes well above the mineralised horizon. $\delta^{34}S$ values greater than about 14‰ (CDT), the regional background from aerosol fallout, indicate the presence of locally-derived sulfides and sulfates. Relatively high $^{87}Sr/^{86}Sr$ ratios, well above background values may indicate alteration related to mineralisation.

Isotope hydrogeochemistry represents an exciting new exploration technology and has the potential to offer a cost-effective exploration technique applicable at both the local and regional scale. Despite only limited testing, isotope methods could be usefully added to conventional hydrogeochemical surveys. The isotopic composition of Pb and S provide robust targets that are significantly independent of the style of mineralization being sought.

Sulfur and Sr isotope analyses will have application in regional target definition, whereas Pb will have application in prospect scale evaluation.

6 ACKNOWLEDGEMENTS

Aberfoyle Resources, Denehurst Limited, North Limited and Pasminco supported this research through AMIRA. Andrew Bryce, Steve Craven, Lesley Dotter, Barbara Gardner and Andrew Todd carried out the analyses.

7 REFERENCES

Andrew, A.S., Carr, G.R., Giblin, A.M. and Whitford, D.J. (1998) Isotope hydrogeochemistry in exploration for buried and blind mineralization. *Aust. J. Earth Sci.* (in press).

Carr, G.R., Dean, J.A., Suppel, D.J. and Heithersay, P.S. (1995) Precise lead isotope fingerprinting of hydrothermal activity associated with Ordovician to Carboniferous metallogenic events in the Lachlan Fold Belt of New South Wales. *Economic Geology* 90: 1467 - 1505.

Chivas, A.R., Andrew, A.S., Lyons, W.B., Bird, M.I. and Donnelly, T.H. (1991) Isotopic constraints on the origin of salts in Australian playas. I. Sulphur. *Palaeogeog. Palaeoclim. Palaeoecol.* 84: 309-332.

Giblin, A. (1997) Geochemistry of groundwaters in the vicinity of Stawell, Clunes, Ararat and Ballarat gold deposits. In "The AusIMM 1997 Annual Conference", Ballarat, 12-15 March, 1997 (Australasian Institute of Mining and Metallurgy, Carlton, 1997) Publ. Ser. 1/97, 181-191.

Spectral characterisation of the hydrothermal alteration at Hishikari, Japan

K. Yang, J. F. Huntington & K. M. Scott
CSIRO Exploration and Mining, North Ryde, N.S.W., Australia

ABSTRACT: In this study, the field infrared spectroscopic method has been applied to characterise the hydrothermal alteration system of the Hishikari Au-Ag deposit, where hydrothermal alteration about quartz-adularia vein systems grades out from chlorite-sericite, mixed layer clay-chlorite, quartz-smectite to cristobalite-smectite. The spectral data can be used to not only identify minerals, but also index the relative abundance and cation composition of certain key alteration minerals, such as chlorite and white mica. The spectral data have proved very useful in characterising the alteration zoning and identifying possible mineralisation-related mineralogical features.

1 INTRODUCTION

The pervasive hydrothermal alteration associated with the Hishikari Au-Ag deposit is characterised mainly by development of chlorite, white mica, mixed layer clays, smectite and kaolinite. Distribution of these phyllosilicate minerals is zoned with respect to Au and Ag-bearing vein systems. Mapping out the zonation, therefore, is of direct significance for exploration.

Field spectroscopic work using the portable infrared mineral analyser (PIMA-II) provides not only rapid identification in the field of the alteration minerals, but also valuable information about the relative abundance and cation composition of the minerals.

The spectral work reported in this paper is part of the AMIRA (Australian Mineral Industries Research Association Ltd) Project 435, *Mineral Mapping with Field Spectroscopy for Exploration*. In this work, we have aimed to document and understand the spectral signatures of a series of hydrothermal alteration minerals, and to search for possible spectral features that may be indicative of gold mineralisation.

2 GEOLOGY

Details of the geology, hydrothermal alteration and mineralisation at Hishikari have been described by Suzuki and Ibaragi (1987), Izawa and others (1990), Matsuhisa and Aoki (1994) and Nakayama (1995).

The Hishikari epithermal Au-Ag deposit is located in northeastern Kagoshima Prefecture, Kyushu, about 60 km north of Kagoshima city on Kyushu Island, Japan. Rocks in the mine area comprise the Cretaceous sedimentary basement (Shimanto Supergroup: sandstone and shale) and the overlying Pleistocene volcanic rocks (Hishikari Lower Andesite and Kurozonsan Dacite).

Mineralised quartz-adularia veins strike 45°-70°, and dip 70°-90° to the north. The veins commonly occur near the unconformity between Shimanto Supergroup and the Pleistocene volcanic rocks. Three major vein systems, Honko, Yamada and Sanjin, are identified. All the samples analysed in this study were derived from the Yamada system, where quartz-adularia veins were developed mainly in the lower andesite of the Pleistocene volcanic rocks.

Local uplifting of the sedimentary basement at Hishikari has resulted in a dome-like structure. Mineralised quartz-adularia veins are located near the top of the dome. Hydrothermal alteration in the Cretaceous basement is structure-controlled and restricted to the contact zone with the quartz-adularia veins. Hydrothermal alteration in the overlying Pleistocene volcanic rocks is much more extensive and zoned with respect to the quartz-adularia veins (Figure 1). The alteration zonation is, from deep (inner) to shallow levels (outer), chlorite-sericite (zone IV), mixed-layer clay (zone III), quartz-smectite (zone II) and cristobalite-smectite (zone I) (Figure 1). In places, a mixed layer clay-chlorite sub-zone (IIIa) may be present in zone III.

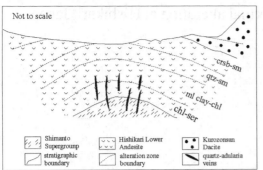

Figure 1. Schematic cross-section of the alteration system at Hishikari (generalised from Izawa et al, 1990). Alteration: crsb-sm = cristobalite-smectite zone, qtz-sm = quartz-smectite zone, ml clay-chl = mixed layer clay-chlorite zone, chl-ser = chlorite-sericite zone.

3 METHODS

218 core samples from six diamond drill-holes were measured with a PIMA-II portable infrared spectrometer. The PIMA-II measures the reflectance in the short-wave infrared (SWIR) region between 1300 nm and 2500 nm. Minerals with molecular vibrations active in this wavelength region include phyllosilicates, sulfates and carbonates, *i.e.* those commonly present in hydrothermal alteration systems. No sample preparation is required. The method, therefore, is particularly useful for in situ alteration studies.

Spectral data were processed using the software package X-SPECTRA, developed by the Mineral Mapping Technologies Group, CSIRO Exploration and Mining.

To confirm the spectral interpretation, XRD and microprobe analyses were carried out on a small number of samples.

4 RESULTS AND DISCUSSION

4.1 *Mineral zoning*

Many alteration minerals may have one or more absorption features at distinct wavelengths of the SWIR region. By selecting and quantifying the diagnostic absorption features of various alteration minerals, the alteration zonation can be spectrally characterised. The main features used in this study are those present at 2200-2225 nm (for white mica), 2250-2260 nm (for chlorite), 2160-2180 nm (for kaolinite), and 2200-2208 nm (for smectite). Other features were considered when the main features of two phases (*e.g.* white mica and smectite) overlapped. Computer-aided feature deconvolution is very helpful in dealing with compound features derived from two or more minerals.

Using those spectral features, the revealed vertical variations in phyllosilicate minerals at Hishikari are from the upper smectite-rich assemblages, through a kaolinite-rich transition, to the lower chlorite-rich, and finally the chlorite + white mica assemblage close to the mineralised veins (Figure 2).

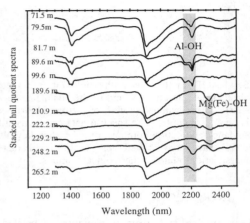

Figure 2. Stacked hull quotient spectra of selected DDH141 samples. Alteration zonation is revealed by the spectral variations, which indicates the mineralogical change with increasing depth from smectite ± kaolinite (71.5 m & 79.5 m), kaolinite ± smectite (81.7-99.6 m), through chlorite ± smectite (189.6-229.2 m), to white mica + chlorite (248.2 m & 265.2 m). The approximate wavelength regions of the identified absorptions are marked by the shaded areas.

On a broad scale, the zoning is spectrally displayed by an upper sequence of intense Al-OH absorption and a lower sequence of intense Mg(Fe)-OH absorption features (Figures 3 & 4). The Al-OH absorption in the upper sequence is caused by smectite and/or kaolinite, whereas the subordinate Al-OH absorption within the lower chlorite-dominated part is due to white mica. Chlorite is responsible for the observed Mg(Fe)-OH absorption, and its abundance tends to increase gradually with depth toward the mineralisation.

Mixed layer clays are present between the upper smectite and the lower chlorite-rich intervals. Based on spectral responses, chlorite-smectite appears to be the main type of mixed layer clay, in which chlorite is commonly the main component.

Wavelength of the minor Al-OH absorption near 2160 nm suggests kaolinite, rather than dickite, as the kaolin species. In zone II, kaolinite may locally become enriched to form a sub-zone (Figure 5).

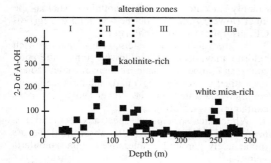

Figure 3. Variation in second derivative (2-D) of the major Al-OH feature through drill-hole DDH141. Increase in the 2-D value corresponds to enrichment of the Al-OH infrared active minerals (kaolinite, white mica or smectite).

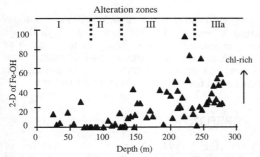

Figure 4. Variation in second derivative (2-D) of the Fe-OH feature through drill-hole DDH141. Increase in the 2-D value corresponds to the increase in chlorite abundance.

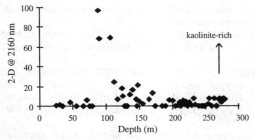

Figure 5. Variation in second derivative (2-D) of the minor Al-OH feature near 2160 nm through drill-hole DDH141. Increase in the 2-D value corresponds to the increase in kaolinite abundance. Kaolinite is spectrally unidentifiable when the 2-D value < 10-15.

4.2 Chlorite composition

Abundant chlorite is present in the inner (lower) alteration zone (Figure 4). Most chlorite is Mg-rich (estimated Mg# > 0.8), as suggested by the short Mg-OH wavelength (2318-2330 nm). In addition, minor amounts of chlorite with relatively longer Mg-OH absorption (2340-2348 nm) formed further out (shallower) in the alteration system, as seen in drill-hole DDH143 (Figure 6). The longer Mg-OH wavelengths suggest relatively more Fe-rich composition (estimated Mg# at 0.5-0.6) for the distal chlorite.

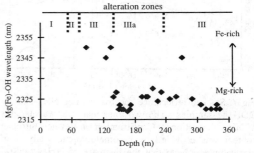

Figure 6. Variation in wavelength of the Mg(Fe)-OH absorption through drill-hole DDH143. Shorter Mg(Fe)-OH wavelength suggests chlorite of more Mg-rich composition.

4.3 White mica composition

The cation composition of white mica can also be spectrally estimated, as the Al-OH wavelength is negatively correlated with the amount of aluminum in octahedral sites (Duke, 1994).

The Al-OH wavelength of white mica at Hishikari varies between 2200 nm and 2225 nm. This wide range indicates a significant compositional variation for white mica, which, based on our unpublished data, corresponds the Al(vi)/Σcations(vi) values between > 0.9 (typical muscovite) and 0.65 (phengite). The compositional change of white mica in the alteration system at Hishikari appears to be random, with no definite relationships to alteration zoning or mineralisation observed.

4.4 Ammonium-bearing mineral

An ammonium-bearing mineral was encountered during this study at Hishikari. The absorptions by NH_4^+ are clearly displayed in the SWIR spectra by two features at around 2015 nm and 2120 nm, with possibly a third minor one near 1560 nm (Krohn & Altaner, 1987) (Figure 7).

The host mineral is not known yet. NH_4^+ could reside in buddingtonite or other phyllosilicates such as chlorite, illite and smectite.

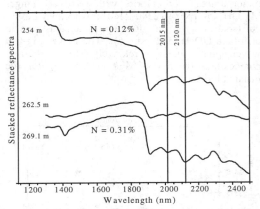

Fig 7. Stacked reflectance spectra of three NH_4^+-bearing DDH141 samples. The two lines mark the wavelengths of NH_4^+ absorptions. Confirmation of elevated nitrogen content was provided by chromatographic analysis of two samples. Note that the intensity of NH_4^+ spectral responses is correlated with the analysed N content.

The ammonium-bearing samples were found in several intervals (up to 20 m thick) within the chlorite-enriched part of the mixed layer clay zone (III) and the mixed layer clay-chlorite zone (IIIa). No ammonium was detected in the upper alteration zones (I and II). Ammonium-bearing minerals were also spectroscopically detected by Sumitomo geologists (Kuriyama, 1997, pers. comm.).

It is not clear whether or not the ammonium-bearing mineral formed in the same alteration event as the other phyllosilicates.

5 CONCLUSIONS

The results show that the hydrothermal alteration system at Hishikari can be spectrally characterised by using field portable instrumentation. Some spectral features may be used to indicate proximity to the mineralised quartz-adularia veins.

In particular, the following observations have been made in this study:
1. Chlorite and smectite are the two most common phyllosilicate minerals. Subordinate minerals are kaolinite and white mica.
2. Spectrally, the outer (upper) alteration zones are characterised by an intense Al-OH absorption (mainly smectite, with some kaolinite), whereas the inner (lower) zones are defined by intense Mg(Fe)-OH absorptions (mainly chlorite).
3. In the alteration system, Mg- and Fe-rich chlorite may occur at the lower (inner alteration zone) and upper volcanic sequence (outer zone), respectively. Au-bearing quartz-adularia veins are commonly located in the lower volcanic rocks containing Mg-chlorite.
4. White mica enrichment, spectrally recorded as a deep zone of intense Al-OH absorption, occurs immediately around mineralised quartz-adularia veins.
5. An NH_4^+-bearing mineral is present in the mixed layer clay-chlorite zone.
6. The revealed zoning of phyllosilicates upward from chlorite + white mica, through chlorite + mixed layer clays to smectite ± kaolinite, indicates a temperature drop with decreasing depth during hydrothermal alteration.

Acknowledgments: We are indebted to Sumitomo Metal Mining Co Ltd for providing the drill-hole samples and related data. Many thanks are due to AMIRA and all the sixteen sponsor companies and government geological surveys for supporting the project and for permission to publish this work.

6 REFERENCES

Duke, E.F. 1994. Near infrared spectra of muscovite, Tschermak substitution, and metamorphic reaction progress: implications for remote sensing. *Geology* 22: 621-624.

Izawa, E., Y. Urashima, K. Ibaraki, R. Suzuki, T. Yokoyama, K. Kawasaki, A. Koga & S. Taguchi 1990. The Hishikari gold deposit: high-grade epithermal veins in Quaternary volcanic rocks of southern Kyushu, Japan. *Journal of Geochemical Exploration* 36: 1-56.

Krohn, M.D. & S.P. Altaner 1987. Near-infrared detection of ammonium minerals. *Geophysics* 7: 924-930.

Matsuhisa, Y. & M. Aoki 1994. Temperature and oxygen isotope variations during formation of the Hishikari epithermal gold-silver veins, Southern Kyushu, Japan. *Economic Geology* 89: 1608-1613.

Nakayama, K. 1995. Case history of the discovery of the Hishikari gold deposit, Japan. In: Mauk, J.L. & J.D. St George (eds), *Proceedings of PACRIM Congress 95*: 429-434. Carlton, Vic: AusIMM.

Suzuki, R. & K. Ibaragi 1987. The Hishikari gold deposit: a case history and recent status of exploration. In *Proceedings of Pacific Rim Congress 87*: 417-422. Parkville, Vic: AusIMM.

Sulfur-isotope geochemistry of Chinkuashih copper-gold deposits, Taiwan: Preliminary results

H.W.Yeh
Institute of Earth Sciences, Academia Sinica, Taipei, Taiwan

L.P.Tan
Department of Geology, National Taiwan University, Taipei, Taiwan

M.Kusakabe
Institute for Study of the Earth's Interior, Okayama University, Misasa, Japan

ABSTRACT: The Pleistocene, hydrothermal copper-gold deposits of Chinkuashih district are located to the northeast corner of Taiwan. Diverse sulfide and sulfate minerals are known to occur in association with the deposits.

$\delta^{34}S_{CDT}$ values of various sulfide and barite samples we studied are highly variable but are all, with one exception of sulfide sample, significantly above 0.0 ‰ and +20 ‰ for sulfide and barite, respectively. The isotopic temperature estimated with Ohmoto & Lasaga's (1982) equation ranges from 180 to 260℃.

It appears that seawater sulfate is a major component of the hydrothermal sulfur. Magmatic sulfur is probably another major component and the relative contribution of these two components to the hydrothermal sulfur fluctuated in time and in space. Similarly, the temperature of the hydrothermal fluid responsible for formation of the sulfide and sulfate minerals fluctuated with a magnitude of greater than 80℃.

1. INTRODUCTION

Chinkuashih, an area of approximately 10 km², is situated at the north-east corner of Taiwan (Fig. 1). The area is mainly covered by the Miocene clastic sediments with several Pleistocene dacite intrusives (Wang, 1953; Hwang & Meyer, 1983). The most important copper-gold deposits known in Taiwan are located here and a total of ca. 100 tons of gold and 100,000 tons of copper were produced from the deposits (Tan et al., 1991; Tan, 1991). The ore bodies generally occur as irregular lenses or pipes in the silicified Miocene sandstone or the Pleistocene dacite near the contacts with the sandstone (Huang, 1955). The deposits are obviously of hydrothermal origin and probably belong to the class of epithermal, in terms of depth below surface, ores (Huang, 1955). The hydrothermal activities responsible for the copper-gold deposits appear to be a consequence of emplacements of the dacite intrusives (Huang, 1955; Juan et al., 1959). Thanks to the economic significance of the copper-gold deposits, they have been relatively well studied (See Tan, 1991 for a review). There are, however, some important aspects regarding the origin of the deposits which are still poorly understood. For example, the sources(s) and the isotopic geochemistry of sulfur are not well-understood. This, inspite of the fact that sulfur might have played an important role in transporting, depositing and accumulating gold during the formation of the deposits (Boyle, 1987; Crocket, 1974). This is a study of sulfur-isotope geochemistry of the deposits. One purpose of this study is to better define the source(s) of the sulfur. Folinsbee et al. (1972) collected four sulfur-mineral samples from two close-by veins in the underground workings south of Penshan (Fig. 1). They reported six $\delta^{34}S_{CDT}$ values, two for pyrite, enargite and barite each but did not really address the isotopic geochemistry of the deposits.

2. METHODS AND MATERIALS

The sulfur-isotope-geochemistry method was used in this study to obtain information on the source(s) of the sulfur and the temperature of the hydrothermal fluid at the time the sulfur minerals were formed.

This is mainly based on the fact that the $\delta^{34}S_{CDT}$ values of potential reservoirs of sulfur are often distinctive and that the difference in $\delta^{34}S_{CDT}$ values between coexisting sulfur minerals at isotopic equilibrium is a function of temperature of their formation (Ohmoto, 1986). The $\delta^{34}S_{CDT}$ values of a suite of samples selected from the deposits were measured for the aforementioned purposes. The sampling sites cover most of the important ore bodies and the minerals include barite, sphalerite, luzonite, pyrite, enargite and cinnabar (Fig. 1). The Sulfide samples were all mono-mineralic and were converted to $BaSO_4$ from which SO_2 for mass spectrometric measurement was prepared. Natural barite was reprecipitated as $BaSO_4$ after Na_2CO_3 dissolution. These are well-established procedures for determining their $\delta^{34}S_{CDT}$ values. The reproducibility of the results is better than ±0.2 ‰ at 95 % level of confidence.

3. RESULTS AND DISCUSSIONS

The $\delta^{34}S_{CDT}$ values range from +29.5 to +36.0 and from -2.9 to +7.2 ‰ for barite and sulfides, respectively. For comparision, the values reported in Folinsbee et al. (1972) range from +23.4 to +26.3 and from +0.7 to +2.1 ‰ for barite and sulfide samples, respectively. There are significant differences in both the values proper and the ranges of the values between these two sets of data. These differences will lead to very different conclusion regarding, for example, the source(s) of the sulfur and underscore the importance of having a set of representative samples to study. We will include the data of Folinsbee et al. (1972) in the final discussion.

Except sample CKY-Lu-1 (luzonite) of which the $\delta^{34}S_{CDT}$ value is -2.6, the $\delta^{34}S_{CDT}$ values of the rest of sulfide samples are all significantly above 0.0 ‰. The $\delta^{34}S_{CDT}$ values of the barite samples are all significantly larger than +20.0 ‰. Assuming that the sulfide/sulfate ratio is close to 1, the results indicate that the $\delta^{34}S_{CDT}$ values of the total sulfur (ΣS) of the hydrothermal system is probably close to +15 ‰. This value is necessarily subjected to a large error because the uncertainties in estimating the ratio as well as the $\delta^{34}S_{CDT}$ values of the sulfide and sulfate. The ratis was estimated by counting the number of reported occurrences of the minerals, and is obviously an approximation. Nonetheless, judging from all the available facts, it is highly unlikely that $\delta^{34}S_{\Sigma S}$ value is less than +12‰. This in turn suggests that sea-water sulfate was probably a primary end member of the source(s) of the sulfur.

The variations in $\delta^{34}S_{CDT}$ values are obvious not only for all the sulfide samples but also for the same mineral samples (Table 1). The ranges, the averages and the Standard deviations of the $\delta^{34}S_{CDT}$ values of the same mineral samples are given in Table 1. The variations most likely reflect the heterogeneity of the isotopic composition of the dissolved sulfur in the hydrothermal fluids. This aspect will be discussed elsewhere with information on the paragenesis of the sulfide minerals. We want to point out, though, that the averaged $\delta^{34}S_{CDT}$ values in Table 1 provided further support to the above mentioned supposition that the seawater sulfate was probably a major source of sulfur of the hydrothermal fluids.

Table 1. The ranges of $\delta^{34}S_{CDT}$ values of barite and some sulfite samples from Chinkuashih copper-gold-deposits

mineral	$\delta^{34}S_{CDT}$ (‰)		δ	No. of samples
	Range	Average		
Barite	+23.4 to +36.0	+29.5	4.9	5[a]
Enargite	+1.8 to +3.4	+2.4	0.7	4[a]
Pyrite	+0.7 to +5.4	+2.8	2.0	5[a]
Sphalerite	+4.7 to +7.2	+5.8	1.3	3

a: Two each from Folinsbee et al. (1972)

The isotopic temperatures can be obtained in several ways employing sulfur-isotope data. For example, assuming that the averaged $\delta^{34}S_{CDT}$ values of barite and of a sulfide mineral can be used to approximate the $\delta^{34}S_{CDT}$ value of parent dissolved sulfate and sulfide, then the isotopic temperature can be obtained using the SO_4/H_2S fractionation factors such as of Ohmoto & Lasaga (1982). Temperatures estimated in this way are ca. 224, 229 and 260℃. for sphalerite, pyrite and enargite, respectively. Alternatively, the $\delta^{34}S_{CDT}$ of a coexisting mineral pair may be used to estimate the temperature of their formation using similar fractionation factors. Thus temperatures of formation of luzonite-barite and sphalerite-barite are estimated to be ca. 180 and 195 ℃ respectively (Table 2). These were estimated from the isotopic data of the mineral-pairs CKY-Ba-

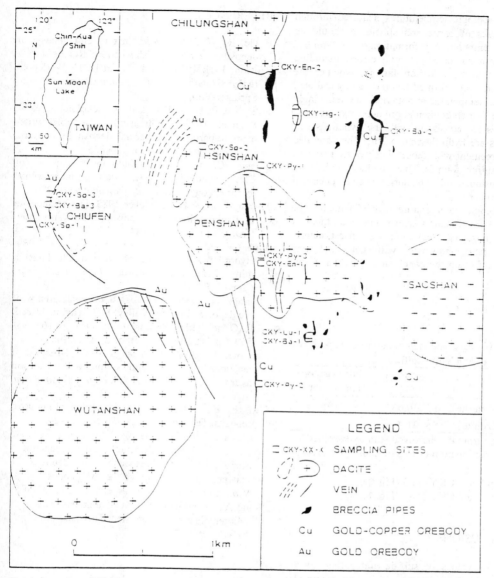

Fig. 1: Location of Chinkuashih, Taiwan and the sketch of distribution of the dacite and Orebodies, and of sample localities.

1/CKY-LU-1 and CKY-Ba-3/CKY-Sp-3 (Fig.1). These temperatures will be meaningful only if isotopic equilibrium was reached between the dissolved sulfide and sulfate, in addition to the requirement that the $\delta^{34}S_{CDT}$ values of the dissolved species and the equivalent minerals are essentially the same. It is believed that isotopic equilibrium between the dissolved species is reached in hydothermal fluid. However, even if under isotopic equilibrium, the sulfide minerals precipitanted from the same hydothermal fluid usually have different $\delta^{34}S_{CDT}$ values from each other and from the dissolved sulfide. This is because isotopic fractionation exists between the dissolved sulfide and the sulfide minerals and the fractionation factor is different for different sulfide mineral. It appears that sphalerite and the dissolved sulfide has a negligible isotopic fractionation at 150℃ and above.

Hence, the isotopic temperatures acquired through $\delta^{34}S_{CDT}$ values of barite and sphalerite should be close to the temperature of formation of the samples. Actually, with so many potential errors, including uncertainties in fractionation factors, whether the minerals were really formed together, analytical etc., the isotopic temperatures obtained are amazingly "reasonable" for an epithemal ore deposits. All these temperatures are consistent with the hypothsis that the deposits are hydrothermal origin. They are also in good agreement with those obtained from means other than sulfur isotopes. These include methods of fluid inclusion and sphalerite/galena chemical geothermometer.

The isotopic temperatures will natualy be different if fractionation factors other than those of Ohmoto & Lasaga (1982) were used in estimation. The differences, however, will not have any significant effect on the conclusions reached in this report.

Table 2. $\triangle\delta^{34}S_{CDT}$ and isotopic temperatures of two pairs of coexisting barite-sulfite.

mineral pair	$\triangle\delta^{34}S_{CDT}$ (‰)	Sulfur-isotope temperature[a]
Barite-Sphalerite[b]	30.6	195
Barite-Luzonite[c]	31.4	180

a: Estimated from the dissolved sulfate-dissolved sulfide fractionation factors of Ohmoto & Lasaga (1982).
b: CKY-Sp-3 and CKY-Ba-3 (Fig. 1).
c: CKY-Lu-1 and CKY-Ba-1 (Fig. 1).

4. CONCLUSIONS

The $\delta^{34}S_{CDT}$ values of sulfide and sulfate of the hydrothermal fluid responsible for origin of Chinkuashih copper-gold deposits are apparently heterogeneous. It appears that sea-water sulfate is a major source of the sulfur of the hydrothermal fluid. Another major source should have $\delta^{34}S_{CDT}$ values close to 0‰ and is probably magmatic sulfur. The relative importance of these two sources probably fluctuated in space as well as time.

The temperature of the hydrothermal fluid probably also fluctuated in space and time. The temperature of sulfur-minerals formed probably fluctuated by more than at least 80°C between 180 to 260°C.

REFERENCES

Boyle, R. W. 1987. *Gold*, history and genesis of deposits. New York: Van Nostrand Reinhold.

Crocket, J. H. 1974. Gold, 79. In K. H. Wedepohl (ed), *Handbook of geochemistry*. Berlin: Springer-Verlag.

Folinsbee, R. E., K. Kirkland, A. Nekolaichur, & V. Smejkal 1972. Chinkuashih - a gold-pyrite-enargite-barite hydrothermal deposit in Taiwan.. *Geol. Soc. Am. Mem.* 135: 323-325.

Huang, C. K. 1955. Gold-copper deposit of Chinkuashih mine, Taiwan, with special reference to mineralogy. *Acta Geol. Taiwanica* 7: 1-20.

Hwang, J. Y. & O. A. Meyer 1983. Dacite-andesite of Chikuashih region, northern Taiwan. *Mem. Geol. Soc. China* 5: 67-84.

Juan, V. C., Y. Wang & S. S. Sun 1959. Hydrothermal alteration of dacite at Chinkuashih mine, Taipeihsien, Taiwan. *Proc. Geol. Soc. China* 2: 73-92.

Ohmoto, H. 1986. Stable isotope geochemistry of ore deposits. In J. W. Valley, H. P. Taylor, Jr. & J. R. O'Neil (eds), *Reviews in mineralogy* 16: 491-559. Chelsa, Michigan: BookCraffers, Inc..

Ohmoto, H. & A. Lasaga 1982. Kinetics of reactions between aqueous sulfates and sulfides in hydrothermal systems. *Geochim. Cosmochim. Acta* 46: 1727-1745.

Tan, L. P. 1991. The Chinkuashih gold-copper deposits, Taiwan. *SEC Newsletter* 7: 21-24.

Tan, L. P., C. H. Chen, H. W. Yeh, T. K. Liu & A. Takeuchi 1991. Tectonic and geochemical characteristic of Pleistocene gold deposits in Taiwan. In J. Cosgrove & M. Jones (eds), *Neotectonics and Resources.* chapter 23: 290-305.

Wang, Y. 1953. Geology of the Chinkuashih and Chinfen districts, Taipeihsien, Taiwan. *Acta Geol. Taiwanica* 5: 47-64.

Hydrogen and oxygen isotopes of water-rock interaction in Dalongshan uranium deposit, Anhui province, China

J.P.Zhai, H.F.Ling & K.Hu
Department of Earth Sciences, Nanjing University, People's Republic of China

ABSTRACT: The δD and $\delta^{18}O$ values for ore-forming fluids of the Dalongshan uranium deposit agree well with the results of water-rock interaction between the melt-derived water, meteoric water and Jurassic sandstone of the Xiangshan Group at various temperatures. The early mineralization of the deposit was related with the melt-derived water and the sandstone, while the major and late mineralization resulted from exchange between the meteoric water and sandstone. The ore-forming uranium was derived from the sandstone of the Xiangshan Group.

1 INTRODUCTION

Uranium deposits associated with crust-derived granites are the main uranium resource in China, whereas granites derived from mixed crust-mantle materials are often associated with deposits of Fe, Cu, Pb and Zn etc. (Du and Rong, 1990). However, some uranium deposits associated with crust-mantle mixture derived granites have been found recently in China. Dalongshan uranium deposit is a representative among them (Zhang et al., 1991). The Dalongshan intrusive body is located at the southwestern edge of the Lu-Zhong volcanic basin in Anhui province, China. The outcrop area of the body is 90 km^2, dominated by quartz-syenite. The wall rocks of the body are mainly Jurassic sandstone of the Xiangshan Group and subordinate Paleozoic strata and Triassic limestone. The genetic characteristics of the body are similar to those of A-type granite. The body has low uranium content (average 7.5 ppm, Zhai, 1989). The uranium deposit is located in the contact zone between the top of the quartz-syenite and the overlying sandstone of the Xiangshan Group. Mineralization is confined in the contact zone. Based on the cements and mineral assemblage and the relationship between the ore bodies, the mineralization can be divided into early (Ⅰ), major (Ⅱ) and late (Ⅲ) stages.

2 SAMPLES AND ANALYTIC METHODS

All the samples for analyses are quartz and carbonate occurred with pitchblende. The samples were pre-heated to 100 °C for 10 minutes to expel absorbed gases and gases released from secondary inclusions before gases in original inclusions were released by decrepitation method at 400 °C and introduced to an ST-04 Micro-Water-chromatography for measuring their compositions with a detection limit of 10^{-5}. Quartz samples were ground to 40-80 meshes in deionized water to extract fluid from inclusions for composition analyses whose detection limits are better than 10^{-9}. Hydrogen and oxgen were extracted through reduction by uranium and oxidation by BrF$_5$ respectively (Shen, 1991). Their isotope compositions were measured with MAT252. The $\delta^{18}O$ values of fluids from which minerals were formed were calculated from $\delta^{18}O$ values of minerals through fractionation equations (Zhang, 1985). δD of the fluids were measured directly from fluid inclusions. All the analyses were fulfilled at the State Key Lab of Mineral Deposit Research in Nanjing University.

3 RESULTS AND DISCUSSION

The gases in fluid inclusions of the Dalongshan

Table 1 Liquid phase composition of fluid inclusions in Dalongshan uranium deposit

Stage	Mineral	pH	Eh(mV)	Liquid composition (10^{-6})								Na^+/K^+	F^-/Cl^-
				K^+	Na^+	Ca^{2+}	Mg^{2+}	F^-	Cl^-	HCO_3^-	SO_4^{2-}		
I	quartz	7.89	-11.8	4.32	2.01	5.88	3.10	2.46	2.57	8.75	0.17	0.46	0.96
	quartz	7.70	-0.53	3.24	1.51	8.40	4.14	3.84	3.50	8.33	0.28	0.47	1.08
II	quartz	7.76	-4.08	0.70	2.64	4.40	3.14	2.34	8.29	8.44	0.33	3.77	0.28
III	quartz	7.63	3.61	0.56	1.71	7.60	2.62	1.26	2.86	8.08	3.00	3.05	0.44

Table 2 Oxygen and hydrogen isotope compositions for Dalongshan uranium deposit

Stage	Mineral	$\delta^{18}O_{Min.}$ (‰)	$\delta^{18}O_{H2O}$ (‰)	δD_{H2O} (‰)	T^* (°C)
I	quartz	15.22	10.70	-73.0	380
	quartz	13.79	8.48	-106	350
	quartz	14.41	6.60	-81.5	275
	quartz	16.41	8.62	-81.5	275
II	quartz	14.62	8.79	-46.3	332
	quartz	14.47	5.97	-48.3	260
	calcite	16.53	7.50	-62.0	210
	dolomite	11.74	2.64	-75.9	225
III	quartz	12.22	0.83	-56.8	205
	clacite	11.12	1.32	-60.0	195
	clacite	12.51	0.91	-72.3	165
	clacite	11.48	-5.61	-66.0	100

T^*, fluid inclusion homogenization temperature.

uranium deposit are predominantly H_2O and CO_2. The salinity of the fluids ranges from 4.5% to 8.2% NaCl equivalent. Ion concentrations in fluids of the three stages are distinctly different (Table 1). The fluid of the early stage has $Na^+/K^+ \approx 0.46$ and $F^-/Cl^- \approx 1$, which agrees with feature of melt-derived hydrothermal fluid (Lu, 1990). The fluids of the major and late stages have low K^+ content and $Na^+/K^+ > 3$, $Ca^+ > Na^+ > K^+$ and $F^-/Cl^- < 1$, which are similar to characteristics of meteoric hydrothermal fluid (Ji and Wang, 1994).

Fig. 1 shows hydrogen and oxygen isotopic curves, calculated by mass balance equation (after Gat, 1996; Taylor, 1977), for fluids supposedly formed by water/rock interaction between meteoric water, melt-derived water and the quartz-syenite (Fig. 1a), or sandstone of the Xiangshan Group (Fig. 1b). The average $\delta^{18}O$ and δD values for the quartz-syenite are 9.5 ‰ and -87 ‰ respectively and for the sandstone 13 ‰ and -80 ‰ respectively (Zhu et al.,1992). The $\delta^{18}O$ and δD values of the melt-derived fluids, which were in equilibrium with crystallizing minerals of quartz, biotite, magnetite and amphibole of the Dalongshan quartz-syenite, were estimated 9 ‰ and -90 ‰ respectively (Zhu et al., 1992). The present-day meteoric water at Dalongshan area has $\delta^{18}O$ of -8.15 ‰ and δD of -52 ‰. According to Zhang (1985), since the coast was more distant from southeast China in the Mesozoic than at present, the $\delta^{18}O$ and δD values of Jurassic and Cretaceous meteoric water in southeast China were about 4 ‰ and 20 ‰ lower than those of the present day meteoric water, respectively. Therefore the $\delta^{18}O$ values used in the calculation for the quartz-syenite, sandstone, meteoric water and melt-derived water are 9.5 ‰, 13 ‰, -12 ‰ and 9 ‰ respectively; δD, -87 ‰, -80 ‰, -72 ‰ and -90 ‰ respectively.

The calculated evolution curves in Fig. 1a do not agree well with the $\delta^{18}O$ and δD values for the fluids of the three mineralization stages (Table 2). Some $\delta^{18}O$ values of the fluids at the early and major mineralization stages are much higher than curves D and E which are calculated for exchange between melt-derived fluid and quartz-syenite at 320 °C and 400 °C respectively. The curves for exchange between meteoric water and quartz-syenite at 200 °C (curve A), 250 °C (curve B) and 300 °C (curve C) seem to agree with the points for the fluid of the late stage. However, the average temperature of the late stage is about 180 °C, lower than those for the calculated curves. Therefore, it is difficult to interpret the genesis of the fluids of the uranium deposit by water-rock interaction between the quartz-syenite and meteoric water or melt-derived water.

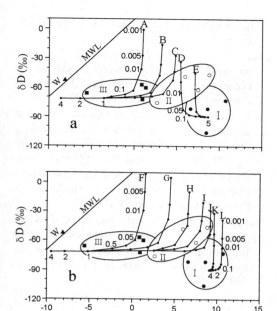

Fig. 1 Oxygen and hydrogen isotope compositions of Dalongshan uranium deposit and evolution curves calculated for exchange between meteoric water and quartz-syenite, melt-derived water and quartz-syenite (a), meteoric water and sandstone, melt-derived water and sandstone (b). Numbers corresponding to the small dots on curves are water/rock ratios. W: modern meteoric water in Dalongshan area; MWL: meteoric water line (Sheppard, 1977); I, II, III: Fluids of early, major and late mineralization stages in the uranium deposit. In Fig. 1a, curves A, B and C represent exchange between meteoric water and quartz-syenite at of 200 °C, 250 °C, and 300 °C respectively, and curves D and E represent exchange between melt-derived water and quartz-syenite at 320 °C and 400 °C. In Fig. 1b, curves F, G, H, I and J represent exchange between meteoric water and sandstone of the Xiangshan Group at 150 °C, 200 °C, 250 °C, 300 °C, and 350 °C respectively, and curves K and L represent exchange between melt-derived water and the sandstone at 350 °C and 400 °C. The $\delta^{18}O$ values used in the calculation for the quartz-syenite, sandstone, meteoric water, melt-derived water are 9.5‰, 13‰, -12‰ and 9‰, respectively; δD, -87‰, -80‰, -72‰ and -90‰, respectively (see text).

In Fig. 1b, curves F, G, H, I and J represent exchange between meteoric water and sandstone of the Xiangshan Group at temperatures of 150 °C, 200 °C, 250 °C, 300 °C, and 350 °C respectively, and curves K and L represent exchange between melt-derived water and the sandstone at 350 °C and 400 °C. These curves agree better with the $\delta^{18}O$ and δD values for the fluids of the three mineralization stages than the curves in Fig. 1a. The average homogenization temperatures of the inclusions of the early, major and late stages are about 350 °C, 280 °C and 180 °C respectively. The points for the early stage are near or among the curves for the exchanges between meteoric water and the sandstone at 350 °C and between melt-derived water and the sandstone at 350 °C and 400 °C. The points for the major and late stages are close to the curves for the exchanges between meteoric water and the sandstone at 250~350 °C and at 150~200 °C respectively. From early to late stage, the water/rock ratios were increasing. In a word, the $\delta^{18}O$ and δD values of the fluids of the uranium deposit indicate that the fluids were very likely formed by exchange between melt-derived water and the sandstone at early stage, and interactions between meteoric water and the sandstone at the major and late stages.

The source of the ore-forming material can be inferred from the water-rock exchange of the deposit. The quantity of uranium from the melt-derived fluid was likely small since the intrusive body itself has low uranium content. Therefore the uranium of the early stage was mainly from the sandstone through interaction between melt-derived fluid and the sandstone. The heat released from the cooling and crystallization of the melt would drive deep convection of meteoric water in the sandstone for a long time. The meteoric water convection would leach large quantity of uranium from the sandstone and deposit it at the contact zone between the intrusive body and the sandstone, forming the major and late stage mineralization. Based on lead isotopes, Zhu et al. (1992) estimated uranium loss or gain of rocks possibly related with formation of the Dalongshan uranium deposit. The initial and present uranium contents of the igneous rocks in the area are similar. In other words, the igneous rocks had no loss or gain of uranium during the formation process of the uranium deposit. The initial uranium content of the Xiangshan Group sandstone, particularly those affected by heat from emplacement of the Dalongshan quartz-syenite, was greater than the present uranium content. The loss of uranium in the sandstone could be 25% to 86%. Thus the study by Zhu et al. (1992) also suggested that the Xiangshan Group sandstone might provide uranium to the Dalongshan uranium deposit.

4 ACKNOWLEDGEMENTS

We thank the reviewers for their detailed review and many constructive suggestions. Thanks also go to Mr. Y.-C. Xu, Mr. J. Tu and Mr. G.-P. Xu for their help.

5 REFERENCES

Du, L.T. and Rong, J.S., 1990. The regional distribution of granite-type uranium deposit in China and their criteria of prognosis and evalution, Uranium provinces in China, Collection of papers presented by Chinese participants to IAEA'TC-Meeting on Uranium Provinces in Asia and the Pacific, Beijing.

Gat, J.R., 1996. Oxygen and hydrogen isotopes in the hydrologic cycle. Annu Rew Earth Planet Sci, 24: 225~262.

Ji, K.J. and Wang, L.P., 1994. The significant research progress of the source of hydrothermal solution and "triple-source" metasomatic hydrothermal metallogeny. Earth Science Frontiers (in Chinese). 1 (4): 126 — 132.

Lu, H.Z., Li, B.L., Shen, K. et al., 1990. Fluid inclusion geochemistry. Geology Press, Beijing (in Chinese): 102-154.

Shen, W.Z., 1991. Characteristic of isotopic geichemistry of U-bearing granites in S. China. Atomic Energy Press, Beijing (in Chinese): 123-137.

Sheppard, S.M.F., 1977. The Cornubian batholith, SW England: D/H and $^{18}O/^{16}O$ studies of kaolinite and other alteration minerals. Jour Geol Soc Lond, 133: 573 — 591.

Taylor, H.P., 1977. Water /rock interacterions and the origin of H_2O in grantic batholith. Jour Geol Soc, 133 (6): 509 — 558.

Zhai, J.P., 1989. Sr isotopic characteristics and geneses of Kunshan, Chengshan and Dalongshan igneous bodies. Geochimica (in Chinese), 18: 202 — 209.

Zhai, J.P., Ling, H.F. and Zhang, H., 1997. A study on geochemistry of ore-forming fluid of the Dalongshan uranium deposit. J. Nanjing University (in Chinese). 33: 94 — 100.

Zhang, L.G., 1985. Applcations of stable isotopes in geosciences. Shanxi Science and Technology Press, Xi'an (in Chinese): 23 — 28.

Zhang, Z.H., Zhang, B.T. and Zhai, J.P. et al., 1991. On the uranium-bearing granites and their related uranium deposits in south China. Atomic Energy Press, Bejing (in Chinese): 13 — 25.

Zhu, J.C., Zheng, M.G. and Ying, J.L. et al., 1992. A study on geological characteristics of stable isotopes in Dalongshan and Kunshan uranium deposits. Uranium Geology (in Chinese), 8: 338 — 347.

10 Geothermal fluids and gases

Boron isotopes in geothermal and ground waters in New Zealand

J.K.Aggarwal
Institute of Geological and Nuclear Sciences, Lower Hutt, New Zealand

ABSTRACT: Boron isotope ratios have been determined in a number of geothermal waters and ground waters from the North Island of New Zealand. Where the waters emerge with temperatures greater than 40°C and there is no influence of seawater B, e.g. Ngawha geothermal system, the B isotopic composition of the waters is consistent with other studies of subaerial hydrothermal systems ($\delta^{11}B$ between -10 and +5‰). Rotokautuku on the east coast shows a heavy B isotope signature (+43.2‰), and this reflects the influence of fluids from dewatering of the subducting Pacific plate. Warm hydrothermal fluids show $\delta^{11}B$ values reflecting the loss of ^{10}B to secondary minerals. Ground waters show variable B isotope ratios as a result of mixing of different endmembers, fractionation due to the effects secondary minerals and the effects of anthropogenic discharges.

1 INTRODUCTION

Boron has two stable isotopes (^{10}B and ^{11}B) with abundances of about 20% and 80% respectively. B isotope ratios ($^{11}B/^{10}B$) can vary from 4.25 to 3.9 in nature (Barth 1993). Boron is a highly mobile element and is strongly partitioned into the aqueous phase during water-rock interaction at high temperature.

In geothermal systems it is possible to identify three potential sources of B. The water recharging the geothermal system may inherently contain B, especially if seawater is involved. Additional B may be derived from the leaching of B from the rocks at high temperatures. In the case of geothermal systems, B may also be derived from the heat source that drives the geothermal system, e.g. a magmatic gas/juvenile source.

At temperatures > 40°C, B is leached from rocks with no apparent fractionation, however, at lower temperatures (<40°C) there is significant amounts of fractionation resulting in the fluid becoming enriched in the heavy isotope (Aggarwal et al in prep).

B isotope ratios have been determined in a number of subaerial geothermal systems; Yellowstone (Palmer and Sturchio 1990); Iceland (Aggarwal et al, 1992); Japan (Oi et al 1996); Israel (Vengosh et al 1994). This study is a pilot study of a much larger programme with the aim of determining B isotope ratios in a highly complex tectonic environment as exists in the North Island of New Zealand.

2 GEOLOGY

The geology of New Zealand is highly complex, with the Pacific plate being subducted under the Australian plate (Figure 1). This gives rise to the Taupo Volcanic Zone (TVZ) in the North Island, which shows andesitic and rhyolitic volcanism typical of an island arc to the east, and extensional volcanism in the west (Giggenbach pers. comms.). The east coast of the North Island shows many features typical of an accretionary prism e.g. mud volcanoes. To the north of the North Island is a waning geothermal province (Ngawha) which lies in a dormant volcanic complex.

3 METHODS

Hydrothermal fluids were obtained from collections at the Institute of Geological and Nuclear Sciences, and ground waters were collected specifically for this study. Figure 2 shows the sites that were sampled. Samples were filtered through 0.4µ filters before preparation for analysis.

Waters with Si/B < 1 were concentrated by evaporation on a hot plate at 40°C, to ensure that a 2µl aliquot contained 1-5ng B. Those with high B concentrations were diluted with sub-boiling distilled water. Removal of Si in samples with Si/B >1 was carried out by the addition of sufficient HF and B-free seawater, and digestion on a warm plate at 40°C. This allowed the removal of Si without volatilisation of B to enable large and stable signals to be produced in the mass spectrometer. Once the samples had been taken to dryness, they were loaded with 1µl 1M HCl onto Re filaments and dried with the aid of an infra-red lamp.

respectively. Results are reported relative to N.B.S. 951 and calculated as;

$$\delta^{11}B\ (\text{‰}) = \left[\left(\frac{^{11}B/^{10}B_{sample}}{^{11}B/^{10}B_{standard}}\right) - 1\right] \times 1000$$

Typical errors in $\delta^{11}B$ are 0.7 ‰.

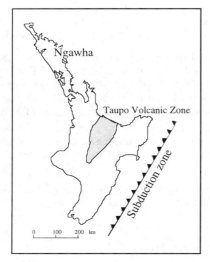

Figure 1. Map showing North Island, New Zealand with Taupo Volcanic Zone and East Coast Subduction Zone.

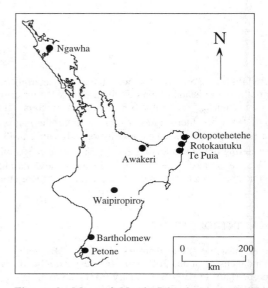

Figure 2. Map of North Island, New Zealand showing the location of some ground water and geothermal water samples studied.

Determination of boron isotopes was carried out by negative ion thermal ionization mass spectrometry of BO_2^- on a Micromass 30B at the Lamont-Doherty Earth Observatory in New York. Isobaric interference of CNO^- on $^{10}BO_2^-$ resulted in low $^{11}B/^{10}B$ ratios, and this was monitored by examination of CN^- at mass 26. At vacuums better than 10^{-8} torr using a liquid N_2 cold finger, isobaric interference was significantly reduced. Samples were heated to 950°C (as determined by an optical pyrometer) to produce a stable ion beam. At temperatures above 950°C the ion beams appeared to fractionate and at less than 950°C the signal intensity was usually too small to obtain precise isotope ratios. A block of five ratio measurements was taken when there was no change in isotope ratio.

NBS 951 and seawater standards were run multiple times and the ratios obtained are 4.002 and 4.159

4 RESULTS AND DISCUSSION

Boron concentrations and $\delta^{11}B$ values are shown in Table 1.

4.1 *Ngawha Geothermal Field*

The B isotope data in Table 1 falls into 2 categories; those with a $\delta^{11}B$ between -5 and +1‰ and those with a $\delta^{11}B$ > 10‰. Samples from Ngawha geothermal system, show high B concentrations and B isotope ratios between 0.6 and -4.4‰. Silica geothermometry indicates that the fluids equilibrated with quartz at between 230 and 150°C. Both the Cl/B and the $\delta^{11}B$ ratios indicate that no seawater has been involved in these systems and that the B must be derived from either the host rock or from a gaseous

Table 1. Boron composition of groundwaters and geothermal wtares from North Island, New Zealand. * Jubliee Pool lies within the Ngawha geothermal field.

Location	temp (°C)	SiO_2 (ppm)	B (ppm)	Cl / B (wt/wt)	$\delta^{11}B$
Ngawha well 4	>100	500	1100	1.5	-3.3
Ngawha well 12	>100	440	980	1.5	+0.6
Ngawha well 13	>100	410	1110	1.5	-0.9
Jubilee pool *	~43	153	880	1.4	-4.4
Waipiropiro	40	25	62	29	+4.6
Te Puia	62	48	78	113	-4.1
Awakeri	52	58	2.5	18.4	+21.2
Otopehetehe	22	10	136	5	+10.6
Rotokautuku	16	19	14	183	+43.6
Bartholomew	7	ND	05	3680	+18.9
Petone	7		0.047	510	+11.9
Seawater			4.6	4130	+39.5

(juvenile) source or both. Geothermal fluids from Ngawha are characterised by their high CO_2 and Hg contents (Mongillo, 1985), which support the influence of a juvenile source into the geothermal system.

Exceedingly high B concentrations at Ngawha are likely to be derived from water rock reactions primarily with some additional B from the magmatic source. To account for such high B concentrations, water/rock ratios must be low.

The B isotope ratio of Ngawha can be deduced to reflect the B isotope ratio of the rock and the juvenile source (assuming that there is fractionation of B during water-rock reaction). $\delta^{11}B$ of Ngawha show similar ratios to those from other non-marine subaerially hosted geothermal systems; Iceland -6.7 to -1.5‰ (Aggarwal et al 1992), Yellowstone (USA) -9.3 to -2.2‰ (Palmer and Sturchio 1990). These 3 geothermal systems show similar $\delta^{11}B$, despite existing in different geologic environments. This data together with that of Barth (1993) who has summarised B isotope ratios from a suite of different rock types, show that crustal rocks can not be differentiated based on their B isotopic composition.

4.2   East Coast hydrothermal systems

Of the systems sampled on the East Coast and Waipiropiro, a relationship can be seen between the temperature of the hydrothermal systems and the B isotope ratio. In general, where the fluids are colder, the B isotope ratios show heavier isotopic signatures, as a result of the adsorption of the $^{10}B(OH)_4^-$ onto secondary minerals. Consequently the resultant fluid becomes heavier. Study by Aggarwal et al (in prep) from the Eastern Lowlands in Iceland has shown that at temperatures below ~40°C, adsorption takes place. It is therefore likely that the waters from Otopotehetehe do not show B isotope ratios reflecting their source, but show isotope ratios that are somewhat heavier. Otopotehetehe's original source is likely to have a B isotopic composition similar to that at Te Puia.

Waipiropiro in the centre of North Island, and Te Puia on the east coast show similar B isotope ratios to that of Ngawha, reflecting the similar crustal influence into the geothermal system. Fluids from the dewatering subducting slab are likely to influence the springs on the east coast and give rise to heavier B isotope ratios (You et al 1993).

Rotokautuku shows a particularly high $\delta^{11}B$ which is similar to that of seawater, however, Cl/B ratios indicate that there is no/little marine influence. It is possible that B may be released from the subducting slab and carried through the accretionary prism to subaerial springs on the East Coast. You et al (1993) shows similar $\delta^{11}B$ for fluids from an accretionary prism in the Japan Sea to that at Rotokautuku. In both cases it is likely that immediately after desorption the source fluids had a lower B isotope ratio, but that the ratio has been modified by the later adsorption of the lighter isotope at a lower temperature.

4.3 Awakeri

The B isotope ratio at Awakeri is difficult to interpret as it shows a relatively high $\delta^{11}B$ and a low Cl/B ratio. It can not be explained on the grounds of low temperature adsorption, nor the influence of seawater. Further data are required to interpret the B isotope signature from Awakeri.

4.4 Ground waters

Ground waters from Petone and Bartholomew show relatively high B isotope ratios which are consistent with the adsorption model. The high Cl/B ratio of Bartholomew and the relatively high $\delta^{11}B$ imply the mixing of seawater and a low $\delta^{11}B$ freshwater endmember. Using the mixing model proposed by Aggarwal et al (in prep), this suggests that the two endmembers would be seawater and a water with a high B concentration and low $\delta^{11}B$. There is no evidence to suggest that this water exists, indicating that the mixing model cannot be applied to the Bartholomew ground water. The Petone ground water is derived from the Petone aquifer, which is recharged from the Hutt River. As with the Bartholomew water, this water can not be explained using the mixing model but may be explained in part by the effects of isotopic fractionation due to adsorption onto secondary minerals and the effects of anthropogenic discharges into the aquifers. More samples and data are required to determine the exact cause of the variation between the Petone and Bartholomew ground waters.

5 CONCLUSIONS

B isotope ratios from Ngawha, Te Puia and Waipiropiro are typical of any subaerial, meteoric water dominated geothermal system. Otopotehetehe on the east coast shows signatures indicative of a source of B from the dewatering subducting Pacific slab. Other low temperature geothermal systems are difficult to interpret due to the fractionation of B onto secondary minerals. Ground waters appear to show variable B isotope ratios which may reflect the influence of anthropogenic B into the water.

REFERENCES

Aggarwal, J. K., Palmer, M. R., Giggenbach, W. F., Arnorsson, S., Gunlaugsson, E., Ragnarsdottir, K. V., & Kristmannsdottir, H. (in prep). The Boron Isotope Systematics of Icelandic Hydrothermal Systems.

Aggarwal, J. K., Palmer, M. R., & Ragnarsdottir, K. V., 1992. Boron isotopic composition of Icelandic hydrothermal systems. In proc.WRI-7, Park City, Utah. Eds: Kharaka, Y.F., Maest, K.A. 893 - 895.

Barth, S., 1993. Boron isotope variations in nature: a synthesis. Geol Rundsch, 82, 640-651.

Mongillo, M. A., 1985. The Ngawha Geothermal field; an introduction and summary. In: The Ngawha Geothermal field; new and updated scientific investigations. Ed. M. A. Mongillo, DSIR report, Wellington, New Zealand.

Palmer M.R, Sturchio N.C. 1990. The boron isotope systematics of the Yellowstone National Park (Wyoming) hydrothermal system: A reconnaissance. Geochim. Cosmochim. 54: 2811-2815.

Oi, T., Ikeda, K., Nakano, M., Ossako, T., & Ossako, J., 1996. Boron isotope geochemistry of hot spring waters in Ibusuki and adjacent areas, Kagoshima, Japan. Geochem. J. 30, 273-287.

Vengosh, A., Starinsky, A., Kolodny, Y., & Chivas, A. R. 1994. Boron isotope geochemistry of thermal springs from the northern Rift Valley, Israel. J. Hydrol., 162, 155-169.

You, C.-F., Spivack, A., Smith, J. M. and Gieskes, J., 1993. Mobilization of boron in convergent margins: implications for the boron geochemical cycle. Geology, 21, 207-210.

Geochemistry of natural waters in Skagafjördur, N-Iceland: I.Chemistry

A.Andrésdóttir, S.Arnórsson & Á.E.Sveinbjörnsdóttir
Science Institute, University of Iceland, Reykjavík, Iceland

ABSTRACT: Geothermal activity is widespread in the valley of Skagafjördur, N-Iceland. Temperatures range from ambient to about 90°C. The geothermal waters are very low in dissolved solids (80-330 ppm), yet higher than surface- and soil waters. The pH of waters at 10-25°C issuing from the bedrock is about 10 but it declines a little with rising temperature above about 25°C. The waters appear to have closely approached equilibrium with secondary minerals at ≥30°C for all major components except Cl and SO_4.

1 INTRODUCTION

Warm and hot springs are widespread in the valley of Skagafjördur in N-Iceland (Fig. 1). Temperatures range from the mean annual temperature (5°C) to about 90°C. The bedrock in the area is for the most part a monotonous pile of Miocene flood basalts. The bedrock appears to have very low permeability except where it has been fractured by recent earth movements. There is a continuum from non-thermal groundwaters to thermal waters permitting evaluation of how early water-rock interaction changes the chemistry of surface waters as they seep into the ground and gain heat. On the basis of their composition and geology the waters in the area have been divided into three groups: (1) surface waters, (2) waters in the soil that is high in organic matter and (3) waters in the bedrock, i.e. groundwater, both thermal and non-thermal. Kristmannsdóttir et al. (1984) and Karlsdóttir et al. (1991) have given detailed accounts of the geothermal manifestations in the area.

In this contribution preliminary results of ongoing studies of natural waters in Skagafjördur are presented.

2 WATER COMPOSITIONS

The surface waters in the Skagafjördur area contain 30-120 ppm dissolved solids but most values lie between 40 and 80 ppm. Soil waters are higher in dissolved solids, largely due to higher Ca and HCO_3 contents. In both these water types the main anions are Cl and HCO_3 and the main cation Na, but also Ca in the case of soil waters. The dissolved solids content of the geothermal waters is higher than that of all non-thermal waters, 80-330 ppm, and it increases rather smoothly with rising temperature, or by about

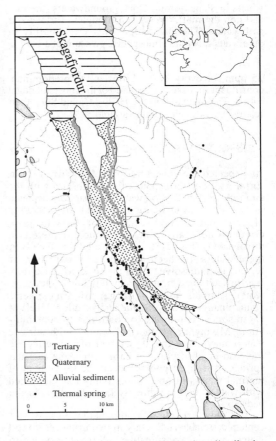

Fig. 1. The Skagafjördur area showing distribution of thermal springs (dots).

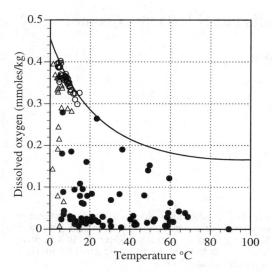

Fig. 2. Dissolved O_2 versus tempererature in natural waters in Skagafjördur. Dots: groundwaters (thermal and non-thermal), circles and triangles: surface and soil waters. The curve represents O_2 solubility in pure water at a total atmospheric pressure of 1 bar.

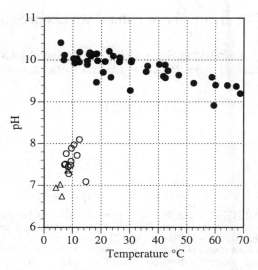

Fig. 3. pH of natural waters in Skagafjördur. Symbols have the same notation as in Fig. 2.

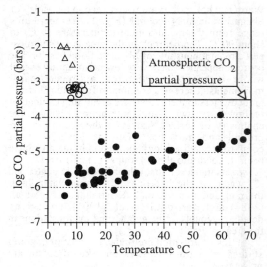

Fig. 4. CO_2 partial pressures in natural waters in Skagafjördur. Symbols have the same notation as in Fig. 2.

25 ppm for every 10°C. Silica is the predominant constituent in these waters and Na the most abundant cation. The main anions are HCO_3, Cl and SO_4, their relative masses being quite variable.

3 DISSOLVED O_2 AND H_2S

As expected the concentrations of dissolved O_2 in surface waters corresponds to equilibrium with the atmosphere (Fig. 2). Soil waters are variably depleted in O_2, the reason being its uptake by decaying organic matter and oxidation of Fe^{+2} to Fe^{+3}. Groundwaters are generally much depleted in dissolved O_2 (Fig. 2). It is considered that the O_2 in these waters has been consumed by oxidation of ferrous iron, dissolved from the rock, to ferric iron. It is difficult to determine accurately low O_2 levels in water samples due to danger of atmospheric contamination. It may be that analysed low levels of O_2 in some samples are too high due to contamination. Mixing of geothermal waters with surface waters may cause the O_2 content of the former to increase or absorption in cases when the thermal springs form open pools. Indeed the O_2 content of thermal waters in open pools is high as compared with water emerging directly from fractures in the bedrock.

With few exceptions waters below 30°C do not contain detectable H_2S. Only in two instances do soil waters contain measureable H_2S, doubtless due to strongly reducing conditions created by decay of organic matter.

4 pH AND CO_2 PARTIAL PRESSURES

The pH of waters in rivers and streams is most often in the range of 7-8. Non-thermal waters emerging in organic soil tend to have lower pH (Fig. 3). On the other hand, waters issuing from the bedrock, both thermal and non-thermal, have considerably higher pH. It is around 10 in waters below 25°C but it decreases somewhat with temperature above about 25°C (Fig. 3).

In waters emerging in organic soil the CO_2 partial pressure is considerably higher than the atmospheric one (Fig. 4). The cause is little doubt supply to the water of CO_2 from decaying organic matter. Waters in rivers and streams have CO_2 partial pressures slightly higher than that of the atmosphere whereas they are lower in groundwaters (Fig. 4). The slightly elevated CO_2 partial pressures in surface waters are accounted for by biological processes. The low CO_2 partial pressure of the groundwaters is to be correlated with their high pH and limited supply to the water of CO_2 once it has been isolated from the atmosphere.

The pH of surface waters is considered to be controlled by steady state conditions involving equal rate of two counteracting reactions, proton consumption by rock dissolution and proton generation by CO_2 uptake from the atmosphere (Gíslason and Eugster, 1987). Soil waters have lower pH due to additional supply of CO_2 from decaying organic matter. More than one process affects the pH of the groundwaters. They include dissolution of minerals from the rock that act like strong bases, dissolution of silica from the rock that produces protons through ionization at the high pH of these waters and possible precipitation of OH-bearing minerals such as smectite. The first process tends to raise the water pH whereas the latter two tend to lower it. Natural waters are undersaturated with most minerals when initially seeping into the ground so during this stage rock dissolution predominates causing the pH to increase. After a certain amount of dissolution the water pH has become sufficiently high for significant silica ionization and for making the water saturated with some secondary minerals. It is to be expected that the rate of secondary mineral precipitation will increase with temperature. It seems likely that such precipitation, involving OH-bearing minerals, is responsible for the progressive decrease in water pH above 25°C.

5 MINERAL SATURATION AND CATION-PROTON RATIOS

Surface and soil waters are strongly calcite undersaturated. On the other hand, waters issuing from the bedrock, both thermal and non-thermal are close to saturation. About half of the groundwaters are close to chalcedony saturation, the other half being supersaturated, as are all soil waters and some of the surface waters. Other surface waters are close to chalcedony saturation. The chalcedony supersaturation at the temperature of thermal spring discharges may, at least in some cases, be caused by cooling of the water in the upflow. All waters are strongly anhydrite undersaturated.

Most of the river waters are undersaturated with respect to both low-albite and microcline (the stable K-feldspar at low temperatures). Other waters, when below about 30°C, are close to saturation or somewhat supersaturated. Geothermal waters above

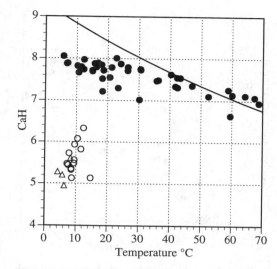

Fig. 5. Calcium/proton activity ratios in natural waters from Skagafjördur. CaH symbolizes the log of the activity of $\sqrt{Ca^{+2}}$ over the activity of H^+. The curve indicates equilibrium ratios for Icelandic geothermal waters according to Arnórsson et al. (1983). Symbols have the same notation as in Fig. 2.

30°C are close to equilibrium with both low-albite and microcline.

The interaction between water and the dark minerals of basalt (olivine, pyroxene and ore) is appropriately described as a titration process where the aqueous solution plays the role of the acid and the minerals the role of the base. Mineral dissolution releases cations to the solution consuming protons in the process. As a result such dissolution tends to raise aqueous cation/proton ratios. It has been shown that cation/proton ratios tend to attain a particular value at equilibrium at a given temperature (Arnórsson et al., 1995). In Skagafjördur surface and soil waters have similar cation/proton ratios and they are much lower than those of the groundwaters. This is exemplified for $\sqrt{Ca^+}/H^+$ ratios in Fig. 5. The pattern for other major cations (Na^+, K^+ and Mg^{+2}) is very similar to that for Ca^{+2}. The geothermal waters, when above 30-40°C have ratios very similar to equilibrium values (Fig. 5). At lower temperatures the ratios are lower. It has not yet been explored whether these waters have not equilibrated or if the "equilibrium" curve in Fig. 5, which is based on drillhole data in the range 50-250°C, is valid at these low temperatures.

6 CHLORINE AND BORON

Arnórsson and Andrésdóttir (1995) have demonstrated that Cl and B act as mobile components in the water-rock environment in Iceland and how these

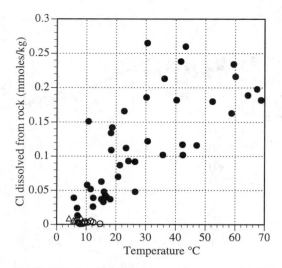

Fig. 6. Relationship between temperature and rock derived Cl in natural waters from Skagafjördur. Symbols have the same signature as in Fig. 2.

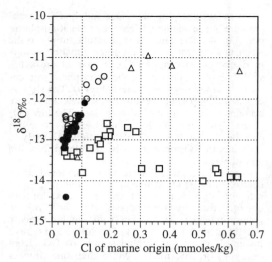

Fig. 7. Relationship bewteen $\delta^{18}O$ and Cl of marine origin. Circles and triangles denote surface and soil waters, respectively. Dots and squares represent groundwaters on the east and west side of the valley of Skagafjördur, respectively.

elements can be used to trace the origin of the water.

Fig. 6 shows how the calculated amount of Cl dissolved from the bedrock in the Skagafjördur waters increases with temperature. The rock derived concentrations of B also increase with the temperature of the water. The pattern for marine derived Cl is very different. Waters in the western part of the area contain the highest marine derived Cl concentrations and in some instances they are significantly higher than the Cl concentrations in local precipitation. When plotting marine derived Cl against $\delta^{18}O$ it is seen that the data points fall into two populations (Fig. 7). One corresponds with increasing Cl with increasing $\delta^{18}O$ and can be accounted for by altitude effect. The other population shows a negative correlation between these two parameters (squares in Fig. 7). This correlation can be explained by assuming the geothermal water to consist of two components, one corresponding with today's local precipitation and the other with water relatively high in Cl of marine origin but depleted in $\delta^{18}O$. Following Arnórsson et al. (1993) it is considered that this latter component represents a mixture and seawater and "iceage" water that sept into the bedrock at the end of the last glaciation, when the area was transgressed by the ocean, or even earlier.

REFERENCES

Arnórsson, S. & A. Andrésdóttir 1995. Processes controlling the distribution of boron and chlorine in natural waters in Iceland. *Geochim. Cosmochim. Acta* 59: 4125-4146.

Arnórsson, S., E. Gunnlaugsson & H. Svavarsson 1983. The chemistry of geothermal waters in Iceland. II. Mineral equilibria and independent variabls controlling water compositions. *Geochim. Cosmochim. Acta* 47: 547-566.

Arnórsson, S., Á.E. Sveinbjörnsdóttir & A. Andrésdóttir 1993. The distribution of Cl, B, δD and $\delta^{18}O$ in natural waters in the Southern Lowlands of Iceland. *Geofluids '93. Contribution to an international conference on fluid evolution, migration and interaction in rocks,* Torquay, England, 313-318.

Arnórsson, S., S.R. Gíslason & A. Andrésdóttir 1995. Processes controlling the pH of geothermal waters. *Proc. World Geothermal Congress, Florence, Italy* 2: 957-962.

Gíslason, S.R. & H.P. Eugster 1987a. Meteoric water-basalt interactions. II. A field study in nothern Iceland. *Geochim. Cosmochim. Acta* 51: 2841-2855.

Karlsdóttir, R., G.I. Haraldsson, A. Ingimarsdóttir, Á. Gudmundsson & Th.H. Hafstad 1991. Skagafjördur: Geology, geothermal activity, fresh water and exploration drillings. *Unpubl. report of the Natiional Energy Authority, OS-91047/JHD-08*, 96 p.

Kristmannsdóttir, H., M.J. Gunnarsdóttir, R. Karlsdóttir, G.I. Haraldsson & H. Jóhannesson 1984. Geothermal activity in Skagafjördur. *Unpubl. report of the National Energy Authority, OS-84050/JHD-09*, 107 p.

Organic gas in Öxarfjördur, NE Iceland

Halldór Ármannsson, Magnús Ólafsson & Gudmundur Ómar Fridleifsson
Orkustofnun, Reykjavík, Iceland

W.George Darling
British Geological Survey, Wallingford, UK

Troels Laier
Geological Survey of Denmark, Copenhagen, Denmark

ABSTRACT: Gas found in the Öxarfjörður geothermal area is mostly nitrogen but with a significant hydrocarbon contribution of which 3.6-17% are C_2-C_6 hydrocarbons. It is as yet unclear whether this is oil-associated gas or gas derived from thermal breakdown of lignite.

1 INTRODUCTION

Organic gases have been discovered in geothermal boreholes in Öxarfjörður, Northeast Iceland. Hitherto methane was the only hydrocarbon that had been confirmed in geothermal gases in Iceland and natural emissions, e.g from Lake Lagarfljót.

Three active northward trending fissure swarms, the Theistareykir, Krafla and Fremri námar fissure swarms, cross the Öxarfjörður region which is at the junction between the NE-SW Iceland spreading zone and the Tjörnes fracture zone, a right-lateral transform zone within which the thickest known sedimentary sequence in Iceland is found on- and off-shore, covering an area 140 km long (N-S) and 40 km wide (E-W), up to 4 km thick and possibly accumulating since the Miocene. The uppermost sediments in the Öxarfjörður area are due to a glacial river delta and formed during the last 10,000 years.

Geothermal surface manifestations are chiefly located within the fissure swarms. The highest surface temperature, 100°C is found in the Skógalón hot spring area (Figure 1). A 10 km² low-resistivity anomaly, underlain by high resistivity, as is common in Icelandic high-temperature areas, is located further to the south at Bakkahlaup Two smaller low-resistivity areas, one at Skógalón, have been observed. Thus an active high-temperature area extending 10 km² and elongated N15°W is located within the Krafla fissure swarm in Öxarfjörður.

The first analyses suggested that the gases might be oil-associated. This aroused considerable interest and it was decided to drill a deeper well that would penetrate the 700 m deep sediment formation known

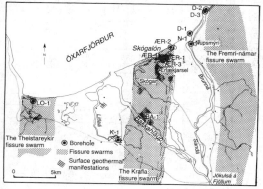

Figure 1. Fissure swarms, geothermal manifestations and boreholes in Öxarfjörður.

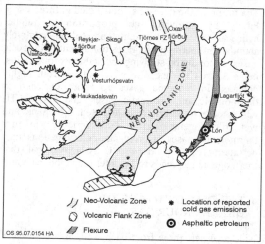

Figure 2. Volcanic zones of Iceland. Principal sites of cold methane emissions and asphaltic petroleum.

Table 1. Examples of gas composition of methane containing natural gas from different sources in Iceland. (Volume %).

Location	Type	CO_2	H_2S	H_2	CH_4	O_2+Ar	N_2
Krafla, well 7	h.t.	86.96	4.64	7.33	0.09	0.04	0.93
Lake Urriðavatn, well 7	l.t.	0.25			0.24	2.08	95.40
Lake Lagarfljót	c.e.	1.24			96.65	0.59	1.52

h.t: high-temperature geothermal area. l.t. low-temperature geothermal area. c.e. cold emission.

to exist in the area. Due to drilling problems and financial constraints it terminated at 450 m depth. The following items were to be studied:
  a) Analyses of the type, quantity and maturation of the organic remains.
  b) Gas analyses
  c) Sedimentary analyses

In this communication the emphasis will be on the results of the gas analyses which will be put in context with the results for the organic remains and the sediment analyses as well as compared with hydrocarbon gases from other locations. The results were described in detail by Ólafsson et al. (1993).

2 HYDROCARBON GASES IN ICELAND

Methane is found in high temperature well and fumarole fluids and in some low-temperature well waters. Cold emissions are known too (Table 1). The locations of the principal sites of cold emissions are shown in Figure 2.

3 CRITERIA FOR CLASSIFICATION OF HYDROCARBONS

The origin of the gases may be studied in three ways. Firstly by studying the proportion of higher hydrocarbons (C_{2+}) which tend to be relatively abundant in thermogenic gases but absent in biogenic gases. Secondly by studying the stable isotope ratios, i.e. $\delta^{13}C$ in CH_4 and C_{2+} which range from highly depleted in biogenic gases to more enriched in thermogenic and magmatic gases, and δD in CH_4 which tends to be relatively depleted in biogenic and wet thermogenic gases but enriched in dry thermogenic gases. Thirdly age-determinations may be attempted, e.g. using ^{14}C, which reveal that thermogenic gas tends to be older than biogenic gas.

4 THE COMPOSITON OF THE ÖXARFJÖRÐUR GAS

Nine analyses, three from each well, are reported in Table 2. If the hydrocarbons only are considered C_{2+} varies from 3.6 to 17% clearly suggesting a thermogenic, probably wet gas (Figure 3). There is a

considerable variation in the $\delta^{13}C_{CH4}$ values. None of the results are depleted enough for a biogenic gas but they can be divided into two groups, one with values close to -40‰ (PDB) and another with values ranging from -22.5 to -31.9‰. The two lower values of the first group were observed in gas from the deepest well, ÆR-04 and the highest value was observed in gas from the shallowest well, ÆR-01.

The δD values are somewhat ambiguous but suggest that there may be two types of gas, one shallow with δD_{CH4} values of the order of -150 ‰(SMOW) and a deeper one with a δD_{CH4} value below -200 ‰(SMOW) suggesting a relationship with the $\delta^{13}C_{CH4}$ results. The values for $\delta^{13}C_{C2H6}$, $\delta^{13}C_{C3H8}$ and $\delta^{13}C_{CO2}$ were found to be -26.9, -25.4 and -9.3 ‰ (PDB) respecively in sample No. 94-200 from well ÆR-04. The values for ethane and propane are typical for thermogenic gases. The value for CO_2 suggests a geothermal gas. The variation in $\delta^{13}C_{CH4}$ suggests that there may be a mixture of geothermal and thermogenic methane present. The idea has been put forward that the thermogenic gas is a result of breakdown of lignite at a fairly high temperature.

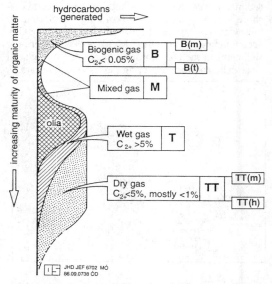

Figure 3. Classification of natural hydrocarbon gases (Schoell 1980).

Table 2. The chemical composition of gas from boreholes at Skógalón, Öxarfjörður (Volume %).

Well No	ÆR-04	ÆR-04	ÆR-04	ÆR-03	ÆR-03	ÆR-03	ÆR-01	ÆR-01	ÆR-01
Sample No.	94-200	91-209	91-189	90-237	89-087	88-149	88-088	88-211	87-119
Sampling mode	Sep.	Sep.	Direct	200 m depth	Direct	Direct	Direct	Direct	Direct
N_2	83.80	86.20	92.80	81.00	93.40	95.90	92.80	93.80	94.60
O_2+Ar	4.26	0.10	0.10	18.40	1.10	1.80	1.20	1.70	2.70
H_2		0.07	0.25	0.06	0.02	0.02	0.07	0.23	0.00
He			0.00	0.00	0.00	<0.01	0.00	0.03	0.00
CO_2	3.51	7.88	0.91	0.05	0.02	<0.05	0.00	0.05	0.04
CH_4	4.33	5.70	5.80	0.45	5.20	3.60	5.60	4.00	2.22
C_2H_6	0.30	0.082	0.120	0.0015	0.230	0.150	0.300	0.200	0.218
C_3H_8	0.085	0.074	0.083	0.002	0.047	0.050	0.057	0.092	0.156
i-C_4H_{10}	0.0107	0.0088	0.011		0.0066	0.0088	0.0076	0.0164	0.0251
n-C_4H_{10}	0.0169	0.024	0.031	0.0034	0.0093	0.0110	0.0100	0.0208	0.0280
neo-C_5H_{12}					0.0002		0.0030		
i-C_5H_{12}	0.0021	0.0019	0.0052	0.0002	0.0016	0.0035	0.0027	0.0123	0.0147
n-C_5H_{12}	0.0031	0.0042	0.0170		0.0018	0.0058	0.0036	0.0152	0.0132
i-C_6H_{14}		<0.0001	0.0007	0.0002	trace		trace		
n-C_6H_{14}		0.0010	0.006						
C_6H_6	0.0080	0.0195	0.0044						
ΣC_{2+}	0.426	0.196	0.274	0.021	0.297	0.229	0.384	0.357	0.455
$\delta^{13}C_{CH4}$ ‰ (PDB)	-28.6	-38.9	-40.0	-26.6	-31.9	-29.6	-29.0	-29.0	-22.5
δD_{CH4} ‰ (SMOW)	-222				-138		-154		

Sep. Steam separated with separator. Direct: Direct from wellhead

Lignite has not been found in Öxarfjörður but a few layers are known to the west of the area, in Tjörnes (Figure 2). One hypothesis is that, due to faulting, sediments of the same origin as those at Tjörnes are preserved below the present drilling area. $\delta^{13}C$ has been determined in three samples of lignite from the Tjörnes sediments (Table 3). The only oil that has been found in Iceland is an asphaltic petroleum found in Skyndidalur in Lón, Southeast Iceland (Figure 2). It was interpreted to be formed by thermal breakdown of lignite (Jakobsson and Friðleifsson 1989). The $\delta^{13}C$ of the lignites from the two areas, one on each side of the neo-volcanic zone (Figure 2) and that of the petroleum are strikingly similar (Table 3). Therefore it is likely that lignite in the area has a $\delta^{13}C$ value in the range -27 to -28 ‰ (PDB) and by comparison with experiments reported by DesMarais et al. (1988) most values obtained for gas from the wells studied could be due to gas produced by such

Table 3. $\delta^{13}C$ of lignites from Tjörnes and Lón and of asphaltic petroleum from Lón.

Sample type and location	Lignite Tjörnes (mean)	Lignite Lón	Asphaltic petroleum Lón
$\delta^{13}C$ ‰ (PDB)	-27.8	-27.1	-27.7

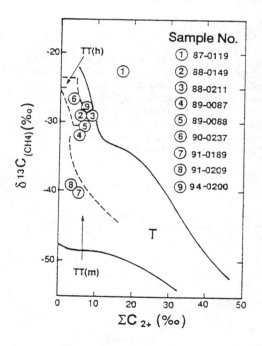

Figure 4. $\delta^{13}C_{CH4}$ vs. ΣC_{2+} in different methanes. See Figure 3 for explanations.

pyrolysis. The depleted values of around -40 ‰ for early samples from ÆR-04 and the enriched value of 22.5‰ obtained for a sample from well ÆR-01 suggest different origins. The $\delta^{13}C_{C2H6}$ and $\delta^{13}C_{C3H8}$ values obtained for gas from well ÆR-04 in 1994 are, however, quite compatible with an asphaltic origin. The $\delta^{13}C_{CH4}$ of geothermal gas in Iceland has been found to vary from -17.8 to -40.4 ‰ (Ármannsson et al. 1989, Sano et al.1985, S. Arnórsson, pers. comm.). Ármannsson et al. (1989) argued that in Krafla, which is on the same fissure swarm as the study area, methane with relatively low $\delta^{13}C_{CH4}$ is derived from the decomposition of organic matter whereas that with a higher $\delta^{13}C_{CH4}$ could be derived from basalt at magmatic conditions.

No alkenes have been detected so the gas is probably not an inorganic gas produced by reaction at high temperature between CO_2 and H_2 (Gunter and Musgrave 1971). The relationships $\delta^{13}C_{CH4}$ vs. ΣC_{2+} (Figure 4) and δD_{CH4} vs. $\delta^{13}C_{CH4}$ (Figure 5) suggest a thermogenic gas bordering on dry and wet gas.

Significant amounts of organic carbon were not detected in the sediments above 450 m. At least the uppermost 350 m are less than 10,000 year old. ^{14}C dating of the gas suggested that it was more than 20,000 year old. It is therefore suggested that the gas originates in deep sediments hitherto not drilled into.

5 CONCLUSIONS

Organic gases were encountered during drilling into sediments in NE Iceland. They are older than the upper sediments that are deltaic and less than 10,000 year old. The composition of the gases is consistent with an origin in breakdown of lignite although oil-associated gases from relatively old sediments at depth are not ruled out. Deep drilling in the area should resolve that question.

6 REFERENCES

Ármannsson, H., Benjamínsson, J. and Jeffrey, A.W.A. 1989. Gas changes in the Krafla geothermal system, Iceland. *Chem. Geol.* 76: 175-196.

Des Marais, D.J., Stallard, M.L., Nehring, N.L. and Truesdell, A.H. 1988. *Chem. Geol.* 71: 159-167.

Jakobsson, S.P. and Friðleifsson, G.Ó. 1989. Asphalt in amygdales in Skyndidalur, Lón (In Icelandic with English abstract). *Náttúrufræðingurinn* 59: 169-188.

Gunter, B.D. and Musgrave, B.C. 1971. New evidence on the origin of methane in hydrothermal gases. *Geochim. cosmochim. Acta* 35: 113-118.

Ólafsson, M., Friðleifsson, G. Ó., Eiríksson, J., Sigvaldason, H. and Ármannsson, H.1993. *On the origin of organic gas in Öxarfjörður, NE Iceland.* Orkustofnun report OS - 93015/ JHD-05, Reykjavík: 76 pp.

Sano, Y., Urabe, A., Wakita, H., Chiba, H. and Sakai, H. 1985. Chemical and isotopic compositions of gases in geothermal fluids in Iceland. *Geochem. J.* 19: 135-148.

Schoell, M. 1980: The hydrogen and carbon isotopic composition of methane from natural gases of various origins. *Geochim. cosmochim. Acta* 44: 649-661.

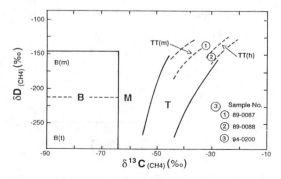

Figure 5. δD_{CH4} vs. $\delta^{13}C_{CH4}$ in different methanes. See Figure 3 for explanations.

Gas chemistry of the Krafla Geothermal Field, Iceland

S.Arnórsson, Th.Fridriksson & I.Gunnarsson
Science Institute, University of Iceland, Reykjavík, Iceland

ABSTRACT: Aqueous CO_2, H_2S and H_2 concentrations in Krafla aquifer fluids are controlled by temperature dependent mineral equilibria and, in the case of CO_2, also by supply from the magma heat source. The upper Krafla system is sub-boiling. In the lower system steam fractions are 0-1% by wt. Excess enthalpy of wells is caused by both phase separation and heat extraction from the rock. N_2 and Ar are lower in the reservoir fluid at present than in air saturated water due to progressive boiling of that fluid and limited recharge into the reservoir.

1 INTRODUCTION

The Krafla geothermal field is located within a central volcanic complex in the active zone of rifting and volcanism in NE-Iceland. In the period 1975-1984 rising basaltic magma developed chambers in the roots of the geothermal system. Degassing of the magma had a profound effect on the gas content of the geothermal fluid.

In all 29 wells have been drilled at Krafla (1138-2222 m deep). Maximum temperatures reach 350°C. Some of the wells discharge almost dry steam. Others have liquid enthalpy. The Krafla reservoir has been divided into an upper system of about 210°C and a lower system which is two phase (Ármannsson et al., 1987). Permeability is generally poor resulting in low average steam yield of wells. The reservoir fluid is low in dissolved solids (1000-1500 ppm). Gas concentrations far exceed those of the solids in the hottest wells.

This contribution focuses on the gas chemistry of the geothermal fluids at Krafla, both well discharges and fumarole steam.

2 GAS CHEMISTRY

Shortly after the eruptive episode, which started in the Krafla central volcano at the end of 1975, the CO_2 content of the only producing well in the area at that time rose about 100 times. Later a change in the gas content of the fumarole steam was observed. This was attributed to the degassing of new magma in the roots of the geothermal system. Many of the wells drilled in the following years were initially high in gas. Since they have declined, but to variable extent. CO_2 is the dominant gas, its concentrations in steam from wells at atmospheric pressure (samples from 1996) range from 15 to 670 mmoles/kg (Table 1). In

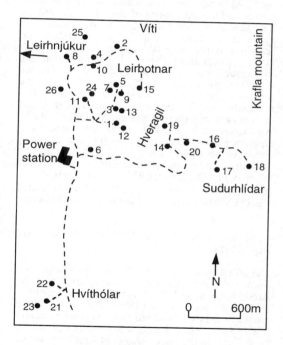

Fig. 1 Map showing location of wells at Krafla.

fumaroles the corresponding numbers are 191 to 1808 mmoles/kg. Concentration variations in fumaroles and well discharges are much higher among the reactive gases (CO_2, H_2S, H_2 and CH_4), particularly H_2, than the unreactive ones (N_2 and Ar). N_2/Ar ratios range from 44 to about 70. They tend to be lower in fumaroles than in well discharges, frequently quite similar to that of air saturated water.

3 STEAM GEOTHERMOMETERS

Temperature equations have been derived for aqueous CO_2, H_2S and H_2 concentrations. Above 230°C they are based on the assumption of equilibrium between the respective gases and specific mineral buffers. The gas-mineral buffer reactions are:

CO_2: 2clinozoisite + 2calcite + 3quartz + $2H_2O$ =
3prehnite + $2CO_{2,aq}$

H_2S: pyrite + pyrrhotite + 2prehnite + $2H_2O$ =
2epidote + $3H_2S_{aq}$

H_2: 4pyrrhotite + 2prehnite + $2H_2O$ =
2epidote + 2pyrite + $3H_{2,aq}$

All these minerals have been identified in drill chips from Krafla. Epidote is not stable below 230°C. Below this temperature the gas-temperature equations are based on relation between measured aquifer temperature and the gas content of discharges of wells in Iceland with the first level of boiling in the well.

The thermodynamic data on the minerals in the reactions above were taken from Helgeson et al. (1978). It has been criticized that their data are systematically in error with respect to all Al-silicates (Hemingway et al., 1982). However, for the reactions above such an error is cancelled. The thermodynamic data on the dissolved gases are based on a review of experimental gas solubility data by Arnórsson et al. (1996). Equations describing aqueous gas concentrations as a function of temperature are shown in Table 2.

Temperature equations have also been derived for the ratios of CO_2, H_2S and H_2 to Ar assuming the Ar concentrations to be equal to that of air saturated water at 5°C (average annual temperature in Iceland).

4 INITIAL STEAM FRACTION IN PRODUCING AQUIFERS

The initial steam fraction in producing aquifers has been estimated by the method described by Arnórsson et al. (1990) using the gas concentration-temperature equations in Table 2. The steam fraction calculates to be 0-1 % of the reservoir fluid by weight (Table 3). At 300°C 1 weight % corresponds to 13% by volume. In the upper system at Krafla the initial steam fraction calculates to be close to zero. This conforms with the fact that the upper system is sub-boiling. In the lower system steam fractions lie in the range 0.16 to 0.95% by weight for individual wells. They are highest in the Sudurhlídar wellfield but lowest in the Hvítholaklif wellfield.

The gas content of the Krafla wells indicates that their "excess" enthalpy is partly caused by separation of flowing water and steam in the producing aquifers, the steam flows into the wells but the water remains partly immobile. Partly the "excess" enthalpy is produced by heat extracted from the rock in the zone of depressurization around discharging wells (X_e in Table 3). Such heat extraction enhances evaporation of the water in this zone.

TABLE 1. The gas content of steam at 1 bar from Krafla wells. Concentrations in mmoles/kg.

well	CO_2	H_2S	H_2	CH_4	N_2	Ar
5	15.6	2.21	0.22	0.052	1.662	0.026
5	14.8	2.46	0.20	0.044	0.700	0.014
9	19.9	2.20	0.24	0.030	0.629	0.013
11	90.1	9.74	2.76	0.050	0.877	0.018
12	213.1	18.22	6.60	0.090	0.712	0.014
13	137.6	16.12	18.20	0.038	8.048	0.101
14	246.1	23.87	17.91	0.058	6.890	0.087
15	186.3	25.86	11.34	0.013	1.834	0.033
17	121.0	18.94	12.54	0.077	0.810	0.018
19	416.0	26.91	19.20	0.054	6.712	0.085
20	670.4	31.82	15.95	0.146	1.128	0.023
20	667.2	28.02	15.30	0.131	11.768	0.149
21	87.1	14.41	7.75	0.211	4.158	0.060
24	14.8	2.86	0.09	0.065	1.153	0.026
24	21.9	1.03	0.12	0.073	8.462	0.114
25	28.2	3.49	0.16	0.152	2.674	0.056
26	240.5	27.15	9.31	0.134	1.219	0.023

5 GEOTHERMOMETRY RESULTS

Results of water (quartz and Na-K) and steam geothermometers according to the equations in Table 2 are similar for wells (5, 9, 24 and 25) producing from the upper system at Krafla (Table 4). For some of the wells producing from the lower system the water and the H_2S geothermometer are also similar, but the CO_2 and H_2 geothermometers indicate higher aquifer temperatures. For other wells producing from the lower system the water geothermometers indicate lower aquifer temperatures than the steam geothermometers. Discrepancy between the steam geothermometers is considered to be due to the presence of reservoir steam in the aquifer. For CO_2 non-equilibrium conditions may also contribute. They are caused by rapid inflow of this gas into the geothermal system from the underlying magma body.

Temperature logging during the heating-up of wells and subsequent stabilization has permitted estimation of the temperature in the producing aquifers. For four of the Krafla wells with aquifers in the lower system the steam geothermometers indicate temperatures similar to those inferred from temperature logging. On the other hand, the water geothermometers indicate lower temperatures. The reason for this dicrepancy is considered to be that these wells extract fluid from two principal horizons, largely water from a shallow one (upper system) and largely steam from a deep one (lower system). The water geothermometers reflect approximately the temperature of the shallow aquifer and the steam geothermometers that of the deep aquifer.

TABLE 2. Temperature equations for steam geothermometers

Gas - gas ratio   Temperature equation (T K)

RESERVOIR WATER
Gas	Equation	Valid range
CO_2	$\log CO_2 = k_o + 5.520 - 412.50/T + 0.0144*T - 5.029*\log T$	(>230°C)
	$\log CO_2 = 6.236 - 4606.63/T - 0.00537*T + 2.386*\log T$	(>100°C)
H_2S	$\log H_2S = k_o + 16.451 - 3635.08/T + 0.00839*T - 6.074*\log T$	(>150°C)
H_2	$\log H_2 = k_o + 17.266 - 4587.39/T + 0.00547*T - 5.356*\log T$	(>150°C)

The value of k_o is determined by the composition (activity) of the minerals with which the respective gas equilibrates. Here it is taken to be -0.523 for the CO_2 reaction and +0.104 for the H_2S and H_2 reactions. These correpond with clinozoisite and epidote activities of 0.3 and 0.7, respectively, but unit activity of all other minerals and water.

STEAM AT 100°C
Gas	Equation
CO_2	$4.724*Q^3 - 11.068*Q^2 + 72.012*Q + 121.8$ (valid for $a_{ep} = 0.3$, i.e. $k_o = -0.155$)[a]
	$10.103*Q^3 - 21.946*Q^2 + 88.299*Q + 129.2$ (valid for $a_{ep} = 0.7$, i.e. $k_o = -0.523$)[a]
H_2S	$4.811*Q^2 + 66.152*Q + 177.6$
H_2	$6.630*Q^3 + 5.836*Q^2 + 56.168*Q + 227.1$
CO_2/N_2	$1.739*Q^3 + 7.599*Q^2 + 48.751*Q + 173.2$
H_2S/Ar	$4.108*Q^2 + 42.265*Q + 137.6$
H_2/Ar	$0.640*Q^2 + 43.260*Q + 170.0$

Q = log mmoles/kg of gas or log of molal gas ratio. [a]The first equation may be more appropriate than the second one at Krafla, at least, in those cases where CO_2 discharge from magma into the geothermal system is high.

TABLE 3. Initial steam fraction in the aquifer fluid on individual Krafla wells.

well	Discharge enthalpy kg/kJ	Aquifer temp.°C	Inflow temp.°C	Sampling pressure bars abs.	Y_{HS}	X_e	V_f	V_r^l	X_c	$X_{es}\%$
5	859	212	170	3.8	-0.03	0.01	0.66	-0.33	0.12	10
5	859	212	170	3.8	-0.04	0.01	0.73	-0.26	0.12	4
9	1033	238	188	8.0	-0.06	0.09	0.32	-0.60	0.15	56
11	1496	261	188	5.6	0.08	0.20	1.01	0.20	0.40	49
12	1956	284	212	13.8	0.28	0.35	1.17	0.52	0.58	61
13	1704	239	212	13.2	0.66	0.14	4.48	3.62	0.45	30
14	2336	246	212	12.3	0.45	0.12	7.65	6.77	0.77	16
15	1899	288	212	13.0	0.47	0.27	1.38	0.65	0.55	48
17	2181	284	212	15.0	0.88	0.43	1.38	0.80	0.69	62
19	2676	248	212	12.6	0.44	0.10	9.41	8.51	0.95	11
20	2316	288	212	13.8	0.61	0.33	2.17	1.50	0.76	43
20	2316	288	212	14.8	0.71	0.38	1.91	1.29	0.76	50
21	2224	251	224	16.0	0.30	0.45	3.62	3.06	0.71	63
24	1014	200	170	4.0	-0.04	0.03	1.94	0.96	0.19	13
24	1014	200	170	4.5	-0.01	0.10	0.69	-0.20	0.18	57
25	935	220	170	3.3	-0.06	0.02	0.87	-0.11	0.17	10
26	1913	299	198	10.0	0.33	0.31	1.00	0.31	0.57	54

Y_{HS} Initial steam fraction (by weight) in reservoir fluid calculated by the method of Arnórsson et al. (1990).
X_e Mass fraction of steam in total well discharge formed by heat extraction from rock.
V_f Relative mass (to well discharge) of reservoir fluid which has boiled to form steam flowing into well.
V_r^l Relative mass (to well discharge) of boiled water retained in aquifer.
X_c Steam fraction in discharge at sampling pressure.
$X_{es}\%$ Percentage of steam in discharge that has formed by heat extraction from rock.

TABLE 4. Geothermometry results (°C) and calculated concentrations (mmoles/kg) of N_2 and Ar in the aquifer fluid of Krafla wells.

well	Qtz	Na-K	H_2S	CO_2	H_2	N_2	Ar	%[a]
5	218	207	217	215	204	0.31	0.005	74
5	218	207	217	213	200	0.12	0.002	88
9	238	237	231	237	215	0.30	0.006	68
11	259	263	271	283	275	0.35	0.007	63
12	277	290	300	330	315	0.35	0.007	64
13	240	238	282	292	344	0.81	0.010	47
14	243	249	287	307	331	0.70	0.009	54
15	290	285	305	314	331	0.73	0.013	30
17	273	273	291	325	369	0.41	0.009	53
19	229	268	288	327	330	0.67	0.009	55
20	279	296	309	377	342	0.39	0.008	57
20	279	296	308	383	346	4.66	0.059	b
21	250	252	293	291	324	0.81	0.012	38
24	213	186	222	211	180	0.11	0.003	86
24	213	186	204	237	198	2.25	0.030	b
25	220	220	227	229	193	0.51	0.011	44
26	297	301	310	330	325	0.69	0.013	31

[a]Percentage degassing of air saturated water with respect to N_2. [b]Sample air contaminated.

The H_2S and H_2S/Ar geothermometers give the most reliable subsurface temperature estimates from fumarole data. H_2S is quite soluble in water and its concentrations in the reservoir fluid are, therefore, less affected by the presence of reservoir steam than the concentrations of the less soluble CO_2 and H_2.

H_2S and H_2S/Ar temperatures are highest in the NE part of the field around Víti and along Hveragil to the south, in the range 280-310°C. At Leirhnúkur and Hvíthólaklif they are lower, 260-280°C.

6 NITROGEN AND ARGON

The annual average air temperature in the Krafla area is around 5°C. The concentration of N_2 and Ar in rain in equilibrium with the atmosphere at this temperature and 1 bar total pressure is close to 0.71 and 0.0193 mmoles/kg, respectively. The calculated concentrations of these gases in producing aquifers of Krafla wells is lower (Table 4). This is taken to indicate that the reservoir water, which boils by depressurization in producing aquifers, has already been partly degassed. It is accordingly envisaged that steam extraction from the reservoir involves progressive boiling of the reservoir water and limited recharge. The N_2/Ar ratios in the well discharges are somewhat higher than in air saturated water. This is attributed to dissolution of trace amounts of N_2 from the rock. Magmatic supply of N_2 is insignificant, if existing.

Condensation of steam beneath fumaroles has been estimated from N_2 and Ar data by a method described by Arnórsson et al. (1995). Condensation is highest in the northern part of the field, or >40%. It diminishes southwards and is insignificant at the southern end of Hveragil. Steam condensation in upflow zones is also limited at Leirhnúkur and Hvíthólaklif, <20%. The steam condensation is accompanied by simultaneous increase in N_2 and Ar concentrations and a decrease in N_2/Ar ratios. Increasing gas concentrations and decreasing N_2/Ar ratios indicate that the steam condenses in cooler water of meteoric origin and dissolves N_2 and Ar from it in the process.

7 CONCEPTUAL MODEL

The three dimensional temperature distribution in the Krafla reservoir, initial reservoir steam fractions and steam condensation in upflow zones have been combined to delineate a model for the Krafla geothermal system. Rising steam from the lower two phase system heats up a groundwater current from the north, thus forming the upper system. Where this steam flow is limited due to limited permeability in the deep system all the rising steam may condense. On the other hand high flow rate of rising steam, such as at Hveragil and Leirhnúkur, may heat the groundwater current to boiling and produce fumaroles at the surface.

REFERENCES

Ármannsson, H., Á. Gudmundsson & B.S. Steingrímsson 1987. Exploration and development of the Krafla geothermal area. *Jökull* 37: 13-30.

Arnórsson, S., S. Björnsson, Z.W. Muna & S.B. Ojiambo 1990. The use of gas chemistry to evaluate boiling processes and initial steam fractions in geothermal reservoirs with an example from the Olkaria field, Kenya. *Geothermics* 19: 497-514.

Arnórsson, S., K. Geirsson, Th. Fridriksson & I. Gunnarsson 1995. Krafla and Námafjall. Gas in wells and fumaroles. *National Power Company rept.*, 14p.

Arnórsson, S., K. Geirsson, A. Andrésdóttir & S. Sigurdsson 1996. Compilation and evaluation of thermodynamic data on aqueous species and dissociational equilibria in aqueous solution. I. The solubility of CO_2, H_2S, H_2, CH_4, N_2, O_2 and Ar in pure water. *Science Institute rept. RH-17-96*, 20p.

Helgeson, H.C., J.M. Delany, H.W. Nesbitt & D.K. Bird 1978. Summary and critique of the thermodynamic properties of rock-forming minerals. *Amer. J. Sci.* 278A: 1-229.

Hemingway, B.S., J.L. Haas & G.R. Robinson Jr. 1982. Thermodynamic properties of selected minerals in the system Al_2O_3-CaO-SiO_2-H_2O at 298.15 K and 1 bar (10^5 Pascals) pressure and at higher temperatures. *U. S. Geol. Surv. Bull. 1544*, 70p.

Precious metals in deep geothermal fluids at the Ohaaki geothermal field

K.L. Brown
Geology Department, University of Auckland, New Zealand

J.G. Webster
ESR Environmental, Auckland, New Zealand

ABSTRACT: A special downhole sampler has been built to retrieve deep geothermal reservoir fluids for precious metal analysis. During a shutdown period, samples were collected from two wells at the Ohaaki geothermal field. The fluids were analyzed by quantitative ICP-MS for gold, silver, copper and platinum group elements. In the deep geothermal reservoir, gold concentrations of 0.50 - 1.16 ppb and silver concentrations of 5.7 - 18.0 ppb were observed. These results agree with previous approximate estimates of reservoir gold and silver concentrations, derived from measurements made on geothermal fluids at the surface. They are, however, lower than expected from thermodynamic predictions of gold solubility as a gold-bisulphide complex.

1. INTRODUCTION

The link between epithermal ore deposits and geothermal systems has been well proven now for some time (e.g., Henley and Ellis, 1983; Brown, 1986). In particular, analysis of orifice plates from Br22 and Br27 showed the presence of significant gold and silver concentrations at the Ohaaki geothermal field (Brown, 1986), and from these observations, a deep reservoir gold concentration of 1.5 ppb was estimated.

Later, pilot plant extraction of gold and silver was attempted at the Kawerau and Rotokawa geothermal fields (Brown and Roberts, 1988). Gold recovery was of the order of 0.5 - 1 ppb which is very similar to that estimated (1.5 ppb) from deposition on to an orifice plate at the surface (Brown, 1986). Gold and silver are transported in geothermal fluids primarily as bisulphide complexes: $Au(HS)_2^-$ and $Ag(HS)_2^-$. Gold solubility, for example, is given by the reaction:

$$Au + H_2S + HS^- = Au(HS)_2^- + 0.5 H_2 \quad (1)$$

After pressure reduction, as occurs at an orifice plate for instance, the H_2S that was previously dissolved in the geothermal liquid becomes distributed primarily into the vapour phase and the gold is deposited.

Thermodynamic calculations for equation (1) have shown that the expected solubility of gold due to the formation of the bisulphide complex was about 10 ppb. Consequently, it appeared that the amount recovered at Kawerau and Rotokawa was an order of magnitude less than that theoretically available in the deep geothermal fluid. However, there was no way to tell whether the gold was undersaturated in the geothermal fluid at depth, or being deposited in the well casing or at points other than in the pilot plant vessel, or whether the thermodynamic calculations were in fact based on incorrect data. The order of magnitude difference between the recovery of gold in the pilot plant and the calculated amount in the deep geothermal reservoir has a large impact on the economic viability of any geothermal gold extraction process and also has ramifications for the genesis of epithermal ore deposits. An accurate determination of the gold concentration in the deep reservoir fluid would help identify the source of this apparent discrepancy.

The aim of this study was to measure, as accurately as possible, the gold concentration in the deep geothermal reservoir at Ohaaki geothermal field. Such measurement is fraught with difficulty. Normal downhole chemical samplers are constructed of stainless steel, which could contain significant trace impurities of gold and silver. Also, when the sampler is brought to the surface, the temperature reduction and loss of dissolved gases, forces gold and silver out of solution and it is deposited inside the sampler. To overcome this problem, the downhole sampler was redesigned specifically to measure precious metal concentrations. Although still based on the standard Klyen sampler, it was constructed in a number of different sections. The main body of the sampler was a mild steel tube, but the sections of this tube to be exposed to the deep geothermal fluids were coated with a ceramic lining. The valve assemblies at either end of the sampler were constructed of titanium, a surface which is more inert than mild steel. The normal stainless steel non-return valve assembly was replaced with a synthetic high temperature EDPM in order to contain and collect any gas from the sampler.

A sample was collected at depth in the normal manner, then the sampler was rinsed with aqua regia to redissolve gold deposited on to the walls of the sample chamber during cooling and gas loss. Both the original geothermal fluid and the aqua regia rinse were analysed for precious metals, and the total gold and silver concentration of the deep geothermal fluid calculated from these analyses.

2. EXPERIMENTAL METHOD

2.1 *Well Selection*

An investigation of the feed depths and chloride concentrations of the Ohaaki wells was carried out in order to determine appropriate wells for sampling. The strictest criterion was the availability of deep geothermal waters without dilution by surface derived waters. As well, a purely liquid feed was required in the well - both to give a good downhole sample, and to ensure that the deep fluid was not diluted with condensed steam. On this basis two wells were chosen: Br20 and Br9. Both of these wells are on the west bank of the Ohaaki field.

2.2 *Chemical Preparation*

Aqua regia (AR) was freshly prepared in the laboratory, and a preliminary washing of the sampler was carried out using AR and DDI (distilled and deionised) water. The first AR wash was discarded and then two AR washes (60ml each), each followed by a DDI water wash (60ml), were kept as an AR "sampler blank". All sample bottles were HDPE Nalgene and were initially prepared by washing in concentrated nitric acid. The bottled were then soaked overnight in 5N nitric acid, and then washed thoroughly with DDI water. Sample bottles to be used for the untreated downhole sampler contents had 20 ml of "Aristar" conc nitric acid added to them prior to use, to preserve the sample. For a "field blank", 625ml DDI water and 20 ml Aristar nitric acid were placed into a sample bottle, and the lid opened at the BR9 field site for ten minutes in a light breeze.

2.3 *Well Sampling & Analysis*

The WHP at BR9 was approximately 4.5Bg and the well had been shut for a few days. The main feed in this well is at 760m CHF, and all samples were collected at this depth. The ceramic sampler was assembled, a sample collected and the sampler allowed to cool. Gas samples were collected into evacuated glass flasks, then the water contents of the sampler were transferred to a Nalgene measuring cylinder that had previously been washed with AR and DDI water. Exactly 500ml of downhole water were recovered in this case (G6 "sampler"). However, it was noted that the viton seals on the upper and lower compartments had fractured on disassembling the sampler, and small pieces of these o-rings were present in this sample.

After the deep reservoir water sample had been recovered from the sampler, the sampler was dismantled to reveal the ceramic chamber. The chamber was washed as thoroughly as possible using two successive washed of 100ml of AR followed by 100ml of DDI water. The 400 ml of combined water and AR washings were collected (G6 "washings"). Two further unsuccessful attempts were made to sample Br9, before another sample was collected (sample G8).

The WHP at Br20 was 14.0Bg, and the well had liquid to the top of the well when sampled. The middle of the large feed zone at 965m CHF was chosen as the sampling site. Samples G9 and G10 were collected, using the same sampling regime as for G6 and G8, before the wells had to be placed back on production making further downhole sampling impossible. Webre separator samples for major ion analysis were collected from both Br9 and Br20, after the wells were flowing again.

The samples were analysed by quantitative ICP-MS for gold, silver, copper and platinum group elements (PGE), as well as semiquantitative ICP-MS analysis for a suite of other elements. Webre samples were analysed for major ions by AAS and wet chemistry, at the Wairakei laboratories of the Institute of Geological and Nuclear Sciences (GNS). The gas samples were analysed by wet chemistry and gas chromatography by GNS.

3. RESULTS AND DISCUSSION

3.1 *Platinum Group Elements and Copper*

All of the PGE concentrations were below the detection limits in all samples. Detection limits, allowing for all of the concentration factors, are:

ruthenium (0.52 - 0.65 ppb), rhodium (0.26 - 0.32ppb), palladium (5.2 - 6.5 ppb) iridium (0.10 - 0.13 ppb) and platinum (0.26 - 0.32 ppb). Therefore, the potential for PGE extraction from geothermal fluids at Ohaaki, would appear to be very limited and almost certainly uneconomic.

The field blank concentrations of copper were comparable to the AR blank concentrations. However, AR "sampler blanks" (washes of the sample chamber prior to use) did have significant copper concentrations, suggesting that copper is being leached from the ceramic coating. Consequently calculations of deep geothermal fluid copper concentrations from copper in the "sampler" and "washing" components of the sample would have been misleading and have not been undertaken.

3.2 Precious metals

The field blank concentrations of gold and silver were comparable to the AR blank concentrations of these metals. This implies that there has been little or no contamination of the sample by the AR, or by AR leaching of the ceramic coating. Deep geothermal fluid concentrations of gold and silver have been obtained for the 4 downhole samples, recalculated from combined "sampler" and "washing" analytical data (Table 1).

Table 1. Gold and silver concentrations (ppb) in deep geothermal fluids at Ohaaki.

Sample	Well	Gold	Silver
G6	Br 9	1.03	18.9
G8	Br 9	0.86	12.7
G9	Br 20	0.50	5.74
G10	Br 20	1.16	12.15

The gold concentrations obtained here are in remarkable agreement with the pilot plant results and the very approximate value calculated for Br 22 (Brown, 1986). It appears that the gold concentration in the geothermal reservoir at Ohaaki is of the order of 1 ppb. Silver concentrations were more variable than those of gold, but were similar, if generally slightly higher, than previously estimated for Br22 (8.0 ppb). Chloride analyses of the downhole samples collected during the shutdown suggested that the runs at Br9 were sampling a fluid diluted with respect to the fluid sampled later by the Webre sampling, while those at Br20 appeared to be only slightly diluted with respect to the later Webre samples. Although there seems to be little difference in the measured gold concentrations between wells Br9 and Br20, possible dilution affects may be reflected in the variable silver concentrations.

3.3 Thermodynamic considerations

Using major ion and gas species concentrations, measured on Webre and gas samples, it is possible to calculate an equilibrium concentration of gold from the measured equilibrium constants for reaction (1). Such a calculation is likely to be more relevant for Br20 than for Br9. Br9 has a discharge enthalpy which indicates that the well does not have a single liquid feed. With this caveat, the two wells yield an expected solubility of gold of 32.5 (or 14.2) ppb for Br 9 and 10.5 (or 4.5) ppb for Br20, with the initial values calculated from data from Seward (1973), and the values in parentheses calculated from Shenberger and Barnes (1989) in each case. For Br 20, thermodynamic predictions of gold solubility are significantly greater than the measured values (Table 1). Notably, predictions using the more recent data of Shenberger and Barnes (1989) are closer to the measured concentrations. For Br 9 there appears to be an even greater divergence from equilibrium solubility.

Under the chemical conditions of the deep Ohaaki fluid, the predominant dissolved silver species present is likely to also be a bisulphide complex: $Ag(HS)_2^-$. Thermodynamic data for the equilibrium reaction with acanthite (Ag_2S):

$$0.5\ Ag_2S + 0.5H_2S + HS^- = Ag(HS)_2^- \qquad (2)$$

have been measured (Gammons and Barnes, 1989). Calculation of silver solubilities using this data yields silver concentrations of 2.9 for Br20 ppb and 3.2 ppb for Br9, values which are considerably less than those measured (Table 1). The $Ag(HS)_2^-$ complex could also be in equilibrium with native silver and a reaction similar to (1) can be written:

$$Ag + H_2S + HS^- = Ag(HS)_2^- + 0.5\ H_2 \qquad (3)$$

Calculation of silver solubilities using Gammons and Barnes (1989) data for reaction (3) yields silver concentrations of 3.9 ppb for Br20 and 12.0 ppb for Br9. These values do appear to be more consistent with measured silver concentrations in Br9 and Br20.

The silver/gold ratio calculated from the thermodynamic data is much lower than the reservoir measurements. For both Br20 and Br9, the calculated

Ag/Au ratio is <1, compared with measured ratios of >10.

We have assumed that gold and silver are transported as the bisulphide complexes, but other complexes or solubility-limiting processes may need to be considered before the differences between predicted and measured gold concentrations can be explained. For example, cyanide complexes of both gold and silver are very strong. Even though a cyanide analysis of the deep Ohaaki geothermal fluid indicated levels of < 0.01 ppm (i.e., below detection limit), trace levels of CN may still influence gold and silver solubility. Thermodynamic calculations comparing the formation constants of $Au(HS)_2^-$ with those of $Au(CN)_2^-$ (Vlassopoulos and Wood, 1990; Skibsted and Bjerrum, 1977) show that, at 25°C, CN^- concentrations of only 1/200th those of HS^- are required to obtain the same gold concentration in solution. Assuming that the temperature dependence of the formation constants is similar, then a concentration of only ca. 0.01ppm of CN^- is required to provide the gold concentrations measured at Ohaaki. Consequently the possibility of gold- or silver-cyanide complexing cannot be ruled out until a more accurate, low-level determination of cyanide concentration can be made.

4. CONCLUSIONS

Gold concentrations measured in the deep geothermal fluids at Ohaaki were considerably lower than thermodynamic predictions of gold solubility. Consequently, commercial extraction of gold is unlikely to be economically viable at the Ohaaki geothermal field. However, the measured concentrations still allow for the transport of 1 million ounces of gold in less than 7000 years, which is well within the typical lifetime of a geothermal systems.

The platinum group elements were not observed at concentrations above the detection limit of the ICPMS. Consequently, the extraction of these elements is also unlikely to be economically viable.

The measured reservoir silver concentrations seem to be closer than those of gold to the theoretical values calculated from thermodynamic data.

5. ACKNOWLEDGEMENTS

We would like to thank Contact Energy for financial support, the opportunity to take these samples, and permission to publish the data. We thank Chris Morris for fitting the sampling into the busy Ohaaki shutdown schedule, Tom Gould for help above and beyond the call of duty in running the winch late into the evening, Lew Klyen for performing the acrobatic ballet necessary to trigger the sampler, the GNS laboratory at Wairakei, for putting up with days of aqua regia pong, and ESR for cyanide analyses. Part of this work was funded by Contract C0 5508 from FRST.

6. REFERENCES

Brown, K.L. (1986) "Gold Deposition from Geothermal Discharges in New Zealand", *Economic Geology*, **81**, 979-983.

Brown, K.L. and Roberts, P.R. (1988) "Extraction of gold and silver from Geothermal Fluid". Proceedings of International Symposium on Geothermal Energy, Kumamoto/Beppu Japan.

Gammons, C.H., and Barnes, H.L. (1989). "The solubility of Ag_2S in near-neutral aqueous sulphide solutions at 25 to 300 °C" *Geochim et Cosmochim Acta*, **53**, 279 - 290.

Henley, R.W. and Ellis, A.J. (1983) "Geothermal Systems, Ancient and Modern" *Earth Science Reviews*, **19**, 1-50.

Seward, T.M. (1973). "Thio complexes of gold and the transport of gold in hydrothermal ore solutions", *Geochim et Cosmochim Acta*, **37**, 379 - 399.

Shenberger, D.M. and Barnes, H.L. (1989). "Solubility of gold in aqueous sulfide solutions from 150 to 350°C" *Geochim et Cosmochim Acta*, **53**, 269 - 278.

Skibsted, L.H. and Bjerrum, J. (1977). "Studies on gold complexes (III) - the standard electrode potentials of aqua gold ions" *Acta Chemica Scand.* **A31**, 155 - 156.

Vlassopoulos, D. and Wood, S.A., (1990). "Gold speciation in natural waters: (I)-solubility and hydrolysis reactions of gold in aqueous solution", *Geochim et Cosmochim Acta*, **54**, 3 - 12.

New data on the chemical composition of waters in the Paratunka hydrothermal system, Kamchatka

O.V.Chudaev
Far East Geological Institute, Vladivostok, Russia

V.A.Chudaeva
Pacific Institute of Geography, Vladivostok, Russia

P.Shand & W.M.Edmunds
British Geological Survey, Hydrogeology Research Group, Wallingford, UK

ABSTRACT: Thermal waters are widespread features of the volcanically active Kamchatka Peninsula of the Russian Far East. The chemical composition of thermal waters in the Paratunka hydrothermal system have been studied as part of a large project aimed at characterizing the thermal features of this region. The thermal waters are of Na-SO$_4$ type with high concentrations of halogens and alkali earth elements. Geochemical and isotopic data (oxygen, hydrogen, sulphur, carbon, and helium) have been used to construct a model for the chemical evolution of waters of the Paratunka hydrothermal system.

1 INTRODUCTION

The Paratunka geothermal area is located in the southern part of the Kamchatka Peninsula and is one of four main geothermal areas in Eastern Kamchatka (Averyev, 1966; Figure 1).

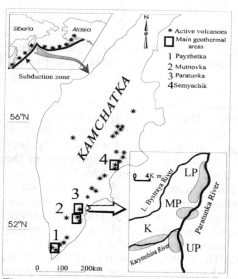

Figure 1. Location map for the main geothermal areas of Kamchatka and the sampled springs.

The high temperature hydrothermal systems of the Payzetka and Semyachik geothermal areas are concentrated in volcanic depressions. The Paratunka and Mutnovka geothermal areas are composed of strongly dislocated volcanic and sedimentary rocks. These areas were actively studied during the 1970's and 1980's when many boreholes were drilled to depth down to 2000m and monitoring and resource evaluation were carried out. However, during the last decades little research has been completed. The Paratunka geothermal area, described below, is one of the largest reservoirs of thermal waters in Russia with total discharge of 780 l/sec (for waters with T's of 70-80^{0}C). This area is located near the city of Petropavlovsk-Kamchatskii and is very important for Kamchatka (horticulture, central heating, resort and so on). The Paratunka geothermal area includes: the Paratunka hydrothermal system (PGS), which is located along the Paratunka river, the Nachiki group of thermal waters, the Bolshoi-Banny, and the Maly-Banny thermal waters. The Karymshinskii hot springs, located along the Karymshinskaya river, a tributary of the Paratunka river, are also included in this system. In this paper we present data on the Paratunka hydrothermal system which is formed by three distinct thermalanomalies:

Upper Paratunka (UP) and Karymshinskii (K), Middle Paratunka (MP), and Low Paratunka (LP), Figure 1. This system was described by many Russian scientists and hydrogeologists from the Hydrogeological Expedition of Kamchatka. The complete data were published by Manukhin and Vorozeikina (1976), Trukhin and Petrova (1976), and Sereznikov and Zimin (1976). The objective of this paper is to present new chemical and isotopic data on the geochemistry of the thermal and cold waters of Paratunka.

2 TECHNIQUES

Water samples were collected from thermal and cold springs and boreholes of the Karymshinskii group of waters (K), Upper Paratunka, Middle Paratunka, and Low Paratunka groups. Field determination were made of SEC, pH, DO, and T°C; and HCO_3 by titration. Major and trace cations were determined in the laboratory by ICP-OES and anions by automated colorimetry. The stable isotopes δ^2H, $\delta^{18}O$, and $\delta^{13}C$ data were measured by mass spectrometry.

3 GEOLOGICAL SETTING

The discharge area of the PGS is concentrated in a graben structure with NE orientation. This graben consists of a few blocks dipping to the North and filled by deposits of the Paratunka unit. This unit is presented by tuffs of basaltic and andesitic composition as well as sedimentary rocks, and there are many dykes. Diorites were intruded in this unit in the Miocene. The thickness of the unit is approximately 2000m. The Paratunka unit is covered by Quaternary deposits represented by argillaceous rocks, gravels, and conglomerates. In the Paratunka valley the Quaternary clays form a confining layer. For the Karymshinskii hot springs, the confining layer is the rocks of Paratunka unit. As a result of hydrothermal activity the host rocks have undergone propylitization. In the altered rocks albite, adularia, epidote, chlorite, clay minerals, and zeolites are widespread. The age of the buried PGS is > 250 000 years (Manukhin, Vorozeikina, 1976). The distribution of fractures in the host rocks and their alteration are main factors determining the transport and geochemistry of the thermal waters.

4 RESULTS

A comparison of the chemical data of the different groups of waters (K, UP, MP, LP) shows that there are not many differences among groups. These waters have high TDS (SEC up to 1.8mS cm^{-1}) and high concentrations of Na and SO_4 (Figure 2).

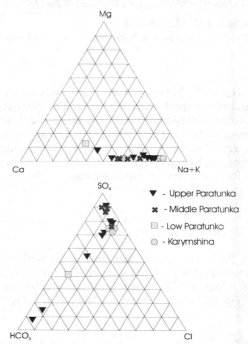

Figure 2. Trilinear diagrams of thermal and cold waters of the Paratunka hydrothermal area.

The contents of K, Mg, Ca, are higher in the LP waters. The boreholes waters generally have higher mineralization and temperature than the springs, where waters have been diluted by fresh shallow groundwaters. The temperatures of the thermal waters range from 31 to 75°C and the cold waters from 2.1 to 10°C. The cold waters circulate mainly in the Quaternary deposits. The waters of the PGS have an alkaline character. The values of pH in boreholes and warm springs increases from the UP to LP groups. The K group has lower pH than the UP group. The cold waters have a narrow range of pH, from 7.2 to 7.4. Redox potential (Eh) in the boreholes is often less than -100mV, but for warm springs Eh can vary up to +200mV. In the cold springs, Eh can be more than +350mV. In general, SEC in the hot waters is greater than 1000 μs/cm. Only in warm springs does TDS decrease to c. 500 μs/cm. For the cold springs SEC is generally low (39 to 190 μs/cm). The hot waters of the Paratunka hydrothermal system have a Na-SO4 composition. The concentration of Na is more than 200mg/l. In the cold springs concentrations of Ca are higher than Na and overlap with the surface waters of the Paratunka catchment. Potassium concentrations in the thermal waters vary from 1.2 to 9mg/l and increase from the UP to LP groups. The concentrations of Mg are very low in the hot springs (<0.04mg/l). Only in the LP area do Mg concentrations reach 3.8 mg/l, similar to concentrations in the cold springs. Concentration of Si are relatively high in the thermal waters (up to 29 mg/l) and decrease in the warm springs. The surface waters of the Paratunka river have variable Si (3.6-8.7 mg/l). The concentrations of Sr in the hot waters are typically high (up to 1440 μg/l) and there is a close correlation with Ca. The contents of Li in the hot waters vary from 200 to 800 μg/l, whereas in the fresh waters they only reach 22 μg/l. Li decreases from K to LP. The concentrations of B are also high (up to 6000 μg/l) in comparison with surface waters (22 μg/l). Boron correlates well with Li and both decrease from K to LP groups. The concentration of Br in the thermal waters range from 98 μg/l to 307 μg/l and in the cold waters the minimum value is 5 μg/l. Br/Cl ratios are similar to seawater. The behavior of I is the same as Br and the maximum content is 59.6 μg/l in K group. In the LP area the value of I is 17.8 μg/l. The anions in the thermal waters are dominated by SO_4 which reaches 90% of the sum of main anions. Bicarbonate concentrations are not high and generally close to, or lower than, 50 mg/l. The concentrations of Cl in the thermal waters range from 30mg/l to 115 mg/l and decrease to 2.5 mg/l in the fresh waters. The molar ratio SO_4/Cl is relatively high in the thermal waters (>1.4) and shows a variation from the upper to lower parts of the catchment away from the main volcanic centers. There is a close correlation between Cl and Br (Figure 3).

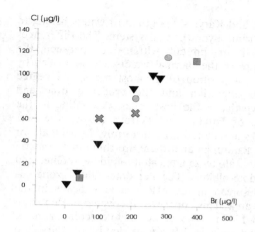

Figure 3. Cl versus Br in the waters of the Paratunka hydrothermal area. Symbols as Fig.2.

The values of $\delta^{18}O$ and δ^2H of the Paratunka system lie along the world meteoric line, however, the cold waters have a heavier signature ($\delta^{18}O$ = -12.9 to -14‰) than do the thermal waters ($\delta^{18}O$ = -14 to -15,1 ‰) forming distinct groups (Figure 4).

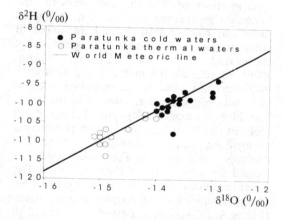

Figure 4. Plot of $\delta^{18}O$ versus δ^2H for the Paratunka hydrothermal area.

The hottest temperature thermal waters all have $\delta^{18}O$ close to -15 ‰ and the variation in the warm waters is most likely a consequence of mixing between a high-T source and shallow ground water with a heavier signature.

5 DISSCUSION

On the basis of the presented data Paratunka

area and Karymshina group of waters belong to the same hydrothermal system. The differences which are present within and between the groups of the thermal waters are controlled by the composition of the host rocks, the degree their alteration and the residence time and flowpaths in the host rocks. According to the data of Trukhin and Petrova (1976), the host rocks are altered non-uniformly. In general, for the Paratunka area there are three metasomatic zones: albite-zeolite; albite-epidote-zeolite, and epidote-albite. The epidote-albite zone is widespread in the MP area and locally in the LP area. This zone was discovered at depths of 115-1540m. The albite-epidote-zeolite zone is found only in LP are at depths of 150-620m (Trukhin, Petrova, 1976). Using data on the fresh and altered basalt of the studied area (Trukhin, Petrova, 1976) we suggest that during alteration the host rocks lost Ca, Si, K, Na, Sr, and Li, but gained Mg during water-rock interaction. The experiments by Bishoff & Dixon, 1975) support this idea. The low magnesium is a consequence of chlorite stability in the hydrothermal system. The concentrations of B in the host rocks of the Paratunka geothermal area are high in altered rocks and reach $1.8*10^{-3}$% and the lowest B concentrations correspond to the geochemical background of this area (Trukhin, Petrova, 1976). In general, the origin of the Paratunka hydrothermal system can be described in terms of circulation of meteoric waters in the crust. The high fracture density and dislocation in rocks of the Paratunka unit favour penetration of meteoric water to depths of 2-4km. The discharge area for the Paratunka system is located along the valley of Paratunka river. The meteoric source of the water is supported by stable isotopic data of oxygen and hydrogen (Figure 4). Chloride, I, Br and B are added to the water from exhalations, which are typical for the volcanic area of Kamchatka. The data on the isotope of $\delta^{34}S$ (Vinogradov, 1970) indicate two sources for the sulphur: deep and from marine deposits. Our data on $\delta^{13}C=7.0$ °/$_{oo}$ in the gas phase of LP springs is too heavy to be from marine carbonates and too light to be of biogenic origin and more consistent with a mantle origin for the carbon. Most mantle values of $\delta^{13}C$ fall within a range of -5 to -8 °/$_{oo}$, from which CO_2 outgases can acquire values as enriched as -2.5 °/$_{oo}$ depending on the degree of out gassing (Javoy, 1986). Polyak et al. (1979) have reported $^3He/^4He$ ratios in the Paratunka thermal waters from $6.9*10^{-6}$ to $8.7*10^{-6}$ which corresponds to a mantle source. In conclusion, meteoric water is heated within hydrothermal systems associated with the volcanic centers and interacts with the host rocks with consequent gains Ca, Si, K, Na, Sr, and Li, and losses in Mg. Chlorine, I, Br, B, and part of the S are added to the water from sub-volcanic exhalation.

6 ACKNOWLEDGEMENT

We acknowledge financial support from INTAS, contract 94-1592 in support of this work.

7 REFERENCES

Averyev V.V. 1966. Hydrothermal processes in the volcanic areas and their relations with mgmatic activity. In: Modern volcanism, Moscow, Nauka, p. 80-95.

Bishoff J. & Dixon F. 1975. Seawater-basalt interactions at 200°C and 500 bars. Earth and Planetary Science Letters, 25, p.385-397

Manukhin Yi. F., Vorozeikina L. A. 1976. Hydrogeology of the Paratunka hydrothermal system and conditions of its origin. In: Hydrothermal system and thermal areas of Kamchatka. Vladivostok, p.143-178.

Javoy, M., Pineati, F. & Delone H. 1986. Carbon and nitrogen isotopes in the mantle. Chem. Geol., 57, p. 41-62.

Polyak B.G., Tolstikhin I.N., Yakuzeni V.P. 1979. Helium isotopes and heat flow and geophysical aspects of tectogenesis. Geotectonics #5 p.3- 23.

Sereznikov A.I., Zimin V.M. 1976. Geological setting of the Paratunka geothermal area. In: Hydrothermal systems and thermal areas of Kamchatka. Vladivostok, p. 115-142.

Trukhin Yu.P., Petrova V.V. 1976. Some regulaririesof the modern hydrothermal processes. Moscow, Nauka, 135p.

Vinogradov V.I. 1970. Isotopic composition of sulpher in thermal waters of Kamchatka and Kuril islands and its genetic interpretation. In: Essay of geochemistry of mercury, molybdenum, and sulphur in the hydrothermal process. Moscow, Nauka, p. 258-271.

Thermal fluids and scalings in the geothermal power plant of Kizildere, Turkey

L.B.Giese, A.Pekdeger & E.Dahms
Institute for Environmental Geology, Free University of Berlin, Germany

ABSTRACT: The Kizildere geothermal power plant is located in western Anatolia, Turkey. Large amounts of scalings have been observed in the production wells, separators and waste water systems. The upwelling fluid looses 33% of its CO_2 and 2% of water. The initial degassing starts at 600 m below the well head. Gases enrich towards the head of the well and cause pressure anomalies. The loss of CO_2 leads to increasing pH, oversaturation with respect to $CaCO_3$ and precipitation of scalings. The scalings are mainly composed of calcite, in the upper part of the well aragonite is dominating. Calculations with SOLMINEQ.88 confirmed a non monotonous development of the precipitation rates within the production well.

1 INTRODUCTION

The Kizildere geothermal power station is located in western Anatolia, Turkey, about 25 km west of the famous sinter terraces of Pamukkale, Denizli. It is producing 10 MW of electric power by the separation of steam from eight production wells. The production of the thermal fluid causes several problems, i.e. (i) formation of carbonate scalings - these scalings appear especially at depths between 550 and 70 m in the detailed investigated production well KD 22 (Giese, 1997) -, (ii) pressure dropdown in the reservoir and (iii) contamination of the Büyük Menderes River by the input of the thermal waste water.

A district heating system for the province capitol Denizli and a re-injection of the thermal waste waters is planned. The heat exchange systems and the re-injection system will also be threatened by the formation of scalings (compare Lindal & Kristmannsdottir, 1989). Problems with carbonate and also silica scalings are expected. The first aim of this investigation was the understanding of the processes in the production well during the ascent of the thermal fluid. The chemical and mineralogic type of the precipitated scalings will be discussed.

2 EXPERIMENTAL

All thermal fluids of Kizildere and the adjacent areas were sampled. The composition of all fluids within the separator systems, the thermal waste water systems and also natural outcrop waters were of primary interest. Mineral samples were taken from production well scalings, separator scalings, and waste water channel scalings as well as from natural sinters for comparison.

Physico-chemical parameters and unstable constituents were determined in the liquid phases on-site. For the chemical analyses AAS, FES, CFS-photometry, cuvette-photometry, titration and ion selective electrode methods were used. The chemical composition of solid phases were determined by cold HCl degestions. Mineral phases were determined semiquantitatively by XRD.

P/T-logs (pressure/temperature) of the upwelling fluid in the well KD 22 were used to calculate the fluid density, contents and composition of the released gases as a function of depth and the determination of the initial degassing point (IDP). Therefore it was possible to calculate a total mass and energy balance for the production system. Characteristics of all phases, fluids as well as mineral phases, appearing before and after the separator system had to be taken into account.

The last methodical step was the computer based calculation of changing three-phase-equilibria within the production well. The program SOLMINEQ.88 (Kharaka et al., 1988) was used. Starting with a calcite saturated fluid the solution was boiled in steps of 100 m under measured P/T-conditions. In this procedure the solution was resaturated in every step with respect to calcite. From the mass of precipitated

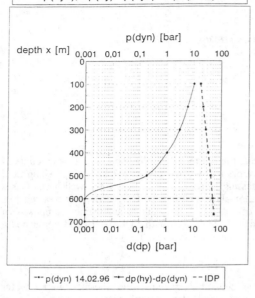

Figure 1. Anomaly of pressure decrease in the production well KD 22. The value of d(dp) is calculated as difference between the normal hydrostatic development and measured change of pressure.

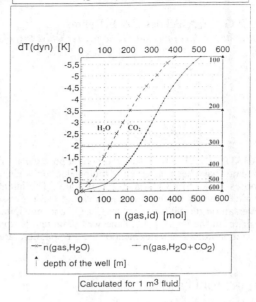

Figure 2. Calculated molar amounts of degassed H_2O and CO_2 from the fluid as function of temperature decrease. The upper curve shows the sum of both gases, the lower curve represents H_2O.

calcium the volume of $CaCO_3$ scaling was calculated. The scaling development was estimated for 11 months of utilisation.

3 RESULTS

Figure 1 shows the anomaly of pressure decrease in the production well KD 22. The difference between the regular pressure loss in an incompressible liquid and the measured values was calculated. The anomaly of pressure decrease depends on developed gas bubbles. The initial depth (IDP) is 600 m below the well head.

What happens in the well above this depth? The first significant appearance of carbonate scalings is observed at a depth of 550 m below the well head. A temperature decrease indicates the dominance of a heat transforming process, which was identified as boiling of water. Due to the low partial pressure of steam this process is associated with the degassing of carbon dioxide. Figure 2 shows calculated molar amounts (n) of degassed H_2O and CO_2 in relation to the temperature decrease (dT). Contents of gases were calculated using (i) the ideal gas law and (ii) a temperature dependent partial pressure. In the beginning mainly CO_2 degasses. Degassing seems to be completed at 300 m below the well head with nearly 6 g CO_2 per kg liquid phase (33% of the dissolved amount). The molar amount of developed steam correlates with the loss of temperature as expected. The thermodynamically allowed maximum amount of degassed water was calculated to be 2.2%. A comparison of both methods indicates, that in the main part of the well, the measured values of steam content in the pressure profile reach only 50% of the ones documented in the thermic profile (dT = 9.5 K). In the upper part of the well it is vice versa. These facts indicate an enrichment of gases near the head of the well because of the higher migration velocity of gases in two-phase systems.

The mineralogical investigations showed that scalings at depths between 550 m and 200 m contain nearly 90% calcite and only 10% aragonite. In the upper part of the well the content of aragonite

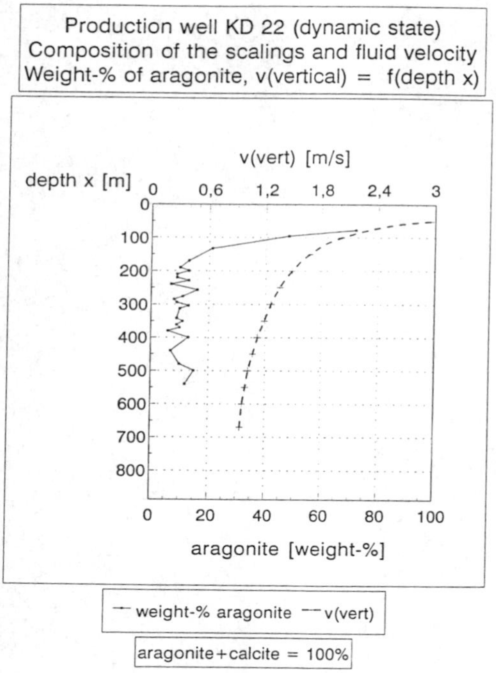

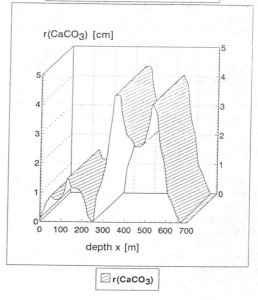

Figure 3. Relation of the aragonite contents in the scaling and the fluid velocity as function of depth.

Figure 4. Calculated radial scaling profile estimated for 11 months of utilisation (production well KD 22).

increases to nearly 70%. Figure 3 shows the relation of aragonite contents in the scaling and the fluid velocity. Due to degassing the velocity is increasing. When the velocity exceeds 1.5 m/s at depths between 200 and 70 m, aragonite is the preferably precipitated mineral phase.

In calcite crystal selective ion ratios increase for Mg/Ca from 0.03 to 0.14 from bottom to top. Due to preferential Ca precipitation the ion ratio in the solution is enriched with Mg. High Mg/Ca ratios result in an kinetical inhibition of calcite formation. The content of strontium in the aragonite crystal seems to react more sensitive to the velocity of precipitating than the Mg-calcite system. High contents are correlated with fast precipitation. The molar ratio of Sr/Ca in aragonite varies between 0.05 to 0.7.

The effects of boiling and degassing processes in the well were recalculated with the thermodynamic program SOLMINEQ.88 (Kharaka et al., 1988). Kinetic effects were postulated as fast enough to be ignored. The calculated profile in the well showed (Figure 4), that saturation indices of scaling bearing carbonate minerals are developing not constantly in the fluid during ascent. This depends on the effects of temperature, pressure, and pH within the carbon species system. The amount of scaling formation reaches its maximum at depths between 550 m and 350 m below the well head, in the uppermost part of the well the precipitation is decreasing to zero. In detail there are several collapses of the precipitation rate which can be detected in chemical signatures of the scaling (foreign components) and by the estimated scaling thickness during the annual clean up of the wells.

4 CONCLUSIONS

The physico-chemical development of ascending fluids and the precipitated scalings could be characterised and quantified.

The upwelling fluid gets oversaturated with carbon dioxide under decreasing hydrostatic pressure. This leads to the development of a gas phase within the fluid. Partial boiling of water causes an adiabatic heat conversion (90% of temperature decrease). The rest of the energy is lost by conductive cooling. The progressive degassing and expansion during ascent reduces the density in the two-phase system. This results in (1) increasing fluid velocities, (2) an increasing amount of heat loss, (3) an increasing deviation from the behaviour of an incompressible fluid, (4) a change of pH to more alcalic values, and (5) the precipitation of scalings.

Therefore, the scaling properties depend on the production mode. Increasing production rates lead to a sinking pressure level in the well. With a sinking IDP the scaling problems move to depth, possibly down to the reservoir level. The secondary effect could be an enrichment of colder, CO_2-rich waters, which also leads to earlier degassing processes. A practical conclusion is to use moderate treatments for the production process in order to keep the problems with carbonate scalings on an acceptable level.

ACKNOWLEDGEMENTS

We would like to thank our technical supporters B. Alberts, A. Bartels, A. Gallo, and U. Müller for their excellent work, also thanks to V. Neumann. This study was realized in cooperation with MTA/Ankara, TEAS/Ankara, and the Dokuz Eylül University/ Izmir. Financial supports were given by DFG (Deutsche Forschungsgemeinschaft, project numbers Pe 362/9-1 and Pe 362/9-2) and DAAD (Deutscher Akademischer Austauschdienst).

REFERENCES

Giese, L.B., 1997. Geotechnische und umweltgeologische Aspekte bei der Förderung und Reinjektion von Thermalfluiden zur Nutzung geothermischer Energie am Beispiel des Geothermalfeldes Kizildere und des Umfeldes, W-Anatolien/ Türkei. Ph.D. thesis FU Berlin: 201 p., in press.

Kharaka, Y.K., Gunter, W.D., Aggarwal, P.K., Perkins, E.H. & Debraal, J.D., 1988. SOLMINEQ.88: A computer program for geochemical modeling of water-rock interactions. *U.S. Geol. Surv. Water Res. Inv. Rep.* 88-4227: 420 p..

Lindal, B. & Kristmannsdottir, H., 1989. The scaling properties of the effluent water from Kizildere power station, Turkey, and recommendation for a pilot plant in view of district heating applications. *Geothermics* 18: 217-223.

Correlations between B/Cl ratios and other chemical and isotopic components of Taupo Volcanic Zone, NZ geothermal fluids – Evidence for water-rock interaction as the major source of boron and gas

J.R. Hulston
Institute of Geological and Nuclear Sciences Limited, Lower Hutt, New Zealand (Presently: Isotope Consulting Limited)

ABSTRACT: Boron to chloride ratios generally correlate with increases in carbon dioxide, methane, nitrogen and hydrogen contents of geothermal discharges and decreases in $\delta^{13}C(CO_2)$ from west to east across the Taupo Volcanic Zone (TVZ), New Zealand. These correlations are largely attributed to interaction of the ascending mantle fluids with the metasedimentary basement, particularly at the boundaries of the TVZ, rather than originating from oceanic sediments carried down by the subducting Pacific plate.

1 INTRODUCTION

Early research work on the geothermal fields of TVZ (see Figure 2 for location) showed distinctive differences in the chemistry between fields, the most notable being in carbon dioxide content which is of importance in geothermal power station design. Later studies (see Ellis and Mahon, 1977) showed a variation of other components of the fluid, particularly boron (which is detrimental to some horticultural plants) while others such as cesium are less variable.

Various attempts had been made to characterise these variations and to elucidate their source (Figure 1). Ellis and Mahon (1967) suggested water-rock interaction as the source of these chemicals and conducted laboratory experiments which showed significant release of Cl, F, B, Na and other solutes from volcanic and greywacke rocks at temperatures above 200°C, which they considered sufficient to explain the results available at that time. Since then further areas have been studied and additional measurements have been made, particularly of helium isotopes. Giggenbach (1995) summarised variations reported up to that time and explained the observed variations in terms of interaction of volatiles released from the subducted sediments with material of the mantle wedge to form a volatile-charged, high alumina basalt arising above the volcanic front in the east with the volatile-depleted residue arising beneath the western part of TVZ. Hulston and Lupton (1996) however found from their study of helium isotopes in the TVZ, a trend of decreasing $^3He/^4He$ and increasing $^{40}Ar/^{36}Ar$ ratio with increasing B/Cl in the TVZ (plotted as increases in $^3He/^4He$ and $^{40}Ar/^{36}Ar$ in Figure 3), suggesting that the decrease in helium isotopes is resulting from the addition of radiogenic helium in conjunction with boron from the greywacke metasedimentary rocks, on the eastern (SE) boundary of TVZ (Figure 2). A similar result was observed for some areas near the western (NW) boundary of TVZ. This paper considers other isotopic and geochemical evidence which supports this hypothesis.

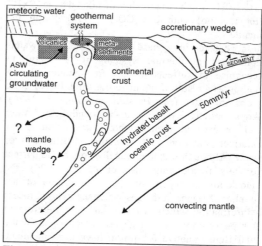

Figure 1 - Diagram of the sources of fluids in a geothermal system (modified from Giggenbach, 1995 and Hulston and Lupton, 1996).

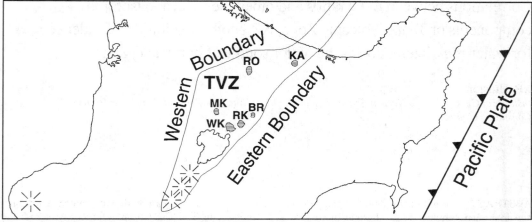

Figure 2 - Map showing the location of the TVZ and the subducting Pacific plate. The geothermal fields shown are Mokai (MK), Wairakei (WK), Rotokawa (RK), Ohaaki-Broadlands (BR), Kawerau (KA) and Rotorua (RO).

2 BACKGROUND

In order to discuss the origin of the geothermal fluids it is necessary to consider the whole process from subduction to groundwater infiltration. ^{3}He degassing from the mantle is considered one of the most useful tools in determining the role of mantle gases in crustal systems. When mantle gases enter a volcanic system (Figure 1) ^{3}He provides a useful tracer through the ascending system. It is therefore frequently used as a control variable in considering the composition of other components. One such ratio is $CO_2/^3$He which is generally considered to be ~3×10^9 (molar ratio) with $\delta^{13}C(CO_2)$ = -6±1 ‰ in the mantle.

In considering these components in water-dominated geothermal systems it is useful to recognise the effects of gas fractionation during boiling as a result of the differing solubility of these gases in water (see Hulston and Lupton, 1996). Argon (strictly ^{36}Ar) from air-saturated surface water can be used to monitor this fractionation as air saturated water at 15°C contains ~0.3µmAr/m of water. Although it is theoretically possible to correct for this fractionation (Mazor et al., 1990), the calculation is dependent on a knowledge of the number of stages in the gas-water separation process - information which is rarely known. Therefore in this paper only those analyses where argon in the total fluid is above 0.1 µm/m have been included. This restriction ensures that the most fractionated samples, which come from gas depleted waters are eliminated from the plots. The technique of Giggenbach (1995) of using the non-atmospheric component, N_{2C} has been followed in the correlation plots below. N_{2C} is calculated by deducting the atmospheric nitrogen component using the argon content as a reference measure of the atmospheric component. Since the effect of radiogenic ^{40}Ar is generally less than 15% in TVZ it has been neglected in this correction.

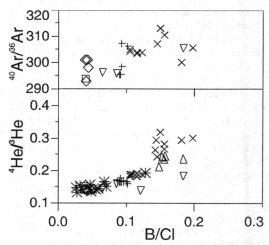

Figure 3 - Increase of radiogenic ^{4}He and ^{40}Ar with B/Cl molar ratio (Hulston and Lupton, 1996; ^{4}He/^{3}He values relative to air).

3. CORRELATIONS OBSERVED IN TVZ

3.1 *The relationship between gases and B/Cl.*

The data set used for this study is a combination of the TVZ results reported by Hulston and Lupton (1996) and those of Giggenbach (1995). Figure 4 shows ratios of carbon dioxide, nitrogen, methane and hydrogen relative to ^{3}He plotted against the B/Cl ratio for all TVZ samples in the database where $Ar/H_2O_{TD} > 0.1$ µm/m.

Although there is some scatter in the points, there is a clear trend for all these ratios to increase with increasing B/Cl ratio. Figure 3 shows that this increase is also correlated with increasing radiogenic ^{4}He and ^{40}Ar. Thus the CO_2, CH_4, N_{2C} and H_2 appear to be coming from a radiogenic rich source. One such source could be the Mesozoic metasedimentary greywacke basement which crops out at edges of the TVZ.

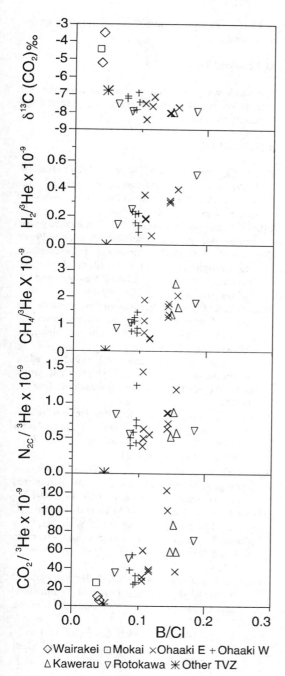

◇Wairakei □Mokai ×Ohaaki E +Ohaaki W
△Kawerau ▽Rotokawa ✳Other TVZ

Figure 4 - Variation of geochemical components with B/Cl

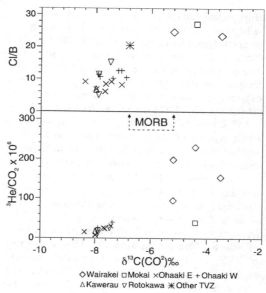

◇Wairakei □Mokai ×Ohaaki E +Ohaaki W
△Kawerau ▽Rotokawa ✳Other TVZ

Figure 5 - Variation of carbon isotopic composition of CO2 with Cl/B and ^{3}He/CO_2

3.2 *Carbon isotopes in TVZ*

The data set of carbon isotope composition of carbon dioxide has been obtained from results by Hulston and McCabe (1962), Lyon and Hulston (1984), Hulston (1986) and Giggenbach (1995). In contrast to the linear type relationships with B/Cl found above, a plot of $\delta^{13}C(CO_2)$ against B/Cl (Figure 4) shows an asymptotic relationship. In order to demonstrate a more linear relationship, the inverse functions Cl/B and $^3He/CO_2$ have been plotted against $\delta^{13}C(CO_2)$ in Figure 5 (again using the condition that $Ar/H_2O_{TD} > 0.1$ um/m). Despite the scatter in these results, a clear trend of increase in $\delta^{13}C(CO_2)$ with increases in Cl/B and $^3He/CO_2$ is evident, suggesting mixing between a source with low CO_2 content and $\delta^{13}C(CO_2) > -2‰$ and a source of much higher CO_2 content with $\delta^{13}C(CO_2)$ of $-9±1‰$. It should be noted that the range of $CO_2/^3He$ and $\delta^{13}C(CO_2)$ values observed in the TVZ is similar to that reported by Sano and Marty (1995) for the Unzen, Kusatsu-Shirane and Ontake volcanic area in Japan, but no similar correlation was found there. In both cases, the $CO_2/^3He$ ratio is several times that of MORB, which explains why the MORB carbon isotopic signature does not predominate.

4 DISCUSSION AND CONCLUSIONS

Hulston and Lupton (1986) showed that increases in radiogenic helium and argon in TVZ fluids are associated with increases in the B/Cl ratios of these fluids, and, in particular with boron in the system. The question is: do these components come from the metasedimentary greywacke basement or from organics in ocean sediments carried down by the subducting plate? The above results indicate that in the TVZ most of the boron in the ocean sediments is either scraped off in the accretionary wedge or is volatilised before reaching the depth where magmatic fluids begin to rise. Sano and Marty (1995) discussed their $\delta^{13}C(CO_2)$ results in terms of a mixture of CO_2 produced by decarbonation of subducted marine limestone and carbonate of $\delta^{13}C$ ~0‰ and of organic sediments of $\delta^{13}C < -20‰$ however their results fell within the narrower range of -9‰ to -1‰, similar to that of TVZ. In the case of TVZ, the low CO_2 -2‰ component could be explained by a mixture of two-thirds CO_2 from slab carbonate and one-third MORB CO_2. This slab component could also explain much of the chloride in the fluid (and in the volcanic rocks). The intercept at B/Cl ~0.03 in the Figure 4 would then be due to a residual boron component carried down by the slab.

The high CO_2 -9‰ component found in these TVZ fluids is similar to those found in calcite from a Torlesse greywacke sample (S. Woldemichael, *pers comm*). Thus it is concluded that the metasedimentary greywacke basement rocks are the most likely source of the B, CO_2, CH_4, H_2 and non-atmospheric N_2 in TVZ geothermal fluids. It should be noted however that there are likely to be many systems in the world where these conclusions are not applicable.

ACKNOWLEDGEMENTS

This work was supported by funding from the NZ Foundation for Research Science and Technology Research contract C05607.

REFERENCES

Ellis A.J. and Mahon W.A.J. 1967. Natural hydrothermal systems and experimental hot water/rock interactions (Part II) *Geochim. Cosmochim. Acta* **31**: 519-538.

Ellis A.J. and Mahon W.A.J. 1977. *Chemistry and geothermal systems.* New York: Academic Press.

Giggenbach W.F. 1995. Variations in the chemical and isotopic composition of fluids discharged from the Taupo Volcanic Zone, New Zealand. *J. Volcan. Geotherm. Res.*, **68**, 89-116.

Hulston J.R. 1986: Further isotopic evidence on the origin of methane in geothermal systems. pp. 270-273 In: IAGC WRI-5 Extended abstracts. Orkustofnun; Reykjavik.

Hulston J.R. and McCabe W.J. 1962. Mass spectrometer measurements in the thermal areas of New Zealand. *Geochim. Cosmochim. Acta* 26: 383-410.

Hulston, J.R. and Lupton, J.E. 1996. Helium isotope studies of geothermal fields in the Taupo Volcanic Zone, New Zealand. *J. Volcan. Geotherm. Res.*, 74:297-321.

Lyon G.L. and Hulston J.R. 1984. Carbon and hydrogen isotopic compositions of geothermal gases. *Geochim. Cosmochim. Acta*, **48**, 169-173.

Mazor E., Bosch A., Stewart M.K. and Hulston J.R. 1990. The geothermal system of Wairakei, New Zealand: physical processes and age estimates inferred from noble gases. *Appl. Geochem.* **5**, 605-624.

Poreda R.J. and Arnorsson S. 1992. Helium isotopes in Icelandic geothermal systems: II. Helium-heat relationships. *Geochim. et Cosmochim. Acta*, **56**, 4229-4235.

Sano Y. and Marty B. 1995. Origin of carbon in fumarolic gas from island arcs. *Chem. Geol.* **119**:265-274.

Chemical and isotopic features of gas manifestations at Phlegrean Fields and Ischia island, Italy

S. Inguaggiato & G. Pecoraino
Istituto di Geochimica dei Fluidi, CNR, Palermo, Italy

ABSTRACT: Fumarolic gases have been collected in Phlegrean Fields and Ischia Island with the aim to characterize their isotopic compositions and to calculate the temperatures of the deep hydrothermal reservoir feeding these fluids. The deep temperatures estimated range between 240° and 270°C for Phlegrean Fields and between 270° and 330°C for Ischia Island. Helium isotope ratios indicate that these gases have a mantle origin that underwent crustal contamination processes. $\delta^{13}C$ of CO_2 values range between -1.4 and -5 ‰ vs PDB, similar to those observed in other volcanic active areas of Italy, more positive in respect to the classic magmatic values (-8 to -5‰ vs PDB). Therefore, the more negative isotopic values of $\delta^{13}C$ of CO_2 reflect contamination processes between deep fluids and shallow hydrothermal fluids.

1 INTRODUCTION

Ischia, Phlegrean Fields and Vesuvius are part of the classic active volcanic area of the alkali-potassic Quaternary Roman Province of central-southern Italy (Figure 1). Their volcanism is related to Plio-Quaternary extensional tectonics that caused the horst and graben structures, with NW-SE and NE-SW trends along the Tyrrhenian margin of southern Apennine chain. The NE-SW tectonic trend marks the alignment of the volcanic areas of Phlegrean Fields and Ischia that were formed by many eruptive centres and are characterized by the presence of many thermal springs, fumaroles, frequent earthquakes and bradyseismic phenomena (Vezzoli et al., 1988).

1.1 *Phlegrean fields*

Phlegrean Fields make up the floor of an impressive caldera, about 12 km in diameter, that was formed approx. 35000 years ago by emission of more than 80 km³ of pyroclastic products. The recent volcanic history is characterized by the occurrence of very explosive eruptions (Rosi et al., 1983). Erupted products cover the whole compositional range: trachybasalts to latites, trachytes and alkali-trachytes, phonolities and peralkaline trachites. The most recent historical eruption in this area, which created Mt Nuovo, took place in 1538. Since then, the Phegrean Fields have been characterized by moderately intense hydrothermal activity and also by bradyseismic phenomena (Carapezza et al., 1984).

This area has continually undergone alternating phases of uplift and subsidence of the ground. According to both historical reports and evidences left by lithodomi on Roman ruins (Serapeo Temple in Pozzuoli), these phenomena date back to at least 2000 years ago. Before the last eruption in 1538 the ground rose up about 7 meters. During the last

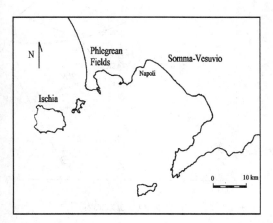

Figure 1: Map of investigated areas.

century there was a gradual subsidence, that in the summer of 1969 reached 170 cm (Osservatorio Vesuviano, 1983). Between 1972 and 1974 the ground subsided again by about 20 cm and thereafter, up to 1982, the situation remained fairly stable. Beginning in the summer of 1982, the ground rose again, gaining another meter by January 1984. Therefore, between 1969 and January 1984 there was a total rise of about 2.5 meters. This phenomenon began with a significant variations in gas concentrations like H_2Ovap, H_2S, CH_4, H_2, reducing capacity (Carapezza et al., 1984, Inguaggiato, 1986) and with increase in local seismic activity (Corrado et al., 1983). In particular, the observed increase in CH_4 contents are explained by Carapezza et al. (1984) as an increase of pressure in the deep hydrothermal reservoir. The present fumarolic activity is mainly concentrated in the southeastern part of the area inside the Solfatara of Pozzuoli. In the last 20 years, the highest temperature was about 160°C in Bocca Grande fumarole, while the other steam vents are characterized by discharge temperatures close to 100°C.

1.2 Ischia Island

The island of Ischia has a surface of 46 km^2 and is part of a volcanic field that was larger that the present island. Mt. Epomeo, an active volcano-tectonic horst (787 m a.s.l.), is the central and the highest part of the island, and is the only part of the island involved in the tectonic uplift (Vezzoli 1988). Almost all the rocks of the island are volcanic, mostly alkali-trachytes with subordinate trachybasalts, latites and phonolites. The last eruption on the island took place in 1301 A.D. at Arso. Several archaeological, geological and geodetic data testify to vertical ground movements during recent times (Vezzoli, 1988). The archeological data include the finding out of the Roman level (C. Romana and S. Resistita necropolis etc) on the sea bottom (5 - 8 metres deep). The geological evidence include the existence of several springs and thermal springs, previously located inside the island but today are in the offshore area (e.g. Citara) and the discovery of fossiliferous shore deposits outcrop at 10 - 40 metres asl. Geothermal research was carried out on the island in the 1940's (Penta & Conforto, 1951). Several wells were drilled there and the hottest temperature (232°C) was measured in the deepest well at 1150 m (Penta, 1963). Many thermal springs (T≤90°C) and fumaroles (T~100°C), and shallow earthquakes (the last one in 1881, Mercalli, 1884) testify to the persistent state of activity of the magmatic system.

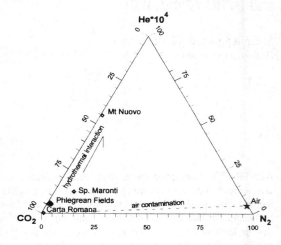

Figure 2: He*10^4-CO$_2$-N$_2$ diagram.

2 RESULTS AND CONCLUSIONS

Gas manifestations in these areas are sampled and analyzed with the aim to characterize their chemical and isotopic compositions and to make an estimation of the temperature of deep reservoirs that feed the two hydrothermal systems (table 1). The sampled gases show a CO$_2$-matrix without significant atmospheric pollution (figure 2).
All samples show a very low He/CO$_2$ ratio, except

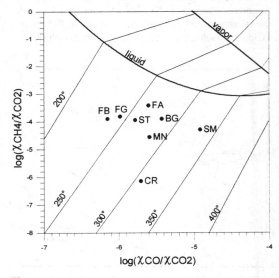

Figure 3: Evaluation of CH$_4$,CO and CO$_2$ equilibration conditions.

Table 1: Chemical and isotopic composition of sampled gases. The chemical compositions are expressed in %vol. The isotopic values of $\delta^{13}C$ of CO_2 are expressed in ‰ vs PDB. The helium isotopic values are expressed as R/Ra (Ra = atmospheric $^3He/^4He$ ratio =$1.4 * 10^{-6}$).

	data	T°C	He	H2	O2	N2	CO	CH4	CO2	d13C	3He/4He
Phlegrean F											
Bocca G.	Apr-93	157	5.50E-04	0.18	0.11	0.53	3.70E-04	1.25E-02	98.9	-1.50	2.89
Stufe	apr-93	99	5.80E-04	0.05	0.47	2.00	1.60E-04	1.10E-02	97.0	n.a.	n.a.
Fratt. B	apr-93	96	6.50E-04	0.05	0.19	0.91	7.00E-05	1.25E-02	99.0	n.a.	n.a.
Fangaia	apr-93	26	5.00E-04	0.06	0.44	2.34	1.00E-04	1.48E-02	96.6	1.53	n.a.
Fratt. A	apr-93	105	5.20E-04	0.26	0.30	1.50	2.40E-04	3.64E-02	96.0	-1.40	n.a.
Ischia											
C. Romana	oct-95	16	n.d.	n.d.	0.40	0.97	1.90E-04	7.00E-05	96.4	-2.49	n.a.
Sp.Maronti	oct-95	101	1.39E-03	0.06	3.20	10.10	1.06E-03	4.50E-03	88.3	-3.26	n.a.
Mt.Nuovo	oct-95	99	1.13E-02	2.46	0.40	1.70	2.30E-04	2.48E-03	89.5	-5.00	3.56

for the Mt Nuovo sample characterized by high He contents (> 0.01 %Vol). This behaviour can be explained by interaction processes between deep gases and shallow hydrothermal groundwater. This process causes the removal by selective dissolution of CO_2 enriching helium in the gas phase (Capasso et al., 1997). To use a graphic method proposed by Giggenbach (1991), based on CO/CO_2 and CH_4/CO_2 ratios, we have estimated the temperature of the deep reservoirs feeding these two volcanic systems. The obtained temperatures range between 240° and 270°C for Phlegrean Fields and between 270° and 330°C for Ischia Island. Helium isotopic ratios (R/Ra) indicate that these gases have a mantle origin that underwent crustal contamination processes. $\delta^{13}C$ of CO_2 values range between -1.4 and -5 ‰ vs PDB, similar to those observed in other volcanic active areas in Italy (ranging between -3 and 0‰). Such values are more positive in respect to the classic magmatic ones indicated for MORB (-8 to -5 ‰, according to Taylor, 1967, Deines, 1970). The more positive isotopic values of these samples in respect to the typical MORB compositions, rather than reflecting shallow contamination processes by CO_2 derived from thermal metamorphism of carbonate sediments of the basement of the volcano, would suggest that the source of magmatic CO_2 is already contaminated (Marty et al. 1994, Hoernle et al., 1995). Such a hypothesis is supported by the results obtained from recent volcanological studies in the Mediterranean area (Capasso et al., 1996, 1997; D'Alessandro et al.,1997; Giammanco & Inguaggiato, 1996, Giammanco et al., 1997, Inguaggiato & Italiano this volume). Helium and $\delta^{13}C$ of CO_2 isotope compositions are close to the typical MORB values, but do not fall into the MORB compositional field because their source shows a marked crustal contamination. Therefore, the more negative isotopic values of $\delta^{13}C$ of CO_2, up to –5‰ vs PDB, observed in the gases of Ischia Island are the result of the interaction processes between the deep CO2-rich fluids and the shallow hydrothermal fluids as proposed by Capasso et al. (1997) in the other volcanic areas.

3 ACKNOWLEDGEMENTS

The authors wish to thank the colleagues Rocco Favara and Franco Italiano for their support to isotopic analyses and comments on the manuscript.

4 REFERENCES

Capasso G., Favara R., and Inguaggiato S. (1996) Chemical and isotopic characterization of gaseous emissions on Vulcano Island (Aeolian Islands, Italy). *Proc. 4th Internl. Symp. on the Geochemistry of the Earth's Surface*, 554-558.

Capasso G., Favara R., and Inguaggiato S. (1997) Chemical features and isotopic composition of gaseous manifestations on Vulcano Island (Aeolian Islands, Italy): an interpretative model of fluid circulation. *Geochim. Cosmochim. Acta*, vol 61, 16, 3425-3440.

Carapezza M., Nuccio P.M., Valenza M., 1984. Geochemical sourveillance of the Solfatara of Pozzuoli (Phlegrean Fields) during 1983. *Bull. Volcanol.,* vol. 47-2: 303-311.

Corrado G., Guerra I., Lo Bascio A., Luongo G. and Rampolli R., 1983. Inflation and microearthqhake activity of Phlegrean Fields, Italy. *Bull. Volcanol.,* vol. 40-3: 1-20.

D'Alessandro W., Giammanco S., Inguaggiato S., and Parello F., (1997) Carbon isotopic composition of CO_2 in the Mt. Etna area. *Proc. 2nd Intl Workshop on European Laboratory Volcanoes.*

Deines P. (1970) The carbon and oxygen isotopic composition of carbonates from the Oka Carbonatite complex, Quebec, Canada. *Geochim. Cosmochim. Acta*, vol 34, 1199-1225.

Giammanco S. and Inguaggiato S. (1996) Soil gas emissions on Mount Etna (Sicily, Italy): geochemical characterization and volcanic influences. *Proc. 4th Internl. Symp. on the Geochemistry of the Earth's Surface*, 569-573

Giammanco S., Inguaggiato S. & Valenza M. (1997) Soil and Fumarole Gases of Mount Etna: Geochemistry and Relations with Volcanic Activity. *In press on J. Volcanol. Geotherm. Res.*

Giggenbach W.F. (1991) Chemical techniques in geothermal exploration. In F.D'Amore (ed), *Application of geochemistry in geothermal reservoir development :* Rome. 119-144.

Hoernle K., Yu-Shen Z., and Graham D. (1995) Seismic and geochemical evidence for large-scale mantle upwelling beneath the eastern Atlantic and western and central Europe. *Nature* vol. 374, 34-39.

Inguaggiato S. (1986) Sorveglianza geochimica Campi Flegrei – Parte prima. *Thesis, Università di Palermo.*

Inguaggiato S. and Italiano F. (1998) Helium and carbon isotopes in the submarine gases from the Aeolian arc (Italy). *This volume*

Marty B., Trull T., Lussiez P., Basile I., Tanguy J. C. (1994) He, Ar, O, Sr and Nd isotope constraints on the origin and evolution of Mount Etna magmatism, *Earth Planet. Sci. Lett.,* vol. 126, 23-39.

Penta F., 1963. Ricerche e studi sui fenomeni esalativo-idrotermali ed il problema delle forze endogene. *Ann. Geof.,* vol. 8:1-94.

Penta F. & Conforto B., 1951. Risultati di sondaggi e ricerche geominerarie nell'isola di Ischia dal 1939 al 1943 nel campo del vapore, delle acque termali e delle forze endogene in generale. *Ann. Geof.,* vol. 4: 1-33.

Rosi M., Sbrana A., Principe C., 1983. The Phlegrean Fields: structural evolution, volcanic history and eruptive mechanism. *Jour. Volcanol. Geoth. Res.* vol. 17, ¼: 273-288.

Taylor H. P., Frechen J., and Degens E. T. (1967) Oxygen and carbon isotope studies of carbonatites from the Laacher See district, West Germany and the Alnö district, Sweden. *Geochim. Cosmochim. Acta* vol. 31, 407-430.

Vezzoli L., 1988. Island of Ischia. In C.N.R. *Quaderni della Ricerca Scientifica* n. 114, Progetto Finalizzato Geodinamica, Monografie Finali, vol. 10: 133pp.

Fluid chemistry and water-rock interaction in a CO_2-rich geothermal area, Northern Portugal

J.M. Marques, L. Aires-Barros, R.C. Graça, M.J. Matias & M.J. Basto
Instituto Superior Técnico, Laboratory of Mineralogy and Petrology, Technical University of Lisbon, Portugal

ABSTRACT: The results of geochemical and isotopic studies on the nature of hot (76°C) and cold (17°C) HCO_3-Na-CO_2-rich mineral waters and water-granite interaction processes are present. The upflow zone of these waters is restricted to one of major regional NNE-trending faults located in the northern part of the Portuguese mainland. The high mineralization of the cold waters seems to be mainly related with water-rock interaction controlled by the CO_2 rather than by a high temperature environment. Mineropetrografic observations of cores from an exploration drillhole indicate that the granite underwent a pervasive alteration stage which is very widespread. A vein alteration stage was also detected along major fracture zones, where the granite is almost completely altered into a mineral assemblage where quartz and white mica are the most important phases. The low W/R ratios suggest that we are in the presence of a local rock-dominated environment.

1 INTRODUCTION

During the past few years, special emphasis has been given to the development of the circulation model of Chaves geothermal waters (Aires-Barros et al. 1995, 1997). The present paper deals mainly with the characterisation of mineropetrographic and isotopic changes in granitic cores, from an exploration drillhole (230m depth), in order to predict the water-rock interaction processes occurring at depth. We also try to interpret the influence of the CO_2 in the geochemistry of the hot and cold mineral waters.

2 GEOLOGICAL SETTING

At Vilarelho da Raia, Chaves, Vidago and Pedras Salgadas areas (Figure 1), the youngest formations are sedimentary of Cenozoic age, having their maximum thickness along the central axis of the NNE-SSW depression. The oldest formations (Ante-Ordovician) consist of a schisto-graywacke complex. At Ordovician and Silurian times, quartzites and schists were formed, being metamorphised at the end of Paleozoic near the contact of the Hercynian granitic intrusion. Later on, these rocks were subjected to the activity of Alpine Orogeny, and this is responsible for the formation of the several hydrothermal systems.

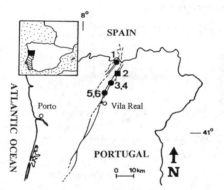

Figure 1. Location map of studied springs. (1) Vilarelho da Raia, (2) Chaves, (3) Campilho, (4) Vidago, (5) Sabroso, (6) Pedras Salgadas.

3 FLUID CHEMISTRY

The studied waters can be divided into "north" and "south" waters for geographical (Figure 1), physico-chemical and isotopic reasons (Table 1). The "north" group consists of the Chaves hot waters (76 °C) and the Vilarelho da Raia cold waters (16°C). These waters are characterised by high mineralization (TDS ≈ 1800 mg/l) and free CO_2 contents between 350 to 1100 mg/l. The "south" group, enclosing Campilho, Vidago, Sabroso and Pedras Salgadas waters, is characterised by higher concentrations of Mg^{2+} and Ca^{2+} and higher free CO_2 contents (up to 2500 mg/l).

Table 1. Main analytical data on water samples. Data from April 1991.

	(a)	(b)	(c)	(d)	(e)	(f)
T(°C)	15.7	75.4	18.4	17.8	17.1	15.6
pH	6.69	7.24	6.19	6.80	6.44	6.45
Na	649.3	588.3	372.6	832.6	559.1	9.8
K	25.0	65.7	30.2	105.6	28.9	1.0
Ca	31.4	22.4	66.7	214.3	190.5	5.6
Mg	6.6	4.35	11.85	40.95	29.70	1.75
HCO_3	1895	1671	1293	4984	2134	12
SO_4	1.7	23.9	11.8	4.0	10.4	8.8
Cl	20.1	38.2	18.5	67.5	35.8	10.1
SiO_2	49.8	75.4	55.8	52.6	70.4	17.0
$\delta^{18}O$	-8.14	-8.15	-6.80	-6.85	-7.32	-7.87
δD	-54.8	-54.5	-46.9	-50.9	-43.9	-50.5
^3H	n.d.	0.69	3.84	n.d.	2.19	7.00

(a) Vilarelho da Raia; (b) Chaves; (c) Vidago AC16; (d) Vidago AC18; (e) Pedras Salgadas AC17; (f) recharge water collected at a high altitude site (± 1000 m a.s.w.l.). Values are in mg/l. $\delta^{18}O$ and δD are in ‰ vs SMOW. ^3H is in T.U. n. d. is below detection limits.

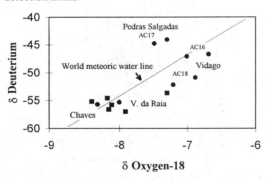

Figure 2. $\delta^{18}O$ and δD relationship in hot (■) and cold (●) mineral waters of the studied region.

$\delta^{18}O$ and δD values of the hot and cold mineral waters (analyses performed at the ITN Laboratory of Environmental Isotopes / Portugal) indicate their meteoric origin. $\delta^{18}O$ and δD values of the "north" waters are more negative than those of the "south" waters (Figure 2) possibly as the result of different recharge altitudes. The presence of ^3H in the "south" mineral waters could be explained by the influx of recent meteoric waters on the margins of the Vidago and Pedras Salgadas basins. The $\delta^{13}C_{(TDC)}$ measurements (Marques et al. 1997) indicate a deep-seated (mantle) origin for the most of CO_2.

4 WATER MINERALIZATION

In the logarithmic Ca^{2+} - HCO_3^- diagram of Figure 3, all the studied hot and cold CO_2-rich mineral waters plot far from the line of electroneutrality $[HCO_3^-] = 2[Ca^{2+}]$, indicating that most HCO_3^- should be related to the dissolution of CO_2 in water. This is very important as the most mineralised water in the system (Vidago AC18) is a cold water showing the highest TDS (4500 mg/l) and HCO_3^- (4984 mg/l) values, more than twice the values of the hot mineral waters in the area. It seems that in Vidago and Pedras Salgadas water-rock interaction is enhanced by the presence of CO_2 in a low temperature environment. This trend seems to explain not only the higher concentrations of alkaline earth metals Ca^{2+} and Mg^{2+} in Vidago and Pedras Salgadas waters but also the high TDS of the cold waters compared with the hot waters.

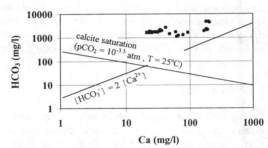

Figure 3. Logarithmic Ca^{2+} - HCO_3^- diagram for the studied hot (■) and cold (●) mineral waters.

5 Rb-Sr ISOTOPE SYSTEMATICS

Rb and Sr concentrations and isotopic ratios ($^{87}Sr/^{86}Sr$ and $^{87}Rb/^{86}Sr$) have been measured in the hot and cold mineral waters of the region and in two granitic whole-rock samples, representative of the deep Rb and Sr contributors. From the comparison of the $^{87}Sr/^{86}Sr$ ratios in the minerals from the rock samples (microcline 0.76359 and 0.75644; plagioclase 0.72087 and 0.71261; muscovite 0.84459; biotite 4.18370 and 2.43928 and quartz 0.70948) with the $^{87}Sr/^{86}Sr$ ratios of the dry residues (from 0.71530 to 0.72804), the origin for the $^{87}Sr/^{86}Sr$ ratios in waters has been interpreted as dominated by dissolution of plagioclase (Aires-Barros et al. 1997).

The whole-rock, K-feldspar and plagioclase fractions give an isochron, corresponding to an age of 310×10^6 years for the Vilarelho da Raia granite

(Figure 4) and of 290×10^6 years for Vidago granite, that is in good agreement with the values presented by Mendes (1968). In the case of Vilarelho da Raia granite, if the muscovite data is incorporated in the calculations, a Rb-Sr isochron age of about 108×10^6 years can be estimated. This absolute age could be in error owing to gain or loss of Rb and Sr after rock crystallisation. Chemical exchange with circulating water seems to be a reasonable explanation.

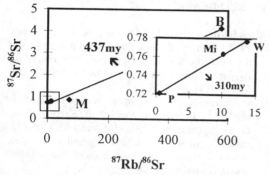

Figure 4. Rb-Sr diagram for the Vilarelho da Raia granite. (W) stands for whole-rock, (Mi) for microcline, (P) for plagioclase, (M) for muscovite and (B) for biotite. Profitable and disturbed isochrons are given. The ages were calculated using λ (^{87}Rb) = 1.39×10^{-11} yr^{-1}.

On the other hand, the Rb-Sr data of biotite from both rock samples shifts the whole-rock, K-feldspar and plagioclase isochrons to an age of about 437×10^6 years (Figure 4), but its significance is doubtful. The scatter of the biotite data could be due to somewhat different initial Sr compositions.

6 ALTERATION AND MINERALOGY

Mineropetrographic observations indicate a pervasive alteration stage which has affected the whole drilled massif. Transformation of biotite and plagioclase has given way to chlorite and sericite, respectively. Microcline rarely includes any secondary minerals. A vein alteration stage was also observed along major fracture zones where the granite is altered into a quartz and white mica (mainly muscovite 2M$_1$) mineral paragenesis, as revealed by IRS and XRD data. Illite, halloysite, chlorite, vermiculite, siderite and iron oxides are also found in some samples. Vein alteration zones are one to several centimetres thick and quite visible in cores. Preliminary studies on the chemistry of micas occurring along vein alteration zones, performed at the LURE (Laboratoire pour l'Utilisation du Rayonnement Electromagnétique / Orsay / France) by non-destructive XRF analysis using a synchrotron radiation microprobe, indicate that: (a) compared to white micas, the green micas show higher contents of Fe, Ti and Mn; (b) the white micas are characterised by higher contents of K, Rb and Ga.

7 WATER/ROCK RATIOS

For a closed system, water/rock (W/R) ratios can be estimated by the equation proposed by Taylor (1978):

$$W/R_{closed} = \frac{\delta^{18}O_{rock}^{final} - \delta^{18}O_{rock}^{initial}}{\delta^{18}O_{fluid}^{initial} - \delta^{18}O_{fluid}^{final}}$$

where W and R are the atom percentages in the fluid (W) and in the rock (R). In an open system, where the heated water is lost from the system by escape to the surface, we have (Taylor, 1978):

$$W/R_{open} = ln\ (W/R_{closed} + 1)$$

The initial δ^{18}O values for the rock (+11.41 and +11.47°/oo) were obtained by analysing the Vilarelho da Raia granite in two different sites outside the springs area. These values fall within Taylor's (1978) "high -^{18}O granites" group. The final δ^{18}O values for the rock (+10.10 to +10.91°/oo) are the values measured, on core samples, in bands displaying the above mentioned vein alteration characteristics. For core samples where more than one isotopic analysis has been performed, the δ^{18}O values decrease with decreasing distance from the vein. The δ^{18}O and δD values of whole-rocks were performed at Delta Isotopes / The Netherlands and XRAL / Canada Laboratories. In our case, the main questions regarding the application of this methodology are related with the initial and final composition of the fluid. Assuming the meteoric origin of the water responsible for the rock alteration, the initial water composition was calculated from the D/H ratio of the alteration assemblage (along veins) and the world meteoric water equation ($\delta D=8\delta^{18}O+10$). The values obtained (from -8.25 to -6.50 °/oo) do not seem unrealistic, since they are similar to those of the present-day local meteoric waters (see Figure 2). Only one sample indicates a lighter δ^{18}O - value (-11.33 °/oo) for the initial fluid. The final water composition was estimated by the O'Neil and Taylor (1967) plagioclase-water isotope fractionation equation:

$$1000 \ln \alpha_{\text{plag-water}} = -3.52 + 2.68 \, (10^6 / T^2)$$

assuming that $\delta^{18}O_{\text{plag}} \approx \delta^{18}O_{\text{whole-rock}}$ (final composition), because plagioclase is an abundant mineral in the rocks and exhibits the greatest rate of ^{18}O exchange with an external fluid phase. In our case, water-rock interaction temperatures (T) were estimated by the stability fields of the alteration minerals. As indicated by several authors (Genter, 1990; Petrucci et al. 1993; Lutz et al. 1996), the observed mineral assemblages indicate formation temperatures between 150°C and 250°C. Considering these temperatures, the W/R ratios obtained for the open system range between 0.056 and 0.24 and between 0.038 and 0.12, respectively. These values suggest that we are in the presence of a local rock-dominated environment. A slight ^{18}O depletion (10.10 ‰ ⇐ 11.47‰) in the rock induces a high positive ^{18}O shift (-6.5 ‰ ⇒ -1.4 ‰) in the water. The lowest W/R ratios (0.038 and 0.056) are related with core samples characterised by the lack of visible natural fractures, but displaying macro and microscopic aspects similar to vein alteration zones. In those cases, one may think on the existence of metasomatic substitutions of pneumatolitic origin, more than hydrothermal fluids circulation.

8 CONCLUDING REMARKS

The low temperature and high ^3H activity of Vidago and Pedras Salgadas waters should correspond to shallow flowpaths, being the high TDS of these cold waters controlled by the presence of CO_2 in a low temperature environment. The higher Na^+ contents of the hot and cold mineral waters compared with the K^+ contents of the same waters could be due (a) to the dominated dissolution of plagioclases, or (b) to K-feldspar↔Na-feldspar equilibrium. The dominant presence of white mica along vein alteration zones seems to be responsible for the conservation of large amounts of K^+ by the solid phase. The lack of an ^{18}O-shift in the present-day Chaves thermal waters indicates that these waters could be considered as "mature" discharges of an hydrothermal system with small change in isotopic composition, as the result of relatively short circulation time.

9 ACKNOWLEDGEMENTS

The authors would like to thank Águas de Carvalhelhos Company for supplying the well cores.

This work was funded by the PRAXIS Project "Aquatransfer" under the Contract No. 3/3.1/ /CEG/2664/95 and by the EC TMR (Training and Mobility of Researchers) Programme. Dr. J. R. Hulston and an anonymous reviewer gave very helpful comments and suggestions and we thank them both.

10 REFERENCES

Aires-Barros, L., J. M. Marques & R. C. Graça 1995. Elemental and isotopic geochemistry in the hydrothermal area of Chaves/Vila Pouca de Aguiar (northern Portugal). *Environmental Geology*. 25 (4): 232-238.

Aires-Barros, L., J. M. Marques, R. C. Graça, M. J. Matias, C. H. van der Weijden, R. Kreulen & H. G. M. Eggenkamp 1997. Hot and cold CO_2-rich mineral waters in Chaves geothermal area (northern Portugal). Paper accepted for publication in *Geothermics*. 28.

Genter, A. 1990. Géothermie roches chaudes sèches: le granite de Soultz-sous-Forêts (Bas-Rhin, France). Fracturation naturelle, altérations hydrothermales et interaction eau-roche. *Document du BRGM*. 85: 193 pp.

Lutz, S. J., J. N. Moore & D. Benoit 1996. Alteration mineralogy of the Dixie Valley geothermal system, Nevada. *Geothermal Resources Council Transactions*. 20: 353-362.

Marques, J. M., P. M. Carreira, L. Aires-Barros, R. C. Graça & A. Monge Soares 1997. O $\delta^{13}C$ das águas gasocarbónicas frias de Vilarelho da Raia, Vidago e Pedras Salgadas como indicador da origem do CO_2. *Actas da X Semana de Geoquímica - IV Congresso de Geoquímica dos Países de Língua Portuguesa*: 571 - 574.

Mendes, F. 1968. Contribuition à l'étude géochronologique, par la méthode au strontium, des formations cristallines du Portugal. *Boletim do Museu e Laboratório Mineralógico e Geológico da Faculdade de Ciências*. 11 (1): 3-157.

O'Neil, J. R. & P. Taylor Jr. 1967. The oxygen isotope and cation exchange chemistry of feldspars. *Am. Mineral*. 52: 1415-1437.

Petrucci, E., S. M. F. Sheppard & B. Turi. 1993. Water/rock interaction in the Larderello geothermal field. *Journal of Volcanology and Geothermal Research*. 59: 145-160.

Taylor, Jr., H. P. 1978. Oxygen and hydrogen isotope studies of plutonic granitic rocks. *Earth Planet. Sci. Lett.*. 38: 177-210.

Sulfur redox chemistry and the origin of thiosulfate in hydrothermal waters of Yellowstone National Park

D.K. Nordstrom, K.M. Cunningham & J.W. Ball
US Geological Survey, Water Resources Division, Boulder, Colo., USA

Y. Xu & M.A.A. Schoonen
Department of Geosciences, SUNY at Stony Brook, N.Y., USA

ABSTRACT: Thiosulfate ($S_2O_3^{2-}$), polythionate ($S_xO_6^{2-}$), dissolved sulfide (H_2S), and sulfate (SO_4^{2-}) concentrations in 39 alkaline and acidic springs in Yellowstone National Park were determined. The analyses were conducted *on site*, using ion chromatography for thiosulfate, polythionate, and sulfate, and using colorimetry for dissolved sulfide. Thiosulfate concentrations ranged from <0.01 to 8 ppm. Polythionate was detected only in Cinder Pool, Norris Geyser basin, at concentrations up to 1.8 ppm, with an average sulfur-chain-length from 4.1 to 4.9 sulfur atoms. The results indicate that no thiosulfate occurs in the deeper parts of the hydrothermal system. Thiosulfate may form from (1) hydrolysis of native sulfur by hydrothermal solutions in the shallower parts (<50 m) of the system, (2) oxidation of dissolved sulfide upon mixing of a deep hydrothermal water with aerated shallow groundwater, and (3) the oxidation of dissolved sulfide by air. Air oxidation of H_2S-containing hot springs proceeds very rapidly.

1. INTRODUCTION

The occurrence of thiosulfate in hydrothermal waters has been investigated for many active hydrothermal systems throughout the world (e.g., Allen and Day, 1935; Boulegue, 1978; Ivanov, 1982; Subzhiyeva and Volkov, 1982; Webster, 1987; Veldeman et al., 1991; Migdisov et al., 1992). The interest in thiosulfate in hydrothermal solutions stems from the notion that thiosulfate may be of importance as a complexer of precious metals, and as an intermediate in sulfur isotope exchange between sulfate and dissolved sulfide.

As shown in laboratory studies, thiosulfate may easily form as a result of incomplete oxidation of dissolved sulfide (defined operationally as that measured using the methylene blue colorimetric method and includes H_2S, HS^-, and S_x^{2-}; e.g., Chen and Morris, 1972; O'Brien and Birkner, 1977; Zhang and Millero, 1993). Therefore, it is not clear whether the thiosulfate detected in surficial hot waters is representative of deep hydrothermal solutions. Furthermore, sampling and preservation may significantly affect sulfur speciation in hot spring waters between the time of sampling and analysis. Oxidation of dissolved sulfide species to thiosulfate occurs rapidly upon brief exposure to oxygen and thiosulfate decomposes slowly over longer periods of time. Consequently, thiosulfate concentrations reported for hydrothermal waters will be affected by changes after sampling.

As part of a comprehensive investigation into the geochemistry of intermediate sulfoxyanions in hydrothermal solutions, the goal of this study is to evaluate the abundance and origin of thiosulfate in surficial hydrothermal waters in Yellowstone National Park (YNP). To minimize the formation or decomposition of thiosulfate after sampling, the analyses were conducted *on site* in a mobile laboratory operated and maintained by the U.S. Geological Survey. Detailed studies were conducted at Azure Spring and Ojo Caliente Spring in Lower Geyser Basin, Cinder Pool in Norris Geyser Basin, Angel Terrace Spring of Mammoth Hot Springs, and Frying Pan Spring located near Norris Geyser Basin.

2. SAMPLING AND ANALYSES

Sampling and analyzing waters containing thiosulfate and polythionates is challenging. The storage time was kept as short as practically possible by analyzing the samples on site. For small and non-violent springs, waters were drawn into a 60 mL plastic syringe through Teflon tubing (0.31 cm ID). The sample was immediately filtered using a syringe filter (0.45 mm) and collected in a 30 mL sampling bottle. For large and violent features, samples were collected using a stainless steel thermos can attached to an aluminum pole. Samples for thiosulfate determination were collected in a 30 mL amber glass bottle containing 1mL of a 1M $ZnCl_2$ solution. The bottles were filled to the rim in order to avoid trapping atmospheric oxygen. The bottles were kept on ice until analysis. For analysis, an aliquot of water was drawn into a 3 mL syringe and injected into a Dionex 2010i/2000i ion chromatograph through a luer-lock syringe filter (0.45 mm pore-size). Using two AG4A guard columns in series, in combination with a standard $NaHCO_3/Na_2CO_3$ eluent, it was possible to determine thiosulfate within 3 minutes. Using a conductivity detector, the detection limit for thiosulfate is 0.01 ppm. Samples for polythionate analysis were collected in 30 mL amber glass bottles containing 1mL of 1M $ZnCl_2$, 1mL of 1M NaOH, and 1mL of 1M KCN solutions. Polythionate concentrations and average chain lengths were determined using alkaline cyanolysis modified from a method by Kelley et al. (1969), Moses et al. (1984), and Schoonen (1989). The detection limit of thiocyanate was 0.05 ppm, and the retention time is about 6 minutes.

Concentrations of dissolved sulfide were determined on site using a portable HACH DR-2000 UV-Vis absorption spectrometer. The HACH method # 8131, based on the methylene blue method, was used. The detection range is 0.666 - 0.001 ppm, and some samples had to be diluted. Dissolved oxygen was measured using the Winkler method, the fixed samples were titrated typically within 2 hours of collection.

3. RESULTS AND CONCLUSIONS

The $S_2O_3^{2-}$, $S_nO_6^{2-}$, H_2S and SO_4^{2-} concentrations for 6 selected hot springs of YNP are given in Table 1.

Ojo Caliente, representative for low-thiosulfate and high-sulfide springs, was studied in detail. Ojo Caliente consists of a pool and a well-defined drainage system (photos of this spring and others studied in detail are accessible via the WWW, *http://sbmp97.ess.sunysb.edu/*). The water in the pool contains 1 ppm sulfide, but only 0.27 ppm thiosulfate and 0.1 ppm dissolved oxygen. As the water discharges into the drainage, the dissolved oxygen content increases rapidly (Fig. 1A). As a result, sulfide is oxidized to thiosulfate and sulfate. Sulfide decreases at an average rate of about 0.05 ppm s^{-1}, while thiosulfate forms at an average rate of 0.01 ppm s^{-1}. Sulfate increases at a rate about 0.02 ppm s^{-1} after an initial lag period, suggesting that sulfate is not the initial product of the oxidation of sulfide. It is important to note that only 25% of the discharged dissolved sulfide was oxidized to thiosulfate and sulfate, the remainder volatilizes to the atmosphere.

Rapid formation of thiosulfate from air oxidation of dissolved H_2S is illustrated by the results from Angel Terrace. In the orifice of this spring, the thiosulfate is below the detection limit (<0.01ppm), yet within 1 m of the orifice ($\leq$ 1 second travel time down a waterfall), the thiosulfate concentration increased to 0.5 ppm along with a rapid increase of dissolved oxygen concentration (see Fig. 1B).

Azure Spring is a hot spring which discharges alkaline bicarbonate-chloride water at a maximum flow rate of 2L/sec. Temperature fluctuating between 60° to 87°C have been observed. Fig. 2 shows changes in thiosulfate and dissolved sulfide concentrations along the drainage of Azure Spring. In contrast to the drainage profiles of Ojo Caliente and Angel Terrace, oxidation of dissolved sulfide after discharge does not contribute much to the thiosulfate concentrations along the Azure drainage. Instead, thiosulfate concentrations decreased slightly along the drainage. Notice that H_2S is much lower

Table 1. Sulfur species, chloride, and dissolved oxygen in selected hot springs.

Spring	T °C	pH	D.O. ppm	$S_2O_3^{2-}$ ppm	$S_nO_6^{2-}$ ppm	H_2S ppm	SO_4^{2-} ppm	Cl$^-$ ppm
Ojo Caliente Spring	93.0	7.8	0.1	0.27	NA	1.09	20.7	324
Azure Spring	84.2	8.7	0.0	6.20	NA	0.36	41.0	306
Cinder Pool	92.0	4.22	0.5	4.20	1.8	1.60	96.0	601
Cistern Spring		6.10		7.93	<0.05		89.5	431
Frying Pan Spring	78.2	2.34	2.4	0.63	<0.05	1.03	437	9
Angel Terrace Spring	72.2	6.43	0.5	<0.01	<0.05	3.01	547	165

NA: data not analyzed.

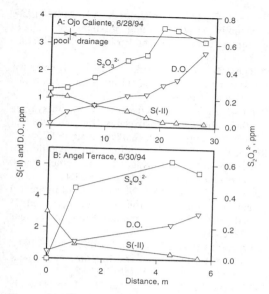

Figure 1. Concentration of sulfur in the form of thiosulfate, dissolved sulfide and dissolved oxygen vs distance along the drainage of (Fig. 1A) Ojo Caliente and (Fig. 1B) Angel Terrace Spring.

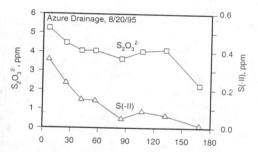

Figure 2. Thiosulfate and sulfide concentrations with distance along the drainage of Azure Spring.

and $S_2O_3^{2-}$ much higher at Azure than at Ojo Caliente or Angel Terrace.

The highest thiosulfate concentration (8 ppm) of all springs sampled in this study was obtained in Cistern Spring, a slightly acidic (pH 6.1) sulfate-chloride spring. The only sulfate-chloride spring with pH<5 that contains a significant amount of thiosulfate is Cinder Pool. Polythionate was found only in Cinder Pool. The concentrations varied from below detection (0.05 ppm) to 1.8 ppm. The average length of the sulfur chain of the polythionate was 4.1 to 4.9. Since Cinder Pool has not only a unique speciation of aqueous sulfoxyanions, but also contained floating sulfur spherules and a subsurface molten sulfur layer (White et al., 1988). Additional laboratory experiments were conducted to investigate the effects of the presence of the molten sulfur pool and sulfur spherules on the aqueous sulfur speciation. The experiments demonstrated that thiosulfate was readily generated from the hydrolysis of elemental sulfur at the conditions found at the bottom of Cinder Pool (120°C, pH 4). Furthermore, the polythionate in Cinder Pool appears to be formed by air oxidation of thiosulfate catalyzed by pyrite in the sulfur spherules.

The enrichment of thiosulfate in slightly acidic to alkaline chloride waters which are enriched in sulfate suggests that thiosulfate formation is part of the processes that lead to the formation of acidic sulfate-chloride waters. In laboratory experiments, thiosulfate has been observed to be an intermediate product in: 1) H_2S oxidation by dissolved oxygen (O'Brien and Birkner, 1977; Zhang and Millero, 1993); 2) H_2S - SO_3^{2-} interaction in aqueous solutions (Zhang and Millero, 1993; Xu et al., 1997); and 3) hydrolysis of elemental sulfur at elevated temperature (Robinson, 1973; Giggenbach 1974a, 1974b; Rafal'sky et al., 1983; Dadze and Sorokin, 1993; Xu et al., 1997). Each of these processes will be briefly discussed.

The fact that many springs have no significant amount of thiosulfate in their orifice or pools despite high dissolved sulfide concentration in some of these springs indicates that the aeration of dissolved sulfide in hot springs before discharging at the surface is not an effective mechanism for thiosulfate formation. While thiosulfate formation via oxidation of sulfide is of importance in the drainages, it plays little or no significant role in the springs and pools where the water first emerges.

There is little evidence to support a significant discharge of magmatic SO_2 gas into the hydrothermal system at YNP (Zinder and Brock, 1977).

One process which could result in local enrichment of thiosulfate in hydrothermal waters is the hydrolysis of elemental sulfur by ascending hydrothermal solutions. It is well known that in aqueous solutions at elevated temperatures, S_8 opens and disproportionation reactions occur to form sulfide and sulfoxyanions (Dadze and Sorokin, 1993; Xu, 1997). The composition of the oxyanions formed during the disproportionation depends on the temperature and pH of the solution.

Sulfur hydrolysis as the predominant source for thiosulfate in Norris Geyser Basin is consistent with the geochemical and sulfur-isotopic characteristics of waters in the hydrothermal system at Norris. When the ascending neutral hydrothermal solution comes in contact with locally buried sulfur deposits at shallow depth in Norris Basin, the hydrolysis of elemental sulfur can produce H_2S and thiosulfate.

REFERENCES

Allen, E.T. & A. Day 1935. Hot Springs of the Yellowstone National Park, Carnegie Institute of Washington, Pub. 466, 525 pp.

Boulegue, J. 1978. Metastable sulfur species and trace metals (Mn, Fe, Cu, Zn, Cd, Pb) in hot brines from the French Dogger. *Am. J. Sci.* 278:1394-1411.

Chen, K.Y. & J.C. Morris 1972. Kinetics of oxidation of aqueous sulfide by O_2. *Environ. Sci. Tech.* 6:529-537.

Dadze, T.P. & V.I. Sorokin 1993. Experimental determination of the concentrations of H_2S, HSO_4, $SO_{2(aq)}$, $H_2S_2O_3$, $S^0_{(aq)}$, and S_{tot} in the aqueous phase in the $S-H_2O$ system at elevated temperatures. *Geochemistry International* 30:36-51.

Giggenbach, W.F. 1974a. Equilibria involving polysulfide ions in aqueous sulfide solutions up to 240 °C. *Inorganic Chem.* 13:1724-17303.

Giggenbach, W.F. 1974b. Kinetics of the polysulfide-thiosulfate disproportionation up to 240 °C. *Inorganic Chem.* 13:1731-1734.

Ivanov, B.V. 1982. Recent hydrothermal activity in the region of the Karym group of volcanos. In: *Hydrothermal mineral-forming solutions in the areas of active volcanism*, ed. Naboka S. I., 39-48, Nauka Publisher, Novosibirsk.

Kelley, D.P. L.A. Chambers & P.A. Trudinger 1969. Cyanolysis and spectrophotometric estimation of trithionate in mixtures with thiosulfate and tetrathionate. *Anal. Chem.* 41:891-901.

Migdisov, A. A., T.P. Dadze & L.A. Gerasimovskaya 1992. Forms of sulfur in the ore-bearing hydrothermal solutions of Uzon Caldera (Kamchatka). *Geochem. Inter.* 29:17-27.

Moses, C.O., D.K. Nordstrom & A.L. Mills 1984. Sampling and analyzing mixtures of sulfate, sulfite, thiosulfate and polythionate. *Talanta.* 31:331-339.

O'Brien, D.J. & F.G. Birkner 1977. Kinetics of oxygenation of reduced sulfur species in aqueous solutions. *Environ. Sci. Tech.* 11:1114-1120.

Rafal'skiy, R.P., N.I. Medvedeva & N.I. Prisyagina 1983. Interaction between sulfur and water at elevated temperatures. *Geokhimia* 5:665-676.

Robinson, B. W. 1973. Sulfur isotope equilibrium during sulfur hydrolysis at high temperatures. *Earth and Planet. Sci. Lett.* 18:443-450.

Schoonen, M.A.A. 1989. Mechanisms of pyrite and marcasite formation from solutions between 25 and 300°C. Ph.D. dissertation, Pennsylvania State University.

Subzhiyeva, T. & I.I. Volkov 1982. Thiosulfates and sulfites in thermal and hydrothermal waters. *Geochemical International* 19:94-98.

Veldeman, E., L. Van't Dac, R. Gijbels & E. Pentcheva 1991. Sulfur species and associated trace elements in south-west Bulgarian thermal waters. *Appl. Geochem.* 6:49-62.

Webster, J.G. 1987. Thiosulfate in surficial geothermal waters, North Island, New Zealand. *Appl. Geochem.* 2:579-584.

White, D.E., R.A. Hutchinson & T.E.C. Keith 1988. The geology and remarkable thermal activity of Norris Geyser Basin, Yellowstone National Park, Wyoming. *U.S. Geol. Survey Prof. Paper* 1456.

Xu, Y. 1997. Kinetics of the redox transformations of aqueous sulfur species: the role of intermediate sulfur oxyanions and mineral surfaces, Ph. D dissertation, SUNY, Stony Brook

Xu, Y., M.A.A. Schoonen, D.K. Nordstrom, K.M. Cunningham & J.W. Ball 1997. Sulfur Geochemistry of Hydrothermal Waters in Yellowstone National Park: II The geochemical cycle of sulfur in Cinder Pool, in review.

Zhang, J.Z. & F.J. Millero 1993. Kinetics of oxidation of hydrogen sulfide in natural waters. In *Environmental geochemistry of sulfide oxidation.* ed C.A. Alpers & D.W. Blowes. 393-40., Am. Chem. Soc. Symp. Series 550, Washington, D.C.

Zinder, S. & T.D. Brock 1977. Sulfur dioxide in geothermal waters and gases. *Geochim. Cosmochim. Acta* 41:73-79.

Hydrogeochemical and isotope geochemical features of the thermal waters of Kizildere, Salavatli, and Germencik in the rift zone of the Büyük Menderes, western Anatolia, Turkey: Preliminary studies

N. Özgür & A. Pekdeger
Freie Universität Berlin, FR Rohstoff- und Umweltgeologie, Germany

M. Wolf, W. Stichler & K. P. Seiler
GSF-Institute of Hydrology, Neuherberg, Germany

M. Satir
Eberhard-Karls-Universität Tübingen, Lehrstuhl für Geochemie, Germany

ABSTRACT: Hydrogeochemical and isotope geochemical features of the geothermal waters of Kizildere, Salavatli, and Germencik in the rift zone of Büyük Menderes were investigated. The metamorphic and sedimentary rocks in the thermal fields are altered intensively. The hydrothermal alteration is distinguished by phyllic, argillic, silicic ± haematitization, and carbonatization zones. The surface temperatures of the waters range from 30 to 99 °C whereas the geochemical thermometers indicate a reservoir temperature between 162 and 232 °C. The $NaSO_4$-$NaCO_3$ type thermal waters show meteoric origin due to isotopic values of δ^2H and $\delta^{18}O$.

1 INTRODUCTION

The geothermal fields of Kizildere, Salavatli, and Germencik are located along the rift zone of the Büyük Menderes from Denizli in the eastern part to Kusadasi at the Aegean Sea coastal area in the western Part within the Menderes Massif (Figs. 1 and 2). The Menderes Massif was uplifted due to the compressional tectonics in the Middle Miocene (Özgür et al., 1997). Subsequently, the rift zones in the crystalline massif formed by extensional tectonics, one of which is the rift zone of the Büyük Menderes. It is represented by a great number of geothermal hot springs (Figs. 1 and 2). We have investigated the thermal waters of Kizildere, Salavatli, Germencik, and their environs (Fig. 2). These geothermal waters are related to faults which strike NW-SE and NE-SW, diagonal to general strike of the Büyük Menderes rift zone; they are generated by compressional tectonic stresses and uplift between two extentional rift zones (Özgür & Pekdeger, 1995; Özgür et al., 1997). In addition to earthquake activity and heat flow anomalies in the rift zone of Büyük Menderes (Alptekin et al., 1990), we have mapped the localities of a great number of calcalkaline basic towards acidic volcanic rocks in three rift zones which range in age from Middle Miocene to 20.000 a (Özgür et al., 1997; Özgür, 1997). These volcanic rocks are of products of continental crust based on isotope analyses of $^{87}Sr/^{86}Sr$ and $^{143}Nd/^{144}Nd$ and considered as heat source for the heating of thermal fluids in the rift zones within the Menderes Massif (Özgür, 1997). Additionally, the geothermal gradients can contribute to heat the thermal waters in the reservoirs which consist of limestone, conglomerate, marble, quartzite, and gneiss.

The aim of this paper is to give an overview and to present hydrogeochemical and isotope geochemical data of the thermal waters in the rift zone of the Büyük Menderes in combination with interpretation the origin and the evolution of these waters.

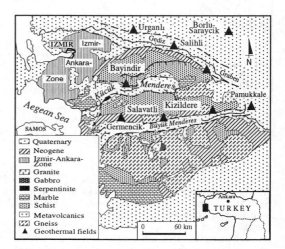

Fig. 1 Geologic setting of the Menderes Massif and location map of the investigated thermal fields of Kizildere, Salavatli, and Germencik.

2 GEOLOGIC SETTING

The geothermal field of Kizildere with eight (8) production wells to a depth down to 1242 m is located in the north of the eastern part of the Büyük Menderes rift zone (Figs. 1 and 2) where there is a 20 MW geothermal power plant. Here, the Paleozoic basement rocks are overlain by Pliocene sedimentary rocks which have good permeability. The Sazak formation, which consists of altered limestone, forms the first shallow reservoir of thermal water. The Igdecik formation composed of quartzite, marble, and mica schist is of second deep thermal water reservoir. The geothermal field of Salavatli with two (2) production wells to a depth down to 1510 m is situated in the north of middle part of the Büyük Menderes rift zone (Figs. 1 and 2) and consists of Paleozoic metamorphic rocks and sedimentary rocks in age from Miocene to Pliocene. Marble, quartzite, and mica schist form the thermal water reservoir. The thermal field of Germencik with nine (9) production wells to a depth down to 2398 m is located in the north of the western part of the Büyük Menderes rift zone (Figs. 1 and 2) and consists of Paleozoic metamorphic rocks and Miocene to Pliocene conglomerates, marble, quartzite, and mica schist form in Germencik the geothermal reservoir.

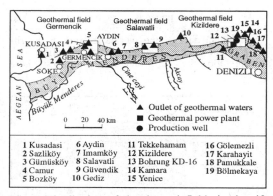

Fig. 2 Distribution of the thermal fields in the rift zone of Büyük Menderes.

3 HYDROTHERMAL ALTERATION

At the surface, the sedimentary rocks at Kizildere, Salavatli, and Germencik are intensively hydrothermally altered recognized by a distinct color change of the rocks in the thermal fields and investigations of rock microscopy on thin sections and x-ray diffractometry. The hydrothermal alteration is noticeable at the surface and distinguished by phyllic, argillic, silicic ± haematitization, and carbonatization alteration zones. The phyllic and argillic alteration zones occur in the Paleozoic metamorphic rocks during the important alteration zone of carbonatization can be observed in sedimentary rocks only (Özgür et al. 1998, this volume). Due to high temperature, pressure, and CO_2 content in the thermal water reservoir of the investigated area, high temperature acid water infiltrates the soil formations and dissolves $CaCO_3$. The solution aims to get in an equilibrium constantly, but does not reach the suitable parameters of the corresponding environment. Thus, the solution loses the temperature, pressure, and CO_2 on its way to the surface, for which reason the pH values increase up to 9.2. Consequently, the geothermal fluid is unable to keep $CaCO_3$ in solution, but it precipitates and occurs in sedimentary rocks near the surface. The Silicic alteration zone can be observed in both Paleozoic metamorphic and Miocene to Pliocene sedimentary rocks (Özgür & Pekdeger, 1995).

4 HYDROGEOCHEMISTRY OF THE THERMAL WATERS

During the present study the outflow of 10 thermal springs and 13 production wells has been sampled in different seasons since 1992. Additionally, we have collected about 240 rock samples.

These thermal waters flow out of clastic sediments of Miocene to Pliocene age. The discharge of thermal waters is linked to faults which strike in E-W direction dip vertically. The surface temperatures of the thermal waters are 99 °C in Kizildere, 45 to 95 °C in Salavatli, and 30 to 92 °C in Germencik whereas the geochemical thermometers show reservoir temperatures of 200 to 212 °C in Kizildere, 162 to 180 °C in Salavatli, and 216 to 232 °C in Germencik. The total dissolved solids of the thermal waters ranges from 1800 to 6000 mg/l. At the surface the scale formations are generated by decreasing temperature and pressure. The scale formations consist of carbonates with high Sr anomalies and minor contents of silicates and sulfides. These minerals are mostly aragonite and calcite.

Hydrogeochemically, the thermal waters can be considered as Na-(Cl)-(SO_4)-HCO_3 type (Fig. 3) and are distinguished by an increase and a decrease of dominantly cations, anions, and trace elements from marine towards continental environment. Trace elements which indicate intensive high-temperature water-rock interaction such as $B^{(III)}$ and F^-, are found in high concentrations in these waters. With exception of As, Sb, and Hg, the thermal waters in the investigated fields are very poor in heavy metals. As, Sb, and Hg are known to precipitate mostly at the surface.

There is a close correlation between F and B in depending on the increasing Na contents and the decreasing Ca contents in high temperature waters (Özgür & Pekdeger, 1995). The origin of $B^{(III)}$ can be attributed to mineral phases, i.e. tourmaline, biotite in metamorphic rocks, and boron deposits at depth. Fluorine can be lead either to a volcanic activity or to the lower Ca contents in the fluids.

5 ISOTOPE GEOCHEMISTRY

Samples from hot springs, production wells, and groundwaters in the rift zone of the Büyük Menderes have been analyzed for their oxygen-18, deuterium and tritium contents. Additionally, the thermal waters of Kizildere were considered due to contents of ^{13}C and ^{14}C.

The stable isotope contents of δ^2H and δ^{18}O are represented in Fig. 4. The above mentioned rainwaters and mixed groundwater-thermal water systems lie along the meteoric water line whereas the high temperature deep groundwater systems deviate from the MWL suggesting a fluid-rock interaction under high temperature conditions. The differences in the degree of isotope shift from Kizildere to Germencik indicate a mixing between three thermal waters which are characterized by δ^2H, δ^{18}O, δ^{13}C, ^{3}H, and ^{14}C values.

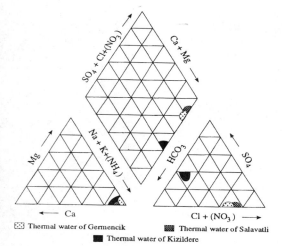

Fig. 3: Geothermal waters of the thermal fields of Kizildere, Salavatli, and Germencik in the rift zone of the Büyük Menderes.

The tritium measurements reveal that (i) the production wells of Kizildere, Salavatli, and Germencik do not contain any measurable tritium and (ii) the mineralized groundwater and groundwater-thermal water systems contain atmospheric and anthropogenic tritium. Therefore, the tritium is not suitable for the age determination of the thermal waters of Kizildere, Salavatli, and Germencik. According to ^{13}C data, the origin of CO_2 might be attributed to magmatic activity as well as to reaction of thermal waters with carbonate rocks at depth (Özgür, 1997). These thermal waters show ^{14}C contents under detection limit because of the dilution of ^{14}C by high contents of CO_2. Under the consideration of this fact, the thermal waters of Kizildere, Salavatli, and Germencik might be interpreted as old waters with a relatively age of 10.000 to 30.000 a.

The isotopic data are well correlated with the results of the chemical analyses which also indicate intensive fluid-rock interaction and reactions with silicates. Due to high temperature and intensive water-rock interaction, the waters are chemically modified by the reaction with crystalline rocks. Good correlations are given with $NaSO_4$-$NaCO_3$ type waters and indicator elements such as Li, B, and F.

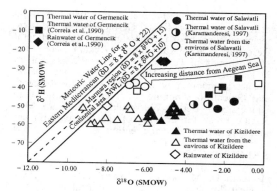

Fig. 4 Plot of δ^2H versus δ^{18}O of thermal waters of Kizildere, Salavatli, and Germencik in the rift zone of Büyük Menderes.

6 CONCLUSIONS

The investigated thermal waters of Kizildere, Salavatli, and Germencik are of meteoric origin. These meteoric waters percolate above the permeable clastic sediments in the unpermeable metamorphic rocks (mica schist, quartzite) heated by cooling magmatic melting. The heating of the thermal waters by a subvolcanic activity are proven by the distinctly enrichment of mantle helium in geothermal fluids of Kizildere (Gülec, 1988; Ercan et al., 1995) which might be interpreted as ^{3}He surplus in comparison to pure continental crustal fluids. This high value of mantle helium might be interpreted that basic volcanic rocks from the earth mantle show an interaction with the geothermal fluids in the environs of the thermal field of Kizildere. Besides some intrusive rocks in the rift zones of Menderes Massif, calkalcaline basic to intermediate volcanic rocks exist which are generated from Middle Miocene to recent. For example, the basalts of Kula 150 km northern part of the investigated area, occur in a young age of 20.000 a. Thereby, the presence of magma chamber at depth of the rift zone of Büyük Menderes is not to exclude.

Additionally, it is noticeable, that the geothermal gradients form second possibility to heat the meteoric waters in the area which correspond to the intensity of heat flow and earthquake activity. Nevertheless, the waters at depth react with heated rocks which leads to water-rock interaction. The hydrothermal convection cells bring the heated waters to the surface due to their lower density. In the rift zone of the Büyük Menderes, the thermal waters occur in the faults of NW-SE and NE-SW directions in terms of steams and hot springs consequently.

7 ACKNOWLEDGMENTS

Sampling of the thermal waters in production wells of Kizildere, Salavatli, and Germencik was made by technical support of General Directorate of Mineral Research and Exploration, Ankara and the establishment of Turkish Electricity Works, Ankara in Kizildere, Denizli. This project has been partly financial supported by the Commission for Research and Scientific Training of New Recruits, Free University of Berlin. We would like to thank Mrs. Renate Erbas for drawing of the figures with her indefatigable patience and feeling.

8 REFERENCES

Alptekin, Ö., M. Ilkisik, Ü. Ezen & S.B. Ücer 1990. Heat flow, seismicity and the crustal structure of western Anatolia. In: *Proc. Internat. Earth Sc. Congr. on Aegean Regions, Izmir/Turkey*, M.Y Savascin & A.H. Eronat (eds.), v. II, p. 1-12.

Correia, H., C. Escobar & B. Gauthier 1990. Germencik geothermal field feasibility report: part two: reservoir and well test analysis: *internal report, Turkish Republic, Ministry of Energy and Natural Resources, Turkey*, 80 p.

Ercan, T., J.-I. Matsuda, K. Nagao & I. Kita 1995. Noble gas isotopic compositions in gas and water samples from Anatolia. In: *Geology of the Black Sea Region: Proc. Internat. Symp. on the Geology of the Black Sea Region, September 7-11, 1992, Ankara, Turkey*, A. Erler, T. Ercan, E. Bingöl & S. Örcen (eds.), p. 197-206.

Gülec, N. 1988. Helium-3 distribution in western Turkey. *MTA Bull.* 108: 35-42.

Karamanderesi, I.H. 1997. Geology and hydrothermal alteration processes in the Salavatli (Aydin) geothermal field. *Ph.D. Thesis, Dokuz Eylül Üniversitesi, Izmir, Turkey*, 240p.

Özgür, N. & A. Pekdeger 1995. Active geothermal systems in the rift zones of the Menderes Massif, western Anatolia, Turkey. In: *Proc. Internat. 8th Symp. on Water-Rock Interaction, Vladivostok, Russia*, : Y. K. Kharaka & O.V. Chudaev (eds.), p. 529-532.

Özgür, N., P. Halbach, A. Pekdeger, C. Sommer-v. Jarmersted, N. Sönmez, O.Ö. Dora, D.-S. Ma, M. Wolf, W. Stichler 1997. Epithermal antimony, mercury, and gold deposits in the continental rift zone of the Kücük Menderes, western Anatolia, Turkey: preliminary studies. in: *Mineral Deposits: Research and Exploration (Where do They Meet?*, H. Papunen (ed.), 4th Biennial SGA Meeting, August 11-13, 1997, Turku, Finland, p. 269-272.

Özgür, N. 1997. Aktive und fossile Geothermalsysteme in den Riftzonen des Menderes-Massives, W-Anatolien/Türkei. *Habilitationsschrift, Freie Universität Berlin* (in prep.).

Özgür, N., M. Vogel & A. Pekdeger 1998. A new type of hydrothermal alteration in the geothermal field of Kizildere in the rift zone of the Büyük Menderes, western Anatolia, Turkey. *this volume*.

Precious and base metal deposition in an active hydrothermal system, La Primavera, Mexico

R.M. Prol-Ledesma & F. Juárez-Sánchez – *Institut de Geofísica, UNAM Cd. Universitaria, México D.F., México*

R. Lozano-Sta. Cruz – *Institut de Geología, UNAM Cd. Universitaria, México, D.F., México*

E. Alcalá-Montiel, V.A. Cruz-Casas & S. Hernández-Lombardini – *Facultade de Ingeniería, UNAM Cd. Universitaria, México, D.F., México*

A. Canals – *Dep. Crist. Miner. Dep. Min., Universitat de Barcelona, Spain*

E. Cardellach – *Dep. Geol., Universitat Autónoma de Barcelona, Bellaterra, Spain*

ABSTRACT: The chemical study of La Primavera geothermal field in México was undertaken after high arsenic concentrations were recorded in the discharged water. Major and trace elements concentration was determined on water and rock samples. High gold concentrations was observed in water and rock: 59 and 263 ppb respectively. Anomalies of Zn and Ba were frequently observed within the geothermal reservoir samples. Shallow samples are enriched in Zn, Ba, Nb and Y, and depleted in Ni, Co, Cr, V, Sr, Rb, Zr, Ga. On the other hand, deep samples show an enrichment in Ni, Co, Cr and V, and a depletion in Nb, Sr, Rb, and Zr.

1. INTRODUCTION

Trace element distribution at depth in a geothermal system may be useful in understanding the processes that take place during the formation of an epithermal ore deposit. The distribution of trace elements in hydrothermal waters has been extensively studied (Ellis and Mahon, 1977; Weissberg et al., 1979; Henley et al., 1984; Mann, et al., 1986; Philpotts, et al., 1987; Hinkley et al., 1988); however, the distribution of trace elements in the rocks of a geothermal reservoir has not been studied in such detail. Trace element concentration has been determined in surface deposits related with hot springs, hot pools (White, 1967; Weissberg, 1969), scales from geothermal wells (Brown, 1986), in samples from exploration wells (Cole and Ravinsky, 1984; Goff and Gardner, 1994; Reyes and Vickridge, 1996), and in sediments related to active volcanoes (Glasby, 1975; Giggenbach and Glasby, 1977). These studies have been mostly focused on the determination of precious metals and related elements (As, Sb, Hg). Also, analyses have been made on silica deposited in the shallow parts of the systems, and on pipe deposits (Henley et al., 1984).

The available data on trace element concentration in geothermal reservoirs, as in the Kawerau geothermal field (Reyes and Vickridge, 1996), in the Beowawe geothermal field (Cole and Ravinsky, 1984) and in the Valles Caldera (Goff and Gardner, 1994), do not provide a definite pattern for their distribution with depth in active hydrothermal systems. The most complete studies about chemical changes due to hydrothermal alteration have been reported on the interaction of typical MORB with ocean water (Hart, 1969; Verma, 1981; Alt et al., 1986; Bienvenu et al., 1990; Verma, 1992; Jochum and Verma, 1996). The studies that include trace elements show that as a result of that interaction the elements: Th, Sn, Sb, Tl, Cs, Pb, Rb and Ba present a significant enrichment in comparison to the fresh MORB samples, and that the less altered basalt samples have the lowest trace element concentrations.

A detailed sampling of the available geothermal fluid, core, cuttings and surface rock samples from La Primavera geothermal field in México was undertaken. In addition to the chemical analyses of the samples, a petrographic study was carried out which included observations in the petrographic microscope and X-ray diffraction for the identification of clay minerals. The chemical characterisation of altered rocks in La Primavera has been undertaken in order to determine the conditions of deposition of base and precious metals in an active hydrothermal environment.

2. LOCAL GEOLOGY AND HYDROTHERMAL ALTERATION

La Primavera geothermal field is located in the western section of the Mexican Volcanic Belt (Fig. 1). It is contained within a volcanic centre that is formed by a caldera structure. The geology (Fig. 2) and eruptive history of La Primavera has been reported by Mahood (1980a, b). The generalised lithologic column of the area is based on data from exploration wells. At shallow depths it is dominated by rocks of rhyolitic composition; however, the deep reservoir is mostly contained in andesitic strata interlayered with lithic tuffs (Gutiérrez-Negrín, 1991). The studies on hydrogeology show the presence of two geothermal reservoirs: a deep high temperature (>200°C) reservoir, and a shallow dilute low temperature (<150°C) reservoir (Cerriteño, 1991).

Samples from 11 wells were available for sampling, 10 of them are located in the central part of the geothermal field (Fig. 1) above the main upflow of the system, and one well was drilled on a lateral outflow plume in the peripheral area.

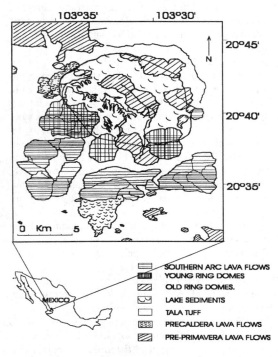

Fig. 1 Surface geology and location of La Primavera geothermal field.

Hydrothermal minerals identified in thin section include: quartz, tridymite, adularia, calcite, epidote, chlorite, sericite, pyrite, hematite, chalcopyrite.
Petrographic studies also included X-ray diffraction analysis which was useful to identify the small grained minerals: montmorillonite, illite, inter-layered clays (illite-smectite, chlorite-smectite), and kaolinite.

The information on alteration minerals from the analysed samples was complemented with the reported alteration (JICA, 1989) to define the generalised alteration pattern shown in Fig. 2. The pattern presents a typical grading from argillic to propylitic alteration with depth.

3. GEOCHEMISTRY OF ALTERED SAMPLES.

Water was sampled from the discharge of warm springs and exploration wells (PR-1 and PR-11) which produce condensate through a valve. Samples were analysed for major ions and trace elements (Cu, Pb, Ni, Tl, Ba, Au, As and Zn). Gold concentrations in the geothermal water attains values of 59 ppb and 2 ppm for arsenic. Core, cuttings and surface rock samples were analysed to determine their chemical composition. The results for major and trace elements were obtained by X-ray fluorescence, precious metals, and thallium were analysed using absorption spectrometry.

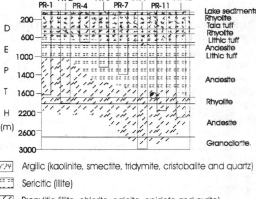

Fig. 2 Distribution of hydrothermal alteration at depth in La Primavera geothermal field.

Enrichment or depletion of trace elements was determined in relation to measured and mean values for fresh rock (Rosler and Lange, 1972; Govett, 1983) and the standard deviation of the data (Cole and Ravinsky, 1984). The only element that occurs in concentrations higher than the mean value in all samples (independently of depth and lithology) is zinc. Anomalously high values for precious and base metals are presented in the section shown in Figure 3. Silver concentrations are low and homogeneous, from 0 to less than 50 ppb. On the other hand, gold concentration varies from 1 to more than 200 ppb.

Thallium is an element typically associated with precious metal mineralization in epithermal deposits; however, in La Primavera there is no direct correlation between Tl and Au concentration anomalies, measured thallium concentrations are very low: less than 15 ppb; however Tl shows slightly higher values below and above the gold anomalous value at 1050 m depth.

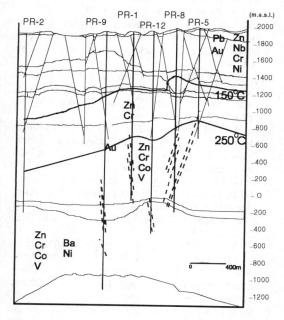

Fig. 3 Cross section showing areas with anomalous concentrations of metals (denoted by element symbols).

4. DISCUSSION AND CONCLUSIONS

The variations in the chemistry of all altered samples with respect to unaltered samples indicates depletion of manganese, and the alkalis; and enrichment in iron and magnesium. Most samples show an enrichment in aluminium and titanium, and depletion in silica and calcium. Trace elements follow different trends at various depths, shallow depths are more favourable for deposition of the analysed trace elements than the surface or the deep part of the reservoir. High concentrations of gold in La Primavera are observed in surface or shallow samples where the most drastic changes in the physical and chemical characteristics of the fluid take place. The only sample enriched in gold at depth (1050 m) is related to the transition from neutral-pH propylitic alteration to slightly acid-pH phyllic alteration, and with high Cu, Pb and Zn values. The conditions in the zones of gold anomalies indicate that possible oxidation of the geothermal fluid is related with gold deposition, as has been suggested by observations in some epithermal deposits and geothermal systems (Barnes, 1979).

In La Primavera samples, the frequent enrichment of Zn, Pb, Cu, Ni, Ba and Nb in altered rocks was observed, similarly to what has been reported for altered MORB samples. Zn and Ni anomalies may be related to their entrapment in the clay structure (Eslinger and Pevear, 1988). These results disagree with the trace element concentrations reported in altered rocks in the Kawerau geothermal field, which indicate depletion in Zn and Pb and immobile behaviour for elements like Nb, Al, Ti and Sr (Reyes and Vickridge, 1996). In the Beowawe geothermal field, the anomalously high metal concentrations are related spatially with the occurrence of diabase dikes, and in Broadlands they are restricted to the deep and the peripheral areas of the field (Cole and Ravinsky, 1984). All anomalous values in La Primavera samples are located in reported fluid loss zones related to faults intersecting the wells or in the vicinities of lithologic changes; therefore, a sudden increase in permeability or the chemical reaction with the host rock triggers metal deposition. Also, anomalous values of metals are common in the shallow aquifer where geothermal fluid is mixed with cold ground water. At shallow depths an enrichment in lead is observed, contrary to what is expected from some studies (e.g. Weissberg, 1969) and deep Pb enrichment patterns observed in epithermal deposits. Also, depletion in Cu and Zn was expected in the shallow samples; however, an enrichment in Zn was observed in wide areas of the shallow reservoir, and no systematic enrichment pattern at depth was observed for Cu. This shows that base metals are highly mobile and they will be transported by various geothermal fluids unless drastic changes in the physical and chemical conditions trigger metal deposition. Shallow samples are depleted in Ni, Co, Cr, V, Sr, Rb, Zr, Ga, while deep samples show an enrichment in Ni, Co, Cr, V, and Ba, and a depletion in Nb, Sr, Rb, and Zr.

ACKNOWLEDGEMENTS

Financial support was provided by the European Commission through the contract CI1*-CT94-0075 (DG 12 HSMU). CFE kindly provided samples and information.

REFERENCES

Alt, J.C., Honnorez, J., Laverne, C. and Emmermann, R., (1986). Hydrothermal alteration of a section through the upper oceanic crust, DSDP Hole 504B: the mineralogy, chemistry and evolution of sea water-basalt interaction. J. Geophys. Res., Vol. 91, 10309-10335.

Barnes, H.L., (1979). Solubilities of ore minerals. In: Barnes, H.L. (ed) Geochemistry of Hydrothermal Ore Deposits. Wiley Int., 404-460.

Bienvenu, P., Bougault, H., Joron, J.L., Treuil, M. and Dmitriev, L., (1990). MORB alteration: rare-earth element/non-rare-earth hygromagmaphile element fractionation. Chem. Geol., , Vol. 82, 1-14.

Brown, K.L., (1986). Gold deposition from geothermal discharges in New Zealand. Econ. Geol., Vol. 81, 979-983.

Cerriteño, O., (1991). Características Hidrodinámicas del Acuífero Somero de La Primavera. CFE Internal Report 12/91.

Cole, D.R. and Ravinsky, L.I., (1984). Hydrothermal alteration zoning in the Beowave geothermal system, Eureka and Lander counties, Nevada. Econ. Geol. Vol. 79, 759-767.

Ellis, A.J. and Mahon, W.A.J. (1977). Chemistry and Geothermal Systems. Academic Press. New York. 392 pp.

Eslinger, E., and Pevear, D., (1988). Clay Minerals for petroleum geologists and engineers. SEPM short course No. 22, Soc. Econ. Paleont. Mineral. USA. .

Giggenbach, W.F. and Glasby, G.P., (1977). The Influence of thermal activity on the trace metal distribution in marine sediments around White Island, New Zealand. Geochemistry 1977, DSIR Bulletin 218, 121-126.

Glasby, G.P., (1975). Geochemical dispersion patterns associated with submarine geothermal activity in the Bay of Plenty, New Zealand. Geochem. Jour., Vol. 9, 125-138.

Goff, F., and Gardner, J.N., (1994). Evolution of a mineralized geothermal system, Valles Caldera, New Mexico. Econ. Geol., Vol. 89. 1803-1832.

Govett, G.J.S., (1983). Handbook of Exploration Geochemistry. Vol. 3. Rock Geochemistry in Mineral Exploration. Elsevier. 461 pp.

Gutiérrez-Negrín, L.C.A., (1991). Recursos Geotérmicos en La Primavera, Jalisco. Ciencia y Desarrollo, Vol. 16 (96), 57-69.

Hart, S.R., (1969). K, Rb, Cs contents and K/Rb, K/Cs ratios of fresh and altered submarine basalts. Earth Planet. Sci. Lett., Vol. 6, 295-303.

Henley, R.W., Truesdell, A.H. and Barton Jr., P.B., (1984). Fluid-mineral equilibria in hydrothermal systems. Rev. in Econ. Geol. Vol. 1, 115-127.

Hinkley, T.K., Seeley, J.L. and Tatsumoto, M., (1988). Major- and Minor-metal composition of three distinct solid material fractions associated with Juan de Fuca hydrothermal fluids (Northeast Pacific), and calculation of dilution fractions of fluid samples. Chemical Geol., Vol. 70, 235-248.

JICA, (1989). Evaluación del Yacimiento Geotérmico en La Primavera. 122 pp.

Jochum, K.P. and Verma, S.P., (1996). Extreme enrichment of Sb, Tl and other trace elements in altered MORB. Chem. Geol., Vol. 130, 289-299.

Mahood, G.A., (1980a). Geological Evolution of a Pleistocene Rhyolitic Center: Sierra La Primavera, Jalisco, México. J. Volcanol. Geotherm. Res., Vol. 8, 199-230.

Mahood, G.A., (1980b). Chemical Evolution of a Pleistocene Rhyolitic Center: Sierra La Primavera, Jalisco, México. Contrib. Mineral. Petrol., Vol. 77, 129-149.

Mann, A.W., Fardy, J.J. and Hedenquist, J.W., (1986). Major and trace element concentrations in some New Zealand Geothermal Waters. Int. Volc. Cong. Proceedings, Auckland, New Zealand. 61-66.

Philpotts, J.A., Aruscavage, P.J. and Von Damm, K.L., (1987). Uniformity and diversity in the composition of mineralizing fluids from hydrothermal vents on the southern Juan de Fuca Ridge. J. Geophys. Res., Vol. 92, 11387-11399.

Reyes, A.G., and Vickridge, I.C., (1996). Distribution of lithium, boron and chloride between fresh and altered rocks in the Kawerau geothermal system, New Zealand. Proceedings of the 18th NZ Geothermal Workshop, 121-126.

Rosler, H.J., and Lange, H., (1972). Geochemical Tables. Elsevier. 457 pp.

Verma, S.P., (1981). Seawater alteration effects on 87Sr/86Sr, K, Rb, Cs, Ba and Sr in oceanic igneous rocks. Chem. Geol., Vol. 34, 81-89.

Verma, S.P., (1992). Seawater alteration effects on REE, K, Rb, Cs, Sr, U, Th, Pb and Sr-Nd-Pb isotope systematics of Mid-Ocean Ridge Basalt. Geochem. J., 159-177.

Weissberg, B.G., (1969). Gold- Silver ore-grade precipitates from New Zealand Thermal Waters. Econ. Geol., Vol. 64, 95-108.

Weissberg, B.G., Browne, P.R.L. and Seward, T.M., (1979). Ore metals in Active Geothermal Systems. In: Barnes, H.L. (ed.) Geochemistry of Hydrothermal Ore Deposits (2nd Edition). Wiley Int., 738-780.

White, D.E., (1967). Mercury and base metal deposits with associated thermal and mineral waters. In Barnes, H.L. (ed.) Geochemistry of hydrothermal ore deposits. Holt, Rinehart and Winston, New York. 575-631.

Geochemistry of natural waters in Skagafjördur, N-Iceland: II. Isotopes

Á.E. Sveinbjörnsdóttir & Stefán Arnórsson
Science Institute, University of Iceland, Reykjavík, Iceland

Jan Heinemeier & Elisabetta Boaretto
AMS ^{14}C Dating Laboratory, Institute of Physics and Astronomy, University of Aarhus, Denmark

ABSTRACT: $\delta^{18}O$ and δD of the Skagafjördur natural waters follow the meteoric line, apart from the isotopically lightest and hottest samples, which show a slight oxygen shift (< 1‰). An attempt was made to date the waters by tritium and radiocarbon measurements. The tritium results for surface and soil waters range from 8 to 12 TU, whereas tritium is not detected in waters with temperatures >40°C. Boron concentrations in the waters were used to correct the ^{14}C concentration for the contribution of rock-derived carbon in the groundwater and thus calculate the initial (undiluted) ^{14}C concentration. When this correction is applied most of the waters lie in the range 30 to 140 pMC (percentage of Modern Carbon). Other waters have higher pMC and for these waters, which may be of mixed origin or have absorbed carbon from decaying organic matter, the boron based model overestimates the ^{14}C-dilution.

1 INTRODUCTION

The deuterium content of the mean annual precipitation in Iceland has been mapped, allowing the deuterium to be used as a tracer to delineate regional movement of groundwater, including geothermal water (Árnason, 1976, 1977).

From stable isotope measurements and general hydrological considerations, Árnason (1976) concluded that the water in cold groundwater systems is relatively young (an average of a few years to a few decades old), whereas hot water is of varying ages and, in most cases, much older than the cold water. Tritium can be used to date groundwaters in the limited time interval back to the beginning of the atmospheric nuclear bomb testing in the late 1950's and the ^{14}C radioisotope, with its half-life of about 5700 years, offers the potential of dating groundwater up to about 40.000 years.

In this contribution we report the first results of geochemical studies on natural waters within the Skagafjördur region, N-Iceland, especially aimed at evaluating origin and age of the groundwater.

2 WATER CHEMISTRY AND ISOTOPES

2.1 Stable isotopes of oxygen and hydrogen

The Skagafjördur region in N-Iceland is geothermally active on both sites of the valley, where the water temperatures range from the annual mean temperature (about 5°C) to 90°C (Arnórsson et al., this issue). Fig. 1 shows the $\delta^{18}O$ and δD relationship of natural waters in Skagafjördur and for comparison the meteoric line from Craig (1961) is also shown. Apart from the isotopically lightest waters, which are geothermal and show a slight oxygen shift (< 1‰), the waters follow approximately the meteoric line. It

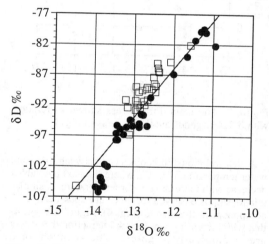

Figure 1. $\delta^{18}O$ and δD relationship of natural waters in Skagafjördur, where waters from the western part of the valley are represented by points and from the eastern part by squares. For comparison the meteoric line from Craig (1961) is also shown.

is, however, striking that many of the water samples from the eastern part of the valley plot above the meteoric line whereas the samples from the western part follow the line more closely. This is further demonstrated in Fig. 2 where the deuterium excess (d), where d is defined as $\delta D - 8\,\delta^{18}O$, is shown in relation with $\delta^{18}O$. It is suggested that the difference between the d-values in the western and eastern part of the Skagafjördur valley can be accounted for by different origin of the moisture, as precipitation originating from the seas north of Iceland is expected to be more common in the high mountains east of the Skagafjördur valley.

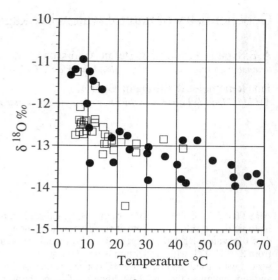

Figure 3. The relation of $\delta^{18}O$ and water temperature. Legends are the same as in Fig. 1.

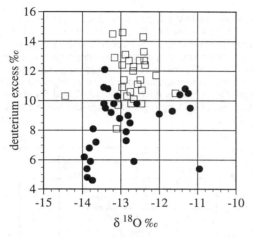

Figure 2. The deuterium excess shown in relation with $\delta^{18}O$. Legends are the same as in Fig. 1.

A clear relationship is observed between $\delta^{18}O$ and water temperature (Fig. 3), where the isotopic values decrease with increasing water temperature. A similar relationship is observed for $\delta^{18}O$ and rock-derived Cl component in the waters (Fig. 4). Figure 4 suggests that with increasing water-rock interaction the waters are isotopically lighter. The conventional interpretation of these results suggests that the hottest waters, which also are isotopically lightest, originate from the highlands, farthest away from the sampling point, and have therefore been interacting with the rock longer than waters with lower temperatures. This interpretation is, however, not valid for all geothermal waters in Skagafjördur, as it has been suggested (Arnórsson et al., this issue) that the isotopically lightest waters are a mixture of at least three components: (1) seawater that has infiltrated the Skagafjördur region at the end of last glaciation (2) precipitation also from that time, which is isotopically light due to the cold climate at that time and (3) precipitation from the present climate regime.

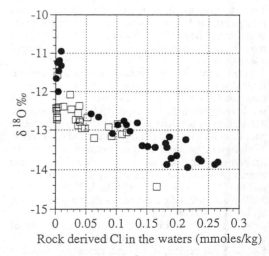

Figure 4. The relation of $\delta^{18}O$ and rock-derived Cl component in the waters. Legends are the same as in Fig. 1.

2.2 Tritium results

The tritium results for surface and soil waters (i.e. cold springs from the soil zone) range from 8 to 12 TU, whereas tritium is not detected in waters with temperatures >40°C (Fig. 5). Thus the warm waters are at least older than the beginning of the atmospheric nuclear bomb testing in the late 1950's. In some instances tritium is detected in warm waters with temperature <40°C, possibly because of mixing with younger surface waters. The largest glacier rivers contain less tritium than normal surface waters, due to their component of tritium-poor icemelt of old glaciers.

the procedures described in Sveinbjörnsdóttir et al. (1995) the concentration of leached CO_2 in the water samples were calculated. Unfortunately, data on boron and carbon in Icelandic basalts are very limited. In Sveinbjörnsdóttir et al. (1995) data on tholeiites from eastern Iceland was used (1.2 ppm B and 0.05 % (wt) CO_2) As no data is yet available for the

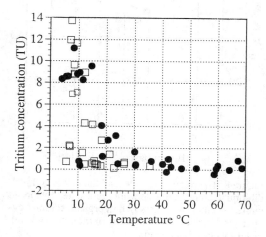

Figure 5. The relation of tritium and water temperature Legends are the same as in Fig. 1.

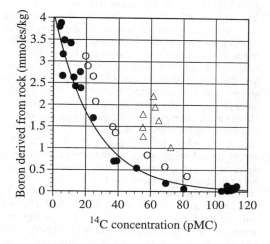

Figure 6. Correlation between rock-derived boron and the ^{14}C concentration. Open circles and triangles represent waters where the boron-based correction model overestimates the ^{14}C dilutuion.

2.3 Carbon isotopes

Carbon in groundwater may be derived from several sources with different ^{14}C concentration and $\delta^{13}C$. In Iceland at least three different sources of carbon supply to the water are expected: atmospheric CO_2, soil CO_2 of organic origin and CO_2 leached from the rock through which the water seeps. Addition of carbon from the rock dilutes the ^{14}C concentration of the water and in this way yields a high apparent ^{14}C age. Because of the multiple character of the carbon supply to the water, a simple two-component isotopic dilution model (Mook, 1980) is inadequate to correct ^{14}C measurements for the Icelandic samples. Sveinbjörnsdóttir et al. (1995) showed that by assuming that the ratio of boron to CO_2 is the same in rock and groundwater, it is possible to correct for the contribution of rock-derived carbon to the groundwater. Fig. 6 shows the correlation between the boron concentration and the ^{14}C concentration in pMC (percentage of Modern Carbon) observed for the Skagafjördur samples. The fit is exponential, as expected if the boron concentration is a measure of dilution with ^{14}C-dead CO_2 from the rock. It is, therefore, proposed that the boron concentration may be taken as a measure of the amount of dead CO_2 in the groundwater leached from the rock. Following

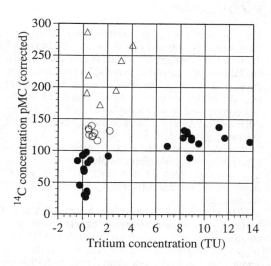

Figure 7. The relationship between the tritium concentration of the waters and the boron-based corrected ^{14}C concentration. Symbols have the same notation as in Fig. 6.

Skagafjördur region the above numbers were also applied in the present study. These numbers do however not characterize well the Skagafjördur region and thus the results on ^{14}C dilution for the Skagafjördur waters are still preliminary and give minimum ^{14}C ages.

The calculations showed that a substantial fraction of the carbon entering the water has precipitated, probably as calcite. Subsequently the ^{14}C dilution was calculated by assuming the average CO_2 composition of surface waters as the initial CO_2 concentration of the water. The relative success of this boron-based correction is shown in Fig. 7, where the relation between boron-based corrected pMC values and the tritium meaurements is shown. The figure shows that the corrected values for most of the samples lie in the range 10.500 BP to modern. However, the boron-based model gives too high ^{14}C concentration for some of the water samples (triangles and open circles). It is suggested that these samples are of mixed origin or have absorbed carbon from decaying organic matter. In these cases the boron based model overestimates the ^{14}C-dilution.

The $\delta^{13}C$ in cold surface waters ranges from -1 to -7 ‰ whereas in soil waters the $\delta^{13}C$ is much lighter and ranges from -16 to -22 ‰. In the geothermal waters the $\delta^{13}C$ lies most commonly in the range -6 to -13 ‰. Fig. 8 suggests a linear relationship for the $\delta^{13}C$ of geothermal water and the caluated ^{14}C dilution, except where the thermal waters emerge in thick peat soil explaining their low $\delta^{13}C$ values (open circles).

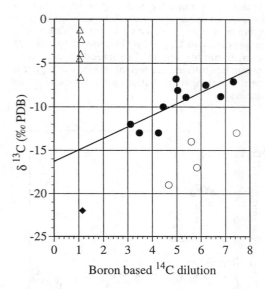

Figure 8. The relation of $\delta^{13}C$ and the calculated ^{14}C dilution. Geothermal waters are represented by closed circles, geothermal waters rich in organic carbon by open circles, surface rivers by open triangles and soil water by closed diamond.

3 CONCLUSIONS

The conventional interpretation of deuterium measurements suggests that the source of supply of the water to the thermal systems is precipitation that has fallen in the interior highlands. This interpretation does not fit for all geothermal waters in Skagafjördur as some of the waters consist of three components, one corresponding with today's precipitation and the others with "ice age" water relatively high in Cl of marine origin but depleted in $\delta^{18}O$.

It is suggested that the difference in deuterium excess values (d) of water samples from the eastern and western part of the Skagafjördur valley respectively, may be explained by different origin of the moisture.

No tritium was detected in water samples > 40°C and radiocarbon measurements suggest that most of these waters range in age from 10.500 BP to 500 BP, after applying the boron correction procedure. In some cases the boron based correction overestimates the ^{14}C dilution.

4 ACKNOWLEDGMENTS

This project is financed by the Icelandic Research Council and the Danish Natural Science Research Council.

5 REFERENCES

Árnason, B. 1976. Groundwater systems in Iceland traced by deuterium. Societas Scientiarum Islandica 42: 236p.

Árnason, B. 1977. Hydrothermal systems in Iceland traced by deuterium. Geothermics 5: 125-151.

Craig, H. 1961. Isotope variations in meteoric waters. Science 153: 10702-10703.

Mook, W.G. 1980. Carbon-14 in hydrological studies. In P. Fritz & J. Ch.Fontes (eds), Handbook of Environmental Isotope Geochemistry. The Terrestrial Environment. Amsterdam, Elsevier: 49-74.

Sveinbjörnsdóttir, A.E., J. Heinemeier & S. Arnórsson. 1995. Origin of ^{14}C in Icelandic Groundwater. Proc. 15th Radiocarbon Conference, edited by G.T. Cook, D.D. Harkness, B. F. Millerand & E. M. Scott. Radiocarbon 37: 551-565.

Gas geochemistry in the Yangbajing geothermal field, Tibet

Zhao Ping, Jin Jian & Zhang Haizheng
Institute of Geology, Chinese Academy of Sciences, Beijing, People's Republic of China

Duo Ji & Liang Tingli
Geothermal Geology Team, Tibet Bureau of Geology and Mineral Resources

ABSTRACT: Gases of the deep and shallow fluids in the Yangbajing geothermal field, Tibet, are distinct in chemical composition, concentration and $\delta^{34}S\text{-}H_2S$ value. No difference in $\delta^{13}C\text{-}CO_2$ value (-11.33 ~ -7.72‰, PDB) is observed. The ^4He enrichment of the gases (R/Ra=0.144~0.390) shows the thermal fluid has a deep circulation in the crust. Argon mainly comes from air or air-saturated water with small addition of radiogenic ^{40}Ar. There is non-atmospheric nitrogen in the deep thermal fluids. The CO_2 and H_2S are considered to derive from the Nyainquentanglha complex heating by a magmatic bogy in the region. The drawdown in the field will effect the gas concentration of steam in production wells.

1 INTRODUCTION

The Yangbajing geothermal field is one of the high-temperature fields in the Himalayan Geothermal Belt in Tibet. It is situated at 94km northwest of Lhasa. The Yangbajing basin stretches from NE to SW and is bordered by the Nyainquentanglha Range on the northwest and the Tang Range on the southeast. Faults controlling thermal fluid flow are mainly NW-SE trend in the region.

The China-Nepal highway divides the field into two parts, the southern and northern. The southern part is covered by widespread alluvium and altered granite. The thickness of Quaternary alluvium is in the 100 ~300m range. Opal and calcite precipitate from the thermal fluid in the pore space of alluvium. The northern part is mostly occupied by altered granite that is mainly replaced by kaolinite. Magneto-telluric survey, carried by UN/DDSMS experts in 1995, revealed resistivity anomaly at 5km depth in the northern part, which has been explained as a magmatic body.

The host rocks in the shallow reservoir are Quaternary alluvium in the southern and the Himalayan granite in the northern. The shallow reservoir is at a depth of 100m to 500m. Temperature inside is relative uniform, 150~165°C, decreasing from northwest to southeast. Pressures in production wells dropped quickly in the recent years after 20 years' exploitation.

Two deep wells were drilled to look for the deep geothermal resources in the northern part. The first deep well ZK4002 recorded 329°C at 1850m depth on May 8, 1994. The discharge in the well was mainly steam. The cuttings of well ZK4002 were mostly altered granites. The second deep well ZK4001 was completed in November 1996. The average temperature is 248°C for deep feed zones, at 1100~1300m depth. The flow-rate is about 300 tons/h. There appears to be potential of deep resource in the northern part.

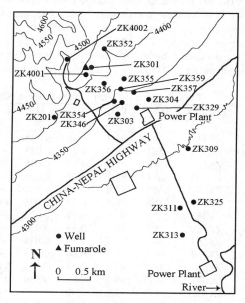

Figure 1. Map of the Yangbajing geothermal field showing the sampling locations.

2 SAMPLING AND ANALYSIS

The sampling method is based on the procedure used in the National Energy Authority of Iceland (Olafsson, 1988). This includes collection of dry gas, condensate and NaOH absorption solution. A vacuum pump and a cooling device are brought in the field. Gas samples prepared for helium isotope measurements are stored in alloy aluminum tubes with two steel valves on one end; these are vacuumed prior to the sampling. Carbon dioxide for isotope analysis is precipitated as $SrCO_3$ from the NaOH absorption solution. HgS deposits when steam enters $Hg(CH_3COO)_2$ solution. Adding a few drops of AgCl solution, HgS is replaced by Ag_2S that is prepared for sulfur isotope analysis.

Gas compositions are measured by a gas chromatograph. The $^3He/^4He$, $^{13}C/^{12}C$ and $^{34}S/^{32}S$ ratios are analyzed on VG5400, MAT252 and δS mass spectrometers, respectively. Figure 1 gives the sampling locations in the Yangbajing geothermal field.

3 DISCUSSION

3.1 *Gas chemical composition*

The carbon dioxide is a dominant non-condensable gas (dry gas) in geothermal systems (Mahon et al., 1980). It normally comprises more than 85% in volume in the Yangbajing geothermal field. Other gases are N_2, H_2S, H_2, CH_4, Ar, He and O_2 with variable contents (Table 1). CO and NH_3 are not measured due to technical reasons. The gas/steam ratios in wells ZK357 and ZK359 are less than 0.1 liter/kg so that gas-sampling is easily contaminated by air. All the samples have been plotted in the He-Ar-N_2 triangular diagram (Giggenbach, 1987) (Fig. 2). The data have not been corrected for air contamination. There are two possible origins for the samples near 10He corner; that is either a deep crustal or a local basaltic origin (Giggenbach, 1992). Magmatism in the region are generally acidic to intermediate. The helium component in the Yangbajing gases seems to originate in the crust. The samples could be classified into two groups. The first group includes the samples W12, W13, F14 (Table 1). The gases in the wells ZK4002 and ZK4001 come from the deep reservoir. The ratios of N_2/Ar are higher than 84, meaning the part of nitrogen is non-atmospheric origin. The He/H_2 ratios are relatively high also. Fumarole R1 has a chemical composition similar to that of the deep wells. Its steam composition is similar to that of well ZK4001 except H_2 (Table 2). The second group includes the samples collected in the production wells. There is no methane detected in the samples except sample W6. The thermal fluid produced a small

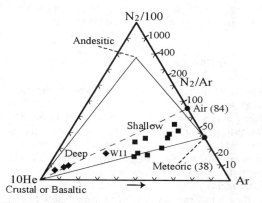

Figure 2. The relative N_2, He and Ar contents of non-condensable gases.

Table 1. Chemical composition of non-condensate gases in the Yangbajing geothermal field.

Location	Sample no.	CO_2 %	H_2S^a %	H_2 %	CH_4 %	N_2 %	He %	Ar %	O_2 %	N_2/Ar	He/H_2
ZK311	W1	95.2	0.40	0.025	---	4.06	0.0095	0.10	1.06	41	0.38
ZK313	W2	81.3	0.43	0.035	---	15.6	0.0052	0.23	1.96	68	0.15
ZK325	W3	92.5	0.22	0.035	---	6.17	0.0078	0.16	0.59	39	0.22
ZK309	W4	85.7	0.27	0.028	---	11.7	0.0044	0.21	2.28	56	0.16
ZK304	W5	93.8	0.33	0.041	---	4.67	0.0104	0.14	0.47	33	0.25
ZK329	W6	90.4	0.30	0.038	0.08	7.70	0.0091	0.12	2.03	64	0.24
ZK355	W7	91.0	0.27	0.058	---	7.10	0.0052	0.14	1.23	51	0.09
ZK359	W8	77.2	0.35	0.028	---	17.2	0.0104	0.31	4.21	55	0.37
ZK303	W9	92.7	0.23	0.034	---	5.02	0.0145	0.14	0.92	36	0.43
ZK354	W10	91.9	0.60	0.053	---	9.10	0.0143	0.14	0.07	65	0.27
ZK201	W11	87.6	0.28	0.065	0.06	11.2	0.0377	0.17	1.81	66	0.58
ZK4002	W12	89.2	0.52	0.036	0.15	8.12	0.1167	0.06	1.38	135	3.24
ZK4001	W13	91.3	0.40	0.017	0.08	5.86	0.0601	0.06	0.74	98	3.54
R1	F14	94.5	0.18	0.058	0.10	5.38	0.0472	0.06	0.56	90	0.81

"a", field data. "---", lower than the detected limit of gas chromatograph.

Table 2. Steam composition in the wells and a fumarole (corrected to 1 atm, in mmol/kg steam).

Location	Sample no.	Date	CO_2	H_2S	H_2	CH_4
ZK311	W1	95-09-16	137.7	0.390	0.0353	---
ZK313	W2-1	95-09-12	21.6	0.595	0.0053	---
ZK313	W2-2	96-09-21	77.4	1.91	0.0307	---
ZK325	W3	95-08-18	21.4	0.021	0.0044	---
ZK309	W4	96-09-21	34.0	0.510	0.0083	---
ZK304	W5-1	95-09-12	26.4	0.119	0.0095	---
ZK304	W5-2	96-09-17	46.6	0.278	0.0171	---
ZK329	W6-1	95-08-26	28.5	0.435	0.0106	0.022
ZK329	W6-2	96-09-19	42.8	0.593	0.0162	---
ZK355	W7-1	95-09-13	28.1	0.259	0.0056	---
ZK355	W7-2	96-09-17	30.2	0.342	0.0147	---
ZK303	W9	95-09-13	66.1	0.366	0.0227	---
ZK354	W10-1	95-08-21	30.6	0.400	0.0095	---
ZK354	W10-2	96-09-20	27.4	0.585	0.0145	---
ZK201	W11	96-09-20	308	1.62	0.23	0.208
ZK4001	W13	96-11-13	315.5	3.07	0.057	0.266
R1	F14	95-08-22	368	4.63	0.23	0.396

quantity of gas. Most of the N_2/Ar ratios are between 38 and 84. N_2 and Ar components are thought to mostly come from air or air-saturated water. Well ZK201 is a damaged exploration well. There is a steam escaping from the well. Sample W11 is similar in composition to the first group.

3.2 *Gas isotopic composition and its origin*

The $^3He/^4He$ ratio, expressed as R/Ra, is in the 0.144~0.390 range and lower than the air ratio in the Yangbajing field. The ^4He enrichment reflects the thermal fluid has a deep circulation in the crust and an addition of radiogenic ^4He. Most of the $^{40}Ar/^{36}Ar$ ratios are slight higher than the atmospheric ratio (~295). The samples have a small contribution of radiogenic ^{40}Ar ($^{40}Ar/^{36}Ar > 300$). The $^4He/^{20}Ne$ ratios are in the range of 7.0 to 22.5, higher than 0.318 in the air. The isotope measurements reflect the inert components in the thermal fluids with different sources.

Tong et al.(1981) analyzed sulfur isotopic composition in natural sulfur deposits. The average $\delta^{34}S$ of three samples was 1.0‰ (CDT). During this survey, the $\delta^{34}S$ value in hydrogen sulfide are 0.2‰, 1.5‰ and 1.8‰ for the samples W12, W13 and F14, respectively and are in the range 1.1‰ to 8.3‰ for the others. The $\delta^{34}S$ value of natural sulfur is 2.1‰ at a vent of the fumarole and is 5.2‰ for sulfur near well ZK4002. The $\delta^{34}S$ in sulphate is 18.9‰ for the deep fluid and is from 7.5‰ to 9.5‰ for the shallow. Sulfur isotope equilibrium between hydrogen sulfide and sulphate is not approached in either the deep or shallow reservoir. The $\delta^{34}S-H_2S$ value increases and the $\delta^{34}S-SO_4$ value decreases along NW-SE trend faults.

The chloride and boron in thermal waters shows that the shallow thermal water comes from a combination of deep and cold ground water. The H_2S content in the total discharge of well ZK4001 is about 22ppm, higher than that in the shallow wells (~2ppm), but the SO_4 content is in reverse, 25.9ppm in the deep hot water and 34~43ppm in the shallow ones. The part of H_2S component is expected to be oxidized in SO_4 as the deep fluids' ascent. The $\delta^{34}S-H_2S$ value is affected by water/rock interaction and possible biological activities in the alluvium. The H_2S of the deep fluids is thought to be a deep crustal origin.

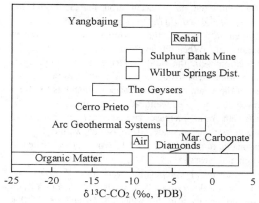

Figure 3. Diagram showing $\delta^{13}C-CO_2$ of the Yangbajing geothermal field compared to values from others (modified from Goff et al.,1995).

No distinct difference in the isotopic composition of carbon dioxide is observed between the deep and shallow thermal fluids. The $\delta^{13}C-CO_2$ values are in the range of -11.33‰ to -7.72‰ (PDB)(Fig. 3). Most samples are more depleted ^{13}C than the CO_2 comes from mantle and the samples in the Rehai high-temperature geothermal field, West Yunnan (Tong et al., 1989). The $\delta^{13}C$-CO_2 value is similar to those found in the Sulphur Bank Mine and Wilbur Springs, California (Goff et al., 1995). The CO_2 is inferred to derive from the Nyainquentanglha complex that is heated by the magmas at >5km depth.

3.3 Gas concentration changes during production

Chemical monitoring has proved to be an important tool for the geothermal production (Kristmannsdottir et al., 1996). No long-term monitoring program has being carried out in the Yangbajing geothermal field. The collection of thermal fluids is only executed at the wellheads due to lack of the down-hole sampling equipment. During 1995-1996, a change of steam compositions was observed (Table 2), which was very useful for understanding the state of reservoirs.

Well ZK311 of the southern field was drilled in 1980, 81.82m at depth. The maximum value of recorded temperature was 157°C in the well. Sample W1 was collected in September 1995. The gas concentration of the steam was much higher than the wells' nearby. The thermal water was acid with a low salinity at the sampling time. The total discharge was depleted in ^{18}O and D. Pressure drop and drawdown caused the well to stop producing in 1996. Two samples were collected in well ZK313, not far from well ZK311. The thermal water appeared alkaline in 1995, changed to acid in 1996 while the gas concentration increased and the salinity decreased. The reason was the long-term exploitation made the boiling level dropped continuously. When the level was below the bottom of a production well, the fluids entering the borehole were vapor-dominated, losing heat in the well resulted in the part of the steam condensated back to liquid as the fluid flowed up to the wellhead. This process led the gas/steam ratio increase and the pH of hot water, mixing with condensate, decrease at wellhead. It is reasonable to predict well ZK313 would be exhausted in the coming years.

The chemical composition variation of gases and brines were also observed in wells ZK329 and ZK304. It did not change a lot for wells ZK354 and ZK355 that were closer to the upflowing channel in the field. As mentioned-above, there is an internal relation between the deep and shallow thermal fluids. Putting well ZK4001 into electricity generation would accelerate the exhaustion of the shallow resources. It is important to re-evaluate the potential of the geothermal resources prior to installing a new generating unit. Chemical monitoring should be kept in the Yangbajing geothermal field.

4 CONCLUSIONS

On the basis of the He-Ar-N_2 diagram, gas concentration in the steam and $\delta^{34}S$-H_2S value, all the samples could be classified in two groups that come from the deep and shallow reservoirs, respectively. The $^3He/^4He$ ratio indicates the helium is mainly the crust origin. Most of the argon comes from air or air-saturated water. The part of nitrogen in the deep reservoir is non-atmospheric origin. The magmas in the region heats the Nyainquentanglha complex to generate CO_2 and H_2S. Gas concentration monitoring is necessary for understanding the state of reservoirs.

5 ACKNOWLEDGMENTS

Financial support received from National Natural Science Foundation of China (Contract No. 49402039 and 49672168) and International Atomic Energy Agency (Contract No. 9095) is gratefully acknowledged.

6 REFERENCES

Giggenbach, W.F. 1987. Redox processes governing the chemistry of fumarolic gas discharges from White Island, New Zealand. *Appl. Geochem.*, 2, 143-161.

Giggenbach, W.F. 1992. The composition of gases in geothermal and volcanic systems as a function of tectonic setting. *Proc. Seventh Internat. Symp. Water-Rock Interaction*, 873-878.

Goff, F., Janik, C.J. and Stimac, J.A. 1995. Sulphur Bank Mine, California: an example of a magmatic rather than metamorphic hydrothermal system? *Proceedings of the World Geothermal Congress*. Florence, Italy, 1105-1110.

Kristmannsdottir, H. and Armannsson, H. 1996. Chemical monitoring of Icelandic geothermal fields during production. *Geothermics*, 25(3), 349-364.

Mahon, W.A.J., McDowell, G.D. and Finlayson, J.B. 1980. Carbon dioxide: its role in geothermal systems. *New Zealand Journal of Science*, 23, 133-148.

Olafsson, M. 1988. Sampling methods for geothermal fluids and gases. *Orkustofnun, Geothermal Division OS-88041/JHD-06*, Reykjavik.

Tong Wei and Zhang Mingtao. 1981. *Geothermics in Tibet* (in Chinese). Science Press, Beijing, 1-170.

Tong Wei and Zhang Mingtao. 1989. *Geothermics in Tengchong* (in Chinese). Science Press, Beijing, 1-262.

11 Geothermal general

Pliocene to present-day water-rock interaction processes at 3.5 km depth within a 3.8 Ma old Larderello monzogranite

G. Cavarretta
CNR, Centro di Studio per il Quaternario e l'Evoluzione Ambientale, Dipartimento di Scienze della Terra, Roma, Italy

M. Puxeddu
CNR, Istituto Internazionale Ricerche Geotermiche, Pisa, Italy

ABSTRACT: The auto-metasomatic interaction of late-magmatic, early-hydrothermal fluids with a F-Li-B-rich two-mica S-type monzogranite body produced albitisation of feldspars and phengitisation of muscovite. The circulation of later meteoric fluids, possibly still active, leads to chloritisation of biotite. Albitisation took place in the innermost, partially molten, parts of the body, far from the intruded rocks. Early formed muscovite crystals indicate temperatures lower than 480°-600°C. Later magmatic to hydrothermal muscovite indicate an almost continuous temperature range down to about 350°C. Temperatures inferred from chlorite composition are consistent with fluid inclusion data and present-day in-hole measured values (330°C).

1 INTRODUCTION

In the western part of the Larderello geothermal field (Tuscany, Italy) at 3.5 km depth, the MV7 well (Monteverdi area) crossed fine grained two-mica monzogranite bodies, probably apophyses of the main Larderello batholith. Biotite yielded a K-Ar age of 3.8 ±0.1 Ma (Villa et al., 1987). The high F contents of muscovite, biotite, apatite and titanite, the occurrence of tourmaline veins and of LiCl-rich brines within the fluid inclusions, the presence of cordierite, andalusite and sporadically sillimanite and hercynite allow to attribute the MV7 rock to the F-Li-B-rich peraluminous two-mica S-granites.

2 PETROGRAPHY AND MINERAL CHEMISTRY

The MV7 rock shows fine to medium grain size, inequigranular, hypidiomorphic, microporphyritic texture and the following modal contents: quartz 32%, plagioclase 28%, K-feldspar 26%, biotite 9%, muscovite 4%, accessories 1% (Villa et al., 1987). Muscovite appears as subhedral primary crystals and late secondary fine grained alteration product of feldspars. In the Al-M^{2+}-Si diagram of Monier and Robert (1986), drawn for 2 kb pressure, early formed muscovite crystals indicate temperatures of 480°-600°C; at lower pressure, as in the case of MV7, such values should be even lower. Later magmatic to hydrothermal muscovite indicate an almost continuous temperature range down to 300°C. Biotite is present as euhedral to anhedral crystals often altered into aggregates of chlorite, titanite and rutile. Plagioclase appears with two generations: early calcic (An$_{20-57}$) and late almost pure albite. K-feldspar also includes early crypto-perthitic hyper-solvus phenocrysts and late fine grained crystals. Textural relationships, the grain size decrease and the transition from euhedral to anhedral shapes observed for titanite and apatite indicate that both minerals crystallised in a wide temperature range. Microprobe analyses are reported in Table 1.

Tab. 1. Representative microprobe analyses of F-rich apatite, titanite, muscovite, phengite, biotite from the Monteverdi 7 monzogranite. The composition of late-hydrothermal chlorite is also reported.

	Apat.	Titan.	Mus.	Phen.	Bioti.	Chlo.
SiO2	0.17	32.00	46.44	43.35	35.95	25.56
Al2O3	0.07	12.40	34.17	29.17	18.66	20.97
TiO2	0	21.78	0.38	0	2.97	0.06
FeO	0.47	0.14	1.46	7.20	19.60	27.36
MnO	0.43	0.07	0	0.11	0.16	0.44
MgO	0	0.12	0.96	3.74	7.36	12.52
CaO	54.32	29.25	0	0	0	0.02
Na2O	0.10	0.01	0.57	0.15	0.21	0.02
K2O	0.09	0.01	10.32	10.62	9.34	0.01
P2O5	41.21	n.d.	n.d.	n.d.	n.d.	n.d.
F	3.80	4.66	0.59	0.95	1.24	0
Cl	0.07	0.01	n.d.	0.11	n.d.	n.d.
Total	99.12	98.49	94.70	94.99	94.97	86.97

3 WATER-ROCK INTERACTION

3.1 *Albitisation*

Water-rock interaction processes affected the MV7 monzogranite in a extremely heterogeneous way as suggested by the remarkably different alkali contents between two rock samples: MV7$_F$ yields K$_2$O=5.00 wt%, Na$_2$O=3.02% and MV7$_3$ K$_2$O=1.73% and Na$_2$O=5.60%. The strong variation of the Na/K ratio and the replacement of K-feldspar by late albite, also observed in thin section, indicate that some parts of the monzogranite body underwent a very strong albitisation process with replacement of K by Na. These albitised parts were probably located far from the wall rocks and within the still partially molten core of the monzogranite body near pockets filled by the last Na-enriched residual melts and related fluids. The albitisation process in a F-Li-B-rich magma can be explained according to Kovalenko & Kovalenko (1984) with a strong Na enrichment of the residual melt within the apical part of a magma chamber, due to the strong increase of fluorine content in the same melt as crystallisation proceeds. On the other hand, the albitisation process should have been much less active at *subsolidus* temperatures as indicated by high NaCl content of post-magmatic, early hydrothermal fluid inclusions in quartz (450°-500°C, Valori et al., 1992). In fact Shinohara & Fujimoto (1994) showed that the HCl/NaCl ratio strongly decreases with decreasing temperature and with pressure increase in the reaction

1/2 Al$_2$SiO$_5$ + 5/2 SiO$_2$ + 1/2 H$_2$O + NaCl = NaAlSi$_3$O$_8$ + HCl

In our case the retrograde increase of fluid pressure synergistic with temperature decrease in a solidified, mainly closed system, should have inhibited further albite formation.

Fluid inclusion studies also showed that, in the early stage, the Larderello field was characterised by two distinct systems: a deep one, below 2.5-3 km depth, with lithostatic pressures, high salinities and 350°-600°C; a shallower one with hydrostatic pressures, low salinities and 200°-350°C. These two systems were intermittently connected by the opening of new fissures and reactivation of old fractures after hydraulic or tectonic activity.

3.2 *Chloritisation*

Within turbid albitised volumes, biotite appears strongly chloritised probably because of an increasing interaction with late meteoric fluids, whose penetration in the wall rock is strongly enhanced by the presence of a fine scale channel network, revealed by the micropores observed within turbid albite (see Finch and Walker, 1991). In the unaltered volumes of early crystallised granite near the contact with country-rock, biotite is perfectly preserved as indicated by its remarkably homogeneous composition and by the abundance of fluorine always in excess of 0.5%, which is the minimum value for fresh biotite according to Nash (1976). In fact no fluorine was detected in strongly chloritised biotite crystals. According to the relation between the octahedral occupancy in chlorite and temperature as reported by Cathelineau and Nieva (1985), the MV7 chlorite analyses indicate an average temperature of 344°C. This value is consistent with fluid inclusions data (Th 342°-362°C) obtained by Valori et al. (1992) on quartz of MV7 granite and is comparable with MV7 bottom-hole measured temperature (330°C) suggesting that the observed hydrothermal alteration, due to circulation of meteoric fluids, could be still active.

3.3 *Muscovite behaviour*

Muscovite behaves as a more retentive mineral, as shown by its relatively high fluorine content, even after interaction with meteoric fluids. This was observed also by Petrucci et al. (1994) on the basis of δ^{18}O data of silicates; while muscovite was not altered by meteoric fluids, biotite showed evidence of interaction with ^{18}O-depleted fluids, likely of meteoric origin, down to 3.8 km depth.

In the diagram Al$_{iv}$ vs. (Al$_{vi}$-1) + 2 Ti (Leroy & Cathelineau, 1981), most analyses of early magmatic muscovite crystals plot between Fe^{3+}/Fe$_{tot}$ = 0-0.05, with few values in the range 0.05-0.10. The more phengitic compositions of late magmatic-hydrothermal crystals show a continuous variation of Fe^{3+} from 0.05 up to 0.50 apfu (Fig. 1). The relatively high fO$_2$ values of the hydrothermal-magmatic fluid should be tentatively related to hydrogen leakage produced by sudden and intermittent opening of new fissures in overlying low-permeability rocks, as already mentioned.

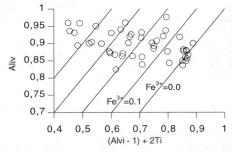

Fig. 1. Al$_{iv}$ vs. Al$_{vi}$ - 1 + 2Ti diagram. Early magmatic muscovites plot along the Fe^{3+}/Fe$_{tot}$=0.0 line. Crystals of later generation are spread towards more oxidising conditions.

Most "primary-looking" muscovite blades show the following chemical differences in comparison with the two certain magmatic crystals: higher phengite contents, with values that Monier & Robert (1986) considered typical of secondary muscovites, simultaneous decrease of Mg/Fe ratio and remarkable F content often slightly higher than those of magmatic crystals. These data confirm that, as a consequence of magma cooling and retrograde increase of fluid pressure, the phengite content increases in late-magmatic muscovites as observed in metamorphic ones (Guidotti, 1978). The chemical features just described can be explained by a thorough long-lasting interaction between magmatic muscovite and a late-magmatic hydrothermal fluid characterised by high F and Fe contents. The abundance of fluorine should be a valid indicator of the "non-meteoric" origin of such fluid; its circulation inserted cleavages, fractures and crystal defects, continuously changing the original muscovite composition, except in small less permeable micro environments. In fact, within the microprobe traverse on one of the two magmatic crystals, reported in Fig. 2, one spot revealed very high phengite contents, between two nearby spots (about 10 μm apart) having phengite-poor compositions.

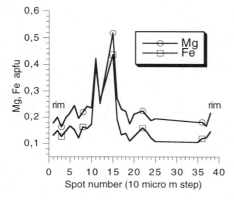

Fig. 2. Mg and Fe chemical profiles across a muscovite crystal. The peaks show phengitic compositions and are also related to SiO_2 increase.

Very small sericite crystals (only some tens of μm size) found along fractures and cleavages mainly in early crystallised feldspars show negligible or undetectable F content (<0.1%) and high phengite contents comparable to those measured within late-magmatic muscovite crystals.

4 CONCLUSIONS

The following chronological sequence was observed for white mica:
1) early magmatic muscovite with high Ti, moderately high F and low phengite contents;
2) late magmatic muscovite with low or undetectable Ti, relatively high F and high phengite contents;
3) hydrothermal acicular sericite with low to undetectable F and Ti and high phengite contents.

Hydrothermal sericite, together with chlorite, should be related to present day circulation, as indicated by the comparison of in-hole measured temperatures with fluid inclusion data and chlorite geothermometer data here reported.

Such multiple stage evolution is in agreement with the fluid inclusion study of Valori et al. (1992) who found as much as 9 generations of different fluid compositions.

5 ACKNOWLEDGEMENTS

Samples were provided by ENEL (Italian Electricity Board) under an IIRG-ENEL agreement. M. Serracino (CSQEA) made electron probe analyses.

6 REFERENCES

Cathelineau, M. & Nieva, D. 1985. A chlorite solid solution geothermometer. The Los Azufres (Mexico) geothermal system. *Contrib. Mineral. Petrol.*, 91: 235-244.
Finch, A.A. & Walker, F.D.L. 1991. A cathodoluminescence and microporosity in alkali felspars from the Blå Måne Sø perthosite, S Greenland. *Mineral. Mag.*, 55: 583-589.
Guidotti, C.V. 1978. Muscovite and K-feldspar from two-mica adamellite in northwestern Maine: composition and petrogenetic implications. *Am. Mineral.* 63: 750-753.
Kovalenko, V.I. & Kovalenko, N.I. 1984. Problems of the origin, ore-bearing and evolution of rare-metal granitoids. *Phys. Earth Planet. Interiors*, 35: 51-62.
Leroy, J. & Cathelineau, M. 1981. Les minéraux philliteux dans les gisements isothermaux d'uranium. 1-cristallochimie des micas hérités et néoformés. *Bull. Minéral.*, 105: 99-109.
Monier, G. & Robert, J.L. 1986. Muscovite solid solutions in the system K_2O-MgO-FeO-Al_2O_3-SiO_2-H_2O: an experimental study at 2 kbar P_{H2O} and comparison with natural Li-free white micas. *Mineral. Mag.*, 50: 257-266.
Nash, W.P. 1976. Fluorine, chlorine, and OH-bearing minerals in the Skaergaard intrusion. *Am. J. Sci.* 276: 546-557.
Petrucci, E., Gianelli, G., Puxeddu, M. & Iacumin, P. 1994. An oxygen isotope study of silicates in the Larderello geothermal field, Italy. *Geothermics*, 23: 327-337.
Shinohara, H. & Fujimoto, K. 1994. Experimental study in the system albite-andalusite-quartz-NaCl-HCl-H_2O at 600°C and 400 to 2000 bars. *Geochim. Cosmochim. Acta*, 58: 4857-4866.
Valori, A., Cathelineau, M. & Marignac, C. 1992. Early fluid migration in a deep part of the Larderello geothermal field: a fluid inclusion study of the granite sill from well Monteverdi 7. *J. Volcanol. Geothermal Res.*, 51: 115-131.
Villa, I.M., Gianelli, G., Puxeddu, M., Bertini, G. & Pandeli, E. 1987. Granitic dykes of 3.8 Ma age from a 3.5 km deep geothermal well at Larderello (Italy). *Soc. Ital. Mineral. Petrol. Proceedings of the Conference "Granites and their surroundings"*, Verbania, 28 sept.-3 oct. 1987, Italy: 163-164.

Geothermal system in Tapi rift basin, Northern Deccan Province, India

D. Chandrasekharam
Department of Earth Sciences, IIT, Bombay, India & CNR, Centro di Studio per la Minerogenesi e la Geochimica Applicata, Firenze, Italy

S. R. Prasad
Department of Earth Sciences, ITT, Bombay, India

ABSTRACT: Jalgaon thermal springs of the Deccan volcanic province are controlled by E-W trending faluts. The chemistry of the springs together with the subsurface geology and structure of this region indicate that these springs are attaining chemical equilibrium with the Precambrian crystallines and suggest a deeper thermal circulatory system.

1 INTRODUCTION

The Narmada-Tapi rift double graben, with the Satpura horst in between, has been reactivated during the Deccan volcanic episode (Sheth and Chandrasekharam, 1997) and hosts several thermal springs. These springs issue through the Deccan basalts near Jalgaon while at Tattapani they issue through the Precambrian crystallines (Chandrasekharam and Antu, 1995). As a part of our continuing investigation on the Indian thermal springs, we report new data on three thermal springs occurring near Jalgaon and discuss their evolution in terms of their chemistry and tectono magmatic regime using field, chemical signatures and published geophysical data The surface temperature of the thermal springs varies from 35 °C to 59 °C and the flow rate is about 50 l/m. The location of these thermal springs is shown in figure 1.

2 MAJOR ION CHEMISTRY

Water samples from the thermal springs, river and groundwater were analysed (Table 1) and plotted on the trilinear diagram (Piper, 1944) shown in figure 2. Data on the thermal springs from western Deccan volcanic province, Tattapani and Puttur reported by Chandrasekharam and Antu (1995) are also plotted in figure 2 for comparison.

The Jalgaon springs, though flowing through basic volcanics, plot in the Na-HCO$_3$ field (Fig. 2) and thus are similar to those flowing through the Precambrian crystallines. The Precambrian crystallines, as reported earlier (Chandrasekharam et.al., 1992; Chandrasekharam and Antu, 1995) include granite pegmatite and granite gneisses. Thus the chemical signature of these springs indicate chemical reaction with rocks other than basic volcanics.

3 GEOLOGY AND TECTONIC SETTING

The Deccan volcanics are exposed on the northern and southern parts of the area and are covered by a thick cover of alluvium along the Tapi rift basin through which the thermal springs are issuing. The thickness of the alluvium varies from 100 to 400m. Bore hole logs and resistivity surveys indicate a number of faults in the basin extending to a depth beyond 200m (Fig. 3). Occurrence of several E-W

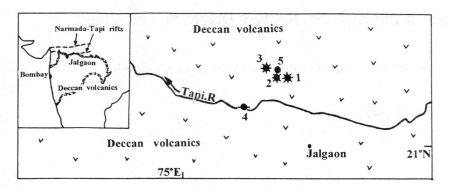

Figure 1. Geological and sample location map of Jalgaon area.
✸ Thermal springs: 1. Unapdeo, 2. Ramtalab, 3. Nazerdeo. ⬤ 4. Tapi river. ● 5. Groundwater.

trending faults over the surface and below the alluvium indicate intense tectonic activity in this region during and after the main Deccan volcanic episode. Neotectonic activity has also been reported along the rift basin (Ravisankar, 1987).

The tectonic fabric of Tapi basin was developed due to plume-rift interaction and lithospheric extension along the Narmada (Sheth and Chandrasekharam, 1997), which resulted in uplift of continental crustal blocks (Chandrasekharam and Viladkar, 1989) and alkaline magmatism along the Narmada rift (Sheth and Chandrasekharam, 1997). Magnetotelluric investigation across the Tapi basin, near the study area (Rao et.al., 1995), indicates presence of a complex granitic body between 2 and 10 km below the surface covered by thick Gondwana sedimentary members. Thus the geophysical and field data indicate presence of Gondwana sediments and granite below the Deccan volcanics.

4 DISCUSSION AND CONCLUSIONS

The Na-HCO_3 character of the thermal springs indicate that the waters must be equilibrating with formations rich in Na rather than formations rich Ca, such as the Deccan volcanics. The chemical plots of the surface and subsurface waters in figure 2 suggest that meteoric water is the main

Table 1. Chemical analyses of water samples (ppm).

		pH	T°C	Na	K	Ca	Mg	HCO_3	Cl	SO_4
1	Unapdeo	9	59	67.5	1.2	8.8	3.4	70.4	4.8	31.8
2	Ramtalab	8	36	57.5	nd	24.8	nd	48.4	3.9	26.9
3	Nazerdeo	7	35	65	2	13.6	2.4	44.4	3.9	26.9
4	Tapi river	7	33	46.2	2.5	30.4	10.7	88.7	1.4	16.9
5	Groundwater	7	32	63.7	nd	40.8	1.4	116.9	1.2	3.4

nd: not determined

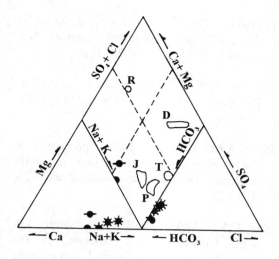

Figure 2. Trilinear diagram. ✹ Thermal springs. ● Tapi river, ○ Rain. ● Groundwater.
D: Deccan volcanic springs. J: Jalgaon springs.
P: Puttur springs. T: Tattapani springs

source for the thermal springs and thus defines a simple circulatory system in this region. The estimated reservoir temperatures, using Na-K geothermometers (Fournier, 1979), vary from 105 °C (Unapdeo) to 133 °C (Nazardeo) springs.

Assuming a geothermal gradient of 60 °C/km (Ravisankar, 1988) for this region, the estimated depth of the reservoir appears to be at about 2 km. The presence of Gondwana sediments and granite at such depths (Rao et.al., 1995) and the chemical signature of the thermal springs clearly indicate the reservoir to be either the Gondwana sedimentary members or the Precambrian crystallines.

Thus these springs, though issuing through the Deccan volcanics covered by recent alluvium, are circulating to depths varying from 2 km and below thus indicating a complex geothermal system in this region which appears to be similar to that existing at Tattapani geothermal field (Chandrasekharam and Antu, 1995).

The presence of ancient sedimentary sequence over the Precambrians coupled with a high geothermal gradient makes the area suitable for further exploration for geothermal energy resources. DSS profiles across the Narmada rift demonstrate that the faults across the rift extend up to the upper mantle depths (Kailasm, 1979). Investigation on rare gas chemistry is being carried out on these springs to determine

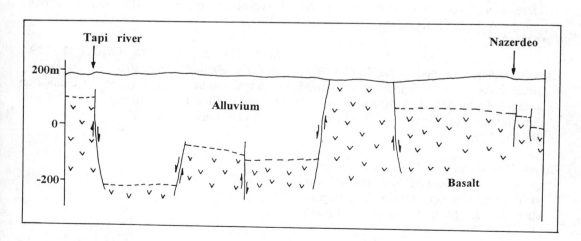

Figure 3. Subsurface structure along Tapi-Nazerdeo traverse (modified after Dubey and Saxena, 1987).

the possible involvement of mantle in the evolution of the geothermal system along the Narmada rift in general and Jalgaon thermal springs in particular.

5 ACKNOWLEDGEMENTS

The authors thank CNR (Italian National Research Council, Florence) and I.C.T.P. (International Centre for Theoretical Physics, Trieste) for providing necessary facilities for writing this paper.

6 REFERENCES

Chandrasekharam, D and Viladkar, S. 1989. Constitution of the Precambrian crust below the Deccan volcanics- Evidence from a composite dyke of Mandaleswar, Deccan volcanic province. In: *Precambrian Granitoids- petrogenesis, geochemistry and metallogeny,* I. Happala and Y. Kahkonen (eds), *Geol. Sur. Finland, Sp. Pap.,* 8.

Chandrasekharam, D., Ramanathan, A and Selvakumar, R.L. 1992. Thermal springs in the Precambrian crystalline of the western continental margin of India: Field and experimental results. *Water-Rock Interaction 7,* (eds) Y.K. Kharaka and A. Maest, A.A. Balkema Pub., Rotterdam, 1271-1274.

Chandrasekharam,D and Antu, M.C. 1995. Geochemistry of Tattapani thermal springs, Madhya Pradesh, India- Field and experimental investigations. *Geothermics,* 24, 553-559.

Dubey, R and Saxena, R.K. 1987. Thermal springs and associated tectonics of Tapi basin area, Districts Dhule and Jalgaon, Maharashtra. *Geol. Sur. India, Records,* 115, 150-174.

Fournier, R.O. 1979. A revised equation for the Na/K geothermometer. *Geother. Res. Council Bull.,* 3, 221-224.

Kailasm, L.N. 1979. Plateau uplift in Peninsular India. *Tectonophy.,* 61, 243-269.

Piper, A.M. 1944. A graphic procedure in the geochemical interpretation of water analysis. *Am. Geophy. Union, Trans.,* 25, 914-923.

Rao, C.K., Gokarn, S.G. and Singh, B.P. 1995. Upper crustal structure in the Torni-Purnad region, central India using magnetotelluric studies. *J. Geomag. Geoelect.,* 47, 411-420.

Ravisankar. 1987. Neotectonic activity along the Tapi-Satpura lineament in central India. *Indian Minerals,* 41, 19-30.

Ravisankar. 1988. Heat flow map of India and discussion on its geological and economic significance. *Indian Minerals,* 42, 89-110.

Sheth, H.C. and Chandrasekharam, D. 1997. Plume-rift interaction in the Deccan volcanic province. *Phy. Earth. Planet. Inter.,* 99, 179-187.

Sheth, H.C. and Chandrasekharam, D. 1997. Early alkaline magmatism in the Deccan traps: Implications for plume incubation and lithospheric rifting. *Phy. Earth. Planet. Inter.* (in press).

Thermal and chemical evolution of the Tiwi Geothermal System, Philippines

Joseph N. Moore – *Energy and Geoscience Institute, University of Utah, Salt Lake City, Utah, USA*
Thomas S. Powell – *Unocal Geothermal and Power Operations, Santa Rosa, Calif., USA*
Carol J. Bruton – *Lawrence Livermore National Laboratory, Geosciences Division, Calif., USA*
David I. Norman & Matthew T. Heizler – *Dept. of Geosciences, New Mexico Tech, Socorro, N.Mex., USA*

ABSTRACT: The Tiwi geothermal field is related to young volcanic activity on the southern coast of Luzon, Philippines. Nine stages of alteration and vein mineralization have been documented in core from the western part of the Matalibong sector where measured temperatures are close to 270°C. The earliest stage is characterized by chalcedony and clays (stage 1). Stage 2 veins are filled with sericite and chlorite. Stage 3, 5, and 7 veins are characterized by quartz ± epidote ± adularia ± wairakite whereas stage 4, 6, and 8 veins consist of anhydrite and/or calcite. Illite ± chlorite were deposited during stage 9. The maximum fluid-inclusion temperatures of stage 4, 5, and 6 vein minerals are close to the boiling point and typically exceeded 300°C while the minimum temperatures are more than 75°C below the present measured values. $^{40}Ar/^{39}Ar$ spectrum dating indicates that stage 5 adularia was deposited at about 0.30 Ma. The results of the spectrum dating combined with fluid-inclusion data and mineral geothermometers indicate that mineralization during stages 2 to 8 reflects repeated cycles of boiling, pressure drawdown and the incursion of cooler fluids between 0.20 and >0.30 Ma. Stage 9 mineralization appears to be related to renewed heating and boiling that occurred in response to intrusive activity during the last 0.05 Ma. Fluid-inclusion gas compositions and salinities indicate that the hydrothermal fluids were mainly mixtures of fresh and seawater modified by interactions with crustal and possibly magmatic gases. Only stage 9 minerals are in equilibrium with the present-day fluids.

INTRODUCTION

The Tiwi geothermal field, with an installed capacity of 330 MWe, is a major producer of electricity in the Philippines (Gambill and Beraquit, 1993). Although development of the field began in the 1960's, it was not until 1992 when Unocal drilled Matalibong-25 that continuous core from the reservoir became available (Fig.1). The well was drilled in the western part of the field where temperatures are currently close to 270°C. Core was taken from 789 to 2439 m. In this paper, we describe the chemical and thermal evolution of the veins encountered in the cored portion of the well.

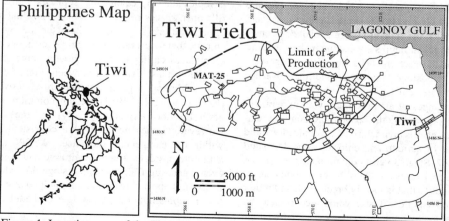

Figure 1. Location map of the Tiwi Geothermal Field.

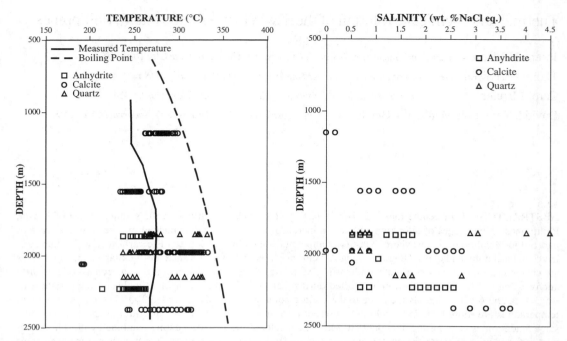

Figure 2. Distribution of fluid-inclusion homogenization temperatures and salinities (in weight percent NaCl equivalent) in vein minerals as a function of depth. For comparison, the measured temperatures and the boiling-point curve for a 2.0 weight percent NaCl solution are also shown on the left.

GEOLOGY AND HYDROTHERMAL ALTERATION

Tiwi is located on the eroded flanks of Mt. Malinao, which was active between 0.5 and 0.06 Ma (Gambill and Beraquit, 1993). Above a depth of 1601 m, the rocks in Matalibong-25 are mainly andesite and basaltic andesite flows, flow breccias and volcaniclastics deposited in a subaerial environment. Between 1601 and 2012 m, poorly-sorted sandstones that are intercalated with claystones, andesite breccias, and lava flows mark the transition to a subaqueous environment. At greater depths, sandstone makes up more than 90% of the section. The underlying basement consists of mudstone and limestone that overlies quartz-muscovite schist.

Nine stages of alteration and vein mineralization were recognized on the basis of crosscutting relationships. Stage 1 is characterized by chalcedony and clays. Veins of sericite and chlorite were deposited during stage 2. Stage 3 veins contain quartz ± pyrite ± epidote. Stage 4 veins consist of blocky, often coarse-grained anhydrite that is locally brecciated, and minor rhombohedral calcite. These veins are crosscut by veins dominated by epidote or quartz ± adularia (stage 5). Veins deposited during stage 5 are the most diverse, with mineral abundances varying markedly even among individual fractures in the same thin section. Rarely, veins assigned to this stage consist of quartz and bladed calcite. Stage 6 veins are characterized by rhombohedral calcite and minor anhydrite. Veins deposited during stage 7 contain wairakite ± epidote ± quartz. However, wairakite-bearing veins are present only in the upper part of the well. Stage 8 is represented by calcite that replaces wairakite and fills open spaces in stage 7 veins. The final stage of mineralization (stage 9) is characterized by illite ± chlorite that forms radiating aggregates in open spaces or replaces earlier calcite, although some late-stage sericite was also deposited between stages 6 and 7.

Minerals typical of temperatures above 300°C (Henley and Ellis, 1983) are not common and only actinolite occurs sporadically in these rocks. However, it does not occupy a consistent position within the paragenetic sequence. At 996 m depth, actinolite postdates stage 7 wairakite and epidote and even later vug-filling calcite (stage 8), whereas at 2135 m, it appears to be contemporaneous with stage 5 veins containing quartz, epidote, and adularia.

FLUID-INCLUSION TEMPERATURES AND SALINITIES

The results of heating and freezing measurements on more than 575 liquid-rich, mostly secondary fluid inclusions hosted in vein calcite, quartz, and anhydrite are shown in Figure 2. Only two-phase inclusions were observed, but both vapor- and liquid-rich types are present. The presence of vapor-rich inclusions provides clear, independent evidence of boiling.

Homogenization temperatures of the inclusions ranged from 332° to 191°C while salinities varied from 4.5 to 0.0 weight percent NaCl equivalent. In general, the maximum temperatures are close to the boiling point to depth curve, whereas the minimum temperatures below a depth of about 1524 m are nearly constant at 230°C.

There is an overall increase in the maximum salinities of the fluid inclusions with depth. Below 1600 m, the maximum salinities ranged from 3.1 to 4.5 weight percent NaCl equivalent and are close to or greater than that of seawater. In contrast, fluid inclusions from shallower depths have salinities that do not exceed 1.7 weight percent NaCl equivalent. For comparison, the reservoir fluid produced by Matalibong-25 has a salinity of 1.2 weight percent NaCl equivalent.

FLUID-INCLUSION GAS COMPOSITIONS

The gas compositions of fluid inclusions in 15 samples of vein calcite, quartz, and anhydrite from depths ranging from 1113 to 2428 m were measured by quadrapole mass spectrometry after being liberated by crushing or thermal decrepitation (Norman and Sawkins, 1987; Norman et al., 1997). The principal component of the inclusions was water, which accounted for 70 to 99.8 mole percent of the analysis. CO_2, CH_4, H_2, N_2, Ar, H_2S, SO_2, and hydrocarbons (C_{2-7}) were also detected, with CO_2 and CH_4 dominating. Most samples have N_2/Ar ratios ranging from air-saturated water (36) to 232. N_2/Ar ratios up to about 80 can be explained by boiling (Norman et al., 1997). Higher N_2 contents imply contributions from either crustal rocks beneath the reservoir or a diluted magmatic fluid. The high CH_4 contents are markedly different from the present-day fluids which are CH_4 poor. This argues that the fluid inclusions trapped gases generated during the early stages of the system's evolution.

DISCUSSION

The mineralogic and fluid-inclusion data demonstrate that veins in Matalibong-25 formed under a variety of conditions and flow regimes. Bruton et al. (1997) determined the saturation state of the vein minerals with respect to the modern reservoir fluids and simulated the effects of boiling, cooling, heating, and the addition of CO_2-rich steam on their deposition. The models suggest that assemblages consisting of quartz ± epidote ± adularia ± wairakite (stages 3, 5, and 7) formed in response to boiling, but that monomineralic veins of wairakite could also have formed by conductive cooling. The only process that favors deposition of calcite + anhydrite (stages 4, 6, and 8) is conductive heating. Crystals of calcite in these veins typically have a blocky habit whereas calcite deposited as a result of boiling is often bladed and intergrown with quartz. Taken together, the vein parageneses document periods of upflow that oscillated with the incursion of cooler fluids during stages 4, 6, and 8. Fluid-inclusion temperatures, however, demonstrate that the maximum temperatures during stages 4, 5, and 6 exceeded 300°C throughout most of the well. Even though we were unable to obtain fluid-inclusion data on minerals clearly related to stages 3 and 7, the presence of epidote in these veins indicates that temperatures were above 240°C while post-stage 8 actinolite at 996 m provides evidence of temperatures above 300°C (Henley and Ellis, 1983).

Late-stage sericite (stage 9) is commonly associated with corroded calcite. Bruton et al. (1977) demonstrated that the modern reservoir fluid produced by Matalibong-25 is slightly acidic with a pH of 5.3 compared to a neutral pH of 5.6 at 270°C. The pH of this fluid, which may reflect the addition of CO_2-rich steam, favors the deposition of sericite and can lead to corrosion of calcite. Thus, the occurrence of sericite as the latest alteration phase is consistent with the chemistry of the modern fluids. A similar mechanism may account for the formation of the stage 1 sericite veins.

$^{40}Ar/^{39}Ar$ spectrum dating of adularia from depths of 1809, 1849, and 1852 m has yielded an age of approximately 0.30 Ma for stage 5 mineralization. Fluid inclusions in quartz intergrown with adularia, from a depth of 1849 m, demonstrate that the temperatures during this mineralization were close to 330°C. Modeling of the age spectra combined with Ar kinetic data for the Tiwi adularia leads to three important conclusions regarding the timing of subsequent events. First, the models indicate that temperatures could have remained above 300°C until about 0.2 Ma or that the rocks were heated above 300°C at this time. Fluid-inclusion homogenization temperatures above 300°C in stage 6 minerals and the presence of post-stage 8 actinolite constrain the age of the main mineralization (stages 2-8) to more than 0.2 Ma. Secondly, the models indicate that significant cooling to temperatures below 250°C must have occurred between 0.2 Ma and the recent past. This cooling is reflected in fluid-inclusion temperatures that are as low as 191°C. Finally, the models show that the present-day temperatures of 270°C could not have been maintained for more than the last 0.05 Ma. Thus, the data record a relatively long period of cooling between the initial magmatic pulse that generated the hydrothermal system and recent, renewed magmatic activity.

CONCLUSIONS

Petrologic studies of core from Matalibong-25, combined with $^{40}Ar/^{39}Ar$ dating provide a unique glimpse into the development of a long lived, volcanic-hosted hydrothermal system. During the main stage of the system's evolution, a complex pattern of fluid circulation developed. Three major cycles, each consisting of a period of upwelling and boiling followed by recharge and the influx of cooler fluids are reflected in alternating sequences of silicate and carbonate and/or sulphate mineralization. The incursion of cooler waters during the second phase of each cycle may have occurred in response to the development of pressure sinks that formed as the fluids boiled off. $^{40}Ar/^{39}Ar$ spectrum dating indicates that the second of these cycles occurred at 0.30 Ma and that the third cycle is more than 0.20 my old.

The hydrothermal fluids were mixtures of fresh water and seawater with seawater dominating at depths below about 1800 m. Gas ratios of fluid inclusions document the presence of a significant crustal component in the fluids at all depths.

The present geothermal system represents a period of renewed heating during the last 0.05 Ma that is probably related to emplacement of subvolcanic intrusions. Measured downhole temperatures are more than 75°C higher than the minimum fluid-inclusion temperatures, reaching a maximum of 270°C at about 1829 m depth. Deposition of sericite as open-space fillings and the corrosion of calcite appear to be modern features of the geothermal system that are related to recent boiling.

ACKNOWLEDGMENTS

This study would not have been possible without the support of the management and staff of the Philippine National Power Corporation, Philippine Geothermal, Inc., and Unocal Geothermal and Power Operations. We would like to thank D. Nielson for helping collect the samples, J. Hulen for helpful discussions on the mineralogy of the well, and R. Turner and R. Wilson, who drafted the illustrations and helped put the paper together. Funding for JNM was provided by the U.S. Department of Energy, under contract no. DE-ACO7-95ID13274. Work by CJB was performed under the auspices of the U. S. Department of Energy under contract W-7405-Eng-48.

REFERENCES

Bruton, C. J., Moore, J. N., and Powell, T. S., 1997, Geochemical analysis of fluid-mineral relations in the Tiwi geothermal field, Philippines: Twenty-second Workshop on Geothermal Reservoir Engineering, Stanford University, p. 457-463.

Gambill, D. T. and Beraquit, D. B., 1993, Development history of the Tiwi geothermal field, Philippines: Geothermics, v. 22, p. 403-416.

Henley, R. W., and Ellis, A. J., 1983, Geothermal systems ancient and modern: a geochemical review: Earth Science Reviews, v. 19, p. 1-50.

Norman, D. I., and Sawkins, F. J., 1987, Analysis of gases in fluid inclusions by mass spectrometer: Chemical Geology, v. 61, p. 110-121.

Norman D. I., Moore J. N., and Musgrave, J., 1997, More on the use of fluid inclusion gaseous species as tracers in geothermal systems: Twenty-second Workshop on Geothermal Reservoir Engineering, Stanford University, p. 419-426.

Low-temperature alteration of basalts from the Tangihua Complex, New Zealand

K. N. Nicholson & P. M. Black
Department of Geology, University of Auckland, New Zealand

ABSTRACT: Low-temperature hydrothermal alteration of the Tangihua Complex of Northland, New Zealand, is charactered by zeolite minerals in veins, filling open spaces and in the rock. The alteration assemblages can be divided into three main stages based primarily on temperature but also influenced by the water/rock ratios. The initial stage of alteration was characterised by Na-rich zeolites: analcime, stilbite and thomsonite with lesser amounts of natrolite and mesolite, while the groundmass and phenocryst alteration minerals include chlorite, actinolite, epidote and pumpellyite. Cooling of the system resulted in the transitional stage of alteration which was characterized by lower temperature minerals including mixed layer chlorite-smectite clays and Ca-rich zeolites with minor precipitation of both Na- and K-rich end members. Finally the system reached temperatures <50°C with high water/rock ratios. During this stage of alteration both K^+ and Ca^{2+} started to precipitate as apophyllite and calcite in veins and open spaces and as calcite, smectite and smectite-chlorite in the groundmass. Eventually the Ca^{2+} became dominant resulting in the overprinting of earlier phases by secondary calcite.

INTRODUCTION

The interaction between seawater and basalt occurs under conditions ranging from low-temperature seafloor alteration (≈2°) to hydrothermal alteration within the oceanic crust (≤350°). There have been many studies of the high-temperature hydrothermal alteration processes of basalts which include the Ocean Drilling Project (eg. Jeffery et al., 1986) and oceanic ridge hot springs (eg. Edmonds et al., 1979 a, b; Campbell et al., 1988; Bowers et al., 1988; Massoth et al., 1989). In contrast there is little data available on either the fluid or the solid phases involved in low-temperature alteration processes.

Most data regarding mechanisms operating at low temperatures are derived from experimental studies which show that low-temperature basalt alteration produces a slight but progressive loss of Mg^{2+}, Na^+ and K^+ and an enrichment of Ca^{2+} and SiO_2 in seawater while pH decreases slowly (Seyfried and Bischoff, 1979). More recent studies based on field relations and mineral assemblages (ie. Bednarz and Schmincke,1989; Guy et al., 1992) support earlier results. Bednarz and Schmincke (1989) reported an increase in K^+ and Mg^{2+} and a decrease in Na^+ in rocks in a cold seawater alteration zone (<20°C) where as in their low-temperature hydrothermal alteration zone (<170°C) only Ca^{2+} is depleted and both Na^+ and K^+ are enriched.

Similar results were obtained by Guy et al. (1992) who showed that low-temperature alteration of basalts by seawater is a function of the rock/water ratio (R/W). As R/W increases (up to 70g/l) there is an almost total removal of Mg^{2+} and K^+ and an enrichment of Ca^{2+}, Sr^{2+} and Si^{4+} from the interstitial waters. This process is accompanied by a decrease in pH and Eh and a depletion of ^{18}O and D. Increasing the R/W above 70 g/l results in a reversal of the Mg-Ca exchange and an enrichment of both ^{18}O and D. Their study attributes both the chemical and isotopic reversals to the transformation of early precipitated Mg-rich saponite to a Ca-Na-rich saponite.

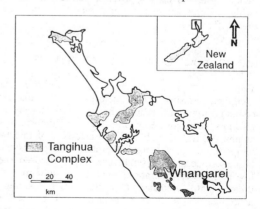

Figure 1. Outline maps showing the location and distribution of the ophiolite massifs of the Tangihua Complex, Northland, New Zealand.

The ophiolites of the Tangihua Complex in New Zealand (Figure 1) provide an excellent opportunity to study the effects of low-temperature hydrothermal alteration on basalts. The present study is based on new sampling, petrography, XRD analyses and the reinterpretation of previous work. The findings of this study have been used in combination with other investigations of low-temperature alteration of basalts in order to determine the alteration history of the Tangihua Complex.

GEOLOGICAL SETTING

The ophiolitic igneous complexes in the northern regions of the North Island, New Zealand, are collectively called the Tangihua Complex and informally known as the Northland Ophiolite. It is one of a number of ophiolites on the southwest Pacific rim with an early Tertiary emplacement age. The others include the Papuan Ultramafic Belt and the Massif du Sud ultramafics and the Poya terrane basalts of New Caledonia (Davies, 1968; Prinzhofer et al., 1980; Malpas et al., 1992). The ophiolites in New Caledonia and Papua New Guinea include major belts of ultramafic and plutonic rocks which represent deep levels of oceanic lithosphere where as the Tangihua Complex is dominated by rocks from the uppermost levels of the oceanic crust (Malpas et al., 1992).

Tangihua Complex ophiolites occur as structurally discrete, rootless bodies which range in area from a few square metres to high standing massifs of up to 900 km^2. In general the complex consists of massive and pillowed basaltic lava sequences with intercalated sediments and lesser gabbro, microgabbro and basaltic intrusives. Chemically, the lavas are dominantly tholeiitic basalts with N-MORB affinities. The complex also includes minor felsic derivatives, younger alkalic intrusives (which exhibit within plate characteristics) and rare ultramafic rocks.

MINERALOGY OF SECONDARY PHASES

There were two distinct periods of alteration in the Tangihua massifs. The majority of the massifs contain low-temperature hydrothermal/diagenetic minerals such as zeolites. Some of the massifs exhibit mineral assemblages characteristic of higher-temperature hydrothermal alteration. In places the higher-temperature assemblages have obscured the low-temperature alteration, however, in some massifs there is no evidence that the low-temperature assemblages were ever present. The massifs containing the higher-temperature assemblages are associated with either cogenetic intrusive activity or much younger intrusions which have been responsible for the higher grade alteration. The secondary phases produced by high-temperature hydrothermal alteration in the Tangihua Complex are described by Leitch (1966), Mason (1973) and Farnell (1973).

The lower temperature alteration mineral assemblages in the Tangihua basalts can be divided into three stages: an early, higher-temperature, stage (Table 1), a transitional stage (Table 2) and late stage alteration (Table 3). The early higher-temperature stage of alteration is accompanied by the formation of chlorite, pumpellyite, actinolite, epidote and quartz both in the groundmass and replacing phenocrysts. In the later alteration stages only chlorite-smectite, smectite and rare actinolite occur in the groundmass. The majority of the secondary minerals occur in veins and filling open spaces, not in the groundmass. The restricted pervasiveness of the alteration reflects both the chemistry of the fluids and especially the lack of fracture related permeability.

Minor secondary opaque minerals were found disseminated throughout the groundmass and in veins. The dominant opaque mineral was pyrite which formed during both the early, higher-temperature stage and the transitional stage of alteration. There does not appear to be any relationship between the alteration stages and the precipitation of pyrite.

The initial higher-temperature stage of alteration was characterised by the precipitation of sodic zeolites, such as analcime, stilbite and thomsonite, in veins and open spaces (Table 1). This stage is similar to the type I, high-temperature hydrothermal alteration zone of Bednarz and Schmincke (1989) and to the transitional zone of Gillis and Robinson (1988).

Table 1. Early higher-temperature alteration minerals	
Vein and open space filling	Groundmass and phenocryst alteration
analcime: Na$_{16}$(Al$_{16}$Si$_{32}$O$_{96}$) · 16H$_2$O	chlorite: (Mg,Al,Fe)$_3$(Si,Al)$_4$O$_{10}$(OH)$_2$ · (Mg,Al,Fe)$_3$(OH)$_6$
stilbite: NaCa$_4$(Al$_9$Si$_{27}$O$_{72}$)	actinolite: Ca$_2$(Mg,Fe^{2+})$_5$Si$_8$O$_{22}$(OH)$_2$
thomsonite: Na$_4$Ca$_8$(Al$_{20}$Si$_{20}$O$_{80}$) · 24H$_2$O	epidote: Ca$_2$Fe^{3+}Al$_2$O(Si$_2$O$_7$)(SiO$_4$)OH
natrolite: Na$_{16}$(Al$_{16}$Si$_{24}$O$_{80}$) · 16H$_2$O	pumpellyite: Ca$_2$MgAl$_2$(SiO$_4$)(Si$_2$O$_7$)(OH)$_2$ · H$_2$O
mesolite: Na$_{16}$Ca$_{16}$(Al$_{48}$Si$_{72}$O$_{240}$) · 64H$_2$O	opaques
opaques	

During the reaction between seawater and basalts, at temperatures ranging from 70° to 425°C and a variety of water/rock ratios, Mg^{2+} is highly reactive. The precipitation of Mg-bearing clays (such as chlorite) and Mg-bearing minerals results in the depletion of Mg^{2+} in the fluids (Bischoff and Dickson, 1975; Seyfried and Mottl, 1982) which generates high acidity. This generally results in the leaching of Ca^{2+} and K^+ from the rocks. The enrichment of Na in the fluids is thought to be the result of a combination of lower water/rock ratios and the higher-temperatures. These findings are in agreement with Bednarz and Schmincke (1989). Bednarz and Schmincke (1989) believe that the marked uptake of K^+ accompanied by a characteristic change from Na_2O loss to Na_2O uptake in the fluids, indicates increasing temperature and/or lower water/rock ratios (eg. Mottl et al., 1979; Seyfried and Bischoff, 1981; Mottl 1983).

Secondary minerals in the groundmass and phenocrysts suggest a temperature for hydrothermal alteration of approximately 250°-300°C: given that chlorite forms at temperatures up to 275°C, actinolite at >250°C and epidote at >260°C. Precipitation of the cogenetic zeolite assemblage in veins and open spaces is possible within this temperature range. Thomsonite has a moderately well constrained temperature of formation of between 150° and 275°C where as stilbite generally precipitates at temperatures between 70° and 200°C.

During the transitional alteration stage the water/rock ratio remained low. At low water/rock ratios the pH of the fluids is reduced to near-neutral and the fluids become enriched in Mg^{2+}, where as Ca^{2+} and to a lesser extent K^+ are removed (Guy et al., 1992). The transitional stage of alteration also represents the result of cooling within the system.

Table 2. Transitional stage alteration minerals	
Vein and open space filling	Groundmass and phenocryst alteration
chabazite: $Ca_2(Al_4Si_8O_{24}) \cdot 12H_2O$	chlorite-smectite and/or smectite
heulandite: $(Na,K)Ca_4(Al_9Si_{27}O_{72}) \cdot 24H_2O$	calcite: $CaCO_3$
clinoptilolite: $(Na,K)_6(Al_6Si_{30}O_{72}) \cdot 20H_2O$	prehnite: $Ca_2Al(AlSi_3O_{10})(OH)_2$
laumontite: $Ca_4(Al_8Si_{16}O_{48}) \cdot 16H_2O$	opaques
opaques	

The secondary phases seen in the groundmass are minor but include calcite, mixed layer chlorite-smectite (150°-180°C) and lesser amounts of prehnite where as the precipitating zeolites range from Ca-rich to K-rich end members (Table 2). Hence, a cooler system with low water/rock ratios resulted in a Ca-rich fluid which precipitated predominantly Ca-rich zeolites, where as the formation of clay minerals (such as chlorite-smectite) depleted the fluid of what little Mg^{2+} it contained.

The latest stage of alteration in these rocks was a low-temperature (<50°C) stage characterised by the precipitation of Ca- and minor K-rich minerals in the veins and open spaces (Table 3). Plagioclase phenocrysts are almost totally replaced by smectite and calcite which appears to overprint earlier alteration of plagioclase to K-feldspar. Calcite and apophyllite occur in veins and open spaces. Most other primary phenocrysts and groundmass assemblages have been altered to smectite and lesser amounts of both calcite and smectite-chlorite.

Table 3. Late stage alteration minerals	
Vein and open space filling	Groundmass and phenocryst alteration
apophyllite: $KCa_4Si_8O_{20}(F,OH) \cdot 8H_2O$	smectite and/or smectite-chlorite
calcite: $CaCO_3$	calcite: $CaCO_3$

The secondary mineral assemblage associated with the late stage alteration suggests temperatures of less than 50°C and high water/rock ratios such that the rock becomes enriched in K^+ while the fluids are enriched in Na^+ and Ca^{2+} (Gillis and Robinson, 1988; Guy et al., 1992). This stage of alteration corresponds to the cold seawater alteration zone of Bednarz and Schmincke (1989). At some stage the fluids become oversaturated with respect to Ca^+ resulting in the precipitation of calcite as the final mineral phase in the paragenetic sequence. Calcite precipitation may have completely overprinted earlier K-rich assemblages which are typically characteristic of this type of alteration (Gillis and Robinson, 1988; Guy et al., 1992), however, there is little evidence of an early K-rich assemblage.

CONCLUSIONS

Alteration in the Tangihua Complex of Northland, New Zealand, can be divided into three main stages based primarily on temperature but is also influenced by the water/rock ratios. The earliest stage of alteration occurred at high temperatures (250°-300°C) and low water/rock ratios and is characterised by Na-rich zeolites, such as analcime, stilbite and thomsonite, which are found filling veins and open spaces. Cooling of the system results in a transitional stage of alteration which was charactered by lower temperature minerals including mixed layer chlorite-smectite clays and Ca-rich zeolites with minor precipitation of both Na- and K-rich end members. The water/rock ratio remained low during the transitional stage. The last stage of alteration in the system occurs at temperatures <50°C with high water/rock ratios. While both K^+ and Ca^{2+} start to precipitate, the system must have become Ca-oversaturated as calcite is the late precipitating phase. The lack of characteristic groundmass alteration minerals may be due to low permeability in the rock.

The result of this study are in agreement with previous experimental and field based studies regarding low-temperature alteration of basalts (Seyfried and Bischoff, 1979; Gillis and Robinson, 1988; Bednarz and Schmincke, 1989; Guy et al., 1992). The zeolite minerals appear to be the most responsive minerals to changes in the water/rock ratio, fluid chemistries and temperature. Hence, it is possible to use the precipitated zeolite mineral assemblages to characterise each stage of low temperature alteration in basalts from the Tangihua Complex, Northland, New Zealand.

REFERENCES

Bednarz, U. and Schmincke, H.-U. 1989. Mass transfer during sub-seafloor alteration of the upper Troodos crust (Cyprus). *Contributions to Mineralogy and Petrology* 102: 93-101.

Bischoff, J.L. and Dickson, F.W. 1975. Seawater-basalt interaction at 200°C and 500 bars: implications for origin of sea-floor heavy metal deposits and regulation of seawater chemistry. *Earth and Planetary Science Letters* 25: 385-397.

Bowers, T.S., Campbell, A.C., Measures, C.I., Spivack, A., Khadem, M. and Edmond, J.M. 1988. Chemical controls on the composition of vent fluids at 13°-11°N and 21°N, East Pacific Rise. *Journal of Geophysical Research* 93: 4522-4536.

Campbell A.C., Bowers, T.S., Measures, C.I., Falkner, K., Khadem, M. and Edmond, J.M. 1988. A time series of various fluid composition from 21°N East Pacific Rise (1979, 1981, 1985) and the Guyamas basin, gulf of California (1982, 1985). *Journal of Geophysical Research* 93: 4537-4549.

Davies, H. L., 1968, Papuan Ultramafic Belt: 23rd International Geological Congress, Section 1, p. 209-220.

Edmonds J.M., Measures, C., McDuff, R.E., Chan, L., Collier, R., Grant, B., Gordon, L.I. and Corliss, J. 1979a. Ridge crest hydrothermal activity and the balances of the major and minor elements in the ocean: The Galapagos data, *Earth and Planetary Science Letters* 46:1-18.

Edmonds, J.M., Measures, C., Mangum, B., Grant, B., Sclater, F.R., Collier, R., Hudson, A., Gordon, L.I. and Corliss, J. 1979b. On the formation of metal-rich deposits at ridge crests. *Earth and Planetary Science Letters* 46:19-30.

Farnell, E.J. 1973. Geology of the Cape Reinga area, Northland. Unpublished MSc. Thesis, University of Auckland.

Gillis, K.M. and Robinson, P.T. 1988. Distribution of alteration zones in the upper oceanic crust. *Geology* 16: 262-266.

Guy, C., Schott, J., Destrigneville, C. and Chiappini, R. 1992. Low-temperature alteration of basalt by interstitial seawater, Mururoa, French Polynesia. *Geochimica et Cosmochimica Acta* 56: 4169-4189.

Jeffery, C.A., Honorez, J., Laverne, C. And Emmermann, R. 1986. Hydrothermal alteration of a 1 km section through the upper oceanic crust, deep sea drilling project hole 504b: Mineralogy, chemistry and evolution of seawater-basalt interactions. *Journal of Geophysical Research* 91:309-355.

Leitch, E.C. 1966. The geology of the North Cape area, northern most New Zealand. Unpublished MSc. thesis, University of Auckland.

Malpas, J., Spörli, K. B., Black, P. M. and Smith, I. E. M. 1992. Northland ophiolite, New Zealand, and implications for plate-tectonic evolution of the southwest Pacific. *Geology* 20: 149-152.

Malpas, J., Smith, I. E. M. And Williams, D., 1994, Comparative genesis and tectonic setting of ophiolitic rocks of the South and North Islands of New Zealand: Proceedings of the 29th International Geological Congress, Part D, p. 29-46.

Mason, D.O. 1973. Geology of the Parakao, Pakotai and Pupuke copper deposits, Northland. Unpublished MSc., University of Auckland.

Massoth, G.J., Butterfield, D.A., Lupton, J.E., McDuff, R.E., Lilley, M.D. and Jonasson, I.R. 1989. Submarine venting of phase-separated hydrothermal fluids at Axial Volcano, Juan de Fuca Ridge. *Nature* 340:702-705.

Mottl, M.J. 1983. Metabasalts axial hot springs and the structure of hydrothermal systems at mid-ocean ridges. *Geological Society of America Bulletin* 94: 161-180.

Mottl, M.J., Holland, H. and Corr, R.F. 1979. Chemical exchange during hydrothermal alteration of basalts by seawater - II: experimental results for Fe, Mn, and sulfur species. *Geochimica et Cosmochimica Acta* 43: 869-884.

Prinzhofer, A., Nicolas, A., Cassard, D., Moutte, J., Leblanc, M., Pars, P and Rabinovitch, M., Structures in the New Caledonia Peridotites-Gabbros: Implications for Oceanic Mantle and Crust, *Tectonophysics,* 69: 85-112.

Saccocia, P.J. and Gillis, K.M. 1995. Hydrothermal upflow zones in the oceanic crust. *Earth and Planetary Science Letters* 136: 1-16.

Seyfried, W.E.Jr. and Bischoff, J.L. 1979. Low temperature basalt alteration by seawater: An experimental study at 70°C and 150°C. *Geochimica et Cosmochimica Acta* 43:1937-1947.

Seyfried, W.E.Jr. and Mottl, M.J. 1982. Hydrothermal alteration of basalt by seawater under seawater-dominated conditions. *Geochimica et Cosmochimica Acta* 46: 985-1002.

A new type of hydrothermal alteration at the Kizildere geothermal field in the rift zone of the Büyük Menderes, western Anatolia, Turkey

N.Özgür, M.Vogel & A.Pekdeger
Freie Universität Berlin, FR Rohstoff- und Umweltgeologie, Germany

ABSTRACT: In the geothermal field of Kizildere, high temperature acid water infiltrates the sedimentary rocks and dissolves $CaCO_3$. The fluids aim still to get in an equilibrium constantly, but does not reach suitable parameters of the corresponding environment. Consequently, the fluids lose T, P, and CO_2 on ist way to the surface, for which reason pH increases up to 9.2. Therefore, the fluid is unable to keep $CaCO_3$ in solution, but it precipitates and occurs in sedimentary rocks near the surface. The Kolonkaya and Tosunlar formations in Kizildere show typical color of alteration which is considered as a new alteration type of carbonatization.

1 INTRODUCTION

Turkey is a country with a big potential for the exploitation of geothermal energy. Therefore, we have also investigated hydrogeochemical and isotope geochemical features of geothermal waters in Kizildere and its environs.

The subduction of the Africa plate under the Eurasian plate causes compressional tectonic features, which result in the lifting of the Menderes Massif. From the Middle Miocene to Pleistocene, the continental rift zone of Büyük Menderes was formed because of extensional tectonic features, which are represented by a great number of hot springs. The thermal waters are related to faults which strike preferentially in NW-SE and NE-SW directions, located diagonal to the general strike of the rift zone of Büyük Menderes, and are generated by compressional tectonic stress revealing the deformation of uplift between two extentional rift zones (Özgür et al., 1997).

We have selected to investigate the Kizildere thermal field with a geothermal power plant of 20 MW. For comparison, the other hot springs in the environs have been studied. The aim of this paper to present a new type of hydrothermal alteration of carbonatization in the geothermal field of Kizildere and hydrogeochemical features of the thermal waters in combination with the origin and evolution of the thermal waters.

A research scheme was carried out in 1993/1995, divided into two main fields:
(i) geological and geochemical investigations based on detailed geological mapping and rock sampling and (ii) comprehensive hydrogeological and hydrogeochemical investigations with sampling of groundwater and thermal waters.

2 GEOLOGIC SETTING

The Kizildere geothermal field is located in the northern part of the Büyük Menderes rift zone, which represents an important tectonic structure of Quaternary age within the Menderes Massif (Fig. 1). The Massif is one of the oldest basements in Turkey and consists of (i) a gneiss-core surrounded by a schist and marble envelope, (ii) a less metamorphosed cover series and (iii) an intensely deformed volcano-sedimentary sequence with incipient HP/LT metamorphism. The rift zones within the massif are the results of an extension which is believed to be closely related to the northward movement of the Arabian plate in the east pushing Anatolia westwards through the North Anatolian and East Anatolian Faults. The southerly bending of the North Anatolian Fault in the northern Aegean and Greece prevents the escape of the Anatolian plate further westwards placing the system in a locking geometry (Dewey & Sengör, 1979). This creates an E-W compression in the Menderes Massif which is relieved by N-S extension. The driving force of extension in the Aegean is believed to be the subduction along the Hellenic Trench (McKenzie, 1978). For the timing of initiation of extension an age within a range from 26 Ma (Spakman, 1989) about 12 Ma (Le Pichon & Angelier, 1979) to 5 Ma (Patton, 1992) is given. In the thermal field of Kizildere, the basement is comprised by the Paleozoic metamorphics which consist of gneiss, schist and the igdecik formation with intercalations of quartzites, mica schists, and marbles (Simsek, 1985; Vogel, 1997).

The basement rocks are overlain by Pliocene sediments. These sediments show fluvial and lacustrine characters and consist of (i) the 200-m-thick Kizilburun formation, (ii) the Sazak formation with a thickness from 100 to 250 m, (iii) the Kolonkaya formation in a range of thickness from 350 to 500 m, and (iv) the 500-m-thick Tosunlar formation. The Kizilburun formation is an intercalation of red and brown conglomerates, sandstones, shales, and lignites, the Sazak formation is composed of intercalated grey limestones, marls, and siltstones. The Kolonkaya formation contains yellowish green marls, siltstones, and sandstones, the Tosunlar formation is composed of an alternation of conglomerates, sandstones and mudstones with fossiliferous clay units. Quaternary alluvium overlies all of the units and reaches a maximum thickness of some hundred meters.

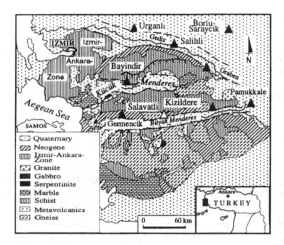

Fig. 1 Geologic setting of the Menderes Massif and location map of the geothermal field of Kizildere (open triangle: the geothermal field of Kizildere; closed triangles: other geothermal fields).

The thermal field is regionally controlled by E-W trending faults. Locally, NW-SE or NE-SW trending faults are active in the field. The development of these trending faults lead to a compression, which was generated by the extension during the formation of the rift zone of Büyük Menderes (Özgür et al., 1997). The northern and southern flanks of the metamorphic basement are affected by uplifting and dissected into a great number of step faults.

3 HYDROTHERMAL ALTERATION

In the geothermal field of Kizildere, the metamorphic and sedimentary rocks are distinguished by intensely hydrothermal alteration. Besides phyllic, argillic and silicic alteration ± haematitization, a new alteration type of carbonatization is recognized in sedimentary rocks especially. Silicification ± haematitization is an additional overprinting type of alteration and does not depend upon the rock chemistry. A noticeable result of silicification is the increase of density and hardness. Hydrothermal fluids will be able to precipitate quartz with haematite in veins, pores, and/or hollows in all kind of rocks.

The precondition for the formation of phyllic and argillic alteration is the presence of feldspar in the rocks. The metamorphic rocks show typical potassium metasomatism and plagioclase. Over a wide area, white clay minerals, such as montmorillonite and kaolinite and fine-grained white micas as sericite and illite, displace K-feldspar and biotite by hydrolysis. In these metasomatic reactions hydrogen ions will be consumed while potassium will be released. The macroscopic effect is a white powdery texture of the involved rocks, which loses its texture.

The Pliocene sedimentary rocks of the Kizildere geothermal field show already from distance intense coloured zones. Near the surface, the marls and limestones of the Kolonkaya and Tosunlar formations vary from pale white to yellow, orange and brown to dark red colored shapes. Responsible for the colors are iron mineralizations, but the fading is a result of the phenomena of carbonatization (Fig. 2). By the process of hydrothermal alteration, Ca^{2+} is usually heavily depleted except by the type of carbonatization (Meyer & Hemley, 1967).

Fig. 2 Soil carbonatization by the infiltration of high temperature acid water in marls and limestones of the Kolonkaya formation in Kizildere (Scale: 1:100).

In this process, the carbonic acid in geothermal waters determines whether these waters will dissolve carbonates (limestones and marls) or precipitate them as follows:

$$CaCO_3 + H_2CO_3 \underset{\text{deposition}}{\overset{\text{erosion}}{\Leftrightarrow}} Ca^{2+} + 2HCO_3^-$$
Solid solution   solution solution

In contrast to other types of hydrothermal alteration, it is a formation of carbonate assemblage by introduction of CO_3^{2-} which is able to fix metal ions such as Mg^{2+}, Fe^{3+}, Ca^{2+}, and Mn^{2+}. Therefore, many different kinds of carbonatization can be discriminated. Dolomitization of limestones by magnesium metasomatism is a simple base-exchange process and one of them. Two or more different types of carbonates can occur in the same alteration zone depending on the chemical composition of the rocks and the solutions. In the investigated area of Kizildere, it seems to be obvious that the carbonate is dissolved in one place and precipitated at another by hydrothermal solutions. This can be explained by the changes of physicochemical parameters of hydrothermal fluids.

In general, strongly altered rocks occur along the tectonicaly weakened zones because of the better possibility of intense hydrothermal water rock interaction. In spite of this the domain of hydrothermal carbonatization shows the greatest horizontal extension in the investigated area. For the formation of an alteration type of carbonatization, an immediate contact with high temperature thermal waters is not necessary.

4 GEOCHEMISTRY

Within the present study, the outflow of 10 thermal springs, 8 drill holes, 1 groundwater spring were sampled in different seasons from 1992 to 1995. For hydrogeochemical and isotopic comparison, we have taken a rain water sample from the geothermal area of Kizildere. Additionally, 120 rock samples and 48 precipitations were collected in the investigated area. At the surface, the metamorphic and sedimentary rocks are distinguished by an enrichment of the metals of Hg, Sb, As, Tl, and Ag in connection with the intensity of hydrothermal alteration which geochemically can be compared with the fields of epithermal ore deposits (Özgür et al., 1997).

Generally, Hg values are high in the less altered metamorphic and sedimentary rocks. The rocks of Igdecik, Sazak and Kolonkaya formations show extremely high Hg values, between 800 and 865 ppm, which might be attributed to the geothermal activity in the investigated area. The Tosunlar formation and travertine have Hg values ranging from 1.0 to 1.3 ppm. The less altered metamorphic and sedimentary rocks show Sb values from 0.2 to 1.0 ppm. Besides the Sb values up to 39 ppm in travertine, the Sazak formation, as a reservoir rock, shows the highest Sb concentrations up to 106 ppm which might be also attributed to hydrothermal activity. The less altered metamorphic rocks are distinguished by As contents of 1.0 ppm. The rocks of Kizilburun, Sazak and Tosunlar formations show As concentrations of about 100 ppm. The Kolonkaya formation reaches As values of 265 ppm. Travertine has As contents up to 420 ppm. The high As concentrations in the altered sedimentary rocks are linked to the hydrothermal activity. High boron contents, up to 900 ppm, are found in gneiss which can be led to the mineral phases of tourmaline and biotite. In the sedimentary formations, there are B contents ranging from 2 to 20 ppm, but up to 680 ppm in the Kolonkaya formation and 2850 ppm in travertine. Moreover, the high concentrations of B might be attributed to a B deposit at depth in sedimentary rocks. Fluorine shows values up to 1000 ppm in gneiss and up to 2125 in the sedimentary rocks and travertine which may be attributed to a hydrothermal activity.

The surface temperatures of the Kizildere thermal waters in drill holes differ from 95 to 100 °C whereas the outflows of 10 thermal springs in the surrounding area show temperatures from 40 to 73 °C. In the thermal field of Kizildere, the geochemical thermometers indicate a reservoir temperature of 220-230 °C (Na-K), 230-260 °C (Na-K-Ca), and 260 °C (SiO_2). The Kizildere thermal waters are distinguished by pH values of 9.2, an Eh value of -150 mV, a mean value of conductivity of 5300 µS/cm, and total dissolved solids (TDS) of 4500 mg/l. In comparison, the waters in the vicinity have pH values from 6.1 to 7.1, Eh values in a range from -137 to 311 mV, conductivity values from 2360 to 3660 µS/cm, and TDS from 1780 to 3838 mg/l.

Hydrogeochemically, the thermal waters of Kizildere and its environs can be classified as Na-(SO_4)-HCO_3-type, during the waters of Pamukkale and Karahayit show Ca-Na-(SO_4)-HCO_3-type (Fig. 3). This can be explained by mixing of a thermal sodium bicarbonate component, a cold calcium sulfate component and a cold calcium bicarbonate component in different proportions (Guidi et al., 1990); alternatively, the sodium bicarbonate thermal water could dissolve calcium sulfate from the Neogene sediments and mix with cold calcium bicarbonate water to originate these intermediate waters. The shallow waters, located near the village of Kizildere, exhibit sodium magnesium sulfate compositions and TDS of 244 mg/l.

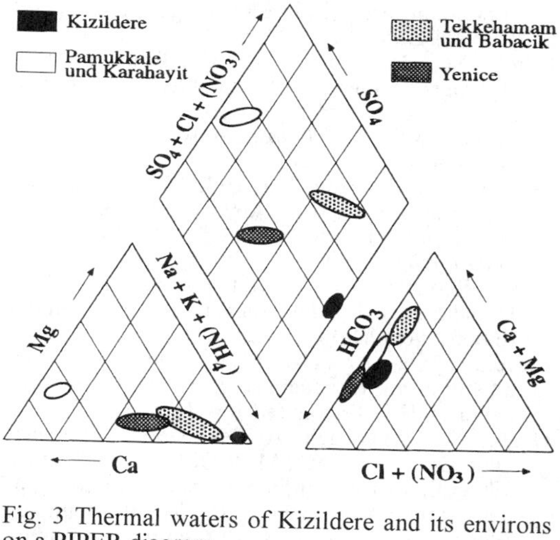

Fig. 3 Thermal waters of Kizildere and its environs on a PIPER diagram.

The genetic mechanism of these waters implies the infiltration of sodium bicarbonate thermal waters into shallow environments where leaching of Neogene sedimentary rocks and calcium carbonate precipitation takes place. Trace elements which indicate intensive high temperature water-rock interaction such as B (35 mg/l) and F (35 mg/l) are found in high concentrations in these waters (Özgür et al., 1997). As heavy metals, there are high concentrations of As (1 mg/l) and Sb (0.1 mg/l) in the waters of Kizildere. The contents of Au and Ag together with base metals (Cu, Pb, Zn) lie under the detection limits, which could be probably precipitated in deep environments under suitable pH, Eh and temperature conditions.

5 CONCLUSIONS

Due to high temperature, pressure, and high content of CO_2 in the thermal waters of Kizildere, high temperature acid water infiltrates the soil formations and dissolves $CaCO_3$. The solution aims to get in an equilibrium constantly, but does not reach the suitable parameters of the corresponding environment. Thus, the solution loses temperature, pressure and (CO_2) gas on its way to the surface, for which reason the pH values increase up to pH 9,2. Consequently, the geothermal fluid is unable to keep $CaCO_3$ in solution, but it precipitates and occurs in the sedimentary rocks near the surface. Especially in the S and SE of the Kizildere geothermal field, the marls and limestones of the Pliocene Kolonkaya and Tosunlar formations show typical values in color and structure which can be considered as an alteration type of carbonatization.

6 REFERENCES

Dewey, J.F. & A.M.C. Sengör, 1979. Aegean and surrounding regions: complex multiplate and continuum tectonics in a convergent zone. *Geol. Soc. Am. Bull. Part I*, 90: 84-92.

Guidi, M., L. Marini, & C. Principe, 990. Hydrogeochemistry of Kizildere geothermal system and nearby region. *Geothermal Resources Council Transactions*, Vol. 14, Part II: 901-908.

Le Pichon, X. & J. Angelier, 1979. The Hellenic arc and trench system: a key to the neotectonic evolution of the eastern Mediterranean. *Tectonophysics* 60: 1-42.

Mckenzie, D.P., 1978. Active tectonics of the Alpin-Himalayan belt: the Aegean and surrounding regions: geophys. *J. R.Astr. Soc.* 55: 217-254.

Meyer, C. & J.J. HEMLEY, 1967. Wall-rock alteration. in: Geochemistry of hydrothermal ore deposits, H.L. Barnes (ed.): p. 166-235.

Özgür, N., P. Halbach, A. Pekdeger, N. Sönmez, O.Ö. Dora, D.-S. Ma, M. WOLF, & W. Stichler, 1997, Epithermal antimony, mercury, and gold deposits in the rift zone of the Kücük Menderes, Western Anatolia, Turkey: preliminary studies: in: *Mineral Deposits: Research and Exploration, Where do They Meet?*, H. Papunen (ed.): 4th Biennnial SGA Meeting, Turku, Finland, August 11-13, 1997, p. 269-272.

Patton, S., 1992. The relationship between extension and volcanism in western Turkey, the Aegean Sea and Central Greece. *Ph.D. thesis, University of Cambridge*, 300 p.

Simsek, S. 1985. Geothermal model of Denizli, Sararyköy-Buldan area: *Geothermics* 14: 393-417.

Spakman, W., 1989. Tomographic mapping of the upper mantle structure beneath the Alpine collision belt. In: *Crust-mantle recycling at convergence zones*, S.R. Hart & L. Gülen (eds.):. NATO ASI series, C258, Kluwer, Dodrecht.

Vogel, M. 1997. Zur Geologie und Hydrogeochemie des Kizildere Geothermalfeldes und seiner Umgebung in der Riftzone des Büyük Menderes, W-Anatolien/Türkei: *M.Sc. Thesis, Freie Universität Berlin* (in prep.).

I-S series in geothermal fields: Comparison with diagenetic I-S series

P. Patrier
ESIP-MC, Poitiers, France

H. Traineau
Bureau de Recherches Géologiques et Minières, Orléans, France

P. Papanagiotou, E. Turgné & D. Beaufort
CNRS-UMR 6532, Université de Poitiers, France

ABSTRACT: The composition of dioctahedral smectitic material in geothermal environments is mainly beidellitic. The significance of these smectites is rather kinetic than thermic. Their crystallization is related to discharge of hydrothermal fluids (in reservoir or emergence zones). With time, these smectites are transformed into illite via I/S mixed-layers. Considering their composition and their conditions of formation, the transposition of information (as paleotemperatures) obtained from mixed layers in a diagenetic environment cannot be applicable to geothermal systems.

1 INTRODUCTION

In geothermal systems, clay minerals have received considerable attention during the last twenty years. In these environments, illite/smectite mixed-layers (I/S) are widespread and therefore have been frequently used for geothermometry, by comparison with the conversion series observed in diagenetic environment (Harvey & Browne, 1991 for example). However, can I/S mixed layers in active geothermal field be compared with diagenetic ones? Indeed, the conditions of formation (kinetic control on conversion of mixed layers, hydrodynamic constraints on clay nucleation...) will be very different in these two geological environments. If smectitic material in diagenetic sequences consists essentially of montmorillonite, the nature of the smectitic end member in active geothermal fields needs to be clarified. The aim of this paper, based on the study of two active geothermal systems: the field of Chipilapa (El Salvador) and Bouillante (Guadeloupe, Lesser Antilles) is to define precisely the nature of this smectitic end member.

2 GEOLOGICAL BACKGROUND

Both fields are developed in an andesistic volcanic arc context.

2.1 *Chipilapa geothermal field.*

This field is located on the eastern margin of the Ahuachapan geothermal system, 80 km west of San Salvador. It emplaced in a Plio-Quaternary volcanic chain as a response to subduction of the Cocos plate under the Carribean plate. The magmatism consists of calcalkaline series (pyroclastics and lava flows) of dacitic to andesitic composition. The geothermal activity postdated the last magmatic manifestations within the Late Pleistocene. The present day thermal anomaly of Chipilapa is centered on an area delimited by connected faults and presents a typical mushroom shape. Two reservoir zones have been evidenced: a shallow zone (500-650 m) which is interpreted as a vapor cap containing fluids at a temperature higher than 185 °C and a deep zone which consists of a liquid dominated reservoir with fluids at temperatures higher than 200°C.

2.2 *Bouillante geothermal field.*

The hydrothermal activity is related to fissural volcanic activity which is thought to be correlated to the existence of a major regional tectonic lineament oriented NW-SE (the Montserrat-Marie Galante fault) (Traineau et al., 1997). The field is located on the southern margin of this regional fault. The basement of the field is made of intercalations of submarine volcanoclastic and subaerial formations (pyroclastics and lavas flows).

3 OCCURRENCE OF I/S

3.1 *Chipilapa geothermal field.*

In lowly permeable rocks, the I/S conversion grades from nearly pure smectite (97-100 % S) to I/S mixed layers with less than 5 % of smectite at depth of 1750 m via regularly ordered R1 structure (35 % of smectite) (Fig. 1).

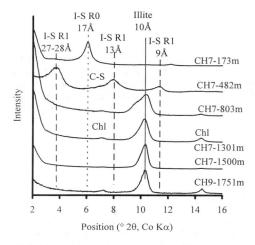

Fig. 1 Characterisitics X ray diffractograms showing the general evolution of I/S versus increasing depth (Chipilapa geothermal field).

In permeable rocks, the smectitic content of the I/S mixed-layers is higher than in the surrounding less permeable formations (pure smectite or regularly ordered I/S).

Therefore, I/S with ≈100% smectite crystallized in two different contexts :
- within the surficial zone (down to 350m depth) for a moderate temperature (<130 °C) in low permeable rocks
- at greater depth, in the present vapor dominated reservoir where the temperatures are higher (>180 °C) and the rocks highly permeable.

In the reservoir zone, I/S are close to the smectite end member (90 % of smectite, « high temperature smectite ») and exhibit beidellitic characteristics (XRD, FTIR).

In surficial levels, the dioctahedral smectites (« moderate temperature smectites ») are more heterogeneous. They consist of an association of beidellitic (± nontronite) and montmorillonitic layers as suggested by XRD diffractograms obtained after ethylene-glycol (EG) saturation and lithium

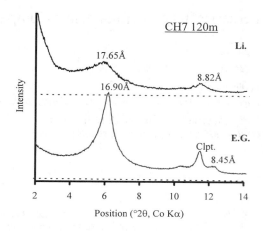

Fig. 2 X Ray diffractograms of mixed bedeillite-montmorillonite after EG and Li saturation according to the Hoffman-Klemen test (Chipilapa geothermal field). Clpt. : clinoptilolite.

saturation (Hoffman-Klemen Li test) (Hoffman & Klemen, 1950)(Fig. 2).

This identification is confirmed by chemical compositions obtained for these dioctahedral clays (table 1).

Table 1. Structural analyses of I/S from pure smectite to nearly pure illite (Chipilapa geothermal field). Basis : 22 oxygens.

	beidellite	I/S with decreasing smectite content			
Si	7.23	6.94	6.65	6.53	6.41
AlIV	0.77	1.06	1.35	1.47	1.59
AlVI	3.79	3.31	3.65	3.47	3.53
Fe^{3+}	0.08	0.40	0.20	0.26	0.33
Mg	0.17	0.32	0.16	0.31	0.15
Ti	0.00	0.00	0.01	0.01	0.00
Mn	0.01	0.00	0.00	0.00	0.00
Oct	4.04	4.03	4.02	4.05	4.01
Ca	0.38	0.20	0.03	0.03	0.02
Na	0.00	0.00	0.00	0.00	0.00
K	0.06	0.88	1.38	1.55	1.68

The compositions of I/S developed in lowly permeable zones plotted on the coordinates MR3-3R2-2R3 (Velde, 1985) underline the bedeillitic nature of the smectitic end member of the conversion sequence developed in low permeability rocks (Fig. 3). Compositions range along the illite-beideillite compositional line according to the smectite content. There is no evidence of a montmorillonite to illite trend.

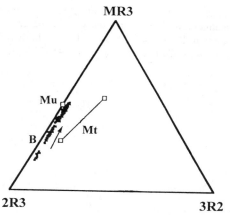

Fig. 3 Structural formula of I/S plotted on the coordinates MR3-3R2-2R3 (Chipilapa geothermal field). B: beidellite, Mt: montmorillonite, Mu: muscovite.

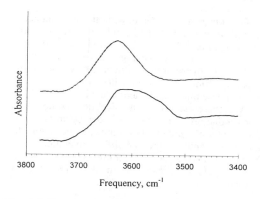

Fig. 4 Infrared spectra of the smectitic component sampled in the discharge zones of the Bouillante geothermal area.

3.2 Bouillante geothermal field.

Samples have been collected at the emergence of thermal sources in surficial (subaerial and submarine) environments. Fluid temperatures range from 40 to 90°C. The fluid consists of a mixing of sea water and meteoric water (and/or vapour condensate) with rather alkaline pH (>7).

The clay content of these samples differs totally of the weathering products which are halloysite and/or kaolinite/smectite mixed layers. Such products have been already described by Delvaux et al. (1990) in volcanic ash soils from Cameroon.

The collected samples are mainly composed of a smectitic material and zeolites (± calcite, ± quartz, ± gypsum).

The deposit of smectite in the emergence of thermal source is associated to a discharge of thermal fluids.

These smectites are dioctahedral. Their 060 reflection underlines a small ferric component (nontronite). According to the X ray diffractograms obtained after ethylene-glycol (EG) saturation and lithium saturation (Hoffman-Klemen Li test), the dioctahedral component appears to be of bedeillitic nature. Such an identification is confirmed by FTIR. The band close to 3630 cm^{-1} is typical of dioctahedral smectites. The broadening towards the low frequency agrees with the occurrence of nontronite and correspond to Fe-Fe band (Fig. 4).

4 INTERPRETATION

In the two studied geothermal areas, smectites are of beidellitic composition. So, they differ from the smectitic end member of the I/S sequence developed in diagenetic environments which is mainly of montmorillonitic composition (Velde and Brusewitz, 1986). In a geothermal environment, montmorillonite appears only in low temperature levels, associated with beidellite.

Although the chemical composition of rocks (andesitic to dacitic) will favour the development of montmorillonite, this smectite does not constitute the smectitic end member of the I/S conversion series observed in the Chipilapa and Bouillante geothermal areas. Fe and Mg are mainly incorporated in trioctahedral clays within the chlorite to smectite conversion series which develops simultaneously.

So, what can explain the occurrence of beidellite in a geothermal context for high to moderate temperatures?

The metastable state of montmorillonite at high temperature has been demonstrated by synthesis experiments. With time, this smectite is transformed into an association of beidellite and saponite (Yamada & Nakazawa, 1993). Such a transformation is also described by Sato et al. (1996) in a diagenetic environment, just before the illitization stage.

So according to these studies, beidellite nucleation seems to prevail in a high energy environment. However, a beidellitic composition probably favours a rapid conversion to illite. Indeed, this mineral characterized by tetrahedral substitutions appears much closer to illite than montmorillonite where the electric charge is located in octahedral sites.

Our observations agree with these conclusions. Indeed, in the two studied fields : 1) the nucleation of beidellite occurs, for moderate to high temperature, in high energy environment where boiling of hydrothermal fluids allows an abrupt supersaturation ; 2) this smectite is only identified as a pure phase in presently active level.

All these results (on smectite composition, transformation process...) demonstrate that transposition cannot be realized between the diagenetic and geothermal I/S conversion series. Moreover, it must be pointed out that randomly I/S mixed-layers are absent from the illite/bedeillite sequence.

REFERENCES

Delvaux, B., Herbillon, A.J., Vielvoye, L. and M.M. Mestdagh 1990. Surface properties and clay mineralogy of hydrated halloysitic soil clays. II : evidence for the presence of halloysite/smectite (H/Sm) mixed-layer clays. *Clay Miner.* 25: 141-160.

Harvey, C.C. & P.R.L. Browne 1991. Mixed-layer clay geothermometry in the Wairakei geothermal field, New Zealand. *Clays and Clay Miner.* 39: 614-621.

Hoffman, U. & R. Klemen 1950. Verlust des austauschfahigkeit von lithiumionen an bentonite durch. *Z. Anorg. Allg. Chem.* 262: 95-99.

Sato, T., Murakami, T. and T. Watanabe 1996. Change in layer charge of smectites and smectite layers in illite/smectite during diagenetic alteration. *Clays and Clay Miner.* 44: 460-469.

Traineau, H., Sanjuan, B., Beaufort, D., Brach, M., Castaing, C., Correia, H., Genter, A. and B. Herbrich 1997. The Bouillante geothermal field revisited : new data on the fractured geothermal reservoir in light of a future stimulation experiment in a low productive well. *22nd Workshop on geothermal reservoir engineering.* Stanford University. SGP-TR-155.

Velde B. 1985. *Clay minerals : a physico-chemical explanation of their occurrence.* Elsevier, Amsterdam.

Velde, B. & A.M. Brusewitz 1986. Compositional variation in component layers in natural illite/smectite. *Clays and Clay Miner.* 34: 651-657.

Woldegabriel, G. & F. Goff 1992. K/Ar dates of hydrothermal clays from core hole VC-2B, Valles Caldera, New Mexico and their relation to alteration in large hydrothermal system. *J. Volcanol. Geotherm. Res.* 50: 207-230.

Yamada, H. & H. Nakazawa 1993. Isothermal treatments of regularly interstratified montmorillonite-beidellite at hydrothermal conditions. *Clays and Clay Miner.* 41: 726-730.

Geothermal resource development at Tattapani in Madhaya Pradesh, India

S.K. Sharma & J. Tikku
KDM Institute of Petroleum Exploration, ONGC, Dehra Dun, India

ABSTRACT : The Tattapani geothermal field, situated in the Son river basin in Madhaya Pradesh, forms one of the most important geothermal manifestations in Narmada-Son-Tapti mega lineament zone. This mega lineament inhabits about 46 geothermally active localities and the Tattapani anomaly covering an area of about 0.2 Km^2 is the strongest of all. Two distinct lithological sequences belonging to Proterozoic and Gondwana supergroup are mappable in the area. The area experiences geothermal gradients of over 5-6 times the normal and the heat flow value is inferred to be over 2-4 times the normal heat flow. Geochemically computed based temperature are of the order of $160° \pm 16°C$. Chemical analysis of water from the boreholes indicates alkaline water of $Na-HCO_3$ type. Variation in silica content in the boreholes could be due to the mixing of cold water at shallow level in varying proportion whereas the elevated level of fluorine is attributed to the leaching of fluoride from the underlying hornblende gneiss of crystalline basement. The presence of hydrothermal alteration minerals like zeolites, quartz, calcite and clays along fracture planes in 26 boreholes drilled in an area of about 9 km^2 reflects the aerial extent of one time geothermal boundary which has now been retreated inwards to form an anomalous area of about 0.2 km^2 where five boreholes are situated yielding a cumulative geofluid flow of 1800 lpm at $112°C$. Causes of this boundary retreat and use of geothermal fluid are presented in the paper.

1. INTRODUCTION

The energy requirments of India are increasing by leaps and bounds, averaging 6% to 8% per annum. To keep pace with the demand and supply dynamics especially with respect to petroleum sector, thrust is being given to tap the alternative energy sources specially geothermal to bridge the gap.

In India, a total of 340 geothermal manifestations, including 113 hot springs, have been identified by the Geological Survey of India, holding a total stored heat potential of 41×10^{12} mega calories, equivalent to 28000 million barrels of oil (Krishnaswami and Ravishanker, 1982). Based on paleogeothermal activity noticed in the form of white encrustation in soil and silica sinter along fractures and cracks towards west of the geothermal field in an area of about 9 Km^2, a total of 26 wells including 5 wells, namely, GW/TAT/6, GW/TAT/23, GW/TAT/24, GW/TAT/25 and GW/TAT/26 in the strongest anomalous area of 0.2 km^2 have been drilled to develop the geothermal resource at Tattapani. Considering the well test as well as the geochemical geothermometry data, an attempt has been made to model the flow pattern of hot water for explaining the inward retreat of geothermal boundary in the area.

2. GEOLOGICAL SETTING

Rocks of the Proterozoic age comprising granitic gneiss, phyllite, biotite schsist, hornblende gneiss and amphibolite, overlain by the Gondwanas, consisting of brown to reddish brown coarse sandstone, grey shale, coal seams with plant fossils and green splintery shale, and siltstone are exposed in the area. Low angle reverse faults have also been observed in drill holes. Two major lineaments F and F-2 trending ENE-WSW are present in the area (Fig. 1). Apart from these lineaments, several cross faults are also present. The area is reported to be tectonically active (Ravishanker, 1987) which is mainly responsible for the geothermal activity. The study of borehole core indicates that medium to coarse sandstone of Gondwana group is encountered to depths of approximately 120 m. in boreholes GW/TAT/6, GW/TAT/23 and GW/TAT/24 wheras the Gondwana sequence is met at the depth of 55 m. in borehole GW/TAT/25 and 77 m. in GW/TAT/26.

2.1 *Geohydrological characterstics.*

Geohydrological studies around the Tattapani thermal springs have delineated three sub basins (Fig. 1). Basins I and II show flow of ground water towards the Kanhar river in an easterly direction. The gradients in Basin III area towards west feeding Sendur river. Both the rivers flowing towards north ultimately feed the Son river. The ground water in Gondwana occurs in confined conditions while the wells in Proterozoic rocks area tap groundwater under water table conditions. The depth of water level ranges from 6 to 10 m below ground level. Seasonal variations in water table are quite high in basement rocks whereas the dug wells in Gondwana rocks show slight variations. Boreholes No. GW/TAT/6, 23, 24, 25 and 26 show free flow of water at the well heads. The meteoric water, due to its deep circulation, collects heat and rises to the surface through the conduits provided by highly fractured fault zones. The Gondwana groundwater reservoir is a shallow aquifer zone responsible for recharge and cold water mixing at shallow levels.

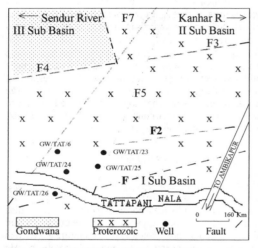

Fig. 1. Geological Map of Tattapani Hot Spring Area

2.2 Geothermal Geochemistry

Chemical analysis of water from hot springs and boreholes of the Tattapani geothermal field indicates two types of water : (a) Type 1 - $NaCl-SO_4$ and (b) Type 2 - $NaHCO_3$. Most of the thermal waters fall under Type 2. The cold water wells in the vecinity of hot springs or fault breccia indicate mixing of hot and cold waters. The comparative water analysis from different boreholes show slightly alkaline pH (8.3 to 8.6) nature. Results of the chemical analysis are summarised in Table 1 indicating low Ca (4.00 ppm), K (7 to 13 ppm), CO_3 (16 to 25 ppm) contents and low hardness, and moderate contents of SO_4 (60 to 62 ppm), Cl (68 to 72 ppm), TDS (492 to 541 ppm) and HCO_3 (87 to 109 ppm). Silica content varies from borehole to borehole (104 to 116 ppm) which could be due to mixing of cold water at shallow level in varying proportions. The Boron content (0.5 ppm) from all the boreholes is same indicating common source for the thermal waters. The fluorine content is rather high (15 to 20 ppm) which may be due to leaching of fluoride from hornblende gneiss. The boreholes/dug wells within Gondwana Terrain do not show the presence of hot water. Hence the geothermal reservoir appears to be restricted to basement rocks/granite gneiss over which Gondwana sediments act as a cap rock. Based on tritium studies, the thermal water has more than 30-40 years of residency period (Thussu, 1987).

2.3 Hydrothermal alteration mineralogy

The study of borehole cores in 26 wells covering an area of about 9 km² has indicated deposition of stilbite, laumonite (zeolites), calcite, quartz, montmorillonite and pyrite along the fracture planes. Hydrothermal deposition is commonly seen around the vents of hot springs and hot water pools in the marshy land. Calcite is major mineral observed up to a depth of 250 m. Stilbite mostly occurs below depth of 120m. The mineral assemblages found in hydrothermal deposits suggest temperature of 125° C to 150° C of thermal fluid (Thussu, Prasad and Saxena, 1987). Based on silica geothermometry, the reservoir temperature is estimated to be 150°C to 160° C. The hydrothermal deposit at depth of around 200-400 m analyses 20 ppm of Cu and up to 100 ppm As. The hydrothermal encrustations inside the discharge pipe of borehole GW/TAT/24 analysed 4 ppm of Ag also (Sarolkar, 1993).

2.4 Heat flow and thermal gradient

Initally, shallow and intermediate depth boreholes were drilled by the Geological Survey of India for measuring thermal gradients. A conductive temperature gradient of 0.1° C/m was measured with maximum borehole temperature of 112° C. The bore hole GW/TAT/3 drilled on the Tattapani fault recorded a temperature of 102°C at 78 m depth (gradient being 0.87° C/m). Instead of about 9 km² an area of 0.2 Km² around Tattapani has been found to be anomalous, this indicates an inward retreat of geothermal boundary from an early large system. An attempt was made by Ravishanker in 1987 to compute heat flow by using available thermal conductivity values. The average heat flow value for the Tattapani area is computed to be 290 ± 50° mW/m² (Ravishanker, 1987) which is 5-6 times the global average. It is the highest amongst 46 other geothermally active areas on the mega lineament.

3. CHEMICAL GEOTHERMOMETRY

Based on the following chemical analysis, some of the important geothermometers have been calculated and used to study the reservoir temperatures. From this table it is concluded that,

i) There is negligible steam loss as the fluid is ascending because the no steam loss geothermometer is giving a uniform temperature and is suited to springs at sub-boiling temperature and as such maximum steam loss geothermometer is also comparable.

ii) Since the Na-K geothermometer is least affected by steam separation, it could be reflecting the reservoir temperature and probably be of the order of 200° C.

iii) It has been reported by Thussu, 1987 that a sub surface temperature of 162° C could be existing based on Na-K, Na-K-Ca geothermometers and chloride enthalpy plot. A minimum of 140° C hot water is also suggested based on silica geothermometry.

4. MODELLING THE TATTAPANI FIELD

An attempt towards geological modeling based on observations for the flow pattern of hot water in the Tattapani geothermal field is made to understand the reservoir. The data used is the geochemical geothermometry data (Table 1) and the well test data (Table 2). Some observations drawn are,

i) Ascending fluid is hot water with negligible amount of steam and is cooling down as its ascending.

ii) In the shallow depths, the convective zone has been identified to be about 100m thick. This zone of constant temperature may actually be a secondary reservoir. If the deeper probing is unable to reach the primary reservoir then this be the only producing zone available.

Table 1 : Results of Water Analysis and Chemical Geothermometry

Well No. GW/TAT	Concentration in ppm												Chemical (°C) Geothermometres				
	CO_3	HCO_3	Cl	SO_4	Ca	Mg	Ca-CO_3	Na	K	TDS	SiO_2	B	F	NSL	MSL	K-Mg	Na-K-Ca	Na-K
6	25	87	72	62	4	1	12	122	13	506	116	0.50	15	146	140	102	159	222
23	19	101	70	60	4	1	14	122	10	494	107	0.50	15	141	136	95	147	200
24	22	92	72	61	4	1	12	123	11	492	104	0.50	20	140	135	97	151	208
25	16	109	68	61	4	1	16	142	7	541	108	0.50	20	141	136	85	133	163

* pH = 8.3 - 8.6 (Alkaline) in all the wells   * NSL = No Steam Loss.   * MSL = Maximum Steam Loss.

Table 2. Summary of downhole testing

Well No. GW/TAT	Drilled Depth (m)	Convective Zone (m)	Temperature (°C) Static	Temperature (°C) Flowing	Well Head Temperature (°C)	Well Head Pressure Kg/Cm²	Bottomhole Pressure	Temperature Reversal Depth (m)	Flow Rate lpm	Remarks
6	506	120-220	109.1	110.7	93	0.81	Both Static & flowing pressures, linearly rising with depth.	Below 110	304	Static pressure is higher than hydrostatic head, hence well will flow.
23	353	200-310	111.9	111.9	106	0.81	--- do ---	155	415	----- do -----
24	244	-	110.7	-	100	0.81	--- do ---	-	389	----- do -----
25	350	235 and below	110.6	110.8	89	0.93	--- do ---	95-135	425	----- do -----
26	239	200 and below	110.5	111.6	95	0.81	--- do ---	110-150	268	----- do -----

The interference test conducted in well no. 23 suggests that the flow in other wells is affecting well no. 23 also. During injection test, the temperature profiles taken during cold water injection into the wells has helped in identifying the feeder zones at different intervals.

iii) Since all wells are giving differing flow rates, there is difference in lateral fracture permeability.
iv) The thickness of the convective zone is largest in Well No. 23 and is reducing laterally. Hence an upflow zone around well No. 23 is inferred.
v) The temperature inversions in the wells suggest mixing with waters of lower temperatures.
vi) Static pressures in all the wells is higher than the hydrostatic head and as a result all the wells are going to flow.

Fig. 2 shows the relationship between the upflow zone, outflow zone and mixing zones for a fractured area within Tattapani sedimentary basin. The horizontal static temperature gradients of wells with respect to well No. GW/TAT/23, below the casing have been calculated, using (GW/TAT/23 - GW/TAT/x)/Distance($^\circ$C/m) formula where GW/TAT/x is a well. It is observed that,

i) Horizontal gradient is minimum (0.005) in well No. 24's direction.
ii) It is greater (0.02) towards well No. 6 and is even larger (0.03) towards well No. 26.

In all probability well No. 23 has an upflow zone under it and laterally its flow is more directed towards well No. 26 rather than others, and is spreading out and cooling down due to mixing.

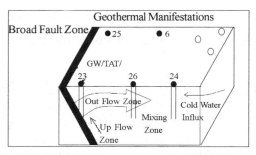

Fig 2 : Schematic Diagram of Tattapani Sedimentary Gothermal System.

The elevated anomalous manifastations at Tattapani in an area of about 0-2 km^2 appears to be the result of well documented upper-mantle and intracrustal disturbances during the Late Tertiary-Quaternary times resulting in the emplacements of several large plutonic bodies, along well defined zones at variable depth within the crust, which may still be cooling and acting as heat source under favourable circumstances (Ravishanker, 1987). Moving westwards from Tattapani, the intensity of surface geothermal activity is greatly reduced due to the masking effect of actual deep geothermal regime by the thick pile of Deccan Traps (Ravishanker, 1987) resulting in an inward retreat of geothermal boundary from the original large geothermal system.

5. RESOURCE EVALUATION AT TATTAPANI

The downhole testing in five wells, namely, GW/TAT/6, 23, 24, 25 & 26 has been done for (i) Static and flowing temperature profiles, (ii) Static and flowing pressure profiles, (iii) Interference, injection and build-up tests. These results are compiled and presented in Table 2. An electric power generating potential of 200-300 Kwe has been established with the present cumulative flow rate of 1800 lpm with the reservoir temperature of 112° C and taking 87° C an end limit of binary power plant.

6. DISCUSSION AND CONCLUSIONS

The presence of pyrite, formation of stilbite and quartz in association with montmorillonite in the alteration mineral assemblages all suggest change in temperature, pressure, boiling zone and chemical composition of Geothermal fluids with passage of time in an area of about 9 km^2. This mineral assemblage is in equilibrium with temperature range of 100° C + 10° C which is actually recorded in boreholes. The conceptual model indicates an upflow zone along the permeable intersecting fault under well No. GW/TAT/23. Different flow rates indicate difference in lateral fracture permeability. An anomalous area of nearly 9 Km^2 where paleogeothermal manifestations could still be noticed at Tattapani has now been reduced to an area of 0.2 Km^2 due to masking effect of most of the anomalous area / geothermal regime by the Deccan Trap activity thus explaining the inward retreat of geothermal boundary.

The area promises an electric generation potential of 200 to 300 Kwe.

ACKNOWLEDGEMENT

The authers are grateful to Shri Kuldeep Chandra, R.D., KDM Institute of Petroleum Exploration, Dehra Dun for granting permission to publish the paper.

REFERENCES

Krishnaswami, V.S. and Ravishanker 1982. Scope of Development, Exploration and Preliminary Assessment of the Geothermal Resource Potentials of India, *Records of GSI*, Vol. III, Pt. 2

Ravishanker 1987. Status of Geothermal Exploration in Maharashtra and Madhaya Pradesh (C.R.), *GSI Records*, Vol. 115, part 6, pp 7-29

Sarolkar, P.B., 1993. Subsurface geological studies in Tattapani Geothermal Field, M.P., *GSI Records* Vol.126, part 6, pp36.

Thussu, J.L., Prasad, J.M. and Saxena, R.K. 1987. Geothermal Energy Resources Potential of Tattapani Hot Spring Belt, Distt. Surguja, M.P., *GSI Records* Vol. 115, Pt. 6, pp 30-55.

Illite, illite-smectite and smectite occurrences in the Broadlands-Ohaaki geothermal system and their implications for clay mineral geothermometry

S. F. Simmons & P. R. L. Browne
Geothermal Institute, University of Auckland, New Zealand

ABSTRACT: The distribution of smectite, interlayer illite-smectite and illite in altered rocks of the Broadlands-Ohaaki geothermal system broadly correlate with the measured reservoir temperatures. Interlayer clays in cores of volcanic rocks from wells Br-6 and Br-7 are predominantly illite-rich and intensely altered, but predominantly smectite-rich samples are slightly altered. Calibration of a clay geothermometer based on the proportions of illite present in these interlayer clays has an uncertainty of about ±50 °C. While clay mineral distribution patterns can be used to interpret thermal trends, the formulation of a reliable and general clay geothermometer on the basis of interlayer clay compositions alone appears to be still elusive.

1 INTRODUCTION

The occurrence of illite, interlayer illite/smectite and smectite clays has been correlated with temperature gradients in active geothermal systems (e.g. Steiner, 1968; Muffler and White, 1969; Browne and Ellis, 1970; Esslinger and Savin, 1973; Browne, 1978; Yau et al., 1988; Harvey and Browne, 1991) and where sediments have undergone diagenesis (e.g. Burst, 1959; Perry and Hower, 1970; Jennings and Thompson, 1986). Based on these studies, the generally recognised upper limit of the thermal stabilities for smectite, interlayer illite/smectite and illite, are <120°, <220° C and 300° C, respectively; illite transforms to muscovite above 300° C. Systematic changes in the nature of compositional interlayering of illite/smectite with temperature are also thought to exist (e.g. Steiner, 1968) and this feature was used for geothermometry in a study of the Creede epithermal deposit. (Horton, 1986). However, the reliability of illite/smectite mineral geothermometer has recently been questioned by Essene and Peacor (1995). Here we give data of illite-smectite occurrences in the Broadlands-Ohaaki geothermal system that indicate limitations in the application of this geothermometer.

2 THE BROADLANDS-OHAAKI GEOTHERMAL SYSTEM

The Broadlands-Ohaaki geothermal system is one of about twenty, liquid-dominated geothermal systems located in the Taupo Volcanic Zone. Exploration drilling in this system (1964-1984) totalled 52 wells, ranging from 370 to 2600 m in depth. The cores, cuttings and fluid samples have been the subject of numerous studies concerning their hydrothermal minerals, fluid and isotopes compositions, and fluid inclusions (see Simmons and Christenson, 1994).

The stratigraphy comprises a sequence of young (<1.6 Ma), predominantly rhyolitic, pyroclastic deposits and lavas that unconformably overlie a block-faulted Mesozoic greywacke basement (Grindley and Browne, 1968; Wilson et al., 1995). Faults, fractures, lithologic contacts and permeable units form the main fluid conduits. In these zones pressures are close to those of hydrostatic boiling down to about 2.5 km depth.

The hydrothermal parent water is near neutral in pH and contains about 0.1 wt % Cl and 2.6 wt % CO_2 (Hedenquist, 1990). This deeply derived water reacts with host rocks to form a typical assemblage of quartz, albite, adularia, illite, calcite, chlorite, pyrite, and rarer epidote and wairakite (Browne and Ellis, 1970). As this parent liquid ascends, CO_2 exsolves from it through boiling. Above the zone of boiling, the separated CO_2 and steam are absorbed into the shallow ground waters to form CO_2-rich steam-heated waters (Hedenquist 1990) that descend down the margins of the upflow zone and locally dilute the rising waters. Clays (illite, illite-smectite, smectite, kaolinite), calcite and siderite are the main hydrothermal minerals associated with the CO_2-rich steam-heated waters (Hedenquist, 1990; Simmons and Christenson, 1994).

3 TECHNIQUES AND RESULTS

The mineralogy of core samples was determined by thin section petrography and bulk powder X-ray diffraction. The results of clay studies reported here are based on core samples from two spatially marginal wells (Br-6 and Br-7) associated with the CO_2-rich steam-heated waters.

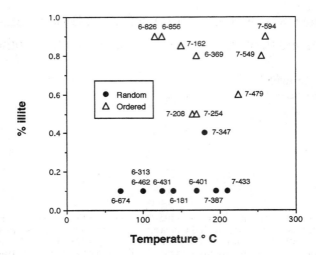

Figure 1. Percentage of illite in interlayered illite-smectite clays in core samples from geothermal wells Br-6 and Br-7 versus *in situ* temperature; the labels give the well number-depth meters. Br-6 samples of Ohaaki Rhyolite, Broadlands Rhyolite and Rautiwiri Breccia were taken from cores between 181 and 890 m depth; Br-7 samples of Broadlands Dacite were taken from cores between 162 and 594 m depth. The distinction between random and ordered samples was determined based on the criteria given in Reynolds (1980).

The samples were gently crushed with motar and pestle, and then dispersed in a water slurry using a blender for 5 minutes. The <0.2 and 0.2-2.0 µm fractions were obtained by gravitational settling and centrifuging, and then mounted and air dried on glass slides. The mineralogy of the oriented clay samples (dry and glycolated) was determined by X-ray diffraction using CuK alpha radiation and a graphite monochrometer. The ordering and percentage of illite present in the interlayer clays were determined using calculated diffractograms (Reynolds and Hower, 1970; Reynolds, 1980); the proportions of illite layers are accurate to ±10%.

Additional data were obtained from interpretation of XRD patterns undertaken during core logging (Browne, 1971; Wood, 1983) and stable isotope studies (Esslinger and Savin, 1973). Pre-exploitation temperature gradients were deduced from down-hole measurements (Hedenquist, 1990), and the *in situ* temperatures used here are believed to be accurate to better than ±10 °C.

3.1 *Results*

The clay minerals in core samples from the two wells are dominated by illite, illite-smectite and smectite; chlorite or kaolinite were found in only a few samples. There is no difference in the mineralogy of the <0.2 and 0.2-2.0 µm size fractions. With few exceptions, smectite-rich interlayer clays occur in little altered rocks where fresh volcanic glass commonly exists, but illite-rich interlayer clays occur in rocks that are intensely altered. Figure 1 shows that, in general, the illite content of clay minerals correlates poorly with the measured bore temperature.

In order to see how these results appear field wide, we plot the distributions of illite, illite-smectite and smectite along two cross-sections in Figure 2. These are the same cross-sections used by Hedenquist (1990) and Simmons and Christenson (1994) to portray the hydrological structure and the distribution of carbonate minerals, respectively. As might be expected, smectite gives way to illite with increasing temperature and depth, with the occurrences of interlayer illite/smectite representing a transition between the two zones. The upper boundary of the illite zone has a shape similar to that of the 200° C isotherm. In most wells, however, the 150° and 200° C isotherms cross cut these clay zones, indicating that clay occurrence is also influenced by factors other than temperature. Only in the upper 500 m of Br-6 does the distribution of clays closely match the position of temperature reversals and outflow zones; here illite-rich clays (80% illlite) correspond to the hottest temperatures, ~170° C.

4 DISCUSSION

Our results indicate that calibration of an illite-smectite geothermometer based on the proportions of interlayer illite leads to uncertainties of the order of ±50 °C. Nevertheless, the overall pattern of the smectite, illite-smectite and illite distributions corresponds with increasing temperature depth and position with respect to the center of the upflow zones.

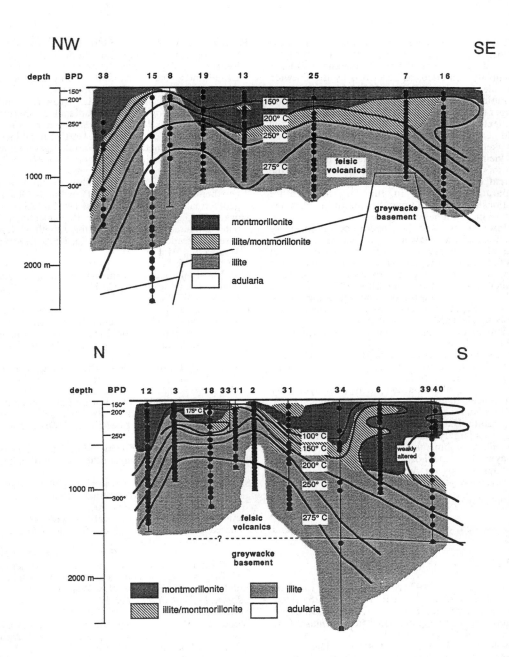

Figure 2. Distribution of clay minerals in Broadlands-Ohaaki geothermal wells (Browne, 1971; Wood, 1982).

There are two reasons that could explain these trends: 1) the thermal gradients have recently changed; 2) the conversion of smectite to illite is kinetically contolled, as described by the Ostwald step rule (Essene and Peacor, 1995).

Fluid inclusion evidence indicates that higher temperatures (near to the boiling point for depth gradient) once existed in the southeast part of the field near Br-16 and Br-7 (Hedenquist, 1990), and this likely accounts for the abundant illite occurring in this vicinity at shallow depths. There is no indication that this illite is converting back to smectite in the current thermal regime. Therefore some of the disparaties apparent in the clay-temperature relationship (Figure 1) may be due to the complex thermal history it has undergone.

However, much of the smectite occurs in glassy rhyolitic rocks with low alteration intensity; i.e, there is a high proportion of primary phases as yet unaffected by hydrothermal alteration. Smectite is commonly the first phase to replace glass (e.g. Steiner, 1968; Browne and Ellis, 1970). In these rocks, smectite more likely represents a metastable phase of reaction progress (Essene and Peacor, 1995) involving glass and steam-heated water.

There are only four occurrences of interlayer clays of intermediate composition (Figure 1), and they derive from the upper part of the Broadlands Dacite in Br-7. Here the dacite is intensely altered with a clay content that exceeds 50%.

These results contrast with interlayer clay occurrences in the Wairakei geothermal system, whose compositions more closely correlate with temperature (Harvey and Browne, 1991), though that study was largely made on poorly permeable, fine-grained lake sediments of the Huka Falls Formation. It may be, then, that host rock permeability/reactivity provides a basis for recognising sequences where interlayer clay geothermometry may apply.

5 ACKNOWLEDGEMENTS

This research was partially funded by AMIRA and FRST.

6 REFERENCES

Browne, P.R.L. 1971. Petrological logs of Broadlands drillholes BR 1 to BR 25. *New Zealand* Browne, P.R.L. 1978. Hydrothermal alteration in active geothermal fields. *Annual Rev. Earth Planet. Sci.*6:229-250.

Browne, P. R. L. & A. J. Ellis 1970. The Ohaki-Broadlands geothermal area, New Zealand: mineralogy and related geochemistry. *American Journal of Science.*269:97-131.

Burst, J. F. Jr.1959. Postdiagenetic clay mineral environmental relationships in the Gulf Coast Eocene. *Clays Clay Min., proc. 6th Natl. Conf.,*: 327-341.

Essene, E. J. & D. R. Peacor 1995. Clay mineral geothermometry-A critical review. *Clays and Clay Minerals.* 43:540-553.

Esslinger, E. V., & S. M. Savin 1973. Mineralogy and oxygen isotope geochemistry of the hydrothermally altered rocks of the Ohaki-Broadlands, New Zealand geothermal area. *American Journal of Science.*273: 240-267.

Grindley, G. W. & P. R. L. Browne 1968. Subsurface geology of the Broadlands geothermal field. *New Zealand Geological Survey Report* 34.

Harvey, C.C. & P.R.L. Browne 1991. Mixed layer clay geothermometry in the Wairakei geothermal field, New Zealand. *Clays and Clay Minerals.* 39:614-621.

Hedenquist, J. W. 1990. The thermal and geochemical structure of the Broadlands-Ohaaki geothermal system. *Geothermics.*19:151-185.

Horton, D. G. 1985. Mixed-layer illite/smectite as a paleotemperature indicator in the Amethyst vein system, Creede district, Colorado, USA. *Contrib. Min. Petrol.* 91:171-179.

Jennings, S. & G. R. Thompson 1986. Diagenesis of Plio-Pleistocene sediments of the Colorado River delta, southern California. *Jour. Sed. Petrol.*56: 89-98.

Muffler, L. J. P. & D. E. White 1969. Active metamorphism of Upper Cenozoic sediments in the Salton Sea geothermal field and the Salton trough, southeastern California. *Geol. Soc. Am. Bull.* 80: 157-182.

Perry, E. & J. Hower 1970. Burial diagenesis in Gulf Coast pelitic sediments. *Clays Clay Min.*18:165-177.

Reynolds, R. C. 1980. Interstratified clay minerals: in G.W. Brindley and G. Brown (eds) *Crystal Structures of Clay Minerals and Their X-ray Identification*:: 249-303

Reynolds, R. C. & J. Hower 1970. The nature of interlayering in mixed-layer illite-montmorillonites. *Clays Clay Min.*18: 25-36.

Simmons, S. F. & B. W. Christenson 1994, Origins of calcite in a boiling geothermal system. *American Journal of Science* 294:361-400.

Steiner, A. 1968. Clay minerals in hydrothermally altered rocks at Wairakei, New Zealand. *Clays and Clay Minerals.* 16:193-213.

Wilson, C. J. N., B. F. Houghton, M. O. McWilliams, M. A. Lanphere, S. D. Weaver & R. M. Briggs 1995, Volcanic and structural evolution of Taupo Volcanic Zone, New Zealand: A review. *Journal of Volcanology and Geothermal Research* 68:1-28.

Wood, C. P.1983. Petrological logs of drillholes BR 26 to BR 40, Broadlands geothermal field. *New Zealand Geological Survey Rept.* 108.

Yau, Y., D. R. Peacor, R. E. Beane, E. J. Essene, and S. D. McDowell 1988. Microstructures, formation mechanisms, and depth-zoning of phyllosilicates in geothermally altered shales, Salton Sea, California. *Clays Clay Min.*36:1-10.

Gas behavior at some geothermal fields in Japan, revealed by Laser Raman Microprobe analysis of fluid inclusions

S. Taguchi
Fukuoka University, Japan

H. Takagi & H. Maeda
West Japan Engineering Consultants, Fukuoka, Japan

K. Sanada
New Japan Environmental Measurements, Fukuoka, Japan

M. Hayashi
Kyushu Sangyo University, Fukuoka, Japan

M. Sasada & T. Sawaki
Geological Survey of Japan, Tsukuba, Japan

T. Uchida & T. Fujino
New Energy and Industrial Technology Development Organization, (NEDO), Tokyo, Japan

ABSTRACT: Laser Raman microprobe (LRM) was applied to fluid inclusion analysis during a development of prospective technology to explore fractured-type geothermal reservoirs, which was promoted by NEDO. LRM is a powerful tool to get useful information on dynamic fluid behavior and geothermal structures even in a geothermal system with very dilute gas contents such as Japanese one.

1 INTRODUCTION

Major gas components in geothermal fluids are usually CO_2, and H_2S with minor amounts of CH_4. These gases are usually separated by boiling from deeper fluid, and migrate to shallower depth, and form steam heated waters. Such CO_2-rich steam-heated waters, were reported from the Broadlands system in New Zealand (Hedenquist and Stewart, 1985). The CO_2-rich water in the Broadlands form over the top of the system and then drain to deeper levels on the margins.

The occurrence and distribution of such CO_2-rich steam heated waters was also revealed by fluid inclusion research. Moore et al. (1992) reported the distribution of CO_2-clathrate in inclusions at Los Azufres in Mexico and Zunil in Guatemala. The clathrate occurs at the upper and marginal parts of the systems forming an umbrella-shaped cap around the main zones of upwelling of geothermal fluids. They call this " clathrate cap".

The gas distribution in a geothermal system is closely associated with the geothermal structure. If we can determine the gas distribution in a geothermal system using fluid inclusions, it will become a great advantage for geothermal development. However, since most of the fluids in Japanese geothermal systems are very low in CO_2 contents (usually less than 0.1 - 0.2%), it will be hard to detect CO2-clathrate from ice-melting experiments. Therefore, New Energy and Industrial Technology Development Organization (NEDO) introduced laser Raman microprobe (LRM) to fluid inclusion analysis during a development of prospective technology to explore fractured-type geothermal reservoirs with low gas contents.

As the LRM has recently been improved, its detection limit of gas in fluid inclusions from low gas geothermal fields like Japan can be revealed.

2 ANALYTICAL METHODS

Measurements were made on samples from three geothermal fields in Japan; the Hohi in Kyushu, the Kakkonda in Northern Honshu, and the Ogiri in Kyushu, Japan. The fluid inclusion samples used for LRM analysis are the same for the microthermometric measurements. The LRM used is a RAMANOR T64000 by Jobin Ybon. The detection limits of the instrument at room temperarure, when thickness of the silica tube is 20 microns, are 0.12, 0.15, 0.04 and 0.07bars for N_2, CO_2, CH_4 and H_2S, respectively (Maeda et al., 1998). These detection limits are improved one order of magnitude compared to older type.

3 RESULTS AND DISCUSSION

At Hohi, CO_2 and H_2S gases were detected at only one place in Well YT-2, in the upper part of the boiling zone and the under the advanced argillic alteration zone. Tm ice of the gas-rich inclusion is -0.4° C, lower than that of liquid-rich, gas poor inclusions of -0.2°C. The difference in the Tm ice is equivalent to 0.1mol/kg. H_2O of CO_2. These gases suggest that they are separated by boiling and migrated to shallow level under the advanced argillic alteration zone, i.e." gas cap" was formed under the advanced argillic alteration zone.

At Kakkonda, gases detected are CO_2 and CH_4, and rarely N_2. These gases are also occur in the upper part of the boiling zone in the upper Kunimitoge Formation (Neogene sediments), and in the Torigoeno-taki Dacite intrusion. The gases are accumulated beneath the dacite intrusion, suggesting that the dacite acted as a cap rock for the gas components.

At Ogiri, CO_2 and CH_4 gases were also detected at shallower levels as shown in Fig.1 Although the boiling, evidenced by fluid inclusion observation, started -500m sea level at Well KE1-4, the gases were not detected in the lower part of the boiling zone. The gases started to accumulate in the inclusions at least above -250m sea level. The most gas accumulated parts are below the argillic alteration zone which act as cap rock, and close to the Ginyu Fault (Fig.2). The Ginyu fault is the main production fracture of the field (Gokou, 1995), and it reaches to the surface forming steaming ground, Ginyu Jigoku.

At the subsurface, the gases are accumulated dominantly in the hanging-wall side of the fault as shown in Fig.2; CO_2 gas absorbed largely at depths

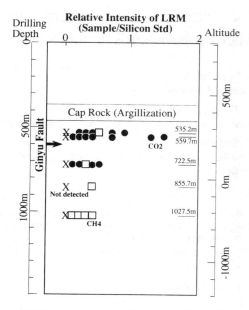

Fig.2 Relation among the relative intensity of gas components in vapor phase of fluid inclusions detected by LRM, depth, and fault location at Well KZ-2.

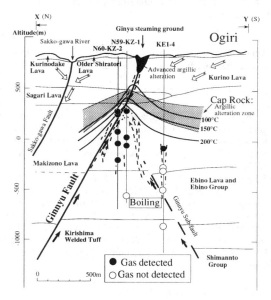

Fig.1 Distribution of gas detected fluid inclusions and the geothermal structure of the Ogiri geothermal field.

of 535 and 559 meters of Well KZ-2, above the Ginyu fault. The highest CO_2 intensity of vapor phase in fluid inclusions in Fig.2 is estimated to be 5 to 6bars at room temperarure (Maeda et al., 1998). These facts suggest that the CO_2 and CH_4 gases were separated by boiling from less gassy geothermal fluid at deeper levels, and migrated to shallow levels, and resulted in the formation of argillic alteration zone. Further accumulation of the gases were occurred after the formation of the argillic alteration zone, especially near the fractures along the Ginyu Fault.

As mentioned above, gas analysis of individual fluid inclusions by laser Raman microprobe give us useful information on dynamic fluid behavior and geothermal structures, and relation between geothermal fluid and altered rocks.

4 REFERENCES

Gokou, K. (1995) Geological analysis and evaluation of the Ogiri geothermal structure in the Kirishima Geothermal area. Resource Geology, Vol.45.41-52 (in Japanese with English abstract).

Hedenquist, J.W. and Stewart, M.K. (1985) Natural CO_2-rich steam-heated waters at Broadlands, New Zealand: Their chemistry, distribution and corrosive nature. Geoth. Res. Council, Transactions Pt.2, 245-250.

Maeda, S., Taguchi, S., Takagi, H., Sanada, K., Hayashi, M., Sasada, M., Sawaki, T., Fujino, T. and Uchida, T. (1998) An experimental measurement of CO_2 gas contents in liquid rich fluid inclusions by laser Raman microanalysis. (In this volume).

Moore, J.N., Adams, M.C. and Lemieux, M.M.(1992) The formation and distribution of CO_2-enriched fluid inclusions in epithermal environment. Geochim. Cosmochim. Acta, Vol.56, 121 - 135

Chemical stability of the hydrothermal silicates at the Los Azufres geothermal field, Mexico

I.S.Torres-Alvarado
Centro de Investigación en Energía, UNAM, Temixco, Morelos, Mexico

ABSTRACT: The equilibrium state between present day fluids and hydrothermal mineral phases was examined at the Los Azufres geothermal field. Ion activity ratios, calculated from geothermal fluid compositions, were used to construct activity diagrams for the most important silicate phases observed in the field. There is a tendency for the reservoir rocks to approach equilibrium with the hydrothermal fluids. Physical reservoir processes, specifically those controlling the CO_2 content of the fluid are very important in governing the type of hydrothermal mineralogy formed in the field.

1 INTRODUCTION

The Los Azufres Geothermal Field is located in central Mexico, approximately 200 km Northwest of Mexico City. It has been extensively investigated and developed since 1970. With a production of 98 MW it is the second most important geothermal field in Mexico (Quijano León and Gutiérrez Negrín, 1995).

Studies of hydrothermal alteration at Los Azufres have been carried out by different authors (see for example Cathelineau et al., 1985; Torres-Alvarado, 1996). These studies have shown that partial to complete hydrothermal metamorphism has occurred, with mineral +paragenesis similar to those for greenschist to amphibolite facies (Cathelineau et al., 1991).

The purpose of this study is to examine the thermodynamic relations between the fluids and the hydrothermal mineral phases in the reservoir. It is hoped to shed light on the question, whether the present day fluids are in equilibrium with the phases observed in the cores and cuttings from field wells.

This paper is focused on the equilibrium thermodynamic relations among silicate phases. This is because silicates are the most important hydrothermal minerals in the field and due to the lack of reliable chemical analyses of metal cations (besides Fe^{2+} or Mg^{++}) from the fluids of Los Azufres.

2 GEOLOGICAL SETTING AND METHODOLOGY

Los Azufres is one of a number of Pleistocene silicic volcanic centres with active geothermal systems that lie in the Mexican Volcanic Belt. This belt extends from the Gulf of Mexico to the Pacific Coast, and comprises Late Tertiary to Quaternary volcanics represented by cinder cones, domes, calderas and stratovolcanoes, along a nearly east-west axis (Aguilar y Vargas and Verma, 1987). The volcanism at Los Azufres is made up of two principal units (Fig. 1): (1) a silicic sequence of rhyodacites, rhyolites and dacites with ages between 1.0 and 0.15 m.y. and a thickness up to 1000 m, and (2) a 2700 m thick interstratification of lava flows and pyroclastic rocks, of andesitic to basaltic composition with ages up to 18 m.y., forming the local basement (Dobson, 1984).

The thermal fluids are sodium-chloride rich waters

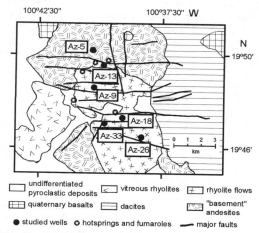

Fig. 1. Geologic map of the Los Azufres area, simplified after Dobson (1984).

Table 1. Some chemical components (mg/kg) and ion activity ratios of the total discharge composition for some wells in Los Azufres geothermal field.

Well	Cl	SiO$_2$	CO$_2$	T$_{Na-K}$*	log (a_{Na+}/a_{H+})	log (a_{K+}/a_{H+})	log ($a_{Ca2+}/[a_{H+}]^2$)	log ($a_{Mg2+}/[a_{H+}]^2$)
Az-5	799	275	17049	302	4.36	3.57	6.87	4.81
Az-9	1677	375	3774	313	5.20	4.34	7.87	5.76
Az-13	1193	313	9212	299	5.12	4.30	8.06	5.69
Az-18	1167	336	43221	296	4.43	3.59	7.75	5.16
Az-26	2085	406	15038	291	5.30	4.43	8.56	5.94
Az-33	330	72	27167	273	3.74	2.82	7.51	4.94

* Temperature in (°C) calculated with the Na-K geothermometer (Nieva and Nieva, 1987)

with high CO$_2$ and H$_2$S contents, and pH 7-8. These fluids percolate through fractures and faults mainly in the mafic units, reaching occasionally to the surface. In this interaction rocks are hydrothermaly altered in incipient to complete form (building a wide variety of minerals and textures) and the chemistry of the geothermal fluids is consequently changed.

The chemical data used for this study was the geothermal fluid composition from the wells Az-5, 9, 13, 18, 26 and 33 (Izquierdo Montalvo, 1988; Moreno Ochoa, 1989). These wells build a near NW-SE section through all the field (Fig. 1). Total discharge compositions were calculated using the methodology proposed by Henley et al. (1984; Table 1). Ion activities for these fluids were calculated using the software package The Geochemist's Workbench (Bethke, 1992) at the well bottom temperature (Table 1). Al^{3+}-content was assumed to be 0.1 mg/kg for all the studied wells, due to the lack of any Al-analysis of the fluids. This value is according to Nicholson (1993) common in other geothermal fields and should be considered as a minimum value for Los Azufres.

Activity diagrams (Figs. 2 and 3) were calculated for the chemical characteristics of the fluids from Los Azufres, under quartz excess conditions and at 300°C, which could be considered the reservoir temperature for the majority of the studied wells. These diagrams were calculated with the software program *The Geochemist's Workbench* (Bethke, 1992), whose thermodynamic data base is mainly supported on that created at the Lawrence Livermore National Laboratory (Delany and Lundeen, 1990). The activity ratios of the different geothermal fluids were also plotted on the activity diagrams of Figs. 2 and 3. Changes in the activity of the chemical components due to temperature differences (for instance in well Az-33) have little effect on the position of the point in the diagram, considering the point size and precision of the phase boundaries.

As a simplification and due to the lack of mineral chemistry data, mineral activities were assumed to be one, although this is not always true, as shown by Torres-Alvarado (1996) for feldspar and epidote solid solutions. For brevity the following discussion will be concentrated on those intensive variables, which are most relevant in establishing those phases characteristic of the reservoir.

3 DISCUSSION

Fig. 2(a) shows the stability fields of important mineral phases in Los Azufres, as a function of the

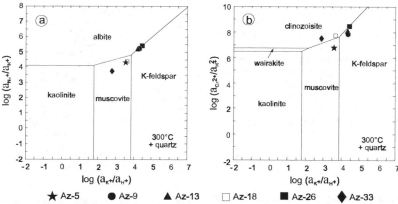

Fig. 2. Activity diagrams for important mineral phases at Los Azufres geothermal field. Fluid activity ratios calculated for some wells are also ploted.

activity ratios K^+/H^+ and Na^+/H^+. In the absence of a sodium bearing hydro-mica in the system, a separate field for paragonite has not been included in this diagram. Well discharges lie near to the triple point albite/K-mica/K-feldspar reflecting alteration of primary plagioclase, according to reactions like: Na-Plag+K^++H^+ = K-mica+Na^+. The principal phases controlling the alkali metals are feldspars, K-mica (mostly in form of sericite) and to some extent K-clay, specially illite (Izquierdo et al., 1995). This is consistent with petrographical observations and supports the applicability of cation exchange geothermometer in the field.

Like in other geothermal fields (Hedenquist, 1991; Hedenquist and Browne, 1989) the discharge compositions tend to fall along an extension of the albite/K-feldspar boundary into the muscovite field. The fact that some discharges plot inside the muscovite (sericite/illite) field indicates that this is the stable phase in the system. The trend may be interpreted as the effects of dissolution of primary feldspars in the reservoir with an equilibrium condition between albite and K-feldspar and may be a metastable condition. It is interesting to note that wells Az-9, 13 and 26 have discharges with the lowest CO_2 contents, suggesting the presence of isoenthalpic cooling and thus shifting the fluid composition from the muscovite field to positions along the albite/K-feldspar join.

The activity diagram K^+/H^+ vs. $Ca^{2+}/(H^+)^2$ is shown in Fig. 2(b). Most total discharge compositions plotted near to the triple point clinozoisite/muscovite/K-feldspar, suggesting the important role of the reaction 3K-feldspar+H_2O+2Ca^{2+} = clinozoisite+ +6Qtz+3K^++H^+ in the system. Az-18 and Az-33 plott in the clinozoisite field, possibly reflecting their high CO_2 activity, which would favour epidote formation (Henley et al., 1984). The low Ca^{2+} activity may explain the position of Az-5 in the muscovite field.

The stability field of clinozoisite was drawn for a clinozoisite activity equal to one. However the natural epidote group phases in the systems are pistacites with up to 60% clinozoisite component (Torres-Alvarado, 1996). According to Bird and Helgeson (1980) the stability field of epidote solid solution increases with decreasing clinozoisite component. This effect of the epidote composition (which has not been calculated for Los Azufres yet) suggests that the stability field for epidote solid solution could be larger, showing the epidote group as the stable mineral phase at this temperature for most wells. This is in agreement with petrographic observations. Although the hegemony of pistacite over clinozoisite and the occurrence of hematite in the field shows the presence of oxidising conditions, more detailed studies concerning the redox state of the fluids are necessary for a better understanding of the stability of these phases.

Fig. 3(a) shows the stability fields of system relevant calc-silicates, as a function of the activity ratios $Ca^{2+}/(H^+)^2$ and $SiO_{2(aq)}$. As shown in Fig. 2(b) clinozoisite is the most important phase. The proximity of most well discharges to the wairakite field could suggest the importance for the system of the following reaction: 3Wai+Ca^{2+} = 2Clinozoisite+ +4H_2O+6Qtz+2H^+. Although wairakite has been only rarely observed, this agrees with the paragenesis epidote+wairakite+quartz found in the field (Torres-Alvarado, 1996). Only Az-26 plotted in the grossular field, possibly reflecting its high Ca^{2+} activity. Margarite has not been yet petrographically observed. The presence of some well discharges in the margarite field in Fig. 3(a) however, reinforces the hypothesis that this mineral should play a more important role than thought before, as shown by the saturation indices calculated for this mineral from geothermal fluid compositions (Torres-Alvarado, 1996).

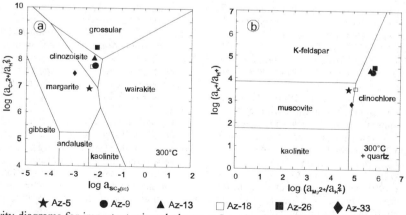

Fig. 3. Activity diagrams for important mineral phases at Los Azufres geothermal field. Fluid activity ratios calculated for some wells are also ploted.

Equilibria involving the clinochlore and the potassium bearing phases is portrayed in Fig. 3(b) as a function of K^+/H^+ vs. $Ca^{2+}/(H^+)^2$ activities. Chlorites appear to control magnesium activity in Los Azufres. This agrees with the mineralogical evidence on the field (Torres-Alvarado, 1996). The existence of two well discharges and the proximity of another one to the clinochlore/muscovite boundary agrees with the observation made by Giggenbach (1986), that chlorite involving reactions (at Los Azufres specifically the following reaction $3\text{Clinochl}+2K^++28H^+ = 2\text{Musc}+ +24H_2O+3Qtz+15Mg^{2+}$) may be very important controlling Mg^{2+} and K^+ in hot water systems.

4 CONCLUSIONS

Activity diagrams compared to textural relations among hydrothermal minerals at Los Azufres show that there is a tendency for the reservoir rocks to approach a state of chemical equilibrium with the hydrothermal fluids percolating through them.

Although temperature plays an important role controlling phase stability, other physical processes, specifically those controlling the CO_2 content of the fluid, are very important (even dominant in some places) governing the type of hydrothermal mineralogy formed in the field.

ACKNOWLEDGEMENTS

This work was supported by the foundations Alfried-Krupp-von-Bohlen-and-Halbach and Hans-Böckler. The final version of the paper benefited from helpful comments from S. P. Verma and P. Browne.

REFERENCES

Aguilar y Vargas, V.H. & S.P.Verma 1987. Composición química (elementos mayores) de los magmas en el Cinturón Volcánico Mexicano. *Geof.Int.*26:195-272.

Bethke, C. 1992. *The Geochemist's Workbench. A users guide to Rxn, Act2, Tact, React and Gtplot.* University of Illinois.

Bird, D.K. & H.C.Helgeson 1980. Chemical interaction of aqueous solutions with epidote-feldspar mineral assemblages in geologic systems. I. Thermodynamic analysis of phase relations in the system $CaO-FeO-Fe_2O_3-Al_2O_3-SiO_2-H_2O-CO_2$. *Am.J.Sci.*280:907-941.

Cathelineau, M., G.Izquierdo, G.R.Vázquez & M.Guevara 1991. Deep geothermal wells in the Los Azufres (Mexico) caldera: Volcanic basement stratigraphy based on major element analysis. *J.Volcanol.Geotherm.Res.*47:149-159.

Cathelineau, M., R.Oliver, A.Garfias, & D.Nieva 1985. Mineralogy and distribution of hydrothermal mineral zones in the Los Azufres (Mexico) geothermal field. *Geothermics.*14:49-57.

Delany, J.M. & S.R.Lundeen 1990. *The LLNL thermochemical database.* Lawrence Livermore National Laboratory.

Dobson, P.F. 1984. Volcanic stratigraphy and geochemistry of the Los Azufres Geothermal Center, Mexico. M.Sc. Thesis, Stanford University.

Giggenbach, W.F. 1986. Graphical techniques for the evaluation of water/rock equilibration conditions by use of Na, K, Mg and Ca contents of discharge waters. *8th. N.Z. Geothermal Workshop:* 37-44.

Hedenquist, J.W. 1991. Boiling and dilution in the shallow portion of the Waiotapu geothermal system, New Zealand. *Geochim.Cosmochim. Acta.*55: 2753-2765.

Hedenquist, J.W. & P.R.L.Browne 1989. The evolution of the Waiotapu geothermal system, New Zealand, based on the chemical and isotopic composition of its fluids, minerals and rocks. *Geochim.Cosmochim.Acta.*53:2235-2257.

Henley, R.W., A.H.Truesdell, P.B.Barton Jr., & J.A.Whitney 1984. Fluid-mineral equilibria in hydrothermal systems. *Reviews in Ec.Geol.* 1:268.

Izquierdo, G., M.Cathelineau & A.García 1995. Clay minerals, fluid inclusions and stabilized temperature estimation in two wells from Los Azufres Geothermal Field, Mexico. *Proceedings World Geothermal Congress:*1083-1086.

Izquierdo Montalvo, G. 1988. Caracterización de yacimientos geotérmicos por medio de la determinación de parámetros fisico-químicos. Unpublished report, IIE, Mexico.

Moreno Ochoa, J. 1989. Hidrogeoquímica de la región de Los Azufres, Michoacán. Unpublished report, CFE, Mexico.

Nicholson, K. 1993. *Geothermal fluids. Chemistry and exploration techniques.* Springer.

Nieva, D. & R.Nieva 1987. A cationic composition geothermometer for prospection of geothermal resources. *Developments in geothermal energy in Mexico.*7:243-258.

Quijano León, J.L. & L.C.A.Gutiérrez Negrín 1995. Present situation of geothermics in Mexico. *Proceedings World Geothermal Congress:*245-250.

Torres-Alvarado, I.S. 1996. Wasser/Gesteins-Wechselwirkung im geothermischen Feled von Los Azufres, Mexiko: Mineralogische, thermochemische und isotopengeochemische Untersuchungen. *TuebingerGeowiss.Arbeiten,* E2.

Evaluation of geothermal activity using thermally stimulated and radiation storage processes of quartz

N.Tsuchiya, T.Suzuki & K.Nakatsuka
Department of Geoscience and Technology, Graduate School of Engineering, Tohoku University, Sendai, Japan

ABSTRACT: Thermoluminescence of quartz in pyroclastic and volcanic rocks distributed in the geothermal field was investigated. Thermoluminescence emission was depleted when approaching the central area of geothermal activity and is decreased with depth. The kinetic equations including thermally stimulated and radiation storage processes were described and a temperature-time trajectory was calculated using those rate expressions under constant annual dose condition. Thermoluminescence behavior is applicable as a new research tool to evaluate the prediction of paleo-temperature history in geothermal systems.

1 INTRODUCTION

Thermoluminescence phenomena have been applicable to as an useful and effective dating method for geological material which were younger than about 1 Ma (Aitken, 1985; McKeever, 1985). The equation for calculating the age of a specimen using thermoluminescence is that natural thermoluminescence is divided by the natural dose rate. Here the natural thermoluminescence is the thermoluminescence accumulated by the specimen during its lifetime and the thermoluminescence per unit dose is sensitivity of that specimen for exhibiting thermoluminescence following a given absorbed dose of radiation.

The age determined by the thermoluminescence dating method can indicate absolute age of the material since its formation and/or last heating event, at least in principle. However, if the temperature condition in nature had been changed, the state of natural thermoluminescence in a given rock should reflect the thermal and radiation history of the rock.

Ypma and Hochman(1991) applied thermoluminescence behavior to evaluate paleo-temperature history of sedimentary basin. Natural thermoluminescence phenomena in some active geothermal areas is described to analyze temperature influence due to geothermal activity. We represent a possibility of a geothermometry and/or a geospeedmetry using thermoluminescence phenomena of quartz.

2 EXPERIMENTAL

Rock samples were collected from outcrop in the Kakkonda geothermal field (Iwate Prefecture) and borehole cores drilled in the Minase geothermal field (Akita Prefecture) in Northeast Japan. These samples were crushed and quartz grains were picked out. The 74-250 μm fraction of quartz was obtained by sieving, and this was followed by etch in HF to reduce the alpha dose.

Thermoluminescence pulses were detected with a photomultiplier tube coupled to an interference filter of which maximum transmittance is at 620 nm and half height width is 2 nm.

Quartz samples were heated up to 320℃ to release the trapped electrons for evaluating trap parameters of the storage processes. After annealing, quartz was irradiated by ^{60}Co gamma ray.

3 THERMOLUMINESCENCE IN GEOTHERMAL FIELDS

Fig.1 shows the distribution of the integrated intensity of thermoluminescence in the Kakkonda geothermal field and an isopleth of the emission is also described (Tsuchiya et al., 1994). In this field, The Tertiary formations are exposed as a fenster near the Kakkonda geothermal power plant(KGPP)and the Quaternary pyroclastic rocks are distributed around the Tertiary formations.

Paleodose of older rocks is higher than that of younger samples on the basis of the concept of thermoluminescence dating, however, integrated emission of thermoluminescence is no relation to the stratigraphic boundaries. Approaching KGPP, integrated emission intensity decreases. The general trend of distribution of isoplethes of TL emission is in a concentric configuration of which the center is located near KGPP. No thermoluminescence can be observed near KGPP, and 5% isopleth is located at a point about 2km away from KGPP in a northwest to a southeast direction along the Kakkonda river, and 1 to 2 km in a northeast to a southwest direction.

Fig.2 shows stratigraphical column and a

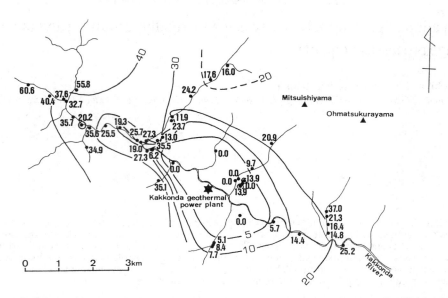

Fig.1 Map showing isopleth o ralative integrated thermoluminescence intensity of quartz in the Kakkonda geothermal field. KGPP: Kakkonda geothermal power plant (after Tsuchiya *et al.*, 1994)

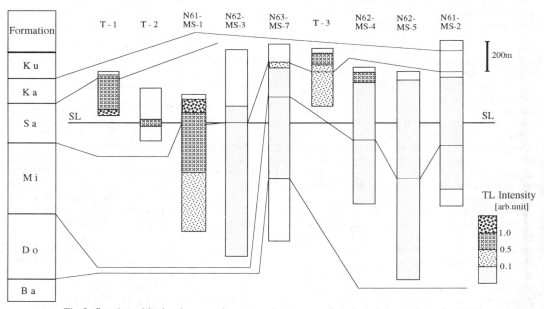

Fig.2 Stratigraphical column and a comparison among borehole intersections for relative thermoluminescence intensity of quartz in the Minase geothermal field. Ba Basement Do Doroyu F., Mi Minase F., Sa Sanzugawa F., Ka Kabutoyama F., Ku Kurikomayama volcanics, SL sea level.(after Nambu *et al.*, 1996).

comparison among borehole intersections for thermoluminescence intensity of the Minase geothermal field (Nambu *et al.*, 1996). TL intensity in arbitrary unit is represented as four zones. This area is composed mainly of the Doroyu, Minasegawa, and Sanzugawa Formations which are Neogene in age and Quaternary volcanics. TL intensities decrease with increasing depth, and no obvious changes can be recognized through boreholes except T-1 boreholes which is located at the northwest boundary of the geothermal area. Thus, there is no clear relation between vertical variations of thermoluminescence intensity and the stratigraphical unit in the central geothermal area. These facts indicate that thermoluminescence intensity doesn't depend on geological age of the rock. The TL intensity decreases in accordance with the present temperature in the boreholes.

The temperature gradient in the geothermal field is extremely high compared with the other areas, so that the trapped electrons have been released by the geothermal activity.

4 ENERGITICS OF THERMOLUMINESCENCE

The theoretical treatment for a thermoluminescence was given by Randall and Wilkins (1945), who assumed no retrapping and a TL intensity proportional to the rate of charge of the concentration of trapped electrons. The fundamental equation of the thermoluminescence mechanism including storage and thermally stimulated processes is described as follows:

$$I(T) = \frac{dn_1}{dt} = -n_1 s \exp(-\frac{E}{kT}) + \beta n(N - n_1)$$

where I is the TL intensity, s is a "escape-frequency factor" or sometimes just the "pre-exponential factor", N, n and n_1 are the concentrations of traps in question, of free electrons and of trapped electrons, respectively. E is the activation energy which is the same as the trap depth, T is the absolute temperature, k is Boltzmann's constant, β is the recombination probability. The first term in the right side represents the thermally stimulated processes and the second term is the storage processes of radiation energy.

The concentration of the trapped electrons (n_1) is very low compared with N at the early stage of the irradiation after zeroset, then the second term becomes λrN. The concentration of trapped electrons through the storage processes is given by

$$n_1 = \lambda N D_a$$
$$D_a = rt$$

where λ is a proportionality constant relating to the recombination probability and irradiation efficiency, r is the irradiation dose rate and D_a is the dose. If the thermoluminescence glow-curve is divided into plural peaks, the trap parameters such as s, E, n_{10} (the initial concentration of trapped electrons), λ and N of decomposed glow-curve are determined independently. The initial rise method and another estimation technique proposed by Balarin and Zetzsche (1962) were applied to evaluate the trap parameters of s, E and n_{10} (Tsuchiya et al., 1997). λ and N are obtained by the artificial irradiation of gamma rays.

Fig.3 shows the concentration of trapped electrons as a function of irradiation dose under 1.1 Gy/hr dose rate condition. L, M and H peaks represent the lower, middle and higher temperature glow-curve, respectively. The concentration of trapped electrons increases linearly with increasing dose. The gradient of the approximation line indicates λN of each glow-curve. Fig.4 shows relationship between dose and integrated thermoluminescence intensity. TL intensity increases linearly until 2000 Gy, however, it increases slowly when approaching the saturated concentration of trapped electrons in the high dose region.

5 T-t TRAJECTORY

Tsuchiya et al. (1997) showed that the thermoluminescence glow-curve of quartz in pyroclastic and volcanic rocks in the Kakkonda geothermal field could be divided into three peaks on the basis of the artificial annealing and peak analysis. The results of peak deconvolution indicate that an isolated trap depth is recognized and three kinetic equations including

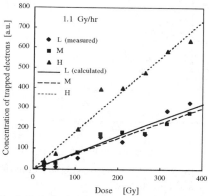

Fig.3 Relationship between irradiation dose and concentarions of trapped electrons.

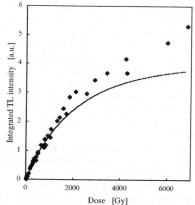

Fig.4 Relationship between irradiation dose and integrated thermoluminescence intensity.

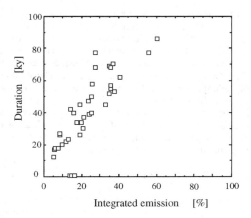

Fig.5 Relationship between integrated emission and duration of heating.

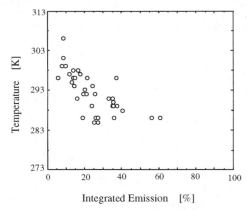

Fig.6 Relationship between integrated emission and temperature

thermally stimulated and radiation storage processes are obtained as a function of time and temperature. Hence the solutions among independent kinetic equations suggest an optimum T-t pair of rocks.

Relationship between thermoluminescence emission and temperature calculated using kinetic equations of L and H peaks of quartz in the Kakkonda geothermal field is shown in Fig.5, and Fig.6 shows duration of heating as a function of integrated emission. The calculation was carried out under 1.2 mSv/year annual dose condition which was determined by *in situ* natural radiation measurement using TL dosimeter (Mg_2SiO_4:Tb).

With increasing integrated thermoluminescence intensity, the temperature is decreasing and the duration of heating is increasing. These facts suggest that duration of relatively high temperature is shorter than low temperature period. These *T- t* pairs indicate equivalent thermal effect due to geothermal activity.

6 CONCLUSION

The thermoluminescence intensities of quartz in the Miocene and the Quaternary rocks show no relation to the stratigraphic boundaries. The integrated thermoluminescence emission is depleted in the central part of the Kakkonda geothermal field and is decreasing with depth in the Minase geothermal field.

Thermoluminescence glow-curve of quartz collected from the Kakkonda geothermal field could be divided into three peaks on the basis of the results of artificial annealing and peak analysis. Independent kinetic equations of decomposed glow-curves indicate temperature and time trajectories which suggests equivalent thermal effect geothermal activity

Thermoluminescence has been developed as the dating technique for geological materials, however, the paleodose in quartz of geothermal field was reduced by natural annealing associated with geothermal activity. Thermoluminescence behavior is applicable as a new research tool to evaluate the prediction paleo-temperature history in the geothermal systems.

REFERENCES

Aitken, M.J. (1985) Thermoluminescence Dating. Academic Press, 359pp.
Balarin, M. and Zetzsche, A.(1962) Bestimmung der Aktivierungsenergie für die Beweglichkeit von Gitterdefekten durch zeitlineares Aufheizen. *Phys.Stat. Sol.* 2, 1670-1682 .
MaKeever, S.W.S.(1985) Thermoluminescence of solids. Cambridge University Press,376p.
Nambu, M., Mikami, K., Tsuchiya, N. and Nakatsuka, K. (1996) Thermoluminescence of quartz in the borehole cores from the Minase geothermal area, Akita prefecture, Japan. *J. Geotherm. Res. Soc. Japan.* 18,39-49. (in Japanese with English abstract)
Randall, J.T. and Wilkins, M.H.F. (1945) Phosphorescence and electron traps I. The study of trap distributions. *Proc.Roy. Soc. London* A184, 366-389.
Tsuchiya, N., Yamamoto, A. and Nakatsuka, K. (1994) Thermoluminescence of quartz in volcanic and pyroclastic rocks from the Kakkonda geothermal area, Northeast Japan - Preliminary study of thermoluminescence geothermometer. *J. Geotherm. Res. Soc. Japan,* 16, 57-70. (in Japanese with English abstract)
Tsuchiya, N., Suzuki, T. and Nakatsuka, K. (1997) Thermoluminescence as a new research tool for evaluation of geothermal activity of the Kakkonda geothermal field, Northeast Japan. *Geothermics. submitted.*
Ypma, P.J. and Hochman, M.B.M. (1991) Thermoluminescence geothermometry- A case study of the Otway Basin. *APEA Jour.* 31(1), 312-324.

Water-rock interaction at the boundary of Wairakei geothermal field

C.P.Wood
Institute of Geological and Nuclear Sciences, Wairakei, New Zealand

ABSTRACT: Drilling through the sharp NE boundary of Wairakei geothermal field has provided evidence that stratigraphy and structure constrain fluid movement, that hydrothermal alteration has not produced an extensive zone of self-sealing, and that the position of the boundary has remained essentially static with time. Interlayered permeable tuffs and impermeable siltstones channel hot and cold flows through the boundary, and a postulated, buried, listric caldera fault may control deep hot fluid movement. The distribution of temperature-sensitive hydrothermal minerals such as epidote and clays suggests the boundary has remained in its present position throughout much of the life of the field, but production-induced pressure decline is now allowing cold water to invade, and the boundary to retreat inwards.

1 INTRODUCTION

The Taupo volcanic Zone (TVZ) of New Zealand contains many high-temperature geothermal systems, located mainly in the rhyolite-dominated central part of the zone, where at least eight major calderas have been active in the past 1.5 million years (Wood, 1995). Tuffs and lavas interbedded with fluvial and lacustrine sediments fill the TVZ to a depth of 2.5 km or more, and comprise the host rocks for hydrothermal plumes generated at depths down to about 8 km (Bibby et al. 1995).

Wairakei is a typical water-dominated TVZ system, that has been continuously explored and developed for the past 50 years. Despite this, very little is known about the boundary structure of the field, and many important questions remain to be answered. To what extent does the geological structure and stratigraphy control the movement of fluids? Has hydrothermal alteration and mineral deposition created an impermeable barrier (the self-sealing effect) which defines the extent of the boundary? Has the boundary migrated with time?

In recent years, the operators of Wairakei power station have investigated the reinjection potential of areas along the banks of the Waikato River (Fig.1) where the resistivity boundary is the most abrupt seen in any TVZ geothermal field (Risk & Bibby, 1997). In 1995, Contact Energy Ltd drilled deviated wells into and through a segment of this sharp boundary, providing a unique opportunity to study its structure. This paper discusses the geology, hydrothermal alteration and geothermal history of this sector of the Wairakei field boundary.

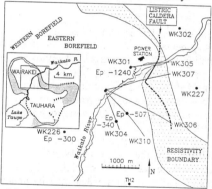

Fig.1 Map of NE boundary of Wairakei Geothermal Field. Deviated wells are labelled at end of track (eg. WK305), vertical at wellhead. Ep is first appearance of epidote in metres relative to sea level. Resistivity boundary is from Risk and Bibby (1997).

2 BOUNDARY STRATIGRAPHY

The geological cross-section in the plane of wells WK301 and WK305 summarises the stratigraphy of the boundary area (Fig. 2A). The Huka Falls Formation (HFF) locally controls the hydrology, and comprises a permeable, water-laid pumice breccia enveloped between impermeable lacustrine siltstones.

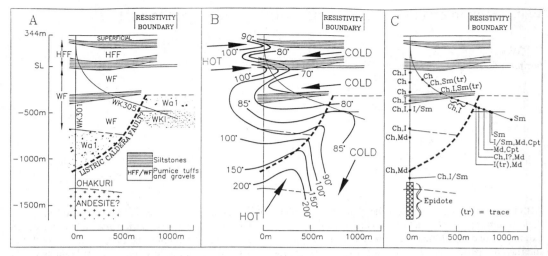

Fig. 2: Cross-section through drillholes WK301 (vertical) and WK305 (deviated). Vertical scale is relative to sea level (SL). Resistivity Boundary is the position interpreted from shallow Schlumberger resistivity data as mapped at the surface by Risk and Bibby (1997). A shows the stratigraphy and fault structure. HFF = Huka Falls Formation; WF = Waiora Formation; Wa1 is the Waiora 1 ignimbrite sequence; WKI = Wairakei Ignimbrite. B shows an interpretation of the thermal structure with isotherms constructed from downhole temperature measurements. Water flow directions shown by large arrows are the vectors in the plane of the cross-section only. C shows the location of XRD analysed samples: Ch = chlorite; I = illite; Sm = smectite; I/Sm = interlayered illite/smectite; md = mordenite; Cpt = clinoptilolite.

Beneath the HFF is a varied sequence of volcanic and sedimentary rocks belonging to the Waiora Formation (WF), that consists mostly of permeable tuffs, gravels and sands, but also includes an interbedded siltstone aquiclude, which occurs in infield wells, but cannot be traced as a distinct unit across the boundary to the outfield wells WK227 and WK302.

The Waiora Formation Member 1 (Wa1) ignimbrites, and the underlying Wairakei Ignimbrite (WKI) are the key to interpreting the boundary structure. The two ignimbrites may be genetically related (Wood, 1994), and belong to the voluminous ~0.33Ma Whakamaru-group ignimbrites eruptive episode (Wilson et al. 1986).

Drillhole data suggest the Wa1 deposits ponded in a deep depression centred beneath the Waikato River, extending south through wells WK310 and TH2. Outside the field, WK227 penetrates through the base of 200m of Wa1 tuffs (Wood 1994) into WKI at -503 mRL (m relative to sea level).

The surface of the Wairakei Ignimbrite slopes at 15° to the SE from a high of -200 mRL beneath the Western Borefield to deeper than -430 mRL in the Eastern Borefield. In well WK301, the ignimbrite is missing, and is represented by just a few metres of gravels lying between Wa1 and Ohakuri Group rocks at -1110mRL. It appears again in the resistivity boundary region less than 800m from WK301, at about -500 mRL in wells WK305 and WK307.

Dacite lavas and tuffs of the oldest drilled formation at Wairakei, the Ohakuri Group, occur only in WK301 where they overlie propylitised andesite. Andesite was also recovered from the bottom of WK306.

3 BOUNDARY STRUCTURE

A structural interpretation must explain why the Wa1/WKI contact occurs at about -500mRL outside the field, but immediately infield WKI is missing beneath thick Wa1 down to -1100mRL. Caldera collapse at the time of the WKI and Wa1 eruptions offers a possible explanation (Wood, 1995; Wood et al., 1997). Thus, following the emplacement of WKI from the NW, caldera collapse caused rotational displacement on a listric fault (Fig. 2A). Wa1 ignimbrites then filled the caldera and further collapse occurred. No movement has occurred since. The fault is speculatively shown in Fig.1 as a scalloped curve, typical of caldera ring faults.

4 TEMPERATURE AND PRESSURE

The pumice breccias of the middle HFF and upper WF form discrete aquifers (Fig. 2B). In the vicinity of wells WK301 and WK305, each aquifer forms a thermal couplet (temperature differential ~25°C) in which denser cold water flows inwards from the overpressured outfield area, while buoyant warmer water moves outwards trapped beneath a siltstone aquiclude. The warm flow does not extend far outside the resistivity boundary (eg WK302 is cold throughout). Although the layered HFF/WF stratigraphy occurs in all wells drilled along the left bank of the Waikato River, the thermal structure seen in WK301/WK305 does not occur in wells to the SW: thus it appears that thermal cross flows are evident only where the boundary is narrowest.

Deep temperatures near the field boundary are known only from WK301, where there is an overall gradient of ~11°C/100m between the base of the WF siltstones and the postulated listric fault. Below the fault, the temperature rises more sharply at 20°C/100m reaching ~260°C at well bottom. Fig. 2B shows a speculative interpretation of the deep thermal structure in which hot fluid rises upwards and outwards along the line of the listric fault into the region dominated by cold groundwater.

Since 1958, fluid withdrawal has caused a uniform pressure decline up to 25 bars across the entire Wairakei reservoir (ECNZ 1992). Across the resistivity boundary there is a steep pressure rise of about 20 bars. The pressure control points in wells WK305 and WK307 are in ignimbrites on the upthrow side of the caldera boundary fault, and these wells, along with outfield wells WK227 and WK302 fall on a pressure curve 20 bars higher than the infield wells. This is a strong indication that the hydrological field boundary is related to the buried caldera fault in this area.

5 HYDROTHERMAL ALTERATION

The life and times of a geothermal system can be measured by the hydrothermal alteration of its host rocks. A hot plume generated at depth will expand upwards and outwards, balanced by the surrounding cold groundwater, and constrained by geological permeability. In time, the alteration mineralogy will reach equilibrium with the prevailing temperature and fluids, and there will be a transition from altered to fresh rocks across the boundary (assuming a pristine initial state). If the heat supply dwindles the system will cool and contract inwards leaving altered rocks stranded outside the retreating field boundary. Subsequent erosion may expose the now cold relic as a "fossil" geothermal field.

Bibby et al (1995) used resistivity data to suggest that hydrothermal plumes in the TVZ are stable for 200 000 years or more, and are surrounded by unaltered, cold areas. A long, stable life combined with continual deposition may explain why there are 17 extant systems but only one exposed fossil system in the TVZ, but also requires a hydrothermal plume to remain active and in situ even when drastic changes are imposed on the aquifers by volcano-tectonic activity. The drillholes through the Wairakei boundary provide an ideal opportunity to test the hypothesis.

The most useful minerals for estimating past hydrothermal conditions are epidote, clays, and zeolites. Epidote is common in a wide range of rock and mineral associations at Wairakei, where Steiner (1977) noted it formed above 235°C, and hence should be a guide to paleotemperature changes in the hotter parts of the reservoir.

Fig. 1 shows the occurrence of epidote in some wells. It does not occur in any boundary-penetrating well (even in WK306 at -845mRL), but is present deep in WK301 and at progressively higher elevations to the south towards Tauhara. To the east, epidote occurs in the Eastern Borefield (>235°C at 90mRL). Epidote distribution suggests there has always been a steep interface between hot and cold water in the area of the resistivity boundary, and points to a 35-45°C decline in the inner boundary region. Production has cooled the Western Borefield by about 30°C (ECNZ, 1992), but this effect has not extended southwards to WK226 and TH2. Either the hydrothermal plume has cooled naturally in the inner boundary region, or epidote formed at a lower temperature than Steiner suggested.

Fig. 2C shows clay mineral and zeolite distribution. The middle HFF aquifer is unaltered, and has not been subjected to higher temperatures than at present. By contrast, the WF deposits are weakly to moderately altered and dominated by discrete chlorite or chlorite+illite clay assemblages, with traces only of smectite or interlayered illite/smectite ($\sim I_{0.7}/Sm$). Harvey and Browne (1991) reported that I (or I/Sm with $I \geq 0.8$) forms at $\geq 180°C$, at Wairakei, in agreement with my own observations in many TVZ fields. They suggested that the deposition of discrete I+Ch as opposed to I/Sm sequences occurred where there was channel flow (high permeability). The WF is the main production aquifer at Wairakei (ECNZ, 1992) so good initial permeability may explain the lack of interlayered clays in the shallowest WF aquifer where there is thermal indication of flowing waters. However, the deeper WF deposits were impermeable when drilled.

The clay geothermometry suggests the WF aquifers adjacent to the resistivity boundary were once 80-100°C hotter than now. Direct invasion of cold water through the boundary in response to the 20 bars production drawdown seems the best explanation for this cooling.

In WK301 the Wa1 ignimbrites are only weakly altered, and contain chlorite and mordenite. WK301 and WK305 present very different aspects of hydrothermal alteration on the older/upthrow side of the caldera boundary fault. Across the fault in WK301 there is a change from weakly altered pumice lithic tuff (Wa1), to intensely altered dacite (Ohakuri Group) replaced by quartz, albite, calcite, pyrite and $I_{>.8}$/Sm: beneath the dacite are tuffs and intensely propylitised andesite.

In WK305, the alteration rank decreases sharply across the fault, and the Ch+I assemblage found in the Wa2 tuffs is replaced by a zeolite-bearing (mordenite+clinoptilolite) association with traces of interlayered I/Sm clays. WK305 terminates in welded Wairakei Ignimbrite, unaltered except for weak smectite development.

6 CONCLUSIONS

Provisional answers to the introductory questions can now be given. The geology does constrain the boundary hydrology. Cold water enters and hot water leaves the field via discrete thermal flow couplets channelled in tuff aquifers layered between siltstone aquicludes. A postulated listric caldera fault may provide vertical egress for deep hot fluid to rise and mix with cold groundwater impounded on the upthrow side of the fault.

There is no strong evidence that hydrothermal alteration or mineral deposition has created a boundary by sealing. However, clay deposition in tuffs banked against the caldera fault may have reduced their permeability and slowed the ingress of cold water.

The boundary has remained essentially static through time, but may now be migrating inwards as cold water penetrates in response to the pressure decline caused by production.

7 ACKNOWLEDGEMENTS

Contact Energy Ltd are thanked for providing access to drill core and cuttings, and supplying temperature and pressure data. Research was funded by FRST Contract C05605.

8 REFERENCES

Bibby, H.M., T.G.Caldwell, F.J.Davey, and T.H.Webb, 1995. Geophysical evidence on the structure of the Taupo Volcanic Zone and its hydrothermal circulation. *J. Vol. Geoth. Res.* 68: 29-58.

ECNZ 1992. *Resource consent application for reinjection: Wairakei geothermal field*. Electricity Corporation of New Zealand. Wairakei.

Harvey, C.C. & P.R.L.Browne 1991. Mixed-layer clay geothermometry in the Wairakei geothermal field, New Zealand. *Clays and Clay Mins.* 39: 614-621.

Risk, G.F & H.M.Bibby 1997. Eastern boundary of Wairakei geothermal field: resistivity measurements. Submitted to *Proc. 19th N. Z. Geoth. Workshop 1994*. Geothermal Institute, University of Auckland.

Steiner, A. 1977. The Wairakei Geothermal Area, North Island, New Zealand. *NZ Geol. Surv. Bull. 90*, DSIR, Wellington.

Wilson, C.J.N., B.F.Houghton, and E.F.Lloyd, 1986. Volcanic history and evolution of the Maroa-Taupo area, Central North Island. In: *Late Cenozoic Volcanism in New Zealand*, I.E.M. Smith (ed), Roy. Soc. NZ Bull. 23, 194-223.

Wood, C.P. 1994. Waiora Formation geothermal aquifer, Taupo Volcanic Zone, New Zealand. *Proc. 16th N. Z. Geoth. Workshop 1994*. Geothermal Institute, University of Auckland: 121-126.

Wood, C.P. 1995. Calderas and geothermal systems in the Taupo Volcanic Zone, New Zealand. *Proc. World Geoth. Cong. Florence 1995*. vol.2: 1331-1336.

Wood, C.P., E.K. Mroczek & B.S. Carey 1997. The boundary of Wairakei geothermal field: geology and chemistry. Submitted to *Proc. 19th N. Z. Geoth. Workshop 1994*. Geothermal Institute, University of Auckland.

12 Oceanic

Alteration of basalts from the Ninetyeast Ridge, Indian Ocean (ODP data)

A.V.Artamonov, V.B.Kurnosov & B.P.Zolotarev
Geological Institute, Moscow, Russia

ABSTRACT: Volcanic rocks recovered during DSDP Leg 26 and ODP Leg 121 on the Ninetyeast Ridge are weakly differentiated tholeiitic basalts. All studied basalts have been affected by low-temperature hydrothermal alteration. Two type of alteration were documented: "oxidative" and "non-oxidative" ones. The oxidative alteration of basalts is characterized by the presence of brown and greenish-yellow smectites, Fe-oxides and hydroxides, and calcite. Green smectite dominates among the products of non-oxidative alteration of basalts. Changes in oxidizing and non-oxidizing conditions result in drastic redistribution of some elements. Unlike the mid-ocean ridge, oxidative and non-oxidative types of alteration take place simultaneously in volcanic edifices. Alteration of basalts occurs during the waning stage of volcanic activity under variable temperatures. The temperature of alteration depends on distance from eruptive centers. Age of basalts from the Ninetyeast Ridge has relatively little effect on their degree of alteration.

1 INTRODUCTION

Aseismic ridges, plateaus, and seamounts are important units of ocean floor with distinctive morphology, composition, thermal history, and alteration characteristics which differ from those of mid-ocean ridges. Alteration of basalts from volcanic edifices (aseismic ridges, plateaus, and seamounts) has been studied in less detail than that of basalts from mid-ocean ridges.

We studied altered basalts recovered during DSDP Leg 26 (Hole 254) and ODP Leg 121 (Holes 756D, 757C, and 758A) on Ninetyeast Ridge. The aseismic Ninetyeast Ridge is located in the eastern sector of the Indian Ocean and extends over a distance about 5000 km in the longitudinal direction (90°E). Holes 254 and 756D are located in the southern segment of the Ninetyeast Ridge. Holes 757C and 758A are located in the central and northern segments of the Ninetyeast Ridge, respectively (Fig.1). The age of basalts increases northward along the Ninetyeast Ridge from 38 Ma (Sites 254 and 756) to 58 Ma (Site 757) and to slightly more than 82 Ma for Site 758 (von der Boch, Sclater, et al., 1974; Duncan, 1978, 1991).

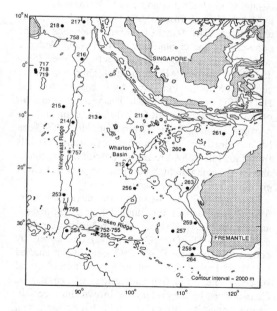

Fig. 1. Location map of the Indian Ocean showing DSDP and ODP sites.

2 GEOCHEMISTRY

The geochemical data provide evidence that the volcanic rocks of the Ninetyeast Ridge are weakly differentiated tholeiitic basalts. Olivine-bearing aphyric basalts prevail among the studied rocks. Plagioclase-phryric basalts were observed only in Hole 757C. Trends of the REE distribution suggest the presence of normal and enriched varieties (N-MORB and E-MORB) among the rocks of the Ninetyeast Ridge. Concentrations of the REE in basalts of Ninetyeast Ridge are higher then those in the mid-ocean ridge basalts. The geochemical data suggest that these basalts are derived from mantle melts generated at different depths. These melts were probably mixed in various proportions.

3 ALTERATION

All basalts have been unevenly affected by hydrothermal alteration at low temperatures. Alteration degree is indicated by H_2O^+ content. The basalts from Hole 757C (6.2 wt % H_2O^+) are most altered. The basalts from Hole 756D are least altered (H_2O^+ content up to 2 wt %). H_2O^+ content in the basalts from Holes 254 and 758A are from 1.7 wt % to 4 wt % and from 0.88 wt % to 3.6 wt %, respectively. The extent of alteration depends on the permeability and crystallinity of the basalts. Aphyric basalts are less altered. The basalts from Holes 756D and 757C suffered hydrothermal alteration under similar conditions, but non-aphyric basalts from the Hole 757C are more altered. The weaker alteration of basalts from the Hole 758A as compared to petrographically similar basalts from Hole 757C can be explained by low seawater-rock ratio. Basalts from the Hole 758A are massive, with rare vesicles, and do not containe high permeable zones that allow circulation of heated seawater.

Two types of low-temperature alteration were observed in each of the studied holes: "oxidative" and "non-oxidative" ones. The boundary between these types ($Fe_2O_3/FeO = 1.60$) is taken by convention, because Fe_2O_3/FeO ratio in the studied basalts is higher than that in typical non-oxidized basalts. The oxidative alteration of basalts is characterized by the presence of brown and greenish-yellow smectite, Fe-oxides and hydroxides, and calcite. K-feldspar occurs in the basalts from Hole 757C. Green smectites in groundmass, fractures, and vesicles prevail among the products of non-oxidative alteration. In some cases, clay minerals with "chloritic" structures (mixed-layer smectite-chlorite, swelling chlorite, chlorite, and hydromica) formed together with smectite. We have identified zeolites in basalts from Holes 757C and 758A and sulfides in basalts from Hole 758A. Quartz and chalcedony rarely occur in the basalts from Hole 758A. In general, smectites dominate among secondary minerals in the products of both types of alteration.

We have calculated the mass balance of major and minor elements in the studied basalts. For the mass balance calculation we have used the method of atomic-volume recalculation of chemical analyses, taking into account the porosity of the rocks (Kazitzin & Rudnik, 1968). The results show the redistribution of chemical elements during alteration of basalts from Ninetyeast ridge. Calculations of mobility of major and minor elements during the alteration of basalts indicate the loss (in various degree) of Si, Ca, Fe, Mn, Na, P, REE, Sc, Co, Cu, Zn, Y, Nb, Sr, Zr, and Ba and gain of Mg under non-oxidizing environment. In oxidizing conditions, the basalts are depleted in Si, Ca, Mg and are enriched in K, Fe, Mn, Na, P, REE, Sc, V, Y, Cu, Zn, Nb, Rb, Sr, Zr, and Ba. The decrease of Si and Ca in amounts with the increase of the alteration degree is most illustrating (Fig. 2).

4 RESULTS AND DISCUSSION

Oxidative alteration occurs in basalts erupted in shallow-water and subaerial environments, along the margins of separate lava flows (especially in the vesicular upper part of flows), in brecciated zones, in fractures, and in vesicles. Oxidizing environments prevail on flanks of volcanic edifices (Kurnosov, 1986; Kurnosov & Murdmaa, 1996). The absence of highly permeable zones results in low seawater-rock ratios and does not favor an oxidizing environment. Therefore, the massive basalts from Hole 758A did not undergo oxidative alteration. These basalts were erupted in a deep-sea environment.

Frey et al. (1991) and Saunders et al. (1991) hypothesized that oxidative alteration follows non-oxidative alteration. We suggest that, unlike the mid-ocean ridge, both oxidative and non-oxidative alteration take place in volcanic edifices simultaneously. Seawater permeates into volcanic edifices along the margins of lava flows, brecciated zones, and fractures. Basalts interact with oxygenated seawater under high seawater-rock ratio. The seawater infiltrates into the flow interiors, but oxidizing conditions only occur near permeable zones. The process is expressed in the presence of both oxidized zones (about 10 cm) in the marginal part of flow units and in haloes around calcite-filled veins. The non-oxidizing environment prevails within remaining volume of lava flows which are altered under low seawater-rock ratio. The formation of new fractures can lead to recurrence of this process. Previous zones of alteration can be intersected and overprinted by secondary veins with mineral assemblages of the oxidizing environment.

We have not determined any vertical zonation in distribution of secondary minerals in any of the studied holes. Nevertheless, we have registered clay minerals with "chloritic" structure within thick units throghout the studied sections. This suggest the idea on the subhorizontal water flow on the flanks of volcanic edifices. Similar results were obtained by study of altered basalts from Suiko Seamount (Emperor Seamount Chain) and Allison Guyot (West Pacific Guyots) drilled during DSDP Leg 55 and ODP Leg 143, respectively (Kurnosov, 1986; Kurnosov et al., 1995).

Alteration of basalts occurs during the waning stage of volcanic activity under different temperatures. The temperature of alteration depends on the distance from eruptive centers. During oxidative alteration temperatures vary from 10°C to about 80°C. The temperature of non-oxidative alteration is similar to that of oxidative alteration, but typically somewhat higher. For example, natrolite from Hole 757C, where both alteration processes occured was formed in the zeolite zone at about 120°C (Saunders et al., 1991). Temperature differences of oxidative and non-oxidative alteration can be explain by cooling due to high seawater-rock ratio. Temperature of calcite generation increase from 10-40°C to 41-71°C in Holes 756D and 758A, respectively (Lawrence, 1991). This may reflect the distance of these holes from eruptive centers. In Hole 758A, where sulfides occur, the temperature of non-oxidative alteration reached 200°C. We suggest that the basalts from Hole 758A were altered by hot hydrothermal flow under non-oxidizing conditions.

The age of the basalts from the Ninetyeast Ridge had practically do no effect on the degree of their alteration. When the volcanic edifices cool down completely, the basalts are generally not subjected to hydrothermal alteration. Very weak alteration of basalts can take place when the cold-water circulation is driven by a temperature difference between the cold bottom seawater and the warmer seamount interior, due to the elevated geothermal gradient of the oceanic crust under seamounts (Kurnosov & Murdmaa, 1996).

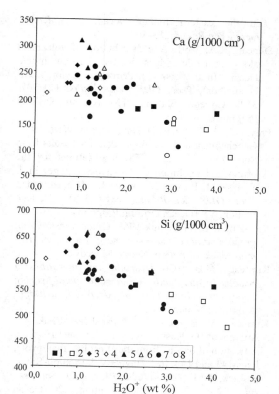

Fig. 2. The decrease of Si and Ca in amounts (g/1000 cm^3) with the increase of alteration degree (H_2O^+, wt %) in massive and sparsely vesicular basalts from Legs 26 and 121: 1 - "non-oxidized" basalts from Hole 254; 2 - oxidized basalts from Hole 254; 3 - "non-oxidized" basalts from Hole 756D; 4 - oxidized basalts from Hole 756D; 5 - "non-oxidized" basalts from Hole 757C; 6 - oxidized basalts from Hole 757C; 7 - "non-oxidized" basalts from Hole 758A; 8 - oxidized basalts from Hole 758A.

5 ACKNOWLEDGMENTS

This study was financially supported by the Russian Foundation for Fundamental Research (grants 95-05-15379 and 95-05-14368).

6 REFERENCES

Duncan, R.A. 1978. Geochronology of basalts from the Ninetyeast Ridge and continental dispersion in the eastern Indian Ocean. *J. Volcanol. Geotherm. Res.*, 4:283-305.

Duncan, R.A. 1991. Age distribution of volcanism along aseismic ridges in the Eastern Indian Ocean. In J. Weissel, J. Prirce, E. Taylor, J. Alt et al. (eds), *Proc. ODP, Sci. Results*, 121:507-517. College Station, TX (Ocean Drilling Program).

Frey, F.A., W.B. Jones, H. Davies, & D. Weis, 1991. Geochemical and petrologic data for basalts from Sites 756, 757, and 758: implications for the origin and evolution of Ninetyeast Ridge. In J. Weissel, J. Prirce, E. Taylor, J. Alt et al. (eds), *Proc. ODP, Sci. Results*, 121:611-659. College Station, TX (Ocean Drilling Program).

Kazitzin, Y. & V. Rudnik 1968. *Guide to estimation of mass balance and inner energy during metasomatic rock formation.* Nedra: Moscow.

Kurnosov, V.B. 1986. *Hydrothermal alterations of basalts in the Pacific Ocean and metal-bearing deposits(using data of deep-sea drilling).* Moscow : Nauka.

Kurnosov, V., B. Zolotarev, V. Eroshchev-Shak, A. Artamonov, G. Kashinzev, & I. Murdmaa 1995. Alteration of basalts from the West Pacific Guyots, Legs 143 and 144. In J. Haggerty, I. Premoli-Silva, F. Rack, & M. McNutt (eds), *Proc. ODP, Sci. Results*, 144:475-491. College Station, TX (Ocean Drilling Program).

Kurnosov, V. & I. Murdmaa 1996. Hydrothermal and cold-water circulation within the intraplate seamounts: effects on rock alteration. In R. Williams & H. Sloan (eds), *The Oceanic Lithosphere & Scientific Drilling into the 21st Century*: 87-88. Woods Hole, MA (ODP - InterRidge - IAVCEI Workshop).

Lawrence, J.R. 1991. Stable isotopic composition of pore waters and calcite veins. In J. Weissel, J. Prirce, E. Taylor, J. Alt et al. (eds), *Proc. ODP, Sci. Results*, 121:447-455. College Station, TX (Ocean DrillingProgram).

Saunders, A.D., M. Storey, I.L. Gibson, P. Leat, J. Hergt, & R.N. Thompson 1991. Chemical and isotopic constraints on the origin of basalts from Ninetyeast Ridge, Indian Ocean: results from DSDP Legs 22 and 26 and ODP Leg 121. In J. Weissel, J. Prirce, E. Taylor, J. Alt et al. (eds). *Proc. ODP, Sci. Results*, 121:559-590. College Station, TX Ocean Drilling Program).

von der Borch, C., J.G. Sclater et al. 1974. *Init. Repts. DSDP*, 22. Washington (U.S. Govt. Printing Office).

Modelling the halmyrolytic formation of palygorskite from serpentinite

C. M. Destrigneville, A. M. Karpoff & D. Charpentier
EOST, Centre de Géochimie de la Surface, ULP-CNRS, Strasbourg, France

ABSTRACT: Uplifted serpentinized peridotites of Galicia Bank (NE Atlantic) were exposed to halmyrolysis since late Albian times and transformed into palygorskite and smectite. From thermodynamic modelling we deduce that palygorskite is an alteration product of serpentine at common sea-floor conditions. The observed prevalence of palygorskite over smectite involves a long term interaction which could have operated since the late Albian peridotites uplift. Mg is lost to sea water as a result of the replacement of serpentine by palygorskite.

1 SETTLING AND OBJECTIVE

The authigenesis of palygorskite as a result of marine alteration at low temperature of a mantellic material was first described by Karpoff et al. (1989). In the pelagic realm, the origin of palygorskite was previously related to detrital, diagenetic or hydrothermal processes.

In diagenetic sediments, the associated phases are Mg-smectites and zeolites. Hydrothermal Mg-silicates occur at mid-oceanic ridges and along transform faults within sedimentary layers containing smectites, clinoptilolite, even dolomite and oxyhydroxides (Bowles et al., 1971; Bonatti et al., 1983, among others). Invoked temperatures are low, below 70°C (McKenzie et al., 1990).

1.1 Site and sample

The process of the upwelling and outcropping of mantle bodies on seafloor is now established. On the west Galicia margin (NE Atlantic), at the boundary of the thinned continental and oceanic crusts, a ridge of peridotites was uplifted and serpentinized during the rifting (140 to 110 Ma) (references in Karpoff et al., 1989).

During the Galinaute cruise of the submersible *Nautile*, a site on the "Hill 5100" mount provided the *in situ* sampling of the altered facies of serpentinized peridotites. The last stage of serpentinization is expressed by a low-temperature (10°C) calcite infilling fractures (Féraud et al., 1988, and Agrinier et al., 1989).

The salmon colored matrix of the collected clayey sample contains centimetric, reddish to yellowish brown, patches of residual serpentinites surrounded by goethite and poorly-crystallized MnO_2. The bulk facies consists of palygorskite, smectite, Al-serpentine, scarce pyroxene and quartz. The later could form the tubes of burrowing organisms found at the surface. The <2μm fraction consists of palygorskite and dioctaedral smectite. SEM and TEM observations establish the prevalence of the authigenetic palygorskite. No genetic relationship between the palygorskite fibers and the flaky smectite particles was observed.

Chemical compositions (Table 1) show greater Fe, Mn, Ni and Cr contents in the bulk facies, related to the residual phases and oxides. The clay fraction has a composition of a marine palygorskite, with the Al/Al+Fe+Mg ratio of 0.41.

1.2 Objective

At the Galicia Bank site which has been exposed to sea water since Aptian times, the palygorskite is not sedimentary and so, neither detrital nor diagenetic. The facies is similar as those ascribed to hydrothermalism but differs by the lack of zeolites.

In fact, the last hydrothermal event recorded by the serpentinites is related to a low-temperature calcite.

Palygorskite or sepiolite formation is possible in oceanic conditions, but these minerals were always considered to be precipitated from fluids. The formation and stability of palygorskite require high silica and magnesium activities and alkaline pH.

Table 1: The chemical composition of the bulk rock sample (BR) and of the extracted <2μm fraction (FF) LOI: Loss On Ignition.

wt%	BR	FF	ppm	BR	FF
SiO_2	47.5	51.2	Mn	8136	4104
Al_2O_3	13.8	14.5	Ba	224	190
MgO	6.9	5.8	Sr	88	89
CaO	1.7	1.9	Ni	1438	918
Fe_2O_3	11.8	10.4	Co	124	59
Na_2O	2.5	0.3	Cr	1792	641
K_2O	3.3	3.5	Zn	288	252
LOI	10.8	9.9	Cu	167	128

The textural relationship between the bed rock and the studied clay is obvious, and the later could originate from the halmyrolysis of the serpentinites by a dissolution-reprecipitation process.

In order to establish the thermodynamic validity of such a process, a modelling of the water-rock reaction is now tested.

2 MODELLING

The alteration of serpentinite by sea water was modelled using the thermodynamic option of the KINDIS computer code (Madé et al., 1990).

2.1 Principles

The dissolution of the reactant is modelled by adding progressively its chemical constituents to 1 kg of solution. At each step, the thermodynamic equilibrium of the system is calculated by precipitating the secondary minerals reaching saturation. The evolution of the fluid chemistry and of the secondary mineral assemblage is given as a function of the mass of the destroyed reactant.

2.2 Conditions

Modelling was performed at constant temperature (25°C), in a closed system and with constant oxygen and carbon dioxide fugacities, in order to test the serpentinite alteration with a dominating oceanic influence. The initial fluid was standard sea water with a pH of 7 and alcalinity and oxygen solubility values corresponding to present deep North Atlantic Ocean conditions (Millero, 1974, and Chester, 1989). A representative chemical composition of serpentine was chosen from analyses of rock samples cored in the same site (Agrinier et al., 1988). Its structural formula is : $Si_{2.036}Mg_{2.148}Al_{0.318}Fe_{0.263}Ca_{0.032}O_5(OH)_4$.

Iron oxides, carbonates, silica and smectites were considered as possible alteration products. The representative end-members of tri- and di-octahedral smectites which were chosen in the available database comprised a Mg- and Fe-rich saponite, an Al- and Mg- rich montmorillonite, a beidellite and a nontronite.

Three compositions of palygorskites were chosen:
1. The pure Mg end-member.
2. Two compositions obtained from STEM analyses on the fine fraction from samples collected *in situ* and calculated by taking the iron ion either divalent or trivalent.

The thermodynamic constants of the palygorskites were calculated using Tardy and Garrels's method (1974). The structural formulae of palygorskites and their thermodynamic constants at 25°C are given in Table 2.

Table 2. Structural formulae of Mg- [1], Fe^{2+}- [2] and Fe^{3+}-palygorskite [3] and their log(K) at 25°C.

[1] $Si_8Mg_5O_{16}(OH)_{10}$	37.4
[2] $(Si_{6.39}Al_{1.61})(Mg_{2.49}Al_{1.61}Fe_{0.9})O_{16}(OH)_{10}$	-31.5
[3] $(Si_{5.74}Al_{2.26})(Mg_{2.74}Al_{1.27}Fe_{0.99})O_{16}(OH)_{10}$	-36.7

3 RESULTS

Calculations were performed where palygorskites are either allowed or prevented from precipitation. The calculations with palygorskites prevented were run in order to test their influence on the resulting secondary minerals distribution and nature, and fluid chemistry.

Only results with palygorskite precipitation allowed are illustrated.

3.1 Secondary minerals

The alteration of serpentine by sea water at 25°C leads to the precipitation of goethite (G), dolomite (Dol), quartz (Qtz), beidellite (Bd), saponite (Sp) and palygorskites (Pl) (Figure 1).

Both calculations give the same results as long as the fluid is undersaturated with respect to palygorskite (below 73 g of destroyed serpentine): goethite first precipitates followed by beidellite, and then quartz and then dolomite. As saponite reaches saturation, it becomes rapidly the major secondary phase prevailing over quartz and consuming Mg and limiting the dolomite precipitation.

The first palygorskite type to precipitate is the ferric one and its occurrence coincides with the beginning of the dissolution of beidellite (Table 3). The precipitation of Mg-palygorskite at 806 g of destroyed serpentine corresponds to a stabilization in the amount of saponite produced (Table 3) and to its decrease in the secondary assemblage (Figure 1).

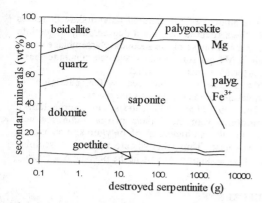

Figure 1: The evolution of the secondary assemblage: cumulative weight % of secondary minerals.

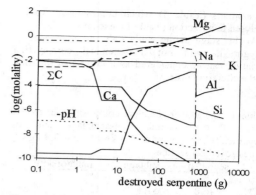

Figure 2: The evolution of the fluid chemistry.

Table 3: The mass of secondary minerals as a function of the mass of destroyed serpentine (DS).

DS	0.1	1.3	12.	106.	580.	806.	3950.
G	0.005	0.09	0.86	7.3	40.	56.	228.
Dol	0.04	0.86	2.	4.1	14.	19.	84.5
Qtz	0.02	0.37					
Bd	0.02	0.33	1.55				
Sp			7.3	70.2	384.	536.	536.
Pl				13.4	73.	119.	2544.
Tt	0.085	1.65	11.7	95.	511.	730.	3392.5

Tt: total produced.

The same trend for saponite was observed in runs without palygorskite which is then replaced by a 45% quartz and 65% beidellite assemblage.

3.2 The fluid

The evolution of the fluid chemistry is exactly the same for most major elements modelling the alteration either with or without palygorskite.

pH increases from 7 to 9.4 as cations are released in the fluid due to the serpentine dissolution.

As alteration proceeds, Ca, Na, Si and Fe concentrations decrease in the fluid when Al, C and Mg concentrations increase (Figure 2, except for Fe whose concentration lies below 10^{-17} mol/kgH$_2$O). The Ca concentration is mainly controlled by dolomite until saponite precipitates and takes over, controlling the Na concentration at the same time.

The formation of dioctahedral smectites, quartz and palygorskite control Al and Si concentrations. Their evolution in the fluid depends on the palygorskite composition and on the increasing proportion of palygorskite against saponite. When palygorskite is not allowed to precipitate, Si concentration is controlled by the quartz saturation and Al reaches a lower concentration in equilibrium with beidellite (10^{-8} mol/kgH$_2$O).

The Mg concentration and the alcalinity keep on increasing as serpentine undergo alteration despite the precipitation of magnesian carbonates and silicates. As the CO$_2$ fugacity remains constant, the thermodynamic equilibrium in the fluid leads to (MgCO$_3$)° as the prevalent aqueous species.

4 DISCUSSION

Thermodynamic modelling predicts the precipitation of iron oxides and carbonates as the minor phases and saponite and palygorskite as the major phases, and all these minerals were observed *in situ*.

Palygorskite precipitates when Al and Si concentrations are high enough in the fluid and when pH is sufficiently high (~8.5), which corresponds to an advanced serpentine alteration degree.

The precipitation of palygorskite does not seem to be correlated to that of saponite since the saponite stability in the secondary assemblage occurs whether palygorskite precipitates or not. Saponite consequently cannot be considered as a palygorskite precursor.

The resulting global chemical compositions of the secondary assemblage and of palygorskites alone (Table 4) show a good agreement with those analysed (Table 1) considering the limited choice in the chemical compositions of the palygorskites (9 to 23 of MgO wt%) and the two saponites (Mg- and Fe-rich available end-members) allowed to precipitate, as well as the possible contamination of analysed minerals by iron oxide, which could explain their high Fe content.

As alteration exceeds 1.3 kg of destroyed serpentine in 1 kg of sea water, the proportion of palygorskite increases and its global composition evolves to a higher Mg content and to lower Al and

Fe contents (Table 4) which comes closer to the analysed fine fraction (Table 1), except for Mg.

At this stage of the study only the abundance of palygorskite and not its chemical composition variation can be considered as a criteria of duration of the alteration process.

Table 4: The global chemical composition in oxides wt% of the secondary assemblage and of palygorskite for a mass of 1.3 kg (*) and 3.9 kg (°) of destroyed serpentine.

	secondary assemblage		palygorskite	
	*	°	*	°
SiO_2	46.8	51.4	43.9	51.2
Al_2O_3	14.6	11.5	21.0	12.4
MgO	21.7	21.9	16.5	20.5
CaO	0.	0.	0.	0.
Fe_2O_3	4.8	4.4	9.2	5.6
Na_2O	2.2	0.9	0.	0.
K_2O	0.	0.	0.	0.
H_2O	9.8	9.9	9.6	10.

For an easier comparison, composition are calculated with a 10wt% H_2O content.

The observation of the replacement of serpentine by a prevalent proportion of palygorskite over smectite involves a long term interaction which is consistent with the long exposition of uplifted serpentinites to sea water since their late Albian uplift.

However as long as thermodynamic modelling is performed and no kinetics are considered due to the lack of data on palygorskite, nothing about absolute time can be given.

Finally modelling suggests a Mg release in the fluid which accounts for the mass difference between destroyed serpentine and secondary minerals although modeled secondary minerals are richer in Mg than observed ones. Serpentine alteration can have a considering influence on the bulk Mg budget in the ocean. And depending on the surrounding geological formations, this Mg released as the ($MgCO_3$) aqueous species could be potentially involved in a dolomitization process (McKenzie et al., 1990).

5 CONCLUSIONS

Thermodynamic simulations indicate that palygorskite precipitates when serpentine undergoes alteration by sea water at low temperature.

Saponite does not appear as a precursor for palygorskite whose saturation mainly depends on the Al and Si high concentrations and on the high pH in the fluid.

The amount of precipitated palygorskite increases with the serpentine alteration degree. As palygorskite is the major phase of the secondary assemblage of the Galicia Margin serpentinites, we can conclude that a long-term interaction occurred.

As a result of the sea water alteration of serpentine, Mg is released to the ocean.

The *in situ* halmyrolysis of serpentinized peridotites from the Galicia margin defines a new way for the authigenesis of palygorskite in marine environment. Thus, this silicate as well as sepiolite may play an important role in the oceanic Mg cycle.

REFERENCES

Agrinier, P., C. Mevel & J. Girardeau 1988. Hydrothermal alteration of the peridotites cored at the ocean/continent boundary of the iberian margin: petrologic and stable isotope evidence. In G. Boillot, E.L. Winterer et al., *Proc. ODP Sci. Results* 103:225-233. College Station TX.

Bonatti, E., E. Craig Simmons, D. Breger, P.R. Hamlyn & J. Lawrence 1983. Ultramafic rock/seawater interaction in the oceanic crust: Mg-silicate (sepiolite) deposit from the Indian Ocean Floor. *Earth Planet. Sci. Lett.* 62:229-238.

Bowles, F.A., E.A. Angino, J.W. Hosterman & O.K. Galk 1971. Precipitation of deep-sea palygorskite and sepiolite. *Earth Planet. Sci. Lett.* 11:324-332.

Chester, R. 1989. *Marine Geochemistry.* London: Unwin Hyman Ltd.

Féraud, G., J. Girardeau, M.O. Beslier & G. Boillot 1988. Datation ^{39}Ar-^{40}Ar de la mise en place des péridotites bordant la marge de Galice (Espagne). *C. R. Acad. Sci.* 307:49-55.

Karpoff, A.M., Y. Lagabrielle, G. Boillot & J. Girardeau 1989. L'authigenèse océanique de palygorskite par halmyrolyse de péridotites serpentinisées (Marge de Galice): ses implications géodynamiques. *C. R. Acad. Sci. Paris* 308:647-654.

Madé, B., Clément, A. & B.Fritz 1990. Modélisation cinetique et thermodynamique de l'altération: le modèle géochimique KINDIS. *C. R. Acad. Sci. Paris* 310:31-36.

McKenzie, J.A., A. Isern, A.M. Karpoff & P.K. Swart 1990. Basal dolomitic sediments, Tyrrhenian Sea, ODP Leg 107. In K. Kastens, J. Mascle et al., *Proc. ODP Sci. Results* 107:141-152.

Millero, F.J. 1974. Sea water as a multicomponent electrolyte solution. In E.D.Goldberg (ed), *The Sea* 5:3-80. New-York: Wiley-Interscience.

Tardy, Y. & R.M. Garrels 1974. A method of estimating the Gibbs energies of formation of layer silicates. *Geochim. Cosmochim. Acta* 38:1101-1116.

The underwater eruption in the Academia Nauk caldera (Kamchatka) and its consequences

S.M. Fazlullin, S.V. Ushakov, R.A. Shuvalov, A.G. Nikolaeva & E.G. Lupikina
Institute of Volcanology, Far East Division, Russian Academy of Sciences, Petropavlovsk-Kamchatsky, Russia

Masahiro Aoki
Mineral and Fuel Resources Department, Geological Survey of Japan, Tsukuba, Japan

ABSTRACT: An underwater eruption in the Karymskoe fresh-water lake (in the Academia Nauk caldera) - was studied by scientists from the Institute of Volcanology. Geomorphologic, hydrological, hydrochemical and hydrobiological investigations were conducted. This publication is the first report of the general results of the investigation of this phenomenon.

1 INTRODUCTION

In early 1996 a simultaneous eruption from two volcanic craters, located six kilometers from one another (on Kamchatka), was observed. One of them was in the crater of Karymsky volcano, while the other was in the northern part of Karymskoe caldera lake. As a result of the eruption an accumulated volcanic structure was formed in the northern part of the lake.

Investigations permitted the determination of the characteristic peculiarities of the lake's post-eruption morphometry. A map of depth-contours in the lake and a new crater, formed in the lake's northern part, was constructed (Fig. 1). Major element composition was determined in the chemical Laboratory of the Institute of Volcanology. Trace element composition analysis of water samples was conducted under the leadership of Dr. Aoki of the Geological Survey of Japan.

2 DESCRIPTION OF THE SUBMARINE ERUPTION

The eruption was preceded by an increase in seismologic activity, beginning in April, 1995. On December 31, 1995 at 19:26 local time an earthquake with magnitude 5.8 occurred in Kronotsky Bay, which is located 50-60 km north-east of Karymsky volcano. On January 1, 1996 at 20:57 an earthquake of magnitude 5.2 was registered near Karymsky volcano and at 21:57 an earthquake of magnitude 6.9 was registered 25 km to the south of Karymsky volcano (Fedotov, 1996). Over the next few days an earthquake swarm with events of magnitude > 5.0 was recorded in this zone. Near midnight on January 2, 1996 an eruption began in the crater of Karymsky volcano, and after 14-15 hours an underwater eruption commenced in the northern part of Karymskoe lake (6 km south from the Karymsky volcano). According to available information (Fazlullin et.al., 1996, Karpov et.al., 1996a) the underwater eruption in the Karymskoe lake began on January 2, 1996 between 14:00 and 15:40. The surface of the lake was free of ice at midday on January 2, while one week earlier an ice layer had been observed.

Aerial observations were conducted at a distance of 5-7 km between 15:40 and 16:20. A gas-ash plume rose up to 3 km and spread to the south-east, in the direction of the Pacific ocean. Ash fall from the plume covered hundreds of square kilometers. Simultaneously, in the northern part of the Academia Nauk caldera, steam and gas mixed with a dark material (presumably, volcanic ash and lake sediment) rose in columns above the lake with a frequency averaging once every 5-6 minutes (Karpov et.al., 1996a). In the process of the volcanic expulsions a cupola-like cloud with a diameter of 200-300 m formed. Over the course of several seconds the cloud changed its form. A basic wave in the form of a taurus-shaped cloud spread to all sides from the eruptive center. Its top edge was approximately 100m high. At the same time, a white gas-steam plume extended upward 7-8 km from the eruptive center. Each series of underwater explosions was accompanied by the formation of waves on the surface of the lake, some of which reached several meters in height. The energy of one explosion, as calculated by Fedotov (1996) based on the maximum height of the gas-ash clouds, reached

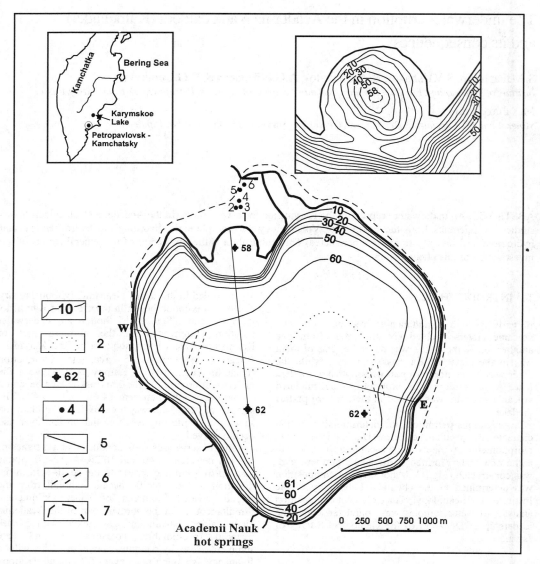

Fig. 1. Scheme of depth distribution in Karymsky lake and the new crater as shown by data from echo sounding profiling (constructed by S.V. Ushakov and Dr. Ya.D. Muravyev, September 1996). 1 - isobaths; 2 - 61 m isobath; 3 - points with maximum depths; 4 - small explosive vents at the surface of volcanic edifice; 5 - echo sounding profile; 6 - small grubens presumably related to fault zones; 7 - approximate lake boundary prior to the eruption;
Inset map at upper right shows details of the newly formed crater at the north of the lake.

10^{14} J. As a result of the eruption, a flood occurred in Karymsky River, which flows out of the lake. A significant part of the caldera of Karymsky volcano, adjacent to the river valley, was covered by a mud flow.

On January 3, explosive activity in the lake stopped. According to our calculations, the activity lasted a total of 12-14 hours. The entire surface of the lake steamed, and the steam reached a height of 800-1000m. In the northern part of the lake, a beach of erupted material was formed, creating a new shore line (Table 1). Plumes of fumarolic gas rose from several vents on the surface of this beach.

Table 1. Characteristics of the lake, based on observations of Sept., 1996

Length, m	4050
Maximum width, m	3750
Average depth, m	47
Maximum depth, m	61-62
Area of the water surface, km^2	9.8
Shore line, km	12.5
Volume of the lake, million. m^3	460.6
Volume of the new crater, million. m^3	5.1
Volume of the new volcanic structure in the north part of Karymskoe lake, million m^3	60.1
Angle of the lake sides, degree	12-15
Average angle of the crater, degree	12

Water level in the lake roused by several meters. The flow of the Karymsky river from the lake was blocked by a dam, formed from erupted material. Several tens of square kilometers of land around the lake were covered by mad, formed from lake water and eruption products. At a distance of several hundred meters from the center of the underwater eruption, impressions of various size, formed by falling bombs and other material, were noted in the snow (Karpov et.al., 1996b). The waves created by the eruption removed the snow cover on the shoreline, in places up to 15-20m above the lake surface. In many places along the shore, land slides and an absence of top soil were noted. The inner slope of the Academia Nauk caldera, adjacent to the zone of the eruption, was most effected. Tens of thousands of cubic meters of soil were eroded by water. The underwater eruption damaged plant cover over an area of several square kilometers, covering everything with a layer of mud. The underwater eruption was a catastrophe for the lake's biota. The enormous waves, along with the major changes in chemical composition and water temperature, lead to the death of practically all life in the lake.

3 AN ESTIMATE OF HEAT RELEASE

In order to calculate the amount of heat energy released during the course of the underwater eruption, we used the vertical temperature distribution and hydrochemical characteristics of the lake in early May, 1996. Persistent stratification existed in the lake, excluding the new, water-filled crater. An analysis of the vertical temperature distribution and composition in the lake showed that the water layer below the thermocline formed as a result of the January eruption, and has undergone little change since then. According to our calculations, the underwater eruption released more than 11.8×10^{15} J of heat over the first two days of the eruption.

4 CHANGES IN THE LAKE WATER COMPOSITION

As a result of the underwater eruption, the hydrochemical characteristics of Karymsky Lake changed drastically over the course of several hours (Table 2).

Table 2. Water composition in Karymsky Lake in various years (before (5-7) and after (1-4) the eruption) in mg/l

	1	2	3	4	5	6	7
H^+	0,58	0,6	0,62	0,64	--	--	--
NH_4^+	0,2	0,2	0,2	0,2	0,0	0,0	0,9
Na^+	65,3	64,4	73,0	64,6	14,2	10,4	14,0
K^+	6,6	6,7	6,6	6,6	1,7	1,6	0,84
Ca^{2+}	64,0	64,0	63,2	63,2	3,6	1,6	6,0
Mg^{2+}	18,9	19,3	18,4	18,4	3,16	0,5	0,6
Mn^{2+}	1,94	1,94	1,92	1,92	n.d.	n.d.	0,006
Fe^{3+}	6,1	6,1	6,1	6,1	n.d.	n.d.	0,002
Al^{3+}	4,9	4,9	4,9	4,9	n.d.	n.d	0,0002
Cl^-	40,5	39,0	38,3	37,6	12,0	8,5	21,3
SO_4^{2-}	374,0	374,0	374,0	374,0	5,4	3,8	4,8
HCO_3^-	--	--	--	--	34,2	35,1	22,0
F^-	1,7	1,76	1,74	1,84	n.d.	n.d.	0.08
H_3BO_3	2.42	2.29	2.00	2.17	1,2	--	<0,5
H_4SiO_4	190,0	194,0	192,0	192,0	57,6	45,0	48,0
pH	3,28	3,25	3,22	3,20	7,2	7,05	7,45

notes: water samples from the lake (1- surface, 2- bottom) and in the new crater (3- surface, 4- bottom) taken in May 1996, 5- water from the lake (09.10.85), 6 - (19.06.85), 7- (20.08.93);
In samples 1- 4 Fe^{2+} concentration was less than 0.5 mg/l, and in the other samples none was detected.
(n.d. - not detected, d.l. - detection limit)

The lake became one of the largest reservoirs of acid water in the world. The general salt content in the lake water increased significantly. We calculated the change in overall salt reserves in the lake, using data on water composition before and after the eruption (Table 3).

Table 3. Total reserve of chemical elements in the water of Karymsky Lake before and after the eruption. (in tons).

Year	Cl	SO₄	HCO₃	Na	K	Ca	Mg	Al	Fe	Mn	Zn	Cu
1993	11228	2530	11598	7380	442	3163	316	0.1	1.17	0.3	2.32	0.01
1996	24460	216730	none	40910	8350	38412	8381	3232	5629	992	32.6	26.8

6 CONCLUSION

The underwater eruption in the Academia Nauk Caldera Jan.2-3, 1996 was a unique event, seen by specialists for the first time. As a result of this natural event, the hydrological regime of the lake, along with its composition, changed completely. This can be classified as a natural ecological catastrophe. The results of the first year's investigations into the consequences of the eruption allow us to construct the long-term evolution process of the lake back to its original state.

7 ACKNOWLEDGEMENTS

We would like to think Ya.D. Muravyev, V.I. Andreev, I.A. Markov for taking regular samples, as well as monitoring the temperature and level in the lake.

Our work was supported by the Russian Government Committee on Science and Technology, the Russian Fund for Fundamental Research, and the INTAS fund (project 94-3129).

REFERENCES

Fazlullin S.M., Shuvalov R.A., Karpov G.A., Ushakov S.V. 1996. Subaqueous Eruption from Karymsky Lake (Kamchatka) and its Effects: *Crater Lakes, Terrestrial Degassing and Hyperacid Fluids in the Environment.* Chapman Conference. Abstracts. September 4-9, 1996 Crater Lake, Oregon. P. 15.

Fedotov, S.A. 1996. The simultaneous eruption of two volcanoes on Kamchatka in January 1996. *«Zemlya i vselennya»*. N3. P.60-65. (in Russian)

Karpov G.A., Muraviev Ya.D., Shuvalov R.A., Fazlullin S.M., Chebrov V.N. 1996a. More details about the early January eruption. *Bulletin of Global Volkanism Network.* 21: 9-10.

Karpov G.A., Muraviev Ya.D., Shuvalov R.A., Fazlullin S.M., Chebrov V.N. 1996b. A subaqueous eruption from the caldera of Akademii Nauk volcano on January 2-3 1996. *Newsletter of the IAVCEI Commission on Volcanic Lakes.* 9:14-17.

Halide systematics in sedimentary hydrothermal systems, Escanaba Trough – ODP Leg 169

J.M.Gieskes & C.Mahn
Scripps Institution of Oceanography, La Jolla, Calif., USA

R.James
Department of Geology, University of Bristol, UK

J.Ishibashi
Institute for Earthquake Chemistry, University of Tokyo, Japan

ABSTRACT: Interstitial water samples were obtained from two sites in the Escanaba Trough, off the coast of Northern California. Chloride concentrations in the drill site most affected by the presence of hydrothermal fluids, show values well above and below average sea water. These profiles can be understood best in terms of supercritical phase separation processes occurring at depth in the hydrothermal system at the base of the sediment column. Phase separation is associated with phase segregation in which low chloride fluids exit the hydrothermal zone through sandy horizons, and the fluids with elevated chloride emanate into the bottom waters through cracks and fissures. Although values of δ^3He indicate the importance of interactions with basalts, the increased values of the Br/Cl ratio, as well as the presence of iodide, indicate that pore fluid-sediment interactions also must have played an important role. In this paper we investigate the halide systematics of the pore fluids of both a low heat-flow reference site (Site 1037) and of a hydrothermal site (Site 1038).

INTRODUCTION

Studies of the halide systematics of hydrothermal fluids have revealed both high and low chloride contents, which usually has been considered to be the result of phase separation processes in the subsurface of the hydrothermal systems. The halide systematics have been the subject of a detailed enquiry by You et al. (1994), who concluded that constant Br/Cl ratios in mid-ocean hydrothermal systems (MOR) can be understood in terms of super- and/or subcritical phase separation processes taking place at depth. Deviations from the average sea water molar Br/Cl ratio of $1.54*10^{-3}$ are relatively small above a chloride content of 250 mM/kg, but below this value lower ratios may prevail (Oosting and Von Damm, 1995). In sedimented hydrothermal systems (SDR), however, much higher ratios of Br/Cl prevail, due to the release of bromide during the diagenesis of organic matter. In addition, in sedimentary hydrothermal systems the regeneration of iodide as a result of organic matter diagenesis will lead to increased values of iodide in the hydrothermal fluids (Campbell and Edmond, 1989; Magenheim and Gieskes, 1992; Campbell et al., 1994). SDR produce fluids that have signatures of hydrothermal interactions both with basalts and sediments.

In this paper we extend the data base on the systematics of Cl, Br, I, and F in sedimentary hydrothermal systems, by considering information on the recent drill sites in the Escanaba Trough off the coast of Northern California. These sites offer the opportunity to investigate both a reference site, affected by relatively low temperatures (Site 1037) and a site characterized by hydrothermal alteration processes (Site 1038).

DATA BASE

Data on the distribution of Cl, Br, and I for Site 1037 and Hole 1038I in the Escanaba Trough are presented in Figures 1 A and B. Chloride variations in Site 1037 are within 4 % of the sea water value and represent variations due both to variations in oceanic chloride contents with time and as a result of reactions or advective flow in the sediments. The concentrations of iodide and bromide show a complicated pattern, with the deposits below 300 mbsf being characterized by greater releases of I and Br (Klamath river derived deposits). Hole 1038I indicates a complex pattern in the distribution of Cl, I, and Br. The low chloride fluids in the sandy horizons between 90 and 250 mbsf (meters below sea floor) have been interpreted in terms of an advective flow of low chloride fluids through the sandy horizons. Iodide and bromide contents of these fluids are also relatively low. In the depth horizons around 50 mbsf and 350 mbsf, zones that are not thought to be affected by the hydrothermal processes at this drill site, Cl contents are similar to those of sea water, and I and Br show maxima as a result of in situ generation of these components from organic carbon decomposition. Just below a sill intrusion in 1038I, high chlorides do occur, which are due either to interaction with the basaltic sill (water uptake in alteration products) or are potentially due to advective input of high chloride fluids during the sill intrusion.

Data for fluoride concentrations for Site 1038 are presented in Figure 2. Whereas F concentrations drop to low values in most holes, large increases occur in the fluids of Holes 1038H and 1038G.

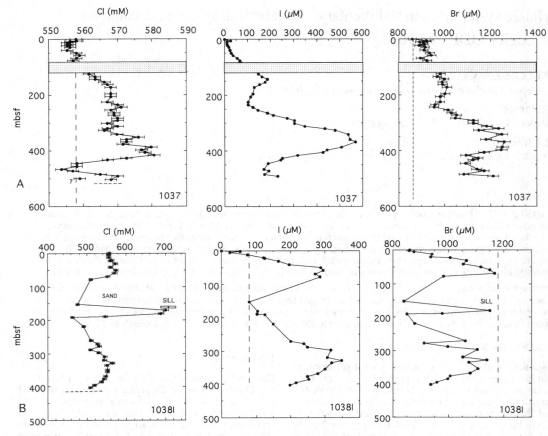

Figure 1 A: Chloride, iodide, and bromide concentration depth profiles at Site 1037 - Escanaba Trough Reference Site. Hatched areas are sandy horizons.

Figure 1 B: Chloride, iodide, and bromide concentration depth profiles in Hole 1038I - Escanaba Trough, Central Hill drillhole.

DISCUSSION

Of particular interest are the observations of both high and low chloride fluids in the pore fluids of the various holes drilled at Site 1038. Campbell et al. (1994) found that the chloride concentrations of the hydrothermal vents at the sea floor of Escanaba Trough had concentrations of about 670 mM, well above the bottom water value of about 556 mM. No low chloride fluids were detected in these vents. Data on Cl in several drill holes in the vicinity of the hydrothermal vents show the presence of fluids with chloride concentrations well above those of the vent fluids (up to 850 mM in Hole 1038G). On the other hand, low chloride fluids are observed in the sandy horizons of Holes 1038A, 1038H, and 1038I, reaching concentrations as low as 350 mM in Hole 1038A. This phenomenon is demonstrated in Figure 3. The most logical explanation for these observations is supercritical phase separation that formed both a low chloride phase and a high chloride phase during high temperature interaction of the fluids with basalts.

Observations of elevated values of the δ^3He of dissolved helium in the pore fluids (J. Ishibashi, pers.com.) clearly indicate that hydrothermal interaction with the basalts underlying the sediment column must have occurred. After phase separation, a process of phase segregation must have occurred resulting in the advective flow of high chloride fluids to the sea floor through cracks and fissures and the advection of the low chloride endmember through more diffuse flow via the sandy horizons at depth. This is the first time that such a process has

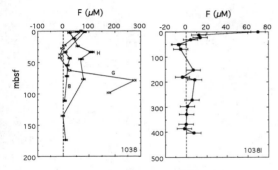

Figure 2. Fluoride concentrations in Sites 1037 and 1038, Escanaba Trough. Letters refer to separate drillholes.

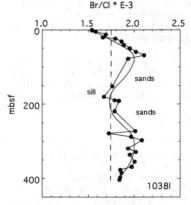

Figure 4. Br/Cl ratio Hole 1038I, Central Hill

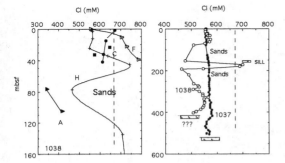

Figure 3. Chloride concentrations in Site 1038, Escanaba Trough. Note background values for hydrothermal vent fluids (dashed lines).

been demonstrated in a sedimented ridge hydrothermal system. This phase segregation process has been postulated previously by Fox (1992) in the subsurface of a basalt hosted hydrothermal system on the Juan de Fuca Ridge.

The hydrothermal fluids of the Escanaba Trough hydrothermal field have been shown to have Br/Cl ratios of ~ $1.76*10^{-3}$, i.e., elevated above the sea water value of $1.54*10^{-3}$ (Campbell et al., 1994). The Br/Cl ratios in many of the pore fluid samples exceed the vent ratio, probably because the hydrothermal fluids may flow more slowly through the sediments and may pick up additional Br from organic carbon diagenesis.

Of interest here is the distribution of Br/Cl in Hole 1038I (Figure 4). It is evident that in the zone of low chloride fluids the Br/Cl ratio is close to that of the vent fluids, but in zones above and below the sand layers values well above the vent fluid ratios are observed. Br/Cl ratios are quite similar to the ratios observed near the Br maximum at Site 1037, suggesting that at both sites diagenetic processes are similar eventhough there are large differences in temperatures.

Fluoride concentrations in basalt hosted hydrothermal systems are low, though not all fluoride appears to be removed (Von Damm et al., 1985; Oosting and Von Damm, personal communication). In Holes 1038B, the upper part of 1038G, and almost the entire section of 1038I, fluoride appears to be removed almost completely. However, in the deeper parts of Holes 1038G and 1038H, large increases in F occur, indicating mobilization of fluoride. No concurrent increases in phosphate have been detected suggesting that dissolution of fluoroapatite is not controlling F content. At this stage it is not certain what processes are involved in the fluoride mobilization.

CONCLUSIONS

The most important result from the study of the halide systematics of drill sites in the Escanaba Trough is the evidence for phase separation in the deeper part of the hydrothermal system. Helium isotopic data show that alteration of basalts containing mantle derived He is involved in these hydrothermal interactions. High chloride fluids seem to escape towards the sea floor through cracks and fissures, resulting in relatively rapid expulsion of these fluids. On the other hand, a phase segregated low chloride fluid escapes into deeper lying sandy horizons, as is evident particularly in the sandy horizons of Hole 1038I on the Central Hill of the Escanaba Trough hydrothermal system.

Data on the distribution of iodide and bromide indicate production from organic carbon diagenesis, leading to Br/Cl ratios well above those observed previously in hydrothermal vents in this area (Campbell et al., 1994). It is not certain whether the enrichments in bromide and iodide occur prior to the phase separation process. However, since there is no evidence for a change in the Br/Cl ratio during the phase separation process, increases in the Br/Cl ratio over those observed in the vent fluids (Campbell et al., 1994) are interpreted as a result of alteration of organic matter during the ascent of the hydrothermal fluids through the sediments.

Fluoride concentrations generally decrease with depth in the hydrothermal Site 1038, but there are local large increases. Processes responsible for these observations have as yet not been identified.

REFERENCES

Campbell, A.C. and Edmond, J.M. , Halide systematics of submarine hydrothermal vents. *Nature*, 342: 514-519, 1989

Campbell, A.C., German, C.R., Pamer, M.R., Gamo, T., and Edmond, J.M., Chemistry of Hydrothermal Fluids from Escanaba Trough, Gorda Ridge. In: Geologic, Hydrothermal, and Biological Studies at Escanaba Trough, Gorda Ridge, Offshore Northern California. *U.S. Geological Survey, Bulletin* 2022: 201-221

Fox, C.G., Consequences of phase separation on the distribution of hydrothermal fluids at Ashes Vent Field, Axial Volcano, Juan de Fuca Ridge, *J. Geophys. Res.*, 95: 12,923-12,960, 1990

Magenheim, A. and Gieskes, J.M., Hydrothermal discharge and alteration in near-surface sedimentsfrom the Guaymas Basin, Gulf of California. *Geochim. Cosmichim. Acta*, 56: 2329-2338

Oosting, S.E. and Von Damm, K.L., Bromide/chloride fractionation in seafloor hydrothermal fluids from 9-10°N East Pacific Rise. *Earth and Planetary Science Letters*, 144: 133-145

Von Damm,K.L., Edmond, J.M., Grant, B., Measures, C.I., Walden, B., and Weiss, R.F., Chemistry of submarine hydrothermal solutions at 21°N, East Pacific Rise. *Geochim. Cosmochim. Acta*, 49: 2197-2220, 1985

You, C.-F., Butterfield, D.A., Spivack, A.J., Gieskes, J.M., Gamo, T., and Campbell, A.J., Boron and halide systmeatics in submarine hydrothermal systyems: Effects of phase separation and sedimentary contributions. *Earth and Planetary Sci. Letters*, 123: 227-238

Helium and carbon isotopes in submarine gases from the Aeolian arc (Southern Italy)

S. Inguaggiato & F. Italiano
Istituto di Geochimica dei Fluidi, CNR, Palermo, Italy

ABSTRACT: Hydrothermal fluids sampled from submarine exhalations along the Aeolian arc are enriched in magmatic components. The helium isotopic ratios, in the range of 2-6 Ra, exibit values lower than other volcanic arcs. Both the He and C isotopes, as well as the chemical composition, indicate that reactions between hydrothermal fluids and CO_2 - rich gases control the final chemical and isotopic composition of the fluids. Helium is sometimes enriched in the gas phase and its isotopic composition is mainly modified by the addition of shallow fluids. Fractionation accompanying chemical equilibria decreasing the $\delta^{13}C$ values of the hydrothermal fluids by some δ units.

1 INTRODUCTION

The volcanic arc of the Aeolian islands consists of seven volcanic islands (Vulcano, Lipari, Salina, Panarea, Stromboli, Filicudi and Alicudi; fig. 1) and some sea-mounts developed during the quaternary age. Two islands are still active volcanoes: Vulcano, which is in a state of solphataric activity after the last eruption (1888-1890) and Stromboli which historicallly has erupts regular low-energy eruptions. The rest of

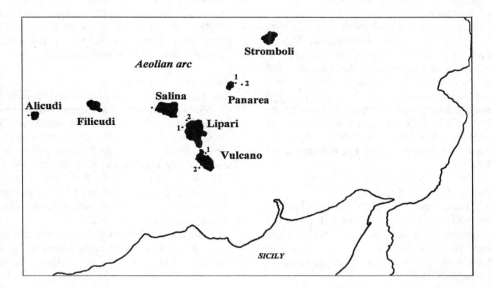

Figure 1 - Sketch map of the Aeolian arc (13 miles off the Northern coasts of Sicily, Southern Italy). The black spots show the sampling sites. Two samples from subaerial vents have been taken at the craters of Stromboli and Vulcano as references.

the islands are in waning state showing only traces of the old volcanic activity. Weak exhalative activity occurs at Lipari and Panarea islands, but no in-land degassing manifestations are recognizable on the rest of the islands.
Nevertheless, the volcanism is considered as a typical arc-volcanism (Barberi et al., 1974), however the helium and carbon isotopic signatures differ from those from other arcs in the world.
Recent investigations carried on by research vessels (NR Urania) documented the existence of hydrothermal exhalations off the coasts of all the islands (Italiano and Nuccio, 1990; Francofonte et al., 1996). Hydrothermal fluids venting at shallow depths have been investigated by scubadiving, but deep exhalations were also identified to depths of about 800m. Gas samples were taken by diving directly at the emission points to avoid gas bubbling into the sea-water and possible fractionation phenomena during the sample collection.
The chemical composition has been determined by gas-chromatography (Perkin Elmer 8500),while the carbon and helium isotopic ratios were measured by mass-spectometry (Finnigan Delta-S for carbon and VG5400 split flight tube for helium)

2 RESULTS AND DISCUSSION
Gas composition
All the sampled gases are dominated by CO_2 with O_2/N_2 ratios well below atmospheric. H_2S is present with concentrations up to 5%, H_2, CO and CH_4 concentrations are in the range of 1-100 ppm vol. He is always present with concentrations ranging from 5 to 20 ppm. The chemical composition is interpreted to result from reequilibration of the gas phase in hydrothermal aquifers at temperatures as high as 150-300°C (Italiano and Nuccio, 1991; Francofonte et al., 1996).

Helium and Carbon isotopes
The results of the He and C isotopic determinations are listed in Table 1. The isotopic composition of helium in the sampled gases ranges between about 2 and 6.4 Ra; the carbon isotopic composition shows values between -0.8 and -4‰ vs PDB.
The commonly accepted opinion that high helium isotopic ratios are from active volcanism and low ratios are from waning volcanoes seems to be controversial at the Aeolian Islands arc. The $^3He/^4He$ ratios from Stromboli volcano (tab. 1) are quite low, despite the fact that it is an active edifice with frequent emission of magma ejecta, while the highest R/Ra ratio was recorded at Alicudi where no in-land exhalative activity exists All the carbon isotopic ratios are more positive with respect to those reported for volcanic gases (Deines, 1970; Taylor et al., 1967) in agreement with the helium isotope results.

Magmatic helium contribution
Poreda and Craig (1989) showed that more than 75% of helium is mantle-derived in most arc volcanoes, implying that the mantle wedge is the main source of helium and only about 25% of helium is derived from sources, such as the subducting slab. Although some of the Aeolian volcanoes are in a waning stage (i.e.-Alicudi, Filicudi, Salina), and the youngest products of Alicudi are older than 34 Ka, their fluids show helium isotope ratios comparable or higher than those of the active volcano at Vulcano (tab. 1). Slightly lower values are displayed by fluids sampled at Lipari, which volcanic activity lasted up to 580 A.D.
The recorded relatively low helium isotope ratios are consistent with either the derivation from a mantle source anomalously enriched by a crustal component, or the assimilation of crustal material by mantle-derived magma during ascent and eruption. Similar hypotheses were formulated by Hilton et al. (1992) to explain the low helium isotope ratio in the east Sunda arc/Banda arcs.

Shallow depth interactions
The high helium concentration (Francofonte et al., 1996) can be the consequence of interactions between CO_2-rich fluids and shallow CO_2 undersaturated fluids. The helium enrichment in the gas phase is related to the large difference in the solubility coefficients: 8.75 ml/l for He and 759 ml/l for CO_2 (T=25°C, P = 1 atm; D'Amore and Truesdell, 1988). The pH of the water table can also affect the concentration of helium due to the influence on CO_2 solubility.
The interaction of hydrothermal systems at depth with the shallow water tables beneath the sea bottom, could result in carbon isotope fractionation during due to degassing as the fluid rise.
The graph of Figure 2 shows the relationship between the He to Ne ratios and the $\delta^{13}C$ values.

Table 1 - Results of the sampled gases from Aeolian islands. The helium isotopic values are expressed as R/Ra (Ra = atmospheric $^3He/^4He$ ratio = 1.4×10^{-6}) The three components percentages (A = atmospheric; C = crustal; M = magmatic) are calculated following the method proposed by Sano and Wakita (1985). The $\delta^{13}C$ (CO_2) are expressed in ‰ vs PDB.

Sample	Date	$\delta^{13}C$	He/Ne	R/Ra	%A	%C	%M
Panarea 1	Aug. 1996	-0.82	12.2	2.1	3	67	30
Panarea 2	Aug. 1996	-0.79	37.5	2.2	1	67	32
Vulcano 1	Aug. 1996	-3.05	4.7	2.3	7	62	31
Vulcano 2	Aug. 1996	-2.41	8.1	2.1	4	67	29
Vulcano 3	Aug. 1996	-1.76	13.4	2.1	4	67	29
Vulcano crater	Sep. 1996	0.0	130.2	5.2	1	33	66
Salina	Aug. 1996	-2.21	7.2	3.0	5	53	42
Lipari 1	Aug. 1996	-3.0	3.6	2.6	10	56	34
Lipari 2	Aug. 1996	-4.0	0.7	1.8	44	43	13
Stromboli crater	June 1993	-2.0	2.6	3.2	13	53	34
Alicudi	June 1993	-3.67	36.1	6.4	2	18	80
Filicudi	June 1993	n.d	31.0	4.9	2	37	61

The sample taken at the active crater of Vulcano island is plotted as reference. Its $\delta^{13}C$ value, which ranges around 0‰, has been interpreted as characteristic of the feeding source of the high-temperature volcanic gases (Capasso et al., 1997). This value is considered as representative of the deep CO_2 of Vulcano island, and it can be reasonably taken as the marker of all the deep, magmatic, gases of the Aeolian arc. The trend shows that lower He/Ne ratio, correlate with lower $\delta^{13}C$ values. Because the He/Ne ratio in volcanic gases will be lowered by addiction of atmospheric neon (1000 is considered as a typical ratio for volcanic gases, while the atmospheric one is 0.318), the observed trend is here interpreted as the consequence of increased interactions with shallow fluids that lowers the He/Ne ratio as the consequence of addiction of shallow, atmospheric, components, and decrease the $\delta^{13}C$ values because of kinetic carbon fractionations during the gas uprising through shallow hydrothermal aquifers by several δ per mil as proposed by Capasso et al. (1997).
The Alicudi sample falls off the trend, a possible consequence of a higher fractionation under low-CO_2 flux (because of the low fluxes of the submarine exhalations), or of a different mantle source beneath the island.

3 CONCLUSIONS

The carbon and helium isotopic compositions of some submarine gases collected at the Aeolian islands can be considered anomalous with respect of the typical values recorded at other volcanic arcs around the world.
All the sampled gases show the presence of a magmatic component up to 80%, but their helium

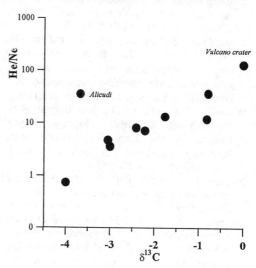

Figure 2 - He/Ne vs $\delta^{13}C$ graph.

isotopic ratios (expressed as R/Ra values) are lower than those reported for other active or waning volcanoes. The $\delta^{13}C$ values range from 0.8 to -4.0‰, are hevier than those measured in other volcanic gases. The results are consistent with the existence of a mantle source enriched in shallow crustal components beneath the Aeolian arc. Interactions of the uprising magmatic gases with shallow fluids can add atmospheric compounds and decrease the $\delta^{13}C$ values because of kinetic carbon fractionation.

4 REFERENCES

Barberi F., Civetta L., Gasparini F., Innocenti F. and Scandone R., 1974. Evolution of a section of the Africa-Europe plate bondary: paleomanetic and volcanological evidence from Sicily. *Eart and Planet. Sci Lett.*, 22, 123-133

Capasso G., Favara R., Inguaggiato S., 1997. Chemical features and isotopic composition of gaseous manifestations on Vulcano island (Aeolian islands, Italy): an interpretative model of fluid circulation. *Geoch. Cosmoch. Acta, 61*, 3425-3440

D'Amore F. and Truesdell A.H., 1988. A review of equilibrium constants for gaseous species of geothermal interest. *Sci. Géol. Bull.* 41, 309-332.

Deines P., 1970. The carbn and oxygen isotopic composition of carbonates from the Oka carbonatite complex, Quebec, Canada. *Geoch. Cosmoch. Acta, 34*, 1199-1225

Francofonte S., Inguaggiato S., Italiano F., Nuccio P.M., 1996. Ricerca di sistemi magmatici attivi delle isole Eolie tramite la geochimica delle esalazioni sottomarine. *XII congr AIOL*, Vulcano, 19-21 Settembre 1996.

Hilton D., Hoogewerff J.A., Van Bergen M.J., Hamershmidt K., 1992. Mapping magma sources in the East Sunda-Banda arcs, Indonesia: constrains for helium isotopes. *Geoch. Cosm. Acta*, 56, 851-859

Italiano F., Nuccio P.M., 1990. Submarine emissions at the Aeolian arc, *GNV-CNR Ann. Scient. Meeting*, Ravenna, December 1990

Italiano F. and Nuccio P.M., 1991. Geochemical investigations of submarine volcanic exhalations to the East of Panarea, Aeolian Islands, Italy. *J. Volcanol. Geotherm. Res.* 46, 125-141

Poreda R.J. and Craig H., 1989. Helium isotope ratios in circum-Pacific volcanic arcs. *Nature*, 338, 473-477.

Sano Y. and Wakita H., 1985. Geographical distribution of the $^3He/^4He$ ratios in Japan: Implications for arc tectonics and incipient magmatism. *Jour. Geoph. Res.*, 90, B10, 8729-8741

Taylor H.P., Frechen J. and Degens E.T., 1967. Oxygen and carbon isotope studies of carbonatites from the Laacher See district, West Germany and the Alno district, Sweden. *Geoch. Cosm. Acta.* 31, 407-430

Fluid chemistry of sediment-rich hydrothermal systems on the continental margin and the mid-oceanic ridge

J. Ishibashi – *Faculty of Science, University of Tokyo, Japan*
U. Tsunogai – *Institute for Hydrospheric-Atmospheric Sciences, Nagoya University, Japan*
T. Gamo – *Ocean Research Institute, University of Tokyo, Japan*
H. Chiba – *Faculty of Science, Kyushu University, Japan*

ABSTRACT: From data compilation, we discuss possible factors controlling chemical composition of high temperature fluid venting from sediment-rich hydrothermal systems. Organic species derived from fluid-sediment interaction provide a strong buffer compared with the rock-buffer in sediment-poor hydrothermal systems. Relationship among major cations and Cl of the sediment-rich fluids suggests enhanced leaching. These factors would act as favorable factors for hydrothermal mineralization in sediment-rich environment.

1 INTRODUCTION

Because the chemical composition of high temperature fluid of seafloor hydrothermal systems is quite different from that of recharging seawater, many geochemical studies have attempted to elucidate the factors that modify and control fluid composition. In the case of sediment-poor hydrothermal systems on the mid-oceanic ridge (MOR), chemical equilibria between hydrothermal fluid and some alteration mineral assemblages (so called rock-buffer) and phase separation have been invoked to explain the wide range of chemical composition of the fluids venting from seafloor (e.g. Von Damm, 1995).

Hydrothermal fluid from sediment-rich systems has distinct composition compared with the MOR fluids, as a result of contribution from fluid-sediment interaction. In this report, we compile and discuss geochemical data on sediment-rich hydrothermal systems (Table 1). Three active hydrothermal sites in the Mid-Okinawa Trough backarc basin on the continental margin provide an important data set that allows comparing of fluid composition in different tectonic setting. The pore fluid studies conducted during ODP Leg139 & 169 provide new insights into fluid interaction beneath the seafloor.

Table 1 Hydrothermal systems in sediment-rich environment

Area	Site	depth	tectonic setting	fluid temp.	Reference for fluid chemistry
Guaymas Basin	South Field	2000m	ridge	308°C	Campbell et al. (1988)
	East Hill	2000m	ridge	288°C	Campbell et al. (1988)
Middle Valley	Dead Dog	2420m	ridge	276°C	Butterfield et al. (1994b)
	ODP Mound	2460m	ridge	265°C	Butterfield et al. (1994b)
Escanaba Trough	Central Hill	3250m	ridge	217°C	Campbell et al. (1994)
Okinawa Trough	Izena Cauldron	1340m	backarc rifting	320°C	Sakai et al. (1990)
	Minami-Ensei	710m	backarc rifting	278°C	Chiba et al. (1993)
	Iheya North	1000m	backarc rifting	238°C	Chiba et al. (1996)
Endeavour segment	Endeavour	2200m	ridge*	353°C	Butterfield et al. (1994a)

* sediment is not obvious on the seafloor, but fluid chemistry shows evidence for interaction with sediment

2 ORGANIC-DERIVED SPECIES

Hydrothermal fluid from sediment-rich system is enriched in organic-derived species, such as NH_3, CO_2, CH_4, B, halogens, and alkalinity, which are derived from thermogenic degradation of organic matter in sediments. Involvement of these species during the recharge stage of fluid circulation is suggested by composition of the borehole samples collected from Hole858G during ODP Leg169, because the fluid comes directly from basaltic basement below the borehole casing which isolates the sediment (Shipboard Scientific Party, in press).

The addition of organic-derived species causes fluid pH and alkalinity to be higher. Dissociation of ammonium ion (NH_3-NH_4^+) has a strong buffer capacity under high temperature fluid conditions. Calcite dissolution equilibrium should be an another important factor that controls fluid pH. Calcite precipitation induces dissociation of carbonic acid in high temperature fluid according to

$$Ca^{2+} + H_2CO_3 = CaCO_3 + 2H^+,$$

which results in lower pH.

A combination of the ammonium dissociation and calcite dissolution could keep fluid pH in narrow range. Moreover, this combination controls pH of fluid even at lower temperature, because ion dissociation in solution is quick reaction. This is in contract to the case of sediment-poor system where a combination of mineral dissolution equilibria controls fluid pH. This pH control may be an important factor for occurrence of massive hydrothermal mineralization in sediment-rich environment.

3 MAJOR ELEMENT COMPOSITION

Fig. 1 illustrates the major element composition of hydrothermal fluids collected from the sediment-rich hydrothermal systems. Plots of the MOR fluids exhibit a simple relationship both in K vs Cl, and Ca vs Cl diagrams. This is one of evidence for control of their fluid chemistry by the same factors.

The fluids from sediment-rich systems show deviation from the MOR relationship. Guaymas, Escanaba, and Okinawa show enrichment in K, while Middle Valley and Endeavour show enrichment in Ca. It is interesting to note that no fluids show enrichment both in K and Ca and that all the sediment-rich fluids show depletion in Na to compensate charge balance.

It is assumed that major cations (K and Ca) are derived by leaching of primary minerals during recharge stage of fluid circulation, these data imply enhanced leaching in sediment-rich system. The complete depletion of K observed in altered sediment from the Dead Dog hydrothermal field (Goodfellow & Peter, 1994) would support this idea. The high Ca concentration of the Dead Dog fluid could be attributed to Ca leaching after complete K uptake. High Water/Rock ratio during fluid-sediment interaction, and/or high acidity caused by addition of organic-derived CO_2 are possible factors which contribute to this enhanced leaching.

Although it is arguable, metal elements in hydrothermal fluid are assumed to be accumulated during fluid circulation by leaching of primary minerals. The enrichment in cations in fluid from sediment-rich systems implies enhanced metal elements accumulation in the hydrothermal fluid.

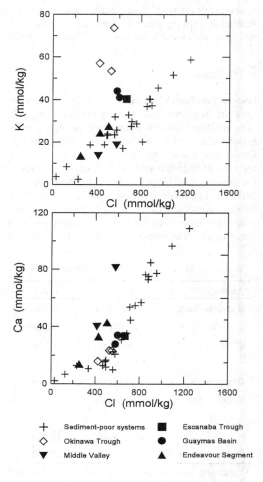

Fig. 1 Relationship of K vs Cl and Ca vs Cl of hydrothermal fluids. Data sources are written in Table 1.

4 REFERENCES

Butterfield, D.A., R. E. McDuff, M. J. Mottl, M.D. Lilley, J. E. Lupton & G. J. Massoth 1994a. Gradients in the composition of hydrothermal fluids from the Endeavour segment vent field: Phase separation and brine loss. *J. Geophys. Res.* 99: 9561-9583.

Butterfield, D.A., R. E. McDuff, J. Franklin & C. G. Wheat 1994b. Geochemistry of hydrothermal vent fluids from Middle Valley, Juan de Fuca Ridge. In M. J. Mottl, E. E. Davis, A. T. Fisher & J. F. Slack (eds), *Proc. ODP, Sci. Results*, 139: 395-410. College Station, TX: Ocean Drilling Program.

Campbell, A. C., T. S. Bowers, C. I. Measures, K. K. Falkner, M. Khadem & J. M. Edmond 1988. A time series of vent fluid compositions from 21°N, East Pacific Rise (1979, 1981, 1985), and the Guaymas Basin, Gulf of California (1982, 1985). *J. Geophys. Res.* 93: 4537-4549.

Campbell, A.C., C. R. German, M. R. Palmer, T. Gamo & J. M. Edmond 1994. Chemistry of hydrothermal fluids from Escanaba Trough, Gorda Ridge. In J. L. Morton, R. A. Zierenberg & C. A. Reiss (eds), *Geologic, Hydrothermal, and Biologic Studies at Escanaba Trough, Gorda Ridge, Offshore Northern California*: 201-221. U.S. Geol. Surv. Bull. vol. 2022.

Chiba, H., K. Nakashima, T. Gamo, J. Ishibashi, T. Tsunogai & H. Sakai 1993. Hydrothermal activity at the Minami-Ensei knoll, Okinawa Trough: Chemical characteristics of hydrothermal solutions. *Proc. JAMSTEC Symp. Deep Sea Res.* 9: 271-282.

Chiba, H., J. Ishibashi, H. Ueno, T. Oomori, N. Uchiyama, T. Takeda, C. Takamine, J. Ri, & A. Itomitsu 1996. Seafloor hydrothermal systems at north knoll, Iheya Ridge, Okinawa Trough. *JAMSTEC Deep Sea Res.* 12: 211-219.

Goodfellow, W. & J. M. Peter 1994. Geochemistry of hydrothermally altered sediment, Middle Valley, Northern Juan de Fuca Ridge. In M. J. Mottl, E. E. Davis, A. T. Fisher & J. F. Slack (eds), *Proc. ODP, Sci. Results*, 139: 207-289. College Station, TX: Ocean Drilling Program.

Sakai, H., T. Gamo, E.-S. Kim, K. Shitashima, F. Yanagisawa, M. Tsutsumi, J. Ishibashi, Y. Sano, H. Wakita, T. Tanaka, M. Matsumoto, T. Naganuma & K. Mitsuzawa 1990. Unique chemistry of the hydrothermal solution in the mid-Okinawa Trough backarc basin. *Geophys. Res. Lett.* 17: 2133-2136.

Shipboard scientific party (in press) In Y. Fouquet, R.A. Zierenberg, D.J. Miller and others (eds) *Proc. ODP, Init. Repts.* 169: College Station, TX: Ocean Drilling Program.

Von Damm, K. L. 1995 Controls on the chemistry and temporal variability of seafloor hydrothermal fluids. In (J. Lupton, L. Mullineaux & R. Zierenberg (eds), *Physical, Chemical, Biological, and Geological Interactions within Hydrothermal Systems:* 222-247. Washington DC: Monogr. v.91 Am. Geophys. Un.

Fluid chemistry of seafloor magmatic hydrothermal system in the Manus Basin, PNG*

J. Ishibashi & H. Takahashi – *Faculty of Science, University of Tokyo, Japan*
T. Gamo & K. Okamura – *Ocean Research Institute, University of Tokyo, Japan*
T. Yamanaka & H. Chiba – *Faculty of Science, Kyushu University, Japan*
J.-L. Charlou – *IFREMER, Centre de Brest, France*
K. Shitashima – *Central Research Institute of Electric Power Industry, Abiko, Japan*

ABSTRACT: The hydrothermal system located at a depth of 1950m of the DESMOS submarine caldera in the Manus backarc basin, is interpreted to be an analogue of high-sulfidation epithermal systems. Venting fluid from a white smoker has a temperature around 100°C and shows atypical chemistry for seafloor hydrothermal systems (pH <2, alkalinity <-9.1mM, SO_4 >Seawater value). This unique fluid chemistry is interpreted to result from direct involvement of magmatic fluid during seawater circulation.

1 INTRODUCTION

Hydrothermal activities have been widely discovered in the western Pacific region as well as in the mid-oceanic ridge (MOR). They occur in association with submarine magmatic activities in arc-backarc systems and have rather shallow seawater circulation cells caused by near-surface magmatic body (Ishibashi and Urabe, 1995). This is contrast to hydrothermal systems in the MOR, rather is similar to geothermal systems on arc islands. Because seawater evolves to hydrothermal fluid during hydrothermal circulation, fluids from arc-backarc systems are expected to be more influenced by magmatic contributions (Ishibashi and Urabe, 1995).

In hydrothermal system located on the northwest slope (depth = 1950m) of the DESMOS submarine caldera in the East Manus backarc basin (Fig. 1), milky venting fluid with sulfur-rich particles was observed (Gamo et al., 1997). We report fluid composition and discuss its similarities to "high sulfidation epithermal systems", which are closely associated with magmatic activity.

2 SUBMERSIBLE SURVEYS

Surveys and fluid sampling were conducted during two dive missions using manned submersibles belonging to JAMSTEC (Japan Marine Science and Technology Center); ManusFlux cruise using SHINKAI 6500 in 1995 (Auzende et al., 1996) and Bioaccess cruise using SHINKAI 2000 in 1996

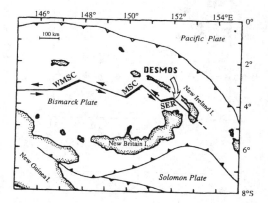

Fig.1 Tectonic map of the Manus backarc basin, indicating location of DESMOS site (Gamo et al., 1997)

(Auzende et al., in press; Ohta et al., in press). Fluid samples were collected using the ORI-type pump sampler (Ishibashi et al, 1994).

3 FLUID CHEMISTRY

In Fig. 2, major element composition of the collected samples is illustrated as relationship between SiO_2 concentration which was determined by molybdate colorimetry. Because Mg concentration does not show large variation, SiO_2 concentration is chosen as the hydrothermal component indicator. The sample of 3K#308 was ambient seawater collected outside of the

* Shipboard scientific party of Bioaccess cruise

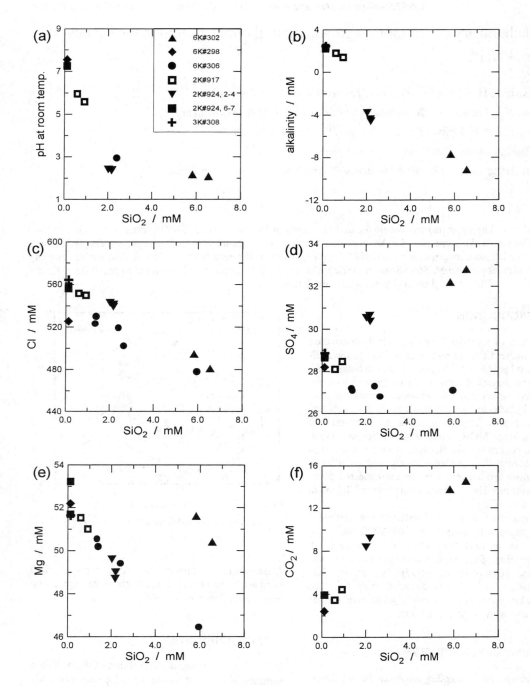

Fig. 2 Relationships between SiO_2 and (a) pH, (b) alkalinity, (c) Cl, (d) SO_4 (e) Mg, and (f) CO_2 of collected samples. The samples of 6K#302 & 2K#924,2-4 were collected from a white milky vent at the center of the hydrothermal field. The sample of 3K#308 was ambient seawater collected from outside of the hydrothermal field.

hydrothermal field.

Significant deviation from ambient seawater is obvious in the concentration of some species, although fluid temperature is only around 100°C. Extremely low pH (<2.1) and alkalinity (<-9.1mM) is notable. Gas species such as CO_2 (up to 14.5mM), H_2S (up to 8.9mM) show significant enrichment, while major elements such as Mg and Cl do not show large variations in concentration.

These data suggest that involvement of a magmatic fluid enriched in volatile species causes modification of fluid chemistry (Gamo et al., 1997), while high-temperature fluid-rock interaction does not appear to have a major effect. Concentration of Cl lower than seawater value could be explained by involvement of magmatic water.

Another strong indication of direct involvement of magmatic fluid is high SO_4 concentration (up to 32.8mM) detected in the fluid venting from a white milky smoker at the center of the hydrothermal field. Geochemical studies of high sulfidation epithermal system (e.g. Giggenbach, 1992) have revealed that magmatic fluid is responsible for acidic, sulfur-rich fluid. High temperature magmatic fluid contains SO_2 as the major S species, which disproportionates according to

$$4H_2O + 4SO_2 = 3H_2SO_4 + H_2S$$

when involved in seawater circulation. The resultant sulfuric acid causes low pH and low alkalinity. Isotope data on sulfur species support this reaction in DESMOS field. The $\delta^{34}S(H_2S)$ value of -5.5 to -5.7 ‰ and the $\delta^{34}S(SO_4)$ value of +19.8 ‰ can be interpreted as the product by isotope fractionation during the disproportionation at temperature of 250°C (Gamo et al., 1997).

Volatile composition of the magmatic fluid (e.g. total-C/total-S ratio) can be estimated from the vent fluid data, and is in the range of volcanic gases from island arcs. Isotopic composition of helium ($R/R_A = 8$) and $CO_2/^3He$ ratio of the DESMOS fluid are also in the range of island arc. Moreover, the fluid from the PACMANUS hydrothermal field, 20km west of the DESMOS field, show very close values in these ratios. These data indicate that the volatile composition of a magma beneath the DESMOS field is not atypical, and cannot be responsible for the unique fluid chemistry.

High sulfidation signature is considered as an indication of magmatic fluid involvement into water circulation system at high temperature, because major S specie in a magmatic fluid shifts from SO_2 to H_2S as temperature decreases (Giggenbach, 1992). Therefore, the unique fluid chemistry of the DESMOS field would represent vigorous magmatic activity associated with this hydrothermal system.

4 REFERENCES

Auzende, J.-M., Urabe, T., & Scientific Party 1996. Hydrothermal vents of the Manus Basin explored by Shinkai6500. *EOS (Trans. Am. Geophys. Union)* 77: 244.

Auzende, J.-M., Hashimoto J., Fiala-Medioni, A., Ohta, S. & Scientific Party (in press). In situ geological and biological study of two hydrothermal zones in the Manus Basin (Papua New Guinea). *Comptes Rendus Acad. Sc. Paris* in press.

Gamo, T., Okamura, K., Charlou, J.-L., Auzende, J.-M., Ishibashi, J., Shitashima, K., Chiba, H. & Shipboard Scientific Party 1997. Acidic and sulfate-rich hydro-thermal fluid from the Manus back-arc basin, Papua New Guinea. *Geology* 25: 139-142.

Giggenbach, W. F. 1992 Magma degassing and mineral deposition in hydrothermal systems along convergent plate boundaries. *Econ. Geol.* 87: 1927-1944.

Ishibashi, J., Grimaud, D., Nojiri, Y., Auzende, J.-M. & Urabe, T. 1994. Fluctuation of chemical compositions of the phase-separated hydrothermal fluid from the North Fiji Basin Ridge. *Mar. Geol.* 116: 215-226.

Ishibashi, J. & T. Urabe 1995. Hydrothermal activity related to arc-backarc magmatism in the Western Pacific. In B. Taylor (ed), Backarc Basins: Tectonics and Magmatism: 451-495. New York: Plenum.

Ohta, S., Hashimoto, J., & Participants of BIOACCESS-Manus96 Cruise. 1997. Hydrothermal vent fields and vent-associated biological communities in the DESMOS Cauldron, eastern Manus Basin. *JAMSTEC J. Deep Sea Res.* 13: in press (in Japanese with English abstract).

Alkali element and B geochemistry of sedimented hydrothermal systems

Rachael H. James & Martin R. Palmer
Department of Geology, Bristol University, UK

ABSTRACT: A series of porefluid samples have been collected during ODP Leg 169 from within the sedimentary sequence which hosts a massive sulphide deposit at the Escanaba Trough, Gorda Ridge. The Cl content of the fluids ranges from 300 to 800 mM. The only way to explain this variation is that the porefluids contain a hydrothermal component which has undergone phase separation at depth. The alkali element and boron characteristics of the porefluids suggest that following phase separation, the high-temperature fluids have interacted extensively with the overlying sediment sequence. The data indicate that secondary uptake processes are probably important for Cs and Rb, and ^{6}Li is preferentially retained in secondary mineral phases during sediment alteration.

1 INTRODUCTION

Drilling of massive sulphide deposits at the Escanaba Trough during ODP Leg 169 provided the unprecedented opportunity to sample fluids circulating within the clastic sedimentary sequences that host these deposits; the chemical composition of these fluids provides key information for mineral formation processes (Humphris et al., 1995; James et al., 1996). Models for the formation of these modern day deposits have previously relied heavily upon samples recovered from the mound surface (Magenheim and Gieskes, 1994); information regarding the three dimensional sub-surface nature of ore deposits has only recently begun to emerge, largely through the efforts of deep-sea drilling (e.g. ODP Leg 158; Humphris et al., 1995).

The Cl content of porefluids recovered from within sediments recovered from the Escanaba Trough ranges from 300 to 800 mM (Fig. 1). This wide variation is inconsistent with thermal alteration of mineral phases, or interaction with sill complexes (e.g. Gieskes et al., 1982). The only way to explain this variation is that the interstitial fluids contain a hydrothermal component which has undergone phase separation with subsequent re-mixing of the Cl-rich and Cl-poor phases in variable proportions with background porefluids. In order to fully interpret these data, the hydrothermal and background porefluid signatures need to be separated. In this connection, boron and the alkali elements are highly mobile during water-rock reactions and are the most sensitive elements to hydrothermal alteration. This enables their concentrations and isotope compositions to be used to fingerprint sources (e.g. secondary mineral formation) operating to control element distributions in hydrothermal systems.

This paper discusses the alkali element and boron geochemistry of porefluids and their host sediment recovered from nine holes (Holes 1038A-I) drilled in the Escanaba Trough hydrothermal area. Results are then compared with samples recovered from a nearby reference site (Site 1037), which is located away from the zone of hydrothermal upflow.

2 RESULTS

Results of porefluids analyses are given in Figs. 2-6. Hole 1038I is discriminated as it is the deepest and best recovered core from Site 1038. Hole 1038I is located between an area of high temperature venting and an area of diffuse venting in the Escanaba Trough hydrothermal zone. Basalt fragments, interpreted to be a sill, were recovered from this hole at a depth of ~160 mbsf. The remaining holes drilled at Site 1038 are located within a few meters of the high temperature or diffuse vents.

2.1 *Li and Li isotopes*

Lithium is clearly enriched in porefluids from the upper ~100 m of Site 1038, and around the sill at

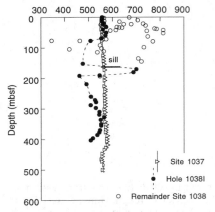

Fig. 1: Cl profile in porefluids from the Escanaba Trough.

Hole 1038I compared to samples from Site 1037 (Fig. 2a). Furthermore, the concentrations of Li measured at Site 1038 are generally higher than have previously been recorded in high-temperature hydrothermal fluids from the Escanaba Trough (~1300 μM; Campbell et al., 1994). Sediments recovered from Site 1038 are slightly depleted in Li compared to Site 1037 (particularly samples from around the sill at Hole 1038I; Fig. 2b).

With the exception of near-surface samples from Hole 1038I, the Li isotopic composition of porefluids from Site 1038 is lighter (more positive) than porefluids from Site 1037 (Fig. 3a). In contrast, the Li isotopic composition of the conjugate sediment samples shows a similar range of values at both sites (Fig. 3b).

2.2 B and B isotopes

To date, a limited number of B isotope analyses have been performed on porefluids collected from Site 1038. A preliminary interpretation of the data suggests that they form a mixing line between a pure hydrothermal endmember (as defined by zero Mg) and seawater (Fig. 4a). The hydrothermal endmember has a $\delta^{11}B$ value that is only slightly elevated over the value of unaltered sediments, suggesting that its boron isotope signature is dominated by boron leached from the sediments. A plot of B concentration versus $\delta^{11}B$ values (Fig. 4b) gives preliminary suggestions of two different hydrothermal systems; one with relatively high B concentrations and the other with slightly lower B concentrations that also includes local hydrothermal vent fluids. However, more data are required to confirm this.

2.3 Cs and Rb

The Cs and Rb content of porefluids from the near-surface of Site 1038, and from around the sill at Hole 1038I is clearly enriched relative to Site 1037 (Figs. 5a and 6a respectively). With the exception of one sample (which was recovered adjacent to a sill at Hole 1038H), the concentrations of Cs and Rb at Site 1038 are less than in vent fluids from the Escanaba Trough (600-7700 nM Cs and 80-105 μM Rb; Campbell et al., 1994).

In the sediment phase, concentrations of Cs and Rb are variable at both sites (Figs. 5b and 6b respectively). There is a clear depletion in Cs and Rb in sediments adjacent to the sill at Hole 1038I.

3 DISCUSSION

Chemical analyses of porefluids from the Escanaba Trough hydrothermal area indicate diffuse venting of high-temperature fluids in the shallow sub-surface (0-80 m) and close to basalt intrusions. This hydrothermal component appears to have undergone phase separation at depth with subsequent segregation of the conjugate Cl-poor and Cl-rich phases.

The alkali element and boron characteristics of the porefluids suggest that following phase separation, the vent fluids have reacted extensively with the sediment sequence resulting in increases in porefluid Li, B, and δ^6Li, and reductions in Cs, Rb, and $\delta^{11}B$, relative to the vent fluids. In the conjugate sediment phase, there are corresponding reductions in Li, Cs, and Rb, while δ^6Li increases.

Of note is that we do not see nearly complete depletion in Cs and Rb in the mineral phases in altered sediments from Site 1038. This means that secondary uptake processes are probably important for Cs and Rb; this is consistent with the porefluid analyses which have lower concentrations of Cs and Rb than vent fluids.

Hydrothermal vent fluids from the Escanaba Trough have δ^6Li values in the range -9 to -5.6‰. Several porefluid samples from Site 1038 have a lighter δ^6Li value than measured in the vent fluids which suggests they have interacted extensively with altered sediments; secondary mineral phases preferentially retain ^6Li (Chan et al., 1994). Furthermore, high porefluid Li samples have higher δ^6Li values in the sediment phase, consistent with preferential retention of ^6Li during sediment alteration.

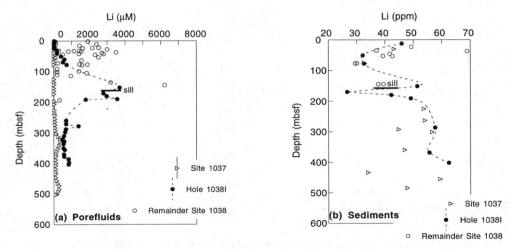

Fig. 2: Li concentration of a) porefluids, and b) sediments from the Escanaba Tough.

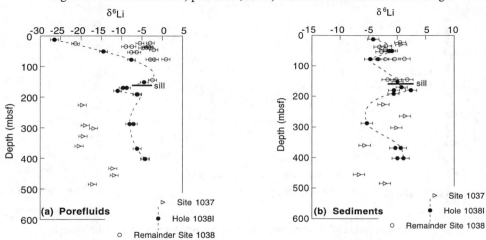

Fig. 3: Li isotopic composition of a) porefluids, and b) sediments from the Escanaba Trough. Li isotopic ratios are expressed as δ^6Li relative to the NBS L-SVEC standard (see Chan et al., 1994).

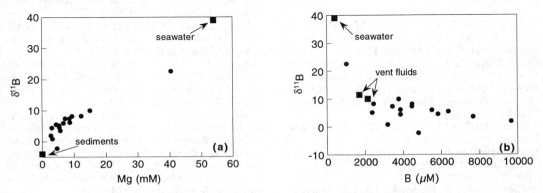

Fig. 4: B isotope composition of porefluids from Site 1038 as a function of a) Mg, and b) B. B isotopic ratios are expressed as δ^{11}B relative to the NBS SRM 951 standard.

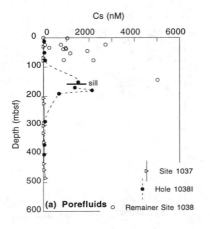
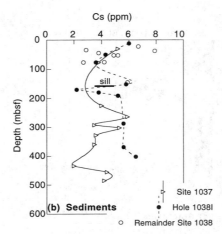

Fig. 5: Cs concentration of a) porefluids, and b) sediments from the Escanaba Trough.

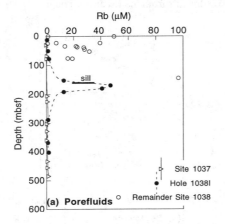
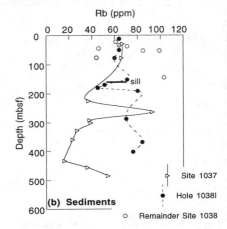

Fig. 6: Rb concentration of a) porefluids, and b) sediments from the Escanaba Trough.

REFERENCES

Campbell, A.C., C.R. German, M.R. Palmer, T. Gamo & J.M. Edmond 1994. Chemistry of hydrothermal fluids from the Escanaba Trough, Gorda Ridge. In J.L. Morton, R.A. Zierenberg & C.A. Reiss (eds), *Geologic, Hydrothermal, and Biologic Studies at Escanaba Trough, Gorda Ridge, Offshore Northern California*. U.S. Geol. Surv. Bull. 2022:201-221.

Chan, L.H., J.M. Gieskes, C.F. You & J.M. Edmond 1994. Lithium isotope geochemistry of sediments and hydrothermal fluids of the Guaymas basin, Gulf of California. *Geochim. Cosmochim. Acta* 58:4443-4454.

Gieskes, J.M., H. Elderfield, J.R. Lawerence, J. Johnson, B. Meyers & A. Campbell 1982. Geochemistry of interstitial waters and sediments, Leg 64, Gulf of California. In J.R. Curray, D.G. Moore et al., *Init. Repts. DSDP*, 64:675-694.

Humphris, S.E., P.M. Herzig, D.J. Miller et al. 1995. The internal structure of an active sea-floor massive sulfide deposit. *Nature*, 377:713-716.

James, R.H. & H. Elderfield 1996. Chemistry of ore-forming fluids and mineral formation rates in an active hydrothermal sulfide deposit on the Mid-Atlantic Ridge. *Geology*, 24:1147-1150.

Magenheim, A.J. & J.M. Gieskes 1994. Evidence for hydrothermal fluid flow through surficial sediments, Escanaba Trough. In J.L. Morton, R.A. Zierenberg & C.A. Reiss (eds.), *Geologic, Hydrothermal, and Biologic Studies at Escanaba Trough, Gorda Ridge, Offshore Northern California*. U.S. Geol. Surv. Bull. 2022:241-255.

Formation of clay minerals in the sedimentary sequence of Middle Valley, Juan de Fuca Ridge – ODP Leg 169

K.S.Lackschewitz, R.Botz, D.Garbe-Schönberg, P.Stoffers & K.Horz
Geological-Palaeontological Institute, University of Kiel, Germany

A.Singer
Seagram Center for Soil and Water Sciences, Hebrew University of Jerusalem, Israel

D.Ackermand
Mineralogical and Petrographical Institute, University of Kiel, Germany

ABSTRACT: Mineralogical and chemical characteristics of clays from two holes drilled in the Middle Valley at the northern end of Juan de Fuca Ridge were studied using X-ray powder diffraction (XRD), differential thermal analysis (DTA), infrared spectroscopy (IRS), electron microscopy (SEM) and transmission electron microscopy (TEM), X-ray fluorescense (XRF) and ICP mass spectrometry. Oxygen isotope measurements on authigenic clay minerals provide a record of mineral formation temperatures. At Hole 1035D, the upper 50mbsf are composed of a mixture of detrital and authigenic minerals in various proportions. Nearly pure chlorite is dominant below a depth of 100mbsf. That section was classified as the high-temperature alteration zone, for which the formation temperature of chlorite was determined to be 277-292°C. Smectite and some mixed-layer chlorite/smectite are dominant phyllosilicates in the upper part of Hole 1036A. Nearly monomineralic mixed-layer chlorite/smectite mainly occur between 20 and 35mbsf. Temperatures of formation of these mixed-layers were calculated at 110-340°C. In the deepest unit, chlorite is the dominant phyllosilicate coexisting with quartz.

1 INTRODUCTION

Middle Valley on the northern Juan de Fuca Ridge is one of a few well-studied sediment-covered spreading centers and was the subject of detailed investigations during Ocean Drilling Programs Leg 139 and Leg 169 (Figure 1). Two main areas about two km apart were drilled during Leg 169, the Dead Dog vent field in the area of active venting (Site 1036), and the Bent Hill Massive Sulfide Deposit (BHMS) in the Bent Hill area (Site 1035, Figure 2). BHMS is a 35m high and 100m wide sulfide mound near the southern margin of the Bent Hill Mound. Hole 1035D was drilled 75m east of the BHMS mound. Sediments consists of clastic sulfides, unaltered to altered hemipelagic sediments and massive to semi-massive sulfides (FOUQUET, ZIERENBERG, MILLER et al., in press). The Dead Dog area is characterized by several active vents with temperatures up to 268°C when last measured (BUTTERFIELD at al. 1994). The vent fluid composition indicates significant interaction of hydrothermal fluid with sediment. Surface sediments consist of anhydrite rubble with minor Mg-smectites and sulfide minerals, whereas the deeper sequences are characterized by unaltered to altered hemipelagic sediments interbedded with turbidites. Clay minerals from the Middle Valley area were described by BUATIER et al. (1994) as Mg-smectite, corrensite and chlorite.

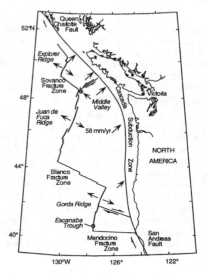

Figure 1. Location map showing the tectonic setting of the northern Juan de Fuca Ridge and the sedimented Middle Valley rift. Arrows indicate plate movement (after DAVIES, MOTTL, FISHER et al. 1992).

743

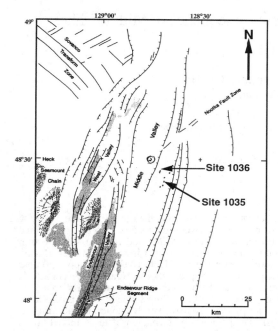

Figure 2. Tectonic interpretation of the Middle Valley with names shown for each of the major tectonic elements. Arrows referring to sites 1036, 858 and 1035. Stippled areas show recent volcanic flows. (modified after DAVIS and VILLINGER 1992).

The objective of our study is to discuss the clay mineral genesis within hydrothermal deposits drilled at Middle Valley. Lateral and vertical variations in the distribution of authigenic clay minerals and changes in their morphology, crystallography, and chemical composition will be studied to derive insights into changing physico-chemical parameters.

The mineral composition of the clay fraction from sediments sampled at Sites 1035 and 1036 (Leg 169) from the Middle Valley area of the northern Juan de Fuca Ridge were examined using X-ray diffraction (XRD), transmission electron microscopy (TEM), scanning electron microscopy (SEM), differential thermal analysis (DTA) and infrared-spectroscopy (IRS). The chemical composition of the samples were determined by X-ray fluorescence analysis (XRF) and ICP mass spectrometry. In order to calculate the formation temperature we have determined the oxygen isotope composition of monomineralic clay samples.

2 DISCUSSION

At both drill sites, the clay fractions are composed of a mixture of detrital and authigenic minerals. Quartz, feldspar, illite and chlorite are the major components in the upper parts of Hole 1035D whereas nearly pure chlorite (by means of XRD, DTA, TEM, SEM, XRF and electron-microprobe) occurs in the lower part of the hole as the dominant component (Figures 3 and 4). The geochemistry of these chlorites, i.e. the high contents of Fe (up to 30%), Zn (200-300ppm) and Cu (130-180ppm), clearly indicates a hydrothermal origin. Furthermore, they are characterized by very low contents of Rb (<0,4ppm), Sr (<1,2ppm) and Ba (<5,5ppm). Finally, the presence of a significant Eu depletion in these secondary chlorites reflects a high temperature hydrothermal alteration of their precursor minerals (e.g. plagioclase and clinopyroxene). MICHARD & ALBAREDE (1986) have documented that hydrothermal solution on the East Pacific Rise are strongly enriched in Eu. This has been interpreted to be related to reduction of Eu^{3+} to Eu^{2+}, associated with high temperatures. These results are confirmed by a formation temperature of

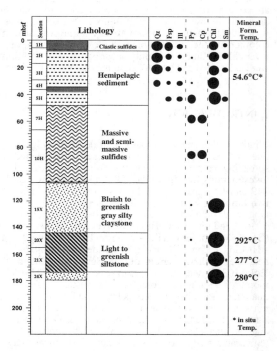

Figure 3. Core description, mineral assemblages and mineral formation temperatures of Hole 1035D. The size of the black circles indicate relative variations of the mineral assemblages. Qz-Quartz, Fsp-Feldspar, Ill-Illite, Py-Pyrite, Cp-Chalcopyrite, Chl-Chlorite, Sm-Smectite. (*in situ temperature from FOUQUET, ZIERENBERG, MILLER et al., in press)

Figure 4. SEM photograph of the clay fraction of sample 169-1035D-24XCC-34.5 showing chlorite minerals with typical platelet morphology.

277°C-290°C at 150-180mbsf for the pure chlorites calculated from oxygen isotope data. These calculated high formation temperatures for the pure chlorite samples are similiar to the calculated temperatures of 275°C for pure chlorites of the Nesjavellir geothermal field on Iceland (SCHIFFMANN & FRIDLEIFSSON, 1991).

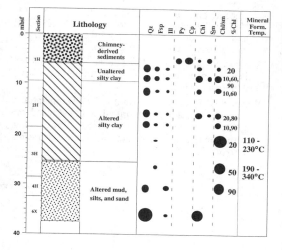

Figure 5. Core description, mineral assemblages and mineral formation temperatures of Hole 1036A. The size of the black circles indicate relative variations of the mineral assemblages. Qz-Quartz, Fsp-Feldspar, Ill-Illite, Py-Pyrite, Cp-Chalcopyrite, Chl-Chlorite, Sm-Smectite, Corr-Corrensite.

Hole 1036A shows a distinct change in silicate mineralogy with depth (Figure 5). The upper 20mbsf are composed of a mixture of detrital quartz, illite, feldspar with smectite and two or three types of mixed-layer smectite/chlorite. In contrast the clay fraction between 20 and 35mbsf is dominated by mixed-layer chlorite/smectite showing an increasing chlorite content with depth. The IR spectrum shows an absorption peak at 3690 cm^{-1} which may arise from the hydroxide sheet of a trioctahedral smectite. Reflections from (060) peak indicate that the mixed-layer chlorite/smectite is an interstratification of trioctahedral chlorite with trioctahedral smectite. An ordered mixed-layer chlorite/smectite of 50% chlorite and 50% smectite (corrensite) was identified by XRD and SEM (Figure 6). These nearly pure mixed-layers are characterized chemically by high contents of Mg (28-30%) and Zn (220-330ppm) suggesting that both are hydrothermal products. While the corrensite at 27.8 mbsf shows a negative Eu anomaly, the smectite-rich mixed-layer at 21.8 mbsf displays a positive europium anomaly. This difference may be due to a thermal gradient related to hydrothermal fluids.

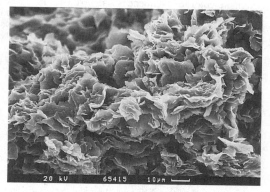

Figure 6. SEM photograph of the clay fraction of Sample 169-1036A-3H06-129 showing corrensite with typical corn-flake morphology.

Because an oxygen isotope fractionation equation for mixed-layer chlorite/smectite has not been established, the oxygen isotope equilibration temperatures were calculated for mixed-layer chlorite/smectite using the chlorite - water oxygen isotope equation after WENNER & TAYLOR (1971) and the smectite - water oxygen isotope fraction equation after SHEPPARD & GILG (1996). The formation temperatures calculated from oxygen isotope data lie for the smectite-rich mixed-layer sample between 110 and 230°C (mean 170°C) and for the corrensite between 190 and 340°C (mean 265°C). SCHIFFMAN & FRIDLEIFSSON (1991) have calculated that smectites transform into smectite-rich, randomly interstratified mixed-layer chlorite/smectite between 200 and 240°C, whereas corrensite occurs between 240 and 265°C.

In the deepest unit of Hole 1036A, chlorite is the dominant phyllosilicate coexisting with quartz indicating temperatures similiar to the pure chlorites of Hole 1035D.

3 CONCLUSIONS

Sediments from Hole 1035D, located approximately 50m east of the BHMS, contain a record of hemipelagic and turbiditic sedimentation in which hydrothermal activity produced an alteration zone dominated by nearly pure chlorite downward until the base of the hole. The hydrothermal origin is established by a temperature of formation of approximately 280°C calculated from oxygen isotope data.

In Hole 1036A the mineral assemblages change with depth. Quartz, feldspar, illite, smectite, chlorite and some mixed-layer chlorite/smectite reflect the mineralogical mixture of detrital and hyrothermal phases above 20mbsf. An increase in alteration below 20mbsf is associated with the disappearance of smectite and the appearance of nearly pure mixed-layer chlorite/smectite. At depths below 35mbsf chlorite is the dominant phyllosilicate. The observed clay mineral sequence of Hole 1036A is similar to the smectite-corrensite-chlorite sequences described in other geothermal fields. The mineralogical sequence and oxygen isotope data obtained from hydrothermal clays indicate high mineral formation temperatures below 20mbsf (approximately 170-265°C).

Microprobe analyses, XRD, and EDX analyses point to a more Fe-rich (up to 30% Fe_2O_3) nature of hydrothermal clays from the Bent Hill Massive Sulfide Deposit (Hole 1035D) whereas clays from the Dead Dog area (Hole 1036A) reflect more Mg-rich phases (up to 30% MgO). The shipboard-measured pore-water geochemistry from Hole 1036A shows evidence for intense lateral flow in the interval 20-30mbsf which is characterized by a high Mg concentration (FOUQUET, ZIERENBERG, MILLER *et al.*, in press). Thus, the variations in Fe- and Mg-content may be related to the different chemical compositions of the fluids whereas the change in mineralogic sequence downhole probably reflects a prevailing high temperature gradient.

4 ACKNOWLEDGEMENTS

The authors acknowledge the valuable comments of two unknown reviewers. We thank U. Schuldt for photographic assistance. This work was financially supported by the Deutsche Forschungsgemeinschaft grant Sto 110/26.

5 REFERENCES

BUATIER, M.D., KARPOFF, A.-M., BONI, M., FRÜH-GREEN, G.L. & J.A. MCKENZIE 1994. Mineralogic and petrographic records of sedimend-fluid interaction in the sedimentary sequence at Middle Valley, Juan de Fuca Ridge, Leg 139. In: Mottl, M.J., Davies, E.E., Fisher, A.T. & J.F. Slack (Eds.), *Proc. ODP, Sci. Res.*, **139**; College Station, TX (Ocean Drilling Program), 133-154.

BUTTERFIELD, D.A., MCDUFF, R.E., FRANKLIN, J. & C.G. WHEAT 1994. Geochemistry of hydrothermal vent fluids from Middle Valley, Juan de Fuca Ridge. In: Mottl, M.J., Davies, E.E., Fisher, A.T. & J.F. Slack (Eds.), *Proc. ODP, Sci. Res.*, **139**; College Station, TX (Ocean Drilling Program), 395-410.

DAVIES, E.A. & H. VILLINGER 1992. Tectonic and thermal structure of the Middle Valley sedimented rift, northern Juan de Fuca Ridge. In: Davies, E.E., Mottl, M.J., Fisher, A.T. & J.F. Slack (Eds.), *Proc. ODP, Sci. Res.*, **139**, ; College Station, TX (Ocean Drilling Program), 9-41.

DAVIES, E.A., MOTTL, M.J., FISHER, A.T., *et al.* 1992. *Proc. ODP Init. Rept.* **139**; College Station, TX (Ocean Drilling Program), 1-1026.

FOUQUET, Y., ZIERENBERG, R.A., MILLER, D.J., *et al.* in press. *Proc. Ocean Drilling Program. Init. Rept.* **169**; College Station, TX (Ocean Drilling Program).

MICHARD, A. & F. ALBAREDE 1986. The REE content of some hydrothermal fluids. *Chem. Geol.*, **55**, 51-60.

SCHIFFMAN, P. & G.O. FRIDLEIFSSON 1991. The smectite-chlorite transition in drillhole NJ-15, Nesjavellir geothermal field, Iceland: XRD, BSE and electron microprobe investigations. *J. metamorphic Geol.*, **9**, 679-696.

SHEPPARD, S.M.F. & H.A. GILG 1996. Stable isotope geochemistry of clay minerals. *Clay Minerals*, **31**, 1-24.

WENNER, D.B. & H.P. TAYLOR 1971. Temperatures of serpentinization of ultramafic rocks based on O^{18}/O^{16} fractionation between coexisting serpentine and magnetite. *Contrib. Mineral. Petrol.*, **32**, 165-185.

Hydrothermal basalt alteration at the surface of the TAG active mound, MAR26°N

H. Masuda, M. Nakamura & K. Tanaka – *Department of Geosciences, Osaka City University, Japan*
H. Chiba – *Department of Earth and Planetary Science, Kyushu University, Fukuoka, Japan*
T. Gamo – *Ocean Research Institute, University of Tokyo, Japan*
K. Fujioka – *JAMSTEC, Yokosuka, Japan*

ABSTRACT: Samples of rubble of hydrothermally altered basalt and hydrothermal deposits were studied to determine their mineralogy and compostions and relationship to the cementation of massive sulfide deposits within the shallow hydrothermal mound at TAG, Mid Atlantic Ridge (MAR) 26°N. The basalt was hydrothermally altered by continuously changing hydrothermal fluids; an initial low temperature, white smoker fluid and a latter high temperature, black smoker fluid. At the latest stage of the hydrothermal alteration, Si and Al were leached from the basalt into the hydrothermal fluid. Since fragmental basalts are distributed in the shallow mound, elements leached from these rocks may contribute to cementation of the massive sulfide deposits during low temperature alteration by diffusive flow of hydrothermal fluid.

1 INTRODUCTION

The TAG active mound, located in the eastern edge of the rift valley of the Mid-Atlantic Ridge at 26°08'N (Fig. 1), 200m in diameter and 50m high, and is one of the largest known seafloor active hydrothermal mound in the world. Hydrothermal activity of the mound is estimated to have been intermittently active during 20,000 yrs (Lalou et al., 1993). The mound is covered by Fe-Cu-Zn rich massive sulfide aggregates, derived from collapsed black smoker chimneys and consolidated by diagenetic and low temperature alteration, and is thought to be a modern analogue of massive sulfide deposits associated with ophiolites.

In this study the mineralogy and composition of hydrothermally altered basalts and massive sulfide aggregates collected by the submersible "*Shinkai 6500*" in 1994 were studied. Based on these data, processes of alteration of basalts at shallow levels within the mound and the consolidation of massive sulfide aggregates are discussed.

2 SAMPLES AND ANALYTICAL METHODS

A few pieces of basalt rubble were collected from the talus on the northeastern slope of the mound. One of the fragmental basalt has alteration halo of about 2.5 cm thick (Fig. 2), and was cut into twelve slices to sample accross the alteration. Mineralogy and chemistry were analyzed by X-ray diffractometry and ICP-AES and AAS, respectively. Whole rock oxygen isotope ratios of these slices were also analyzed.

Massive sulfide aggregates were widely distributed on the surface of the active mound. The sample used in this study was collected from the eastern area of the upper terrace, where diffusive black smoker and white smoker fluids were emanating. It was examined by XRD and SEM equipped EDX.

3 RESULTS

3.1 *Hydrothermally altered basalt*

X-ray diffraction patterns of twelve slices of the basalt and molar ratios of major chemical compositions relative to Ti are shown in Figures 3 and 4, respectively. The oxygen isotope ratios are also shown at the bottom of Figure 4. Location of the sampling traverse (least alteration) to alteration rim (#12) is shown by Line L1. Even in the least altered innermost slice (No. 1), smectite was identified by X-ray diffractometry. The oxygen isotope ratio of the innermost three slices is 4.1 ‰ relative to SMOW and is lower than those reported for fresh MORB (about 6 ‰. e. g., Muehlenbachs and Clayton, 1972). Thus, the studied basalt has been altered to form smectite, although anorthite still remains in the inner part.

The alteration halos were distinguished by their color; the inner halo (Slice Nos. 2 to 5) was green gray and the outer halo (Nos. 6 to 12) was gray.

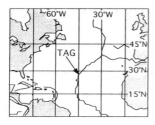

Figure 1 Locality of the TAG hydrothermal field

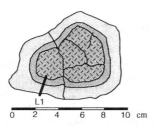

Fig. 2 Sketch of the alteration basalt having two alteration halos taken from the TAG hydrothermal mound.

The difference of the color corresponds to differences in mineralogy and chemistry. In the inner zone, anorthite remains, however, it was completely decomposed in the outer zone. In the inner zone, smectite is the major alteration silicate, but chlorite is the major alteration phase in the outer zone. Major element chemical composition changes drastically in the inner halo, however, it changes rather smoothly in the outer halo. From the inner to the outer halo, Si, Ca, Na and K decrease, but Fe and Mg increase. Copper is enriched in the outer halo, while Zn is enriched in the inner. Copper enrichment is caused by presence of chalcopyrite and Zn is probably due to sphalerite. In the outermost portions (Nos. 11 and 12), Fe is enriched due to the presence of pyrite and Al and Mg are depleted.

Oxygen isotope ratios of slices become lower from the inside (4.1‰) to the outside (1.1‰), concordant with the formation of Al-silicate alteration minerals, chlorite and smectite. In the outermost slice (No. 12, 2.3‰), the oxygen isotope ratio becomes higher, probably due to cooling of the altering fluid mixed with seawater.

3.2 *Distribution of elements in the cement of massive sulfide aggregates*

Figure 4 shows the distribution of major elements in the cement of massive sulfide aggregates. Si, O and minor Al and Mg are distributed in the same area, indicative of quartz and clay minerals. Cu occupies the same area as O and Mg, but is not associated with S or Fe. This indicates that primary chalcopyrite has been oxidized in the cement, though it is the major Cu-bearing mineral in the active black smoker chimneys. XRD examinations suggests that chalcopyrite was oxidized to form atacamite and melanterite on the mound surface. Fe and S distri-butions suggest that pyrite is still present in the cements.

4   DISCUSSION

4.1 *Processes of hydrothermal alteration of basalt at the shallow part of the active mound*

In the inner alteration halo of the analyzed basalt fragment, anorthite remained unaltered. It indicates that the hydrothermal alteration was not complete in the inner halo, probably due to the lower temperature and/or short period of hydrothermal alteration. Smectite is the major alteration mineral in the inner halo and is indicative of low temperature alteration. On the other hand, chlorite is the major alteration mineral in the slices of the outer halo. Thus, the outer halo seemed to have been altered at higher temperature than the inner halo after initial low temperature alteration of the whole crust. Alteration temperature of the outer halo can be estimated by oxygen isotope geothermometer. Assuming that the whole rock $\delta^{18}O$ values are reflected of chlorite ($\delta^{18}O$ values of 1.9‰ for Nos. 6 and 7), the oxygen isotope equilibrium temperature is calculated to be about 290°C using the fractionation factor between chlorite and water (Wenner and Taylor, 1971) and $\delta^{18}O$ value of the black smoker fluid (1.7‰, Shanks et al., 1995).

The maximum concentration of Cu occurs in the outer halo. This Cu enrichment is due to chalcopyrite precipitation in the alteration zone. On the other hand, The maximum concentration of Zn occurs in the inner halo. However, presence of sphalerite was not confirmed by X-ray diffractometry, even though it is the likely Zn phase. The distributions of Cu and Zn between two halos are consistent with the difference in the alteration temperature indicated by alteration minerals. Cu precipitates as chalcopyrite in high temperature environment, but Zn precipitates later at lower temperatures (e.g. Hannington, 1993). Temperature control on Cu and Zn compositions of the venting fluids at the active mound is also observed. High temperature black smoker fluids have high Cu

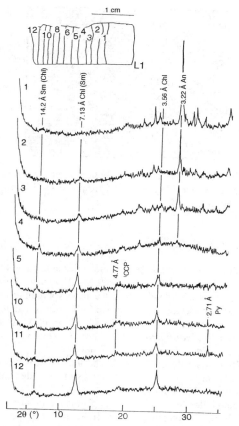

Figure 3 X-ray diffraction patterns of the sliced basalt. The inserted figure indicates the sample slices along the L1 in the Figure 2.

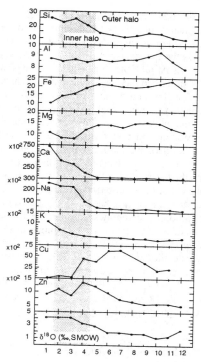

Figure 4 Major element and oxygen isotope compositions of the sliced basalt samples. Sample Numbers are the same as those in Figure 3, and the chemical composition is represented by the molar ratio relative to Ti concentration.

concentrations, while Zn was concent-ated in the lower temperature white smoker fluids (e.g., Edmond et al., 1995). This suggests that the studied basalt was altered by continuously changing hydrothermal fluid; by the low temperature white smoker-type fluid at an early stage of hydrothermal infiltration and the later by high temperature black smoker-type fluid.

Decrease in Al and Mg in the outermost part may be related to the enhanced pyrite precipitation, which is indicated by the higher Fe concentration in the outermost part. If pyrite precipitates due to the cooling of and/or mixing of seawater with the high temperature hydrothermal fluid. The pH of the fluid decreases (Edmond et al., 1995). Mixing of seawater in the venting fluid was indicated by the study of REE of the white smoker fluids (Mills R. A. and Elderfield H., 1995), which are more acidic compared to the black smoker fluid. If such a more acidic fluid was present at the end of the outer halo forming process, Al and Mg would be more effectively leached from the altered basalt. Decrease of fluid temperature at the end of outer halo formation is also consistent with the higher $\delta^{18}O$ value (2.3‰) of the outermost slice of the altered basalt.

4.2 Effect of basalt alteration within the mound on the cementation of massive sulfide aggregates

The altered basalt sample discussed above was recovered from the talus, 25m above the surrounding basaltic seafloor. It is plausible that the studied basalt resided in the shallow part of the mound before it became a part of rubble deposit. ODP Leg. 158 revealed the presence of altered basalt fragments at shallow levels within the hydrothermal mound (Humphris et al., 1995). Although the original location of the studied basalt is not clear at the present, similar hydrothermally altered basalt may be abundant within the mound. If so, it may supply materials necessary for the cementation of massive sulfide aggregate during its alteration process.

Silica (mainly quartz) is the dominant mineral in the cement. Minor amount of Mg and Al occur in the same area as the quartz cement (Fig. 5). Thus, smectite containing Al and Mg appearently coexists with quartz. As described before, chalcopyrite, which is one of the major sulfide minerals in active black smoker chimneys, was mostly decomposed to form atacamite and melanterite in the cement. Such an oxidation reaction must have occurred at the surface or shallow level of the mound where oxidative seawater could react with chalcopyrite. Mixing of seawater with the black smoker fluid causes a decrease in the pH of the fluid and probably promotes the release of Al and Mg from altering basalt, as discussed in the previous section. Al supply within the shallow level of the mound would elevate its concentration in the fluid and promote smectite formation in the cement. The basalt alteration within the mound may contribute to the consolidation of the sulfide aggregates by supplying Al and SiO_2, though most of these elements would be supplied from the deep reaction zone by the high temperature hydrothermal fluid.

5 SUMMARY AND CONCLUSIONS

The altered basalt sample collected from the talus of the active mound was found to be hydrothermally altered by continuously changing hydrothermal fluids: the low temperature white smoker-type fluid at the earlier stage and high temperature black smoker-type fluid at the later. Two distinct alteration halos of the altered basalt correspond to the alteration by these two types of fluid. The inner halo was altered by the low temperature fluid and was characterized by the presence of smectite and enrichment of Zn. The high temperature alteration by the black smoker-type fluid was overprinted in the outer halo. It is characterized by the presence of chlorite and enrichment of Cu.

In the outermost part of the altered basalt, Si, Mg and Al were leached compared to the other part. This may be due to the interaction with highly acidic fluid formed by the seawater mixing into the hydrothermal fluid and/or simple cooling of the hydrothermal fluid causing pyrite precipitation, which release H^+. One of these processes took place at the final stage of alteration of this sample. Elements released by the basalt alteration within the mound may contribute to the formation of hydrothermal alteration products, such as quartz and smectite, in the cement of massive sulfide deposits that are distributed widely on and in the shallow mound.

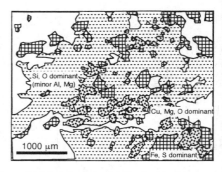

Figure 5 Major element distribution in the cement of massive sulfide aggregates from the TAG active mound.

REFERENCES

Edmond J.M. et al. 1995. Time-series studies of vent fluids from the TAG and MARK sites (1986, 1990) Mid-Atlantic Ridge: a new solution chemistry model and a mechanism for Cu/Zn zonation in massive sulphide orebodies, In ParsonL.M. et al. (eds), *Hydrothermal vents and Processes: Geol. Soc. Spec. Publ.*No. 86: 77-886.

Humphris S.E. et al. 1995. The internal structure of an active sea-floor massive sulphide deposits. *Nature* 377: 713-719.

Hannington M.D. 1993. The formation of atacamite during weathering of sulfides on the modern seafloor. *Can. Mineral.* 31: 945-956.

Lalou C. et al. 1993. New age data for Mid-Atlantic Ridge hydrothermal sites : TAG and Snakepit chronology revisited. *Jour. Geophys. Res.* 98: 9705-9713.

Mills R.A. & Elderfield H. 1995. Rare earth element geochemistry of hydrothermal deposits from the active TAG mound, 26°N Mid-Atlantic Ridge. *Geochim. Cosmochim. Acta* 59: 3511-3524.

Muehlenbachs K. and Clayton R.N. 1972. Oxygen isotope studies of fresh and weathered submarine basalts. *Can. Jour. Earth Sci.* 8: 1591-1594.

Shanks III W.C.et al. 1995. Stable isotopes in Mid-Ocean Ridge hydrothermal systems: Interactions between fluids, minerals, and oraganisms. In Humphris S.E. et al. (eds) *Seafloor Hydrothermal systems: Physical, chemical, giological, and geological interactions. Geophysical Monograph* 91: 194-221.

Wenner D.B. & Taylor Jr. H.P. 1971. Temperatures of serpentinization of ultramafic rocks based on O^{18}/O^{16} fractionation between coexisting serpentine and magnetite. *Contrib. Mineral. Petrol.* 32: 165-185.

Feeder zones of massive sulfide deposits: Constraints from Bent Hill, Juan de Fuca Ridge – ODP Leg 169*

P. Nehlig
BRGM-DR, Orléans, France

L. Marquez
Department of Geological Sciences, Northwestern University, Evanston, Ill., USA

ABSTRACT : Drilling through the turbidite-hosted Bent Hill (N. Juan de Fuca Ridge) sulfide deposit during ODP Leg 169 provides an unprecedented insight into the geometry and genesis of a large massive sulfide feeder zone. Vein- and impregnation-dominated hydrothermal mineralization extend for as much as 100 meters below massive sulfide mineralization, which is also 100 m thick in the core of the deposit. With a few minor exceptions all hydrothermal veins are extension veins. Their width is typically a few millimeters, and most show only one episode of extension and mineral filling. They have a strong tendency to be subvertical although subhorizontal cracks are common along preexisting mechanical anisotropies. Vein frequency is typically very high, on the order of 20 veins/m and decreases downward. In the absence of faults, the formation of the stockwork is explained by an upward moving hydraulic cracking front within the fine-grained sediments, associated with localized lateral flow in the coarser-grained sediments.

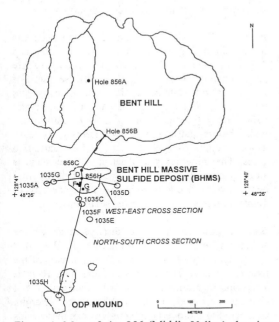

Figure 1. Map of site 856 (Middle Valley) showing the location of the Bent Hill and the two sulfide mounds to the south (modified from Goodfellow and Peter, 1994). Note the location of the two cross sections shown in figure 2.

1 INTRODUCTION

Studies of fossil and active hydrothermal systems have shown that massive-sulfide deposits are rooted within stockworks, which mark the paleo-subsurface path of the ascending ore fluids that deposited massive-sulfides (Solomon, 1984; Davis et al., 1992; Large, 1992; Nehlig et al., 1994; Tivey et al., 1995). However, few studies have attempted a geometric characterization of this zone. In this paper we examine the veining processes. We then discuss the mechanisms involved in fracture genesis. The work was carried out by studying the material filling veins, their continuity, spacing and orientation, and by review of existing seismic, heat flow, side-scan sonar and geophysical logging data.

2 GEOLOGY

An array of 7 holes was drilled during Leg 169 in addition to the 8 holes drilled during Leg 139. The holes are sited on a NS and EW transect centered on the Bent Hill Massive Sulfide Deposit (BHMSD) (Figure 1).

Although Middle Valley is a failed rift (Currie and Davis, 1994), it is evident from seismic studies (Rohr and Schmidt, 1994) that tectonic and volcanic

* Leg 169, Scientific party

activity has been active throughout the 7 km wide zone during the late Pleistocene and possibly the Holocene. However, seismic profiles around Bent Hill show no major tectonic feature directly associated with Bent Hill, the BHMSD or the active venting site to the south (Rohr and Schmidt, 1994).

Based on stratigraphic data, Mottl et al. (1994) infer that the BHMSD was deposited during the Pleistocene (~140-220 ka), whereas the high level laccolithic intrusion (Bent Hill) occurred during the Holocene, possibly within the past few thousand years.

3 THE HYDROTHERMAL VEIN SYSTEM

While all drill-holes recovered massive sulfide, whether as cemented or loose chimney clasts or as extensive impregnations, only Holes 856H, 1035F and 1035H crosscut a major stringer zone underlying the massive sulfide deposit. Away from the stringer zones the vein densities are typically very low (Figure 2). The transition between the Bent Hill massive sulfide deposit and the underlying stockwork is typically very sharp. Below we provide a detailed summary of the veining within Hole 856H (Figures 1 and 2).

3.1. Description of veining in Hole 856H

Massive sulfide deposit : The massive sulfides (0 to 104 mbsf) recovered from Hole 856H during Leg 139 and Leg 169 are composed predominantly of pyrite and pyrrhotite with subordinate amounts of chalcopyrite and sphalerite and are generally not veined.

Stockwork : Directly below the massive sulfide deposit (104 to 143 mbsf) the hydrothermal vein system is composed of millimeter-wide subvertical chalcopyrite, pyrrhotite, isocubanite and pyrite veins which coalesce. These veins occasionally feed small sulfide impregnation pockets of a few, probably more permeable, sediment layers. This network of predominantly subvertical coalesced veins continues downhole. Coalesced copper-iron sulfide veins have maximum thicknesses of several millimeters and sometimes contain relict sedimentary clasts. These composite-veins typically exhibit more than ten less than 1 mm-wide veins over a 1 cm wide interval, are subvertical, and include relicts of pale gray altered sediments. The coalesced nature of the veins suggests that they formed by multiple increments of cracking and mineral filling indicative of periodic fluid over-pressuring.

Sediments : Below 143 mbsf, this dense stockwork disappears and gives way to sediments with only a few 1 mm wide sulfide veinlets and sulfide impregnations. The sediments are less-indurated and are only crosscut by a few millimeter-wide veins and sulfide impregnations. The veins are Cu-sulfide-bearing, and subperpendicular to the bedding plane (vein density within this interval decreases to less than 2 m^{-1}), while the impregnations are concordant with the layering and appear to follow the more permeable layers.

Deep Copper Zone : Below (201.0-210.6 mbsf), the sediments are very extensively mineralized by bedding-parallel Cu-Fe sulfide impregnations (>30%), but are almost devoid of hydrothermal veins (less than 1 m^{-1}).

Interbedded turbidites and hemipelagic sediments : Downhole, the number of hydrothermal veins decreases significantly and the density of hydrothermal veins drops to less than 1 m^{-1} of subvertical, less than 1mm wide, anhydrite veins.

Basaltic sill complex : Abundant veining reappears in an alternating sequence of basaltic sills and sediments (431.7-434.3 mbsf). The sediments are predominantly crosscut by a diffuse set of millimeter-wide quartz-sulfide veins (density less than 5 m^{-1}). The basaltic sills have a much greater vein density (greater than 10 m^{-1}) of millimeter-wide chlorite-quartz and vuggy quartz-pyrite veins. Trace amounts of epidote occur in the quartz-rich veins. The mineralogy and density of the veins is strongly controlled by lithology: abundant chlorite-bearing veins occur in sills and lower density of quartz-sulfide veins occur in the sediments.

Basaltic flows : With the disappearance of the sediments and the appearance of the first hyaloclastites (479.9-489.5 mbsf), the density of the hydrothermal veins increases to more than 10 m^{-1}. The veins are mainly chloritic, with trace amounts of chalcopyrite, and have a chloritized margin. Compared to the overlying vein network these veins have a sinuous and discontinuous geometry which is interpreted as mineral filling of thermal contraction cracks in the pillow-lava basalts.

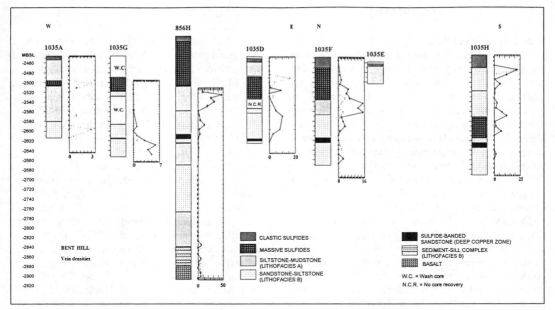

Figure 2. Diagram showing the density of veins expressed as m^{-1} (veins per meter ; sulfide veins as filled symbols and solid line and non-sulfide veins as unfilled symbols and dashed line) within the different drill holes. Note that the horizontal scales of the vein density diagrams are different. For location of WE and NS cross sections see figure 1

3.2. Width, spacing and orientation of the hydrothermal vein system

With a few minor exceptions all hydrothermal veins are extension veins. Their width is typically a few millimeters (Figure 3), and they show only one episode of extension and mineral filling.

These observations and crosscutting relationships imply that the stockwork was formed by multiple increments of cracking and mineral infilling. The veins have a strong tendency to be subvertical although subhorizontal cracks are common (Figure 4) along preexisting mechanical anisotropies.

The vein azimuths are as yet unknown, however a few observations, in less disrupted cores, show coherent azimuths over long intervals suggesting that the veins may have a preferred orientation.

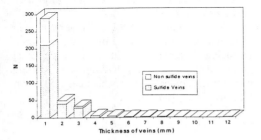

Figure 3. Histogram showing the modal distribution of the vein thickness (mm) of the sulfide and non-sulfide bearing hydrothermal veins.

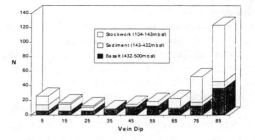

Figure 4. Histogram showing the modal distribution of the dip of the veins within Hole 856H for the major lithological units.

4 DISCUSSION AND CONCLUSION

Beds dipping more than 10° have been measured in a few cores and in a few limited intervals within cores sampled during Leg 169. These high bedding dips have always been related to slumping and do not mark in-situ rotated fault blocks. Furthermore, no fault gouge was recovered. This, along with the near-absence of shear structures suggests that faults are absent in the Bent Hill area and that fluid flow is controlled by 1) the preexisting high permeability in the siltstone and sandstone turbidite layers and 2) fluid overpressure-assisted extension veining within the less permeable, indurated mudstone layers or in cemented coarser-grained layers.

In the absence of a major fault, it is not clear why the hydrothermal fluid discharge remained focused. It can, however, be speculated that coarse-grained sandstone lenses act not only as major porous-flow conduits, but also as major hydrothermal fluid reservoirs. Because mudstones have a low permeability, the fluid flux and advective transport of heat through the seafloor is limited. However, the mudstones gain strength through hydrothermal alteration and thermal induration which allows the formation of contraction cracks and hydrothermal flow. The formation of the stockwork and of the initial high flow permeabiliy in the upper hydrothermal system can thus be explained by an upward moving cracking front within the fine-grained sediments, associated with important lateral flow in the coarser-grained sediments.

This accounts for the general observation of smaller vein densities in the coarse layers, which are characterized by mainly horizontal porous flow, and for the high subvertical vein densities in the mudstone layers. It also accounts for the observation that sulfides are more enriched within the coarser parts of the turbidites, in both lithified and non-lithified sediments.

The basaltic sills interlayered within the sediment unit have caused limited increased induration, hydrothermal alteration and fracturing of the surrounding sediments. They are highly fractured and veined, and compared to the surrounding sediments, they do not act as barriers to fluid flow.

These observations indicate that the permeability structure and evolution of the Bent Hill sediment-hosted hydrothermal system is primarily controlled by the lithology and by the fluid pressure-induced veining.

5 REFERENCES

Curie R.G. and Davis E.E., 1994. Low crustal magnetization of the Middle Valley sedimented rift inferred from sea-surface magnetic anomalies. In Mottl, M.J., Davis, E.E., Fisher, A.T., Slack, J.F. Eds. Proc. ODP, Sc. Results 139, College Station TX (Ocean Drilling Program): 19-27.

Davis, E.E., Mottl, M.J. and Fischer A.T. et al., 1992. Proc. ODP, Init. Repts., 139, College Station TX (Ocean Drilling Program).

Goodfellow W.D. and Peter J.M., 1994. Geochemistry of hydrothermally altered sediment, Middle Valley, northern Juan de Fuca Ridge. In Mottl, M.J., Davis, E.E., Fisher, A.T., Slack, J.F. Eds. Proc. ODP, Sc. Results 139, College Station TX (Ocean Drilling Program): 207-290.

Large, R.R., 1992. Australian volcanic-hosted massive sulfide deposits: features, styles, and genetic models, Econ. Geol., 87, pp 477-510

Mottl M., Wheat C.G. and Boulegue J., 1994. Timing of ore deposition and sill intrusion at site 856 : Evidence from stratigraphy, alteration and sediment pore-water composition. In Mottl, M.J., Davis, E.E., Fisher, A.T., Slack, J.F. Eds. Proc. ODP, Sc. Results 139, College Station TX (Ocean Drilling Program): 679-693.

Nehlig, P., Juteau, T., Bendel, V. and Cotten, J., 1994. The root zones of oceanic hydrothermal systems: constraints from the Semail ophiolite (Oman). Jour Geophys. Res., 99, B3, 4703-4713.

Rohr K.M.M. and Schmidt U., 1994. Seismic structure of Middle Valley near sites 855-858, Leg 139, Juan de Fuca Ridge. In Mottl, M.J., Davis, E.E., Fisher, A.T., Slack, J.F. Eds. Proc. ODP, Sc. Results 139, College Station TX (Ocean Drilling Program): 3-17.

Solomon, M., 1984. The formation of stockworks beneath massive-sulphide deposits (abstract), IGC Abst., 27, 6, 308.

Tivey, M.K., Humphris, S.E., Thompson, G., Hannington, M.D. and Rona, P.A., 1995. Deducing patterns of fluid flow and mixing within the active TAG hydrothermal mound using mineralogical and geochemical data. Jour. Geophys. Res., 100, 12527.

Trace elements in hydrothermal fluids at the Manus Basin, Papua New Guinea*

K. Shitashima
Central Research Institute of Electric Power Industry, Abiko, Japan

T. Gamo & K. Okamura
Ocean Research Institute, University of Tokyo, Japan

J. Ishibashi
Faculty of Science, University of Tokyo, Japan

ABSTRACT: Total (dissolved plus particulate) concentration of trace elements was determined in hydrothermal fluid samples collected from three hydrothermally active sites at the Manus Basin in 1995 and 1996. The DESMOS fluids have higher concentrations of Al and lower concentrations of other trace elements than those of the other two sites. In particular, the concentrations of Fe, Cu and Zn are extremely low. Owing to the sulfidation of trace elements may predominate in the DESMOS fluid, Mn, Fe, Ni, Cu, Zn, Cd and Pb precipitate as sulfide and scavenged from the hydrothermal fluid before being discharged from the hydrothermal vent. In contrast, Al is not scavenged from the hydrothermal fluid because water soluble aluminum sulfate is formed during the hydrothermal circulation. Therefore, Al is discharged at a high concentration from the vent into the ambient sea water. Furthermore, low pH of the DESMOS fluid limits formation of insoluble aluminum hydroxide.

1 INTRODUCTION

Deep-sea hydrothermal systems play an important role as significant sources or sinks of trace metals in the ocean, so hydrothermal activity has a major impact on the chemistry of the ocean. Many trace elements are extracted from the basalt by the vent fluid during hydrothermal circulation. Consequently, trace elements are introduced into the deep sea via the hydrothermal plume at high concentrations. The hydrothermal systems in the back-arc basin have been a major subject of study in recent years because the comparison of the chemical flux on a global scale between back-arc ridge systems and mid ocean ridge systems is important.

Manus Basin is a fast-spreading back-arc basin in the Bismarck Sea, western Pacific. The basin consists of three spreading (or stretching) ridge segments. These are the western Manus spreading center (WMSC), Manus spreading centers (MSC) and the southeastern ridges (SER) (Martinez and Taylor, 1996). Recently, three active hydrothermal sites, Vienna Woods, PACMANUS and DESMOS, were described by Lisitsyn et al. (1993), Binns and Scott, (1993) and Gamo et al. (1993), respectively. During the KH90-3 cruise of R/V Hakuho Maru (ORI, Univ. Tokyo), hydrothermal plumes that displayed water column anomalies for pH, CH_4, Mn and Al were detected over the submarine caldera at the DESMOS site in the Manus Basin (Gamo et al., 1993). The highest concentration of Al over the DESMOS site at KH90-3 is 1560 nM/L, 150 times greater than the Al concentration in ambient sea water. Shitashima et al. (1995) measured a maximum Al concentration of 18nM/L in buoyant plumes over the southern East Pacific Rise (S-EPR), which is only 3 times higher than the ambient Al concentration. Gamo et al. (1993) calculated the dilution factor of the hydrothermal plume at the KH90-3 to be approximately 10,000 times diluted by ambient sea water, based on the potential temperature anomaly. They estimated the endmember Al concentration to be about 10 mM/L. Resing et al. (1992) reported that lava-sea water interactions at high temperature extract extremely high concentrations of Al into the water phase from the molten rock. Furthermore, Gamo et al. (1993) and Resing and Sansone (1996) suggested that the hydrothermal plumes with large Al anomalies at KH90-3 originate from the contact of lava and sea water that is associated with episodic submarine eruption events. This interpretation for the large Al anomaly over the DESMOS site was made before fluid samples were collected directly from the hydrothermal vents at the DESMOS site.

Hydrothermal fluid samples were collected from the three hydrothermally active sites (Vienna Woods, PACMANUS and DESMOS Site) in the Manus Basin during two cruises using manned submersibles belongs to JAMSTEC; ManusFlux Cruises by "Shinkai 6500" in 1995 and Bioaccess Cruise by "Shinkai 2000" in 1996. This paper examines the reasons for the high Al concentration in the hydrothermal plumes at the DESMOS site by using the results of trace element analysis of hydrothermal fluid samples from the Manus Basin.

2 OBSERVATIONS

Hydrothermal fluid samples were collected from the three hydrothermal active sites in the Manus Basin using a manifold-pump sampling system (Sakai et al. 1990) attached to the submersible during the two cruises in 1995 and 1996. Figure 1 shows a geodynamic sketch of the Manus Basin and the hydrothermal sites for sample collection during the two cruises. Major cations and anions (ion chromatography), pH (with a pH meter), alkalinity and H_2S (potentiometric titration) were measured on board the mother vessels Yokosuka and Natsushima. Total carbonate (TCO_2) was determined using the coulometric titration system (UIC Inc., Carbon Coulometer model 5011) described by Johnson et al. (1985) with the modified CO_2 extraction system

---

* Shipboard scientific parties of ManusFlux cruise and Bioaccess cruise

described by Shitashima et al. (1996). Unfiltered fluid samples for trace elements analysis were acidified to pH 1 with ultra-pure HNO_3 and kept at room temperature until the samples were analyzed on land. Total (dissolved plus particulate, acid leachable sulfide and hydroxide) concentrations of trace elements (Al, Mn, Fe, Ni, Co, Cu, Zn, Cd, Pb) were determined using graphite furnace atomic absorption spectrophotometery (Perkin-Elmer, Model 4100ZL) with Zeeman background correction by the direct injection method of 1/2 diluted samples calibrated against standards made up in the same matrix as the samples. During the KH90-3 cruise, the Al in dissolved form was determined by fluorescence spectrometry using Al-lumogallion complex in a similar way to Hydes and Liss (1976).

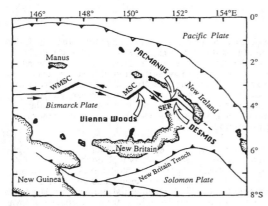

Figure-1: Geodynamic sketch of the Manus Basin as well as locations of PACMANUS, DESMOS and VIENNA WOODS hydrothermal sites.

3 RESULTS AND DISCUSSION

The chemical characteristics of the Manus Basin hydrothermal fluid taken from the three sites are completely different (Gamo et al. 1996, 1997). Figure 2 shows Mg vs. pH, total carbonate and H_2S relationships for the hydrothermal fluid samples taken from the three hydrothermal sites in the Manus Basin. The Vienna Woods fluids (over 100°C), erupting from anhydrite-sulfide chimneys on a basaltic basement, have similar characteristics to those from mid-ocean ridges. The PACMANUS fluids (black smoker; 268°C), erupting as black smokers from sulfide chimneys on a dacite basement, have lower pH (4-2.5) and they are more depleted in Ca and enriched in the other elements and total carbonate relative to those of the Vienna Woods. The DESMOS fluids (white smoker; 88°C) show unique chemical characteristics of extremely low pH (2.1), alkalinity (-9.2mM) and Si concentration and high H_2S (8.9mM) and SO_4^{2-} (32.8mM) concentrations. Native sulfur was deposited around the vents at the DESMOS site, suggesting strong magmatic contributions at this site.

Relationships between Mg and Al, Ni, Fe, Cu, Zn, Pb, Mn and Cd for the hydrothermal fluids collected at the PACMANUS, DESMOS and Vienna Woods hydrothermal sites during the cruise are shown in Figure 3. The DESMOS fluids have higher concentrations of Al and lower concentrations of the other trace elements

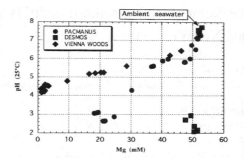
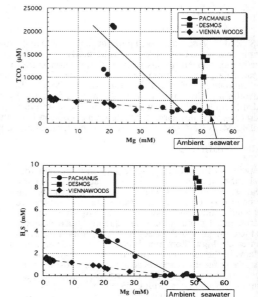

Figure-2: Mg vs. pH, total carbonate and H_2S relationships for the hydrothermal fluid samples taken from three hydrothermal sites in the Manus Basin.

than those of the other two sites. In particular, the concentrations of Fe, Cu and Zn are extremely low. The concentrations of trace elements in the DESMOS fluids are 2 to 3 times greater in Al and 1/2 times, 1/5 times, 1/3 times and 1/1000 times lower in Fe, Mn, Zn and Ni, respectively, as compared with those in the S-EPR fluids (Shitashima et al., 1995). In general, the solubilities of trace elements increase under low pH conditions. Therefore, the concentration of trace elements in the DESMOS fluid are expected to be high. The solubility of Al steeply increases as pH decreases in acidic solution (Bourcier et al., 1993). Gamo et al. (1997) suggested that the high Mg concentration (51.2mM) in the DESMOS fluid could be attributed to dissolution of Mg-silicate minerals by the highly acidic (pH ~1) fluid. However, the DESMOS fluid has low concentrations of all trace elements except Al. The sulfidation of trace elements may predominate in the DESMOS fluid

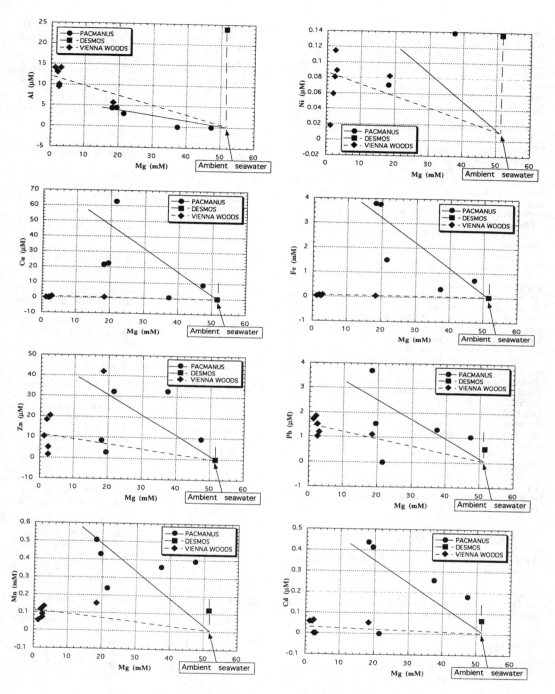

Figure-3: Relationships between Mg and Al, Ni, Fe, Cu, Zn, Pb, Mn and Cd for hydrothermal fluids collected at PACMANUS, DESMOS and VIENNA WOODS hydrothermal sites during the cruise.

because Cu, Cd and Pb easily form sulfides and Fe, Co and Zn form sulfide and hydroxide. If this is the case, then Fe, Ni, Cu, Zn, Cd and Pb precipitate as sulfides and scavenged from the hydrothermal fluid before being discharged from the hydrothermal vent. The concentration of Mn in the DESMOS fluid does not become as low as Fe because Mn forms sulfides less easily than Fe.

Because aluminum sulfide (Al_2S_3) is hydrolyzed to aluminum hydroxide ($Al(OH)_3$) and H_2S in the aqueous phase, aluminum sulfide is absent in the DESMOS fluid. In ordinary hydrothermal systems, it is considered that the aluminum hydroxide is transported to the ambient sea water in the diffused hydrothermal plume as colloidal particles, amorphous aluminum hydroxide ($[(Al(OH)_3)]_n$), or aluminosilicate (Elderfield et al., 1993; Shitashima et al., 1995). However, the Al anomalies at the DESMOS site detected during KH90-3 cruise were not derived from colloidal particle forms of Al because the lumogallion method can only detect dissolved forms. In the DESMOS fluid, aluminum hydroxide colloidal particles cannot form because the hydrolysis of aluminum hydroxide may not proceed due to the extremely low pH of the fluid. Furthermore, the high concentration of SO_4^{2-} in the fluid enhances the formation of aluminum sulfate ($Al_2(SO_4)_3$) from aluminum hydroxide and low Si concentration of the fluid restricts aluminosilicate formation. The water soluble aluminum sulfate therefore is discharged to the ambient sea water without being scavenged from the fluid. It is concluded that the reason for the high Al concentration in the fluid and hydrothermal plume at the DESMOS site is the formation of the water soluble aluminum sulfate in the fluid and diffusion of Al in that form to the ambient sea water. The low pH in the DESMOS fluid further limits the hydrolysis of the insoluble aluminum hydroxide. During the mixing between the fluid and ambient sea water, the aluminum sulfate may be gradually hydrolyzed with increasing pH (from pH ~6-8).

In order to define the behavior of trace elements in Manus Basin fluids and hydrothermal plumes, in the future we plan to carry out a comprehensive survey including long-term monitoring of hydrothermal activity at the DESMOS site under the New STERMAR project.

4 ACKNOWLEDGEMENTS

This study was funded by the RidgeFlux project of Science and Technology Agency of Japan. We thank the captains, crew, Shinkai team and scientists aboard the cruises for their kind help in the sampling and onboard analysis. We also thank S. Kraines for helpful comments and critical reading of our manuscript.

5 REFERENCE

Binns, R. A. and Scott, S. D. (1993) Actively forming polymetallic sulfide deposits associated with felsic volcanic rocks in the eastern Manus back-arc basin, Papua new Guinea. Econ. Geol., 88, 2226-2236.

Bourcier, W. L., Knauss, K. G. and Jackson, K. J. (1993) Aluminum hydrolysis constants to 250°C from boehmite solubility measurements. Geochim. Cosmochim. Acta, 57, 747-762.

Elderfield, H., Greaves, M. J. and Rucnicki, M. D. (1993) Aluminum Reactivity in Hydrothermal Plumes at the Mid-Atlantic Ridge. J. Geophys. Res., 98, B6, 9667-9670.

Gamo, T., Sakai, H., Ishibashi J., Nakayama E., Isshiki K., Matsumoto H., Shitashima K., Takeuchi K. and Ohota S. (1993) Hydrothermal Plumes in the Eastern Manus Basin, Bismarck Sea: CH_4, Mn, Al and pH Anomalies. Deep-Sea Res. I, 40, 2335-2349.

Gamo, T., Ishibashi J., Shitashima K. (1996) Unique hydrothermal fluid the DESMOS caldera, Manus Basin: reply to comments by J. A. Resing and F. J. Sansone. Deep-Sea Res. I, 43, 1873-1875.

Gamo, T., Okamura K., Charlou J-L., Urabe T., Auzende J-M., Ishibashi J., Shitashima K., Chiba H. and Shipboard Scientific Party of the ManusFlux Cruise (1997) Acidic and Sulfate-rich hydrothermal fluids from the Manus back-arc basin, Papua New Guinea. Geology., 25, (2), 139-142.

Hydes, D. J. and Liss P. S. (1997) Fluorimetric method for the determination of low concentrations of dissolved aluminium in natural waters. Analyst, 101, 922-931.

Johnson, K. M., King A. E., and Sieburth J. M. (1985) Coulometric TCO_2 analysis for marine studies: An introduction. Mar. Chem. 16, 61-82.

Lisitsyn, A. P., Crook, K. A. W., Bogdanov, Yu. A., Zonenshayn, L. P., Murav' Yev, K. G., Tufar, W., Gurvich, Ye. G., Gordeyev, V. V. and Lvanov, G. V. (1993) A hydrothermal fluid in the rift zone of the Manus Basin, Bismarck Sea. Internat. Geol. rev., 35, 105-126.

Martinez, F. and Taylor B. (1996) Fast backarc spreading, rifting and microplate rotation between transform faults in the Manus Basin, Bismarck Sea. Mar. Geophys. Res., Spec. Iss. (edited by J-M. Auzende and J-Y. Collot), 18, 1-3.

Resing, J., Sansone F. J., Wheat C. G., Measures C. I., McMurtry G. M., Sedwick P. N. and Massoth G. J. (1992) The extraction of aluminum from hot rock: the elemental signature of an eruptive vs classical hydrothermal system. EOS, 73, 254.

Resing, J. and Sansone F. J. (1996) Al and pH anomalies in the Manus Basin. Deep-Sea Res. I, 43, 1865-1870.

Sakai, H., Gamo T., Kim E.-S., Shitashima K., Yanagisawa F., Tsutsumi M., Ishibashi J., Sano Y., Wakita H., Tanaka T., Matsumoto T., Naganuma T. and Mitsuzawa K. (1990) Unique chemistry of the hydrothermal solution in the mid-Okinawa Trough backarc basin. Geophys. Res. Lett., 17, 2133-2136.

Shitashima, K., Sonoda A, Feely R A and Butterfield D A (1995) Hydrothermal Fluxes of Trace Elements (Al, Fe and Mn) from the Southern East Pacific Rise. EOS, 76, F321.

Shitashima, K., Tshumune D. and Kraines S. (1996) C.7 Distribution of oceanic CO_2, pH and TA, In Cruise Report for P8S -A hydrographic Section along 130°E- by R/V KAIYO from 17 June to 2 July 1996, 44-49.

Mineralogy and chemical composition of clay minerals, TAG hydrothermal mound

A.A.Sturz & M.J.T.Itoh
Marine Science, University of San Diego, Calif., USA

S.E.Smith
Department of Geosciences, University of Houston, Tex., USA

ABSTRACT: This study examines distribution and chemical composition of clay minerals in rocks from the Trans-Atlantic Geotraverse (TAG) hydrothermal mound, at 26° N latitude, Mid-Atlantic Ridge. Chemical analyses of clay mineral separates and bulk rock basalt alteration rims indicate that chlorite and smectite have similar major-element compositions and that the clay minerals have lower SiO_2, and higher Fe_2O_3, MgO and Zn than bulk basalt. Spatial distributions through the mound of SiO_2/Fe_2O_3, MgO, and Zn in clay minerals and alteration rims suggest that fluids circulating through the TAG mound originate from two sources. 1) A shallowly circulating fluid that is nearer to unaltered seawater chemical composition has affected the southeasterly and shallow central portions of the mound. 2) A deeply circulating hydrothermal end-member fluid has affected the northwesterly and deeper central portions of the mound.

1 INTRODUCTION

Rocks and minerals analyzed during this study were recovered during Ocean Drilling Program Leg 158. An important objective of ODP Leg 158 was to investigate subsurface mineral distribution and modes of alteration at the Trans-Atlantic Geotraverse (TAG) hydrothermal mound, 26°N latitude, Mid-Atlantic Ridge. This study examines the origin and distribution of subsurface fluids using evidence from clay mineral data.

Work reported prior to Leg 158 (e.g. Becker et al., 1993; 1994; Fujioka et al., 1994; Humphris et al., 1996; Tivey et al., 1994; and references there in), indicates that the TAG mound is circular, about 200 m in diameter, with its highest part about 50 m above the adjacent basalt plain. The highest part of the mound is occupied by black smoker chimneys that emit >360°C fluids (Campbell et al., 1988; Von Damm, 1995). West of the black smoker complex is a zone of relatively low heat flow, considered by Becker et al. (1993) to be a region of seawater entrainment into the mound. Southeast of the black smoker complex are white smokers venting fluids with temperatures between 260° and 300° C. Shipboard interpretation of rocks recovered during Leg 158 suggests the following stratigraphic sequence. 1) The upper few meters are dominated by pyrite breccias derived from crusts, chimney talus, and near-surface precipitates. 2) Below the pyrite breccia is an anhydrite-rich zone that may contribute to mound growth by subsurface deposition related to entrainment of seawater through the shallow portions of the mound. 3) Below the anhydrite-rich zone is a zone of quartz mineralization dominated by silicified pyrite breccias, grading to chloritized basalt.

2 RESULTS

Details of methods used during this study are given in Sturz et al. in Leg 158 Scientific Results (Humphris, Herzig, Miller et al, in press). This paper focuses on relative differences in iron, silica, and magnesium contents in clay minerals and altered rims. Results are given in Table 1.

Smectite recovered from TAG-4, the northwesterly side of the mound, (sample k, Table 1), contains SiO_2, (48.4 wt%) and Fe_2O_3 (19.8 wt%), closest in content to that of the unaltered basalt (50.38 wt% and 10.96 wt%, respectively). Smectite from the overlying breccia (sample j) has lower SiO_2, (38.9 wt%) and higher Fe_2O_3 (43.7 wt%). Bulk-rock altered red rim (sample m) and green rims (samples l and n) contain SiO_2, (48.3, 47.8, and 42.7 wt%, respectively) and Fe_2O_3 (12.8, 13.3, and 19.2 wt%, respectively), all very close to that of the basalt. Chlorites associated with massive pyrite breccia (samples e, f and g,) and red alteration rims (samples h and i) are found at relatively shallow depth at TAG-2, the southeasterly side of the mound. These chlorites contain SiO$_2$, (34.5, 35.4, and 39.9 wt%, respectively) and Fe$_2$O$_3$ (32.4, 30.1, and 23.9 wt%, respectively), lower silica and higher iron than observed in smectites at TAG-4 and in the basalt. The red rims contain SiO$_2$, (35.13 and 35.41 wt%, respectively) and Fe$_2$O$_3$ (17.92 and 25.29 wt%, respectively), also lower silica and, in sample i, higher iron than at TAG-4 and in the basalt. Smectites recovered from pyrite silica breccia at relatively shallow depth at TAG-1 (samples a and b) contain SiO$_2$, (35.5 and 37.2 wt%, respectively) and Fe$_2$O$_3$ (31.3 and 29.8 wt%, respectively), similar to the chlorites from TAG-4. Chlorites recovered from the chloritized basalt, from below 115 mbsf at TAG-1, contain SiO$_2$, (38.5 and 40.9 wt%, respectively) and Fe$_2$O$_3$ (39.9 and 40.4 wt%, respectively), similar to the smectites from TAG-2.

With the exception of one red rim (bulk rock sample i), magnesium contents in the TAG-4 samples and TAG-1 samples recovered from below 115 mbsf are lower (8.59 to 17.2 wt%) than the magnesium contents in the TAG-2 samples and TAG-1 samples recovered above 50 mbsf (24.4 to 29.0 wt%). The smectites and alteration rims from the TAG-4 region have a narrow range of MgO contents, only slightly higher than the basalt. The MgO contents of chlorites from the TAG-2 region also have a narrow range, but are much higher than the basalt. Altered rims from the TAG-2 region have a larger range of MgO contents (16.4-26.6 wt%). The smectites from the shallower region of TAG-1 are higher (27.2 and 24.4 wt% MgO) than the basalt, more similar to the MgO contents of the chlorites from TAG-2. The values of chlorites from the deeper region of TAG-1 (17.2 and 12.8 wt% MgO) are closer to that of the basalt and more similar to the smectites from TAG-4.

3 DISCUSSION

Formation of Fe-Mg-Zn-bearing clay minerals, such as those observed in the TAG rocks, obviously requires an abundant source of these elements plus SiO$_2$. Because the local bottom water contains less than detectable concentrations of dissolved Fe, Zn and dissolved SiO$_2$ near 30 µM (Humphris, Herzig, Miller, et al., 1995), the primary source of Fe, Zn and SiO$_2$ must be the basalt. Sources of Mg can be either or both the basalt and seawater. Mg is quantitatively removed from seawater early in its hydrothermal evolution history (e.g. Bischoff and Seyfried, 1978; Von Damm, 1988). A deeply circulating end-member hydrothrmal fluid is expected to have a different major-element compostion that a fluid dominated by local bottom water. Clay minerals formed in association with fluids of different circulation histories may reflect this difference.

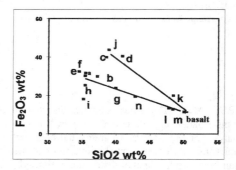

Figure 1. Weight percent SiO$_2$ and Fe$_2$O$_3$ in clay minerals and alteration rims.

Data shown in Figure 1 suggest two distinct trends of Fe$_2$O$_3$/SiO$_2$ in clay minerals. Fe$_2$O$_3$/SiO$_2$ of the unaltered basalt is at the intersection of the two trend lines. Smectites associated with the pyrite breccia, found at relatively shallow depths in the central mound (TAG-1) region plot along the more steeply sloping line. Chlorites found below 115 mbsf at the TAG-1 region, associated with chloritized basalt, plot along the less steeply sloping line. Samples from the western portion of the mound (TAG-4) cluster along the same trend as the chlorites found deeper at TAG-1. Samples from the southeasterly portion of the mound (TAG-2) cluster along the same trend as the smectites from the shallow portion of TAG-1.

The Fe$_2$O$_3$/SiO$_2$ relationship observed in the alteration rims is less clear. The Fe$_2$O$_3$/SiO$_2$ data shown in Figure 1 for the alteration rims plot along the more shallowly sloping line. However, the alteration rim analyses were performed on bulk-rock samples. The presence of quartz intermixed with the clay minerals would dilute iron concentrations and enrich silica concentrations measured relative to the clay mineral composition, thus moving the point plotted down and to the right on the diagram. The different trends between Fe$_2$O$_3$/SiO$_2$ and basalt composition, suggest that different regions of the mound have been influenced by fluids of different Fe$_2$O$_3$/SiO$_2$ values.

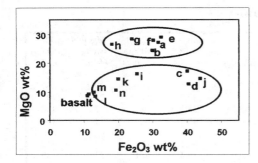

Figure 2. Weight percent Fe2O3 and MgO in clay minerals and alteration rims.

Figure 2 suggests two fields of Fe_2O_3/MgO relative ratio. Samples recovered from TAG-2 and from the shallow portion of TAG-1 plot in a field of relatively high MgO. Samples recovered from the deeper portion of TAG-1 and from TAG-4 plot in the field of relatively lower MgO. One sample from TAG-2, plots in the field of relatively lower MgO. However, MgO and Fe_2O_3 concentrations of that sample were measured on bulk rock from a red alteration rim and may be diluted by quartz.

Fluids circulating through the southeasterly side of the mound (TAG-2) and the upper portion the central mound (shallow portion of TAG-1) produced clay minerals with relatively high MgO contents, suggesting a fluid near to normal sea water composition during the episode(s) of clay mineral formation. Fluids circulating through the westerly (TAG-4) and deep central portion of the mound at TAG-1 produced clay minerals with relatively low MgO contents, suggesting that the host fluid is end-member hydrothermal fluid.

Another important finding is that Zn concentrations in smectite and chlorite are higher around the peripheral regions of the mound (TAG-2 and TAG-4) than in central portion of the mound and that the red alteration rims have lower Zn content than the green rims (Table 1). The highest Zn concentrations were measured in the TAG-4 area. These regions with Zn-enriched clay minerals are associated with the pyrite breccias, interpreted to be chimney talus (Humphris et al., 1996). Zinc in the chlorites and smectites from the TAG-1 region are relatively low, with the highest Zn concentrations in this region found in the shallowest samples. The variability in Zn concentrations associated with clay minerals may represent differences in dissolved Zn in circulating fluids. The variation in Zn associated with different colors of the alteration rims also implies specific fluid zones.

4 CONCLUSIONS

The observed distributions of Fe_2O_3/SiO_2, MgO, and Zn in chlorites and smectites from the TAG mound are consistent with two fluid sources: 1) end-member hydrothermal fluid originating from a high-temperature reaction zone, and 2) local bottom water entrained into shallow portions of the mound. These data suggest that mixing of the two fluids within the mound are not uniformly distributed. Locally derived seawater has a larger influence on the clay minerals formed in the southeastern TAG-2 and shallow central TAG-1 regions than in other parts of the mound. This is consistent with measured exit temperatures of present-day fluids (260°-300°C) at the TAG-2 region. Clay minerals recovered from the deeper central portion of the mound were formed in association with a greater proportion of end-member hydrothermal fluids. This is consistent with the fluids presently exiting with temperature in excess of 360°C from the apex of the mound at the black smoker complex. Based on the clay-mineral composition in the shallow portions of TAG-1, fluids presently exiting at the black smoker complex must pass rapidly through the talus mound with little loss of dissolved components into clay minerals. The clay minerals from the shallow regions of TAG-1 were apparently derived from an episode of clay formation from fluids more similar to those presently exiting from the TAG-2 region (that is, fluids derived from mixing a greater proportion of local bottom water).

Interpretation of the TAG-4 region data is more ambiguous. Becker et al. (1993) suggest that the northwesterly side of the mound is a region of relatively low heat flow, perhaps a zone of bottom-water entrainment into the mound. However, the clay mineral data reflect authigenic mineral formation from end-member hydrothermal fluid, perhaps similar to that presently exiting from the black smoker complex. Also, Zn contents in the clay minerals and green alteration rims from the TAG-4 region are more consistent with a fluid relatively high in dissolved Zn. This suggests that the Zn may be acquired from a different composition fluid, through ion exchange at some time after the mineral has formed and implies multiple episodes of fluid circulation.

Table 1. Chemical composition of clay minerals and altered bulk rock.

Location	Sample	symbols shown in figures 1 & 2	depth mbsf	lithology	SiO2 (wt%) +/- 3.4	Fe2O3 (wt%) +/- 0.7	MgO (wt%) +/- 1.3	Zn (mg/kg) +/- 17
TAG-1	158-957C-13N-02,24-27	a	37.4	2	35.5	31.6	27.2	267
	158-957C-16N-01,28-32	b	46.5	2	37.2	29.8	24.4	199
	158-957E-17R-01,01-04	c	116.1	3	38.5	39.9	17.2	188
	158-957E-18R-01,33-35	d	121.1	3	40.9	40.4	12.8	180
TAG-2	158-957B-04R-01,28-30	e	20.1	1	34.5	32.4	29.0	418
	158-957B-04R-01,39-41	f	20.4	1	35.4	30.1	27.9	349
	158-957B-04R-01,55-57	g	20.6	1	39.9	23.9	28.4	403
	158-957B-04R-01,25-27	h		4	35.13	17.92	26.61	330
	158-957B-05R-01,04-09	i		4	35.41	25.29	16.41	207
	158-957B-4R-01,55-62*	basalt	20.6	4	49.97	11.27	9.34	77.8
TAG-4	158-957M-09R-01,25-27	j	42.6	1	38.9	43.7	14.7	1106
	158-957M-10R-01,96-100	k	47.2	4	48.4	19.8	14.5	627
	158-957M-09R-01,67-71	l		4	47.77	13.25	8.59	1572
	158-957M-10R-01,79-74	m		4	48.29	12.75	9.88	345
	158-957M-10R-02,11-14	n		4	42.77	19.24	10.63	4316
	158-957M-09R-01,38-43*	basalt	42.8	4	50.38	10.96	8.72	98.98
	local bottom water				544	<0.1	54.4	
	TAG black smoker fluid+				659	1.64	0	

1 = massive pyrite breccia; 2 = pyrite silica breccia; 3 = chloritized basalt; 4 = basalt

5 REFERENCES

Becker, Von Herzen, and Rona (1993) Conductive heat flow measurements using Alvin at the TAG hydrothermal mound, EOS, 74:99.

Bischoff and Seyfried (1978) Hydrothermal chemistry of seawater from 25°C to 350°C. Amer. Jour. Sci., 278, 838-860.

Campbell, Klinkhammer, Palmer, Bowers, Edmond, Lawrence, Casey, Thompson, Humphris, Rona, and Karson (1988) Chemistry of hot springs on the Mid-Atlantic Ridge: TAG and MARK sites. Nature, 335:514-519.

Fujioka, Von Herzen, and the Shipboard Scientific Party (1994) Shinkai-Yokosuka MODE '94 Leg 2 Cruise Summary: Studies of an active hydrothermal mound at the TAG area on the Mid-Atlantic Ridge. Interridge News, 3:21-22.

Humphris, Herizg, Miller, et al. (1996) Initial Reports, Proceedings of the Ocean Drilling Program, Volume 158.

Humphris, Herizg, Miller, et al. (in press) Scientific Results, Proceedings of the Ocean Drilling Program, Volume 158.

Humphris, Herizg, Miller, et al.(1996) Internal structure of an active sea-floor massive sulphide deposit. Nature, 377, 713-716.

Smith and Humphris (in press) In Humphris, Herizg, Miller, et al. Scientific Results, Proceedings of the Ocean Drilling Program, Volume 158.

Sturz, Itoh, and Smith (in press) In Humphris, Herizg, Miller, et al. Scientific Results, Proceedings of the Ocean Drilling Program, Volume 158.

Tivey, Humphris, Thompson, et al. (1994) Deducing patterns of fluid flow and mixing within the active TAG mound using mineralogical and geochemical data. Jour. Geophys. Res., 100, 12527-12555.

Von Damm (1988) Systematics of and postulated controls on submarine hydrothermal solution chemistry. Jour. Geophys. Res., 93, 4551-4561.

Von Damm (1995) Controls on the chemistry and temporal variability of seafloor hydrothermal fluids. In Seafloor Hydrothermal System, Humphris, Zierenberg, Mullineaux and Thompson, editors. Geophysical Monograph

13 Fluids and tectonics

Rock-exchanged fluid oxygen isotope ratios in active collisional mountain belts, Pakistan and New Zealand

D. Craw & P.O. Koons
Geology Department, University of Otago, Dunedin, New Zealand

C.P. Chamberlain & M. Poage
Department of Earth Sciences, Dartmouth College, Hanover, N.H., USA

ABSTRACT: Rock-exchanged oxygen isotope ratios for vein-forming fluids have been estimated using laser ablation oxygen isotope data on quartz veins from two active mountain belts. Meteoric water penetrating into a tectonic thermal anomaly at Nanga Parbat, Pakistan, is fully equilibrated with the host rocks because of enhanced isotopic exchange between low density dry steam and hot rock. Isotopically exchanged fluids in the Southern Alps, New Zealand rise almost to the surface at the crest of the mountains. Rise of these deep-sourced fluids is due to enhanced permeability resulting from strain partitioning and dilatation during oblique collisional tectonics.

1 INTRODUCTION

Oxygen isotope ratios of vein minerals in ancient metamorphic belts are almost invariably the same as the same minerals in the host rocks. This is because rocks contain a high proportion of oxygen (ca. 50%) which readily exchanges with hot fluids forming veins. Irrespective of the original source of the fluid, water-rock interaction ensures that isotopic exchange results in a relatively uniform system. Hence, without some additional constraints, oxygen isotope studies generally provide little useful information.

In active mountain belts, direct and indirect observations of structure, geometry and fluid chemistry provide constraints on sources and migration of fluids. This additional information can then be combined with oxygen isotope data to deduce the nature of processes in the middle to shallow crust in tectonically-driven hydrothermal systems. This paper outlines two examples where oxygen isotopes, in conjunction with other data, can be used to identify and quantify active crustal processes.

2 NANGA PARBAT, PAKISTAN

2.1 Fluids and oxygen isotopes

Nanga Parbat is an 8000 m mountain of Indian basement rocks which is currently undergoing uplift of at least 5 mm/year (Zeitler et al. 1993). The rapid uplift results in an elevated geothermal gradient beneath the mountain, and surface hot spring activity due to topographically driven circulation of meteoric water through this thermal anomaly (Craw et al. 1994; Winslow et al. 1994). The thermal anomaly also ensures that the brittle-ductile transition is elevated to lie as close as 6 km below surface (Craw et al. 1994). Post-metamorphic quartz veins formed as the rocks became brittle during uplift right across the massif. The veins record, in fluid inclusions, the nature of the fluid which filled the fractures during vein formation.

There are three main vein-forming fluids: water-rich, carbon dioxide-rich, and mixtures of these two (Fig. 1). The carbon dioxide-rich fluid occurs in veins in relatively low grade lower amphibolite facies rocks around the north, east and south sides of Nanga Parbat. The water-rich fluid occurs in veins in the highest grade rocks in the core of the massif. Mixtures between these fluids occur in veins in rocks between the core and the rim of the massif (Fig. 1). Water-carbon dioxide mixtures were also trapped in fluid inclusions in metamorphic rocks (not veins) in the core of the massif (Winslow et al. 1994). The carbon dioxide-rich fluid is metamorphic and/or magmatic in origin (Craw et al. 1997).

Water-rich fluid inclusions are almost exclusively low density dry steam trapped at 350-410°C with densities between 0.3 and 0.07 g/cm^3 (Craw et al. 1997). This water has a meteoric origin and has penetrated deep into the basement rocks driven by the high topography (Chamberlain et al. 1995; Craw et al. 1997). Meteoric water penetration along shear zones has resulted in lowering of oxygen isotopic ratios of host rocks (Chamberlain et al. 1995).

A set of oxgen isotope ratios for quartz veins cutting host rock schistosity has been obtained in a near-complete transect across the massif from west to east. Isotope analyses were obtained using the laser

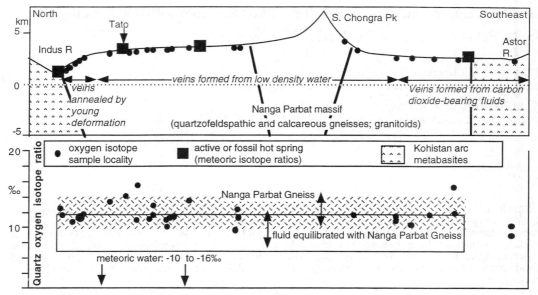

Fig. 1. Cross section through the Nanga Parbat massif, Pakistan, showing relief, hot springs, and sampled quartz vein localities. Oxygen isotope analyses of quartz from these localities are indicated in the lower portion of the diagram, with calculated fluid isotopic composition (fractionations after Matsuhisa et al. 1979).

ablation line at Dartmouth College, NH, USA and NBS-28 as standard. Analysed values have an error of ±0.2 per mil.

Oxygen isotope ratios for quartz veins are remarkably constant across the massif, and are similar to ratios for the host rocks (Fig. 1). There is no discernable difference between veins formed from dry steam in the core of the massif and veins formed from carbon dioxide-rich fluids on the massif margins. The meteoric fluid had fully equilibrated isotopically with the host rocks, and is isotopically indistinguishable from the carbon dioxide-bearing deeper crustal fluid.

2.2 Low density fluid isotopic exchange

Such complete isotopic exchange is a result of interaction between rock and low density dry steam. Calculated water/rock ratios (Ohmoto & Rye 1974) are distinctly lower if a density correction is used for the water. For a given mass ratio of, e.g., 1, low density water will exhibit much more pronounced isotopic exchange than higher density water (Craw et al. 1997). The actual amounts of water passing through the system may have been larger than implied by these calculations due to kinetic effects (Blattner & Lassey 1989), but the principle of differing water densities is still significant.

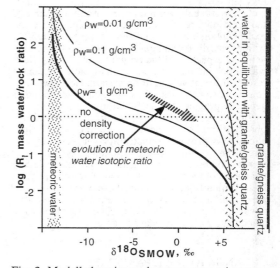

Fig. 2. Modelled exchange between meteoric water and hypothetical host rocks (after Ohmoto & Rye 1974; heavy line), using quartz =+10 per mil as representative of Nanga Parbat basement. Lighter lines show exchange with low density water; 0.1 g/cm^3 is representative of Nanga Parbat dry steam.

3 SOUTHERN ALPS, NEW ZEALAND

The Southern Alps are forming at the obliquely

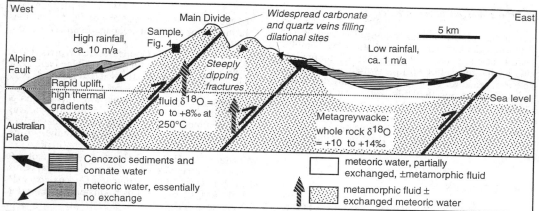

Fig. 3. Cross section through the Southern Alps, New Zealand, showing the topography and major structures (partly after Koons 1989; Cox & Findlay 1995). Approximate distribution of principal fluid zones and their direction of movement are indicated.

converging Pacific-Australian plate boundary. The structure varies along strike, but the main features can be depicted in cross section (Fig. 3). Uplift at about 8 mm/a is occurring adjacent to the Alpine Fault (Simpson et al. 1994), and erosion is keeping pace with uplift (Koons 1989). Uplift and erosion decrease eastwards.

Fluids passing through the western portion of the mountain belt are dominated by topography-driven meteoric water ($\delta^{18}O$= ca. -9‰, possibly down to the brittle-ductile transition (Jenkin et al. 1994). East of the main mountain belt, meteoric water and connate water form veins at shallow to moderate depths in Cenozoic sediments and adjacent basement rocks, particularly in fault zones (Templeton et al. 1997).

3.1 Main Divide region

The Main Divide region is cut by numerous generations of moderate to steeply dipping fractures and faults. This region is undergoing some dilatancy due to extension parallel to strike, driven by the obliquity of the collision (Koons et al. 1997). This dilatational region has focussed fluid flow from depth. Veins have formed in the fractures zones from below 5 km to near-surface regions (Cox et al. 1997). All veins formed in this zone have strongly positive $\delta^{18}O$, typically >10 per mil (Cox et al. 1997). Calculated fluid isotopic compositions depend strongly on temperatures of formation. Fluid inclusion estimates of about 250°C for many veins except the shallowest ones imply that the fluid had $\delta^{18}O$ near 0 to +8 per mil (Fig. 3).

Laser ablation isotopic analyses were conducted on a single quartz crystal with fluid inclusion temperatures well constrained by boiling (Craw 1997), to provide some more rigorous constraints on mineralization temperature (Fig. 4). These data confirm that the vein-forming fluid which formed this crystal, at about 500 m depth, had $\delta^{18}O$ of +1 to +5 per mil.

These positive $\delta^{18}O$ ratios are significant because they demonstrate that the hydrological system is dominated by rising hot fluids from depth. These hot, isotopically exchanged fluids can apparently penetrate almost to the surface. This is in strong contrast to the hydrological system only a few kilometres to the west where unexchanged or partially exchanged meteoric water dominates to substantial depths.

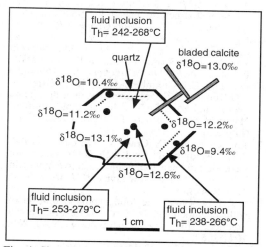

Fig. 4. Sketch cross section through a single quartz crystal from a cavity near the Main Divide (Fig. 3), showing the locations of oxygen isotope analyses, and fluid inclusion homogenization temperatures.

3.2 Unexchanged meteoric water

In both the active mountain systems discussed above, meteoric water can penetrate to shallow levels in the crust, heat up in the conductive thermal anomalies, and discharge as hot springs which still have a meteoric signature. This is almost certainly due to the lower temperatures that this water reached in the Southern Alps (<150°C, Allis & Shi 1995), and the same reason may apply at Nanga Parbat. Isotopic exchange is inhibited by unfavourable kinetics below about 150°C (Truesdell & Hulston, 1980)), in agreement with our empirical observations.

4 REFERENCES

Allis, R. G.& Y. Shi 1995: New insights to temperature and pressure beneath the central Southern Alps, New Zealand. *N.Z. J. Geol. Geophys.38*: 585-592.

Blattner, P. & K.R.Lassey 1989. Stable isotope exchange fronts, Damköhler Numbers and fluid to rock ratios. *Chem. Geol., 78*: 381-392.

Chamberlain, C.P., P.K.Zeitler, D.E.Barnett, D.Winslow, S.R.Poulson, T. Leahy & J.E. Hammer 1995. Active hydrothermal systems during the recent uplift of Nanga Parbat. *J. Geophys. Res., 100*: 439-453

Cox, S.C & R.H.Findlay 1995. The Main Divide Fault Zone and its role in formation of the Southern Alps, New Zealand. *N.Z. J. Geol. Geophys.38*: 489-499.

Cox, S.C., D.Craw & C.P. Chamberlain, 1997. Structure and fluid migration in a late Cenozoic duplex system forming the Main Divide in the central Southern Alps, New Zealand.*N.Z. J. Geol. Geophys.40* : 359-373

Craw, D. 1997: Fluid inclusion evidence for geothermal structure beneath the Southern Alps, New Zealand. *N.Z. J. Geol. Geophys.40* : 43-52.

Craw, D., P.O.Koons, D.Winslow, C.P. Chamberlain & P.K.Zeitler 1994. Boiling fluids in a region of rapid uplift, Nanga Parbat massif, Pakistan. *Earth Plan. Sci. Letters, 128*: 169-182.

Craw, D., C.P.Chamberlain, P.K. Zeitler & P.O.Koons 1997. Geochemistry of a dry steam geothermal zone formed during rapid uplift of Nanga Parbat, northern Pakistan. *Chem. Geol.* (in press).

Jenkin, G. R. T., D. Craw & A.E.Fallick 1994. Stable isotopic and fluid inclusion evidence for meteoric fluid penetration into an active mountain belt, Alpine Schist, New Zealand. *J. Meta. Geol. 12*: 429-444.

Koons, P.O. 1989. The topographic evolution of collisional mountain belts: a numerical look at the Southern Alps, New Zealand. *Amer. J. Sci. 289*: 1041-1069.

Matsuhisa, Y., J.R. Goldsmith & R.N. Clayton 1979. Oxygen isotopic fractionatin in the system quartz-albite-anorthite-water.
Geochim.Cosmochim. Acta 43: 1131-1140.

Ohmoto, H. & R.O.Rye 1974. Hydrogen and oxygen isotopic compositions of fluid inclusions in the Kuroko deposits, Japan. *Econ. Geol., 69*: 947-953.

Simpson, G. D., A.F.Cooper & R.J.Norris 1994. Late Quaternary evolution of the Alpine Fault Zone at Paringa, South Westland, New Zealand. *N.Z. J. Geol. Geophys. 37*: 49-58.

Templeton, A.S., C.P.Chamberlain, P.O.Koons & D. Craw 1997. Isotopic evidence for mixing between metamorphic fluids and surface-derived waters during Recent uplift of the Southern Alps, New Zealand *Earth Plan. Sci. Letters* (in press)

Truesdell, A. H.& J.R.Hulston 1980. Isotopic evidence on environments of geothermal systems. In: *Handbook of environmental Geochemistry* (P. Fritz & J. C. Fontes, eds), Amsterdam, Elsevier. Volume 1: 179-226.

Winslow, D.M., P.K.Zeitler, C.P.Chamberlain, & L.S.Hollister 1994. Direct evidence for a steep geotherm under conditions of rapid denudation, Western Himalaya, Pakistan. *Geology, 22*:: 1075-1078.

Zeitler, P. K., C.P. Chamberlain & H.A.Smith 1993. Synchronous anatexis, metamorphism, and rapid denudation at Nanga Parbat (Pakistan Himalaya) *Geology, 21*: 347-350.

Underpressured paleofluids and future fluid flow in the host rocks of a planned radioactive waste repository

L.W. Diamond
Institute of Mineralogy and Petrology, University of Bern, Switzerland

ABSTRACT: Low-permeability marls at Wellenberg, Central Helvetic Alps, are the host rocks of a planned radioactive waste repository for Switzerland. Fluid inclusions in the marls indicate that, during Oligo-Miocene Alpine metamorphism, two-phase CH_4-H_2O-NaCl fluids formed in-situ with pressures that evolved from lithostatic to subhydrostatic. The underpressures resulted from porosity being created by nappe folding under hydrodynamically closed conditions. Free fluids in boreholes exhibit the same two-phase state, the same compositions, and locally the same underpressured (now subhydrostatic) regime as the paleofluids. Whereas other workers have explained the modern underpressures as transient effects of post-glacial rebound, the new correlations suggest the pressure anomalies are relics of Alpine tectonics. The implications are that zones of the marls have remained hydrodynamically closed for ~20 Ma, and that therefore they are likely to remain closed into the distant future, potentially providing a suitable environment to isolate waste.

1 INTRODUCTION

The proposed host rocks of Switzerland's final underground repository for low- and intermediate-level radioactive waste are highly impermeable carbonaceous marls of the Palfris Formation at Wellenberg, Central Helvetic Alps. The marls were metamorphosed at low grade during the Oligo-Miocene Alpine orogeny, and concurrent folding and thrust imbrication thickened the unit to over 5 times its original sedimentary thickness.

As part of a broad geological assessment of the site by Nagra (Swiss National Cooperative for the Disposal of Radioactive Waste), boreholes have been drilled into the Palfris formation to sample present-day fluids. The holes encountered immiscible methane gas and various weakly saline aqueous solutions (Mazurek et al., 1988). Surprisingly, these fluids are under subhydrostatic pressures at certain depths within the unit (Fig.1). Preliminary modelling by Vinard et al. (1993) suggests that the observed underpressures could be due to transient poro-elastic dilation induced by post-glacial unloading several thousand years ago, but the input parameters for the calculations are not well constrained. The origin of the down-hole fluids and the period over which their underpressures are likely to recover are questions relevant to the future isolation of radioactive waste at Wellenberg.

In order to elucidate the history of fluid activity in the marls, fluid inclusions have been analysed in quartz-calcite veins and vugs recovered from the boreholes (Diamond & Marshall, 1994). This paper reports the results of this study; it proposes an origin

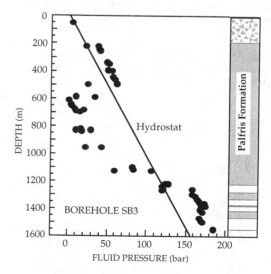

Figure 1. Present-day fluid pressures measured in borehole SB3 (from data in Vinard et al., 1993).

for the immiscible fluids and an alternative mechanism to explain the underpressures in the boreholes. These results lead to conclusions regarding the possible duration of the present hydraulic regime in the underpressured zones of the marls.

2 RESULTS

Several generations of veins are found in the Palfris-Formation. Early ankerite and grey calcite veins are strongly deformed by Alpine folding and truncated by tectonic stylolites composed of illite-smectite and mature kerogens. Lineations in the plane of the stylolites allow the local stress axes during folding to be identified. Younger veins and open dilatant vugs rimmed with euhedral quartz and calcite are arrayed in two conjugate sets perpendicular to the plane of the stylolites (parallel to the axis of compression). The vugs are up to several cm wide and several dm long.

Vein quartz and calcite contain many generations of primary and secondary fluid inclusions, but they all represent the same two contemporaneous, immiscible fluids: (1) a weakly saline, methane-saturated aqueous "liquid" and (2) a conjugate, water-saturated methane "vapour". Table 1 shows analyses of the two end-members based on microthermometry, laser-Raman spectroscopy, cryo-SEM-EDS analysis, and thermodynamic modelling. The relative amounts of the two compositional types of inclusions vary enormously between samples.

Because the inclusions formed from a two-phase parent fluid, entrapment conditions can be deduced directly from the homogenisation temperatures and homogenisation pressures of inclusions that contain

Table 1. Typical bulk compositions, densities and molar volumes of coexisting aqueous and methane-rich fluid inclusions.

		Aqueous inclusions	Methane inclusions
H_2O	(Mol%)	97.92	3.2
NaCl	(Mol%)	0.62	–
KCl	(Mol%)	<0.01	–
$CaCl_2$	(Mol%)	<0.01	–
CH_4	(Mol%)	1.46	95.4
C_2H_6	(Mol%)	–	<0.6
C_3H_8	(Mol%)	–	<0.13
CO_2	(Mol%)	<0.1	1.4
N_2	(Mol%)	–	<1.5
H_2S	(Mol%)	–	<0.15
Density (g/cm³)		0.95	0.36
Molar volume (cm³/mol)		19	46

either of the pure end-member fluids. Thus the veins are seen to have formed between 190-250 °C over a remarkably wide range of pressures, from 2500 to 400 bars (e.g. Fig. 2). All generations of inclusions show approximately the same range in temperatures. However, the earliest inclusions, which lie in crack-seal traces, formed at the highest pressures, whereas later primary and secondary inclusions in undeformed quartz and calcite formed at progressively lower pressures.

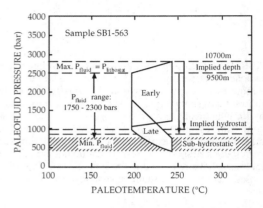

Figure 2. Reconstructed P-T conditions of formation of early and late fluid inclusions in sample SB1-563.

3 ORIGIN OF METAMORPHIC FLUIDS

The observation that all the aqueous fluid inclusions are saturated with respect to CH_4 is strong evidence that the two parent fluids in the veins shared a common origin, or at least a common history of intimate mixing during migration. The latter possibility, that the marls were flushed with a well mixed fluid of external origin, can be excluded on the grounds that: (1) CO_2 contents of the methane "vapour" inclusions vary between samples (i.e. between levels in the boreholes) but not within samples; (2) similar sample-specific variations are visible in the NaCl contents of the aqueous inclusions; (3) all chemical components present in the vein minerals and fluids are also present in the adjacent wall rocks; (4) no metasomatic reactions are present at the vein/wall rock contacts. All these features point to the veins having formed in highly localized, geochemically closed systems.

The source of the CH_4-H_2O-NaCl vein fluids which best fits the available evidence is the immediately adjacent marl wall rocks. CH_4 was produced by progressive thermal decomposition of wall rock carbonaceous matter during metamorphic heating, whereas the NaCl-H_2O components appear to have derived from sea water originally trapped the in rock pores during sedimentation and later diluted by H$_2$O from metamorphic dehydration of matrix clay minerals (Diamond and Marshall, 1994). These conclusions are well supported by analyses of stable and radiogenic isotopes, vitrinite reflectance and kerogen fluorescence (Ballantine et al., 1994; Mazurek et al., 1998).

The state of immiscibility in the Wellenberg veins was a simple consequence of the sub-solvus P-T conditions of in-situ fluid production and mineral precipitation during Alpine metamorphism.

4 ORIGIN OF FLUID P-T REGIME

The highest vein formation temperatures are consistent with the local peak temperatures of Alpine metamorphism, as indicated by kerogen maturity and mineral stabilities (Mazurek et al., 1998). Given the heating rates involved in the regional metamorphism, the 60°C span observed in vein formation temperatures is interpreted to reflect a prolonged period of fluid migration into the veins, preceding and outlasting the local metamorphic peak at ~20 Ma.

Whereas the metamorphic thermal regime in the wall rocks apparently buffered the vein fluid temperatures at all times, the wide spread in vein fluid pressures reflects an evolving hydrodynamic regime during progressive folding, cavity opening and mineral deposition. The nature of this regime is indicated by three points of evidence: (1) The small amounts of quartz and calcite precipitated on the vein and vug walls can be accounted for by the volume reduction caused by pressure solution in the stylolites. This, together with the fact that quartz and calcite are excess phases in the marl matrix, implies that only tiny amounts of fluid were present in the rocks; (2) As the earliest fluid inclusions lie in crack-seal traces, the fluid pressure during incipient veining can be assumed to have been close to lithostatic (~2500 bar). If this lithostat is used to calculate the depth of veining (9.5-10.7 km), then the fluid pressures recorded by later fluid inclusions (as low as 400 bar) are seen to be nominally subhydrostatic in some cases (e.g. Fig. 2); (3) When pressures deduced from inclusions of the same late generation are plotted as a function of depth (e.g. Fig. 3), it is evident that fluid pressure does not increase with depth.

Clearly, fluid pressure during vug opening was independent of the lithostatic load. Short-circuits in flow paths could be imagined in deforming units to explain the apparently subhydrostatic pressures as true hydrostatic pressures, but the compositional properties of the fluids argue against the presence of long drainage routes leading outside the marls. Fluid pressure thus appears to have been independent of the hydrostatic head as well. Fluid pressures in the vugs are therefore interpreted to have evolved in

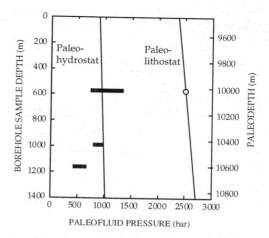

Figure 3. Late-stage paleofluid pressures (black bars) in vugs from borehole SB1 at ~220 °C. The lithostat is drawn through the maximum pressure of early fluid inclusions (circle), as shown in Fig. 2.

highly localized, hydrodynamically closed systems within the marls.

The process of cavity opening and mineral deposition can be reconstructed in the following simplified model (Fig. 4). Thermal decomposition of kerogen near the metamorphic peak generated high methane pressures that eventually exceeded the lithostat by amounts equivalent to the tensile strength of the marls, leading to hydrofracturing parallel to the axis of compression, σ_1 (Fig. 4a). The fractures were initially sealed by quartz and calcite dissolved from the adjacent stylolites (Fig. 4b). Subsequent extension of the marls (parallel to σ_3 in Fig. 4) opened the existing fractures into macroscopic cavities, but two-phase fluid flow into these voids was too meagre to maintain lithostatic pressures. Hence fluid pressures in the cavities continued to drop as more space opened due to gradual extension, and the final pressures were dictated by the balance between fluid influx and available space (Fig. 4c).

5 ORIGIN OF BOREHOLE FLUIDS AND PRESENT-DAY UNDERPRESSURES

The compositions of the immiscible metamorphic fluids represented by the Alpine fluid inclusions match closely the compositions of free fluids found in the deeper segments of the boreholes: dilute, aqueous NaCl solutions compare well with the aqueous "liquid" phase in the fluid inclusions, while the free CO$_2$-bearing methane gas has its analogue in the conjugate "vapour" phase of the inclusions. Even the isotopic characteristics of the borehole fluids are

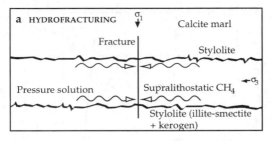
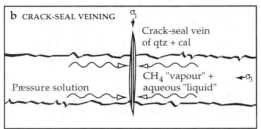
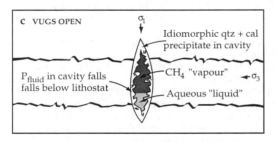
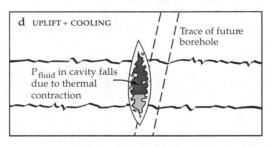

Figure 4. Simplified reconstruction of vug opening and the generation of sublithostatic fluid pressures. (a-c) Sequence of events during the peak of Alpine metamorphism (190-230 °C) in the Palfris Formation, at ~20 Ma. (d) Post-metamorphic cooling and uplift to present-day situation.

similar to those of the fluid inclusions (Mazurek et al., 1988). These correlations strongly suggest that the present-day fluids are simply relict Alpine metamorphic fluids, which have been released from vugs in the marls by drilling (Diamond & Marshall, 1994).

In the light of these compositional correlations it seems highly likely that the modern underpressures in the borehole fluids are also inherited from paleo-underpressures in the Alpine vugs. The much lower absolute pressures measured today in the boreholes are presumably the consequence of quasi-isochoric thermal contraction of the vug fluids upon post-metamorphic cooling of the Palfris Formation (Fig. 4d).

6   CONCLUSIONS

Alpine metamorphism and deformation some 20 Ma ago generated anomalously underpressured, two-phase CH_4-H_2O-NaCl fluids in the Palfris marls. The absence of any fluids of external origin, and the fact that underpressures could be generated at all in the vugs, attest to the extreme impermeability and low fluid content of the marls. Boreholes in the deeper sections of the marls have tapped the same relict metamorphic fluids in the vugs, and so it is concluded that the marls have remained hydrodynamically closed for at least the last 20 Ma.

Although transient processes such as post-glacial rebound may have contributed to the present-day fluid underpressures, the borehole observations can be fully accounted for by inheritance from the Alpine tectono-metamorphic event. There is therefore good reason to expect that the underpressured zones of the Palfris Formation will remain hydrodynamically closed well into the future, and hence they may provide a suitably isolated environment for the disposal of radioactive waste.

7   REFERENCES

Ballentine, C.J., Mazurek, M. & Gautschi, A. 1994. Thermal constraints on crustal rare gas release and migration: Evidence from Alpine fluid inclusions. *Geochim. Cosmochim. Acta* 58:4333-4348.

Diamond L.W. & Marshall D.D. 1994. Fluid inclusions in vein samples from boreholes at Wellenberg, Switzerland. *Nagra Interner Bericht* 94-12. 88 pp.

Mazurek, M., Waber, H.N., & Gautschi, A. 1998. Hydrocarbon gases and fluid evolution in very low-grade metamorphic terranes: A case study from the Central Swiss Alps. *Proc. 9th Int. Symp. Water-Rock Interaction* (this volume).

Vinard P., Blümling P., McCord J.P., & Aristonenas G. 1993. Evaluation of hydraulic underpressures at Wellenberg, Switzerland. *Ind. J. Rock. Mech. Mining Sci.* 30:1143-1150.

Soil gas emissions and tectonics in volcanic areas of Italy and Hawaii

S. Gurrieri, S. De Gregorio, I. S. Diliberto & S. Giammanco
Istituto di Geochimica dei Fluidi, CNR, Palermo, Italy

M. Valenza
Istituto di Mineralogia, Petrografia e Geochimica, Università di Palermo, Italy

ABSTRACT: Measurements of CO_2 and He concentration in the soil along with diffuse CO_2 flux were carried out in some active and extinct volcanic areas of southern Italy and Hawaii, in order to highlight the main anomalous zones and to correlate them with the local tectonic framework. Anomalous soil degassing was preferentially found in areas subject to tectonic strain rather than along eruptive fissures. This confirms that faults are preferential pathways for migration of deep gases toward the surface. The methodology used in the present work also proves useful in highlighting buried active faults, thus helping constrain the structural models of the studied areas.

1 INTRODUCTION

Soil gas emissions occur preferentially along tectonic structures as observed in both volcanically and seismically active areas (e.g., Irwin and Barnes, 1980; Klusman, 1993). This phenomenon was already studied in order to detect faults in areas where they are buried or hidden (Giammanco et al., 1997). In this study we present data relevant to different volcanic and seismic areas of Sicily (Mount Etna volcano and Naftia Lake) and Hawaii (Kilauea volcano). The aims of the investigation were to:
1. Study existing faults as regards soil degassing.
2. Highlight buried faults.
3. Develop structural models based on the geochemical data.
4. Verify whether this prospecting method is applicable in different geological contexts.

2 SAMPLING AND ANALYTICAL METHODS

The concentrations of CO_2 and He, as well as the flux of CO_2 were measured in the soil at a depth of 50 cm. CO_2 fluxes were determined in the field using the method of Gurrieri & Valenza (1988), while CO_2 and He concentrations were respectively measured with a IR spectrophotometer (Riken Keiki 550A) and a modified leak detector (Leybold). The ^{13}C (CO_2) isotopic compositions of some fumarolic and soil gas samples were analysed in the laboratory using a mass spectrometer for stable isotopes (Finnigan Delta S). Soil gases were measured along lines with a sampling interval ranging between 20 and 200 m, depending on logistics and detail of information required. This allowed us to complete each line in few hours, therefore variations in the degassing activity and/or in atmospheric conditions could be assumed negligible. Most of the sampling lines were almost orthogonal to the main tectonic structures. In the case of Naftia Lake the measurements were carried out on a grid of points in order to obtain maps of concentrations and fluxes. A threshold of $4 \cdot 10^{-5}$ $cm^3 cm^{-2} s^{-1}$, corresponding to 1.44 $lm^{-2} h^{-1}$, was chosen to highlight the CO_2 flux anomalies. This value represents the highest CO_2 flux due to microbial activity in the soil (Kanemasu et al., 1974).

3 RESULTS

Given the very large number of measurements (about 1400 sites), a table with all the data is not displayed. Some representative lines (Etna and Kilauea areas) and a contour map of CO_2 flux (Naftia Lake zone) are shown instead.

3.1 *MT. ETNA (East Sicily, Italy).* Measurements were carried out in about 1000 sites distributed along 26 sampling lines in different areas of the volcano. On the NE flank (box E in Fig.1) anomalous fluxes of CO_2 were found on the Pernicana Fault, along directions parallel to it and

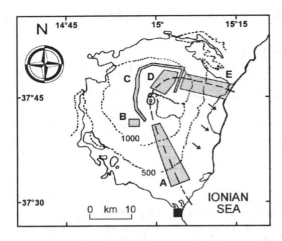

Figure 1. Location of investigated areas on Etna: A) SE rift, B) Milia, C) Forest road, D) NE rift, E) Pernicana. Arrows show the direction of the collapsing sector (dashed lines).

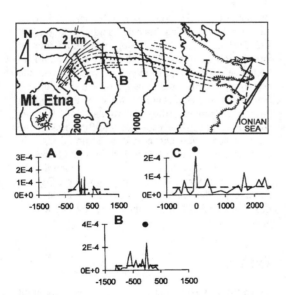

Figure 2. Map of the structural model of NE rift and Pernicana fault (boxes D and E in fig. 1) based on soil gas data (upper part of the figure). Ticked solid lines = Pernicana fault, dashed lines = hypothesised buried faults, solid lines = NE rift. The location of the sampling lines is also shown. The diagrams below the map show some profiles of CO_2 fluxes. In each profile abscissas indicate distances (in metres) from the Pernicana fault plane (solid circle) and the horizontal lines indicate the threshold ($4 \cdot 10^{-5}$ $cm^3 cm^{-2} s^{-1}$, see text).

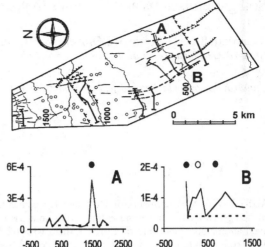

Figure 3. The upper part of the figure shows the structural model of the SE-rift (box A in figure 1) based on soil gas data. Ticked solid lines = known faults (dashed when uncertain), solid lines = eruptive fissures, open circles = parasitic cones, dashed lines = hypotised buried faults. The location of the sampling lines is also shown. The diagrams below the map show two lines of CO_2 flux and the intercepted faults (filled circles) and eruptive fissures (open circles). The abscissas indicate distances (in metres) from the beginning of each line. The horizontal dotted lines indicate the threshold (see text).

also parallel to the "NE rift" (box D in Fig.1). Some anomalies are located in zones with no surface trace of tectonic structures, thus suggesting the presence of a number of buried faults (Fig. 2). These latters, together with the main one, would make a complex system, which extends for about 18km from the Ionian coast up to 2500m a.s.l. on the northern slopes of Etna (Fig. 2). The complex structure seems to have quite deep roots as suggested by the isotopic composition of CO_2 (Giammanco et al., 1997).

The above hypothesised system should represent the northern boundary of a large eastward collapsing sector of the volcanic edifice, as proposed by Borgia et al. (1993). The "SE rift" represents the southern boundary of this sliding sector (box A in Fig.1). The SE rift is another complex structure consisting of many faults, both visible and buried, as highlighted with the geochemical surveys (Fig.3).

The lack of anomalous degassing along the fissures of the "NE rift" would suggest that, after each eruptive episode, the cooled dikes seal the fissures, thus preventing the deep gases from being drained towards the surface. In general, this low permeability condition can be reverted if tectonic strain occurs. According to our data, this does not seem to be the case of the NE rift.

Similar results were obtained in the other areas of Etna: soil CO_2 degassing takes place preferentially along active faults, even though buried, rather than along eruptive fissures. In fact, anomalous soil degassing was found along 95% of the investigated faults and along 42% of the eruptive fissures. The degassing eruptive structures found in these areas should be connected to active faults. This is also supported by the agreement between the orientation of the exhaling eruptive fissures and the direction of the main structural trends in the region.

3.2 NAFTIA LAKE (South East Sicily, Italy).

This area is located about 50km South East of Etna. It was characterised by active volcanism until 1.7 Ma ago (Barberi et al., 1974). In historical times, this area was marked by strong seismicity and by very high soil CO_2 emissions. The highest exhaling site is actually exploited for commercial uses (Mofeta dei Palici plant). No fault surface traces are known in this zone, but volcanic eruptions in neighbour areas were driven by lithospheric faults oriented about N70E (Carbone et al., 1987). This structural direction belongs to an important regional trend, which affects Etna as well.

Measurements were carried out on a grid of 180 sampling points distributed over a surface of about 10 km^2. The areal distributions of He and CO_2 concentrations as well as that of CO_2 fluxes is characterised by the same pattern, and suggest the presence of a structural direction well oriented as the main regional trend. Furthermore, the highest He content (100 ppm by volume) was measured in coincidence with the largest anomaly of CO_2 flux and of CO_2 concentration. The isotopic composition of He from the highest degasssing site (6.6 R/Ra, after Polyak et al., 1979) indicates a deep origin, likely mantle, of the emitted gases.

The estimated total amount of CO_2 output from this area is about 2400 tons/day. This value was calculated assuming a linear model of continuity as suggested by the variogram (Klusman, 1993) and it

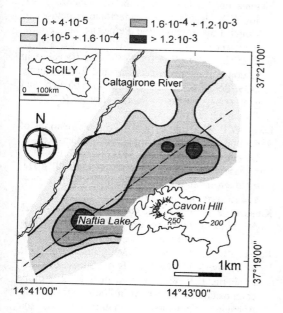

Figure 4. Areal distribution of CO_2 fluxes in the Naftia Lake area. The dashed line represents the hypothetical structural trend. Values expressed in cm/s.

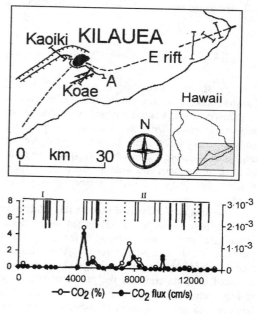

Figure 5. Location of surveyed lines on Kilauea (top of figure). Ticked solid lines = main structural trends, dashed lines = rift zones. The diagram shows values of CO_2 flux and concentrations measured along line A, as well as the intercepted faults (solid lines) and eruptive fissures (dashed lines). Vertical thick lines = eruptive fissures, vertical thin lines = faults (dashed when uncertain), I = Kilauea caldera area, II = East rift area.

does not include the amount of CO_2 (few hundred tons/day) trapped by the Mofeta dei Palici plant.

3.3 KILAUEA (Hawaii, USA). 220 measurements of He and CO_2 concentrations as well as of CO_2 fluxes in the soils were carried out on Kilauea. Sampling lines were performed across the summit caldera, the upper East rift zone, the Koae fault system and across the lowest portion of the East rift zone (Fig. 5). Some measurements were also carried out on Mauna Loa over the Kaoiki fault system that borders Kilauea to the northwest. The highest He and CO_2 concentrations were found near the summit crater of Kilauea and just to the NE of it in an area of steam and sulfur vents. As for Etna, fewer soil gas anomalies were found along eruptive fissures than along purely tectonic structures. However, the tectonic structures located in the western and northwestern parts of Kilauea's caldera showed gas emissions much lower than those in the southern flank (line A of figure 5). Buried faults were detected on the lower part of the East rift area. Anomalous soil degassing was found on the Kaoiki faults as well.

4 CONCLUSIONS

Soil gas studies carried out in southern Italy and Hawaii confirm that tectonic structures in general act as pathways for the upward migration of crustal and sub-crustal gases. Different levels of soil degassing between faults and eruptive fissures suggest that any degassing structure is one subject to strain. Actually, tectonic phenomena are able to keep open those fissures that were previously sealed by magmatic dikes.

The prospecting method used in this study also allowed us to highlight several buried faults in the investigated areas, thus providing further information on the local structural trends. The tectonic framework inferred from the geochemical data acquired on Etna confirms the presence of a huge sliding sector of the volcano towards the east. On Kilauea, soil gas data seem to indicate extensional tectonic forces acting mainly in the southern sector of the volcano and along the rift zones.

The same prospecting method applied to a different geological context proved useful, too. In fact, around Naftia Lake, an extinct volcanic area lacking in surface fault traces, soil gas anomalies suggest a structural direction oriented in agreement with the main regional trend.

5 ACKNOLEDGEMENTS

Work partially supported by financial contribution of Gruppo Nazionale per la Vulcanologia - C.N.R. (Italy).

6 REFERENCES

Barberi, F., Civetta, L., Gasparini, P., Innocenti, F., Scandone, R. & Villari, L. 1974. Evolution of a section of the Africa - Europe plate boundary: paleomagnetic and volcanology evidence from Sicily. *Earth Planet. Sci. Lett.*, 22:123 - 132.

Borgia, A., Ferrari, L. & G. Pasquarè 1992. Importance of gravitational spreading in the tectonic and volcanic evolution of Mt. Etna. *Nature* 357:231-234.

Carbone, S., Grasso, M. & Lentini, F. 1987. Lineamenti geologici del plateau ibleo (Sicilia S.E.). Presentazione delle carte geologiche della Sicilia sud-orientale. *Mem. Soc. Geol. It.*, 38:127 - 135.

D'Alessandro, W., Giammanco, S., Parello, F. & M. Valenza 1997. CO_2 output and $\delta^{13}C(CO_2)$ from Mount Etna as indicators of degassing of shallow astenosphere. *Bull. Volcanol.* 58:455-458.

Giammanco, S., Gurrieri, S. & M. Valenza 1997. Soil CO_2 degassing along tectonic structures of Mount Etna (Sicily): the Pernicana fault. *Appl. Geochem.* 13.

Gurrieri, S. & M. Valenza 1988. Gas transport in natural porous mediums: a method for measuring CO_2 flows from the ground in volcanic and geothermal areas. *Rend. Soc. It. Min. Petrog.* 43:1151-1158.

Irwin, W.P. & I. Barnes 1980. Tectonic relations of Carbon Dioxide discharges and earthquakes. *J. Geoph. Res.* 85(B6):3115-3121.

Kanemasu, E.T., Powers, W.L. & J.W. Sij 1974. Field chamber measurements of CO_2 flux from soil surface. *Soil Sci.*,118: 233-237.

Klusman, R.W. 1993. *Soil gas and related methods for natural resource exploration*. New York: John Wiley and Sons.

Polyak, B.G., Prasolov, E.M., Buachidze, G.I., Kononov, V.I., Mamyrin, B.A., Surovtseva, I.I., Khabarin, L.V. & V.S. Yudenich 1979. Isotope composition of He and Ar in fluids of the Alpine-Apennine region and its relation with volcanic activity. *Doklady Akad. Nauk SSSR* 247:1220-1225.

Mineral-water interactions and stress: Pressure solution of halite aggregates

R. Hellmann, J. P. Gratier & T. Chen
Crustal Fluids Group, CNRS, Observatoire de Grenoble, Université Joseph Fourier à Grenoble, LGIT-IRIGM, France

ABSTRACT: The deformation of synthetic halite aggregates in the presence of circulating fluids was investigated in order to better understand the mechanisms of pressure solution reactions. Halite samples were subjected to differential stresses ranging from 1-5 MPa over periods of time from several weeks to several months. The samples were deformed in the presence of various fluids at 25°C: NaCl-saturated, salt brine, H_2O, and ethanol. The axial deformation rates of the samples varied as a function of the differential stress, the initial state of the samples (dry vs. wet), the chemical nature of the circulating fluids, and the total amount of deformation. The relationship between axial strain rate and differential stress was essentially linear. The chemistry of the fluids was of critical importance to the deformation process; in the presence of ethanol, in which halite is only sparingly soluble, deformation rates and total deformation were significantly less than was the case in aqueous fluids.

1 INTRODUCTION

When considering cataclasis, plastic deformation, and pressure solution, the three major mechanisms of rock deformation in the upper crust, only pressure solution is based on a chemical reaction mechanism between a solution and the constituent grains of a rock. Rock deformation at pressures and temperatures ranging up to greenschist facies conditions is thought to take place primarily by solution mass transport, in which pressure solution takes a prominent role (Evans & Kohlstedt, 1995). Pressure solution reactions play an important role in diagenesis reactions associated with sediment burial, compaction and lithification, the development of pressure solution cleavage, and mineral segregation during low grade metamorphism.

The overall pressure solution process is based on dissolution reactions at grain-to-grain boundaries containing an intergranular fluid, aqueous transport of dissolved species, and reprecipitation in intergranular void spaces. The dissolution rate of a given chemical species at a grain-solution interface is a function of the difference in chemical potentials between the surface and the fluid. Such factors as the applied effective stress and the surface free energy (with contributions from elastic, plastic, and interfacial energies) determine the chemical potential of the solid (see details in Paterson, 1973). Measured strain rates, which are dependent on dissolution rates at grain-fluid boundaries under differential stress, are not only dependent on chemical driving forces, but also on intergranular diffusion rates and grain dimensions (for a general review, see Evans & Kohlstedt, 1995; and references therein).

The study of pressure solution reactions applied to the deformation of salt aggregates is of particular interest in understanding mass transfer and flow associated with the geological evolution of salt deposits and diapirs. Salt pressure solution studies also have certain industrial applications in the petroleum and mining fields. In addition, there is also interest in the use of salt mines as potential nuclear and chemical waste repository sites. Applications such as these require precise knowledge of salt deformation behavior in the presence of fluids.

There have been numerous studies published on the deformation of salt by pressure solution (e.g., Urai et al., 1986; Spiers et al., 1990; Spiers & Brzesowsky, 1993; and references therein). Other studies have concentrated on grain indentation and the precise nature of grain-to-grain contacts to better define the operative mechanisms associated with pressure solution (e.g. Tada & Siever, 1986; Hickman & Evans, 1991; Gratier, 1993). The present study emphasizes the deformation behavior of halite in the presence of various fluids over extended

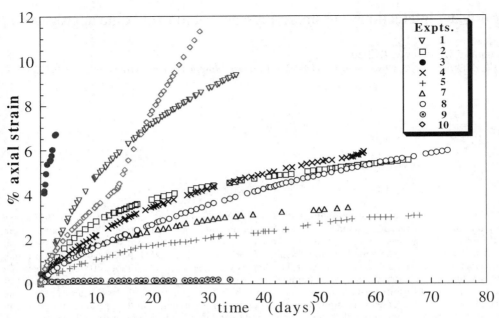

Fig. 1 Measured axial strain as a function of time. All of the experiments were conducted in NaCl - saturated solutions, with the exception of expts. 8 (water), 9 (ethanol), and 10 (salt brine). The differential stress ($\sigma_1 - \sigma_3$) was on the order of 3 MPa, except for expts. 1 (5 MPa), 2 (1 MPa), and 10 (1, 3 MPa). Steady-state strain rates were generally reached within 1 month. The break in the slope of data from expt. 10 is due to an increase in the axial stress during the experiment.

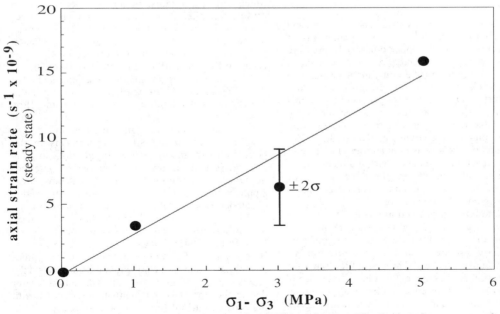

Fig. 2 Axial strain rate as a function of the applied differential stress. The data are from one set of experiments with approximately the same grain size ($\approx$ 50-60 µm). The relative linearity of the data is indicative of Newtonian behavior.

periods of time, as well as a complete characterization of the time-dependent evolution of permeability and porosity using various techniques.

2 EXPERIMENTAL METHODS

The deformation behavior of salt aggregates in the presence of fluids was studied over periods of time ranging up to 70 days. This ensured that true steady-state deformational flow of salt was achieved, free from the effects of purely mechanical deformation associated with grain diminution, grain sliding, and partial pore collapse. This study used a triaxial compaction cell, where the axial stress (σ_1), the confining stress (σ_3), and the fluid pressure (P_f) were independent. We used an open compaction cell; this enabled us to measure fluid flow and permeability as a function of time during the experiments. The main experimental parameters were the applied differential stress ($\sigma_1 - \sigma_3$) and the chemical composition of the fluids. The chemical nature of the fluids was varied (NaCl-saturated, salt brine, water, ethanol) to test the importance of dissolution rates of halite with respect to the overall rates of measured deformation. Acoustic velocities and x-ray tomographic images were used to characterize the density of the samples before, during, and after deformation in order to monitor the changing permeability and to understand the nature of fluid flow within the samples.

3 RESULTS AND DISCUSSION

3.1. *Axial strain rates and effective stress*

Measured axial strains as a function of time for all of the experiments are shown in Fig. 1. Steady-state axial strain rates ranged up to about 1.6×10^{-8} s^{-1}. A plot of steady-state axial strain rates vs. effective axial stress (i.e. axial stress (σ_1) - confining stress (σ_3)) reveals an approximate linear relationship (Fig. 2). This is indicative of a Newtonian rheology for these particular stress conditions. The definition for the effective axial stress is normally ($\sigma_1 - P_f$). However, based on SEM images, we have assumed that the intergranular fluid films have the following approximate pressures:

$P_{f1} \approx \sigma_1$ (fluid film $\perp \sigma_1$); $P_{f3} \approx \sigma_3$ (fluid film $\perp \sigma_3$). The effect of fluids in microchannels, where $P \approx P_f$, is probably of minor importance to the overall, steady-state deformation process.

3.2. *Chemistry of the fluids*

The importance of dissolution reactions with respect to the overall deformation process is well illustrated in these experiments. The use of fluids that were good solvents (polar solvents with elevated dielectric constants: NaCl-saturated fluids, salt brine, and H_2O) vs. ethanol, which is a non-solvent (non-polar, low dielectric constant), showed that the chemical nature of the fluid plays a critical role in determining the rate of deformation. The recorded total deformation, as well as the deformation rate, of the salt in the presence of ethanol was only a fraction of that recorded for samples deformed in the presence of aqueous solutions (Fig. 2). This can be explained by the fact that the salt-ethanol reaction has a smaller chemical affinity than the corresponding affinities of salt-aqueous solution reactions. At conditions close to equilibrium, a smaller chemical affinity translates to a slower rate of dissolution. This is proof that chemical reactions play a dominant role in wet-salt deformation under these differential stress conditions.

3.3 *Permeability*

The permeability of the samples continually evolved during the course of the experiments. This is evidence that the permeability of a rock or an aggregate is not an intrinsic, fixed property, but rather continually evolves during deformation due to reactions between permeating fluids and mineral grains. Plots of cumulative mass of fluid collected as a function of time revealed, in almost all cases, sigmoidal behavior. This type of behavior is the result of samples that were initially impermeable to fluid flow, followed by a rapid increase in permeability. Fluid flow was apparently channelized in the samples, as shown by x-ray tomography images of samples before and after deformation. Eventually, however, most of the samples became impermeable to further fluid flow. Porosimetry measurements before and after each experiment revealed that pore dimensions decreased, due mostly to a destruction of pores with diameters between 1-10μm. The decrease in pore volumes is in accord with the samples becoming impermeable to fluid flow. The net transfer of material from grain-to-grain contacts subject to stress to pore voids is the principal reason for the decrease in permeability. Acoustic measurements of samples during the course of experiments showed a significant increase in velocities, this being evidence for an increase in the overall density of the samples, and concomitantly, a decrease in void spaces.

4 REFERENCES

Evans B. & D.L. Kohlstedt 1995. Rheology of rocks. In T.J. Ahrens (ed), Rock Physics & Phase

Relations A Handbook of Physical Constants AGU Reference Shelf 3: 148-165. Washington, D.C.: A.G.U.

Gratier, J.P. 1993. Experimental pressure solution of halite by an indenter technique. *Geophys. Res. Lett.* 20: 1647-1650.

Hickman, S.H. & B. Evans 1991. Experimental pressure solution in halite 1: The effect of grain/interphase structure. *J. Geol. Soc. London* 148: 549-560.

Paterson M.S. 1973. Nonhydrostatic thermodynamics and its geologic applications. *Rev. Geophys. Space Phys.* 11: 355-389.

Spiers C.J. & R.H. Brzesowsky 1993. Densification behaviour of wet granular salt: Theory versus experiment. *Seventh Symposium on Salt*, 1: 82-92. London: Elsevier Science Publishers.

Spiers C.J., P.M.T.M. Schutjens, R. Brzesowsky, C.J. Peach, J.L. Liezenberg, & H.J. Zwart 1990. Experimental determination of constitutive parameters governing creep of rocksalt by pressure solution. In R. J. Knipe & E. H. Rutter (eds), *Deformation Mechanisms, Rheology and Tectonics, Geological Society Special Publication No. 54*: 215-227. London: The Geological Society.

Tada, R. & R. Siever 1986. Experimental knife-edge pressure solution of halite. *Geochim. Cosmochim. Acta* 50: 29-36.

Urai J.L., C.J. Spiers, H.J. Zwart, & G.S. Lister 1986. Weakening of rock salt by water during long-term creep. *Nature* 324: 554-557.

Fluids and faults: The chemistry, origin and interactions of fluids associated with the San Andreas fault system, California, USA

Y. K. Kharaka, J. J. Thordsen & W. C. Evans
US Geological Survey, Menlo Park, Calif., USA

B. M. Kennedy
Lawrence Berkeley Laboratory, University of California, Calif., USA

ABSTRACT: Heat flow and stress orientation measurements indicate that the San Andreas fault is anomalously weak; the most plausible explanation for this weakness relates to the distribution of fluid pressures. To investigate the role of fluids in the dynamics of this fault, we carried out detailed chemical and isotopic analyses of water and associated gases from 41 thermal and saline springs and wells located near the fault between San Francisco and Los Angeles. Results indicate that the chemical compositions of water and gases are controlled mainly by the enclosing rock types. The δD and $\delta^{18}O$ values establish that waters are predominantly of meteoric or connate origins, with no significant contributions from mantle or metamorphic sources. Chemical geothermometry gives reservoir temperatures of up to 150°C, indicating water circulation depths of up to 6 km. However, composition and isotope abundance of noble gases and $\delta^{13}C$ values indicate a high (up to 50%) mantle component for volatiles, favoring the continuous deep flow model as the origin of high pore pressure, especially at seismogenic depths > 6 km. Numerical simulations indicate that the CO_2 fluxes of mantle and deep crustal origins are sufficient to generate lithostatic fluid pressures, possibly initiating fault rupture, in time scales comparable to those of earthquake recurrence.

1 INTRODUCTION

It is generally accepted that faults and shear zones provide conduits focusing fluid flow in the upper crust and that water and gases play a critical role in a variety of faulting processes, including earthquakes (for a recent discussion, see Hickman et al., 1995, and other contributions to that section). Fault-hosted gold-quartz and other metalliferous mineral veins and associated hydrothermal rock alteration provide good evidence for large fluxes of deeply sourced fluids along faults (Wintsch et al., 1995). The role of fluid pressure in controlling the strength of faults has been recognized for ~50 years; the complete formulation of fault failure is generally represented (Lachenbruch & Sass, 1992) by a criterion of Coulomb form:

$$\tau_s = C + \mu\sigma_n' = C + \mu(\sigma_n - P_f) \qquad (1)$$

where τ_s is the frictional shear strength, C is the cohesive strength of fault gouge, μ is the static coefficient of friction (with values of 0.6-0.9), σ_n' and σ_n are the effective and normal stresses on the fault, respectively, and P_f is the fluid pressure. Equation (1) indicates that fault strength would be reduced drastically, if fluid pressure increased from hydrostatic to lithostatic.

There are many theoretical and laboratory studies indicating that the chemical effects of fluids on fault zone rheology are probably as important as the physical role (Wintsch et al., 1995). Specific hydrothermal water-rock interactions, including pressure solution, crack growth and healing, and clay formation through retrograde reactions, have been identified that could reduce or increase fault strengths principally by modifying the cohesion strength and coefficient of friction of its gouge (Kharaka et al., 1997).

A group of scientists have proposed a program that includes drilling a 10 km deep hole in the San Andreas fault zone to improve our understanding of the physics and geochemistry of faulting at seismogenic depths. As part of this effort, we carried out detailed chemical and isotope analyses of water, solutes and associated gases for 41 thermal/mineral springs and wells located on the San Andreas fault system (Fig. 1). In this summary, we emphasize results and conclusions based on the chemical and isotopic compositions of gases, especially CO_2, and discuss the implications of our results to the dynamics of the San Andreas fault.

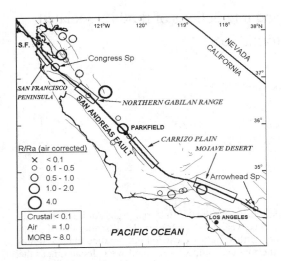

Figure 1. Location map of the study area showing the distribution of ^{3}He/^{4}He in fluids associated with the San Andreas fault system and the four candidate segments for the deep drilling.

1.1 The San Andreas Fault System

The San Andreas fault system refers to the network of active faults with mostly right-lateral offset that together have accommodated ~350 km of relative motion between the North American and Pacific plates in the last 30 Ma (Wallace, 1990). Intensive geological and geophysical studies, following the great San Francisco earthquake of 1906, have shown that the geology and tectonic evolution of California are extremely complex, controlled mainly by interaction between the continental American plate and oceanic Pacific and related plates. This interaction has included plate convergence leading to the underthrusting of oceanic crust and the sweeping of island arcs and other crustal material into the zone of interaction, as well as lateral plate translation leading to fragmentation, slicing and movement of large crustal fragments. These processes have led to the juxtaposition of "terranes" of variable lithology, metamorphic facies, origin and age (e.g., Wallace, 1990).

Recent sophisticated geodynamic models (e.g., Rice, 1992; Byerlee, 1993) describe the spatial and temporal behavior of this fault. Most fault models accept the conclusion, based on field determinations of heat flow and stress orientations, that the San Andreas fault is anomalously weak, both in an absolute sense and relative to the adjoining crust (Zoback et al., 1987). The fault weakness (shear strength ~10 MPa instead of ~100 MPa) cannot be attributed to the presence of serpentine, clay or other authigenic minerals with low coefficients of friction, because laboratory friction coefficients obtained on these minerals are too high at seismogenic conditions (Moore et al., 1996).

The most plausible explanation for the weakness of this fault relates to the distribution of fluid pressures within the fault zone and in the surrounding crust. Byerlee (1993) and Rice (1992) proposed models that require lithostatic fluid pressures in the fault zone, but hydrostatic pressures in the adjoining rocks, to satisfy both the heat flow and stress orientation constraints. The main differences between these models relate to the origin and maintenance of the high pore pressures. In the continuous flow model of Rice (1992), the high fluid pressures in the fault zone are maintained by flow of pressurized fluids from deeper crust or mantle. Byerlee (1993) proposed an episodic model in which meteoric water from the adjacent country rock is trapped in the fault zone following an earthquake. Fluid pressure is increased by compaction in compartments sealed by mineral precipitation as the gouge is sheared. The pore pressure increase continues until a critical fluid pressure causes seal rupture, thereby triggering an earthquake and starting another earthquake cycle.

2 RESULTS AND DISCUSSION

Data from literature and our estimates of water discharges indicate that thermal and mineral springs with high water flows ($\geq$2 L/s) are relatively few in the study area, and are especially rare on the trace of the San Andreas fault (Fig. 1). On the San Andreas fault, relatively high total discharges are present only at Arrowhead (~10 L/s) and Congress (~3 L/s) springs at the southern and northern ends of the study area (Fig. 1). This lack of high discharge springs even in mountainous regions of significant relief is likely caused mainly by the low annual precipitation, generally <50 cm/yr. It implies little fluid circulation in the shallow crust, and that water circulation is not responsible for the absense of a high heat flow anomaly across the fault (Lachenbruch & Sass, 1992).

2.1 Water and gas compositions

The chemical composition of water is highly variable, with salinity ranging from fresh water with <1,000 mg/L total dissolved solids to brine with salinity of ~50,000 mg/L (Kharaka et al., 1997). The chemical composition of water is determined primarily by the regional geology, especially the enclosing rock types. In general, water samples from the Salinian block granitoids have relatively low salinity and high pH and are of Na-SO$_4$ type; waters from the sedimentary Great Valley sequence have the highest salinity and are dominantly of Na-Cl type; and samples from the Franciscan assemblage have a low to moderate salinity, are Na-HCO$_3$ type, but contain significant concentrations of Mg, B, NH$_3$ and CO$_2$.

The δD and δ^{18}O values indicate that waters are predominantly of meteoric and connate origins, with no significant contribution from mantle or metamorphic sources (Kharaka et al., 1997). Chemical geothermometry yields reservoir temperatures of 70-150°C, indicating moderate water circulation depths of up to 6 km.

The gas compositions are also controlled mainly by geology and rock types, with samples from the Salinian block granitoids being composed mainly of atmospheric N$_2$ and minor CH$_4$ and CO$_2$. Samples from the Great Valley sequence are dominated by CH$_4$ of thermogenic origin (δ^{13}C = -36 to -50‰). Gas samples from Franciscan rocks are CO$_2$-rich but also have relatively high concentrations of CH$_4$ and N$_2$. The δ^{13}C values of bicarbonate and CO$_2$ range from -20 to +23‰, spanning the known sources of carbon, including organic, marine carbonate and mantle. Samples with high Ra (Ra is the ^{3}He/^{4}He ratio in air) values and/or those obtained from CO$_2$-rich Franciscan rocks have δ^{13}C range of -6 to -11‰, which are closer to that expected from a mantle source (δ^{13}C = -4 to -10‰).

The fluid samples are enriched in total He relative to its concentration in air saturated water: ^{4}He/^{36}Ar enrichment factors, range from ~1 to 4,300. The He isotopic compositions (Fig. 1) have Ra values from 0.06 to 4.0, indicating a range from crustal values (<0.1 Ra), to high enrichment in ^{3}He of mantle (~8 Ra in MORB) origin. In Kennedy et al. (1997), we discuss the main controls on these isotopes; our conclusion, however, is that the high ^{3}He/^{4}He ratios indicate a pervasive mantle degassing throughout the San Andreas fault system that does not correlate with rock provenance, with water types, or with the age or distance of nearby volcanic rocks (Kharaka et al., 1997).

2.2 Implications for the San Andreas Fault

Our model calculations indicate that the total flux of ^{3}He for the entire San Andreas fault system, as defined by Wallace (1990), is a modest ~6 moles/a (Kennedy et al., 1997). More importantly, mantle ^{3}He is coupled to other more abundant volatiles, especially CO$_2$ and water. Mid-oceanic ridge basalts originating in the mantle, for instance, have a relatively constant CO$_2$/^{3}He ratio of ~10^{10} (Trull et al., 1993). Using this ratio and the ^{3}He flux, yields a very high CO$_2$ flux of ~10^{11} moles/a for the entire San Andreas fault system. Because the density and viscosity of CO$_2$ are comparable to those of water at seismogenic depths, CO$_2$ flux into the fault zone would generate high pore pressures, thus lowering the effective stress and weakening the fault as originally proposed by Irwin and Barnes (1980), as calculated by Bredehoeft & Ingebritsen (1990) and as postulated in the Rice (1992) model. Bredehoeft & Ingebritsen (1990), used computer simulations to show that global CO$_2$ fluxes, focused along seismically active areas, could generate high pore pressures in the crust in time scales of 10^4-10^6a.

Generation of high fluid pressure by CO$_2$ flux into the San Andreas fault was modeled with the same simulation package and parameters used by Bredehoeft and Ingebritsen (1990), but with temperature and CO$_2$ flux data obtained from our study area. Results indicate that for reasonable geologic and hydrologic parameters (Fig. 2) operating at a 10 km depth, fluid pressures could increase from hydrostatic (~74 MPa) to lithostatic (250 MPa) in a short period of only ~200 years, using the highest CO$_2$ flux; a CO$_2$ flux, that is lower by a factor of 10, yields a period of ~2000 years.

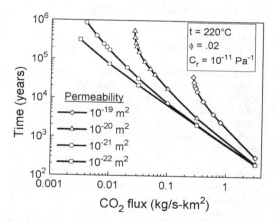

Figure 2. Time required to generate lithostatic fluid pressures with CO$_2$ flux at 10 km depth (ϕ is porosity; C$_r$ is rock compressibility).

The highest CO$_2$ flux used in the model adds unrealistic high advective heat to the fault system. However, the stress relieved by an earthquake is of the order of ~10 MPa (Lachenbruch & Sass, 1992), the fluid pressure at the start of the earthquake cycle could be close to lithostatic, and time periods required to reach a fluid pressure that would cause fault rupture and trigger an earthquake are probably of the order of 1/10 of the values shown in Fig. 2. The model is being refined further, especially to account for the advective heat transport by CO$_2$ flux and heat loss by CO$_2$ expansion. Preliminary results, however, indicate that CO_2 flux alone could be responsible for generation of high fluid pressures at time scales of earthquake recurrence.

CO_2 may also play an important role in water-rock interactions that impact the mechanical behavior of the San Andreas fault. Thermochemical data indicate that CO_2, present mainly as H_2CO_3 at high temperatures, becomes much more reactive with enclosing rocks at temperatures $\leq 300°C$. Addition of CO_2 to subsurface waters generally results in lowering of pH values, in redissolution of carbonate minerals and in an increase of mineral dissolution rates. In addition, low pH solutions in the presence of dissolved Mg could result in the formation of chlorite or biotite, and thus cause "fault softening." Increased fluid pressure and "fault softening" reactions could lead to earthquake triggering. On the other hand, loss of CO_2 from water following a fluid pressure drop induced by an earthquake, would result in higher pH, in precipitation of calcite and other minerals and would shift water-rock interactions to "fault hardening" reactions (e.g., Wintsch et al., 1995). Mineral precipitation would lead to fault crack healing, to sealing of fluid flow paths and to the start of a new earthquake cycle.

2.3 *Concluding remarks*

Fluid-rock interactions in the seismogenic zone are probably more complex than those described above. Calculations show that the measured total C (H_2CO_3, HCO_3^-, CO_3^{2-}, and other C species) in our samples is higher, at times by >1000, than that supportable by the measured mantle ^3He. The high $C/^3He$ could be due to selective He loss from some fluids prior to sampling, and/or to the addition of CO_2 from non mantle sources. The additional C is probably mainly from decarbonation of organic matter as indicated by the low $\delta^{13}C$ (up to -20‰) values, but some may be derived from the rare marine carbonate units ($\delta^{13}C$ ~0‰). Unlike the ^3He, however, ^{13}C isotopes do not provide an unambiguous tracer for the source of dissolved C because of mixing of C from different sources and isotopic exchange between dissolved C and carbonate minerals (Kharaka et al., 1997).

The role of any water associated with mantle ^3He is also problematic. The measured δD and $\delta^{18}O$ values indicate that water is meteoric or connate in origin with no significant mantle component. Also, thermochemical data indicate that water present with mantle He and CO_2, likely would become incorporated in clays and other minerals, rendering lower crust 'dry'. These results indicate that meteoric water would be dominant at shallower depths, and, likely would also play an important role in the earthquake cycle as postulated by the episodic model of Byerlee (1993).

These results and discussions indicate that there are many unanswered questions. A deep drilling program would allow for direct observation of the fault at seismogenic depths and greatly improve our understanding of the role of fluids in the dynamics of this important plate boundary fault.

REFERENCES

Bredehoeft, J.D. & Ingebritsen, S.E. 1990. Degassing of carbon dioxide as a possible source of high pore pressure in the crust: In *The Role of Fluids in Crustal Processes*. National Academy Press, Washington D.C. p.158-164.

Byerlee, J.D. 1993. A model for episodic flow of high pressure water in fault zones before earthquakes. *Geology*, 21, 303-306.

Hickman, S.H., Sibson, R. & Bruhn, R. 1995. Introduction to special section: Mechanical involvement of fluids in faulting. *J. Geophys. Res.*, 100B, 13,113-13,132.

Irwin, W.P. & Barnes, I. 1980. Tectonic relations of carbon dioxide discharges and earthquakes. *J. Geophys. Res.*, 85, 3115-3121.

Kennedy, B.M., Kharaka, Y.K., Evans, W.C., & others 1997. The San Andreas fault system, California: Fault zone fluid chemistry and evidence for mantle derived fluids. Submitted to *Science*.

Kharaka, Y.K., Kennedy, B.M., Thordsen, J.J. & Evans, W. C. 1997. Role of fluids in the dynamics of the San Andreas fault system, California, USA. *Proc. Geofluids II '97*, Belfast, N. Ireland, p.107-110.

Lachenbruch, A.H. & Sass, J.H. 1992. Heat flow from Cajon Pass, fault strength, and tectonic implications. *J.Geophys. Res.*, 97, 4995-5015.

Moore, D.E., Lockner, D.A., Summers, R. & others 1996. Strength of chrysotile-serpentinite gouge under hydrothermal conditions: Can it explain a weak San Andreas fault? *Geology*, 24, 1041-1044.

Rice, J.R. 1992. Fault stress states, pore pressure distributions, and the weakness of the San Andreas fault: In B. Evans & Wong T.-F. (eds) *Fault Mechanics and Transport Properties of Rocks*, Academic Press, San Diego. p. 475-503.

Trull, T., Nadeau, S., Pineau, F., Polve, M., & Javoy, M. 1993. C-He systematics in hotspot xenoliths: implications for mantle carbon contents and C recycling. *Earth Planet. Sci. Lett.*, 118, 43-64.

Wallace, R.E. 1990. General features: In Wallace, R.E. (ed) *The San Andreas Fault System, California*, U.S. Geol.. Surv. Prof. Paper 1515. p.3-12.

Wintsch, R.P., Christofferson, R. & Kronenberg, A.K. 1995. Fluid-rock reaction weakening of fault zones. *J. Geophys. Res.*, 100B, 13,021-13,032.

Zoback, M.D., Zoback, M.L., Mount, V.S., & others 1987. New evidence on the state of stress of the San Andreas fault system. *Science*, 238, 1105-1111.

Fluid flow during folding and thrusting in carbonates: 2-D patterns of Sr and O isotope alteration

A.M. McCaig & J.G. Kirby
Department of Earth Sciences, Leeds University, UK

ABSTRACT: Sr and O isotope data are presented from Cretaceous carbonates affected by thrusting and folding beneath the Gavarnie Thrust (GT) in the central Pyrenees. Detailed sampling with a rock drill allows 2-D contour maps to be drawn for both isotopes. These show that limited mass-transfer occurred across the unconformity between Cretaceous carbonates and Triassic redbeds, probably largely by diffusion. Advection occurred along a discrete thrust fault cutting the carbonates, but was very limited compared to advection along a mylonite zone beneath a larger thrust. The most important factor in determining integrated fluid flux was probably the timescale over which deformation took place. Our data suggest that active deformation, probably involving brittle deformation or veining, is required to significantly enhance permeability, and hence long-lived structures with large displacements focus fluid flow to a much greater extent than minor faults.

1 INTRODUCTION

It is well established that faults and shear zones focus fluid flow in the middle and upper crust (e.g. Oliver, 1996), and that fluid flow occurs during deformation as evidenced by syntectonic veins and alteration (Knipe and McCaig, 1994). This must mean that the permeability of rocks is enhanced by deformation, but the precise mechanisms by which this occurs, and the extent of permeability enhancement, remain uncertain. In this paper we present a combined Sr and O isotope study of carbonates which have been affected by both ductile folding and brittle faulting beneath the GT in the Central Pyrenees.

2 GEOLOGICAL SETTING.

The data presented comes from the Pic de Port Vieux, on the French-Spanish border 10 km north of

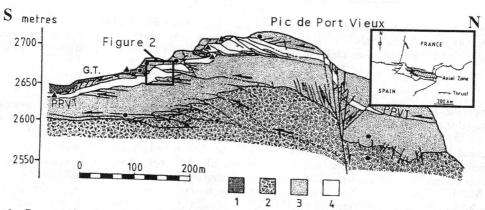

Figure 1. Cross section through the Pic de Port Vieux. Circles and triangles denote fluid inclusion analyses from Banks et al 1991. PPVT: Pic de Port Vieux Thrust; GT: Gavarnie Thrust; 1: Silurian graphitic schists; 2: Cambro-Ordovician basement; 3: Cretaceous carbonates; 4: Triassic sandstones.

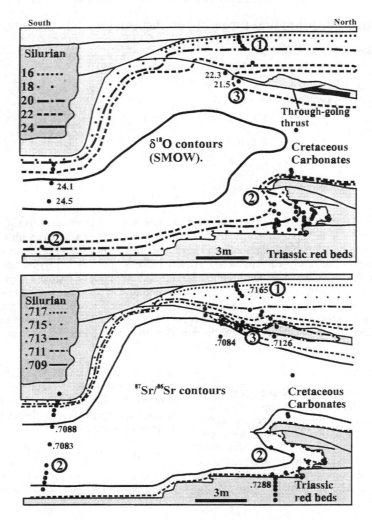

Figure 2. $\delta^{18}O$ of calcite separates and $^{87}Sr/^{86}Sr$ of whole rock samples from the Lower Fold Structure of the Pic de Port Vieux Thrust system. See figure 1 for location of sections. Solid circles denote sample locations. Circled numbers refer to points raised in results section. Values of selected samples have been added for clarity.

Bielsa in the central Pyrenees. The local geology and structural sequence have been described in detail by Grant (1990). Figure 1 shows that the highest structural unit is the Gavarnie thrust sheet, which contains Silurian graphitic schists and Devonian phyllites and marbles, and was thrust 10-15 km southwards in the Eocene. The footwall of the repeated by the Pic de Port Vieux Thrust (PPVT), which has a displacement of about 1 km and branches onto the GT. In the later stages of thrusting, minor thrusts formed beneath the PPVT and folded both it and the GT into a series of southward-verging folds. Strain within the thrust sheets was accommodated by locally intense cleavage in redbeds and carbonates, particularly in antiformal hinge zones, by complex networks of contractional and extensional faults, and by abundant syntectonic veins. Mylonites in the carbonates beneath the GT and the PPVT were folded during this event and therefore formed early in the deformation history. Because the PPVT branched onto the GT, it is likely that the 1 m mylonite zone seen beneath the PPVT actually formed beneath the GT.

McCaig et al. (1995) undertook a regional study of Sr isotope variations in the carbonates beneath the GT, supported by limited stable isotope data. Undeformed carbonates show isotopic values consistent with precipitation from Cretaceous seawater with minor modification due to diagenesis and radioactive decay ($^{87}Sr/^{86}Sr$ 0.708 to 0.710; $\delta^{18}O$ +25 to +28 ‰). Carbonate mylonites and veins typically show higher values of $^{87}Sr/^{86}Sr$ (up to 0.720) and lower values of $\delta^{18}O$ (down to +15‰). Gradients in isotopic composition southwards along the GT mylonite zone were inferred to reflect southward flow of fluid stored in Triassic redbeds and underlying fractured basement rocks, and expelled during thrusting. The fluid was a hypersaline calcic brine containing up to 2000 ppm Sr, and probably represents a formation water from Upper Triassic evaporites, eroded away in the Pic de Port Vieux area before the mid-Cretaceous unconformity formed (Banks et al., 1991; Yardley et al., 1993).

A key feature of the previous dataset is the discontinuity in Sr and O isotope composition which exists at the unconformity between Triassic redbeds and unconformably overlying Cretaceous carbonates. If fluid flowed across this boundary from redbeds into the carbonates, the isotopic discontinuity should have been displaced by advective mass-transport in the direction of fluid flow (Bickle, 1992; McCaig et al., 1995), the distance moved being proportional to the integrated fluid flux and the fluid-rock partition coefficient for the isotope in question. Some smoothing of the discontinuity by diffusive mass-transport is also to be expected. The rationale of the present study is that by contouring isotopic compositions in folded, cleaved and faulted carbonates, the fluid flow pathways through the carbonate layer can be established, and the effectiveness of different deformation mechanisms in controlling permeability and mass-transport constrained.

3 ISOTOPE DATA

Sr isotopes on whole rock carbonates and silicates were measured at Leeds on a fully automated VG-54E double-collector thermal ionisation mass spectrometer (TIMS), and Rb contents were measured on a VG micromass 30 single collector solid source TIMS after standard dissolution and ion exchange (see McCaig et al, 1995 for details). Stable isotope data were collected using the sealed vessel method (Swart et al 1991). Calcite gas was collected by reaction with phosphoric acid at 25°C for 30 minutes, and was analysed automatically on a VG SIRA 12 (Middlewich, Cheshire) dual-inlet, triple-collector mass spectrometer.

Figure 2 presents Sr and O isotope data around the lower fold structure of Grant (1990). The data set shows the following features (the points raised are highlighted by circled numbers in figure 2):

1) The most altered carbonates in terms of both $^{87}Sr/^{86}Sr$ and $\delta^{18}O$ occur within and below the mylonite zone beneath the PPVT on the upper limb of the fold structure. A gradient for both isotopes occurs over about 2 m, with the most altered samples immediately beneath the thrust. South of the fold structure, the gradient is much steeper and the maximum deviation from unaltered values is less.

2) South of the fold structure, alteration dies out upwards within 1 m of the unconformity for Sr and 3 m for O. In the syncline, adjacent to the folded and thrust Triassic, the corresponding distances measured parallel to the axial plane of the fold are 2 m for $^{87}Sr/^{86}Sr$ and ≥ 4 m for $\delta^{18}O$. No values of $^{87}Sr/^{86}Sr$ >0.713 are seen anywhere close to the unconformity.

3) Higher values of $^{87}Sr/^{86}Sr$ (up to 0.714) are seen in carbonates along the minor thrust within the Cretaceous on the top limb of the fold. The 0.711 contour is displaced about 1.5 m along this fault from the cutoff of the unconformity, compared with a maximum of about 30 cm displacement away from the unconformity anywhere else. Less displacement appears to be present in the $\delta^{18}O$ contours, although more data are required.

4 DISCUSSION

Our data show that the isotopic anomaly penetrated twice as far into the carbonates in the intensely cleaved core of the fold structure than in relatively undeformed carbonates south of the fold structure. Mass-transport distances along the minor thrust fault are two to three times greater again. The data also show that the fluid which carried the isotopic anomaly within the mylonites beneath the PPVT must have originated well south of the fold structure. We follow McCaig et al. (1995) in interpreting this anomaly to have formed during larger scale thrust-parallel fluid movement largely before formation of the fold structures. The steep gradient in isotopic composition downwards into the less deformed carbonates beneath the mylonite probably reflects diffusive modification of the advectively produced anomaly across the side of the conduit, although dispersive effects cannot be ruled out (Bowman et al, 1994).

The sub-Cretaceous unconformity was present throughout the history of mass transfer, and the gradient in Sr and O isotope values across this may have been produced largely by diffusion. It is difficult to be sure whether the larger mass-transport distance within the highly cleaved carbonates in the syncline core resulted from enhanced diffusion along the cleavage, or from advection. In any case, the scale of advection and the integrated flux were very limited (< 1 m^3/m^2, cf. McCaig et al., 1995).

The minor thrust fault within the Cretaceous carbonates on the upper limb of the fold is thought to be a relatively early structure, because of the thinning of the Cretaceous sequence between it and the PPVT. Fluid flow appears to have been focused along this fault, although to a much lesser extent than along the mylonite beneath the PPVT. This fault is a discrete structure, whereas the mylonite must have been active over a long period of time. We conclude that it is active movement on a fault which enhances permeability, to a greater extent than the presence of a discontinuity or fault rock.

5 CONCLUSIONS

1. In weakly deformed rocks, fluid flux across the Cretaceous/Triassic unconformity was negligible, and was low even where intense cleavage is present
2. Higher fluxes occurred along a discrete minor thrust fault, but these were still at least an order of magnitude lower than fluxes along a 1 m thick mylonite zone. Abundant deformed veins within the mylonite may be the main conduits for flow.
3. Permeability is evidently enhanced by deformation, but continuing deformation (large displacements) is required to produce large fluxes.

6 REFERENCES

Banks, D. A., Davies, G. R., Yardley, B. W. D., McCaig, A. M., & Grant, N. T. (1991) The chemistry of brines from an Alpine thrust system in the Pyrenees: An application of fluid inclusion analysis to the study of fluid behaviour in orogenesis. *Geochim. Cosmochim. Acta:* 55, 1021-1030.

Bickle M. J. (1992) Transport mechanisms by fluid flow in metamorphic rocks: Oxygen and strontium decoupling in the Trois Seigneurs Massif - a consequence of kinetic dispersion? *Am. J. Sci:* 292, 289-316

Bowman, J. R., Willett, S. D. & Cook, S. J. (1994) Oxygen isotopic transport and exchange during fluid flow: one-D models and applications. *Am. J. Sci:* 294, 1-55.

Grant, N.T. (1990) Episodic discrete and distributed deformation: consequences and controls in a thrust culmination from the central Pyrenees, J. Struct. Geol. 12, 835-850.

Knipe, R. J. & McCaig, A. M. (1994) Microstructural and microchemical consequences of fluid flow in deforming rocks, in Geofluids: *Origin, Migration, and Evolution of fluids in Sedimentary Basins,* (ed Parnell, J.), Geol. Soc. Spec. Pub: 78, 99-111.

McCaig, A. M., Wayne, D. M., Marshall, D. M., Banks, J. D. & Henderson, I. (1995) Isotopic and fluid inclusion studies of fluid movement along the Gavarnie Thrust, Central Pyrenees: Reaction fronts in carbonate mylonites. *Am. J. Sci:* 295, 309-343.

Oliver, N. H. S. (1996) Review and classification of structural controls on fluid flow during regional metamorphism. *J Met. Geol:* 14, 477-492.

Swart, K. K., Burns, S. J. & Leder, J. J. (1991) Fractionation of the stable isotopes of oxygen and carbon in carbon-dioxide as a function of temperature and technique. *Chem. Geol:* 86, 89-96.

Yardley, B. W. D., Banks, D. A. & Munz, I. A. (1993). Fluid penetration into crystalline crust: evidence from the halogen chemistry of inclusion fluids. In: J. Parnell, A. H. Ruffell & N. R. Moles (Editors) *Geofluids '93.* Extended abst, 350-353.

The formation of albite veins in high-pressure terrains: Examples from Corsica and Zermatt-Saas, Switzerland

J.A. Miller & I. Cartwright
Department of Earth Sciences and VIEPS, Monash University, Melbourne, Vic., Australia

A.C. Barnicoat
Department of Earth Sciences, The University of Leeds, UK

ABSTRACT: Blueschist and eclogite facies ophiolites from Alpine Corsica, France, and Zermatt-Saas, Switzerland, contain centimetre- to decimetre-wide albite veins with greenschist alteration haloes. Oxygen isotope data suggest the vein-forming fluids were derived from the ophiolite pile and may have been produced from the breakdown of hydrous minerals (eg lawsonite) during early decompression. The precipitation of albite is favoured by a reduction in pressure. Albite precipitation in veins in the Corsican and Zermatt-Saas metabasalts may have been driven by the exhumation of these terrains across the blueschist-greenschist transition. Alteration haloes of chlorite + albite + tremolite/actinolite + clinozoisite/epidote were probably produced by fluids exsolved from the crystallising veins.

1 INTRODUCTION

Albite veins are a common feature of high-pressure terrains. They are often associated with distinct alteration haloes and are generally thought to reflect the presence of fluids. However, the volume of fluid involved in vein formation is not clear. Different models of vein formation have suggested that veins may form through large-scale fluid flow (e.g. Ferry & Dipple, 1991) as well as by localised fluid flow often coupled with diffusion (e.g. Cartwright et al., 1994). If albite veins represent the pathways of large-scale fluid flow in high-pressure tectonic settings, the source of this fluid is problematic given that high-pressure rocks have normally been substantially dehydrated. It is also unclear what influences the precipitation of albite in veins under high-pressure conditions. While a number of studies have looked at particular components of albite vein formation, few have tried to integrate all aspects. In this contribution, we attempt to address these questions and put forward one possible model for the formation of albite veins in high-pressure rocks from Alpine Corsica and the Zermatt-Saas ophiolite zone.

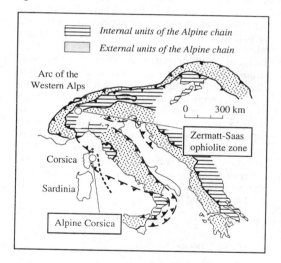

Fig. 1: Geology of the western Alps and the location of the two areas discussed in this study

2 GEOLOGICAL SETTING

Alpine Corsica (France) and the Zermatt-Saas ophiolite (Switzerland) are part of the Pennine zone of the internal units of the Alpine chain (Fig. 1). Both areas comprise ophiolitic and sedimentary material that have undergone high-pressure metamorphism during the Alpine orogeny to produce lawsonite-blueschist, blueschist, and eclogite facies mineralogies. On Corsica two high grade metamorphic episodes have been identified: 1) early eclogite facies metamorphism (1.2 GPa, 530 °C); and 2) pervasive blueschist facies metamorphism (0.7-0.9 GPa, 400 °C) (Gibbons et al., 1986). In the Zermatt-Saas ophiolite, the highest grade metamorphic event reached 1.7 - 2.0 GPa and 500 - 600 °C (Barnicoat & Fry, 1986) at 50-70 Ma (Bowtell et al., 1994) producing eclogitic mineral assemblages and other cofacial parageneses. Both areas also show evidence for a late stage greenschist facies overprint that is often spatially related to the distribution of albite veins. On Corsica and around Zermatt-Saas, this event is thought to have occurred at 0.5 GPa, 450 °C and 0.8 GPa, 450 °C respectively (Fournier et al., 1991; Barnicoat & Fry, 1986).

3 PETROLOGY

Albite veins from Alpine Corsica typically occur in blueschist facies metabasalts that preserve evidence of an early eclogitic mineralogy. The metabasalts are composed of lawsonite, omphacite, garnet, albite glaucophane and clinozoisite and are locally retrogressed around laterally discordant, near-monomineralic coarse-grained albite veins that may reach 20 cm in width. All the albite veins have alteration haloes up to 1m wide, characterised by the formation of albite + chlorite + clinozoisite/epidote + tremolite/actinolite at the expense of glaucophane and lawsonite, suggesting that the alteration haloes formed under greenschist facies conditions.

In comparison, the higher-grade Zermatt-Saas metabasalts are composed of paragonite, omphacite, garnet, glaucophane, clinozoisite, late albite and minor lawsonite pseudomorphed by clinozoisite and paragonite. Albite-rich veins (>90% albite) with minor chlorite and ankerite, and up to 10% clinozoisite (Barnicoat, 1988) occur both as narrow (<10 cm) conjugate shear-veins and as broader (> 20 cm) en echelon dilatant fractures that bisect the shear-veins. They locally have alteration haloes up to 2 m wide but more commonly < 1m, with similar mineralogies to the Corsican albite veins.

There are several reactions that may have generated the mineral assemblages in the vein alteration haloes, including the breakdown of glaucophane and lawsonite:

$$gl + lws + H_2O = act + alb + chl + qtz \quad (1)$$

or glaucophane and garnet:

$$gl + qtz + gnt + H_2O = alb + chl + tr. \quad (2)$$

The H_2O required to drive reactions (1) and (2) was probably sourced from fluids exsolved from the crystallising veins. The formation of the alteration haloes indicate that the vein-forming fluid was not in equilibrium with the immediate host rock and may have been externally derived.

4 STABLE ISOTOPE DATA

In contrast to the mineralogical evidence, oxygen isotope data suggests the fluids were derived from within the ophiolite although not necessarily from the pillow lava units. Figure 2 illustrates the change in $\delta^{18}O$ values along a traverse through the unaltered wallrock (10.3-9.0‰), the altered wallrock (9.0-7.6‰) and the albite vein (11.7-12.5‰), for a Corsican albite vein. $\delta^{18}O$ values for the albite vein indicate that the fluid was either derived from within the ophiolite complex or from a fluid that equilibrated isotopically with the ophiolite rocks. In the second case the fluids may have been derived from the overlying Schistes Lustrés. However, given that the Schistes Lustrés on Corsica are dominated by calcareous metasediments containing both calcite and quartz veins, it is unlikely that the albite veins within the ophiolite units were formed from fluids derived from these units. The more likely scenario is that the fluids involved in vein formation were derived from the ophiolitic rocks since, in addition to being similar isotopically to the albite veins, they are also compositionally more compatible as ophiolitic rocks are often enriched in Na by early seafloor hydrothermal alteration prior to subduction. The slight lowering of $\delta^{18}O$ values in the vicinity of the vein suggests either that the fluid had a lower $\delta^{18}O$ value than the host rock or that the fluid was derived from "hotter" rocks.

Both of these explanations are consistent with the fluids being derived from units lower down in the ophiolite pile, for example the gabbros or

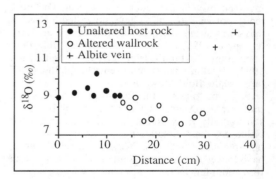

Fig 2: Variation in oxygen isotopes in Corsican metabasalt with distance from an albite vein

Table 1. $\delta^{18}O$ for albite veins and associated wallrock from the Zermatt-Saas region

Sample #	Albite Vein	Altered Wallrock	Unaltered Wallrock
96A6-11	7.7	7.8	6.6
96A30-32	8.1	6.2	6.1
96A37-39	7.4	6.8	8.4
96A48-51	8.8	7.5	7.3

serpentinites. Overall, the oxygen isotope data suggest that large scale fluid flow between different Corsican Alpine nappes did not occur. A similar pattern is observed for the Zermatt-Saas veins and associated wallrock (Table 1).

5 FLUID SOURCES IN H-P TERRAINS

Hydrated oceanic crust may transport significant volumes of fluid into subduction zones. Fluids may be derived through compaction and dewatering at high crustal levels and through dehydration reactions at deeper crustal levels. The extent and timing of dehydration of the subducting slab depends upon the P-T path it follows within the subduction zone (Peacock, 1990). Rocks that follow a fairly cold P-T path within the subduction zone may not substantially dehydrate, producing a lawsonite-blueschist facies rock that potentially contains around 6 wt% H_2O, primarily bound in hydrous minerals such as lawsonite and clinozoisite. These rocks may then be an important source of fluid at deeper levels in the subduction zone.

5.1 Lawsonite as a Fluid Reservoir During High-Pressure Metamorphism

Lawsonite is stable over a wide range of pressures at fairly low temperatures (Pawley, 1994). Given that lawsonite contains 11.5 wt% H_2O, it has the potential to transport considerable volumes of water to deep levels within subduction zones. Lawsonite is abundant (up to 20%) in Corsican blueschist facies metabasalts. Its presence in garnet porphyroblasts indicates that it was present prior to the peak of high-pressure metamorphism. During high-pressure metamorphism lawsonite may breakdown via reactions such as:

$$omp + lws = cz + gnt + pa + qtz + H_2O \quad (3)$$

or at lower pressures via:

$$lws + gl = ab + chl + cz + qtz + H_2O \quad (4)$$

Both of these reactions release substantial amounts of H_2O into the surrounding rocks. Dehydration reactions similar to (3) and (4) may be an important source of fluid for the formation of albite veins during retrogression.

5.2 Other Fluid Sources in High-Pressure Terrains

Clearly albite veins are not restricted to lawsonite-bearing rocks. The oceanic crust is variably hydrated during early hydrothermal alteration on the seafloor so that some units of the ophiolite are more important fluid reservoirs than others. In the previous discussion we have focussed on the basaltic units as a fluid reservoir, however, peridotites are often substantially hydrated to form serpentinites, which may also be an important source of fluids during subduction. Rocks from the Zermatt-Saas ophiolite followed a warmer P-T path during subduction and underwent a greater degree of dehydration than the Corsican metabasalts to form eclogites, considered to be fairly anhydrous. In these rocks water tends to bound up in minerals such as clinozoisite and white micas and it is unclear whether these minerals have the potential to provide sufficient fluid to form the albite veins. However, additional fluid may have been sourced from the serpentinites through the dehydration of antigorite which is stable to depths in excess of 80 kms.

5.3 Fluid Mobility Under High-Pressures

Fluids generated through dehydration reactions during high-pressure metamorphism may escape from the subducting slab and migrate back up the subduction complex through the accretionary wedge or flux into the overlying mantle wedge. However, there is evidence to suggest that some fluid may remain in the subducting slab (e.g. Thompson & Connolly, 1990). Whether or not a fluid will escape from the rocks depends on the rocks permeability. If the rate of fluid production is greater than the rate of fluid escape, fluids may be trapped in the rocks until sufficiently high fluid pressures are built up to generate the opening of microcracks. If the fluid pressure becomes sufficiently high then the rocks may produce discrete fractures that act to focus fluids out of the rocks. As discussed above, the retrograde transition from lawsonite-blueschist to greenschist facies involves substantial dehydration as greenschist facies rocks generally contain half the fluid content of lawsonite-blueschists (Peacock, 1990). The generation of fluids during this transition has the potential to substantially increase the fluid pressure which may lead to the formation of fractures in which albite may be precipitated.

6 PRECIPITATION OF ALBITE

The formation of albite veins requires that fluids associated with vein formation contain aqueous Na, Al, and Si. In order to transfer chemical components from one part of a rock to another, a driving force such as a chemical potential gradient, pressure or temperature gradient is required. In the case of albite veins, since the solubility of Al is pressure sensitive (e.g. Lin & Manning, 1994), albite precipitation is likely to be driven by a pressure gradient. This suggests that fluids may dissolve Al at medium to high pressures such as those associated with lawsonite breakdown via reactions (3) and (4).

The precipitation of albite may occur via:

$$Na^+ + 3SiO_2 + Al(OH)_4^- = NaAlSi_3O_8 + 2H_2O \quad (5)$$

with the precipitation of albite favoured at lower pressures. Two factors may facilitate albite precipitation; 1) the pressure drop associated with the blueschist to greenschist transition during uplift; and 2) the drop in fluid pressures associated with the opening of fractures. While it is not clear whether a drop in fluid pressure is sufficient to cause albite precipitation, uplift of the Corsican alpine nappes may be related to extensional exhumation in a core complex (Fournier et al., 1991). The associated pressure drop may be sufficient to drive albite precipitation.

In the Corsican metabasalts albite may be produced during prograde metamorphism, and hence reaction (5) may be an important source of aqueous Na, Si and Al. The breakdown of lawsonite at medium to high pressures via reaction (4) produces coexisting albite and H_2O and at these pressures albite dissolution will be favoured. If reaction (4) begins at the onset of decompression, assuming that the reaction does not go to completion, the fluids may precipitate albite into veins leaving a water-rich fluid that forms the alteration haloes around the veins by reaction (1). While albite precipitation is pressure sensitive, it is unclear what magnitude of pressure change is required to precipitate albite. However, albite precipitation by reaction (5) probably requires the pressure change to be overstepped, it is unlikely that albite precipitation will occur until a significant amount of decompression has taken place.

In the Zermatt-Saas metabasalts albite is generally not stable with respect to the higher pressure phases jadeite and omphacite. However, similar dissolution reactions are possible for other Na, Al silicates produced by reactions involving the breakdown of high-pressure hydrous minerals, such as reaction (3). These dissolution reactions would then give rise to a fluid saturated with respect to Na, Al and Si which may precipitate albite at lower pressures according to reaction (5).

7 CONCLUSIONS

While albite veins are common in many high-pressure terrains, the processes that lead to their formation are complex. The devolatilisation of hydrous minerals such as lawsonite at medium to high pressures may provide an important fluid source for the formation of albite veins. This is also supported by stable isotope data that indicates fluids involved in the formation of albite veins from Corsican and Zermatt-Saas metabasalts were probably derived from within the ophiolites, suggesting that there has been no large-scale fluid flow between different nappes. Similar processes may operate in other terrains. The precipitation of albite from a fluid is favoured by a decrease in pressure. Alteration haloes around the veins are characterised by greenschist facies minerals suggesting that albite precipitation may be driven by the drop in pressure associated with the blueschist to greenschist transition possibly during exhumation and core complex formation.

8 REFERENCES

Barnicoat, A.C., 1988. The mechanism of veining and retrograde alteration of Alpine eclogites. *J. Metamorphic Geol.*, 6: 545-558.

Barnicoat, A.C. & N. Fry, 1986. High-pressure metamorphism of the Zermatt-Saas ophiolite zone, Switzerland. *J. Geol. Soc. London*, 143, 603-618

Bowtell, S.A., R.A. Cliff and A.C. Barnicoat, 1994. Sm-Nd isotopic evidence on the age of eclogitisation in the Zermatt-Saas ophiolite. *J. Metamorphic Geol.*, 12:187-196.

Cartwright, I., Power, W.L., Oliver, N.H.S., Valenta, R.K., & McLatchie, G.S., 1994. Fluid migration and vein formation during deformation and greenschist-facies metamorphism at Ormiston Gorge, Australia. *J. Metamorphic Geol.*, 12, 373-386.

Ferry, J.M. & Dipple, G.M., 1991. Fluid flow, mineral reactions, and metasomatism. *Geology*, 19, 211-214.

Fournier, M., Jolivet, L., Goffé, B., & Dubois, R., 1991. Alpine Corsica metamorphic core complex. *Tectonics*, 10, 1173-1186.

Gibbons, W., Waters, C. & Warburton, J., 1986. The blueschist facies Schistes Lustrés of Alpine Corsica: A Review. *Geol. Soc. Am. Mem.*, 164, 301-311.

Lin, H.A. & C.E. Manning, 1994. Solubility of Jadeite + Quartz in H_2O at 12-20 kbar, 400-600 °C. *EOS*, 75(44) Suppl, p. 693.

Pawley, A.R., 1994. The pressure and temperature stability limits of lawsonite: implications for H2O recycling in subduction zones. *Contrib. Mineral. Petrol.*, 118, 99-108

Peacock, 1990. Numerical simulation of metamorphic pressure-temperature-time paths and fluid production in subducting slabs. *Tectonics*, 9, 1197-1211.

Thompson, A.B. & Connolly, J.A.D., 1990. Metamorphic fluids and anomalous porosities in the lower crust. *Tectonphysics*, 182, 47-55.

Lateral variations in mylonite thickness as influenced by fluid/rock interactions in a shear zone in Africa

U. Ring
Institut für Geowissenschaften, Johannes Gutenberg-Universität, Mainz, Germany

ABSTRACT: The transcurrent Mkamasa River shear zone, northern Malawi, east-central Africa, has been studied to explore the relationship between shear-zone thickness, volume strain and fluid flow. Mylonite thickness varies over a distance of about 2.5 km along strike from almost 50 m in the E to ~4 m in the W. Shear-zone formation is characterized by the breakdown of feldspar and biotite and the formation of sillimanite, quartz and water. Based on the degree of mylonitisation, fluid/rock ratio and volume strain, mylonite I and mylonite II have been distinguished in the shear zone. In the E, only mylonite I is developed, whereas in the W, ~3 m of mylonite II developed in the centre of the shear zone. Mass-balance calculations indicate approximately 50% volume loss in mylonite II. Fluid/rock ratios estimated from calculated depletions of silica are up to 400. In mylonite I, volume loss (≤9%) and fluid/rock ratios (<15) are less pronounced. The initiation and activity of the shear zone was accompanied by large amounts of fluids that infiltrated the shear zone. These fluids caused an almost complete dealkalisation within the shear zone leading to the development of almost "dry" and sheet-silicate-free mylonite II. It is proposed that the associated destabilisation of micas destroyed the pathways for fluid circulation and ultimately caused the cessation of shearing.

1 INTRODUCTION

Retrograde shear zones are ubiquitous features in continental basement complexes and are associated with the excavation of metamorphic terrains. They are sites of enhanced metamorphic reactions and conduits through which large quantities of crustal fluids circulate. Therefore, retrograde shear zones are considered a fundamental feature of metamorphic flow pattern. Mid-crustal shear zones are commonly associated with profound changes in chemical composition reflecting extensive fluid infiltration usually causing high volume loss on the order of up to 60-70%.

2 SETTING AND PETROGRAPHY

The sinistral Mkamasa River shear zone is part of the Mugesse shear belt of northern Malawi (Fig.1) and developed in medium-grained (2-4 mm) paragneiss which shows a metamorphic banding into leuco- and melanocratic layers. The dark layers consist mainly of large biotite flakes with rare sillimanite, muscovite, garnet and hornblende. The leucocratic bands are composed of quartz, potassium feldspar and plagioclase. Based on changes in grain size, mineralogy and composition, two types of mylonite have been distinguished within the shear zone: (1) Mylonite I, characterized by grain-size reduction and minor compositional changes and; (2) mylonite II, which shows pronounced grain-size reduction and a dra-matic change in mineralogy and composition. In the W, the shear zone varies in width from 3 m to almost 10 m over a distance of approximately 1 km. In this area, a variably thick zone of mylonite II is developed within the centre of the shear zone. Further E, the shear zone widens and shear-zone thickness increases to about 30-60 m. In the eastern locality, no mylonite II developed within the shear zone.

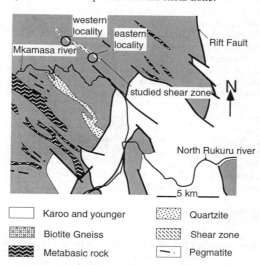

Fig.1 Regional geologic map of the Mkamasa River shear zone in northern Malawi.

Mylonitisation is petrographically characterized by the incipient breakdown of feldspar, biotite and muscovite. Garnet is occasionally marginally replaced by sillimanite, biotite and plagioclase. The latter two and potassium feldspar are converted into sillimanite, quartz and opaques. The development of mylonite II is characterized by the almost complete breakdown of feldspar and biotite to form sillimanite, small new garnet, quartz and opaques. Almost all OH-bearing minerals (mainly biotite, but also muscovite and hornblende) have been transformed to sillimanite, garnet, quartz and opaques.

Pre-mylonitisation P-T conditions were between 710 and 770°C (grt-hbl thermometer), with pressure ranging from 5.7-6.7 kbar (for 725°C; GRAIL barometer). In mylonite I, thermometry supplied 600-630°C and pressure estimates are 5.3-5.7 kbar (for 610°C). Due to largely unsuitable petro-graphy, no reliable P-T estimates could be obtained in mylonite II. In one sample, the paragenesis grt+rt +sil+il+qtz has been observed. The application of the GRAIL barometer yielded 4.4-5.5 kbar (for 600°C). The results show that mylonitisation took place as both pressure and temperature decreased.

3 VOLUME LOSS

The chemical data show that Ca, K, Na, Si, Rb and Sr, as well as P and Ba decrease systematically from wallrock to mylonite II. The decrease in the alkaline elements and Ca correlates well with the breakdown of feldspar and biotite. Ti, Fe, Al, Mg, Mn, Nb and Zr and also Co, Cr and V are enriched. The relative increase in Ti, Fe, Al, Mg and Mn can be linked to the increase in the modal abundance of sillimanite, opaques and garnet. Al-though quartz increased in modal abundance, the total amount of Si in the rock decreased. A decrease in structurally bound H_2O is inferred from the decreased loss on ignition and the virtual absence of hydrous phases in mylonite II. Overall, the chemical and mineralogical variation trends in the shear zone indicate that deformation was accompanied by volume loss. The variable thickness along strike of the Mkamasa River shear zone is manifested quantitatively in different mineralogical and chemical changes, although qualitatively these trends are similar. In the western part of the shear zone, development of mylonite I was largely an isochemical process, whereas the switch from mylonite I to mylonite II is characterized by a considerable volume loss of 46%. In the eastern part of the shear zone, the relative enrichment in immobile elements is small, consequently the estimated volume loss in mylonite I is modest (9%).

4 VOLUME OF FLUIDS

Assuming that the chemical changes reflect fluid/rock interaction during negative dilatation, fluid/rock ratios and the total volume of fluids infiltrating the shear zones can be estimated. The volume of fluid that passed through the shear zone has been estimated by mulitplying the fluid/rock ratio by columns of rock with unit cross-sectional area and height equal to the thickness of the fault zone prior to deformation (Newman & Mitra, 1993). The amount of water that can be released by the described retrograde dehydration reactions has been calculated from the total amount of H_2O that is structurally bound in the water-bearing minerals in the wallrock.

Mass-balance equations indicate that the bulk of the volume loss was achieved by progressive depletion of silica towards the centre of the shear zone. Assuming a 90% silica-saturated fluid and quartz solubility of 1 wt%, a fluid/rock ratio of ~400 in mylonite II is estimated. The volume of fluid that passed through the above defined column of mylonite II was on the order of 2400 m^3. The volume of structurally-bound water in the wallrock is 1.8 wt%, which results in a total volume of water in the rock column of ~0.1 m^3. The estimated fluid/rock ratio in mylonite I is ~100, and the volume of fluid that infiltrated the mylonite column is ~5000 m^3. The volume of structurally-bound water in the wallrock is 2.2 wt%, creating 1.2 m^3 of water that could be potentially released within the considered rock column.

Comparing the two outcrops in the Mkamasa River shear zone shows that the fluid/rock ratio in the centre of the shear zone was about four times higher in the relatively thin mylonite zone in the W, but the volume of infiltrating fluid was only about double as much in the more than 10 times thicker mylonite zone in the E.

5 DEFORMATION AND METAMORPHISM

Deformation/metamorphism relationships and their progressive change with increasing deformation are very similar in both parts of the studied shear zone. The development of the schistosity in the wallrock is expressed by a flattening of feldspar and the preferred alignment of biotite parallel to the flattened planes of feldspar. Radiating aggregates of fibrolitic sillimanite overgrew feldspar/biotite boundaries at the margin of the shear zones. The grain boundaries of biotite and feldspar meet the prismatic faces of fibrolite at random angles, suggesting that the fibrolitic sillimanite grew after the final positioning of the boundaries between the other minerals.

In mylonite, biotite is partly replaced by fibrolitic sillimanite. The long axes of both minerals are sub-parallel to each other. Sillimanite occasionally defines S-C-type structures. Feldspar and, where present, cordierite have also been replaced by fibrolitic sillimanite. Feldspar between the biotite/sillimanite layers recrystallised synkinematically to form core-and-mantle structures. Potassium feldspar developed myrmekite structures where crystal faces are parallel to foliation planes.

With increasing deformation the total amount of fibrolite, garnet and opaques, and to a lesser degree also quartz, in the rock increased, mainly at the expense of biotite and feldspar. Bunches of fibrolite partly enclose relics of biotite. Between fibrolite layers, ribbon quartz with aspect ratios of up to 14 developed. The grain boundaries of quartz meet the fibrolite bundles approximately at 90° angles, suggesting that fibrolitic sillimanite grew before or during the development of the quartz ribbons. Quartz between fibrolite bundles commonly shows microcracks which formed at a high angle to the fibrolite bundles. The material in the cracks appears to be fibrolite and phyllosilicate.

To explore the relationship between water-releasing biotite breakdown, the related enrichment of sil+qtz and deformation, the sillimanite/biotite ratio has been plotted against the aspect ratio of quartz in XZ sections (Fig.2). The diagrams show: (1) A pronounced increase in strain (as represented by the aspect ratios of quartz) at very low sillimanite/biotite ratios (~<0.3), and (2) the increase in strain slows down, but gets very high, as biotite is intensively replaced by sillimanite.

The notable increase in strain at low sillimanite/biotite ratios is probably solely due to deformation since the sillimanite-producing reaction had not yet proceeded significantly. Deformation at this stage is characterized by sliding and local boudinage of biotite along (001) planes and the formation of deformation lamallae, undulose extinction and subgrain structures in quartz. These processes tend to increase the total surface area of biotite and quartz which leads to high internal free energy which in turn triggers mineral reactions or enhance rates of mineral reactions and therefore may have ultimately caused the pronounced increase in the sillimanite/biotite ratio.

The number of microcracks shows a positive correlation with the sillimanite/biotite ratio. The development of such microcracks in amphibolite-facies rocks is due to a reduction in mean stress as a consequence of increased fluid pressure. The distinct spatial relationship between the mircocracks, the fibrolite bundles and the sillimanite/biotite ratio suggest that biotite breakdown to sillimanite and quartz caused an increased fluid pressure at the reaction sites.

6 DISCUSSION

6.1 *Fluid sources*

Estimates for the volume of infiltrating fluid and structurally-bound water in the wallrock indicates that the infiltrating fluid was not generated in situ. Ring (1993) showed that the pattern of contractional and transcurrently displacing amphibolite-facies shear zones in northern Malawi occurred during a subduction/collision event, that is, in a scenario in which material in the lower plate may still undergo prograde metamorphic reactions whereas rocks in the upper plate undergo decompression and retrograde metamorphism. Dehydration reactions during prograde metamorphism create large volumes of fluid. As shown by Ague (1994), prograde metamorphism may cause volume loss on the order of 30% in amphibolite-facies rocks. The most obvious chemical change during such volume loss is the progressive removal of silica with increasing metamorphic grade. This suggests that the liberated fluids were highly saturated with respect to silica.

The problem with deriving fluids from deeper crustal levels is that these upward, i.e. down regional PT gradients, migrating fluids are not likely to dissolve large amounts of silica. In a contractional setting, however, thrusts usually place higher isotherms over lower ones and therefore create inverted thermal gradients thus permitting silica to be dissolved by the fluid upon passage through the shear zone. Possible shear heating would increase this effect.

6.2 *Retrogressive shearing and fluid flow*

The initiation of the shear zone is intimately associated with the growth of fibrolite and quartz from biotite and feldspar. The formation of fibrolite and quartz in high-strain zones appears to be a consequence of dehydration reactions and dealkalisation accompanying mylonitisation, both of which caused

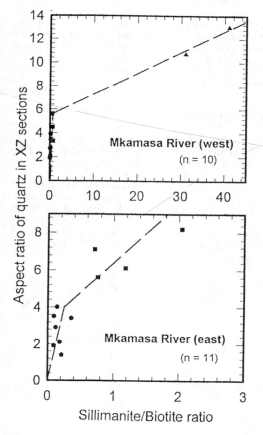

Fig.2 Aspect ratio of quartz ribbons in XZ sections plotted against sillimanite/biotite ratio.

the breakdown of biotite. Although the total amount of water liberated by this dehydration reaction is negligible with respect to the total volume of fluid that infiltrated the shear zone, the relationship between the sillimanite/biotite ratio and strain suggests that the released water had a catalytic influence on the deformation behaviour of quartz. The dealkalisation within the shear zone caused the destabilisation of micas and also prevented the growth of new micas in the shear zone. The end product of progressive shearing is almost "dry", sheet slilicate-free mylonite II. The weak bonding between (001) planes in layer silicates results in higher porosity than in strongly bonded framework and ortho silicates. The almost complete breakdown of biotite destroyed the pathways for fluid circulation in the mylonite which ultimately resulted in strain hardening and the cessation of shearing.

6.3 *Lateral variations in shear-zone thickness*

A remarkable feature of the Mkamasa River shear zone is the lateral variation of its overall structure. High volume loss, high fluid/rock ratio and stretching lineations oriented subperpendicular to the shear direction in the thin mylonite zone in the W compare to relatively little volume loss, a lower fluid/rock ratio and a "normal", i.e. transport-parallel, orientation of the stretching lineations in the thick mylonite zone in the E. In the western part of the Mkamasa River shear zone, wallrock and mylonite I consist of relatively little biotite (11% and 6%, respectively) as compared to the shear zone in the east (20% and 13% biotite). The amount of framework and ortho silicates in the W increases towards the centre of the shear zone (75%→79%→89%), whereas in the E the amount of framework and ortho silicates is reduced in mylonite I (76%→71%). As discussed above, the weak bonding between (001) planes in layer silicates results in higher porosity than in strongly bonded framework and ortho silicates. Accordingly, the relatively high amount of biotite relative to framework and ortho silicates aided fluids to infiltrate a larger volume of rock in the E. In the western part of the Mkamasa River shear zone, the infiltration of fluid was more difficult, hence a relatively thin zone with a high fluid/rock ratio developed. This high fluid/rock ratio aided diffusion processes at grain boundaries and may have caused dissolution and transport of silica causing the large volume loss in the W.

Newman & Mitra (1993) compared varitions in fluid flow along the Linville Falls fault zone to the grain-boundary model of islands and channels. Channels are represented by thin zones with a high fluid/rock ratio and islands are characterised by thick zones with low fluid/rock ratios. The structural data indicate that the degree of coaxial deformation was higher in the thin western part (channel) of the Mkamasa River shear zone than in the broad eastern part (island). The formation of the channels may have been aided by an increased amount of shear-zone perpendicular contraction. On a larger scale, such a fluid-flow and strain partitioning ultimately results in an anastomosing pattern of shear zones in the Mugesse shear belt.

The transport-normal orientation of the stretching lineations in the western part of the Mkamasa River shear zone indicates that the maximum principal elongation became the axis of the rotational component of deformation. This orientation of the stretching lineations was controlled by the pure-shear component of deformation because fabric elements can rotate faster in pure shear than in simple shear.

7 CONCLUSIONS

I believe that the following conclusions can be drawn from this study:

(1) The studied retrograde strike-slip shear zone is characterised by the progressive breakdown of feldspar and biotite to form sillimanite, quartz and water. The shear zones developed under amphibolite-facies P-T conditions below the "wet" granite solidus and are characterized by high fluid/rock ratios and considerably negative volume strains. The bulk of the volume loss was achieved by loss of silica.

(2) A dramatic consequence of the high fluid flow was an almost complete dealkalisation within the shear zone. The associated destabilisation of micas destroyed the pathways for fluid circulation, "dried out" the shear zone and ultimately caused the cessation of shearing.

8 REFERENCES

Ague, J.J. 1994. Mass transfer during Barrovian metamorphism of pelites, south-central Connecticut. I: Evidence for changes in composition and volume. *Am. J. Sci.* 294: 989-1057.

Newman, J. & G. Mitra, Lateral variations in mylonite zone thickness as influenced by fluid-rock interactions, Linville Falls fault, North Carolina. *J. Struc. Geol.* 15: 849-863, 1993.

Ring, U. 1993. Aspects of the kinematic history and mechanism of superposition of the Proterozoic mobile belts of eastern Central Africa. *Precambr. Res.* 61: 207-226.

Subsurface horst features beneath the geothermal reservoirs in Taupo Volcanic Zone (TVZ)

S.Tamanyu
Geological Survey of Japan, Tsukuba, Japan

ABSTRACT: The subsurface structures related to geothermal reservoirs have been reviewed at Wairakei, Ohaaki (Broadlands) and Rotokawa fields which are located at the eastern part of Taupo Volcanic Zone (TVZ). Stratigraphic correlations between the drill holes indicate that the subsurface geological structures are characterized by horst and graben structures, or caldera structures in TVZ. Temperature logging data show that the subsurface isothermal structures are characterized by relatively higher temperatures on the horst blocks. These features are consistent with residual gravity and resistivity maps. It can be concluded that potential reservoirs would occur at shallow depth above horst blocks in TVZ geothermal areas.

1. INTRODUCTION

The TVZ is defined by Quaternary vent positions and caldera structural boundaries, and regarded as a rifted arc about 300 km long and up to 60 km wide (Wilson et al., 1995). It is characterized by voluminous rhyolite volcanism and by high rates of magma production. Associated with this volcanism are numerous high-temperature (>250°C) geothermal systems through which a total natural heat output of 4200±500 MW is channelled (Bibby et al., 1995). Most geothermal fields are contained within the area of rhyolite volcanism. Boundaries of calderas are often speculative, but of 20 geothermal systems considered, 15 occur on or next to a caldera margin where there is enhanced deep permeability (Wood, 1995). Three geothermal power plants operate at Wairakei, Ohaaki (Broadlands) and Kawerau, and two other new plants have been constructed and operated since 1997 at Poihipi (southwestern margin of Wairakei) and Rotokawa, respectively.

2. GEOTHERMAL STRUCTURES REVEALED BY DRILL HOLES

Three exploited geothermal fields, Wairakei, Ohaaki (Broadlands) and Rotokawa are good model fields to examine the subsurface structures because their geothermal models have been based on stratigraphic correlation. These areas are located in the eastern part of the central TVZ (Fig. 1).

2.1 Wairakei

The geology of this field was investigated in detail, and was summarized as a 1: 63,360 geologic map (Grindley, 1961) and New Zealand Geological Survey Bulletin (Grindley, 1965). The dominant structural features of Wairakei area are two north-

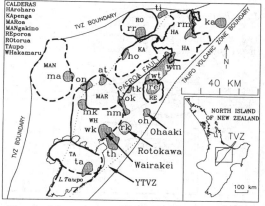

Fig. 1: Map of the central TVZ showing caldera boundary and geothermal fields (Wood, 1995). Caldera boundaries are shown as heavy solid or dashed lines (where less certain). YTVZ=Young Taupo Volcanic Zone. Geothermal fields are stipples areas labelled: at=Atiamuri, ho=Horohoro, ka=Kawerau, ma=Mangakino, mk=Mokai, nm=Ngatamariki, oh=Ohaaki, or=Orakeikorako, on=Ongaroto, re=Reporoa, rk=Rotokawa, rm=Rotoma, rr=Rotorua, ta=Taupo, th=Tauhara, ti=Tikitere, tk=Te Kopia, wk=Wairakei, wm=Waimangu, wt=Waiotapu.

east trending basins - Taupo-Reporoa Basin and Taupo-Rotorua Basin - separated by a complex, fault-bounded horst, the Wairakei Block. Cross sections indicate the following subsurface structures in the borefield: (a) throws of the faults bounding major blocks, increase with depth; (b) faults appear to have steep dips (80-88°); (c) there are fewer major faults shown at depth than in the shallow strata; (d) presence of a large, buried, non-active, (caldera?) fault-scarp was identified. The latest interpretation of the subsurface structure in this field is more focused on caldera structure recognition (Wood, 1995).

The reservoir characteristics of the field are illustrated in Fig. 2 (Grant, 1988). The cross sections show the geology, isotherms in the natural pre-exploitation state and in the 1980's after 25 years production (Fig. 2). The principal permeability from which production wells derive fluid is the contact between the Waiora Formation and Wairakei Ignimbrite. Some shallow wells also draw on the Huka Falls Formation - Waiora Formation contacts, and outlying wells (not in current use) on the the Waiora Formation - Karapiti Rhyolite contact. On the reservoir scale, the uniformity of drawdown across the field shows that good horizontal permeability exists across Wairakei and may be continuous to SSE extension, Tauhara field (Fig. 1). These data suggest that deep geothermal fluid ascends through faults and/or formation boundaries from the subsurface horst block-faulted region and then spreads out horizontally.

2.2 Ohaaki (Broadlands)

The subsurface geology in this field has been investigated using the recently drilled deep production wells with pre-existing drill hole data from the viewpoint of basement geology by Wood (1996) (Fig. 3). The geological structure is reviewed in this paper as follows. A 3D view of the basement shows how the surface lower from SE to NW over the buried scarps of two major basement faults (Fault A and B) which comprise further part of the Kaingaroa Fault Zone. Fault A is further northwest out of the cross section. Fault B trends SW-NE parallel to the TVZ boundary, but the downthrown side of Fault A is defined only by BR15, and its SW-NE orientation is presumed. Fault C is poorly defined, but also has a westerly downthrow. BR16 stratigraphy suggests that Fault D lowers the basement in the SE corner of the field down to the SE by at least 200 m, in a contrary sense to the general westward downthrow into the TVZ. All the faults appear to dislocate the Rangitaiki Ignimbrite, and possibly the overlying Rautawiri Breccia. Younger formations are not noticeably affected, suggesting that most movement occurred prior to about 0.3 Ma.

The graywacke has a poor production record. Circulation losses while drilling in the basement were recorded in BR10, BR42 and BR43 (at 72 m, 185 m and 20 m below basement surface respectively), but only BR43 could sustain production. None-the-less, the basement is hot and potentially productive. Fig. 3 is a geologic cross-section from BR16 to BR34, overlaid on a thermal cross section. A thermal plume rising into the volcanics from the basement in the vicinity of Faults B and C is clearly shown. These data suggest that deep geothermal fluid ascends through faults and/or formation boundaries from the subsurface horst region and then spreads out horizontally. In this respect, fluid behaviour is the same as at Wairakei.

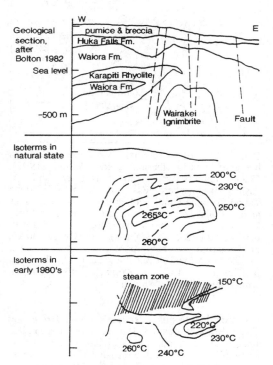

Fig. 2: West-East cross section of Wairakei geothermal field (Grant, 1988).

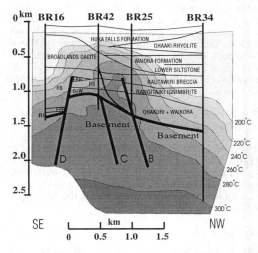

Fig. 3: Southeast-Northwest cross section of Ohaaki (Broadlands) area (Wood, 1996). Fault A is further northwest out of the cross section. EBR: East Broadlands Rhyolite, RI: Rangitaiki Ignimbrite.

2.3 Rotokawa

The subsurface stratigraphy of the Rotokawa field has been explored by shallow and deep drilling. Interbedded silicic pyroclastic rocks, lacustrine sediments and rhyolite lava overlie thick (>1,100 m) andesite lava flows resting unconformably upon the Mesozoic graywacke and argillite basement rocks of the region (Grindley et al., 1985).

Collar and Browne (1985) identified a series of eruption vents and associated ejecta deposited over a period between 3,400 and 20,000 years ago. At least 8 major hydrothermal eruptions have occurred at Rotokawa in the last 20,000 years, the biggest about 6,000 years ago, but there is no evidence for any in the past 3,400 years.

Deep (>2 km) drillholes in this field penetrated a sequence of Quaternary silicic volcanic rocks and andesite lava, and Mesozoic graywacke basement (Browne, et al., 1992, Arehart et al., 1997 in press). A north-south geological cross section is shown on Fig. 4 (Arehart et al., 1997 in press). In accordance with Grindley (1961), the block penetrated by RK4 should be regarded as a horst structure in this area. Subsurface isothermal structure is estimated from temperature logging data (e.g., Collar, 1985 in Stokes, 1994), and indicates that geothermal fluids rise up from the horst block in a similar way to fluid behaviour at Wairakei and Ohaaki (Broadlands).

3. GEOPHYSICAL EVIDENCE OF SUBSURFACE HORSTS

Electrical resistivity techniques have proved to be the most effective geophysical tool for identifying the near-surface extent of geothermal waters. Available data measured with the 500-m Schlumberger array has been compiled, and geothermal fields are

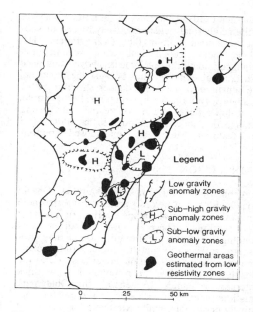

Fig. 5: Residual gravity anomalies overlaid on low resistivity anomalies (partly revised from Bibby et al., 1995).

delineated as low-resistivity zones (e.g. Bibby et al., 1995). Gravity studies have also proved to be a useful technique to estimate the subsurface density distribution. The residual gravity anomalies indicate that a broad area of low gravity, almost equivalent to the structurally-defined TVZ exists, and that it may be sub-divided into a number of smaller areas of even lower gravity and sub-high gravity anomaly zones. These smaller-scale low gravity features are suggested to relate to caldera collapse structures, subsequently infilled with low-density material. Sub-high gravity anomaly zones are assumed to be associated with high-level Mesozoic graywacke or lowermost Cenozoic andesite basement (e.g. Bibby et al., 1995). An overlay of apparent resistivity and residual gravity (Fig. 5) shows that most geothermal areas occur in the eastern TVZ within the area designated "young TVZ" (Wilson, et al., 1995), as defined by the distribution of volcanic vents of 0.34 Ma age and younger. Many of the geothermal areas are located in close relation to sub-high gravity anomaly zones within the low gravity anomaly of the TVZ. This suggests that most geothermal fields are associated with sub-horst structures in graywacke and/or the lowermost Quaternary andesite, and these sub-horsts provide relatively high temperature conditions because conductive heat transfer is predominant within them, in contrast to the overlying water-saturated Quaternary volcanics. Although the sub-horsts shown in Fig. 2, 3 and 4 are of smaller scale than the sub-high gravity anomaly zones, the faults developed within and around the sub-horsts seem to be preferred locations for the upwelling of

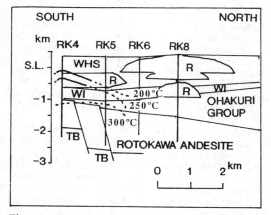

Fig. 4: Geothermal model for Rotokawa area (Arehart et al., 1997 in press). R: Rhyolite lavas, TB: Torlesse basement (greywacke), WHS: Tuffs and sediments of the Falls/superficial formations, WI: Wairakei Ignimbrite (330 ka)

geothermal fluids. Therefore, the author emphasises that the geothermal fields are closely related to sub-horst structures within the TVZ graben. The available other geophysical data, seismic and aeromagnetic data do not indicate above mentioned subsurface sub-horst features because these data are more sensitive to the more deeper structure rather than shallow (<3 km depth) structures. For example, the sub-horst features can not be delineated using aeromagnetic data which respond better to deep (>3 km) features such as magnetised plutons (Soengkono, 1995).

4. CONCLUSION

The geothermal fluid flows in Wairakei, Ohaaki (Broadlands) and Rotokawa areas are inferred here from isothermal contour maps overlaid on geologic cross sections. These diagrams indicate that ascending fluid flows are strongly controlled by sub-horst structures within graben structures of Mesozoic graywacke basement and/or lowermost Cenozoic andesite, which have been covered with thick Quaternary volcanics. In general, the Quaternary formations (except lacustrine sediments) are relatively permeable, while Mesozoic basement and the Cenozoic andesite are impermeable except for fault zones. The top surface of the Mesozoic basement or the andesite is generally consistent with isothermal contours because conductive heat transfer is dominant in these horizons. In contrast, fluid convection is dominant in overlying Quaternary volcanics. Potential geothermal fields can therefore be expected above sub-horsts at shallow depth rather than deeper grabens. These sub-horst structures play the role at heat source regions for overlying geothermal reservoirs. This idea has been developed from the conceptual modelling of the Hohi geothermal area in the Beppu-Kuju graben, Japan, where the subsurface geothermal structure has been clarified by many deep drill holes (e.g., Tamanyu, 1994).

ACKNOWLEDGEMENTS

The author expresses his appreciation to C.P. Wood and M. Rosenberg for their critical readings and comments on the original manuscript, and also to C.P. Wood for his offer of some original figures.

REFERENCES

Arehart, G.B., Wood, C.P., Christenson, B.W., Browne, P.R.L. and K.A. Foland, 1997 in press. Timing of volcanism and geothermal activity at Ngatamariki and Rotokawa, New Zealand. *Proc. 19th N. Z. Geotherm. Workshop. 1997.* University of Auckland:

Bibby, H.M., Caldwell, T.G., Davey, F.J. and T.H.Webb, 1995. Geophysical evidence on the structure of the Taupo Volcanic Zone and its hydrothermal circulation. *J. Volc. Geotherm. Res.* 68: 29-58.

Bolton, R.S. 1982. Fracturing in the Wairakei geothermal field. *Proc. Geotherm. Resources Council Workshop,* Hawaii, 27-28. Aug. 1982.

Browne, P.R.L., Graham, I.J., Parker, R.J. and C.P. Wood, 1992. Subsurface andesite lavas and plutonic rocks in the Rotokawa and Ngatamariki geothermal systems, Taupo Volcanic Zone, New Zealand. *J. Volc. Geotherm. Res.* 51: 199-215.

Collar, R.J. 1985. Hydrothermal eruptions in the Rotokawa Geothermal System, Taupo Volcanic Zone, New Zealand. Unpubl. Geothermal Institute Report, University of Auckland.

Collar, R.J. & P.R.L. Browne, 1985. Hydrothermal eruptions at the Rotokawa geothermal field, Taupo Volcanic Zone, New Zealand. *Proc. 7th N. Z. Geotherm. Workshop, 1985.* University of Auckland: 171-175.

Grindley, G.W. 1961. *Sheet N94 Taupo, Geological Map of New Zealand* 1:63,360, Wellington, DSIR.

Grindley, G.W. 1965. The Geology, structure, and exploitation of the Wairakei Geothermal Field, Taupo, New Zealand. *N. Z. Geol. Surv., Bull.* 75, 131P.

Grindley, G. W., Browne, P. R. L. and M. W. Gardner 1985. Geologic report. In, Mongillo, M.A., compiler, Rotokawa geothermal field: A preliminary report of the scientific investigations. Unpubl. DSIR Report.

Grant, M.A. 1988. Reservoir engineering of Wairakei geothermal field. *Proc. 13th Workshop Geotherm. Reservoir Engineering, Stanford, California,* 191-198.

Soengkono, S. 1995. A magnetic model for deep plutonic bodies beneath the central Taupo Volcanic Zone, North Island, New Zealand. *J. Volc. Geotherm. Res.,* 68: 193-207.

Stokes, E. 1994. *Rotokawa Geothermal Area, some historical perspectives.* Department of Geography, University of Waikato, Hamilton, New Zealand. 177P.

Tamanyu, S. 1994. Structural controlled reservoirs within the horst and graben structure at the Hohi geothermal area, Japan. *Geothermal Resources Council Transactions,* 18: 611-615.

Wilson, C.J.N, Houghton, B.F., McWilliams, M.O., Lanphere, M.A., Weaver, S.D. and R.M. Briggs. 1995. Volcanic and structural evolution of Taupo Volcanic Zone, New Zealand: a review. *J. Volc. Geotherm. Res.,* 68: 1-28.

Wood, C.P. 1994. The Waiora formation geothermal aquifer, Taupo Volcanic Zone, New Zealand. *Proc. 16th N. Z. Geotherm. Workshop, 1994.* University of Auckland: 121-126.

Wood, C.P. 1995. Calderas and geothermal systems in the Taupo volcanic zone. New Zealand. *Proc. World Geotherm. Cong. Florence 1995.* vol. 2: 1331-1336.

Wood, C.P. 1996. Basement geology and structure of TVZ geothermal fields, New Zealand. *Proc. 18th N. Z. Geotherm. Workshop, 1996.* University of Auckland: 157-162.

Ar/Ar dating and uplift rate of hydrothermal minerals in the Southern Alps, New Zealand

Damon A. H. Teagle & Chris M. Hall
Geological Sciences, University of Michigan, Ann Arbor, Mich., USA

Simon C. Cox & Dave Craw
Department of Geology, University of Otago, Dunedin, New Zealand

ABSTRACT: The rapid uplift of the Southern Alps, New Zealand, along the Alpine Fault has resulted in high geothermal gradients and vigorous hydrothermal circulation in the upper crust. The dating of minerals associated with these geothermal systems by Ar-Ar techniques have been unsuccessful to date. A procedure has been developed to remove inherited Ar from vein minerals by liberating fluid inclusions prior to final analysis for dating. Useful ages are provided by adularia and biotite from the central Southern Alps of 880 kyr (minimum) and 1.26±0.02 Myr respectively. Our data constrain the uplift rate of the Almer Ridge mineralized zone to be <1 mm/yr. This rate is very slow compared to uplift rates of ca. 8 mm/yr along the Alpine Fault only 10 km away. Similar variations in uplift rates should be expected in all active collisional mountain chains and in exhumed metamorphic terranes which formed beneath such mountain belts.

1 INTRODUCTION

The Southern Alps are an actively rising mountain belt formed along an obliquely convergent portion of the Australian–Pacific plate boundary. The western margin of the belt is the Alpine Fault, which is a dextral oblique-slip fault accommodating most of the nearly 40 mm/year oblique convergence. The Southern Alps are underlain by meta-greywackes variably reconstituted during Mesozoic metamorphism. This ancient metamorphic belt was partially exhumed between 160-120 million years ago, exposing prehnite-pumpellyite facies rocks along the Main Divide of the Southern Alps. The rise of the mountains since the Pliocene has exposed pumpellyite-actinolite, greenschist and amphibolite facies rocks according to the total amount of uplift, with higher grades rocks cropping out at lower altitudes progressively closer to the present day trace of the Alpine Fault.

The uplift rate immediately east of the Alpine Fault in the central portion of the mountain belt is about 8 mm/year (based on uplifted Quaternary deposits; Simpson et al., 1994), and uplift keeps pace with erosion due to very high rainfall (Koons, 1990). Consequently, although this part of the mountain belt is near sea level, there has been nearly 20 km of uplift and erosion over the 2-3 million year life of the belt (Cooper, 1980). Uplift rate is only about 1 mm/year on the east side of the mountains (based on tilted lake beaches; Wellman, 1979), and in that rain shadow area erosion does not keep pace with uplift.

The observable mountain crests and ridges indicate the approximate total amount of uplift. Between the eastern and western slopes described above, uplift rate decreases generally from west to east across the drainage divide (Main Divide). Uplift rate is difficult to quantify in this rugged region because there are no reliable Quaternary markers. The present study attempts to quantify uplift rate at two localities to the west of the Main Divide using Ar/Ar dating and fluid inclusion data from hydrothermal minerals.

The rapid uplift of collisional mountain belts results in pronounced thermal anomalies at shallow levels in the crust (Koons, 1987) and the high geothermal gradients encourage vigorous fluid (Koons & Craw, 1991). Fluid flow is driven by topography and strain, and involves both meteoric and deep crustal waters (Koons & Craw, 1991). These fluids deposit vein minerals, mainly quartz and calcite, throughout the mountain belt, and are associated with widespread minor gold mineralization (Cox & Craw, 1995). The hydrothermal veins crosscut structures associated with the rise of the Southern Alps and must be late Cenozoic in age, but previous attempts at absolute dating have been unsuccessful. This study provides the first absolute dates, on distinctive vein generations seen throughout the western slopes of the high mountains.

Previous endeavours using Ar systems in the Southern Alps were fraught with the recurrent problem of excess inherited argon (Wellman & Cooper, 1971; Adams, 1981; Cooper et al., 1987). To address this problem a procedure has been developed to remove inherited Ar from minerals by liberating fluid inclusions prior to final analysis for dating.

Minerals from hydrothermal veins from adjacent valleys in the Franz Josef to Fox Glacier region of the central Southern Alps have been investigated. These samples are hosted by greenschist facies schists at an altitude of 1500–1800 m above sea level (Craw, 1997). The most promising results come from Ar-Ar determinations of the timing of precipitation of hydrothermal adularia and late metamorphic biotite.

We have attempted to date samples of adularia, muscovite and actinolite from near Almer Hut on the northern flank of the Franz Josef glacier. Fibrous actinolite occurs with quartz in early veins synchronous with folded metamorphic veins. Adularia occurs with quartz and calcite as large euhedral crystals that fill and protrude into open fissures. These veins display a variety of orientations but are most commonly steeply dipping and fill fractures that cut schistosity and upright folds associated with Alpine Fault deformation (Cooper et al., 1987; Craw et al., 1994). Unique to this locality, late stage muscovite occurs coating the prismatic minerals and this is the only reported occurrence of muscovite in this type of vein. Porphyroblastic biotite forming a new foliation in garnet zone schist from near the junction of the Victoria Valley and Fox glacier has also been investigated. This locality is approximately 10 km south west of the Almer Ridge and is closer to the Alpine Fault.

Adularia from the Almer Ridge mineralized zone contains numerous large (up to 100 µm) fluid inclusions. Most of these are round and irregularly distributed through the crystals and are of primary origin, but some cross-cutting secondary inclusion trails occur also. Primary inclusions homogenize at 240-280°C, and secondary inclusions typically homogenize at lower temperatures (220-230°C) but up to 250°C. Ice-melting temperatures are -0.8 to -2.0°C.

Adularia with bladed calcite commonly indicate mineral deposition at or near a hydrothermal boiling zone (Browne, 1978; Simmons & Christenson, 1994), and vapour-rich fluid inclusions have been found in adularia and calcite in similar veins a few km to the northeast (Craw, 1997). This means that the adularia homogenization temperatures reflect true trapping temperatures, and the fluid pressure (35 to 65 bars) can be estimated from the liquid-vapour curve in pressure-temperature space (Fisher, 1976). This implies mineralization depths of about 400-700 metres assuming hydrostatic pressure and overlying water density of 0.9 to 1.0 g/cm^3 (Craw, 1997; Cox et al., 1997). Similar veins in the Douglas Valley about 30 km southwest of Almer Ridge formed at higher temperatures (300-350°C) and therefore greater depths (down to 2 km; Cox et al., 1997).

2 AR-AR DATING OF HYDROTHERMAL MINERALS

Mineral samples were irradiated, along with packets of interlaboratory standard biotite (FCT-3, 27.99 Ma; Hall & Farrell, 1995) in the Phoenix-Ford Memorial Nuclear Reactor of the University of Michigan. Multiple grains of actinolite, biotite and muscovite were transferred into sample wells for conventional laser step heating. Adularia grains were treated in a two stage process in order to liberate fluids trapped in inclusions that pilot studies had shown to contain a significant proportion of inherited Ar. Single grains of adularia were heated with a diffuse laser beam in two increments to burst fluid inclusions and liberate all the trapped gas, but without significantly disturbing lattice bound Ar. The vacuum was then broken and an outgassed flux bead was added to each sample well. The single grains of the pale, translucent adularia were melted with the aid of the flux bead, with a focussed high power laser beam.

2.1 Adularia - Almer Ridge

63 single grains of adularia have been analyzed from two samples from the same outcrop of different veins of the same generation. The analytical data for the two samples are indistinguishable, and all data will be dealt with together. The results of the low-temperature heating steps and the high-temperature fusions with flux beads are plotted on Fig. 1. There is a clear separation between the low and high temperature data sets with the low temperature points clustering at very low $^{39}Ar/^{40}Ar$ ratios (<0.03) and displaying $^{40}Ar/^{36}Ar$ spanning the range from atmosphere to extremely radiogenic values (>60,000). The high temperature points do not define an isochron and scatter in an elongate data cloud with a gentle negative slope (Fig. 1). Age constraints can be placed on these data by assuming that the analyses record mixing in a 3 component system bounded by atmosphere, an inherited Ar component as defined by the low temperature data, and Ar due to in situ radioactive decay. Reference lines have been calculated for mixing between atmosphere and in situ decay for 800, 900, and 1000 kyr. No samples yield ages less than 800 kyr and the youngest bounding line that will contain all data is ca. 880 kyr (Fig. 1), with older ages reflecting slight contamination by highly radiogenic inherited Ar as liberated from fluid inclusions in the low temperature analyses. The bounding line at 880 kyr represents a minimum age but due to the asymmetric nature of the data and the irregular contamination by inherited Ar precise error estimates cannot be assigned. The adularia could

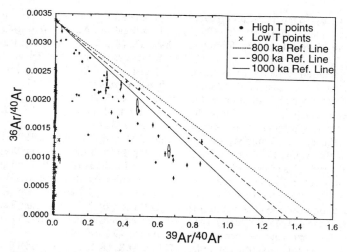

Fig. 1: Ar-Ar analyses of single grains of adularia from Almer Ridge, Southern Alps, New Zealand.

conceivably be younger than 880 kyr if the samples that define the uppermost bounding line still contain inherited ^{40}Ar. However, after 63 analyses there is as yet no evidence for younger ages. Whilst adularia is a common host for Ar-rich fluid inclusions it appears that Ar is not incorporated into the crystalline lattice during growth of the mineral.

2.2 Actinolite and Muscovite - Almer Ridge

Step heating of actinolite yielded an irregular U-shaped pattern with no clearly defined plateau and shows strong effects of having formed from a chlorine-rich brine. The low K content of actinolite means that the Ar is most probably trapped within inclusions, and not from *in situ* radioactive decay.

Ar-Ar laser step-heating analyses of muscovite from the Almer Ridge mineralized fissures do not define isochrons or provide geologically reasonable gas ages. However, the data are consistent with the contemporaneous precipitation of adularia and muscovite, strongly contaminated with inherited Ar of similar isotopic compositions to that released from fluid inclusions in the adularia.

2.3 Biotite - Confluence of Fox & Victoria Glaciers

Conventional laser step-heating of late metamorphic biotite from garnet schists from the Fox Glacier yields an isochron age of 1.26±0.02 Myr (MSWD=0.54; n=7). The isochron has a near atmospheric intercept indicating only a very minor inherited radiogenic component.

3 DISCUSSION

Useful ages are provided by multiple fusions of adularia and a step-heating analysis of biotite of 880 kyr (minimum) and 1.26±0.02 Myr, respectively. High concentrations of highly radiogenic, ancient Ar are present in the fluids circulating through the deep level rocks of the Southern Alps and excess Ar can plausibly be invoked to explain large age dis-crepancies in previous studies (e.g., Adams, 1981; Cooper et al., 1987) as well as analyses of muscovite and actinolite in this study.

The similarity between the Ar blocking temperature (200–250°C; e.g., Lovera et al., 1993) and the fluid inclusion homogenization temperature (Craw, 1997) suggests that the minimum age of 880 kyr yielded for the adularia samples most probably records the time of formation and hydrothermal activity in the Almer Ridge mineralized zone. In contrast, the biotite isochron age most probably records the passage of the host garnet schist through the biotite Ar retention temperature (350–400°C for cooling >200°C/Myr; Harrison et al., 1985).

A pressure-temperature path for the uplift of the central Southern Alps has been estimated from fluid inclusions in metamorphic and hydrothermal vein minerals (Holm et al., 1989). By combining this information with a range of Ar blocking temperatures for biotite (350–400°C) an uplift rate of between 4.8–8.8 mm/yr is estimated for the garnet schists from the Fox Glacier. This rate is rapid and similar to determinations based upon uplifted marine terraces (5–20 mm/yr; Bull & Cooper, 1986) and fission track dating of zircons (1-8 mm/yr; Tippet & Kamp, 1993). Bull & Cooper (1986) and Simpson et al. (1994) estimate the present rate of uplift to be near 8 mm/yr.

The exposure of greenschist facies rocks from beneath metagreywackes to the east of the Alpine Fault implies uplift of the order of 5–10 km over the 2 to 3 Myr of orogeny (Cooper, 1980). Hence the expected uplift rates for this portion of the Alpine schists is expected to be ca. 1.6–5 mm/yr, significantly lower than near the Alpine Fault (see above). For the mineralized adularia veins at Almer Ridge the depth of formation (400–700 m) as estimated from fluid inclusions (Craw, 1997) and the minimum age determined by Ar-Ar dating (880 kyr) constrain the uplift rate to be in the range of 0.45–0.8 mm/yr, and certainly an uplift rate less than 1mm/yr seems to be inescapable. Hence, our new data quantifies an order of magnitude change in uplift rate over ca. 10 km in the Southern Alps. Similar variations in uplift rates should be expected in all active collisional mountain belts and in exhumed metamorphic belts which formed beneath such mountain belts.

4 REFERENCES

Adams, C.J., 1981. Uplift rates and thermal structure in the Alpine fault zone and Alpine schists, Southern Alps, New Zealand. *Geol. Soc. Lond. Spec. Pub.* 9:211-222.

Browne, P.R.L., 1978. Hydrothermal alteration in active geothermal fields. *Ann. Rev. Earth Planet. Sci.* 6:229-250.

Bull, W.J., & A.F. Cooper, 1986. Uplifted marine terraces along the Alpine Fault, New Zealand. *Science*, 234:1225-1228.

Cooper, A.F., 1980. Retrograde alteration of chromian kyanite in metachert and amphibolite whiteschist from the Southern Alps, New Zealand, with implications for uplift on the Alpine Fault. *Contrib. Mineral. Petrol.* 75:153-164.

Cooper, A.F., B.A. Barreiro, D.L. Kimbrough, J.M. Mattinson, 1987. Lamprophyre dike intrusion and the age of the Alpine Fault, New Zealand. *Geology*, 15:941-944.

Cox, S.C., & D. Craw, 1995. Structural permeability in the Southern Alps hydrothermal system, New Zealand. *Proc. PACRIM 95*:157-162.

Cox, S.C., D. Craw, and C.P. Chamberlain, 1997. Structure and fluid migration in a late Cenozoic duplex system forming the Main Divide in the central Southern Alps, New Zealand. *N.Z. J. Geol. Geophys.*, In press.

Craw, D., 1997. Fluid inclusion evidence for geothermal structure beneath the Southern Alps, New Zealand. *N.Z. J. Geol. Geophys.* 40:43-52.

Craw, D., M.S. Rattenbury, and R.D. Johnstone, 1994. Structures within the greenschist facies Alpine Schist, central Southern Alps, New Zealand. *N.Z. J. Geol. Geophys.* 37:101-111.

Fisher, J.R., 1976. The volumetric properties of H_2O: a graphical portrayal. *U.S. Geol. Surv. J. Res.* 4:189-193.

Hall, C.M. and J.W. Farrell, 1995. Laser $^{40}Ar/^{39}Ar$ ages of tephra from Indian Ocean deep-sea sediments; tie points for the astronomical and geomagnetic polarity time scales. *Earth Planet. Sci. Lett.*, 133:327-338.

Harrison, T.M., I. Duncan, and I. McDougall, 1985. Diffusion of ^{40}Ar in biotite: Temperature, pressure and compositional effects. *Geochim. Cosmochim. Acta*, 49:2461-2468.

Holm, D.K., R.J. Norris, and D. Craw, 1989. Brittle and ductile deformation in a zone of rapid uplift: Central Southern Alps, New Zealand. *Tectonics*, 8:153-168.

Lovera, O.M., M. Heizler, and T.M. Harrison, 1993. Argon diffusion domains in K-feldspar; II, Kinetic properties of MH-10. *Contrib. Mineral. Petrol.* 113:381-393.

Koons, P.O., 1987. Some thermal and mechanical consequences of rapid uplift: an example from the Southern Alps, New Zealand. *Earth Planet. Sci. Lett.* 86:307-319.

Koons, P.O., 1990. Two-sided orogen: Collision and erosion from the sandbox to the Southern Alps, New Zealand. *Geology*, 18:679-682.

Koons, P.O., & D. Craw, 1991. Gold mineral-isation as a consequence of continental collision: an example from the Southern Alps, New Zealand. *Earth Planet. Sci. Lett.* 103:1-9.

Simmons, S.F., & B.W. Christenson, 1994. Origins of calcite in a boiling geothermal system. *Am. J. Sci.* 294:361-400.

Simpson, G.D., A.F. Cooper, and R.J. Norris, 1994. Late Quaternary evolution of the Alpine Fault Zone at Paringa, South Westland, New Zealand. *N.Z. J. Geol. Geophys.* 37:49-58.

Tippett, J. M., & P.J.J. Kamp, 1993. Fission track analysis of the late Cenozoic vertical kinematics of continental Pacific crust, South Island, New Zealand. *J. Geophys. Res.* 98:16119-16148.

Wellman, H. W. 1979. An uplift map for the South Island of New Zealand and a model for uplift of the Southern Alps. *Royal Soc. N.Z. Bull.* 18:13-20.

Wellman, P., & A.F. Cooper, 1971 Potassium-argon age of some New Zealand lamprophyre dikes near the Alpine Fault. *N.Z. J. Geol. Geophys.* 14:341-350.

Trap integrity and fluid migration: Coupled mechanical/fluid flow models

P. Upton & K. Baxter
CSIRO, Division of Exploration and Mining, Nedlands, W.A. & the Australian Geodynamics Cooperative Research Centre, Australia

G.W. O'Brien
Australian Geological Survey Organisation, Canberra, A.C.T. & the Australian Geodynamics Cooperative Research Centre, Australia

ABSTRACT: This study presents fully coupled mechanical/fluid flow numerical models which consider the mechanics of fault reactivation and subsequent fluid flow and the application of these to oil and gas accumulations in the Timor Sea off Australia. The ability to predict regions of volume change with deformation and associated effects on the fluid migration represents significant advancement in the understanding of basins worldwide. By integrating observations of charge history and tectonic framework with a mechanically constrained model, we can move towards the generation of a truly predictive reservoir modelling tool.

1 INTRODUCTION

A recent integrated study of fluid flow, hydrocarbon charge and thermal histories of a series of traps in the Timor Sea, off the North West Shelf of Australia (Fig. 1), has suggested the presence of a fluid flow event in the latest Miocene/Early Pliocene (O'Brien et al., 1996). This event, which has been linked to tectonism associated with the collision of the Australian and Eurasian plates, involved the flow of hot, saline brines up major faults and through the Mesozoic and Tertiary sequences with associated breaking of seals/traps and loss of hydrocarbons from Triassic and Jurassic reservoirs in reactivated, charged Mesozoic traps. The biodegradation of the leaked hydrocarbons has been associated with the development of the Hydrocarbon-Related Diagenetic Zones (HRDZ) as recognised by O'Brien and Woods (1995). The HRDZs are defined by zones of localised, intense carbonate cementation within otherwise poorly cemented sands and are seismically resolvable. Their size and seismic velocity are a function of the amount of hydrocarbon which has passed through the aquifer. Trap integrity is an important factor in controlling the distribution of oil and gas accumulations in the North West Shelf (O'Brien & Lisk, 1995; O'Brien et al. 1996). Here we describe models aimed at investigating factors which may determine the integrity of a trap such as; fault rheology and permeability; seal rheology and permeability; and the presence of overpressured units within the basin.

There appears to be a relationship between the type of trap and the angle between the deep Mesozoic rift faults and the Mio-Pliocene faulting in the area. The Mio-Pliocene faults associated with high integrity traps (HIT) are aligned with the deeper Mesozoic rift faults whereas in the moderate integrity traps (MIT) and low integrity traps (LIT) the later faults are usually divergent from the earlier Mesozoic rift faults (5-20° for MITs and 30-50° for LITs).

To date these studies have been based on geometries found in the North West Shelf of Australia and in the Timor Sea in particular. In this study we have used fully coupled mechanical/fluid flow numerical

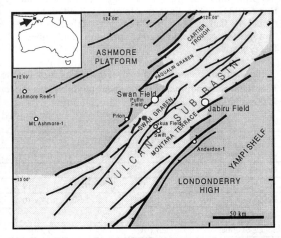

Fig. 1 Location map of the Vulcan Sub-basin showing key structural elements, exploration wells and the location of the Swan Graben.

modelling to consider the effects that the mechanics of the system may have had upon the fault reactivation and the resultant fluid flow directions. By using a fully coupled mechanical/fluid flow approach we are able to consider the effects of rheology, tectonic boundary conditions, fault and rock strengths on the system.

The ability to predict the reactivation of faults and the associated effects on fluid migration and accumulations represents significant steps in understanding not only the evolution of the Timor Sea area, but also the development of characteristic pressure compartments apparent in numerous basins worldwide. By integrating the observations of charge history and structural history with a mechanically constrained model we develop the first steps in generating a truly predictive reservoir modelling tool.

2 MODELS OF THE SWAN GRABEN

We have developed models in two dimensions using finite difference continuum codes (Cundall & Board 1998, ITASCA 1995) in order to investigate the mechanics of fault reactivation and to consider the mechanical and hydrodynamic conditions that would result in fluid flow up the reactivated Mesozoic faults.

A number of different material rheologies are available to the modeller and in this study those used included an isotropic elastic material; a non-associated elastic-plastic material which uses a Mohr-Coulomb yield condition and flow law as well as an elastic-plastic Mohr-Coulomb material which exhibits a well-defined strength anisotropy (Vermeer & de Borst 1984). The later rheology has been used to model the fault zones. Fluid flow is modelled as flow through a permeable/porous medium and obeys Darcy's law.

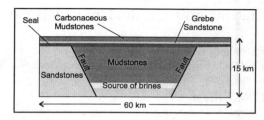

Fig. 2 The geometry and lithologic units used in our depth profile of a simple graben system, based on the Swan Graben.

The aim of the modelling approach was to isolate those processes which have contributed to reservoir breaching giving an improved understanding of the interaction between fault reactivation and fluid flow. This was undertaken in an interactive manner using those constraints which are known (amount of extension, structural and stratigraphic history) and testing the less well-constrained parameters such as varying fluid pressure, rock strength and fault strength. Pore fluid pressure has also been increased within discrete layers in order to test the effects of overpressuring on reservoir breaching.

We have created a two-dimensional depth profile of a graben system with similar geometry and stratigraphy to the Swan Graben in the southern Vulcan Sub-basin (Fig. 2). The model was run initially to isolate the fundamental effects and the geometry is thus fairly simplistic. This model consists of two Mesozoic faults dipping at 60° and bounding a basin consisting of mudstones and a deep source of high salinity brines. The upthrown sides of the faults consist of reservoir sandstones. These units are overlain by a seal and the Grebe Sandstone, which represents the unit in which the HRDZs are found,

Table 1. Input material and hydrologic properties for the lithologic units of the models described in Figure 1. K is the bulk modulus, G is the shear modulus, t is the tension limit, c is the cohesion, ϕ is the friction angle, ψ is the dilation angle and k is the permeability.

	K (Pa)	G (Pa)	t (Pa)	c (Pa)	ϕ	ψ	k (m^2)
Carbonaceous mudstone	2.26×10^{10}	1.11×10^{10}	1×10^3	6×10^3	10	15	10^{-15}
Grebe Sandstone	2.68×10^{10}	7×10^9	1.17×10^3	2.7×10^4	28	15	10^{-14}
Seal	1.56×10^{10}	1.08×10^{10}	1.17×10^3	3.84×10^3	14.4	10	10^{-16}
Reservoir Sandstone	2.68×10^{10}	7×10^9	1.17×10^3	2.7×10^4	28	15	10^{-14}
Basin fill mudstone	1.56×10^{10}	1.08×10^{10}	1.17×10^3	3.84×10^3	14.4	10	10^{-15}
Source of brines	2.3×10^8	1.5×10^8	1×10^7	1×10^8	5	1	10^{-15}
faults	2×10^8	2×10^8	1×10^7	0	20	10	10^{-13} ($10^{-12}/10^{-16}$)#
fault joints*			0	0	5	2	

* angled along the fault plane
# In some models this parameter varies between the bracketed values

and a carbonaceous mudstone in about 200m water depth. Properties of the different units are listed in Table 1. In some models, the layer within the basin which contains the high salinity brines has been overpressured slightly as a possible source for fluids moving up the faults.

A number of models were run which looked at the effects of: varying the rheology and permeability structure of the deep Mesozoic rift faults; the presence of an overpressured region of high salinity brines at depth within the basin; and changing the permeability structure of the seal and how failure of the seal effects the fluid flow regime beneath it.

We find that under an extensional regime with a low permeability seal, the deformation/fluid flow pattern is stable with the faults reactivating from the bottom up and fluids moving up the faults (Fig. 3A). The presence of a region of overpressure increases the flow rates. If the seal is allowed to fail, enhancing fluid movement through it, a transient pattern of deformation and fluid flow develops (Fig. 3B).

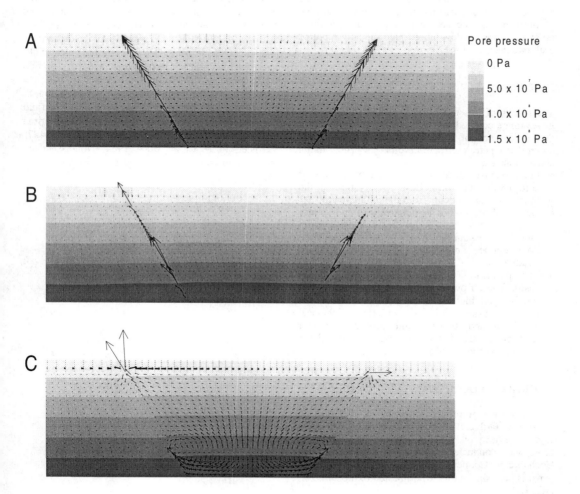

Fig.3 Pore pressure contours and fluid flow vectors for two of the models of the swan graben. A: fluid moves up the faults which are continually failing and providing a high permeability conduit for fluid flow. B: fluid moves up faults where they are failing and where the seal fails, as it is above the left fault, fluid moves through the seal into the sandstones above. C: The faults are low permeability barriers to fluid flow. Fluids move upward through the basin and overpressure leads to seal rupture and fluid migration into the units above.

If the faults are sealed, and are a barrier to flow rather than providing a conduit for fluid flow, the pattern of failure and fluid flow is significantly different, however seal rupture may still result (Fig. 3C). Most of the material movement is taken up along the faults, which are zones of weakness. This leads to deformation within the seal, particularly above the fault tips. Fluids move up through the basin from the region of overpressured brines at depth. Initially, the fluids are trapped within the basin, between the low permeability seal and the two low permeability faults. This leads to overpressuring within the basin. As the amount of movement along the faults increases and the pore pressure within the basin increases, the seal begins to fail in tension, allowing fluids to move up through the now ruptured seal, into the sandstones above.

3 SUMMARY

The results of this modelling are the initial steps in generating a coupled mechanical/fluid flow model for the prediction of reservoir breaching in fault controlled petroleum reservoirs. The initial conclusions are:
- In an extensional regime with a stable pattern of deformation and fluid flow, reactivation of the Mesozoic faults occurs from the base upward.
- Failure of the seal, if associated with an increase in the permeability of the seal and hence with fluid flow, produces a transient deformation/fluid flow regime which does not evolve towards a steady state.
- Low permeability faults trap the fluids in the basin. These flow upward from the overpressured saline layer, increasing the pore pressure at the top of the basin. This leads to failure of the seal in shear and tension and provides another scenario for seal rupture and lose of hydrocarbons into the overlying units.

4 ACKNOWLEDGMENTS

This paper was improved following internal reviews by Alison Ord and Yanhua Zhang in CSIRO and external reviews by WRI-9 reviewers. Work reported here was conducted as part of the Australian Geodynamic Cooperative Research Centre and this paper is published with the permission of the Director, AGCRC.

5 REFERENCES

Cundall, P. & M. Board 1988. A microcomputer program for modelling large-strain plasticity problems: In Swododa, C. (ed) *Numerical Methods in Geomechanics. Proceedings of the 6th International Conference on Numerical Methods in Geomechanics. Innsbruck, Austria,* Balkemapp, Rotterdam, p2101-2108.

Itasca, 1995, FLAC Fast Lagrangian Analysis of Continua Version 3.3, Minneapolis, USA.

ITASCA, 1996, FLAC Fast Lagrangian Analysis of Continua in 3 Dimensions Version 1.1, Minneapolis, USA.

O'Brien, G.W. & M. Lisk 1995. Determination of hydrocarbon charge history and trap integrity from the recognition of hydrocarbon-related diagenesis on the North-West Shelf. AOGC Conference, Adelaide, SA.

O'Brien, G.W. & E.P. Woods 1995. Hydrocarbon-related diagenetic zones (HRDZs) in the Vulcan Sub-basin, Timor Sea: recognition and exploration implications. *APPEA Journal*, 35, 220-252.

O'Brien, G.W., M. Lisk, I. Duddy, P.J., Eadington, S. Cadman & M. Fellows 1996. Late Tertiary fluid migration in the Timor Sea: A key control on thermal and diagenetic histories *APPEA Journal*, 36, 399-427.

Vermeer, P. & R. de Borst 1984, Non-associated plasticity for soils, concrete and rock. *Heron*. 29, 1-64.

Monitoring of thermal and mineral waters in the frame of READINESS

H. Woith, C. Milkereit & J. Zschau
GeoForschungsZentrum Potsdam, Germany

U. Maiwald & A. Pekdeger
FU Berlin, FB Geowissenschaften, Germany

ABSTRACT: Within the frame of the earthquake research project READINESS thermal and mineral waters are studied in the eastern Mediterranean. Since 1994 continuously monitoring field laboratories were installed at springs in Armenia and Turkey to detect temporal changes of the fluid compositions. The site selection is based on hydrogeological mapping. In this paper examples from the North-Anatolian Fault Zone are presented, where most systems are influenced by mixing of a deep fluid with a shallow water component. The mixing leads to intensive ion-exchange processes with the surrounding rocks and to a cooling of the geothermal waters. A spectral analysis of the available time series revealed that at least 2 of the monitored ground water systems are sensitive to earth tidal strain. This means a strain sensitivity down to 10^{-8} for periodic strain variations. We think that those sites are especially favourable as natural stress/strain sensors.

1 INTRODUCTION

The interdisciplinary earthquake research initiative READINESS ("REAltime Data Information Network in Earth ScienceS") aims at the understanding of interactions between large scale fault systems. The idea of triggering distant earthquakes goes back to the very beginning of seismology. Recently large scale interactions of earthquake regions were observed in the western US, when the M=7.3 Landers earthquake of June 28, 1992 triggered a series of smaller events up to a distance of 1200 km (Mori et al. 1992, Schwarzschild 1993). Co-seismic water level changes were observed at distances up to 1100 km (Roeloffs et al., 1995).

Large spatial dimensions have also been proposed for the preparation zones of major earthquakes. This theory is supported by applying the SEISMOLAP-algorithm to various earthquake catalogues and showing that large earthquakes are preceded by relative seismic quiescence in areas of up to more than 500 km around the future event (Zschau, 1996). Monitoring and detection of critical stress build-up within such earthquake preparation zones is a further goal of READINESS.

2 INVESTIGATION AREA

The investigation area is the eastern Mediterranean. Of special interest are the earthquakes occurring along the plate boundaries between the Anatolian, the Arabian and the Eurasian plate (Fig. 1). They mainly occur in the collision zone of the Caucasus and along prominent faults like the North-Anatolian fault (NAF), the North-East-Anatolian fault (NEAF), the East-Anatolian fault (EAF) and the Dead-Sea fault (DSF). In the first phase the project concentrated on monitoring the NAF and the Caucasian region. The relative northward movement of the Arabian plate causes (i) faulting and uplift in the Caucasus region and (ii) a westward migration of the Anatolian block. Recent GPS surveys showed that this "escape" movement is realised by a rotation of the Anatolian micro plate with velocities up to 25 mm per year (Oral, 1994). Seismicity along the NAF is characterised by alternating phases of activity and relative quiescence lasting up to 150 years (Ambraseys 1970, Sengör 1979). The last activity cycle started in December 1939 with a magnitude M=8.0 earthquake near Erzincan (near the intersection of NAF and NEAF). A series of strong earthquakes migrated westward until the province of Adapazari (W of site "6" in Fig. 1) was reached in 1967.

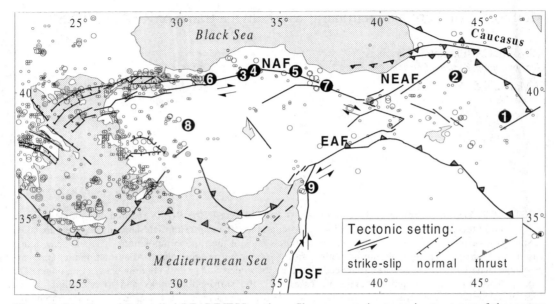

Figure 1. Location of 9 installed READINESS stations. Shown are major tectonic structures of the eastern Mediterranean (after Oral 1994) as well as earthquakes with magnitudes above M=6 for the time period from 500 B.C. until 1992 (NEIC catalogue). NAF: North-Anatolian Fault, EAF: East-Anatolian Fault, NEAF: North-East-Anatolian Fault, DFS: Dead Sea Fault.

For the second phase it is intended to extend the project to the EAF and the DSF. According to the NEIC catalogue, only a few earthquakes above magnitude 6 occurred along the DSF during the past 2500 years. From paleo-seismological studies north of the Red Sea, Zilberman et al. (1996) obtained a recurrence period for earthquakes above M>6 of less than 1000 years. The Gulf of Eilat itself was considered aseismic during the first half of this century (see Salamon et al., 1996). On November 22, 1995, a M_w=7.2 earthquake occurred in the Gulf of Eilat possibly marking the beginning of a new earthquake cycle in this region.

3 EXPERIMENTAL

The READINESS concept requires (i) standardised equipment and quality control to enable a comparison of different observation clusters and (ii) online-availability of all relevant data. A central data bank at the GeoForschungsZentrum Potsdam stores the incoming data.

READINESS started in October/November 1994 mapping and monitoring physico-chemical parameters of deep ground waters along the North Anatolian Fault in Turkey and two major fault systems in Armenia (Fig. 1, Table 1).

Table 1: Geographic coordinates of the READINESS stations. Sites 1 and 2 are located on the territory of Armenia, the following ones are in Turkey.

No	ID	location	°E	°N
1	KAT	Katjaran	46.20	39.16
2	ACH	Achurik	43.75	40.75
3	CAV	Cavundur Aci Su	33.18	40.83
4	KAZ	Kazanci Aci Su	33.70	41.01
5	HAM	Hamamayagi Kaplica	35.79	40.97
6	CEP	Cepni Aci Su	31.52	40.67
7	RES	Resadiye Kaplica	37.33	40.39
8	AFY	Afyon Gecek Kaplica	30.42	38.85
9	SAR	Sarkhamam	36.60	36.37

Meanwhile, 9 monitoring stations are recording radon, water temperature, pH, and the specific electrical conductivity in 10-minute intervals. Additionally, meteorological parameters like air temperature, air humidity, air pressure and precipitation are recorded. Every hour the data are sent to Potsdam, Ankara and Yerevan via satellite (METEOSAT). With this set of parameters we should be able to detect ground water anomalies, especially those caused by a mixing of normally separated aquifer systems. Mixing of different water

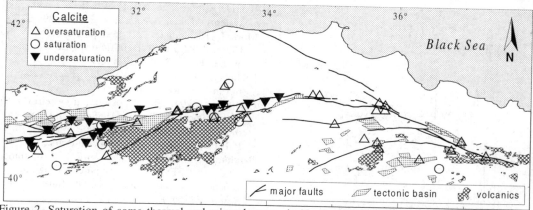

Figure 2. Saturation of some thermal and mineral waters in Northern Anatolia with respect to the mineral calcite. Saturation indices were calculated using the program PHREEQE (Parkhurst et al., 1980).

types is the most accepted model to explain seismo-tectonically induced anomalies in ground water time series (Thomas, 1988).

The site selection was based on hydrogeological mapping of the mineral and thermal waters. According to our present knowledge we think that a "good" monitoring site is characterised by the following features:
- tectonic aspects
  - located on an active fault
  - located near geometrical singularities
- hydrogeological aspects
  - deep water (no tritium, mantle components ^{3}He)
  - no surface water (no mixing, no seasons)
  - strain sensitive (Earth tides)
  - artesian systems (no pumping)
- not used by humans
- technical limits
  - water temperature below 60°C
  - contact to METEOSAT
  - flood safe

Deep waters with such properties are extremely rare in nature. Therefore, we accept simple two-component systems - e.g. 1 deep water mixing with 1 shallow water - as a second choice for a monitoring site, so that we have the chance to quantify the contributions of each component.

4 RESULTS

Generally, the deep ground waters along the North Anatolian Fault Zone are relatively cold. More than 75% of the investigated waters are cooler than 35 °C at surface. The content of total dissolved solids varies between 300 mg/l and 12 g/l, with Ca-HCO$_3$, Mg-HCO$_3$ and Na-HCO$_3$ waters as dominant groundwater types.

Most of these waters are influenced by ion exchange and mixing processes: (i) Na-HCO$_3$ type waters may originate from rising Na-Cl type thermal waters intruding into shallow aquifers with ground waters of Ca-HCO$_3$ type. (ii) undersaturation with respect to the mineral calcite (Fig. 2) may be caused by rising CO$_2$ gas mixing with shallow ground waters. (iii) calculated geothermometer temperatures are much higher than the measured temperatures (Fig. 3). These observations can be indicators for a mixing of deep fluids - gas or water - with shallow

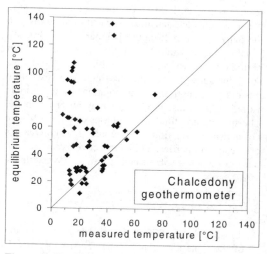

Figure 3. Equilibrium temperatures calculated with the Chalcedony geothermometer versus measured temperatures.

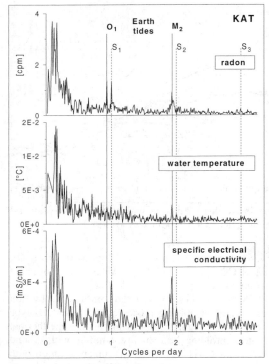

Figure 4. Power spectra calculated for KAT show the presence of earth tides (O_1 and M_2) in the signals of radon, water temperature and conductivity.

ground waters (Arnorsson, 1985; Nicholson, 1993).

A spectral analysis of the available time series revealed that at least 2 of the monitored ground water systems are sensitive to earth tidal strain (see the daily O_1 and half-daily M_2 lunar tides in Fig. 4). This means a strain sensitivity down to 10^{-8} for periodic strain variations. We think that those sites are especially favourable as natural stress/strain sensors, although the extrapolation from periodic variations to transient changes may be problematic. The spectral amplitudes and phases of O_1 and M_2 are not constant over time. This could indicate changes of crustal properties, i.e. modified elastic modules due to changes of effective stress (Westerhaus, 1997).

5 REFERENCES

Ambraseys, N.N. 1970. Some characteristic features of the Anatolian Fault Zone. *Tectonophysics*, 9: 143-165.

Arnorsson, S. 1985. The use of mixing models and chemical geothermometers for estimating underground temperatures in geothermal systems.- *J. Volc. Geoth. Res.*, 23, 3-4: 299-235.

Mori, J., K. Hudnut, L. Jones, E. Hauksson & K. Hutton 1992. Rapid scientific response to Landers quake.- *EOS*, 73, 39: 417-418.

Nicholson, K. 1993. *Geothermal fluids.*- 263 p., Berlin-Heidelberg (Springer).

Oral, M.B. 1994. Global positioning system (GPS) measurements in Turkey (1988-1992): kinematics of the Africa-Arabia-Eurasia plate collision zone.- PhD thesis, 344 p., MIT, Massachusetts.

Parkhurst, D.L., D.C. Thorstenson & L.N. Plummer 1980. PHREEQE - A computer program for geochemical calculations.- USGS Surv. *Water-Resour. Invest. Rep.*, 80-96: 210 p.

Roeloffs, E., W.R. Danskin, C.D. Farrar, D.L. Galloway, S.N. Hamlin, E.G. Quilty, H.M. Quinn, D.H. Schaefer, M.L. Sorey & D.E. Woodcock 1995. Hydrologic effects associated with the June 28, 1992 Landers, California, earthquake sequence.- Open-File Report - USGS: 68 p.

Salamon, A., A. Hofstetter, Z. Garfunkel & H. Ron 1996. Seismicity of the eastern Mediterranean region; prespective from the Sinai Subplate.- *Tectonophysics*, 263, 1-4: 293-305.

Schwarzschild, B. 1993. Earthquake yields first real evidence of remote triggering.- *Physics Today*, September 1993: 17-19.

Sengör, A.M.C. 1979. The North Anatolian transform fault: its age offset and tectonic significance.- *J. Geol. Soc. London*, 136: 269-282.

Thomas, D. 1988. Geochemical precursors to seismic activity.- *Pure Appl. Geophy.*, 126, 2-4: 241-266.

Westerhaus, M. 1997. Tidal tilt modification along an active fault.- In: Tidal Phenomena, Lecture Notes in Earth Sciences, H. Wilhelm, W. Zürn, H.-G. Wenzel (Eds.), pp. 311-339, Berlin-Heidelberg (Springer).

Zilberman, E., R. Amit, Y. Enzel, N. Porat, D. Wachs & H. Wust 1996. Large scale paleoearthquakes (M>6) and seismic hazards along the southern Dead Sea transform, southern Arava Valley, Israel.- Proc. XXV ESC, Reykjavik, Iceland.

Zschau, J. 1996. SEISMOLAP - Ein Schritt in Richtung Erdbebenvorhersage.- *Geowissenschaften*, 14: 1-7.

Large temperature fluctuations recorded in veins in the Victory gold deposit, Western Australia: A consequence of episodic fluid influx during progressive deformation?

Y. Xu & J.M. Palin
Ore Genesis Group, Research School of Earth Sciences, The Australian National University, Canberra, A.C.T., Australia

ABSTRACT: Mineralised shear zones in the Victory gold deposit of Western Australia exhibit extensive hydrothermal quartz vein development. Oxygen isotope analysis of mineral pairs (mainly quartz-albite and quartz-scheelite) from the veins indicate that precipitation temperatures varied between 290°C and 540°C. Such large fluctuations in fluid temperature are considered to be a consequence of episodic influx of deeply derived fluid during crack-seal vein growth within the progressively deforming shear zones.

1 INTRODUCTION

Faults and shear zones have long been known as favorable sites for hydrothermal mineral deposits. Over the past few years, there has been increasing interest in understanding fluid infiltration and migration along faults and shear zones on a variety of scales. Detailed structural studies have revealed the involvement of dynamic (episodic) faulting processes in hydrothermal vein formation (e.g. Cox et al., 1995; Robert et al., 1995). Associated cyclic variations in fluid pressure within active faults have also been recognised (Sibson, 1989). However, little attention has been given to fluctuations in fluid temperature which may accompany such episodic fluid influx.

In this paper, we present results of a detailed oxygen isotope study of hydrothermal veins within a mineralised shear zone in the Victory gold deposit of Western Australia. Temperatures of vein precipitation along this feature exhibit a large range. These results contrast with the assumption of nearly constant fluid temperature or smooth spatial and/or temporal temperature variations usually made in studies of mineralised vein systems. In addition, this approach provides important information on the fluid history of the shear zone.

2 GEOLOGIC SETTING

The Victory gold deposit is located 20 km south of Kambalda in the Norseman-Wiluna greenstone belt of the Yilgarn Craton. The geology of Kambalda area and of the Victory gold deposit have been previously described by Clark et al. (1986; 1989) and Roberts and Elias (1990).

The stratigraphy of the Kambalda area is dominated by metamorphosed Archean mafic and ultramafic volcanics and felsic granitoids. Although gold mineralisation is found in every rock type within the Victory complex, iron-rich cherty sediments and differentiated gabbros are the most important host rocks (Fig. 1).

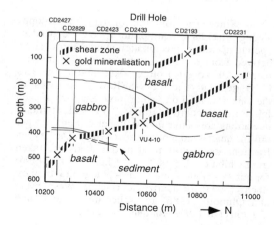

Fig. 1 NW-SE geologic cross section through the Victory gold deposit.

Gold mineralisation is mainly controlled by NNW-trending and E-dipping shear zones and reverse faults that splay from the major Boulder-Lefroy fault system and appear to have formed relatively late in the deformation history. Gold mineralisation has been dated by the U-Pb method on rutile at 2627±7 Ma (Clark et al., 1989) and by the $^{40}Ar/^{39}Ar$ method on actinolite at 2627±14 Ma (Kent, 1994).

3 RESULTS AND DISCUSSION

Hydrothermal vein samples were collected from the most important shear zone within the mine and cover a horizontal distance of 1 km and a depth range of about 300m (Fig. 1). The veins range from vein breccias to single veinlets depending upon the local density of vein development (Clark et al., 1986). Extension veins occur adjacent to the major veins within the shear zone and can be observed in the underground workings.

The veins contain mainly quartz with lesser amounts of dolomite and albite and minor pyrite, scheelite, calcite, and chlorite. The texture and morphology of vein minerals suggest that quartz, albite and dolomite were generally coprecipitated. However, in some samples, secondary dolomite infills pre-existing quartz and albite. Chlorite, muscovite, biotite, and fine-grained albite and carbonate typically occur in the altered wallrocks immediately adjacent to the veins. The mineralogy of the altered wallrock is dependent to a significant degree on original composition (Clark et al., 1989).

Since fractionation of oxygen isotopes among minerals at equilibrium is solely dependent upon temperature, precipitation temperatures can be derived from oxygen isotope analysis of mineral pairs. All oxygen isotopic data reported here as $\delta^{18}O$ are relative to the V-SMOW scale. Most of the data are based on two or more analyses of each mineral, especially albite. Typical standard deviations are better than 0.05 per mil and 0.1 per mil for quartz and albite, respectively. The oxygen fractionation factors used in this study are taken from Chiba et al. (1989) and Chacko et al. (1996) for quartz - albite - muscovite - phlogopite (biotite); Zheng (1993) for quartz - tourmaline; Wesolowski and Ohmoto (1986) for scheelite - water; and Clayton et al. (1972) for quartz - water.

Particular importance is placed on quartz-albite pairs because these minerals are the most common silicate minerals coexisting in the hydrothermal veins at Victory, and the quartz-albite fractionation factor is well known. Other mineral pairs are given lesser emphasis because of uncertainties in paragenesis or poorer knowledge of fractionation factors.

Temperatures of hydrothermal vein precipitation calculated from the $\delta^{18}O$ values of coexisting mineral pairs exhibit a relatively large range within the same shear zone, from 275°C to 590°C (Fig. 2). These temperatures do not correlate with distance, depth, or wallrock lithology. Temperatures estimated solely from the quartz-albite pairs range from 350°C to 540°C. The highest of these temperatures is within the range of estimated peak metamorphic temperatures for the Kambalda area (525±25°C, Bavington, 1979), while the lowest is consistent with temperatures estimated for gold-associated hydrothermal alteration in the north-central portion of the Victory deposit (390±40°C, Clark et al., 1989). Quartz-scheelite pairs from this area of the deposit give lower temperatures, ranging from approximately 290°C to 350°C. Direct, comparison of temperatures calculated from these two mineral pairs is hindered by the lack of experimental calibration of the quartz-scheelite fractionation factor and the absence to date of coexisting quartz-albite-scheelite vein assemblages.

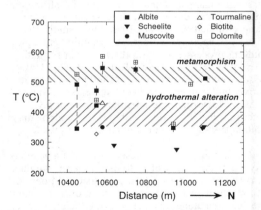

Fig. 2 Vein precipitation temperatures based on oxygen isotope fractionation between quartz and coexisting minerals from the major shear zone in the Victory gold deposit. Temperature ranges for peak metamorphism and hydrothermal alteration associated with gold mineralisation are from previous work (see text).

An important question arises as to whether temperatures estimated from the quartz-albite pairs are accurate given the common occurrence of isotopic disequilibrium between these minerals in hydrothermally altered granites. Petrographic examination shows that quartz and albite coexist in apprent textural equilibrium and that albite is mineralogically unaltered. Electron microprobe analysis further demonstrates albite to be pure. Moreover, although quartz is the predominant mineral in the veins, the range of quartz $\delta^{18}O$ values is greater than that for albite. This would not be the case had quartz and albite undergone significant oxygen isotopic exchange with each other after precipitation or had albite exchanged isotopically while quartz retained its original value. In addition, since quartz is more abundant and is generally more resistant to isotopic exchange than albite, it is unlikely that quartz was altered while albite remained intact. Hence, we conclude that the temperature estimates are accurate and that large fluctuations in fluid temperature occurred during vein precipitation.

The $\delta^{18}O$ values of vein quartz vary from 11.0 to 12.8 with an average of 11.8±0.5 (1 s.d., n=19), and are consistent with previous work by Golding et al. (1988). If all the veins had precipitated from a similar fluid with temperature varying smoothly with depth, the quartz values would correlate with depth. However, when we plot the quartz $\delta^{18}O$ values verses depth (Fig. 3), there is no systematic trend. This requires the occurrence of highly variable fluid isotopic compositions and/or temperatures within the shear zone during the course of vein precipitation.

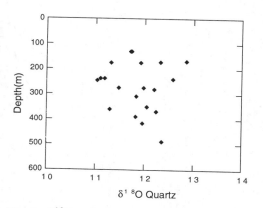

Fig. 3 $\delta^{18}O$ values of quartz in veins exhibit no correlation with depth.

A number of studies of hydrothermal systems associated with faults and shear zones have suggested that vein growth in such environments is episodic, involving numerous cycles of stress buildup and earthquake rupture each leading to fluid influx into the active fault zone (Sibson, 1989). Following fault rupture, hot ascending fluid is channeled into newly openned void space, and dissolved constituents precipitate upon reaching or exceeding saturation. If the precipitation rate is faster than the wall separation rate, the fault becomes sealed and a vein is formed. When fluid pressure builds up high enough under such a fault "valve", the fracture is reopened and a new episode of fluid influx takes place. This cycle repeats itself many times during the lifetime of a hydrothermal system, although each new fracture surface can occupy various locations. Rock fracture can repeatedly take place in the interior of a major vein, irregularly through vein and wallrock, or along an entirely new surface in the rock mass. As a consequence, vein systems in shear zones occur in two principal forms: 1) thick, massive veins (including vein breccias and laminated veins) produced by multiple opening and fluid infiltration events, and 2) thin veinlets or extension veins generating by single fracture and fluid influx events.

The spatial association of gold mineralisation and hydrothermal alteration with shear zones in the Victory deposit suggests that these were the main channels for deeply-derived hydrothermal fluids. Open space filling textures observed in some extensional veins indicate that these fractures remained open for relatively long periods and were never reopened after initial sealing. Conversely, brecciated and laminated veins provide evidence of repeated crack-seal events. Thus, fluid migration in the shear zones was spatially and temporally variable. Temperatures recorded by the veins represent both single fluid influx events in the case of single extensional veinlets or multiple episodes of fluid influx in the case of large vein breccias or laminated veins. For example, whereas most of the replicate analysis on quartz-albite pairs gave reproducible results, repeated analyses of a vein from DDH CD2423 gave two groups of temperatures, one around 350°C and the other around 500°C (Fig. 2). This is most likely due to real temperature fluctuations during at least two episodes of fluid influx and vein precipitation.

4 CONCLUSIONS

Hydrothermal veins within mineralized shear zones in the Victory gold deposit provide an unique opportunity to document dynamic fluid fluxes during deformation. The results of our oxygen isotope study demonstrate that large fluctuations in the temperatures of vein precipitation accompanied crack-seal vein-forming processes. Although conclusions regarding the composition and source of the ore-forming fluid pend the results of ongoing study, the ore-components must have been introduced during the extensive deformational and hydrothermal history of the shear zones.

5 ACKNOWLEDGEMENTS

We acknowledge Western Mining Corporation for their generous support, and in particular former Senior Research Geologist Dr Kim A.A. Hein and the Victory mine staff for guidance with the geology and sampling. We thank Ian Campbell, Majid Ghaderi, and Bob Loucks for their assistance in the field and helpful discussions. This work is part of the "Crustal Fluids" program of the RSES Ore Genesis and Petrophysics Groups.

REFERENCES

Bavington, O. A. 1979. Interflow sedimentary rocks from the Kambalda ultramafic sequence: their geochemistry, metamorphism and genesis. Ph.D. thesis, ANU (unpubl.).

Chacko, T., Mayeda, T., Clayton, R. N. & Goldsmith, J. 1996. Oxygen isotope fractionations in muscovite, phlogopite, and rutile. *Geochim. Cosmochim. Acta* : 60, 2595-2608.

Chiba, H., Chacko, T., Clayton, R. N. & Goldsmith, J. R. 1989. Oxygen isotope fractionations involving diopside, forsterite, magnetite, and calcite: Application to geothermometry. *Geochim. Cosmochim. Acta* : 53, 2985-2995.

Clark, M. E., Archibald, N. J. & Hodgson, C. J. 1986. The structure and metamorphic setting of the Victory gold mine, Kambalda, Western Australia. Gold'86, An International Symposium on the Geology of Gold Deposits, Proceedings Volume. Toronto.

Clark, M. E., Carmichael, D. M., Hodgson, C. J. & Fu, M. 1989. Wall-rock alteration, Victory Gold Mine, Kambalda, Western Australia: Processesand P-T-XCO2 conditions of metasomatism. Econmic Geology Monograph 6, pp. 445-459.

Clayton, R. N. & Mayeda, T. K. 1963. The use of bromine pentafluoride in the extraction of oxygen from oxides and silicates for isotopic analysis. *Geochim. et Cosmochim. Acta* : 27, 43-52.

Clayton R. N., O'Neil J. R. & Mayeda T. K. 1972. Oxygen isotope exchange between quartz and water. *Geochim. et Cosmochim. Acta* : 77, 3057-3067.

Cox, S. F., Sun, S.-S., Etheridge, M. A., Wall, V. J. & Potter, T. F. 1995. Structural and geochemical controls on the development of turbidite-hosted gold quartz vein deposits, Wattle Gully Mine, Central Victoria, Australia. *Econ. Geol.* : 90, 1722-1746.

Golding, S. D., McNaughton, N. J., Barley, M. E., Groves, D. I., Ho, S. E. Rock, N. M. S.& Turner, J.W. 1988. Archean Carbon and Oxygen Reservoirs: Their Significance for Fluid Sources and Circulation Paths for Archean Mesothermal Gold Deposits of the Norseman-Wiluna Belt, Western Australia. Econ. Geol. Monograph 6, 376-388.

Kent, A. J. 1994 Geochronological constraints on the timing of Achaean gold mineralisation in the Yilgarn Craton, WA. Ph.D. thesis, ANU (unpubl.).

Robert, F., Boullier, A., & Firdaous, K. 1995. Gold-quartz veins in metamorphic terranes and their bearing on the role fluids in faulting. *J. Geop. Res.*: 100, 12861-12879.

Roberts, D. E. & Elias, M. 1990. Gold deposits of the Kambalda-St Ives region. In Hughes, F. E. ed. *Geology of the Mineral Deposits of Australia and Paupa New Guinea*, 1, pp. 479-491.

Sibson, R. 1989. Earthquake faulting as a structural process. *J. Struct. Geol.*: 11, 11-14.

Wesolowski, D. & Ohmoto, H. 1986. Calculated oxygen isotope fractionation factors between water and the minerals sheelite and powellite. *Econ. Geol.*: 81, 471-477.

Zheng, Y. 1993. Calculation of oxygen isotope fractionation in hydroxyl-bearing silicates. *Chem. Geol.*: 120, 247-263.

14 Experimental

Pitzer specific ion interaction parameters for Ag-Cl from solubility measurements in the system AgCl-HCl-H$_2$O to 275°C

J.J. Bao & D.A. Polya
Department of Earth Sciences, The University of Manchester, UK

ABSTRACT: Solubilities of silver chloride in aqueous hydrochloric acid solutions were measured from 40°C to 275°C at initial concentrations of HCl from 0.015 to 4.8 mol kg^{-1} at pressures on the liquid-vapour curve. The set of 87 measurements, which show similar trends to the data of Ruaya and Seward (1987) but are consistently somewhat lower, were found to be best fitted by the following model Pitzer specific ion interaction parameters for Ag-Cl and the silver chloride solubility product over the temperature range 40°C to 200°C:

$\beta^{(0)}_{AgCl}$ (T) = -1.108517 + 0.000000 ln (T/K) + 0.002333 (T/K)
$\beta^{(1)}_{AgCl}$ (T) = -38.84957 + 6.186868 ln (T/K) - 0.010770 (T/K)
$\beta^{(2)}_{AgCl}$ (T) = -7857.574 + 1484.878 ln (T/K) - 3.047264 (T/K)
K_{sp} (T) = exp(-361.26184 + 65.287458 ln (T/K) - 0.11636088 (T/K))

with no requirement being found for non-zero C^{ϕ}_{AgCl}, ψ_{AgHCl} or θ_{AgH} terms.
The large negative $\beta^{(2)}_{AgCl}$ values are consistent with the known large degrees of ion association between Ag$^+$ and Cl$^-$ in solution.

1 INTRODUCTION

The Pitzer specific ion interaction approach to modelling activity-composition relations in aqueous fluids (Pitzer, 1973) has had great success in predicting the thermodynamic behaviour of soluble salts in highly saline low temperature waters, such as seawater (Harvie and Weare, 1980). The application of this approach to the modeling of transition metals in geothermal or hydrothermal systems, however, is hampered, in part, by a lack of experimental data and , in part, by the theoretical complications arising from the extensive complexing or ion pairing of transition metals, particularly in saline solutions.

This study reports use of the Pitzer specific ion interaction approach to preliminary modelling of the experimentally determined solubilities of a transition metal chloride, silver chloride, in synthetic hydrothermal solutions up to 275°C. The system AgCl-HCl-H$_2$O was chosen for this study because (i) it has been previously studied experimentally by, amongst others, Ruaya and Seward (1987), (ii) Pitzer specific ion interaction parameters for HCl have previously been determined (Simonson et al., 1990) and (iii) silver was not expected to undergo any significant changes in oxidation state under the experimental conditions adopted. The extraction of Pitzer specific ion interaction parameters for AgCl was carried out using a self-consistent set of experimental solubility data obtained as part of this study.

2 EXPERIMENTAL STUDY

Solubility experiments were conducted after the method of Seward (1976) in sealed silica glass tubes (10 cm length x 0.6 cm inside diameter) containing AnalaR or equivalent hydrochloric acid, standardised by titration with sodium carbonate, and pressed SpecPure silver chloride pellets (0.48 cm diameter), prepared by pressing the powder in a hydraulic pressure under 0.5 kbar pressure. The initial degree of liquid fill was around 50 volume % for the all runs. The experiments were conducted at 40°C, 60°C and 80°C using water-filled polypropylene jackets in thermostated water baths and at 125°C, 175°C, 200°C, 250° and 275°C using aluminum heating blocks placed in large volume fluidised alumina baths. Temporal and spatial variations in the water baths and fluidised baths were less than 2°C. The duration of the experiments was 2 days for the 40°C - 80°C runs and 12 hours for the 125°C - 275°C runs, these times having been shown by Ruaya and Seward (1987) to be sufficient to achieve equilibrium. It was ensured that the silver pellet was in contact with the hydrochloric acid solution during the experimental runs, at the end of which the tubes, which were constructed with a mid-point constriction, were inverted to separate the solution from the silver chloride pellet. The solution plus any silver chloride identified morphologically as a quench product was diluted/dissolved with a 0.1 mol kg^{-1} sodium thiosulfate/0.1 mol kg^{-1} ammonia solution and then with deionised water. These diluted solutions were analysed by ICP-AES with an estimated analytical precision of better than 5 %. XRD, SEM and XPS analysis confirmed that silver chloride was the only major crystalline solid product of the experiments and that no significant photolytic reduction of silver had taken place. Measured silver chloride solubilities in aqueous HCl solutions are presented in Fig. 1.

3 THERMODYNAMIC MODEL AND DERIVATION OF PITZER SPECIFIC ION INTERACTION PARAMETERS

We model the thermodynamic behaviour of the AgCl-HCl-H$_2$O system by explicitly considering the the formation of AgCl (s), and the formation of the HCl (aq) ion pair, with a dissociation constant calculated from Frantz and Marshall (1984) using equations for the vapour pressure and saturated liquid density of pure water from Saul and Wagner (1987). Ion association between Ag and Cl is not explicitly considered but is implicitly reflected in calculated values of the $\beta^{(2)}_{AgCl}$ parameter (cf. Pitzer, 1987). Activity coefficients for aqueous Ag$^+$, H$^+$, Cl$^-$ and HCl^0 are calculated through the equations of Pitzer (1987) for ln(γ_M), ln(γ_X) ln(γ_N) from the Pitzer specific ion interaction parameters $\beta^{(0)}_{HCl}$, $\beta^{(1)}_{HCl}$, $C^{(\phi)}_{HCl}$ and $\lambda_{HCl,HCl}$ from the AI model of Simonson et al. (1990) and $\beta^{(0)}_{AgCl}$, $\beta^{(1)}_{AgCl}$, $\beta^{(2)}_{AgCl}$ and $C^{(\phi)}_{AgCl}$, which four parameters were derived iteratively in this study, together with the solubility product, K_{sp}, of silver chloride, in simple temperature-dependent functions of the form:

f (T) = A + B ln (T/K) + C (T/K)

This was achieved by iteratively minimising the objective function:

$F = \Sigma_i \; [(S^e_{Ag} - S^c_{Ag})/S^e_{Ag}]^2$

where S^e_{Ag} and S^c_{Ag} are the experimental and calculated solubility of silver chloride and F is a function of the A, B & C terms for $\beta^{(0)}_{AgCl}, \beta^{(1)}_{AgCl}$, $\beta^{(2)}_{AgCl}$, $C^{(\phi)}_{AgCl}$ and K_{sp}. The ψ_{AgHCl} and θ_{AgH} parameters were set to zero and no significant fit improvement could be made by adopting non-zero values of $C^{(\phi)}_{AgCl}$. Calculated best fit model parameters aregiven in the caption to Fig. 1, in which model solubilities calculated using these parameters are compared to the experimentally obtained solubilities.

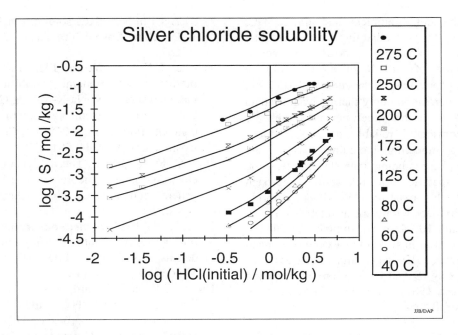

Figure 1. Silver chloride solubility in the system AgCl-HCl-H$_2$O at saturated vapor pressure. Experimental data (this study) indicated by symbols. Solid lines indicate calculated solubilities using the Pitzer specific ion interaction parameters and silver chloride solubility products derived in this study, viz.

$\beta^{(0)}_{AgCl}(T) = -1.108517 + 0.000000 \ln(T/K) + 0.002333 (T/K)$

$\beta^{(1)}_{AgCl}(T) = -38.84957 + 6.186868 \ln(T/K) - 0.010770 (T/K)$

$\beta^{(2)}_{AgCl}(T) = -7857.574 + 1484.878 \ln(T/K) - 3.047264 (T/K)$

$K_{sp}(T) = \exp(-361.26184 + 65.287458 \ln(T/K) - 0.11636088 (T/K))$

4 DISCUSSION & CONCLUSIONS

The trends of increasing solubility of silver chloride with increasing temperature and increasing initial HCl concentration found in this study mirror those found by Ruaya and Seward (1987), however the magnitude of the solubilities are consistently lower than those of Ruaya and Seward (1987).

Cognisant of the correction that Gammons et al. (1993) report for H$_2$O and HCl loss to the vapor phase in similar experiments, it is speculated that uncorrected differential partitioning of H$_2$O and HCl between the liquid and vapour phases, particularly at run temperatures above 200°C may explain the differences in the solubilities measured by Ruaya and Seward (1987) and those determined in this study, particularly as both our and Ruaya and Seward's (1987) experiments (T.M. Seward pers. comm.) had degrees of fill sufficiently low to ensure the presence of vapor phase during all runs. Estimates of such partitioning, based on the data of Saul and Wagner (1987) for H$_2$O and Simonson and Palmer (1993) for HCl, suggests that of our experiments, for those run at temperatures below 250°C the HCl molality of the liquid phase at run temperature was less than 3 % different from the initial ambient temperature HCl molality. It is noted that this error is less than estimated analytical error. However, corrections of 10 % or more would be required for some of the runs at 250°C and 275°C, correspondingly thermodynamic data derived from the experimental studies are only considered reliable to 200°C. More rigorous modeling including loss to the vapour phase is being done Nevertheless, the Pitzer specific ion interaction approach models the experimental silver chloride solubility data with a mean relative error of less than 10 % even though Ag-Cl ion association has not explictly been considered. Known high degrees of complexing (Seward, 1976) are, as expected (cf. Pitzer, 1987), reflected in large negative values of $\beta^{(2)}_{AgCl}$.

ACKNOWLEDGEMENTS

We thank Terry Seward for discussions and advice, Kathy England for XPS and Barbara Fenlon and Cathy Davies for considerable assistance in preparation and analysis. We gratefully acknowledge the support of a British Council PhD studentship to JJB and a NERC grant GR3/6817 to DAP.

REFERENCES

Bao, J.J., 1992. Thermodynamic behaviour of zinc in hydrothermal fluids. Unpub. PhD thesis, The University of Manchester, 441pp.

Frantz, J.D. and Marshall, W.L., 1984. Electrical conductances and ionization constants of acids and bases in supercritical aqueous fluids. I. Hydrochloric acid from 100° to 700°C at pressures to 4000 bars. American J. Sci., 284, 651-667.

Gammons, C.H., Yu, Y. and Bloom, M.S., 1993. The solubility of Ag-Pd alloy + AgCl in NaCl/HCl solutions at 300°C. Geochim. Cosmochim. Acta, 57, 2469-2479.

Harvie, C.E. and Weare, J.H., 1980. The prediction of mineral solubilities in natural waters: the Na-K-Mg-Ca-Cl-SO$_4$-H$_2$O system from zero to high concentration at 25°C Geochim. Cosmochim. Acta, 44, 981-997.

Pitzer, K.S., 1973. Thermodynamics of electrolytes. I. Theoretical basis and general equations. J. Phys. Chem., 77, 268-277.

Pitzer, K.S., 1987. A thermodynamic model for aqueous solutions of liquid-like density IN Ribbe, P.H. (Ed.) Rev. Mineralogy, 17, 97-142.

Ruaya, J.R. and Seward, T.M., 1987. The ion-pair constant and other thermodynamic properties of HCl up to 300°C. Geochim. Cosmochim. Acta, 51, 121-130.

Saul, A. and Wagner, W., 1987. International equations for the saturation properties of ordinary water substance. J. Phys. Chem. Ref. Data, 16, 893-901.

Seward, T.M. 1976, The stability of chloride complexes of silver in hydrothermal solutions up to 350°C. Geochim. Cosmochim. Acta, 40, 1329-1341.

Simonson, J.M., Holmes, H.F., Busey, R.H. and Mesmer, R.E., 1990. Modeling of the Thermodynamic of Electrolyte Solutions to High Temperatures. Application to Hydrochloric Acid, J. Phys. Chem., 94, 7675-7681.

Simonson, J.M. and Palmer, D.A., 1993. Liquid-vapor partitioning of HCl(aq) to 350°C. Geochim. Cosmochim. Acta, 57, 1-7.

Dissolution of sanidine up to 300°C near equilibrium at approximately neutral pH

G. Berger
Paul Sabatier University, CNRS, UMR 5563, Toulouse, France

D. Beaufort
Poitiers University, CNRS, UMR 6532, Poitiers, France

J.-C. Lacharpagne
Elf Exploration Production, Pau, France

ABSTRACT: The reaction rate/affinity relationship for alkali feldspar dissolution is discussed by considering solely the competition of parallel hydrolysis reactions of network-former oxides. This approach raises the problem of the equilibrium state. Batch experiments conducted at 100 and 300°C to approach equilibrium between sanidine and near neutral solutions saturated with respect to quartz (typical natural conditions) suggest that the solutions do not equilibrate with the bulk mineral, although they reach a steady-state composition. An amorphous and K-depleted layer is clearly identified at the surface of the samples altered at 300°C, but not at 100°C. If an altered layer exists at low temperature, its thickness is less than 50 nm. The protective effect of such a layer and/or the saturation of the network-former oxides (Si and Al) likely prevent a full and reversible equilibrium between the fresh mineral and the bulk solution.

1 INTRODUCTION

Most mineral reactions in natural water-rock systems progress at conditions close to chemical equilibrium. The kinetics of these reactions, in particular the dissolution rate of the primary minerals, is a major constraint on the numerical modelling of diagenetic and hydrothermal processes. In the case of aluminosilicates, recent experimental studies have pointed out that more detailed information is necessary to understand the elementary reactions controlling the dissolution process. Far from equilibrium, it has been established that the bulk dissolution rate is related to the surface charge of the mineral (see Walther, 1996, and references herein). This suggests a competitive effect of Si and Al detachment, depending on the specific charge of the relevant surface sites. When the reaction approaches chemical equilibrium, the dependence of dissolution rate on the chemical affinity can be derived in terms of the Transition State Theory (TST; Lasaga, 1981 and references therein) which assumes that the dissolution rate is controlled by the desorption kinetics of an activated complex formed at the surface. However, the extrapolation of this concept from homogeneous to heterogeneous kinetics requires a clear definition of the elementary molecular reactions controlling the overall rate. In the case of simple oxides, it may be assumed that the rupture of the metal-oxygen bond is the only reaction controlling the overall rate and that the activated complex is the same for dissolution and for precipitation. The rate-affinity relation then takes a simple form (first-order law) by balancing the forward and reverse reactions:

$$r = k_+ (1 - (Q/K)) \quad (1)$$

where r denotes the overall dissolution rate (mole/surface/time), k_+ is the rate constant for dissolution at a given temperature and pH, Q/K is the degree of saturation of the solution with respect to the dissolving phase. At a given pH and temperature, Eq.(1) expresses a linear relation between the rate/k_+ ratio and the saturation index of the solid. This simple relation has been established, for example, for quartz dissolution in pure water (Rimstidt and Barnes, 1980; Berger et al., 1994). However, for complex silicates, such a simple relation does not describe accurately the dependence of dissolution rate on the chemical affinity of reaction. In the case of alkali feldspars, several models have been proposed to account for the observed dissolution rate/chemical affinity relationships:

- Oelkers et al. (1994) and Gautier et al. (1995) observed an inhibitor effect of dissolved aluminum on the dissolution rate of kaolinite and albite at 150°C in acidic or alkaline solutions. These authors assumed that the reaction rate is controlled by silica-rich aluminum-deficient precursor complexes on the mineral surface, and proposed the following rate law for alkaline feldspar dissolution:

$$r = kS \ [1/(K° + (a_{Al(OH)_4^-} \cdot a_{H^+})^{1/3})] \cdot [1-(Q/K)^{1/3}] \quad (2)$$

where a_{H^+} and $a_{Al(OH)_4^-}$ are activities of H+ and Al(OH)$_4^-$ in solution and k and $K°$ are constants.

- For sanidine and albite at 300°C and pH 9, Alekseyev et al. (1997) proposed an empirical rate law derived from Eq.(1), where p and q are constants depending on the mineral:

$$r = k \ S \ [1 - (Q/K)^p]^q \quad (3)$$

- For sanidine at near neutral pH far from

equilibrium, Berger (1995) found that the dissolution rate decreases as the aqueous silica concentration increases, without any correlation with the dissolved aluminum concentration. However, when the solution approaches saturation with respect to quartz, the rate becomes dependent on the aqueous aluminum concentration. These results were interpreted by Berger (1995) as a competition effect between several controlling reactions: far from equilibrium, the hydrolysis of the Si-O bonds controls the bulk rate; in silica rich solutions, this reaction is inhibited and the bulk rate is dependent on the Al-O breakdown reaction.

This latter approach can be applied to literature data in acid or basic media. Fig 1 shows an example of the dependence of rate on various aqueous aluminum concentrations in basic media. The nearly linear dependence can be interpreted as a chemical affinity effect on Al-O bond hydrolysis according to Eq.(1). Like in neutral media, an alternation of the controlling reaction is suspected when the solution composition changes: when the Al-O bonds are inhibited, the dissolution rate likely reflects the hydrolysis of the Si-O bonds. However, this model raises the question of the equilibrium state. It predicts that dissolution of the network will cease when all the network-former oxides are saturated, whatever the saturation state of the bulk solid (the alkali are not considered here).

2 EXPERIMENTAL PROTOCOL

In order to address this problem, sanidine was reacted with near-neutral solutions in a closed system at 100°, 200° and 300°C. At 100°C we used different potassium concentrations. The starting solutions were previously equilibrated with quartz. These conditions are chosen to prevent or minimize the precipitation of secondary clays which form in agressive solutions. We assume that, under these conditions, the dissolution of the mineral is controlled by the rupture of the Al-O bonds, because the Si-O bonds were already "saturated". The sample was ultrasonically cleaned, graded for grain size and leached in deionized water to dissolve surface defects resulting from grinding. The experiments were conducted in a titanium reactor at 200° and 300°C and

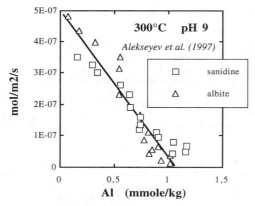

Fig.1: Example of dependence dissolution rate on the aqueous aluminum concentration.

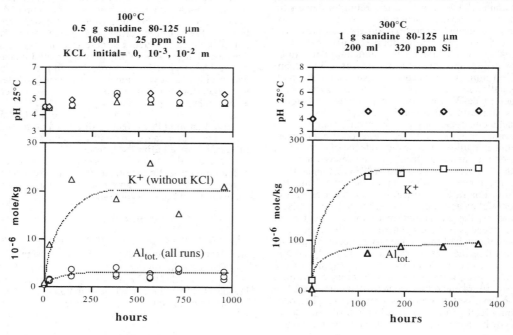

Fig.2: Dissolution data of sanidine batch reactor experiments. At 100°C several KCl concentrations were used.

in teflon reactors at 100°C. Aliqots of solution were sampled during the course of the runs and analysed for Si, Al, K and pH. At the end of the experiments, the altered samples were cleaned in deionized water and observed by scanning (SEM) and transmission (TEM) electronic microscopy.

3 RESULTS

At all temperatures, the solution chemistry evolves towards a steady-state composition and reflects a preferential release of potassium from the sanidine samples. Examples are shown in Fig.2. The excess of potassium in solution, with respect to aluminum, suggests the presence of a K-leached layer of 19 and 180 nm in depth at 100 and 300°C, respectively.

When potassium is added to the starting solution (100°C runs), the aqueous aluminum reaches the same steady-state concentration. This suggests that the solution chemistry does not reflect an equilibrium with sanidine. It remains to elucidate the nature of the phase that controls the solution composition.

At 300°C, no secondary mineral can be identified by SEM at the surface of the altered sample. Nevertheless, chemical analyses and diffraction data collected by TEM on ultrathin sections show the existence of an altered layer at the surface of the altered grains. This layer is amorphous, while microanalyses show a progressive loss in K and enrichment in Al (Fig.3B, points 1-4). The Si/Al ratio decreases from 3 in the fresh feldspar to 1 at point 4.

At 100°C, neither secondary products nor an altered layer can be observed after 400 hours of reaction. However, a full equilibrium between the solution and fresh sanidine is unlikely because the steady state concentration of aqueous aluminum is independent on the potassium concentrations, while pH and Si are constant in all the 100°C runs. If a protective, modified layer exists, its thickness is less than 50 nm (the resolution of microanalyses by TEM), according to the solution data.

4 DISCUSSION AND CONCLUSION

The formation of a leached layer at the surface of feldspars is often observed when the mineral is dissolved far from equilibrium, especially in acid solutions (Hellmann et al., 1997, among others). Our results show that, under typical natural conditions (quartz saturation, neutral pH), a strongly modified and altered layer develops rapidly at high temperatures. The chemical composition and the low crystallinity of this phase suggests that its formation is promoted by the preferential release of potassium from the mineral. This process can be compared with the hydrothermal alteration of silicate glasses (Berger et al., 1987). At lower temperatures (100°C), such a modified layer is not observed by TEM, according to the solution data which indicate a very thin K-leached layer (19 nm).

In our experiments, the solutions did not equilibrate with the fresh sanidine, although they reached a steady state composition. A first assumption is that a surficial altered layer acts as a diffusional barrier between the fresh mineral and the bulk solution; the solution composition reflects the solubility of the altered layer. A full equilibrium may possibly only be attained after a long period of time. Another assumption is that the network ceases to dissolve when all the network-former oxides are saturated, whatever the saturation state of the bulk solid. Note that this second scenario does not exclude the development of a leached layer (release of modifier cations), which may promote the formation of a silicate gel under appropriate conditions.

Our experiments do not allow identification of the exact controlling process, but strongly suggest that the approach to equilibrium in water-rock reactions is not governed solely by the solubility of the bulk mineral. Furthermore, equilibrium implies reversibility. Unfortunately, the lack of detailed investigations on the precipitation of feldspar near equilibrium prevents any definitive conclusionto be drawn.

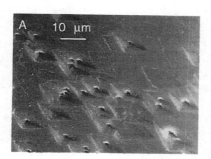
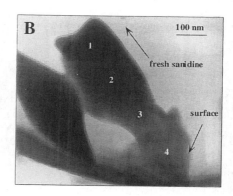

Fig.3: Sanidine reacted at 300°C. A=SEM; B= SEM

REFERENCES

Alekseyev, V.A., L.S.Medvedeva, N.I.Prisyagina, S.S.Meshalkin & I.A.Balabin 1997. Change in the dissolution rates of alkali feldspars as a result of secondary mineral precipitation and approach to equilibrium. Geochim. Cosmochim. Acta 61: 1125-1142.

Berger, G., J.Schott & M.Loubet 1987. Fundamental process controlling the first stage of alteration of a basalt glass by seawater: an experimental study between 200°C and 320°C. Earth and Planet. Sci. Let. 84: 431-445.

Berger, G. 1995. The dissolution rate of sanidine at near neutral pH as a function of silica activity at 100, 200, and 300°C. In Water-Rock-Interaction, Y.K Kharaka and O. Chudaev (ed.), Balkema, Rotterdam: 141-144.

Gautier, J.M., E.H.Oelkers & J.Schott 1994. Experimental study of K-feldspar dissolution rates as a function of chemical affinity at 150°C and pH 9. Geochim. Cosmochim. Acta 58: 4549-4560.

Hellmann, R., J.C.Dran & G.Della Mea 1997. The albite-water system: Part III. Characterization of leached and hydrogen-enriched layers formed at 300°C using MeV ion beam techniques. Geochim. Cosmochim. Acta 61: 1575-1594.

Lasaga, A.C. 1981. Transition State Theory. In Kinetics of Geochemical Process, A.C. Lasaga & R.J. Kirkpatrick Ed., Rev. Mineral. 8, Mineral Soc. Am., Washington, D.C, 169-195.

Oelkers, E.H., J.Schott & J.L.Devidal 1994. The effect of aluminum, pH, and chemical affinity on the rates of aluminosilicate dissolution reactions. Geochim. Cosmochim. Acta 58: 2011-2024.

Rimstidt, J. D. & H.L.Barnes 1980. The kinetics of silica-water reactions. Geochim. Cosmochim. Acta 44: 1683-1699.

Walther, J. 1996. Relation between rates of aluminosilicate mineral dissolution, pH, temperature, and surface charge. Amer. J. Sci. 296: 693-728.

Stable isotope exchange equilibria and kinetics in mineral-fluid systems

David R. Cole, Lee R. Riciputi & Juske Horita
Chemical and Analytical Sciences Division, Oak Ridge National Laboratory, Tenn., USA

Tom Chacko
Department of Earth and Atmospheric Sciences, University of Alberta, Edmonton, Alb., Canada

ABSTRACT: Critical to the interpretation of stable isotope distributions in natural systems is the knowledge of the equilibrium fractionations and rates of exchange between minerals and fluids. Oxygen isotope exchange rates between either metal carbonates or sheet silicates and water have been measured at 300 and 350°C, respectively. The dominant mechanism of exchange in these systems is recrystallization. For a given mineral group, a linear relationship is observed between the rates and the lattice energies of the minerals. Preliminary oxygen isotope fractionation data are also presented for the system chlorite-water for the temperature range of 170 to 350°C. A new approach to the determination of equilibrium hydrogen isotope fractionation using the ion microprobe is described with application to the epidote-water system.

1 INTRODUCTION

Equilibrium isotope fractionation factors and rates of isotope exchange (via diffusion or mineral transformation mechanisms) form the cornerstones for interpretation of stable isotope data from natural systems. Over the past three decades, a tremendous amount of effort has been expended in determining the equilibrium isotope fractionations of O, H, C, and S among fluid and mineral phases at high temperatures (>350°C), from experimental exchange studies, theoretical calculations (from spectroscopic data and statistical mechanical methods), and studies of natural materials (O'Neil, 1986). These isotope fractionation factors have been used to make reasonable estimates of the temperatures of fluid-rock interaction and the origin of the fluids. Despite the success of the equilibrium approach, a number of studies of natural systems indicate that isotope heterogeneity and disequilibrium may be more widespread than previously realized. Efforts at ORNL have focused on the measurement of both fractionation factors and rates of isotope exchange in a variety of mineral-fluid systems at temperatures and pressures appropriate for sedimentary and geothermal systems. A number of representative examples of this effort are presented below.

2 RATES AND MINERAL CHEMISTRY

Our knowledge of the rates and mechanisms of isotope exchange controlled by surface reactions comes principally from high temperature (300°C) - high pressure (≥1 kbar) partial isotope exchange experiments that were initially far from both chemical and isotope equilibrium resulting in recrystallization. Within a given mineral group there appears to be a systematic relationship between rate and mineral chemistry (Cole and Ohmoto, 1986). During the recrystallization of a phase, the structure of the reassembled mineral is determined by the tendency of atoms to take up positions whereby their total potential energy is reduced to a minimum. We proposed that the greater the lattice energy the greater the energy required to break up the crystal into its constituent ions, and hence the more sluggish the rate of both chemical and isotopic exchange. An empirical model describing the kinetics of oxygen isotope exchange between water and either silicates or carbonates recrystallizing under hydrothermal conditions was proposed for temperatures between 300 and 600°C (Cole, 1994). This model describes a relationship between rate and both temperaure and the mineral's electrostatic lattice energy (from Smyth, 1989; Smyth and Bish, 1988) normalized per number of cations. It is important to point out that (a) both cation and oxygen site potentials are used to estimate the total electrostatic attractive lattice energy for any given phase, and (b) the much more subordinate repulsive energies have been ignored. Hence, the more negative the U' value, the greater the "lattice" energy.

2.1 Carbonate-fluid exchange

We tested our empirical model directly by conducting exchange experiments in two systems: carbonates and sheet silicates. Strontianite ($SrCO_3$), witherite ($BaCO_3$) and calcite were reacted at 300°C with either pure water or NaCl solutions (1, 3, 5 molal) at fluid/solid mass ratios of 3-4 for durations of 271, 695, and 1390 hrs. At 300°C, the "salt effect" on fractionation can be appreciable depending on the concentration (~1 per mil at 5 m), so the equilibrium fractionation factors were corrected with data given by Horita et al. (1995). Diffusion at 300°C is negligible, so the fraction of exchange (F) values are related only to a recrystallization mechanism. Increases in mean grain diameter were monitored by SEM images, and recast as surface areas from an empirical relationship between measured specific surface areas and grain size. The F values for strontianite and witherite exhibit the characteristic rapid increase early on, then a flattening with time. Calcite data, on the other hand, show a much less pronounced increase with time. An increase in salinity clearly has a major effect on the rates. The fits of these data to various rate models differed very little, so for comparison purposes we chose the pseudo-first order rate model that has adequately described the higher temperature calcite data (Cole, 1992), given as $R=[-\ln(1-F)n_{fl}^o n_c^o]/(n_{fl}^o + n_c^o) \cdot At$ where n_{fl}^o and n_c^o represent the total moles of O in the fluid and solid, respectively, A is the total surface area of the solid (m^2), t is time (s), and R is in units of moles O m^{-2}s^{-1}. Rates for the carbonate-pure water systems are plotted against U' values in Fig. 1. This plot demonstrates a clear linear relationship between ln R and U', where rates increase (BaCO$_3$>SrCO$_3$>CaCO$_3$) with decreasing electrostatic attractive lattice energy.

2.2 Sheet silicate-water exchange

In order to test the empirical relationship between rates and mineral chemistry further, we elected to investigate a silicate system. One of the underlying themes in all of our isotope experimental work is to better constrain exchange rates, as well as equilibrium fractionation at low temperatures. We had already completed a series of oxygen isotope exchange experiments between chlorite and three isotopically different waters (-24.79, 0.0, +21.50 per mil), in an attempt to measure the equilibrium fractionation factor at 350°C, 250 bars. To augment these results, we have completed exchange experiments between biotite-H$_2$O and muscovite-H$_2$O at 350°C and a pressure of ~250 bars. The biotite and muscovite were reacted with only the -24.79 per mil water. The fluid/solid ratios averaged about 3 and run durations of 132 (168), 669, 1384, and 3282 hr. were used. In the case of the chlorite-H$_2$O experiments, three isotopically different waters were used for all four run times. Detailed SEM observations and N$_2$ BET surface area measurements were used to characterize each starting material. The chlorite has a mean grain diameter of about 10 μm and ΣFe/Mg mole ratio of ~1. The biotite, on the other hand, is much finer grained (mean grain diameter ~ 2.8 μm) and much richer in Fe (ΣFe/Mg mole ratio ~ 2.5). The muscovite is a relatively pure pegmatitic variety with a mean grain diameter of about 5.7 μm.

The F versus time results from these experiments exhibit similar patterns as the carbonates; modest increase early on then a flattening of the slopes. Overall, the fractions of exchange are much lower, less than 0.3 for the longest run time. However, there is clear evidence from SEM images that all three phases experienced some degree of recrystallization. Surface areas were estimated from an empirical relation between N$_2$ BET specific surface areas and mean grain diameters of well characterized sheet silicates in a manner similar to our carbonate studies. The best fit to the data was obtained from a psuedo-first rate order model that yielded ln R's of -23.99±0.3, -23.10±0.32, and 22.31±0.38 moles O m^{-2} sec^{-1} for muscovite, biotite and chlorite, respectively. There was no need to correct the F values for a possible contribution by diffusion because oxygen diffusion in micas is very slow at 350°C (Fortier and Giletti, 1991). The rates are shown in Figure 1b plotted against the electrostatic lattice energies estimated from the combined cation (Smyth and Bish, 1988) and oxygen site potentials (Smyth, 1989). As with the carbonates, we observe an increase in isotope exchange rate, chl>bio>musc, with decreasing lattice energy. This order is consistent with our model and reflects, in part, an increase in the iron content for this particular mineral group. The order is also consistent with the order in the rate of weathering of micas, wherein Fe-rich varieties react much faster than their more alkali-rich counter parts (Brady and Walther, 1989).

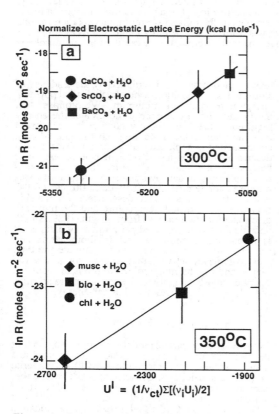

Figure 1. Oxygen isotope rates for metal carbonates (a) and sheet silicates (b) reacted with H$_2$O versus normalized electrostatic lattice energies (U'). ΣU' represents the sum of electrostatic site potential for each cation and anion site, υ$_i$ is the stoichiometric value for each site in the crystal, and υ$_{ct}$ is the number of cations per formula unit in each phase.

3 CHLORITE-WATER FRACTIONATION

We have completed a series of Northrop-Clayton type exchange experiments with chlorite. Three different isotopic waters were reacted with chlorite for four different durations (132 - 3282 hr) at 350°C and 250 bars. The percents of exchange determined for the four times from shortest to longest are 4.4, 6.5, 8.0, and 11.9. The fractionation factors calculated from the Northrop-Clayton (1966) method are in modest aggreement for the four run durations: 0.13, 0.26 -0.46 and -0.55 per mil (Fig. 2). However, the errors for each of these fractionation factors are quite large because of the small percents of exchange. Clearly, the value determined for the longest run of ~20 weeks is probably the most reliable of the group. Interestingly, this value compares very closely with a value estimated from Wenner and Taylor (1971) based on natural samples, -0.7 per mil. Agreement is also surprisingly good with the estimate described by Savin and Lee (1988), -1.3 per mil, using their type c equation for an Fe-Mg chlorite based the bond-type approach. Reduced partition function estimates for chlorite, made by Onuma et al (1972), when combined with similar data on H_2O yield a fractionation factor of ~ -0.5 per mil at 347°C, in good agreement with our value at 350°C.

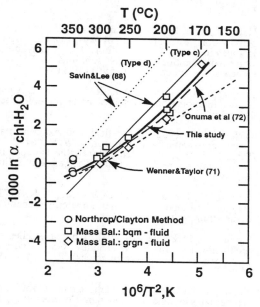

Figure 2. Oxygen isotope fractionation factors estimated from both Northrop/Clayton-type experiments (350°C) and from chlorite-water exchange in fluid-granite hydrothermal experiments (170- 300°C). bqm refers to biotite quartz monzonite; grgn refers to granite gneiss.

Additional fractionation data have been estimated from hydrothermal granite-fluid experiments where biotites altered to chlorites (Cole et al. 1992). New chemical, SEM and isotopic results have been obtained that augment an original data set reported by Cole (1985). Variations in the oxygen isotopic composition of biotites altered to chlorite have been monitored as a function of time from 16 hydrothermal granite-fluid experiments conducted at the following conditions: T = 170 to 300°C, P = 100 - 300 b, NaCl molality = 0.1 - 1.0, fluid/biotite mass ratios = 1-60, run durations = 200 - 840 hr. Detailed thin section, SEM and XRD studies demonstrate that biotite is altered exclusively to chlorite in 11 of the 16 experiments. The amount of chlorite, quantified through point counting, increased with increasing temperature as well as time. For times exceeding 600 hours, the weight percent of biotite altered to chlorite averaged approximately 90 -100, 60-80, and 35-50 for temperatures of 300, 250, and 200°C, respectively. The isotopic compositions of chlorite were calculated from mass balance, and compared with final measured $\delta^{18}O$ of the fluids. The 1000 ln α values average 0.2, 1.2 and 3.2 per mil for 300, 250, and 200°C, respectively (Fig. 2). The errors become progressively larger with decreasing temperature because the calculated final chlorite isotope value is more sensitive to errors in point counting as the extent of reaction becomes less. We estimate that the errors are at least ±0.5 per mil at 250 and 300°C, and as much as ±1 per mil at 200°C. Despite the drawbacks to the use of empirical estimates, our data are in good agreement with the curves given by Wenner and Taylor (1971), Savin and Lee (1988) and Onuma et al (1972). However, below about 200°C, our new data tend to predict consistently larger fractionations compared to several other studies.

4 EPIDOTE-WATER FRACTIONATION

There is presently little consensus on hydrogen isotope fractionation factors for many important mineral-water systems. To resolve this issue, we have developed a novel experimental technique (Fig. 3) that uses large single crystals rather than fine powders reacted with

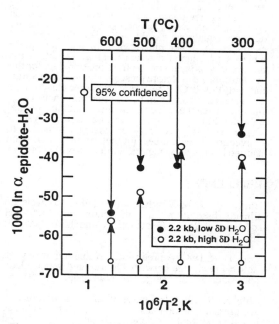

Figure 3. Hydrogen isotope fractionations in the system epidote- H_2O at 2.2 kbars determined with the ion microprobe. The arrows indicate the direction of isotope reversibility from the initial epidote-H_2O fractionations.

labeled waters, and analyzed these with the ion microprobe. The advantage of this method over conventional approaches is that it allows fractionation factors to be determined in experiments in which effectively all the isotope exchange takes place by a diffusional process. This makes for an ideal exchange experiment in that the only thermodynamic driving force for exchange is the free energy change associated with the isotope exchange reaction itself. To test this approach, we conducted experiments in which a large excess of water was exchanged with museum-quality crystals of epidote ($Ps_{32.5}$; δD=-40.7‰) at 300-600°C, 2.2 kbar. Two experiments were run at each temperature, identical in all respects except for containing isotopically different waters (δD=+24.4, -104.9‰). The epidote run products were analyzed on the ion microprobe following the procedure of Deloule et al. (1991). Analytical precision was on the order of 4-9‰ (95% confidence). Despite initial epidote-H_2O fractionations far from equilibrium, the experiments showed a close approach towards equilibrium fractionation as low as 300°C (Fig. 3). Fractionation factors (ΔD[ep-H_2O]) range from -36±8‰ at 300°C to -54±8‰ at 600°C.

5 SUMMARY

Oxygen isotope exchange rate data on carbonates and sheet silicates confirm our empirical model that describes, for a particular mineral group, a linear relationship between rates of exchange and the lattice energies of the minerals. Although the results may seem intuitive, it is important to point out that by establishing an unambiguous relationship between rate, temperature and lattice energy, one can easily develop equations capable of predicting rates of exchange for other phases of a mineral group for which experimental data are lacking. Additionally, preliminary oxygen isotope fractionation data for the system chlorite-water are presented for the temperature range, 170-350°C. These new data agree closely with estimates based on a number of other approaches, e.g., increment method, bond-strength method, natural chlorites. Finally, the ion microprobe D/H analysis of single crystals provides a promising alternative to conventional techniques for determining hydrogen isotope fractionation factors in mineral-water systems.

REFERENCES

Brady, P. V. & Walther, J. V. 1989, Controls on silicate dissolution rates in neutral and basic pH solutions at 25°C. Geochim. Cosmochim. Acta 53: 2823-2830.

Cole, D. R. 1985. A preliminary evaluation of oxygen isotopic exchange between chlorite and water. Geol. Soc.Amer. Annual Mtg, Abs. with Prog. 17: 550.

Cole, D. R. 1992. Influence of solution composition and pressure on the rates of oxygen isotope exchange in the system: calcite-H_2O-NaCl at elevated temperatures. Chem. Geol. 102: 199-216.

Cole, D. R. 1994. Oxygen isotope exchange rates in mineral-fluid systems: correlations and predictions. Min. Mag. 58A:189-190.

Cole, D. R. & Ohmoto, H. 1986. Kinetics of isotopic exchange at elevated temperatures and pressures. In J. W.Valley, H. P. Taylor, Jr., & J. R. O'Neil (eds) Stable Isotopes in High Temperature Geological Process. Rev. Min.16: 41-90.

Cole, D. R., Ohmoto, H. & Jacobs, G. K. 1992. Isotope exchange in mineral-fluid systems. III. Rates and mechanisms of oxygen isotope exchange in the systems: granite-H_2O-NaCl-KCl at hydrothermal conditions. Geochim. Cosmochim. Acta. 56: 445-466.

Deloule, E., France-Lanford, C., and Allegre, F. 1991. D/H analysis of minerals by ion microprobe. In H. P. Taylor, Jr., J. R. O'Neil & I. R. Kaplan (eds) Stable Isotope Geochem., Geochem. Soc.. Pub. 3: 53-62.

Fortier, S. M. & Giletti, B. J. 1991. Volume self-diffusion of oxygen in biotite, muscovite, and phlogopite micas. Geochim. Cosmochim. Acta 55: 1319-1330.

Horita, J., Cole, D. R. & Wesolowski D. J. 1995. The activity-composition relations of oxygen and hydrogen isotopes in aqueous salt solutions. III. Vapor-liquid water equilibration of NaCl solutions to 350°C. Geochim. Cosmochim. Acta 59: 1139-1151.

Northrop, D. A. & Clayton, R. N. 1966. Oxygen isotope fractionation in systems containing dolomite. J. Geol 64: 174-196.

O'Neil, J. R. 1986. Theoretical and experimental aspects of isotopic fractionation. In J. W. Valley, H. P. Taylor, Jr., & J. R. O'Neil (eds) Stable Isotopes in High Temperature Geological Processes. Rev. Min. 16: 1-40.

Onuma, N., Clayton, R. N. & Mayeda, T. K. 1972. Oxygen isotope cosmothermometer. Geochim. Cosmochim. Acta. 36: 169-188.

Savin, S. M. & Lee, M. 1988. Isotopic studies of phyllosilicates. In S. W. Bailey (ed.) Hydrous Phyllosilicates (exclusive of micas). Rev. Min. 19: 189-223.

Smyth, J. R. 1989. Electrostatic characterization of oxygen sites in minerals. Geochim. Cosmochim. Acta 53: 1101-1110.

Smyth, J. R. & Bish, D. L. 1988. Crystal Structures and Cation Sites in Rock-Forming Minerals. Allen & Unwin, 386p.

Wenner, D. B. & Taylor, H. P., Jr. 1971. Temperatures of serpentinization of ultramafic rocks based on O^{18}/O^{16} fractionation between coexisting serpentine and magnetite. Contrib. Min. Petrol. 32: 165-185.

Reviews by J. R. Hulston and G. L. Lyon are gratefully acknowledged. Research sponsored by the Geosciences Program of the Office of Basic Energy Sciences and the Geothermal Technology Program (EERE), U. S. Dept. of Energy under contract no. DE-AC05-96OR22464 to Lockheed Martin Energy Research Corporation.

Solubility and potentiometric studies of REE complexation with simple carboxylate (acetate, oxalate) ligands from 25° to 80°C

Ru Ding & S.A.Wood
Department of Geology and Geological Engineering, University of Idaho, Moscow, Idaho, USA

C.H.Gammons
Department of Geological Engineering, Montana Technical University, Butte, Mont., USA

ABSTRACT: The stability constants of La^{3+} and Nd^{3+} complexes with acetate have been measured potentiometrically in NaCl media at constant ionic strengths of 0.1, 0.2, 1.0 and 2.0 molal and temperatures from 25° to 70°C. Under the conditions of the study, only the first complex ($LnAc^{2+}$) was detected. The stability constants exhibit a weak but steady increase with temperature and agree well with accepted literature values at 25°C. The solubility of $Yb_2(C_2O_4)_3 \cdot yH_2O$ was investigated as a function of oxalate concentration, pH, ionic strength and temperature (25-80°C). At low oxalate concentrations there is evidence for $Yb(C_2O_4)^+$ and at high oxalate concentrations $Yb(C_2O_4)_2^-$ forms. Also, at high oxalate and NaCl concentrations and low temperatures, a double salt becomes the stable phase. Solubility at low oxalate concentrations exhibits a relatively weak temperature dependence.

1 INTRODUCTION

REE mass transfer is of concern to geologists studying processes in sedimentary basins, and the genesis of associated petroleum and base-metal deposits (Scherer & Seitz, 1980; Banner et al., 1988; Dorobek & Filby, 1988; Kontak et al., 1995). The mobility of the REE in aqueous solutions also is relevant environmentally. The REE are produced during the fission of U and Pu (Rard 1988), and they function as analogues of the actinides (Choppin 1983). The REE and actinides are especially mobile in the presence of organic chelating agents, e.g., oxalate, citrate, EDTA (Means et al., 1978).

Oxalate occurs naturally in oil field brines (up to 500 ppm; MacGowan & Surdam, 1988) and soils (Graustein et al., 1977), is a product of the thermal degradation of more complex organic matter (Surdam et al., 1984), is used as a reagent in the separation and/or chelation of the REE and forms relatively strong chelates with the REE. Acetate occurs in high concentrations in oil-field brines (up to 10,000 ppm, Carothers & Kharaka, 1978) and in sediment-hosted seafloor hydrothermal fluids (Martens, 1990), is a product of thermal degradation of more complex organic matter (Surdam et al., 1984) and persists metastably in aqueous solution at elevated temperatures (Kharaka et al., 1983; Palmer & Drummond, 1986). Complexes between the REE and acetate or oxalate may therefore be important agents of mass transfer. However, thermodynamic data for REE-oxalate complexes are poorly known at any temperature and for REE-acetate complexes there are few data at elevated temperatures (see Wood, 1993 for a review).

2 EXPERIMENTAL

2.1 Potentiometric measurements

The stability constants of La^{3+}, Nd^{3+}, Gd^{3+} and Yb^{3+} complexes with acetate were measured potentiometrically at ionic strengths of 0.1, 0.2, 1.0 and 2.0 molal (NaCl medium) and temperatures of 25, 40, 50, 60 and 70°C. The experiments were conducted in jacketed reaction vessels, the temperature of which was controlled by a circulating water bath. Argon was continually bubbled through the solutions to exclude CO_2. The pH values of the test solutions as a function of titrant concentration were recorded with a glass combination electrode, calibrated against reference solutions with the same temperature and ionic strength as the test solutions. The experimental data were fitted to various speciation models using non-linear least squares regression and stability constants were derived from the simplest model providing the best fit to the data.

2.2 Solubility measurements.

The solubility of solid ytterbium oxalate was measured from 25° to 80°C in solutions with ionic strengths of 0.05, 0.10 and 0.20. NaCl was employed as an inert electrolyte to maintain constant ionic strength. The free oxalate concentration was varied both by adjusting the total oxalate concentration and the pH of the solutions. The starting solid was $Yb_2(C_2O_4)_3 \cdot yH_2O$, but this apparently transformed during the experiments to the double salt $NaYb(C_2O_4)_2 \cdot xH_2O$ at high oxalate and NaCl concentrations (see below). Grenthe et al. (1969) encountered similar double salts.

The pH was measured using a combination glass electrode calibrated against buffer solutions at the same ionic strengths and temperatures as the experiments. Concentrations of Yb and oxalate were determined using ICP-AES and ion chromatography, respectively. Approximately 50 g of each experimental solution and 300 mg of Yb oxalate were loaded into cylindrical glass tubes. The tubes were placed on a rack in a reciprocal shaking constant temperature (±0.1°C) water bath. Samples for Yb and oxalate analyses were withdrawn via syringe and forced through a 0.45 μm filter. Samples for Yb analysis were acidified with HCl. A separate series of experiments demonstrated attainment of equilibrium within 3 days.

3 RESULTS AND DISCUSSION

Preliminary data obtained thus far for acetate complexes are presented in the table. These data represent averages of 2-4 replicate titrations and are "raw", i.e., unsmoothed values. They correspond to the first cumulative stability constant (β_1) for the reaction:

$$Ln^{3+} + Ac^- = LnAc^{2+}$$

where Ln^{3+} represents any trivalent REE ion and Ac^- represents the acetate ion and β_1 is defined as:

$$\beta_1 = \frac{[LnAc^{2+}]}{[Ln^{3+}][Ac^-]}$$

REE	I (molal)	t(°C)	log β_1
La	0.1	25	1.916±0.064
		30	2.152±0.010
		40	2.116±0.045
		50	2.226±0.043
		60	2.272±0.030
		70	2.377±0.054
	0.2	25	1.910±0.143
		40	1.735±0.028
		50	2.051±0.059
		60	2.066±0.035
		70	2.177±0.041
	1.0	25	1.687±0.097
		40	1.517±0.073
		50	1.544±0.109
		60	1.766±0.041
		70	1.985±0.130
	2.0	25	1.412±0.078
		40	1.024±0.062
		50	1.174±0.074
		60	1.499±0.100
		70	2.209±0.031
Nd	0.1	25	2.191±0.010
		40	2.311±0.069
		50	2.363±0.048
		60	2.514±0.058
		70	2.562±0.023
	0.2	25	2.071±0.017
		40	2.100±0.055
		50	2.283±0.019
		60	2.401±0.005
		70	2.482±0.014

The data show a slight, positive temperature dependence over the range investigated, and the trends with respect to temperature and ionic strength are relatively smooth. Our data at room temperature are in excellent agreement with stability constants reported in the literature for $NaClO_4$ media (Kolat & Powell, 1962; Kovar & Powell, 1966). The similarity of REE-acetate complex stability constants in chloride and perchlorate media suggest that REE-chloride complexes are very weak under the conditions investigated. Our data for La are also consistent with data recently reported by Deberdt et al. (1997) from 25° to 80°C. Under the conditions of our experiments, we did not detect complexes higher than the 1:1 complex $LnAc^{2+}$.

Representative data from the Yb oxalate solubility study are shown in Figures 1 and 2. In all cases we have obtained U-shaped solubility curves indicative of a change from positively charged to negatively charged Yb-oxalate complexes with increasing oxalate concentration. At 80°C $Yb_2(C_2O_4)_3 \cdot 10H_2O$ is the stable phase over the entire range of oxalate concentrations (Figure 1). At 40°C, there are two branches to the solubility curve at high oxalate concentrations. The lower branch corresponds to the double salt $NaYb(C_2O_4)_2 \cdot xH_2O$ and the upper branch corresponds to the metastable simple oxalate $Yb_2(C_2O_4)_3 \cdot 10H_2O$. There is evidence for the formation of the first Yb-oxalate complex $Yb(C_2O_4)^+$ at low oxalate concentrations, and transformation to $Yb(C_2O_4)_2^-$ at higher oxalate concentrations. In our initial experiments we did not attain low enough free oxalate concentrations to detect the uncomplexed Yb^{3+} ion and so cannot derive the first stability constants for the complex $Yb(C_2O_4)^+$. Experiments at low pH and high ionic strength are currently in progress to obtain these data. However, we are able to derive equilibrium constants for the reaction:

$$1/2 Yb_2(C_2O_4)_3(s) = Yb(C_2O_4)^+ + 1/2 C_2O_4^{2-}$$

and we have obtained the following values of log K at 0.05 molal ionic strength: -6.80 (40°C), -6.74 (60°C) and -6.69 (80°C). These values compare with log K = -7.12 derived by Crouthamel & Martin (1950) at 25°C and infinite dilution.

A significant outcome of this study is that the solubility of ytterbium oxalate in pure water and solutions containing < 10^{-4} molal free oxalate is virtually independent of temperature in the range 25 to 80°C. There is a slight prograde temperature dependence at very high oxalate, which may be due to the thermal decomposition of the double salt. In very high salinity brines, such as those occurring in sedimentary basins, it is possible that some double REE oxalate salts may occur as natural minerals, owing to the very low solubilities of these phases.

Work on both the REE acetate and oxalate systems continues in our laboratories. Data such as those presented in this paper should lead to a better understanding of the behavior of the REE in sedimentary basins and in mixed radioactive-organic waste sites.

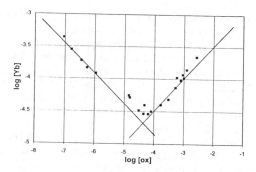

Figure 1: Solubility data as a function of oxalate at 80°C and I = 0.05 molal NaCl. The stable solid phase is $Yb_2(C_2O_4)_3 \cdot 10H_2O$ over the range of conditions shown. The straight lines with slopes -1/2 and plus 1/2 indicate the predominance of $Yb(C_2O_4)^+$ at low oxalate concentrations, and $Yb(C_2O_4)_2^-$ at high oxalate concentrations, respectively.

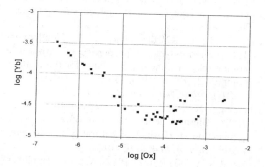

Figure 2: Solubility data as a function of oxalate at 40°C and I = 0.05 molal NaCl. At low oxalate concentrations $Yb_2(C_2O_4)_3 \cdot 10H_2O$ is the stable phase. At high oxalate concentrations the upper branch of the solubility curve corresponds to the metastable $Yb_2(C_2O_4)_3 \cdot 10H_2O$ and the lower branch corresponds to the double salt $NaYb(C_2O_4)_2 \cdot xH_2O$.

REFERENCES

Banner, J.L., G.N. Hanson, & W.J. Meyers 1988. Rare earth element and Nd isotopic variations in regionally extensive dolomites from the Burlington-Keokuk formation (Mississippian): Implications for REE mobility during carbonate diagenesis. *J. Sed. Petrol.* 58: 415-432.

Carothers, W.W. & Y.K. Kharaka 1978. Aliphatic acid anions in oil-field waters and their implications for the origin of natural gas. *Bull. Amer. Assoc. Pet. Geol.* 62: 2441-2453.

Choppin, G.R. 1983. Comparison of the solution chemistry of the actinides and lanthanides. *Jour. Less-Common Metals* 93: 323-330.

Crouthamel, C.E. & D.S. Martin, Jr. 1950. The solubility of ytterbium oxalate and complex ion formation in oxalate solutions. *J. Amer. Chem. Soc.* 72: 1382-1386.

Deberdt, S., S. Castet, J.-L. Dandurand, J. Schott, C. Harrichoury, & I. Louiset 1997. Experimental study of La(OH)$_3$ and Gd(OH)$_3$ solubilities (25 to 150°C) and La-acetate complexing (25 to 80°C). *Chem. Geol.* (submitted).

Dorobek, S.L. & R.H. Filby 1988. Origin of dolomites in a downslope biostrome, Jefferson formation (Devonian), central Idaho: evidence from REE patterns, stable isotopes, and petrography. *Bull. Can. Petrol. Geol.* 36: 202-215.

Graustein, W.C., K. Cromack, Jr., & P. Sollins 1977. Calcium oxalate: Occurrence in soils and effect on nutrient and geochemical cycles. *Science* 198: 1252.

Grenthe, I., G. Gardhammar, & E. Hundcrantz 1969. Thermodynamic properties of rare earth complexes. VI. Stability constants for the oxalate complexes of Ce(III), Eu(III), Tb(III) and Lu(III). *Acta Chem. Scand.* 23: 93-108.

Kharaka, Y.K., W.W. Carothers, & R.J. Rosenbauer 1983. Thermal decarboxylation of acetic acid: Implications for origin of natural gas. *Geochim. Cosmochim. Acta* 47: 397-402.

Kolat, R.S. & J.E. Powell 1962. Acetate complexes of the rare earth and several transition metal ions. *Inorg. Chem.* 1: 293-296.

Kontak, D.J. and S. Jackson 1995. Laser-ablation ICP-MS micro-analysis of calcite cement from a Mississippi-Valley-type Zn-Pb deposit, Nova Scotia: dramatic variability in REE content on macro- and micro-scales. *Can. Min.* 33: 445-468.

Kovar, L.E. & J.E. Powell 1966. Stability constants of rare earths with some weak carboxylic acids. Report TID-4500, Ames Laboratory, Iowa State University, 54 p.

MacGowan, D.B. & R.C. Surdam 1988. Difunctional carboxylic acid anions in oil field waters. *Org. Geochem.* 12: 245-259.

Martens, C. S. 1990. Generation of short chain organic acid anions in hydrothermally altered sediments of the Guayamas Basin, Gulf of California. *Appl. Geochem.* 5: 71-76.

Means, J.L., D.A. Crerar, and J.O. Duguid 1978. Migration of radioactive wastes: Radionuclide mobilization by complexing agents. *Science*, 200: 1477-1481.

Palmer, D.A. & S.E. Drummond 1986. Thermal decarboxylation of acetate. Part I. The kinetics and mechanism of reaction in aqueous solution. *Geochim. Cosmochim. Acta* 50: 813-823.

Rard, J.A. 1988. Aqueous solubilities of praseodymium, europium and lutetium sulfates. *Jour. Sol. Chem.* 17: 499-517.

Scherer, M. & H. Seitz 1980. Rare-earth element distribution in Holocene and Pleistocene corals and their redistribution during diagenesis. *Chem. Geol.* 28: 279-289.

Surdam, R.C., S.W. Boese, and L.J. Crossey 1984. The chemistry of secondary porosity. *Amer. Assoc. Petrol. Geol. Mem.* 37: 127-149.

Wood, S.A. 1993. The aqueous geochemistry of the rare earth elements: critical stability constants for complexes with simple carboxylic acids at 25°C and 1 bar and their application to nuclear waste management. *J. Geol. Engineering* 34: 229-259.

Two-dimensional measurement of natural radioactivity of rocks by photostimulated luminescence

M. Hareyama, N. Tsuchiya & M. Takebe
Department of Geoscience and Technology, Graduate School of Engineering, Tohoku University, Sendai, Japan

ABSTRACT: Radiation images of small volume specimens such as a thin sections can be obtained by an imaging plate consisting of a storage film coated with photostimulated phosphor ($BaFBr:Eu^{2+}$). This plate shows distinctive performance compared with X-ray film and other two-dimensional radiation detectors. Irradiation image of rocks is inhomogeneous and intensity of photostimulated luminescence (PSL) per unit area which is principally converted into integrated dose is increasing with increasing potassium content in rocks. These fact indicate that beta-disintegration of ^{40}K is dominant radiation source of rocks.

In situ measurement of natural radiation was carried out in the 250m level drift of the Kamaishi Mine, which is applicable to characterize radioactive properties and to evaluate the distribution of radioactive nuclei within rocks in the field.

1 INTRODUCTION

Barium fluorobromide doped with Eu^{2+} ($BaFBr:Eu^{2+}$) is a useful storage-phosphor in which ionizing radiation-produced images are recorded as a latent image. Since an Imaging Plate, IP, has striking performances of radiation detection, such as a simple usage, a high position resolution, a large detection area, a high detection sensitivity with good signal-to-noise ratio, long time dose accumulation, good dose linearity, extremely wide dynamic range of dose and erasing for reuse (Takebe et al., 1995), it is extensively used in various fields.

The imaging plate has a possibility to obtain a high quality two-dimensional natural irradiation pattern of rocks without any skilled techniques compared with conventional methods. This study describes an application of the imaging plate to measure extremely low level natural radiation of rocks, and *in situ* two-dimensional measurement of environmental radiation with respect to mass transfer through fractures is also discussed.

2 IMAGING PLATE

The imaging plate (IP) is a storage film coated with photostimulated phosphor ($BaFBr:Eu^{2+}$) and shows distinctive performance of a bi-dimensional position radiation detector such as a high position resolution of 25μm, a high detection sensitivity of approximately 10^3 times that of X-ray films, long time dose accumulation, good dose linearity, extremely wide dynamic range of dose of 10^5.

The photostimulated luminescence (PSL) mechanism of $BaFBr:Eu^{2+}$ was described by Takahashi et al. (1984) and Iwabuchi et al. (1994). The electrons excited to the conduction band by radiation are caught by F$^+$ centers (halogen ion vacancies) to form two types of F centers, i.e., F (Br$^-$) and F (F$^-$), while holes in the valence band are trapped by Eu^{2+} and converted into Eu^{3+}. The latent image of ionizing radiation was stored as distribution of concentration of trapped electrons in the imaging plate. By visible light irradiation the trapped electrons are again liberated to the conduction band and recombined with holes associated with Eu^{3+} ions. Thus, the Eu^{2+} luminescence is observed as a photostimulated luminescence. The Energy level scheme of $BaFBr:Eu^{2+}$ and the PSL processes are schematically illustrated in Fig.1.

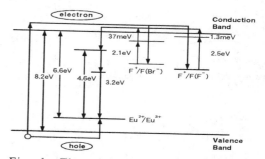

Fig. 1 The energy term scheme and the photostimulated luminescence processes in $BaFBr:Eu^{2+}$ (modified after Iwabuchi et al., 1994).

3 MEASUREMENTS OF NATURAL RADIATION OF ROCKS

The samples used in this study were collected from drifts, quarries and outcrops in Northeast Japan. The description of the samples and whole rock chemical composition determined using XRF is represented in Table 1.

Polished slabs of samples (1 to 2cm thickness) and glass backing uncovered thin-section samples (30mm thickness) were laid on the imaging plate (BAS-SR Fuji Photo Film Co., Ltd.) . The exposure duration in air at room temperature is 72 hours for slab samples and is 168 hours for the thin section samples in a shield box made from radioactive-free iron.

In situ measurement of natural radiation was carried out on the 250m level drift of the Kamaishi Mine. The imaging plate was placed against the side and roof walls including several kinds of veins which are filled with prehnite and chlorite. The exposure duration *in situ* is 24 hours.

The 390 nm photo stimulated luminescence of Eu^{2+} was measured by scanning with a 633 nm He-Ne laser (BAS-5000, Fuji Photo Film Co., Ltd.), and is reconstructed as a two-dimensional dot image. The specifications of the BAS-SR imaging plate and its imaging system used in this work are as follows; detection area: 20×25 cm^2, position resolution: $25\mu m$, digital image : 16 bits, dynamic range of dose: 10^4. Fig. 2 shows the principle of the radiation imaging system.

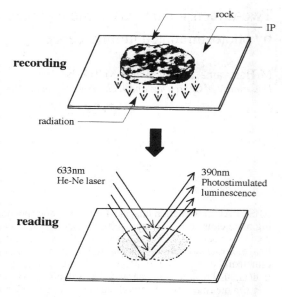

Fig. 2 The principles detection and imaging of radiation in rocks using the imaging plate.

Table 1 Representative analyses of whole rock chemical composition of samples used in this study.

Rock type	Granitic	rocks	Pyroclastic rocks	Volcanic rocks	Metamorphic rocks
	granodiorite	granodiorite	tuff	andesite	pelitic hornfels
Locality	Abukuma Plateau quarry	Kamaishi Mine 250m Level drift	Yuzawa outcrop	Kakkonda outcrop	Kamaishi Mine 250m Level drift
SiO_2 [wt%]	68.39	62.87	68.90	56.63	51.07
TiO_2	0.36	0.73	0.63	0.69	1.04
Al_2O_3	15.77	13.71	14.18	15.87	20.45
Fe_2O_3	2.87	7.77	5.56	8.50	10.36
MnO	0.06	0.13	0.10	0.14	0.20
MgO	1.24	3.45	3.14	6.15	1.61
CaO	3.29	5.36	1.29	8.63	4.81
Na_2O	3.93	2.94	3.93	2.32	4.90
K_2O	2.11	2.34	1.96	0.72	3.38
P_2O_5	0.08	0.17	0.09	0.06	0.51
total	98.1	99.51	99.76	99.69	98.32

4 RESULTS & DISCUSSION

The distribution of natural radioactivity in the Kurihashi granodiorite measured using the imaging plate is represented in Fig.3. The sample contains a greenish vein which is composed mainly of prehnite and chlorite, and a reddish altered zone is recognized around the vein (Fig.3(a)). It is very obvious that the distribution of radioactive sources within the rock is heterogeneous. The greenish vein itself is free of radioactivity. However, relatively strong PSL intensities are observed in the reddish altered zone around the greenish vein.

Fig. 4 Autoradiography of the thin section of the Iidate granodiorite.

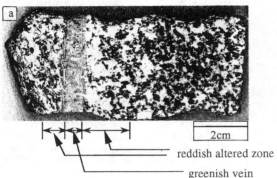

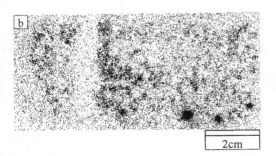

Fig. 3 Autoradiography of the Kurihashi granodiorite (a) Optical photograph of polished surface (b) Pseudo level representation of the imaging plate.

Fig.4 shows an autoradiograph of the thin section of the Iidate granite. The volume of thin section that is a radioactive source is remarkably small. However, irradiation images could be detected by the imaging plate because of its high detection sensitivity. Some distinctive spots are recognized and zircon is identified by optical microscopy in these spots. Thus, the dark spots are related to the location of zircon and are caused mainly by alpha particles from zircon grains.

Fig. 5 shows the relationship between integrated PSL intensity per unit area and potassium content in whole rocks, where PSL represents the intensity of photostimulated luminescence in arbitrary units. PSL increases with increasing potassium content regardless of rock type. Potassium content is considerably greater than that of other radioactive elements (i.e., U,Th and Rb), thus potassium is the main beta and gamma source in the natural rocks. However, PSL values of the granitic rocks are scattered compared with those of other rock types. This scattering is considered to depend on the number of alpha emitters in the rocks.

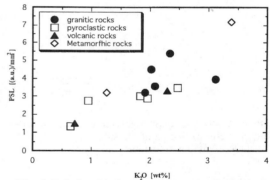

Fig. 5 Relationship between photostimulated luminescence, PSL, and potassium content.

Fig. 6 shows the result of *in situ* measurement of the natural radiation of granitic side wall including the same type of vein shown in Fig.3. The PSL intensity of the reddish altered zone distribution along the greenish vein is relatively high, which is the same radioactive characteristics determined in the laboratory measurement shown in Fig.3. Therefore *in situ* measurement of the natural radiation of rocks using the imaging plate is applicable to characterize radioactive properties and to evaluate the distribution of radioactive nuclei within rocks in the field.

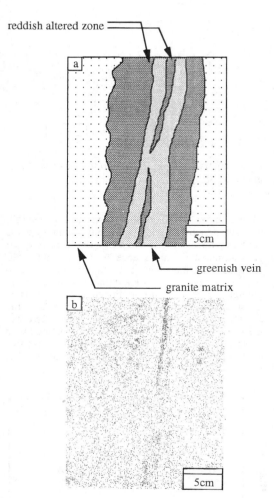

Fig. 6 Autoradiography of granitic side wall including the vein (a) Sketch of the side wall (b) Pseudo level representation of the imaging plate.

samples correspond to zircon that contents radioactive nuclei such as U and Th as an alpha-emitter.

In situ measurement using the imaging plate at the drift in granite rock body is successful in obtaining a two-dimensional natural radiation image, which is applicable to obtain precise environmental radiation images in the field within a relatively short time.

The imaging plate is a new research tool to characterize environmental radiation caused by geological materials and to evaluate geochemical transport phenomena of radioactive nuclei in the Earth's crust.

REFERENCES

Iwabuchi Y., Mori N., Takahashi K., Matsuda T., and Shinoya S. (1994) Mechanism of photostimulated luminescence process in BaFBr:Eu^{2+} phosphors. Jpn. J. Appl. Phys. 33, 178-185.

Takahashi K., Kohda K., Miyahara J., Kanemura Y., Amitani K. and Shionoya S. (1984) Mechanism of photostimulated luminescence in BaFBr:Eu^{2+} (X=Cl, Br) phosphors. J. Luminescence, 31&32, 266-268.

Takebe M., Abe K., Souda M., Satou Y. and Kondo Y. (1995) A particle energy determination with an imaging plate. Nucl. Instrum. & Methods, A359, 625-627.

Takebe M., Abe K., Souda M., Satou Y. and Kondo Y. (1995) A wide range of electron energy determination with an imaging plate. Nucl. Instrum. & Methods, A359, 625-627.

5 CONCLUSION

The imaging plate, which is a new sensitive radiation detector, was applied to autoradiography of natural rocks. Since the imaging plate has a very high sensitivity, it is possible to obtain a two-dimensional radiation image of small volume rock specimens such as thin sections, as well as the slab samples. The irradiation image of rocks is heterogeneous and PSL intensity per unit area is increasing with increasing potassium content in rocks. These facts indicate that potassium is the main beta and gamma source of natural rocks. The highly PSL spots observed in the granite

Leaching experiments with acid cation-exchange resin as a new tool to estimate element availabilities in geological samples

W. Irber
Geology Department, University of Auckland, New Zealand

P. Möller
GeoForschungsZentrum Potsdam, Germany

W. Bach
Woods Hole Oceanographic Institution, Mass., USA

ABSTRACT: A newly developed experimental leaching method using a cation-exchange resin provides insight into the relative availability of major and trace elements in fresh and altered rocks. The method does not intend to simulate natural alteration processes, but it allows the quantification of effects owing to fluid-rock interaction. Examples from two different geological settings are given: (i) the leachability of Rb and Sr in Hercyninan S-type granites and (ii) the availability of REE in fresh and altered mid-ocean ridge basalts. In the granites, a mineralogically controlled, systematic leaching behaviour of Rb and Sr as well as their isotopes could be detected. It changes parallel with differentiation and might explain systematic shifts of Rb and Sr whole-rock isochrons during alteration. In the basalts, the leaching results revealed the interstitial material (formerly residual melt) as the major REE source which may act as an important REE source for crustal fluids during alteration.

1 INTRODUCTION

Fluid-rock interaction describes element exchange between fluids and solid phases. Whereas the behaviour of ions in the fluid system is well known, the total effect of alteration on the release or the fixation of elements in the multi-phase system "rock" is yet poorly understood. Until today, the evaluation of possible element mobility is mostly done with respect to the mineralogical assemblage of the fresh sample, and not of the first altered equivalent where *internal* redistribution of elements has already occured.

Of major interest therefore is in what way does the overall availability of elements change as rocks alter and how does this correspond to changes in mineralogical composition?

A simple leaching method was developed to study this topic. The method uses a cation-exchange resin to collect leached ions from the solution, which also prevents re-adsorption of released ions and precipitation of secondary minerals. The method does not intend to simulate natural alteration processes, but to determine and quantify the effects that occurred owing to natural alteration, e.g. the redistribution of elements from primary to secondary minerals. This redistribution is traced by changes between fresh and altered rocks both in total leachability and the relative leaching behaviour of major and trace elements. To demonstrate the method two examples of common interest are given.

2 EXPERIMENTAL METHOD

The leaching experiments are carried out under acidic conditions (pH ~3) at 70°C (Möller & Giese 1997). Three batches (1-2 g rock powder) of each sample are leached in individual runs with 50 ml deionised water for about 2, 5 and 20 hrs, respectively (water/rock ratio=25). A cation-exchange resin (BIORAD AG50W-X8) collects all leached cations. The pH drops to 3 after about 1 hr; the working temperature of 70°C is limited by the thermal stability of the resin. The leaching solutions obtained after elution of the resin were analysed by ICP-AES for major elements and by ICP-MS for trace elements. The element concentrations were divided by the whole-rock concentration and multiplied by the water-rock ratio and the factor 100 (=leached whole-rock fraction in per cent). In repeated experiments, the reproducibility of the leached fractions was found to be better than 12% (Irber, 1996; Möller and Giese, 1997).

3 EXAMPLES

3.1 Leachability of Rb and Sr (^{87}Sr, ^{86}Sr) and implications for age dating of granites

Whole-rock Rb-Sr age dating of highly evolved granitic rocks is often disputed as these ages tend to be too young with respect to those obtained by U-Pb of zircons or monazites, or by K-Ar of micas. Since highly evolved granites are variably affected by alteration owing to the release of a late magmatic fluid (Strong 1985), these alteration processes close to the intrusion event are believed to be the cause for distorted Rb-Sr systems (Hradetzky & Lippolt 1993).

In order to investigate this question and to test if individual leaching behaviour of Rb and Sr (including isotopes) exists during alteration we performed the leaching experiments on differently evolved granitic rock samples (Irber et al. 1996).

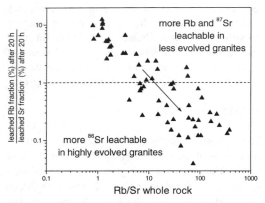

Figure 1 Leachable fractions of Rb (^{87}Sr) and Sr (^{86}Sr) in differently evolved Hercynian S-type granites from the Erzgebirge and the Fichtelgebirge in Germany. Rb is highly leachable in less evolved, biotite-rich granites, Sr in highly evolved, apatite-rich samples.

The results show a systematic tendency in the release of Rb and Sr that is strongly dependant on the mineralogical composition of the granite or the degree of differentiation (Fig. 1). In less evolved biotite granites, a large fraction of Rb and ^{87}Sr can be easily mobilised from the biotite's interlayer sites. With increasing degrees of differentiation, the available fractions of Rb and ^{87}Sr decrease and both elements become increasingly less leachable owing to their incorporation in less soluble hosts such as Li-rich micas, muscovite and/or K feldspar. Interestingly, the behaviour of Rb (and ^{87}Sr) develops towards that of the non-radiogenic Sr during granite evolution. Whereas plagioclase retains Sr in less evolved granites, in the more highly evolved rock types up to 80% of the Sr whole-rock content can be easily leached by selective dissolution of apatite, a major and often underestimated Sr host. Analyses of apatites in less to more highly evolved granites have shown a range in Sr concentrations from <1000 to more than 10000 ppm.

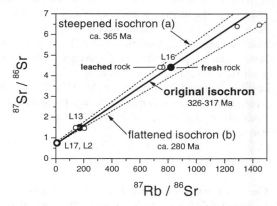

Figure 2 Combination of leaching data and theoretical calculations of alteration dependant element mobility: strong albitization (model case a) or sericitization (model case b) reveals the possibility of systematic shifts of Rb-Sr isochrons

Investigations on alteration sequences revealed that during alteration highly accessible Rb and ^{87}Sr fractions in fresh biotite-bearing rocks were immediately released, but during ongoing alteration only Rb was fixed again in less soluble secondary micas. The Rb whole-rock content was maintained or even increased during alteration. The ^{86}Sr, however, which is mainly hosted in the less soluble plagioclase in less evolved granites, became more and more leachable during alteration as well as the increasing degree of differentiation. This is caused by alteration of the primary feldspar and the Ca and Sr uptake by secondary apatite and carbonate. Parallel with the increase in Sr leachability, an overall loss of Ca and Sr in the naturally altered rocks was observed.

If sericitization is the dominant style of alteration within a granitic pluton, a selective removal of ^{87}Sr (and of Sr in general) occurs. Theoretical calculations based on the experimental leaching results show that the combination of availability and the following control of element mobility by secondary minerals can result in a systematic decrease of the slope of the Rb-Sr whole-rock isochron and a lowering of the Rb-Sr whole-rock age. During strong albitization, the opposite leaching behaviour of Rb and Sr is observed, which results in a steepening of the slope of the Rb-Sr whole-rock isochron and an increase of the Rb-Sr whole-rock age.

In summary, the leaching experiments point to a surprising coherence between the degree of differentiation and the mobility of Rb and Sr during fluid-rock interaction, which in turn could explain a systematic shift of the Rb-Sr isochrons.

3.2 Leachability and source of REE in fresh and altered basaltic rocks

During recent investigations on smoker fluids at midocean ridges, the source of REE in crustal fluids and the REE mobility in basalts became a subject of increasing interest (e.g. Klinkhammer et al. 1994, Wood & Williams-Jones 1994).

In order to investigate the relative REE availability in basalts under experimental conditions and natural alteration, we performed leaching experiments on a weakly (diabase) and a heavily altered (patch) MORB of the ODP 504B drill core close to the Costa-Rica rift (Alt et al. 1996; Bach et al. 1996). Although the style of alteration is similar in both samples, it differs markedly in intensity. A pronounced loss of "immobile" elements (e.g. the REE) is observed in the heavily altered sample, whereas the major elements do not show a significant change in abundances.

The results of the leaching experiments revealed that in both samples the release of REE is independent from the dissolution of the major minerals (Fig. 3; Bach & Irber subm). The REE are dominantly leached not from the rock-forming minerals but from a so far unspecified REE rich source. Among the REE, only Eu shows a leachability which is controlled both by the release from the above mentioned unspecified source *and* the dissolution of plagioclase. And because of the slow dissolution rate of plagioclase the overall REE

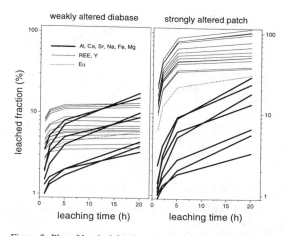

Figure 3 Plot of leached fractions vs. time. The similarities or differences in the leaching curves allow the identification of elements either released from the dissolution of the rock-forming minerals or in the case of REE, Y, and Eu from another unspecified source. For simplification only three element groups are distinguished.

pattern of the solution is characterised by a marked negative Eu anomaly (Fig. 4).

We calculated via Rayleigh fractionation the REE contents of the major minerals to quantify the theoretical amount of REE that can be released from the dissolution of major minerals. The dissolved mineral fractions were derived with respect to the solution's major element chemistry by matrix calculations (Bach & Irber, subm.).

The calculations revealed that the large quantity of REE extracted in our experiments can be only explained if about 10 per cent of residual melt during crystallisation is assumed. The residual melt is strongly enriched in REE (preferably in LREE), but depleted in Eu, as the major minerals discriminate against the REE during crystallisation (except Eu).

The alteration caused a increase in the REE's leachability, but did not change the leached REE pattern at all. We suggest that the alteration only changed the mineralogical composition of the the interstitial material (former residual melt?) and made the REE more susceptible to removal by interacting fluids. This may explain the unexpected and marked loss of the "nominally" immobile elements in the heavily altered sample.

In summary, the leaching experiments demonstrate that in the studied basalts the REE are

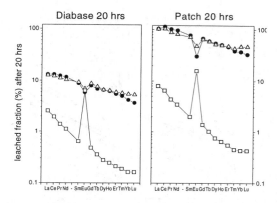

Figure 4 The experimentally leached REE after 20 hrs *(solid black)* are compared to the fraction of REE theoretically leached from the dissolution of major minerals *(open squares)* and the dissolution of major minerals together with interstitial material and apatite *(open triangles)*. Theoretical dissolution of the major minerals only does not explain the leached REE fractions (see text).

dominantly hosted in an unspecified phase (= former residual melt?) which is more soluble than the major rock-forming minerals. We propose that interstitial material in basaltic rocks may in general act as a major easily available REE reservoir for circulating fluids in the oceanic crust.

4 CONCLUSIONS

The leaching experiment provides detailed information about the relative leachability of major and trace elements in whole-rock samples before and after alteration. The experiments do not simulate natural alteration, however, the results allow us to trace the alteration-caused internal redistribution of elements from primary to secondary minerals and to emphasize the individual participation of elements during alteration. The data obtained give new insights for the interpretation of alteration processes and support the quantification of effects that occur during fluid-rock interaction. The given examples demonstrate that the leaching data in combination with classic methods in geosciences are an useful tool to gain new insight into the complex process of element mobilisation.

5 REFERENCES

Alt, J.C., Laverne, C., Vanko, D., Tartarotti, P., Teagle, D.A.H., Bach, W., Zuleger, E., Erzinger, J. and Honnorez, J. (1996) Hydrothermal alteration of a section of upper oceanic crust in the eastern equatorial Pacific: A synthesis of results from DSDP/ODP Legs 69, 70, 83, 111, 137, 140 and 148 at Site 504B. In: Alt, J.C., Kinoshita, H., Stokking, L.B. and Michael, P.J., Eds., *Proc. ODP*, Sci. Results, 148: 417-434.

Bach, W., Erzinger, J., Alt, J.C. and Teagle, D.A.H., 1996. Chemistry of the lower sheeted dike complex, ODP Hole 504B, Leg 148. The influence of magmatic differentiation and hydrothermal alteration. In: Alt, J.C., Kinoshita, H., Stokking, L.B. and Michael, P.J., Eds., *Proc. ODP*, Sci. Results, 148: 39-55.

Bach, W. & Irber, W. (subm.) REE mobility in the oceanic lower sheeted dyke complex: Evidence from geochemical data and leaching experiments. *Chem. Geol.*

Hradetzky, H., Lippolt, H. J. (1993) Generation and distortion of Rb/Sr whole-rock isochrons - effects of metamorphism and alteration. *Eur. J. Mineral.* 5: 1175-1193

Irber, W. (1996) Laugungsexperimente an peraluminischen Graniten als Sonde für Alterationsprozesse im finalen Stadium der Granitkristallisation mit Anwendung auf das Rb-Sr-Isotopensystem. Ph.D. Thesis, FU Berlin, Germany.

Irber, W., Siebel, W., Möller, P., Teufel, S. (1996) Leaching of Rb and Sr ($87Sr-/86Sr$) of Hercynian peraluminous granites with application to age determination. *Journal of Conference Abstracts*, 1(1): 281

Klinkhammer, G.P., Elderfield, H., Edmond, J.M. and Mitra, A. (1994) Geochemical implications of rare earth element patterns in hydrothermal fluids from mid-ocean ridges. *Geochim. Cosmochim. Acta*, 58: 5105-5113.

Möller, P. & Giese, U. (1997) Determination of easily accessible metal fractions in rocks by batch leaching with acid cation exchange resin. *Chem. Geol.*, 137, 1-2: 41-55.

Strong, D. F. (1985) A review and model for granite-related mineral deposits.- In: Recent advances in the geology of granite-related mineral deposits, proceedings of the CIM conference on granite-related mineral deposits. 39: 424-445.

Wood, S.A. and Williams-Jones, A.E. (1994) The aqueous geochemistry of the rare-earth elements and yttrium, 4. Monacite solubility and REE mobility in exhalative massive sulfide-depositing environments. *Chem. Geol.*, 115: 47-60.

Quantitative analysis of high density fluid inclusions

P. Knoll & M. Pressl
Institute of Experimental Physics, Karl-Franzens University, Graz, Austria

R. Abart & R. A. Kaindl
Institute of Mineralogy-Crystallography and Petrology, Karl-Franzens University, Graz, Austria

ABSTRACT: A combined microthermometric and micro-Raman spectroscopic method for the quantitative analysis of fluid inclusions is presented. This method avoids systematic errors due to nonlinear Raman behavior. The improvement on the accuracy of fluid-composition determination is demonstrated in the N_2-CH_4 system. At fluid densities higher than 0.008 mole/cm^3 the accuracy is significantly increased.

1 INTRODUCTION

Many fluid inclusions of geological interest belong to the C-O-H-N chemical system. If present at room temperature, the vapor phase of such fluid inclusions is dominated by the molecular species CO_2, CH_4, and N_2. The quantitative determination of these species is a major issue in metamorphic and applied geology. On the one hand, knowledge of compositions from individual inclusions provides independent information for the reconstruction of the complex P-T evolution of metamorphic terranes, and on the other hand, the chemical characterization of paleofluids is a major concern in the investigation of mineralization. From the chemical compositions of fluid inclusions, paleo-redox states and the dielectric properties of the paleofluid and thus mineral-fluid equilibria and the transport properties of the paleofluid can be derived (Dubessy et al. 1989).

The determination of the proportions of CO_2, CH_4 and N_2 in individual fluid inclusions by microthermometric measurements may be hampered by the fact that the observable phase transitions are too few and that the topologies of the corresponding phase diagrams are not known well enough. In this case, analysis by micro-Raman spectroscopy may be the only option to obtain quantitative compositional information from individual fluid inclusions. Quantitative Raman analyses are commonly based on the linear superposition of the individual fluid components based on the assumption of a linear relation between Raman signal and fluid density. A widely used method of deriving the relative species abundance from Raman signals has been introduced by Placzek (1934).

However, inclusions in minerals are known over a wide density range (0.003 to 0.025 mole/cm^3) and even pure gases of one component are reported to show nonlinear Raman response with density. This is due to non negligible molecule-molecule interactions. Seitz et al. (1993, 1996) have reported, that in CH_4 the Raman signal is nonlinear over the pressure range from 1 to about 700 bars at room temperature (Fig.1). Similar effects occur in CO_2 (especially around the critical density) and even in N_2 (Fig.2), which at room temperature mostly behaves like an ideal gas (Wang et al., 1973). In order to compensate for such effects, sophisticated calibration methods have been used which are based on a direct comparison of an artificial mixture of the same components at a density comparable to that of the individual inclusion (Seitz 1993, 1996; Kerkhof 1993). Such methods are demanding with regard to the preparation of mixtures at several densities and restrict the standard application of a quantitative micro-Raman analysis. Furthermore, it is impossible to produce all the artificial mixtures at exactly the same conditions as found in natural inclusions; this causes an additional systematic error.

In our contribution, we improve Placzek's method by including nonlinear Raman behavior. We present an algorithm that combines micro-Raman investigations with micro-thermometric analyses. Our algorithm requires a parametrization of the binary subsystems that has to be performed only once at any laboratory and a simple standard calibration of the Raman spectrometer. For the N_2-CH_4 system we demonstrate how the accuracy of compositional determinations of fluid inclusions can be improved.

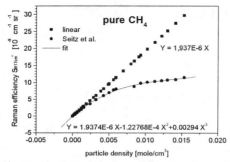

Fig.1: Absolute Raman efficiencies for methane: Idealized linear behavior (Schrötter 1979) and experimental data from Seitz (1993, 1996), scaled to absolute values.

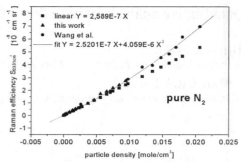

Fig.2: Absolute Raman efficiencies for nitrogen. Idealized linear behavior from Schrötter (1979). Experimental data from Wang (1973) scaled to absolute efficiencies.

2 COMBINED RAMAN SPECTROSCOPY AND MICRO-THERMOMETRY

Raman spectroscopy is based on the inelastic scattering of photons and can be taken as a fingerprint of molecules. Each type of molecule (species a) causes a Raman signal at a characteristic frequency ω_a (energy difference between incoming and scattered photon, $\omega_L - \omega_S$). In the case of a linear approximation, the relative concentration $x = n_a/(n_a+n_b)$ of the two Raman active species, a and b, in the same phase of a fluid inclusion, may be obtained from measurements of the corresponding peak areas, A_a and A_b, of the Raman bands according to the relation originally derived by Placzek (1934).

We develop an algorithm that accounts for nonlinear Raman behavior within a consistent framework using only the measured Raman efficiency $S_a(\omega_a)$, the differential Raman scattering cross section of a molecule $\sigma_a(\omega_a)$, and the number of molecules per volume n_a. The intrinsic microscopic quantity of the Raman process is the differential Raman scattering cross section $\sigma_a(\omega_a)$ of a molecule which is defined like any differential cross section as the ratio of the number of inelastic scattered photons per time and solid angle, and the number of incoming photons per area and time; the units are $cm^2 sr^{-1}$. The corresponding value for the Q branch of N_2 at standard condition (760 Torr, 298 K) is $\sigma_{N2}(2331 cm^{-1}) = 43 \cdot 10^{-32} cm^2 sr^{-1}$ (for 514,5nm laser excitation, Schrötter 1979). Values for other gases can be obtained by comparison with N_2: $\sigma_{CH4}(2917 cm^{-1}) = 3,2 \cdot 10^{-30} cm^2 sr^{-1}$ (Schrötter 1979).

In a real experiment we do not see a single molecule but a certain volume with a large number of molecules. As long as there is no interaction between the molecules the Raman signal increases linearly with the molecule density n_a. Then, the relation of the Raman efficiency (units of $cm^{-1} sr^{-1}$) $S_a(\omega_a) = n_a \sigma_a(\omega_a)$ will be valid for all densities (Shown in Fig.1 and Fig.2). However, molecule-molecule interactions will disturb this simple behavior and we may write for a gas mixture a power series in n_a, n_b, n_c, ... etc.:

$$S_a(\omega_a) = \sum_b f_{ab}(n_a, n_b) n_b + ... \text{ or as a matrix}$$

equation $\vec{S}(\vec{\omega}) = \vec{\vec{f}}_{ab}(n_a, n_b) \vec{n} + ...$ (1)

If we restrict molecule-molecule interactions to two species only, the matrix elements f_{ab} are only power series in two particle numbers. Then, it is a straight forward calculation to invert the matrix f_{ab} and apply an iterative algorithm to obtain the concentrations $n_{a,b,c,...}$ of the individual species from this nonlinear equation. The components of the matrix are polynominyls in the pairs of concentrations n_a, n_b and have to be determined from an artificial gas mixture. Without any interactions, the off-diagonal elements of the matrix f_{ab} vanish and the diagonal elements reduce to the corresponding $\sigma_a(\omega_a)$ values. The experimental determination of the Raman efficiencies $S_a(\omega_a)$ of a Raman band (ω_a) is a standard procedure in Raman spectroscopy. This quantity may be obtained by integrating over the specific Raman band (peak areas) of the proper corrected Raman spectrum.

$S_a(\omega_a) =$

$$\frac{n(\omega_L)}{1-R(\omega_L)} \int_{\omega_a-\Delta}^{\omega_a+\Delta} \frac{n(\omega_s)}{1-R(\omega_s)}(I(\omega)-B(\omega))\eta(\omega_s)d\omega$$

This expression corresponds to the equation given by Pressl et al. (1996) with the exception of neglecting optical absorption $\alpha_L = \alpha_S = 0$, the integration over the Raman band and the explicit consideration of the whole Raman system sensitivity $\eta(\omega_s)$ at the frequency of the scattered light $\omega_s = \omega_L - \omega$. This sensitivity can be obtained by measuring a source with known photon emission, e.g., standard tungsten lamps, black body radiation, etc. At our lab we use the OSRAM WI17G tungsten standard lamp with additional spectral calibration. A scaling in absolute efficiencies (in units of cm^{-1} sr^{-1}) as it is done in this work can be achieved by further comparison with already calibrated signals - as, e.g., the Q branch of N_2 (Schrötter 1979) - but is not required for our analysis. However, in order to compare the experimental data of Fig.1 and 2 with the linear behavior of the Schrötter data we have scaled to absolute efficiencies so that the slopes at low densities merge. The halfwidth Δ of the integration range and the background signal $B(\omega)$ have to be chosen so that all contributions to the Raman signal $I(\omega)$ from one vibration are considered but no contributions from other parts of the spectrum are included.

Eq.2 can be simplified. As a first approximation, the factors which correct for reflectivity $R(\omega_{L,s})$ and refractive index $n(\omega_{L,s})$ can be neglected for gases. If one assumes that the dispersion of the Raman system's sensitivity is negligible within the integration limits of the Raman bands one derives the simple expression:

$$S_a(\omega_a) = \eta_a \int_{\omega_a - \Delta}^{\omega_a + \Delta} (I(\omega) - B(\omega)) d\omega = \eta_a A_a \quad (3)$$

As a sensitivity calibration with a standard lamp is much simpler than the calibration for the sensitivities η_a for each individual Raman band at ω_a we use Eq.2 instead of the simpler form. (We also do not use the method given by Wopenka and Pasteris (1987) who combine the instrumental sensitivity η_a with relative Raman cross sections to the so called quantification factor F as this requires a calibration on each Raman instrument for all individual Raman bands and all polarization directions in order to make the results comparable between different laboratories.)

Absolute Raman efficiencies can not be obtained from fluid inclusions in general. The exact determination of Raman efficiencies in inclusions with complicated surrounding conditions (shape of the inclusion, refractive index, etc.) is far too difficult and is not applicable for standard investigations in geological and mineralogical sciences. Therefore, we can not deal with Eq.1 but have to consider that the efficiencies are only known by a common factor g.

$$\vec{S}(\vec{\omega}) = g \vec{f}_{ab}(n_a, n_b) \vec{n} \quad (4)$$

This additional factor will not affect the determination of the ratio $x = n_a/(n_a+n_b)$ in one iteration step. But the matrix coefficients f_{ab} are functions of the absolute particle concentrations and additional informations as, e.g., Raman frequency shifts, line shapes, etc. are required in order to solve Equ.4. However, a rather simple algorithm can be achieved if we consider the homogenization temperature T_h (observed in micro-

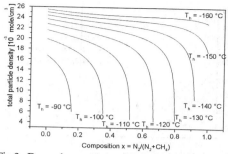

Fig.3: Dependence of the density on the composition for several homogenization temperatures T_h in the binary subsystem N_2-CH_4. (Data from Kerkhof 1988)

thermometry) which relates the relative composition x_i to the total density ρ.

$$\rho = \sum_i n_i = d(x_i, T_h) \quad , \quad x_i = \frac{n_i}{\sum_i n_i} \quad (5)$$

For the binary subsystem N_2-CH_4 the relations $d(x, T_h)$ (Kerkhof, 1988) according to Eq.5 are shown for several homogenization temperatures in Fig.3 and have to be included into the iterative method in order to solve Eq.4. First the relative composition x is determined by Raman analysis with a first guess of the absolute concentrations n_i of the inclusion. (It is assumed that the Raman investigations are performed at a temperature above the homogenization temperature with the same density of the inclusion as at T_h. For most cases this will be true for room-temperature Raman analyses.) Secondly, the homogenization temperature is determined by micro-thermometric measurements. With the help of the relations $d(x_i, T_h)$ for the specific system the density ρ of the inclusion is determined. Then the absolute concentrations $n_i = \rho x_i$ are obtained and enter into the next iteration step for the Raman analysis.

3 GEOLOGICAL APPLICATIONS

The relevant question for the application of this method to geological problems is how the accuracy can be improved. The accuracy of the Raman analysis is determined by two factors: First, the systematic error of nonlinear Raman response in a gas mixture with molecule-molecule interaction, and second, the accuracy of the Raman cross sections of the individual molecules at standard conditions. A comparison of such Raman cross sections at standard conditions for a collection of different gases has been given by Schrötter and Klöckner (1979) with an accuracy of 10%. In a binary system this standard error of 10%

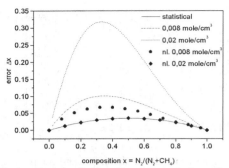

Fig.4: Error of the Raman analysis of fluid compositions in the N_2-CH_4 binary system: Statistical limit (full line), Standard Raman analysis (broken lines), our nonlinear method (symbols).

produces a statistical error in the determination of the composition as plotted in Fig.4 (full line).

In order to estimate the error of the Raman analysis with respect to nonlinear Raman behavior we consider the N_2-CH_4 subsystem. We neglect off diagonal elements in the f_{ab} matrix and for the diagonal elements we take the functions obtained from the fits of the experimental data shown in Fig.1 and Fig.2. First, we calculate the systematic error Δx if the standard Raman method with a linear Raman response is applied instead of our non linear method.

In Fig.4 the sum of the systematic error and the statistical error is plotted for two different densities. At rather low densities the deviation of the assumed linear behavior from the actual one is small and, therefore, the total error in standard Raman analysis is only increased by a small amount. At higher densities the situation changes dramatically and the total error is dominated by the non-linearity effects. If we apply our nonlinear method we avoid this systematic error but we have to consider an additional statistical error in the homogenization temperature T_h, which we assume to 1°C. We calculate the total error as shown in Fig.4 for two different densities (diamonds and full circles). At high densities the total error of our non linear method matches the statistical limit of a binary system and improves the standard Raman analyses significantly. At low densities the additional statistical error in T_h grows up dramatically because of the steep slopes in Fig.3 and the resulting huge inaccuracy of the total density.

4 REFERENCES

Dubessy, J., B. Poty and C. Ramboz 1989. Advances in C-O-H-N-S Fluid Geochemistry Based on Micro-Raman Spectrometric Analysis of Fluid Inclusions. *Eur. J. Mineral.* 1: 517-534.

Kerkhof, A.M.v.d. 1988. *The System CO2/CH4-N2 in Fluid Inclusions: Theoretical Modelling and geological Applications.* Theses, Free University Press, Amsterdam.

Kerkhof, A.M.v.d. & H.J. Kisch 1993. CH_4-rich inclusions from quartz veins in the Valley-and-Ridge province and the anthracite fields of the Pennsylvania Appalachians - Reply. *Amer. Min.* 78: 220-224.

Placzek, G. 1934. Rayleigh-Streuung und Ramaneffekt. In Marx, E. (eds), *Handbuch der Radiologie*, Leipzig, Akademische Verlagsgesellschaft, p. 205-374.

Pressl, M., Mayer, M., Knoll, P., Lo, S. Hohenester, U. and E. Holzinger-Schweiger 1996. Magnetic Raman Scattering in Undoped and Doped Antiferromagnets. *J. Raman Spec.* 27: 343-349.

Schrötter, H.W. & H.W. Klöckner 1979. Raman Scattering Cross-Sections in Gases and Liquids. In Weber, A. (eds), *Raman Spectroscopy of Gases and Liquids*, p. 123-166.

Seitz, J.C., J.D. Pasteris and I.M. Chou 1993. Raman spectroscopic characterization of gas mixtures. 1. Quantitative composition and pressure determination of CH_4, N_2 and their mixtures. *Am. J. Sci.* 293: 297-321.

Seitz, J.C., J.D. Pasteris and I.M. Chou 1996. Raman spectroscopic characterization of gas mixtures. 2. Quantitative composition and pressure determination of the CO_2-CH_4 system. *Am. J. Sci.* 296: 577-600.

Wang, C.H. & R.B. Wright 1973. Effect of density on the Raman scattering of molecular fluids. I. A detailed study of the scattering polarization, intensity, frequency shift and spectral shape of gaseous N_2. *J. Chem. Phys.* 59: 1706-1712.

Wopenka, B., J.D. Pasteris and J.J. Freeman 1987. Raman intensities and detection limits of geochemically relevant gas mixtures for a Laser Raman micro-probe. *Analyt. Chem.* 59: 2165-2170.

Modified set-up for column experiments to improve the comparability of water-rock interaction data: Column cap and hydraulic control system

D. Lazik
UFZ Centre for Environmental Research Leipzig-Halle, Department of Hydrogeology, Germany

ABSTRACT: Column experiments are a commonly used tool for the characterisation of hydrogeological or geochemical processes. However, it is crucial that the experimental data obtained be comparable with natural systems. Furthermore, it is difficult to compare column experiments between different laboratories. To overcome this problem, a special column cap was developed which enables both the utilization of undisturbed samples and the standardization of column size as a criterion for comparability.
In order to understand the hydraulic behaviour of different phases (groundwater, oil and gas) and the combined transport-interaction behaviour of migrants in deep aquifer systems, certain processes must be simulated separately at the laboratory scale. However, as the special deep aquifer properties are difficult to simulate using active technical control systems, a passive hydraulic control system was developed.

1 INTRODUCTION

The dynamic behaviour of hydrogeological systems with known solid and liquid phase composition can be simply tested in column experiments, for instance by changing boundary conditions. Consequently, column experiments are a widespread tool for characterising hydrogeological or geochemical processes. One critical aspect is ensuring the comparability of the experimental data with natural systems. Simple column equipment was developed to overcome this difficulty.

2 COLUMN CAP

Column experiments are usually prepared by sampling sediments, transferring them to the lab and compressing them layer by layer into the columns. However, this procedure upsets the structure, texture and biogeochemistry of natural sediment-water systems (e.g. Reynolds, 1993, Williams, & Higgo, 1994). In order to side-step this problem, recent research has focused on replacing disturbed column structures with undisturbed ones. Obviously, in the case of undisturbed column experiments, simulation gets as close as possible to site-specific natural conditions, and parameter ranges and model sensitivity can thus be estimated similar to specific site conditions. Previously, however, filling columns with undisturbed sediment cores presented an insoluble problem, especially when drilling was used.
The conventional procedure comprises the following steps. At the sampling site the drill cores ("liners") are temporarily sealed by caps; after transferring them to the laboratory the liners are manipulated again (e.g. by pushing the soil cores into prepared columns and flushing them with gas or water). Soil labs have developed special column-handling procedures based on their own experience. Hence, columns differ in size, diameter and length. Differently shaped column walls and water dispersion systems have been used in order to implement 1D flow through the sediment and to avoid bypass-flow at the column walls. Therefore comparing column experiments performed by different laboratories is difficult.
A water- and air-proof column cap (Fig. 1) was developed in order to support both the utilization of undisturbed samples and the standardization of column size as criteria for comparability between aquifers and the lab, as well as among different labs (Lazik et al. 1996).

2.1 CONSTRUCTION OF THE CAP

Fig. 1 shows the column cap. It is divided into a counter-ring (5) and a cap (7) which have to be fixed on the liner (2) by means of a seal ring (4) by tightening the screws (6). A porous sheet (3) is positioned between the soil sample (1) and the disperser (9). A small seal ring (4) between the porous sheet (3) and the disperser (9) can also be used to avoid soil losses. Fittings (8) for fluid entry are situated on the cap forehead and if required on the cap range. Inert materials are used for all segments of the column cap which come in contact with the sediment-water system. Porous sheets with

Legend:
1) Soil sample
2) Liner [a]
3) Porous sheet [b]
4) Seal ring [d]
5) Counter ring [a]
6) Screw/foot [c]
7) Cap [a]
8) Fitting [a]
9) Disperser [a]

Materials:
a) PVC, Teflon, stainless steel
b) Teflon, PP, PE, glass, etc.
c) Stainless steel
d) Rubber, etc.

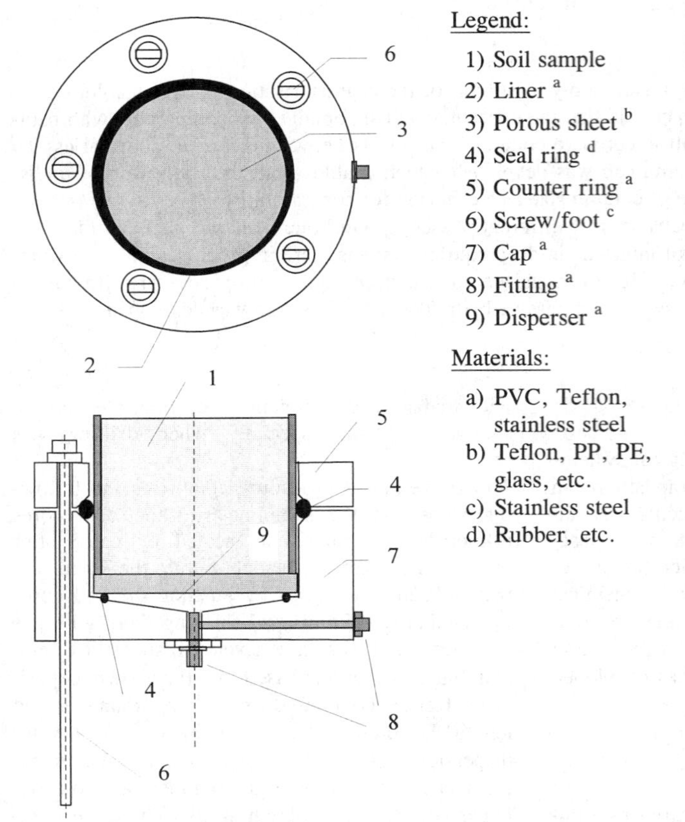

Figure 1. Construction of column cap.

different pore sizes reduce clogging effects or losses of fine-grained sediment. Different surface properties of the sheets in combination with the pore size control the phases' movement through the column in multiphase flow experiments. For example Teflon as a sheet material is very hydrophobic, polypropylene (PP) is slightly hydrophobic, and glass is hydrophilic.

2.2 USAGE AND EXPERIENCE

The column cap is generally adaptable to any pipe diameter. Therefore the cap can also be used for lysimeters. In particular the simple, fast and safe construction of columns from undisturbed drilling cores enables on-line handling with drilling in the field. Because drilling only uses a few liner sizes, the heterogeneity of column construction and size can be reduced to typical liners. Moreover, the field construction of columns means that the sediment is only subject to the disturbance caused by drilling work in the field; all other mechanical and chemical effects normally suffered by the samples via manipulation in the laboratory are eliminated. Additional interactions between the sample and the air are minor and can be further reduced by flushing the column with gas (N_2, Ar). (Lazik et al. 1996)

The study of leaching behaviour of dangerous waste is important for safety assessment, e.g. for radioactive deposits, mine tailings or chemical industry dumps. The method described here enables observation to be carried out without risk for personnel or the lab and thus usually without the need for involved lab-safety regulations. As the columns are erected at the deposit, toxic material is encapsulated throughout the entire duration of transfer and observation.

The high-pressure usage of erected columns (for instance to estimate the hydraulic conductivity of clays or to save the handling of toxic material) is possible simply by tightening both caps between the screws with an elastic band.

The column caps can be reused once the experiments are complete. They are reasonably priced, simple to fit, and improve the safety of sampling and column handling.

Nearly 50 columns augmented by the cap developed and fitted with 100 μm (pore-size) PE-filter sheets have been used for a year in a long-term study of the saturated and unsaturated coupled hydrogeochemical behaviour of two lignite dump profiles. The sample profiles contain fine-grained sands with a high silt and clay mineral content. So far no clogging of the columns or sediment loss has been observed. Another experiment uses hand hand-cut liners to observe the sorption-desorption behaviour of organics on lignite-rich sediments. Cutting the drill cores by hand means the liners have a very rough face. Nevertheless, despite the rough edges the columns remain watertight and in this experiment too were found to function smoothly.

3 HYDRAULIC CONTROL SYSTEM

Deep groundwaters contain high concentrations of dissolved substances, have high temperatures, and owing to the enormous hydrostatic pressure high gas levels (especially CO_2) are typical. The equilibrium of calcite with carbonic acid plays an important role in the geochemistry of such waters. Sampling groundwater upwards causes CO_2 to escape and calcite to be precipitated. The redox potential and pH change rapidly. Moreover, hydroxides are precipitated in contact with air. This means the chemistry of the water changes rapidly after sampling and chemical analysis is difficult or even impossible. Multiphase movement for different (p,T)-conditions is a second unsolved problem affecting for example oil/gas exploration, resource management or the long-term contamination capacity of deep aquifers by DNPLA (dense non-aqueous-phase liquids). Therefore, in order to understand the hydraulic behaviour of different phases and/or the combined hydrogeochemical behaviour of migrants in deep aquifer systems, certain processes must be simulated separately at laboratory scale. However, the special deep aquifer properties such as high hydraulic potential, low hydraulic gradient and long-term flow stability are difficult and expensive to simulate using active technical control systems (usually used in low potential column experiments). Therefore a passive hydraulic control system (Fig. 2) was developed (Lazik & Schneider, 1996).

3.1 CONSTRUCTION OF THE HYDRAULIC CONTROL SYSTEM

The system consists of two potential bottles, A and B, which are connected by both a fluid line and a pressure (or potential) line. The bottles are fixed at different levels in order to create a hydraulic gradient $\Delta h/l$ over the column (length: l) which is arranged in the fluid line. Both bottles are identical in construction but can differ in size. The sample bottle B is rotated by 180° so that the potential line which is connected for example to a pressure-gas bottle generates the same high potential in the closed system. Consequently its potential at the height of the bubble level of the tube h_A in bottle A is equal to that at the drop formation level h_B in bottle B; as $\Delta h=0$, the fluid phases cannot move (Lazik & Schneider, 1996). A number of valves are used to couple the system, or to change the fluids in the bottles or the elements of the system.

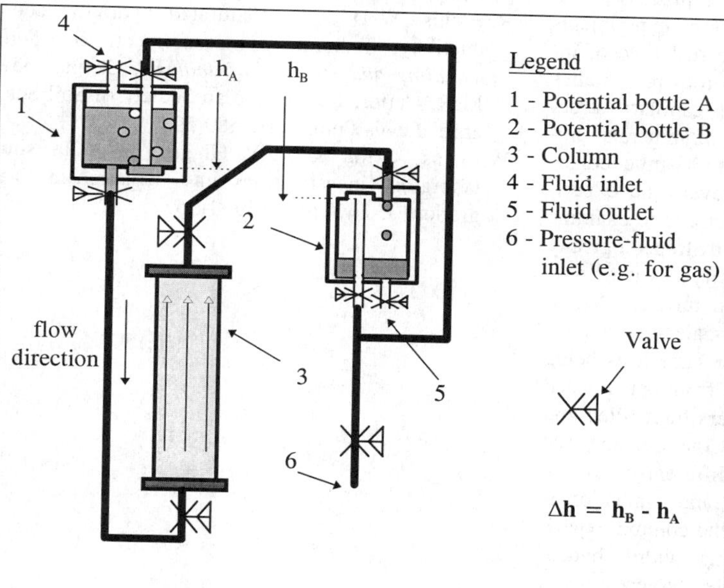

Figure 2. Hydraulic control system.

Legend
1 - Potential bottle A
2 - Potential bottle B
3 - Column
4 - Fluid inlet
5 - Fluid outlet
6 - Pressure-fluid inlet (e.g. for gas)

$\Delta h = h_B - h_A$

3.2 APPLICATION

Because of the uncoupled regulation mechanisms of both the hydraulic potential (by a pressure fluid) and the hydraulic gradient (by different levels in the potential bottles A and B), the control system is very precise. The only restriction is that both fluids must be immiscible. Of special interest is that the content of dissolved gases (e.g. CO_2) can be chosen by the composition of the pressure fluid with respect to the pressure used.

The hydraulic control system features a simple design, is easy to use, reasonably priced, and is highly accurate irrespective of external changes (e.g. air pressure and temperature).

4 SUMMARY

Over the past few years, the sphere of usage of column experiments has both widened and shifted. In addition to gaining a fundamental grasp of phenomena, rising importance is being attached to balancing and understanding complex processes and interactions, meaning that column experiments have become closely linked to the real field of investigation. Therefore, not only porous rock bodies but also other properties such as capillary structure, the geochemical behaviour of minerals, microbiology and the actual thermodynamic conditions must be reproduced. However, the experimental and technical apparatus required to examine samples in a natural state under hydrogeologically relevant conditions in the laboratory has until now been largely unavailable. This in turn has led to enormous variety concerning the scale and methods of experiments, with virtually the same tests being carried out using set-ups ranging from pencil-sized columns to columns several meters high filled and compacted in different ways. Consequently, the important findings obtained by different groups of researchers cannot be compared and evaluation is impossible. The significance of the column experiment suffers from this dilemma within hydrogeological research: although urgently necessary to help assess environmental issues, its application to is only being slowly recognized by state authorities.

The low-cost, user-friendly experimental aids we developed raised the quality of the experimental equipment in our laboratory to the high standards required by scientific research. Moreover, the possibility of obtaining high-grade rock body almost completely unaffected in pedomechanical, geochemical or hydrogeological terms by combining the proven coring technique with the column cap resulted in an experiment whose performance is closely related to the relatively uniformly dimensioned core-boring techniques. We are thus convinced that the obvious advantages of using the column cap will definitely result in the standardization of column experiments.

5 REFERENCES

Lazik, D., M. Sander, E. Seifert, Th. Schneider, 1996. Säulenverschlußkappe und Verfahren zur Vorbereitung und Durchführung von Säulenversuchen. *International Patent Application PCT / EP 97 / 02884,* European Patent Office DG 1 - Berlin sub-office, 04. 06. 96.

Lazik, D. & Th. Schneider, 1996 Anlage und Verfahren zur Durchführung von Säulenuntersuchungen unter hydrostatischem Druck. *International Patent Application PCT / EP 97 / 04085*, European Patent Office, DG 1 - Berlin sub-office, 01. 08. 96.

Reynolds, W.D. 1993. Saturated hydraulic conductivity: Laboratory measurement. In: *Soil sampling and methods of analysis*, Chapter 55, M.R. Carter, Ed., Canadian Society of Soil Science. Lewis Publishers. 589-598.

Williams, G.M. & J.J.W.Higgo, 1994. In situ laboratory investigations into contaminant migration. *J. Hydology* 159, 1-25.

Solubility of Platinum in aqueous fluids buffered by manganese oxides

G.G. Likhoidov, L.P. Plyusnina & J.A. Scheka
Far East Geological Institute, Vladivostok, Russia

ABSTRACT: The solubility of metallic Pt in water and aqueous chloride solutions was determined at 300 to 500°C, and 1 kb total pressure at oxidation states buffered by the $MnO-Mn_3O_4$, $Mn_3O_4-Mn_2O_3$, and $Mn_2O_3-MnO_2$ mixtures. The HM and NNO traditional buffers were used for comparison. The total Pt content in quenched solutions varied from 90 ppb to 320 ppm. The Pt solubility increases with rise in fO_2 most of all, compared to the increase with both temperature and acidity. At studied temperatures, the measured high solubility of platinum at $logfO_2$ exceeding about -12 (in chloride free solutions) may be connected with a stable existence of Pt(IV) species. So, the highly-oxidized fluids buffered by an appropriate Mn assemblage, for example, are capable of transporting a significant amount of platinum.

The presence of marked Pt concentrations in ferromanganese crusts (Halbach et al., 1984, Govorov et al., 1995), unconformity-related uranium deposits (Jaireth, 1992), and uranium-bearing coal deposits (Smariovich et al., 1992) confirm low temperature mobility of Pt in some geological settings. Most probably, in all the cases Pt was mobilized during water-rock interaction. The last is confirmed by values of the Ni/Pt ratio in seamount crusts and underlying basalt (Halbach et al., 1984, Govorov et al., 1995). To better understand possible hydrothermal mobilization of Pt we have studied its solubility at 300-500°C (1 kb total pressure) in the presence of some mixed manganese oxides. Mn, as a transition metal, is capable of existing in a number of different oxidation states ($2^+, 3^+, 4^+, 6^+$, and 7^+). The 2^+ and $4+$ oxidation states are the lowest and highest that have been observed usually in nature (Brookins, 1987). Hence, the most appropriate for study are MnO, Mn_3O_4, and Mn_2O_3.

Experiments were conducted in externally heated autoclaves produced of Ti alloys using a furnace with an accuracy better than ±5°C. Internal pressure equal to 1 kb was mainted by adding appropriate amounts of distilled water, in accord with P-V-T relations. As a source of Pt we used Pt ampoules (9·80·0.2 mm) welded after charging. The charge included starting solution (water, 1mNaCl, 0.1mHCl) along with 200-250 mg of oxygen buffer (powdered reagent grade: $MnO-Mn_3O_4$ (Mn1), $Mn_3O_4-Mn_2O_3$ (Mn2), and $Mn_2O_3-MnO_2$ (Mn3)) per run. Some runs were carried out in the presence of traditional $Fe_2O_3-Fe_3O_4$ (HM) and Ni-NiO (NNO) fO_2-buffers for comparison. Thus, an immediate contact between the Pt, solution, and buffer mix was achieved in the runs. X-ray was used for control over the buffer end composition. The quenched solutions were diluted with 0.1mHCl immediately after opening, carefully filtered, and specially adapted to the analysis. The Pt content was measured using atomic absorption spectroscopy with an accuracy about ±20%. Experimental results as well as run duration are given in Table 1. The runs duration was defined by a special kinetic series (Fig.1).

X-ray analysis of resultant Mn1 phase composition has revealed that, at 300°C, MnO is oxidized to Mn_3O_4 with admixture of β-MnOOH (the main d/n - 4.63, 2.66, 2.36 Å) in water, and of $Mn_2(OH)_3Cl$ (d/n - 5.75, 5.40, 4.44, 2.98, 2.91, 2.39 Å) - in chloride solutions.

Their formation can be described as:
$4MnO + 0.5H_2O + 0.75O_2 = Mn_3O_4 + \beta MnOOH$,
and $2MnO + HCl + H_2O = Mn_2(OH)_3Cl$.

Thus, the Mn1 buffer transforms to Mn2 one through the intermediate composition. The transformation is accompanied by a rise in Pt solubility about 1-2 orders of magnitude, and by precipitation of tiny platinum globules (Pt_m) upon quenching, identified optically and by the X-ray. In run products, the globules are present in a mixture with Mn_3O_4, β-MnOOH, and $Mn_2(OH)_3Cl$ (see Table 1). The metastable character of β-MnOOH at experimental study of Mn(II) oxidation was established formerly (Murray et al., 1985). Kinetic runs at 300°C herein also support its metastable formation. Thus, metastable coprecipitation of the Pt_m with Mn hydroxides is obvious. Similar processes are pointed out on the ocean floor, where the metastable formation of Mn-bearing hydroxides is usual (Novikov, 1996). Widespread hydrohausmannite was found to be a mixture of Mn_3O_4 and β-MnOOH rather than a single phase, and generation of the latter is accompanied by an essential increase in alkalinity of the solutions (up to pH about 10) and concomitant jump of Eh (Bricker, 1965).

At 500°C, the MnO_2 is unstable, being reduced to Mn_2O_3 in 72 hours. So, Mn_3O_4-Mn_2O_3 mix is the more appropriate buffer for all studied temperatures. However, values of total platinum concentrations (Pt_{aq}) in the runs with starting Mn2 buffer are somewhat less in comparison with the equilibrium ones (Fig.2). In a study of the Au-Mn-Cl-H_2O system, an analogous phenomenon was described (Zotov et al., 1990).

On this evidence, the transition of Mn1 into Mn2 buffer, at 300°C, and of Mn3 into Mn2, at 500°C, can be used as a control of approach to the equilibrium position from below and above, correspondingly.

So, the approach to the equilibrium is confirmed by results of kinetic runs, similarity in the measured values of Pt solubility at f_{O2} equivalent to the Mn1 and HM buffers, and by regular variations of the C_{Pt} values in dependence on f_{O2}, including runs in the presence of the NNO buffer (Fig.1).

Table 1 illustrates a slight increase in Pt solubility in the chloride solutions in comparison with the chloride-free ones, more pronounced in the presence of 0.1mHCl. At 500°C, the maximum value of Pt_{aq} (320 ppm) in such solution buffered by the Mn3 is fixed. The Pt contents obtained in aqueous solutions reveal a visible difference between reduced and oxidized conditions. As a result, an intersection of the solubility isotherms in the $\log C_{Pt}$-$\log f_{O2}$ diagram at $\log f_{O2}$ about -12 is observed (Fig.2). The intersection may be connected with change of oxidation state from Pt(II) to Pt(IV) species

Table 1. The Pt concentrations (C_{Pt}, gr/l) in the quenched solutions

Initial solution	$\log C_{Pt}$	N	Duration, hrs	Buffer
300°C				
H_2O	-1.57	2	168	Mn1*
H_2O	-3.64	4	330	Mn1**
H_2O	-3.58	2	330	Mn2*
H_2O	-3.32	3	248	Mn3
H_2O	-5.19	12	168-330	NNO
H_2O	-4.03	4	125-248	HM
1mNaCl	-3.26	2	331	Mn1**
0.1mHCl	-1.56	3	168	Mn1*
0.1mHCl	-3.26	2	331	Mn1
0.1mHCl	-3.19	2	331	Mn2
0.1mHCl	-1.83	3	331	Mn3
400°C				
H_2O	-3.59	3	188	Mn1
H_2O	-3.43	2	188	Mn2
H_2O	-2.50	3	168	Mn3
H_2O	-4.50	9	168-175	NNO
H_2O	-3.53	4	168	HM
1mNaCl	-3.41	3	175	Mn1
1mNaCl	-1.39	2	175	Mn3
0.1mHCl	-3.17	3	175	Mn1
0.1mHCl	-3.37	4	175	Mn2
500°C				
H_2O	-3.81	3	108	Mn1
H_2O	-3.67	2	108	Mn2
H_2O	-2.05	3	72	Mn3
H_2O	-4.57	8	72-108	NNO
H_2O	-3.84	2	168	HM
1mNaCl	-3.39	2	108	Mn1
1mNaCl	-3.70	2	108	Mn2
1mNaCl	-1.38	3	72	Mn3
0.1mHCl	-3.22	3	120	Mn1
0.1mHCl	-2.30	2	120	Mn2
0.1mHCl	-0.495	2	72	Mn3

Newly formed phases: * - $Mn_3O_4 + \beta MnOOH \pm Mn_2(OH)_3Cl + Pt_m$, ** - the same without Pt_m, N - number of runs.

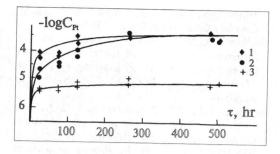

Fig.1 Plot of Pt contents (gr/l) in aqueous solutions as a function of time (P_{tot} = 1 kb). In the presence of - HM (1), Mn1 (2), at 400°C, and of NNO (3), at 300°C.

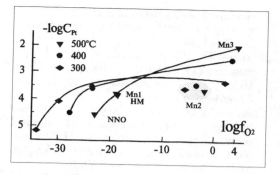

Fig.2 f_{O_2}-T diagram of the Pt_{aq} content (gr/l) in the quenched chlorine-free solutions.

The log f_{O_2} values were calculated using (Chou I-Ming, 1987); the region of invalid Mn2 buffer is shaded.

in the reduced, and oxidized media, respectively. In a series of publications (Mountain & Wood, 1988, Wood et al., 1992) thermodynamic calculations have been presented on the solubility, transport and deposition of Pt in aqueous solutions. They suppose that hydroxide Pt(II) complexes may be dominant species in oxidized surficial waters such as seawater. According to Mountain and Wood (1988) a relatively oxidizing and acidic fluid would be required to transport 10 ppb or 1 ppm of Pt as chloride complexes near 300°C. In our runs, essentially higher Pt contents were obtained in chloride solutions buffered by the MnO_2-Mn_2O_3 assemblage (Table 1). This discrepancy may be explained by Pt(IV) complexing under these conditions. Jaireth (1992) predicted that Pt(IV) species may dominate in highly oxygenated solutions especially under sea-floor weathering regime, where the Mn oxide minerals (e.g. Mn_3O_4/Mn_2O_3) buffer oxidation state at a higher level. According to data of Gammons et al. (1992) the Pt(IV) chloride complexes are stable with respect to Pt(II) ones over a range of f_{O_2}-pH conditions which narrows quickly with increase in T. So, at 300°C, the location of the aqueous Pt(II)/Pt(IV) boundary in neutral solutions is defined by $logf_{O_2}$ values about -6. The experimental study of Gammons (1995) indicates that natural occurrence of the Pt(IV) chloride complexes will be restricted to highly oxidized conditions in the T range from 25 to approximately 150°C. Our study points out that high oxygen fugacity stabilize the Pt(IV) species towards higher T. According to

experimental data (Gammons, 1995, Plyusnina et al., 1995), relatively high solubility of Pt in hydrothermal fluids occurs at highly oxidized (>Mn_3O_4/Mn_2O_3) and high acidity (pH<4) conditions. Pt dissolved as chloride complexes may be deposited in response to reduction, increase in pH, and/or temperature decrease. Of these, reduction is the most effective mechanism and explains the common association of anomalous Pt with strongly reduced horizons within sedimentary basins, ones enriched in organic matter, and sea-floor deposits.

Pt enrichment in ferromanganese nodules is readily explained as due to Pt(IV) species existence (Brookins, 1987). The occurence of hydrothermal manganese plumes (Klinkhammer et al., 1985, Horibe et all., 1986) confirms the hydrothermal origin of some Mn compounds.

ACKNOWLEDGMENTS

The authors would like to thank the anonymous reviewer for constructive critical comments, T.B.Aphanas'ieva for careful X-ray analysis. This research was supported by the Russian Fond of Fundamental Researches (Grant 96-05-64088).

REFERENCES

Bricker O. 1965. Some stability relations in the

system Mn-O_2-H_2O at 25°C and one atmosphere total pressure. *Amer. Mineral.* 20:1296-1354.

Brookins D. 1987. Platinoid element Eh-pH diagrams (25°C, 1 bar) in the systems M-O-H-S with geochemical applications. *Chem. Geol.* 64:17-24.

Chou I-Ming. 1987. Oxygen buffer and hydrogen sensor techniques at elevated pressures and temperatures. In H.P.Eugster (ed.) *Hydrothermal experimental techniques.* 61-99. J.Wiley & Sons. New York.

Gammons C.H. 1995. Experimental investigations of the hydrothermal geochemistry of platinum and palladium: IV. The stoichiometry of Pt(IV) and Pt(II) chloride complexes at 100 to 300°C. *Geocim. Cosmochim. Acta.* 59:1655-1667

Gammons C.H., M.S. Bloom, and Y.Yu 1992. Experimental investigation of the hydrothermal geochemistry of platinum and palladium. *Geochim. Cosmochim. Acta.* 57: 2469-2479.

Govorov I.N., K.F.Simanenko, V.P.Simanenko 1995. Pacific Co-Pt-rich manganese crusts and auriferous ores as a result of water-rock interaction. In Kharaka & Chudaev (eds.), *WRI-8 Water-Rock Interaction*: 707-709. Rotterdam: Balkema.

Halbach P., D.Pitenaus, & F.Manheim 1984. Platinum concentration in ferromanganese crusts from the Central Pacific. *Naturwissenchaten.* 71:577-579.

Horibe Y., K.R.Kim, & H.Craig 1986. Hydrothermal plumes in the Mariana back-arc spreading center. *Nature.* 324:131-133.

Jaireth S. 1992. The calculated solubility of platinum and gold in oxygen-saturated fluids and the genesis of platinum-palladium and gold mineralization in the unconformity-related uranium deposits. *Miner. Deposita.* 27:42-54.

Klinkhammer G., R.Rona, M.Greaves, & M.Elderfield 1985. Hydrothermal Mn plumes in the Mid-Atlantic Ridge rift valley. *Nature.* 314:727-731.

Mountain B.W. & S.A. Wood 1988. Chemical controls on the solubility, transport and deposition of platinum and palladium in hydrothermal solutions: a thermodynamic approach. *Econ. Geol..* 83: 492-510

Murray J, J.Dollard, R.Givanoli, H.Moers, & V.Stumm 1985. Oxidation of Mn(II): initial mineralogy, oxidation state and aging. *Geochim. Cosmochim. Acta.* 49:463-470.

Novikov G.V. 1996. Oceanic Fe-Mn formations - sorbents of metal ions: chemi-mineralogical aspects. *Proceedings of Russian Mineralog. Soc.* Pt.CXXV:No3:38-49. (in Russian).

Plyusnina L.P., I.Ya.Nekrasov, and J.A.Scheka 1995. Experimental study of platinum solubility in aqueous-chloride solutions at 300-500°C and 1 kb. In *Reports of Russian Academy.* 340: No 4: 525-527. (in Russian).

Smariovich E.M., B.I.Natalchenko, K.G. Brovnin et al. 1992. The behavior of noble metals in ore-forming infiltration processes. *Izvestia AN, geol.ser.* 9:121-132. (in Russian).

Wood S.A., B.W. Mountain & P. Pan 1992. The aqueous geochemistry of platinum, palladium and gold: recent experimental constraints and a re-evaluation oh theoretical predictions. *Canad. Miner.* 30: 955-982.

Zotov A.V., N.N.Baranova, T.G.Dar'ina, & L.N.Bannykh 1985. Stability of Au(OH)$°_{aq}$ specie in water at 300-500°C and pressures from 500 to 1500 bar. *Geochimia.* 1:105-109. (in Russian).

Semiquantitative measurements of CO_2 gas in liquid-rich inclusions by laser Raman microspectroscopy

S. Maeda & H. Takagi
West Japan Engineering Consultants (WestJEC), Fukuoka, Japan

S. Taguchi
Fukuoka University, Japan

K. Sanada
New Japan Environmental Measurements, Fukuoka, Japan

M. Hayashi
Kyusyu Sangyo University, Fukuoka, Japan

M. Sasada & T. Sawaki
Geological Survey of Japan, Tsukuba, Japan

T. Fujino & T. Uchida
New Energy and Industrial Technology Development Organization (NEDO), Tokyo, Japan

ABSTRACT: The detection limits and F-values for N_2, CO_2, CH_4 and H_2S gases in the laser Raman microspectrometry have been determined utilizing the RAMANOR T64000 microprobe. The major factors which affect the spectrum intensity were also examined using a 25 mol. % CO_2 standard, synthesized inclusion. Very wide fluctuations in the spectrum intensity of the gas were reduced down to relative error 14 % as a result of normalizing the CO_2 using the H_2O water intensity in the same fluid inclusion. This method can be applied to liquid-rich fluid inclusions for the semiquantitative measurements of CO_2 gas.

1. INTRODUCTION

Fluid inclusions are commonly found in hydrothermal minerals formed in conduits of geothermal fluids. The homogenization temperature of liquid-rich inclusions has been used to estimate underground temperature in almost all the geothermal fields in Japan.

On the other hand, it is difficult to detect the kind and the content of gases in fluid inclusions, which are useful for a better understanding of the nature and the evolution of a geothermal system. The discovery of Raman effect by C. V. Raman and K. S. Krishnan (1928) made detection practically possible. NEDO (New Energy and Industrial Technology Development Organization in Japan) has been carrying out on extensive project to develop prospective technologies for exploring "fractured-type geothermal reservoirs" (Taguchi et al., 1996; Horikoshi et al., 1996). In this project we carried out a study of the laser Raman microscopy of fluid inclusions.

2. ANALYTICAL TECHNIQUE

The laser Raman microprobe used for this study is the RAMANOR T64000 by Jobin Yvon. A laser beam focuses down to one micrometer precision with high Raman intensity reflection. The laser produces 200 mW power. A spectrum intensity is the mean of three measurements of 180 seconds.

The detection limits of the instrument are 0.12, 0.15, 0.04 and 0.07 bars for N_2, CO_2, CH_4 and H_2S.

These values are close to the detection limits reported by Wopenka and Pasteris(1987); 0.85, 1.33, 0.34 and 0.35 atm for RAMANOR U-1000. F-values for N_2 gas standard were determined with a glass tube of twenty micro meters thickness. The F-vales of pure gases are 7.10±0.08, 1.42±0.02 and 7.11±0.02 for H_2S, CO_2 and CH_4. The F-values of mixed-gases are 7.60±1.3, 1.48±0.04 and 8.10±0.97 for H_2S, CO_2 and CH_4. The F-vales of the pure gases are consistent within the error ranges with those of the mixed-gases.

3. RAMAN INTENSITY IN STANDARD CO_2 GAS

The intensity of the CO_2 band fluctuated widely among fluid inclusions, even in the CO_2 25 mol. % standard synthesized inclusion (Sterner and Bodnar, 1984) with a relative standard deviation of about 120 % (Fig. 1). Such a wide range was not observed in the tube experiments for F-values and detection limits.

Accordingly, the wide range of intensity may be caused by the optical paths in a host mineral of the fluid inclusion and by the conditions of the fluid inclusion itself.

There are many factors that affect the spectrum intensity:

(1) The size of the gas bubble in a fluid inclusion is roughly proportional to the Raman intensity, in spite of the small beam diameter.

(2) The depth from the polished surface to a fluid

inclusion is roughly inversely proportional to the spectrum intensity: Intensities of Raman reflections are different when measured from opposite sides of the polished mineral.

(3) Raman intensity changes according to the optical orientation of a host mineral. The intensity of a diagonal position in a host mineral is stronger than that measured at extinction position. The four intensities measured at diagonal position are not always the same.

(4) Raman intensity in the vapor phase of fluid phase inclusion changes depending on the analyzed spot.

These major factors, inducing intensity differences in the laser Raman analysis, were found to be dependent on the size of fluid inclusions, the distance between the mineral surface to fluid inclusions, the optical orientation of a host mineral, the form of fluid inclusions, etc..

4. NORMALIZATION OF CO_2 LASER RAMAN INTENSITY

Although the spectrum intensity of a substance with a given composition fluctuates widely, such as the CO_2 gas in the standard, the Raman spectrum intensity can be expected to behave similarly among all the substances in one fluid inclusion (Bodnar et al., 1996). Besides, the H_2O in liquid-rich inclusions from geothermal fields is regarded as virtually pure water in most cases. Accordingly, the CO_2 gas intensity can be normalized by the H_2O intensity. As a result, its normalized intensity varied in a narrow range of 4 to 9 with the mean of 6.49 (Fig. 2) and the standard deviation of 0.92 (14 relative %).

5. DISCUSSION AND CONCLUSIONS

The detection limits of the T64000 were low enough to detect gases at low concentration in liquid-rich fluid inclusions from some Japanese geothermal fields. Such gases as CO_2, H_2S and CH_4 were detected in fluid inclusions formed in the upper part of reservoir, where boiling begins (Taguchi et al., 1997).

The normalization of the Raman gas intensity using the H_2O Raman intensity in liquid-rich inclusions made it possible to semiquantitatively determine the gas content. An application of the method to the Ogiri geothermal field in Japan showed that the maximum CO_2 partial pressure in the KZ-2 well is 5-6 bar. One disadvantage of this method is that vapor-rich inclusions with no visible water, cannot be analyzed. In addition, the volume ratio of gas to liquid in a fluid inclusion probably influences its gas content. This method, however, will be a step toward success of the quantitative Raman microspectroscopy.

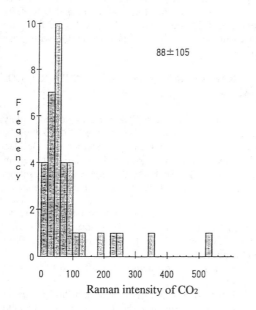

Fig. 1 Histogram of the Raman intensities in the standard CO_2 gas.

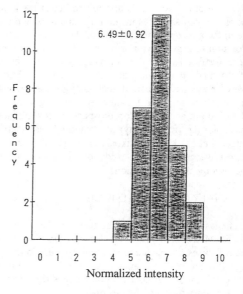

Fig. 2 Histogram of the normalized intensities in the standard CO_2 gas.

6. REFERENCES

Bodnar, R. J., Szabo, Cs., Shilobreeva, S.N. and S. Newman. 1996. Quantitative Analysis of H_2O-CO_2 Fluid Inclusions by Raman Spectroscopy. *PACROFI VI, Program and Abstracts, 14.*

Horikoshi, T., Uchida, T., Oketani, Y., Hisatani, K. and N. Nagahama. 1996. NEDO's development of fracture-type reservoir exploration technology. *Proc. 18th NZ Geothermal Workshope, Geothermal Institute, U of Auckland. 95-102.*

Raman, C. V. and K. S. Krishnan 1928. The optical analogue of the Compton effect. *Nature. 121; 711.*

Sterner, S. M. and R. J. Bodnar. 1984. Synthetic fluid inclusions in natural quartz. I. Compositional types synthesized and applications to experimental geochemistry. *Geochim. et Cosmochim. Acta 48; 2659-2668.*

Taguchi, S., Fujino, T., Takagi, H., Sanada, K., Hayashi, M., Sasada, M., Sawaki, T. and T. Uchida. 1996. Application of Laser Raman Microprobe (RAMANOR T64000) to gas analysis of fluid inclusions from active geothermal fields in Japan: Separation and migration of gas components. *30th IGC Abstracts, 1996, Beijing, China. 3:2481.*

Taguchi, S., Takagi, H., Maeda S., Sanada, K., Hayashi, M., Sasada, M., Sawaki, T., Uchida, T. and T. Fujino. 1997. Gas behavior at some geothermal fields in Japan, revealed by fluid inclusions using laser Raman Microprobe: *(In this volume).*

Wopenka, B. and J. D. Pasteris. 1987. Raman intensities and detection limits of geochemically relevant gas mixtures for a laser Raman microprobe. *Anal. Chemist. 59:2165-2170.*

An autoradiographic method for studying irradiation-induced luminescence in feldspars

M. Siitari-Kauppi & S. Pinnioja
Department of Radiochemistry, University of Helsinki, Finland

A. Lindberg
Geological Survey of Finland, Finland

ABSTRACT: Irradiation-induced luminescence in feldspars depends on the elemental composition of the mineral. Autoradiography and optical densitometry, with application of digital image processing, was found to be a suitable method for determining the intensity of luminescence in feldspar minerals. The elemental composition was studied by energy dispersive X-ray spectrometer analysis (EDS). A correlation was found between optical densities of autoradiographs and elemental composition of feldspars. Increase of sodium concentration and decrease of potassium concentration decrease the luminescence intensity of potassium feldspar. Likewise, decrease of sodium concentration reduces the luminescence intensity of plagioclase. Increase in FeO concentration in feldspars of both types is accompanied by a decrease in the luminescence intensity.

1 INTRODUCTION

For the assessment of migration processes of reactants and reaction products through a compact rock matrix in geological formations, information is needed about the interaction between rock and water. Retention properties of rock-forming minerals are affected by mineral composition. In the case of feldspars, the sorption capacities are dependent on the elemental composition of the minerals and the degree of alteration.

Ionizing radiation induces a luminescence in minerals which is characteristic in both intensity and spectral shape (McKeever, 1985). Thermoluminescence (TL) of feldspars is widely exploited in archaeological age determinations of geological materials. Potassium feldspar exhibits more intensive TL than plagioclase, which consists of a solid-solution series having albite ($NaAlSi_3O_8$) and anorthite ($CaAl_2Si_2O_8$) as the end members. In view of the characteristic TL of minerals it was recently suggested that TL might be used even for the identification of feldspar composition (Poolton et al., 1996).

Autoradiography is a useful technique for measuring the luminescence emission of rock matrices. We describe a new method to differentiate feldspars on polished rock and mineral surfaces. The intensity of the luminescence is great enough to expose an autoradiographic film (Pinnioja et al., in press). The method is based on autoradiography and optical densitometry, with application of a digital image processing technique combined with SEM/EDS analysis.

2 EXPERIMENTAL

2.1 Mineralogy

The outcrop sample of Rapakivi granite (2-YSP-83, Elimäki, Finland) contained potassium feldspar (48.8%), quartz (29.6%), plagioclase (7.3%) and hornblende (3.5%). The feldspar ovoids in the medium-grained groundmass of the rock were mantled with a plagioclase rim. The sample studied was a polished, sawn rock piece 2 x 3.5 x 0.5 cm^3 in size.

2.2 Irradiation

The sample was irradiated with a 500 Ci ^{60}Co irradiation source. The doses were 100 Gy, 500 Gy, 1 kGy, 5 kGy and 10 kGy and the dose rate was 1.9 kGy/h. Irradiation was performed in air at ambient temperature.

2.3 Autoradiography

After irradiation and 18 hours delay in dark the polished rapakivi granite rock surface was exposed on a β film (Amersham, Hyperfilm-βmax) and on an X-ray film (Kodak X Omat MA). The exposure times were 5 hours and 23 hours, respectively. Quantitative interpretation of the results was based on digital image analysis, which began with digitizing of the autoradiograph into pixels (84x84 µm). All the intensities of the subdomains were then converted into corresponding optical densities, which are dependent on the luminescence intensity of the different minerals. Because the responses of the image source (table scanner Ricoh FS 2, maximum optical resolution 600 dpi) and the amplifier of the digital image analyser are linear, the digitized grey levels of the film can be handled as intensities. Intensity here means the light intensity coming through the autoradiographic film. Optical densities were derived from intensities as follows:

$$OD = \log\left(\frac{I_0}{I}\right) \quad (1),$$

where OD is the optical density, I_0 the intensity of the background and I the intensity of the sample. The digitized areas in the autoradiographs were 1 mm^2 and the scanner resolution was 300 dpi. The measured areas were the same as the areas for SEM/EDS analysis. Quantification of the autoradiographs by digital image analysis has been described in detail earlier by Hellmuth et al. (1993, 1994).

2.4 Electron microscopy measurements

Scanning electron microscopy and energy dispersive X-ray spectrometer analysis (SEM/EDS) were carried out to determine the element composition of feldspars in the rock sample. The analyses were done on the same parts of the minerals as the autoradiographic measurements. Sawn and polished rock surfaces were carbon-coated before the analysis. Electron microscopic analysis of polished rock sections was done with a Zeiss DSM 962 and Link ISIS, with a UTW Si(Li) detector operated in backscattered electron image (BSE) mode (Department of Electron Microscopy, University of Helsinki).

3 RESULTS

A photograph of a polished rock surface and the corresponding autoradiograph of the irradiated rock are displayed in Figure 1. Digital image analyses and SEM/EDS analyses were done from plagioclase grains at points A to D and from K feldspar grains at points E to J (see the autoradiograph in Fig. 1). Major elements, given as oxides, are listed in Table 1. Figure 2 shows the optical densities in plagioclase and K feldspar grains of the rapakivi granite sample as a function of irradiation dose. The intensities at points A and B in the autoradiographs were near background when irradiation doses were low and exposure time was five hours, and these points are not included in Figure 2.

The luminescence intensity expressed as optical densities of the autoradiographs showed clear dependence on the irradiation dose. The luminescence intensity could be divided into three classes according to the minerals. First class: OD was at background level in areas where mafic minerals were identified and in parts of the plagioclase rims. Second class: OD remained at low level even when irradiation doses were high. These mineral areas were mainly plagioclase. Third class: OD increased strongly with increase in the irradiation dose. K feldspar in the form of perthite was identified in these areas. Clear differences in optical densities were observed, especially with the highest irradiation dose.

Table 1. Composition of plagioclase and perthite grains measured by scanning electron microscopy and energy dispersive X-ray spectrometry (SEM/EDS) (beam 1µm, parameters: voltage 20 kV, current 10 nA, quantitative method: ZAF)

	SiO$_2$	Al$_2$O$_3$	K$_2$O	Na$_2$O	CaO	FeO
A	62.15	20.80	1.78	12.41	2.59	0.61
B	61.90	21.38	1.51	12.09	3.44	0.48
C	60.52	24.26	2.84	10.52	1.39	0.53
D	59.89	24.16	1.22	9.76	4.80	0.26
F	65.99	17.51	11.90	4.65	0.02	0.14
G	67.34	15.20	15.32	2.94	0.01	0.15
H	67.91	15.17	12.41	5.09	0.04	0.16
I	65.91	17.95	14.31	2.10	0.02	0.08
J	66.43	17.74	15.17	1.16	0.01	0.07

Figure 1. Rapakivi granite rock sample (left), autoradiograph (right) of the surface after irradiation to a dose of 10 kGy. Autoradiographic and EDS analyses are from points A to J (see Table 1 and Fig. 2).

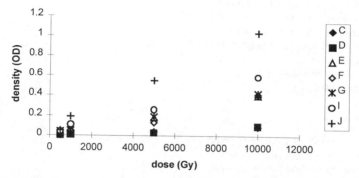

Figure 2. Luminescence intensity of measured points (see Fig. 1) as a function of irradiation dose, determined by the quantitative autoradiographic method.

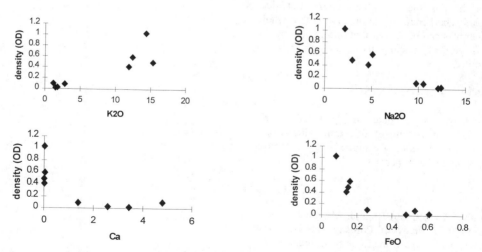

Figure 3. Luminescence expressed as optical densities of autoradiographs shown as a function of K_2O (upper left), Na_2O (upper right), CaO (down left) and FeO (down right) content (wt-%) of feldspars in irradiated rapakivi granite.

Luminescences expressed as optical densities (OD) of autoradiographs as a function of potassium (a), sodium (b), calcium (c) and iron (d) content of feldspars are illustrated in Figure 3. The OD values were measured from the autoradiograph, which was recorded after 10 kGy irradiation and 23 hours exposure on X-ray film (see Fig. 1). In SEM/EDS analyses the first four mineral areas (points A to D) represented plagioclase, which is close to the albite end of the plagioclase (oligoclase) series, and the rest (points F to J) were K feldspar in the form of perthite.

Increase in potassium concentration in perthite increased the luminescence intensity, whereas increase in sodium concentration decreased the luminescence. Potassium concentration in plagioclase did not show any trend, while decrease of sodium was observed to reduce the luminescence intensity. Even a small calcium content decreased the luminescence of plagioclase. Increased FeO concentration in both potassium feldspar and plagioclase was accompanied by decrease in the luminescence.

4 CONCLUSIONS

Autoradiography and optical densitometry with application of digital image processing technique was found to be a suitable method for determining the mineral specific differences in the luminescence effect in rock samples. Quantitative autoradiography combined with SEM/EDS analysis makes it possible to characterize individual feldspars. The benefit of the method is that whole rock samples can be analysed at once, allowing the tedious work of crushing and separating grains to be avoided.

5 ACKNOWLEDGEMENTS

The authors would like to thank Mr. Jyrki Juhanoja, Department of Electron Microscopy, University of Helsinki, for supervising the electron microscopy measurements.

6 REFERENCES

Hellmuth, K.H., Siitari-Kauppi, M. and Lindberg, A. (1993) Study of Porosity and Migration Pathways in Crystalline Rock by Impregnation with ^{14}C-polymethylmethacrylate. Journal of Contaminant Hydrology, 13: 403-418.

Hellmuth, K.H., Lukkarinen, S. and Siitari-Kauppi, M. (1994) Rock Matrix Studies with Carbon-14-Polymethylmethacrylate (PMMA); Method Development and Applications. Isotopenpraxis Environ. Health Stud., 30: 47-60.

McKeever, S. W. S. (1985) Thermoluminescence of Solids. Cambridge University Press.

Poolton, N. J. R., Bøtter-Jensen and Johnsen, O. (1996) On the Relationship Between Luminescence Excitation Spectra and Feldspar Mineralogy, Radiation Measurements 26 (1), 93.

Pinnioja, S., Siitari-Kauppi, M. and Lindberg, A. (1997) Effect of Feldspar Composition on Luminescence in Minerals Separated from Food, Radiat. Phys. Chem. in press

A Raman spectroscopic study of thio-arsenite and arsenite species in low-temperature aqueous solutions

S.A.Wood
Department of Geology and Geological Engineering, University of Idaho, Moscow, Idaho, USA

C.D.Tait & D.R.Janecky
Los Alamos National Laboratory, N.Mex., USA

ABSTRACT: The Raman spectra of thio-arsenite and arsenite species in aqueous solution were obtained at room temperature. Two series of solutions at constant $\Sigma As + \Sigma S$ of 0.1 and 0.5 molal were prepared with various S/As ratios (0.1-9.0) and pH values (7-13.2). Our data suggest that the speciation of As under the conditions investigated is complicated. The Raman measurements offer evidence for at least five independent S-bearing As species whose principal bands are centered near 370, 383, 400, 412 and 420 cm^{-1}. In general, the relative proportions of these species are dependent on total As concentration, S/As ratio and pH. At very low S/As ratios we also observe Raman bands attributable to the dissociation products of H_3AsO_3(aq).

1 INTRODUCTION

Thio-arsenite complexes are thought to be responsible for the mass transfer of As in low-temperature, sulfide-rich, ore-forming fluids, but the stoichiometries and thermodynamics of these complexes continue to be disputed. Spycher and Reed (1989; 1990a,b) reviewed the available literature at the time and proposed that the trimers, $H_3As_3S_6^0$, $H_2As_3S_6^-$ and $HAs_3S_6^{2-}$, are predominant. Krupp (1990a,b) favored dimers, ($H_2As_2S_4$(aq), $HAs_2S_4^-$, or $As_2S_4^{2-}$), by analogy with Sb. Recent experimental solubility studies have been interpreted both in terms of trimers (Webster, 1990; Eary, 1992) and dimers (Mironova et al., 1990). Using EXAFS, Raman spectroscopy and *ab initio* molecular orbital calculations, Helz et al. (1995) concluded that, in solutions undersaturated with respect to either orpiment or amorphous As_2S_3, monomeric species (i.e., $H_2AsS_3^-$ and $HAsS_3^{2-}$, actually formulated by Helz et al. as $AsS(SH)_2^-$ and $AsS_2(SH)^{2-}$) are predominant. However, Helz et al. (1995) reinterpreted previous solubility experiments and suggested that at saturation the predominant species are the trimer $As_3S_4(SH)_2^-$ and the monomer $AsS(OH)(SH)^-$.

Further work is required to completely resolve the nature and thermodynamics of thio-arsenite species. We present preliminary results of Raman spectroscopic measurements conducted on thio-arsenite and arsenite species at 25°C in near-neutral to alkaline solutions which shed additional light on the issue of thio-arsenite speciation.

2 METHODS

Solution preparation and manipulation were conducted in a low-oxygen, nitrogen-filled glove bag. Solutions were prepared for which the S/As ratio varied but the sum of the total arsenic and sulfide concentrations was held constant. A NaHS stock solution was prepared by bubbling H_2S gas through a solution of known NaOH concentration. A sodium arsenite solution was prepared by dissolving reagent-grade As_2O_3(s) in a NaOH solution. These two stock solutions were then mixed in various proportions to obtain solutions with S/As ratios of 9, 5.25, 4, 3.17, 1.94, 1.00, 0.515, 0.25 and 0.11. Two different sets of experiments were carried out: one in which $\Sigma As + \Sigma S = 0.1$ molal, and one in which $\Sigma As + \Sigma S = 0.5$ molal. The pH of each of the series of solutions was adjusted to a constant value using NaOH or HCl. A pH range from 13.2 to 8.4 was investigated. Lower pH values caused precipitation of As-sulfide. An aliquot of each solution to be tested was placed in a glass NMR tube, and the tube was capped and sealed with parafilm. The Raman spectrum of each solution was acquired within 1-2 hours after being sealed in the

NMR tube. The Raman instrumentation has been described in detail by Marley et al. (1988) and Tait et al. (1991).

3 RESULTS

All spectra exhibit a broad, asymmetric, relatively intense band from approximately 325 to 550 cm^{-1}. Most of the intensity of this band can be attributed to Raman signals from the glass NMR tube, with a smaller contribution from librational motions of liquid water. Less intense bands centered near 625 and 800 cm^{-1} are observed in deionized water, and these are also attributable to water or the glass tube. Bands for thio-arsenite species are superimposed on this broad background band.

At $\Sigma S + \Sigma As = 0.1$ molal, pH = 13.2 and a S/As ratio of 9, a prominent, narrow band centered at approximately 383 cm^{-1} is evident, together with a low intensity shoulder on its high wavenumber side (Fig. 1). Upon a decrease in S/As ratio to 5.25, another narrow band, centered near 400 cm^{-1} emerges at the expense of the band at 383 cm^{-1}. At a S/As ratio of 4, the band at 383 cm^{-1} disappears. Between S/As = 4 and S/As = 0.25, the band at 400 cm^{-1} loses intensity, while a band centered near 420 cm^{-1} becomes relatively more prominent, until it too loses intensity. Meanwhile, broad bands near 600 and 800 cm^{-1} increase in intensity. At S/As = 0.11, the bands in the 350-450 cm^{-1} range disappear, leaving only the broad band observed in the background spectrum, and bands of increased intensity near 600 and 800 cm^{-1}. The broad bands observed at 600 and 800 cm^{-1} may be assigned to $HAsO_3^{2-}$ and AsO_3^{3-} according to the previous Raman measurements of Loehr and Plane (1968). The bands observed in the region 350-450 cm^{-1} can be attributed to As-S bonds (cf. Rogstad, 1972; Mikenda et al., 1982; Helz et al., 1995). These species may be pure thioarsenites, or mixed species such as AsO_2S^{3-}, for example.

In solutions with $\Sigma S + \Sigma As = 0.1$ molal and pH = 12.6, only one prominent band is noted in the spectral region 350-450 cm^{-1}, but the maximum of this band appears to shift from about 383 cm^{-1} to 396 cm^{-1} as the S/As ratio varies from 9 to 0.25 (Fig. 2). The band is asymmetric to the high wavenumber side, and there may also be some intensity above background on the low wavenumber side. All of these observations suggest the presence of more than one component to the band (at least one band at 383 cm^{-1} and one at 400 cm^{-1}). Similarly to pH = 13.2, as the S/As ratio decreases, bands

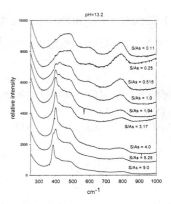

Figure 1: Raman spectra for solutions with $\Sigma S + \Sigma As = 0.1$ molal and pH = 13.2 at various S/As ratios.

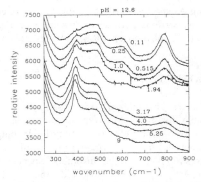

Figure 2: Raman spectra for solutions with $\Sigma S + \Sigma As = 0.1$ molal and pH = 12.6 at various S/As ratios.

near 600 and 800 cm^{-1} increase in relative intensity, suggesting the formation of arsenite species.

At $\Sigma S + \Sigma As = 0.1$ molal, pH = 10.5 and S/As = 9.0, there are two prominent bands in the low wavenumber region: one with a peak near 383 cm^{-1} and one near 412 cm^{-1} (Fig. 3). There is also a low intensity band near 325 cm^{-1}. As S/As ratio decreases, the intensity of the bands at 325 and 412 cm^{-1} increase at the expense of that at 383 cm^{-1}. With further decrease in S/As ratio, a very weak band appears near 420 cm^{-1} and then disappears, and we also observe a broad, intense band at about 790 cm^{-1}, which can be attributed to an arsenite species ($HAsO_3^{2-}$). For solutions with $\Sigma S + \Sigma As = 0.1$ molal and pH = 8.4-8.5 (Fig. 4), the low wavenumber portion of the spectrum behaves very much like that at pH = 10.5. However, in the region 600-700 cm^{-1} the spectrum closely resembles that attributed by

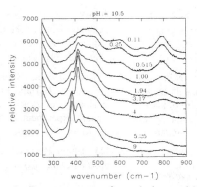

Figure 3: Raman spectra for solutions with ΣS + ΣAs = 0.1 molal and pH = 10.5 at various S/As ratios.

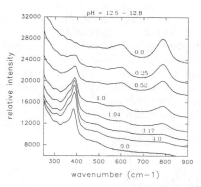

Figure 5: Raman spectra for solutions with ΣS + ΣAs = 0.5 molal and pH = 12.5-12.8 at various S/As ratios.

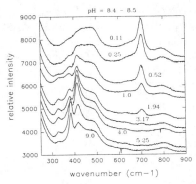

Figure 4: Raman spectra for solutions with ΣS + ΣAs = 0.1 molal and pH = 8.4-8.5 at various S/As ratios.

Loehr & Plane (1968) to H_3AsO_3(aq).

If ΣS + ΣAs is increased to 0.5 at pH = 12.5-12.8 and S/As = 9.0, we observe a main asymmetric band in the low wavenumber region centered near 383 cm^{-1}, with a tail toward the low wavenumber side (Fig. 5). There is also a low intensity band centered near 310 cm^{-1}. With decreasing S/As ratio a shoulder appears on the 383 cm^{-1} band near 370 cm^{-1}. We have not observed previously any bands centered near 370 cm^{-1} at any pH or S/As ratio at ΣS + ΣAs = 0.1 molal. Thus, we conclude that the species responsible for the shoulder at 370 cm^{-1} must be yet another thio-arsenite species containing a greater number of As atoms than any species previously detected. As S/As ratio decreases from 4.0 to 0.25, the relative intensities of the 383 cm^{-1} band and the 370 cm^{-1} shoulder do not appear to change greatly, although the intensities of both decrease relative to bands attributable to arsenite species at 600 and 800 cm^{-1}.

4 DISCUSSION

Our Raman measurements provide evidence for the existence of at least five different S-bearing As species under the conditions of our experiments. We observe at least five independent Raman bands in the S-As stretching region of the Raman spectrum, centered near 370, 383, 400, 412 and 420 cm^{-1}. Additional species may exist but we cannot confirm their existence from our data. Also, other species may exist in more acidic solutions which were not studied here. Our results suggest that some of the confusion in the literature regarding speciation may be due to the existence of a series of thio-arsenites with differing degrees of polymerization. We also observe the conversion of thio-arsenites to arsenites as S/As ratio of the solution is lowered. Additional Raman and solubility measurements, required to establish the exact stoichiometries of the S-bearing As species, are in progress in our laboratories.

Helz et al. (1995) studied a solution containing 1 M NaHS and 0.08 M As at pH 8.2 and 12 at 25°C. They observed only three bands over the range 150-750 cm^{-1} at 325, 382 and 412 cm^{-1}. Upon an increase in pH from 8.2 to 12, the relative intensity of the band at 382 cm^{-1} increased at the expense of the two bands at 325 and 412 cm^{-1}. Based on *ab initio* calculations, Helz et al. (1995) attributed the bands at 382 cm^{-1} to the species $AsS_2(HS)^{2-}$ and the bands at 325 and 412 cm^{-1} to the species $AsS(HS)_2^-$, which can be related to one another through a simple protonation reaction. Like Helz et al. (1995), we observe two main bands in the pH range 8.5-10.5, at about 383 and 412 cm^{-1}, and we also observe a relative increase in the intensity of the former band with increasing pH. However, we further observe that, at constant pH, the intensity of the band at 383 cm^{-1} decreases relative to that at 412 cm^{-1} with decreasing S/As ratio. This observation indicates the existence of two species with different S/As ratios, not just differing degrees of protonation. Thus, either two species exist under these conditions which differ both in S/As ratio and degree of protonation or several different species exist, at least some of which exhibit closely overlapping Raman bands.

REFERENCES

Eary, L.E. 1992. The solubility of amorphous As_2S_3 from 25 to 90°. *Geochim. Cosmochim. Acta* 56: 2267-2280.

Helz, G.R., J.A.Tossell, J.M.Charnock, R.A.D. Pattrick, D.J. Vaughan, & C.D.Garner 1995. Oligomerization in As(III) sulfide solutions: Theoretical constraints and spectroscopic evidence. *Geochim. Cosmochim Acta* 59: 4591-4604.

Krupp, R.E. 1990a. Comment on "As(III) and Sb(III) sulfide complexes: An evaluation of stoichiometry and stability from existing experimental data" by N.F. Spycher and M.H. Reed. *Geochim. Cosmochim. Acta* 54: 3239-3240.

Krupp, R.E. 1990b. Response to the Reply by N.F. Spycher and M.H. Reed. *Geochim. Cosmochim. Acta* 54: 3245.

Loehr, T.M. & R.A. Plane 1968. Raman spectra and structures of arsenious acid and arsenites in aqueous solution. *Inorg. Chem.* 7: 1708-1714.

Marley, N.A., et al. 1988. High temperature and pressure system for laser Raman spectroscopy of aqueous solutions. *Rev. Sci. Instrum.* 59: 2247-2253.

Mikenda, W., H. Steidl, & A. Preisinger 1982. Raman spectra of $Na_3AsS_4·8(D,H)_2O$ and $Na_3SbS_4·9(D,H)_2O$ and O - D(H)...S bonds in salt hydrates. *Jour. Raman Spectr.* 12: 217-221.

Mironova, G.D., A. Zotov, & N.I. Gul'ko 1990. The solubility of orpiment in sulfide solutions at 25-150°C and the stability of arsenic sulfide complexes. *Geochem. Int.* 27: 61-73.

Rogstad, A. 1972. High-temperature laser Raman spectroscopic study on mixtures of arsenic and sulphur vapours. *Jour. Mol. Struct.* 14: 421-426.

Spycher, N.F. & M.H. Reed 1989. As(III) and Sb(III) sulfide complexes: An evaluation of stoichiometry and stability from existing experimental data. *Geochim. Cosmochim. Acta* 53: 2185-2194.

Spycher, N.F. and M.H. Reed 1990a. Reply to comments by R.E. Krupp on "As(III) and Sb(III) sulfide complexes: An evaluation of stoichiometry and stability from existing experimental data". *Geochim. Cosmochim. Acta* 54: 3241-3243.

Spycher, N.F. and Reed, M.H., 1990b. Response to the Response of R.E. Krupp. *Geochim. Cosmochim. Acta* 54: 3246.

Tait, C.D., D.R. Janecky, & P.S.Z. Rogers 1991. Speciation of aqueous palladium(II) chloride solutions using optical spectroscopies. *Geochim. Cosmochim. Acta* 55: 1253-1264.

Webster, J. 1990. The solubility of As_2S_3 and speciation of As in dilute and sulphide-bearing fluids at 25 and 90°C. *Geochim. Cosmochim. Acta* 54: 1009-1017.

Study of electrical conductivity of H_2O at 0.21-4.18GPa and 20-350°C

Zheng Haifei
Department of Geology, Peking University, Beijing, People's Republic of China

Xie Hongsen, Xu Yousheng, Song Maoshuang, Guo Jie & Zhang Yueming
Institute of Geochemistry, Chinese Academy of Sciences, Guiyang, People's Republic of China

ABSTRACT: Measurement of the electrical conductivity of H_2O at 0.21-4.18GPa and 20-350°C shows that the conductivity of H_2O increases with temperature and pressure, but there are discontinuities at pressures between 0.57 and 0.9GPa, and between 2.11 and 2.58GPa, which are consistent with the formation of polymorphs of ice (ice$_V$, ice$_{VI}$ and ice$_{VII}$), indicating that H_2O at different pressures has quite different electronic properties of which will affect the interaction between rocks/minerals and water.

1 INTRODUCTION

Water is the most important component in the Earth's lithosphere. A large number of studies indicate that water plays an important role in the origin of magmas, mineral deposits and the evolution of the earth's crust. Therefore, understanding the fundamental properties of water at high temperature and high pressure will aid study of the geological phenomena and geochemistry.

The study of electrical conductivity of water at high temperature and high pressure is one of the important methods for obtaining information on water properties. Since 1960, many investigators have carried out such investigations. However, the data obtained are all from the experiments at pressures below 1.0GPa (Holzapfel, 1969; Marshall & Franck, 1981; Tanger & Pitzer, 1989; Ho, Palmer & Mesmer, 1994). Recently, we have carried out the measurement of the electrical conductivity of H_2O at pressures up to 5.0GPa and temperatures to 350°C. We present a portion of the results in this paper.

2 EXPERIMENTAL METHOD

The experiments were carried out in a YJ-3000 ton press fitted with a wedge-type cubic anvil. The samples were pressed using pyrophyllite as the pressure medium, and a tube made of a piece of stainless steel was used to heat up the sample resistively. The temperatures were measured with PtRh-Pt thermocouples and the measured pressures were calibrated by the melting points of water at high pressure.

The sample cell for the experiment was made of Teflon, and platinum metal was used as electrodes. Since the sample cell would be changed at high temperature and high pressure, we assumed that the forces acting on the sample from different directions are the same and modified the cell constant using the Redlich-Kwong equation of state (Holland & Powell, 1991).

3 RESULT

In general, the electrical conductivity increases with temperature and pressure. The plot of Log κ vs. 1/K is basically linear, but the slopes of the straight lines are different at various pressure(Fig. 1). In addition, it can be seen that there are discontinuities in the electrical conductivity(Fig.2). A very obvious discontinuity exists between 2.11GPa and 2.58GPa, and the slope of the logarithm of electrical conductivity vs. pressure is larger at 1.27-2.11GPa than at 2.58-4.18GPa. It can also be seen that, by comparing the electrical conductivities between 2.11 and 2.58GPa, the latter increases suddenly. Another discontinuous point is between 0.57 and 0.9GPa, that is, the electrical conductivity decreases suddenly when the pressure increases from 0.57 to 0.9GPa. The discontinuities of water are

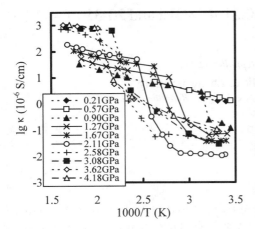

Figure 1 Plot of the relationship between Log κ and 1/K at different pressures.

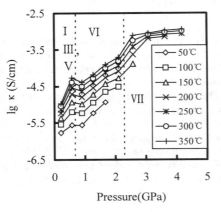

Figure 2 Plot of the relationship between Log κ and pressure at different temperatures (the Roman numerals represent polymorphs of ice).

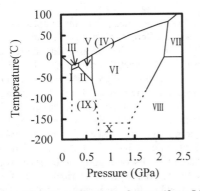

Figure 3 Phase diagram of water (from Lin-gun Liu, 1986).

consistent with those of ice(Fig.3). This reflects that the water at high pressures has basic microcosmic structure of ice at the corresponding pressure.

4 DISCUSSION

In general, the effects on the electrical conductivity are: (1)the properties of water or aqueous solutions, such as concentration of ions, ionic mobility, viscosity and dielectric constant. (2)external conditions, such as the influence of temperature and pressure on the concentration of ions, viscosity and dielectric constant. They mainly depend on the method of transport of electric charge and the influence of temperature and pressure on the concentration of ions. For example, increasing pressure will increase viscosity and dielectric constant, but these properties have opposite effects on the electrical conductivity. That is, increasing viscosity will decrease electrical conductivity, whereas the increase of the dielectric constant reflects the increase of dipole moment of molecule, thus making hydrogen easier remove from water molecules to form H_3O^+ and OH^- ions and increases the average concentration of ions in water. Therefore, the relation between electrical conductivity and temperature /pressure can provide basic information on the properties of water or solution, such as the information of dielectric constant, ionization.

It can be seen from our experimental results that over the experimental range of pressure, electrical conductivity increases with temperature. This finding proves that decreasing viscosity and increasing ionic mobility are the main influence on the electrical conductivity. At different pressures, electrical conductivity increases with pressure, which is consistent with increase of the dielectric constant with pressure, but at two points of pressure (0.57-0.9GPa and 2.11-2.58GPa), there may be sudden changes in the dielectric constant. that is, when the pressure increases from 0.57 to 0.9GPa and from 2.11 to 2.58GPa the dielectric constant suddenly decreases and increases, respectively. This can probably be explained by microcosm that increasing pressure will cause the bond distance of O-H to short and the bond distance of the hydrogen bond (H ⋯ O) to elonge (decreasing the energy of hydrogen bond), thus causing the electrical conductivity to increase. Since the relative distance of O-H and H ⋯ O bonds is related to the structure of ice/ and the potential barrier to structural transformations, the change of structure will result in the sudden change of electrical conductivity. Although water in liquid is different from ice, it has a short rang order similar to that of ice, and shows sudden changes of electrical conductivity at various pressures.

Based on the available information of geophysics, the pressure range of the layer with high electrical conductance in the low crust basically corresponds to water $_{III}$ and water $_V$. This indicates that the layer is related to the water property. Even though the electrical conductivity of water is not obviously higher than that of adjacent phase of water, it can be noticed from figure 1 and 2 that the slope of Log κ vs. 1/K is larger at 0.57 GPa than at higher pressure. Therefore, the electrical conductivity at higher temperatures will be much higher than those at pressures above 0.9 GPa. In addition, it will be possible to make the conductance at the pressure of 0.57 GPa increase notably when ionic concentration increases (possibly solving of minerals or rocks). This should be further investigated by carrying out an experimental study on the electrical conductivity of water-rock systems at high temperature and high pressure.

In most areas around the would, the top boundary of the asthenosphere in the upper mantle is at ~100km. For example, the low velocity zone exists at a depth of 82-93km in the upper mantle of north China and a depth of 80km below the Mariana Islands trench. The pressure corresponding to these depths is consistent with the stable pressure of water $_{VII}$. These phenomena could not simply be a coincidence. They probably imply that the properties of water $_{VII}$ are obviously different with those of water $_{VI}$ and the difference probably is an important reason that the melting point at higher pressure (above 2.58 GPa) is lower than that at lower pressure (below 2.11 GPa). This explanation is consistent with the melting experiment results of rocks containing about 2% water at high temperature and high pressure (Zheng Haifei et al., 1995). A detailed understanding of the microscopic mechanism of melting reactions requires studies of the dielectric constant and infrared spectroscopy of water at high temperature and high pressure.

5 ACKNOWLEDGEMENTS

The authors thank Prof. Tu Guangzhi of the Institute of Geochemistry of the Chinese Academy of Sciences for support and discussion. This work was supported by the National Science Foundation of China (Grant No. 49573192, 49603049)

6 REFERENCES

Holzapfel, W.B. 1969. Effect of pressure and temperature on the conductivity and ionic dissociation of water up to 10 Kbar and 1000℃. J. Chem. Phys. 50: 4424-4428.

Marshall, W.L. and Franck, E.U.1981. Ion product of water substance at 0-1000℃, 1-10000bars, new international formulation and its background. Phys. Chem. Ref. Data. 10: 295-304.

Tanger, J.C. and Pitzer, K.S.1989. Calculation of the ionization constant of H_2O to 2273K and 500MPa. AICHE J. 35:1631-1638.

Ho PC. Palmer DA. and Mesmer RE. 1994. Electrical condictivity measurements of aqueous sodium chloride solutions to 600℃ and 300MPa, Journal of Solution Chemistry. 23(9): 997-1018.

Holland, T. and Powell, R. 1991. Compensated-Redlich-Kwong(CORK) equation for volunes and fugacities of CO2 and H2O in the range 1 bar to 50 kbar and 100-1600℃. Contrib. Mineral. Petrol. 109:265-273.

Lin-gun Liu and Willliam A.Bassett, Elements, Oxides, Silicates: High-Presspre Phases with Implications for the Earth's Interior. 1986. Oxford University Press, New York and Clarendon, Oxford: 92-94.

Zheng Haifei, Xie Hongsen, Xu Yousheng et al. 1995. Experimental study of the influence of hydrous minerals on the melting behaviour of rocks at high temperatures and pressures. Acta Geologica Sinica. Vol.9(2):157-167.

15 Modelling

Enhancements to the geochemical model PHREEQC – 1D transport and reaction kinetics

C.A.J. Appelo
Faculty of Earth Sciences, Free University, Amsterdam, Netherlands

D.L. Parkhurst
US Geological Survey, Denver Federal Center, Colo., USA

ABSTRACT: The geochemical equilibrium model PHREEQC (Parkhurst, 1995) has been extended with 1D transport capabilities and a BASIC interpreter for kinetic rate equations. The transport part comprises a complete 1D advection and dispersion/diffusion model; dual porosity media can be simulated with the first order exchange model or with explicit diffusive mixing of stagnant and mobile cells. The BASIC interpreter allows flexible definition of rate equations in the input file in the form of BASIC statements. A selection of kinetic rate equations from the literature is presented to demonstrate the variety of forms which are used for describing kinetic geochemical reactions. Model calculations for kinetic dissolution of calcite in a batch experiment and in a column illustrate the flexibility and applicability of PHREEQC 2.0 for modeling geochemical systems. PHREEQC 2.0 will be available via ftp (brrcrftp.cr.usgs.gov) and web site (http://water.usgs.gov).

1 INTRODUCTION

The application of geochemical equilibrium models such as PHREEQC, MINTEQ, and EQ3/6 is hindered by at least two drawbacks. The first is that many geochemical systems are transient, and transport of solutes plays a dominant role. Transport cannot simply be calculated with a model that is meant for batch reactions. The second, even more serious drawback, is that minerals often do not react to equilibrium in the timeframe of an experiment or a model period. Kinetic rate equations have been published for numerous minerals, and for various conditions of temperature, pressure, and solution composition. However, the rate equations are manifold, and if the formulas are hard coded in the model, incorporation of each new equation requires modification and recompilation of the source code.

PHREEQC 2.0 tackles these difficulties. A complete 1D transport algorithm which includes dispersion and/or diffusion, and dual porosity exchange has been included in the code. The problem of defining kinetic rate equations has been circumvented in a novel way. PHREEQC 2.0 allows kinetic rate equations to be defined with BASIC language statements in the input file. The BASIC statements are interpreted by PHREEQC 2.0 without need to recompile the code. The BASIC statements can define kinetic reactions in a completely general way, with access to all parameters (activities, molalities, saturation indexes, and other calculated properties) of the aqueous model. We present some often used kinetic rate equations, and give example calculations for calcite dissolution according to the Plummer-Wigley-Parkhurst rate equation (Plummer et al., 1978).

2 SELECTED RATE EQUATIONS

The overall rate for a kinetic reaction of minerals and other solids is:

$$R_k = r_k \, A/V \, (m_k/m_{0k})^n, \quad (1)$$

where R_k is the overall rate for solid k (mol/dm^3/s), r_k is the specific rate (mol/dm^2/s), A is the initial surface area of the solid (dm^2), V is the solution volume (dm^3), m_{0k} are the initial moles of solid, m_k are the moles of solid at a given time, and $(m_k/m_{0k})^n$ is a factor to account for changes in A/V during dissolution and also for selective dissolution and aging of the solid. For uniformly dissolving spheres and cubes $n = 2/3$.

Equations for the specific rates r may have various forms, largely depending on the completeness of the experimental information. When information is lacking, an often applied, simple rate equation is:

$$r_k = k_t \, (1 - (IAP/K_k)^\sigma). \quad (2)$$

Here, k_t is an empirical constant, and IAP/K_k is the saturation ratio (SR) for mineral k. This rate equation can be derived from transition state theory. The coefficient σ is related to the stoichiometry of the reaction when an activated complex is formed (Aagaard and Helgeson, 1982; Delany et al., 1986). Often, $\sigma = 1$. An advantage of this rate equation is that it applies for both super- and subsaturation, and the rate is zero at equilibrium. The rate is linear if it is applied far from equilibrium ($IAP/K < 0.1$).

The specific rate may also be based on the saturation index (SI):

$$r_k = k_t\, \sigma\, \ln(IAP/K_k). \tag{3}$$

This rate equation has been applied with success to dissolution of dolomite (Appelo et al., 1984).

For many biochemical reactions the Monod, or Michaelis-Menten, rate equation is applicable. This rate equation has the form:

$$r_k = r_{max}\, m_i / (K_m + m_i), \tag{4}$$

where r_{max} is the maximum reaction rate, m_i is the concentration of chemical i which limits the reaction, and K_m is the concentration of i where $r_k = \tfrac{1}{2} r_{max}$. This equation gives a rate which shows first order dependence on i at low concentrations, when $m_i \ll K_m$. At high concentrations of i, when $m_i \gg K_m$, the reaction becomes independent of the concentration of i. When the concentration of i is of the order of K_m, $m_i \approx K_m$, a mixed kinetics rate is calculated.

The Monod rate equation can be used for simulating the sequential steps that occur in the oxidation of organic matter (Van Cappellen and Wang, 1996). In line with the energy yield of the oxidant, first O_2 is consumed, then NO_3^-, and successively other, more slowly operating oxidants such as Fe(III) oxides and SO_4^{2-}. The coefficients in the Monod equation can be related to commonly applied first order rate equations for the individual processes. For degradation of organic matter (C) in soils it is:

$$dC/dt = k_1\, C, \tag{5}$$

where C is organic carbon content (mol/kg soil), and k_1 is the first order decay constant (/s). The value of k_1 is approximately $k_1 \approx 0.025/$ yr in a temperate climate and in aerobic soils (Russell, 1973), while in sandy aquifers in The Netherlands, where NO_3^- is the oxidant, $k_1 \approx 5e-4/yr$ (Van Beek, pers. comm.). Concentrations of up to 3 μM O_2 are found in groundwater even outside the redox-domain of organic degradation by O_2, and 3 μM O_2 may be considered the concentration where aerobic degradation equals denitrification. The combination ($k_1 = 5e-4$ for 3 μM O_2, and $k_1 = 0.025/yr$ for 0.3 mM O_2 in aerobic soils) yields $r_{max} = 1.57e-9/s$ and $K_m = 294$ μM for aerobic degradation. A similar estimate for denitrification can be based on $k_1 = 5e-4/yr$ for $m_{NO_3^-} = 3$ mM and $k_1 = 1e-5/yr$ for 3 μM, and yields $r_{max} = 1.67e-11/s$ and $K_m = 155$ μM. The combined rate for aquifer sediments is then:

$$r_C = -\left(\frac{1.6e-9\, m_{O_2}}{2.9e-4 + m_{O_2}} + \frac{1.7e-11\, m_{NO_3^-}}{1.5e-4 + m_{NO_3^-}}\right) C. \tag{6}$$

Still other rate equations are based on detailed measurements in solutions with varying concentrations of the components which influence the rate. For oxidation of pyrite the rate is (Williamson and Rimstidt, 1994):

$$r_{Pyrite} = 10^{-10.19}\, (m_{O_2})^{0.5}\, (m_{H^+})^{-0.11}, \tag{7}$$

which shows a square root dependence on the molality of oxygen, and a small increase of the rate with increase in pH. This rate equation is applicable while the solution contains oxygen, and for the dissolution reaction only. It is not suitable when oxygen is depleted, and the solution approaches equilibrium.

An example of a more complete rate equation that applies for both dissolution and precipitation is the rate equation for calcite of Plummer et al. (1978):

$$r_{Calcite} = k_1\,[H^+] + k_2\,[H_2CO_3] + k_3\,[H_2O] - k_4\,[Ca^{2+}][HCO_3^-] \tag{8a}$$

or

$$r_{Calcite} = r_f - r_b, \tag{8b}$$

in which the square brackets indicate activity for the ions or molecules and k_1, k_2 and k_3 are temperature dependent constants given by Plummer et al. (1978). The rate contains a forward part r_f, and a backward part r_b. The value of k_4 in the backward rate is variable, but it must be such that the reaction rate is zero at equilibrium. In a pure water/calcite system, the backward term can be approximated as:

$$r_b = k_4[Ca^{2+}][HCO_3^-] \approx 2\, k_4[Ca^{2+}]^2. \tag{9}$$

At equilibrium, the net reaction rate $r_{Calcite} = 0$, and $[Ca^{2+}]$ is the activity at saturation, $[Ca^{2+}]_s$. Therefore:

$$2\, k_4 = r_f / [Ca^{2+}]_s^2. \tag{10}$$

This gives:

$$r_{Calcite} = r_f \left\{ 1 - \left(\frac{[Ca^{2+}]}{[Ca^{2+}]_s} \right)^2 \right\}. \quad (11)$$

In a pure Ca-CO$_2$ system at constant CO$_2$ pressure, the ion activity product IAP is:

$$IAP_{Calcite} = [Ca^{2+}][HCO_3^-]^2 / P_{CO_2}$$

$$\approx 4\,[Ca^{2+}]^3 / P_{CO_2}. \quad (12)$$

The rate for calcite can thus be approximated by:

$$r_{Calcite} \approx r_f\,(\,1 - (IAP/K_{Calcite})^{2/3}\,), \quad (13)$$

where r_f contains the terms given in Eqn. (8).

3 KINETICS IN PHREEQC 2.0

The rates can be specified in the input file of PHREEQC 2.0 in the form of statements in the BASIC computer language. For example, the kinetic dissolution of calcite is defined as shown in Fig. 1.

```
RATES
Calcite
    -start
1   rem  parm(1) = A/V, 1/cm
10  si_cc = si("Calcite")
20  if (m <= 0 and si_cc < 0) then goto 200
30  k1 = 10^(0.198 - 444.0 / (273.16 + tc))
40  k2 = 10^(2.84 - 2177.0 / (273.16 + tc))
50  k3 = 10^(-5.86 - 317.0 / (273.16 + tc))
60  r1 = k1 * act("H+")
70  r2 = k2 * act("CO2")
80  r3 = k3 * act("H2O")
90  moles = parm(1) * (r1 + r2 + r3) * time
100 moles = moles * (1 - 10^(2/3*si_cc))
110 if (moles > m) then moles = m
200 save moles
    -end
```

Fig. 1 Kinetic dissolution of calcite: rate defined in BASIC language in PHREEQC 2.0.

Access is given to parameters of the aqueous model, e.g. `si("Calcite")` in line 10 gives the saturation index for calcite, `m` in line 20 indicates the remaining moles of calcite, `tc` in line 30 is the temperature in °C, `act("H+")` in line 60 gives the activity of H$^+$. Variable names can be defined at will, e.g. `si_cc` in line 10, `k1` in line 30, and others. The rate equations are normally part of a database. Other constants for a specific simulation are defined under the keyword KINETICS in the input file. This keyword calls the kinetics module with a number of parameters, e.g. the error tolerance in the integration, and parm(1), the ratio of surface area A to solution volume V.

The rate R_k in eqn. (1) corresponds to a change in concentration of an aqueous species i:

$$d\,m_i\,/\,d\,t = c_{i,k}\,R_k, \quad (14)$$

where $c_{i,k}$ is the stoichiometric coefficient of species i in the kinetic reaction. The rate is integrated with a Runge-Kutta-Fehlberg scheme with 4th and 5th order pair formulas. The local error is of order $O(h^6)$ and is used to automatically adapt the step size h.

The results of PHREEQC 2.0 are compared with experimental results in Figure 2.

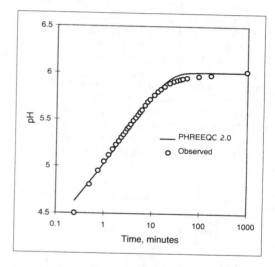

Fig. 2 Change of pH with time during kinetic dissolution of calcite in water with constant P_{CO_2} = 1 atm: comparison of experiment 8 by Plummer and Wigley (1976) and PHREEQC 2.0 calculations.

4 TRANSPORT IN PHREEQC 2.0

A one-dimensional transport algorithm is included in PHREEQC 2.0 for simulating hydrogeochemical transport processes including the effects of diffusion or dispersion and interaction with stagnant zones (dual porosity media). The algorithm integrates the terms of the advection-reaction-dispersion equation with an operator splitting technique. Central differences in time and space are used for dispersion and an upstream scheme is used for advection calculations.

A major advantage of operator splitting is that

numerical accuracy and stability can be obtained by adjusting timestep to grid size for the individual parts of the equation, cf. Chapter 9 in Appelo and Postma (1993). The time step for advective transport is split into a number of equal time intervals for calculating dispersion; the number depends on the ratio of dispersivity and cell-length. Reactions are calculated after both the advective and the dispersive steps.

Results for different discretizations show generally excellent agreement when cell-length is equal or smaller than dispersivity, $\Delta x \leq \alpha$. Dual porosity media (structured soils with mobile/immobile pores, or fractured rocks) can be simulated with the first order exchange approximation, or diffusive mixing can be defined explicitly among mobile and stagnant zones.

As an example, we show the calculated composition of effluent from a column which contains calcite with $A/V = 15$/m, which is sufficiently low that equilibrium with calcite is not attained. The column is initially filled with distilled water that has equilibrated with calcite. Water with $P_{CO_2} = 1$ atm is pumped into the column where the flow velocity $v = 1$ cm/hr. The column is 8 cm long and has a dispersivity $\alpha = 0.2$ cm. The calculated composition of the effluent is shown in Figure 3. Two discretizations have been used, one with $\Delta x = \alpha$ (symbols), the other for a three times smaller grid with $\Delta x = \alpha/3$ (lines). The results of the two simulations are indiscernible at the scale of the plot.

The effluent composition in Figure 3 shows the displacement of the distilled water with a low Ca^{2+} concentration by the CO_2 charged water which attains a higher Ca^{2+} concentration by greater dissolution of calcite. The $SI_{Calcite}$ is initially zero; due to mixing of resident and injected water in the dispersive front it increases slightly above zero, then decreases markedly, and rises again to the final $SI_{Calcite} = -0.618$ in the effluent. The final $SI_{Calcite}$ results solely from the kinetic reaction of calcite with the injected, CO_2 charged water as it travels through the column.

REFERENCES

Aagaard, P. and Helgeson, H.C., 1982. Thermodynamic and kinetic constraints on reaction rates among minerals and aqueous solutions-I. theoretical considerations. Am. J. Sci. 282, 237-285.

Appelo, C.A.J., Beekman, H.E. and Oosterbaan, A.W.A., 1984. Hydrochemistry of springs from dolomite reefs in the southern Alps of Northern Italy. IAHS Pub. 150, 125-138.

Appelo, C.A.J. and Postma, D., 1993. Geochemistry, groundwater and pollution. Balkema, Rotterdam, 536 p.

Delany, J.M., Puigdomenech, I. and Wolery, T.J., 1986. Precipitation kinetics option for the EQ6 geochemical reaction path code. LLNL, Univ. California, Livermore, 44 p.

Parkhurst, D.L., 1995. User's guide to PHREEQC-A computer program for speciation, reaction-path, advective transport, and inverse geochemical calculations. US Geol. Surv. Water Resour. Inv. 95-4227, 143 p.

Plummer, L.N. and Wigley, T.M.L., 1976. The dissolution of calcite in CO_2-saturated solutions at 25°C and 1 atmosphere total pressure. Geochim. Cosmochim. Acta 40, 191-202.

Plummer, L.N., Wigley, T.M.L. and Parkhurst, D.L., 1978. The kinetics of calcite dissolution in CO_2 water systems at 5° to 60°C and 0.0 to 1.0 atm CO_2. Am. J. Sci. 278, 179-216.

Russell, E.W., 1973. Soil conditions and plant growth. Longman, London, 849 pp.

Van Cappellen, P. and Wang, Y., 1996. Cycling of iron and manganese in surface sediments. Am. J. Sci. 296, 197-243.

Williamson, M.A. and Rimstidt, J.D., 1994. The kinetics and electrochemical rate-determining step of aqueous pyrite oxidation. Geochim. Cosmochim. Acta 58, 5443-5454.

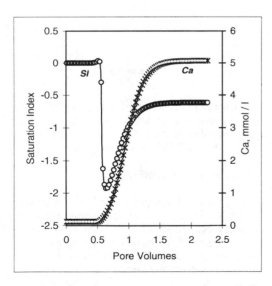

Fig. 3 Simulated effluent composition from a column with kinetic reaction of calcite. The column contains initially distilled water that has equilibrated with calcite; water with $P_{CO_2} = 1$ atm is injected.

Physicochemical model of water-atmosphere-coal system

O.V. Avchenko
Far East Geological Institute, Vladivostok, Russia

ABSTRACT: With the program complex "SELEKTOR-C", physicochemical modelling of the water-atmosphere-coal system has been done under the conditions of air excess (model A) and absent (model B). In model A, water acidity increases sharply through the supply of sulphur from coal and formation of SO_4^{-2} anion. In model B (at $Eh<0$), water strongly increases alkalinity through the formation of NH_4^+ cation and sharp lowering of SO_4^{-2} anion concentration. Quantity of the atmosphere in the water-atmosphere-coal system is a vigorous factor in the water composition evolution. The impact of P_{CO2} in the atmosphere on pH magnitude of the surface and rain waters is discussed.

1 INTRODUCTION

The program complex "SELEKTOR-C" and underlying algorithm of searching for the minimum of Gibbs free energy in heterogenous systems is a powerful tool for investigation of many physicochemical problems (Karpov et.al, 1995). A significant feature of "SELEKTOR-C" is the possibility to construct the many-reactor models that permit observation of the successive change of the composition of a fluid, gas, or water when they pass through rock. One of such many-reactor model is discussed in this paper.

2 FORMULATION OF THE PROBLEM

The problem of the physicochemical modelling of the water-atmosphere-coal system has been formulated in two stages: (1) to consider the chemical equilibrium of water and atmosphere ($P=1$ bar; $T=25^0C$) and (2) to study the condition of the chemical equilibrium of water, coal, and atmosphere ($P=1$ bar; $T=25^0C$) and to reveal the main features of the water composition change through water-coal interaction under the conditions of air excess and absent. In both stages, the problem was described in terms of 13 independent components: C, O, H, N, K, Na, Ca, Mg, Si, F, B, Cl, S. The model of the water-atmosphere or water-atmosphere-coal system system included 104 water species (cations, anions, and neutral compounds) 20 gas components, and 13 solid phases, that provided completeness of thermodynamic description of the model.

3 MODEL OF THE WATER - ATMOSPHERE SYSTEM

When considering the water-atmosphere system, we used the analyses of water from Chistovodnye source (Chudaeva et.al. 1995). The water of the source was in contact with the atmosphere. In modelling the air excess was prescribed or the water system was considered a system opened relative to the atmosphere. In other words, the components of the air - C, O, H, gave water their chemical potentials. On this approach it is obvious that the water pH formation (pH=log[H+]) is influenced by the quantity of CO_2 solubility in water and subsequent formation of carbonic acid. The value of water pH of the Chistovodnoye source is 7.69 from field measurements. This pH value in the model is obtained for the water composition available at $P_{CO2} = 3.77 \cdot 10^{-4}$ bar in the atmosphere. Data (Keeling et.al., 1994) showed that P_{CO2} in the Earth's atmosphere increases with time gradually and in 1996 it will approximate $3.6 \cdot 10^{-4}$ bar. Thus, P_{CO2} value obtained over Chistovodnoye source is in gratifying agreement with P_{CO2} estimates in the Earth's atmosphere. Alternatively, P_{CO2} value in the air would be prescribed according to the reference book (Voitkevich et.al., 1990), where P_{CO2} estimate in the atmosphere equals $3 \cdot 10^{-4}$ bar, then pH of the given water will be 7.8 and not agree with the experimental determination. Besides pH, there is good agreement between the Chistovodnoye natural water composition determined by analysis and its modelled analogue on other components (Table 1).

Table 1. Natural composition (mg/l) of water (0), its modelled analogue in equilibrium with the atmosphere (1), and water compositions in line A (reactors 2-5) and B (reactors 6-9).

Component	0	1	2	3	4	5	6	7	8	9
Na^+	2.40	2.39	2.39	2.39	2.39	2.39	2.39	2.39	2.39	2.39
K^+	0.54	0.54	0.54	0.54	0.54	0.54	0.54	0.54	0.54	0.54
Ca^{+2}	4.46	4.41	4.407	4.38	4.35	4.31	4.41	4.20	4.20	1.77
Mg^{+2}	0.74	0.69	0.70	0.72	0.73	0.73	0.73	0.70	0.67	0.025
$Ca(HCO_3)^+$	n.d.	0.03	0.028	0.012	-	-	0.028	0.397	0.32	0.07
NH_4^+	n.d.	-	-	-	-	-	-	74.46	59.53	39.44
HCO_3^-	20.1	17.93	15.14	6.59	0.006	0.002	15.47	269.25	210.3	108.7
SO_4^{-2}	3.15	3.11	5.39	12.22	21.20	31.94	5.39	0.115	0.04	0.000
$NaSO_4^-$	n.d.	-	0.004	0.008	0.014	0.021	0.004	0.0001	-	-
HSO_4^-	n.d.	-	-	-	0.147	0.83	-	-	-	-
NO_3^-	<0.2	0.07	0.057	0.02	-	-	0.001	-	-	-
CO_3^{-2}	n.d.	0.04	0.03	0.005	-	-	0.001	0.15	0.59	6.76
F^-	0.09	0.089	0.089	0.089	0.081	0.063	0.089	0.089	0.089	0.089
HS^-	n.d.	-	-	-	-	-	-	2.13	6.13	11.11
$H_2CO_3^*$	n.d.	0.83	0.89	1.02	1.08	1.15	27.77	58.00	9.05	0.21
$Mg(OH)_2^*$	n.d.	0.09	0.06	0.008	-	-	-	0.0034	0.086	1.68
CO_2^*	n.d.	0.55	0.60	0.68	0.73	0.77	18.60	38.84	6.06	0.14
$CaSO_4^*$	n.d.	0.05	0.09	0.197	0.337	0.473	0.09	0.0013	0.005	-
$MgCO_3^*$	n.d.	0.001	-	-	-	-	-	0.0035	0.014	0.006
HF^*	n.d.	-	-	-	-	0.028	-	-	-	-
$CaCO_3^*$	n.d.	-	-	-	-	-	-	0.03	0.138	0.70
NH_4OH^*	n.d.	-	-	-	-	-	-	0.54	2.22	33.43
NH_3^*	n.d.	-	-	-	-	-	-	0.22	1.56	23.51
H_2S^*	n.d.	-	-	-	-	-	-	1.96	1.13	0.09
CH_4^*	n.d.	-	-	-	-	-	-	0.007	0.04	1.85
O_2^*	8.2	8.26	8.26	8.26	8.26	8.26	7.64	-	-	-
solid phases	-	-	-	-	-	-	-	C	C	C, Cc
wt.% gas phase	-	83.4	83.4	83.4	83.4	83.4	1	0.034	0.004	-
water/coal ratio	-	∞	10000	3330	2500	2000	10000	3330	2500	2000
pH	7.69	7.69	7.59	7.17	4.1	3.52	6.1	7.00	7.71	9.06
Eh	-	0.762	0.768	0.793	0.974	1.00	0.855	-0.23	-0.28	-0.39

n.d. -not determined, * -neutral compounds, C -graphite, Cc - calcite.

There are some differences for HCO_3^- anion, $H_4SiO_4^*$ compound, and in addition, in the modelled water composition the following components are present: $Ca(HCO_3)^+$, CO_3^{-2}, $H_2CO_3^*$, $Mg(OH)_2^*$, CO_2^*, $CaSO_4^*$, N_2^*, SiO_2^*, that have not been analysed chemically. Slightly lower concentration of HCO_3^- anion in the model as compared to the analysis is caused by insignificant lack of cations in the analysis, as electroneutrality of water in the natural water sample from Chistovodnyae source is -3.17%. Difference in $H_4SiO_4^*$ results from the absence of analysis for SiO_2^*. The model (table 1) shows the equlibrium of this water with the atmosphere, at least in regard to oxygen, as O_2^* concentration in water is 8.2 mg/l in the analysis and 8.26 mg/l in the model. Unfortunately, in the water analysis given, N_2^* and CO_2^* are absent, so we cannot strongly state the equilibrium of water and atmosphere on these components. However, such an equilibrium, as it is in regard to O_2^*, is quit reasonable. According to the calculations with "SELECTOR-C", rain waters in equilibrium with the atmosphere at $P_{CO2}= 4 \cdot 10^{-4}$ give pH = 5.6. Apparently, "SELEKTOR-C" described quite correctly the equlibrium of the surface and rain waters and the atmosphere, that allows complication of the model and consideration of the equlibrium of water-coal-atmosphere system.

4 MANY-REACTOR PHYSICO-CHEMICAL MODEL WATER - ATMOSPHERE - COAL

The model consists of two lines: in the first line water reacts with coal under the air excess (model A), in the second line water interacts with coal under the air absent (model B). Model A relates to the natural interaction of the surface waters with coal in mines, adits, coal quarries and beds having contact with the atmosphere. Model B is true in the cases when the air access to the place of water-coal interactis is hampered, that is quite usual when the infiltrating ground water reach the coal beds at a depth. Reactor 1 is common for both models A and B. It includes one ton of water (Chistovodnoye source) and five tons of air being in equlibrium with this water. From reactor 1 water goes in two directions: 2-5 (model A) and 6-9 (model B). Each of 2-5 reactors contains brown coal (wt.%): C=61.4%, H=6.1%, O=30.9%, N=0.75%, S=0.77% and five tons of air. Coal amount increases from 2 to 5 reactors: 100, 300, 400, and 500 g . In reactors 6-9 there are coal of the some composition and amount, but the air quantity in them another: 10 (6), 0.5 (7), 0.035 (8), and 0.01 kg (9). Thus, the water in models A and B went successively through the reactors and reacted with increased amounts of coal, that modelled the water infiltration to a coal bed under the air excess (model A) and absent (model B). Coal and air components were prescribed immobile. Table 1 shows all peculiarities of the water composition change in both models. In model A sulphur comes to water from coal and under the oxidation conditions (or air excess) gives SO_4^{-2} anion that results in water acidity increase (pH decrease). In parallel to acidity and Eh increase, HCO_3^- content decreases sharply to practically zero value, in spite of the fact that water reacts with increasing coal amount. In model B the process is much more complicated. While oxygen is present in the water, water-coal reaction results in the water acidity increase (6, table 1), but when dissolved oxygen disappear, water starts to increase its alkalinity mainly at the expense of NH_4^+ cation formation and decrease of SO_4^{-2} concentration. At the same time, with increase of NH_4^+ concentration in water, HCO_3^- concentration increases sharply compensating NH_4^+ cation positive charge (7, table 1). Under reduction conditions (Eh<0), sulphure supply from coal to water does not result in the SO_4^- anion formed but HS^- anion concentration increases gradually (7-9, table 1). Besides, under reduction conditions water is enriched in NH_4OH^*, NH_3^*, and CH_4^* and quckly reaches saturation on carbon that results in graphite precipitation. With pH increase, carbonate barrier is reached and calcite begins to precipitate from water (9, table 1). It is interesting that, according to the calculations, several hundreds grams of coal are enough to change cardinally the composition of one ton of slightly mineralized water , similar in composition to 1 (table 1), and turn it to acid (model A) or alkaline (model B) water.

Consideration of models A and B shows what a vigorous factor in the water composition evolution during their reaction with coals is the atmosphere quantity. F. I. Tutunova (1987) points to acidic nature of waters draining coal quarries and spoil banks. At the same time , the lower water-bearing horizons of the exploited coal deposits show higher values of pH. This noticed regularity is fully illustrated by the many-reactor model considered above.

5 ACKNOWLEDGMENTS

This study was financially supported by the Russian Fundamental Researching Fund (Project 97-05-65818). We would like to express our thankfulness to E.I. Nasdratenko (governor of Primorye region) for his support.

6 REFERENCES

Chudaeva, V.A., T.N.Lutsenko, O.V.Chudaev, A.N.Chelnokov, W.M.Edmunds, P.Shand, 1995 Thermal waters of the Primorye region, Eastern Russia. In: *Water-Rock Interaction WRI-8*, Y.K. Karaka & O.V.Chudaev (eds.), Proc. 8-th Int. Symp., Vladivostok, Russia. Rotterdam, Balkema, 375-378.

Karpov, I.K., K.V.Chudnenko, D.A.Kulik, 1995. Minimization of thermodynamic potentials in geochemical modelling: State of art. In: *Water-Rock Interaction WRI-8*, Y.K. Karaka & O.V.Chudaev (eds), Proc. 8-th Int. Symp., Vladivostok, Russia. Rotterdam, Balkema, 733-735.

Keeling, C.D. & T.P.Whorf 1994. Atmospheric CO records from site in the SIO air sampling network. In: *Trends '93: A Compendium of Data on Global Change*, T.A.Boden, D.P.Kaiser, R.J.Sepanski, and F.W.Stoss (eds), ORNL/CDIAC-65. Carbon Dioxide Information Analysis Center, Oak Ridge National Laboratory, Oak Ridge, Tenn., U.S.A.

Tutunova, F.I. 1987. *Hydrogeochemistry of the technogenes*. Moskov, Nauka, (In Russian).

Modeling the metasomatism of marbles

V. N. Balashov
Institute of Experimental Mineralogy of Russian Academy of Science, Chernogolovka, Moscow District, Russia

B. W. D. Yardley
Department of Earth Sciences, University of Leeds, UK

ABSTRACT: We have modelled infiltration and metasomatism of marble allowing for the effects of decarbonation on the flow patterns. Rates of porosity creation by reaction and porosity loss by creep have been calculated for calcite - quartz - wollastonite marble, and show that porosity generation is likely to outstrip collapse of porosity. The reaction of calcite + quartz to wollastonite in response to influx of water has been investigated further using a 1D finite difference model. We find that fluid released by decarbonation creates a back-flow, so that further advance of the reaction front is a result of diffusion of water against the Darcy flux. The effect of creep is especially to reduce the secondary porosity developed at the reaction front. In a 3D situation, the porous zone of reacted marble becomes a conduit for layer-parallel advection, and is then infilled by calc-silicate minerals due to silica metasomatism.

1 INTRODUCTION

The significance of calc-silicate rocks and skarns, as markers of fluid flow during metamorphism, has been recognised for many years, and their development has been linked to the creation of secondary porosity during decarbonation. Deformation also influences fluid flow processes (Knipe and McCaig, 1994), most simply by the closure of pore space in a ductile rock media due to the effective confining pressure: $\sigma_{eff} = \sigma_3 - P_f$.

In this paper, we model metasomatic processes that may lead to skarn formation in marbles, making allowance for fluid produced by decarbonation, enhancement of porosity by reaction and reduction of porosity by effective pressure creep.

2 THE MODEL FOR REACTION AND FLOW

Consider an isothermal - isobaric system in which a generalized mixed volatile reaction takes place. The independent component transport equations for the barycentric reference system, which define the balance between diffusive and advective flow, and have the form (Balashov and Lebedeva, 1996)

$$\frac{\partial \phi C_1}{\partial t} = \nabla \cdot (D^0 \frac{M_2}{\overline{V}_2} \frac{F}{\rho} \nabla C_1) - \nabla \cdot (C_1 \mathbf{q}) + \nu_1 j^r$$

$$\frac{\partial \phi C_2}{\partial t} = \nabla \cdot (D^0 \frac{M_1}{\overline{V}_1} \frac{F}{\rho} \nabla C_2) - \nabla \cdot (C_2 \mathbf{q}) + \nu_2 j^r \quad (1)$$

Where C_k - component concentration in the fluid, M_k - molecular mass, $\overline{V}_k$ - partial molar volume of component, ϕ - porosity, D^0 - counterdiffusion coefficient, F - reciprocal Archie's law formation factor, j^r - reaction rate. The Darcy velocity, $\mathbf{q}$ is given by:

$$\mathbf{q} = -\frac{k^p}{\eta}(\nabla P_f + \mathbf{i}_y \rho g) \quad (2)$$

where k^p - permeability of the porous medium, η - fluid viscosity, P_f - fluid pressure, ρ - fluid density, $\mathbf{i}_y$ - unit vector for vertical direction, g - acceleration due to gravity.

The bulk fluid transport equation is then:

$$\frac{\partial \phi \rho}{\partial t} = -\nabla \cdot (\rho \mathbf{q}) + j^r(\nu_2 M_2 - \nu_1 M_1) \quad (3)$$

Overall, the change of porosity with time in a rock undergoing reaction and compaction is given by:

$$\frac{\partial \phi}{\partial t} = -j^r \Delta V_s^0 + \dot{\phi}^{cr} \quad (4)$$

where ΔV_s^0 is the solid volume effect of reaction. The first term of equation 4 defines the change in porosity due to reaction, and the second term, $\dot{\phi}^{cr}$, is the rate of porosity loss due to creep.

3 APPLICATION TO A NATURAL METAMORPHIC REACTION

In this study we have taken reaction:

$$Cal + Qtz \xrightarrow{k^+} Wo + CO_2 \qquad (5)$$

occurring in marble at 600°C and 200 MPa. Data are available to evaluate both the reaction rate and the creep rate (Tanner et al. 1985, Walker et al.,1990, Zhang et al., 1994, Paterson, 1995).

Consider a marble at an initial time t^0. For a given initial porosity, any degree of reaction overstepping will lead to a rate of porosity generation which may be compensated by creep due to the stress, σ_{eff}. The opposing effects of chemical reaction and compaction have been combined in Figure 1, which shows conditions for equal porosity generation and porosity loss at an initial time t^0 along a line which we term here the Critical Initial Stress Curve. The position of this curve is itself a function of porosity, and it marks the threshold between a field in which infiltration-driven reactions are likely to be self-accelerating and a field in which they will come to a halt.

The results shown in Figure 1 are remarkable in that they show that appreciable porosities (several percent) may be maintained during reaction even when stresses are of the order of several hundred bars. This is in good accord with textural studies of calc-silicate rocks which illustrate growth into open porosity (Yardley and Lloyd, 1989).

4 COUPLED CREEP AND REACTION FLOW MODEL

Calc-silicate zones often develop at the edges of marble units, due to diffusion or infiltration of aqueous fluid from adjacent siliceous rocks into the marble. As a next step to understanding this process, we developed a 1-D model for infiltration-driven reaction, which includes the effects of reaction and creep on fluid budget and permeability.

4.1 *Formulation of the problem*

Consider a quartz-bearing marble in a vertical layer 10 m wide, within which wollastonite can grow due to horizontal infiltration of $CO_2 + H_2O$ under isothermal conditions of 600°C at 200 MPa. The boundary conditions are ($t \geq 0$)

$$x = 0 \begin{cases} X_1 = X_1^0 \\ P_f = P_f^0 \end{cases}, \quad x = L \begin{cases} \frac{\partial X_1}{\partial x} = 0 \text{ and } P_f = P_f^L \end{cases} \quad (6)$$

The initial mole fraction of CO_2 in the pore fluid throughout the marble is the equilibrium value for reaction (5): $X_{CO2} = 0.184$. The initial profile of fluid pressure through the layer is taken to be linear with boundary fluid pressures defined to support a pressure difference of 0.2 MPa, comparable to a lithostatic gradient in P_f.

The problem was solved for three possible stress regimes: $\sigma=0$, $\sigma=1$ MPa and $\sigma=2.5$ MPa and for two values of initial marble porosity: 1% and 0.1%. The initial calcite mode (volume fraction) was always 90%; the quartz mode was thus 9% or 9.9% according to the porosity value. The sizes of calcite and quartz grains were 0.1 and 0.01 cm respectively. The mutual diffusion coefficient in an $H_2O - CO_2$ fluid D^0, and viscosity η, were taken as $7.1 \cdot 10^{-4}$ [cm^2/s] and $7.3 \cdot 10^{-5}$ [Pa·s] respectively (Labotka, 1991). The permeability of the porous medium due to the interconnected porosity was taken to be

$$k^p = a^p \phi^3 \qquad (7)$$

(Zhang et. al., 1994, Zaraisky and Balashov, 1995), where a^p is a constant with the value 10^{-8} cm^2.

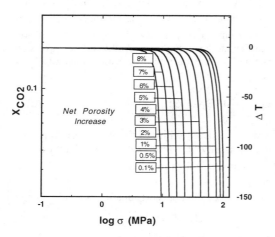

Figure 1. Family of Critical Initial Stress Curves showing the stress value (σ) needed for porosity collapse to balance the rate of porosity generation at any given degree of overstepping of reaction (5). Curves correspond to different initial porosities, as labelled. The vertical axes provide alternative measures of degree of overstepping from equilibrium ($\Delta T = 0$, $X_{CO2} = 0.184$)

4.2 Fluid Pressure and Flow

Reaction (5) yields CO_2 which then contributes to the Darcy flux of fluid. Figure 2a illustrates the distribution of fluid pressure inside the marble layer at $\phi^0=1\%$, $\sigma=0$ MPa. The initial reaction generates sufficient fluid to reverse the flow behind the front, developing a maximum in fluid pressure at the front (inset to Figure 2a).

The reaction front advances with time, but is driven by diffusion: the diffusive flux of water, less the Darcy flux of water, behind the reaction front, is equal to the Darcy flux of water ahead of the reaction front. With time, the intensity of the back-flow decreases, disappearing after 17.5a. The effect of reducing the initial porosity 10 times results in a c.100-fold increase in the maximum of fluid pressure, because it decreases the fluid flux through the marble layer. For this reason, where the initial porosity is very low, metamorphic reaction within the layer is dominated by reversed fluid flow. Applying a stress so that marble compacts by creep leads to more complex patterns of fluid pressure in space and time, (Figure 2b). A fluid pressure maximum exists at the reaction front, but it increases in magnitude with time due to the compaction.

5 IMPLICATIONS FOR NATURAL FLOW AND SKARN FORMATION

The main constraint on the fluid flux through the 1D model considered here is the low permeability of the unreacted marble ahead of the advancing front. In a natural, 3D situation, gradients in hydraulic head along the boundaries of the marble layer are inevitable, and will result in large, layer-parallel flows through the porous reacted margin of the marble, while the advance of the reacted zone itself into the marble remains diffusion-controlled. In the presence of a large flux of fluid derived from quartz-bearing lithologies, further production of wollastonite from calcite may take place in the permeable zone behind the reaction "front" (which is now strictly a "side") through reaction:

$$Cal + SiO^0_{2(aq)} \rightarrow Wo + CO_2 \quad (8)$$

Unlike reaction (5), reaction (8) leads to a reduction in porosity equivalent to 7.5% of the volume of the calcite consumed. In calc-silicate skarn layers at the edges of marbles, it is commonly the case that no carbonate remains, and hence the porosity generated by quartz-based decarbonation reactions, such as (5), must have been sufficient to permit complete

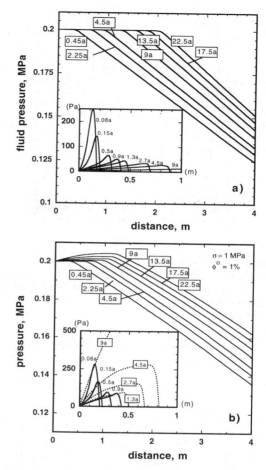

Figure 2. Evolution of the fluid pressure gradient across the infiltrated marble layer through time. Reaction front advances from the left. a) In the absence of stress, reacted rock behind the front has a fluid pressure close to the boundary value (0.2 MPa above the 200 MPa system pressure), with a small overpressure at the site of reaction which gives rise to backflow. This is shown in the inset (enlarged vertical scale gives excess pressure above the boundary value). b) For $\sigma=1$ MPa, collapse of porosity results in greater overpressure at the reaction site. The inset contrasts overpressures developed under stress (dotted lines) with those from part a (solid lines).

carbonate breakdown by metasomatic silica introduction, as in reaction (8), assuming no further creation of porosity by calcite dissolution.

For our model composition of marble, the amount

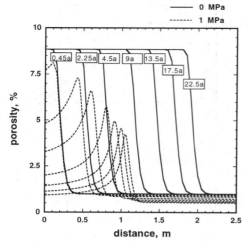

Figure 3. Evolution of the porosity profile across the reaction front through time. In the presence of stress, the maximum porosity is restricted to the site of reaction and decreases in magnitude as the front advances.

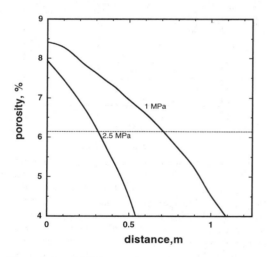

Figure 4. Locus of the maximum porosity attained at any time across the infiltrated marble for σ_{eff}=1 Mpa (from Figure 3) and 2.5 Mpa. Above the horizontal line, complete conversion of calcite to wollastonite by silica metasomatism is possible.

of porosity that must develop to permit total conversion of the remaining calcite to wollastonite by silica metasomatism, is 6.14%. We have plotted the locus of the maximum porosity attained as a function of distance (Figure 3), for stresses of 1MPa and 2.5MPa in Figure 4. Assuming that the layer parallel flow is sufficiently extensive for reaction 8 not to be limited by silica supply, it can be seen that pure wollastonite skarns will be restricted to the outermost 10's of cms of the reacting layer. In the higher stress case it is clear that only the outermost metre of marble is likely to be metasomatised at all, which is in good accord with observation.

REFERENCES

Balashov, V.N. and Lebedeva, M.I., 1996, On self-consistent model of chemical transport at metamorphism of rocks. *Doklady Russian Acad. Sci.* (in press).

Knipe, R.J., and McCaig, A.M., 1994, Microstructural and microchemical consequences of fluid flow in deforming rocks: In *Geofluids; origin, migration and evolution of fluids in sedimentary basins*. (ed,. J. Parnell) Geol. Soc. Spec. Publ. 78, pp.99-111.

Labotka, T.C., 1991, Chemical and physical properties of fluids. *Reviews in Mineralogy*, 26, 43-104.

Paterson, M.S., 1995, A theory for granular flow accommodated by material transfer via an intergranular fluid. *Tectonophysics*, **245**, 135-151.

Tanner, S.B., Kerrick, D.M., and Lasaga, A.C., 1985, Experimental kinetic study of the reaction: calcite + quartz = wollastonite + carbon dioxide, from 1 to 3 kilobars and 500 to 850°C. *Am. Jl Sci.*, 285, 577-620.

Walker, A.N., Rutter, E.H. and Brodie, K.H., 1990, Experimental study of grain-size sensitive flow of synthetic, hot-pressed calcite rocks. In *Deformation Mechanisms: Rheology and Tectonics*. (eds. R.J. Knipe and E.H. Rutter) Geol. Soc. Spec. Publ. 54, pp 259-282.

Yardley, B.W.D., and Lloyd, G.E., 1989, An application of cathodoluminescence microscopy to the study of textures and reactions in high-grade marbles from Connemara, Ireland. *Geological Magazine*, 126, 333-337.

Zaraisky, G.P., and Balashov, V.N., 1995, Thermal decompaction of rocks. In *Fluids in the Crust*. (eds. K.I. Schmulovich, B.W.D. Yardley and G.G. Gonchar) Chapman & Hall, pp.253-284.

Zhang, S., Paterson, M.S., and Cox, S.F., 1994, Porosity and permeability evolution during hot isostatic pressing of calcite aggregates. *Jl. Geophys. Res.*, 99, 15741-15760.

Forward modelling of complex water evolution – Soda waters in Northland, New Zealand

Franz May
Lehrstuhl Geodynamik-Physik der Lithosphäre, Rheinische Friedrich-Wilhelms-Universität, Bonn, Germany

ABSTRACT: Several geochemical sources and processes contribute to the formation of Northlands' soda waters. Their composition can be simulated by a stepwise batch reaction sequence. Dissolution of ascending magmatic CO_2 in ground water and wallrock alteration are modelled as irreversible reactions. Thermodynamic equilibrium is assumed for the precipitation of secondary phases. Mixing with saline ground water of marine origin is reflected by Br/Cl ratios. CO_2-degasing and Fe precipitation are kinetically controlled reactions within the free flowing springs. The calculated main element concentrations agree well with Northland HCO_3 waters. Some trace element concentrations are sensitive to the selection of the minor phase assemblage and parameter uncertainties.

1 INTRODUCTION

Mineral waters are the final product of water-rock interactions along subterranean paths from recharge areas to springs. Numerical simulations help to identify possible reactions contributing to mineral water genesis. Different numerical strategies can be used, involving inverse simulations, forward models and reaction kinetics coupled to fluid dynamic models. Forward models are best suited to study multi-stage processes. Water can participate in a variety of reactions with different rocks at different times and places within an aquifer. The identification and understanding of individual geochemical processes is essential for the design of numerical strategies, especially when reactions with slow kinetics are involved, e.g. dissolution of silicates in cold ground waters. Equilibrium assumptions depend on the relative flow- and reaction velocities. Irreversible reactions can be simulated using velocity constants or reaction progress (mass transfer) models. Forward models developed for HCO_3 in the Rhenish Massif (May 1996) are extended to simulate the evolution of Northlands' soda waters.

The code PHREEQC (Parkhurst 1995) has been used for the simulations. When CO_2 enters ground waters at shallow crustal depths, carbonic acid forms, which reacts with wallrocks. The basic process producing the Northland 'soda waters' is the neutralisation of carbonic acid by wallrocks. The waters are not in equilibrium with the rocks' primary minerals and their composition reflects several processes of water evolution.

2 NUMERICAL REALISATION

Different evolution steps revealed by the interpretation of water analyses lead to concepts for the numerical models, including:

1) initial ground water composition
2) diagenetic alteration of sea water
3) mixing of ground water and altered sea water
4) CO_2 dissolution
5) mineral composition of aquifers
6) wallrock alteration
7) oxidation by atmospheric oxygen.

The mineral waters originate from local meteoric recharge, as indicated by stable isotopes (Sheppard and Giggenbach 1985). The initial ground water can be a significant source only for Cl and S which are scarce in wallrocks. However, altered sea water contributes most of these elements. Thus, the initial composition is no major element source and mean values of local ground waters represent the initial fluid compositions. Br/Cl ratios indicate a sea water origin of these elements. Low Mg and SO_4 contents in high concentrated NaCl rich waters probably result from early diagenetic SO_4 reduction. Oxidation of organic matter in marine sediments by ascending CH_4 in the anoxic zone a few meters below the sea floor occurs widespread in shelf sequences. It is often associated with pyrite formation and carbonate precipitation. The complex redox reactions during early diagenesis can not be reconstructed from the Northland soda waters. Thus, a simplified geochemical model, assuming equilibrium between average sea water, hematite, CH_4, dolomite and iron-monosulfide, is included in the numerical strategy, despite arbitrary assumptions about CH_4 partial pressures that affect the resulting SO_4 concentrations. Since the halogenides do not participate in this reaction, the proportions of altered sea water and ground water are calculated from the Cl concentrations in the mineral waters.

Somewhere along the flow path water comes into contact with CO_2. The gas-water ratios determine the waters alteration potential. Wallrock minerals can dissolve as long as the water contains CO_2 that can be converted to HCO_3. The initial CO_2-water ratio can not be derived from the waters total C content, because of carbonate precipitation during alteration and subsequent dilution by fresh ground waters. Two extreme assumptions can be made however, to constrain the initial CO_2-water ratio. 1) The initial CO_2 content is estimated by trial and error parameter variation. A model with just enough CO_2 for carbonate precipitation and the dissolved carbon species represents the fact of dilution, but it can underestimate the H_3O^+ concentration at the deep reaction site. 2) Equilibrium between water and CO_2 at likely aquifer depths depends on temperature and pressure. Equilibrium models take into account, that the reactivity of the carbonic acid solutions is highest at the gas-water interface, and that this interface is a likely site of reactions, provided that unaltered wallrock is present. However, the resulting CO_2 and H_3O^+ concentrations are higher than measured. The main uncertainty of this model is the depth of the gas-water interface, because of the pressure and temperature dependence of CO_2 solubility and H_2CO_3 dissociation. At a geothermal gradient of 30 K/km and hydrostatic pressures, the pH in a pure H_2O-CO_2 system is nearly constant (3±0.1) between 0.4 and 4.5 km depth. Thus the variation of the initial CO_2 content with depth is of minor importance for the overall water evolution and both assumptions produce reasonable results.

Altered aquifer rocks are not accessible in Northland, but in the Rhenish Massif (Central Europe), aquifer rocks from wells producing HCO_3 waters similar to those from Northland show various degrees of alteration. Chlorite and feldspars have been transformed to kaolinite and carbonates (May et al. 1996). The clastic aquifers in the Rhenish Massif and in the Northland Waipapa terrane are made up of thick sequences of greywackes, sandstones and argillites. Average values for the Waipapa basement (Cox 1985) are used in the simulations. Illite, the major K source in the Rhenish rocks is not altered, thus only 10 wt % of the total K is assumed to be contained in soluble minerals. Quartz is also sparingly soluble, but it is the most abundant mineral in the siliciclastic rocks. It makes up about 30 wt % of the average rocks. The actual amount of the remaining Si contained in soluble minerals is not known precisely, but it is sufficient for the formation of secondary Al-silicates and the precipitation of excess silica, since the Al/Si ratio of kaolinite is higher than the respective ratio in the primary rock-forming minerals. The Si concentration in solution is then only a function of temperature and the silica polymorph precipitating. Minor amounts of spilites, pillow basalts and carbonates among the basement rocks have little effect on the Waipapa bulk rock average, but they can be an important sources of minor species for the fluids passing through these rocks. The Cl, F, and B values (12, 280, 33 ppm) given by Ellis and Mahon (1964) and an average value for Li in New Zealand greywacke (46 ppm, Reyes & Vickridge 1997) have been used for the reaction progress simulations. Many trace elements are higher concentrated in fine clastic lithologies than in sandstones. Average B concentrations in pelagic clays are around 140 ppm. The redox reactions between basement rocks and water and the rocks S content are not known. The production of SO_4 or sulfide minerals by wallrock alteration has not been considered in the models. S(VI) is the predominant calculated redox state in solution and the measured SO_4 concentrations are mostly similar to those in fresh ground waters. Thus the wallrocks do not appear to be a major S source for the mineral waters. Fe and Mn in the basement rocks are assumed to occur entirely in the valence states II and IV respectively. Variable chlorite-kaolinite ratios in Rhenish rocks and linear correlations between the major cations and HCO_3 concentrations indicate that the rock alteration is initially almost congruent dissolution and the state of reaction progress achieved is probably limited by kinetic factors. The reaction rate limiting step is the decomposition of primary minerals, but values are unknown. Simulations of progressive rock alteration and evolution of Rhenish HCO_3 waters (May 1996) are adapted to the aquifer rocks in Northland. The alteration reaction is simulated as successive irreversible addition of the soluble portion of average Waipapa basement rocks to the carbonated water. The following irreversible reaction was formulated, using the above approximations and the mean rock analysis:

$Si_{582}Al_{312}Fe_{61.4}Mn_{1.41}Mg_{37.2}Ca_{41.2}Li_{0.66}Na_{117.8}K_{6.58}P_{2.96}$ $B_{0.315}Cl_{0.034}F_{1.47}N_{2.24}H_{213.665}O_{1947.075}$ + 393.71 H_2O + 1.504 e^- + 1336.82 H^+ ↔ 582 H_4SiO_4 + 312 Al^{3+} + 61.4 Fe^{2+} + 1.41 Mn^{2+} + 41.2 Ca^{2+} + 0.66 Li^+ + 117.8 Na^+ + 6.58 K^+ + 37.2 Mg^{2+} + 2.96 PO_4^{-3} + 0.315 H_3BO_3 + 0.034 Cl^- + 1.47 F^- + 2.24 NH_4^+

The precipitation of the secondary phases proceeds in equilibrium with the changing fluid compositions. The main secondary phases are assumed to be the same as those found in the Rhenish rocks: quartz, kaolinite and Fe-rich carbonates. The Ca-Mg-Fe carbonates form solid solutions, which have been approximated by pure siderite and dolomite, and by $Mg_{0.1}Fe_{0.9}$-siderite, using a linear mixing model between siderite and magnesite to calculate the thermodynamic parameters.

Iron-oxihydrate precipitates in and around some Northland springs due to the contact with atmospheric oxygen. This final water evolution step assumes equilibrium between water, amorphous $Fe(OH)_3$ and O_2 at atmospheric partial pressure and spring temperature. The free CO_2 contents depend on the springs degasing kinetics (May 1997). Loss of CO_2 increases the waters H_3O^+ concentrations, resulting in a redistribution of pH dependent species, but it does not affect the measured element concentrations. Thus, this evolution step as well as subsequent formation of travertine or silica sinter are not modelled.

3 SIMULATION RESULTS

The results of numerical simulations are compared with mineral water analyses from two representative Northland springs. The calculated concentrations of many elements agree well with the measured values. The simulation conditions for Mangamuka (fig. 1) are: 0.1 vol. % altered sea water; equilibration with CO_2 at 23 °C and 25 bar pCO_2 (40 g/l initial CO_2); 9 g whole-rock alteration (including inert quartz and illite). The measured K, Al and SO_4 concentrations are below the simulated values. Their concentrations are probably limited by alunite, a minor phase in altered rocks from the Rhenish Massif. The ability of alunite to form a wide range of solid solutions with the alkali elements can be the reason for the low Li values in the Mangamuka water as well. The thermodynamic values for alunite given in the PHREEQC data base, do not indicate supersaturation however. The measured Fe concentration is lower than the calculated value for siderite-Fe(II) equilibrium at reducing conditions, but higher than the value calculated for additional Fe(III)-goethite equilibrium at atmospheric O_2 partial pressure. This reflects the Fe precipitation kinetics within the spring.

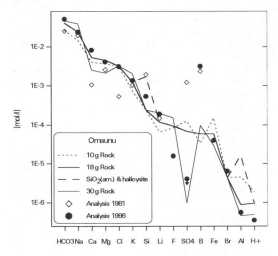

Figure 2. Comparison of Omaunu analyses with simulated concentrations.

The simulation parameters for Omaunu (fig. 2) are 0.5 vol. % sea water; 5.8 g/l initial CO_2; 23 °C reaction temperature; 10 - 30 g whole-rock alteration. The measured B concentrations exceed the simulated ones. High B concentrations are a general feature in many Northland mineral waters. Intensive leaching of B and Cl from basement rocks prior to carbonic acid neutralisation is a likely process explaining the high B values in geothermal waters from Ngawha (May 1997). The main variable affecting the simulated element concentrations is the amount of rock alteration (fig. 2). The concentrations of conservative elements, not included in secondary phases (Na, K, Li, B), are proportional to the amount of rock alteration. Water-rock ratios can be obtained for the wallrock alteration process from the multi-element plots, omitting species affected by special reactions, such as B, SO_4, H^+, CO_2, and Eh. The analytical variation of different elements, is too large to calculate 'precise' ratios and various parameter sets can produce

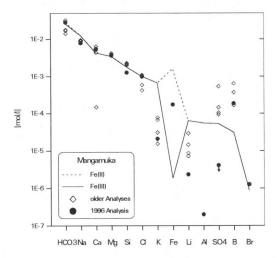

Figure 1. Comparison of Mangamuka analyses with simulated concentrations.

comparable results. The second equally important parameter is the set of secondary minerals. The Ca, Mg, and Fe concentrations depend on the assumed carbonate compositions. Mineralogical studies of carbonate from well samples are necessary to improve the calculations for these elements. Most element concentrations are insensitive to moderate changes of the reaction temperature. The Si concentrations are temperature dependent within the pH range of HCO_3 waters. Provided the corresponding silica species is known, Si can be used for geothermometry. The main cations often used in geothermometers only reflect the wallrock composition and the degree of reaction progress attained. Al concentrations are sensitive to the concentrations of the complexing ligands OH^- and F^-. This way the formation of F-apatite can significantly change the Al concentrations. Including further secondary phases (calcite, montmorillonite, pyrite, halloysite instead of kaolinite, or different silica species) improves the models for individual springs. These minerals will be minor phases and their actual presence in altered aquifer rocks should be checked with mineralogical analyses.

The relative amounts of the calculated secondary phases is in agreement with their abundance in altered rocks from Rhenish wells. Kaolinite is the predominant alteration product before quartz, siderite and dolomite. The mass-balance is -3.1 wt % for the Mangamuka simulation and +1.1 wt % for Omaunu. At a low degree of rock alteration, the Mangamuka water is still undersaturated with respect to dolomite, while the precipitation of siderite and dolomite is likely to reduce permeability in the hydrothermal system at Omaunu.

4 CONCLUSIONS

The conceptual sequence of reaction steps is a simplification of the continuous water evolution along flow paths in space and time. Among other methods, numerical models are suitable tools to test and to quantify concepts of multi-stage rock alteration and the associated water evolution, often involving processes of unknown kinetics, and to investigate the effects of parameter uncertainties, provided a comprehensive conceptual model is based on geochemical investigations of mineral waters and associated aquifer rocks.

ACKNOWLEDGEMENTS

This research has been supported by the German Science Foundation (DFG) and by the New Zealand Institute of Geological and Nuclear Sciences.

REFERENCES

Cox, M. E. 1985: *Geochemical examination of the active hydrothermal system at Ngawha, Northland, New Zealand: hydrochemical model, element distribution and geological setting.* Thesis Univ. Auckland, New Zealand.

Ellis, A. J. & Mahon, W. A. J. 1964: Natural hydrothermal systems and experimental hot-water/rock interactions. *Geochim. Cosmochim. Acta* 28: 1323-1357.

May, F. 1996: Equilibrium versus kinetics in CO_2 dominated hydrothermal alteration. In: Simmons, S. F.; Rahman, M. M.; Watson, A. eds. *Proc. 18th New Zealand Geothermal Workshop*, Auckland, New Zealand: 337-344.

May, F.; Hoernes, S.; Neugebauer, H. J. 1996: Genesis and distribution of mineral waters as a consequence of recent lithospheric dynamics: The Rhenish Massif, Central Europe. *Geologische Rundschau* 85: 782-799.

May, F. 1997 Geochemical models of mineral and thermal water genesis in Northland, New Zealand. *New Zealand J. Geol. Geophys.* (submitted).

Parkhurst, D. L. 1995: User's guide to PHREEQC - A computer program for speciation, reaction-path, advective-transport, and inverse geochemical calculations. *USGS Water-Resources Inv. Rep.* 95-4227: 143 p.

Reyes, A. G.; Vickridge, I. C. 1997: Lithium, boron and chloride in volcanics and greywackes in Northland, Auckland and the Taupo Volcanic Zone. *IGNS Sc. Rep.* (in press).

Sheppard, D. S.; Giggenbach, W. F. 1985: Ngawha well fluid compositions. In: Mongillo M.A. ed. The Ngawha geothermal field. *DSIR Geothermal Rep.* 8: 103-119.

Water-Rock Interaction, Arehart & Hulston (eds) © 1998 Balkema, Rotterdam, ISBN 90 5410 942 4

Chemical and isotopic features and flow path modelling of thermal fluids of the Abano system, Italy

C. Panichi, L. Bellucci, S. Caliro, F. Gherardi & G. Volpi
Istituto Internazionale per le Ricerche Geotermiche, Pisa, Italy
G. Magro & M. Pennisi
Istituto di Geocronologia e Geochimica Isotopica, Pisa, Italy

ABSTRACT: δD and $\delta^{18}O$ in waters, $\delta^{13}C$ in CO_2 and in CH_4, $^3He/^4He$ and $^{40}Ar/^{36}Ar$ ratios in natural gases were used to define the origin of water, CO_2 and CH_4. A good correspondence were obtained between measured chemical compositions of thermal springs emerging at Berici Hills and in the exploited spa areas, and those computed by mean a numerical simulation.

1 INTRODUCTION

Because of its high profile as a popular spa, where the thermal waters are also utilised for space and greenhouse-heating, the Abano region has received considerable geoscientific attention. Over the years, a series of geochemical surveys have produced a large set of data. A thorough inventory of the chemical analyses of cold and thermal waters and of the gas mixtures was carried out by Mameli & Carretta (1954) and Piccoli et al. (1973).

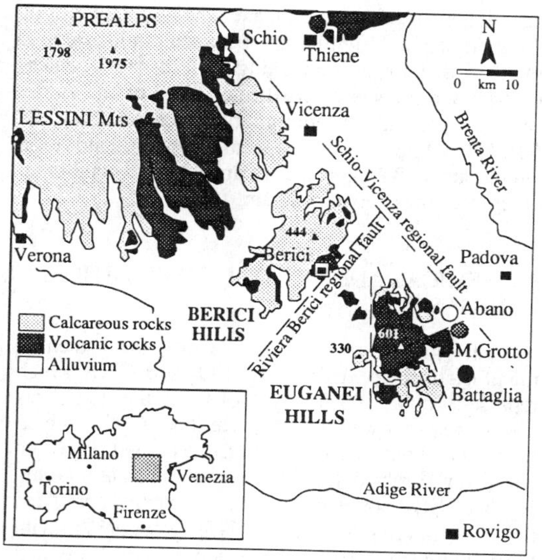

Figure 1. Geological sketch map of the Abano thermal area.

Circulation of waters in the Abano region is interpreted to be the result of meteoric water infiltration into outcrops of Permian and Mesozoic aquifers in the Prealps, which lie north of the thermal area. These aquifers, mainly constituted by limestone, dolomite and evaporite-bearing formations, have southerly dips and extend under the Abano region at depths between 0.5 and 2.5 km. This particular aquifer geometry is conductive to force convection both of aqueous ions and thermal energy along flow paths which extend from the outcrops in the source region to depths of 2 km below the Abano region and upward along high angle faults to thermal springs. On the basis of chemical composition of major ions and of the ^{18}O contents of water samples, Norton & Panichi (1978) suggest that the measured solute contents were the result of water/rock interactions with the bulk mineralogical assemblage of the main litologies crossed by the thermal fluids.

Moving from this result, this paper is aimed to give a comprehensive modelling of fluids flow path of the Abano thermal system. Old data are integrated by new chemical data relative to gaseous phase associated to water samples, and by ^{18}O, D and ^{13}C measurements of gas and water samples. To evaluate the reliability of field data with the hypothesised evolutionary path, a numerical simulation constrained by geological, physical and chemical parameters, is done. In addition, the availability of a large set of data permits to explain the observed variations in the fluids emerging in the intensely exploited spa areas of Battaglia, Montegrotto and Abano.

2 ISOTOPIC AND CHEMICAL COMPOSITION OF WATERS

The scatterplot of δD, δ^{18}O and Cl data given in Figure 2 provides a chemical and isotopic picture of the thermal waters emerging in the Abano area.

The δ values have been used (a) to infer the origin and the area of infiltration of the meteoric waters that recharge the hydrothermal system, and (b) to define the mixing between deep fluids and shallow groundwaters. Parent deep waters reaching the surface in Berici, Abano, Montegrotto and Battaglia springs and wells show similar δD values, ranging from -83 to -86‰. The oxygen-18 and chloride contents are however more variable, and reveal significative differences between the fluids delivered in each area.

Emergences of deep hot waters are aligned in a NW-SE direction in Abano, Montegrotto and Battaglia areas, and in an ENE-WSW direction near the Berici manifestations (Figure 1). Because these alignments are sub-parallel to the trends of the Schio-Vicenza and Riviera dei Berici regional faults, respectively, it appears reasonable to assume that the deep circulation is controlled by these main tectonic discontinuities.

Moving from a common source region (Prealps), the deep 'parent' waters emerge after about 70 km in the Berici area, where their chloride contents reach

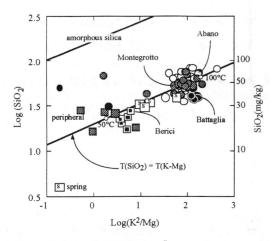

Figure 3. Log (SiO$_2$) vs Log (K^2)/(Mg) diagram.

≈0.1 g Cl/Kg, the minimum value for thermal waters in the Abano region. Waters from the same source that reach the surface in the Battaglia, Montegrotto and Abano areas 35 km away, exhibit a progressive enrichment in salinity, but maintain the same deuterium contents. Between Battaglia and Abano, the fluids also become enriched in oxygen-18. The data plot on mixing lines between the deep component and shallow waters, which are characterised by Cl contents of about 5 mg/kg and δ^{18}O values ranging between -9 and -8‰.

The diagram in Figure 3 shows silica concentrations versus K^2/Mg ratios of solutes, and has been used to evaluate the attainment of thermodynamic equilibrium between fluids and a mineral assemblage containing K-feldspar, K-mica, chlorite and a generic Al-silicate. Most of the Berici samples indicate re-equilibration temperatures of 60±5°C, while the Montegrotto, Abano, and to a slight extent, Battaglia waters appear to re-equilibrate at temperatures between 80 and 100°C.

3 CHEMICAL AND ISOTOPIC COMPOSITION OF GASES

The gases associated with the thermal manifestations are mainly N$_2$, CO$_2$ and CH$_4$, which represent about 95% (by vol.) of the total gas discharge, followed by Ar, He and C$_2$H$_6$, which represent about 1.3% (by vol.). H$_2$ and H$_2$S are absent or barely detectable by gas chromatography. N$_2$ and CO$_2$ concentrations are strictly correlated [(%N$_2$)+1,4414×(%CO$_2$)=95,595; R^2=0,969] with their abundances ranging between 65 and 95% and 0.5 and 20.5% (by vol.), respectively. Chemical features of the gas samples are grouped as a function of the sampling area. N$_2$ concentration, in particular,

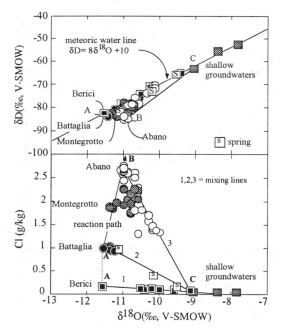

Figure 2. δD, δ^{18}O and Cl relationships in shallow cold waters and thermal springs in the Abano region.

decreases moving from colder marginal towards hotter spa areas, and viceversa for CO_2.

A ternary plot of N_2, CH_4 and CO_2 (Figure 4) shows the origin and the chemical evolution of gases within the thermal region. Data points show a CO_2/CH_4 ratio of about 2 and also that CO_2 and CH_4 concentrations become progressively higher as N_2 concentration decreases moving from the marginal areas to Montegrotto and Abano.

Figure 4 indicates that vapour loss at depth at the estimated temperature of 100°C cannot explain the chemical differentiation observed at the surface as it did for the situation between Battaglia and Abano. The most plausible explanation of the observed trend is the mixing between a deep component rich in CO_2 and CH_4 and a shallow component rich in N_2.

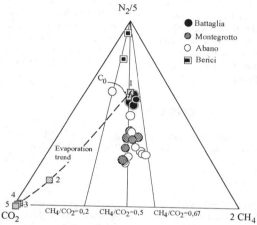

Figure 4. Relative concentrations (in mol/mol) of CO_2, N_2 and CH_4 in the gas mixtures. The evaporation curve (dashed) describes the variation of gas composition produced by a multistage vapour separation from 100 to 95°C.

Carbon dioxide has almost constant $\delta^{13}C$ values of -6.5±2‰. This signature is characteristic of mantle CO_2 (-5±2‰; Hoefs, 1997), but the helium isotopic composition having crustal values ($^3He/^4He$ between 0,07 and 0,09 Ra) exclude any mantle contribution. Probably the most important source of CO_2 is sedimentary and represented by hydrothermal alteration of the carbonatic and dolomitic rocks of the reservoir. Isotopic fractionation during hydrolysis at low temperatures can account for the measured values, which are somewhat lower than are typical of calcareous limestones (0±3‰, Hoefs, 1997).

The $\delta^{13}C$ values of CH_4 (-40±2‰) are higher than those of bacterial origin ($\delta^{13}C$<-60‰; Hoefs, 1997), and are lower than in geothermal fields (-27±3‰; Welhan, 1988). Considering the geological setting of area, these data suggest that the methane derives from the thermogenic breakdown of organic matter in reservoir calcareous rocks crossed by the geothermal fluids. This is further corroborated by the hydrogen composition of CH_4 (δD=-164±14.4‰), as it appears that the isotopic signatures of our samples fit into the standard classification schemes for thermogenic methanes proposed in the literature.

Comparing the $\delta^{13}C$ values of methane with those of CO_2, we obtain ε values between 34.2 and 35.1‰, which correspond to re-equilibration temperatures of 195±5°C, which are much higher than those obtained by solute geothermometers (Figure 3). The most logical explanation is that CO_2 and CH_4 have not reached isotopic equilibrium.

From the chemical viewpoint there is no equilibrium between CO_2 and CH_4, since the CH_4/CO_2 molecular ratios range between 0.2 and 0.67, corresponding to temperatures of 203.8 and 179.3°C, respectively. These data confirm the different mechanisms of formation of the two species and the lack of subsequent re-equilibration.

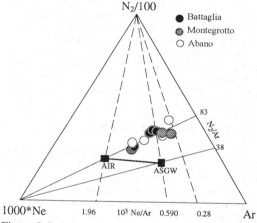

Figure 5. Relative concentrations of N_2, Ar and Ne.

The data points of Figure 5 lie within a narrow zone that shifts slightly towards the N_2 corner near the air-saturated groundwater component. As indicated in the literature (Giggenbach, 1990), Ne of geothermal continental systems is considered almost completely of atmospheric origin. In the Euganean gaseous mixtures, the Ar isotopic composition ($^{40}Ar/^{36}Ar$=295±2) also indicates an atmospheric origin. In this context, the N_2/Ar molecular ratio becomes an important genetic tracer. The N_2/Ar values ranging between 55.5 and 96.7, coupled with the [10^3*(Ne/Ar)] ratio, ranging between 0.278 and 1.012, clearly indicate a nitrogen excess. In other words, N_2 is principally atmospheric, but a relatively small percentage is probably of sedimentary origin and is produced via thermocatalitic alteration of sedimentary organic matter.

Table 1. Estimated and measured solute contents of natural springs in the Abano thermal area.

	T	P CO$_2$	NaCl	S.I. 1	S.I. 2	Ca	Mg	Na	K	Cl	HCO$_3$	SO$_4$	pH	Comments
	S.I. 3	S.I. 4	S.I. 5	S.I. 6										
1	25	0.006	0.001	0	-0.5	55	8	23	1	35	179	23	7.4	Estimated A
	0	0	1	-2.5										("Parent" meteoric fluid)
2	60	0.006	0.001	0	0	106	37	114	18	177	135	320	7	Estimated B
	0	0	0.4	-1.1	85	45	113	17	174	136	298	7		Berici (spring)
3	100	-	0.029	0	-0.6	146	31	620	39	1027	125	331	6.4	Estimated C
	0	0	0.55	-0.55	155	31	641	44	1040	160	320	6.5		Battaglia (spring)
4	100	-	0.079	0	-0.7	355	64	1494	76	2483	140	861	6.3	Estimated D
	0	0	0.55	0.05	392	59	1473	80	2517	131	830	6.6		Abano (spring)

S.I.= saturation index; 1= calcite; 2= dolomite; 3=chalcedony; 4= K-smectite; 5= kaolinite; 6=anhydrite.
P(CO2) input values of steps 3 and 4 coincide with the calculated value at equilibrium for step 2.
(T, P(CO$_2$), NaCl are expressed in °C, bar and mol/kg, respectively; solute contents are expressed in mg/l)

4 FLOW PATH SIMULATION

Water/rock interaction processes occurring all along the underground circulation from the Prealps to the Abano spa areas are summarised in Table 1.

Assuming an infiltrating water of meteoric origin with atmospheric P(CO$_2$) content, after reaction at 25°C with a calcite-dolomite-anhydrite-K-smectite-chalcedony and kaolinite mineral assemblage we obtain "Estimated A". This water is saturated in calcite and chalcedony, but remains slightly undersaturated in dolomite (SI=-0.5), because the upper part of the circuit in the fluids interact with a carbonatic aquifer. At the same time the lack of anhydrite leads to a strong undersaturation (SI=-2.5). Argillaceous materials are thought to be widespread in the limestone.

This bicarbonate water, while circulating through deeper horizons, is heated (up to 60°C, as estimated by geothermometers in the middle part of the circuit, see Figure 3), and tends to reach equilibrium conditions (SI=0) with the same mineral assemblage, except for anhydrite, only as a consequence of the increases in temperature and reaction times. This "Estimated B" coincides very well with the deep 'parent' water emerging in the Berici area. The presence of the Riviera dei Berici regional fault (Figure 1) facilitates the emergence of the deep waters and a simultaneous deepening of the circulation through the Permian evaporites at 2.5 km depth. Water circulation at this depth permits a progressive water/rock interaction at higher temperature. Steps 3 and 4 of Table 1 describe the chemical variations in the deep solution during its flow towards the thermal manifestations in the Battaglia and Abano areas. Estimated C and D solutions almost coincide with spring waters and indicate that the fluid emerging at Abano represents the last evolutionary step of the deep solution, which attained saturation conditions with the mineral assemblage described in Table 1 at temperatures of about 100°C. The observed variations in solute contents in the thermal springs inside the spa zones are ascribed to different reaction times occurring in the deep solution between Battaglia and Abano areas. The Schio-Vicenza regional fault (Figure 1) constrains the underground circulation at the eastern boundary of the thermal areas and provides an ascent route for the fluids to the natural springs.

ACKNOWLEDGEMENTS

The authors are indebted to D.Hayba and J.R.Hulston for careful reviews of the manuscript.

REFERENCES

Giggenbach, W.F. 1990. Water and gas chemistry of Lake Nyos and its bearing on the eruptive process. *J.Volcanol. Geotherm. Res.* 42: 337-362.

Hoefs, J. 1997. *Stable Isotope Geochemistry*. 4th Edition. 201pp, Springer Ed.

Mameli, E. & L.Carretta 1954. Due secoli di indagini fisiche e chimiche sulle acque minerali ipertermali, sui fanghi e sui gas Euganei. *Memorie dell'Accademia Patavina di Scienze, Lettere ed Arti*, 66: 1-146.

Norton, D. & C.Panichi 1978. Determination of the sources and circulation paths of thermal fluids: the Abano region, northern Italy. *Geoch. Cosm. Acta*. 42: 1283-1294.

Piccoli, G., Dal Prà, A., Sedea, R., Bellati, R., Di Lallo, E., Cataldi, R., Baldi, P. & G.C Ferrara. 1973. Contributo alla conoscenza del sistema idrotermale Euganeo-Berico. *Atti Acc. Naz. Lincei, Memorie Classe Sc. Fis. Mat. Nat.*, VIII,. 11: 103-133.

Welhan, J.A. 1988. Origins of methane in hydrothermal systems. *Chem. Geol.* 71: 183-198.

Trace element speciation in hydrotherms due to the influence of CO_2 on genetic 'silicate rock-thermal fluid' processes

E. N. Pentcheva
Bulgarian Academy of Sciences, Geological Institute, Sofia, Bulgaria

L. Van't Dack & R. Gijbels
University of Antwerp (UIA), Department of Chemistry, Antwerp-Wilrijk, Belgium

ABSTRACT: The present complex investigation (experimental and thermodynamical) elucidates: 1) the consequences of the influence of CO_2 on genetic processes in the "water - gas - silicate rock" dynamic system (in its hydrogeochemical evolution) leading to a modification of the existence and migration forms of trace elements (in solution and in suspension, as well as during formation of deposits at the emergence); 2) the differentiation of the trace element speciation in N_2-bearing thermal fluids and the influence exercised by CO_2 during the complete evolution: rock --> solution --> suspension --> deposit; 3) the present and future equilibrium state between the different phases and species which determines the final hydrochemical characteristics and activity of CO_2- and N_2-bearing thermal waters; and 4) the possibilities for the optimum utilisation and forecasting the evolution of the studied hydrothermal systems.

1. INTRODUCTION

Earlier investigations (Pentcheva, 1984; Pentcheva et al., 1995) proved that:
1. the secondary influence of CO_2, linked to more recent volcanic phenomena, on N_2-bearing thermal waters in granitic rocks (in Southwest Bulgaria), can be shown as an important metamorphosing action of the solute and gaseous composition;
2. the modification of the rich primary composition, typical for alkaline N_2-bearing deep fluids hosted in silicate rocks (Pentcheva, 1984) is a reflection of the processes acting on the fluids during the ascent of the water.

The aim of the present study was to identify the results of the influence of CO_2 on the formation, existence and migration of trace element species, both in solution and suspension, in N_2-bearing and especially in CO_2-influenced hydrothermal systems. This allows to elucidate the present and future state of the dynamic equilibrium between the different phases and species which determine the definitive "image" and activity of the thermal water.

2. METHODOLOGIES

Analytical data on major, minor and trace elements, in solution, suspension (as retained on a 0.4 μm pore-size membrane filter) and in the deposits found at the emergence, as well as for experimental water-rock interaction solutions, were obtained by a variety of techniques such as INAA, ICP-MS, etc. (Pentcheva et al., 1983; Veldeman, 1991). The speciation of major and trace elements and the mineral phases controlling their solubility were calculated by means of the WATEQ4F programme (Ball et al., 1987) together with concepts developed by Pentcheva (1984). Solid deposits were also studied by means of electron probe microanalysis (EPMA) (Pentcheva et al., 1992).

3. RESULTS AND DISCUSSION

1. The various water-rock interaction experiments (at temperatures up to 150°C) in the presence of CO_2 showed:
(a) the influence of CO_2 enrichment imposes an important deforming action on the "water - rock" system, with reactions displaced towards the enhanced solubility (with an inhibition of secondary sorption processes). The relative enrichment, due to the aggressivity of CO_2, of the Na-HCO_3 leachates in trace elements typical for pure granitic N_2-bearing alkaline thermal waters, with respect to the host rock is established (for ex. Fig. 1). However, the modelling solutions remain poorer in these elements relative to the natural fluids. This is probably due to secondary

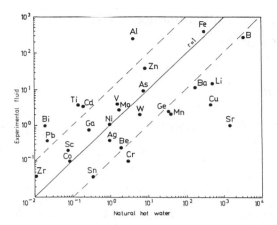

Figure 1. Trace element content of the CO_2-influenced water (connected with gneisses at depth) compared to the experimental Na-HCO_3 model solution.

metamorphisation processes limiting the migration in the aqueous thermal phase (sorption and precipitation barriers, etc.);

(b) an important genetic difference between the water rock interaction processes in the studied hydrotherms and those observed in the endogenic-CO_2 geothermal area in volcanic rocks of the French Massif Central (Pentcheva and Mercier, 1994) is established. The aqueous composition reflects another characteristic type of definitive microcomposition formation and specificity and not a simple deformation (Pentcheva et al., 1995).

2. (a) In the CO_2-influenced hydrothermal systems (Pentcheva et al., 1997), the transport in the suspended matter appears to be quite significant for a number of trace elements such as Mo, Ga, Au, Ag, Sc, Se (Fig. 2), and also for those elements having an unfavourable migration behaviour in alkaline waters, e.g. Zr, V, Se, heavy metals, Ce and U. The higher Eh and calcium content are suggesting a solubility diminution. In juxtaposition, the colloidal state is favoured in the alkaline medium of N_2-bearing waters for elements such as Tl, Cd, Ni, Sn, Zn and Ba (see Pentcheva et al., 1995). The EPMA study of the chemical composition and morphology of single particles of the suspended matter of CO_2-bearing water reveals the presence of aluminosilicates, calcium- and other -carbonates, silica polymorphs. They dominate the sulphide containing and mixed Fe-S-Si-Al particles characteristic for the suspended matter found in N_2-bearing alkaline thermal waters;

(b) under atmospheric conditions, the deposits formed at the emergence of CO_2-bearing waters appear to be very rich in trace elements, extracting (through coprecipitation, sorption, occlusion, etc.) not only elements transported in suspension but also the quite soluble rare alkali elements (Fig. 2). The relative enrichment (2 to 3 orders of magnitude) of the deposits with respect to the suspended matter attests to the mixed origin of this phase. This is probably due to the prevalence of direct precipitation out of solution due to CO_2 degassing. The evolution of trace elements in CO_2-influenced thermal waters from rock --> water --> deposit, leading to a significant loss of constituents, is a distinct disadvantage for the exploitation and consumption of these waters with respect to those of the unmetamorphosed, slightly mineralised N_2-bearing waters from crystalline rocks.

3. (a) In CO_2-influenced waters, the migration in solution of a lot of trace elements is favoured and their speciation is proving to be significantly different in juxtaposition to N_2-bearing hydrotherms (for ex. Figs. 3 and 4). It is evident that the hydrogeochemical deformation due to the action of CO_2 is leading to an abundance of bicarbonate, carbonate and sulphate (and biphosphate) species displacing the hydroxide, carbonate and sulphide forms (and simple cations) found in alkaline waters.

(b) The lower pH (< 7) of the CO_2-influenced waters is not so favourable for a good migration of W, Ga, Ge, Mo, V, Be and Zr, which are present in exceptionally stable anionic form in the "pure" alkaline N_2-bearing thermal waters (W: WO_4^{2-} and mixed species with OH^- and/or F^-; Ge: anions and neutral forms such as $H_2GeO_3^0$ and $H_4GeO_4^0$; Ga: as hydroxycomplexes such as $Ga(OH)_4^-$, $Ga(OH)_3^0$, $Ga(OH)_5^{2-}$; Mo: as anion or sulphide complexes; V as anions of V^{3+}, V^{4+} and V^{5+}; Be: as bicarbonate and carbonate complexes; U: as bicarbonate complexes of UO_2^{2+}) (Pentcheva, 1984). The present results show however that due to complexation (e.g. with HCO_3^- and SO_4^{2-}, and to a minor extent with CO_3^{2-} and OH^-) the mobility of these elements is not altered very much. The good migration ability of HCO_3^- species ensures the presence in these waters of elements specific for thermal fluids from crystalline rocks notwithstanding the elevated mineralisation and all the consequences of the secondary deformation (see Pentcheva et al., 1995).

4. The migration in N_2- and CO_2-bearing waters is controlled by a number of mineral species of different importance for both types (Fig. 5), more or less near to equilibrium. For the "deformed" CO_2-bearing waters, chalcedony, barite, sepiolite (c), magnesite, etc. are nearest to equilibrium. For the N_2-bearing thermal waters (from the same geological environment) the most important mineral phases controlling the solubility are fluorite, aragonite, covellite, (a,L)SiO_2, chalcedony, etc.

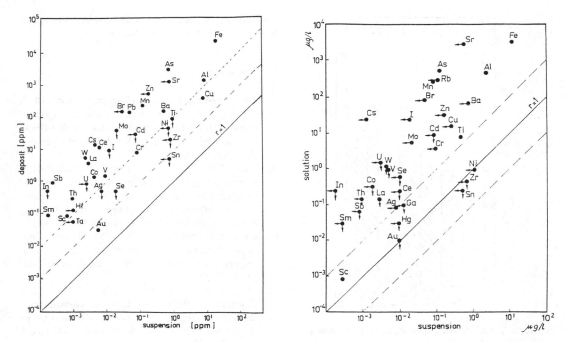

Figure 2. Juxtaposition of suspension to either solution or deposit composition for CO_2-influenced thermal water.

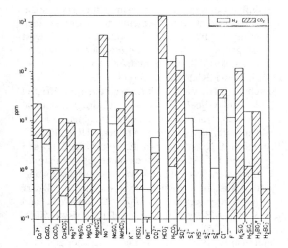

Figure 3. Speciation of major components in CO_2- and N_2-bearing thermal water.

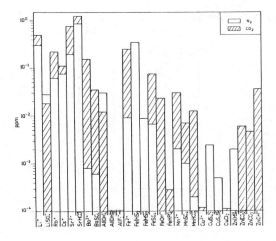

Figure 4. Trace element speciation in CO_2- and N_2-bearing thermal waters.

4. CONCLUSIONS

The processes leading to the definitive microchemical composition, in its complexity and dynamics, and, especially, of the trace element speciation can be considered as a reflection of complicated "water - gas - rock" processes and their geochemical evolution. The trace elements play a decisive role for the favourable or unfavourable action of the water on the organisms

The results of the present experimental and theoretical study differentiate the trace element migration in CO_2- and N_2-bearing thermal fluids. This

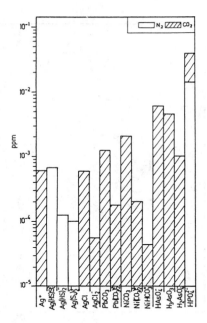

Figure 5. Trace element speciation in CO_2- and N_2-bearing thermal waters.

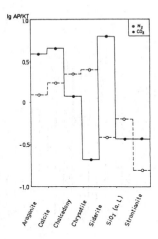

Figure 6. Some mineral phases controlling the solubility.

completes the knowledge regarding the natural dynamic system "rock - gas - water - suspension - solid deposits". These and the interphase geochemical processes are discussed from a primary and secondary genetic point of view.

The information obtained allows to forecast the evolution of the studied hydrothermal systems and to determine their optimum utilisation.

5. REFERENCES

Ball J.W., Nordstrom D.K. and Zachmann D.W., 1987. WATEQ4F - a personal computer Fortran translation of the geochemical model WATEQ2 with revised data base. *U.S. Geol. Surv. Open-file Rep.*, 87-50, 107 p.

Pentcheva E.N., 1984. Formation et évolution de la spécificité microgéochimique indicatrice des hydrothermes azotées de roches silicatées (système de modélisation expérimentale). Doct. Habil. Thesis, Geol. Inst. Acad. Sci. Sofia, 254 p.

Pentcheva E.N. and Mercier F., 1994. L'interaction "eau thermale - roche volcanique" et son importance génétique. *C.R. Acad. Bulg. Sci.*, **47**, no. 2, 69-72.

Pentcheva E.N., Swenters K., Van't dack L. and Gijbels R., 1983. Recherches comparatives microchimiques (SME and AAN) d'hydrothermes des granites de la Bulgarie du Sud. *C.R. Acad. Bulg. Sci.*, **37**, 509-512.

Pentcheva E.N., Van't dack L. and Gijbels R., 1995. Influence of recent volcanism on the geochemical behaviour of trace elements and gases in deep granitic hydrothermal systems, southwest Bulgaria. in *"Water-Rock Interaction - WRI-8"* (Kharaka Y.K. and Chudaev O.V., eds.), A.A. Balkema, Rotterdam, 383-387.

Pentcheva E.N., Van't dack L., Veldeman E., Hristov V. and Gijbels R., 1997. Hydrogeochemical characteristics of geothermal systems in South Bulgaria. University of Antwerp (UIA), 121 p.

Pentcheva E.N., Veldeman E., Van't dack L. and Gijbels R., 1992. Trace element geochemistry of the system "rock-thermal water-suspended matter-deposits" in a granitic environment. in *"Water Rock Interanction - WRI-7"* (Kharaka Y.K. and Maest A.S., eds.), A.A. Balkema, Rotterdam, 1321-1325.

Veldeman E., 1991. Geochemical characterization of hydrothermal systems in a granitic environment. Ph.D. thesis, Univ. Antwerp., Belgium, 335 p.

PHOX: Automated calculation of mineral stability and aqueous species predominance fields in Eh (or log (fO_2) or pε)-pH space

D.A. Polya
Department of Earth Sciences, The University of Manchester, UK

ABSTRACT: The Turbo-Pascal program PHOX (Version 1.0) presented here enables the calculation and presentation of Eh-pH, pε-pH or log (fO_2)-pH aqueous species predominance, mineral stability and relative mineral stability diagrams for any 5 component chemical system of the form M-L-O_2-H^+-H_2O, where M and L may be **any** chemical elements, but most usefully the major elements other than oxygen and hydrogen found in geochemically important metal and ligand species respectively. The program requires the input of (i) temperature, (ii) analytical concentrations of the 'metal' and 'ligand' species and (iii) standard Gibbs free energies of formation and chemical formulae for all species to be considered. Diagrams for the As-S-O_2-H^+-H_2O and S-O_2-H^+-H_2O systems illustrate the utility of the code in calculating complex equilibria involving over 25 species (including polynuclear and hydrated species) and 300 equilibrium boundaries.

1 INTRODUCTION

The redox state and acidity of an aqueous fluid as characterised by Eh (or log (fO_2) or pε) and pH are two of the most important chemical parameters controlling the speciation of dissolved elements, mineral solubility and surface adsorption processes. Consequently, diagrams which illustrate, as a function of Eh and pH, (i) the predominance fields of the aqueous species, (ii) the solubility of the mineral species and (iii) the relative stability fields of the various minerals species of a given component have great utility in enabling the visualisation of the behaviour of metals and other solutes in natural and synthetic aqueous solutions at ambient and elevated temperatures. Such diagrams are widely used in one form or another by geochemists, corrosion chemists and electrochemists. Pourbaix (1966) and Brookins (1988), notably, have published atlases of such diagrams, but the construction of further such diagrams can be a laborious and time-consuming task prone to error, particularly for systems in which a large number of species need to be considered. Reported here is a Turbo-Pascal computer program, PHOX, designed to accurately and robustly calculate and construct such diagrams at any temperature and pressure for which suitable data exist for 5 component chemical systems of the type M-L-O_2-H^+-H_2O (cf. M-L-O-H-(e^-)), where M and L might typically be metal and ligand components respectively.

2 DESCRIPTION OF CODE

2.1 *Input*

The program PHOX requires input of (i) temperature, (ii) analytical concentrations of

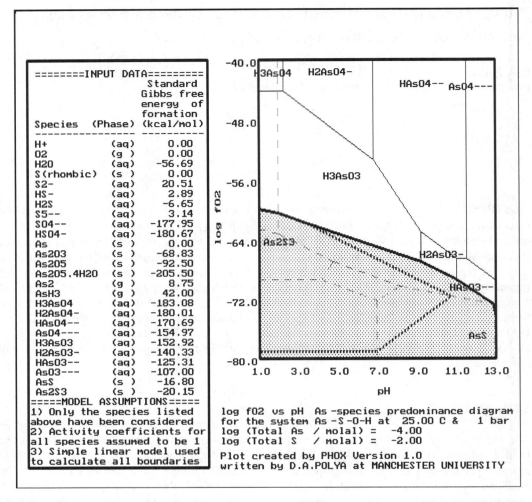

Figure 1. Example of aqueous species predominance and relative mineral stability diagram created by PHOX Version 1.0. Chemical system and analytical concentrations of 'metal' and 'ligand' as specified, together with the list of the explicitly considered chemical species, ΔG_f° data for which are from Garrels and Christ (1965) and Pourbaix and Pourbaix (1992). If log (fO_2)-pH conditions exist for which a solid 'ligand' species, such as rhombic sulfur, is supersaturated for the specified analytical concentration of the 'ligand', then for these conditions the analytical concentration of the 'ligand' is limited to that required for saturation of that solid 'ligand' species. Aqueous species predominance boundaries are indicated by light continuous lines. Mineral solubility boundaries are indicated by heavy continuous lines. Relative mineral stability boundaries are indicated by heavy dotted lines. Regions of mineral supersaturation are shaded. Light dashed lines indicate species predominance boundaries in the 'ligand'-O_2-H^+-H_2O system.

the 'metal' and 'ligand' species and (iii) the standard Gibbs free energies of formation and chemical formulae of the species to be considered, necessarily including O_2 (g), H^+ (aq) and H_2O (aq).

2.2 Assumptions

The assumptions or approximations below are made implicitly or explicitly by the PHOX code:(i) the species for which data have been input constitute a complete set of species to be considered; (ii) equilibrium exists in the solutions modelled; (iii) activity coefficients for all aqueous and solid species as well as for the solvent are unity; and (iv) the concentration of any aqueous species of a given component within that species' predominance field is equal to the analytical concentration of that component, provided that the solution is not saturated with respect to a solid phase containing that component. The last assumption constrains all constructed boundaries in Eh-pH space to be linear (this is clearly only an approximation, particularly near species predominance boundaries). These assumptions are sufficient to enable the determination of the species predominance fields and relative and absolute mineral stabilities from suitable data without resorting to iterative techniques.

2.3 Method of calculation

The methods of calculation are outlined only briefly here as space does not permit a rigorous treatment. For systems of the type $M-O_2-H^+-H_2O$, firstly enummeration is made of all possible equilibria between pairs of M-bearing chemical species, balanced using some combination of O_2 (g), H^+ (aq) and H_2O (aq). All such equilibria are balanced, the standard Gibbs free energy of reaction calculated by:

$$\Delta G_R° = \sum_i \nu_i \Delta G_{f,i}°$$

Where each of the pair of species are aqueous then a species predominance boundary in log (fO_2)-pH space is calculated. Where one species is aqueous and the other solid, a mineral stability (i.e. solubility) boundary) is calculated. Where both species are solid, a realtive mineral stability boundary is calculated.

All possible univariant points are then calculated and then eliminations made of points that are found within the fields of more predominant aqueous species or more stable solid species not involved in the definition of that univariant point. An algorithm based on Le Chatelier's Principle is used to establish the position of stability or predominance fields relative to calculated boundaries. For $M-L-O_2-H^+-H_2O$ systems, a more complex calculation procedure follows the calculation of equilibria in the $L-O_2-H^+-H_2O$ system. If required, calculations of equilibria in Eh-pH or pε-pH spaced are derived using well established relations between Eh, pε, log (fO_2) and pH from the equivalent log (fO_2)-pH equilibria.

2.4 Executing the PHOX code

The PHOX code has been run and executed on DOS and Windows based Pentium PCs. User defined thermodynamic data are listed in a suitably formatted text-file, whilst the selection of chemical system and the specification of analytical concentrations of the 'metal' and, if required, ' ligand' components are input interactively. A post-processing code, PHPLOT, enables rapid plotting of calculated boundaries using user-defined scales and axes (i.e. Eh, pε, log (fO_2)). Output is to VDU or file.

2.5 Examples of Output

Figure 1 shows an example of a diagram generated by PHOX in under 30 seconds using a 133 MHz Pentium processor. It illustrate the utility of the code in calculating complex equilibria involving up to 26 species and nearly 300 equilibrium boundaries

3 USES & LIMITATIONS

The PHOX code enables, amongst other uses, a rapid assessment to be made of (i) the most likely predominant aqueous species of a given component as a function temperature, Eh, pH and the analytical concentrations of the component and a 'ligand' component; and the (ii) sensitivity of predominance and stability boundaries to variations in values of Gibbs free energies of formation of individual species. In the modeling systems where the formation of certain mineral or aqueous species is kinetically inhibited, phase equilibria may be rapidly recalculated using PHOX simply by omitting or suppressing these species in the user-defined database.

Inherent limitations of the code include that (i) significant curvature may exist for boundaries in regions where more than two species of the component considered are significantly present, that (ii) the accuracy of the calculations is reduced where analytical concentration of the 'metal' component approach that of the 'ligand' component.

As for any code calculating chemical equilibria, any meaningful application to real systems requires an assessment of the validity of model used, in particular with regard to the accuracy of the thermodynamic data input & the activity-composition model, the validity or otherwise of the assumption of equilibrium and the significance of ignoring species and components not explicitly included in the calculations.

Clearly, for more accurate modelling, particularly of systems containing more than 5 significant components, the use of packages such as EQ3, SOLMINEQ.88, PHREEQE or Geochemists Workbench (cf. Plummer (1992), Nordstrom and Munoz (1994) and Bethke (1996) and references therein) rather than PHOX is indicated.

4 CONCLUSIONS

The PHOX (Version 1.0) code reported here enables automated calculation and construction of species predominance, mineral stability and relative mineral stability boundaries in Eh (or $\log(fO_2)$ or $p\varepsilon$) -pH space for $M-L-O_2-H^+-H_2O$ systems at any temperature and pressure for which suitable thermodynamic data exist.

ACKNOWLEDGEMENTS

Thanks are due to participants at the MDSG meeting in Manchester in 1995 for their comments on a previous version of the code; to Jeff Brown for β-testing; and to John Walshe for introducing us to mineral-fluid equilibria in his graduate classes at the University of Tasmania.

REFERENCES

Bethke, C.M., 1996. *Geochemical Reaction Modeling*. Oxford University Press, New York, 397pp.

Brookins, D.G., 1988, *Eh-pH diagrams for geochemists*. Springer-Verlag.

Garrels, R.M. and Christ, C.M., 1965, *Solutions, minerals and equilibria*. Freeman and Cooper, 450p.

Nordstrom, D.K. and Munoz, J.L., 1994, *Geochemical Thermodynamics*, Blackwell Scientific Publications, 477p.

Plummer, L.N.,1992. Geochemical modeling of water-rock interaction. In Kharaka, Y.K. and Maest, A.S.(Eds). Proceedings WRI-7, Park City, Utah, Balkema, 1, 23-33.

Pourbaix, M., 1966, *Atlas of electrochemical equilibrium in aqueous solutions*, Pergamon Press, 644p.

Pourbaix, M. and Pourbaix, A., 1992. Potential - pH diagrams for the system S-H_2O from 25 to 150°C: Influence of access to oxygen in sulphide solutions. *Geochim. Cosmochim. Acta*, 56, 3157-3178.

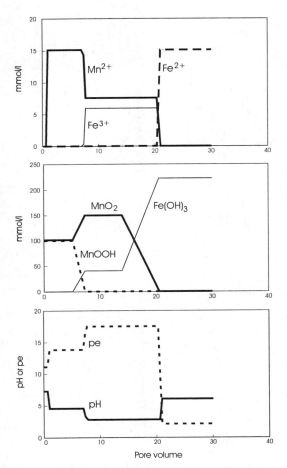

Fig. 3 PHREEQC modeling results of transporting a 15 mM $FeCl_2$ through a column containing equal amounts of MnOOH and MnO_2.

the breakthrough of Fe^{2+} occurs at around 20 pore volumes.

The presence of MnOOH could give rise to two possible reactions within the column. First reaction between Fe^{2+} and MnOOH:

$$Fe^{2+} + MnOOH \rightarrow FeOOH + Mn^{2+} \quad (2)$$

The other possible reaction is of MnOOH with the protons produced by reaction (1)

$$2MnOOH + 2H^+ \rightarrow MnO_2 + Mn^{2+} + 2H_2O \quad (3)$$

To evaluate how reactions (1-3) relate to the experimentally observed breakthrough curves of Mn^{2+} and Fe^{2+} (Fig. 3) some solute transport modeling was carried out.

4 MODELING

Solute transport modeling was carried out using the code PHREEQC (Parkhurst, 1995) with extended facilities by Appelo and Parkhurst (this volume). A model flowtube is dimensioned equivalent to the column of experiment 2. The water in the flowtube is equilibrated with birnessite, manganite and $Fe(OH)_3$. Equal amounts of MnO_2 and MnOOH are initially present but the initial amount of $Fe(OH)_3$ is neglectable. A 15 mM $FeCl_2$ solution is transported through the flowtube at a rate of 3 ml/hr. Dispersion and diffusion are at this stage neglected.

Some modeling results are displayed in Figure 3. The processes occurring downstream in the flowtube can best be understood by reading the changes in effluent composition from higher numbered pore volumes to lower numbered pore volumes, i.e. from right to left in Figure 3. The first reaction front occurs near 20 pore volumes where Fe^{2+} reacts with MnO_2 according to reaction (1). Part of the Fe^{3+} precipitates as $Fe(OH)_3$ and the pH drops. However, at pH values around 3, some of the ferric iron may remain in solution and in reality Fe-oxide has here a pH buffering effect. The second reaction front is found at about 7 pore volumes and here the protons produced at the Fe^{2+}/MnO_2 reaction front react with MnOOH according to reaction (3). Thus the result is disintegration of MnOOH into Mn^{2+} and MnO_2. Fe^{3+} reaching the $H^+/$MnOOH front will precipitate as $Fe(OH)_3$ due to the increase in pH caused by the proton consumption of reaction (3). Note that Fe^{2+} never reaches MnOOH and reaction (2) is therefore not possible. The reason why the two reaction fronts become separate is that the MnO_2 production at the $H^+/$MnOOH front slows down the following Fe^{2+}/MnO_2 front.

Comparison of the model results (Fig. 3) with the experimental results (Fig. 2), shows that the model simulates the overall behavior of Fe^{2+} and Mn^{2+}. It explains the breakthrough curve for Mn^{2+} with two plateaus, where the highest plateau is due to the disintegration of MnOOH and the second plateau corresponds to the reaction between Fe^{2+} and MnO_2. In fact the experimentally observed breakthrough curves for Fe^{2+} and Mn^{2+} at the Fe^{2+}/MnO_2 reaction front perfectly matches the 2:1 stoichiometry predicted by Eqn (1). However the experimentally observed Fe^{2+}/MnO_2 reaction front (Fig. 2) is more gradual than predicted by the equilibrium model (Fig. 3), indicating that reaction kinetics are of importance as well. Furthermore, additional processes must be taken into consideration to explain the experimental results. First of all, ion exchange plays a role as indicated by the Ca^{2+}

breakthrough curve. Also, the appearance of Al^{3+} suggests that part of the pH buffering role of $Fe(OH)_3$ is replaced by $Al(OH)_3$, which may explain the discrepancy between the observed and modelled pH. Work is in progress to include these processes in the model.

REFERENCES

Appelo, C.A.J. and Parkhurst, D.L., 1998, Enhancements to the geochemical model PHREEQC- 1D transport and reaction kinetics, WRI-9, this volume.

Giovanoli, R., 1980a, Vernadite is random stacked birnessite. Mineral. Deposita 15: 251-253.

Giovanoli, R., 1980b, On natural and synthetic manganese nodules. In Varentsov, I.M. and Grasselly, G. (Eds) Geology and Geochemistry of Manganese, v. 1, p. 159-202.

Murray, J.W., Balistrieri, L.S. and Paul, B., 1984, The oxidation state of manganese in marine sediments and ferromanganese nodules. Geochim. Cosmochim. Acta 48: 1237-1247.

Parkhurst, D.L., 1995, User's guide to PHREEQC- A computer program for speciation, reaction-path, advective transport, and inverse geochemical calculations. US Geol. Surv. Water Resour. Inv. 95-4227.

Postma, D., 1981, Formation of siderite and vivianite and the pore-water composition of a recent bog sediment in Denmark, Chem. Geol 31: 255-244.

Postma, D., 1985, Concentration of Mn and separation from Fe in sediments-I. Kinetics and stoichiometry of the reaction between birnessite and dissolved Fe(II) at 10°C. Geochim. Cosmochim. Acta 49: 1023-1033.

Hydrogeochemical processes at the fracture/matrix boundary of fractured sandstones

M. Sauter & R. Liedl
Applied Geology, University of Tübingen, Germany

ABSTRACT: The quantification of hydrogeochemical dissolution processes at the boundary between fracture and matrix of fractured porous sandstones is important for the understanding of weathering processes and for the assessment of the acid-neutralisation capacity of the rock material, resulting from e.g. the impact of acid rain or mining activities. Using laboratory reactive diffusion experiments the parameters determining the migration of the reaction front into the rock matrix were determined. The advance of the front was simulated using an analytical model based on the "shrinking core" concept. Because of the temporal change in parameters (diffusion coefficient, porosity, hydraulic conductivity) a finite-difference model was developed to investigate the coupled reactive transport and to analyse the nature of the effective diffusion coefficient determined in the laboratory. Using this numerical model, a simple relationship could be derived which allows the determination of the effective diffusion coefficient for a retreating front from independent rock parameters. Data from a natural analogue were used to validate the model.

1 INTRODUCTION

Groundwater quality of fractured rock aquifers is mainly determined by the hydrochemical interaction between the rock matrix and the groundwater flowing in the fractures. These processes are responsible for changes in the groundwater composition as well as for chemical-mineralogical and physical changes in the rock matrix (porosity, permeability). The understanding of these processes is a prerequisite for the understanding of groundwater quality, especially in the context of the assessment of weathering processes, the effect of acid rain or mine workings (new installations, flooding, tailings).

In many cases, the calcitic cement of the sandstone matrix is the only buffering mechanism for the acid waters. According to our present knowledge, transport of ions from the reaction front into the rock matrix is mainly determined by diffusion and advection in the outer higher permeability zone of the matrix seam, created as a result of the removal of soluble minerals. The relative importance of each of these processes depends on the hydraulic and hydrogeochemical parameters as well as the hydraulic gradient in the fracture. In this study the role of carbonate cement of sandstones in the long-term buffering of proton input is analysed and, in particular, it is examined how the effective diffusion coefficient, which changes with time can be determined from independent measurements.

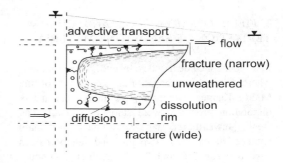

Figure 1. Conceptual model of migration of reaction front into the rock matrix

2 CONCEPT AND PREVIOUS WORK

The dissolution process of carbonate cement in a fractured porous sandstone is visualised as shown in Figure 1. Calcium-ions are transported from a dissolution front in the matrix into the mobile water in the fracture. The rate of advance of this front is determined by the properties of the carbonate-depleted solution zone (solution rim), i.e. porosity and permeability, the fracture geometry, the hydraulic potential distribution, and the reaction kinetics of the carbonate system. The rate of advance increases with the flow rate of the fracture water, which provides the upkeep of a concentration gradient between dissolution front and mobile fracture water. In narrower fractures, this concentration gradient is lower and therefore the rate of advance of the dissolution front is reduced.

The chemical processes occuring in the fractures and the adjacent pore space not only affect the water quality but also the hydraulic properties of the rock. Such types of processes have been studied for example within the context of the genesis of uranium deposits and their subsequent leaching during mineral exploitation (e.g. Bennett et al., 1992). For the risk assessment of nuclear repositories, the complex chemical processes, coupled with groundwater flow and solute transport have been studied extensively (e.g. Tsang, 1987). There are a number of theoretical and modelling studies, evaluating the effect of the chemical leaching of fractured porous systems (e.g. Read et al., 1993).

In this paper, a comprehensive study is presented comprising field, experimental laboratory and modelling investigations for this coupled fracture/matrix dissolution process.

3 FIELD OBSERVATIONS AND CHARACTERISATION OF SANDSTONE MATERIAL

In an abandoned quarry (Gschwend, SW Germany), where formerly carbonate-rich sandstones of the Middle Keuper formation were mined, the matrix dissolution processes were studied in detail. The fresh, unweathered sandstones have carbonate contents of ca. 40% (weight) and the porosities vary between 2% and 3.4%. Hydraulic conductivities between $1.8*10^{-11}$ and $3*10^{-12}$ m/s were measured. The fractures examined in the field display spacings of between a few and approximately ten meters. Near the surface they show apertures of ca. 0.15 m, which are reduced with depth to only a few centimeters. Contrary to our expectations based on the conceptual model, the dissolution rims increase with increasing depth (Figure 2). This observation can be explained by the disintegration of the rock matrix after removal of the carbonate cement. Residual carbonate contents vary between 15% and 20% (weight). The porosities in the solution rim vary between 18% and 33% and hydraulic conductivities fluctuate between ca. 10^{-4} and 10^{-5} m/s.

4 DIFFUSION EXPERIMENTS

It is believed that diffusion processes contribute to a major degree to the transport of ions between dis-

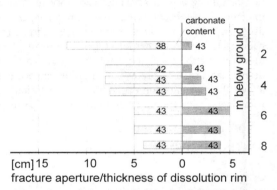

Figure 2. Changes in fracture aperture and dissolution rims with depth

solution front and mobile fracture water, especially if the dissolution rim is further advanced into the rock matrix. In order to investigate diffusion with a retreating boundary and the concomitant changes in the diffusion coefficient, laboratory experiments were conducted using a diffusion cell as shown in Figure 3, with samples of a diameter of 4.6 cm and thicknesses varying between 0.5 cm and 1.0 cm. One of the reservoirs contained hydrochloric acid of pH 1 and pH 2 for two types of experiments, the other contained an isomolar solution of potassium iodide in order to prevent osmotic effects. During the experiment, the reaction front moves from right to left and the hydraulic conductivity, porosity and the diffusion coefficient in the affected zone increase with time. The changes in the mass of Ca^{2+} and K^+ in the respective cells were monitored and effective

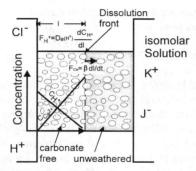

Figure 3. Diffusion cell used for determination of effective diffusion coefficient

diffusion coefficients were determined using an analytical solution, based on the "shrinking core" conceptual model (Hinsenveld, 1993). The approach is based on the assumption that the diffusive flux of H^+ to the dissolution front equals the flux of protons consumed for calcite dissolution in the space interval dl at the front during the time dt.

$$m_{Ca^{2+}}(t) = \sqrt{\frac{2D_e^{H^+} C_H \cdot C_0^2 t}{\beta}} \quad (1)$$

The Calcium concentration in the rock (C_o) in mole/m³, the acid neutralisation capacity of the rock (β, [mole/m³]), i.e. the number of moles acid required to dissolve the calcite in one cubic meter of rock, and the proton concentration (C_{H+}, [mole/m³], pH) can be determined independently. As apparent from equation (1), the effective diffusion coefficient D_e^{H+} can be determined by plotting the cumulative mass of Ca^{2+} (m_{Ca2+}) against the square root of time (Einsele et al., 1995). An average effective diffusion coefficient of $4.3*10^{-10}$ m²/s was calculated with values varying between $2.9*10^{-10}$ and $5.5*10^{-10}$ m²/s. The dimensonless diffusion coefficient (D', D' = D_e/D_{aq}, D_{aq}^{H+} = $9.3*10^{-10}$ m²/s) can be calculated from the effective diffusion coefficient determined experimentally and is independent of the dissolved substance. There was no dependence of the diffusion coefficient on the pH. These data allowed the calculation of the migration of the reaction front into the matrix for longer time periods.

5 EFFECTIVE DIFFUSION COEFFICIENT

Generally, diffusion coefficients in porous rocks can be determined analogously to Archie's law using D' = an^b from the porosity (n) without using diffusion experiments. The coefficients a and b are determined empirically, with a generally assumed to be unity and b varying between 1.5 and 2.5, depending on the rock material (Grathwohl, 1995). The change of D_e with time and the associated non-linearity make the independent determination of the effective diffusion coefficient for the above described process very difficult.

In order to investigate the important and relevant parameters for the above non-linear diffusion process, a numerical finite-difference model was developed, simulating the flow, transport and reaction processes in a fractured porous system. The model uses an implicit procedure for flow and an explicit procedure for transport simulation. It is assumed that the carbonate dissolution process is very rapid relative to the flow and transport processes, i.e. kinetic effects are assumed to be negligible. The model was verified for a constant diffusion coefficient by comparison with the analytical solution after Hinsenveld (1993). For further details regarding the verification see Sauter & Liedl (1996).

The model was used to show the relationship between the effective diffusion coefficient D_e, Archie's law exponent b, the volume fraction f of calcite in the sandstone and the initial porosity n_{ini}, i.e. the porosity before the start of the dissolution process. Sensitivity analyses showed that a linear relationship exists between log D_e and b and that the slope of this relationship did not depend on f (Figure 4). Therefore an approach of the form

$$\log \frac{D_e}{D_{aq}} = \alpha + \beta \log n_{ini} + \gamma b + \delta \log f + \epsilon b \log f \quad (2)$$

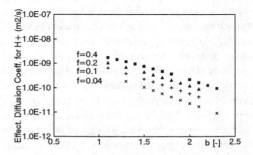

Figure 4. Relationship between effective diffusion coefficient D_e, fraction of carbonate f in sandstone and parameter b

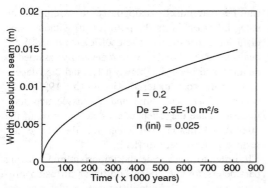

Figure 5. Development of dissolution seam with time, calculated after Hinsenveld (1993)

was chosen, with the coefficients $\alpha = 2.14355$, $\beta = 0.862729$, $\gamma = -0.98702$, $\delta = 0.69327$ and $\epsilon = 0.12559$. The parameter b mainly depends on the type of sandstone. For the type of sandstone used during the described experiments, a value of $b = 1.7$ was determined. For a sandstone with clayey cement, Mauch (1993) determined a parameter b of 2.5, i.e the resulting diffusion coefficients are lower as expected.

6 INTERPRETATION OF FIELD DATA

Based on the diffusion coefficient found for the sandstone investigated by using equation (2), the development of the reaction rim was calculated, based on the analytical solution after Hinsenveld (1993). Because only ca. 50% of the calcite cement was removed f was assumed to be 0.2. With $b = 1.7$ and $n_{ini} = 0.025$, an effective diffusion coefficient of $2.5*10^{-10}$ m²/s was calculated. With large fracture apertures and therefore high flow rates maximal concentration gradients between reaction front and mobile fracture water can be assumed. Figure 5 shows that after geological time periods dissolution rims of only a few centimeters develop. Similar thicknesses were measured in the field, which supports the use of the above described approach. The circulation system in the Gschwend quarry can be assumed to have been active over a period of approximately 1 Million years (Poppe, pers. comm.). The dissolution rims measured in the field are generally thicker, which can be explained by higher initial porosities or lower calcite contents of the original, unweathered rock. Especially at the base of the profile shown in Figure 2, where the rims show thicknesses of around 4 cm, lower calcite contents were determined.

7 CONCLUSIONS

It could be shown, using field investigations, laboratory experiments and model simulations that the dissolution of carbonate cement in a porous sandstone can be described by a relatively simple analytical solution. The results are important for the assessment of groundwater quality in mining areas and fractured aquifers affected by groundwater acidification. Presently, the role of the advective transport on the development of the dissolution rim is being examined.

REFERENCES

Bennett, D.G., Read, D., Lawless, T.A. & Sims, R.J. (1992): The migration of Uranium through sandstone.- *U.K. Dep. Environ. London*, Rep. No. DOE /HMIP/RR/92.101.

Einsele, G., Sauter, M., Clemens, T., Boehme, M. & Poppe, R. (1995): Carbonate dissolution along fractures of sandstone aquifers: field observations and modelling.- *IAHS Publ.* no. 225, 71-78.

Grathwohl, P. (1995): *Diffusion limited sorption and desorption of organic contaminants in soils and sediments.-* Habilitationsschrift, Geowissenschaftliche Fakultät, Universität Tübingen, 152p.

Hinsenveld, M. (1993): A sound and practical method to determine the quality of stabilization.- In: Arendt, F., Annokkee, G.J., Bosman, R. & van den Brink, W.J (eds): *Contaminated Soil '93*. Kluwer, Netherlands, 1519-1528.

Mauch, S. (1993): *Ermittlung von Tracer-Diffusionskoeffizienten (Jodid) in Sandsteinen.-* unveröffentlichte Diplomarbeit, Geowissenschaftliche Fakultät der Universität Tübingen, 103 S.

Read, D., Lawless, T.A., Sims, R.J. & Butter, K.R. (1993): Uranium migration through intact sandstone cores.- *J. Cont. Hydrology*, **13**, 277-289.

Sauter, M & R. Liedl 1996. Quantifizierung hydrogeochemischer Austauschvorgänge an der Grenze Kluft-Matrix. In: Merkel, B. *Grundwasser und Rohstoffgewinnung*, Geocongress, 2, Sven non Loga, Köln.

Tsang, C.-F., ed, (1987): *Coupled Processes Associated with Nuclear Waste repositories*. Academic Press, Orlando.

Calculations of fluid-ternary solid solution equilibria: An application of the Wilson equation to fluid-(Fe,Mn,Mg)TiO$_3$ equilibria

Y. Shibue
Geoscience Institute, Hyogo University of Teacher Education, Kato-gun, Japan

ABSTRACT: The Wilson equation is applied to (Fe,Mn,Mg)TiO$_3$ solid solutions. The mixing property of the ternary solid solution is obtained from the element partition data between (Fe,Mn)Cl$_{2(aq)}$ and (Fe,Mn)TiO$_3$, (Mn,Mg)Cl$_{2(aq)}$ and (Mn,Mg)TiO$_3$, and (Fe,Mg)Cl$_{2(aq)}$ and (Fe,Mg)TiO$_3$ at 600°C and 1kb. Then, the compositions of (Fe,Mn,Mg)Cl$_{2(aq)}$ in equilibria with the ternary solid solution are predicted. The predicted results are in good agreement with the experimental values.

1 MARGULES EQUATION AND WILSON EQUATION

Margules equation has been used for describing the mixing properties of mineral solid solutions. By means of Margules equation, excess Gibbs free energy (G^{ex}) of ternary solid solution is written as follows (e.g., Mukhopadhyay et al., 1993):

$$G^{ex} = X_1 X_2 (X_2 W_{12} + X_1 W_{21}) + X_1 X_3 (X_3 W_{13}$$
$$+ X_1 W_{31}) + X_2 X_3 (X_3 W_{23} + X_2 W_{32})$$
$$+ X_1 X_2 X_3 C_{123} \qquad (1)$$

where X and W stand for the mole fraction of the subscript and Margules parameter for the subscript pair, respectively. C_{123} is the ternary interaction parameter, which is independent of the mole fractions and Margules parameters for binary solid solution. If the ternary system consists of only symmetric regular binaries, the ternary interaction parameter is equal to 0. Otherwise, there are no reasons for assuming $C_{123} = 0$ without experimental confirmation. Therefore, we need the experimental data for obtaining the mixing property of the ternary solid solution. Consequently, we can not carry out the equilibrium calculations between fluid–ternary solid solution without obtaining the ternary interaction parameter.

Mixing properties of multicomponent solutions have been described by using the Wilson equation (Wilson, 1964). In contrast to Margules equation, the Wilson equation has the basis of molecular thermodynamics for its expression. The excess Gibbs free energy is written as follows (Prausnitz et al., 1986):

$$G^{ex} = -RT \sum_{i=1}^{3} X_i \ln\left(\sum_{j=1}^{3} \Lambda_{ij} X_j\right) \qquad (2)$$

where R is the universal gas constant, T is the absolute temperature, and Λ_{ij} stands for the Wilson parameter for the pair of component i and component j. By definition, $\Lambda_{ii} = 1$ in Eq. (2). For a binary solution, Eq. (2) is reduced into the following equation.

$$G^{ex} = -RT\{X_1 \ln(X_1 + \Lambda_{12} X_2)$$
$$+ X_2 \ln(\Lambda_{21} X_1 + X_2)\} \qquad (3)$$

If $\Lambda_{12} = \Lambda_{21}$, then the binary solution shows the symmetric regular solution behavior.

By an appropriate differentiation of Eq. (2), the activity coefficient of component i (γ_i) can be expressed as follows.

$$\ln \gamma_i = 1 - \ln\left(\sum_{j=1}^{3} X_j \Lambda_{ij}\right) - \sum_{k=1}^{3} \left(\frac{X_k \Lambda_{ki}}{\sum_{j=1}^{3} X_j \Lambda_{kj}}\right) \qquad (4)$$

We can compute the activity coefficients of the components in the ternary solution by combining the Wilson parameters for the three binary solutions whose components constitute the ternary system. Although the Wilson equation is more complex than Margules equation, the main advantage of the Wilson equation is its applicability to multicomponent solutions without the ternary interaction parameter.

This paper aims to apply the Wilson equation to mineral solid solutions. As an illustrative example, this study analyzes the experimental results on fluid–$(Fe,Mn,Mg)TiO_3$ equilibria (Kubo et al., 1992). Based on the results of the three binary cation-exchange experiments, this study obtains the mixing property of the ternary solid solution and computes the compositions of $(Fe,Mn,Mg)Cl_{2(aq)}$ in equilibria with $(Fe,Mn,Mg)TiO_3$. Then, the computed results are compared with the experimental results.

2 FLUID–$(Fe,Mn)TiO_3$, FLUID–$(Mn,Mg)TiO_3$, AND FLUID–$(Fe,Mg)TiO_3$ EQUILIBRIA

Kubo et al. (1992) carried out the cation–exchange experiments on the following systems at 600°C and 1 kb.

$$FeTiO_3 + MnCl_{2(aq)} = MnTiO_3 + FeCl_{2(aq)} \quad (5)$$

$$MnTiO_3 + MgCl_{2(aq)} = MgTiO_3 + MnCl_{2(aq)} \quad (6)$$

$$MgTiO_3 + FeCl_{2(aq)} = FeTiO_3 + MgCl_{2(aq)} \quad (7)$$

Total chloride concentrations were 2.0 molar in all the experimental runs.

By using Eq. (4), Gibbs free energy of reaction (5) can be expressed as follows.

$$\frac{\Delta G^o_{(5)}}{RT} = -\ln \frac{X_{Mn} \gamma_{Mn} Y_{Fe}}{X_{Fe} \gamma_{Fe} Y_{Mn}}$$

$$= -\ln \frac{X_{Mn} Y_{Fe}}{X_{Fe} Y_{Mn}} + \ln \frac{\Lambda_{MnFe} X_{Fe} + X_{Mn}}{X_{Fe} + \Lambda_{FeMn} X_{Mn}}$$

$$+ \frac{\Lambda_{FeMn} X_{Fe} - X_{Fe}}{X_{Fe} + \Lambda_{FeMn} X_{Mn}} + \frac{X_{Mn} - \Lambda_{MnFe} X_{Mn}}{\Lambda_{MnFe} X_{Fe} + X_{Mn}} \quad (8)$$

where Y_{Fe} and Y_{Mn} designate the following ratios.

Table 1. ΔG^o for reactions (5)–(7) and the Wilson parameters for $(Fe,Mn)TiO_3$, $(Mn,Mg)TiO_3$, and $(Fe,Mg)TiO_3$ solid solutions.

$$\frac{\Delta G^o_{(5)}}{RT} = 1.462(0.042)$$

$$\frac{\Delta G^o_{(6)}}{RT} = -0.975(0.036)$$

$$\frac{\Delta G^o_{(7)}}{RT} = -0.487(0.035)$$

$\Lambda_{MnFe} = 0.585(0.220)$, $\Lambda_{FeMn} = 1.314(0.370)$

$\Lambda_{MgMn} = 0.371(0.087)$, $\Lambda_{MnMg} = 0.393(0.073)$

$\Lambda_{FeMg} = 0.962(0.213)$, $\Lambda_{MgFe} = 0.406(0.138)$

Values in parentheses indicate the standard errors.

$$Y_{Fe} = \frac{m_{FeCl_{2(aq)}}}{m_{FeCl_{2(aq)}} + m_{MnCl_{2(aq)}} + m_{MgCl_{2(aq)}}} \quad (9)$$

$$Y_{Mn} = \frac{m_{MnCl_{2(aq)}}}{m_{FeCl_{2(aq)}} + m_{MnCl_{2(aq)}} + m_{MgCl_{2(aq)}}} \quad (10)$$

ΔG^o for reaction (6) or (7) can be obtained by substituting Mg for Fe or Mg for Mn as the subscript in Eq. (8). Y_{Mg} in the resultant equations denotes the following ratio.

$$Y_{Mg} = \frac{m_{MgCl_{2(aq)}}}{m_{FeCl_{2(aq)}} + m_{MnCl_{2(aq)}} + m_{MgCl_{2(aq)}}} \quad (11)$$

For the binary cation-exchange reactions, one of the metal chloride concentrations is equal to 0 molal. This study ignores the ionic species and assumes that the activity coefficients of the neutral aqueous species are equal to 1.

By the non-linear least-square regression of the experimental data of Kubo et al. (1992), the Wilson parameters and ΔG^o values are obtained for reactions (5)–(7). During the regression analyses, this study considers the following constraints.

$$\Delta G^o_{(5)} = -(\Delta G^o_{(6)} + \Delta G^o_{(7)}) \quad (12)$$

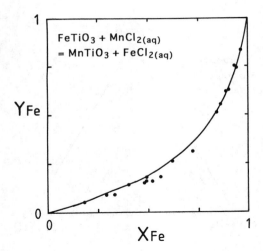
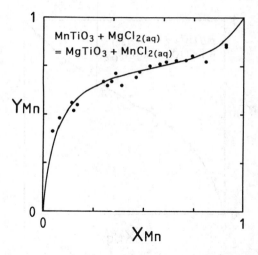

Fig. 1. The computed cation-exchange isotherm and the experimental data points (Kubo et al., 1992) for the reaction $FeTiO_3 + MnCl_{2(aq)} = MnTiO_3 + FeCl_{2(aq)}$.

Fig. 2. The computed cation-exchange isotherm and the experimental data points (Kubo et al., 1992) for the reaction $MnTiO_3 + MgCl_{2(aq)} = MgTiO_3 + MnCl_{2(aq)}$.

$$\Lambda_{MgFe} = \frac{\Lambda_{MgMn}\Lambda_{MnFe}\Lambda_{FeMg}}{\Lambda_{MnMg}\Lambda_{FeMn}} \quad (13)$$

The latter constraint was derived by Hala (1982). By imposing these constraints, the obtained $\Delta G°$ values as well as the Wilson parameters are internally consistent (Table 1).

The obtained Wilson parameters indicate that $(Fe,Mn)TiO_3$ and $(Fe,Mg)TiO_3$ are asymmetric solid solutions while $(Mn,Mg)TiO_3$ is a symmetric solid solution.

From Eqs. (8)–(10) and the equivalent expressions for reactions (6) and (7), we get the following equations:

$$\frac{Y_{Fe}}{Y_{Mn}} = \exp\left\{\left(\frac{-\Delta G°_{(5)}}{RT}\right) - \ln\left(\frac{X_{Mn}\gamma_{Mn}}{X_{Fe}\gamma_{Fe}}\right)\right\} \equiv Z_1 \quad (14)$$

for reaction (5),

$$\frac{Y_{Mn}}{Y_{Mg}} = \exp\left[\left(\frac{-\Delta G°_{(6)}}{RT}\right) - \ln\left(\frac{X_{Mg}\gamma_{Mg}}{X_{Mn}\gamma_{Mn}}\right)\right] \equiv Z_2 \quad (15)$$

for reaction (6), and

$$\frac{Y_{Mg}}{Y_{Fe}} = \exp\left[\left(\frac{-\Delta G°_{(7)}}{RT}\right) - \ln\left(\frac{X_{Fe}\gamma_{Fe}}{X_{Mg}\gamma_{Mg}}\right)\right] \equiv Z_3 \quad (16)$$

for reaction (7). When the solid composition is fixed, then the activity coefficients of the components in the solid phase are determined by Eq. (4). By using the activity coefficients at specific solid compositions and the $\Delta G°$ value for the relevant reactions, cation-exchange isotherms for reactions (5), (6), and (7) are computed from Eqs. (14), (15), and (16), respectively (Figs. 1–3). The computed isotherms show the good agreement with the experimental results of Kubo et al. (1992).

3 CALCULATIONS OF FLUID–$(Fe,Mn,Mg)TiO_3$ EQUILIBRIA

Using the obtained $\Delta G°$ values and the Wilson parameters, the compositions of $(Fe,Mn,Mg)Cl_{2(aq)}$ in equilibria with $(Fe,Mn,Mg)TiO_3$ are computed by Eqs. (14)–(16). The solid compositions are taken from the experimental results of Kubo et al. (1992). After solving Eqs. (14)–(16), this study computes the weight-averaged values of fluid compositions following the method of Shallcross et al. (1988). The computed fluid compositions and the experimental results obtained by Kubo et al. (1992) are compared in Fig. 4. The agreement is generally good although the computed compositions deviate from the experimental compositions at low Y_{Fe}

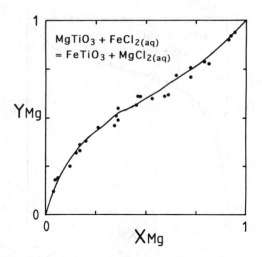
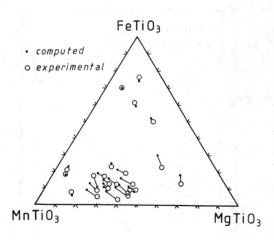

Fig. 3. The computed cation-exchange isotherm and the experimental data points (Kubo et al., 1992) for the reaction $MgTiO_3+FeCl_{2(aq)}=FeTiO_3+MgCl_{2(aq)}$.

Fig. 4. Fluid compositions in equilibria with $(Fe,Mn,Mg)TiO_3$ solid solution. Fluid compositions tied by lines correspond to the same solid compositions. The tie lines are omitted when the computed compositions closely agree with the experimental compositions. The experimental compositions of solid and fluid phases are taken from Kubo et al. (1992).

region. The deviations of the computed compositions from the experimental compositions are within ±0.08 in terms of Y value for any component. It can be concluded that fluid compositions in equilibria with the ternary solid solution may be predicted from the binary cation-exchange experiments.

4 CONCLUSIONS

This study obtains the mixing property of the ternary solid solution of $(Fe,Mn,Mg)TiO_3$ by means of the Wilson equation. The derived property is based on the analyses of the element partition data between $(Fe,Mn)Cl_{2(aq)}-(Fe,Mn)TiO_3$, $(Mn,Mg)Cl_{2(aq)}-(Mn,Mg)TiO_3$, and $(Fe,Mg)Cl_{2(aq)}-(Fe,Mg)TiO_3$ at 600°C and 1 kb. By using the obtained $\Delta G°$ values and the Wilson parameters, this study computes the fluid compositions of $(Fe,Mn,Mg)Cl_{2(aq)}$ in equilibria with $(Fe,Mn,Mg)TiO_3$. The computed fluid compositions agree well with the experimental fluid compositions when the solid phase compositions are known.

5 REFERENCES

Hala, E. 1972. Note to Bruin-Prausnitz one-parameter local composition equation. *Ind. Chem. Process Des. Develop.* 11: 638.

Kubo, T., E. Uchida, Y. Furukawa & N. Imai 1992. Experimental study on ion exchange equilibria between solid solution (Fe^{2+}, Mn^{2+},Mg^{2+})TiO_3 and aqueous (Fe^{2+},Mn^{2+},Mg^{2+})Cl_2 solution. *Kobutsugaku Zasshi* 21: 59-67.

Mukhopadhyay, B., S. Basu & M. J. Holdway 1993. A discussion of Margules-type formulation for multicomponent solutions with a generalized approach. *Geochim. Cosmochim. Acta* 57: 277-283.

Prausnitz, J. M., R. N. Lichtenthaler & E. G. Azevedo 1986. *Molecular Thermodynamics of Fluid-phase Equilibria*. 2nd. ed. New York: Prentice Hall.

Shallcross, D. C., C. C. Herrmann & B. J. McCoy 1988. An improved model for the prediction of multicomponent ion exchange equilibria. *Chem. Eng. Sci.* 43: 279-288.

Wilson, G. M. 1964. Vapor–liquid equilibrium. XI. A new expression for the excess free energy mixing. *J. Amer Chem. Soc.* 86: 127-130.

Competitive pool growth model and numerical simulation for morphological diversity of hot-spring mineral deposits

Hiroshi Shigeno
Geological Survey of Japan, Tsukuba, Ibaraki, Japan

ABSTRACT: A competitive pool growth model was proposed, and two-dimensional numerical simulations were conducted concerning formation mechanism for diverse macroscopic morphology of hot-spring mineral deposits. Simulation results show that dome, typical terracettes, and slope types of deposits were formed through middle stage during full development of the deposits according to mineral formation reaction rate in water. The spectacular terracettes deposits, such as those at Pamukkale and Mammoth Hot Springs, have probably been developing in fairly stable environments with intermediate parameter value combinations.

1 INTRODUCTION

Sub-aerial hot springs produce various kinds of mineral deposits of diverse morphology through abiological and biological processes (e.g. Kitano, 1963; Bargar, 1978; Ford & Pedley, 1992). Their macroscopic features could be grouped into types such as dome, typical terracettes, slope and massive terrace, according to their cross sections. The typical terracettes (rimstone pools) systems, which are probably the most spectacular of these deposits, are observed, for example, at Pamukkale, western Turkey (e.g. Shigeno, 1995) (see Fig. 1), Mammoth Hot Springs, Yellowstone National Park, U.S.A. (e.g. Bargar, 1978; Fournier et al., 1992), and White and Pink Terraces, New Zealand (e.g. Simmons et al., 1993). These deposits are composed of $CaCO_3$ or SiO_2.

In the present paper, a competitive pool growth model, as a formation mechanism for the macroscopic features of the deposits, will be proposed. Two-dimensional numerical simulation results based on the model will be discussed, especially for formation conditions of the terracettes systems.

2 GROWTH MODEL

On the surfaces of active hot-spring mineral deposits, tiny to small (e.g. 1 to 10 cm wide) shallow pool structures generally elongated to the traverse direction of slopes are often observed. I estimate that these small pools could be developed from rough surfaces of the slopes, and would develop gradually to the large (e.g. 0.5 to 5 m wide) terracettes systems, which are composed of rimstones, shallow pools and bottom deposits, if conditions allow.

Concerning the formation mechanism of the macroscopic deposit features, especially for the typical terracettes systems, I propose the competitive pool growth model as follows: The pools grow upward if the rim tops grow faster than the pool bottoms. During the growth processes, a pool located upstream merges to an adjacent lower pool, if the latter grows sufficiently faster than the former, forming a united larger pool.

The faster upward growth of the rim tops, compared with the pool bottoms, could be explained by rim top accumulation of the minerals that are formed at surfaces and in waters of the pools, in addition to direct mineral deposition on the rim tops. The upward rim top growth may be faster in deep pools, where total mineral production rate from the pool water per unit bottom area tends to be larger, than in shallow pools.

Figure 1. Spectacular hot-spring terracettes system, composed of $CaCO_3$, at Pamukkale, Turkey.

3 METHOD OF NUMERICAL SIMULATION

Fig. 2 shows the simple setting of the two-dimensional model used for analyzing the development processes of hot-spring deposits by numerical simulation. Major assumptions are as follows (see Fig. 2):

(1) The mineral deposition occurs on an initially plane slope of a length (L, m), a unit width (W, m), and an angle (Ang, °). Inflow water steadily coming up through a fault flows down the slope, and discharges to a river.

(2) Flow rate (Fr, m^3/s) and initial over-saturated concentration for a mineral (C0, mg/l) of the inflow water are constant. The over-saturated mineral concentration (C, mg/l) changes with time (t, s) according to a first-order mineral formation reaction with a kinetic constant (k, s^{-1}), as follows:

$$\ln C = -k*t + \ln C0 \quad (1)$$

(3) The deposit develops following the competitive pool growth model mentioned above. Initially, a number (NP0) of small rugged features, in the ranges of the mutual distances (Pd0, m) and the heights (Ph0, m), across the slope are randomly distributed. The rimstones develop on these seeds.

(4) As a hydrological constraint of the inflow water, the upward growth of the deposit is limited to a level (Ymax, m) from the top of the initial slope. The inflow water channel migrates downward to the outer bottom of a terracette, when its growth reaches to the limit.

On the basis of the above setting, flow and residence of the water, change of the over-saturated mineral concentration of the water, deposition of the mineral, and growth of the pool at each terracette (n) and each time step (ts) are represented by the following difference equations (2) to (10) (refer to Fig. 3). Masses deposited on the rimstones were neglected on the assumption that these are much smaller than the masses deposited on the pool bottoms.

$$Tr(n, ts) = Pv(n, ts)/Fr(n, ts) \quad (2)$$
$$Cin(n, ts) = Cout(n-1, ts) \quad (3)$$
$$Cout(n, ts) = Cin(n, ts)*\exp(-0.693*Tr(n, ts)/T1/2) \quad (4)$$
$$Mw(n, ts) = (Cin(n, ts) - Cout(n, ts))*Fr(n, ts)*\Delta ts/1000 \quad (5)$$
$$Mv(n, ts) = Mw(n, ts)/Rmin/1000 \quad (6)$$
$$Ub(n, ts) = Mv(n, ts)/Ab(n, ts) \quad (7)$$
$$Ur(n, ts) = Ub(n, ts)*Frim \quad (8)$$
$$Pv(n, ts+1) = Pv(n, ts) + A(n, ts)*Ur(n, ts) - Ab(n, ts)*Ub(n, ts) \quad (9)$$
$$Pd(n, ts+1) = Pd(n, ts) + Ur(n, ts) - Ub(n, ts) \quad (10)$$

Symbols are as follows:
A: Area of pool surface (m^2)
Ab: Area of pool bottom (m^2)
Cin: Over-saturated mineral concentration of inflow water (mg/l)
Cout: Over-saturated mineral concentration of

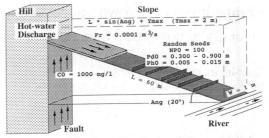

Figure 2. Two-dimensional simulation model for development processes of hot-spring mineral deposits: Assumptions for initial topography and mode of water flow. See the text for the symbols.

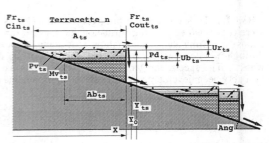

Figure 3. Two-dimensional simulation model for development processes of hot-spring mineral deposits: Cross section for flow and residence of water, mineral deposition, pool growth at each terracette (n) and each time step (ts). See the text for the symbols.

outflow water (mg/l)
Frim: Height increase ratio of rim top to pool bottom (-, const.)
Mv: Volume of new deposit (m^3)
Mw: Weight of newly deposited mineral (kg)
Pd: Average depth of pool (m)
Pv: Volume of pool (m^3)
Rmin: Density of deposit (g/cm^3, const.)
T1/2: Half-life of mineral formation reaction (s, 0.693/k, const.)
Tr: Average residence time of water in pool (s)
Ub: Height increase of pool bottom (m)
Ur: Height increase of rim top (m)
Δts: Span of time step (s, const.)

Numerical simulations were conducted mainly using Apple Macintosh LC520 and Microsoft QuickBasic.

4 RESULTS AND DISCUSSION

Table 1 summarizes selected simulation results based on the above method with changed parameter values, Ang and T1/2, and Δts values adjusted to growth speed of the deposits. Figs. 4 and 5 show temporal changes of the two evaluation functions: number ratio of remaining terracettes to the initial seed,

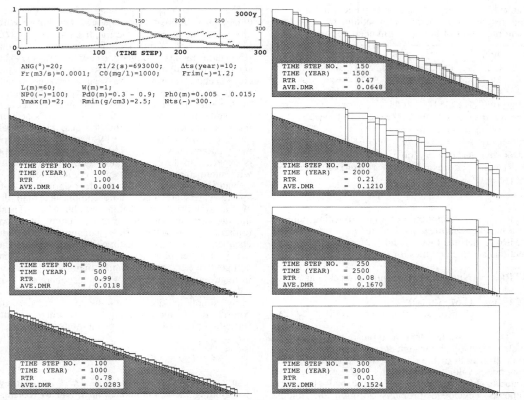

Figure 4. Results of the two-dimensional simulation, Run 1, for development processes of hot-spring mineral deposits. Top left figure shows temporal changes of the two evaluation functions, RTR(ts) and DMR(ts), by large and small circles, respectively. Following seven figures show temporal changes of the cross section of the deposit, at the years 100, 500, 1000, 1500, 2000, 2500 and 3000 from the start, respectively. Two horizontal lines of each terracette are surface and bottom of the pool. See the text and Table 1 for parameters.

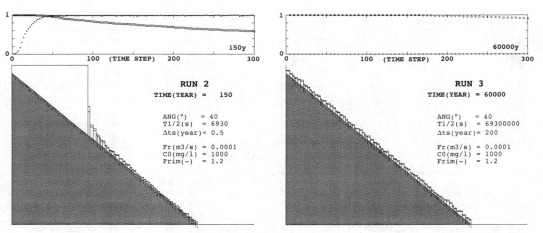

Figure 5. Results of the two-dimensional simulation, Runs 2 and 3, for development processes of hot-spring mineral deposits. Top figures show temporal changes of RTR(ts) and DMR(ts) by large and small circles, and bottom figures show cross sections at the years from the start, respectively. See the text and Table 1.

Table 1. Summary of parameter value combinations and results of numerical simulations for development processes of hot-spring mineral deposits.

Run	Ang (°)	$T_{1/2}$ (s)	Δts (year)	Result type of deposit
1	20	$0.693*10^6$	10	Terracettes[#1]
2	40	$0.693*10^4$	0.5	Dome[#2]
3	40	$0.693*10^8$	200	Slope

Constants: $Fr=0.0001 m^3/s$; $C0=1000 mg/l$; $Frim=1.2$; $L=60m$; $W=1m$; $NP0=100$; $Pd0=0.3-0.9m$; $Ph0=0.005-0.015m$; $Ymax=2m$; $Rmin=2.5g/cm^3$; $Nts=300$.

[#1] Slope -> Terracettes -> Massive Terrace
[#2] (Slope ->) Dome

RTR(ts), and weight ratio of total depositing mineral to total over-saturated mineral in the inflow water, DMR(ts), defined by (11) and (12), and typical cross sections at selected times for the above cases.

RTR(ts) = (Number of remaining terracettes at ts) / (Number of initial seeds)   (11)

DMR(ts) = $(\Sigma Mw(n, ts))$ / $(Cin(0, ts)*Fr(0, ts)*\Delta ts/1000)$ (12)

Fig. 4 indicates that a typical terracettes-type system developed through the middle stage during the full growth of the deposit, where an appropriate parameter value combination was selected. In contrast, Fig. 5 shows that dome and slope-type deposits developed probably through the middle stage, where the $T_{1/2}$ value was much smaller or larger than that in Run 1. The development of the typical terracettes type is characterized by steadily decreasing RTR values, and gradually increasing (but moderate) DMR values through most of the growth period (compare Figs. 4 and 5).

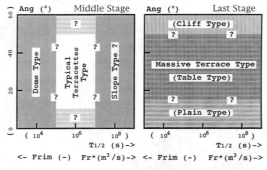

Figure 6. Summary of the two-dimensional simulation results for development processes of hot-spring mineral deposits. Left and right figures show the middle and last stages of macroscopic morphological changes of the deposits, respectively. * indicates the condition of $Fr*C0 = const$. See the text and Table 1 for parameters and deposit types.

Fig. 6 conceptually summarizes the two-dimensional deposit type changes developed with time, with the effects of the major parameters, $T_{1/2}$, Fr^*, Frim and Ang, including some additional simulation results (Shigeno, 1995). At the last stage of the full development of the deposits, the cross section features will be massive terraces, which might be called plain, table and cliff types, depending on the Ang and Ymax values. However, through the middle stage, more variations of the cross section features: dome, typical terracettes and slope types develop. Those are caused by the combinations of many parameter values. The typical terracettes type is expected to develop in fairly stable environments with intermediate parameter value combinations.

Natural conditions of mineral deposition from hot-spring waters are much more complicated than the above simplified simulation conditions. These include three-dimensional effects, setting of initial topography and water inflow, changes of flow and chemistry of waters with time, and weathering and decay of developing deposit (Shigeno, 1995). The competitive pool growth model may be hypothetical in most environments, for example, where the $T_{1/2}$ value is very small or large. However, the above fairly simple model and simulation results could contribute to understanding the essence of the formation processes of hot-spring mineral deposits.

ACKNOWLEDGMENT

I am very grateful to Ali Kocak (M.T.A., Turkey) for guiding me at Pamukkale area, and to Koichiro Fujimoto (G.S.J.) and Kohei Akaku (NEDO) for commenting on manuscript. I thank Dan Hayba and John R. Hulston for kindly reviewing the paper.

REFERENCES

Bargar, K.E. 1978. Geology and thermal history of Mammoth Hot Springs, Yellowstone National Park, Wyoming. *U.S. Geol. Surv. Bull.*, 1444.

Ford, T.D. & Pedley, H.M. 1992. Tufa deposits of the world. *Jour. Speleol. Soc. Japan* 17: 46-63.

Fournier, R.O., Christiansen, R.L., Hutchinson, R.A. & Pierce, K.L. 1992. *WRI-7 Yellowstone National Park Field Trip*. 7th Internat. Symp. on Water-Rock Interaction, Park City 1992.

Kitano, Y. 1963. Geochemistry of calcareous deposits found in hot springs. *Jour. Earth Sci., Nagoya Univ.*, no. 11: 68-100.

Shigeno, H. 1995. Mechanism and numerical simulation of terracettes (rimstone pools) formation by mineral deposition from hot-spring waters. *Jour. Japan Geothermal Energy Assoc.* 32: 317-336. (in Japanese with English Abstract)

Simmons, S.F., Keywood, M., Scott, B.J. & Keam, R.F. 1993. Irreversible change of the Rotomahana-Waimangu hydrothermal system (New Zealand) as a consequence of a volcanic eruption. *Geology* 21: 643-646.

16 Mineral surfaces

Colloidal interactions of precipitated Zn carbonates with clay minerals

H. B. Bradl
Bilfinger + Berger Bauaktiengesellschaft, Mannheim, Germany

ABSTRACT: The influence of precipitated Zn carbonates on the surface charge of clays has been investigated by means of polyelectrolyte titration (PET). Clay suspensions have been mixed with solutions of $ZnCl_2$ and Na_2CO_3 in different concentrations. The distribution of the particular ion species and their concentration has been calculated with the help of the geochemical program PHREEQE and then been compared to the data measured. A process called charge blocking could be observed. The precipitation of solids in the presence of a clay suspension results in a considerable reduction of surface charge due to a film formation on the clay surface.

1 INTRODUCTION

The interaction of clay surfaces with other substances has major environmental, sedimentological and geological consequences. Clays are of special interest due to their origin, mineralogical structures and surface properties. Especially their ion sorption capacity is of utmost importance for a wide variety of applications. The sorption capacity of clays is caused by the permanent negative charge of the clay surfaces. Many investigations were carried out on the sorption of charged particles onto clays (Czurda & Wagner, 1981). Only little information is available on the interaction of precipitated substances with the clay surface. Such reactions occur in a variety of natural and artifical environments e.g. aquifers, sedimentary basins or waste deposits. It is a well known fact in colloid chemistry that chemically inactive substances like gold sols (Thiessen, 1942) can be adsorbed onto clays. The system water - clay - precipitate can be looked upon as a socalled mixed colloidal system. The properties of such systems are determined by the surface charge of the suspended particles and the interactions between them (Gherardi & Matijevic, 1986). It is clear that the surface charge of the clay particle is a very important parameter. It is possible to determine the surface charge of suspended clay particles in a very exact way with the help of the socalled polyelectrolyte titration or PET (Kliesch, 1989).

2 MATERIALS AND METHODS

For the titration experiments the fraction $< 2\mu m$ of two natural clays from Mainflingen near Aschaffenburg, Bavaria and Woellstein near Bad Kreuznach (Rhineland-Palatinate) and two artificially activated Ca-Bentonites (Montigel F, Calcigel, Südchemie Corp., Munich) were used. The clays from Woellstein contain carbonate and are of Lower Oligocene (Rupelian) age. The carbonate free clays from Mainflingen are of Upper Pliocene age. Samples of these clays were oven dried at 105° C, ground carefully and then sieved. The fraction $< 2\mu m$ was obtained from the fraction $< 125\ \mu m$ by sedimentation.

The two bentonites are very pure bentonites with a montmorillonite content of 70-75 %. Montigel F is activated by Sodium ions. The clays could be used directly without further processing. The bentonites are decomposition products of acid volcanic glass tuffs from the Bavarian deposits of late Upper Miocene age as previously described by Hofmann (1973).

2.1 X-Ray Diffraction Investigations

The differentiation of clay minerals has been executed with the help of X-Ray diffraction. The semiquantitative mineral content of the clays can be estimated from the peak intensities. A Philips X-ray diffractometer with Cu-K_α-radiation was used. The evaluation was made according to Thorez (1975). Kaolinite, illite and smectite could be found in different quantities with the help of XRD (table 1).

Table 1. Semiquantitative clay mineral associations.

Clay	%K	%I	%M
Woellstein brown	27	30	43
Woellstein grey	21	48	31
Mainflingen yellow	49	46	5
Mainflingen black	26	74	0
Mainflingen grey	42	48	10
	100	100	100

K = Kaolinite; I = Illite; M = Montmorillonite

2.2 Polyelectrolyte Titration (PET)

Polyelectrolytes are macromolecules with dissociative groups like e.g. -COOH and -SO_3H (anionic) or -NH_2, -$N(R_3)^+$ X^- (cationic) (Krause et al., 1984). The polyelectrolytes used were the cationic CATFLOC (polydimethyldiallylammoniumchloride) or the anionic KPVS (potassiumpolyvinylphosphate). The amount of polyelectrolyte consumed is the measured parameter for the charge density of the clay suspension. If the specific surface of the clay is known, the active net charge of the clay can be calculated (Bradl, 1995).

2.3 Titration experiments with clay suspensions

A clay suspension was prepared from the dried fraction < 2µm by adding bidistilled water. For preparation procedure see Bradl (1995). First of all the influence of pH on the surface charge of the uninfluenced suspensions was investigated by adding HNO_3 or NH_3. Then the polyelectrolyte consumption was measured with the help of the SCD apparatus. To investigate the influence of heavy metal carbonates which precipitate in the presence of the clay suspension on the surface charge the procedure was as follows:

10^{-2} m and 10^{-3} m solutions were made from solid $ZnCl_2$ and Na_2CO_3 and bidistilled water. The following solutions have been prepared:

Concentration 10/10: 10^{-2} m $ZnCl_2$ solution/10^{-2} m Carbonate solution
Concentration 1/10: 10^{-3} m $ZnCl_2$ solution/10^{-2} m Carbonate solution
Concentration 10/1: 10^{-2} m $ZnCl_2$ solution/10^{-3} m Carbonate solution
Concentration 1/1: 10^{-3} m $ZnCl_2$ solution /10^{-3} m Carbonate solution

20 ml clay suspension were mixed with aliquot parts of the heavy metal and the carbonate solutions, shaken gently and then pH was adjusted to the desired pH-values. Then the sample was investigated with the SCD apparatus.

3 RESULTS

3.1 Precipitation Behavior of Zinc

Precipitation products were produced according to the procedure described above. The sediment was filtered and the heavy metal concentration of the solutions was then analysed by atomic adsorption spectroscopy. For the four concentrations used the distribution of the theoretically possible ion species and their concentrations was calculated with the help of the geochemical program PHREEQE (Parkhurst et al., 1980). To compare the different concentrations the values are given in percent Zn as carbonate and are not given in absolute numbers.

Table 2 shows the percentage of precipitated heavy metal carbonate (here as $ZnCO_3$) as a function of solution pH for a concentration of 10^{-3} m $ZnCl_2$ and 10^{-3} m Na_2CO_3 in the blank solution calculated by PHREEQE compared to the values measured in the suspension with Montigel F.

Table 2: Comparison of measured vs. calculated data.

pH	Montigel F (%)	Blank (%)
2	5	0
3	7	0
4	10	0
5	12	10
6	38	40
7	41	72
8	82	83
9	95	99

For the calculation procedure of the percents of precipitated substance see Bradl (1995).

The observation of different clay suspensions revealed the following results:

1. The most important parameter for the precipitation reaction of Zn carbonates are pH and concentration used. For pH between 8 and 9 the Zn ions exist completely as solids of low solubility.

2. The existence of a clay suspension causes a superposition of precipitation and sorption mechanisms in medium pH ranges. The amount of sorption depends on the mineralogical composition of the clays. The sorption process which took place in the clay suspension influences the sorption capacity of the clays, especially of the highly swellable bentonites.

3.2 Influence of pH on the surface charge

Suspensions of the fraction < 2μm of the different clays were prepared. The concentration was 1 g clay in 1 l bidistilled water. These suspensions are called "uninfluenced suspensions". Figure 1 shows the influence of the change of pH on the surface charge of the clay particles. If the consumption of cationic CATFLOC is plotted versus the solution pH, characteristic curves are obtained which reveal informations on the processes at the clay mineral surface. The higher the CATFLOC consumption of a clay suspension, the higher is the negative surface charge of the clay and its adsorption capacity.

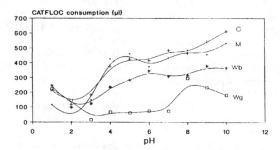

Figure 1: CATFLOC consumption vs. solution pH. C = Calcigel; M = Montigel; Wb = Woellstein brown; Wg = Woellstein grau

The measured curves show a common tendency. They reach their maximum values at a high pH between 8 and 10. The maximum value is on different levels according to the clay used. Consumption decreases with decreasing pH to a minimum value to rise again at very low pH values. The behavior at high pH is controlled by the pH dependent dissociation of Hydroxyl ions. The higher pH, the more H_3O^+ ions will solve and the higher the negative surface charge (Mitchell, 1976). At decreasing pH the negative surface charge decreases as well. At the socalled point of zero charge the surface is uncharged.

For the suspension with Montigel F this point of zero charge happens to be at pH 2 precisely. The surface is uncharged and CATFLOC consumption is zero. The point of zero charge is known for different materials by titration with acid. Bartoli et al. (1987) determined the point of zero charge for a pure kaolinite suspension of pH 3,4. This value is in good agreement with the minima of the curves of the kaolinitic clays from Mainflingen. The slight rise which all curves show at very low pH can be explained by dissolution phenomena at low pH (Veith & Schwerdtmann, 1979).

The exchange of permanent charge in the crystal lattice against H_3O^+ ions leads to the release of structurally bounded Aluminium and Silicium ions (Weiss, 1958a, b). One can see that the rise at low pH is more inferior for the kaolinitic Mainflingen clays than for Montigel F and Calcigel and the Lower Oligocene clays from Woellstein which have a certain amount of smectites.

3.3 Influenced suspensions

To investigate the influence of Zn carbonates which precipitate in the presence of clay suspensions precipitation products were produced. These suspensions are called "influenced suspensions". Then the polyelectrolyte consumption of the sample was measured. Figures 2 and 3 show the results for a clay with high smectite content (Calcigel) and for a clay with lower smectite content (Woellstein brown).

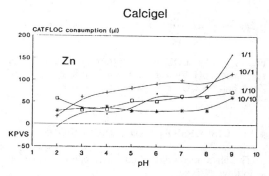

Figure 2: CATFLOC consumption as a function of suspension pH for different concentrations of $ZnCl_2$ and Na_2CO_3, clay: Calcigel.

All curves show a maximum at high pH and decrease more or less steeply with decreasing pH. If one looks at the absolute value of the polyelectrolyte consumption then one can see a considerable reduction compared to the uninfluenced suspension. For the Calcigel suspension the value at pH 9 amounts to 50 μl in comparison to 600 μl for the uninfluenced suspension. This means that the surface charge is reduced to one twelfth of its origin value due to the precipitation process of zinc carbonate. For the lowest concentration 1/1 the surface charge becomes positive at pH 2. The suspension consumes anionic KPVS.

This phenomenon can be explained by the formation of zinc-hexaquocomplexes which form at excess zinc concentrations in the solution and act as polyelectrolytes.

A pure $ZnCl_2 \cdot Na_2CO_3$ solution (a socalled „blank") of this concentration consumed 148 µl of KPVS.

In contrast to the curve in figure 2 the curve in figure 3 shows a more heterogeneous structure. The reduction of surface charge can be observed as well but the amount of reduction is much lower than for the clays with high smectite content. The clay shows a certain resistance at very low pH ranges due to the buffering effect of carbonate.

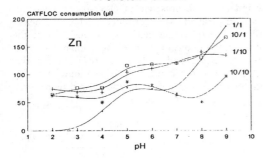

Fig. 3 : CATFLOC consumption as a function of suspension pH for different concentration ratios of $ZnCl_2$ and Na_2CO_3, clay: Woellstein brown.

4 DISCUSSION AND CONCLUSIONS

The phenomena described above can be explained as follows:
At the pH range > 6 Zn carbonates exist as solids in the solution. When pH decreases, only Zn ions respectively their chloro- and hydroxocomplexes exist. In mixed-colloidal systems the reduction of surface charge can be caused by film formation of heavy metal carbonate on the clay mineral surfaces („charge blocking"). Especially the surface charge of clays with high smectite contents is affected by this charge blocking effect while clays containing non-swellable phases like kaolinite are not affected to such a degree.

Similar film formations have been observed by Gherardi & Matijevic (1986) when investigating the precipitation of hematite in the presence of a finely dispersed titania sol. This mechanism depends on pH and concentration and is more effective with higher concentrations. A similar mechanism has been recently observed by Booth et al. (1997) when investigating the dissolution of calcite in aqueous solution at pH below 5-6. Exposure of calcite to sulphuric acid lead to the rapid formation of an overlayer of Calcium Sulphate (gypsum) which passivates the surface against further dissolution.

5 REFERENCES

Bartoli, F. & R. Philippy 1987. The colloidal stability of variable-charge mineral suspensions. *Clay Min.* 22 (1): 93-107.

Booth, J., Hong, Q., Compton R.G., Prout, K. & R.M. Payne 1997. Gypsum overgrowths passivate Calcite to acid attacks. *J. Colloid Interface Sci.* (submitted).

Bradl, H.B., 1995. New aspects regarding failure of clays in barrier systems. Proc. 5th Int. Landfill Symposium, Cagliari, Italy II: 213-224.

Czurda, K.A. & J.F. Wagner 1988. Transfer and sorption of heavy metals in clayey barrier rocks. *Schr. Angew. Geol. Karlsruhe* 4: 225-245.

Gherardi, P. & E. Matijevic 1986. Interactions of precipitated Hematite with preformed colloidal Titania dispersions. *J. Colloid Interface Sci.* 109 (1): 543-546.

Hofmann, B. 1973. Erläuterungen zur geologischen Karte von Bayern 1:25.000 Blatt 7439 Landshut Ost. 133 pp., Munich.

Kliesch, K. 1989. Der Einfluß des Porenwassers auf das Quellverhalten tertiärer Tone: Analyse mittels Polyelektrolyttitration. Ber. 7. Nat. Tag. Ing.-Geol., 69-76, Bensheim.

Krause, T., Schempp, W. & P. Hess 1984. Die Polyelektrolyttitration und ihre Anwendung in der Papierchemie. *Dtsche Papierwirtschaft* 3: 127-130.

Mitchell, J.K. 1976. Fundamentals of Soil Behaviour. New York: John Wiley and Sons.

Parkhurst, D.L., Thorstenson, D.C. & L.N. Plummer 1980. PHREEQE - a computer program for geochemical calculations. US Geological Survey Water Resour. Inv. 80-96, 210 pp., Washington.

Thiessen, P.A. 1942. Wechselseitige Adsorption von Kolloiden. *Z. Elektrochemie* 48: 675-681.

Thorez, J. 1975. Phyllosilicates and clay minerals. Ed. G. Lelotte, 578 pp., Dison, Belgique.

Veith, J. & U. Schwerdtmann 1979. Reaktionen von Ca-Montmorilloniten und Ca-Vermiculit mit Kohlensäure. *Z. Pflanzenernaehr. Dueng. Bodenkd.* 131 (1): 21-37.

Weiss, A. 1958a. Über das Kationenaustauschvermögen der Tonminerale. I. Vergleich der Untersuchungsmethoden. *Z. Anorg. Allg. Chem.* 297: 232-256.

Weiss, A. 1958b. Über das Kationenaustauschvermögen der Tonminerale. II. Der Kationenaustausch bei den Mineralen der Glimmer-, Vermiculit- und Montmorillonitgruppe. *Z. Anorg. Allg. Chem.* 297: 257-286.

'Natural' schwertmannite formed in a lake from waters draining pyritic deposits

C.W. Childs – *School of Chemical and Physical Sciences, Victoria University, Wellington, New Zealand*
K. Inoue & C. Mizota – *Faculty of Agriculture, Iwate University, Morioka, Japan*
M. Soma – *Graduate School of Nutritional and Environmental Sciences, University of Shizuoka, Japan*
B.K.G. Theng – *Manaaki-Whenua Landcare Research NZ Ltd, Palmerston North, New Zealand*

ABSTRACT: Schwertmannite, a recently described iron hydroxysulfate ('ideal' formula $Fe_8O_8(OH)_6SO_4$), forms naturally as a fluffy brownish-yellow precipitate in Lake Matsuo-Goshikinuma, Iwate Prefecture, Japan, during part of the distinctive annual geochemical cycle of the lake. The only significant inflow to the lake, other than precipitation, is anoxic ground water which drains disseminated pyritic deposits. The lake water typically has pH~3 and contains ~125 mg/l soluble-S and ~30 mg/l soluble-Fe. From approximately each September to June the lake circulates completely, dissolved oxygen levels are relatively high, iron-oxidizing bacteria (*Thiobacillus ferrooxidans*) are abundant, and the lake appears turbid yellowish-brown as schwertmannite is formed. Samples have been identified and characterized by X-ray powder diffraction, X-ray fluorescence analysis, extraction with oxalate and dithionite reagents, and X-ray photoelectron spectroscopy. Freshly precipitated samples are predominantly schwertmannite; samples that have remained in contact with the lake water for months or more show partial transformation to goethite, but the process is incomplete even after 30 years.

1 INTRODUCTION

Schwertmannite is a recently recognized iron hydroxysulfate mineral that occurs as a brownish-yellow oxidation product of anoxic acid (pH~3) waters rich in sulfate and iron (Bigham et al, 1994). Although iron-oxidizing bacteria (esp. *Thiobacillus ferrooxidans*) are often associated with its formation in nature, the mineral can be synthesized abiotically in the laboratory (Bigham et al, 1990). The 'ideal' formula is $Fe_8O_8(OH)_6SO_4$ and the proposed structure is analogous to that of akaganéite with sulfate, instead of chloride, occupying the tunnels present in the structural FeO(OH) array.

Most occurrences of schwertmannite, recorded to date, are attributed to mining or other anthropogenic activities associated with the surface or near-surface oxidation of iron sulfides (Bigham et al, 1994). Schwertmann et al (1995) recently described an entirely 'natural' occurrence in a small stream draining a pyritic schist. Lake Matsuo-Goshikinuma and its immediate environment are natural and the formation of schwertmannite here does not appear to have been influenced by anthropogenic activities.

To our knowledge, this is the first report on the natural occurrence of schwertmannite in Japan, and only the second one from anywhere.

2 LAKE MATSUO-GOSHIKINUMA

Lake Matsuo-Goshikinuma is a nearly round lake, with diameter 45 m and maximum depth 13 m. It lies at an altitude of 890 m about 35 km north-west of Morioka City, in the Towada-Hachimantai National Park. The name 'Goshikinuma' means 'lake of five colors' and refers to regular seasonal changes in color and appearance. Descriptions of the lake, its seasonal changes, chemistry and microbiological activity, are documented in the Japanese scientific literature (Yoshida et al, 1978; Wakao et al, 1980; Wakao & Yoshida, 1985; Maki, 1995a, 1995b).

The only significant inflow to Lake Matsuo-Goshikinuma, other than precipitation, is anoxic ground water entering though a vent in the lake floor, and deriving from the dissolution of disseminated pyritic deposits associated with Quaternary volcanism. The flow rate through the vent, measured in March 1977, was 30 l/s. There is a small surface overflow outlet from the lake into an adjacent shallow lake and swamp known as Gozaishonuma. The pH of Lake Matsuo-Goshikinuma is within the range 3.0-3.2 throughout the year and there is no significant variation with depth. The soluble-S concentration (predominantly SO_4^{2-}) ranges from 120 to 130 mg/l, and the soluble-Fe concentration from 20 to 40 mg/l throughout most years. The lake cycles regularly through three major annual phases. From September to June the lake water circulates completely and continually and contains abundant dissolved oxygen (~50% or more saturation). Any incoming Fe^{2+} and H_2S are oxidized, counts of iron-oxidizing bacteria (*Thiobacillus ferrooxidans*) in the water are relatively high, and the lake has a turbid yellowish-brown appearance. In this phase, Lake Matsuo-Goshikinuma has been likened to a continuous culture system. In July, circulation ceases and there is a progressive decrease in the level of dissolved oxygen and microbiological activity. Eventually there is insufficient oxygen to oxidize incoming Fe^{2+} though any H_2S is oxidized (mainly to S), and the lake's appearance evolves through greyish green to bluish grey to slightly turbid pale blue. In August, the whole lake is depleted in oxygen (<10% saturation), apart from a shallow surface stratum of ~0.5 m. Oxidation then essentially ceases, and the lake becomes clear blue with high visibility (~10 m).

In some years, the lake takes on a relatively clear, slightly turbid bluish-green appearance during March to May even though circulation continues and the lake water is oxygenated. This supplementary phase appears to be due, at least in part, to dilution caused by relatively high inputs of water from the spring thaw period. Heavier rainfalls and stronger winds than usual are also thought to occasionally perturb the normal annual cycle.

3 CHARACTERIZATION OF SAMPLES

Two samples, MG1004 and MG1118, of 'fluffy' freshly precipitated iron oxide were collected from the surface of Lake Matsuo-Goshikinuma in October/November 1995, soon after the onset of lake circulation, mixing and aeration. Sample MG1004 was treated with boiling 5% H_2O_2 to remove fine organic debris and oven-dried; sample MG1118 was collected relatively free of organic material and was dried only at 40°C. Munsell colors of the dried samples are 7.5YR6/8 (reddish yellow) and 10YR7/8 (yellow) respectively.

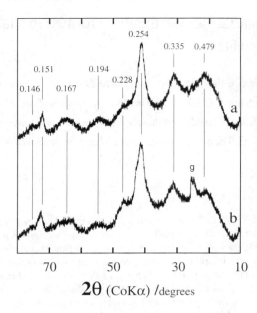

Figure 1. X-ray powder diffraction patterns for MG1118 (a) and MG1004 (b). (d-spacings in nm; goethite, g)

X-ray diffraction (XRD) patterns are shown in Figure 1. There is excellent agreement between the positions of the eight broad peaks and the distinctive overall diffraction envelopes with those previously reported for schwertmannite (Bigham et al, 1990; Bigham et al, 1994). A small amount of goethite is present in MG1004 though it is scarcely detectable in MG1118.

Major oxide analysis of MG1118 by X-ray fluorescence (XRF) yielded (wt%, oven-dry) Fe_2O_3 66.47, MnO 0.01, TiO_2 0.07, CaO 0.19, K_2O 0.25, SO_3 0.49, P_2O_5 0.10, SiO_2 2.91, Al_2O_3 0.94, MgO 0.04, Na_2O 0.48; LOI 27.94; Sum 99.89. The SO_3 value here represents only the S remaining in the sample after ignition. XRF analysis on a pressed oven-dried powder yielded (wt%, oven-dry) S 4.06, As 0.07 and V 0.08. Arsenate and vanadate anions may occupy a small proportion of the structural 'sulfate' sites within the samples or they may merely be adsorbed on the schwertmannite surface.

The Fe/S atomic ratio for MG1118 is 6.6. The ratio expected for $Fe_8O_8(OH)_6(SO_4)$ is 8 though smaller values (ranging to <5) are usually found and are attributable to adsorption of excess sulfate on the mineral surface (Bigham et al, 1990; Schwertmann et al, 1995).

The extraction of Fe by acid-ammonium-oxalate (AAO) reagent (0.15 M, pH3.5, 2 h shaking in dark) and by dithionite-citrate (DC) reagent (Holmgren, 1967) is useful in characterizing iron oxides formed in surficial environments. Schwertmannite samples have been found to be completely soluble in AAO reagent (Fe_o/Fe_t~1) and DC reagent (Fe_o/Fe_d~1) (Bigham et al, 1990). In contrast, goethite and other more crystalline iron oxides are essentially insoluble in AAO reagent though are usually soluble in DC reagent (Fe_o/Fe_t~0, Fe_d/Fe_t~1). Both MG1004 and MG1118 gave Fe_o/Fe_t~1 consistent with schwertmannite.

The 'surface' Fe/S ratio of the samples has been determined by X-ray photoelectron spectroscopy (XPS). XPS 'sees' the sample surfaces to a depth of approximately a few nm - the escape depth of electrons liberated by soft X-rays (Paterson & Swaffield, 1994). The Fe/S atomic ratios by XPS for MG1004 and MG1118 are 11.1 and 4.8 respectively; that for bulk MG1118 by XRF is 6.6. The apparent surface depletion of S for MG1004 is probably due to the removal of a S-rich surface phase during the H_2O_2 treatment. The lower Fe/S value for surface MG1118 relative to bulk is attributable to adsorption of excess sulfate on surface sites. Electron binding energies, also determined by XPS, for Fe $2p_{3/2}$ and O $1s$ are similar to those for ferrihydrite (Soma et al, 1996) though the O $1s$ line-shape is modified by the additional presence of sulfate-O in schwertmannite.

4 TRANSFORMATION TO GOETHITE

Schwertmannite is metastable with respect to goethite (Schwertmann et al, 1995) and eventually transforms to it if left in contact with the waters in which it initially forms (Bigham et al, 1996). Additional samples, varying in estimated age, were examined. Two samples of bottom sediments from the lake (both with Fe_o/Fe_d=0.10) showed goethite and no other iron oxide by XRD. These two samples represent material accumulated in the lake over a long period and the goethite is likely to be a transformation product of initially precipitated schwertmannite. Three samples (Fe_o/Fe_d 0.88-0.92), collected in July 1996 and considered to represent material precipitated during the previous September to June, all showed dominant schwertmannite with significant goethite by XRD. Two samples from an exposed margin of the lake near a previous outlet, and probably of the order of 30 y since precipitation, showed strong XRD peaks for goethite. Any reflections for schwertmannite were effectively obscured though the Fe_o/Fe_d values (0.39, 0.44) are consistent with its presence.

5 ACKNOWLEDGEMENTS

This work was supported by the award of a Senior Fellowship of the Japan Society for the Promotion of Science to CWC, and by the NZ Foundation for Research, Science and Technology through Manaaki Whenua Landcare Research.

6 REFERENCES

Bigham, J.M., U.Schwertmann, L.Carlson & E. Murad 1990. A poorly crystallized oxyhydroxysulfate of iron formed by bacterial oxidation of Fe(II) in acid mine waters. *Geochim. Cosmochim. Acta* 54: 2743-2758.

Bigham, J.M., L.Carlson & E.Murad 1994. Schwertmannite, a new iron oxyhydroxysulfate from Pyäsalmi, Finland, and other localities. *Mineral. Mag.* 58: 641-648.

Bigham, J.M., U.Schwertmann, S.J.Traina, R.L. Winland & M.Wolf 1996. Schwertmannite and the chemical modeling of iron in acid sulfate waters. *Geochim. Cosmochim. Acta* 60: 2111-2121.

Holmgren, G.G.S. 1967. A rapid citrate-dithionite extractable iron procedure. *Soil Sci. Soc. Am. Proc.* 31: 210-211.

Maki, Y. 1995a. Contribution of the microbial activities oxidizing ferrous iron or elemental sulfur to seasonal color change in the water of Lake Matsuo-Goshikinuma in northern Honshu, Japan. *Jap. J. Limnol.* 56: 183-193. (In Japanese with English summary.)

Maki, Y. 1995b. A possibility that heavy rainfalls with strong winds altered the seasonal color change in the water of Lake Matsuo-Goshikinuma in northern Honshu, Japan. *Jap. J. Limnol.* 56: 227-231. (In Japanese with English summary.)

Paterson, E. & R.Swaffield 1994. X-ray photelectron spectroscopy. In M.J.Wilson (ed), *Clay Mineralogy: Spectroscopic and Chemical Determinative Methods:* 226-259. London: Chapman & Hall.

Schwertmann, U., J.M. Bigham & E. Murad 1995. The first occurrence of schwertmannite in a natural stream environment. *Europ. J. Mineral.* 7: 547-552.

Soma, M., H.Seyama, N.Yoshinaga, B.K.G.Theng &

C.W.Childs 1996. Bonding state of silicon in natural ferrihydrites by X-ray photoelectron spectroscopy. *Clay Sci.* 9: 385-391.

Wakao, N. & M.Yoshida 1985. Seasonal changes in total bacterial number, iron-oxidizing activity and color of Lake Matsuo-Goshikinuma, Iwate Prefecture. *J. Fac. Agric. Iwate Univ.* 17: 333-341. (In Japanese with English summary.)

Wakao, N., M.Yoshida, Y.Sakurai & H.Shiota 1980. Microbial iron oxidation and color change of Lake Matsuo-Goshikinuma having a character of continuous culture system. *J. Fac. Agric. Iwate Univ.* 15: 29-47. (In Japanese with English summary.)

Yoshida, M., N.Wakao, & K.Inoue 1978. Seasonal change in color of Lake Matsuo-Goshikinuma, Iwate Prefecture. *Jap. J. Limnol.* 39: 170-175. (In Japanese with English summary.)

Attachment features between an aerobic *Pseudomonas sp.* bacteria and hematite observed with atomic-force microscopy

J. Forsythe & P. Maurice
Department of Geology, Kent State University, USA

L. Hersman
Los Alamos National Laboratories, USA

ABSTRACT: Although microorganisms have been shown to play important roles in the weathering of Fe(III)-(hydr)oxide minerals, little is known regarding microbial-mineral interactions in aerobic environments. Our research is focused on determining the rates and mechanisms by which obligate aerobic *Pseudomonas* bacteria obtain Fe from geologic materials. In the study reported here, we have grown colonies of *Pseudomonas* species in suspensions of hematite (α-Fe$_2$O$_3$) in which the mineral materials are the only sources of Fe. Atomic-force microscopy (AFM) and scanning electron microscopy (SEM) show that bacteria colonize mineralogic aggregates, forming networks of fiber-like attachment features intertwined through amorphous-looking 'ooze'. In addition, we have found close associations between bacterial flagella and growth-medium salts that precipitate on drying, which suggest that the flagella are hydrophilic. Our SEM and AFM images concur with, and further enhance, our understanding of microbial adhesion.

1 INTRODUCTION

Fe(III)-(hydr)oxides, though often present as only trace constituents, are some of the most abundant of the soil minerals. They generally occur as small, high-surface area particles and as coatings on the surfaces of other grains; hence, they help to define soil characteristics (e.g. adsorptivity, mobility of metals and organic pollutants). Additionally, Fe is nutritionally important and a metabolic requirement of living systems.

Microorganisms play a dominant role in biogeochemical transformations of minerals within terrestrial systems. Surface soil horizons, where microbes abound, often are aerobic, neutral environments. Fe is very insoluble under these conditions. So how, then, do the microbes living in these environments acquire sufficient quantities of Fe for survival? We speculate that some organisms may create a microenvironment conducive to dissolution through the use of extracellular materials. It is well documented in the literature that surfaces in general are well colonized by bacteria. However, it is not clear whether attachment is necessary for Fe acquisition, or whether growth on surfaces is simply preferred by microorganisms.

Previous studies have acknowledged the importance of microbial sorption to surfaces, including initial reversible sorption followed by irreversible sorption (Marshall et al., 1971). Irreversible sorption may be associated with the production of extracellular fibrils that serve to attach the cell firmly to the surface (Marshall and Cruickshank, 1973). Production of amorphous, granular, and fibrous holdfast materials (Hirsch and Pankratz, 1970) and involvement of polysaccharide capsules and extracellular slime (Humphrey et al., 1979) also have been described in the literature. Additionally, Meadows (1971) and Sjoblad and Doetsch (1982) demonstrated the involvement of flagella in attachment of motile bacteria to surfaces.

In the study reported here, we describe a first approach to this problem of Fe acquisition, concentrating on the methods of attachment used by an aerobic *Pseudomonas sp.* This approach includes the use of tapping-mode atomic-force microscopy (TMAFM) to image microbial attachment features that form upon reaction of an obligate aerobe *Pseudomonas sp.* with hematite.

2 MATERIALS AND METHODS

2.1 Mineralogic samples

Hematite (α-Fe$_2$O$_3$) was synthesized according to Hematite Sample Preparation Method 4, described in Schwertmann and Cornell (1991). Qualitative XRD (Rigaku X-ray powder diffractometer D/MAX-2 TBX θ/θ) analysis showed that the sample was monomineralic. TMAFM and TEM analysis concurred with these results and showed the hematite particles to be very small (50-100 nm diameter), rhombohedral to hexagonal shaped particles (Figs. 1 and 2). Two batches were produced, and BET measurements performed found the surface areas of the hematite batches to be 32.3 and 29.5 m^2 g^{-1}.

2.2 Bacterial sample

The microorganism used for our experiments is a soil bacterium isolated from surface soil samples collected at the Nevada Test Site. Dr. Myron Sasser (Microbial

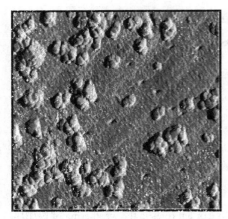

Figure 1. Synthetic hematite particles on pc membrane filter (note pores). Because the particles are very small, tip-sample interaction occurs which causes rounding of particle edges. Ex-situ TMAFM; amplitude data type. Scale: 3.00 μm on a side.

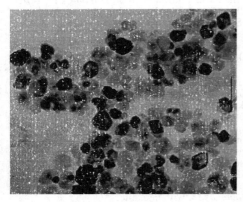

Figure 2. Synthetic hematite particles analyzed by TEM. The particle rhombohedral to hexagonal shapes are much clearer by this method of analysis. Scale: 608 nm wide x 523 nm high.

ID, Inc., Newark, Del.) determined by fatty acid analysis that this bacterium is most similar to *Pseudomonas stutzeri*. It tested positive for catalase, oxidase, and nitrate reduction (assimilatory), but negative for fermentation and dissimilatory nitrate, iron, and sulfate reduction. Hence, this microbe is an obligate aerobe, not capable of using iron as a terminal electron acceptor for oxidative phosphorylation.

2.3. *Microbial dissolution experiments*

For these experiments, the following growth medium was used: 0.091 g K_2HPO_4, 1.0 g NH_4Cl, 0.0625 g $MgSO_4 \cdot 7H_2O$, 0.05 g $CaCl_2$, 1.5 g succinic acid, and 125 mL trace elements. pH was adjusted to 6.8. The medium purposely contained too little Fe to support rich microbial growth so that in order for the microbes to obtain Fe, they must promote the dissolution of the hematite. The hematite was added in varying amounts (0.1067, 0.5400, 2.693, 13.46 (reflected in the data presented here), and 67.29 m^2/L) to acid-washed 125-mL Teflon flasks containing 30 mL of the succinate medium. Following sterilization by autoclave, the flasks were inoculated with 20 μL of an overnight culture of *Pseudomonas* sp. having an optical density of 0.2 adsorbance units at 600 nm wave number. These cultures were then incubated at room temperature, in the dark, while shaking at 50 rpm. Controls lacking Fe particles or microbes were prepared for comparison. Growth measurements, determined by adsorbance at 600 nm, were recorded on days 1, 2, 3, 4, 6, 8, and 10. After the last day, those particles having attached bacterial growth, as determined by inspection, were removed from the flasks using a sterile 1.0 mL pipette. The particle/bacteria aggregates were deposited on a pc membrane filter (25 mm diameter, 0.1 μm pore size), rinsed with sterile succinate medium, and air-dried. Uninoculated Fe particles from the control flasks also were placed on separate pc filters and dried. The pc filters provide a relatively smooth surface with pores serving as reference indicators when imaging with TMAFM. Additionally, we found that the particles and microbes attached well to this substrate.

2.4. *TMAFM*

A Digital Instruments, Inc. NanoScope III AFM was used under ambient (air) conditions to analyze the resulting relationships between the microbes and the particles. A description of AFM and its application to similar experiments can be found in Maurice et al. (1995). Briefly, AFM works by rastering a sample mounted on a piezoelectric tube under a sharp tip that is part of a cantilever system. In the case of TMAFM, the cantilever is oscillated at a resonant frequency while the analysis is taking place. This results in a decrease in the lateral frictional forces that normally occur with conventional contact-mode AFM. We collect two data types simultaneously: Height mode and Amplitude mode. While Height mode gives more quantitative data (i.e. height of sample's z axis), Amplitude mode gives data that are essentially first derivative of the Height mode, resulting in an image that is more easily discernible with the human eye. The Height mode image represents directly the microtopography of a sample surface, as it is a record of lateral changes in the z-axis piezoelectric cylinder. Amplitude mode images, on the other hand, give a record of lateral changes in the photodiode voltage. This reflects the magnitude of lag between the tip's deflection by a topographic feature and the movement of the z-axis piezoelectrode to maintain a constant force between the tip and the sample.

2.5. *SEM/EDS*

A Princeton Gamma Tech ISIABT SEM (model SX-40A) was used to check the TMAFM images against a more conventional methodology and to gain better insight into the larger-scale structure of the microbe-mineral interface. EDS analysis was used to analyze the chemical constituents of the larger-scale particles viewed by TMAFM and SEM. Samples were sputtered with either Au or C following AFM analysis and prior to SEM analysis.

3 RESULTS AND DISCUSSION

3.1. Microbial growth rates

Figure 3 depicts the growth curves for the microbes reacted with the iron oxide particles. Growth of bacteria in solution is monitored; these growth curves do not include bacteria attached to mineral surfaces. As observed by Hersman et al. (1996), the increased bacterial growth in the presence of hematite (sample) relative to the control suggests that the bacteria are removing Fe from the hematite. These data reflect the samples prepared with 13.46m^2 hematite surface area per L, as well as a hematite-free control. The decrease in microbial concentrations after 3 days may be related to depletion of the growth medium.

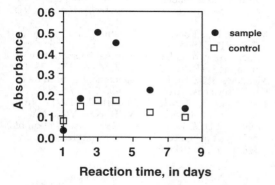

Figure 3. Bacterial growth as measured by adsorbance, in sample (bacteria + hematite + growth medium) versus control (bacteria + growth media) cultures.

3.2. Microbial attachment structures

Reaction blanks (hematite in growth media with no bacteria present) were identical to Figure 1. In bacterial-hematite cultures, we observed macroscopically the formation of microbial-mineral aggregates after several days of reaction. Figure 4 shows an SEM micrograph of the extracellular material, 'ooze', dried onto the surface of a pc membrane filter. Figure 5 shows the more detailed observations of the same material using TMAFM. These images reveal that the material is bound together by a loose network of fibers, each fiber approximately 20 nm in width and several microns in length. Other images show that this material tends to coat particle aggregates. Hence, it appears that at least some fraction of microbes form colonies on Fe(III)-(hydr)oxide aggregates using holdfast fibers intertwined through mineralogic particles and amorphous-looking 'ooze'. We do not know at this time whether the fibers are the original bacterial flagella or additional flagella-like structures formed upon attachment. In addition to the aggregates containing 'ooze' material, there were numerous aggregates and single particles of hematite that did not appear to contain microbial attachment features, and we also observed numerous single bacteria, not attached to surfaces.

In addition, several TMAFM images show a distinct relationship between the microbial flagella and small particles. As shown in Figure 6, <200 nm particles were observed attached to flagella like pearls on a string. Subsequent SEM/EDS analysis revealed that these particles were composed primarily of Na and Cl. Apparently, these particles are salts that precipitated from the growth medium upon drying. These salt crystals were not observed on control samples (hematite in growth media with no bacteria present). We believe that this may be due to pH increases in solution in the presence of bacteria. The close association between the flagella and the salt crystals suggests to us that the flagella are hydrophilic. Hence, the salt solution tends to remain associated with the flagella upon filtration and drying. Experiments are currently being conducted in order to remove and/or rinse the growth media to prevent this salt-crystal formation.

Our work to date has been performed ex-situ. One problem with this method is that we cannot be certain that the drying process has not affected the structures and microbe-mineral relationships. Recently, we developed a method for coating glass slides with thin layers of goethite needles. As the needles tend to lie flat, they can be imaged successfully by TMAFM. These goethite-coated slides will eventually be reacted with the microbes and analyzed using both ex-situ and in-situ TMAFM. The results of these studies should show even more conclusively the relationship between the microbes, their attachment structures, and the iron (hydr)oxide particles.

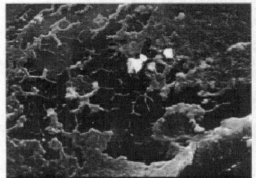

Figure 4. SEM micrograph of microbial sample reacted with hematite particles. This micrograph shows the large-scale structure of microbial extracellular material. Scale: 123 μm wide x 90 μm high.

4 CONCLUSIONS

We observed microbial growth on hematite, in aerobic Fe-limited conditions, indicating that bacteria are capable of extracting Fe for metabolic activity. Attachment features consisting of an extracellular, fibrous material were found coating the hematite particles. We also detected a relationship between bacterial flagella and salt particles, indicating that the flagella are most probably hydrophilic. These observations support the statements by Sjoblad and Doetsch (1982) that though relatively unrecognized, flagellar interactions with surfaces may play important roles in the microbial ecology of interfaces in numerous habitats, and that bacterial flagella should be recognized as attachment organelles, not solely as propulsion organelles.

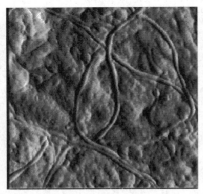

Figure 5. Ex-situ TMAFM image (Amplitude data type) of extracellular polymer. Here, the detailed structure of the 'ooze' tied together by fibrils can be seen. Scale: 1.25 µm on a side.

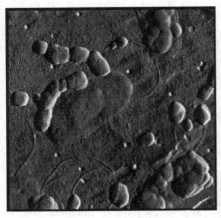

Figure 6. "String of pearls" relationship between the bacterial flagella and the salt crystals. Ex-situ TMAFM; Amplitude image. Scale: 3.66 µm on a side.

5 ACKNOWLEDGEMENTS

Funding for this research was provided by Department of Energy grant #DEFG02-96ER14668.

6 REFERENCES

Hersman, L., P.Maurice, and G.Sposito, 1996. Iron acquisition from hydrous Fe(III) oxides by an aerobic *Pseudomonas* sp.*Chem. Geol.* 132: 25-31.

Hirsch, P. & S.H.Pankratz, 1970. Study of bacterial populations in natural environments by use of submerged electron microscope grids. *Z. allg. Mikrobiol.* 10: 589-605.

Humphrey, B.A., M.R.Dickson, and K.C.Marshall, 1979. Physiochemical and in situ observations on the adhesion of gliding bacteria to surfaces. *Arch. Microbiol.* 120: 231-238.

Marshall, K.C., R.Stout, and R.Mitchell, 1971. Mechanism of the initial events in the sorption of marine bacteria to surfaces. *J. Gen. Microbiol.* 68: 337-348.

Marshall, K.C. & R.H.Cruickshank, 1973. Cell surface hydrophobicity and the orientation of certain bacteria at interfaces. *Arch. Mikrobiol.* 91: 29-40.

Maurice, P., J.Forsythe, L.Hersman, and G.Sposito, 1995. Application of atomic-force microscopy to studies of microbial interactions with hydrous Fe(III)-oxides. *Chem. Geol.* 132: 33-43.

Meadows, P.S., 1971. The attachment of bacteria to solid surfaces. *Arch. Microbiol.* 75: 374-381.

Schwertmann, U. & R.M.Cornell, 1991. Iron oxides in the laboratory. VCH, New York, NY.

Sjoblad, R.D. & R.N.Doetsch, 1982. Adsorption of polarly flagellated bacteria to surfaces. *Curr. Microbiol.* 7: 191-194.

Influence of temperature on the sorption isotherm of potassium on a montmorillonite

E.C.Gaucher
Commissariat à l'Energie Atomique, DCC/DESD/ SESD/LIRE, Saclay, Gif-sur-Yvette & Laboratoire de Géochimie des Eaux, Université Paris VII, France

L.Claude, H.Pitsch & J.Ly
Commissariat à l'Energie Atomique, DCC/DESD/ SESD/LIRE, Saclay, Gif-sur-Yvette, France

ABSTRACT : Sorption of potassium on Wyoming montmorillonite is studied. Two types of sorption sites are put in evidence and characterised by their exchange capacities and their selectivity coefficients for the exchange reaction between H^+ and K^+ in 2 mM KCl solutions. The experiments are conducted at two temperatures : 25°C and 80°C. No modification of the adsorption capacities of the sorption sites are induced by the temperature. On the contrary, the selectivity coefficients can be strongly affected by this parameter.

1-INTRODUCTION

This work takes place in a more general research on the control of the chemical composition of groundwaters by clays. The purpose of this global study is to describe the dissolution-precipitation reactions of clays in connection with ion exchange equilibria (Gaucher et al, 1997).

In the present paper, we focus our attention on the influence of temperature on the sorption of potassium on a montmorillonite. In the literature, this type of phenomenon is very poorly documented. Laudelout (1980) presents the inversion in the order of affinities of K^+ and Ca^{2+} for a smectite induced by temperature. At low temperature Ca^{2+} is more strongly adsorbed on a smectite than K^+, but at elevated temperature, the contrary is observed. Inoue (1983) and Cuevas et al (1994) show that the amount of K^+ fixed by clays increases with increasing temperatures.

Batch experiments are realised to construct the sorption isotherm of K^+ as a function of pH at two temperatures : 25°C and 80°C. The experimental procedures are derived from Gorgeon (1994) and Rytwo et al (1996) and include two steps. In the first one, the mineral adsorbs a certain amount of K^+ in a solution of known pH. In a second step, a solution of an organic cation of large binding affinity (crystal violet) is used to displace the adsorbed potassium. The analytical data are processed using a multisite ion exchange model developed by Ly (1991). This model enables to discriminate the different types of sorption sites and to characterise them by their intrinsic capacity and the selectivity coefficients for pairs of cations. These sites are located either on the basal surface of the clay platelets, or on the edges of the sheets.

2-ION EXCHANGE MODEL

In this study, the cation exchange between two monovalent ions is considered (K^+ and H^+). For the exchange reaction (1) the thermodynamic equilibrium constant is represented by expression (2), and (3) gives the expression of the selectivity coefficient. The existence of sites of different natures is considered, the equations (1) to (5) are given for a particular type of sites noted (i).

$$\{Xi^-K^+\} + H^+ \Leftrightarrow \{Xi^-H^+\} + K^+ \quad (1)$$

$$K_{K/H}^{°i} = \frac{[Xi^-H^+][K^+]}{[Xi^-K^+][H^+]} \frac{f_{X^-H^+} \cdot y_K}{f_{X^-K^+} \cdot y_H} \quad (2)$$

$$K_{K^+/H^+}^{*i} = \frac{[Xi^-H^+][K^+]}{[Xi^-K^+][H^+]} \quad (3)$$

[ ] : concentration of species either in solution or adsorbed.
y : activity coefficient of species in solution.
f : activity coefficient of adsorbed species

As experiments are carried out at low ionic strength (10 to 20 mM), the ratio of the activity coefficients of the monocharged cations in the aqueous phase is practically equal to 1. So the difference between the selectivity coefficient and the equilibrium constant is reduced to the ratio of the activity coefficients of the adsorbed cations, which is not computable from our data. Nevertheless, in our experimental conditions, we can assume that this ratio will not be very different from unity.

The exchange capacity of a type of sites (i) EC_i can be expressed by equation (4), as K^+ and H^+ are the unique cations presented in the system.

$$EC_i = [X_i^-K^+] + [X_i^-H^+] \quad (4)$$

So, the concentration of adsorbed K^+ can be expressed by equation (5).

$$[X_i^-K^+] = \frac{EC_i}{1 + \frac{K_{K/H}^{*i} \cdot [H^+]}{[K^+]}} \quad (5)$$

K^+ can be adsorbed on several sites of different nature, so the total concentration of adsorbed K^+ (C_K) will be the sum (6).

$$C_K = \sum_{i=1}^{n}[X_i^-K^+] = \sum_{i=1}^{n} \frac{EC_i}{1 + \frac{K_{K/H}^{*i} \cdot [H^+]}{[K^+]}} \quad (6)$$

This last equation (6) is used for a non-linear regression by the least squares method using the experimental data : [K^+] and pH of the solution, and C_K. (i) pairs of parameters (EC_i and $K_{K/H}^{*i}$) are unknown.

3-EXPERIMENTAL PART

Purified and homoionic K-montmorillonite is prepared using a Wyoming montmorillonite, by sequential immersion in KCl solutions and washing with water and alcohol. From chemical analysis of treated clay samples, the structural formula of the clay is determined using the classical model of Bain and Smith (1987) :

$K_{0.7}(Si_8)(Al_{3.11}Fe^{3+}_{0.269}Mg_{0.471}Fe^{2+}_{0.108})O_{20}(OH)_4$

Batch experiments are conducted in polycarbonate centrifuge tubes. In the first step (adsorption), clay suspensions are prepared by dispersing 0.2 g of K-montmorillonite in 25 ml of KCl 10 mM, HCl 10 mM solutions. Appropriate amounts of KOH 10^{-1}M are added to the clay suspension to obtain a wide range of pH (2-12). After 12 h of incubation at the selected temperature in a shaker, the final pH is measured at the same temperature in a water-bath. Then the suspension is centrifuged at 8000 rpm for 90 min. The supernatant (A) is separated and an aliquot is filtered for analysis. By a precise weighing, the residual amount of solution (A) present in the sedimented clay is calculated.

The second step (desorption) begins with the dispersion of the sedimented clay in 25 ml of a 10 mM crystal violet solution ($C_{25}H_{30}N_3Cl$). The suspension of clay in the dye solution is shaken vigorously for 72 h. Then the suspension is centrifuged at 8000 rpm for 90 min. The supernatant (B) is separated and an aliquot is filtered for analysis. Potassium concentrations are determined by capillary electrophoresis (Quanta 4000 Waters).

The total amount of K^+ in the solution (B) corresponds to the amount of K^+ adsorbed on a precise weigh of clay and to the amount of K^+ present in the interstitial water. The volume of interstitial water is determined between the two steps and the concentration of K^+ in this water corresponds to the concentration of the solution A. So, by subtraction of this minor source of potassium to the solution (B), a precise value of K^+ adsorbed on the clay can be obtained.

4-RESULTS AND DISCUSSION

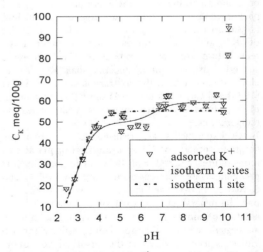

Figure 1 : Adsorption of K^+ onto montmorillonite versus pH at 80°C. Comparison of the exchange model between K^+ and H^+ for 1 or 2 sites of adsorption.

The experimental data, obtained at 80°C, are presented on figure 1 (adsorbed K^+ versus pH). The adsorption of K^+ on the montmorillonite increases in a non-linear manner with pH.

In a first approach the data are processed with an exchange model assuming a single type of sites (Figure 1). But, in this case, the residuals are not randomly distributed (Figure 2) : two groups of points are observed (α and β). The lack of homogeneity indicates that the model with one site is biased. On the contrary, the exchange model with 2 sites shows an homogenous distribution of the residuals (Figure 3). So, the best fit of the experimental data is obtained for an exchange model with two types of sorption sites.

indicates the saturation of the first type of sites. The second type of sites is converted to the K^+ form between pH 6 and 8 and the second plateau indicates that both type of sites are saturated, with a total capacity of 59 meq/100g. Beyond pH 10, the two points indicate that a third K^+/H^+ exchange reaction is occurring on a third type of sites, but the experimental data are insufficient for a proper modelling. Nevertheless, they are consistent with the total cation exchange capacity measured by adsorption of Cs^+ (93 meq/100g).

The comparison between the two sets of data obtained at 25°C and 80°C is presented in figure 4. The exchange model with two sites is applied with success for the two temperatures. But an important shift of the position of the second plateau can be observed.

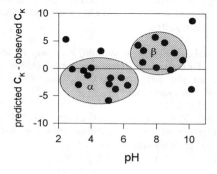

Figure 2 : Residuals for exchange model with 1 site.

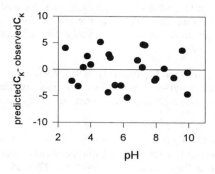

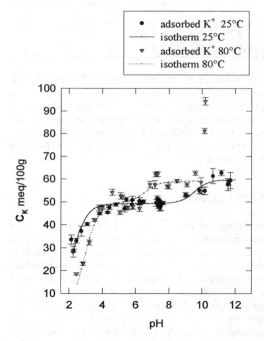

Figure 4 : Adsorption of K^+ onto montmorillonite versus pH at 25°C and 80°C.

Figure 3 : Residuals for exchange model with 2 sites.

The adsorption isotherm (Figure 1) shows that at low pH (<2), sorption of K^+ is negligible; between pH 2 and 4, mobile protons from the first type of sites of the clay surface are replaced by potassium ions. Between pH 4 and 6, the plateau at 49 meq/100g

Table 1 Cation Exchange Capacities of the two sites.

	25°C	80°C
EC1 meq/100g	49.2 ± 0.4	49.8 ± 1.5
EC2 meq/100g	10.4 ± 0.9	9.3 ± 1.9
CEC meq/100g	59.6 ± 1	59.1 ± 2.4

The parameters calculated from non-linear regression are presented in the tables 1 and 2.

Table 1 gives the Cation Exchange Capacity related to each type of sites obtained by calculation at both temperatures. The CEC is the sum of the capacities of the two sites. The value of 59 meq/100g for the total potassium adsorption capacity of this Wyoming montmorillonite is within the classical range. The values determined in this study are close to the values of Goulding and Talibudeen (1980) for the two sites they have defined (40 and 16.4 meq/100g).

No difference can be observed between the results obtained at the two temperatures. Indeed the adsorption capacity of a site is a mineralogical parameter, which is not modified by the temperature.

Table 2 Selectivity coefficient for K^+/H^+ exchange.

	25°C	80°C
log $K_{K/H}^{*1}$	0.37 ± 0.04	1.10 ± 0.08
log $K_{K/H}^{*2}$	7.8 ± 0.2	4.9 ± 0.5

On the contrary, the selectivity coefficients ($K_{K/H}^{*i}$), are affected by temperature (Table 2). While the selectivity coefficient for the first kind of sites is moderately affected by the increase of temperature, the selectivity coefficient for the second type of sites decreases by a factor 850 when the temperature increases from 25°C to 80°C.

5-CONCLUSION

Modelling of the sorption isotherm of K^+ on a montmorillonite puts in evidence the existence of three types of ion exchange sites and allows to fully characterise two of them. Assumptions on the nature of these sites can be made. In fact, the first sites with a high exchange capacity (49.5 meq/100g) are probably on the basal surface. The origin of this first type of sites is the substitution of cations in the octahedral polyhedrons of the clay sheet. This internal origin of the permanent charge of clay probably explains the little influence of a temperature modification on the K^+/H^+ exchange. On the contrary, the selectivity coefficient of the second types of sites is largely influenced by temperature. The capacity of this type of sites (10 meq/100g) and the high variability of $K_{K/H}^{*2}$ suggest that they may be sites of sheet edge.

Goulding and Talibudeen (1980), using microcalorimetric measurements, distinguish between different groups of cations exchange sites, based on regions of constant differential enthalpy of exchange. They distinguish two types of sites for the Wyoming montmorillonite. In this study, these two types of sites are confirmed by a totally different method.

REFERENCES

Bain D.C. and Smith B.F.L.,1987. Chemical analysis. 248-274, in: *A handbook of determinative methods in clay mineralogy*. Wilson M.J. eds, Blackie & Son Ltd.

Cuevas J., Leguey S., Pusch R., 1994. Hydrothermal stability of saponitic clays from Madrid Basin. *Applied Clay Science*, vol 8, 467-484.

Gaucher E.C., Beaucaire C., Ly J., and Michard G., 1997. Stability of minerals having ion exchange capacity or how clays control the chemical composition of waters. *Mudrocks at the basin scale, London, Jan. 1997*.

Gorgeon L., 1994. Contribution à la modélisation physico-chimique de la rétention de radioéléments à vie longue par des matériaux argileux. *Thèse de doctorat. Université Paris VI*.

Goulding K.W.T. and Talibudeen O., 1980. Heterogeneity of cation-exchange sites for K-Ca, exchange in aluminosilicates. *J. of Colloid and Interface science*, vol 78, No 1, 15-24.

Inoue A., 1983. Potassium fixation by clay minerals during hydrothermal treatment. *Clays and Clay Minerals*, vol 31, No 2, 81-91.

Laudelout H., 1980. L'échange d'ions dans les argiles. in: *Géochimie des interactions entre les eaux, les minéraux et les roches*, Y. Tardy ed., Eléments, Tarbes, 7-26.

Ly J., Stammose D., Pitsch H., 1991. Description of actinides sorption onto clays by ion exchange mechanisms. *Migration'91, Jerez de la Frontera, Spain*.

Rytwo G., Banin A., and Nir S., 1996. Exchange reactions in the Ca-Mg-Na Montmorillonite system. *Clays and Clay Minerals*, vol 44, No 2, 276-285,

Mineral surfaces and the sorption of bacteria in groundwater

J. S. Herman, A. L. Mills & E. P. Knapp
Program of Interdisciplinary Research in Contaminant Hydrogeology, Department of Environmental Sciences, University of Virginia, Charlottesville, Va., USA

ABSTRACT

Among the demonstrated processes influencing the transport of bacteria through aquifers, the deposition of cells on mineral surfaces is one of the most important. Bacterial sorption to clean, quartz sand yielded equilibrium, linear, isotherms, with greatest sorption at highest groundwater ionic strength. When iron-oxyhydroxide-coated sand was used, all of the added bacteria were sorbed up to a threshold beyond which no further sorption occurred. Bacterial transport was evaluated in experiments through columns of quartz sand in which three distributions of Fe(III)-oxyhydroxide-coated sand, present as 10% of the mass, were prepared. A large fraction of cells was retained: 14.7-15.8% of the cells were recovered after three pore volumes had eluted through clean quartz sand, and only 2.1-4.0% were recovered from the Fe(III)-oxyhydroxide-coated sand mixtures regardless of the spatial distribution of iron coatings.

INTRODUCTION

The presence of bacteria in groundwater can be viewed as harmful, as in the contamination of water supply by pathogens, or as beneficial, as in the biodegradation of noxious organic compounds. Elucidation of factors influencing the transport of bacteria is needed, and understanding the processes involved in bacterial migration is especially critical for conditions appropriate to the hydrogeological environment.

A variety of complex biological, physical, and chemical phenomena are involved in determining bacterial attachment to mineral surfaces (Harvey, 1991). The general physicochemical concept used to describe the adsorption of charged solutes has been found to successfully describe the partitioning of bacterial cells between an aqueous suspension and solid phase (Daniels, 1980), and the distribution of mass between aqueous and solid phases can be expressed by a sorption isotherm.

Long-range electrostatic interactions are critical elements of bacterial adhesion to soil mineral surfaces (van Loosdrecht et al., 1989). The sign and magnitude of the surface charge of the minerals have been found to control the attachment of bacteria to surfaces. Bacteria, usually negatively charged, interact weakly with quartz which has a negative surface charge in contact with natural waters (pH_{zpc}, zero point of charge, is 2.0) but adsorb strongly to $Fe(OH)_3$ which has a positive surface charge (pH_{zpc} is 8.5). The effect of Fe(III)-oxyhydroxide distribution on the transport of bacteria has not been thoroughly elucidated.

We investigated the interaction of bacteria with mineral surfaces in laboratory experiments. Understanding the interaction between bacterial cells and mineral surfaces is essential to our attempts to quantify and predict the transport of microbes in groundwater.

METHODS

Preparation of sand, water, and bacteria

Fine quartz sand (0.5-0.6 mm diameter) was cleaned by rinsing sequentially in HNO_3, NaOH, and deionized water (DIW). Some sand was coated by adding it to a flask containing an acidic solution prepared from $FeCl_3 \cdot H_2O$. The contents of the flask were titrated with 0.5 N NaOH. The slurry at final pH around 5.0 was then allowed to shake for 24 h to ensure complete coating of the quartz sand by Fe(III) oxyhydroxide. The newly coated sand was rinsed repeatedly with DIW and dried at 90°C.

The solution used for experiments was an artificial groundwater (AGW) of total ionic strength 57.9 mm. Other ionic strengths were obtained by dilution. The bacterial strains used were isolated from groundwater samples collected from a fresh, surficial aquifer on the Coastal Plain of Virginia, U.S.A. Both strains W8 and S1 are non-motile rods (Mills et al., 1994). For experiments, cultured cells were suspended in filtered AGW and placed on a rotary shaker for a period of 24 h to ensure that they had entered a resting stage where growth ceased.

Batch sorption experiments

The experiments were generally conducted by adding 26 mL of a bacterial stock suspension to each of three Erlenmeyer flasks. Immediately, 1 mL of suspension from each flask was extracted with a pipette and counted using the acridine orange direct count (AODC). Sand (12.5 g) was then added to each flask, and the flasks were allowed to swirl gently for 3 h, a time chosen to ensure equilibration. The number of bacteria in initial sand-free bacterial suspensions and in the final, equilibrium suspensions were counted, and the number sorbed was determined by difference.

Column transport experiments

The transport experiments were conducted in glass chromatography columns (4.8-cm inside diameter). A total of 175 mL of water was poured into each column. Then, 440 g of sand, comprised of 90% clean quartz sand and 10% Fe(III)-coated sand, was introduced. The water was drained from each column until it reached the top the sand. The final length of the sand column was 14 cm. The pore volume of the columns was 105 mL, and the porosity was 0.35. Flow of AGW was regulated by a variable flow peristaltic pump placed at the outlet of the column. The flow rate used was 1.0 pore volume h^{-1} (14 cm h^{-1}). A 5-mL pulse of 10^8 bacterial cells mL^{-1} was introduced to the top of the columns. Cell concentrations in eluted groundwater were determined by AODC.

RESULTS

Batch sorption isotherms

The adsorption isotherms in Fig. 1 show a clear effect of ionic strength on sorption of bacteria to clean quartz sand. The isotherm slopes (K_d) increased with increasing ionic strength. An effect of bacterial strain was also seen. At the lowest ionic strength, the K_d values were similar for both strains (0.55 mL g^{-1} for S1 and 0.56 mL g^{-1} for W8). At the highest ionic strength, the difference in K_d values had increased substantially, with strain W8 being the more strongly sorbed ($K_d = 6.11$) as opposed to S1 ($K_d = 3.72$).

The effect of mineral surface coatings (surface charge) was apparent. Using the negatively charged uncoated quartz, the maximum amount of cells adsorbed for the most sorption-favorable conditions of highest ionic strength AGW and strain W8 was 75% of the initial bacterial suspension concentration. The least amount of adsorption was approximately 28% with strain S1 and lowest ionic strength AGW. When using Fe(III)-coated quartz, practically all of the cells adsorbed to the positively charged mineral surface.

Additional batch experiments were performed with Fe(III)-coated sand using higher initial cell suspension concentrations to determine an isotherm appropriate for the coated sand grains. Apparent saturation of the sorption sites on the sand was reached at a sorbed concentration of 6.93 x 10^8 cells g^{-1}. Once that level was reached, no additional cells sorbed (Fig. 2). When samples of the sand were placed in AGW of the same ionic strength for 48 h, no desorption of cells above the background could be detected.

Column transport

Breakthrough curves for bacteria for all the columns showed the peak of the bacterial breakthrough for each of the columns in the experiment was at approximately one pore volume (Fig. 3). The concentration of cells at the peak breakthrough from the clean quartz columns was considerably greater than from the columns with a 10% mixture of Fe(III)-oxyhydroxide-coated sand as was total mass recovery over three pore volumes (Table 1). The average mass recovery for the clean columns was 15.25%, and the average for all the Fe(III) columns was 3.0%, with the percentage recovery of bacteria through each of the three physical arrangements of chemical heterogeneity being similar. The agreement across the three Fe(III) treatments in the timing and magnitude of the peaks along with the similarity of the overall shapes of the curves indicated that there was no significant difference among them.

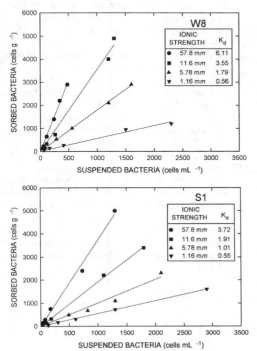

Fig. 1. Sorption isotherms for bacterial strains W8 and S1 to clean quartz sand. Each curve represents a different ionic strength. Slopes (K_d, the distribution coefficient in mL g^{-1}) were obtained by linear regression.

DISCUSSION

Sorption isotherms

The isotherms are linear and, with zero intercepts, satisfy the model for K_d. The range of ionic strengths used is reasonable in terms of dilute ground water, and the impact of solution composition is significant. The mechanistic basis of the observed ionic strength effect is related to the shrinking of the double layer as the ionic strength increases, allowing bacteria to approach mineral surfaces more closely (*e.g.*, van Loosdrecht *et al.*, 1989; Harvey, 1991). Thus, in isotherms of bacterial adsorption, the K_d value is higher at higher ionic strengths (Fig. 1).

The sorption of bacteria to Fe(III)-coated sand is not reversible. Nearly complete uptake of cells to the surface occurs up to a threshold beyond which no further cells are attracted to the mineral surface (Fig. 2). Subsequent exposure of the bacteria-laden Fe(III)-coated sand grains to bacteria-free, dilute AGW does not promote desorption. The sorption of negatively charged bacterial cells to positively charged Fe(III)-oxyhydroxide coating is clearly favored by electrostatic attraction.

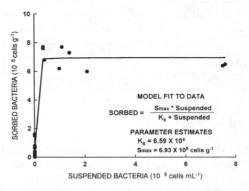

Fig. 2. Saturation plot for strain W8 and Fe(III)-coated sand. The Langmuir isotherm was fitted simply to obtain a value for maximum cell coverage.

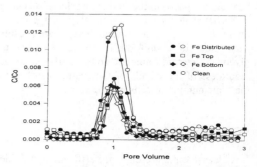

Fig. 3. The breakthrough of bacteria for each of the experimental treatments. The plot shows the reduced concentration of bacteria (C/C_o) vs. number of pore volumes eluted.

Table 1. The center of mass and the percentage recovery for each of the experimental treatments.

	Center of Mass	
	Column A	Column B
Fe distributed	1.29	1.38
Fe top	1.36	1.18
Fe bottom	1.18	1.11
AVERAGE	1.25 (std. dev. 0.10)	
Clean	1.05	1.06
AVERAGE	1.055 (std. dev. 0.005)	

	Percentage Recovery	
	Column A	Column B
Fe distributed	4.0	2.1
Fe top	3.4	2.8
Fe bottom	2.1	3.6
AVERAGE	3.0 (std. dev. 0.73)	
Clean	14.7	15.8
AVERAGE	15.25 (std. dev. 0.55)	

Column transport

The presence of relatively small amounts of Fe(III)-oxyhydroxide coatings on sand grains has a profound effect on the transport of bacteria. Our mass-retention results show a five-fold decrease in percentage of eluted cells when only 10% of the sand was coated with Fe(III)-oxyhydroxide compared to clean quartz sand. The results from each of the three arrangements of chemical heterogeneity are similar, indicating that distribution of Fe(III)-coated sand within the flow path does not affect the transport of bacteria. The column results show that, in a uniform hydraulic environment, given a certain amount of Fe(III) coatings, regardless of the details of the spatial distribution, the same number of cells will be sorbed.

CONCLUSIONS

It is clear that the presence of Fe(III)-oxyhydroxide-coated sand increases the retention of bacterial cells relative to an Fe(III)-free system. The expected impact of bacterial sorption to the Fe(III) oxides ubiquitous in shallow sedimentary aquifers can be confounded by a number of factors. Complex interactions among aquifer mineralogy, geochemistry, hydrology, and microbial ecology point to the need for more study prior to the effective prediction of bacterial migration in groundwater systems.

REFERENCES

Daniels S.L. (1980) Mechanisms involved in sorption of microorganisms to solid surfaces. In: Bitton G. and Marshall K.C. (Eds), Adsorption of Microorganisms to Surfaces. John Wiley and Sons, New York pp 7-58

Harvey R.W. (1991) Parameters involved in modeling movement of bacteria in groundwater. In: Hurst C.J. (Ed.) Modeling the Environmental Fate of Microorganisms. Washington, D.C., American Society for Microbiology pp 89-114

Marshall K.C. (1980) Adsorption of microorganisms to soils and sediments. In: Bitton G. and Marshall K.C. (Eds.), Bacterial Adhesion. New York, Plenum Press pp 58-65

Mills A.L., Herman J.S., Hornberger G.M., and DeJesus T.H. (1994) Effect of solution ionic strength and iron coatings on minerals grains on the sorption of bacterial cells to quartz sand. *Applied and Environmental Microbiology* **60**: 3300-3306.

van Loosdrecht M.C.M., Lyklema J., Norde W., and Zehnder A.J.B. (1989) Bacterial adhesion: A physicochemical approach. *Microbial Ecology* **17**:1-15.

Lead adsorption onto aquifer gravel using batch experiments and XPS

C. Hinton & M. E. Close
Institute of Environmental Science and Research Ltd (ESR), Christchurch, New Zealand

ABSTRACT: Lead adsorption onto gravel taken from an aquifer located at Burnham, Canterbury, New Zealand was considered. The gravel often bears a natural coating, which is up to 13 μm thick and composed largely of iron oxides. The characteristics of the coated gravel and those of non-coated gravel and fine material (≤ 2 mm) were determined by bulk extractions and surface-sensitive XPS. The role of the natural oxide in adsorption processes was addressed. Lead is more rapidly adsorbed by the coated gravel than by the non-coated, however at equilibrium there is little difference in the extent of adsorption. Adsorption is shown to be described by a Langmuir isotherm over the range 1×10^{-6} to 1×10^{-4} M Pb^{2+}. The pH-adsorption edge shows a large amount of adsorption of 1×10^{-5} M lead even at low pH. XPS allows *in situ* examination of the adsorbed lead, and initial measurements show quantification is possible.

1 INTRODUCTION

Groundwater serves as the source of drinking water for 37% of New Zealanders, and agriculture and industry require large volumes of groundwater. Groundwater supplies may be threatened by heavy metal contamination through mining and other industrial activities, and seepage from landfills. Such contamination is very difficult to remove, and impacts on water supply in future decades.

The ability to predict the movement and fate of dissolved species is fundamental to the reliable prediction and better management of heavy metal contamination of groundwater. This requires an understanding of the adsorption reactions of metal ions with aquifer solids (Coston et al., 1995).

The present study was undertaken to assess heavy metal adsorption onto material from an aquifer of fluvioglacial outwash gravels. The gravel often bears an oxide coating, the role of which in adsorption processes is investigated. This paper describes the properties of the aquifer material, and the adsorption of lead onto the alluvial gravels as a function of adsorbate concentration and solution pH. Bulk properties were determined by extraction, and surface properties ascertained primarily by X-ray photoelectron spectroscopy (XPS). Adsorption was investigated by batch experiments, and the use of XPS, a technique well suited to adsorption studies as it provides an *in situ* measure of surface coverage and yields information on the state of sorbate and sorbent (Martin & Smart, 1987), is illustrated.

2 MATERIALS AND METHODS

The aquifer material was taken from a field site at Burnham, Canterbury, New Zealand. The site is on a formation of well sorted, little weathered fluvioglacial outwash gravels with varying proportions of sand, silt and clay (Pang et al., 1997). The composite material was sorted into three fractions: coated gravel, non-coated gravel, and fine material passing through a 2 mm sieve. Gravels were washed with tap water, which closely approximates Burnham groundwater, and air dried before use.

Glassware and plastic vessels were acid washed before use. All reagents were analytical grade, and solutions prepared with MilliQ water. Metal ion concentrations were determined by flame or flameless atomic absorption spectroscopy (AAS).

For acid extractions, a 1:1 ratio (w/v) of gravel to extractant was used, 1:100 for the fine material. Standard methods for dithionite-citrate (Mehra & Jackson, 1960), ammonium oxalate (McKeague & Day, 1966), and hydroxylamine hydrochloride (Chao, 1972) extractions were used as described for fine material, and adapted for use on the gravels. Reagent blanks were analysed for all extractions.

Cation exchange capacities were measured at pH 7 using a leaching method (Blakemore et al., 1987) modified for the gravel samples.

Batch adsorption experiments were performed in sealed HDPE jars held on an orbital shaker. A 1:1 ratio (w/v) of gravel to 0.1 M $NaNO_3$ solution was used at $20 \pm 2°C$. Gravels were soaked for 36 hours at pH 6 ± 0.5. The 0.1 M $NaNO_3$ was replaced, the pH adjusted to that desired, and the solution degassed, before commencement of the experiment. Readjustment of the pH was required after degassing and after Pb^{2+} addition. A nitrogen atmosphere was maintained.

Kinetic experiments were performed at pH 6, and samples taken over 2 days. For the concentration dependence a pH of 6.5 and equilibration period of 4 hours was used. pH dependence experiments were begun at pH 3, and incrementally increased by 0.5 pH units to pH 7. A sample was collected and the equilibrium pH measured after 4 hours at each pH. In all cases blanks were run to determine the initial Pb^{2+} concentration and as a check for precipitation.

Thin sections 30 μm thick were prepared from two coated gravel samples, mounted on a slide, and examined under a microscope.

Gravel samples for XPS analysis were prepared by cutting and grinding the gravel to a 1x1 cm square of thickness ≤ 5 mm. H_2O was used as lubricant and the upper (analysis) gravel surface protected from abrasion. Iron oxide compounds were prepared as finely crushed powders, and all samples were mounted on double sided electrically conductive tape. XPS was performed on a Kratos XSAM800 spectrometer using an Mg X-ray source operated at 14 kV and 12 mA. A pass energy of 65 eV was used for wide scans, and 20 eV for region scans.

3 RESULTS

3.1 *Bulk properties of aquifer components*

Table 1 lists for each extractant the amount of iron and manganese as a percentage of metal extracted per gram of each aquifer component.

Table 1 shows that for all iron extracts, a greater amount of iron is removed from the coated gravel than for the non-coated gravel. The manganese extractants show the same is true for this metal. The coating of the alluvial gravel thus consists of iron oxide(s), with some manganese oxide associated with it. Following extraction by oxalate, the coated gravel retains a yellow covering, believed to be goethite. This covering is removed by subsequent treatment with dithionite-citrate, which alone removes the coating in its entirety.

Though the iron content of the two gravel types does differ, the distinction is not as great as visual inspection suggests. Moreover, the bulk gravel as a whole, and not just the coated gravel, has a significant iron content, as established by multi-element analysis of nitric acid extracts. On the other hand, Table 1 shows that manganese is more exclusively associated with the oxide coating. On a weight basis, the fine material contains significantly more iron and manganese than the gravel fractions.

3.2 *Surface properties of aquifer components*

The cation exchange capacities (CEC), of the coated gravel, non-coated gravel and fine material are 0.19, 0.18 and 2.47 cmol(+)/kg respectively. The gravels have essentially the same CEC, whilst that of the fine material is greater. Given that a typical CEC for a New Zealand topsoil is 30 cmol(+)/kg, the values for all aquifer components are extremely low.

Thin sections of the coated gravel reveal the oxide thickness to vary from about 6 to 13 μm. The gravel surface has a distinct peak and valley profile, the peaks having little coverage, the valleys being oxide rich. The oxide coating is present on the surface either as a continuous uniform layer, or in semi-spherical nodules. No crystal structure is apparent under the microscope.

XPS affords information on the uppermost 0.5 to 2 nm of a surface. The elements present and their relative abundance are determined from a wide scan of the surface. Such spectra showed the three aquifer components to have a similar surface composition. O, Si and Al are the major surface constituents, comprising over 70% of the total. Iron constitutes only 1-2% of the surface. The coated gravel has an elevated Fe content relative to the non-coated gravel. Notably the coated gravel is the only component to contain Mn, suggesting it to be specific to the oxide coating. The level of iron for the fine material is similar to that of the non-coated gravel.

Peak positions determined from detailed XPS spectra collected over narrow energy windows allow identification of chemical form. The coating was characterised by comparing the iron spectra of the alluvial gravel with those of the standard iron

Table 1. Concentration of iron and manganese removed from aquifer fractions by extraction.

Extractant	Iron extracted (% x 100)			Manganese extracted (% x 100)		
	Coated gravel	Non-coated gravel	Fine material	Coated gravel	Non-coated gravel	Fine material
Dithionite-citrate	1.7 ± 0.1	1.3 ± 0.1	26 ± 4.0	0.10 ± 0.02	0.034 ± 0.005	0.54 ± 0.01
Ammonium oxalate	0.33 ± 0.01	0.19 ± 0.03	8.0 ± 1.0	-	-	-
Hydroxylamine-HCl	-	-	-	0.035	0.0070	0.19

oxides: hydrous ferric oxide (HFO); goethite and ferric oxide. The Fe $2p^{1/2}$ and $2p^{3/2}$ peak positions of the coated gravel and their separation coincide well with those of HFO, and are distinguishable from those of the other oxides. Peak broadening indicates this is not the only, but rather the predominant form of iron present.

Elemental composition and speciation data were used to assess the characteristics of the surface of the aquifer components. It was found that:
- the coating composition is essentially homogeneous as a function of depth over 14 nm.
- the coating composition is homogeneous over the surface of a single coated gravel sample.
- reproducibility of the surface composition is evident between multiple samples for each of the three aquifer components.

3.3 *Lead adsorption onto aquifer gravels*

Monitoring the adsorption of 1×10^{-5} M Pb^{2+} as a function of time, it was found that both the coated and non-coated gravels adsorb lead. Adsorption is initially more rapid for the coated gravel, but at equilibrium the extent of adsorption (~98%) is similar for the two gravel types. Equilibrium is achieved in approximately 4 hours, and the smooth kinetic curve is suggestive of a single, rapid adsorption process for lead onto the aquifer gravels.

Adsorption isotherms are developed by measuring metal ion partitioning between the solid and solution phase over a range of total metal concentrations. For coated and non-coated gravels, the initial Pb^{2+} concentration was varied from 1×10^{-6} to 1×10^{-4} M, to generate the isotherms shown in Figure 1, a plot of the sorbed lead concentration S in mg/g versus the equilibrium solution concentration C in mg/L.

Figure 1 shows lead adsorption to increase with increasing solution concentration. For a given equilibrium concentration, slightly more lead will be adsorbed by the coated gravel than the non-coated. By taking samples as a function of time, it was also found that the approach to equilibrium is slowed with increasing initial lead concentration.

In Figure 2 are shown the pH-adsorption edges for 1×10^{-5} M Pb^{2+} onto aquifer gravels. Even at low pH there is significant lead adsorption, and it increases to about 98% with increasing pH. Up to pH 5.1 there is slightly more lead adsorption onto the coated gravel than the non-coated gravel for a given pH.

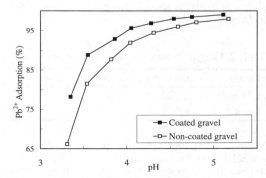

Figure 2. pH-adsorption edges for the adsorption of lead onto coated and non-coated aquifer gravels.

4 DISCUSSION

Natural oxides often serve as very effective sorbents, and it was thought that the aquifer gravel coating might be important in heavy metal adsorption. The observed coating is a largely amorphous mixture of iron and manganese oxides up to 13 μm thick. It has however been shown by bulk extraction and by surface analysis that the difference in iron content of the coated and non-coated gravels is not dramatic. The ratio of iron from non-coated to coated gravel is 0.77 by both dithionite-citrate extraction and XPS. More so than iron, manganese contents were shown to distinguish the coated and non-coated gravel surfaces. Further, XPS and extracts have shown that the elemental composition of both gravels, and the fine material, is quite similar. In general, metal ion sorption in natural systems is dominated by surface reactions with iron and aluminium oxyhydroxides (Coston *et al.*, 1995). Given the bulk and surface similarities in iron content of the aquifer fractions, there should be the potential for all components to adsorb metal ions.

Taking bulk gravel samples, it has been shown that coated and non-coated gravels alike adsorb lead. Sorption is quite rapid, and the smooth kinetic curve suggests a single-step. Adsorption onto the coated gravel is faster than for the non-coated gravel, however at equilibrium the amount of lead adsorbed is quite similar.

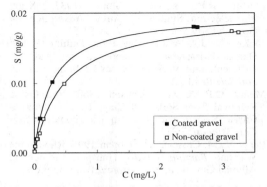

Figure 1. Isotherms for the adsorption of lead onto coated and non-coated aquifer gravels at pH 6.5.

The lead adsorption isotherms for the aquifer gravels (Figure 1) are well fit by the Langmuir equation (1)

$$S = \frac{Q^\circ k C}{1 + k C} \qquad (1)$$

where Q° is the maximum sorptive capacity of the surface and k a measure of bond strength. The optimised fit for the coated and non-coated gravels is shown in Figure 1, and the derived parameters listed in Table 2.

Table 2. Langmuir parameters for aquifer gravels.

Parameter	Coated gravel	Non-coated gravel
Q° (mg/g)	0.020	0.020
k (mL/g)	3.6	2.1

Table 2 shows that the gravels do not differ in potential sorptive capacity for lead. Rather the binding of lead to the coated gravel is stronger than for the non-coated gravel. Concentration studies showed that the approach to equilibrium was slowed with increasing lead concentration. This is believed to be a consequence of surface charge effects.

The pH adsorption curves (Figure 2) show coated and non-coated gravels to be very effective lead sorbents even at low pH. At pH ~3.3, the coated and non-coated gravels sorb 78% and 66% respectively of the initial 1×10^{-5} M Pb^{2+}. At pH 5.1 adsorption is complete (98%). Coston *et al.* (1995) found a natural Al and Fe-bearing aquifer sand to be a similarly effective sorbent: adsorption of 1×10^{-6} M lead increased from 50% to 100% from pH 4.5 to 6. Up to pH 5.1 adsorption onto the two gravel types is distinguishable, the coated gravel adsorbing slightly more lead than the non-coated gravel for a given pH.

The described batch experiments derive adsorption data by difference. XPS can be used to observe directly the adsorbed lead *in situ*. Figure 3 shows the lead signal from a coated gravel sample pre-equilibrated with 1×10^{-4} M Pb^{2+}. The peak appears as a doublet due to the Pb $4f^{7/2}$ and Pb $4f^{5/2}$ electrons. In XPS adsorption studies the ratio of sorbate to sorbent is used for quantification, and in Table 3 is listed this ratio for the systems considered.

Table 3. XPS Pb/Fe ratios for aquifer gravels.

Sorbate	1×10^{-4} M Pb^{2+}		1×10^{-5} M Pb^{2+}
Sorbent	Coated gravel	Non-coated gravel	Coated gravel
Pb/Fe	0.071, 0.084	0.059	0.039

Two coated gravel samples pre-equilibrated with 1×10^{-4} M Pb^{2+} show good reproducibility of the Pb/Fe ratio. Differences between the coated and non-coated gravel samples are measurable. The varied lead concentration is also reflected in the ratio. These data indicate XPS is reliable for the quantitative determination of lead adsorption onto the aquifer gravels.

5 REFERENCES

Blakemore, L.C., P.L. Searle & B.K. Daly 1987. Methods for Chemical Analysis of Soils, *NZ Soil Bureau Scientific Report 80*, Lower Hutt.

Chao, T.T. 1972. Selective Dissolution of Manganese Oxides from Soils and Sediments with Acidified Hydroxylamine Hydrochloride. *Soil Sci. Soc. Amer. Proc.* 36:764-768.

Coston, J.A., C.C. Fuller & J.A. Davis 1995. Pb^{2+} and Zn^{2+} Adsorption by a Natural Aluminum- and Iron-Bearing Surface Coating on an Aquifer Sand. *Geochim. Cosmochim. Acta* 59:3535-3547.

Martin, R.R. & R.S. Smart 1987. X-ray Photoelectron Studies of Anion Adsorption on Goethite. *Soil Sci. Soc. Am. J.* 51:54-56.

McKeague, J.A. & J.H. Day 1966. Dithionite- and Oxalate-Extractable Fe and Al as Aids in Differentiating Various Classes of Soils. *Can. J. Soil Sci.* 46:13-22.

Mehra, O.P. & M.L. Jackson 1960. Iron Oxide Removal from Soils and Clays by a Dithionite-Citrate System with Sodium Bicarbonate. *Clays Clay Mineral.* 7:317-327.

Pang, L., M. Close & M. Noonan 1997. Rhodamine WT and *Bacillus subtilis* Transport through an Alluvial Gravel Aquifer. *Ground Water* in press.

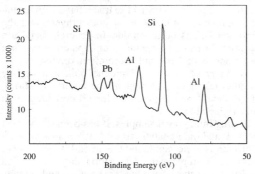

Figure 3. That region of the XPS spectrum showing the presence of lead on coated gravel.

Water-rock interaction and sorption of redox-sensitive elements: Experiments on olivine and uranium

J. Suksi & M. Upero
Laboratory of Radiochemistry, University of Helsinki, Finland

A. Adriaens
Department of Chemistry, University of Antwerp, Belgium

K.-H. Hellmuth
Finnish Centre for Radiation and Nuclear Safety, Helsinki, Finland

ABSTRACT: The formation of alteration products on the surface of high-FeO olivine was studied by secondary ion mass spectrometry (SIMS) and the interaction of U(VI) (i) during water-rock interaction and (ii) with preformed alteration products at 20, 120 and 180°C applying selective leaching methods. Uranium sorption was associated with Fe(III)-rich secondary products undergoing transitions from loosely to more strongly bound states with increasing temperature and time.

1 INTRODUCTION

Fe(II)-rich minerals in the rock matrix dominate the control of the redox state of groundwaters in almost all water-rock systems. Their surfaces determine the retardation potential of rocks towards radionuclides from nuclear waste repositories. Numerous laboratory studies on the dissolution behaviour of iron silicate minerals mainly under weathering conditions have been conducted (see e.g. Schott, J., Berner, E.A., 1985 and Wogelius, R.A., Walther, J.V., 1992). In this work the formation of alteration products on high-FeO olivine and their interaction with redox-sensitive elements, in this case uranium(VI), was studied at various temperatures. In contrast to work reported in the literature, this work, simulating the evolution of repository conditions, involved the transition from initially oxidizing to reducing and high-pH conditions. In such kind of experiments the application of selective leaching methods, known mainly from soil science, has, to our knowledge, not been reported before. The final goal of the investigations is (i) the validation of laboratory sorption experiments and (ii) the simulation of long-term processes to be expected in a deep nuclear waste repository. This study is focussed on (i) the behaviour of uranium(VI) during initial stages of water-rock interaction and (ii) the interaction of uranium(VI) with preformed olivine alteration products. Simulations of repository conditions start from (i) initially oxygenated (near field) or (ii) reducing (far field) conditions. Due to the difficulties to simulate reducing conditions representative of deep geological formations in the laboratory the latter will be subject of a separate project.

2 EXPERIMENTAL

The mineral content of the ultrabasic rock from Lovasjärvi (SE-Finland) was: olivine (65%; fayalite content in olivine: 39–58 mol-%), plagioclase (20%), magnetite (8%), pyroxene (4%) and serpentine (3%)

Table 1. Sequential extraction technique applied for studying the sorbed U-236 tracer.

Target phase of U in the samples	Applied reagents	Leaching conditions
Physically and electrostatically (ion exchange) adsorbed fraction (A)	1M ammonium acetate buffered to pH 4.8 with acetic acid	20° C, shaking, 15 min, liquid-solid ratio 5–10
Fraction specifically adsorbed on Fe(III)-rich phases (oxyhydroxides) (T)	Mixture of 0.175M ammonium oxalate and 0.1M oxalic acid (Tamm's reagent)	20° C, pH 3.5, shaking 2 h in the dark, liquid-solid ratio 5–10
Fractions fixed more permanently, possibly incorporated into the structure of crystalline phases; chemical substitution (HCl)	3M HCl (Extraction with HCl was taken to replace Mehra & Jackson reagent developed for dissolving crystalline Fe oxides)	Boiling, 90 min

(total FeO content, 28%). Crushed samples (size fraction 0.063–0.50 mm; median about 0.25 mm; sample size 0.5–1 g) were brought into contact with uranium(VI) (uranium-236; initial U concentration 6×10^{-6} M in double-distilled water) in titanium autoclaves equipped with copper or teflon seal (Hellmuth et al., 1994). The autoclaves were filled under oxygenated conditions. Reaction temperatures were 20, 120 and 180°C, reaction times varied between 1.5 hours and 107 days. After reaction the autoclaves were handled in an Ar-filled glove bag or in a nitrogen glove box.

Uranium sorption and its distribution in the sorbed phase was examined using sequential extraction techniques (Table 1) and liquid scintillation counting. Sequential extraction has long been adapted in evaluating the behaviour of heavy metals in soils and sub-surface environments. Here, the choice of Tamm's oxalate as an extractant was based on existing knowledge of the Fe(III)-rich alteration products of olivine (Hellmuth et al., 1994; Rodrigues et al., 1997).

The formation of thin, homogeneous layers of alteration products suitable for SIMS profiling was conducted using polished (1 μm diamond paste) rock disks of 14 mm diameter at temperatures between 140 and 270°C and reaction times between 2 and 71 hours. The water-rock ratio (2.5–12 g/g; normally 2.5 g/g; double distilled water or simulated granite groundwater) and the amount of air in the head space (0.5–12 ml; normally 0.5 ml) were varied. The SIMS measurements on 15 samples were carried out using a Cameca IMS 4f instrument (8 keV; 200 nA negative oxygen ion beam; no surface coating).

3 RESULTS AND DISCUSSION

The SIMS profiles of the altered layers on olivine showed the same principal pattern independent of the conditions. A typical example is shown in Fig. 1 together with a profile measured on unaltered olivine. At short reaction times at lower temperatures (140°C) the alteration layer was hardly observable under the stereomicroscope (false colours due to light refraction) while at temperatures above 200°C and times above 40–70 hours the progress of the alteration front became visibly heterogeneous following grain boundaries, cracks, microfissures and weakness zones, providing unfavourable conditions for SIMS profiling. Due to experimental difficulties (adjusting the beam onto the center of small olivine grains; finding the sputtered craters in the profilometer) and principal limitations of the SIMS method (calibration of the signals for quantitative analysis difficult) only qualitative conclusions could be drawn. It seems that the alteration mechanism did not change significantly within the range of conditions applied. The most obvious feature was the enrichment of iron, in the average the signal increased by a factor of about seven, as Fe(III) at the olivine surface as inferred from the deep red-brown colour and recent photoelectron spectroscopy (ESCA) measurements (Rodrigues et al., 1997). Also Ca and Al were enriched following a similar pattern indicating involvement of cations leached from plagioclase. Because of the very sensitive detection of these ions, in particular of Al, the concentrations were low. Mg was often very slightly depleted at the surface. The thickness of the alteration layer seemed to increase at 180°C between 2 and 22 hours in a way which would imply a diffusion-controlled mechanism. However, no sufficient consistency was found when comparing results from other conditions. The alteration layer thickness was mostly in the range 100–1400 nm (one minute of sputter time corresponded to very roughly 10 nm depth). The profiling reached in most cases the unaltered rock matrix. There were considerable differences in details such as the thickness of a Fe-depleted layer formed on top of the Fe enrichment and the depth of the Fe enrichment. The decrease of the Fe concentration towards the depth as well as the surface cannot be due to consumption of the residual oxygen, because with crushed rock providing larger reactive surface area and 0.5 ml air in the head space negative redox potential and a high pH was measured first after 15–20 d reaction time at 180°C.

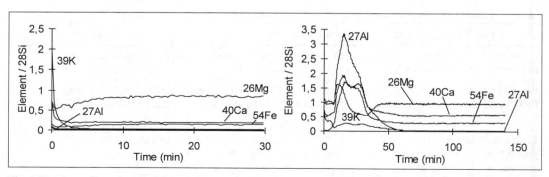

Fig. 1 Depth profiles measured by SIMS on fresh (left) and altered (180°C, 22 h, pure water) (right) olivine.

The transition to reducing, high-pH conditions was not achieved at the lower reaction temperatures (20 and 120°C) and short reaction times. The pH remained between 6 and 7 (influenced by residual CO_2). After 15 d at 180°C the pH was 9–10, rising to about 11.5 after 20 d. The lowest redox potential measured with a Pt electrode was –170 mV. Although at pH values over 9 weakly sorbing uranyl carbonate complex would dominate, the progress of the serpentinization reaction might limit carbonate to very low values. For more information the liquid phase should be analyzed, too.

Water-rock interaction removed U from the liquid phase efficiently. At 20°C and after twenty days contact time over 80 % of the added U was sorbed, the loosely-bound form being the dominating fraction (Fig. 2). After 20 d interaction the system obviously was still in a state of transition. The seven days distribution pattern with preformed alteration products appeared to approach that of twenty days

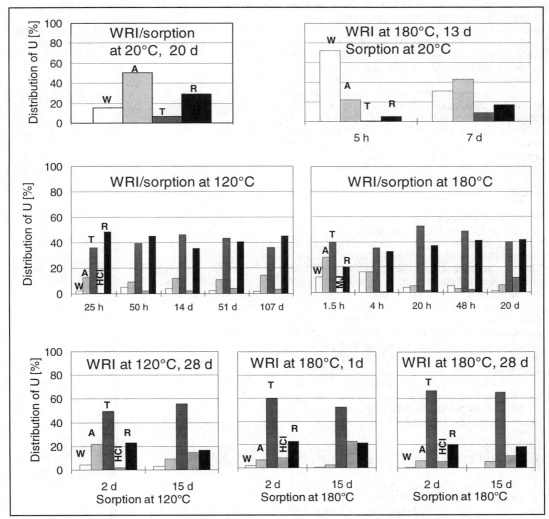

Figure 2. Uranium (U-236) distribution between liquid and various sorbed phases (determined by sequential extraction techniques (see Table 1)) in the course of, and, after a certain water-olivine interaction (WRI) time. W = components left in the water phase (non-sorbed U; due to strong sorption difficult to measure in some cases), A = adsorbed components, T = components associated mainly to the Fe(III)-rich phase, MJ and HCl = components incorporated in crystalline Fe(III)-phases, and R = residual U (added U minus non-sorbed U and sorbed U removed in extractions). WRI/sorption denotes sorption in the course of water-olivine interaction; otherwise sorption occurred on preformed secondary products.

simultaneous interaction, suggesting a kinetic control for sorption. At 120 and 180°C fixation of U was fast. In both cases components associated with a Fe(III)-rich phase dominated together with an even more strongly bound fraction. At 180°C the loosely bound fraction decreased rapidly with increasing time. On preformed alteration products the residual fraction was less pronounced, but a clear, increasing trend between 2 and 15 d sorption time was evident in the HCl fraction at both 120 and 180°C. Probably, the Fe(III)-rich phase formed in the early stage of water-rock interaction prevented U from reaching the more strongly bound states. The clearly larger Tamm's fractions in the case of preformed alteration products might support this view. Aging of the alteration products leading to a loss of sorption sites might be inferred from the slight decrease of sorption observed between 25 and 50 h at 120°C and 1.5 and 4 h at 180°C. This aspect needs further experimental data including the influence of other components in the liquid phase. Although the chemical state of fixed U remained unclear, the distribution pattern of sorbed U obtained, however, offers a useful tool to consider the sorption behaviour of U in more detail.

4 CONCLUSIONS

Sorption of redox-sensitive uranium on Fe(II)-rich mineral surfaces undergoes transitions from loosely to more strongly bound states connected with formation and changes of secondary, Fe(III)-rich products. This is corroborated by SIMS profiling and ESCA measurements (Rodrigues et al., 1997). By ESCA even a reduction of U(VI) to U(IV) on the olivine surface was observed. The evaluation of the retardation capability of rocks based on "standard" parameters such as distribution coefficients and the extrapolation to varying conditions and geological time scales obviously needs further investigations. This work will be continued by characterizing the oxidation state of U in the liquid as well as in various sorbed phases. Information from natural analogues (U profiles in weathering/alteration rims) will be used to support the laboratory results.

5 REFERENCES

Hellmuth, K-H., Siitari-Kauppi, M., Rauhala, E., Johanson, B., Zilliacus, R., Gijbels, R., Adriaens, A., 1994. *Reactions of high-FeO olivine rock with groundwater and redox-sensitive elements studied by surface-analytical methods and autoradiography.* In: R. van Konynenburg & A. Barkatt (eds.), Scientific Basis for Nuclear Waste Management XVII, Mat. Res. Soc. Symp. Proc. 333: 947-953.

Rodrigues, E., El Aamrani, FZ., Gimenez, J., Casas, I., Torrero, ME., de Pablo, J., Duro, L., Hellmuth, K-H., 1997. *Surface characterization of olivine-rock by X-ray photoelectron spectroscopy (XPS). Leaching and U(VI)-sorption experiments.* MRS-Symp. Scientific Basis for Nuclear Waste Management, Davos (submitted).

Schott, J., Berner, E.A., 1985. *Dissolution mechanisms of pyroxenes and olivines during weathering.* In: Drever, J.I. (ed.), The Chemistry of Weathering. D. Reidel Publ. Comp., Dordrecht, p. 35-53.

Wogelius, R.A., Walther, J.V., 1992. *Olivine dissolution kinetics at near-surface conditions.* Chemical Geology, 97: 101-112.

Arsenic removal from geothermal bore waters: The effect of mono-silicic acid

P.J. Swedlund & J.G. Webster
Institute of Environmental Science and Research, Auckland, New Zealand

ABSTRACT: Adsorption onto a hydrous ferric oxide (HFO) floc is a viable option for the removal of arsenic (As) from geothermal bore waters. However, As(III) and As(V) adsorption was considerably less efficient from bore water than from a 0.1m $NaNO_3$ electrolyte, or than calculated using the diffuse layer model and literature surface complexation constants. By determining the effect of individual bore water components on the adsorption of As(III) and As(V) from 0.1m $NaNO_3$, mono-silicic acid (H_4SiO_4) was identified as the component inhibiting As adsorption. Bore water H_4SiO_4 was considered to be both adsorbing and polymerising on the HFO surface. Surface complexation constants were derived for As(III), As(V) and H_4SiO_4 adsorption which improved model accuracy. However, the effect of H_4SiO_4 on As adsorption continued to be underestimated, presumably because the model does not include the surface polymerisation of H_4SiO_4, which may inhibit As adsorption by increased site occupation, steric or surface charge effects.

1. INTRODUCTION

Arsenic concentrations in New Zealand's longest and most intensely utilised waterway, the Waikato River, typically vary between 10 and 80 μgL^{-1}, exceeding the revised WHO drinking water guideline. The largest source of As entering the Waikato River is the discharge of spent bore water, containing approximately 4,000 μgL^{-1} As, from the Wairakei Geothermal Power Station. Although hydrous ferric oxide (HFO) flocculation can be an effective As(V) removal process for bore water at pH 4, laboratory experiments demonstrated that As adsorption from bore water was significantly inhibited when compared to adsorption from a 0.1m $NaNO_3$ electrolyte solution (Webster and Webster 1995). This inhibition of As adsorption was not predicted by diffuse layer model calculations using literature (Dzombak and Morel 1990) thermodynamic data. In fact, significant enhancement of As(V) adsorption above pH 8 was predicted, though inhibition was observed. The aim of this study was to determine the effect of individual bore water components on the HFO adsorption of As from 0.1m $NaNO_3$ and improve model accuracy for this system.

2. METHODS AND MATERIALS

HFO was synthesised by rapidly raising the pH of a 0.1m $NaNO_3$ and 1.0×10^{-3}m $Fe(NO_3)_3$ solution from 2.0 to 7.0 and ageing the resulting suspension for 24 hr. For adsorption from bore water, the 0.1m $NaNO_3$ from above the settled floc was replaced with bore water taken from Flash Plant No. 10 at the Wairakei Geothermal Power Station. Bore water As was >90% As(III). For As(V) adsorption experiments, 5 mL of 30% H_2O_2 was added per 1L of bore water. Synthetic solutions were prepared by adding As (5.3×10^{-5}m) and one, or more, other bore water component(s) to the HFO suspension. These components included: Cl (6.2×10^{-2}m); SO_4 (4.3×10^{-3}m); K (5.3×10^{-3}m); Ca (2.6×10^{-4}m); H_3BO_3 (2.7×10^{-3}m); CO_3/HCO_3 (1.1×10^{-3}m); and H_4SiO_4 (1.8×10^{-3}m or 2.2×10^{-3}m). The suspension pH was raised to 12 with NaOH and then lowered sequentially with HNO_3 in steps of 1-1.5 pH units. The pH was remeasured and the suspension sampled 30 minutes after each pH adjustment. The sample was filtered and analysed for As. Adsorption experiments performed under N_2 indicated that atmospheric CO_2 did not influence adsorption. Batch experiments to determine adsorption kinetics and equilibrium adsorption for As and H_4SiO_4 were undertaken in HDPE bottles agitated on a side-to-side shaker at 25°C.

Samples were analysed for As (total As, and As(III) for selected samples) by Hydride Generation Atomic Adsorption Spectrophotometry (Aggett and Aspell 1976). Total silica was measured by N_2O/C_2H_2 AAS, after adding 1mL of 1m NaOH and 10,000ppm EDTA per 10 mL sample. Molybdate-reactive silica was measured by the beta-silicomolybdate method (Iler 1979). Colour development in both bore water and sodium silicate solutions was complete within 90s and all molybdate reactive silica was considered to be mono-silicic acid (Hansen et al. 1994).

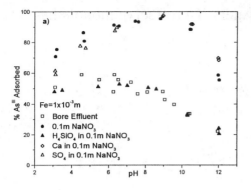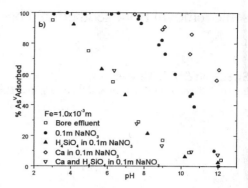

Figure 1. Adsorption after 30 min. of As(III) (a) and As(V) (b) from bore water (4.9×10^{-5}m As), 0.1m NaNO$_3$ (5.3×10^{-5}m As), and 0.1m NaNO$_3$ with: SO$_4$ (4.3×10^{-3}m), H$_4$SiO$_4$ (2.2×10^{-3}m) and/or Ca (2.6×10^{-4}m).

3. RESULTS

3.1 *Effect of individual bore water components*

The bore water components having the most significant effect on As adsorption were H$_4$SiO$_4$ and Ca (Figures 1a and b). Fresh hot Wairakei geothermal bore water contained 8.3×10^{-3}m H$_4$SiO$_4$, which polymerises to form a silica colloid (SiO$_{2(am)}$) upon cooling. After storing bore water at room temperature for 1 week, the H$_4$SiO$_4$ concentration stabilised at 2.0×10^{-3}m, rising to 2.2×10^{-3}m at the high pH (>10) used at the start of the adsorption experiments. Sodium silicate added to synthetic solutions was initially all present as H$_4$SiO$_4$. The adsorption of As(III) and As(V) from 0.1m NaNO$_3$ to which 2.2×10^{-3}m sodium silicate was added was essentially identical to that observed from bore water. Adsorption of As(V) from 0.1m NaNO$_3$ at pH>8 was significantly enhanced by 2.6×10^{-4}m Ca^{2+}, however, this was not observed from bore water or from 0.1m NaNO$_3$ with 2.6×10^{-4}m Ca^{2+} *and* 2.2×10^{-3}m H$_4$SiO$_4$.

3.2 *Kinetics and Equilibrium Adsorption*

For bore water at pH<10 and synthetic solutions As adsorption stabilised to within ± 2% within 24 hr, whereas H$_4$SiO$_4$ adsorption from 0.1m NaNO$_3$ required up to 150 hr. Total filterable SiO$_2$ decreased over 70 hr from 8.3×10^{-3}m to 5.7×10^{-3}m at pH 4.2, 6.4×10^{-3}m at pH 7.0 and 6.8×10^{-3}m at pH 9.9. Adsorption of H$_4$SiO$_4$ could account for up to 0.5×10^{-3}m of this decrease and the remainder was considered to be due to the interaction between the SiO$_{2(am)}$ and HFO colloidal particles.

Arsenic adsorption after 24 hr was measured in bore water and 0.1m NaNO$_3$ with and without 1.8×10^{-3}m H$_4$SiO$_4$ (Figures 2a and b). Again As adsorption from 0.1m NaNO$_3$ with 1.8×10^{-3}m H$_4$SiO$_4$ was essentially identical to adsorption from bore water. Equilibrium As(V) adsorption from 0.1m NaNO$_3$ was also measured with higher As/Fe ratios, to allow derivation of all As(V) adsorption constants. A 240 hr equilibration time was used for the higher As/Fe ratios (Fuller et al. 1993). The adsorption edges measured for 1.8×10^{-3}m and 1.0×10^{-3}m H$_4$SiO$_4$ on 1.0×10^{-3}m HFO exhibit adsorption densities (Γ_{H4SiO4}) up to 0.46 and 0.33 respectively, which was inconsistent with simple adsorption. To derive H$_4$SiO$_4$ constants, adsorption was measured using a lower Si/Fe ratio; 1.0×10^{-3}m H$_4$SiO$_4$ and 3.7×10^{-2}m HFO.

4.0 DISCUSSION

Adsorption was considered in terms of the diffuse layer model and the geochemical speciation programme MINTEQA2 (Allison et al. 1991) was used. The values for HFO surface area (600m^2/g), adsorption site densities (0.2mol/mol Fe), surface acidity constants (pK$_{A1}$=7.29;pK$_{A2}$=8.93) and adsorption reactions were taken from Dzombak and Morel (1990). FITEQL3.2 (Herbelin and Westall 1996) was used to determine the best fit equilibrium constants presented in Table 1.

4.1 *Arsenic*

The best fit equilibrium constants for As(III) and As(V) and the weighted averages for the As(V) constants (using the method of Dzombak and Morel 1990), are given in Table 1. Adsorption edges modelled using these equilibrium constants are shown, together with experimental data, in Figures 2a and 2b. The Dzombak and Morel (1990) adsorption constants for As(V), derived predominantly from Pierce and Moore(1982) data, are considerably smaller than those derived in this work. However, unlike Pierce and Moore (1982) the constants from this work are in good agreement with Leckie et al. (1980) and Fuller et al.

Table 1 Adsorption constants for As and H_4SiO_4 (standard deviations in parentheses). Weighted average equilibrium constants for As(V) are also shown, with the 95% uncertainty level (*in italics in parentheses*).

System	$logK_1$	$logK_2$	$logK_3$	$logK_4$	WSOS/DF
As(V) Species	$\equiv FeH_2AsO_4$	$\equiv FeHAsO_4^-$	$\equiv FeAsO_4^{2-}$	$\equiv FeOHAsO_4^{3-}$	
As/Fe=0.053	-	-	-0.58 (0.04)	-8.92 (0.04)	3.28
As/Fe=0.107	11.72 (0.3)	5.30 (0.90)	-0.25 (0.05)	-8.85 (0.1)	0.71
As/Fe=0.214	11.43 (0.07)	6.21 (0.08)	-0.84 (0.16)	-8.50 (0.09)	0.57
Weighted Average					
This study	11.46 (*1.0*)	6.14 (*2.0*)	-0.48 (*0.4*)	-8.80 (*0.33*)	
Dzombak & Morel (1990)	8.61 (*1.0*)	2.81 (*0.19*)	-	-10.2 (*0.4*)	
As(III) Species	$\equiv FeH_2AsO_3$	$\equiv FeHAsO_3^-$	$\equiv Fe(OH)HAsO_3^{2-}$		
As/Fe=0.053	5.68 (0.04)	-2.70 (0.4)	-10.62 (0.05)	-	5.12
Dzombak & Morel (1990)	5.41 (*0.15*)	-	-	-	
H_4SiO_4	$\equiv FeH_3SiO_4$	$\equiv FeH_2SiO_4^-$	$\equiv Fe(OH)H_2SiO_4^{2-}$		
Si/Fe=0.027	4.08 (0.06)	-3.81 (1.4)	-11.5 (0.3)		0.21
Hansen et al (1994)	3.62	-	-		

(1993) data. The $logK_1^{INT}$ value for As(III) from this work is in reasonable agreement with the literature value, also derived from Pierce and Moore (1982).

4.2 *Mono-silicic Acid*

The high Γ_{H4SiO4} observed in this study (with Fe of 1.0×10^{-3}) implied that significant H_4SiO_4 polymerisation had occurred at the HFO surface (Herbillon & Tran Vinh An 1969). Because polymerisation occurred at H_4SiO_4 concentrations well below saturation with respect to bulk phase silica (in solutions with 1.0×10^{-3} m H_4SiO_4) the HFO surface can be considered to act as a template for polymerisation. Polymerisation of H_4SiO_4 has a significant effect on the HFO surface charge. As the number of siloxane linkages increases, the acidity of the remaining silanol groups increases, therefore, although H_4SiO_4 has pK_{A1} of 9.86, the PZC of pure silica is ca. 2 (Iler 1979). Schwertmann and Fechter (1982) and Anderson and Benjamin (1985) have measured HFO PZC decreasing 2-3 pH units as the Γ_{H4SiO4} increased from 0.0 to 0.3.

On the other hand, with 1.0×10^{-3} m H_4SiO_4 and 3.7×10^{-2} m Fe, the maximum Γ_{H4SiO4} was 0.027, consistent with adsorption. The value for $logK_1^{INT}$ (Table 1) is a little larger than that derived by Hansen et al. (1994) for H_4SiO_4 adsorption. This could be due to the long dialysis time (9 days) used by Hansen et al. (1994). Fuller et al. (1993) have shown that long HFO ageing times decreased As(V) adsorption.

4.3 *Arsenic and H_4SiO_4*

The modelled effect of H_4SiO_4 on As(III) and As(V) adsorption using the derived adsorption constants (Tables 1) is shown, with the experimental data, in Figures 2a and b. The model underestimates the effect of H_4SiO_4 on As(III) and As(V) adsorption below a pH of ca. 9. This is likely to be because the model does not take into account the polymerisation of H_4SiO_4, which would presumably inhibit As adsorption by increased site occupation, steric or surface charge effects. Although there is limited experimental data at pH>9 it appears that the modelled edges converge with the experimental, which may reflect that H_4SiO_4 polymerisation is less favoured as the pH approaches the pK_{A1} of H_4SiO_4.

Anderson and Benjamin (1985) found that adsorption edges for SeO_3^{2-} shifted to lower pH as the Γ_{H4SiO4} increased from 0.0 to 0.3, but the edges converged when the pH was normalised to the PZC (ie. pH-PZC) determined for each Γ_{H4SiO4}. This suggested that the effect of the particulate Si on the HFO surface potential was the only important mechanism inhibiting anion adsorption. Changes in PZC due to polymerisation of H_4SiO_4 clearly can not explain the effect of H_4SiO_4 on As(III) adsorption at pH<9 in this work. For As(V), the adsorption edge shift to lower pH does approximately correspond to the estimated decrease in PZC, but from diffuse layer calculations H_4SiO_4 competition for adsorption sites is apparently a significant factor.

4.4 *Calcium*

The adsorption of alkaline earth metals occurs over a wider pH range, is more sensitive to ionic strength and has a lower stoichiometry of proton release than is typical for transition or post transition metals.

A divalent surface species is invoked to model this behaviour (Dzombak and Morel 1990). Thus, Ca has a more significant effect on the HFO surface potential than non-alkaline earth metals. The increase in As(V) adsorption due to 2.6×10^{-4} m Ca in the bore water components experiments (Figure 1b), although not at equilibrium, is consistent with the MINTEQA2 modelled effect of Ca^{2+}, (using the equations of Dzombak and Morel 1990) where adsorption

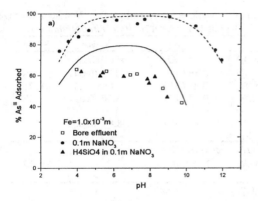
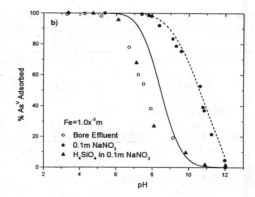

Figure 2. Experimentally determined and modelled As(III) (a) and As(V) (b) equilibrium adsorption from bore water, and from 0.1m $NaNO_3$ with and without H_4SiO_4 (1.8×10^{-3}m). The dashed line shows modelled adsorption from 0.1m $NaNO_3$, and the solid line modelled adsorption from 0.1m $NaNO_3$ containing H_4SiO_4.

decreased gradually above pH 8 to ca. 80% at pH 12.

As(V) adsorption was not enhanced by Ca^{2+} in the presence of H_4SiO_4, neither in bore water nor in a 0.1m $NaNO_3$ solution (Figure 1b). Consistent with this result, the diffuse layer model predicted that Ca^{2+} would not have a significant effect on As(V) adsorption in the presence of 1.8×10^{-3}m H_4SiO_4. Modelling indicated that, not only would Ca^{2+} adsorption decrease, due to H_4SiO_4 competition for surface sites, but adsorption of the negative H_4SiO_4 surface species decreased the electrostatic changes due to Ca^{2+} adsorption.

REFERENCES

Aggett, J. & Aspell, A.C. 1976. The determination of arsenic (III) and total arsenic by atomic-adsorption spectroscopy. *The Analyst* 10:341-347.

Allison, J.D., Brown, D.S. & Novo-Gradac, K.J. 1991. *MINTEQA2/PRODEFA2, A Geochemical Assessment Model for Environmental Systems*. EPA/600/3-91/021, USEPA, Athens, Georgia.

Anderson, P.R. & Benjamin, M.M. 1985. Effects of silicon on the crystallisation and adsorption properties of ferric oxides. *Environ. Sci. Technol.* 19:1048-1053.

Dzombak, D.A. & Morel F.M.M. 1990. *Surface Complexation Modelling; Hydrous Ferric Oxide*, John Wiley & Sons: New York.

Fuller, C., Davis, J.A. & Waychunas, G.A. 1993. Surface chemistry of ferrihydrite: Part 2. Kinetics of arsenate adsorption and co-precipitation. *Geochim. Cosmo.Acta* 57:2271-2282.

Hansen, H.C.B., Wethe, T.P., Raulund-Rasmussen K. & Borggaard, O.K. 1994. Stability constants for silicate adsorbed to ferrihydrite. *Clay Minerals* 29: 341-350.

Herbelin, A.L. & Westall, J.C. 1996. *FITEQL: A Computer Program for Determination of Chemical Equilibrium Constants;* Report 96-01;Chem. Dept., Oregon State University: Corvallis, OR.

Herbillon, A.J. & Tran Vinh An, J., 1969. Heterogeneity in silicon-iron mixed hydroxides. *Journal of Soil Science* 20:223-235.

Iler, R.K., 1979. *The Chemistry of Silica;* John Wiley & Sons: New York.

Leckie, J.O.; Benjamin, M.M.; Hayes, K.; Kaufman, G.; Altmann, S. *Adsorption/Coprecipitation of Trace Elements From Water with Iron Oxyhydroxide;* Research Project 910-1; EPRI: Palo Alto, California, 1980.

Pierce, M.T. & Moore, C.B. 1982. Adsorption of arsenite and arsenate on amorphous iron hydroxide. *Water Res.* 16: 1247-1253.

Schwertmann, U.; Fechter, H. 1982. *Clay Minerals*, 17, 471 - 476.

Webster, J.G. & Webster, K.S. 1995. Arsenic adsorption from geothermal water. *Proceedings of 16th Annual PNOC Geothermal Conference* 35-42. Manila, Philippines.

Trace metal adsorption onto acid mine drainage iron oxide

J.G.Webster, P.J.Swedlund & K.S.Webster
Institute of Environmental Science and Research, Auckland, New Zealand

ABSTRACT: Trace metal (Cu, Pb, Zn & Cd) adsorption onto pure hydrous iron oxide has been compared to adsorption onto a natural sulphate-rich iron oxide from an acid mine drainage system, and onto synthetic schwertmannite (a poorly crystallised oxyhydroxysulphate of iron). For Cu, Zn and Cd, the natural oxide is a more efficient adsorbent than pure iron oxide, enhancing adsorption through the formation of Cu and Zn ternary complexes with SO_4 at the oxide surface, and an additional, possibly bacterially mediated, mechanism.

For Pb, natural schwertmannite is a less efficient adsorbent than pure iron oxide. Although there is evidence for the formation of Pb ternary surface complexes with SO_4, adsorption on the natural oxide appears to be adversely affected by the relatively low surface area, and reduced availability of binding sites.

1 INTRODUCTION

In aquatic environments hydrous oxides of iron play an important role in the transport of trace metals, and in regulating dissolved trace metal concentrations through adsorption. The degree of trace metal adsorption onto pure amorphous hydrous ferric oxide (HFO) and onto goethite has previously been determined (eg., Benjamin & Leckie, 1980; Balestieri & Murray, 1982) and the results used to derive surface complexation constants (Dzombak & Morel.,1990) and model metal adsorption in natural systems (eg., MINTEQA2: Allison et al., 1991). However, natural oxides are rarely pure. The ochreous oxides precipitated in acid mine drainage systems (AMD), for example, include poorly crystallised oxyhydroxysulphates of iron which contain up to 14% SO_4 (Bigham et al., 1980). This sulphate-rich oxide has been formally recognised and named as the mineral "schwertmannite" (Bigham et al, 1994), which has a general unit cell formula of $Fe_{16}O_{16}(OH)_y(SO_4)_z \cdot nH_2O$ where $16-y = 2z$, and z may vary between 2.5 and 3.0.

The aim of this study was to quantify trace metal adsorption onto natural AMD oxide and onto synthetic schwertmannite, through the determination of adsorption edges (adsorption as a function of pH). Metal adsorption onto pure HFO was also determined and compared to adsorption onto the sulphate-rich oxide phases.

2 EXPERIMENTAL METHOD & MATERIALS

2.1 Sorbent synthesis & characterisation

Natural sulphate-rich AMD oxides were collected from the drainage below the tailings dam of the former Tui mine near Te Aroha at Coromandel, New Zealand. The oxide was washed in deionised water and wet sieved through a 85 μm nylon mesh, and then stored as an acidic slurry prior to use. Pure HFO and synthetic schwertmannite were synthesised by raising the pH of a $Fe(NO_3)_3 \cdot 9H_2O$ solution. For synthetic schwertmannite, the solution also contained 2000mg/kg SO_4, and pH was raised slowly to 4.0-5.0 over a period of 30 hrs. HFO formed a red/brown loose gelatinous precipitate, whereas schwertmannite formed an ochreous yellow/brown oxide which adhered to the precipitation vessel. The oxides were aged in solution for 24hrs before use.

The concentration of SO_4 in the oxides was determined by high performance ion chromatography (HPIC), on a HCl acid digestion from which Fe had been removed by extraction into diethylether. The concentration of SO_4 adsorbed at the oxide surface was determined by the addition of $Ba(NO_3)_2$ to oxide suspensions (after Bigham et al, 1990). Adsorbed SO_4 made up ca. 60% of the total SO_4 content of both the natural oxide and the synthetic schwertmannite.

Synthetic schwertmannite contained 8.2-13.3 wt% SO_4 ($Fe_{16}O_{16}(OH)_{12.2-9.0}(SO_4)_{1.9-3.5}$), depending on the degree of SO_4 adsorption at the oxide surface. The natural oxide averaged 11.3% SO_4 but had very low levels of other impurities such as Cl, NO_3, Al, Si, Mn and TOC.

Over 90% of the Fe in natural oxide and synthetic schwertmannite was oxalate-soluble. XRD and IR analysis indicated detectable levels of crystalline goethite (α-FeOOH) in the natural oxide, but only schwertmannite in the synthetic phase. XRD could not confirm or deny the presence of schwertmannite in the natural oxide, as schwertmannite has a relatively low intensity spectra, and this would be obscured by that of any goethite present. Viewed under SEM, the natural oxide had a larger particle size than synthetic schwertmannite, and a more angular morphology than HFO. The surface area of the dried natural oxide was ca. 90 m^2/g, compared with ca. 120 m^2/g for the synthetic schwertmannite, and ca. 300 m^2/g for pure HFO (as determined by *para*-nitrophenol adsorption: Theng, 1996).

2.2 Adsorption edge measurement

Adsorption edges for Cu, Pb, Zn and Cd were measured in a 0.1m $NaNO_3$ medium, with sorbent oxide concentrations of 1.0 ($\pm$ 0.1) x 10^{-3}m Fe. Trace metals were added from nitrate salt stock solutions to achieve concentrations of 0.5mg/kg. Reactions times of 2hr at each pH were allowed and, at least two measurements on the steeper part of each adsorption edge were repeated after 24 hr to ensure that there had been no further change in metal adsorption. Aliquots were extracted, filtered (0.45µm) and acidified, prior to analysis by AAS or, for low-level Pb, ICP-mass spectrometry. Sorbent concentrations were measured as Fe on an unfiltered, acidified aliquot of the solution, using UV/VIS spectrometry (as red $Fe(SCN)_3$ at 450nm).

"Control" experiments without an oxide sorbent present were also undertaken to determine the extent of metal adsorption onto the HDPE container walls, and of metal-bearing salt precipitation.

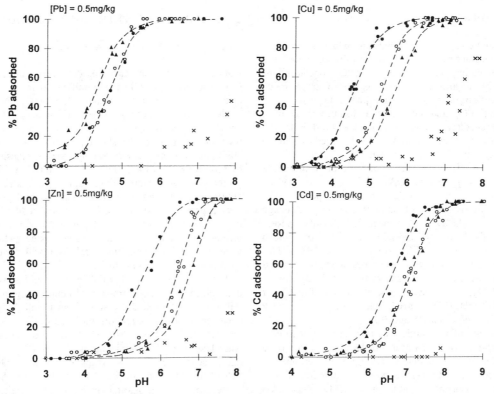

Figure 1. The adsorption edges for Pb, Cu, Zn and Cd adsorption onto natural AMD oxide (•), synthetic schwertmannite (°), pure HFO (▲), and in the absence of any oxide surface (×).

3. RESULTS

Lead adsorption was inhibited onto natural oxide and synthetic schwertmannite, relative to adsorption onto HFO at pH <5.5 (Figure 1). Natural oxide and synthetic schwertmannite adsorption edges were similar and were displaced relative to the HFO adsorption edge by ca. 0.5 pH unit.

Copper and Zn adsorption onto the natural oxide and synthetic schwertmannite, was enhanced relative to adsorption onto pure HFO, at pH <6.5 and pH<7.5 respectively (Figure 1). Copper and Zn adsorption onto the natural oxide was also considerably greater than onto synthetic schwertmannite at this pH. Adsorption edges were displaced by ca. 0.3 - 0.5 pH units for synthetic schwertmannite, and by ca. 1.2 pH units for the natural oxide, relative to HFO.

Cadmium adsorption onto the natural oxide was enhanced relative to adsorption onto pure HFO and synthetic schwertmannite at pH <8 (Figure 1). The difference between the adsorption edges for HFO (and the coincident synthetic schwertmannite edge) and the natural oxide was 0.5 pH unit, somewhat less than that observed for Cu and Zn adsorption.

Adsorption edges were also measured for Pb, Cu and Zn adsorption onto natural oxide or synthetic schwertmannite, from which surface SO_4 had been removed by ageing the oxides at pH 6.5-8.0, or by $Ba(NO_3)_2$ addition. These adsorption edges moved to a higher pH (see Figure 2) and, for Cu and Zn adsorption on synthetic schwertmannite, became indistinguishable from those for HFO.

In the presence of high solution SO_4 concentrations (2000mg/kg), adsorption edges determined for Pb, Cu and Zn adsorption onto pure HFO shifted to lower pH (Figure 2). For Cu and Zn, the adsorption edges shifted closer to those observed for adsorption onto synthetic schwertmannite. For Pb, however, the adsorption edges for synthetic and the natural oxide lie in the opposite direction.

4.0 DISCUSSION

4.1 *The influence of SO_4 on metal adsorption.*

Metal adsorption was evidently enhanced in the presence of solution (i.e, co-adsorbing) SO_4, and decreased when surface-adsorbed SO_4 was removed from both the natural AMD oxide and synthetic schwertmannite. There are at least three ways in which adsorbed SO_4 could have encouraged metal adsorption: by altering electrostatic conditions at the oxide surface, through the formation of a ternary surface complexes (eg., $\equiv FeOHCuSO_4$) or by surface precipitation of a metal-SO_4 salt.

MINTEQA2, using a diffuse layer model to calculate the degree of metal adsorption onto pure HFO in the presence of 2000 mg/kg SO_4, indicated that the changes in surface charge incurred by SO_4 adsorption are small and can not account for the increased adsorption observed. Surface precipitation at the oxide surface appears to have been a possibility only for Pb, which approaches saturation with respect to anglesite in the 2000mg/kg SO_4 solution.

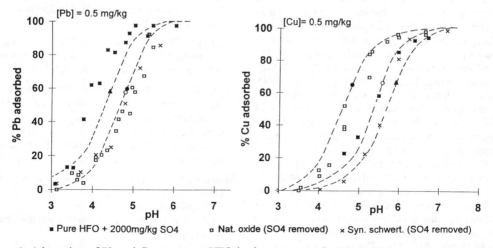

Figure 2. Adsorption of Pb and Cu onto pure HFO in the presence of solution SO_4, and onto natural AMD oxide and synthetic schwertmannite from which surface-adsorbed SO_4 has been removed. Dashed curves indicate position of original adsorption edges shown in Figure 1.

Therefore, the most likely mechanism by which pre-adsorbed or co-adsorbed SO_4 may enhance metal adsorption is through the formation of ternary complexes between the oxide surface, SO_4, and the metal ion. This mechanism has recently been proposed to account for enhanced Cu adsorption onto goethite in the presence of SO_4 (Ali & Dzombak, 1996), and can now be used to account for enhanced adsorption of Cu and Zn on synthetic schwertmannite. Cd adsorption occurred at pH >6, where SO_4 desorption from the schwertmannite surface would appear to have precluded any major interaction between Cd and SO_4 at the oxide surface.

For Pb adsorption on synthetic schwertmannite, however, the positive influence of ternary surface complex formation was evidently overwhelmed by some other factor, possibly the limitations of low surface area. MINTEQA2 modelled Pb adsorption onto HFO decreased by 10-20% over the pH range of 3.5 - 5.5, as surface area was arbitrarily decreased from >300 m^2/g to 100 m^2/g. There is good agreement with observed Pb adsorption on the natural AMD oxide and synthetic schwertmannite of similarly low surface area.

4.2 *Adsorption on natural schwertmannite*

There is a significant and as yet unexplained difference between Cu, Zn and Cd adsorption onto the natural oxide, and onto pure HFO or synthetic schwertmannite. The observed enhanced adsorption of Cu, Zn and Cd is not consistent with the lower surface area of the natural oxide. Chemical characterisation of the natural oxide indicated that there are unlikely to be other major cations or anions interacting with these trace metals on the oxide surface, or inorganic phases acting as additional adsorbing surfaces. The organic content of the natural schwertmannite was very low (TOC = 0.5%). Although adsorbed organic acids could potentially influence oxide surface characteristics, experimental Cu adsorption edges determined after removal of humic acids from the natural oxide, or after addition of humic acids to synthetic schwertmannite, remained unchanged (Webster et al., 1997).

Iron-oxidizing bacteria such as *Leptosprillum sp.* were present in the AMD waters which precipitated the natural oxide (pers comm., Julie Williamson, Landcare). Likewise, *Thiobacillus* growth was observed on plates of enrichment media smeared with the natural AMD oxide sample (pers. comm. Renate Domel, ANSTO). The presence of bacterial cells in the natural oxide sample may well have increased its ability to adsorb or precipitate selected heavy metals (eg., Beveridge & Fyfe, 1985).

We conclude that the degree of Cu, Pb, Zn and Cd adsorption onto natural iron oxides in AMD systems can not be reliably predicted using adsorption data for pure HFO.

REFERENCES

Ali M.A.& Dzombak D.A. 1996. Interactions of copper, organic ligands and sulfate in goethite suspensions. *Geochim. Cosmo. Acta* 60:5045-5055.

Allison J.D., Brown D.S. & Novo-Gradac K.J. 1991. MINTEQA2/PRODEFA2, a geochemical model for environmental systems. *USEPA Report*: EPA/600/3-91/021.

Balistrieri L.S. & Murray J.W. 1982. The adsorption of Cu, Pb, Zn and Cd on goethite from major ion seawater. *Geochim. Cosmo. Acta* 46:1253-1265.

Benjamin M.M. & Leckie J.O. 1980. Multiple-site adsorption of Cd, Cu, Zn and Pb on amorphous iron oxide. *J. Colloid Interfac. Sci.* 79:209-221.

Beveridge T.J. & Fyfe W.S. 1985. Metal fixation by bacterial cell walls.*Can.J.Earth Sci.* 22:1893-1898.

Bigham J.M., Schwertmann U., Carlson L.& Murad E. 1990. A poorly crystallised oxyhydroxysulfate of iron formed by bacterial oxidation of Fe(II) in acid mine waters. *Geochim. Cosmo. Acta* 54: 2743-2758.

Bigham J.M., Carlson L. & Murad E. 1994 Schwertmannite, a new iron oxyhydroxysulphate from Pyhasalmi, Finland and other localities. *Min. Mag.* 58:641-648.

Dzombak, D.A. & Morel, F.M.M.1990. *Surface Complexation Modelling*, Wiley Interscience Publications, Wiley & Sons Inc., USA. 393pp.

Theng B.K.G. 1996 (in press). On measuring the specific surface area of clays and soils by adsorption of *para*-nitrophenol: use and limitations. In *Clays: Controlling the Environment* (Eds: Churchman, Fitzpatrick & Eggleton), CSIRO, Melbourne.

Webster J.G., Swedlund P.J. and Webster K.S. 1997 (in press) Trace metal adsorption onto acid mine drainage ferric oxide: the influence of co-adsorbed sulphate and organic acids. *Environmental Science & Technology*.

Kinetics of calcite precipitation: Molar measurements and molecular descriptions

Pierpaolo Zuddas & Giovanni De Giudici
Laboratoire de Géochimie des Eaux, Institut de Physique du Globe et Université Paris 7, France

Alfonso Mucci
Department of Earth and Planetary Sciences, McGill University, Montreal, Que., Canada

ABSTRACT: The mechanism of calcite precipitation in seawater has been evaluated using the results of a set of experiments in which the influence of the solution composition and specific components were measured. The rate data obtained by these measurements give a macroscopic representation of the overall reaction, whereas reaction mechanisms require a molecular scale description. To properly describe the molecular-scale mechanisms, microscopic and nanoscopic studies must be carried out.

1 INTRODUCTION

To precisely describe $CaCO_3$ crystal growth and dissolution reactions in seawater solutions, accurate reaction rates, rate laws, and mechanisms must be identified. The classical approach has been to follow the evolution of individual solution components during a reaction and determine their influence on the observed rates of reaction. However, these observations are based on measurements carried out at the molar scale whereas reaction mechanisms occur at a molecular scale. This scaling leads to much speculation in the elaboration of the reaction mechanism and should be recognized when developing geokinetic operative models. In this paper, we compare the results obtained from kinetics studies at the molar scale to the meso and nano-scale observations of calcite precipitation from seawater solutions.

2. A MOLAR MECHANISTIC APPROACH

Calcite precipitation from seawater is assumed to be dominated by the following reaction:

$$Ca^{2+} + CO_3^{2-} = CaCO_3 \quad (1)$$

The overall rate of this reaction can be expressed by the difference between the forward and backward reaction rates according to:

$$R = R_f - R_b = k_f(a_{Ca})^{na}(a_{CO3})^{nb} - k_b(a_{CaCO3})^{nc} \quad (2)$$

where R is the net calcite precipitation rate, R_f, R_b, and k_f, k_b are respectively, the forward and backward reaction rates and rate constants for the overall reaction, and a_i and n_i are, respectively, the activities and partial reaction orders for the species involved in the reaction. Since the activity of relatively pure solids is conventionally assigned a value of one, Eqn (2) can be reduced to:

$$R = k_f(aCa)^{na}(aCO_3^{2-})^{nb} - k_b \quad (3)$$

Given the constancy of the calcium concentration in natural seawater, Eqn (3) can be further reduced to:

$$R = k_f'[CO_3^{2-}]^{nb} - k_b \quad (4)$$

where $k_f' = k_f(aCa)^{na}(\gamma CO_3^{2-})^{nb} \quad (4a)$

Classically, the partial reaction order with respect to the carbonate ion concentration is the empirical parameter that fixing the form of the kinetic equation gives information on the mechanism of the reaction (Laidler, 1987).

2.1 Experimental results

The influence of solution composition on the mechanism of calcite crystal growth from seawater solutions was investigated by carrying out seeded-growth experiments using the 'constant addition system' (Zhong and Mucci, 1993). This experimental technique allows one to evaluate the reaction rate while maintaining nearly invariant the composition of the reacting solution.

The parameterisation of the model expressed by Eqn (4) for seawater solutions gives:

$$\text{Log } R = 3 \text{ Log } [CO_3^{2-}] + 3.6 \qquad (5)$$

Eqn (5) suggests that calcite precipitation from seawater is a complex reaction and mechanistically different from the precipitation reaction from simple and dilute solutions where the partial reaction order with respect to the carbonate ion concentration was found to be equal to 1 (Chou et al. 1989).

The influence of the major seawater constituents and known reaction inhibitors such a Mg^{2+} and SO_4^{2-} was evaluated from the results of a set of experiments carried out in solutions having the same ionic strength as seawater (i.e. 0.7 m) in the absence of individual seawater components and known inhibitors.

Since growth rates in both seawater and NaCl-CaCl$_2$ solutions yield similar rate equations (same empirical reaction order with respect to the carbonate ion) it was suggested that the precipitation mechanism may not change significantly as a result of the adsorption and incorporation of Mg and SO$_4$ in calcite precipitated from seawater.

Similarly, the presence of dissolved humic acid, which inhibits calcite growth, does not affect the partial reaction order with respect to the carbonate ion concentration in these strong electrolyte solutions (Zuddas et al. 1998).

In contrast, variations of the total ionic strength of the solution induces a change of both the rate constants and the reaction order with respect to the carbonate ion concentration. Growth rate measurements in both NaCl- CaCl$_2$ and seawater solutions indicate that when the ionic strength is increased from 0.10 to 0.93m, the partial reaction order with respect to the CO_3^{2-} concentration increases from 1 to 3 and the forward reaction rate constant, k_f, increases by several orders of magnitude (Zuddas and Mucci, 1997)

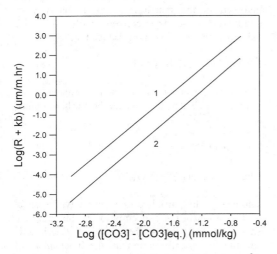
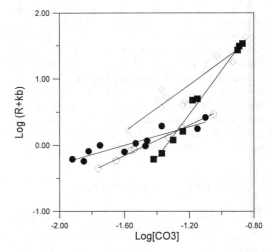

Figure 1. Log (R + k_{b*}) as function of Log ([CO_3^{2-}] - [CO_3^{2-}]$_{eq.}$) for calcite growth at 298.15K and 3.14 x 10^2 Pa. (1) from a 0.7 m NaCl-CaCl$_2$ solution (Zuddas and Mucci, 1994), (2) from a phosphate-free seawater solution (Zhong and Mucci, 1993)

Figure 2. Log (R + k_b) (μmol. m^{-2} hr^{-1}) as a function of Log[CO_3^{2-}] (mmol/kg) for calcite growth in Na-Cl-CaCl$_2$ solutions at 298.15 K and 3.14 x 10^2 Pa, at various ionic strengths I: solid circles at I = 0.10 m, open circles at I = 0.34 m, rhombs at I = 0.54, solid squares at I = 0.93 m.

3 LIMITATIONS OF MACROSCOPIC DESCRIPTIONS

Molar measurements allow one to describe Eqn (1) according to a mass budget. If Eqn (1) represents also a reaction mechanism, we must assume that the amount of energy required to achieve the reaction (kJ/mol) is homogeneously distributed on all the reactive sites of the crystal surface. The free energy at the surface of the crystal, however, varies with the crystallographic orientation of the surface (Desjonquères and Spanjaard, 1993). The distribution of the energy during the calcite growth is therefore, a priori, anisotropic. The change of scale is theoretically incorrect and molecular-scale measurements are required to properly describe the calcite growth reaction.

4 MICROSCOPIC OBSERVATIONS AT MESO AND NANO-SCALE

Transmission Electron Microscopy studies of calcite growth from seawater indicate that, at the mesoscale (μm), the presence of foreign ions like Mg, favours the development of irregular surfaces along specific edges and corners of the calcite crystal, even at high seawater saturation states (Paquette et al. 1996). The ability to directly observe mineral growth at a mineral-solution interface in real time (Binning et al. 1986) may improve our understanding of reaction mechanisms. Atomic Force Microscopy (AFM) studies by Hillner et al. (1992a, b) showed that calcite growth on a $(10\bar{1}4)$ cleavage plane occurs through spreading of monomolecular steps (0.3 +/- 0.1 nm high) oriented along one crystallographic direction. Dove and Hochella (1993) also observed that during the early stages of exposure to a supersaturated solution, calcite growth occurs by a polynuclear birth and spread mechanism, with nuclei coalescing to form a flat surface. Spiral growth was observed only after long reaction periods or in solutions close to equilibrium. AFM images at molecular scale are, however, influenced by tip geometry and interference with the growing crystal so that images are not equivocally interpreted from in situ liquid cell measurements (Stipp et al. 1994, Cunault, 1997).

We examined the influence of the ionic strength on the crystal growth mechanism on a fresh cleaved $(10\bar{1}4)$ face of calcite by AFM under liquid cell conditions. Our images indicate that at equivalent saturation states (i.e. $\Omega = 5$) calcite grows by spreading of a 2 nm thick film or step at the ionic strength of 0.1 m, whereas by 'nuclei' of 10 nm high at the ionic strength of 0.9m.

5 CONCLUSION

Observations based on measurements carried out at the molar scale suggest that the variation of the total ionic strength mainly controls the change of the mechanism of calcite crystal growth from seawater. This could be tentatively associated to molecular scale observations where it is observed that calcite grows by spreading of a 2nm thick film at low ionic strength whereas by 'nuclei' of 10 nm high at high ionic strength

REFERENCES

Binning G., Quate C.F., Gerber Ch. (1986). Atomic force microscope. Phys. Rev. Lett. 56, 930-933.

Chou L., Garrels R.M., Wollast R. (1989). Comparative study of the kinetic and mechanisms of dissolution of carbonate minerals. Chem. Geol. 78, 269-282.

Culnaut J.B. (1997). Changement d'échelle dans la réactivité de la calcite. Memoire D.E.A.Université Paris 7.

Desjonquères M.C. and Spanjaard D. (1993). Concepts in surface physics. Springer series in surface sciences 30. Springer-Verlag.

Dove P.M. and Hochella M.F. Jr. (1993) Calcite precipitation mechanisms and inhibition by orthophosphate: In situ observations by Scanning Force Microscopy. Geochim. Cosmochim. Acta 57, 705-714.

Hillner P.O., Manne S., Gratz A.J. and Hansma P.K. (1992). AFM images of dissolution and growth on a calcite crystal. Ultramicrosc. 42-44, 1387-1393.

Hillner P.E., Gratz A.J., Manne S., and Hansma P.K. (1992). Atomic-scale imaging of calcite growth and dissolution in real time. Geology 20, 359-362.

Paquette J., Vali H., Mucci A. (1996). TEM study of Pt-C replicas of calcite overgrowths precipitated from electrolyte solutions. Geochim. Cosmochim. Acta 60, 4689-4699.

Stipp S.L., Eggleston C.M., and Nielsen B.S. (1994). Calcite surface structure observed at microtopographic and molecular scale with the atomic force microscopy (AFM). Geochim. Cosmochim. Acta 58, 3023-3033.

Zhong S. and Mucci A. (1993). Calcite precipitation from seawater using a constant addition technique: a new overall reaction kinetic expression. Geochim. Cosmochim. Acta 57, 1409-1417.

Zuddas P. and Mucci A. (1994). Kinetics of calcite precipitation from seawater: I. A classical chemical kinetics description for strong electrolyte solutions. Geochim. Cosmochim. Acta 58, 4353-4362.

Zuddas P. and Mucci A. (1997). Kinetics of calcite precipitation from seawater: II. The influence of the ionic strength. Geochim. Cosmochim. Acta (in press)

Zuddas P., Descostes M., Prévot F. (1998)
The influence of the dissolved organic matter on the kinetics of calcite precipitation from seawater. Chem. Geol. (to be submitted)

17 Waste storage and disposal

Geochemical modelling of groundwater/bentonite interaction for waste disposal systems

D. Arcos, J. Bruno & L. Duro
QuantiSci, S.L. Parc Technològic del Vallès, Cerdanyola del Vallès, Spain

ABSTRACT: Bentonite is widely used as engineered barrier in nuclear and toxic waste disposal systems, as a hydrological and geochemical barrier. A geochemical model showing the chemical evolution of groundwater through this barrier was performed using the PHREEQC code. Bentonite impurities, as calcite and pyrite, are the main pH and oxygen fugacity buffers. Groundwater composition controls the Na-Ca replacement in bentonite, exerting an indirect control on pH due to the calcite dissolution-precipitation process driven by the Ca concentration in the porewater.

1 INTRODUCTION

Most underground designs for nuclear and toxic waste repositories include bentonite as a physico-chemical barrier. The interaction between groundwater and bentonite is a key factor in order to assess the stability of the barrier and its consequences for the waste. This has been done by using recently developed conceptual models for the stability of bentonite (Wanner et al. 1992; Wieland et al. 1994) in combination with the PHREEQC geochemical code package (Parkhurst 1995).

2 BENTONITE/GROUNDWATER MODEL

2.1 System composition and parameter selection

MX-80 Wyoming bentonite has been proposed by Swedish Waste Management Co. (Lajudie et al. 1994; Wieland et al. 1994) as a possible engineered barrier for high-level nuclear waste repositories. The MX-80 bentonite is mainly composed of Na-montmorillonite (up to 75 wt. %), with minor contents of calcite (1.4 wt. %), pyrite (0.3 wt. %), quartz (<15.0 wt. %), feldspar (<9.0 wt. %) and traces of other impurities such as gypsum and NaCl.

According to the design specifications (SR 95, 1996), the density of highly compacted bentonite is of 1.6 g/cm³, the porewater/bentonite ratio at repository conditions is 0.33, and the estimated barrier width around nuclear waste is of 0.35 m.

Two Swedish granitic groundwaters, with different chemical compositions, were selected (Table 1) in order to test the effect of water composition on the evolution of the selected system. Main differences between the two selected groundwaters are the high salinity and a high calcium content in Äspö groundwater, compared to that of Gideå.

Table 1. Selected groundwater compositions.

Component	Äspö (mols/l)	Gideå (mols/l)
Na	$9.13 \cdot 10^{-2}$	$4.57 \cdot 10^{-3}$
K	$2.05 \cdot 10^{-4}$	$5.13 \cdot 10^{-5}$
Ca	$4.73 \cdot 10^{-2}$	$5.25 \cdot 10^{-4}$
Fe total	$4.30 \cdot 10^{-6}$	$8.95 \cdot 10^{-7}$
Mg	$1.73 \cdot 10^{-3}$	$4.11 \cdot 10^{-5}$
Mn	$5.28 \cdot 10^{-6}$	$1.82 \cdot 10^{-7}$
Al	$3.70 \cdot 10^{-7}$	$7.00 \cdot 10^{-7}$
HCO_3^-	$1.64 \cdot 10^{-4}$	$2.95 \cdot 10^{-4}$
Cl	$1.81 \cdot 10^{-1}$	$5.01 \cdot 10^{-3}$
F	$7.89 \cdot 10^{-5}$	$1.68 \cdot 10^{-4}$
SiO_2	$1.46 \cdot 10^{-4}$	$1.67 \cdot 10^{-4}$
S total	$5.83 \cdot 10^{-3}$	$1.04 \cdot 10^{-6}$
pH	7.7	9.3
pe	-5.21/-1.24	-3.41/-1.01
Data source	SR 95 (1996)	Laurent (1983)

The basic conceptual model has been developed by Wanner et al. (1992), where the chemistry of the montmorillonite surface is coupled to that of the impurities (calcite, pyrite, etc.). We used the

PHREEQC code (Parkhurst, 1995) to simulate the chemical evolution of the system. Fresh groundwater was considered to completely replace the bentonite porewater in successive cycles every 12,000 years, and we considered that the first porewater was the result of equilibrating distilled water with bentonite.

The model was calibrated against experimental results reported by Werme (1992). We selected the best set of exchange (Olin et al. 1995) and surface acidity constants (Wieland et al. 1994), reported in Table 2, from the comparison with the experimental data (Fig. 1).

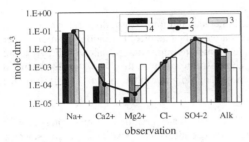

Figure 1. Comparative graph of calculations using different exchange and surface acidity constants reported by different authors and the experiments of Werme (1992). 1. Wanner et al. (1992); 2. Wieland et al. (1994); 3. This study using constants from Olin et al. (1995); 4. This study using the same constants as in Wieland et al. (1994); and 5. Experiments reported by Werme (1992).

Table 2. Set of parameters included in the model.

Cation Exchange Capacity (CEC)= 85.0 meq/100 g	
Na: 81.7 % Mg: 3.9 % Ca: 14.1 % K: 0.3 %	
Equilibria (Olin et al. 1995)	
$2NaX + Ca^{2+} = CaX_2 + 2Na^+$	log K = 2.10
$2NaX + Mg^{2+} = MgX_2 + 2Na^+$	log K = 1.50
$NaX + K^+ = KX + Na^+$	log K = 0.27
Acidity surface sites (OH-groups)= 2.8 meq/100 g	
Equilibria (Wieland et al. 1994)	
$>SOH + H^+ = >SOH_2^+$	log K = 5.40
$>SOH = >SO^- + H^+$	log K = -6.70

2.2 Results

We performed the simulations by using the selected groundwaters from Äspö and Gideå (Table 1), and assuming that they are in equilibrium with atmospheric oxygen pressure (0.21 bar), calcite, quartz, and pyrite. Gypsum and halite impurities were assumed to totally dissolve during the first interaction.

For the calculations the bentonite barrier was divided in five cells, each of 7 cm thick, perpendicular to the water flow direction. The first cell is the one in contact with the host rock (granite) and the last is the one in contact with the canister containing the waste. In the simulation, fresh groundwater replaces bentonite porewater in the first cell, while porewater of this cell replaces that of the next cell and so on, up to 500 replacements in each cell. This is equivalent to total evolution time of about 1.1 Ma.

We performed the calculations performed at two different oxygen contents in the inflowing groundwater (0.21 bar and 0.6 bar) in order to test the effect of pyrite dissolution on the redox state of the system. The obtained results are shown in Figure 2. It can be seen that as pyrite totally dissolves the pe rises to a value close to 13.

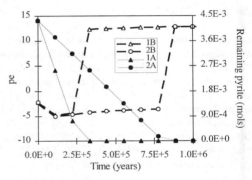

Figure 2. Effect of the total pyrite dissolution on the redox state of the system as a function of the oxygen content of groundwater. 1B and 1A. pe evolution and remaining pyrite in bentonite respectively, for groundwater with partial pressure of oxygen of 0.6 bar; 2B and 2A the same for partial pressure of oxygen of 0.21 bar.

The dissolution of calcite depends strongly on the availability of calcium in groundwater. This is in turn reflected on the Ca-Na exchange in the bentonite. This behaviour is shown in Figure 3. In the case of the interaction with the Äspö groundwater, the dissolution of calcite stops as the replacement of Ca for Na is completed (according to equilibria reported in Table 2). In the case of Gideå groundwater, calcite dissolution is the source of calcium for the Na-Ca exchange. This is due to the low initial Ca^{2+} concentration in groundwater. When calcite is totally dissolved, there is no more calcium available to complete the Na-Ca replacement, even if the initial amount of calcite is much larger than in the previous case.

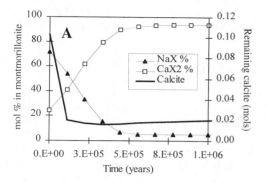

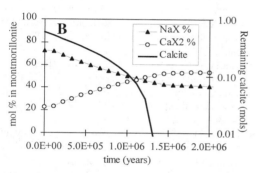

Figure 3. Relationship between calcite dissolution and Na-Ca replacement in montmorillonite in the case of Äspö (A) and Gideå (B). Note than in case B the time-scale is larger (up to 2Ma) than in case A (to 1 Ma), and the initial calcite content is also larger in case B than in case A.

The results of the interaction with the Ca-rich Äspö groundwater are shown in Figure 4. The pH is mainly controlled by calcite, pyrite and montmorillonite surface. Total substitution of Ca for Na in montmorillonite is achieved through all the bentonite barrier after 0.5 Ma. This is due to the high calcium content of groundwater, which in turn avoids total calcite dissolution in the studied time-scale. Pyrite equilibrium buffers the Eh, maintaining oxygen fugacities at very low values (log $fO_2 \approx$ -66 bar), as pyrite does not totally dissolve.

In the case of low-salinity Gideå groundwater/bentonite interaction (Fig. 5), the main difference respect to the previous case is the very slow Na-Ca exchange in the montmorillonite, which is even slower once calcite is totally dissolved. This behaviour is due to the low initial calcium content of the Gideå groundwater as shown in Figure 3. After calcite dissolution, the pH is mainly controlled by the inflowing groundwater (pH= 9.3), although small pH changes are due to pyrite dissolution.

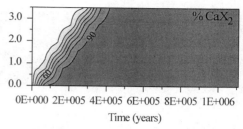

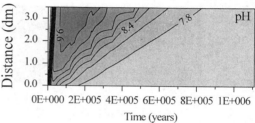

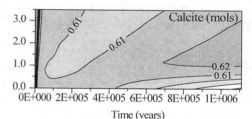

Figure 4. Time-distance graphs showing the evolution of the main parameters in the interaction model using Äspö groundwater.

In general, it is worth noting that the alkalinity buffering ability of bentonite is readily consumed by low-salinity groundwaters, as that of Gideå. This is illustrated by the total dissolution of calcite impurities during the interaction with this groundwater. Nevertheless, the buffering redox effect exerted by pyrite is due to the low solubility of this phase independently of groundwater composition.

Analytical redox measurements are rather unreliable determinations as they do not establish which redox pairs, if any, are in equilibrium The Eh values measured in groundwaters may not be indicative of the actual redox state of the system. If we assume that Äspö and Gideå groundwaters are equilibrated with their respective host rocks, they should be equilibrated with pyrite and, therefore, pyrite contained in bentonite will not dissolve as readily as calculated here. This implies that the redox buffering capacity exerted by bentonite will extend to longer time periods.

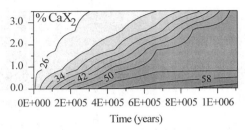

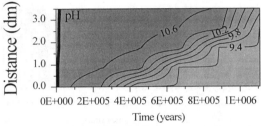

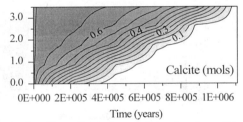

Figure 5. Time-distance graphs showing the evolution of the main parameters in the interaction model using Gideå groundwater.

3 CONCLUSIONS

The main processes affecting the chemical stability of the bentonite/groundwater system are: a) pH buffering due to calcite dissolution-precipitation and proton exchange reactions at the montmorillonite surface; and b) redox buffering mainly due to the oxidation of pyrite impurities contained in bentonite.

These chemical buffering capacities are not largely affected by the replacement of granitic groundwater for bentonite porewater. However, subsequent replacement of the porewater causes the slow depletion of calcite and pyrite impurities. This depletion is faster when low salinity groundwaters (Gideå) interact with bentonite than when groundwater flowing through the buffer has larger salt contents (Äspö). The salinity difference also affects the Na-Ca exchange in the montmorillonite, which is faster and complete when the calcium content of the groundwater is larger and slower and incomplete in the case of lower calcium contents.

4 REFERENCES

Lajudie, A., Raynal, J., Petit, J.C., and Toulhoat, P., 1995. Clay-based materials for engineered barriers: A review. *MRS Symp. Proc.* 353: 221-230.

Laurent, S., 1983. Analysis of groundwater from deep boreholes in Gideå. *Swedish Environmental Research Institute SKBF-KBS Tech. Rep.* 83-17.

Olin, M., Lehikoinen, J., and Muurinen, A., 1995. Coupled chemical and diffusion model for compacted bentonite. *MRS Symp. Proc.* 353: 253-260.

Parkhurst, D.L., 1995. User's guide to PHREEQC - A computer program for speciation, reaction-path, advective-transport, and inverse geochemical calculations. *U.S. Geol. Surv. Water-Resources Invest. Rep.* 95-4227.

SKI 1996. Deep Repository Performance Assesment Project. *Statens Kärnkraftinspektion (Swedish Nuclear Power Inspectorate) SKI Rep.* 96:36.

SR 95, 1996. Template for safety reports with descriptive example. *Swedish Nuclear Fuel and Waste Management Co. SKB Tech. Rep.* 96-05.

Wanner, H., Wersin, P., and Sierro, N., 1992. Thermodynamic modelling of bentonite-groundwater interaction and implications for near field chemistry in a repository for spent fuel. *Swedish Nuclear Fuel and Waste Management Co. SKB Tech. Rep.* 92-37.

Werme, L., 1992. Data in: Wanner H., Wersin, P., and Sierro, N., 1992. Thermodynamic modelling of bentonite-groundwater interaction and implications for near field chemistry in a repository for spent fuel. *Swedish Nuclear Fuel and Waste Management Co. SKB Tech. Rep.* 92-37.

Wieland, E., Wanner, H., Albinsson, Y., Wersin, P., and Karnland, O., 1994. A surface chemical model of the bentonite-water interface and its implications for modelling the near field chemistry in a repository for spent fuel. *Swedish Nuclear Fuel and Waste Management Co. SKB Tech. Rep.* 94-26.

Uranium series isotopic data of fracture infill materials from the potential underground laboratory site in the Vienne granitoids, France

J. Casanova
BRGM, Department of Hydrology, Geochemistry and Transfers, Orléans, France

J.-F. Aranyossy
ANDRA, Research and Development Department, Châtenay-Malabry, France

ABSTRACT: A mineralogical and uranium series isotopic study of fracture infill materials from the Vienne granitoids was performed on samples from deep cores drilled in the Charroux-Civray region. Occurrences of ^{234}U preferentially mobilized relative to ^{238}U are presented and discussed according to two main processes of uranium mobility: preferential solution related to oxidation paleoevents in the upper zone (0 - 400m) of the site and alpha-recoil induced solution in the lower zone (400 - 900m).

1 INTRODUCTION

Underground facilities provide the opportunity for research, technical development and demonstration in a realistic setting that are required for eventual geologic disposal of nuclear waste. As part of the preliminary geological characterization program to assess the feasibility of an underground laboratory in granitic rock, 17 boreholes, continuously cored and oriented (down to a maximum depth of 900 m) have been drilled in the Charroux-Civray region (Vienne district). The concerned batholith, located below a 150 m thick Jurassic sedimentary cover, is mainly composed of dioritic and tonalitic to granodioritic plutons (Cathelineau et al., 1997).

Owing to the chemistry of these rocks and the development of paleo-hydrothermal systems, most of the fractures are filled with Ca-Mg-Fe minerals. One of the main objectives of the on-going investigations is to develop and test a conceptual model of the groundwater flow system. A mineralogical and uranium series isotopic study on fracture infillings was undertaken in order to characterize the eventual traces of recent fluid circulation in the granite basement (Casanova et al., 1997).

2 METHODS

A total of 65 samples of fracture fillings was selected along the CHA112, CHA212 and CHA312 boreholes according to the following criteria: (i) location in water-bearing zones, (ii) presence of open spaces, drusy or dissolution cavities, (iii) presence of loose material in the core section, (iv) fracture surfaces covered by poorly crystallized clay minerals or Fe-oxy-hydroxides. The nature of fracture infillings and their paragenetic order of crystallization have been determined to select the samples for U-series analyses (measured by alpha-spectrometry and mass spectrometry). Both small-sized samples (0.2 to 2 gr), corresponding to separated mineralogies, and aliquots from bulk samples (10 to 40 gr) of fracture filling materials were analysed. The uranium content and $^{234}U/^{238}U$ isotopic ratio were also measured (TIMS) in shallow groundwaters from boreholes in the two sedimentary aquifers (Dogger and Infra Toarcian) and in deep groundwaters from boreholes in the granitoid basement.

3 RESULTS

Textural and fluid inclusion data show that most of the fracture filling minerals crystallized during activity of the geothermal system (Cathelineau et al., 1997). Five main phases of crystallization sequences have been identified: (1) adularia(I) - hematite ; (2) dolomite, with frequent dissolution cavities ; (3) quartz, baryte(I), fluorite, pyrite, chalcopyrite, clays, calcite ; (4) adularia(II), clays, Fe-oxy-hydroxides, baryte(II), rare-earth minerals, framboidal pyrite, tabular to lanceolate gypsum; (5) needle-shaped gypsum. Small crystals of adularia, baryte and pyrite (sometimes framboidal) are commonly associated with Fe-oxy-hydroxides precipitated within finely crystallized clay minerals. The small-sized, late, poorly to non-crystallized phases, are assumed to be related to the most recent groundwater circulations.

The $^{234}U/^{238}U$ ratio distribution from fracture infill materials (Fig. 1) defines two distinct zones according to depth.

3.1 Upper zone (0 - 400 m)

In the upper 200 m of the granitoid basement (level 160-360 m), the uranium contents vary from 420 ppb to 23.85 ppm and the highest concentrations are found in the Fe-oxy-hydroxide samples. The $^{234}U/^{238}U$ ratios range from 0.84 to 1.94. The observed ^{234}U excess is mainly carried by Fe-oxy-hydroxides and is interpreted as an uranium re-mobilization related to oxygenated, low temperature, paleowater circulations. Preferential removal of ^{234}U in this zone is also indicated by some $^{230}Th/^{234}U$ activity ratios higher than unity.

The direct calculation of uranium residence time was possible on few samples, devoid of ^{232}Th. Because of mixing of small-sized minerals of various ages, it is unrealistic to expect to be able to analyse coexisting minerals with varying amounts of juvenile and inherited material in order to estimate an initial $^{230}Th/^{2320}Th$ ratio. Although the sample selection tends to minimize the effect of initial ^{230}Th, the calculated ages must be considered to be maxima. These samples are characterized by $^{230}Th/^{234}U$ activity ratios lower than unity indicating "recent" (<350 ka) uranium uptake from the fracture fluids. The Fe-oxy-hydroxide located at 207 m has a moderate uranium content ([U]=9.4 ppm) with a $^{234}U/^{238}U$ ratio in equilibrium and yields an age of 102 ± 5 ka. The Fe-oxy-hydroxide located at 277 m contains 6.9 ppm of uranium and has a high $^{234}U/^{238}U$ ratio of 1.7; it yields an age of 173 ± 15 ka. The Fe-oxy-hydroxide located at 300 m is rich in uranium (~ 24 ppm U) and has a deficiency of ^{234}U relative to ^{238}U, but its $^{230}Th/^{234}U$ activity ratio of 0.82 indicates recent (181 ± 10 ka) uptake of uranium. An illite/chlorite-bearing sample, located at 390 m, contains 6.3 ppm of uranium with a $^{234}U/^{238}U$ ratio in equilibrium and provides an age of 177 ± 17 ka. These preliminary results will be completed by further characterization of uranium behavior in specific minerals with respect to bulk samples from the same fracture. As a matter of fact, some fractures (i.e. levels 207 m & 396 m) exhibit minerals characterized by ^{234}U preferentially mobilized relative to ^{238}U, while bulk analyses show total isotopic equilibrium.

3.2 Lower zone (400 - 900 m)

In the lower part of the granitoid basement, the uranium contents vary from 7.8 ppb to 11.3 ppm with the notable exception of the baryte sample, which contains 27.5 ppm of uranium. The $^{234}U/^{238}U$ ratios range from 0.67 to 1.07. Most of the samples plot on or close to the secular equilibrium line and the observed isotopic disequilibriums are mainly related to a ^{234}U deficiency relative to ^{238}U. This is probably related to an alpha-recoil effect in finely and

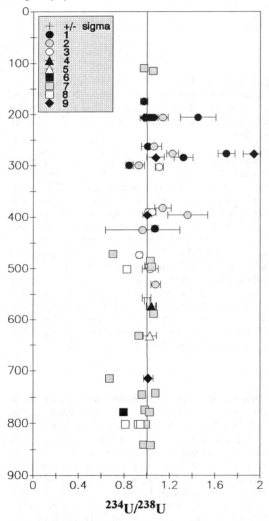

Fig. 1. Depth versus uranium isotopic ratio for fracture infillings analysed by alpha spectrometry: 1 Fe-oxy-hydroxides ; 2 carbonates ; 3 chlorites ; 4 smectites ; 5 barytes ; 9 bulk material, or by TIMS: 6 Fe-oxy-hydroxides ; 7 carbonates ; 8 chlorites (CHA112, CHA212 and CHA312 cores).

poorly crystallized phases (Andrews et al., 1989; Griffault et al., 1993).

Serial samples with identical mineralogy and sampled from the same fracture, show decreasing ^{234}U deficit with increasing sample weight. The ^{234}U/^{238}U ratio of a bulk sample located at 715 m is unity whereas a small-sized carbonate sample (0.70 gr, [U] = 600 ppb) from the same fracture shows a ^{234}U/^{238}U ratio of 0.67. This suggests that such ^{234}U/^{238}U disequilibrium observed in the lower zone may be a consequence of alpha-recoil induced solution of ^{234}U from the mineral surfaces to the fracture fluids. Large-sized bulk samples are, indeed, more likely to readsorb recoil-ejected ^{234}U atoms, taking into account the reducing water chemistry at depth.

A correlative distribution of ^{234}U/^{238}U activity ratio in grounwaters partly reflects these two main processes of uranium mobilization: preferential solution and alpha-recoil induced solution. The ^{234}U/^{238}U activity ratio for uranium dissolved in groundwater is greater than unity (Fig. 2).

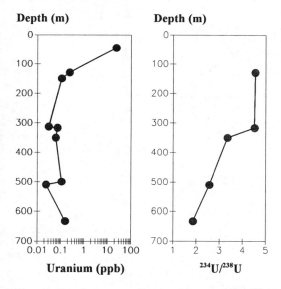

Fig. 2. Depth versus dissolved U and ^{234}U/^{238}U isotopic ratio for the Vienne groundwaters (CHA212 and CHA312 boreholes).

The small to moderate disequilibrium (1.87 < ^{234}U/^{238}U < 3.36) observed in the deep groundwater, together with very low uranium concentration, is characteristic of reducing conditions (Andrews et al. 1989, Osmond & Ivanovich 1992, Osmond & Cowart 1992). It is very likely that deep groundwaters with uranium contents as low as 0.022 ppb, underwent activity ratio increase due to the alpha-recoil process, while shallow groundwaters, with uranium content close to 24 ppb, result from preferential ^{234}U-solution processes.

4 CONCLUSIONS

The results of the uranium series isotopic study of fracture infill materials from the Vienne granitoids are presently interpreted as the occurrence of two different zones according to depth. In the upper zone (0 - 400 m), an excess of ^{234}U relative to ^{238}U shown by small-sized minerals, and especially Fe-oxyhydroxides, suggests that preferential ^{234}U-solution processes occurred some 102 to 181 ka ago, assuming a closed-system behaviour of these iron oxides. In the lower zone (400 - 900 m), a deficiency of ^{234}U relative to ^{238}U, more likely evokes an alpha-recoil induced solution of ^{234}U from the mineral surfaces to the fracture fluids.

5 REFERENCES

Andrews, J.N., D.J.Ford, N.Hussain, D.Trivedi and M.J.Youngman, 1989. Natural radioelement solution by circulating groundwaters in the Stripa granite. *Geochimica et Cosmochimica Acta.* 53: 1791-1802.

Casanova, J., P.Negrel and A.Bourguignon, 1997. Etude des cristallisations de fractures en milieu granitique. Approche géochimique et géochronologique. Etude des forages Sud-Vienne CHA112-CHA212-CHA312. Rapport BRP0ANT 97-021/B, 90 p. ANDRA, Chatenay-Malabry.

Cathelineau, M., M.Cuney, Y.Coulibaly, M.A. Ougougdal and M.C.Boiron, 1997. Circulations récentes à actuelles dans les granites. Cristallisations dans les fractures. Rapport BRP O CRE 97, 78 p., CREGU, Nancy.

Griffault L.Y., M.Gascoyne, C.Kamineni, R.Kerrich and T.T.Vandergraaf, 1993 Actinide and rare earth element characteristics of deep fracture zones in the Lac du Bonnet granitic batholith, Manitoba, Canada. *Geochimica et Cosmochimica Acta.*, 57: 1181-1202.

Osmond, J.K. & M.Ivanovich, 1992. Uranium-series mobilization and surface hydrology. In M. Ivanovich & R.S.Harmon (eds), *Uranium series disequilibrium: application to environmental problems*, 259-289, Clarendon Press, Oxford.

Osmond, J.K. & J.B.Cowart, 1992. Ground water. In M. Ivanovich & R.S.Harmon (eds), *Uranium series disequilibrium: application to environmental problems*, 290-333, Clarendon Press, Oxford.

Influence of mine watering on groundwater quality at Monteponi, Sardinia, Italy

R. Cidu & L. Fanfani
Dipartimento di Scienze della Terra, Cagliari, Italy

ABSTRACT: Dewatering has been carried out at the Monteponi mine for a century. A marked increase in salinity and metal concentration of deep groundwater occurred as the water table level was lowered with time. Closure of the mine and a stop of dewatering calls for an assessment of the contamination risk for shallow groundwater which supplies the town of Iglesias. Because the volume of water pumped out in six months has been reduced, the water table level has raised ~ 60 m. This has caused the deep water to mix with the shallow aquifer and has led to an increase in salinity and dissolved metal concentration in the shallow groundwater.

1 INTRODUCTION

The Cambrian carbonate formations (Pillola, 1991) in the Iglesias district (SW Sardinia) host important aquifers due to intense fracturing and karstification. This district has been an important mining area for centuries. Lead-zinc ores hosted in Cambrian limestone-dolomite have been exploited by some 40 mines spread out over about 50 km^2, an area known as the Metalliferous ring (Fig.1).

Since the last century, drainage systems, which have been improved with time, have been used to avoid flooding at the mine stopes. When mining reached the lowest depth in 1990, the quality of groundwater being drained deteriorated significantly due to a marked increase in marine salts and toxic metals. However, an important water supply to the town of Iglesias still derives from the relatively shallow, fresh, and uncontaminated wells in the Campo Pisano mine. This resource could be affected by mixing with the deeper component due to the rise in the water table caused by the stop of pumping at Monteponi (the main draining station from 1910), as a consequence of the mine closure for its non-economicity.

The extent of the watering and its potential effect can be better appreciated when one considers that the whole western and southern part of the Metalliferous ring has been affected by the mine drainage. Watering operations gradually started in January 1997 and are still in progress. This study was undertaken to monitor the water quality before and during watering operations to assess the risk of contamination of shallow groundwater.

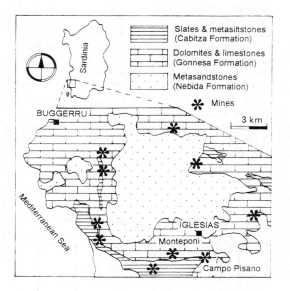

Fig. 1. Location of the most important mines.

2 HISTORICAL RECORDS OF DEWATERING

Most mines in the Iglesias district reached the groundwater level in the XIII century. Since then several draining works were developed to keep the galleries dry. At Monteponi, records from 1870 indicate that the water table was lowered from 70 m to 45 m a.s.l. (Civita et al., 1983 and references therein). In 1910, a 6 km long drainage gallery was completed from the Monteponi mine to the sea and the water table settled at 14 m a.s.l.. Pumping stations were successively installed at increasing depths: in 1928 at −15 m; in 1936 at -60 m; in 1956 at -100 m; in 1990 at -200 m below sea level (Bellé and Cherchi, 1996; Bellé et al., 1996). At Monteponi, 2.5×10^9 m^3 of water were pumped out from 1890 to 1990 (Bonato et al., 1992). Fig. 2 shows the changes in the water table with time in the mining area at different sites.

Under dewatering conditions the Metalliferous ring receives water supplies in different quantity and of a different quality. A small amount comes from rainfall infiltration and shows low salinity (<1 g/l) and chemical composition dominated by Ca-Mg-HCO$_3$, as observed in a few springs in the NE part of the area (Bertorino et al., 1981), and in some shallow mine waters that preserve these characteristics. The waters pumped off from the deepest levels at the mines at Monteponi and San Giovanni show high salinity (>10 g/l) and Na-Cl composition, suggesting sources other than rainfall. This is supported by hydrological studies showing that the average effective infiltration is extremely small compared to the volumes pumped out of mines (Civita et al., 1983).

The results from structural, remote-sensing, hydrological and isotopic studies in the Iglesias district support the hypothesis of contamination of the water system by sea water infiltration into deep fractures far from the coast (Civita et al, 1983). Fig. 3 shows increasing concentrations of Cl as the water table level was lowered with time, while the drainage flow was increased.

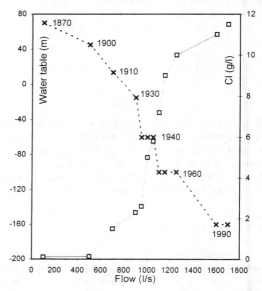

Fig. 3. Variations of Cl (□) and water table (x) versus flow from the 1870 to 1990 at Monteponi.

3 SAMPLING AND METHODS

Due to the partial closure of the mines and the beginning of the watering process, not many sampling sites could be easily accessed. On the basis of previous records, water samples from Campo Pisano (No.3, mixing of 4 pumped outputs) and San Marco (No. 5, well water) are considered to reach aquifers fed essentially by rainfall, while in the Monteponi mine, water samples from Monsignore (No.6) and Satira (No.8) wells derive from aquifers largely fed by saline waters. The water from the drainage gallery (No.1) is the bulk water pumped off from −160 m below sea level. Since the well waters show relevant stratification, they were sampled at about 15 m below the water table level. The first sampling was carried out under dewatering conditions in July 1996. Sampling continued in March and April 1997 while watering operations were in progress.

At the sampling site water samples were filtered (0.4 μm) and acidified, and temperature, pH, redox potential, conductivity and alkalinity were measured.

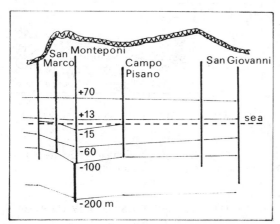

Fig. 2. Scheme of changes in the water table in the Iglesias district (modified after Bonato et al., 1992).

Metals were determined by atomic absorption spectroscopy (AAS), inductively coupled plasma optical emission spectroscopy (ICP-OES) and inductively coupled plasma mass spectrometry (ICP-MS). Mercury, arsenic and antimony were determined by ICP-MS after Hg-vapour or hydride generation (As and Sb were close to the detection limit). The detection limit for chemical components was calculated at five times the standard deviation of blank solutions. Accuracy was evaluated using standard reference materials (NIST 1643 and NRC CASS-3). Speciation and equilibrium calculations were carried out using the WATEQP (Appelo, 1988) computer program. The saturation index with respect to a mineral phase is defined as Log IAP/K (IAP is the ionic activity product and K the thermodynamic constant of equilibrium for a given temperature).

4 RESULTS AND DISCUSSION

Since this is an ongoing study, only the main results achieved so far will be presented. Table 1 shows some selected data for the most representative samples collected in July 1996 and April 1997. During the second sampling, watering operations were already in progress and differences in composition reflecting this fact are discussed in the next paragraph. The pH is near-neutral as expected in a carbonate environment. The redox potential is always >0.4 V indicating an oxidizing environment.

As expected, Campo Pisano (No. 3) and San Marco (No. 5) waters contain low total dissolved solids (TDS ≤1 g/l), CaMg-HCO$_3$ composition with Ca/Mg molar ratios close to 1 due to interaction with dolomite formations. Equilibrium with respect to calcite and dolomite is attained both in the less and most saline samples.

Mixing of shallow groundwater with deep water (e.g. No.1, 6, 8) is marked by increasing TDS, decreasing Ca/Mg molar ratios, Na-Cl dominant character, and higher concentrations of Br, B and Sr; the Na/Cl and Br/Cl molar ratios are close to values observed in seawater; Cl, Na, K, Br, B and Sr show high mutual correlation coefficients; all these data support a mixing of deep water with diluted seawater. Sulfate concentrations are higher than expected from a dilution process involving seawater. The excess SO$_4$ amount (non-marine SO$_4$ was calculated from the Cl/SO$_4$ ratio in seawater) can be related to the interaction of water with the buried mineral deposits or with the mineralised waste material released at depth. The same process also brings metals such as Zn, Pb (Fig.4), Cd, and Ag into solution. Metal concentration is high despite the near-neutral environment. Metals are mostly present in solution as complexes, the free ion being usually <20% of the total concentration. Zinc and Pb are preferably complexed by the CO$_3^{-2}$ ligand, either at low or high TDS conditions; the CdCl$^+$ and CdCl$_2^0$ species represent about 80% of total Cd; while Ag is only present as AgCl$_3^{-2}$ and AgCl$_2^-$ complexes.

Table 1. Selected results of waters from the Iglesias district.

No.	3	3	5	5	6	6	8	8	1	1
Date	Jul.96	Apr.97	Jul.96	Apr.97	Jul.96	Apr.97	Jul.96	Apr.97	Jul.96	Apr.97
T°C	20	21	21	19	23	18	23	20	23	22
pH	6.4	7.3	7.3	7.3	7.3	7.1	7.0	7.1	7.5	7.6
TDS g/l	0.88	0.99	0.79	1.10	4.73	11.26	4.37	7.82	20.20	23.09
Ca mg/l	105	110	87	98	477	800	339	296	533	556
Mg	69	84	53	63	222	462	228	260	692	764
Na	106	116	122	190	898	2420	851	2127	6000	6730
K	5.4	5.4	5.9	7.7	47	90	33	68	225	252
Cl	180	214	242	396	1476	4700	1634	4240	11000	12900
Alk	430	427	368	352	193	317	390	384	328	317
SO$_4$	200	240	82	158	1485	2600	1014	594	1600	1725
Br	0.8	0.8	0.9	1.4	6.7	16.4	6.2	14.5	38.1	42.5
B µg/l	60	56	20	59	240	600	290	440	1750	2000
Sr	170	179	141	177	710	1700	700	1400	3900	4900
Zn	430	483	820	1200	7400	35500	4000	1860	1820	3000
Cd	0.5	0.3	2.5	2.3	29	254	8	9	7	20.7
Pb	7	9	30	99	1000	1485	165	181	88	210
Hg	1.8	2.3	1.0	6.6	42	69	300	28	50	25
Ag	<0.1	0.3	<0.1	0.8	0.1	4	7	29	46	42

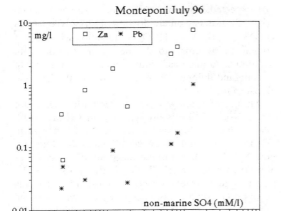

Fig. 4. Plot of Zn and Pb versus non-marine SO_4.

The concentration of Hg appears apparently unrelated to any other chemical component and at times does not increase with increasing TDS as the water table level decreases. On the basis of speciation computation, mercury is transported as $HgCl_4^{-2}$, $HgCl_2^0$ and $HgCl_3^-$ species.

5 ENVIRONMENTAL IMPLICATIONS AND CONCLUSIONS

Gradual watering of the mine aims to avoid contamination of the Campo Pisano water by deep saline water, and Hg.

Concerning the first objective we may observe that during the first sampling campaign in 1996 about 1800 l/s were pumped off Monteponi to keep the water table level at −160 m below sea level. The flow was gradually reduced to about 800 l/s in April 1997, and in six months the water table level was raised by about 60 m to reach the −100 m level. On comparing the chemical composition of waters from the last sampling with the composition of the waters prior to watering operations, a general increase in TDS is observed (about 15% at the drain). The salinity increase is greater in the central part of the Monteponi mine, such as at sample sites No.s 6 and 8 close to the pumping station, while at the peripheral wells (i.e. No.s 3 and 5) mixing with deep water is of a lesser entity. An increase in metals as the deep saline water rises has also been observed.

At this stage of the study we may conclude that a certain degree of salinization will occur in fresh water such as at the sampling site at Campo Pisano

and San Marco due to mixing. However, a stratification process, which will allow the saline water to settle at depth, is expected with time.

The Hg concentration is linked to the availability of leacheable Hg, either from the mineralization or from clays which have been accumulating Hg through weathering processes (Amat et al., 1996), and to the dissolution power of chlorine by forming Hg-Cl-complexes. The reduction in water circulation at depth with a reduction in pumping can explain the decrease in Hg content at sampling sites 1 and 8, while the input of chlorine-rich waters explains Hg enrichment with salinity at Campo Pisano and San Marco.

6 REFERENCES

Amat, P., E.Contini, R.Enne, C.Garbarino, R.Sarritzu, S.Tocco & R.Virdis 1996. Il ciclo del mercurio nel sistema idraulico carsico associato alle mineralizzazioni piombo-zincifere della sinclinale di Iglesias: implicazioni *Mem. Associazione Mineraria Sarda*: 69-85.

Appelo, C.A.J. 1988. WATEQP A computer program for equilibrium calculations of water analyses. Inst. Earth Sciences, Free University of Amsterdam.

Bellè, O. & F.Cherchi 1996. Il bacino idrogeologico dell'Anello Metallifero. *Mem. Associazione Mineraria Sarda*: 51-66.

Bellè, O., F.Cherchi & I.Salvadori 1996. The aquifer of the Cambrian carbonate basin in the Iglesiente district and the mining industry, mutual effects. *Proc. 1st Int. Conf. The Impact of Industry on Groundwater Resources Cernobbio, Como, Italy, 22-24 May 1996*: 295-305.

Bertorino, G., R.Caboi, A.M.Caredda, L.Fanfani, R.Cidu, R.Sitzia, A.R.Zanzari & P.Zuddas 1981. Le manifestazioni termali del Sulcis (Sardegna sud-occidentale) *Per. Miner.* 50:233-255.

Bonato, M., M.Congiu, E.Fioravanti & D.Lipari 1992. Impianto di eduzione −200 di Monteponi. *Boll. Ass. Min. Subalpina*, 29: 313-321.

Civita, M., T.Cocozza, P.Forti, G.Perna & B.Turi 1983. Idrogeologia del bacino minerario dell'Iglesiente (Sardegna sud occidentale). *Mem. Ist. It. Speleologia, ser.II*, 2: 7-137.

Pillola, G.L. 1991. Trilobites du Cambrien inferieur du SW de la Sardaigne, Italie. *Palaeontographia Italica* 78: 1-174.

Ferricrete provides record of natural acid drainage, New World District, Montana

G. Furniss & N.W. Hinman
Department of Geology, University of Montana, Missoula, Mont., USA

ABSTRACT: Ancient deposits of ferricrete (stratified iron oxyhydroxide or clastic sediment cemented by iron oxyhydroxide) are present in headwater streams draining a historic mining district in the Rocky Mountains, U.S.A. These deposits, which form under conditions of low pH and high metal concentrations, are present in streams draining many other mineralized areas. Radiocarbon dating of wood fragments entombed within the ferricrete indicates dates of deposition as old as 8,840 ± 50 years before present (B.P.). Chemical analysis of the ferricrete deposits provide evidence that natural, metal-rich acid rock drainage has been occurring in this mineralized region for thousands of years as a result of weathering and oxidation of exposed or near-surface massive sulfide ore deposits. The dating and chemical composition of the ferricrete deposits have applications to the environmental remediation of historic mining sites and to environmental issues related to natural background chemistry.

1 INTRODUCTION

Natural deposits of ferricrete are common in mineralized areas where oxidation of sulfide minerals results in acidic drainage (Runnells et al., 1992; Miller & McHugh, 1994; Furniss & Hinman, 1996; Logsdon et al., 1996). At the historic New World Mining District in south-central Montana, U.S.A. (Fig. 1), acidic drainage and natural deposits of ferricrete are present. The New World Mining District covers an area of 130 km² (Lovering, 1929) and straddles three alpine headwater stream valleys within an uplift in the Middle Rocky Mountains. The streams within the study area drain portions of mineralized Henderson, Fisher, and Scotch Bonnet Mountains (Fig. 1). Prominent ferricretes occur along each creek.

Economic deposits of precious and base metals have been identified within an area approximately 3 km by 8 km in the New World Mining District. Sulfide ore zones are associated with vertical hydrothermal breccia pipes and high-angle faults, and may contain high percentages of pyrite. Hydrothermal fluids associated with dacite igneous intrusions selectively replaced Paleozoic carbonate strata with sulfide minerals near feeder fracture and pipe systems.

Mining in the New World Mining District began in the late 1800s and ended in 1953 (Glidden, 1982). During the early history of mining, the natural formation of ferricrete deposits within the Fisher Creek drainage was documented by U.S. Geological Survey geologist T.S. Lovering (Lovering, 1929) as follows: "Talus breccias cemented by limonite cover many acres near the headwaters of the Clarks Fork of the Yellowstone. Large pyritic deposits occur near by, and both surface water and ground water move from the sulphides to the breccia, where deposition of iron hydroxide is going on actively."

In the present-day streams draining the New World Mining District, iron oxyhydroxide coatings form on the stream sediments as dissolved Fe^{2+} oxidizes to Fe^{3+} and precipitates, especially in riffles which aerate the flow. Herein, a comparison is made between the modern and ancient stream precipitates.

2 SAMPLING AND ANALYTICAL METHODS

Reconnaissance mapping and sampling were conducted of: 1) recent iron oxyhydroxide precipitates and coatings on the Fisher and Daisy Creek stream beds, and 2) ancient stratified ferricrete outcrops located along the Fisher, Daisy, and Miller Creek drainages. Samples were collected over three field seasons (1994-1996).

The ancient ferricrete samples were collected from stream terraces that have been incised and exposed by modern streams adjusting to lower erosional base levels. Samples were collected from pits excavated into the surface of these tabular, layered, terrace deposits or from the outer exposed

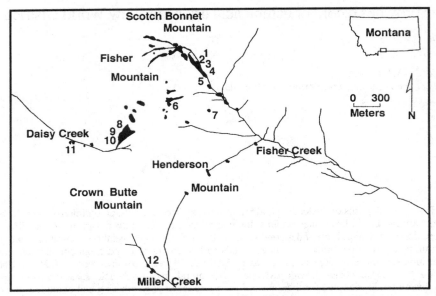

Figure 1. Location map of the New World Mining District, Montana. Dark areas indicate areal extent of mapped ferricrete deposits. Numbers indicate ferricrete sample locations with radiometric dates.

surfaces. The modern iron oxyhydroxides that are currently forming in the streams were collected from submerged cobbles in the streams at low flow.

Wood samples from the ferricrete strata were collected and shipped to Beta Analytic Inc. for ^{14}C measurement and age calculation. Samples were analyzed using accelerator mass spectrophotometry or beta decay counting (Sheppard, 1975).

The iron oxyhydroxide in the ancient ferricrete and fresh precipitates was separated from cobbles and finer sediment clasts using non-metallic tools under a binocular microscope. The samples were analyzed for metals by strong acid leach digestion (nitric acid/hydrogen peroxide; US EPA Method 3050; US EPA, 1987) and inductively-coupled plasma emission spectrophotometry (Energy Laboratories). X-ray diffraction (XRD) and scanning electron microscopy (SEM) were performed at Montana State University.

3 RESULTS AND DISCUSSION

Remnant deposits of ancient ferricrete (Fig. 1) are present along the stream drainages as terraces, 0.1 to 2 meters thick, and extend downstream along the first 1.2 km of the headwater streams.

The ferricrete is similar in morphology and mineralogy to the iron oxyhydroxide coatings currently forming in the modern streams and acidic springs of the New World District. The ancient ferricrete and the modern iron oxyhydroxides both occur as laminated deposits within and on stream sediment. XRD analysis of the modern iron oxyhydroxide coatings identifies them as amorphous or poorly crystalline ferrihydrite, whereas XRD analysis of ancient ferricrete shows them to be goethite. This difference in mineralogy suggests that the amorphous iron oxyhydroxide transforms over time to a crystalline mineralogical phase, as described by Langmuir and Whittemore (1971). SEM analysis of the modern iron oxyhydroxides and the ancient ferricrete show rod-shaped structures within the matrix, which may be from iron-sheathed bacteria.

3.1 Radiocarbon Dates

The rapid deposition of the iron oxyhydroxides that form the ferricrete results in the cementation of plant debris dispersed within the matrix. Radiocarbon dating of 22 wood samples that were tightly cemented within the ferricrete reveals ^{14}C ages ranging from 30 ± 50 years B.P. to $8,840 \pm 50$ years B.P. (Table 1). Neither the locations of the deposits nor the systematic radiocarbon stratigraphy of the ferricrete are compatible with the possibility of recent cementation of ancient wood fragments.

Field observations of current conditions show that wood fragments are periodically transported from the forested areas on the steep mountain slopes to the stream drainages by winter snow avalanches. These materials become cemented into the modern streams during low flow when metal concentrations and precipitation rates are highest.

At four locations, samples were collected from the vertical profiles of ferricrete documenting the succession of deposition through time. For example, five samples collected from a 1.2-meter-thick terrace deposit on Fisher Creek had ^{14}C dates from 5,810 ± 80 years B.P. at the uppermost surface to 7,170 ± 70 years B.P. at the base (Fig. 1, site #3). Five samples collected from a thick ferricrete apron which is still forming within the incised stream channel of Fisher Creek had ^{14}C dates from 30 ± 50 years B.P. at the surface to 890 ± 70 years B.P. at depth within a 0.2 m interval (Fig. 1, site #5).

In addition to stream deposits, large ferricrete aprons occur around several acidic springs in unmined portions of the district. One of these aprons on the east flank of Fisher Mountain, a few hundred feet above Fisher Creek, contains wood dated at 4,000 ± 60 years B.P. (Fig. 1, site #6).

3.2 Chemical Analysis

The mean and ranges for chemical concentrations of both the ancient and modern iron oxyhydroxide coatings are shown in Table 2. Similarities between elemental relationships would indicate similar depositional environments in modern and ancient streams. Such similarities between compositions of ancient ferricrete and modern iron oxyhydroxide coatings are observed. Concentrations of iron range from 21 to 450 mg/g in the ferricrete (n = 30), compared to 69 to 390 mg/g in the modern iron oxyhydroxides (n = 5). Concentrations of copper range from 0.081 to 13 mg/g in the ferricrete, and from 0.075 to 21 mg/g in the modern iron oxyhydroxides. Chemical conditions within the drainages have persisted since the exposure of the ore body. Arsenic, antimony, and molybdenum were not analyzed but do occur in ferricretes and oxyhydroxide collected from other mining districts.

4 CONCLUSIONS

Natural acid rock drainage and ferricrete deposition have occurred for at least nine thousand years in the New World Mining District of Montana as a result of oxidative weathering of near-surface sulfide ore deposits. The dating of wood fragments contained within the deposits of ferricrete provides evidence that acidic rock drainage in the New World Mining District predates the earliest mining activity of the past century. The well-preserved ferricrete deposits lend themselves to the investigation of natural background chemistry and environmental remediation in mineralized areas.

Table 1. Radiocarbon dates for wood collected from ferricrete deposits (in stratigraphic order where more than one date shown).

Sample Location	Analytical Method	Radiocarbon Date (yr. B.P.)
1	B	6,880 ± 70
2	B	8,690 ± 80
3	B	5,810 ± 80
	B	6,920 ± 80
	B	7,030 ± 60
	B	7170 ± 70
	B	7170 ± 70
4	B	5,970 ± 150
	B	8270 ± 70
5	B	30 ± 50
	B	60 ± 70
	B	100 ± 100
	B	550 ± 80
	B	890 ± 70
6	A	4000 ± 60
7	B	1670 ± 40
8	A	8620 ± 60
9	B	310 ± 110
10	A	8700 ± 50
	A	8840 ± 50
11	A	6490 ± 60
12	B	2050 ± 50

A = accelerator mass spectrometer. B = beta decay.

Table 2. Composition (mg/g) of iron oxyhydroxide matrix collected from ancient ferricrete deposits and modern iron oxyhydroxide stream cements.

Element	Mean Ancient Samples	Mean Modern Samples	Range Ancient Samples	Range Modern Samples
S	6.9	32	0.8 -- 17.4 (n=30)	1.6 -- 49.8 (n=4)
Al	10.8	63	0.33--55.8 (n=30)	0.75--141 (n=5)
Cu	2.58	5.8	0.08--12.60 (n=30)	0.075--21.1 (n=5)
Fe	239	236	21.8--446 (n=30)	69.1-- 394 (n=5)
Pb	0.13	0.14	0.01 -- 1.04 (n=17)	0.12 -- 0.16 (n=3)
Mg	1.75	1.95	0.16 -- 4.10 (n=11)	1 -- 2.9 (n=2)
Mn	1.10	0.27	0.01-- 8.01 (n=15)	0.06 -- 0.64 (n=4)
P	1.80	1.20	0.16--10.6 (n=25)	0.76 -- 2 (n=4)
K	1.24	3.20	0.23 -- 2.6 (n=9)	1.3 -- 5 (n=2)
Zn	0.20	0.27	0.01 -- 2.1 (n=25)	0.001-- 0.75 (n=4)

Samples analyzed by strong acid leach digestion (nitric acid/hydrogen peroxide) and inductively-coupled plasma emission spectrophotometry (ICP)(US EPA Method 3050).

5 REFERENCES

Furniss, G. & N.W. Hinman, 1996. Iron oxide deposits resulting from natural acid drainage records Holocene environment of Paymaster Creek, Heddleston District, Lewis and Clark County, Montana, *Geol. Soc. Am. Abstr.* 28 (7) A-98.

Glidden, R., 1982. *Exploring the Yellowstone High Country, A History of the Cooke City Area.* (Cooke City Store, Montana).

Langmuir, D. & D.O. Whittemore, 1971. Variations in the stability of precipitated ferric oxyhydroxides, in *Nonequilibrium Systems in Natural Water Chemistry*, J.D. Hem Ed. (Advances in Chemistry Series 106, American Chemical Society, Washington, DC), pp. 208-234.

Logsdon, M.J., S. Miller, & E. Swanson, 1996. Pre-mining water quality in historically mined area: iron bogs and ferricretes in the San Juan Mountains, Colorado, *Geol. Soc. Am. Abstr.* 28 (7) A-518.

Lovering, T.S., 1929. The New World or Cooke City Mining District, Park County, Montana, *U.S. Geol. Surv. Bull.* 811A.

Miller, W.R. & J.B. McHugh, 1994. Natural Acid Drainage from Altered Areas Within and Adjacent to the Upper Alamosa River Basin, Colorado, *U.S. Geol. Surv. OFR* 94-144.

Runnells, D.D., T.A. Shepherd, & E.E. Angino, 1992. Estimation of Natural Background Concentrations of Metals in Water in Mineralized Regions after Disturbance by Mining, *Environ. Sci. Tech.* 26, 2316-2323.

Sheppard, J.C., 1975. Radiocarbon Dating Primer. *College of Engineering Bulletin 338*, Washington State University, Washington.

US EPA, 1987. U.S. Environmental Protection Agency *Test Methods for Evaluating Solid Waste, Physical/Chemical Methods,* Third Edition. Office of Solid Waste and Emergency Response.

Landfill leachate – Chalk rock interactions: The fate of nitrogen and sulphur species

N.C. Ingrey
Environment Agency, Thames Region, North East Area, Hatfield, UK

J.D. Mather
Royal Holloway, University of London, UK

ABSTRACT: Porewaters analysed from Chalk cores taken from directly beneath the base of a 'dilute and disperse' landfill provide evidence of attenuating and retarding processes affecting both nitrogen and sulphur species. The variation within the data sets may reflect different processes operating beneath mature 40 year-old wastes in borehole HL5 as compared to more recent waste material in borehole HL6. Alternatively, the variation may reflect part of a continuous sequence which is being sampled at two periods within its history. Well defined anaerobic and aerobic zones are demonstrated in both HL5 and HL6 beneath the landfill. These zones play a key role in leachate/rock interactions within the unsaturated zone.

1 INTRODUCTION

It has been widely debated in the UK (Rees 1981, Mather 1989, Palmer and Young 1992) whether or not the unsaturated zone of the Cretaceous Chalk provides a buffer against contamination by landfill leachates. If such a buffering capacity exists it is unlikely that unlined landfill sites will cause significant groundwater contamination, providing the unsaturated zone is of sufficient thickness to maintain optimum conditions for geochemical and biological processes to be effective.

In the UK there are several thousand closed and operational unlined landfills - a legacy of waste management practices prior to the 1980's. Standard monitoring results (from observation boreholes outside the landfill boundary) indicate only a small percentage present a significant risk to the environment and even fewer are creating significant groundwater contamination problems (National Rivers Authority 1995). However, unsaturated zones are rarely monitored beneath these landfills, and so the water quality directly below the landfill itself is unknown. Thus the potential environmental impact posed by these landfills in the future remains a subject of debate.

Results from Foxhill Landfill (Suffolk, UK) demonstrate that significant time can elapse (up to twenty years) between the initial landfilling activity and the onset of leachate generation (Palmer and Young 1992). Coupling this with the attenuating / retarding effects of the Chalk unsaturated zone, it is conceivable that landfill leachate contaminants have yet to fully penetrate the unsaturated zone and thus remain undetected. The attenuating mechanisms of the chalk unsaturated zone with respect to landfill leachate are poorly understood, and it is possible that a legacy of contamination exists in the unsaturated zone as has been demonstrated for nitrate (Foster 1982).

It has been suggested that all landfills constructed within the last 30 years that have no leachate management systems must be considered as a serious potential threat to groundwater and surface water quality, unless benefiting from extensive natural containment by geological clays (Palmer and Young 1992). This has serious implications especially in the light of the European Groundwater Directive and recent UK legislation. Are comments such as these purely scare mongering or is there a costly pollution legacy to pay for 'dilute and disperse' landfill sites in the United Kingdom?

2 NUTTINGSWOOD LANDFILL, BUCKS, UK

This 40 hectare landfill site has been in operation since the early 1950's accepting both domestic and industrial wastes including liquid wastes until 1980. Waste thickness varies from 3-30 metres and reflects the different waste handling techniques progressively adopted during the last 40 years.

In 1992 two boreholes were drilled through the landfill and the unsaturated and the saturated zones beneath the Chalk. Continuous core material was removed and its chemical and physical properties determined. Borehole HL5 is located in waste from the 1950's, whereas borehole HL6 is in waste from the 1980's.

Porewater analysis (Ingrey and Mather, 1995) defined sharp vertical profiles of individual species with depth. The data normalised to chloride provided similar Na/Cl and Ca/Cl profiles in both boreholes, suggesting that the dominant control was attenuation in the unsaturated zone and not other variables such as waste type or depth. The profiles also indicated that processes other than dilution were taking place and that the Total Organic Carbon (TOC) front had advanced deeper than the chloride front - perhaps as a result of methane gas dissolution in porewaters.

3 RESULTS

In HL5, drilled through 40 year-old waste, the presence of NH_4-rich porewaters provides evidence of strongly reducing anaerobic conditions directly beneath the landfill base to a depth of around 5 metres. Nitrate concentrations in the porewaters are well below 0.5 mg/l until the zone of water table fluctuation is reached at approximately 40 metres, where NO_3 concentrations of 50-70 mg/l exist. This exceeds the typical local groundwater conditions of 15-20 mg/l, and suggests an anthropogenic source (figure 1a).

The depth profile of NH_4 is very similar to that of K, with species achieving background conditions approximately 5 m below the landfill base. This is in contrast to Cl and TOC which penetrate to 15 m (Ingrey and Mather 1995).

A totally different picture is seen in HL6 which penetrates relatively recent waste. Ammonium concentrations are much lower and approach background levels at approximately 2.5 metres below the landfill (figure 1b); whereas both Cl and TOC have penetrated to around 12 metres (Ingrey and Mather, 1995).

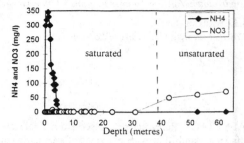

Figure 1a, borehole HL5, NH_4 and NO_3 (mg/l) vs depth beneath the base of landfill.

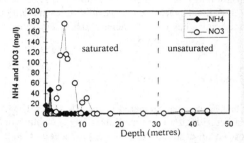

Figure 1b, borehole HL6, NH_4 and NO_3 (mg/l) vs depth beneath the base of the landfill.

The correlation with the K profile is much weaker, but the real difference lies in the increase in NO_3 which is evident in the profile beneath 3 metres below the landfill base (figure 1b). Concentrations increase to over 100 mg/l between 4 and 7 metres and then return rapidly to background levels. This reflects very different behaviour of the nitrogen species in the upper parts of the two profiles.

The picture with respect to sulphate (figures 3a and 3b) is also very different within the two profiles. Beneath the mature waste in HL5 (figure 3a), SO_4 concentrations lie between 100 and 700 mg/l within the first metre of the unsaturated Chalk. There is then a sharp reduction to concentrations approximately 5 mg/l increasing to 11 mg/l at a depth of 5 metres and remaining around a background level of approximately 25 mg/l from 7.5 metres to the zone of water table fluctuation at about 40 metres below the landfill base.

In HL6 high sulphate concentrations of over 1000 mg/l occur within the first 3 metres of the unsaturated Chalk. This reduces rapidly to background concentrations of 20-35 mg/l at about 6 metres below the base of the landfill (figure 3b). In both boreholes concentrations within the unsaturated zone porewaters are higher at 20-35 mg/l than porewaters from the saturated Chalk which have concentrations below 15 mg/l.

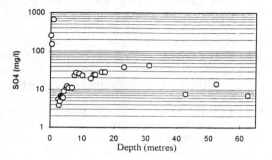

Figure 3a, borehole HL5, SO_4 (mg/l) vs depth beneath the base of the landfill.

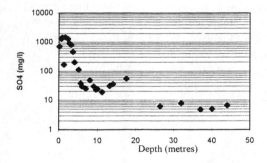

Figure 3b, borehole HL6, SO_4 (mg/l) vs depth beneath the base of the landfill.

4 DISCUSSION

The very different picture within the porewater profiles from the two boreholes may be the result of different processes operating beneath the more degraded waste in borehole HL5 compared with the younger material in the landfill above borehole HL6. On the other hand the profiles may be part of a continuous sequence which is being sampled at two periods in its history. As there is little evidence of differences in waste composition or geology at the two locations, the latter appears to be a more likely explanation.

In borehole HL5 NH_4-rich porewaters have migrated to a depth of around 5 metres (figure 1a). There is a strong correlation between the NH_4 profile and the K profile which is consistent with a scenario of NH_4 removal by ion exchange as has been suggested by Cezan, et al. (1989). Below 5 metres nitrogen species more or less disappear from the profile until the zone of water table fluctuation is reached.

The strongly anaerobic conditions immediately below the landfill are not reflected by the SO_4 data, particularly by the high SO_4 concentrations in the porewaters in the upper metre of the unsaturated Chalk (figure 3b). Orange and brown precipitates scraped from the fracture surfaces at 3 and 4 metres beneath the landfill and analysed by XRD show evidence of the formation of iron and copper sulphides. The biogenic formation of iron sulphides in semi-saline conditions at low temperature and neutral pH is well documented (Rickards, 1969) and it is thought that the sulphides have been formed in situ beneath the landfill. Despite detailed investigations no other sulphides have been detected from core material from HL6 or from a control borehole drilled outside the landfill. For some reason sulphate reduction is not taking place in the first metre immediately beneath the landfill. This may be because the environment is too hostile for the microbes which enable mineralisation of sulphate to sulphide.

In borehole HL6 the NH_4 front has penetrated to only 2.5 metres below the landfill base and high concentrations of nitrate are present below this. The data shows that nitrogen species are present in the profile to a depth of 12 metres below the landfill base close to the depth reached by both Cl and TOC. The possibility therefore exists that below 2.5 metres NH_4 has been oxidised to be replaced by NO_3 in the profile. The mechanism by which this oxidation is achieved is unclear and needs further investigation involving analysis of possible intermediate species such as N_2O and NO_2.

The SO_4 profile in HL6 (figure 3b) shows no evidence of reduction as in HL5. However the delay in the migration of the SO_4 front relative to Cl demonstrates that some attenuation process is active, possibly involving the precipitation of gypsum ($CaSO_4.2H_2O$). However despite detailed XRD work no evidence of gypsum precipitation has been demonstrated.

From this review of the unsaturated zone porewater profiles involving nitrogen and sulphur species beneath both the mature and young wastes it is suggested that the following processes occur:

1) Removal of NH_4 by cation exchange but also possibly by oxidation. Beneath immature waste this oxidation proceeds as far as yielding NO_3 but it is possible that beneath old waste nitrogen species are lost from the profile as N_2.

2) SO_4 is attenuated beneath immature waste possibly by the precipitation of gypsum and beneath mature waste by reduction to sulphides.

5 CONCLUSIONS

In the introduction to this paper the question was raised as to whether or not dilute and disperse landfill sites on the Chalk pose a serious pollution threat to groundwaters. The evidence from the present work suggests that even beneath mature waste there has been very little migration of either nitrogen or sulphur species. This work has been carried out on porewaters and it is possible that some of the leachate has bypassed the Chalk matrix and migrated more rapidly to the water-table via fracture systems. Figure 1a may support this argument, as nitrate concentrations in the porewaters taken from below the water table are significantly higher than those in the surrounding groundwaters. However evidence of such by-pass flow was not detected at the water-table in either of the boreholes drilled.

6 ACKNOWLEDGMENTS

This paper is presented by permission of Buckinghamshire County Council who operate the landfill, and the Environment Agency UK sponsoring the research.

The views expressed in this paper are those of the authors and do not necessarily reflect the views of the Environment Agency or any other Authority.

7 REFERENCES

Cezan,M.L., Thurman,E.M., and Smith,R.L.,(1989) Retardation of ammonium and potassium transport through a contaminated sand and gravel aquifer: The role of cation exchange. *Environ. Sci. Technol.* 23:p.1402-1408.

Foster, S.S.D., Cripps, A.C., and Smith-Carrington, A.K., (1982), Nitrate leaching to groundwaters. *Phil. Tran. Royal. Soc. Lond.* 296: p.477-489.

Ingrey, N.C., and Mather, J.D., (1995), Leachate/Rock interactions beneath a landfill site in the Cretaceous Chalk aquifer in Bucks, UK. In: *Water Rock Interactions (WRI-8)*, Eds. Kharaka Y.F. and Chudaev O.V. Rotterdam: AA Balkema.p. 445-448.

Mather, J.D., (1989), The attenuation of the organic component of landfill leachate in the unsaturated zone, a review. *Quart. Jour. Eng. Geol.* 22, p.241-246.

National Rivers Authority, (1995), The effect of old landfill sites on groundwater quality - phase 1. *R&D Project No.* 569/5/A.

Palmer, C., and Young, P.J., (1992), Leachate contamination from closed landfills - predicting impact. In: *AEA Waste Management Symposium*, Harwell, UK.

Rees, J.F., (1981), Landfill attenuation in the Lower Chalk - the role of microbial processes. *AERE Harwell, Report No.* AERE-R-10271.

Rickards, D.T., (1969), Chemistry of iron sulphide formation at low temperatures. *Stockholm Contrib. Geol.* 20: p.67-95.

Hydrogeochemical characteristics of deep groundwater in Korea for geological disposal of radioactive waste

J.U. Lee, H.T. Chon & Y.W. John
Department of Mineral and Petroleum Engineering, College of Engineering, Seoul National University, Korea

ABSTRACT: The main purpose of this study is to construct basic geochemical data for future application to the water-rock interaction around the disposal site of radioactive waste in Korea. This study reports a series of results from hydrogeochemical analysis of deep groundwater in the various type of rock in Korea.

1 INTRODUCTION

In Korea, a systematic long-term project has been set up for the geological disposal of radioactive waste which requires high levels of safely for long time. One of the most important factor is a safe control of geological barrier which is highly dependent on the geochemistry of groundwater. In this study, a variety of geochemical data has been collected to provide a basic data for future research on the water-rock interaction in such system in Korea.

2 SAMPLING AND CHEMICAL ANALYSIS

Groundwaters were sampled at wells, springs, deep tunnels, deep mines and boreholes. At each sampling site, pH, Eh, electrical conductivity and temperature were measured. Cations were determined using ICP and AAS, and anions by IC in laboratory. Alkalinity was measured by titration method. Quality control was performed for each analysis. Electroneutrality of each sample was calculated and samples above 30% were abandoned (Nordstrom et al., 1990). Some samples nitrate ion contents which were above 45mg/l were also omitted from the dataset because there might be due to some contamination by anthropogenic origin through agricultural activity.

3 RESULT AND DISCUSSION

3.1 Granitic groundwater

In granitic region, the contents of total dissolved solids increase statistically and progressively from shallow to deep groundwater, which indicates the presence of more concentrated water with depth due to water-rock interaction. Other, most dissolved, constituents and pH and temperature show the same trend as TDS. However, the contents of K^+ and Ca^{2+} are unchanged and that of Mg^{2+}, NO_3^- and Eh, on the contrary, decreases with depth.

The behaviour of dissolved ions and saturation indices indicate that congruent dissolution of calcite present at soil layers and in veins, fissures and joints mainly controls chemical composition of shallow groundwater. However, when it moves to intermediate depth, calcite is precipitated as solid phase and incongruent dissolution of plagioclase become major reaction at deep groundwater. This change leads the fact that the chemical composition of groundwater evolves from initial $Ca^{2+}+(Cl^-+SO_4^{2-})$ or $Ca^{2+}+HCO_3^-$ type to final $Na^++HCO_3^-$ or $Na^++(Cl^-+SO_4^{2-})$ type, via $Ca^{2+}+HCO_3^-$ type (Figure 1).

At intermediate and deep depths, another important reaction seems to occur which controls the behaviour of dissolved silica. Equilibrium reactions between kaolinite-smectite and/or kaolinite-illite is thought to play a significant role for this control. The decrease of Mg^{2+} and K^+ concentration is presumably due to the formation of these clay minerals as well as low solubility of Mg- and K-minerals.

3.2 Metasedimentary groundwater

Chemical characteristics of metasedimentary ground-water are somewhat different from that of granitic groundwater. They consistently show

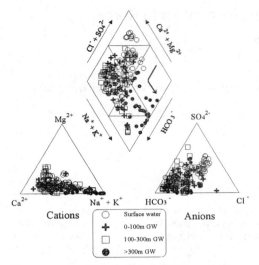

Figure 1. Chemical compositions of water samples in granitic region on the Piper's diagram. GW: groundwater

$Ca^{2+}+HCO_3^-$ type at depth and low sodium concentrations compared with granitic groundwater, which is due to the lack of plagioclase as Na^+ supplier in their aquifer rock. Therefore chemical composition of metasedimentary groundwater is controlled predominantly by undersaturated calcite dissolution, irrespective of depth.

According to factor analysis, chemical characteristics of metasedimentary groundwater at 100-300m depth are simplified as following factors (Table 1):

1. Dissolution of calcite and Mg-carbonates
2. Transformation of kaolinite to illite
3. Presence of sodium as not product of plagioclase dissolution but product of artificial salt dissolution.

Discriminant analysis between granitic and metasedimentary groundwaters derives following unstandardized discriminant function (Eq.3.1). It shows good discriminating ability with 81.8%

Z_{ust} = 1.0918 logF^- + 1.5120 logMg^{2+} + 0.1523 $SiO_{2(aq)}$ + 1.8888 logHCO_3^- - 3.1341 logNa^+ - 1.3482 logCa^{2+} - 0.9657 log Cl^- - 0.4851
(3.1)

All groundwater samples of volcanic aquifer, which has abundant plagioclase, are included in the granitic group according to Eq.3.1. This confirms the fact that the presence of plagioclase is main discriminator between granitic and metasedimentary groundwaters.

Table 1. Varimax factor matrix of chemical constituents and factor scores for the groundwater samples from 100~300m depth in metasedimentary region.

Variable	Factor 1	Factor 2	Factor 3	Communality
log HCO_3^-	0.931			0.9162
log SO_4^{2-}	0.922			0.9043
log Ca^{2+}	0.911			0.9585
log Mg^{2+}	0.902			0.9524
$SiO_{2(aq)}$		0.902		0.9236
log K^+		0.851		0.8785
log F^-		0.667	0.667	0.8903
log Na^+			0.905	0.9281
log Cl^-			0.836	0.7476
Eigenvalues	3.70	2.86	1.54	
Percent of variance explained by factor	41.1	31.7	17.1	
Cumulative percent of variance	41.1	72.9	90.0	

* Factor loadings less than 0.4 are omitted.

3.3 Spa water

Studied spa waters have isotopically meteoric origin. Thus it seems to be meaningful to compare with cool groundwater in order to anticipate the chemical changes of groundwaters near disposal site due to radioactive decay heat.

Spa water samples show the result of active water-rock interaction due to high temperature (mean: 38.5℃). The depth of studied spa waters are corresponding to intermediate depth, but they show the same characteristics as that of deep granitic ground- water. Furthermore, Na^+, F^-, SO_4^{2-} and Sr are enriched and Mg^{2+} is depleted, compared with deep granitic groundwater (Figure 2).

3.4 Equilibrium constants of kaolinite-smectites

The distribution of solubility quotients for kaolinite-smectite reactions shows the trend of reaching at equilibrium with Ca- and Mg-smectite for deep groundwater. The values of $10^{-15.25}$ and $10^{-16.25}$ are proposed as equilibrium constants between kaolinite and Ca- and Mg-smectite end members, respectively. On stability diagrams drawn based on these values, most of deep groundwaters are plotted at equilibrium boundaries between

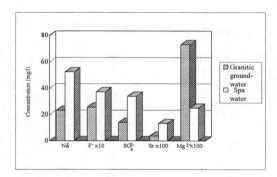

Figure 2. Median concentration of some dissolved major ions for the groundwater from depth deeper than 300m in granitic region and some spa waters.

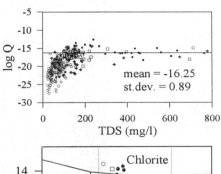

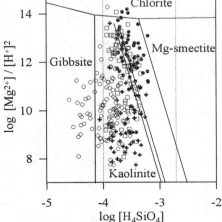

Figure 3. Stability relationships among some minerals in the system $MgO-Al_2O_3-SiO_2-H_2O$ at 298K and 1atm. The boundaries between kaolinite and Mg-smectite are by Helgeson(1969), this study and Nesbitt(1977) from the left to the right. Q means solubility for reaction of kaolinite and Mg-smectite. Symbols have the same meaning as that of Figure 1.

Table 2. The mixing ratios in models obtained by mass balance modelling for two tunnel seepage water sampled before and after waterproof operations (MW-1 → MW-2). SW-2 and SW-3 denote granite and limestone groundwater, respectively. The models which provide the best fit to the measured data are indicated in bold-face type (adapted from Schramke et al., 1996).

Model No.	Initial samples		NO_3^- (mg/l)	Cl^- (mg/l)
Measured values at sample MW-1			0.25	4.75
	SW-2	SW-3		
1	80.9 %	19.1 %	3.73	5.21
2	92.0 %	8.0 %	2.75	4.96
3	12.0 %	88.0 %	9.79	6.73
4	**98.0 %**	**2.0 %**	**2.23**	**4.83**
5	66.7 %	33.3 %	4.98	5.52
Measured values at sample MW-2			8.89	6.13
	SW-2	SW-3		
1, 2, 4, 5	67.7 %	32.3 %	4.89	5.50
3	8.8 %	91.2 %	10.06	6.80
6, 7, 8, 9, 10, 11, 12, 18, 19, 21, 22	61.1 %	38.9 %	5.47	5.64
13, 14, 15, 16	71.9 %	28.1 %	4.52	5.40
17	**19.7 %**	**80.3 %**	**9.11**	**6.56**
20	**38.0 %**	**62.0 %**	**7.50**	**6.16**

stability fields of kaolinite and smectites or on stability fields of smectites and illite (Figure 3).

Compared with rock-forming minerals of granitic rock, those of basic rock dissolve rapidly (Eggleton, 1986; Drever, 1988) and supply much cations into solution. Thus groundwaters in basic rock aquifer are plotted at the area of deep granitic groundwater on stability diagrams.

3.5 *Modelling of water-rock interaction based on mass balance calculation*

The reaction modellings based on mass balance calculation for two local study areas, i.e. Youngcheon and Donghae in Korea proved to be reliable because they are in accordance with the results of studies on the behaviour of dissolved constituents and thermodynamic consideration.

Chemical composition of deep groundwater in the vicinity of Youngcheon Tunnel area is thought to be resulted from following reaction (Eq.3.2):

'surface water' + 0.31 kaolinite + 0.58 plagioclase + 0.09 K-feldspar + 0.27 calcite + 0.08 pyrite + 0.34 CO_2 = 'deep groundwater' + 0.45 Ca-smectite + 0.16 illite + 0.04 hematite (unit : mmol/l)

(3.2)

Another mass balance calculation for seepage water in the tunnel at Donghae area, which is located in the border of granite and limestone, shows that the proportion of granitic groundwater decrease from 98% to 20~38% as a result of waterproof operations (Table 2).

REFERENCES

Drever, J.I., 1988, *The geochemistry of natural waters* (2nd ed), New Jersey: Prentice Hall.

Eggleton, R.A., 1986, The relation between crystal structure and silicate weathering rates, In S.M. Colman & D.P.Dethier (eds), *Rates of chemical weathering of rocks and minerals*:21-40, Florida: Academic Press.

Helgeson, H.C., 1969, Thermodynamics of hydrothermal systems at elevated temperatures and pressures, *Amer. J. Sci.* 267:729-804.

Kim, K.H., 1991, *Isotope geology (in Korean)*, Seoul: Mineumsa.

Nesbitt, H.W., 1977, Estimation of the thermodynamic properties of Na-, Ca- and Mg-beidellites, *Canadian Mineralogist* 15:22-30.

Nordstrom, D.K., I.Puigdomènech & R.H.McNutt, 1990, Geochemical modelling of water-rock interactions at the Osamu Utsumi mine and Morro do Ferro analogue study sites, Poços de Caldas, Brazil, *SKB Technical Report 90-23*.

Schramke, J.A., E.M.Murphy & B.D.Wood, 1996, The use of geochemical mass-balance and mixing models to determine groundwater sources, *Applied Geochem.* 11:523-539.

Alteration of cold crucible melter Ti/Zr-based ceramics

G. Leturcq, T. Advocat & A. Bonnetier
CEA Valrho DCC/DRRV, SCD-BP171, Bagnols/Cèze, France

G. Berger
Paul Sabatier University, CNRS, UMR 5563, Toulouse, France

ABSTRACT: Three Ti/Zr-based ceramics were synthesized in an induction-heated cold crucible at laboratory scale (1 kg) from an oxide mixture: a zirconia, a batch in zirconolite stoichiometry and a synroc. The chemical durability of both melted materials, as determined by static leaching of powder samples in initially pure water at 90°C with an high SA/V ratio. Cerium, used in this investigation to simulate the presence of tri- and tetravalent actinides, was found in steady-state concentrations on the order of 1 ppb. These results are encouraging for the investigation to develop specific containment of actinides.

1 INTRODUCTION

Current research on advanced chemical separation processes for the long-lived actinides (Np, Am, Pu, Cm) and the possible option of geological disposal if transmutation is not adopted, have prompted investigation to develop specific containment matrices for these radioelements. The ability of Synroc titanate-based ceramic phases (primarily zirconolite) to incorporate substantial quantities of actinides (e.g. 10% PuO_2), their resistance to aqueous leaching and to significant irradiation doses without any major reduction in their chemical durability has been demonstrated by several studies (Vance et al., 1995; Jostsons et al., 1995; Blackford et al.; 1992). Another possibility is incorporation of actinides in ZrO_2. The objective of this study was to confirm the technical feasability of producing these ceramics by cold crucible melting (Jouan et al., 1995) and to verify their chemical durability.

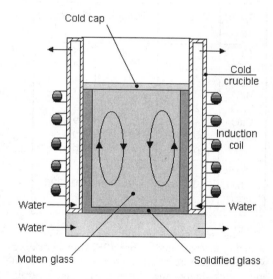

Figure 1. Induction-heated cold crucible schematic

2 EXPERIMENTAL METHODS

2.1 Melting Principle (Jouan et al., 1995)

The technique is illustrated in Figure 1. The material to be melted (1 kg) is placed inside the sectorized crucible where it is heated by Joule effect resulting from the electric currents induced by an electromagnetic field. A solidified layer of material forms on contact with the water-cooled crucible wall, allowing refractory materials to be melted at high temperatures without damaging the melter.

Three ceramics were fabricated by melting an oxide mixture in a laboratory-scale melter : a zirconia enriched in Ce, a batch in zirconolite stoichiometry and a synroc named Rocsynmar. The characteristics of the melting facilities are indicated in Table 1 and the compositions of the samples are

Table 1. Melting facilities used to fabricate samples

Crucible diameter (mm)	Crucible height (mm)	Crucible material	Induction generator (kW)	Induction frequency (kHz)	Initial material
60	250	Stainless steel	20	3000	Oxide mixture

reported in Table 2. The ceramics were fabricated experimentally from elementary oxide powder samples (Table 2). Induction was initiated by adding about 1 wt% titanium metal. One kilogram of material was completely melted, then cooled by reducing the induction power after one hour. Cerium oxide is used here to simulate the tri- and tetravalent actinides. After melting, the material was recovered in the form of an ingot about 12 cm high.

Table 2. Calculated compositions (oxide wt%) of samples and BET specific surface area of 40-50 μm powders (m^2/g).

Oxide	Zirconia	Zirconolite	Rocsynmar
TiO_2		44.8	65.3
ZrO_2	38.0	27.5	5.6
Al_2O_3			4.8
BaO			4.8
Li_2O	2.0	2.0	
CaO		15.7	9.6
CeO_2	60.0	10.0	9.9
Total	100.0	100.0	100.0
specific surface area	0.3787	0.1348	0.1296

2.2 Characterization Methods

Scanning electron microscope (SEM) examinations were performed on polished cross sections obtained after embedding the sample in epoxy resin (*Cambridge* SEM with an attached *Link* system). The mineral phases were also identified with X-ray diffraction on powder samples (*Siemens* D500, 0.01° steps, 2 seconds per step, JCPDS file).

Samples were leached in static mode at 90°C ± 1°C in initially ultrapure water (MillQPlus system) at SA/V ratios of 8000 to 20000 m^{-1} according to a test protocol described by Vernaz et al. (1989). The materials were ground and screened to obtain a grain size between 40 and 50 μm. Melted ceramics powder specimens containing 0.29 g each were placed in contact with 5 ml of ultrapure water in 7 ml *Savillex* PFA containers, which were then sealed in glass flasks maintained in a temperature-regulated oven and stirred continuously. Tests were conducted in triplicate for periods of 7 days to one year. Leachates were ultra-filtered at 10000 D. Ti, Ba, Al and Ce concentrations in the leachate were then measured by ICP–MS (*Perkin-Elmer* Elan 5000).

This leaching protocol is widely used to investigate the long-term behavior of nuclear glass. It is designed to simulate a very high degree of alteration reaction progress in the laboratory by quickly reaching saturation conditions, and to assess the long-term evolution of the alterability of materials (French AFNOR standard X30-407).

3 CHARACTERIZATION RESULTS

3.1 Zirconia

Two phases were identified by X-ray diffraction:
- $Zr_{0.4}Ce_{0.6}O_2$ (JCPDS 38-1439) observed in SEM examination of polished cross sections (white crystals in Figure 2a),
- Li_2ZrO_3 (JCPDS 38-0843) observed in SEM as interstitial crystals (black in Figure 2a).

Li was added initially to determine dissolution rates but it crystallizes separately. So we can not use it as a dissolution tracer.

3.2 Zirconolite

Three phases were identified by X-ray diffraction:
- perovskite (JCPDS 42-0423) dark gray in SEM image (Figure 2b) with dendrites containing Zr (white).
- zirconolite (JCPDS 17-0495) : xenomorphous crystals measuring about 50 to 100 μm (white and pale gray in figure 2b)
- $Li_2Ti_3O_7$ (JCPDS 34-0393), interstitial phase (black in Figure 2b).

We note than Li forms its own phase, so we can not use it as dissolution tracer, and crystallization of zirconolite is accompanied by crystallization of perovskite.

3.3 Rocsynmar

Four types of phases characteristic of hot-pressed Synroc (Reeve et al., 1984) were identified by X-ray diffraction in this sample of melted-oxide Synroc: rutile (JCPDS 21-1276) dendritic strips observed in Figure 2c (black), hollandite (JCPDS 33-0133) xenomorphous crystals measuring about 50 μm (dark gray in Figure 2c), perovskite (JCPDS 42-0423) subautomorphous crystals (dark gray in Figure 2c) measuring up to 50 μm and often exhibiting zoning with exchanges between Ca and Ce, and zirconolite (JCPDS 17-0495) star-shaped crystals (pale gray in Figure 2c) measuring up to 30 μm. See also Advocat *et al.*(1997).

4 LEACH TEST RESULTS

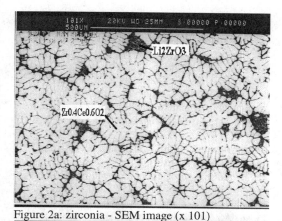

Figure 2a: zirconia - SEM image (x 101)

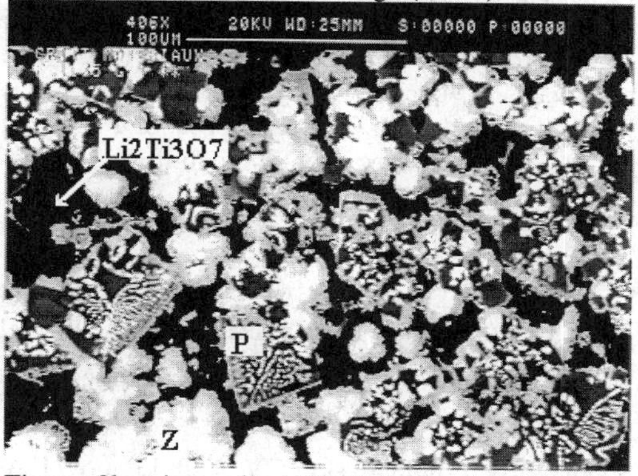

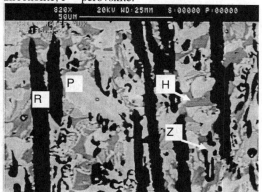

Figure 2b: zirconolite- SEM image (x 406) Z = zirconolite; P = perovskite.

Figure 2c: rocsynmar. Detail view of zirconolite (Z), perovskite (P), hollandite (H) and rutile (R) crystals.

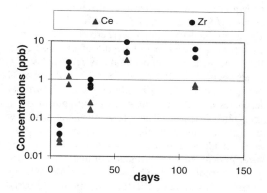

Figure 3a: Zirconia. Element concentrations (static leach test at 90°C and SA/V ratio of 20000 m^{-1})

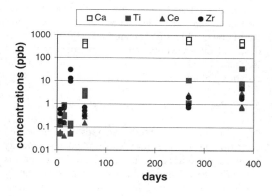

Figure 3b: Zirconolite. Element concentrations (static leach test at 90°C and SA/V ratio of 8000 m^{-1})

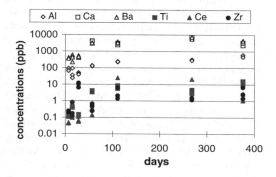

Figure 3c: Rocsynmar. Element concentrations (static leach test at 90°C and SA/V ratio of 8000 m^{-1})

The variations in time of the dissolved species concentrations in the ultra-filtered leachates are shown in Figure 3 for the three materials. pHs were about 10.5, 10.5 and 7.5 at 90°C for zirconia, zirconolite and rocsynmar respectively. For the two first, basic pH are due to leaching of Li phases. For the three materials, concentrations reached steady-states about 1 and 10 ppb of Ce and Zr, 10 ppb of Ti concentrations of zirconolite and rocsynmar leachates and 100 ppb of Al and few ppm of Ba in rocsynmar leachates. These low concentrations show the good chemical durability of such matrices: if we consider Ca as a matrix dissolution tracer, we can determine the dissolved fraction of material after one year of leaching at 90°C and SA/V = 8000 m^{-1}. We obtain 0.01% and 0.2% for the zirconolite-based ceramic and the melted synroc (rocsynmar) respectively. Solubility data (logK) are needed to confirm saturation states conditions of the leachates with respect to the initial crystals or eventually secondary phases. The low Ce concentrations measured in the leachates prove its good confinement in such materials. Finally results observed on rocsynmar are in agreement with Van Iseghem et al. (1996) with respect to chemical durability of hot-pressed synroc C.

5 CONCLUSION

The use of Ti/Zr-based ceramics for dedicated conditioning of the rare earth elements (mainly cerium), considered as chemical simulants of the actinides, was found to be feasible at a laboratory scale by cold crucible melting of the basic component oxides. This process considerably reduces the number and complexity of the steps required to produce hot-pressed Synroc. Although the thermal scenario was not optimized, the zirconolite, perovskite, hollandite and rutile phases crystallized as the melt was cooled. These tests confirmed the results of earlier work by other authors (Sobolev et al., 1993) on Synroc-type ceramics with similar concentrations for the major oxides (TiO$_2$, ZrO$_2$, CaO, Al$_2$O$_3$, BaO) but with lower concentrations of rare earth oxides (mainly CeO$_2$) simulating the presence of actinides. Crystallization of zirconolite is accompanied by crystallization of perovskite. It is also possible to obtain Ce-enriched zirconia by this way. Another test objective was to verify the long-term chemical durability of synthetic materials rich in rare earth elements by aqueous leach testing at high SA/V ratios. The preliminary results are encouraging, and confirm the intrinsic low solubility of the Ti and Zr phases. For the three tested materials, cerium, simulating the actinides, was found at steady-state concentrations on the order of 1 ppb.

6 REFERENCES

AFNOR X30-407. Méthodologie pour la détermination du comportement à long terme. 1995.

Advocat T., Leturcq G., Lacombe J., Berger G, Day R.A., Hart K., Vernaz E. and Bonnetier A. 1997 Ateration of cold crucible melter titanate-based ceramics: comparison with hot-pressed titanate based ceramic. Mat. Res. Soc. Symp. Proc. XX, ed. Mc Gray P.

Blackford M.G., Smith K.L. and Hart K.P. 1992 Microstructure, partitioning and dissolution behaviour of synroc containing actinides. Mat. Res. Soc. Symp. Proc. Vol. 257, ed Sombret C.G., Pittsburgh, PA, 243-249.

Jouan A., Moncouyoux J.P. and Merlin S. The development of nuclear waste vitrification in France *Waste Management 95*, Tuscon 26 February - 2 March 1995, section 17-54.

Jostsons A., Vance E.R., Mercer D.J. and Oversby V.M. 1995 Synroc for immobilising excess weapons plutonium. Mat. Res. Soc. Symp. Proc. Vol. 353, eds Murakami T. and Ewing R.C., 775-781.

Reeve K. D., Levins D. M., Wolfrey J. L. and Ramm E. J. 1984 Immobilization of high-level radioactive waste in Synroc. Nuclear Waste Management, ed by Wicks G. and Ross W., 200-208.

Sobolev I.A., Stefanovsky S.V. and Lifanov F.A. 1993 Radiochemistry, 35,99.

Vance E.R., Begg B.D., Day R.A. and Ball C.J. 1995 Zirconolite-rich ceramics for actinide wastes. Mat. Res. Soc. Symp. Proc. Vol. 353, eds Murakami T. and Ewing R.C., 767-774.

Van Iseghem P., Jiang W. and Blanchaert M. 1996 The interaction between synroc-C and pure water or Boom clay. Mat. Res. Soc. Symp. Proc. Vol. 412, eds Murphy W.M. and Knecht D.A., Boston, MA, 305-312.

Vernaz E. , Advocat T. , Dussossoy J.L. 1989 in Ceramic Transactions, vol. 9, Nuclear Waste Management III, ed. G.B. Mellinger, 175-185.

Remediation of a sandstone aquifer following chemical mining of uranium in the Stráž deposit, Czech Republic

J. Novák, R. Smetana & J. Šlosar
DIAMO s.p., Stráž pod Ralskem, Czech Republic

ABSTRACT: Chemical mining of uranium in situ causes serious ecological problems. Remediation following the closure of the mining operation requires selection of the optimal strategy, appropriate technologies and evaluation of alternative scenarios with sophisticated mathematical models. For simulation of the remediation course the three dimensional transport-reactive model has been developed. This finite element model enables simultaneous simulation of the transport and chemical processes in underground.

1 INTRODUCTION

The uranium deposit Stráž, Northern part of the Czech Republic was exploited by underground acidic leaching between 1968 and 1996. More than 14 000 tons of uranium were produced during this period. More than 4 million tons of H_2SO_4, 300 thousand tons of HNO_3, 120 thousand tons of NH_3 and other chemicals were injected in Cenomanian sandstones. The mining has resulted in large contamination of the ground waters. This requires intensive remediation activities.

2 CHEMICAL LEACHING OF URANIUM

The chemical leaching is done by forced circulation of the leaching solution in a closed cycle (Beneš 1993). After the separation of uranium necessary reagents are added into the exhausted leachate and this renovated leaching solution is again injected into the aquifer. The acidic leaching using sulfuric acid is not selective. The solution is reacting with minerals of the rock environment and hydrogen ions H^+ are replaced by other cations leached from the rock. Besides uranium, other elements are leached from the rock, especially Fe and Al. The separation of uranium by classical ion exchange technology requires high selectivity because the quality of the uranium concentrate is negatively influenced by minor components. Al and Fe sulfates and other products of the reactions remain in the solutions and the content of sulfates gradually increases.

In addition to sulfuric acid, remains of chemicals used in surface technologies were added into the leaching solution. Nitric acid, used for uranium desorption from the ionex, served also as an oxidizing reagent in leaching. Ammonia remaining after the precipitation of the yellow cake (ammonium diuranate) served no technological purpose in the underground processes. It is, however, the most serious ecological pollutant.

3 HYDROGEOLOGY

The Stráž tectonic block, where the North Bohemian uranium deposits are situated, is part of the Bohemian Cretaceous basin. The basin represents an important reservoir of drinking water of European significance. The Stráž block is, to a large extent, an

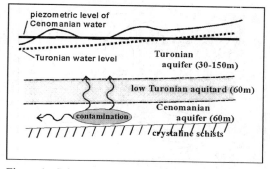

Figure 1. Schematic hydrogeological cross-section of the Stráž deposit with possible pathways of contamination

independent hydrogeological unit, the boundary of which is formed by important tectonic lines. The block consists of Upper Cretaceous sediments, mainly sandstones and siltstones with dykes and pipes of tertiary basaltic volcanites and the underlying crystalline rocks (Fiedler & Slezák 1993).

In the Stráž block two aquifers exist (Fig.1). The first, Middle Turonian aquifer with free water level, is the main regional water supply. A layer of siltstones and wackestones of the Lower Turonian stratum forms the aquitard with a thickness of about 60 meters. The aquitard separates the first Turonian aquifer from the second Cenomanian aquifer with the confined water level. Original water of the Cenomanian aquifer in the deposit area was characterised by high radioactivity and was not utilised.

4 UNDERGROUND CONTAMINATION

The volume of contaminated waters in the Cenomanian aquifer is about 190 million m^3 with 1 to 100 g/l of total dissolved solids (TDS). The main components of the solutions are SO_4^{2-}, Al, Fe and NH_4^+. Table 1 shows the range of concentrations of the main components of concentrated solutions in the leaching area.

Table 1. Range of concentrations of the main components of solutions in the leaching area [g/l]

SO_4^{2-}	H_2SO_4	Al	Fe	NH_4^+	NO_3^-
33-80	10-25 *	4-6	1-2	0.8-1.5	0.6-1.4

* in SO_4^{2-}

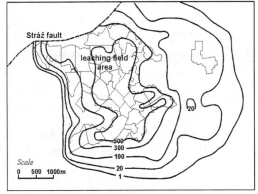

Figure 2. Contamination of the Cenomanian aquifer in 1996 - total dissolved solids (TDS) in kg/m^2

Distribution of the contaminants is very uneven. Fig.2 shows the amount of TDS in the solutions falling on 1 m^2 of the deposit area. In the central part of the leaching field area the square density of TDS is higher than 500 kg/m^2, in the large periphery area it is less than 100 kg/m^2. The total contaminated area is about 28 km^2. The dispersion of solutions outside the leaching field area has been caused by a very complicated hydrogeological situation as a consequence of the neighbouring deep mining activities.

Tab.2 shows the quantity and proportions of the main components in the solutions.

Table 2. Quantity of the main components in solution [thousand tons]

TDS	SO_4^{2-}	H_2SO_4	Al^{3+}
4773.2	3802.8	933.2	412.5
Fe	NH_4^+	Ca_2^+	NO_3^-
126.7	91.4	31.6	27.3

Contamination of the Cenomanian aquifer does not represent an urgent threat to the environment but it involves two potential risks. The first one is a threat to Elbe river. This river drains the Cretaceous sediments in the distance of 40 km from the contaminated area. Modelling shows that the contaminated solutions could reach this line in 1500 years and the concentrations would then increase for another 3500-8500 years.

The second potential risk is possible inter-aquifer migration of the contaminants from the Cenomanian to the Turonian aquifer. Natural communication has not been observed but almost 15000 wells in the deposit area have reduced isolating ability of the Lower Turonian aquitard layer. The quality of Turonian water would be affected after several decades. Drinking water supply in the region could be devastated and the river Ploučnice could be endangered after 100-500 years.

5 REMEDIATION PROCEDURE

In view of these risks, most of the contaminants will have to be withdrawn onto surface. Only such quantity of contaminants will be left in the ground water, which makes risks as low as possible. This process will require the building of new high-end technological devices (evaporating station with a multi-

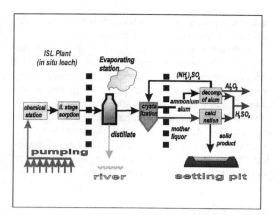

Figure 3. Scheme of remediation technology.

stage evaporator and crystallizing unit, membrane units, etc). The technology and the remediation procedure are described in detail in Šlosar & Ondrušek (1995) and Beneš (1995). The schematic for the process is shown in Fig.3. The removal of the contaminants will continue for approximately 30 years.

Geochemical interaction between the solutions and rocks will change. At present, in the leaching fields area, where the most concentrated solutions are located, the major part of reactive minerals is decomposed and the sorption capacity is minimized. Consequently, the content of free H_2SO_4 is decreasing very slowly in the central area (about 8% per year with an average concentration of acid of about 10 g/l). More significant changes are observed only in the peripheral areas.

After the remediation is finished, the solutions will move to unaffected rocks of the Cenomanian and/or Turonian stratum. Neutralising capacity of the rock, influence of reductive environment and sorption capacity undamaged by acid environment will all play the important role here. The pH of solutions will change from present values of about 1 to circum neutral pH. The oxidation-reduction potential will decrease and some of the components will be removed from the solutions by adsorbtion.

Besides natural neutralization, immobilization of part of the residual acid solutions in situ by injection of lime milk or sodium bicarbonate is also considered. It would cause precipitation of secondary solid phases ($CaSO_4$, Fe and Al hydroxides) in the pores of sandstones. Lowered permeability of the rocks will decelerate dispersion of the remaining pollutants.

6 MODELLING

In connection with the remediation a set of mathematical models have been developed (Novák et al. 1995, Novák & Smetana 1996). These models enable a better understanding of the on-going processes, their prognosis, control and optimization. The most complete model is the 3D transport-reactive model for simulation of pumping and simultaneously running chemical processes in underground.

The model calculates a flow by the finite element method. The shape of the elements is a vertical trilateral prism with a general slope of bases. This shape of the elements respects the complicated shape of a layer boundary as well as geological non-homogeneities (bodies of volcanic rock) and technological objects (boreholes). The resulting solution of the mathematical formulation of the problem is in a form of balanced volume of liquid which is transferred between elements.

The transport is solved in time steps by an inter-element transfer of the solution with dissolved solids. After the transfer, the balance of solids as well as the calculation of concentrations of the solution components are done. The chemical processes in a particular time steps are solved inside the element independent of the situation in surrounding elements. This procedure is described in detail in Štrof (1997). First, the thermodynamic equilibrium of the new solution is calculated. It corresponds with quick reactions of components in a liquid phase. The next step is calculation of the oxidation-reduction potential of the solution, to determine the solution-rock interaction.

As opposed to calculation of the solution composition, where all components are evaluated simultaneously, the interaction is solved step by step for single minerals. It begins with evaluation of the saturation index for all minerals. Depending on its value, it is followed by calculation of the kinetics of precipitation or dissolution. The sorption processes are again done separately for all components of the solid phase. The results of this procedure are new concentrations both in the solution and in the rock in a single element. These concentrations are an input for the next calculation step.

The initial conditions for the calculation are concentrations of the solution and rock components. The model used at the present time calculates with 10 components of the solution, 14 components of the solid phase and 5 sorbed components.

Calibration of the model is based on gradual decrease in the difference between the model result and

the behaviour of a real object. It can be done by one of the following steps:
– evaluation of completeness of the thermodynamic system
– change of kinetic parameters of some component
– change of the initial concentrations of components whose distribution in the underground is not well-known
– adjustment of the sorption capacity of minerals and the sorption affinity of cations

The parameters are verified by laboratory and field experiments.

The course of calculation is driven by remediation scenarios. Time schedule of the basic variant includes re-injection of the concentrate from the evaporator to the aquifer by the end of 2002. After the year 2003, a set of surface technologies for the treatment of solutions will be in operation. The technologies include handling of alum, production of H_2SO_4 and Al_2O_3, and disposal of mother liquor. Starting in 2014, the first station reverse electrodialysis will be in operation, and the second and third station will follow.

The result of the calculations is, on the one hand, a description of global situation in the underground; on the other hand, it is a measure of removed solids and the quantity of products of remediation in time. In compliance with an actual opinion the target concentration will be 3 g/l. To achieve this target, it is necessary to remove 3.9 million tons of solids which can be converted to 0.6 million tons of Al_2O_3 and more than 3 million tons of H_2SO_4. This will be achieved in 2034.

The variants of the final state in the underground after the remediation will be transformed to the initial conditions for the long-term models. These models will be used for evaluation of the remediation effect for protection of the environment. The models will calculate vertical transfer of the contaminants into the Turonian aquifer and long-distance transport of the contaminants in the Cenomanian aquifer.

7 CONCLUSIONS

Large contamination of the Cenomanian aquifer in the area of the uranium deposit Stráž represents one of the most serious ecological problems in the Czech Republic. At present, the situation is stabilized due to the operation of the evaporating station. It causes a hydraulic depression in the uranium deposit and prevents dispersion of the solutions in the aquifer. Environmental risk assessment is presently in progress and the construction of technological devices for handling of the contaminants is under preparation.

Mathematical modelling is used to define the limits of the residual contamination, evaluate the different procedures and optimize the remediation process. Two types of models are used: thermodynamical and kinetic models for detailed solution of the chemical reactions in underground and 3D transport-reactive models for global description of the process in the whole region for simulation of cleaning the aquifer and for balancing the volume of pumped water and removed substances.

8 REFERENCES

Beneš V. 1993. In-situ leaching of uranium in Northern Bohemia. *Proc. IAEA TCM Uranium In-Situ Leaching 1992.* IAEA Vienna, Austria.

Beneš V. 1995. Station for contaminated solutions liquidation, I. and II. stage. *Proc. Workshop on In Situ Leach Mining 1995.* DIAMO&IWACO (Netherlands), Harrachov, Czech Republic.

Fiedler J. & Slezák J. 1993. Some experience with the co-existence of classical deep mining and in-situ leaching of uranium in Northern Bohemia. *Proc. IAEA TCM Uranium In-Situ Leaching 1992.* IAEA Vienna, Austria.

Novák J. & Smetana R. 1996. Modeling of the ISL remediation in the Stráž deposit. *Proc. IAEA TCM Computer Application in Uranium Exploration and Production 1994.* IAEA Vienna, Austria.

Novák J., Mareček P. and Smetana R. 1995. System of the model solutions of the physical-chemical processes in the Stráž deposit used for rock environment restoration. *Proc. Conference and Workshop Uranium-Mining and Hydrogeology 1995.* Technische Universitat Freiberg, Germany.

Šlosar J. & Ondrušek M. 1995. Cleaning of acidic solutions in Stráž deposit using membrane technology. *Proc. Workshop on In Situ Leach Mining 1995.* DIAMO&IWACO, Harrachov, Czech Republic

Štrof P., Ira P., Emmer J., Novák J., Gombos L. and Pačes T. 1997. Alkaline leaching of uranium ore from the North Bohemian Cretaceous (Czech Republic). *Proc. 9th International Symposium on Water-Rock Interaction 1997.*

Molecular characterization of manganese oxides and trace metals in stream sediments from a mining-contaminated site

Peggy A. O'Day & Katherine E. Geiger
Geology Department, Arizona State University, Tempe, Ariz., USA

Christopher C. Fuller
Water Resources Division, US Geological Survey, Menlo Park, Calif., USA

ABSTRACT: In streams contaminated by mining activities, the mobility of hazardous trace metals is controlled by reactions between waters and surface coatings of secondary minerals and amorphous phases on detrital sediments. At Pinal Creek, Arizona, the precipitation of Mn-oxide surface coatings, both inorganic and biologically-mediated, is a primary control on the uptake from solution of trace transition metals. In this study we used synchrotron radiation X-ray absorption spectroscopy (SR-XAS), electron microprobe, and laser-ablation ICP-MS to characterize Mn-oxidation states, the bonding sites of trace concentrations of Zn, and the distribution of metals in contaminated stream sediments.

1 BACKGROUND

Contamination of the stream bed and alluvial aquifer from elevated metal concentrations has occurred from copper mining activities over the last century in the Pinal Creek Basin in central Arizona. Studies by the U.S. Geological Survey, begun in 1984 and continuing, indicate that metals have been mobilized by acidic discharges, primarily from a former unlined waste pond near the head of Miami Wash into the perennial stream and the alluvial aquifer (Eychaner 1991; Hem et al., 1989). At present, a 15-km-long acidic plume (pH ≈ 4) is present in the alluvium along Miami Wash and Lower Pinal Creek. Reaction of acidic groundwater with aquifer sediments neutralizes and retards the advancement of the plume. Neutralized contaminated groundwater (pH ≈ 5.5-7) is discharged down-gradient into the perennial reach of Lower Pinal Creek and contains elevated levels of dissolved solids (Ca, SO_4^{2-}) and metals (primarily Mn, Ni, Co, and Zn). High concentrations of Mn^{2+}(aq) (≈ -70 mg/L) in the upper stream reaches result in precipitation of Mn(III,IV) oxide minerals on the surfaces of streambed sediments in the hyporheic zone to a depth of about 10 cm. Extraction studies of Mn, Ni, Co, Zn, and Cu from sediments collected from the upper 4 cm of the stream bed indicate that metal uptake is coupled to the oxidation/precipitation of Mn-oxide phases in the hyporheic zone of the stream, which is in turn influenced by both mineral substrate and microbial activity (Lind and Hem, 1993; Fuller and Harvey, 1994). Owing to the arid climate, there is little evidence for organic or colloidal control of metal mobility (Fuller and Harvey, 1994). Central questions related to the long-term fate of metals in this system are: what mineral phase(s) is each metal associated with, and what is the nature of the metal uptake (i.e., surface adsorbed, interlayer or interior adsorbed, co-precipitated)? Quantitative modeling of metal partitioning between aqueous and solid phases requires molecular-level information combined with kinetic and thermodynamic constraints in order to predict the long-term release of metals and to assess the potential for "self-remediation" of the stream In this study we report initial results of X-ray absorption spectroscopy, electron microprobe, and laser-ablation ICP-MS studies of Mn-oxide surface coatings and trace metal distribution in Pinal Creek stream sediments.

2 ANALYTICAL TECHNIQUES

Synchrotron radiation X-ray absorption spectra for sediments were collected at the Stanford Synchrotron Radiation laboratory (SSRL) on wiggler beamlines 4-1 and 4-3 under dedicated conditions (3 GeV, 40-99 mA) using an unfocused beam. Stream sediment samples were separated into magnetic and non-magnetic fractions and size-fractioned, but otherwise were uncrushed and untreated. Samples were loaded into 1 mm thick Teflon sample holders and sealed with Mylar film. Both wet and dried samples were studied at both ambient and cryogenic temperatures (10-15 K) to compare changes in the absorption spectra. For some samples, stream sediments were equilibrated for several days with a synthetic stream water solution containing excess dissolved metal (either Co, Ni, or Zn) to study sorbed metal complexes. Total concentrations of Co, Ni, and Zn in the stream sediments were very low, in the range of 10-200 µg/g sand. Sorption of excess aqueous metal was about equal to that of the original bulk concentration, so very low total concentrations were examined in all cases. In contrast, total Mn concentrations were much higher, generally about 1,000-5,000 µg/g sand.

K-edge XAS spectra were collected using either a Si(111) or Si(220) monochromator crystal and element foils were used for energy calibration. Element fluorescence was detected using either a 13-element solid-state Ge-fluorescence array detector or a Stern-Heald-type fluorescence detector. Harmonic rejection was achieved by detuning the incoming beam by 30-50% of maximum intensity. Depending on absorber concentration and background noise, 3-48 scans were collected and averaged for each sample. Spectra for solid reference compounds were collected in order to calibrate fit parameters for theoretically calculated phase shift and amplitude functions (using FEFF6) and to estimate errors in fit results (see O'Day et al., 1994, for details of data analysis).

Electron microscopy and X-ray element dot maps were collected using a JEOL JXA- 8600 with a Noran 5500/5600 Energy Dispersive Spectrometer (EDS) at Arizona State University. Laser ablation ICP-MS was done with a sector-field inductively coupled plasma mass spectrometer (Finnigan MAT 'Element') at Washington University, St. Louis, MO. Loose sediment grains were mounted in epoxy, cut, and polished to expose cross-sections of detrital grains and surface overgrowths.

3 RESULTS AND DISCUSSION

3.1 SR-XAS

X-ray absorption Mn K-edge spectra for Pinal Creek sediment samples are compared in Fig. 1 with solid Mn-oxide reference compounds with Mn in three different oxidation states (Mn(II, III, IV)). The energy position and spectral features of the Mn K-absorption edge are sensitive to Mn oxidation state and coordination. The stream sediments (labeled A-E) represent stream material collected at different times from different parts of the stream, and two sediments with sorbed aqueous Zn or Co. Table 1 summarizes the attributes of sample collection and XAS data collection. The top three spectra in Fig. 1 are all roughly similar; slight differences in edge features may be attributed to the sorption of excess Zn or Co for B and C, respectively. The edge features of D and E are markedly different from those of the other sediments. Comparison with the reference compounds suggests mixed oxidation states, with a component of Mn(III) in sample D and a component of Mn(II) in sample E. The absorption spectrum for sample D was measured within a week of sample collection from Pinal Creek. The sample was taken from a part of the stream in which spring floods had ripped out old Mn crusts and new Mn precipitates were actively forming on fresh mineral surfaces. The sample was preserved in stream water and kept refrigerated before XAS data collection at ambient temperature. Sample E was a sediment sample that was collected and preserved in stream water. Shortly after collection, nutrients were added and the native microbial population was incubated for several weeks. The presence of Mn(III) in sample D suggests the formation of a mixed Mn(III, IV) oxide in newly precipitated material. Older samples from further downsteam (A-C) are more similar to pyrolusite (Mn(IV)O_2) and appear to contain mostly Mn(IV). These field observations are consistent with laboratory studies and XAS data which have proposed that Mn(III) in mixed Mn(III,IV)-oxide minerals (birnessite or buserite) disproportionates with time into Mn(IV) oxides and Mn^{2+}(aq) (Friedl et al., 1997). Sample E, however, indicates that microbial activity also plays an important role in controlling Mn

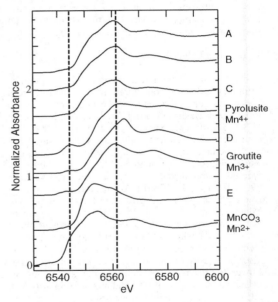

Fig. 1. Mn K-edge spectra of Pinal Creek sediment samples (A-E) and Mn reference compounds.

Table 1. Sample conditions for spectra in Fig. 1

Sample	State	T_{XAS}(°C)	Treatment
A	dry	cryogenic	unreacted
B	wet	ambient	Co^{2+}(aq)
C	wet	ambient	Zn^{2+}(aq)
D	wet	ambient	fresh
E	wet	ambient	biotic

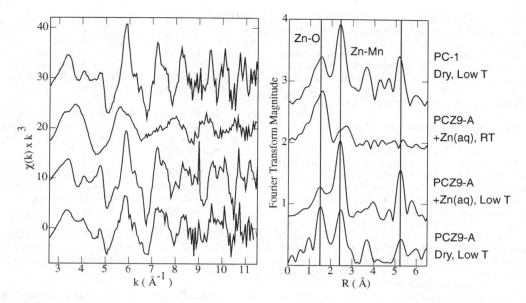

Fig. 2. Zinc K-edge EXAFS and Fourier transforms of sediment samples. Spectra were collected for Pinal Creek sediments at two sites (PC-1 and PCZ9-A) at low temperature (Low T = 10-15 K) or ambient room temperature (RT) with and without excess adsorbed Zn^{2+}(aq).

oxidation state and coordination. The edge position in sample E suggests the presence of a fraction of reduced Mn(II), which may be either sorbed or associated with carbonate, as has been observed in reduced lake sediments (Friedl et al., 1997). Analysis is underway to quantify the relative proportions of different Mn oxidation states in the stream sediments, and to use spatially-resolved methods to identify Mn phases and their distributions within surface coatings (see below).

The extended X-ray absorption fine structure (EXAFS) spectra for Zn in two stream sediment samples (PCZ9-A and PC-1) are shown in Fig. 2. Three different spectra for sample PCZ9-A are shown: dried, untreated, low temperature; with excess Zn^{2+}(aq) sorbed, frozen, low temperature; with excess Zn^{2+}(aq) sorbed, wet, ambient temperature. Sample PC-1 was dried and collected at low temperature. Spectra were collected at cryogenic temperatures to reduce thermal disorder and to enhance the fine structure at high photon energy. All four spectra have prominent peaks at ≈ 1.5 Å, 2.4 Å, and 5.2 Å (distances (R) shown in Fig. 2 are uncorrected for backscatterer phase shifts, and are ≈ 0.4-0.5 Å shorter than true distances). Backscatterer amplitudes are reduced significantly in the ambient temperature spectra at high R due to thermal disorder. Comparison of spectra with and without sorbed Zn at low temperature shows an increase in second-neighbor peak amplitude, but no significant change in distance.

This suggests that surface sorption sites that bind Zn in the stream are not saturated with respect to Zn uptake, and that these same sites bind Zn^{2+}(aq) during short-term equilibration. Changes in the shape of the second-neighbor peaks at ≈ 2.4 Å in wet vs. dry samples suggest some changes in Zn sorption sites that might be cause by freezing the sample to cyrogenic temperatures.

Least-squares fits of the Zn-EXAFS with reference functions indicate that first-neighbor backscatterers are oxygen, and that more distant backscatterers are probably Mn. Because Mn and Fe differ in atomic number by one, the difference between Mn and Fe as backscatterers cannot be determined directly from fitting. Because Mn-oxides form thick surface coatings, and because the magnetic fraction was removed in these samples (leaving the bulk concentration of Mn much greater than that of Fe), the assumption of Mn as the primary backscatterer is probably reasonable. Note in Fig. 2 that the ambient-temperature wet sample has broader first- and second-neighbor peaks compared with peaks of the cryogenic samples. Least-squares fitting shows that at ambient and cryogenic temperatures, the first O neighbors and the second Mn neighbors can be well modeled with two sets of interatomic distances for both Zn-O and Zn-Mn. Fitting of the second-neighbor peak assuming Mn backscattering gives Zn-Mn distances of 2.9-3.0 Å. This range of interatomic distances requires that Zn and Mn octahedra share polyhedral edges. Corner-sharing of Zn tetrahedra or octahedra and Mn octahedra results in significantly longer Zn-Mn distances (≈ 3.5 Å). This suggests that Zn is octahedrally coordinated by oxygen and may bond by edge-sharing to octahedral Mn. The two Zn-Mn distances might be explained by the presence of both Mn(III) and Mn(IV), each of which have different cation sizes and polyhedral distortions.

3.2 Electron microprobe and laser-ablation ICP-MS

Examination of grain cross-sections by EM and LA-ICP-MS shows that surface coatings are not homogenous Mn-oxides. Although some coatings are dominantly Mn, others contain Fe-rich layers or domains. Analysis of trace elements suggests variations in composition related to Mn- and Fe-rich areas. The detrital sediments are primary quartz and feldspar; minor phases include magnetic and ilmenite. Some of the grains have a layer of Fe coating rimming the primary grain, and then layers of Mn-oxides outward from the Fe-rim. These observations suggest a history of weathering for different minerals that may influence the microbial population, and thus the type of Mn coating that precipitates and its trace metal signature. On-going studies are aimed at identifying the overgrowth phases and the mineralogy of Mn-oxides.

4 SUMMARY

Our initial results point out the heterogeneity of sediment surface coatings that control the uptake of metals in this contaminated system. Results from SR-XAS suggest that secondary Mn-oxides are mostly Mn(IV), but that Mn(III) and Mn(II) states are found in some samples. It appears that oxidation states change as a function of time and microbial activity. Our results show support for disproportionation of Mn(III) oxyhydroxide into Mn(IV) oxides and Mn^{2+}(aq) with time as previously proposed in laboratory experiments. Likewise, the distribution of trace Zn in this system depends on available sorption sites on or in Mn-oxides, and may include both adsorption of surface complexes and substitution in oxides phases. Molecular characterization is useful for identifying the primary modes of metal uptake in natural samples, which greatly simplifies the number of partitioning reactions needed for geochemical modeling. This information is especially useful in coupled chemical speciation and hydrologic transport models that attempt to predict the long-term fate and transport of hazardous metals.

5 ACKNOWLEDGMENTS

Funding for this work was provided by the U.S. National Science Foundation (EAR-9629276). SSRL is supported by the Department of Energy, Office of Basic Energy Sciences.

6 REFERENCES

Eychaner, J.H. 1991. The Globe, Arizona, research site—contaminants related to copper mining in a hydrologically integrated environment. In G. E. Mallard & D. A. Aronson (eds.), USGS Water Resour. Invest. Rept. 91-4034: 439-447.

Friedl, G., B.Wehrli, and A.Manceau, 1997. Solid phase cycling of manganese in eutrophic lakes: New insights from EXAFS spectroscopy. *Geochim. Cosmochim. Acta* 61: 2811-2822.

Fuller, C.C. & J.W.Harvey 1994. Effect of surface chemical reactions in the hyporheic zone on trace metal partitioning, Pinal Creek, Arizona, EOS (abstr.), Am. Geophys. Union, 261.

Hem, J.D., C.J.Lind, and C.E.Roberson, 1989. Coprecipitation and redox reactions of manganese oxides with copper and nickel. *Geochim. Cosmochim. Acta* 53: 2811-2822.

Lind C.J. & J.D.Hem 1993. Manganese minerals and associated fine particulates in the streambed of Pinal Creek, Arizona, U.S.A.; a mining-related acid drainage problem. *Applied Geochem.* 8 : 67-80.

O'Day P.A., J.J.Rehr, S.I.Zabinsky, and G.E.Brown, Jr., 1994. Extended X-ray absorption fine structure (EXAFS) analysis of disorder and multiple-scattering in complex crystalline solids. *J. Am. Chem. Soc.* 116: 2938-2949.

Determination of background chemistry of water at mining and milling sites, Salt Lake Valley, Utah, USA

D. D. Runnells & D. P. Dupon
Shepherd Miller, Inc., Fort Collins, Colo., USA

R. L. Jones
Kennecott Utah Copper Corporation, Combined Labs, Magna, Utah, USA

D. J. Cline
Kennecott Utah Copper Corporation, Bingham Canyon, Utah, USA

ABSTRACT: In order to establish meaningful remediation goals and monitoring programs in areas disturbed by mining and associated industries, it is desirable to determine the natural concentrations of metals that existed in the area prior to disturbance. In mineralized areas, multiple natural populations of metals and associated components may be present. Probability distribution diagrams were used to identify multiple natural and anthropogenic background populations of dissolved components in surface and ground waters adjacent to major mining, milling, and smelting facilities in Salt Lake Valley, Utah.

1 INTRODUCTION

Estimation of natural background concentrations in water from mineralized areas can provide realistic goals for subsequent remediation of mining, milling, and smelting sites. The objective of this study is to present a methodology which can be used to characterize natural background populations in mineralized or disturbed mining, milling, and smelting sites.

2 METHODS FOR DETERMINING BACKGROUND CHEMISTRY

There are at least five possible approaches to estimating natural background chemistry, including: (1) up-gradient and cross-gradient sampling (U.S. EPA, 1992), (2) historical water quality data (Runnells et al., 1992), (3) extrapolation from similar geochemical environments (Miller and McHugh, 1994), (4) geochemical modeling (Nordstrom et al., 1996), and (5) statistical methods (Stackelberg and Siwiec, 1993).

Probability distribution diagrams represent a simple statistical approach for summarizing large amounts of data in a manner that is concise, comprehensive, and easily understood. The diagrams can yield natural background concentrations in the form of mean concentrations and standard deviations for distinct geochemical families of waters, as well as various "threshold" values.

3 PROBABILITY DISTRIBUTION DIAGRAMS

The use of probability distribution diagrams has found widespread use and acceptance in the field of geochemical exploration (Sinclair, 1976). This paper uses the methods of Sinclair (1976) to identify natural background or baseline concentrations of dissolved components in water from mineralized and disturbed areas.

The computer program PROBPLOT (Stanley, 1987) was used for the construction and interpretation of probability distribution diagrams for the studies presented in this paper. Populations on a probability graph can be identified by inflections in the lines that represent in the distribution of data. The program PROBPLOT (Stanley, 1987) can then be used to determine summary statistics (mean and standard deviations) for each contributing geochemical family, in either arithmetic or geometric units.

To overcome the bias introduced into PROBPLOT by non-detect (ND) values, and to handle multiple detection limits, a second probability approach can be utilized, as described by (Helsel, 1990). A computer program, MDL (Multiple Detection Limits) (Helsel, 1990), facilitates application of the fill-in method and calculates summary statistics for normal and log-normal distributions.

4 EXAMPLES OF THE USE OF PROBABILITY DISTRIBUTION DIAGRAMS

Probability distribution diagrams were utilized in recent studies for characterizing background concentrations in surface and ground water in the North and South Operational Areas of Kennecott Utah Copper (Kennecott), near Salt Lake City, Utah. The study incorporated chemical analyses of water from nine surface and ground water hydrogeologic units in the Kennecott North Area and five ground water hydrogeologic units in the South Area. The total number of individual chemical values exceeded 100,000.

4.1 Multiple natural populations.

One of the nine distinct hydrogeologic units in the North Area is surface water, represented by 24 sampling sites in proximity to the Kennecott tailings impoundment. A total of 235 surface water analyses were available. Figure 1 presents a probability graph for 233 concentrations of dissolved Cl. The probability graph shows that the observed distribution of Cl concentrations is made up of two populations, comprising 90% and 10% of the data, with mean arithmetic concentrations of 1600 and 9600 mg L^{-1}, respectively. In this example, both populations are of natural origin, representing the influence of the highly saline brines of the Great Salt Lake immediately to the northwest.

A portion of the principal ground water hydrogeologic unit in the Kennecott North Area is the Principal (Deep Lacustrine) Aquifer. Figure 2 shows a probability graph representing 47 concentrations of Cl from this aquifer. The three natural populations (logarithmic) which best describe the data for Cl have corresponding arithmetic mean values of 570, 4,000, and 15,000 mg L^{-1}. An almost identical distribution was observed for Na in this aquifer (data not shown).

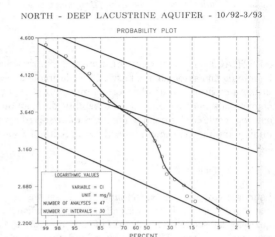

Figure 2. Probability graph of 47 Cl concentrations in shallow ground water, Kennecott North Area.

4.2 Contaminant populations.

A second hydrogeologic unit in the Kennecott North Area, the Alluvial Aquifer, is composed of alluvial deposits of coarse-grained sediments. Shown in Figures 3 and 4 are probability graphs for pH and Mg in this aquifer, respectively. Two distinct populations are identified for each parameter, with arithmetic mean concentrations of 7.08 and 2.04 for pH, and 71 and 2100 mg L^{-1} for dissolved Mg. The lower population of pH and the upper population of Mg are most pronounced in wells related to historical effects from past smelting and refining operations.

This example shows the distinct separation that can be achieved for natural versus man-influenced populations. It also shows the utility of magnesium as an indicator of historical acid impacts at this site.

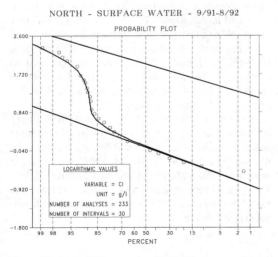

Figure 1. Probability graph of 233 Cl concentrations from surface water samples, Kennecott North Area.

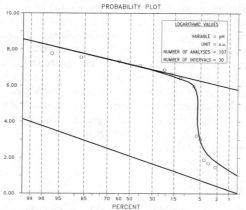

Figure 3. Probability graph for pH values, Kennecott North Area.

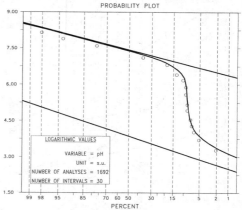

Figure 5. Probability graph for 1692 pH values in ground water, Kennecott South Area.

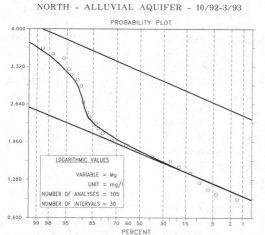

Figure 4. Probability graph for Mg concentrations, Kennecott North Area.

4.3 Identifying previously unknown contaminant populations.

Shown in Figure 5 are 1692 pH values representing ground water in one of the important aquifers of the Kennecott South Area, adjacent to the Bingham copper mine. A historic plume of acidic, metal-bearing water is responsible for the lower population of values, comprising 7 percent of the analyses. The two populations shown in Figure 5 have arithmetic mean pH values of 3.82 and 7.31.

To clarify the probability distributions in another important hydrogeologic unit in the South Area, the Principal Alluvial Volcanic Aquifer, the data were separated into eastern and western areas according to their proximity to former industrial evaporation ponds and unlined agricultural irrigation canals. This was done for two reasons: (1) to separate the potential natural and anthropogenic sources contributing to the aquifer and (2) to isolate surface irrigation canals as a possible source of elevated Cl. A subset of 105 Cl values for this aquifer is shown in Figure 6. All three geochemical populations (arithmetic means of 130, 250, and 670 mg L^{-1}) in this eastern area were determined to represent contributions from distinct ground water sources based on evaluation of well screen depths and geographic locations. The irrigation canals and related surface irrigation were identified as significant sources of Cl (and SO_4), as represented by the middle population in Figure 6. The highest population in Figure 6 represents waters of geothermal origin, and the lowest population represents normal ground water in that area.

4.4 Trace metal populations.

Figure 7 shows a well-defined single natural distribution of As concentrations for the Shallow Lacustrine Aquifer of the Kennecott North Area. The arithmetic mean concentration for this natural population is 0.068 mg L^{-1}. From similar occurrences of As in the Shallow Lacustrine Aquifer throughout the greater Salt Lake Valley, it is clear that this natural population, with a mean value of As that

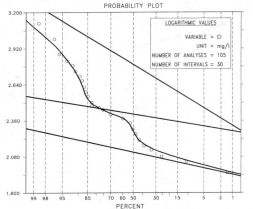

Figure 6. Probability graph for Cl concentrations in ground water, Kennecott South Area.

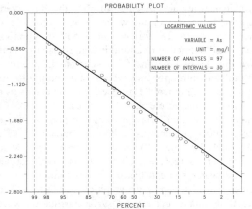

Figure 7. Probability graph showing one natural population of As for 97 ground water samples, Kennecott North Area.

exceeds the drinking water standard (0.05 mg L^{-1}), is caused by natural As-enriched natural sediments of the Salt Lake Valley. Elevated natural As concentrations are common in closed desert basins in the western U.S., in the complete absence of mining and milling activity.

5 CONCLUSIONS

Probability distribution graphs offer a simple and useful method for examining and classifying large sets of geochemical data in natural and/or highly disturbed areas. Such graphs can be used to identify multiple natural background populations in mineralized and non-mineralized areas. Anthropogenic effects can also be recognized in the form of distinct populations and separated from natural background populations. The method allows realistic and meaningful values to be established for baseline studies and site remediation. In the absence of valid estimates of natural background concentrations, standards or goals may be set for remediation that are unrealistically low.

REFERENCES

Helsel, D.R., 1990, Less than obvious-statistical treatment of data below the detection limit. *Environ. Sci. Technol.* 24, 1766-1744.

Miller, W.R, and McHugh, J.B., 1994, Natural acid drainage from altered areas within and adjacent to the Upper Alamosa River Basin, Colorado: U.S. Geol. Survey Open File Report 94-144, 47p.

Nordstrom, D.K., Alpers, C.N. and Wright, W.G., 1996, Geochemical methods for estimating pre-mining and background water-quality conditions in mineralized areas: Geological Soc. of America, Abstracts with Programs, 1996 Annual Meeting, Denver, CO, Oct. 28-31, p. A-465.

Runnells, D.D., Shepherd, T.A., and Angino, E.E., 1992, Determining natural background concentrations in mineralized areas. *Environ. Sci. Technol.* 26, 2316-2322.

Sinclair, A.J., 1976, *Applications of Probability Graphs in Mineral Exploration*. Association of Exploration Geochemists, Special Volume 4, Rexdale, Ontario, Canada and Denver, Colorado.

Stackelberg, P.E. and Siwiec, S.F., 1993, Evaluation of statistical models to predict chemical quality of shallow ground water in the Pine Barrens of Suffolk County, Long Island, New York: U.S. Geol. Survey Water-Resources Invests. Report 92-4100, 26p.

Stanley, C.R., 1987, PROBPLOT - A computer program to fit mixtures of normal (or log normal) distributions with maximum likelihood optimization procedures. Association of Exploration Geochemists, Special Volume 14.

U.S. EPA (U.S. Environmental Protection Agency), 1992, General ground-water monitoring requirements: Code of Federal Regulations, 40 CFR, Chapter 1, Parts 260-299.

Attenuation of leachate contaminants in an engineered wetland

M. Sartaj & L. Fernandes
Civil Engineering Department, University of Ottawa, Ont., Canada

ABSTRACT: Ground water is the most important source of potable water, but increasingly threatened by human activities such as landfilling. The primary concern is the production of leachate. The Huneault Landfill, located in Ontario, has been receiving industrial, commercial and institutional (IC&I) waste as well as construction/demolition waste since 1971. A constructed wetland system consisting of a vertical flow subsurface system (VFSS) followed by three free water surface (FWS) wetlands was selected as the treatment method for leachate. Results obtained confirm that artificial wetlands can provide a sound wastewater treatment alternative for managing landfill leachate. Removal efficiencies for some of the selected parameters were 90% for boron, 59% for TOC and TDS, 95% for ammonia and at least 45% for BOD_5. With the addition of a peat filter, the large land surface area generally associated with artificial wetlands can be significantly reduced.

1. INTRODUCTION

Waste disposal facilities (landfill) are potential threats to groundwater quality (Howard, 1997). The primary concern associated with landfills is the production of leachate. Leachate contains a variety of pollutants and is highly toxic, and could impact aquatic life and degrade water quality (Cameron & Koch, 1980).

Wetlands have been identified as a promising approach for treatment of a variety of waste waters including landfill leachate. Constructed wetlands are categorized into two main groups: free water surface (FWS) and subsurface flow systems (SFS). A third type of design, which is presently evolving to allow manipulation of the flow, is vertical flow subsurface systems (VFSS). In VFSS wetlands, water flows vertically down through the porous media and is collected at the base.

Performance of artificial wetlands is measured by removal efficiencies. Wetland systems can significantly reduce 5-day biological oxygen demand (BOD_5), suspended solids (SS), nitrogen, trace organics and pathogens as well as heavy metals. BOD_5 and SS removal efficiencies in the range of 70-90%, nitrogen removals of 60-86%, and the removal of Cu, Zn and Cd at rates of 99%, 97% and 99%, respectively, have been observed (Pries, 1994; Gersberg et al., 1985).

This paper discusses the design and application of a constructed wetland for attenuation of leachate produced at the Huneault Landfill, Ontario. The physical characteristics of the landfill site, physico-chemical characteristics of the leachate produced, and the design of the system as well as its effectiveness are presented and discussed.

2. CHARACTERISTICS OF THE SITE

The Huneault Waste Management site, occupying an area of 40 hectares located in Ontario, Canada, has been operating since 1971 under a certificate of approval from the Ministry of Environment and Energy (MOEE) to receive construction/demolition waste, industrial/commercial/institutional (IC&I) waste, as well as miscellaneous inert material. In 1996, a total of 177,769 tonnes of materials entered the landfill site of which 50.5% were landfilled and the rest were either converted or used as final cover. The average annual precipitation for the region is 917 millimeters. The average temperatures vary from a low of -10.6°C in January to 20.7°C in July, with an annual average of 6 °C.

3. CHARACTERISTICS OF THE LEACHATE

Leachate generated at this site has been monitored since 1990. This landfill is unique in the sense that the collected leachate is distinctly different from most other landfill sites in Ontario, with most analytes having much lower concentrations. This is attributed to the type of waste material being landfilled at this site. Concentrations of some selected contaminants (Table 1) were on average 16.5 mg/l of ammonia, 40 mg/l of BOD_5, 10.2 mg/l of boron, 2.8 mg/l of iron, 0.03 mg/l of lead, 0.03 mg/l of zinc and 51 mg/l of total suspended solids.

4. PEAT FILTER AND WETLAND SYSTEM

The design of the system consists of four cells, the first one comprising the peat filter serving as vertical flow subsurface system and the other three cells serving as free surface water wetlands, as shown in Figure 1. Leachate is collected in the two existing ponds and then pumped and distributed over the peat filter by way of spray irrigation. The filtrate is then collected at the bottom of the peat filter by a network of subsurface drains that flow into the engineered wetlands as free surface water. The peat filter has an average length and width of 103 and 57 m, respectively, providing a surface area of 5580 m^2. The depth of the peat filter is 1.4 m. The hydraulic conductivity of the peat measured at 15 observation wells installed in the peat filter ranges from 1.1×10^{-5} to 2.0×10^{-3} cm/sec and averages 3.0×10^{-4} m/sec. The upper layer of the peat generally has a higher hydraulic conductivity averaging 7.65×10^{-4} cm/sec.

The three engineered FWS wetland cells operate in series with a total surface area of 4374 m^2. The total length of free water surface wetlands is 268 m and the average channel width is 12.7 m. The cells which make up the FWS range in size from 0.84 to 1.18 acres for a total of 3.0 acres.

Table 1: Characteristics of leachate at Huneault Landfill and average concentrations in other Ontario landfills

Parameter	Huneault Landfill			Ontario
	Min.	Max.	Avg.	Avg.
NH_3	0.1	31	16.5	175
BOD_5	2	173	39.9	4975
Cu	0.002	4.5	0.43	0.045
B	0.4	16.9	10.2	10.4
Fe	0.003	8.09	2.8	58
TDS	1920	2572	2234	4327
Na	54	514	369	935
Mn	0.13	8.56	1.67	3.5
COD	69	600	289	7855
TSS	2.0	130.0	51.2	445
Cl	76.5	533	366	270
Pb	0.0	0.28	0.03	0.07
Zn	0.01	0.13	0.03	1.42

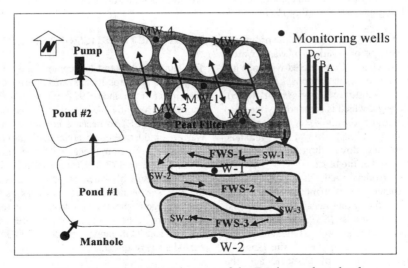

Figure 1: Schematic Diagram of the Engineered wetland

Capillary barriers for the surface sealings of landfills

Nico von der Hude & Frank Huppert
BILFINGER + BERGER Bauaktiengesellschaft, Department of Foundation and Environmental Engineering, Mannheim, Germany

ABSTRACT: The surface seals of landfills are engineering structures which, depending on prevailing local conditions, have to withstand particularly heavy loading, resulting from total or differential settlement, desiccation, penetration of roots and ageing processes. Compared to other alternatives, capillary barriers are outstandingly efficient sealing systems.

A capillary barrier consists of two sloping layers. Fine-grained sand is used in the capillary layer, which overlies a coarse-grained layer (the capillary block). The water is kept above the interface and runs off laterally using the fact that in unsaturated fine sand layers, capillary forces counteract the power of gravity and prevent water in the capillary layer from seeping down into the capillary block. In suitable applications and with a correct choice of materials a capillary barrier is able to provide an efficient surface seal. Several pilot projects covering slopes of landfills in Germany were set up in order to demonstrate that large-scale capillary barriers would be a feasible technical solution for the sealing of landfills.

1 INTRODUCTION

The main problem with landfills and remediation sites is that precipitation can infiltrate into the waste body and cause the migration of perched leachate. A barrier layer at the base of the landfill enables the collection and treatment of seepage water before reaching the underlying earth. The next, highly important step is the construction of a final cover which reduces water seepage and encapsulates the waste body. If the base of the landfill or remediation site is not sealed the only protection that exists is provided by the surface seal. Total and differential settlement, temperature extremes, wet and dry cycles, penetration by roots, animals or insects, long-term moisture changes and alterations caused by gas derived from decomposition will pose unpredictable loads on all liners in a final cover. Investigations of compacted clay liners indicate that they are subject to desiccation. Laboratory and test-field measurements have shown that capillary barriers can be used as an alternative type of sealing structure.

2 STRUCTURE OF A CAPILLARY BARRIER

A capillary barrier (Fig. 1) consists of a capillary layer of fine sand on top of a coarse-grained material (i.e. the capillary block). If unsaturated fine-grained soil overlies an unsaturated coarse-grained soil along a sloping contact area, infiltrating water can, in appropriate conditions, be diverted away from the coarser material. The water flows downdip above the plane of contact between the fine- and the coarse-grained layer. The fine-grained soil, which will consist mostly of fine sand, is called the capillary

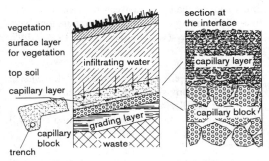

Fig. 1: Cross-sectional representation of a capillary barrier

layer, while the coarse-grained material is known as the capillary block.

Used in a sloping landfill (min.incline ≈1:6), it consists in its simpliest form of the following layers:

- the surface and protection layer, consisting of reafforestable topsoil which stores precipitation and supports evapotranspiring vegetation
- the capillary layer, consisting of fine-grained (0/1mm) sand with a steeply ascending grain-size distribution curve
- the capillary block, consisting of coarse material (1/4mm) with a steeply ascending grain-size distribution curve
- a grading layer overlying the body of waste

3 HOW A CAPILLARY BARRIER WORKS

A capillary barrier exploits the relationships between water content, capillary pressure and unsaturated hydraulic conductivity. At the boundary between two layers with different pore-size distributions, the capillary pressure on both sides of the interface comes to be the same. Fig. 2 shows in simplified form the capillary pressure and the hydraulic conductivity of a coarse-grained (1/4mm) and a fine-grained material(0/1mm) as functions of the degree of saturation. The unsaturated hydraulic conductivity k_u is much more dependent on water content than on pore size. Thus, when the capillary pressures in fine-grained sand are low, the level of saturation in this sand will be high and the level of unsaturated conductivity (k_{uf}) will also be relatively high. The degree of saturation of the coarse-grained sand, on the other hand, will be low, and its conductivity (k_{ug}) will be several powers of ten less than that of the fine-grained sand.

The functioning of a capillary barrier depends on:

- the existence of a distinct difference in the pore sizes of the fine-grained and the coarse-grained materials across the entire interface between the two layers (steep particle-size distribution curve for both materials).
- filter stability of the two materials.
- the materials used, the thickness of the layers, the degree of incline and the field length in that virtually all seeping water migrates laterally via the capillary seam, and no water is able to percolate down into the layer of coarse-grained sand.

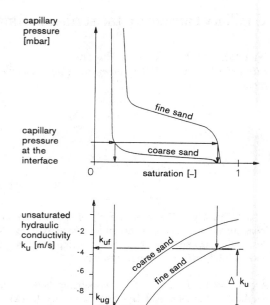

Fig. 2: Capillary pressure and unsaturated hydraulic conductivity k_u as functions of the degree of saturation S (qualitative)

4 LABORATORY TESTS AND TESTS UNDER FIELD CONDITIONS

Laboratory experiments involving the use of two 8-metre-long tilted flumes have been in progress at the Institute of Water-Resources Engineering at the Technical University of Darmstadt since 1990.

These experiments have thrown considerable light on how capillary barriers function. Application limits and the materials that can be used in various fine- and coarse-grained material combinations have been tested sites (v. d. Hude 1993, 1994, 1995, Kämpf 1995).

Trials have been in progress in test fields outside Marburg, Germany since April 1992, and in January 1994 four capillary barrier systems were brought into operation in the "Monte Scherbelino" landfill near Frankfurt (Jelinek, 1997).

In practical applications, the capillary barrier will be part of a cover-sealing system. Fig. 3 shows the